Commercial Renovation Costs with RSMeans data

Thomas Lane, Senior Editor

GORDIAN®

2022
43rd annual edition

Vice President, Data
Chris Gaudreau

Vice President, Product
Ted Kail

Principal Engineer
Bob Mewis

Engineers: Architectural Divisions
1, 3, 4, 5, 6, 7, 8, 9, 10, 11, 12, 13, 41
Matthew Doheny, Manager
Sam Babbitt
Sergeleen Edouard
Richard Goldin
Scott Keller
Thomas Lane
Giles Munyard

Engineers: Civil Divisions and Wages
2, 31, 32, 33, 34, 35, 44, 46
Derrick Hale, PE, Manager
Christopher Babbitt
Stephen Bell
Michael Lynch
Elisa Mello
David Yazbek

Engineers: Mechanical, Electrical, Plumbing & Conveying Divisions
14, 21, 22, 23, 25, 26, 27, 28, 48
Joseph Kelble, Manager
Brian Adams
Michelle Curran
Antonio D'Aulerio

Thomas Lyons
Jake MacDonald

Contributing Engineers
John Melin, PE, Manager
Paul Cowan
Barry Hutchinson
Gerard Lafond, PE
Matthew Sorrentino

Technical Project Lead
Kevin Souza

Production
Debra Panarelli, Manager
Jonathan Forgit
Edward Green
Sharon Larsen
Sheryl Rose
Janice Thalin

Data Quality
Euricka Thomas, Manager
David Byars
Audrey Considine
Ellen D'Amico

Cover Design
Blaire Collins

Innovation
Chris Gaudreau, Vice President
Kedar Gaikwad
Todd Glowac
Srini Narla
Joseph Woughter

RSMeans data from Gordian
Construction Publishers & Consultants
30 Patewood Dr., Suite 350
Greenville, SC 29615
United States of America
1.800.448.8182
RSMeans.com

Copyright 2021 by The Gordian Group Inc.
All rights reserved.

Cover photo © iStock.com/Pulvret80

Printed in the United States of America
ISSN 2161-4628
ISBN 978-1-955341-02-8

0042 $330.00 per copy (in United States)
Price is subject to change without prior notice.

Related Data and Services

Our engineers recommend the following products and services to complement *Commercial Renovation Costs with RSMeans data:*

Annual Cost Data Books
2022 Facilities Construction Costs with RSMeans data
2022 Facilities Maintenance & Repair Costs with RSMeans data

Reference Books
Estimating Building Costs
RSMeans Estimating Handbook
Green Building: Project Planning & Estimating
How to Estimate with RSMeans data
Plan Reading & Material Takeoff
Project Scheduling & Management for Construction
Universal Design for Style, Comfort & Safety

Virtual, Instructor-led & On-site Training Offerings
Facilities Construction Estimating
Training for our online estimating solution
Practical Project Management for Construction Professionals
Mechanical & Electrical Estimating
Maintenance & Repair Estimating for Facilities

RSMeans data
For access to the latest cost data, an intuitive search, and an easy-to-use estimate builder, take advantage of the time savings available from our online application.

To learn more visit: **RSMeans.com/online**

Enterprise Solutions
Building owners, facility managers, building product manufacturers, and attorneys across the public and private sectors engage with RSMeans data Enterprise to solve unique challenges where trusted construction cost data is critical.

To learn more visit: **RSMeans.com/Enterprise**

Custom Built Data Sets
Building and Space Models: Quickly plan construction costs across multiple locations based on geography, project size, building system component, product options, and other variables for precise budgeting and cost control.

Predictive Analytics: Accurately plan future builds with custom graphical interactive dashboards, negotiate future costs of tenant build-outs, and identify and compare national account pricing.

Consulting
Building Product Manufacturing Analytics: Validate your claims and assist with new product launches.

Third-Party Legal Resources: Used in cases of construction cost or estimate disputes, construction product failure vs. installation failure, eminent domain, class action construction product liability, and more.

API
For resellers or internal application integration, RSMeans data is offered via API. Deliver Unit, Assembly, and Square Foot Model data within your interface. To learn more about how you can provide your customers with the latest in localized construction cost data visit: **RSMeans.com/API**

Table of Contents

Foreword

The Value of RSMeans data from Gordian

Since 1942, RSMeans data has been the industry-standard materials, labor, and equipment cost information database for contractors, facility owners and managers, architects, engineers, and anyone else that requires the latest localized construction cost information. More than 75 years later, the objective remains the same: to provide facility and construction professionals with the most current and comprehensive construction cost database possible.

With the constant influx of new construction methods and materials, in addition to ever-changing labor and material costs, last year's cost data is not reliable for today's designs, estimates, or budgets. Gordian's cost engineers apply real-world construction experience to identify and quantify new building products and methodologies, adjust productivity rates, and adjust costs to local market conditions across the nation. This adds up to more than 22,000 hours in cost research annually. This unparalleled construction cost expertise is why so many facility and construction professionals rely on RSMeans data year over year.

About Gordian

Gordian is the leading provider of facility and construction cost data, software, and services for all phases of the building lifecycle. From planning to design, procurement, construction, and operations, Gordian delivers groundbreaking solutions to contractors, architects, engineers, business and financial officers, and facility owners in local, state, and federal governments, education, healthcare, and other industries. A pioneer of Job Order Contracting (JOC), Gordian's offerings also include our proprietary RSMeans data and Facility Intelligence solutions. We develop and maintain the largest collection of labor, material, and equipment data and associated costs for all areas of construction. Gordian's solutions are accessed through our innovative software platforms and supported by a team of industry experts proven to help clients maximize efficiency, optimize cost savings, and increase building quality.

Our Commitment

At Gordian, we do more than talk about the quality of our data and the usefulness of its application. We stand behind all of our RSMeans data—from historical cost indexes to construction materials and techniques—to craft current costs and predict future trends. If you have any questions about our products or services, please call us toll-free at 800.448.8182 or visit our website at gordian.com.

How the Cost Data Is Built: An Overview

Unit Prices*

All cost data have been divided into 50 divisions according to the MasterFormat® system of classification and numbering.

Assemblies*

The cost data in this section have been organized in an "Assemblies" format. These assemblies are the functional elements of a building and are arranged according to the 7 elements of the UNIFORMAT II classification system. For a complete explanation of a typical "Assembly", see "RSMeans data: Assemblies—How They Work."

Residential Models*

Model buildings for four classes of construction—economy, average, custom, and luxury—are developed and shown with complete costs per square foot.

Commercial/Industrial/Institutional Models*

This section contains complete costs for 77 typical model buildings expressed as costs per square foot.

Green Commercial/Industrial/Institutional Models*

This section contains complete costs for 25 green model buildings expressed as costs per square foot.

References*

This section includes information on Equipment Rental Costs, Crew Listings, Historical Cost Indexes, City Cost Indexes, Location Factors, Reference Tables, and Change Orders, as well as a listing of abbreviations.

- **Equipment Rental Costs:** Included are the average costs to rent and operate hundreds of pieces of construction equipment.
- **Crew Listings:** This section lists all the crews referenced in the cost data. A crew is composed of more than one trade classification and/or the addition of power equipment to any trade classification. Power equipment is included in the cost of the crew. Costs are shown both with bare labor rates and with the installing contractor's overhead and profit added. For each, the total crew cost per eight-hour day and the composite cost per labor-hour are listed.

Unit Cost data

Assembly Cost data

Square Foot Models

- **Historical Cost Indexes:** These indexes provide you with data to adjust construction costs over time.
- **City Cost Indexes:** All costs in this data set are U.S. national averages. Costs vary by region. You can adjust for this by CSI Division to over 730 cities in 900+ 3-digit zip codes throughout the U.S. and Canada by using this data.
- **Location Factors:** You can adjust total project costs to over 730 cities in 900+ 3-digit zip codes throughout the U.S. and Canada by using the weighted number, which applies across all divisions.
- **Reference Tables:** At the beginning of selected major classifications in the Unit Prices are reference numbers indicators. These numbers refer you to related information in the Reference Section. In this section, you'll find reference tables, explanations, and estimating information that support how we develop the unit price data, technical data, and estimating procedures.
- **Change Orders:** This section includes information on the factors that influence the pricing of change orders.

- **Abbreviations:** A listing of abbreviations used throughout this information, along with the terms they represent, is included.

Index (printed versions only)

A comprehensive listing of all terms and subjects will help you quickly find what you need when you are not sure where it occurs in MasterFormat®.

Conclusion

This information is designed to be as comprehensive and easy to use as possible.

The Construction Specifications Institute (CSI) and Construction Specifications Canada (CSC) have produced the 2018 edition of MasterFormat®, a system of titles and numbers used extensively to organize construction information.

All unit prices in the RSMeans cost data are now arranged in the 50-division MasterFormat® 2018 system.

---

*   Not all information is available in all data sets

Note: The material prices in RSMeans cost data are "contractor's prices." They are the prices that contractors can expect to pay at the lumberyards, suppliers'/distributors' warehouses, etc. Small orders of specialty items would be higher than the costs shown, while very large orders, such as truckload lots, would be less. The variation would depend on the size, timing, and negotiating power of the contractor. The labor costs are primarily for new construction or major renovation rather than repairs or minor alterations. With reasonable exercise of judgment, the figures can be used for any building work.

Estimating with RSMeans data: Unit Prices

Following these steps will allow you to complete an accurate estimate using RSMeans data: Unit Prices.

1. Scope Out the Project

- Think through the project and identify the CSI divisions needed in your estimate.
- Identify the individual work tasks that will need to be covered in your estimate.
- The Unit Price data have been divided into 50 divisions according to CSI MasterFormat® 2018.
- In printed versions, the Unit Price Section Table of Contents on page 1 may also be helpful when scoping out your project.
- Experienced estimators find it helpful to begin with Division 2 and continue through completion. Division 1 can be estimated after the full project scope is known.

2. Quantify

- Determine the number of units required for each work task that you identified.
- Experienced estimators include an allowance for waste in their quantities. (Waste is not included in our Unit Price line items unless otherwise stated.)

3. Price the Quantities

- Use the search tools available to locate individual Unit Price line items for your estimate.
- Reference Numbers indicated within a Unit Price section refer to additional information that you may find useful.
- The crew indicates who is performing the work for that task. Crew codes are expanded in the Crew Listings in the Reference Section to include all trades and equipment that comprise the crew.
- The Daily Output is the amount of work the crew is expected to complete in one day.
- The Labor-Hours value is the amount of time it will take for the crew to install one unit of work.
- The abbreviated Unit designation indicates the unit of measure upon which the crew, productivity, and prices are based.
- Bare Costs are shown for materials, labor, and equipment needed to complete the Unit Price line item. Bare costs do not include waste, project overhead, payroll insurance, payroll taxes, main office overhead, or profit.
- The Total Incl O&P cost is the billing rate or invoice amount of the installing contractor or subcontractor who performs the work for the Unit Price line item.

4. Multiply

- Multiply the total number of units needed for your project by the Total Incl O&P cost for each Unit Price line item.
- Be careful that your take off unit of measure matches the unit of measure in the Unit column.
- The price you calculate is an estimate for a completed item of work.
- Keep scoping individual tasks, determining the number of units required for those tasks, matching each task with individual Unit Price line items, and multiplying quantities by Total Incl O&P costs.
- An estimate completed in this manner is priced as if a subcontractor, or set of subcontractors, is performing the work. The estimate does not yet include Project Overhead or Estimate Summary components such as general contractor markups on subcontracted work, general contractor office overhead and profit, contingency, and location factors.

5. Project Overhead

- Include project overhead items from Division 1-General Requirements.
- These items are needed to make the job run. They are typically, but not always, provided by the general contractor. Items include, but are not limited to, field personnel, insurance, performance bond, permits, testing, temporary utilities, field office and storage facilities, temporary scaffolding and platforms, equipment mobilization and demobilization, temporary roads and sidewalks, winter protection, temporary barricades and fencing, temporary security, temporary signs, field engineering and layout, final cleaning, and commissioning.
- Each item should be quantified and matched to individual Unit Price line items in Division 1, then priced and added to your estimate.
- An alternate method of estimating project overhead costs is to apply a percentage of the total project cost—usually 5% to 15% with an average of 10% (see General Conditions).
- Include other project related expenses in your estimate such as:
  - ☐ Rented equipment not itemized in the Crew Listings
  - ☐ Rubbish handling throughout the project (see section 02 41 19.19)

6. Estimate Summary

- Include sales tax as required by laws of your state or county.
- Include the general contractor's markup on self-performed work, usually 5% to 15% with an average of 10%.
- Include the general contractor's markup on subcontracted work, usually 5% to 15% with an average of 10%.
- Include the general contractor's main office overhead and profit:
  - ☐ RSMeans data provides general guidelines on the general contractor's main office overhead (see section 01 31 13.60 and Reference Number R013113-50).
  - ☐ Markups will depend on the size of the general contractor's operations, projected annual revenue, the level of risk, and the level of competition in the local area and for this project in particular.
- Include a contingency, usually 3% to 5%, if appropriate.
- Adjust your estimate to the project's location by using the City Cost Indexes or the Location Factors in the Reference Section:
  - ☐ Look at the rules in "How to Use the City Cost Indexes" to see how to apply the Indexes for your location.
  - ☐ When the proper Index or Factor has been identified for the project's location, convert it to a multiplier by dividing it by 100, then multiply that multiplier by your estimated total cost. The original estimated total cost will now be adjusted up or down from the national average to a total that is appropriate for your location.

Editors' Note:
We urge you to spend time reading and understanding the supporting material. An accurate estimate requires experience, knowledge, and careful calculation. The more you know about how we at RSMeans developed the data, the more accurate your estimate will be. In addition, it is important to take into consideration the reference material such as Equipment Listings, Crew Listings, City Cost Indexes, Location Factors, and Reference Tables.

How to Use the Cost Data: The Details

What's Behind the Numbers? The Development of Cost Data

RSMeans data engineers continually monitor developments in the construction industry to ensure reliable, thorough, and up-to-date cost information. While overall construction costs may vary relative to general economic conditions, price fluctuations within the industry are dependent upon many factors. Individual price variations may, in fact, be opposite to overall economic trends. Therefore, costs are constantly tracked and complete updates are performed yearly. Also, new items are frequently added in response to changes in materials and methods.

Costs in U.S. Dollars

All costs represent U.S. national averages and are given in U.S. dollars. The City Cost Index (CCI) with RSMeans data can be used to adjust costs to a particular location. The CCI for Canada can be used to adjust U.S. national averages to local costs in Canadian dollars. No exchange rate conversion is necessary because it has already been factored in.

G The processes or products identified by the green symbol in our publications have been determined to be environmentally responsible and/or resource-efficient solely by RSMeans data engineering staff. The inclusion of the green symbol does not represent compliance with any specific industry association or standard.

Material Costs

RSMeans data engineers contact manufacturers, dealers, distributors, and contractors all across the U.S. and Canada to determine national average material costs. If you have access to current material costs for your specific location, you may wish to make adjustments to reflect differences from the national average. Included within material costs are fasteners for a normal installation. RSMeans data engineers use manufacturers' recommendations, written specifications, and/or standard construction practices for the sizing and spacing of fasteners. Adjustments to material costs may be required for your specific application or location. The manufacturer's warranty is assumed. Extended warranties are not included in the material costs. **Material costs do not include sales tax.**

Labor Costs

Labor costs are based upon a mathematical average of trade-specific wages in 30 major U.S. cities. The type of wage (union, open shop, or residential) is identified on the inside back cover of printed publications or selected by the estimator when using the electronic products. Markups for the wages can also be found on the inside back cover of printed publications and/or under the labor references found in the electronic products.

- If wage rates in your area vary from those used, or if rate increases are expected within a given year, labor costs should be adjusted accordingly.

Labor costs reflect productivity based on actual working conditions. In addition to actual installation, these figures include time spent during a normal weekday on tasks, such as material receiving and handling, mobilization at the site, site movement, breaks, and cleanup.

Productivity data is developed over an extended period so as not to be influenced by abnormal variations and reflects a typical average.

Equipment Costs

Equipment costs include not only rental but also operating costs for equipment under normal use. The operating costs include parts and labor for routine servicing, such as the repair and replacement of pumps, filters, and worn lines. Normal operating expendables, such as fuel, lubricants, tires, and electricity (where applicable), are also included. Extraordinary operating expendables with highly variable wear patterns, such as diamond bits and blades, are excluded. These costs are included under materials. Equipment rental rates are obtained from industry sources throughout North America—contractors, suppliers, dealers, manufacturers, and distributors.

Rental rates can also be treated as reimbursement costs for contractor-owned equipment. Owned equipment costs include depreciation, loan payments, interest, taxes, insurance, storage, and major repairs.

Equipment costs do not include operators' wages.

Equipment Cost/Day—The cost of equipment required for each crew is included in the Crew Listings in the Reference Section (small tools that are considered essential everyday tools are not listed out separately). The Crew Listings itemize specialized tools and heavy equipment along with labor trades. The daily cost of itemized equipment included in a crew is based on dividing the weekly bare rental rate by 5 (number of working days per week), then adding the hourly operating cost times 8 (the number of hours per day). This Equipment Cost/Day is shown in the last column of the Equipment Rental Costs in the Reference Section.

Mobilization, Demobilization—The cost to move construction equipment from an equipment yard or rental company to the job site and back again is not included in equipment costs. Mobilization (to the site) and demobilization (from the site) costs can be found in the Unit Price Section. If a piece of equipment is already at the job site, it is not appropriate to utilize mobilization or demobilization costs again in an estimate.

Overhead and Profit

Total Cost including O&P for the installing contractor is shown in the last column of the Unit Price and/or Assemblies. This figure is the sum of the bare material cost plus 10% for profit, the bare labor cost plus total overhead and profit, and the bare equipment cost plus 10% for profit. Details for the calculation of overhead and profit on labor are shown on the inside back cover of the printed product and in the Reference Section of the electronic product.

General Conditions

Cost data in this data set are presented in two ways: Bare Costs and Total Cost including O&P (Overhead and Profit). General Conditions, or General Requirements, of the contract should also be added to the Total Cost including O&P when applicable. Costs for General Conditions are listed in Division 1 of the Unit Price Section and in the Reference Section.

General Conditions for the installing contractor may range from 0% to 10% of the Total Cost including O&P. For the general or prime contractor, costs for General Conditions may range from 5% to 15% of the Total Cost including O&P, with a figure of 10% as the most typical allowance. If applicable, the Assemblies and Models sections use costs that include the installing contractor's overhead and profit (O&P).

Factors Affecting Costs

Costs can vary depending upon a number of variables. Here's a listing of some factors that affect costs and points to consider.

Quality—The prices for materials and the workmanship upon which productivity is based represent sound construction work. They are also in line with industry standard and manufacturer specifications and are frequently used by federal, state, and local governments.

Overtime—We have made no allowance for overtime. If you anticipate premium time or work beyond normal working hours, be sure to make an appropriate adjustment to your labor costs.

Productivity—The productivity, daily output, and labor-hour figures for each line item are based on an eight-hour work day in daylight hours in moderate temperatures and up to a 14' working height unless otherwise indicated. For work that extends beyond normal work hours or is performed under adverse conditions, productivity may decrease.

Size of Project—The size, scope of work, and type of construction project will have a significant impact on cost. Economies of scale can reduce costs for large projects. Unit costs can often run higher for small projects.

Location—Material prices are for metropolitan areas. However, in dense urban areas, traffic and site storage limitations may increase costs. Beyond a 20-mile radius of metropolitan areas, extra trucking or transportation charges may also increase the material costs slightly. On the other hand, lower wage rates may be in effect. Be sure to consider both of these factors when preparing an estimate, particularly if the job site is located in a central city or remote rural location. In addition, highly specialized subcontract items may require travel and per-diem expenses for mechanics.

Other Factors—

- season of year
- contractor management
- weather conditions
- local union restrictions
- building code requirements
- availability of:
  - adequate energy
  - skilled labor
  - building materials
- owner's special requirements/restrictions
- safety requirements
- environmental considerations
- access

Unpredictable Factors—General business conditions influence "in-place" costs of all items. Substitute materials and construction methods may have to be employed. These may affect the installed cost and/or life cycle costs. Such factors may be difficult to evaluate and cannot necessarily be predicted on the basis of the job's location in a particular section of the country. Thus, where these factors apply, you may find significant but unavoidable cost variations for which you will have to apply a measure of judgment to your estimate.

Rounding of Costs

In printed publications only, all unit prices in excess of $5.00 have been rounded to make them easier to use and still maintain adequate precision of the results.

How Subcontracted Items Affect Costs

A considerable portion of all large construction jobs is usually subcontracted. In fact, the percentage done by subcontractors is constantly increasing and may run over 90%. Since the workers employed by these companies do nothing else but install their particular products, they soon become experts in that line. As a result, installation by these firms is accomplished so efficiently that the total in-place cost, even with the general contractor's overhead and profit, is no more, and often less, than if the principal contractor had handled the installation. Companies that deal with construction specialties are anxious to have their products perform well and, consequently, the installation will be the best possible.

Contingencies

The allowance for contingencies generally provides for unforeseen construction difficulties. On alterations or repair jobs, 20% is not too much. If drawings are final and only field contingencies are being considered, 2% or 3% is probably sufficient and often nothing needs to be added. Contractually, changes in plans will be covered by extras. The contractor should consider inflationary price trends and possible material shortages during the course of the job. These escalation factors are dependent upon both economic conditions and the anticipated time between the estimate and actual construction. If drawings are not complete or approved, or a budget cost is wanted, it is wise to add 5% to 10%. Contingencies, then, are a matter of judgment.

Important Estimating Considerations

The productivity, or daily output, of each craftsman or crew assumes a well-managed job where tradesmen with the proper tools and equipment, along with the appropriate construction materials, are present. Included are daily set-up and cleanup time, break time, and plan layout time. Unless otherwise indicated, time for material movement on site (for items that can be transported by hand) of up to 200' into the building and to the first or second floor is also included. If material has to be transported by other means, over greater distances, or to higher floors, an additional allowance should be considered by the estimator.

While horizontal movement is typically a sole function of distances, vertical transport introduces other variables that can significantly impact productivity. In an occupied building, the use of elevators (assuming access, size, and required protective measures are acceptable) must be understood at the time of the estimate. For new construction, hoist wait and cycle times can easily be 15 minutes and may result in scheduled access extending beyond the normal work day. Finally, all vertical transport will impose strict weight limits likely to preclude the use of any motorized material handling.

The productivity, or daily output, also assumes installation that meets manufacturer/designer/ standard specifications. A time allowance for quality control checks, minor adjustments, and any task required to ensure proper function or operation is also included. For items that require connections to services, time is included for positioning, leveling, securing the unit, and making all the necessary connections (and start up where applicable) to ensure a complete installation. Estimating of the services themselves (electrical, plumbing, water, steam, hydraulics, dust collection, etc.) is separate.

In some cases, the estimator must consider the use of a crane and an appropriate crew for the installation of large or heavy items. For those situations where a crane is not included in the assigned crew and as part of the line item cost,

then equipment rental costs, mobilization and demobilization costs, and operator and support personnel costs must be considered.

Labor-Hours

The labor-hours expressed in this publication are derived by dividing the total daily labor-hours for the crew by the daily output. Based on average installation time and the assumptions listed above, the labor-hours include: direct labor, indirect labor, and nonproductive time. A typical day for a craftsman might include but is not limited to:

- Direct Work
  - ☐ Measuring and layout
  - ☐ Preparing materials
  - ☐ Actual installation
  - ☐ Quality assurance/quality control
- Indirect Work
  - ☐ Reading plans or specifications
  - ☐ Preparing space
  - ☐ Receiving materials
  - ☐ Material movement
  - ☐ Giving or receiving instruction
  - ☐ Miscellaneous
- Non-Work
  - ☐ Chatting
  - ☐ Personal issues
  - ☐ Breaks
  - ☐ Interruptions (i.e., sickness, weather, material or equipment shortages, etc.)

If any of the items for a typical day do not apply to the particular work or project situation, the estimator should make any necessary adjustments.

Final Checklist

Estimating can be a straightforward process provided you remember the basics. Here's a checklist of some of the steps you should remember to complete before finalizing your estimate.

Did you remember to:

- factor in the City Cost Index for your locale?
- take into consideration which items have been marked up and by how much?
- mark up the entire estimate sufficiently for your purposes?
- read the background information on techniques and technical matters that could impact your project time span and cost?
- include all components of your project in the final estimate?
- double check your figures for accuracy?
- call RSMeans data engineers if you have any questions about your estimate or the data you've used? Remember, Gordian stands behind all of our products, including our extensive RSMeans data solutions. If you have any questions about your estimate, about the costs you've used from our data, or even about the technical aspects of the job that may affect your estimate, feel free to call the Gordian RSMeans editors at 1.800.448.8182.

Using Minimum Labor/Equipment Charges for Small Quantities

Estimating small construction or repair tasks often creates situations in which the quantity of work to be performed is very small. When this occurs, the labor and/or equipment costs to perform the work may be too low to allow for the crew to get to the job, receive instructions, find materials, get set up, perform the work, clean up, and get to the next job. In these situations, the estimator should compare the developed labor and/or equipment costs for performing the work (e.g., quantity × labor and/or equipment costs) with the *minimum labor/equipment charge* within the Unit Price section of the data set. (These minimum labor/equipment charge line items appear only in *Facilities Construction Costs with RSMeans data* and *Commercial Renovation Costs with RSMeans data*.)

If the labor and/or equipment costs developed by the estimator are LOWER THAN the *minimum labor/equipment charge* listed at the bottom of specific sections of Unit Price costs, the estimator should adjust the developed costs upward to the *minimum labor/equipment charge*. The proper use of a *minimum labor/equipment charge* results in having enough money in the estimate to cover the contractor's higher cost of performing a very small amount of work during a partial workday.

A *minimum labor/equipment charge* should be used only when the task being estimated is the only task the crew will perform at the job site that day. **If, however, the crew will be able to perform other tasks at the job site that day, the use of a *minimum labor/equipment charge* is not appropriate.**

08 52 Wood Windows

08 52 10 – Plain Wood Windows

08 52 10.40 Casement Window		Crew	Daily Output	Labor-Hours	Unit	Material	2022 Bare Costs Labor	Equipment	Total	Total Incl O&P
0010	**CASEMENT WINDOW**, including frame, screen and grilles									
0100	2'-0" x 3'-0" H, double insulated glass [G]	1 Carp	10	.800	Ea.	320	45		365	420
0150	Low E glass [G]		10	.800		360	45		405	465
0200	2'-0" x 4'-6" high, double insulated glass [G]		9	.889		425	50		475	545
0250	Low E glass [G]		9	.889		425	50		475	545
9000	Minimum labor/equipment charge	↓	3	2.667	Job		150		150	239

Example:

Establish the bid price to install two casement windows. Assume installation of 2' × 4'-6" metal clad windows with insulating glass [Unit Price line number 08 52 10.40 0250] and that this is the only task this crew will perform at the job site that day.

Solution:

Step One – Develop the Bare Labor Cost for this task:
Bare Labor Cost = 2 windows @ $50 each = $100.00

Step Two – Evaluate the *minimum labor/equipment charge* for this Unit Price section against the developed Bare Labor Cost for this task:
Minimum labor/equipment charge = $150.00 (compare with $100.00)

Step Three – Choose to adjust the developed labor cost upward to the *minimum labor/equipment charge*.

Step Four – Develop the bid price for this task (including O&P):
Add together the marked-up Bare Material Cost for this task and the marked-up *minimum labor/equipment charge* for this Unit Price section.

2 × ($425.00 + 10%) + ($150.00 + 58.9%)

=   2 × ($425.00 + $42.50) + ($150.00 + $88.80)

=   2 × ($467.50) + $238.80

= $935.00 + $238.80

= $1,173.80

ANSWER: $1,173.80 is the correct bid price to use. This sum takes into consideration the material cost (with 10% for profit) for these two windows, plus the *minimum labor/equipment charge* (with O&P included) for this section of the Unit Price.

Unit Price Section

Table of Contents

Table of Contents (cont.)

Table of Contents (cont.)

RSMeans data: Unit Prices— How They Work

All RSMeans data: Unit Prices are organized in the same way.

03 30 Cast-In-Place Concrete

03 30 53 – Miscellaneous Cast-In-Place Concrete

	03 30 53.40 Concrete In Place		Crew	Daily Output	Labor-Hours	Unit	Material	2022 Bare Costs Labor	Equipment	Total	Total Incl O&P
0010	**CONCRETE IN PLACE**	R033105-10									
0020	Including forms (4 uses), Grade 60 rebar, concrete (Portland cement	R033105-20									
0050	Type I), placement and finishing unless otherwise indicated	R033105-70									
0300	Beams (3500 psi), 5 kip/L.F., 10' span	R033105-85	C-14A	15.62	12.804	C.Y.	485	720	50	1,255	1,750
0350	25' span		"	18.55	10.782		510	605	42	1,157	1,575
0500	Chimney foundations (5000 psi), over 5 C.Y.		C-14C	32.22	3.476		193	187	.87	380.87	510
0510	(3500 psi), under 5 C.Y.		"	23.71	4.724		230	254	1.18	485.18	655
0700	Columns, square (4000 psi), 12" x 12", up to 1% reinforcing by area		C-14A	11.96	16.722		535	940	65.50	1,540.50	2,150
3540	Equipment pad (3000 psi), 3' x 3' x 6" thick		C-14H	45	1.067	Ea.	66.50	58.50	.62	125.62	167
3550	4' x 4' x 6" thick			30	1.600		101	88	.93	189.93	251
3560	5' x 5' x 8" thick			18	2.667		180	147	1.55	328.55	430
3570	6' x 6' x 8" thick			14	3.429		245	189	1.99	435.99	570
3580	8' x 8' x 10" thick			8	6		510	330	3.48	843.48	1,075
3590	10' x 10' x 12" thick			5	9.600		895	530	5.55	1,430.55	1,825

It is important to understand the structure of RSMeans data: Unit Prices so that you can find information easily and use it correctly.

---

1 Line Numbers

Line Numbers consist of 12 characters, which identify a unique location in the database for each task. The first 6 or 8 digits conform to the Construction Specifications Institute MasterFormat® 2018. The remainder of the digits are a further breakdown in order to arrange items in understandable groups of similar tasks. Line numbers are consistent across all of our publications, so a line number in any of our products will always refer to the same item of work.

2 Descriptions

Descriptions are shown in a hierarchical structure to make them readable. In order to read a complete description, read up through the indents to the top of the section. Include everything that is above and to the left that is not contradicted by information below. For instance, the complete description for line 03 30 53.40 3550 is "Concrete in place, including forms (4 uses), Grade 60 rebar, concrete (Portland cement Type 1), placement and finishing unless otherwise indicated; Equipment pad (3,000 psi), 4' × 4' × 6" thick."

3 RSMeans data

When using **RSMeans data**, it is important to read through an entire section to ensure that you use the data that most closely matches your work. Note that sometimes there is additional information shown in the section that may improve your price. There are frequently lines that further describe, add to, or adjust data for specific situations.

4 Reference Information

Gordian's RSMeans engineers have created **reference** information to assist you in your estimate. **If** there is information that applies to a section, it will be indicated at the start of the section. The Reference Section is located in the back of the data set.

5 Crews

Crews include labor and/or equipment necessary to accomplish each task. In this case, Crew C-14H is used. Gordian's RSMeans staff selects a crew to represent the workers and equipment that are typically used for that task. In this case, Crew C-14H consists of one carpenter foreman (outside), two carpenters, one rodman, one laborer, one cement finisher, and one gas engine vibrator. Details of all crews can be found in the Reference Section.

Crews - Renovation

Crew No.	Bare Costs		Incl. Subs O & P		Cost Per Labor-Hour	
Crew C-14H	Hr.	Daily	Hr.	Daily	Bare Costs	Incl. O&P
1 Carpenter Foreman (outside)	$58.30	$466.40	$92.60	$740.80	$55.06	$86.98
2 Carpenters	56.30	900.80	89.45	1431.20		
1 Rodman (reinf.)	60.70	485.60	94.90	759.20		
1 Laborer	45.60	364.80	72.45	579.60		
1 Cement Finisher	53.15	425.20	83.05	664.40		
1 Gas Engine Vibrator		27.87		30.65	.58	.64
48 L.H., Daily Totals		$2670.67		$4205.85	$55.64	$87.62

6 Daily Output

The **Daily Output** is the amount of work that the crew can do in a normal 8-hour workday, including mobilization, layout, movement of materials, and cleanup. In this case, crew C-14H can install thirty 4' × 4' × 6" thick concrete pads in a day. Daily output is variable and based on many factors, including the size of the job, location, and environmental conditions. RSMeans data represents work done in daylight (or adequate lighting) and temperate conditions.

7 Labor-Hours

The figure in the **Labor-Hours** column is the amount of labor required to perform one unit of work—in this case the amount of labor required to construct one 4' × 4' equipment pad. This figure is calculated by dividing the number of hours of labor in the crew by the daily output (48 labor-hours divided by 30 pads = 1.6 hours of labor per pad). Multiply 1.6 times 60 to see the value in minutes: 60 × 1.6 = 96

minutes. Note: the labor-hour figure is not dependent on the crew size. A change in crew size will result in a corresponding change in daily output, but the labor-hours per unit of work will not change.

8 Unit of Measure

All RSMeans data: Unit Prices include the typical **Unit of Measure** used for estimating that item. For concrete-in-place the typical unit is cubic yards (C.Y.) or each (Ea.). For installing broadloom carpet it is square yard and for gypsum board it is square foot. The estimator needs to take special care that the unit in the data matches the unit in the take-off. Unit conversions may be found in the Reference Section.

9 Bare Costs

Bare Costs are the costs of materials, labor, and equipment that the installing contractor pays. They represent the cost, in U.S. dollars, for one unit of work. They do not include any markups for profit or labor burden.

10 Bare Total

The **Total column** represents the total bare cost for the installing contractor in U.S. dollars. In this case, the sum of $101 for material + $88 for labor + $.93 for equipment is $189.93.

11 Total Incl O&P

The **Total Incl O&P column** is the total cost, including overhead and profit, that the installing contractor will charge the customer. This represents the cost of materials plus 10% profit, the cost of labor plus labor burden and 10% profit, and the cost of equipment plus 10% profit. It does not include the general contractor's overhead and profit. Note: See the inside back cover of the printed product or the Reference Section of the electronic product for details on how the labor burden is calculated.

National Average

The RSMeans data in our print publications represent a "national average" cost. This data should be modified to the project location using the **City Cost Indexes** *or* **Location Factors** *tables found in the Reference Section. Use the Location Factors to adjust estimate totals if the project covers multiple trades. Use the City Cost Indexes (CCI) for single trade*

projects or projects where a more detailed analysis is required. All figures in the two tables are derived from the same research. The last row of data in the CCI—the weighted average—is the same as the numbers reported for each location in the location factor table.

Project Name: Pre-Engineered Steel Building				Architect: As Shown				
Location:	**Anywhere, USA**						**01/01/22**	**R+R**
Line Number	**Description**	**Qty**	**Unit**	**Material**	**Labor**	**Equipment**	**SubContract**	**Estimate Total**
03 30 53.40 3940	Strip footing, 12" x 24", reinforced	15	C.Y.	$3,255.00	$1,890.00	$8.70	$0.00	
03 30 53.40 3950	Strip footing, 12" x 36", reinforced	34	C.Y.	$7,072.00	$3,400.00	$15.98	$0.00	
03 11 13.65 3000	Concrete slab edge forms	500	L.F.	$230.00	$1,430.00	$0.00	$0.00	
03 22 11.10 0200	Welded wire fabric reinforcing	150	C.S.F.	$4,800.00	$4,725.00	$0.00	$0.00	
03 31 13.35 0300	Ready mix concrete, 4000 psi for slab on grade	278	C.Y.	$45,592.00	$0.00	$0.00	$0.00	
03 31 13.70 4300	Place, strike off & consolidate concrete slab	278	C.Y.	$0.00	$5,699.00	$141.78	$0.00	
03 35 13.30 0250	Machine float & trowel concrete slab	15,000	S.F.	$0.00	$10,650.00	$750.00	$0.00	
03 15 16.20 0140	Cut control joints in concrete slab	950	L.F.	$38.00	$446.50	$85.50	$0.00	
03 39 23.13 0300	Sprayed concrete curing membrane	150	C.S.F.	$2,032.50	$1,155.00	$0.00	$0.00	
Division 03	**Subtotal**			**$63,019.50**	**$29,395.50**	**$1,001.96**	**$0.00**	**$93,416.96**
08 36 13.10 2650	Manual 10' x 10' steel sectional overhead door	8	Ea.	$12,400.00	$4,000.00	$0.00	$0.00	
08 36 13.10 2860	Insulation and steel back panel for OH door	800	S.F.	$4,720.00	$0.00	$0.00	$0.00	
Division 08	**Subtotal**			**$17,120.00**	**$4,000.00**	**$0.00**	**$0.00**	**$21,120.00**
13 34 19.50 1100	Pre-Engineered Steel Building, 100' x 150' x 24'	15,000	SF Flr.	$0.00	$0.00	$0.00	$412,500.00	
13 34 19.50 6050	Framing for PESB door opening, 3' x 7'	4	Opng.	$0.00	$0.00	$0.00	$2,600.00	
13 34 19.50 6100	Framing for PESB door opening, 10' x 10'	8	Opng.	$0.00	$0.00	$0.00	$10,800.00	
13 34 19.50 6200	Framing for PESB window opening, 4' x 3'	6	Opng.	$0.00	$0.00	$0.00	$3,870.00	
13 34 19.50 5750	PESB door, 3' x 7', single leaf	4	Opng.	$2,940.00	$792.00	$0.00	$0.00	
13 34 19.50 7750	PESB sliding window, 4' x 3' with screen	6	Opng.	$2,910.00	$678.00	$68.10	$0.00	
13 34 19.50 6550	PESB gutter, eave type, 26 ga., painted	300	L.F.	$2,580.00	$927.00	$0.00	$0.00	
13 34 19.50 8650	PESB roof vent, 12" wide x 10' long	15	Ea.	$750.00	$3,705.00	$0.00	$0.00	
13 34 19.50 6900	PESB insulation, vinyl faced, 4" thick	27,400	S.F.	$14,522.00	$10,686.00	$0.00	$0.00	
Division 13	**Subtotal**			**$23,702.00**	**$16,788.00**	**$68.10**	**$429,770.00**	**$470,328.10**
			Subtotal	$103,841.50	$50,183.50	$1,070.06	$429,770.00	$584,865.06
Division 01	**General Requirements @ 7%**			**7,268.91**	**3,512.85**	**74.90**	**30,083.90**	
			Estimate Subtotal	$111,110.41	$53,696.35	$1,144.96	$459,853.90	$584,865.06
			Sales Tax @ 5%	5,555.52		57.25	11,496.35	
			Subtotal A	116,665.93	53,696.35	1,202.21	471,350.25	
			GC O & P	11,666.59	31,251.27	120.22	47,135.02	
			Subtotal B	128,332.52	84,947.62	1,322.43	518,485.27	$733,087.84
			Contingency @ 5%					36,654.39
			Subtotal C					$769,742.23
			Bond @ $12/1000 +10% O&P					10,160.60
			Subtotal D					$779,902.83
			Location Adjustment Factor		113.10			102,167.27
			Grand Total					**$882,070.10**

This estimate is based on an interactive spreadsheet. You are free to download it and adjust it to your methodology.
A copy of this spreadsheet is available at **RSMeans.com/2022books.**

Sample Estimate

This sample demonstrates the elements of an estimate, including a tally of the RSMeans data lines and a summary of the markups on a contractor's work to arrive at a total cost to the owner. The Location Factor with RSMeans data is added at the bottom of the estimate to adjust the cost of the work to a specific location.

---

❶ Work Performed

The body of the estimate shows the RSMeans data selected, including the line number, a brief description of each item, its take-off unit and quantity, and the bare costs of materials, labor, and equipment. This estimate also includes a column titled "SubContract." This data is taken from the column "Total Incl O&P" and represents the total that a subcontractor would charge a general contractor for the work, including the sub's markup for overhead and profit.

❷ Division 1, General Requirements

This is the first division numerically but the last division estimated. Division 1 includes project-wide needs provided by the general contractor. These requirements vary by project but may include temporary facilities and utilities, security, testing, project cleanup, etc. For small projects a percentage can be used—typically between 5% and 15% of project cost. For large projects the costs may be itemized and priced individually.

❸ Sales Tax

If the work is subject to state or local sales taxes, the amount must be added to the estimate. Sales tax may be added to material costs, equipment costs, and subcontracted work. In this case, sales tax was added in all three categories. It was assumed that approximately half the subcontracted work would be material cost, so the tax was applied to 50% of the subcontract total.

❹ GC O&P

This entry represents the general contractor's markup on material, labor, equipment, and subcontractor costs. Our standard markup on materials, equipment, and subcontracted work is 10%. In this estimate, the markup on the labor performed by the GC's workers uses "Skilled Workers Average" shown in Column F on the table "Installing Contractor's Overhead & Profit," which can be found on the inside back cover of the printed product or in the Reference Section of the electronic product.

❺ Contingency

A factor for contingency may be added to any estimate to represent the cost of unknowns that may occur between the time that the estimate is performed and the time the project is constructed. The amount of the allowance will depend on the stage of design at which the estimate is done and the contractor's assessment of the risk involved. Refer to section 01 21 16.50 for contingency allowances.

❻ Bonds

Bond costs should be added to the estimate. The figures here represent a typical performance bond, ensuring the owner that if the general contractor does not complete the obligations in the construction contract the bonding company will pay the cost for completion of the work.

❼ Location Adjustment

Published prices are based on national average costs. If necessary, adjust the total cost of the project using a location factor from the "Location Factor" table or the "City Cost Index" table. Use location factors if the work is general, covering multiple trades. If the work is by a single trade (e.g., masonry) use the more specific data found in the "City Cost Indexes."

Estimating Tips
01 20 00 Price and Payment Procedures

- Allowances that should be added to estimates to cover contingencies and job conditions that are not included in the national average material and labor costs are shown in Section 01 21.
- When estimating historic preservation projects (depending on the condition of the existing structure and the owner's requirements), a 15–20% contingency or allowance is recommended, regardless of the stage of the drawings.

01 30 00 Administrative Requirements

- Before determining a final cost estimate, it is good practice to review all the items listed in Subdivisions 01 31 and 01 32 to make final adjustments for items that may need customizing to specific job conditions.
- Requirements for initial and periodic submittals can represent a significant cost to the General Requirements of a job. Thoroughly check the submittal specifications when estimating a project to determine any costs that should be included.

01 40 00 Quality Requirements

- All projects will require some degree of quality control. This cost is not included in the unit cost of construction listed in each division. Depending upon the terms of the contract, the various costs of inspection and testing can be the responsibility of either the owner or the contractor. Be sure to include the required costs in your estimate.

01 50 00 Temporary Facilities and Controls

- Barricades, access roads, safety nets, scaffolding, security, and many more requirements for the execution of a safe project are elements of direct cost. These costs can easily be overlooked when preparing an estimate. When looking through the major classifications of this subdivision, determine which items apply to each division in your estimate.
- Construction equipment rental costs can be found in the Reference Section in Section 01 54 33. Operators' wages are not included in equipment rental costs.
- Equipment mobilization and demobilization costs are not included in equipment rental costs and must be considered separately.

- The cost of small tools provided by the installing contractor for his workers is covered in the "Overhead" column on the "Installing Contractor's Overhead and Profit" table that lists labor trades, base rates, and markups. Therefore, it is included in the "Total Incl. O&P" cost of any unit price line item.

01 70 00 Execution and Closeout Requirements

- When preparing an estimate, thoroughly read the specifications to determine the requirements for Contract Closeout. Final cleaning, record documentation, operation and maintenance data, warranties and bonds, and spare parts and maintenance materials can all be elements of cost for the completion of a contract. Do not overlook these in your estimate.

Reference Numbers

Reference numbers are shown at the beginning of some major classifications. These numbers refer to related items in the Reference Section. The reference information may be an estimating procedure, an alternate pricing method, or technical information.

Note: Not all subdivisions listed here necessarily appear. ∎

Same Data. Simplified.

Enjoy the convenience and efficiency of accessing your costs anywhere:

- **Skip the multiplier** by setting your location
- **Quickly search,** edit, favorite and share costs
- **Stay on top of price changes** with automatic updates

Discover more at rsmeans.com/online

01 11 Summary of Work

01 11 31 – Professional Consultants

01 11 31.10 Architectural Fees		Crew	Daily Output	Labor-Hours	Unit	Material	2022 Bare Costs Labor	Equipment	Total	Total Incl O&P
0010	**ARCHITECTURAL FEES**	R011110-10								
0020	For new construction									
0060	Minimum				Project				4.90%	4.90%
0090	Maximum								16%	16%
0100	For alteration work, to $500,000, minimum								7.35%	7.35%
0125	Maximum								24%	24%
0150	For alteration work, over $500,000, minumum								6.13%	6.13%
0175	Maximum				↓				20%	20%

01 11 31.20 Construction Management Fees										
0010	**CONSTRUCTION MANAGEMENT FEES**									
0060	For work to $100,000				Project				10%	10%
0070	To $250,000								9%	9%
0090	To $1,000,000								6%	6%
0100	To $5,000,000								5%	5%
0110	To $10,000,000				↓				4%	4%

01 11 31.30 Engineering Fees										
0010	**ENGINEERING FEES**	R011110-30								
0020	Educational planning consultant, minimum				Project				.50%	.50%
0100	Maximum				"				2.50%	2.50%
0400	Elevator & conveying systems, minimum				Contrct				2.50%	2.50%
0500	Maximum								5%	5%
1000	Mechanical (plumbing & HVAC), minimum								4.10%	4.10%
1100	Maximum				↓				10.10%	10.10%
1200	Structural, minimum				Project				1%	1%
1300	Maximum				"				2.50%	2.50%

01 11 31.75 Renderings										
0010	**RENDERINGS** Color, matted, 20" x 30", eye level,									
0020	1 building, minimum				Ea.	2,400			2,400	2,650
0050	Average					3,475			3,475	3,825
0100	Maximum					5,675			5,675	6,250
1000	5 buildings, minimum					4,200			4,200	4,600
1100	Maximum				↓	8,375			8,375	9,225

01 21 Allowances

01 21 16 – Contingency Allowances

01 21 16.50 Contingencies										
0010	**CONTINGENCIES**, Add to estimate									
0020	Conceptual stage				Project				20%	20%
0050	Schematic stage								15%	15%
0100	Preliminary working drawing stage (Design Dev.)				↓				10%	10%
0150	Final working drawing stage				Project				3%	3%

01 21 53 – Factors Allowance

01 21 53.50 Factors										
0010	**FACTORS** Cost adjustments	R012153-10								
0100	Add to construction costs for particular job requirements									
0500	Cut & patch to match existing construction, add, minimum				Costs	2%	3%			
0550	Maximum					5%	9%			
0800	Dust protection, add, minimum					1%	2%			
0850	Maximum					4%	11%			
1100	Equipment usage curtailment, add, minimum				↓	1%	1%			

10

01 21 53 – Factors Allowance

01 21 53.50 Factors		Crew	Daily Output	Labor-Hours	Unit	Material	2022 Bare Costs Labor	Equipment	Total	Total Incl O&P
1150	Maximum				Costs	3%	10%			
1400	Material handling & storage limitation, add, minimum					1%	1%			
1450	Maximum					6%	7%			
1700	Protection of existing work, add, minimum					2%	2%			
1750	Maximum					5%	7%			
2000	Shift work requirements, add, minimum						5%			
2050	Maximum						30%			
2300	Temporary shoring and bracing, add, minimum					2%	5%			
2350	Maximum				▼	5%	12%			

01 21 53.60 Security Factors

		Crew	Daily Output	Labor-Hours	Unit	Material	Labor	Equipment	Total	Total Incl O&P
0010	**SECURITY FACTORS** R012153-60									
0100	Additional costs due to security requirements									
0110	Daily search of personnel, supplies, equipment and vehicles									
0120	Physical search, inventory and doc of assets, at entry				Costs		30%			
0130	At entry and exit						50%			
0140	Physical search, at entry						6.25%			
0150	At entry and exit						12.50%			
0160	Electronic scan search, at entry						2%			
0170	At entry and exit						4%			
0180	Visual inspection only, at entry						.25%			
0190	At entry and exit						.50%			
0200	ID card or display sticker only, at entry						.12%			
0210	At entry and exit				▼		.25%			
0220	Day 1 as described below, then visual only for up to 5 day job duration									
0230	Physical search, inventory and doc of assets, at entry				Costs		5%			
0240	At entry and exit						10%			
0250	Physical search, at entry						1.25%			
0260	At entry and exit				▼		2.50%			
0270	Electronic scan search, at entry				Costs		.42%			
0280	At entry and exit				"		.83%			
0290	Day 1 as described below, then visual only for 6-10 day job duration									
0300	Physical search, inventory and doc of assets, at entry				Costs		2.50%			
0310	At entry and exit						5%			
0320	Physical search, at entry						.63%			
0330	At entry and exit						1.25%			
0340	Electronic scan search, at entry						.21%			
0350	At entry and exit				▼		.42%			
0360	Day 1 as described below, then visual only for 11-20 day job duration									
0370	Physical search, inventory and doc of assets, at entry				Costs		1.25%			
0380	At entry and exit						2.50%			
0390	Physical search, at entry						.31%			
0400	At entry and exit						.63%			
0410	Electronic scan search, at entry						.10%			
0420	At entry and exit				▼		.21%			
0430	Beyond 20 days, costs are negligible									
0440	Escort required to be with tradesperson during work effort				Costs		6.25%			

01 21 53.65 Infectious Disease Precautions

		Crew	Daily Output	Labor-Hours	Unit	Material	Labor	Equipment	Total	Total Incl O&P
0010	**INFECTIOUS DISEASE PRECAUTIONS** R012153-65									
0100	Additional costs due to infectious disease precautions									
0120	Daily temperature checks				Costs		1%			
0130	Donning & doffing masks and gloves						1%			
0140	Washing hands				▼		1%			

01 21 Allowances

01 21 53 – Factors Allowance

01 21 53.65 Infectious Disease Precautions	Crew	Daily Output	Labor-Hours	Unit	Material	2022 Bare Costs Labor	Equipment	Total	Total Incl O&P
0150	Informational meetings				Costs		2%			
0160	Maintaining social distance						1%			
0170	Disinfecting tools or equipment				↓		3%			
0200	N95 rated masks				Ea.	1.71			1.71	1.88
0210	Surgical masks				"	.30			.30	.33
0220	Black nitrile disposable gloves				Pr.	.36			.36	.39
0250	Hand washing station & 2 x weekly service				Week				158	174

01 21 55 – Job Conditions Allowance

01 21 55.50 Job Conditions

		Crew	Daily Output	Labor-Hours	Unit	Material	2022 Bare Costs Labor	Equipment	Total	Total Incl O&P
0010	**JOB CONDITIONS** Modifications to applicable									
0020	cost summaries									
0100	Economic conditions, favorable, deduct				Project				2%	2%
0200	Unfavorable, add								5%	5%
0300	Hoisting conditions, favorable, deduct				↓				2%	2%
0400	Unfavorable, add				Project				5%	5%
0500	General contractor management, experienced, deduct								2%	2%
0600	Inexperienced, add								10%	10%
0700	Labor availability, surplus, deduct								1%	1%
0800	Shortage, add								10%	10%
0900	Material storage area, available, deduct								1%	1%
1000	Not available, add								2%	2%
1100	Subcontractor availability, surplus, deduct								5%	5%
1200	Shortage, add								12%	12%
1300	Work space, available, deduct								2%	2%
1400	Not available, add				↓				5%	5%

01 21 63 – Taxes

01 21 63.10 Taxes

			Crew	Daily Output	Labor-Hours	Unit	Material	2022 Bare Costs Labor	Equipment	Total	Total Incl O&P
0010	**TAXES**	R012909-80									
0020	Sales tax, State, average					%	5.08%				
0050	Maximum	R012909-85					7.50%				
0200	Social Security, on first $118,500 of wages							7.65%			
0300	Unemployment, combined Federal and State, minimum	R012909-86						.60%			
0350	Average							9.60%			
0400	Maximum					↓		12%			

01 31 Project Management and Coordination

01 31 13 – Project Coordination

01 31 13.20 Field Personnel

		Crew	Daily Output	Labor-Hours	Unit	Material	2022 Bare Costs Labor	Equipment	Total	Total Incl O&P
0010	**FIELD PERSONNEL**									
0020	Clerk, average				Week		495		495	790
0160	General purpose laborer, average	1 Clab	.20	40			1,825		1,825	2,900
0180	Project manager, minimum						2,175		2,175	3,475
0200	Average						2,500		2,500	4,000
0220	Maximum						2,850		2,850	4,575
0240	Superintendent, minimum						2,125		2,125	3,400
0260	Average						2,325		2,325	3,725
0280	Maximum				↓		2,650		2,650	4,225

For customer support on your Commercial Renovation Costs with RSMeans data, call 800.448.8182.

01 31 13 – Project Coordination

01 31 13.30 Insurance

01 31 13.30 Insurance		Crew	Daily Output	Labor-Hours	Unit	Material	2022 Bare Costs Labor	Equipment	Total	Total Incl O&P
0010	**INSURANCE** R013113-40									
0020	Builders risk, standard, minimum				Job				.24%	.24%
0050	Maximum R013113-50								.80%	.80%
0200	All-risk type, minimum								.25%	.25%
0250	Maximum R013113-60				↓				.62%	.62%
0400	Contractor's equipment floater, minimum				Value				.50%	.50%
0450	Maximum				"				1.50%	1.50%
0600	Public liability, average				Job				2.02%	2.02%
0810	Workers' compensation & employer's liability									
2000	Range of 35 trades in 50 states, excl. wrecking & clerical, min.				Payroll		1.37%			
2100	Average						10.60%			
2200	Maximum				↓		120.29%			

01 31 13.40 Main Office Expense

01 31 13.40 Main Office Expense		Crew	Daily Output	Labor-Hours	Unit	Material	Labor	Equipment	Total	Total Incl O&P
0010	**MAIN OFFICE EXPENSE** Average for General Contractors									
0020	As a percentage of their annual volume									
0030	Annual volume to $300,000, minimum				% Vol.				20%	
0040	Maximum								30%	
0060	To $500,000, minimum								17%	
0070	Maximum								22%	
0080	To $1,000,000, minimum								16%	
0090	Maximum								19%	
0110	To $3,000,000, minimum								14%	
0120	Maximum								16%	
0130	To $5,000,000, minimum								8%	
0140	Maximum				↓				10%	

01 31 13.50 General Contractor's Mark-Up

01 31 13.50 General Contractor's Mark-Up		Crew	Daily Output	Labor-Hours	Unit	Material	Labor	Equipment	Total	Total Incl O&P
0010	**GENERAL CONTRACTOR'S MARK-UP** on Change Orders									
0200	Extra work, by subcontractors, add				%				10%	10%
0250	By General Contractor, add								15%	15%
0400	Omitted work, by subcontractors, deduct all but								5%	5%
0450	By General Contractor, deduct all but								7.50%	7.50%
0600	Overtime work, by subcontractors, add								15%	15%
0650	By General Contractor, add				↓				10%	10%

01 31 13.80 Overhead and Profit

01 31 13.80 Overhead and Profit		Crew	Daily Output	Labor-Hours	Unit	Material	Labor	Equipment	Total	Total Incl O&P
0010	**OVERHEAD & PROFIT** Allowance to add to items in this R013113-50									
0020	book that do not include Subs O&P, average				%				25%	
0100	Allowance to add to items in this book that R013113-55									
0110	do include Subs O&P, minimum				%				5%	5%
0150	Average								10%	10%
0200	Maximum								15%	15%
0290	Typical, by size of project, under $50,000								40%	
0310	$50,000 to $100,000								35%	
0320	$100,000 to $500,000								25%	
0330	$500,000 to $1,000,000				↓				20%	

01 31 13.90 Performance Bond

01 31 13.90 Performance Bond		Crew	Daily Output	Labor-Hours	Unit	Material	Labor	Equipment	Total	Total Incl O&P
0010	**PERFORMANCE BOND**									
0020	For buildings, minimum				Job				.60%	.60%
0100	Maximum				Job				2.50%	2.50%

For customer support on your Commercial Renovation Costs with RSMeans data, call 800.448.8182.

13

01 31 Project Management and Coordination

01 31 14 – Facilities Services Coordination

01 31 14.20 Lock Out/Tag Out

	01 31 14.20 Lock Out/Tag Out	Crew	Daily Output	Labor-Hours	Unit	Material	2022 Bare Costs Labor	Equipment	Total	Total Incl O&P
0010	**LOCK OUT/TAG OUT**									
0020	Miniature circuit breaker lock out device	1 Elec	220	.036	Ea.	27.50	2.41		29.91	33.50
0030	Miniature pin circuit breaker lock out device		220	.036		23	2.41		25.41	28.50
0040	Single circuit breaker lock out device		220	.036		24	2.41		26.41	29.50
0050	Multi-pole circuit breaker lock out device (15 to 225 Amp)		210	.038		22	2.52		24.52	28
0060	Large 3 pole circuit breaker lock out device (over 225 Amp)		210	.038		24.50	2.52		27.02	30.50
0080	Square D I-Line circuit breaker lock out device		210	.038		37	2.52		39.52	45
0090	Lock out disconnect switch, 30 to 100 Amp		330	.024		13.15	1.60		14.75	16.90
0100	100 to 400 Amp		330	.024		13.15	1.60		14.75	16.90
0110	Over 400 Amp		330	.024		13.15	1.60		14.75	16.90
0120	Lock out hasp for multiple lockout tags		200	.040		6.15	2.65		8.80	10.85
0130	Electrical cord plug lock out device		220	.036		11.55	2.41		13.96	16.40
0140	Electrical plug prong lock out device (3-wire grounding plug)		220	.036		7.65	2.41		10.06	12.10
0150	Wall switch lock out		200	.040		29	2.65		31.65	35.50
0160	Fire alarm pull station lock out	1 Stpi	200	.040		21.50	2.81		24.31	28.50
0170	Sprinkler valve tamper and flow switch lock out device	1 Skwk	220	.036		19.30	2.14		21.44	24.50
0180	Lock out sign		330	.024		20.50	1.43		21.93	25.50
0190	Lock out tag		440	.018		5.50	1.07		6.57	7.75

01 41 Regulatory Requirements

01 41 26 – Permit Requirements

01 41 26.50 Permits

	01 41 26.50 Permits	Crew	Daily Output	Labor-Hours	Unit	Material	2022 Bare Costs Labor	Equipment	Total	Total Incl O&P
0010	**PERMITS**									
0020	Rule of thumb, most cities, minimum				Job				.50%	.50%
0100	Maximum				"				2%	2%

01 45 Quality Control

01 45 23 – Testing and Inspecting Services

01 45 23.50 Testing

	01 45 23.50 Testing	Crew	Daily Output	Labor-Hours	Unit	Material	2022 Bare Costs Labor	Equipment	Total	Total Incl O&P
0010	**TESTING** and Inspecting Services									
1800	Compressive test, cylinder, delivered to lab, ASTM C 39				Ea.				36	40
1900	Picked up by lab, 30 minute round trip, add to above	1 Skwk	16	.500			29.50		29.50	46.50
1950	1 hour round trip, add to above		8	1			59		59	93
2000	2 hour round trip, add to above		4	2			118		118	186
2200	Compressive strength, cores (not incl. drilling), ASTM C 42								95	105
2300	Patching core holes, to 12" diameter	1 Cefi	22	.364		36	19.35		55.35	70
4730	Soil testing									
4750	Moisture content, ASTM D 2216				Ea.				18	20
4780	Permeability test, double ring infiltrometer								500	550
4800	Permeability, var. or constant head, undist., ASTM D 2434								264	290
4850	Recompacted								250	275
4900	Proctor compaction, 4" standard mold, ASTM D 698								230	253
4950	6" modified mold								110	120
5100	Shear tests, triaxial, minimum								410	450
5150	Maximum								545	600
5550	Technician for inspection, per day, earthwork								556	612
5650	Bolting								556	612
5750	Roofing								556	612
5790	Welding								556	612

01 45 Quality Control

01 45 23 – Testing and Inspecting Services

01 45 23.50 Testing	Crew	Daily Output	Labor-Hours	Unit	Material	2022 Bare Costs Labor	Equipment	Total	Total Incl O&P
5820	Non-destructive metal testing, dye penetrant				Day				556	612
5840	Magnetic particle								556	612
5860	Radiography								556	612
5880	Ultrasonic				▼				556	612
6000	Welding certification, minimum				Ea.				91	100
6100	Maximum				"				364	400
7000	Underground storage tank									
7520	Volumetric tightness, °= 12,000 gal.				Ea.	475			475	525
7530	12,000 – 29,999 gal.					595			595	655
7540	= 30,000 gal.				▼	675			675	740
7600	Vadose zone (soil gas) sampling, 10-40 samples, min.				Day				1,375	1,500
7610	Maximum				"				2,275	2,500
7700	Ground water monitoring incl. drilling 3 wells, min.				Total				4,550	5,000
7710	Maximum				"				6,375	7,000
8000	X-ray concrete slabs				Ea.				182	200

01 51 Temporary Utilities

01 51 13 – Temporary Electricity

01 51 13.80 Temporary Utilities	Crew	Daily Output	Labor-Hours	Unit	Material	2022 Bare Costs Labor	Equipment	Total	Total Incl O&P
0010	**TEMPORARY UTILITIES**									
0350	Lighting, lamps, wiring, outlets, 40,000 S.F. building, 8 strings	1 Elec	34	.235	CSF Flr	3.21	15.60		18.81	27.50
0360	16 strings	"	17	.471		6.40	31		37.40	55
0400	Power for temp lighting only, 6.6 KWH, per month								.92	1.01
0430	11.8 KWH, per month								1.65	1.82
0450	23.6 KWH, per month				▼				3.30	3.63
0600	Power for job duration incl. elevator, etc., minimum				CSF Flr				53	58
0650	Maximum				"				110	121
0675	Temporary cooling				Ea.	1,075			1,075	1,175

01 52 Construction Facilities

01 52 13 – Field Offices and Sheds

01 52 13.20 Office and Storage Space	Crew	Daily Output	Labor-Hours	Unit	Material	2022 Bare Costs Labor	Equipment	Total	Total Incl O&P
0010	**OFFICE AND STORAGE SPACE**									
0020	Office trailer, furnished, no hookups, 20' x 8', buy	2 Skwk	1	16	Ea.	10,200	940		11,140	12,700
0250	Rent per month					194			194	214
0300	32' x 8', buy	2 Skwk	.70	22.857		15,200	1,350		16,550	18,900
0350	Rent per month					247			247	271
0400	50' x 10', buy	2 Skwk	.60	26.667		30,900	1,575		32,475	36,500
0450	Rent per month					360			360	395
0500	50' x 12', buy	2 Skwk	.50	32		26,200	1,875		28,075	31,800
0550	Rent per month					460			460	505
0700	For air conditioning, rent per month, add				▼	53.50			53.50	59
0800	For delivery, add per mile				Mile	12.15			12.15	13.35
1200	Storage boxes, 20' x 8', buy	2 Skwk	1.80	8.889	Ea.	5,300	525		5,825	6,650
1250	Rent per month					119			119	131
1300	40' x 8', buy	2 Skwk	1.40	11.429		7,500	675		8,175	9,325
1350	Rent per month				▼	139			139	153

For customer support on your Commercial Renovation Costs with RSMeans data, call 800.448.8182.

15

01 54 Construction Aids

01 54 16 – Temporary Hoists

01 54 16.50 Weekly Forklift Crew

		Crew	Daily Output	Labor-Hours	Unit	Material	2022 Bare Costs Labor	Equipment	Total	Total Incl O&P
0010	**WEEKLY FORKLIFT CREW**									
0100	All-terrain forklift, 45' lift, 35' reach, 9000 lb. capacity	A-3P	.20	40	Week		2,275	3,150	5,425	7,000

01 54 19 – Temporary Cranes

01 54 19.50 Daily Crane Crews

		Crew	Daily Output	Labor-Hours	Unit	Material	Labor	Equipment	Total	Total Incl O&P
0010	**DAILY CRANE CREWS** for small jobs, portal to portal									
0100	12-ton truck-mounted hydraulic crane	A-3H	1	8	Day		505	1,575	2,080	2,525
0200	25-ton	A-3I	1	8			505	1,700	2,205	2,650
0300	40-ton	A-3J	1	8			505	1,825	2,330	2,800
0400	55-ton	A-3K	1	16			940	2,025	2,965	3,700
0500	80-ton	A-3L	1	16	↓		940	2,175	3,115	3,875
0900	If crane is needed on a Saturday, Sunday or Holiday									
0910	At time-and-a-half, add				Day		50%			
0920	At double time, add				"		100%			

01 54 19.60 Monthly Tower Crane Crew

		Crew	Daily Output	Labor-Hours	Unit	Material	Labor	Equipment	Total	Total Incl O&P
0010	**MONTHLY TOWER CRANE CREW**, excludes concrete footing									
0100	Static tower crane, 130' high, 106' jib, 6200 lb. capacity	A-3N	.05	176	Month		11,200	39,300	50,500	60,500

01 54 23 – Temporary Scaffolding and Platforms

01 54 23.60 Pump Staging

		Crew	Daily Output	Labor-Hours	Unit	Material	Labor	Equipment	Total	Total Incl O&P
0010	**PUMP STAGING**, Aluminum R015423-20									
0200	24' long pole section, buy				Ea.	395			395	435
0300	18' long pole section, buy					315			315	345
0400	12' long pole section, buy					200			200	220
0500	6' long pole section, buy					132			132	146
0600	6' long splice joint section, buy					78.50			78.50	86
0700	Pump jack, buy					179			179	196
0900	Foldable brace, buy					65.50			65.50	72
1000	Workbench/back safety rail support, buy					75.50			75.50	83
1100	Scaffolding planks/workbench, 14" wide x 24' long, buy					625			625	685
1200	Plank end safety rail, buy					99.50			99.50	109
1250	Safety net, 22' long, buy				↓	435			435	480
1300	System in place, 50' working height, per use based on 50 uses	2 Carp	84.80	.189	C.S.F.	5.95	10.60		16.55	23.50
1400	100 uses		84.80	.189		2.98	10.60		13.58	20
1500	150 uses	↓	84.80	.189	↓	2	10.60		12.60	19.10

01 54 23.70 Scaffolding

		Crew	Daily Output	Labor-Hours	Unit	Material	Labor	Equipment	Total	Total Incl O&P
0010	**SCAFFOLDING** R015423-10									
0015	Steel tube, regular, no plank, labor only to erect & dismantle									
0090	Building exterior, wall face, 1 to 5 stories, 6'-4" x 5' frames	3 Carp	8	3	C.S.F.		169		169	268
0200	6 to 12 stories	4 Carp	8	4			225		225	360
0301	13 to 20 stories	5 Clab	8	5			228		228	360
0460	Building interior, wall face area, up to 16' high	3 Carp	12	2			113		113	179
0560	16' to 40' high		10	2.400	↓		135		135	215
0800	Building interior floor area, up to 30' high	↓	150	.160	C.C.F.		9		9	14.30
0900	Over 30' high	4 Carp	160	.200	"		11.25		11.25	17.90
0906	Complete system for face of walls, no plank, material only rent/mo				C.S.F.	83			83	91.50
0908	Interior spaces, no plank, material only rent/mo				C.C.F.	5.50			5.50	6.05
0910	Steel tubular, heavy duty shoring, buy									
0920	Frames 5' high 2' wide				Ea.	60			60	66.50
0925	5' high 4' wide					71.50			71.50	78.50
0930	6' high 2' wide					72			72	79.50
0935	6' high 4' wide				↓	86.50			86.50	95
0940	Accessories									

01 54 23.70 Scaffolding		Crew	Daily Output	Labor-Hours	Unit	Material	2022 Bare Costs Labor	Equipment	Total	Total Incl O&P
0945	Cross braces				Ea.	21.50			21.50	24
0950	U-head, 8" x 8"					13.75			13.75	15.15
0955	J-head, 4" x 8"				▼	16.90			16.90	18.60
0960	Base plate, 8" x 8"				Ea.	18.30			18.30	20
0965	Leveling jack				"	25.50			25.50	28
1000	Steel tubular, regular, buy									
1100	Frames 3' high 5' wide				Ea.	66.50			66.50	73
1150	5' high 5' wide					79			79	87
1200	6'-4" high 5' wide					88			88	97
1350	7'-6" high 6' wide					158			158	174
1500	Accessories, cross braces					21			21	23
1550	Guardrail post					24.50			24.50	27
1600	Guardrail 7' section					10.55			10.55	11.60
1650	Screw jacks & plates					26			26	29
1700	Sidearm brackets					35			35	38.50
1750	8" casters					45			45	49.50
1800	Plank 2" x 10" x 16'-0"					67.50			67.50	74.50
1900	Stairway section					290			290	320
1910	Stairway starter bar					33			33	36
1920	Stairway inside handrail					56			56	61.50
1930	Stairway outside handrail					93			93	102
1940	Walk-thru frame guardrail				▼	42.50			42.50	47
2000	Steel tubular, regular, rent/mo.									
2100	Frames 3' high 5' wide				Ea.	10.05			10.05	11.05
2150	5' high 5' wide					13.90			13.90	15.30
2200	6'-4" high 5' wide					5.85			5.85	6.40
2250	7'-6" high 6' wide					10.05			10.05	11.05
2500	Accessories, cross braces					3.35			3.35	3.69
2550	Guardrail post					3.30			3.30	3.63
2600	Guardrail 7' section					11.25			11.25	12.40
2650	Screw jacks & plates					4.35			4.35	4.79
2700	Sidearm brackets					8.70			8.70	9.55
2750	8" casters					7.85			7.85	8.65
2800	Outrigger for rolling tower					11.10			11.10	12.20
2850	Plank 2" x 10" x 16'-0"					10.05			10.05	11.05
2900	Stairway section					32.50			32.50	36
2940	Walk-thru frame guardrail				▼	8.35			8.35	9.20
3000	Steel tubular, heavy duty shoring, rent/mo.									
3250	5' high 2' & 4' wide				Ea.	8.55			8.55	9.40
3300	6' high 2' & 4' wide				"	8.55			8.55	9.40
3500	Accessories, cross braces				Ea.	.92			.92	1.01
3600	U-head, 8" x 8"					2.52			2.52	2.77
3650	J-head, 4" x 8"					2.52			2.52	2.77
3700	Base plate, 8" x 8"					.92			.92	1.01
3750	Leveling jack					2.50			2.50	2.75
5700	Planks, 2" x 10" x 16'-0", labor only to erect & remove to 50' H	3 Carp	72	.333			18.75		18.75	30
5800	Over 50' high	4 Carp	80	.400	▼		22.50		22.50	36

01 54 23 – Temporary Scaffolding and Platforms

01 54 23.75 Scaffolding Specialties	Crew	Daily Output	Labor-Hours	Unit	Material	2022 Bare Costs Labor	2022 Bare Costs Equipment	Total	Total Incl O&P
0010 **SCAFFOLDING SPECIALTIES**									
1200 Sidewalk bridge, heavy duty steel posts & beams, including									
1210 parapet protection & waterproofing (material cost is rent/month)									
1220 8' to 10' wide, 2 posts	3 Carp	15	1.600	L.F.	49.50	90		139.50	198
1230 3 posts	"	10	2.400	"	72.50	135		207.50	295
1500 Sidewalk bridge using tubular steel scaffold frames including									
1510 planking (material cost is rent/month)	3 Carp	45	.533	L.F.	9.90	30		39.90	58.50
1600 For 2 uses per month, deduct from all above					50%				
1700 For 1 use every 2 months, add to all above					100%				
1900 Catwalks, 20" wide, no guardrails, 7' span, buy				Ea.	140			140	154
2000 10' span, buy					201			201	221
3720 Putlog, standard, 8' span, with hangers, buy					68.50			68.50	75.50
3730 Rent per month					14.25			14.25	15.70
3750 12' span, buy					88			88	96.50
3755 Rent per month					17.80			17.80	19.55
3760 Trussed type, 16' span, buy					268			268	295
3770 Rent per month					21.50			21.50	23.50
3790 22' span, buy					320			320	350
3795 Rent per month					28.50			28.50	31
4000 7 step					670			670	740
4100 Rolling towers, buy, 5' wide, 7' long, 10' high					1,275			1,275	1,425
4200 For additional 5' high sections, to buy					201			201	221
4300 Complete incl. wheels, railings, outriggers,									
4350 21' high, to buy				Ea.	2,175			2,175	2,375
4400 Rent/month = 5% of purchase cost				"	109			109	119
5000 Motorized work platform, mast climber									
5050 Base unit, 50' W, less than 100' tall, rent/mo				Ea.	3,250			3,250	3,575
5100 Less than 200' tall, rent/mo					3,275			3,275	3,600
5150 Less than 300' tall, rent/mo					4,000			4,000	4,400
5200 Less than 400' tall, rent/mo				Ea.	4,575			4,575	5,050
5250 Set up and demob, per unit, less than 100' tall	B-68F	16.60	1.446	C.S.F.		86	24.50	110.50	163
5300 Less than 200' tall		25	.960			57	16.15	73.15	108
5350 Less than 300' tall		30	.800			47.50	13.45	60.95	90.50
5400 Less than 400' tall		33	.727			43.50	12.25	55.75	82
5500 Mobilization (price includes freight in and out) per unit	B-34N	1	16	Ea.		910	705	1,615	2,200

01 54 23.80 Staging Aids

01 54 23.80 Staging Aids	Crew	Daily Output	Labor-Hours	Unit	Material	2022 Bare Costs Labor	2022 Bare Costs Equipment	Total	Total Incl O&P
0010 **STAGING AIDS** and fall protection equipment									
0100 Sidewall staging bracket, tubular, buy				Ea.	74.50			74.50	82
0110 Cost each per day, based on 250 days use				Day	.30			.30	.33
0200 Guard post, buy				Ea.	56.50			56.50	62
0210 Cost each per day, based on 250 days use				Day	.23			.23	.25
0300 End guard chains, buy per pair				Pr.	45			45	49.50
0310 Cost per set per day, based on 250 days use				Day	.30			.30	.33
1010 Cost each per day, based on 250 days use				"	.05			.05	.06
1100 Wood bracket, buy				Ea.	37			37	40.50
1110 Cost each per day, based on 250 days use				Day	.15			.15	.16
2010 Cost per pair per day, based on 250 days use					.64			.64	.70
3010 Cost each per day, based on 250 days use					.23			.23	.25
3100 Aluminum scaffolding plank, 20" wide x 24' long, buy				Ea.	885			885	975
3110 Cost each per day, based on 250 days use				Day	3.54			3.54	3.90
4000 Nylon full body harness, lanyard and rope grab				Ea.	182			182	200
4010 Cost each per day, based on 250 days use				Day	.73			.73	.80

01 54 Construction Aids

01 54 23 – Temporary Scaffolding and Platforms

01 54 23.80 Staging Aids	Crew	Daily Output	Labor-Hours	Unit	Material	2022 Bare Costs Labor	Equipment	Total	Total Incl O&P
4100 Rope for safety line, 5/8" x 100' nylon, buy				Ea.	56			56	61.50
4110 Cost each per day, based on 250 days use				Day	.22			.22	.25
4200 Permanent U-Bolt roof anchor, buy				Ea.	87			87	95.50
4300 Temporary (one use) roof ridge anchor, buy				"	7.30			7.30	8.05
5000 Installation (setup and removal) of staging aids									
5010 Sidewall staging bracket	2 Carp	64	.250	Ea.		14.10		14.10	22.50
5020 Guard post with 2 wood rails	"	64	.250			14.10		14.10	22.50
5030 End guard chains, set	1 Carp	64	.125			7.05		7.05	11.20
5100 Roof shingling bracket		96	.083			4.69		4.69	7.45
5200 Ladder jack		64	.125			7.05		7.05	11.20
5300 Wood plank, 2" x 10" x 16'	2 Carp	80	.200			11.25		11.25	17.90
5310 Aluminum scaffold plank, 20" x 24'	"	40	.400			22.50		22.50	36
5410 Safety rope	1 Carp	40	.200			11.25		11.25	17.90
5420 Permanent U-Bolt roof anchor (install only)	2 Carp	40	.400	Ea.		22.50		22.50	36
5430 Temporary roof ridge anchor (install only)	1 Carp	64	.125	"		7.05		7.05	11.20

01 54 26 – Temporary Swing Staging

01 54 26.50 Swing Staging	Crew	Daily Output	Labor-Hours	Unit	Material	2022 Bare Costs Labor	Equipment	Total	Total Incl O&P
0010 SWING STAGING, 500 lb. cap., 2' wide to 24' long, hand operated									
0020 Steel cable type, with 60' cables, buy				Ea.	6,750			6,750	7,400
0030 Rent per month				"	675			675	740
0600 Lightweight (not for masons) 24' long for 150' height,									
0610 manual type, buy				Ea.	13,900			13,900	15,200
0620 Rent per month					1,375			1,375	1,525
0700 Powered, electric or air, to 150' high, buy					35,800			35,800	39,400
0710 Rent per month					2,500			2,500	2,750
0780 To 300' high, buy					36,400			36,400	40,100
0800 Rent per month					2,550			2,550	2,800
1000 Bosun's chair or work basket 3' x 3'6", to 300' high, electric, buy					14,300			14,300	15,700
1010 Rent per month					1,000			1,000	1,100
2200 Move swing staging (setup and remove)	E-4	2	16	Move		1,000	74	1,074	1,675

01 54 36 – Equipment Mobilization

01 54 36.50 Mobilization	Crew	Daily Output	Labor-Hours	Unit	Material	2022 Bare Costs Labor	Equipment	Total	Total Incl O&P
0010 MOBILIZATION (Use line item again for demobilization) R015436-50									
0015 Up to 25 mi. haul dist. (50 mi. RT for mob/demob crew)									
1200 Small equipment, placed in rear of, or towed by pickup truck	A-3A	4	2	Ea.		114	45	159	228
1300 Equipment hauled on 3-ton capacity towed trailer	A-3Q	2.67	3			171	95	266	370
1400 20-ton capacity	B-34U	2	8			440	230	670	950
1500 40-ton capacity	B-34N	2	8			455	355	810	1,100
1600 50-ton capacity	B-34V	1	24			1,400	1,125	2,525	3,425
1700 Crane, truck-mounted, up to 75 ton (driver only)	1 Eqhv	4	2			127		127	198
1800 Over 75 ton (with chase vehicle)	A-3E	2.50	6.400			375	72	447	665
2400 Crane, large lattice boom, requiring assembly	B-34W	.50	144			8,025	7,950	15,975	21,500
2500 For each additional 5 miles haul distance, add						10%	10%		
3000 For large pieces of equipment, allow for assembly/knockdown									
3100 For mob/demob of micro-tunneling equip, see Section 33 05 07.36									
3200 For mob/demob of pile driving equip, see Section 31 62 19.10									

For customer support on your Commercial Renovation Costs with RSMeans data, call 800.448.8182.

19

01 55 Vehicular Access and Parking

01 55 23 – Temporary Roads

01 55 23.50 Roads and Sidewalks		Crew	Daily Output	Labor-Hours	Unit	Material	2022 Bare Costs Labor	Equipment	Total	Total Incl O&P
0010	**ROADS AND SIDEWALKS** Temporary									
0050	Roads, gravel fill, no surfacing, 4" gravel depth	B-14	715	.067	S.Y.	4.96	3.21	.46	8.63	11.05
0100	8" gravel depth	B-14	615	.078	S.Y.	9.90	3.73	.53	14.16	17.40

01 56 Temporary Barriers and Enclosures

01 56 13 – Temporary Air Barriers

01 56 13.60 Tarpaulins		Crew	Daily Output	Labor-Hours	Unit	Material	2022 Bare Costs Labor	Equipment	Total	Total Incl O&P
0010	**TARPAULINS**									
0020	Cotton duck, 10-13.13 oz./S.Y., 6' x 8'				S.F.	.73			.73	.80
0050	30' x 30'					.34			.34	.37
0100	Polyvinyl coated nylon, 14-18 oz., minimum					1.12			1.12	1.23
0150	Maximum					1.18			1.18	1.30
0200	Reinforced polyethylene 3 mils thick, white					.05			.05	.06
0300	4 mils thick, white, clear or black					.08			.08	.09
0400	5.5 mils thick, clear					.08			.08	.09
0500	White, fire retardant					.48			.48	.53
0600	12 mils, oil resistant, fire retardant					.39			.39	.43
0700	8.5 mils, black					.31			.31	.34
0720	Steel reinforced polyethylene, 20 mils thick					.43			.43	.47
0730	Polyester reinforced w/integral fastening system, 11 mils thick					.16			.16	.18
0740	Polyethylene, reflective, 23 mils thick					1.03			1.03	1.13

01 56 13.90 Winter Protection		Crew	Daily Output	Labor-Hours	Unit	Material	2022 Bare Costs Labor	Equipment	Total	Total Incl O&P
0010	**WINTER PROTECTION**									
0100	Framing to close openings	2 Clab	500	.032	S.F.	.71	1.46		2.17	3.10
0200	Tarpaulins hung over scaffolding, 8 uses, not incl. scaffolding		1500	.011		.19	.49		.68	.98
0250	Tarpaulin polyester reinf. w/integral fastening system, 11 mils thick		1600	.010		.19	.46		.65	.93
0300	Prefab fiberglass panels, steel frame, 8 uses		1200	.013		3.14	.61		3.75	4.42

01 56 16 – Temporary Dust Barriers

01 56 16.10 Dust Barriers, Temporary		Crew	Daily Output	Labor-Hours	Unit	Material	2022 Bare Costs Labor	Equipment	Total	Total Incl O&P
0010	**DUST BARRIERS, TEMPORARY**, erect and dismantle									
0020	Spring loaded telescoping pole & head, to 12', erect and dismantle	1 Clab	240	.033	Ea.		1.52		1.52	2.42
0025	Cost per day (based upon 250 days)				Day	.32			.32	.35
0030	To 21', erect and dismantle	1 Clab	240	.033	Ea.		1.52		1.52	2.42
0035	Cost per day (based upon 250 days)				Day	.75			.75	.82
0040	Accessories, caution tape reel, erect and dismantle	1 Clab	480	.017	Ea.		.76		.76	1.21
0045	Cost per day (based upon 250 days)				Day	.35			.35	.38
0060	Foam rail and connector, erect and dismantle	1 Clab	240	.033	Ea.		1.52		1.52	2.42
0065	Cost per day (based upon 250 days)				Day	.10			.10	.11
0070	Caution tape	1 Clab	384	.021	C.L.F.	2.61	.95		3.56	4.38
0080	Zipper, standard duty		60	.133	Ea.	11	6.10		17.10	22
0090	Heavy duty		48	.167	"	11.10	7.60		18.70	24.50
0100	Polyethylene sheet, 4 mil	1 Clab	37	.216	Sq.	4.40	9.85		14.25	20.50
0110	6 mil	"	37	.216	"	6	9.85		15.85	22.50
1000	Dust partition, 6 mil polyethylene, 1" x 3" frame	2 Carp	2000	.008	S.F.	.37	.45		.82	1.12
1080	2" x 4" frame	"	2000	.008	"	.55	.45		1	1.32
1085	Negative air machine, 1800 CFM				Ea.	900			900	990
1090	Adhesive strip application, 2" width	1 Clab	192	.042	C.L.F.	10.95	1.90		12.85	15.05
4000	Dust & infectious control partition, adj. to 10' high, obscured, 4' panel	2 Carp	90	.178	Ea.	580	10		590	655
4010	3' panel		90	.178		550	10		560	620
4020	2' panel		90	.178		430	10		440	490

01 56 Temporary Barriers and Enclosures

01 56 16 – Temporary Dust Barriers

01 56 16.10 Dust Barriers, Temporary

		Crew	Daily Output	Labor-Hours	Unit	Material	2022 Bare Costs Labor	Equipment	Total	Total Incl O&P
4030	1' panel	1 Carp	90	.089	Ea.	285	5		290	325
4040	6" panel	"	90	.089		265	5		270	300
4050	2' panel with HEPA filtered discharge port	2 Carp	90	.178		500	10		510	565
4060	3' panel with 32" door		90	.178		895	10		905	1,000
4070	4' panel with 36" door		90	.178		995	10		1,005	1,125
4080	4'-6" panel with 44" door		90	.178		1,200	10		1,210	1,350
4090	Hinged corner		80	.200		185	11.25		196.25	222
4100	Outside corner		80	.200		150	11.25		161.25	183
4110	T post		80	.200		150	11.25		161.25	183
4120	Accessories, ceiling grid clip		360	.044		7.45	2.50		9.95	12.20
4130	Panel locking clip		360	.044		5.15	2.50		7.65	9.65
4140	Panel joint closure strip		360	.044		8	2.50		10.50	12.80
4150	Screw jack		360	.044		6.65	2.50		9.15	11.30
4160	Digital pressure difference gauge					275			275	305
4180	Combination lockset	1 Carp	13	.615		205	34.50		239.50	280
4185	Sealant tape, 2" wide	1 Clab	192	.042	C.L.F.	10.95	1.90		12.85	15.05
4190	System in place, including door and accessories									
4200	Based upon 25 uses	2 Carp	51	.314	L.F.	9.85	17.65		27.50	39
4210	Based upon 50 uses		51	.314		4.94	17.65		22.59	33.50
4230	Based upon 100 uses		51	.314		2.47	17.65		20.12	30.50

01 56 23 – Temporary Barricades

01 56 23.10 Barricades

		Crew	Daily Output	Labor-Hours	Unit	Material	2022 Bare Costs Labor	Equipment	Total	Total Incl O&P
0010	**BARRICADES**									
0020	5' high, 3 rail @ 2" x 8", fixed	2 Carp	20	.800	L.F.	8.75	45		53.75	81
0150	Movable	"	30	.533	"	7.45	30		37.45	55.50
0300	Stock units, 58" high, 8' wide, reflective, buy				Ea.	199			199	218
0350	With reflective tape, buy				"	242			242	266
0400	Break-a-way 3" PVC pipe barricade									
0410	with 3 ea. 1' x 4' reflectorized panels, buy				Ea.	175			175	192
0500	Barricades, plastic, 8" x 24" wide, foldable				Ea.	95			95	104
0800	Traffic cones, PVC, 18" high					7.75			7.75	8.55
0850	28" high					13.75			13.75	15.15
0900	Barrels, 55 gal., with flasher	1 Clab	96	.083		93	3.80		96.80	108
1000	Guardrail, wooden, 3' high, 1" x 6" on 2" x 4" posts	2 Carp	200	.080	L.F.	1.82	4.50		6.32	9.15
1100	2" x 6" on 4" x 4" posts	"	165	.097		3.37	5.45		8.82	12.35
1200	Portable metal with base pads, buy					13.50			13.50	14.85
1250	Typical installation, assume 10 reuses	2 Carp	600	.027		2.25	1.50		3.75	4.87
1300	Barricade tape, polyethylene, 7 mil, 3" wide x 300' long roll	1 Clab	128	.063	Ea.	7.80	2.85		10.65	13.15
3000	Detour signs, set up and remove									
3010	Reflective aluminum, MUTCD, 24" x 24", post mounted	1 Clab	20	.400	Ea.	3.37	18.25		21.62	32.50
4000	Roof edge portable barrier stands and warning flags, 50 uses	1 Rohe	9100	.001	L.F.	.08	.03		.11	.14
4010	100 uses	"	9100	.001	"	.04	.03		.07	.10

01 56 26 – Temporary Fencing

01 56 26.50 Temporary Fencing

		Crew	Daily Output	Labor-Hours	Unit	Material	2022 Bare Costs Labor	Equipment	Total	Total Incl O&P
0010	**TEMPORARY FENCING**									
0020	Chain link, 11 ga., 4' high	2 Clab	400	.040	L.F.	2.40	1.82		4.22	5.55
0100	6' high		300	.053		3.79	2.43		6.22	8.05
0200	Rented chain link, 6' high, to 1000' (up to 12 mo.)		400	.040		2.65	1.82		4.47	5.80
0250	Over 1000' (up to 12 mo.)		300	.053		3.13	2.43		5.56	7.30
0350	Plywood, painted, 2" x 4" frame, 4' high	A-4	135	.178		7.90	9.50		17.40	24
0400	4" x 4" frame, 8' high	"	110	.218		15.15	11.65		26.80	35
0500	Wire mesh on 4" x 4" posts, 4' high	2 Carp	100	.160		12.80	9		21.80	28.50

For customer support on your Commercial Renovation Costs with RSMeans data, call 800.448.8182.

21

01 56 Temporary Barriers and Enclosures

01 56 26 – Temporary Fencing

01 56 26.50 Temporary Fencing	Crew	Daily Output	Labor-Hours	Unit	Material	2022 Bare Costs Labor	Equipment	Total	Total Incl O&P
0550	8' high	2 Carp	80	.200	L.F.	19.20	11.25		30.45	39

01 56 29 – Temporary Protective Walkways

01 56 29.50 Protection

		Crew	Daily Output	Labor-Hours	Unit	Material	Labor	Equipment	Total	Total Incl O&P
0010	**PROTECTION**									
0020	Stair tread, 2" x 12" planks, 1 use	1 Carp	75	.107	Tread	8.25	6		14.25	18.60
0100	Exterior plywood, 1/2" thick, 1 use		65	.123		2.28	6.95		9.23	13.50
0200	3/4" thick, 1 use		60	.133		2.83	7.50		10.33	15.05
2200	Sidewalks, 2" x 12" planks, 2 uses		350	.023	S.F.	1.37	1.29		2.66	3.55
2300	Exterior plywood, 2 uses, 1/2" thick		750	.011		.38	.60		.98	1.37
2400	5/8" thick		650	.012		.41	.69		1.10	1.55
2500	3/4" thick		600	.013		.47	.75		1.22	1.71

01 57 Temporary Controls

01 57 33 – Temporary Security

01 57 33.50 Watchman

		Crew	Daily Output	Labor-Hours	Unit	Material	Labor	Equipment	Total	Total Incl O&P
0010	**WATCHMAN**									
0020	Service, monthly basis, uniformed person, minimum				Hr.				27.65	30.40
0100	Maximum								56	61.62
0200	Person and command dog, minimum								31	34
0300	Maximum								60	65

01 58 Project Identification

01 58 13 – Temporary Project Signage

01 58 13.50 Signs

		Crew	Daily Output	Labor-Hours	Unit	Material	Labor	Equipment	Total	Total Incl O&P
0010	**SIGNS**									
0020	High intensity reflectorized, no posts, buy				Ea.	13.30			13.30	14.65

01 66 Product Storage and Handling Requirements

01 66 19 – Material Handling

01 66 19.10 Material Handling

		Crew	Daily Output	Labor-Hours	Unit	Material	Labor	Equipment	Total	Total Incl O&P
0010	**MATERIAL HANDLING**									
0020	Above 2nd story, via stairs, per C.Y. of material per floor	2 Clab	145	.110	C.Y.		5.05		5.05	8
0030	Via elevator, per C.Y. of material		240	.067			3.04		3.04	4.83
0050	Distances greater than 200', per C.Y. of material per each addl 200'		300	.053			2.43		2.43	3.86

01 71 Examination and Preparation

01 71 23 – Field Engineering

01 71 23.13 Construction Layout

		Crew	Daily Output	Labor-Hours	Unit	Material	Labor	Equipment	Total	Total Incl O&P
0010	**CONSTRUCTION LAYOUT**									
1100	Crew for layout of building, trenching or pipe laying, 2 person crew	A-6	1	16	Day		905	36.50	941.50	1,475
1200	3 person crew	A-7	1	24			1,475	36.50	1,511.50	2,350
1400	Crew for roadway layout, 4 person crew	A-8	1	32			1,900	36.50	1,936.50	3,025

01 74 Cleaning and Waste Management

01 74 13 – Progress Cleaning

01 74 13.20 Cleaning Up	Crew	Daily Output	Labor-Hours	Unit	Material	2022 Bare Costs Labor	2022 Bare Costs Equipment	Total	Total Incl O&P
0010 **CLEANING UP**									
0020 After job completion, allow, minimum				Job				.30%	.30%
0040 Maximum				"				1%	1%
0052 Cleanup of floor area, continuous, per day, during const.	A-5	16	1.125	M.S.F.	3.30	52	5.30	60.60	92
0100 Final by GC at end of job	"	11.50	1.565	"	3.49	72	7.40	82.89	127

01 76 Protecting Installed Construction

01 76 13 – Temporary Protection of Installed Construction

01 76 13.20 Temporary Protection	Crew	Daily Output	Labor-Hours	Unit	Material	2022 Bare Costs Labor	2022 Bare Costs Equipment	Total	Total Incl O&P
0010 **TEMPORARY PROTECTION**									
0020 Flooring, 1/8" tempered hardboard, taped seams	2 Carp	1500	.011	S.F.	.32	.60		.92	1.30
0030 Peel away carpet protection	1 Clab	3200	.003	"	.11	.11		.22	.30

01 91 Commissioning

01 91 13 – General Commissioning Requirements

01 91 13.50 Building Commissioning	Crew	Daily Output	Labor-Hours	Unit	Material	2022 Bare Costs Labor	2022 Bare Costs Equipment	Total	Total Incl O&P
0010 **BUILDING COMMISSIONING**									
0100 Systems operation and verification during turnover				%				.25%	.25%
0150 Including all systems subcontractors								.50%	.50%
0200 Systems design assistance, operation, verification and training								.50%	.50%
0250 Including all systems subcontractors								1%	1%

01 93 Facility Maintenance

01 93 09 – Facility Equipment

01 93 09.50 Moving Equipment	Crew	Daily Output	Labor-Hours	Unit	Material	2022 Bare Costs Labor	2022 Bare Costs Equipment	Total	Total Incl O&P
0010 **MOVING EQUIPMENT**, Remove and reset, 100' distance,									
0020 No obstructions, no assembly or leveling unless noted									
0100 Annealing furnace, 24' overall	B-67	4	4	Ea.		235	83	318	455
0200 Annealing oven, small		14	1.143			67	23.50	90.50	130
0240 Very large		1	16			940	330	1,270	1,825
0400 Band saw, small		12	1.333			78.50	27.50	106	153
0440 Large		8	2			118	41.50	159.50	228
0500 Blue print copy machine		7	2.286			134	47.50	181.50	260
0600 Bonding mill, 6"		7	2.286			134	47.50	181.50	260
0620 12"		6	2.667			157	55.50	212.50	305
0640 18"		4	4			235	83	318	455
0660 24"		2	8			470	166	636	915
0700 Boring machine (jig)	B-68	7	3.429			204	47.50	251.50	365
0800 Bridgeport mill, standard	B-67	14	1.143			67	23.50	90.50	130
1000 Calibrator, 6 unit	"	14	1.143			67	23.50	90.50	130
1100 Comparitor, bench top	2 Clab	14	1.143			52		52	83
1140 Floor mounted	B-67	7	2.286			134	47.50	181.50	260
1200 Computer, desk top	2 Clab	25	.640			29		29	46.50
1300 Copy machine	"	25	.640			29		29	46.50
1500 Deflasher	B-67	14	1.143			67	23.50	90.50	130
1600 Degreaser, small		14	1.143			67	23.50	90.50	130
1640 Large 24' overall		1	16			940	330	1,270	1,825

For customer support on your Commercial Renovation Costs with RSMeans data, call 800.448.8182.

23

01 93 09.50 Moving Equipment		Crew	Daily Output	Labor-Hours	Unit	Material	2022 Bare Costs		Total	Total Incl O&P
							Labor	Equipment		
1700	Desk with chair	2 Clab	25	.640	Ea.		29		29	46.50
1725	Desk only	"	22	.727			33		33	52.50
1750	Chair only	1 Clab	300	.027			1.22		1.22	1.93
1800	Dial press	B-67	7	2.286			134	47.50	181.50	260
1900	Drafting table	2 Clab	14	1.143			52		52	83
2000	Drill press, bench top	"	14	1.143			52		52	83
2040	Floor mounted	B-67	14	1.143			67	23.50	90.50	130
2080	Industrial radial	"	7	2.286			134	47.50	181.50	260
2100	Dust collector, portable	2 Clab	25	.640			29		29	46.50
2140	Stationary, small	B-67	7	2.286			134	47.50	181.50	260
2180	Stationary, large	"	2	8			470	166	636	915
2300	Electric discharge machine	B-68	7	3.429			204	47.50	251.50	365
2400	Environmental chamber walls, including assembly	4 Clab	18	1.778	L.F.		81		81	129
2600	File cabinet	2 Clab	25	.640	Ea.		29		29	46.50
2800	Grinder/sander, pedestal mount	B-67	14	1.143			67	23.50	90.50	130
3000	Hack saw, power	2 Clab	24	.667			30.50		30.50	48.50
3100	Hydraulic press	B-67	14	1.143			67	23.50	90.50	130
3500	Laminar flow tables	"	14	1.143			67	23.50	90.50	130
3600	Lathe, bench	2 Clab	14	1.143			52		52	83
3640	6"	B-67	14	1.143			67	23.50	90.50	130
3680	10"		13	1.231			72.50	25.50	98	140
3720	12"		12	1.333			78.50	27.50	106	153
4000	Milling machine		8	2			118	41.50	159.50	228
4100	Molding press, 25 ton		5	3.200			188	66.50	254.50	365
4140	60 ton		4	4			235	83	318	455
4180	100 ton		2	8			470	166	636	915
4220	150 ton		1.50	10.667			625	221	846	1,225
4260	200 ton		1	16			940	330	1,270	1,825
4300	300 ton		.75	21.333			1,250	445	1,695	2,425
4700	Oil pot stand		14	1.143			67	23.50	90.50	130
5000	Press, 10 ton		14	1.143			67	23.50	90.50	130
5040	15 ton		12	1.333			78.50	27.50	106	153
5080	20 ton		10	1.600			94	33	127	183
5120	30 ton		8	2			118	41.50	159.50	228
5160	45 ton		6	2.667			157	55.50	212.50	305
5200	60 ton		4	4			235	83	318	455
5240	75 ton		2.50	6.400			375	133	508	730
5280	100 ton		2	8			470	166	636	915
5500	Raised floor, including assembly	2 Carp	250	.064	S.F.		3.60		3.60	5.70
5600	Rolling mill, 6"	B-67	7	2.286	Ea.		134	47.50	181.50	260
5640	9"		6	2.667			157	55.50	212.50	305
5680	12"		4	4			235	83	318	455
5720	13"		3.50	4.571			269	95	364	520
5760	18"		2	8			470	166	636	915
5800	25"		1	16			940	330	1,270	1,825
6000	Sander, floor stand		14	1.143			67	23.50	90.50	130
6100	Screw machine		7	2.286			134	47.50	181.50	260
6200	Shaper, 16"		14	1.143			67	23.50	90.50	130
6300	Shear, power assist		4	4			235	83	318	455
6400	Slitter, 6"		14	1.143			67	23.50	90.50	130
6440	8"		13	1.231			72.50	25.50	98	140
6480	10"		12	1.333			78.50	27.50	106	153
6520	12"		11	1.455			85.50	30	115.50	166

For customer support on your Commercial Renovation Costs with RSMeans data, call 800.448.8182.

01 93 Facility Maintenance

01 93 09 – Facility Equipment

01 93 09.50 Moving Equipment		Crew	Daily Output	Labor-Hours	Unit	Material	2022 Bare Costs Labor	2022 Bare Costs Equipment	Total	Total Incl O&P
6560	16"	B-67	10	1.600	Ea.		94	33	127	183
6600	20"		8	2			118	41.50	159.50	228
6640	24"		6	2.667			157	55.50	212.50	305
6800	Snag and tap machine		7	2.286			134	47.50	181.50	260
6900	Solder machine (auto)		7	2.286			134	47.50	181.50	260
7000	Storage cabinet metal, small	2 Clab	36	.444			20.50		20.50	32
7040	Large		25	.640			29		29	46.50
7100	Storage rack open, small		14	1.143			52		52	83
7140	Large		7	2.286			104		104	166
7200	Surface bench, small	B-67	14	1.143			67	23.50	90.50	130
7240	Large	"	5	3.200			188	66.50	254.50	365
7300	Surface grinder, large wet	B-68	5	4.800			285	66.50	351.50	515
7500	Time check machine	2 Clab	14	1.143			52		52	83
8000	Welder, 30 kVA (bench)		14	1.143			52		52	83
8100	Work bench with chair		25	.640			29		29	46.50
8900	For moving distance to 200', add						25%			

Division Notes

	CREW	DAILY OUTPUT	LABOR-HOURS	UNIT	BARE COSTS				TOTAL INCL O&P
					MAT.	LABOR	EQUIP.	TOTAL	

Estimating Tips
02 30 00 Subsurface Investigation
In preparing estimates on structures involving earthwork or foundations, all information concerning soil characteristics should be obtained. Look particularly for hazardous waste, evidence of prior dumping of debris, and previous stream beds.

02 40 00 Demolition and Structure Moving
The costs shown for selective demolition do not include rubbish handling or disposal. These items should be estimated separately using RSMeans data or other sources.

- Historic preservation often requires that the contractor remove materials from the existing structure, rehab them, and replace them. The estimator must be aware of any related measures and precautions that must be taken when doing selective demolition and cutting and patching. Requirements may include special handling and storage, as well as security.

- In addition to Subdivision 02 41 00, you can find selective demolition items in each division. Example: Roofing demolition is in Division 7.
- Absent of any other specific reference, an approximate demolish-in-place cost can be obtained by halving the new-install labor cost. To remove for reuse, allow the entire new-install labor figure.

02 40 00 Building Deconstruction
This section provides costs for the careful dismantling and recycling of most low-rise building materials.

02 50 00 Containment of Hazardous Waste
This section addresses on-site hazardous waste disposal costs.

02 80 00 Hazardous Material Disposal/Remediation
This subdivision includes information on hazardous waste handling, asbestos remediation, lead remediation, and mold remediation. See reference numbers

R028213-20 and R028319-60 for further guidance in using these unit price lines.

02 90 00 Monitoring Chemical Sampling, Testing Analysis
This section provides costs for on-site sampling and testing hazardous waste.

Reference Numbers
Reference numbers are shown at the beginning of some major classifications. These numbers refer to related items in the Reference Section. The reference information may be an estimating procedure, an alternate pricing method, or technical information.

Note: Not all subdivisions listed here necessarily appear. ■

02 21 Surveys

02 21 13 – Site Surveys

02 21 13.09 Topographical Surveys

		Crew	Daily Output	Labor-Hours	Unit	Material	2022 Bare Costs Labor	Equipment	Total	Total Incl O&P
0010	**TOPOGRAPHICAL SURVEYS**									
0020	Topographical surveying, conventional, minimum	A-7	3.30	7.273	Acre	45	445	11.05	501.05	755
0050	Average	"	1.95	12.308		67.50	750	18.70	836.20	1,275
0100	Maximum	A-8	.60	53.333	↓	90	3,175	61	3,326	5,150

02 21 13.13 Boundary and Survey Markers

		Crew	Daily Output	Labor-Hours	Unit	Material	2022 Bare Costs Labor	Equipment	Total	Total Incl O&P
0010	**BOUNDARY AND SURVEY MARKERS**									
0300	Lot location and lines, large quantities, minimum	A-7	2	12	Acre	33	735	18.25	786.25	1,200
0320	Average	"	1.25	19.200		66	1,175	29	1,270	1,950
0400	Small quantities, maximum	A-8	1	32	↓	66	1,900	36.50	2,002.50	3,100

02 31 Geophysical Investigations

02 31 23 – Electromagnetic Investigations

02 31 23.10 Ground Penetrating Radar

		Crew	Daily Output	Labor-Hours	Unit	Material	2022 Bare Costs Labor	Equipment	Total	Total Incl O&P
0010	**GROUND PENETRATING RADAR**									
0100	Ground Penetrating Radar				Hr.		210		210	235

02 32 Geotechnical Investigations

02 32 13 – Subsurface Drilling and Sampling

02 32 13.10 Boring and Exploratory Drilling

		Crew	Daily Output	Labor-Hours	Unit	Material	2022 Bare Costs Labor	Equipment	Total	Total Incl O&P
0010	**BORING AND EXPLORATORY DRILLING**									
0020	Borings, initial field stake out & determination of elevations	A-6	1	16	Day		905	36.50	941.50	1,475
0100	Drawings showing boring details				Total		335		335	425
0200	Report and recommendations from P.E.						775		775	970
0300	Mobilization and demobilization	B-55	4	6	↓		283	191	474	660
0350	For over 100 miles, per added mile		450	.053	Mile		2.52	1.70	4.22	5.85
0600	Auger holes in earth, no samples, 2-1/2" diameter		78.60	.305	L.F.		14.40	9.75	24.15	33.50
0650	4" diameter		67.50	.356			16.80	11.35	28.15	39
0800	Cased borings in earth, with samples, 2-1/2" diameter		55.50	.432		14.75	20.50	13.80	49.05	64
0850	4" diameter		32.60	.736		14.45	35	23.50	72.95	97.50
1000	Drilling in rock, "BX" core, no sampling	B-56	34.90	.458			23.50	45.50	69	87
1050	With casing & sampling		31.70	.505		14.75	26	50	90.75	112
1200	"NX" core, no sampling		25.92	.617	↓		31.50	61.50	93	118
1250	With casing and sampling		25	.640	↓	10.90	33	63.50	107.40	134
1400	Borings, earth, drill rig and crew with truck mounted auger	B-55	1	24	Day		1,125	765	1,890	2,650
1450	Rock using crawler type drill	B-56	1	16	"		820	1,600	2,420	3,050
1500	For inner city borings, add, minimum								10%	10%
1510	Maximum								20%	20%

02 41 Demolition

02 41 13 – Selective Site Demolition

02 41 13.15 Hydrodemolition	Crew	Daily Output	Labor-Hours	Unit	Material	2022 Bare Costs Labor	2022 Bare Costs Equipment	Total	Total Incl O&P
0010 **HYDRODEMOLITION**									
0015 Hydrodemolition, concrete pavement									
0120 Includes removal but excludes disposal of concrete									
0130 2" depth	B-5E	1000	.064	S.F.		3.23	3	6.23	8.40
0410 4" depth		800	.080			4.04	3.75	7.79	10.55
0420 6" depth		600	.107			5.40	5	10.40	14
0520 8" depth	↓	300	.213	↓		10.75	10	20.75	28

02 41 13.16 Hydroexcavation	Crew	Daily Output	Labor-Hours	Unit	Material	2022 Bare Costs Labor	2022 Bare Costs Equipment	Total	Total Incl O&P
0010 **HYDROEXCAVATION**									
0015 Hydroexcavation									
0105 Assumes onsite disposal									
0110 Mobilization or Demobilization	B-6D	8	2.500	Ea.		130	165	295	385
0130 Normal Conditions		48	.417	B.C.Y.		21.50	27.50	49	64
0140 Adverse Conditions		30	.667	"		34.50	44	78.50	103
0160 Minimum labor/equipment charge	↓	4	5	Ea.		260	330	590	775

02 41 13.17 Demolish, Remove Pavement and Curb	Crew	Daily Output	Labor-Hours	Unit	Material	2022 Bare Costs Labor	2022 Bare Costs Equipment	Total	Total Incl O&P
0010 **DEMOLISH, REMOVE PAVEMENT AND CURB** R024119-10									
5010 Pavement removal, bituminous roads, up to 3" thick	B-38	690	.058	S.Y.		2.97	1.94	4.91	6.80
5050 4"-6" thick	"	420	.095	"		4.88	3.20	8.08	11.20
5100 Bituminous driveways	B-38	640	.063	S.Y.		3.20	2.10	5.30	7.35
5200 Concrete to 6" thick, hydraulic hammer, mesh reinforced		255	.157			8.05	5.25	13.30	18.45
5300 Rod reinforced		200	.200	↓		10.25	6.70	16.95	23.50
5400 Concrete, 7"-24" thick, plain		33	1.212	C.Y.		62	40.50	102.50	143
5500 Reinforced		24	1.667	"		85.50	56	141.50	196
5590 Minimum labor/equipment charge	↓	6	6.667	Job		340	224	564	785
5600 With hand held air equipment, bituminous, to 6" thick	B-39	1900	.025	S.F.		1.21	.17	1.38	2.10
5700 Concrete to 6" thick, no reinforcing		1600	.030			1.43	.21	1.64	2.50
5800 Mesh reinforced		1400	.034			1.64	.24	1.88	2.86
5900 Rod reinforced	↓	765	.063	↓		3	.43	3.43	5.25
5990 Minimum labor/equipment charge	B-38	6	6.667	Job		340	224	564	785
6000 Curbs, concrete, plain	B-6	360	.067	L.F.		3.29	.91	4.20	6.20
6100 Reinforced		275	.087			4.31	1.20	5.51	8.10
6200 Granite		360	.067			3.29	.91	4.20	6.20
6300 Bituminous		528	.045	↓		2.25	.62	2.87	4.23
6390 Minimum labor/equipment charge		6	4	Job		198	55	253	370

02 41 13.23 Utility Line Removal	Crew	Daily Output	Labor-Hours	Unit	Material	2022 Bare Costs Labor	2022 Bare Costs Equipment	Total	Total Incl O&P
0010 **UTILITY LINE REMOVAL**									
0015 No hauling, abandon catch basin or manhole	B-6	7	3.429	Ea.		169	47	216	320
0020 Remove existing catch basin or manhole, masonry		4	6			296	82	378	555
0030 Catch basin or manhole frames and covers, stored		13	1.846			91	25.50	116.50	172
0040 Remove and reset	↓	7	3.429	↓		169	47	216	320
0900 Hydrants, fire, remove only	B-21A	5	8			450	114	564	830
0950 Remove and reset	"	2	20	↓		1,125	286	1,411	2,075
0990 Minimum labor/equipment charge	2 Plum	2	8	Job		560		560	860
2900 Pipe removal, sewer/water, no excavation, 12" diameter	B-6	175	.137	L.F.		6.75	1.88	8.63	12.75
2930 15"-18" diameter	B-12Z	150	.160			8.25	8.95	17.20	23
2960 21"-24" diameter		120	.200			10.30	11.15	21.45	28.50
3000 27"-36" diameter	↓	90	.267			13.75	14.90	28.65	38
3200 Steel, welded connections, 4" diameter	B-6	160	.150			7.40	2.06	9.46	13.95
3300 10" diameter		80	.300	↓		14.80	4.11	18.91	28
3390 Minimum labor/equipment charge	↓	3	8	Job		395	110	505	745

For customer support on your Commercial Renovation Costs with RSMeans data, call 800.448.8182.

29

02 41 Demolition

02 41 13 – Selective Site Demolition

02 41 13.30 Minor Site Demolition

	02 41 13.30 Minor Site Demolition	Crew	Daily Output	Labor-Hours	Unit	Material	2022 Bare Costs Labor	Equipment	Total	Total Incl O&P
0010	**MINOR SITE DEMOLITION** R024119-10									
0100	Roadside delineators, remove only	B-80	175	.183	Ea.		9.15	3.10	12.25	17.90
0110	Remove and reset	"	100	.320	"		16.05	5.40	21.45	31.50
0400	Minimum labor/equipment charge	B-6	4	6	Job		296	82	378	555
0800	Guiderail, corrugated steel, remove only	B-80A	100	.240	L.F.		10.95	3.77	14.72	21.50
0850	Remove and reset	"	40	.600	"		27.50	9.45	36.95	54
0860	Guide posts, remove only	B-80B	120	.267	Ea.		12.90	4.75	17.65	25.50
0870	Remove and reset	B-55	50	.480	"		22.50	15.30	37.80	53
0890	Minimum labor/equipment charge	2 Clab	4	4	Job		182		182	290
1000	Masonry walls, block, solid	B-5	1800	.031	C.F.		1.56	.84	2.40	3.38
1200	Brick, solid		900	.062			3.12	1.68	4.80	6.75
1400	Stone, with mortar		900	.062			3.12	1.68	4.80	6.75
1500	Dry set		1500	.037			1.87	1.01	2.88	4.06
1600	Median barrier, precast concrete, remove and store	B-3	430	.112	L.F.		5.70	6.30	12	15.95
1610	Remove and reset	"	390	.123	"		6.25	6.95	13.20	17.55
1650	Minimum labor/equipment charge	A-1	4	2	Job		91	27.50	118.50	176
4000	Sidewalk removal, bituminous, 2" thick	B-6	350	.069	S.Y.		3.39	.94	4.33	6.40
4010	2-1/2" thick		325	.074			3.65	1.01	4.66	6.85
4050	Brick, set in mortar		185	.130			6.40	1.78	8.18	12.05
4100	Concrete, plain, 4"		160	.150			7.40	2.06	9.46	13.95
4110	Plain, 5"		140	.171			8.45	2.35	10.80	15.95
4120	Plain, 6"		120	.200			9.90	2.74	12.64	18.60
4200	Mesh reinforced, concrete, 4"		150	.160			7.90	2.19	10.09	14.85
4210	5" thick		131	.183			9.05	2.51	11.56	17
4220	6" thick		112	.214			10.60	2.94	13.54	19.95
4290	Minimum labor/equipment charge	B-39	12	4	Job		191	27.50	218.50	335

02 41 13.33 Railtrack Removal

	02 41 13.33 Railtrack Removal	Crew	Daily Output	Labor-Hours	Unit	Material	Labor	Equipment	Total	Total Incl O&P
0010	**RAILTRACK REMOVAL**									
3500	Railroad track removal, ties and track	B-13	330	.170	L.F.		8.40	6.35	14.75	20.50
3600	Ballast	B-14	500	.096	C.Y.		4.59	.66	5.25	7.95
3700	Remove and re-install, ties & track using new bolts & spikes		50	.960	L.F.		46	6.60	52.60	80
3800	Turnouts using new bolts and spikes		1	48	Ea.		2,300	330	2,630	3,975
3890	Minimum labor/equipment charge		5	9.600	Job		460	66	526	800

02 41 13.60 Selective Demolition Fencing

	02 41 13.60 Selective Demolition Fencing	Crew	Daily Output	Labor-Hours	Unit	Material	Labor	Equipment	Total	Total Incl O&P
0010	**SELECTIVE DEMOLITION FENCING** R024119-10									
1600	Fencing, barbed wire, 3 strand	2 Clab	430	.037	L.F.		1.70		1.70	2.70
1650	5 strand	"	280	.057			2.61		2.61	4.14
1700	Chain link, posts & fabric, 8'-10' high, remove only	B-6	445	.054			2.66	.74	3.40	5

02 41 16 – Structure Demolition

02 41 16.13 Building Demolition

	02 41 16.13 Building Demolition	Crew	Daily Output	Labor-Hours	Unit	Material	Labor	Equipment	Total	Total Incl O&P
0010	**BUILDING DEMOLITION** Large urban projects, incl. 20 mi. haul R024119-10									
0011	No foundation or dump fees, C.F. is vol. of building standing									
0020	Steel	B-8	21500	.003	C.F.		.16	.22	.38	.50
0050	Concrete	B-8	15300	.004	C.F.		.22	.31	.53	.70
0080	Masonry		20100	.003			.17	.24	.41	.52
0100	Mixture of types		20100	.003			.17	.24	.41	.52
0500	Small bldgs, or single bldgs, no salvage included, steel	B-3	14800	.003			.17	.18	.35	.46
0600	Concrete		11300	.004			.22	.24	.46	.60
0650	Masonry		14800	.003			.17	.18	.35	.46
0700	Wood		14800	.003			.17	.18	.35	.46
0750	For buildings with no interior walls, deduct								30%	30%

02 41 Demolition

02 41 16 – Structure Demolition

02 41 16.13 Building Demolition

		Crew	Daily Output	Labor-Hours	Unit	Material	2022 Bare Costs Labor	2022 Bare Costs Equipment	Total	Total Incl O&P
1000	Demolition single family house, one story, wood 1600 S.F.	B-3	1	48	Ea.		2,450	2,725	5,175	6,875
1020	3200 S.F.		.50	96			4,875	5,425	10,300	13,700
1200	Demolition two family house, two story, wood 2400 S.F.		.67	71.964			3,650	4,075	7,725	10,300
1220	4200 S.F.		.38	128			6,525	7,250	13,775	18,300
1300	Demolition three family house, three story, wood 3200 S.F.		.50	96			4,875	5,425	10,300	13,700
1320	5400 S.F.	▼	.30	160			8,150	9,050	17,200	22,900
5000	For buildings with no interior walls, deduct								30%	30%

02 41 16.17 Building Demolition Footings and Foundations

		Crew	Daily Output	Labor-Hours	Unit	Material	2022 Bare Costs Labor	2022 Bare Costs Equipment	Total	Total Incl O&P
0010	**BUILDING DEMOLITION FOOTINGS AND FOUNDATIONS** R024119-10									
0200	Floors, concrete slab on grade,									
0240	4" thick, plain concrete	B-13L	5000	.003	S.F.		.20	.52	.72	.89
0280	Reinforced, wire mesh		4000	.004			.25	.65	.90	1.11
0300	Rods		4500	.004			.23	.58	.81	.98
0400	6" thick, plain concrete		4000	.004			.25	.65	.90	1.11
0420	Reinforced, wire mesh		3200	.005			.32	.81	1.13	1.38
0440	Rods		3600	.004	▼		.28	.72	1	1.23
1000	Footings, concrete, 1' thick, 2' wide		300	.053	L.F.		3.38	8.65	12.03	14.75
1080	1'-6" thick, 2' wide		250	.064			4.06	10.35	14.41	17.70
1120	3' wide		200	.080			5.05	12.95	18	22
1140	2' thick, 3' wide	▼	175	.091			5.80	14.80	20.60	25.50
1200	Average reinforcing, add								10%	10%
1220	Heavy reinforcing, add				▼				20%	20%
2000	Walls, block, 4" thick	B-13L	8000	.002	S.F.		.13	.32	.45	.56
2040	6" thick		6000	.003			.17	.43	.60	.74
2080	8" thick		4000	.004			.25	.65	.90	1.11
2100	12" thick	▼	3000	.005			.34	.86	1.20	1.48
2200	For horizontal reinforcing, add								10%	10%
2220	For vertical reinforcing, add								20%	20%
2400	Concrete, plain concrete, 6" thick	B-13L	4000	.004			.25	.65	.90	1.11
2420	8" thick	"	3500	.005	▼		.29	.74	1.03	1.26
2440	10" thick	B-13L	3000	.005	S.F.		.34	.86	1.20	1.48
2500	12" thick	"	2500	.006			.41	1.04	1.45	1.77
2600	For average reinforcing, add								10%	10%
2620	For heavy reinforcing, add				▼				20%	20%
9000	Minimum labor/equipment charge	A-1	2	4	Job		182	55	237	350

02 41 16.19 Minor Building Deconstruction

			Crew	Daily Output	Labor-Hours	Unit	Material	2022 Bare Costs Labor	2022 Bare Costs Equipment	Total	Total Incl O&P
0010	**MINOR BUILDING DECONSTRUCTION** R024119-10										
0011	For salvage, avg house										
0225	2 story, pre 1970 house, 1400 S.F., labor for all salv mat, min	G	6 Clab	25	1.920	SF Flr.		87.50		87.50	139
0230	Maximum	G	"	15	3.200	"		146		146	232
0235	Salvage of carpet, tackless	G	2 Clab	2400	.007	S.F.		.30		.30	.48
0240	Wood floors, incl. denailing and packaging	G	3 Clab	270	.089	"		4.05		4.05	6.45
0260	Wood doors and trim, standard	G	1 Clab	16	.500	Ea.		23		23	36
0280	Base or cove mouldings, incl. denailing and packaging	G		500	.016	L.F.		.73		.73	1.16
0320	Closet shelving and trim, incl. denailing and packaging	G		18	.444	Set		20.50		20.50	32
0340	Kitchen cabinets, uppers and lowers, prefab type	G	2 Clab	24	.667	L.F.		30.50		30.50	48.50
0360	Kitchen cabinets, uppers and lowers, built-in	G		12	1.333	"		61		61	96.50
0370	Bath fixt, incl. toilet, tub, vanity, and med cabinet	G	▼	3	5.333	Set		243		243	385

For customer support on your Commercial Renovation Costs with RSMeans data, call 800.448.8182.

31

02 41 Demolition

02 41 19 – Selective Demolition

02 41 19.13 Selective Building Demolition	Crew	Daily Output	Labor-Hours	Unit	Material	2022 Bare Costs Labor	2022 Bare Costs Equipment	Total	Total Incl O&P
0010 **SELECTIVE BUILDING DEMOLITION**									
0020 Costs related to selective demolition of specific building components									
0025 are included under Common Work Results (XX 05)									
0030 in the component's appropriate division.									

02 41 19.16 Selective Demolition, Cutout

02 41 19.16 Selective Demolition, Cutout	Crew	Daily Output	Labor-Hours	Unit	Material	2022 Bare Costs Labor	2022 Bare Costs Equipment	Total	Total Incl O&P
0010 **SELECTIVE DEMOLITION, CUTOUT** R024119-10									
0020 Concrete, elev. slab, light reinforcement, under 6 C.F.	B-9	65	.615	C.F.		28.50	5.10	33.60	50.50
0050 Light reinforcing, over 6 C.F.		75	.533	"		24.50	4.43	28.93	44
0200 Slab on grade to 6" thick, not reinforced, under 8 S.F.		85	.471	S.F.		21.50	3.91	25.41	39
0250 8-16 S.F.		175	.229	"		10.50	1.90	12.40	18.80
0255 For over 16 S.F. see Line 02 41 16.17 0400									
0600 Walls, not reinforced, under 6 C.F.	B-9	60	.667	C.F.		30.50	5.55	36.05	54.50
0650 6-12 C.F.	"	80	.500	"		23	4.15	27.15	41
0655 For over 12 C.F. see Line 02 41 16.17 2500									
1000 Concrete, elevated slab, bar reinforced, under 6 C.F.	B-9	45	.889	C.F.		41	7.40	48.40	73
1050 Bar reinforced, over 6 C.F.		50	.800	"		37	6.65	43.65	66
1200 Slab on grade to 6" thick, bar reinforced, under 8 S.F.		75	.533	S.F.		24.50	4.43	28.93	44
1250 8-16 S.F.		150	.267	"		12.25	2.21	14.46	22
1255 For over 16 S.F. see Line 02 41 16.17 0440									
1400 Walls, bar reinforced, under 6 C.F.	B-9	50	.800	C.F.		37	6.65	43.65	66
1450 6-12 C.F.	"	70	.571	"		26.50	4.74	31.24	47
1455 For over 12 C.F. see Lines 02 41 16.17 2500 and 2600									
2000 Brick, to 4 S.F. opening, not including toothing									
2040 4" thick	B-9	30	1.333	Ea.		61.50	11.05	72.55	110
2060 8" thick		18	2.222			102	18.45	120.45	183
2080 12" thick		10	4			184	33	217	330
2400 Concrete block, to 4 S.F. opening, 2" thick		35	1.143			52.50	9.50	62	94
2420 4" thick		30	1.333			61.50	11.05	72.55	110
2440 8" thick		27	1.481			68	12.30	80.30	122
2460 12" thick		24	1.667			76.50	13.85	90.35	137
2600 Gypsum block, to 4 S.F. opening, 2" thick		80	.500			23	4.15	27.15	41
2620 4" thick		70	.571			26.50	4.74	31.24	47
2640 8" thick		55	.727			33.50	6.05	39.55	59.50
2800 Terra cotta, to 4 S.F. opening, 4" thick		70	.571			26.50	4.74	31.24	47
2840 8" thick		65	.615			28.50	5.10	33.60	50.50
2880 12" thick		50	.800			37	6.65	43.65	66
4000 For toothing masonry, see Section 04 01 20.50									
6000 Walls, interior, not including re-framing,									
6010 openings to 5 S.F.									
6100 Drywall to 5/8" thick	1 Clab	24	.333	Ea.		15.20		15.20	24
6200 Paneling to 3/4" thick		20	.400			18.25		18.25	29
6300 Plaster, on gypsum lath		20	.400			18.25		18.25	29
6340 On wire lath		14	.571			26		26	41.50
7000 Wood frame, not including re-framing, openings to 5 S.F.									
7200 Floors, sheathing and flooring to 2" thick	1 Clab	5	1.600	Ea.		73		73	116
7310 Roofs, sheathing to 1" thick, not including roofing		6	1.333			61		61	96.50
7410 Walls, sheathing to 1" thick, not including siding		7	1.143			52		52	83
8500 Minimum labor/equipment charge		4	2	Job		91		91	145

02 41 19 – Selective Demolition

02 41 19.18 Selective Demolition, Disposal Only		Crew	Daily Output	Labor-Hours	Unit	Material	2022 Bare Costs Labor	2022 Bare Costs Equipment	Total	Total Incl O&P
0010	**SELECTIVE DEMOLITION, DISPOSAL ONLY** R024119-10									
0015	Urban bldg w/salvage value allowed									
0020	Including loading and 5 mile haul to dump									
0200	Steel frame	B-3	430	.112	C.Y.		5.70	6.30	12	15.95
0300	Concrete frame		365	.132			6.70	7.45	14.15	18.80
0400	Masonry construction		445	.108			5.50	6.10	11.60	15.40
0500	Wood frame	B-3	247	.194	C.Y.		9.90	11	20.90	28

02 41 19.19 Selective Demolition

02 41 19.19 Selective Demolition		Crew	Daily Output	Labor-Hours	Unit	Material	2022 Bare Costs Labor	2022 Bare Costs Equipment	Total	Total Incl O&P
0010	**SELECTIVE DEMOLITION**, Rubbish Handling R024119-10									
0020	The following are to be added to the demolition prices									
0050	The following are components for a complete chute system									
0100	Top chute circular steel, 4' long, 18" diameter	B-1C	15	1.600	Ea.	279	74	24.50	377.50	450
0102	23" diameter		15	1.600		305	74	24.50	403.50	480
0104	27" diameter		15	1.600		325	74	24.50	423.50	505
0106	30" diameter		15	1.600		350	74	24.50	448.50	530
0108	33" diameter		15	1.600		370	74	24.50	468.50	555
0110	36" diameter		15	1.600		395	74	24.50	493.50	580
0112	Regular chute, 18" diameter		15	1.600		210	74	24.50	308.50	375
0114	23" diameter		15	1.600		233	74	24.50	331.50	400
0116	27" diameter		15	1.600		256	74	24.50	354.50	425
0118	30" diameter		15	1.600		268	74	24.50	366.50	440
0120	33" diameter		15	1.600		305	74	24.50	403.50	480
0122	36" diameter		15	1.600		325	74	24.50	423.50	505
0124	Control door chute, 18" diameter		15	1.600		395	74	24.50	493.50	580
0126	23" diameter		15	1.600		420	74	24.50	518.50	605
0128	27" diameter		15	1.600		440	74	24.50	538.50	630
0130	30" diameter		15	1.600		465	74	24.50	563.50	655
0132	33" diameter		15	1.600		490	74	24.50	588.50	680
0134	36" diameter		15	1.600		510	74	24.50	608.50	710
0136	Chute liners, 14 ga., 18"-30" diameter		15	1.600		214	74	24.50	312.50	380
0138	33"-36" diameter		15	1.600		268	74	24.50	366.50	440
0140	17% thinner chute, 30" diameter		15	1.600		233	74	24.50	331.50	400
0142	33% thinner chute, 30" diameter		15	1.600		174	74	24.50	272.50	335
0144	Top chute cover	1 Clab	24	.333		247	15.20		262.20	296
0146	Door chute cover	"	24	.333		146	15.20		161.20	185
0148	Top chute trough	2 Clab	12	1.333		495	61		556	640
0150	Bolt down frame & counter weights, 250 lb.	B-1	4	6		4,300	278		4,578	5,175
0152	500 lb.		4	6		6,350	278		6,628	7,425
0154	750 lb.		4	6		10,800	278		11,078	12,300
0156	1000 lb.		2.67	8.989		10,600	415		11,015	12,300
0158	1500 lb.		2.67	8.989		13,300	415		13,715	15,400
0160	Chute warning light system, 5 stories	B-1C	4	6		8,175	278	92	8,545	9,550
0162	10 stories	"	2	12		15,200	555	184	15,939	17,800
0164	Dust control device for dumpsters	1 Clab	8	1		93	45.50		138.50	175
0166	Install or replace breakaway cord		8	1		27.50	45.50		73	103
0168	Install or replace warning sign		16	.500		10	23		33	47
0600	Dumpster, weekly rental, 1 dump/week, 6 C.Y. capacity (2 tons)				Week	415			415	455
0700	10 C.Y. capacity (3 tons)					480			480	530
0725	20 C.Y. capacity (5 tons) R024119-20					565			565	625
0800	30 C.Y. capacity (7 tons)					730			730	800
0840	40 C.Y. capacity (10 tons)					775			775	850
0900	Alternate pricing for dumpsters									

For customer support on your Commercial Renovation Costs with RSMeans data, call 800.448.8182.

33

02 41 19.19 Selective Demolition

		Crew	Daily Output	Labor-Hours	Unit	Material	2022 Bare Costs Labor	Equipment	Total	Total Incl O&P
0910	Delivery, average for all sizes				Ea.	75			75	82.50
0920	Haul, average for all sizes					235			235	259
0930	Rent per day, average for all sizes					20			20	22
0940	Rent per month, average for all sizes					80			80	88
0950	Disposal fee per ton, average for all sizes				Ton	88			88	97
2000	Load, haul, dump and return, 0'-50' haul, hand carried	2 Clab	24	.667	C.Y.		30.50		30.50	48.50
2005	Wheeled		37	.432			19.70		19.70	31.50
2040	0'-100' haul, hand carried		16.50	.970			44		44	70.50
2045	Wheeled		25	.640			29		29	46.50
2050	Forklift	A-3R	25	.320			18.25	11.35	29.60	41
2080	Haul and return, add per each extra 100' haul, hand carried	2 Clab	35.50	.451			20.50		20.50	32.50
2085	Wheeled		54	.296			13.50		13.50	21.50
2120	For travel in elevators, up to 10 floors, add		140	.114			5.20		5.20	8.30
2130	0'-50' haul, incl. up to 5 riser stairs, hand carried		23	.696			31.50		31.50	50.50
2135	Wheeled		35	.457			21		21	33
2140	6-10 riser stairs, hand carried		22	.727			33		33	52.50
2145	Wheeled		34	.471			21.50		21.50	34
2150	11-20 riser stairs, hand carried		20	.800			36.50		36.50	58
2155	Wheeled		31	.516			23.50		23.50	37.50
2160	21-40 riser stairs, hand carried		16	1			45.50		45.50	72.50
2165	Wheeled		24	.667			30.50		30.50	48.50
2170	0'-100' haul, incl. 5 riser stairs, hand carried		15	1.067			48.50		48.50	77.50
2175	Wheeled		23	.696			31.50		31.50	50.50
2180	6-10 riser stairs, hand carried		14	1.143			52		52	83
2185	Wheeled		21	.762			34.50		34.50	55
2190	11-20 riser stairs, hand carried		12	1.333			61		61	96.50
2195	Wheeled		18	.889			40.50		40.50	64.50
2200	21-40 riser stairs, hand carried		8	2			91		91	145
2205	Wheeled		12	1.333			61		61	96.50
2210	Haul and return, add per each extra 100' haul, hand carried		35.50	.451			20.50		20.50	32.50
2215	Wheeled	2 Clab	54	.296	C.Y.		13.50		13.50	21.50
2220	For each additional flight of stairs, up to 5 risers, add		550	.029	Flight		1.33		1.33	2.11
2225	6-10 risers, add		275	.058			2.65		2.65	4.22
2230	11-20 risers, add		138	.116			5.30		5.30	8.40
2235	21-40 risers, add		69	.232			10.55		10.55	16.80
3000	Loading & trucking, including 2 mile haul, chute loaded	B-16	45	.711	C.Y.		34	17.05	51.05	73
3040	Hand loading truck, 50' haul	"	48	.667			32	16	48	68.50
3080	Machine loading truck	B-17	120	.267			13.40	6	19.40	27.50
3120	Wheeled 50' and ramp dump loaded	2 Clab	24	.667			30.50		30.50	48.50
5000	Haul, per mile, up to 8 C.Y. truck	B-34B	1165	.007			.36	.66	1.02	1.30
5100	Over 8 C.Y. truck	"	1550	.005			.27	.50	.77	.98

02 41 19.20 Selective Demolition, Dump Charges

		Crew	Daily Output	Labor-Hours	Unit	Material	2022 Bare Costs Labor	Equipment	Total	Total Incl O&P
0010	**SELECTIVE DEMOLITION, DUMP CHARGES** R024119-10									
0020	Dump charges, typical urban city, tipping fees only									
0100	Building construction materials				Ton	74			74	81
0200	Trees, brush, lumber					63			63	69.50
0300	Rubbish only					63			63	69.50
0500	Reclamation station, usual charge					74			74	81

02 41 Demolition

02 41 19 – Selective Demolition

02 41 19.21 Selective Demolition, Gutting

02 41 19.21 Selective Demolition, Gutting	Crew	Daily Output	Labor-Hours	Unit	Material	2022 Bare Costs Labor	2022 Bare Costs Equipment	Total	Total Incl O&P
0010 **SELECTIVE DEMOLITION, GUTTING** R024119-10									
0020 Building interior, including disposal, dumpster fees not included									
0500 Residential building									
0560 Minimum	B-16	400	.080	SF Flr.		3.84	1.92	5.76	8.20
0580 Maximum	"	360	.089	"		4.26	2.13	6.39	9.15
0900 Commercial building									
1000 Minimum	B-16	350	.091	SF Flr.		4.38	2.19	6.57	9.35
1020 Maximum		250	.128	"		6.15	3.07	9.22	13.15
3000 Minimum labor/equipment charge		4	8	Job		385	192	577	820

02 41 19.25 Selective Demolition, Saw Cutting

02 41 19.25 Selective Demolition, Saw Cutting	Crew	Daily Output	Labor-Hours	Unit	Material	2022 Bare Costs Labor	2022 Bare Costs Equipment	Total	Total Incl O&P
0010 **SELECTIVE DEMOLITION, SAW CUTTING** R024119-10									
0012 For concrete saw cutting, see Section 03 81									
0015 Asphalt, up to 3" deep	B-89	1050	.015	L.F.	.10	.82	.60	1.52	2.06
0020 Each additional inch of depth	"	1800	.009		.03	.48	.35	.86	1.17
1200 Masonry walls, hydraulic saw, brick, per inch of depth	B-89B	300	.053		.03	2.86	3.58	6.47	8.50
1220 Block walls, solid, per inch of depth	"	250	.064		.03	3.44	4.30	7.77	10.15
2000 Brick or masonry w/hand held saw, per inch of depth	A-1	125	.064		.05	2.92	.88	3.85	5.65
5000 Wood sheathing to 1" thick, on walls	1 Carp	200	.040			2.25		2.25	3.58
5020 On roof	"	250	.032			1.80		1.80	2.86
9000 Minimum labor/equipment charge	A-1	2	4	Job		182	55	237	350

02 41 19.27 Selective Demolition, Torch Cutting

02 41 19.27 Selective Demolition, Torch Cutting	Crew	Daily Output	Labor-Hours	Unit	Material	2022 Bare Costs Labor	2022 Bare Costs Equipment	Total	Total Incl O&P
0010 **SELECTIVE DEMOLITION, TORCH CUTTING** R024119-10									
0020 Steel, 1" thick plate	E-25	333	.024	L.F.	2.22	1.54	.04	3.80	4.96
0040 1" diameter bar	"	600	.013	Ea.	.37	.85	.02	1.24	1.80
1000 Oxygen lance cutting, reinforced concrete walls									
1040 12"-16" thick walls	1 Clab	10	.800	L.F.		36.50		36.50	58
1080 24" thick walls	"	6	1.333	"		61		61	96.50
1090 Minimum labor/equipment charge	E-25	2	4	Job		256	6.70	262.70	415
1100 See Section 05 05 21.10									

02 42 Removal and Diversion of Construction Materials

02 42 10 – Building Deconstruction

02 42 10.10 Estimated Salvage Value or Savings

02 42 10.10 Estimated Salvage Value or Savings	Crew	Daily Output	Labor-Hours	Unit	Material	2022 Bare Costs Labor	2022 Bare Costs Equipment	Total	Total Incl O&P
0010 **ESTIMATED SALVAGE VALUE OR SAVINGS**									
0015 Excludes material handling, packaging, container costs and									
0020 transportation for salvage or disposal									
0050 All items in Section 02 42 10.10 are credit deducts and not costs									
0100 Copper wire salvage value Ⓖ				Lb.	1.63			1.63	1.79
0200 Brass salvage value Ⓖ					1.20			1.20	1.32
0250 Bronze					1.39			1.39	1.53
0300 Steel Ⓖ					.05			.05	.06
0304 304 Stainless Steel					.29			.29	.32
0316 316 Stainless Steel					.43			.43	.47
0400 Cast Iron Ⓖ					.06			.06	.07
0600 Aluminum siding Ⓖ					.32			.32	.35
0610 Aluminum wire					.45			.45	.50

For customer support on your Commercial Renovation Costs with RSMeans data, call 800.448.8182.

35

02 42 10.20 Deconstruction of Building Components		Crew	Daily Output	Labor-Hours	Unit	Material	2022 Bare Costs Labor	Equipment	Total	Total Incl O&P
0010	**DECONSTRUCTION OF BUILDING COMPONENTS**									
0012	Buildings one or two stories only									
0015	Excludes material handling, packaging, container costs and									
0020	transportation for salvage or disposal									
0050	Deconstruction of plumbing fixtures									
0100	Wall hung or countertop lavatory	G	2 Clab	16	1	Ea.		45.50		45.50	72.50
0110	Single or double compartment kitchen sink	G		14	1.143			52		52	83
0120	Wall hung urinal	G		14	1.143			52		52	83
0130	Floor mounted	G		8	2			91		91	145
0140	Floor mounted water closet	G		16	1			45.50		45.50	72.50
0150	Wall hung	G		14	1.143			52		52	83
0160	Water fountain, free standing	G		16	1			45.50		45.50	72.50
0170	Wall hung or deck mounted	G		12	1.333			61		61	96.50
0180	Bathtub, steel or fiberglass	G	2 Clab	10	1.600	Ea.		73		73	116
0190	Cast iron	G		8	2			91		91	145
0200	Shower, single	G		6	2.667			122		122	193
0210	Group	G		7	2.286			104		104	166
0300	Deconstruction of electrical fixtures									
0310	Surface mounted incandescent fixtures	G	2 Clab	48	.333	Ea.		15.20		15.20	24
0320	Fluorescent, 2 lamp	G		32	.500			23		23	36
0330	4 lamp	G		24	.667			30.50		30.50	48.50
0340	Strip fluorescent, 1 lamp	G		40	.400			18.25		18.25	29
0350	2 lamp	G		32	.500			23		23	36
0400	Recessed drop-in fluorescent fixture, 2 lamp	G		27	.593			27		27	43
0410	4 lamp	G		18	.889			40.50		40.50	64.50
0500	Deconstruction of appliances									
0510	Cooking stoves	G	2 Clab	26	.615	Ea.		28		28	44.50
0520	Dishwashers	G	"	26	.615	"		28		28	44.50
0600	Deconstruction of millwork and trim									
0610	Cabinets, wood	G	2 Carp	40	.400	L.F.		22.50		22.50	36
0620	Countertops	G		100	.160	"		9		9	14.30
0630	Wall paneling, 1" thick	G		500	.032	S.F.		1.80		1.80	2.86
0640	Ceiling trim	G		500	.032	L.F.		1.80		1.80	2.86
0650	Wainscoting	G		500	.032	S.F.		1.80		1.80	2.86
0660	Base, 3/4"-1" thick	G		600	.027	L.F.		1.50		1.50	2.39
0700	Deconstruction of doors and windows									
0710	Doors, wrap, interior, wood, single, no closers	G	2 Carp	21	.762	Ea.	2	43		45	70
0720	Double	G		13	1.231		4	69.50		73.50	114
0730	Solid core, single, exterior or interior	G		10	1.600		2	90		92	145
0740	Double	G		8	2		4	113		117	183
0810	Windows, wrap, wood, single									
0812	with no casement or cladding	G	2 Carp	21	.762	Ea.	2	43		45	70
0820	with casement and/or cladding	G	"	18	.889	"	2	50		52	81.50
0900	Deconstruction of interior finishes									
0910	Drywall for recycling	G	2 Clab	1775	.009	S.F.		.41		.41	.65
0920	Plaster wall, first floor	G		1775	.009			.41		.41	.65
0930	Second floor	G		1330	.012			.55		.55	.87
1000	Deconstruction of roofing and accessories									
1010	Built-up roofs	G	2 Clab	570	.028	S.F.		1.28		1.28	2.03
1020	Gutters, fascia and rakes	G	"	1140	.014	L.F.		.64		.64	1.02
2000	Deconstruction of wood components									
2010	Roof sheeting	G	2 Clab	570	.028	S.F.		1.28		1.28	2.03

02 42 10.20 Deconstruction of Building Components

		Crew	Daily Output	Labor-Hours	Unit	Material	2022 Bare Costs Labor	2022 Bare Costs Equipment	Total	Total Incl O&P
2020	Main roof framing G	2 Clab	760	.021	L.F.		.96		.96	1.53
2030	Porch roof framing G	↓	445	.036			1.64		1.64	2.60
2040	Beams 4" x 8" G	B-1	375	.064			2.96		2.96	4.70
2050	4" x 10" G		300	.080			3.70		3.70	5.90
2055	4" x 12" G		250	.096			4.44		4.44	7.05
2060	6" x 8" G		250	.096			4.44		4.44	7.05
2065	6" x 10" G		200	.120			5.55		5.55	8.80
2070	6" x 12" G		170	.141			6.55		6.55	10.40
2075	8" x 12" G		126	.190			8.80		8.80	14
2080	10" x 12" G	↓	100	.240			11.10		11.10	17.65
2100	Ceiling joists G	2 Clab	800	.020			.91		.91	1.45
2150	Wall framing, interior G		1230	.013	↓		.59		.59	.94
2160	Sub-floor G		2000	.008	S.F.		.36		.36	.58
2170	Floor joists G		2000	.008	L.F.		.36		.36	.58
2200	Wood siding (no lead or asbestos) G		1300	.012	S.F.		.56		.56	.89
2300	Wall framing, exterior G		1600	.010	L.F.		.46		.46	.72
2400	Stair risers G		53	.302	Ea.		13.75		13.75	22
2500	Posts G	↓	800	.020	L.F.		.91		.91	1.45
3000	Deconstruction of exterior brick walls									
3010	Exterior brick walls, first floor G	2 Clab	200	.080	S.F.		3.65		3.65	5.80
3020	Second floor G		64	.250	"		11.40		11.40	18.10
3030	Brick chimney G	↓	100	.160	C.F.		7.30		7.30	11.60
4000	Deconstruction of concrete									
4010	Slab on grade, 4" thick, plain concrete G	B-9	500	.080	S.F.		3.68	.66	4.34	6.60
4020	Wire mesh reinforced G		470	.085			3.91	.71	4.62	7
4030	Rod reinforced G		400	.100			4.60	.83	5.43	8.20
4110	Foundation wall, 6" thick, plain concrete G		160	.250			11.50	2.08	13.58	20.50
4120	8" thick G		140	.286			13.15	2.37	15.52	23.50
4130	10" thick G	↓	120	.333	↓		15.35	2.77	18.12	27.50
9000	Deconstruction process, support equipment as needed									
9010	Daily use, portal to portal, 12-ton truck-mounted hydraulic crane crew G	A-3H	1	8	Day		505	1,575	2,080	2,525
9020	Daily use, skid steer and operator G	A-3C	1	8			455	420	875	1,175
9030	Daily use, backhoe 48 HP, operator and labor G	"	1	8	↓		455	420	875	1,175

02 42 10.30 Deconstruction Material Handling

		Crew	Daily Output	Labor-Hours	Unit	Material	2022 Bare Costs Labor	2022 Bare Costs Equipment	Total	Total Incl O&P
0010	**DECONSTRUCTION MATERIAL HANDLING**									
0012	Buildings one or two stories only									
0100	Clean and stack brick on pallet G	2 Clab	1200	.013	Ea.		.61		.61	.97
0200	Haul 50' and load rough lumber up to 2" x 8" size G		2000	.008	"		.36		.36	.58
0210	Lumber larger than 2" x 8" G		3200	.005	B.F.		.23		.23	.36
0300	Finish wood for recycling stack and wrap per pallet G		8	2	Ea.	16	91		107	163
0350	Light fixtures		6	2.667		29	122		151	225
0375	Windows		6	2.667		27	122		149	223
0400	Miscellaneous materials	↓	8	2	↓	8	91		99	154
1000	See Section 02 41 19.19 for bulk material handling									

02 43 Structure Moving

02 43 13 – Structure Relocation

02 43 13.13 Building Relocation		Crew	Daily Output	Labor-Hours	Unit	Material	2022 Bare Costs Labor	Equipment	Total	Total Incl O&P
0010	**BUILDING RELOCATION**									
0011	One day move, up to 24' wide									
0020	Reset on existing foundation				Total				11,500	11,500
0040	Wood or steel frame bldg., based on ground floor area	G	B-4	185	.259	S.F.		12.25	2.78	15.03	22.50
0060	Masonry bldg., based on ground floor area	G	"	137	.350			16.55	3.75	20.30	30.50
0200	For 24'-42' wide, add									15%	15%

02 58 Snow Control

02 58 13 – Snow Fencing

02 58 13.10 Snow Fencing System		Crew	Daily Output	Labor-Hours	Unit	Material	2022 Bare Costs Labor	Equipment	Total	Total Incl O&P
0010	**SNOW FENCING SYSTEM**									
7001	Snow fence on steel posts 10' OC, 4' high	B-1	500	.048	L.F.	3.67	2.22		5.89	7.55

02 65 Underground Storage Tank Removal

02 65 10 – Underground Tank and Contaminated Soil Removal

02 65 10.30 Removal of Underground Storage Tanks

02 65 10.30 Removal of Underground Storage Tanks			Crew	Daily Output	Labor-Hours	Unit	Material	2022 Bare Costs Labor	Equipment	Total	Total Incl O&P
0010	**REMOVAL OF UNDERGROUND STORAGE TANKS**	R026510-20									
0011	Petroleum storage tanks, non-leaking										
0100	Excavate & load onto trailer										
0110	3,000 gal. to 5,000 gal. tank	G	B-14	4	12	Ea.		575	82	657	1,000
0120	6,000 gal. to 8,000 gal. tank	G	B-3A	3	13.333			645	330	975	1,400
0130	9,000 gal. to 12,000 gal. tank		"	2	20			970	495	1,465	2,075
0190	Known leaking tank, add					%				100%	100%
0200	Remove sludge, water and remaining product from bottom										
0201	of tank with vacuum truck										
0300	3,000 gal. to 5,000 gal. tank	G	A-13	5	1.600	Ea.		91	154	245	310
0310	6,000 gal. to 8,000 gal. tank	G		4	2			114	192	306	390
0320	9,000 gal. to 12,000 gal. tank	G		3	2.667			152	257	409	520
0390	Dispose of sludge off-site, average					Gal.				6.25	6.80
0400	Insert inert solid CO_2 "dry ice" into tank										
0401	For cleaning/transporting tanks (1.5 lb./100 gal. cap)	G	1 Clab	500	.016	Lb.	1.26	.73		1.99	2.55
0503	Disconnect and remove piping	G	1 Plum	160	.050	L.F.		3.49		3.49	5.40
0603	Transfer liquids, 10% of volume	G	"	1600	.005	Gal.		.35		.35	.54
0703	Cut accessway into underground storage tank	G	1 Clab	5.33	1.501	Ea.		68.50		68.50	109
0813	Remove sludge, wash and wipe tank, 500 gal.	G	1 Plum	8	1			70		70	108
0823	3,000 gal.	G		6.67	1.199			83.50		83.50	129
0833	5,000 gal.	G		6.15	1.301			91		91	140
0843	8,000 gal.	G		5.33	1.501			105		105	162
0853	10,000 gal.	G		4.57	1.751			122		122	188
0863	12,000 gal.	G		4.21	1.900			133		133	204
1020	Haul tank to certified salvage dump, 100 miles round trip										
1023	3,000 gal. to 5,000 gal. tank					Ea.				760	830
1026	6,000 gal. to 8,000 gal. tank									880	960
1029	9,000 gal. to 12,000 gal. tank									1,050	1,150
1100	Disposal of contaminated soil to landfill										
1110	Minimum					C.Y.				145	160
1111	Maximum					"				400	440
1120	Disposal of contaminated soil to										
1121	bituminous concrete batch plant										

02 65 Underground Storage Tank Removal

02 65 10 – Underground Tank and Contaminated Soil Removal

02 65 10.30 Removal of Underground Storage Tanks		Crew	Daily Output	Labor-Hours	Unit	Material	2022 Bare Costs Labor	Equipment	Total	Total Incl O&P
1130	Minimum				C.Y.				80	88
1131	Maximum				"				115	125
1203	Excavate, pull & load tank, backfill hole, 8,000 gal. +	G	B-12C	.50	32	Ea.		1,750	1,950	3,700	4,875
1213	Haul tank to certified dump, 100 miles rt, 8,000 gal. +	G	B-34K	1	8			425	890	1,315	1,650
1223	Excavate, pull & load tank, backfill hole, 500 gal.	G	B-11C	1	16			850	330	1,180	1,675
1233	Excavate, pull & load tank, backfill hole, 3,000-5,000 gal.	G	B-11M	.50	32			1,700	805	2,505	3,550
1243	Haul tank to certified dump, 100 miles rt, 500 gal.	G	B-34L	1	8			455	340	795	1,075
1253	Haul tank to certified dump, 100 miles rt, 3,000-5,000 gal.	G	B-34M	1	8	▼		455	375	830	1,125
2010	Decontamination of soil on site incl poly tarp on top/bottom										
2011	Soil containment berm and chemical treatment										
2020	Minimum	G	B-11C	100	.160	C.Y.	7.90	8.50	3.29	19.69	25.50
2021	Maximum	G	"	100	.160		10.20	8.50	3.29	21.99	28
2050	Disposal of decontaminated soil, minimum									135	150
2055	Maximum					▼				400	440

02 81 Transportation and Disposal of Hazardous Materials

02 81 20 – Hazardous Waste Handling

02 81 20.10 Hazardous Waste Cleanup/Pickup/Disposal

		Crew	Daily Output	Labor-Hours	Unit	Material	2022 Bare Costs Labor	Equipment	Total	Total Incl O&P
0010	**HAZARDOUS WASTE CLEANUP/PICKUP/DISPOSAL**									
0100	For contractor rental equipment, i.e., dozer,									
0110	Front end loader, dump truck, etc., see 01 54 33 Reference Section									
1000	Solid pickup									
1100	55 gal. drums				Ea.				240	265
1120	Bulk material, minimum				Ton				190	210
1130	Maximum				"				595	655
1200	Transportation to disposal site									
1220	Truckload = 80 drums or 25 C.Y. or 18 tons									
1260	Minimum				Mile				3.95	4.45
1270	Maximum				"				7.25	7.98
3000	Liquid pickup, vacuum truck, stainless steel tank									
3100	Minimum charge, 4 hours									
3110	1 compartment, 2200 gallon				Hr.				140	155
3120	2 compartment, 5000 gallon				"				200	225
3400	Transportation in 6900 gallon bulk truck				Mile				7.95	8.75
3410	In teflon lined truck				"				10.20	11.25
5000	Heavy sludge or dry vacuumable material				Hr.				140	155
6000	Dumpsite disposal charge, minimum				Ton				140	155
6020	Maximum				"				415	455

02 82 Asbestos Remediation

02 82 13 – Asbestos Abatement

02 82 13.39 Asbestos Remediation Plans and Methods

		Crew	Daily Output	Labor-Hours	Unit	Material	2022 Bare Costs Labor	Equipment	Total	Total Incl O&P
0010	**ASBESTOS REMEDIATION PLANS AND METHODS**									
0100	Building Survey-Commercial Building				Ea.				2,200	2,400
0200	Asbestos Abatement Remediation Plan				"				1,350	1,475

For customer support on your Commercial Renovation Costs with RSMeans data, call 800.448.8182.

39

02 82 13 – Asbestos Abatement

02 82 13.41 Asbestos Abatement Equipment	Crew	Daily Output	Labor-Hours	Unit	Material	2022 Bare Costs Labor	Equipment	Total	Total Incl O&P
0010 **ASBESTOS ABATEMENT EQUIPMENT** R028213-20									
0011 Equipment and supplies, buy									
0200 Air filtration device, 2000 CFM				Ea.	1,100			1,100	1,200
0250 Large volume air sampling pump, minimum					435			435	480
0260 Maximum					460			460	505
0300 Airless sprayer unit, 2 gun					2,600			2,600	2,850
0350 Light stand, 500 watt					48			48	52.50
0400 Personal respirators									
0410 Negative pressure, 1/2 face, dual operation, minimum				Ea.	28			28	31
0420 Maximum					22			22	24.50
0450 P.A.P.R., full face, minimum					710			710	780
0460 Maximum					1,275			1,275	1,400
0470 Supplied air, full face, including air line, minimum				Ea.	211			211	233
0480 Maximum					845			845	930
0500 Personnel sampling pump					215			215	236
1500 Power panel, 20 unit, including GFI					565			565	620
1600 Shower unit, including pump and filters					960			960	1,050
1700 Supplied air system (type C)					3,525			3,525	3,875
1750 Vacuum cleaner, HEPA, 16 gal., stainless steel, wet/dry					1,450			1,450	1,600
1760 55 gallon					1,550			1,550	1,700
1800 Vacuum loader, 9-18 ton/hr.					113,000			113,000	124,000
1900 Water atomizer unit, including 55 gal. drum					325			325	355
2000 Worker protection, whole body, foot, head cover & gloves, plastic					10.50			10.50	11.55
2500 Respirator, single use					28			28	30.50
2550 Cartridge for respirator					4.24			4.24	4.66
2570 Glove bag, 7 mil, 50" x 64"					11.90			11.90	13.05
2580 10 mil, 44" x 60"					9.15			9.15	10.10
6000 Disposable polyethylene bags, 6 mil, 3 C.F.					1.20			1.20	1.32
6300 Disposable fiber drums, 3 C.F.					20			20	22
6400 Pressure sensitive caution labels, 3" x 5"					3.57			3.57	3.93
6450 11" x 17"					13.10			13.10	14.40
6500 Negative air machine, 1800 CFM					900			900	990

02 82 13.42 Preparation of Asbestos Containment Area

	Crew	Daily Output	Labor-Hours	Unit	Material	2022 Bare Costs Labor	Equipment	Total	Total Incl O&P
0010 **PREPARATION OF ASBESTOS CONTAINMENT AREA**									
0100 Pre-cleaning, HEPA vacuum and wet wipe, flat surfaces	A-9	12000	.005	S.F.	.02	.34		.36	.55
0200 Protect carpeted area, 2 layers 6 mil poly on 3/4" plywood	"	1000	.064		2.67	4.04		6.71	9.35
0300 Separation barrier, 2" x 4" @ 16", 1/2" plywood ea. side, 8' high	2 Carp	400	.040		3.48	2.25		5.73	7.40
0310 12' high		320	.050		3.39	2.82		6.21	8.20
0320 16' high		200	.080		3.35	4.50		7.85	10.85
0400 Personnel decontam. chamber, 2" x 4" @ 16", 3/4" ply ea. side		280	.057		3.76	3.22		6.98	9.25
0450 Waste decontam. chamber, 2" x 4" studs @ 16", 3/4" ply ea. side		360	.044		3.76	2.50		6.26	8.10
0500 Cover surfaces with polyethylene sheeting									
0501 Including glue and tape									
0550 Floors, each layer, 6 mil	A-9	8000	.008	S.F.	.06	.50		.56	.87
0551 4 mil		9000	.007		.05	.45		.50	.76
0560 Walls, each layer, 6 mil		6000	.011		.06	.67		.73	1.13
0561 4 mil		7000	.009		.05	.58		.63	.96
0570 For heights above 14', add						20%			
0575 For heights above 20', add						30%			
0580 For fire retardant poly, add					100%				
0590 For large open areas, deduct					10%	20%			
0600 Seal floor penetrations with foam firestop to 36 sq. in.	2 Carp	200	.080	Ea.	16.05	4.50		20.55	25

02 82 13 – Asbestos Abatement

02 82 13.42 Preparation of Asbestos Containment Area

		Crew	Daily Output	Labor-Hours	Unit	Material	2022 Bare Costs Labor	Equipment	Total	Total Incl O&P
0610	36 sq. in. to 72 sq. in.	2 Carp	125	.128	Ea.	32	7.20		39.20	47
0615	72 sq. in. to 144 sq. in.		80	.200		64	11.25		75.25	88.50
0620	Wall penetrations, to 36 sq. in.		180	.089		16.05	5		21.05	25.50
0630	36 sq. in. to 72 sq. in.		100	.160		32	9		41	50
0640	72 sq. in. to 144 sq. in.		60	.267		64	15		79	94.50
0800	Caulk seams with latex	1 Carp	230	.035	L.F.	.18	1.96		2.14	3.31
0900	Set up neg. air machine, 1-2k CFM/25 M.C.F. volume	1 Asbe	4.30	1.860	Ea.		117		117	185
0950	Set up and remove portable shower unit	2 Asbe	4	4	"		252		252	400

02 82 13.43 Bulk Asbestos Removal

		Crew	Daily Output	Labor-Hours	Unit	Material	2022 Bare Costs Labor	Equipment	Total	Total Incl O&P
0010	**BULK ASBESTOS REMOVAL**									
0020	Includes disposable tools and 2 suits and 1 respirator filter/day/worker									
0100	Beams, W 10 x 19	A-9	235	.272	L.F.	.86	17.20		18.06	28
0110	W 12 x 22		210	.305		.96	19.25		20.21	31.50
0120	W 14 x 26		180	.356		1.12	22.50		23.62	36.50
0130	W 16 x 31		160	.400		1.26	25		26.26	41.50
0140	W 18 x 40		140	.457		1.44	29		30.44	47
0150	W 24 x 55		110	.582		1.84	36.50		38.34	60
0160	W 30 x 108		85	.753		2.38	47.50		49.88	77.50
0170	W 36 x 150		72	.889		2.81	56		58.81	91.50
0200	Boiler insulation		480	.133	S.F.	.56	8.40		8.96	13.90
0210	With metal lath, add				%				50%	50%
0300	Boiler breeching or flue insulation	A-9	520	.123	S.F.	.39	7.75		8.14	12.70
0310	For active boiler, add				%				100%	100%
0400	Duct or AHU insulation	A-10B	440	.073	S.F.	.23	4.59		4.82	7.50
0500	Duct vibration isolation joints, up to 24 sq. in. duct	A-9	56	1.143	Ea.	3.61	72		75.61	118
0520	25 sq. in. to 48 sq. in. duct		48	1.333		4.21	84		88.21	138
0530	49 sq. in. to 76 sq. in. duct		40	1.600		5.05	101		106.05	166
0600	Pipe insulation, air cell type, up to 4" diameter pipe		900	.071	L.F.	.22	4.49		4.71	7.35
0610	4" to 8" diameter pipe		800	.080		.25	5.05		5.30	8.30
0620	10" to 12" diameter pipe		700	.091		.29	5.75		6.04	9.40
0630	14" to 16" diameter pipe		550	.116		.37	7.35		7.72	12
0650	Over 16" diameter pipe		650	.098	S.F.	.31	6.20		6.51	10.15
0700	With glove bag up to 3" diameter pipe		200	.320	L.F.	3.20	20		23.20	35.50
1000	Pipe fitting insulation up to 4" diameter pipe	A-9	320	.200	Ea.	.63	12.60		13.23	20.50
1100	6" to 8" diameter pipe		304	.211		.66	13.30		13.96	21.50
1110	10" to 12" diameter pipe		192	.333		1.05	21		22.05	34
1120	14" to 16" diameter pipe		128	.500		1.58	31.50		33.08	51.50
1130	Over 16" diameter pipe		176	.364	S.F.	1.15	23		24.15	38
1200	With glove bag, up to 8" diameter pipe		75	.853	L.F.	8.05	54		62.05	94
2000	Scrape foam fireproofing from flat surface		2400	.027	S.F.	.08	1.68		1.76	2.75
2100	Irregular surfaces		1200	.053		.17	3.37		3.54	5.50
3000	Remove cementitious material from flat surface		1800	.036		.11	2.24		2.35	3.66
3100	Irregular surface		1000	.064		.14	4.04		4.18	6.55
4000	Scrape acoustical coating/fireproofing, from ceiling		3200	.020		.06	1.26		1.32	2.06
5000	Remove VAT and mastic from floor by hand		2400	.027		.08	1.68		1.76	2.75
5100	By machine	A-11	4800	.013		.04	.84	.01	.89	1.39
5150	For 2 layers, add				%				50%	50%
6000	Remove contaminated soil from crawl space by hand	A-9	400	.160	C.F.	.51	10.10		10.61	16.50
6100	With large production vacuum loader	A-12	700	.091	"	.29	5.75	1.10	7.14	10.65
7000	Radiator backing, not including radiator removal	A-9	1200	.053	S.F.	.17	3.37		3.54	5.50
8000	Cement-asbestos transite board and cement wall board	2 Asbe	1000	.016		.21	1.01		1.22	1.82
8100	Transite shingle siding	A-10B	750	.043		.29	2.70		2.99	4.58

02 82 Asbestos Remediation

02 82 13 – Asbestos Abatement

02 82 13.43 Bulk Asbestos Removal	Crew	Daily Output	Labor-Hours	Unit	Material	2022 Bare Costs Labor	2022 Bare Costs Equipment	Total	Total Incl O&P
8200	Shingle roofing	A-10B	2000	.016	S.F.	.10	1.01		1.11	1.71
8250	Built-up, no gravel, non-friable	B-2	1400	.029		.10	1.31		1.41	2.20
8260	Bituminous flashing	1 Rofc	300	.027		.10	1.32		1.42	2.39
8300	Asbestos millboard, flat board and VAT contaminated plywood	2 Asbe	1000	.016	▼	.09	1.01		1.10	1.69
9000	For type B (supplied air) respirator equipment, add				%				10%	10%

02 82 13.44 Demolition In Asbestos Contaminated Area

		Crew	Daily Output	Labor-Hours	Unit	Material	2022 Bare Costs Labor	2022 Bare Costs Equipment	Total	Total Incl O&P
0010	**DEMOLITION IN ASBESTOS CONTAMINATED AREA**									
0200	Ceiling, including suspension system, plaster and lath	A-9	2100	.030	S.F.	.10	1.92		2.02	3.15
0210	Finished plaster, leaving wire lath		585	.109		.35	6.90		7.25	11.30
0220	Suspended acoustical tile		3500	.018		.06	1.15		1.21	1.88
0230	Concealed tile grid system		3000	.021		.07	1.35		1.42	2.20
0240	Metal pan grid system		1500	.043		.13	2.69		2.82	4.40
0250	Gypsum board		2500	.026	▼	.08	1.62		1.70	2.64
0260	Lighting fixtures up to 2' x 4'	▼	72	.889	Ea.	2.81	56		58.81	91.50
0400	Partitions, non load bearing									
0410	Plaster, lath, and studs	A-9	690	.093	S.F.	.73	5.85		6.58	10.05
0450	Gypsum board and studs	"	1390	.046	"	.15	2.91		3.06	4.75
9000	For type B (supplied air) respirator equipment, add				%				10%	10%

02 82 13.45 OSHA Testing

		Crew	Daily Output	Labor-Hours	Unit	Material	2022 Bare Costs Labor	2022 Bare Costs Equipment	Total	Total Incl O&P
0010	**OSHA TESTING**									
0100	Certified technician, minimum				Day		340		340	340
0110	Maximum						670		670	670
0121	Industrial hygienist, minimum						385		385	385
0130	Maximum				▼		755		755	755
0200	Asbestos sampling and PCM analysis, NIOSH 7400, minimum	1 Asbe	8	1	Ea.	18	63		81	119
0210	Maximum		4	2		24	126		150	226
1000	Cleaned area samples		8	1		18	63		81	119
1100	PCM air sample analysis, NIOSH 7400, minimum		8	1		18	63		81	119
1110	Maximum	▼	4	2		24	126		150	226
1200	TEM air sample analysis, NIOSH 7402, minimum								80	106
1210	Maximum				▼				360	450

02 82 13.46 Decontamination of Asbestos Containment Area

		Crew	Daily Output	Labor-Hours	Unit	Material	2022 Bare Costs Labor	2022 Bare Costs Equipment	Total	Total Incl O&P
0010	**DECONTAMINATION OF ASBESTOS CONTAINMENT AREA**									
0100	Spray exposed substrate with surfactant (bridging)									
0200	Flat surfaces	A-9	6000	.011	S.F.	.44	.67		1.11	1.54
0250	Irregular surfaces		4000	.016	"	.47	1.01		1.48	2.12
0300	Pipes, beams, and columns		2000	.032	L.F.	.63	2.02		2.65	3.88
1000	Spray encapsulate polyethylene sheeting		8000	.008	S.F.	.46	.50		.96	1.31
1100	Roll down polyethylene sheeting		8000	.008	"		.50		.50	.80
1500	Bag polyethylene sheeting		400	.160	Ea.	1.05	10.10		11.15	17.10
2000	Fine clean exposed substrate, with nylon brush		2400	.027	S.F.		1.68		1.68	2.66
2500	Wet wipe substrate		4800	.013			.84		.84	1.33
2600	Vacuum surfaces, fine brush	▼	6400	.010	▼		.63		.63	1
3000	Structural demolition									
3100	Wood stud walls	A-9	2800	.023	S.F.		1.44		1.44	2.28
3500	Window manifolds, not incl. window replacement		4200	.015			.96		.96	1.52
3600	Plywood carpet protection	▼	2000	.032	▼		2.02		2.02	3.19
4000	Remove custom decontamination facility	A-10A	8	3	Ea.	16.35	190		206.35	320
4100	Remove portable decontamination facility	3 Asbe	12	2	"	15	126		141	216
5000	HEPA vacuum, shampoo carpeting	A-9	4800	.013	S.F.	.12	.84		.96	1.46
9000	Final cleaning of protected surfaces	A-10A	8000	.003	"		.19		.19	.30

02 82 Asbestos Remediation

02 82 13 – Asbestos Abatement

02 82 13.47 Asbestos Waste Pkg., Handling, and Disp.	Crew	Daily Output	Labor-Hours	Unit	Material	2022 Bare Costs Labor	Equipment	Total	Total Incl O&P
0010 ASBESTOS WASTE PACKAGING, HANDLING, AND DISPOSAL									
0100 Collect and bag bulk material, 3 C.F. bags, by hand	A-9	400	.160	Ea.	1.20	10.10		11.30	17.25
0200 Large production vacuum loader	A-12	880	.073		1.01	4.59	.87	6.47	9.30
1000 Double bag and decontaminate	A-9	960	.067		1.20	4.21		5.41	7.95
2000 Containerize bagged material in drums, per 3 C.F. drum	A-9	800	.080	Ea.	20	5.05		25.05	30
3000 Cart bags 50' to dumpster	2 Asbe	400	.040	"		2.52		2.52	3.98
5000 Disposal charges, not including haul, minimum				C.Y.				61	67
5020 Maximum				"				355	395
9000 For type B (supplied air) respirator equipment, add				%				10%	10%

02 82 13.48 Asbestos Encapsulation With Sealants

	Crew	Daily Output	Labor-Hours	Unit	Material	2022 Bare Costs Labor	Equipment	Total	Total Incl O&P
0010 ASBESTOS ENCAPSULATION WITH SEALANTS									
0100 Ceilings and walls, minimum	A-9	21000	.003	S.F.	.54	.19		.73	.89
0110 Maximum		10600	.006		.58	.38		.96	1.24
0200 Columns and beams, minimum		13300	.005		.54	.30		.84	1.07
0210 Maximum		5325	.012		.65	.76		1.41	1.92
0300 Pipes to 12" diameter including minor repairs, minimum		800	.080	L.F.	.65	5.05		5.70	8.70
0310 Maximum		400	.160	"	1.27	10.10		11.37	17.35

02 83 Lead Remediation

02 83 19 – Lead-Based Paint Remediation

02 83 19.21 Lead Paint Remediation Plans and Methods

	Crew	Daily Output	Labor-Hours	Unit	Material	2022 Bare Costs Labor	Equipment	Total	Total Incl O&P
0010 LEAD PAINT REMEDIATION PLANS AND METHODS									
0100 Building Survey-Commercial Building				Ea.				2,050	2,250
0200 Lead Abatement Remediation Plan								1,225	1,350
0300 Lead Paint Testing, AAS Analysis								51	56
0400 Lead Paint Testing, X-Ray Fluorescence								51	56

02 83 19.22 Preparation of Lead Containment Area

	Crew	Daily Output	Labor-Hours	Unit	Material	2022 Bare Costs Labor	Equipment	Total	Total Incl O&P
0010 PREPARATION OF LEAD CONTAINMENT AREA									
0020 Lead abatement work area, test kit, per swab	1 Skwk	16	.500	Ea.	3.40	29.50		32.90	50
0025 For dust barriers see Section 01 56 16.10									
0050 Caution sign	1 Skwk	48	.167	Ea.	4.44	9.80		14.24	20.50
0100 Pre-cleaning, HEPA vacuum and wet wipe, floor and wall surfaces	3 Skwk	5000	.005	S.F.	.02	.28		.30	.47
0105 Ceiling, 6'-11' high		4100	.006		.10	.34		.44	.66
0108 12'-15' high		3550	.007		.09	.40		.49	.73
0115 Over 15' high		3000	.008		.12	.47		.59	.88
0500 Cover surfaces with polyethylene sheeting									
0550 Floors, each layer, 6 mil	3 Skwk	8000	.003	S.F.	.06	.18		.24	.35
0560 Walls, each layer, 6 mil	"	6000	.004	"	.06	.24		.30	.44
0570 For heights above 14', add						20%			
2400 Post abatement cleaning of protective sheeting, HEPA vacuum & wet wipe	3 Skwk	5000	.005	S.F.	.02	.28		.30	.47
2450 Doff, bag and seal protective sheeting		12000	.002		.01	.12		.13	.20
2500 Post abatement cleaning, HEPA vacuum & wet wipe		5000	.005		.02	.28		.30	.47

02 83 19.23 Encapsulation of Lead-Based Paint

	Crew	Daily Output	Labor-Hours	Unit	Material	2022 Bare Costs Labor	Equipment	Total	Total Incl O&P
0010 ENCAPSULATION OF LEAD-BASED PAINT									
0020 Interior, brushwork, trim, under 6"	1 Pord	240	.033	L.F.	2.34	1.59		3.93	5.10
0030 6"-12" wide		180	.044		3.16	2.12		5.28	6.85
0040 Balustrades		300	.027		2.14	1.27		3.41	4.37
0050 Pipe to 4" diameter		500	.016		1.31	.76		2.07	2.65
0060 To 8" diameter		375	.021		1.55	1.02		2.57	3.32
0070 To 12" diameter		250	.032		2.33	1.53		3.86	4.98

For customer support on your Commercial Renovation Costs with RSMeans data, call 800.448.8182.

43

02 83 Lead Remediation

02 83 19 – Lead-Based Paint Remediation

02 83 19.23 Encapsulation of Lead-Based Paint

	Crew	Daily Output	Labor-Hours	Unit	Material	2022 Bare Costs Labor	Equipment	Total	Total Incl O&P
0080	To 16" diameter	1 Pord	170	.047	L.F.	3.46	2.25		5.71	7.35
0090	Cabinets, ornate design		200	.040	S.F.	3.29	1.91		5.20	6.65
0100	Simple design		250	.032	"	2.54	1.53		4.07	5.20
0110	Doors, 3' x 7', both sides, incl. frame & trim									
0120	Flush	1 Pord	6	1.333	Ea.	29.50	63.50		93	134
0130	French, 10-15 lite		3	2.667		6.65	127		133.65	209
0140	Panel		4	2		33.50	95.50		129	188
0150	Louvered		2.75	2.909		31.50	139		170.50	255
0160	Windows, per interior side, per 15 S.F.									
0170	1-6 lite	1 Pord	14	.571	Ea.	23	27.50		50.50	68
0180	7-10 lite		7.50	1.067		23	51		74	106
0190	12 lite		5.75	1.391		34	66.50		100.50	142
0200	Radiators		8	1		72.50	48		120.50	155
0210	Grilles, vents		275	.029	S.F.	2.10	1.39		3.49	4.51
0220	Walls, roller, drywall or plaster		1000	.008		.67	.38		1.05	1.34
0230	With spunbonded reinforcing fabric		720	.011		.82	.53		1.35	1.74
0240	Wood		800	.010		.71	.48		1.19	1.54
0250	Ceilings, roller, drywall or plaster		900	.009		.82	.42		1.24	1.57
0260	Wood		700	.011		.93	.55		1.48	1.88
0270	Exterior, brushwork, gutters and downspouts		300	.027	L.F.	2.09	1.27		3.36	4.32
0280	Columns		400	.020	S.F.	1.41	.96		2.37	3.06
0290	Spray, siding		600	.013	"	1.08	.64		1.72	2.20
0300	Miscellaneous									
0310	Electrical conduit, brushwork, to 2" diameter	1 Pord	500	.016	L.F.	1.31	.76		2.07	2.65
0320	Brick, block or concrete, spray		500	.016	S.F.	1.35	.76		2.11	2.70
0330	Steel, flat surfaces and tanks to 12"		500	.016		1.54	.76		2.30	2.90
0340	Beams, brushwork		400	.020		1.41	.96		2.37	3.06
0350	Trusses		400	.020		1.41	.96		2.37	3.06

02 83 19.26 Removal of Lead-Based Paint

	Crew	Daily Output	Labor-Hours	Unit	Material	2022 Bare Costs Labor	Equipment	Total	Total Incl O&P
0010	**REMOVAL OF LEAD-BASED PAINT** R028319-60									
0011	By chemicals, per application									
0050	Baseboard, to 6" wide	1 Pord	64	.125	L.F.	.85	6		6.85	10.40
0070	To 12" wide	"	32	.250	"	1.53	11.95		13.48	20.50
0200	Balustrades, one side	1 Pord	28	.286	S.F.	1.49	13.65		15.14	23
1400	Cabinets, simple design		32	.250		1.41	11.95		13.36	20.50
1420	Ornate design		25	.320		1.70	15.30		17	26
1600	Cornice, simple design		60	.133		1.57	6.35		7.92	11.85
1620	Ornate design		20	.400		5.50	19.10		24.60	36
2800	Doors, one side, flush		84	.095		1.58	4.55		6.13	8.95
2820	Two panel		80	.100		1.37	4.78		6.15	9.05
2840	Four panel		45	.178		1.40	8.50		9.90	15
2880	For trim, one side, add		64	.125	L.F.	.81	6		6.81	10.35
3000	Fence, picket, one side		30	.267	S.F.	1.37	12.75		14.12	21.50
3200	Grilles, one side, simple design		30	.267		1.37	12.75		14.12	21.50
3220	Ornate design		25	.320		1.46	15.30		16.76	25.50
3240	Handrails		90	.089	L.F.	1.37	4.25		5.62	8.20
4400	Pipes, to 4" diameter		90	.089		1.58	4.25		5.83	8.45
4420	To 8" diameter		50	.160		3.16	7.65		10.81	15.60
4440	To 12" diameter		36	.222		4.74	10.60		15.34	22
4460	To 16" diameter		20	.400		6.30	19.10		25.40	37
4500	For hangers, add		40	.200	Ea.	2.52	9.55		12.07	17.85
4800	Siding		90	.089	S.F.	1.37	4.25		5.62	8.20

02 83 Lead Remediation

02 83 19 – Lead-Based Paint Remediation

02 83 19.26 Removal of Lead-Based Paint	Crew	Daily Output	Labor-Hours	Unit	Material	2022 Bare Costs Labor	Equipment	Total	Total Incl O&P
5000	Trusses, open	1 Pord	55	.145	SF Face	2.11	6.95		9.06	13.30
6200	Windows, one side only, double-hung, 1/1 light, 24" x 48" high		4	2	Ea.	27.50	95.50		123	182
6220	30" x 60" high		3	2.667		34.50	127		161.50	240
6240	36" x 72" high		2.50	3.200		39.50	153		192.50	285
6280	40" x 80" high		2	4		52	191		243	355
6400	Colonial window, 6/6 light, 24" x 48" high		2	4		55	191		246	360
6420	30" x 60" high		1.50	5.333		69	255		324	480
6440	36" x 72" high		1	8		96.50	380		476.50	710
6480	40" x 80" high		1	8		96.50	380		476.50	710
6600	8/8 light, 24" x 48" high		2	4		50	191		241	355
6620	40" x 80" high		1	8		96.50	380		476.50	710
6800	12/12 light, 24" x 48" high		1	8		96.50	380		476.50	710
6820	40" x 80" high		.75	10.667		138	510		648	955
6840	Window frame & trim items, included in pricing above									
7000	Hand scraping and HEPA vacuum, less than 4 S.F.	1 Pord	8	1	Ea.	1.20	48		49.20	77
8000	Collect and bag bulk material, 3 C.F. bags, by hand		30	.267	"	1.20	12.75		13.95	21.50
9000	Minimum labor/equipment charge		3	2.667	Job		127		127	202

02 87 Biohazard Remediation

02 87 13 – Mold Remediation

02 87 13.16 Mold Remediation Preparation and Containment

		Crew	Daily Output	Labor-Hours	Unit	Material	2022 Bare Costs Labor	Equipment	Total	Total Incl O&P
0010	**MOLD REMEDIATION PREPARATION AND CONTAINMENT**									
0015	Mold remediation plans and methods									
0020	Initial inspection, areas to 2,500 S.F.				Total				268	295
0030	Areas to 5,000 S.F.								440	485
0032	Areas to 10,000 S.F.								440	485
0040	Testing, air sample each								126	139
0050	Swab sample								115	127
0060	Tape sample								125	140
0070	Post remediation air test								126	139
0080	Mold abatement plan, area to 2,500 S.F.								1,225	1,325
0090	Areas to 5,000 S.F.								1,625	1,775
0095	Areas to 10,000 S.F.								2,550	2,800
0100	Packup & removal of contents, average 3 bedroom home, excl. storage								8,150	8,975
0110	Average 5 bedroom home, excl. storage								15,300	16,800
0600	For demolition in mold contaminated areas, see Section 02 87 13.33									
0610	For personal protection equipment, see Section 02 82 13.41									
6010	Preparation of mold containment area									
6100	Pre-cleaning, HEPA vacuum and wet wipe, flat surfaces	A-9	12000	.005	S.F.	.02	.34		.36	.55
6300	Separation barrier, 2" x 4" @ 16", 1/2" plywood ea. side, 8' high	2 Carp	400	.040		5.20	2.25		7.45	9.30
6310	12' high		320	.050		6.15	2.82		8.97	11.25
6320	16' high		200	.080		2.54	4.50		7.04	9.95
6400	Personnel decontam. chamber, 2" x 4" @ 16", 3/4" ply ea. side		280	.057		5.15	3.22		8.37	10.80
6450	Waste decontam. chamber, 2" x 4" studs @ 16", 3/4" ply each side		360	.044		20.50	2.50		23	27
6500	Cover surfaces with polyethylene sheeting									
6501	Including glue and tape									
6550	Floors, each layer, 6 mil	A-9	8000	.008	S.F.	.06	.50		.56	.87
6551	4 mil		9000	.007		.05	.45		.50	.76
6560	Walls, each layer, 6 mil		6000	.011		.06	.67		.73	1.13
6561	4 mil		7000	.009		.05	.58		.63	.96
6570	For heights above 14', add						20%			

For customer support on your Commercial Renovation Costs with RSMeans data, call 800.448.8182.

45

02 87 Biohazard Remediation

02 87 13 – Mold Remediation

02 87 13.16 Mold Remediation Preparation and Containment

	02 87 13.16 Mold Remediation Preparation and Containment	Crew	Daily Output	Labor-Hours	Unit	Material	2022 Bare Costs Labor	Equipment	Total	Total Incl O&P
6575	For heights above 20', add						30%			
6580	For fire retardant poly, add					100%				
6590	For large open areas, deduct					10%	20%			
6600	Seal floor penetrations with foam firestop to 36 sq. in.	2 Carp	200	.080	Ea.	16.05	4.50		20.55	25
6610	36 sq. in. to 72 sq. in.	2 Carp	125	.128	Ea.	32	7.20		39.20	47
6615	72 sq. in. to 144 sq. in.		80	.200		64	11.25		75.25	88.50
6620	Wall penetrations, to 36 sq. in.		180	.089		16.05	5		21.05	25.50
6630	36 sq. in. to 72 sq. in.		100	.160		32	9		41	50
6640	72 sq. in. to 144 sq. in.		60	.267		64	15		79	94.50
6800	Caulk seams with latex caulk	1 Carp	230	.035	L.F.	.18	1.96		2.14	3.31
6900	Set up neg. air machine, 1-2k CFM/25 M.C.F. volume	1 Asbe	4.30	1.860	Ea.		117		117	185

02 87 13.33 Removal and Disposal of Materials With Mold

		Crew	Daily Output	Labor-Hours	Unit	Material	2022 Bare Costs Labor	Equipment	Total	Total Incl O&P
0010	**REMOVAL AND DISPOSAL OF MATERIALS WITH MOLD**									
0015	Demolition in mold contaminated area									
0200	Ceiling, including suspension system, plaster and lath	A-9	2100	.030	S.F.	.10	1.92		2.02	3.15
0210	Finished plaster, leaving wire lath		585	.109		.35	6.90		7.25	11.30
0220	Suspended acoustical tile		3500	.018		.06	1.15		1.21	1.88
0230	Concealed tile grid system		3000	.021		.07	1.35		1.42	2.20
0240	Metal pan grid system		1500	.043		.13	2.69		2.82	4.40
0250	Gypsum board		2500	.026		.08	1.62		1.70	2.64
0255	Plywood		2500	.026		.08	1.62		1.70	2.64
0260	Lighting fixtures up to 2' x 4'		72	.889	Ea.	2.81	56		58.81	91.50
0400	Partitions, non load bearing									
0410	Plaster, lath, and studs	A-9	690	.093	S.F.	.73	5.85		6.58	10.05
0450	Gypsum board and studs		1390	.046		.15	2.91		3.06	4.75
0465	Carpet & pad		1390	.046		.15	2.91		3.06	4.75
0600	Pipe insulation, air cell type, up to 4" diameter pipe		900	.071	L.F.	.22	4.49		4.71	7.35
0610	4" to 8" diameter pipe		800	.080		.25	5.05		5.30	8.30
0620	10" to 12" diameter pipe		700	.091		.29	5.75		6.04	9.40
0630	14" to 16" diameter pipe		550	.116		.37	7.35		7.72	12
0650	Over 16" diameter pipe		650	.098	S.F.	.31	6.20		6.51	10.15
9000	For type B (supplied air) respirator equipment, add				%				10%	10%

02 87 19 – Bio-Hazard and Infectious Disease Control Remediation

02 87 19.10 Temporary Barriers, Including Ppe

		Crew	Daily Output	Labor-Hours	Unit	Material	2022 Bare Costs Labor	Equipment	Total	Total Incl O&P
0010	**TEMPORARY BARRIERS, including PPE**									
0100	Infectious control partition, adjustable to 10' high, obscured, 4' panel	2 Carp	75	.213	Ea.	580	12		592	660
0110	3' panel		75	.213		550	12		562	625
0120	2' panel		75	.213		430	12		442	495
0130	1' panel	1 Carp	75	.107		285	6		291	325
0140	6" panel	"	75	.107		265	6		271	300
0150	2' panel with HEPA filtered discharge port	2 Carp	75	.213		500	12		512	570
0160	3' panel with 32" door	"	75	.213		895	12		907	1,000
0170	4' panel with 36" door	2 Carp	75	.213	Ea.	995	12		1,007	1,125
0180	4'-6" panel with 44" door		75	.213		1,200	12		1,212	1,350
0190	Hinged corner		68	.235		186	13.25		199.25	225
0200	Outside corner		68	.235		151	13.25		164.25	187
0210	T post		68	.235		151	13.25		164.25	187
0220	Accessories, ceiling grid clip		300	.053		7.65	3		10.65	13.15
0230	Panel locking clip		300	.053		5.35	3		8.35	10.65
0240	Panel joint closure strip		300	.053		8.20	3		11.20	13.75
0250	Screw jack		300	.053		6.85	3		9.85	12.30
0260	Digital pressure difference guage					275			275	305

02 87 19 – Bio-Hazard and Infectious Disease Control Remediation

02 87 19.10 Temporary Barriers, Including Ppe

		Crew	Daily Output	Labor-Hours	Unit	Material	2022 Bare Costs Labor	2022 Bare Costs Equipment	Total	Total Incl O&P
0280	Combination lockset	1 Carp	11	.727	Ea.	207	41		248	293
0285	Sealant tape, 2" wide	1 Clab	165	.048	C.L.F.	11.10	2.21		13.31	15.75
0290	System in place, including door and accessories									
0300	Based upon 25 uses	2 Carp	43	.372	L.F.	11.25	21		32.25	46
0310	Based upon 50 uses		43	.372		6.30	21		27.30	40.50
0330	Based upon 100 uses		43	.372		3.84	21		24.84	37.50
0400	Separation barrier, 2" x 4" @ 16" O.C., 1/2" plywood each side, 8' high		340	.047	S.F.	3.65	2.65		6.30	8.20
0410	12' high		275	.058		3.61	3.28		6.89	9.15
0420	16' high		170	.094		3.70	5.30		9	12.45
0430	5/8" FR GWB each side, 8' high		290	.055		2.57	3.11		5.68	7.75
0440	12' high		235	.068		2.49	3.83		6.32	8.85
0450	16' high		144	.111		2.45	6.25		8.70	12.65

02 87 19.20 Temporary Protection, Including Ppe

		Crew	Daily Output	Labor-Hours	Unit	Material	2022 Bare Costs Labor	2022 Bare Costs Equipment	Total	Total Incl O&P
0010	**TEMPORARY PROTECTION, including PPE**									
0100	Infectious control protection, 6 mil plastic sheeting, applied to floors	1 Skwk	1900	.004	S.F.	.06	.25		.31	.46
0110	Applied to walls		1400	.006		.06	.34		.40	.60
0120	Applied to ceilings		1000	.008		.06	.47		.53	.82
0130	Sealant tape, 2" wide	1 Clab	165	.048	C.L.F.	11.10	2.21		13.31	15.75
0140	Floor covering, 1/8" tempered hardboard	2 Carp	878	.018	S.F.	.51	1.03		1.54	2.20
0150	1/2" CDX		878	.018		.95	1.03		1.98	2.68
0160	3/4" CDX		755	.021		1.14	1.19		2.33	3.15

02 87 19.30 Infectious Control Equipment and Supplies

		Crew	Daily Output	Labor-Hours	Unit	Material	2022 Bare Costs Labor	2022 Bare Costs Equipment	Total	Total Incl O&P
0010	**INFECTIOUS CONTROL EQUIPMENT AND SUPPLIES**									
0100	Entry/exit decontamination chamber, portable, 10 uses	A-10C	12	2	Ea.	228	126	28.50	382.50	480
0105	Negative air machine, 1800 CFM					900			900	990
0110	Disposable polyethylene bags, 6 mil, 3 C.F.					1.20			1.20	1.32
0130	Paper towels, carton of 12, 350' rolls					77.50			77.50	85
0140	Heavy duty wipes, 160 per box					28			28	30.50
0200	Tent, including frame, weights, roof, walls, 5' x 5'	2 Skwk	32	.500	Ea.	2,150	29.50		2,179.50	2,425
0205	5' x 10'		24	.667		2,500	39.50		2,539.50	2,800
0210	10' x 10'		24	.667		3,125	39.50		3,164.50	3,475
0220	13' x 26'		12	1.333		9,350	78.50		9,428.50	10,400
0230	13' x 26', flat roof infirmary		12	1.333		7,600	78.50		7,678.50	8,475
0240	13' x 26', drive through screening		12	1.333		8,800	78.50		8,878.50	9,800
0242	LED light kit for tents, 15 watt	1 Elec	48	.167		199	11.05		210.05	236
0244	Halogen heating kit, 1500 watt	"	48	.167		500	11.05		511.05	565
0250	Medical partition system, including frame, drapes, 5' x 5'	2 Skwk	48	.333		1,300	19.65		1,319.65	1,450

02 87 19.40 Infectious Control Cleaning, Includes Ppe

		Crew	Daily Output	Labor-Hours	Unit	Material	2022 Bare Costs Labor	2022 Bare Costs Equipment	Total	Total Incl O&P
0010	**INFECTIOUS CONTROL CLEANING, Includes PPE**									
0100	Infectious control cleaning, including PPE and antimicrobial spray, floors	1 Skwk	3600	.002	S.F.	.03	.13		.16	.24
0110	Walls		3000	.003		.03	.16		.19	.28
0120	Ceilings		2700	.003		.03	.17		.20	.31
0130	Countertops		3300	.002		.03	.14		.17	.26
0140	Other surfaces		2600	.003		.03	.18		.21	.32
0150	Plumbing fixtures		14	.571	Ea.	2.31	33.50		35.81	56
0160	Bathroom accessories		30	.267		1.06	15.70		16.76	26
0170	Kitchen appliance		9	.889		3.88	52.50		56.38	87.50
0180	Electrical switches and receptacles		140	.057		.23	3.37		3.60	5.60
0190	Work station		9	.889		4.08	52.50		56.58	87.50
0200	Door and window hardware		88	.091		.36	5.35		5.71	8.85
0250	Supply / return ductwork, up to 2 S.F. cross section		240	.033	L.F.	.24	1.96		2.20	3.38
0260	2 to 4 S.F. cross section		192	.042		.35	2.45		2.80	4.27

02 87 19 – Bio-Hazard and Infectious Disease Control Remediation

02 87 19.40 Infectious Control Cleaning, Includes Ppe	Crew	Daily Output	Labor-Hours	Unit	Material	2022 Bare Costs Labor	Equipment	Total	Total Incl O&P
0270	4 to 8 S.F. cross section	1 Skwk	144	.056	L.F.	.44	3.27		3.71	5.70
0280	8 SF cross section		102	.078	↓	.67	4.62		5.29	8.05
0300	Duct coil		5	1.600	Ea.	6.70	94		100.70	156
0310	Duct damper and vane set		77	.104		.78	6.10		6.88	10.55
0320	Fan coil unit or VAV box		8	1		4.09	59		63.09	97.50
0330	Air handling unit and coil		2	4		16.35	236		252.35	395
0340	Grill / diffuser or register	↓	32	.250	↓	1.16	14.75		15.91	25

Estimating Tips
General

- Carefully check all the plans and specifications. Concrete often appears on drawings other than structural drawings, including mechanical and electrical drawings for equipment pads. The cost of cutting and patching is often difficult to estimate. See Subdivision 03 81 for Concrete Cutting, Subdivision 02 41 19.16 for Cutout Demolition, Subdivision 03 05 05.10 for Concrete Demolition, and Subdivision 02 41 19.19 for Rubbish Handling (handling, loading, and hauling of debris).
- Always obtain concrete prices from suppliers near the job site. A volume discount can often be negotiated, depending upon competition in the area. Remember to add for waste, particularly for slabs and footings on grade.

03 10 00 Concrete Forming and Accessories

- A primary cost for concrete construction is forming. Most jobs today are constructed with prefabricated forms. The selection of the forms best suited for the job and the total square feet of forms required for efficient concrete forming and placing are key elements in estimating concrete construction. Enough forms must be available for erection to make efficient use of the concrete placing equipment and crew.
- Concrete accessories for forming and placing depend upon the systems used. Study the plans and specifications to ensure that all special accessory requirements have been included in the cost estimate, such as anchor bolts, inserts, and hangers.
- Included within costs for forms-in-place are all necessary bracing and shoring.

03 20 00 Concrete Reinforcing

- Ascertain that the reinforcing steel supplier has included all accessories, cutting, bending, and an allowance for lapping, splicing, and waste. A good rule of thumb is 10% for lapping, splicing, and waste. Also, 10% waste should be allowed for welded wire fabric.
- The unit price items in the subdivisions for Reinforcing In Place, Glass Fiber Reinforcing, and Welded Wire Fabric include the labor to install accessories such as beam and slab bolsters, high chairs, and bar ties and tie wire. The material cost for these accessories is not included; they may be obtained from the Accessories Subdivisions.

03 30 00 Cast-In-Place Concrete

- When estimating structural concrete, pay particular attention to requirements for concrete additives, curing methods, and surface treatments. Special consideration for climate, hot or cold, must be included in your estimate. Be sure to include requirements for concrete placing equipment and concrete finishing.
- For accurate concrete estimating, the estimator must consider each of the following major components individually: forms, reinforcing steel, ready-mix concrete, placement of the concrete, and finishing of the top surface. For faster estimating, Subdivision 03 30 53.40 for Concrete-In-Place can be used; here, various items of concrete work are presented that include the costs of all five major components (unless specifically stated otherwise).

03 40 00 Precast Concrete
03 50 00 Cast Decks and Underlayment

- The cost of hauling precast concrete structural members is often an important factor. For this reason, it is important to get a quote from the nearest supplier. It may become economically feasible to set up precasting beds on the site if the hauling costs are prohibitive.

Reference Numbers

Reference numbers are shown at the beginning of some major classifications. These numbers refer to related items in the Reference Section. The reference information may be an estimating procedure, an alternate pricing method, or technical information.

Note: Not all subdivisions listed here necessarily appear. ■

Same Data. Simplified.

Enjoy the convenience and efficiency of accessing your costs anywhere:

- **Skip the multiplier** by setting your location
- **Quickly search,** edit, favorite and share costs
- **Stay on top of price changes** with automatic updates

Discover more at rsmeans.com/online

03 01 30.62 Concrete Patching		Crew	Daily Output	Labor-Hours	Unit	Material	2022 Bare Costs Labor	Equipment	Total	Total Incl O&P
0010	**CONCRETE PATCHING**									
0100	Floors, 1/4" thick, small areas, regular grout	1 Cefi	170	.047	S.F.	1.34	2.50		3.84	5.40
0150	Epoxy grout	"	100	.080	"	11.60	4.25		15.85	19.40
2000	Walls, including chipping, cleaning and epoxy grout									
2100	1/4" deep	1 Cefi	65	.123	S.F.	4.55	6.55		11.10	15.20
2150	1/2" deep	1 Cefi	50	.160	S.F.	9.10	8.50		17.60	23.50
2200	3/4" deep		40	.200	"	13.65	10.65		24.30	31.50
9000	Minimum labor/equipment charge		4.50	1.778	Job		94.50		94.50	148

03 01 30.71 Concrete Crack Repair		Crew	Daily Output	Labor-Hours	Unit	Material	2022 Bare Costs Labor	Equipment	Total	Total Incl O&P
0010	**CONCRETE CRACK REPAIR**									
1000	Structural repair of concrete cracks by epoxy injection (ACI RAP-1)									
1001	suitable for horizontal, vertical and overhead repairs									
1010	Clean/grind concrete surface(s) free of contaminants	1 Cefi	400	.020	L.F.		1.06		1.06	1.66
1015	Rout crack with v-notch crack chaser, if needed	C-32	600	.027		.04	1.32	.27	1.63	2.40
1020	Blow out crack with oil-free dry compressed air (1 pass)	C-28	3000	.003			.14	.01	.15	.23
1030	Install surface-mounted entry ports (spacing = concrete depth)	1 Cefi	400	.020	Ea.	.95	1.06		2.01	2.71
1040	Cap crack at surface with epoxy gel (per side/face)		400	.020	L.F.	.19	1.06		1.25	1.86
1050	Snap off ports, grind off epoxy cap residue after injection		200	.040	"		2.13		2.13	3.32
1100	Manual injection with 2-part epoxy cartridge, excludes prep									
1110	Up to 1/32" (0.03125") wide x 4" deep	1 Cefi	160	.050	L.F.	.09	2.66		2.75	4.25
1120	6" deep		120	.067		.14	3.54		3.68	5.70
1130	8" deep		107	.075		.19	3.97		4.16	6.40
1140	10" deep		100	.080		.23	4.25		4.48	6.90
1150	12" deep		96	.083		.28	4.43		4.71	7.20
1210	Up to 1/16" (0.0625") wide x 4" deep		160	.050		.19	2.66		2.85	4.35
1220	6" deep		120	.067		.28	3.54		3.82	5.85
1230	8" deep		107	.075		.37	3.97		4.34	6.60
1240	10" deep		100	.080		.47	4.25		4.72	7.15
1250	12" deep		96	.083		.56	4.43		4.99	7.50
1310	Up to 3/32" (0.09375") wide x 4" deep		160	.050		.28	2.66		2.94	4.46
1320	6" deep		120	.067		.42	3.54		3.96	6
1330	8" deep		107	.075		.56	3.97		4.53	6.80
1340	10" deep		100	.080		.70	4.25		4.95	7.40
1350	12" deep		96	.083		.84	4.43		5.27	7.80
1410	Up to 1/8" (0.125") wide x 4" deep		160	.050		.37	2.66		3.03	4.56
1420	6" deep		120	.067		.56	3.54		4.10	6.15
1430	8" deep		107	.075		.74	3.97		4.71	7
1440	10" deep		100	.080		.93	4.25		5.18	7.65
1450	12" deep		96	.083		1.12	4.43		5.55	8.15
1500	Pneumatic injection with 2-part bulk epoxy, excludes prep									
1510	Up to 5/32" (0.15625") wide x 4" deep	C-31	240	.033	L.F.	.30	1.77	1.38	3.45	4.62
1520	6" deep		180	.044		.45	2.36	1.84	4.65	6.20
1530	8" deep		160	.050		.60	2.66	2.07	5.33	7.10
1540	10" deep	C-31	150	.053	L.F.	.76	2.83	2.21	5.80	7.70
1550	12" deep		144	.056		.91	2.95	2.30	6.16	8.15
1610	Up to 3/16" (0.1875") wide x 4" deep		240	.033		.36	1.77	1.38	3.51	4.69
1620	6" deep		180	.044		.54	2.36	1.84	4.74	6.30
1630	8" deep		160	.050		.73	2.66	2.07	5.46	7.25
1640	10" deep		150	.053		.91	2.83	2.21	5.95	7.85
1650	12" deep		144	.056		1.09	2.95	2.30	6.34	8.35
1710	Up to 1/4" (0.25") wide x 4" deep		240	.033		.48	1.77	1.38	3.63	4.82
1720	6" deep		180	.044		.73	2.36	1.84	4.93	6.50

03 01 30 – Maintenance of Cast-In-Place Concrete

03 01 30.71 Concrete Crack Repair		Crew	Daily Output	Labor-Hours	Unit	Material	2022 Bare Costs Labor	Equipment	Total	Total Incl O&P
1730	8" deep	C-31	160	.050	L.F.	.97	2.66	2.07	5.70	7.50
1740	10" deep		150	.053		1.21	2.83	2.21	6.25	8.20
1750	12" deep	↓	144	.056	↓	1.45	2.95	2.30	6.70	8.75
2000	Non-structural filling of concrete cracks by gravity-fed resin (ACI RAP-2)									
2001	suitable for individual cracks in stable horizontal surfaces only									
2010	Clean/grind concrete surface(s) free of contaminants	1 Cefi	400	.020	L.F.		1.06		1.06	1.66
2020	Rout crack with v-notch crack chaser, if needed	C-32	600	.027		.04	1.32	.27	1.63	2.40
2030	Blow out crack with oil-free dry compressed air (1 pass)	C-28	3000	.003			.14	.01	.15	.23
2040	Cap crack with epoxy gel at underside of elevated slabs, if needed	1 Cefi	400	.020		.19	1.06		1.25	1.86
2050	Insert backer rod into crack, if needed		400	.020		.07	1.06		1.13	1.74
2060	Partially fill crack with fine dry sand, if needed		800	.010		.02	.53		.55	.85
2070	Apply two beads of sealant alongside crack to form reservoir, if needed	↓	400	.020	↓	.27	1.06		1.33	1.96
2100	Manual filling with squeeze bottle of 2-part epoxy resin									
2110	Full depth crack up to 1/16" (0.0625") wide x 4" deep	1 Cefi	300	.027	L.F.	.07	1.42		1.49	2.29
2120	6" deep		240	.033		.11	1.77		1.88	2.89
2130	8" deep		200	.040		.15	2.13		2.28	3.48
2140	10" deep		185	.043		.18	2.30		2.48	3.79
2150	12" deep		175	.046		.22	2.43		2.65	4.04
2210	Full depth crack up to 1/8" (0.125") wide x 4" deep		270	.030		.15	1.57		1.72	2.62
2220	6" deep		215	.037		.22	1.98		2.20	3.33
2230	8" deep		180	.044		.29	2.36		2.65	4.01
2240	10" deep		165	.048		.37	2.58		2.95	4.43
2250	12" deep		155	.052		.44	2.74		3.18	4.77
2310	Partial depth crack up to 3/16" (0.1875") wide x 1" deep		300	.027		.05	1.42		1.47	2.27
2410	Up to 1/4" (0.250") wide x 1" deep		270	.030		.07	1.57		1.64	2.54
2510	Up to 5/16" (0.3125") wide x 1" deep		250	.032		.09	1.70		1.79	2.76
2610	Up to 3/8" (0.375") wide x 1" deep	↓	240	.033	↓	.11	1.77		1.88	2.89

03 01 30.72 Concrete Surface Repairs

		Crew	Daily Output	Labor-Hours	Unit	Material	Labor	Equipment	Total	Total Incl O&P
0010	**CONCRETE SURFACE REPAIRS**									
1000	Spall repairs using low-pressure spraying (ACI RAP-3)									
1001	Suitable for vertical and overhead repairs up to 3" thick									
1010	Sound the concrete surface to locate delaminated areas	1 Cefi	2000	.004	S.F.		.21		.21	.33
1020	Remove concrete in repair areas to fully expose reinforcing bars	B-9E	160	.100			4.94	.18	5.12	8
1030	Mark the perimeter of each repair area	1 Cefi	2000	.004	↓		.21		.21	.33
1040	Saw cut the perimeter of each repair area down to reinforcing bars	B-89C	160	.050	L.F.	.03	2.66	.42	3.11	4.65
1050	If corroded rebar is exposed, remove concrete to 3/4" below corroded rebars									
1051	Single layer of #4 rebar	B-9E	64	.250	S.F.		12.35	.46	12.81	19.95
1052	#5 rebar		53.30	.300			14.80	.55	15.35	24
1053	#6 rebar		45.70	.350			17.30	.64	17.94	27.50
1054	#7 rebar		40	.400			19.75	.73	20.48	32
1055	#8 rebar		35.60	.449			22	.82	22.82	36
1056	Double layer of #4 rebar		35.60	.449			22	.82	22.82	36
1057	#5 rebar		32	.500			24.50	.92	25.42	40
1058	#6 rebar		29.10	.550			27	1.01	28.01	44
1059	#7 rebar		26.70	.599			29.50	1.10	30.60	47.50
1060	#8 rebar	↓	24.60	.650			32	1.19	33.19	52
1100	See section 03 21 11.60 for placement of supplemental reinforcement bars									
1104	See section 02 41 19.19 for handling & removal of debris									
1106	See section 02 41 19.20 for disposal of debris									
1110	Final cleaning by sandblasting	E-11	1100	.029	S.F.	.57	1.46	.28	2.31	3.32
1120	By high pressure water	C-29	500	.016			.73	.20	.93	1.38
1130	Blow off dust and debris with oil-free dry compressed air	C-28	16000	.001	↓		.03		.03	.04

03 01 Maintenance of Concrete

03 01 30 – Maintenance of Cast-In-Place Concrete

03 01 30.72 Concrete Surface Repairs

		Crew	Daily Output	Labor-Hours	Unit	Material	2022 Bare Costs Labor	Equipment	Total	Total Incl O&P
1140	Mix up repair material with concrete mixer	C-30	200	.040	C.F.	10.55	1.82	.70	13.07	15.25
1150	Place repair material by low-pressure spray in lifts up to 3" thick	C-9	200	.160	"		8.05	2.14	10.19	15
1160	Screed and initial float of repair material by hand	C-10	2000	.012	S.F.		.61		.61	.95
1170	Float and steel trowel finish repair material by hand	"	1265	.019			.96		.96	1.51
1180	Cure with sprayed membrane curing compound	2 Cefi	6200	.003	↓	.14	.14		.28	.36
2000	Surface repairs using form-and-pour techniques (ACI RAP-4)									
2001	Suitable for walls, columns, beam sides/bottoms, and slab soffits									
2010	Sound the concrete surface to locate delaminated areas	1 Cefi	2000	.004	S.F.		.21		.21	.33
2020	Remove concrete in repair areas to fully expose reinforcing bars	B-9E	160	.100	↓		4.94	.18	5.12	8
2030	Mark the perimeter of each repair area	1 Cefi	2000	.004	↓		.21		.21	.33
2040	Saw cut the perimeter of each repair area down to reinforcing bars	B-89C	160	.050	L.F.	.03	2.66	.42	3.11	4.65
2050	If corroded rebar is exposed, remove concrete to 3/4" below corroded rebars									
2051	Single layer of #4 rebar	B-9E	64	.250	S.F.		12.35	.46	12.81	19.95
2052	#5 rebar		53.30	.300			14.80	.55	15.35	24
2053	#6 rebar		45.70	.350			17.30	.64	17.94	27.50
2054	#7 rebar		40	.400			19.75	.73	20.48	32
2055	#8 rebar		35.60	.449			22	.82	22.82	36
2056	Double layer of #4 rebar		35.60	.449			22	.82	22.82	36
2057	#5 rebar		32	.500			24.50	.92	25.42	40
2058	#6 rebar		29.10	.550			27	1.01	28.01	44
2059	#7 rebar		26.70	.599			29.50	1.10	30.60	47.50
2060	#8 rebar	↓	24.60	.650	↓		32	1.19	33.19	52
2100	See section 03 21 11.60 for placement of supplemental reinforcement bars									
2105	See section 02 41 19.19 for handling & removal of debris									
2110	See section 02 41 19.20 for disposal of debris									
2120	For slab soffit repairs, drill 2" diam. holes to 6" deep for conc placement	B-89A	16.50	.970	Ea.	.30	50.50	7.20	58	89
2130	For each additional inch of slab thickness in same hole, add	"	1080	.015		.05	.77	.11	.93	1.40
2140	Drill holes for expansion shields, 3/4" diameter x 4" deep	1 Carp	45	.178	↓		10		10	15.90
2150	Final cleaning by sandblasting	E-11	1100	.029	S.F.	.57	1.46	.28	2.31	3.32
2160	By high pressure water	C-29	500	.016			.73	.20	.93	1.38
2170	Blow off dust and debris with oil-free dry compressed air	C-28	16000	.001	↓		.03		.03	.04
2180	Build wood forms including chutes where needed for concrete placement	C-2	960	.050	SFCA	4.52	2.74		7.26	9.35
2190	Insert expansion shield, coil rod, coil tie and plastic cone in each hole	1 Carp	32	.250	Ea.	5.10	14.10		19.20	28
2200	Install wood forms and fasten with coil rod and coil nut	C-2	960	.050	SFCA	.57	2.74		3.31	4.99
2210	Mix up bagged repair material by hand with a mixing paddle	1 Cefi	30	.267	C.F.	10.55	14.15		24.70	33.50
2220	By concrete mixer	C-30	200	.040	↓	10.55	1.82	.70	13.07	15.25
2230	Place repair material by hand and consolidate	C-6A	200	.080	↓		4.25	.23	4.48	6.90
2240	Remove wood forms	C-2	1600	.030	SFCA		1.65		1.65	2.61
2250	Chip/grind off extra concrete at fill chutes	B-9E	16	1	Ea.	.14	49.50	1.83	51.47	80
2260	Break ties and patch voids	1 Cefi	270	.030	"	.07	1.57		1.64	2.54
2270	Cure with sprayed membrane curing compound	2 Cefi	6200	.003	S.F.	.14	.14		.28	.36
3000	Surface repairs using form-and-pump techniques (ACI RAP-5)									
3001	Suitable for walls, columns, beam sides/bottoms, and slab soffits									
3010	Sound the concrete surface to locate delaminated areas	1 Cefi	2000	.004	S.F.		.21		.21	.33
3020	Remove concrete in repair areas to fully expose reinforcing bars	B-9E	160	.100	S.F.		4.94	.18	5.12	8
3030	Mark the perimeter of each repair area	1 Cefi	2000	.004	"		.21		.21	.33
3040	Saw cut the perimeter of each repair area down to reinforcing bars	B-89C	160	.050	L.F.	.03	2.66	.42	3.11	4.65
3050	If corroded rebar is exposed, remove concrete to 3/4" below corroded rebars									
3051	Single layer of #4 rebar	B-9E	64	.250	S.F.		12.35	.46	12.81	19.95
3052	#5 rebar		53.30	.300			14.80	.55	15.35	24
3053	#6 rebar		45.70	.350			17.30	.64	17.94	27.50
3054	#7 rebar		40	.400			19.75	.73	20.48	32
3055	#8 rebar	↓	35.60	.449	↓		22	.82	22.82	36

03 01 Maintenance of Concrete

03 01 30 – Maintenance of Cast-In-Place Concrete

03 01 30.72 Concrete Surface Repairs		Crew	Daily Output	Labor-Hours	Unit	Material	2022 Bare Costs Labor	Equipment	Total	Total Incl O&P
3056	Double layer of #4 rebar	B-9E	35.60	.449	S.F.		22	.82	22.82	36
3057	#5 rebar		32	.500			24.50	.92	25.42	40
3058	#6 rebar		29.10	.550			27	1.01	28.01	44
3059	#7 rebar		26.70	.599			29.50	1.10	30.60	47.50
3060	#8 rebar		24.60	.650			32	1.19	33.19	52
3100	See section 03 21 11.60 for placement of supplemental reinforcement bars									
3104	See section 02 41 19.19 for handling & removal of debris									
3106	See section 02 41 19.20 for disposal of debris									
3110	Drill holes for expansion shields, 3/4" diameter x 4" deep	1 Carp	45	.178	Ea.		10		10	15.90
3120	Final cleaning by sandblasting	E-11	1100	.029	S.F.	.57	1.46	.28	2.31	3.32
3130	By high pressure water	C-29	500	.016			.73	.20	.93	1.38
3140	Blow off dust and debris with oil-free dry compressed air	C-28	16000	.001			.03		.03	.04
3150	Build wood forms and drill holes for injection ports	C-2	960	.050	SFCA	4.52	2.74		7.26	9.35
3160	Assemble PVC injection ports with ball valves and install in forms	1 Plum	8	1	Ea.	98	70		168	216
3170	Insert expansion shield, coil rod, coil tie and plastic cone in each hole	1 Carp	32	.250	"	5.10	14.10		19.20	28
3180	Install wood forms and fasten with coil rod and coil nut	C-2	960	.050	SFCA	.57	2.74		3.31	4.99
3190	Mix up self-consolidating repair material with concrete mixer	C-30	200	.040	C.F.	10.90	1.82	.70	13.42	15.60
3200	Place repair material by pump and pressurize	C-9	200	.160	"		8.05	2.14	10.19	15
3210	Remove wood forms	C-2	1600	.030	SFCA		1.65		1.65	2.61
3220	Chip/grind off extra concrete at injection ports	B-9E	64	.250	Ea.	.14	12.35	.46	12.95	20
3230	Break ties and patch voids	1 Cefi	270	.030	"	.07	1.57		1.64	2.54
3240	Cure with sprayed membrane curing compound	2 Cefi	6200	.003	S.F.	.14	.14		.28	.36
4000	Vertical & overhead spall repair by hand application (ACI RAP-6)									
4001	Generally recommended for thin repairs that are cosmetic in nature									
4010	Remove concrete in repair areas to fully expose reinforcing bars	B-9E	160	.100	S.F.		4.94	.18	5.12	8
4020	Saw cut the perimeter of each repair area to 1/2" depth	B-89C	160	.050	L.F.	.02	2.66	.42	3.10	4.63
4030	If corroded rebar is exposed, remove concrete to 3/4" below corroded rebars									
4031	Single layer of #4 rebar	B-9E	64	.250	S.F.		12.35	.46	12.81	19.95
4032	#5 rebar		53.30	.300			14.80	.55	15.35	24
4033	#6 rebar		45.70	.350			17.30	.64	17.94	27.50
4034	#7 rebar		40	.400			19.75	.73	20.48	32
4035	#8 rebar		35.60	.449			22	.82	22.82	36
4036	Double layer of #4 rebar		35.60	.449			22	.82	22.82	36
4037	#5 rebar		32	.500			24.50	.92	25.42	40
4038	#6 rebar		29.10	.550			27	1.01	28.01	44
4039	#7 rebar		26.70	.599			29.50	1.10	30.60	47.50
4040	#8 rebar		24.60	.650			32	1.19	33.19	52
4080	See section 03 21 11.60 for placement of supplemental reinforcement bars									
4090	See section 02 41 19.19 for handling & removal of debris									
4095	See section 02 41 19.20 for disposal of debris									
4100	Final cleaning by sandblasting	E-11	1100	.029	S.F.	.57	1.46	.28	2.31	3.32
4110	By high pressure water	C-29	500	.016			.73	.20	.93	1.38
4120	Blow off dust and debris with oil-free dry compressed air	C-28	16000	.001			.03		.03	.04
4130	Brush on bonding material	1 Cefi	320	.025		.11	1.33		1.44	2.20
4140	Mix up bagged repair material by hand with a mixing paddle	"	30	.267	C.F.	64	14.15		78.15	92.50
4150	By concrete mixer	C-30	200	.040		64	1.82	.70	66.52	74
4160	Place repair material by hand in lifts up to 2" deep for vertical repairs	C-10	48	.500			25.50		25.50	40
4170	Up to 1" deep for overhead repairs		36	.667			34		34	53
4180	Screed and initial float of repair material by hand		2000	.012	S.F.		.61		.61	.95
4190	Float and steel trowel finish repair material by hand		1265	.019			.96		.96	1.51
4200	Cure with sprayed membrane curing compound	2 Cefi	6200	.003		.14	.14		.28	.36
5000	Spall repair of horizontal concrete surfaces (ACI RAP-7)									
5001	Suitable for structural slabs, slabs-on-grade, balconies, interior floors									

03 01 30.72 Concrete Surface Repairs		Crew	Daily Output	Labor-Hours	Unit	Material	2022 Bare Costs Labor	Equipment	Total	Total Incl O&P
5010	Sound the concrete surface to locate delaminated areas	1 Cefi	2000	.004	SF Flr.		.21		.21	.33
5020	Mark the perimeter of each repair area	"	2000	.004	"		.21		.21	.33
5030	Saw cut the perimeter of each repair area down to reinforcing bars	B-89	1060	.015	L.F.	.03	.81	.59	1.43	1.97
5040	Remove concrete in repair areas to fully expose reinforcing bars	B-9E	160	.100	S.F.		4.94	.18	5.12	8
5050	If corroded rebar is exposed, remove concrete to 3/4" below corroded rebars									
5051	Single layer of #4 rebar	B-9E	64	.250	S.F.		12.35	.46	12.81	19.95
5052	#5 rebar	"	53.30	.300	"		14.80	.55	15.35	24
5053	#6 rebar	B-9E	45.70	.350	S.F.		17.30	.64	17.94	27.50
5054	#7 rebar		40	.400			19.75	.73	20.48	32
5055	#8 rebar		35.60	.449			22	.82	22.82	36
5056	Double layer of #4 rebar		35.60	.449			22	.82	22.82	36
5057	#5 rebar		32	.500			24.50	.92	25.42	40
5058	#6 rebar		29.10	.550			27	1.01	28.01	44
5059	#7 rebar		26.70	.599			29.50	1.10	30.60	47.50
5060	#8 rebar	▼	24.60	.650	▼		32	1.19	33.19	52
5100	See section 03 21 11.60 for placement of supplemental reinforcement bars									
5110	See section 02 41 19.19 for handling & removal of debris									
5115	See section 02 41 19.20 for disposal of debris									
5120	Final cleaning by sandblasting	E-11	1100	.029	S.F.	.57	1.46	.28	2.31	3.32
5130	By high pressure water	C-29	500	.016			.73	.20	.93	1.38
5140	Blow off dust and debris with oil-free dry compressed air	C-28	16000	.001			.03		.03	.04
5160	Brush on bonding material	1 Cefi	320	.025	▼	.11	1.33		1.44	2.20
5170	Mix up bagged repair material by hand with a mixing paddle	"	30	.267	C.F.	64	14.15		78.15	92.50
5180	By concrete mixer	C-30	200	.040		64	1.82	.70	66.52	74
5190	Place repair material by hand, consolidate and screed	C-6A	200	.080	▼		4.25	.23	4.48	6.90
5200	Bull float, manual float & broom finish	C-10	1850	.013	S.F.		.66		.66	1.03
5210	Bull float, manual float & trowel finish	"	1265	.019			.96		.96	1.51
5220	Cure with sprayed membrane curing compound	2 Cefi	6200	.003	▼	.14	.14		.28	.36
6000	Spall repairs by the Preplaced Aggregate method (ACI RAP-9)									
6001	Suitable for vertical, overhead, and underwater repairs									
6010	Sound the concrete surface to locate delaminated areas	1 Cefi	2000	.004	S.F.		.21		.21	.33
6020	Remove concrete in repair areas to fully expose reinforcing bars	B-9E	160	.100			4.94	.18	5.12	8
6030	Mark the perimeter of each repair area	1 Cefi	2000	.004	▼		.21		.21	.33
6040	Saw cut the perimeter of each repair area down to reinforcing bars	B-89C	160	.050	L.F.	.03	2.66	.42	3.11	4.65
6050	If corroded rebar is exposed, remove concrete to 3/4" below corroded rebars									
6051	Single layer of #4 rebar	B-9E	64	.250	S.F.		12.35	.46	12.81	19.95
6052	#5 rebar		53.30	.300			14.80	.55	15.35	24
6053	#6 rebar		45.70	.350			17.30	.64	17.94	27.50
6054	#7 rebar		40	.400			19.75	.73	20.48	32
6055	#8 rebar		35.60	.449			22	.82	22.82	36
6056	Double layer of #4 rebar		35.60	.449			22	.82	22.82	36
6057	#5 rebar		32	.500			24.50	.92	25.42	40
6058	#6 rebar	▼	29.10	.550	▼		27	1.01	28.01	44
6059	#7 rebar	B-9E	26.70	.599	S.F.		29.50	1.10	30.60	47.50
6060	#8 rebar	"	24.60	.650	"		32	1.19	33.19	52
6100	See section 03 21 11.60 for placement of supplemental reinforcement bars									
6104	See section 02 41 19.19 for handling & removal of debris									
6106	See section 02 41 19.20 for disposal of debris									
6110	Drill holes for expansion shields, 3/4" diameter x 4" deep	1 Carp	45	.178	Ea.		10		10	15.90
6120	Final cleaning by sandblasting	E-11	1100	.029	S.F.	.57	1.46	.28	2.31	3.32
6130	By high pressure water	C-29	500	.016			.73	.20	.93	1.38
6140	Blow off dust and debris with oil-free dry compressed air	C-28	16000	.001	▼		.03		.03	.04
6150	Build wood forms and drill holes for injection ports	C-2	960	.050	SFCA	4.52	2.74		7.26	9.35

03 01 30 – Maintenance of Cast-In-Place Concrete

03 01 30.72 Concrete Surface Repairs	Crew	Daily Output	Labor-Hours	Unit	Material	2022 Bare Costs Labor	Equipment	Total	Total Incl O&P
6160	Assemble PVC injection ports with ball valves and install in forms	1 Plum	8	1	Ea.	98	70		168	216
6170	Insert expansion shield, coil rod, coil tie and plastic cone in each hole	1 Carp	32	.250	"	5.10	14.10		19.20	28
6180	Install wood forms and fasten with coil rod and coil nut	C-2	960	.050	SFCA	.57	2.74		3.31	4.99
6185	Install pre-washed gap-graded aggregate as forms are installed	B-24	160	.150	C.F.	.88	7.75		8.63	13.20
6190	Mix up pre-bagged grout material with concrete mixer	C-30	200	.040		13.50	1.82	.70	16.02	18.50
6200	Place grout by pump from bottom up	C-9	200	.160	↓		8.05	2.14	10.19	15
6210	Remove wood forms	C-2	1600	.030	SFCA		1.65		1.65	2.61
6220	Chip/grind off extra concrete at injection ports	B-9E	64	.250	Ea.	.14	12.35	.46	12.95	20
6230	Break ties and patch voids	1 Cefi	270	.030	."	.07	1.57		1.64	2.54
6240	Cure with sprayed membrane curing compound	2 Cefi	6200	.003	S.F.	.14	.14		.28	.36
7000	Leveling & reprofiling of vertical & overhead surfaces (ACI RAP-10)									
7001	Used to provide an acceptable surface for aesthetic or protective coatings									
7010	Clean the surface by sandblasting, leaving an open profile	E-11	1100	.029	S.F.	.57	1.46	.28	2.31	3.32
7020	By high pressure water	C-29	500	.016			.73	.20	.93	1.38
7030	Blow off dust and debris with oil-free dry compressed air	C-28	16000	.001	↓		.03		.03	.04
7040	Mix up bagged repair material by hand with a mixing paddle	1 Cefi	30	.267	C.F.	64	14.15		78.15	92.50
7050	Place repair material by hand with steel trowel to fill pores & honeycombs	C-10	1265	.019	S.F.		.96		.96	1.51
7060	Cure with sprayed membrane curing compound	2 Cefi	6200	.003	"	.14	.14		.28	.36
8000	Concrete repair by shotcrete application (ACI RAP-12)									
8001	Suitable for vertical and overhead surfaces									
8010	Sound the concrete surface to locate delaminated areas	1 Cefi	2000	.004	S.F.		.21		.21	.33
8020	Remove concrete in repair areas to fully expose reinforcing bars	B-9E	160	.100			4.94	.18	5.12	8
8030	Mark the perimeter of each repair area	1 Cefi	2000	.004	↓		.21		.21	.33
8040	Saw cut the perimeter of each repair area down to reinforcing bars	B-89C	160	.050	L.F.	.03	2.66	.42	3.11	4.65
8050	If corroded rebar is exposed, remove concrete to 3/4" below corroded rebars									
8051	Single layer of #4 rebar	B-9E	64	.250	S.F.		12.35	.46	12.81	19.95
8052	#5 rebar		53.30	.300			14.80	.55	15.35	24
8053	#6 rebar		45.70	.350			17.30	.64	17.94	27.50
8054	#7 rebar		40	.400			19.75	.73	20.48	32
8055	#8 rebar		35.60	.449			22	.82	22.82	36
8056	Double layer of #4 rebar		35.60	.449			22	.82	22.82	36
8057	#5 rebar		32	.500			24.50	.92	25.42	40
8058	#6 rebar		29.10	.550			27	1.01	28.01	44
8059	#7 rebar		26.70	.599			29.50	1.10	30.60	47.50
8060	#8 rebar	↓	24.60	.650	↓		32	1.19	33.19	52
8100	See section 03 21 11.60 for placement of supplemental reinforcement bars									
8110	Final cleaning by sandblasting	E-11	1100	.029	S.F.	.57	1.46	.28	2.31	3.32
8120	By high pressure water	C-29	500	.016			.73	.20	.93	1.38
8130	Blow off dust and debris with oil-free dry compressed air	C-28	16000	.001	↓		.03		.03	.04
8140	Mix up repair material with concrete mixer	C-30	200	.040	C.F.	10.55	1.82	.70	13.07	15.25
8150	Place repair material by wet shotcrete process to full depth of repair	C-8C	200	.240	"		11.90	5.10	17	24.50
8160	Carefully trim excess material and steel trowel finish without overworking	C-10	1265	.019	S.F.		.96		.96	1.51
8170	Cure with sprayed membrane curing compound	2 Cefi	6200	.003	"	.14	.14		.28	.36
9000	Surface repair by Methacrylate flood coat (ACI RAP-13)									
9010	Suitable for healing and sealing horizontal surfaces only									
9020	Large cracks must previously have been repaired or filled									
9100	Shotblast entire surface to remove contaminants	A-1A	4000	.002	S.F.		.12	.05	.17	.25
9200	Blow off dust and debris with oil-free dry compressed air	C-28	16000	.001			.03		.03	.04
9300	Flood coat surface w/Methacrylate, distribute w/broom/squeegee, no prep.	3 Cefi	8000	.003		.87	.16		1.03	1.21
9400	Lightly broadcast even coat of dry silica sand while sealer coat is tacky	1 Cefi	8000	.001	↓	.01	.05		.06	.10

For customer support on your Commercial Renovation Costs with RSMeans data, call 800.448.8182.

55

03 05 Common Work Results for Concrete

03 05 05 – Selective Demolition for Concrete

03 05 05.10 Selective Demolition, Concrete

		Crew	Daily Output	Labor-Hours	Unit	Material	2022 Bare Costs Labor	Equipment	Total	Total Incl O&P
0010	**SELECTIVE DEMOLITION, CONCRETE** R024119-10									
0012	Excludes saw cutting, torch cutting, loading or hauling									
0050	Break into small pieces, reinf. less than 1% of cross-sectional area	B-9	24	1.667	C.Y.		76.50	13.85	90.35	137
0060	Reinforcing 1% to 2% of cross-sectional area		16	2.500			115	21	136	206
0070	Reinforcing more than 2% of cross-sectional area		8	5			230	41.50	271.50	410
0150	Remove whole pieces, up to 2 tons per piece	E-18	36	1.111	Ea.		69	22.50	91.50	135
0160	2-5 tons per piece		30	1.333			82.50	27	109.50	162
0170	5-10 tons per piece		24	1.667			103	34	137	203
0180	10-15 tons per piece		18	2.222			138	45.50	183.50	270
0250	Precast unit embedded in masonry, up to 1 C.F.	D-1	16	1			50		50	80.50
0260	1-2 C.F.		12	1.333			67		67	107
0270	2-5 C.F.		10	1.600			80		80	128
0280	5-10 C.F.		8	2			100		100	161
0990	For hydrodemolition see Section 02 41 13.15									
1910	Minimum labor/equipment charge	B-9	2	20	Job		920	166	1,086	1,625

03 05 13 – Basic Concrete Materials

03 05 13.20 Concrete Admixtures and Surface Treatments

		Crew	Daily Output	Labor-Hours	Unit	Material	2022 Bare Costs Labor	Equipment	Total	Total Incl O&P
0010	**CONCRETE ADMIXTURES AND SURFACE TREATMENTS**									
0040	Abrasives, aluminum oxide, over 20 tons				Lb.	1.47			1.47	1.62
0050	1 to 20 tons					1.53			1.53	1.68
0070	Under 1 ton					1.59			1.59	1.75
0100	Silicon carbide, black, over 20 tons					2.27			2.27	2.50
0110	1 to 20 tons					2.36			2.36	2.60
0120	Under 1 ton					2.45			2.45	2.70
0200	Air entraining agent, .7 to 1.5 oz. per bag, 55 gallon drum				Gal.	17.70			17.70	19.45
0220	5 gallon pail					32			32	35.50
0300	Bonding agent, acrylic latex, 250 S.F./gallon, 5 gallon pail					27			27	29.50
0320	Epoxy resin, 80 S.F./gallon, 4 gallon case					66			66	72.50
0400	Calcium chloride, 50 lb. bags, T.L. lots				Ton	950			950	1,050
0420	Less than truckload lots				Bag	19.95			19.95	22
0500	Carbon black, liquid, 2 to 8 lb. per bag of cement				Lb.	11.50			11.50	12.65
0600	Concrete admixture, integral colors, dry pigment, 5 lb. bag				Ea.	29.50			29.50	32.50
0610	10 lb. bag					51.50			51.50	57
0620	25 lb. bag					80			80	88
0920	Dustproofing compound, 250 S.F./gal., 5 gallon pail				Gal.	11.40			11.40	12.55
1010	Epoxy based, 125 S.F./gal., 5 gallon pail				"	30.50			30.50	33.50
1100	Hardeners, metallic, 55 lb. bags, natural (grey)				Lb.	1.66			1.66	1.82
1200	Colors					1.42			1.42	1.57
1300	Non-metallic, 55 lb. bags, natural grey					.37			.37	.41
1320	Colors					1.06			1.06	1.16
1550	Release agent, for tilt slabs, 5 gallon pail				Gal.	17.15			17.15	18.85
1570	For forms, 5 gallon pail				Gal.	13.95			13.95	15.35
1590	Concrete release agent for forms, 100% biodegradable, zero VOC, 5 gal. pail G					20.50			20.50	22.50
1595	55 gallon drum G					15.50			15.50	17.05
1600	Sealer, hardener and dustproofer, epoxy-based, 125 S.F./gal., 5 gallon unit					30.50			30.50	33.50
1620	3 gallon unit					33.50			33.50	36.50
1630	Sealer, solvent-based, 250 S.F./gal., 55 gallon drum					22			22	24.50
1640	5 gallon pail					30			30	33
1650	Sealer, water based, 350 S.F., 55 gallon drum					27			27	30
1660	5 gallon pail					25			25	27.50
1900	Set retarder, 100 S.F./gal., 1 gallon pail					6.15			6.15	6.80
2000	Waterproofing, integral 1 lb. per bag of cement				Lb.	1.36			1.36	1.50

03 05 Common Work Results for Concrete

03 05 13 – Basic Concrete Materials

03 05 13.20 Concrete Admixtures and Surface Treatments		Crew	Daily Output	Labor-Hours	Unit	Material	2022 Bare Costs Labor	Equipment	Total	Total Incl O&P
2100	Powdered metallic, 40 lbs. per 100 S.F., standard colors				Lb.	4.19			4.19	4.61
2120	Premium colors				▼	5.90			5.90	6.50
3000	For colored ready-mix concrete, add to prices in section 03 31 13.35									
3100	Subtle shades, 5 lb. dry pigment per C.Y., add				C.Y.	29.50			29.50	32.50
3400	Medium shades, 10 lb. dry pigment per C.Y., add					51.50			51.50	57
3700	Deep shades, 25 lb. dry pigment per C.Y., add				▼	80			80	88

03 05 13.85 Winter Protection

		Crew	Daily Output	Labor-Hours	Unit	Material	2022 Bare Costs Labor	Equipment	Total	Total Incl O&P
0010	**WINTER PROTECTION**									
0012	For heated ready mix, add				C.Y.	8.50			8.50	9.35
0100	Temporary heat to protect concrete, 24 hours	2 Clab	50	.320	M.S.F.	184	14.60		198.60	225
0200	Temporary shelter for slab on grade, wood frame/polyethylene sheeting									
0201	Build or remove, light framing for short spans	2 Carp	10	1.600	M.S.F.	485	90		575	675
0210	Large framing for long spans	"	3	5.333	"	585	300		885	1,125
0710	Electrically heated pads, 15 watts/S.F., 20 uses				S.F.	.64			.64	.71

03 11 Concrete Forming

03 11 13 – Structural Cast-In-Place Concrete Forming

03 11 13.20 Forms In Place, Beams and Girders

			Crew	Daily Output	Labor-Hours	Unit	Material	2022 Bare Costs Labor	Equipment	Total	Total Incl O&P
0010	**FORMS IN PLACE, BEAMS AND GIRDERS**	R031113-40									
0500	Exterior spandrel, job-built plywood, 12" wide, 1 use	R031113-60	C-2	225	.213	SFCA	4.82	11.70		16.52	24
0650	4 use			310	.155		1.57	8.50		10.07	15.20
1000	18" wide, 1 use			250	.192		4.38	10.55		14.93	21.50
1150	4 use			315	.152		1.44	8.35		9.79	14.90
1500	24" wide, 1 use			265	.181		4.02	9.95		13.97	20
1650	4 use			325	.148		1.31	8.10		9.41	14.30
2000	Interior beam, job-built plywood, 12" wide, 1 use			300	.160		5.55	8.80		14.35	20
2150	4 use		C-2	377	.127	SFCA	1.81	7		8.81	13.10
2500	24" wide, 1 use			320	.150		4.11	8.25		12.36	17.55
2650	4 use			395	.122		1.33	6.65		7.98	12.05
3000	Encasing steel beam, hung, job-built plywood, 1 use			325	.148		4.60	8.10		12.70	17.90
3150	4 use			430	.112		1.50	6.10		7.60	11.40
9000	Minimum labor/equipment charge		2 Carp	2	8	Job		450		450	715

03 11 13.25 Forms In Place, Columns

			Crew	Daily Output	Labor-Hours	Unit	Material	2022 Bare Costs Labor	Equipment	Total	Total Incl O&P
0010	**FORMS IN PLACE, COLUMNS**	R031113-40									
0500	Round fiberglass, 4 use per mo., rent, 12" diameter		C-1	160	.200	L.F.	23	10.75		33.75	42.50
0550	16" diameter	R031113-60		150	.213		27	11.45		38.45	47.50
0600	18" diameter			140	.229		29.50	12.25		41.75	52
0650	24" diameter			135	.237		37	12.70		49.70	60.50
0700	28" diameter			130	.246		41.50	13.20		54.70	66.50
0800	30" diameter			125	.256		43.50	13.75		57.25	70
0850	36" diameter			120	.267		57.50	14.30		71.80	85.50
1500	Round fiber tube, recycled paper, 1 use, 8" diameter	G		155	.206		2.48	11.05		13.53	20.50
1550	10" diameter	G		155	.206		4.03	11.05		15.08	22
1600	12" diameter	G		150	.213		5	11.45		16.45	23.50
1650	14" diameter	G		145	.221		7.15	11.85		19	26.50
1700	16" diameter	G		140	.229		8.35	12.25		20.60	28.50
1720	18" diameter	G		140	.229		9.75	12.25		22	30
1750	20" diameter	G		135	.237		11.65	12.70		24.35	33
1800	24" diameter	G		130	.246		13.45	13.20		26.65	36
1850	30" diameter	G		125	.256		19.60	13.75		33.35	43.50

For customer support on your Commercial Renovation Costs with RSMeans data, call 800.448.8182.

57

03 11 13 – Structural Cast-In-Place Concrete Forming

03 11 13.25 Forms In Place, Columns

		Crew	Daily Output	Labor-Hours	Unit	Material	2022 Bare Costs Labor	Equipment	Total	Total Incl O&P
1900	36" diameter	G C-1	115	.278	L.F.	28.50	14.90		43.40	54.50
1950	42" diameter	G	100	.320		52.50	17.15		69.65	85
2000	48" diameter	G ↓	85	.376	↓	55.50	20		75.50	93.50
2200	For seamless type, add					15%				
3000	Round, steel, 4 use per mo., rent, regular duty, 14" diameter	G C-1	145	.221	L.F.	18.45	11.85		30.30	39.50
3050	16" diameter	G	125	.256		18.80	13.75		32.55	42.50
3100	Heavy duty, 20" diameter	G	105	.305		21	16.35		37.35	49
3150	24" diameter	G	85	.376		22.50	20		42.50	57
3200	30" diameter	G	70	.457		26	24.50		50.50	67.50
3250	36" diameter	G	60	.533		28	28.50		56.50	76
3300	48" diameter	G	50	.640		41.50	34.50		76	100
3350	60" diameter	G ↓	45	.711	↓	51	38		89	117
4500	For second and succeeding months, deduct					50%				
5000	Job-built plywood, 8" x 8" columns, 1 use	C-1	165	.194	SFCA	3.76	10.40		14.16	20.50
5050	2 use	C-1	195	.164	SFCA	2.16	8.80		10.96	16.35
5100	3 use		210	.152		1.50	8.15		9.65	14.65
5150	4 use		215	.149		1.23	8		9.23	14.05
5500	12" x 12" columns, 1 use		180	.178		3.76	9.55		13.31	19.30
5550	2 use		210	.152		2.07	8.15		10.22	15.25
5600	3 use		220	.145		1.50	7.80		9.30	14.05
5650	4 use		225	.142		1.22	7.65		8.87	13.45
6000	16" x 16" columns, 1 use		185	.173		3.83	9.30		13.13	18.95
6050	2 use		215	.149		2.03	8		10.03	14.95
6100	3 use		230	.139		1.54	7.45		8.99	13.55
6150	4 use		235	.136		1.25	7.30		8.55	13
6500	24" x 24" columns, 1 use		190	.168		4.39	9.05		13.44	19.15
6550	2 use		216	.148		2.41	7.95		10.36	15.25
6600	3 use		230	.139		1.75	7.45		9.20	13.80
6650	4 use		238	.134		1.43	7.20		8.63	13
7000	36" x 36" columns, 1 use		200	.160		3.71	8.60		12.31	17.75
7050	2 use		230	.139		2.08	7.45		9.53	14.15
7100	3 use	↓	245	.131	↓	1.48	7		8.48	12.80
7150	4 use		250	.128		1.20	6.85		8.05	12.25
7400	Steel framed plywood, based on 50 uses of purchased									
7420	forms, and 4 uses of bracing lumber									
7500	8" x 8" column	C-1	340	.094	SFCA	2.39	5.05		7.44	10.65
7550	10" x 10"		350	.091		2.12	4.90		7.02	10.15
7600	12" x 12"		370	.086		1.80	4.64		6.44	9.35
7650	16" x 16"		400	.080		1.40	4.29		5.69	8.35
7700	20" x 20"		420	.076		1.24	4.09		5.33	7.85
7750	24" x 24"		440	.073		.87	3.90		4.77	7.15
7755	30" x 30"		440	.073		1.11	3.90		5.01	7.40
7760	36" x 36"	↓	460	.070	↓	.98	3.73		4.71	7.05
9000	Minimum labor/equipment charge	2 Carp	2	8	Job		450		450	715

03 11 13.35 Forms In Place, Elevated Slabs

		Crew	Daily Output	Labor-Hours	Unit	Material	2022 Bare Costs Labor	Equipment	Total	Total Incl O&P
0010	**FORMS IN PLACE, ELEVATED SLABS** R031113-40									
1000	Flat plate, job-built plywood, to 15' high, 1 use	C-2	470	.102	S.F.	5.75	5.60		11.35	15.25
1150	4 use R031113-60		560	.086		1.87	4.70		6.57	9.50
1500	15' to 20' high ceilings, 4 use		495	.097		2.01	5.30		7.31	10.65
2000	Flat slab, drop panels, job-built plywood, to 15' high, 1 use		449	.107		6.75	5.85		12.60	16.75
2150	4 use		544	.088		2.20	4.84		7.04	10.10
2250	15' to 20' high ceilings, 4 use	↓	480	.100	↓	3.64	5.50		9.14	12.70

03 11 13.35 Forms In Place, Elevated Slabs

		Crew	Daily Output	Labor-Hours	Unit	Material	2022 Bare Costs Labor	Equipment	Total	Total Incl O&P
3500	Floor slab, with 1-way joist pans, 1 use	C-2	415	.116	S.F.	10.55	6.35		16.90	21.50
3650	4 use	C-2	500	.096	S.F.	5.95	5.25		11.20	14.90
4500	With 2-way waffle domes, 1 use		405	.119		11.15	6.50		17.65	22.50
4550	4 use		470	.102		6.55	5.60		12.15	16.10
5000	Box out for slab openings, over 16" deep, 1 use		190	.253	SFCA	5.25	13.85		19.10	28
5050	2 use		240	.200	"	2.89	10.95		13.84	20.50
5500	Shallow slab box outs, to 10 S.F.		42	1.143	Ea.	17.70	62.50		80.20	119
5550	Over 10 S.F. (use perimeter)		600	.080	L.F.	2.37	4.39		6.76	9.55
6000	Bulkhead forms for slab, with keyway, 1 use, 2 piece		500	.096		2.39	5.25		7.64	11
6100	3 piece (see also edge forms)		460	.104		2.73	5.70		8.43	12.10
6200	Slab bulkhead form, 4-1/2" high, exp metal, w/keyway & stakes G	C-1	1200	.027		1.18	1.43		2.61	3.57
6210	5-1/2" high G		1100	.029		1.36	1.56		2.92	3.98
6215	7-1/2" high G		960	.033		1.68	1.79		3.47	4.69
6220	9-1/2" high G		840	.038		1.83	2.04		3.87	5.25
6500	Curb forms, wood, 6" to 12" high, on elevated slabs, 1 use		180	.178	SFCA	2.67	9.55		12.22	18.10
6550	2 use		205	.156		1.47	8.35		9.82	14.90
6600	3 use		220	.145		1.07	7.80		8.87	13.55
6650	4 use		225	.142		.87	7.65		8.52	13.05
7000	Edge forms to 6" high, on elevated slab, 4 use		500	.064	L.F.	.31	3.43		3.74	5.80
7070	7" to 12" high, 1 use		162	.198	SFCA	2.02	10.60		12.62	19.05
7080	2 use		198	.162		1.11	8.65		9.76	14.95
7090	3 use		222	.144		.81	7.75		8.56	13.20
7101	4 use		350	.091		.66	4.90		5.56	8.50
7500	Depressed area forms to 12" high, 4 use		300	.107	L.F.	1.17	5.70		6.87	10.40
7550	12" to 24" high, 4 use		175	.183		1.59	9.80		11.39	17.35
8000	Perimeter deck and rail for elevated slabs, straight		90	.356		19.60	19.05		38.65	52
8050	Curved		65	.492		27	26.50		53.50	71.50
8500	Void forms, round plastic, 8" high x 3" diameter G		450	.071	Ea.	1.45	3.81		5.26	7.65
8550	4" diameter G		425	.075		2.16	4.04		6.20	8.80
8600	6" diameter G		400	.080		3.55	4.29		7.84	10.70
8650	8" diameter G		375	.085		6.35	4.58		10.93	14.25
9000	Minimum labor/equipment charge	2 Carp	2	8	Job		450		450	715

03 11 13.40 Forms In Place, Equipment Foundations

		Crew	Daily Output	Labor-Hours	Unit	Material	2022 Bare Costs Labor	Equipment	Total	Total Incl O&P
0010	**FORMS IN PLACE, EQUIPMENT FOUNDATIONS** R031113-40									
0020	1 use	C-2	160	.300	SFCA	4.14	16.45		20.59	30.50
0050	2 use R031113-60		190	.253		2.28	13.85		16.13	24.50
0100	3 use		200	.240		1.66	13.15		14.81	23
0150	4 use		205	.234		1.35	12.85		14.20	22
9000	Minimum labor/equipment charge	1 Carp	3	2.667	Job		150		150	239

03 11 13.45 Forms In Place, Footings

		Crew	Daily Output	Labor-Hours	Unit	Material	2022 Bare Costs Labor	Equipment	Total	Total Incl O&P
0010	**FORMS IN PLACE, FOOTINGS** R031113-40									
0020	Continuous wall, plywood, 1 use	C-1	375	.085	SFCA	7.85	4.58		12.43	15.90
0150	4 use R031113-60	"	485	.066	"	2.57	3.54		6.11	8.40
1500	Keyway, 4 use, tapered wood, 2" x 4"	1 Carp	530	.015	L.F.	.35	.85		1.20	1.73
1550	2" x 6"	"	500	.016	"	.48	.90		1.38	1.96
3000	Pile cap, square or rectangular, job-built plywood, 1 use	C-1	290	.110	SFCA	4.40	5.90		10.30	14.25
3150	4 use		383	.084		1.43	4.48		5.91	8.65
5000	Spread footings, job-built lumber, 1 use		305	.105		3.13	5.65		8.78	12.40
5150	4 use		414	.077		1.02	4.15		5.17	7.70
9000	Minimum labor/equipment charge	1 Carp	3	2.667	Job		150		150	239

For customer support on your Commercial Renovation Costs with RSMeans data, call 800.448.8182.

59

03 11 Concrete Forming

03 11 13 − Structural Cast-In-Place Concrete Forming

03 11 13.47 Forms In Place, Gas Station Forms

		Crew	Daily Output	Labor-Hours	Unit	Material	2022 Bare Costs Labor	Equipment	Total	Total Incl O&P
0010	**FORMS IN PLACE, GAS STATION FORMS**									
0050	Curb fascia, with template, 12 ga. steel, left in place, 9" high G	1 Carp	50	.160	L.F.	17.15	9		26.15	33
1000	Sign or light bases, 18" diameter, 9" high G		9	.889	Ea.	108	50		158	199
1050	30" diameter, 13" high G		8	1	"	172	56.50		228.50	279
1990	Minimum labor/equipment charge		2	4	Job		225		225	360
2000	Island forms, 10' long, 9" high, 3'-6" wide G	C-1	10	3.200	Ea.	480	172		652	805
2050	4' wide G		9	3.556		495	191		686	850
2500	20' long, 9" high, 4' wide G		6	5.333		795	286		1,081	1,325
2550	5' wide G		5	6.400		830	345		1,175	1,450
9000	Minimum labor/equipment charge	1 Carp	3	2.667	Job		150		150	239

03 11 13.50 Forms In Place, Grade Beam

		Crew	Daily Output	Labor-Hours	Unit	Material	2022 Bare Costs Labor	Equipment	Total	Total Incl O&P
0010	**FORMS IN PLACE, GRADE BEAM** R031113-40									
0020	Job-built plywood, 1 use	C-2	530	.091	SFCA	3.90	4.97		8.87	12.20
0150	4 use R031113-60	"	605	.079	"	1.27	4.35		5.62	8.30
9000	Minimum labor/equipment charge	2 Carp	2	8	Job		450		450	715

03 11 13.65 Forms In Place, Slab On Grade

		Crew	Daily Output	Labor-Hours	Unit	Material	2022 Bare Costs Labor	Equipment	Total	Total Incl O&P
0010	**FORMS IN PLACE, SLAB ON GRADE** R031113-40									
1000	Bulkhead forms w/keyway, wood, 6" high, 1 use	C-1	510	.063	L.F.	1.41	3.37		4.78	6.90
1400	Bulkhead form for slab, 4-1/2" high, exp metal, incl. keyway & stakes R031113-60		1200	.027		1.18	1.43		2.61	3.57
1410	5-1/2" high G		1100	.029		1.36	1.56		2.92	3.98
1420	7-1/2" high G		960	.033		1.68	1.79		3.47	4.69
1430	9-1/2" high G		840	.038		1.83	2.04		3.87	5.25
2000	Curb forms, wood, 6" to 12" high, on grade, 1 use		215	.149	SFCA	3.56	8		11.56	16.60
2150	4 use		275	.116	"	1.16	6.25		7.41	11.15
3000	Edge forms, wood, 4 use, on grade, to 6" high		600	.053	L.F.	.46	2.86		3.32	5.05
3050	7" to 12" high		435	.074	SFCA	1.22	3.95		5.17	7.60
3060	Over 12"		350	.091	"	1.45	4.90		6.35	9.40
3500	For depressed slabs, 4 use, to 12" high	C-1	300	.107	L.F.	1.19	5.70		6.89	10.40
3550	To 24" high		175	.183		1.57	9.80		11.37	17.35
4000	For slab blockouts, to 12" high, 1 use		200	.160		1.27	8.60		9.87	15.05
4050	To 24" high, 1 use		120	.267		1.62	14.30		15.92	24.50
4100	Plastic (extruded), to 6" high, multiple use, on grade		800	.040		11.80	2.15		13.95	16.35
5000	Screed, 24 ga. metal key joint, see Section 03 15 16.30									
5020	Wood, incl. wood stakes, 1" x 3"	C-1	900	.036	L.F.	.92	1.91		2.83	4.04
5050	2" x 4"		900	.036	"	1.11	1.91		3.02	4.25
6000	Trench forms in floor, wood, 1 use		160	.200	SFCA	2.88	10.75		13.63	20
6150	4 use		185	.173	"	.94	9.30		10.24	15.80
8760	Void form, corrugated fiberboard, 4" x 12", 4' long G		3000	.011	S.F.	4.65	.57		5.22	6
8770	6" x 12", 4' long		3000	.011		5.05	.57		5.62	6.45
8780	1/4" thick hardboard protective cover for void form	2 Carp	1500	.011		.87	.60		1.47	1.91
9000	Minimum labor/equipment charge	1 Carp	2	4	Job		225		225	360

03 11 13.85 Forms In Place, Walls

		Crew	Daily Output	Labor-Hours	Unit	Material	2022 Bare Costs Labor	Equipment	Total	Total Incl O&P
0010	**FORMS IN PLACE, WALLS** R031113-10									
0100	Box out for wall openings, to 16" thick, to 10 S.F.	C-2	24	2	Ea.	34	110		144	211
0150	Over 10 S.F. (use perimeter) R031113-40	"	280	.171	L.F.	3.07	9.40		12.47	18.30
0250	Brick shelf, 4" W, add to wall forms, use wall area above shelf									
0260	1 use R031113-60	C-2	240	.200	SFCA	3.49	10.95		14.44	21.50
0350	4 use		300	.160	"	1.40	8.80		10.20	15.50
0500	Bulkhead, wood with keyway, 1 use, 2 piece		265	.181	L.F.	3.40	9.95		13.35	19.55
0600	Bulkhead forms with keyway, 1 piece expanded metal, 8" wall G	C-1	1000	.032		1.68	1.72		3.40	4.58
0610	10" wall G		800	.040		1.83	2.15		3.98	5.40

03 11 Concrete Forming

03 11 13 – Structural Cast-In-Place Concrete Forming

03 11 13.85 Forms In Place, Walls

		Crew	Daily Output	Labor-Hours	Unit	Material	2022 Bare Costs Labor	Equipment	Total	Total Incl O&P
0620	12" wall ⒢	C-1	525	.061	L.F.	2.20	3.27		5.47	7.60
0700	Buttress, to 8' high, 1 use	C-2	350	.137	SFCA	5.80	7.50		13.30	18.35
0850	4 use		480	.100	"	1.91	5.50		7.41	10.80
1000	Corbel or haunch, to 12" wide, add to wall forms, 1 use		150	.320	L.F.	3.68	17.55		21.23	32
1150	4 use		180	.267	"	1.20	14.65		15.85	24.50
2000	Wall, job-built plywood, to 8' high, 1 use		370	.130	SFCA	4.16	7.10		11.26	15.85
2050	2 use		435	.110		2.66	6.05		8.71	12.55
2100	3 use		495	.097		1.93	5.30		7.23	10.60
2150	4 use		505	.095		1.57	5.20		6.77	10.05
2400	Over 8' to 16' high, 1 use		280	.171		4.58	9.40		13.98	20
2450	2 use		345	.139		1.70	7.65		9.35	13.95
2500	3 use		375	.128		1.22	7		8.22	12.50
2550	4 use	↓	395	.122	↓	.99	6.65		7.64	11.70
2700	Over 16' high, 1 use	C-2	235	.204	SFCA	3.52	11.20		14.72	21.50
2750	2 use		290	.166		1.94	9.10		11.04	16.55
2800	3 use		315	.152		1.41	8.35		9.76	14.85
2850	4 use		330	.145		1.15	8		9.15	13.90
4000	Radial, smooth curved, job-built plywood, 1 use		245	.196		2.78	10.75		13.53	20
4150	4 use		335	.143		.90	7.85		8.75	13.50
4200	Below grade, job-built plywood, 1 use		225	.213		3.55	11.70		15.25	22.50
4210	2 use		225	.213		1.96	11.70		13.66	21
4220	3 use		225	.213		1.66	11.70		13.36	20.50
4230	4 use		225	.213		1.16	11.70		12.86	19.85
4600	Retaining wall, battered, job-built plyw'd, to 8' high, 1 use		300	.160		2.52	8.80		11.32	16.70
4750	4 use		390	.123		.82	6.75		7.57	11.60
4900	Over 8' to 16' high, 1 use		240	.200		2.81	10.95		13.76	20.50
5050	4 use	↓	320	.150	↓	.91	8.25		9.16	14.05
7500	Lintel or sill forms, 1 use	1 Carp	30	.267		4.99	15		19.99	29.50
7560	4 use	"	37	.216	↓	1.62	12.15		13.77	21
7800	Modular prefabricated plywood, based on 20 uses of purchased									
7820	forms, and 4 uses of bracing lumber									
7860	To 8' high	C-2	800	.060	SFCA	1.40	3.29		4.69	6.80
8060	Over 8' to 16' high		600	.080		1.48	4.39		5.87	8.60
8600	Pilasters, 1 use		270	.178		4.82	9.75		14.57	21
8660	4 use	↓	385	.125	↓	1.57	6.85		8.42	12.55
9010	Steel framed plywood, based on 50 uses of purchased									
9020	forms, and 4 uses of bracing lumber									
9060	To 8' high	C-2	600	.080	SFCA	.66	4.39		5.05	7.70
9260	Over 8' to 16' high		450	.107		.66	5.85		6.51	10.05
9460	Over 16' to 20' high	↓	400	.120	↓	.66	6.60		7.26	11.20
9475	For elevated walls, add						10%			
9480	For battered walls, 1 side battered, add					10%	10%			
9485	For battered walls, 2 sides battered, add					15%	15%			
9900	Minimum labor/equipment charge	2 Carp	2	8	Job		450		450	715

03 11 16 – Architectural Cast-in-Place Concrete Forming

03 11 16.13 Concrete Form Liners

		Crew	Daily Output	Labor-Hours	Unit	Material	2022 Bare Costs Labor	Equipment	Total	Total Incl O&P
0010	**CONCRETE FORM LINERS**									
5750	Liners for forms (add to wall forms), ABS plastic									
5800	Aged wood, 4" wide, 1 use	1 Carp	256	.031	SFCA	4.54	1.76		6.30	7.80
5820	2 use	"	256	.031	"	2.50	1.76		4.26	5.55
5830	3 use	1 Carp	256	.031	SFCA	1.82	1.76		3.58	4.80
5840	4 use	↓	256	.031	↓	1.48	1.76		3.24	4.42

03 11 Concrete Forming

03 11 16 – Architectural Cast-in-Place Concrete Forming

03 11 16.13 Concrete Form Liners

		Crew	Daily Output	Labor-Hours	Unit	Material	2022 Bare Costs Labor	Equipment	Total	Total Incl O&P
5900	Fractured rope rib, 1 use	1 Carp	192	.042	SFCA	6.60	2.35		8.95	11
5925	2 use		192	.042		3.62	2.35		5.97	7.70
5950	3 use		192	.042		2.64	2.35		4.99	6.65
6000	4 use		192	.042		2.14	2.35		4.49	6.10
6100	Ribbed, 3/4" deep x 1-1/2" OC, 1 use		224	.036		6.60	2.01		8.61	10.45
6125	2 use		224	.036		3.62	2.01		5.63	7.20
6150	3 use		224	.036		2.64	2.01		4.65	6.10
6200	4 use		224	.036		2.14	2.01		4.15	5.55
6300	Rustic brick pattern, 1 use		224	.036		4.54	2.01		6.55	8.20
6325	2 use		224	.036		2.50	2.01		4.51	5.95
6350	3 use		224	.036		1.82	2.01		3.83	5.20
6400	4 use		224	.036		1.48	2.01		3.49	4.81
6500	3/8" striated, random, 1 use		224	.036		4.54	2.01		6.55	8.20
6525	2 use		224	.036		2.50	2.01		4.51	5.95
6550	3 use		224	.036		1.82	2.01		3.83	5.20
6600	4 use		224	.036		1.48	2.01		3.49	4.81

03 15 Concrete Accessories

03 15 05 – Concrete Forming Accessories

03 15 05.70 Shores

			Crew	Daily Output	Labor-Hours	Unit	Material	2022 Bare Costs Labor	Equipment	Total	Total Incl O&P
0010	**SHORES**										
0020	Erect and strip, by hand, horizontal members										
0500	Aluminum joists and stringers	G	2 Carp	60	.267	Ea.		15		15	24
0600	Steel, adjustable beams	G		45	.356			20		20	32
0700	Wood joists			50	.320			18		18	28.50
0800	Wood stringers			30	.533			30		30	47.50
1000	Vertical members to 10' high	G		55	.291			16.40		16.40	26
1050	To 13' high	G		50	.320			18		18	28.50
1100	To 16' high			45	.356			20		20	32
1500	Reshoring	G		1400	.011	S.F.	.82	.64		1.46	1.92
1600	Flying truss system	G	C-17D	9600	.009	SFCA		.52	.13	.65	.96
1760	Horizontal, aluminum joists, 6-1/4" high x 5' to 21' span, buy	G				L.F.	21.50			21.50	23.50
1770	Beams, 7-1/4" high x 4' to 30' span	G				"	25.50			25.50	28
1810	Horizontal, steel beam, W8x10, 7' span, buy	G				Ea.	81.50			81.50	89.50
1830	10' span	G					110			110	121
1920	15' span	G					158			158	173
1940	20' span	G					190			190	209
1970	Steel stringer, W8x10, 4' to 16' span, buy	G				L.F.	10.85			10.85	11.95
3000	Rent for job duration, aluminum joist @ 2' OC, per mo.	G				SF Flr.	.53			.53	.59
3050	Steel W8x10	G					.27			.27	.30
3060	Steel adjustable	G					.28			.28	.30
3500	#1 post shore, steel, 5'-7" to 9'-6" high, 10,000# cap., buy	G				Ea.	209			209	230
3550	#2 post shore, 7'-3" to 12'-10" high, 7800# capacity	G					242			242	266
3600	#3 post shore, 8'-10" to 16'-1" high, 3800# capacity	G					265			265	291
5010	Frame shoring systems, steel, 12,000#/leg, buy										
5040	Frame, 2' wide x 6' high	G				Ea.	72			72	79.50
5250	X-brace	G					21.50			21.50	24
5550	Base plate	G					18.30			18.30	20
5600	Screw jack	G					25.50			25.50	28
5650	U-head, 8" x 8"	G					13.75			13.75	15.15

03 15 Concrete Accessories

03 15 05 – Concrete Forming Accessories

03 15 05.85 Stair Tread Inserts

		Crew	Daily Output	Labor-Hours	Unit	Material	2022 Bare Costs Labor	Equipment	Total	Total Incl O&P
0010	**STAIR TREAD INSERTS**									
0105	Cast nosing insert, abrasive surface, pre-drilled, includes screws									
0110	Aluminum, 3" wide x 3' long	1 Cefi	32	.250	Ea.	43.50	13.30		56.80	68.50
0120	4' long		31	.258		67	13.70		80.70	95
0130	5' long		30	.267		85.50	14.15		99.65	116
0135	Extruded nosing insert, black abrasive strips, continuous anchor									
0140	Aluminum, 3" wide x 3' long	1 Cefi	64	.125	Ea.	34.50	6.65		41.15	48.50
0150	4' long		60	.133		45	7.10		52.10	60.50
0160	5' long		56	.143		53.50	7.60		61.10	71
0165	Extruded nosing insert, black abrasive strips, pre-drilled, incl. screws									
0170	Aluminum, 3" wide x 3' long	1 Cefi	32	.250	Ea.	43.50	13.30		56.80	68.50
0180	4' long		31	.258		55	13.70		68.70	82
0190	5' long		30	.267		67	14.15		81.15	95.50
9000	Minimum labor/equipment charge	1 Carp	4	2	Job		113		113	179

03 15 13 – Waterstops

03 15 13.50 Waterstops

		Crew	Daily Output	Labor-Hours	Unit	Material	2022 Bare Costs Labor	Equipment	Total	Total Incl O&P
0010	**WATERSTOPS**, PVC and Rubber									
0020	PVC, ribbed 3/16" thick, 4" wide	1 Carp	155	.052	L.F.	1.75	2.91		4.66	6.55
0050	6" wide		145	.055		2.76	3.11		5.87	8
0500	With center bulb, 6" wide, 3/16" thick		135	.059		2.87	3.34		6.21	8.45
0550	3/8" thick		130	.062		3.32	3.46		6.78	9.15
0600	9" wide x 3/8" thick		125	.064		4.39	3.60		7.99	10.55

03 15 16 – Concrete Construction Joints

03 15 16.20 Control Joints, Saw Cut

		Crew	Daily Output	Labor-Hours	Unit	Material	2022 Bare Costs Labor	Equipment	Total	Total Incl O&P
0010	**CONTROL JOINTS, SAW CUT**									
0100	Sawcut control joints in green concrete									
0120	1" depth	C-27	2000	.008	L.F.	.03	.43	.08	.54	.78
0140	1-1/2" depth		1800	.009		.04	.47	.09	.60	.89
0160	2" depth		1600	.010		.05	.53	.10	.68	1
0180	Sawcut joint reservoir in cured concrete									
0182	3/8" wide x 3/4" deep, with single saw blade	C-27	1000	.016	L.F.	.04	.85	.16	1.05	1.55
0184	1/2" wide x 1" deep, with double saw blades		900	.018		.08	.94	.18	1.20	1.76
0186	3/4" wide x 1-1/2" deep, with double saw blades		800	.020		.16	1.06	.20	1.42	2.06
0190	Water blast joint to wash away laitance, 2 passes	C-29	2500	.003			.15	.04	.19	.27
0200	Air blast joint to blow out debris and air dry, 2 passes	C-28	2000	.004			.21	.02	.23	.35
0300	For backer rod, see Section 07 91 23.10									
0340	For joint sealant, see Section 03 15 16.30 or 07 92 13.20									
0900	For replacement of joint sealant, see Section 07 01 90.81									

03 15 16.30 Expansion Joints

			Crew	Daily Output	Labor-Hours	Unit	Material	2022 Bare Costs Labor	Equipment	Total	Total Incl O&P
0010	**EXPANSION JOINTS**										
0020	Keyed, cold, 24 ga., incl. stakes, 3-1/2" high	G	1 Carp	200	.040	L.F.	1.05	2.25		3.30	4.74
0050	4-1/2" high	G		200	.040		1.18	2.25		3.43	4.88
0100	5-1/2" high	G		195	.041		1.36	2.31		3.67	5.15
0150	7-1/2" high	G		190	.042		1.68	2.37		4.05	5.60
0160	9-1/2" high	G		185	.043		1.83	2.43		4.26	5.90
0300	Poured asphalt, plain, 1/2" x 1"		1 Clab	450	.018		.94	.81		1.75	2.32
0350	1" x 2"			400	.020		3.74	.91		4.65	5.55
0500	Neoprene, liquid, cold applied, 1/2" x 1"			450	.018		2.13	.81		2.94	3.63
0550	1" x 2"			400	.020		8.50	.91		9.41	10.80
0700	Polyurethane, poured, 2 part, 1/2" x 1"			400	.020		1.45	.91		2.36	3.05
0750	1" x 2"			350	.023		5.80	1.04		6.84	8.05

For customer support on your Commercial Renovation Costs with RSMeans data, call 800.448.8182.

63

03 15 16 – Concrete Construction Joints

03 15 16.30 Expansion Joints		Crew	Daily Output	Labor-Hours	Unit	Material	2022 Bare Costs Labor	Equipment	Total	Total Incl O&P
0900	Rubberized asphalt, hot or cold applied, 1/2" x 1"	1 Clab	450	.018	L.F.	.55	.81		1.36	1.90
0950	1" x 2"		400	.020		2.28	.91		3.19	3.96
1100	Hot applied, fuel resistant, 1/2" x 1"		450	.018		.83	.81		1.64	2.20
1150	1" x 2"		400	.020		3.42	.91		4.33	5.20
2000	Premolded, bituminous fiber, 1/2" x 6"	1 Carp	375	.021		.41	1.20		1.61	2.36
2050	1" x 12"		300	.027		2.04	1.50		3.54	4.63
2140	Concrete expansion joint, recycled paper and fiber, 1/2" x 6" G		390	.021		.47	1.15		1.62	2.35
2150	1/2" x 12" G		360	.022		.94	1.25		2.19	3.03
2250	Cork with resin binder, 1/2" x 6"		375	.021		1.26	1.20		2.46	3.30
2300	1" x 12"		300	.027		3.53	1.50		5.03	6.25
2500	Neoprene sponge, closed cell, 1/2" x 6"		375	.021		3.15	1.20		4.35	5.40
2550	1" x 12"		300	.027		9.55	1.50		11.05	12.90
2750	Polyethylene foam, 1/2" x 6"	1 Carp	375	.021	L.F.	.74	1.20		1.94	2.72
2800	1" x 12"		300	.027		3.93	1.50		5.43	6.70
3000	Polyethylene backer rod, 3/8" diameter		460	.017		.13	.98		1.11	1.70
3050	3/4" diameter		460	.017		.17	.98		1.15	1.75
3100	1" diameter		460	.017		.30	.98		1.28	1.89
3500	Polyurethane foam, with polybutylene, 1/2" x 1/2"		475	.017		1.43	.95		2.38	3.08
3550	1" x 1"		450	.018		3.64	1		4.64	5.60
3750	Polyurethane foam, regular, closed cell, 1/2" x 6"		375	.021		1.06	1.20		2.26	3.08
3800	1" x 12"		300	.027		3.79	1.50		5.29	6.55
4000	Polyvinyl chloride foam, closed cell, 1/2" x 6"		375	.021		2.88	1.20		4.08	5.10
4050	1" x 12"		300	.027		9.95	1.50		11.45	13.30
4250	Rubber, gray sponge, 1/2" x 6"		375	.021		2.42	1.20		3.62	4.57
4300	1" x 12"		300	.027		8.70	1.50		10.20	12
4400	Redwood heartwood, 1" x 4"		400	.020		3.28	1.13		4.41	5.40
4450	1" x 6"		375	.021		4.80	1.20		6	7.20
5000	For installation in walls, add						75%			
5250	For installation in boxouts, add						25%			

03 15 19 – Cast-In Concrete Anchors

03 15 19.05 Anchor Bolt Accessories

0010	**ANCHOR BOLT ACCESSORIES**									
0015	For anchor bolts set in fresh concrete, see Section 03 15 19.10									
8150	Anchor bolt sleeve, plastic, 1" diameter bolts	1 Carp	60	.133	Ea.	9.20	7.50		16.70	22
8500	1-1/2" diameter		28	.286		14.40	16.10		30.50	41.50
8600	2" diameter		24	.333		16.55	18.75		35.30	48
8650	3" diameter		20	.400		42	22.50		64.50	82

03 15 19.10 Anchor Bolts

0010	**ANCHOR BOLTS**									
0015	Made from recycled materials									
0025	Single bolts installed in fresh concrete, no templates									
0030	Hooked w/nut and washer, 1/2" diameter, 8" long G	1 Carp	132	.061	Ea.	2.23	3.41		5.64	7.85
0040	12" long G		131	.061		2.48	3.44		5.92	8.20
0050	5/8" diameter, 8" long G		129	.062		3.06	3.49		6.55	8.90
0060	12" long G		127	.063		3.76	3.55		7.31	9.80
0070	3/4" diameter, 8" long G		127	.063		3.76	3.55		7.31	9.80
0080	12" long G		125	.064		4.70	3.60		8.30	10.85
0090	2-bolt pattern, including job-built 2-hole template, per set									
0100	J-type, incl. hex nut & washer, 1/2" diameter x 6" long G	1 Carp	21	.381	Set	8.80	21.50		30.30	43.50
0110	12" long G		21	.381		9.80	21.50		31.30	45
0120	18" long G		21	.381		11.30	21.50		32.80	46.50
0130	3/4" diameter x 8" long G	1 Carp	20	.400	Set	12.40	22.50		34.90	49.50

03 15 Concrete Accessories

03 15 19 – Cast-In Concrete Anchors

03 15 19.10 Anchor Bolts		Crew	Daily Output	Labor-Hours	Unit	Material	2022 Bare Costs Labor	2022 Bare Costs Equipment	Total	Total Incl O&P
0140	12" long	G 1 Carp	20	.400	Set	14.25	22.50		36.75	51.50
0150	18" long	G	20	.400		17.10	22.50		39.60	55
0160	1" diameter x 12" long	G	19	.421		33	23.50		56.50	73.50
0170	18" long	G	19	.421		38.50	23.50		62	80
0180	24" long	G	19	.421		46	23.50		69.50	88
0190	36" long	G	18	.444		61	25		86	108
0200	1-1/2" diameter x 18" long	G	17	.471		52.50	26.50		79	100
0210	24" long	G	16	.500		62	28		90	113
0300	L-type, incl. hex nut & washer, 3/4" diameter x 12" long	G	20	.400		19.75	22.50		42.25	57.50
0310	18" long	G	20	.400		24	22.50		46.50	62.50
0320	24" long	G	20	.400		28	22.50		50.50	67
0330	30" long	G	20	.400		34.50	22.50		57	74
0340	36" long	G	20	.400		38.50	22.50		61	78.50
0350	1" diameter x 12" long	G	19	.421		30	23.50		53.50	70.50
0360	18" long	G	19	.421		36	23.50		59.50	77.50
0370	24" long	G	19	.421		43.50	23.50		67	85.50
0380	30" long	G	19	.421		50.50	23.50		74	93
0390	36" long	G	18	.444		57	25		82	103
0400	42" long	G	18	.444		68	25		93	115
0410	48" long	G	18	.444		76	25		101	124
0420	1-1/4" diameter x 18" long	G	18	.444		50.50	25		75.50	95.50
0430	24" long	G	18	.444		59	25		84	105
0440	30" long	G	17	.471		68	26.50		94.50	117
0450	36" long	G	17	.471		76.50	26.50		103	126
1000	4-bolt pattern, including job-built 4-hole template, per set									
1100	J-type, incl. hex nut & washer, 1/2" diameter x 6" long	G 1 Carp	19	.421	Set	12.80	23.50		36.30	51.50
1110	12" long	G	19	.421		14.80	23.50		38.30	54
1120	18" long	G	18	.444		17.75	25		42.75	59.50
1130	3/4" diameter x 8" long	G	17	.471		19.90	26.50		46.40	64
1140	12" long	G	17	.471		23.50	26.50		50	68
1150	18" long	G	17	.471		29.50	26.50		56	74
1160	1" diameter x 12" long	G	16	.500		61	28		89	112
1170	18" long	G	15	.533		72.50	30		102.50	127
1180	24" long	G	15	.533		87	30		117	144
1190	36" long	G	15	.533		118	30		148	177
1200	1-1/2" diameter x 18" long	G	13	.615		100	34.50		134.50	165
1210	24" long	G	12	.667		119	37.50		156.50	191
1300	L-type, incl. hex nut & washer, 3/4" diameter x 12" long	G	17	.471		34.50	26.50		61	80
1310	18" long	G 1 Carp	17	.471	Set	43	26.50		69.50	89.50
1320	24" long	G	17	.471		51.50	26.50		78	98.50
1330	30" long	G	16	.500		64	28		92	115
1340	36" long	G	16	.500		72.50	28		100.50	124
1350	1" diameter x 12" long	G	16	.500		55	28		83	105
1360	18" long	G	15	.533		67.50	30		97.50	122
1370	24" long	G	15	.533		82	30		112	138
1380	30" long	G	15	.533		96	30		126	154
1390	36" long	G	15	.533		109	30		139	168
1400	42" long	G	14	.571		132	32		164	196
1410	48" long	G	14	.571		147	32		179	213
1420	1-1/4" diameter x 18" long	G	14	.571		96.50	32		128.50	157
1430	24" long	G	14	.571		114	32		146	176
1440	30" long	G	13	.615		131	34.50		165.50	199
1450	36" long	G	13	.615		148	34.50		182.50	218

For customer support on your Commercial Renovation Costs with RSMeans data, call 800.448.8182.

65

03 15 Concrete Accessories

03 15 19 – Cast-In Concrete Anchors

03 15 19.45 Machinery Anchors		Crew	Daily Output	Labor-Hours	Unit	Material	2022 Bare Costs Labor	Equipment	Total	Total Incl O&P
0010	**MACHINERY ANCHORS**, heavy duty, incl. sleeve, floating base nut,									
0020	lower stud & coupling nut, fiber plug, connecting stud, washer & nut.									
0030	For flush mounted embedment in poured concrete heavy equip. pads.									
0200	Stud & bolt, 1/2" diameter	G	E-16	40	.400	Ea.	70	25	3.68	98.68	122
0300	5/8" diameter	G		35	.457		82	29	4.21	115.21	141
0500	3/4" diameter	G		30	.533		99	33.50	4.91	137.41	168
0600	7/8" diameter	G		25	.640		111	40.50	5.90	157.40	194
0800	1" diameter	G		20	.800		121	50.50	7.35	178.85	222
0900	1-1/4" diameter	G	↓	15	1.067	↓	161	67	9.80	237.80	296

03 21 Reinforcement Bars

03 21 11 – Plain Steel Reinforcement Bars

03 21 11.60 Reinforcing In Place

03 21 11.60 Reinforcing In Place		Crew	Daily Output	Labor-Hours	Unit	Material	2022 Bare Costs Labor	Equipment	Total	Total Incl O&P
0010	**REINFORCING IN PLACE**, 50-60 ton lots, A615 Grade 60 R032110-10									
0020	Includes labor, but not material cost, to install accessories									
0030	Made from recycled materials									
0102	Beams & Girders, #3 to #7	G	4 Rodm	3200	.010	Lb.	.73	.61		1.34	1.75
0152	#8 to #18	G		5400	.006		.73	.36		1.09	1.36
0202	Columns, #3 to #7	G		3000	.011		.73	.65		1.38	1.81
0252	#8 to #18	G		4600	.007		.73	.42		1.15	1.46
0402	Elevated slabs, #4 to #7	G		5800	.006		.73	.33		1.06	1.32
0502	Footings, #4 to #7	G		4200	.008		.73	.46		1.19	1.52
0552	#8 to #18	G		7200	.004		.73	.27		1	1.22
0602	Slab on grade, #3 to #7	G	↓	4200	.008	↓	.73	.46		1.19	1.52
0702	Walls, #3 to #7	G	4 Rodm	6000	.005	Lb.	.73	.32		1.05	1.31
0752	#8 to #18	G	"	8000	.004	"	.73	.24		.97	1.18
0900	For other than 50-60 ton lots									
1000	Under 10 ton job, #3 to #7, add						25%	10%			
1010	#8 to #18, add						20%	10%			
1050	10-50 ton job, #3 to #7, add						10%				
1060	#8 to #18, add						5%				
1100	60-100 ton job, #3 to #7, deduct						5%				
1110	#8 to #18, deduct						10%				
1150	Over 100 ton job, #3 to #7, deduct						10%				
1160	#8 to #18, deduct						15%				
2000	Unloading & sorting, add to above		C-5	100	.560	Ton		34	21	55	76
2200	Crane cost for handling, 90 picks/day, up to 1.5 tons/bundle, add to above			135	.415			25	15.50	40.50	56
2210	1.0 ton/bundle			92	.609			37	23	60	82.50
2220	0.5 ton/bundle		↓	35	1.600	↓		96.50	60	156.50	217
2400	Dowels, 2 feet long, deformed, #3	G	2 Rodm	520	.031	Ea.	.60	1.87		2.47	3.58
2410	#4	G		480	.033		1.07	2.02		3.09	4.33
2420	#5	G		435	.037		1.67	2.23		3.90	5.30
2430	#6	G		360	.044	↓	2.40	2.70		5.10	6.85
2450	Longer and heavier dowels, add	G		725	.022	Lb.	.80	1.34		2.14	2.97
2500	Smooth dowels, 12" long, 1/4" or 3/8" diameter	G		140	.114	Ea.	1.02	6.95		7.97	12
2520	5/8" diameter	G		125	.128		1.79	7.75		9.54	14.10
2530	3/4" diameter	G	↓	110	.145	↓	2.22	8.85		11.07	16.25
2600	Dowel sleeves for CIP concrete, 2-part system									
2610	Sleeve base, plastic, for 5/8" smooth dowel sleeve, fasten to edge form		1 Rodm	200	.040	Ea.	.57	2.43		3	4.43
2615	Sleeve, plastic, 12" long, for 5/8" smooth dowel, snap onto base		↓	400	.020	↓	1.41	1.21		2.62	3.45

03 21 Reinforcement Bars

03 21 11 – Plain Steel Reinforcement Bars

03 21 11.60 Reinforcing In Place

		Crew	Daily Output	Labor-Hours	Unit	Material	2022 Bare Costs Labor	Equipment	Total	Total Incl O&P
2620	Sleeve base, for 3/4" smooth dowel sleeve	1 Rodm	175	.046	Ea.	.57	2.77		3.34	4.97
2625	Sleeve, 12" long, for 3/4" smooth dowel		350	.023		1.41	1.39		2.80	3.72
2630	Sleeve base, for 1" smooth dowel sleeve		150	.053		.96	3.24		4.20	6.10
2635	Sleeve, 12" long, for 1" smooth dowel		300	.027		1.48	1.62		3.10	4.16
2700	Dowel caps, visual warning only, plastic, #3 to #8	2 Rodm	800	.020		.64	1.21		1.85	2.60
2720	#8 to #18		750	.021		1.11	1.29		2.40	3.24
2750	Impalement protective, plastic, #4 to #9		800	.020		1.60	1.21		2.81	3.66
9000	Minimum labor/equipment charge	1 Rodm	4	2	Job		121		121	190

03 21 21 – Composite Reinforcement Bars

03 21 21.11 Glass Fiber-Reinforced Polymer Reinf. Bars

		Crew	Daily Output	Labor-Hours	Unit	Material	2022 Bare Costs Labor	Equipment	Total	Total Incl O&P
0010	**GLASS FIBER-REINFORCED POLYMER REINFORCEMENT BARS**									
0020	Includes labor, but not material cost, to install accessories									
0050	#2 bar, .043 lb./L.F.	4 Rodm	9500	.003	L.F.	.43	.20		.63	.79
0100	#3 bar, .092 lb./L.F.		9300	.003		.70	.21		.91	1.10
0150	#4 bar, .160 lb./L.F.		9100	.004		1	.21		1.21	1.43
0200	#5 bar, .258 lb./L.F.		8700	.004		1.23	.22		1.45	1.70
0250	#6 bar, .372 lb./L.F.		8300	.004		2.15	.23		2.38	2.74
0300	#7 bar, .497 lb./L.F.		7900	.004		2.65	.25		2.90	3.30
0350	#8 bar, .620 lb./L.F.		7400	.004		2.41	.26		2.67	3.06
0400	#9 bar, .800 lb./L.F.		6800	.005		4.37	.29		4.66	5.25
0450	#10 bar, 1.08 lb./L.F.		5800	.006		3.97	.33		4.30	4.89
0500	For bends, add per bend				Ea.	1.67			1.67	1.84

03 22 Fabric and Grid Reinforcing

03 22 11 – Plain Welded Wire Fabric Reinforcing

03 22 11.10 Plain Welded Wire Fabric

			Crew	Daily Output	Labor-Hours	Unit	Material	2022 Bare Costs Labor	Equipment	Total	Total Incl O&P
0010	**PLAIN WELDED WIRE FABRIC** ASTM A185	R032205-30									
0020	Includes labor, but not material cost, to install accessories										
0030	Made from recycled materials										
0050	Sheets										
0100	6 x 6 - W1.4 x W1.4 (10 x 10) 21 lb./C.S.F.	G	2 Rodm	35	.457	C.S.F.	22	28		50	68
0200	6 x 6 - W2.1 x W2.1 (8 x 8) 30 lb./C.S.F.	G		31	.516		32	31.50		63.50	84
0300	6 x 6 - W2.9 x W2.9 (6 x 6) 42 lb./C.S.F.	G		29	.552		43	33.50		76.50	100
0400	6 x 6 - W4 x W4 (4 x 4) 58 lb./C.S.F.	G		27	.593		52	36		88	113
0500	4 x 4 - W1.4 x W1.4 (10 x 10) 31 lb./C.S.F.	G		31	.516		35.50	31.50		67	88
0600	4 x 4 - W2.1 x W2.1 (8 x 8) 44 lb./C.S.F.	G		29	.552		44.50	33.50		78	102
0650	4 x 4 - W2.9 x W2.9 (6 x 6) 61 lb./C.S.F.	G		27	.593		71	36		107	134
0700	4 x 4 - W4 x W4 (4 x 4) 85 lb./C.S.F.	G		25	.640		88.50	39		127.50	158
0750	Rolls										
0800	2 x 2 - #14 galv., 21 lb./C.S.F., beam & column wrap	G	2 Rodm	6.50	2.462	C.S.F.	64	149		213	305
0900	2 x 2 - #12 galv. for gunite reinforcing	G	"	6.50	2.462	"	89	149		238	330
9000	Minimum labor/equipment charge		1 Rodm	4	2	Job		121		121	190

03 22 13 – Galvanized Welded Wire Fabric Reinforcing

03 22 13.10 Galvanized Welded Wire Fabric

		Crew	Daily Output	Labor-Hours	Unit	Material	2022 Bare Costs Labor	Equipment	Total	Total Incl O&P
0010	**GALVANIZED WELDED WIRE FABRIC**									
0100	Add to plain welded wire pricing for galvanized welded wire				Lb.	.29			.29	.31

For customer support on your Commercial Renovation Costs with RSMeans data, call 800.448.8182.

67

03 22 Fabric and Grid Reinforcing

03 22 16 – Epoxy-Coated Welded Wire Fabric Reinforcing

03 22 16.10 Epoxy-Coated Welded Wire Fabric	Crew	Daily Output	Labor-Hours	Unit	Material	2022 Bare Costs Labor	2022 Bare Costs Equipment	Total	Total Incl O&P
0010 **EPOXY-COATED WELDED WIRE FABRIC**									
0100 Add to plain welded wire pricing for epoxy-coated welded wire				Lb.	.57			.57	.62

03 24 Fibrous Reinforcing

03 24 05 – Reinforcing Fibers

03 24 05.30 Synthetic Fibers

	Crew	Daily Output	Labor-Hours	Unit	Material	Labor	Equipment	Total	Total Incl O&P
0010 **SYNTHETIC FIBERS**									
0100 Synthetic fibers, add to concrete				Lb.	6.25			6.25	6.85
0110 1-1/2 lb./C.Y.				C.Y.	9.65			9.65	10.60

03 24 05.70 Steel Fibers

		Crew	Daily Output	Labor-Hours	Unit	Material	Labor	Equipment	Total	Total Incl O&P
0010 **STEEL FIBERS**										
0140 ASTM A850, Type V, continuously deformed, 1-1/2" long x 0.045" diam.										
0150 Add to price of ready mix concrete	G				Lb.	1.22			1.22	1.34
0205 Alternate pricing, dosing at 5 lb./C.Y., add to price of RMC	G				C.Y.	6.10			6.10	6.70
0210 10 lb./C.Y.	G					12.20			12.20	13.40
0215 15 lb./C.Y.	G					18.30			18.30	20
0220 20 lb./C.Y.	G					24.50			24.50	27
0225 25 lb./C.Y.	G					30.50			30.50	33.50
0230 30 lb./C.Y.	G					36.50			36.50	40.50
0235 35 lb./C.Y.	G					42.50			42.50	47
0240 40 lb./C.Y.	G					49			49	53.50
0250 50 lb./C.Y.	G					61			61	67
0275 75 lb./C.Y.	G					91.50			91.50	101
0300 100 lb./C.Y.	G					122			122	134

03 30 Cast-In-Place Concrete

03 30 53 – Miscellaneous Cast-In-Place Concrete

03 30 53.40 Concrete In Place

		Crew	Daily Output	Labor-Hours	Unit	Material	Labor	Equipment	Total	Total Incl O&P
0010 **CONCRETE IN PLACE**	R033105-10									
0020 Including forms (4 uses), Grade 60 rebar, concrete (Portland cement	R033105-20									
0050 Type I), placement and finishing unless otherwise indicated	R033105-70									
0300 Beams (3500 psi), 5 kip/L.F., 10' span	R033105-85	C-14A	15.62	12.804	C.Y.	485	720	50	1,255	1,750
0350 25' span		"	18.55	10.782		510	605	42	1,157	1,575
0500 Chimney foundations (5000 psi), over 5 C.Y.		C-14C	32.22	3.476		193	187	.87	380.87	510
0510 (3500 psi), under 5 C.Y.		"	23.71	4.724		230	254	1.18	485.18	655
0700 Columns, square (4000 psi), 12" x 12", up to 1% reinforcing by area		C-14A	11.96	16.722		535	940	65.50	1,540.50	2,150
0720 Up to 2% reinforcing by area			10.13	19.743		835	1,100	77	2,012	2,750
0740 Up to 3% reinforcing by area			9.03	22.148		1,250	1,250	86.50	2,586.50	3,450
0800 16" x 16", up to 1% reinforcing by area			16.22	12.330		425	695	48	1,168	1,625
0820 Up to 2% reinforcing by area			12.57	15.911		715	895	62	1,672	2,275
0840 Up to 3% reinforcing by area			10.25	19.512		1,100	1,100	76.50	2,276.50	3,000
0900 24" x 24", up to 1% reinforcing by area			23.66	8.453		360	475	33	868	1,175
0920 Up to 2% reinforcing by area			17.71	11.293		635	635	44	1,314	1,750
0940 Up to 3% reinforcing by area			14.15	14.134		1,000	795	55.50	1,850.50	2,400
1000 36" x 36", up to 1% reinforcing by area			33.69	5.936		320	335	23	678	905
1020 Up to 2% reinforcing by area			23.32	8.576		560	485	33.50	1,078.50	1,425
1040 Up to 3% reinforcing by area			17.82	11.223		940	630	44	1,614	2,075
1100 Columns, round (4000 psi), tied, 12" diameter, up to 1% reinforcing by area			20.97	9.537		505	535	37.50	1,077.50	1,450

68

03 30 Cast-In-Place Concrete

03 30 53 – Miscellaneous Cast-In-Place Concrete

03 30 53.40 Concrete In Place	Crew	Daily Output	Labor-Hours	Unit	Material	2022 Bare Costs Labor	Equipment	Total	Total Incl O&P
1120 Up to 2% reinforcing by area	C-14A	15.27	13.098	C.Y.	805	735	51	1,591	2,125
1140 Up to 3% reinforcing by area		12.11	16.515		1,200	930	64.50	2,194.50	2,875
1200 16" diameter, up to 1% reinforcing by area		31.49	6.351		455	355	25	835	1,100
1220 Up to 2% reinforcing by area		19.12	10.460		760	590	41	1,391	1,800
1240 Up to 3% reinforcing by area		13.77	14.524		1,125	815	57	1,997	2,600
1300 20" diameter, up to 1% reinforcing by area		41.04	4.873		440	274	19.05	733.05	940
1320 Up to 2% reinforcing by area		24.05	8.316		715	470	32.50	1,217.50	1,550
1340 Up to 3% reinforcing by area		17.01	11.758		1,100	660	46	1,806	2,325
1400 24" diameter, up to 1% reinforcing by area		51.85	3.857		400	217	15.10	632.10	800
1420 Up to 2% reinforcing by area		27.06	7.391		700	415	29	1,144	1,450
1440 Up to 3% reinforcing by area		18.29	10.935		1,075	615	43	1,733	2,200
1500 36" diameter, up to 1% reinforcing by area		75.04	2.665		410	150	10.40	570.40	705
1520 Up to 2% reinforcing by area		37.49	5.335		680	300	21	1,001	1,250
1540 Up to 3% reinforcing by area		22.84	8.757		1,050	495	34	1,579	1,975
1900 Elevated slab (4000 psi), flat slab with drops, 125 psf Sup. Load, 20' span	C-14B	38.45	5.410		395	305	20.50	720.50	940
1950 30' span		50.99	4.079		420	229	15.35	664.35	835
2100 Flat plate, 125 psf Sup. Load, 15' span		30.24	6.878		360	385	26	771	1,025
2150 25' span		49.60	4.194		375	235	15.75	625.75	795
2300 Waffle const., 30" domes, 125 psf Sup. Load, 20' span		37.07	5.611		405	315	21	741	970
2350 30' span		44.07	4.720		375	265	17.75	657.75	855
2500 One way joists, 30" pans, 125 psf Sup. Load, 15' span		27.38	7.597		510	425	28.50	963.50	1,275
2550 25' span		31.15	6.677		470	375	25	870	1,150
2700 One way beam & slab, 125 psf Sup. Load, 15' span		20.59	10.102		390	565	38	993	1,375
2750 25' span		28.36	7.334		370	410	27.50	807.50	1,075
2900 Two way beam & slab, 125 psf Sup. Load, 15' span		24.04	8.652		375	485	32.50	892.50	1,225
2950 25' span		35.87	5.799		325	325	22	672	900
3100 Elevated slabs, flat plate, including finish, not									
3110 including forms or reinforcing									
3150 Regular concrete (4000 psi), 4" slab	C-8	2613	.021	S.F.	2.15	1.07	.29	3.51	4.38
3200 6" slab		2585	.022		3.16	1.09	.29	4.54	5.50
3250 2-1/2" thick floor fill		2685	.021		1.40	1.05	.28	2.73	3.50
3300 Lightweight, 110 #/C.F., 2-1/2" thick floor fill		2585	.022		1.34	1.09	.29	2.72	3.50
3400 Cellular concrete, 1-5/8" fill, under 5000 S.F.		2000	.028		.92	1.40	.38	2.70	3.63
3450 Over 10,000 S.F.		2200	.025		.88	1.28	.34	2.50	3.36
3500 Add per floor for 3 to 6 stories high		31800	.002			.09	.02	.11	.17
3520 For 7 to 20 stories high		21200	.003			.13	.04	.17	.25
3540 Equipment pad (3000 psi), 3' x 3' x 6" thick	C-14H	45	1.067	Ea.	66.50	58.50	.62	125.62	167
3550 4' x 4' x 6" thick		30	1.600		101	88	.93	189.93	251
3560 5' x 5' x 8" thick		18	2.667		180	147	1.55	328.55	430
3570 6' x 6' x 8" thick		14	3.429		245	189	1.99	435.99	570
3580 8' x 8' x 10" thick		8	6		510	330	3.48	843.48	1,075
3590 10' x 10' x 12" thick		5	9.600		895	530	5.55	1,430.55	1,825
3800 Footings (3000 psi), spread under 1 C.Y.	C-14C	28	4	C.Y.	256	215	1	472	620
3825 1 C.Y. to 5 C.Y.		43	2.605		315	140	.65	455.65	570
3850 Over 5 C.Y.		75	1.493		287	80.50	.37	367.87	440
3900 Footings, strip (3000 psi), 18" x 9", unreinforced	C-14L	40	2.400		195	126	.70	321.70	415
3920 18" x 9", reinforced	C-14C	35	3.200		230	172	.80	402.80	525
3925 20" x 10", unreinforced	C-14L	45	2.133		189	112	.62	301.62	385
3930 20" x 10", reinforced	C-14C	40	2.800		218	151	.70	369.70	480
3935 24" x 12", unreinforced	C-14L	55	1.745		188	92	.51	280.51	355
3940 24" x 12", reinforced	C-14C	48	2.333		217	126	.58	343.58	440
3945 36" x 12", unreinforced	C-14L	70	1.371		181	72	.40	253.40	315
3950 36" x 12", reinforced	C-14C	60	1.867		208	100	.47	308.47	390

03 30 53 – Miscellaneous Cast-In-Place Concrete

03 30 53.40 Concrete In Place		Crew	Daily Output	Labor-Hours	Unit	Material	2022 Bare Costs Labor	Equipment	Total	Total Incl O&P
4000	Foundation mat (3000 psi), under 10 C.Y.	C-14C	38.67	2.896	C.Y.	315	156	.72	471.72	595
4200	Wall, free-standing (3000 psi), 8" thick, 8' high	C-14D	45.83	4.364		244	244	17.05	505.05	670
4250	14' high		27.26	7.337		289	410	28.50	727.50	1,000
4260	12" thick, 8' high		64.32	3.109		221	174	12.15	407.15	530
4270	14' high		40.01	4.999		236	279	19.55	534.55	725
4300	15" thick, 8' high		80.02	2.499		214	140	9.75	363.75	470
4350	12' high		51.26	3.902		214	218	15.25	447.25	600
4500	18' high	▼	48.85	4.094	▼	241	229	16	486	650
4520	Handicap access ramp (4000 psi), railing both sides, 3' wide	C-14H	14.58	3.292	L.F.	435	181	1.91	617.91	770
4525	5' wide		12.22	3.928		450	216	2.28	668.28	840
4530	With 6" curb and rails both sides, 3' wide		8.55	5.614		450	310	3.26	763.26	990
4535	5' wide	▼	7.31	6.566	▼	455	360	3.81	818.81	1,075
4650	Slab on grade (3500 psi), not including finish, 4" thick	C-14E	60.75	1.449	C.Y.	191	79.50	.46	270.96	335
4700	6" thick	"	92	.957	"	183	52.50	.31	235.81	285
4701	Thickened slab edge (3500 psi), for slab on grade poured									
4702	monolithically with slab; depth is in addition to slab thickness;									
4703	formed vertical outside edge, earthen bottom and inside slope									
4705	8" deep x 8" wide bottom, unreinforced	C-14L	2190	.044	L.F.	5.35	2.31	.01	7.67	9.55
4710	8" x 8", reinforced	C-14C	1670	.067		8.80	3.61	.02	12.43	15.35
4715	12" deep x 12" wide bottom, unreinforced	C-14L	1800	.053		10.85	2.81	.02	13.68	16.40
4720	12" x 12", reinforced	C-14C	1310	.086		17.10	4.60	.02	21.72	26
4725	16" deep x 16" wide bottom, unreinforced	C-14L	1440	.067		18.25	3.51	.02	21.78	25.50
4730	16" x 16", reinforced	C-14C	1120	.100		25.50	5.40	.03	30.93	37
4735	20" deep x 20" wide bottom, unreinforced	C-14L	1150	.083		27.50	4.39	.02	31.91	37.50
4740	20" x 20", reinforced	C-14C	920	.122		37	6.55	.03	43.58	51.50
4745	24" deep x 24" wide bottom, unreinforced	C-14L	930	.103		39	5.45	.03	44.48	51.50
4750	24" x 24", reinforced	C-14C	740	.151	▼	51	8.15	.04	59.19	69.50
4751	Slab on grade (3500 psi), incl. troweled finish, not incl. forms									
4760	or reinforcing, over 10,000 S.F., 4" thick	C-14F	3425	.021	S.F.	2.09	1.07	.01	3.17	3.99
4820	6" thick		3350	.021		3.05	1.09	.01	4.15	5.10
4840	8" thick		3184	.023		4.17	1.15	.01	5.33	6.40
4900	12" thick		2734	.026		6.25	1.34	.01	7.60	9
4950	15" thick	▼	2505	.029	▼	7.85	1.46	.01	9.32	10.95
5000	Slab on grade (3000 psi), incl. broom finish, not incl. forms									
5001	or reinforcing, 4" thick	C-14G	2873	.019	S.F.	2.07	.98	.01	3.06	3.83
5010	6" thick		2590	.022		3.24	1.09	.01	4.34	5.30
5020	8" thick	▼	2320	.024	▼	4.23	1.21	.01	5.45	6.55
5200	Lift slab in place above the foundation, incl. forms, reinforcing,									
5210	concrete (4000 psi) and columns, over 20,000 S.F./floor	C-14B	2113	.098	S.F.	9.25	5.55	.37	15.17	19.30
5250	10,000 S.F. to 20,000 S.F./floor		1650	.126		10.05	7.10	.47	17.62	23
5300	Under 10,000 S.F./floor	▼	1500	.139	▼	10.80	7.80	.52	19.12	25
5500	Lightweight, ready mix, including screed finish only,									
5510	not including forms or reinforcing									
5550	1:4 (2500 psi) for structural roof decks	C-14B	260	.800	C.Y.	127	45	3.01	175.01	214
5600	1:6 (3000 psi) for ground slab with radiant heat	C-14F	92	.783		119	40	.31	159.31	194
5650	1:3:2 (2000 psi) with sand aggregate, roof deck	C-14B	260	.800		113	45	3.01	161.01	198
5700	Ground slab (2000 psi)	C-14F	107	.673		113	34	.26	147.26	178
5900	Pile caps (3000 psi), incl. forms and reinf., sq. or rect., under 10 C.Y.	C-14C	54.14	2.069		265	111	.52	376.52	470
5950	Over 10 C.Y.		75	1.493		246	80.50	.37	326.87	395
6000	Triangular or hexagonal, under 10 C.Y.		53	2.113		193	114	.53	307.53	395
6050	Over 10 C.Y.		85	1.318		216	71	.33	287.33	350
6200	Retaining walls (3000 psi), gravity, 4' high see Section 32 32	C-14D	66.20	3.021		214	169	11.80	394.80	515
6250	10' high	▼	125	1.600	▼	206	89.50	6.25	301.75	375

For customer support on your Commercial Renovation Costs with RSMeans data, call 800.448.8182.

03 30 Cast-In-Place Concrete

03 30 53 – Miscellaneous Cast-In-Place Concrete

03 30 53.40 Concrete In Place

		Crew	Daily Output	Labor-Hours	Unit	Material	2022 Bare Costs Labor	Equipment	Total	Total Incl O&P
6300	Cantilever, level backfill loading, 8' high	C-14D	70	2.857	C.Y.	228	160	11.15	399.15	515
6350	16' high	↓	91	2.198	↓	222	123	8.60	353.60	450
6800	Stairs (3500 psi), not including safety treads, free standing, 3'-6" wide	C-14H	83	.578	LF Nose	8.65	32	.34	40.99	60.50
6850	Cast on ground		125	.384	"	7.10	21	.22	28.32	41.50
7000	Stair landings, free standing		200	.240	S.F.	6.95	13.20	.14	20.29	29
7050	Cast on ground	↓	475	.101	"	5.40	5.55	.06	11.01	14.80
9000	Minimum labor/equipment charge	2 Carp	1	16	Job		900		900	1,425

03 31 Structural Concrete

03 31 13 – Heavyweight Structural Concrete

03 31 13.25 Concrete, Hand Mix

		Crew	Daily Output	Labor-Hours	Unit	Material	2022 Bare Costs Labor	Equipment	Total	Total Incl O&P
0010	**CONCRETE, HAND MIX** for small quantities or remote areas									
0050	Includes bulk local aggregate, bulk sand, bagged Portland									
0060	cement (Type I) and water, using gas powered cement mixer									
0125	2500 psi	C-30	135	.059	C.F.	4.21	2.70	1.04	7.95	10.10
0130	3000 psi		135	.059		4.54	2.70	1.04	8.28	10.45
0135	3500 psi		135	.059		4.78	2.70	1.04	8.52	10.70
0140	4000 psi		135	.059		5.05	2.70	1.04	8.79	11
0145	4500 psi		135	.059		5.40	2.70	1.04	9.14	11.40
0150	5000 psi	↓	135	.059	↓	5.70	2.70	1.04	9.44	11.70
0300	Using pre-bagged dry mix and wheelbarrow (80-lb. bag = 0.6 C.F.)									
0340	4000 psi	1 Clab	48	.167	C.F.	11.75	7.60		19.35	25

03 31 13.30 Concrete, Volumetric Site-Mixed

		Crew	Daily Output	Labor-Hours	Unit	Material	2022 Bare Costs Labor	Equipment	Total	Total Incl O&P
0010	**CONCRETE, VOLUMETRIC SITE-MIXED**									
0015	Mixed on-site in volumetric truck									
0020	Includes local aggregate, sand, Portland cement (Type I) and water									
0025	Excludes all additives and treatments									
0100	3000 psi, 1 C.Y. mixed and discharged				C.Y.	220			220	242
0110	2 C.Y.					162			162	178
0120	3 C.Y.					147			147	161
0130	4 C.Y.					129			129	142
0140	5 C.Y.				↓	126			126	139
0200	For truck holding/waiting time past first 2 on-site hours, add				Hr.	94			94	103
0210	For trip charge beyond first 20 miles, each way, add				Mile	3.72			3.72	4.09
0220	For each additional increase of 500 psi, add				Ea.	4.78			4.78	5.25

03 31 13.35 Heavyweight Concrete, Ready Mix

		Crew	Daily Output	Labor-Hours	Unit	Material	2022 Bare Costs Labor	Equipment	Total	Total Incl O&P
0010	**HEAVYWEIGHT CONCRETE, READY MIX,** delivered									
0012	Includes local aggregate, sand, Portland cement (Type I) and water									
0015	Excludes all additives and treatments R033105-20									
0020	2000 psi				C.Y.	152			152	167
0100	2500 psi					155			155	170
0150	3000 psi					159			159	175
0200	3500 psi					161			161	177
0300	4000 psi					164			164	180
0350	4500 psi					135			135	148
0400	5000 psi					139			139	153
0411	6000 psi					143			143	157
0412	8000 psi					150			150	165
0413	10,000 psi					158			158	174
0414	12,000 psi				↓	166			166	182

03 31 Structural Concrete

03 31 13 – Heavyweight Structural Concrete

03 31 13.35 Heavyweight Concrete, Ready Mix

		Crew	Daily Output	Labor-Hours	Unit	Material	2022 Bare Costs Labor	Equipment	Total	Total Incl O&P
1000	For high early strength (Portland cement Type III), add				C.Y.	10%				
1300	For winter concrete (hot water), add					8.50			8.50	9.35
1410	For mid-range water reducer, add					4.27			4.27	4.70
1420	For high-range water reducer/superplasticizer, add					9			9	9.90
1430	For retarder, add					5.10			5.10	5.60
1440	For non-Chloride accelerator, add					8.80			8.80	9.70
1450	For Chloride accelerator, per 1%, add					4			4	4.40
1460	For fiber reinforcing, synthetic (1 lb./C.Y.), add					8.40			8.40	9.25
1500	For Saturday delivery, add				▼	10			10	11
1510	For truck holding/waiting time past 1st hour per load, add				Hr.	115			115	127
1520	For short load (less than 4 C.Y.), add per load				Ea.	125			125	138
2000	For all lightweight aggregate, add				C.Y.	45%				

03 31 13.70 Placing Concrete

		Crew	Daily Output	Labor-Hours	Unit	Material	2022 Bare Costs Labor	Equipment	Total	Total Incl O&P
0010	**PLACING CONCRETE** R033105-70									
0020	Includes labor and equipment to place, level (strike off) and consolidate									
0050	Beams, elevated, small beams, pumped	C-20	60	1.067	C.Y.		52	13.50	65.50	97
0100	With crane and bucket	C-7	45	1.600			79	52.50	131.50	182
0200	Large beams, pumped	C-20	90	.711			34.50	9	43.50	64.50
0250	With crane and bucket	C-7	65	1.108			54.50	36.50	91	126
0400	Columns, square or round, 12" thick, pumped	C-20	60	1.067			52	13.50	65.50	97
0450	With crane and bucket	C-7	40	1.800			88.50	59	147.50	205
0600	18" thick, pumped	C-20	90	.711			34.50	9	43.50	64.50
0650	With crane and bucket	C-7	55	1.309			64.50	43	107.50	150
0800	24" thick, pumped	C-20	92	.696			34	8.80	42.80	63
0850	With crane and bucket	C-7	70	1.029			50.50	34	84.50	117
1000	36" thick, pumped	C-20	140	.457			22	5.80	27.80	41.50
1050	With crane and bucket	C-7	100	.720			35.50	23.50	59	82
1200	Duct bank, direct chute	C-6	155	.310			14.60	.36	14.96	23.50
1400	Elevated slabs, less than 6" thick, pumped	C-20	140	.457			22	5.80	27.80	41.50
1450	With crane and bucket	C-7	95	.758			37.50	25	62.50	86.50
1500	6" to 10" thick, pumped	C-20	160	.400			19.45	5.05	24.50	36.50
1550	With crane and bucket	C-7	110	.655			32	21.50	53.50	74.50
1600	Slabs over 10" thick, pumped	C-20	180	.356			17.30	4.50	21.80	32.50
1650	With crane and bucket	C-7	130	.554			27.50	18.25	45.75	63
1900	Footings, continuous, shallow, direct chute	C-6	120	.400			18.90	.46	19.36	30.50
1950	Pumped	C-20	150	.427			21	5.40	26.40	39
2000	With crane and bucket	C-7	90	.800			39.50	26.50	66	91
2100	Footings, continuous, deep, direct chute	C-6	140	.343			16.20	.40	16.60	26
2150	Pumped	C-20	160	.400			19.45	5.05	24.50	36.50
2200	With crane and bucket	C-7	110	.655			32	21.50	53.50	74.50
2400	Footings, spread, under 1 C.Y., direct chute	C-6	55	.873			41	1.01	42.01	66
2450	Pumped	C-20	65	.985			48	12.45	60.45	89
2500	With crane and bucket	C-7	45	1.600			79	52.50	131.50	182
2600	Over 5 C.Y., direct chute	C-6	120	.400			18.90	.46	19.36	30.50
2650	Pumped	C-20	150	.427			21	5.40	26.40	39
2700	With crane and bucket	C-7	100	.720			35.50	23.50	59	82
2900	Foundation mats, over 20 C.Y., direct chute	C-6	350	.137			6.45	.16	6.61	10.45
2950	Pumped	C-20	400	.160			7.80	2.03	9.83	14.55
3000	With crane and bucket	C-7	300	.240			11.80	7.90	19.70	27.50
3200	Grade beams, direct chute	C-6	150	.320			15.10	.37	15.47	24.50
3250	Pumped	C-20	180	.356			17.30	4.50	21.80	32.50
3300	With crane and bucket	C-7	120	.600			29.50	19.75	49.25	68

03 31 13.70 Placing Concrete	Crew	Daily Output	Labor-Hours	Unit	Material	2022 Bare Costs Labor	Equipment	Total	Total Incl O&P
3500	High rise, for more than 5 stories, pumped, add per story	C-20	2100	.030	C.Y.		1.48	.39	1.87	2.76
3510	With crane and bucket, add per story	C-7	2100	.034			1.69	1.13	2.82	3.90
3700	Pile caps, under 5 C.Y., direct chute	C-6	90	.533			25	.62	25.62	40.50
3750	Pumped	C-20	110	.582			28.50	7.35	35.85	52.50
3800	With crane and bucket	C-7	80	.900			44.50	29.50	74	103
3850	Pile cap, 5 C.Y. to 10 C.Y., direct chute	C-6	175	.274			12.95	.32	13.27	21
3900	Pumped	C-20	200	.320			15.55	4.05	19.60	29
3950	With crane and bucket	C-7	150	.480			23.50	15.80	39.30	55
4000	Over 10 C.Y., direct chute	C-6	215	.223			10.55	.26	10.81	17
4050	Pumped	C-20	240	.267			12.95	3.38	16.33	24
4100	With crane and bucket	C-7	185	.389			19.15	12.80	31.95	44
4300	Slab on grade, up to 6" thick, direct chute	C-6	110	.436			20.50	.51	21.01	33
4350	Pumped	C-20	130	.492			24	6.25	30.25	45
4400	With crane and bucket	C-7	110	.655			32	21.50	53.50	74.50
4600	Over 6" thick, direct chute	C-6	165	.291			13.75	.34	14.09	22
4650	Pumped	C-20	185	.346			16.85	4.38	21.23	31.50
4700	With crane and bucket	C-7	145	.497			24.50	16.35	40.85	56.50
4900	Walls, 8" thick, direct chute	C-6	90	.533			25	.62	25.62	40.50
4950	Pumped	C-20	100	.640			31	8.10	39.10	58
5000	With crane and bucket	C-7	80	.900			44.50	29.50	74	103
5050	12" thick, direct chute	C-6	100	.480			22.50	.56	23.06	36.50
5100	Pumped	C-20	110	.582			28.50	7.35	35.85	52.50
5200	With crane and bucket	C-7	90	.800			39.50	26.50	66	91
5300	15" thick, direct chute	C-6	105	.457			21.50	.53	22.03	34.50
5350	Pumped	C-20	120	.533			26	6.75	32.75	48.50
5400	With crane and bucket	C-7	95	.758	▼		37.50	25	62.50	86.50
5600	Wheeled concrete dumping, add to placing costs above									
5610	Walking cart, 50' haul, add	C-18	32	.281	C.Y.		12.90	3.72	16.62	24.50
5620	150' haul, add		24	.375			17.20	4.95	22.15	33
5700	250' haul, add	▼	18	.500			23	6.60	29.60	44
5800	Riding cart, 50' haul, add	C-19	80	.113			5.15	1.75	6.90	10.10
5810	150' haul, add		60	.150			6.85	2.33	9.18	13.45
5900	250' haul, add	▼	45	.200	▼		9.15	3.10	12.25	17.95
6000	Concrete in-fill for pan-type metal stairs and landings. Manual placement									
6010	includes up to 50' horizontal haul from point of concrete discharge.									
6100	Stair pan treads, 2" deep									
6110	Flights in 1st floor level up/down from discharge point	C-8A	3200	.015	S.F.		.73		.73	1.15
6120	2nd floor level		2500	.019			.93		.93	1.47
6130	3rd floor level		2000	.024			1.16		1.16	1.84
6140	4th floor level	▼	1800	.027	▼		1.29		1.29	2.04
6200	Intermediate stair landings, pan-type 4" deep									
6210	Flights in 1st floor level up/down from discharge point	C-8A	2000	.024	S.F.		1.16		1.16	1.84
6220	2nd floor level		1500	.032			1.55		1.55	2.45
6230	3rd floor level		1200	.040			1.94		1.94	3.06
6240	4th floor level	▼	1000	.048	▼		2.33		2.33	3.67
9000	Minimum labor/equipment charge	C-6	2	24	Job		1,125	28	1,153	1,825

For customer support on your Commercial Renovation Costs with RSMeans data, call 800.448.8182.

73

03 35 Concrete Finishing

03 35 13 – High-Tolerance Concrete Floor Finishing

03 35 13.30 Finishing Floors, High Tolerance

		Crew	Daily Output	Labor-Hours	Unit	Material	2022 Bare Costs Labor	Equipment	Total	Total Incl O&P
0010	**FINISHING FLOORS, HIGH TOLERANCE**									
0012	Finishing of fresh concrete flatwork requires that concrete									
0013	first be placed, struck off & consolidated									
0015	Basic finishing for various unspecified flatwork									
0100	Bull float only	C-10	4000	.006	S.F.		.30		.30	.48
0125	Bull float & manual float		2000	.012			.61		.61	.95
0150	Bull float, manual float & broom finish, w/edging & joints		1850	.013			.66		.66	1.03
0200	Bull float, manual float & manual steel trowel	↓	1265	.019	↓		.96		.96	1.51
0210	For specified Random Access Floors in ACI Classes 1, 2, 3 and 4 to achieve									
0215	Composite Overall Floor Flatness and Levelness values up to FF35/FL25									
0250	Bull float, machine float & machine trowel (walk-behind)	C-10C	1715	.014	S.F.		.71	.05	.76	1.17
0300	Power screed, bull float, machine float & trowel (walk-behind)	C-10D	2400	.010			.51	.07	.58	.88
0350	Power screed, bull float, machine float & trowel (ride-on)	C-10E	4000	.006	↓		.30	.08	.38	.57
0352	For specified Random Access Floors in ACI Classes 5, 6, 7 and 8 to achieve									
0354	Composite Overall Floor Flatness and Levelness values up to FF50/FL50									
0356	Add for two-dimensional restraightening after power float	C-10	6000	.004	S.F.		.20		.20	.32
0358	For specified Random or Defined Access Floors in ACI Class 9 to achieve									
0360	Composite Overall Floor Flatness and Levelness values up to FF100/FL100									
0362	Add for two-dimensional restraightening after bull float & power float	C-10	3000	.008	S.F.		.41		.41	.64
0364	For specified Superflat Defined Access Floors in ACI Class 9 to achieve									
0366	Minimum Floor Flatness and Levelness values of FF100/FL100									
0368	Add for 2-dim'l restraightening after bull float, power float, power trowel	C-10	2000	.012	S.F.		.61		.61	.95
9100	Minimum labor/equipment charge	"	2	12	Job		610		610	955

03 35 29 – Tooled Concrete Finishing

03 35 29.60 Finishing Walls

		Crew	Daily Output	Labor-Hours	Unit	Material	2022 Bare Costs Labor	Equipment	Total	Total Incl O&P
0010	**FINISHING WALLS**									
0020	Break ties and patch voids	1 Cefi	540	.015	S.F.	.04	.79		.83	1.27
0050	Burlap rub with grout	"	450	.018		.04	.94		.98	1.52
0300	Bush hammer, green concrete	B-39	1000	.048			2.30	.33	2.63	3.99
0350	Cured concrete	"	650	.074			3.53	.51	4.04	6.15
0500	Acid etch	1 Cefi	575	.014		.11	.74		.85	1.29
0700	Sandblast, light penetration	E-11	1100	.029		.57	1.46	.28	2.31	3.32
0750	Heavy penetration	"	375	.085	↓	1.14	4.28	.83	6.25	9.15
0850	Grind form fins flush	1 Clab	700	.011	L.F.		.52		.52	.83
9000	Minimum labor/equipment charge	C-10	2	12	Job		610		610	955

03 35 33 – Stamped Concrete Finishing

03 35 33.50 Slab Texture Stamping

		Crew	Daily Output	Labor-Hours	Unit	Material	2022 Bare Costs Labor	Equipment	Total	Total Incl O&P
0010	**SLAB TEXTURE STAMPING**									
0050	Stamping requires that concrete first be placed, struck off, consolidated,									
0060	bull floated and free of bleed water. Decorative stamping tasks include:									
0100	Step 1 - first application of dry shake colored hardener	1 Cefi	6400	.001	S.F.	.57	.07		.64	.72
0110	Step 2 - bull float		6400	.001			.07		.07	.10
0130	Step 3 - second application of dry shake colored hardener	↓	6400	.001		.28	.07		.35	.41
0140	Step 4 - bull float, manual float & steel trowel	3 Cefi	1280	.019			1		1	1.56
0150	Step 5 - application of dry shake colored release agent	1 Cefi	6400	.001		.10	.07		.17	.21
0160	Step 6 - place, tamp & remove mats	3 Cefi	2400	.010		1.14	.53		1.67	2.09
0170	Step 7 - touch up edges, mat joints & simulated grout lines	1 Cefi	1280	.006			.33		.33	.52
0300	Alternate stamping estimating method includes all tasks above	4 Cefi	800	.040		2.09	2.13		4.22	5.60
0400	Step 8 - pressure wash @ 3000 psi after 24 hours	1 Cefi	1600	.005			.27		.27	.42
0500	Step 9 - roll 2 coats cure/seal compound when dry	"	800	.010	↓	.64	.53		1.17	1.54

03 35 43.10 Polished Concrete Floors	Crew	Daily Output	Labor-Hours	Unit	Material	2022 Bare Costs Labor	Equipment	Total	Total Incl O&P
0010 **POLISHED CONCRETE FLOORS**									
0015 Processing of cured concrete to include grinding, honing,									
0020 and polishing of interior floors with 22" segmented diamond									
0025 planetary floor grinder (2 passes in different directions per grit)									
0100 Removal of pre-existing coatings, dry, with carbide discs using									
0105 dry vacuum pick-up system, final hand sweeping									
0110 Glue, adhesive or tar	J-4	1.60	15	M.S.F.	22	760	144	926	1,375
0120 Paint, epoxy, 1 coat		3.60	6.667		22	340	64	426	625
0130 2 coats	↓	1.80	13.333	↓	22	675	128	825	1,225
0200 Grinding and edging, wet, including wet vac pick-up and auto									
0205 scrubbing between grit changes									
0210 40-grit diamond/metal matrix	J-4A	1.60	20	M.S.F.	37.50	990	263	1,290.50	1,875
0220 80-grit diamond/metal matrix		2	16		37.50	790	210	1,037.50	1,525
0230 120-grit diamond/metal matrix		2.40	13.333		37.50	660	175	872.50	1,250
0240 200-grit diamond/metal matrix	↓	2.80	11.429		37.50	565	150	752.50	1,100
0300 Spray on dye or stain (1 coat)	1 Cefi	16	.500		202	26.50		228.50	264
0400 Spray on densifier/hardener (2 coats)	"	8	1		330	53		383	450
0410 Auto scrubbing after 2nd coat, when dry	J-4B	16	.500	↓		23	11.90	34.90	49
0500 Honing and edging, wet, including wet vac pick-up and auto									
0505 scrubbing between grit changes									
0510 100-grit diamond/resin matrix	J-4A	2.80	11.429	M.S.F.	37.50	565	150	752.50	1,100
0520 200-grit diamond/resin matrix	"	2.80	11.429	"	37.50	565	150	752.50	1,100
0530 Dry, including dry vacuum pick-up system, final hand sweeping									
0540 400-grit diamond/resin matrix	J-4A	2.80	11.429	M.S.F.	37.50	565	150	752.50	1,100
0600 Polishing and edging, dry, including dry vac pick-up and hand									
0605 sweeping between grit changes									
0610 800-grit diamond/resin matrix	J-4A	2.80	11.429	M.S.F.	37.50	565	150	752.50	1,100
0620 1500-grit diamond/resin matrix		2.80	11.429		37.50	565	150	752.50	1,100
0630 3000-grit diamond/resin matrix	↓	2.80	11.429		37.50	565	150	752.50	1,100
0700 Auto scrubbing after final polishing step	J-4B	16	.500	↓		23	11.90	34.90	49

03 37 13.30 Gunite (Dry-Mix)

	Crew	Daily Output	Labor-Hours	Unit	Material	Labor	Equipment	Total	Total Incl O&P
0010 **GUNITE (DRY-MIX)**									
0020 Typical in place, 1" layers, no mesh included	C-16	2000	.028	S.F.	.52	1.40	.21	2.13	3.02
0300 Typical in place, including mesh, 2" thick, flat surfaces		1000	.056		1.94	2.81	.41	5.16	7
0350 Curved surfaces		500	.112		1.94	5.60	.83	8.37	11.90
0500 4" thick, flat surfaces		750	.075		2.98	3.75	.55	7.28	9.80
0550 Curved surfaces	↓	350	.160		2.98	8.05	1.18	12.21	17.25
0900 Prepare old walls, no scaffolding, good condition	C-10	1000	.024			1.22		1.22	1.91
0950 Poor condition	"	275	.087			4.42		4.42	6.95
1100 For high finish requirement or close tolerance, add						50%			
1150 Very high				↓		110%			
9000 Minimum labor/equipment charge	C-10	1	24	Job		1,225		1,225	1,900

03 37 13.60 Shotcrete (Wet-Mix)

	Crew	Daily Output	Labor-Hours	Unit	Material	Labor	Equipment	Total	Total Incl O&P
0010 **SHOTCRETE (WET-MIX)**									
0020 Wet mix, placed @ up to 12 C.Y./hour, 3000 psi	C-8C	80	.600	C.Y.	169	30	12.70	211.70	247
0100 Up to 35 C.Y./hour	C-8E	240	.200	"	152	9.80	4.87	166.67	189
1010 Fiber reinforced, 1" thick	C-8C	1740	.028	S.F.	1.24	1.37	.58	3.19	4.17

03 37 Specialty Placed Concrete

03 37 13 – Shotcrete

03 37 13.60 Shotcrete (Wet-Mix)		Crew	Daily Output	Labor-Hours	Unit	Material	2022 Bare Costs Labor	Equipment	Total	Total Incl O&P
1020	2" thick	C-8C	900	.053	S.F.	2.48	2.65	1.13	6.26	8.15
1030	3" thick	↓	825	.058		3.72	2.89	1.23	7.84	10
1040	4" thick	↓	750	.064	↓	4.97	3.18	1.35	9.50	11.95

03 39 Concrete Curing

03 39 13 – Water Concrete Curing

03 39 13.50 Water Curing

		Crew	Daily Output	Labor-Hours	Unit	Material	2022 Bare Costs Labor	Equipment	Total	Total Incl O&P
0010	**WATER CURING**									
0015	With burlap, 4 uses assumed, 7.5 oz.	2 Clab	55	.291	C.S.F.	14.90	13.25		28.15	37.50
0100	10 oz.	"	55	.291	"	27	13.25		40.25	50.50
0400	Curing blankets, 1" to 2" thick, buy				S.F.	.64			.64	.70
9000	Minimum labor/equipment charge	1 Clab	5	1.600	Job		73		73	116

03 39 23 – Membrane Concrete Curing

03 39 23.13 Chemical Compound Membrane Concrete Curing

		Crew	Daily Output	Labor-Hours	Unit	Material	2022 Bare Costs Labor	Equipment	Total	Total Incl O&P
0010	**CHEMICAL COMPOUND MEMBRANE CONCRETE CURING**									
0300	Sprayed membrane curing compound	2 Clab	95	.168	C.S.F.	13.55	7.70		21.25	27
0700	Curing compound, solvent based, 400 S.F./gal., 55 gallon lots				Gal.	22			22	24.50
0720	5 gallon lots					30			30	33
0800	Curing compound, water based, 250 S.F./gal., 55 gallon lots					27			27	30
0820	5 gallon lots				↓	25			25	27.50

03 39 23.23 Sheet Membrane Concrete Curing

		Crew	Daily Output	Labor-Hours	Unit	Material	2022 Bare Costs Labor	Equipment	Total	Total Incl O&P
0010	**SHEET MEMBRANE CONCRETE CURING**									
0200	Curing blanket, burlap/poly, 2-ply	2 Clab	70	.229	C.S.F.	24.50	10.40		34.90	43.50

03 41 Precast Structural Concrete

03 41 13 – Precast Concrete Hollow Core Planks

03 41 13.50 Precast Slab Planks

		Crew	Daily Output	Labor-Hours	Unit	Material	2022 Bare Costs Labor	Equipment	Total	Total Incl O&P
0010	**PRECAST SLAB PLANKS** R034105-30									
0020	Prestressed roof/floor members, grouted, solid, 4" thick	C-11	2400	.030	S.F.	8.20	1.84	.99	11.03	13.10
0050	6" thick		2800	.026		8.35	1.58	.85	10.78	12.60
0100	Hollow, 8" thick		3200	.023		10.65	1.38	.75	12.78	14.80
0150	10" thick		3600	.020		9.70	1.23	.66	11.59	13.35
0200	12" thick	↓	4000	.018	↓	11.40	1.10	.60	13.10	15

03 41 23 – Precast Concrete Stairs

03 41 23.50 Precast Stairs

		Crew	Daily Output	Labor-Hours	Unit	Material	2022 Bare Costs Labor	Equipment	Total	Total Incl O&P
0010	**PRECAST STAIRS**									
0020	Precast concrete treads on steel stringers, 3' wide	C-12	75	.640	Riser	173	36	26.50	235.50	276
0300	Front entrance, 5' wide with 48" platform, 2 risers		16	3	Flight	675	168	124	967	1,150
0350	5 risers		12	4		1,250	224	165	1,639	1,900
0500	6' wide, 2 risers		15	3.200		720	179	132	1,031	1,225
0550	5 risers	↓	11	4.364		1,375	244	180	1,799	2,100
1200	Basement entrance stairwell, 6 steps, incl. steel bulkhead door	B-51	22	2.182		1,850	102	15.45	1,967.45	2,225
1250	14 steps	"	11	4.364	↓	3,350	204	31	3,585	4,025

03 41 Precast Structural Concrete

03 41 33 – Precast Structural Pretensioned Concrete

03 41 33.10 Precast Beams

		Crew	Daily Output	Labor-Hours	Unit	Material	2022 Bare Costs Labor	2022 Bare Costs Equipment	Total	Total Incl O&P
0010	**PRECAST BEAMS** R034105-30									
0011	L-shaped, 20' span, 12" x 20"	C-11	32	2.250	Ea.	4,675	138	74.50	4,887.50	5,450
0060	18" x 36"		24	3		6,450	184	99.50	6,733.50	7,500
0100	24" x 44"		22	3.273		7,750	201	108	8,059	8,950
0150	30' span, 12" x 36"		24	3		9,125	184	99.50	9,408.50	10,500
0200	18" x 44"		20	3.600		10,900	221	119	11,240	12,500
0250	24" x 52"		16	4.500		13,100	276	149	13,525	15,000
0400	40' span, 12" x 52"		20	3.600		13,800	221	119	14,140	15,700
0450	18" x 52"		16	4.500		15,400	276	149	15,825	17,600
0500	24" x 52"		12	6		17,500	370	199	18,069	20,000
1200	Rectangular, 20' span, 12" x 20"		32	2.250		4,050	138	74.50	4,262.50	4,775
1250	18" x 36"		24	3		5,900	184	99.50	6,183.50	6,900
1300	24" x 44"		22	3.273		6,875	201	108	7,184	8,025
1400	30' span, 12" x 36"		24	3		6,725	184	99.50	7,008.50	7,775
1450	18" x 44"		20	3.600		9,350	221	119	9,690	10,800
1500	24" x 52"		16	4.500		11,400	276	149	11,825	13,100
1600	40' span, 12" x 52"		20	3.600		11,200	221	119	11,540	12,800
1650	18" x 52"		16	4.500		13,200	276	149	13,625	15,100
1700	24" x 52"		12	6		15,200	370	199	15,769	17,500
2000	"T" shaped, 20' span, 12" x 20"		32	2.250		5,650	138	74.50	5,862.50	6,525
2050	18" x 36"		24	3		7,425	184	99.50	7,708.50	8,575
2100	24" x 44"		22	3.273		7,625	201	108	7,934	8,850
2200	30' span, 12" x 36"		24	3		10,100	184	99.50	10,383.50	11,500
2250	18" x 44"		20	3.600		12,800	221	119	13,140	14,600
2300	24" x 52"		16	4.500		12,500	276	149	12,925	14,300
2500	40' span, 12" x 52"		20	3.600		16,500	221	119	16,840	18,600
2550	18" x 52"		16	4.500		15,500	276	149	15,925	17,700
2600	24" x 52"		12	6		16,600	370	199	17,169	19,100

03 41 33.15 Precast Columns

		Crew	Daily Output	Labor-Hours	Unit	Material	2022 Bare Costs Labor	2022 Bare Costs Equipment	Total	Total Incl O&P
0010	**PRECAST COLUMNS** R034105-30									
0020	Rectangular to 12' high, 16" x 16"	C-11	120	.600	L.F.	266	37	19.90	322.90	375
0050	24" x 24"	"	96	.750	"	365	46	25	436	500

03 45 Precast Architectural Concrete

03 45 13 – Faced Architectural Precast Concrete

03 45 13.50 Precast Wall Panels

		Crew	Daily Output	Labor-Hours	Unit	Material	2022 Bare Costs Labor	2022 Bare Costs Equipment	Total	Total Incl O&P
0010	**PRECAST WALL PANELS**									
2200	Fiberglass reinforced cement with urethane core									
2210	R20, 8' x 8', 5" plain finish	E-2	750	.075	S.F.	30.50	4.57	3.31	38.38	44.50
2220	Exposed aggregate or brick finish	"	600	.093	"	50.50	5.70	4.13	60.33	69

For customer support on your Commercial Renovation Costs with RSMeans data, call 800.448.8182.

77

03 48 Precast Concrete Specialties

03 48 43 – Precast Concrete Trim

03 48 43.40 Precast Lintels	Crew	Daily Output	Labor-Hours	Unit	Material	2022 Bare Costs Labor	Equipment	Total	Total Incl O&P
0010 **PRECAST LINTELS**, smooth gray, prestressed, stock units only									
0800 4" wide x 8" high x 4' long	D-10	28	1.143	Ea.	39.50	63	20.50	123	166
0850 8' long		24	1.333		90	73.50	24	187.50	242
1000 6" wide x 8" high x 4' long		26	1.231		66.50	68	22	156.50	206
1050 10' long		22	1.455		174	80.50	26	280.50	350
1200 8" wide x 8" high x 4' long		24	1.333		64.50	73.50	24	162	214
1250 12' long		20	1.600		223	88.50	28.50	340	420
1275 For custom sizes, types, colors, or finishes of precast lintels, add					150%				

03 48 43.90 Precast Window Sills

03 48 43.90 Precast Window Sills	Crew	Daily Output	Labor-Hours	Unit	Material	2022 Bare Costs Labor	Equipment	Total	Total Incl O&P
0010 **PRECAST WINDOW SILLS**									
0600 Precast concrete, 4" tapers to 3", 9" wide	D-1	70	.229	L.F.	5.80	11.45		17.25	24.50
0650 11" wide	"	60	.267	"	6.45	13.35		19.80	28.50

03 51 Cast Roof Decks

03 51 13 – Cementitious Wood Fiber Decks

03 51 13.50 Cementitious/Wood Fiber Planks	Crew	Daily Output	Labor-Hours	Unit	Material	2022 Bare Costs Labor	Equipment	Total	Total Incl O&P
0010 **CEMENTITIOUS/WOOD FIBER PLANKS**									
0050 Plank, beveled edge, 1" thick	2 Carp	1000	.016	S.F.	3.68	.90		4.58	5.50
0100 1-1/2" thick		975	.016		5.55	.92		6.47	7.55
0150 T&G, 2" thick		950	.017		3.85	.95		4.80	5.75
0200 2-1/2" thick		925	.017		4.36	.97		5.33	6.35
1000 Bulb tee, sub-purlin and grout, 6' span, add	E-1	5000	.005		2.01	.29	.03	2.33	2.71
1100 8' span	"	4200	.006		2.01	.35	.04	2.40	2.81

03 52 Lightweight Concrete Roof Insulation

03 52 16 – Lightweight Insulating Concrete

03 52 16.13 Lightweight Cellular Insulating Concrete

03 52 16.13 Lightweight Cellular Insulating Concrete		Crew	Daily Output	Labor-Hours	Unit	Material	2022 Bare Costs Labor	Equipment	Total	Total Incl O&P
0010 **LIGHTWEIGHT CELLULAR INSULATING CONCRETE**	R035216-10									
0020 Portland cement and foaming agent	G	C-8	50	1.120	C.Y.	136	56	15.10	207.10	254

03 52 16.16 Lightweight Aggregate Insulating Concrete

03 52 16.16 Lightweight Aggregate Insulating Concrete		Crew	Daily Output	Labor-Hours	Unit	Material	2022 Bare Costs Labor	Equipment	Total	Total Incl O&P
0010 **LIGHTWEIGHT AGGREGATE INSULATING CONCRETE**	R035216-10									
0100 Poured vermiculite or perlite, field mix,										
0110 1:6 field mix	G	C-8	50	1.120	C.Y.	330	56	15.10	401.10	470
0200 Ready mix, 1:6 mix, roof fill, 2" thick	G		10000	.006	S.F.	1.84	.28	.08	2.20	2.55
0250 3" thick	G		7700	.007	"	2.77	.36	.10	3.23	3.72

03 53 Concrete Topping

03 53 16 – Iron-Aggregate Concrete Topping

03 53 16.50 Floor Topping	Crew	Daily Output	Labor-Hours	Unit	Material	2022 Bare Costs Labor	Equipment	Total	Total Incl O&P
0010 **FLOOR TOPPING**									
0400 Integral topping/finish, on fresh concrete, using 1:1:2 mix, 3/16" thick	C-10B	1000	.040	S.F.	.13	1.94	.31	2.38	3.56
0450 1/2" thick		950	.042		.34	2.05	.33	2.72	3.97
0500 3/4" thick		850	.047		.51	2.29	.37	3.17	4.58
0600 1" thick		750	.053		.68	2.59	.42	3.69	5.30
0800 Granolithic topping, on fresh or cured concrete, 1:1:1-1/2 mix, 1/2" thick		590	.068		.38	3.30	.53	4.21	6.20
0820 3/4" thick		580	.069		.57	3.35	.54	4.46	6.55
0850 1" thick		575	.070		.77	3.38	.55	4.70	6.80

For customer support on your Commercial Renovation Costs with RSMeans data, call 800.448.8182.

03 53 Concrete Topping

03 53 16 – Iron-Aggregate Concrete Topping

03 53 16.50 Floor Topping	Crew	Daily Output	Labor-Hours	Unit	Material	2022 Bare Costs Labor	Equipment	Total	Total Incl O&P
0950	2" thick	C-10B	500	.080	S.F.	1.53	3.89	.63	6.05	8.50

03 54 Cast Underlayment

03 54 13 – Gypsum Cement Underlayment

03 54 13.50 Poured Gypsum Underlayment

		Crew	Daily Output	Labor-Hours	Unit	Material	2022 Bare Costs Labor	Equipment	Total	Total Incl O&P
0010	**POURED GYPSUM UNDERLAYMENT**									
0400	Underlayment, gypsum based, self-leveling 2500 psi, pumped, 1/2" thick	C-8	24000	.002	S.F.	.19	.12	.03	.34	.42
0500	3/4" thick		20000	.003		.28	.14	.04	.46	.57
0600	1" thick		16000	.004		.38	.18	.05	.61	.74
1400	Hand placed, 1/2" thick	C-18	450	.020		.19	.92	.26	1.37	1.96
1500	3/4" thick	"	300	.030		.28	1.37	.40	2.05	2.93

03 54 16 – Hydraulic Cement Underlayment

03 54 16.50 Cement Underlayment

		Crew	Daily Output	Labor-Hours	Unit	Material	2022 Bare Costs Labor	Equipment	Total	Total Incl O&P
0010	**CEMENT UNDERLAYMENT**									
2510	Underlayment, P.C. based, self-leveling, 4100 psi, pumped, 1/4" thick	C-8	20000	.003	S.F.	2.06	.14	.04	2.24	2.53
2520	1/2" thick		19000	.003		4.12	.15	.04	4.31	4.80
2530	3/4" thick		18000	.003		6.20	.16	.04	6.40	7.10
2540	1" thick		17000	.003		8.25	.17	.04	8.46	9.35
2550	1-1/2" thick		15000	.004		12.35	.19	.05	12.59	13.95
2560	Hand placed, 1/2" thick	C-18	450	.020		4.12	.92	.26	5.30	6.30
2610	Topping, P.C. based, self-leveling, 6100 psi, pumped, 1/4" thick	C-8	20000	.003		2.19	.14	.04	2.37	2.66
2620	1/2" thick		19000	.003		4.37	.15	.04	4.56	5.10
2630	3/4" thick		18000	.003		6.55	.16	.04	6.75	7.50
2660	1" thick		17000	.003		8.75	.17	.04	8.96	9.90
2670	1-1/2" thick		15000	.004		13.10	.19	.05	13.34	14.80
2680	Hand placed, 1/2" thick	C-18	450	.020		4.37	.92	.26	5.55	6.55

03 63 Epoxy Grouting

03 63 05 – Grouting of Dowels and Fasteners

03 63 05.10 Epoxy Only

		Crew	Daily Output	Labor-Hours	Unit	Material	2022 Bare Costs Labor	Equipment	Total	Total Incl O&P
0010	**EPOXY ONLY**									
1500	Chemical anchoring, epoxy cartridge, excludes layout, drilling, fastener									
1530	For fastener 3/4" diam. x 6" embedment	2 Skwk	72	.222	Ea.	4.23	13.10		17.33	25
1535	1" diam. x 8" embedment		66	.242		6.35	14.30		20.65	29.50
1540	1-1/4" diam. x 10" embedment		60	.267		12.70	15.70		28.40	39
1545	1-3/4" diam. x 12" embedment		54	.296		21	17.45		38.45	51
1550	14" embedment		48	.333		25.50	19.65		45.15	59
1555	2" diam. x 12" embedment		42	.381		34	22.50		56.50	72.50
1560	18" embedment		32	.500		42.50	29.50		72	93

For customer support on your Commercial Renovation Costs with RSMeans data, call 800.448.8182.

79

03 81 Concrete Cutting

03 81 13 – Flat Concrete Sawing

03 81 13.50 Concrete Floor/Slab Cutting

		Crew	Daily Output	Labor-Hours	Unit	Material	2022 Bare Costs Labor	Equipment	Total	Total Incl O&P
0010	**CONCRETE FLOOR/SLAB CUTTING**									
0050	Includes blade cost, layout and set-up time									
0300	Saw cut concrete slabs, plain, up to 3" deep	B-89	1060	.015	L.F.	.10	.81	.59	1.50	2.04
0320	Each additional inch of depth		3180	.005		.03	.27	.20	.50	.69
0400	Mesh reinforced, up to 3" deep		980	.016		.11	.88	.64	1.63	2.20
0420	Each additional inch of depth		2940	.005		.04	.29	.21	.54	.73
0500	Rod reinforced, up to 3" deep		800	.020		.14	1.07	.78	1.99	2.70
0520	Each additional inch of depth		2400	.007	↓	.05	.36	.26	.67	.90
0590	Minimum labor/equipment charge	↓	2	8	Job		430	315	745	1,025

03 81 13.75 Concrete Saw Blades

		Crew	Daily Output	Labor-Hours	Unit	Material	2022 Bare Costs Labor	Equipment	Total	Total Incl O&P
0010	**CONCRETE SAW BLADES**									
3000	Blades for saw cutting, included in cutting line items									
3020	Diamond, 12" diameter				Ea.	207			207	228
3040	18" diameter					325			325	360
3080	24" diameter					460			460	510
3120	30" diameter					820			820	900
3160	36" diameter					1,150			1,150	1,275
3200	42" diameter				↓	1,750			1,750	1,925

03 81 16 – Track Mounted Concrete Wall Sawing

03 81 16.50 Concrete Wall Cutting

		Crew	Daily Output	Labor-Hours	Unit	Material	2022 Bare Costs Labor	Equipment	Total	Total Incl O&P
0010	**CONCRETE WALL CUTTING**									
0750	Includes blade cost, layout and set-up time									
0800	Concrete walls, hydraulic saw, plain, per inch of depth	B-89B	250	.064	L.F.	.03	3.44	4.30	7.77	10.15
0820	Rod reinforcing, per inch of depth		150	.107	"	.05	5.75	7.15	12.95	16.95
0890	Minimum labor/equipment charge	↓	2	8	Job		430	540	970	1,275

03 82 Concrete Boring

03 82 13 – Concrete Core Drilling

03 82 13.10 Core Drilling

		Crew	Daily Output	Labor-Hours	Unit	Material	2022 Bare Costs Labor	Equipment	Total	Total Incl O&P
0010	**CORE DRILLING**									
0015	Includes bit cost, layout and set-up time									
0020	Reinforced concrete slab, up to 6" thick									
0100	1" diameter core	B-89A	17	.941	Ea.	.22	49	7	56.22	86
0150	For each additional inch of slab thickness in same hole, add		1440	.011		.04	.58	.08	.70	1.05
0200	2" diameter core		16.50	.970		.30	50.50	7.20	58	89
0250	For each additional inch of slab thickness in same hole, add		1080	.015		.05	.77	.11	.93	1.40
0300	3" diameter core		16	1		.42	52.50	7.45	60.37	91.50
0350	For each additional inch of slab thickness in same hole, add		720	.022		.07	1.16	.17	1.40	2.10
0500	4" diameter core		15	1.067		.56	55.50	7.95	64.01	98
0550	For each additional inch of slab thickness in same hole, add		480	.033		.09	1.74	.25	2.08	3.13
0700	6" diameter core		14	1.143		.89	59.50	8.50	68.89	105
0750	For each additional inch of slab thickness in same hole, add		360	.044		.15	2.32	.33	2.80	4.20
0900	8" diameter core		13	1.231		1.30	64.50	9.15	74.95	114
0950	For each additional inch of slab thickness in same hole, add		288	.056		.22	2.90	.41	3.53	5.30
1100	10" diameter core		12	1.333		1.60	69.50	9.95	81.05	123
1150	For each additional inch of slab thickness in same hole, add		240	.067		.27	3.48	.50	4.25	6.35
1300	12" diameter core		11	1.455		1.83	76	10.85	88.68	134
1350	For each additional inch of slab thickness in same hole, add		206	.078		.30	4.06	.58	4.94	7.45
1500	14" diameter core		10	1.600		2.26	83.50	11.90	97.66	149
1550	For each additional inch of slab thickness in same hole, add	↓	180	.089	↓	.38	4.64	.66	5.68	8.50

03 82 Concrete Boring

03 82 13 – Concrete Core Drilling

03 82 13.10 Core Drilling

		Crew	Daily Output	Labor-Hours	Unit	Material	2022 Bare Costs Labor	Equipment	Total	Total Incl O&P
1700	18" diameter core	B-89A	9	1.778	Ea.	2.94	93	13.25	109.19	165
1750	For each additional inch of slab thickness in same hole, add		144	.111		.49	5.80	.83	7.12	10.65
1754	24" diameter core		8	2		4.67	105	14.90	124.57	188
1756	For each additional inch of slab thickness in same hole, add		120	.133		.78	6.95	.99	8.72	13
1760	For horizontal holes, add to above						20%	20%		
1770	Prestressed hollow core plank, 8" thick									
1780	1" diameter core	B-89A	17.50	.914	Ea.	.29	48	6.80	55.09	83.50
1790	For each additional inch of plank thickness in same hole, add		3840	.004		.04	.22	.03	.29	.42
1794	2" diameter core		17.25	.928		.40	48.50	6.90	55.80	85
1796	For each additional inch of plank thickness in same hole, add		2880	.006		.05	.29	.04	.38	.56
1800	3" diameter core		17	.941		.57	49	7	56.57	86.50
1810	For each additional inch of plank thickness in same hole, add		1920	.008		.07	.44	.06	.57	.84
1820	4" diameter core		16.50	.970		.75	50.50	7.20	58.45	89.50
1830	For each additional inch of plank thickness in same hole, add		1280	.013		.09	.65	.09	.83	1.24
1840	6" diameter core		15.50	1.032		1.19	54	7.70	62.89	95.50
1850	For each additional inch of plank thickness in same hole, add		960	.017		.15	.87	.12	1.14	1.68
1860	8" diameter core		15	1.067		1.73	55.50	7.95	65.18	99
1870	For each additional inch of plank thickness in same hole, add		768	.021		.22	1.09	.16	1.47	2.14
1880	10" diameter core		14	1.143		2.13	59.50	8.50	70.13	106
1890	For each additional inch of plank thickness in same hole, add		640	.025		.27	1.31	.19	1.77	2.56
1900	12" diameter core		13.50	1.185		2.44	62	8.85	73.29	110
1910	For each additional inch of plank thickness in same hole, add		548	.029		.30	1.53	.22	2.05	3
1999	Drilling, core, minimum labor/equipment charge		2	8	Job		420	59.50	479.50	730
3000	Bits for core drilling, included in drilling line items									
3010	Diamond, premium, 1" diameter				Ea.	87.50			87.50	96
3020	2" diameter					120			120	132
3030	3" diameter					170			170	186
3040	4" diameter					226			226	249
3060	6" diameter					355			355	395
3080	8" diameter					520			520	570
3110	10" diameter					640			640	705
3120	12" diameter					730			730	805
3140	14" diameter					905			905	995
3180	18" diameter					1,175			1,175	1,300
3240	24" diameter					1,875			1,875	2,050

03 82 16 – Concrete Drilling

03 82 16.10 Concrete Impact Drilling

		Crew	Daily Output	Labor-Hours	Unit	Material	2022 Bare Costs Labor	Equipment	Total	Total Incl O&P
0010	**CONCRETE IMPACT DRILLING**									
0020	Includes bit cost, layout and set-up time, no anchors									
0050	Up to 4" deep in concrete/brick floors/walls									
0100	Holes, 1/4" diameter	1 Carp	75	.107	Ea.	.05	6		6.05	9.60
0150	For each additional inch of depth in same hole, add		430	.019		.01	1.05		1.06	1.67
0200	3/8" diameter		63	.127		.05	7.15		7.20	11.40
0250	For each additional inch of depth in same hole, add		340	.024		.01	1.32		1.33	2.11
0300	1/2" diameter		50	.160		.04	9		9.04	14.35
0350	For each additional inch of depth in same hole, add		250	.032		.01	1.80		1.81	2.87
0400	5/8" diameter		48	.167		.07	9.40		9.47	15
0450	For each additional inch of depth in same hole, add		240	.033		.02	1.88		1.90	3
0500	3/4" diameter		45	.178		.10	10		10.10	16
0550	For each additional inch of depth in same hole, add		220	.036		.03	2.05		2.08	3.28
0600	7/8" diameter		43	.186		.13	10.45		10.58	16.80
0650	For each additional inch of depth in same hole, add		210	.038		.03	2.14		2.17	3.44

For customer support on your Commercial Renovation Costs with RSMeans data, call 800.448.8182.

81

03 82 Concrete Boring

03 82 16 – Concrete Drilling

03 82 16.10 Concrete Impact Drilling		Crew	Daily Output	Labor-Hours	Unit	Material	2022 Bare Costs Labor	Equipment	Total	Total Incl O&P
0700	1" diameter	1 Carp	40	.200	Ea.	.19	11.25		11.44	18.10
0750	For each additional inch of depth in same hole, add		190	.042		.05	2.37		2.42	3.82
0800	1-1/4" diameter		38	.211		.21	11.85		12.06	19.10
0850	For each additional inch of depth in same hole, add		180	.044		.05	2.50		2.55	4.04
0900	1-1/2" diameter		35	.229		.32	12.85		13.17	21
0950	For each additional inch of depth in same hole, add		165	.048		.08	2.73		2.81	4.43
1000	For ceiling installations, add						40%			

Estimating Tips

04 05 00 Common Work Results for Masonry

- The terms mortar and grout are often used interchangeably—and incorrectly. Mortar is used to bed masonry units, seal the entry of air and moisture, provide architectural appearance, and allow for size variations in the units. Grout is used primarily in reinforced masonry construction and to bond the masonry to the reinforcing steel. Common mortar types are M (2500 psi), S (1800 psi), N (750 psi), and O (350 psi), and they conform to ASTM C270. Grout is either fine or coarse and conforms to ASTM C476, and in-place strengths generally exceed 2500 psi. Mortar and grout are different components of masonry construction and are placed by entirely different methods. An estimator should be aware of their unique uses and costs.

- Mortar is included in all assembled masonry line items. The mortar cost, part of the assembled masonry material cost, includes all ingredients, all labor, and all equipment required. Please see reference number RO040513-10.

- Waste, specifically the loss/droppings of mortar and the breakage of brick and block, is included in all unit cost lines that include mortar and masonry units in this division. A factor of 25% is added for mortar and 3% for brick and concrete masonry units.

- Scaffolding or staging is not included in any of the Division 4 costs. Refer to Subdivision 01 54 23 for scaffolding and staging costs.

04 20 00 Unit Masonry

- The most common types of unit masonry are brick and concrete masonry. The major classifications of brick are building brick (ASTM C62), facing brick (ASTM C216), glazed brick, fire brick, and pavers. Many varieties of texture and appearance can exist within these classifications, and the estimator would be wise to check local custom and availability within the project area. For repair and remodeling jobs, matching the existing brick may be the most important criteria.

- Brick and concrete block are priced by the piece and then converted into a price per square foot of wall. Openings less than two square feet are generally ignored by the estimator because any savings in units used are offset by the cutting and trimming required.

- It is often difficult and expensive to find and purchase small lots of historic brick. Costs can vary widely. Many design issues affect costs, selection of mortar mix, and repairs or replacement of masonry materials. Cleaning techniques must be reflected in the estimate.

- All masonry walls, whether interior or exterior, require bracing. The cost of bracing walls during construction should be included by the estimator, and this bracing must remain in place until permanent bracing is complete. Permanent bracing of masonry walls is accomplished by masonry itself, in the form of pilasters or abutting wall corners, or by anchoring the walls to the structural frame. Accessories in the form of anchors, anchor slots, and ties are used, but their supply and installation can be by different trades. For instance, anchor slots on spandrel beams and columns are supplied and welded in place by the steel fabricator, but the ties from the slots into the masonry are installed by the bricklayer. Regardless of the installation method, the estimator must be certain that these accessories are accounted for in pricing.

Reference Numbers

Reference numbers are shown at the beginning of some major classifications. These numbers refer to related items in the Reference Section. The reference information may be an estimating procedure, an alternate pricing method, or technical information.

Note: Not all subdivisions listed here necessarily appear. ■

04 01 20.20 Pointing Masonry

		Crew	Daily Output	Labor-Hours	Unit	Material	2022 Bare Costs Labor	Equipment	Total	Total Incl O&P
0010	**POINTING MASONRY**									
0300	Cut and repoint brick, hard mortar, running bond	1 Bric	80	.100	S.F.	.83	5.55		6.38	9.75
0320	Common bond		77	.104		.83	5.75		6.58	10.10
0360	Flemish bond		70	.114		.87	6.30		7.17	11.05
0400	English bond		65	.123		.87	6.80		7.67	11.85
0600	Soft old mortar, running bond		100	.080		.83	4.42		5.25	8
0620	Common bond		96	.083		.83	4.60		5.43	8.30
0640	Flemish bond		90	.089		.87	4.91		5.78	8.80
0680	English bond		82	.098		.87	5.40		6.27	9.60
0700	Stonework, hard mortar		140	.057	L.F.	1.10	3.16		4.26	6.25
0720	Soft old mortar		160	.050	"	1.10	2.76		3.86	5.65
1000	Repoint, mask and grout method, running bond		95	.084	S.F.	1.10	4.65		5.75	8.65
1020	Common bond		90	.089		1.10	4.91		6.01	9.05
1040	Flemish bond		86	.093		1.16	5.15		6.31	9.55
1060	English bond		77	.104		1.16	5.75		6.91	10.50
2000	Scrub coat, sand grout on walls, thin mix, brushed		120	.067		4.96	3.68		8.64	11.35
2020	Troweled		98	.082		6.95	4.51		11.46	14.80
9000	Minimum labor/equipment charge		3	2.667	Job		147		147	236

04 01 20.30 Pointing CMU

		Crew	Daily Output	Labor-Hours	Unit	Material	2022 Bare Costs Labor	Equipment	Total	Total Incl O&P
0010	**POINTING CMU**									
0300	Cut and repoint block, hard mortar, running bond	1 Bric	190	.042	S.F.	.34	2.33		2.67	4.10
0310	Stacked bond		200	.040		.34	2.21		2.55	3.91
0600	Soft old mortar, running bond		230	.035		.34	1.92		2.26	3.45
0610	Stacked bond		245	.033		.34	1.80		2.14	3.26

04 01 20.41 Unit Masonry Stabilization

		Crew	Daily Output	Labor-Hours	Unit	Material	2022 Bare Costs Labor	Equipment	Total	Total Incl O&P
0010	**UNIT MASONRY STABILIZATION**									
0100	Structural repointing method									
0110	Cut/grind mortar joint	1 Bric	240	.033	L.F.		1.84		1.84	2.95
0120	Clean and mask joint		2500	.003		.13	.18		.31	.43
0130	Epoxy paste and 1/4" FRP rod		240	.033		2.28	1.84		4.12	5.45
0132	3/8" FRP rod		160	.050		3.40	2.76		6.16	8.15
0140	Remove masking		14400	.001			.03		.03	.05
0300	Structural fabric method									
0310	Primer	1 Bric	600	.013	S.F.	1.33	.74		2.07	2.64
0320	Apply filling/leveling paste		720	.011		1.01	.61		1.62	2.09
0330	Epoxy, glass fiber fabric		720	.011		11.60	.61		12.21	13.75
0340	Carbon fiber fabric		720	.011		22	.61		22.61	25

04 01 20.50 Toothing Masonry

		Crew	Daily Output	Labor-Hours	Unit	Material	2022 Bare Costs Labor	Equipment	Total	Total Incl O&P
0010	**TOOTHING MASONRY**									
0500	Brickwork, soft old mortar	1 Clab	40	.200	V.L.F.		9.10		9.10	14.50
0520	Hard mortar		30	.267			12.15		12.15	19.30
0700	Blockwork, soft old mortar		70	.114			5.20		5.20	8.30
0720	Hard mortar		50	.160			7.30		7.30	11.60
9000	Minimum labor/equipment charge		4	2	Job		91		91	145

04 01 20.52 Cleaning Masonry

		Crew	Daily Output	Labor-Hours	Unit	Material	2022 Bare Costs Labor	Equipment	Total	Total Incl O&P
0010	**CLEANING MASONRY**									
0200	By chemical, brush and rinse, new work, light construction dust	D-1	1000	.016	S.F.	.06	.80		.86	1.35
0220	Medium construction dust		800	.020		.09	1		1.09	1.71
0240	Heavy construction dust, drips or stains		600	.027		.12	1.34		1.46	2.27
0260	Low pressure wash and rinse, light restoration, light soil		800	.020		.13	1		1.13	1.75
0270	Average soil, biological staining		400	.040		.19	2.01		2.20	3.42
0280	Heavy soil, biological and mineral staining, paint		330	.048		.26	2.43		2.69	4.17

04 01 Maintenance of Masonry

04 01 20 – Maintenance of Unit Masonry

04 01 20.52 Cleaning Masonry

		Crew	Daily Output	Labor-Hours	Unit	Material	2022 Bare Costs Labor	Equipment	Total	Total Incl O&P
0300	High pressure wash and rinse, heavy restoration, light soil	D-1	600	.027	S.F.	.10	1.34		1.44	2.25
0310	Average soil, biological staining		400	.040		.15	2.01		2.16	3.37
0320	Heavy soil, biological and mineral staining, paint		250	.064		.20	3.21		3.41	5.35
0400	High pressure wash, water only, light soil	C-29	500	.016			.73	.20	.93	1.38
0420	Average soil, biological staining		375	.021			.97	.27	1.24	1.85
0440	Heavy soil, biological and mineral staining, paint		250	.032			1.46	.40	1.86	2.76
0800	High pressure water and chemical, light soil		450	.018		.15	.81	.22	1.18	1.71
0820	Average soil, biological staining		300	.027		.23	1.22	.34	1.79	2.55
0840	Heavy soil, biological and mineral staining, paint		200	.040		.31	1.82	.50	2.63	3.80
1200	Sandblast, wet system, light soil	J-6	1750	.018		.38	.91	.18	1.47	2.04
1220	Average soil, biological staining		1100	.029		.57	1.44	.28	2.29	3.21
1240	Heavy soil, biological and mineral staining, paint		700	.046		.76	2.27	.44	3.47	4.89
1400	Dry system, light soil		2500	.013		.38	.63	.12	1.13	1.56
1420	Average soil, biological staining		1750	.018		.57	.91	.18	1.66	2.25
1440	Heavy soil, biological and mineral staining, paint		1000	.032		.76	1.59	.31	2.66	3.67
1800	For walnut shells, add					1.36			1.36	1.50
1820	For corn chips, add					.29			.29	.32
2000	Steam cleaning, light soil	A-1H	750	.011			.49	.11	.60	.89
2020	Average soil, biological staining		625	.013			.58	.13	.71	1.07
2040	Heavy soil, biological and mineral staining		375	.021			.97	.21	1.18	1.78
4000	Add for masking doors and windows	1 Clab	800	.010		.07	.46		.53	.80
4200	Add for pedestrian protection				Job				10%	10%
9000	Minimum labor/equipment charge	D-4	2	16	"		810	69.50	879.50	1,350

04 01 20.70 Brick Washing

		Crew	Daily Output	Labor-Hours	Unit	Material	2022 Bare Costs Labor	Equipment	Total	Total Incl O&P
0010	**BRICK WASHING**									
0012	Acid cleanser, smooth brick surface	1 Bric	560	.014	S.F.	.07	.79		.86	1.33
0050	Rough brick		400	.020		.09	1.11		1.20	1.87
0060	Stone, acid wash		600	.013		.11	.74		.85	1.30
1000	Muriatic acid, price per gallon in 5 gallon lots				Gal.	13.25			13.25	14.55

04 05 Common Work Results for Masonry

04 05 05 – Selective Demolition for Masonry

04 05 05.10 Selective Demolition

		Crew	Daily Output	Labor-Hours	Unit	Material	2022 Bare Costs Labor	Equipment	Total	Total Incl O&P
0010	**SELECTIVE DEMOLITION** R024119-10									
0200	Bond beams, 8" block with #4 bar	2 Clab	32	.500	L.F.		23		23	36
0300	Concrete block walls, unreinforced, 2" thick		1200	.013	S.F.		.61		.61	.97
0310	4" thick		1150	.014			.63		.63	1.01
0320	6" thick		1100	.015			.66		.66	1.05
0330	8" thick		1050	.015			.69		.69	1.10
0340	10" thick		1000	.016			.73		.73	1.16
0360	12" thick		950	.017			.77		.77	1.22
0380	Reinforced alternate courses, 2" thick		1130	.014			.65		.65	1.03
0390	4" thick		1080	.015			.68		.68	1.07
0400	6" thick		1035	.015			.70		.70	1.12
0410	8" thick		990	.016			.74		.74	1.17
0420	10" thick		940	.017			.78		.78	1.23
0430	12" thick		890	.018			.82		.82	1.30
0440	Reinforced alternate courses & vertically 48" OC, 4" thick		900	.018			.81		.81	1.29
0450	6" thick		850	.019			.86		.86	1.36
0460	8" thick		800	.020			.91		.91	1.45
0480	10" thick		750	.021			.97		.97	1.55

For customer support on your Commercial Renovation Costs with RSMeans data, call 800.448.8182.

85

04 05 05.10 Selective Demolition	Crew	Daily Output	Labor-Hours	Unit	Material	2022 Bare Costs Labor	Equipment	Total	Total Incl O&P
0490	12" thick	2 Clab	700	.023	S.F.		1.04		1.04	1.66
1000	Chimney, 16" x 16", soft old mortar	1 Clab	55	.145	C.F.		6.65		6.65	10.55
1020	Hard mortar		40	.200			9.10		9.10	14.50
1030	16" x 20", soft old mortar		55	.145			6.65		6.65	10.55
1040	Hard mortar		40	.200			9.10		9.10	14.50
1050	16" x 24", soft old mortar		55	.145			6.65		6.65	10.55
1060	Hard mortar		40	.200			9.10		9.10	14.50
1080	20" x 20", soft old mortar		55	.145			6.65		6.65	10.55
1100	Hard mortar		40	.200			9.10		9.10	14.50
1110	20" x 24", soft old mortar		55	.145			6.65		6.65	10.55
1120	Hard mortar		40	.200			9.10		9.10	14.50
1140	20" x 32", soft old mortar		55	.145			6.65		6.65	10.55
1160	Hard mortar		40	.200			9.10		9.10	14.50
1200	48" x 48", soft old mortar		55	.145			6.65		6.65	10.55
1220	Hard mortar	▼	40	.200	▼		9.10		9.10	14.50
1250	Metal, high temp steel jacket, 24" diameter	E-2	130	.431	V.L.F.		26.50	19.10	45.60	63
1260	60" diameter	"	60	.933			57	41.50	98.50	137
1280	Flue lining, up to 12" x 12"	1 Clab	200	.040			1.82		1.82	2.90
1282	Up to 24" x 24"		150	.053			2.43		2.43	3.86
2000	Columns, 8" x 8", soft old mortar		48	.167			7.60		7.60	12.10
2020	Hard mortar		40	.200			9.10		9.10	14.50
2060	16" x 16", soft old mortar		16	.500			23		23	36
2100	Hard mortar		14	.571			26		26	41.50
2140	24" x 24", soft old mortar		8	1			45.50		45.50	72.50
2160	Hard mortar		6	1.333			61		61	96.50
2200	36" x 36", soft old mortar		4	2			91		91	145
2220	Hard mortar		3	2.667	▼		122		122	193
2230	Alternate pricing method, soft old mortar		30	.267	C.F.		12.15		12.15	19.30
2240	Hard mortar	▼	23	.348	"		15.85		15.85	25
3000	Copings, precast or masonry, to 8" wide									
3020	Soft old mortar	1 Clab	180	.044	L.F.		2.03		2.03	3.22
3040	Hard mortar	"	160	.050	"		2.28		2.28	3.62
3100	To 12" wide									
3120	Soft old mortar	1 Clab	160	.050	L.F.		2.28		2.28	3.62
3140	Hard mortar	"	140	.057	"		2.61		2.61	4.14
4000	Fireplace, brick, 30" x 24" opening									
4020	Soft old mortar	1 Clab	2	4	Ea.		182		182	290
4040	Hard mortar		1.25	6.400			292		292	465
4100	Stone, soft old mortar		1.50	5.333			243		243	385
4120	Hard mortar		1	8	▼		365		365	580
5000	Veneers, brick, soft old mortar		140	.057	S.F.		2.61		2.61	4.14
5020	Hard mortar		125	.064			2.92		2.92	4.64
5050	Glass block, up to 4" thick		500	.016			.73		.73	1.16
5100	Granite and marble, 2" thick		180	.044			2.03		2.03	3.22
5120	4" thick		170	.047			2.15		2.15	3.41
5140	Stone, 4" thick		180	.044			2.03		2.03	3.22
5160	8" thick		175	.046	▼		2.08		2.08	3.31
5400	Alternate pricing method, stone, 4" thick		60	.133	C.F.		6.10		6.10	9.65
5420	8" thick		85	.094	"		4.29		4.29	6.80
9000	Minimum labor/equipment charge	▼	2	4	Job		182		182	290

04 05 13 – Masonry Mortaring

04 05 13.10 Cement	Crew	Daily Output	Labor-Hours	Unit	Material	2022 Bare Costs Labor	Equipment	Total	Total Incl O&P
0010 **CEMENT**									
0100 Masonry, 70 lb. bag, T.L. lots				Bag	11			11	12.10
0150 L.T.L. lots					11.65			11.65	12.85
0200 White, 70 lb. bag, T.L. lots					17.10			17.10	18.80
0250 L.T.L. lots				↓	12.50			12.50	13.75

04 05 13.20 Lime

	Crew	Daily Output	Labor-Hours	Unit	Material	Labor	Equipment	Total	Total Incl O&P
0010 **LIME**									
0020 Masons, hydrated, 50 lb. bag, T.L. lots				Bag	13.55			13.55	14.90
0050 L.T.L. lots					14.90			14.90	16.40
0200 Finish, double hydrated, 50 lb. bag, T.L. lots					14.75			14.75	16.25
0250 L.T.L. lots				↓	16.25			16.25	17.85

04 05 13.23 Surface Bonding Masonry Mortaring

	Crew	Daily Output	Labor-Hours	Unit	Material	Labor	Equipment	Total	Total Incl O&P
0010 **SURFACE BONDING MASONRY MORTARING**									
0020 Gray or white colors, not incl. block work	1 Bric	540	.015	S.F.	.14	.82		.96	1.47

04 05 13.30 Mortar

	Crew	Daily Output	Labor-Hours	Unit	Material	Labor	Equipment	Total	Total Incl O&P
0010 **MORTAR**									
0020 With masonry cement									
0100 Type M, 1:1:6 mix	1 Brhe	143	.056	C.F.	7	2.52		9.52	11.75
0200 Type N, 1:3 mix		143	.056		6	2.52		8.52	10.65
0300 Type O, 1:3 mix		143	.056		10	2.52		12.52	15.05
0400 Type PM, 1:1:6 mix, 2500 psi		143	.056		7.95	2.52		10.47	12.80
0500 Type S, 1/2:1:4 mix	↓	143	.056	↓	10.30	2.52		12.82	15.35
2000 With Portland cement and lime									
2100 Type M, 1:1/4:3 mix	1 Brhe	143	.056	C.F.	14.35	2.52		16.87	19.85
2200 Type N, 1:1:6 mix, 750 psi		143	.056		10.60	2.52		13.12	15.75
2300 Type O, 1:2:9 mix (Pointing Mortar)		143	.056		10	2.52		12.52	15.05
2400 Type PL, 1:1/2:4 mix, 2500 psi		143	.056		9.60	2.52		12.12	14.60
2600 Type S, 1:1/2:4 mix, 1800 psi	↓	143	.056		10.70	2.52		13.22	15.80
2650 Pre-mixed, type S or N					8.55			8.55	9.40
2700 Mortar for glass block	1 Brhe	143	.056	↓	13.70	2.52		16.22	19.10
2900 Mortar for fire brick, dry mix, 10 lb. pail				Ea.	30			30	33

04 05 13.91 Masonry Restoration Mortaring

	Crew	Daily Output	Labor-Hours	Unit	Material	Labor	Equipment	Total	Total Incl O&P
0010 **MASONRY RESTORATION MORTARING**									
0020 Masonry restoration mix				Lb.	.52			.52	.57
0050 White				"	.45			.45	.50

04 05 13.93 Mortar Pigments

	Crew	Daily Output	Labor-Hours	Unit	Material	Labor	Equipment	Total	Total Incl O&P
0010 **MORTAR PIGMENTS**, 50 lb. bags (2 bags per M bricks)									
0020 Color admixture, range 2 to 10 lb. per bag of cement, light colors				Lb.	8.10			8.10	8.95
0050 Medium colors					14.60			14.60	16.10
0100 Dark colors				↓	28.50			28.50	31.50

04 05 13.95 Sand

	Crew	Daily Output	Labor-Hours	Unit	Material	Labor	Equipment	Total	Total Incl O&P
0010 **SAND**, screened and washed at pit									
0020 For mortar, per ton				Ton	33.50			33.50	37
0050 With 10 mile haul					33.50			33.50	37
0100 With 30 mile haul					27.50			27.50	30.50
0200 Screened and washed, at the pit				C.Y.	47			47	51.50
0250 With 10 mile haul					47			47	51.50
0300 With 30 mile haul				↓	38.50			38.50	42.50

For customer support on your Commercial Renovation Costs with RSMeans data, call 800.448.8182.

87

04 05 13 – Masonry Mortaring

04 05 13.98 Mortar Admixtures

		Crew	Daily Output	Labor-Hours	Unit	Material	2022 Bare Costs Labor	Equipment	Total	Total Incl O&P
0010	**MORTAR ADMIXTURES**									
0020	Waterproofing admixture, per quart (1 qt. to 2 bags of masonry cement)				Qt.	7.50			7.50	8.25

04 05 16 – Masonry Grouting

04 05 16.30 Grouting

		Crew	Daily Output	Labor-Hours	Unit	Material	2022 Bare Costs Labor	Equipment	Total	Total Incl O&P
0010	**GROUTING**									
0011	Bond beams & lintels, 8" deep, 6" thick, 0.15 C.F./L.F.	D-4	1480	.022	L.F.	.80	1.09	.09	1.98	2.72
0020	8" thick, 0.2 C.F./L.F.		1400	.023		1.23	1.16	.10	2.49	3.31
0050	10" thick, 0.25 C.F./L.F.		1200	.027		1.33	1.35	.12	2.80	3.73
0060	12" thick, 0.3 C.F./L.F.		1040	.031		1.59	1.56	.13	3.28	4.37
0200	Concrete block cores, solid, 4" thk., by hand, 0.067 C.F./S.F. of wall	D-8	1100	.036	S.F.	.36	1.86		2.22	3.37
0210	6" thick, pumped, 0.175 C.F./S.F.	D-4	720	.044		.93	2.25	.19	3.37	4.80
0250	8" thick, pumped, 0.258 C.F./S.F.		680	.047		1.37	2.38	.20	3.95	5.50
0300	10" thick, pumped, 0.340 C.F./S.F.		660	.048		1.80	2.45	.21	4.46	6.10
0350	12" thick, pumped, 0.422 C.F./S.F.		640	.050		2.24	2.53	.22	4.99	6.70
0500	Cavity walls, 2" space, pumped, 0.167 C.F./S.F. of wall		1700	.019		.89	.95	.08	1.92	2.58
0550	3" space, 0.250 C.F./S.F.		1200	.027		1.33	1.35	.12	2.80	3.73
0600	4" space, 0.333 C.F./S.F.		1150	.028		1.77	1.41	.12	3.30	4.31
0700	6" space, 0.500 C.F./S.F.		800	.040		2.65	2.02	.17	4.84	6.35
0800	Door frames, 3' x 7' opening, 2.5 C.F. per opening		60	.533	Opng.	13.25	27	2.31	42.56	60
0850	6' x 7' opening, 3.5 C.F. per opening		45	.711	"	18.60	36	3.08	57.68	81
2000	Grout, C476, for bond beams, lintels and CMU cores		350	.091	C.F.	5.30	4.62	.40	10.32	13.65
9000	Minimum labor/equipment charge	1 Bric	2	4	Job		221		221	355

04 05 19 – Masonry Anchorage and Reinforcing

04 05 19.05 Anchor Bolts

		Crew	Daily Output	Labor-Hours	Unit	Material	2022 Bare Costs Labor	Equipment	Total	Total Incl O&P
0010	**ANCHOR BOLTS**									
0015	Installed in fresh grout in CMU bond beams or filled cores, no templates									
0020	Hooked, with nut and washer, 1/2" diameter, 8" long	1 Bric	132	.061	Ea.	2.23	3.35		5.58	7.80
0030	12" long		131	.061		2.48	3.37		5.85	8.15
0040	5/8" diameter, 8" long		129	.062		3.06	3.43		6.49	8.85
0050	12" long		127	.063		3.76	3.48		7.24	9.70
0060	3/4" diameter, 8" long		127	.063		3.76	3.48		7.24	9.70
0070	12" long		125	.064		4.70	3.54		8.24	10.80

04 05 19.16 Masonry Anchors

		Crew	Daily Output	Labor-Hours	Unit	Material	2022 Bare Costs Labor	Equipment	Total	Total Incl O&P
0010	**MASONRY ANCHORS**									
0020	For brick veneer, galv., corrugated, 7/8" x 7", 22 ga.	1 Bric	10.50	.762	C	13.15	42		55.15	82
0100	24 ga.		10.50	.762		12.10	42		54.10	81
0150	16 ga.		10.50	.762		20.50	42		62.50	90
0200	Buck anchors, galv., corrugated, 16 ga., 2" bend, 8" x 2"		10.50	.762		50.50	42		92.50	123
0250	8" x 3"		10.50	.762		60.50	42		102.50	134
0300	Adjustable, rectangular, 4-1/8" wide									
0350	Anchor and tie, 3/16" wire, mill galv.									
0400	2-3/4" eye, 3-1/4" tie	1 Bric	1.05	7.619	M	460	420		880	1,175
0500	4-3/4" tie		1.05	7.619		500	420		920	1,225
0520	5-1/2" tie		1.05	7.619		540	420		960	1,275
0550	4-3/4" eye, 3-1/4" tie		1.05	7.619		505	420		925	1,225
0570	4-3/4" tie		1.05	7.619		555	420		975	1,300
0580	5-1/2" tie		1.05	7.619		650	420		1,070	1,400
0660	Cavity wall, Z-type, galvanized, 6" long, 1/8" diam.		10.50	.762	C	23.50	42		65.50	93.50
0670	3/16" diameter		10.50	.762		29.50	42		71.50	99.50
0680	1/4" diameter		10.50	.762		38.50	42		80.50	110
0850	8" long, 3/16" diameter		10.50	.762		26.50	42		68.50	97

04 05 19.16 Masonry Anchors		Crew	Daily Output	Labor-Hours	Unit	Material	2022 Bare Costs Labor	Equipment	Total	Total Incl O&P
0855	1/4" diameter	1 Bric	10.50	.762	C	48.50	42		90.50	121
1000	Rectangular type, galvanized, 1/4" diameter, 2" x 6"		10.50	.762		76.50	42		118.50	152
1050	4" x 6"		10.50	.762		87.50	42		129.50	164
1100	3/16" diameter, 2" x 6"		10.50	.762		52	42		94	125
1150	4" x 6"		10.50	.762		53	42		95	126
1200	Mesh wall tie, 1/2" mesh, hot dip galvanized									
1400	16 ga., 12" long, 3" wide	1 Bric	9	.889	C	88	49		137	175
1420	6" wide		9	.889		129	49		178	221
1440	12" wide		8.50	.941		203	52		255	310
1500	Rigid partition anchors, plain, 8" long, 1" x 1/8"		10.50	.762		183	42		225	270
1550	1" x 1/4"		10.50	.762		315	42		357	415
1580	1-1/2" x 1/8"		10.50	.762		217	42		259	305
1600	1-1/2" x 1/4"		10.50	.762		345	42		387	450
1650	2" x 1/8"		10.50	.762		320	42		362	420
1700	2" x 1/4"		10.50	.762		380	42		422	490
2000	Column flange ties, wire, galvanized									
2300	3/16" diameter, up to 3" wide	1 Bric	10.50	.762	C	77	42		119	153
2350	To 5" wide		10.50	.762		84	42		126	160
2400	To 7" wide		10.50	.762		90.50	42		132.50	168
2600	To 9" wide		10.50	.762		96	42		138	174
2650	1/4" diameter, up to 3" wide		10.50	.762		107	42		149	186
2700	To 5" wide		10.50	.762		127	42		169	208
2800	To 7" wide		10.50	.762		143	42		185	225
2850	To 9" wide		10.50	.762		155	42		197	238
2900	For hot dip galvanized, add					35%				
4000	Channel slots, 1-3/8" x 1/2" x 8"									
4100	12 ga., plain	1 Bric	10.50	.762	C	244	42		286	335
4150	16 ga., galvanized	"	10.50	.762	"	137	42		179	219
4200	Channel slot anchors									
4300	16 ga., galvanized, 1-1/4" x 3-1/2"				C	61.50			61.50	67.50
4350	1-1/4" x 5-1/2"					71.50			71.50	78.50
4400	1-1/4" x 7-1/2"					65.50			65.50	72
4500	1/8" plain, 1-1/4" x 3-1/2"					137			137	151
4550	1-1/4" x 5-1/2"					146			146	161
4600	1-1/4" x 7-1/2"					160			160	176
4700	For corrugation, add					74			74	81
4750	For hot dip galvanized, add					35%				
5000	Dowels									
5100	Plain, 1/4" diameter, 3" long				C	47			47	51.50
5150	4" long					52			52	57
5200	6" long					47.50			47.50	52
5300	3/8" diameter, 3" long					59			59	65
5350	4" long					69			69	76
5400	6" long					89			89	98
5500	1/2" diameter, 3" long					90.50			90.50	99.50
5550	4" long					107			107	118
5600	6" long					50.50			50.50	55.50
5700	5/8" diameter, 3" long					124			124	137
5750	4" long					153			153	168
5800	6" long					210			210	231
6000	3/4" diameter, 3" long					154			154	170
6100	4" long					188			188	207
6150	6" long					280			280	310

04 05 19 – Masonry Anchorage and Reinforcing

04 05 19.16 Masonry Anchors	Crew	Daily Output	Labor-Hours	Unit	Material	2022 Bare Costs Labor	Equipment	Total	Total Incl O&P
6300 For hot dip galvanized, add				C	35%				

04 05 19.26 Masonry Reinforcing Bars

04 05 19.26 Masonry Reinforcing Bars	Crew	Daily Output	Labor-Hours	Unit	Material	2022 Bare Costs Labor	Equipment	Total	Total Incl O&P
0010 **MASONRY REINFORCING BARS**									
0015 Steel bars A615, placed horiz., #3 & #4 bars	1 Bric	450	.018	Lb.	.73	.98		1.71	2.37
0020 #5 & #6 bars		800	.010		.73	.55		1.28	1.69
0050 Placed vertical, #3 & #4 bars		350	.023		.73	1.26		1.99	2.82
0060 #5 & #6 bars		650	.012		.73	.68		1.41	1.89
0200 Joint reinforcing, regular truss, to 6" wide, mill std galvanized		30	.267	C.L.F.	26.50	14.75		41.25	53
0250 12" wide		20	.400		30	22		52	68.50
0400 Cavity truss with drip section, to 6" wide		30	.267		26.50	14.75		41.25	53
0450 12" wide		20	.400		31.50	22		53.50	70.50
0500 Joint reinforcing, ladder type, mill std galvanized									
0600 9 ga. sides, 9 ga. ties, 4" wall	1 Bric	30	.267	C.L.F.	16	14.75		30.75	41
0650 6" wall		30	.267		23	14.75		37.75	49
0700 8" wall		25	.320		27	17.70		44.70	58
0750 10" wall		20	.400		30	22		52	68.50
0800 12" wall		20	.400		22	22		44	59.50
1000 Truss type									
1100 9 ga. sides, 9 ga. ties, 4" wall	1 Bric	30	.267	C.L.F.	21.50	14.75		36.25	47.50
1150 6" wall		30	.267		22.50	14.75		37.25	48
1200 8" wall		25	.320		23	17.70		40.70	54
1250 10" wall		20	.400		23.50	22		45.50	61.50
1300 12" wall		20	.400		24.50	22		46.50	62.50
1500 3/16" sides, 9 ga. ties, 4" wall		30	.267		21.50	14.75		36.25	47.50
1550 6" wall		30	.267		23	14.75		37.75	48.50
1600 8" wall		25	.320		23.50	17.70		41.20	54
1650 10" wall		20	.400		23.50	22		45.50	61.50
1700 12" wall		20	.400		25	22		47	63
2000 3/16" sides, 3/16" ties, 4" wall		30	.267		22	14.75		36.75	47.50
2050 6" wall		30	.267		23	14.75		37.75	48.50
2100 8" wall		25	.320		24	17.70		41.70	54.50
2150 10" wall		20	.400		24.50	22		46.50	62.50
2200 12" wall		20	.400		25	22		47	63
2500 Cavity truss type, galvanized									
2600 9 ga. sides, 9 ga. ties, 4" wall	1 Bric	25	.320	C.L.F.	50.50	17.70		68.20	84
2650 6" wall		25	.320		50.50	17.70		68.20	84
2700 8" wall		20	.400		52	22		74	92.50
2750 10" wall		15	.533		53	29.50		82.50	105
2800 12" wall		15	.533		53	29.50		82.50	106
3000 3/16" sides, 9 ga. ties, 4" wall		25	.320		40	17.70		57.70	72.50
3050 6" wall		25	.320		29	17.70		46.70	60.50
3100 8" wall		20	.400		53.50	22		75.50	94.50
3150 10" wall		15	.533		38	29.50		67.50	89
3200 12" wall		15	.533		38	29.50		67.50	89
3500 For hot dip galvanizing, add				Ton	570			570	625

04 05 23 – Masonry Accessories

04 05 23.13 Masonry Control and Expansion Joints

04 05 23.13 Masonry Control and Expansion Joints	Crew	Daily Output	Labor-Hours	Unit	Material	2022 Bare Costs Labor	Equipment	Total	Total Incl O&P
0010 **MASONRY CONTROL AND EXPANSION JOINTS**									
0020 Rubber, for double wythe 8" minimum wall (Brick/CMU)	1 Bric	400	.020	L.F.	5.15	1.11		6.26	7.45
0025 "T" shaped		320	.025		1.12	1.38		2.50	3.44
0030 Cross-shaped for CMU units		280	.029		1.35	1.58		2.93	4.02
0050 PVC, for double wythe 8" minimum wall (Brick/CMU)		400	.020		1.50	1.11		2.61	3.42

04 05 23 – Masonry Accessories

04 05 23.13 Masonry Control and Expansion Joints	Crew	Daily Output	Labor-Hours	Unit	Material	2022 Bare Costs Labor	Equipment	Total	Total Incl O&P
0120	"T" shaped	1 Bric	320	.025	L.F.	.72	1.38		2.10	3
0160	Cross-shaped for CMU units	↓	280	.029	↓	.73	1.58		2.31	3.33

04 05 23.19 Masonry Cavity Drainage, Weepholes, and Vents

		Crew	Daily Output	Labor-Hours	Unit	Material	Labor	Equipment	Total	Total Incl O&P
0010	**MASONRY CAVITY DRAINAGE, WEEPHOLES, AND VENTS**									
0020	Extruded aluminum, 4" deep, 2-3/8" x 8-1/8"	1 Bric	30	.267	Ea.	15.15	14.75		29.90	40
0050	5" x 8-1/8"		25	.320		21.50	17.70		39.20	52
0100	2-1/4" x 25"		25	.320		62.50	17.70		80.20	97.50
0150	5" x 16-1/2"		22	.364		87	20		107	128
0175	5" x 24"		22	.364		69.50	20		89.50	109
0200	6" x 16-1/2"		22	.364		75	20		95	115
0250	7-3/4" x 16-1/2"	↓	20	.400		7.65	22		29.65	44
0400	For baked enamel finish, add					35%				
0500	For cast aluminum, painted, add					60%				
1000	Stainless steel ventilators, 6" x 6"	1 Bric	25	.320		168	17.70		185.70	214
1050	8" x 8"		24	.333		108	18.40		126.40	149
1100	12" x 12"		23	.348		111	19.20		130.20	153
1150	12" x 6"		24	.333		77.50	18.40		95.90	115
1200	Foundation block vent, galv., 1-1/4" thk, 8" high, 16" long, no damper	↓	30	.267	↓	19.95	14.75		34.70	45.50
1250	For damper, add					3.14			3.14	3.45
1450	Drainage and ventilation fabric	2 Bric	1450	.011	S.F.	.84	.61		1.45	1.90

04 05 23.95 Wall Plugs

		Crew	Daily Output	Labor-Hours	Unit	Material	Labor	Equipment	Total	Total Incl O&P
0010	**WALL PLUGS** (for nailing to brickwork)									
0020	25 ga., galvanized, plain	1 Bric	10.50	.762	C	60	42		102	134
0050	Wood filled	"	10.50	.762	"	125	42		167	205

04 21 Clay Unit Masonry

04 21 13 – Brick Masonry

04 21 13.13 Brick Veneer Masonry

		Crew	Daily Output	Labor-Hours	Unit	Material	Labor	Equipment	Total	Total Incl O&P
0010	**BRICK VENEER MASONRY**, T.L. lots, excl. scaff., grout & reinforcing R042110-20									
0015	Material costs incl. 3% brick and 25% mortar waste									
2000	Standard, sel. common, 4" x 2-2/3" x 8" (6.75/S.F.) R042110-50	D-8	230	.174	S.F.	7.75	8.90		16.65	23
2020	Red, 4" x 2-2/3" x 8", running bond		220	.182		5.05	9.30		14.35	20.50
2050	Full header every 6th course (7.88/S.F.)		185	.216		5.90	11.05		16.95	24
2100	English, full header every 2nd course (10.13/S.F.)		140	.286		7.60	14.60		22.20	32
2150	Flemish, alternate header every course (9.00/S.F.)		150	.267		6.75	13.65		20.40	29.50
2200	Flemish, alt. header every 6th course (7.13/S.F.)		205	.195		5.35	10		15.35	22
2250	Full headers throughout (13.50/S.F.)		105	.381		10.10	19.50		29.60	42
2300	Rowlock course (13.50/S.F.)		100	.400		10.10	20.50		30.60	44
2350	Rowlock stretcher (4.50/S.F.)		310	.129		3.40	6.60		10	14.30
2400	Soldier course (6.75/S.F.)		200	.200		5.05	10.25		15.30	22
2450	Sailor course (4.50/S.F.)		290	.138		3.40	7.05		10.45	15.05
2600	Buff or gray face, running bond (6.75/S.F.)		220	.182		5.35	9.30		14.65	21
2700	Glazed face brick, running bond		210	.190		16.10	9.75		25.85	33.50
2750	Full header every 6th course (7.88/S.F.)		170	.235		18.80	12.05		30.85	40
3000	Jumbo, 6" x 4" x 12" running bond (3.00/S.F.)		435	.092		5.40	4.70		10.10	13.50
3050	Norman, 4" x 2-2/3" x 12" running bond (4.5/S.F.)		320	.125		6.75	6.40		13.15	17.65
3100	Norwegian, 4" x 3-1/5" x 12" (3.75/S.F.)		375	.107		7.70	5.45		13.15	17.25
3150	Economy, 4" x 4" x 8" (4.50/S.F.)		310	.129		4.72	6.60		11.32	15.75
3200	Engineer, 4" x 3-1/5" x 8" (5.63/S.F.)		260	.154		3.98	7.85		11.83	16.95
3250	Roman, 4" x 2" x 12" (6.00/S.F.)	↓	250	.160		8.15	8.20		16.35	22

For customer support on your Commercial Renovation Costs with RSMeans data, call 800.448.8182.

91

04 21 Clay Unit Masonry

04 21 13 – Brick Masonry

04 21 13.13 Brick Veneer Masonry

		Crew	Daily Output	Labor-Hours	Unit	Material	2022 Bare Costs Labor	Equipment	Total	Total Incl O&P
3300	S.C.R., 6" x 2-2/3" x 12" (4.50/S.F.)	D-8	310	.129	S.F.	6.15	6.60		12.75	17.30
3350	Utility, 4" x 4" x 12" (3.00/S.F.)	↓	360	.111		5.55	5.70		11.25	15.20
3360	For less than truck load lots, add					.10%				
3400	For cavity wall construction, add						15%			
3450	For stacked bond, add						10%			
3500	For interior veneer construction, add						15%			
3550	For curved walls, add				↓		30%			
9000	Minimum labor/equipment charge	D-1	2	8	Job		400		400	640

04 21 13.14 Thin Brick Veneer

		Crew	Daily Output	Labor-Hours	Unit	Material	2022 Bare Costs Labor	Equipment	Total	Total Incl O&P
0010	**THIN BRICK VENEER**									
0015	Material costs incl. 3% brick and 25% mortar waste									
0020	On & incl. metal panel support sys, modular, 2-2/3" x 5/8" x 8", red	D-7	92	.174	S.F.	14.05	8.35		22.40	28.50
0100	Closure, 4" x 5/8" x 8"		110	.145		9.60	7		16.60	21.50
0110	Norman, 2-2/3" x 5/8" x 12"		110	.145		9.95	7		16.95	22
0120	Utility, 4" x 5/8" x 12"		125	.128		9.30	6.15		15.45	19.80
0130	Emperor, 4" x 3/4" x 16"		175	.091		10.50	4.39		14.89	18.40
0140	Super emperor, 8" x 3/4" x 16"	↓	195	.082	↓	11.55	3.94		15.49	18.85
0150	For L shaped corners with 4" return, add				L.F.	9.25			9.25	10.20
0200	On masonry/plaster back-up, modular, 2-2/3" x 5/8" x 8", red	D-7	137	.117	S.F.	8.90	5.60		14.50	18.55
0210	Closure, 4" x 5/8" x 8"		165	.097		4.45	4.65		9.10	12.15
0220	Norman, 2-2/3" x 5/8" x 12"		165	.097		4.82	4.65		9.47	12.55
0230	Utility, 4" x 5/8" x 12"		185	.086		4.16	4.15		8.31	11.10
0240	Emperor, 4" x 3/4" x 16"		260	.062		5.40	2.95		8.35	10.50
0250	Super emperor, 8" x 3/4" x 16"	↓	285	.056	↓	6.40	2.69		9.09	11.25
0260	For L shaped corners with 4" return, add				L.F.	10.10			10.10	11.10
0270	For embedment into pre-cast concrete panels, add				S.F.	14.40			14.40	15.85

04 21 13.15 Chimney

		Crew	Daily Output	Labor-Hours	Unit	Material	2022 Bare Costs Labor	Equipment	Total	Total Incl O&P
0010	**CHIMNEY**, excludes foundation, scaffolding, grout and reinforcing									
0100	Brick, 16" x 16", 8" flue	D-1	18.20	.879	V.L.F.	43.50	44		87.50	118
0150	16" x 20" with one 8" x 12" flue		16	1		61.50	50		111.50	148
0200	16" x 24" with two 8" x 8" flues		14	1.143		84	57.50		141.50	185
0250	20" x 20" with one 12" x 12" flue		13.70	1.168		75.50	58.50		134	177
0300	20" x 24" with two 8" x 12" flues		12	1.333		97	67		164	214
0350	20" x 32" with two 12" x 12" flues	↓	10	1.600	↓	129	80		209	270

04 21 13.18 Columns

		Crew	Daily Output	Labor-Hours	Unit	Material	2022 Bare Costs Labor	Equipment	Total	Total Incl O&P
0010	**COLUMNS**, solid, excludes scaffolding, grout and reinforcing									
0050	Brick, 8" x 8", 9 brick/V.L.F.	D-1	56	.286	V.L.F.	7.05	14.30		21.35	31
0100	12" x 8", 13.5 brick/V.L.F.		37	.432		10.55	21.50		32.05	46
0200	12" x 12", 20 brick/V.L.F.		25	.640		15.60	32		47.60	68.50
0300	16" x 12", 27 brick/V.L.F.		19	.842		21	42		63	90.50
0400	16" x 16", 36 brick/V.L.F.		14	1.143		28	57.50		85.50	123
0500	20" x 16", 45 brick/V.L.F.		11	1.455		35	73		108	156
0600	20" x 20", 56 brick/V.L.F.		9	1.778		43.50	89		132.50	191
0700	24" x 20", 68 brick/V.L.F.		7	2.286		53	115		168	243
0800	24" x 24", 81 brick/V.L.F.		6	2.667		63.50	134		197.50	284
1000	36" x 36", 182 brick/V.L.F.	↓	3	5.333	↓	142	267		409	585
9000	Minimum labor/equipment charge		2	8	Job		400		400	640

04 21 13.30 Oversized Brick

		Crew	Daily Output	Labor-Hours	Unit	Material	2022 Bare Costs Labor	Equipment	Total	Total Incl O&P
0010	**OVERSIZED BRICK**, excludes scaffolding, grout and reinforcing									
0100	Veneer, 4" x 2.25" x 16"	D-8	387	.103	S.F.	4.89	5.30		10.19	13.85
0102	8" x 2.25" x 16", multicell		265	.151		17.90	7.70		25.60	32
0105	4" x 2.75" x 16"	↓	412	.097		5.90	4.97		10.87	14.40

04 21 13.30 Oversized Brick

		Crew	Daily Output	Labor-Hours	Unit	Material	2022 Bare Costs Labor	Equipment	Total	Total Incl O&P
0107	8" x 2.75" x 16", multicell	D-8	295	.136	S.F.	17.85	6.95		24.80	31
0110	4" x 4" x 16"		460	.087		3.90	4.45		8.35	11.45
0120	4" x 8" x 16"		533	.075		5.60	3.84		9.44	12.35
0122	4" x 8" x 16" multicell		327	.122		16	6.25		22.25	27.50
0125	Loadbearing, 6" x 4" x 16", grouted and reinforced		387	.103		12.05	5.30		17.35	21.50
0130	8" x 4" x 16", grouted and reinforced		327	.122		12.05	6.25		18.30	23.50
0132	10" x 4" x 16", grouted and reinforced		327	.122		28.50	6.25		34.75	41
0135	6" x 8" x 16", grouted and reinforced		440	.091		15.55	4.65		20.20	24.50
0140	8" x 8" x 16", grouted and reinforced		400	.100		16.35	5.10		21.45	26
0145	Curtainwall/reinforced veneer, 6" x 4" x 16"		387	.103		17.05	5.30		22.35	27
0150	8" x 4" x 16"		327	.122		19.85	6.25		26.10	32
0152	10" x 4" x 16"		327	.122		31.50	6.25		37.75	44.50
0155	6" x 8" x 16"		440	.091		21.50	4.65		26.15	31
0160	8" x 8" x 16"		400	.100		28.50	5.10		33.60	39.50
0200	For 1 to 3 slots in face, add					15%				
0210	For 4 to 7 slots in face, add					25%				
0220	For bond beams, add					20%				
0230	For bullnose shapes, add					20%				
0240	For open end knockout, add					10%				
0250	For white or gray color group, add					10%				
0260	For 135 degree corner, add					250%				

04 21 13.35 Common Building Brick

						Material	Labor	Equipment	Total	Total Incl O&P
0010	**COMMON BUILDING BRICK**, C62, T.L. lots, material only R042110-20									
0020	Standard				M	970			970	1,075
0050	Select				"	1,025			1,025	1,125

04 21 13.40 Structural Brick

		Crew	Daily Output	Labor-Hours	Unit	Material	Labor	Equipment	Total	Total Incl O&P
0010	**STRUCTURAL BRICK** C652, Grade SW, incl. mortar, scaffolding not incl.									
0100	Standard unit, 4-5/8" x 2-3/4" x 9-5/8"	D-8	245	.163	S.F.	4.57	8.35		12.92	18.45
0120	Bond beam		225	.178		4.69	9.10		13.79	19.70
0140	V cut bond beam		225	.178		4.58	9.10		13.68	19.60
0160	Stretcher quoin, 5-5/8" x 2-3/4" x 9-5/8"		245	.163		7.65	8.35		16	22
0180	Corner quoin		245	.163		8.35	8.35		16.70	22.50
0200	Corner, 45 degree, 4-5/8" x 2-3/4" x 10-7/16"		235	.170		9.35	8.70		18.05	24.50

04 21 13.45 Face Brick

						Material	Labor	Equipment	Total	Total Incl O&P
0010	**FACE BRICK** Material Only, C216, T.L. lots R042110-20									
0300	Standard modular, 4" x 2-2/3" x 8"				M	635			635	695
0450	Economy, 4" x 4" x 8"					920			920	1,000
0510	Economy, 4" x 4" x 12"					1,450			1,450	1,600
0550	Jumbo, 6" x 4" x 12"					1,650			1,650	1,825
0610	Jumbo, 8" x 4" x 12"					1,650			1,650	1,825
0650	Norwegian, 4" x 3-1/5" x 12"					1,900			1,900	2,075
0710	Norwegian, 6" x 3-1/5" x 12"					1,600			1,600	1,750
0850	Standard glazed, plain colors, 4" x 2-2/3" x 8"					2,225			2,225	2,450
1000	Deep trim shades, 4" x 2-2/3" x 8"					2,400			2,400	2,650
1080	Jumbo utility, 4" x 4" x 12"					1,700			1,700	1,875
1120	4" x 8" x 8"					1,975			1,975	2,175
1140	4" x 8" x 16"					6,950			6,950	7,650
1260	Engineer, 4" x 3-1/5" x 8"					590			590	645
1350	King, 4" x 2-3/4" x 10"					580			580	640
1770	Standard modular, double glazed, 4" x 2-2/3" x 8"					3,025			3,025	3,325
1850	Jumbo, colored glazed ceramic, 6" x 4" x 12"					2,825			2,825	3,125
2050	Jumbo utility, glazed, 4" x 4" x 12"					5,825			5,825	6,400

04 21 13 – Brick Masonry

04 21 13.45 Face Brick	Crew	Daily Output	Labor-Hours	Unit	Material	2022 Bare Costs Labor	Equipment	Total	Total Incl O&P
2100	4" x 8" x 8"				M	6,850			6,850	7,525
2150	4" x 16" x 8"					7,300			7,300	8,050
3050	Used brick					500			500	550
3150	Add for brick to match existing work, minimum					5%				
3200	Maximum					50%				

04 21 26 – Glazed Structural Clay Tile Masonry

04 21 26.10 Structural Facing Tile

		Crew	Daily Output	Labor-Hours	Unit	Material	Labor	Equipment	Total	Total Incl O&P
0010	**STRUCTURAL FACING TILE**, std. colors, excl. scaffolding, grout, reinforcing									
0020	6T series, 5-1/3" x 12", 2.3 pieces per S.F., glazed 1 side, 2" thick	D-8	225	.178	S.F.	10	9.10		19.10	25.50
0100	4" thick		220	.182		13.25	9.30		22.55	29.50
0150	Glazed 2 sides		195	.205		17.60	10.50		28.10	36
0250	6" thick		210	.190		18.90	9.75		28.65	36.50
0300	Glazed 2 sides		185	.216		22.50	11.05		33.55	42.50
0400	8" thick		180	.222		23	11.35		34.35	43.50
0500	Special shapes, group 1		400	.100	Ea.	7.65	5.10		12.75	16.60
0550	Group 2		375	.107		13.70	5.45		19.15	24
0600	Group 3		350	.114		17.20	5.85		23.05	28.50
0650	Group 4		325	.123		35.50	6.30		41.80	49
0700	Group 5		300	.133		42.50	6.80		49.30	58
0750	Group 6		275	.145		63.50	7.45		70.95	81.50
1000	Fire rated, 4" thick, 1 hr. rating		210	.190	S.F.	18.70	9.75		28.45	36
1050	2 hr. rating		170	.235		28.50	12.05		40.55	51
1300	Acoustic, 4" thick		210	.190		40	9.75		49.75	59.50
1310	6" thick		170	.235		41.50	12.05		53.55	65.50
1320	8" thick		145	.276		44	14.10		58.10	71
1350	8T acoustic, 4" thick		265	.151		20	7.70		27.70	34.50
1360	6" thick		256	.156		29.50	8		37.50	45.50
1370	8" thick		208	.192		31.50	9.85		41.35	51
2000	8W series, 8" x 16", 1.125 pieces per S.F.									
2050	2" thick, glazed 1 side	D-8	360	.111	S.F.	9.65	5.70		15.35	19.70
2100	4" thick, glazed 1 side		345	.116		13.65	5.95		19.60	24.50
2150	Glazed 2 sides		325	.123		14.55	6.30		20.85	26
2200	6" thick, glazed 1 side		330	.121		31.50	6.20		37.70	44.50
2210	Glazed 2 sides		250	.160		31.50	8.20		39.70	48
2250	8" thick, glazed 1 side		310	.129		29.50	6.60		36.10	43
2260	Glazed 2 sides		220	.182		35.50	9.30		44.80	54
2500	Special shapes, group 1		300	.133	Ea.	16.05	6.80		22.85	28.50
2550	Group 2		280	.143		21.50	7.30		28.80	35
2600	Group 3		260	.154		23	7.85		30.85	37.50
2650	Group 4		250	.160		54.50	8.20		62.70	73
2700	Group 5		240	.167		47.50	8.55		56.05	65.50
2750	Group 6		230	.174		104	8.90		112.90	129
2900	Fire rated, 2" thick, glazed 1 side		360	.111	S.F.	12.80	5.70		18.50	23
3000	4" thick, glazed 1 side		345	.116		16	5.95		21.95	27
3020	Glazed 2 sides		345	.116		24	5.95		29.95	36
3040	6" thick, glazed 1 side		330	.121		24	6.20		30.20	36.50
3045	Glazed 2 sides		250	.160		25.50	8.20		33.70	41
3047	8" thick, glazed 1 side		310	.129		30.50	6.60		37.10	44
3048	Glazed 2 sides		220	.182		32	9.30		41.30	50
3050	Special shapes, group 1, fire rated		300	.133		21	6.80		27.80	34
3060	Group 2		280	.143		29.50	7.30		36.80	44
3070	Group 3		260	.154		30.50	7.85		38.35	46

04 21 Clay Unit Masonry

04 21 26 – Glazed Structural Clay Tile Masonry

04 21 26.10 Structural Facing Tile

		Crew	Daily Output	Labor-Hours	Unit	Material	2022 Bare Costs Labor	Equipment	Total	Total Incl O&P
3080	Group 4	D-8	250	.160	S.F.	53.50	8.20		61.70	72
3090	Group 5		240	.167		53	8.55		61.55	72
3100	Acoustic, 4" thick	↓	345	.116	↓	22.50	5.95		28.45	34.50
3120	4W series, 8" x 8", 2.25 pieces per S.F.									
3125	2" thick, glazed 1 side	D-8	360	.111	S.F.	9.30	5.70		15	19.30
3130	4" thick, glazed 1 side		345	.116		12.80	5.95		18.75	23.50
3135	Glazed 2 sides		325	.123		15.95	6.30		22.25	27.50
3140	6" thick, glazed 1 side		330	.121		16.35	6.20		22.55	28
3145	Glazed 2 sides		250	.160		22	8.20		30.20	37
3150	8" thick, glazed 1 side		310	.129	↓	24.50	6.60		31.10	37.50
3155	Special shapes, group 1		300	.133	Ea.	7.70	6.80		14.50	19.40
3160	Group 2	↓	280	.143	"	8.95	7.30		16.25	21.50
3200	For designer colors, add					25%				
3300	For epoxy mortar joints, add				S.F.	1.90			1.90	2.09
9000	Minimum labor/equipment charge	D-1	2	8	Job		400		400	640

04 21 29 – Terra Cotta Masonry

04 21 29.10 Terra Cotta Masonry Components

		Crew	Daily Output	Labor-Hours	Unit	Material	2022 Bare Costs Labor	Equipment	Total	Total Incl O&P
0010	**TERRA COTTA MASONRY COMPONENTS**									
0020	Coping, split type, not glazed, 9" wide	D-1	90	.178	L.F.	11.75	8.90		20.65	27.50
0100	13" wide		80	.200		16.85	10.05		26.90	34.50
0200	Coping, split type, glazed, 9" wide		90	.178		20.50	8.90		29.40	37
0250	13" wide	↓	80	.200	↓	25	10.05		35.05	43.50
0500	Partition or back-up blocks, scored, in C.L. lots									
0700	Non-load bearing 12" x 12", 3" thick, special order	D-8	550	.073	S.F.	21	3.72		24.72	29
0750	4" thick, standard		500	.080		7	4.09		11.09	14.25
0800	6" thick		450	.089		8.50	4.55		13.05	16.65
0850	8" thick		400	.100		10.45	5.10		15.55	19.70
1000	Load bearing, 12" x 12", 4" thick, in walls		500	.080		6.50	4.09		10.59	13.70
1050	In floors		750	.053		6.50	2.73		9.23	11.50
1200	6" thick, in walls		450	.089		9.90	4.55		14.45	18.20
1250	In floors		675	.059		9.90	3.03		12.93	15.75
1400	8" thick, in walls		400	.100		11	5.10		16.10	20.50
1450	In floors		575	.070		11	3.56		14.56	17.80
1600	10" thick, in walls, special order		350	.114		28.50	5.85		34.35	41
1650	In floors, special order		500	.080		28.50	4.09		32.59	38
1800	12" thick, in walls, special order		300	.133		26.50	6.80		33.30	40.50
1850	In floors, special order	↓	450	.089		26.50	4.55		31.05	37
2000	For reinforcing with steel rods, add to above				↓	15%	5%			
2025	Placed vertical, #3 & #4 bars	1 Bric	350	.023	Lb.	.73	1.26		1.99	2.82
2100	For smooth tile instead of scored, add				S.F.	2.97			2.97	3.27
2200	For L.C.L. quantities, add				"	10%	10%			
9000	Minimum labor/equipment charge	D-1	2	8	Job		400		400	640

04 21 29.20 Terra Cotta Tile

		Crew	Daily Output	Labor-Hours	Unit	Material	2022 Bare Costs Labor	Equipment	Total	Total Incl O&P
0010	**TERRA COTTA TILE**, on walls, dry set, 1/2" thick									
0100	Square, hexagonal or lattice shapes, unglazed	1 Tilf	135	.059	S.F.	5.25	3.15		8.40	10.70
0300	Glazed, plain colors		130	.062		8.20	3.27		11.47	14.10
0400	Intense colors	↓	125	.064	↓	9.55	3.40		12.95	15.85

For customer support on your Commercial Renovation Costs with RSMeans data, call 800.448.8182.

95

04 22 10 – Concrete Masonry Units

04 22 10.11 Autoclave Aerated Concrete Block		Crew	Daily Output	Labor-Hours	Unit	Material	2022 Bare Costs Labor	Equipment	Total	Total Incl O&P
0010	**AUTOCLAVE AERATED CONCRETE BLOCK**, excl. scaffolding, grout & reinforcing									
0050	Solid, 4" x 8" x 24", incl. mortar G	D-8	600	.067	S.F.	3.09	3.41		6.50	8.85
0060	6" x 8" x 24" G		600	.067		5.15	3.41		8.56	11.10
0070	8" x 8" x 24" G		575	.070		3.54	3.56		7.10	9.60
0080	10" x 8" x 24" G		575	.070		4.25	3.56		7.81	10.35
0090	12" x 8" x 24" G		550	.073		5.10	3.72		8.82	11.55

04 22 10.14 Concrete Block, Back-Up										
0010	**CONCRETE BLOCK, BACK-UP**, C90, 2000 psi									
0020	Normal weight, 8" x 16" units, tooled joint 1 side									
0050	Not-reinforced, 2000 psi, 2" thick	D-8	475	.084	S.F.	1.58	4.31		5.89	8.65
0200	4" thick		460	.087		1.89	4.45		6.34	9.25
0300	6" thick		440	.091		4.01	4.65		8.66	11.85
0350	8" thick		400	.100		2.60	5.10		7.70	11.05
0400	10" thick		330	.121		3.84	6.20		10.04	14.15
0450	12" thick	D-9	310	.155		4.48	7.75		12.23	17.35
1000	Reinforced, alternate courses, 4" thick	D-8	450	.089		2.05	4.55		6.60	9.55
1100	6" thick		430	.093		4.17	4.76		8.93	12.20
1150	8" thick		395	.101		2.77	5.20		7.97	11.35
1200	10" thick		320	.125		4.01	6.40		10.41	14.65
1250	12" thick	D-9	300	.160		4.66	8		12.66	17.95
9000	Minimum labor/equipment charge	D-1	2	8	Job		400		400	640

04 22 10.16 Concrete Block, Bond Beam										
0010	**CONCRETE BLOCK, BOND BEAM**, C90, 2000 psi									
0020	Not including grout or reinforcing									
0125	Regular block, 6" thick	D-8	584	.068	L.F.	3.37	3.50		6.87	9.30
0130	8" high, 8" thick	"	565	.071		4.72	3.62		8.34	11
0150	12" thick	D-9	510	.094		6.10	4.72		10.82	14.25
0525	Lightweight, 6" thick	D-8	592	.068		3.22	3.46		6.68	9.10
0530	8" high, 8" thick	"	575	.070		3.79	3.56		7.35	9.85
0550	12" thick	D-9	520	.092		5.10	4.63		9.73	13
2000	Including grout and 2 #5 bars									
2100	Regular block, 8" high, 8" thick	D-8	300	.133	L.F.	7.50	6.80		14.30	19.20
2150	12" thick	D-9	250	.192		9.55	9.60		19.15	26
2500	Lightweight, 8" high, 8" thick	D-8	305	.131		6.60	6.70		13.30	18
2550	12" thick	D-9	255	.188		8.55	9.45		18	24.50
9000	Minimum labor/equipment charge	D-1	2	8	Job		400		400	640

04 22 10.19 Concrete Block, Insulation Inserts										
0010	**CONCRETE BLOCK, INSULATION INSERTS**									
0100	Styrofoam, plant installed, add to block prices									
0200	8" x 16" units, 6" thick				S.F.	1.63			1.63	1.79
0250	8" thick					2.10			2.10	2.31
0300	10" thick					1.71			1.71	1.88
0350	12" thick					2.18			2.18	2.40
0500	8" x 8" units, 8" thick					2.08			2.08	2.29
0550	12" thick					2.06			2.06	2.27

04 22 10.23 Concrete Block, Decorative										
0010	**CONCRETE BLOCK, DECORATIVE**, C90, 2000 psi									
0020	Embossed, simulated brick face									
0100	8" x 16" units, 4" thick	D-8	400	.100	S.F.	3.23	5.10		8.33	11.75
0200	8" thick		340	.118		3.39	6		9.39	13.35
0250	12" thick		300	.133		4.92	6.80		11.72	16.35
0400	Embossed both sides									

04 22 10.23 Concrete Block, Decorative		Crew	Daily Output	Labor-Hours	Unit	Material	2022 Bare Costs Labor	Equipment	Total	Total Incl O&P
0500	8" thick	D-8	300	.133	S.F.	5.75	6.80		12.55	17.25
0550	12" thick	"	275	.145	"	6.95	7.45		14.40	19.55
1000	Fluted high strength									
1100	8" x 16" x 4" thick, flutes 1 side,	D-8	345	.116	S.F.	4.28	5.95		10.23	14.20
1150	Flutes 2 sides		335	.119		4.85	6.10		10.95	15.15
1200	8" thick		300	.133		5.45	6.80		12.25	16.95
1250	For special colors, add					.69			.69	.76
1400	Deep grooved, smooth face									
1450	8" x 16" x 4" thick	D-8	345	.116	S.F.	2.68	5.95		8.63	12.45
1500	8" thick	"	300	.133	"	4.68	6.80		11.48	16.10
2000	Formblock, incl. inserts & reinforcing									
2100	8" x 16" x 8" thick	D-8	345	.116	S.F.	4.64	5.95		10.59	14.60
2150	12" thick	"	310	.129	"	6	6.60		12.60	17.15
2500	Ground face									
2600	8" x 16" x 4" thick	D-8	345	.116	S.F.	2.84	5.95		8.79	12.65
2650	6" thick		325	.123		5.10	6.30		11.40	15.70
2700	8" thick		300	.133		4.11	6.80		10.91	15.45
2725	10" thick		280	.143		5.85	7.30		13.15	18.10
2750	12" thick	D-9	265	.181		4.82	9.10		13.92	19.85
2900	For special colors, add, minimum					15%				
2950	For special colors, add, maximum					45%				
4000	Slump block									
4100	4" face height x 16" x 4" thick	D-1	165	.097	S.F.	5.50	4.86		10.36	13.85
4150	6" thick		160	.100		6.90	5		11.90	15.65
4200	8" thick		155	.103		9.85	5.15		15	19.15
4250	10" thick		140	.114		12.70	5.75		18.45	23
4300	12" thick		130	.123		13.90	6.15		20.05	25
4400	6" face height x 16" x 6" thick		155	.103		5.25	5.15		10.40	14.10
4450	8" thick		150	.107		7.95	5.35		13.30	17.30
4500	10" thick		130	.123		15.20	6.15		21.35	26.50
4550	12" thick	"	120	.133	"	15.10	6.70		21.80	27.50
5000	Split rib profile units, 1" deep ribs, 8 ribs									
5100	8" x 16" x 4" thick	D-8	345	.116	S.F.	5.05	5.95		11	15.05
5150	6" thick		325	.123		5.85	6.30		12.15	16.55
5200	8" thick		300	.133		7.10	6.80		13.90	18.75
5225	10" thick		300	.133		7.25	6.80		14.05	18.90
5250	12" thick	D-9	275	.175		8.10	8.75		16.85	23
5400	For special deeper colors, 4" thick, add					1.49			1.49	1.64
5450	12" thick, add					1.44			1.44	1.58
5600	For white, 4" thick, add					1.49			1.49	1.64
5650	6" thick, add					1.44			1.44	1.58
5700	8" thick, add					1.55			1.55	1.70
5750	12" thick, add					1.49			1.49	1.64
6000	Split face									
6100	8" x 16" x 4" thick	D-8	350	.114	S.F.	3.96	5.85		9.81	13.70
6150	6" thick		325	.123		4.54	6.30		10.84	15.10
6200	8" thick		300	.133		5.80	6.80		12.60	17.30
6225	10" thick		275	.145		6.10	7.45		13.55	18.60
6250	12" thick	D-9	270	.178		6.45	8.90		15.35	21.50
6300	For scored, add					.41			.41	.45
6400	For special deeper colors, 4" thick, add					.69			.69	.76
6450	6" thick, add					.87			.87	.96
6500	8" thick, add					.75			.75	.82

For customer support on your Commercial Renovation Costs with RSMeans data, call 800.448.8182.

97

04 22 10.23 Concrete Block, Decorative	Crew	Daily Output	Labor-Hours	Unit	Material	2022 Bare Costs Labor	Equipment	Total	Total Incl O&P
6550	12" thick, add				S.F.	.87			.87	.96
6650	For white, 4" thick, add					1.49			1.49	1.64
6700	6" thick, add					1.34			1.34	1.48
6750	8" thick, add					1.47			1.47	1.62
6800	12" thick, add					1.44			1.44	1.58
7000	Scored ground face, 2 to 5 scores									
7100	8" x 16" x 4" thick	D-8	340	.118	S.F.	2.80	6		8.80	12.75
7150	6" thick		310	.129		10.55	6.60		17.15	22
7200	8" thick		290	.138		4.03	7.05		11.08	15.75
7250	12" thick	D-9	265	.181		4.72	9.10		13.82	19.75
8000	Hexagonal face profile units, 8" x 16" units									
8100	4" thick, hollow	D-8	340	.118	S.F.	4.29	6		10.29	14.35
8200	Solid		340	.118		5.45	6		11.45	15.65
8300	6" thick, hollow		310	.129		4.35	6.60		10.95	15.35
8350	8" thick, hollow		290	.138		4.31	7.05		11.36	16.05
8500	For stacked bond, add						26%			
8550	For high rise construction, add per story	D-8	67.80	.590	M.S.F.		30		30	48.50
8600	For scored block, add					10%				
8650	For honed or ground face, per face, add				Ea.	1.24			1.24	1.37
8700	For honed or ground end, per end, add				"	1.60			1.60	1.77
8750	For bullnose block, add					10%				
8800	For special color, add					13%				
9000	Minimum labor/equipment charge	D-1	2	8	Job		400		400	640

04 22 10.24 Concrete Block, Exterior

		Crew	Daily Output	Labor-Hours	Unit	Material	2022 Bare Costs Labor	Equipment	Total	Total Incl O&P
0010	**CONCRETE BLOCK, EXTERIOR**, C90, 2000 psi									
0020	Reinforced alt courses, tooled joints 2 sides									
0100	Normal weight, 8" x 16" x 6" thick	D-8	395	.101	S.F.	2.43	5.20		7.63	10.95
0200	8" thick		360	.111		4.58	5.70		10.28	14.15
0250	10" thick		290	.138		4.36	7.05		11.41	16.10
0300	12" thick	D-9	250	.192		5	9.60		14.60	21
9000	Minimum labor/equipment charge	D-1	2	8	Job		400		400	640

04 22 10.26 Concrete Block Foundation Wall

		Crew	Daily Output	Labor-Hours	Unit	Material	2022 Bare Costs Labor	Equipment	Total	Total Incl O&P
0010	**CONCRETE BLOCK FOUNDATION WALL**, C90/C145									
0050	Normal-weight, cut joints, horiz joint reinf, no vert reinf.									
0200	Hollow, 8" x 16" x 6" thick	D-8	455	.088	S.F.	4.52	4.50		9.02	12.20
0250	8" thick		425	.094		3.13	4.81		7.94	11.15
0300	10" thick		350	.114		4.38	5.85		10.23	14.15
0350	12" thick	D-9	300	.160		5.05	8		13.05	18.40
0500	Solid, 8" x 16" block, 6" thick	D-8	440	.091		3.95	4.65		8.60	11.80
0550	8" thick	"	415	.096		4.21	4.93		9.14	12.55
0600	12" thick	D-9	350	.137		4.20	6.85		11.05	15.60
1000	Reinforced, #4 vert @ 48"									
1100	Hollow, 8" x 16" block, 4" thick	D-8	455	.088	S.F.	3.07	4.50		7.57	10.60
1125	6" thick		445	.090		5.60	4.60		10.20	13.50
1150	8" thick		415	.096		4.64	4.93		9.57	13
1200	10" thick		340	.118		6.35	6		12.35	16.60
1250	12" thick	D-9	290	.166		7.40	8.30		15.70	21.50
1500	Solid, 8" x 16" block, 6" thick	D-8	430	.093		3.95	4.76		8.71	11.95
1600	8" thick	"	405	.099		4.21	5.05		9.26	12.75
1650	12" thick	D-9	340	.141		4.20	7.10		11.30	15.95
9000	Minimum labor/equipment charge	D-1	2	8	Job		400		400	640

04 22 10 – Concrete Masonry Units

04 22 10.30 Concrete Block, Interlocking

		Crew	Daily Output	Labor-Hours	Unit	Material	2022 Bare Costs Labor	Equipment	Total	Total Incl O&P
0010	**CONCRETE BLOCK, INTERLOCKING**									
0100	Not including grout or reinforcing									
0200	8" x 16" units, 2,000 psi, 8" thick	D-1	245	.065	S.F.	2.95	3.27		6.22	8.50
0300	12" thick	"	220	.073		3.90	3.65		7.55	10.15
0400	Including grout & reinforcing, 8" thick	D-4	245	.131		8.85	6.60	.57	16.02	21
0450	12" thick		220	.145		10.10	7.35	.63	18.08	23.50
0500	16" thick		185	.173		12.55	8.75	.75	22.05	28.50
9000	Minimum labor/equipment charge	D-1	2	8	Job		400		400	640

04 22 10.32 Concrete Block, Lintels

		Crew	Daily Output	Labor-Hours	Unit	Material	2022 Bare Costs Labor	Equipment	Total	Total Incl O&P
0010	**CONCRETE BLOCK, LINTELS**, C90, normal weight									
0100	Including grout and horizontal reinforcing									
0200	8" x 8" x 8", 1 #4 bar	D-4	300	.107	L.F.	5.10	5.40	.46	10.96	14.65
0250	2 #4 bars		295	.108		5.40	5.50	.47	11.37	15.15
0400	8" x 16" x 8", 1 #4 bar		275	.116		4.39	5.90	.50	10.79	14.75
0450	2 #4 bars		270	.119		4.72	6	.51	11.23	15.30
1000	12" x 8" x 8", 1 #4 bar		275	.116		6.65	5.90	.50	13.05	17.25
1100	2 #4 bars		270	.119		7	6	.51	13.51	17.80
1150	2 #5 bars		270	.119		7.35	6	.51	13.86	18.20
1200	2 #6 bars		265	.121		7.80	6.10	.52	14.42	18.80
1500	12" x 16" x 8", 1 #4 bar		250	.128		7.75	6.45	.55	14.75	19.40
1600	2 #3 bars		245	.131		7.80	6.60	.57	14.97	19.65
1650	2 #4 bars		245	.131		8.05	6.60	.57	15.22	20
1700	2 #5 bars		240	.133		8.45	6.75	.58	15.78	20.50
9000	Minimum labor/equipment charge	D-1	2	8	Job		400		400	640

04 22 10.33 Lintel Block

		Crew	Daily Output	Labor-Hours	Unit	Material	2022 Bare Costs Labor	Equipment	Total	Total Incl O&P
0010	**LINTEL BLOCK**									
3481	Lintel block 6" x 8" x 8"	D-1	300	.053	Ea.	1.51	2.67		4.18	5.95
3501	6" x 16" x 8"		275	.058		2.11	2.92		5.03	7
3521	8" x 8" x 8"		275	.058		1.46	2.92		4.38	6.30
3561	8" x 16" x 8"		250	.064		1.62	3.21		4.83	6.95

04 22 10.34 Concrete Block, Partitions

		Crew	Daily Output	Labor-Hours	Unit	Material	2022 Bare Costs Labor	Equipment	Total	Total Incl O&P
0010	**CONCRETE BLOCK, PARTITIONS**, excludes scaffolding									
1000	Lightweight block, tooled joints, 2 sides, hollow									
1100	Not reinforced, 8" x 16" x 4" thick	D-8	440	.091	S.F.	1.94	4.65		6.59	9.60
1150	6" thick		410	.098		1.75	4.99		6.74	9.90
1200	8" thick		385	.104		3.15	5.30		8.45	11.95
1250	10" thick		370	.108		3.14	5.55		8.69	12.30
1300	12" thick	D-9	350	.137		3.80	6.85		10.65	15.20
1500	Reinforced alternate courses, 4" thick	D-8	435	.092		2.06	4.70		6.76	9.80
1600	6" thick		405	.099		1.92	5.05		6.97	10.20
1650	8" thick		380	.105		3.35	5.40		8.75	12.35
1700	10" thick		365	.110		3.36	5.60		8.96	12.70
1750	12" thick	D-9	345	.139		3.96	6.95		10.91	15.50
4000	Regular block, tooled joints, 2 sides, hollow									
4100	Not reinforced, 8" x 16" x 4" thick	D-8	430	.093	S.F.	1.79	4.76		6.55	9.55
4150	6" thick		400	.100		3.91	5.10		9.01	12.50
4200	8" thick		375	.107		2.50	5.45		7.95	11.50
4250	10" thick		360	.111		3.74	5.70		9.44	13.20
4300	12" thick	D-9	340	.141		4.38	7.10		11.48	16.15
4500	Reinforced alternate courses, 8" x 16" x 4" thick	D-8	425	.094		1.95	4.81		6.76	9.85
4550	6" thick		395	.101		4.07	5.20		9.27	12.80

For customer support on your Commercial Renovation Costs with RSMeans data, call 800.448.8182.

99

04 22 Concrete Unit Masonry

04 22 10 – Concrete Masonry Units

04 22 10.34 Concrete Block, Partitions

		Crew	Daily Output	Labor-Hours	Unit	Material	2022 Bare Costs Labor	Equipment	Total	Total Incl O&P
4600	8" thick	D-8	370	.108	S.F.	2.67	5.55		8.22	11.80
4650	10" thick	↓	355	.113		4.41	5.75		10.16	14.10
4700	12" thick	D-9	335	.143	↓	4.56	7.20		11.76	16.50
9000	Minimum labor/equipment charge	D-1	2	8	Job		400		400	640

04 22 10.38 Concrete Brick

		Crew	Daily Output	Labor-Hours	Unit	Material	2022 Bare Costs Labor	Equipment	Total	Total Incl O&P
0010	**CONCRETE BRICK**, C55, grade N, type 1									
0100	Regular, 4" x 2-1/4" x 8"	D-8	660	.061	Ea.	.71	3.10		3.81	5.75
0125	Rusticated, 4" x 2-1/4" x 8"		660	.061		.80	3.10		3.90	5.85
0150	Frog, 4" x 2-1/4" x 8"		660	.061		.77	3.10		3.87	5.80
0200	Double, 4" x 4-7/8" x 8"	↓	535	.075	↓	1.24	3.82		5.06	7.50

04 22 10.44 Glazed Concrete Block

		Crew	Daily Output	Labor-Hours	Unit	Material	2022 Bare Costs Labor	Equipment	Total	Total Incl O&P
0010	**GLAZED CONCRETE BLOCK** C744									
0100	Single face, 8" x 16" units, 2" thick	D-8	360	.111	S.F.	12.80	5.70		18.50	23
0200	4" thick		345	.116		13.15	5.95		19.10	24
0250	6" thick		330	.121		15.50	6.20		21.70	27
0300	8" thick		310	.129		17.20	6.60		23.80	29.50
0350	10" thick	↓	295	.136		17.65	6.95		24.60	30.50
0400	12" thick	D-9	280	.171		18.95	8.60		27.55	35
0700	Double face, 8" x 16" units, 4" thick	D-8	340	.118		17.80	6		23.80	29.50
0750	6" thick		320	.125		22.50	6.40		28.90	35.50
0800	8" thick		300	.133	↓	24	6.80		30.80	37.50
1000	Jambs, bullnose or square, single face, 8" x 16", 2" thick		315	.127	Ea.	21.50	6.50		28	34
1050	4" thick		285	.140	"	22	7.20		29.20	36
1200	Caps, bullnose or square, 8" x 16", 2" thick		420	.095	L.F.	21	4.87		25.07	31.50
1250	4" thick		380	.105	"	24.50	5.40		29.90	35.50
1256	Corner, bullnose or square, 2" thick		280	.143	Ea.	25	7.30		32.30	38.50
1258	4" thick		270	.148		28.50	7.60		36.10	43
1260	6" thick		260	.154		31.50	7.85		39.35	47
1270	8" thick		250	.160		41	8.20		49.20	58.50
1280	10" thick		240	.167		33.50	8.55		42.05	50.50
1290	12" thick		230	.174	↓	33	8.90		41.90	50.50
1500	Cove base, 8" x 16", 2" thick		315	.127	L.F.	11.25	6.50		17.75	23
1550	4" thick		285	.140		11.35	7.20		18.55	24
1600	6" thick		265	.151		12.10	7.70		19.80	25.50
1650	8" thick	↓	245	.163	↓	12.65	8.35		21	27.50
9000	Minimum labor/equipment charge	D-1	2	8	Job		400		400	640

04 23 Glass Unit Masonry

04 23 13 – Vertical Glass Unit Masonry

04 23 13.10 Glass Block

		Crew	Daily Output	Labor-Hours	Unit	Material	2022 Bare Costs Labor	Equipment	Total	Total Incl O&P
0010	**GLASS BLOCK**									
0100	Plain, 4" thick, under 1,000 S.F., 6" x 6"	D-8	115	.348	S.F.	26.50	17.80		44.30	58
0150	8" x 8"		160	.250		24.50	12.80		37.30	47
0160	end block		160	.250		52	12.80		64.80	77.50
0170	90 degree corner		160	.250		60.50	12.80		73.30	87
0180	45 degree corner		160	.250		56	12.80		68.80	82
0200	12" x 12"		175	.229		29	11.70		40.70	50.50
0210	4" x 8"		160	.250		26.50	12.80		39.30	50
0220	6" x 8"		160	.250		18.50	12.80		31.30	41
0300	1,000 to 5,000 S.F., 6" x 6"	↓	135	.296	↓	26	15.15		41.15	53

For customer support on your Commercial Renovation Costs with RSMeans data, call 800.448.8182.

04 23 Glass Unit Masonry

04 23 13 – Vertical Glass Unit Masonry

	04 23 13.10 Glass Block	Crew	Daily Output	Labor-Hours	Unit	Material	2022 Bare Costs Labor	Equipment	Total	Total Incl O&P
0350	8" x 8"	D-8	190	.211	S.F.	24	10.75		34.75	43.50
0400	12" x 12"		215	.186		28	9.50		37.50	46.50
0410	4" x 8"		215	.186		26	9.50		35.50	44
0420	6" x 8"		215	.186		18.10	9.50		27.60	35
0500	Over 5,000 S.F., 6" x 6"		145	.276		25.50	14.10		39.60	50.50
0700	For solar reflective blocks, add					100%				
1000	Thinline, plain, 3-1/8" thick, under 1,000 S.F., 6" x 6"	D-8	115	.348	S.F.	23.50	17.80		41.30	54.50
1050	8" x 8"		160	.250		13.65	12.80		26.45	35.50
1250	8" x 8"		215	.186		20.50	9.50		30	38
1400	For cleaning block after installation (both sides), add		1000	.040		.21	2.05		2.26	3.51
9000	Minimum labor/equipment charge	D-1	2	8	Job		400		400	640

04 27 Multiple-Wythe Unit Masonry

04 27 10 – Multiple-Wythe Masonry

04 27 10.10 Cornices

		Crew	Daily Output	Labor-Hours	Unit	Material	Labor	Equipment	Total	Total Incl O&P
0010	**CORNICES**									
0110	Face bricks, 12 brick/S.F.	D-1	30	.533	SF Face	5.30	26.50		31.80	49
0150	15 brick/S.F.		23	.696	"	6.35	35		41.35	63
9000	Minimum labor/equipment charge		1.50	10.667	Job		535		535	855

04 27 10.30 Brick Walls

		Crew	Daily Output	Labor-Hours	Unit	Material	Labor	Equipment	Total	Total Incl O&P
0010	**BRICK WALLS**, including mortar, excludes scaffolding									
0020	Estimating by number of brick									
0140	Face brick, 4" thick wall, 6.75 brick/S.F.	D-8	1.45	27.586	M	745	1,400		2,145	3,075
0150	Common brick, 4" thick wall, 6.75 brick/S.F.		1.60	25		1,100	1,275		2,375	3,250
0204	8" thick, 13.50 brick/S.F.		1.80	22.222		1,100	1,125		2,225	3,050
0250	12" thick, 20.25 brick/S.F.		1.90	21.053		1,125	1,075		2,200	2,950
0304	16" thick, 27.00 brick/S.F.		2	20		1,125	1,025		2,150	2,875
0500	Reinforced, face brick, 4" thick wall, 6.75 brick/S.F.		1.40	28.571		765	1,450		2,215	3,200
0520	Common brick, 4" thick wall, 6.75 brick/S.F.		1.55	25.806		1,125	1,325		2,450	3,350
0550	8" thick, 13.50 brick/S.F.		1.75	22.857		1,125	1,175		2,300	3,125
0600	12" thick, 20.25 brick/S.F.		1.85	21.622		1,150	1,100		2,250	3,025
0650	16" thick, 27.00 brick/S.F.		1.95	20.513		1,150	1,050		2,200	2,925
0790	Alternate method of figuring by square foot									
0800	Face brick, 4" thick wall, 6.75 brick/S.F.	D-8	215	.186	S.F.	5	9.50		14.50	21
0850	Common brick, 4" thick wall, 6.75 brick/S.F.	R042110-50	240	.167		7.35	8.55		15.90	22
0900	8" thick, 13.50 brick/S.F.		135	.296		15	15.15		30.15	41
1000	12" thick, 20.25 brick/S.F.		95	.421		22.50	21.50		44	59.50
1050	16" thick, 27.00 brick/S.F.		75	.533		30.50	27.50		58	77
1200	Reinforced, face brick, 4" thick wall, 6.75 brick/S.F.		210	.190		5.20	9.75		14.95	21.50
1220	Common brick, 4" thick wall, 6.75 brick/S.F.		235	.170		7.50	8.70		16.20	22
1250	8" thick, 13.50 brick/S.F.		130	.308		15.30	15.75		31.05	42
1260	8" thick, 2.25 brick/S.F.		130	.308		1.85	15.75		17.60	27
1300	12" thick, 20.25 brick/S.F.		90	.444		23	22.50		45.50	62
1350	16" thick, 27.00 brick/S.F.		70	.571		31	29		60	81
9000	Minimum labor/equipment charge	D-1	2	8	Job		400		400	640

04 27 10.40 Steps

		Crew	Daily Output	Labor-Hours	Unit	Material	Labor	Equipment	Total	Total Incl O&P
0010	**STEPS**									
0012	Entry steps, select common brick	D-1	.30	53.333	M	1,025	2,675		3,700	5,400

For customer support on your Commercial Renovation Costs with RSMeans data, call 800.448.8182.

101

04 41 Dry-Placed Stone

04 41 10 – Dry Placed Stone

04 41 10.10 Rough Stone Wall

		Crew	Daily Output	Labor-Hours	Unit	Material	2022 Bare Costs Labor	2022 Bare Costs Equipment	Total	Total Incl O&P
0011	**ROUGH STONE WALL**, Dry									
0012	Dry laid (no mortar), under 18" thick	G D-1	60	.267	C.F.	14.15	13.35		27.50	37
0100	Random fieldstone, under 18" thick	G D-12	60	.533		14.15	27		41.15	59
0150	Over 18" thick	G "	63	.508	↓	17	25.50		42.50	59.50
0500	Field stone veneer	G D-8	120	.333	S.F.	13.10	17.05		30.15	42
0510	Valley stone veneer	G	120	.333		13.10	17.05		30.15	42
0520	River stone veneer	G ↓	120	.333	↓	13.10	17.05		30.15	42
0600	Rubble stone walls, in mortar bed, up to 18" thick	G D-11	75	.320	C.F.	17.15	16.80		33.95	46
9000	Minimum labor/equipment charge	D-1	2	8	Job		400		400	640

04 43 Stone Masonry

04 43 10 – Masonry with Natural and Processed Stone

04 43 10.05 Ashlar Veneer

		Crew	Daily Output	Labor-Hours	Unit	Material	2022 Bare Costs Labor	2022 Bare Costs Equipment	Total	Total Incl O&P
0011	**ASHLAR VENEER** +/- 4" thk, random or random rectangular									
0150	Sawn face, split joints, low priced stone	D-8	140	.286	S.F.	12.50	14.60		27.10	37.50
0200	Medium priced stone		130	.308		15	15.75		30.75	41.50
0300	High priced stone		120	.333		18.10	17.05		35.15	47.50
0600	Seam face, split joints, medium price stone		125	.320		17.85	16.35		34.20	45.50
0700	High price stone		120	.333		17.30	17.05		34.35	46.50
1000	Split or rock face, split joints, medium price stone		125	.320		10.10	16.35		26.45	37
1100	High price stone	↓	120	.333	↓	17.65	17.05		34.70	47

04 43 10.10 Bluestone

		Crew	Daily Output	Labor-Hours	Unit	Material	2022 Bare Costs Labor	2022 Bare Costs Equipment	Total	Total Incl O&P
0010	**BLUESTONE**, cut to size									
0500	Sills, natural cleft, 10" wide to 6' long, 1-1/2" thick	D-11	70	.343	L.F.	13	18		31	43.50
0550	2" thick		63	.381		14.25	20		34.25	47.50
0600	Smooth finish, 1-1/2" thick		70	.343		13.95	18		31.95	44.50
0650	2" thick		63	.381		16.50	20		36.50	50
0800	Thermal finish, 1-1/2" thick		70	.343		12.30	18		30.30	42.50
0850	2" thick	↓	63	.381		10.55	20		30.55	43.50
1000	Stair treads, natural cleft, 12" wide, 6' long, 1-1/2" thick	D-10	115	.278		15.05	15.35	4.97	35.37	46.50
1050	2" thick		105	.305		16.40	16.85	5.45	38.70	51
1100	Smooth finish, 1-1/2" thick		115	.278		14.10	15.35	4.97	34.42	45.50
1150	2" thick		105	.305		14.90	16.85	5.45	37.20	49.50
1300	Thermal finish, 1-1/2" thick		115	.278		14.50	15.35	4.97	34.82	46
1350	2" thick	↓	105	.305	↓	13.80	16.85	5.45	36.10	48
9000	Minimum labor/equipment charge	D-1	2.50	6.400	Job		320		320	515

04 43 10.45 Granite

		Crew	Daily Output	Labor-Hours	Unit	Material	2022 Bare Costs Labor	2022 Bare Costs Equipment	Total	Total Incl O&P
0010	**GRANITE**, cut to size									
0050	Veneer, polished face, 3/4" to 1-1/2" thick									
0150	Low price, gray, light gray, etc.	D-10	130	.246	S.F.	24	13.60	4.40	42	53
0180	Medium price, pink, brown, etc.		130	.246		27.50	13.60	4.40	45.50	57
0220	High price, red, black, etc.	↓	130	.246	↓	41.50	13.60	4.40	59.50	72
0300	1-1/2" to 2-1/2" thick, veneer									
0350	Low price, gray, light gray, etc.	D-10	130	.246	S.F.	28	13.60	4.40	46	57.50
0500	Medium price, pink, brown, etc.	"	130	.246	"	32.50	13.60	4.40	50.50	62.50
2500	Steps, copings, etc., finished on more than one surface									
2550	Low price, gray, light gray, etc.	D-10	50	.640	C.F.	89	35.50	11.45	135.95	166
2575	Medium price, pink, brown, etc.		50	.640		115	35.50	11.45	161.95	196
2600	High price, red, black, etc.	↓	50	.640		142	35.50	11.45	188.95	225
9000	Minimum labor/equipment charge	D-1	2	8	Job		400		400	640

04 43 Stone Masonry

04 43 10 – Masonry with Natural and Processed Stone

04 43 10.55 Limestone		Crew	Daily Output	Labor-Hours	Unit	Material	2022 Bare Costs Labor	Equipment	Total	Total Incl O&P
0010	**LIMESTONE**, cut to size									
0020	Veneer facing panels									
0500	Texture finish, light stick, 4-1/2" thick, 5' x 12'	D-4	300	.107	S.F.	25.50	5.40	.46	31.36	37
0750	5" thick, 5' x 14' panels	D-10	275	.116		27	6.45	2.08	35.53	42
1000	Sugarcube finish, 2" thick, 3' x 5' panels		275	.116		30	6.45	2.08	38.53	45.50
1050	3" thick, 4' x 9' panels		275	.116		31	6.45	2.08	39.53	46.50
1200	4" thick, 5' x 11' panels		275	.116		33.50	6.45	2.08	42.03	49.50
1400	Sugarcube, textured finish, 4-1/2" thick, 5' x 12'		275	.116		37	6.45	2.08	45.53	53
1450	5" thick, 5' x 14' panels		275	.116		37.50	6.45	2.08	46.03	54
2000	Coping, sugarcube finish, top & 2 sides		30	1.067	C.F.	64.50	59	19.05	142.55	185
2100	Sills, lintels, jambs, trim, stops, sugarcube finish, simple		20	1.600		64.50	88.50	28.50	181.50	242
2150	Detailed		20	1.600		64.50	88.50	28.50	181.50	242
2300	Steps, extra hard, 14" wide, 6" rise		50	.640	L.F.	28	35.50	11.45	74.95	99
3000	Quoins, plain finish, 6" x 12" x 12"	D-12	25	1.280	Ea.	53	65		118	162
3050	6" x 16" x 24"	"	25	1.280	"	70.50	65		135.50	182
9000	Minimum labor/equipment charge	D-1	2	8	Job		400		400	640

04 43 10.60 Marble

04 43 10.60 Marble		Crew	Daily Output	Labor-Hours	Unit	Material	2022 Bare Costs Labor	Equipment	Total	Total Incl O&P
0011	**MARBLE**, ashlar, split face, +/- 4" thick, random									
0040	Lengths 1' to 4' & heights 2" to 7-1/2", average	D-8	175	.229	S.F.	16.25	11.70		27.95	36.50
0100	Base, polished, 3/4" or 7/8" thick, polished, 6" high	D-10	65	.492	L.F.	10.85	27	8.80	46.65	64.50
0300	Carvings or bas-relief, from templates, simple design		80	.400	S.F.	151	22	7.15	180.15	209
0350	Intricate design		80	.400	"	335	22	7.15	364.15	415
1000	Facing, polished finish, cut to size, 3/4" to 7/8" thick									
1050	Carrara or equal	D-10	130	.246	S.F.	34	13.60	4.40	52	64
1100	Arabescato or equal		130	.246		35.50	13.60	4.40	53.50	66
1300	1-1/4" thick, Botticino Classico or equal		125	.256		22	14.15	4.57	40.72	51.50
1350	Statuarietto or equal		125	.256		36.50	14.15	4.57	55.22	67.50
1500	2" thick, Crema Marfil or equal		120	.267		46.50	14.75	4.76	66.01	80
1550	Cafe Pinta or equal		120	.267		64	14.75	4.76	83.51	99.50
1700	Rubbed finish, cut to size, 4" thick									
1740	Average	D-10	100	.320	S.F.	40	17.65	5.70	63.35	78.50
1780	Maximum	"	100	.320	"	75	17.65	5.70	98.35	117
2200	Window sills, 6" x 3/4" thick	D-1	85	.188	L.F.	12.65	9.45		22.10	29
2500	Flooring, polished tiles, 12" x 12" x 3/8" thick									
2510	Thin set, Giallo Solare or equal	D-11	90	.267	S.F.	14.25	14		28.25	38
2600	Sky Blue or equal		90	.267		10.95	14		24.95	34.50
2700	Mortar bed, Giallo Solare or equal		65	.369		14.35	19.40		33.75	47
2740	Sky Blue or equal		65	.369		10.95	19.40		30.35	43
2780	Travertine, 3/8" thick, Sierra or equal	D-10	130	.246		11.10	13.60	4.40	29.10	38.50
2790	Silver or equal	"	130	.246		29.50	13.60	4.40	47.50	59
2800	Patio tile, non-slip, 1/2" thick, flame finish	D-11	75	.320		12.95	16.80		29.75	41.50
2900	Shower or toilet partitions, 7/8" thick partitions									
3050	3/4" or 1-1/4" thick stiles, polished 2 sides, average	D-11	75	.320	S.F.	46.50	16.80		63.30	78.50
3201	Soffits, add to above prices				"	20%	100%			
3210	Stairs, risers, 7/8" thick x 6" high	D-10	115	.278	L.F.	14.40	15.35	4.97	34.72	46
3360	Treads, 12" wide x 1-1/4" thick	"	115	.278	"	44.50	15.35	4.97	64.82	79
3500	Thresholds, 3' long, 7/8" thick, 4" to 5" wide, plain	D-12	24	1.333	Ea.	33.50	67.50		101	145
3550	Beveled		24	1.333	"	73.50	67.50		141	189
3700	Window stools, polished, 7/8" thick, 5" wide		85	.376	L.F.	18.95	19.05		38	51.50
9000	Minimum labor/equipment charge	D-1	2	8	Job		400		400	640

For customer support on your Commercial Renovation Costs with RSMeans data, call 800.448.8182.

103

04 43 10 – Masonry with Natural and Processed Stone

04 43 10.75 Sandstone or Brownstone	Crew	Daily Output	Labor-Hours	Unit	Material	2022 Bare Costs Labor	Equipment	Total	Total Incl O&P
0011 **SANDSTONE OR BROWNSTONE**									
0100 Sawed face veneer, 2-1/2" thick, to 2' x 4' panels	D-10	130	.246	S.F.	21	13.60	4.40	39	49.50
0150 4" thick, to 3'-6" x 8' panels		100	.320		21	17.65	5.70	44.35	57.50
0300 Split face, random sizes		100	.320		14.50	17.65	5.70	37.85	50.50
9000 Minimum labor/equipment charge	D-1	2.50	6.400	Job		320		320	515

04 43 10.80 Slate

	Crew	Daily Output	Labor-Hours	Unit	Material	2022 Bare Costs Labor	Equipment	Total	Total Incl O&P
0010 **SLATE**									
0040 Pennsylvania - blue gray to black									
0050 Vermont - unfading green, mottled green & purple, gray & purple									
0100 Virginia - blue black									
3100 Stair landings, 1" thick, black, clear	D-1	65	.246	S.F.	20	12.35		32.35	42
3200 Ribbon	"	65	.246	"	22	12.35		34.35	44.50
3500 Stair treads, sand finish, 1" thick x 12" wide									
3550 Under 3 L.F.	D-10	85	.376	L.F.	22	21	6.70	49.70	65
3600 3 L.F. to 6 L.F.	"	120	.267	"	24	14.75	4.76	43.51	55.50
3700 Ribbon, sand finish, 1" thick x 12" wide									
3750 To 6 L.F.	D-10	120	.267	L.F.	20	14.75	4.76	39.51	51
4000 Stools or sills, sand finish, 1" thick, 6" wide	D-12	160	.200		11.55	10.15		21.70	29
4100 Honed finish		160	.200		11.05	10.15		21.20	28.50
4200 10" wide		90	.356		17.85	18		35.85	48.50
4250 Honed finish		90	.356		16.60	18		34.60	47.50
4400 2" thick, 6" wide		140	.229		18.60	11.55		30.15	39
4450 Honed finish		140	.229		17.75	11.55		29.30	38
4600 10" wide		90	.356		29	18		47	61
4650 Honed finish		90	.356		27.50	18		45.50	59.50
4800 For lengths over 3', add					25%				
9000 Minimum labor/equipment charge	D-1	2.50	6.400	Job		320		320	515

04 43 10.85 Window Sill

	Crew	Daily Output	Labor-Hours	Unit	Material	2022 Bare Costs Labor	Equipment	Total	Total Incl O&P
0010 **WINDOW SILL**									
0020 Bluestone, thermal top, 10" wide, 1-1/2" thick	D-1	85	.188	S.F.	11.70	9.45		21.15	28
0050 2" thick		75	.213	"	18.10	10.70		28.80	37
0100 Cut stone, 5" x 8" plain		48	.333	L.F.	12.35	16.70		29.05	40.50
0200 Face brick on edge, brick, 8" wide		80	.200		3.57	10.05		13.62	20
0400 Marble, 9" wide, 1" thick		85	.188		8.80	9.45		18.25	25
0900 Slate, colored, unfading, honed, 12" wide, 1" thick		85	.188		9	9.45		18.45	25
0950 2" thick		70	.229		10.90	11.45		22.35	30.50
9000 Minimum labor/equipment charge	1 Bric	2	4	Job		221		221	355

04 54 Refractory Brick Masonry

04 54 10 – Refractory Brick Work

04 54 10.10 Fire Brick

	Crew	Daily Output	Labor-Hours	Unit	Material	2022 Bare Costs Labor	Equipment	Total	Total Incl O&P
0010 **FIRE BRICK**									
0012 Low duty, 2000°F, 9" x 2-1/2" x 4-1/2"	D-1	.60	26.667	M	2,525	1,325		3,850	4,925
0050 High duty, 3000°F	"	.60	26.667	"	2,975	1,325		4,300	5,425

04 54 10.20 Fire Clay

	Crew	Daily Output	Labor-Hours	Unit	Material	2022 Bare Costs Labor	Equipment	Total	Total Incl O&P
0010 **FIRE CLAY**									
0020 Gray, high duty, 100 lb. bag				Bag	28.50			28.50	31.50
0050 100 lb. drum, premixed (400 brick per drum)				Drum	44.50			44.50	49

04 57 Masonry Fireplaces

04 57 10 – Brick or Stone Fireplaces

04 57 10.10 Fireplace

		Crew	Daily Output	Labor-Hours	Unit	Material	2022 Bare Costs Labor	Equipment	Total	Total Incl O&P
0010	**FIREPLACE**									
0100	Brick fireplace, not incl. foundations or chimneys									
0110	30" x 29" opening, incl. chamber, plain brickwork	D-1	.40	40	Ea.	605	2,000		2,605	3,875
0200	Fireplace box only (110 brick)	"	2	8	"	156	400		556	810
0300	For elaborate brickwork and details, add					35%	35%			
0400	For hearth, brick & stone, add	D-1	2	8	Ea.	206	400		606	865
0410	For steel, damper, cleanouts, add		4	4		17.40	201		218.40	340
0600	Plain brickwork, incl. metal circulator		.50	32		1,025	1,600		2,625	3,725
0800	Face brick only, standard size, 8" x 2-2/3" x 4"		.30	53.333	M	590	2,675		3,265	4,925
0900	Stone fireplace, fieldstone, add				SF Face	8.75			8.75	9.65
1000	Cut stone, add				"	9.55			9.55	10.50
9000	Minimum labor/equipment charge	D-1	2	8	Job		400		400	640

04 71 Manufactured Brick Masonry

04 71 10 – Simulated or Manufactured Brick

04 71 10.10 Simulated Brick

		Crew	Daily Output	Labor-Hours	Unit	Material	2022 Bare Costs Labor	Equipment	Total	Total Incl O&P
0010	**SIMULATED BRICK**									
0020	Aluminum, baked on colors	1 Carp	200	.040	S.F.	4.57	2.25		6.82	8.65
0050	Fiberglass panels		200	.040		10	2.25		12.25	14.60
0100	Urethane pieces cemented in mastic		150	.053		8.35	3		11.35	13.95
0150	Vinyl siding panels		200	.040		11.20	2.25		13.45	15.90

04 72 Cast Stone Masonry

04 72 10 – Cast Stone Masonry Features

04 72 10.10 Coping

		Crew	Daily Output	Labor-Hours	Unit	Material	2022 Bare Costs Labor	Equipment	Total	Total Incl O&P
0010	**COPING**, stock units									
0050	Precast concrete, 10" wide, 4" tapers to 3-1/2", 8" wall	D-1	75	.213	L.F.	7.20	10.70		17.90	25
0100	12" wide, 3-1/2" tapers to 3", 10" wall		70	.229		7.80	11.45		19.25	27
0110	14" wide, 4" tapers to 3-1/2", 12" wall		65	.246		10.50	12.35		22.85	31.50
0150	16" wide, 4" tapers to 3-1/2", 14" wall		60	.267		11.35	13.35		24.70	34
0300	Limestone for 12" wall, 4" thick		90	.178		16.30	8.90		25.20	32.50
0350	6" thick		80	.200		23	10.05		33.05	41
0500	Marble, to 4" thick, no wash, 9" wide		90	.178		10.05	8.90		18.95	25.50
0550	12" wide		80	.200		17.85	10.05		27.90	35.50
0700	Terra cotta, 9" wide		90	.178		7.85	8.90		16.75	23
0750	12" wide		80	.200		7.90	10.05		17.95	25
0800	Aluminum, for 12" wall		80	.200		2.31	10.05		12.36	18.60
9000	Minimum labor/equipment charge		2	8	Job		400		400	640

04 72 20 – Cultured Stone Veneer

04 72 20.10 Cultured Stone Veneer Components

		Crew	Daily Output	Labor-Hours	Unit	Material	2022 Bare Costs Labor	Equipment	Total	Total Incl O&P
0010	**CULTURED STONE VENEER COMPONENTS**									
0110	On wood frame and sheathing substrate, random sized cobbles, corner stones	D-8	70	.571	V.L.F.	13.75	29		42.75	62
0120	Field stones		140	.286	S.F.	9.15	14.60		23.75	33.50
0130	Random sized flats, corner stones		70	.571	V.L.F.	8.95	29		37.95	57
0140	Field stones		140	.286	S.F.	9.50	14.60		24.10	34
0150	Horizontal lined ledgestones, corner stones		75	.533	V.L.F.	13.75	27.50		41.25	58.50
0160	Field stones		150	.267	S.F.	10.35	13.65		24	33.50
0170	Random shaped flats, corner stones		65	.615	V.L.F.	13.75	31.50		45.25	65.50
0180	Field stones		150	.267	S.F.	10.35	13.65		24	33.50

For customer support on your Commercial Renovation Costs with RSMeans data, call 800.448.8182.

105

04 72 Cast Stone Masonry

04 72 20 - Cultured Stone Veneer

04 72 20.10 Cultured Stone Veneer Components	Crew	Daily Output	Labor-Hours	Unit	Material	2022 Bare Costs Labor	Equipment	Total	Total Incl O&P
0190	Random shaped/textured face, corner stones	D-8	65	.615	V.L.F.	13.75	31.50		45.25	65.50
0200	Field stones		130	.308	S.F.	10.35	15.75		26.10	36.50
0210	Random shaped river rock, corner stones		65	.615	V.L.F.	13.75	31.50		45.25	65.50
0220	Field stones		130	.308	S.F.	10.35	15.75		26.10	36.50
0240	On concrete or CMU substrate, random sized cobbles, corner stones		70	.571	V.L.F.	12.65	29		41.65	61
0250	Field stones		140	.286	S.F.	8.55	14.60		23.15	33
0260	Random sized flats, corner stones		70	.571	V.L.F.	7.85	29		36.85	55.50
0270	Field stones		140	.286	S.F.	8.95	14.60		23.55	33.50
0280	Horizontal lined ledgestones, corner stones		75	.533	V.L.F.	12.65	27.50		40.15	57.50
0290	Field stones		150	.267	S.F.	9.80	13.65		23.45	33
0300	Random shaped flats, corner stones		70	.571	V.L.F.	12.65	29		41.65	61
0310	Field stones		140	.286	S.F.	9.80	14.60		24.40	34.50
0320	Random shaped/textured face, corner stones		65	.615	V.L.F.	12.65	31.50		44.15	64.50
0330	Field stones		130	.308	S.F.	9.80	15.75		25.55	36
0340	Random shaped river rock, corner stones		65	.615	V.L.F.	12.65	31.50		44.15	64.50
0350	Field stones	▼	130	.308	S.F.	9.80	15.75		25.55	36
0360	Cultured stone veneer, #15 felt weather resistant barrier	1 Clab	3700	.002	Sq.	4.43	.10		4.53	5.05
0370	Expanded metal lath, diamond, 2.5 lb./S.Y., galvanized	1 Lath	85	.094	S.Y.	5.05	5.25		10.30	13.75
0390	Water table or window sill, 18" long	1 Bric	80	.100	Ea.	15.30	5.55		20.85	25.50

04 73 Manufactured Stone Masonry

04 73 20 - Simulated or Manufactured Stone

04 73 20.10 Simulated Stone

		Crew	Daily Output	Labor-Hours	Unit	Material	Labor	Equipment	Total	Total Incl O&P
0010	**SIMULATED STONE**									
0100	Insulated fiberglass panels, 5/8" ply backer	L-4	200	.120	S.F.	10.10	6.45		16.55	21.50

Estimating Tips
05 05 00 Common Work Results for Metals

- Nuts, bolts, washers, connection angles, and plates can add a significant amount to both the tonnage of a structural steel job and the estimated cost. As a rule of thumb, add 10% to the total weight to account for these accessories.

- Type 2 steel construction, commonly referred to as "simple construction," consists generally of field-bolted connections with lateral bracing supplied by other elements of the building, such as masonry walls or x-bracing. The estimator should be aware, however, that shop connections may be accomplished by welding or bolting. The method may be particular to the fabrication shop and may have an impact on the estimated cost.

05 10 00 Structural Steel

- Steel items can be obtained from two sources: a fabrication shop or a metals service center. Fabrication shops can fabricate items under more controlled conditions than crews in the field can. They are also more efficient and can produce items more economically. Metal service centers serve as a source of long mill shapes to both fabrication shops and contractors.

- Most line items in this structural steel subdivision, and most items in 05 50 00 Metal Fabrications, are indicated as being shop fabricated. The bare material cost for these shop fabricated items is the "Invoice Cost" from the shop and includes the mill base price of steel plus mill extras, transportation to the shop, shop drawings and detailing where warranted, shop fabrication and handling, sandblasting and a shop coat of primer paint, all necessary structural bolts, and delivery to the job site. The bare labor cost and bare equipment cost for these shop fabricated items are for field installation or erection.

- Line items in Subdivision 05 12 23.40 Lightweight Framing, and other items scattered in Division 5, are indicated as being field fabricated. The bare material cost for these field fabricated items is the "Invoice Cost" from the metals service center and includes the mill base price of steel plus mill extras, transportation to the metals service center, material handling, and delivery of long lengths of mill shapes to the job site. Material costs for structural bolts and welding rods should be added to the estimate. The bare labor cost and bare equipment cost for these items are for both field fabrication and field installation or erection, and include time for cutting, welding, and drilling in the fabricated metal items. Drilling into concrete and fasteners to fasten field fabricated items to other work is not included and should be added to the estimate.

05 20 00 Steel Joist Framing

- In any given project the total weight of open web steel joists is determined by the loads to be supported and the design. However, economies can be realized in minimizing the amount of labor used to place the joists. This is done by maximizing the joist spacing and therefore minimizing the number of joists required to be installed on the job. Certain spacings and locations may be required by the design, but in other cases maximizing the spacing and keeping it as uniform as possible will keep the costs down.

05 30 00 Steel Decking

- The takeoff and estimating of a metal deck involve more than the area of the floor or roof and the type of deck specified or shown on the drawings. Many different sizes and types of openings may exist. Small openings for individual pipes or conduits may be drilled after the floor/roof is installed, but larger openings may require special deck lengths as well as reinforcing or structural support. The estimator should determine who will be supplying this reinforcing. Additionally, some deck terminations are part of the deck package, such as screed angles and pour stops, and others will be part of the steel contract, such as angles attached to structural members and cast-in-place angles and plates. The estimator must ensure that all pieces are accounted for in the complete estimate.

05 50 00 Metal Fabrications

- The most economical steel stairs are those that use common materials, standard details, and most importantly, a uniform and relatively simple method of field assembly. Commonly available A36/A992 channels and plates are very good choices for the main stringers of the stairs, as are angles and tees for the carrier members. Risers and treads are usually made by specialty shops, and it is most economical to use a typical detail in as many places as possible. The stairs should be pre-assembled and shipped directly to the site. The field connections should be simple and straightforward enough to be accomplished efficiently, and with minimum equipment and labor.

Reference Numbers
Reference numbers are shown at the beginning of some major classifications. These numbers refer to related items in the Reference Section. The reference information may be an estimating procedure, an alternate pricing method, or technical information.

Note: Not all subdivisions listed here necessarily appear. ∎

Same Data. Simplified.

Enjoy the convenience and efficiency of accessing your costs anywhere:

- **Skip the multiplier** by setting your location
- **Quickly search,** edit, favorite and share costs
- **Stay on top of price changes** with automatic updates

Discover more at rsmeans.com/online

05 05 05 – Selective Demolition for Metals

05 05 05.10 Selective Demolition, Metals		Crew	Daily Output	Labor-Hours	Unit	Material	2022 Bare Costs Labor	Equipment	Total	Total Incl O&P
0010	**SELECTIVE DEMOLITION, METALS**	R024119-10								
0015	Excludes shores, bracing, cutting, loading, hauling, dumping									
0020	Remove nuts only up to 3/4" diameter	1 Sswk	480	.017	Ea.		1.03		1.03	1.66
0030	7/8" to 1-1/4" diameter		240	.033			2.06		2.06	3.32
0040	1-3/8" to 2" diameter		160	.050			3.09		3.09	4.98
0060	Unbolt and remove structural bolts up to 3/4" diameter		240	.033			2.06		2.06	3.32
0070	7/8" to 2" diameter		160	.050			3.09		3.09	4.98
0140	Light weight framing members, remove whole or cut up, up to 20 lb.		240	.033			2.06		2.06	3.32
0150	21-40 lb.	2 Sswk	210	.076			4.71		4.71	7.60
0160	41-80 lb.	3 Sswk	180	.133			8.25		8.25	13.30
0170	81-120 lb.	4 Sswk	150	.213			13.20		13.20	21.50
0230	Structural members, remove whole or cut up, up to 500 lb.	E-19	48	.500			30.50	17	47.50	67
0240	1/4-2 tons	E-18	36	1.111			69	22.50	91.50	135
0250	2-5 tons	E-24	30	1.067			65.50	70	135.50	182
0260	5-10 tons	E-20	24	2.667			163	91	254	360
0270	10-15 tons	E-2	18	3.111			191	138	329	455
0340	Fabricated item, remove whole or cut up, up to 20 lb.	1 Sswk	96	.083			5.15		5.15	8.30
0350	21-40 lb.	2 Sswk	84	.190			11.80		11.80	18.95
0360	41-80 lb.	3 Sswk	72	.333			20.50		20.50	33
0370	81-120 lb.	4 Sswk	60	.533			33		33	53
0380	121-500 lb.	E-19	48	.500			30.50	17	47.50	67
0390	501-1000 lb.	"	36	.667			40.50	22.50	63	89.50
0500	Steel roof decking, uncovered, bare	B-2	5000	.008	S.F.		.37		.37	.58
2950	Minimum labor/equipment charge	E-19	2	12	Job		730	410	1,140	1,625

05 05 19 – Post-Installed Concrete Anchors

05 05 19.10 Chemical Anchors

		Crew	Daily Output	Labor-Hours	Unit	Material	Labor	Equipment	Total	Total Incl O&P
0010	**CHEMICAL ANCHORS**									
0020	Includes layout & drilling									
1430	Chemical anchor, w/rod & epoxy cartridge, 3/4" diameter x 9-1/2" long	B-89A	27	.593	Ea.	7.70	31	4.41	43.11	62.50
1435	1" diameter x 11-3/4" long		24	.667		15.10	35	4.97	55.07	77
1440	1-1/4" diameter x 14" long		21	.762		29.50	40	5.70	75.20	102
1445	1-3/4" diameter x 15" long		20	.800		70	42	5.95	117.95	150
1450	18" long		17	.941		84	49	7	140	178
1455	2" diameter x 18" long		16	1		123	52.50	7.45	182.95	227
1460	24" long		15	1.067		162	55.50	7.95	225.45	275

05 05 19.20 Expansion Anchors

			Crew	Daily Output	Labor-Hours	Unit	Material	Labor	Equipment	Total	Total Incl O&P
0010	**EXPANSION ANCHORS**										
0100	Anchors for concrete, brick or stone, no layout and drilling										
0200	Expansion shields, zinc, 1/4" diameter, 1-5/16" long, single	G	1 Carp	90	.089	Ea.	.35	5		5.35	8.35
0300	1-3/8" long, double	G		85	.094		.67	5.30		5.97	9.15
0400	3/8" diameter, 1-1/2" long, single	G		85	.094		.70	5.30		6	9.15
0500	2" long, double	G		80	.100		1.47	5.65		7.12	10.55
0600	1/2" diameter, 2-1/16" long, single	G		80	.100		1.22	5.65		6.87	10.30
0700	2-1/2" long, double	G		75	.107		2.51	6		8.51	12.30
0800	5/8" diameter, 2-5/8" long, single	G		75	.107		2.77	6		8.77	12.60
0900	2-3/4" long, double	G		70	.114		3.54	6.45		9.99	14.10
1000	3/4" diameter, 2-3/4" long, single	G		70	.114		4.39	6.45		10.84	15.05
1100	3-15/16" long, double	G		65	.123		6.45	6.95		13.40	18.05
2100	Hollow wall anchors for gypsum wall board, plaster or tile										
2300	1/8" diameter, short	G	1 Carp	160	.050	Ea.	.27	2.82		3.09	4.77
2400	Long	G		150	.053		.31	3		3.31	5.10
2500	3/16" diameter, short	G		150	.053		.52	3		3.52	5.35

05 05 19 – Post-Installed Concrete Anchors

05 05 19.20 Expansion Anchors

		Crew	Daily Output	Labor-Hours	Unit	Material	2022 Bare Costs Labor	2022 Bare Costs Equipment	Total	Total Incl O&P
2600	Long	G 1 Carp	140	.057	Ea.	.60	3.22		3.82	5.75
2700	1/4" diameter, short	G	140	.057		.65	3.22		3.87	5.80
2800	Long	G	130	.062		.79	3.46		4.25	6.35
3000	Toggle bolts, bright steel, 1/8" diameter, 2" long	G	85	.094		.25	5.30		5.55	8.70
3100	4" long	G	80	.100		.29	5.65		5.94	9.25
3200	3/16" diameter, 3" long	G	80	.100		.33	5.65		5.98	9.30
3300	6" long	G	75	.107		.46	6		6.46	10.05
3400	1/4" diameter, 3" long	G	75	.107		.38	6		6.38	9.95
3500	6" long	G	70	.114		.57	6.45		7.02	10.85
3600	3/8" diameter, 3" long	G	70	.114		1.01	6.45		7.46	11.30
3700	6" long	G	60	.133		1.56	7.50		9.06	13.65
3800	1/2" diameter, 4" long	G	60	.133		2.26	7.50		9.76	14.45
3900	6" long	G	50	.160		2.57	9		11.57	17.15
4000	Nailing anchors									
4100	Nylon nailing anchor, 1/4" diameter, 1" long	1 Carp	3.20	2.500	C	17.70	141		158.70	244
4200	1-1/2" long		2.80	2.857		20	161		181	278
4300	2" long		2.40	3.333		25	188		213	325
4400	Metal nailing anchor, 1/4" diameter, 1" long	G	3.20	2.500		15.80	141		156.80	241
4500	1-1/2" long	G	2.80	2.857		20	161		181	278
4600	2" long	G	2.40	3.333		24	188		212	325
8000	Wedge anchors, not including layout or drilling									
8050	Carbon steel, 1/4" diameter, 1-3/4" long	G 1 Carp	150	.053	Ea.	.75	3		3.75	5.60
8100	3-1/4" long	G	140	.057		.98	3.22		4.20	6.20
8150	3/8" diameter, 2-1/4" long	G	145	.055		.43	3.11		3.54	5.40
8200	5" long	G	140	.057		.75	3.22		3.97	5.95
8250	1/2" diameter, 2-3/4" long	G	140	.057		1.13	3.22		4.35	6.35
8300	7" long	G	125	.064		1.93	3.60		5.53	7.80
8350	5/8" diameter, 3-1/2" long	G	130	.062		2.11	3.46		5.57	7.80
8400	8-1/2" long	G	115	.070		4.49	3.92		8.41	11.15
8450	3/4" diameter, 4-1/4" long	G	115	.070		3.41	3.92		7.33	9.95
8500	10" long	G	95	.084		7.75	4.74		12.49	16.10
8550	1" diameter, 6" long	G	100	.080		5.60	4.50		10.10	13.30
8575	9" long	G	85	.094		7.25	5.30		12.55	16.40
8600	12" long	G	75	.107		7.85	6		13.85	18.15
8650	1-1/4" diameter, 9" long	G 1 Carp	70	.114	Ea.	33	6.45		39.45	46.50
8700	12" long	G "	60	.133	"	42.50	7.50		50	58.50
8750	For type 303 stainless steel, add					350%				
8800	For type 316 stainless steel, add					450%				
8950	Self-drilling concrete screw, hex washer head, 3/16" diam. x 1-3/4" long	G 1 Carp	300	.027	Ea.	.21	1.50		1.71	2.62
8960	2-1/4" long	G	250	.032		.29	1.80		2.09	3.18
8970	Phillips flat head, 3/16" diam. x 1-3/4" long	G	300	.027		.20	1.50		1.70	2.61
8980	2-1/4" long	G	250	.032		.31	1.80		2.11	3.20
9000	Minimum labor/equipment charge		4	2	Job		113		113	179

05 05 21 – Fastening Methods for Metal

05 05 21.10 Cutting Steel

		Crew	Daily Output	Labor-Hours	Unit	Material	2022 Bare Costs Labor	2022 Bare Costs Equipment	Total	Total Incl O&P
0010	**CUTTING STEEL**									
0020	Hand burning, incl. preparation, torch cutting & grinding, no staging									
0050	Steel to 1/4" thick	E-25	400	.020	L.F.	.64	1.28	.03	1.95	2.81
0100	1/2" thick		320	.025		1.04	1.60	.04	2.68	3.76
0150	3/4" thick		260	.031		1.61	1.97	.05	3.63	5
0200	1" thick		200	.040		2.22	2.56	.07	4.85	6.65
9000	Minimum labor/equipment charge		2	4	Job		256	6.70	262.70	415

05 05 21.15 Drilling Steel

		Crew	Daily Output	Labor-Hours	Unit	Material	2022 Bare Costs Labor	Equipment	Total	Total Incl O&P
0010	**DRILLING STEEL**									
1910	Drilling & layout for steel, up to 1/4" deep, no anchor									
1920	Holes, 1/4" diameter	1 Sswk	112	.071	Ea.	.06	4.42		4.48	7.15
1925	For each additional 1/4" depth, add		336	.024		.06	1.47		1.53	2.44
1930	3/8" diameter		104	.077		.07	4.76		4.83	7.75
1935	For each additional 1/4" depth, add		312	.026		.07	1.59		1.66	2.63
1940	1/2" diameter		96	.083		.08	5.15		5.23	8.40
1945	For each additional 1/4" depth, add		288	.028		.08	1.72		1.80	2.86
1950	5/8" diameter		88	.091		.12	5.60		5.72	9.20
1955	For each additional 1/4" depth, add		264	.030		.12	1.87		1.99	3.15
1960	3/4" diameter		80	.100		.16	6.20		6.36	10.10
1965	For each additional 1/4" depth, add		240	.033		.16	2.06		2.22	3.49
1970	7/8" diameter		72	.111		.19	6.85		7.04	11.25
1975	For each additional 1/4" depth, add		216	.037		.19	2.29		2.48	3.89
1980	1" diameter		64	.125		.28	7.75		8.03	12.75
1985	For each additional 1/4" depth, add		192	.042		.28	2.58		2.86	4.45
1990	For drilling up, add						40%			
2000	Minimum labor/equipment charge	1 Carp	4	2	Job		113		113	179

05 05 21.90 Welding Steel

		Crew	Daily Output	Labor-Hours	Unit	Material	2022 Bare Costs Labor	Equipment	Total	Total Incl O&P
0010	**WELDING STEEL**, Structural R050521-20									
0020	Field welding, 1/8" E6011, cost per welder, no operating engineer	E-14	8	1	Hr.	6.40	64	18.40	88.80	131
0200	With 1/2 operating engineer	E-13	8	1.500		6.40	92.50	18.40	117.30	175
0300	With 1 operating engineer	E-12	8	2		6.40	121	18.40	145.80	220
0500	With no operating engineer, 2# weld rod per ton	E-14	8	1	Ton	6.40	64	18.40	88.80	131
0600	8# E6011 per ton	"	2	4		25.50	256	73.50	355	520
0800	With one operating engineer per welder, 2# E6011 per ton	E-12	8	2		6.40	121	18.40	145.80	220
0900	8# E6011 per ton	"	2	8		25.50	485	73.50	584	875
1200	Continuous fillet, down welding									
1300	Single pass, 1/8" thick, 0.1#/L.F.	E-14	150	.053	L.F.	.32	3.41	.98	4.71	6.95
1400	3/16" thick, 0.2#/L.F.		75	.107		.64	6.80	1.96	9.40	13.85
1500	1/4" thick, 0.3#/L.F.		50	.160		.96	10.25	2.95	14.16	20.50
1610	5/16" thick, 0.4#/L.F.		38	.211		1.28	13.45	3.88	18.61	27
1800	3 passes, 3/8" thick, 0.5#/L.F.		30	.267		1.60	17.05	4.91	23.56	34.50
2010	4 passes, 1/2" thick, 0.7#/L.F.		22	.364		2.24	23.50	6.70	32.44	47.50
2200	5 to 6 passes, 3/4" thick, 1.3#/L.F.		12	.667		4.16	42.50	12.30	58.96	86.50
2400	8 to 11 passes, 1" thick, 2.4#/L.F.		6	1.333		7.65	85.50	24.50	117.65	172
2600	For vertical joint welding, add						20%			
2700	Overhead joint welding, add						300%			
2900	For semi-automatic welding, obstructed joints, deduct						5%			
3000	Exposed joints, deduct						15%			
4000	Cleaning and welding plates, bars, or rods									
4010	to existing beams, columns, or trusses	E-14	12	.667	L.F.	1.60	42.50	12.30	56.40	84
9000	Minimum labor/equipment charge	"	4	2	Job		128	37	165	247

05 05 23.30 Lag Screws

			Crew	Daily Output	Labor-Hours	Unit	Material	2022 Bare Costs Labor	Equipment	Total	Total Incl O&P
0010	**LAG SCREWS**										
0020	Steel, 1/4" diameter, 2" long	G	1 Carp	200	.040	Ea.	.14	2.25		2.39	3.73
0200	1/2" diameter, 3" long	G		130	.062		1.22	3.46		4.68	6.85
0300	5/8" diameter, 3" long	G		120	.067		1.80	3.75		5.55	7.95

05 05 Common Work Results for Metals

05 05 23 – Metal Fastenings

05 05 23.50 Powder Actuated Tools and Fasteners		Crew	Daily Output	Labor- Hours	Unit	Material	2022 Bare Costs Labor	Equipment	Total	Total Incl O&P
0010	**POWDER ACTUATED TOOLS & FASTENERS**									
0020	Stud driver, .22 caliber, single shot				Ea.	90			90	99
0100	.27 caliber, semi automatic, strip				"	630			630	690
0300	Powder load, single shot, .22 cal, power level 2, brown				C	8.25			8.25	9.10
0400	Strip, .27 cal, power level 4, red					12.75			12.75	14.05
0600	Drive pin, .300 x 3/4" long	G	1 Carp	4.80	1.667		6.40	94		100.40	156
0700	.300 x 3" long with washer	G	"	4	2		17.95	113		130.95	199

05 05 23.70 Structural Blind Bolts

		Crew	Daily Output	Labor- Hours	Unit	Material	Labor	Equipment	Total	Total Incl O&P
0010	**STRUCTURAL BLIND BOLTS**									
0100	1/4" diameter x 1/4" grip	G	1 Sswk	240	.033	Ea.	.88	2.06		2.94	4.29
0150	1/2" grip	G		216	.037		.93	2.29		3.22	4.71
0200	3/8" diameter x 1/2" grip	G		232	.034		1.71	2.13		3.84	5.30
0250	3/4" grip	G		208	.038		1.73	2.38		4.11	5.75
0300	1/2" diameter x 1/2" grip	G		224	.036		3.87	2.21		6.08	7.80
0350	3/4" grip	G		200	.040		3.87	2.47		6.34	8.25
0400	5/8" diameter x 3/4" grip	G		216	.037		6.05	2.29		8.34	10.35
0450	1" grip	G		192	.042		6.05	2.58		8.63	10.80

05 05 23.87 Weld Studs

		Crew	Daily Output	Labor- Hours	Unit	Material	Labor	Equipment	Total	Total Incl O&P
0010	**WELD STUDS**									
0020	1/4" diameter, 2-11/16" long	G	E-10	1120	.014	Ea.	.43	.90	.47	1.80	2.44
0100	4-1/8" long	G		1080	.015		.40	.93	.49	1.82	2.47
0200	3/8" diameter, 4-1/8" long	G		1080	.015		.79	.93	.49	2.21	2.90
0300	6-1/8" long	G		1040	.015		1.03	.97	.50	2.50	3.24
9000	Minimum labor/equipment charge		1 Sswk	2	4	Job		247		247	400
0020	1" thick		E-25	.50	16	S.F.	2.22	1,025	26.50	1,053.72	1,675
0022	1" thick			.30	26.667		2.22	1,700	44.50	1,746.72	2,800
0024	1" thick			1	8		2.22	510	13.35	525.57	840
0026	1" thick			1.75	4.571		2.22	292	7.65	301.87	480

05 12 Structural Steel Framing

05 12 23 – Structural Steel for Buildings

05 12 23.05 Canopy Framing

		Crew	Daily Output	Labor- Hours	Unit	Material	Labor	Equipment	Total	Total Incl O&P
0010	**CANOPY FRAMING**									
0020	6" and 8" members, shop fabricated	G	E-4	3000	.011	Lb.	1.96	.67	.05	2.68	3.28
9000	Minimum labor/equipment charge		1 Sswk	1	8	Job		495		495	795

05 12 23.10 Ceiling Supports

		Crew	Daily Output	Labor- Hours	Unit	Material	Labor	Equipment	Total	Total Incl O&P
0010	**CEILING SUPPORTS**									
1000	Entrance door/folding partition supports, shop fabricated	G	E-4	60	.533	L.F.	33	33.50	2.46	68.96	92
1100	Linear accelerator door supports	G		14	2.286		149	143	10.55	302.55	405
1200	Lintels or shelf angles, hung, exterior hot dipped galv.	G		267	.120		22.50	7.45	.55	30.50	37
1250	Two coats primer paint instead of galv.	G		267	.120		19.35	7.45	.55	27.35	34
1400	Monitor support, ceiling hung, expansion bolted	G		4	8	Ea.	520	500	37	1,057	1,425
1450	Hung from pre-set inserts	G		6	5.333		560	335	24.50	919.50	1,175
1600	Motor supports for overhead doors	G		4	8		264	500	37	801	1,125
1700	Partition support for heavy folding partitions, without pocket	G		24	1.333	L.F.	74.50	83	6.15	163.65	223
1750	Supports at pocket only	G		12	2.667		149	166	12.30	327.30	445
2000	Rolling grilles & fire door supports	G		34	.941		64	58.50	4.34	126.84	169
2100	Spider-leg light supports, expansion bolted to ceiling slab	G		8	4	Ea.	213	249	18.45	480.45	655
2150	Hung from pre-set inserts	G	E-4	12	2.667	Ea.	229	166	12.30	407.30	535
2400	Toilet partition support	G		36	.889	L.F.	74.50	55.50	4.10	134.10	176

For customer support on your Commercial Renovation Costs with RSMeans data, call 800.448.8182.

111

05 12 23.10 Ceiling Supports		Crew	Daily Output	Labor-Hours	Unit	Material	2022 Bare Costs Labor	Equipment	Total	Total Incl O&P
2500	X-ray travel gantry support	E-4	12	2.667	L.F.	255	166	12.30	433.30	565

05 12 23.17 Columns, Structural

		Crew	Daily Output	Labor-Hours	Unit	Material	Labor	Equipment	Total	Total Incl O&P
0010	COLUMNS, STRUCTURAL R051223-10									
0015	Made from recycled materials									
0020	Shop fab'd for 100-ton, 1-2 story project, bolted connections									
0800	Steel, concrete filled, extra strong pipe, 3-1/2" diameter	E-2	660	.085	L.F.	54	5.20	3.76	62.96	72
0830	4" diameter		780	.072		60	4.40	3.18	67.58	76.50
0890	5" diameter		1020	.055		71.50	3.36	2.43	77.29	87
0930	6" diameter		1200	.047		95	2.86	2.07	99.93	112
0940	8" diameter		1100	.051		95	3.12	2.25	100.37	112
1500	Steel pipe, extra strong, no concrete, 3" to 5" diameter	G	16000	.004	Lb.	1.64	.21	.16	2.01	2.31
3300	Structural tubing, square, A500GrB, 4" to 6" square, light section	G	11270	.005	"	1.64	.30	.22	2.16	2.53
8090	For projects 75 to 99 tons, add				%	10%				
8092	50 to 74 tons, add					20%				
8094	25 to 49 tons, add					30%	10%			
8096	10 to 24 tons, add					50%	25%			
8098	2 to 9 tons, add					75%	50%			
8099	Less than 2 tons, add					100%	100%			
9000	Minimum labor/equipment charge	1 Sswk	1	8	Job		495		495	795

05 12 23.18 Corner Guards

		Crew	Daily Output	Labor-Hours	Unit	Material	Labor	Equipment	Total	Total Incl O&P
0010	CORNER GUARDS									
0020	Steel angle w/anchors, 1" x 1" x 1/4", 1.5#/L.F.	2 Carp	160	.100	L.F.	8	5.65		13.65	17.75
0100	2" x 2" x 1/4" angles, 3.2#/L.F.	"	150	.107	"	20.50	6		26.50	32
9000	Minimum labor/equipment charge	1 Carp	2	4	Job		225		225	360

05 12 23.20 Curb Edging

		Crew	Daily Output	Labor-Hours	Unit	Material	Labor	Equipment	Total	Total Incl O&P
0010	CURB EDGING									
0020	Steel angle w/anchors, shop fabricated, on forms, 1" x 1", 0.8#/L.F.	G E-4	350	.091	L.F.	1.99	5.70	.42	8.11	11.85
0300	4" x 4" angles, 8.2#/L.F.	G	275	.116	"	17.15	7.25	.54	24.94	31
9000	Minimum labor/equipment charge		4	8	Job		500	37	537	845

05 12 23.40 Lightweight Framing

		Crew	Daily Output	Labor-Hours	Unit	Material	Labor	Equipment	Total	Total Incl O&P
0010	LIGHTWEIGHT FRAMING									
0015	Made from recycled materials									
0200	For load-bearing steel studs see Section 05 41 13.30									
0400	Angle framing, field fabricated, 4" and larger	G E-3	440	.055	Lb.	.95	3.41	.33	4.69	6.90
0450	Less than 4" angles	G	265	.091		.98	5.65	.56	7.19	10.80
0600	Channel framing, field fabricated, 8" and larger	G	500	.048		.98	3	.29	4.27	6.25
0650	Less than 8" channels	G	335	.072		.98	4.48	.44	5.90	8.75
1000	Continuous slotted channel framing system, shop fab, simple framing	G 2 Sswk	2400	.007	Lb.	5.05	.41		5.46	6.25
1200	Complex framing	"	1600	.010		5.75	.62		6.37	7.30
1250	Plate & bar stock for reinforcing beams and trusses	G				1.80			1.80	1.98
1300	Cross bracing, rods, shop fabricated, 3/4" diameter	G E-3	700	.034		1.96	2.14	.21	4.31	5.85
1310	7/8" diameter	G	850	.028		1.96	1.77	.17	3.90	5.20
1320	1" diameter	G	1000	.024		1.96	1.50	.15	3.61	4.74
1330	Angle, 5" x 5" x 3/8"	G	2800	.009		1.96	.54	.05	2.55	3.08
1350	Hanging lintels, shop fabricated	G	850	.028		1.96	1.77	.17	3.90	5.20
1380	Roof frames, shop fabricated, 3'-0" square, 5' span	G E-2	4200	.013		1.96	.82	.59	3.37	4.11
1400	Tie rod, not upset, 1-1/2" to 4" diameter, with turnbuckle	G 2 Sswk	800	.020		2.13	1.24		3.37	4.33
1420	No turnbuckle	G	700	.023		2.05	1.41		3.46	4.53
1500	Upset, 1-3/4" to 4" diameter, with turnbuckle	G	800	.020		2.13	1.24		3.37	4.33
1520	No turnbuckle	G	700	.023		2.05	1.41		3.46	4.53
9000	Minimum labor/equipment charge		2	8	Job		495		495	795

05 12 23 – Structural Steel for Buildings

05 12 23.45 Lintels

			Crew	Daily Output	Labor-Hours	Unit	Material	2022 Bare Costs Labor	2022 Bare Costs Equipment	Total	Total Incl O&P
0010	**LINTELS**										
0015	Made from recycled materials										
0020	Plain steel angles, shop fabricated, under 500 lb.	G	1 Bric	550	.015	Lb.	1.26	.80		2.06	2.68
0100	500 to 1,000 lb.	G		640	.013		1.23	.69		1.92	2.46
0200	1,000 to 2,000 lb.	G		640	.013		1.19	.69		1.88	2.42
0300	2,000 to 4,000 lb.	G	↓	640	.013		1.16	.69		1.85	2.39
0500	For built-up angles and plates, add to above	G					1.64			1.64	1.80
0700	For engineering, add to above						.16			.16	.18
0900	For galvanizing, add to above, under 500 lb.						.38			.38	.42
0950	500 to 2,000 lb.						.35			.35	.38
1000	Over 2,000 lb.					↓	.33			.33	.36
2000	Steel angles, 3-1/2" x 3", 1/4" thick, 2'-6" long	G	1 Bric	47	.170	Ea.	17.70	9.40		27.10	34.50
2100	4'-6" long	G		26	.308		32	17		49	62
2500	3-1/2" x 3-1/2" x 5/16", 5'-0" long	G		18	.444		47	24.50		71.50	91.50
2600	4" x 3-1/2", 1/4" thick, 5'-0" long	G		21	.381		40.50	21		61.50	78
2700	9'-0" long	G		12	.667		73	37		110	140
2800	4" x 3-1/2" x 5/16", 7'-0" long	G		12	.667		70.50	37		107.50	137
2900	5" x 3-1/2" x 5/16", 10'-0" long	G		8	1	↓	114	55.50		169.50	214
9000	Minimum labor/equipment charge		↓	4	2	Job		111		111	177

05 12 23.60 Pipe Support Framing

			Crew	Daily Output	Labor-Hours	Unit	Material	2022 Bare Costs Labor	2022 Bare Costs Equipment	Total	Total Incl O&P
0010	**PIPE SUPPORT FRAMING**										
0020	Under 10#/L.F., shop fabricated	G	E-4	3900	.008	Lb.	2.19	.51	.04	2.74	3.27
0200	10.1 to 15#/L.F.	G		4300	.007		2.16	.46	.03	2.65	3.17
0400	15.1 to 20#/L.F.	G		4800	.007		2.13	.42	.03	2.58	3.04
0600	Over 20#/L.F.	G	↓	5400	.006	↓	2.10	.37	.03	2.50	2.92

05 12 23.65 Plates

			Crew	Daily Output	Labor-Hours	Unit	Material	2022 Bare Costs Labor	2022 Bare Costs Equipment	Total	Total Incl O&P
0010	**PLATES**										
0015	Made from recycled materials										
0020	For connections & stiffener plates, shop fabricated										
0050	1/8" thick (5.1 lb./S.F.)	G				S.F.	8.35			8.35	9.20
0100	1/4" thick (10.2 lb./S.F.)	G					16.70			16.70	18.35
0300	3/8" thick (15.3 lb./S.F.)	G					25			25	27.50
0400	1/2" thick (20.4 lb./S.F.)	G					33.50			33.50	36.50
0450	3/4" thick (30.6 lb./S.F.)	G					50			50	55
0500	1" thick (40.8 lb./S.F.)	G				↓	67			67	73.50
2000	Steel plate, warehouse prices, no shop fabrication										
2100	1/4" thick (10.2 lb./S.F.)	G				S.F.	10.65			10.65	11.70
2210	1/4" steel plate, welded in place		E-18	528	.076		10.65	4.69	1.54	16.88	21
2220	1/2" steel plate, welded in place			480	.083		21.50	5.15	1.70	28.35	33.50
2230	3/4" steel plate, welded in place			384	.104		32	6.45	2.12	40.57	47.50
2240	1" steel plate, welded in place			320	.125		42.50	7.75	2.55	52.80	62
2250	1-1/2" steel plate, welded in place			256	.156		64	9.70	3.18	76.88	89.50
2260	2" steel plate, welded in place		↓	192	.208		85.50	12.90	4.25	102.65	119

05 12 23.75 Structural Steel Members

				Crew	Daily Output	Labor-Hours	Unit	Material	2022 Bare Costs Labor	2022 Bare Costs Equipment	Total	Total Incl O&P
0010	**STRUCTURAL STEEL MEMBERS**		R051223-10									
0015	Made from recycled materials											
0020	Shop fab'd for 100-ton, 1-2 story project, bolted connections											
0102	Beam or girder, W 6 x 9	R051223-15	G	E-2	600	.093	L.F.	16.20	5.70	4.13	26.03	31.50
0302	W 8 x 10		G		600	.093		18	5.70	4.13	27.83	33.50
0502	x 31		G		550	.102		56	6.25	4.51	66.76	76.50
0702	W 10 x 22		G		600	.093		39.50	5.70	4.13	49.33	57

05 12 23.75 Structural Steel Members

			Crew	Daily Output	Labor-Hours	Unit	Material	2022 Bare Costs Labor	Equipment	Total	Total Incl O&P
0902	x 49	G	E-2	550	.102	L.F.	88	6.25	4.51	98.76	112
1102	W 12 x 16	G		880	.064		29	3.90	2.82	35.72	41
1302	x 22	G		880	.064		39.50	3.90	2.82	46.22	53
1502	x 26	G		880	.064		47	3.90	2.82	53.72	61
1902	W 14 x 26	G		990	.057		47	3.46	2.51	52.97	60
2102	x 30	G		900	.062		54	3.81	2.76	60.57	68.50
2702	W 16 x 26	G		1000	.056		47	3.43	2.48	52.91	59.50
2902	x 31	G	↓	900	.062		56	3.81	2.76	62.57	70.50
8490	For projects 75 to 99 tons, add						10%				
8492	50 to 74 tons, add						20%				
8494	25 to 49 tons, add						30%	10%			
8496	10 to 24 tons, add						50%	25%			
8498	2 to 9 tons, add						75%	50%			
8499	Less than 2 tons, add					↓	100%	100%			
9000	Minimum labor/equipment charge		E-2	2	28	Job		1,725	1,250	2,975	4,100

05 12 23.77 Structural Steel Projects

			Crew	Daily Output	Labor-Hours	Unit	Material	2022 Bare Costs Labor	Equipment	Total	Total Incl O&P
0010	**STRUCTURAL STEEL PROJECTS**										
0015	Made from recycled materials										
0020	Shop fab'd for 100-ton, 1-2 story project, bolted connections										
0700	Offices, hospitals, etc., steel bearing, 1 to 2 stories R050521-20	G	E-5	10.30	7.767	Ton	3,275	480	255	4,010	4,650
1300	Industrial bldgs., 1 story, beams & girders, steel bearing	G		12.90	6.202		3,275	380	204	3,859	4,425
1400	Masonry bearing R051223-10	G		10	8		3,275	495	263	4,033	4,675
1600	1 story with roof trusses, steel bearing	G		10.60	7.547		3,875	465	248	4,588	5,275
1700	Masonry bearing R051223-15	G	↓	8.30	9.639		3,875	595	315	4,785	5,550
5390	For projects 75 to 99 tons, add						10%				
5392	50 to 74 tons, add						20%				
5394	25 to 49 tons, add						30%	10%			
5396	10 to 24 tons, add						50%	25%			
5398	2 to 9 tons, add						75%	50%			
5399	Less than 2 tons, add					↓	100%	100%			

05 12 23.78 Structural Steel Secondary Members

		Unit	Material	2022 Bare Costs Labor	Equipment	Total	Total Incl O&P
0010	**STRUCTURAL STEEL SECONDARY MEMBERS**						
0015	Made from recycled materials						
0020	Shop fabricated for 20-ton girt/purlin framing package, materials only						
0100	Girts/purlins, C/Z-shapes, includes clips and bolts						
0110	6" x 2-1/2" x 2-1/2", 16 ga., 3.0 lb./L.F.	L.F.	4.42			4.42	4.86
0115	14 ga., 3.5 lb./L.F.		5.15			5.15	5.65
0120	8" x 2-3/4" x 2-3/4", 16 ga., 3.4 lb./L.F.		5			5	5.50
0125	14 ga., 4.1 lb./L.F.		6.05			6.05	6.65
0130	12 ga., 5.6 lb./L.F.		8.25			8.25	9.10
0135	10" x 3-1/2" x 3-1/2", 14 ga., 4.7 lb./L.F.		6.90			6.90	7.60
0140	12 ga., 6.7 lb./L.F.		9.85			9.85	10.85
0145	12" x 3-1/2" x 3-1/2", 14 ga., 5.3 lb./L.F.		7.80			7.80	8.60
0150	12 ga., 7.4 lb./L.F.	↓	10.90			10.90	12
0200	Eave struts, C-shape, includes clips and bolts						
0210	6" x 4" x 3", 16 ga., 3.1 lb./L.F.	L.F.	4.57			4.57	5
0215	14 ga., 3.9 lb./L.F.		5.75			5.75	6.30
0220	8" x 4" x 3", 16 ga., 3.5 lb./L.F.		5.15			5.15	5.65
0225	14 ga., 4.4 lb./L.F.		6.50			6.50	7.15
0230	12 ga., 6.2 lb./L.F.		9.15			9.15	10.05
0235	10" x 5" x 3", 14 ga., 5.2 lb./L.F.		7.65			7.65	8.45
0240	12 ga., 7.3 lb./L.F.	↓	10.75			10.75	11.85

114

05 12 Structural Steel Framing

05 12 23 – Structural Steel for Buildings

05 12 23.78 Structural Steel Secondary Members	Crew	Daily Output	Labor-Hours	Unit	Material	2022 Bare Costs Labor	Equipment	Total	Total Incl O&P
0245	12" x 5" x 4", 14 ga., 6.0 lb./L.F.				L.F.	8.85			8.85	9.70
0250	12 ga., 8.4 lb./L.F.				↓	12.40			12.40	13.60
0300	Rake/base angle, excludes concrete drilling and expansion anchors									
0310	2" x 2", 14 ga., 1.0 lb./L.F.	2 Sswk	640	.025	L.F.	1.47	1.55		3.02	4.11
0315	3" x 2", 14 ga., 1.3 lb./L.F.		535	.030		1.92	1.85		3.77	5.10
0320	3" x 3", 14 ga., 1.6 lb./L.F.		500	.032		2.36	1.98		4.34	5.80
0325	4" x 3", 14 ga., 1.8 lb./L.F.	↓	480	.033	↓	2.65	2.06		4.71	6.25
0600	Installation of secondary members, erection only									
0610	Girts, purlins, eave struts, 16 ga., 6" deep	E-18	100	.400	Ea.		25	8.15	33.15	48.50
0615	8" deep		80	.500			31	10.20	41.20	60.50
0620	14 ga., 6" deep		80	.500			31	10.20	41.20	60.50
0625	8" deep		65	.615			38	12.55	50.55	75
0630	10" deep		55	.727			45	14.80	59.80	88.50
0635	12" deep		50	.800			49.50	16.30	65.80	97.50
0640	12 ga., 8" deep		50	.800			49.50	16.30	65.80	97.50
0645	10" deep		45	.889			55	18.10	73.10	108
0650	12" deep	↓	40	1	↓		62	20.50	82.50	122
0900	For less than 20-ton job lots									
0905	For 15 to 19 tons, add				%	10%				
0910	For 10 to 14 tons, add					25%				
0915	For 5 to 9 tons, add					50%	50%	50%		
0920	For 1 to 4 tons, add					75%	75%	75%		
0925	For less than 1 ton, add				↓	100%	100%	100%		
0990	Minimum labor/equipment charge	E-18	1	40	Job		2,475	815	3,290	4,875

05 15 Wire Rope Assemblies

05 15 16 – Steel Wire Rope Assemblies

05 15 16.70 Temporary Cable Safety Railing

		Crew	Daily Output	Labor-Hours	Unit	Material	Labor	Equipment	Total	Total Incl O&P
0010	**TEMPORARY CABLE SAFETY RAILING**, Each 100' strand incl.									
0020	2 eyebolts, 1 turnbuckle, 100' cable, 2 thimbles, 6 clips									
0025	Made from recycled materials									
0100	One strand using 1/4" cable & accessories	G	2 Sswk	4	4	C.L.F.	84	247		331	495
0200	1/2" cable & accessories	G	"	2	8	"	162	495		657	975

05 21 Steel Joist Framing

05 21 13 – Deep Longspan Steel Joist Framing

05 21 13.50 Deep Longspan Joists

		Crew	Daily Output	Labor-Hours	Unit	Material	Labor	Equipment	Total	Total Incl O&P
0010	**DEEP LONGSPAN JOISTS**									
3010	DLH series, 40-ton job lots, bolted cross bridging, shop primer									
3015	Made from recycled materials									
3040	Spans to 144' (shipped in 2 pieces)	G	E-7	13	6.154	Ton	2,400	380	213	2,993	3,475
3500	For less than 40-ton job lots									
3502	For 30 to 39 tons, add				%	10%				
3504	20 to 29 tons, add					20%				
3506	10 to 19 tons, add					30%				
3507	5 to 9 tons, add					50%	25%			
3508	1 to 4 tons, add					75%	50%			
3509	Less than 1 ton, add				↓	100%	100%			
4010	SLH series, 40-ton job lots, bolted cross bridging, shop primer									

For customer support on your Commercial Renovation Costs with RSMeans data, call 800.448.8182.

115

05 21 Steel Joist Framing

05 21 13 – Deep Longspan Steel Joist Framing

05 21 13.50 Deep Longspan Joists

		Crew	Daily Output	Labor-Hours	Unit	Material	2022 Bare Costs Labor	Equipment	Total	Total Incl O&P
4040	Spans to 200' (shipped in 3 pieces) **G**	E-7	13	6.154	Ton	2,475	380	213	3,068	3,575
6100	For less than 40-ton job lots									
6102	For 30 to 39 tons, add				%	10%				
6104	20 to 29 tons, add					20%				
6106	10 to 19 tons, add					30%				
6107	5 to 9 tons, add					50%	25%			
6108	1 to 4 tons, add					75%	50%			
6109	Less than 1 ton, add					100%	100%			

05 21 16 – Longspan Steel Joist Framing

05 21 16.50 Longspan Joists

		Crew	Daily Output	Labor-Hours	Unit	Material	2022 Bare Costs Labor	Equipment	Total	Total Incl O&P
0010	**LONGSPAN JOISTS**									
2000	LH series, 40-ton job lots, bolted cross bridging, shop primer									
2015	Made from recycled materials									
2040	Longspan joists, LH series, up to 96' **G**	E-7	13	6.154	Ton	2,100	380	213	2,693	3,150
2600	For less than 40-ton job lots									
2602	For 30 to 39 tons, add				%	10%				
2604	20 to 29 tons, add					20%				
2606	10 to 19 tons, add					30%				
2607	5 to 9 tons, add					50%	25%			
2608	1 to 4 tons, add					75%	50%			
2609	Less than 1 ton, add					100%	100%			

05 21 19 – Open Web Steel Joist Framing

05 21 19.10 Open Web Joists

		Crew	Daily Output	Labor-Hours	Unit	Material	2022 Bare Costs Labor	Equipment	Total	Total Incl O&P
0010	**OPEN WEB JOISTS**									
0015	Made from recycled materials									
0050	K series, 40-ton lots, horiz. bridging, spans to 30', shop primer **G**	E-7	12	6.667	Ton	1,850	410	231	2,491	2,950
0440	K series, 30' to 50' spans **G**	"	17	4.706	"	2,625	290	163	3,078	3,525
0800	For less than 40-ton job lots									
0802	For 30 to 39 tons, add				%	10%				
0804	20 to 29 tons, add					20%				
0806	10 to 19 tons, add					30%				
0807	5 to 9 tons, add					50%	25%			
0808	1 to 4 tons, add					75%	50%			
0809	Less than 1 ton, add					100%	100%			
1010	CS series, 40-ton job lots, horizontal bridging, shop primer									
1040	Spans to 30' **G**	E-7	12	6.667	Ton	1,925	410	231	2,566	3,025
1500	For less than 40-ton job lots									
1502	For 30 to 39 tons, add				%	10%				
1504	20 to 29 tons, add					20%				
1506	10 to 19 tons, add					30%				
1507	5 to 9 tons, add					50%	25%			
1508	1 to 4 tons, add					75%	50%			
1509	Less than 1 ton, add					100%	100%			
6400	Individual steel bearing plate, 6" x 6" x 1/4" with J-hook **G**	1 Bric	160	.050	Ea.	9.80	2.76		12.56	15.25
9000	Minimum labor and equipment, bar joists	F-6	1	40	Job		2,150	1,975	4,125	5,550

05 31 Steel Decking

05 31 13 – Steel Floor Decking

05 31 13.50 Floor Decking

		Crew	Daily Output	Labor-Hours	Unit	Material	2022 Bare Costs Labor	Equipment	Total	Total Incl O&P
0010	**FLOOR DECKING**	R053100-10								
0015	Made from recycled materials									
5100	Non-cellular composite decking, galvanized, 1-1/2" deep, 16 ga.	G	E-4	3500	.009	S.F.	4.57	.57	.04	5.18	6
5120	18 ga.	G		3650	.009		2.82	.55	.04	3.41	4.02
5140	20 ga.	G		3800	.008		3.33	.53	.04	3.90	4.55
5200	2" deep, 22 ga.	G		3860	.008		2.25	.52	.04	2.81	3.35
5300	20 ga.	G		3600	.009		3.20	.55	.04	3.79	4.46
5400	18 ga.	G		3380	.009		2.94	.59	.04	3.57	4.23
5500	16 ga.	G		3200	.010		4.52	.62	.05	5.19	6.05
5700	3" deep, 22 ga.	G		3200	.010		2.45	.62	.05	3.12	3.75
5800	20 ga.	G		3000	.011		3.52	.67	.05	4.24	4.99
5900	18 ga.	G		2850	.011		2.94	.70	.05	3.69	4.42
6000	16 ga.	G		2700	.012		5.15	.74	.05	5.94	6.90
9000	Minimum labor/equipment charge		1 Sswk	1	8	Job		495		495	795

05 31 23 – Steel Roof Decking

05 31 23.50 Roof Decking

		Crew	Daily Output	Labor-Hours	Unit	Material	Labor	Equipment	Total	Total Incl O&P
0010	**ROOF DECKING**									
0015	Made from recycled materials									
2100	Open type, 1-1/2" deep, Type B, wide rib, galv., 22 ga., under 50 sq.	G	E-4	4500	.007	S.F.	2.54	.44	.03	3.01	3.54
2400	Over 500 squares	G	"	5100	.006	"	1.83	.39	.03	2.25	2.67

05 31 33 – Steel Form Decking

05 31 33.50 Form Decking

		Crew	Daily Output	Labor-Hours	Unit	Material	Labor	Equipment	Total	Total Incl O&P
0010	**FORM DECKING**									
0015	Made from recycled materials									
6100	Slab form, steel, 28 ga., 9/16" deep, Type UFS, uncoated	G	E-4	4000	.008	S.F.	1.79	.50	.04	2.33	2.80
6200	Galvanized	G	"	4000	.008	"	1.58	.50	.04	2.12	2.58

05 35 Raceway Decking Assemblies

05 35 13 – Steel Cellular Decking

05 35 13.50 Cellular Decking

		Crew	Daily Output	Labor-Hours	Unit	Material	Labor	Equipment	Total	Total Incl O&P
0010	**CELLULAR DECKING**									
0015	Made from recycled materials									
0200	Cellular units, galv, 1-1/2" deep, Type BC, 20-20 ga., over 15 squares	G	E-4	1460	.022	S.F.	11.20	1.37	.10	12.67	14.60
0400	3" deep, Type NC, galvanized, 20-20 ga.	G		1375	.023		12.30	1.45	.11	13.86	16
1000	4-1/2" deep, Type JC, galvanized, 18-20 ga.	G		1100	.029		11.85	1.81	.13	13.79	16.05
1500	For acoustical deck, add						15%				
1900	For multi-story or congested site, add							50%			

05 41 Structural Metal Stud Framing

05 41 13 – Load-Bearing Metal Stud Framing

05 41 13.05 Bracing

		Crew	Daily Output	Labor-Hours	Unit	Material	Labor	Equipment	Total	Total Incl O&P
0010	**BRACING**, shear wall X-bracing, per 10' x 10' bay, one face										
0015	Made of recycled materials										
0120	Metal strap, 20 ga. x 4" wide	G	2 Carp	18	.889	Ea.	19.35	50		69.35	101
0130	6" wide	G		18	.889		31	50		81	114
0160	18 ga. x 4" wide	G		16	1		32	56.50		88.50	125
0170	6" wide	G		16	1		47	56.50		103.50	141
0410	Continuous strap bracing, per horizontal row on both faces										

05 41 13 – Load-Bearing Metal Stud Framing

05 41 13.05 Bracing

			Crew	Daily Output	Labor-Hours	Unit	Material	2022 Bare Costs Labor	2022 Bare Costs Equipment	Total	Total Incl O&P
0420	Metal strap, 20 ga. x 2" wide, studs 12" OC	G	1 Carp	7	1.143	C.L.F.	55	64.50		119.50	163
0430	16" OC	G		8	1		55	56.50		111.50	150
0440	24" OC	G		10	.800		55	45		100	132
0450	18 ga. x 2" wide, studs 12" OC	G		6	1.333		77	75		152	204
0460	16" OC	G		7	1.143		77	64.50		141.50	187
0470	24" OC	G	▼	8	1	▼	77	56.50		133.50	174

05 41 13.10 Bridging

			Crew	Daily Output	Labor-Hours	Unit	Material	2022 Bare Costs Labor	2022 Bare Costs Equipment	Total	Total Incl O&P
0010	**BRIDGING**, solid between studs w/1-1/4" leg track, per stud bay										
0015	Made from recycled materials										
0200	Studs 12" OC, 18 ga. x 2-1/2" wide	G	1 Carp	125	.064	Ea.	.93	3.60		4.53	6.70
0210	3-5/8" wide	G	"	120	.067	"	1.11	3.75		4.86	7.15
0220	4" wide	G	1 Carp	120	.067	Ea.	1.04	3.75		4.79	7.10
0230	6" wide	G		115	.070		1.53	3.92		5.45	7.90
0240	8" wide	G		110	.073		1.88	4.09		5.97	8.55
0300	16 ga. x 2-1/2" wide	G		115	.070		1.15	3.92		5.07	7.45
0310	3-5/8" wide	G		110	.073		1.36	4.09		5.45	8
0320	4" wide	G		110	.073		1.52	4.09		5.61	8.15
0330	6" wide	G		105	.076		1.90	4.29		6.19	8.90
0340	8" wide	G		100	.080		2.34	4.50		6.84	9.75
1200	Studs 16" OC, 18 ga. x 2-1/2" wide	G		125	.064		1.19	3.60		4.79	7
1210	3-5/8" wide	G		120	.067		1.42	3.75		5.17	7.50
1220	4" wide	G		120	.067		1.34	3.75		5.09	7.40
1230	6" wide	G		115	.070		1.96	3.92		5.88	8.35
1240	8" wide	G		110	.073		2.41	4.09		6.50	9.15
1300	16 ga. x 2-1/2" wide	G		115	.070		1.47	3.92		5.39	7.80
1310	3-5/8" wide	G		110	.073		1.75	4.09		5.84	8.40
1320	4" wide	G		110	.073		1.95	4.09		6.04	8.65
1330	6" wide	G		105	.076		2.44	4.29		6.73	9.50
1340	8" wide	G		100	.080		3	4.50		7.50	10.45
2200	Studs 24" OC, 18 ga. x 2-1/2" wide	G		125	.064		1.72	3.60		5.32	7.60
2210	3-5/8" wide	G		120	.067		2.05	3.75		5.80	8.20
2220	4" wide	G		120	.067		1.93	3.75		5.68	8.10
2230	6" wide	G		115	.070		2.84	3.92		6.76	9.30
2240	8" wide	G		110	.073		3.49	4.09		7.58	10.35
2300	16 ga. x 2-1/2" wide	G		115	.070		2.12	3.92		6.04	8.55
2310	3-5/8" wide	G		110	.073		2.53	4.09		6.62	9.30
2320	4" wide	G		110	.073		2.82	4.09		6.91	9.60
2330	6" wide	G		105	.076		3.53	4.29		7.82	10.70
2340	8" wide	G	▼	100	.080	▼	4.34	4.50		8.84	11.95
3000	Continuous bridging, per row										
3100	16 ga. x 1-1/2" channel thru studs 12" OC	G	1 Carp	6	1.333	C.L.F.	51.50	75		126.50	176
3110	16" OC	G		7	1.143		51.50	64.50		116	159
3120	24" OC	G		8.80	.909		51.50	51		102.50	139
4100	2" x 2" angle x 18 ga., studs 12" OC	G		7	1.143		78	64.50		142.50	188
4110	16" OC	G		9	.889		78	50		128	166
4120	24" OC	G		12	.667		78	37.50		115.50	146
4200	16 ga., studs 12" OC	G		5	1.600		98	90		188	251
4210	16" OC	G		7	1.143		98	64.50		162.50	210
4220	24" OC	G		10	.800	▼	98	45		143	180

05 41 13 – Load-Bearing Metal Stud Framing

05 41 13.25 Framing, Boxed Headers/Beams		Crew	Daily Output	Labor-Hours	Unit	Material	2022 Bare Costs Labor	Equipment	Total	Total Incl O&P
0010	**FRAMING, BOXED HEADERS/BEAMS**									
0015	Made from recycled materials									
0200	Double, 18 ga. x 6" deep	G	2 Carp	220	.073	L.F.	5.45	4.09		9.54	12.50
0210	8" deep	G		210	.076		5.65	4.29		9.94	13
0220	10" deep	G		200	.080		7.25	4.50		11.75	15.10
0230	12" deep	G		190	.084		7.90	4.74		12.64	16.25
0300	16 ga. x 8" deep	G		180	.089		6.55	5		11.55	15.15
0310	10" deep	G		170	.094		8.10	5.30		13.40	17.30
0320	12" deep	G		160	.100		8.80	5.65		14.45	18.60
0400	14 ga. x 10" deep	G		140	.114		9.55	6.45		16	20.50
0410	12" deep	G		130	.123		10.45	6.95		17.40	22.50
1210	Triple, 18 ga. x 8" deep	G		170	.094		8.55	5.30		13.85	17.80
1220	10" deep	G		165	.097		10.30	5.45		15.75	20
1230	12" deep	G		160	.100		11.30	5.65		16.95	21.50
1300	16 ga. x 8" deep	G		145	.110		9.90	6.20		16.10	21
1310	10" deep	G		140	.114		11.60	6.45		18.05	23
1320	12" deep	G		135	.119		12.65	6.65		19.30	24.50
1400	14 ga. x 10" deep	G		115	.139		13	7.85		20.85	27
1410	12" deep	G		110	.145		14.40	8.20		22.60	29

05 41 13.30 Framing, Stud Walls

05 41 13.30 Framing, Stud Walls		Crew	Daily Output	Labor-Hours	Unit	Material	2022 Bare Costs Labor	Equipment	Total	Total Incl O&P
0010	**FRAMING, STUD WALLS** w/top & bottom track, no openings,									
0020	Headers, beams, bridging or bracing									
0025	Made from recycled materials									
4100	8' high walls, 18 ga. x 2-1/2" wide, studs 12" OC	G	2 Carp	54	.296	L.F.	8	16.70		24.70	35.50
4110	16" OC	G		77	.208		6.45	11.70		18.15	25.50
4120	24" OC	G		107	.150		4.88	8.40		13.28	18.75
4130	3-5/8" wide, studs 12" OC	G		53	.302		9.55	17		26.55	37.50
4140	16" OC	G		76	.211		7.70	11.85		19.55	27.50
4150	24" OC	G		105	.152		5.85	8.60		14.45	20
4160	4" wide, studs 12" OC	G		52	.308		9.75	17.30		27.05	38
4170	16" OC	G		74	.216		7.80	12.15		19.95	28
4180	24" OC	G		103	.155		5.85	8.75		14.60	20.50
4190	6" wide, studs 12" OC	G		51	.314		12.70	17.65		30.35	42
4200	16" OC	G		73	.219		10.25	12.35		22.60	31
4210	24" OC	G		101	.158		7.80	8.90		16.70	23
4220	8" wide, studs 12" OC	G		50	.320		16.60	18		34.60	47
4230	16" OC	G		72	.222		13.35	12.50		25.85	34.50
4240	24" OC	G		100	.160		10.10	9		19.10	25.50
4300	16 ga. x 2-1/2" wide, studs 12" OC	G		47	.340		10.50	19.15		29.65	42
4310	16" OC	G		68	.235		8.30	13.25		21.55	30
4320	24" OC	G		94	.170		6.10	9.60		15.70	22
4330	3-5/8" wide, studs 12" OC	G		46	.348		12.05	19.60		31.65	44.50
4340	16" OC	G		66	.242		9.55	13.65		23.20	32
4350	24" OC	G		92	.174		7.05	9.80		16.85	23.50
4360	4" wide, studs 12" OC	G		45	.356		12.95	20		32.95	46.50
4370	16" OC	G		65	.246		10.20	13.85		24.05	33
4380	24" OC	G		90	.178		7.45	10		17.45	24
4390	6" wide, studs 12" OC	G		44	.364		16.45	20.50		36.95	50.50
4400	16" OC	G		64	.250		13.05	14.10		27.15	37
4410	24" OC	G		88	.182		9.70	10.25		19.95	27
4420	8" wide, studs 12" OC	G		43	.372		19.75	21		40.75	55
4430	16" OC	G		63	.254		15.70	14.30		30	40

For customer support on your Commercial Renovation Costs with RSMeans data, call 800.448.8182.

119

05 41 13 – Load-Bearing Metal Stud Framing

05 41 13.30 Framing, Stud Walls		Crew	Daily Output	Labor-Hours	Unit	Material	2022 Bare Costs Labor	Equipment	Total	Total Incl O&P
4440	24" OC	G	2 Carp	86	.186	L.F.	11.65	10.45		22.10	29.50
5100	10' high walls, 18 ga. x 2-1/2" wide, studs 12" OC	G		54	.296		9.55	16.70		26.25	37
5110	16" OC	G		77	.208		7.60	11.70		19.30	27
5120	24" OC	G		107	.150		5.65	8.40		14.05	19.65
5130	3-5/8" wide, studs 12" OC	G		53	.302		11.40	17		28.40	39.50
5140	16" OC	G		76	.211		9.10	11.85		20.95	29
5150	24" OC	G		105	.152		6.75	8.60		15.35	21
5160	4" wide, studs 12" OC	G		52	.308		11.70	17.30		29	40.50
5170	16" OC	G		74	.216		9.25	12.15		21.40	29.50
5180	24" OC	G		103	.155		6.85	8.75		15.60	21.50
5190	6" wide, studs 12" OC	G		51	.314		15.10	17.65		32.75	44.50
5200	16" OC	G		73	.219		12.05	12.35		24.40	33
5210	24" OC	G		101	.158		9	8.90		17.90	24
5220	8" wide, studs 12" OC	G		50	.320		19.90	18		37.90	50.50
5230	16" OC	G		72	.222		15.80	12.50		28.30	37.50
5240	24" OC	G		100	.160		11.75	9		20.75	27
5300	16 ga. x 2-1/2" wide, studs 12" OC	G		47	.340		12.65	19.15		31.80	44.50
5310	16" OC	G		68	.235		9.95	13.25		23.20	32
5320	24" OC	G		94	.170		7.20	9.60		16.80	23
5330	3-5/8" wide, studs 12" OC	G		46	.348		14.50	19.60		34.10	47
5340	16" OC	G		66	.242		11.40	13.65		25.05	34
5350	24" OC	G		92	.174		8.30	9.80		18.10	24.50
5360	4" wide, studs 12" OC	G		45	.356		15.70	20		35.70	49.50
5370	16" OC	G		65	.246		12.25	13.85		26.10	35.50
5380	24" OC	G		90	.178		8.85	10		18.85	25.50
5390	6" wide, studs 12" OC	G		44	.364		19.80	20.50		40.30	54.50
5400	16" OC	G		64	.250		15.60	14.10		29.70	39.50
5410	24" OC	G	2 Carp	88	.182	L.F.	11.35	10.25		21.60	29
5420	8" wide, studs 12" OC	G		43	.372		24	21		45	59.50
5430	16" OC	G		63	.254		18.75	14.30		33.05	43
5440	24" OC	G		86	.186		13.70	10.45		24.15	31.50
6190	12' high walls, 18 ga. x 6" wide, studs 12" OC	G		41	.390		17.55	22		39.55	54.50
6200	16" OC	G		58	.276		13.90	15.55		29.45	40
6210	24" OC	G		81	.198		10.25	11.10		21.35	29
6220	8" wide, studs 12" OC	G		40	.400		23	22.50		45.50	61.50
6230	16" OC	G		57	.281		18.25	15.80		34.05	45
6240	24" OC	G		80	.200		13.35	11.25		24.60	32.50
6390	16 ga. x 6" wide, studs 12" OC	G		35	.457		23	25.50		48.50	66.50
6400	16" OC	G		51	.314		18.15	17.65		35.80	48
6410	24" OC	G		70	.229		13.05	12.85		25.90	35
6420	8" wide, studs 12" OC	G		34	.471		28	26.50		54.50	72.50
6430	16" OC	G		50	.320		22	18		40	52.50
6440	24" OC	G		69	.232		15.70	13.05		28.75	38
6530	14 ga. x 3-5/8" wide, studs 12" OC	G		34	.471		22.50	26.50		49	66.50
6540	16" OC	G		48	.333		17.35	18.75		36.10	49
6550	24" OC	G		65	.246		12.45	13.85		26.30	35.50
6560	4" wide, studs 12" OC	G		33	.485		23.50	27.50		51	69.50
6570	16" OC	G		47	.340		18.30	19.15		37.45	50.50
6580	24" OC	G		64	.250		13.15	14.10		27.25	37
6730	12 ga. x 3-5/8" wide, studs 12" OC	G		31	.516		30.50	29		59.50	79.50
6740	16" OC	G		43	.372		23.50	21		44.50	59.50
6750	24" OC	G		59	.271		16.65	15.25		31.90	43
6760	4" wide, studs 12" OC	G		30	.533		32	30		62	83

05 41 Structural Metal Stud Framing

05 41 13 – Load-Bearing Metal Stud Framing

05 41 13.30 Framing, Stud Walls

			Crew	Daily Output	Labor-Hours	Unit	Material	2022 Bare Costs Labor	Equipment	Total	Total Incl O&P
6770	16" OC	G	2 Carp	42	.381	L.F.	25	21.50		46.50	61.50
6780	24" OC	G		58	.276		17.55	15.55		33.10	44
7390	16' high walls, 16 ga. x 6" wide, studs 12" OC	G		33	.485		30	27.50		57.50	76.50
7400	16" OC	G		48	.333		23	18.75		41.75	55.50
7410	24" OC	G		67	.239		16.45	13.45		29.90	39.50
7420	8" wide, studs 12" OC	G		32	.500		36	28		64	84
7430	16" OC	G		47	.340		28	19.15		47.15	61
7440	24" OC	G		66	.242		19.75	13.65		33.40	43
7560	14 ga. x 4" wide, studs 12" OC	G		31	.516		30.50	29		59.50	79.50
7570	16" OC	G		45	.356		23.50	20		43.50	58
7580	24" OC	G		61	.262		16.55	14.75		31.30	42
7590	6" wide, studs 12" OC	G		30	.533		38.50	30		68.50	89.50
7600	16" OC	G		44	.364		29.50	20.50		50	65
7610	24" OC	G	2 Carp	60	.267	L.F.	21	15		36	47
7760	12 ga. x 4" wide, studs 12" OC	G		29	.552		42	31		73	95.50
7770	16" OC	G		40	.400		32	22.50		54.50	71.50
7780	24" OC	G		55	.291		22.50	16.40		38.90	50.50
7790	6" wide, studs 12" OC	G		28	.571		54	32		86	111
7800	16" OC	G		39	.410		41.50	23		64.50	82
7810	24" OC	G		54	.296		29	16.70		45.70	58.50
8590	20' high walls, 14 ga. x 6" wide, studs 12" OC	G		29	.552		47	31		78	101
8600	16" OC	G		42	.381		36	21.50		57.50	74
8610	24" OC	G		57	.281		25.50	15.80		41.30	53
8620	8" wide, studs 12" OC	G		28	.571		52.50	32		84.50	109
8630	16" OC	G		41	.390		40.50	22		62.50	79.50
8640	24" OC	G		56	.286		28.50	16.10		44.60	56.50
8790	12 ga. x 6" wide, studs 12" OC	G		27	.593		67	33.50		100.50	127
8800	16" OC	G		37	.432		51	24.50		75.50	94.50
8810	24" OC	G		51	.314		35	17.65		52.65	67
8820	8" wide, studs 12" OC	G		26	.615		79	34.50		113.50	142
8830	16" OC	G		36	.444		60.50	25		85.50	107
8840	24" OC	G		50	.320		41.50	18		59.50	74.50
9000	Minimum labor/equipment charge			4	4	Job		225		225	360

05 42 Cold-Formed Metal Joist Framing

05 42 13 – Cold-Formed Metal Floor Joist Framing

05 42 13.05 Bracing

			Crew	Daily Output	Labor-Hours	Unit	Material	2022 Bare Costs Labor	Equipment	Total	Total Incl O&P
0010	**BRACING**, continuous, per row, top & bottom										
0015	Made from recycled materials										
0120	Flat strap, 20 ga. x 2" wide, joists at 12" OC	G	1 Carp	4.67	1.713	C.L.F.	57.50	96.50		154	217
0130	16" OC	G		5.33	1.501		55.50	84.50		140	195
0140	24" OC	G		6.66	1.201		53.50	67.50		121	166
0150	18 ga. x 2" wide, joists at 12" OC	G		4	2		76.50	113		189.50	263
0160	16" OC	G		4.67	1.713		75	96.50		171.50	236
0170	24" OC	G		5.33	1.501		73.50	84.50		158	215

05 42 13.10 Bridging

			Crew	Daily Output	Labor-Hours	Unit	Material	2022 Bare Costs Labor	Equipment	Total	Total Incl O&P
0010	**BRIDGING**, solid between joists w/1-1/4" leg track, per joist bay										
0015	Made from recycled materials										
0230	Joists 12" OC, 18 ga. track x 6" wide	G	1 Carp	80	.100	Ea.	1.53	5.65		7.18	10.65
0240	8" wide	G		75	.107		1.88	6		7.88	11.60
0250	10" wide	G		70	.114		2.39	6.45		8.84	12.85

For customer support on your Commercial Renovation Costs with RSMeans data, call 800.448.8182.

121

05 42 13 – Cold-Formed Metal Floor Joist Framing

05 42 13.10 Bridging

			Crew	Daily Output	Labor-Hours	Unit	Material	2022 Bare Costs Labor	Equipment	Total	Total Incl O&P
0260	12" wide	G	1 Carp	65	.123	Ea.	2.66	6.95		9.61	13.95
0330	16 ga. track x 6" wide	G		70	.114		1.90	6.45		8.35	12.30
0340	8" wide	G		65	.123		2.34	6.95		9.29	13.60
0350	10" wide	G		60	.133		3.02	7.50		10.52	15.30
0360	12" wide	G		55	.145		3.47	8.20		11.67	16.80
0440	14 ga. track x 8" wide	G		60	.133		3.05	7.50		10.55	15.30
0450	10" wide	G		55	.145		3.71	8.20		11.91	17.10
0460	12" wide	G		50	.160		4.29	9		13.29	19
0550	12 ga. track x 10" wide	G		45	.178		5.45	10		15.45	22
0560	12" wide	G		40	.200		5.75	11.25		17	24
1230	16" OC, 18 ga. track x 6" wide	G		80	.100		1.96	5.65		7.61	11.10
1240	8" wide	G		75	.107		2.41	6		8.41	12.20
1250	10" wide	G		70	.114		3.07	6.45		9.52	13.60
1260	12" wide	G		65	.123		3.42	6.95		10.37	14.75
1330	16 ga. track x 6" wide	G		70	.114		2.44	6.45		8.89	12.90
1340	8" wide	G		65	.123		3	6.95		9.95	14.30
1350	10" wide	G		60	.133		3.88	7.50		11.38	16.20
1360	12" wide	G		55	.145		4.46	8.20		12.66	17.90
1440	14 ga. track x 8" wide	G		60	.133		3.91	7.50		11.41	16.25
1450	10" wide	G		55	.145		4.75	8.20		12.95	18.25
1460	12" wide	G		50	.160		5.50	9		14.50	20.50
1550	12 ga. track x 10" wide	G		45	.178		6.95	10		16.95	23.50
1560	12" wide	G		40	.200		7.35	11.25		18.60	26
2230	24" OC, 18 ga. track x 6" wide	G		80	.100		2.84	5.65		8.49	12.05
2240	8" wide	G		75	.107		3.49	6		9.49	13.40
2250	10" wide	G		70	.114		4.44	6.45		10.89	15.10
2260	12" wide	G		65	.123		4.94	6.95		11.89	16.45
2330	16 ga. track x 6" wide	G		70	.114		3.53	6.45		9.98	14.10
2340	8" wide	G		65	.123		4.34	6.95		11.29	15.80
2350	10" wide	G		60	.133		5.60	7.50		13.10	18.10
2360	12" wide	G		55	.145		6.45	8.20		14.65	20
2440	14 ga. track x 8" wide	G		60	.133		5.65	7.50		13.15	18.15
2450	10" wide	G		55	.145		6.85	8.20		15.05	20.50
2460	12" wide	G		50	.160		7.95	9		16.95	23
2550	12 ga. track x 10" wide	G		45	.178		10.05	10		20.05	27
2560	12" wide	G		40	.200		10.65	11.25		21.90	29.50

05 42 13.25 Framing, Band Joist

			Crew	Daily Output	Labor-Hours	Unit	Material	2022 Bare Costs Labor	Equipment	Total	Total Incl O&P
0010	**FRAMING, BAND JOIST** (track) fastened to bearing wall										
0015	Made from recycled materials										
0220	18 ga. track x 6" deep	G	2 Carp	1000	.016	L.F.	1.25	.90		2.15	2.80
0230	8" deep	G		920	.017		1.53	.98		2.51	3.25
0240	10" deep	G		860	.019		1.95	1.05		3	3.81
0320	16 ga. track x 6" deep	G		900	.018		1.55	1		2.55	3.30
0330	8" deep	G	2 Carp	840	.019	L.F.	1.91	1.07		2.98	3.80
0340	10" deep	G		780	.021		2.47	1.15		3.62	4.54
0350	12" deep	G		740	.022		2.84	1.22		4.06	5.05
0430	14 ga. track x 8" deep	G		750	.021		2.49	1.20		3.69	4.65
0440	10" deep	G		720	.022		3.02	1.25		4.27	5.30
0450	12" deep	G		700	.023		3.50	1.29		4.79	5.90
0540	12 ga. track x 10" deep	G		670	.024		4.43	1.34		5.77	7
0550	12" deep	G		650	.025		4.68	1.39		6.07	7.35

05 42 Cold-Formed Metal Joist Framing

05 42 13 – Cold-Formed Metal Floor Joist Framing

05 42 13.30 Framing, Boxed Headers/Beams

		Crew	Daily Output	Labor-Hours	Unit	Material	2022 Bare Costs Labor	Equipment	Total	Total Incl O&P
0010	**FRAMING, BOXED HEADERS/BEAMS**									
0015	Made from recycled materials									
0200	Double, 18 ga. x 6" deep G	2 Carp	220	.073	L.F.	5.45	4.09		9.54	12.50
0210	8" deep G		210	.076		5.65	4.29		9.94	13
0220	10" deep G		200	.080		7.25	4.50		11.75	15.10
0230	12" deep G		190	.084		7.90	4.74		12.64	16.25
0300	16 ga. x 8" deep G		180	.089		6.55	5		11.55	15.15
0310	10" deep G		170	.094		8.10	5.30		13.40	17.30
0320	12" deep G		160	.100		8.80	5.65		14.45	18.60
0400	14 ga. x 10" deep G		140	.114		9.55	6.45		16	20.50
0410	12" deep G		130	.123		10.45	6.95		17.40	22.50
0500	12 ga. x 10" deep G		110	.145		12.55	8.20		20.75	27
0510	12" deep G		100	.160		13.90	9		22.90	29.50
1210	Triple, 18 ga. x 8" deep G		170	.094		8.55	5.30		13.85	17.80
1220	10" deep G		165	.097		10.30	5.45		15.75	20
1230	12" deep G		160	.100		11.30	5.65		16.95	21.50
1300	16 ga. x 8" deep G		145	.110		9.90	6.20		16.10	21
1310	10" deep G		140	.114		11.60	6.45		18.05	23
1320	12" deep G		135	.119		12.65	6.65		19.30	24.50
1400	14 ga. x 10" deep G		115	.139		13.75	7.85		21.60	27.50
1410	12" deep G		110	.145		15.15	8.20		23.35	29.50
1500	12 ga. x 10" deep G		90	.178		18.30	10		28.30	36
1510	12" deep G		85	.188		20.50	10.60		31.10	39.50

05 42 13.40 Framing, Joists

		Crew	Daily Output	Labor-Hours	Unit	Material	2022 Bare Costs Labor	Equipment	Total	Total Incl O&P
0010	**FRAMING, JOISTS**, no band joists (track), web stiffeners, headers,									
0020	Beams, bridging or bracing									
0025	Made from recycled materials									
0030	Joists (2" flange) and fasteners, materials only									
0220	18 ga. x 6" deep G				L.F.	1.84			1.84	2.02
0230	8" deep G					1.94			1.94	2.14
0240	10" deep G					2.31			2.31	2.54
0320	16 ga. x 6" deep G				L.F.	1.98			1.98	2.18
0330	8" deep G					2.43			2.43	2.67
0340	10" deep G					2.77			2.77	3.05
0350	12" deep G					3.13			3.13	3.44
0430	14 ga. x 8" deep G					3.06			3.06	3.36
0440	10" deep G					3.52			3.52	3.87
0450	12" deep G					4			4	4.40
0540	12 ga. x 10" deep G					5.10			5.10	5.60
0550	12" deep G					5.80			5.80	6.40
1010	Installation of joists to band joists, beams & headers, labor only									
1220	18 ga. x 6" deep	2 Carp	110	.145	Ea.		8.20		8.20	13
1230	8" deep		90	.178			10		10	15.90
1240	10" deep		80	.200			11.25		11.25	17.90
1320	16 ga. x 6" deep		95	.168			9.50		9.50	15.05
1330	8" deep		70	.229			12.85		12.85	20.50
1340	10" deep		60	.267			15		15	24
1350	12" deep		55	.291			16.40		16.40	26
1430	14 ga. x 8" deep		65	.246			13.85		13.85	22
1440	10" deep		45	.356			20		20	32
1450	12" deep		35	.457			25.50		25.50	41
1540	12 ga. x 10" deep		40	.400			22.50		22.50	36

For customer support on your Commercial Renovation Costs with RSMeans data, call 800.448.8182.

123

05 42 13 – Cold-Formed Metal Floor Joist Framing

05 42 13.40 Framing, Joists		Crew	Daily Output	Labor-Hours	Unit	Material	2022 Bare Costs Labor	Equipment	Total	Total Incl O&P
1550	12" deep	2 Carp	30	.533	Ea.		30		30	47.50
9000	Minimum labor/equipment charge	↓	4	4	Job		225		225	360

05 42 13.45 Framing, Web Stiffeners

	05 42 13.45 Framing, Web Stiffeners		Crew	Daily Output	Labor-Hours	Unit	Material	2022 Bare Costs Labor	Equipment	Total	Total Incl O&P
0010	**FRAMING, WEB STIFFENERS** at joist bearing, fabricated from										
0020	Stud piece (1-5/8" flange) to stiffen joist (2" flange)										
0025	Made from recycled materials										
2120	For 6" deep joist, with 18 ga. x 2-1/2" stud	G	1 Carp	120	.067	Ea.	.78	3.75		4.53	6.80
2130	3-5/8" stud	G		110	.073		.93	4.09		5.02	7.50
2140	4" stud	G		105	.076		.97	4.29		5.26	7.85
2150	6" stud	G		100	.080		1.22	4.50		5.72	8.50
2160	8" stud	G		95	.084		1.63	4.74		6.37	9.35
2220	8" deep joist, with 2-1/2" stud	G		120	.067		1.05	3.75		4.80	7.10
2230	3-5/8" stud	G		110	.073		1.25	4.09		5.34	7.85
2240	4" stud	G		105	.076		1.30	4.29		5.59	8.25
2250	6" stud	G		100	.080		1.63	4.50		6.13	8.95
2260	8" stud	G		95	.084		2.18	4.74		6.92	9.95
2320	10" deep joist, with 2-1/2" stud	G		110	.073		1.29	4.09		5.38	7.90
2330	3-5/8" stud	G		100	.080		1.54	4.50		6.04	8.85
2340	4" stud	G		95	.084		1.61	4.74		6.35	9.30
2350	6" stud	G		90	.089		2.03	5		7.03	10.20
2360	8" stud	G		85	.094		2.71	5.30		8.01	11.40
2420	12" deep joist, with 2-1/2" stud	G		110	.073		1.56	4.09		5.65	8.20
2430	3-5/8" stud	G		100	.080		1.86	4.50		6.36	9.20
2440	4" stud	G		95	.084		1.94	4.74		6.68	9.70
2450	6" stud	G		90	.089		2.44	5		7.44	10.65
2460	8" stud	G		85	.094		3.26	5.30		8.56	12
3130	For 6" deep joist, with 16 ga. x 3-5/8" stud	G		100	.080		1.24	4.50		5.74	8.50
3140	4" stud	G		95	.084		1.37	4.74		6.11	9.05
3150	6" stud	G		90	.089		1.69	5		6.69	9.80
3160	8" stud	G		85	.094		2.02	5.30		7.32	10.60
3230	8" deep joist, with 3-5/8" stud	G		100	.080		1.66	4.50		6.16	9
3240	4" stud	G		95	.084		1.84	4.74		6.58	9.55
3250	6" stud	G		90	.089		2.26	5		7.26	10.45
3260	8" stud	G		85	.094		2.71	5.30		8.01	11.40
3330	10" deep joist, with 3-5/8" stud	G		85	.094		2.06	5.30		7.36	10.65
3340	4" stud	G		80	.100		2.27	5.65		7.92	11.45
3350	6" stud	G		75	.107		2.81	6		8.81	12.65
3360	8" stud	G		70	.114		3.35	6.45		9.80	13.90
3430	12" deep joist, with 3-5/8" stud	G		85	.094		2.48	5.30		7.78	11.15
3440	4" stud	G		80	.100		2.74	5.65		8.39	11.95
3450	6" stud	G		75	.107		3.38	6		9.38	13.25
3460	8" stud	G		70	.114		4.04	6.45		10.49	14.65
4230	For 8" deep joist, with 14 ga. x 3-5/8" stud	G		90	.089		2.20	5		7.20	10.35
4240	4" stud	G		85	.094		2.29	5.30		7.59	10.90
4250	6" stud	G		80	.100		2.91	5.65		8.56	12.15
4260	8" stud	G		75	.107		3.20	6		9.20	13.05
4330	10" deep joist, with 3-5/8" stud	G		75	.107		2.72	6		8.72	12.55
4340	4" stud	G		70	.114		2.84	6.45		9.29	13.30
4350	6" stud	G		65	.123		3.60	6.95		10.55	14.95
4360	8" stud	G		60	.133		3.97	7.50		11.47	16.30
4430	12" deep joist, with 3-5/8" stud	G		75	.107		3.28	6		9.28	13.15
4440	4" stud	G		70	.114		3.42	6.45		9.87	13.95

05 42 Cold-Formed Metal Joist Framing

05 42 13 – Cold-Formed Metal Floor Joist Framing

05 42 13.45 Framing, Web Stiffeners

		Crew	Daily Output	Labor-Hours	Unit	Material	2022 Bare Costs Labor	Equipment	Total	Total Incl O&P
4450	6" stud	G 1 Carp	65	.123	Ea.	4.34	6.95		11.29	15.75
4460	8" stud	G	60	.133		4.78	7.50		12.28	17.20
5330	For 10" deep joist, with 12 ga. x 3-5/8" stud	G	65	.123		3.88	6.95		10.83	15.25
5340	4" stud	G	60	.133		4.05	7.50		11.55	16.40
5350	6" stud	G	55	.145		5.25	8.20		13.45	18.75
5360	8" stud	G 1 Carp	50	.160	Ea.	6.20	9		15.20	21
5430	12" deep joist, with 3-5/8" stud	G	65	.123		4.68	6.95		11.63	16.15
5440	4" stud	G	60	.133		4.88	7.50		12.38	17.30
5450	6" stud	G	55	.145		6.30	8.20		14.50	19.95
5460	8" stud	G	50	.160		7.45	9		16.45	22.50

05 42 23 – Cold-Formed Metal Roof Joist Framing

05 42 23.05 Framing, Bracing

		Crew	Daily Output	Labor-Hours	Unit	Material	2022 Bare Costs Labor	Equipment	Total	Total Incl O&P
0010	**FRAMING, BRACING**									
0015	Made from recycled materials									
0020	Continuous bracing, per row									
0100	16 ga. x 1-1/2" channel thru rafters/trusses @ 16" OC	G 1 Carp	4.50	1.778	C.L.F.	51.50	100		151.50	216
0120	24" OC	G	6	1.333		51.50	75		126.50	176
0300	2" x 2" angle x 18 ga., rafters/trusses @ 16" OC	G	6	1.333		78	75		153	205
0320	24" OC	G	8	1		78	56.50		134.50	176
0400	16 ga., rafters/trusses @ 16" OC	G	4.50	1.778		98	100		198	267
0420	24" OC	G	6.50	1.231		98	69.50		167.50	218

05 42 23.10 Framing, Bridging

		Crew	Daily Output	Labor-Hours	Unit	Material	2022 Bare Costs Labor	Equipment	Total	Total Incl O&P
0010	**FRAMING, BRIDGING**									
0015	Made from recycled materials									
0020	Solid, between rafters w/1-1/4" leg track, per rafter bay									
1200	Rafters 16" OC, 18 ga. x 4" deep	G 1 Carp	60	.133	Ea.	1.34	7.50		8.84	13.40
1210	6" deep	G	57	.140		1.96	7.90		9.86	14.70
1220	8" deep	G	55	.145		2.41	8.20		10.61	15.65
1230	10" deep	G	52	.154		3.07	8.65		11.72	17.15
1240	12" deep	G	50	.160		3.42	9		12.42	18.05
2200	24" OC, 18 ga. x 4" deep	G	60	.133		1.93	7.50		9.43	14.10
2210	6" deep	G	57	.140		2.84	7.90		10.74	15.65
2220	8" deep	G	55	.145		3.49	8.20		11.69	16.85
2230	10" deep	G	52	.154		4.44	8.65		13.09	18.65
2240	12" deep	G	50	.160		4.94	9		13.94	19.75

05 42 23.50 Framing, Parapets

		Crew	Daily Output	Labor-Hours	Unit	Material	2022 Bare Costs Labor	Equipment	Total	Total Incl O&P
0010	**FRAMING, PARAPETS**									
0015	Made from recycled materials									
0100	3' high installed on 1st story, 18 ga. x 4" wide studs, 12" OC	G 2 Carp	100	.160	L.F.	4.89	9		13.89	19.70
0110	16" OC	G	150	.107		4.17	6		10.17	14.15
0120	24" OC	G	200	.080		3.44	4.50		7.94	10.95
0200	6" wide studs, 12" OC	G	100	.160		6.60	9		15.60	21.50
0210	16" OC	G	150	.107		5.65	6		11.65	15.80
0220	24" OC	G	200	.080		4.75	4.50		9.25	12.35
1100	Installed on 2nd story, 18 ga. x 4" wide studs, 12" OC	G	95	.168		4.89	9.50		14.39	20.50
1110	16" OC	G 2 Carp	145	.110	L.F.	4.17	6.20		10.37	14.45
1120	24" OC	G	190	.084		3.44	4.74		8.18	11.35
1200	6" wide studs, 12" OC	G	95	.168		6.60	9.50		16.10	22.50
1210	16" OC	G	145	.110		5.65	6.20		11.85	16.10
1220	24" OC	G	190	.084		4.75	4.74		9.49	12.75
2100	Installed on gable, 18 ga. x 4" wide studs, 12" OC	G	85	.188		4.89	10.60		15.49	22.50
2110	16" OC	G	130	.123		4.17	6.95		11.12	15.60

05 42 23 – Cold-Formed Metal Roof Joist Framing

05 42 23.50 Framing, Parapets		Crew	Daily Output	Labor-Hours	Unit	Material	2022 Bare Costs Labor	Equipment	Total	Total Incl O&P
2120	24" OC	G 2 Carp	170	.094	L.F.	3.44	5.30		8.74	12.20
2200	6" wide studs, 12" OC	G	85	.188		6.60	10.60		17.20	24
2210	16" OC	G	130	.123		5.65	6.95		12.60	17.25
2220	24" OC	G	170	.094		4.75	5.30		10.05	13.60

05 42 23.60 Framing, Roof Rafters										
0010	**FRAMING, ROOF RAFTERS**									
0015	Made from recycled materials									
0100	Boxed ridge beam, double, 18 ga. x 6" deep	G 2 Carp	160	.100	L.F.	5.45	5.65		11.10	14.95
0110	8" deep	G	150	.107		5.65	6		11.65	15.75
0120	10" deep	G	140	.114		7.25	6.45		13.70	18.15
0130	12" deep	G	130	.123		7.90	6.95		14.85	19.70
0200	16 ga. x 6" deep	G	150	.107		5.70	6		11.70	15.85
0210	8" deep	G	140	.114		6.55	6.45		13	17.40
0220	10" deep	G	130	.123		8.10	6.95		15.05	19.90
0230	12" deep	G	120	.133		8.80	7.50		16.30	21.50
1100	Rafters, 2" flange, material only, 18 ga. x 6" deep	G				1.84			1.84	2.02
1110	8" deep	G				1.94			1.94	2.14
1120	10" deep	G				2.31			2.31	2.54
1130	12" deep	G				2.66			2.66	2.92
1200	16 ga. x 6" deep	G				1.98			1.98	2.18
1210	8" deep	G				2.43			2.43	2.67
1220	10" deep	G				2.77			2.77	3.05
1230	12" deep	G				3.13			3.13	3.44
2100	Installation only, ordinary rafter to 4:12 pitch, 18 ga. x 6" deep	2 Carp	35	.457	Ea.		25.50		25.50	41
2110	8" deep		30	.533			30		30	47.50
2120	10" deep		25	.640			36		36	57.50
2130	12" deep		20	.800			45		45	71.50
2200	16 ga. x 6" deep		30	.533			30		30	47.50
2210	8" deep		25	.640			36		36	57.50
2220	10" deep		20	.800			45		45	71.50
2230	12" deep		15	1.067			60		60	95.50
8100	Add to labor, ordinary rafters on steep roofs						25%			
8110	Dormers & complex roofs						50%			
8200	Hip & valley rafters to 4:12 pitch						25%			
8210	Steep roofs						50%			
8220	Dormers & complex roofs						75%			
8300	Hip & valley jack rafters to 4:12 pitch						50%			
8310	Steep roofs						75%			
8320	Dormers & complex roofs						100%			
9000	Minimum labor/equipment charge	2 Carp	4	4	Job		225		225	360

05 42 23.70 Framing, Soffits and Canopies										
0010	**FRAMING, SOFFITS & CANOPIES**									
0015	Made from recycled materials									
0130	Continuous ledger track @ wall, studs @ 16" OC, 18 ga. x 4" wide	G 2 Carp	535	.030	L.F.	.89	1.68		2.57	3.66
0140	6" wide	G	500	.032		1.31	1.80		3.11	4.30
0150	8" wide	G	465	.034		1.61	1.94		3.55	4.85
0160	10" wide	G	430	.037		2.05	2.09		4.14	5.60
0230	Studs @ 24" OC, 18 ga. x 4" wide	G	800	.020		.85	1.13		1.98	2.73
0240	6" wide	G	750	.021		1.25	1.20		2.45	3.28
0250	8" wide	G	700	.023		1.53	1.29		2.82	3.73
0260	10" wide	G	650	.025		1.95	1.39		3.34	4.35
1000	Horizontal soffit and canopy members, material only									

05 42 Cold-Formed Metal Joist Framing

05 42 23 – Cold-Formed Metal Roof Joist Framing

05 42 23.70 Framing, Soffits and Canopies

			Crew	Daily Output	Labor-Hours	Unit	Material	2022 Bare Costs Labor	Equipment	Total	Total Incl O&P
1030	1-5/8" flange studs, 18 ga. x 4" deep	G				L.F.	1.16			1.16	1.28
1040	6" deep	G					1.46			1.46	1.61
1050	8" deep	G					1.96			1.96	2.15
1140	2" flange joists, 18 ga. x 6" deep	G					2.10			2.10	2.31
1150	8" deep	G					2.22			2.22	2.44
1160	10" deep	G					2.64			2.64	2.90
4030	Installation only, 18 ga., 1-5/8" flange x 4" deep		2 Carp	130	.123	Ea.		6.95		6.95	11
4040	6" deep			110	.145			8.20		8.20	13
4050	8" deep			90	.178			10		10	15.90
4140	2" flange, 18 ga. x 6" deep			110	.145			8.20		8.20	13
4150	8" deep			90	.178			10		10	15.90
4160	10" deep			80	.200			11.25		11.25	17.90
6010	Clips to attach fascia to rafter tails, 2" x 2" x 18 ga. angle	G	1 Carp	120	.067		.92	3.75		4.67	6.95
6020	16 ga. angle	G	"	100	.080		1.16	4.50		5.66	8.40
9000	Minimum labor/equipment charge		2 Carp	4	4	Job		225		225	360

05 44 Cold-Formed Metal Trusses

05 44 13 – Cold-Formed Metal Roof Trusses

05 44 13.60 Framing, Roof Trusses

			Crew	Daily Output	Labor-Hours	Unit	Material	2022 Bare Costs Labor	Equipment	Total	Total Incl O&P
0010	**FRAMING, ROOF TRUSSES**										
0015	Made from recycled materials										
0020	Fabrication of trusses on ground, Fink (W) or King Post, to 4:12 pitch										
0120	18 ga. x 4" chords, 16' span	G	2 Carp	12	1.333	Ea.	54.50	75		129.50	179
0130	20' span	G		11	1.455		68	82		150	205
0140	24' span	G		11	1.455		81.50	82		163.50	220
0150	28' span	G		10	1.600		95	90		185	248
0160	32' span	G		10	1.600		109	90		199	263
0250	6" chords, 28' span	G		9	1.778		120	100		220	291
0260	32' span	G		9	1.778		137	100		237	310
0270	36' span	G		8	2		154	113		267	350
0280	40' span	G		8	2		171	113		284	365
1120	5:12 to 8:12 pitch, 18 ga. x 4" chords, 16' span	G		10	1.600		62	90		152	212
1130	20' span	G		9	1.778		77.50	100		177.50	245
1140	24' span	G		9	1.778		93	100		193	261
1150	28' span	G		8	2		109	113		222	299
1160	32' span	G		8	2		124	113		237	315
1250	6" chords, 28' span	G		7	2.286		137	129		266	355
1260	32' span	G		7	2.286		156	129		285	375
1270	36' span	G		6	2.667		176	150		326	430
1280	40' span	G		6	2.667		195	150		345	455
2120	9:12 to 12:12 pitch, 18 ga. x 4" chords, 16' span	G		8	2		77.50	113		190.50	265
2130	20' span	G		7	2.286		97	129		226	310
2140	24' span	G		7	2.286		116	129		245	330
2150	28' span	G		6	2.667		136	150		286	390
2160	32' span	G		6	2.667		155	150		305	410
2250	6" chords, 28' span	G		5	3.200		171	180		351	475
2260	32' span	G		5	3.200		195	180		375	500
2270	36' span	G		4	4		220	225		445	600
2280	40' span	G		4	4		244	225		469	630
4900	Minimum labor/equipment charge			4	4	Job		225		225	360
5120	Erection only of roof trusses, to 4:12 pitch, 16' span		F-6	48	.833	Ea.		44.50	41.50	86	116

For customer support on your Commercial Renovation Costs with RSMeans data, call 800.448.8182.

127

05 44 Cold-Formed Metal Trusses

05 44 13 – Cold-Formed Metal Roof Trusses

05 44 13.60 Framing, Roof Trusses

		Crew	Daily Output	Labor-Hours	Unit	Material	2022 Bare Costs Labor	Equipment	Total	Total Incl O&P
5130	20' span	F-6	46	.870	Ea.		46.50	43	89.50	121
5140	24' span		44	.909			48.50	45	93.50	127
5150	28' span		42	.952			51	47	98	133
5160	32' span		40	1			53.50	49.50	103	139
5170	36' span		38	1.053			56.50	52	108.50	147
5180	40' span		36	1.111			59.50	55	114.50	155
5220	5:12 to 8:12 pitch, 16' span		42	.952			51	47	98	133
5230	20' span		40	1			53.50	49.50	103	139
5240	24' span		38	1.053			56.50	52	108.50	147
5250	28' span		36	1.111			59.50	55	114.50	155
5260	32' span		34	1.176			63	58.50	121.50	164
5270	36' span		32	1.250			67	62	129	174
5280	40' span		30	1.333			71.50	66	137.50	186
5320	9:12 to 12:12 pitch, 16' span		36	1.111			59.50	55	114.50	155
5330	20' span		34	1.176			63	58.50	121.50	164
5340	24' span		32	1.250			67	62	129	174
5350	28' span		30	1.333			71.50	66	137.50	186
5360	32' span		28	1.429			76.50	71	147.50	199
5370	36' span		26	1.538			82	76.50	158.50	214
5380	40' span		24	1.667			89	82.50	171.50	232
9000	Minimum labor/equipment charge		2	20	Job		1,075	990	2,065	2,800

05 51 Metal Stairs

05 51 13 – Metal Pan Stairs

05 51 13.50 Pan Stairs

			Crew	Daily Output	Labor-Hours	Unit	Material	2022 Bare Costs Labor	Equipment	Total	Total Incl O&P
0010	**PAN STAIRS**, shop fabricated, steel stringers										
0015	Made from recycled materials										
0200	Metal pan tread for concrete in-fill, picket rail, 3'-6" wide	G	E-4	35	.914	Riser	570	57	4.21	631.21	725
0300	4'-0" wide	G		30	1.067		565	66.50	4.92	636.42	730
0350	Wall rail, both sides, 3'-6" wide	G		53	.604		490	37.50	2.78	530.28	605
1500	Landing, steel pan, conventional	G		160	.200	S.F.	75.50	12.45	.92	88.87	104

05 51 16 – Metal Floor Plate Stairs

05 51 16.50 Floor Plate Stairs

			Crew	Daily Output	Labor-Hours	Unit	Material	2022 Bare Costs Labor	Equipment	Total	Total Incl O&P
0010	**FLOOR PLATE STAIRS**, shop fabricated, steel stringers										
0015	Made from recycled materials										
0400	Cast iron tread and pipe rail, 3'-6" wide	G	E-4	35	.914	Riser	540	57	4.21	601.21	690
0450	4'-0" wide	G		30	1.067		565	66.50	4.92	636.42	735
0475	5'-0" wide	G		28	1.143		625	71.50	5.25	701.75	805
0500	Checkered plate tread, industrial, 3'-6" wide	G		28	1.143		310	71.50	5.25	386.75	460
0550	Circular, for tanks, 3'-0" wide	G		33	.970		350	60.50	4.47	414.97	485
0600	For isolated stairs, add							100%			
0800	Custom steel stairs, 3'-6" wide, economy	G	E-4	35	.914		475	57	4.21	536.21	615
0810	Medium priced	G		30	1.067		625	66.50	4.92	696.42	800
0900	Deluxe	G		20	1.600		785	100	7.40	892.40	1,025
1100	For 4' wide stairs, add							5%	5%		
1300	For 5' wide stairs, add							10%	10%		

05 51 Metal Stairs

05 51 19 – Metal Grating Stairs

05 51 19.50 Grating Stairs		Crew	Daily Output	Labor-Hours	Unit	Material	2022 Bare Costs Labor	Equipment	Total	Total Incl O&P
0010	**GRATING STAIRS**, shop fabricated, steel stringers, safety nosing on treads									
0015	Made from recycled materials									
0020	Grating tread and pipe railing, 3'-6" wide	G E-4	35	.914	Riser	370	57	4.21	431.21	505
0100	4'-0" wide	G	30	1.067	"	365	66.50	4.92	436.42	515
9000	Minimum labor/equipment charge		2	16	Job		1,000	74	1,074	1,675

05 51 23 – Metal Fire Escapes

05 51 23.25 Fire Escapes

		Crew	Daily Output	Labor-Hours	Unit	Material	Labor	Equipment	Total	Total Incl O&P
0010	**FIRE ESCAPES**, shop fabricated									
0200	2' wide balcony, 1" x 1/4" bars 1-1/2" OC, with railing	G 2 Sswk	10	1.600	L.F.	55.50	99		154.50	220
0400	1st story cantilevered stair, standard, with railing	G	.50	32	Ea.	2,225	1,975		4,200	5,625
0500	Cable counterweighted, with railing	G	.40	40	"	2,125	2,475		4,600	6,325
0700	36" x 40" platform & fixed stair, with railing	G	.40	40	Flight	985	2,475		3,460	5,050
0900	For 3'-6" wide escapes, add to above					100%	150%			

05 51 23.50 Fire Escape Stairs

		Crew	Daily Output	Labor-Hours	Unit	Material	Labor	Equipment	Total	Total Incl O&P
0010	**FIRE ESCAPE STAIRS**, portable									
0100	Portable ladder				Ea.	126			126	139

05 51 33 – Metal Ladders

05 51 33.13 Vertical Metal Ladders

		Crew	Daily Output	Labor-Hours	Unit	Material	Labor	Equipment	Total	Total Incl O&P
0010	**VERTICAL METAL LADDERS**, shop fabricated									
0015	Made from recycled materials									
0020	Steel, 20" wide, bolted to concrete, with cage	G E-4	50	.640	V.L.F.	122	40	2.95	164.95	202
0100	Without cage	G	85	.376		54	23.50	1.74	79.24	99.50
0300	Aluminum, bolted to concrete, with cage	G	50	.640		119	40	2.95	161.95	199
0400	Without cage	G	85	.376		53.50	23.50	1.74	78.74	99
9000	Minimum labor/equipment charge		2	16	Job		1,000	74	1,074	1,675

05 51 33.16 Inclined Metal Ladders

		Crew	Daily Output	Labor-Hours	Unit	Material	Labor	Equipment	Total	Total Incl O&P
0010	**INCLINED METAL LADDERS**, shop fabricated									
0015	Made from recycled materials									
3900	Industrial ships ladder, steel, 24" W, grating treads, 2 line pipe rail	G E-4	30	1.067	Riser	239	66.50	4.92	310.42	375
4000	Aluminum	G "	30	1.067	"	405	66.50	4.92	476.42	555

05 51 33.23 Alternating Tread Ladders

		Crew	Daily Output	Labor-Hours	Unit	Material	Labor	Equipment	Total	Total Incl O&P
0010	**ALTERNATING TREAD LADDERS**, shop fabricated									
0015	Made from recycled materials									
0800	Alternating tread ladders, 68° angle of incline									
0810	8' vertical rise, steel, 149 lb., standard paint color	B-68G	3	5.333	Ea.	2,075	330	134	2,539	2,975
0820	Non-standard paint color		3	5.333		2,475	330	134	2,939	3,400
0830	Galvanized		3	5.333		2,500	330	134	2,964	3,425
0840	Stainless		3	5.333		3,600	330	134	4,064	4,650
0850	Aluminum, 87 lb.		3	5.333		2,650	330	134	3,114	3,575
1010	10' vertical rise, steel, 181 lb., standard paint color		2.75	5.818		2,575	360	147	3,082	3,600
1020	Non-standard paint color		2.75	5.818		2,950	360	147	3,457	3,975
1030	Galvanized		2.75	5.818		3,025	360	147	3,532	4,075
1040	Stainless		2.75	5.818		4,350	360	147	4,857	5,525
1050	Aluminum, 103 lb.		2.75	5.818		3,175	360	147	3,682	4,250
1210	12' vertical rise, steel, 245 lb., standard paint color		2.50	6.400		3,025	395	161	3,581	4,125
1220	Non-standard paint color		2.50	6.400		3,425	395	161	3,981	4,575
1230	Galvanized		2.50	6.400		3,525	395	161	4,081	4,675
1240	Stainless		2.50	6.400		5,075	395	161	5,631	6,400
1250	Aluminum, 103 lb.		2.50	6.400		3,725	395	161	4,281	4,900
1410	14' vertical rise, steel, 281 lb., standard paint color		2.25	7.111		3,475	440	179	4,094	4,725

05 51 Metal Stairs

05 51 33 – Metal Ladders

05 51 33.23 Alternating Tread Ladders		Crew	Daily Output	Labor-Hours	Unit	Material	2022 Bare Costs Labor	Equipment	Total	Total Incl O&P
1420	Non-standard paint color	B-68G	2.25	7.111	Ea.	3,900	440	179	4,519	5,200
1430	Galvanized		2.25	7.111		4,050	440	179	4,669	5,375
1440	Stainless		2.25	7.111		5,825	440	179	6,444	7,300
1450	Aluminum, 136 lb.		2.25	7.111		4,250	440	179	4,869	5,575
1610	16' vertical rise, steel, 317 lb., standard paint color		2	8		3,925	495	202	4,622	5,350
1620	Non-standard paint color		2	8		4,400	495	202	5,097	5,850
1630	Galvanized		2	8		4,575	495	202	5,272	6,050
1640	Stainless		2	8		6,550	495	202	7,247	8,250
1650	Aluminum, 153 lb.		2	8		4,800	495	202	5,497	6,300

05 52 Metal Railings

05 52 13 – Pipe and Tube Railings

05 52 13.50 Railings, Pipe

			Crew	Daily Output	Labor-Hours	Unit	Material	2022 Bare Costs Labor	Equipment	Total	Total Incl O&P
0010	**RAILINGS, PIPE**, shop fab'd, 3'-6" high, posts @ 5' OC										
0015	Made from recycled materials										
0020	Aluminum, 2 rail, satin finish, 1-1/4" diameter	G	E-4	160	.200	L.F.	54	12.45	.92	67.37	80
0030	Clear anodized	G		160	.200		63.50	12.45	.92	76.87	91
0040	Dark anodized	G		160	.200		72	12.45	.92	85.37	100
0080	1-1/2" diameter, satin finish	G		160	.200		60.50	12.45	.92	73.87	88
0090	Clear anodized	G		160	.200		68.50	12.45	.92	81.87	96
0100	Dark anodized	G		160	.200		75	12.45	.92	88.37	104
0140	Aluminum, 3 rail, 1-1/4" diam., satin finish	G		137	.234		69	14.55	1.08	84.63	101
0150	Clear anodized	G		137	.234		95.50	14.55	1.08	111.13	130
0160	Dark anodized	G		137	.234		94.50	14.55	1.08	110.13	129
0200	1-1/2" diameter, satin finish	G		137	.234		90.50	14.55	1.08	106.13	124
0210	Clear anodized	G		137	.234		93	14.55	1.08	108.63	127
0220	Dark anodized	G		137	.234		114	14.55	1.08	129.63	150
0500	Steel, 2 rail, on stairs, primed, 1-1/4" diameter	G		160	.200		38.50	12.45	.92	51.87	63.50
0520	1-1/2" diameter	G		160	.200		32.50	12.45	.92	45.87	57
0540	Galvanized, 1-1/4" diameter	G		160	.200		49.50	12.45	.92	62.87	75.50
0560	1-1/2" diameter	G	E-4	160	.200	L.F.	58.50	12.45	.92	71.87	85.50
0580	Steel, 3 rail, primed, 1-1/4" diameter	G		137	.234		54.50	14.55	1.08	70.13	84.50
0600	1-1/2" diameter	G		137	.234		60	14.55	1.08	75.63	90.50
0620	Galvanized, 1-1/4" diameter	G		137	.234		76.50	14.55	1.08	92.13	109
0640	1-1/2" diameter	G		137	.234		92	14.55	1.08	107.63	126
0700	Stainless steel, 2 rail, 1-1/4" diam., #4 finish	G		137	.234		155	14.55	1.08	170.63	195
0720	High polish	G		137	.234		251	14.55	1.08	266.63	300
0740	Mirror polish	G		137	.234		310	14.55	1.08	325.63	370
0760	Stainless steel, 3 rail, 1-1/2" diam., #4 finish	G		120	.267		232	16.65	1.23	249.88	284
0770	High polish	G		120	.267		385	16.65	1.23	402.88	455
0780	Mirror finish	G		120	.267		470	16.65	1.23	487.88	545
0900	Wall rail, alum. pipe, 1-1/4" diam., satin finish	G		213	.150		25.50	9.35	.69	35.54	44
0905	Clear anodized	G		213	.150		35.50	9.35	.69	45.54	55
0910	Dark anodized	G		213	.150		41.50	9.35	.69	51.54	61.50
0915	1-1/2" diameter, satin finish	G		213	.150		31.50	9.35	.69	41.54	50.50
0920	Clear anodized	G		213	.150		39.50	9.35	.69	49.54	59.50
0925	Dark anodized	G		213	.150		49.50	9.35	.69	59.54	70.50
0930	Steel pipe, 1-1/4" diameter, primed	G		213	.150		22	9.35	.69	32.04	40
0935	Galvanized	G		213	.150		32	9.35	.69	42.04	51
0940	1-1/2" diameter	G		176	.182		18.65	11.35	.84	30.84	39.50
0945	Galvanized	G		213	.150		32	9.35	.69	42.04	51

05 52 Metal Railings

05 52 13 – Pipe and Tube Railings

05 52 13.50 Railings, Pipe

			Crew	Daily Output	Labor-Hours	Unit	Material	2022 Bare Costs Labor	Equipment	Total	Total Incl O&P
0955	Stainless steel pipe, 1-1/2" diam., #4 finish	G	E-4	107	.299	L.F.	124	18.65	1.38	144.03	168
0960	High polish	G		107	.299		252	18.65	1.38	272.03	310
0965	Mirror polish	G		107	.299		298	18.65	1.38	318.03	360
2000	2-line pipe rail (1-1/2" T&B) with 1/2" pickets @ 4-1/2" OC,										
2005	attached handrail on brackets										
2010	42" high aluminum, satin finish, straight & level	G	E-4	120	.267	L.F.	415	16.65	1.23	432.88	485
2050	42" high steel, primed, straight & level	G	"	120	.267		157	16.65	1.23	174.88	200
4000	For curved and level rails, add							10%	10%		
4100	For sloped rails for stairs, add							30%	30%		
9000	Minimum labor/equipment charge		1 Sswk	2	4	Job		247		247	400

05 52 16 – Industrial Railings

05 52 16.50 Railings, Industrial

			Crew	Daily Output	Labor-Hours	Unit	Material	2022 Bare Costs Labor	Equipment	Total	Total Incl O&P
0010	**RAILINGS, INDUSTRIAL**, shop fab'd, 3'-6" high, posts @ 5' OC										
0020	2 rail, 3'-6" high, 1-1/2" pipe	G	E-4	255	.125	L.F.	43	7.80	.58	51.38	60
0200	For 4" high kick plate, 10 ga., add	G						6.05		6.05	6.65
0500	For curved level rails, add							10%	10%		
0550	For sloped rails for stairs, add						L.F.	30%	30%		
9000	Minimum labor/equipment charge		1 Sswk	2	4	Job		247		247	400

05 55 Metal Stair Treads and Nosings

05 55 13 – Metal Stair Treads

05 55 13.50 Stair Treads

			Crew	Daily Output	Labor-Hours	Unit	Material	2022 Bare Costs Labor	Equipment	Total	Total Incl O&P
0010	**STAIR TREADS**, stringers and bolts not included										
3000	Diamond plate treads, steel, 1/8" thick										
3005	Open riser, black enamel										
3010	9" deep x 36" long	G	2 Sswk	48	.333	Ea.	114	20.50		134.50	158
3020	42" long	G		48	.333		120	20.50		140.50	165
3030	48" long	G		48	.333		125	20.50		145.50	171
3040	11" deep x 36" long	G		44	.364		120	22.50		142.50	168
3050	42" long	G		44	.364		127	22.50		149.50	176
3060	48" long	G		44	.364		135	22.50		157.50	185
3110	Galvanized, 9" deep x 36" long	G		48	.333		180	20.50		200.50	231
3120	42" long	G		48	.333		191	20.50		211.50	243
3130	48" long	G		48	.333		203	20.50		223.50	256
3140	11" deep x 36" long	G		44	.364		186	22.50		208.50	241
3150	42" long	G		44	.364		203	22.50		225.50	259
3160	48" long	G		44	.364		211	22.50		233.50	268
3200	Closed riser, black enamel										
3210	12" deep x 36" long	G	2 Sswk	40	.400	Ea.	138	24.50		162.50	191
3220	42" long	G		40	.400		149	24.50		173.50	204
3230	48" long	G		40	.400		157	24.50		181.50	213
3240	Galvanized, 12" deep x 36" long	G		40	.400		217	24.50		241.50	279
3250	42" long	G		40	.400		237	24.50		261.50	300
3260	48" long	G		40	.400		248	24.50		272.50	315
4000	Bar grating treads										
4005	Steel, 1-1/4" x 3/16" bars, anti-skid nosing, black enamel										
4010	8-5/8" deep x 30" long	G	2 Sswk	48	.333	Ea.	56	20.50		76.50	94.50
4020	36" long	G		48	.333		65.50	20.50		86	105
4030	48" long	G		48	.333		101	20.50		121.50	144
4040	10-15/16" deep x 36" long	G		44	.364		72.50	22.50		95	116

131

05 55 Metal Stair Treads and Nosings

05 55 13 – Metal Stair Treads

05 55 13.50 Stair Treads		Crew	Daily Output	Labor-Hours	Unit	Material	2022 Bare Costs Labor	Equipment	Total	Total Incl O&P
4050	48" long	G	2 Sswk	44	.364	Ea.	102	22.50		124.50	149
4060	Galvanized, 8-5/8" deep x 30" long	G		48	.333		64	20.50		84.50	104
4070	36" long	G		48	.333		76	20.50		96.50	117
4080	48" long	G		48	.333		112	20.50		132.50	156
4090	10-15/16" deep x 36" long	G		44	.364		89	22.50		111.50	134
4100	48" long	G		44	.364		115	22.50		137.50	163
4200	Aluminum, 1-1/4" x 3/16" bars, serrated, with nosing										
4210	7-5/8" deep x 18" long	G	2 Sswk	52	.308	Ea.	54	19.05		73.05	90
4220	24" long	G		52	.308		62	19.05		81.05	98.50
4230	30" long	G		52	.308		71.50	19.05		90.55	110
4240	36" long	G		52	.308		185	19.05		204.05	235
4250	8-13/16" deep x 18" long	G		48	.333		70.50	20.50		91	111
4260	24" long	G		48	.333		117	20.50		137.50	161
4270	30" long	G		48	.333		117	20.50		137.50	161
4280	36" long	G		48	.333		204	20.50		224.50	257
4290	10" deep x 18" long	G		44	.364		136	22.50		158.50	186
4300	30" long	G		44	.364		181	22.50		203.50	235
4310	36" long	G		44	.364		220	22.50		242.50	278
5000	Channel grating treads										
5005	Steel, 14 ga., 2-1/2" thick, galvanized										
5010	9" deep x 36" long	G	2 Sswk	48	.333	Ea.	102	20.50		122.50	146
5020	48" long	G		48	.333	"	136	20.50		156.50	183
9000	Minimum labor/equipment charge			4	4	Job		247		247	400

05 55 19 – Metal Stair Tread Covers

05 55 19.50 Stair Tread Covers for Renovation

		Crew	Daily Output	Labor-Hours	Unit	Material	Labor	Equipment	Total	Total Incl O&P
0010	**STAIR TREAD COVERS FOR RENOVATION**									
0205	Extruded tread cover with nosing, pre-drilled, includes screws									
0210	Aluminum with black abrasive strips, 9" wide x 3' long	1 Carp	24	.333	Ea.	104	18.75		122.75	145
0220	4' long		22	.364		137	20.50		157.50	183
0230	5' long		20	.400		185	22.50		207.50	240
0240	11" wide x 3' long		24	.333		147	18.75		165.75	191
0250	4' long		22	.364		200	20.50		220.50	253
0260	5' long		20	.400		239	22.50		261.50	298
0305	Black abrasive strips with yellow front strips									
0310	Aluminum, 9" wide x 3' long	1 Carp	24	.333	Ea.	127	18.75		145.75	169
0320	4' long		22	.364		169	20.50		189.50	219
0330	5' long		20	.400		217	22.50		239.50	274
0340	11" wide x 3' long		24	.333		161	18.75		179.75	207
0350	4' long		22	.364		205	20.50		225.50	259
0360	5' long		20	.400		267	22.50		289.50	330
0405	Black abrasive strips with photoluminescent front strips									
0410	Aluminum, 9" wide x 3' long	1 Carp	24	.333	Ea.	160	18.75		178.75	206
0420	4' long		22	.364		181	20.50		201.50	233
0430	5' long		20	.400		255	22.50		277.50	315
0440	11" wide x 3' long		24	.333		181	18.75		199.75	230
0450	4' long	1 Carp	22	.364	Ea.	242	20.50		262.50	299
0460	5' long		20	.400	"	300	22.50		322.50	370
9000	Minimum labor/equipment charge		4	2	Job		113		113	179

05 56 Metal Castings

05 56 13 – Metal Construction Castings

05 56 13.50 Construction Castings		Crew	Daily Output	Labor-Hours	Unit	Material	2022 Bare Costs Labor	Equipment	Total	Total Incl O&P
0010	**CONSTRUCTION CASTINGS**									
0020	Manhole covers and frames, see Section 33 44 13.13									
0100	Column bases, cast iron, 16" x 16", approx. 65 lb.	G	E-4	46	.696	Ea.	126	43.50	3.21	172.71	212
0200	32" x 32", approx. 256 lb.	G		23	1.391	"	455	87	6.40	548.40	645
0600	Miscellaneous C.I. castings, light sections, less than 150 lb.	G		3200	.010	Lb.	8.20	.62	.05	8.87	10.05
1300	Special low volume items	G		3200	.010	"	11.25	.62	.05	11.92	13.40

05 58 Formed Metal Fabrications

05 58 13 – Column Covers

05 58 13.05 Column Covers

			Crew	Daily Output	Labor-Hours	Unit	Material	2022 Bare Costs Labor	Equipment	Total	Total Incl O&P
0010	**COLUMN COVERS**										
0015	Made from recycled materials										
0020	Excludes structural steel, light ga. metal framing, misc. metals, sealants										
0100	Round covers, 2 halves with 2 vertical joints for backer rod and sealant										
0110	Up to 12' high, no horizontal joints										
0120	12" diameter, 0.125" aluminum, anodized/painted finish	G	2 Sswk	32	.500	V.L.F.	30.50	31		61.50	83.50
0130	Type 304 stainless steel, 16 gauge, #4 brushed finish	G		32	.500		44	31		75	98.50
0140	Type 316 stainless steel, 16 gauge, #4 brushed finish	G		32	.500		53	31		84	109
0150	18" diameter, aluminum	G		32	.500		45.50	31		76.50	100
0160	Type 304 stainless steel	G		32	.500		66	31		97	123
0170	Type 316 stainless steel	G		32	.500		79.50	31		110.50	138
0180	24" diameter, aluminum	G		32	.500		60.50	31		91.50	117
0190	Type 304 stainless steel	G		32	.500		88	31		119	147
0200	Type 316 stainless steel	G		32	.500		106	31		137	167
0210	30" diameter, aluminum	G		30	.533		76	33		109	137
0220	Type 304 stainless steel	G		30	.533		110	33		143	174
0230	Type 316 stainless steel	G		30	.533		133	33		166	199
0240	36" diameter, aluminum	G		30	.533		91	33		124	153
0250	Type 304 stainless steel	G		30	.533		132	33		165	198
0260	Type 316 stainless steel	G		30	.533		159	33		192	228
0400	Up to 24' high, 2 stacked sections with 1 horizontal joint										
0410	18" diameter, aluminum	G	2 Sswk	28	.571	V.L.F.	48	35.50		83.50	110
0450	Type 304 stainless steel	G	2 Sswk	28	.571	V.L.F.	69.50	35.50		105	134
0460	Type 316 stainless steel	G		28	.571		83.50	35.50		119	149
0470	24" diameter, aluminum	G		28	.571		63.50	35.50		99	127
0480	Type 304 stainless steel	G		28	.571		92.50	35.50		128	159
0490	Type 316 stainless steel	G		28	.571		112	35.50		147.50	180
0500	30" diameter, aluminum	G		24	.667		79.50	41		120.50	154
0510	Type 304 stainless steel	G		24	.667		116	41		157	194
0520	Type 316 stainless steel	G		24	.667		140	41		181	220
0530	36" diameter, aluminum	G		24	.667		95.50	41		136.50	172
0540	Type 304 stainless steel	G		24	.667		139	41		180	220
0550	Type 316 stainless steel	G		24	.667		167	41		208	251

05 58 23 – Formed Metal Guards

05 58 23.90 Window Guards

			Crew	Daily Output	Labor-Hours	Unit	Material	2022 Bare Costs Labor	Equipment	Total	Total Incl O&P
0010	**WINDOW GUARDS**, shop fabricated										
0015	Expanded metal, steel angle frame, permanent	G	E-4	350	.091	S.F.	21	5.70	.42	27.12	32.50
0025	Steel bars, 1/2" x 1/2", spaced 5" OC	G	"	290	.110	"	14.65	6.90	.51	22.06	28
0030	Hinge mounted, add	G				Opng.	42.50			42.50	46.50
0040	Removable type, add	G				"	27			27	29.50

133

For customer support on your Commercial Renovation Costs with RSMeans data, call 800.448.8182.

05 58 Formed Metal Fabrications

05 58 23 – Formed Metal Guards

05 58 23.90 Window Guards

		Crew	Daily Output	Labor-Hours	Unit	Material	2022 Bare Costs Labor	Equipment	Total	Total Incl O&P
0050	For galvanized guards, add				S.F.	35%				
0070	For pivoted or projected type, add					105%	40%			
0100	Mild steel, stock units, economy	G	E-4	405	.079		5.75	4.93	.36	11.04	14.70
0200	Deluxe	G		405	.079		11.70	4.93	.36	16.99	21.50
0400	Woven wire, stock units, 3/8" channel frame, 3' x 5' opening	G		40	.800	Opng.	156	50	3.69	209.69	256
0500	4' x 6' opening	G		38	.842		249	52.50	3.88	305.38	365
0800	Basket guards for above, add	G					222			222	245
1000	Swinging guards for above, add	G					76			76	83.50
9000	Minimum labor/equipment charge		1 Sswk	2	4	Job		247		247	400

05 58 25 – Formed Lamp Posts

05 58 25.40 Lamp Posts

		Crew	Daily Output	Labor-Hours	Unit	Material	2022 Bare Costs Labor	Equipment	Total	Total Incl O&P
0010	**LAMP POSTS**										
0020	Aluminum, 7' high, stock units, post only	G	1 Carp	16	.500	Ea.	94	28		122	148
0100	Mild steel, plain	G		16	.500	"	58.50	28		86.50	109
9000	Minimum labor/equipment charge			4	2	Job		113		113	179

05 71 Decorative Metal Stairs

05 71 13 – Fabricated Metal Spiral Stairs

05 71 13.50 Spiral Stairs

		Crew	Daily Output	Labor-Hours	Unit	Material	2022 Bare Costs Labor	Equipment	Total	Total Incl O&P
0010	**SPIRAL STAIRS**										
1805	Shop fabricated, custom ordered										
1810	Aluminum, 5'-0" diameter, plain units	G	E-4	45	.711	Riser	625	44.50	3.28	672.78	760
1820	Fancy units	G	"	45	.711	"	1,025	44.50	3.28	1,072.78	1,200
1900	Cast iron, 4'-0" diameter, plain units	G	E-4	45	.711	Riser	750	44.50	3.28	797.78	900
1920	Fancy units	G		25	1.280		1,625	80	5.90	1,710.90	1,900
2000	Steel, industrial checkered plate, 4' diameter	G		45	.711		590	44.50	3.28	637.78	725
2200	6' diameter	G		40	.800		635	50	3.69	688.69	785
3100	Spiral stair kits, 12 stacking risers to fit exact floor height										
3110	Steel, flat metal treads, primed, 3'-6" diameter	G	2 Carp	1.60	10	Flight	1,275	565		1,840	2,300
3120	4'-0" diameter	G		1.45	11.034		1,375	620		1,995	2,475
3130	4'-6" diameter	G		1.35	11.852		1,650	665		2,315	2,850
3140	5'-0" diameter	G		1.25	12.800		1,825	720		2,545	3,150
3210	Galvanized, 3'-6" diameter	G		1.60	10		2,050	565		2,615	3,150
3220	4'-0" diameter	G		1.45	11.034		2,150	620		2,770	3,350
3230	4'-6" diameter	G		1.35	11.852		2,425	665		3,090	3,725
3240	5'-0" diameter	G		1.25	12.800		2,725	720		3,445	4,150
3310	Checkered plate tread, primed, 3'-6" diameter			1.45	11.034		1,525	620		2,145	2,650
3320	4'-0" diameter			1.35	11.852		1,625	665		2,290	2,825
3330	4'-6" diameter			1.25	12.800		1,900	720		2,620	3,225
3340	5'-0" diameter			1.15	13.913		2,100	785		2,885	3,550
3410	Galvanized, 3'-6" diameter			1.45	11.034		2,325	620		2,945	3,525
3420	4'-0" diameter			1.35	11.852		2,425	665		3,090	3,700
3430	4'-6" diameter			1.25	12.800		2,625	720		3,345	4,025
3440	5'-0" diameter			1.15	13.913		3,050	785		3,835	4,600
3510	Red oak covers on flat metal treads, 3'-6" diameter			1.35	11.852		1,750	665		2,415	2,975
3520	4'-0" diameter			1.25	12.800		1,900	720		2,620	3,225
3530	4'-6" diameter			1.15	13.913		2,150	785		2,935	3,625
3540	5'-0" diameter			1.05	15.238		2,550	860		3,410	4,200

05 73 Decorative Metal Railings

05 73 16 – Wire Rope Decorative Metal Railings

05 73 16.10 Cable Railings		Crew	Daily Output	Labor-Hours	Unit	Material	2022 Bare Costs Labor	Equipment	Total	Total Incl O&P
0010	**CABLE RAILINGS**, with 316 stainless steel 1 x 19 cable, 3/16" diameter									
0015	Made from recycled materials									
0100	1-3/4" diameter stainless steel posts x 42" high, cables 4" OC	G 2 Sswk	25	.640	L.F.	47	39.50		86.50	116

05 73 23 – Ornamental Railings

05 73 23.50 Railings, Ornamental		Crew	Daily Output	Labor-Hours	Unit	Material	2022 Bare Costs Labor	Equipment	Total	Total Incl O&P
0010	**RAILINGS, ORNAMENTAL**, 3'-6" high, posts @ 6' OC									
0020	Bronze or stainless, hand forged, plain	G 2 Sswk	24	.667	L.F.	247	41		288	340
0100	Fancy	G	18	.889		495	55		550	635
0200	Aluminum, panelized, plain	G	24	.667		16.15	41		57.15	84.50
0300	Fancy	G	18	.889		23.50	55		78.50	115
0400	Wrought iron, hand forged, plain	G 2 Sswk	24	.667	L.F.	69	41		110	143
0500	Fancy	G	18	.889		218	55		273	330
0550	Steel, panelized, plain	G	24	.667		21	41		62	89.50
0560	Fancy	G	18	.889		23	55		78	114
0600	Composite metal/wood/glass, plain		18	.889		144	55		199	247
0700	Fancy		12	1.333		287	82.50		369.50	450
9000	Minimum labor/equipment charge	1 Sswk	2	4	Job		247		247	400

05 75 Decorative Formed Metal

05 75 13 – Columns

05 75 13.10 Aluminum Columns		Crew	Daily Output	Labor-Hours	Unit	Material	2022 Bare Costs Labor	Equipment	Total	Total Incl O&P
0010	**ALUMINUM COLUMNS**									
0015	Made from recycled materials									
0020	Aluminum, extruded, stock units, no cap or base, 6" diameter	G E-4	240	.133	L.F.	20	8.30	.61	28.91	36
0100	8" diameter	G "	170	.188	"	24	11.75	.87	36.62	46.50
0460	Caps, ornamental, plain	G			Set	293			293	320
0470	Fancy	G			"	1,650			1,650	1,800
0500	For square columns, add to column prices above				L.F.	50%				

135

For customer support on your Commercial Renovation Costs with RSMeans data, call 800.448.8182.

Division Notes

	CREW	DAILY OUTPUT	LABOR-HOURS	UNIT	BARE COSTS				TOTAL INCL O&P
					MAT.	LABOR	EQUIP.	TOTAL	

Estimating Tips
06 05 00 Common Work Results for Wood, Plastics, and Composites

- Common to any wood-framed structure are the accessory connector items such as screws, nails, adhesives, hangers, connector plates, straps, angles, and hold-downs. For typical wood-framed buildings, such as residential projects, the aggregate total for these items can be significant, especially in areas where seismic loading is a concern. For floor and wall framing, the material cost is based on 10 to 25 lbs. of accessory connectors per MBF. Hold-downs, hangers, and other connectors should be taken off by the piece.

  Included with material costs are fasteners for a normal installation. Gordian's RSMeans engineers use manufacturers' recommendations, written specifications, and/or standard construction practice for the sizing and spacing of fasteners. Prices for various fasteners are shown for informational purposes only. Adjustments should be made if unusual fastening conditions exist.

06 10 00 Carpentry

- Lumber is a traded commodity and therefore sensitive to supply and demand in the marketplace. Even with "budgetary" estimating of wood-framed projects, it is advisable to call local suppliers for the latest market pricing.

- The common quantity unit for wood-framed projects is "thousand board feet" (MBF). A board foot is a volume of wood—1" x 1' x 1' or 144 cubic inches. Board-foot quantities are generally calculated using nominal material dimensions—dressed sizes are ignored. Board foot per lineal foot of any stick of lumber can be calculated by dividing the nominal cross-sectional area by 12. As an example, 2,000 lineal feet of 2 x 12 equates to 4 MBF by dividing the nominal area, 2 x 12, by 12, which equals 2, and multiplying that by 2,000 to give 4,000 board feet. This simple rule applies to all nominal dimensioned lumber.

- Waste is an issue of concern at the quantity takeoff for any area of construction. Framing lumber is sold in even foot lengths, i.e., 8', 10', 12', 14', 16', and depending on spans, wall heights, and the grade of lumber, waste is inevitable. A rule of thumb for lumber waste is 5–10% depending on material quality and the complexity of the framing.

- Wood in various forms and shapes is used in many projects, even where the main structural framing is steel, concrete, or masonry. Plywood as a back-up partition material and 2x boards used as blocking and cant strips around roof edges are two common examples. The estimator should ensure that the costs of all wood materials are included in the final estimate.

06 20 00 Finish Carpentry

- It is necessary to consider the grade of workmanship when estimating labor costs for erecting millwork and an interior finish. In practice, there are three grades: premium, custom, and economy. The RSMeans daily output for base and case moldings is in the range of 200 to 250 L.F. per carpenter per day. This is appropriate for most average custom-grade projects. For premium projects, an adjustment to productivity of 25–50% should be made, depending on the complexity of the job.

Reference Numbers

Reference numbers are shown at the beginning of some major classifications. These numbers refer to related items in the Reference Section. The reference information may be an estimating procedure, an alternate pricing method, or technical information.

Note: Not all subdivisions listed here necessarily appear. ∎

Same Data. Simplified.

Enjoy the convenience and efficiency of accessing your costs anywhere:

- **Skip the multiplier** by setting your location
- **Quickly search,** edit, favorite and share costs
- **Stay on top of price changes** with automatic updates

Discover more at rsmeans.com/online

06 05 Common Work Results for Wood, Plastics, and Composites

06 05 05 – Selective Demolition for Wood, Plastics, and Composites

06 05 05.10 Selective Demolition Wood Framing	Crew	Daily Output	Labor-Hours	Unit	Material	2022 Bare Costs Labor	Equipment	Total	Total Incl O&P
0010 **SELECTIVE DEMOLITION WOOD FRAMING** R024119-10									
0100 Timber connector, nailed, small	1 Clab	96	.083	Ea.		3.80		3.80	6.05
0110 Medium		60	.133			6.10		6.10	9.65
0120 Large		48	.167			7.60		7.60	12.10
0130 Bolted, small		48	.167			7.60		7.60	12.10
0140 Medium		32	.250			11.40		11.40	18.10
0150 Large		24	.333			15.20		15.20	24
2958 Beams, 2" x 6"	2 Clab	1100	.015	L.F.		.66		.66	1.05
2960 2" x 8"		825	.019			.88		.88	1.41
2965 2" x 10"		665	.024			1.10		1.10	1.74
2970 2" x 12"		550	.029			1.33		1.33	2.11
2972 2" x 14"		470	.034			1.55		1.55	2.47
2975 4" x 8"	B-1	413	.058			2.69		2.69	4.27
2980 4" x 10"		330	.073			3.37		3.37	5.35
2985 4" x 12"		275	.087			4.04		4.04	6.40
3000 6" x 8"		275	.087			4.04		4.04	6.40
3040 6" x 10"		220	.109			5.05		5.05	8
3080 6" x 12"		185	.130			6		6	9.55
3120 8" x 12"		140	.171			7.95		7.95	12.60
3160 10" x 12"		110	.218			10.10		10.10	16.05
3162 Alternate pricing method	B-1	1.10	21.818	M.B.F.		1,000		1,000	1,600
3170 Blocking, in 16" OC wall framing, 2" x 4"	1 Clab	600	.013	L.F.		.61		.61	.97
3172 2" x 6"		400	.020			.91		.91	1.45
3174 In 24" OC wall framing, 2" x 4"		600	.013			.61		.61	.97
3176 2" x 6"		400	.020			.91		.91	1.45
3178 Alt method, wood blocking removal from wood framing		.40	20	M.B.F.		910		910	1,450
3179 Wood blocking removal from steel framing		.36	22.222	"		1,025		1,025	1,600
3180 Bracing, let in, 1" x 3", studs 16" OC		1050	.008	L.F.		.35		.35	.55
3181 Studs 24" OC		1080	.007			.34		.34	.54
3182 1" x 4", studs 16" OC		1050	.008			.35		.35	.55
3183 Studs 24" OC		1080	.007			.34		.34	.54
3184 1" x 6", studs 16" OC		1050	.008			.35		.35	.55
3185 Studs 24" OC		1080	.007			.34		.34	.54
3186 2" x 3", studs 16" OC		800	.010			.46		.46	.72
3187 Studs 24" OC		830	.010			.44		.44	.70
3188 2" x 4", studs 16" OC		800	.010			.46		.46	.72
3189 Studs 24" OC		830	.010			.44		.44	.70
3190 2" x 6", studs 16" OC		800	.010			.46		.46	.72
3191 Studs 24" OC		830	.010			.44		.44	.70
3192 2" x 8", studs 16" OC		800	.010			.46		.46	.72
3193 Studs 24" OC		830	.010			.44		.44	.70
3194 "T" shaped metal bracing, studs at 16" OC		1060	.008			.34		.34	.55
3195 Studs at 24" OC		1200	.007			.30		.30	.48
3196 Metal straps, studs at 16" OC		1200	.007			.30		.30	.48
3197 Studs at 24" OC		1240	.006			.29		.29	.47
3199 2" x 6", studs 16" OC		480	.017	S.F.		.76		.76	1.21
3200 Columns, round, 8' to 14' tall		40	.200	Ea.		9.10		9.10	14.50
3202 Dimensional lumber sizes	2 Clab	1.10	14.545	M.B.F.		665		665	1,050
3250 Blocking, between joists	1 Clab	320	.025	Ea.		1.14		1.14	1.81
3252 Bridging, metal strap, between joists		320	.025	Pr.		1.14		1.14	1.81
3254 Wood, between joists		320	.025	"		1.14		1.14	1.81
3260 Door buck, studs, header & access., 8' high 2" x 4" wall, 3' wide		32	.250	Ea.		11.40		11.40	18.10
3261 4' wide		32	.250			11.40		11.40	18.10

06 05 05 – Selective Demolition for Wood, Plastics, and Composites

06 05 05.10 Selective Demolition Wood Framing		Crew	Daily Output	Labor-Hours	Unit	Material	2022 Bare Costs Labor	Equipment	Total	Total Incl O&P
3262	5' wide	1 Clab	32	.250	Ea.		11.40		11.40	18.10
3263	6' wide		32	.250			11.40		11.40	18.10
3264	8' wide		30	.267			12.15		12.15	19.30
3265	10' wide		30	.267			12.15		12.15	19.30
3266	12' wide		30	.267			12.15		12.15	19.30
3267	2" x 6" wall, 3' wide	1 Clab	32	.250	Ea.		11.40		11.40	18.10
3268	4' wide		32	.250			11.40		11.40	18.10
3269	5' wide		32	.250			11.40		11.40	18.10
3270	6' wide		32	.250			11.40		11.40	18.10
3271	8' wide		30	.267			12.15		12.15	19.30
3272	10' wide		30	.267			12.15		12.15	19.30
3273	12' wide		30	.267			12.15		12.15	19.30
3274	Window buck, studs, header & access., 8' high 2" x 4" wall, 2' wide		24	.333			15.20		15.20	24
3275	3' wide		24	.333			15.20		15.20	24
3276	4' wide		24	.333			15.20		15.20	24
3277	5' wide		24	.333			15.20		15.20	24
3278	6' wide		24	.333			15.20		15.20	24
3279	7' wide		24	.333			15.20		15.20	24
3280	8' wide		22	.364			16.60		16.60	26.50
3281	10' wide		22	.364			16.60		16.60	26.50
3282	12' wide		22	.364			16.60		16.60	26.50
3283	2" x 6" wall, 2' wide		24	.333			15.20		15.20	24
3284	3' wide		24	.333			15.20		15.20	24
3285	4' wide		24	.333			15.20		15.20	24
3286	5' wide		24	.333			15.20		15.20	24
3287	6' wide		24	.333			15.20		15.20	24
3288	7' wide		24	.333			15.20		15.20	24
3289	8' wide		22	.364			16.60		16.60	26.50
3290	10' wide		22	.364			16.60		16.60	26.50
3291	12' wide		22	.364			16.60		16.60	26.50
3360	Deck or porch decking		825	.010	L.F.		.44		.44	.70
3400	Fascia boards, 1" x 6"		500	.016			.73		.73	1.16
3440	1" x 8"		450	.018			.81		.81	1.29
3480	1" x 10"		400	.020			.91		.91	1.45
3490	2" x 6"		450	.018			.81		.81	1.29
3500	2" x 8"		400	.020			.91		.91	1.45
3510	2" x 10"		350	.023			1.04		1.04	1.66
3610	Furring, on wood walls or ceiling		4000	.002	S.F.		.09		.09	.14
3620	On masonry or concrete walls or ceiling		1200	.007	"		.30		.30	.48
3800	Headers over openings, 2 @ 2" x 6"		110	.073	L.F.		3.32		3.32	5.25
3840	2 @ 2" x 8"		100	.080			3.65		3.65	5.80
3880	2 @ 2" x 10"		90	.089			4.05		4.05	6.45
3885	Alternate pricing method		.26	30.651	M.B.F.		1,400		1,400	2,225
3920	Joists, 1" x 4"		1250	.006	L.F.		.29		.29	.46
3930	1" x 6"	1 Clab	1135	.007	L.F.		.32		.32	.51
3940	1" x 8"		1000	.008			.36		.36	.58
3950	1" x 10"		895	.009			.41		.41	.65
3960	1" x 12"		765	.010			.48		.48	.76
4200	2" x 4"	2 Clab	1000	.016			.73		.73	1.16
4230	2" x 6"		970	.016			.75		.75	1.20
4240	2" x 8"		940	.017			.78		.78	1.23
4250	2" x 10"		910	.018			.80		.80	1.27
4280	2" x 12"		880	.018			.83		.83	1.32

For customer support on your Commercial Renovation Costs with RSMeans data, call 800.448.8182.

139

06 05 05.10 Selective Demolition Wood Framing	Crew	Daily Output	Labor-Hours	Unit	Material	2022 Bare Costs Labor	2022 Bare Costs Equipment	Total	Total Incl O&P
4281	2" x 14"	2 Clab	850	.019	L.F.		.86		.86	1.36
4282	Composite joists, 9-1/2"		960	.017			.76		.76	1.21
4283	11-7/8"		930	.017			.78		.78	1.25
4284	14"		897	.018			.81		.81	1.29
4285	16"		865	.019			.84		.84	1.34
4290	Wood joists, alternate pricing method		1.50	10.667	M.B.F.		485		485	775
4500	Open web joist, 12" deep		500	.032	L.F.		1.46		1.46	2.32
4505	14" deep		475	.034			1.54		1.54	2.44
4510	16" deep		450	.036			1.62		1.62	2.58
4520	18" deep		425	.038			1.72		1.72	2.73
4530	24" deep		400	.040			1.82		1.82	2.90
4550	Ledger strips, 1" x 2"	1 Clab	1200	.007			.30		.30	.48
4560	1" x 3"		1200	.007			.30		.30	.48
4570	1" x 4"		1200	.007			.30		.30	.48
4580	2" x 2"		1100	.007			.33		.33	.53
4590	2" x 4"		1000	.008			.36		.36	.58
4600	2" x 6"		1000	.008			.36		.36	.58
4601	2" x 8 or 2" x 10"		800	.010			.46		.46	.72
4602	4" x 6"		600	.013			.61		.61	.97
4604	4" x 8"		450	.018			.81		.81	1.29
5400	Posts, 4" x 4"	2 Clab	800	.020			.91		.91	1.45
5405	4" x 6"		550	.029			1.33		1.33	2.11
5410	4" x 8"		440	.036			1.66		1.66	2.63
5425	4" x 10"		390	.041			1.87		1.87	2.97
5430	4" x 12"		350	.046			2.08		2.08	3.31
5440	6" x 6"		400	.040			1.82		1.82	2.90
5445	6" x 8"		350	.046			2.08		2.08	3.31
5450	6" x 10"		320	.050			2.28		2.28	3.62
5455	6" x 12"		290	.055			2.52		2.52	4
5480	8" x 8"		300	.053			2.43		2.43	3.86
5500	10" x 10"		240	.067			3.04		3.04	4.83
5660	T&G floor planks	2 Clab	2	8	M.B.F.		365		365	580
5682	Rafters, ordinary, 16" OC, 2" x 4"		880	.018	S.F.		.83		.83	1.32
5683	2" x 6"		840	.019			.87		.87	1.38
5684	2" x 8"		820	.020			.89		.89	1.41
5685	2" x 10"		820	.020			.89		.89	1.41
5686	2" x 12"		810	.020			.90		.90	1.43
5687	24" OC, 2" x 4"		1170	.014			.62		.62	.99
5688	2" x 6"		1117	.014			.65		.65	1.04
5689	2" x 8"		1091	.015			.67		.67	1.06
5690	2" x 10"		1091	.015			.67		.67	1.06
5691	2" x 12"		1077	.015			.68		.68	1.08
5795	Rafters, ordinary, 2" x 4" (alternate method)		862	.019	L.F.		.85		.85	1.34
5800	2" x 6" (alternate method)		850	.019			.86		.86	1.36
5840	2" x 8" (alternate method)		837	.019			.87		.87	1.38
5855	2" x 10" (alternate method)		825	.019			.88		.88	1.41
5865	2" x 12" (alternate method)		812	.020			.90		.90	1.43
5870	Sill plate, 2" x 4"	1 Clab	1170	.007			.31		.31	.50
5871	2" x 6"		780	.010			.47		.47	.74
5872	2" x 8"		586	.014			.62		.62	.99
5873	Alternate pricing method		.78	10.256	M.B.F.		470		470	745
5885	Ridge board, 1" x 4"	2 Clab	900	.018	L.F.		.81		.81	1.29
5886	1" x 6"		875	.018			.83		.83	1.32

140

For customer support on your Commercial Renovation Costs with RSMeans data, call 800.448.8182.

06 05 05.10 Selective Demolition Wood Framing	Crew	Daily Output	Labor-Hours	Unit	Material	2022 Bare Costs Labor	Equipment	Total	Total Incl O&P
5887	1" x 8"	2 Clab	850	.019	L.F.		.86		.86	1.36
5888	1" x 10"		825	.019			.88		.88	1.41
5889	1" x 12"		800	.020			.91		.91	1.45
5890	2" x 4"		900	.018			.81		.81	1.29
5892	2" x 6"		875	.018			.83		.83	1.32
5894	2" x 8"		850	.019			.86		.86	1.36
5896	2" x 10"		825	.019			.88		.88	1.41
5898	2" x 12"		800	.020			.91		.91	1.45
5900	Hip & valley rafters, 2" x 6"		500	.032			1.46		1.46	2.32
5940	2" x 8"		420	.038			1.74		1.74	2.76
6050	Rafter tie, 1" x 4"		1250	.013			.58		.58	.93
6052	1" x 6"		1135	.014			.64		.64	1.02
6054	2" x 4"		1000	.016			.73		.73	1.16
6056	2" x 6"		970	.016			.75		.75	1.20
6070	Sleepers, on concrete, 1" x 2"	1 Clab	4700	.002			.08		.08	.12
6075	1" x 3"		4000	.002			.09		.09	.14
6080	2" x 4"		3000	.003			.12		.12	.19
6085	2" x 6"	1 Clab	2600	.003	L.F.		.14		.14	.22
6086	Sheathing from roof, 5/16"	2 Clab	1600	.010	S.F.		.46		.46	.72
6088	3/8"		1525	.010			.48		.48	.76
6090	1/2"		1400	.011			.52		.52	.83
6092	5/8"		1300	.012			.56		.56	.89
6094	3/4"		1200	.013			.61		.61	.97
6096	Board sheathing from roof		1400	.011			.52		.52	.83
6100	Sheathing, from walls, 1/4"		1200	.013			.61		.61	.97
6110	5/16"		1175	.014			.62		.62	.99
6120	3/8"		1150	.014			.63		.63	1.01
6130	1/2"		1125	.014			.65		.65	1.03
6140	5/8"		1100	.015			.66		.66	1.05
6150	3/4"		1075	.015			.68		.68	1.08
6152	Board sheathing from walls		1500	.011			.49		.49	.77
6158	Subfloor/roof deck, with boards		2200	.007			.33		.33	.53
6159	Subfloor/roof deck, with tongue & groove boards		2000	.008			.36		.36	.58
6160	Plywood, 1/2" thick		768	.021			.95		.95	1.51
6162	5/8" thick		760	.021			.96		.96	1.53
6164	3/4" thick		750	.021			.97		.97	1.55
6165	1-1/8" thick		720	.022			1.01		1.01	1.61
6166	Underlayment, particle board, 3/8" thick	1 Clab	780	.010			.47		.47	.74
6168	1/2" thick		768	.010			.48		.48	.75
6170	5/8" thick		760	.011			.48		.48	.76
6172	3/4" thick		750	.011			.49		.49	.77
6200	Stairs and stringers, straight run	2 Clab	40	.400	Riser		18.25		18.25	29
6240	With platforms, winders or curves	"	26	.615	"		28		28	44.50
6300	Components, tread	1 Clab	110	.073	Ea.		3.32		3.32	5.25
6320	Riser		80	.100	"		4.56		4.56	7.25
6390	Stringer, 2" x 10"		260	.031	L.F.		1.40		1.40	2.23
6400	2" x 12"		260	.031			1.40		1.40	2.23
6410	3" x 10"		250	.032			1.46		1.46	2.32
6420	3" x 12"		250	.032			1.46		1.46	2.32
6590	Wood studs, 2" x 3"	2 Clab	3076	.005			.24		.24	.38
6600	2" x 4"		2000	.008			.36		.36	.58
6640	2" x 6"		1600	.010			.46		.46	.72
6720	Wall framing, including studs, plates and blocking, 2" x 4"	1 Clab	600	.013	S.F.		.61		.61	.97

06 05 05 – Selective Demolition for Wood, Plastics, and Composites

06 05 05.10 Selective Demolition Wood Framing		Crew	Daily Output	Labor-Hours	Unit	Material	2022 Bare Costs Labor	2022 Bare Costs Equipment	Total	Total Incl O&P
6740	2" x 6"	1 Clab	480	.017	S.F.		.76		.76	1.21
6750	Headers, 2" x 4"		1125	.007	L.F.		.32		.32	.52
6755	2" x 6"		1125	.007	"		.32		.32	.52
6760	2" x 8"	1 Clab	1050	.008	L.F.		.35		.35	.55
6765	2" x 10"		1050	.008			.35		.35	.55
6770	2" x 12"		1000	.008			.36		.36	.58
6780	4" x 10"		525	.015			.69		.69	1.10
6785	4" x 12"		500	.016			.73		.73	1.16
6790	6" x 8"		560	.014			.65		.65	1.04
6795	6" x 10"		525	.015			.69		.69	1.10
6797	6" x 12"		500	.016			.73		.73	1.16
7000	Trusses									
7050	12' span	2 Clab	74	.216	Ea.		9.85		9.85	15.65
7150	24' span	F-3	66	.606			35	30	65	88.50
7200	26' span		64	.625			36	31	67	91
7250	28' span		62	.645			37	32	69	94
7300	30' span		58	.690			40	34	74	101
7350	32' span		56	.714			41	35.50	76.50	104
7400	34' span		54	.741			43	36.50	79.50	108
7450	36' span		52	.769			44.50	38	82.50	112
8000	Soffit, T&G wood	1 Clab	520	.015	S.F.		.70		.70	1.11
8010	Hardboard, vinyl or aluminum	"	640	.013			.57		.57	.91
8030	Plywood	2 Carp	315	.051			2.86		2.86	4.54
9000	Minimum labor/equipment charge	1 Clab	4	2	Job		91		91	145
9500	See Section 02 41 19.19 for rubbish handling									

06 05 05.20 Selective Demolition Millwork and Trim

06 05 05.20 Selective Demolition Millwork and Trim		Crew	Daily Output	Labor-Hours	Unit	Material	2022 Bare Costs Labor	2022 Bare Costs Equipment	Total	Total Incl O&P
0010	SELECTIVE DEMOLITION MILLWORK AND TRIM R024119-10									
1000	Cabinets, wood, base cabinets, per L.F.	2 Clab	80	.200	L.F.		9.10		9.10	14.50
1020	Wall cabinets, per L.F.	"	80	.200	"		9.10		9.10	14.50
1060	Remove and reset, base cabinets	2 Carp	18	.889	Ea.		50		50	79.50
1070	Wall cabinets		20	.800			45		45	71.50
1072	Oven cabinet, 7' high		11	1.455			82		82	130
1074	Cabinet door, up to 2' high	1 Clab	66	.121			5.55		5.55	8.80
1076	2' - 4' high	"	46	.174			7.95		7.95	12.60
1100	Steel, painted, base cabinets	2 Clab	60	.267	L.F.		12.15		12.15	19.30
1120	Wall cabinets		60	.267	"		12.15		12.15	19.30
1200	Casework, large area		320	.050	S.F.		2.28		2.28	3.62
1220	Selective		200	.080	"		3.65		3.65	5.80
1500	Counter top, straight runs		200	.080	L.F.		3.65		3.65	5.80
1510	L, U or C shapes		120	.133			6.10		6.10	9.65
1550	Remove and reset, straight runs	2 Carp	50	.320			18		18	28.50
1560	L, U or C shape	"	40	.400			22.50		22.50	36
2000	Paneling, 4' x 8' sheets	2 Clab	2000	.008	S.F.		.36		.36	.58
2100	Boards, 1" x 4"		700	.023			1.04		1.04	1.66
2120	1" x 6"		750	.021			.97		.97	1.55
2140	1" x 8"		800	.020			.91		.91	1.45
3000	Trim, baseboard, to 6" wide		1200	.013	L.F.		.61		.61	.97
3040	Greater than 6" and up to 12" wide		1000	.016			.73		.73	1.16
3080	Remove and reset, minimum	2 Carp	400	.040			2.25		2.25	3.58
3090	Maximum	"	300	.053			3		3	4.77
3100	Ceiling trim	2 Clab	1000	.016			.73		.73	1.16
3120	Chair rail		1200	.013			.61		.61	.97

06 05 Common Work Results for Wood, Plastics, and Composites

06 05 05 – Selective Demolition for Wood, Plastics, and Composites

06 05 05.20 Selective Demolition Millwork and Trim	Crew	Daily Output	Labor-Hours	Unit	Material	2022 Bare Costs Labor	Equipment	Total	Total Incl O&P
3140	Railings with balusters	2 Clab	240	.067	L.F.		3.04		3.04	4.83
3160	Wainscoting	↓	700	.023	S.F.		1.04		1.04	1.66
9000	Minimum labor/equipment charge	1 Clab	4	2	Job		91		91	145

06 05 23 – Wood, Plastic, and Composite Fastenings

06 05 23.10 Nails

		Crew	Daily Output	Labor-Hours	Unit	Material	2022 Bare Costs Labor	Equipment	Total	Total Incl O&P
0010	**NAILS**, material only, based upon 50# box purchase									
0020	Copper nails, plain				Lb.	13			13	14.30
0400	Stainless steel, plain					12.65			12.65	13.90
0500	Box, 3d to 20d, bright					3.10			3.10	3.41
0520	Galvanized					3.95			3.95	4.35
0600	Common, 3d to 60d, plain					1.67			1.67	1.84
0700	Galvanized					1.76			1.76	1.94
0800	Aluminum					12.80			12.80	14.10
1000	Annular or spiral thread, 4d to 60d, plain					3.33			3.33	3.66
1200	Galvanized					4.40			4.40	4.84
1400	Drywall nails, plain					2.90			2.90	3.19
1600	Galvanized					3			3	3.30
1800	Finish nails, 4d to 10d, plain					2			2	2.20
2000	Galvanized					3.30			3.30	3.63
2100	Aluminum					12.35			12.35	13.60
2300	Flooring nails, hardened steel, 2d to 10d, plain					3.51			3.51	3.86
2400	Galvanized					5.90			5.90	6.50
2500	Gypsum lath nails, 1-1/8", 13 ga. flathead, blued					3.20			3.20	3.52
2600	Masonry nails, hardened steel, 3/4" to 3" long, plain					3.49			3.49	3.84
2700	Galvanized					4.56			4.56	5
2900	Roofing nails, threaded, galvanized					3.54			3.54	3.89
3100	Aluminum					8.65			8.65	9.55
3300	Compressed lead head, threaded, galvanized					4.20			4.20	4.62
3600	Siding nails, plain shank, galvanized					2.53			2.53	2.78
3800	Aluminum					13			13	14.30
5000	Add to prices above for cement coating					.19			.19	.21
5200	Zinc or tin plating					.24			.24	.26
5500	Vinyl coated sinkers, 8d to 16d					1.45			1.45	1.60

06 05 23.50 Wood Screws

		Crew	Daily Output	Labor-Hours	Unit	Material	2022 Bare Costs Labor	Equipment	Total	Total Incl O&P
0010	**WOOD SCREWS**									
0020	#8, 1" long, steel				C	6.45			6.45	7.10
0100	Brass					15.50			15.50	17.05
0200	#8, 2" long, steel					11			11	12.10
0300	Brass					28			28	31
0400	#10, 1" long, steel					4.50			4.50	4.95
0500	Brass					19.85			19.85	22
0600	#10, 2" long, steel					7.25			7.25	7.95
0700	Brass					34			34	37.50
0800	#10, 3" long, steel					12.10			12.10	13.30
1000	#12, 2" long, steel					10.70			10.70	11.80
1100	Brass					44			44	48.50
1500	#12, 3" long, steel					15.55			15.55	17.10
2000	#12, 4" long, steel					27.50			27.50	30
2050	Composite trim screws									
2100	Stainless steel, trim head, #7 x 1-5/8" long				C	11.55			11.55	12.70
2200	#7 x 2" long					11.45			11.45	12.55
2300	#7 x 2-1/2" long					25			25	27.50

For customer support on your Commercial Renovation Costs with RSMeans data, call 800.448.8182.

143

06 05 Common Work Results for Wood, Plastics, and Composites

06 05 23 – Wood, Plastic, and Composite Fastenings

06 05 23.50 Wood Screws	Crew	Daily Output	Labor-Hours	Unit	Material	2022 Bare Costs Labor	Equipment	Total	Total Incl O&P
2400	#7 x 3" long				C	18.65			18.65	20.50
2500	#9 x 3-1/2" long					30			30	33
2600	#9 x 4" long					32.50			32.50	36
3000	Stainless steel, flat head, #6 x 1-5/8" long					12.05			12.05	13.25
3100	#8 x 2" long					12.70			12.70	13.95
3200	#8 x 2-1/2" long					22			22	24
3300	#8 x 3" long					17.75			17.75	19.55
3400	#10 x 2" long					17.45			17.45	19.15
3500	#10 x 3-1/2" long					38			38	42
3600	#12 x 3" long					43.50			43.50	48
3700	#12 x 4" long					48			48	53
3800	#12 x 6" long					121			121	133

06 05 23.60 Timber Connectors

06 05 23.60 Timber Connectors	Crew	Daily Output	Labor-Hours	Unit	Material	2022 Bare Costs Labor	Equipment	Total	Total Incl O&P
0010	**TIMBER CONNECTORS**									
0020	Add up cost of each part for total cost of connection									
0100	Connector plates, steel, with bolts, straight	2 Carp	75	.213	Ea.	25.50	12		37.50	47
0110	Tee, 7 ga.		50	.320		68.50	18		86.50	104
0120	T-Strap, 14 ga., 12" x 8" x 2"		50	.320		68.50	18		86.50	104
0150	Anchor plates, 7 ga., 9" x 7"		75	.213		25.50	12		37.50	47
0200	Bolts, machine, sq. hd. with nut & washer, 1/2" diameter, 4" long	1 Carp	140	.057		.78	3.22		4	5.95
0300	7-1/2" long		130	.062		1.41	3.46		4.87	7.05
0500	3/4" diameter, 7-1/2" long		130	.062		3.84	3.46		7.30	9.70
0610	Machine bolts, w/nut, washer, 3/4" diameter, 15" L, HD's & beam hangers		95	.084		7.20	4.74		11.94	15.45
0800	Drilling bolt holes in timber, 1/2" diameter		450	.018	Inch		1		1	1.59
0900	1" diameter		350	.023	"		1.29		1.29	2.04
1100	Framing anchor, angle, 3" x 3" x 1-1/2", 12 ga.		175	.046	Ea.	3.20	2.57		5.77	7.60
1150	Framing anchors, 18 ga., 4-1/2" x 2-3/4"		175	.046		3.20	2.57		5.77	7.60
1160	Framing anchors, 18 ga., 4-1/2" x 3"		175	.046		3.20	2.57		5.77	7.60
1170	Clip anchors plates, 18 ga., 12" x 1-1/8"		175	.046		3.20	2.57		5.77	7.60
1250	Holdowns, 3 ga. base, 10 ga. body		8	1		69.50	56.50		126	166
1260	Holdowns, 7 ga. 11-1/16" x 3-1/4"		8	1		69.50	56.50		126	166
1270	Holdowns, 7 ga. 14-3/8" x 3-1/8"		8	1		69.50	56.50		126	166
1275	Holdowns, 12 ga. 8" x 2-1/2"		8	1		69.50	56.50		126	166
1300	Joist and beam hangers, 18 ga. galv., for 2" x 4" joist		175	.046		1.09	2.57		3.66	5.30
1400	2" x 6" to 2" x 10" joist		165	.048		1.95	2.73		4.68	6.50
1600	16 ga. galv., 3" x 6" to 3" x 10" joist		160	.050		6.95	2.82		9.77	12.10
1700	3" x 10" to 3" x 14" joist		160	.050		7.35	2.82		10.17	12.50
1800	4" x 6" to 4" x 10" joist		155	.052		5.90	2.91		8.81	11.05
1900	4" x 10" to 4" x 14" joist		155	.052		7.90	2.91		10.81	13.30
2000	Two-2" x 6" to two-2" x 10" joists		150	.053		6.20	3		9.20	11.60
2100	Two-2" x 10" to two-2" x 14" joists		150	.053		5.60	3		8.60	10.90
2300	3/16" thick, 6" x 8" joist		145	.055		51	3.11		54.11	61
2400	6" x 10" joist		140	.057		53	3.22		56.22	63.50
2500	6" x 12" joist		135	.059		90.50	3.34		93.84	105
2700	1/4" thick, 6" x 14" joist		130	.062		104	3.46		107.46	120
2900	Plywood clips, extruded aluminum H clip, for 3/4" panels					.13			.13	.14
3000	Galvanized 18 ga. back-up clip					.16			.16	.18
3200	Post framing, 16 ga. galv. for 4" x 4" base, 2 piece	1 Carp	130	.062		16.25	3.46		19.71	23.50
3300	Cap	"	130	.062		49.50	3.46		52.96	60
3500	Rafter anchors, 18 ga. galv., 1-1/2" wide, 5-1/4" long	1 Carp	145	.055	Ea.	.56	3.11		3.67	5.55
3600	10-3/4" long		145	.055		2.28	3.11		5.39	7.45
3800	Shear plates, 2-5/8" diameter		120	.067		5.55	3.75		9.30	12.05

144

06 05 23 – Wood, Plastic, and Composite Fastenings

06 05 23.60 Timber Connectors		Crew	Daily Output	Labor-Hours	Unit	Material	2022 Bare Costs Labor	Equipment	Total	Total Incl O&P
3900	4" diameter	1 Carp	115	.070	Ea.	4.64	3.92		8.56	11.30
4000	Sill anchors, embedded in concrete or block, 25-1/2" long		115	.070		3.14	3.92		7.06	9.65
4100	Spike grids, 3" x 6"		120	.067		2.20	3.75		5.95	8.35
4400	Split rings, 2-1/2" diameter		120	.067		5.70	3.75		9.45	12.25
4500	4" diameter		110	.073		6.05	4.09		10.14	13.15
4550	Tie plate, 20 ga., 7" x 3-1/8"		110	.073		6.05	4.09		10.14	13.15
4560	5" x 4-1/8"		110	.073		6.05	4.09		10.14	13.15
4575	Twist straps, 18 ga., 12" x 1-1/4"		110	.073		6.05	4.09		10.14	13.15
4580	16" x 1-1/4"		110	.073		6.05	4.09		10.14	13.15
4600	Strap ties, 20 ga., 2-1/16" wide, 12-13/16" long		180	.044		1.67	2.50		4.17	5.80
4700	16 ga., 1-3/8" wide, 12" long		180	.044		1.67	2.50		4.17	5.80
4800	1-1/4" wide, 21-5/8" long		160	.050		2.44	2.82		5.26	7.15
5000	Toothed rings, 2-5/8" or 4" diameter		90	.089		2.87	5		7.87	11.10
5200	Truss plates, nailed, 20 ga., up to 32' span	↓	17	.471		20	26.50		46.50	64
5400	Washers, 2" x 2" x 1/8"					.59			.59	.65
5500	3" x 3" x 3/16"				↓	1.57			1.57	1.73
9000	Minimum labor/equipment charge	1 Carp	4	2	Job		113		113	179

06 05 23.80 Metal Bracing

		Crew	Daily Output	Labor-Hours	Unit	Material	2022 Bare Costs Labor	Equipment	Total	Total Incl O&P
0010	**METAL BRACING**									
0302	Let-in, "T" shaped, 22 ga. galv. steel, studs at 16" OC	1 Carp	580	.014	L.F.	1.26	.78		2.04	2.62
0402	Studs at 24" OC		600	.013		1.26	.75		2.01	2.58
0502	Steel straps, 16 ga. galv. steel, studs at 16" OC		600	.013		1.95	.75		2.70	3.34
0602	Studs at 24" OC	↓	620	.013	↓	1.95	.73		2.68	3.30

06 11 Wood Framing

06 11 10 – Framing with Dimensional, Engineered or Composite Lumber

06 11 10.01 Forest Stewardship Council Certification

		Crew	Daily Output	Labor-Hours	Unit	Material	2022 Bare Costs Labor	Equipment	Total	Total Incl O&P
0010	**FOREST STEWARDSHIP COUNCIL CERTIFICATION**									
0020	For Forest Stewardship Council (FSC) cert dimension lumber, add	G				65%				

06 11 10.02 Blocking

		Crew	Daily Output	Labor-Hours	Unit	Material	2022 Bare Costs Labor	Equipment	Total	Total Incl O&P
0010	**BLOCKING**									
2600	Miscellaneous, to wood construction									
2620	2" x 4"	1 Carp	.17	47.059	M.B.F.	985	2,650		3,635	5,275
2625	Pneumatic nailed		.21	38.095		990	2,150		3,140	4,500
2660	2" x 8"		.27	29.630		1,075	1,675		2,750	3,825
2665	Pneumatic nailed	↓	.33	24.242	↓	1,075	1,375		2,450	3,375
2720	To steel construction									
2740	2" x 4"	1 Carp	.14	57.143	M.B.F.	985	3,225		4,210	6,175
2780	2" x 8"		.21	38.095	"	1,075	2,150		3,225	4,575
9000	Minimum labor/equipment charge	↓	4	2	Job		113		113	179

06 11 10.04 Wood Bracing

		Crew	Daily Output	Labor-Hours	Unit	Material	2022 Bare Costs Labor	Equipment	Total	Total Incl O&P
0010	**WOOD BRACING**									
0012	Let-in, with 1" x 6" boards, studs @ 16" OC	1 Carp	150	.053	L.F.	.98	3		3.98	5.85
0202	Studs @ 24" OC	"	230	.035	"	.98	1.96		2.94	4.18

06 11 10.06 Bridging

		Crew	Daily Output	Labor-Hours	Unit	Material	2022 Bare Costs Labor	Equipment	Total	Total Incl O&P
0010	**BRIDGING**									
0012	Wood, for joists 16" OC, 1" x 3"	1 Carp	130	.062	Pr.	.76	3.46		4.22	6.35
0017	Pneumatic nailed		170	.047		.83	2.65		3.48	5.10
0102	2" x 3" bridging		130	.062		.81	3.46		4.27	6.40
0107	Pneumatic nailed	↓	170	.047	↓	.82	2.65		3.47	5.10

For customer support on your Commercial Renovation Costs with RSMeans data, call 800.448.8182.

145

06 11 10 - Framing with Dimensional, Engineered or Composite Lumber

06 11 10.06 Bridging		Crew	Daily Output	Labor-Hours	Unit	Material	2022 Bare Costs Labor	Equipment	Total	Total Incl O&P
0302	Steel, galvanized, 18 ga., for 2" x 10" joists at 12" OC	1 Carp	130	.062	Pr.	2.04	3.46		5.50	7.75
0352	16" OC		135	.059		.02	3.34		3.36	5.30
0402	24" OC		140	.057		3.14	3.22		6.36	8.55
0602	For 2" x 14" joists at 16" OC		130	.062		2.71	3.46		6.17	8.50
0902	Compression type, 16" OC, 2" x 8" joists		200	.040		1.61	2.25		3.86	5.35
1002	2" x 12" joists		200	.040		2.14	2.25		4.39	5.95

06 11 10.10 Beam and Girder Framing		Crew	Daily Output	Labor-Hours	Unit	Material	2022 Bare Costs Labor	Equipment	Total	Total Incl O&P
0010	**BEAM AND GIRDER FRAMING** R061110-30									
1000	Single, 2" x 6"	2 Carp	700	.023	L.F.	.97	1.29		2.26	3.11
1005	Pneumatic nailed		812	.020		.97	1.11		2.08	2.83
1020	2" x 8"		650	.025		1.43	1.39		2.82	3.78
1025	Pneumatic nailed		754	.021		1.44	1.19		2.63	3.48
1040	2" x 10"		600	.027		1.99	1.50		3.49	4.57
1045	Pneumatic nailed		696	.023		2	1.29		3.29	4.25
1060	2" x 12"		550	.029		2.77	1.64		4.41	5.65
1065	Pneumatic nailed		638	.025		2.78	1.41		4.19	5.30
1080	2" x 14"		500	.032		5.30	1.80		7.10	8.65
1085	Pneumatic nailed		580	.028		5.30	1.55		6.85	8.30
1100	3" x 8"		550	.029		5.55	1.64		7.19	8.70
1120	3" x 10"		500	.032		6.85	1.80		8.65	10.35
1140	3" x 12"		450	.036		8.20	2		10.20	12.20
1160	3" x 14"		400	.040		7	2.25		9.25	11.30
1170	4" x 6"	F-3	1100	.036		4.15	2.10	1.80	8.05	9.85
1180	4" x 8"		1000	.040		5.60	2.31	1.98	9.89	12.05
1200	4" x 10"		950	.042		7.05	2.43	2.09	11.57	13.90
1220	4" x 12"		900	.044		8.65	2.57	2.20	13.42	16.05
1240	4" x 14"		850	.047		10.15	2.72	2.33	15.20	18.05
1250	6" x 8"		525	.076		11.45	4.40	3.78	19.63	23.50
1260	6" x 10"	F-3	500	.080	L.F.	14.70	4.62	3.97	23.29	28
1290	8" x 12"	"	300	.133		32	7.70	6.60	46.30	55
2000	Double, 2" x 6"	2 Carp	625	.026		1.94	1.44		3.38	4.42
2005	Pneumatic nailed		725	.022		1.95	1.24		3.19	4.11
2020	2" x 8"		575	.028		2.87	1.57		4.44	5.65
2025	Pneumatic nailed		667	.024		2.88	1.35		4.23	5.30
2040	2" x 10"		550	.029		3.97	1.64		5.61	6.95
2045	Pneumatic nailed		638	.025		3.99	1.41		5.40	6.65
2060	2" x 12"		525	.030		5.55	1.72		7.27	8.85
2065	Pneumatic nailed		610	.026		5.55	1.48		7.03	8.50
2080	2" x 14"		475	.034		10.60	1.90		12.50	14.65
2085	Pneumatic nailed		551	.029		10.60	1.63		12.23	14.25
3000	Triple, 2" x 6"		550	.029		2.91	1.64		4.55	5.80
3005	Pneumatic nailed		638	.025		2.92	1.41		4.33	5.45
3020	2" x 8"		525	.030		4.30	1.72		6.02	7.45
3025	Pneumatic nailed		609	.026		4.32	1.48		5.80	7.10
3040	2" x 10"		500	.032		5.95	1.80		7.75	9.40
3045	Pneumatic nailed		580	.028		6	1.55		7.55	9.05
3060	2" x 12"		475	.034		8.30	1.90		10.20	12.15
3065	Pneumatic nailed		551	.029		8.35	1.63		9.98	11.80
3080	2" x 14"		450	.036		16.75	2		18.75	21.50
3085	Pneumatic nailed		522	.031		15.90	1.73		17.63	20
9000	Minimum labor/equipment charge	1 Carp	2	4	Job		225		225	360

06 11 Wood Framing

06 11 10 – Framing with Dimensional, Engineered or Composite Lumber

06 11 10.12 Ceiling Framing		Crew	Daily Output	Labor-Hours	Unit	Material	2022 Bare Costs Labor	Equipment	Total	Total Incl O&P
0010	**CEILING FRAMING**									
6000	Suspended, 2" x 3"	2 Carp	1000	.016	L.F.	.46	.90		1.36	1.94
6050	2" x 4"		900	.018		.66	1		1.66	2.31
6100	2" x 6"		800	.020		.97	1.13		2.10	2.86
6150	2" x 8"	↓	650	.025	↓	1.43	1.39		2.82	3.78
9000	Minimum labor/equipment charge	1 Carp	4	2	Job		113		113	179

06 11 10.14 Posts and Columns		Crew	Daily Output	Labor-Hours	Unit	Material	2022 Bare Costs Labor	Equipment	Total	Total Incl O&P
0010	**POSTS AND COLUMNS**									
0100	4" x 4"	2 Carp	390	.041	L.F.	2.32	2.31		4.63	6.20
0150	4" x 6"		275	.058		4.15	3.28		7.43	9.75
0200	4" x 8"		220	.073		5.60	4.09		9.69	12.70
0250	6" x 6"		215	.074		9.50	4.19		13.69	17.10
0300	6" x 8"		175	.091		11.45	5.15		16.60	21
0350	6" x 10"	↓	150	.107	↓	14.70	6		20.70	25.50
9000	Minimum labor/equipment charge	1 Carp	2	4	Job		225		225	360

06 11 10.18 Joist Framing		Crew	Daily Output	Labor-Hours	Unit	Material	2022 Bare Costs Labor	Equipment	Total	Total Incl O&P
0010	**JOIST FRAMING** R061110-30									
2000	Joists, 2" x 4"	2 Carp	1250	.013	L.F.	.66	.72		1.38	1.86
2005	Pneumatic nailed	2 Carp	1438	.011	L.F.	.66	.63		1.29	1.73
2100	2" x 6"		1250	.013		.97	.72		1.69	2.21
2105	Pneumatic nailed		1438	.011		.97	.63		1.60	2.07
2150	2" x 8"		1100	.015		1.43	.82		2.25	2.88
2155	Pneumatic nailed		1265	.013		1.44	.71		2.15	2.71
2200	2" x 10"		900	.018		1.99	1		2.99	3.77
2205	Pneumatic nailed		1035	.015		2	.87		2.87	3.57
2250	2" x 12"		875	.018		2.77	1.03		3.80	4.69
2255	Pneumatic nailed		1006	.016		2.78	.90		3.68	4.48
2300	2" x 14"		770	.021		5.30	1.17		6.47	7.65
2305	Pneumatic nailed		886	.018		5.30	1.02		6.32	7.45
2350	3" x 6"		925	.017		4.18	.97		5.15	6.15
2400	3" x 10"		780	.021		6.85	1.15		8	9.35
2450	3" x 12"		600	.027		8.20	1.50		9.70	11.40
2500	4" x 6"		800	.020		4.15	1.13		5.28	6.35
2550	4" x 10"		600	.027		7.05	1.50		8.55	10.15
2600	4" x 12"		450	.036		8.65	2		10.65	12.75
2605	Sister joist, 2" x 6"		800	.020		.97	1.13		2.10	2.86
2606	Pneumatic nailed		960	.017		.97	.94		1.91	2.56
2610	2" x 8"		640	.025		1.43	1.41		2.84	3.82
2611	Pneumatic nailed		768	.021		1.44	1.17		2.61	3.44
2615	2" x 10"		535	.030		1.99	1.68		3.67	4.86
2616	Pneumatic nailed		642	.025		2	1.40		3.40	4.42
2620	2" x 12"		455	.035		2.77	1.98		4.75	6.20
2625	Pneumatic nailed		546	.029	↓	2.78	1.65		4.43	5.70
3000	Composite wood joist 9-1/2" deep		.90	17.778	M.L.F.	4,250	1,000		5,250	6,275
3010	11-1/2" deep		.88	18.182		4,475	1,025		5,500	6,550
3020	14" deep		.82	19.512		5,150	1,100		6,250	7,425
3030	16" deep		.78	20.513		7,525	1,150		8,675	10,100
4000	Open web joist 12" deep		.88	18.182	↓	7,600	1,025		8,625	9,975
4002	Per linear foot		880	.018	L.F.	7.60	1.02		8.62	10
4004	Treated, per linear foot		880	.018	"	8.65	1.02		9.67	11.15
4010	14" deep		.82	19.512	M.L.F.	8,250	1,100		9,350	10,800
4012	Per linear foot		820	.020	L.F.	8.25	1.10		9.35	10.80

For customer support on your Commercial Renovation Costs with RSMeans data, call 800.448.8182.

147

06 11 Wood Framing

06 11 10 – Framing with Dimensional, Engineered or Composite Lumber

06 11 10.18 Joist Framing

		Crew	Daily Output	Labor-Hours	Unit	Material	2022 Bare Costs Labor	2022 Bare Costs Equipment	Total	Total Incl O&P
4014	Treated, per linear foot	2 Carp	820	.020	L.F.	9.50	1.10		10.60	12.15
4020	16" deep		.78	20.513	M.L.F.	11,800	1,150		12,950	14,800
4022	Per linear foot		780	.021	L.F.	11.80	1.15		12.95	14.80
4024	Treated, per linear foot		780	.021	"	13.20	1.15		14.35	16.35
4030	18" deep		.74	21.622	M.L.F.	9,150	1,225		10,375	12,000
4032	Per linear foot	2 Carp	740	.022	L.F.	9.15	1.22		10.37	12
4034	Treated, per linear foot		740	.022	"	10.75	1.22		11.97	13.75
6000	Composite rim joist, 1-1/4" x 9-1/2"		.90	17.778	M.L.F.	2,700	1,000		3,700	4,575
6010	1-1/4" x 11-1/2"		.88	18.182		4,200	1,025		5,225	6,225
6020	1-1/4" x 14-1/2"		.82	19.512		5,450	1,100		6,550	7,750
6030	1-1/4" x 16-1/2"		.78	20.513		6,275	1,150		7,425	8,750
9000	Minimum labor/equipment charge	1 Carp	4	2	Job		113		113	179

06 11 10.24 Miscellaneous Framing

		Crew	Daily Output	Labor-Hours	Unit	Material	2022 Bare Costs Labor	2022 Bare Costs Equipment	Total	Total Incl O&P
0010	**MISCELLANEOUS FRAMING**									
2000	Firestops, 2" x 4"	2 Carp	780	.021	L.F.	.66	1.15		1.81	2.55
2005	Pneumatic nailed		952	.017		.66	.95		1.61	2.23
2100	2" x 6"		600	.027		.97	1.50		2.47	3.46
2105	Pneumatic nailed		732	.022		.97	1.23		2.20	3.03
5000	Nailers, treated, wood construction, 2" x 4"		800	.020		.86	1.13		1.99	2.73
5005	Pneumatic nailed		960	.017		.86	.94		1.80	2.44
5100	2" x 6"		750	.021		1.21	1.20		2.41	3.24
5105	Pneumatic nailed		900	.018		1.21	1		2.21	2.92
5120	2" x 8"		700	.023		1.61	1.29		2.90	3.81
5125	Pneumatic nailed		840	.019		1.61	1.07		2.68	3.47
5200	Steel construction, 2" x 4"		750	.021		.86	1.20		2.06	2.85
5220	2" x 6"		700	.023		1.21	1.29		2.50	3.37
5240	2" x 8"		650	.025		1.61	1.39		3	3.97
7000	Rough bucks, treated, for doors or windows, 2" x 6"		400	.040		1.21	2.25		3.46	4.91
7005	Pneumatic nailed		480	.033		1.21	1.88		3.09	4.31
7100	2" x 8"		380	.042		1.61	2.37		3.98	5.55
7105	Pneumatic nailed		456	.035		1.61	1.98		3.59	4.91
8000	Stair stringers, 2" x 10"		130	.123		1.99	6.95		8.94	13.20
8100	2" x 12"		130	.123		2.77	6.95		9.72	14.05
8150	3" x 10"		125	.128		6.85	7.20		14.05	18.95
8200	3" x 12"		125	.128		8.20	7.20		15.40	20.50
8870	Laminated structural lumber, 1-1/4" x 11-1/2"		130	.123		4.19	6.95		11.14	15.60
8880	1-1/4" x 14-1/2"		130	.123		5.40	6.95		12.35	16.95
9000	Minimum labor/equipment charge	1 Carp	4	2	Job		113		113	179

06 11 10.26 Partitions

		Crew	Daily Output	Labor-Hours	Unit	Material	2022 Bare Costs Labor	2022 Bare Costs Equipment	Total	Total Incl O&P
0010	**PARTITIONS**									
0020	Single bottom and double top plate, no waste, std. & better lumber									
0180	2" x 4" studs, 8' high, studs 12" OC	2 Carp	80	.200	L.F.	7.25	11.25		18.50	26
0185	12" OC, pneumatic nailed		96	.167		7.25	9.40		16.65	23
0200	16" OC		100	.160		5.90	9		14.90	21
0205	16" OC, pneumatic nailed	2 Carp	120	.133	L.F.	5.95	7.50		13.45	18.50
0300	24" OC		125	.128		4.60	7.20		11.80	16.50
0305	24" OC, pneumatic nailed		150	.107		4.62	6		10.62	14.65
0380	10' high, studs 12" OC		80	.200		8.55	11.25		19.80	27.50
0385	12" OC, pneumatic nailed		96	.167		8.60	9.40		18	24.50
0400	16" OC		100	.160		6.90	9		15.90	22
0405	16" OC, pneumatic nailed		120	.133		6.95	7.50		14.45	19.55
0500	24" OC		125	.128		5.25	7.20		12.45	17.25

06 11 10.26 Partitions

		Crew	Daily Output	Labor-Hours	Unit	Material	2022 Bare Costs Labor	Equipment	Total	Total Incl O&P
0505	24" OC, pneumatic nailed	2 Carp	150	.107	L.F.	5.30	6		11.30	15.35
0580	12' high, studs 12" OC		65	.246		9.85	13.85		23.70	33
0585	12" OC, pneumatic nailed		78	.205		9.90	11.55		21.45	29.50
0600	16" OC		80	.200		7.90	11.25		19.15	26.50
0605	16" OC, pneumatic nailed		96	.167		7.90	9.40		17.30	23.50
0700	24" OC		100	.160		5.90	9		14.90	21
0705	24" OC, pneumatic nailed		120	.133		5.95	7.50		13.45	18.50
0780	2" x 6" studs, 8' high, studs 12" OC		70	.229		10.65	12.85		23.50	32
0785	12" OC, pneumatic nailed		84	.190		10.70	10.70		21.40	29
0800	16" OC		90	.178		8.70	10		18.70	25.50
0805	16" OC, pneumatic nailed		108	.148		8.75	8.35		17.10	23
0900	24" OC		115	.139		6.80	7.85		14.65	19.90
0905	24" OC, pneumatic nailed		138	.116		6.80	6.55		13.35	17.85
0980	10' high, studs 12" OC		70	.229		12.60	12.85		25.45	34.50
0985	12" OC, pneumatic nailed		84	.190		12.65	10.70		23.35	31
1000	16" OC		90	.178		10.15	10		20.15	27
1005	16" OC, pneumatic nailed		108	.148		10.20	8.35		18.55	24.50
1100	24" OC		115	.139		7.75	7.85		15.60	21
1105	24" OC, pneumatic nailed		138	.116		7.80	6.55		14.35	18.90
1180	12' high, studs 12" OC		55	.291		14.55	16.40		30.95	42
1185	12" OC, pneumatic nailed		66	.242		14.60	13.65		28.25	37.50
1200	16" OC		70	.229		11.60	12.85		24.45	33.50
1205	16" OC, pneumatic nailed		84	.190		11.70	10.70		22.40	30
1300	24" OC		90	.178		8.70	10		18.70	25.50
1305	24" OC, pneumatic nailed		108	.148		8.75	8.35		17.10	23
1400	For horizontal blocking, 2" x 4", add		600	.027		.66	1.50		2.16	3.11
1500	2" x 6", add		600	.027		.97	1.50		2.47	3.46
1600	For openings, add		250	.064			3.60		3.60	5.70
1702	Headers for above openings, material only, add				B.F.	1.19			1.19	1.31
9000	Minimum labor/equipment charge	1 Carp	4	2	Job		113		113	179

06 11 10.28 Porch or Deck Framing

		Crew	Daily Output	Labor-Hours	Unit	Material	2022 Bare Costs Labor	Equipment	Total	Total Incl O&P
0010	**PORCH OR DECK FRAMING**									
0100	Treated lumber, posts or columns, 4" x 4"	2 Carp	390	.041	L.F.	1.63	2.31		3.94	5.45
0110	4" x 6"		275	.058		2.83	3.28		6.11	8.30
0120	4" x 8"		220	.073		6.75	4.09		10.84	13.95
0130	Girder, single, 4" x 4"		675	.024		1.63	1.33		2.96	3.92
0140	4" x 6"		600	.027		2.83	1.50		4.33	5.50
0150	4" x 8"		525	.030		6.75	1.72		8.47	10.20
0160	Double, 2" x 4"		625	.026		1.73	1.44		3.17	4.19
0170	2" x 6"		600	.027		2.44	1.50		3.94	5.05
0180	2" x 8"		575	.028		3.25	1.57		4.82	6.05
0190	2" x 10"		550	.029		3.77	1.64		5.41	6.75
0200	2" x 12"		525	.030		5.95	1.72		7.67	9.30
0210	Triple, 2" x 4"		575	.028		2.60	1.57		4.17	5.35
0220	2" x 6"		550	.029		3.66	1.64		5.30	6.60
0230	2" x 8"		525	.030		4.87	1.72		6.59	8.10
0240	2" x 10"		500	.032		5.65	1.80		7.45	9.05
0250	2" x 12"		475	.034		8.95	1.90		10.85	12.85
0260	Ledger, bolted 4' OC, 2" x 4"		400	.040		1.15	2.25		3.40	4.85
0270	2" x 6"		395	.041		1.50	2.28		3.78	5.25
0280	2" x 8"		390	.041		1.89	2.31		4.20	5.75
0290	2" x 10"		385	.042		2.15	2.34		4.49	6.10

For customer support on your Commercial Renovation Costs with RSMeans data, call 800.448.8182.

149

06 11 10 – Framing with Dimensional, Engineered or Composite Lumber

06 11 10.28 Porch or Deck Framing		Crew	Daily Output	Labor-Hours	Unit	Material	2022 Bare Costs Labor	Equipment	Total	Total Incl O&P
0300	2" x 12"	2 Carp	380	.042	L.F.	3.23	2.37		5.60	7.35
0310	Joists, 2" x 4"		1250	.013		.89	.72		1.61	2.12
0320	2" x 6"		1250	.013		1.25	.72		1.97	2.52
0330	2" x 8"		1100	.015		1.67	.82		2.49	3.13
0340	2" x 10"		900	.018		1.94	1		2.94	3.72
0350	2" x 12"		875	.018		2.33	1.03		3.36	4.20
0360	Railings and trim, 1" x 4"	1 Carp	300	.027		.63	1.50		2.13	3.09
0370	2" x 2"		300	.027		.80	1.50		2.30	3.27
0380	2" x 4"		300	.027		.86	1.50		2.36	3.33
0390	2" x 6"		300	.027		1.21	1.50		2.71	3.72
0400	Decking, 1" x 4"		275	.029	S.F.	2.66	1.64		4.30	5.55
0410	2" x 4"		300	.027		2.91	1.50		4.41	5.60
0420	2" x 6"		320	.025		2.61	1.41		4.02	5.10
0430	5/4" x 6"		320	.025		2.86	1.41		4.27	5.40
0440	Balusters, square, 2" x 2"	2 Carp	660	.024	L.F.	.81	1.36		2.17	3.06
0450	Turned, 2" x 2"		420	.038		1.07	2.14		3.21	4.59
0460	Stair stringer, 2" x 10"		130	.123		1.94	6.95		8.89	13.15
0470	2" x 12"		130	.123		2.33	6.95		9.28	13.55
0480	Stair treads, 1" x 4"	2 Carp	140	.114	L.F.	2.67	6.45		9.12	13.15
0490	2" x 4"		140	.114		.89	6.45		7.34	11.20
0500	2" x 6"		160	.100		1.33	5.65		6.98	10.40
0510	5/4" x 6"		160	.100		1.35	5.65		7	10.45
0520	Turned handrail post, 4" x 4"		64	.250	Ea.	16.70	14.10		30.80	41
0530	Lattice panel, 4' x 8', 1/2"		1600	.010	S.F.	1.36	.56		1.92	2.39
0535	3/4"		1600	.010	"	2.02	.56		2.58	3.12
0540	Cedar, posts or columns, 4" x 4"		390	.041	L.F.	5.05	2.31		7.36	9.20
0550	4" x 6"		275	.058		8.20	3.28		11.48	14.20
0560	4" x 8"		220	.073		12.60	4.09		16.69	20.50
0800	Decking, 1" x 4"		550	.029		4.59	1.64		6.23	7.65
0810	2" x 4"		600	.027		9.50	1.50		11	12.85
0820	2" x 6"		640	.025		14.30	1.41		15.71	18
0830	5/4" x 6"		640	.025		11.35	1.41		12.76	14.75
0840	Railings and trim, 1" x 4"		600	.027		4.59	1.50		6.09	7.45
0860	2" x 4"		600	.027		9.50	1.50		11	12.85
0870	2" x 6"		600	.027		14.30	1.50		15.80	18.15
0920	Stair treads, 1" x 4"		140	.114		4.59	6.45		11.04	15.25
0930	2" x 4"		140	.114		9.50	6.45		15.95	20.50
0940	2" x 6"		160	.100		14.30	5.65		19.95	24.50
0950	5/4" x 6"		160	.100		11.35	5.65		17	21.50
0980	Redwood, posts or columns, 4" x 4"		390	.041		12.35	2.31		14.66	17.25
0990	4" x 6"		275	.058		19.65	3.28		22.93	26.50
1000	4" x 8"		220	.073		37	4.09		41.09	47
1240	Decking, 1" x 4"	1 Carp	275	.029	S.F.	11.25	1.64		12.89	14.95
1260	2" x 6"		340	.024		20.50	1.32		21.82	25
1270	5/4" x 6"		320	.025		13.10	1.41		14.51	16.65
1280	Railings and trim, 1" x 4"	2 Carp	600	.027	L.F.	3.29	1.50		4.79	6
1310	2" x 6"		600	.027		20.50	1.50		22	25.50
1420	Alternative decking, wood/plastic composite, 5/4" x 6" [G]		640	.025		2.31	1.41		3.72	4.78
1440	1" x 4" square edge fir		550	.029		2.68	1.64		4.32	5.55
1450	1" x 4" tongue and groove fir		450	.036		1.98	2		3.98	5.35
1460	1" x 4" mahogany		550	.029		2.12	1.64		3.76	4.94
1462	5/4" x 6" PVC		550	.029		3.30	1.64		4.94	6.25
1465	Framing, porch or deck, alt deck fastening, screws, add	1 Carp	240	.033	S.F.		1.88		1.88	2.98

06 11 Wood Framing

06 11 10 – Framing with Dimensional, Engineered or Composite Lumber

06 11 10.28 Porch or Deck Framing

		Crew	Daily Output	Labor-Hours	Unit	Material	2022 Bare Costs Labor	Equipment	Total	Total Incl O&P
1470	Accessories, joist hangers, 2" x 4"	1 Carp	160	.050	Ea.	1.09	2.82		3.91	5.65
1480	2" x 6" through 2" x 12"	↓	150	.053		1.95	3		4.95	6.90
1530	Post footing, incl excav, backfill, tube form & concrete, 4' deep, 8" diam.	F-7	12	2.667	↓	17.90	136		153.90	236
1540	10" diameter	F-7	11	2.909	Ea.	29	148		177	267
1550	12" diameter	"	10	3.200	"	39	163		202	300

06 11 10.30 Roof Framing

		Crew	Daily Output	Labor-Hours	Unit	Material	2022 Bare Costs Labor	Equipment	Total	Total Incl O&P
0010	**ROOF FRAMING** R061110-30									
1900	Rough fascia, 2" x 6"	2 Carp	250	.064	L.F.	.97	3.60		4.57	6.75
2000	2" x 8"		225	.071		1.43	4		5.43	7.95
2100	2" x 10"		180	.089		1.99	5		6.99	10.15
2200	2" x 12"		180	.089		2.77	5		7.77	11
5000	Rafters, to 4 in 12 pitch, 2" x 6", ordinary		1000	.016		.97	.90		1.87	2.50
5060	2" x 8", ordinary		950	.017		1.43	.95		2.38	3.09
5120	2" x 10", ordinary		630	.025		1.99	1.43		3.42	4.45
5180	2" x 12", ordinary		575	.028		2.77	1.57		4.34	5.55
5250	Composite rafter, 9-1/2" deep		575	.028		4.25	1.57		5.82	7.15
5260	11-1/2" deep		575	.028		4.48	1.57		6.05	7.40
5300	Hip and valley rafters, 2" x 6", ordinary		760	.021		.97	1.19		2.16	2.95
5360	2" x 8", ordinary		720	.022		1.43	1.25		2.68	3.57
5420	2" x 10", ordinary		570	.028		1.99	1.58		3.57	4.69
5480	2" x 12", ordinary		525	.030		2.77	1.72		4.49	5.80
5540	Hip and valley jacks, 2" x 6", ordinary		600	.027		.97	1.50		2.47	3.46
5600	2" x 8", ordinary		490	.033		1.43	1.84		3.27	4.50
5660	2" x 10", ordinary		450	.036		1.99	2		3.99	5.35
5720	2" x 12", ordinary	↓	375	.043	↓	2.77	2.40		5.17	6.85
5761	For slopes steeper than 4 in 12, add						30%			
5780	Rafter tie, 1" x 4", #3	2 Carp	800	.020	L.F.	.63	1.13		1.76	2.48
5790	2" x 4", #3		800	.020		.66	1.13		1.79	2.51
5800	Ridge board, #2 or better, 1" x 6"		600	.027		.98	1.50		2.48	3.46
5820	1" x 8"		550	.029		1.36	1.64		3	4.09
5840	1" x 10"		500	.032		1.85	1.80		3.65	4.90
5860	2" x 6"		500	.032		.97	1.80		2.77	3.93
5880	2" x 8"		450	.036		1.43	2		3.43	4.76
5900	2" x 10"		400	.040		1.99	2.25		4.24	5.75
5920	Roof cants, split, 4" x 4"		650	.025		2.32	1.39		3.71	4.75
5940	6" x 6"		600	.027		9.50	1.50		11	12.85
5960	Roof curbs, untreated, 2" x 6"		520	.031		.97	1.73		2.70	3.82
5980	2" x 12"		400	.040		2.77	2.25		5.02	6.65
6000	Sister rafters, 2" x 6"		800	.020		.97	1.13		2.10	2.86
6020	2" x 8"		640	.025		1.43	1.41		2.84	3.82
6040	2" x 10"		535	.030		1.99	1.68		3.67	4.86
6060	2" x 12"	↓	455	.035	↓	2.77	1.98		4.75	6.20
9000	Minimum labor/equipment charge	1 Carp	4	2	Job		113		113	179

06 11 10.32 Sill and Ledger Framing

		Crew	Daily Output	Labor-Hours	Unit	Material	2022 Bare Costs Labor	Equipment	Total	Total Incl O&P
0010	**SILL AND LEDGER FRAMING**									
0020	Extruded polystyrene sill sealer, 5-1/2" wide	1 Carp	1600	.005	L.F.	.18	.28		.46	.65
2002	Ledgers, nailed, 2" x 4"	2 Carp	755	.021		.66	1.19		1.85	2.62
2052	2" x 6"		600	.027		.97	1.50		2.47	3.46
2102	Bolted, not including bolts, 3" x 6"		325	.049		4.16	2.77		6.93	9
2152	3" x 12"		233	.069		8.15	3.87		12.02	15.15
2602	Mud sills, redwood, construction grade, 2" x 4"		895	.018		6.25	1.01		7.26	8.45
2622	2" x 6"	↓	780	.021	↓	9.40	1.15		10.55	12.20

For customer support on your Commercial Renovation Costs with RSMeans data, call 800.448.8182.

151

06 11 Wood Framing

06 11 10 - Framing with Dimensional, Engineered or Composite Lumber

06 11 10.32 Sill and Ledger Framing

		Crew	Daily Output	Labor-Hours	Unit	Material	2022 Bare Costs Labor	Equipment	Total	Total Incl O&P
4002	Sills, 2" x 4"	2 Carp	600	.027	L.F.	.65	1.50		2.15	3.10
4052	2" x 6"		550	.029		.96	1.64		2.60	3.65
4082	2" x 8"		500	.032		1.42	1.80		3.22	4.42
4101	2" x 10"		450	.036		1.96	2		3.96	5.35
4121	2" x 12"		400	.040		2.75	2.25		5	6.60
4202	Treated, 2" x 4"		550	.029		.85	1.64		2.49	3.53
4222	2" x 6"		500	.032		1.19	1.80		2.99	4.17
4242	2" x 8"		450	.036		1.59	2		3.59	4.93
4261	2" x 10"		400	.040		1.84	2.25		4.09	5.60
4281	2" x 12"		350	.046		2.93	2.57		5.50	7.30
4402	4" x 4"		450	.036		1.60	2		3.60	4.94
4422	4" x 6"		350	.046		2.78	2.57		5.35	7.15
4462	4" x 8"		300	.053		6.70	3		9.70	12.10
4481	4" x 10"		260	.062		7.75	3.46		11.21	14
9000	Minimum labor/equipment charge	1 Carp	4	2	Job		113		113	179

06 11 10.34 Sleepers

		Crew	Daily Output	Labor-Hours	Unit	Material	2022 Bare Costs Labor	Equipment	Total	Total Incl O&P
0010	**SLEEPERS**									
0100	On concrete, treated, 1" x 2"	2 Carp	2350	.007	L.F.	.47	.38		.85	1.12
0150	1" x 3"		2000	.008		.48	.45		.93	1.25
0200	2" x 4"		1500	.011		.91	.60		1.51	1.96
0250	2" x 6"		1300	.012		1.32	.69		2.01	2.56
9000	Minimum labor/equipment charge	1 Carp	4	2	Job		113		113	179

06 11 10.36 Soffit and Canopy Framing

		Crew	Daily Output	Labor-Hours	Unit	Material	2022 Bare Costs Labor	Equipment	Total	Total Incl O&P
0010	**SOFFIT AND CANOPY FRAMING**									
1002	Canopy or soffit framing, 1" x 4"	2 Carp	900	.018	L.F.	.63	1		1.63	2.28
1021	1" x 6"		850	.019		.98	1.06		2.04	2.75
1042	1" x 8"		750	.021		1.36	1.20		2.56	3.40
1102	2" x 4"		620	.026		.66	1.45		2.11	3.03
1121	2" x 6"		560	.029		.97	1.61		2.58	3.63
1142	2" x 8"		500	.032		1.43	1.80		3.23	4.44
1202	3" x 4"		500	.032		2.31	1.80		4.11	5.40
1221	3" x 6"	2 Carp	400	.040	L.F.	4.18	2.25		6.43	8.20
1242	3" x 10"	"	300	.053	"	6.85	3		9.85	12.25
9000	Minimum labor/equipment charge	1 Carp	4	2	Job		113		113	179

06 11 10.38 Treated Lumber Framing Material

		Crew	Daily Output	Labor-Hours	Unit	Material	2022 Bare Costs Labor	Equipment	Total	Total Incl O&P
0010	**TREATED LUMBER FRAMING MATERIAL**									
0100	2" x 4"				M.B.F.	1,275			1,275	1,400
0110	2" x 6"					1,200			1,200	1,300
0120	2" x 8"					1,200			1,200	1,300
0130	2" x 10"					1,100			1,100	1,225
0140	2" x 12"					1,475			1,475	1,600
0200	4" x 4"					1,200			1,200	1,325
0210	4" x 6"					1,400			1,400	1,525
0220	4" x 8"					2,500			2,500	2,750

06 11 10.40 Wall Framing

		Crew	Daily Output	Labor-Hours	Unit	Material	2022 Bare Costs Labor	Equipment	Total	Total Incl O&P
0010	**WALL FRAMING** R061110-30									
0100	Door buck, studs, header, access, 8' high, 2" x 4" wall, 3' wide	1 Carp	32	.250	Ea.	27.50	14.10		41.60	52.50
0110	4' wide		32	.250		29	14.10		43.10	54.50
0120	5' wide		32	.250		36	14.10		50.10	62
0130	6' wide		32	.250		39	14.10		53.10	65
0140	8' wide		30	.267		53.50	15		68.50	83
0150	10' wide		30	.267		77.50	15		92.50	110

152

For customer support on your Commercial Renovation Costs with RSMeans data, call 800.448.8182.

06 11 10 – Framing with Dimensional, Engineered or Composite Lumber

06 11 10.40 Wall Framing		Crew	Daily Output	Labor-Hours	Unit	Material	2022 Bare Costs Labor	Equipment	Total	Total Incl O&P
0160	12' wide	1 Carp	30	.267	Ea.	150	15		165	189
0170	2" x 6" wall, 3' wide		32	.250		37.50	14.10		51.60	63.50
0180	4' wide		32	.250		39	14.10		53.10	65.50
0190	5' wide		32	.250		46	14.10		60.10	73
0200	6' wide		32	.250		49	14.10		63.10	76
0210	8' wide		30	.267		63.50	15		78.50	94
0220	10' wide		30	.267		87.50	15		102.50	121
0230	12' wide		30	.267		160	15		175	200
0240	Window buck, studs, header & access, 8' high 2" x 4" wall, 2' wide		24	.333		29	18.75		47.75	62
0250	3' wide		24	.333		34	18.75		52.75	67
0260	4' wide		24	.333		37	18.75		55.75	71
0270	5' wide		24	.333		44	18.75		62.75	78.50
0280	6' wide		24	.333		48.50	18.75		67.25	83.50
0290	7' wide		24	.333		61	18.75		79.75	97
0300	8' wide		22	.364		67.50	20.50		88	107
0310	10' wide		22	.364		93.50	20.50		114	136
0320	12' wide		22	.364		170	20.50		190.50	220
0330	2" x 6" wall, 2' wide		24	.333		41.50	18.75		60.25	76
0340	3' wide		24	.333		46.50	18.75		65.25	81
0350	4' wide		24	.333		50.50	18.75		69.25	85.50
0360	5' wide	1 Carp	24	.333	Ea.	57.50	18.75		76.25	93.50
0370	6' wide		24	.333		63.50	18.75		82.25	99.50
0380	7' wide		24	.333		77	18.75		95.75	115
0390	8' wide		22	.364		84	20.50		104.50	125
0400	10' wide		22	.364		111	20.50		131.50	155
0410	12' wide		22	.364		189	20.50		209.50	241
2002	Headers over openings, 2" x 6"	2 Carp	360	.044	L.F.	.97	2.50		3.47	5.05
2007	2" x 6", pneumatic nailed		432	.037		.97	2.09		3.06	4.38
2052	2" x 8"		340	.047		1.43	2.65		4.08	5.80
2057	2" x 8", pneumatic nailed		408	.039		1.44	2.21		3.65	5.10
2101	2" x 10"		320	.050		1.99	2.82		4.81	6.65
2106	2" x 10", pneumatic nailed		384	.042		2	2.35		4.35	5.90
2152	2" x 12"		300	.053		2.77	3		5.77	7.80
2157	2" x 12", pneumatic nailed		360	.044		2.78	2.50		5.28	7.05
2180	4" x 8"		260	.062		5.60	3.46		9.06	11.70
2185	4" x 8", pneumatic nailed		312	.051		5.65	2.89		8.54	10.80
2191	4" x 10"		240	.067		7.05	3.75		10.80	13.70
2196	4" x 10", pneumatic nailed		288	.056		7.05	3.13		10.18	12.70
2202	4" x 12"		190	.084		8.65	4.74		13.39	17.10
2207	4" x 12", pneumatic nailed		228	.070		8.70	3.95		12.65	15.85
2241	6" x 10"		165	.097		14.70	5.45		20.15	25
2246	6" x 10", pneumatic nailed		198	.081		14.70	4.55		19.25	23.50
2251	6" x 12"		140	.114		16.85	6.45		23.30	29
2256	6" x 12", pneumatic nailed		168	.095		16.90	5.35		22.25	27
5002	Plates, untreated, 2" x 3"		850	.019		.46	1.06		1.52	2.19
5007	2" x 3", pneumatic nailed		1020	.016		.46	.88		1.34	1.91
5022	2" x 4"		800	.020		.66	1.13		1.79	2.51
5027	2" x 4", pneumatic nailed		960	.017		.66	.94		1.60	2.22
5041	2" x 6"		750	.021		.97	1.20		2.17	2.98
5046	2" x 6", pneumatic nailed		900	.018		.97	1		1.97	2.66
5122	Studs, 8' high wall, 2" x 3"		1200	.013		.46	.75		1.21	1.70
5127	2" x 3", pneumatic nailed		1440	.011		.46	.63		1.09	1.50
5142	2" x 4"		1100	.015		.65	.82		1.47	2.01

06 11 Wood Framing

06 11 10 – Framing with Dimensional, Engineered or Composite Lumber

06 11 10.40 Wall Framing

		Crew	Daily Output	Labor-Hours	Unit	Material	2022 Bare Costs Labor	Equipment	Total	Total Incl O&P
5147	2" x 4", pneumatic nailed	2 Carp	1320	.012	L.F.	.66	.68		1.34	1.81
5162	2" x 6"		1000	.016		.97	.90		1.87	2.50
5167	2" x 6", pneumatic nailed		1200	.013		.97	.75		1.72	2.26
5182	3" x 4"		800	.020		2.31	1.13		3.44	4.33
5187	3" x 4", pneumatic nailed	↓	960	.017	↓	2.32	.94		3.26	4.04
8200	For 12' high walls, deduct						5%			
8220	For stub wall, 6' high, add						20%			
8240	3' high, add						40%			
8250	For second story & above, add						5%			
8300	For dormer & gable, add						15%			
9000	Minimum labor/equipment charge	1 Carp	4	2	Job		113		113	179

06 11 10.42 Furring

		Crew	Daily Output	Labor-Hours	Unit	Material	2022 Bare Costs Labor	Equipment	Total	Total Incl O&P
0010	**FURRING**									
0012	Wood strips, 1" x 2", on walls, on wood	1 Carp	550	.015	L.F.	.30	.82		1.12	1.63
0015	On wood, pneumatic nailed		710	.011		.30	.63		.93	1.34
0300	On masonry		495	.016		.33	.91		1.24	1.82
0400	On concrete		260	.031		.33	1.73		2.06	3.12
0600	1" x 3", on walls, on wood		550	.015		.46	.82		1.28	1.81
0605	On wood, pneumatic nailed		710	.011		.46	.63		1.09	1.52
0700	On masonry		495	.016		.51	.91		1.42	2.01
0800	On concrete		260	.031		.51	1.73		2.24	3.31
0850	On ceilings, on wood		350	.023		.46	1.29		1.75	2.55
0855	On wood, pneumatic nailed		450	.018		.46	1		1.46	2.10
0900	On masonry		320	.025		.51	1.41		1.92	2.80
0950	On concrete		210	.038	↓	.51	2.14		2.65	3.97
9000	Minimum labor/equipment charge	↓	4	2	Job		113		113	179

06 11 10.44 Grounds

		Crew	Daily Output	Labor-Hours	Unit	Material	2022 Bare Costs Labor	Equipment	Total	Total Incl O&P
0010	**GROUNDS**									
0020	For casework, 1" x 2" wood strips, on wood	1 Carp	330	.024	L.F.	.30	1.36		1.66	2.50
0100	On masonry		285	.028		.33	1.58		1.91	2.88
0200	On concrete		250	.032		.33	1.80		2.13	3.23
0400	For plaster, 3/4" deep, on wood		450	.018		.30	1		1.30	1.92
0500	On masonry		225	.036		.33	2		2.33	3.55
0600	On concrete		175	.046		.33	2.57		2.90	4.46
0700	On metal lath		200	.040		.33	2.25		2.58	3.95
9000	Minimum labor/equipment charge	↓	4	2	Job		113		113	179

06 12 Structural Panels

06 12 10 – Structural Insulated Panels

06 12 10.10 OSB Faced Panels

			Crew	Daily Output	Labor-Hours	Unit	Material	2022 Bare Costs Labor	Equipment	Total	Total Incl O&P
0010	**OSB FACED PANELS**										
0100	Structural insul. panels, 7/16" OSB both faces, EPS insul., 3-5/8" T	G	F-3	2075	.019	S.F.	6.50	1.11	.96	8.57	9.95
0110	5-5/8" thick	G		1725	.023		4.54	1.34	1.15	7.03	8.35
0120	7-3/8" thick	G		1425	.028		5.65	1.62	1.39	8.66	10.35
0130	9-3/8" thick	G		1125	.036		6.15	2.05	1.76	9.96	11.95
0140	7/16" OSB one face, EPS insul., 3-5/8" thick	G	↓	2175	.018	↓	5.25	1.06	.91	7.22	8.45
0150	5-5/8" thick	G	F-3	1825	.022	S.F.	5.45	1.27	1.09	7.81	9.15
0160	7-3/8" thick	G		1525	.026		6.25	1.51	1.30	9.06	10.70
0170	9-3/8" thick	G		1225	.033		6.65	1.88	1.62	10.15	12.10
0190	7/16" OSB - 1/2" GWB faces, EPS insul., 3-5/8" T	G		2075	.019		3.81	1.11	.96	5.88	7

06 12 Structural Panels

06 12 10 – Structural Insulated Panels

06 12 10.10 OSB Faced Panels

			Crew	Daily Output	Labor-Hours	Unit	Material	2022 Bare Costs Labor	Equipment	Total	Total Incl O&P
0200	5-5/8" thick	G	F-3	1725	.023	S.F.	4.93	1.34	1.15	7.42	8.80
0210	7-3/8" thick	G		1425	.028		5.65	1.62	1.39	8.66	10.35
0220	9-3/8" thick	G		1125	.036		5.40	2.05	1.76	9.21	11.15
0240	7/16" OSB - 1/2" MRGWB faces, EPS insul., 3-5/8" T	G		2075	.019		4.67	1.11	.96	6.74	7.95
0250	5-5/8" thick	G		1725	.023		5.70	1.34	1.15	8.19	9.65
0260	7-3/8" thick	G		1425	.028		5.80	1.62	1.39	8.81	10.50
0270	9-3/8" thick	G	▼	1125	.036		5.65	2.05	1.76	9.46	11.40
0300	For 1/2" GWB added to OSB skin, add	G					1.60			1.60	1.76
0310	For 1/2" MRGWB added to OSB skin, add	G					1.77			1.77	1.95
0320	For one T1-11 skin, add to OSB-OSB	G					2.09			2.09	2.30
0330	For one 19/32" CDX skin, add to OSB-OSB	G				▼	1.72			1.72	1.89
0500	Structural insulated panel, 7/16" OSB both sides, straw core										
0510	4-3/8" T, walls (w/sill, splines, plates)	G	F-6	2400	.017	S.F.	8.65	.89	.83	10.37	11.85
0520	Floors (w/splines)	G		2400	.017		8.65	.89	.83	10.37	11.85
0530	Roof (w/splines)	G		2400	.017		8.65	.89	.83	10.37	11.85
0550	7-7/8" T, walls (w/sill, splines, plates)	G		2400	.017		12.25	.89	.83	13.97	15.75
0560	Floors (w/splines)	G		2400	.017		12.25	.89	.83	13.97	15.75
0570	Roof (w/splines)	G	▼	2400	.017	▼	12.25	.89	.83	13.97	15.75

06 12 19 – Composite Shearwall Panels

06 12 19.10 Steel and Wood Composite Shearwall Panels

		Crew	Daily Output	Labor-Hours	Unit	Material	2022 Bare Costs Labor	Equipment	Total	Total Incl O&P
0010	**STEEL & WOOD COMPOSITE SHEARWALL PANELS**									
0020	Anchor bolts, 36" long (must be placed in wet concrete)	1 Carp	150	.053	Ea.	51.50	3		54.50	61.50
0030	On concrete, 2" x 4" & 2" x 6" walls, 7'-10' high, 360 lb. shear, 12" wide	2 Carp	8	2		630	113		743	875
0040	715 lb. shear, 15" wide		8	2		785	113		898	1,050
0050	1860 lb. shear, 18" wide		8	2		860	113		973	1,125
0060	2780 lb. shear, 21" wide		8	2		945	113		1,058	1,200
0070	3790 lb. shear, 24" wide		8	2		995	113		1,108	1,275
0080	2" x 6" walls, 11'-13' high, 1180 lb. shear, 18" wide		6	2.667		950	150		1,100	1,300
0090	1555 lb. shear, 21" wide		6	2.667		1,450	150		1,600	1,850
0100	2280 lb. shear, 24" wide		6	2.667	▼	1,125	150		1,275	1,500
0110	For installing above on wood floor frame, add									
0120	Coupler nuts, threaded rods, bolts, shear transfer plate kit	1 Carp	16	.500	Ea.	85.50	28		113.50	139
0130	Framing anchors, angle (2 required)	"	96	.083	"	3.20	4.69		7.89	10.95
0140	For blocking see Section 06 11 10.02									
0150	For installing above, first floor to second floor, wood floor frame, add									
0160	Add stack option to first floor wall panel				Ea.	94			94	103
0170	Threaded rods, bolts, shear transfer plate kit	1 Carp	16	.500		99	28		127	154
0180	Framing anchors, angle (2 required)	"	96	.083	▼	3.20	4.69		7.89	10.95
0190	For blocking see section 06 11 10.02									
0200	For installing stacked panels, balloon framing									
0210	Add stack option to first floor wall panel				Ea.	94			94	103
0220	Threaded rods, bolts kit	1 Carp	16	.500	"	56	28		84	106

For customer support on your Commercial Renovation Costs with RSMeans data, call 800.448.8182.

155

06 13 Heavy Timber Construction

06 13 23 – Heavy Timber Framing

06 13 23.10 Heavy Framing	Crew	Daily Output	Labor-Hours	Unit	Material	2022 Bare Costs Labor	Equipment	Total	Total Incl O&P
0010 **HEAVY FRAMING**									
0020 Beams, single 6" x 10"	2 Carp	1.10	14.545	M.B.F.	2,225	820		3,045	3,750
0100 Single 8" x 16"		1.20	13.333	"	3,600	750		4,350	5,175
0202 Built from 2" lumber, multiple 2" x 14"		900	.018	B.F.	2.26	1		3.26	4.07
0212 Built from 3" lumber, multiple 3" x 6"		700	.023		2.77	1.29		4.06	5.10
0222 Multiple 3" x 8"		800	.020		2.76	1.13		3.89	4.83
0232 Multiple 3" x 10"		900	.018		2.72	1		3.72	4.58
0242 Multiple 3" x 12"		1000	.016		2.72	.90		3.62	4.42
0252 Built from 4" lumber, multiple 4" x 6"		800	.020		2.06	1.13		3.19	4.06
0262 Multiple 4" x 8"		900	.018		2.10	1		3.10	3.90
0272 Multiple 4" x 10"		1000	.016		2.10	.90		3	3.74
0282 Multiple 4" x 12"		1100	.015		2.15	.82		2.97	3.67
0292 Columns, structural grade, 1500 fb, 4" x 4"		450	.036	L.F.	3.31	2		5.31	6.80
0302 6" x 6"		225	.071		8.50	4		12.50	15.70
0402 8" x 8"		240	.067		15.45	3.75		19.20	23
0502 10" x 10"		90	.178		24	10		34	42.50
0602 12" x 12"		70	.229		35	12.85		47.85	59
0802 Floor planks, 2" thick, T&G, 2" x 6"		1050	.015	B.F.	3.25	.86		4.11	4.94
0902 2" x 10"		1100	.015		4.25	.82		5.07	6
1102 3" thick, 3" x 6"		1050	.015		4.25	.86		5.11	6.05
1202 3" x 10"		1100	.015		4.25	.82		5.07	6
1402 Girders, structural grade, 12" x 12"		800	.020		2.90	1.13		4.03	4.98
1502 10" x 16"		1000	.016		5	.90		5.90	6.95
2302 Roof purlins, 4" thick, structural grade		1050	.015		2.10	.86		2.96	3.67
9000 Minimum labor/equipment charge	1 Carp	2	4	Job		225		225	360

06 15 Wood Decking

06 15 16 – Wood Roof Decking

06 15 16.10 Solid Wood Roof Decking

	Crew	Daily Output	Labor-Hours	Unit	Material	2022 Bare Costs Labor	Equipment	Total	Total Incl O&P
0010 **SOLID WOOD ROOF DECKING**									
0350 Cedar planks, 2" thick	2 Carp	350	.046	S.F.	8.10	2.57		10.67	13.05
0400 3" thick		320	.050		12.15	2.82		14.97	17.85
0500 4" thick		250	.064		16.25	3.60		19.85	23.50
0550 6" thick		200	.080		24.50	4.50		29	34
0650 Douglas fir, 2" thick		350	.046		5.35	2.57		7.92	10
0700 3" thick		320	.050		8.05	2.82		10.87	13.30
0800 4" thick		250	.064		10.70	3.60		14.30	17.50
0850 6" thick		200	.080		16.10	4.50		20.60	25
0950 Hemlock, 2" thick		350	.046		5.45	2.57		8.02	10.10
1000 3" thick		320	.050		8.20	2.82		11.02	13.45
1100 4" thick		250	.064		10.90	3.60		14.50	17.70
1150 6" thick		200	.080		16.35	4.50		20.85	25
1250 Western white spruce, 2" thick		350	.046		3.48	2.57		6.05	7.90
1300 3" thick		320	.050		5.25	2.82		8.07	10.20
1400 4" thick		250	.064		6.95	3.60		10.55	13.35
1450 6" thick		200	.080		10.45	4.50		14.95	18.65
9000 Minimum labor/equipment charge	1 Carp	2	4	Job		225		225	360

For customer support on your Commercial Renovation Costs with RSMeans data, call 800.448.8182.

06 15 Wood Decking

06 15 23 – Laminated Wood Decking

06 15 23.10 Laminated Roof Deck	Crew	Daily Output	Labor-Hours	Unit	Material	2022 Bare Costs Labor	Equipment	Total	Total Incl O&P
0010 **LAMINATED ROOF DECK**									
0020 Pine or hemlock, 3" thick	2 Carp	425	.038	S.F.	9.60	2.12		11.72	13.90
0100 4" thick		325	.049		11	2.77		13.77	16.50
0300 Cedar, 3" thick		425	.038		16.30	2.12		18.42	21.50
0400 4" thick		325	.049		22	2.77		24.77	28.50
0600 Fir, 3" thick		425	.038		12.80	2.12		14.92	17.40
0700 4" thick		325	.049		16.25	2.77		19.02	22.50
9000 Minimum labor/equipment charge	1 Carp	3	2.667	Job		150		150	239

06 16 Sheathing

06 16 13 – Insulating Sheathing

06 16 13.10 Insulating Sheathing		Crew	Daily Output	Labor-Hours	Unit	Material	2022 Bare Costs Labor	Equipment	Total	Total Incl O&P
0010 **INSULATING SHEATHING**										
0020 Expanded polystyrene, 1#/C.F. density, 3/4" thick, R2.89	G	2 Carp	1400	.011	S.F.	.37	.64		1.01	1.42
0030 1" thick, R3.85	G		1300	.012		.44	.69		1.13	1.58
0040 2" thick, R7.69	G		1200	.013		.72	.75		1.47	1.98
0050 Extruded polystyrene, 15 psi compressive strength, 1" thick, R5	G		1300	.012		.74	.69		1.43	1.91
0060 2" thick, R10	G		1200	.013		.91	.75		1.66	2.19
0070 Polyisocyanurate, 2#/C.F. density, 3/4" thick	G	2 Carp	1400	.011	S.F.	.81	.64		1.45	1.91
0080 1" thick	G		1300	.012		1	.69		1.69	2.20
0090 1-1/2" thick	G		1250	.013		1.19	.72		1.91	2.45
0100 2" thick	G		1200	.013		1.52	.75		2.27	2.86

06 16 23 – Subflooring

06 16 23.10 Subfloor		Crew	Daily Output	Labor-Hours	Unit	Material	2022 Bare Costs Labor	Equipment	Total	Total Incl O&P
0010 **SUBFLOOR**	R061636-20									
0011 Plywood, CDX, 1/2" thick		2 Carp	1500	.011	SF Flr.	.76	.60		1.36	1.79
0015 Pneumatic nailed			1860	.009		.76	.48		1.24	1.61
0102 5/8" thick			1350	.012		.82	.67		1.49	1.96
0107 Pneumatic nailed			1674	.010		.82	.54		1.36	1.75
0202 3/4" thick			1250	.013		.94	.72		1.66	2.18
0207 Pneumatic nailed			1550	.010		.94	.58		1.52	1.96
0302 1-1/8" thick, 2-4-1 including underlayment			1050	.015		3.62	.86		4.48	5.35
0440 With boards, 1" x 6", S4S, laid regular			900	.018		2.11	1		3.11	3.91
0452 1" x 8", laid regular			1000	.016		2.16	.90		3.06	3.80
0462 Laid diagonal			850	.019		2.16	1.06		3.22	4.05
0502 1" x 10", laid regular			1100	.015		2.32	.82		3.14	3.85
0602 Laid diagonal			900	.018		2.32	1		3.32	4.14
8990 Subfloor adhesive, 3/8" bead		1 Carp	2300	.003	L.F.	.10	.20		.30	.42
9000 Minimum labor/equipment charge		"	4	2	Job		113		113	179

06 16 26 – Underlayment

06 16 26.10 Wood Product Underlayment		Crew	Daily Output	Labor-Hours	Unit	Material	2022 Bare Costs Labor	Equipment	Total	Total Incl O&P
0010 **WOOD PRODUCT UNDERLAYMENT**	R061636-20									
0015 Plywood, underlayment grade, 1/4" thick		2 Carp	1500	.011	S.F.	1.85	.60		2.45	2.99
0018 Pneumatic nailed			1860	.009		1.85	.48		2.33	2.81
0030 3/8" thick			1500	.011		2.06	.60		2.66	3.22
0070 Pneumatic nailed			1860	.009		2.06	.48		2.54	3.04
0102 1/2" thick			1450	.011		2.75	.62		3.37	4.02
0107 Pneumatic nailed			1798	.009		2.75	.50		3.25	3.83
0202 5/8" thick			1400	.011		2.81	.64		3.45	4.11

06 16 Sheathing

06 16 26 – Underlayment

06 16 26.10 Wood Product Underlayment

		Crew	Daily Output	Labor-Hours	Unit	Material	2022 Bare Costs Labor	2022 Bare Costs Equipment	Total	Total Incl O&P
0207	Pneumatic nailed	2 Carp	1736	.009	S.F.	2.81	.52		3.33	3.91
0302	3/4" thick		1300	.012		3.08	.69		3.77	4.49
0306	Pneumatic nailed		1612	.010		3.08	.56		3.64	4.28
0502	Particle board, 3/8" thick		1500	.011		.58	.60		1.18	1.59
0507	Pneumatic nailed		1860	.009		.58	.48		1.06	1.41
0602	1/2" thick		1450	.011		1	.62		1.62	2.09
0607	Pneumatic nailed		1798	.009		1	.50		1.50	1.90
0802	5/8" thick		1400	.011		1.03	.64		1.67	2.15
0807	Pneumatic nailed	2 Carp	1736	.009	S.F.	1.03	.52		1.55	1.95
0902	3/4" thick		1300	.012		1.06	.69		1.75	2.27
0907	Pneumatic nailed		1612	.010		1.06	.56		1.62	2.06
1100	Hardboard, underlayment grade, 4' x 4', .215" thick [G]		1500	.011		.87	.60		1.47	1.91
9000	Minimum labor/equipment charge	1 Carp	4	2	Job		113		113	179

06 16 33 – Wood Board Sheathing

06 16 33.10 Board Sheathing

		Crew	Daily Output	Labor-Hours	Unit	Material	2022 Bare Costs Labor	2022 Bare Costs Equipment	Total	Total Incl O&P
0009	**BOARD SHEATHING**									
0010	Roof, 1" x 6" boards, laid horizontal	2 Carp	725	.022	S.F.	2.11	1.24		3.35	4.29
0020	On steep roof		520	.031		2.11	1.73		3.84	5.05
0040	On dormers, hips, & valleys		480	.033		2.11	1.88		3.99	5.30
0050	Laid diagonal		650	.025		2.11	1.39		3.50	4.52
0070	1" x 8" boards, laid horizontal		875	.018		2.16	1.03		3.19	4.01
0080	On steep roof		635	.025		2.16	1.42		3.58	4.62
0090	On dormers, hips, & valleys		580	.028		2.21	1.55		3.76	4.90
0100	Laid diagonal		725	.022		2.16	1.24		3.40	4.34
0110	Skip sheathing, 1" x 4", 7" OC	1 Carp	1200	.007		.78	.38		1.16	1.46
0120	1" x 6", 9" OC		1450	.006		.97	.31		1.28	1.56
0180	T&G sheathing/decking, 1" x 6"		1000	.008		2.54	.45		2.99	3.51
0190	2" x 6"		1000	.008		5.90	.45		6.35	7.15
0200	Walls, 1" x 6" boards, laid regular	2 Carp	650	.025		2.11	1.39		3.50	4.52
0210	Laid diagonal		585	.027		2.11	1.54		3.65	4.77
0220	1" x 8" boards, laid regular		765	.021		2.16	1.18		3.34	4.24
0230	Laid diagonal		650	.025		2.16	1.39		3.55	4.57

06 16 36 – Wood Panel Product Sheathing

06 16 36.10 Sheathing

		Crew	Daily Output	Labor-Hours	Unit	Material	2022 Bare Costs Labor	2022 Bare Costs Equipment	Total	Total Incl O&P
0010	**SHEATHING** R061110-30									
0012	Plywood on roofs, CDX									
0032	5/16" thick	2 Carp	1600	.010	S.F.	.91	.56		1.47	1.89
0037	Pneumatic nailed R061636-20		1952	.008		.91	.46		1.37	1.73
0052	3/8" thick		1525	.010		.68	.59		1.27	1.69
0057	Pneumatic nailed		1860	.009		.68	.48		1.16	1.52
0102	1/2" thick		1400	.011		.76	.64		1.40	1.86
0103	Pneumatic nailed		1708	.009		.76	.53		1.29	1.68
0202	5/8" thick		1300	.012		.82	.69		1.51	2
0207	Pneumatic nailed		1586	.010		.82	.57		1.39	1.80
0302	3/4" thick		1200	.013		.94	.75		1.69	2.23
0307	Pneumatic nailed		1464	.011		.94	.62		1.56	2.02
0502	Plywood on walls, with exterior CDX, 3/8" thick		1200	.013		.68	.75		1.43	1.94
0507	Pneumatic nailed	2 Carp	1488	.011	S.F.	.68	.61		1.29	1.71
0602	1/2" thick		1125	.014		.76	.80		1.56	2.11
0607	Pneumatic nailed		1395	.011		.76	.65		1.41	1.87
0702	5/8" thick		1050	.015		.82	.86		1.68	2.26
0707	Pneumatic nailed		1302	.012		.82	.69		1.51	2

06 16 36 – Wood Panel Product Sheathing

06 16 36.10 Sheathing		Crew	Daily Output	Labor-Hours	Unit	Material	2022 Bare Costs Labor	Equipment	Total	Total Incl O&P
0802	3/4" thick	2 Carp	975	.016	S.F.	.94	.92		1.86	2.51
0807	Pneumatic nailed	↓	1209	.013	↓	.94	.75		1.69	2.22
1000	For shear wall construction, add						20%			
1200	For structural 1 exterior plywood, add				S.F.	10%				
3000	Wood fiber, regular, no vapor barrier, 1/2" thick	2 Carp	1200	.013		.73	.75		1.48	1.99
3100	5/8" thick		1200	.013		.69	.75		1.44	1.95
3300	No vapor barrier, in colors, 1/2" thick		1200	.013		.79	.75		1.54	2.06
3400	5/8" thick		1200	.013		.83	.75		1.58	2.10
3600	With vapor barrier one side, white, 1/2" thick		1200	.013		.81	.75		1.56	2.08
3700	Vapor barrier 2 sides, 1/2" thick		1200	.013		.96	.75		1.71	2.25
3800	Asphalt impregnated, 25/32" thick		1200	.013		.44	.75		1.19	1.67
3850	Intermediate, 1/2" thick		1200	.013		.37	.75		1.12	1.60
4500	Oriented strand board, on roof, 7/16" thick G		1460	.011		.47	.62		1.09	1.50
4505	Pneumatic nailed G		1780	.009		.47	.51		.98	1.32
4550	1/2" thick G		1400	.011		.47	.64		1.11	1.54
4555	Pneumatic nailed G		1736	.009		.47	.52		.99	1.34
4600	5/8" thick G		1300	.012		1.07	.69		1.76	2.28
4605	Pneumatic nailed G		1586	.010		1.07	.57		1.64	2.08
4610	On walls, 7/16" thick G		1200	.013		.47	.75		1.22	1.71
4615	Pneumatic nailed G		1488	.011		.47	.61		1.08	1.48
4620	1/2" thick G		1195	.013		.47	.75		1.22	1.72
4625	Pneumatic nailed G		1325	.012		.47	.68		1.15	1.60
4630	5/8" thick G		1050	.015		1.07	.86		1.93	2.54
4635	Pneumatic nailed G		1302	.012		1.07	.69		1.76	2.28
4700	Oriented strand board, factory laminated W.R. barrier, on roof, 1/2" thick G		1400	.011		1.37	.64		2.01	2.53
4705	Pneumatic nailed G		1736	.009		1.37	.52		1.89	2.33
4720	5/8" thick G		1300	.012		2.47	.69		3.16	3.82
4725	Pneumatic nailed G		1586	.010		2.47	.57		3.04	3.62
4730	5/8" thick, T&G G		1150	.014		1.83	.78		2.61	3.25
4735	Pneumatic nailed, T&G G		1400	.011		1.83	.64		2.47	3.03
4740	On walls, 7/16" thick G	↓	1200	.013	↓	1.11	.75		1.86	2.41
4745	Pneumatic nailed G		1488	.011		1.11	.61		1.72	2.18
4750	1/2" thick G	2 Carp	1195	.013	S.F.	1.37	.75		2.12	2.71
4755	Pneumatic nailed G		1325	.012	"	1.37	.68		2.05	2.59
4800	Joint sealant tape, 3-1/2"		7600	.002	L.F.	.30	.12		.42	.52
4810	Joint sealant tape, 6"	↓	7600	.002	"	.57	.12		.69	.82
9000	Minimum labor/equipment charge	1 Carp	2	4	Job		225		225	360

06 16 43 – Gypsum Sheathing

06 16 43.10 Gypsum Sheathing		Crew	Daily Output	Labor-Hours	Unit	Material	2022 Bare Costs Labor	Equipment	Total	Total Incl O&P
0010	**GYPSUM SHEATHING**									
0020	Gypsum, weatherproof, 1/2" thick	2 Carp	1125	.014	S.F.	.52	.80		1.32	1.84
0040	With embedded glass mats	"	1100	.015	"	.62	.82		1.44	1.98

06 17 Shop-Fabricated Structural Wood

06 17 33 – Wood I-Joists

06 17 33.10 Wood and Composite I-Joists	Crew	Daily Output	Labor-Hours	Unit	Material	2022 Bare Costs Labor	Equipment	Total	Total Incl O&P
0010 **WOOD AND COMPOSITE I-JOISTS**									
0100 Plywood webs, incl. bridging & blocking, panels 24" OC									
1200 15' to 24' span, 50 psf live load	F-5	2400	.013	SF Flr.	4.82	.76		5.58	6.50
1300 55 psf live load		2250	.014		5.05	.81		5.86	6.90
1400 24' to 30' span, 45 psf live load		2600	.012		5.85	.70		6.55	7.50
1500 55 psf live load		2400	.013		8.55	.76		9.31	10.60

06 17 53 – Shop-Fabricated Wood Trusses

06 17 53.10 Roof Trusses

	Crew	Daily Output	Labor-Hours	Unit	Material	Labor	Equipment	Total	Total Incl O&P
0010 **ROOF TRUSSES**									
0100 Fink (W) or King post type, 2'-0" OC									
0200 Metal plate connected, 4 in 12 slope									
0210 24' to 29' span	F-3	3000	.013	SF Flr.	3.26	.77	.66	4.69	5.55
0300 30' to 43' span		3000	.013		4.93	.77	.66	6.36	7.40
0400 44' to 60' span		3000	.013		4.75	.77	.66	6.18	7.15
0700 Glued and nailed, add					50%				

06 18 Glued-Laminated Construction

06 18 13 – Glued-Laminated Beams

06 18 13.20 Laminated Framing

	Crew	Daily Output	Labor-Hours	Unit	Material	Labor	Equipment	Total	Total Incl O&P
0010 **LAMINATED FRAMING**									
0020 30 lb., short term live load, 15 lb. dead load									
0200 Straight roof beams, 20' clear span, beams 8' OC	F-3	2560	.016	SF Flr.	3.38	.90	.77	5.05	6
0300 Beams 16' OC		3200	.013		2.48	.72	.62	3.82	4.55
0500 40' clear span, beams 8' OC		3200	.013		6.45	.72	.62	7.79	8.85
0600 Beams 16' OC		3840	.010		5.30	.60	.52	6.42	7.35
0800 60' clear span, beams 8' OC	F-4	2880	.017		11	.95	.79	12.74	14.45
0900 Beams 16' OC	"	3840	.013		8.25	.71	.59	9.55	10.85
1100 Tudor arches, 30' to 40' clear span, frames 8' OC	F-3	1680	.024		14.35	1.37	1.18	16.90	19.20
1200 Frames 16' OC	F-3	2240	.018	SF Flr.	11.20	1.03	.89	13.12	14.90
1400 50' to 60' clear span, frames 8' OC	F-4	2200	.022		15.40	1.25	1.03	17.68	20
1500 Frames 16' OC		2640	.018		13.15	1.04	.86	15.05	17.10
1700 Radial arches, 60' clear span, frames 8' OC		1920	.025		14.45	1.43	1.18	17.06	19.45
1800 Frames 16' OC		2880	.017		11.50	.95	.79	13.24	15
2000 100' clear span, frames 8' OC		1600	.030		14.85	1.71	1.42	17.98	20.50
2100 Frames 16' OC		2400	.020		13.15	1.14	.95	15.24	17.30
2300 120' clear span, frames 8' OC		1440	.033		19.75	1.90	1.58	23.23	26
2400 Frames 16' OC		1920	.025		18.05	1.43	1.18	20.66	23.50
2600 Bowstring trusses, 20' OC, 40' clear span	F-3	2400	.017		8.95	.96	.83	10.74	12.30
2700 60' clear span	F-4	3600	.013		8.05	.76	.63	9.44	10.75
2800 100' clear span		4000	.012		11.40	.69	.57	12.66	14.25
2900 120' clear span		3600	.013		12.10	.76	.63	13.49	15.20
3000 For less than 1000 B.F., add					20%				
3100 For premium appearance, add to S.F. prices					5%				
3300 For industrial type, deduct					15%				
3500 For stain and varnish, add					5%				
3900 For 3/4" laminations, add to straight					25%				
4100 Add to curved					15%				
4300 Alternate pricing method: (use nominal footage of									
4310 components). Straight beams, camber less than 6"	F-3	3.50	11.429	M.B.F.	4,225	660	565	5,450	6,325
4400 Columns, including hardware		2	20		4,525	1,150	990	6,665	7,900

06 18 Glued-Laminated Construction

06 18 13 – Glued-Laminated Beams

06 18 13.20 Laminated Framing

	Crew	Daily Output	Labor-Hours	Unit	Material	2022 Bare Costs Labor	2022 Bare Costs Equipment	Total	Total Incl O&P
4600 Curved members, radius over 32'	F-3	2.50	16	M.B.F.	4,650	925	795	6,370	7,425
4700 Radius 10' to 32'	↓	3	13.333		4,600	770	660	6,030	7,000
4900 For complicated shapes, add maximum					100%				
5100 For pressure treating, add to straight					35%				
5200 Add to curved				↓	45%				
6000 Laminated veneer members, southern pine or western species									
6050 1-3/4" wide x 5-1/2" deep	2 Carp	480	.033	L.F.	6.35	1.88		8.23	10
6100 9-1/2" deep		480	.033		5.40	1.88		7.28	8.95
6150 14" deep		450	.036		8.50	2		10.50	12.55
6200 18" deep	↓	450	.036	↓	11.75	2		13.75	16.10
6300 Parallel strand members, southern pine or western species									
6350 1-3/4" wide x 9-1/4" deep	2 Carp	480	.033	L.F.	5.65	1.88		7.53	9.20
6400 11-1/4" deep		450	.036		6.70	2		8.70	10.55
6450 14" deep	↓	400	.040	↓	9.90	2.25		12.15	14.50
6500 3-1/2" wide x 9-1/4" deep	2 Carp	480	.033	L.F.	27	1.88		28.88	32.50
6550 11-1/4" deep		450	.036		29.50	2		31.50	35.50
6600 14" deep		400	.040		27	2.25		29.25	33
6650 7" wide x 9-1/4" deep		450	.036		50	2		52	58
6700 11-1/4" deep		420	.038		61	2.14		63.14	70.50
6750 14" deep	↓	400	.040	↓	73	2.25		75.25	83.50
9000 Minimum labor/equipment charge	F-3	2.50	16	Job		925	795	1,720	2,325

06 22 Millwork

06 22 13 – Standard Pattern Wood Trim

06 22 13.15 Moldings, Base

	Crew	Daily Output	Labor-Hours	Unit	Material	2022 Bare Costs Labor	2022 Bare Costs Equipment	Total	Total Incl O&P
0010 **MOLDINGS, BASE**									
5100 Classic profile, 5/8" x 5-1/2", finger jointed and primed	1 Carp	250	.032	L.F.	1.71	1.80		3.51	4.74
5105 Poplar		240	.033		1.72	1.88		3.60	4.88
5110 Red oak		220	.036		2.64	2.05		4.69	6.15
5115 Maple		220	.036		3.98	2.05		6.03	7.65
5120 Cherry		220	.036		4.72	2.05		6.77	8.45
5125 3/4" x 7-1/2", finger jointed and primed		250	.032		2.17	1.80		3.97	5.25
5130 Poplar		240	.033		2.75	1.88		4.63	6
5135 Red oak		220	.036		3.84	2.05		5.89	7.50
5140 Maple		220	.036		5.30	2.05		7.35	9.10
5145 Cherry		220	.036		6.30	2.05		8.35	10.20
5150 Modern profile, 5/8" x 3-1/2", finger jointed and primed		250	.032		1.02	1.80		2.82	3.99
5155 Poplar		240	.033		1.10	1.88		2.98	4.19
5160 Red oak		220	.036		1.81	2.05		3.86	5.25
5165 Maple		220	.036		2.80	2.05		4.85	6.35
5170 Cherry		220	.036		3.01	2.05		5.06	6.55
5175 Ogee profile, 7/16" x 3", finger jointed and primed		250	.032		.73	1.80		2.53	3.67
5180 Poplar		240	.033		.76	1.88		2.64	3.82
5185 Red oak		220	.036		1.01	2.05		3.06	4.36
5200 9/16" x 3-1/2", finger jointed and primed		250	.032		1.10	1.80		2.90	4.07
5205 Pine		240	.033		1.20	1.88		3.08	4.30
5210 Red oak		220	.036		1.81	2.05		3.86	5.25
5215 9/16" x 4-1/2", red oak		220	.036		2.71	2.05		4.76	6.25
5220 5/8" x 3-1/2", finger jointed and primed		250	.032		1.10	1.80		2.90	4.07
5225 Poplar		240	.033		1.16	1.88		3.04	4.26
5230 Red oak	↓	220	.036	↓	1.81	2.05		3.86	5.25

06 22 13.15 Moldings, Base

		Crew	Daily Output	Labor-Hours	Unit	Material	2022 Bare Costs Labor	2022 Bare Costs Equipment	Total	Total Incl O&P
5235	Maple	1 Carp	220	.036	L.F.	2.80	2.05		4.85	6.35
5240	Cherry		220	.036		3.01	2.05		5.06	6.55
5245	5/8" x 4", finger jointed and primed	▼	250	.032	▼	1.41	1.80		3.21	4.41
5250	Poplar	1 Carp	240	.033	L.F.	1.50	1.88		3.38	4.63
5255	Red oak		220	.036		2.03	2.05		4.08	5.50
5260	Maple		220	.036		3.13	2.05		5.18	6.70
5265	Cherry		220	.036		3.27	2.05		5.32	6.85
5270	Rectangular profile, oak, 3/8" x 1-1/4"		260	.031		2.01	1.73		3.74	4.96
5275	1/2" x 2-1/2"		255	.031		2.38	1.77		4.15	5.45
5280	1/2" x 3-1/2"		250	.032		3.02	1.80		4.82	6.20
5285	1" x 6"		240	.033		2.84	1.88		4.72	6.10
5290	1" x 8"		240	.033		4.15	1.88		6.03	7.55
5295	Pine, 3/8" x 1-3/4"		260	.031		.52	1.73		2.25	3.33
5300	7/16" x 2-1/2"		255	.031		1.06	1.77		2.83	3.98
5305	1" x 6"		240	.033		1.99	1.88		3.87	5.15
5310	1" x 8"		240	.033		2.99	1.88		4.87	6.25
5315	Shoe, 1/2" x 3/4", primed		260	.031		.56	1.73		2.29	3.37
5320	Pine		240	.033		.66	1.88		2.54	3.71
5325	Poplar		240	.033		.46	1.88		2.34	3.49
5330	Red oak		220	.036		.62	2.05		2.67	3.94
5335	Maple		220	.036		.88	2.05		2.93	4.22
5340	Cherry		220	.036		.92	2.05		2.97	4.27
5345	11/16" x 1-1/2", pine		240	.033		.72	1.88		2.60	3.78
5350	Caps, 11/16" x 1-3/8", pine		240	.033		.71	1.88		2.59	3.76
5355	3/4" x 1-3/4", finger jointed and primed		260	.031		.90	1.73		2.63	3.74
5360	Poplar		240	.033		1.13	1.88		3.01	4.23
5365	Red oak		220	.036		1.33	2.05		3.38	4.72
5370	Maple		220	.036		1.79	2.05		3.84	5.20
5375	Cherry		220	.036		4.25	2.05		6.30	7.95
5380	Combination base & shoe, 9/16" x 3-1/2" & 1/2" x 3/4", pine		125	.064		1.87	3.60		5.47	7.75
5385	Three piece oak, 6" high		80	.100		4.80	5.65		10.45	14.25
5390	Including 3/4" x 1" base shoe		70	.114		5.45	6.45		11.90	16.15
5395	Flooring cant strip, 3/4" x 3/4", pre-finished pine	▼	260	.031		.41	1.73		2.14	3.20
5400	For pre-finished, stain and clear coat, add					1.05			1.05	1.16
5405	Clear coat only, add				▼	.89			.89	.98
9000	Minimum labor/equipment charge	1 Carp	4	2	Job		113		113	179

06 22 13.30 Moldings, Casings

		Crew	Daily Output	Labor-Hours	Unit	Material	2022 Bare Costs Labor	2022 Bare Costs Equipment	Total	Total Incl O&P
0010	**MOLDINGS, CASINGS**									
0085	Apron, 9/16" x 2-1/2", pine	1 Carp	250	.032	L.F.	1.86	1.80		3.66	4.91
0090	5/8" x 2-1/2", pine		250	.032		2.23	1.80		4.03	5.30
0110	5/8" x 3-1/2", pine		220	.036		2.53	2.05		4.58	6.05
0300	Band, 11/16" x 1-1/8", pine	▼	270	.030	▼	.98	1.67		2.65	3.73
0310	11/16" x 1-1/2", finger jointed and primed	1 Carp	270	.030	L.F.	.85	1.67		2.52	3.59
0320	Pine		270	.030		1.31	1.67		2.98	4.09
0330	11/16" x 1-3/4", finger jointed and primed		270	.030		1.05	1.67		2.72	3.81
0350	Pine		270	.030		1.43	1.67		3.10	4.23
0355	Beaded, 3/4" x 3-1/2", finger jointed and primed		220	.036		1.10	2.05		3.15	4.46
0360	Poplar		220	.036		1.29	2.05		3.34	4.67
0365	Red oak		220	.036		1.81	2.05		3.86	5.25
0370	Maple		220	.036		2.80	2.05		4.85	6.35
0375	Cherry		220	.036		3.38	2.05		5.43	6.95
0380	3/4" x 4", finger jointed and primed	▼	220	.036		1.92	2.05		3.97	5.35

06 22 13 – Standard Pattern Wood Trim

06 22 13.30 Moldings, Casings	Crew	Daily Output	Labor-Hours	Unit	Material	2022 Bare Costs Labor	Equipment	Total	Total Incl O&P
0385	Poplar	1 Carp	220	.036	L.F.	1.99	2.05		4.04	5.45
0390	Red oak		220	.036		2.84	2.05		4.89	6.40
0395	Maple		220	.036		4.38	2.05		6.43	8.05
0400	Cherry		220	.036		4.58	2.05		6.63	8.30
0405	3/4" x 5-1/2", finger jointed and primed		200	.040		1.80	2.25		4.05	5.55
0410	Poplar		200	.040		2.23	2.25		4.48	6.05
0415	Red oak		200	.040		3.12	2.25		5.37	7
0420	Maple		200	.040		4.09	2.25		6.34	8.10
0425	Cherry		200	.040		4.70	2.25		6.95	8.75
0430	Classic profile, 3/4" x 2-3/4", finger jointed and primed		250	.032		.96	1.80		2.76	3.92
0435	Poplar		250	.032		1.16	1.80		2.96	4.14
0440	Red oak		250	.032		1.61	1.80		3.41	4.63
0445	Maple		250	.032		2.29	1.80		4.09	5.40
0450	Cherry		250	.032		2.60	1.80		4.40	5.70
0455	Fluted, 3/4" x 3-1/2", poplar		220	.036		1.29	2.05		3.34	4.67
0460	Red oak		220	.036		1.81	2.05		3.86	5.25
0465	Maple		220	.036		2.80	2.05		4.85	6.35
0470	Cherry		220	.036		3.38	2.05		5.43	6.95
0475	3/4" x 4", poplar		220	.036		1.99	2.05		4.04	5.45
0480	Red oak		220	.036		2.84	2.05		4.89	6.40
0485	Maple		220	.036		4.38	2.05		6.43	8.05
0490	Cherry		220	.036		4.58	2.05		6.63	8.30
0495	3/4" x 5-1/2", poplar		200	.040		2.23	2.25		4.48	6.05
0500	Red oak		200	.040		3.12	2.25		5.37	7
0505	Maple		200	.040		4.09	2.25		6.34	8.10
0510	Cherry		200	.040		4.70	2.25		6.95	8.75
0515	3/4" x 7-1/2", poplar		190	.042		2.03	2.37		4.40	6
0520	Red oak		190	.042		5.40	2.37		7.77	9.65
0525	Maple		190	.042		7.40	2.37		9.77	11.90
0530	Cherry	1 Carp	190	.042	L.F.	8.85	2.37		11.22	13.50
0535	3/4" x 9-1/2", poplar		180	.044		5.30	2.50		7.80	9.80
0540	Red oak		180	.044		8.35	2.50		10.85	13.20
0545	Maple		180	.044		11.50	2.50		14	16.65
0550	Cherry		180	.044		12.50	2.50		15	17.75
0555	Modern profile, 9/16" x 2-1/4", poplar		250	.032		1.43	1.80		3.23	4.44
0560	Red oak		250	.032		1.83	1.80		3.63	4.88
0565	11/16" x 2-1/2", finger jointed & primed		250	.032		1.16	1.80		2.96	4.14
0570	Pine		250	.032		2.89	1.80		4.69	6.05
0575	3/4" x 2-1/2", poplar		250	.032		1.01	1.80		2.81	3.97
0580	Red oak		250	.032		1.37	1.80		3.17	4.37
0585	Maple		250	.032		2.07	1.80		3.87	5.15
0590	Cherry		250	.032		2.60	1.80		4.40	5.70
0595	Mullion, 5/16" x 2", pine		270	.030		1.10	1.67		2.77	3.86
0600	9/16" x 2-1/2", finger jointed and primed		250	.032		1.16	1.80		2.96	4.14
0605	Pine		250	.032		1.41	1.80		3.21	4.41
0610	Red oak		250	.032		3.61	1.80		5.41	6.85
0615	1-1/16" x 3-3/4", red oak		220	.036		7.45	2.05		9.50	11.45
0620	Ogee, 7/16" x 2-1/2", poplar		250	.032		1.42	1.80		3.22	4.43
0625	Red oak		250	.032		1	1.80		2.80	3.96
0630	9/16" x 2-1/4", finger jointed and primed		250	.032		.58	1.80		2.38	3.50
0635	Poplar		250	.032		.66	1.80		2.46	3.59
0640	Red oak		250	.032		.90	1.80		2.70	3.85
0645	11/16" x 2-1/2", finger jointed and primed		250	.032		1.16	1.80		2.96	4.14

06 22 13 – Standard Pattern Wood Trim

06 22 13.30 Moldings, Casings		Crew	Daily Output	Labor-Hours	Unit	Material	2022 Bare Costs Labor	Equipment	Total	Total Incl O&P
0700	Pine	1 Carp	250	.032	L.F.	1.44	1.80		3.24	4.45
0701	Red oak		250	.032		3.44	1.80		5.24	6.65
0730	11/16" x 3-1/2", finger jointed and primed		220	.036		1.06	2.05		3.11	4.42
0750	Pine		220	.036		1.78	2.05		3.83	5.20
0755	3/4" x 2-1/2", finger jointed and primed		250	.032		1.77	1.80		3.57	4.81
0760	Poplar		250	.032		1.53	1.80		3.33	4.55
0765	Red oak		250	.032		1.37	1.80		3.17	4.37
0770	Maple		250	.032		2.07	1.80		3.87	5.15
0775	Cherry		250	.032		2.60	1.80		4.40	5.70
0780	3/4" x 3-1/2", finger jointed and primed		220	.036		1.06	2.05		3.11	4.42
0785	Poplar		220	.036		1.27	2.05		3.32	4.65
0790	Red oak		220	.036		1.89	2.05		3.94	5.35
0795	Maple		220	.036		2.93	2.05		4.98	6.50
0800	Cherry		220	.036		3.53	2.05		5.58	7.15
4700	Square profile, 1" x 1", teak		215	.037		2.08	2.09		4.17	5.60
4800	Rectangular profile, 1" x 3", teak	1 Carp	200	.040	L.F.	6.20	2.25		8.45	10.45
9000	Minimum labor/equipment charge	"	4	2	Job		113		113	179

06 22 13.35 Moldings, Ceilings										
0010	**MOLDINGS, CEILINGS**									
0600	Bed, 9/16" x 1-3/4", pine	1 Carp	270	.030	L.F.	.83	1.67		2.50	3.57
0650	9/16" x 2", pine		270	.030		1.32	1.67		2.99	4.11
0710	9/16" x 1-3/4", oak		270	.030		2.73	1.67		4.40	5.65
1200	Cornice, 9/16" x 1-3/4", pine		270	.030		1.11	1.67		2.78	3.87
1300	9/16" x 2-1/4", pine		265	.030		1.55	1.70		3.25	4.40
1350	Cove, 1/2" x 2-1/4", poplar		265	.030		1.38	1.70		3.08	4.21
1360	Red oak		265	.030		1.85	1.70		3.55	4.73
1370	Hard maple		265	.030		2.79	1.70		4.49	5.75
1380	Cherry		265	.030		3.51	1.70		5.21	6.55
2400	9/16" x 1-3/4", pine		270	.030		.79	1.67		2.46	3.52
2500	11/16" x 2-3/4", pine		265	.030		2.62	1.70		4.32	5.60
2510	Crown, 5/8" x 5/8", poplar		300	.027		.52	1.50		2.02	2.97
2520	Red oak		300	.027		.58	1.50		2.08	3.03
2530	Hard maple		300	.027		.78	1.50		2.28	3.25
2540	Cherry		300	.027		.82	1.50		2.32	3.30
2600	9/16" x 3-5/8", pine		250	.032		1.54	1.80		3.34	4.55
2700	11/16" x 4-1/4", pine		250	.032		2.56	1.80		4.36	5.65
2705	Oak		250	.032		3.27	1.80		5.07	6.45
2710	3/4" x 1-3/4", poplar		270	.030		.83	1.67		2.50	3.57
2720	Red oak		270	.030		1.16	1.67		2.83	3.93
2730	Hard maple		270	.030		1.60	1.67		3.27	4.41
2740	Cherry		270	.030		1.88	1.67		3.55	4.72
2750	3/4" x 2", poplar		270	.030		1.08	1.67		2.75	3.84
2760	Red oak		270	.030		.69	1.67		2.36	3.41
2770	Hard maple		270	.030		1.93	1.67		3.60	4.78
2780	Cherry		270	.030		2.15	1.67		3.82	5
2790	3/4" x 2-3/4", poplar		265	.030		1.19	1.70		2.89	4.01
2800	Red oak		265	.030		1.73	1.70		3.43	4.60
2810	Hard maple		265	.030		2.30	1.70		4	5.25
2820	Cherry		265	.030		2.64	1.70		4.34	5.60
2830	3/4" x 3-1/2", poplar		250	.032		1.52	1.80		3.32	4.53
2840	Red oak		250	.032		2.16	1.80		3.96	5.25
2850	Hard maple		250	.032		2.82	1.80		4.62	5.95

06 22 13 – Standard Pattern Wood Trim

06 22 13.35 Moldings, Ceilings		Crew	Daily Output	Labor-Hours	Unit	Material	2022 Bare Costs Labor	Equipment	Total	Total Incl O&P
2860	Cherry	1 Carp	250	.032	L.F.	3.29	1.80		5.09	6.50
2870	FJP poplar		250	.032		1.12	1.80		2.92	4.09
2880	3/4" x 5", poplar	1 Carp	245	.033	L.F.	2.24	1.84		4.08	5.40
2890	Red oak		245	.033		3.14	1.84		4.98	6.35
2900	Hard maple		245	.033		4.11	1.84		5.95	7.45
2910	Cherry		245	.033		4.73	1.84		6.57	8.10
2920	FJP poplar		245	.033		1.55	1.84		3.39	4.62
2930	3/4" x 6-1/4", poplar		240	.033		2.66	1.88		4.54	5.90
2940	Red oak		240	.033		3.81	1.88		5.69	7.15
2950	Hard maple		240	.033		4.97	1.88		6.85	8.45
2960	Cherry		240	.033		5.90	1.88		7.78	9.50
2970	7/8" x 8-3/4", poplar		220	.036		5.30	2.05		7.35	9.10
2980	Red oak		220	.036		8.75	2.05		10.80	12.90
2990	Hard maple		220	.036		11.50	2.05		13.55	15.90
3000	Cherry		220	.036		8.45	2.05		10.50	12.55
3010	1" x 7-1/4", poplar		220	.036		4.34	2.05		6.39	8
3020	Red oak		220	.036		6.75	2.05		8.80	10.65
3030	Hard maple		220	.036		9.30	2.05		11.35	13.50
3040	Cherry		220	.036		10.80	2.05		12.85	15.15
3050	1-1/16" x 4-1/4", poplar		250	.032		2.63	1.80		4.43	5.75
3060	Red oak		250	.032		3.09	1.80		4.89	6.25
3070	Hard maple		250	.032		4.09	1.80		5.89	7.35
3080	Cherry		250	.032		5.85	1.80		7.65	9.25
3090	Dentil crown, 3/4" x 5", poplar		250	.032		2.14	1.80		3.94	5.20
3100	Red oak		250	.032		3.28	1.80		5.08	6.45
3110	Hard maple		250	.032		4.21	1.80		6.01	7.50
3120	Cherry		250	.032		4.94	1.80		6.74	8.30
3130	Dentil piece for above, 1/2" x 1/2", poplar		300	.027		3.13	1.50		4.63	5.85
3140	Red oak		300	.027		3.60	1.50		5.10	6.35
3150	Hard maple		300	.027		4.09	1.50		5.59	6.90
3160	Cherry		300	.027		4.45	1.50		5.95	7.30
9000	Minimum labor/equipment charge		4	2	Job		113		113	179

06 22 13.40 Moldings, Exterior		Crew	Daily Output	Labor-Hours	Unit	Material	2022 Bare Costs Labor	Equipment	Total	Total Incl O&P
0010	**MOLDINGS, EXTERIOR**									
0100	Band board, cedar, rough sawn, 1" x 2"	1 Carp	300	.027	L.F.	.72	1.50		2.22	3.19
0110	1" x 3"		300	.027		1.07	1.50		2.57	3.57
0120	1" x 4"		250	.032		1.43	1.80		3.23	4.44
0130	1" x 6"		250	.032		2.14	1.80		3.94	5.20
0140	1" x 8"		225	.036		2.84	2		4.84	6.30
0150	1" x 10"		225	.036		3.54	2		5.54	7.05
0160	1" x 12"		200	.040		4.24	2.25		6.49	8.25
0240	STK, 1" x 2"		300	.027		.42	1.50		1.92	2.85
0250	1" x 3"	1 Carp	300	.027	L.F.	.46	1.50		1.96	2.90
0260	1" x 4"		250	.032		.92	1.80		2.72	3.88
0270	1" x 6"		250	.032		1.48	1.80		3.28	4.49
0280	1" x 8"		225	.036		2.60	2		4.60	6.05
0290	1" x 10"		225	.036		3.58	2		5.58	7.10
0300	1" x 12"		200	.040		4.93	2.25		7.18	9.05
0310	Pine, #2, 1" x 2"		300	.027		.33	1.50		1.83	2.76
0320	1" x 3"		300	.027		.49	1.50		1.99	2.93
0330	1" x 4"		250	.032		.69	1.80		2.49	3.62
0340	1" x 6"		250	.032		1.03	1.80		2.83	4

For customer support on your Commercial Renovation Costs with RSMeans data, call 800.448.8182.

165

06 22 13 – Standard Pattern Wood Trim

06 22 13.40 Moldings, Exterior		Crew	Daily Output	Labor-Hours	Unit	Material	2022 Bare Costs Labor	Equipment	Total	Total Incl O&P
0350	1" x 8"	1 Carp	225	.036	L.F.	1.41	2		3.41	4.73
0360	1" x 10"		225	.036		1.94	2		3.94	5.30
0370	1" x 12"		200	.040		2.68	2.25		4.93	6.55
0380	D & better, 1" x 2"		300	.027		.74	1.50		2.24	3.21
0390	1" x 3"		300	.027		1.10	1.50		2.60	3.60
0400	1" x 4"		250	.032		1.54	1.80		3.34	4.56
0410	1" x 6"		250	.032		2.85	1.80		4.65	6
0420	1" x 8"		225	.036		3.50	2		5.50	7.05
0430	1" x 10"		225	.036		3.83	2		5.83	7.40
0440	1" x 12"		200	.040		4.18	2.25		6.43	8.20
0450	Redwood, clear all heart, 1" x 2"		300	.027		1.73	1.50		3.23	4.29
0460	1" x 3"		300	.027		2.58	1.50		4.08	5.20
0470	1" x 4"		250	.032		3.35	1.80		5.15	6.55
0480	1" x 6"		252	.032		4.86	1.79		6.65	8.20
0490	1" x 8"		225	.036		5.35	2		7.35	9.10
0500	1" x 10"		225	.036		16.45	2		18.45	21.50
0510	1" x 12"		200	.040		6.35	2.25		8.60	10.60
0530	Corner board, cedar, rough sawn, 1" x 2"		225	.036		.72	2		2.72	3.98
0540	1" x 3"		225	.036		1.07	2		3.07	4.36
0550	1" x 4"		200	.040		1.43	2.25		3.68	5.15
0560	1" x 6"		200	.040		2.14	2.25		4.39	5.95
0570	1" x 8"		200	.040		2.84	2.25		5.09	6.70
0580	1" x 10"		175	.046		3.54	2.57		6.11	8
0590	1" x 12"		175	.046		4.24	2.57		6.81	8.75
0670	STK, 1" x 2"		225	.036		.42	2		2.42	3.64
0680	1" x 3"		225	.036		.46	2		2.46	3.69
0690	1" x 4"		200	.040		.89	2.25		3.14	4.56
0700	1" x 6"		200	.040		1.48	2.25		3.73	5.20
0710	1" x 8"		200	.040		2.60	2.25		4.85	6.45
0720	1" x 10"		175	.046		3.58	2.57		6.15	8.05
0730	1" x 12"	1 Carp	175	.046	L.F.	4.93	2.57		7.50	9.55
0740	Pine, #2, 1" x 2"		225	.036		.33	2		2.33	3.55
0750	1" x 3"		225	.036		.49	2		2.49	3.72
0760	1" x 4"		200	.040		.69	2.25		2.94	4.34
0770	1" x 6"		200	.040		1.03	2.25		3.28	4.72
0780	1" x 8"		200	.040		1.41	2.25		3.66	5.15
0790	1" x 10"		175	.046		1.94	2.57		4.51	6.20
0800	1" x 12"		175	.046		2.68	2.57		5.25	7.05
0810	D & better, 1" x 2"		225	.036		.74	2		2.74	4
0820	1" x 3"		225	.036		1.10	2		3.10	4.39
0830	1" x 4"		200	.040		1.54	2.25		3.79	5.30
0840	1" x 6"		200	.040		2.85	2.25		5.10	6.70
0850	1" x 8"		200	.040		3.50	2.25		5.75	7.45
0860	1" x 10"		175	.046		3.83	2.57		6.40	8.30
0870	1" x 12"		175	.046		4.18	2.57		6.75	8.70
0880	Redwood, clear all heart, 1" x 2"		225	.036		1.73	2		3.73	5.10
0890	1" x 3"		225	.036		2.58	2		4.58	6
0900	1" x 4"		200	.040		3.35	2.25		5.60	7.25
0910	1" x 6"		200	.040		4.86	2.25		7.11	8.95
0920	1" x 8"		200	.040		5.35	2.25		7.60	9.50
0930	1" x 10"		175	.046		16.45	2.57		19.02	22
0940	1" x 12"		175	.046		6.35	2.57		8.92	11.10
0950	Cornice board, cedar, rough sawn, 1" x 2"		330	.024		.72	1.36		2.08	2.97

For customer support on your Commercial Renovation Costs with RSMeans data, call 800.448.8182.

06 22 13 – Standard Pattern Wood Trim

06 22 13.40 Moldings, Exterior		Crew	Daily Output	Labor-Hours	Unit	Material	2022 Bare Costs Labor	Equipment	Total	Total Incl O&P
0960	1" x 3"	1 Carp	290	.028	L.F.	1.07	1.55		2.62	3.65
0970	1" x 4"		250	.032		1.43	1.80		3.23	4.44
0980	1" x 6"		250	.032		2.14	1.80		3.94	5.20
0990	1" x 8"		200	.040		2.84	2.25		5.09	6.70
1000	1" x 10"		180	.044		3.54	2.50		6.04	7.85
1010	1" x 12"		180	.044		4.24	2.50		6.74	8.65
1020	STK, 1" x 2"		330	.024		.42	1.36		1.78	2.63
1030	1" x 3"		290	.028		.46	1.55		2.01	2.98
1040	1" x 4"		250	.032		.92	1.80		2.72	3.88
1050	1" x 6"		250	.032		1.48	1.80		3.28	4.49
1060	1" x 8"		200	.040		2.60	2.25		4.85	6.45
1070	1" x 10"		180	.044		3.58	2.50		6.08	7.90
1080	1" x 12"		180	.044		4.93	2.50		7.43	9.45
1500	Pine, #2, 1" x 2"		330	.024		.33	1.36		1.69	2.54
1510	1" x 3"		290	.028		.34	1.55		1.89	2.84
1600	1" x 4"		250	.032		.69	1.80		2.49	3.62
1700	1" x 6"		250	.032		1.03	1.80		2.83	4
1800	1" x 8"	1 Carp	200	.040	L.F.	1.41	2.25		3.66	5.15
1900	1" x 10"		180	.044		1.94	2.50		4.44	6.10
2000	1" x 12"		180	.044		2.68	2.50		5.18	6.95
2020	D & better, 1" x 2"		330	.024		.74	1.36		2.10	2.99
2030	1" x 3"		290	.028		1.10	1.55		2.65	3.68
2040	1" x 4"		250	.032		1.54	1.80		3.34	4.56
2050	1" x 6"		250	.032		2.85	1.80		4.65	6
2060	1" x 8"		200	.040		3.50	2.25		5.75	7.45
2070	1" x 10"		180	.044		3.83	2.50		6.33	8.20
2080	1" x 12"		180	.044		4.18	2.50		6.68	8.60
2090	Redwood, clear all heart, 1" x 2"		330	.024		1.73	1.36		3.09	4.07
2100	1" x 3"		290	.028		2.58	1.55		4.13	5.30
2110	1" x 4"		250	.032		3.35	1.80		5.15	6.55
2120	1" x 6"		250	.032		4.86	1.80		6.66	8.20
2130	1" x 8"		200	.040		5.35	2.25		7.60	9.50
2140	1" x 10"		180	.044		16.45	2.50		18.95	22
2150	1" x 12"		180	.044		6.35	2.50		8.85	11
2160	3 piece, 1" x 2", 1" x 4", 1" x 6", rough sawn cedar		80	.100		4.31	5.65		9.96	13.70
2180	STK cedar		80	.100		2.83	5.65		8.48	12.05
2200	#2 pine		80	.100		2.05	5.65		7.70	11.20
2210	D & better pine		80	.100		5.15	5.65		10.80	14.60
2220	Clear all heart redwood		80	.100		9.95	5.65		15.60	19.90
2230	1" x 8", 1" x 10", 1" x 12", rough sawn cedar		65	.123		10.55	6.95		17.50	22.50
2240	STK cedar		65	.123		11.05	6.95		18	23
2300	#2 pine		65	.123		5.95	6.95		12.90	17.55
2320	D & better pine		65	.123		11.45	6.95		18.40	23.50
2330	Clear all heart redwood		65	.123		28	6.95		34.95	42
2340	Door/window casing, cedar, rough sawn, 1" x 2"		275	.029		.72	1.64		2.36	3.40
2350	1" x 3"		275	.029		1.07	1.64		2.71	3.78
2360	1" x 4"		250	.032		1.43	1.80		3.23	4.44
2370	1" x 6"		250	.032		2.14	1.80		3.94	5.20
2380	1" x 8"		230	.035		2.84	1.96		4.80	6.25
2390	1" x 10"		230	.035		3.54	1.96		5.50	7
2395	1" x 12"		210	.038		4.24	2.14		6.38	8.10
2410	STK, 1" x 2"		275	.029		.42	1.64		2.06	3.06
2420	1" x 3"		275	.029		.46	1.64		2.10	3.11

06 22 13.40 Moldings, Exterior		Crew	Daily Output	Labor-Hours	Unit	Material	2022 Bare Costs Labor	Equipment	Total	Total Incl O&P
2430	1" x 4"	1 Carp	250	.032	L.F.	.92	1.80		2.72	3.88
2440	1" x 6"		250	.032		1.48	1.80		3.28	4.49
2450	1" x 8"		230	.035		2.60	1.96		4.56	5.95
2460	1" x 10"	1 Carp	230	.035	L.F.	3.58	1.96		5.54	7.05
2470	1" x 12"		210	.038		4.93	2.14		7.07	8.85
2550	Pine, #2, 1" x 2"		275	.029		.33	1.64		1.97	2.97
2560	1" x 3"		275	.029		.49	1.64		2.13	3.14
2570	1" x 4"		250	.032		.69	1.80		2.49	3.62
2580	1" x 6"		250	.032		1.03	1.80		2.83	4
2590	1" x 8"		230	.035		1.41	1.96		3.37	4.66
2600	1" x 10"		230	.035		1.94	1.96		3.90	5.25
2610	1" x 12"		210	.038		2.68	2.14		4.82	6.35
2620	Pine, D & better, 1" x 2"		275	.029		.74	1.64		2.38	3.42
2630	1" x 3"		275	.029		1.10	1.64		2.74	3.81
2640	1" x 4"		250	.032		1.54	1.80		3.34	4.56
2650	1" x 6"		250	.032		2.85	1.80		4.65	6
2660	1" x 8"		230	.035		3.50	1.96		5.46	6.95
2670	1" x 10"		230	.035		3.83	1.96		5.79	7.35
2680	1" x 12"		210	.038		4.18	2.14		6.32	8
2690	Redwood, clear all heart, 1" x 2"		275	.029		1.73	1.64		3.37	4.50
2695	1" x 3"		275	.029		2.58	1.64		4.22	5.45
2710	1" x 4"		250	.032		3.35	1.80		5.15	6.55
2715	1" x 6"		250	.032		4.86	1.80		6.66	8.20
2730	1" x 8"		230	.035		5.35	1.96		7.31	9
2740	1" x 10"		230	.035		16.45	1.96		18.41	21
2750	1" x 12"		210	.038		6.35	2.14		8.49	10.40
3500	Bellyband, pine, 11/16" x 4-1/4"		250	.032		2.98	1.80		4.78	6.15
3610	Brickmold, pine, 1-1/4" x 2"		200	.040		2.19	2.25		4.44	6
3620	FJP, 1-1/4" x 2"		200	.040		1.17	2.25		3.42	4.87
5100	Fascia, cedar, rough sawn, 1" x 2"		275	.029		.72	1.64		2.36	3.40
5110	1" x 3"		275	.029		1.07	1.64		2.71	3.78
5120	1" x 4"		250	.032		1.43	1.80		3.23	4.44
5200	1" x 6"		250	.032		2.14	1.80		3.94	5.20
5300	1" x 8"		230	.035		2.84	1.96		4.80	6.25
5310	1" x 10"		230	.035		3.54	1.96		5.50	7
5320	1" x 12"		210	.038		4.24	2.14		6.38	8.10
5400	2" x 4"		220	.036		1.46	2.05		3.51	4.86
5500	2" x 6"		220	.036		2.19	2.05		4.24	5.65
5600	2" x 8"		200	.040		2.91	2.25		5.16	6.80
5700	2" x 10"		180	.044		3.62	2.50		6.12	7.95
5800	2" x 12"		170	.047		8.60	2.65		11.25	13.65
6120	STK, 1" x 2"		275	.029		.42	1.64		2.06	3.06
6130	1" x 3"	1 Carp	275	.029	L.F.	.46	1.64		2.10	3.11
6140	1" x 4"		250	.032		.92	1.80		2.72	3.88
6150	1" x 6"		250	.032		1.48	1.80		3.28	4.49
6160	1" x 8"		230	.035		2.60	1.96		4.56	5.95
6170	1" x 10"		230	.035		3.58	1.96		5.54	7.05
6180	1" x 12"		210	.038		4.93	2.14		7.07	8.85
6185	2" x 2"		260	.031		.91	1.73		2.64	3.75
6190	Pine, #2, 1" x 2"		275	.029		.33	1.64		1.97	2.97
6200	1" x 3"		275	.029		.49	1.64		2.13	3.14
6210	1" x 4"		250	.032		.69	1.80		2.49	3.62
6220	1" x 6"		250	.032		1.03	1.80		2.83	4

06 22 13 – Standard Pattern Wood Trim

06 22 13.40 Moldings, Exterior		Crew	Daily Output	Labor-Hours	Unit	Material	2022 Bare Costs Labor	Equipment	Total	Total Incl O&P
6230	1" x 8"	1 Carp	230	.035	L.F.	1.41	1.96		3.37	4.66
6240	1" x 10"		230	.035		1.94	1.96		3.90	5.25
6250	1" x 12"		210	.038		2.68	2.14		4.82	6.35
6260	D & better, 1" x 2"		275	.029		.74	1.64		2.38	3.42
6270	1" x 3"		275	.029		1.10	1.64		2.74	3.81
6280	1" x 4"		250	.032		1.54	1.80		3.34	4.56
6290	1" x 6"		250	.032		2.85	1.80		4.65	6
6300	1" x 8"		230	.035		3.50	1.96		5.46	6.95
6310	1" x 10"		230	.035		3.83	1.96		5.79	7.35
6312	1" x 12"		210	.038		4.18	2.14		6.32	8
6330	Southern yellow, 1-1/4" x 5"		240	.033		3.25	1.88		5.13	6.55
6340	1-1/4" x 6"		240	.033		2.64	1.88		4.52	5.90
6350	1-1/4" x 8"		215	.037		3.52	2.09		5.61	7.20
6360	1-1/4" x 12"		190	.042		5.65	2.37		8.02	10
6370	Redwood, clear all heart, 1" x 2"		275	.029		1.73	1.64		3.37	4.50
6380	1" x 3"		275	.029		3.35	1.64		4.99	6.30
6390	1" x 4"		250	.032		3.35	1.80		5.15	6.55
6400	1" x 6"		250	.032		4.86	1.80		6.66	8.20
6410	1" x 8"		230	.035		5.35	1.96		7.31	9
6420	1" x 10"		230	.035		16.45	1.96		18.41	21
6430	1" x 12"		210	.038		6.35	2.14		8.49	10.40
6440	1-1/4" x 5"		240	.033		5.10	1.88		6.98	8.60
6450	1-1/4" x 6"		240	.033		6.10	1.88		7.98	9.70
6460	1-1/4" x 8"		215	.037		8.05	2.09		10.14	12.20
6470	1-1/4" x 12"		190	.042		9.50	2.37		11.87	14.20
6580	Frieze, cedar, rough sawn, 1" x 2"		275	.029		.72	1.64		2.36	3.40
6590	1" x 3"		275	.029		1.07	1.64		2.71	3.78
6600	1" x 4"		250	.032		1.43	1.80		3.23	4.44
6610	1" x 6"		250	.032		2.14	1.80		3.94	5.20
6620	1" x 8"	1 Carp	250	.032	L.F.	2.84	1.80		4.64	6
6630	1" x 10"		225	.036		3.54	2		5.54	7.05
6640	1" x 12"		200	.040		4.18	2.25		6.43	8.20
6650	STK, 1" x 2"		275	.029		.42	1.64		2.06	3.06
6660	1" x 3"		275	.029		.46	1.64		2.10	3.11
6670	1" x 4"		250	.032		.92	1.80		2.72	3.88
6680	1" x 6"		250	.032		1.48	1.80		3.28	4.49
6690	1" x 8"		250	.032		2.60	1.80		4.40	5.70
6700	1" x 10"		225	.036		3.58	2		5.58	7.10
6710	1" x 12"		200	.040		4.93	2.25		7.18	9.05
6790	Pine, #2, 1" x 2"		275	.029		.33	1.64		1.97	2.97
6800	1" x 3"		275	.029		.49	1.64		2.13	3.14
6810	1" x 4"		250	.032		.69	1.80		2.49	3.62
6820	1" x 6"		250	.032		1.03	1.80		2.83	4
6830	1" x 8"		250	.032		1.41	1.80		3.21	4.41
6840	1" x 10"		225	.036		1.94	2		3.94	5.30
6850	1" x 12"		200	.040		2.68	2.25		4.93	6.55
6860	D & better, 1" x 2"		275	.029		.74	1.64		2.38	3.42
6870	1" x 3"		275	.029		1.10	1.64		2.74	3.81
6880	1" x 4"		250	.032		1.54	1.80		3.34	4.56
6890	1" x 6"		250	.032		2.85	1.80		4.65	6
6900	1" x 8"		250	.032		3.50	1.80		5.30	6.70
6910	1" x 10"		225	.036		3.83	2		5.83	7.40
6920	1" x 12"		200	.040		4.18	2.25		6.43	8.20

06 22 13.40 Moldings, Exterior		Crew	Daily Output	Labor-Hours	Unit	Material	2022 Bare Costs Labor	Equipment	Total	Total Incl O&P
6930	Redwood, clear all heart, 1" x 2"	1 Carp	275	.029	L.F.	1.73	1.64		3.37	4.50
6940	1" x 3"		275	.029		2.58	1.64		4.22	5.45
6950	1" x 4"		250	.032		3.35	1.80		5.15	6.55
6960	1" x 6"		250	.032		4.86	1.80		6.66	8.20
6970	1" x 8"		250	.032		5.35	1.80		7.15	8.75
6980	1" x 10"		225	.036		16.45	2		18.45	21.50
6990	1" x 12"		200	.040		6.35	2.25		8.60	10.60
7000	Grounds, 1" x 1", cedar, rough sawn		300	.027		.38	1.50		1.88	2.80
7010	STK		300	.027		.27	1.50		1.77	2.68
7020	Pine, #2		300	.027		.21	1.50		1.71	2.62
7030	D & better		300	.027		.46	1.50		1.96	2.89
7050	Redwood		300	.027		1.05	1.50		2.55	3.54
7060	Rake/verge board, cedar, rough sawn, 1" x 2"		225	.036		.72	2		2.72	3.98
7070	1" x 3"		225	.036		1.07	2		3.07	4.36
7080	1" x 4"		200	.040		1.43	2.25		3.68	5.15
7090	1" x 6"		200	.040		2.14	2.25		4.39	5.95
7100	1" x 8"	1 Carp	190	.042	L.F.	2.84	2.37		5.21	6.90
7110	1" x 10"		190	.042		3.54	2.37		5.91	7.65
7120	1" x 12"		180	.044		4.24	2.50		6.74	8.65
7130	STK, 1" x 2"		225	.036		.42	2		2.42	3.64
7140	1" x 3"		225	.036		.46	2		2.46	3.69
7150	1" x 4"		200	.040		.92	2.25		3.17	4.60
7160	1" x 6"		200	.040		1.48	2.25		3.73	5.20
7170	1" x 8"		190	.042		2.60	2.37		4.97	6.65
7180	1" x 10"		190	.042		3.58	2.37		5.95	7.70
7190	1" x 12"		180	.044		4.93	2.50		7.43	9.45
7200	Pine, #2, 1" x 2"		225	.036		.33	2		2.33	3.55
7210	1" x 3"		225	.036		.49	2		2.49	3.72
7220	1" x 4"		200	.040		.69	2.25		2.94	4.34
7230	1" x 6"		200	.040		1.03	2.25		3.28	4.72
7240	1" x 8"		190	.042		1.41	2.37		3.78	5.30
7250	1" x 10"		190	.042		1.94	2.37		4.31	5.90
7260	1" x 12"		180	.044		2.68	2.50		5.18	6.95
7340	D & better, 1" x 2"		225	.036		.74	2		2.74	4
7350	1" x 3"		225	.036		1.10	2		3.10	4.39
7360	1" x 4"		200	.040		1.54	2.25		3.79	5.30
7370	1" x 6"		200	.040		2.85	2.25		5.10	6.70
7380	1" x 8"		190	.042		3.50	2.37		5.87	7.60
7390	1" x 10"		190	.042		3.83	2.37		6.20	8
7400	1" x 12"		180	.044		4.18	2.50		6.68	8.60
7410	Redwood, clear all heart, 1" x 2"		225	.036		1.73	2		3.73	5.10
7420	1" x 3"		225	.036		2.58	2		4.58	6
7430	1" x 4"		200	.040		3.35	2.25		5.60	7.25
7440	1" x 6"		200	.040		4.86	2.25		7.11	8.95
7450	1" x 8"		190	.042		5.35	2.37		7.72	9.65
7460	1" x 10"		190	.042		16.45	2.37		18.82	22
7470	1" x 12"		180	.044		6.35	2.50		8.85	11
7480	2" x 4"		200	.040		6.30	2.25		8.55	10.55
7490	2" x 6"		182	.044		9.45	2.47		11.92	14.35
7500	2" x 8"		165	.048		12	2.73		14.73	17.55
7630	Soffit, cedar, rough sawn, 1" x 2"	2 Carp	440	.036		.72	2.05		2.77	4.05
7640	1" x 3"		440	.036		1.07	2.05		3.12	4.43
7650	1" x 4"		420	.038		1.43	2.14		3.57	4.99

For customer support on your Commercial Renovation Costs with RSMeans data, call 800.448.8182.

06 22 13 – Standard Pattern Wood Trim

06 22 13.40 Moldings, Exterior		Crew	Daily Output	Labor-Hours	Unit	Material	2022 Bare Costs Labor	Equipment	Total	Total Incl O&P
7660	1" x 6"	2 Carp	420	.038	L.F.	2.14	2.14		4.28	5.75
7670	1" x 8"		420	.038		2.84	2.14		4.98	6.55
7680	1" x 10"		400	.040		3.54	2.25		5.79	7.45
7690	1" x 12"	2 Carp	400	.040	L.F.	4.24	2.25		6.49	8.25
7700	STK, 1" x 2"		440	.036		.42	2.05		2.47	3.71
7710	1" x 3"		440	.036		.46	2.05		2.51	3.76
7720	1" x 4"		420	.038		.92	2.14		3.06	4.43
7730	1" x 6"		420	.038		1.48	2.14		3.62	5.05
7740	1" x 8"		420	.038		2.60	2.14		4.74	6.25
7750	1" x 10"		400	.040		3.58	2.25		5.83	7.50
7760	1" x 12"		400	.040		4.93	2.25		7.18	9.05
7770	Pine, #2, 1" x 2"		440	.036		.33	2.05		2.38	3.62
7780	1" x 3"		440	.036		.49	2.05		2.54	3.79
7790	1" x 4"		420	.038		.69	2.14		2.83	4.17
7800	1" x 6"		420	.038		1.03	2.14		3.17	4.55
7810	1" x 8"		420	.038		1.41	2.14		3.55	4.96
7820	1" x 10"		400	.040		1.94	2.25		4.19	5.70
7830	1" x 12"		400	.040		2.68	2.25		4.93	6.55
7840	D & better, 1" x 2"		440	.036		.74	2.05		2.79	4.07
7850	1" x 3"		440	.036		1.10	2.05		3.15	4.46
7860	1" x 4"		420	.038		1.54	2.14		3.68	5.10
7870	1" x 6"		420	.038		2.85	2.14		4.99	6.55
7880	1" x 8"		420	.038		3.50	2.14		5.64	7.25
7890	1" x 10"		400	.040		3.83	2.25		6.08	7.80
7900	1" x 12"		400	.040		4.18	2.25		6.43	8.20
7910	Redwood, clear all heart, 1" x 2"		440	.036		1.73	2.05		3.78	5.15
7920	1" x 3"		440	.036		2.58	2.05		4.63	6.10
7930	1" x 4"		420	.038		3.35	2.14		5.49	7.10
7940	1" x 6"		420	.038		4.86	2.14		7	8.75
7950	1" x 8"		420	.038		5.35	2.14		7.49	9.30
7960	1" x 10"		400	.040		16.45	2.25		18.70	21.50
7970	1" x 12"		400	.040		6.35	2.25		8.60	10.60
8050	Trim, crown molding, pine, 11/16" x 4-1/4"	1 Carp	250	.032		4.68	1.80		6.48	8
8060	Back band, 11/16" x 1-1/16"		250	.032		.93	1.80		2.73	3.88
8070	Insect screen frame stock, 1-1/16" x 1-3/4"		395	.020		.45	1.14		1.59	2.31
8080	Dentils, 2-1/2" x 2-1/2" x 4", 6" OC		30	.267		1.24	15		16.24	25.50
8100	Fluted, 5-1/2"		165	.048		.40	2.73		3.13	4.78
8110	Stucco bead, 1-3/8" x 1-5/8"		250	.032		2.41	1.80		4.21	5.50
9000	Minimum labor/equipment charge		4	2	Job		113		113	179

06 22 13.45 Moldings, Trim

0010	**MOLDINGS, TRIM**									
0200	Astragal, stock pine, 11/16" x 1-3/4"	1 Carp	255	.031	L.F.	1.33	1.77		3.10	4.27
0250	1-5/16" x 2-3/16"	"	240	.033	"	4.89	1.88		6.77	8.40
0800	Chair rail, stock pine, 5/8" x 2-1/2"	1 Carp	270	.030	L.F.	1.53	1.67		3.20	4.33
0900	5/8" x 3-1/2"		240	.033		4.16	1.88		6.04	7.55
1000	Closet pole, stock pine, 1-1/8" diameter		200	.040		1.28	2.25		3.53	4.99
1100	Fir, 1-5/8" diameter		200	.040		1.93	2.25		4.18	5.70
3300	Half round, stock pine, 1/4" x 1/2"		270	.030		.46	1.67		2.13	3.16
3350	1/2" x 1"		255	.031		.62	1.77		2.39	3.49
3400	Handrail, fir, single piece, stock, hardware not included									
3450	1-1/2" x 1-3/4"	1 Carp	80	.100	L.F.	2.74	5.65		8.39	11.95
3470	Pine, 1-1/2" x 1-3/4"		80	.100		2.16	5.65		7.81	11.35

06 22 13.45 Moldings, Trim

		Crew	Daily Output	Labor-Hours	Unit	Material	2022 Bare Costs Labor	Equipment	Total	Total Incl O&P
3500	1-1/2" x 2-1/2"	1 Carp	76	.105	L.F.	3.36	5.95		9.31	13.10
3600	Lattice, stock pine, 1/4" x 1-1/8"		270	.030		.47	1.67		2.14	3.17
3700	1/4" x 1-3/4"		250	.032		.62	1.80		2.42	3.54
3800	Miscellaneous, custom, pine, 1" x 1"		270	.030		.39	1.67		2.06	3.08
3850	1" x 2"		265	.030		.79	1.70		2.49	3.57
3900	1" x 3"		240	.033		1.18	1.88		3.06	4.28
4100	Birch or oak, nominal 1" x 1"		240	.033		1.12	1.88		3	4.21
4200	Nominal 1" x 3"		215	.037		3.35	2.09		5.44	7
4400	Walnut, nominal 1" x 1"		215	.037		1.53	2.09		3.62	5
4500	Nominal 1" x 3"		200	.040		4.58	2.25		6.83	8.65
4700	Teak, nominal 1" x 1"		215	.037		6.15	2.09		8.24	10.15
4800	Nominal 1" x 3"		200	.040		18.50	2.25		20.75	24
4900	Quarter round, stock pine, 1/4" x 1/4"		275	.029		.49	1.64		2.13	3.14
4950	3/4" x 3/4"		255	.031		.68	1.77		2.45	3.56
5600	Wainscot moldings, 1-1/8" x 9/16", 2' high, minimum		76	.105	S.F.	14.50	5.95		20.45	25.50
5700	Maximum		65	.123	"	21.50	6.95		28.45	34.50
9000	Minimum labor/equipment charge	▼	4	2	Job		113		113	179

06 22 13.50 Moldings, Window and Door

		Crew	Daily Output	Labor-Hours	Unit	Material	2022 Bare Costs Labor	Equipment	Total	Total Incl O&P
0010	**MOLDINGS, WINDOW AND DOOR**									
2800	Door moldings, stock, decorative, 1-1/8" wide, plain	1 Carp	17	.471	Set	48.50	26.50		75	95.50
2900	Detailed		17	.471	"	117	26.50		143.50	170
2960	Clear pine door jamb, no stops, 11/16" x 4-9/16"		240	.033	L.F.	6.40	1.88		8.28	10.05
3150	Door trim set, 1 head and 2 sides, pine, 2-1/2" wide		12	.667	Opng.	24.50	37.50		62	86
3170	3-1/2" wide		11	.727	"	30	41		71	98
3250	Glass beads, stock pine, 3/8" x 1/2"		275	.029	L.F.	.41	1.64		2.05	3.05
3270	3/8" x 7/8"		270	.030		.53	1.67		2.20	3.23
4850	Parting bead, stock pine, 3/8" x 3/4"		275	.029		.62	1.64		2.26	3.28
4870	1/2" x 3/4"		255	.031		.53	1.77		2.30	3.39
5000	Stool caps, stock pine, 11/16" x 3-1/2"	1 Carp	200	.040	L.F.	2.82	2.25		5.07	6.70
5100	1-1/16" x 3-1/4"		150	.053	"	4.22	3		7.22	9.40
5300	Threshold, oak, 3' long, inside, 5/8" x 3-5/8"		32	.250	Ea.	16	14.10		30.10	40
5400	Outside, 1-1/2" x 7-5/8"	▼	16	.500	"	65	28		93	116
5900	Window trim sets, including casings, header, stops,									
5910	stool and apron, 2-1/2" wide, FJP	1 Carp	13	.615	Opng.	36	34.50		70.50	95
5950	Pine		10	.800		42.50	45		87.50	119
6000	Oak		6	1.333	▼	86	75		161	214
9000	Minimum labor/equipment charge		4	2	Job		113		113	179

06 22 13.60 Moldings, Soffits

		Crew	Daily Output	Labor-Hours	Unit	Material	2022 Bare Costs Labor	Equipment	Total	Total Incl O&P
0010	**MOLDINGS, SOFFITS** R061110-30									
0200	Soffits, pine, 1" x 4"	2 Carp	420	.038	L.F.	.62	2.14		2.76	4.10
0210	1" x 6"		420	.038		.97	2.14		3.11	4.48
0220	1" x 8"		420	.038		1.35	2.14		3.49	4.89
0230	1" x 10"		400	.040		1.84	2.25		4.09	5.60
0240	1" x 12"		400	.040		2.58	2.25		4.83	6.40
0250	STK cedar, 1" x 4"		420	.038		.86	2.14		3	4.36
0260	1" x 6"		420	.038		1.42	2.14		3.56	4.97
0270	1" x 8"		420	.038		2.54	2.14		4.68	6.20
0280	1" x 10"		400	.040		3.49	2.25		5.74	7.40
0290	1" x 12"		400	.040		4.84	2.25		7.09	8.90
1000	Exterior AC plywood, 1/4" thick		400	.040	S.F.	1.70	2.25		3.95	5.45
1050	3/8" thick		400	.040		2.06	2.25		4.31	5.85
1100	1/2" thick	▼	400	.040	▼	2.75	2.25		5	6.60

06 22 Millwork

06 22 13 - Standard Pattern Wood Trim

06 22 13.60 Moldings, Soffits		Crew	Daily Output	Labor-Hours	Unit	Material	2022 Bare Costs Labor	Equipment	Total	Total Incl O&P
1150	Polyvinyl chloride, white, solid	1 Carp	230	.035	S.F.	1.56	1.96		3.52	4.83
1160	Perforated	"	230	.035	↓	1.71	1.96		3.67	4.99
1170	Accessories, "J" channel 5/8"	2 Carp	700	.023	L.F.	.66	1.29		1.95	2.77
9000	Minimum labor/equipment charge	"	5	3.200	Job		180		180	286

06 25 Prefinished Paneling

06 25 13 - Prefinished Hardboard Paneling

06 25 13.10 Paneling, Hardboard

			Crew	Daily Output	Labor-Hours	Unit	Material	2022 Bare Costs Labor	Equipment	Total	Total Incl O&P
0010	**PANELING, HARDBOARD**										
0050	Not incl. furring or trim, hardboard, tempered, 1/8" thick	G	2 Carp	500	.032	S.F.	.32	1.80		2.12	3.21
0100	1/4" thick	G		500	.032		.70	1.80		2.50	3.63
0300	Tempered pegboard, 1/8" thick	G		500	.032		.53	1.80		2.33	3.44
0400	1/4" thick	G		500	.032		.96	1.80		2.76	3.92
0600	Untempered hardboard, natural finish, 1/8" thick	G		500	.032		.35	1.80		2.15	3.25
0700	1/4" thick	G		500	.032		.59	1.80		2.39	3.51
0900	Untempered pegboard, 1/8" thick	G	↓	500	.032	↓	.63	1.80		2.43	3.55
1000	1/4" thick	G	2 Carp	500	.032	S.F.	.71	1.80		2.51	3.64
1200	Plastic faced hardboard, 1/8" thick	G		500	.032		.54	1.80		2.34	3.45
1300	1/4" thick	G		500	.032		1.05	1.80		2.85	4.02
1500	Plastic faced pegboard, 1/8" thick	G		500	.032		.81	1.80		2.61	3.75
1600	1/4" thick	G		500	.032		1.09	1.80		2.89	4.06
1800	Wood grained, plain or grooved, 1/8" thick	G		500	.032	↓	.90	1.80		2.70	3.85
1900	1/4" thick	G		425	.038		1.70	2.12		3.82	5.25
2100	Moldings, wood grained MDF			500	.032	L.F.	.52	1.80		2.32	3.43
2200	Pine		↓	425	.038	"	1.43	2.12		3.55	4.94
9000	Minimum labor/equipment charge		1 Carp	2	4	Job		225		225	360

06 25 16 - Prefinished Plywood Paneling

06 25 16.10 Paneling, Plywood

			Crew	Daily Output	Labor-Hours	Unit	Material	2022 Bare Costs Labor	Equipment	Total	Total Incl O&P
0010	**PANELING, PLYWOOD**	R061636-20									
2400	Plywood, prefinished, 1/4" thick, 4' x 8' sheets										
2410	with vertical grooves. Birch faced, economy		2 Carp	500	.032	S.F.	.88	1.80		2.68	3.83
2420	Average			420	.038		1.53	2.14		3.67	5.10
2430	Custom			350	.046		2	2.57		4.57	6.30
2600	Mahogany, African			400	.040		3.30	2.25		5.55	7.20
2700	Philippine (Lauan)			500	.032		.53	1.80		2.33	3.44
2900	Oak			500	.032		2.10	1.80		3.90	5.15
3000	Cherry			400	.040		3.22	2.25		5.47	7.10
3200	Rosewood			320	.050		3.92	2.82		6.74	8.80
3400	Teak			400	.040		4.16	2.25		6.41	8.15
3600	Chestnut			375	.043		6.85	2.40		9.25	11.35
3800	Pecan			400	.040		3.20	2.25		5.45	7.10
3900	Walnut, average			500	.032		3.38	1.80		5.18	6.60
3950	Custom			400	.040		3.13	2.25		5.38	7
4000	Plywood, prefinished, 3/4" thick, stock grades, economy			320	.050		1.90	2.82		4.72	6.55
4100	Average			224	.071		6.40	4.02		10.42	13.45
4300	Architectural grade, custom			224	.071		6.90	4.02		10.92	14
4400	Luxury			160	.100		7.10	5.65		12.75	16.75
4600	Plywood, "A" face, birch, VC, 1/2" thick, natural			450	.036		2.81	2		4.81	6.25
4700	Select			450	.036		2.42	2		4.42	5.85
4900	Veneer core, 3/4" thick, natural		↓	320	.050	↓	2.89	2.82		5.71	7.65

06 25 Prefinished Paneling

06 25 16 – Prefinished Plywood Paneling

06 25 16.10 Paneling, Plywood	Crew	Daily Output	Labor-Hours	Unit	Material	2022 Bare Costs Labor	Equipment	Total	Total Incl O&P
5000	Select	2 Carp	320	.050	S.F.	3.40	2.82		6.22	8.20
5200	Lumber core, 3/4" thick, natural		320	.050		3.81	2.82		6.63	8.65
5500	Plywood, knotty pine, 1/4" thick, A2 grade		450	.036		2.08	2		4.08	5.45
5600	A3 grade		450	.036		2.40	2		4.40	5.80
5800	3/4" thick, veneer core, A2 grade	2 Carp	320	.050	S.F.	3.09	2.82		5.91	7.85
5900	A3 grade		320	.050		2.81	2.82		5.63	7.55
6100	Aromatic cedar, 1/4" thick, plywood		400	.040		2.94	2.25		5.19	6.80
6200	1/4" thick, particle board		400	.040		1.48	2.25		3.73	5.20
9000	Minimum labor/equipment charge	1 Carp	2	4	Job		225		225	360

06 25 26 – Panel System

06 25 26.10 Panel Systems

		Crew	Daily Output	Labor-Hours	Unit	Material	2022 Bare Costs Labor	Equipment	Total	Total Incl O&P
0010	**PANEL SYSTEMS**									
0100	Raised panel, eng. wood core w/wood veneer, std., paint grade	2 Carp	300	.053	S.F.	14.50	3		17.50	20.50
0110	Oak veneer		300	.053		28	3		31	36
0120	Maple veneer		300	.053		39	3		42	48
0130	Cherry veneer		300	.053		38.50	3		41.50	47
0300	Class I fire rated, paint grade		300	.053		16.75	3		19.75	23
0310	Oak veneer		300	.053		37.50	3		40.50	46
0320	Maple veneer		300	.053		42.50	3		45.50	51.50
0330	Cherry veneer		300	.053		61	3		64	72
0510	Beadboard, 5/8" MDF, standard, primed		300	.053		9.60	3		12.60	15.30
0520	Oak veneer, unfinished		300	.053		17.15	3		20.15	23.50
0530	Maple veneer, unfinished		300	.053		24.50	3		27.50	32
0610	Rustic paneling, 5/8" MDF, standard, maple veneer, unfinished		300	.053		21	3		24	28

06 26 Board Paneling

06 26 13 – Profile Board Paneling

06 26 13.10 Paneling, Boards

		Crew	Daily Output	Labor-Hours	Unit	Material	2022 Bare Costs Labor	Equipment	Total	Total Incl O&P
0010	**PANELING, BOARDS**									
6400	Wood board paneling, 3/4" thick, knotty pine	2 Carp	300	.053	S.F.	2.33	3		5.33	7.35
6500	Rough sawn cedar		300	.053		5.15	3		8.15	10.40
6700	Redwood, clear, 1" x 4" boards		300	.053		14.05	3		17.05	20
6900	Aromatic cedar, closet lining, boards		275	.058		2.48	3.28		5.76	7.95
9000	Minimum labor/equipment charge	1 Carp	2	4	Job		225		225	360

06 43 Wood Stairs and Railings

06 43 13 – Wood Stairs

06 43 13.20 Prefabricated Wood Stairs

		Crew	Daily Output	Labor-Hours	Unit	Material	2022 Bare Costs Labor	Equipment	Total	Total Incl O&P
0010	**PREFABRICATED WOOD STAIRS**									
0100	Box stairs, prefabricated, 3'-0" wide									
0110	Oak treads, up to 14 risers	2 Carp	39	.410	Riser	1,500	23		1,523	1,675
0600	With pine treads for carpet, up to 14 risers	"	39	.410	"	1,000	23		1,023	1,125
1100	For 4' wide stairs, add				Flight	25%				
1550	Stairs, prefabricated stair handrail with balusters	1 Carp	30	.267	L.F.	82.50	15		97.50	115
1700	Basement stairs, prefabricated, pine treads									
1710	Pine risers, 3' wide, up to 14 risers	2 Carp	52	.308	Riser	1,000	17.30		1,017.30	1,125
4000	Residential, wood, oak treads, prefabricated		1.50	10.667	Flight	28,000	600		28,600	31,800
4200	Built in place		.44	36.364	"	2,725	2,050		4,775	6,250
4400	Spiral, oak, 4'-6" diameter, unfinished, prefabricated,									

06 43 13 – Wood Stairs

06 43 13.20 Prefabricated Wood Stairs	Crew	Daily Output	Labor-Hours	Unit	Material	2022 Bare Costs Labor	Equipment	Total	Total Incl O&P
4500	incl. railing, 9' high	2 Carp	1.50	10.667	Flight	2,325	600		2,925	3,500
9000	Minimum labor/equipment charge	"	3	5.333	Job		300		300	475

06 43 13.40 Wood Stair Parts

		Crew	Daily Output	Labor-Hours	Unit	Material	2022 Bare Costs Labor	Equipment	Total	Total Incl O&P
0010	**WOOD STAIR PARTS**									
0020	Pin top balusters, 1-1/4", oak, 34"	1 Carp	96	.083	Ea.	7.10	4.69		11.79	15.25
0030	38"		96	.083		7.75	4.69		12.44	16
0040	42"		96	.083		8.15	4.69		12.84	16.45
0050	Poplar, 34"		96	.083		5.30	4.69		9.99	13.25
0060	38"		96	.083		5.70	4.69		10.39	13.70
0070	42"		96	.083		5.95	4.69		10.64	14
0080	Maple, 34"		96	.083		16.35	4.69		21.04	25.50
0090	38"		96	.083		18.80	4.69		23.49	28
0100	42"		96	.083		20	4.69		24.69	30
0130	Primed, 34"		96	.083		7.70	4.69		12.39	15.90
0140	38"		96	.083		8.65	4.69		13.34	16.95
0150	42"		96	.083		9.60	4.69		14.29	18
0180	Box top balusters, 1-1/4", oak, 34"		60	.133		8.35	7.50		15.85	21
0190	38"		60	.133		8.50	7.50		16	21.50
0200	42"		60	.133		8.95	7.50		16.45	22
0210	Poplar, 34"		60	.133		6.25	7.50		13.75	18.80
0220	38"		60	.133		6.25	7.50		13.75	18.80
0230	42"		60	.133		6.50	7.50		14	19.10
0240	Maple, 34"		60	.133		18.25	7.50		25.75	32
0250	38"		60	.133		28.50	7.50		36	43.50
0260	42"		60	.133		17.20	7.50		24.70	31
0290	Primed, 34"		60	.133		8.10	7.50		15.60	21
0300	38"		60	.133		8.85	7.50		16.35	21.50
0310	42"		60	.133	▼	9.85	7.50		17.35	23
0340	Square balusters, cut from lineal stock, pine, 1-1/16" x 1-1/16"		180	.044	L.F.	6.35	2.50		8.85	10.95
0350	1-5/16" x 1-5/16"		180	.044		2.14	2.50		4.64	6.35
0360	1-5/8" x 1-5/8"		180	.044	▼	8.85	2.50		11.35	13.75
0370	Turned newel, oak, 3-1/2" square, 48" high		8	1	Ea.	91.50	56.50		148	191
0380	62" high		8	1		140	56.50		196.50	244
0390	Poplar, 3-1/2" square, 48" high		8	1		58.50	56.50		115	154
0400	62" high	▼	8	1	▼	96	56.50		152.50	196
0410	Maple, 3-1/2" square, 48" high	1 Carp	8	1	Ea.	58.50	56.50		115	154
0420	62" high		8	1		153	56.50		209.50	258
0430	Square newel, oak, 3-1/2" square, 48" high		8	1		119	56.50		175.50	221
0440	58" high		8	1		114	56.50		170.50	216
0450	Poplar, 3-1/2" square, 48" high		8	1		58.50	56.50		115	154
0460	58" high		8	1		118	56.50		174.50	220
0470	Maple, 3" square, 48" high		8	1		124	56.50		180.50	227
0480	58" high		8	1	▼	156	56.50		212.50	261
0490	Railings, oak, economy		96	.083	L.F.	19.65	4.69		24.34	29
0500	Average		96	.083		18.10	4.69		22.79	27.50
0510	Custom		96	.083		16.90	4.69		21.59	26
0520	Maple, economy		96	.083		11.25	4.69		15.94	19.80
0530	Average		96	.083		21.50	4.69		26.19	31.50
0540	Custom		96	.083		23	4.69		27.69	33
0550	Oak, for bending rail, economy		48	.167		27.50	9.40		36.90	45
0560	Average		48	.167		57	9.40		66.40	77.50
0570	Custom	▼	48	.167	▼	35	9.40		44.40	53.50

06 43 13.40 Wood Stair Parts		Crew	Daily Output	Labor-Hours	Unit	Material	2022 Bare Costs Labor	Equipment	Total	Total Incl O&P
0580	Maple, for bending rail, economy	1 Carp	48	.167	L.F.	69	9.40		78.40	90.50
0590	Average		48	.167		32.50	9.40		41.90	50.50
0600	Custom		48	.167	↓	63.50	9.40		72.90	85
0610	Risers, oak, 3/4" x 8", 36" long		80	.100	Ea.	44.50	5.65		50.15	58
0620	42" long		70	.114		52	6.45		58.45	67
0630	48" long		63	.127		59.50	7.15		66.65	77
0640	54" long		56	.143		67	8.05		75.05	86.50
0650	60" long		50	.160		74.50	9		83.50	96
0660	72" long		42	.190		89	10.70		99.70	115
0670	Poplar, 3/4" x 8", 36" long		80	.100		33.50	5.65		39.15	46
0680	42" long		71	.113		39	6.35		45.35	53
0690	48" long		63	.127		44.50	7.15		51.65	60.50
0700	54" long		56	.143		50.50	8.05		58.55	68.50
0710	60" long		50	.160		56	9		65	76
0720	72" long		42	.190		67	10.70		77.70	91
0730	Pine, 1" x 8", 36" long		80	.100		4.05	5.65		9.70	13.40
0740	42" long		70	.114		4.72	6.45		11.17	15.40
0750	48" long		63	.127		5.40	7.15		12.55	17.30
0760	54" long		56	.143		6.05	8.05		14.10	19.50
0770	60" long		50	.160		6.75	9		15.75	21.50
0780	72" long		42	.190		8.10	10.70		18.80	26
0790	Treads, oak, no returns, 1-1/32" x 11-1/2" x 36" long	↓	32	.250	↓	47	14.10		61.10	74
0800	42" long	1 Carp	32	.250	Ea.	54.50	14.10		68.60	82.50
0810	48" long		32	.250		62.50	14.10		76.60	91.50
0820	54" long		32	.250		70.50	14.10		84.60	100
0830	60" long		32	.250		78	14.10		92.10	109
0840	72" long		32	.250		94	14.10		108.10	126
0850	Mitred return one end, 1-1/32" x 11-1/2" x 36" long		24	.333		77	18.75		95.75	115
0860	42" long		24	.333		90	18.75		108.75	129
0870	48" long		24	.333		103	18.75		121.75	143
0880	54" long		24	.333		116	18.75		134.75	157
0890	60" long		24	.333		129	18.75		147.75	172
0900	72" long		24	.333		154	18.75		172.75	200
0910	Mitred return two ends, 1-1/32" x 11-1/2" x 36" long		12	.667		68.50	37.50		106	135
0920	42" long		12	.667		80	37.50		117.50	148
0930	48" long		12	.667		91.50	37.50		129	161
0940	54" long		12	.667		103	37.50		140.50	173
0950	60" long		12	.667		114	37.50		151.50	186
0960	72" long		12	.667		137	37.50		174.50	211
0970	Starting step, oak, 48", bullnose		8	1		182	56.50		238.50	290
0980	Double end bullnose		8	1	↓	255	56.50		311.50	370
1030	Skirt board, pine, 1" x 10"		55	.145	L.F.	1.84	8.20		10.04	15.05
1040	1" x 12"		52	.154	"	2.58	8.65		11.23	16.60
1050	Oak landing tread, 1-1/16" thick		54	.148	S.F.	8.85	8.35		17.20	23
1060	Oak cove molding		96	.083	L.F.	1.93	4.69		6.62	9.55
1070	Oak stringer molding		96	.083	"	3.74	4.69		8.43	11.55
1090	Rail bolt, 5/16" x 3-1/2"		48	.167	Ea.	3.47	9.40		12.87	18.70
1100	5/16" x 4-1/2"		48	.167		3.13	9.40		12.53	18.35
1120	Newel post anchor		16	.500		30.50	28		58.50	78
1130	Tapered plug, 1/2"		240	.033		.93	1.88		2.81	4
1140	1"		240	.033	↓	.38	1.88		2.26	3.40
9000	Minimum labor/equipment charge	↓	3	2.667	Job		150		150	239

06 43 Wood Stairs and Railings

06 43 16 – Wood Railings

06 43 16.10 Wood Handrails and Railings	Crew	Daily Output	Labor-Hours	Unit	Material	2022 Bare Costs Labor	Equipment	Total	Total Incl O&P
0010 **WOOD HANDRAILS AND RAILINGS**									
0020 Custom design, architectural grade, hardwood, plain	1 Carp	38	.211	L.F.	11.30	11.85		23.15	31.50
0100 Shaped		30	.267		58	15		73	88
0300 Stock interior railing with spindles 4" OC, 4' long		40	.200		61	11.25		72.25	85.50
0400 8' long		48	.167		61	9.40		70.40	82.50
9000 Minimum labor/equipment charge		3	2.667	Job		150		150	239

06 44 Ornamental Woodwork

06 44 19 – Wood Grilles

06 44 19.10 Grilles

	Crew	Daily Output	Labor-Hours	Unit	Material	2022 Bare Costs Labor	Equipment	Total	Total Incl O&P
0010 **GRILLES** and panels, hardwood, sanded									
0020 2' x 4' to 4' x 8', custom designs, unfinished, economy	1 Carp	38	.211	S.F.	56.50	11.85		68.35	81.50
0050 Average		30	.267		65	15		80	95.50
0100 Custom		19	.421		76	23.50		99.50	121
9000 Minimum labor/equipment charge		2	4	Job		225		225	360

06 44 33 – Wood Mantels

06 44 33.10 Fireplace Mantels

	Crew	Daily Output	Labor-Hours	Unit	Material	2022 Bare Costs Labor	Equipment	Total	Total Incl O&P
0010 **FIREPLACE MANTELS**									
0015 6" molding, 6' x 3'-6" opening, plain, paint grade	1 Carp	5	1.600	Opng.	480	90		570	675
0100 Ornate, oak		5	1.600		665	90		755	875
0300 Prefabricated pine, colonial type, stock, deluxe		2	4		2,225	225		2,450	2,800
0400 Economy		3	2.667		785	150		935	1,100
9000 Minimum labor/equipment charge		3	2.667	Job		150		150	239

06 44 33.20 Fireplace Mantel Beam

	Crew	Daily Output	Labor-Hours	Unit	Material	2022 Bare Costs Labor	Equipment	Total	Total Incl O&P
0010 **FIREPLACE MANTEL BEAM**									
0020 Rough texture wood, 4" x 8"	1 Carp	36	.222	L.F.	11.45	12.50		23.95	32.50
0100 4" x 10"		35	.229	"	12.85	12.85		25.70	34.50
0300 Laminated hardwood, 2-1/4" x 10-1/2" wide, 6' long		5	1.600	Ea.	113	90		203	267
0400 8' long		5	1.600	"	190	90		280	350
0600 Brackets for above, rough sawn		12	.667	Pr.	12.30	37.50		49.80	73
0700 Laminated		12	.667	"	93	37.50		130.50	163
9000 Minimum labor/equipment charge		4	2	Job		113		113	179

06 44 39 – Wood Posts and Columns

06 44 39.10 Decorative Beams

	Crew	Daily Output	Labor-Hours	Unit	Material	2022 Bare Costs Labor	Equipment	Total	Total Incl O&P
0010 **DECORATIVE BEAMS**									
0020 Rough sawn cedar, non-load bearing, 4" x 4"	2 Carp	180	.089	L.F.	1.97	5		6.97	10.10
0100 4" x 6"		170	.094		1.77	5.30		7.07	10.35
0200 4" x 8"		160	.100		2.35	5.65		8	11.55
0300 4" x 10"		150	.107		4.32	6		10.32	14.30
0400 4" x 12"		140	.114		5.05	6.45		11.50	15.75
0500 8" x 8"		130	.123		4.84	6.95		11.79	16.30
9000 Minimum labor/equipment charge	1 Carp	3	2.667	Job		150		150	239

06 44 39.20 Columns

	Crew	Daily Output	Labor-Hours	Unit	Material	2022 Bare Costs Labor	Equipment	Total	Total Incl O&P
0010 **COLUMNS**									
0050 Aluminum, round colonial, 6" diameter	2 Carp	80	.200	V.L.F.	11.60	11.25		22.85	30.50
0100 8" diameter		62.25	.257		17.60	14.45		32.05	42.50
0200 10" diameter		55	.291		27.50	16.40		43.90	56.50
0250 Fir, stock units, hollow round, 6" diameter		80	.200		35.50	11.25		46.75	57.50
0300 8" diameter		80	.200		32.50	11.25		43.75	53.50

06 44 39.20 Columns		Crew	Daily Output	Labor-Hours	Unit	Material	2022 Bare Costs Labor	2022 Bare Costs Equipment	Total	Total Incl O&P
0350	10" diameter	2 Carp	70	.229	V.L.F.	50	12.85		62.85	75.50
0360	12" diameter	2 Carp	65	.246	V.L.F.	73	13.85		86.85	103
0400	Solid turned, to 8' high, 3-1/2" diameter		80	.200		12	11.25		23.25	31
0500	4-1/2" diameter		75	.213		14.95	12		26.95	35.50
0600	5-1/2" diameter		70	.229		20	12.85		32.85	42.50
0800	Square columns, built-up, 5" x 5"		65	.246		35	13.85		48.85	60.50
0900	Solid, 3-1/2" x 3-1/2"		130	.123		11.80	6.95		18.75	24
1600	Hemlock, tapered, T&G, 12" diam., 10' high		100	.160		49	9		58	68.50
1700	16' high		65	.246		84.50	13.85		98.35	115
1900	14" diameter, 10' high		100	.160		116	9		125	141
2000	18' high		65	.246		101	13.85		114.85	133
2200	18" diameter, 12' high		65	.246		177	13.85		190.85	216
2300	20' high		50	.320		129	18		147	170
2500	20" diameter, 14' high		40	.400		243	22.50		265.50	305
2600	20' high	▼	35	.457		196	25.50		221.50	257
2800	For flat pilasters, deduct				▼	33%				
3000	For splitting into halves, add				Ea.	137			137	151
4000	Rough sawn cedar posts, 4" x 4"	2 Carp	250	.064	V.L.F.	4.87	3.60		8.47	11.05
4100	4" x 6"		235	.068		8.20	3.83		12.03	15.10
4200	6" x 6"		220	.073		15	4.09		19.09	23
4300	8" x 8"	▼	200	.080	▼	25	4.50		29.50	34.50
9000	Minimum labor/equipment charge	1 Carp	3	2.667	Job		150		150	239

06 48 Wood Frames

06 48 13 – Exterior Wood Door Frames

06 48 13.10 Exterior Wood Door Frames and Accessories

		Crew	Daily Output	Labor-Hours	Unit	Material	2022 Bare Costs Labor	2022 Bare Costs Equipment	Total	Total Incl O&P
0010	**EXTERIOR WOOD DOOR FRAMES AND ACCESSORIES**									
0400	Exterior frame, incl. ext. trim, pine, 5/4 x 4-9/16" deep	2 Carp	375	.043	L.F.	7.20	2.40		9.60	11.75
0420	5-3/16" deep		375	.043		8.15	2.40		10.55	12.80
0440	6-9/16" deep		375	.043		8.70	2.40		11.10	13.35
0600	Oak, 5/4 x 4-9/16" deep		350	.046		10.75	2.57		13.32	15.95
0620	5-3/16" deep		350	.046		11.85	2.57		14.42	17.10
0640	6-9/16" deep		350	.046		24.50	2.57		27.07	31
1000	Sills, 8/4 x 8" deep, oak, no horns		100	.160		18.35	9		27.35	34.50
1020	2" horns		100	.160		22	9		31	39
1040	3" horns		100	.160		22	9		31	39
1100	8/4 x 10" deep, oak, no horns		90	.178		7.45	10		17.45	24
1120	2" horns		90	.178		29	10		39	47.50
1140	3" horns		90	.178	▼	29	10		39	47.50
2000	Wood frame & trim, ext., colonial, 3' opng., fluted pilasters, flat head	▼	22	.727	Ea.	525	41		566	640
2010	Dentil head	2 Carp	21	.762	Ea.	655	43		698	795
2020	Ram's head		20	.800		630	45		675	765
2100	5'-4" opening, in-swing, fluted pilasters, flat head		17	.941		560	53		613	700
2120	Ram's head		15	1.067		1,375	60		1,435	1,625
2140	Out-swing, fluted pilasters, flat head		17	.941		500	53		553	635
2160	Ram's head		15	1.067		1,400	60		1,460	1,650
2400	6'-0" opening, in-swing, fluted pilasters, flat head		16	1		500	56.50		556.50	640
2420	Ram's head		10	1.600		1,575	90		1,665	1,875
2460	Out-swing, fluted pilasters, flat head		16	1		500	56.50		556.50	640
2480	Ram's head		10	1.600	▼	1,600	90		1,690	1,900
2600	For two sidelights, flat head, add		30	.533	Opng.	355	30		385	440

06 48 Wood Frames

06 48 13 – Exterior Wood Door Frames

06 48 13.10 Exterior Wood Door Frames and Accessories	Crew	Daily Output	Labor-Hours	Unit	Material	2022 Bare Costs Labor	Equipment	Total	Total Incl O&P
2620	Ram's head, add	2 Carp	20	.800	Opng.	885	45		930	1,050
2700	Custom birch frame, 3'-0" opening		16	1	Ea.	269	56.50		325.50	385
2750	6'-0" opng.		16	1		430	56.50		486.50	565
2900	Exterior, modern, plain trim, 3' opng., in-swing, FJP		26	.615		54.50	34.50		89	115
2920	Fir		24	.667		63	37.50		100.50	129
2940	Oak		22	.727		72.50	41		113.50	145

06 48 16 – Interior Wood Door Frames

06 48 16.10 Interior Wood Door Jamb and Frames

		Crew	Daily Output	Labor-Hours	Unit	Material	2022 Bare Costs Labor	Equipment	Total	Total Incl O&P
0010	INTERIOR WOOD DOOR JAMB AND FRAMES									
3000	Interior frame, pine, 11/16" x 3-5/8" deep	2 Carp	375	.043	L.F.	5.80	2.40		8.20	10.20
3020	4-9/16" deep		375	.043		6.70	2.40		9.10	11.20
3200	Oak, 11/16" x 3-5/8" deep		350	.046		2.54	2.57		5.11	6.90
3220	4-9/16" deep		350	.046		6.40	2.57		8.97	11.15
3240	5-3/16" deep		350	.046		19.20	2.57		21.77	25
3400	Walnut, 11/16" x 3-5/8" deep		350	.046		15.15	2.57		17.72	21
3420	4-9/16" deep		350	.046		12.50	2.57		15.07	17.85
3440	5-3/16" deep		350	.046		12.10	2.57		14.67	17.40
3600	Pocket door frame		16	1	Ea.	69	56.50		125.50	166
3800	Threshold, oak, 5/8" x 3-5/8" deep		200	.080	L.F.	2.75	4.50		7.25	10.20
3820	4-5/8" deep		190	.084		4.50	4.74		9.24	12.50
3840	5-5/8" deep		180	.089		7.15	5		12.15	15.85
9000	Minimum labor/equipment charge	1 Carp	4	2	Job		113		113	179

06 49 Wood Screens and Exterior Wood Shutters

06 49 19 – Exterior Wood Shutters

06 49 19.10 Shutters, Exterior

		Crew	Daily Output	Labor-Hours	Unit	Material	2022 Bare Costs Labor	Equipment	Total	Total Incl O&P
0010	SHUTTERS, EXTERIOR									
0012	Aluminum, louvered, 1'-4" wide, 3'-0" long	1 Carp	10	.800	Pr.	320	45		365	420
0200	4'-0" long	1 Carp	10	.800	Pr.	340	45		385	445
0300	5'-4" long		10	.800		435	45		480	545
0400	6'-8" long		9	.889		525	50		575	660
1000	Pine, louvered, primed, each 1'-2" wide, 3'-3" long		10	.800		251	45		296	350
1100	4'-7" long		10	.800		330	45		375	435
1250	Each 1'-4" wide, 3'-0" long		10	.800		262	45		307	360
1350	5'-3" long		10	.800		420	45		465	530
1500	Each 1'-6" wide, 3'-3" long		10	.800		305	45		350	410
1600	4'-7" long		10	.800		410	45		455	520
1620	Cedar, louvered, 1'-2" wide, 5'-7" long		10	.800		390	45		435	500
1630	Each 1'-4" wide, 2'-2" long		10	.800		205	45		250	297
1640	3'-0" long		10	.800		252	45		297	350
1650	3'-3" long		10	.800		279	45		324	375
1660	3'-11" long		10	.800		325	45		370	430
1670	4'-3" long		10	.800		350	45		395	455
1680	5'-3" long		10	.800		470	45		515	590
1690	5'-11" long		10	.800		465	45		510	580
1700	Door blinds, 6'-9" long, each 1'-3" wide		9	.889		615	50		665	755
1710	1'-6" wide		9	.889		725	50		775	880
1720	Cedar, solid raised panel, each 1'-4" wide, 3'-3" long		10	.800		315	45		360	420
1730	3'-11" long		10	.800		370	45		415	475
1740	4'-3" long		10	.800		395	45		440	505

179

For customer support on your Commercial Renovation Costs with RSMeans data, call 800.448.8182.

06 49 Wood Screens and Exterior Wood Shutters

06 49 19 – Exterior Wood Shutters

06 49 19.10 Shutters, Exterior		Crew	Daily Output	Labor-Hours	Unit	Material	2022 Bare Costs Labor	Equipment	Total	Total Incl O&P
1750	4'-7" long	1 Carp	10	.800	Pr.	420	45		465	530
1760	4'-11" long		10	.800		445	45		490	560
1770	5'-11" long		10	.800		525	45		570	645
1800	Door blinds, 6'-9" long, each 1'-3" wide		9	.889		555	50		605	690
1900	1'-6" wide		9	.889		655	50		705	800
2500	Polystyrene, solid raised panel, each 1'-4" wide, 3'-3" long		10	.800		102	45		147	184
2600	3'-11" long		10	.800		108	45		153	191
2700	4'-7" long		10	.800		123	45		168	207
2800	5'-3" long		10	.800		140	45		185	226
2900	6'-8" long		9	.889		179	50		229	277
4500	Polystyrene, louvered, each 1'-2" wide, 3'-3" long		10	.800		35	45		80	110
4600	4'-7" long		10	.800		70	45		115	149
4750	5'-3" long		10	.800		82	45		127	162
4850	6'-8" long		9	.889		103	50		153	194
6000	Vinyl, louvered, each 1'-2" x 4'-7" long		10	.800		101	45		146	183
6200	Each 1'-4" x 6'-8" long		9	.889		134	50		184	228
9000	Minimum labor/equipment charge	1 Carp	4	2	Job		113		113	179

06 51 Structural Plastic Shapes and Plates

06 51 13 – Plastic Lumber

06 51 13.10 Recycled Plastic Lumber

			Crew	Daily Output	Labor-Hours	Unit	Material	2022 Bare Costs Labor	Equipment	Total	Total Incl O&P
0010	**RECYCLED PLASTIC LUMBER**										
4000	Sheeting, recycled plastic, black or white, 4' x 8' x 1/8"	G	2 Carp	1100	.015	S.F.	1.35	.82		2.17	2.79
4010	4' x 8' x 3/16"	G		1100	.015		1.76	.82		2.58	3.24
4020	4' x 8' x 1/4"	G		950	.017		1.87	.95		2.82	3.57
4030	4' x 8' x 3/8"	G		950	.017		3.44	.95		4.39	5.30
4040	4' x 8' x 1/2"	G		900	.018		4.57	1		5.57	6.65
4050	4' x 8' x 5/8"	G		900	.018		7.95	1		8.95	10.35
4060	4' x 8' x 3/4"	G		850	.019		8.90	1.06		9.96	11.50
4070	Add for colors	G				Ea.	5%				
8500	100% recycled plastic, var colors, NLB, 2" x 2"	G				L.F.	2.16			2.16	2.38
8510	2" x 4"	G					4.46			4.46	4.91
8520	2" x 6"	G					6.70			6.70	7.40
8530	2" x 8"	G					9.90			9.90	10.85
8540	2" x 10"	G					14			14	15.40
8550	5/4" x 4"	G					4.70			4.70	5.15
8560	5/4" x 6"	G					6.55			6.55	7.20
8570	1" x 6"	G					3.22			3.22	3.54
8580	1/2" x 8"	G					3.80			3.80	4.18
8590	2" x 10" T&G	G					14			14	15.40
8600	3" x 10" T&G	G					24			24	26.50
8610	Add for premium colors	G					20%				

06 51 13.12 Structural Plastic Lumber

		Crew	Daily Output	Labor-Hours	Unit	Material	2022 Bare Costs Labor	Equipment	Total	Total Incl O&P
0010	**STRUCTURAL PLASTIC LUMBER**									
1320	Plastic lumber, posts or columns, 4" x 4"	2 Carp	390	.041	L.F.	17.40	2.31		19.71	23
1325	4" x 6"		275	.058		16.10	3.28		19.38	23
1330	4" x 8"		220	.073		23.50	4.09		27.59	32.50
1340	Girder, single, 4" x 4"		675	.024		17.40	1.33		18.73	21.50
1345	4" x 6"		600	.027		16.10	1.50		17.60	20
1350	4" x 8"		525	.030		23.50	1.72		25.22	28.50
1352	Double, 2" x 4"		625	.026		16.70	1.44		18.14	20.50

06 51 Structural Plastic Shapes and Plates

06 51 13 – Plastic Lumber

06 51 13.12 Structural Plastic Lumber		Crew	Daily Output	Labor-Hours	Unit	Material	2022 Bare Costs Labor	2022 Bare Costs Equipment	Total	Total Incl O&P
1354	2" x 6"	2 Carp	600	.027	L.F.	25	1.50		26.50	29.50
1356	2" x 8"		575	.028		32	1.57		33.57	37.50
1358	2" x 10"		550	.029		39.50	1.64		41.14	45.50
1360	2" x 12"		525	.030		48	1.72		49.72	55.50
1362	Triple, 2" x 4"		575	.028		25	1.57		26.57	30
1364	2" x 6"		550	.029		37	1.64		38.64	43.50
1366	2" x 8"	2 Carp	525	.030	L.F.	48	1.72		49.72	55.50
1368	2" x 10"		500	.032		59	1.80		60.80	68
1370	2" x 12"		475	.034		72	1.90		73.90	82.50
1372	Ledger, bolted 4' OC, 2" x 4"		400	.040		8.65	2.25		10.90	13.10
1374	2" x 6"		550	.029		12.45	1.64		14.09	16.25
1376	2" x 8"		390	.041		16.30	2.31		18.61	21.50
1378	2" x 10"		385	.042		19.90	2.34		22.24	25.50
1380	2" x 12"		380	.042		24.50	2.37		26.87	30.50
1382	Joists, 2" x 4"		1250	.013		8.40	.72		9.12	10.35
1384	2" x 6"		1250	.013		12.40	.72		13.12	14.80
1386	2" x 8"		1100	.015		16.05	.82		16.87	18.95
1388	2" x 10"		500	.032		19.70	1.80		21.50	24.50
1390	2" x 12"		875	.018		24	1.03		25.03	28
1392	Railings and trim, 5/4" x 4"	1 Carp	300	.027		5.45	1.50		6.95	8.40
1394	2" x 2"		300	.027		4.42	1.50		5.92	7.25
1396	2" x 4"		300	.027		8.35	1.50		9.85	11.55
1398	2" x 6"		300	.027		12.35	1.50		13.85	16

06 52 Plastic Structural Assemblies

06 52 10 – Fiberglass Structural Assemblies

06 52 10.30 Fiberglass Grating

		Crew	Daily Output	Labor-Hours	Unit	Material	2022 Bare Costs Labor	2022 Bare Costs Equipment	Total	Total Incl O&P
0010	**FIBERGLASS GRATING**									
0100	Molded, green (for mod. corrosive environment)									
0140	1" x 4" mesh, 1" thick	2 Sswk	400	.040	S.F.	5.85	2.47		8.32	10.45
0180	1-1/2" square mesh, 1" thick		400	.040		12.90	2.47		15.37	18.20
0220	1-1/4" thick		400	.040		6.10	2.47		8.57	10.70
0260	1-1/2" thick		400	.040		18.80	2.47		21.27	24.50
0300	2" square mesh, 2" thick		320	.050		18.30	3.09		21.39	25
1000	Orange (for highly corrosive environment)									
1040	1" x 4" mesh, 1" thick	2 Sswk	400	.040	S.F.	11.85	2.47		14.32	17
1080	1-1/2" square mesh, 1" thick		400	.040		16.35	2.47		18.82	22
1120	1-1/4" thick		400	.040		13.75	2.47		16.22	19.10
1160	1-1/2" thick		400	.040		16.20	2.47		18.67	22
1200	2" square mesh, 2" thick		320	.050		14.45	3.09		17.54	21
3000	Pultruded, green (for mod. corrosive environment)									
3040	1" OC bar spacing, 1" thick	2 Sswk	400	.040	S.F.	11.05	2.47		13.52	16.15
3080	1-1/2" thick		320	.050		12.75	3.09		15.84	19.05
3120	1-1/2" OC bar spacing, 1" thick		400	.040		11.50	2.47		13.97	16.65
3160	1-1/2" thick		400	.040		12.80	2.47		15.27	18.05
4000	Grating support legs, fixed height, no base					Ea.	35			35	38.50
4040	With base					Ea.	28.50			28.50	31.50
4080	Adjustable to 60"					"	42.50			42.50	47

181

06 52 Plastic Structural Assemblies

06 52 10 – Fiberglass Structural Assemblies

06 52 10.40 Fiberglass Floor Grating	Crew	Daily Output	Labor-Hours	Unit	Material	2022 Bare Costs Labor	Equipment	Total	Total Incl O&P
0010 **FIBERGLASS FLOOR GRATING**									
0100 Reinforced polyester, fire retardant, 1" x 4" grid, 1" thick	E-4	510	.063	S.F.	10.55	3.91	.29	14.75	18.20
0200 1-1/2" x 6" mesh, 1-1/2" thick		500	.064		12.15	3.99	.30	16.44	20
0300 With grit surface, 1-1/2" x 6" grid, 1-1/2" thick	↓	500	.064	↓	12.40	3.99	.30	16.69	20.50

06 63 Plastic Railings

06 63 10 – Plastic (PVC) Railings

06 63 10.10 Plastic Railings	Crew	Daily Output	Labor-Hours	Unit	Material	2022 Bare Costs Labor	Equipment	Total	Total Incl O&P
0010 **PLASTIC RAILINGS**									
0100 Horizontal PVC handrail with balusters, 3-1/2" wide, 36" high	1 Carp	96	.083	L.F.	25.50	4.69		30.19	35.50
0150 42" high		96	.083		24.50	4.69		29.19	34.50
0200 Angled PVC handrail with balusters, 3-1/2" wide, 36" high		72	.111		25.50	6.25		31.75	38
0250 42" high		72	.111		29.50	6.25		35.75	42.50
0300 Post sleeve for 4 x 4 post		96	.083	↓	27.50	4.69		32.19	38
0400 Post cap for 4 x 4 post, flat profile		48	.167	Ea.	13	9.40		22.40	29
0450 Newel post style profile		48	.167		36.50	9.40		45.90	55.50
0500 Raised corbeled profile		48	.167		35	9.40		44.40	53.50
0550 Post base trim for 4 x 4 post	↓	96	.083	↓	18.35	4.69		23.04	27.50

06 65 Plastic Trim

06 65 10 – PVC Trim

06 65 10.10 PVC Trim, Exterior	Crew	Daily Output	Labor-Hours	Unit	Material	2022 Bare Costs Labor	Equipment	Total	Total Incl O&P
0010 **PVC TRIM, EXTERIOR**									
0100 Cornerboards, 5/4" x 6" x 6"	1 Carp	240	.033	L.F.	10.05	1.88		11.93	14.10
0110 Door/window casing, 1" x 4"		200	.040		2.47	2.25		4.72	6.30
0120 1" x 6"		200	.040		3.50	2.25		5.75	7.45
0130 1" x 8"		195	.041		3.90	2.31		6.21	7.95
0140 1" x 10"		195	.041		4.98	2.31		7.29	9.10
0150 1" x 12"		190	.042		6.40	2.37		8.77	10.80
0160 5/4" x 4"		195	.041		3.24	2.31		5.55	7.25
0170 5/4" x 6"		195	.041		4.04	2.31		6.35	8.10
0180 5/4" x 8"		190	.042		4.79	2.37		7.16	9
0190 5/4" x 10"		190	.042		5.95	2.37		8.32	10.30
0200 5/4" x 12"		185	.043		6.90	2.43		9.33	11.45
0210 Fascia, 1" x 4"		250	.032		2.47	1.80		4.27	5.55
0220 1" x 6"		250	.032		3.50	1.80		5.30	6.70
0230 1" x 8"		225	.036		3.90	2		5.90	7.45
0240 1" x 10"		225	.036		4.98	2		6.98	8.65
0250 1" x 12"	↓	200	.040	↓	6.40	2.25		8.65	10.65
0260 5/4" x 4"	1 Carp	240	.033	L.F.	3.24	1.88		5.12	6.55
0270 5/4" x 6"		240	.033		4.04	1.88		5.92	7.40
0280 5/4" x 8"		215	.037		4.79	2.09		6.88	8.60
0290 5/4" x 10"		215	.037		5.95	2.09		8.04	9.90
0300 5/4" x 12"		190	.042		6.90	2.37		9.27	11.35
0310 Frieze, 1" x 4"		250	.032		2.47	1.80		4.27	5.55
0320 1" x 6"		250	.032		3.50	1.80		5.30	6.70
0330 1" x 8"		225	.036		3.90	2		5.90	7.45
0340 1" x 10"		225	.036		4.98	2		6.98	8.65
0350 1" x 12"		200	.040	↓	6.40	2.25		8.65	10.65

For customer support on your Commercial Renovation Costs with RSMeans data, call 800.448.8182.

06 65 Plastic Trim

06 65 10 - PVC Trim

06 65 10.10 PVC Trim, Exterior	Crew	Daily Output	Labor-Hours	Unit	Material	2022 Bare Costs Labor	Equipment	Total	Total Incl O&P
0360	5/4" x 4"	1 Carp	240	.033	L.F.	3.24	1.88		5.12	6.55
0370	5/4" x 6"		240	.033		4.04	1.88		5.92	7.40
0380	5/4" x 8"		215	.037		4.79	2.09		6.88	8.60
0390	5/4" x 10"		215	.037		5.95	2.09		8.04	9.90
0400	5/4" x 12"		190	.042		6.90	2.37		9.27	11.35
0410	Rake, 1" x 4"		200	.040		2.47	2.25		4.72	6.30
0420	1" x 6"		200	.040		3.50	2.25		5.75	7.45
0430	1" x 8"		190	.042		3.90	2.37		6.27	8.05
0440	1" x 10"		190	.042		4.98	2.37		7.35	9.20
0450	1" x 12"		180	.044		6.40	2.50		8.90	11.05
0460	5/4" x 4"		195	.041		3.24	2.31		5.55	7.25
0470	5/4" x 6"		195	.041		4.04	2.31		6.35	8.10
0480	5/4" x 8"		185	.043		4.79	2.43		7.22	9.10
0490	5/4" x 10"		185	.043		5.95	2.43		8.38	10.40
0500	5/4" x 12"		175	.046		6.90	2.57		9.47	11.70
0510	Rake trim, 1" x 4"		225	.036		2.47	2		4.47	5.90
0520	1" x 6"		225	.036		3.50	2		5.50	7.05
0560	5/4" x 4"		220	.036		3.24	2.05		5.29	6.80
0570	5/4" x 6"	↓	220	.036		4.04	2.05		6.09	7.70
0610	Soffit, 1" x 4"	2 Carp	420	.038		2.47	2.14		4.61	6.10
0620	1" x 6"		420	.038		3.50	2.14		5.64	7.25
0630	1" x 8"		420	.038		3.90	2.14		6.04	7.70
0640	1" x 10"		400	.040		4.98	2.25		7.23	9.05
0650	1" x 12"		400	.040		6.40	2.25		8.65	10.65
0660	5/4" x 4"		410	.039		3.24	2.20		5.44	7.05
0670	5/4" x 6"		410	.039		4.04	2.20		6.24	7.95
0680	5/4" x 8"		410	.039		4.79	2.20		6.99	8.75
0690	5/4" x 10"		390	.041		5.95	2.31		8.26	10.20
0700	5/4" x 12"	↓	390	.041	↓	6.90	2.31		9.21	11.25

06 73 Composite Decking

06 73 13 - Composite Decking

06 73 13.10 Woodgrained Composite Decking

		Crew	Daily Output	Labor-Hours	Unit	Material	Labor	Equipment	Total	Total Incl O&P
0010	**WOODGRAINED COMPOSITE DECKING**									
0100	Woodgrained composite decking, 1" x 6"	2 Carp	640	.025	L.F.	3.24	1.41		4.65	5.80
0110	Grooved edge		660	.024		3.42	1.36		4.78	5.95
0120	2" x 6"		640	.025		4.69	1.41		6.10	7.40
0130	Encased, 1" x 6"		640	.025		5.40	1.41		6.81	8.20
0140	Grooved edge		660	.024		5.60	1.36		6.96	8.30
0150	2" x 6"	↓	640	.025	↓	6.65	1.41		8.06	9.55

For customer support on your Commercial Renovation Costs with RSMeans data, call 800.448.8182.

183

06 81 Composite Railings

06 81 10 – Encased Railings

06 81 10.10 Encased Composite Railings	Crew	Daily Output	Labor-Hours	Unit	Material	2022 Bare Costs Labor	Equipment	Total	Total Incl O&P
0010 **ENCASED COMPOSITE RAILINGS**									
0100 Encased composite railing, 6' long, 36" high, incl. balusters	1 Carp	16	.500	Ea.	180	28		208	243
0110 42" high, incl. balusters		16	.500		241	28		269	310
0120 8' long, 36" high, incl. balusters		12	.667		248	37.50		285.50	335
0130 42" high, incl. balusters		12	.667		365	37.50		402.50	460
0140 Accessories, post sleeve, 4" x 4", 39" long		32	.250		51.50	14.10		65.60	79
0150 96" long		24	.333		81.50	18.75		100.25	120
0160 6" x 6", 39" long		32	.250		60.50	14.10		74.60	89.50
0170 96" long		24	.333		166	18.75		184.75	212
0180 Accessories, post skirt, 4" x 4"		96	.083		5.45	4.69		10.14	13.45
0190 6" x 6"		96	.083		6.35	4.69		11.04	14.45
0200 Post cap, 4" x 4", flat		48	.167		9.80	9.40		19.20	25.50
0210 Pyramid		48	.167		8.50	9.40		17.90	24.50
0220 Post cap, 6" x 6", flat		48	.167		12.70	9.40		22.10	29
0230 Pyramid		48	.167		16.25	9.40		25.65	33

Estimating Tips

07 10 00 Dampproofing and Waterproofing

- Be sure of the job specifications before pricing this subdivision. The difference in cost between waterproofing and dampproofing can be great. Waterproofing will hold back standing water. Dampproofing prevents the transmission of water vapor. Also included in this section are vapor retarding membranes.

07 20 00 Thermal Protection

- Insulation and fireproofing products are measured by area, thickness, volume, or R-value. Specifications may give only what the specific R-value should be in a certain situation. The estimator may need to choose the type of insulation to meet that R-value.

07 30 00 Steep Slope Roofing
07 40 00 Roofing and Siding Panels

- Many roofing and siding products are bought and sold by the square. One square is equal to an area that measures 100 square feet.

  This simple change in unit of measure could create a large error if the estimator is not observant. Accessories necessary for a complete installation must be figured into any calculations for both material and labor.

07 50 00 Membrane Roofing
07 60 00 Flashing and Sheet Metal
07 70 00 Roofing and Wall Specialties and Accessories

- The items in these subdivisions compose a roofing system. No one component completes the installation, and all must be estimated. Built-up or single-ply membrane roofing systems are made up of many products and installation trades. Wood blocking at roof perimeters or penetrations, parapet coverings, reglets, roof drains, gutters, downspouts, sheet metal flashing, skylights, smoke vents, and roof hatches all need to be considered along with the roofing material. Several different installation trades will need to work together on the roofing system. Inherent difficulties in the scheduling and coordination of various trades must be accounted for when estimating labor costs.

07 90 00 Joint Protection

- To complete the weather-tight shell, the sealants and caulkings must be estimated. Where different materials meet—at expansion joints, at flashing penetrations, and at hundreds of other locations throughout a construction project—caulking and sealants provide another line of defense against water penetration. Often, an entire system is based on the proper location and placement of caulking or sealants. The detailed drawings that are included as part of a set of architectural plans show typical locations for these materials. When caulking or sealants are shown at typical locations, this means the estimator must include them for all the locations where this detail is applicable. Be careful to keep different types of sealants separate, and remember to consider backer rods and primers if necessary.

Reference Numbers

Reference numbers are shown at the beginning of some major classifications. These numbers refer to related items in the Reference Section. The reference information may be an estimating procedure, an alternate pricing method, or technical information.

Note: Not all subdivisions listed here necessarily appear. ■

07 01 50 – Maintenance of Membrane Roofing

07 01 50.10 Roof Coatings	Crew	Daily Output	Labor-Hours	Unit	Material	2022 Bare Costs Labor	Equipment	Total	Total Incl O&P
0010 **ROOF COATINGS**									
0012 Asphalt, brush grade, material only				Gal.	8.60			8.60	9.45
0200 Asphalt base, fibered aluminum coating Ⓖ					10.40			10.40	11.40
0300 Asphalt primer, 5 gal.					9.70			9.70	10.65
0600 Coal tar pitch, 200 lb. barrels				Ton	1,325			1,325	1,450
0700 Tar roof cement, 5 gal. lots				Gal.	13.20			13.20	14.55
0800 Glass fibered roof & patching cement, 5 gal.				"	10.40			10.40	11.40
0900 Reinforcing glass membrane, 450 S.F./roll				Ea.	52.50			52.50	57.50
1000 Neoprene roof coating, 5 gal., 2 gal./sq.				Gal.	30			30	33
1100 Roof patch & flashing cement, 5 gal.				Gal.	7.85			7.85	8.65
1200 Roof resaturant, glass fibered, 3 gal./sq.					8.15			8.15	8.95
1600 Reflective roof coating, white, elastomeric, approx. 50 S.F./gal. Ⓖ					14.10			14.10	15.50

07 01 90 – Maintenance of Joint Protection

07 01 90.81 Joint Sealant Replacement

	Crew	Daily Output	Labor-Hours	Unit	Material	2022 Bare Costs Labor	Equipment	Total	Total Incl O&P
0010 **JOINT SEALANT REPLACEMENT**									
0050 Control joints in concrete floors/slabs									
0100 Option 1 for joints with hard dry sealant									
0110 Step 1: Sawcut to remove 95% of old sealant									
0112 1/4" wide x 1/2" deep, with single saw blade	C-27	4800	.003	L.F.	.01	.18	.03	.22	.33
0114 3/8" wide x 3/4" deep, with single saw blade		4000	.004		.02	.21	.04	.27	.40
0116 1/2" wide x 1" deep, with double saw blades		3600	.004		.05	.24	.04	.33	.47
0118 3/4" wide x 1-1/2" deep, with double saw blades		3200	.005		.09	.27	.05	.41	.57
0120 Step 2: Water blast joint faces and edges	C-29	2500	.003			.15	.04	.19	.27
0130 Step 3: Air blast joint faces and edges	C-28	2000	.004			.21	.02	.23	.35
0140 Step 4: Sand blast joint faces and edges	E-11	2000	.016			.80	.16	.96	1.48
0150 Step 5: Air blast joint faces and edges	C-28	2000	.004			.21	.02	.23	.35
0200 Option 2 for joints with soft pliable sealant									
0210 Step 1: Plow joint with rectangular blade	B-62	2600	.009	L.F.		.46	.10	.56	.83
0220 Step 2: Sawcut to re-face joint faces									
0222 1/4" wide x 1/2" deep, with single saw blade	C-27	2400	.007	L.F.	.02	.35	.07	.44	.64
0224 3/8" wide x 3/4" deep, with single saw blade		2000	.008		.03	.43	.08	.54	.78
0226 1/2" wide x 1" deep, with double saw blades		1800	.009		.06	.47	.09	.62	.91
0228 3/4" wide x 1-1/2" deep, with double saw blades		1600	.010		.12	.53	.10	.75	1.08
0230 Step 3: Water blast joint faces and edges	C-29	2500	.003			.15	.04	.19	.27
0240 Step 4: Air blast joint faces and edges	C-28	2000	.004			.21	.02	.23	.35
0250 Step 5: Sand blast joint faces and edges	E-11	2000	.016			.80	.16	.96	1.48
0260 Step 6: Air blast joint faces and edges	C-28	2000	.004			.21	.02	.23	.35
0290 For saw cutting new control joints, see Section 03 15 16.20									
8910 For backer rod, see Section 07 91 23.10									
8920 For joint sealant, see Section 03 15 16.30 or 07 92 13.20									

186

For customer support on your Commercial Renovation Costs with RSMeans data, call 800.448.8182.

07 05 05 – Selective Demolition for Thermal and Moisture Protection

07 05 05.10 Selective Demo., Thermal and Moist. Protection		Crew	Daily Output	Labor-Hours	Unit	Material	2022 Bare Costs Labor	Equipment	Total	Total Incl O&P
0010	**SELECTIVE DEMO., THERMAL AND MOISTURE PROTECTION**									
0020	Caulking/sealant, to 1" x 1" joint R024119-10	1 Clab	600	.013	L.F.		.61		.61	.97
0120	Downspouts, including hangers	1 Clab	350	.023	L.F.		1.04		1.04	1.66
0220	Flashing, sheet metal		290	.028	S.F.		1.26		1.26	2
0420	Gutters, aluminum or wood, edge hung		240	.033	L.F.		1.52		1.52	2.42
0520	Built-in		100	.080	"		3.65		3.65	5.80
0620	Insulation, air/vapor barrier		3500	.002	S.F.		.10		.10	.17
0670	Batts or blankets		1400	.006	C.F.		.26		.26	.41
0720	Foamed or sprayed in place	2 Clab	1000	.016	B.F.		.73		.73	1.16
0770	Loose fitting	1 Clab	3000	.003	C.F.		.12		.12	.19
0870	Rigid board		3450	.002	B.F.		.11		.11	.17
1120	Roll roofing, cold adhesive		12	.667	Sq.		30.50		30.50	48.50
1170	Roof accessories, adjustable metal chimney flashing		9	.889	Ea.		40.50		40.50	64.50
1171	Selective demo, firstopping		20	.400			18.25		18.25	29
1325	Plumbing vent flashing		32	.250			11.40		11.40	18.10
1375	Ridge vent strip, aluminum		310	.026	L.F.		1.18		1.18	1.87
1620	Skylight to 10 S.F.		8	1	Ea.		45.50		45.50	72.50
2120	Roof edge, aluminum soffit and fascia		570	.014	L.F.		.64		.64	1.02
2170	Concrete coping, up to 12" wide	2 Clab	160	.100			4.56		4.56	7.25
2220	Drip edge	1 Clab	1000	.008			.36		.36	.58
2270	Gravel stop		950	.008			.38		.38	.61
2370	Sheet metal coping, up to 12" wide		240	.033			1.52		1.52	2.42
2470	Roof insulation board, over 2" thick	B-2	7800	.005	B.F.		.24		.24	.37
2520	Up to 2" thick	"	3900	.010	S.F.		.47		.47	.75
2620	Roof ventilation, louvered gable vent	1 Clab	16	.500	Ea.		23		23	36
2670	Remove, roof hatch	G-3	15	2.133			121		121	189
2675	Rafter vents	1 Clab	960	.008			.38		.38	.60
2720	Soffit vent and/or fascia vent		575	.014	L.F.		.63		.63	1.01
2775	Soffit vent strip, aluminum, 3" to 4" wide		160	.050			2.28		2.28	3.62
2820	Roofing accessories, shingle moulding, to 1" x 4"		1600	.005			.23		.23	.36
2870	Cant strip	B-2	2000	.020			.92		.92	1.46
2920	Concrete block walkway	1 Clab	230	.035			1.59		1.59	2.52
3070	Roofing, felt paper, #15		70	.114	Sq.		5.20		5.20	8.30
3125	#30 felt		30	.267	"		12.15		12.15	19.30
3170	Asphalt shingles, 1 layer	B-2	3500	.011	S.F.		.53		.53	.84
3180	2 layers		1750	.023	"		1.05		1.05	1.67
3370	Modified bitumen		26	1.538	Sq.		71		71	112
3420	Built-up, no gravel, 3 ply		25	1.600			73.50		73.50	117
3470	4 ply		21	1.905			87.50		87.50	139
3620	5 ply	B-2	1600	.025	S.F.		1.15		1.15	1.83
3720	5 ply, with gravel		890	.045			2.07		2.07	3.28
3725	Loose gravel removal		5000	.008			.37		.37	.58
3730	Embedded gravel removal		2000	.020			.92		.92	1.46
3870	Fiberglass sheet		1200	.033			1.53		1.53	2.44
4120	Slate shingles		1900	.021			.97		.97	1.54
4170	Ridge shingles, clay or slate		2000	.020	L.F.		.92		.92	1.46
4320	Single ply membrane, attached at seams		52	.769	Sq.		35.50		35.50	56
4370	Ballasted		75	.533			24.50		24.50	39
4420	Fully adhered		39	1.026			47		47	75
4550	Roof hatch, 2'-6" x 3'-0"	1 Clab	10	.800	Ea.		36.50		36.50	58
4670	Wood shingles	B-2	2200	.018	S.F.		.84		.84	1.33
4820	Sheet metal roofing	"	2150	.019			.86		.86	1.36
4970	Siding, horizontal wood clapboards	1 Clab	380	.021			.96		.96	1.53

07 05 05 – Selective Demolition for Thermal and Moisture Protection

07 05 05.10 Selective Demo., Thermal and Moist. Protection	Crew	Daily Output	Labor-Hours	Unit	Material	2022 Bare Costs Labor	Equipment	Total	Total Incl O&P
5025	Exterior insulation finish system	1 Clab	120	.067	S.F.		3.04		3.04	4.83
5070	Tempered hardboard, remove and reset	1 Carp	380	.021			1.19		1.19	1.88
5120	Tempered hardboard sheet siding	"	375	.021			1.20		1.20	1.91
5170	Metal, corner strips	1 Clab	850	.009	L.F.		.43		.43	.68
5225	Horizontal strips		444	.018	S.F.		.82		.82	1.31
5320	Vertical strips		400	.020			.91		.91	1.45
5520	Wood shingles		350	.023			1.04		1.04	1.66
5620	Stucco siding		360	.022			1.01		1.01	1.61
5670	Textured plywood		725	.011			.50		.50	.80
5720	Vinyl siding		510	.016			.72		.72	1.14
5770	Corner strips		900	.009	L.F.		.41		.41	.64
5870	Wood, boards, vertical		400	.020	S.F.		.91		.91	1.45
5880	Steel siding, corrugated/ribbed		402.50	.020	"		.91		.91	1.44
5920	Waterproofing, protection/drain board	2 Clab	3900	.004	B.F.		.19		.19	.30
5970	Over 1/2" thick		1750	.009	S.F.		.42		.42	.66
6020	To 1/2" thick		2000	.008	"		.36		.36	.58
9000	Minimum labor/equipment charge	1 Clab	2	4	Job		182		182	290

07 11 Dampproofing

07 11 13 – Bituminous Dampproofing

07 11 13.10 Bituminous Asphalt Coating

		Crew	Daily Output	Labor-Hours	Unit	Material	2022 Bare Costs Labor	Equipment	Total	Total Incl O&P
0010	**BITUMINOUS ASPHALT COATING**									
0030	Brushed on, below grade, 1 coat	1 Rofc	665	.012	S.F.	.22	.60		.82	1.27
0100	2 coat		500	.016		.43	.79		1.22	1.84
0300	Sprayed on, below grade, 1 coat		830	.010		.22	.48		.70	1.06
0400	2 coat	1 Rofc	500	.016	S.F.	.42	.79		1.21	1.83
0500	Asphalt coating, with fibers				Gal.	10.40			10.40	11.40
0600	Troweled on, asphalt with fibers, 1/16" thick	1 Rofc	500	.016	S.F.	.45	.79		1.24	1.87
0700	1/8" thick		400	.020		.80	.99		1.79	2.59
1000	1/2" thick		350	.023		2.60	1.13		3.73	4.80
9000	Minimum labor/equipment charge		3	2.667	Job		132		132	228

07 11 16 – Cementitious Dampproofing

07 11 16.20 Cementitious Parging

		Crew	Daily Output	Labor-Hours	Unit	Material	2022 Bare Costs Labor	Equipment	Total	Total Incl O&P
0010	**CEMENTITIOUS PARGING**									
0020	Portland cement, 2 coats, 1/2" thick	D-1	250	.064	S.F.	.35	3.21		3.56	5.55
0100	Waterproofed Portland cement, 1/2" thick, 2 coats	"	250	.064	"	2.66	3.21		5.87	8.10

07 12 Built-up Bituminous Waterproofing

07 12 13 – Built-Up Asphalt Waterproofing

07 12 13.20 Membrane Waterproofing

		Crew	Daily Output	Labor-Hours	Unit	Material	2022 Bare Costs Labor	Equipment	Total	Total Incl O&P
0010	**MEMBRANE WATERPROOFING**									
0012	On slabs, 1 ply, felt, mopped	G-1	3000	.019	S.F.	.40	.87	.18	1.45	2.13
0015	On walls, 1 ply, felt, mopped		3000	.019		.40	.87	.18	1.45	2.13
0100	On slabs, 1 ply, glass fiber fabric, mopped		2100	.027		.46	1.24	.26	1.96	2.92
0105	On walls, 1 ply, glass fiber fabric, mopped		2100	.027		.46	1.24	.26	1.96	2.92
0300	On slabs, 2 ply, felt, mopped		2500	.022		.81	1.04	.22	2.07	2.92
0305	On walls, 2 ply, felt, mopped		2500	.022		.81	1.04	.22	2.07	2.92
0400	On slabs, 2 ply, glass fiber fabric, mopped		1650	.034		1.04	1.57	.33	2.94	4.22
0405	On walls, 2 ply, glass fiber fabric, mopped		1650	.034		1.04	1.57	.33	2.94	4.22

07 12 Built-up Bituminous Waterproofing

07 12 13 – Built-Up Asphalt Waterproofing

07 12 13.20 Membrane Waterproofing	Crew	Daily Output	Labor-Hours	Unit	Material	2022 Bare Costs Labor	Equipment	Total	Total Incl O&P
0600	On slabs, 3 ply, felt, mopped	G-1	2100	.027	S.F.	1.21	1.24	.26	2.71	3.75
0605	On walls, 3 ply, felt, mopped		2100	.027		1.21	1.24	.26	2.71	3.75
0700	On slabs, 3 ply, glass fiber fabric, mopped		1550	.036		1.37	1.67	.36	3.40	4.78
0705	On walls, 3 ply, glass fiber fabric, mopped		1550	.036		1.37	1.67	.36	3.40	4.78
0710	Asphaltic hardboard protection board, 1/8" thick	2 Rofc	500	.032		1.24	1.59		2.83	4.09
0715	1/4" thick		450	.036		1.56	1.76		3.32	4.75
1000	EPS membrane protection board, 1/4" thick		3500	.005		.37	.23		.60	.80
1050	3/8" thick		3500	.005		.41	.23		.64	.84
1060	1/2" thick		3500	.005		.44	.23		.67	.87
1070	Fiberglass fabric, black, 20/10 mesh		116	.138	Sq.	18.05	6.85		24.90	31.50
9000	Minimum labor/equipment charge		2	8	Job		395		395	685

07 13 Sheet Waterproofing

07 13 53 – Elastomeric Sheet Waterproofing

07 13 53.10 Elastomeric Sheet Waterproofing and Access.	Crew	Daily Output	Labor-Hours	Unit	Material	2022 Bare Costs Labor	Equipment	Total	Total Incl O&P
0010	ELASTOMERIC SHEET WATERPROOFING AND ACCESS.									
0090	EPDM, plain, 45 mils thick	2 Rofc	580	.028	S.F.	.79	1.37		2.16	3.21
0100	60 mils thick		570	.028		.91	1.39		2.30	3.40
0300	Nylon reinforced sheets, 45 mils thick		580	.028		.97	1.37		2.34	3.42
0400	60 mils thick		570	.028		1.13	1.39		2.52	3.64
0600	Vulcanizing splicing tape for above, 2" wide				C.L.F.	63			63	69.50
0700	4" wide				"	355			355	390
0900	Adhesive, bonding, 60 S.F./gal.				Gal.	42			42	46
1000	Splicing, 75 S.F./gal.				"	37			37	40.50
1200	Neoprene sheets, plain, 45 mils thick	2 Rofc	580	.028	S.F.	1.31	1.37		2.68	3.79
1300	60 mils thick		570	.028		2.03	1.39		3.42	4.63
1500	Nylon reinforced, 45 mils thick		580	.028		2.63	1.37		4	5.25
1600	60 mils thick		570	.028		4.81	1.39		6.20	7.70
1800	120 mils thick		500	.032		8.10	1.59		9.69	11.65
1900	Adhesive, splicing, 150 S.F./gal. per coat				Gal.	37			37	40.50
2100	Fiberglass reinforced, fluid applied, 1/8" thick	2 Rofc	500	.032	S.F.	1.58	1.59		3.17	4.47
2200	Polyethylene and rubberized asphalt sheets, 60 mils thick		550	.029		.70	1.44		2.14	3.25
2400	Polyvinyl chloride sheets, plain, 10 mils thick		580	.028		.17	1.37		1.54	2.54
2500	20 mils thick		570	.028		.23	1.39		1.62	2.65
2700	30 mils thick		560	.029		.23	1.42		1.65	2.69
3000	Adhesives, trowel grade, 40-100 S.F./gal.				Gal.	28.50			28.50	31.50
3100	Brush grade, 100-250 S.F./gal.				"	25.50			25.50	28.50
3300	Bitumen modified polyurethane, fluid applied, 55 mils thick	2 Rofc	665	.024	S.F.	1.17	1.19		2.36	3.34
9000	Minimum labor/equipment charge	"	2	8	Job		395		395	685

For customer support on your Commercial Renovation Costs with RSMeans data, call 800.448.8182.

189

07 16 Cementitious and Reactive Waterproofing

07 16 16 – Crystalline Waterproofing

07 16 16.20 Cementitious Waterproofing		Crew	Daily Output	Labor-Hours	Unit	Material	2022 Bare Costs Labor	Equipment	Total	Total Incl O&P
0010	**CEMENTITIOUS WATERPROOFING**									
0020	1/8" application, sprayed on	G-2A	1000	.024	S.F.	.67	1.06	.66	2.39	3.25
0050	4 coat cementitious metallic slurry	1 Cefi	1.20	6.667	C.S.F.	81.50	355		436.50	645

07 19 Water Repellents

07 19 19 – Silicone Water Repellents

07 19 19.10 Silicone Based Water Repellents

		Crew	Daily Output	Labor-Hours	Unit	Material	2022 Bare Costs Labor	Equipment	Total	Total Incl O&P
0010	**SILICONE BASED WATER REPELLENTS**									
0020	Water base liquid, roller applied	2 Rofc	7000	.002	S.F.	.45	.11		.56	.69
0200	Silicone or stearate, sprayed on CMU, 1 coat	1 Rofc	4000	.002		.42	.10		.52	.63
0300	2 coats		3000	.003		.84	.13		.97	1.16
9000	Minimum labor/equipment charge		3	2.667	Job		132		132	228

07 21 Thermal Insulation

07 21 13 – Board Insulation

07 21 13.10 Rigid Insulation

			Crew	Daily Output	Labor-Hours	Unit	Material	2022 Bare Costs Labor	Equipment	Total	Total Incl O&P
0010	**RIGID INSULATION**, for walls										
0040	Fiberglass, 1.5#/C.F., unfaced, 1" thick, R4.1	G	1 Carp	1000	.008	S.F.	.46	.45		.91	1.23
0060	1-1/2" thick, R6.2	G		1000	.008		.56	.45		1.01	1.34
0080	2" thick, R8.3	G		1000	.008		.62	.45		1.07	1.40
0120	3" thick, R12.4	G		800	.010		.75	.56		1.31	1.72
0370	3#/C.F., unfaced, 1" thick, R4.3	G		1000	.008		.73	.45		1.18	1.52
0390	1-1/2" thick, R6.5	G		1000	.008		.93	.45		1.38	1.74
0400	2" thick, R8.7	G		890	.009		1.47	.51		1.98	2.42
0420	2-1/2" thick, R10.9	G		800	.010		1.32	.56		1.88	2.34
0440	3" thick, R13	G		800	.010		2.08	.56		2.64	3.18
0520	Foil faced, 1" thick, R4.3	G		1000	.008		1.02	.45		1.47	1.84
0540	1-1/2" thick, R6.5	G		1000	.008		1.41	.45		1.86	2.27
0560	2" thick, R8.7	G		890	.009		1.77	.51		2.28	2.75
0580	2-1/2" thick, R10.9	G		800	.010		2.23	.56		2.79	3.34
0600	3" thick, R13	G		800	.010		2.36	.56		2.92	3.49
1600	Isocyanurate, 4' x 8' sheet, foil faced, both sides										
1610	1/2" thick	G	1 Carp	800	.010	S.F.	.37	.56		.93	1.30
1620	5/8" thick	G		800	.010		.79	.56		1.35	1.76
1630	3/4" thick	G		800	.010		.56	.56		1.12	1.51
1640	1" thick	G		800	.010		.70	.56		1.26	1.66
1650	1-1/2" thick	G		730	.011		.83	.62		1.45	1.89
1660	2" thick	G		730	.011		1.11	.62		1.73	2.20
1670	3" thick	G		730	.011		1.94	.62		2.56	3.11
1680	4" thick	G		730	.011		2.66	.62		3.28	3.91
1700	Perlite, 1" thick, R2.77	G		800	.010		.59	.56		1.15	1.54
1750	2" thick, R5.55	G		730	.011		1.06	.62		1.68	2.15
1900	Extruded polystyrene, 25 psi compressive strength, 1" thick, R5	G		800	.010		.73	.56		1.29	1.69
1940	2" thick, R10	G		730	.011		1.45	.62		2.07	2.58
1960	3" thick, R15	G		730	.011		2.18	.62		2.80	3.38
2100	Expanded polystyrene, 1" thick, R3.85	G		800	.010		.28	.56		.84	1.20
2120	2" thick, R7.69	G		730	.011		.56	.62		1.18	1.60
2140	3" thick, R11.49	G		730	.011		.84	.62		1.46	1.90
9000	Minimum labor/equipment charge			4	2	Job		113		113	179

07 21 Thermal Insulation

07 21 13 – Board Insulation

07 21 13.13 Foam Board Insulation		Crew	Daily Output	Labor-Hours	Unit	Material	2022 Bare Costs Labor	Equipment	Total	Total Incl O&P
0010	**FOAM BOARD INSULATION**									
0600	Polystyrene, expanded, 1" thick, R4	G 1 Carp	680	.012	S.F.	.28	.66		.94	1.36
0700	2" thick, R8	G 1 Carp	675	.012	S.F.	.56	.67		1.23	1.68
9000	Minimum labor/equipment charge	"	4	2	Job		113		113	179

07 21 16 – Blanket Insulation

07 21 16.10 Blanket Insulation for Floors/Ceilings

		Crew	Daily Output	Labor-Hours	Unit	Material	2022 Bare Costs Labor	Equipment	Total	Total Incl O&P
0010	**BLANKET INSULATION FOR FLOORS/CEILINGS**									
0020	Including spring type wire fasteners									
2000	Fiberglass, blankets or batts, paper or foil backing									
2100	3-1/2" thick, R13	G 1 Carp	700	.011	S.F.	.73	.64		1.37	1.82
2150	6-1/4" thick, R19	G	600	.013		.99	.75		1.74	2.28
2210	9-1/2" thick, R30	G	500	.016		1.14	.90		2.04	2.68
2220	12" thick, R38	G	475	.017		1.44	.95		2.39	3.09
3000	Unfaced, 3-1/2" thick, R13	G	600	.013		.54	.75		1.29	1.78
3010	6-1/4" thick, R19	G	500	.016		.78	.90		1.68	2.29
3020	9-1/2" thick, R30	G	450	.018		1.13	1		2.13	2.83
3030	12" thick, R38	G	425	.019		1.66	1.06		2.72	3.51
9000	Minimum labor/equipment charge		4	2	Job		113		113	179

07 21 16.20 Blanket Insulation for Walls

		Crew	Daily Output	Labor-Hours	Unit	Material	2022 Bare Costs Labor	Equipment	Total	Total Incl O&P
0010	**BLANKET INSULATION FOR WALLS**									
0020	Kraft faced fiberglass, 3-1/2" thick, R11, 15" wide	G 1 Carp	1350	.006	S.F.	.53	.33		.86	1.11
0030	23" wide	G	1600	.005		.53	.28		.81	1.03
0060	R13, 11" wide	G	1150	.007		.60	.39		.99	1.28
0080	15" wide	G	1350	.006		.60	.33		.93	1.19
0100	23" wide	G	1600	.005		.60	.28		.88	1.11
0110	R15, 11" wide	G	1150	.007		.89	.39		1.28	1.60
0120	15" wide	G	1350	.006		.89	.33		1.22	1.51
0130	23" wide	G	1600	.005		.89	.28		1.17	1.43
0140	6" thick, R19, 11" wide	G	1150	.007		.86	.39		1.25	1.57
0160	15" wide	G	1350	.006		.86	.33		1.19	1.48
0180	23" wide	G	1600	.005		.86	.28		1.14	1.40
0182	R21, 11" wide	G	1150	.007		.82	.39		1.21	1.52
0184	15" wide	G	1350	.006		.82	.33		1.15	1.43
0186	23" wide	G	1600	.005		.82	.28		1.10	1.35
0188	9" thick, R30, 11" wide	G	985	.008		1.14	.46		1.60	1.98
0200	15" wide	G	1150	.007		1.14	.39		1.53	1.87
0220	23" wide	G	1350	.006		1.14	.33		1.47	1.78
0230	12" thick, R38, 11" wide	G	985	.008		1.44	.46		1.90	2.31
0240	15" wide	G	1150	.007		1.44	.39		1.83	2.20
0260	23" wide	G	1350	.006		1.44	.33		1.77	2.11
0410	Foil faced fiberglass, 3-1/2" thick, R13, 11" wide	G	1150	.007		.70	.39		1.09	1.39
0420	15" wide	G	1350	.006		.70	.33		1.03	1.30
0440	23" wide	G 1 Carp	1600	.005	S.F.	.70	.28		.98	1.22
0442	R15, 11" wide	G	1150	.007		.60	.39		.99	1.28
0444	15" wide	G	1350	.006		.60	.33		.93	1.19
0446	23" wide	G	1600	.005		.60	.28		.88	1.11
0448	6" thick, R19, 11" wide	G	1150	.007		.97	.39		1.36	1.69
0460	15" wide	G	1350	.006		.97	.33		1.30	1.60
0480	23" wide	G	1600	.005		.97	.28		1.25	1.52
0482	R21, 11" wide	G	1150	.007		.84	.39		1.23	1.54
0484	15" wide	G	1350	.006		.84	.33		1.17	1.45

07 21 16 – Blanket Insulation

07 21 16.20 Blanket Insulation for Walls

		Crew	Daily Output	Labor-Hours	Unit	Material	2022 Bare Costs Labor	Equipment	Total	Total Incl O&P
0486	23" wide	G	1 Carp	1600	.005	S.F.	.84	.28		1.12	1.37
0488	9" thick, R30, 11" wide	G		985	.008		1.54	.46		2	2.42
0500	15" wide	G		1150	.007		1.54	.39		1.93	2.31
0550	23" wide	G		1350	.006		1.54	.33		1.87	2.22
0560	12" thick, R38, 11" wide	G		985	.008		1.44	.46		1.90	2.31
0570	15" wide	G		1150	.007		1.44	.39		1.83	2.20
0580	23" wide	G		1350	.006		1.44	.33		1.77	2.11
0620	Unfaced fiberglass, 3-1/2" thick, R13, 11" wide	G		1150	.007		.54	.39		.93	1.21
0820	15" wide	G		1350	.006		.54	.33		.87	1.12
0830	23" wide	G		1600	.005		.54	.28		.82	1.04
0832	R15, 11" wide	G		1150	.007		.75	.39		1.14	1.45
0836	23" wide	G		1600	.005		.75	.28		1.03	1.28
0838	6" thick, R19, 11" wide	G		1150	.007		.78	.39		1.17	1.48
0860	15" wide	G		1150	.007		.78	.39		1.17	1.48
0880	23" wide	G		1350	.006		.78	.33		1.11	1.39
0882	R21, 11" wide	G		1150	.007		.97	.39		1.36	1.69
0886	15" wide	G		1350	.006		.97	.33		1.30	1.60
0888	23" wide	G		1600	.005		.97	.28		1.25	1.52
0890	9" thick, R30, 11" wide	G		985	.008		1.13	.46		1.59	1.97
0900	15" wide	G		1150	.007		1.13	.39		1.52	1.86
0920	23" wide	G		1350	.006		1.13	.33		1.46	1.77
0930	12" thick, R38, 11" wide	G		985	.008		1.66	.46		2.12	2.56
0940	15" wide	G		1000	.008		1.66	.45		2.11	2.55
0960	23" wide	G		1150	.007		1.66	.39		2.05	2.45
1300	Wall or ceiling insulation, mineral wool batts										
1320	3-1/2" thick, R15	G	1 Carp	1600	.005	S.F.	1.04	.28		1.32	1.59
1340	5-1/2" thick, R23	G		1600	.005		1.63	.28		1.91	2.25
1380	7-1/4" thick, R30	G		1350	.006		2.15	.33		2.48	2.90
1700	Non-rigid insul., recycled blue cotton fiber, unfaced batts, R13, 16" wide	G		1600	.005		.86	.28		1.14	1.40
1710	R19, 16" wide	G		1600	.005		1.29	.28		1.57	1.87
1850	Friction fit wire insulation supports, 16" OC		1 Carp	960	.008	Ea.	.13	.47		.60	.89
9000	Minimum labor/equipment charge		"	4	2	Job		113		113	179

07 21 19 – Foamed In Place Insulation

07 21 19.10 Masonry Foamed In Place Insulation

		Crew	Daily Output	Labor-Hours	Unit	Material	2022 Bare Costs Labor	Equipment	Total	Total Incl O&P
0010	**MASONRY FOAMED IN PLACE INSULATION**										
0100	Amino-plast foam, injected into block core, 6" block	G	G-2A	6000	.004	Ea.	.22	.18	.11	.51	.66
0110	8" block	G		5000	.005		.26	.21	.13	.60	.80
0120	10" block	G		4000	.006		.33	.26	.17	.76	.98
0130	12" block	G		3000	.008		.44	.35	.22	1.01	1.32
0140	Injected into cavity wall	G		13000	.002	B.F.	.08	.08	.05	.21	.29
0150	Preparation, drill holes into mortar joint every 4 V.L.F., 5/8" diameter		1 Clab	960	.008	Ea.		.38		.38	.60
0160	7/8" diameter			680	.012			.54		.54	.85
0170	Patch drilled holes, 5/8" diameter			1800	.004		.04	.20		.24	.36
0180	7/8" diameter			1200	.007		.05	.30		.35	.54
9000	Minimum labor/equipment charge		G-2A	2	12	Job		530	330	860	1,250
9010	Minimum labor/equipment charge		1 Clab	4	2	"		91		91	145

07 21 23 – Loose-Fill Insulation

07 21 23.10 Poured Loose-Fill Insulation

		Crew	Daily Output	Labor-Hours	Unit	Material	2022 Bare Costs Labor	Equipment	Total	Total Incl O&P
0010	**POURED LOOSE-FILL INSULATION**										
0020	Cellulose fiber, R3.8 per inch	G	1 Carp	200	.040	C.F.	.83	2.25		3.08	4.50
0021	4" thick	G		1000	.008	S.F.	.20	.45		.65	.94
0022	6" thick	G		800	.010	"	.34	.56		.90	1.26

07 21 Thermal Insulation

07 21 23 – Loose-Fill Insulation

07 21 23.10 Poured Loose-Fill Insulation

		Crew	Daily Output	Labor-Hours	Unit	Material	2022 Bare Costs Labor	Equipment	Total	Total Incl O&P
0080	Fiberglass wool, R4 per inch	G 1 Carp	200	.040	C.F.	.71	2.25		2.96	4.37
0081	4" thick	G	600	.013	S.F.	.24	.75		.99	1.46
0082	6" thick	G	400	.020	"	.35	1.13		1.48	2.17
0100	Mineral wool, R3 per inch	G	200	.040	C.F.	.67	2.25		2.92	4.32
0101	4" thick	G	600	.013	S.F.	.22	.75		.97	1.43
0102	6" thick	G	400	.020	"	.34	1.13		1.47	2.16
0300	Polystyrene, R4 per inch	G	200	.040	C.F.	1.78	2.25		4.03	5.55
0301	4" thick	G	600	.013	S.F.	.59	.75		1.34	1.84
0302	6" thick	G	400	.020	"	.89	1.13		2.02	2.77
0400	Perlite, R2.78 per inch	G	200	.040	C.F.	3.15	2.25		5.40	7.05
0401	4" thick	G	1000	.008	S.F.	1.05	.45		1.50	1.87
0402	6" thick	G	800	.010	"	1.58	.56		2.14	2.62
9000	Minimum labor/equipment charge		4	2	Job		113		113	179

07 21 23.20 Masonry Loose-Fill Insulation

		Crew	Daily Output	Labor-Hours	Unit	Material	2022 Bare Costs Labor	Equipment	Total	Total Incl O&P
0010	**MASONRY LOOSE-FILL INSULATION**, vermiculite or perlite									
0100	In cores of concrete block, 4" thick wall, .115 C.F./S.F.	G D-1	4800	.003	S.F.	.36	.17		.53	.67
0200	6" thick wall, .175 C.F./S.F.	G "	3000	.005	"	.55	.27		.82	1.04
0300	8" thick wall, .258 C.F./S.F.	G D-1	2400	.007	S.F.	.81	.33		1.14	1.43
0400	10" thick wall, .340 C.F./S.F.	G	1850	.009		1.07	.43		1.50	1.87
0500	12" thick wall, .422 C.F./S.F.	G	1200	.013		1.33	.67		2	2.53
0600	Poured cavity wall, vermiculite or perlite, water repellent	G	250	.064	C.F.	3.15	3.21		6.36	8.60
0700	Foamed in place, urethane in 2-5/8" cavity	G G-2A	1035	.023	S.F.	1.84	1.02	.64	3.50	4.45
0800	For each 1" added thickness, add	G "	2372	.010	"	.70	.45	.28	1.43	1.83

07 21 26 – Blown Insulation

07 21 26.10 Blown Insulation

		Crew	Daily Output	Labor-Hours	Unit	Material	2022 Bare Costs Labor	Equipment	Total	Total Incl O&P
0010	**BLOWN INSULATION** Ceilings, with open access									
0020	Cellulose, 3-1/2" thick, R13	G G-4	5000	.005	S.F.	.29	.22	.11	.62	.79
0030	5-3/16" thick, R19	G	3800	.006		.42	.29	.14	.85	1.09
0050	6-1/2" thick, R22	G	3000	.008		.54	.37	.18	1.09	1.39
0100	8-11/16" thick, R30	G	2600	.009		.73	.43	.21	1.37	1.71
0120	10-7/8" thick, R38	G	1800	.013		.93	.62	.30	1.85	2.34
1000	Fiberglass, 5.5" thick, R11	G	3800	.006		.24	.29	.14	.67	.89
1050	6" thick, R12	G	3000	.008		.35	.37	.18	.90	1.17
1100	8.8" thick, R19	G	2200	.011		.43	.50	.25	1.18	1.55
1200	10" thick, R22	G	1800	.013		.50	.62	.30	1.42	1.86
1300	11.5" thick, R26	G	1500	.016		.60	.74	.36	1.70	2.24
1350	13" thick, R30	G	1400	.017		.69	.79	.39	1.87	2.45
1450	16" thick, R38	G	1145	.021		.88	.97	.48	2.33	3.03
1500	20" thick, R49	G	920	.026		1.17	1.21	.59	2.97	3.86
9000	Minimum labor/equipment charge		4	6	Job		278	136	414	590

07 21 29 – Sprayed Insulation

07 21 29.10 Sprayed-On Insulation

		Crew	Daily Output	Labor-Hours	Unit	Material	2022 Bare Costs Labor	Equipment	Total	Total Incl O&P
0010	**SPRAYED-ON INSULATION**									
0020	Fibrous/cementitious, finished wall, 1" thick, R3.7	G G-2	2050	.012	S.F.	.46	.56	.07	1.09	1.46
0100	Attic, 5.2" thick, R19	G	1550	.015	"	.54	.74	.09	1.37	1.86
0200	Fiberglass, R4 per inch, vertical	G	1600	.015	B.F.	.26	.71	.09	1.06	1.51
0210	Horizontal	G	1200	.020	"	.26	.95	.12	1.33	1.92
0300	Closed cell, spray polyurethane foam, 2 lb./C.F. density	G								
0310	1" thick	G G-2A	6000	.004	S.F.	.70	.18	.11	.99	1.19
0320	2" thick	G	3000	.008		1.41	.35	.22	1.98	2.38
0330	3" thick	G	2000	.012		2.11	.53	.33	2.97	3.57

07 21 Thermal Insulation

07 21 29 – Sprayed Insulation

07 21 29.10 Sprayed-On Insulation		Crew	Daily Output	Labor-Hours	Unit	Material	2022 Bare Costs Labor	Equipment	Total	Total Incl O&P
0335	3-1/2" thick	G	G-2A	1715	.014	S.F.	2.46	.62	.39	3.47	4.17
0340	4" thick	G		1500	.016		2.81	.71	.44	3.96	4.75
0350	5" thick	G		1200	.020		3.51	.88	.55	4.94	5.95
0355	5-1/2" thick	G		1090	.022		3.86	.97	.61	5.44	6.55
0360	6" thick	G	G-2A	1000	.024	S.F.	4.22	1.06	.66	5.94	7.15
9000	Minimum labor/equipment charge		G-2	2	12	Job		570	69.50	639.50	980

07 21 53 – Reflective Insulation

07 21 53.10 Reflective Insulation Options

			Crew	Daily Output	Labor-Hours	Unit	Material	Labor	Equipment	Total	Total Incl O&P
0010	**REFLECTIVE INSULATION OPTIONS**										
0020	Aluminum foil on reinforced scrim	G	1 Carp	19	.421	C.S.F.	19.05	23.50		42.55	58.50
0100	Reinforced with woven polyolefin	G		19	.421		30	23.50		53.50	71
0500	With single bubble air space, R8.8	G		15	.533		33.50	30		63.50	84.50
0600	With double bubble air space, R9.8	G		15	.533		40.50	30		70.50	92
9000	Minimum labor/equipment charge			4	2	Job		113		113	179

07 22 Roof and Deck Insulation

07 22 16 – Roof Board Insulation

07 22 16.10 Roof Deck Insulation

			Crew	Daily Output	Labor-Hours	Unit	Material	Labor	Equipment	Total	Total Incl O&P
0010	**ROOF DECK INSULATION**, fastening excluded										
0016	Asphaltic cover board, fiberglass lined, 1/8" thick		1 Rofc	1400	.006	S.F.	.50	.28		.78	1.04
0018	1/4" thick			1400	.006		1.02	.28		1.30	1.61
0020	Fiberboard low density, 1/2" thick, R1.39	G		1300	.006		.38	.31		.69	.95
0030	1" thick, R2.78	G		1040	.008		.78	.38		1.16	1.52
0080	1-1/2" thick, R4.17	G		1040	.008		1.18	.38		1.56	1.96
0100	2" thick, R5.56	G		1040	.008		1.13	.38		1.51	1.90
0110	Fiberboard high density, 1/2" thick, R1.3	G		1300	.006		.41	.31		.72	.98
0120	1" thick, R2.5	G		1040	.008		.67	.38		1.05	1.40
0130	1-1/2" thick, R3.8	G		1040	.008		1.42	.38		1.80	2.22
0200	Fiberglass, 3/4" thick, R2.78	G		1300	.006		.64	.31		.95	1.23
0400	15/16" thick, R3.70	G		1300	.006		.88	.31		1.19	1.50
0460	1-1/16" thick, R4.17	G		1300	.006		1.23	.31		1.54	1.88
0600	1-5/16" thick, R5.26	G		1300	.006		1.53	.31		1.84	2.21
0650	2-1/16" thick, R8.33	G		1040	.008		1.72	.38		2.10	2.55
0700	2-7/16" thick, R10	G		1040	.008		1.84	.38		2.22	2.68
0800	Gypsum cover board, fiberglass mat facer, 1/4" thick			1400	.006		.52	.28		.80	1.06
0810	1/2" thick			1300	.006		.69	.31		1	1.29
0820	5/8" thick			1200	.007		.68	.33		1.01	1.32
0830	Primed fiberglass mat facer, 1/4" thick			1400	.006		.63	.28		.91	1.18
0840	1/2" thick			1300	.006		.71	.31		1.02	1.31
0850	5/8" thick			1200	.007		.73	.33		1.06	1.37
1650	Perlite, 1/2" thick, R1.32	G		1365	.006		.31	.29		.60	.84
1655	3/4" thick, R2.08	G		1040	.008		.40	.38		.78	1.10
1660	1" thick, R2.78	G		1040	.008		.62	.38		1	1.34
1670	1-1/2" thick, R4.17	G	1 Rofc	1040	.008	S.F.	.88	.38		1.26	1.63
1680	2" thick, R5.56	G		910	.009		1.10	.44		1.54	1.96
1685	2-1/2" thick, R6.67	G		910	.009		1.41	.44		1.85	2.30
1690	Tapered for drainage	G		1040	.008	B.F.	1.21	.38		1.59	1.99
1700	Polyisocyanurate, 2#/C.F. density, 3/4" thick	G		1950	.004	S.F.	.65	.20		.85	1.07
1705	1" thick	G		1820	.004		.84	.22		1.06	1.30
1715	1-1/2" thick	G		1625	.005		1.03	.24		1.27	1.55

07 22 Roof and Deck Insulation

07 22 16 – Roof Board Insulation

07 22 16.10 Roof Deck Insulation		Crew	Daily Output	Labor-Hours	Unit	Material	2022 Bare Costs Labor	Equipment	Total	Total Incl O&P
1725	2" thick	G 1 Rofc	1430	.006	S.F.	1.36	.28		1.64	1.98
1735	2-1/2" thick	G	1365	.006		1.72	.29		2.01	2.39
1745	3" thick	G	1300	.006		1.13	.31		1.44	1.77
1755	3-1/2" thick	G	1300	.006		2.55	.31		2.86	3.34
1765	Tapered for drainage	G	1820	.004	B.F.	1.03	.22		1.25	1.51
1900	Extruded polystyrene									
1910	15 psi compressive strength, 1" thick, R5	G 1 Rofc	1950	.004	S.F.	.58	.20		.78	.99
1920	2" thick, R10	G	1625	.005		.75	.24		.99	1.25
1930	3" thick, R15	G	1300	.006		1.51	.31		1.82	2.19
1932	4" thick, R20	G	1300	.006		2.03	.31		2.34	2.76
1934	Tapered for drainage	G	1950	.004	B.F.	.52	.20		.72	.92
1940	25 psi compressive strength, 1" thick, R5	G	1950	.004	S.F.	.79	.20		.99	1.22
1942	2" thick, R10		1625	.005		1.50	.24		1.74	2.07
1944	3" thick, R15		1300	.006		2.29	.31		2.60	3.05
1946	4" thick, R20		1300	.006		3.16	.31		3.47	4.01
1948	Tapered for drainage	G	1950	.004	B.F.	.59	.20		.79	1
1950	40 psi compressive strength, 1" thick, R5	G	1950	.004	S.F.	.88	.20		1.08	1.32
1952	2" thick, R10	G	1625	.005		1.67	.24		1.91	2.26
1954	3" thick, R15	G	1300	.006		2.42	.31		2.73	3.19
1956	4" thick, R20	G	1300	.006		3.17	.31		3.48	4.01
1958	Tapered for drainage	G	1820	.004	B.F.	.89	.22		1.11	1.36
1960	60 psi compressive strength, 1" thick, R5	G	1885	.004	S.F.	1.10	.21		1.31	1.57
1962	2" thick, R10	G	1560	.005		2.09	.25		2.34	2.74
1964	3" thick, R15	G	1270	.006		3.41	.31		3.72	4.29
1966	4" thick, R20	G	1235	.006		4.24	.32		4.56	5.20
1968	Tapered for drainage	G	1820	.004	B.F.	1.01	.22		1.23	1.49
2010	Expanded polystyrene, 1#/C.F. density, 3/4" thick, R2.89	G	1950	.004	S.F.	.21	.20		.41	.58
2020	1" thick, R3.85	G	1950	.004		.28	.20		.48	.66
2100	2" thick, R7.69	G	1625	.005		.56	.24		.80	1.04
2110	3" thick, R11.49	G	1625	.005		.84	.24		1.08	1.34
2120	4" thick, R15.38	G	1625	.005		1.12	.24		1.36	1.65
2130	5" thick, R19.23	G 1 Rofc	1495	.005	S.F.	1.40	.27		1.67	2
2140	6" thick, R23.26	G	1495	.005	"	1.68	.27		1.95	2.31
2150	Tapered for drainage	G	1950	.004	B.F.	.46	.20		.66	.86
2400	Composites with 2" EPS									
2410	1" fiberboard	G 1 Rofc	1325	.006	S.F.	1.59	.30		1.89	2.27
2420	7/16" oriented strand board	G	1040	.008		1.41	.38		1.79	2.21
2430	1/2" plywood	G	1040	.008		3.28	.38		3.66	4.27
2440	1" perlite	G	1040	.008		1.07	.38		1.45	1.84
2450	Composites with 1-1/2" polyisocyanurate									
2460	1" fiberboard	G 1 Rofc	1040	.008	S.F.	1.23	.38		1.61	2.01
2470	1" perlite	G	1105	.007		1.02	.36		1.38	1.74
2480	7/16" oriented strand board	G	1040	.008		2.75	.38		3.13	3.69
3000	Fastening alternatives, coated screws, 2" long		3744	.002	Ea.	.05	.11		.16	.24
3010	4" long		3120	.003		.11	.13		.24	.34
3020	6" long		2675	.003		.24	.15		.39	.52
3030	8" long		2340	.003		.29	.17		.46	.61
3040	10" long		1872	.004		.53	.21		.74	.94
3050	Pre-drill and drive wedge spike, 2-1/2"		1248	.006		.64	.32		.96	1.25
3060	3-1/2"		1101	.007		1.42	.36		1.78	2.18
3070	4-1/2"		936	.009		1.99	.42		2.41	2.92
3075	3" galvanized deck plates		7488	.001		.23	.05		.28	.34
3080	Spot mop asphalt	G-1	295	.190	Sq.	5.50	8.80	1.87	16.17	23.50

For customer support on your Commercial Renovation Costs with RSMeans data, call 800.448.8182.

195

07 22 Roof and Deck Insulation

07 22 16 - Roof Board Insulation

07 22 16.10 Roof Deck Insulation		Crew	Daily Output	Labor-Hours	Unit	Material	2022 Bare Costs Labor	Equipment	Total	Total Incl O&P
3090	Full mop asphalt	G-1	192	.292	Sq.	11	13.50	2.87	27.37	39
3110	Low-rise polyurethane adhesive, 12" OC beads	1 Rofc	45	.178		97	8.80		105.80	122
3120	6" OC beads		32	.250		194	12.40		206.40	235
3130	4" OC beads		30	.267		294	13.25		307.25	350
9000	Minimum labor/equipment charge		3.25	2.462	Job		122		122	210

07 24 Exterior Insulation and Finish Systems

07 24 13 - Polymer-Based Exterior Insulation and Finish System

07 24 13.10 Exterior Insulation and Finish Systems

			Crew	Daily Output	Labor-Hours	Unit	Material	2022 Bare Costs Labor	Equipment	Total	Total Incl O&P
0010	**EXTERIOR INSULATION AND FINISH SYSTEMS**										
0095	Field applied, 1" EPS insulation	G	J-1	390	.103	S.F.	.40	5.05	.29	5.74	8.75
0100	With 1/2" cement board sheathing	G		268	.149		1.13	7.35	.42	8.90	13.35
0105	2" EPS insulation	G		390	.103		1.59	5.05	.29	6.93	10.05
0110	With 1/2" cement board sheathing	G		268	.149		1.41	7.35	.42	9.18	13.65
0115	3" EPS insulation	G		390	.103		.96	5.05	.29	6.30	9.40
0120	With 1/2" cement board sheathing	G		268	.149		1.69	7.35	.42	9.46	13.95
0125	4" EPS insulation	G		390	.103		1.24	5.05	.29	6.58	9.70
0130	With 1/2" cement board sheathing	G	J-1	268	.149	S.F.	2.70	7.35	.42	10.47	15.10
0140	Premium finish add		"	1265	.032		.45	1.56	.09	2.10	3.07
0145	Drainage and ventilation cavity, add		2 Plas	1450	.011		.84	.57		1.41	1.82
0150	Heavy duty reinforcement add		J-1	914	.044		.60	2.15	.12	2.87	4.21
0160	2.5#/S.Y. metal lath substrate add		1 Lath	75	.107	S.Y.	5.05	5.95		11	14.85
0170	3.4#/S.Y. metal lath substrate add		"	75	.107	"	5.15	5.95		11.10	15
0180	Color or texture change		J-1	1265	.032	S.F.	.76	1.56	.09	2.41	3.41
0190	With substrate leveling base coat		1 Plas	530	.015		.87	.78		1.65	2.19
0210	With substrate sealing base coat		1 Pord	1224	.007		.14	.31		.45	.64
0370	V groove shape in panel face					L.F.	.74			.74	.81
0380	U groove shape in panel face					"	.89			.89	.98
0433	Crack repair, acrylic rubber, fluid applied, 20 mils thick		1 Plas	350	.023	S.F.	1.60	1.17		2.77	3.62
0437	50 mils thick, reinforced		"	200	.040	"	2.85	2.06		4.91	6.40
0440	For higher than one story, add							25%			

07 25 Weather Barriers

07 25 10 - Weather Barriers or Wraps

07 25 10.10 Weather Barriers

		Crew	Daily Output	Labor-Hours	Unit	Material	2022 Bare Costs Labor	Equipment	Total	Total Incl O&P
0010	**WEATHER BARRIERS**									
0400	Asphalt felt paper, #15	1 Carp	37	.216	Sq.	4.43	12.15		16.58	24
0401	Per square foot	"	3700	.002	S.F.	.04	.12		.16	.24
0450	Housewrap, exterior, spun bonded polypropylene									
0470	Small roll	1 Carp	3800	.002	S.F.	.14	.12		.26	.35
0480	Large roll	"	4000	.002	"	.06	.11		.17	.25
2100	Asphalt felt roof deck vapor barrier, class 1 metal decks	1 Rofc	37	.216	Sq.	19.85	10.70		30.55	40.50
2200	For all other decks	"	37	.216		15.40	10.70		26.10	35.50
2800	Asphalt felt, 50% recycled content, 15 lb., 4 sq./roll	1 Carp	36	.222		5.70	12.50		18.20	26
2810	30 lb., 2 sq./roll	"	36	.222		12.75	12.50		25.25	34
3000	Building wrap, spun bonded polyethylene	2 Carp	8000	.002	S.F.	.18	.11		.29	.38
9960	Minimum labor/equipment charge	1 Carp	2	4	Job		225		225	360

07 26 13 – Above-Grade Vapor Retarders

07 26 13.10 Vapor Retarders	Crew	Daily Output	Labor-Hours	Unit	Material	2022 Bare Costs Labor	Equipment	Total	Total Incl O&P
0010 **VAPOR RETARDERS**									
0020 Aluminum and kraft laminated, foil 1 side [G]	1 Carp	37	.216	Sq.	12	12.15		24.15	32.50
0100 Foil 2 sides [G]		37	.216		17.55	12.15		29.70	38.50
0600 Polyethylene vapor barrier, standard, 2 mil [G]		37	.216		2	12.15		14.15	21.50
0700 4 mil [G]		37	.216		4.40	12.15		16.55	24
0900 6 mil [G]		37	.216		6	12.15		18.15	26
1200 10 mil [G]	1 Carp	37	.216	Sq.	8.15	12.15		20.30	28.50
1300 Clear reinforced, fire retardant, 8 mil [G]		37	.216		25.50	12.15		37.65	47.50
1350 Cross laminated type, 3 mil [G]		37	.216		12.30	12.15		24.45	33
1400 4 mil [G]		37	.216		13.65	12.15		25.80	34.50
1800 Reinf. waterproof, 2 mil polyethylene backing, 1 side		37	.216		11.50	12.15		23.65	32
1900 2 sides		37	.216		14.50	12.15		26.65	35.50
2400 Waterproofed kraft with sisal or fiberglass fibers		37	.216		18.05	12.15		30.20	39
9950 Minimum labor/equipment charge		4	2	Job		113		113	179

07 27 Air Barriers

07 27 13 – Modified Bituminous Sheet Air Barriers

07 27 13.10 Modified Bituminous Sheet Air Barrier

	Crew	Daily Output	Labor-Hours	Unit	Material	Labor	Equipment	Total	Total Incl O&P
0010 **MODIFIED BITUMINOUS SHEET AIR BARRIER**									
0100 SBS modified sheet laminated to polyethylene sheet, 40 mils, 4" wide	1 Carp	1200	.007	L.F.	.32	.38		.70	.95
0120 6" wide		1100	.007		.43	.41		.84	1.12
0140 9" wide		1000	.008		.60	.45		1.05	1.38
0160 12" wide		900	.009		.77	.50		1.27	1.64
0180 18" wide	2 Carp	1700	.009	S.F.	.72	.53		1.25	1.63
0200 36" wide	"	1800	.009		.70	.50		1.20	1.57
0220 Adhesive for above	1 Carp	1400	.006			.32		.32	.51

07 27 26 – Fluid-Applied Membrane Air Barriers

07 27 26.10 Fluid Applied Membrane Air Barrier

	Crew	Daily Output	Labor-Hours	Unit	Material	Labor	Equipment	Total	Total Incl O&P
0010 **FLUID APPLIED MEMBRANE AIR BARRIER**									
0100 Spray applied vapor barrier, 25 S.F./gallon	1 Pord	1375	.006	S.F.	.02	.28		.30	.46

07 31 Shingles and Shakes

07 31 13 – Asphalt Shingles

07 31 13.10 Asphalt Roof Shingles

	Crew	Daily Output	Labor-Hours	Unit	Material	Labor	Equipment	Total	Total Incl O&P
0010 **ASPHALT ROOF SHINGLES**									
0100 Standard strip shingles									
0150 Inorganic, class A, 25 year	1 Rofc	5.50	1.455	Sq.	96	72		168	230
0155 Pneumatic nailed		7	1.143		96	56.50		152.50	204
0200 30 year		5	1.600		126	79.50		205.50	276
0205 Pneumatic nailed		6.25	1.280		126	63.50		189.50	248
0250 Standard laminated multi-layered shingles									
0300 Class A, 240-260 lb./square	1 Rofc	4.50	1.778	Sq.	123	88		211	287
0305 Pneumatic nailed		5.63	1.422		123	70.50		193.50	256
0350 Class A, 250-270 lb./square		4	2		123	99		222	305
0355 Pneumatic nailed		5	1.600		123	79.50		202.50	272
0400 Premium, laminated multi-layered shingles									
0450 Class A, 260-300 lb./square	1 Rofc	3.50	2.286	Sq.	160	113		273	370
0455 Pneumatic nailed	1 Rofc	4.37	1.831	Sq.	160	91		251	330
0500 Class A, 300-385 lb./square		3	2.667		415	132		547	690

For customer support on your Commercial Renovation Costs with RSMeans data, call 800.448.8182.

197

07 31 Shingles and Shakes

07 31 13 – Asphalt Shingles

07 31 13.10 Asphalt Roof Shingles		Crew	Daily Output	Labor-Hours	Unit	Material	2022 Bare Costs Labor	Equipment	Total	Total Incl O&P
0505	Pneumatic nailed	1 Rofc	3.75	2.133	Sq.	415	106		521	640
0800	#15 felt underlayment		64	.125		4.43	6.20		10.63	15.50
0825	#30 felt underlayment		58	.138		8.70	6.85		15.55	21.50
0850	Self adhering polyethylene and rubberized asphalt underlayment		22	.364		66	18.05		84.05	104
0900	Ridge shingles		330	.024	L.F.	2.87	1.20		4.07	5.25
0905	Pneumatic nailed		412.50	.019	"	2.87	.96		3.83	4.82
1000	For steep roofs (7 to 12 pitch or greater), add						50%			
9000	Minimum labor/equipment charge	1 Rofc	3	2.667	Job		132		132	228

07 31 16 – Metal Shingles

07 31 16.10 Aluminum Shingles

		Crew	Daily Output	Labor-Hours	Unit	Material	2022 Bare Costs Labor	Equipment	Total	Total Incl O&P
0010	**ALUMINUM SHINGLES**									
0020	Mill finish, .019" thick	1 Carp	5	1.600	Sq.	283	90		373	455
0100	.020" thick	"	5	1.600		340	90		430	520
0300	For colors, add					25.50			25.50	28
0600	Ridge cap, .024" thick	1 Carp	170	.047	L.F.	3.75	2.65		6.40	8.35
0700	End wall flashing, .024" thick		170	.047		3.20	2.65		5.85	7.75
0900	Valley section, .024" thick		170	.047		4.20	2.65		6.85	8.85
1000	Starter strip, .024" thick		400	.020		3.20	1.13		4.33	5.30
1200	Side wall flashing, .024" thick		170	.047		3.83	2.65		6.48	8.40
1500	Gable flashing, .024" thick		400	.020		2.43	1.13		3.56	4.46
9000	Minimum labor/equipment charge		3	2.667	Job		150		150	239

07 31 16.20 Steel Shingles

		Crew	Daily Output	Labor-Hours	Unit	Material	2022 Bare Costs Labor	Equipment	Total	Total Incl O&P
0010	**STEEL SHINGLES**									
0012	Galvanized, 26 ga.	1 Rots	2.20	3.636	Sq.	293	180		473	630
0200	24 ga.	"	2.20	3.636		435	180		615	790
0300	For colored galvanized shingles, add					61.50			61.50	67.50
9000	Minimum labor/equipment charge	1 Rots	3	2.667	Job		132		132	228

07 31 26 – Slate Shingles

07 31 26.10 Slate Roof Shingles

			Crew	Daily Output	Labor-Hours	Unit	Material	2022 Bare Costs Labor	Equipment	Total	Total Incl O&P
0010	**SLATE ROOF SHINGLES**										
0100	Buckingham Virginia black, 3/16" - 1/4" thick	G	1 Rots	1.75	4.571	Sq.	690	227		917	1,150
0200	1/4" thick	G		1.75	4.571		690	227		917	1,150
0900	Pennsylvania black, Bangor, #1 clear	G		1.75	4.571		620	227		847	1,075
1200	Vermont, unfading, green, mottled green	G		1.75	4.571		630	227		857	1,075
1300	Semi-weathering green & gray	G		1.75	4.571		505	227		732	945
1400	Purple	G		1.75	4.571		620	227		847	1,075
1500	Black or gray	G		1.75	4.571		540	227		767	985
2500	Slate roof repair, extensive replacement		1 Rots	1	8	Sq.	755	395		1,150	1,525
2600	Repair individual pieces, scattered			19	.421	Ea.	8.90	21		29.90	46
2700	Ridge shingles, slate			200	.040	L.F.	12.70	1.98		14.68	17.35
9000	Minimum labor/equipment charge			3	2.667	Job		132		132	228

07 31 29 – Wood Shingles and Shakes

07 31 29.13 Wood Shingles

			Crew	Daily Output	Labor-Hours	Unit	Material	2022 Bare Costs Labor	Equipment	Total	Total Incl O&P
0010	**WOOD SHINGLES**	R061110-30									
0012	16" No. 1 red cedar shingles, 5" exposure, on roof		1 Carp	2.50	3.200	Sq.	465	180		645	795
0015	Pneumatic nailed			3.25	2.462		465	139		604	730
0200	7-1/2" exposure, on walls			2.05	3.902		310	220		530	690
0205	Pneumatic nailed			2.67	2.996		310	169		479	610
0300	18" No. 1 red cedar perfections, 5-1/2" exposure, on roof			2.75	2.909		465	164		629	770
0305	Pneumatic nailed			3.57	2.241		465	126		591	710
0500	7-1/2" exposure, on walls			2.25	3.556		340	200		540	695

07 31 Shingles and Shakes

07 31 29 – Wood Shingles and Shakes

07 31 29.13 Wood Shingles

		Crew	Daily Output	Labor-Hours	Unit	Material	2022 Bare Costs Labor	2022 Bare Costs Equipment	Total	Total Incl O&P
0505	Pneumatic nailed	1 Carp	2.92	2.740	Sq.	340	154		494	620
0600	Resquared and rebutted, 5-1/2" exposure, on roof		3	2.667		360	150		510	635
0605	Pneumatic nailed		3.90	2.051		360	115		475	580
0900	7-1/2" exposure, on walls		2.45	3.265		264	184		448	580
0905	Pneumatic nailed	▼	3.18	2.516	▼	264	142		406	515
1000	Add to above for fire retardant shingles					72			72	79
1060	Preformed ridge shingles	1 Carp	400	.020	L.F.	4.85	1.13		5.98	7.15
2000	White cedar shingles, 16" long, extras, 5" exposure, on roof		2.40	3.333	Sq.	160	188		348	475
2005	Pneumatic nailed		3.12	2.564		160	144		304	405
2050	5" exposure on walls		2	4		160	225		385	535
2055	Pneumatic nailed		2.60	3.077		160	173		333	450
2100	7-1/2" exposure, on walls		2	4		114	225		339	485
2105	Pneumatic nailed		2.60	3.077		114	173		287	400
2150	"B" grade, 5" exposure on walls		2	4		180	225		405	560
2155	Pneumatic nailed		2.60	3.077		180	173		353	475
2300	For #15 organic felt underlayment on roof, 1 layer, add		64	.125		4.43	7.05		11.48	16.05
2400	2 layers, add	▼	32	.250	▼	8.85	14.10		22.95	32.50
2600	For steep roofs (7/12 pitch or greater), add to above						50%			
2700	Panelized systems, No.1 cedar shingles on 5/16" CDX plywood									
2800	On walls, 8' strips, 7" or 14" exposure	2 Carp	700	.023	S.F.	9.30	1.29		10.59	12.25
3500	On roofs, 8' strips, 7" or 14" exposure	1 Carp	3	2.667	Sq.	705	150		855	1,025
3505	Pneumatic nailed		4	2	"	705	113		818	960
9000	Minimum labor/equipment charge	▼	3	2.667	Job		150		150	239

07 31 29.16 Wood Shakes

		Crew	Daily Output	Labor-Hours	Unit	Material	2022 Bare Costs Labor	2022 Bare Costs Equipment	Total	Total Incl O&P
0010	**WOOD SHAKES**									
1100	Hand-split red cedar shakes, 1/2" thick x 24" long, 10" exp. on roof	1 Carp	2.50	3.200	Sq.	500	180		680	835
1105	Pneumatic nailed		3.25	2.462		500	139		639	770
1110	3/4" thick x 24" long, 10" exp. on roof		2.25	3.556		500	200		700	870
1115	Pneumatic nailed		2.92	2.740		500	154		654	795
1200	1/2" thick, 18" long, 8-1/2" exp. on roof		2	4		415	225		640	815
1205	Pneumatic nailed		2.60	3.077		415	173		588	730
1210	3/4" thick x 18" long, 8-1/2" exp. on roof		1.80	4.444		415	250		665	855
1215	Pneumatic nailed		2.34	3.419		415	192		607	760
1255	10" exposure on walls		2	4		400	225		625	800
1260	10" exposure on walls, pneumatic nailed	▼	2.60	3.077		400	173		573	715
1700	Add to above for fire retardant shakes, 24" long					72			72	79
1800	18" long				▼	72			72	79
1810	Ridge shakes	1 Carp	350	.023	L.F.	5.40	1.29		6.69	8

07 32 Roof Tiles

07 32 13 – Clay Roof Tiles

07 32 13.10 Clay Tiles

		Crew	Daily Output	Labor-Hours	Unit	Material	2022 Bare Costs Labor	2022 Bare Costs Equipment	Total	Total Incl O&P
0010	**CLAY TILES**, including accessories									
0300	Flat shingle, interlocking, 15", 166 pcs./sq., fireflashed blend	3 Rots	6	4	Sq.	560	198		758	955
0500	Terra cotta red		6	4		660	198		858	1,075
0600	Roman pan and top, 18", 102 pcs./sq., fireflashed blend	▼	5.50	4.364		520	216		736	940
0640	Terra cotta red	1 Rots	2.40	3.333		605	165		770	950
1100	Barrel mission tile, 18", 166 pcs./sq., fireflashed blend	3 Rots	5.50	4.364		435	216		651	845
1140	Terra cotta red		5.50	4.364		440	216		656	855
1700	Scalloped edge flat shingle, 14", 145 pcs./sq., fireflashed blend		6	4		1,125	198		1,323	1,600
1800	Terra cotta red		6	4		1,125	198		1,323	1,575

For customer support on your Commercial Renovation Costs with RSMeans data, call 800.448.8182.

199

07 32 Roof Tiles

07 32 13 – Clay Roof Tiles

07 32 13.10 Clay Tiles	Crew	Daily Output	Labor-Hours	Unit	Material	2022 Bare Costs Labor	Equipment	Total	Total Incl O&P
3010 #15 felt underlayment	1 Rofc	64	.125	Sq.	4.43	6.20		10.63	15.50
3020 #30 felt underlayment		58	.138		8.70	6.85		15.55	21.50
3040 Polyethylene and rubberized asph. underlayment		22	.364		66	18.05		84.05	104
9010 Minimum labor/equipment charge	1 Rots	2	4	Job		198		198	340

07 32 16 – Concrete Roof Tiles

07 32 16.10 Concrete Tiles	Crew	Daily Output	Labor-Hours	Unit	Material	2022 Bare Costs Labor	Equipment	Total	Total Incl O&P
0010 **CONCRETE TILES**									
0020 Corrugated, 13" x 16-1/2", 90 per sq., 950 lb./sq.									
0050 Earthtone colors, nailed to wood deck	1 Rots	1.35	5.926	Sq.	119	294		413	635
0150 Blues		1.35	5.926		107	294		401	620
0200 Greens		1.35	5.926		118	294		412	635
0250 Premium colors		1.35	5.926		118	294		412	635
0500 Shakes, 13" x 16-1/2", 90 per sq., 950 lb./sq.									
0600 All colors, nailed to wood deck	1 Rots	1.50	5.333	Sq.	174	265		439	645
1500 Accessory pieces, ridge & hip, 10" x 16-1/2", 8 lb. each	"	120	.067	Ea.	3.78	3.31		7.09	9.85
1700 Rake, 6-1/2" x 16-3/4", 9 lb. each					3.78			3.78	4.16
1800 Mansard hip, 10" x 16-1/2", 9.2 lb. each					3.78			3.78	4.16
1900 Hip starter, 10" x 16-1/2", 10.5 lb. each					10.50			10.50	11.55
2000 3 or 4 way apex, 10" each side, 11.5 lb. each					12			12	13.20
9000 Minimum labor/equipment charge	1 Rots	3	2.667	Job		132		132	228

07 32 19 – Metal Roof Tiles

07 32 19.10 Metal Roof Tiles	Crew	Daily Output	Labor-Hours	Unit	Material	2022 Bare Costs Labor	Equipment	Total	Total Incl O&P
0010 **METAL ROOF TILES**									
0020 Accessories included, .032" thick aluminum, mission tile	1 Carp	2.50	3.200	Sq.	805	180		985	1,175
0200 Spanish tiles		3	2.667	"	570	150		720	865
9000 Minimum labor/equipment charge		3	2.667	Job		150		150	239

07 33 Natural Roof Coverings

07 33 63 – Vegetated Roofing

07 33 63.10 Green Roof Systems		Crew	Daily Output	Labor-Hours	Unit	Material	2022 Bare Costs Labor	Equipment	Total	Total Incl O&P
0010 **GREEN ROOF SYSTEMS**										
0020 Soil mixture for green roof 30% sand, 55% gravel, 15% soil										
0100 Hoist and spread soil mixture 4" depth up to 5 stories tall roof	G	B-13B	4000	.014	S.F.	.26	.69	.57	1.52	2.01
0150 6" depth	G		2667	.021		.39	1.04	.85	2.28	3.01
0200 8" depth	G		2000	.028		.52	1.39	1.13	3.04	4.01
0250 10" depth	G		1600	.035		.65	1.74	1.42	3.81	5
0300 12" depth	G		1335	.042		.78	2.08	1.70	4.56	6
0310 Alt. man-made soil mix, hoist & spread, 4" deep up to 5 stories tall roof	G		4000	.014		1.78	.69	.57	3.04	3.68
0350 Mobilization 55 ton crane to site	G	1 Eqhv	3.60	2.222	Ea.		141		141	219
0355 Hoisting cost to 5 stories per day (Avg. 28 picks per day)	G	B-13B	1	56	Day		2,775	2,275	5,050	6,875
0360 Mobilization or demobilization, 100 ton crane to site driver & escort	G	A-3E	2.50	6.400	Ea.		375	72	447	665
0365 Hoisting cost 6-10 stories per day (Avg. 21 picks per day)	G	B-13C	1	56	Day		2,775	1,400	4,175	5,925
0370 Hoist and spread soil mixture 4" depth 6-10 stories tall roof	G		4000	.014	S.F.	.26	.69	.35	1.30	1.78
0375 6" depth	G		2667	.021		.39	1.04	.53	1.96	2.65
0380 8" depth	G		2000	.028		.52	1.39	.70	2.61	3.53
0385 10" depth	G		1600	.035		.65	1.74	.88	3.27	4.42
0390 12" depth	G		1335	.042		.78	2.08	1.05	3.91	5.30
0400 Green roof edging treated lumber 4" x 4", no hoisting included	G	2 Carp	400	.040	L.F.	1.63	2.25		3.88	5.40
0410 4" x 6"	G		400	.040		2.83	2.25		5.08	6.70
0420 4" x 8"	G		360	.044		6.75	2.50		9.25	11.45

07 33 Natural Roof Coverings

07 33 63 – Vegetated Roofing

07 33 63.10 Green Roof Systems

		Crew	Daily Output	Labor-Hours	Unit	Material	2022 Bare Costs Labor	Equipment	Total	Total Incl O&P
0430	4" x 6" double stacked	[G] 2 Carp	300	.053	L.F.	5.65	3		8.65	11
0500	Green roof edging redwood lumber 4" x 4", no hoisting included	[G]	400	.040		12.35	2.25		14.60	17.20
0510	4" x 6"	[G]	400	.040		19.65	2.25		21.90	25
0520	4" x 8"	[G]	360	.044		37	2.50		39.50	44.50
0530	4" x 6" double stacked	[G]	300	.053		39.50	3		42.50	48
0550	Components, not including membrane or insulation:									
0560	Fluid applied rubber membrane, reinforced, 215 mil thick	[G] G-5	350	.114	S.F.	.27	5.15	.48	5.90	9.65
0570	Root barrier	[G] 2 Rofc	775	.021		.99	1.02		2.01	2.85
0580	Moisture retention barrier and reservoir	[G] "	900	.018		3.97	.88		4.85	5.90
0600	Planting sedum, light soil, potted, 2-1/4" diameter, 2 per S.F.	[G] 1 Clab	420	.019		15.95	.87		16.82	18.95
0610	1 per S.F.	[G] "	840	.010		8	.43		8.43	9.45
0630	Planting sedum mat per S.F. including shipping (4000 S.F. min)	[G] 4 Clab	4000	.008		19.15	.36		19.51	21.50
0640	Installation sedum mat system (no soil required) per S.F. (4000 S.F. min)	[G] "	4000	.008		26.50	.36		26.86	30
0645	Note: pricing of sedum mats shipped in full truck loads (4000-5000 S.F.)									

07 41 Roof Panels

07 41 13 – Metal Roof Panels

07 41 13.10 Aluminum Roof Panels

		Crew	Daily Output	Labor-Hours	Unit	Material	2022 Bare Costs Labor	Equipment	Total	Total Incl O&P
0010	**ALUMINUM ROOF PANELS**									
0020	Corrugated or ribbed, .0155" thick, natural	G-3	1200	.027	S.F.	.79	1.51		2.30	3.24
0300	Painted		1200	.027		1.16	1.51		2.67	3.65
0400	Corrugated, .018" thick, on steel frame, natural finish		1200	.027		1.37	1.51		2.88	3.88
0600	Painted		1200	.027		1.70	1.51		3.21	4.24
0700	Corrugated, on steel frame, natural, .024" thick		1200	.027		1.98	1.51		3.49	4.55
0800	Painted		1200	.027		2.26	1.51		3.77	4.86
0900	.032" thick, natural		1200	.027		3.22	1.51		4.73	5.90
1200	Painted		1200	.027		5.25	1.51		6.76	8.15
9000	Minimum labor/equipment charge	1 Rofc	3	2.667	Job		132		132	228

07 41 13.20 Steel Roofing Panels

		Crew	Daily Output	Labor-Hours	Unit	Material	2022 Bare Costs Labor	Equipment	Total	Total Incl O&P
0010	**STEEL ROOFING PANELS**									
0012	Corrugated or ribbed, on steel framing, 30 ga. galv	G-3	1100	.029	S.F.	1.19	1.65		2.84	3.89
0100	28 ga.		1050	.030		1.62	1.72		3.34	4.49
0300	26 ga.		1000	.032		2	1.81		3.81	5.05
0400	24 ga.		950	.034		3.15	1.91		5.06	6.45
0600	Colored, 28 ga.		1050	.030		1.97	1.72		3.69	4.88
0700	26 ga.		1000	.032		2.23	1.81		4.04	5.30
0710	Flat profile, 1-3/4" standing seams, 10" wide, standard finish, 26 ga.		1000	.032		4.10	1.81		5.91	7.35
0715	24 ga.		950	.034		4.78	1.91		6.69	8.25
0720	22 ga.		900	.036		5.90	2.01		7.91	9.65
0725	Zinc aluminum alloy finish, 26 ga.		1000	.032		3.27	1.81		5.08	6.45
0730	24 ga.		950	.034		3.86	1.91		5.77	7.25
0735	22 ga.		900	.036		4.41	2.01		6.42	8
0740	12" wide, standard finish, 26 ga.		1000	.032		6.55	1.81		8.36	10.10
0745	24 ga.		950	.034		5.50	1.91		7.41	9.05
0750	Zinc aluminum alloy finish, 26 ga.		1000	.032		4.64	1.81		6.45	7.95
0755	24 ga.		950	.034		4.97	1.91		6.88	8.45
0840	Flat profile, 1" x 3/8" batten, 12" wide, standard finish, 26 ga.		1000	.032		3.60	1.81		5.41	6.80
0845	24 ga.		950	.034		6.35	1.91		8.26	9.95
0850	22 ga.		900	.036		5.10	2.01		7.11	8.75
0855	Zinc aluminum alloy finish, 26 ga.		1000	.032		3.47	1.81		5.28	6.65
0860	24 ga.		950	.034		5.70	1.91		7.61	9.30

07 41 13 – Metal Roof Panels

07 41 13.20 Steel Roofing Panels		Crew	Daily Output	Labor-Hours	Unit	Material	2022 Bare Costs Labor	Equipment	Total	Total Incl O&P
0865	22 ga.	G-3	900	.036	S.F.	4.45	2.01		6.46	8.05
0870	16-1/2" wide, standard finish, 24 ga.		950	.034		5.60	1.91		7.51	9.20
0875	22 ga.		900	.036		4.70	2.01		6.71	8.30
0880	Zinc aluminum alloy finish, 24 ga.		950	.034		5.05	1.91		6.96	8.55
0885	22 ga.		900	.036		4.14	2.01		6.15	7.70
0890	Flat profile, 2" x 2" batten, 12" wide, standard finish, 26 ga.		1000	.032		4.14	1.81		5.95	7.40
0895	24 ga.		950	.034		5.30	1.91		7.21	8.85
0900	22 ga.		900	.036		6.50	2.01		8.51	10.30
0905	Zinc aluminum alloy finish, 26 ga.		1000	.032		4.19	1.81		6	7.45
0910	24 ga.		950	.034		6.35	1.91		8.26	9.95
0915	22 ga.		900	.036		5.15	2.01		7.16	8.85
0920	16-1/2" wide, standard finish, 24 ga.		950	.034		4.60	1.91		6.51	8.05
0925	22 ga.		900	.036		5.35	2.01		7.36	9
0930	Zinc aluminum alloy finish, 24 ga.		950	.034		4.14	1.91		6.05	7.55
0935	22 ga.		900	.036		4.72	2.01		6.73	8.35
9000	Minimum labor/equipment charge	1 Rofc	2	4	Job		198		198	340

07 41 33 – Plastic Roof Panels

07 41 33.10 Fiberglass Panels

		Crew	Daily Output	Labor-Hours	Unit	Material	Labor	Equipment	Total	Total Incl O&P
0010	**FIBERGLASS PANELS**									
0012	Corrugated panels, roofing, 8 oz./S.F.	G-3	1000	.032	S.F.	3.44	1.81		5.25	6.60
0100	12 oz./S.F.		1000	.032		4.78	1.81		6.59	8.10
0300	Corrugated siding, 6 oz./S.F.		880	.036		2.79	2.06		4.85	6.30
0400	8 oz./S.F.		880	.036		3.44	2.06		5.50	7
0500	Fire retardant		880	.036		4.11	2.06		6.17	7.75
0600	12 oz. siding, textured		880	.036		3.88	2.06		5.94	7.50
0700	Fire retardant		880	.036		4.98	2.06		7.04	8.75
0900	Flat panels, 6 oz./S.F., clear or colors		880	.036		3.14	2.06		5.20	6.70
1100	Fire retardant, class A		880	.036		3.73	2.06		5.79	7.35
1300	8 oz./S.F., clear or colors		880	.036		2.47	2.06		4.53	5.95
9000	Minimum labor/equipment charge	1 Rofc	2	4	Job		198		198	340

07 42 Wall Panels

07 42 13 – Metal Wall Panels

07 42 13.10 Mansard Panels

		Crew	Daily Output	Labor-Hours	Unit	Material	Labor	Equipment	Total	Total Incl O&P
0010	**MANSARD PANELS**									
0600	Aluminum, stock units, straight surfaces	1 Shee	115	.070	S.F.	4.85	4.70		9.55	12.65
0700	Concave or convex surfaces, add		75	.107	"	2.63	7.20		9.83	14.10
0800	For framing, to 5' high, add		115	.070	L.F.	3.96	4.70		8.66	11.65
0900	Soffits, to 1' wide		125	.064	S.F.	3.05	4.32		7.37	10.10
9000	Minimum labor/equipment charge		2.50	3.200	Job		216		216	335

07 42 13.20 Aluminum Siding Panels

		Crew	Daily Output	Labor-Hours	Unit	Material	Labor	Equipment	Total	Total Incl O&P
0010	**ALUMINUM SIDING PANELS**									
0012	Corrugated, on steel framing, .019" thick, natural finish	G-3	775	.041	S.F.	1.97	2.34		4.31	5.85
0100	Painted		775	.041		1.90	2.34		4.24	5.75
0400	Farm type, .021" thick on steel frame, natural		775	.041		1.80	2.34		4.14	5.65
0600	Painted		775	.041		1.90	2.34		4.24	5.75
0700	Industrial type, corrugated, on steel, .024" thick, mill		775	.041		2.47	2.34		4.81	6.40
0900	Painted		775	.041		2.63	2.34		4.97	6.55
1000	.032" thick, mill		775	.041		5.25	2.34		7.59	9.45
1200	Painted		775	.041		5.80	2.34		8.14	10

07 42 Wall Panels

07 42 13 – Metal Wall Panels

07 42 13.20 Aluminum Siding Panels

		Crew	Daily Output	Labor-Hours	Unit	Material	2022 Bare Costs Labor	Equipment	Total	Total Incl O&P
1300	V-Beam, on steel frame, .032" thick, mill	G-3	775	.041	S.F.	4.57	2.34		6.91	8.70
1500	Painted		775	.041		5.05	2.34		7.39	9.20
1600	.040" thick, mill		775	.041		2.97	2.34		5.31	6.95
1800	Painted	↓	775	.041		4.21	2.34		6.55	8.30
3800	Horizontal, colored clapboard, 8" wide, plain	2 Carp	515	.031		3.21	1.75		4.96	6.30
3810	Insulated		515	.031		3.10	1.75		4.85	6.20
3830	8" embossed, painted		515	.031		3.86	1.75		5.61	7.05
3840	Insulated		515	.031		3.38	1.75		5.13	6.50
3860	12" painted, smooth		600	.027		3.65	1.50		5.15	6.40
3870	Insulated		600	.027		3.31	1.50		4.81	6.05
3890	12" embossed, painted		600	.027		3.17	1.50		4.67	5.90
3900	Insulated		515	.031		3.23	1.75		4.98	6.35
4000	Vertical board & batten, colored, non-insulated	↓	515	.031		2.70	1.75		4.45	5.75
4200	For simulated wood design, add				↓	.18			.18	.20
4300	Corners for above, outside	2 Carp	515	.031	V.L.F.	4.13	1.75		5.88	7.30
4500	Inside corners	"	515	.031	"	1.85	1.75		3.60	4.82
9000	Minimum labor/equipment charge	1 Carp	3	2.667	Job		150		150	239

07 42 13.30 Steel Siding

		Crew	Daily Output	Labor-Hours	Unit	Material	2022 Bare Costs Labor	Equipment	Total	Total Incl O&P
0010	**STEEL SIDING**									
0020	Beveled, vinyl coated, 8" wide	1 Carp	265	.030	S.F.	1.95	1.70		3.65	4.85
0050	10" wide	"	275	.029		1.91	1.64		3.55	4.70
0080	Galv, corrugated or ribbed, on steel frame, 30 ga.	G-3	800	.040		1.21	2.26		3.47	4.88
0100	28 ga.		795	.040		1.31	2.28		3.59	5
0300	26 ga.		790	.041		1.70	2.29		3.99	5.45
0400	24 ga.		785	.041		1.71	2.31		4.02	5.50
0600	22 ga.		770	.042		1.86	2.35		4.21	5.75
0700	Colored, corrugated/ribbed, on steel frame, 10 yr. finish, 28 ga.		800	.040		2.38	2.26		4.64	6.15
0900	26 ga.		795	.040		2.09	2.28		4.37	5.85
1000	24 ga.		790	.041		2.34	2.29		4.63	6.15
1020	20 ga.	↓	785	.041	↓	2.98	2.31		5.29	6.90
9000	Minimum labor/equipment charge	1 Carp	3	2.667	Job		150		150	239

07 46 Siding

07 46 23 – Wood Siding

07 46 23.10 Wood Board Siding

		Crew	Daily Output	Labor-Hours	Unit	Material	2022 Bare Costs Labor	Equipment	Total	Total Incl O&P
0010	**WOOD BOARD SIDING**									
3200	Wood, cedar bevel, A grade, 1/2" x 6"	1 Carp	295	.027	S.F.	4.60	1.53		6.13	7.50
3300	1/2" x 8"		330	.024		7.85	1.36		9.21	10.75
3500	3/4" x 10", clear grade		375	.021		7.65	1.20		8.85	10.30
3600	"B" grade		375	.021		4.56	1.20		5.76	6.90
3800	Cedar, rough sawn, 1" x 4", A grade, natural		220	.036		9.15	2.05		11.20	13.30
3900	Stained		220	.036		9.10	2.05		11.15	13.25
4100	1" x 12", board & batten, #3 & Btr., natural		420	.019		5.05	1.07		6.12	7.25
4200	Stained		420	.019		5.50	1.07		6.57	7.75
4400	1" x 8" channel siding, #3 & Btr., natural		330	.024		5.45	1.36		6.81	8.15
4500	Stained		330	.024		6.25	1.36		7.61	9.05
4700	Redwood, clear, beveled, vertical grain, 1/2" x 4"		220	.036		5.20	2.05		7.25	8.95
4750	1/2" x 6"		295	.027		5.65	1.53		7.18	8.65
4800	1/2" x 8"		330	.024		5.75	1.36		7.11	8.50
5000	3/4" x 10"		375	.021		5.60	1.20		6.80	8.05
5200	Channel siding, 1" x 10", B grade	↓	375	.021		5.10	1.20		6.30	7.50

For customer support on your Commercial Renovation Costs with RSMeans data, call 800.448.8182.

203

07 46 23 – Wood Siding

07 46 23.10 Wood Board Siding	Crew	Daily Output	Labor-Hours	Unit	Material	2022 Bare Costs Labor	Equipment	Total	Total Incl O&P
5250	Redwood, T&G boards, B grade, 1" x 4"	1 Carp	220	.036	S.F.	7.30	2.05		9.35	11.25
5270	1" x 8"		330	.024		9.35	1.36		10.71	12.45
5400	White pine, rough sawn, 1" x 8", natural		330	.024		2.51	1.36		3.87	4.93
5500	Stained		330	.024		2.51	1.36		3.87	4.93
9000	Minimum labor/equipment charge		2	4	Job		225		225	360

07 46 29 – Plywood Siding

07 46 29.10 Plywood Siding Options

		Crew	Daily Output	Labor-Hours	Unit	Material	Labor	Equipment	Total	Total Incl O&P
0010	**PLYWOOD SIDING OPTIONS**									
0900	Plywood, medium density overlaid, 3/8" thick	2 Carp	750	.021	S.F.	1.50	1.20		2.70	3.56
1000	1/2" thick		700	.023		1.98	1.29		3.27	4.22
1100	3/4" thick		650	.025		2.62	1.39		4.01	5.10
1600	Texture 1-11, cedar, 5/8" thick, natural		675	.024		2.69	1.33		4.02	5.10
1700	Factory stained		675	.024		2.98	1.33		4.31	5.40
1900	Texture 1-11, fir, 5/8" thick, natural		675	.024		1.47	1.33		2.80	3.74
2000	Factory stained		675	.024		2.10	1.33		3.43	4.43
2050	Texture 1-11, S.Y.P., 5/8" thick, natural		675	.024		1.66	1.33		2.99	3.95
2100	Factory stained		675	.024		1.60	1.33		2.93	3.88
2200	Rough sawn cedar, 3/8" thick, natural		675	.024		1.33	1.33		2.66	3.58
2300	Factory stained		675	.024		1.60	1.33		2.93	3.88
2500	Rough sawn fir, 3/8" thick, natural		675	.024		1.06	1.33		2.39	3.29
2600	Factory stained		675	.024		1.18	1.33		2.51	3.42
2800	Redwood, textured siding, 5/8" thick		675	.024		2.14	1.33		3.47	4.47
3000	Polyvinyl chloride coated, 3/8" thick		750	.021		1.23	1.20		2.43	3.26
9000	Minimum labor/equipment charge	1 Carp	2	4	Job		225		225	360

07 46 33 – Plastic Siding

07 46 33.10 Vinyl Siding

		Crew	Daily Output	Labor-Hours	Unit	Material	Labor	Equipment	Total	Total Incl O&P
0010	**VINYL SIDING**									
3995	Clapboard profile, woodgrain texture, .048 thick, double 4	2 Carp	495	.032	S.F.	1.15	1.82		2.97	4.16
4000	Double 5		550	.029		1.15	1.64		2.79	3.87
4005	Single 8		495	.032		1.96	1.82		3.78	5.05
4010	Single 10		550	.029		2.35	1.64		3.99	5.20
4015	.044 thick, double 4		495	.032		1.14	1.82		2.96	4.15
4020	Double 5		550	.029		1.17	1.64		2.81	3.89
4025	.042 thick, double 4		495	.032		1.25	1.82		3.07	4.27
4030	Double 5		550	.029		1.89	1.64		3.53	4.68
4035	Cross sawn texture, .040 thick, double 4		495	.032		.80	1.82		2.62	3.77
4040	Double 5		550	.029		.77	1.64		2.41	3.45
4045	Smooth texture, .042 thick, double 4		495	.032		1.91	1.82		3.73	4.99
4050	Double 5		550	.029		.85	1.64		2.49	3.54
4055	Single 8		495	.032		.77	1.82		2.59	3.74
4060	Cedar texture, .044 thick, double 4		495	.032		1.21	1.82		3.03	4.22
4065	Double 6		600	.027		1.56	1.50		3.06	4.11
4070	Dutch lap profile, woodgrain texture, .048 thick, double 5		550	.029		1.28	1.64		2.92	4.01
4075	.044 thick, double 4.5		525	.030		1.15	1.72		2.87	4
4080	.042 thick, double 4.5		525	.030		.96	1.72		2.68	3.79
4085	.040 thick, double 4.5		525	.030		.79	1.72		2.51	3.60
4100	Shake profile, 10" wide		400	.040		3.93	2.25		6.18	7.90
4105	Vertical pattern, .046 thick, double 5		550	.029		2.10	1.64		3.74	4.91
4110	.044 thick, triple 3		550	.029		1.80	1.64		3.44	4.58
4115	.040 thick, triple 4		550	.029		1.82	1.64		3.46	4.60
4120	.040 thick, triple 2.66		550	.029		2.01	1.64		3.65	4.81
4125	Insulation, fan folded extruded polystyrene, 1/4"		2000	.008		.32	.45		.77	1.07

07 46 Siding

07 46 33 – Plastic Siding

07 46 33.10 Vinyl Siding

		Crew	Daily Output	Labor-Hours	Unit	Material	2022 Bare Costs Labor	Equipment	Total	Total Incl O&P
4130	3/8"	2 Carp	2000	.008	S.F.	.35	.45		.80	1.11
4135	Accessories, J channel, 5/8" pocket		700	.023	L.F.	.68	1.29		1.97	2.78
4140	3/4" pocket		695	.023		.76	1.30		2.06	2.89
4145	1-1/4" pocket		680	.024		1.28	1.32		2.60	3.50
4150	Flexible, 3/4" pocket		600	.027		3.50	1.50		5	6.25
4155	Under sill finish trim		500	.032		.73	1.80		2.53	3.66
4160	Vinyl starter strip		700	.023		.79	1.29		2.08	2.90
4165	Aluminum starter strip		700	.023		.32	1.29		1.61	2.39
4170	Window casing, 2-1/2" wide, 3/4" pocket		510	.031		2.04	1.77		3.81	5.05
4175	Outside corner, woodgrain finish, 4" face, 3/4" pocket		700	.023		2.74	1.29		4.03	5.05
4180	5/8" pocket		700	.023		2.11	1.29		3.40	4.36
4185	Smooth finish, 4" face, 3/4" pocket		700	.023		2.78	1.29		4.07	5.10
4190	7/8" pocket		690	.023		2.01	1.31		3.32	4.28
4195	1-1/4" pocket		700	.023		1.59	1.29		2.88	3.79
4200	Soffit and fascia, 1' overhang, solid		120	.133		5.80	7.50		13.30	18.35
4205	Vented		120	.133		5.80	7.50		13.30	18.35
4207	18" overhang, solid		110	.145		6.85	8.20		15.05	20.50
4208	Vented		110	.145		6.85	8.20		15.05	20.50
4210	2' overhang, solid		100	.160		7.90	9		16.90	23
4215	Vented		100	.160		7.90	9		16.90	23
4217	3' overhang, solid		100	.160		9.95	9		18.95	25.50
4218	Vented		100	.160		9.95	9		18.95	25.50
4220	Colors for siding and soffits, add				S.F.	.16			.16	.18
4225	Colors for accessories and trim, add				L.F.	.30			.30	.33
9000	Minimum labor/equipment charge	1 Carp	3	2.667	Job		150		150	239

07 46 33.20 Polypropylene Siding

		Crew	Daily Output	Labor-Hours	Unit	Material	2022 Bare Costs Labor	Equipment	Total	Total Incl O&P
0010	**POLYPROPYLENE SIDING**									
4090	Shingle profile, random grooves, double 7	2 Carp	400	.040	S.F.	3.70	2.25		5.95	7.65
4092	Cornerpost for above	1 Carp	365	.022	L.F.	16.40	1.23		17.63	20
4095	Triple 5	2 Carp	400	.040	S.F.	3.88	2.25		6.13	7.85
4097	Cornerpost for above	1 Carp	365	.022	L.F.	13.15	1.23		14.38	16.40
5000	Staggered butt, double 7"	2 Carp	400	.040	S.F.	4.33	2.25		6.58	8.35
5002	Cornerpost for above	1 Carp	365	.022	L.F.	15.25	1.23		16.48	18.70
5010	Half round, double 6-1/4"	2 Carp	360	.044	S.F.	4.50	2.50		7	8.95
5020	Shake profile, staggered butt, double 9"	"	510	.031	"	4.01	1.77		5.78	7.20
5022	Cornerpost for above	1 Carp	365	.022	L.F.	12.25	1.23		13.48	15.45
5030	Straight butt, double 7"	2 Carp	400	.040	S.F.	4.50	2.25		6.75	8.55
5032	Cornerpost for above	1 Carp	365	.022	L.F.	15.30	1.23		16.53	18.75
6000	Accessories, J channel, 5/8" pocket	2 Carp	700	.023		.68	1.29		1.97	2.78
6010	3/4" pocket		695	.023		.76	1.30		2.06	2.89
6020	1-1/4" pocket		680	.024		1.28	1.32		2.60	3.50
6030	Aluminum starter strip		700	.023		.32	1.29		1.61	2.39

07 46 46 – Fiber Cement Siding

07 46 46.10 Fiber Cement Siding

		Crew	Daily Output	Labor-Hours	Unit	Material	2022 Bare Costs Labor	Equipment	Total	Total Incl O&P
0010	**FIBER CEMENT SIDING**									
0020	Lap siding, 5/16" thick, 6" wide, 4-3/4" exposure, smooth texture	2 Carp	415	.039	S.F.	1.94	2.17		4.11	5.60
0025	Woodgrain texture		415	.039		1.94	2.17		4.11	5.60
0030	7-1/2" wide, 6-1/4" exposure, smooth texture		425	.038		1.77	2.12		3.89	5.30
0035	Woodgrain texture		425	.038		1.77	2.12		3.89	5.30
0040	8" wide, 6-3/4" exposure, smooth texture		425	.038		1.90	2.12		4.02	5.45
0045	Rough sawn texture		425	.038		1.90	2.12		4.02	5.45
0050	9-1/2" wide, 8-1/4" exposure, smooth texture		440	.036		1.77	2.05		3.82	5.20

07 46 46 – Fiber Cement Siding

07 46 46.10 Fiber Cement Siding

07 46 46.10 Fiber Cement Siding	Crew	Daily Output	Labor-Hours	Unit	Material	2022 Bare Costs Labor	Equipment	Total	Total Incl O&P
0055	Woodgrain texture	2 Carp	440	.036	S.F.	1.77	2.05		3.82	5.20
0060	12" wide, 10-3/8" exposure, smooth texture		455	.035		1.67	1.98		3.65	4.99
0065	Woodgrain texture		455	.035		1.67	1.98		3.65	4.99
0070	Panel siding, 5/16" thick, smooth texture		750	.021		1.33	1.20		2.53	3.37
0075	Stucco texture		750	.021		1.33	1.20		2.53	3.37
0080	Grooved woodgrain texture		750	.021		1.33	1.20		2.53	3.37
0088	Shingle siding, 48" x 15-1/4" panels, 7" exposure		700	.023		2.50	1.29		3.79	4.79
0090	Wood starter strip		400	.040	L.F.	.43	2.25		2.68	4.05

07 46 73 – Soffit

07 46 73.10 Soffit Options

07 46 73.10 Soffit Options	Crew	Daily Output	Labor-Hours	Unit	Material	2022 Bare Costs Labor	Equipment	Total	Total Incl O&P
0010	**SOFFIT OPTIONS**									
0012	Aluminum, residential, .020" thick	1 Carp	210	.038	S.F.	1.69	2.14		3.83	5.25
0100	Baked enamel on steel, 16 or 18 ga.		105	.076		4.94	4.29		9.23	12.25
0300	Polyvinyl chloride, white, solid		230	.035		1.56	1.96		3.52	4.83
0400	Perforated		230	.035		1.71	1.96		3.67	4.99
0500	For colors, add					.11			.11	.12
9000	Minimum labor/equipment charge	1 Carp	3	2.667	Job		150		150	239

07 51 Built-Up Bituminous Roofing

07 51 13 – Built-Up Asphalt Roofing

07 51 13.10 Built-Up Roofing Components

07 51 13.10 Built-Up Roofing Components	Crew	Daily Output	Labor-Hours	Unit	Material	2022 Bare Costs Labor	Equipment	Total	Total Incl O&P
0010	**BUILT-UP ROOFING COMPONENTS**									
0012	Asphalt saturated felt, #30, 2 sq./roll	1 Rofc	58	.138	Sq.	8.70	6.85		15.55	21.50
0200	#15, 4 sq./roll, plain or perforated, not mopped		58	.138		4.43	6.85		11.28	16.60
0300	Roll roofing, smooth, #65		15	.533		10.35	26.50		36.85	57
0500	#90		12	.667		43	33		76	105
0520	Mineralized		12	.667		43	33		76	105
0540	D.C. (double coverage), 19" selvage edge		10	.800		50	39.50		89.50	124
0580	Adhesive (lap cement)				Gal.	11.20			11.20	12.35
0800	Steep, flat or dead level asphalt, 10 ton lots, packaged				Ton	915			915	1,000
9000	Minimum labor/equipment charge	1 Rofc	4	2	Job		99		99	171

07 51 13.13 Cold-Applied Built-Up Asphalt Roofing

07 51 13.13 Cold-Applied Built-Up Asphalt Roofing	Crew	Daily Output	Labor-Hours	Unit	Material	2022 Bare Costs Labor	Equipment	Total	Total Incl O&P
0010	**COLD-APPLIED BUILT-UP ASPHALT ROOFING**									
0020	3 ply system, installation only (components listed below)	G-5	50	.800	Sq.		36	3.33	39.33	65.50
0100	Spunbond poly. fabric, 1.35 oz./S.Y., 36"W, 10.8 sq./roll				Ea.	173			173	191
0500	Base & finish coat, 3 gal./sq., 5 gal./can				Gal.	8.60			8.60	9.45
0600	Coating, ceramic granules, 1/2 sq./bag				Ea.	30.50			30.50	33.50
0700	Aluminum, 2 gal./sq.				Gal.	19.80			19.80	22
0800	Emulsion, fibered or non-fibered, 4 gal./sq.				"	7.40			7.40	8.15

07 51 13.20 Built-Up Roofing Systems

07 51 13.20 Built-Up Roofing Systems	Crew	Daily Output	Labor-Hours	Unit	Material	2022 Bare Costs Labor	Equipment	Total	Total Incl O&P
0010	**BUILT-UP ROOFING SYSTEMS**									
0120	Asphalt flood coat with gravel/slag surfacing, not including									
0140	Insulation, flashing or wood nailers									
0200	Asphalt base sheet, 3 plies #15 asphalt felt, mopped	G-1	22	2.545	Sq.	109	118	25	252	350
0350	On nailable decks		21	2.667		101	124	26	251	355
0500	4 plies #15 asphalt felt, mopped		20	2.800		134	130	27.50	291.50	400
0550	On nailable decks		19	2.947		118	137	29	284	395
0700	Coated glass base sheet, 2 plies glass (type IV), mopped		22	2.545		111	118	25	254	355
0850	3 plies glass, mopped		20	2.800		133	130	27.50	290.50	400
0950	On nailable decks		19	2.947		126	137	29	292	405

07 51 Built-Up Bituminous Roofing

07 51 13 – Built-Up Asphalt Roofing

07 51 13.20 Built-Up Roofing Systems

	Crew	Daily Output	Labor-Hours	Unit	Material	2022 Bare Costs Labor	Equipment	Total	Total Incl O&P
1100	4 plies glass fiber felt (type IV), mopped	G-1	20	2.800	Sq.	164	130	27.50	321.50	435
1150	On nailable decks		19	2.947		149	137	29	315	430
1200	Coated & saturated base sheet, 3 plies #15 asph. felt, mopped		20	2.800		110	130	27.50	267.50	375
1250	On nailable decks		19	2.947		102	137	29	268	380
1300	4 plies #15 asphalt felt, mopped	↓	22	2.545	↓	127	118	25	270	370
2000	Asphalt flood coat, smooth surface									
2200	Asphalt base sheet & 3 plies #15 asphalt felt, mopped	G-1	24	2.333	Sq.	103	108	23	234	325
2400	On nailable decks		23	2.435		95	113	24	232	325
2600	4 plies #15 asphalt felt, mopped		24	2.333		120	108	23	251	345
2700	On nailable decks	↓	23	2.435	↓	112	113	24	249	345
2900	Coated glass fiber base sheet, mopped, and 2 plies of									
2910	glass fiber felt (type IV)	G-1	25	2.240	Sq.	105	104	22	231	320
3100	On nailable decks		24	2.333		99	108	23	230	320
3200	3 plies, mopped		23	2.435		127	113	24	264	360
3300	On nailable decks		22	2.545		120	118	25	263	365
3800	4 plies glass fiber felt (type IV), mopped		23	2.435		150	113	24	287	385
3900	On nailable decks		22	2.545		142	118	25	285	390
4000	Coated & saturated base sheet, 3 plies #15 asph. felt, mopped		24	2.333		104	108	23	235	325
4200	On nailable decks		23	2.435		96	113	24	233	325
4300	4 plies #15 organic felt, mopped	↓	22	2.545	↓	121	118	25	264	365
4500	Coal tar pitch with gravel/slag surfacing									
4600	4 plies #15 tarred felt, mopped	G-1	21	2.667	Sq.	222	124	26	372	485
4800	3 plies glass fiber felt (type IV), mopped	"	19	2.947	"	182	137	29	348	470
5000	Coated glass fiber base sheet, and 2 plies of									
5010	glass fiber felt (type IV), mopped	G-1	19	2.947	Sq.	185	137	29	351	470
5300	On nailable decks		18	3.111		162	144	30.50	336.50	460
5600	4 plies glass fiber felt (type IV), mopped		21	2.667		253	124	26	403	520
5800	On nailable decks	↓	20	2.800	↓	230	130	27.50	387.50	505

07 51 13.30 Cants

	Crew	Daily Output	Labor-Hours	Unit	Material	2022 Bare Costs Labor	Equipment	Total	Total Incl O&P
0010	**CANTS**									
0012	Lumber, treated, 4" x 4" cut diagonally	1 Rofc	325	.025	L.F.	1.96	1.22		3.18	4.26
0300	Mineral or fiber, trapezoidal, 1" x 4" x 48"		325	.025		.37	1.22		1.59	2.51
0400	1-1/2" x 5-5/8" x 48"	↓	325	.025	↓	.49	1.22		1.71	2.64
9000	Minimum labor/equipment charge	↓	4	2	Job		99		99	171

07 51 13.40 Felts

	Crew	Daily Output	Labor-Hours	Unit	Material	2022 Bare Costs Labor	Equipment	Total	Total Incl O&P
0010	**FELTS**									
0012	Glass fibered roofing felt, #15, not mopped	1 Rofc	58	.138	Sq.	9.95	6.85		16.80	22.50
0300	Base sheet, #80, channel vented		58	.138		48.50	6.85		55.35	65
0400	#70, coated		58	.138		19.25	6.85		26.10	33
0500	Cap, #87, mineral surfaced		58	.138		73.50	6.85		80.35	93
0600	Flashing membrane, #65		16	.500		10.35	25		35.35	54
0800	Coal tar fibered, #15, no mopping		58	.138		19.15	6.85		26	33
0900	Asphalt felt, #15, 4 sq./roll, no mopping		58	.138		4.43	6.85		11.28	16.60
1100	#30, 2 sq./roll		58	.138		8.70	6.85		15.55	21.50
1200	Double coated, #33		58	.138		12.35	6.85		19.20	25.50
1400	#40, base sheet		58	.138		11.35	6.85		18.20	24
1450	Coated and saturated		58	.138		12.45	6.85		19.30	25.50
1500	Tarred felt, organic, #15, 4 sq. rolls		58	.138		16.60	6.85		23.45	30
1550	#30, 2 sq. roll	↓	58	.138		27.50	6.85		34.35	42
1700	Add for mopping above felts, per ply, asphalt, 24 lb./sq.	G-1	192	.292		11	13.50	2.87	27.37	39
1800	Coal tar mopping, 30 lb./sq.		186	.301		19.80	13.95	2.96	36.71	49.50
1900	Flood coat, with asphalt, 60 lb./sq.	↓	60	.933		27.50	43.50	9.20	80.20	115

07 51 Built-Up Bituminous Roofing

07 51 13 – Built-Up Asphalt Roofing

07 51 13.40 Felts

		Crew	Daily Output	Labor-Hours	Unit	Material	2022 Bare Costs Labor	Equipment	Total	Total Incl O&P
2000	With coal tar, 75 lb./sq.	G-1	56	1	Sq.	49.50	46.50	9.85	105.85	145
9000	Minimum labor/equipment charge	1 Rofc	4	2	Job		99		99	171

07 51 13.50 Walkways for Built-Up Roofs

		Crew	Daily Output	Labor-Hours	Unit	Material	2022 Bare Costs Labor	Equipment	Total	Total Incl O&P
0010	**WALKWAYS FOR BUILT-UP ROOFS**									
0020	Asphalt impregnated, 3' x 6' x 1/2" thick	1 Rofc	400	.020	S.F.	1.74	.99		2.73	3.63
0100	3' x 3' x 3/4" thick		400	.020	"	5	.99		5.99	7.20
0600	100% recycled rubber, 3' x 4' x 3/8" [G]		400	.020	L.F.	7.20	.99		8.19	9.65
0610	3' x 4' x 1/2" [G]		400	.020		8.05	.99		9.04	10.55
0620	3' x 4' x 3/4" [G]		400	.020	↓	8.45	.99		9.44	11
9000	Minimum labor/equipment charge	↓	2.75	2.909	Job		144		144	248

07 52 Modified Bituminous Membrane Roofing

07 52 13 – Atactic-Polypropylene-Modified Bituminous Membrane Roofing

07 52 13.10 APP Modified Bituminous Membrane

		Crew	Daily Output	Labor-Hours	Unit	Material	2022 Bare Costs Labor	Equipment	Total	Total Incl O&P
0010	**APP MODIFIED BITUMINOUS MEMBRANE** R075213-30									
0020	Base sheet, #15 glass fiber felt, nailed to deck	1 Rofc	58	.138	Sq.	11.95	6.85		18.80	25
0030	Spot mopped to deck	G-1	295	.190		15.45	8.80	1.87	26.12	34
0040	Fully mopped to deck	"	192	.292		21	13.50	2.87	37.37	49.50
0050	#15 organic felt, nailed to deck	1 Rofc	58	.138		6.40	6.85		13.25	18.80
0060	Spot mopped to deck	G-1	295	.190		9.95	8.80	1.87	20.62	28
0070	Fully mopped to deck	"	192	.292	↓	15.40	13.50	2.87	31.77	43.50
2100	APP mod., smooth surf. cap sheet, poly. reinf., torched, 160 mils	G-5	2100	.019	S.F.	.75	.86	.08	1.69	2.40
2150	170 mils		2100	.019		.68	.86	.08	1.62	2.32
2200	Granule surface cap sheet, poly. reinf., torched, 180 mils		2000	.020		.87	.90	.08	1.85	2.60
2250	Smooth surface flashing, torched, 160 mils		1260	.032		.75	1.43	.13	2.31	3.44
2300	170 mils		1260	.032		.68	1.43	.13	2.24	3.36
2350	Granule surface flashing, torched, 180 mils	↓	1260	.032		.87	1.43	.13	2.43	3.57
2400	Fibrated aluminum coating	1 Rofc	3800	.002	↓	.10	.10		.20	.29
2450	Seam heat welding	"	205	.039	L.F.	.10	1.94		2.04	3.44

07 52 16 – Styrene-Butadiene-Styrene Modified Bituminous Membrane Roofing

07 52 16.10 SBS Modified Bituminous Membrane

		Crew	Daily Output	Labor-Hours	Unit	Material	2022 Bare Costs Labor	Equipment	Total	Total Incl O&P
0010	**SBS MODIFIED BITUMINOUS MEMBRANE**									
0080	Mod. bit. rfng., SBS mod, gran surf. cap sheet, poly. reinf.									
0650	120 to 149 mils thick	G-1	2000	.028	S.F.	1.27	1.30	.28	2.85	3.93
0750	150 to 160 mils		2000	.028		1.96	1.30	.28	3.54	4.69
1600	Smooth surface cap sheet, mopped, 145 mils		2100	.027		.85	1.24	.26	2.35	3.36
1620	Lightweight base sheet, fiberglass reinforced, 35 to 47 mil		2100	.027		.28	1.24	.26	1.78	2.73
1625	Heavyweight base/ply sheet, reinforced, 87 to 120 mil thick	↓	2100	.027		.87	1.24	.26	2.37	3.38
1650	Granulated walkpad, 180 to 220 mils	1 Rofc	400	.020		1.75	.99		2.74	3.63
1700	Smooth surface flashing, 145 mils	G-1	1260	.044		.85	2.06	.44	3.35	4.97
1800	150 mils		1260	.044		.49	2.06	.44	2.99	4.57
1900	Granular surface flashing, 150 mils		1260	.044		.65	2.06	.44	3.15	4.75
2000	160 mils	↓	1260	.044		.73	2.06	.44	3.23	4.83
2010	Elastomeric asphalt primer	1 Rofc	2600	.003		.16	.15		.31	.44
2015	Roofing asphalt, 30 lb./square	G-1	19000	.003		.14	.14	.03	.31	.42
2020	Cold process adhesive, 20 to 30 mils thick	1 Rofc	750	.011		.20	.53		.73	1.13
2025	Self adhering vapor retarder, 30 to 45 mils thick	G-5	2150	.019	↓	1.01	.84	.08	1.93	2.64
2050	Seam heat welding	1 Rofc	205	.039	L.F.	.10	1.94		2.04	3.44

208

07 53 Elastomeric Membrane Roofing

07 53 16 – Chlorosulfonate-Polyethylene Roofing

07 53 16.10 Chlorosulfonated Polyethylene Roofing

07 53 16.10 Chlorosulfonated Polyethylene Roofing	Crew	Daily Output	Labor-Hours	Unit	Material	2022 Bare Costs Labor	Equipment	Total	Total Incl O&P
0010 **CHLOROSULFONATED POLYETHYLENE ROOFING**									
0800 Chlorosulfonated polyethylene (CSPE)									
0900 45 mils, heat welded seams, plate attachment	G-5	35	1.143	Sq.	266	51.50	4.75	322.25	385
1100 Heat welded seams, plate attachment and ballasted		26	1.538		278	69.50	6.40	353.90	430
1200 60 mils, heat welded seams, plate attachment		35	1.143		365	51.50	4.75	421.25	495
1300 Heat welded seams, plate attachment and ballasted		26	1.538		375	69.50	6.40	450.90	540

07 53 23 – Ethylene-Propylene-Diene-Monomer Roofing

07 53 23.20 Ethylene-Propylene-Diene-Monomer Roofing

07 53 23.20 Ethylene-Propylene-Diene-Monomer Roofing	Crew	Daily Output	Labor-Hours	Unit	Material	2022 Bare Costs Labor	Equipment	Total	Total Incl O&P
0010 **ETHYLENE-PROPYLENE-DIENE-MONOMER ROOFING (EPDM)**									
3500 Ethylene-propylene-diene-monomer (EPDM), 45 mils, 0.28 psf									
3600 Loose-laid & ballasted with stone (10 psf)	G-5	51	.784	Sq.	101	35.50	3.26	139.76	176
3700 Mechanically attached		35	1.143		95	51.50	4.75	151.25	198
3800 Fully adhered with adhesive		26	1.538		90	69.50	6.40	165.90	225
4500 60 mils, 0.40 psf									
4600 Loose-laid & ballasted with stone (10 psf)	G-5	51	.784	Sq.	118	35.50	3.26	156.76	194
4700 Mechanically attached		35	1.143		110	51.50	4.75	166.25	215
4800 Fully adhered with adhesive		26	1.538		105	69.50	6.40	180.90	242
4810 45 mil, 0.28 psf, membrane only				Sq.	64.50			64.50	71
4820 60 mil, 0.40 psf, membrane only				"	78.50			78.50	86.50
4850 Seam tape for membrane, 3" x 100' roll				Ea.	42			42	46
4900 Batten strips, 10' sections					5.70			5.70	6.30
4910 Cover tape for batten strips, 6" x 100' roll					156			156	172
4930 Plate anchors				M	105			105	115
4970 Adhesive for fully adhered systems, 60 S.F./gal.				S.F.	.70			.70	.77

07 53 29 – Polyisobutylene Roofing

07 53 29.10 Polyisobutylene Roofing

07 53 29.10 Polyisobutylene Roofing	Crew	Daily Output	Labor-Hours	Unit	Material	2022 Bare Costs Labor	Equipment	Total	Total Incl O&P
0010 **POLYISOBUTYLENE ROOFING**									
7500 Polyisobutylene (PIB), 100 mils, 0.57 psf									
7600 Loose-laid & ballasted with stone/gravel (10 psf)	G-5	51	.784	Sq.	219	35.50	3.26	257.76	305
7700 Partially adhered with adhesive		35	1.143		262	51.50	4.75	318.25	385
7800 Hot asphalt attachment		35	1.143		250	51.50	4.75	306.25	370
7900 Fully adhered with contact cement		26	1.538		276	69.50	6.40	351.90	430

07 54 Thermoplastic Membrane Roofing

07 54 16 – Ketone Ethylene Ester Roofing

07 54 16.10 Ketone Ethylene Ester Roofing

07 54 16.10 Ketone Ethylene Ester Roofing	Crew	Daily Output	Labor-Hours	Unit	Material	2022 Bare Costs Labor	Equipment	Total	Total Incl O&P
0010 **KETONE ETHYLENE ESTER ROOFING**									
0100 Ketone ethylene ester roofing, 50 mil, fully adhered	G-5	26	1.538	Sq.	215	69.50	6.40	290.90	365
0120 Mechanically attached		35	1.143		143	51.50	4.75	199.25	251
0140 Ballasted with stone		51	.784		149	35.50	3.26	187.76	229
0160 50 mil, fleece backed, adhered w/hot asphalt	G-1	26	2.154		160	100	21	281	370
0180 Accessories, pipe boot	1 Rofc	32	.250	Ea.	26	12.40		38.40	50
0200 Pre-formed corners		32	.250	"	9.75	12.40		22.15	32.50
0220 Ketone clad metal, including up to 4 bends		330	.024	S.F.	4.11	1.20		5.31	6.60
0240 Walkway pad	2 Rofc	800	.020	"	4.27	.99		5.26	6.40
0260 Stripping material	1 Rofc	310	.026	L.F.	.97	1.28		2.25	3.27

For customer support on your Commercial Renovation Costs with RSMeans data, call 800.448.8182.

209

07 54 Thermoplastic Membrane Roofing

07 54 19 – Polyvinyl-Chloride Roofing

07 54 19.10 Polyvinyl-Chloride Roofing (PVC)	Crew	Daily Output	Labor-Hours	Unit	Material	2022 Bare Costs Labor	Equipment	Total	Total Incl O&P
0010 **POLYVINYL-CHLORIDE ROOFING (PVC)**									
8200 Heat welded seams									
8700 Reinforced, 48 mils, 0.33 psf									
8750 Loose-laid & ballasted with stone/gravel (12 psf)	G-5	51	.784	Sq.	124	35.50	3.26	162.76	201
8800 Mechanically attached		35	1.143		118	51.50	4.75	174.25	224
8850 Fully adhered with adhesive		26	1.538		110	69.50	6.40	185.90	246
8860 Reinforced, 60 mils, 0.40 psf									
8870 Loose-laid & ballasted with stone/gravel (12 psf)	G-5	51	.784	Sq.	135	35.50	3.26	173.76	213
8880 Mechanically attached	"	35	1.143	"	129	51.50	4.75	185.25	236
8890 Fully adhered with adhesive	G-5	26	1.538	Sq.	120	69.50	6.40	195.90	258

07 54 23 – Thermoplastic-Polyolefin Roofing

07 54 23.10 Thermoplastic Polyolefin Roofing (T.P.O.)

	Crew	Daily Output	Labor-Hours	Unit	Material	2022 Bare Costs Labor	Equipment	Total	Total Incl O&P
0010 **THERMOPLASTIC POLYOLEFIN ROOFING (T.P.O.)**									
0100 45 mil, loose laid & ballasted with stone (1/2 ton/sq.)	G-5	51	.784	Sq.	90.50	35.50	3.26	129.26	165
0120 Fully adhered		25	1.600		79	72	6.65	157.65	218
0140 Mechanically attached		34	1.176		78.50	53	4.89	136.39	183
0160 Self adhered		35	1.143		155	51.50	4.75	211.25	265
0180 60 mil membrane, heat welded seams, ballasted		50	.800		106	36	3.33	145.33	182
0200 Fully adhered		25	1.600		93.50	72	6.65	172.15	234
0220 Mechanically attached		34	1.176		99	53	4.89	156.89	205
0240 Self adhered		35	1.143		186	51.50	4.75	242.25	298

07 55 Protected Membrane Roofing

07 55 10 – Protected Membrane Roofing Components

07 55 10.10 Protected Membrane Roofing Components

	Crew	Daily Output	Labor-Hours	Unit	Material	2022 Bare Costs Labor	Equipment	Total	Total Incl O&P
0010 **PROTECTED MEMBRANE ROOFING COMPONENTS**									
0100 Choose roofing membrane from 07 50									
0120 Then choose roof deck insulation from 07 22									
0130 Filter fabric	2 Rofc	10000	.002	S.F.	.05	.08		.13	.20
0140 Ballast, 3/8" - 1/2" in place	G-1	36	1.556	Ton	25	72	15.30	112.30	168
0150 3/4" - 1-1/2" in place	"	36	1.556	"	25	72	15.30	112.30	168
0200 2" concrete blocks, natural	1 Clab	115	.070	S.F.	3.70	3.17		6.87	9.10
0210 Colors	"	115	.070	"	2.42	3.17		5.59	7.70

07 56 Fluid-Applied Roofing

07 56 10 – Fluid-Applied Roofing Elastomers

07 56 10.10 Elastomeric Roofing

	Crew	Daily Output	Labor-Hours	Unit	Material	2022 Bare Costs Labor	Equipment	Total	Total Incl O&P
0010 **ELASTOMERIC ROOFING**									
0020 Acrylic, 44% solids, 2 coats, on corrugated metal	2 Rofc	2400	.007	S.F.	.70	.33		1.03	1.34
0025 On smooth metal		3000	.005		.56	.26		.82	1.07
0030 On foam or modified bitumen		1500	.011		1.12	.53		1.65	2.14
0035 On concrete		1500	.011		1.12	.53		1.65	2.14
0040 On tar and gravel		1500	.011		1.12	.53		1.65	2.14
0045 36% solids, 2 coats, on corrugated metal		2400	.007		.70	.33		1.03	1.34
0050 On smooth metal		3000	.005		.56	.26		.82	1.07
0055 On foam or modified bitumen		1500	.011		1.12	.53		1.65	2.14
0060 On concrete		1500	.011		1.12	.53		1.65	2.14
0065 On tar and gravel		1500	.011		1.12	.53		1.65	2.14

07 56 Fluid-Applied Roofing

07 56 10 – Fluid-Applied Roofing Elastomers

07 56 10.10 Elastomeric Roofing	Crew	Daily Output	Labor-Hours	Unit	Material	2022 Bare Costs Labor	Equipment	Total	Total Incl O&P
0070	Primer if required, 2 coats on corrugated metal	2 Rofc	2400	.007	S.F.	1.40	.33		1.73	2.11
0075	On smooth metal		3000	.005	↓	1.40	.26		1.66	2
0080	On foam or modified bitumen	2 Rofc	1500	.011	S.F.	.83	.53		1.36	1.82
0085	On concrete		1500	.011		.62	.53		1.15	1.59
0090	On tar & gravel/rolled roof	↓	1500	.011		.83	.53		1.36	1.82
0110	Acrylic rubber, fluid applied, 20 mils thick	G-5	2000	.020		2.65	.90	.08	3.63	4.56
0120	50 mils, reinforced		1200	.033		.64	1.50	.14	2.28	3.43
0130	For walking surface, add	↓	900	.044		1.70	2	.18	3.88	5.50
0300	Neoprene, fluid applied, 20 mil thick, not reinforced	G-1	1135	.049		1.35	2.29	.49	4.13	5.95
0600	Non-woven polyester, reinforced		960	.058		1.51	2.70	.57	4.78	6.95
0700	5 coat neoprene deck, 60 mil thick, under 10,000 S.F.		325	.172		4.50	8	1.70	14.20	20.50
0900	Over 10,000 S.F.	↓	625	.090	↓	4.50	4.15	.88	9.53	13.05
9000	Minimum labor/equipment charge	1 Rofc	2	4	Job		198		198	340

07 57 Coated Foamed Roofing

07 57 13 – Sprayed Polyurethane Foam Roofing

07 57 13.10 Sprayed Polyurethane Foam Roofing (S.P.F.)

		Crew	Daily Output	Labor-Hours	Unit	Material	2022 Bare Costs Labor	Equipment	Total	Total Incl O&P
0010	**SPRAYED POLYURETHANE FOAM ROOFING (S.P.F.)**									
0100	Primer for metal substrate (when required)	G-2A	3000	.008	S.F.	.59	.35	.22	1.16	1.48
0200	Primer for non-metal substrate (when required)		3000	.008		.22	.35	.22	.79	1.07
0300	Closed cell spray, polyurethane foam, 3 lb./C.F. density, 1", R6.7		15000	.002		.83	.07	.04	.94	1.08
0400	2", R13.4		13125	.002		1.66	.08	.05	1.79	2.02
0500	3", R18.6		11485	.002		2.49	.09	.06	2.64	2.94
0550	4", R24.8		10080	.002		3.31	.11	.07	3.49	3.90
0700	Spray-on silicone coating	↓	2500	.010	↓	.02	.42	.26	.70	1.02

07 58 Roll Roofing

07 58 10 – Asphalt Roll Roofing

07 58 10.10 Roll Roofing

		Crew	Daily Output	Labor-Hours	Unit	Material	2022 Bare Costs Labor	Equipment	Total	Total Incl O&P
0010	**ROLL ROOFING**									
0100	Asphalt, mineral surface									
0200	1 ply #15 organic felt, 1 ply mineral surfaced									
0300	Selvage roofing, lap 19", nailed & mopped	G-1	27	2.074	Sq.	72	96	20.50	188.50	267
0400	3 plies glass fiber felt (type IV), 1 ply mineral surfaced									
0500	Selvage roofing, lapped 19", mopped	G-1	25	2.240	Sq.	124	104	22	250	340
0600	Coated glass fiber base sheet, 2 plies of glass fiber									
0700	Felt (type IV), 1 ply mineral surfaced selvage									
0800	Roofing, lapped 19", mopped	G-1	25	2.240	Sq.	133	104	22	259	350
0900	On nailable decks	"	24	2.333	"	122	108	23	253	345
1000	3 plies glass fiber felt (type III), 1 ply mineral surfaced									
1100	Selvage roofing, lapped 19", mopped	G-1	25	2.240	Sq.	124	104	22	250	340

For customer support on your Commercial Renovation Costs with RSMeans data, call 800.448.8182.

211

07 61 Sheet Metal Roofing

07 61 13 – Standing Seam Sheet Metal Roofing

07 61 13.10 Standing Seam Sheet Metal Roofing, Field Fab.

		Crew	Daily Output	Labor-Hours	Unit	Material	2022 Bare Costs Labor	Equipment	Total	Total Incl O&P
0010	**STANDING SEAM SHEET METAL ROOFING, FIELD FABRICATED**									
0400	Copper standing seam roofing, over 10 squares, 16 oz., 125 lb./sq.	1 Shee	1.30	6.154	Sq.	1,775	415		2,190	2,600
0600	18 oz., 140 lb./sq.		1.20	6.667		1,975	450		2,425	2,875
0700	20 oz., 150 lb./sq.		1.10	7.273		2,150	490		2,640	3,125
1200	For abnormal conditions or small areas, add					25%	100%			
1300	For lead-coated copper, add					25%				
1400	Lead standing seam roofing, 5 lb./S.F.	1 Shee	1.30	6.154		2,600	415		3,015	3,500
1500	Zinc standing seam roofing, .020" thick		1.20	6.667		750	450		1,200	1,525
1510	.027" thick		1.15	6.957		895	470		1,365	1,725
1520	.032" thick		1.10	7.273		920	490		1,410	1,800
1530	.040" thick		1.05	7.619		1,300	515		1,815	2,225
9000	Minimum labor/equipment charge		2	4	Job		270		270	420

07 61 16 – Batten Seam Sheet Metal Roofing

07 61 16.10 Batten Seam Sheet Metal Roofing, Field Fab.

		Crew	Daily Output	Labor-Hours	Unit	Material	2022 Bare Costs Labor	Equipment	Total	Total Incl O&P
0010	**BATTEN SEAM SHEET METAL ROOFING, FIELD FABRICATED**									
0012	Copper batten seam roofing, over 10 sq., 16 oz., 130 lb./sq.	1 Shee	1.10	7.273	Sq.	2,250	490		2,740	3,250
0020	Lead batten seam roofing, 5 lb./S.F.		1.20	6.667		2,725	450		3,175	3,675
0100	Zinc/copper alloy batten seam roofing, .020" thick		1.20	6.667		785	450		1,235	1,575
0200	Copper roofing, batten seam, over 10 sq., 18 oz., 145 lb./sq.		1	8		2,500	540		3,040	3,600
0300	20 oz., 160 lb./sq.		1	8		2,725	540		3,265	3,850
0500	Stainless steel batten seam roofing, type 304, 28 ga.		1.20	6.667		495	450		945	1,250
0600	26 ga.		1.15	6.957		615	470		1,085	1,400
0800	Zinc, copper alloy roofing, batten seam, .027" thick		1.15	6.957		935	470		1,405	1,750
0900	.032" thick		1.10	7.273		965	490		1,455	1,825
1000	.040" thick		1.05	7.619		1,350	515		1,865	2,300
9000	Minimum labor/equipment charge		2	4	Job		270		270	420

07 61 19 – Flat Seam Sheet Metal Roofing

07 61 19.10 Flat Seam Sheet Metal Roofing, Field Fabricated

		Crew	Daily Output	Labor-Hours	Unit	Material	2022 Bare Costs Labor	Equipment	Total	Total Incl O&P
0010	**FLAT SEAM SHEET METAL ROOFING, FIELD FABRICATED**									
0900	Copper flat seam roofing, over 10 squares, 16 oz., 115 lb./sq.	1 Shee	1.20	6.667	Sq.	1,650	450		2,100	2,500
0950	18 oz., 130 lb./sq.		1.15	6.957		1,850	470		2,320	2,750
1000	20 oz., 145 lb./sq.		1.10	7.273		2,000	490		2,490	2,975
1008	Zinc flat seam roofing, .020" thick		1.20	6.667		670	450		1,120	1,425
1010	.027" thick		1.15	6.957		800	470		1,270	1,600
1020	.032" thick		1.12	7.143		820	485		1,305	1,650
1030	.040" thick		1.05	7.619		1,150	515		1,665	2,075
1100	Lead flat seam roofing, 5 lb./S.F.		1.30	6.154		2,325	415		2,740	3,200
9000	Minimum labor/equipment charge		2.75	2.909	Job		197		197	305

07 62 Sheet Metal Flashing and Trim

07 62 10 – Sheet Metal Trim

07 62 10.10 Sheet Metal Cladding

		Crew	Daily Output	Labor-Hours	Unit	Material	2022 Bare Costs Labor	Equipment	Total	Total Incl O&P
0010	**SHEET METAL CLADDING**									
0100	Aluminum, up to 6 bends, .032" thick, window casing	1 Carp	180	.044	S.F.	4.24	2.50		6.74	8.65
0200	Window sill		72	.111	L.F.	4.24	6.25		10.49	14.60
0300	Door casing		180	.044	S.F.	4.24	2.50		6.74	8.65
0400	Fascia		250	.032		4.24	1.80		6.04	7.50
0500	Rake trim		225	.036		4.24	2		6.24	7.85
0700	.024" thick, window casing		180	.044		1.55	2.50		4.05	5.70
0800	Window sill		72	.111	L.F.	1.55	6.25		7.80	11.65

07 62 Sheet Metal Flashing and Trim

07 62 10 – Sheet Metal Trim

07 62 10.10 Sheet Metal Cladding	Crew	Daily Output	Labor-Hours	Unit	Material	2022 Bare Costs Labor	2022 Bare Costs Equipment	Total	Total Incl O&P
0900	Door casing	1 Carp	180	.044	S.F.	1.55	2.50		4.05	5.70
1000	Fascia		250	.032		1.55	1.80		3.35	4.57
1100	Rake trim		225	.036		1.55	2		3.55	4.89
1200	Vinyl coated aluminum, up to 6 bends, window casing		180	.044		2.04	2.50		4.54	6.20
1300	Window sill		72	.111	L.F.	2.04	6.25		8.29	12.20
1400	Door casing		180	.044	S.F.	2.04	2.50		4.54	6.20
1500	Fascia		250	.032		2.04	1.80		3.84	5.10
1600	Rake trim		225	.036		2.04	2		4.04	5.40

07 65 Flexible Flashing

07 65 10 – Sheet Metal Flashing

07 65 10.10 Sheet Metal Flashing and Counter Flashing

		Crew	Daily Output	Labor-Hours	Unit	Material	2022 Bare Costs Labor	2022 Bare Costs Equipment	Total	Total Incl O&P
0010	**SHEET METAL FLASHING AND COUNTER FLASHING**									
0011	Including up to 4 bends									
0020	Aluminum, mill finish, .013" thick	1 Rofc	145	.055	S.F.	1.33	2.74		4.07	6.15
0030	.016" thick		145	.055		2.59	2.74		5.33	7.55
0060	.019" thick		145	.055		5.20	2.74		7.94	10.45
0100	.032" thick		145	.055		4.24	2.74		6.98	9.35
0200	.040" thick		145	.055		4.47	2.74		7.21	9.65
0300	.050" thick		145	.055		6.25	2.74		8.99	11.55
0325	Mill finish 5" x 7" step flashing, .016" thick		1920	.004	Ea.	.17	.21		.38	.55
0350	Mill finish 12" x 12" step flashing, .016" thick		1600	.005	"	.62	.25		.87	1.11
0400	Painted finish, add				S.F.	.32			.32	.35
1000	Mastic-coated 2 sides, .005" thick	1 Rofc	330	.024		1.91	1.20		3.11	4.17
1100	.016" thick		330	.024		2.11	1.20		3.31	4.39
1600	Copper, 16 oz. sheets, under 1000 lb.		115	.070		14.15	3.45		17.60	21.50
1700	Over 4000 lb.	1 Rofc	155	.052	S.F.	14.15	2.56		16.71	19.95
1900	20 oz. sheets, under 1000 lb.		110	.073		17.15	3.61		20.76	25
2000	Over 4000 lb.		145	.055		16.30	2.74		19.04	22.50
2200	24 oz. sheets, under 1000 lb.		105	.076		20.50	3.78		24.28	29
2300	Over 4000 lb.		135	.059		19.35	2.94		22.29	26.50
2500	32 oz. sheets, under 1000 lb.		100	.080		27.50	3.97		31.47	37
2600	Over 4000 lb.		130	.062		26	3.05		29.05	34
2700	W shape for valleys, 16 oz., 24" wide		100	.080	L.F.	18.40	3.97		22.37	27
5800	Lead, 2.5 lb./S.F., up to 12" wide		135	.059	S.F.	6.45	2.94		9.39	12.15
5900	Over 12" wide		135	.059		4.44	2.94		7.38	9.95
8900	Stainless steel sheets, 32 ga.		155	.052		3.63	2.56		6.19	8.40
9000	28 ga.		155	.052		8.30	2.56		10.86	13.55
9100	26 ga.		155	.052		6.45	2.56		9.01	11.50
9200	24 ga.		155	.052		7.75	2.56		10.31	12.90
9290	For mechanically keyed flashing, add					40%				
9320	Steel sheets, galvanized, 20 ga.	1 Rofc	130	.062	S.F.	3	3.05		6.05	8.55
9322	22 ga.		135	.059		1.69	2.94		4.63	6.90
9324	24 ga.		140	.057		.99	2.83		3.82	5.95
9326	26 ga.		148	.054		2.30	2.68		4.98	7.15
9328	28 ga.		155	.052		1.98	2.56		4.54	6.60
9340	30 ga.		160	.050		1.67	2.48		4.15	6.10
9400	Terne coated stainless steel, .015" thick, 28 ga.		155	.052		8.35	2.56		10.91	13.55
9500	.018" thick, 26 ga.		155	.052		9.30	2.56		11.86	14.65
9600	Zinc and copper alloy (brass), .020" thick		155	.052		9.25	2.56		11.81	14.55
9700	.027" thick		155	.052		10.90	2.56		13.46	16.35

07 65 Flexible Flashing

07 65 10 – Sheet Metal Flashing

07 65 10.10 Sheet Metal Flashing and Counter Flashing

07 65 10.10 Sheet Metal Flashing and Counter Flashing	Crew	Daily Output	Labor-Hours	Unit	Material	2022 Bare Costs Labor	2022 Bare Costs Equipment	Total	Total Incl O&P
9800	.032" thick	1 Rofc	155	.052	S.F.	13.65	2.56		16.21	19.40
9900	.040" thick		155	.052		19.90	2.56		22.46	26.50
9950	Minimum labor/equipment charge	↓	3	2.667	Job		132		132	228

07 65 12 – Fabric and Mastic Flashings

07 65 12.10 Fabric and Mastic Flashing and Counter Flashing

		Crew	Daily Output	Labor-Hours	Unit	Material	Labor	Equipment	Total	Total Incl O&P
0010	**FABRIC AND MASTIC FLASHING AND COUNTER FLASHING**									
1300	Asphalt flashing cement, 5 gallon				Gal.	11.75			11.75	12.95
4900	Fabric, asphalt-saturated cotton, specification grade	1 Rofc	35	.229	S.Y.	3.40	11.35		14.75	23
5000	Utility grade		35	.229		1.48	11.35		12.83	21
5300	Close-mesh fabric, saturated, 17 oz./S.Y.		35	.229		2.26	11.35		13.61	22
5500	Fiberglass, resin-coated		35	.229	↓	1.17	11.35		12.52	21
8500	Shower pan, bituminous membrane, 7 oz.	↓	155	.052	S.F.	1.84	2.56		4.40	6.45

07 65 13 – Laminated Sheet Flashing

07 65 13.10 Laminated Sheet Flashing

		Crew	Daily Output	Labor-Hours	Unit	Material	Labor	Equipment	Total	Total Incl O&P
0010	**LAMINATED SHEET FLASHING**, Including up to 4 bends									
0500	Aluminum, fabric-backed 2 sides, mill finish, .004" thick	1 Rofc	330	.024	S.F.	1.62	1.20		2.82	3.85
0700	.005" thick		330	.024		1.90	1.20		3.10	4.16
0750	Mastic-backed, self adhesive		460	.017		3.49	.86		4.35	5.30
0800	Mastic-coated 2 sides, .004" thick		330	.024		1.62	1.20		2.82	3.85
2800	Copper, paperbacked 1 side, 2 oz.		330	.024		2.20	1.20		3.40	4.49
2900	3 oz.		330	.024		3.21	1.20		4.41	5.60
3100	Paperbacked 2 sides, 2 oz.		330	.024		2.44	1.20		3.64	4.75
3150	3 oz.		330	.024		2.28	1.20		3.48	4.58
3200	5 oz.		330	.024		3.97	1.20		5.17	6.45
3250	7 oz.		330	.024		6.25	1.20		7.45	8.95
3400	Mastic-backed 2 sides, copper, 2 oz.		330	.024		2.32	1.20		3.52	4.62
3500	3 oz.		330	.024		2.63	1.20		3.83	4.96
3700	5 oz.		330	.024		4	1.20		5.20	6.45
3800	Fabric-backed 2 sides, copper, 2 oz.		330	.024		2.09	1.20		3.29	4.37
4000	3 oz.		330	.024		2.95	1.20		4.15	5.30
4100	5 oz.		330	.024		4.13	1.20		5.33	6.60
4300	Copper-clad stainless steel, .015" thick, under 500 lb.		115	.070		7.05	3.45		10.50	13.70
4400	Over 2000 lb.		155	.052		6.95	2.56		9.51	12.05
4600	.018" thick, under 500 lb.		100	.080		8.10	3.97		12.07	15.75
4700	Over 2000 lb.		145	.055		8.35	2.74		11.09	13.85
8550	Shower pan, 3 ply copper and fabric, 3 oz.		155	.052		4.22	2.56		6.78	9.05
8600	7 oz.		155	.052		5.10	2.56		7.66	10
9300	Stainless steel, paperbacked 2 sides, .005" thick	↓	330	.024	↓	4.14	1.20		5.34	6.60

07 65 19 – Plastic Sheet Flashing

07 65 19.10 Plastic Sheet Flashing and Counter Flashing

		Crew	Daily Output	Labor-Hours	Unit	Material	Labor	Equipment	Total	Total Incl O&P
0010	**PLASTIC SHEET FLASHING AND COUNTER FLASHING**									
7300	Polyvinyl chloride, black, 10 mil	1 Rofc	285	.028	S.F.	.28	1.39		1.67	2.71
7400	20 mil		285	.028		.27	1.39		1.66	2.70
7600	30 mil		285	.028		.34	1.39		1.73	2.77
7700	60 mil		285	.028		.90	1.39		2.29	3.39
7900	Black or white for exposed roofs, 60 mil	↓	285	.028	↓	4.89	1.39		6.28	7.80
8060	PVC tape, 5" x 45 mils, for joint covers, 100 L.F./roll				Ea.	189			189	208
8850	Polyvinyl chloride, 30 mil	1 Rofc	160	.050	S.F.	1.06	2.48		3.54	5.45

07 65 Flexible Flashing

07 65 23 – Rubber Sheet Flashing

07 65 23.10 Rubber Sheet Flashing and Counter Flashing	Crew	Daily Output	Labor-Hours	Unit	Material	2022 Bare Costs Labor	Equipment	Total	Total Incl O&P
0010	**RUBBER SHEET FLASHING AND COUNTER FLASHING**									
4810	EPDM 90 mils, 1" diameter pipe flashing	1 Rofc	32	.250	Ea.	22	12.40		34.40	45.50
4820	2" diameter	1 Rofc	30	.267	Ea.	4.15	13.25		17.40	27.50
4830	3" diameter		28	.286		4.65	14.15		18.80	29.50
4840	4" diameter		24	.333		8.35	16.55		24.90	37.50
4850	6" diameter		22	.364		9.65	18.05		27.70	41.50
8100	Rubber, butyl, 1/32" thick		285	.028	S.F.	2.35	1.39		3.74	4.99
8200	1/16" thick		285	.028		4.91	1.39		6.30	7.80
8300	Neoprene, cured, 1/16" thick		285	.028		2.35	1.39		3.74	4.99
8400	1/8" thick		285	.028		6.20	1.39		7.59	9.20

07 65 26 – Self-Adhering Sheet Flashing

07 65 26.10 Self-Adhering Sheet or Roll Flashing	Crew	Daily Output	Labor-Hours	Unit	Material	2022 Bare Costs Labor	Equipment	Total	Total Incl O&P
0010	**SELF-ADHERING SHEET OR ROLL FLASHING**									
0020	Self-adhered flashing, 25 mil cross laminated HDPE, 4" wide	1 Rofc	960	.008	L.F.	.27	.41		.68	1.01
0040	6" wide		896	.009		.41	.44		.85	1.21
0060	9" wide		832	.010		.62	.48		1.10	1.50
0080	12" wide		768	.010		.82	.52		1.34	1.79

07 71 Roof Specialties

07 71 16 – Manufactured Counterflashing Systems

07 71 16.10 Roof Drain Boot

		Crew	Daily Output	Labor-Hours	Unit	Material	2022 Bare Costs Labor	Equipment	Total	Total Incl O&P
0010	**ROOF DRAIN BOOT**									
0100	Cast iron, 4" diameter	1 Shee	125	.064	L.F.	127	4.32		131.32	146
0300	4" x 3"		125	.064		121	4.32		125.32	140
0400	5" x 4"		125	.064		248	4.32		252.32	280

07 71 16.20 Pitch Pockets, Variable Sizes

		Crew	Daily Output	Labor-Hours	Unit	Material	2022 Bare Costs Labor	Equipment	Total	Total Incl O&P
0010	**PITCH POCKETS, VARIABLE SIZES**									
0100	Adjustable, 4" to 7", welded corners, 4" deep	1 Rofc	48	.167	Ea.	26.50	8.25		34.75	43.50
0200	Side extenders, 6"	"	240	.033	"	3.59	1.65		5.24	6.80

07 71 19 – Manufactured Gravel Stops and Fasciae

07 71 19.10 Gravel Stop

		Crew	Daily Output	Labor-Hours	Unit	Material	2022 Bare Costs Labor	Equipment	Total	Total Incl O&P
0010	**GRAVEL STOP**									
0020	Aluminum, .050" thick, 4" face height, mill finish	1 Shee	145	.055	L.F.	4.26	3.73		7.99	10.50
0080	Duranodic finish		145	.055		4.26	3.73		7.99	10.50
0100	Painted		145	.055		4.26	3.73		7.99	10.50
0300	6" face height		135	.059		9.30	4		13.30	16.50
0350	Duranodic finish		135	.059		10.80	4		14.80	18.15
0400	Painted		135	.059		9.50	4		13.50	16.70
0600	8" face height		125	.064		8.90	4.32		13.22	16.55
0650	Duranodic finish		125	.064		10.80	4.32		15.12	18.60
0700	Painted		125	.064		12	4.32		16.32	19.95
0900	12" face height, .080" thick, 2 piece		100	.080		13.10	5.40		18.50	23
0950	Duranodic finish		100	.080		12.70	5.40		18.10	22.50
1000	Painted		100	.080		17.05	5.40		22.45	27
1200	Copper, 16 oz., 3" face height	1 Shee	145	.055	L.F.	41.50	3.73		45.23	51.50
1300	6" face height		135	.059		50	4		54	61.50
1350	Galv steel, 24 ga., 4" leg, plain, with continuous cleat, 4" face		145	.055		8.55	3.73		12.28	15.20
1360	6" face height		145	.055		7.80	3.73		11.53	14.40
1500	Polyvinyl chloride, 6" face height		135	.059		6.90	4		10.90	13.85

For customer support on your Commercial Renovation Costs with RSMeans data, call 800.448.8182.

215

07 71 Roof Specialties

07 71 19 – Manufactured Gravel Stops and Fasciae

07 71 19.10 Gravel Stop

		Crew	Daily Output	Labor-Hours	Unit	Material	2022 Bare Costs Labor	Equipment	Total	Total Incl O&P
1600	9" face height	1 Shee	125	.064	L.F.	8.25	4.32		12.57	15.85
1800	Stainless steel, 24 ga., 6" face height		135	.059		20	4		24	29
1900	12" face height		100	.080		26.50	5.40		31.90	37.50
2100	20 ga., 6" face height		135	.059		25	4		29	34
2200	12" face height		100	.080		31.50	5.40		36.90	43
9000	Minimum labor/equipment charge		3.50	2.286	Job		154		154	240

07 71 19.30 Fascia

		Crew	Daily Output	Labor-Hours	Unit	Material	2022 Bare Costs Labor	Equipment	Total	Total Incl O&P
0010	**FASCIA**									
0100	Aluminum, reverse board and batten, .032" thick, colored, no furring incl.	1 Shee	145	.055	S.F.	8.85	3.73		12.58	15.55
0200	Residential type, aluminum	1 Carp	200	.040	L.F.	2.61	2.25		4.86	6.45
0220	Vinyl	"	200	.040	"	3.45	2.25		5.70	7.40
0300	Steel, galv and enameled, stock, no furring, long panels	1 Shee	145	.055	S.F.	6.90	3.73		10.63	13.40
0600	Short panels		115	.070	"	6.50	4.70		11.20	14.45
9000	Minimum labor/equipment charge		4	2	Job		135		135	210

07 71 23 – Manufactured Gutters and Downspouts

07 71 23.10 Downspouts

		Crew	Daily Output	Labor-Hours	Unit	Material	2022 Bare Costs Labor	Equipment	Total	Total Incl O&P
0010	**DOWNSPOUTS**									
0020	Aluminum, embossed, .020" thick, 2" x 3"	1 Shee	190	.042	L.F.	1.40	2.84		4.24	5.95
0100	Enameled		190	.042		1.40	2.84		4.24	5.95
0300	.024" thick, 2" x 3"		180	.044		1.40	3		4.40	6.20
0400	3" x 4"		140	.057		2.90	3.86		6.76	9.20
0600	Round, corrugated aluminum, 3" diameter, .020" thick		190	.042		4.80	2.84		7.64	9.75
0700	4" diameter, .025" thick		140	.057		8.40	3.86		12.26	15.20
0900	Wire strainer, round, 2" diameter		155	.052	Ea.	2.14	3.49		5.63	7.80
1000	4" diameter		155	.052		3.79	3.49		7.28	9.60
1200	Rectangular, perforated, 2" x 3"		145	.055		3.06	3.73		6.79	9.15
1300	3" x 4"		145	.055		4.62	3.73		8.35	10.90
1500	Copper, round, 16 oz., stock, 2" diameter		190	.042	L.F.	9.65	2.84		12.49	15.10
1600	3" diameter		190	.042		9.65	2.84		12.49	15.10
1800	4" diameter		145	.055		13.05	3.73		16.78	20
1900	5" diameter		130	.062		18.85	4.16		23.01	27.50
2100	Rectangular, corrugated copper, stock, 2" x 3"		190	.042		10.55	2.84		13.39	16.05
2200	3" x 4"		145	.055		15.20	3.73		18.93	22.50
2400	Rectangular, plain copper, stock, 2" x 3"	1 Shee	190	.042	L.F.	8.95	2.84		11.79	14.30
2500	3" x 4"		145	.055	"	9.90	3.73		13.63	16.70
2700	Wire strainers, rectangular, 2" x 3"		145	.055	Ea.	22	3.73		25.73	30.50
2800	3" x 4"		145	.055		18.30	3.73		22.03	26
3000	Round, 2" diameter		145	.055		12.20	3.73		15.93	19.20
3100	3" diameter		145	.055		14.10	3.73		17.83	21.50
3300	4" diameter		145	.055		16.30	3.73		20.03	24
3400	5" diameter		115	.070		21	4.70		25.70	30.50
3600	Lead-coated copper, round, stock, 2" diameter		190	.042	L.F.	35	2.84		37.84	43
3700	3" diameter		190	.042		33.50	2.84		36.34	41
3900	4" diameter		145	.055		37	3.73		40.73	47
4000	5" diameter, corrugated		130	.062		18.85	4.16		23.01	27.50
4200	6" diameter, corrugated		105	.076		25	5.15		30.15	35.50
4300	Rectangular, corrugated, stock, 2" x 3"		190	.042		19.75	2.84		22.59	26
4500	Plain, stock, 2" x 3"		190	.042		34.50	2.84		37.34	42.50
4600	3" x 4"		145	.055		55	3.73		58.73	66.50
4800	Steel, galvanized, round, corrugated, 2" or 3" diameter, 28 ga.		190	.042		3.45	2.84		6.29	8.25
4900	4" diameter, 28 ga.		145	.055		4.70	3.73		8.43	10.95
5100	5" diameter, 26 ga.		130	.062		5.15	4.16		9.31	12.10

07 71 Roof Specialties

07 71 23 – Manufactured Gutters and Downspouts

07 71 23.10 Downspouts

		Crew	Daily Output	Labor-Hours	Unit	Material	2022 Bare Costs Labor	Equipment	Total	Total Incl O&P
5400	6" diameter, 28 ga.	1 Shee	105	.076	L.F.	5.90	5.15		11.05	14.50
5500	26 ga.		105	.076		5.95	5.15		11.10	14.50
5700	Rectangular, corrugated, 28 ga., 2" x 3"		190	.042		3.10	2.84		5.94	7.85
5800	3" x 4"		145	.055		3.86	3.73		7.59	10.05
6000	Rectangular, plain, 28 ga., galvanized, 2" x 3"		190	.042		3.10	2.84		5.94	7.85
6100	3" x 4"		145	.055		3.80	3.73		7.53	10
6300	Epoxy painted, 24 ga., corrugated, 2" x 3"		190	.042		5.80	2.84		8.64	10.85
6400	3" x 4"		145	.055		7.10	3.73		10.83	13.65
6600	Wire strainers, rectangular, 2" x 3"		145	.055	Ea.	25	3.73		28.73	33.50
6700	3" x 4"		145	.055		25.50	3.73		29.23	34
6900	Round strainers, 2" or 3" diameter		145	.055		6.25	3.73		9.98	12.65
7000	4" diameter		145	.055		9.45	3.73		13.18	16.20
9000	Minimum labor/equipment charge		4	2	Job		135		135	210

07 71 23.20 Downspout Elbows

		Crew	Daily Output	Labor-Hours	Unit	Material	2022 Bare Costs Labor	Equipment	Total	Total Incl O&P
0010	**DOWNSPOUT ELBOWS**									
0020	Aluminum, embossed, 2" x 3", .020" thick	1 Shee	100	.080	Ea.	1.41	5.40		6.81	9.95
0100	Enameled		100	.080		2.49	5.40		7.89	11.15
0200	Embossed, 3" x 4", .025" thick		100	.080		1.41	5.40		6.81	9.95
0300	Enameled		100	.080		4.39	5.40		9.79	13.25
0400	Embossed, corrugated, 3" diameter, .020" thick	1 Shee	100	.080	Ea.	4.61	5.40		10.01	13.45
0500	4" diameter, .025" thick		100	.080		8.25	5.40		13.65	17.50
0600	Copper, 16 oz., 2" diameter		100	.080		12.40	5.40		17.80	22
0700	3" diameter		100	.080		14.10	5.40		19.50	24
0800	4" diameter		100	.080		18.10	5.40		23.50	28.50
1000	Rectangular, 2" x 3" corrugated		100	.080		10.95	5.40		16.35	20.50
1100	3" x 4" corrugated		100	.080		16.90	5.40		22.30	27
1300	Vinyl, 2-1/2" diameter, 45 or 75 degree bend		100	.080		5.90	5.40		11.30	14.85
1400	Tee Y junction		75	.107		14.65	7.20		21.85	27.50
9000	Minimum labor/equipment charge		4	2	Job		135		135	210

07 71 23.30 Gutters

		Crew	Daily Output	Labor-Hours	Unit	Material	2022 Bare Costs Labor	Equipment	Total	Total Incl O&P
0010	**GUTTERS**									
0012	Aluminum, stock units, 5" K type, .027" thick, plain	1 Shee	125	.064	L.F.	2.40	4.32		6.72	9.40
0100	Enameled		125	.064		2.35	4.32		6.67	9.35
0300	5" K type, .032" thick, plain		125	.064		2.70	4.32		7.02	9.70
0400	Enameled		125	.064		2.67	4.32		6.99	9.70
0700	Copper, half round, 16 oz., stock units, 4" wide		125	.064		8.30	4.32		12.62	15.85
0900	5" wide		125	.064		8.30	4.32		12.62	15.85
1000	6" wide		118	.068		11.15	4.58		15.73	19.40
1200	K type, 16 oz., stock, 5" wide		125	.064		11.75	4.32		16.07	19.70
1300	6" wide		125	.064		10.95	4.32		15.27	18.80
1500	Lead coated copper, 16 oz., half round, stock, 4" wide		125	.064		19.40	4.32		23.72	28.50
1600	6" wide		118	.068		19.55	4.58		24.13	28.50
1800	K type, stock, 5" wide		125	.064		20.50	4.32		24.82	29.50
1900	6" wide		125	.064		11.75	4.32		16.07	19.70
2100	Copper clad stainless steel, K type, 5" wide		125	.064		8.15	4.32		12.47	15.70
2200	6" wide		125	.064		4.22	4.32		8.54	11.40
2400	Steel, galv, half round or box, 28 ga., 5" wide, plain		125	.064		3.63	4.32		7.95	10.75
2500	Enameled		125	.064		3.60	4.32		7.92	10.70
2700	26 ga., stock, 5" wide		125	.064		3.63	4.32		7.95	10.75
2800	6" wide		125	.064		4.90	4.32		9.22	12.15
3000	Vinyl, O.G., 4" wide	1 Carp	115	.070		2.13	3.92		6.05	8.55
3100	5" wide		115	.070		2.13	3.92		6.05	8.55

For customer support on your Commercial Renovation Costs with RSMeans data, call 800.448.8182.

217

07 71 Roof Specialties

07 71 23 – Manufactured Gutters and Downspouts

07 71 23.30 Gutters		Crew	Daily Output	Labor-Hours	Unit	Material	2022 Bare Costs Labor	Equipment	Total	Total Incl O&P
3200	4" half round, stock units	1 Carp	115	.070	L.F.	1.49	3.92		5.41	7.85
3250	Joint connectors				Ea.	4.88			4.88	5.35
3300	Wood, clear treated cedar, fir or hemlock, 3" x 4"	1 Carp	100	.080	L.F.	10.75	4.50		15.25	19
3400	4" x 5"	"	100	.080	"	18.95	4.50		23.45	28
5000	Accessories, end cap, K type, aluminum 5"	1 Shee	625	.013	Ea.	.81	.86		1.67	2.24
5010	6"	1 Shee	625	.013	Ea.	1.44	.86		2.30	2.93
5020	Copper, 5"		625	.013		3.77	.86		4.63	5.50
5030	6"		625	.013		3.97	.86		4.83	5.70
5040	Lead coated copper, 5"		625	.013		14.45	.86		15.31	17.25
5050	6"		625	.013		15.45	.86		16.31	18.30
5060	Copper clad stainless steel, 5"		625	.013		3.93	.86		4.79	5.65
5070	6"		625	.013		3.93	.86		4.79	5.65
5080	Galvanized steel, 5"		625	.013		1.73	.86		2.59	3.25
5090	6"		625	.013		3.58	.86		4.44	5.30
5100	Vinyl, 4"	1 Carp	625	.013		2.88	.72		3.60	4.31
5110	5"	"	625	.013		3.19	.72		3.91	4.65
5120	Half round, copper, 4"	1 Shee	625	.013		5.10	.86		5.96	6.95
5130	5"		625	.013		5.80	.86		6.66	7.70
5140	6"		625	.013		9.40	.86		10.26	11.70
5150	Lead coated copper, 5"		625	.013		16.30	.86		17.16	19.30
5160	6"		625	.013		24	.86		24.86	28
5170	Copper clad stainless steel, 5"		625	.013		5.55	.86		6.41	7.45
5180	6"		625	.013		5.05	.86		5.91	6.95
5190	Galvanized steel, 5"		625	.013		3.68	.86		4.54	5.40
5200	6"		625	.013		3.84	.86		4.70	5.55
5210	Outlet, aluminum, 2" x 3"		420	.019		.98	1.29		2.27	3.08
5220	3" x 4"		420	.019		1.15	1.29		2.44	3.27
5230	2-3/8" round		420	.019		.73	1.29		2.02	2.80
5240	Copper, 2" x 3"		420	.019		8.55	1.29		9.84	11.45
5250	3" x 4"		420	.019		9.65	1.29		10.94	12.60
5260	2-3/8" round		420	.019		5.40	1.29		6.69	7.95
5270	Lead coated copper, 2" x 3"		420	.019		29	1.29		30.29	33.50
5280	3" x 4"		420	.019		32.50	1.29		33.79	38
5290	2-3/8" round		420	.019		29	1.29		30.29	34
5300	Copper clad stainless steel, 2" x 3"		420	.019		8.05	1.29		9.34	10.85
5310	3" x 4"		420	.019		9.15	1.29		10.44	12.05
5320	2-3/8" round		420	.019		5.40	1.29		6.69	7.95
5330	Galvanized steel, 2" x 3"		420	.019		4.18	1.29		5.47	6.60
5340	3" x 4"		420	.019		8	1.29		9.29	10.80
5350	2-3/8" round		420	.019		5.45	1.29		6.74	8
5360	K type mitres, aluminum		65	.123		7.15	8.30		15.45	21
5370	Copper		65	.123		19.35	8.30		27.65	34.50
5380	Lead coated copper		65	.123		61.50	8.30		69.80	80.50
5390	Copper clad stainless steel		65	.123		30.50	8.30		38.80	46.50
5400	Galvanized steel	1 Shee	65	.123	Ea.	31	8.30		39.30	47
5420	Half round mitres, copper		65	.123		78	8.30		86.30	99
5430	Lead coated copper		65	.123		100	8.30		108.30	123
5440	Copper clad stainless steel		65	.123		63	8.30		71.30	82
5450	Galvanized steel		65	.123		37	8.30		45.30	54
5460	Vinyl mitres and outlets		65	.123		12	8.30		20.30	26
5470	Sealant		940	.009	L.F.	.01	.57		.58	.90
5480	Soldering		96	.083	"	.40	5.65		6.05	9.20
9000	Minimum labor/equipment charge		3.75	2.133	Job		144		144	224

07 71 Roof Specialties

07 71 23 – Manufactured Gutters and Downspouts

07 71 23.35 Gutter Guard

		Crew	Daily Output	Labor-Hours	Unit	Material	2022 Bare Costs Labor	Equipment	Total	Total Incl O&P
0010	**GUTTER GUARD**									
0020	6" wide strip, aluminum mesh	1 Carp	500	.016	L.F.	2.32	.90		3.22	3.98
0100	Vinyl mesh		500	.016	"	1.29	.90		2.19	2.85
9000	Minimum labor/equipment charge		4	2	Job		113		113	179

07 71 26 – Reglets

07 71 26.10 Reglets and Accessories

		Crew	Daily Output	Labor-Hours	Unit	Material	2022 Bare Costs Labor	Equipment	Total	Total Incl O&P
0010	**REGLETS AND ACCESSORIES**									
0020	Reglet, aluminum, .025" thick, in parapet	1 Carp	225	.036	L.F.	2.25	2		4.25	5.65
0300	16 oz. copper		225	.036		8.65	2		10.65	12.75
0400	Galvanized steel, 24 ga.		225	.036		2.16	2		4.16	5.55
0600	Stainless steel, .020" thick		225	.036		4.88	2		6.88	8.55
0900	Counter flashing for above, 12" wide, .032" aluminum	1 Shee	150	.053		3.41	3.60		7.01	9.35
1200	16 oz. copper		150	.053		12.20	3.60		15.80	19
1300	Galvanized steel, 26 ga.		150	.053		1.95	3.60		5.55	7.75
1500	Stainless steel, .020" thick		150	.053		8.25	3.60		11.85	14.65
9000	Minimum labor/equipment charge	1 Carp	3	2.667	Job		150		150	239

07 71 29 – Manufactured Roof Expansion Joints

07 71 29.10 Expansion Joints

		Crew	Daily Output	Labor-Hours	Unit	Material	2022 Bare Costs Labor	Equipment	Total	Total Incl O&P
0010	**EXPANSION JOINTS**									
0300	Butyl or neoprene center with foam insulation, metal flanges									
0400	Aluminum, .032" thick for openings to 2-1/2"	1 Rofc	165	.048	L.F.	15.20	2.40		17.60	21
0600	For joint openings to 3-1/2"		165	.048		14	2.40		16.40	19.55
0610	For joint openings to 5"		165	.048		17.30	2.40		19.70	23
0620	For joint openings to 8"		165	.048		35	2.40		37.40	42.50
0700	Copper, 16 oz. for openings to 2-1/2"		165	.048		29.50	2.40		31.90	36.50
0900	For joint openings to 3-1/2"		165	.048		23	2.40		25.40	29.50
0910	For joint openings to 5"		165	.048		29	2.40		31.40	35.50
0920	For joint openings to 8"		165	.048		31.50	2.40		33.90	38.50
1000	Galvanized steel, 26 ga. for openings to 2-1/2"		165	.048		9.25	2.40		11.65	14.35
1200	For joint openings to 3-1/2"	1 Rofc	165	.048	L.F.	10.10	2.40		12.50	15.30
1210	For joint openings to 5"		165	.048		12.50	2.40		14.90	17.90
1220	For joint openings to 8"		165	.048		19.35	2.40		21.75	25.50
1300	Lead-coated copper, 16 oz. for openings to 2-1/2"		165	.048		42.50	2.40		44.90	50.50
1500	For joint openings to 3-1/2"		165	.048		46.50	2.40		48.90	55
1600	Stainless steel, .018", for openings to 2-1/2"		165	.048		14.55	2.40		16.95	20
1800	For joint openings to 3-1/2"		165	.048		16.75	2.40		19.15	22.50
1810	For joint openings to 5"		165	.048		19.05	2.40		21.45	25
1820	For joint openings to 8"		165	.048		26	2.40		28.40	33
1900	Neoprene, double-seal type with thick center, 4-1/2" wide		125	.064		17.10	3.17		20.27	24.50
1950	Polyethylene bellows, with galv steel flat flanges		100	.080		9.15	3.97		13.12	16.95
1960	With galvanized angle flanges		100	.080		9.25	3.97		13.22	17.05
2000	Roof joint with extruded aluminum cover, 2"	1 Shee	115	.070		40	4.70		44.70	51.50
2100	Expansion, stainless flange, foam center, standard	1 Rofc	100	.080		16.30	3.97		20.27	25
2200	Large	"	100	.080		22	3.97		25.97	31
2500	Roof to wall joint with extruded aluminum cover	1 Shee	115	.070		36	4.70		40.70	47
2700	Wall joint, closed cell foam on PVC cover, 9" wide	1 Rofc	125	.064		7.40	3.17		10.57	13.55
2800	12" wide	"	115	.070		8.35	3.45		11.80	15.15
9000	Minimum labor/equipment charge	1 Shee	3	2.667	Job		180		180	280

07 71 Roof Specialties

07 71 43 – Drip Edge

	07 71 43.10 Drip Edge, Rake Edge, Ice Belts	Crew	Daily Output	Labor-Hours	Unit	Material	2022 Bare Costs Labor	Equipment	Total	Total Incl O&P
0010	**DRIP EDGE, RAKE EDGE, ICE BELTS**									
0020	Aluminum, .016" thick, 5" wide, mill finish	1 Carp	400	.020	L.F.	.69	1.13		1.82	2.55
0100	White finish		400	.020		.70	1.13		1.83	2.56
0200	8" wide, mill finish		400	.020		.84	1.13		1.97	2.71
0300	Ice belt, 28" wide, mill finish		100	.080		6.20	4.50		10.70	13.95
0310	Vented, mill finish		400	.020		2.24	1.13		3.37	4.25
0320	Painted finish		400	.020		2.97	1.13		4.10	5.05
0400	Galvanized, 5" wide		400	.020		.96	1.13		2.09	2.85
0500	8" wide, mill finish		400	.020		.99	1.13		2.12	2.88
0510	Rake edge, aluminum, 1-1/2" x 1-1/2"		400	.020		.34	1.13		1.47	2.16
0520	3-1/2" x 1-1/2"		400	.020		.49	1.13		1.62	2.33
9000	Minimum labor/equipment charge		4	2	Job		113		113	179

07 72 Roof Accessories

07 72 23 – Relief Vents

07 72 23.10 Roof Vents

		Crew	Daily Output	Labor-Hours	Unit	Material	2022 Bare Costs Labor	Equipment	Total	Total Incl O&P
0010	**ROOF VENTS**									
0020	Mushroom shape, for built-up roofs, aluminum	1 Rofc	30	.267	Ea.	89	13.25		102.25	121
0100	PVC, 6" high	1 Rofc	30	.267	Ea.	29	13.25		42.25	55
9000	Minimum labor/equipment charge	"	2.75	2.909	Job		144		144	248

07 72 23.20 Vents

		Crew	Daily Output	Labor-Hours	Unit	Material	2022 Bare Costs Labor	Equipment	Total	Total Incl O&P
0010	**VENTS**									
0100	Soffit or eave, aluminum, mill finish, strips, 2-1/2" wide	1 Carp	200	.040	L.F.	.56	2.25		2.81	4.20
0200	3" wide		200	.040		.54	2.25		2.79	4.17
0300	Enamel finish, 3" wide		200	.040		.61	2.25		2.86	4.25
0400	Mill finish, rectangular, 4" x 16"		72	.111	Ea.	1.99	6.25		8.24	12.15
0500	8" x 16"		72	.111		2.90	6.25		9.15	13.15
2420	Roof ventilator	Q-9	16	1		58.50	61		119.50	159
2500	Vent, roof vent	1 Rofc	24	.333		29.50	16.55		46.05	61

07 72 26 – Ridge Vents

07 72 26.10 Ridge Vents and Accessories

		Crew	Daily Output	Labor-Hours	Unit	Material	2022 Bare Costs Labor	Equipment	Total	Total Incl O&P
0010	**RIDGE VENTS AND ACCESSORIES**									
0100	Aluminum strips, mill finish	1 Rofc	160	.050	L.F.	2.53	2.48		5.01	7.05
0150	Painted finish		160	.050	"	3.75	2.48		6.23	8.40
0200	Connectors		48	.167	Ea.	4.44	8.25		12.69	19.15
0300	End caps		48	.167	"	2.04	8.25		10.29	16.50
0400	Galvanized strips		160	.050	L.F.	5	2.48		7.48	9.75
0430	Molded polyethylene, shingles not included		160	.050	"	2.62	2.48		5.10	7.15
0440	End plugs		48	.167	Ea.	2.04	8.25		10.29	16.50
0450	Flexible roll, shingles not included		160	.050	L.F.	2.55	2.48		5.03	7.05
2300	Ridge vent strip, mill finish	1 Shee	155	.052	"	2.75	3.49		6.24	8.50

07 72 33 – Roof Hatches

07 72 33.10 Roof Hatch Options

		Crew	Daily Output	Labor-Hours	Unit	Material	2022 Bare Costs Labor	Equipment	Total	Total Incl O&P
0010	**ROOF HATCH OPTIONS**									
0500	2'-6" x 3', aluminum curb and cover	G-3	10	3.200	Ea.	1,125	181		1,306	1,525
0520	Galvanized steel curb and aluminum cover		10	3.200		1,050	181		1,231	1,425
0540	Galvanized steel curb and cover		10	3.200		835	181		1,016	1,200
0600	2'-6" x 4'-6", aluminum curb and cover		9	3.556		1,400	201		1,601	1,850
0800	Galvanized steel curb and aluminum cover		9	3.556		1,500	201		1,701	1,975

07 72 Roof Accessories

07 72 33 - Roof Hatches

07 72 33.10 Roof Hatch Options

	07 72 33.10 Roof Hatch Options	Crew	Daily Output	Labor-Hours	Unit	Material	2022 Bare Costs Labor	Equipment	Total	Total Incl O&P
0900	Galvanized steel curb and cover	G-3	9	3.556	Ea.	1,100	201		1,301	1,550
1100	4' x 4' aluminum curb and cover		8	4		2,250	226		2,476	2,825
1120	Galvanized steel curb and aluminum cover		8	4		2,350	226		2,576	2,925
1140	Galvanized steel curb and cover		8	4		1,225	226		1,451	1,700
1200	2'-6" x 8'-0", aluminum curb and cover		6.60	4.848		2,250	274		2,524	2,900
1400	Galvanized steel curb and aluminum cover		6.60	4.848		2,425	274		2,699	3,100
1500	Galvanized steel curb and cover	↓	6.60	4.848		2,025	274		2,299	2,650
1800	For plexiglass panels, 2'-6" x 3'-0", add to above				↓	630			630	695
9000	Minimum labor/equipment charge	2 Carp	2	8	Job		450		450	715

07 72 36 - Smoke Vents

07 72 36.10 Smoke Hatches

					Unit	Material	Labor	Equipment	Total	Total Incl O&P
0010	**SMOKE HATCHES**									
0200	For 3'-0" long, add to roof hatches from Section 07 72 33.10				Ea.	25%	5%			
0250	For 4'-0" long, add to roof hatches from Section 07 72 33.10					20%	5%			
0300	For 8'-0" long, add to roof hatches from Section 07 72 33.10				↓	10%	5%			

07 72 36.20 Smoke Vent Options

		Crew	Daily Output	Labor-Hours	Unit	Material	Labor	Equipment	Total	Total Incl O&P
0010	**SMOKE VENT OPTIONS**									
0100	4' x 4' aluminum cover and frame	G-3	13	2.462	Ea.	2,475	139		2,614	2,950
0200	Galvanized steel cover and frame		13	2.462		1,925	139		2,064	2,325
0300	4' x 8' aluminum cover and frame		8	4		3,750	226		3,976	4,475
0400	Galvanized steel cover and frame	↓	8	4	↓	3,100	226		3,326	3,775
9000	Minimum labor/equipment charge	2 Carp	2	8	Job		450		450	715

07 72 53 - Snow Guards

07 72 53.10 Snow Guard Options

		Crew	Daily Output	Labor-Hours	Unit	Material	Labor	Equipment	Total	Total Incl O&P
0010	**SNOW GUARD OPTIONS**									
0100	Slate & asphalt shingle roofs, fastened with nails	1 Rofc	160	.050	Ea.	14.15	2.48		16.63	19.80
0200	Standing seam metal roofs, fastened with set screws		48	.167		20	8.25		28.25	36.50
0300	Surface mount for metal roofs, fastened with solder		48	.167	↓	8.80	8.25		17.05	24
0400	Double rail pipe type, including pipe	↓	130	.062	L.F.	40.50	3.05		43.55	50.50

07 72 80 - Vent Options

07 72 80.30 Vent Options

		Crew	Daily Output	Labor-Hours	Unit	Material	Labor	Equipment	Total	Total Incl O&P
0010	**VENT OPTIONS**									
0020	Plastic, for insulated decks, 1 per M.S.F.	1 Rofc	40	.200	Ea.	23	9.90		32.90	42.50
0100	Heavy duty		20	.400		59.50	19.85		79.35	99.50
0300	Aluminum	↓	30	.267		21	13.25		34.25	46
0800	Polystyrene baffles, 12" wide for 16" OC rafter spacing	1 Carp	90	.089		.48	5		5.48	8.50
0900	For 24" OC rafter spacing		110	.073	↓	.83	4.09		4.92	7.40
9000	Minimum labor/equipment charge	↓	3	2.667	Job		150		150	239

07 76 Roof Pavers

07 76 16 – Roof Decking Pavers

07 76 16.10 Roof Pavers and Supports

	07 76 16.10 Roof Pavers and Supports	Crew	Daily Output	Labor-Hours	Unit	Material	2022 Bare Costs Labor	Equipment	Total	Total Incl O&P
0010	ROOF PAVERS AND SUPPORTS									
1000	Roof decking pavers, concrete blocks, 2" thick, natural	1 Clab	115	.070	S.F.	3.70	3.17		6.87	9.10
1100	Colors		115	.070	"	2.42	3.17		5.59	7.70
1200	Support pedestal, bottom cap		960	.008	Ea.	2.73	.38		3.11	3.60
1300	Top cap		960	.008		5	.38		5.38	6.10
1400	Leveling shims, 1/16"		1920	.004		1.24	.19		1.43	1.66
1500	1/8"		1920	.004		2.05	.19		2.24	2.56
1600	Buffer pad		960	.008		4.63	.38		5.01	5.70
1700	PVC legs (4" SDR 35)	1 Clab	2880	.003	Inch	.18	.13		.31	.40
2000	Alternate pricing method, system in place	"	101	.079	S.F.	8.20	3.61		11.81	14.80

07 81 Applied Fireproofing

07 81 16 – Cementitious Fireproofing

07 81 16.10 Sprayed Cementitious Fireproofing

	07 81 16.10 Sprayed Cementitious Fireproofing	Crew	Daily Output	Labor-Hours	Unit	Material	2022 Bare Costs Labor	Equipment	Total	Total Incl O&P
0010	SPRAYED CEMENTITIOUS FIREPROOFING									
0050	Not including canvas protection, normal density									
0100	Per 1" thick, on flat plate steel	G-2	3000	.008	S.F.	.59	.38	.05	1.02	1.29
0200	Flat decking		2400	.010		.59	.48	.06	1.13	1.46
0400	Beams		1500	.016		.59	.76	.09	1.44	1.95
0500	Corrugated or fluted decks		1250	.019		.88	.91	.11	1.90	2.54
0700	Columns, 1-1/8" thick		1100	.022		.66	1.04	.13	1.83	2.51
0800	2-3/16" thick		700	.034		1.66	1.63	.20	3.49	4.64
0900	For canvas protection, add		5000	.005		.11	.23	.03	.37	.51
1000	Not including canvas protection, high density									
1100	Per 1" thick, on flat plate steel	G-2	3000	.008	S.F.	2.35	.38	.05	2.78	3.24
1110	On flat decking		2400	.010		2.35	.48	.06	2.89	3.41
1120	On beams		1500	.016		2.35	.76	.09	3.20	3.90
1130	Corrugated or fluted decks		1250	.019		2.35	.91	.11	3.37	4.16
1140	Columns, 1-1/8" thick		1100	.022		2.64	1.04	.13	3.81	4.70
1150	2-3/16" thick		1100	.022		5.30	1.04	.13	6.47	7.60
1170	For canvas protection, add		5000	.005		.11	.23	.03	.37	.51
1200	Not including canvas protection, retrofitting									
1210	Per 1" thick, on flat plate steel	G-2	1500	.016	S.F.	.53	.76	.09	1.38	1.90
1220	On flat decking		1200	.020		.53	.95	.12	1.60	2.23
1230	On beams		750	.032		.53	1.52	.18	2.23	3.21
1240	Corrugated or fluted decks		625	.038		.80	1.83	.22	2.85	4.02
1250	Columns, 1-1/8" thick		550	.044		.60	2.08	.25	2.93	4.24
1260	2-3/16" thick		500	.048		1.20	2.29	.28	3.77	5.25
1400	Accessories, preliminary spattered texture coat		4500	.005		.05	.25	.03	.33	.48
1410	Bonding agent	1 Plas	1000	.008		.11	.41		.52	.77
9000	Minimum labor/equipment charge	G-2	3	8	Job		380	46	426	655

07 84 13 – Penetration Firestopping

07 84 13.10 Firestopping		Crew	Daily Output	Labor-Hours	Unit	Material	2022 Bare Costs Labor	Equipment	Total	Total Incl O&P
0010	**FIRESTOPPING**	R078413-30								
0100	Metallic piping, non insulated									
0110	Through walls, 2" diameter	1 Carp	16	.500	Ea.	11.50	28		39.50	57
0120	4" diameter	"	14	.571	"	20	32		52	73
0130	6" diameter	1 Carp	12	.667	Ea.	29.50	37.50		67	92
0140	12" diameter		10	.800		25	45		70	99
0150	Through floors, 2" diameter		32	.250		5.75	14.10		19.85	29
0160	4" diameter		28	.286		10.10	16.10		26.20	36.50
0170	6" diameter		24	.333		14.75	18.75		33.50	46.50
0180	12" diameter		20	.400		12.95	22.50		35.45	50.50
0190	Metallic piping, insulated									
0200	Through walls, 2" diameter	1 Carp	16	.500	Ea.	20	28		48	66.50
0210	4" diameter		14	.571		29.50	32		61.50	83.50
0220	6" diameter		12	.667		39.50	37.50		77	103
0230	12" diameter		10	.800		67.50	45		112.50	146
0240	Through floors, 2" diameter		32	.250		10.10	14.10		24.20	33.50
0250	4" diameter		28	.286		14.75	16.10		30.85	42
0260	6" diameter		24	.333		19.80	18.75		38.55	52
0270	12" diameter		20	.400		34	22.50		56.50	73
0280	Non metallic piping, non insulated									
0290	Through walls, 2" diameter	1 Carp	12	.667	Ea.	70	37.50		107.50	137
0300	4" diameter		10	.800		129	45		174	214
0310	6" diameter		8	1		247	56.50		303.50	360
0330	Through floors, 2" diameter		16	.500		44.50	28		72.50	93
0340	4" diameter		6	1.333		81	75		156	208
0350	6" diameter		6	1.333		147	75		222	280
0370	Ductwork, insulated & non insulated, round									
0380	Through walls, 6" diameter	1 Carp	12	.667	Ea.	39.50	37.50		77	103
0390	12" diameter		10	.800		67.50	45		112.50	146
0400	18" diameter		8	1		82	56.50		138.50	180
0410	Through floors, 6" diameter		16	.500		19.80	28		47.80	66.50
0420	12" diameter		14	.571		14.75	32		46.75	67.50
0430	18" diameter		12	.667		17.80	37.50		55.30	79
0440	Ductwork, insulated & non insulated, rectangular									
0450	With stiffener/closure angle, through walls, 6" x 12"	1 Carp	8	1	Ea.	52	56.50		108.50	147
0460	12" x 24"		6	1.333		104	75		179	233
0470	24" x 48"		4	2		207	113		320	405
0480	With stiffener/closure angle, through floors, 6" x 12"		10	.800		10.50	45		55.50	83
0490	12" x 24"		8	1		52	56.50		108.50	147
0500	24" x 48"		6	1.333		104	75		179	233
0510	Multi trade openings									
0520	Through walls, 6" x 12"	1 Carp	2	4	Ea.	79	225		304	445
0530	12" x 24"	1 Carp	1	8	Ea.	213	450		663	950
0540	24" x 48"	2 Carp	1	16		645	900		1,545	2,125
0550	48" x 96"	"	.75	21.333		2,575	1,200		3,775	4,725
0560	Through floors, 6" x 12"	1 Carp	2	4		48.50	225		273.50	415
0570	12" x 24"	"	1	8		64.50	450		514.50	785
0580	24" x 48"	2 Carp	.75	21.333		139	1,200		1,339	2,050
0590	48" x 96"	"	.50	32		325	1,800		2,125	3,200
0600	Structural penetrations, through walls									
0610	Steel beams, W8 x 10	1 Carp	8	1	Ea.	74.50	56.50		131	172
0620	W12 x 14		6	1.333		108	75		183	238
0630	W21 x 44		5	1.600		166	90		256	325

For customer support on your Commercial Renovation Costs with RSMeans data, call 800.448.8182.

223

07 84 13.10 Firestopping		Crew	Daily Output	Labor-Hours	Unit	Material	2022 Bare Costs Labor	Equipment	Total	Total Incl O&P
0640	W36 x 135	1 Carp	3	2.667	Ea.	235	150		385	500
0650	Bar joists, 18" deep		6	1.333		43.50	75		118.50	167
0660	24" deep		6	1.333		61	75		136	187
0670	36" deep		5	1.600		92	90		182	244
0680	48" deep		4	2		133	113		246	325
0690	Construction joints, floor slab at exterior wall									
0700	Precast, brick, block or drywall exterior									
0710	2" wide joint	1 Carp	125	.064	L.F.	11.10	3.60		14.70	17.90
0720	4" wide joint	"	75	.107	"	17.10	6		23.10	28.50
0730	Metal panel, glass or curtain wall exterior									
0740	2" wide joint	1 Carp	40	.200	L.F.	12.60	11.25		23.85	32
0750	4" wide joint	"	25	.320	"	14.30	18		32.30	44.50
0760	Floor slab to drywall partition									
0770	Flat joint	1 Carp	100	.080	L.F.	14.50	4.50		19	23
0780	Fluted joint		50	.160		16.85	9		25.85	33
0790	Etched fluted joint		75	.107		20.50	6		26.50	32
0800	Floor slab to concrete/masonry partition									
0810	Flat joint	1 Carp	75	.107	L.F.	16.90	6		22.90	28
0820	Fluted joint	"	50	.160	"	18.40	9		27.40	34.50
0830	Concrete/CMU wall joints									
0840	1" wide	1 Carp	100	.080	L.F.	20.50	4.50		25	29.50
0850	2" wide		75	.107		32	6		38	44.50
0860	4" wide		50	.160		36	9		45	54.50
0870	Concrete/CMU floor joints									
0880	1" wide	1 Carp	200	.040	L.F.	21.50	2.25		23.75	27.50
0890	2" wide		150	.053		25	3		28	32.50
0900	4" wide		100	.080		36	4.50		40.50	47

07 91 13.10 Compression Seals

		Crew	Daily Output	Labor-Hours	Unit	Material	Labor	Equipment	Total	Total Incl O&P
0010	**COMPRESSION SEALS**									
4900	O-ring type cord, 1/4"	1 Bric	472	.017	L.F.	.43	.94		1.37	1.97
4910	1/2"		440	.018		1.14	1		2.14	2.86
4920	3/4"		424	.019		2.29	1.04		3.33	4.19
4930	1"		408	.020		4.59	1.08		5.67	6.80
4940	1-1/4"		384	.021		11.05	1.15		12.20	14
4950	1-1/2"		368	.022		13.55	1.20		14.75	16.85
4960	1-3/4"		352	.023		15.85	1.26		17.11	19.40
4970	2"		344	.023		27.50	1.28		28.78	32.50

07 91 16.10 Joint Gaskets

		Crew	Daily Output	Labor-Hours	Unit	Material	Labor	Equipment	Total	Total Incl O&P
0010	**JOINT GASKETS**									
4400	Joint gaskets, neoprene, closed cell w/adh, 1/8" x 3/8"	1 Bric	240	.033	L.F.	.22	1.84		2.06	3.19
4500	1/4" x 3/4"		215	.037		.76	2.06		2.82	4.13
4700	1/2" x 1"		200	.040		.73	2.21		2.94	4.34
4800	3/4" x 1-1/2"		165	.048		1.42	2.68		4.10	5.85

224

For customer support on your **Commercial Renovation Costs** with RSMeans data, call 800.448.8182.

07 91 Preformed Joint Seals

07 91 23 – Backer Rods

07 91 23.10 Backer Rods	Crew	Daily Output	Labor-Hours	Unit	Material	2022 Bare Costs Labor	Equipment	Total	Total Incl O&P
0010 **BACKER RODS**									
0032　Backer rod, polyethylene, 1/4" diameter	1 Bric	460	.017	L.F.	.07	.96		1.03	1.62
0052　　　　1/2" diameter		460	.017		.16	.96		1.12	1.72
0072　　　　3/4" diameter		460	.017		.17	.96		1.13	1.73
0092　　　　1" diameter		460	.017		.30	.96		1.26	1.87

07 91 26 – Joint Fillers

07 91 26.10 Joint Fillers	Crew	Daily Output	Labor-Hours	Unit	Material	2022 Bare Costs Labor	Equipment	Total	Total Incl O&P
0010 **JOINT FILLERS**									
4360　Butyl rubber filler, 1/4" x 1/4"	1 Bric	290	.028	L.F.	.28	1.52		1.80	2.75
4365　　　　1/2" x 1/2"		250	.032		1.13	1.77		2.90	4.07
4370　　　　1/2" x 3/4"		210	.038		1.69	2.10		3.79	5.25
4375　　　　3/4" x 3/4"		230	.035		2.54	1.92		4.46	5.85
4380　　　　1" x 1"		180	.044		3.38	2.46		5.84	7.65
4390　　　　For coloring, add					12%				
4980　Polyethylene joint backing, 1/4" x 2"	1 Bric	2.08	3.846	C.L.F.	13	213		226	355
4990　　　　1/4" x 6"		1.28	6.250	"	30	345		375	590
5600　Silicone, room temp vulcanizing foam seal, 1/4" x 1/2"		1312	.006	L.F.	.31	.34		.65	.88
5610　　　　1/2" x 1/2"		656	.012		.61	.67		1.28	1.75
5620　　　　1/2" x 3/4"		442	.018		.92	1		1.92	2.61
5630　　　　3/4" x 3/4"		328	.024		1.38	1.35		2.73	3.68
5640　　　　1/8" x 1"		1312	.006		.31	.34		.65	.88
5650　　　　1/8" x 3"	1 Bric	442	.018	L.F.	.92	1		1.92	2.61
5670　　　　1/4" x 3"		295	.027		1.84	1.50		3.34	4.42
5680　　　　1/4" x 6"		148	.054		3.68	2.99		6.67	8.80
5690　　　　1/2" x 6"		82	.098		7.35	5.40		12.75	16.75
5700　　　　1/2" x 9"		52.50	.152		11.05	8.40		19.45	25.50
5710　　　　1/2" x 12"		33	.242		14.70	13.40		28.10	37.50

07 92 Joint Sealants

07 92 13 – Elastomeric Joint Sealants

07 92 13.10 Masonry Joint Sealants	Crew	Daily Output	Labor-Hours	Unit	Material	2022 Bare Costs Labor	Equipment	Total	Total Incl O&P
0010 **MASONRY JOINT SEALANTS**, 1/2" x 1/2" joint									
0050　Re-caulk only, oil base	1 Bric	225	.036	L.F.	.62	1.96		2.58	3.83
0100　　　　Acrylic latex		205	.039		.39	2.16		2.55	3.88
0200　　　　Polyurethane		200	.040		.75	2.21		2.96	4.36
0300　　　　Silicone		195	.041		.61	2.27		2.88	4.30
1000　Cut out and re-caulk, oil base		145	.055		.62	3.05		3.67	5.55
1050　　　　Acrylic latex		130	.062		.39	3.40		3.79	5.90
1100　　　　Polyurethane		125	.064		.75	3.54		4.29	6.45
1150　　　　Silicone		120	.067		.61	3.68		4.29	6.55
9000　Minimum labor/equipment charge		4	2	Job		111		111	177

07 92 13.20 Caulking and Sealant Options	Crew	Daily Output	Labor-Hours	Unit	Material	2022 Bare Costs Labor	Equipment	Total	Total Incl O&P
0010 **CAULKING AND SEALANT OPTIONS**									
0050　Latex acrylic based, bulk				Gal.	31			31	34
0055　　　　Bulk in place 1/4" x 1/4" bead	1 Bric	300	.027	L.F.	.10	1.47		1.57	2.47
0060　　　　1/4" x 3/8"		294	.027		.16	1.50		1.66	2.59
0065　　　　1/4" x 1/2"		288	.028		.22	1.53		1.75	2.70
0075　　　　3/8" x 3/8"		284	.028		.24	1.56		1.80	2.76
0080　　　　3/8" x 1/2"		280	.029		.33	1.58		1.91	2.89
0085　　　　3/8" x 5/8"		276	.029		.41	1.60		2.01	3.02

For customer support on your Commercial Renovation Costs with RSMeans data, call 800.448.8182.

225

07 92 13 – Elastomeric Joint Sealants

07 92 13.20 Caulking and Sealant Options

		Crew	Daily Output	Labor-Hours	Unit	Material	2022 Bare Costs Labor	Equipment	Total	Total Incl O&P
0095	3/8" x 3/4"	1 Bric	272	.029	L.F.	.49	1.63		2.12	3.14
0100	1/2" x 1/2"		275	.029		.44	1.61		2.05	3.05
0105	1/2" x 5/8"		269	.030		.54	1.64		2.18	3.23
0110	1/2" x 3/4"		263	.030		.65	1.68		2.33	3.41
0115	1/2" x 7/8"		256	.031		.76	1.73		2.49	3.61
0120	1/2" x 1"		250	.032		.87	1.77		2.64	3.79
0125	3/4" x 3/4"		244	.033		.98	1.81		2.79	3.98
0130	3/4" x 1"		225	.036		1.31	1.96		3.27	4.59
0135	1" x 1"		200	.040		1.74	2.21		3.95	5.45
0190	Cartridges				Gal.	54.50			54.50	60
0200	11 fl. oz. cartridge				Ea.	4.68			4.68	5.15
0500	1/4" x 1/2"	1 Bric	288	.028	L.F.	.38	1.53		1.91	2.88
0600	1/2" x 1/2"	"	275	.029	"	.76	1.61		2.37	3.41
0800	3/4" x 3/4"	1 Bric	244	.033	L.F.	1.72	1.81		3.53	4.79
0900	3/4" x 1"		225	.036		2.29	1.96		4.25	5.65
1000	1" x 1"		200	.040		2.87	2.21		5.08	6.70
1400	Butyl based, bulk				Gal.	42			42	46.50
1500	Cartridges				"	42			42	46.50
1700	1/4" x 1/2", 154 L.F./gal.	1 Bric	288	.028	L.F.	.27	1.53		1.80	2.76
1800	1/2" x 1/2", 77 L.F./gal.	"	275	.029	"	.55	1.61		2.16	3.17
2300	Polysulfide compounds, 1 component, bulk				Gal.	86			86	94.50
2400	Cartridges				"	175			175	192
2600	1 or 2 component, in place, 1/4" x 1/4", 308 L.F./gal.	1 Bric	300	.027	L.F.	.28	1.47		1.75	2.67
2700	1/2" x 1/4", 154 L.F./gal.		288	.028		.56	1.53		2.09	3.07
2900	3/4" x 3/8", 68 L.F./gal.		272	.029		1.26	1.63		2.89	3.99
3000	1" x 1/2", 38 L.F./gal.		250	.032		2.26	1.77		4.03	5.30
3200	Polyurethane, 1 or 2 component				Gal.	56			56	61.50
3300	Cartridges				"	66.50			66.50	73.50
3500	Bulk, in place, 1/4" x 1/4"	1 Bric	300	.027	L.F.	.18	1.47		1.65	2.56
3655	1/2" x 1/4"		288	.028		.36	1.53		1.89	2.86
3800	3/4" x 3/8"		272	.029		.82	1.63		2.45	3.51
3900	1" x 1/2"		250	.032		1.45	1.77		3.22	4.43
4100	Silicone rubber, bulk				Gal.	67.50			67.50	74
4200	Cartridges				"	45.50			45.50	50.50
4300	Bulk in place, 1/4" x 1/2", 154 L.F./gal.	1 Bric	235	.034	L.F.	.44	1.88		2.32	3.49

07 92 16 – Rigid Joint Sealants

07 92 16.10 Rigid Joint Sealants

		Crew	Daily Output	Labor-Hours	Unit	Material	Labor	Equipment	Total	Total Incl O&P
0010	**RIGID JOINT SEALANTS**									
5802	Tapes, sealant, PVC foam adhesive, 1/16" x 1/4"				L.F.	.04			.04	.04
5902	1/16" x 1/2"					.09			.09	.10
5952	1/16" x 1"					.15			.15	.17
6002	1/8" x 1/2"					.09			.09	.10

07 92 19 – Acoustical Joint Sealants

07 92 19.10 Acoustical Sealant

		Crew	Daily Output	Labor-Hours	Unit	Material	Labor	Equipment	Total	Total Incl O&P
0010	**ACOUSTICAL SEALANT**									
0020	Acoustical sealant, elastomeric, cartridges				Ea.	8.25			8.25	9.10
0025	In place, 1/4" x 1/4"	1 Bric	300	.027	L.F.	.34	1.47		1.81	2.73
0030	1/4" x 1/2"		288	.028		.67	1.53		2.20	3.20
0035	1/2" x 1/2"		275	.029		1.35	1.61		2.96	4.05
0040	1/2" x 3/4"		263	.030		2.02	1.68		3.70	4.92
0045	3/4" x 3/4"		244	.033		3.04	1.81		4.85	6.25
0050	1" x 1"		200	.040		5.40	2.21		7.61	9.50

07 95 Expansion Control

07 95 13 – Expansion Joint Cover Assemblies

07 95 13.50 Expansion Joint Assemblies	Crew	Daily Output	Labor-Hours	Unit	Material	2022 Bare Costs Labor	Equipment	Total	Total Incl O&P
0010 **EXPANSION JOINT ASSEMBLIES**									
0200 Floor cover assemblies, 1" space, aluminum	1 Sswk	38	.211	L.F.	29	13		42	53
0300 Bronze		38	.211		54	13		67	80
0500 2" space, aluminum		38	.211		40.50	13		53.50	66
0600 Bronze		38	.211		62.50	13		75.50	89.50
0800 Wall and ceiling assemblies, 1" space, aluminum		38	.211		17.40	13		30.40	40
0900 Bronze		38	.211		55	13		68	81.50
1100 2" space, aluminum		38	.211		25.50	13		38.50	49
1200 Bronze		38	.211		53	13		66	79
1400 Floor to wall assemblies, 1" space, aluminum		38	.211		25	13		38	48.50
1500 Bronze or stainless		38	.211		64.50	13		77.50	92
1700 Gym floor angle covers, aluminum, 3" x 3" angle		46	.174		21.50	10.75		32.25	41
1800 3" x 4" angle		46	.174		25.50	10.75		36.25	45.50
2000 Roof closures, aluminum, flat roof, low profile, 1" space		57	.140		18.90	8.70		27.60	35
2100 High profile		57	.140		26	8.70		34.70	42.50
2300 Roof to wall, low profile, 1" space		57	.140		25	8.70		33.70	41.50
2400 High profile		57	.140		33	8.70		41.70	50
9000 Minimum labor/equipment charge		2	4	Job		247		247	400

For customer support on your Commercial Renovation Costs with RSMeans data, call 800.448.8182.

227

Division Notes

	CREW	DAILY OUTPUT	LABOR-HOURS	UNIT	BARE COSTS				TOTAL INCL O&P
					MAT.	LABOR	EQUIP.	TOTAL	

Estimating Tips
08 10 00 Doors and Frames

All exterior doors should be addressed for their energy conservation (insulation and seals).

- Most metal doors and frames look alike, but there may be significant differences among them. When estimating these items, be sure to choose the line item that most closely compares to the specification or door schedule requirements regarding:
  - □ type of metal
  - □ metal gauge
  - □ door core material
  - □ fire rating
  - □ finish
- Wood and plastic doors vary considerably in price. The primary determinant is the veneer material. Lauan, birch, and oak are the most common veneers. Other variables include the following:
  - □ hollow or solid core
  - □ fire rating
  - □ flush or raised panel
  - □ finish
- Door pricing includes bore for cylindrical locksets and mortise for hinges.

08 30 00 Specialty Doors and Frames

- There are many varieties of special doors, and they are usually priced per each. Add frames, hardware, or operators required for a complete installation.

08 40 00 Entrances, Storefronts, and Curtain Walls

- Glazed curtain walls consist of the metal tube framing and the glazing material. The cost data in this subdivision is presented for the metal tube framing alone or the composite wall. If your estimate requires a detailed takeoff of the framing, be sure to add the glazing cost and any tints.

08 50 00 Windows

- Steel windows are unglazed and aluminum can be glazed or unglazed. Some metal windows are priced without glass. Refer to 08 80 00 Glazing for glass pricing. The grade C indicates commercial grade windows, usually ASTM C-35.
- All wood windows and vinyl are priced preglazed. The glazing is insulating glass. Add the cost of screens and grills if required and not already included.

08 70 00 Hardware

- Hardware costs add considerably to the cost of a door. The most efficient method to determine the hardware requirements for a project is to review the door and hardware schedule together. One type of door may have different hardware, depending on the door usage.
- Door hinges are priced by the pair, with most doors requiring 1-1/2 pairs per door. The hinge prices do not include installation labor because it is included in door installation.

Hinges are classified according to the frequency of use, base material, and finish.

08 80 00 Glazing

- Different openings require different types of glass. The most common types are:
  - □ float
  - □ tempered
  - □ insulating
  - □ impact-resistant
  - □ ballistic-resistant
- Most exterior windows are glazed with insulating glass. Entrance doors and window walls, where the glass is less than 18" from the floor, are generally glazed with tempered glass. Interior windows and some residential windows are glazed with float glass.
- Coastal communities require the use of impact-resistant glass, dependent on wind speed.
- The insulation or 'u' value is a strong consideration, along with solar heat gain, to determine total energy efficiency.

Reference Numbers

Reference numbers are shown at the beginning of some major classifications. These numbers refer to related items in the Reference Section. The reference information may be an estimating procedure, an alternate pricing method, or technical information.

Note: Not all subdivisions listed here necessarily appear. ■

Same Data. Simplified.

Enjoy the convenience and efficiency of accessing your costs anywhere:

- ■ **Skip the multiplier** by setting your location
- ■ **Quickly search,** edit, favorite and share costs
- ■ **Stay on top of price changes** with automatic updates

Discover more at rsmeans.com/online

---

08 01 Operation and Maintenance of Openings

08 01 11 – Operation and Maintenance of Metal Doors and Frames

08 01 11.10 Door and Window Maintenance

	08 01 11.10 Door and Window Maintenance	Crew	Daily Output	Labor-Hours	Unit	Material	2022 Bare Costs Labor	Equipment	Total	Total Incl O&P
0010	**DOOR & WINDOW MAINTENANCE**									
1130	Install door	2 Carp	13.07	1.224	Ea.		69		69	110
1140	Remove deadbolt	1 Carp	13.85	.578			32.50		32.50	51.50
1150	Remove panic bar		7.69	1.040			58.50		58.50	93
1420	Remove door, plane to fit, rail, reinstall door	↓	9.60	.833	↓		47		47	74.50

08 01 14 – Operation and Maintenance of Wood Doors

08 01 14.15 Door and Window Maintenance

	08 01 14.15 Door and Window Maintenance	Crew	Daily Output	Labor-Hours	Unit	Material	2022 Bare Costs Labor	Equipment	Total	Total Incl O&P
0010	**DOOR & WINDOW MAINTENANCE**									
0050	Rehang single door	1 Carp	32	.250	Ea.		14.10		14.10	22.50
0100	Rehang double door		16	.500	Pr.		28		28	44.50
0350	Install window sash cord	↓	16	.500	Ea.	4.93	28		32.93	50

08 01 53 – Operation and Maintenance of Plastic Windows

08 01 53.81 Solid Vinyl Replacement Windows

	08 01 53.81 Solid Vinyl Replacement Windows		Crew	Daily Output	Labor-Hours	Unit	Material	2022 Bare Costs Labor	Equipment	Total	Total Incl O&P
0010	**SOLID VINYL REPLACEMENT WINDOWS**	R085313-20									
0020	Double-hung, insulated glass, up to 83 united inches	G	2 Carp	8	2	Ea.	261	113		374	465
0040	84 to 93	G	"	8	2	"	264	113		377	470
0060	94 to 101	G	2 Carp	6	2.667	Ea.	345	150		495	620
0080	102 to 111	G		6	2.667		320	150		470	595
0100	112 to 120	G		6	2.667	↓	350	150		500	625
0120	For each united inch over 120, add	G		800	.020	Inch	3.89	1.13		5.02	6.05
0140	Casement windows, one operating sash, 42 to 60 united inches	G		8	2	Ea.	380	113		493	600
0160	61 to 70	G		8	2		365	113		478	585
0180	71 to 80	G		8	2		630	113		743	875
0200	81 to 96	G		8	2		515	113		628	745
0220	Two operating sash, 58 to 78 united inches	G		8	2		880	113		993	1,150
0240	79 to 88	G		8	2		995	113		1,108	1,275
0260	89 to 98	G		8	2		1,275	113		1,388	1,600
0280	99 to 108	G		6	2.667		1,050	150		1,200	1,400
0300	109 to 121	G		6	2.667		1,150	150		1,300	1,525
0320	Two operating, one fixed sash, 73 to 108 united inches	G		8	2		835	113		948	1,100
0340	109 to 118	G		8	2		910	113		1,023	1,175
0360	119 to 128	G		6	2.667		1,025	150		1,175	1,400
0380	129 to 138	G		6	2.667		1,050	150		1,200	1,400
0400	139 to 156	G		6	2.667		1,275	150		1,425	1,650
0420	Four operating sash, 98 to 118 united inches	G		8	2		1,775	113		1,888	2,125
0440	119 to 128	G		8	2		1,850	113		1,963	2,225
0460	129 to 138	G		6	2.667		1,900	150		2,050	2,350
0480	139 to 148	G		6	2.667		1,950	150		2,100	2,400
0500	149 to 168	G		6	2.667		2,100	150		2,250	2,550
0520	169 to 178	G		6	2.667		2,175	150		2,325	2,625
0560	Fixed picture window, up to 63 united inches	G		8	2		243	113		356	445
0580	64 to 83	G		8	2		260	113		373	465
0600	84 to 101	G		8	2	↓	315	113		428	530
0620	For each united inch over 101, add	G	↓	900	.018	Inch	4.54	1		5.54	6.60
0800	Cellulose fiber insulation, poured into sash balance cavity	G	1 Carp	36	.222	C.F.	.83	12.50		13.33	21
0820	Silicone caulking at perimeter	G	"	800	.010	L.F.	.15	.56		.71	1.05
2000	Impact resistant replacement windows										
2005	Laminated glass, 120 MPH rating, measure in united inches										
2010	Installation labor does not cover any rework of the window opening										
2020	Double-hung, insulated glass, up to 101 united inches		2 Carp	8	2	Ea.	590	113		703	830
2025	For each united inch over 101, add			80	.200	Inch	4.68	11.25		15.93	23
2100	Casement windows, impact resistant, up to 60 united inches		↓	8	2	Ea.	365	113		478	580

08 01 Operation and Maintenance of Openings

08 01 53 – Operation and Maintenance of Plastic Windows

08 01 53.81 Solid Vinyl Replacement Windows		Crew	Daily Output	Labor-Hours	Unit	Material	2022 Bare Costs Labor	Equipment	Total	Total Incl O&P
2120	61 to 70	2 Carp	8	2	Ea.	515	113		628	745
2130	71 to 80		8	2		565	113		678	800
2140	81 to 100		8	2		325	113		438	535
2150	For each united inch over 100, add		80	.200	Inch	7.10	11.25		18.35	25.50
2200	Awning windows, impact resistant, up to 60 united inches		8	2	Ea.	350	113		463	565
2220	61 to 70		8	2		505	113		618	735
2230	71 to 80		8	2		560	113		673	795
2240	For each united inch over 80, add		80	.200	Inch	6.70	11.25		17.95	25.50
2300	Picture windows, impact resistant, up to 63 united inches		8	2	Ea.	405	113		518	625
2320	63 to 83		8	2		520	113		633	750
2330	84 to 101		8	2		560	113		673	800
2340	For each united inch over 101, add		80	.200	Inch	4.50	11.25		15.75	23

08 05 Common Work Results for Openings

08 05 05 – Selective Demolition for Openings

08 05 05.10 Selective Demolition Doors

		Crew	Daily Output	Labor-Hours	Unit	Material	2022 Bare Costs Labor	Equipment	Total	Total Incl O&P
0010	**SELECTIVE DEMOLITION DOORS** R024119-10									
0200	Doors, exterior, 1-3/4" thick, single, 3' x 7' high	1 Clab	16	.500	Ea.		23		23	36
0202	Doors, exterior, 3' - 6' wide x 7' high		12	.667			30.50		30.50	48.50
0210	3' x 8' high		10	.800			36.50		36.50	58
0215	Double, 3' x 8' high		6	1.333			61		61	96.50
0220	Double, 6' x 7' high		12	.667			30.50		30.50	48.50
0500	Interior, 1-3/8" thick, single, 3' x 7' high		20	.400			18.25		18.25	29
0520	Double, 6' x 7' high		16	.500			23		23	36
0700	Bi-folding, 3' x 6'-8" high		20	.400			18.25		18.25	29
0720	6' x 6'-8" high		18	.444			20.50		20.50	32
0900	Bi-passing, 3' x 6'-8" high		16	.500			23		23	36
0940	6' x 6'-8" high		14	.571			26		26	41.50
0960	Interior metal door 1-3/4" thick, 3'-0" x 6'-8"		18	.444			20.50		20.50	32
0980	Interior metal door 1-3/4" thick, 3'-0" x 7'-0"		18	.444			20.50		20.50	32
1000	Interior wood door 1-3/4" thick, 3'-0" x 6'-8"		20	.400			18.25		18.25	29
1020	Interior wood door 1-3/4" thick, 3'-0" x 7'-0"		20	.400			18.25		18.25	29
1100	Door demo, floor door	2 Sswk	5	3.200			198		198	320
1500	Remove and reset, hollow core	1 Carp	8	1			56.50		56.50	89.50
1520	Solid		6	1.333			75		75	119
2000	Frames, including trim, metal		8	1			56.50		56.50	89.50
2200	Wood	2 Carp	32	.500			28		28	44.50
2201	Alternate pricing method	1 Carp	200	.040	L.F.		2.25		2.25	3.58
2205	Remove door hardware	"	45.70	.175	Ea.		9.85		9.85	15.65
3000	Special doors, counter doors	2 Carp	6	2.667	Ea.		150		150	239
3100	Double acting		10	1.600			90		90	143
3200	Floor door (trap type), or access type		8	2			113		113	179
3300	Glass, sliding, including frames		12	1.333			75		75	119
3400	Overhead, commercial, 12' x 12' high		4	4			225		225	360
3440	up to 20' x 16' high		3	5.333			300		300	475
3445	up to 35' x 30' high		1	16			900		900	1,425
3500	Residential, 9' x 7' high		8	2			113		113	179
3540	16' x 7' high		7	2.286			129		129	204
3600	Remove and reset, small		4	4			225		225	360
3620	Large		2.50	6.400			360		360	570
3700	Roll-up grille		5	3.200			180		180	286

08 05 05 – Selective Demolition for Openings

08 05 05.10 Selective Demolition Doors		Crew	Daily Output	Labor-Hours	Unit	Material	2022 Bare Costs Labor	Equipment	Total	Total Incl O&P
3800	Revolving door, no wire or connections	2 Carp	2	8	Ea.		450		450	715
3900	Storefront swing door	↓	3	5.333			300		300	475
3902	Cafe/bar swing door	2 Clab	8	2			91		91	145
4000	Residential lockset, exterior	1 Carp	28	.286			16.10		16.10	25.50
4224	Pocket door, no frame		8	1			56.50		56.50	89.50
4300	Remove and reset deadbolt		8	1			56.50		56.50	89.50
5500	Remove door closer	1 Clab	10	.800			36.50		36.50	58
5520	Remove push/pull plate		20	.400			18.25		18.25	29
5540	Remove door bolt		30	.267			12.15		12.15	19.30
5560	Remove kick/mop plate		20	.400			18.25		18.25	29
5585	Remove panic device		10	.800			36.50		36.50	58
5590	Remove mail slot		45	.178			8.10		8.10	12.90
5595	Remove peep hole	↓	45	.178			8.10		8.10	12.90
5604	Remove 9' x 7' garage door	2 Carp	9.50	1.684			95		95	151
5624	16' x 7' garage door		8	2			113		113	179
5626	9' x 7' swing up garage door		7	2.286			129		129	204
5628	16' x 7' swing up garage door	↓	7	2.286			129		129	204
5630	Garage door demolition, remove garage door track	1 Carp	6	1.333			75		75	119
5644	Remove overhead door opener	1 Clab	5	1.600			73		73	116
5774	Remove heavy gage sectional door, 20 ga., 8' x 8'	2 Carp	7	2.286			129		129	204
5784	10' x 10'		6	2.667			150		150	239
5794	12' x 12'		5	3.200			180		180	286
5805	14' x 14'	↓	4	4			225		225	360
6384	Remove French door unit	1 Clab	8	1	↓		45.50		45.50	72.50
7100	Remove double swing pneumatic doors, openers and sensors	2 Skwk	.50	32	Opng.		1,875		1,875	2,975
7570	Remove shock absorbing door	2 Sswk	1.90	8.421	"		520		520	840
9000	Minimum labor/equipment charge	1 Carp	4	2	Job		113		113	179

08 05 05.20 Selective Demolition of Windows

08 05 05.20 Selective Demolition of Windows		Crew	Daily Output	Labor-Hours	Unit	Material	2022 Bare Costs Labor	Equipment	Total	Total Incl O&P
0010	**SELECTIVE DEMOLITION OF WINDOWS** R024119-10									
0200	Aluminum, including trim, to 12 S.F.	1 Clab	16	.500	Ea.		23		23	36
0240	To 25 S.F.		11	.727			33		33	52.50
0280	To 50 S.F.		5	1.600			73		73	116
0320	Storm windows/screens, to 12 S.F.		27	.296			13.50		13.50	21.50
0360	To 25 S.F.		21	.381			17.35		17.35	27.50
0400	To 50 S.F.		16	.500	↓		23		23	36
0600	Glass, up to 10 S.F./window		200	.040	S.F.		1.82		1.82	2.90
0620	Over 10 S.F./window		150	.053	"		2.43		2.43	3.86
1000	Steel, including trim, to 12 S.F.		13	.615	Ea.		28		28	44.50
1020	To 25 S.F.		9	.889			40.50		40.50	64.50
1040	To 50 S.F.		4	2			91		91	145
2000	Wood, including trim, to 12 S.F.		22	.364			16.60		16.60	26.50
2020	To 25 S.F.		18	.444			20.50		20.50	32
2060	To 50 S.F.		13	.615			28		28	44.50
2065	To 180 S.F.	↓	8	1	↓		45.50		45.50	72.50
4400	Remove skylight, prefabricated glass block with metal frame	G-3	180	.178	S.F.		10.05		10.05	15.80
4410	Remove skylight, plstc domes, flush/curb mtd	2 Carp	395	.041	"		2.28		2.28	3.62
4420	Remove skylight, plstc/glass up to 2' x 3'	1 Carp	15	.533	Ea.		30		30	47.50
4440	Remove skylight, plstc/glass up to 4' x 6'	2 Carp	10	1.600			90		90	143
4480	Remove roof window up to 3' x 4'	1 Carp	8	1			56.50		56.50	89.50
4500	Remove roof window up to 4' x 6'	2 Carp	6	2.667			150		150	239
5020	Remove and reset window, up to a 2' x 2' window	1 Carp	6	1.333			75		75	119
5040	Up to a 3' x 3' window	↓	4	2			113		113	179

08 05 Common Work Results for Openings

08 05 05 – Selective Demolition for Openings

08 05 05.20 Selective Demolition of Windows	Crew	Daily Output	Labor-Hours	Unit	Material	2022 Bare Costs Labor	Equipment	Total	Total Incl O&P
5080	Up to a 4' x 5' window	1 Carp	2	4	Ea.		225		225	360
6000	Screening only	1 Clab	4000	.002	S.F.		.09		.09	.14
9000	Minimum labor/equipment charge	"	4	2	Job		91		91	145

08 11 Metal Doors and Frames

08 11 16 – Aluminum Doors and Frames

08 11 16.10 Entrance Doors

		Crew	Daily Output	Labor-Hours	Unit	Material	2022 Bare Costs Labor	Equipment	Total	Total Incl O&P
0010	**ENTRANCE DOORS** and frame, aluminum, narrow stile									
0011	Including standard hardware, clear finish, no glass									
0012	Top and bottom offset pivots, 1/4" beveled glass stops, threshold									
0013	Dead bolt lock with inside thumb screw, standard push pull									
0020	3'-0" x 7'-0" opening	2 Sswk	2	8	Ea.	1,125	495		1,620	2,025
0025	Anodizing aluminum entr. door & frame, add					110			110	121
0030	3'-6" x 7'-0" opening	2 Sswk	2	8		880	495		1,375	1,775
0100	3'-0" x 10'-0" opening, 3' high transom		1.80	8.889		1,375	550		1,925	2,375
0200	3'-6" x 10'-0" opening, 3' high transom		1.80	8.889		1,500	550		2,050	2,525
0280	5'-0" x 7'-0" opening		2	8		1,600	495		2,095	2,575
0300	6'-0" x 7'-0" opening		1.30	12.308		1,100	760		1,860	2,425
0301	6'-0" x 7'-0" opening		1.30	12.308	Pr.	1,100	760		1,860	2,425
0400	6'-0" x 10'-0" opening, 3' high transom		1.10	14.545	"	1,650	900		2,550	3,250
0520	3'-0" x 7'-0" opening, wide stile		2	8	Ea.	930	495		1,425	1,825
0540	3'-6" x 7'-0" opening		2	8		1,025	495		1,520	1,925
0560	5'-0" x 7'-0" opening		2	8		1,525	495		2,020	2,500
0580	6'-0" x 7'-0" opening		1.30	12.308	Pr.	1,675	760		2,435	3,075
0600	7'-0" x 7'-0" opening		1	16	"	2,375	990		3,365	4,200
1200	For non-standard size, add				Leaf	80%				
1250	For installation of non-standard size, add						20%			
1300	Light bronze finish, add				Leaf	36%				
1400	Dark bronze finish, add					25%				
1500	For black finish, add					40%				
1600	Concealed panic device, add					1,125			1,125	1,225
1700	Electric striker release, add				Opng.	335			335	370
1800	Floor check, add				Leaf	720			720	790
1900	Concealed closer, add				"	575			575	635
9000	Minimum labor/equipment charge	2 Carp	4	4	Job		225		225	360

08 11 63 – Metal Screen and Storm Doors and Frames

08 11 63.23 Aluminum Screen and Storm Doors and Frames

		Crew	Daily Output	Labor-Hours	Unit	Material	2022 Bare Costs Labor	Equipment	Total	Total Incl O&P
0010	**ALUMINUM SCREEN AND STORM DOORS AND FRAMES**									
0020	Combination storm and screen									
0420	Clear anodic coating, 2'-8" wide	2 Carp	14	1.143	Ea.	243	64.50		307.50	370
0440	3'-0" wide	"	14	1.143	"	197	64.50		261.50	320
0500	For 7'-0" door height, add					8%				
1020	Mill finish, 2'-8" wide	2 Carp	14	1.143	Ea.	254	64.50		318.50	380
1040	3'-0" wide	"	14	1.143		274	64.50		338.50	400
1100	For 7'-0" door, add					8%				
1520	White painted, 2'-8" wide	2 Carp	14	1.143		266	64.50		330.50	395
1540	3'-0" wide	"	14	1.143		300	64.50		364.50	430
1600	For 7'-0" door, add					8%				
2000	Wood door & screen, see Section 08 14 33.20									
9000	Minimum labor/equipment charge	1 Carp	4	2	Job		113		113	179

For customer support on your Commercial Renovation Costs with RSMeans data, call 800.448.8182.

233

08 12 13.13 Standard Hollow Metal Frames		Crew	Daily Output	Labor-Hours	Unit	Material	2022 Bare Costs Labor	Equipment	Total	Total Incl O&P
0010	**STANDARD HOLLOW METAL FRAMES**									
0020	16 ga., up to 5-3/4" jamb depth									
0025	3'-0" x 6'-8" single	G	2 Carp	16	1	Ea.	216	56.50		272.50	330
0028	3'-6" wide, single	G		16	1		213	56.50		269.50	325
0030	4'-0" wide, single	G		16	1		226	56.50		282.50	340
0040	6'-0" wide, double	G		14	1.143		236	64.50		300.50	360
0045	8'-0" wide, double	G		14	1.143		275	64.50		339.50	405
0100	3'-0" x 7'-0" single	G		16	1		175	56.50		231.50	283
0110	3'-6" wide, single			16	1		184	56.50		240.50	292
0112	4'-0" wide, single			16	1		184	56.50		240.50	292
0140	6'-0" wide, double			14	1.143		261	64.50		325.50	390
0145	8'-0" wide, double			14	1.143		260	64.50		324.50	390
1000	16 ga., up to 4-7/8" deep, 3'-0" x 7'-0" single	G		16	1		195	56.50		251.50	305
1140	6'-0" wide, double	G		14	1.143		240	64.50		304.50	365
1200	16 ga., 8-3/4" deep, 3'-0" x 7'-0" single	G		16	1		325	56.50		381.50	450
1240	6'-0" wide, double	G		14	1.143		310	64.50		374.50	440
2800	14 ga., up to 3-7/8" deep, 3'-0" x 7'-0" single	G		16	1		219	56.50		275.50	330
2840	6'-0" wide, double	G		14	1.143		305	64.50		369.50	435
3000	14 ga., up to 5-3/4" deep, 3'-0" x 6'-8" single	G		16	1		175	56.50		231.50	282
3002	3'-6" wide, single	G		16	1		226	56.50		282.50	340
3005	4'-0" wide, single	G		16	1		226	56.50		282.50	340
3600	up to 5-3/4" jamb depth, 4'-0" x 7'-0" single			15	1.067		251	60		311	370
3620	6'-0" wide, double			12	1.333		235	75		310	380
3640	8'-0" wide, double			12	1.333		385	75		460	545
3700	8'-0" high, 4'-0" wide, single			15	1.067		287	60		347	410
3740	8'-0" wide, double			12	1.333		390	75		465	550
4000	6-3/4" deep, 4'-0" x 7'-0" single	G		15	1.067		300	60		360	425
4020	6'-0" wide, double	G		12	1.333		315	75		390	465
4040	8'-0" wide, double	G		12	1.333		315	75		390	465
4100	8'-0" high, 4'-0" wide, single	G		15	1.067		345	60		405	475
4140	8'-0" wide, double	G		12	1.333		505	75		580	680
4400	8-3/4" deep, 4'-0" x 7'-0", single			15	1.067		380	60		440	515
4440	8'-0" wide, double			12	1.333		495	75		570	665
4500	4'-0" x 8'-0", single			15	1.067		425	60		485	565
4540	8'-0" wide, double		▼	12	1.333		535	75		610	710
4900	For welded frames, add						124			124	136
5380	Steel frames, KD, 14 ga., "B" label, to 5-3/4" throat, to 3'-0" x 7'-0"		2 Carp	15	1.067		268	60		328	390
5400	14 ga., "B" label, up to 5-3/4" deep, 4'-0" x 7'-0" single	G		15	1.067		310	60		370	435
5440	8'-0" wide, double	G		12	1.333		425	75		500	590
5800	6-3/4" deep, 7'-0" high, 4'-0" wide, single	G		15	1.067		187	60		247	300
5840	8'-0" wide, double	G		12	1.333		390	75		465	550
6200	8-3/4" deep, 4'-0" x 7'-0" single	G		15	1.067		435	60		495	575
6240	8'-0" wide, double	G	▼	12	1.333	▼	445	75		520	610
6300	For "A" label use same price as "B" label										
6400	For baked enamel finish, add						30%	15%			
6500	For galvanizing, add						20%				
6600	For hospital stop, add					Ea.	284			284	315
6620	For hospital stop, stainless steel, add					"	156			156	172
7900	Transom lite frames, fixed, add		2 Carp	155	.103	S.F.	60	5.80		65.80	75.50
8000	Movable, add		"	130	.123	"	80.50	6.95		87.45	99.50
9000	Minimum labor/equipment charge		1 Carp	4	2	Job		113		113	179

234

08 12 13.15 Borrowed Lites	Crew	Daily Output	Labor-Hours	Unit	Material	2022 Bare Costs Labor	Equipment	Total	Total Incl O&P
0010 **BORROWED LITES**									
0100 Hollow metal section 20 ga., 3-1/2" with glass stop	2 Carp	100	.160	L.F.	23.50	9		32.50	40.50
0110 3-3/4" with glass stop		100	.160		8.55	9		17.55	23.50
0120 4" with glass stop		100	.160		11.05	9		20.05	26.50
0130 4-5/8" with glass stop		100	.160		11.45	9		20.45	27
0140 4-7/8" with glass stop		100	.160		11.65	9		20.65	27
0150 5" with glass stop		100	.160		12.65	9		21.65	28.50
0160 5-3/8" with glass stop		100	.160		14.90	9		23.90	30.50
0300 Hollow metal section 18 ga., 3-1/2" with glass stop		80	.200		11.75	11.25		23	31
0310 3-3/4" with glass stop		80	.200		12	11.25		23.25	31
0320 4" with glass stop		80	.200		16.65	11.25		27.90	36
0330 4-5/8" with glass stop		80	.200		12.75	11.25		24	32
0340 4-7/8" with glass stop		80	.200		13.05	11.25		24.30	32.50
0350 5" with glass stop		80	.200		13.20	11.25		24.45	32.50
0360 5-3/8" with glass stop		80	.200		13.20	11.25		24.45	32.50
0370 6-5/8" with glass stop		80	.200		14.70	11.25		25.95	34
0380 6-7/8" with glass stop		80	.200		14.90	11.25		26.15	34.50
0390 7-1/4" with glass stop		80	.200		15.55	11.25		26.80	35
0500 Mullion section 18 ga., 3-1/2" with double stop	2 Carp	50	.320	L.F.	27.50	18		45.50	58.50
0510 3-3/4" with double stop		50	.320		28	18		46	59
0530 4-5/8" with double stop		50	.320		30	18		48	61.50
0540 4-7/8" with double stop		50	.320		30.50	18		48.50	62
0550 5" with double stop		50	.320		30.50	18		48.50	62
0560 5-3/8" with double stop		50	.320		31.50	18		49.50	63
0570 6-5/8" with double stop		50	.320		34.50	18		52.50	66
0580 6-7/8" with double stop		50	.320		35	18		53	67
0590 7-1/4" with double stop		50	.320		35.50	18		53.50	67.50
1000 Assembled frame 2'-0" x 2'-0" 16 ga., 3-3/4" with glass stop		10	1.600	Ea.	258	90		348	425
1010 3'-0" x 2'-0"		10	1.600		258	90		348	425
1020 4'-0" x 2'-0"		10	1.600		259	90		349	430
1030 5'-0" x 2'-0"		10	1.600		345	90		435	520
1040 6'-0" x 2'-0"		8	2		385	113		498	605
1050 7'-0" x 2'-0"		8	2		385	113		498	605
1060 8'-0" x 2'-0"		8	2		385	113		498	605
1100 2'-0" x 7'-0"		8	2		385	113		498	605
1110 3'-0" x 7'-0"		8	2		385	113		498	605
1120 4'-0" x 7'-0"		8	2		385	113		498	605
1130 5'-0" x 7'-0"		8	2		385	113		498	605
1140 6'-0" x 7'-0"		6	2.667		420	150		570	700
1150 7'-0" x 7'-0"		6	2.667		420	150		570	700
1160 8'-0" x 7'-0"		6	2.667		420	150		570	700
1200 Assembled frame 2'-0" x 2'-0" 16 ga., 4-7/8" with glass stop		10	1.600		259	90		349	430
1210 3'-0" x 2'-0"		10	1.600		259	90		349	430
1220 4'-0" x 2'-0"		8	2		259	113		372	465
1230 5'-0" x 2'-0"		8	2		385	113		498	605
1240 6'-0" x 2'-0"		8	2		385	113		498	605
1250 7'-0" x 2'-0"		8	2		385	113		498	605
1260 8'-0" x 2'-0"		8	2		385	113		498	605
1300 2'-0" x 7'-0"		8	2		385	113		498	605
1310 3'-0" x 7'-0"		8	2		385	113		498	605
1320 4'-0" x 7'-0"		8	2		385	113		498	605
1330 5'-0" x 7'-0"		6	2.667		385	150		535	665

For customer support on your Commercial Renovation Costs with RSMeans data, call 800.448.8182.

235

08 12 Metal Frames

08 12 13 – Hollow Metal Frames

08 12 13.15 Borrowed Lites

		Crew	Daily Output	Labor-Hours	Unit	Material	2022 Bare Costs Labor	Equipment	Total	Total Incl O&P
1340	6'-0" x 7'-0"	2 Carp	6	2.667	Ea.	420	150		570	700
1350	7'-0" x 7'-0"		6	2.667		370	150		520	645
1360	8'-0" x 7'-0"		6	2.667		370	150		520	645
1400	Assembled frame 2'-0" x 3'-0" 16 ga., 6-1/8" with glass stop		6	2.667		260	150		410	525
1410	3'-0" x 3'-0"		8	2		259	113		372	465
1420	4'-0" x 3'-0"		8	2		259	113		372	465
1430	5'-0" x 3'-0"		6	2.667		385	150		535	665
1440	6'-0" x 3'-0"		6	2.667		370	150		520	650
1450	7'-0" x 3'-0"		6	2.667		385	150		535	665
1460	8'-0" x 3'-0"		10	1.600		385	90		475	570

08 12 13.25 Channel Metal Frames

			Crew	Daily Output	Labor-Hours	Unit	Material	2022 Bare Costs Labor	Equipment	Total	Total Incl O&P
0010	**CHANNEL METAL FRAMES**										
0020	Steel channels with anchors and bar stops										
0100	6" channel @ 8.2#/L.F., 3' x 7' door, weighs 150#	G	E-4	13	2.462	Ea.	295	153	11.35	459.35	585
0200	8" channel @ 11.5#/L.F., 6' x 8' door, weighs 275#	G		9	3.556		540	222	16.40	778.40	970
0300	8' x 12' door, weighs 400#	G		6.50	4.923		785	305	22.50	1,112.50	1,375
0400	10" channel @ 15.3#/L.F., 10' x 10' door, weighs 500#	G		6	5.333		980	335	24.50	1,339.50	1,625
0500	12' x 12' door, weighs 600#	G		5.50	5.818		1,175	365	27	1,567	1,925
0600	12" channel @ 20.7#/L.F., 12' x 12' door, weighs 825#	G		4.50	7.111		1,625	445	33	2,103	2,525
0700	12' x 16' door, weighs 1000#	G		4	8		1,975	500	37	2,512	3,000
0800	For frames without bar stops, light sections, deduct						15%				
0900	Heavy sections, deduct						10%				
9000	Minimum labor/equipment charge		E-4	4	8	Job		500	37	537	845

08 13 Metal Doors

08 13 13 – Hollow Metal Doors

08 13 13.13 Standard Hollow Metal Doors

			Crew	Daily Output	Labor-Hours	Unit	Material	2022 Bare Costs Labor	Equipment	Total	Total Incl O&P
0010	**STANDARD HOLLOW METAL DOORS**	R081313-20									
0015	Flush, full panel, hollow core										
0017	When noted doors are prepared but do not include glass or louvers										
0020	1-3/8" thick, 20 ga., 2'-0" x 6'-8"	G	2 Carp	20	.800	Ea.	525	45		570	645
0040	2'-8" x 6'-8"	G		18	.889		485	50		535	615
0060	3'-0" x 6'-8"	G		17	.941		535	53		588	675
0100	3'-0" x 7'-0"			17	.941		540	53		593	680
0120	For vision lite, add						310			310	340
0140	For narrow lite, add						42			42	46
0320	Half glass, 20 ga., 2'-0" x 6'-8"	G	2 Carp	20	.800		860	45		905	1,025
0340	2'-8" x 6'-8"	G		18	.889		900	50		950	1,075
0360	3'-0" x 6'-8"	G		17	.941		860	53		913	1,025
0400	3'-0" x 7'-0"	G		17	.941		870	53		923	1,050
0410	1-3/8" thick, 18 ga., 2'-0" x 6'-8"	G		20	.800		705	45		750	845
0420	3'-0" x 6'-8"	G		17	.941		420	53		473	550
0425	3'-0" x 7'-0"	G		17	.941		550	53		603	690
0450	For vision lite, add						310			310	340
0452	For narrow lite, add				Ea.		42			42	46
0460	Half glass, 18 ga., 2'-0" x 6'-8"	G	2 Carp	20	.800		955	45		1,000	1,125
0465	2'-8" x 6'-8"	G		18	.889		990	50		1,040	1,150
0470	3'-0" x 6'-8"	G		17	.941		965	53		1,018	1,125
0475	3'-0" x 7'-0"	G		17	.941		805	53		858	970
0500	Hollow core, 1-3/4" thick, full panel, 20 ga., 2'-8" x 6'-8"	G		18	.889		550	50		600	685
0520	3'-0" x 6'-8"	G		17	.941		550	53		603	690

08 13 13.13 Standard Hollow Metal Doors

		Crew	Daily Output	Labor-Hours	Unit	Material	2022 Bare Costs Labor	Equipment	Total	Total Incl O&P
0640	3'-0" x 7'-0" G	2 Carp	17	.941	Ea.	560	53		613	700
0680	4'-0" x 7'-0" G		15	1.067		760	60		820	930
0700	4'-0" x 8'-0" G		13	1.231		915	69.50		984.50	1,100
1000	18 ga., 2'-8" x 6'-8" G		17	.941		650	53		703	800
1020	3'-0" x 6'-8" G		16	1		665	56.50		721.50	825
1120	3'-0" x 7'-0" G		17	.941		665	53		718	820
1150	3'-6" x 7'-0"	1 Carp	14	.571		830	32		862	960
1180	4'-0" x 7'-0"	2 Carp	14	1.143		780	64.50		844.50	955
1200	4'-0" x 8'-0" G	"	17	.941		1,150	53		1,203	1,325
1212	For vision lite, add					310			310	340
1214	For narrow lite, add					42			42	46
1230	Half glass, 20 ga., 2'-8" x 6'-8" G	2 Carp	20	.800		775	45		820	920
1240	3'-0" x 6'-8" G		18	.889		735	50		785	885
1260	3'-0" x 7'-0" G		18	.889		870	50		920	1,025
1280	Embossed panel, 1-3/4" thick, poly core, 20 ga., 3'-0" x 7'-0" G		18	.889		980	50		1,030	1,150
1290	Half glass, 1-3/4" thick, poly core, 20 ga., 3'-0" x 7'-0" G		18	.889		890	50		940	1,050
1320	18 ga., 2'-8" x 6'-8" G		18	.889		830	50		880	990
1340	3'-0" x 6'-8" G		17	.941		820	53		873	985
1360	3'-0" x 7'-0" G		17	.941		835	53		888	1,000
1380	4'-0" x 7'-0" G		15	1.067		1,050	60		1,110	1,250
1400	4'-0" x 8'-0" G		14	1.143		1,125	64.50		1,189.50	1,350
1500	Flush full panel, 16 ga., steel hollow core									
1520	2'-0" x 6'-8" G	2 Carp	20	.800	Ea.	865	45		910	1,025
1530	2'-8" x 6'-8" G		20	.800		865	45		910	1,025
1540	3'-0" x 6'-8" G		20	.800		865	45		910	1,025
1560	2'-8" x 7'-0" G		18	.889		910	50		960	1,075
1570	3'-0" x 7'-0" G		18	.889		910	50		960	1,075
1580	3'-6" x 7'-0" G		18	.889		1,050	50		1,100	1,225
1590	4'-0" x 7'-0" G		18	.889		1,175	50		1,225	1,375
1600	2'-8" x 8'-0" G		18	.889		1,100	50		1,150	1,275
1620	3'-0" x 8'-0" G		18	.889		1,200	50		1,250	1,400
1630	3'-6" x 8'-0" G	2 Carp	18	.889	Ea.	1,275	50		1,325	1,500
1640	4'-0" x 8'-0" G		18	.889		1,400	50		1,450	1,600
1650	1-13/16", 14 ga., 2'-8" x 7'-0" G		10	1.600		975	90		1,065	1,225
1670	3'-0" x 7'-0" G		10	1.600		975	90		1,065	1,225
1690	3'-6" x 7'-0" G		10	1.600		1,075	90		1,165	1,350
1700	4'-0" x 7'-0" G		10	1.600		1,175	90		1,265	1,450
1720	Insulated, 1-3/4" thick, full panel, 18 ga., 3'-0" x 6'-8" G		15	1.067		720	60		780	890
1740	2'-8" x 7'-0" G		16	1		660	56.50		716.50	815
1760	3'-0" x 7'-0" G		15	1.067		560	60		620	710
1800	4'-0" x 8'-0" G		13	1.231		550	69.50		619.50	715
1820	Half glass, 18 ga., 3'-0" x 6'-8" G		16	1		835	56.50		891.50	1,000
1840	2'-8" x 7'-0" G		17	.941		865	53		918	1,050
1860	3'-0" x 7'-0" G		16	1		930	56.50		986.50	1,125
1900	4'-0" x 8'-0" G		14	1.143		800	64.50		864.50	980
2000	For vision lite, add					310			310	340
2010	For narrow lite, add					42			42	46
8100	For bottom louver, add					440			440	485
8110	For baked enamel finish, add					30%	15%			
8120	For galvanizing, add					20%				
8190	Commercial door, flush steel, fire door see section 08 13 13 15									
8270	For dutch door shelf, add to standard door				Ea.	300			300	330
9000	Minimum labor/equipment charge	1 Carp	4	2	Job		113		113	179

For customer support on your Commercial Renovation Costs with RSMeans data, call 800.448.8182.

237

08 13 13.15 Metal Fire Doors

		Crew	Daily Output	Labor-Hours	Unit	Material	2022 Bare Costs Labor	Equipment	Total	Total Incl O&P
0010	**METAL FIRE DOORS** R081313-20									
0015	Steel, flush, "B" label, 90 minutes									
0020	Full panel, 20 ga., 2'-0" x 6'-8"	2 Carp	20	.800	Ea.	565	45		610	690
0040	2'-8" x 6'-8"		18	.889		565	50		615	705
0060	3'-0" x 6'-8"		17	.941		570	53		623	710
0080	3'-0" x 7'-0"		17	.941		590	53		643	735
0140	18 ga., 3'-0" x 6'-8"		16	1		785	56.50		841.50	950
0160	2'-8" x 7'-0"		17	.941		830	53		883	995
0180	3'-0" x 7'-0"		16	1		690	56.50		746.50	845
0200	4'-0" x 7'-0"		15	1.067		1,225	60		1,285	1,450
0220	For "A" label, 3 hour, 18 ga., use same price as "B" label									
0240	For vision lite, add				Ea.	232			232	255
0300	Full panel, 16 ga., 2'-0" x 6'-8"	2 Carp	20	.800		830	45		875	980
0310	2'-8" x 6'-8"		18	.889		850	50		900	1,025
0320	3'-0" x 6'-8"		17	.941		615	53		668	760
0350	2'-8" x 7'-0"	2 Carp	17	.941	Ea.	885	53		938	1,050
0360	3'-0" x 7'-0"		16	1		885	56.50		941.50	1,075
0370	4'-0" x 7'-0"		15	1.067		1,125	60		1,185	1,325
0520	Flush, "B" label, 90 minutes, egress core, 20 ga., 2'-0" x 6'-8"		18	.889		915	50		965	1,075
0540	2'-8" x 6'-8"		17	.941		920	53		973	1,100
0560	3'-0" x 6'-8"		16	1		925	56.50		981.50	1,125
0580	3'-0" x 7'-0"		16	1		965	56.50		1,021.50	1,150
0640	Flush, "A" label, 3 hour, egress core, 18 ga., 3'-0" x 6'-8"		15	1.067		1,000	60		1,060	1,200
0660	2'-8" x 7'-0"		16	1		1,050	56.50		1,106.50	1,250
0680	3'-0" x 7'-0"		15	1.067		555	60		615	705
0700	4'-0" x 7'-0"		14	1.143		1,275	64.50		1,339.50	1,500
9000	Minimum labor/equipment charge	1 Carp	4	2	Job		113		113	179

08 13 13.20 Residential Steel Doors

			Crew	Daily Output	Labor-Hours	Unit	Material	2022 Bare Costs Labor	Equipment	Total	Total Incl O&P
0010	**RESIDENTIAL STEEL DOORS**										
0020	Prehung, insulated, exterior										
0030	Embossed, full panel, 2'-8" x 6'-8"	G	2 Carp	17	.941	Ea.	480	53		533	615
0040	3'-0" x 6'-8"	G		15	1.067		495	60		555	635
0060	3'-0" x 7'-0"	G		15	1.067		485	60		545	630
0070	5'-4" x 6'-8", double	G		8	2		1,100	113		1,213	1,400
0220	Half glass, 2'-8" x 6'-8"	G		17	.941		455	53		508	585
0240	3'-0" x 6'-8"	G		16	1		455	56.50		511.50	590
0260	3'-0" x 7'-0"	G		16	1		550	56.50		606.50	695
0270	5'-4" x 6'-8", double	G		8	2		1,300	113		1,413	1,625
1320	Flush face, full panel, 2'-8" x 6'-8"	G		16	1		475	56.50		531.50	610
1340	3'-0" x 6'-8"	G		15	1.067		480	60		540	625
1360	3'-0" x 7'-0"	G		15	1.067		420	60		480	560
1380	5'-4" x 6'-8", double	G		8	2		1,075	113		1,188	1,375
1420	Half glass, 2'-8" x 6'-8"	G		17	.941		470	53		523	600
1440	3'-0" x 6'-8"	G		16	1		470	56.50		526.50	605
1460	3'-0" x 7'-0"	G		16	1		635	56.50		691.50	785
1480	5'-4" x 6'-8", double	G		8	2		915	113		1,028	1,175
1500	Sidelight, full lite, 1'-0" x 6'-8" with grille	G					605			605	665
1510	1'-0" x 6'-8", low E	G					515			515	570
1520	1'-0" x 6'-8", half lite	G					415			415	455
1530	1'-0" x 6'-8", half lite, low E	G					405			405	445
2300	Interior, residential, closet, bi-fold, 2'-0" x 6'-8"	G	2 Carp	16	1		120	56.50		176.50	222
2330	3'-0" wide	G		16	1		380	56.50		436.50	510

For customer support on your Commercial Renovation Costs with RSMeans data, call 800.448.8182.

08 13 Metal Doors

08 13 13 – Hollow Metal Doors

08 13 13.20 Residential Steel Doors

		Crew	Daily Output	Labor-Hours	Unit	Material	2022 Bare Costs Labor	Equipment	Total	Total Incl O&P
2360	4'-0" wide	G 2 Carp	15	1.067	Ea.	395	60		455	530
2400	5'-0" wide	G	14	1.143		505	64.50		569.50	655
2420	6'-0" wide	G	13	1.231		400	69.50		469.50	550
9000	Minimum labor/equipment charge	1 Carp	4	2	Job		113		113	179

08 13 13.25 Doors Hollow Metal

			Crew	Daily Output	Labor-Hours	Unit	Material	2022 Bare Costs Labor	Equipment	Total	Total Incl O&P
0010	**DOORS HOLLOW METAL**										
0500	Exterior, commercial, flush, 20 ga., 1-3/4" x 7'-0" x 2'-6" wide	G	2 Carp	15	1.067	Ea.	555	60		615	710
0530	2'-8" wide	G		15	1.067		555	60		615	710
0560	3'-0" wide	G		14	1.143		560	64.50		624.50	715
0590	3'-6" wide	G		14	1.143		710	64.50		774.50	880
1000	18 ga., 1-3/4" x 7'-0" x 2'-6" wide	G		15	1.067		635	60		695	795
1030	2'-8" wide	G		15	1.067		635	60		695	795
1060	3'-0" wide	G		14	1.143		635	64.50		699.50	800
1500	16 ga., 1-3/4" x 7'-0" x 2'-6" wide	G		15	1.067		750	60		810	920
1530	2'-8" wide	G		15	1.067		750	60		810	920
1560	3'-0" wide	G		14	1.143		645	64.50		709.50	810
1590	3'-6" wide	G		14	1.143		865	64.50		929.50	1,050
2900	Fire door, "A" label, 18 gauge, 1-3/4" x 2'-6" x 7'-0"	G		15	1.067		685	60		745	850
2930	2'-8" wide	G		15	1.067		685	60		745	850
2960	3'-0" wide	G		14	1.143		780	64.50		844.50	960
2990	3'-6" wide	G		14	1.143		835	64.50		899.50	1,025
3100	"B" label, 2'-6" wide	G		15	1.067		610	60		670	765
3130	2'-8" wide	G		15	1.067		610	60		670	765
3160	3'-0" wide	G		14	1.143		610	64.50		674.50	770

08 13 16 – Aluminum Doors

08 13 16.10 Commercial Aluminum Doors

		Crew	Daily Output	Labor-Hours	Unit	Material	2022 Bare Costs Labor	Equipment	Total	Total Incl O&P
0010	**COMMERCIAL ALUMINUM DOORS**, flush, no glazing									
5000	Flush panel doors, pair of 2'-6" x 7'-0"	2 Sswk	2	8	Pr.	1,300	495		1,795	2,225
5050	3'-0" x 7'-0", single		2.50	6.400	Ea.	670	395		1,065	1,375
5100	Pair of 3'-0" x 7'-0"		2	8	Pr.	1,450	495		1,945	2,375
5150	3'-6" x 7'-0", single		2.50	6.400	Ea.	880	395		1,275	1,600

08 14 Wood Doors

08 14 13 – Carved Wood Doors

08 14 13.10 Types of Wood Doors, Carved

		Crew	Daily Output	Labor-Hours	Unit	Material	2022 Bare Costs Labor	Equipment	Total	Total Incl O&P
0010	**TYPES OF WOOD DOORS, CARVED**									
3000	Solid wood, 1-3/4" thick stile and rail									
3020	Mahogany, 3'-0" x 7'-0", six panel	2 Carp	14	1.143	Ea.	1,350	64.50		1,414.50	1,575
3030	With two lites		10	1.600		1,525	90		1,615	1,825
3040	3'-6" x 8'-0", six panel		10	1.600		1,050	90		1,140	1,300
3050	With two lites		8	2		2,925	113		3,038	3,400
3100	Pine, 3'-0" x 7'-0", six panel		14	1.143		1,350	64.50		1,414.50	1,575
3110	With two lites		10	1.600		950	90		1,040	1,200
3120	3'-6" x 8'-0", six panel		10	1.600		1,100	90		1,190	1,350
3130	With two lites		8	2		1,675	113		1,788	2,025
3200	Red oak, 3'-0" x 7'-0", six panel		14	1.143		390	64.50		454.50	525
3210	With two lites		10	1.600		2,550	90		2,640	2,950
3220	3'-6" x 8'-0", six panel		10	1.600		2,650	90		2,740	3,075
3230	With two lites		8	2		3,450	113		3,563	3,975
4000	Hand carved door, mahogany									

For customer support on your Commercial Renovation Costs with RSMeans data, call 800.448.8182.

239

08 14 Wood Doors

08 14 13 – Carved Wood Doors

08 14 13.10 Types of Wood Doors, Carved		Crew	Daily Output	Labor-Hours	Unit	Material	2022 Bare Costs Labor	Equipment	Total	Total Incl O&P
4020	3'-0" x 7'-0", simple design	2 Carp	14	1.143	Ea.	1,800	64.50		1,864.50	2,100
4030	Intricate design		11	1.455		3,575	82		3,657	4,050
4040	3'-6" x 8'-0", simple design		10	1.600		3,250	90		3,340	3,725
4050	Intricate design		8	2		3,775	113		3,888	4,350
4400	For custom finish, add					560			560	615
4600	Side light, mahogany, 7'-0" x 1'-6" wide, 4 lites	2 Carp	18	.889		1,150	50		1,200	1,325
4610	6 lites		14	1.143		3,025	64.50		3,089.50	3,425
4620	8'-0" x 1'-6" wide, 4 lites		14	1.143		2,125	64.50		2,189.50	2,450
4630	6 lites		10	1.600		2,350	90		2,440	2,725
4640	Side light, oak, 7'-0" x 1'-6" wide, 4 lites		18	.889		1,475	50		1,525	1,700
4650	6 lites		14	1.143		2,450	64.50		2,514.50	2,800
4660	8'-0" x 1'-6" wide, 4 lites		14	1.143		1,400	64.50		1,464.50	1,650
4670	6 lites		10	1.600		2,450	90		2,540	2,850

08 14 16 – Flush Wood Doors

08 14 16.09 Smooth Wood Doors

		Crew	Daily Output	Labor-Hours	Unit	Material	2022 Bare Costs Labor	Equipment	Total	Total Incl O&P
0010	**SMOOTH WOOD DOORS**									
0015	Flush, interior, hollow core									
0025	Lauan face, 1-3/8", 3'-0" x 6'-8"	2 Carp	17	.941	Ea.	101	53		154	195
0030	4'-0" x 6'-8"		16	1		156	56.50		212.50	261
0080	1-3/4", 2'-0" x 6'-8"		17	.941		75	53		128	167
0085	3'-0" x 6'-8"	1 Carp	16	.500		101	28		129	156
0108	3'-0" x 7'-0"	2 Carp	16	1		254	56.50		310.50	370
0112	Pair of 3'-0" x 7'-0"		9	1.778	Pr.	173	100		273	350
0140	Birch face, 1-3/8", 2'-6" x 6'-8"		17	.941	Ea.	110	53		163	205
0180	3'-0" x 6'-8"		17	.941		121	53		174	218
0200	4'-0" x 6'-8"		16	1		197	56.50		253.50	305
0202	1-3/4", 2'-0" x 6'-8"		17	.941		102	53		155	196
0204	2'-4" x 7'-0"		16	1		173	56.50		229.50	280
0206	2'-6" x 7'-0"		16	1		186	56.50		242.50	295
0210	3'-0" x 7'-0"		16	1		200	56.50		256.50	310
0214	Pair of 3'-0" x 7'-0"	2 Carp	9	1.778	Pr.	400	100		500	600
0220	Oak face, 1-3/8", 2'-0" x 6'-8"		17	.941	Ea.	153	53		206	253
0280	3'-0" x 6'-8"		17	.941		201	53		254	305
0300	4'-0" x 6'-8"		16	1		203	56.50		259.50	315
0305	1-3/4", 2'-6" x 6'-8"		17	.941		77	53		130	169
0310	3'-0" x 7'-0"		16	1		325	56.50		381.50	450
0320	Walnut face, 1-3/8", 2'-0" x 6'-8"		17	.941		285	53		338	400
0340	2'-6" x 6'-8"		17	.941		267	53		320	375
0380	3'-0" x 6'-8"		17	.941		278	53		331	390
0400	4'-0" x 6'-8"		16	1		288	56.50		344.50	405
0430	For 7'-0" high, add					37.50			37.50	41
0440	For 8'-0" high, add					62			62	68
0480	For prefinishing, clear, add					78.50			78.50	86.50
0500	For prefinishing, stain, add					86.50			86.50	95
1320	M.D. overlay on hardboard, 1-3/8", 2'-0" x 6'-8"	2 Carp	17	.941		166	53		219	266
1340	2'-6" x 6'-8"		17	.941		183	53		236	286
1380	3'-0" x 6'-8"		17	.941		168	53		221	269
1400	4'-0" x 6'-8"		16	1		280	56.50		336.50	400
1420	For 7'-0" high, add					25.50			25.50	28
1440	For 8'-0" high, add					57.50			57.50	63
1720	H.P. plastic laminate, 1-3/8", 2'-0" x 6'-8"	2 Carp	16	1		310	56.50		366.50	435
1740	2'-6" x 6'-8"		16	1		305	56.50		361.50	425

08 14 16 – Flush Wood Doors

08 14 16.09 Smooth Wood Doors		Crew	Daily Output	Labor-Hours	Unit	Material	2022 Bare Costs Labor	Equipment	Total	Total Incl O&P
1780	3'-0" x 6'-8"	2 Carp	15	1.067	Ea.	325	60		385	450
1800	4'-0" x 6'-8"	↓	14	1.143		425	64.50		489.50	570
1820	For 7'-0" high, add					25			25	28
1840	For 8'-0" high, add					34			34	37.50
2020	Particle core, lauan face, 1-3/8", 2'-6" x 6'-8"	2 Carp	15	1.067		131	60		191	240
2040	3'-0" x 6'-8"		14	1.143		102	64.50		166.50	214
2080	3'-0" x 7'-0"		13	1.231		235	69.50		304.50	370
2085	4'-0" x 7'-0"		12	1.333		200	75		275	340
2110	1-3/4", 3'-0" x 7'-0"		13	1.231		254	69.50		323.50	390
2120	Birch face, 1-3/8", 2'-6" x 6'-8"		15	1.067		168	60		228	281
2140	3'-0" x 6'-8"		14	1.143		198	64.50		262.50	320
2180	3'-0" x 7'-0"		13	1.231		192	69.50		261.50	320
2200	4'-0" x 7'-0"		12	1.333		219	75		294	360
2205	1-3/4", 3'-0" x 7'-0"		13	1.231		260	69.50		329.50	395
2220	Oak face, 1-3/8", 2'-6" x 6'-8"		15	1.067		192	60		252	310
2240	3'-0" x 6'-8"	2 Carp	14	1.143	Ea.	180	64.50		244.50	300
2280	3'-0" x 7'-0"		13	1.231		201	69.50		270.50	330
2300	4'-0" x 7'-0"		12	1.333		247	75		322	390
2305	1-3/4", 3'-0" x 7'-0"		13	1.231		286	69.50		355.50	425
2320	Walnut face, 1-3/8", 2'-0" x 6'-8"		15	1.067		170	60		230	283
2340	2'-6" x 6'-8"		14	1.143		191	64.50		255.50	310
2380	3'-0" x 6'-8"		13	1.231		220	69.50		289.50	350
2400	4'-0" x 6'-8"	↓	12	1.333		282	75		357	430
2440	For 8'-0" high, add					61			61	67
2460	For 8'-0" high walnut, add					50			50	55
2720	For prefinishing, clear, add					58			58	63.50
2740	For prefinishing, stain, add					73.50			73.50	80.50
3320	M.D. overlay on hardboard, 1-3/8", 2'-6" x 6'-8"	2 Carp	14	1.143		235	64.50		299.50	360
3340	3'-0" x 6'-8"		13	1.231		266	69.50		335.50	405
3380	3'-0" x 7'-0"		12	1.333		270	75		345	415
3400	4'-0" x 7'-0"	↓	10	1.600		365	90		455	545
3440	For 8'-0" height, add					56			56	62
3460	For solid wood core, add					95			95	105
3720	H.P. plastic laminate, 1-3/8", 2'-6" x 6'-8"	2 Carp	13	1.231		232	69.50		301.50	365
3740	3'-0" x 6'-8"		12	1.333		247	75		322	390
3780	3'-0" x 7'-0"		11	1.455		263	82		345	420
3800	4'-0" x 7'-0"	↓	8	2		595	113		708	835
3840	For 8'-0" height, add					56			56	62
3860	For solid wood core, add					56.50			56.50	62.50
4000	Exterior, flush, solid core, birch, 1-3/4" x 2'-6" x 7'-0"	2 Carp	15	1.067		199	60		259	315
4020	2'-8" wide		15	1.067		288	60		348	410
4040	3'-0" wide	↓	14	1.143		225	64.50		289.50	350
4045	3'-0" x 8'-0"	1 Carp	8	1		365	56.50		421.50	490
4100	Oak faced 1-3/4" x 2'-6" x 7'-0"	2 Carp	15	1.067		268	60		328	390
4120	2'-8" wide		15	1.067		345	60		405	475
4140	3'-0" wide	↓	14	1.143		286	64.50		350.50	415
4160	Walnut faced, 1-3/4" x 3'-0" x 6'-8"	1 Carp	17	.471		670	26.50		696.50	775
4180	3'-6" wide	"	17	.471		760	26.50		786.50	875
4200	Walnut faced, 1-3/4" x 2'-6" x 7'-0"	2 Carp	15	1.067		445	60		505	585
4220	2'-8" wide		15	1.067		485	60		545	625
4240	3'-0" wide	↓	14	1.143		425	64.50		489.50	565
4300	For 6'-8" high door, deduct from 7'-0" door					23.50			23.50	26
5000	Wood doors, for vision lite, add					310			310	340

08 14 Wood Doors

08 14 16 – Flush Wood Doors

08 14 16.09 Smooth Wood Doors	Crew	Daily Output	Labor-Hours	Unit	Material	2022 Bare Costs Labor	Equipment	Total	Total Incl O&P
5010 Wood doors, for narrow lite, add				Ea.	42			42	46
5015 Wood doors, for bottom (or top) louver, add				↓	440			440	485
9000 Minimum labor/equipment charge	1 Carp	4	2	Job		113		113	179

08 14 16.10 Wood Doors Decorator	Crew	Daily Output	Labor-Hours	Unit	Material	2022 Bare Costs Labor	Equipment	Total	Total Incl O&P
0010 **WOOD DOORS DECORATOR**									
1800 Exterior, flush, solid wood core, birch 1-3/4" x 2'-6" x 7'-0"	2 Carp	15	1.067	Ea.	218	60		278	335
1820 2'-8" wide		15	1.067		227	60		287	345
1840 3'-0" wide		14	1.143		585	64.50		649.50	745
1900 Oak faced, 1-3/4" x 2'-6" x 7'-0"		15	1.067		222	60		282	340
1920 2'-8" wide		15	1.067		560	60		620	715
1940 3'-0" wide		14	1.143		655	64.50		719.50	820
2100 Walnut faced, 1-3/4" x 2'-6" x 7'-0"		15	1.067		455	60		515	595
2120 2'-8" wide		15	1.067		445	60		505	585
2140 3'-0" wide	↓	14	1.143	↓	405	64.50		469.50	545

08 14 16.20 Wood Fire Doors	Crew	Daily Output	Labor-Hours	Unit	Material	2022 Bare Costs Labor	Equipment	Total	Total Incl O&P
0010 **WOOD FIRE DOORS**									
0020 Particle core, 7 face plys, "B" label,									
0040 1 hour, birch face, 1-3/4" x 2'-6" x 6'-8"	2 Carp	14	1.143	Ea.	515	64.50		579.50	665
0080 3'-0" x 6'-8"		13	1.231		460	69.50		529.50	615
0090 3'-0" x 7'-0"		12	1.333		475	75		550	645
0100 4'-0" x 7'-0"		12	1.333		800	75		875	1,000
0140 Oak face, 2'-6" x 6'-8"		14	1.143		635	64.50		699.50	795
0180 3'-0" x 6'-8"		13	1.231		635	69.50		704.50	810
0190 3'-0" x 7'-0"		12	1.333		655	75		730	840
0200 4'-0" x 7'-0"		12	1.333		750	75		825	945
0240 Walnut face, 2'-6" x 6'-8"		14	1.143		600	64.50		664.50	760
0280 3'-0" x 6'-8"		13	1.231		610	69.50		679.50	780
0290 3'-0" x 7'-0"		12	1.333		445	75		520	605
0300 4'-0" x 7'-0"		12	1.333		880	75		955	1,100
0440 M.D. overlay on hardboard, 2'-6" x 6'-8"		15	1.067		410	60		470	545
0480 3'-0" x 6'-8"		14	1.143		450	64.50		514.50	595
0490 3'-0" x 7'-0"		13	1.231		480	69.50		549.50	640
0500 4'-0" x 7'-0"		12	1.333		520	75		595	695
0740 90 minutes, birch face, 1-3/4" x 2'-6" x 6'-8"		14	1.143		515	64.50		579.50	665
0780 3'-0" x 6'-8"		13	1.231		485	69.50		554.50	640
0790 3'-0" x 7'-0"		12	1.333		500	75		575	670
0800 4'-0" x 7'-0"		12	1.333		665	75		740	850
0840 Oak face, 2'-6" x 6'-8"		14	1.143		445	64.50		509.50	590
0880 3'-0" x 6'-8"		13	1.231		500	69.50		569.50	660
0890 3'-0" x 7'-0"		12	1.333		535	75		610	705
0900 4'-0" x 7'-0"		12	1.333		670	75		745	855
0940 Walnut face, 2'-6" x 6'-8"		14	1.143		575	64.50		639.50	730
0980 3'-0" x 6'-8"		13	1.231		460	69.50		529.50	615
0990 3'-0" x 7'-0"		12	1.333		635	75		710	820
1000 4'-0" x 7'-0"		12	1.333		770	75		845	970
1140 M.D. overlay on hardboard, 2'-6" x 6'-8"		15	1.067		505	60		565	650
1180 3'-0" x 6'-8"		14	1.143		490	64.50		554.50	640
1190 3'-0" x 7'-0"		13	1.231		530	69.50		599.50	695
1200 4'-0" x 7'-0"	↓	12	1.333		615	75		690	795
1240 For 8'-0" height, add					95.50			95.50	105
1260 For 8'-0" height walnut, add					106			106	117
2200 Custom architectural "B" label, flush, 1-3/4" thick, birch,									

08 14 Wood Doors

08 14 16 – Flush Wood Doors

08 14 16.20 Wood Fire Doors		Crew	Daily Output	Labor-Hours	Unit	Material	2022 Bare Costs Labor	Equipment	Total	Total Incl O&P
2210	Solid core									
2220	2'-6" x 7'-0"	2 Carp	15	1.067	Ea.	415	60		475	555
2260	3'-0" x 7'-0"		14	1.143		540	64.50		604.50	695
2300	4'-0" x 7'-0"		13	1.231		620	69.50		689.50	795
2420	4'-0" x 8'-0"		11	1.455		475	82		557	650
2480	For oak veneer, add					50%				
2500	For walnut veneer, add					75%				
9000	Minimum labor/equipment charge	1 Carp	4	2	Job		113		113	179

08 14 33 – Stile and Rail Wood Doors

08 14 33.10 Wood Doors Paneled

		Crew	Daily Output	Labor-Hours	Unit	Material	2022 Bare Costs Labor	Equipment	Total	Total Incl O&P
0010	**WOOD DOORS PANELED**									
0020	Interior, six panel, hollow core, 1-3/8" thick									
0040	Molded hardboard, 2'-0" x 6'-8"	2 Carp	17	.941	Ea.	108	53		161	203
0060	2'-6" x 6'-8"		17	.941		119	53		172	215
0070	2'-8" x 6'-8"		17	.941		91.50	53		144.50	185
0080	3'-0" x 6'-8"		17	.941		106	53		159	200
0140	Embossed print, molded hardboard, 2'-0" x 6'-8"		17	.941		88	53		141	181
0160	2'-6" x 6'-8"		17	.941		88	53		141	181
0180	3'-0" x 6'-8"		17	.941		106	53		159	200
0540	Six panel, solid, 1-3/8" thick, pine, 2'-0" x 6'-8"		15	1.067		212	60		272	330
0560	2'-6" x 6'-8"		14	1.143		227	64.50		291.50	350
0580	3'-0" x 6'-8"		13	1.231		258	69.50		327.50	395
1020	Two panel, bored rail, solid, 1-3/8" thick, pine, 1'-6" x 6'-8"		16	1		325	56.50		381.50	445
1040	2'-0" x 6'-8"		15	1.067		435	60		495	575
1060	2'-6" x 6'-8"		14	1.143		480	64.50		544.50	625
1340	Two panel, solid, 1-3/8" thick, fir, 2'-0" x 6'-8"	2 Carp	15	1.067	Ea.	213	60		273	330
1360	2'-6" x 6'-8"		14	1.143		287	64.50		351.50	415
1380	3'-0" x 6'-8"		13	1.231		520	69.50		589.50	685
1740	Five panel, solid, 1-3/8" thick, fir, 2'-0" x 6'-8"		15	1.067		385	60		445	520
1760	2'-6" x 6'-8"		14	1.143		505	64.50		569.50	655
1780	3'-0" x 6'-8"		13	1.231		505	69.50		574.50	665
4190	Exterior, Knotty pine, paneled, 1-3/4", 3'-0" x 6'-8"		16	1		735	56.50		791.50	900
4195	Double 1-3/4", 3'-0" x 6'-8"		16	1		1,475	56.50		1,531.50	1,725
4200	Ash, paneled, 1-3/4", 3'-0" x 6'-8"		16	1		555	56.50		611.50	700
4205	Double 1-3/4", 3'-0" x 6'-8"		16	1		1,125	56.50		1,181.50	1,325
4210	Cherry, paneled, 1-3/4", 3'-0" x 6'-8"		16	1		1,500	56.50		1,556.50	1,750
4215	Double 1-3/4", 3'-0" x 6'-8"		16	1		2,975	56.50		3,031.50	3,375
4230	Ash, paneled, 1-3/4", 3'-0" x 8'-0"		16	1		450	56.50		506.50	585
4235	Double 1-3/4", 3'-0" x 8'-0"		16	1		895	56.50		951.50	1,075
4240	Hard maple, paneled, 1-3/4", 3'-0" x 8'-0"		16	1		495	56.50		551.50	635
4245	Double 1-3/4", 3'-0" x 8'-0"		16	1		990	56.50		1,046.50	1,200
4250	Cherry, paneled, 1-3/4", 3'-0" x 8'-0"		16	1		915	56.50		971.50	1,100
4255	Double 1-3/4", 3'-0" x 8'-0"		16	1		1,825	56.50		1,881.50	2,125
9000	Minimum labor/equipment charge	1 Carp	4	2	Job		113		113	179

08 14 33.20 Wood Doors Residential

		Crew	Daily Output	Labor-Hours	Unit	Material	2022 Bare Costs Labor	Equipment	Total	Total Incl O&P
0010	**WOOD DOORS RESIDENTIAL**									
0200	Exterior, combination storm & screen, pine									
0260	2'-8" wide	2 Carp	10	1.600	Ea.	325	90		415	505
0280	3'-0" wide		9	1.778		340	100		440	535
0300	7'-1" x 3'-0" wide		9	1.778		365	100		465	560
0400	Full lite, 6'-9" x 2'-6" wide		11	1.455		350	82		432	515
0420	2'-8" wide		10	1.600		340	90		430	520

For customer support on your Commercial Renovation Costs with RSMeans data, call 800.448.8182.

243

08 14 33.20 Wood Doors Residential	Crew	Daily Output	Labor-Hours	Unit	Material	2022 Bare Costs Labor	Equipment	Total	Total Incl O&P
0440	3'-0" wide	2 Carp	9	1.778	Ea.	350	100		450	545
0500	7'-1" x 3'-0" wide		9	1.778		375	100		475	575
0700	Dutch door, pine, 1-3/4" x 2'-8" x 6'-8", 6 panel		12	1.333		1,175	75		1,250	1,400
0720	Half glass		10	1.600		1,025	90		1,115	1,300
0800	3'-0" wide, 6 panel		12	1.333		650	75		725	840
0820	Half glass		10	1.600		1,150	90		1,240	1,425
1000	Entrance door, colonial, 1-3/4" x 6'-8" x 2'-8" wide		16	1		630	56.50		686.50	785
1020	6 panel pine, 3'-0" wide		15	1.067		169	60		229	282
1100	8 panel pine, 2'-8" wide		16	1		740	56.50		796.50	905
1120	3'-0" wide		15	1.067		680	60		740	840
1200	For tempered safety glass lites (min. of 2), add					144			144	158
1300	Flush, birch, solid core, 1-3/4" x 6'-8" x 2'-8" wide	2 Carp	16	1		490	56.50		546.50	625
1320	3'-0" wide		15	1.067		505	60		565	650
1350	7'-0" x 2'-8" wide		16	1		140	56.50		196.50	244
1360	3'-0" wide		15	1.067		585	60		645	740
1380	For tempered safety glass lites, add					119			119	131
1420	6'-8" x 3'-0" wide, fir	2 Carp	16	1		640	56.50		696.50	795
1720	Carved mahogany 3'-0" x 6'-8"		15	1.067		1,700	60		1,760	1,975
1800	Lauan, solid core, 1-3/4" x 7'-0" x 2'-4" wide		16	1		147	56.50		203.50	251
1810	2'-6" wide		15	1.067		154	60		214	265
1820	2'-8" wide		9	1.778		192	100		292	370
1830	3'-0" wide		16	1		209	56.50		265.50	320
1840	3'-4" wide		16	1		276	56.50		332.50	395
1850	Pair of 3'-0" wide		15	1.067	Pr.	415	60		475	555
2700	Interior, closet, bi-fold, w/hardware, no frame or trim incl.									
2720	Flush, birch, 2'-6" x 6'-8"	2 Carp	13	1.231	Ea.	172	69.50		241.50	299
2740	3'-0" wide		13	1.231		190	69.50		259.50	320
2760	4'-0" wide		12	1.333		299	75		374	450
2780	5'-0" wide		11	1.455		330	82		412	495
2800	6'-0" wide		10	1.600		365	90		455	545
2820	Flush, hardboard, primed, 6'-8" x 2'-6" wide		13	1.231		146	69.50		215.50	271
2840	3'-0" wide		13	1.231		161	69.50		230.50	287
2860	4'-0" wide		12	1.333		335	75		410	485
2880	5'-0" wide		11	1.455		292	82		374	450
2900	6'-0" wide		10	1.600		166	90		256	325
3000	Raised panel pine, 6'-6" or 6'-8" x 2'-6" wide		13	1.231		540	69.50		609.50	705
3020	3'-0" wide		13	1.231		575	69.50		644.50	745
3040	4'-0" wide		12	1.333		930	75		1,005	1,150
3060	5'-0" wide		11	1.455		425	82		507	595
3080	6'-0" wide		10	1.600		465	90		555	655
3200	Louvered, pine, 6'-6" or 6'-8" x 2'-6" wide		13	1.231		171	69.50		240.50	298
3220	3'-0" wide		13	1.231		270	69.50		339.50	405
3240	4'-0" wide		12	1.333		930	75		1,005	1,150
3260	5'-0" wide		11	1.455		277	82		359	435
3280	6'-0" wide		10	1.600		340	90		430	515
4400	Bi-passing closet, incl. hardware and frame, no trim incl.									
4420	Flush, lauan, 6'-8" x 4'-0" wide	2 Carp	12	1.333	Opng.	192	75		267	330
4440	5'-0" wide		11	1.455		258	82		340	415
4460	6'-0" wide		10	1.600		191	90		281	355
4600	Flush, birch, 6'-8" x 4'-0" wide		12	1.333		285	75		360	435
4620	5'-0" wide		11	1.455		244	82		326	400
4640	6'-0" wide		10	1.600		400	90		490	585
4800	Louvered, pine, 6'-8" x 4'-0" wide		12	1.333		490	75		565	660

08 14 33 – Stile and Rail Wood Doors

08 14 33.20 Wood Doors Residential		Crew	Daily Output	Labor-Hours	Unit	Material	2022 Bare Costs Labor	Equipment	Total	Total Incl O&P
4820	5'-0" wide	2 Carp	11	1.455	Opng.	585	82		667	770
4840	6'-0" wide		10	1.600		855	90		945	1,075
5000	Paneled, pine, 6'-8" x 4'-0" wide		12	1.333		545	75		620	720
5020	5'-0" wide		11	1.455		860	82		942	1,075
5040	6'-0" wide		10	1.600		1,000	90		1,090	1,250
5042	8'-0" wide	▼	12	1.333	▼	1,100	75		1,175	1,325
6100	Folding accordion, closet, including track and frame									
6120	Vinyl, 2 layer, stock	2 Carp	10	1.600	Ea.	675	90		765	885
6200	Rigid PVC	"	10	1.600	"	127	90		217	283
7310	Passage doors, flush, no frame included									
7320	Hardboard, hollow core, 1-3/8" x 6'-8" x 1'-6" wide	2 Carp	18	.889	Ea.	43	50		93	127
7330	2'-0" wide		18	.889		48	50		98	132
7340	2'-6" wide		18	.889		54.50	50		104.50	140
7350	2'-8" wide		18	.889		56.50	50		106.50	142
7360	3'-0" wide		17	.941		56.50	53		109.50	147
7420	Lauan, hollow core, 1-3/8" x 6'-8" x 1'-6" wide		18	.889		85	50		135	173
7440	2'-0" wide		18	.889		73.50	50		123.50	161
7450	2'-4" wide		18	.889		84	50		134	172
7460	2'-6" wide		18	.889		84	50		134	172
7480	2'-8" wide		18	.889		96	50		146	186
7500	3'-0" wide		17	.941		72	53		125	163
7700	Birch, hollow core, 1-3/8" x 6'-8" x 1'-6" wide		18	.889		101	50		151	192
7720	2'-0" wide		18	.889		115	50		165	207
7740	2'-6" wide		18	.889		109	50		159	200
7760	2'-8" wide		18	.889		124	50		174	217
7780	3'-0" wide		17	.941		117	53		170	213
8000	Pine louvered, 1-3/8" x 6'-8" x 1'-6" wide		19	.842		201	47.50		248.50	297
8020	2'-0" wide		18	.889		223	50		273	325
8040	2'-6" wide		18	.889		239	50		289	345
8060	2'-8" wide		18	.889		274	50		324	380
8080	3'-0" wide	▼	17	.941	▼	245	53		298	355
8300	Pine paneled, 1-3/8" x 6'-8" x 1'-6" wide	2 Carp	19	.842	Ea.	222	47.50		269.50	320
8320	2'-0" wide		18	.889		320	50		370	430
8330	2'-4" wide		18	.889		320	50		370	430
8340	5'-6" wide		18	.889		345	50		395	460
8360	2'-8" wide		18	.889		345	50		395	455
8380	3'-0" wide		17	.941		370	53		423	495
8804	Pocket door, 6 panel pine, 2'-6" x 6'-8" with frame		10.50	1.524		340	86		426	510
8814	2'-8" x 6'-8"		10.50	1.524		505	86		591	690
8824	3'-0" x 6'-8"		10.50	1.524		380	86		466	550
9000	Passage doors, flush, no frame, birch, solid core, 1-3/8" x 2'-4" x 7'-0"		16	1		165	56.50		221.50	272
9020	2'-8" wide		16	1		175	56.50		231.50	282
9040	3'-0" wide		16	1		208	56.50		264.50	320
9060	3'-4" wide		15	1.067	▼	370	60		430	505
9080	Pair of 3'-0" wide		9	1.778	Pr.	370	100		470	565
9100	Lauan, solid core, 1-3/8" x 7'-0" x 2'-4" wide		16	1	Ea.	222	56.50		278.50	335
9120	2'-8" wide		16	1		198	56.50		254.50	310
9140	3'-0" wide		16	1		262	56.50		318.50	380
9160	3'-4" wide		15	1.067	▼	271	60		331	395
9180	Pair of 3'-0" wide		9	1.778	Pr.	530	100		630	745
9200	Hardboard, solid core, 1-3/8" x 7'-0" x 2'-4" wide		16	1	Ea.	219	56.50		275.50	330
9220	2'-8" wide		16	1		224	56.50		280.50	335
9240	3'-0" wide		16	1		231	56.50		287.50	345

For customer support on your Commercial Renovation Costs with RSMeans data, call 800.448.8182.

245

08 14 Wood Doors

08 14 33 – Stile and Rail Wood Doors

08 14 33.20 Wood Doors Residential		Crew	Daily Output	Labor-Hours	Unit	Material	2022 Bare Costs Labor	Equipment	Total	Total Incl O&P
9260	3'-4" wide	2 Carp	15	1.067	Ea.	480	60		540	625
9900	Minimum labor/equipment charge	1 Carp	4	2	Job		113		113	179

08 14 35 – Torrified Doors

08 14 35.10 Torrified Exterior Doors

		Crew	Daily Output	Labor-Hours	Unit	Material	2022 Bare Costs Labor	Equipment	Total	Total Incl O&P
0010	**TORRIFIED EXTERIOR DOORS**									
0020	Wood doors made from torrified wood, exterior									
0030	All doors require a finish be applied, all glass is insulated									
0040	All doors require pilot holes for all fasteners									
0100	6 panel, paint grade poplar, 1-3/4" x 3'-0" x 6'-8"	2 Carp	12	1.333	Ea.	1,325	75		1,400	1,575
0120	Half glass 3'-0" x 6'-8"	"	12	1.333		1,425	75		1,500	1,675
0200	Side lite, full glass, 1-3/4" x 1'-2" x 6'-8"					785			785	865
0220	Side lite, half glass, 1-3/4" x 1'-2" x 6'-8"					765			765	840
0300	Raised face, 2 panel, paint grade poplar, 1-3/4" x 3'-0" x 7'-0"	2 Carp	12	1.333		1,550	75		1,625	1,850
0320	Side lite, raised face, half glass, 1-3/4" x 1'-2" x 7'-0"					890			890	980
0500	6 panel, Fir, 1-3/4" x 3'-0" x 6'-8"	2 Carp	12	1.333	↓	1,875	75		1,950	2,175
0520	Half glass 3'-0" x 6'-8"	2 Carp	12	1.333	Ea.	1,700	75		1,775	2,000
0600	Side lite, full glass, 1-3/4" x 1'-2" x 6'-8"					795			795	875
0620	Side lite, half glass, 1-3/4" x 1'-2" x 6'-8"					825			825	905
0700	6 panel, Mahogany, 1-3/4" x 3'-0" x 6'-8"	2 Carp	12	1.333		2,475	75		2,550	2,850
0800	Side lite, full glass, 1-3/4" x 1'-2" x 6'-8"					900			900	990
0820	Side lite, half glass, 1-3/4" x 1'-2" x 6'-8"				↓	910			910	1,000

08 14 40 – Interior Cafe Doors

08 14 40.10 Cafe Style Doors

		Crew	Daily Output	Labor-Hours	Unit	Material	2022 Bare Costs Labor	Equipment	Total	Total Incl O&P
0010	**CAFE STYLE DOORS**									
6520	Interior cafe doors, 2'-6" opening, stock, panel pine	2 Carp	16	1	Ea.	430	56.50		486.50	565
6540	3'-0" opening	"	16	1	"	445	56.50		501.50	575
6550	Louvered pine									
6560	2'-6" opening	2 Carp	16	1	Ea.	340	56.50		396.50	465
8000	3'-0" opening		16	1		335	56.50		391.50	460
8010	2'-6" opening, hardwood		16	1		380	56.50		436.50	510
8020	3'-0" opening	↓	16	1	↓	430	56.50		486.50	560
9000	Minimum labor/equipment charge	1 Carp	4	2	Job		113		113	179

08 16 Composite Doors

08 16 13 – Fiberglass Doors

08 16 13.10 Entrance Doors, Fibrous Glass

			Crew	Daily Output	Labor-Hours	Unit	Material	2022 Bare Costs Labor	Equipment	Total	Total Incl O&P
0010	**ENTRANCE DOORS, FIBROUS GLASS**										
0020	Exterior, fiberglass, door, 2'-8" wide x 6'-8" high	G	2 Carp	15	1.067	Ea.	1,125	60		1,185	1,325
0040	3'-0" wide x 6'-8" high	G		15	1.067		281	60		341	405
0060	3'-0" wide x 7'-0" high	G		15	1.067		525	60		585	675
0080	3'-0" wide x 6'-8" high, with two lites	G		15	1.067		720	60		780	890
0100	3'-0" wide x 8'-0" high, with two lites	G		15	1.067		550	60		610	700
0110	Half glass, 3'-0" wide x 6'-8" high	G		15	1.067		485	60		545	630
0120	3'-0" wide x 6'-8" high, low E	G		15	1.067		530	60		590	680
0130	3'-0" wide x 8'-0" high	G		15	1.067		615	60		675	775
0140	3'-0" wide x 8'-0" high, low E	G	↓	15	1.067		695	60		755	860
0150	Side lights, 1'-0" wide x 6'-8" high	G					285			285	315
0160	1'-0" wide x 6'-8" high, low E	G					280			280	310
0180	1'-0" wide x 6'-8" high, full glass	G					375			375	410
0190	1'-0" wide x 6'-8" high, low E	G				↓	395			395	430

08 16 Composite Doors

08 16 14 – French Doors

08 16 14.10 Exterior Doors With Glass Lites

08 16 14.10 Exterior Doors With Glass Lites	Crew	Daily Output	Labor-Hours	Unit	Material	2022 Bare Costs Labor	Equipment	Total	Total Incl O&P
0010 **EXTERIOR DOORS WITH GLASS LITES**									
0020 French, Fir, 1-3/4", 3'-0" wide x 6'-8" high	2 Carp	12	1.333	Ea.	1,250	75		1,325	1,500
0025 Double	"	12	1.333	"	2,525	75		2,600	2,900
0030 Maple, 1-3/4", 3'-0" wide x 6'-8" high	2 Carp	12	1.333	Ea.	745	75		820	935
0035 Double		12	1.333		1,475	75		1,550	1,750
0040 Cherry, 1-3/4", 3'-0" wide x 6'-8" high		12	1.333		865	75		940	1,075
0045 Double		12	1.333		1,725	75		1,800	2,025
0100 Mahogany, 1-3/4", 3'-0" wide x 8'-0" high		10	1.600		1,250	90		1,340	1,550
0105 Double		10	1.600		2,525	90		2,615	2,925
0110 Fir, 1-3/4", 3'-0" wide x 8'-0" high		10	1.600		555	90		645	755
0115 Double		10	1.600		1,100	90		1,190	1,375
0120 Oak, 1-3/4", 3'-0" wide x 8'-0" high		10	1.600		600	90		690	805
0125 Double	↓	10	1.600	↓	1,200	90		1,290	1,475

08 17 Integrated Door Opening Assemblies

08 17 13 – Integrated Metal Door Opening Assemblies

08 17 13.10 Hollow Metal Doors and Frames

08 17 13.10 Hollow Metal Doors and Frames		Crew	Daily Output	Labor-Hours	Unit	Material	2022 Bare Costs Labor	Equipment	Total	Total Incl O&P
0010 **HOLLOW METAL DOORS AND FRAMES**										
0100 Prehung, flush, 18 ga., 1-3/4" x 6'-8" x 2'-8" wide		1 Carp	16	.500	Ea.	780	28		808	900
0120 3'-0" wide			15	.533		1,550	30		1,580	1,750
0130 3'-6" wide			13	.615		955	34.50		989.50	1,100
0140 4'-0" wide			10	.800		1,050	45		1,095	1,225
0300 Double, 1-3/4" x 3'-0" x 6'-8"			5	1.600		1,175	90		1,265	1,450
0350 3'-0" x 8'-0"		↓	4	2	↓	2,500	113		2,613	2,925

08 17 13.20 Stainless Steel Doors and Frames

08 17 13.20 Stainless Steel Doors and Frames		Crew	Daily Output	Labor-Hours	Unit	Material	2022 Bare Costs Labor	Equipment	Total	Total Incl O&P
0010 **STAINLESS STEEL DOORS AND FRAMES**										
0020 Stainless steel (304) prehung 24 ga. 2'-6" x 6'-8" door w/16 g frame	G	2 Carp	6	2.667	Ea.	2,150	150		2,300	2,625
0025 2'-8" x 6'-8"	G		6	2.667		2,175	150		2,325	2,650
0030 3'-0" x 6'-8"	G		6	2.667		2,175	150		2,325	2,625
0040 3'-0" x 7'-0"	G		6	2.667		2,250	150		2,400	2,725
0050 4'-0" x 7'-0"	G		5	3.200		3,975	180		4,155	4,650
0100 Stainless steel (316) prehung 24 ga. 2'-6" x 6'-8" door w/16 g frame	G		6	2.667		4,775	150		4,925	5,500
0110 2'-8" x 6'-8"	G		6	2.667		2,575	150		2,725	3,100
0120 3'-0" x 6'-8"	G		6	2.667		4,875	150		5,025	5,625
0150 3'-0" x 7'-0"	G		6	2.667		4,275	150		4,425	4,950
0160 4'-0" x 7'-0"	G		6	2.667		5,125	150		5,275	5,900
0300 Stainless steel (304) prehung 18 ga. 2'-6" x 6'-8" door w/16 g frame	G		6	2.667		3,650	150		3,800	4,275
0310 2'-8" x 6'-8"	G		6	2.667		3,700	150		3,850	4,300
0320 3'-0" x 6'-8"	G		6	2.667		2,525	150		2,675	3,025
0350 3'-0" x 7'-0"	G		6	2.667		2,650	150		2,800	3,150
0360 4'-0" x 7'-0"	G		5	3.200		3,650	180		3,830	4,300
0500 Stainless steel, prehung door, foam core, 14 ga., 3'-0" x 7'-0"	G	↓	5	3.200	↓	4,475	180		4,655	5,200
0600 Stainless steel, prehung double door, foam core, 14 ga., 3'-0" x 7'-0"	G	2 Carp	4	4	Ea.	7,400	225		7,625	8,500

08 17 23 – Integrated Wood Door Opening Assemblies

08 17 23.10 Pre-Hung Doors

08 17 23.10 Pre-Hung Doors	Crew	Daily Output	Labor-Hours	Unit	Material	2022 Bare Costs Labor	Equipment	Total	Total Incl O&P
0010 **PRE-HUNG DOORS**									
0300 Exterior, wood, comb. storm & screen, 6'-9" x 2'-6" wide	2 Carp	15	1.067	Ea.	262	60		322	385
0320 2'-8" wide		15	1.067		350	60		410	480
0340 3'-0" wide	↓	15	1.067		330	60		390	460
0360 For 7'-0" high door, add				↓	46			46	50.50

For customer support on your Commercial Renovation Costs with RSMeans data, call 800.448.8182.

247

08 17 Integrated Door Opening Assemblies

08 17 23 – Integrated Wood Door Opening Assemblies

08 17 23.10 Pre-Hung Doors		Crew	Daily Output	Labor-Hours	Unit	Material	2022 Bare Costs Labor	Equipment	Total	Total Incl O&P
1600	Entrance door, flush, birch, solid core									
1620	4-5/8" solid jamb, 1-3/4" x 6'-8" x 2'-8" wide	2 Carp	16	1	Ea.	375	56.50		431.50	505
1640	3'-0" wide		16	1		445	56.50		501.50	580
1642	5-5/8" jamb		16	1		410	56.50		466.50	545
1680	For 7'-0" high door, add					25.50			25.50	28
2000	Entrance door, colonial, 6 panel pine									
2020	4-5/8" solid jamb, 1-3/4" x 6'-8" x 2'-8" wide	2 Carp	16	1	Ea.	755	56.50		811.50	920
2040	3'-0" wide	"	16	1		645	56.50		701.50	800
2060	For 7'-0" high door, add					64.50			64.50	71
2200	For 5-5/8" solid jamb, add					49.50			49.50	54.50
2230	French style, exterior, 1 lite, 1-3/4" x 3'-0" x 6'-8"	1 Carp	14	.571		805	32		837	935
2235	9 lites	"	14	.571		815	32		847	945
2245	15 lites	2 Carp	14	1.143		930	64.50		994.50	1,125
2250	Double, 15 lites, 2'-0" x 6'-8", 4'-0" opening		7	2.286	Pr.	1,475	129		1,604	1,800
2260	2'-6" x 6'-8", 5'-0" opening		7	2.286		1,625	129		1,754	2,000
2280	3'-0" x 6'-8", 6'-0" opening		7	2.286		1,925	129		2,054	2,325
2430	3'-0" x 7'-0", 15 lites		14	1.143	Ea.	895	64.50		959.50	1,075
2432	Two 3'-0" x 7'-0"		7	2.286	Pr.	1,875	129		2,004	2,250
2435	3'-0" x 8'-0"		14	1.143	Ea.	895	64.50		959.50	1,075
2437	Two, 3'-0" x 8'-0"		7	2.286	Pr.	1,875	129		2,004	2,250
4000	Interior, passage door, 4-5/8" solid jamb									
4350	Paneled, primed, hollow core, 2'-8" wide	2 Carp	17	.941	Ea.	150	53		203	249
4360	3'-0" wide		17	.941		216	53		269	320
4400	Lauan, flush, solid core, 1-3/8" x 6'-8" x 2'-6" wide		17	.941		223	53		276	330
4420	2'-8" wide		17	.941		223	53		276	330
4440	3'-0" wide		16	1		225	56.50		281.50	340
4600	Hollow core, 1-3/8" x 6'-8" x 2'-6" wide		17	.941		227	53		280	335
4620	2'-8" wide		17	.941		150	53		203	249
4640	3'-0" wide		16	1		225	56.50		281.50	340
4700	For 7'-0" high door, add				Ea.	48.50			48.50	53.50
5000	Birch, flush, solid core, 1-3/8" x 6'-8" x 2'-6" wide	2 Carp	17	.941		315	53		368	430
5020	2'-8" wide		17	.941		244	53		297	350
5040	3'-0" wide		16	1		345	56.50		401.50	470
5200	Hollow core, 1-3/8" x 6'-8" x 2'-6" wide		17	.941		215	53		268	320
5220	2'-8" wide		17	.941		242	53		295	350
5240	3'-0" wide		16	1		251	56.50		307.50	365
5280	For 7'-0" high door, add					41.50			41.50	45.50
5500	Hardboard paneled, 1-3/8" x 6'-8" x 2'-6" wide	2 Carp	17	.941		175	53		228	277
5520	2'-8" wide		17	.941		184	53		237	286
5540	3'-0" wide		16	1		185	56.50		241.50	294
6000	Pine paneled, 1-3/8" x 6'-8" x 2'-6" wide		17	.941		310	53		363	430
6020	2'-8" wide		17	.941		330	53		383	450
6040	3'-0" wide		16	1		340	56.50		396.50	465
8200	Birch, flush, solid core, 1-3/4" x 6'-8" x 2'-4" wide	1 Carp	17	.471		243	26.50		269.50	310
8220	2'-6" wide		17	.471		235	26.50		261.50	300
8240	2'-8" wide		17	.471		220	26.50		246.50	284
8260	3'-0" wide		16	.500		229	28		257	297
8280	3'-6" wide		15	.533		400	30		430	490
8500	Pocket door frame with lauan, flush, hollow core, 1-3/8" x 3'-0" x 6'-8"		17	.471		335	26.50		361.50	410
9000	Minimum labor/equipment charge		4	2	Job		113		113	179

For customer support on your Commercial Renovation Costs with RSMeans data, call 800.448.8182.

08 31 Access Doors and Panels

08 31 13 – Access Doors and Frames

08 31 13.10 Types of Framed Access Doors

	Crew	Daily Output	Labor-Hours	Unit	Material	2022 Bare Costs Labor	2022 Bare Costs Equipment	Total	Total Incl O&P
0010 TYPES OF FRAMED ACCESS DOORS									
1000 Fire rated door with lock									
1100 Metal, 12" x 12"	1 Carp	10	.800	Ea.	172	45		217	261
1150 18" x 18"		9	.889		340	50		390	455
1200 24" x 24"		9	.889		345	50		395	460
1250 24" x 36"		8	1		345	56.50		401.50	470
1300 24" x 48"		8	1		410	56.50		466.50	540
1350 36" x 36"		7.50	1.067		510	60		570	655
1400 48" x 48"		7.50	1.067		665	60		725	825
1600 Stainless steel, 12" x 12"		10	.800		335	45		380	440
1650 18" x 18"		9	.889		370	50		420	485
1700 24" x 24"		9	.889		545	50		595	680
1750 24" x 36"	↓	8	1	↓	650	56.50		706.50	805
2000 Flush door for finishing									
2100 Metal 8" x 8"	1 Carp	10	.800	Ea.	38	45		83	114
2150 12" x 12"	"	10	.800	"	38.50	45		83.50	114
3000 Recessed door for acoustic tile									
3100 Metal, 12" x 12"	1 Carp	4.50	1.778	Ea.	48.50	100		148.50	213
3150 12" x 24"		4.50	1.778		111	100		211	281
3200 24" x 24"		4	2		117	113		230	310
3250 24" x 36"	↓	4	2	↓	163	113		276	360
4000 Recessed door for drywall									
4100 Metal 12" x 12"	1 Carp	6	1.333	Ea.	75.50	75		150.50	202
4150 12" x 24"		5.50	1.455		118	82		200	260
4200 24" x 36"	↓	5	1.600	↓	188	90		278	350
6000 Standard door									
6100 Metal, 8" x 8"	1 Carp	10	.800	Ea.	31.50	45		76.50	107
6150 12" x 12"		10	.800		284	45		329	380
6200 18" x 18"		9	.889		35	50		85	118
6250 24" x 24"		9	.889		425	50		475	550
6300 24" x 36"		8	1		175	56.50		231.50	282
6350 36" x 36"		8	1		99	56.50		155.50	199
6500 Stainless steel, 8" x 8"		10	.800		80	45		125	160
6550 12" x 12"		10	.800		94	45		139	175
6600 18" x 18"		9	.889		170	50		220	267
6650 24" x 24"		9	.889	↓	259	50		309	365
9000 Minimum labor/equipment charge	↓	4	2	Job		113		113	179

08 31 13.20 Bulkhead/Cellar Doors

	Crew	Daily Output	Labor-Hours	Unit	Material	2022 Bare Costs Labor	2022 Bare Costs Equipment	Total	Total Incl O&P
0010 BULKHEAD/CELLAR DOORS									
0020 Steel, not incl. sides, 44" x 62"	1 Carp	5.50	1.455	Ea.	715	82		797	920
0100 52" x 73"		5.10	1.569		895	88.50		983.50	1,125
0500 With sides and foundation plates, 57" x 45" x 24"		4.70	1.702		890	96		986	1,125
0600 42" x 49" x 51"		4.30	1.860	↓	605	105		710	830
9000 Minimum labor/equipment charge	↓	2	4	Job		225		225	360

08 31 13.30 Commercial Floor Doors

	Crew	Daily Output	Labor-Hours	Unit	Material	2022 Bare Costs Labor	2022 Bare Costs Equipment	Total	Total Incl O&P
0010 COMMERCIAL FLOOR DOORS									
0020 Aluminum tile, steel frame, one leaf, 2' x 2' opng.	2 Sswk	3.50	4.571	Opng.	760	283		1,043	1,300
0050 3'-6" x 3'-6" opening		3.50	4.571		1,625	283		1,908	2,225
0500 Double leaf, 4' x 4' opening		3	5.333		1,650	330		1,980	2,350
0550 5' x 5' opening		3	5.333	↓	2,750	330		3,080	3,550
9000 Minimum labor/equipment charge	↓	2	8	Job		495		495	795

For customer support on your Commercial Renovation Costs with RSMeans data, call 800.448.8182.

249

08 31 Access Doors and Panels

08 31 13 – Access Doors and Frames

08 31 13.35 Industrial Floor Doors	Crew	Daily Output	Labor-Hours	Unit	Material	2022 Bare Costs Labor	Equipment	Total	Total Incl O&P
0010 **INDUSTRIAL FLOOR DOORS**									
0020 Steel 300 psf L.L., single leaf, 2' x 2', 175#	2 Sswk	6	2.667	Opng.	1,800	165		1,965	2,250
0050 3' x 3' opening, 300#	2 Sswk	5.50	2.909	Opng.	1,400	180		1,580	1,850
0300 Double leaf, 4' x 4' opening, 455#		5	3.200		2,675	198		2,873	3,250
0350 5' x 5' opening, 645#		4.50	3.556		3,150	220		3,370	3,825
1000 Aluminum, 300 psf L.L., single leaf, 2' x 2', 60#		6	2.667		735	165		900	1,075
1050 3' x 3' opening, 100#		5.50	2.909		890	180		1,070	1,275
1500 Double leaf, 4' x 4' opening, 160#		5	3.200		3,200	198		3,398	3,850
1550 5' x 5' opening, 235#		4.50	3.556		4,125	220		4,345	4,875
9000 Minimum labor/equipment charge		2	8	Job		495		495	795

08 32 Sliding Glass Doors

08 32 13 – Sliding Aluminum-Framed Glass Doors

08 32 13.10 Sliding Aluminum Doors

08 32 13.10 Sliding Aluminum Doors	Crew	Daily Output	Labor-Hours	Unit	Material	2022 Bare Costs Labor	Equipment	Total	Total Incl O&P
0010 **SLIDING ALUMINUM DOORS**									
0350 Aluminum, 5/8" tempered insulated glass, 6' wide									
0400 Premium	2 Carp	4	4	Ea.	1,500	225		1,725	2,000
0450 Economy		4	4		860	225		1,085	1,300
0500 8' wide, premium		3	5.333		1,725	300		2,025	2,375
0550 Economy		3	5.333		1,525	300		1,825	2,150
0600 12' wide, premium		2.50	6.400		3,000	360		3,360	3,875
0650 Economy		2.50	6.400		1,525	360		1,885	2,250
4000 Aluminum, baked on enamel, temp glass, 6'-8" x 10'-0" wide		4	4		1,150	225		1,375	1,625
4020 Insulating glass, 6'-8" x 6'-0" wide		4	4		950	225		1,175	1,400
4040 8'-0" wide		3	5.333		1,125	300		1,425	1,725
4060 10'-0" wide		2	8		1,300	450		1,750	2,150
4080 Anodized, temp glass, 6'-8" x 6'-0" wide		4	4		465	225		690	875
4100 8'-0" wide		3	5.333		585	300		885	1,125
4120 10'-0" wide		2	8		690	450		1,140	1,475
5000 Aluminum sliding glass door system									
5010 Sliding door 4' wide opening single side	2 Carp	2	8	Ea.	6,375	450		6,825	7,725
5015 8' wide opening single side		2	8		1,700	450		2,150	2,600
5020 Telescoping glass door system, 4' wide opening biparting		2	8		5,475	450		5,925	6,750
5025 8' wide opening biparting		2	8		6,375	450		6,825	7,725
5030 Folding glass door, 4' wide opening biparting		2	8		8,500	450		8,950	10,100
5035 8' wide opening biparting		2	8		5,100	450		5,550	6,325
5040 ICU-CCU sliding telescoping glass door, 4' x 7', single side opening		2	8		3,250	450		3,700	4,300
5045 8' x 7', single side opening		2	8		4,950	450		5,400	6,150
7000 Electric swing door operator and control, single door w/sensors	1 Carp	4	2		2,975	113		3,088	3,450
7005 Double door w/sensors		2	4		6,000	225		6,225	6,925
7010 Electric folding door operator and control, single door w/sensors		4	2		7,050	113		7,163	7,925
7015 Bi-folding door		4	2		4,875	113		4,988	5,525
7020 Electric swing door operator and control, single door		4	2		4,025	113		4,138	4,600

08 32 19 – Sliding Wood-Framed Glass Doors

08 32 19.10 Sliding Wood Doors

	Crew	Daily Output	Labor-Hours	Unit	Material	2022 Bare Costs Labor	Equipment	Total	Total Incl O&P
0010 **SLIDING WOOD DOORS**									
0020 Wood, tempered insul. glass, 6' wide, premium	2 Carp	4	4	Ea.	1,500	225		1,725	2,000
0100 Economy		4	4		1,175	225		1,400	1,650
0150 8' wide, wood, premium		3	5.333		1,925	300		2,225	2,600
0200 Economy		3	5.333		1,425	300		1,725	2,025

For customer support on your Commercial Renovation Costs with RSMeans data, call 800.448.8182.

08 32 Sliding Glass Doors

08 32 19 – Sliding Wood-Framed Glass Doors

08 32 19.10 Sliding Wood Doors		Crew	Daily Output	Labor-Hours	Unit	Material	2022 Bare Costs Labor	Equipment	Total	Total Incl O&P
0235	10' wide, wood, premium	2 Carp	2.50	6.400	Ea.	2,825	360		3,185	3,675
0240	Economy		2.50	6.400		2,400	360		2,760	3,225
0250	12' wide, wood, premium		2.50	6.400		3,100	360		3,460	4,000
0300	Economy		2.50	6.400		2,575	360		2,935	3,425
9000	Minimum labor/equipment charge	1 Carp	2	4	Job		225		225	360

08 32 19.15 Sliding Glass Vinyl-Clad Wood Doors

			Crew	Daily Output	Labor-Hours	Unit	Material	Labor	Equipment	Total	Total Incl O&P
0010	**SLIDING GLASS VINYL-CLAD WOOD DOORS**										
0020	Glass, sliding vinyl-clad, insul. glass, 6'-0" x 6'-8"	G	2 Carp	4	4	Opng.	1,575	225		1,800	2,075
0025	6'-0" x 6'-10" high	G		4	4		1,725	225		1,950	2,250
0030	6'-0" x 8'-0" high	G		4	4		2,125	225		2,350	2,675
0050	5'-0" x 6'-8" high	G		4	4		1,650	225		1,875	2,175
0100	8'-0" x 6'-10" high	G		4	4		2,250	225		2,475	2,825
0150	8'-0" x 8'-0" high	G		4	4		2,425	225		2,650	3,025
0500	4 leaf, 9'-0" x 6'-10" high	G		3	5.333		3,275	300		3,575	4,075
0550	9'-0" x 8'-0" high	G		3	5.333		3,925	300		4,225	4,775
0600	12'-0" x 6'-10" high	G		3	5.333		4,250	300		4,550	5,150
0650	12'-0" x 8'-0" high	G		3	5.333		4,250	300		4,550	5,150
9000	Minimum labor/equipment charge		1 Carp	4	2	Job		113		113	179

08 33 Coiling Doors and Grilles

08 33 13 – Coiling Counter Doors

08 33 13.10 Counter Doors, Coiling Type		Crew	Daily Output	Labor-Hours	Unit	Material	Labor	Equipment	Total	Total Incl O&P
0010	**COUNTER DOORS, COILING TYPE**									
0020	Manual, incl. frame and hardware, galv. stl., 4' roll-up, 6' long	2 Carp	2	8	Opng.	1,350	450		1,800	2,200
0300	Galvanized steel, UL label		1.80	8.889		1,350	500		1,850	2,275
0600	Stainless steel, 4' high roll-up, 6' long		2	8		2,250	450		2,700	3,200
0700	10' long		1.80	8.889		2,725	500		3,225	3,775
2000	Aluminum, 4' high, 4' long		2.20	7.273		1,875	410		2,285	2,725
2020	6' long		2	8		2,525	450		2,975	3,500
2040	8' long		1.90	8.421		2,400	475		2,875	3,375
2060	10' long	2 Carp	1.80	8.889	Opng.	2,375	500		2,875	3,425
2080	14' long		1.40	11.429		3,650	645		4,295	5,050
2100	6' high, 4' long		2	8		2,575	450		3,025	3,550
2120	6' long		1.60	10		1,875	565		2,440	2,950
2140	10' long		1.40	11.429		2,475	645		3,120	3,750
9000	Minimum labor/equipment charge	1 Carp	2	4	Job		225		225	360

08 33 16 – Coiling Counter Grilles

08 33 16.10 Coiling Grilles		Crew	Daily Output	Labor-Hours	Unit	Material	Labor	Equipment	Total	Total Incl O&P
0010	**COILING GRILLES**									
2020	Aluminum, manual operated, mill finish	2 Sswk	82	.195	S.F.	31	12.05		43.05	53.50
2040	Bronze anodized		82	.195	"	56.50	12.05		68.55	82
2060	Steel, manual operated, 10' x 10' high		1	16	Opng.	2,725	990		3,715	4,600
2080	15' x 8' high		.80	20	"	3,650	1,225		4,875	6,000
3000	For safety edge bottom bar, electric, add				L.F.	51			51	56.50
8000	For motor operation, add	2 Sswk	5	3.200	Opng.	1,575	198		1,773	2,075
9000	Minimum labor/equipment charge	"	1	16	Job		990		990	1,600

For customer support on your Commercial Renovation Costs with RSMeans data, call 800.448.8182.

251

08 33 23.10 Coiling Service Doors		Crew	Daily Output	Labor-Hours	Unit	Material	2022 Bare Costs Labor	Equipment	Total	Total Incl O&P
0010	**COILING SERVICE DOORS** Steel, manual, 20 ga., incl. hardware									
0050	8' x 8' high	2 Sswk	1.60	10	Ea.	535	620		1,155	1,575
0130	12' x 12' high, standard		1.20	13.333		1,500	825		2,325	2,975
0160	10' x 20' high, standard		.50	32		1,600	1,975		3,575	4,950
2000	Class A fire doors, manual, 20 ga., 8' x 8' high		1.40	11.429		1,875	705		2,580	3,225
2100	10' x 10' high		1.10	14.545		3,100	900		4,000	4,850
2200	20' x 10' high		.80	20		5,300	1,225		6,525	7,825
2300	12' x 12' high		1	16		4,075	990		5,065	6,100
2400	20' x 12' high		.80	20		5,700	1,225		6,925	8,275
2500	14' x 14' high		.60	26.667		4,450	1,650		6,100	7,550
2600	20' x 16' high		.50	32		6,175	1,975		8,150	9,975
2700	10' x 20' high		.40	40		5,625	2,475		8,100	10,200
3000	For 18 ga. doors, add				S.F.	1.59			1.59	1.75
3300	For enamel finish, add				"	2.97			2.97	3.27
3600	For safety edge bottom bar, pneumatic, add				L.F.	29.50			29.50	32
4000	For weatherstripping, extruded rubber, jambs, add					2.21			2.21	2.43
4100	Hood, add					9.50			9.50	10.45
4200	Sill, add					9.65			9.65	10.60
4500	Motor operators, to 14' x 14' opening	2 Sswk	5	3.200	Ea.	970	198		1,168	1,400
4700	For fire door, additional fusible link, add				"	80.50			80.50	88.50
9000	Minimum labor/equipment charge	2 Sswk	1	16	Job		990		990	1,600

08 34 13.10 Doors for Cold Area Storage		Crew	Daily Output	Labor-Hours	Unit	Material	2022 Bare Costs Labor	Equipment	Total	Total Incl O&P
0010	**DOORS FOR COLD AREA STORAGE**									
0020	Single, 20 ga. galvanized steel									
0300	Horizontal sliding, 5' x 7', manual operation, 3.5" thick	2 Carp	2	8	Ea.	3,300	450		3,750	4,350
0400	4" thick		2	8		3,850	450		4,300	4,950
0500	6" thick		2	8		3,800	450		4,250	4,900
0800	5' x 7', power operation, 2" thick		1.90	8.421		5,475	475		5,950	6,750
0900	4" thick		1.90	8.421		6,300	475		6,775	7,675
1000	6" thick		1.90	8.421		7,175	475		7,650	8,650
1300	9' x 10', manual operation, 2" insulation		1.70	9.412		4,775	530		5,305	6,100
1400	4" insulation		1.70	9.412		5,400	530		5,930	6,800
1500	6" insulation		1.70	9.412		5,450	530		5,980	6,850
1800	Power operation, 2" insulation		1.60	10		7,525	565		8,090	9,175
1900	4" insulation		1.60	10		8,575	565		9,140	10,300
2000	6" insulation		1.70	9.412		9,075	530		9,605	10,800
2300	For stainless steel face, add					25%				
3000	Hinged, lightweight, 3' x 7'-0", 2" thick	2 Carp	2	8	Ea.	1,375	450		1,825	2,250
3050	4" thick		1.90	8.421		1,725	475		2,200	2,650
3300	Polymer doors, 3' x 7'-0"		1.90	8.421		1,475	475		1,950	2,375
3350	6" thick		1.40	11.429		2,575	645		3,220	3,875
3600	Stainless steel, 3' x 7'-0", 4" thick		1.90	8.421		1,950	475		2,425	2,900
3650	6" thick		1.40	11.429		3,175	645		3,820	4,525
3900	Painted, 3' x 7'-0", 4" thick		1.90	8.421		1,475	475		1,950	2,350
3950	6" thick		1.40	11.429		2,600	645		3,245	3,875
5000	Bi-parting, electric operated									
5010	6' x 8' opening, galv. faces, 4" thick for cooler	2 Carp	.80	20	Opng.	7,975	1,125		9,100	10,600

08 34 Special Function Doors

08 34 13 – Cold Storage Doors

08 34 13.10 Doors for Cold Area Storage

		Crew	Daily Output	Labor-Hours	Unit	Material	2022 Bare Costs Labor	Equipment	Total	Total Incl O&P
5050	For freezer, 4" thick	2 Carp	.80	20	Opng.	8,775	1,125		9,900	11,500
5300	For door buck framing and door protection, add		2.50	6.400		710	360		1,070	1,350
6000	Galvanized batten door, galvanized hinges, 4' x 7'		2	8		2,100	450		2,550	3,025
6050	6' x 8'		1.80	8.889		2,750	500		3,250	3,825
6500	Fire door, 3 hr., 6' x 8', single slide		.80	20		9,450	1,125		10,575	12,200
6550	Double, bi-parting		.70	22.857		12,100	1,275		13,375	15,400
9000	Minimum labor/equipment charge	1 Carp	2	4	Job		225		225	360

08 34 36 – Darkroom Doors

08 34 36.10 Various Types of Darkroom Doors

		Crew	Daily Output	Labor-Hours	Unit	Material	2022 Bare Costs Labor	Equipment	Total	Total Incl O&P
0010	**VARIOUS TYPES OF DARKROOM DOORS**									
0015	Revolving, standard, 2 way, 36" diameter	2 Carp	3.10	5.161	Opng.	3,800	291		4,091	4,625
0020	41" diameter		3.10	5.161		3,525	291		3,816	4,325
0050	3 way, 51" diameter		1.40	11.429		5,350	645		5,995	6,925
1000	4 way, 49" diameter		1.40	11.429		4,100	645		4,745	5,525
2000	Hinged safety, 2 way, 41" diameter		2.30	6.957		4,250	390		4,640	5,300
2500	3 way, 51" diameter		1.40	11.429		5,575	645		6,220	7,150
3000	Pop out safety, 2 way, 41" diameter		3.10	5.161		4,325	291		4,616	5,200
4000	3 way, 51" diameter		1.40	11.429		6,250	645		6,895	7,900
5000	Wheelchair-type, pop out, 51" diameter		1.40	11.429		6,275	645		6,920	7,925
5020	72" diameter		.90	17.778		11,800	1,000		12,800	14,600

08 34 53 – Security Doors and Frames

08 34 53.20 Steel Door

		Crew	Daily Output	Labor-Hours	Unit	Material	2022 Bare Costs Labor	Equipment	Total	Total Incl O&P
0010	**STEEL DOOR** flush with ballistic core and welded frame both 14 ga.									
0120	UL 752 Level 3, 1-3/4", 3'-0" x 7'-0"	2 Carp	1.50	10.667	Opng.	2,600	600		3,200	3,800
0125	1-3/4", 3'-6" x 7'-0"		1.50	10.667		2,700	600		3,300	3,925
0130	1-3/4", 4'-0" x 7'-0"		1.20	13.333		2,800	750		3,550	4,275
0150	UL 752 Level 8, 1-3/4", 3'-0" x 7'-0"		1.50	10.667		6,450	600		7,050	8,050
0155	1-3/4", 3'-6" x 7'-0"		1.50	10.667		6,600	600		7,200	8,200
0160	1-3/4", 4'-0" x 7'-0"		1.20	13.333		6,700	750		7,450	8,550
1000	Safe room sliding door and hardware, 1-3/4", 3'-0" x 7'-0" UL 752 Level 3		.50	32		19,000	1,800		20,800	23,800
1050	Safe room swinging door and hardware, 1-3/4", 3'-0" x 7'-0" UL 752 Level 3		.50	32		22,800	1,800		24,600	28,000

08 34 53.30 Wood Ballistic Doors

		Crew	Daily Output	Labor-Hours	Unit	Material	2022 Bare Costs Labor	Equipment	Total	Total Incl O&P
0010	**WOOD BALLISTIC DOORS** with frames and hardware									
0050	Wood, 1-3/4", 3'-0" x 7'-0" UL 752 Level 3	2 Carp	1.50	10.667	Opng.	3,425	600		4,025	4,725

08 34 56 – Security Gates

08 34 56.10 Gates

		Crew	Daily Output	Labor-Hours	Unit	Material	2022 Bare Costs Labor	Equipment	Total	Total Incl O&P
0010	**GATES**									
0015	Driveway gates include mounting hardware									
0500	Wood, security gate, driveway, dual, 10' wide	H-4	.80	25	Opng.	3,525	1,325		4,850	5,975
0505	12' wide		.80	25		3,825	1,325		5,150	6,300
0510	15' wide		.80	25		4,425	1,325		5,750	6,975
0600	Steel, security gate, driveway, single, 10' wide		.80	25		2,700	1,325		4,025	5,075
0605	12' wide		.80	25		3,250	1,325		4,575	5,675
0620	Steel, security gate, driveway, dual, 12' wide		.80	25		2,800	1,325		4,125	5,200
0625	14' wide		.80	25		3,275	1,325		4,600	5,700
0630	16' wide		.80	25		2,350	1,325		3,675	4,675
0700	Aluminum, security gate, driveway, dual, 10' wide		.80	25		1,200	1,325		2,525	3,425
0705	12' wide		.80	25		1,300	1,325		2,625	3,525
0710	16' wide		.80	25		4,475	1,325		5,800	7,025
1000	Security gate, driveway, opener 12 VDC				Ea.	2,375			2,375	2,600
1010	Wireless					1,925			1,925	2,125

For customer support on your Commercial Renovation Costs with RSMeans data, call 800.448.8182.

253

08 34 Special Function Doors

08 34 56 – Security Gates

08 34 56.10 Gates		Crew	Daily Output	Labor-Hours	Unit	Material	2022 Bare Costs Labor	Equipment	Total	Total Incl O&P
1020	Security gate, driveway, opener 24 VDC				Ea.	2,250			2,250	2,500
1030	Wireless					2,175			2,175	2,375
1040	Security gate, driveway, opener 12 VDC, solar panel 10 watt	1 Elec	2	4		2,325	265		2,590	2,975
1050	20 watt	"	2	4		1,250	265		1,515	1,775

08 34 73 – Sound Control Door Assemblies

08 34 73.10 Acoustical Doors

		Crew	Daily Output	Labor-Hours	Unit	Material	Labor	Equipment	Total	Total Incl O&P
0010	**ACOUSTICAL DOORS**									
0020	Including framed seals, 3' x 7', wood, 40 STC rating	2 Carp	1.50	10.667	Ea.	3,250	600		3,850	4,525
0100	Steel, 41 STC rating		1.50	10.667		1,825	600		2,425	2,950
0200	45 STC rating		1.50	10.667		2,275	600		2,875	3,450
0300	48 STC rating		1.50	10.667		2,925	600		3,525	4,175
0400	52 STC rating		1.50	10.667		5,150	600		5,750	6,625
9000	Minimum labor/equipment charge	1 Carp	4	2	Job		113		113	179

08 36 Panel Doors

08 36 13 – Sectional Doors

08 36 13.10 Overhead Commercial Doors

		Crew	Daily Output	Labor-Hours	Unit	Material	Labor	Equipment	Total	Total Incl O&P
0010	**OVERHEAD COMMERCIAL DOORS**									
1000	Stock, sectional, heavy duty, wood, 1-3/4" thick, 8' x 8' high	2 Carp	2	8	Ea.	1,200	450		1,650	2,050
1200	12' x 12' high		1.50	10.667		2,775	600		3,375	4,000
1300	Chain hoist, 14' x 14' high		1.30	12.308		3,775	695		4,470	5,250
1600	20' x 16' high		.65	24.615		7,125	1,375		8,500	10,100
2100	For medium duty custom door, deduct					5%	5%			
2150	For medium duty stock doors, deduct					10%	5%			
2300	Fiberglass and aluminum, heavy duty, sectional, 12' x 12' high	2 Carp	1.50	10.667	Ea.	3,375	600		3,975	4,675
2450	Chain hoist, 20' x 20' high	"	.50	32		7,725	1,800		9,525	11,400
2900	For electric trolley operator, 1/3 HP, to 12' x 12', add	1 Carp	2	4		865	225		1,090	1,325
2950	Over 12' x 12', 1/2 HP, add		1	8		865	450		1,315	1,675
2980	Overhead, for row of clear lites, add		1	8		262	450		712	1,000
9000	Minimum labor/equipment charge	2 Carp	1.50	10.667	Job		600		600	955

08 36 13.20 Residential Garage Doors

		Crew	Daily Output	Labor-Hours	Unit	Material	Labor	Equipment	Total	Total Incl O&P
0010	**RESIDENTIAL GARAGE DOORS**									
0050	Hinged, wood, custom, double door, 9' x 7'	2 Carp	4	4	Ea.	2,300	225		2,525	2,875
0070	16' x 7'		3	5.333		3,900	300		4,200	4,750
0200	Overhead, sectional, incl. hardware, fiberglass, 9' x 7', standard		5	3.200		960	180		1,140	1,325
0220	Deluxe		5	3.200		1,175	180		1,355	1,550
0300	16' x 7', standard		6	2.667		2,175	150		2,325	2,650
0320	Deluxe	2 Carp	6	2.667	Ea.	2,400	150		2,550	2,900
0500	Hardboard, 9' x 7', standard		8	2		730	113		843	980
0520	Deluxe		8	2		930	113		1,043	1,200
0600	16' x 7', standard		6	2.667		1,350	150		1,500	1,725
0620	Deluxe		6	2.667		1,650	150		1,800	2,075
0700	Metal, 9' x 7', standard		8	2		605	113		718	845
0720	Deluxe		6	2.667		1,450	150		1,600	1,850
0800	16' x 7', standard		6	2.667		935	150		1,085	1,275
0820	Deluxe		5	3.200		1,625	180		1,805	2,050
0900	Wood, 9' x 7', standard		8	2		2,300	113		2,413	2,700
0920	Deluxe		8	2		2,375	113		2,488	2,775
1000	16' x 7', standard		6	2.667		3,500	150		3,650	4,100
1020	Deluxe		6	2.667		3,175	150		3,325	3,725

08 36 Panel Doors

08 36 13 – Sectional Doors

08 36 13.20 Residential Garage Doors		Crew	Daily Output	Labor-Hours	Unit	Material	2022 Bare Costs Labor	Equipment	Total	Total Incl O&P
1800	Door hardware, sectional	1 Carp	4	2	Ea.	370	113		483	590
1810	Door tracks only		4	2		173	113		286	370
1820	One side only		7	1.143		123	64.50		187.50	237
4000	For electric operator, economy, add		8	1		440	56.50		496.50	575
4100	Deluxe, including remote control		8	1		640	56.50		696.50	795
4500	For transmitter/receiver control, add to operator				Total	117			117	129
4600	Transmitters, additional				"	68.50			68.50	75.50
6000	Replace section, on sectional door, fiberglass, 9' x 7'	1 Carp	4	2	Ea.	715	113		828	970
6020	16' x 7'		3.50	2.286		750	129		879	1,025
6200	Hardboard, 9' x 7'		4	2		189	113		302	385
6220	16' x 7'		3.50	2.286		248	129		377	475
6300	Metal, 9' x 7'		4	2		226	113		339	425
6320	16' x 7'		3.50	2.286		330	129		459	570
6500	Wood, 9' x 7'		4	2		123	113		236	315
6520	16' x 7'		3.50	2.286		243	129		372	470
7010	Garage doors, row of lites					226			226	249
9000	Minimum labor/equipment charge	1 Carp	2.50	3.200	Job		180		180	286

08 36 19 – Multi-Leaf Vertical Lift Doors

08 36 19.10 Sectional Vertical Lift Doors

		Crew	Daily Output	Labor-Hours	Unit	Material	2022 Bare Costs Labor	Equipment	Total	Total Incl O&P
0010	**SECTIONAL VERTICAL LIFT DOORS**									
0020	Motorized, 14 ga. steel, incl. frame and control panel									
0050	16' x 16' high	L-10	.50	48	Ea.	22,400	3,025	3,975	29,400	33,900
0100	10' x 20' high		1.30	18.462		38,800	1,175	1,525	41,500	46,200
0120	15' x 20' high		1.30	18.462		47,300	1,175	1,525	50,000	55,500
0140	20' x 20' high		1	24		55,500	1,525	1,975	59,000	65,500
0160	25' x 20' high		1	24		56,000	1,525	1,975	59,500	66,000
0170	32' x 24' high		.75	32		52,000	2,025	2,650	56,675	63,000
0180	20' x 25' high		1	24		58,000	1,525	1,975	61,500	68,000
0200	25' x 25' high		.70	34.286		80,500	2,150	2,825	85,475	95,000
0220	25' x 30' high		.70	34.286		84,500	2,150	2,825	89,475	99,500
0240	30' x 30' high		.70	34.286		93,000	2,150	2,825	97,975	108,500
0260	35' x 30' high		.70	34.286		101,500	2,150	2,825	106,475	118,000

08 38 Traffic Doors

08 38 19 – Rigid Traffic Doors

08 38 19.20 Double Acting Swing Doors

		Crew	Daily Output	Labor-Hours	Unit	Material	2022 Bare Costs Labor	Equipment	Total	Total Incl O&P
0010	**DOUBLE ACTING SWING DOORS**									
0020	Including frame, closer, hardware and vision panel									
1000	Polymer, 7'-0" high, 4'-0" wide	2 Carp	4.20	3.810	Pr.	2,025	214		2,239	2,575
1025	6'-0" wide		4	4		2,500	225		2,725	3,100
1050	6'-8" wide		4	4		2,650	225		2,875	3,275
2000	3/4" thick, stainless steel									
2010	Stainless steel, 7' high opening, 4' wide	2 Carp	4	4	Pr.	2,500	225		2,725	3,100
2050	7' wide		3.80	4.211	"	2,700	237		2,937	3,350
9000	Minimum labor/equipment charge		2	8	Job		450		450	715

08 38 19.30 Shock Absorbing Doors

		Crew	Daily Output	Labor-Hours	Unit	Material	2022 Bare Costs Labor	Equipment	Total	Total Incl O&P
0010	**SHOCK ABSORBING DOORS**									
0020	Rigid, no frame, 1-1/2" thick, 5' x 7'	2 Sswk	1.90	8.421	Opng.	1,525	520		2,045	2,525
0100	8' x 8'		1.80	8.889		3,100	550		3,650	4,275
0500	Flexible, no frame, insulated, .16" thick, economy, 5' x 7'		2	8		2,850	495		3,345	3,925

For customer support on your Commercial Renovation Costs with RSMeans data, call 800.448.8182.

255

08 38 Traffic Doors

08 38 19 – Rigid Traffic Doors

08 38 19.30 Shock Absorbing Doors	Crew	Daily Output	Labor-Hours	Unit	Material	2022 Bare Costs Labor	Equipment	Total	Total Incl O&P
0600	Deluxe	2 Sswk	1.90	8.421	Opng.	3,550	520		4,070	4,750
1000	8' x 8' opening, economy		2	8		3,100	495		3,595	4,200
1100	Deluxe		1.90	8.421		3,500	520		4,020	4,700
9000	Minimum labor/equipment charge		2	8	Job		495		495	795

08 41 Entrances and Storefronts

08 41 13 – Aluminum-Framed Entrances and Storefronts

08 41 13.20 Tube Framing

		Crew	Daily Output	Labor-Hours	Unit	Material	2022 Bare Costs Labor	Equipment	Total	Total Incl O&P
0010	**TUBE FRAMING**, For window walls and storefronts, aluminum stock									
0050	Plain tube frame, mill finish, 1-3/4" x 1-3/4"	2 Glaz	103	.155	L.F.	10.95	8.40		19.35	25.50
0150	1-3/4" x 4"		98	.163		15.60	8.80		24.40	31
0200	1-3/4" x 4-1/2"		95	.168		17.20	9.10		26.30	33.50
0250	2" x 6"		89	.180		30.50	9.70		40.20	49
0350	4" x 4"		87	.184		25	9.95		34.95	43.50
0400	4-1/2" x 4-1/2"		85	.188		31.50	10.15		41.65	50.50
0450	Glass bead		240	.067		3.45	3.60		7.05	9.50
1000	Flush tube frame, mill finish, 1/4" glass, 1-3/4" x 4", open header		80	.200		15.65	10.80		26.45	34.50
1050	Open sill	2 Glaz	82	.195	L.F.	13.95	10.55		24.50	32
1100	Closed back header		83	.193		24.50	10.40		34.90	43.50
1150	Closed back sill		85	.188		23.50	10.15		33.65	42
1160	Tube fmg., spandrel cover both sides, alum 1" wide	1 Sswk	85	.094	S.F.	100	5.80		105.80	119
1170	Tube fmg., spandrel cover both sides, alum 2" wide	"	85	.094	"	41	5.80		46.80	54.50
1200	Vertical mullion, one piece	2 Glaz	75	.213	L.F.	21.50	11.50		33	42.50
1250	Two piece		73	.219		23.50	11.85		35.35	44.50
1300	90° or 180° vertical corner post		75	.213		36.50	11.50		48	58.50
1400	1-3/4" x 4-1/2", open header		80	.200		18.40	10.80		29.20	37.50
1450	Open sill		82	.195		15.25	10.55		25.80	33.50
1500	Closed back header		83	.193		21.50	10.40		31.90	40
1550	Closed back sill		85	.188		21.50	10.15		31.65	39.50
1600	Vertical mullion, one piece		75	.213		24	11.50		35.50	45
1650	Two piece		73	.219		25.50	11.85		37.35	47
1700	90° or 180° vertical corner post		75	.213		26	11.50		37.50	47
2000	Flush tube frame, mil fin.,ins. glass w/thml brk, 2" x 4-1/2", open header		75	.213		16.15	11.50		27.65	36
2050	Open sill		77	.208		13.60	11.20		24.80	33
2100	Closed back header		78	.205		15	11.10		26.10	34
2150	Closed back sill		80	.200		14.55	10.80		25.35	33
2200	Vertical mullion, one piece		70	.229		17.15	12.35		29.50	38.50
2250	Two piece		68	.235		18.90	12.70		31.60	41
2300	90° or 180° vertical corner post		70	.229		19.50	12.35		31.85	41
5000	Flush tube frame, mill fin., thermal brk., 2-1/4" x 4-1/2", open header		74	.216		16.70	11.70		28.40	37
5050	Open sill		75	.213		14.45	11.50		25.95	34
5100	Vertical mullion, one piece		69	.232		19.20	12.50		31.70	41
5150	Two piece		67	.239		21	12.90		33.90	44
5200	90° or 180° vertical corner post		69	.232		19.40	12.50		31.90	41.50
5295	Flush tube frame, mill finish, 4-1/2" x 4-1/2"		85	.188		31.50	10.15		41.65	50.50
5300	Plain tube frame, mill finish, 2" x 3" jamb		115	.139		11.55	7.50		19.05	24.50
5310	2" x 4"		115	.139		16.50	7.50		24	30
5320	3" x 5-1/2"		107	.150		25.50	8.05		33.55	41.50
5330	Head, 2" x 3"		200	.080		21.50	4.32		25.82	30.50
5340	Mullion, 2" x 3"		200	.080		22	4.32		26.32	31
5350	2" x 4"		115	.139		28	7.50		35.50	42.50

08 41 Entrances and Storefronts

08 41 13 – Aluminum-Framed Entrances and Storefronts

08 41 13.20 Tube Framing

		Crew	Daily Output	Labor-Hours	Unit	Material	2022 Bare Costs Labor	2022 Bare Costs Equipment	Total	Total Incl O&P
5360	3" x 5-1/2"	2 Glaz	107	.150	L.F.	35	8.05		43.05	51.50
5370	4" corner mullion		90	.178		31	9.60		40.60	49.50
5380	Horizontal, 2" x 3"	↓	115	.139	↓	25.50	7.50		33	40
5390	3" x 5-1/2"	2 Glaz	107	.150	L.F.	29.50	8.05		37.55	45.50
5430	Sill section, 1/8" x 6"		133	.120		36.50	6.50		43	51
5440	1/8" x 7"		133	.120		44	6.50		50.50	59
5450	1/8" x 8-1/2"		133	.120		51.50	6.50		58	67
5460	Column covers, aluminum, 1/8" x 26"		57	.281		71	15.15		86.15	102
5470	1/8" x 34"		53	.302		100	16.30		116.30	136
5480	1/8" x 38"		53	.302		102	16.30		118.30	138
5700	Vertical mullions, clear finish, 1/4" thick glass		528	.030		14.85	1.64		16.49	18.95
5720	3/8" thick		445	.036		13.80	1.94		15.74	18.30
5740	1/2" thick		400	.040		14.65	2.16		16.81	19.55
5760	3/4" thick		344	.047		15.90	2.51		18.41	21.50
5780	1" thick		304	.053		15.50	2.84		18.34	21.50
6980	Door stop (snap in)	↓	380	.042	↓	3.25	2.27		5.52	7.20
7000	For joints, 90°, clip type, add				Ea.	27.50			27.50	30.50
7050	Screw spline joint, add					24.50			24.50	27
7100	For joint other than 90°, add				↓	52			52	57
8000	For bronze anodized aluminum, add					15%				
8020	For black finish, add					30%				
8050	For stainless steel materials, add					350%				
8100	For monumental grade, add					53%				
8150	For steel stiffener, add	2 Glaz	200	.080	L.F.	11.70	4.32		16.02	19.70
8200	For 2 to 5 stories, add per story				Story		8%			
9000	Minimum labor/equipment charge	2 Glaz	2	8	Job		430		430	685

08 41 19 – Stainless-Steel-Framed Entrances and Storefronts

08 41 19.10 Stainless-Steel and Glass Entrance Unit

		Crew	Daily Output	Labor-Hours	Unit	Material	2022 Bare Costs Labor	2022 Bare Costs Equipment	Total	Total Incl O&P
0010	**STAINLESS-STEEL AND GLASS ENTRANCE UNIT**, narrow stiles									
0020	3' x 7' opening, including hardware, minimum	2 Sswk	1.60	10	Opng.	8,000	620		8,620	9,800
0050	Average		1.40	11.429		8,475	705		9,180	10,500
0100	Maximum	↓	1.20	13.333	↓	9,100	825		9,925	11,300
1000	For solid bronze entrance units, statuary finish, add					64%				
1100	Without statuary finish, add					45%				
2000	Balanced doors, 3' x 7', economy	2 Sswk	.90	17.778	Ea.	10,600	1,100		11,700	13,500
2100	Premium		.70	22.857	"	17,300	1,425		18,725	21,300
9000	Minimum labor/equipment charge	↓	2	8	Job		495		495	795

08 41 26 – All-Glass Entrances and Storefronts

08 41 26.10 Window Walls Aluminum, Stock

		Crew	Daily Output	Labor-Hours	Unit	Material	2022 Bare Costs Labor	2022 Bare Costs Equipment	Total	Total Incl O&P
0010	**WINDOW WALLS ALUMINUM, STOCK**, including glazing									
0020	Minimum	H-2	160	.150	S.F.	59.50	7.70		67.20	77.50
0050	Average	"	140	.171	"	82	8.80		90.80	104
0100	Maximum	H-2	110	.218	S.F.	193	11.15		204.15	230
0500	For translucent sandwich wall systems, see Section 07 41 33.10									
0850	Cost of the above walls depends on material,									
0860	finish, repetition, and size of units.									
0870	The larger the opening, the lower the S.F. cost									

08 42 Entrances

08 42 26 – All-Glass Entrances

08 42 26.10 Swinging Glass Doors		Crew	Daily Output	Labor-Hours	Unit	Material	2022 Bare Costs Labor	2022 Bare Costs Equipment	Total	Total Incl O&P
0010	**SWINGING GLASS DOORS**									
0020	Including hardware, 1/2" thick, tempered, 3' x 7' opening	2 Glaz	2	8	Opng.	2,400	430		2,830	3,300
0100	6' x 7' opening		1.40	11.429	"	4,575	615		5,190	6,000
9000	Minimum labor/equipment charge		2	8	Job		430		430	685

08 42 36 – Balanced Door Entrances

08 42 36.10 Balanced Entrance Doors

		Crew	Daily Output	Labor-Hours	Unit	Material	2022 Bare Costs Labor	2022 Bare Costs Equipment	Total	Total Incl O&P
0010	**BALANCED ENTRANCE DOORS**									
0020	Hardware & frame, alum. & glass, 3' x 7', econ.	2 Sswk	.90	17.778	Ea.	7,525	1,100		8,625	10,100
0150	Premium		.70	22.857	"	8,975	1,425		10,400	12,100
9000	Minimum labor/equipment charge		1	16	Job		990		990	1,600

08 43 Storefronts

08 43 13 – Aluminum-Framed Storefronts

08 43 13.10 Aluminum-Framed Entrance Doors and Frames

		Crew	Daily Output	Labor-Hours	Unit	Material	2022 Bare Costs Labor	2022 Bare Costs Equipment	Total	Total Incl O&P
0010	**ALUMINUM-FRAMED ENTRANCE DOORS AND FRAMES**									
0015	Standard hardware and glass stops but no glass									
0020	Entrance door, 3' x 7' opening, clear anodized finish	2 Sswk	7	2.286	Opng.	740	141		881	1,050
0040	Bronze finish		7	2.286		925	141		1,066	1,250
0060	Black finish		7	2.286		900	141		1,041	1,225
0200	3'-6" x 7'-0", mill finish		7	2.286		1,000	141		1,141	1,325
0220	Bronze finish		7	2.286		795	141		936	1,100
0240	Black finish		7	2.286		805	141		946	1,125
0500	6' x 7' opening, clear finish		6	2.667		995	165		1,160	1,375
0520	Bronze finish		6	2.667		1,375	165		1,540	1,800
0600	Door frame for above doors 3'-0" x 7'-0", mill finish		6	2.667		625	165		790	955
0620	Bronze finish		6	2.667		665	165		830	1,000
0640	Black finish		6	2.667		750	165		915	1,100
0700	3'-6" x 7'-0", mill finish		6	2.667		450	165		615	760
0720	Bronze finish		6	2.667		450	165		615	760
0740	Black finish		6	2.667		450	165		615	760
0800	6'-0" x 7'-0", mill finish		6	2.667		450	165		615	760
0820	Bronze finish		6	2.667		455	165		620	765
0840	Black finish	2 Sswk	6	2.667	Opng.	490	165		655	805
1000	With 3' high transom above, 3' x 7' opening, clear finish		5.50	2.909		695	180		875	1,050
1050	Bronze finish		5.50	2.909		690	180		870	1,050
1100	Black finish		5.50	2.909		675	180		855	1,025
1300	3'-6" x 7'-0" opening, clear finish		5.50	2.909		450	180		630	785
1320	Bronze finish		5.50	2.909		475	180		655	810
1340	Black finish		5.50	2.909		480	180		660	820
1500	6' x 7' opening, clear finish		5.50	2.909		735	180		915	1,100
1550	Bronze finish		5.50	2.909		860	180		1,040	1,225
1600	Black finish		5.50	2.909		955	180		1,135	1,350
8000	For 8' high doors, add				Ea.	174			174	174
9000	Minimum labor/equipment charge	2 Sswk	4	4	Job		247		247	400

08 43 13.20 Storefront Systems

		Crew	Daily Output	Labor-Hours	Unit	Material	2022 Bare Costs Labor	2022 Bare Costs Equipment	Total	Total Incl O&P
0010	**STOREFRONT SYSTEMS**, aluminum frame clear 3/8" plate glass									
0020	incl. 3' x 7' door with hardware (400 sq. ft. max. wall)									
0500	Wall height to 12' high, commercial grade	2 Glaz	150	.107	S.F.	25.50	5.75		31.25	37
0600	Institutional grade		130	.123		29.50	6.65		36.15	43
0700	Monumental grade		115	.139		49.50	7.50		57	66.50

08 43 Storefronts

08 43 13 – Aluminum-Framed Storefronts

08 43 13.20 Storefront Systems

		Crew	Daily Output	Labor-Hours	Unit	Material	2022 Bare Costs Labor	Equipment	Total	Total Incl O&P
1000	6' x 7' door with hardware, commercial grade	2 Glaz	135	.119	S.F.	75.50	6.40		81.90	93
1100	Institutional grade		115	.139		43	7.50		50.50	59
1200	Monumental grade		100	.160		122	8.65		130.65	148
1500	For bronze anodized finish, add					15%				
1600	For black anodized finish, add					36%				
1700	For stainless steel framing, add to monumental					78%				
9000	Minimum labor/equipment charge	2 Glaz	1	16	Job		865		865	1,375

08 43 29 – Sliding Storefronts

08 43 29.10 Sliding Panels

		Crew	Daily Output	Labor-Hours	Unit	Material	2022 Bare Costs Labor	Equipment	Total	Total Incl O&P
0010	**SLIDING PANELS**									
0020	Mall fronts, aluminum & glass, 15' x 9' high	2 Glaz	1.30	12.308	Opng.	4,725	665		5,390	6,250
0100	24' x 9' high		.70	22.857		6,625	1,225		7,850	9,250
0200	48' x 9' high, with fixed panels		.90	17.778		11,900	960		12,860	14,600
0500	For bronze finish, add					17%				
9000	Minimum labor/equipment charge	2 Glaz	1	16	Job		865		865	1,375

08 45 Translucent Wall and Roof Assemblies

08 45 10 – Translucent Roof Assemblies

08 45 10.10 Skyroofs

		Crew	Daily Output	Labor-Hours	Unit	Material	2022 Bare Costs Labor	Equipment	Total	Total Incl O&P
0010	**SKYROOFS**									
1200	Skylights, circular, clear, double glazed acrylic									
1230	30" diameter	2 Carp	3	5.333	Ea.	1,775	300		2,075	2,450
1250	60" diameter	2 Carp	3	5.333	Ea.	2,950	300		3,250	3,725
1290	96" diameter	"	2	8	"	6,125	450		6,575	7,450
1300	Skylight Barrel Vault, clear, double glazed, acrylic									
1330	3'-0" x 12'-0"	G-3	3	10.667	Ea.	4,425	605		5,030	5,825
1350	4'-0" x 12'-0"		3	10.667		4,900	605		5,505	6,350
1390	5'-0" x 12'-0"		2	16		6,475	905		7,380	8,550
1400	Skylight Pyramid, aluminum frame, clear low E laminated glass									
1410	The glass is installed in the frame except where noted									
1430	Square, 3' x 3'	G-3	3	10.667	Ea.	4,300	605		4,905	5,675
1440	4' x 4'		3	10.667		4,875	605		5,480	6,325
1450	5' x 5', glass must be field installed		3	10.667		6,875	605		7,480	8,500
1460	6' x 6', glass must be field installed		2	16		9,525	905		10,430	11,900
1550	Install pre-cut laminated glass in aluminum frame on a flat roof	2 Glaz	55	.291	SF Surf		15.70		15.70	25
1560	Install pre-cut laminated glass in aluminum frame on a sloped roof	"	40	.400	"		21.50		21.50	34.50
9000	Minimum labor/equipment charge	2 Carp	8	2	Job		113		113	179

08 51 Metal Windows

08 51 13 – Aluminum Windows

08 51 13.10 Aluminum Sash

		Crew	Daily Output	Labor-Hours	Unit	Material	2022 Bare Costs Labor	Equipment	Total	Total Incl O&P
0010	**ALUMINUM SASH**									
0020	Stock, grade C, glaze & trim not incl., casement	2 Sswk	200	.080	S.F.	43	4.95		47.95	55
0050	Double-hung		200	.080		42	4.95		46.95	54
0100	Fixed casement		200	.080		18.45	4.95		23.40	28.50
0150	Picture window		200	.080		19.75	4.95		24.70	29.50
0200	Projected window		200	.080		38	4.95		42.95	50
0250	Single-hung		200	.080		17.80	4.95		22.75	27.50
0300	Sliding		200	.080		23.50	4.95		28.45	34

08 51 Metal Windows

08 51 13 – Aluminum Windows

08 51 13.10 Aluminum Sash

		Crew	Daily Output	Labor-Hours	Unit	Material	2022 Bare Costs Labor	Equipment	Total	Total Incl O&P
1000	Mullions for above, tubular	2 Sswk	240	.067	L.F.	6.55	4.12		10.67	13.85
9000	Minimum labor/equipment charge	1 Sswk	2	4	Job		247		247	400

08 51 13.20 Aluminum Windows

		Crew	Daily Output	Labor-Hours	Unit	Material	2022 Bare Costs Labor	Equipment	Total	Total Incl O&P
0010	**ALUMINUM WINDOWS**, incl. frame and glazing, commercial grade									
1000	Stock units, casement, 3'-1" x 3'-2" opening	2 Sswk	10	1.600	Ea.	730	99		829	965
1050	Add for storms					129			129	141
1600	Projected, with screen, 3'-1" x 3'-2" opening	2 Sswk	10	1.600		605	99		704	825
1700	Add for storms					125			125	138
2000	4'-5" x 5'-3" opening	2 Sswk	8	2		420	124		544	665
2100	Add for storms					128			128	141
2500	Enamel finish windows, 3'-1" x 3'-2"	2 Sswk	10	1.600		385	99		484	585
2600	4'-5" x 5'-3"	"	8	2	↓	435	124		559	680
3000	Single-hung, 2' x 3' opening, enameled, standard glazed	2 Sswk	10	1.600	Ea.	223	99		322	405
3100	Insulating glass		10	1.600		271	99		370	455
3300	2'-8" x 6'-8" opening, standard glazed		8	2		375	124		499	615
3400	Insulating glass		8	2		495	124		619	745
3700	3'-4" x 5'-0" opening, standard glazed		9	1.778		325	110		435	530
3800	Insulating glass		9	1.778		350	110		460	560
4000	Sliding aluminum, 3' x 2' opening, standard glazed		10	1.600		1,025	99		1,124	1,275
4100	Insulating glass		10	1.600		1,200	99		1,299	1,475
4300	5' x 3' opening, standard glazed		9	1.778		1,200	110		1,310	1,500
4400	Insulating glass		9	1.778		1,350	110		1,460	1,675
4600	8' x 4' opening, standard glazed		6	2.667		1,875	165		2,040	2,350
4700	Insulating glass		6	2.667		2,025	165		2,190	2,500
5000	9' x 5' opening, standard glazed		4	4		550	247		797	1,000
5100	Insulating glass		4	4		840	247		1,087	1,325
5500	Sliding, with thermal barrier and screen, 6' x 4', 2 track		8	2		2,475	124		2,599	2,925
5700	4 track	↓	8	2		905	124		1,029	1,200
6000	For above units with bronze finish, add					15%				
6200	For installation in concrete openings, add					8%				
7000	Double-hung, insulating glass, 2'-0" x 2'-0"	2 Sswk	11	1.455		118	90		208	274
7020	2'-0" x 2'-6"		11	1.455		133	90		223	291
7040	3'-0" x 1'-6"		10	1.600		270	99		369	455
7060	3'-0" x 2'-0"		10	1.600		315	99		414	505
7080	3'-0" x 2'-6"		10	1.600		375	99		474	575
7100	3'-0" x 3'-0"		10	1.600		420	99		519	625
7120	3'-0" x 3'-6"		10	1.600		445	99		544	650
7140	3'-0" x 4'-0"		10	1.600		480	99		579	690
7160	3'-0" x 5'-0"		10	1.600		525	99		624	740
7180	3'-0" x 6'-0"	↓	9	1.778	↓	985	110		1,095	1,250

08 51 13.30 Impact Resistant Aluminum Windows

		Crew	Daily Output	Labor-Hours	Unit	Material	2022 Bare Costs Labor	Equipment	Total	Total Incl O&P
0010	**IMPACT RESISTANT ALUMINUM WINDOWS**, incl. frame and glazing									
0100	Single-hung, impact resistant, 2'-8" x 5'-0"	2 Carp	9	1.778	Ea.	1,175	100		1,275	1,450
0120	3'-0" x 5'-0"		9	1.778		1,400	100		1,500	1,700
0130	4'-0" x 5'-0"		9	1.778		1,950	100		2,050	2,275
0250	Horizontal slider, impact resistant, 5'-5" x 5'-2"	↓	9	1.778	↓	2,225	100		2,325	2,600

08 51 23 – Steel Windows

08 51 23.10 Steel Sash

		Crew	Daily Output	Labor-Hours	Unit	Material	2022 Bare Costs Labor	Equipment	Total	Total Incl O&P
0010	**STEEL SASH** Custom units, glazing and trim not included									
0100	Casement, 100% vented	2 Sswk	200	.080	S.F.	69.50	4.95		74.45	84.50
0200	50% vented		200	.080		56.50	4.95		61.45	70
0300	Fixed		200	.080	↓	29	4.95		33.95	40

08 51 Metal Windows

08 51 23 – Steel Windows

08 51 23.10 Steel Sash

		Crew	Daily Output	Labor-Hours	Unit	Material	2022 Bare Costs Labor	2022 Bare Costs Equipment	Total	Total Incl O&P
1000	Projected, commercial, 40% vented	2 Sswk	200	.080	S.F.	53.50	4.95		58.45	67
1100	Intermediate, 50% vented		200	.080		64	4.95		68.95	78.50
1500	Industrial, horizontally pivoted		200	.080		56	4.95		60.95	69.50
1600	Fixed		200	.080		32.50	4.95		37.45	43.50
2000	Industrial security sash, 50% vented		200	.080		61.50	4.95		66.45	75.50
2100	Fixed		200	.080		49	4.95		53.95	62
2500	Picture window		200	.080		32.50	4.95		37.45	44
3000	Double-hung		200	.080	↓	62	4.95		66.95	76
5000	Mullions for above, open interior face		240	.067	L.F.	10.90	4.12		15.02	18.65
5100	With interior cover	↓	240	.067	"	18.10	4.12		22.22	26.50
5200	Single glazing for above, add	2 Glaz	200	.080	S.F.	7.35	4.32		11.67	14.90
6000	Double glazing for above, add		200	.080		13.30	4.32		17.62	21.50
6100	Triple glazing for above, add	↓	85	.188	↓	12.70	10.15		22.85	30
9000	Minimum labor/equipment charge	1 Sswk	2	4	Job		247		247	400

08 51 23.20 Steel Windows

		Crew	Daily Output	Labor-Hours	Unit	Material	2022 Bare Costs Labor	2022 Bare Costs Equipment	Total	Total Incl O&P
0010	**STEEL WINDOWS** Stock, including frame, trim and insul. glass R085123-10									
1000	Custom units, double-hung, 2'-8" x 4'-6" opening	2 Sswk	12	1.333	Ea.	720	82.50		802.50	925
1100	2'-4" x 3'-9" opening		12	1.333		590	82.50		672.50	785
1500	Commercial projected, 3'-9" x 5'-5" opening		10	1.600		1,250	99		1,349	1,525
1600	6'-9" x 4'-1" opening		7	2.286		1,650	141		1,791	2,050
2000	Intermediate projected, 2'-9" x 4'-1" opening		12	1.333		700	82.50		782.50	905
2100	4'-1" x 5'-5" opening	↓	10	1.600	↓	1,425	99		1,524	1,725
9000	Minimum labor/equipment charge	1 Sswk	3	2.667	Job		165		165	266

08 51 66 – Metal Window Screens

08 51 66.10 Screens

		Crew	Daily Output	Labor-Hours	Unit	Material	2022 Bare Costs Labor	2022 Bare Costs Equipment	Total	Total Incl O&P
0010	**SCREENS**									
0020	For metal sash, aluminum or bronze mesh, flat screen	2 Sswk	1200	.013	S.F.	39	.82		39.82	44
0500	Wicket screen, inside window		1000	.016		7	.99		7.99	9.30
0800	Security screen, aluminum frame with stainless steel cloth		1200	.013		25	.82		25.82	29
0900	Steel grate, painted, on steel frame		1600	.010		14	.62		14.62	16.40
1000	Screens for solar louvers	↓	160	.100	↓	25.50	6.20		31.70	38
4000	See Section 05 58 23.90 for window guards									

08 52 Wood Windows

08 52 10 – Plain Wood Windows

08 52 10.20 Awning Window

		Crew	Daily Output	Labor-Hours	Unit	Material	2022 Bare Costs Labor	2022 Bare Costs Equipment	Total	Total Incl O&P
0010	**AWNING WINDOW**, Including frame, screens and grilles									
0100	34" x 22", insulated glass	1 Carp	10	.800	Ea.	390	45		435	500
0200	Low E glass		10	.800		430	45		475	545
0300	40" x 28", insulated glass		9	.889		415	50		465	535
0400	Low E glass		9	.889		460	50		510	585
0500	48" x 36", insulated glass		8	1		630	56.50		686.50	780
0600	Low E glass	↓	8	1		660	56.50		716.50	820
4000	Impact windows, minimum, add					60%				
4010	Impact windows, maximum, add					160%				
9000	Minimum labor/equipment charge	1 Carp	4	2	Job		113		113	179

261

08 52 10 – Plain Wood Windows

08 52 10.30 Wood Windows

		Crew	Daily Output	Labor-Hours	Unit	Material	2022 Bare Costs Labor	2022 Bare Costs Equipment	Total	Total Incl O&P
0010	**WOOD WINDOWS**, double-hung									
0020	Including frame, double insulated glass, screens and grilles									
0040	Double-hung, 2'-2" x 3'-4" high	2 Carp	15	1.067	Ea.	550	60		610	700
0060	2'-2" x 4'-4"		14	1.143		253	64.50		317.50	380
0080	2'-6" x 3'-4"		13	1.231		250	69.50		319.50	385
0100	2'-6" x 4'-0"		12	1.333		256	75		331	400
0120	2'-6" x 4'-8"		12	1.333		270	75		345	415
0140	2'-10" x 3'-4"		10	1.600		245	90		335	415
0160	2'-10" x 4'-0"		10	1.600		273	90		363	445
0180	3'-7" x 3'-4"		9	1.778		280	100		380	470
0200	3'-7" x 5'-4"		9	1.778		325	100		425	520
0220	3'-10" x 5'-4"		8	2		560	113		673	795
3800	Triple glazing for above, add					25%				

08 52 10.40 Casement Window

			Crew	Daily Output	Labor-Hours	Unit	Material	2022 Bare Costs Labor	2022 Bare Costs Equipment	Total	Total Incl O&P
0010	**CASEMENT WINDOW**, including frame, screen and grilles										
0100	2'-0" x 3'-0" H, double insulated glass	G	1 Carp	10	.800	Ea.	320	45		365	420
0150	Low E glass	G		10	.800		360	45		405	465
0200	2'-0" x 4'-6" high, double insulated glass	G		9	.889		425	50		475	545
0250	Low E glass	G		9	.889		425	50		475	545
0260	Casement 4'-2" x 4'-2" double insulated glass	G		11	.727		815	41		856	960
0270	4'-0" x 4'-0" Low E glass	G		11	.727		815	41		856	965
0290	6'-4" x 5'-7" Low E glass	G		9	.889		1,450	50		1,500	1,675
0300	2'-4" x 6'-0" high, double insulated glass	G		8	1		965	56.50		1,021.50	1,175
0350	Low E glass	G		8	1		900	56.50		956.50	1,075
0522	Vinyl-clad, premium, double insulated glass, 2'-0" x 3'-0"	G		10	.800		365	45		410	470
0524	2'-0" x 4'-0"	G		9	.889		435	50		485	560
0525	2'-0" x 5'-0"	G		8	1		510	56.50		566.50	650
0528	2'-0" x 6'-0"	G		8	1		625	56.50		681.50	780
0600	3'-0" x 5'-0"	G		8	1		470	56.50		526.50	610
0700	4'-0" x 3'-0"	G	1 Carp	8	1	Ea.	935	56.50		991.50	1,125
0710	4'-0" x 4'-0"	G		8	1		765	56.50		821.50	935
0720	4'-8" x 4'-0"	G		8	1		765	56.50		821.50	930
0730	4'-8" x 5'-0"	G		6	1.333		880	75		955	1,075
0740	4'-8" x 6'-0"	G		6	1.333		950	75		1,025	1,175
0750	6'-0" x 4'-0"	G		6	1.333		955	75		1,030	1,175
0800	6'-0" x 5'-0"	G		6	1.333		995	75		1,070	1,225
0900	5'-6" x 5'-6"	G	2 Carp	15	1.067		990	60		1,050	1,200
8190	For installation, add per leaf							15%			
8200	For multiple leaf units, deduct for stationary sash										
8220	2' high					Ea.	30			30	33
8240	4'-6" high						34			34	37.50
8260	6' high						45			45	49.50
8300	Impact windows, minimum, add						60%				
8310	Impact windows, maximum, add						160%				
9000	Minimum labor/equipment charge		1 Carp	3	2.667	Job		150		150	239

08 52 10.50 Double-Hung

			Crew	Daily Output	Labor-Hours	Unit	Material	2022 Bare Costs Labor	2022 Bare Costs Equipment	Total	Total Incl O&P
0010	**DOUBLE-HUNG**, Including frame, screens and grilles										
0100	2'-0" x 3'-0" high, low E insul. glass	G	1 Carp	10	.800	Ea.	152	45		197	239
0200	3'-0" x 4'-0" high, double insulated glass	G		9	.889		291	50		341	400
0300	4'-0" x 4'-6" high, low E insulated glass	G		8	1		255	56.50		311.50	370
8000	Impact windows, minimum, add						60%				
8010	Impact windows, maximum, add						160%				

08 52 10 – Plain Wood Windows

08 52 10.50 Double-Hung	Crew	Daily Output	Labor-Hours	Unit	Material	2022 Bare Costs Labor	Equipment	Total	Total Incl O&P
9000	Minimum labor/equipment charge	1 Carp	3	2.667	Job		150		150	239

08 52 10.55 Picture Window	Crew	Daily Output	Labor-Hours	Unit	Material	Labor	Equipment	Total	Total Incl O&P
0010	**PICTURE WINDOW**, Including frame and grilles									
0100	3'-6" x 4'-0" high, double insulated glass	2 Carp	12	1.333	Ea.	700	75		775	890
0150	Low E glass		12	1.333		590	75		665	770
0200	4'-0" x 4'-6" high, double insulated glass		11	1.455		705	82		787	905
0250	Low E glass		11	1.455		800	82		882	1,000
0300	5'-0" x 4'-0" high, double insulated glass		11	1.455		800	82		882	1,000
0350	Low E glass		11	1.455		795	82		877	1,000
0400	6'-0" x 4'-6" high, double insulated glass		10	1.600		860	90		950	1,100
0450	Low E glass		10	1.600		835	90		925	1,075

08 52 10.65 Wood Sash	Crew	Daily Output	Labor-Hours	Unit	Material	Labor	Equipment	Total	Total Incl O&P
0010	**WOOD SASH**, Including glazing but not trim									
0050	Custom, 5'-0" x 4'-0", 1" double glazed, 3/16" thick lites	2 Carp	3.20	5	Ea.	320	282		602	800
0100	1/4" thick lites		5	3.200		465	180		645	795
0200	1" thick, triple glazed		5	3.200		560	180		740	900
0300	7'-0" x 4'-6" high, 1" double glazed, 3/16" thick lites	2 Carp	4.30	3.721	Ea.	600	209		809	995
0400	1/4" thick lites		4.30	3.721		640	209		849	1,050
0500	1" thick, triple glazed		4.30	3.721		785	209		994	1,200
0600	8'-6" x 5'-0" high, 1" double glazed, 3/16" thick lites		3.50	4.571		815	257		1,072	1,300
0700	1/4" thick lites		3.50	4.571		895	257		1,152	1,400
0800	1" thick, triple glazed		3.50	4.571		835	257		1,092	1,325
0900	Window frames only, based on perimeter length				L.F.	5.10			5.10	5.60
1200	Window sill, stock, per lineal foot					11.85			11.85	13.05
1250	Casing, stock					4.32			4.32	4.75

08 52 10.70 Sliding Windows		Crew	Daily Output	Labor-Hours	Unit	Material	Labor	Equipment	Total	Total Incl O&P
0010	**SLIDING WINDOWS**										
0100	3'-0" x 3'-0" high, double insulated	G	1 Carp	10	.800	Ea.	730	45		775	875
0120	Low E glass	G		10	.800		745	45		790	890
0200	4'-0" x 3'-6" high, double insulated	G		9	.889		950	50		1,000	1,125
0220	Low E glass	G		9	.889		565	50		615	700
0300	6'-0" x 5'-0" high, double insulated	G		8	1		645	56.50		701.50	800
0320	Low E glass	G		8	1		785	56.50		841.50	955
9000	Minimum labor/equipment charge			3	2.667	Job		150		150	239

08 52 13 – Metal-Clad Wood Windows

08 52 13.10 Awning Windows, Metal-Clad	Crew	Daily Output	Labor-Hours	Unit	Material	Labor	Equipment	Total	Total Incl O&P
0010	**AWNING WINDOWS, METAL-CLAD**									
2000	Metal-clad, awning deluxe, double insulated glass, 34" x 22"	1 Carp	9	.889	Ea.	330	50		380	440
2050	36" x 25"		9	.889		370	50		420	485
2100	40" x 22"		9	.889		395	50		445	515
2150	40" x 30"		9	.889		455	50		505	580
2200	48" x 28"		8	1		470	56.50		526.50	605
2250	60" x 36"		8	1		500	56.50		556.50	640

08 52 13.20 Casement Windows, Metal-Clad		Crew	Daily Output	Labor-Hours	Unit	Material	Labor	Equipment	Total	Total Incl O&P
0010	**CASEMENT WINDOWS, METAL-CLAD**										
0100	Metal-clad, deluxe, dbl. insul. glass, 2'-0" x 3'-0" high	G	1 Carp	10	.800	Ea.	370	45		415	475
0120	2'-0" x 4'-0" high	G		9	.889		415	50		465	540
0130	2'-0" x 5'-0" high	G		8	1		445	56.50		501.50	580
0140	2'-0" x 6'-0" high	G		8	1		475	56.50		531.50	615
0300	Metal-clad, casement, bldrs mdl, 6'-0" x 4'-0", dbl. insul. glass, 3 panels		2 Carp	10	1.600		1,325	90		1,415	1,625
0310	9'-0" x 4'-0", 4 panels			8	2		1,700	113		1,813	2,025

For customer support on your Commercial Renovation Costs with RSMeans data, call 800.448.8182.

263

08 52 Wood Windows

08 52 13 – Metal-Clad Wood Windows

08 52 13.20 Casement Windows, Metal-Clad		Crew	Daily Output	Labor-Hours	Unit	Material	2022 Bare Costs Labor	2022 Bare Costs Equipment	Total	Total Incl O&P
0320	10'-0" x 5'-0", 5 panels	2 Carp	7	2.286	Ea.	2,325	129		2,454	2,750
0330	12'-0" x 6'-0", 6 panels	↓	6	2.667		2,925	150		3,075	3,475

08 52 13.30 Double-Hung Windows, Metal-Clad

			Crew	Daily Output	Labor-Hours	Unit	Material	Labor	Equipment	Total	Total Incl O&P
0010	**DOUBLE-HUNG WINDOWS, METAL-CLAD**										
0100	Metal-clad, deluxe, dbl. insul. glass, 2'-6" x 3'-0" high	G	1 Carp	10	.800	Ea.	234	45		279	330
0120	3'-0" x 3'-6" high	G		10	.800		260	45		305	360
0140	3'-0" x 4'-0" high	G		9	.889		267	50		317	375
0160	3'-0" x 4'-6" high	G		9	.889		285	50		335	395
0180	3'-0" x 5'-0" high	G		8	1		315	56.50		371.50	435
0200	3'-6" x 6'-0" high	G	↓	8	1	↓	380	56.50		436.50	505

08 52 13.35 Picture and Sliding Windows Metal-Clad

			Crew	Daily Output	Labor-Hours	Unit	Material	Labor	Equipment	Total	Total Incl O&P
0010	**PICTURE AND SLIDING WINDOWS METAL-CLAD**										
2000	Metal-clad, dlx picture, dbl. insul. glass, 4'-0" x 4'-0" high		2 Carp	12	1.333	Ea.	505	75		580	675
2100	4'-0" x 6'-0" high			11	1.455		755	82		837	960
2200	5'-0" x 6'-0" high			10	1.600		835	90		925	1,075
2300	6'-0" x 6'-0" high		↓	10	1.600		955	90		1,045	1,200
2400	Metal-clad, dlx sliding, dbl. insul. glass, 3'-0" x 3'-0" high	G	1 Carp	10	.800		430	45		475	540
2420	4'-0" x 3'-6" high	G		9	.889		525	50		575	655
2440	5'-0" x 4'-0" high	G		9	.889		630	50		680	770
2460	6'-0" x 5'-0" high	G	↓	8	1		1,025	56.50		1,081.50	1,225
9000	Minimum labor/equipment charge		2 Carp	2.75	5.818	Job		330		330	520

08 52 13.40 Bow and Bay Windows, Metal-Clad

			Crew	Daily Output	Labor-Hours	Unit	Material	Labor	Equipment	Total	Total Incl O&P
0010	**BOW AND BAY WINDOWS, METAL-CLAD**										
0100	Metal-clad, deluxe, dbl. insul. glass, 8'-0" x 5'-0" high, 4 panels		2 Carp	10	1.600	Ea.	1,875	90		1,965	2,200
0120	10'-0" x 5'-0" high, 5 panels			8	2		2,000	113		2,113	2,400
0140	10'-0" x 6'-0" high, 5 panels			7	2.286		2,375	129		2,504	2,800
0160	12'-0" x 6'-0" high, 6 panels			6	2.667		3,775	150		3,925	4,400
0400	Double-hung, bldrs. model, bay, 8' x 4' high, dbl. insul. glass			10	1.600		1,475	90		1,565	1,775
0440	Low E glass			10	1.600		1,600	90		1,690	1,925
0460	9'-0" x 5'-0" high, dbl. insul. glass			6	2.667		1,600	150		1,750	2,025
0480	Low E glass			6	2.667		1,700	150		1,850	2,100
0500	Metal-clad, deluxe, dbl. insul. glass, 7'-0" x 4'-0" high			10	1.600		1,425	90		1,515	1,725
0520	8'-0" x 4'-0" high			8	2		1,475	113		1,588	1,800
0540	8'-0" x 5'-0" high			7	2.286		1,500	129		1,629	1,850
0560	9'-0" x 5'-0" high		↓	6	2.667	↓	1,625	150		1,775	2,025

08 52 16 – Plastic-Clad Wood Windows

08 52 16.10 Bow Window

		Crew	Daily Output	Labor-Hours	Unit	Material	Labor	Equipment	Total	Total Incl O&P
0010	**BOW WINDOW** including frames, screens, and grilles									
0020	End panels operable									
1000	Bow type, casement, wood, bldrs. mdl., 8' x 5' dbl. insul. glass, 4 panel	2 Carp	10	1.600	Ea.	1,700	90		1,790	2,025
1050	Low E glass		10	1.600		1,425	90		1,515	1,725
1100	10'-0" x 5'-0", dbl. insul. glass, 6 panels		6	2.667		1,500	150		1,650	1,900
1200	Low E glass, 6 panels		6	2.667		1,600	150		1,750	2,000
1300	Vinyl-clad, bldrs. model, dbl. insul. glass, 6'-0" x 4'-0", 3 panel		10	1.600		1,100	90		1,190	1,350
1340	9'-0" x 4'-0", 4 panel		8	2		1,575	113		1,688	1,900
1380	10'-0" x 6'-0", 5 panels		7	2.286		2,525	129		2,654	2,975
1420	12'-0" x 6'-0", 6 panels		6	2.667		3,325	150		3,475	3,900
2000	Bay window, 8' x 5', dbl. insul. glass		10	1.600		2,125	90		2,215	2,475
2050	Low E glass		10	1.600		2,550	90		2,640	2,975
2100	12'-0" x 6'-0", dbl. insul. glass, 6 panels		6	2.667		2,625	150		2,775	3,150
2200	Low E glass	↓	6	2.667	↓	3,700	150		3,850	4,325

08 52 Wood Windows

08 52 16 – Plastic-Clad Wood Windows

08 52 16.10 Bow Window

		Crew	Daily Output	Labor-Hours	Unit	Material	2022 Bare Costs Labor	Equipment	Total	Total Incl O&P
2300	Vinyl-clad, premium, dbl. insul. glass, 8'-0" x 5'-0"	2 Carp	10	1.600	Ea.	1,775	90		1,865	2,125
2340	10'-0" x 5'-0"		8	2		2,525	113		2,638	2,950
2380	10'-0" x 6'-0"		7	2.286		3,325	129		3,454	3,850
2420	12'-0" x 6'-0"		6	2.667		3,175	150		3,325	3,750
3300	Vinyl-clad, premium, dbl. insul. glass, 7'-0" x 4'-6"		10	1.600		1,500	90		1,590	1,775
3340	8'-0" x 4'-6"		8	2		1,575	113		1,688	1,900
3380	8'-0" x 5'-0"		7	2.286		1,650	129		1,779	2,000
3420	9'-0" x 5'-0"		6	2.667	↓	1,700	150		1,850	2,100
9000	Minimum labor/equipment charge	↓	2.50	6.400	Job		360		360	570

08 52 16.15 Awning Window Vinyl-Clad

		Crew	Daily Output	Labor-Hours	Unit	Material	2022 Bare Costs Labor	Equipment	Total	Total Incl O&P
0010	**AWNING WINDOW VINYL-CLAD** including frames, screens, and grilles									
0240	Vinyl-clad, 34" x 22"	1 Carp	10	.800	Ea.	355	45		400	460
0280	36" x 28"		9	.889		405	50		455	525
0300	36" x 36"		9	.889		465	50		515	590
0340	40" x 22"		10	.800		395	45		440	505
0360	48" x 28"		8	1		490	56.50		546.50	630
0380	60" x 36"	↓	8	1	↓	675	56.50		731.50	830

08 52 16.20 Half Round, Vinyl-Clad

		Crew	Daily Output	Labor-Hours	Unit	Material	2022 Bare Costs Labor	Equipment	Total	Total Incl O&P
0010	**HALF ROUND, VINYL-CLAD**, double insulated glass, incl. grille									
0800	14" height x 24" base	2 Carp	9	1.778	Ea.	440	100		540	645
1040	15" height x 25" base		8	2		400	113		513	620
1060	16" height x 28" base		7	2.286		460	129		589	710
1080	17" height x 29" base	↓	7	2.286		480	129		609	730
2000	19" height x 33" base	1 Carp	6	1.333		485	75		560	655
2100	20" height x 35" base		6	1.333		490	75		565	660
2200	21" height x 37" base	↓	6	1.333		515	75		590	685
2250	23" height x 41" base	2 Carp	6	2.667		575	150		725	875
2300	26" height x 48" base		6	2.667		600	150		750	900
2350	30" height x 56" base	↓	6	2.667		695	150		845	1,000
3000	36" height x 67" base	1 Carp	4	2		1,225	113		1,338	1,525
3040	38" height x 71" base	2 Carp	5	3.200		1,075	180		1,255	1,475
3050	40" height x 75" base	"	5	3.200		1,425	180		1,605	1,825
5000	Elliptical, 71" x 16"	1 Carp	11	.727		1,075	41		1,116	1,250
5100	95" x 21"	"	10	.800	↓	1,550	45		1,595	1,775

08 52 16.35 Double-Hung Window

			Crew	Daily Output	Labor-Hours	Unit	Material	2022 Bare Costs Labor	Equipment	Total	Total Incl O&P
0010	**DOUBLE-HUNG WINDOW** including frames, screens, and grilles										
0300	Vinyl-clad, premium, double insulated glass, 2'-6" x 3'-0"	G	1 Carp	10	.800	Ea.	340	45		385	445
0305	2'-6" x 4'-0"	G		10	.800		296	45		341	395
0400	3'-0" x 3'-6"	G		10	.800		283	45		328	380
0500	3'-0" x 4'-0"	G		9	.889		320	50		370	435
0600	3'-0" x 4'-6"	G		9	.889		355	50		405	470
0700	3'-0" x 5'-0"	G		8	1		365	56.50		421.50	490
0790	3'-4" x 5'-0"	G		8	1		375	56.50		431.50	500
0800	3'-6" x 6'-0"	G		8	1		415	56.50		471.50	550
0820	4'-0" x 5'-0"	G		7	1.143		440	64.50		504.50	585
0830	4'-0" x 6'-0"	G	↓	7	1.143	↓	535	64.50		599.50	690

08 52 16.40 Transom Windows

		Crew	Daily Output	Labor-Hours	Unit	Material	2022 Bare Costs Labor	Equipment	Total	Total Incl O&P
0010	**TRANSOM WINDOWS**									
0050	Vinyl-clad, premium, dbl. insul. glass, 32" x 8"	1 Carp	16	.500	Ea.	249	28		277	320
0100	36" x 8"		16	.500		224	28		252	292
0110	36" x 12"		16	.500		241	28		269	310
2000	Custom sizes, up to 350 sq. in.	↓	12	.667	↓	310	37.50		347.50	400

For customer support on your Commercial Renovation Costs with RSMeans data, call 800.448.8182.

08 52 Wood Windows

08 52 16 – Plastic-Clad Wood Windows

08 52 16.40 Transom Windows

		Crew	Daily Output	Labor-Hours	Unit	Material	2022 Bare Costs Labor	Equipment	Total	Total Incl O&P
2100	351 to 750 sq. in.	1 Carp	12	.667	Ea.	420	37.50		457.50	525
2200	751 to 1150 sq. in.		11	.727		460	41		501	570
2300	1151 to 1450 sq. in.		11	.727		540	41		581	660
2400	1451 to 1850 sq. in.	2 Carp	12	1.333		700	75		775	885
2500	1851 to 2250 sq. in.		12	1.333		705	75		780	895
2600	2251 to 2650 sq. in.		11	1.455		925	82		1,007	1,150
2700	2651 to 3050 sq. in.		11	1.455		820	82		902	1,025
2800	3051 to 3450 sq. in.		11	1.455		880	82		962	1,100
2900	3451 to 3850 sq. in.		10	1.600		945	90		1,035	1,200
3000	3851 to 4250 sq. in.		10	1.600		1,125	90		1,215	1,400
3100	4251 to 4650 sq. in.		10	1.600		1,150	90		1,240	1,425
3200	4651 to 5050 sq. in.		10	1.600		1,275	90		1,365	1,550
3300	5051 to 5450 sq. in.		9	1.778		1,225	100		1,325	1,500
3400	5451 to 5850 sq. in.		9	1.778		1,075	100		1,175	1,350
3600	6251 to 6650 sq. in.		8	2		1,350	113		1,463	1,650
3700	6651 to 7050 sq. in.		8	2		1,700	113		1,813	2,050

08 52 16.45 Trapezoid Windows

		Crew	Daily Output	Labor-Hours	Unit	Material	2022 Bare Costs Labor	Equipment	Total	Total Incl O&P
0010	**TRAPEZOID WINDOWS**									
0900	20" base x 44" leg x 53" leg	2 Carp	13	1.231	Ea.	440	69.50		509.50	590
1000	24" base x 90" leg x 102" leg		8	2		795	113		908	1,050
3000	36" base x 40" leg x 22" leg		12	1.333		495	75		570	665
3010	36" base x 44" leg x 25" leg		13	1.231		525	69.50		594.50	685
3050	36" base x 26" leg x 48" leg		9	1.778		525	100		625	735
3100	36" base x 42" legs, 50" peak		9	1.778		615	100		715	835
3200	36" base x 60" leg x 81" leg		11	1.455		805	82		887	1,025
4320	44" base x 23" leg x 56" leg		11	1.455		620	82		702	815
4350	44" base x 59" leg x 92" leg		10	1.600		1,000	90		1,090	1,250
4500	46" base x 15" leg x 46" leg		8	2		500	113		613	730
4550	46" base x 16" leg x 48" leg		8	2		525	113		638	760
4600	46" base x 50" leg x 80" leg		7	2.286		790	129		919	1,075
6600	66" base x 12" leg x 42" leg		8	2		680	113		793	925
6650	66" base x 12" legs, 28" peak		9	1.778		530	100		630	740
6700	68" base x 23" legs, 31" peak		8	2		680	113		793	925

08 52 16.70 Vinyl-Clad, Premium, DBL. Insul. Glass

			Crew	Daily Output	Labor-Hours	Unit	Material	2022 Bare Costs Labor	Equipment	Total	Total Incl O&P
0010	**VINYL-CLAD, PREMIUM, DBL. INSUL. GLASS**										
1000	Sliding, 3'-0" x 3'-0"	G	1 Carp	10	.800	Ea.	855	45		900	1,000
1050	4'-0" x 3'-6"	G		9	.889		1,175	50		1,225	1,375
1100	5'-0" x 4'-0"	G		9	.889		1,250	50		1,300	1,450
1150	6'-0" x 5'-0"	G		8	1		1,600	56.50		1,656.50	1,850

08 52 50 – Window Accessories

08 52 50.10 Window Grille or Muntin

		Crew	Daily Output	Labor-Hours	Unit	Material	2022 Bare Costs Labor	Equipment	Total	Total Incl O&P
0010	**WINDOW GRILLE OR MUNTIN**, snap in type									
0020	Standard pattern interior grilles									
2000	Wood, awning window, glass size, 28" x 16" high	1 Carp	30	.267	Ea.	41	15		56	69
2060	44" x 24" high		32	.250		59	14.10		73.10	87.50
2100	Casement, glass size, 20" x 36" high		30	.267		45.50	15		60.50	74
2180	20" x 56" high		32	.250		61.50	14.10		75.60	90
2200	Double-hung, glass size, 16" x 24" high		24	.333	Set	79	18.75		97.75	117
2280	32" x 32" high		34	.235	"	183	13.25		196.25	222
2500	Picture, glass size, 48" x 48" high		30	.267	Ea.	170	15		185	211
2580	60" x 68" high		28	.286	"	268	16.10		284.10	320
2600	Sliding, glass size, 14" x 36" high		24	.333	Set	61.50	18.75		80.25	98

08 52 Wood Windows

08 52 50 – Window Accessories

08 52 50.10 Window Grille or Muntin		Crew	Daily Output	Labor-Hours	Unit	Material	2022 Bare Costs Labor	Equipment	Total	Total Incl O&P
2680	36" x 36" high	1 Carp	22	.364	Set	61	20.50		81.50	99.50
9000	Minimum labor/equipment charge	↓	5	1.600	Job		90		90	143

08 52 66 – Wood Window Screens

08 52 66.10 Wood Screens

		Crew	Daily Output	Labor-Hours	Unit	Material	2022 Bare Costs Labor	Equipment	Total	Total Incl O&P
0010	**WOOD SCREENS**									
0020	Over 3 S.F., 3/4" frames	2 Carp	375	.043	S.F.	6.35	2.40		8.75	10.75
0100	1-1/8" frames		375	.043		11.30	2.40		13.70	16.25
0200	Rescreen wood frame	↓	500	.032	↓	2.23	1.80		4.03	5.30
9000	Minimum labor/equipment charge	1 Carp	4	2	Job		113		113	179

08 52 69 – Wood Storm Windows

08 52 69.10 Storm Windows

			Crew	Daily Output	Labor-Hours	Unit	Material	2022 Bare Costs Labor	Equipment	Total	Total Incl O&P
0010	**STORM WINDOWS**, aluminum residential										
0300	Basement, mill finish, incl. fiberglass screen										
0320	1'-10" x 1'-0" high	G	2 Carp	30	.533	Ea.	49	30		79	102
0340	2'-9" x 1'-6" high	G		30	.533		53.50	30		83.50	107
0360	3'-4" x 2'-0" high	G	↓	30	.533	↓	69.50	30		99.50	124
1600	Double-hung, combination, storm & screen										
1700	Custom, clear anodic coating, 2'-0" x 3'-5" high		2 Carp	30	.533	Ea.	123	30		153	183
1720	2'-6" x 5'-0" high			28	.571		179	32		211	248
1740	4'-0" x 6'-0" high			25	.640		315	36		351	410
1800	White painted, 2'-0" x 3'-5" high			30	.533		150	30		180	213
1820	2'-6" x 5'-0" high			28	.571		254	32		286	330
1840	4'-0" x 6'-0" high			25	.640		390	36		426	490
2000	Clear anodic coating, 2'-0" x 3'-5" high	G		30	.533		136	30		166	198
2020	2'-6" x 5'-0" high	G		28	.571		169	32		201	237
2040	4'-0" x 6'-0" high	G		25	.640		174	36		210	249
2400	White painted, 2'-0" x 3'-5" high	G		30	.533		141	30		171	203
2420	2'-6" x 5'-0" high	G		28	.571		239	32		271	315
2440	4'-0" x 6'-0" high	G		25	.640		158	36		194	232
2600	Mill finish, 2'-0" x 3'-5" high	G		30	.533		113	30		143	172
2620	2'-6" x 5'-0" high	G		28	.571		142	32		174	207
2640	4'-0" x 6'-8" high	G	↓	25	.640	↓	180	36		216	255
4000	Picture window, storm, 1 lite, white or bronze finish										
4020	4'-6" x 4'-6" high		2 Carp	25	.640	Ea.	194	36		230	272
4040	5'-8" x 4'-6" high			20	.800		201	45		246	293
4400	Mill finish, 4'-6" x 4'-6" high			25	.640		183	36		219	259
4420	5'-8" x 4'-6" high		↓	20	.800	↓	233	45		278	330
4600	3 lite, white or bronze finish										
4620	4'-6" x 4'-6" high		2 Carp	25	.640	Ea.	236	36		272	320
4640	5'-8" x 4'-6" high			20	.800		236	45		281	330
4800	Mill finish, 4'-6" x 4'-6" high			25	.640		202	36		238	281
4820	5'-8" x 4'-6" high		↓	20	.800	↓	236	45		281	330
6000	Sliding window, storm, 2 lite, white or bronze finish										
6020	3'-4" x 2'-7" high		2 Carp	28	.571	Ea.	196	32		228	267
6040	4'-4" x 3'-3" high			25	.640		211	36		247	290
6060	5'-4" x 6'-0" high		↓	20	.800	↓	345	45		390	445
8000	PVC framed										
8100	Double-hung, combination, storm & screen										
8120	2'-6" x 3'-5"		2 Carp	15	1.067	Ea.	174	60		234	288
8140	4' x 6'		"	12.50	1.280	"	244	72		316	385
8600	Single lite picture storm										
8620	4'-6" x 4'-6"		2 Carp	12.50	1.280	Ea.	236	72		308	375

For customer support on your Commercial Renovation Costs with RSMeans data, call 800.448.8182.

267

08 52 Wood Windows

08 52 69 – Wood Storm Windows

08 52 69.10 Storm Windows		Crew	Daily Output	Labor-Hours	Unit	Material	2022 Bare Costs Labor	Equipment	Total	Total Incl O&P
8640	5'-8" x 4'-6"	2 Carp	10	1.600	Ea.	261	90		351	430
9000	Interior storm window									
9100	Storm window interior glass	1 Glaz	107	.075	S.F.	7.45	4.04		11.49	14.60
9410	Minimum labor/equipment charge	1 Carp	4	2	Job		113		113	179

08 53 Plastic Windows

08 53 13 – Vinyl Windows

08 53 13.20 Vinyl Single-Hung Windows

	08 53 13.20 Vinyl Single-Hung Windows		Crew	Daily Output	Labor-Hours	Unit	Material	2022 Bare Costs Labor	Equipment	Total	Total Incl O&P
0010	**VINYL SINGLE-HUNG WINDOWS**, insulated glass										
0020	Grids, low E, J fin, extension jambs										
0130	25" x 41"	G	2 Carp	20	.800	Ea.	208	45		253	300
0140	25" x 49"	G		18	.889		215	50		265	315
0150	25" x 57"	G		17	.941		235	53		288	340
0160	25" x 65"	G		16	1		263	56.50		319.50	380
0170	29" x 41"	G		18	.889		209	50		259	310
0180	29" x 53"	G		18	.889		227	50		277	330
0190	29" x 57"	G		17	.941		245	53		298	355
0200	29" x 65"	G		16	1		292	56.50		348.50	410
0210	33" x 41"	G		20	.800		246	45		291	345
0220	33" x 53"	G		18	.889		221	50		271	325
0230	33" x 57"	G		17	.941		292	53		345	405
0240	33" x 65"	G		16	1		299	56.50		355.50	420
0250	37" x 41"	G		20	.800		227	45		272	320
0260	37" x 53"	G		18	.889		262	50		312	370
0270	37" x 57"	G		17	.941		274	53		327	385
0280	37" x 65"	G		16	1		310	56.50		366.50	435

08 53 13.30 Vinyl Double-Hung Windows

	08 53 13.30 Vinyl Double-Hung Windows		Crew	Daily Output	Labor-Hours	Unit	Material	2022 Bare Costs Labor	Equipment	Total	Total Incl O&P
0010	**VINYL DOUBLE-HUNG WINDOWS**, insulated glass										
0100	Grids, low E, J fin, ext. jambs, 21" x 53"	G	2 Carp	18	.889	Ea.	143	50		193	238
0102	21" x 37"	G		18	.889		167	50		217	264
0104	21" x 41"	G		18	.889		243	50		293	345
0106	21" x 49"	G		18	.889		216	50		266	315
0110	21" x 57"	G		17	.941		268	53		321	380
0120	21" x 65"	G		16	1		289	56.50		345.50	405
0128	25" x 37"	G		20	.800		231	45		276	325
0130	25" x 41"	G		20	.800		241	45		286	335
0140	25" x 49"	G		18	.889		246	50		296	350
0145	25" x 53"	G		18	.889		271	50		321	380
0150	25" x 57"	G		17	.941		217	53		270	325
0160	25" x 65"	G		16	1		200	56.50		256.50	310
0162	25" x 69"	G		16	1		235	56.50		291.50	350
0164	25" x 77"	G		16	1		241	56.50		297.50	355
0168	29" x 37"	G		18	.889		240	50		290	345
0170	29" x 41"	G		18	.889		260	50		310	365
0172	29" x 49"	G		18	.889		268	50		318	375
0180	29" x 53"	G		18	.889		275	50		325	380
0190	29" x 57"	G		17	.941		185	53		238	288
0200	29" x 65"	G		16	1		277	56.50		333.50	395
0202	29" x 69"	G		16	1		257	56.50		313.50	370
0205	29" x 77"	G		16	1		229	56.50		285.50	340
0208	33" x 37"	G		20	.800		257	45		302	355

08 53 Plastic Windows

08 53 13 – Vinyl Windows

08 53 13.30 Vinyl Double-Hung Windows

			Crew	Daily Output	Labor-Hours	Unit	Material	2022 Bare Costs Labor	Equipment	Total	Total Incl O&P
0210	33" x 41"	G	2 Carp	20	.800	Ea.	264	45		309	360
0215	33" x 49"	G		20	.800		219	45		264	315
0220	33" x 53"	G		18	.889		188	50		238	287
0230	33" x 57"	G		17	.941		213	53		266	320
0240	33" x 65"	G		16	1		245	56.50		301.50	360
0242	33" x 69"	G		16	1		238	56.50		294.50	350
0246	33" x 77"	G		16	1		222	56.50		278.50	335
0250	37" x 41"	G		20	.800		225	45		270	320
0255	37" x 49"	G		20	.800		239	45		284	335
0260	37" x 53"	G		18	.889		245	50		295	350
0270	37" x 57"	G		17	.941		264	53		317	375
0280	37" x 65"	G		16	1		283	56.50		339.50	400
0282	37" x 69"	G		16	1		293	56.50		349.50	410
0286	37" x 77"	G		16	1		289	56.50		345.50	410
0300	Solid vinyl, average quality, double insulated glass, 2'-0" x 3'-0"	G	1 Carp	10	.800		237	45		282	330
0310	3'-0" x 4'-0"	G		9	.889		189	50		239	288
0330	Premium, double insulated glass, 2'-6" x 3'-0"	G		10	.800		255	45		300	350
0340	3'-0" x 3'-6"	G		9	.889		267	50		317	375
0350	3'-0" x 4'-0"	G		9	.889		273	50		323	380
0360	3'-0" x 4'-6"	G		9	.889		375	50		425	495
0370	3'-0" x 5'-0"	G		8	1		335	56.50		391.50	455
0380	3'-6" x 6'-0"	G		8	1		335	56.50		391.50	455

08 53 13.40 Vinyl Casement Windows

			Crew	Daily Output	Labor-Hours	Unit	Material	2022 Bare Costs Labor	Equipment	Total	Total Incl O&P
0010	**VINYL CASEMENT WINDOWS**, insulated glass										
0015	Grids, low E, J fin, extension jambs, screens										
0100	One lite, 21" x 41"	G	2 Carp	20	.800	Ea.	370	45		415	480
0110	21" x 47"	G		20	.800		380	45		425	490
0120	21" x 53"	G		20	.800		410	45		455	520
0128	24" x 35"	G		19	.842		325	47.50		372.50	430
0130	24" x 41"	G		19	.842		375	47.50		422.50	490
0140	24" x 47"	G		19	.842		400	47.50		447.50	515
0150	24" x 53"	G		19	.842		420	47.50		467.50	535
0158	28" x 35"	G		19	.842		370	47.50		417.50	485
0160	28" x 41"	G		19	.842		335	47.50		382.50	440
0170	28" x 47"	G		19	.842		420	47.50		467.50	535
0180	28" x 53"	G		19	.842		273	47.50		320.50	375
0184	28" x 59"	G		19	.842		320	47.50		367.50	425
0188	Two lites, 33" x 35"	G		18	.889		380	50		430	500
0190	33" x 41"	G		18	.889		370	50		420	485
0200	33" x 47"	G		18	.889		390	50		440	510
0210	33" x 53"	G		18	.889		410	50		460	530
0212	33" x 59"	G		18	.889		430	50		480	555
0215	33" x 72"	G		18	.889		465	50		515	595
0220	41" x 41"	G		18	.889		460	50		510	585
0230	41" x 47"	G		18	.889		460	50		510	585
0240	41" x 53"	G		17	.941		550	53		603	690
0242	41" x 59"	G		17	.941		530	53		583	670
0246	41" x 72"	G		17	.941		545	53		598	685
0250	47" x 41"	G		17	.941		595	53		648	740
0260	47" x 47"	G		17	.941		450	53		503	580
0270	47" x 53"	G		17	.941		495	53		548	630
0272	47" x 59"	G		17	.941		670	53		723	825

08 53 Plastic Windows

08 53 13 – Vinyl Windows

08 53 13.40 Vinyl Casement Windows

			Crew	Daily Output	Labor-Hours	Unit	Material	2022 Bare Costs Labor	Equipment	Total	Total Incl O&P
0280	56" x 41"	G	2 Carp	15	1.067	Ea.	495	60		555	640
0290	56" x 47"	G		15	1.067		495	60		555	640
0300	56" x 53"	G		15	1.067		540	60		600	690
0302	56" x 59"	G		15	1.067		540	60		600	690
0310	56" x 72"	G		15	1.067		700	60		760	865
0340	Solid vinyl, premium, double insulated glass, 2'-0" x 3'-0" high	G	1 Carp	10	.800		330	45		375	435
0360	2'-0" x 4'-0" high	G		9	.889		365	50		415	480
0380	2'-0" x 5'-0" high	G		8	1		415	56.50		471.50	545

08 53 13.50 Vinyl Picture Windows

		Crew	Daily Output	Labor-Hours	Unit	Material	2022 Bare Costs Labor	Equipment	Total	Total Incl O&P
0010	**VINYL PICTURE WINDOWS**, insulated glass									
0120	Grids, low E, J fin, ext. jambs, 47" x 35"	2 Carp	12	1.333	Ea.	229	75		304	370
0130	47" x 41"	2 Carp	12	1.333	Ea.	320	75		395	475
0140	47" x 47"		12	1.333		237	75		312	380
0150	47" x 53"		11	1.455		305	82		387	465
0160	71" x 35"		11	1.455		271	82		353	430
0170	71" x 41"		11	1.455		218	82		300	370
0180	71" x 47"		11	1.455		243	82		325	395

08 53 13.60 Vinyl Half Round Windows

		Crew	Daily Output	Labor-Hours	Unit	Material	2022 Bare Costs Labor	Equipment	Total	Total Incl O&P
0010	**VINYL HALF ROUND WINDOWS**, Including grille, J fin, low E, ext. jambs									
0100	10" height x 20" base	2 Carp	8	2	Ea.	257	113		370	460
0110	15" height x 30" base		8	2		238	113		351	440
0120	17" height x 34" base		7	2.286		310	129		439	545
0130	19" height x 38" base		7	2.286		470	129		599	720
0140	19" height x 33" base		7	2.286		410	129		539	655
0150	24" height x 48" base	1 Carp	6	1.333		340	75		415	495
0160	25" height x 50" base	"	6	1.333		420	75		495	585
0170	30" height x 60" base	2 Carp	6	2.667		470	150		620	755

08 54 Composite Windows

08 54 13 – Fiberglass Windows

08 54 13.10 Fiberglass Single-Hung Windows

			Crew	Daily Output	Labor-Hours	Unit	Material	2022 Bare Costs Labor	Equipment	Total	Total Incl O&P
0010	**FIBERGLASS SINGLE-HUNG WINDOWS**										
0100	Grids, low E, 18" x 24"	G	2 Carp	18	.889	Ea.	375	50		425	490
0110	18" x 40"	G		17	.941		365	53		418	490
0130	24" x 40"	G		20	.800		395	45		440	505
0230	36" x 36"	G		17	.941		405	53		458	530
0250	36" x 48"	G		20	.800		440	45		485	550
0260	36" x 60"	G		18	.889		485	50		535	615
0280	36" x 72"	G		16	1		525	56.50		581.50	665
0290	48" x 40"	G		16	1		525	56.50		581.50	665

08 54 13.30 Fiberglass Slider Windows

			Crew	Daily Output	Labor-Hours	Unit	Material	2022 Bare Costs Labor	Equipment	Total	Total Incl O&P
0010	**FIBERGLASS SLIDER WINDOWS**										
0100	Grids, low E, 36" x 24"	G	2 Carp	20	.800	Ea.	375	45		420	480
0110	36" x 36"	G	"	20	.800	"	380	45		425	485

08 54 13.50 Fiberglass Bay Windows

			Crew	Daily Output	Labor-Hours	Unit	Material	2022 Bare Costs Labor	Equipment	Total	Total Incl O&P
0010	**FIBERGLASS BAY WINDOWS**										
0150	48" x 36"	G	2 Carp	11	1.455	Ea.	1,100	82		1,182	1,350

08 56 Special Function Windows

08 56 46 – Radio-Frequency-Interference Shielding Windows

08 56 46.10 Radio-Frequency-Interference Mesh

		Crew	Daily Output	Labor-Hours	Unit	Material	2022 Bare Costs Labor	2022 Bare Costs Equipment	Total	Total Incl O&P
0010	**RADIO-FREQUENCY-INTERFERENCE MESH**									
0100	16 mesh copper 0.011" wire				S.F.	1.82			1.82	2
0150	22 mesh copper 0.015" wire					6.50			6.50	7.15
0200	100 mesh copper 0.022" wire					3.12			3.12	3.43
0250	100 mesh stainless steel 0.0012" wire				S.F.	7.45			7.45	8.20

08 61 Roof Windows

08 61 13 – Metal Roof Windows

08 61 13.10 Roof Windows

		Crew	Daily Output	Labor-Hours	Unit	Material	Labor	Equipment	Total	Total Incl O&P
0010	**ROOF WINDOWS**, fixed high perf tmpd glazing, metallic framed									
0020	46" x 21-1/2", Flashed for shingled roof	1 Carp	8	1	Ea.	300	56.50		356.50	420
0100	46" x 28"		8	1		335	56.50		391.50	460
0125	57" x 44"		6	1.333		400	75		475	560
0130	72" x 28"		7	1.143		400	64.50		464.50	540
0150	Fixed, laminated tempered glazing, 46" x 21-1/2"		8	1		485	56.50		541.50	620
0175	46" x 28"		8	1		535	56.50		591.50	675
0200	57" x 44"		6	1.333		510	75		585	680
0500	Vented flashing set for shingled roof, 46" x 21-1/2"		7	1.143		460	64.50		524.50	605
0525	46" x 28"		6	1.333		575	75		650	755
0550	57" x 44"		5	1.600		955	90		1,045	1,200
0560	72" x 28"		5	1.600		955	90		1,045	1,200
0575	Flashing set for low pitched roof, 46" x 21-1/2"		7	1.143		480	64.50		544.50	625
0600	46" x 28"		7	1.143		515	64.50		579.50	665
0625	57" x 44"		5	1.600		620	90		710	830
0650	Flashing set for curb, 46" x 21-1/2"		7	1.143		675	64.50		739.50	845
0675	46" x 28"		7	1.143		735	64.50		799.50	910
0700	57" x 44"		5	1.600		910	90		1,000	1,150

08 62 Unit Skylights

08 62 13 – Domed Unit Skylights

08 62 13.10 Domed Skylights

			Crew	Daily Output	Labor-Hours	Unit	Material	Labor	Equipment	Total	Total Incl O&P
0010	**DOMED SKYLIGHTS**										
0020	Skylight, fixed dome type, 22" x 22"	G	G-3	12	2.667	Ea.	123	151		274	370
0030	22" x 46"	G		10	3.200		225	181		406	530
0040	30" x 30"	G		12	2.667		169	151		320	425
0050	30" x 46"	G		10	3.200		320	181		501	640
0110	Fixed, double glazed, 22" x 27"	G		12	2.667		235	151		386	495
0120	22" x 46"	G		10	3.200		225	181		406	530
0130	44" x 46"	G		10	3.200		370	181		551	690
0210	Operable, double glazed, 22" x 27"	G		12	2.667		345	151		496	615
0220	22" x 46"	G		10	3.200		800	181		981	1,175
0230	44" x 46"	G		10	3.200		720	181		901	1,075
9000	Minimum labor/equipment charge			2	16	Job		905		905	1,425

08 62 13.20 Skylights

			Crew	Daily Output	Labor-Hours	Unit	Material	Labor	Equipment	Total	Total Incl O&P
0010	**SKYLIGHTS**, flush or curb mounted										
2120	Ventilating insulated plexiglass dome with										
2130	curb mounting, 36" x 36"	G	G-3	12	2.667	Ea.	495	151		646	775
2150	52" x 52"	G		12	2.667		690	151		841	995
2160	28" x 52"	G		10	3.200		575	181		756	920

271

08 62 Unit Skylights

08 62 13 – Domed Unit Skylights

08 62 13.20 Skylights

		Crew	Daily Output	Labor-Hours	Unit	Material	2022 Bare Costs Labor	Equipment	Total	Total Incl O&P
2170	36" x 52" [G]	G-3	10	3.200	Ea.	635	181		816	985
2180	For electric opening system, add [G]					370			370	405
2210	Operating skylight, with thermopane glass, 24" x 48" [G]	G-3	10	3.200		640	181		821	990
2220	32" x 48" [G]		9	3.556		670	201		871	1,050
2300	Insulated safety glass with aluminum frame [G]		160	.200	S.F.	99.50	11.30		110.80	127
4000	Skylight, solar tube kit, incl. dome, flashing, diffuser, 1 pipe, 10" diam. [G]	1 Carp	2	4	Ea.	325	225		550	720
4010	14" diam. [G]		2	4		400	225		625	805
4020	21" diam. [G]		2	4		500	225		725	910
4030	Accessories for, 1' long x 9" diam. pipe [G]		24	.333		63	18.75		81.75	99
4040	2' long x 9" diam. pipe [G]		24	.333		46.50	18.75		65.25	81.50
4050	4' long x 9" diam. pipe [G]		20	.400		79	22.50		101.50	123
4060	1' long x 13" diam. pipe [G]		24	.333		71.50	18.75		90.25	109
4070	2' long x 13" diam. pipe [G]		24	.333		59.50	18.75		78.25	95
4080	4' long x 13" diam. pipe [G]		20	.400		84	22.50		106.50	129
4090	6.5" turret ext. for 21" diam. pipe [G]		16	.500		143	28		171	203
4100	12' long x 21" diam. flexible pipe [G]		12	.667		88	37.50		125.50	157
4110	45 degree elbow, 10" [G]		16	.500		204	28		232	269
4120	14" [G]		16	.500		88	28		116	142
4130	Interior decorative ring, 9" [G]		20	.400		57.50	22.50		80	99.50
4140	13" [G]		20	.400		58.50	22.50		81	101

08 63 Metal-Framed Skylights

08 63 13 – Domed Metal-Framed Skylights

08 63 13.20 Skylight Rigid Metal-Framed

		Crew	Daily Output	Labor-Hours	Unit	Material	2022 Bare Costs Labor	Equipment	Total	Total Incl O&P
0010	**SKYLIGHT RIGID METAL-FRAMED** skylight framing is aluminum									
0050	Fixed acrylic double domes, curb mount, 25-1/2" x 25-1/2"	G-3	10	3.200	Ea.	210	181		391	515
0060	25-1/2" x 33-1/2"		10	3.200		254	181		435	565
0070	25-1/2" x 49-1/2"		6	5.333		279	300		579	780
0080	33-1/2" x 33-1/2"		8	4		310	226		536	695
0090	37-1/2" x 25-1/2"		8	4		271	226		497	655
0100	37-1/2" x 37-1/2"		8	4		370	226		596	765
0110	37-1/2" x 49-1/2"		6	5.333		445	300		745	960
0120	49-1/2" x 33-1/2"		6	5.333		420	300		720	935
0130	49-1/2" x 49-1/2"		6	5.333		465	300		765	990
1000	Fixed tempered glass, curb mount, 17-1/2" x 33-1/2"		10	3.200		155	181		336	455
1020	17-1/2" x 49-1/2"		6	5.333		176	300		476	670
1030	25-1/2" x 25-1/2"		10	3.200		155	181		336	455
1040	25-1/2" x 33-1/2"		10	3.200		184	181		365	485
1050	25-1/2" x 37-1/2"		8	4		196	226		422	570
1060	25-1/2" x 49-1/2"		6	5.333		520	300		820	1,050
1070	25-1/2" x 73-1/2"		6	5.333		355	300		655	865
1080	33-1/2" x 33-1/2"		8	4		215	226		441	590
2000	Manual vent tempered glass & screen, curb, 25-1/2" x 25-1/2"		10	3.200		400	181		581	725
2020	25-1/2" x 37-1/2"		10	3.200		460	181		641	790
2030	25-1/2" x 49-1/2"		8	4		495	226		721	900
2040	33-1/2" x 33-1/2"		8	4		530	226		756	940
2050	33-1/2" x 49-1/2"		6	5.333		670	300		970	1,200
2060	37-1/2" x 37-1/2"		6	5.333		610	300		910	1,150
2070	49-1/2" x 49-1/2"		6	5.333		785	300		1,085	1,325
3000	Electric vent tempered glass, curb mount, 25-1/2" x 25-1/2"		10	3.200		1,050	181		1,231	1,425
3020	25-1/2" x 37-1/2"		10	3.200		1,125	181		1,306	1,525

08 63 Metal-Framed Skylights

08 63 13 – Domed Metal-Framed Skylights

08 63 13.20 Skylight Rigid Metal-Framed		Crew	Daily Output	Labor-Hours	Unit	Material	2022 Bare Costs Labor	Equipment	Total	Total Incl O&P
3030	25-1/2" x 49-1/2"	G-3	8	4	Ea.	1,150	226		1,376	1,625
3040	33-1/2" x 33-1/2"		8	4		1,175	226		1,401	1,650
3050	33-1/2" x 49-1/2"		6	5.333		1,250	300		1,550	1,850
3060	37-1/2" x 37-1/2"		6	5.333		1,225	300		1,525	1,825
3070	49-1/2" x 49-1/2"		6	5.333		1,350	300		1,650	1,950

08 71 Door Hardware

08 71 13 – Automatic Door Operators

08 71 13.10 Automatic Openers Commercial

		Crew	Daily Output	Labor-Hours	Unit	Material	2022 Bare Costs Labor	Equipment	Total	Total Incl O&P
0010	**AUTOMATIC OPENERS COMMERCIAL**									
0020	Pneumatic door opener incl. motion sens, control box, tubing, compressor									
0050	For single swing door, per opening	2 Skwk	.80	20	Ea.	5,550	1,175		6,725	8,000
0100	Pair, per opening		.50	32	Opng.	7,950	1,875		9,825	11,700
1000	For single sliding door, per opening		.60	26.667		5,925	1,575		7,500	9,000
1300	Bi-parting pair		.50	32		7,925	1,875		9,800	11,700
1420	Electronic door opener incl. motion sens, 12 V control box, motor									
1450	For single swing door, per opening	2 Skwk	.80	20	Opng.	3,975	1,175		5,150	6,250
1500	Pair, per opening		.50	32		7,975	1,875		9,850	11,800
1600	For single sliding door, per opening		.60	26.667		5,225	1,575		6,800	8,225
1700	Bi-parting pair		.50	32		5,850	1,875		7,725	9,400
1750	Handicap actuator buttons, 2, incl. 12 V DC wiring, add	1 Carp	1.50	5.333	Pr.	480	300		780	1,000
2000	Electric panic button for door	2 Skwk	.50	32	Opng.	243	1,875		2,118	3,250

08 71 20.15 Hardware

		Crew	Daily Output	Labor-Hours	Unit	Material	2022 Bare Costs Labor	Equipment	Total	Total Incl O&P
0010	**HARDWARE**									
0020	Average hardware cost									
1000	Door hardware, apartment, interior	1 Carp	4	2	Door	535	113		648	770
1300	Average, door hardware, motel/hotel interior, with access card		4	2		785	113		898	1,050
1500	Hospital bedroom, average quality		4	2		505	113		618	735
2000	High quality		3	2.667		535	150		685	830
2250	School, single exterior, incl. lever, incl. panic device		3	2.667		1,900	150		2,050	2,350
2500	Single interior, regular use, lever included		3	2.667		455	150		605	745
2600	Average, door hdwe. set, school, classroom, ANSI F88, incl. lever		3	2.667		515	150		665	810
2850	Stairway, single interior		3	2.667		510	150		660	800
3100	Double exterior, with panic device		2	4	Pr.	3,625	225		3,850	4,350
6020	Add for fire alarm door holder, electro-magnetic	1 Elec	4	2	Ea.	106	132		238	320

08 71 20.30 Door Closers

		Crew	Daily Output	Labor-Hours	Unit	Material	2022 Bare Costs Labor	Equipment	Total	Total Incl O&P
0010	**DOOR CLOSERS** adjustable backcheck, multiple mounting									
0015	and rack and pinion									
0020	Standard Regular Arm	1 Carp	6	1.333	Ea.	167	75		242	305
0040	Hold open arm		6	1.333		102	75		177	231
0100	Fusible link		6.50	1.231		193	69.50		262.50	320
1500	Concealed closers, normal use, head, pivot hung, interior		5.50	1.455		335	82		417	500
1510	Exterior		5.50	1.455		450	82		532	625
1520	Overhead concealed, all sizes, regular arm		5.50	1.455		200	82		282	350
1525	Concealed arm		5	1.600		515	90		605	710
1530	Concealed in door, all sizes, regular arm		5.50	1.455		1,075	82		1,157	1,300
1535	Concealed arm		5	1.600		271	90		361	440
1560	Floor concealed, all sizes, single acting		2.20	3.636		595	205		800	980
1565	Double acting		2.20	3.636		2,400	205		2,605	2,975
1570	Interior, floor, offset pivot, single acting		3.50	2.286		1,050	129		1,179	1,350

08 71 20.30 Door Closers

		Crew	Daily Output	Labor-Hours	Unit	Material	2022 Bare Costs Labor	Equipment	Total	Total Incl O&P
1590	Exterior	1 Carp	3.50	2.286	Ea.	1,150	129		1,279	1,475
1610	Hold open arm		6	1.333		277	75		352	425
1620	Double acting, standard arm		6	1.333		1,300	75		1,375	1,550
1630	Hold open arm		6	1.333		1,300	75		1,375	1,550
1640	Floor, center hung, single acting, bottom arm		6	1.333		475	75		550	640
1650	Double acting		6	1.333		1,000	75		1,075	1,225
1660	Offset hung, single acting, bottom arm		6	1.333		1,050	75		1,125	1,300
2000	Backcheck and adjustable power, hinge face mount									
5000	For cast aluminum cylinder, deduct				Ea.	41.50			41.50	45.50
5040	For delayed action, add					51			51	56.50
5080	For fusible link arm, add					193			193	212
5120	For shock absorbing arm, add					54.50			54.50	60
5160	For spring power adjustment, add					40			40	44
6000	Closer-holder, hinge face mount, all sizes, exposed arm	1 Carp	6.50	1.231		237	69.50		306.50	370
6500	Electro magnetic closer/holder									
6510	Single point, no detector	1 Carp	4	2	Ea.	950	113		1,063	1,225
6515	Including detector		4	2		830	113		943	1,100
6520	Multi-point, no detector		4	2		1,100	113		1,213	1,400
6524	Including detector		4	2		1,725	113		1,838	2,075
6550	Electric automatic operators									
6555	Operator	1 Carp	4	2	Ea.	2,200	113		2,313	2,600
6570	Wall plate actuator		4	2		299	113		412	510
7000	Electronic closer-holder, hinge facemount, concealed arm		5	1.600		495	90		585	690
7400	With built-in detector		5	1.600		620	90		710	830
9000	Minimum labor/equipment charge		4	2	Job		113		113	179

08 71 20.36 Panic Devices

		Crew	Daily Output	Labor-Hours	Unit	Material	2022 Bare Costs Labor	Equipment	Total	Total Incl O&P
0010	**PANIC DEVICES** R087110-10									
0015	Touch bars various styles									
0040	Single door exit only, rim device, wide stile									
0050	Economy US28	1 Carp	5	1.600	Ea.	315	90		405	490
0060	Standard duty US28		5	1.600		840	90		930	1,075
0065	US26D		5	1.600		925	90		1,015	1,175
0070	US10		5	1.600		980	90		1,070	1,225
0075	US3		5	1.600		1,125	90		1,215	1,375
0080	Night latch, economy US28		5	1.600		330	90		420	510
0085	Standard duty, night latch US28		5	1.600		1,025	90		1,115	1,275
0090	US26D		5	1.600		1,075	90		1,165	1,325
0095	US10		5	1.600		1,200	90		1,290	1,475
0100	US3		5	1.600		1,375	90		1,465	1,675
0500	Single door exit only, rim device, narrow stile									
0540	Economy US28	1 Carp	5	1.600	Ea.	535	90		625	730
0550	Standard duty US28		5	1.600		930	90		1,020	1,175
0565	US26D		5	1.600		1,150	90		1,240	1,400
0570	US10		5	1.600		1,000	90		1,090	1,250
0575	US3		5	1.600		1,250	90		1,340	1,525
0580	Economy with night latch US28		5	1.600		400	90		490	585
0585	Standard duty US28		5	1.600		1,100	90		1,190	1,375
0590	US26D		5	1.600		1,550	90		1,640	1,850
0595	US10	1 Carp	5	1.600	Ea.	1,025	90		1,115	1,275
0600	US32D	"	5	1.600	"	1,450	90		1,540	1,750
1040	Single door exit only, surface vertical rod									
1050	Surface vertical rod, economy US28	1 Carp	4	2	Ea.	610	113		723	850

08 71 Door Hardware

08 71 13 - Automatic Door Operators

08 71 20.36 Panic Devices

		Crew	Daily Output	Labor-Hours	Unit	Material	2022 Bare Costs Labor	Equipment	Total	Total Incl O&P
1060	Surface vertical rod, standard duty US28	1 Carp	4	2	Ea.	1,250	113		1,363	1,550
1065	US26D		4	2		1,375	113		1,488	1,700
1070	US10		4	2		1,350	113		1,463	1,650
1075	US3		4	2		1,400	113		1,513	1,700
1080	US32D		4	2		1,550	113		1,663	1,900
1500	Single door exit, rim, narrow stile, economy, surface vertical rod									
1540	Surface vertical rod, narrow, economy US28	1 Carp	4	2	Ea.	610	113		723	855
2040	Single door exit only, concealed vertical rod									
2042	These devices will require separate and extra door preparation									
2060	Concealed rod, standard duty US28	1 Carp	4	2	Ea.	1,250	113		1,363	1,550
2065	US26D		4	2		1,400	113		1,513	1,700
2070	US10		4	2		1,525	113		1,638	1,850
2560	Concealed rod, narrow, standard duty US28		4	2		1,425	113		1,538	1,750
2565	US26D		4	2		1,625	113		1,738	1,950
2567	US26		4	2		2,350	113		2,463	2,750
2570	US10		4	2		1,475	113		1,588	1,800
2575	US3		4	2		1,725	113		1,838	2,075
3040	Single door exit only, mortise device, wide stile									
3042	These devices must be combined with a mortise lock									
3060	Mortise, standard duty US28	1 Carp	4	2	Ea.	1,325	113		1,438	1,625
3065	US26D		4	2		1,350	113		1,463	1,650
3067	US26		4	2		1,575	113		1,688	1,900
3070	US10		4	2		1,425	113		1,538	1,750
3075	US3		4	2		1,400	113		1,513	1,700
3080	US32D		4	2		1,400	113		1,513	1,725
3500	Single door exit only, mortise device, narrow stile									
3510	These devices must be combined with a mortise lock									
3540	Mortise, economy US28	1 Carp	4	2	Ea.	455	113		568	685

08 71 20.40 Lockset

		Crew	Daily Output	Labor-Hours	Unit	Material	2022 Bare Costs Labor	Equipment	Total	Total Incl O&P
0010	**LOCKSET**, Standard duty									
0020	Non-keyed, passage, w/sect. trim	1 Carp	12	.667	Ea.	133	37.50		170.50	206
0100	Privacy		12	.667		144	37.50		181.50	218
0400	Keyed, single cylinder function		10	.800		178	45		223	268
0420	Hotel (see also Section 08 71 20.15)		8	1		266	56.50		322.50	380
0500	Lever handled, keyed, single cylinder function	1 Carp	10	.800	Ea.	176	45		221	266
1000	Heavy duty with sectional trim, non-keyed, passages		12	.667		295	37.50		332.50	385
1100	Privacy		12	.667		197	37.50		234.50	277
1400	Keyed, single cylinder function		10	.800		400	45		445	510
1420	Hotel		8	1		475	56.50		531.50	615
1600	Communicating		10	.800		261	45		306	360
1690	For re-core cylinder, add					58			58	64
3800	Cipher lockset w/key pad (security item)	1 Carp	13	.615		1,075	34.50		1,109.50	1,225
3900	Cipher lockset with dial for swinging doors (security item)		13	.615		1,900	34.50		1,934.50	2,150
3920	with dial for swinging doors & drill resistant plate (security item)		12	.667		2,300	37.50		2,337.50	2,600
3950	Cipher lockset with dial for safe/vault door (security item)		12	.667		1,650	37.50		1,687.50	1,875
3980	Keyless, pushbutton type									
4000	Residential/light commercial, deadbolt, standard	1 Carp	9	.889	Ea.	300	50		350	410
4010	Heavy duty		9	.889		305	50		355	415
4020	Industrial, heavy duty, with deadbolt		9	.889		375	50		425	495
4030	Key override		9	.889		415	50		465	540
4040	Lever activated handle		9	.889		375	50		425	495
4050	Key override		9	.889		415	50		465	540

275

For customer support on your Commercial Renovation Costs with RSMeans data, call 800.448.8182.

08 71 20.40 Lockset		Crew	Daily Output	Labor-Hours	Unit	Material	2022 Bare Costs Labor	Equipment	Total	Total Incl O&P
4060	Double sided pushbutton type	1 Carp	8	1	Ea.	300	56.50		356.50	420
4070	Key override		8	1		950	56.50		1,006.50	1,150
9000	Minimum labor/equipment charge		6	1.333	Job		75		75	119

08 71 20.41 Dead Locks										
0010	**DEAD LOCKS**									
0011	Mortise heavy duty outside key (security item)	1 Carp	9	.889	Ea.	242	50		292	345
0020	Double cylinder		9	.889		156	50		206	252
0100	Medium duty, outside key		10	.800		121	45		166	205
0110	Double cylinder		10	.800		281	45		326	380
1000	Tubular, standard duty, outside key		10	.800		56.50	45		101.50	134
1010	Double cylinder		10	.800		90	45		135	171
1200	Night latch, outside key		10	.800		156	45		201	244
1203	Deadlock night latch		7.70	1.039		156	58.50		214.50	265
1420	Deadbolt lock, single cylinder		10	.800		98	45		143	180
1440	Double cylinder		10	.800		69	45		114	148

08 71 20.42 Mortise Locksets										
0010	**MORTISE LOCKSETS**, Comm., wrought knobs & full escutcheon trim									
0015	Assumes mortise is cut									
0020	Non-keyed, passage, Grade 3	1 Carp	9	.889	Ea.	237	50		287	340
0030	Grade 1	"	8	1	"	395	56.50		451.50	520
0040	Privacy set, Grade 3	1 Carp	9	.889	Ea.	171	50		221	269
0050	Grade 1		8	1		385	56.50		441.50	515
0100	Keyed, office/entrance/apartment, Grade 2		8	1		255	56.50		311.50	370
0110	Grade 1		7	1.143		325	64.50		389.50	460
0120	Single cylinder, typical, Grade 3		8	1		170	56.50		226.50	277
0130	Grade 1		7	1.143		420	64.50		484.50	560
0200	Hotel, room, Grade 3		7	1.143		192	64.50		256.50	315
0210	Grade 1 (see also Section 08 71 20.15)		6	1.333		560	75		635	735
0300	Double cylinder, Grade 3		8	1		241	56.50		297.50	355
0310	Grade 1		7	1.143		585	64.50		649.50	745
1000	Wrought knobs and sectional trim, non-keyed, passage, Grade 3		10	.800		139	45		184	225
1010	Grade 1		9	.889		430	50		480	555
1040	Privacy, Grade 3		10	.800		163	45		208	251
1050	Grade 1		9	.889		485	50		535	610
1100	Keyed, entrance, office/apartment, Grade 3		9	.889		235	50		285	340
1103	Install lockset		6.92	1.156		235	65		300	360
1110	Grade 1		8	1		555	56.50		611.50	700
1120	Single cylinder, Grade 3		9	.889		240	50		290	345
1130	Grade 1		8	1		510	56.50		566.50	655
2000	Cast knobs and full escutcheon trim									
2010	Non-keyed, passage, Grade 3	1 Carp	9	.889	Ea.	279	50		329	385
2020	Grade 1		8	1		460	56.50		516.50	595
2040	Privacy, Grade 3		9	.889		345	50		395	460
2050	Grade 1		8	1		450	56.50		506.50	585
2120	Keyed, single cylinder, Grade 3		8	1		515	56.50		571.50	655
2123	Mortise lock		6.15	1.301		515	73		588	680
2130	Grade 1		7	1.143		550	64.50		614.50	700
3000	Cast knob and sectional trim, non-keyed, passage, Grade 3		10	.800		222	45		267	315
3010	Grade 1		10	.800		430	45		475	545
3040	Privacy, Grade 3		10	.800		231	45		276	325
3050	Grade 1		10	.800		480	45		525	595
3100	Keyed, office/entrance/apartment, Grade 3		9	.889		262	50		312	370

08 71 Door Hardware

08 71 13 – Automatic Door Operators

08 71 20.42 Mortise Locksets

		Crew	Daily Output	Labor-Hours	Unit	Material	2022 Bare Costs Labor	Equipment	Total	Total Incl O&P
3110	Grade 1	1 Carp	9	.889	Ea.	485	50		535	615
3120	Single cylinder, Grade 3		9	.889		274	50		324	380
3130	Grade 1		9	.889		555	50		605	690
3190	For re-core cylinder, add					59			59	65

08 71 20.44 Anti-Ligature Locksets Grade 1

		Crew	Daily Output	Labor-Hours	Unit	Material	2022 Bare Costs Labor	Equipment	Total	Total Incl O&P
0010	**ANTI-LIGATURE LOCKSETS GRADE 1**									
0100	Anti-ligature cylindrical locksets									
0500	Arch handle passage set, US32D	1 Carp	8	1	Ea.	299	56.50		355.50	420
0510	privacy set		8	1		375	56.50		431.50	500
0520	classroom set		8	1		370	56.50		426.50	495
0530	storeroom set		8	1		370	56.50		426.50	495
0550	Lever handle passage set US32D		8	1		355	56.50		411.50	480
0560	Hospital/privacy set		8	1		375	56.50		431.50	500
0570	Office set		8	1		375	56.50		431.50	505
0580	Classroom set		8	1		365	56.50		421.50	495
0590	Storeroom set		8	1		370	56.50		426.50	495
0600	Exit set		8	1		375	56.50		431.50	500
0610	Entry set		8	1		335	56.50		391.50	455
0620	Asylum set		8	1		360	56.50		416.50	485
0630	Back to back dummy set		8	1		229	56.50		285.50	340
0690	Anti-ligature mortise locksets									
0700	Knob handle privacy set, US32D	1 Carp	8	1	Ea.	450	56.50		506.50	585
0710	Office set		8	1		375	56.50		431.50	505
0720	Classroom set		8	1		475	56.50		531.50	610
0730	Hotel set		8	1		455	56.50		511.50	590
0740	Apartment set		8	1		385	56.50		441.50	515
0750	Institutional privacy set		8	1		585	56.50		641.50	735
0760	Lever handle office set		8	1		435	56.50		491.50	570
0770	Classroom set sectional trim		8	1		435	56.50		491.50	570
0780	Hotel set		8	1		445	56.50		501.50	580
0790	Apartment set		8	1		445	56.50		501.50	580
0800	Institutional privacy set		8	1		485	56.50		541.50	620
0810	Office set US32D		8	1		405	56.50		461.50	535
0820	Classroom set escutcheon trim		8	1		555	56.50		611.50	700
0830	Hotel set escutcheon trim		8	1		585	56.50		641.50	730
0840	Apartment set escutcheon trim		8	1		585	56.50		641.50	730
0850	Institutional privacy set escutcheon trim		8	1		585	56.50		641.50	730

08 71 20.45 Peepholes

		Crew	Daily Output	Labor-Hours	Unit	Material	2022 Bare Costs Labor	Equipment	Total	Total Incl O&P
0010	**PEEPHOLES**									
2010	Peephole	1 Carp	32	.250	Ea.	12.95	14.10		27.05	37
2020	Peephole, wide view	"	32	.250	"	9.55	14.10		23.65	33

08 71 20.50 Door Stops

		Crew	Daily Output	Labor-Hours	Unit	Material	2022 Bare Costs Labor	Equipment	Total	Total Incl O&P
0010	**DOOR STOPS**									
0020	Holder & bumper, floor or wall	1 Carp	32	.250	Ea.	32.50	14.10		46.60	58
1300	Wall bumper, 4" diameter, with rubber pad, aluminum		32	.250		5.65	14.10		19.75	28.50
1600	Door bumper, floor type, aluminum		32	.250		6.20	14.10		20.30	29.50
1620	Brass		32	.250		7.15	14.10		21.25	30.50
1630	Bronze		32	.250		26	14.10		40.10	51
1900	Plunger type, door mounted		32	.250		39.50	14.10		53.60	66
2500	Holder, floor type, aluminum		32	.250		21	14.10		35.10	45.50
2520	Wall type, aluminum		32	.250		34	14.10		48.10	59.50
2530	Overhead type, bronze		32	.250		109	14.10		123.10	142

For customer support on your Commercial Renovation Costs with RSMeans data, call 800.448.8182.

277

08 71 Door Hardware

08 71 13 - Automatic Door Operators

08 71 20.50 Door Stops

		Crew	Daily Output	Labor-Hours	Unit	Material	2022 Bare Costs Labor	Equipment	Total	Total Incl O&P
2540	Plunger type, aluminum	1 Carp	32	.250	Ea.	39.50	14.10		53.60	66
2560	Brass		32	.250		44.50	14.10		58.60	71.50
3000	Electromagnetic, wall mounted, US3		3	2.667		276	150		426	545
3020	Floor mounted, US3		3	2.667		320	150		470	595
4000	Doorstop, ceiling mounted		3	2.667		18.45	150		168.45	260
4030	Doorstop, header		3	2.667		43.50	150		193.50	287
9000	Minimum labor/equipment charge		6	1.333	Job		75		75	119

08 71 20.55 Push-Pull Plates

		Crew	Daily Output	Labor-Hours	Unit	Material	2022 Bare Costs Labor	Equipment	Total	Total Incl O&P
0010	**PUSH-PULL PLATES**									
0090	Push plate, 0.050 thick, 3" x 12", aluminum	1 Carp	12	.667	Ea.	5.50	37.50		43	65.50
0100	4" x 16"		12	.667		9.95	37.50		47.45	70.50
0110	6" x 16"		12	.667		7.40	37.50		44.90	67.50
0120	8" x 16"		12	.667		8.70	37.50		46.20	69
0200	Push plate, 0.050 thick, 3" x 12", brass		12	.667		12.65	37.50		50.15	73.50
0210	4" x 16"		12	.667		15.80	37.50		53.30	77
0220	6" x 16"		12	.667		24.50	37.50		62	86
0230	8" x 16"		12	.667		32	37.50		69.50	94.50
0250	Push plate, 0.050 thick, 3" x 12", satin brass		12	.667		12.85	37.50		50.35	73.50
0260	4" x 16"		12	.667		16	37.50		53.50	77
0270	6" x 16"		12	.667		24.50	37.50		62	86.50
0280	8" x 16"		12	.667		32	37.50		69.50	94.50
0490	Push plate, 0.050 thick, 3" x 12", bronze		12	.667		15.25	37.50		52.75	76.50
0500	4" x 16"		12	.667		23	37.50		60.50	84.50
0510	6" x 16"		12	.667		29	37.50		66.50	91.50
0520	8" x 16"		12	.667		35	37.50		72.50	98
0600	Push plate, antimicrobial copper alloy finish, 3.5" x 15"		13	.615		49.50	34.50		84	110
0610	4" x 16"		13	.615		53.50	34.50		88	114
0620	6" x 16"		13	.615		21	34.50		55.50	78.50
0630	6" x 20"		13	.615		38	34.50		72.50	97
0740	Push plate, 0.050 thick, 3" x 12", stainless steel		12	.667		10.80	37.50		48.30	71.50
0750	4" x 16"		12	.667		15.65	37.50		53.15	77
0760	6" x 16"		12	.667		16.90	37.50		54.40	78
0780	8" x 16"		12	.667		21.50	37.50		59	83
0790	Push plate, 0.050 thick, 3" x 12", satin stainless steel	1 Carp	12	.667	Ea.	6.10	37.50		43.60	66
0810	4" x 16"		12	.667		7.50	37.50		45	68
0820	6" x 16"		12	.667		10.65	37.50		48.15	71
0830	8" x 16"		12	.667		14.30	37.50		51.80	75.50
0980	Pull plate, 0.050 thick, 3" x 12", aluminum		12	.667		23.50	37.50		61	85
1000	4" x 16"		12	.667		26.50	37.50		64	89
1050	Pull plate, 0.050 thick, 3" x 12", brass		12	.667		36	37.50		73.50	99.50
1060	4" x 16"		12	.667		57.50	37.50		95	123
1080	Pull plate, 0.050 thick, 3" x 12", bronze		12	.667		45.50	37.50		83	110
1100	4" x 16"		12	.667		53	37.50		90.50	118
1180	Pull plate, 0.050 thick, 3" x 12", stainless steel		12	.667		41.50	37.50		79	106
1200	4" x 16"		12	.667		55	37.50		92.50	120
1250	Pull plate, 0.050 thick, 3" x 12", chrome		12	.667		41.50	37.50		79	106
1270	4" x 16"		12	.667		43.50	37.50		81	108
1500	Pull handle and push bar, aluminum		11	.727		22.50	41		63.50	89.50
2000	Bronze		10	.800		86.50	45		131.50	167
9800	Minimum labor/equipment charge		5	1.600	Job		90		90	143

08 71 Door Hardware

08 71 13 – Automatic Door Operators

08 71 20.60 Entrance Locks

		Crew	Daily Output	Labor-Hours	Unit	Material	2022 Bare Costs Labor	2022 Bare Costs Equipment	Total	Total Incl O&P
0010	**ENTRANCE LOCKS**									
0015	Cylinder, grip handle deadlocking latch	1 Carp	9	.889	Ea.	184	50		234	282
0020	Deadbolt		8	1		235	56.50		291.50	350
0100	Push and pull plate, dead bolt		8	1		188	56.50		244.50	297
0900	For handicapped lever, add					175			175	193

08 71 20.65 Thresholds

		Crew	Daily Output	Labor-Hours	Unit	Material	2022 Bare Costs Labor	2022 Bare Costs Equipment	Total	Total Incl O&P
0010	**THRESHOLDS**									
0011	Threshold 3' long saddles aluminum	1 Carp	48	.167	L.F.	12.05	9.40		21.45	28
0100	Aluminum, 8" wide, 1/2" thick		12	.667	Ea.	60	37.50		97.50	126
0500	Bronze		60	.133	L.F.	36.50	7.50		44	52.50
0600	Bronze, panic threshold, 5" wide, 1/2" thick		12	.667	Ea.	137	37.50		174.50	211
0700	Rubber, 1/2" thick, 5-1/2" wide		20	.400		35.50	22.50		58	75
0800	2-3/4" wide		20	.400		25.50	22.50		48	64
1950	ADA compliant thresholds									
2000	Threshold, wood oak 3-1/2" wide x 24" long	1 Carp	12	.667	Ea.	26	37.50		63.50	88.50
2010	3-1/2" wide x 36" long		12	.667		34.50	37.50		72	97.50
2020	3-1/2" wide x 48" long		12	.667		35.50	37.50		73	98.50
2030	4-1/2" wide x 24" long		12	.667		12.70	37.50		50.20	73.50
2040	4-1/2" wide x 36" long		12	.667		40	37.50		77.50	104
2050	4-1/2" wide x 48" long		12	.667		38	37.50		75.50	102
2060	6-1/2" wide x 24" long	1 Carp	12	.667	Ea.	27	37.50		64.50	89
2070	6-1/2" wide x 36" long		12	.667		49	37.50		86.50	113
2080	6-1/2" wide x 48" long		12	.667		51	37.50		88.50	116
2090	Threshold, wood cherry 3-1/2" wide x 24" long		12	.667		23.50	37.50		61	85
2100	3-1/2" wide x 36" long		12	.667		32	37.50		69.50	94.50
2110	3-1/2" wide x 48" long		12	.667		35	37.50		72.50	98
2120	4-1/2" wide x 24" long		12	.667		25	37.50		62.50	87
2130	4-1/2" wide x 36" long		12	.667		34	37.50		71.50	97
2140	4-1/2" wide x 48" long		12	.667		38	37.50		75.50	102
2150	6-1/2" wide x 24" long		12	.667		25	37.50		62.50	87
2160	6-1/2" wide x 36" long		12	.667		39.50	37.50		77	103
2170	6-1/2" wide x 48" long		12	.667		56	37.50		93.50	121
2180	Threshold, wood walnut 3-1/2" wide x 24" long		12	.667		23.50	37.50		61	85
2190	3-1/2" wide x 36" long		12	.667		31.50	37.50		69	94
2200	3-1/2" wide x 48" long		12	.667		36	37.50		73.50	99
2210	4-1/2" wide x 24" long		12	.667		25	37.50		62.50	87
2220	4-1/2" wide x 36" long		12	.667		33	37.50		70.50	95.50
2230	4-1/2" wide x 48" long		12	.667		38	37.50		75.50	102
2240	6-1/2" wide x 24" long		12	.667		27	37.50		64.50	89
2250	6-1/2" wide x 36" long		12	.667		42.50	37.50		80	106
2260	6-1/2" wide x 48" long		12	.667		56	37.50		93.50	121
2300	Threshold, aluminum 4" wide x 36" long		12	.667		22	37.50		59.50	84
2310	4" wide x 48" long		12	.667		29.50	37.50		67	92
2320	4" wide x 72" long		12	.667		7.40	37.50		44.90	67.50
2330	5" wide x 36" long		12	.667		28.50	37.50		66	90.50
2340	5" wide x 48" long		12	.667		38	37.50		75.50	101
2350	5" wide x 72" long		12	.667		9.45	37.50		46.95	70
2360	6" wide x 36" long		12	.667		12.05	37.50		49.55	73
2370	6" wide x 48" long		12	.667		48.50	37.50		86	113
2380	6" wide x 72" long		12	.667		72.50	37.50		110	139
2390	7" wide x 36" long		12	.667		45.50	37.50		83	110
2400	7" wide x 48" long		12	.667		60.50	37.50		98	126

08 71 20.65 Thresholds	Crew	Daily Output	Labor-Hours	Unit	Material	2022 Bare Costs Labor	Equipment	Total	Total Incl O&P
2410	7" wide x 72" long	1 Carp	12	.667	Ea.	91	37.50		128.50	160
2500	Threshold, ramp, aluminum or rubber 24" x 24"		12	.667		156	37.50		193.50	231
9000	Minimum labor/equipment charge		4	2	Job		113		113	179

08 71 20.75 Door Hardware Accessories	Crew	Daily Output	Labor-Hours	Unit	Material	Labor	Equipment	Total	Total Incl O&P
0010	**DOOR HARDWARE ACCESSORIES**									
0050	Door closing coordinator, 36" (for paired openings up to 56")	1 Carp	8	1	Ea.	134	56.50		190.50	237
0060	48" (for paired openings up to 84")		8	1		142	56.50		198.50	246
0070	56" (for paired openings up to 96")		8	1		151	56.50		207.50	257
1000	Knockers, brass, standard		16	.500		67.50	28		95.50	119
1100	Deluxe		10	.800		98	45		143	180
2000	Torsion springs for overhead doors									
2050	1-3/4" diam., 32" long, 0.243" spring	1 Carp	8	1	Ea.	59.50	56.50		116	155
2060	1-3/4" diam., 32" long, 0.250" spring		8	1		45	56.50		101.50	139
2070	2" diam., 32" long, 0.243" spring		8	1		45.50	56.50		102	140
2080	2" diam., 32" long, 0.253" spring		8	1		46.50	56.50		103	141
4500	Rubber door silencers		540	.015		.37	.83		1.20	1.74
9000	Minimum labor/equipment charge		6	1.333	Job		75		75	119

08 71 20.80 Hasps	Crew	Daily Output	Labor-Hours	Unit	Material	Labor	Equipment	Total	Total Incl O&P
0010	**HASPS**, steel assembly									
0015	3"	1 Carp	26	.308	Ea.	5.90	17.30		23.20	34
0020	4-1/2"		13	.615		7.25	34.50		41.75	63
0040	6"		12.50	.640		10.50	36		46.50	69

08 71 20.90 Hinges	Crew	Daily Output	Labor-Hours	Unit	Material	Labor	Equipment	Total	Total Incl O&P
0010	**HINGES**									
0012	Full mortise, avg. freq., steel base, USP, 4-1/2" x 4-1/2"				Pr.	34.50			34.50	38
0100	5" x 5", USP					27			27	29.50
0200	6" x 6", USP					56.50			56.50	62.50
0400	Brass base, 4-1/2" x 4-1/2", US10					49.50			49.50	54.50
0500	5" x 5", US10					57.50			57.50	63.50
0600	6" x 6", US10					181			181	199
0800	Stainless steel base, 4-1/2" x 4-1/2", US32					55.50			55.50	61
0900	For non removable pin, add (security item)				Ea.	3.24			3.24	3.56
0910	For floating pin, driven tips, add					3.33			3.33	3.66
0930	For hospital type tip on pin, add					8.65			8.65	9.50
0940	For steeple type tip on pin, add					11.50			11.50	12.65
0950	Full mortise, high frequency, steel base, 3-1/2" x 3-1/2", US26D				Pr.	31.50			31.50	35
1000	4-1/2" x 4-1/2", USP					65			65	71.50
1100	5" x 5", USP					22			22	24.50
1200	6" x 6", USP					113			113	124
1400	Brass base, 3-1/2" x 3-1/2", US4					48			48	52.50
1430	4-1/2" x 4-1/2", US10					60.50			60.50	66.50
1500	5" x 5", US10					124			124	136
1600	6" x 6", US10					144			144	158
1800	Stainless steel base, 4-1/2" x 4-1/2", US32				Pr.	55.50			55.50	61
1810	5" x 4-1/2", US32				"	55.50			55.50	61
1930	For hospital type tip on pin, add				Ea.	14.65			14.65	16.10
1950	Full mortise, low frequency, steel base, 3-1/2" x 3-1/2", US26D				Pr.	8.25			8.25	9.05
2000	4-1/2" x 4-1/2", USP					3.43			3.43	3.77
2100	5" x 5", USP					55			55	60.50
2200	6" x 6", USP					96			96	106
2300	4-1/2" x 4-1/2", US3					20.50			20.50	22.50
2310	5" x 5", US3					41.50			41.50	45.50

08 71 20.90 Hinges		Crew	Daily Output	Labor-Hours	Unit	Material	2022 Bare Costs Labor	Equipment	Total	Total Incl O&P
2400	Brass bass, 4-1/2" x 4-1/2", US10				Pr.	94			94	103
2500	5" x 5", US10					120			120	132
2800	Stainless steel base, 4-1/2" x 4-1/2", US32					80.50			80.50	88.50
8000	Install hinge	1 Carp	34	.235	▼		13.25		13.25	21

08 71 20.91 Special Hinges										
0010	**SPECIAL HINGES**									
0015	Paumelle, high frequency									
0020	Steel base, 6" x 4-1/2", US10				Pr.	138			138	152
0100	Brass base, 5" x 4-1/2", US10				Ea.	236			236	260
0200	Paumelle, average frequency, steel base, 4-1/2" x 3-1/2", US10				Pr.	93.50			93.50	103
0400	Olive knuckle, low frequency, brass base, 6" x 4-1/2", US10				Ea.	154			154	169
1000	Electric hinge with concealed conductor, average frequency									
1010	Steel base, 4-1/2" x 4-1/2", US26D				Pr.	345			345	380
1100	Bronze base, 4-1/2" x 4-1/2", US26D				"	350			350	385
1200	Electric hinge with concealed conductor, high frequency									
1210	Steel base, 4-1/2" x 4-1/2", US26D				Pr.	285			285	315
1600	Double weight, 800 lb., steel base, removable pin, 5" x 6", USP					135			135	149
1700	Steel base-welded pin, 5" x 6", USP					174			174	192
1800	Triple weight, 2000 lb., steel base, welded pin, 5" x 6", USP					114			114	125
2000	Pivot reinf., high frequency, steel base, 7-3/4" door plate, USP					171			171	188
2200	Bronze base, 7-3/4" door plate, US10				▼	257			257	283
3000	Swing clear, full mortise, full or half surface, high frequency,									
3010	Steel base, 5" high, USP				Pr.	150			150	165
3200	Swing clear, full mortise, average frequency									
3210	Steel base, 4-1/2" high, USP				Pr.	133			133	146
4000	Wide throw, average frequency, steel base, 4-1/2" x 6", USP				"	94.50			94.50	104
4200	High frequency, steel base, 4-1/2" x 6", USP				Pr.	129			129	142
4440	US26D				"	94.50			94.50	104
4600	Spring hinge, single acting, 6" flange, steel				Ea.	50.50			50.50	55.50
4700	Brass					97			97	107
4900	Double acting, 6" flange, steel					61			61	67
4950	Brass				▼	134			134	148
5000	T-strap, galvanized, 4"				Pr.	22.50			22.50	25
5010	6"					31			31	34
5020	8"				▼	62			62	68
8000	Continuous hinges									
8010	Steel, piano, 2" x 72"	1 Carp	20	.400	Ea.	28	22.50		50.50	67
8020	Brass, piano, 1-1/16" x 30"		30	.267		12.75	15		27.75	38
8030	Acrylic, piano, 1-3/4" x 12"		40	.200		15.85	11.25		27.10	35.50
8040	Aluminum, door, standard duty, 7'		3	2.667		125	150		275	375
8050	Heavy duty, 7'		3	2.667		150	150		300	405
8060	8'		3	2.667		178	150		328	435
8070	Steel, door, heavy duty, 7'		3	2.667		214	150		364	475
8080	8'		3	2.667		425	150		575	705
8090	Stainless steel, door, heavy duty, 7'		3	2.667		273	150		423	540
8100	8'		3	2.667		305	150		455	575
9000	Continuous hinge, steel, full mortise, heavy duty, 96"		2	4		520	225		745	935
9200	Continuous geared hinge, aluminum, full mortise, standard duty, 83"		3	2.667		68.50	150		218.50	315
9250	Continuous geared hinge, aluminum, full mortise, heavy duty, 83"	▼	3	2.667	▼	139	150		289	390

08 71 Door Hardware

08 71 13 – Automatic Door Operators

08 71 20.95 Kick Plates		Crew	Daily Output	Labor-Hours	Unit	Material	2022 Bare Costs Labor	Equipment	Total	Total Incl O&P
0010	**KICK PLATES**									
0020	Stainless steel, .050", 16 ga., 8" x 28", US32	1 Carp	15	.533	Ea.	53	30		83	106
0030	8" x 30"		15	.533		54.50	30		84.50	107
0040	8" x 34"		15	.533		61.50	30		91.50	115
0050	10" x 28"		15	.533		66	30		96	121
0060	10" x 30"		15	.533		71	30		101	126
0070	10" x 34"		15	.533		80	30		110	136
0080	Mop/Kick, 4" x 28"		15	.533		40.50	30		70.50	92.50
0090	4" x 30"		15	.533		43	30		73	95
0100	4" x 34"		15	.533		49	30		79	102
0110	6" x 28"		15	.533		40	30		70	91.50
0120	6" x 30"		15	.533		42.50	30		72.50	94
0130	6" x 34"		15	.533		48	30		78	101
0500	Bronze, .050", 8" x 28"		15	.533		69.50	30		99.50	124
0510	8" x 30"		15	.533		75	30		105	130
0520	8" x 34"	1 Carp	15	.533	Ea.	85	30		115	142
0530	10" x 28"		15	.533		87.50	30		117.50	144
0540	10" x 30"		15	.533		94	30		124	151
0550	10" x 34"		15	.533		106	30		136	165
0560	Mop/Kick, 4" x 28"		15	.533		39	30		69	90.50
0570	4" x 30"		15	.533		43.50	30		73.50	95
0580	4" x 34"		15	.533		43.50	30		73.50	95.50
0590	6" x 28"		15	.533		55	30		85	108
0600	6" x 30"		15	.533		59	30		89	113
0610	6" x 34"		15	.533		67	30		97	122
1000	Acrylic, .125", 8" x 26"		15	.533		30	30		60	80
1010	8" x 36"		15	.533		41.50	30		71.50	93
1020	8" x 42"		15	.533		48	30		78	101
1030	10" x 26"		15	.533		37.50	30		67.50	88.50
1040	10" x 36"		15	.533		52.50	30		82.50	105
1050	10" x 42"		15	.533		61	30		91	115
1060	Mop/Kick, 4" x 26"		15	.533		17.80	30		47.80	67
1070	4" x 36"		15	.533		25	30		55	75
1080	4" x 42"		15	.533		28.50	30		58.50	79
1090	6" x 26"		15	.533		26.50	30		56.50	76.50
1100	6" x 36"		15	.533		36.50	30		66.50	87.50
1110	6" x 42"		15	.533		43	30		73	94.50
1220	Brass, .050", 8" x 26"		15	.533		76.50	30		106.50	132
1230	8" x 36"		15	.533		76.50	30		106.50	132
1240	8" x 42"		15	.533		93.50	30		123.50	151
1250	10" x 26"		15	.533		77.50	30		107.50	133
1260	10" x 36"		15	.533		113	30		143	172
1270	10" x 42"		15	.533		113	30		143	173
1320	Mop/Kick, 4" x 26"		15	.533		28.50	30		58.50	79
1330	4" x 36"		15	.533		39.50	30		69.50	91
1340	4" x 42"		15	.533		46	30		76	98
1350	6" x 26"		15	.533		42.50	30		72.50	94
1360	6" x 36"		15	.533		52	30		82	105
1370	6" x 42"		15	.533		156	30		186	219
1800	Aluminum, .050", 8" x 26"		15	.533		39	30		69	90.50
1810	8" x 36"		15	.533		46.50	30		76.50	98.50
1820	8" x 42"		15	.533		60	30		90	114

08 71 Door Hardware

08 71 13 – Automatic Door Operators

08 71 20.95 Kick Plates

		Crew	Daily Output	Labor-Hours	Unit	Material	2022 Bare Costs Labor	Equipment	Total	Total Incl O&P
1830	10" x 26"	1 Carp	15	.533	Ea.	36	30		66	87
1840	10" x 36"		15	.533		47	30		77	99
1850	10" x 42"		15	.533		51.50	30		81.50	105
1860	Mop/Kick, 4" x 26"	1 Carp	15	.533	Ea.	28	30		58	78
1870	4" x 36"		15	.533		23	30		53	73
1880	4" x 42"		15	.533		34	30		64	85
1890	6" x 26"		15	.533		26	30		56	76
1900	6" x 36"		15	.533		38.50	30		68.50	90
1910	6" x 42"		15	.533		42	30		72	94
9000	Minimum labor/equipment charge		6	1.333	Job		75		75	119

08 71 21 – Astragals

08 71 21.10 Exterior Mouldings, Astragals

		Crew	Daily Output	Labor-Hours	Unit	Material	2022 Bare Costs Labor	Equipment	Total	Total Incl O&P
0010	**EXTERIOR MOULDINGS, ASTRAGALS**									
0400	One piece, overlapping cadmium plated steel, flat, 3/16" x 2"	1 Carp	90	.089	L.F.	5.10	5		10.10	13.60
0600	Prime coated steel, flat, 1/8" x 3"		90	.089		6.10	5		11.10	14.65
0800	Stainless steel, flat, 3/32" x 1-5/8"		90	.089		15.75	5		20.75	25.50
1000	Aluminum, flat, 1/8" x 2"		90	.089		4.55	5		9.55	12.95
1200	Nail on, "T" extrusion		120	.067		2.17	3.75		5.92	8.35
1300	Vinyl bulb insert		105	.076		2.65	4.29		6.94	9.70
1600	Screw on, "T" extrusion		90	.089		4.68	5		9.68	13.10
1700	Vinyl insert		75	.107		4.10	6		10.10	14.05
2000	"L" extrusion, neoprene bulbs		75	.107		4.39	6		10.39	14.40
2100	Neoprene sponge insert		75	.107		6.80	6		12.80	17.05
2200	Magnetic		75	.107		10.40	6		16.40	21
2400	Spring hinged security seal, with cam		75	.107		6.65	6		12.65	16.90
2600	Spring loaded locking bolt, vinyl insert		45	.178		9.05	10		19.05	26
2800	Neoprene sponge strip, "Z" shaped, aluminum		60	.133		8	7.50		15.50	21
2900	Solid neoprene strip, nail on aluminum strip		90	.089		3.78	5		8.78	12.10
3000	One piece stile protection									
3020	Neoprene fabric loop, nail on aluminum strips	1 Carp	60	.133	L.F.	1.23	7.50		8.73	13.30
3110	Flush mounted aluminum extrusion, 1/2" x 1-1/4"		60	.133		6.65	7.50		14.15	19.25
3140	3/4" x 1-3/8"		60	.133		3.93	7.50		11.43	16.25
3160	1-1/8" x 1-3/4"		60	.133		5.40	7.50		12.90	17.85
3300	Mortise, 9/16" x 3/4"		60	.133		4.47	7.50		11.97	16.85
3320	13/16" x 1-3/8"		60	.133		4.16	7.50		11.66	16.55
3600	Spring bronze strip, nail on type		105	.076		2.36	4.29		6.65	9.40
3620	Screw on, with retainer		75	.107		2.62	6		8.62	12.45
3800	Flexible stainless steel housing, pile insert, 1/2" door		105	.076		6.95	4.29		11.24	14.45
3820	3/4" door		105	.076		7.95	4.29		12.24	15.55
4000	Extruded aluminum retainer, flush mount, pile insert		105	.076		2.73	4.29		7.02	9.80
4080	Mortise, felt insert	1 Carp	90	.089	L.F.	4.55	5		9.55	12.95
4160	Mortise with spring, pile insert		90	.089		3.74	5		8.74	12.05
4400	Rigid vinyl retainer, mortise, pile insert		105	.076		3.06	4.29		7.35	10.15
4600	Wool pile filler strip, aluminum backing		105	.076		3.02	4.29		7.31	10.10
5000	Two piece overlapping astragal, extruded aluminum retainer									
5010	Pile insert	1 Carp	60	.133	L.F.	3.17	7.50		10.67	15.45
5020	Vinyl bulb insert		60	.133		1.72	7.50		9.22	13.85
5040	Vinyl flap insert		60	.133		4.33	7.50		11.83	16.70
5060	Solid neoprene flap insert		60	.133		6.45	7.50		13.95	19.05
5080	Hypalon rubber flap insert		60	.133		6.40	7.50		13.90	19
5090	Snap on cover, pile insert		60	.133		9.60	7.50		17.10	22.50
5400	Magnetic aluminum, surface mounted		60	.133		27	7.50		34.50	41.50

For customer support on your Commercial Renovation Costs with RSMeans data, call 800.448.8182.

283

08 71 Door Hardware

08 71 21 – Astragals

08 71 21.10 Exterior Mouldings, Astragals

		Crew	Daily Output	Labor-Hours	Unit	Material	2022 Bare Costs Labor	Equipment	Total	Total Incl O&P
5500	Interlocking aluminum, 5/8" x 1" neoprene bulb insert	1 Carp	45	.178	L.F.	4.31	10		14.31	20.50
5600	Adjustable aluminum, 9/16" x 21/32", pile insert		45	.178		17.70	10		27.70	35.50
5800	Magnetic, adjustable, 9/16" x 21/32"		45	.178		23	10		33	41.50
6000	Two piece stile protection									
6010	Cloth backed rubber loop, 1" gap, nail on aluminum strips	1 Carp	45	.178	L.F.	4.42	10		14.42	21
6040	Screw on aluminum strips		45	.178		6.45	10		16.45	23
6100	1-1/2" gap, screw on aluminum extrusion		45	.178		5.75	10		15.75	22
6240	Vinyl fabric loop, slotted aluminum extrusion, 1" gap		45	.178		2.34	10		12.34	18.45
6300	1-1/4" gap		45	.178		6.10	10		16.10	22.50

08 71 25 – Door Weatherstripping

08 71 25.10 Mechanical Seals, Weatherstripping

		Crew	Daily Output	Labor-Hours	Unit	Material	2022 Bare Costs Labor	Equipment	Total	Total Incl O&P
0010	**MECHANICAL SEALS, WEATHERSTRIPPING**									
1000	Doors, wood frame, interlocking, for 3' x 7' door, zinc	1 Carp	3	2.667	Opng.	53.50	150		203.50	298
1100	Bronze		3	2.667		69.50	150		219.50	315
1300	6' x 7' opening, zinc		2	4		56.50	225		281.50	425
1400	Bronze		2	4		70	225		295	435
1700	Wood frame, spring type, bronze									
1800	3' x 7' door	1 Carp	7.60	1.053	Opng.	22.50	59.50		82	119
1900	6' x 7' door	"	7	1.143	"	31.50	64.50		96	137
2200	Metal frame, spring type, bronze									
2300	3' x 7' door	1 Carp	3	2.667	Opng.	51	150		201	295
2400	6' x 7' door	"	2.50	3.200	"	52	180		232	345
2500	For stainless steel, spring type, add					133%				
2700	Metal frame, extruded sections, 3' x 7' door, aluminum	1 Carp	3	2.667	Opng.	33.50	150		183.50	276
2800	Bronze	"	3	2.667	"	95.50	150		245.50	345
3100	6' x 7' door, aluminum	1 Carp	1.50	5.333	Opng.	41	300		341	520
3200	Bronze	"	1.50	5.333	"	148	300		448	640
3500	Threshold weatherstripping									
3650	Door sweep, flush mounted, aluminum	1 Carp	25	.320	Ea.	22	18		40	53
3700	Vinyl		25	.320		19.75	18		37.75	50
5000	Garage door bottom weatherstrip, 12' aluminum, clear		14	.571		24	32		56	77.50
5010	Bronze		14	.571		94	32		126	155
5050	Bottom protection, rubber		14	.571		55.50	32		87.50	112
5100	Threshold		14	.571		69	32		101	127
9000	Minimum labor/equipment charge		3	2.667	Job		150		150	239

08 75 Window Hardware

08 75 10 – Window Handles and Latches

08 75 10.10 Handles and Latches

		Crew	Daily Output	Labor-Hours	Unit	Material	2022 Bare Costs Labor	Equipment	Total	Total Incl O&P
0010	**HANDLES AND LATCHES**									
1000	Handles, surface mounted, aluminum	1 Carp	24	.333	Ea.	6.90	18.75		25.65	37.50
1020	Brass		24	.333		7.65	18.75		26.40	38.50
1040	Chrome		24	.333		9.10	18.75		27.85	40
1200	Window handles window crank ADA		24	.333		28	18.75		46.75	61
1500	Recessed, aluminum		12	.667		5.50	37.50		43	65.50
1520	Brass		12	.667		5.95	37.50		43.45	66
1540	Chrome		12	.667		5.15	37.50		42.65	65
2000	Latches, aluminum		20	.400		4.34	22.50		26.84	41
2020	Brass		20	.400		7.40	22.50		29.90	44
2040	Chrome		20	.400		6	22.50		28.50	42.50

08 75 Window Hardware

08 75 10 – Window Handles and Latches

08 75 10.10 Handles and Latches

	Crew	Daily Output	Labor-Hours	Unit	Material	2022 Bare Costs Labor	Equipment	Total	Total Incl O&P
9000 Minimum labor/equipment charge	1 Carp	6	1.333	Job		75		75	119

08 75 10.15 Window Opening Control

	Crew	Daily Output	Labor-Hours	Unit	Material	2022 Bare Costs Labor	Equipment	Total	Total Incl O&P
0010 **WINDOW OPENING CONTROL**									
0015 Window stops									
0020 For double-hung window	1 Carp	35	.229	Ea.	43.50	12.85		56.35	68.50
0030 Cam action for sliding window		35	.229		4.89	12.85		17.74	26
0040 Thumb screw for sliding window	↓	35	.229	↓	2.88	12.85		15.73	23.50
0100 Window guards, child safety bars for single or double-hung									
0110 14" to 17" wide max vert opening 26"	1 Carp	16	.500	Ea.	73.50	28		101.50	126
0120 17" to 23" wide max vert opening 26"		16	.500		84	28		112	137
0130 23" to 36" wide max vert opening 26"		16	.500		95.50	28		123.50	150
0140 35" to 58" wide max vert opening 26"		16	.500		111	28		139	167
0150 58" to 90" wide max vert opening 26"		16	.500		149	28		177	209
0160 73" to 120" wide max vert opening 26"	↓	16	.500	↓	261	28		289	330

08 75 30 – Window Weatherstripping

08 75 30.10 Mechanical Weather Seals

	Crew	Daily Output	Labor-Hours	Unit	Material	2022 Bare Costs Labor	Equipment	Total	Total Incl O&P
0010 **MECHANICAL WEATHER SEALS**, Window, double-hung, 3' x 5'									
0020 Zinc	1 Carp	7.20	1.111	Opng.	19.65	62.50		82.15	121
0100 Bronze		7.20	1.111		39.50	62.50		102	143
0200 Vinyl V strip		7	1.143		4.61	64.50		69.11	107
0500 As above but heavy duty, zinc		4.60	1.739		20	98		118	178
0600 Bronze	↓	4.60	1.739	↓	77	98		175	241
9000 Minimum labor/equipment charge	1 Clab	4.60	1.739	Job		79.50		79.50	126

08 79 Hardware Accessories

08 79 20 – Door Accessories

08 79 20.10 Door Hardware Accessories

	Crew	Daily Output	Labor-Hours	Unit	Material	2022 Bare Costs Labor	Equipment	Total	Total Incl O&P
0010 **DOOR HARDWARE ACCESSORIES**									
0140 Door bolt, surface, 4"	1 Carp	32	.250	Ea.	29.50	14.10		43.60	55
0160 Door latch	"	12	.667	"	29.50	37.50		67	92
0200 Sliding closet door									
0220 Track and hanger, single	1 Carp	10	.800	Ea.	88.50	45		133.50	169
0240 Double		8	1		114	56.50		170.50	215
0260 Door guide, single		48	.167		44	9.40		53.40	63.50
0280 Double		48	.167		3.45	9.40		12.85	18.70
0600 Deadbolt and lock cover plate, brass or stainless steel		30	.267		11.35	15		26.35	36.50
0620 Hole cover plate, brass or chrome		35	.229		12.25	12.85		25.10	34
2240 Mortise lockset, passage, lever handle		9	.889		112	50		162	203
4000 Security chain, standard		18	.444		15.80	25		40.80	57.50
4100 Deluxe	↓	18	.444	↓	62	25		87	109

For customer support on your Commercial Renovation Costs with RSMeans data, call 800.448.8182.

285

08 81 10 – Float Glass

08 81 10.10 Various Types and Thickness of Float Glass		Crew	Daily Output	Labor-Hours	Unit	Material	2022 Bare Costs Labor	Equipment	Total	Total Incl O&P
0010	**VARIOUS TYPES AND THICKNESS OF FLOAT GLASS**									
0020	3/16" plain	2 Glaz	130	.123	S.F.	7.15	6.65		13.80	18.40
0200	Tempered, clear		130	.123		21.50	6.65		28.15	34
0300	Tinted		130	.123		7.70	6.65		14.35	19
0600	1/4" thick, clear, plain		120	.133		10.10	7.20		17.30	22.50
0700	Tinted		120	.133		9.55	7.20		16.75	22
0800	Tempered, clear		120	.133		12.90	7.20		20.10	25.50
0900	Tinted		120	.133		26	7.20		33.20	40
1600	3/8" thick, clear, plain		75	.213		14.95	11.50		26.45	34.50
1700	Tinted		75	.213		20.50	11.50		32	41
1800	Tempered, clear		75	.213		28	11.50		39.50	49
1900	Tinted		75	.213		40.50	11.50		52	63
2200	1/2" thick, clear, plain		55	.291		19.35	15.70		35.05	46.50
2300	Tinted		55	.291		35.50	15.70		51.20	64
2400	Tempered, clear	2 Glaz	55	.291	S.F.	32	15.70		47.70	60.50
2500	Tinted		55	.291		28.50	15.70		44.20	56
2800	5/8" thick, clear, plain		45	.356		35.50	19.20		54.70	69.50
2900	Tempered, clear		45	.356		40.50	19.20		59.70	75
3200	3/4" thick, clear, plain		35	.457		47	24.50		71.50	90.50
3300	Tempered, clear		35	.457		56.50	24.50		81	101
3600	1" thick, clear, plain		30	.533		76	29		105	129
8900	For low emissivity coating for 3/16" & 1/4" only, add to above					18%				
9000	Minimum labor/equipment charge	1 Glaz	2	4	Job		216		216	345

08 81 17 – Fire Glass

08 81 17.10 Fire Resistant Glass

		Crew	Daily Output	Labor-Hours	Unit	Material	2022 Bare Costs Labor	Equipment	Total	Total Incl O&P
0010	**FIRE RESISTANT GLASS**									
0020	Fire glass minimum	2 Glaz	40	.400	S.F.	178	21.50		199.50	231
0030	Mid range		40	.400		236	21.50		257.50	294
0050	High end		40	.400		236	21.50		257.50	294

08 81 20 – Vision Panels

08 81 20.10 Full Vision

		Crew	Daily Output	Labor-Hours	Unit	Material	2022 Bare Costs Labor	Equipment	Total	Total Incl O&P
0010	**FULL VISION**, window system with 3/4" glass mullions									
0020	Up to 10' high	H-2	130	.185	S.F.	70	9.45		79.45	92
0100	10' to 20' high, minimum		110	.218		77	11.15		88.15	103
0150	Average		100	.240		80	12.30		92.30	108
0200	Maximum		80	.300		93	15.35		108.35	128
9000	Minimum labor/equipment charge	1 Glaz	2	4	Job		216		216	345

08 81 25 – Glazing Variables

08 81 25.10 Applications of Glazing

		Crew	Daily Output	Labor-Hours	Unit	Material	2022 Bare Costs Labor	Equipment	Total	Total Incl O&P
0010	**APPLICATIONS OF GLAZING** R088110-10									
0600	For glass replacement, add				S.F.		100%			
0700	For gasket settings, add				L.F.	6.05			6.05	6.65
0900	For sloped glazing, add				S.F.		26%			
2000	Fabrication, polished edges, 1/4" thick				Inch	.59			.59	.65
2100	1/2" thick					1.46			1.46	1.61
2500	Mitered edges, 1/4" thick					1.46			1.46	1.61
2600	1/2" thick					2.42			2.42	2.66

08 81 Glass Glazing

08 81 30 – Insulating Glass

08 81 30.10 Reduce Heat Transfer Glass

08 81 30.10 Reduce Heat Transfer Glass		Crew	Daily Output	Labor-Hours	Unit	Material	2022 Bare Costs Labor	Equipment	Total	Total Incl O&P
0010	**REDUCE HEAT TRANSFER GLASS**									
0015	2 lites 1/8" float, 1/2" thk under 15 S.F.									
0020	Clear	R088110-10 2 Glaz	95	.168	S.F.	28	9.10		37.10	45
0100	Tinted	G	95	.168		30.50	9.10		39.60	48
0200	2 lites 3/16" float, for 5/8" thk unit, 15 to 30 S.F., clear	G	90	.178		17.15	9.60		26.75	34
0400	1" thk, dbl. glazed, 1/4" float, 30 to 70 S.F., clear	G	75	.213		17.50	11.50		29	37.50
0500	Tinted	G 2 Glaz	75	.213	S.F.	29.50	11.50		41	51
2000	Both lites, light & heat reflective	G	85	.188		39.50	10.15		49.65	59.50
2500	Heat reflective, film inside, 1" thick unit, clear	G	85	.188		34.50	10.15		44.65	54
2600	Tinted	G	85	.188		32.50	10.15		42.65	52
3000	Film on weatherside, clear, 1/2" thick unit	G	95	.168		24.50	9.10		33.60	41.50
3100	5/8" thick unit	G	90	.178		21.50	9.60		31.10	39
3200	1" thick unit	G	85	.188		34	10.15		44.15	53.50
3350	Clear heat reflective film on inside	G 1 Glaz	50	.160		13.85	8.65		22.50	29
3360	Heat reflective film on inside, tinted	G	25	.320		14.50	17.30		31.80	43.50
3370	Heat reflective film on the inside, metalized	G	20	.400		21.50	21.50		43	58
5000	Spectrally selective film, on ext., blocks solar gain/allows 70% of light	G 2 Glaz	95	.168		15.25	9.10		24.35	31
9000	Minimum labor/equipment charge	1 Glaz	2	4	Job		216		216	345

08 81 40 – Plate Glass

08 81 40.10 Plate Glass

08 81 40.10 Plate Glass		Crew	Daily Output	Labor-Hours	Unit	Material	Labor	Equipment	Total	Total Incl O&P
0010	**PLATE GLASS** Twin ground, polished,									
0020	3/16" thick	2 Glaz	100	.160	S.F.	6.15	8.65		14.80	20.50
0100	1/4" thick		94	.170		8.40	9.20		17.60	24
0200	3/8" thick		60	.267		14.25	14.40		28.65	38.50
0300	1/2" thick		40	.400		28	21.50		49.50	65

08 81 55 – Window Glass

08 81 55.10 Sheet Glass

08 81 55.10 Sheet Glass		Crew	Daily Output	Labor-Hours	Unit	Material	Labor	Equipment	Total	Total Incl O&P
0010	**SHEET GLASS** (window), clear float, stops, putty bed									
0015	1/8" thick, clear float	2 Glaz	480	.033	S.F.	4.35	1.80		6.15	7.65
0500	3/16" thick, clear		480	.033		6.80	1.80		8.60	10.35
0600	Tinted		480	.033		7.70	1.80		9.50	11.35
0700	Tempered		480	.033		12.65	1.80		14.45	16.75
9000	Minimum labor/equipment charge		5	3.200	Job		173		173	274

08 81 65 – Wire Glass

08 81 65.10 Glass Reinforced With Wire

08 81 65.10 Glass Reinforced With Wire		Crew	Daily Output	Labor-Hours	Unit	Material	Labor	Equipment	Total	Total Incl O&P
0010	**GLASS REINFORCED WITH WIRE**									
0012	1/4" thick rough obscure	2 Glaz	135	.119	S.F.	25.50	6.40		31.90	38
1000	Polished wire, 1/4" thick, diamond, clear		135	.119		33.50	6.40		39.90	47
1500	Pinstripe, obscure		135	.119		54	6.40		60.40	69.50

08 83 Mirrors

08 83 13 – Mirrored Glass Glazing

08 83 13.10 Mirrors

		Crew	Daily Output	Labor-Hours	Unit	Material	2022 Bare Costs Labor	Equipment	Total	Total Incl O&P
0010	**MIRRORS**, No frames, wall type, 1/4" plate glass, polished edge									
0100	Up to 5 S.F.	2 Glaz	125	.128	S.F.	9.55	6.90		16.45	21.50
0200	Over 5 S.F.		160	.100		9.15	5.40		14.55	18.65
0500	Door type, 1/4" plate glass, up to 12 S.F.		160	.100		9.50	5.40		14.90	19
1000	Float glass, up to 10 S.F., 1/8" thick	2 Glaz	160	.100	S.F.	6.70	5.40		12.10	15.90
1100	3/16" thick		150	.107		8.65	5.75		14.40	18.65
1500	12" x 12" wall tiles, square edge, clear		195	.082		2.85	4.43		7.28	10.20
1600	Veined		195	.082		6.35	4.43		10.78	14
2000	1/4" thick, stock sizes, one way transparent		125	.128		23	6.90		29.90	36
2010	Bathroom, unframed, laminated		160	.100		16.40	5.40		21.80	26.50
2500	Tempered		160	.100		18.55	5.40		23.95	29

08 83 13.15 Reflective Glass

		Crew	Daily Output	Labor-Hours	Unit	Material	2022 Bare Costs Labor	Equipment	Total	Total Incl O&P
0010	**REFLECTIVE GLASS**									
0100	1/4" float with fused metallic oxide fixed [G]	2 Glaz	115	.139	S.F.	18.25	7.50		25.75	32
0500	1/4" float glass with reflective applied coating [G]	"	115	.139	"	39.50	7.50		47	55

08 84 Plastic Glazing

08 84 10 – Plexiglass Glazing

08 84 10.10 Plexiglass Acrylic

		Crew	Daily Output	Labor-Hours	Unit	Material	2022 Bare Costs Labor	Equipment	Total	Total Incl O&P
0010	**PLEXIGLASS ACRYLIC**, clear, masked,									
0020	1/8" thick, cut sheets	2 Glaz	170	.094	S.F.	15.50	5.10		20.60	25
0200	Full sheets		195	.082		3.42	4.43		7.85	10.80
0500	1/4" thick, cut sheets		165	.097		11.60	5.25		16.85	21
0600	Full sheets		185	.086		6.75	4.67		11.42	14.80
0900	3/8" thick, cut sheets		155	.103		10.25	5.55		15.80	20
1000	Full sheets		180	.089		11	4.80		15.80	19.70
1300	1/2" thick, cut sheets		135	.119		24.50	6.40		30.90	37
1400	Full sheets		150	.107		15.10	5.75		20.85	26
1700	3/4" thick, cut sheets		115	.139		37	7.50		44.50	53
1800	Full sheets		130	.123		21	6.65		27.65	33.50
2100	1" thick, cut sheets		105	.152		54	8.25		62.25	72.50
2200	Full sheets		125	.128		20.50	6.90		27.40	34
3000	Colored, 1/8" thick, cut sheets		170	.094		19.65	5.10		24.75	29.50
3200	Full sheets		195	.082		11.30	4.43		15.73	19.50
3500	1/4" thick, cut sheets		165	.097		21.50	5.25		26.75	32.50
3600	Full sheets		185	.086		16.15	4.67		20.82	25
4000	Mirrors, untinted, cut sheets, 1/8" thick		185	.086		11.40	4.67		16.07	19.95
4200	1/4" thick		180	.089		20.50	4.80		25.30	30

08 84 20 – Polycarbonate

08 84 20.10 Thermoplastic

		Crew	Daily Output	Labor-Hours	Unit	Material	2022 Bare Costs Labor	Equipment	Total	Total Incl O&P
0010	**THERMOPLASTIC**, clear, masked, cut sheets									
0020	1/8" thick	2 Glaz	170	.094	S.F.	4.96	5.10		10.06	13.50
0500	3/16" thick		165	.097		7.45	5.25		12.70	16.50
1000	1/4" thick		155	.103		8.95	5.55		14.50	18.70
1500	3/8" thick		150	.107		16.25	5.75		22	27
9000	Minimum labor/equipment charge	1 Glaz	2	4	Job		216		216	345

08 85 Glazing Accessories

08 85 10 – Miscellaneous Glazing Accessories

08 85 10.10 Glazing Gaskets	Crew	Daily Output	Labor-Hours	Unit	Material	2022 Bare Costs Labor	Equipment	Total	Total Incl O&P
0010 **GLAZING GASKETS**, Neoprene for glass tongued mullion									
0015 1/4" glass	2 Glaz	200	.080	L.F.	4	4.32		8.32	11.25
0020 3/8"		200	.080		1.14	4.32		5.46	8.10
0040 1/2"		200	.080		1.45	4.32		5.77	8.45
0060 3/4"		180	.089		1.86	4.80		6.66	9.65
0080 1"		180	.089		1.74	4.80		6.54	9.50
1000 Glazing compound, wood	1 Glaz	58	.138			7.45		7.45	11.80
1005 Glazing compound, per window, up to 30 L.F.									
1006 Glazing compound, per window, up to 30 L.F.				Ea.	.12			.12	.13
1020 Metal	1 Glaz	58	.138	L.F.		7.45		7.45	11.80

08 85 20 – Structural Glazing Sealants

08 85 20.10 Structural Glazing Adhesives

	Crew	Daily Output	Labor-Hours	Unit	Material	Labor	Equipment	Total	Total Incl O&P
0010 **STRUCTURAL GLAZING ADHESIVES**									
0050 Structural glazing adhesive, 1/4" x 1/4" joint seal	1 Glaz	200	.040	L.F.	.34	2.16		2.50	3.80

08 87 Glazing Surface Films

08 87 13 – Solar Control Films

08 87 13.10 Solar Films On Glass

	Crew	Daily Output	Labor-Hours	Unit	Material	Labor	Equipment	Total	Total Incl O&P
0010 **SOLAR FILMS ON GLASS** (glass not included)									
1000 Bronze, 20% VLT	H-2	180	.133	S.F.	1.82	6.85		8.67	12.85
1020 50% VLT		180	.133		1.09	6.85		7.94	12.05
1050 Neutral, 20% VLT		180	.133		.72	6.85		7.57	11.65
1100 Silver, 15% VLT		180	.133		.86	6.85		7.71	11.80
1120 35% VLT		180	.133		1.19	6.85		8.04	12.15
1150 68% VLT		180	.133		.67	6.85		7.52	11.60
3000 One way mirror with night vision 5% in and 15% out VLT		180	.133		3.48	6.85		10.33	14.70

08 87 16 – Glass Safety Films

08 87 16.10 Safety Films On Glass

	Crew	Daily Output	Labor-Hours	Unit	Material	Labor	Equipment	Total	Total Incl O&P
0010 **SAFETY FILMS ON GLASS** (glass not included)									
0015 Safety film helps hold glass together when broken									
0050 Safety film, clear, 2 mil	H-2	180	.133	S.F.	1.69	6.85		8.54	12.70
0100 Safety film, clear, 4 mil		180	.133		1.60	6.85		8.45	12.60
0110 Safety film, tinted, 4 mil		180	.133		1.06	6.85		7.91	12
0150 Safety film, black tint, 8 mil		180	.133		1.95	6.85		8.80	13

08 87 26 – Bird Control Film

08 87 26.10 Bird Control Film

	Crew	Daily Output	Labor-Hours	Unit	Material	Labor	Equipment	Total	Total Incl O&P
0010 **BIRD CONTROL FILM**									
0050 Patterned, adhered to glass	2 Glaz	180	.089	S.F.	6.50	4.80		11.30	14.75
0200 Decals small, adhered to glass	1 Glaz	50	.160	Ea.	1.85	8.65		10.50	15.75
0250 Decals large, adhered to glass	1 Glaz	50	.160	Ea.	8.55	8.65		17.20	23
0300 Decals set of 4, adhered to glass		25	.320	Set	7.90	17.30		25.20	36
0400 Bird control film tape		10	.800	Roll	5.30	43		48.30	74.50

08 87 33 – Electrically Tinted Window Film

08 87 33.20 Window Film

	Crew	Daily Output	Labor-Hours	Unit	Material	Labor	Equipment	Total	Total Incl O&P
0010 **WINDOW FILM** adhered on glass (glass not included)									
0015 Film is pre-wired and can be trimmed									
0100 Window film 4 S.F. adhered on glass	1 Glaz	200	.040	S.F.	254	2.16		256.16	282
0120 6 S.F.		180	.044		380	2.40		382.40	425
0160 9 S.F.		170	.047		570	2.54		572.54	630

08 87 Glazing Surface Films

08 87 33 – Electrically Tinted Window Film

08 87 33.20 Window Film		Crew	Daily Output	Labor-Hours	Unit	Material	2022 Bare Costs Labor	Equipment	Total	Total Incl O&P
0180	10 S.F.	1 Glaz	150	.053	S.F.	158	2.88		160.88	179
0200	12 S.F.		144	.056		760	3		763	840
0220	14 S.F.		140	.057		885	3.09		888.09	980
0240	16 S.F.		128	.063		1,025	3.38		1,028.38	1,125
0260	18 S.F.		126	.063		1,150	3.43		1,153.43	1,250
0280	20 S.F.		120	.067		1,275	3.60		1,278.60	1,400

08 87 53 – Security Films On Glass

08 87 53.10 Security Film Adhered On Glass

		Crew	Daily Output	Labor-Hours	Unit	Material	2022 Bare Costs Labor	Equipment	Total	Total Incl O&P
0010	**SECURITY FILM ADHERED ON GLASS** (glass not included)									
0020	Security film, clear, 7 mil	1 Glaz	200	.040	S.F.	2.42	2.16		4.58	6.10
0030	8 mil		200	.040		2.64	2.16		4.80	6.35
0040	9 mil		200	.040		1.27	2.16		3.43	4.83
0050	10 mil		200	.040		2.19	2.16		4.35	5.85
0060	12 mil		200	.040		3.21	2.16		5.37	6.95
0075	15 mil		200	.040		2.59	2.16		4.75	6.30
0100	Security film, sealed with structural adhesive, 7 mil		180	.044		2.43	2.40		4.83	6.50
0110	8 mil		180	.044		2.65	2.40		5.05	6.75
0140	14 mil		180	.044		4.39	2.40		6.79	8.65

08 88 Special Function Glazing

08 88 52 – Prefabricated Glass Block Windows

08 88 52.10 Prefabricated Glass Block Windows

		Crew	Daily Output	Labor-Hours	Unit	Material	2022 Bare Costs Labor	Equipment	Total	Total Incl O&P
0010	**PREFABRICATED GLASS BLOCK WINDOWS**									
0015	Includes frame and silicone seal									
0020	Glass Block Window 16" x 8"	2 Glaz	6	2.667	Ea.	153	144		297	395
0025	16" x 16"		6	2.667		191	144		335	440
0030	16" x 24"		6	2.667		224	144		368	475
0050	16" x 48"		4	4		325	216		541	705
0100	Glass Block Window 24" x 8"		6	2.667		180	144		324	425
0110	24" x 16"		6	2.667		231	144		375	480
0120	24" x 24"		6	2.667		291	144		435	550
0150	24" x 48"		6	2.667		430	144		574	700
0200	Glass Block Window 32" x 8"	2 Glaz	6	2.667	Ea.	181	144		325	425
0210	32" x 16"		6	2.667		264	144		408	520
0220	32" x 24"		4	4		330	216		546	710
0250	32" x 48"		4	4		520	216		736	920
0300	Glass Block Window 40" x 8"		6	2.667		230	144		374	480
0310	40" x 16"		6	2.667		300	144		444	560
0320	40" x 24"		6	2.667		370	144		514	640
0350	40" x 48"		6	2.667		615	144		759	905
0400	Glass Block Window 48" x 8"		6	2.667		272	144		416	525
0410	48" x 16"		6	2.667		355	144		499	620
0420	48" x 24"		6	2.667		430	144		574	700
0450	48" x 48"		6	2.667		695	144		839	995
0500	Glass Block Window 56" x 8"		6	2.667		274	144		418	530
0510	56" x 16"		4	4		405	216		621	790
0520	56" x 24"		4	4		500	216		716	895
0550	56" x 48"		4	4		790	216		1,006	1,225

08 88 Special Function Glazing

08 88 52 – Prefabricated Glass Block Windows

08 88 52.20 Prefab. Glass Block Windows Hurricane Resistant	Crew	Daily Output	Labor-Hours	Unit	Material	2022 Bare Costs Labor	Equipment	Total	Total Incl O&P
0010 **PREFABRICATED GLASS BLOCK WINDOWS HURRICANE RESISTANT**									
0015 Includes frame and silicone seal									
3020 Glass Block Window 16" x 16"	2 Glaz	6	2.667	Ea.	430	144		574	700
3030 16" x 24"		6	2.667		505	144		649	785
3040 16" x 32"		4	4		590	216		806	995
3050 16" x 40"		4	4		625	216		841	1,025
3060 16" x 48"		4	4		660	216		876	1,075
3070 16" x 56"		4	4		820	216		1,036	1,250
3080 Glass Block Window 24" x 24"		6	2.667		585	144		729	870
3090 24" x 32"		4	4		660	216		876	1,075
3100 24" x 40"		4	4		835	216		1,051	1,250
3110 24" x 48"		4	4		865	216		1,081	1,300
3120 24" x 56"		4	4		915	216		1,131	1,350
3130 Glass Block Window 32" x 32"		4	4		740	216		956	1,150
3140 32" x 40"		4	4		565	216		781	970
3150 32" x 48"		4	4		980	216		1,196	1,425
3160 32" x 56"		4	4		870	216		1,086	1,300
3200 Glass Block Window 40" x 40"		4	4		980	216		1,196	1,425
3210 40" x 48"		3	5.333		1,175	288		1,463	1,750
3220 40" x 56"		3	5.333		1,325	288		1,613	1,900
3260 Glass Block Window 48" x 48"		3	5.333		1,375	288		1,663	1,975
3270 48" x 56"		3	5.333		1,550	288		1,838	2,150
3310 Glass Block Window 56" x 56"	2 Glaz	3	5.333	Ea.	1,700	288		1,988	2,325

08 88 52.30 Prefabricated Acrylic Block Casement Windows

	Crew	Daily Output	Labor-Hours	Unit	Material	2022 Bare Costs Labor	Equipment	Total	Total Incl O&P
0010 **PREFABRICATED ACRYLIC BLOCK CASEMENT WINDOWS**									
0015 Windows include frame and nail fin.									
3050 6" x 6" x 1-3/4" Block Casement Window 2 x 2 Pattern	2 Glaz	6	2.667	Ea.	450	144		594	725
3055 2 x 3		6	2.667		470	144		614	745
3060 2 x 4		6	2.667		510	144		654	790
3065 2 x 5		6	2.667		545	144		689	830
3070 2 x 6		6	2.667		600	144		744	890
3075 2 x 7		4	4		710	216		926	1,125
3080 2 x 8		4	4		740	216		956	1,150
3085 2 x 9		4	4		815	216		1,031	1,250
3090 3 x 3		4	4		490	216		706	885
3095 3 x 4		4	4		540	216		756	940
3100 3 x 5		4	4		610	216		826	1,025
3110 3 x 6		4	4		650	216		866	1,050
3115 3 x 7		4	4		735	216		951	1,150
3120 3 x 8		4	4		815	216		1,031	1,250
3125 3 x 9		3	5.333		880	288		1,168	1,425
3130 4 x 4		3	5.333		635	288		923	1,150
3135 4 x 5		3	5.333		675	288		963	1,200
3140 4 x 6		3	5.333		730	288		1,018	1,250
3145 4 x 7		3	5.333		765	288		1,053	1,300
3150 4 x 8		3	5.333		880	288		1,168	1,425
3500 8" x 8" x 1-3/4" Block Casement Window 2 x 2 Pattern		6	2.667		415	144		559	685
3510 2 x 3		6	2.667		460	144		604	735
3515 2 x 4		6	2.667		515	144		659	800
3520 2 x 5		6	2.667		590	144		734	880
3525 2 x 6		4	4		660	216		876	1,075
3530 2 x 7		4	4		690	216		906	1,100

For customer support on your Commercial Renovation Costs with RSMeans data, call 800.448.8182.

291

08 88 Special Function Glazing

08 88 52 – Prefabricated Glass Block Windows

08 88 52.30 Prefabricated Acrylic Block Casement Windows	Crew	Daily Output	Labor-Hours	Unit	Material	2022 Bare Costs Labor	Equipment	Total	Total Incl O&P
3535	2 x 8	2 Glaz	3	5.333	Ea.	755	288		1,043	1,275
3540	2 x 9		3	5.333		800	288		1,088	1,325
3545	2 x 10		3	5.333		795	288		1,083	1,325
3550	2 x 11		3	5.333		820	288		1,108	1,350
3555	3 x 3		4	4		515	216		731	910
3560	3 x 4		4	4		570	216		786	975
3565	3 x 5		4	4		685	216		901	1,100
3570	3 x 6		4	4		745	216		961	1,175
3575	3 x 7		4	4		855	216		1,071	1,300
3580	3 x 8		3	5.333		950	288		1,238	1,500
3585	3 x 9		3	5.333		875	288		1,163	1,425
3590	4 x 4		3	5.333		705	288		993	1,225
3595	4 x 5		3	5.333		860	288		1,148	1,400
3600	4 x 6		3	5.333		925	288		1,213	1,475

08 88 56 – Ballistics-Resistant Glazing

08 88 56.10 Laminated Glass

0010	LAMINATED GLASS									
0020	Clear float .03" vinyl 1/4" thick	2 Glaz	90	.178	S.F.	13.35	9.60		22.95	30
0100	3/8" thick		78	.205		32	11.10		43.10	52.50
0200	.06" vinyl, 1/2" thick		65	.246		28	13.30		41.30	52
1000	5/8" thick		90	.178		39.50	9.60		49.10	59
2000	Bullet-resisting, 1-3/16" thick, to 15 S.F.		16	1		132	54		186	231
2100	Over 15 S.F.		16	1		131	54		185	230
2200	2" thick, to 15 S.F.		10	1.600		150	86.50		236.50	300
2300	Over 15 S.F.		10	1.600		133	86.50		219.50	284
2500	2-1/4" thick, to 15 S.F.		12	1.333		206	72		278	340
2600	Over 15 S.F.		12	1.333		195	72		267	330
2700	Level 2 (.357 magnum), NIJ and UL		12	1.333		56	72		128	176
2750	Level 3A (.44 magnum) NIJ, UL 3		12	1.333		60.50	72		132.50	181
2800	Level 4 (AK-47) NIJ, UL 7 & 8		12	1.333		82.50	72		154.50	205
2850	Level 5 (M-16) UL		12	1.333		81.50	72		153.50	204
2900	Level 3 (7.62 Armor Piercing) NIJ, UL 4 & 5		12	1.333		96.50	72		168.50	220

08 91 Louvers

08 91 16 – Operable Wall Louvers

08 91 16.10 Movable Blade Louvers

0010	MOVABLE BLADE LOUVERS									
0100	PVC, commercial grade, 12" x 12"	1 Shee	20	.400	Ea.	105	27		132	158
0110	16" x 16"		14	.571		123	38.50		161.50	195
0120	18" x 18"		14	.571		126	38.50		164.50	199
0130	24" x 24"		14	.571		181	38.50		219.50	259
0140	30" x 30"		12	.667		217	45		262	310
0150	36" x 36"		12	.667		256	45		301	350
0300	Stainless steel, commercial grade, 12" x 12"		20	.400		325	27		352	395
0310	16" x 16"		14	.571		400	38.50		438.50	500
0320	18" x 18"		14	.571		340	38.50		378.50	435
0330	20" x 20"		14	.571		390	38.50		428.50	490
0340	24" x 24"		14	.571		465	38.50		503.50	570
0350	30" x 30"		12	.667		520	45		565	640
0360	36" x 36"		12	.667		630	45		675	765

08 91 19 - Fixed Louvers

08 91 19.10 Aluminum Louvers	Crew	Daily Output	Labor-Hours	Unit	Material	2022 Bare Costs Labor	Equipment	Total	Total Incl O&P
0010 **ALUMINUM LOUVERS**									
0020 Aluminum with screen, residential, 8" x 8"	1 Carp	38	.211	Ea.	22	11.85		33.85	43
0100 12" x 12"		38	.211		28	11.85		39.85	50
0200 12" x 18"		35	.229		63	12.85		75.85	90
0250 14" x 24"		30	.267		43	15		58	71
0300 18" x 24"		27	.296		34	16.70		50.70	64
0500 24" x 30"		24	.333		100	18.75		118.75	140
0700 Triangle, adjustable, small		20	.400		95.50	22.50		118	141
0800 Large		15	.533		132	30		162	194
2100 Round, aluminum, 3/4" deep, 1" diameter		85	.094		1.37	5.30		6.67	9.90
2150 3" diameter		60	.133		3.24	7.50		10.74	15.50
2200 4" diameter		50	.160		7.05	9		16.05	22
2250 6" diameter		30	.267		5.10	15		20.10	29.50
3000 PVC, commercial grade, 12" x 12"	1 Shee	20	.400		124	27		151	179
3010 12" x 18"		20	.400		140	27		167	196
3020 12" x 24"		20	.400		188	27		215	249
3030 14" x 24"		20	.400		197	27		224	258
3100 Aluminum, commercial grade, 12" x 12"		20	.400		215	27		242	279
3110 24" x 24"		16	.500		297	34		331	380
3120 24" x 36"		16	.500		425	34		459	520
3130 36" x 24"		16	.500		440	34		474	540
3140 36" x 36"		14	.571		550	38.50		588.50	665
3150 36" x 48"		10	.800		605	54		659	750
3160 48" x 36"		10	.800		635	54		689	780
3170 48" x 48"		10	.800		640	54		694	785
3180 60" x 48"		8	1		1,175	67.50		1,242.50	1,400
3190 60" x 60"		8	1		950	67.50		1,017.50	1,150

08 91 19.20 Steel Louvers

	Crew	Daily Output	Labor-Hours	Unit	Material	Labor	Equipment	Total	Total Incl O&P
0010 **STEEL LOUVERS**									
3300 Galvanized steel, fixed blades, commercial grade, 18" x 18"	1 Shee	20	.400	Ea.	206	27		233	269
3310 24" x 24"		20	.400		277	27		304	345
3320 24" x 36"		16	.500		360	34		394	450
3330 36" x 24"		16	.500		370	34		404	460
3340 36" x 36"		14	.571		455	38.50		493.50	560
3350 36" x 48"		10	.800		535	54		589	670
3360 48" x 36"		10	.800		520	54		574	655
3370 48" x 48"		10	.800		645	54		699	795
3380 60" x 48"		10	.800		705	54		759	860
3390 60" x 60"		10	.800		845	54		899	1,025

08 91 26 - Door Louvers

08 91 26.10 Steel Louvers, 18 Gauge, Fixed Blade

	Crew	Daily Output	Labor-Hours	Unit	Material	Labor	Equipment	Total	Total Incl O&P
0010 **STEEL LOUVERS, 18 GAUGE, FIXED BLADE**									
0050 12" x 12", with powder coat	1 Carp	20	.400	Ea.	82.50	22.50		105	127
0055 18" x 12"		20	.400		85.50	22.50		108	130
0060 18" x 18"		20	.400		95	22.50		117.50	141
0065 24" x 12"		20	.400		99	22.50		121.50	145
0070 24" x 18"		20	.400		104	22.50		126.50	151
0075 24" x 24"		20	.400		132	22.50		154.50	181
0100 12" x 12", galvanized		20	.400		56	22.50		78.50	97.50
0105 18" x 12"		20	.400		63	22.50		85.50	106
0115 24" x 12"		20	.400		90.50	22.50		113	136

08 91 Louvers

08 91 26 – Door Louvers

08 91 26.10 Steel Louvers, 18 Gauge, Fixed Blade		Crew	Daily Output	Labor-Hours	Unit	Material	2022 Bare Costs Labor	Equipment	Total	Total Incl O&P
0125	24" x 24"	1 Carp	20	.400	Ea.	138	22.50		160.50	188
0300	6" x 12", painted		20	.400		37.50	22.50		60	77.50
0320	10" x 16"		20	.400		59	22.50		81.50	101
0340	12" x 22"		20	.400		87	22.50		109.50	132
0360	14" x 14"		20	.400		87.50	22.50		110	133
0400	18" x 18"		20	.400		128	22.50		150.50	177
0420	20" x 26"		20	.400		119	22.50		141.50	167
0440	22" x 22"		20	.400		112	22.50		134.50	160
0460	26" x 26"		20	.400		223	22.50		245.50	281
3000	Fire rated									
3010	12" x 12"	1 Carp	10	.800	Ea.	148	45		193	234
3050	18" x 18"		10	.800		216	45		261	310
3100	24" x 12"		10	.800		227	45		272	320
3120	24" x 18"		10	.800		243	45		288	340
3130	24" x 24"		10	.800		242	45		287	340

08 91 26.20 Aluminum Louvers

		Crew	Daily Output	Labor-Hours	Unit	Material	2022 Bare Costs Labor	Equipment	Total	Total Incl O&P
0010	**ALUMINUM LOUVERS**, 18 ga., fixed blade, clear anodized									
0050	6" x 12"	1 Carp	20	.400	Ea.	61	22.50		83.50	103
0060	10" x 10"		20	.400		71	22.50		93.50	114
0065	10" x 14"		20	.400		80	22.50		102.50	124
0080	12" x 22"		20	.400		112	22.50		134.50	159
0090	14" x 14"		20	.400		100	22.50		122.50	146
0100	18" x 10"		20	.400		83.50	22.50		106	128
0110	18" x 18"		20	.400		146	22.50		168.50	197
0120	22" x 22"		20	.400		181	22.50		203.50	235
0130	26" x 26"		20	.400		232	22.50		254.50	291

08 95 Vents

08 95 13 – Soffit Vents

08 95 13.10 Wall Louvers

		Crew	Daily Output	Labor-Hours	Unit	Material	2022 Bare Costs Labor	Equipment	Total	Total Incl O&P
0010	**WALL LOUVERS**									
2340	Baked enamel finish	1 Carp	200	.040	L.F.	6.10	2.25		8.35	10.35
2400	Under eaves vent, aluminum, mill finish, 16" x 4"	1 Carp	48	.167	Ea.	1.71	9.40		11.11	16.80
2500	16" x 8"	"	48	.167	"	1.89	9.40		11.29	17

08 95 16 – Wall Vents

08 95 16.10 Louvers

		Crew	Daily Output	Labor-Hours	Unit	Material	2022 Bare Costs Labor	Equipment	Total	Total Incl O&P
0010	**LOUVERS**									
0020	Redwood, 2'-0" diameter, full circle	1 Carp	16	.500	Ea.	96	28		124	151
0100	Half circle		16	.500		233	28		261	300
0200	Octagonal		16	.500		218	28		246	285
0300	Triangular, 5/12 pitch, 5'-0" at base		16	.500		665	28		693	775
1000	Rectangular, 1'-4" x 1'-3"		16	.500		29	28		57	76.50
1100	Rectangular, 1'-4" x 1'-8"		16	.500		45	28		73	94
1200	1'-4" x 2'-2"		15	.533		51.50	30		81.50	104
1300	1'-9" x 2'-2"		15	.533		90	30		120	147
1400	2'-3" x 2'-2"		14	.571		102	32		134	163
1700	2'-4" x 2'-11"		13	.615		102	34.50		136.50	167
2000	Aluminum, 12" x 16"		25	.320		24.50	18		42.50	55.50
2010	16" x 20"		25	.320		36.50	18		54.50	68.50
2020	24" x 30"		25	.320		65	18		83	100

For customer support on your Commercial Renovation Costs with RSMeans data, call 800.448.8182.

08 95 16.10 Louvers		Crew	Daily Output	Labor-Hours	Unit	Material	2022 Bare Costs Labor	Equipment	Total	Total Incl O&P
2100	6' triangle	1 Carp	12	.667	Ea.	198	37.50		235.50	277
3100	Round, 2'-2" diameter		16	.500		200	28		228	266
7000	Vinyl gable vent, 8" x 8"		38	.211		51.50	11.85		63.35	75.50
7020	12" x 12"		38	.211		33.50	11.85		45.35	55.50
7080	12" x 18"		35	.229		52	12.85		64.85	77.50
7200	18" x 24"		30	.267		62.50	15		77.50	92.50
9000	Minimum labor/equipment charge		3.50	2.286	Job		129		129	204

Division Notes

	CREW	DAILY OUTPUT	LABOR-HOURS	UNIT	BARE COSTS				TOTAL INCL O&P
					MAT.	LABOR	EQUIP.	TOTAL	

Estimating Tips
General
- Room Finish Schedule: A complete set of plans should contain a room finish schedule. If one is not available, it would be well worth the time and effort to obtain one.

09 20 00 Plaster and Gypsum Board
- Lath is estimated by the square yard plus a 5% allowance for waste. Furring, channels, and accessories are measured by the linear foot. An extra foot should be allowed for each accessory miter or stop.
- Plaster is also estimated by the square yard. Deductions for openings vary by preference, from zero deduction to 50% of all openings over 2 feet in width. The estimator should allow one extra square foot for each linear foot of horizontal interior or exterior angle located below the ceiling level. Also, double the areas of small radius work.
- Drywall accessories, studs, track, and acoustical caulking are all measured by the linear foot. Drywall taping is figured by the square foot. Gypsum wallboard is estimated by the square foot. No material deductions should be made for door or window openings under 32 S.F.

09 60 00 Flooring
- Tile and terrazzo areas are taken off on a square foot basis. Trim and base materials are measured by the linear foot. Accent tiles are listed per each. Two basic methods of installation are used. Mud set is approximately 30% more expensive than thin set.

The cost of grout is included with tile unit price lines unless otherwise noted. In terrazzo work, be sure to include the linear footage of embedded decorative strips, grounds, machine rubbing, and power cleanup.
- Wood flooring is available in strip, parquet, or block configuration. The latter two types are set in adhesives with quantities estimated by the square foot. The laying pattern will influence labor costs and material waste. In addition to the material and labor for laying wood floors, the estimator must make allowances for sanding and finishing these areas, unless the flooring is prefinished.
- Sheet flooring is measured by the square yard. Roll widths vary, so consideration should be given to use the most economical width, as waste must be figured into the total quantity. Consider also the installation methods available—direct glue down or stretched. Direct glue-down installation is assumed with sheet carpet unit price lines unless otherwise noted.

09 70 00 Wall Finishes
- Wall coverings are estimated by the square foot. The area to be covered is measured—length by height of the wall above the baseboards—to calculate the square footage of each wall. This figure is divided by the number of square feet in the single roll which is being used. Deduct, in full, the areas of openings such as doors and windows. Where a pattern match is required allow 25–30% waste.

09 80 00 Acoustic Treatment
- Acoustical systems fall into several categories. The takeoff of these materials should be by the square foot of area with a 5% allowance for waste. Do not forget about scaffolding, if applicable, when estimating these systems.

09 90 00 Painting and Coating
- New line items created for cut-ins with reference diagram.
- A major portion of the work in painting involves surface preparation. Be sure to include cleaning, sanding, filling, and masking costs in the estimate.
- Protection of adjacent surfaces is not included in painting costs. When considering the method of paint application, an important factor is the amount of protection and masking required. These must be estimated separately and may be the determining factor in choosing the method of application.

Reference Numbers
Reference numbers are shown at the beginning of some major classifications. These numbers refer to related items in the Reference Section. The reference information may be an estimating procedure, an alternate pricing method, or technical information.

Note: Not all subdivisions listed here necessarily appear. ∎

09 01 70 – Maintenance of Wall Finishes

09 01 70.10 Gypsum Wallboard Repairs	Crew	Daily Output	Labor-Hours	Unit	Material	2022 Bare Costs Labor	Equipment	Total	Total Incl O&P
0010 **GYPSUM WALLBOARD REPAIRS**									
0100 Fill and sand, pin/nail holes	1 Carp	960	.008	Ea.		.47		.47	.75
0110 Screw head pops		480	.017			.94		.94	1.49
0120 Dents, up to 2" square		48	.167		.01	9.40		9.41	14.90
0130 2" to 4" square		24	.333		.04	18.75		18.79	30
0140 Cut square, patch, sand and finish, holes, up to 2" square		12	.667		.03	37.50		37.53	59.50
0150 2" to 4" square		11	.727		.07	41		41.07	65
0160 4" to 8" square		10	.800		.19	45		45.19	71.50
0170 8" to 12" square		8	1		.38	56.50		56.88	90
0180 12" to 32" square		6	1.333		1.36	75		76.36	120
0210 16" by 48"		5	1.600		2.25	90		92.25	145
0220 32" by 48"	1 Carp	4	2	Ea.	3.56	113		116.56	183
0230 48" square		3.50	2.286		4.98	129		133.98	209
0240 60" square		3.20	2.500		7.60	141		148.60	232
0500 Skim coat surface with joint compound		1600	.005	S.F.	.04	.28		.32	.49
0510 Prepare, retape and refinish joints		60	.133	L.F.	.75	7.50		8.25	12.80
9000 Minimum labor/equipment charge		2	4	Job		225		225	360

09 01 90 – Maintenance of Painting and Coating

09 01 90.92 Sanding

	Crew	Daily Output	Labor-Hours	Unit	Material	2022 Bare Costs Labor	Equipment	Total	Total Incl O&P
0010 **SANDING** and puttying interior trim, compared to									
0100 Painting 1 coat, on quality work				L.F.		100%			
0300 Medium work						50%			
0400 Industrial grade						25%			
0500 Surface protection, placement and removal									
0510 Surface protection, placement and removal, basic drop cloths	1 Pord	6400	.001	S.F.		.06		.06	.09
0520 Masking with paper		800	.010		.07	.48		.55	.84
0530 Volume cover up (using plastic sheathing or building paper)		16000	.001			.02		.02	.04

09 01 90.93 Exterior Surface Preparation

	Crew	Daily Output	Labor-Hours	Unit	Material	2022 Bare Costs Labor	Equipment	Total	Total Incl O&P
0010 **EXTERIOR SURFACE PREPARATION**									
0015 Doors, per side, not incl. frames or trim									
0020 Scrape & sand									
0030 Wood, flush	1 Pord	616	.013	S.F.		.62		.62	.98
0040 Wood, detail		496	.016			.77		.77	1.22
0050 Wood, louvered		280	.029			1.37		1.37	2.16
0060 Wood, overhead		616	.013			.62		.62	.98
0070 Wire brush									
0080 Metal, flush	1 Pord	640	.013	S.F.		.60		.60	.95
0090 Metal, detail		520	.015			.74		.74	1.16
0100 Metal, louvered		360	.022			1.06		1.06	1.68
0110 Metal or fibr., overhead		640	.013			.60		.60	.95
0120 Metal, roll up		560	.014			.68		.68	1.08
0130 Metal, bulkhead		640	.013			.60		.60	.95
0140 Power wash, based on 2500 lb. operating pressure									
0150 Metal, flush	A-1H	2240	.004	S.F.		.16	.04	.20	.30
0160 Metal, detail		2120	.004			.17	.04	.21	.31
0170 Metal, louvered		2000	.004			.18	.04	.22	.33
0180 Metal or fibr., overhead		2400	.003			.15	.03	.18	.28
0190 Metal, roll up		2400	.003			.15	.03	.18	.28
0200 Metal, bulkhead		2200	.004			.17	.04	.21	.30
0400 Windows, per side, not incl. trim									
0410 Scrape & sand									
0420 Wood, 1-2 lite	1 Pord	320	.025	S.F.		1.20		1.20	1.89

09 01 90.93 Exterior Surface Preparation

	Crew	Daily Output	Labor-Hours	Unit	Material	2022 Bare Costs Labor	Equipment	Total	Total Incl O&P
0430 Wood, 3-6 lite	1 Pord	280	.029	S.F.		1.37		1.37	2.16
0440 Wood, 7-10 lite		240	.033			1.59		1.59	2.52
0450 Wood, 12 lite		200	.040			1.91		1.91	3.02
0460 Wood, Bay/Bow		320	.025			1.20		1.20	1.89
0470 Wire brush									
0480 Metal, 1-2 lite	1 Pord	480	.017	S.F.		.80		.80	1.26
0490 Metal, 3-6 lite		400	.020			.96		.96	1.51
0500 Metal, Bay/Bow		480	.017			.80		.80	1.26
0510 Power wash, based on 2500 lb. operating pressure									
0520 1-2 lite	A-1H	4400	.002	S.F.		.08	.02	.10	.15
0530 3-6 lite		4320	.002			.08	.02	.10	.15
0540 7-10 lite		4240	.002			.09	.02	.11	.16
0550 12 lite		4160	.002			.09	.02	.11	.16
0560 Bay/Bow		4400	.002			.08	.02	.10	.15
0600 Siding, scrape and sand, light=10-30%, med.=30-70%									
0610 Heavy=70-100% of surface to sand									
0650 Texture 1-11, light	1 Pord	480	.017	S.F.		.80		.80	1.26
0660 Med.		440	.018			.87		.87	1.37
0670 Heavy		360	.022			1.06		1.06	1.68
0680 Wood shingles, shakes, light		440	.018			.87		.87	1.37
0690 Med.		360	.022			1.06		1.06	1.68
0700 Heavy		280	.029			1.37		1.37	2.16
0710 Clapboard, light		520	.015			.74		.74	1.16
0720 Med.		480	.017			.80		.80	1.26
0730 Heavy		400	.020			.96		.96	1.51
0740 Wire brush									
0750 Aluminum, light	1 Pord	600	.013	S.F.		.64		.64	1.01
0760 Med.		520	.015			.74		.74	1.16
0770 Heavy		440	.018			.87		.87	1.37
0780 Pressure wash, based on 2500 lb. operating pressure									
0790 Stucco	A-1H	3080	.003	S.F.		.12	.03	.15	.22
0800 Aluminum or vinyl		3200	.003			.11	.02	.13	.21
0810 Siding, masonry, brick & block		2400	.003			.15	.03	.18	.28
1300 Miscellaneous, wire brush									
1310 Metal, pedestrian gate	1 Pord	100	.080	S.F.		3.82		3.82	6.05
1320 Aluminum chain link, both sides	"	250	.032	"		1.53		1.53	2.42
1400 Existing galvanized surface, clean and prime, prep for painting	1 Pord	380	.021	S.F.	.13	1.01		1.14	1.73
8000 For chemical washing, see Section 04 01 30									
8010 For steam cleaning, see Section 04 01 30.20									

09 01 90.94 Interior Surface Preparation

	Crew	Daily Output	Labor-Hours	Unit	Material	2022 Bare Costs Labor	Equipment	Total	Total Incl O&P
0010 **INTERIOR SURFACE PREPARATION**									
0020 Doors, per side, not incl. frames or trim									
0030 Scrape & sand									
0040 Wood, flush	1 Pord	616	.013	S.F.		.62		.62	.98
0050 Wood, detail		496	.016			.77		.77	1.22
0060 Wood, louvered		280	.029			1.37		1.37	2.16
0070 Wire brush									
0080 Metal, flush	1 Pord	640	.013	S.F.		.60		.60	.95
0090 Metal, detail		520	.015			.74		.74	1.16
0100 Metal, louvered		360	.022			1.06		1.06	1.68
0110 Hand wash									
0120 Wood, flush	1 Pord	2160	.004	S.F.		.18		.18	.28

For customer support on your Commercial Renovation Costs with RSMeans data, call 800.448.8182.

299

09 01 90.94 Interior Surface Preparation	Crew	Daily Output	Labor-Hours	Unit	Material	2022 Bare Costs Labor	Equipment	Total	Total Incl O&P
0130	Wood, detail	1 Pord	2000	.004	S.F.		.19		.19	.30
0140	Wood, louvered		1360	.006			.28		.28	.44
0150	Metal, flush		2160	.004			.18		.18	.28
0160	Metal, detail		2000	.004			.19		.19	.30
0170	Metal, louvered	↓	1360	.006	↓		.28		.28	.44
0400	Windows, per side, not incl. trim									
0410	Scrape & sand									
0420	Wood, 1-2 lite	1 Pord	360	.022	S.F.		1.06		1.06	1.68
0430	Wood, 3-6 lite		320	.025			1.20		1.20	1.89
0440	Wood, 7-10 lite		280	.029			1.37		1.37	2.16
0450	Wood, 12 lite		240	.033			1.59		1.59	2.52
0460	Wood, Bay/Bow	↓	360	.022	↓		1.06		1.06	1.68
0470	Wire brush									
0480	Metal, 1-2 lite	1 Pord	520	.015	S.F.		.74		.74	1.16
0490	Metal, 3-6 lite		440	.018			.87		.87	1.37
0500	Metal, Bay/Bow	↓	520	.015	↓		.74		.74	1.16
0600	Walls, sanding, light=10-30%, medium=30-70%,									
0610	heavy=70-100% of surface to sand									
0650	Walls, sand									
0660	Gypsum board or plaster, light	1 Pord	3077	.003	S.F.		.12		.12	.20
0670	Gypsum board or plaster, medium		2160	.004			.18		.18	.28
0680	Gypsum board or plaster, heavy		923	.009			.41		.41	.66
0690	Wood, T&G, light	↓	2400	.003	↓		.16		.16	.25
0700	Wood, T&G, medium	1 Pord	1600	.005	S.F.		.24		.24	.38
0710	Wood, T&G, heavy	"	800	.010	"		.48		.48	.76
0720	Walls, wash									
0730	Gypsum board or plaster	1 Pord	3200	.003	S.F.		.12		.12	.19
0740	Wood, T&G		3200	.003			.12		.12	.19
0750	Masonry, brick & block, smooth		2800	.003			.14		.14	.22
0760	Masonry, brick & block, coarse	↓	2000	.004	↓		.19		.19	.30
8000	For chemical washing, see Section 04 01 30									
8010	For steam cleaning, see Section 04 01 30.20									
9010	Minimum labor/equipment charge	1 Pord	3	2.667	Job		127		127	202

09 01 90.95 Scrape After Fire Damage

		Crew	Daily Output	Labor-Hours	Unit	Material	Labor	Equipment	Total	Total Incl O&P
0010	**SCRAPE AFTER FIRE DAMAGE**									
0050	Boards, 1" x 4"	1 Pord	336	.024	L.F.		1.14		1.14	1.80
0060	1" x 6"		260	.031			1.47		1.47	2.33
0070	1" x 8"		207	.039			1.85		1.85	2.92
0080	1" x 10"		174	.046			2.20		2.20	3.48
0500	Framing, 2" x 4"		265	.030			1.44		1.44	2.28
0510	2" x 6"		221	.036			1.73		1.73	2.74
0520	2" x 8"		190	.042			2.01		2.01	3.18
0530	2" x 10"		165	.048			2.32		2.32	3.67
0540	2" x 12"		144	.056			2.66		2.66	4.20
1000	Heavy framing, 3" x 4"		226	.035			1.69		1.69	2.68
1010	4" x 4"		210	.038			1.82		1.82	2.88
1020	4" x 6"		191	.042			2		2	3.17
1030	4" x 8"		165	.048			2.32		2.32	3.67
1040	4" x 10"		144	.056			2.66		2.66	4.20
1060	4" x 12"		131	.061	↓		2.92		2.92	4.62
2900	For sealing, light damage		825	.010	S.F.	.12	.46		.58	.86
2920	Heavy damage	↓	460	.017	"	.31	.83		1.14	1.65

09 01 Maintenance of Finishes

09 01 90 – Maintenance of Painting and Coating

09 01 90.95 Scrape After Fire Damage	Crew	Daily Output	Labor-Hours	Unit	Material	2022 Bare Costs Labor	Equipment	Total	Total Incl O&P
3000	For sandblasting, see Section 04 01 30.20									
9000	Minimum labor/equipment charge	1 Pord	3	2.667	Job		127		127	202

09 05 Common Work Results for Finishes

09 05 05 – Selective Demolition for Finishes

09 05 05.10 Selective Demolition, Ceilings

		Crew	Daily Output	Labor-Hours	Unit	Material	2022 Bare Costs Labor	Equipment	Total	Total Incl O&P
0010	**SELECTIVE DEMOLITION, CEILINGS** R024119-10									
0200	Ceiling, gypsum wall board, furred and nailed or screwed	2 Clab	800	.020	S.F.		.91		.91	1.45
0220	On metal frame		760	.021			.96		.96	1.53
0240	On suspension system, including system		720	.022			1.01		1.01	1.61
1000	Plaster, lime and horse hair, on wood lath, incl. lath		700	.023			1.04		1.04	1.66
1020	On metal lath	2 Clab	570	.028	S.F.		1.28		1.28	2.03
1100	Gypsum, on gypsum lath		720	.022			1.01		1.01	1.61
1120	On metal lath		500	.032			1.46		1.46	2.32
1200	Suspended ceiling, mineral fiber, 2' x 2' or 2' x 4'		1500	.011			.49		.49	.77
1250	On suspension system, incl. system		1200	.013			.61		.61	.97
1500	Tile, wood fiber, 12" x 12", glued		900	.018			.81		.81	1.29
1540	Stapled		1500	.011			.49		.49	.77
1580	On suspension system, incl. system		760	.021			.96		.96	1.53
2000	Wood, tongue and groove, 1" x 4"		1000	.016			.73		.73	1.16
2040	1" x 8"		1100	.015			.66		.66	1.05
2400	Plywood or wood fiberboard, 4' x 8' sheets		1200	.013			.61		.61	.97
2500	Remove & refinish textured ceiling	1 Plas	222	.036		.04	1.85		1.89	2.98
9000	Minimum labor/equipment charge	1 Clab	2	4	Job		182		182	290

09 05 05.20 Selective Demolition, Flooring

		Crew	Daily Output	Labor-Hours	Unit	Material	2022 Bare Costs Labor	Equipment	Total	Total Incl O&P
0010	**SELECTIVE DEMOLITION, FLOORING** R024119-10									
0200	Brick with mortar	2 Clab	475	.034	S.F.		1.54		1.54	2.44
0400	Carpet, bonded, including surface scraping		2000	.008			.36		.36	.58
0480	Tackless		9000	.002			.08		.08	.13
0550	Carpet tile, releasable adhesive		5000	.003			.15		.15	.23
0560	Permanent adhesive		1850	.009			.39		.39	.63
0600	Composition, acrylic or epoxy		400	.040			1.82		1.82	2.90
0700	Concrete, scarify skin	A-1A	225	.036			2.09	.96	3.05	4.36
0800	Resilient, sheet goods	2 Clab	1400	.011			.52		.52	.83
0820	For gym floors	"	900	.018			.81		.81	1.29
0850	Vinyl or rubber cove base	1 Clab	1000	.008	L.F.		.36		.36	.58
0860	Vinyl or rubber cove base, molded corner	"	1000	.008	Ea.		.36		.36	.58
0870	For glued and caulked installation, add to labor						50%			
0900	Vinyl composition tile, 12" x 12"	2 Clab	1000	.016	S.F.		.73		.73	1.16
2000	Tile, ceramic, thin set		675	.024			1.08		1.08	1.72
2020	Mud set		625	.026			1.17		1.17	1.85
2200	Marble, slate, thin set		675	.024			1.08		1.08	1.72
2220	Mud set		625	.026			1.17		1.17	1.85
2600	Terrazzo, thin set		450	.036			1.62		1.62	2.58
2620	Mud set		425	.038			1.72		1.72	2.73
2640	Terrazzo, cast in place		300	.053			2.43		2.43	3.86
3000	Wood, block, on end	1 Carp	400	.020			1.13		1.13	1.79
3200	Parquet	"	450	.018			1		1	1.59
3400	Strip flooring, interior, 2-1/4" x 25/32" thick	1 Carp	325	.025	S.F.		1.39		1.39	2.20
3500	Exterior, porch flooring, 1" x 4"		220	.036			2.05		2.05	3.25
3800	Subfloor, tongue and groove, 1" x 6"		325	.025			1.39		1.39	2.20

For customer support on your Commercial Renovation Costs with RSMeans data, call 800.448.8182.

301

09 05 05.20 Selective Demolition, Flooring

		Crew	Daily Output	Labor-Hours	Unit	Material	2022 Bare Costs Labor	Equipment	Total	Total Incl O&P
3820	1" x 8"	1 Carp	430	.019	S.F.		1.05		1.05	1.66
3840	1" x 10"		520	.015			.87		.87	1.38
4000	Plywood, nailed		600	.013			.75		.75	1.19
4100	Glued and nailed		400	.020			1.13		1.13	1.79
4200	Hardboard, 1/4" thick		760	.011			.59		.59	.94
8000	Remove flooring, bead blast, simple floor plan	A-1A	1000	.008		.05	.47	.22	.74	1.05
8100	Complex floor plan		400	.020		.05	1.18	.54	1.77	2.51
8150	Mastic only		1500	.005		.05	.31	.14	.50	.72
9000	Minimum labor/equipment charge	1 Clab	4	2	Job		91		91	145
9050	For grinding concrete floors, see Section 03 35 43.10									

09 05 05.30 Selective Demolition, Walls and Partitions

		Crew	Daily Output	Labor-Hours	Unit	Material	2022 Bare Costs Labor	Equipment	Total	Total Incl O&P
0010	**SELECTIVE DEMOLITION, WALLS AND PARTITIONS** R024119-10									
0020	Walls, concrete, reinforced	B-39	120	.400	C.F.		19.15	2.76	21.91	33.50
0025	Plain	"	160	.300			14.35	2.07	16.42	25
0100	Brick, 4" to 12" thick	B-9	220	.182			8.35	1.51	9.86	14.95
0200	Concrete block, 4" thick		1150	.035	S.F.		1.60	.29	1.89	2.86
0280	8" thick		1050	.038			1.75	.32	2.07	3.13
0300	Exterior stucco 1" thick over mesh		3200	.013			.58	.10	.68	1.02
1000	Gypsum wallboard, nailed or screwed	1 Clab	1000	.008			.36		.36	.58
1010	2 layers		400	.020			.91		.91	1.45
1020	Glued and nailed		900	.009			.41		.41	.64
1500	Fiberboard, nailed		900	.009			.41		.41	.64
1520	Glued and nailed		800	.010			.46		.46	.72
1568	Plenum barrier, sheet lead		300	.027			1.22		1.22	1.93
1600	Glass block		65	.123			5.60		5.60	8.90
2000	Movable walls, metal, 5' high		300	.027			1.22		1.22	1.93
2020	8' high		400	.020			.91		.91	1.45
2200	Metal or wood studs, finish 2 sides, fiberboard	B-1	520	.046			2.14		2.14	3.39
2250	Lath and plaster		260	.092			4.27		4.27	6.80
2300	Gypsum wallboard		520	.046			2.14		2.14	3.39
2350	Plywood		450	.053			2.47		2.47	3.92
2800	Paneling, 4' x 8' sheets	1 Clab	475	.017			.77		.77	1.22
3000	Plaster, lime and horsehair, on wood lath		400	.020			.91		.91	1.45
3020	On metal lath		335	.024			1.09		1.09	1.73
3400	Gypsum or perlite, on gypsum lath	1 Clab	410	.020	S.F.		.89		.89	1.41
3420	On metal lath		300	.027	"		1.22		1.22	1.93
3450	Plaster, interior gypsum, acoustic, or cement		60	.133	S.Y.		6.10		6.10	9.65
3500	Stucco, on masonry		145	.055			2.52		2.52	4
3510	Commercial 3-coat		80	.100			4.56		4.56	7.25
3520	Interior stucco		25	.320			14.60		14.60	23
3753	Remove damaged glass block	B-1	134.62	.178	S.F.		8.25		8.25	13.10
3760	Tile, ceramic, on walls, thin set	1 Clab	300	.027			1.22		1.22	1.93
3765	Mud set		250	.032			1.46		1.46	2.32
3800	Toilet partitions, slate or marble		5	1.600	Ea.		73		73	116
3820	Metal or plastic		8	1	"		45.50		45.50	72.50
5000	Wallcovering, vinyl	1 Pape	700	.011	S.F.		.55		.55	.87
5010	With release agent		1500	.005			.26		.26	.41
5025	Wallpaper, 2 layers or less, by hand		250	.032			1.54		1.54	2.43
5035	3 layers or more		165	.048			2.33		2.33	3.68
9000	Minimum labor/equipment charge	1 Clab	4	2	Job		91		91	145

09 05 71 – Acoustic Underlayment

09 05 71.10 Acoustical Underlayment

		Crew	Daily Output	Labor-Hours	Unit	Material	2022 Bare Costs Labor	Equipment	Total	Total Incl O&P
0010	**ACOUSTICAL UNDERLAYMENT**									
0100	Rubber underlayment, 5/64"	1 Tilf	275	.029	S.F.	.41	1.55		1.96	2.86
0110	1/8" thk.		275	.029		.59	1.55		2.14	3.06
0120	13/64" thk.		270	.030		1	1.57		2.57	3.56
0130	15/64" thk.		270	.030		1.49	1.57		3.06	4.10
0140	23/64" thk.		265	.030		1.82	1.60		3.42	4.50
0150	15/32" thk.		265	.030		1.90	1.60		3.50	4.59
0400	Cork underlayment, 3/32"		275	.029		.77	1.55		2.32	3.26
0410	1/8" thk.		275	.029		.69	1.55		2.24	3.17
0420	5/32" thk.		275	.029		.63	1.55		2.18	3.10
0430	15/64" thk.		270	.030		1.39	1.57		2.96	3.99
0440	15/32" thk.		265	.030		2.09	1.60		3.69	4.80
0600	Rubber cork underlayment, 3/64"		275	.029		.47	1.55		2.02	2.93
0610	13/64" thk.		270	.030		2.80	1.57		4.37	5.55
0620	3/8" thk.		270	.030		3.36	1.57		4.93	6.15
0630	25/64" thk.		270	.030		3.52	1.57		5.09	6.35
0800	Foam underlayment, 5/64"		275	.029		.55	1.55		2.10	3.02
0810	1/8" thk.		275	.029		.83	1.55		2.38	3.32
0820	15/32" thk.		265	.030		.67	1.60		2.27	3.24
4000	Nylon matting 0.4" thick, with carbon black spinerette									
4010	plus polyester fabric, on floor	D-7	1600	.010	S.F.	1.62	.48		2.10	2.53

09 21 Plaster and Gypsum Board Assemblies

09 21 16 – Gypsum Board Assemblies

09 21 16.23 Gypsum Board Shaft Wall Assemblies

		Crew	Daily Output	Labor-Hours	Unit	Material	2022 Bare Costs Labor	Equipment	Total	Total Incl O&P
0010	**GYPSUM BOARD SHAFT WALL ASSEMBLIES**									
0020	Cavity type on 25 ga. J-track & C-H studs, 24" OC									
0030	1" thick coreboard wall liner on shaft side									
0040	2-hour assembly with double layer									
0060	5/8" f.r. gyp bd on rm side, 2-1/2" J-track & C-H studs	2 Carp	220	.073	S.F.	2.88	4.09		6.97	9.65
0065	4" J-track & C-H studs		220	.073		3.10	4.09		7.19	9.90
0070	6" J-track & C-H studs		220	.073		3.41	4.09		7.50	10.25
0100	3-hour assembly with triple layer									
0300	5/8" f.r. gyp bd on rm side, 2-1/2" J-track & C-H studs	2 Carp	180	.089	S.F.	2.62	5		7.62	10.85
0305	4" J-track & C-H studs		180	.089		2.84	5		7.84	11.10
0310	6" J-track & C-H studs		180	.089		3.15	5		8.15	11.40
0400	4-hour assembly, 1" coreboard, 5/8" fire rated gypsum board									
0600	and 3/4" galv. metal furring channels, 24" OC, with									
0700	Dbl. layer 5/8" f.r. gyp bd on rm side, 2-1/2" trk. & C-H studs	2 Carp	110	.145	S.F.	3.41	8.20		11.61	16.75
0705	4" J-track & C-H studs		110	.145		3.10	8.20		11.30	16.40
0710	6" J-track & C-H studs		110	.145		3.41	8.20		11.61	16.75
0900	For taping & finishing, add per side	1 Carp	1050	.008		.05	.43		.48	.74
1000	For insulation, see Section 07 21									
5200	For work over 8' high, add	2 Carp	3060	.005	S.F.		.29		.29	.47
5300	For distribution cost 3 stories and above, add per story	"	6100	.003	"		.15		.15	.23

09 21 16.33 Partition Wall

		Crew	Daily Output	Labor-Hours	Unit	Material	2022 Bare Costs Labor	Equipment	Total	Total Incl O&P
0010	**PARTITION WALL** Stud wall, 8' to 12' high									
0050	1/2", interior, gypsum board, std, tape & finish 2 sides									
0500	Installed on and incl. 2" x 4" wood studs, 16" OC	2 Carp	310	.052	S.F.	1.33	2.91		4.24	6.10
1000	Metal studs, NLB, 25 ga., 16" OC, 3-5/8" wide		350	.046		1.20	2.57		3.77	5.40

For customer support on your Commercial Renovation Costs with RSMeans data, call 800.448.8182.

303

09 21 16.33 Partition Wall		Crew	Daily Output	Labor-Hours	Unit	Material	2022 Bare Costs Labor	Equipment	Total	Total Incl O&P
1200	6" wide	2 Carp	330	.048	S.F.	1.38	2.73		4.11	5.85
1400	Water resistant, on 2" x 4" wood studs, 16" OC		310	.052		1.49	2.91		4.40	6.25
1600	Metal studs, NLB, 25 ga., 16" OC, 3-5/8" wide		350	.046		1.36	2.57		3.93	5.60
1800	6" wide		330	.048		1.54	2.73		4.27	6.05
2000	Fire res., 2 layers, 1-1/2 hr., on 2" x 4" wood studs, 16" OC		210	.076		2.67	4.29		6.96	9.75
2200	Metal studs, NLB, 25 ga., 16" OC, 3-5/8" wide		250	.064		2.54	3.60		6.14	8.50
2400	6" wide		230	.070		2.72	3.92		6.64	9.20
2600	Fire & water res., 2 layers, 1-1/2 hr., 2" x 4" studs, 16" OC		210	.076		2.67	4.29		6.96	9.75
2800	Metal studs, NLB, 25 ga., 16" OC, 3-5/8" wide	2 Carp	250	.064	S.F.	2.54	3.60		6.14	8.50
3000	6" wide	"	230	.070	"	2.72	3.92		6.64	9.20
3200	5/8", interior, gypsum board, standard, tape & finish 2 sides									
3400	Installed on and including 2" x 4" wood studs, 16" OC	2 Carp	300	.053	S.F.	1.35	3		4.35	6.25
3600	24" OC		330	.048		1.19	2.73		3.92	5.65
3800	Metal studs, NLB, 25 ga., 16" OC, 3-5/8" wide		340	.047		1.22	2.65		3.87	5.55
4000	6" wide		320	.050		1.40	2.82		4.22	6
4200	24" OC, 3-5/8" wide		360	.044		1.06	2.50		3.56	5.15
4400	6" wide		340	.047		1.20	2.65		3.85	5.55
4800	Water resistant, on 2" x 4" wood studs, 16" OC		300	.053		1.69	3		4.69	6.65
5000	24" OC		330	.048		1.53	2.73		4.26	6
5200	Metal studs, NLB, 25 ga. 16" OC, 3-5/8" wide		340	.047		1.56	2.65		4.21	5.90
5400	6" wide		320	.050		1.74	2.82		4.56	6.40
5600	24" OC, 3-5/8" wide		360	.044		1.40	2.50		3.90	5.50
5800	6" wide		340	.047		1.54	2.65		4.19	5.90
6000	Fire resistant, 2 layers, 2 hr., on 2" x 4" wood studs, 16" OC		205	.078		2.19	4.39		6.58	9.40
6200	24" OC		235	.068		2.03	3.83		5.86	8.35
6400	Metal studs, NLB, 25 ga., 16" OC, 3-5/8" wide		245	.065		2.19	3.68		5.87	8.25
6600	6" wide		225	.071		2.24	4		6.24	8.80
6800	24" OC, 3-5/8" wide		265	.060		1.90	3.40		5.30	7.50
7000	6" wide		245	.065		2.04	3.68		5.72	8.10
7200	Fire & water resistant, 2 layers, 2 hr., 2" x 4" studs, 16" OC		205	.078		2.19	4.39		6.58	9.40
7400	24" OC		235	.068		2.03	3.83		5.86	8.35
7600	Metal studs, NLB, 25 ga., 16" OC, 3-5/8" wide		245	.065		2.06	3.68		5.74	8.10
7800	6" wide		225	.071		2.24	4		6.24	8.80
8000	24" OC, 3-5/8" wide		265	.060		1.90	3.40		5.30	7.50
8200	6" wide		245	.065		2.04	3.68		5.72	8.10
8600	1/2" blueboard, mesh tape both sides									
8620	Installed on and including 2" x 4" wood studs, 16" OC	2 Carp	300	.053	S.F.	1.49	3		4.49	6.40
8640	Metal studs, NLB, 25 ga., 16" OC, 3-5/8" wide		340	.047		1.36	2.65		4.01	5.70
8660	6" wide		320	.050		1.54	2.82		4.36	6.15
8800	Hospital security partition, 5/8" fiber reinf. high abuse gyp. bd.									
8810	Mtl. studs, NLB, 20 ga., 16" OC, 3-5/8" wide, w/sec. mesh, gyp. bd.	2 Carp	208	.077	S.F.	6.30	4.33		10.63	13.85
9000	Exterior, 1/2" gypsum sheathing, 1/2" gypsum finished, interior,									
9100	including foil faced insulation, metal studs, 20 ga.									
9200	16" OC, 3-5/8" wide	2 Carp	290	.055	S.F.	2.68	3.11		5.79	7.90
9400	6" wide		270	.059		2.45	3.34		5.79	8
9600	Partitions, for work over 8' high, add		1530	.010			.59		.59	.94

09 22 Supports for Plaster and Gypsum Board

09 22 03 – Fastening Methods for Finishes

09 22 03.20 Drilling Plaster/Drywall

		Crew	Daily Output	Labor-Hours	Unit	Material	2022 Bare Costs Labor	Equipment	Total	Total Incl O&P
0010	**DRILLING PLASTER/DRYWALL**									
1100	Drilling & layout for drywall/plaster walls, up to 1" deep, no anchor									
1200	Holes, 1/4" diameter	1 Carp	150	.053	Ea.	.01	3		3.01	4.78
1300	3/8" diameter		140	.057		.01	3.22		3.23	5.10
1400	1/2" diameter		130	.062		.01	3.46		3.47	5.50
1500	3/4" diameter		120	.067		.01	3.75		3.76	5.95
1600	1" diameter		110	.073		.02	4.09		4.11	6.55
1700	1-1/4" diameter		100	.080		.03	4.50		4.53	7.20
1800	1-1/2" diameter		90	.089		.04	5		5.04	8
1900	For ceiling installations, add						40%			

09 22 13 – Metal Furring

09 22 13.13 Metal Channel Furring

		Crew	Daily Output	Labor-Hours	Unit	Material	2022 Bare Costs Labor	Equipment	Total	Total Incl O&P
0010	**METAL CHANNEL FURRING**									
0030	Beams and columns, 7/8" hat channels, galvanized, 12" OC	1 Lath	155	.052	S.F.	.48	2.89		3.37	5.05
0050	16" OC		170	.047		.39	2.64		3.03	4.54
0070	24" OC		185	.043		.26	2.42		2.68	4.07
0100	Ceilings, on steel, 7/8" hat channels, galvanized, 12" OC		210	.038		.44	2.13		2.57	3.81
0300	16" OC		290	.028		.39	1.54		1.93	2.84
0400	24" OC		420	.019		.26	1.07		1.33	1.95
0600	1-5/8" hat channels, galvanized, 12" OC		190	.042		.59	2.36		2.95	4.33
0700	16" OC		260	.031		.53	1.72		2.25	3.27
0900	24" OC		390	.021		.35	1.15		1.50	2.18
0930	7/8" hat channels with sound isolation clips, 12" OC		120	.067		1.61	3.73		5.34	7.55
0940	16" OC		100	.080		1.21	4.48		5.69	8.35
0950	24" OC		165	.048		.80	2.72		3.52	5.10
0960	1-5/8" hat channels, galvanized, 12" OC		110	.073		1.76	4.07		5.83	8.30
0970	16" OC		100	.080		1.32	4.48		5.80	8.45
0980	24" OC		155	.052		.88	2.89		3.77	5.50
1000	Walls, 7/8" hat channels, galvanized, 12" OC		235	.034		.44	1.91		2.35	3.45
1200	16" OC		265	.030		.39	1.69		2.08	3.07
1300	24" OC		350	.023		.26	1.28		1.54	2.29
1500	1-5/8" hat channels, galvanized, 12" OC		210	.038		.59	2.13		2.72	3.98
1600	16" OC		240	.033		.53	1.87		2.40	3.49
1800	24" OC	1 Lath	305	.026	S.F.	.35	1.47		1.82	2.68
1920	7/8" hat channels with sound isolation clips, 12" OC		125	.064		1.61	3.58		5.19	7.35
1940	16" OC		100	.080		1.21	4.48		5.69	8.35
1950	24" OC		150	.053		.80	2.99		3.79	5.55
1960	1-5/8" hat channels, galvanized, 12" OC		115	.070		1.76	3.90		5.66	8.05
1970	16" OC		95	.084		1.32	4.72		6.04	8.80
1980	24" OC		140	.057		.88	3.20		4.08	5.95
3000	Z Furring, walls, 1" deep, 25 ga., 24" OC		350	.023		.35	1.28		1.63	2.38
3010	48" OC		700	.011		.17	.64		.81	1.19
3020	1-1/2" deep, 24" OC		345	.023		1.66	1.30		2.96	3.86
3030	48" OC		695	.012		.83	.64		1.47	1.92
3040	2" deep, 24" OC		340	.024		2	1.32		3.32	4.25
3050	48" OC		690	.012		1	.65		1.65	2.11
3060	1" deep, 20 ga., 24" OC		350	.023		2.36	1.28		3.64	4.60
3070	48" OC		700	.011		1.18	.64		1.82	2.30
3080	1-1/2" deep, 24" OC		345	.023		3.90	1.30		5.20	6.30
3090	48" OC		695	.012		1.95	.64		2.59	3.15
4000	2" deep, 24" OC		340	.024		3.33	1.32		4.65	5.70
4010	48" OC		690	.012		1.67	.65		2.32	2.84

09 22 Supports for Plaster and Gypsum Board

09 22 13 – Metal Furring

09 22 13.13 Metal Channel Furring		Crew	Daily Output	Labor-Hours	Unit	Material	2022 Bare Costs Labor	Equipment	Total	Total Incl O&P
9000	Minimum labor/equipment charge	1 Lath	4	2	Job		112		112	175

09 22 16 – Non-Structural Metal Framing

09 22 16.13 Non-Structural Metal Stud Framing

		Crew	Daily Output	Labor-Hours	Unit	Material	2022 Bare Costs Labor	Equipment	Total	Total Incl O&P
0010	**NON-STRUCTURAL METAL STUD FRAMING**									
1600	Non-load bearing, galv., 8' high, 25 ga. 1-5/8" wide, 16" OC	1 Carp	619	.013	S.F.	.59	.73		1.32	1.81
1610	24" OC		950	.008		.44	.47		.91	1.23
1620	2-1/2" wide, 16" OC		613	.013		.58	.73		1.31	1.81
1630	24" OC		938	.009		.44	.48		.92	1.24
1640	3-5/8" wide, 16" OC		600	.013		.62	.75		1.37	1.88
1650	24" OC		925	.009		.46	.49		.95	1.28
1660	4" wide, 16" OC		594	.013		.76	.76		1.52	2.04
1670	24" OC		925	.009		.56	.49		1.05	1.39
1680	6" wide, 16" OC		588	.014		.82	.77		1.59	2.12
1690	24" OC		906	.009		.61	.50		1.11	1.46
1700	20 ga. studs, 1-5/8" wide, 16" OC		494	.016		.48	.91		1.39	1.98
1710	24" OC		763	.010		.36	.59		.95	1.34
1720	2-1/2" wide, 16" OC		488	.016		.62	.92		1.54	2.15
1730	24" OC		750	.011		.50	.60		1.10	1.50
1740	3-5/8" wide, 16" OC		481	.017		1.21	.94		2.15	2.82
1750	24" OC		738	.011		.88	.61		1.49	1.93
1760	4" wide, 16" OC	1 Carp	475	.017	S.F.	.99	.95		1.94	2.60
1770	24" OC		738	.011		.75	.61		1.36	1.79
1780	6" wide, 16" OC		469	.017		1.02	.96		1.98	2.65
1790	24" OC		725	.011		.83	.62		1.45	1.91
2000	Non-load bearing, galv., 10' high, 25 ga. 1-5/8" wide, 16" OC		495	.016		.56	.91		1.47	2.07
2100	24" OC		760	.011		.41	.59		1	1.39
2200	2-1/2" wide, 16" OC		490	.016		.55	.92		1.47	2.07
2250	24" OC		750	.011		.41	.60		1.01	1.40
2300	3-5/8" wide, 16" OC		480	.017		.59	.94		1.53	2.14
2350	24" OC		740	.011		.43	.61		1.04	1.45
2400	4" wide, 16" OC		475	.017		.73	.95		1.68	2.32
2450	24" OC		740	.011		.53	.61		1.14	1.55
2500	6" wide, 16" OC		470	.017		.78	.96		1.74	2.38
2550	24" OC		725	.011		.57	.62		1.19	1.62
2600	20 ga. studs, 1-5/8" wide, 16" OC		395	.020		.46	1.14		1.60	2.31
2650	24" OC		610	.013		.34	.74		1.08	1.54
2700	2-1/2" wide, 16" OC		390	.021		.57	1.15		1.72	2.46
2750	24" OC		600	.013		.45	.75		1.20	1.68
2800	3-5/8" wide, 16" OC		385	.021		1.16	1.17		2.33	3.14
2850	24" OC		590	.014		.83	.76		1.59	2.13
2900	4" wide, 16" OC		380	.021		.94	1.19		2.13	2.91
2950	24" OC		590	.014		.69	.76		1.45	1.97
3000	6" wide, 16" OC		375	.021		.93	1.20		2.13	2.93
3050	24" OC		580	.014		.74	.78		1.52	2.05
3060	Non-load bearing, galv., 12' high, 25 ga. 1-5/8" wide, 16" OC		413	.019		.54	1.09		1.63	2.32
3070	24" OC		633	.013		.39	.71		1.10	1.56
3080	2-1/2" wide, 16" OC		408	.020		.53	1.10		1.63	2.33
3090	24" OC		625	.013		.38	.72		1.10	1.56
3100	3-5/8" wide, 16" OC		400	.020		.57	1.13		1.70	2.42
3110	24" OC		617	.013		.41	.73		1.14	1.61
3120	4" wide, 16" OC		396	.020		.71	1.14		1.85	2.59
3130	24" OC		617	.013		.51	.73		1.24	1.72

09 22 16 – Non-Structural Metal Framing

09 22 16.13 Non-Structural Metal Stud Framing

		Crew	Daily Output	Labor-Hours	Unit	Material	2022 Bare Costs Labor	Equipment	Total	Total Incl O&P
3140	6" wide, 16" OC	1 Carp	392	.020	S.F.	.75	1.15		1.90	2.65
3150	24" OC		604	.013		.54	.75		1.29	1.78
3160	20 ga. studs, 1-5/8" wide, 16" OC		329	.024		.44	1.37		1.81	2.66
3170	24" OC		508	.016		.32	.89		1.21	1.76
3180	2-1/2" wide, 16" OC		325	.025		.53	1.39		1.92	2.79
3190	24" OC		500	.016		.41	.90		1.31	1.88
3200	3-5/8" wide, 16" OC	↓	321	.025	↓	1.13	1.40		2.53	3.47
3210	24" OC	1 Carp	492	.016	S.F.	.80	.92		1.72	2.33
3220	4" wide, 16" OC		317	.025		.90	1.42		2.32	3.25
3230	24" OC		492	.016		.65	.92		1.57	2.17
3240	6" wide, 16" OC		313	.026		.86	1.44		2.30	3.23
3250	24" OC		483	.017		.67	.93		1.60	2.22
3260	Non-load bearing, galv., 16' high, 25 ga. 4" wide, 12" OC		195	.041		.93	2.31		3.24	4.69
3270	16" OC		275	.029		.73	1.64		2.37	3.40
3280	24" OC		400	.020		.52	1.13		1.65	2.37
3290	6" wide, 12" OC		190	.042		.97	2.37		3.34	4.84
3300	16" OC		280	.029		.76	1.61		2.37	3.40
3310	24" OC		400	.020		.56	1.13		1.69	2.40
3320	20 ga. studs, 2-1/2" wide, 12" OC		180	.044		.65	2.50		3.15	4.69
3330	16" OC		254	.032		.52	1.77		2.29	3.40
3340	24" OC		390	.021		.40	1.15		1.55	2.27
3350	3-5/8" wide, 12" OC		170	.047		1.49	2.65		4.14	5.85
3360	16" OC		251	.032		1.16	1.79		2.95	4.13
3370	24" OC		384	.021		.83	1.17		2	2.78
3380	4" wide, 12" OC		170	.047		1.15	2.65		3.80	5.45
3390	16" OC		247	.032		.91	1.82		2.73	3.90
3400	24" OC		384	.021		.67	1.17		1.84	2.59
3410	6" wide, 12" OC		175	.046		1.02	2.57		3.59	5.20
3420	16" OC		245	.033		.83	1.84		2.67	3.83
3430	24" OC		400	.020		.64	1.13		1.77	2.50
3440	Non-load bearing, galv., 20' high, 25 ga. 6" wide, 12" OC		125	.064		.95	3.60		4.55	6.75
3450	16" OC		220	.036		.74	2.05		2.79	4.07
3460	24" OC		360	.022		.54	1.25		1.79	2.58
3470	20 ga. studs, 4" wide, 12" OC		120	.067		1.12	3.75		4.87	7.20
3480	16" OC		215	.037		.90	2.09		2.99	4.33
3490	6" wide, 12" OC		115	.070		.97	3.92		4.89	7.25
3500	16" OC		215	.037		.78	2.09		2.87	4.19
3510	24" OC	↓	331	.024	↓	.60	1.36		1.96	2.82
5000	For load bearing studs, see Section 05 41 13.30									
9000	Minimum labor/equipment charge	1 Carp	4	2	Job		113		113	179

09 22 26 – Suspension Systems

09 22 26.13 Ceiling Suspension Systems

		Crew	Daily Output	Labor-Hours	Unit	Material	2022 Bare Costs Labor	Equipment	Total	Total Incl O&P
0010	**CEILING SUSPENSION SYSTEMS** for gypsum board or plaster									
8000	Suspended ceilings, including carriers									
8200	1-1/2" carriers, 24" OC with:									
8300	7/8" channels, 16" OC	1 Lath	275	.029	S.F.	.61	1.63		2.24	3.21
8320	24" OC	1 Lath	310	.026	S.F.	.48	1.45		1.93	2.78
8400	1-5/8" channels, 16" OC		205	.039		.75	2.19		2.94	4.23
8420	24" OC	↓	250	.032	↓	.57	1.79		2.36	3.43
8600	2" carriers, 24" OC with:									
8700	7/8" channels, 16" OC	1 Lath	250	.032	S.F.	.66	1.79		2.45	3.53
8720	24" OC	↓	285	.028	↓	.53	1.57		2.10	3.04

For customer support on your Commercial Renovation Costs with RSMeans data, call 800.448.8182.

307

09 22 26 – Suspension Systems

09 22 26.13 Ceiling Suspension Systems	Crew	Daily Output	Labor-Hours	Unit	Material	2022 Bare Costs Labor	Equipment	Total	Total Incl O&P
8800	1-5/8" channels, 16" OC	1 Lath	190	.042	S.F.	.80	2.36		3.16	4.56
8820	24" OC		225	.036		.62	1.99		2.61	3.80

09 22 36 – Lath

09 22 36.13 Gypsum Lath

		Crew	Daily Output	Labor-Hours	Unit	Material	Labor	Equipment	Total	Total Incl O&P
0010	**GYPSUM LATH** R092000-50									
0020	Plain or perforated, nailed, 3/8" thick	1 Lath	85	.094	S.Y.	3.42	5.25		8.67	11.95
0100	1/2" thick		80	.100		2.97	5.60		8.57	12
0300	Clipped to steel studs, 3/8" thick		75	.107		3.42	5.95		9.37	13.05
0400	1/2" thick		70	.114		2.97	6.40		9.37	13.25
1500	For ceiling installations, add		216	.037			2.07		2.07	3.24
1600	For columns and beams, add		170	.047			2.64		2.64	4.11
9000	Minimum labor/equipment charge		4.25	1.882	Job		105		105	164

09 22 36.23 Metal Lath

		Crew	Daily Output	Labor-Hours	Unit	Material	Labor	Equipment	Total	Total Incl O&P
0010	**METAL LATH** R092000-50									
0020	Diamond, expanded, 2.5 lb./S.Y., painted				S.Y.	5.10			5.10	5.60
0100	Galvanized					5.05			5.05	5.55
0300	3.4 lb./S.Y., painted					5.70			5.70	6.25
0400	Galvanized					5.15			5.15	5.70
0600	For #15 asphalt sheathing paper, add					.40			.40	.44
0900	Flat rib, 1/8" high, 2.75 lb., painted					4			4	4.40
1000	Foil backed					3.75			3.75	4.13
1200	3.4 lb./S.Y., painted					4.21			4.21	4.63
1300	Galvanized					4.39			4.39	4.83
1500	For #15 asphalt sheathing paper, add					.40			.40	.44
1800	High rib, 3/8" high, 3.4 lb./S.Y., painted					5.10			5.10	5.60
1900	Galvanized					5.10			5.10	5.60
2400	3/4" high, painted, .60 lb./S.F.				S.F.	.85			.85	.94
2500	.75 lb./S.F.				"	1.78			1.78	1.96
2800	Stucco mesh, painted, 3.6 lb.				S.Y.	5.95			5.95	6.50
3000	K-lath, perforated, absorbent paper, regular					2.29			2.29	2.52
3100	Heavy duty					5.35			5.35	5.90
3300	Waterproof, heavy duty, grade B backing					5.25			5.25	5.75
3400	Fire resistant backing					5.80			5.80	6.35
3600	2.5 lb. diamond painted, on wood framing, on walls	1 Lath	85	.094	S.Y.	5.10	5.25		10.35	13.80
3700	On ceilings		75	.107		5.10	5.95		11.05	14.90
3900	3.4 lb. diamond painted, on wood framing, on walls		80	.100		4.21	5.60		9.81	13.40
4000	On ceilings		70	.114		4.21	6.40		10.61	14.65
4200	3.4 lb. diamond painted, wired to steel framing		75	.107		4.21	5.95		10.16	13.95
4300	On ceilings		60	.133		4.21	7.45		11.66	16.30
4600	Cornices, wired to steel		35	.229		4.21	12.80		17.01	24.50
4800	Screwed to steel studs, 2.5 lb.		80	.100		5.10	5.60		10.70	14.35
4900	3.4 lb.		75	.107		5.70	5.95		11.65	15.55
5100	Rib lath, painted, wired to steel, on walls, 2.5 lb.		75	.107		4	5.95		9.95	13.70
5200	3.4 lb.		70	.114		5.10	6.40		11.50	15.60
5400	4.0 lb.		65	.123		5.95	6.90		12.85	17.30
5500	For self-furring lath, add					.12			.12	.13
5700	Suspended ceiling system, incl. 3.4 lb. diamond lath, painted	1 Lath	15	.533		4.34	30		34.34	51.50
5800	Galvanized	"	15	.533		3.97	30		33.97	51
6000	Hollow metal stud partitions, 3.4 lb. painted lath both sides									
6010	Non-load bearing, 25 ga., w/rib lath, 2-1/2" studs, 12" OC	1 Lath	20.30	.394	S.Y.	16.45	22		38.45	52.50
6300	16" OC		21.10	.379		15.15	21		36.15	49.50
6350	24" OC		22.70	.352		13.85	19.75		33.60	46.50

09 22 Supports for Plaster and Gypsum Board

09 22 36 – Lath

09 22 36.23 Metal Lath

		Crew	Daily Output	Labor-Hours	Unit	Material	2022 Bare Costs Labor	Equipment	Total	Total Incl O&P
6400	3-5/8" studs, 16" OC	1 Lath	19.50	.410	S.Y.	15.55	23		38.55	53
6600	24" OC		20.40	.392		14.10	22		36.10	50
6700	4" studs, 16" OC		20.40	.392		16.80	22		38.80	53
6900	24" OC		21.60	.370		14.95	20.50		35.45	49
7000	6" studs, 16" OC		19.50	.410		17.20	23		40.20	55
7100	24" OC		21.10	.379		15.35	21		36.35	50
7200	L.B. partitions, 16 ga., w/rib lath, 2-1/2" studs, 16" OC		20	.400		14.10	22.50		36.60	50.50
7300	3-5/8" studs, 16 ga.		19.70	.406		15.45	22.50		37.95	52.50
7500	4" studs, 16 ga.		19.50	.410		16.50	23		39.50	54
7600	6" studs, 16 ga.		18.70	.428	↓	19.20	24		43.20	58.50
9000	Minimum labor/equipment charge	↓	4.25	1.882	Job		105		105	164

09 23 Gypsum Plastering

09 23 13 – Acoustical Gypsum Plastering

09 23 13.10 Perlite or Vermiculite Plaster

			Crew	Daily Output	Labor-Hours	Unit	Material	2022 Bare Costs Labor	Equipment	Total	Total Incl O&P
0010	**PERLITE OR VERMICULITE PLASTER**	R092000-50									
0020	In 100 lb. bags, under 200 bags					Bag	11.15			11.15	12.30
0300	2 coats, no lath included, on walls		J-1	92	.435	S.Y.	4.77	21.50	1.23	27.50	40.50
0400	On ceilings		"	79	.506	"	4.77	25	1.43	31.20	46.50
0900	3 coats, no lath included, on walls		J-1	74	.541	S.Y.	6.85	26.50	1.53	34.88	51
1000	On ceilings		"	63	.635		6.85	31	1.80	39.65	59
1700	For irregular or curved surfaces, add to above							30%			
1800	For columns and beams, add to above							50%			
1900	For soffits, add to ceiling prices					↓		40%			
9000	Minimum labor/equipment charge		1 Plas	1	8	Job		410		410	650

09 23 20 – Gypsum Plaster

09 23 20.10 Gypsum Plaster On Walls and Ceilings

			Crew	Daily Output	Labor-Hours	Unit	Material	2022 Bare Costs Labor	Equipment	Total	Total Incl O&P
0010	**GYPSUM PLASTER ON WALLS AND CEILINGS**	R092000-50									
0020	80# bag, less than 1 ton					Bag	10.55			10.55	11.65
0300	2 coats, no lath included, on walls		J-1	105	.381	S.Y.	3.45	18.75	1.08	23.28	34.50
0400	On ceilings			92	.435		3.45	21.50	1.23	26.18	39
0900	3 coats, no lath included, on walls			87	.460		4.89	22.50	1.30	28.69	43
1000	On ceilings	↓		78	.513	↓	4.89	25	1.45	31.34	47
1600	For irregular or curved surfaces, add							30%			
1800	For columns & beams, add							50%			
9000	Minimum labor/equipment charge		1 Plas	1	8	Job		410		410	650

09 24 Cement Plastering

09 24 23 – Cement Stucco

09 24 23.40 Stucco

			Crew	Daily Output	Labor-Hours	Unit	Material	2022 Bare Costs Labor	Equipment	Total	Total Incl O&P
0010	**STUCCO**	R092000-50									
0015	3 coats 7/8" thick, float finish, with mesh, on wood frame		J-2	63	.762	S.Y.	8.90	38.50	1.80	49.20	72
0100	On masonry construction, no mesh incl.		J-1	67	.597		2.81	29.50	1.69	34	51.50
0300	For trowel finish, add		1 Plas	170	.047			2.42		2.42	3.84
0600	For coloring, add		J-1	685	.058		.44	2.87	.17	3.48	5.20
0700	For special texture, add		"	200	.200		1.55	9.85	.57	11.97	17.95
0900	For soffits, add		J-2	155	.310		2.36	15.60	.73	18.69	28
1000	Stucco, with bonding agent, 3 coats, on walls, no mesh incl.		J-1	200	.200		3.67	9.85	.57	14.09	20.50
1200	Ceilings	↓		180	.222	↓	3.95	10.95	.63	15.53	22.50

09 24 Cement Plastering

09 24 23 – Cement Stucco

09 24 23.40 Stucco		Crew	Daily Output	Labor-Hours	Unit	Material	2022 Bare Costs Labor	Equipment	Total	Total Incl O&P
1300	Beams	J-1	80	.500	S.Y.	3.95	24.50	1.42	29.87	45
1500	Columns	▼	100	.400	▼	3.95	19.65	1.13	24.73	36.50
1550	Minimum labor/equipment charge	1 Plas	1	8	Job		410		410	650
1600	Mesh, galvanized, nailed to wood, 1.8 lb.	1 Lath	60	.133	S.Y.	2.77	7.45		10.22	14.70
1800	3.6 lb.		55	.145		5.95	8.15		14.10	19.20
1900	Wired to steel, galvanized, 1.8 lb.		53	.151		2.77	8.45		11.22	16.25
2100	3.6 lb.		50	.160	▼	5.95	8.95		14.90	20.50
9000	Minimum labor/equipment charge	▼	4	2	Job		112		112	175

09 25 Other Plastering

09 25 23 – Lime Based Plastering

09 25 23.10 Venetian Plaster

		Crew	Daily Output	Labor-Hours	Unit	Material	2022 Bare Costs Labor	Equipment	Total	Total Incl O&P
0010	**VENETIAN PLASTER**									
0100	Walls, 1 coat primer, roller applied	1 Plas	950	.008	S.F.	.19	.43		.62	.90
0200	Plaster, 3 coats, incl. sanding	2 Plas	700	.023		.54	1.17		1.71	2.46
0210	For pigment, light colors add per S.F. plaster					.02			.02	.02
0220	For pigment, dark colors add per S.F. plaster					.04			.04	.04
0300	For sealer/wax coat incl. burnishing, add	1 Plas	300	.027	▼	.16	1.37		1.53	2.35

09 26 Veneer Plastering

09 26 13 – Gypsum Veneer Plastering

09 26 13.20 Blueboard

		Crew	Daily Output	Labor-Hours	Unit	Material	2022 Bare Costs Labor	Equipment	Total	Total Incl O&P
0010	**BLUEBOARD** For use with thin coat									
0100	plaster application see Section 09 26 13.80									
1000	3/8" thick, on walls or ceilings, standard, no finish included	2 Carp	1900	.008	S.F.	.32	.47		.79	1.10
1100	With thin coat plaster finish		875	.018		.44	1.03		1.47	2.13
1400	On beams, columns, or soffits, standard, no finish included		675	.024		.37	1.33		1.70	2.52
1450	With thin coat plaster finish		475	.034		.49	1.90		2.39	3.55
3000	1/2" thick, on walls or ceilings, standard, no finish included		1900	.008		.40	.47		.87	1.19
3100	With thin coat plaster finish		875	.018		.52	1.03		1.55	2.21
3300	Fire resistant, no finish included		1900	.008		.40	.47		.87	1.19
3400	With thin coat plaster finish		875	.018		.52	1.03		1.55	2.21
3450	On beams, columns, or soffits, standard, no finish included		675	.024		.46	1.33		1.79	2.63
3500	With thin coat plaster finish		475	.034		.58	1.90		2.48	3.65
3700	Fire resistant, no finish included		675	.024		.46	1.33		1.79	2.63
3800	With thin coat plaster finish		475	.034		.58	1.90		2.48	3.65
5000	5/8" thick, on walls or ceilings, fire resistant, no finish included		1900	.008		.61	.47		1.08	1.42
5100	With thin coat plaster finish		875	.018		.73	1.03		1.76	2.44
5500	On beams, columns, or soffits, no finish included		675	.024		.70	1.33		2.03	2.89
5600	With thin coat plaster finish		475	.034		.82	1.90		2.72	3.92
6000	For high ceilings, over 8' high, add		3060	.005			.29		.29	.47
6500	For distribution costs 3 stories and above, add per story	▼	6100	.003	▼		.15		.15	.23
9000	Minimum labor/equipment charge	1 Carp	2	4	Job		225		225	360

09 26 13.80 Thin Coat Plaster

		Crew	Daily Output	Labor-Hours	Unit	Material	2022 Bare Costs Labor	Equipment	Total	Total Incl O&P
0010	**THIN COAT PLASTER** R092000-50									
0012	1 coat veneer, not incl. lath	J-1	3600	.011	S.F.	.12	.55	.03	.70	1.03
1000	In 50 lb. bags				Bag	16.45			16.45	18.05

310

09 28 Backing Boards and Underlayments

09 28 13 – Cementitious Backing Boards

09 28 13.10 Cementitious Backerboard

		Crew	Daily Output	Labor-Hours	Unit	Material	2022 Bare Costs Labor	Equipment	Total	Total Incl O&P
0010	**CEMENTITIOUS BACKERBOARD**									
0070	Cementitious backerboard, on floor, 3' x 4' x 1/2" sheets	2 Carp	525	.030	S.F.	1	1.72		2.72	3.83
0080	3' x 5' x 1/2" sheets		525	.030		.73	1.72		2.45	3.53
0090	3' x 6' x 1/2" sheets		525	.030		.73	1.72		2.45	3.53
0100	3' x 4' x 5/8" sheets		525	.030		.93	1.72		2.65	3.76
0110	3' x 5' x 5/8" sheets		525	.030		.93	1.72		2.65	3.75
0120	3' x 6' x 5/8" sheets		525	.030		.99	1.72		2.71	3.82
0150	On wall, 3' x 4' x 1/2" sheets		350	.046		1	2.57		3.57	5.20
0160	3' x 5' x 1/2" sheets		350	.046		.73	2.57		3.30	4.89
0170	3' x 6' x 1/2" sheets		350	.046		.73	2.57		3.30	4.89
0180	3' x 4' x 5/8" sheets		350	.046		.93	2.57		3.50	5.10
0190	3' x 5' x 5/8" sheets		350	.046		.93	2.57		3.50	5.10
0200	3' x 6' x 5/8" sheets		350	.046		.99	2.57		3.56	5.20
0250	On counter, 3' x 4' x 1/2" sheets		180	.089		1	5		6	9.05
0260	3' x 5' x 1/2" sheets		180	.089		.73	5		5.73	8.75
0270	3' x 6' x 1/2" sheets		180	.089		.73	5		5.73	8.75
0300	3' x 4' x 5/8" sheets		180	.089		.93	5		5.93	9
0310	3' x 5' x 5/8" sheets		180	.089		.93	5		5.93	8.95
0320	3' x 6' x 5/8" sheets		180	.089		.99	5		5.99	9.05

09 29 Gypsum Board

09 29 10 – Gypsum Board Panels

09 29 10.30 Gypsum Board

		Crew	Daily Output	Labor-Hours	Unit	Material	2022 Bare Costs Labor	Equipment	Total	Total Incl O&P
0010	**GYPSUM BOARD** on walls & ceilings R092910-10									
0100	Nailed or screwed to studs unless otherwise noted									
0110	1/4" thick, on walls or ceilings, standard, no finish included	2 Carp	1330	.012	S.F.	.34	.68		1.02	1.45
0115	1/4" thick, on walls or ceilings, flexible, no finish included		1050	.015		.32	.86		1.18	1.71
0117	1/4" thick, on columns or soffits, flexible, no finish included		1050	.015		.32	.86		1.18	1.71
0130	1/4" thick, standard, no finish included, less than 800 S.F.		510	.031		.34	1.77		2.11	3.18
0150	3/8" thick, on walls, standard, no finish included		2000	.008		.25	.45		.70	1
0200	On ceilings, standard, no finish included		1800	.009		.25	.50		.75	1.08
0250	On beams, columns, or soffits, no finish included		675	.024		.25	1.33		1.58	2.40
0300	1/2" thick, on walls, standard, no finish included		2000	.008		.25	.45		.70	1
0350	Taped and finished (level 4 finish)		965	.017		.30	.93		1.23	1.81
0390	With compound skim coat (level 5 finish)		775	.021		.36	1.16		1.52	2.25
0400	Fire resistant, no finish included		2000	.008		.46	.45		.91	1.23
0450	Taped and finished (level 4 finish)		965	.017		.51	.93		1.44	2.04
0490	With compound skim coat (level 5 finish)	2 Carp	775	.021	S.F.	.57	1.16		1.73	2.48
0500	Water resistant, no finish included		2000	.008		.33	.45		.78	1.08
0550	Taped and finished (level 4 finish)		965	.017		.38	.93		1.31	1.90
0590	With compound skim coat (level 5 finish)		775	.021		.44	1.16		1.60	2.34
0600	Prefinished, vinyl, clipped to studs		900	.018		.47	1		1.47	2.11
0700	Mold resistant, no finish included		2000	.008		.39	.45		.84	1.15
0710	Taped and finished (level 4 finish)		965	.017		.44	.93		1.37	1.97
0720	With compound skim coat (level 5 finish)		775	.021		.50	1.16		1.66	2.40
1000	On ceilings, standard, no finish included		1800	.009		.25	.50		.75	1.08
1050	Taped and finished (level 4 finish)		765	.021		.30	1.18		1.48	2.20
1090	With compound skim coat (level 5 finish)		610	.026		.36	1.48		1.84	2.75
1100	Fire resistant, no finish included		1800	.009		.46	.50		.96	1.31
1150	Taped and finished (level 4 finish)		765	.021		.51	1.18		1.69	2.43
1195	With compound skim coat (level 5 finish)		610	.026		.57	1.48		2.05	2.98

09 29 10.30 Gypsum Board		Crew	Daily Output	Labor-Hours	Unit	Material	2022 Bare Costs Labor	Equipment	Total	Total Incl O&P
1200	Water resistant, no finish included	2 Carp	1800	.009	S.F.	.33	.50		.83	1.16
1250	Taped and finished (level 4 finish)		765	.021		.38	1.18		1.56	2.29
1290	With compound skim coat (level 5 finish)		610	.026		.44	1.48		1.92	2.84
1310	Mold resistant, no finish included		1800	.009		.39	.50		.89	1.23
1320	Taped and finished (level 4 finish)		765	.021		.44	1.18		1.62	2.36
1330	With compound skim coat (level 5 finish)		610	.026		.50	1.48		1.98	2.90
1350	Sag resistant, no finish included		1600	.010		.39	.56		.95	1.32
1360	Taped and finished (level 4 finish)		765	.021		.44	1.18		1.62	2.36
1370	With compound skim coat (level 5 finish)		610	.026		.50	1.48		1.98	2.90
1500	On beams, columns, or soffits, standard, no finish included		675	.024		.29	1.33		1.62	2.44
1550	Taped and finished (level 4 finish)		540	.030		.30	1.67		1.97	2.98
1590	With compound skim coat (level 5 finish)		475	.034		.36	1.90		2.26	3.41
1600	Fire resistant, no finish included		675	.024		.46	1.33		1.79	2.63
1650	Taped and finished (level 4 finish)		540	.030		.51	1.67		2.18	3.21
1690	With compound skim coat (level 5 finish)		475	.034		.57	1.90		2.47	3.64
1700	Water resistant, no finish included		675	.024		.38	1.33		1.71	2.54
1750	Taped and finished (level 4 finish)		540	.030		.38	1.67		2.05	3.07
1790	With compound skim coat (level 5 finish)		475	.034		.44	1.90		2.34	3.50
1800	Mold resistant, no finish included		675	.024		.45	1.33		1.78	2.61
1810	Taped and finished (level 4 finish)		540	.030		.44	1.67		2.11	3.14
1820	With compound skim coat (level 5 finish)		475	.034		.50	1.90		2.40	3.56
1850	Sag resistant, no finish included	2 Carp	675	.024	S.F.	.45	1.33		1.78	2.61
1860	Taped and finished (level 4 finish)		540	.030		.44	1.67		2.11	3.14
1870	With compound skim coat (level 5 finish)		475	.034		.50	1.90		2.40	3.56
2000	5/8" thick, on walls, standard, no finish included		2000	.008		.26	.45		.71	1.01
2050	Taped and finished (level 4 finish)		965	.017		.31	.93		1.24	1.82
2090	With compound skim coat (level 5 finish)		775	.021		.37	1.16		1.53	2.26
2100	Fire resistant, no finish included		2000	.008		.34	.45		.79	1.09
2150	Taped and finished (level 4 finish)		965	.017		.39	.93		1.32	1.91
2195	With compound skim coat (level 5 finish)		775	.021		.45	1.16		1.61	2.35
2200	Water resistant, no finish included		2000	.008		.43	.45		.88	1.19
2250	Taped and finished (level 4 finish)		965	.017		.48	.93		1.41	2.01
2290	With compound skim coat (level 5 finish)		775	.021		.54	1.16		1.70	2.45
2300	Prefinished, vinyl, clipped to studs		900	.018		.99	1		1.99	2.68
2510	Mold resistant, no finish included		2000	.008		.44	.45		.89	1.20
2520	Taped and finished (level 4 finish)		965	.017		.49	.93		1.42	2.02
2530	With compound skim coat (level 5 finish)		775	.021		.55	1.16		1.71	2.46
3000	On ceilings, standard, no finish included		1800	.009		.26	.50		.76	1.09
3050	Taped and finished (level 4 finish)		765	.021		.31	1.18		1.49	2.21
3090	With compound skim coat (level 5 finish)		615	.026		.37	1.46		1.83	2.74
3100	Fire resistant, no finish included		1800	.009		.34	.50		.84	1.17
3150	Taped and finished (level 4 finish)		765	.021		.39	1.18		1.57	2.30
3190	With compound skim coat (level 5 finish)		615	.026		.45	1.46		1.91	2.83
3200	Water resistant, no finish included		1800	.009		.43	.50		.93	1.27
3250	Taped and finished (level 4 finish)		765	.021		.48	1.18		1.66	2.40
3290	With compound skim coat (!level 5 finish)		615	.026		.54	1.46		2	2.93
3300	Mold resistant, no finish included		1800	.009		.44	.50		.94	1.28
3310	Taped and finished (level 4 finish)		765	.021		.49	1.18		1.67	2.41
3320	With compound skim coat (level 5 finish)		615	.026		.55	1.46		2.01	2.94
3500	On beams, columns, or soffits, no finish included		675	.024		.30	1.33		1.63	2.45
3550	Taped and finished (level 4 finish)		475	.034		.36	1.90		2.26	3.40
3590	With compound skim coat (level 5 finish)		380	.042		.43	2.37		2.80	4.24
3600	Fire resistant, no finish included		675	.024		.39	1.33		1.72	2.55

312

09 29 10 – Gypsum Board Panels

09 29 10.30 Gypsum Board		Crew	Daily Output	Labor-Hours	Unit	Material	2022 Bare Costs Labor	Equipment	Total	Total Incl O&P
3650	Taped and finished (level 4 finish)	2 Carp	475	.034	S.F.	.45	1.90		2.35	3.51
3690	With compound skim coat (level 5 finish)		380	.042		.45	2.37		2.82	4.27
3700	Water resistant, no finish included		675	.024		.49	1.33		1.82	2.66
3750	Taped and finished (level 4 finish)	2 Carp	475	.034	S.F.	.54	1.90		2.44	3.61
3790	With compound skim coat (level 5 finish)		380	.042		.55	2.37		2.92	4.38
3800	Mold resistant, no finish included		675	.024		.51	1.33		1.84	2.68
3810	Taped and finished (level 4 finish)		475	.034		.55	1.90		2.45	3.62
3820	With compound skim coat (level 5 finish)		380	.042		.57	2.37		2.94	4.39
4000	Fireproofing, beams or columns, 2 layers, 1/2" thick, incl finish		330	.048		1.02	2.73		3.75	5.45
4010	Mold resistant		330	.048		.88	2.73		3.61	5.30
4050	5/8" thick		300	.053		.78	3		3.78	5.65
4060	Mold resistant		300	.053		.98	3		3.98	5.85
4100	3 layers, 1/2" thick		225	.071		1.54	4		5.54	8.05
4110	Mold resistant		225	.071		1.33	4		5.33	7.80
4150	5/8" thick		210	.076		1.18	4.29		5.47	8.10
4160	Mold resistant		210	.076		1.48	4.29		5.77	8.40
5050	For 1" thick coreboard on columns		480	.033		.73	1.88		2.61	3.78
5100	For foil-backed board, add					.20			.20	.22
5200	For work over 8' high, add	2 Carp	3060	.005			.29		.29	.47
5270	For textured spray, add	2 Lath	1600	.010		.05	.56		.61	.93
5300	For distribution cost 3 stories and above, add per story	2 Carp	6100	.003			.15		.15	.23
5350	For finishing inner corners, add		950	.017	L.F.	.11	.95		1.06	1.63
5355	For finishing outer corners, add		1250	.013		.39	.72		1.11	1.57
5500	For acoustical sealant, add per bead	1 Carp	500	.016		.03	.90		.93	1.47
5550	Sealant, 1 quart tube				Ea.	5.55			5.55	6.10
6000	Gypsum sound dampening panels									
6010	1/2" thick on walls, multi-layer, lightweight, no finish included	2 Carp	1500	.011	S.F.	1.65	.60		2.25	2.77
6015	Taped and finished (level 4 finish)		725	.022		1.70	1.24		2.94	3.84
6020	With compound skim coat (level 5 finish)		580	.028		1.76	1.55		3.31	4.41
6025	5/8" thick on walls, for wood studs, no finish included		1500	.011		1.76	.60		2.36	2.89
6030	Taped and finished (level 4 finish)		725	.022		1.81	1.24		3.05	3.96
6035	With compound skim coat (level 5 finish)		580	.028		1.87	1.55		3.42	4.53
6040	For metal stud, no finish included		1500	.011		1.76	.60		2.36	2.89
6045	Taped and finished (level 4 finish)		725	.022		1.81	1.24		3.05	3.96
6050	With compound skim coat (level 5 finish)		580	.028		1.87	1.55		3.42	4.53
6055	Abuse resist, no finish included		1500	.011		1.70	.60		2.30	2.82
6060	Taped and finished (level 4 finish)		725	.022		1.75	1.24		2.99	3.90
6065	With compound skim coat (level 5 finish)		580	.028		1.81	1.55		3.36	4.47
6070	Shear rated, no finish included	2 Carp	1500	.011	S.F.	1.76	.60		2.36	2.89
6075	Taped and finished (level 4 finish)		725	.022		1.81	1.24		3.05	3.96
6080	With compound skim coat (level 5 finish)		580	.028		1.87	1.55		3.42	4.53
6085	For SCIF applications, no finish included		1500	.011		1.56	.60		2.16	2.67
6090	Taped and finished (level 4 finish)		725	.022		1.61	1.24		2.85	3.74
6095	With compound skim coat (level 5 finish)		580	.028		1.67	1.55		3.22	4.31
6100	1-3/8" thick on walls, THX certified, no finish included		1500	.011		8.45	.60		9.05	10.25
6105	Taped and finished (level 4 finish)		725	.022		8.50	1.24		9.74	11.30
6110	With compound skim coat (level 5 finish)		580	.028		8.60	1.55		10.15	11.90
6115	5/8" thick on walls, score & snap installation, no finish included		2000	.008		1.76	.45		2.21	2.66
6120	Taped and finished (level 4 finish)		965	.017		1.81	.93		2.74	3.47
6125	With compound skim coat (level 5 finish)		775	.021		1.87	1.16		3.03	3.91
7020	5/8" thick on ceilings, for wood joists, no finish included		1200	.013		1.76	.75		2.51	3.13
7025	Taped and finished (level 4 finish)		510	.031		1.81	1.77		3.58	4.80
7030	With compound skim coat (level 5 finish)		410	.039		1.87	2.20		4.07	5.55

09 29 10 – Gypsum Board Panels

09 29 10.30 Gypsum Board

		Crew	Daily Output	Labor-Hours	Unit	Material	2022 Bare Costs Labor	Equipment	Total	Total Incl O&P
7035	For metal joists, no finish included	2 Carp	1200	.013	S.F.	1.76	.75		2.51	3.13
7040	Taped and finished (level 4 finish)		510	.031		1.81	1.77		3.58	4.80
7045	With compound skim coat (level 5 finish)		410	.039		1.87	2.20		4.07	5.55
7050	Abuse resist, no finish included		1200	.013		1.70	.75		2.45	3.06
7055	Taped and finished (level 4 finish)		510	.031		1.75	1.77		3.52	4.74
7060	With compound skim coat (level 5 finish)		410	.039		1.81	2.20		4.01	5.50
7065	Shear rated, no finish included		1200	.013		1.76	.75		2.51	3.13
7070	Taped and finished (level 4 finish)		510	.031		1.81	1.77		3.58	4.80
7075	With compound skim coat (level 5 finish)		410	.039		1.87	2.20		4.07	5.55
7080	For SCIF applications, no finish included		1200	.013		1.56	.75		2.31	2.91
7085	Taped and finished (level 4 finish)		510	.031		1.61	1.77		3.38	4.58
7090	With compound skim coat (level 5 finish)		410	.039		1.67	2.20		3.87	5.35
8010	5/8" thick on ceilings, score & snap installation, no finish included		1600	.010		1.76	.56		2.32	2.83
8015	Taped and finished (level 4 finish)		680	.024		1.81	1.32		3.13	4.09
8020	With compound skim coat (level 5 finish)		545	.029		1.87	1.65		3.52	4.69
9000	Minimum labor/equipment charge	1 Carp	2	4	Job		225		225	360

09 29 10.50 High Abuse Gypsum Board

		Crew	Daily Output	Labor-Hours	Unit	Material	2022 Bare Costs Labor	Equipment	Total	Total Incl O&P
0010	**HIGH ABUSE GYPSUM BOARD**, fiber reinforced, nailed or									
0100	screwed to studs unless otherwise noted									
0110	1/2" thick, on walls, no finish included	2 Carp	1800	.009	S.F.	.67	.50		1.17	1.54
0120	Taped and finished (level 4 finish)	2 Carp	870	.018	S.F.	.72	1.04		1.76	2.44
0130	With compound skim coat (level 5 finish)		700	.023		.78	1.29		2.07	2.90
0150	On ceilings, no finish included		1620	.010		.67	.56		1.23	1.62
0160	Taped and finished (level 4 finish)		690	.023		.72	1.31		2.03	2.86
0170	With compound skim coat (level 5 finish)		550	.029		.78	1.64		2.42	3.46
0210	5/8" thick, on walls, no finish included		1800	.009		1.08	.50		1.58	1.99
0220	Taped and finished (level 4 finish)		870	.018		1.13	1.04		2.17	2.89
0230	With compound skim coat (level 5 finish)		700	.023		1.19	1.29		2.48	3.35
0250	On ceilings, no finish included		1620	.010		1.08	.56		1.64	2.07
0260	Taped and finished (level 4 finish)		690	.023		1.13	1.31		2.44	3.31
0270	With compound skim coat (level 5 finish)		550	.029		1.19	1.64		2.83	3.91
0272	5/8" thick, on roof sheathing for 1-hour rating		1620	.010		1.08	.56		1.64	2.07
0310	5/8" thick, on walls, very high impact, no finish included		1800	.009		.91	.50		1.41	1.80
0320	Taped and finished (level 4 finish)		870	.018		.96	1.04		2	2.71
0330	With compound skim coat (level 5 finish)		700	.023		1.02	1.29		2.31	3.17
0350	On ceilings, no finish included		1620	.010		.91	.56		1.47	1.88
0360	Taped and finished (level 4 finish)		690	.023		.96	1.31		2.27	3.13
0370	With compound skim coat (level 5 finish)		550	.029		1.02	1.64		2.66	3.73
0400	High abuse, gypsum core, paper face									
0410	1/2" thick, on walls, no finish included	2 Carp	1800	.009	S.F.	.77	.50		1.27	1.65
0420	Taped and finished (level 4 finish)		870	.018		.82	1.04		1.86	2.55
0430	With compound skim coat (level 5 finish)		700	.023		.88	1.29		2.17	3.01
0450	On ceilings, no finish included		1620	.010		.77	.56		1.33	1.73
0460	Taped and finished (level 4 finish)		690	.023		.82	1.31		2.13	2.97
0470	With compound skim coat (level 5 finish)		550	.029		.88	1.64		2.52	3.57
0510	5/8" thick, on walls, no finish included		1800	.009		.37	.50		.87	1.21
0520	Taped and finished (level 4 finish)		870	.018		.42	1.04		1.46	2.11
0530	With compound skim coat (level 5 finish)		700	.023		.48	1.29		1.77	2.57
0550	On ceilings, no finish included		1620	.010		.37	.56		.93	1.29
0560	Taped and finished (level 4 finish)		690	.023		.42	1.31		1.73	2.53
0570	With compound skim coat (level 5 finish)		550	.029		.48	1.64		2.12	3.13
1000	For high ceilings, over 8' high, add		2750	.006			.33		.33	.52

For customer support on your Commercial Renovation Costs with RSMeans data, call 800.448.8182.

09 29 Gypsum Board

09 29 10 – Gypsum Board Panels

	09 29 10.50 High Abuse Gypsum Board	Crew	Daily Output	Labor-Hours	Unit	Material	2022 Bare Costs Labor	Equipment	Total	Total Incl O&P
1010	For distribution cost 3 stories and above, add per story	2 Carp	5500	.003	S.F.		.16		.16	.26

09 29 15 – Gypsum Board Accessories

09 29 15.10 Accessories, Gypsum Board

		Crew	Daily Output	Labor-Hours	Unit	Material	2022 Bare Costs Labor	Equipment	Total	Total Incl O&P
0010	**ACCESSORIES, GYPSUM BOARD**									
0020	Casing bead, galvanized steel	1 Carp	2.90	2.759	C.L.F.	34	155		189	285
0100	Vinyl		3	2.667		30.50	150		180.50	273
0300	Corner bead, galvanized steel, 1" x 1"		4	2		29.50	113		142.50	212
0400	1-1/4" x 1-1/4"		3.50	2.286		32	129		161	239
0600	Vinyl		4	2		28	113		141	210
0900	Furring channel, galv. steel, 7/8" deep, standard		2.60	3.077		72.50	173		245.50	355
1000	Resilient		2.55	3.137		52	177		229	340
1100	J trim, galvanized steel, 1/2" wide		3	2.667		42.50	150		192.50	286
1120	5/8" wide		2.95	2.712		31.50	153		184.50	278
1500	Z stud, galvanized steel, 1-1/2" wide		2.60	3.077		78	173		251	360
9000	Minimum labor/equipment charge		3	2.667	Job		150		150	239

09 30 Tiling

09 30 13 – Ceramic Tiling

09 30 13.20 Ceramic Tile Repairs

		Crew	Daily Output	Labor-Hours	Unit	Material	2022 Bare Costs Labor	Equipment	Total	Total Incl O&P
0010	**CERAMIC TILE REPAIRS**									
1000	Grout removal, carbide tipped, rotary grinder	1 Clab	240	.033	L.F.		1.52		1.52	2.42

09 30 13.45 Ceramic Tile Accessories

		Crew	Daily Output	Labor-Hours	Unit	Material	2022 Bare Costs Labor	Equipment	Total	Total Incl O&P
0010	**CERAMIC TILE ACCESSORIES**									
0100	Spacers, 1/8"				C	.69			.69	.76
1310	Sealer for natural stone tile, installed	1 Tilf	650	.012	S.F.	.07	.65		.72	1.09

09 30 29 – Metal Tiling

09 30 29.10 Metal Tile

		Crew	Daily Output	Labor-Hours	Unit	Material	2022 Bare Costs Labor	Equipment	Total	Total Incl O&P
0010	**METAL TILE** 4' x 4' sheet, 24 ga., tile pattern, nailed									
0200	Stainless steel	2 Carp	512	.031	S.F.	25.50	1.76		27.26	31
0400	Aluminized steel	"	512	.031	"	11.65	1.76		13.41	15.65
9000	Minimum labor/equipment charge	1 Carp	4	2	Job		113		113	179

09 30 95 – Tile & Stone Setting Materials and Specialties

09 30 95.10 Moisture Resistant, Anti-Fracture Membrane

		Crew	Daily Output	Labor-Hours	Unit	Material	2022 Bare Costs Labor	Equipment	Total	Total Incl O&P
0010	**MOISTURE RESISTANT, ANTI-FRACTURE MEMBRANE**									
0200	Elastomeric membrane, 1/16" thick	D-7	275	.058	S.F.	1.87	2.79		4.66	6.40

09 31 Thin-Set Tiling

09 31 13 – Thin-Set Ceramic Tiling

09 31 13.10 Thin-Set Ceramic Tile

		Crew	Daily Output	Labor-Hours	Unit	Material	2022 Bare Costs Labor	Equipment	Total	Total Incl O&P
0010	**THIN-SET CERAMIC TILE**									
0700	Cove base, 4-1/4" x 4-1/4"	D-7	128	.125	L.F.	5.65	6		11.65	15.55
1000	6" x 4-1/4" high		137	.117		3.10	5.60		8.70	12.15
1300	Sanitary cove base, 6" x 4-1/4" high		124	.129		5.65	6.20		11.85	15.85
1600	6" x 6" high		117	.137		5.85	6.55		12.40	16.70
2500	Bullnose trim, 4-1/4" x 4-1/4"		128	.125		2.24	6		8.24	11.80
2800	2" x 6"		124	.129		1.44	6.20		7.64	11.25
3300	Ceramic tile, porcelain type, 1 color, color group 2, 1" x 1"	D-7	183	.087	S.F.	6.75	4.19		10.94	13.95

09 31 Thin-Set Tiling

09 31 13 – Thin-Set Ceramic Tiling

09 31 13.10 Thin-Set Ceramic Tile

	09 31 13.10 Thin-Set Ceramic Tile	Crew	Daily Output	Labor-Hours	Unit	Material	2022 Bare Costs Labor	Equipment	Total	Total Incl O&P
3310	2" x 2" or 2" x 1"	D-7	190	.084	S.F.	1.14	4.04		5.18	7.55
3350	For random blend, 2 colors, add					.91			.91	1
3360	4 colors, add					1.30			1.30	1.43
4300	Specialty tile, 4-1/4" x 4-1/4" x 1/2", decorator finish	D-7	183	.087		11.85	4.19		16.04	19.55
4500	Add for epoxy grout, 1/16" joint, 1" x 1" tile		800	.020		.60	.96		1.56	2.16
4600	2" x 2" tile		820	.020		.56	.94		1.50	2.08
4610	Add for epoxy grout, 1/8" joint, 8" x 8" x 3/8" tile, add		900	.018		.56	.85		1.41	1.94
4800	Pregrouted sheets, walls, 4-1/4" x 4-1/4", 6" x 4-1/4"									
4810	and 8-1/2" x 4-1/4", 4 S.F. sheets, silicone grout	D-7	240	.067	S.F.	5.20	3.20		8.40	10.70
5100	Floors, unglazed, 2 S.F. sheets,									
5110	urethane adhesive	D-7	180	.089	S.F.	2.53	4.26		6.79	9.45
5400	Walls, interior, 4-1/4" x 4-1/4" tile		190	.084		1.37	4.04		5.41	7.80
5500	6" x 4-1/4" tile		190	.084		13.25	4.04		17.29	21
5700	8-1/2" x 4-1/4" tile		190	.084		10.45	4.04		14.49	17.80
5800	6" x 6" tile		175	.091		1.83	4.39		6.22	8.85
5810	8" x 8" tile		170	.094		5.30	4.52		9.82	12.90
5820	12" x 12" tile		160	.100		2.82	4.80		7.62	10.60
5830	16" x 16" tile		150	.107		1.05	5.10		6.15	9.15
6000	Decorated wall tile, 4-1/4" x 4-1/4", color group 1		270	.059		8.40	2.84		11.24	13.65
6100	Color group 4		180	.089		62.50	4.26		66.76	75
9300	Ceramic tiles, recycled glass, standard colors, 2" x 2" thru 6" x 6" [G]		190	.084		22.50	4.04		26.54	31
9310	6" x 6" [G]		175	.091		20	4.39		24.39	29
9320	8" x 8" [G]		170	.094		21	4.52		25.52	30
9330	12" x 12" [G]		160	.100		21	4.80		25.80	30.50
9340	Earthtones, 2" x 2" to 4" x 8" [G]		190	.084		25	4.04		29.04	33.50
9350	6" x 6" [G]		175	.091		24.50	4.39		28.89	34
9360	8" x 8" [G]		170	.094		24	4.52		28.52	33.50
9370	12" x 12" [G]		160	.100		24	4.80		28.80	34
9380	Deep colors, 2" x 2" to 4" x 8" [G]		190	.084		28.50	4.04		32.54	37.50
9390	6" x 6" [G]		175	.091		28	4.39		32.39	38
9400	8" x 8" [G]		170	.094		29.50	4.52		34.02	39.50
9410	12" x 12" [G]		160	.100		29.50	4.80		34.30	40

09 31 33 – Thin-Set Stone Tiling

09 31 33.10 Tiling, Thin-Set Stone

		Crew	Daily Output	Labor-Hours	Unit	Material	Labor	Equipment	Total	Total Incl O&P
0010	**TILING, THIN-SET STONE**									
3000	Floors, natural clay, random or uniform, color group 1	D-7	183	.087	S.F.	1.75	4.19		5.94	8.50
3100	Color group 2	"	183	.087	"	2.46	4.19		6.65	9.25

09 32 Mortar-Bed Tiling

09 32 13 – Mortar-Bed Ceramic Tiling

09 32 13.10 Ceramic Tile

		Crew	Daily Output	Labor-Hours	Unit	Material	Labor	Equipment	Total	Total Incl O&P
0010	**CERAMIC TILE**									
0600	Cove base, 4-1/4" x 4-1/4" high	D-7	91	.176	L.F.	5.80	8.45		14.25	19.50
0900	6" x 4-1/4" high		100	.160		3.24	7.70		10.94	15.55
1200	Sanitary cove base, 6" x 4-1/4" high		93	.172		5.80	8.25		14.05	19.30
1500	6" x 6" high		84	.190		6	9.15		15.15	21
2400	Bullnose trim, 4-1/4" x 4-1/4"		82	.195		2.35	9.35		11.70	17.20
2700	2" x 6" bullnose trim		84	.190		1.50	9.15		10.65	15.90
6210	Wall tile, 4-1/4" x 4-1/4", better grade	1 Tilf	50	.160	S.F.	9	8.50		17.50	23
6240	2" x 2"		50	.160		7	8.50		15.50	21

09 32 13 – Mortar-Bed Ceramic Tiling

09 32 13.10 Ceramic Tile

		Crew	Daily Output	Labor-Hours	Unit	Material	2022 Bare Costs Labor	Equipment	Total	Total Incl O&P
6250	6" x 6"	1 Tilf	55	.145	S.F.	8.40	7.75		16.15	21.50
6260	8" x 8"	↓	60	.133		5.05	7.10		12.15	16.60
6600	Crystalline glazed, 4-1/4" x 4-1/4", plain	D-7	100	.160		4.28	7.70		11.98	16.70
6700	4-1/4" x 4-1/4", scored tile		100	.160		5.95	7.70		13.65	18.50
6900	6" x 6" plain		93	.172		1.98	8.25		10.23	15.10
7000	For epoxy grout, 1/16" joints, 4-1/4" tile, add		800	.020		.39	.96		1.35	1.93
7200	For tile set in dry mortar, add		1735	.009			.44		.44	.69
7300	For tile set in Portland cement mortar, add		290	.055	↓	.14	2.65		2.79	4.29
9500	Minimum labor/equipment charge	↓	3.25	4.923	Job		236		236	370

09 32 16 – Mortar-Bed Quarry Tiling

09 32 16.10 Quarry Tile

		Crew	Daily Output	Labor-Hours	Unit	Material	2022 Bare Costs Labor	Equipment	Total	Total Incl O&P
0010	**QUARRY TILE**									
0100	Base, cove or sanitary, to 5" high, 1/2" thick	D-7	110	.145	L.F.	1.50	7		8.50	12.55
0300	Bullnose trim, red, 6" x 6" x 1/2" thick		120	.133		3.53	6.40		9.93	13.90
0400	4" x 4" x 1/2" thick		110	.145		3.74	7		10.74	15
0600	4" x 8" x 1/2" thick, using 8" as edge		130	.123	↓	2.95	5.90		8.85	12.45
0700	Floors, 1,000 S.F. lots, red, 4" x 4" x 1/2" thick		120	.133	S.F.	7.75	6.40		14.15	18.50
0900	6" x 6" x 1/2" thick		140	.114		8.45	5.50		13.95	17.85
1000	4" x 8" x 1/2" thick	↓	130	.123		5.30	5.90		11.20	15
1300	For waxed coating, add					.68			.68	.75
1500	For non-standard colors, add					.45			.45	.50
1600	For abrasive surface, add					.48			.48	.53
1800	Brown tile, imported, 6" x 6" x 3/4"	D-7	120	.133		7.45	6.40		13.85	18.20
1900	8" x 8" x 1"		110	.145		8.50	7		15.50	20.50
2100	For thin set mortar application, deduct		700	.023	↓		1.10		1.10	1.71
2700	Stair tread, 6" x 6" x 3/4", plain	↓	50	.320		5.15	15.35		20.50	29.50
2800	Abrasive	D-7	47	.340	S.F.	7.75	16.35		24.10	34
3000	Wainscot, 6" x 6" x 1/2", thin set, red		105	.152		2.59	7.30		9.89	14.25
3100	Non-standard colors		105	.152	↓	2.67	7.30		9.97	14.35
3300	Window sill, 6" wide, 3/4" thick		90	.178	L.F.	7.50	8.55		16.05	21.50
3400	Corners		80	.200	Ea.	5.45	9.60		15.05	21
9000	Minimum labor/equipment charge	↓	3.25	4.923	Job		236		236	370

09 32 23 – Mortar-Bed Glass Mosaic Tiling

09 32 23.10 Glass Mosaics

		Crew	Daily Output	Labor-Hours	Unit	Material	2022 Bare Costs Labor	Equipment	Total	Total Incl O&P
0010	**GLASS MOSAICS** 3/4" tile on 12" sheets, standard grout									
1020	1" tile on 12" sheets, opalescent finish	D-7	73	.219	S.F.	15.45	10.50		25.95	33.50
1040	1" x 2" tile on 12" sheet, blend		73	.219		20.50	10.50		31	39
1060	2" tile on 12" sheet, blend		73	.219		21.50	10.50		32	40
1080	5/8" x random tile, linear, on 12" sheet, blend		73	.219		20.50	10.50		31	39
1600	Dots on 12" sheet		73	.219		20.50	10.50		31	39
1700	For glass mosaic tiles set in dry mortar, add		290	.055		.40	2.65		3.05	4.57
1720	For glass mosaic tiles set in Portland cement mortar, add		290	.055	↓	.01	2.65		2.66	4.14
1730	For polyblend sanded tile grout	↓	96.15	.166	Lb.	1.97	8		9.97	14.60

317

For customer support on your Commercial Renovation Costs with RSMeans data, call 800.448.8182.

09 34 Waterproofing-Membrane Tiling

09 34 13 – Waterproofing-Membrane Ceramic Tiling

09 34 13.10 Ceramic Tile Waterproofing Membrane	Crew	Daily Output	Labor-Hours	Unit	Material	2022 Bare Costs Labor	Equipment	Total	Total Incl O&P
0010 **CERAMIC TILE WATERPROOFING MEMBRANE**									
0020 On floors, including thinset									
0030 Fleece laminated polyethylene grid, 1/8" thick	D-7	250	.064	S.F.	2.28	3.07		5.35	7.30
0040 5/16" thick	"	250	.064	"	2.60	3.07		5.67	7.65
0050 On walls, including thinset									
0060 Fleece laminated polyethylene sheet, 8 mil thick	D-7	480	.033	S.F.	2.55	1.60		4.15	5.30
0070 Accessories, including thinset									
0080 Joint and corner sheet, 4 mils thick, 5" wide	1 Tilf	240	.033	L.F.	1.26	1.77		3.03	4.15
0090 7-1/4" wide		180	.044		1.71	2.36		4.07	5.55
0100 10" wide		120	.067		2.09	3.54		5.63	7.85
0110 Pre-formed corners, inside		32	.250	Ea.	7.65	13.30		20.95	29
0120 Outside		32	.250		7.65	13.30		20.95	29
0130 2" flanged floor drain with 6" stainless steel grate		16	.500		360	26.50		386.50	435
0140 EPS, sloped shower floor		480	.017	S.F.	15.55	.89		16.44	18.50
0150 Curb		32	.250	L.F.	15.40	13.30		28.70	37.50

09 35 Chemical-Resistant Tiling

09 35 13 – Chemical-Resistant Ceramic Tiling

09 35 13.10 Chemical-Resistant Ceramic Tiling

	Crew	Daily Output	Labor-Hours	Unit	Material	2022 Bare Costs Labor	Equipment	Total	Total Incl O&P
0010 **CHEMICAL-RESISTANT CERAMIC TILING**									
0100 4-1/4" x 4-1/4" x 1/4", 1/8" joint	D-7	130	.123	S.F.	8.95	5.90		14.85	19.05
0200 6" x 6" x 1/2" thick		120	.133		10.55	6.40		16.95	21.50
0300 8" x 8" x 1/2" thick		110	.145		13.40	7		20.40	25.50
0400 4-1/4" x 4-1/4" x 1/4", 1/4" joint		130	.123		9.95	5.90		15.85	20
0500 6" x 6" x 1/2" thick		120	.133		12	6.40		18.40	23
0600 8" x 8" x 1/2" thick		110	.145		14.25	7		21.25	26.50
0700 4-1/4" x 4-1/4" x 1/4", 3/8" joint		130	.123		10.90	5.90		16.80	21
0800 6" x 6" x 1/2" thick		120	.133		13.25	6.40		19.65	24.50
0900 8" x 8" x 1/2" thick		110	.145		15.75	7		22.75	28.50

09 35 16 – Chemical-Resistant Quarry Tiling

09 35 16.10 Chemical-Resistant Quarry Tiling

	Crew	Daily Output	Labor-Hours	Unit	Material	2022 Bare Costs Labor	Equipment	Total	Total Incl O&P
0010 **CHEMICAL-RESISTANT QUARRY TILING**									
0100 4" x 8" x 1/2" thick, 1/8" joint	D-7	130	.123	S.F.	10.75	5.90		16.65	21
0200 6" x 6" x 1/2" thick		120	.133		13.80	6.40		20.20	25
0300 8" x 8" x 1/2" thick		110	.145		9.90	7		16.90	22
0400 4" x 8" x 1/2" thick, 1/4" joint		130	.123		12.35	5.90		18.25	23
0500 6" x 6" x 1/2" thick		120	.133		15.25	6.40		21.65	27
0600 8" x 8" x 1/2" thick		110	.145		10.70	7		17.70	22.50
0700 4" x 8" x 1/2" thick, 3/8" joint		130	.123		13.75	5.90		19.65	24.50
0800 6" x 6" x 1/2" thick		120	.133		16.45	6.40		22.85	28
0900 8" x 8" x 1/2" thick		110	.145		12.25	7		19.25	24.50

09 51 Acoustical Ceilings

09 51 13 – Acoustical Panel Ceilings

09 51 13.10 Ceiling, Acoustical Panel

		Crew	Daily Output	Labor-Hours	Unit	Material	2022 Bare Costs Labor	Equipment	Total	Total Incl O&P
0010	**CEILING, ACOUSTICAL PANEL**									
0100	Fiberglass boards, film faced, 2' x 2' or 2' x 4', 5/8" thick	1 Carp	625	.013	S.F.	4.59	.72		5.31	6.20
0120	3/4" thick		600	.013		4.96	.75		5.71	6.65
0130	3" thick, thermal, R11		450	.018		4.19	1		5.19	6.20

09 51 14 – Acoustical Fabric-Faced Panel Ceilings

09 51 14.10 Ceiling, Acoustical Fabric-Faced Panel

		Crew	Daily Output	Labor-Hours	Unit	Material	2022 Bare Costs Labor	Equipment	Total	Total Incl O&P
0010	**CEILING, ACOUSTICAL FABRIC-FACED PANEL**									
0100	Glass cloth faced fiberglass, 3/4" thick	1 Carp	500	.016	S.F.	3.38	.90		4.28	5.15
0120	1" thick		485	.016		7.25	.93		8.18	9.50
0130	1-1/2" thick, nubby face		475	.017		2.85	.95		3.80	4.65

09 51 23 – Acoustical Tile Ceilings

09 51 23.10 Suspended Acoustic Ceiling Tiles

		Crew	Daily Output	Labor-Hours	Unit	Material	2022 Bare Costs Labor	Equipment	Total	Total Incl O&P
0010	**SUSPENDED ACOUSTIC CEILING TILES**, not including									
0100	suspension system									
1110	Mineral fiber tile, lay-in, 2' x 2' or 2' x 4', 5/8" thick, fine texture	1 Carp	625	.013	S.F.	2.25	.72		2.97	3.62
1115	Rough textured		625	.013		1.17	.72		1.89	2.43
1125	3/4" thick, fine textured		600	.013		2.16	.75		2.91	3.57
1130	Rough textured		600	.013		2.07	.75		2.82	3.47
1135	Fissured		600	.013		2.28	.75		3.03	3.70
1150	Tegular, 5/8" thick, fine textured		470	.017		2.55	.96		3.51	4.33
1155	Rough textured		470	.017		1.92	.96		2.88	3.63
1165	3/4" thick, fine textured		450	.018		4.26	1		5.26	6.30
1170	Rough textured		450	.018		1.42	1		2.42	3.15
1175	Fissured		450	.018		1.77	1		2.77	3.54
1185	For plastic film face, add					.76			.76	.84
1190	For fire rating, add					.55			.55	.61
3720	Mineral fiber, 24" x 24" or 48", reveal edge, painted, 5/8" thick	1 Carp	600	.013		1.55	.75		2.30	2.90
3740	3/4" thick		575	.014		1.77	.78		2.55	3.19
5020	66-78% recycled content, 3/4" thick G		600	.013		6.90	.75		7.65	8.80
5040	Mylar, 42% recycled content, 3/4" thick G		600	.013		4.85	.75		5.60	6.55
6000	Remove & install new ceiling tiles, min fiber, 2' x 2' or 2' x 4', 5/8"thk.		335	.024		2.25	1.34		3.59	4.62
9000	Minimum labor/equipment charge		4	2	Job		113		113	179

09 51 23.30 Suspended Ceilings, Complete

		Crew	Daily Output	Labor-Hours	Unit	Material	2022 Bare Costs Labor	Equipment	Total	Total Incl O&P
0010	**SUSPENDED CEILINGS, COMPLETE**, incl. standard									
0100	suspension system but not incl. 1-1/2" carrier channels									
0600	Fiberglass ceiling board, 2' x 4' x 5/8", plain faced	1 Carp	500	.016	S.F.	6.10	.90		7	8.15
0700	Offices, 2' x 4' x 3/4"		380	.021		6.45	1.19		7.64	9
0800	Mineral fiber, on 15/16" T bar susp. 2' x 2' x 3/4" lay-in board		345	.023		4.10	1.31		5.41	6.60
0810	2' x 4' x 5/8" tile		380	.021		5.30	1.19		6.49	7.75
0820	Tegular, 2' x 2' x 5/8" tile on 9/16" grid		250	.032		5.25	1.80		7.05	8.65
0830	2' x 4' x 3/4" tile		275	.029		5.05	1.64		6.69	8.15
0900	Luminous panels, prismatic, acrylic		255	.031		5.40	1.77		7.17	8.70
1200	Metal pan with acoustic pad, steel		75	.107		5.30	6		11.30	15.35
1300	Painted aluminum		75	.107		4.37	6		10.37	14.35
1500	Aluminum, degreased finish		75	.107		6	6		12	16.15
1600	Stainless steel		75	.107		9.30	6		15.30	19.75
1800	Tile, Z bar suspension, 5/8" mineral fiber tile		150	.053		1.09	3		4.09	5.95
1900	3/4" mineral fiber tile		150	.053		2.08	3		5.08	7.05
2402	For strip lighting, see Section 26 51 13.50									
2500	For rooms under 500 S.F., add				S.F.		25%			
9000	Minimum labor/equipment charge	1 Carp	2	4	Job		225		225	360

For customer support on your Commercial Renovation Costs with RSMeans data, call 800.448.8182.

319

09 51 Acoustical Ceilings

09 51 33 – Acoustical Metal Pan Ceilings

09 51 33.10 Ceiling, Acoustical Metal Pan	Crew	Daily Output	Labor-Hours	Unit	Material	2022 Bare Costs Labor	Equipment	Total	Total Incl O&P
0010 **CEILING, ACOUSTICAL METAL PAN**									
0100 Metal panel, lay-in, 2' x 2', sq. edge	1 Carp	500	.016	S.F.	12.10	.90		13	14.75
0110 Tegular edge		500	.016		14.55	.90		15.45	17.45
0120 2' x 4', sq. edge		500	.016		14.30	.90		15.20	17.15
0130 Tegular edge		500	.016		14.40	.90		15.30	17.30
0140 Perforated alum. clip-in, 2' x 2'		500	.016		3.19	.90		4.09	4.94
0150 2' x 4'		500	.016		11.85	.90		12.75	14.50

09 51 53 – Direct-Applied Acoustical Ceilings

09 51 53.10 Ceiling Tile

	Crew	Daily Output	Labor-Hours	Unit	Material	2022 Bare Costs Labor	Equipment	Total	Total Incl O&P
0010 **CEILING TILE**, stapled or cemented									
0100 12" x 12" or 12" x 24", not including furring									
0600 Mineral fiber, vinyl coated, 5/8" thick	1 Carp	300	.027	S.F.	4.32	1.50		5.82	7.15
0700 3/4" thick		300	.027		4.68	1.50		6.18	7.55
0900 Fire rated, 3/4" thick, plain faced		300	.027		1.89	1.50		3.39	4.47
1000 Plastic coated face		300	.027		1.15	1.50		2.65	3.66
1200 Aluminum faced, 5/8" thick, plain		300	.027		2.42	1.50		3.92	5.05
3000 Wood fiber tile, 1/2" thick		400	.020		1.14	1.13		2.27	3.04
3100 3/4" thick		400	.020		1.61	1.13		2.74	3.56
3300 For flameproofing, add					.11			.11	.12
3400 For sculptured 3 dimensional, add					.42			.42	.46
3900 For ceiling primer, add					.15			.15	.17
4000 For ceiling cement, add					.50		○	.50	.55
9000 Minimum labor/equipment charge	1 Carp	4	2	Job		113		113	179

09 53 Acoustical Ceiling Suspension Assemblies

09 53 23 – Metal Acoustical Ceiling Suspension Assemblies

09 53 23.30 Ceiling Suspension Systems

	Crew	Daily Output	Labor-Hours	Unit	Material	2022 Bare Costs Labor	Equipment	Total	Total Incl O&P
0010 **CEILING SUSPENSION SYSTEMS** for boards and tile									
0050 Class A suspension system, 15/16" T bar, 2' x 4' grid	1 Carp	800	.010	S.F.	1.51	.56		2.07	2.55
0300 2' x 2' grid		650	.012		1.94	.69		2.63	3.23
0310 25% recycled steel, 2' x 4' grid [G]		800	.010		1.24	.56		1.80	2.26
0320 2' x 2' grid [G]		650	.012		1.67	.69		2.36	2.94
0350 For 9/16" grid, add					.14			.14	.15
0360 For fire rated grid, add					.08			.08	.09
0370 For colored grid, add					.20			.20	.22
0400 Concealed Z bar suspension system, 12" module	1 Carp	520	.015		.94	.87		1.81	2.41
0600 1-1/2" carrier channels, 4' OC, add	1 Carp	470	.017	S.F.	.12	.96		1.08	1.65
0700 Carrier channels for ceilings with									
0900 recessed lighting fixtures, add	1 Carp	460	.017	S.F.	.22	.98		1.20	1.80
3000 Seismic ceiling bracing, IBC Site Class D, Occupancy Category II									
3050 For ceilings less than 2500 S.F.									
3060 Seismic clips at attached walls	1 Carp	180	.044	Ea.	.88	2.50		3.38	4.95
3100 For ceilings greater than 2500 S.F., add									
3120 Seismic clips, joints at cross tees	1 Carp	120	.067	Ea.	.88	3.75		4.63	6.90
3140 At cross tees and mains, mains field cut	"	60	.133	"	.88	7.50		8.38	12.90
3200 Compression posts, telescopic, attached to structure above									
3210 To 30" high	1 Carp	26	.308	Ea.	36.50	17.30		53.80	67.50
3220 30" to 48" high		25.50	.314		37.50	17.65		55.15	69.50
3230 48" to 84" high		25	.320		45.50	18		63.50	78.50
3240 84" to 102" high		24.50	.327		52	18.40		70.40	86

For customer support on your Commercial Renovation Costs with RSMeans data, call 800.448.8182.

09 53 Acoustical Ceiling Suspension Assemblies

09 53 23 – Metal Acoustical Ceiling Suspension Assemblies

09 53 23.30 Ceiling Suspension Systems	Crew	Daily Output	Labor-Hours	Unit	Material	2022 Bare Costs Labor	Equipment	Total	Total Incl O&P
3250	102" to 120" high	1 Carp	24	.333	Ea.	74.50	18.75		93.25	112
3260	120" to 144" high	↓	24	.333	↓	83.50	18.75		102.25	122
3300	Stabilizer bars									
3310	12" long	1 Carp	240	.033	Ea.	1.01	1.88		2.89	4.09
3320	24" long		235	.034		.93	1.92		2.85	4.08
3330	36" long		230	.035		.93	1.96		2.89	4.13
3340	48" long	↓	220	.036	↓	.75	2.05		2.80	4.07
3400	Wire support for light fixtures, per L.F. height to structure above									
3410	Less than 10 lb.	1 Carp	400	.020	L.F.	.40	1.13		1.53	2.23
3420	10 lb. to 56 lb.	"	240	.033	"	.79	1.88		2.67	3.85
3500	Retrofit existing suspended ceiling grid to current code									
3510	Less than 2500 S.F. (using clips @ perimeter)	1 Carp	1455	.006	S.F.	.08	.31		.39	.58
3520	Greater than 2500 S.F. (using compression posts and clips)		550	.015		.41	.82		1.23	1.75
4000	Remove and replace ceiling grid, class A, 15/16" T bar, 2' x 2' grid		435	.018		1.94	1.04		2.98	3.78
4010	2' x 4' grid	↓	535	.015	↓	1.51	.84		2.35	3

09 54 Specialty Ceilings

09 54 16 – Luminous Ceilings

09 54 16.10 Ceiling, Luminous

		Crew	Daily Output	Labor-Hours	Unit	Material	2022 Bare Costs Labor	Equipment	Total	Total Incl O&P
0010	**CEILING, LUMINOUS**									
0020	Translucent lay-in panels, 2' x 2'	1 Carp	500	.016	S.F.	1.53	.90		2.43	3.11
0030	2' x 6'	"	500	.016	"	18.45	.90		19.35	22

09 54 23 – Linear Metal Ceilings

09 54 23.10 Metal Ceilings

		Crew	Daily Output	Labor-Hours	Unit	Material	2022 Bare Costs Labor	Equipment	Total	Total Incl O&P
0010	**METAL CEILINGS**									
0015	Solid alum. planks, 3-1/4" x 12', open reveal	1 Carp	500	.016	S.F.	2.44	.90		3.34	4.11
0020	Closed reveal		500	.016		3.13	.90		4.03	4.87
0030	7-1/4" x 12', open reveal		500	.016		4.17	.90		5.07	6
0040	Closed reveal		500	.016		5.30	.90		6.20	7.30
0050	Metal, open cell, 2' x 2', 6" cell		500	.016		9.05	.90		9.95	11.40
0060	8" cell		500	.016		10	.90		10.90	12.45
0070	2' x 4', 6" cell		500	.016		5.95	.90		6.85	8
0080	8" cell	↓	500	.016		5.95	.90		6.85	8

09 54 26 – Suspended Wood Ceilings

09 54 26.10 Wood Ceilings

		Crew	Daily Output	Labor-Hours	Unit	Material	2022 Bare Costs Labor	Equipment	Total	Total Incl O&P
0010	**WOOD CEILINGS**									
1000	4"-6" wood slats on heavy duty 15/16" T-bar grid	2 Carp	250	.064	S.F.	22.50	3.60		26.10	30.50

09 54 33 – Decorative Panel Ceilings

09 54 33.20 Metal Panel Ceilings

		Crew	Daily Output	Labor-Hours	Unit	Material	2022 Bare Costs Labor	Equipment	Total	Total Incl O&P
0010	**METAL PANEL CEILINGS**									
0020	Lay-in or screwed to furring, not including grid									
0100	Tin ceilings, 2' x 2' or 2' x 4', bare steel finish	2 Carp	300	.053	S.F.	2.84	3		5.84	7.90
0120	Painted white finish		300	.053	"	4.36	3		7.36	9.55
0140	Copper, chrome or brass finish		300	.053	L.F.	4.35	3		7.35	9.55
0200	Cornice molding, 2-1/2" to 3-1/2" wide, 4' long, bare steel finish		200	.080	S.F.	2.04	4.50		6.54	9.40
0220	Painted white finish		200	.080		3.19	4.50		7.69	10.65
0240	Copper, chrome or brass finish		200	.080		3.48	4.50		7.98	11
0320	5" to 6-1/2" wide, 4' long, bare steel finish		150	.107		2.30	6		8.30	12.10
0340	Painted white finish		150	.107		3.32	6		9.32	13.20

09 54 Specialty Ceilings

09 54 33 – Decorative Panel Ceilings

09 54 33.20 Metal Panel Ceilings	Crew	Daily Output	Labor-Hours	Unit	Material	2022 Bare Costs Labor	Equipment	Total	Total Incl O&P
0360	Copper, chrome or brass finish	2 Carp	150	.107	S.F.	3.45	6		9.45	13.35
0420	Flat molding, 3-1/2" to 5" wide, 4' long, bare steel finish		250	.064		3.32	3.60		6.92	9.35
0440	Painted white finish		250	.064		3.74	3.60		7.34	9.80
0460	Copper, chrome or brass finish		250	.064		3.71	3.60		7.31	9.80

09 61 Flooring Treatment

09 61 19 – Concrete Floor Staining

09 61 19.40 Floors, Interior

		Crew	Daily Output	Labor-Hours	Unit	Material	2022 Bare Costs Labor	Equipment	Total	Total Incl O&P
0010	**FLOORS, INTERIOR**									
0300	Acid stain and sealer									
0310	Stain, one coat	1 Pord	650	.012	S.F.	.12	.59		.71	1.06
0320	Two coats		570	.014		.24	.67		.91	1.33
0330	Acrylic sealer, one coat		2600	.003		.25	.15		.40	.51
0340	Two coats		1400	.006		.51	.27		.78	.99

09 62 Specialty Flooring

09 62 19 – Laminate Flooring

09 62 19.10 Floating Floor

		Crew	Daily Output	Labor-Hours	Unit	Material	2022 Bare Costs Labor	Equipment	Total	Total Incl O&P
0010	**FLOATING FLOOR**									
8300	Floating floor, laminate, wood pattern strip, complete	1 Clab	133	.060	S.F.	5.45	2.74		8.19	10.35
8310	Components, T&G wood composite strips					4.20			4.20	4.62
8320	Film					.74			.74	.81
8330	Foam					.29			.29	.32
8340	Adhesive					.89			.89	.98
8350	Installation kit					.25			.25	.28
8360	Trim, 2" wide x 3' long				L.F.	8			8	8.80
8370	Reducer moulding				"	7.40			7.40	8.15

09 62 29 – Cork Flooring

09 62 29.10 Cork Tile Flooring

			Crew	Daily Output	Labor-Hours	Unit	Material	2022 Bare Costs Labor	Equipment	Total	Total Incl O&P
0010	**CORK TILE FLOORING**										
2200	Cork tile, standard finish, 1/8" thick	G	1 Tilf	315	.025	S.F.	3.51	1.35		4.86	5.95
2250	3/16" thick	G		315	.025		5.10	1.35		6.45	7.70
2300	5/16" thick	G		315	.025		4.49	1.35		5.84	7.05
2350	1/2" thick	G		315	.025		4.12	1.35		5.47	6.65
2500	Urethane finish, 1/8" thick	G		315	.025		5.70	1.35		7.05	8.40
2550	3/16" thick	G		315	.025		8.35	1.35		9.70	11.30
2600	5/16" thick	G		315	.025		6.90	1.35		8.25	9.70
2650	1/2" thick	G		315	.025		6.25	1.35		7.60	8.95

09 63 Masonry Flooring

09 63 13 – Brick Flooring

09 63 13.10 Miscellaneous Brick Flooring		Crew	Daily Output	Labor-Hours	Unit	Material	2022 Bare Costs Labor	Equipment	Total	Total Incl O&P
0010	**MISCELLANEOUS BRICK FLOORING**									
0020	Acid-proof shales, red, 8" x 3-3/4" x 1-1/4" thick	D-7	.43	37.209	M	625	1,775		2,400	3,475
0050	2-1/4" thick	D-1	.40	40		1,025	2,000		3,025	4,325
0200	Acid-proof clay brick, 8" x 3-3/4" x 2-1/4" thick	[G]	.40	40	↓	1,050	2,000		3,050	4,350
0250	9" x 4-1/2" x 3"	[G]	95	.168	S.F.	4.20	8.45		12.65	18.10
0260	Cast ceramic, pressed, 4" x 8" x 1/2", unglazed	D-7	100	.160		7.50	7.70		15.20	20.50
0270	Glazed		100	.160		10	7.70		17.70	23
0280	Hand molded flooring, 4" x 8" x 3/4", unglazed		95	.168		9.90	8.10		18	23.50
0290	Glazed		95	.168		10.90	8.10		19	24.50
0300	8" hexagonal, 3/4" thick, unglazed		85	.188		9.50	9.05		18.55	24.50
0310	Glazed	↓	85	.188		17.15	9.05		26.20	33
0400	Heavy duty industrial, cement mortar bed, 2" thick, not incl. brick	D-1	80	.200		.91	10.05		10.96	17.05
0450	Acid-proof joints, 1/4" wide	"	65	.246		1.72	12.35		14.07	21.50
0500	Pavers, 8" x 4", 1" to 1-1/4" thick, red	D-7	95	.168		2.16	8.10		10.26	15
0510	Ironspot	"	95	.168		5.40	8.10		13.50	18.55
0540	1-3/8" to 1-3/4" thick, red	D-1	95	.168	↓	2.07	8.45		10.52	15.80
0560	Ironspot	D-1	95	.168	S.F.	5.35	8.45		13.80	19.35
0580	2-1/4" thick, red		90	.178		12.25	8.90		21.15	28
0590	Ironspot		90	.178		22.50	8.90		31.40	39.50
0700	Paver, adobe brick, 6" x 12", 1/2" joint	[G] ↓	42	.381		1.30	19.10		20.40	32
0710	Mexican red, 12" x 12"	[G] 1 Tilf	48	.167		1.80	8.85		10.65	15.85
0720	Saltillo, 12" x 12"	[G] "	48	.167	↓	1.38	8.85		10.23	15.35
0800	For sidewalks and patios with pavers, see Section 32 14 16.10									
0870	For epoxy joints, add	D-1	600	.027	S.F.	3.27	1.34		4.61	5.75
0880	For Furan underlayment, add	"	600	.027		2.70	1.34		4.04	5.10
0890	For waxed surface, steam cleaned, add	A-1H	1000	.008	↓	.21	.36	.08	.65	.90
9000	Minimum labor/equipment charge	1 Bric	2	4	Job		221		221	355

09 63 40 – Stone Flooring

09 63 40.10 Marble

		Crew	Daily Output	Labor-Hours	Unit	Material	2022 Bare Costs Labor	Equipment	Total	Total Incl O&P
0010	**MARBLE**									
0020	Thin gauge tile, 12" x 6", 3/8", white Carara	D-7	60	.267	S.F.	7.65	12.80		20.45	28.50
0100	Travertine		60	.267		8.30	12.80		21.10	29
0200	12" x 12" x 3/8", thin set, floors		60	.267		.81	12.80		13.61	21
0300	On walls		52	.308	↓	.81	14.75		15.56	24
1000	Marble threshold, 4" wide x 36" long x 5/8" thick, white		60	.267	Ea.	.81	12.80		13.61	21
9000	Minimum labor/equipment charge	↓	3	5.333	Job		256		256	400

09 63 40.20 Slate Tile

		Crew	Daily Output	Labor-Hours	Unit	Material	2022 Bare Costs Labor	Equipment	Total	Total Incl O&P
0010	**SLATE TILE**									
0020	Vermont, 6" x 6" x 1/4" thick, thin set	D-7	180	.089	S.F.	7.80	4.26		12.06	15.20
9000	Minimum labor/equipment charge	"	3	5.333	Job		256		256	400

For customer support on your Commercial Renovation Costs with RSMeans data, call 800.448.8182.

323

09 64 Wood Flooring

09 64 16 – Wood Block Flooring

09 64 16.10 End Grain Block Flooring

		Crew	Daily Output	Labor-Hours	Unit	Material	2022 Bare Costs Labor	Equipment	Total	Total Incl O&P
0010	**END GRAIN BLOCK FLOORING**									
0020	End grain flooring, coated, 2" thick	1 Carp	295	.027	S.F.	12	1.53		13.53	15.65
0400	Natural finish, 1" thick, fir		125	.064		4.14	3.60		7.74	10.25
0600	1-1/2" thick, pine		125	.064		4.06	3.60		7.66	10.15
0700	2" thick, pine		125	.064		4.70	3.60		8.30	10.85
9000	Minimum labor/equipment charge		2	4	Job		225		225	360

09 64 23 – Wood Parquet Flooring

09 64 23.10 Wood Parquet

		Crew	Daily Output	Labor-Hours	Unit	Material	2022 Bare Costs Labor	Equipment	Total	Total Incl O&P
0010	**WOOD PARQUET** flooring									
5200	Parquetry, 5/16" thk, no finish, oak, plain pattern	1 Carp	160	.050	S.F.	10.95	2.82		13.77	16.50
5300	Intricate pattern		100	.080		19.95	4.50		24.45	29
5500	Teak, plain pattern		160	.050		7.65	2.82		10.47	12.85
5600	Intricate pattern		100	.080		13.05	4.50		17.55	21.50
5650	13/16" thick, select grade oak, plain pattern	1 Carp	160	.050	S.F.	14.75	2.82		17.57	20.50
5700	Intricate pattern		100	.080		22	4.50		26.50	31.50
5800	Custom parquetry, including finish, plain pattern		100	.080		23.50	4.50		28	32.50
5900	Intricate pattern		50	.160		32	9		41	49.50
6700	Parquetry, prefinished white oak, 5/16" thick, plain pattern		160	.050		11.10	2.82		13.92	16.65
6800	Intricate pattern		100	.080		10.15	4.50		14.65	18.35
7000	Walnut or teak, parquetry, plain pattern		160	.050		17.95	2.82		20.77	24
7100	Intricate pattern		100	.080		27	4.50		31.50	36.50
7200	Acrylic wood parquet blocks, 12" x 12" x 5/16",									
7210	Irradiated, set in epoxy	1 Carp	160	.050	S.F.	14.25	2.82		17.07	20

09 64 29 – Wood Strip and Plank Flooring

09 64 29.10 Wood

		Crew	Daily Output	Labor-Hours	Unit	Material	2022 Bare Costs Labor	Equipment	Total	Total Incl O&P
0010	**WOOD** R061110-30									
0020	Fir, vertical grain, 1" x 4", not incl. finish, grade B & better	1 Carp	255	.031	S.F.	5.30	1.77		7.07	8.65
0100	Grade C & better		255	.031		5.35	1.77		7.12	8.70
0300	Flat grain, 1" x 4", not incl. finish, grade B & better		255	.031		5.15	1.77		6.92	8.45
0400	Grade C & better		255	.031		3.95	1.77		5.72	7.15
4000	Maple, strip, 25/32" x 2-1/4", not incl. finish, select		170	.047		5.35	2.65		8	10.05
4100	#2 & better		170	.047		5.05	2.65		7.70	9.75
4300	33/32" x 3-1/4", not incl. finish, #1 grade		170	.047		8.05	2.65		10.70	13.05
4400	#2 & better		170	.047		6.95	2.65		9.60	11.85
4600	Oak, white or red, 25/32" x 2-1/4", not incl. finish									
4700	#1 common	1 Carp	170	.047	S.F.	4.16	2.65		6.81	8.80
4900	Select quartered, 2-1/4" wide		170	.047		4.38	2.65		7.03	9.05
5000	Clear		170	.047		4.63	2.65		7.28	9.30
6100	Prefinished, white oak, prime grade, 2-1/4" wide		170	.047		4.49	2.65		7.14	9.15
6200	3-1/4" wide		185	.043		3.94	2.43		6.37	8.20
6400	Ranch plank		145	.055		5.90	3.11		9.01	11.45
6500	Hardwood blocks, 9" x 9", 25/32" thick		160	.050		10.30	2.82		13.12	15.80
7400	Yellow pine, 3/4" x 3-1/8", T&G, C & better, not incl. finish		200	.040		3.46	2.25		5.71	7.40
7500	Refinish wood floor, sand, 2 coats poly, wax, soft wood	1 Clab	400	.020		.24	.91		1.15	1.72
7600	Hardwood		130	.062		.24	2.81		3.05	4.73
7800	Sanding and finishing, 2 coats polyurethane		295	.027		.24	1.24		1.48	2.23
7900	Subfloor and underlayment, see Section 06 16									
8015	Transition molding, 2-1/4" wide, 5' long	1 Carp	19.20	.417	Ea.	26.50	23.50		50	67
9000	Minimum labor/equipment charge	"	2	4	Job		225		225	360

09 64 Wood Flooring

09 64 36 – Bamboo Flooring

09 64 36.10 Flooring, Bamboo

		Crew	Daily Output	Labor-Hours	Unit	Material	2022 Bare Costs Labor	Equipment	Total	Total Incl O&P
0010	**FLOORING, BAMBOO**									
8600	Flooring, wood, bamboo strips, unfinished, 5/8" x 4" x 3'	G 1 Carp	255	.031	S.F.	7.85	1.77		9.62	11.45
8610	5/8" x 4" x 4'	G	275	.029		7.90	1.64		9.54	11.30
8620	5/8" x 4" x 6'	G	295	.027		2.96	1.53		4.49	5.70
8630	Finished, 5/8" x 4" x 3'	G	255	.031		8.35	1.77		10.12	12
8640	5/8" x 4" x 4'	G	275	.029		9.10	1.64		10.74	12.60
8650	5/8" x 4" x 6'	G	295	.027	↓	4.11	1.53		5.64	6.95
8660	Stair treads, unfinished, 1-1/16" x 11-1/2" x 4'	G	18	.444	Ea.	32	25		57	75
8670	Finished, 1-1/16" x 11-1/2" x 4'	G	18	.444		80	25		105	128
8680	Stair risers, unfinished, 5/8" x 7-1/2" x 4'	G	18	.444		40	25		65	84
8690	Finished, 5/8" x 7-1/2" x 4'	G	18	.444		35	25		60	78.50
8700	Stair nosing, unfinished, 6' long	G	16	.500		59.50	28		87.50	110
8710	Finished, 6' long	G ↓	16	.500	↓	75	28		103	127

09 65 Resilient Flooring

09 65 10 – Resilient Tile Underlayment

09 65 10.10 Latex Underlayment

		Crew	Daily Output	Labor-Hours	Unit	Material	2022 Bare Costs Labor	Equipment	Total	Total Incl O&P
0010	**LATEX UNDERLAYMENT**									
3600	Latex underlayment, 1/8" thk., cementitious for resilient flooring	1 Tilf	160	.050	S.F.	.60	2.66		3.26	4.81
4000	Liquid, fortified				Gal.	17.45			17.45	19.20

09 65 13 – Resilient Base and Accessories

09 65 13.13 Resilient Base

		Crew	Daily Output	Labor-Hours	Unit	Material	2022 Bare Costs Labor	Equipment	Total	Total Incl O&P
0010	**RESILIENT BASE**									
0690	1/8" vinyl base, 2-1/2" H, straight or cove, standard colors	1 Tilf	315	.025	L.F.	1.36	1.35		2.71	3.61
0700	4" high		315	.025		1.39	1.35		2.74	3.64
0710	6" high		315	.025	↓	2.70	1.35		4.05	5.10
0720	Corners, 2-1/2" high		315	.025	Ea.	3.53	1.35		4.88	6
0730	4" high		315	.025		3.89	1.35		5.24	6.40
0740	6" high		315	.025	↓	4.50	1.35		5.85	7.05
0800	1/8" rubber base, 2-1/2" H, straight or cove, standard colors		315	.025	L.F.	1.49	1.35		2.84	3.75
1100	4" high		315	.025		1.72	1.35		3.07	4
1110	6" high		315	.025	↓	3.52	1.35		4.87	6
1150	Corners, 2-1/2" high		315	.025	Ea.	2.92	1.35		4.27	5.30
1153	4" high		315	.025		3.65	1.35		5	6.15
1155	6" high	↓	315	.025	↓	3.52	1.35		4.87	6
1450	For premium color/finish add					50%				
1500	Millwork profile	1 Tilf	315	.025	L.F.	5.05	1.35		6.40	7.65

09 65 13.23 Resilient Stair Treads and Risers

		Crew	Daily Output	Labor-Hours	Unit	Material	2022 Bare Costs Labor	Equipment	Total	Total Incl O&P
0010	**RESILIENT STAIR TREADS AND RISERS**									
0300	Rubber, molded tread, 12" wide, 5/16" thick, black	1 Tilf	115	.070	L.F.	18.90	3.70		22.60	27
0400	Colors		115	.070		18.90	3.70		22.60	27
0600	1/4" thick, black		115	.070		10.95	3.70		14.65	17.80
0700	Colors		115	.070		10.95	3.70		14.65	17.80
0900	Grip strip safety tread, colors, 5/16" thick		115	.070		22.50	3.70		26.20	30.50
1000	3/16" thick		120	.067	↓	19.25	3.54		22.79	26.50
1200	Landings, smooth sheet rubber, 1/8" thick		120	.067	S.F.	5.05	3.54		8.59	11.10
1300	3/16" thick		120	.067	"	6.35	3.54		9.89	12.55
1500	Nosings, 3" wide, 3/16" thick, black		140	.057	L.F.	10.70	3.04		13.74	16.55
1600	Colors		140	.057		12	3.04		15.04	17.95
1800	Risers, 7" high, 1/8" thick, flat		250	.032		11.30	1.70		13	15.10

09 65 Resilient Flooring

09 65 13 – Resilient Base and Accessories

09 65 13.23 Resilient Stair Treads and Risers

		Crew	Daily Output	Labor-Hours	Unit	Material	2022 Bare Costs Labor	Equipment	Total	Total Incl O&P
1900	Coved	1 Tilf	250	.032	L.F.	7.60	1.70		9.30	11.05
2100	Vinyl, molded tread, 12" wide, colors, 1/8" thick		115	.070		12.35	3.70		16.05	19.35
2200	1/4" thick		115	.070		9.35	3.70		13.05	16
2300	Landing material, 1/8" thick		200	.040	S.F.	8.55	2.13		10.68	12.75
2400	Riser, 7" high, 1/8" thick, coved		175	.046	L.F.	4.36	2.43		6.79	8.60
2500	Tread and riser combined, 1/8" thick		80	.100	"	13.20	5.30		18.50	23
9000	Minimum labor/equipment charge		3	2.667	Job		142		142	221

09 65 13.37 Vinyl Transition Strips

		Crew	Daily Output	Labor-Hours	Unit	Material	2022 Bare Costs Labor	Equipment	Total	Total Incl O&P
0010	**VINYL TRANSITION STRIPS**									
0100	Various mats. to various mats., adhesive applied, 1/4" to 1/8"	1 Tilf	315	.025	L.F.	6	1.35		7.35	8.70
0105	0.08" to 1/8"		315	.025		4.21	1.35		5.56	6.75
0110	0.08" to 1/4"		315	.025		5.25	1.35		6.60	7.85
0115	1/4" to 3/8"		315	.025		7.50	1.35		8.85	10.35
0120	1/4" to 1/2"		315	.025		1.44	1.35		2.79	3.69
0125	1/4" to 0.08"		315	.025		1.51	1.35		2.86	3.77
0200	Vinyl wheeled trans. strips, carpet to var. mats., 1/4" to 1/8" x 2-1/2"		315	.025		6	1.35		7.35	8.70
0205	1/4" to 1/8" x 4"		315	.025		7.25	1.35		8.60	10.05
0210	Various mats. to various mats. 1/4" to 0.08" x 2-1/2"		315	.025		5.25	1.35		6.60	7.85
0215	Carpet to various materials, 1/4" to flush x 2-1/2"		315	.025		5	1.35		6.35	7.60
0220	1/4" to flush x 4"		315	.025		8.45	1.35		9.80	11.40
0225	Various materials to resilient, 3/8" to 1/8" x 2-1/2"		315	.025		5	1.35		6.35	7.60
0230	Carpet to various materials, 3/8" to 1/4" x 2-1/2"		315	.025		6.90	1.35		8.25	9.70
0235	1/4" to 1/4" x 2-1/2"		315	.025		8.45	1.35		9.80	11.40
0240	Various materials to resilient, 1/8" to 1/8" x 2-1/2"		315	.025		5.85	1.35		7.20	8.50
0245	Various materials to var. mats., 1/8" to flush x 2-1/2"		315	.025		4.29	1.35		5.64	6.85
0250	3/8" to flush x 4"	1 Tilf	315	.025	L.F.	8.65	1.35		10	11.60
0255	1/2" to flush x 4"		315	.025		10.55	1.35		11.90	13.70
0260	Various materials to resilient, 1/8" to 0.08" x 2-1/2"		315	.025		4.54	1.35		5.89	7.10
0265	0.08" to 0.08" x 2-1/2"		315	.025		4.72	1.35		6.07	7.30
0270	3/8" to 0.08" x 2-1/2"		315	.025		4.72	1.35		6.07	7.30

09 65 16 – Resilient Sheet Flooring

09 65 16.10 Rubber and Vinyl Sheet Flooring

			Crew	Daily Output	Labor-Hours	Unit	Material	2022 Bare Costs Labor	Equipment	Total	Total Incl O&P
0010	**RUBBER AND VINYL SHEET FLOORING**										
5500	Linoleum, sheet goods	G	1 Tilf	360	.022	S.F.	3.92	1.18		5.10	6.15
5900	Rubber, sheet goods, 36" wide, 1/8" thick			120	.067		1.32	3.54		4.86	7
5950	3/16" thick			100	.080		9.15	4.25		13.40	16.75
6000	1/4" thick			90	.089		1.83	4.72		6.55	9.35
8000	Vinyl sheet goods, backed, .065" thick, plain pattern/colors			250	.032		1.20	1.70		2.90	3.97
8050	Intricate pattern/colors			200	.040		1.20	2.13		3.33	4.64
8100	.080" thick, plain pattern/colors			230	.035		2.79	1.85		4.64	5.95
8150	Intricate pattern/colors			200	.040		5.20	2.13		7.33	9.05
8200	.125" thick, plain pattern/colors			230	.035		3.47	1.85		5.32	6.70
8250	Intricate pattern/colors			200	.040		6.80	2.13		8.93	10.75
8400	For welding seams, add			100	.080	L.F.	.11	4.25		4.36	6.75
8450	For integral cove base, add			175	.046	"	8.70	2.43		11.13	13.35
8700	Adhesive cement, 1 gallon per 200 to 300 S.F.					Gal.	19.50			19.50	21.50
8800	Asphalt primer, 1 gallon per 300 S.F.						15.60			15.60	17.15
8900	Emulsion, 1 gallon per 140 S.F.						48			48	52.50

09 65 Resilient Flooring

09 65 19 – Resilient Tile Flooring

09 65 19.19 Vinyl Composition Tile Flooring

		Crew	Daily Output	Labor-Hours	Unit	Material	2022 Bare Costs Labor	2022 Bare Costs Equipment	Total	Total Incl O&P
0010	**VINYL COMPOSITION TILE FLOORING**									
7000	Vinyl composition tile, 12" x 12", 1/16" thick	1 Tilf	500	.016	S.F.	.77	.85		1.62	2.18
7050	Embossed		500	.016		1.40	.85		2.25	2.87
7100	Marbleized		500	.016		1.20	.85		2.05	2.65
7150	Solid		500	.016		3.27	.85		4.12	4.93
7200	3/32" thick, embossed		500	.016		1.29	.85		2.14	2.75
7250	Marbleized		500	.016		.94	.85		1.79	2.36
7300	Solid		500	.016		1.21	.85		2.06	2.66
7350	1/8" thick, marbleized		500	.016		1.96	.85		2.81	3.49
7400	Solid		500	.016		1.10	.85		1.95	2.54
7450	Conductive	▼	500	.016	▼	8.35	.85		9.20	10.55

09 65 19.23 Vinyl Tile Flooring

		Crew	Daily Output	Labor-Hours	Unit	Material	2022 Bare Costs Labor	2022 Bare Costs Equipment	Total	Total Incl O&P
0010	**VINYL TILE FLOORING**									
7500	Vinyl tile, 12" x 12", 3/32" thick, standard colors/patterns	1 Tilf	500	.016	S.F.	1.36	.85		2.21	2.83
7550	1/8" thick, standard colors/patterns	1 Tilf	500	.016	S.F.	1.17	.85		2.02	2.62
7600	1/8" thick, premium colors/patterns		500	.016		6.10	.85		6.95	8.05
7650	Solid colors		500	.016		1.10	.85		1.95	2.54
7700	Marbleized or Travertine pattern		500	.016		2.99	.85		3.84	4.62
7750	Florentine pattern		500	.016		1.98	.85		2.83	3.51
7800	Premium colors/patterns		500	.016	▼	3.25	.85		4.10	4.91
9500	Minimum labor/equipment charge	▼	4	2	Job		106		106	166

09 65 19.33 Rubber Tile Flooring

		Crew	Daily Output	Labor-Hours	Unit	Material	2022 Bare Costs Labor	2022 Bare Costs Equipment	Total	Total Incl O&P
0010	**RUBBER TILE FLOORING**									
6050	Rubber tile, marbleized colors, 12" x 12", 1/8" thick	1 Tilf	400	.020	S.F.	6.35	1.06		7.41	8.65
6100	3/16" thick		400	.020		7.85	1.06		8.91	10.30
6300	Special tile, plain colors, 1/8" thick		400	.020		7.85	1.06		8.91	10.30
6350	3/16" thick	▼	400	.020	▼	9.25	1.06		10.31	11.85

09 65 33 – Conductive Resilient Flooring

09 65 33.10 Conductive Rubber and Vinyl Flooring

		Crew	Daily Output	Labor-Hours	Unit	Material	2022 Bare Costs Labor	2022 Bare Costs Equipment	Total	Total Incl O&P
0010	**CONDUCTIVE RUBBER AND VINYL FLOORING**									
1700	Conductive flooring, rubber tile, 1/8" thick	1 Tilf	315	.025	S.F.	15.30	1.35		16.65	18.95
1800	Homogeneous vinyl tile, 1/8" thick	"	315	.025	"	4.15	1.35		5.50	6.70

09 65 66 – Resilient Athletic Flooring

09 65 66.10 Resilient Athletic Flooring

		Crew	Daily Output	Labor-Hours	Unit	Material	2022 Bare Costs Labor	2022 Bare Costs Equipment	Total	Total Incl O&P
0010	**RESILIENT ATHLETIC FLOORING**									
1000	Recycled rubber rolled goods, for weight rooms, 3/8" thk.	1 Tilf	315	.025	S.F.	2.64	1.35		3.99	5
1050	Interlocking 2' x 2' squares, rubber, 1/4" thk.		310	.026		13.05	1.37		14.42	16.50
1055	5/16" thk.		310	.026		2.32	1.37		3.69	4.69
1060	3/8" thk.		300	.027		4.39	1.42		5.81	7.05
1065	1/2" thk.		310	.026		4.92	1.37		6.29	7.55
2000	Vinyl sheet flooring, 1/4" thk.	▼	315	.025	▼	5.15	1.35		6.50	7.80

09 66 13 – Portland Cement Terrazzo Flooring

09 66 13.10 Portland Cement Terrazzo	Crew	Daily Output	Labor-Hours	Unit	Material	2022 Bare Costs Labor	Equipment	Total	Total Incl O&P
0010	**PORTLAND CEMENT TERRAZZO**, cast-in-place									
0020	Cove base, 6" high, 16 ga. zinc	1 Mstz	20	.400	L.F.	4.08	21.50		25.58	38
0100	Curb, 6" high and 6" wide		6	1.333		6.70	71		77.70	118
0300	Divider strip for floors, 14 ga., 1-1/4" deep, zinc		375	.021		1.48	1.14		2.62	3.41
0400	Brass		375	.021		2.99	1.14		4.13	5.05
0600	Heavy top strip 1/4" thick, 1-1/4" deep, zinc		300	.027		2.27	1.42		3.69	4.72
1200	For thin set floors, 16 ga., 1/2" x 1/2", zinc		350	.023		1.55	1.22		2.77	3.62
1500	Floor, bonded to concrete, 1-3/4" thick, gray cement	J-3	75	.213	S.F.	4.09	10.45	3.19	17.73	24.50
1600	White cement, mud set		75	.213		4.82	10.45	3.19	18.46	25
1800	Not bonded, 3" total thickness, gray cement		70	.229		5.05	11.20	3.42	19.67	27
1900	White cement, mud set		70	.229		5.80	11.20	3.42	20.42	27.50
9000	Minimum labor/equipment charge	1 Mstz	1	8	Job		425		425	665

09 66 16 – Terrazzo Floor Tile

09 66 16.10 Tile or Terrazzo Base

		Crew	Daily Output	Labor-Hours	Unit	Material	2022 Bare Costs Labor	Equipment	Total	Total Incl O&P
0010	**TILE OR TERRAZZO BASE**									
0020	Scratch coat only	1 Mstz	150	.053	S.F.	.20	2.85		3.05	4.67
0500	Scratch and brown coat only		75	.107	"	.40	5.70		6.10	9.35
9000	Minimum labor/equipment charge		1	8	Job		425		425	665

09 66 16.13 Portland Cement Terrazzo Floor Tile

		Crew	Daily Output	Labor-Hours	Unit	Material	2022 Bare Costs Labor	Equipment	Total	Total Incl O&P
0010	**PORTLAND CEMENT TERRAZZO FLOOR TILE**									
1200	Floor tiles, non-slip, 1" thick, 12" x 12"	D-1	60	.267	S.F.	29	13.35		42.35	53.50
1300	1-1/4" thick, 12" x 12"		60	.267		27	13.35		40.35	51.50
1500	16" x 16"		50	.320		30	16.05		46.05	58.50
1600	1-1/2" thick, 16" x 16"		45	.356		27.50	17.80		45.30	58.50
1800	For Venetian terrazzo, add					8.20			8.20	9.05
1900	For white cement, add					.75			.75	.83

09 66 16.30 Terrazzo, Precast

		Crew	Daily Output	Labor-Hours	Unit	Material	2022 Bare Costs Labor	Equipment	Total	Total Incl O&P
0010	**TERRAZZO, PRECAST**									
0020	Base, 6" high, straight	1 Mstz	70	.114	L.F.	13.25	6.10		19.35	24
0100	Cove		60	.133		18	7.10		25.10	31
0300	8" high, straight		60	.133		16.90	7.10		24	29.50
0400	Cove		50	.160		26	8.55		34.55	42
0600	For white cement, add					.64			.64	.70
0700	For 16 ga. zinc toe strip, add					2.49			2.49	2.74
0900	Curbs, 4" x 4" high	1 Mstz	40	.200		46	10.70		56.70	67
1000	8" x 8" high	"	30	.267		50.50	14.25		64.75	77.50
2400	Stair treads, 1-1/2" thick, non-slip, three line pattern	2 Mstz	70	.229		57	12.20		69.20	82
2500	Nosing and two lines		70	.229		57	12.20		69.20	82
2700	2" thick treads, straight		60	.267		66.50	14.25		80.75	95
2800	Curved		50	.320		89.50	17.10		106.60	125
3000	Stair risers, 1" thick, to 6" high, straight sections		60	.267		17.20	14.25		31.45	41
3100	Cove		50	.320		21.50	17.10		38.60	50
3300	Curved, 1" thick, to 6" high, vertical		48	.333		30	17.80		47.80	61
3400	Cove		38	.421		44.50	22.50		67	84
3600	Stair tread and riser, single piece, straight, smooth surface		60	.267		70.50	14.25		84.75	99.50
3700	Non skid surface		40	.400		91.50	21.50		113	134
3900	Curved tread and riser, smooth surface		40	.400		101	21.50		122.50	146
4000	Non skid surface		32	.500		124	26.50		150.50	179
4200	Stair stringers, notched, 1" thick		25	.640		40.50	34		74.50	98
4300	2" thick		22	.727		47.50	39		86.50	113
4500	Stair landings, structural, non-slip, 1-1/2" thick		85	.188	S.F.	45	10.05		55.05	65
4600	3" thick		75	.213		62.50	11.40		73.90	86.50

09 66 Terrazzo Flooring

09 66 16 – Terrazzo Floor Tile

09 66 16.30 Terrazzo, Precast	Crew	Daily Output	Labor-Hours	Unit	Material	2022 Bare Costs Labor	Equipment	Total	Total Incl O&P
4800	Wainscot, 12" x 12" x 1" tiles	1 Mstz	12	.667	S.F.	10.05	35.50		45.55	66.50
4900	16" x 16" x 1-1/2" tiles	"	8	1		19.85	53.50		73.35	106
9500	Minimum labor/equipment charge	1 Tilf	2	4	Job		213		213	330

09 66 23 – Resinous Matrix Terrazzo Flooring

09 66 23.16 Epoxy-Resin Terrazzo Flooring

		Crew	Daily Output	Labor-Hours	Unit	Material	Labor	Equipment	Total	Total Incl O&P
0010	**EPOXY-RESIN TERRAZZO FLOORING**									
1800	Epoxy terrazzo, 1/4" thick, chemical resistant, granite chips	J-3	200	.080	S.F.	6.45	3.93	1.20	11.58	14.50
1900	Recycled porcelain	"	150	.107	"	10.95	5.25	1.60	17.80	22

09 67 Fluid-Applied Flooring

09 67 13 – Elastomeric Liquid Flooring

09 67 13.13 Elastomeric Liquid Flooring

		Crew	Daily Output	Labor-Hours	Unit	Material	Labor	Equipment	Total	Total Incl O&P
0010	**ELASTOMERIC LIQUID FLOORING**									
0020	Cementitious acrylic, 1/4" thick	C-6	520	.092	S.F.	1.91	4.36	.11	6.38	9.10
0200	Methyl methachrylate, 1/4" thick	C-8A	3000	.016	"	7.90	.78		8.68	9.90

09 67 23 – Resinous Flooring

09 67 23.23 Resinous Flooring

		Crew	Daily Output	Labor-Hours	Unit	Material	Labor	Equipment	Total	Total Incl O&P
0010	**RESINOUS FLOORING**									
1200	Heavy duty epoxy topping, 1/4" thick,									
1300	500 to 1,000 S.F.	C-6	420	.114	S.F.	6.65	5.40	.13	12.18	16
1500	1,000 to 2,000 S.F.		450	.107		5.65	5.05	.12	10.82	14.30
1600	Over 10,000 S.F.		480	.100		5.35	4.72	.12	10.19	13.50

09 67 26 – Quartz Flooring

09 67 26.26 Quartz Flooring

		Crew	Daily Output	Labor-Hours	Unit	Material	Labor	Equipment	Total	Total Incl O&P
0010	**QUARTZ FLOORING**									
0600	Epoxy, with colored quartz chips, broadcast, 3/8" thick	C-6	675	.071	S.F.	3.54	3.36	.08	6.98	9.30
0700	1/2" thick		490	.098		4.75	4.62	.11	9.48	12.70
0900	Troweled, minimum		560	.086		3.80	4.04	.10	7.94	10.70
1000	Maximum		480	.100		6.55	4.72	.12	11.39	14.85

09 68 Carpeting

09 68 05 – Carpet Accessories

09 68 05.11 Flooring Transition Strip

		Crew	Daily Output	Labor-Hours	Unit	Material	Labor	Equipment	Total	Total Incl O&P
0010	**FLOORING TRANSITION STRIP**									
0107	Clamp down brass divider, 12' strip, vinyl to carpet	1 Tilf	31.25	.256	Ea.	4.81	13.60		18.41	26.50
0117	Vinyl to hard surface	"	31.25	.256	"	4.81	13.60		18.41	26.50

09 68 10 – Carpet Pad

09 68 10.10 Commercial Grade Carpet Pad

		Crew	Daily Output	Labor-Hours	Unit	Material	Labor	Equipment	Total	Total Incl O&P
0010	**COMMERCIAL GRADE CARPET PAD**									
9000	Sponge rubber pad, 20 oz./sq. yd.	1 Tilf	150	.053	S.Y.	4.88	2.83		7.71	9.75
9100	40 to 62 oz./sq. yd.		150	.053		8.15	2.83		10.98	13.35
9200	Felt pad, 20 oz./sq. yd.		150	.053		6.35	2.83		9.18	11.40
9300	32 to 56 oz./sq. yd.		150	.053		12.20	2.83		15.03	17.80
9400	Bonded urethane pad, 2.7 density		150	.053		5.95	2.83		8.78	10.95
9500	13.0 density		150	.053		7.95	2.83		10.78	13.15
9600	Prime urethane pad, 2.7 density		150	.053		3.65	2.83		6.48	8.45
9700	13.0 density		150	.053		8.15	2.83		10.98	13.35

For customer support on your Commercial Renovation Costs with RSMeans data, call 800.448.8182.

329

09 68 13 – Tile Carpeting

09 68 13.10 Carpet Tile	Crew	Daily Output	Labor-Hours	Unit	Material	2022 Bare Costs Labor	Equipment	Total	Total Incl O&P
0010 **CARPET TILE**									
0100 Tufted nylon, 18" x 18", hard back, 20 oz.	1 Tilf	80	.100	S.Y.	26.50	5.30		31.80	37.50
0110 26 oz.		80	.100		25	5.30		30.30	36
0200 Cushion back, 20 oz.		80	.100		11.55	5.30		16.85	21
0210 26 oz.		80	.100		36.50	5.30		41.80	49
1100 Tufted, 24" x 24", hard back, 24 oz. nylon		80	.100		32	5.30		37.30	44
1180 35 oz.		80	.100		38.50	5.30		43.80	51
5060 42 oz.		80	.100		43.50	5.30		48.80	56.50
6000 Electrostatic dissapative carpet tile, 24" x 24", 24 oz.		80	.100		38.50	5.30		43.80	51
6100 Electrostatic dissapative carpet tile for access floors, 24" x 24", 24 oz.		80	.100		48.50	5.30		53.80	62

09 68 16 – Sheet Carpeting

09 68 16.10 Sheet Carpet

09 68 16.10 Sheet Carpet	Crew	Daily Output	Labor-Hours	Unit	Material	2022 Bare Costs Labor	Equipment	Total	Total Incl O&P
0010 **SHEET CARPET**									
0700 Nylon, level loop, 26 oz., light to medium traffic	1 Tilf	75	.107	S.Y.	15.90	5.65		21.55	26.50
0720 28 oz., light to medium traffic		75	.107		32	5.65		37.65	44.50
0900 32 oz., medium traffic		75	.107		41.50	5.65		47.15	54.50
1100 40 oz., medium to heavy traffic		75	.107		56	5.65		61.65	70.50
2920 Nylon plush, 30 oz., medium traffic		75	.107		15.15	5.65		20.80	25.50
3000 36 oz., medium traffic		75	.107		38.50	5.65		44.15	51.50
3100 42 oz., medium to heavy traffic		70	.114		20.50	6.05		26.55	32
3200 46 oz., medium to heavy traffic		70	.114		54	6.05		60.05	68.50
3300 54 oz., heavy traffic		70	.114		32	6.05		38.05	45
4110 Wool, level loop, 40 oz., medium traffic		70	.114		117	6.05		123.05	139
4500 50 oz., medium to heavy traffic		70	.114		109	6.05		115.05	130
4700 Patterned, 32 oz., medium to heavy traffic		70	.114		99.50	6.05		105.55	119
4900 48 oz., heavy traffic		70	.114		109	6.05		115.05	130
5000 For less than full roll (approx. 1500 S.F.), add					25%				
5100 For small rooms, less than 12' wide, add						25%			
5200 For large open areas (no cuts), deduct						25%			
5600 For bound carpet baseboard, add	1 Tilf	300	.027	L.F.	2.03	1.42		3.45	4.44
5610 For stairs, not incl. price of carpet, add	"	30	.267	Riser		14.15		14.15	22
5620 For borders and patterns, add to labor						18%			
8950 For tackless, stretched installation, add padding from 09 68 10.10 to above									
9850 For brand-named specific fiber, add				S.Y.	25%				
9900 Carpet cleaning machine, rent				Day				38	42
9910 Minimum labor/equipment charge	1 Tilf	3	2.667	Job		142		142	221

09 68 20 – Athletic Carpet

09 68 20.10 Indoor Athletic Carpet

09 68 20.10 Indoor Athletic Carpet	Crew	Daily Output	Labor-Hours	Unit	Material	2022 Bare Costs Labor	Equipment	Total	Total Incl O&P
0010 **INDOOR ATHLETIC CARPET**									
3700 Polyethylene, in rolls, no base incl., landscape surfaces	1 Tilf	275	.029	S.F.	3.59	1.55		5.14	6.35
3800 Nylon action surface, 1/8" thick		275	.029		4.05	1.55		5.60	6.85
3900 1/4" thick		275	.029		5.85	1.55		7.40	8.80
4000 3/8" thick		275	.029		7.35	1.55		8.90	10.45
5500 Polyvinyl chloride, sheet goods for gyms, 1/4" thick		80	.100		6.10	5.30		11.40	15.05
5600 3/8" thick		60	.133		8.95	7.10		16.05	21

09 69 Access Flooring

09 69 13 – Rigid-Grid Access Flooring

09 69 13.10 Access Floors	Crew	Daily Output	Labor-Hours	Unit	Material	2022 Bare Costs Labor	Equipment	Total	Total Incl O&P
0010 **ACCESS FLOORS**									
0015 Access floor pkg. including panel, pedestal, & stringers 1500 lbs. load									
0100 Package pricing, conc. fill panels, no fin., 6" ht.	4 Carp	750	.043	S.F.	15.20	2.40		17.60	20.50
0105 12" ht.		750	.043		17.45	2.40		19.85	23
0110 18" ht.		750	.043		15.60	2.40		18	21
0115 24" ht.		750	.043		15.80	2.40		18.20	21
0120 Package pricing, steel panels, no fin., 6" ht.		750	.043		17.15	2.40		19.55	22.50
0125 12" ht.		750	.043		18.30	2.40		20.70	24
0130 18" ht.		750	.043		17.75	2.40		20.15	23.50
0135 24" ht.		750	.043		18.10	2.40		20.50	23.50
0140 Package pricing, wood core panels, no fin., 6" ht.		750	.043		12.35	2.40		14.75	17.40
0145 12" ht.		750	.043		12.55	2.40		14.95	17.60
0150 18" ht.		750	.043		12.75	2.40		15.15	17.80
0155 24" ht.		750	.043		12.95	2.40		15.35	18
0160 Pkg. pricing, conc. fill pnls, no stringers, no fin., 6" ht, 1250 lbs. load		700	.046		18.15	2.57		20.72	24
0165 12" ht.		700	.046		18.35	2.57		20.92	24
0170 18" ht.		700	.046		18.55	2.57		21.12	24.50
0175 24" ht.		700	.046		18.80	2.57		21.37	24.50
0250 Panels, 2' x 2' conc. fill, no fin.	2 Carp	500	.032		23	1.80		24.80	28.50
0255 With 1/8" high pressure laminate		500	.032		57	1.80		58.80	66
0260 Metal panels with 1/8" high pressure laminate		500	.032		67.50	1.80		69.30	77.50
0265 Wood core panels with 1/8" high pressure laminate		500	.032		44.50	1.80		46.30	52
0400 Aluminum panels, no fin.		500	.032		35.50	1.80		37.30	42
0600 For carpet covering, add					9.35			9.35	10.30
0700 For vinyl floor covering, add					9.95			9.95	10.95
0900 For high pressure laminate covering, add					8.25			8.25	9.05
0910 For snap on stringer system, add	2 Carp	1000	.016		1.47	.90		2.37	3.05
1000 Machine cutouts after initial installation	1 Carp	50	.160	Ea.	20	9		29	36.50
1050 Pedestals, 6" to 12"	2 Carp	85	.188		9.35	10.60		19.95	27
1100 Air conditioning grilles, 4" x 12"	1 Carp	17	.471		73.50	26.50		100	123
1150 4" x 18"	"	14	.571		101	32		133	162
1200 Approach ramps, steel	2 Carp	60	.267	S.F.	30	15		45	57
1300 Aluminum	"	40	.400	"	35.50	22.50		58	75
1500 Handrail, 2 rail, aluminum	1 Carp	15	.533	L.F.	128	30		158	188

09 72 Wall Coverings

09 72 13 – Cork Wall Coverings

09 72 13.10 Covering, Cork Wall

	Crew	Daily Output	Labor-Hours	Unit	Material	2022 Bare Costs Labor	Equipment	Total	Total Incl O&P
0010 **COVERING, CORK WALL**									
0600 Cork tiles, light or dark, 12" x 12" x 3/16"	1 Pape	240	.033	S.F.	4.04	1.60		5.64	6.95
0700 5/16" thick		235	.034		3.38	1.64		5.02	6.30
0900 1/4" basket weave		240	.033		3.12	1.60		4.72	5.95
1000 1/2" natural, non-directional pattern		240	.033		6.60	1.60		8.20	9.80
1100 3/4" natural, non-directional pattern		240	.033		12.35	1.60		13.95	16.10
1200 Granular surface, 12" x 36", 1/2" thick		385	.021		1.35	1		2.35	3.07
1300 1" thick		370	.022		1.76	1.04		2.80	3.58
1500 Polyurethane coated, 12" x 12" x 3/16" thick		240	.033		4.26	1.60		5.86	7.20
1600 5/16" thick		235	.034		6.10	1.64		7.74	9.30
1800 Cork wallpaper, paperbacked, natural		480	.017		1.65	.80		2.45	3.09
1900 Colors		480	.017		2.14	.80		2.94	3.62

331

09 72 Wall Coverings

09 72 16 – Vinyl-Coated Fabric Wall Coverings

09 72 16.13 Flexible Vinyl Wall Coverings	Crew	Daily Output	Labor-Hours	Unit	Material	2022 Bare Costs Labor	2022 Bare Costs Equipment	Total	Total Incl O&P
0010 **FLEXIBLE VINYL WALL COVERINGS**									
3000 Vinyl wall covering, fabric-backed, lightweight, type 1 (12-15 oz./S.Y.)	1 Pape	640	.013	S.F.	1.57	.60		2.17	2.68
3300 Medium weight, type 2 (20-24 oz./S.Y.)		480	.017		1.23	.80		2.03	2.62
3400 Heavy weight, type 3 (28 oz./S.Y.)		435	.018		1.87	.88		2.75	3.46
3600 Adhesive, 5 gal. lots (18 S.Y./gal.)				Gal.	9.70			9.70	10.70

09 72 16.16 Rigid-Sheet Vinyl Wall Coverings

	Crew	Daily Output	Labor-Hours	Unit	Material	Labor	Equipment	Total	Total Incl O&P
0010 **RIGID-SHEET VINYL WALL COVERINGS**									
0100 Acrylic, modified, semi-rigid PVC, .028" thick	2 Carp	330	.048	S.F.	1.54	2.73		4.27	6.05
0110 .040" thick	"	320	.050	"	2.04	2.82		4.86	6.70

09 72 19 – Textile Wall Coverings

09 72 19.10 Textile Wall Covering

	Crew	Daily Output	Labor-Hours	Unit	Material	Labor	Equipment	Total	Total Incl O&P
0010 **TEXTILE WALL COVERING**, including sizing; add 10-30% waste @ takeoff									
0020 Silk	1 Pape	640	.013	S.F.	6.10	.60		6.70	7.65
0030 Cotton		640	.013		4.75	.60		5.35	6.15
0040 Linen		640	.013		4.36	.60		4.96	5.75
0050 Blend		640	.013		3.28	.60		3.88	4.56
0090 Grass cloth, natural fabric [G]		400	.020		1.82	.96		2.78	3.52
0100 Grass cloths with lining paper [G]		400	.020		1.26	.96		2.22	2.91
0110 Premium texture/color [G]		350	.023		3.47	1.10		4.57	5.55

09 72 20 – Natural Fiber Wall Covering

09 72 20.10 Natural Fiber Wall Covering

	Crew	Daily Output	Labor-Hours	Unit	Material	Labor	Equipment	Total	Total Incl O&P
0010 **NATURAL FIBER WALL COVERING**, including sizing; add 10-30% waste @ takeoff									
0015 Bamboo	1 Pape	640	.013	S.F.	2.72	.60		3.32	3.94
0030 Burlap		640	.013		2.72	.60		3.32	3.94
0045 Jute		640	.013		2.36	.60		2.96	3.55
0060 Sisal		640	.013		1.74	.60		2.34	2.86

09 72 23 – Wallpapering

09 72 23.10 Wallpaper

	Crew	Daily Output	Labor-Hours	Unit	Material	Labor	Equipment	Total	Total Incl O&P
0010 **WALLPAPER** including sizing; add 10-30% waste @ takeoff									
0050 Aluminum foil	1 Pape	275	.029	S.F.	1.07	1.40		2.47	3.39
0100 Copper sheets, .025" thick, vinyl backing		240	.033		5.70	1.60		7.30	8.80
0300 Phenolic backing		240	.033		7.40	1.60		9	10.70
2400 Gypsum-based, fabric-backed, fire resistant									
2500 for masonry walls, 21 oz./S.Y.	1 Pape	800	.010	S.F.	.89	.48		1.37	1.74
2600 Average		720	.011		1.31	.53		1.84	2.28
2700 Small quantities		640	.013		.89	.60		1.49	1.93
3700 Wallpaper, average workmanship, solid pattern, low cost paper		640	.013		.60	.60		1.20	1.61
3900 Basic patterns (matching required), avg. cost paper		535	.015		1.35	.72		2.07	2.63
4000 Paper at $85 per double roll, quality workmanship		435	.018		2.34	.88		3.22	3.97
5990 Wallpaper removal, 1 layer		800	.010		.03	.48		.51	.79
6000 Wallpaper removal, 3 layer		400	.020		.09	.96		1.05	1.62
9000 Minimum labor/equipment charge		2	4	Job		192		192	305

09 74 Flexible Wood Sheets

09 74 16 – Flexible Wood Veneers

09 74 16.10 Veneer, Flexible Wood	Crew	Daily Output	Labor-Hours	Unit	Material	2022 Bare Costs Labor	Equipment	Total	Total Incl O&P
0010	**VENEER, FLEXIBLE WOOD**									
0100	Flexible wood veneer, 1/32" thick, plain woods	1 Pape	100	.080	S.F.	2.55	3.84		6.39	8.90
0110	Exotic woods	"	95	.084	"	3.85	4.05		7.90	10.65

09 77 Special Wall Surfacing

09 77 30 – Fiberglass Reinforced Panels

09 77 30.10 Fiberglass Reinforced Plastic Panels

	09 77 30.10 Fiberglass Reinforced Plastic Panels	Crew	Daily Output	Labor-Hours	Unit	Material	2022 Bare Costs Labor	Equipment	Total	Total Incl O&P
0010	**FIBERGLASS REINFORCED PLASTIC PANELS**, .090" thick									
0020	On walls, adhesive mounted, embossed surface	2 Carp	640	.025	S.F.	1.50	1.41		2.91	3.89
0030	Smooth surface		640	.025		1.97	1.41		3.38	4.41
0040	Fire rated, embossed surface		640	.025		3	1.41		4.41	5.55
0050	Nylon rivet mounted, on drywall, embossed surface		480	.033		1.50	1.88		3.38	4.63
0060	Smooth surface		480	.033		2.06	1.88		3.94	5.25
0070	Fire rated, embossed surface		480	.033		2.50	1.88		4.38	5.75
0080	On masonry, embossed surface		320	.050		1.41	2.82		4.23	6
0090	Smooth surface		320	.050		1.77	2.82		4.59	6.40
0100	Fire rated, embossed surface		320	.050		2.50	2.82		5.32	7.20
0110	Nylon rivet and adhesive mounted, on drywall, embossed surface		240	.067		1.63	3.75		5.38	7.75
0120	Smooth surface		240	.067		1.73	3.75		5.48	7.85
0130	Fire rated, embossed surface		240	.067		2.81	3.75		6.56	9.05
0140	On masonry, embossed surface		190	.084		1.63	4.74		6.37	9.35
0150	Smooth surface		190	.084		1.73	4.74		6.47	9.45
0160	Fire rated, embossed surface		190	.084		2.81	4.74		7.55	10.65
0170	For moldings, add	1 Carp	250	.032	L.F.	.30	1.80		2.10	3.19
0180	On ceilings, for lay in grid system, embossed surface		400	.020	S.F.	1.50	1.13		2.63	3.44
0190	Smooth surface		400	.020		1.97	1.13		3.10	3.96
0200	Fire rated, embossed surface		400	.020		3	1.13		4.13	5.10

09 77 43 – Panel Systems

09 77 43.20 Slatwall Panels and Accessories

	09 77 43.20 Slatwall Panels and Accessories	Crew	Daily Output	Labor-Hours	Unit	Material	2022 Bare Costs Labor	Equipment	Total	Total Incl O&P
0010	**SLATWALL PANELS AND ACCESSORIES**									
0100	Slatwall panel, 4' x 8' x 3/4" T, MDF, paint grade	1 Carp	500	.016	S.F.	1.69	.90		2.59	3.29
0110	Melamine finish		500	.016		2.66	.90		3.56	4.36
0120	High pressure plastic laminate finish		500	.016		3.73	.90		4.63	5.55
0125	Wood veneer		500	.016		2.73	.90		3.63	4.43
0130	Aluminum channel inserts, add					2.84			2.84	3.12
0200	Accessories, corner forms, 8' L				L.F.	4.37			4.37	4.81
0210	T-connector, 8' L					5.75			5.75	6.30
0220	J-mold, 8' L					1.31			1.31	1.44
0230	Edge cap, 8' L					3.12			3.12	3.43
0240	Finish end cap, 8' L					3.62			3.62	3.98
0300	Display hook, metal, 4" L				Ea.	.46			.46	.51
0310	6" L					.51			.51	.56
0320	8" L					.56			.56	.62
0330	10" L					.53			.53	.58
0340	12" L					.50			.50	.55
0350	Acrylic, 4" L					.16			.16	.18
0360	6" L					.18			.18	.20
0400	Waterfall hanger, metal, 12"-16"					4.08			4.08	4.49
0410	Acrylic					11			11	12.10
0500	Shelf bracket, metal, 8"					1.96			1.96	2.16

09 77 Special Wall Surfacing

09 77 43 – Panel Systems

09 77 43.20 Slatwall Panels and Accessories	Crew	Daily Output	Labor-Hours	Unit	Material	2022 Bare Costs Labor	Equipment	Total	Total Incl O&P
0510	10"				Ea.	2.04			2.04	2.24
0520	12"					2.04			2.04	2.24
0530	14"					2.50			2.50	2.75
0540	16"					3.20			3.20	3.52
0550	Acrylic, 8"					4.30			4.30	4.73
0560	10"					4.55			4.55	5
0570	12"					4.96			4.96	5.45
0580	14"					5.70			5.70	6.25
0600	Shelf, acrylic, 12" x 16" x 1/4"					24.50			24.50	27
0610	12" x 24" x 1/4"					39			39	43

09 81 Acoustic Insulation

09 81 13 – Acoustic Board Insulation

09 81 13.10 Acoustic Board Insulation

		Crew	Daily Output	Labor-Hours	Unit	Material	Labor	Equipment	Total	Total Incl O&P
0010	**ACOUSTIC BOARD INSULATION**									
0020	Cellulose fiber board, 1/2" thk.	1 Carp	800	.010	S.F.	.90	.56		1.46	1.88

09 81 16 – Acoustic Blanket Insulation

09 81 16.10 Sound Attenuation Blanket

		Crew	Daily Output	Labor-Hours	Unit	Material	Labor	Equipment	Total	Total Incl O&P
0010	**SOUND ATTENUATION BLANKET**									
0020	Blanket, 1" thick	1 Carp	925	.009	S.F.	.30	.49		.79	1.10
0500	1-1/2" thick		920	.009		.37	.49		.86	1.19
1000	2" thick		915	.009		.46	.49		.95	1.29
1500	3" thick		910	.009		.72	.49		1.21	1.58
2000	Wall hung, STC 18-21, 1" thick, 4' x 20'	2 Carp	22	.727	Ea.	107	41		148	183
2010	10' x 20'	"	19	.842		268	47.50		315.50	370
2020	Wall hung, STC 27-28, 3" thick, 4' x 20'	3 Carp	12	2		620	113		733	865
2030	10' x 20'	"	9	2.667		1,550	150		1,700	1,950
3400	Urethane plastic foam, open cell, on wall, 2" thick	2 Carp	2050	.008	S.F.	3.35	.44		3.79	4.39
3500	3" thick		1550	.010		4.46	.58		5.04	5.85
3600	4" thick		1050	.015		6.25	.86		7.11	8.25
3700	On ceiling, 2" thick		1700	.009		3.34	.53		3.87	4.51
3800	3" thick		1300	.012		4.46	.69		5.15	6
3900	4" thick		900	.018		6.25	1		7.25	8.50
9000	Minimum labor/equipment charge	1 Carp	5	1.600	Job		90		90	143

09 84 Acoustic Room Components

09 84 13 – Fixed Sound-Absorptive Panels

09 84 13.10 Fixed Panels

		Crew	Daily Output	Labor-Hours	Unit	Material	Labor	Equipment	Total	Total Incl O&P
0010	**FIXED PANELS** Perforated steel facing, painted with									
0100	Fiberglass or mineral filler, no backs, 2-1/4" thick, modular									
0200	space units, ceiling or wall hung, white or colored	1 Carp	100	.080	S.F.	11.60	4.50		16.10	19.90
0300	Fiberboard sound deadening panels, 1/2" thick	"	600	.013	"	.31	.75		1.06	1.53
0500	Fiberglass panels, 4' x 8' x 1" thick, with									
0600	glass cloth face for walls, cemented	1 Carp	155	.052	S.F.	9.75	2.91		12.66	15.30
0700	1-1/2" thick, dacron covered, inner aluminum frame,									
0710	wall mounted	1 Carp	300	.027	S.F.	8.55	1.50		10.05	11.85
0900	Mineral fiberboard panels, fabric covered, 30" x 108",									
1000	3/4" thick, concealed spline, wall mounted	1 Carp	150	.053	S.F.	6.15	3		9.15	11.55
9000	Minimum labor/equipment charge	"	4	2	Job		113		113	179

09 91 13.30 Fences

		Crew	Daily Output	Labor-Hours	Unit	Material	2022 Bare Costs Labor	2022 Bare Costs Equipment	Total	Total Incl O&P
0010	**FENCES**									
0100	Chain link or wire metal, one side, water base									
0110	Roll & brush, first coat	1 Pord	960	.008	S.F.	.12	.40		.52	.77
0120	Second coat		1280	.006		.12	.30		.42	.60
0130	Spray, first coat		2275	.004		.12	.17		.29	.41
0140	Second coat		2600	.003		.12	.15		.27	.37
0150	Picket, water base									
0160	Roll & brush, first coat	1 Pord	865	.009	S.F.	.13	.44		.57	.85
0170	Second coat		1050	.008		.13	.36		.49	.73
0180	Spray, first coat		2275	.004		.13	.17		.30	.42
0190	Second coat		2600	.003		.13	.15		.28	.38
0200	Stockade, water base									
0210	Roll & brush, first coat	1 Pord	1040	.008	S.F.	.13	.37		.50	.73
0220	Second coat		1200	.007		.13	.32		.45	.65
0230	Spray, first coat		2275	.004		.13	.17		.30	.42
0240	Second coat		2600	.003		.13	.15		.28	.38
9000	Minimum labor/equipment charge		2	4	Job		191		191	300

09 91 13.42 Miscellaneous, Exterior

		Crew	Daily Output	Labor-Hours	Unit	Material	2022 Bare Costs Labor	2022 Bare Costs Equipment	Total	Total Incl O&P
0010	**MISCELLANEOUS, EXTERIOR**									
0100	Railing, ext., decorative wood, incl. cap & baluster									
0110	Newels & spindles @ 12" OC									
0120	Brushwork, stain, sand, seal & varnish									
0130	First coat	1 Pord	90	.089	L.F.	1.08	4.25		5.33	7.90
0140	Second coat	"	120	.067	"	1.08	3.19		4.27	6.25
0150	Rough sawn wood, 42" high, 2" x 2" verticals, 6" OC									
0160	Brushwork, stain, each coat	1 Pord	90	.089	L.F.	.34	4.25		4.59	7.10
0170	Wrought iron, 1" rail, 1/2" sq. verticals									
0180	Brushwork, zinc chromate, 60" high, bars 6" OC									
0190	Primer	1 Pord	130	.062	L.F.	1.53	2.94		4.47	6.35
0200	Finish coat		130	.062		2.30	2.94		5.24	7.20
0210	Additional coat		190	.042		2.68	2.01		4.69	6.15
0220	Shutters or blinds, single panel, 2' x 4', paint all sides									
0230	Brushwork, primer	1 Pord	20	.400	Ea.	1.54	19.10		20.64	31.50
0240	Finish coat, exterior latex		20	.400		1	19.10		20.10	31
0250	Primer & 1 coat, exterior latex		13	.615		2.18	29.50		31.68	49
0260	Spray, primer		35	.229		2.24	10.95		13.19	19.75
0270	Finish coat, exterior latex		35	.229		2.12	10.95		13.07	19.65
0280	Primer & 1 coat, exterior latex		20	.400		2.37	19.10		21.47	32.50
0290	For louvered shutters, add				S.F.	10%				
0300	Stair stringers, exterior, metal									
0310	Roll & brush, zinc chromate, to 14", each coat	1 Pord	320	.025	L.F.	.77	1.20		1.97	2.73
0320	Rough sawn wood, 4" x 12"									
0330	Roll & brush, exterior latex, each coat	1 Pord	215	.037	L.F.	.15	1.78		1.93	2.97
0340	Trellis/lattice, 2" x 2" @ 3" OC with 2" x 8" supports									
0350	Spray, latex, per side, each coat	1 Pord	475	.017	S.F.	.15	.81		.96	1.43
0450	Decking, ext., sealer, alkyd, brushwork, sealer coat		1140	.007		.15	.34		.49	.70
0460	1st coat		1140	.007		.13	.34		.47	.67
0470	2nd coat		1300	.006		.10	.29		.39	.58
0500	Paint, alkyd, brushwork, primer coat		1140	.007		.12	.34		.46	.66
0510	1st coat		1140	.007		.20	.34		.54	.75
0520	2nd coat		1300	.006		.15	.29		.44	.63
0600	Sand paint, alkyd, brushwork, 1 coat		150	.053		.13	2.55		2.68	4.18

09 91 13.42 Miscellaneous, Exterior	Crew	Daily Output	Labor-Hours	Unit	Material	2022 Bare Costs Labor	2022 Bare Costs Equipment	Total	Total Incl O&P
9000 Minimum labor/equipment charge	1 Pord	2	4	Job		191		191	300

09 91 13.60 Siding Exterior

		Crew	Daily Output	Labor-Hours	Unit	Material	Labor	Equipment	Total	Total Incl O&P
0010	**SIDING EXTERIOR**, Alkyd (oil base)									
0450	Steel siding, oil base, paint 1 coat, brushwork	2 Pord	2015	.008	S.F.	.16	.38		.54	.78
0500	Spray		4550	.004		.24	.17		.41	.54
0800	Paint 2 coats, brushwork		1300	.012		.32	.59		.91	1.28
1000	Spray		2750	.006		.15	.28		.43	.61
1200	Stucco, rough, oil base, paint 2 coats, brushwork		1300	.012		.32	.59		.91	1.28
1400	Roller		1625	.010		.34	.47		.81	1.11
1600	Spray		2925	.005		.36	.26		.62	.80
1800	Texture 1-11 or clapboard, oil base, primer coat, brushwork		1300	.012		.15	.59		.74	1.10
2000	Spray		4550	.004		.15	.17		.32	.44
2400	Paint 2 coats, brushwork		810	.020		.47	.94		1.41	2
2600	Spray		2600	.006		.52	.29		.81	1.04
3400	Stain 2 coats, brushwork		950	.017		.23	.81		1.04	1.52
4000	Spray		3050	.005		.25	.25		.50	.68
4200	Wood shingles, oil base primer coat, brushwork		1300	.012		.14	.59		.73	1.09
4400	Spray		3900	.004		.13	.20		.33	.46
5000	Paint 2 coats, brushwork		810	.020		.39	.94		1.33	1.92
5200	Spray		2275	.007		.37	.34		.71	.94
6500	Stain 2 coats, brushwork		950	.017		.23	.81		1.04	1.52
7000	Spray		2660	.006		.31	.29		.60	.80
8000	For latex paint, deduct					10%				
8100	For work over 12' H, from pipe scaffolding, add						15%			
8200	For work over 12' H, from extension ladder, add						25%			
8300	For work over 12' H, from swing staging, add						35%			
9000	Minimum labor/equipment charge	1 Pord	2	4	Job		191		191	300

09 91 13.62 Siding, Misc.

		Crew	Daily Output	Labor-Hours	Unit	Material	Labor	Equipment	Total	Total Incl O&P
0010	**SIDING, MISC.**, latex paint									
0100	Aluminum siding									
0110	Brushwork, primer	2 Pord	2275	.007	S.F.	.14	.34		.48	.68
0120	Finish coat, exterior latex		2275	.007		.13	.34		.47	.67
0130	Primer & 1 coat exterior latex		1300	.012		.29	.59		.88	1.25
0140	Primer & 2 coats exterior latex		975	.016		.42	.78		1.20	1.70
0150	Mineral fiber shingles									
0160	Brushwork, primer	2 Pord	1495	.011	S.F.	.15	.51		.66	.98
0170	Finish coat, industrial enamel		1495	.011		.42	.51		.93	1.27
0180	Primer & 1 coat enamel		810	.020		.57	.94		1.51	2.12
0190	Primer & 2 coats enamel		540	.030		1	1.42		2.42	3.33
0200	Roll, primer		1625	.010		.17	.47		.64	.93
0210	Finish coat, industrial enamel		1625	.010		.45	.47		.92	1.24
0220	Primer & 1 coat enamel		975	.016		.63	.78		1.41	1.93
0230	Primer & 2 coats enamel		650	.025		1.08	1.18		2.26	3.05
0240	Spray, primer		3900	.004		.13	.20		.33	.46
0250	Finish coat, industrial enamel		3900	.004		.38	.20		.58	.73
0260	Primer & 1 coat enamel		2275	.007		.51	.34		.85	1.09
0270	Primer & 2 coats enamel		1625	.010		.89	.47		1.36	1.72
0280	Waterproof sealer, first coat		4485	.004		.12	.17		.29	.40
0290	Second coat		5235	.003		.12	.15		.27	.36
0300	Rough wood incl. shingles, shakes or rough sawn siding									
0310	Brushwork, primer	2 Pord	1280	.013	S.F.	.31	.60		.91	1.29
0320	Finish coat, exterior latex		1280	.013		.23	.60		.83	1.20

09 91 13.62 Siding, Misc.

		Crew	Daily Output	Labor-Hours	Unit	Material	2022 Bare Costs Labor	Equipment	Total	Total Incl O&P
0330	Primer & 1 coat exterior latex	2 Pord	960	.017	S.F.	.54	.80		1.34	1.85
0340	Primer & 2 coats exterior latex		700	.023		.76	1.09		1.85	2.57
0350	Roll, primer		2925	.005		.41	.26		.67	.87
0360	Finish coat, exterior latex		2925	.005		.27	.26		.53	.71
0370	Primer & 1 coat exterior latex		1790	.009		.69	.43		1.12	1.43
0380	Primer & 2 coats exterior latex		1300	.012		.96	.59		1.55	1.98
0390	Spray, primer		3900	.004		.35	.20		.55	.69
0400	Finish coat, exterior latex		3900	.004		.21	.20		.41	.54
0410	Primer & 1 coat exterior latex		2600	.006		.56	.29		.85	1.09
0420	Primer & 2 coats exterior latex		2080	.008		.77	.37		1.14	1.43
0430	Waterproof sealer, first coat		4485	.004		.22	.17		.39	.51
0440	Second coat		4485	.004		.12	.17		.29	.40
0450	Smooth wood incl. butt, T&G, beveled, drop or B&B siding									
0460	Brushwork, primer	2 Pord	2325	.007	S.F.	.22	.33		.55	.77
0470	Finish coat, exterior latex		1280	.013		.23	.60		.83	1.20
0480	Primer & 1 coat exterior latex		800	.020		.45	.96		1.41	2
0490	Primer & 2 coats exterior latex		630	.025		.68	1.21		1.89	2.66
0500	Roll, primer		2275	.007		.25	.34		.59	.80
0510	Finish coat, exterior latex		2275	.007		.24	.34		.58	.80
0520	Primer & 1 coat exterior latex		1300	.012		.49	.59		1.08	1.47
0530	Primer & 2 coats exterior latex		975	.016		.74	.78		1.52	2.05
0540	Spray, primer		4550	.004		.19	.17		.36	.48
0550	Finish coat, exterior latex		4550	.004		.21	.17		.38	.50
0560	Primer & 1 coat exterior latex		2600	.006		.40	.29		.69	.91
0570	Primer & 2 coats exterior latex		1950	.008		.61	.39		1	1.30
0580	Waterproof sealer, first coat		5230	.003		.12	.15		.27	.36
0590	Second coat		5980	.003		.12	.13		.25	.33
0600	For oil base paint, add				S.F.	10%				
9000	Minimum labor/equipment charge	1 Pord	2	4	Job		191		191	300

09 91 13.70 Doors and Windows, Exterior

		Crew	Daily Output	Labor-Hours	Unit	Material	2022 Bare Costs Labor	Equipment	Total	Total Incl O&P
0010	**DOORS AND WINDOWS, EXTERIOR**									
0100	Door frames & trim, only									
0110	Brushwork, primer	1 Pord	512	.016	L.F.	.14	.75		.89	1.33
0120	Finish coat, exterior latex		512	.016		.12	.75		.87	1.32
0130	Primer & 1 coat, exterior latex		300	.027		.26	1.27		1.53	2.31
0135	2 coats, exterior latex, both sides		15	.533	Ea.	11.15	25.50		36.65	53
0140	Primer & 2 coats, exterior latex		265	.030	L.F.	.39	1.44		1.83	2.71
0150	Doors, flush, both sides, incl. frame & trim									
0160	Roll & brush, primer	1 Pord	10	.800	Ea.	10.65	38		48.65	72
0170	Finish coat, exterior latex		10	.800		9.50	38		47.50	71
0180	Primer & 1 coat, exterior latex		7	1.143		20	54.50		74.50	109
0190	Primer & 2 coats, exterior latex		5	1.600		29.50	76.50		106	154
0200	Brushwork, stain, sealer & 2 coats polyurethane		4	2		36.50	95.50		132	192
0210	Doors, French, both sides, 10-15 lite, incl. frame & trim									
0220	Brushwork, primer	1 Pord	6	1.333	Ea.	5.30	63.50		68.80	107
0230	Finish coat, exterior latex		6	1.333		4.74	63.50		68.24	106
0240	Primer & 1 coat, exterior latex		3	2.667		10.05	127		137.05	213
0250	Primer & 2 coats, exterior latex		2	4		14.50	191		205.50	315
0260	Brushwork, stain, sealer & 2 coats polyurethane		2.50	3.200		13.65	153		166.65	257
0270	Doors, louvered, both sides, incl. frame & trim									
0280	Brushwork, primer	1 Pord	7	1.143	Ea.	10.65	54.50		65.15	98
0290	Finish coat, exterior latex		7	1.143		9.50	54.50		64	97

For customer support on your Commercial Renovation Costs with RSMeans data, call 800.448.8182.

337

09 91 Painting

09 91 13 – Exterior Painting

09 91 13.70 Doors and Windows, Exterior

		Crew	Daily Output	Labor-Hours	Unit	Material	2022 Bare Costs Labor	Equipment	Total	Total Incl O&P
0300	Primer & 1 coat, exterior latex	1 Pord	4	2	Ea.	20	95.50		115.50	173
0310	Primer & 2 coats, exterior latex		3	2.667		29	127		156	234
0320	Brushwork, stain, sealer & 2 coats polyurethane	↓	4.50	1.778	↓	36.50	85		121.50	175
0330	Doors, panel, both sides, incl. frame & trim									
0340	Roll & brush, primer	1 Pord	6	1.333	Ea.	10.65	63.50		74.15	113
0350	Finish coat, exterior latex		6	1.333		9.50	63.50		73	111
0360	Primer & 1 coat, exterior latex		3	2.667		20	127		147	224
0370	Primer & 2 coats, exterior latex		2.50	3.200		29	153		182	274
0380	Brushwork, stain, sealer & 2 coats polyurethane	↓	3	2.667	↓	36.50	127		163.50	243
0400	Windows, per ext. side, based on 15 S.F.									
0410	1 to 6 lite									
0420	Brushwork, primer	1 Pord	13	.615	Ea.	2.10	29.50		31.60	49
0430	Finish coat, exterior latex	1 Pord	13	.615	Ea.	1.87	29.50		31.37	48.50
0440	Primer & 1 coat, exterior latex		8	1		3.97	48		51.97	80
0450	Primer & 2 coats, exterior latex		6	1.333		5.75	63.50		69.25	107
0460	Stain, sealer & 1 coat varnish	↓	7	1.143	↓	5.40	54.50		59.90	92.50
0470	7 to 10 lite									
0480	Brushwork, primer	1 Pord	11	.727	Ea.	2.10	35		37.10	57.50
0490	Finish coat, exterior latex		11	.727		1.87	35		36.87	57
0500	Primer & 1 coat, exterior latex		7	1.143		3.97	54.50		58.47	91
0510	Primer & 2 coats, exterior latex		5	1.600		5.75	76.50		82.25	127
0520	Stain, sealer & 1 coat varnish	↓	6	1.333	↓	5.40	63.50		68.90	107
0530	12 lite									
0540	Brushwork, primer	1 Pord	10	.800	Ea.	2.10	38		40.10	63
0550	Finish coat, exterior latex		10	.800		1.87	38		39.87	62.50
0560	Primer & 1 coat, exterior latex		6	1.333		3.97	63.50		67.47	105
0570	Primer & 2 coats, exterior latex		5	1.600		5.75	76.50		82.25	127
0580	Stain, sealer & 1 coat varnish	↓	6	1.333	↓	5.70	63.50		69.20	107
0590	For oil base paint, add					10%				
9000	Minimum labor/equipment charge	1 Pord	2	4	Job		191		191	300

09 91 13.80 Trim, Exterior

		Crew	Daily Output	Labor-Hours	Unit	Material	2022 Bare Costs Labor	Equipment	Total	Total Incl O&P
0010	**TRIM, EXTERIOR**									
0100	Door frames & trim (see Doors, interior or exterior)									
0110	Fascia, latex paint, one coat coverage									
0120	1" x 4", brushwork	1 Pord	640	.013	L.F.	.04	.60		.64	.99
0130	Roll		1280	.006		.04	.30		.34	.52
0140	Spray		2080	.004		.03	.18		.21	.33
0150	1" x 6" to 1" x 10", brushwork		640	.013		.13	.60		.73	1.10
0160	Roll		1230	.007		.14	.31		.45	.64
0170	Spray		2100	.004		.11	.18		.29	.41
0180	1" x 12", brushwork		640	.013		.13	.60		.73	1.10
0190	Roll		1050	.008		.14	.36		.50	.73
0200	Spray		2200	.004		.11	.17		.28	.39
0210	Gutters & downspouts, metal, zinc chromate paint									
0220	Brushwork, gutters, 5", first coat	1 Pord	640	.013	L.F.	.81	.60		1.41	1.84
0230	Second coat		960	.008		.77	.40		1.17	1.47
0240	Third coat		1280	.006		.62	.30		.92	1.15
0250	Downspouts, 4", first coat		640	.013		.81	.60		1.41	1.84
0260	Second coat		960	.008		.77	.40		1.17	1.47
0270	Third coat	↓	1280	.006	↓	.62	.30		.92	1.15
0280	Gutters & downspouts, wood									
0290	Brushwork, gutters, 5", primer	1 Pord	640	.013	L.F.	.14	.60		.74	1.10

09 91 Painting

09 91 13 – Exterior Painting

09 91 13.80 Trim, Exterior

		Crew	Daily Output	Labor-Hours	Unit	Material	2022 Bare Costs Labor	Equipment	Total	Total Incl O&P
0300	Finish coat, exterior latex	1 Pord	640	.013	L.F.	.11	.60		.71	1.07
0310	Primer & 1 coat exterior latex		400	.020		.26	.96		1.22	1.80
0320	Primer & 2 coats exterior latex		325	.025		.39	1.18		1.57	2.29
0330	Downspouts, 4", primer		640	.013		.14	.60		.74	1.10
0340	Finish coat, exterior latex		640	.013		.11	.60		.71	1.07
0350	Primer & 1 coat exterior latex		400	.020		.26	.96		1.22	1.80
0360	Primer & 2 coats exterior latex		325	.025		.19	1.18		1.37	2.07
0370	Molding, exterior, up to 14" wide									
0380	Brushwork, primer	1 Pord	640	.013	L.F.	.17	.60		.77	1.13
0390	Finish coat, exterior latex		640	.013		.13	.60		.73	1.10
0400	Primer & 1 coat exterior latex		400	.020		.32	.96		1.28	1.86
0410	Primer & 2 coats exterior latex		315	.025		.32	1.21		1.53	2.27
0420	Stain & fill		1050	.008		.14	.36		.50	.73
0430	Shellac		1850	.004		.15	.21		.36	.50
0440	Varnish		1275	.006		.16	.30		.46	.65
9000	Minimum labor/equipment charge		2	4	Job		191		191	300

09 91 13.90 Walls, Masonry (CMU), Exterior

		Crew	Daily Output	Labor-Hours	Unit	Material	2022 Bare Costs Labor	Equipment	Total	Total Incl O&P
0010	**WALLS, MASONRY (CMU), EXTERIOR**									
0360	Concrete masonry units (CMU), smooth surface									
0370	Brushwork, latex, first coat	1 Pord	640	.013	S.F.	.06	.60		.66	1.02
0380	Second coat		960	.008		.05	.40		.45	.68
0390	Waterproof sealer, first coat		736	.011		.48	.52		1	1.35
0400	Second coat		1104	.007		.48	.35		.83	1.08
0410	Roll, latex, paint, first coat		1465	.005		.07	.26		.33	.49
0420	Second coat		1790	.004		.05	.21		.26	.40
0430	Waterproof sealer, first coat		1680	.005		.48	.23		.71	.89
0440	Second coat		2060	.004		.48	.19		.67	.82
0450	Spray, latex, paint, first coat		1950	.004		.06	.20		.26	.37
0460	Second coat		2600	.003		.05	.15		.20	.28
0470	Waterproof sealer, first coat		2245	.004		.48	.17		.65	.80
0480	Second coat		2990	.003		.48	.13		.61	.73
0490	Concrete masonry unit (CMU), porous									
0500	Brushwork, latex, first coat	1 Pord	640	.013	S.F.	.12	.60		.72	1.08
0510	Second coat		960	.008		.06	.40		.46	.70
0520	Waterproof sealer, first coat		736	.011		.48	.52		1	1.35
0530	Second coat		1104	.007		.48	.35		.83	1.08
0540	Roll latex, first coat	1 Pord	1465	.005	S.F.	.09	.26		.35	.51
0550	Second coat		1790	.004		.06	.21		.27	.40
0560	Waterproof sealer, first coat		1680	.005		.48	.23		.71	.89
0570	Second coat		2060	.004		.48	.19		.67	.82
0580	Spray latex, first coat		1950	.004		.07	.20		.27	.38
0590	Second coat		2600	.003		.05	.15		.20	.28
0600	Waterproof sealer, first coat		2245	.004		.48	.17		.65	.80
0610	Second coat		2990	.003		.48	.13		.61	.73
9000	Minimum labor/equipment charge		2	4	Job		191		191	300

09 91 23 – Interior Painting

09 91 23.20 Cabinets and Casework

		Crew	Daily Output	Labor-Hours	Unit	Material	2022 Bare Costs Labor	Equipment	Total	Total Incl O&P
0010	**CABINETS AND CASEWORK**									
1000	Primer coat, oil base, brushwork	1 Pord	650	.012	S.F.	.10	.59		.69	1.04
2000	Paint, oil base, brushwork, 1 coat		650	.012		.20	.59		.79	1.15
3000	Stain, brushwork, wipe off		650	.012		.11	.59		.70	1.06
4000	Shellac, 1 coat, brushwork		650	.012		.13	.59		.72	1.07

339

09 91 23 – Interior Painting

09 91 23.20 Cabinets and Casework

		Crew	Daily Output	Labor-Hours	Unit	Material	2022 Bare Costs Labor	2022 Bare Costs Equipment	Total	Total Incl O&P
4500	Varnish, 3 coats, brushwork, sand after 1st coat	1 Pord	325	.025	S.F.	.40	1.18		1.58	2.31
5000	For latex paint, deduct				↓	10%				
6300	Strip, prep and refinish wood furniture									
6310	Remove paint using chemicals, wood furniture	1 Pord	28	.286	S.F.	1.70	13.65		15.35	23.50
6320	Prep for painting, sanding		75	.107		.28	5.10		5.38	8.35
6350	Stain and wipe, brushwork		600	.013		.11	.64		.75	1.14
6355	Spray applied		900	.009		.10	.42		.52	.78
6360	Sealer or varnish, brushwork		1080	.007		.13	.35		.48	.71
6365	Spray applied		2100	.004		.13	.18		.31	.44
6370	Paint, primer, brushwork		720	.011		.10	.53		.63	.95
6375	Spray applied		2100	.004		.09	.18		.27	.39
6380	Finish coat, brushwork		810	.010		.20	.47		.67	.97
6385	Spray applied		2100	.004	↓	.19	.18		.37	.50
9010	Minimum labor/equipment charge		2	4	Job		191		191	300

09 91 23.33 Doors and Windows, Interior Alkyd (Oil Base)

		Crew	Daily Output	Labor-Hours	Unit	Material	2022 Bare Costs Labor	2022 Bare Costs Equipment	Total	Total Incl O&P
0010	**DOORS AND WINDOWS, INTERIOR ALKYD (OIL BASE)**									
0500	Flush door & frame, 3' x 7', oil, primer, brushwork	1 Pord	10	.800	Ea.	3.66	38		41.66	64.50
1000	Paint, 1 coat		10	.800		6.80	38		44.80	68
1400	Stain, brushwork, wipe off		18	.444		2.39	21		23.39	36
1600	Shellac, 1 coat, brushwork		25	.320		2.64	15.30		17.94	27
1800	Varnish, 3 coats, brushwork, sand after 1st coat		9	.889		8.50	42.50		51	76.50
2000	Panel door & frame, 3' x 7', oil, primer, brushwork		6	1.333	↓	3.71	63.50		67.21	105
2200	Paint, 1 coat	1 Pord	6	1.333	Ea.	6.80	63.50		70.30	108
2600	Stain, brushwork, panel door, 3' x 7', not incl. frame		16	.500		2.39	24		26.39	40.50
2800	Shellac, 1 coat, brushwork		22	.364		2.64	17.40		20.04	30.50
3000	Varnish, 3 coats, brushwork, sand after 1st coat		7.50	1.067	↓	8.50	51		59.50	90
4400	Windows, including frame and trim, per side									
4600	Colonial type, 6/6 lites, 2' x 3', oil, primer, brushwork	1 Pord	14	.571	Ea.	.59	27.50		28.09	43.50
5800	Paint, 1 coat		14	.571		1.07	27.50		28.57	44
6200	3' x 5' opening, 6/6 lites, primer coat, brushwork		12	.667		1.47	32		33.47	52
6400	Paint, 1 coat		12	.667		2.68	32		34.68	53.50
6800	4' x 8' opening, 6/6 lites, primer coat, brushwork		8	1		3.13	48		51.13	79
7000	Paint, 1 coat		8	1		5.70	48		53.70	82
8000	Single lite type, 2' x 3', oil base, primer coat, brushwork		33	.242		.59	11.60		12.19	19
8200	Paint, 1 coat		33	.242		1.07	11.60		12.67	19.55
8600	3' x 5' opening, primer coat, brushwork		20	.400		1.47	19.10		20.57	31.50
8800	Paint, 1 coat		20	.400		2.68	19.10		21.78	33
9200	4' x 8' opening, primer coat, brushwork		14	.571		3.13	27.50		30.63	46.50
9400	Paint, 1 coat		14	.571	↓	5.70	27.50		33.20	49.50
9900	Minimum labor/equipment charge		2	4	Job		191		191	300

09 91 23.35 Doors and Windows, Interior Latex

		Crew	Daily Output	Labor-Hours	Unit	Material	2022 Bare Costs Labor	2022 Bare Costs Equipment	Total	Total Incl O&P
0010	**DOORS & WINDOWS, INTERIOR LATEX**									
0100	Doors, flush, both sides, incl. frame & trim									
0110	Roll & brush, primer	1 Pord	10	.800	Ea.	4.44	38		42.44	65.50
0120	Finish coat, latex		10	.800		2.58	38		40.58	63.50
0130	Primer & 1 coat latex		7	1.143		7	54.50		61.50	94
0140	Primer & 2 coats latex		5	1.600		9.45	76.50		85.95	131
0160	Spray, both sides, primer		20	.400		4.68	19.10		23.78	35
0170	Finish coat, latex		20	.400		2.70	19.10		21.80	33
0180	Primer & 1 coat latex		11	.727		7.40	35		42.40	63
0190	Primer & 2 coats latex		8	1	↓	10	48		58	86.50
0200	Doors, French, both sides, 10-15 lite, incl. frame & trim									

09 91 23 – Interior Painting

09 91 23.35 Doors and Windows, Interior Latex

		Crew	Daily Output	Labor-Hours	Unit	Material	2022 Bare Costs Labor	Equipment	Total	Total Incl O&P
0210	Roll & brush, primer	1 Pord	6	1.333	Ea.	2.22	63.50		65.72	103
0220	Finish coat, latex		6	1.333		1.29	63.50		64.79	102
0230	Primer & 1 coat latex		3	2.667		3.51	127		130.51	206
0240	Primer & 2 coats latex	↓	2	4	↓	4.72	191		195.72	305
0260	Doors, louvered, both sides, incl. frame & trim									
0270	Roll & brush, primer	1 Pord	7	1.143	Ea.	4.44	54.50		58.94	91.50
0280	Finish coat, latex		7	1.143		2.58	54.50		57.08	89.50
0290	Primer & 1 coat, latex		4	2		6.75	95.50		102.25	158
0300	Primer & 2 coats, latex		3	2.667		9.60	127		136.60	213
0320	Spray, both sides, primer		20	.400		4.68	19.10		23.78	35
0330	Finish coat, latex		20	.400		2.70	19.10		21.80	33
0340	Primer & 1 coat, latex		11	.727		7.40	35		42.40	63
0350	Primer & 2 coats, latex	↓	8	1	↓	10.15	48		58.15	86.50
0360	Doors, panel, both sides, incl. frame & trim									
0370	Roll & brush, primer	1 Pord	6	1.333	Ea.	4.68	63.50		68.18	106
0380	Finish coat, latex		6	1.333		2.58	63.50		66.08	104
0390	Primer & 1 coat, latex		3	2.667		7	127		134	210
0400	Primer & 2 coats, latex		2.50	3.200		9.60	153		162.60	253
0420	Spray, both sides, primer		10	.800		4.68	38		42.68	65.50
0430	Finish coat, latex		10	.800		2.70	38		40.70	63.50
0440	Primer & 1 coat, latex		5	1.600		7.40	76.50		83.90	129
0450	Primer & 2 coats, latex	↓	4	2	↓	10.15	95.50		105.65	162
0460	Windows, per interior side, based on 15 S.F.									
0470	1 to 6 lite									
0480	Brushwork, primer	1 Pord	13	.615	Ea.	.88	29.50		30.38	47.50
0490	Finish coat, enamel		13	.615		.51	29.50		30.01	47
0500	Primer & 1 coat enamel		8	1		1.39	48		49.39	77
0510	Primer & 2 coats enamel	↓	6	1.333	↓	1.89	63.50		65.39	103
0530	7 to 10 lite									
0540	Brushwork, primer	1 Pord	11	.727	Ea.	.88	35		35.88	56
0550	Finish coat, enamel		11	.727		.51	35		35.51	55.50
0560	Primer & 1 coat enamel		7	1.143		1.39	54.50		55.89	88
0570	Primer & 2 coats enamel	↓	5	1.600		1.89	76.50		78.39	123
0590	12 lite									
0600	Brushwork, primer	1 Pord	10	.800	Ea.	.88	38		38.88	61.50
0610	Finish coat, enamel		10	.800		.51	38		38.51	61
0620	Primer & 1 coat enamel		6	1.333		1.39	63.50		64.89	103
0630	Primer & 2 coats enamel	↓	5	1.600	↓	1.89	76.50		78.39	123
0650	For oil base paint, add					10%				
9000	Minimum labor/equipment charge	1 Pord	2	4	Job		191		191	300

09 91 23.39 Doors and Windows, Interior Latex, Zero Voc

			Crew	Daily Output	Labor-Hours	Unit	Material	2022 Bare Costs Labor	Equipment	Total	Total Incl O&P
0010	**DOORS & WINDOWS, INTERIOR LATEX, ZERO VOC**										
0100	Doors flush, both sides, incl. frame & trim										
0110	Roll & brush, primer	G	1 Pord	10	.800	Ea.	5.35	38		43.35	66.50
0120	Finish coat, latex	G	"	10	.800	"	12	38		50	73.50
0130	Primer & 1 coat latex	G	1 Pord	7	1.143	Ea.	17.35	54.50		71.85	106
0140	Primer & 2 coats latex	G		5	1.600		28.50	76.50		105	153
0160	Spray, both sides, primer	G		20	.400		5.65	19.10		24.75	36
0170	Finish coat, latex	G		20	.400		12.60	19.10		31.70	44
0180	Primer & 1 coat latex	G		11	.727		18.40	35		53.40	75
0190	Primer & 2 coats latex	G	↓	8	1		30.50	48		78.50	109
0200	Doors, French, both sides, 10-15 lite, incl. frame & trim										

For customer support on your Commercial Renovation Costs with RSMeans data, call 800.448.8182.

341

09 91 Painting

09 91 23 – Interior Painting

09 91 23.39 Doors and Windows, Interior Latex, Zero Voc		Crew	Daily Output	Labor-Hours	Unit	Material	2022 Bare Costs Labor	Equipment	Total	Total Incl O&P
0210	Roll & brush, primer	G	1 Pord	6	1.333	Ea.	2.68	63.50		66.18	104
0220	Finish coat, latex	G		6	1.333		6	63.50		69.50	108
0230	Primer & 1 coat latex	G		3	2.667		8.70	127		135.70	212
0240	Primer & 2 coats latex	G		2	4		14.35	191		205.35	315
0360	Doors, panel, both sides, incl. frame & trim										
0370	Roll & brush, primer	G	1 Pord	6	1.333	Ea.	5.65	63.50		69.15	107
0380	Finish coat, latex	G		6	1.333		12	63.50		75.50	114
0390	Primer & 1 coat, latex	G		3	2.667		17.35	127		144.35	221
0400	Primer & 2 coats, latex	G		2.50	3.200		29.50	153		182.50	275
0420	Spray, both sides, primer	G		10	.800		5.65	38		43.65	66.50
0430	Finish coat, latex	G		10	.800		12.60	38		50.60	74.50
0440	Primer & 1 coat, latex	G		5	1.600		18.40	76.50		94.90	141
0450	Primer & 2 coats, latex	G		4	2		31	95.50		126.50	186
0460	Windows, per interior side, based on 15 S.F.										
0470	1 to 6 lite										
0480	Brushwork, primer	G	1 Pord	13	.615	Ea.	1.06	29.50		30.56	47.50
0490	Finish coat, enamel	G		13	.615		2.37	29.50		31.87	49
0500	Primer & 1 coat enamel	G		8	1		3.43	48		51.43	79.50
0510	Primer & 2 coats enamel	G		6	1.333		5.80	63.50		69.30	107
9000	Minimum labor/equipment charge			2	4	Job		191		191	300

09 91 23.40 Floors, Interior

		Crew	Daily Output	Labor-Hours	Unit	Material	2022 Bare Costs Labor	Equipment	Total	Total Incl O&P
0010	**FLOORS, INTERIOR**									
0100	Concrete paint, latex									
0110	Brushwork									
0120	1st coat	1 Pord	975	.008	S.F.	.15	.39		.54	.79
0130	2nd coat		1150	.007		.10	.33		.43	.64
0140	3rd coat		1300	.006		.08	.29		.37	.56
0150	Roll									
0160	1st coat	1 Pord	2600	.003	S.F.	.20	.15		.35	.45
0170	2nd coat		3250	.002		.12	.12		.24	.32
0180	3rd coat		3900	.002		.09	.10		.19	.26
0190	Spray									
0200	1st coat	1 Pord	2600	.003	S.F.	.17	.15		.32	.42
0210	2nd coat		3250	.002		.09	.12		.21	.29
0220	3rd coat		3900	.002		.08	.10		.18	.24

09 91 23.44 Anti-Slip Floor Treatments

		Crew	Daily Output	Labor-Hours	Unit	Material	2022 Bare Costs Labor	Equipment	Total	Total Incl O&P
0010	**ANTI-SLIP FLOOR TREATMENTS**									
1000	Walking surface treatment, ADA compliant, mop on and rinse									
1100	For tile, terrazzo, stone or smooth concrete	1 Pord	4000	.002	S.F.	.16	.10		.26	.33
1110	For marble		4000	.002		.22	.10		.32	.40
1120	For wood		4000	.002		.21	.10		.31	.38
1130	For baths and showers		500	.016		.18	.76		.94	1.41
2000	Granular additive for paint or sealer, add to paint cost					.02			.02	.02

09 91 23.52 Miscellaneous, Interior

		Crew	Daily Output	Labor-Hours	Unit	Material	2022 Bare Costs Labor	Equipment	Total	Total Incl O&P
0010	**MISCELLANEOUS, INTERIOR**									
2400	Floors, conc./wood, oil base, primer/sealer coat, brushwork	2 Pord	1950	.008	S.F.	.10	.39		.49	.73
2450	Roller		5200	.003		.10	.15		.25	.34
2600	Spray		6000	.003		.10	.13		.23	.31
2650	Paint 1 coat, brushwork		1950	.008		.22	.39		.61	.86
2800	Roller		5200	.003		.23	.15		.38	.48
2850	Spray		6000	.003		.24	.13		.37	.47
3000	Stain, wood floor, brushwork, 1 coat		4550	.004		.11	.17		.28	.40

09 91 23 – Interior Painting

09 91 23.52 Miscellaneous, Interior		Crew	Daily Output	Labor-Hours	Unit	Material	2022 Bare Costs Labor	2022 Bare Costs Equipment	Total	Total Incl O&P
3200	Roller	2 Pord	5200	.003	S.F.	.12	.15		.27	.36
3250	Spray		6000	.003		.12	.13		.25	.33
3400	Varnish, wood floor, brushwork		4550	.004		.13	.17		.30	.42
3450	Roller		5200	.003		.14	.15		.29	.39
3600	Spray		6000	.003		.15	.13		.28	.36
3800	Grilles, per side, oil base, primer coat, brushwork	1 Pord	520	.015		.20	.74		.94	1.37
3850	Spray		1140	.007		.20	.34		.54	.75
3920	Paint 2 coats, brushwork		325	.025		.69	1.18		1.87	2.62
3940	Spray		650	.012		.79	.59		1.38	1.80
4600	Miscellaneous surfaces, metallic paint, spray applied									
4610	Water based, non-tintable, warm silver	1 Pord	1140	.007	S.F.	.57	.34		.91	1.15
4620	Rusted iron		1140	.007		.72	.34		1.06	1.32
4630	Low VOC, tintable		1140	.007		.18	.34		.52	.73
5000	Pipe, 1"-4" diameter, primer or sealer coat, oil base, brushwork	2 Pord	1250	.013	L.F.	.10	.61		.71	1.08
5100	Spray		2165	.007		.10	.35		.45	.66
5350	Paint 2 coats, brushwork		775	.021		.33	.99		1.32	1.92
5400	Spray		1240	.013		.37	.62		.99	1.38
6300	13"-16" diameter, primer or sealer coat, brushwork		310	.052		.40	2.47		2.87	4.34
6450	Spray		540	.030		.75	1.42		2.17	3.07
6500	Paint 2 coats, brushwork		195	.082		1.32	3.92		5.24	7.65
6550	Spray		310	.052		1.46	2.47		3.93	5.50
7000	Trim, wood, incl. puttying, under 6" wide									
7200	Primer coat, oil base, brushwork	1 Pord	650	.012	L.F.	.05	.59		.64	.98
7250	Paint, 1 coat, brushwork		650	.012		.09	.59		.68	1.03
7450	3 coats		325	.025		.26	1.18		1.44	2.14
7500	Over 6" wide, primer coat, brushwork		650	.012		.10	.59		.69	1.04
7550	Paint, 1 coat, brushwork		650	.012		.18	.59		.77	1.13
7650	3 coats		325	.025		.52	1.18		1.70	2.43
8000	Cornice, simple design, primer coat, oil base, brushwork		650	.012	S.F.	.10	.59		.69	1.04
8250	Paint, 1 coat		650	.012		.18	.59		.77	1.13
8350	Ornate design, primer coat		350	.023		.10	1.09		1.19	1.84
8400	Paint, 1 coat		350	.023		.18	1.09		1.27	1.93
8600	Balustrades, primer coat, oil base, brushwork		520	.015		.10	.74		.84	1.27
8900	Trusses and wood frames, primer coat, oil base, brushwork		800	.010		.10	.48		.58	.87
8950	Spray		1200	.007		.10	.32		.42	.61
9220	Paint 2 coats, brushwork		500	.016		.35	.76		1.11	1.59
9240	Spray		600	.013		.39	.64		1.03	1.43
9260	Stain, brushwork, wipe off		600	.013		.11	.64		.75	1.14
9280	Varnish, 3 coats, brushwork		275	.029		.40	1.39		1.79	2.65
9350	For latex paint, deduct					10%				
9900	Minimum labor/equipment charge	1 Pord	2	4	Job		191		191	300

09 91 23.62 Electrostatic Painting

		Crew	Daily Output	Labor-Hours	Unit	Material	2022 Bare Costs Labor	2022 Bare Costs Equipment	Total	Total Incl O&P
0010	**ELECTROSTATIC PAINTING**									
0100	In shop									
0200	Flat surfaces (lockers, casework, elevator doors, etc.)									
0300	One coat	1 Pord	200	.040	S.F.	.22	1.91		2.13	3.27
0400	Two coats	"	120	.067	"	.32	3.19		3.51	5.40
0500	Irregular surfaces (furniture, door frames, etc.)									
0600	One coat	1 Pord	150	.053	S.F.	.22	2.55		2.77	4.28
0700	Two coats	"	100	.080	"	.32	3.82		4.14	6.40
0800	On site									
0900	Flat surfaces (lockers, casework, elevator doors, etc.)									

For customer support on your Commercial Renovation Costs with RSMeans data, call 800.448.8182.

343

09 91 23 – Interior Painting

09 91 23.62 Electrostatic Painting	Crew	Daily Output	Labor-Hours	Unit	Material	2022 Bare Costs Labor	Equipment	Total	Total Incl O&P
1000	One coat	1 Pord	150	.053	S.F.	.22	2.55		2.77	4.28
1100	Two coats	"	100	.080	"	.32	3.82		4.14	6.40
1200	Irregular surfaces (furniture, door frames, etc.)									
1300	One coat	1 Pord	115	.070	S.F.	.22	3.33		3.55	5.50
1400	Two coats	"	70	.114	"	.32	5.45		5.77	9
2000	Anti-microbial coating, hospital application	1 Pord	150	.053	S.F.	.41	2.55		2.96	4.49

09 91 23.72 Walls and Ceilings, Interior

		Crew	Daily Output	Labor-Hours	Unit	Material	2022 Bare Costs Labor	Equipment	Total	Total Incl O&P
0010	**WALLS AND CEILINGS, INTERIOR**									
0100	Concrete, drywall or plaster, latex, primer or sealer coat									
0150	Smooth finish, cut-in by brush	1 Pord	1150	.007	L.F.	.02	.33		.35	.55
0200	Brushwork		1150	.007	S.F.	.07	.33		.40	.60
0240	Roller		1350	.006		.07	.28		.35	.52
0280	Spray		2750	.003	↓	.06	.14		.20	.28
0290	Sand finish, cut-in by brush		975	.008	L.F.	.02	.39		.41	.64
0300	Brushwork		975	.008	S.F.	.07	.39		.46	.69
0340	Roller		1150	.007		.07	.33		.40	.60
0380	Spray		2275	.004	↓	.06	.17		.23	.33
0590	Paint 2 coats, smooth finish, cut-in by brush		680	.012	L.F.	.09	.56		.65	.99
0800	Brushwork		680	.012	S.F.	.36	.56		.92	1.29
0840	Roller		800	.010		.36	.48		.84	1.16
0880	Spray		1625	.005	↓	.34	.24		.58	.74
0890	Sand finish, cut-in by brush		605	.013	L.F.	.09	.63		.72	1.10
0900	Brushwork		605	.013	S.F.	.36	.63		.99	1.40
0940	Roller		1020	.008		.36	.37		.73	.99
0980	Spray		1700	.005		.34	.22		.56	.73
1600	Glaze coating, 2 coats, spray, clear		1200	.007		.56	.32		.88	1.12
1640	Multicolor	↓	1200	.007	↓	.91	.32		1.23	1.50
1660	Painting walls, complete, including surface prep, primer &									
1670	2 coats finish, on drywall or plaster, with roller	1 Pord	325	.025	S.F.	.44	1.18		1.62	2.35
1700	For oil base paint, add					10%				
1800	For ceiling installations, add				↓		25%			
2000	Masonry or concrete block, primer/sealer, latex paint									
2090	Primer, smooth finish, cut-in by brush	1 Pord	1000	.008	L.F.	.07	.38		.45	.67
2100	Brushwork		1000	.008	S.F.	.26	.38		.64	.89
2110	Roller		1150	.007		.12	.33		.45	.66
2180	Spray		2400	.003	↓	.10	.16		.26	.36
2190	Sand finish, cut-in by brush		850	.009	L.F.	.03	.45		.48	.74
2200	Brushwork		850	.009	S.F.	.12	.45		.57	.84
2210	Roller		975	.008		.12	.39		.51	.75
2280	Spray		2050	.004	↓	.10	.19		.29	.41
2590	Primer plus one finish coat, smooth cut-in by brush		525	.015	L.F.	.09	.73		.82	1.25
2800	Brushwork		525	.015	S.F.	.37	.73		1.10	1.56
2810	Roller	↓	615	.013	"	.25	.62		.87	1.26
2880	Spray	1 Pord	1200	.007	S.F.	.22	.32		.54	.75
2890	Sand finish, cut-in by brush		450	.018	L.F.	.06	.85		.91	1.41
2900	Brushwork		450	.018	S.F.	.25	.85		1.10	1.62
2910	Roller		515	.016		.25	.74		.99	1.45
2980	Spray		1025	.008		.22	.37		.59	.84
3600	Glaze coating, 3 coats, spray, clear		900	.009		.80	.42		1.22	1.55
3620	Multicolor		900	.009		1.11	.42		1.53	1.89
4000	Block filler, 1 coat, brushwork		425	.019		.29	.90		1.19	1.74
4100	Silicone, water repellent, 2 coats, spray	↓	2000	.004	↓	.46	.19		.65	.81

09 91 Painting

09 91 23 – Interior Painting

09 91 23 – Interior Painting

09 91 23.72 Walls and Ceilings, Interior

		Crew	Daily Output	Labor-Hours	Unit	Material	2022 Bare Costs Labor	Equipment	Total	Total Incl O&P
4120	For oil base paint, add				S.F.	10%				
8200	For work 8'-15' H, add						10%			
8300	For work over 15' H, add						20%			
8400	For light textured surfaces, add						10%			
8410	Heavy textured, add						25%			
9900	Minimum labor/equipment charge	1 Pord	2	4	Job		191		191	300

09 91 23.74 Walls and Ceilings, Interior, Zero VOC Latex

			Crew	Daily Output	Labor-Hours	Unit	Material	2022 Bare Costs Labor	Equipment	Total	Total Incl O&P
0010	**WALLS AND CEILINGS, INTERIOR, ZERO VOC LATEX**										
0100	Concrete, dry wall or plaster, latex, primer or sealer coat										
0190	Smooth finish, cut-in by brush		1 Pord	1150	.007	L.F.	.02	.33		.35	.55
0200	Brushwork	G		1150	.007	S.F.	.07	.33		.40	.61
0240	Roller	G		1350	.006		.07	.28		.35	.52
0280	Spray	G		2750	.003		.05	.14		.19	.28
0290	Sand finish, cut-in by brush			975	.008	L.F.	.02	.39		.41	.64
0300	Brushwork	G		975	.008	S.F.	.07	.39		.46	.69
0340	Roller	G		1150	.007		.07	.33		.40	.61
0380	Spray	G		2275	.004		.06	.17		.23	.33
0790	Paint 2 coats, smooth finish, cut-in by brush			680	.012	L.F.	.03	.56		.59	.93
0800	Brushwork			680	.012	S.F.	.03	.56		.59	.93
0840	Roller	G		800	.010		.13	.48		.61	.91
0880	Spray	G		1625	.005		.11	.24		.35	.50
0890	Sand finish, cut-in by brush			605	.013	L.F.	.03	.63		.66	1.03
0900	Brushwork	G		605	.013	S.F.	.13	.63		.76	1.14
0940	Roller	G		1020	.008		.13	.37		.50	.74
0980	Spray	G		1700	.005		.11	.22		.33	.49
1800	For ceiling installations, add	G						25%			
8200	For work 8' - 15' H, add							10%			
8300	For work over 15' H, add							20%			
9900	Minimum labor/equipment charge		1 Pord	2	4	Job		191		191	300

09 91 23.75 Dry Fall Painting

		Crew	Daily Output	Labor-Hours	Unit	Material	2022 Bare Costs Labor	Equipment	Total	Total Incl O&P
0010	**DRY FALL PAINTING**									
0100	Sprayed on walls, gypsum board or plaster									
0220	One coat	1 Pord	2600	.003	S.F.	.08	.15		.23	.32
0250	Two coats		1560	.005		.17	.25		.42	.58
0280	Concrete or textured plaster, one coat		1560	.005		.08	.25		.33	.48
0310	Two coats		1300	.006		.17	.29		.46	.66
0340	Concrete block, one coat		1560	.005		.08	.25		.33	.48
0370	Two coats		1300	.006		.17	.29		.46	.66
0400	Wood, one coat		877	.009		.08	.44		.52	.78
0430	Two coats		650	.012		.17	.59		.76	1.12
0440	On ceilings, gypsum board or plaster									
0470	One coat	1 Pord	1560	.005	S.F.	.08	.25		.33	.48
0500	Two coats		1300	.006		.17	.29		.46	.66
0530	Concrete or textured plaster, one coat		1560	.005		.08	.25		.33	.48
0560	Two coats		1300	.006		.17	.29		.46	.66
0570	Structural steel, bar joists or metal deck, one coat		1560	.005		.08	.25		.33	.48
0580	Two coats		1040	.008		.17	.37		.54	.77
9900	Minimum labor/equipment charge		2	4	Job		191		191	300

For customer support on your Commercial Renovation Costs with RSMeans data, call 800.448.8182.

345

09 93 Staining and Transparent Finishing

09 93 23 – Interior Staining and Finishing

09 93 23.10 Varnish		Crew	Daily Output	Labor-Hours	Unit	Material	2022 Bare Costs Labor	Equipment	Total	Total Incl O&P
0010	**VARNISH**									
0012	1 coat + sealer, on wood trim, brush, no sanding included	1 Pord	400	.020	S.F.	.20	.96		1.16	1.73
0020	1 coat + sealer, on wood trim, brush, no sanding included, no VOC		400	.020		.22	.96		1.18	1.75
0100	Hardwood floors, 2 coats, no sanding included, roller		1890	.004		.15	.20		.35	.49
9000	Minimum labor/equipment charge		4	2	Job		95.50		95.50	151

09 96 High-Performance Coatings

09 96 23 – Graffiti-Resistant Coatings

09 96 23.10 Graffiti-Resistant Treatments

09 96 23.10 Graffiti-Resistant Treatments		Crew	Daily Output	Labor-Hours	Unit	Material	2022 Bare Costs Labor	Equipment	Total	Total Incl O&P
0010	**GRAFFITI-RESISTANT TREATMENTS**, sprayed on walls									
0100	Non-sacrificial, permanent non-stick coating, clear, on metals	1 Pord	2000	.004	S.F.	.35	.19		.54	.69
0200	Concrete		2000	.004		.40	.19		.59	.74
0300	Concrete block		2000	.004		.52	.19		.71	.87
0400	Brick		2000	.004		.58	.19		.77	.94
0500	Stone		2000	.004		.58	.19		.77	.94
0600	Unpainted wood		2000	.004		.67	.19		.86	1.04
2000	Semi-permanent cross linking polymer primer, on metals		2000	.004		.41	.19		.60	.76
2100	Concrete		2000	.004		.50	.19		.69	.85
2200	Concrete block		2000	.004		.62	.19		.81	.98
2300	Brick	1 Pord	2000	.004	S.F.	.50	.19		.69	.85
2400	Stone		2000	.004		.50	.19		.69	.85
2500	Unpainted wood		2000	.004		.69	.19		.88	1.06
3000	Top coat, on metals		2000	.004		.51	.19		.70	.86
3100	Concrete		2000	.004		.58	.19		.77	.94
3200	Concrete block		2000	.004		.81	.19		1	1.19
3300	Brick		2000	.004		.67	.19		.86	1.04
3400	Stone		2000	.004		.67	.19		.86	1.04
3500	Unpainted wood		2000	.004		.81	.19		1	1.19
5000	Sacrificial, water based, on metal		2000	.004		.22	.19		.41	.54
5100	Concrete		2000	.004		.22	.19		.41	.54
5200	Concrete block		2000	.004		.22	.19		.41	.54
5300	Brick		2000	.004		.22	.19		.41	.54
5400	Stone		2000	.004		.22	.19		.41	.54
5500	Unpainted wood		2000	.004		.22	.19		.41	.54
8000	Cleaner for use after treatment									
8100	Towels or wipes, per package of 30				Ea.	.49			.49	.54
8200	Aerosol spray, 24 oz. can				"	13			13	14.30

09 96 46 – Intumescent Painting

09 96 46.10 Coatings, Intumescent

09 96 46.10 Coatings, Intumescent		Crew	Daily Output	Labor-Hours	Unit	Material	2022 Bare Costs Labor	Equipment	Total	Total Incl O&P
0010	**COATINGS, INTUMESCENT**, spray applied									
0100	On exterior structural steel, 0.25" d.f.t.	1 Pord	475	.017	S.F.	.51	.81		1.32	1.83
0150	0.51" d.f.t.		350	.023		.51	1.09		1.60	2.29
0200	0.98" d.f.t.		280	.029		.51	1.37		1.88	2.72
0300	On interior structural steel, 0.108" d.f.t.		300	.027		.51	1.27		1.78	2.58
0350	0.310" d.f.t.		150	.053		.51	2.55		3.06	4.59
0400	0.670" d.f.t.		100	.080		.51	3.82		4.33	6.60

09 96 High-Performance Coatings

09 96 53 – Elastomeric Coatings

09 96 53.10 Coatings, Elastomeric	Crew	Daily Output	Labor-Hours	Unit	Material	2022 Bare Costs Labor	Equipment	Total	Total Incl O&P
0010 **COATINGS, ELASTOMERIC**									
0020 High build, water proof, one coat system									
0100 Concrete, brush	1 Pord	650	.012	S.F.	.50	.59		1.09	1.48

09 96 56 – Epoxy Coatings

09 96 56.20 Wall Coatings

	Crew	Daily Output	Labor-Hours	Unit	Material	2022 Bare Costs Labor	Equipment	Total	Total Incl O&P
0010 **WALL COATINGS**									
0100 Acrylic glazed coatings, matte	1 Pord	525	.015	S.F.	.40	.73		1.13	1.59
0200 Gloss		305	.026		.80	1.25		2.05	2.86
0300 Epoxy coatings, solvent based		525	.015		.50	.73		1.23	1.70
0400 Water based		170	.047		.40	2.25		2.65	4
2400 Sprayed perlite or vermiculite, 1/16" thick, solvent based		2935	.003		.29	.13		.42	.53
2500 Water based		640	.013		.92	.60		1.52	1.96
2700 Vinyl plastic wall coating, solvent based		735	.011		.43	.52		.95	1.29
2800 Water based		240	.033		1.07	1.59		2.66	3.70
3000 Urethane on smooth surface, 2 coats, solvent based		1135	.007		.36	.34		.70	.93
3100 Water based		665	.012		.66	.58		1.24	1.64

09 97 Special Coatings

09 97 35 – Dry Erase Coatings

09 97 35.10 Dry Erase Coatings

	Crew	Daily Output	Labor-Hours	Unit	Material	2022 Bare Costs Labor	Equipment	Total	Total Incl O&P
0010 **DRY ERASE COATINGS**									
0020 Dry erase coatings, clear, roller applied	1 Pord	1325	.006	S.F.	2.14	.29		2.43	2.81

Division Notes

	CREW	DAILY OUTPUT	LABOR-HOURS	UNIT	BARE COSTS				TOTAL INCL O&P
					MAT.	LABOR	EQUIP.	TOTAL	

Estimating Tips
General

- The items in this division are usually priced per square foot or each.

- Many items in Division 10 require some type of support system or special anchors that are not usually furnished with the item. The required anchors must be added to the estimate in the appropriate division.

- Some items in Division 10, such as lockers, may require assembly before installation. Verify the amount of assembly required. Assembly can often exceed installation time.

10 20 00 Interior Specialties

- Support angles and blocking are not included in the installation of toilet compartments, shower/dressing compartments, or cubicles. Appropriate line items from Division 5 or 6 may need to be added to support the installations.

- Toilet partitions are priced by the stall. A stall consists of a side wall, pilaster, and door with hardware. Toilet tissue holders and grab bars are extra.

- The required acoustical rating of a folding partition can have a significant impact on costs. Verify the sound transmission coefficient rating of the panel priced against the specification requirements.

- Grab bar installation does not include supplemental blocking or backing to support the required load. When grab bars are installed at an existing facility, provisions must be made to attach the grab bars to a solid structure.

Reference Numbers

Reference numbers are shown at the beginning of some major classifications. These numbers refer to related items in the Reference Section. The reference information may be an estimating procedure, an alternate pricing method, or technical information.

Note: Not all subdivisions listed here necessarily appear. ■

10 11 13 – Chalkboards

10 11 13.13 Fixed Chalkboards

		Crew	Daily Output	Labor-Hours	Unit	Material	2022 Bare Costs Labor	Equipment	Total	Total Incl O&P
0010	**FIXED CHALKBOARDS** Porcelain enamel steel									
3900	Wall hung									
4000	Aluminum frame and chalktrough									
4300	3' x 5'	2 Carp	15	1.067	Ea.	218	60		278	335
4600	4' x 12'	"	13	1.231	"	525	69.50		594.50	685
4700	Wood frame and chalktrough									
4800	3' x 4'	2 Carp	16	1	Ea.	420	56.50		476.50	550
5300	4' x 8'	"	13	1.231	"	370	69.50		439.50	515
5400	Liquid chalk, white porcelain enamel, wall hung									
5420	Deluxe units, aluminum trim and chalktrough									
5450	4' x 4'	2 Carp	16	1	Ea.	169	56.50		225.50	276
5550	4' x 12'	"	12	1.333	"	510	75		585	680
5700	Wood trim and chalktrough									
5900	4' x 4'	2 Carp	16	1	Ea.	595	56.50		651.50	745
6200	4' x 8'		14	1.143	"	815	64.50		879.50	1,000
9000	Minimum labor/equipment charge		3	5.333	Job		300		300	475

10 11 13.43 Portable Chalkboards

		Crew	Daily Output	Labor-Hours	Unit	Material	2022 Bare Costs Labor	Equipment	Total	Total Incl O&P
0010	**PORTABLE CHALKBOARDS**									
0100	Freestanding, reversible									
0120	Economy, wood frame, 4' x 6'									
0140	Chalkboard both sides				Ea.	710			710	785
0160	Chalkboard one side, cork other side				"	625			625	685
0200	Standard, lightweight satin finished aluminum, 4' x 6'									
0220	Chalkboard both sides				Ea.	710			710	785
0240	Chalkboard one side, cork other side				"	410			410	450
0300	Deluxe, heavy duty extruded aluminum, 4' x 6'									
0320	Chalkboard both sides				Ea.	405			405	445
0340	Chalkboard one side, cork other side				"	1,300			1,300	1,425

10 11 16 – Markerboards

10 11 16.53 Electronic Markerboards

		Crew	Daily Output	Labor-Hours	Unit	Material	2022 Bare Costs Labor	Equipment	Total	Total Incl O&P
0010	**ELECTRONIC MARKERBOARDS**									
0100	Wall hung or free standing, 3' x 4' to 4' x 6'	2 Carp	8	2	S.F.	28	113		141	210
0150	5' x 6' to 4' x 8'		8	2	"	26.50	113		139.50	208
0500	Interactive projection module for existing whiteboards		8	2	Ea.	845	113		958	1,100

10 11 23 – Tackboards

10 11 23.10 Fixed Tackboards

		Crew	Daily Output	Labor-Hours	Unit	Material	2022 Bare Costs Labor	Equipment	Total	Total Incl O&P
0010	**FIXED TACKBOARDS**									
2120	Prefabricated, 1/4" cork, 3' x 5' with aluminum frame	2 Carp	16	1	Ea.	99	56.50		155.50	199
2140	Wood frame	"	16	1	"	179	56.50		235.50	286
2300	Glass enclosed cabinets, alum., cork panel, hinged doors									
2600	4' x 7', 3 door	2 Carp	10	1.600	Ea.	525	90		615	720
9000	Minimum labor/equipment charge	"	4	4	Job		225		225	360

10 11 23.20 Control Boards

		Crew	Daily Output	Labor-Hours	Unit	Material	2022 Bare Costs Labor	Equipment	Total	Total Incl O&P
0010	**CONTROL BOARDS**									
1000	Hospital patient display board, 4-color custom design									
1010	Porcelain steel dry erase board, 36" x 24"	2 Carp	7.50	2.133	Ea.	239	120		359	455

For customer support on your Commercial Renovation Costs with RSMeans data, call 800.448.8182.

10 13 Directories

10 13 10 – Building Directories

10 13 10.10 Directory Boards

		Crew	Daily Output	Labor-Hours	Unit	Material	2022 Bare Costs Labor	Equipment	Total	Total Incl O&P
0010	**DIRECTORY BOARDS**									
0050	Plastic, glass covered, 30" x 20"	2 Carp	3	5.333	Ea.	219	300		519	715
0100	36" x 48"		2	8		565	450		1,015	1,325
0900	Outdoor, weatherproof, black plastic, 36" x 24"		2	8		535	450		985	1,300
1000	36" x 36"	↓	1.50	10.667	↓	460	600		1,060	1,450
9000	Minimum labor/equipment charge	1 Carp	1	8	Job		450		450	715

10 14 Signage

10 14 19 – Dimensional Letter Signage

10 14 19.10 Exterior Signs

		Crew	Daily Output	Labor-Hours	Unit	Material	2022 Bare Costs Labor	Equipment	Total	Total Incl O&P
0010	**EXTERIOR SIGNS**									
3900	Plaques, custom, 20" x 30", for up to 450 letters, cast aluminum	2 Carp	4	4	Ea.	1,325	225		1,550	1,800
4000	Cast bronze		4	4		2,325	225		2,550	2,900
4200	30" x 36", up to 900 letters, cast aluminum		3	5.333		2,650	300		2,950	3,400
4300	Cast bronze	↓	3	5.333		4,600	300		4,900	5,525
5100	Exit signs, 24 ga. alum., 14" x 12" surface mounted	1 Carp	30	.267		21.50	15		36.50	48
5200	10" x 7"	"	20	.400		13.80	22.50		36.30	51
6400	Replacement sign faces, 6" or 8"	1 Clab	50	.160	↓	4.65	7.30		11.95	16.70
9000	Minimum labor/equipment charge	1 Carp	4	2	Job		113		113	179

10 14 23 – Panel Signage

10 14 23.13 Engraved Panel Signage

		Crew	Daily Output	Labor-Hours	Unit	Material	2022 Bare Costs Labor	Equipment	Total	Total Incl O&P
0010	**ENGRAVED PANEL SIGNAGE**, interior									
1010	Flexible door sign, adhesive back, w/Braille, 5/8" letters, 4" x 4"	1 Clab	32	.250	Ea.	44.50	11.40		55.90	67
1050	6" x 6"		32	.250		68.50	11.40		79.90	93
1100	8" x 2"		32	.250		19.65	11.40		31.05	39.50
1150	8" x 4"		32	.250		26	11.40		37.40	47
1200	8" x 8"		32	.250		38.50	11.40		49.90	60.50
1250	12" x 2"		32	.250		41.50	11.40		52.90	64
1300	12" x 6"		32	.250		52	11.40		63.40	75.50
1350	12" x 12"		32	.250		155	11.40		166.40	188
1500	Graphic symbols, 2" x 2"		32	.250		18.50	11.40		29.90	38.50
1550	6" x 6"		32	.250		24.50	11.40		35.90	45
1600	8" x 8"	↓	32	.250		30.50	11.40		41.90	51.50
2010	Corridor, stock acrylic, 2-sided, with mounting bracket, 2" x 8"	1 Carp	24	.333		28	18.75		46.75	60.50
2020	2" x 10"		24	.333		28.50	18.75		47.25	61
2050	3" x 8"		24	.333		54.50	18.75		73.25	90
2060	3" x 10"		24	.333		35.50	18.75		54.25	69
2070	3" x 12"		24	.333		48.50	18.75		67.25	83
2100	4" x 8"		24	.333		53.50	18.75		72.25	88.50
2110	4" x 10"		24	.333		53.50	18.75		72.25	89
2120	4" x 12"	↓	24	.333		53.50	18.75		72.25	89

10 14 26 – Post and Panel/Pylon Signage

10 14 26.10 Post and Panel Signage

		Crew	Daily Output	Labor-Hours	Unit	Material	2022 Bare Costs Labor	Equipment	Total	Total Incl O&P
0010	**POST AND PANEL SIGNAGE**, Alum., incl. 2 posts,									
0011	2-sided panel (blank), concrete footings per post									
0025	24" W x 18" H	B-1	25	.960	Ea.	144	44.50		188.50	230
0050	36" W x 24" H		25	.960		1,200	44.50		1,244.50	1,400
0100	57" W x 36" H		25	.960		2,100	44.50		2,144.50	2,400
0150	81" W x 46" H	↓	25	.960	↓	2,575	44.50		2,619.50	2,900

10 14 Signage

10 14 43 – Photoluminescent Signage

10 14 43.10 Photoluminescent Signage

10 14 43.10 Photoluminescent Signage	Crew	Daily Output	Labor-Hours	Unit	Material	2022 Bare Costs Labor	Equipment	Total	Total Incl O&P
0010 **PHOTOLUMINESCENT SIGNAGE**									
0011 Photoluminescent exit sign, plastic, 7"x10"	1 Carp	32	.250	Ea.	32	14.10		46.10	57.50
0022 Polyester		32	.250		20.50	14.10		34.60	45
0033 Aluminum		32	.250		71	14.10		85.10	101
0044 Plastic, 10"x14"		32	.250		71.50	14.10		85.60	101
0055 Polyester		32	.250		74	14.10		88.10	104
0066 Aluminum		32	.250		163	14.10		177.10	203
0111 Directional exit sign, plastic, 10"x14"		32	.250		74	14.10		88.10	104
0122 Polyester		32	.250		74	14.10		88.10	104
0133 Aluminum		32	.250		91	14.10		105.10	123
0153 Kick plate exit sign, 10"x34"		28	.286		375	16.10		391.10	435
0173 Photoluminescent tape, 1"x60'		300	.027	L.F.	.68	1.50		2.18	3.14
0183 2"x60'		290	.028	"	1.08	1.55		2.63	3.66
0211 Photoluminescent directional arrows		42	.190	Ea.	14.75	10.70		25.45	33.50

10 14 53 – Traffic Signage

10 14 53.20 Traffic Signs

	Crew	Daily Output	Labor-Hours	Unit	Material	2022 Bare Costs Labor	Equipment	Total	Total Incl O&P
0010 **TRAFFIC SIGNS**									
0012 Stock, 24" x 24", no posts, .080" alum. reflectorized	B-80	70	.457	Ea.	39.50	23	7.75	70.25	88.50
0100 High intensity		70	.457		28.50	23	7.75	59.25	76.50
0300 30" x 30", reflectorized		70	.457		62	23	7.75	92.75	113
0400 High intensity		70	.457		96.50	23	7.75	127.25	151
0600 Guide and directional signs, 12" x 18", reflectorized		70	.457		37.50	23	7.75	68.25	86
0700 High intensity		70	.457		36.50	23	7.75	67.25	85
0900 18" x 24", stock signs, reflectorized		70	.457		32.50	23	7.75	63.25	81
1000 High intensity		70	.457		51.50	23	7.75	82.25	102
1200 24" x 24", stock signs, reflectorized		70	.457		38.50	23	7.75	69.25	87.50
1300 High intensity		70	.457		54.50	23	7.75	85.25	105
1500 Add to above for steel posts, galvanized, 10'-0" upright, bolted		200	.160		32.50	8	2.71	43.21	51.50
1600 12'-0" upright, bolted		140	.229		48.50	11.45	3.87	63.82	76
1800 Highway road signs, aluminum, over 20 S.F., reflectorized		350	.091	S.F.	20	4.59	1.55	26.14	31
2000 High intensity		350	.091		19.45	4.59	1.55	25.59	30.50
2200 Highway, suspended over road, 80 S.F. min., reflectorized		165	.194		21.50	9.75	3.28	34.53	42.50
2300 High intensity		165	.194		21.50	9.75	3.28	34.53	42.50
9000 Replace directional sign		6	5.333	Ea.	96.50	267	90.50	454	630

10 17 Telephone Specialties

10 17 16 – Telephone Enclosures

10 17 16.10 Commercial Telephone Enclosures

	Crew	Daily Output	Labor-Hours	Unit	Material	2022 Bare Costs Labor	Equipment	Total	Total Incl O&P
0010 **COMMERCIAL TELEPHONE ENCLOSURES**									
0300 Shelf type, wall hung, recessed	2 Carp	5	3.200	Ea.	370	180		550	690
0400 Surface mount		5	3.200	"	277	180		457	590
9000 Minimum labor/equipment charge		4	4	Job		225		225	360

10 21 13 – Toilet Compartments

10 21 13.13 Metal Toilet Compartments

		Crew	Daily Output	Labor-Hours	Unit	Material	2022 Bare Costs Labor	Equipment	Total	Total Incl O&P
0010	**METAL TOILET COMPARTMENTS**									
0110	Cubicles, ceiling hung									
0200	Powder coated steel	2 Carp	4	4	Ea.	470	225		695	875
0500	Stainless steel	"	4	4	"	2,050	225		2,275	2,600
0600	For handicap units, add				Ea.	122			122	134
0900	Floor and ceiling anchored									
1000	Powder coated steel	2 Carp	5	3.200	Ea.	780	180		960	1,150
1300	Stainless steel	"	5	3.200		2,000	180		2,180	2,475
1400	For handicap units, add					385			385	425
1610	Floor anchored									
1700	Powder coated steel	2 Carp	7	2.286	Ea.	1,000	129		1,129	1,300
2000	Stainless steel	"	7	2.286		1,050	129		1,179	1,350
2100	For handicap units, add					365			365	400
2200	For juvenile units, deduct					45.50			45.50	50.50
2450	Floor anchored, headrail braced									
2500	Powder coated steel	2 Carp	6	2.667	Ea.	475	150		625	765
2804	Stainless steel	"	6	2.667		1,200	150		1,350	1,550
2900	For handicap units, add					335			335	365
3000	Wall hung partitions, powder coated steel	2 Carp	7	2.286		565	129		694	830
3300	Stainless steel	"	7	2.286		420	129		549	665
3400	For handicap units, add					335			335	365
4000	Screens, entrance, floor mounted, 58" high, 48" wide									
4200	Powder coated steel	2 Carp	15	1.067	Ea.	208	60		268	325
4500	Stainless steel	"	15	1.067	"	520	60		580	665
4650	Urinal screen, 18" wide									
4704	Powder coated steel	2 Carp	6.15	2.602	Ea.	230	146		376	485
5004	Stainless steel	"	6.15	2.602	"	540	146		686	830
5100	Floor mounted, headrail braced									
5300	Powder coated steel	2 Carp	8	2	Ea.	244	113		357	445
5600	Stainless steel	"	8	2	"	625	113		738	870
5750	Pilaster, flush									
5800	Powder coated steel	2 Carp	10	1.600	Ea.	273	90		363	445
6100	Stainless steel		10	1.600		184	90		274	345
6300	Post braced, powder coated steel		10	1.600		172	90		262	330
6800	Powder coated steel		10	1.600		105	90		195	259
7800	Wedge type, powder coated steel		10	1.600		140	90		230	297
8100	Stainless steel		10	1.600		640	90		730	845
9000	Minimum labor/equipment charge	1 Carp	2.50	3.200	Job		180		180	286

10 21 13.16 Plastic-Laminate-Clad Toilet Compartments

		Crew	Daily Output	Labor-Hours	Unit	Material	2022 Bare Costs Labor	Equipment	Total	Total Incl O&P
0010	**PLASTIC-LAMINATE-CLAD TOILET COMPARTMENTS**									
0110	Cubicles, ceiling hung									
0300	Plastic laminate on particle board	2 Carp	4	4	Ea.	650	225		875	1,075
0600	For handicap units, add				Ea.	122			122	134
0900	Floor and ceiling anchored									
1100	Plastic laminate on particle board	2 Carp	5	3.200	Ea.	690	180		870	1,050
1400	For handicap units, add				"	385			385	425
1610	Floor mounted									
1800	Plastic laminate on particle board	2 Carp	7	2.286	Ea.	745	129		874	1,025
2450	Floor mounted, headrail braced									
2600	Plastic laminate on particle board	2 Carp	6	2.667	Ea.	910	150		1,060	1,250
3400	For handicap units, add					335			335	365
4300	Entrance screen, floor mtd., plas. lam., 58" high, 48" wide	2 Carp	15	1.067		282	60		342	405

For customer support on your Commercial Renovation Costs with RSMeans data, call 800.448.8182.

353

10 21 Compartments and Cubicles

10 21 13 – Toilet Compartments

10 21 13.16 Plastic-Laminate-Clad Toilet Compartments

		Crew	Daily Output	Labor-Hours	Unit	Material	2022 Bare Costs Labor	2022 Bare Costs Equipment	Total	Total Incl O&P
4800	Urinal screen, 18" wide, ceiling braced, plastic laminate	2 Carp	8	2	Ea.	291	113		404	500
5400	Floor mounted, headrail braced		8	2		216	113		329	415
5900	Pilaster, flush, plastic laminate		10	1.600		445	90		535	635
6400	Post braced, plastic laminate	↓	10	1.600	↓	239	90		329	405
6700	Wall hung, bracket supported									
6900	Plastic laminate on particle board	2 Carp	10	1.600	Ea.	98.50	90		188.50	251
7450	Flange supported									
7500	Plastic laminate on particle board	2 Carp	10	1.600	Ea.	116	90		206	271
9000	Minimum labor/equipment charge	1 Carp	2.50	3.200	Job		180		180	286

10 21 13.19 Plastic Toilet Compartments

		Crew	Daily Output	Labor-Hours	Unit	Material	2022 Bare Costs Labor	2022 Bare Costs Equipment	Total	Total Incl O&P
0010	**PLASTIC TOILET COMPARTMENTS**									
0110	Cubicles, ceiling hung									
0250	Phenolic	2 Carp	4	4	Ea.	835	225		1,060	1,275
0260	Polymer plastic	"	4	4		350	225		575	745
0600	For handicap units, add				↓	122			122	134
0900	Floor and ceiling anchored									
1050	Phenolic	2 Carp	5	3.200	Ea.	1,025	180		1,205	1,400
1060	Polymer plastic	"	5	3.200		995	180		1,175	1,375
1400	For handicap units, add				↓	385			385	425
1610	Floor mounted									
1750	Phenolic	2 Carp	7	2.286	Ea.	820	129		949	1,100
1760	Polymer plastic	"	7	2.286		925	129		1,054	1,225
2100	For handicap units, add				↓	365			365	400
2200	For juvenile units, deduct					45.50			45.50	50.50
2450	Floor mounted, headrail braced									
2550	Phenolic	2 Carp	6	2.667	Ea.	995	150		1,145	1,350
3600	Polymer plastic		6	2.667		380	150		530	660
3810	Entrance screen, polymer plastic, flr. mtd., 48"x58"		6	2.667		252	150		402	515
3820	Entrance screen, polymer plastic, flr. to clg pilaster, 48"x58"		6	2.667		252	150		402	515
6110	Urinal screen, polymer plastic, pilaster flush, 18" w		6	2.667		244	150		394	505
7110	Wall hung		6	2.667		138	150		288	390
7710	Flange mounted	↓	6	2.667	↓	221	150		371	480
9000	Minimum labor/equipment charge	1 Carp	2.50	3.200	Job		180		180	286

10 21 13.40 Stone Toilet Compartments

		Crew	Daily Output	Labor-Hours	Unit	Material	2022 Bare Costs Labor	2022 Bare Costs Equipment	Total	Total Incl O&P
0010	**STONE TOILET COMPARTMENTS**									
0100	Cubicles, ceiling hung, marble	2 Marb	2	8	Ea.	1,675	440		2,115	2,550
0600	For handicap units, add					122			122	134
0800	Floor & ceiling anchored, marble	2 Marb	2.50	6.400		1,700	350		2,050	2,450
1400	For handicap units, add					385			385	425
1600	Floor mounted, marble	2 Marb	3	5.333		1,075	294		1,369	1,650
2400	Floor mounted, headrail braced, marble	"	3	5.333		1,225	294		1,519	1,825
2900	For handicap units, add					335			335	365
4100	Entrance screen, floor mounted marble, 58" high, 48" wide	2 Marb	9	1.778		735	98		833	965
4600	Urinal screen, 18" wide, ceiling braced, marble	D-1	6	2.667	↓	770	134		904	1,075
5100	Floor mounted, headrail braced									
5200	Marble	D-1	6	2.667	Ea.	660	134		794	940
5700	Pilaster, flush, marble		9	1.778		865	89		954	1,100
6200	Post braced, marble	↓	9	1.778	↓	845	89		934	1,075
9000	Minimum labor/equipment charge	1 Carp	2.50	3.200	Job		180		180	286

10 21 Compartments and Cubicles

10 21 14 – Toilet Compartment Components

10 21 14.13 Metal Toilet Compartment Components

10 21 14.13 Metal Toilet Compartment Components	Crew	Daily Output	Labor-Hours	Unit	Material	2022 Bare Costs Labor	2022 Bare Costs Equipment	Total	Total Incl O&P
0010 **METAL TOILET COMPARTMENT COMPONENTS**									
0100 Pilasters									
0110 Overhead braced, powder coated steel, 7" wide x 82" high	2 Carp	22.20	.721	Ea.	72	40.50		112.50	144
0120 Stainless steel		22.20	.721		166	40.50		206.50	247
0130 Floor braced, powder coated steel, 7" wide x 70" high		23.30	.687		109	38.50		147.50	182
0140 Stainless steel		23.30	.687		215	38.50		253.50	298
0150 Ceiling hung, powder coated steel, 7" wide x 83" high		13.30	1.203		118	67.50		185.50	238
0160 Stainless steel		13.30	1.203		293	67.50		360.50	430
0170 Wall hung, powder coated steel, 3" wide x 58" high		18.90	.847		83.50	47.50		131	168
0180 Stainless steel	▼	18.90	.847	▼	184	47.50		231.50	279
0200 Panels									
0210 Powder coated steel, 31" wide x 58" high	2 Carp	18.90	.847	Ea.	161	47.50		208.50	253
0220 Stainless steel	"	18.90	.847	"	480	47.50		527.50	605
0230 Powder coated steel, 53" wide x 58" high	2 Carp	18.90	.847	Ea.	198	47.50		245.50	294
0240 Stainless steel		18.90	.847		575	47.50		622.50	710
0250 Powder coated steel, 63" wide x 58" high		18.90	.847		262	47.50		309.50	365
0260 Stainless steel	▼	18.90	.847	▼	690	47.50		737.50	835
0300 Doors									
0310 Powder coated steel, 24" wide x 58" high	2 Carp	14.10	1.135	Ea.	188	64		252	310
0320 Stainless steel		14.10	1.135		430	64		494	575
0330 Powder coated steel, 26" wide x 58" high		14.10	1.135		194	64		258	315
0340 Stainless steel		14.10	1.135		475	64		539	625
0350 Powder coated steel, 28" wide x 58" high		14.10	1.135		196	64		260	320
0360 Stainless steel		14.10	1.135		460	64		524	605
0370 Powder coated steel, 36" wide x 58" high		14.10	1.135		204	64		268	325
0380 Stainless steel	▼	14.10	1.135	▼	550	64		614	705
0400 Headrails									
0410 For powder coated steel, 62" long	2 Carp	65	.246	Ea.	39	13.85		52.85	65
0420 Stainless steel		65	.246		39	13.85		52.85	65
0430 For powder coated steel, 84" long		50	.320		32.50	18		50.50	64.50
0440 Stainless steel		50	.320		34	18		52	65.50
0450 For powder coated steel, 120" long		30	.533		45	30		75	97
0460 Stainless steel	▼	30	.533	▼	45	30		75	97
9000 Minimum labor/equipment charge	1 Carp	4	2	Job		113		113	179

10 21 14.16 Plastic-Laminate Clad Toilet Compartment Components

10 21 14.16 Plastic-Laminate Clad Toilet Compartment Components	Crew	Daily Output	Labor-Hours	Unit	Material	2022 Bare Costs Labor	2022 Bare Costs Equipment	Total	Total Incl O&P
0010 **PLASTIC-LAMINATE CLAD TOILET COMPARTMENT COMPONENTS**									
0100 Pilasters									
0110 Overhead braced, 7" wide x 82" high	2 Carp	22.20	.721	Ea.	111	40.50		151.50	187
0130 Floor anchored, 7" wide x 70" high		23.30	.687		111	38.50		149.50	184
0150 Ceiling hung, 7" wide x 83" high		13.30	1.203		129	67.50		196.50	250
0180 Wall hung, 3" wide x 58" high	▼	18.90	.847	▼	87.50	47.50		135	172
0200 Panels									
0210 31" wide x 58" high	2 Carp	18.90	.847	Ea.	188	47.50		235.50	283
0230 51" wide x 58" high		18.90	.847		245	47.50		292.50	345
0250 63" wide x 58" high	▼	18.90	.847	▼	300	47.50		347.50	405
0300 Doors									
0310 24" wide x 58" high	2 Carp	14.10	1.135	Ea.	175	64		239	295
0330 26" wide x 58" high		14.10	1.135		181	64		245	300
0350 28" wide x 58" high		14.10	1.135		185	64		249	305
0370 36" wide x 58" high	▼	14.10	1.135	▼	226	64		290	350
0400 Headrails									
0410 62" long	2 Carp	65	.246	Ea.	26.50	13.85		40.35	51.50

For customer support on your Commercial Renovation Costs with RSMeans data, call 800.448.8182.

355

10 21 14 – Toilet Compartment Components

10 21 14.16 Plastic-Laminate Clad Toilet Compartment Components

		Crew	Daily Output	Labor-Hours	Unit	Material	2022 Bare Costs Labor	2022 Bare Costs Equipment	Total	Total Incl O&P
0430	84" long	2 Carp	60	.267	Ea.	37	15		52	65
0450	120" long	↓	30	.533	↓	46.50	30		76.50	98.50
9000	Minimum labor/equipment charge	1 Carp	4	2	Job		113		113	179

10 21 14.19 Plastic Toilet Compartment Components

		Crew	Daily Output	Labor-Hours	Unit	Material	2022 Bare Costs Labor	2022 Bare Costs Equipment	Total	Total Incl O&P
0010	**PLASTIC TOILET COMPARTMENT COMPONENTS**									
0100	Pilasters									
0110	Overhead braced, polymer plastic, 7" wide x 82" high	2 Carp	22.20	.721	Ea.	111	40.50		151.50	187
0120	Phenolic		22.20	.721		179	40.50		219.50	262
0130	Floor braced, polymer plastic, 7" wide x 70" high		23.30	.687		190	38.50		228.50	271
0140	Phenolic		23.30	.687		159	38.50		197.50	237
0150	Ceiling hung, polymer plastic, 7" wide x 83" high		13.30	1.203		190	67.50		257.50	315
0160	Phenolic		13.30	1.203		190	67.50		257.50	315
0180	Wall hung, phenolic, 3" wide x 58" high	↓	18.90	.847	↓	105	47.50		152.50	192
0200	Panels									
0203	Polymer plastic, 18" wide x 55" high	2 Carp	18.90	.847	Ea.	300	47.50		347.50	405
0206	Phenolic, 18" wide x 58" high		18.90	.847		335	47.50		382.50	445
0210	Polymer plastic, 31" wide x 55" high		18.90	.847		410	47.50		457.50	525
0220	Phenolic, 31" wide x 58" high		18.90	.847		375	47.50		422.50	485
0223	Polymer plastic, 48" wide x 55" high		18.90	.847		570	47.50		617.50	700
0226	Phenolic, 48" wide x 58" high		18.90	.847		525	47.50		572.50	655
0230	Polymer plastic, 51" wide x 55" high		18.90	.847		575	47.50		622.50	710
0240	Phenolic, 51" wide x 58" high		18.90	.847		525	47.50		572.50	655
0250	Polymer plastic, 63" wide x 55" high		18.90	.847		740	47.50		787.50	890
0260	Phenolic, 63" wide x 58" high	↓	18.90	.847	↓	680	47.50		727.50	820
0300	Doors									
0310	Polymer plastic, 24" wide x 55" high	2 Carp	14.10	1.135	Ea.	272	64		336	400
0320	Phenolic, 24" wide x 58" high		14.10	1.135		380	64		444	520
0330	Polymer plastic, 26" wide x 55" high		14.10	1.135		290	64		354	420
0340	Phenolic, 26" wide x 58" high		14.10	1.135		310	64		374	440
0350	Polymer plastic, 28" wide x 55" high		14.10	1.135		325	64		389	460
0360	Phenolic, 28" wide x 58" high		14.10	1.135		420	64		484	560
0370	Polymer plastic, 36" wide x 55" high		14.10	1.135		390	64		454	530
0380	Phenolic, 36" wide x 58" high	↓	14.10	1.135	↓	460	64		524	605
0400	Headrails									
0410	For polymer plastic, 62" long	2 Carp	65	.246	Ea.	39	13.85		52.85	65
0420	Phenolic		65	.246		39	13.85		52.85	65
0430	For polymer plastic, 84" long		50	.320		39	18		57	71.50
0440	Phenolic		50	.320		40.50	18		58.50	73
0450	For polymer plastic, 120" long		30	.533		56.50	30		86.50	110
0460	Phenolic	↓	30	.533	↓	45	30		75	96.50
9000	Minimum labor/equipment charge	1 Carp	4	2	Job		113		113	179

10 21 23 – Cubicle Curtains and Track

10 21 23.16 Cubicle Track and Hardware

		Crew	Daily Output	Labor-Hours	Unit	Material	2022 Bare Costs Labor	2022 Bare Costs Equipment	Total	Total Incl O&P
0010	**CUBICLE TRACK AND HARDWARE**									
0020	Curtain track, box channel, ceiling mounted	1 Carp	135	.059	L.F.	12	3.34		15.34	18.50
0100	Suspended		100	.080		10	4.50		14.50	18.15
0550	Polyester, antimicrobial, 9' ceiling height		425	.019		17.45	1.06		18.51	21
0560	8' ceiling height	↓	425	.019	↓	13.90	1.06		14.96	17

356

For customer support on your Commercial Renovation Costs with RSMeans data, call 800.448.8182.

10 22 Partitions

10 22 16 – Folding Gates

10 22 16.10 Security Gates

	Crew	Daily Output	Labor- Hours	Unit	Material	2022 Bare Costs Labor	Equipment	Total	Total Incl O&P
0010 **SECURITY GATES**									
0015 For roll up type, see Section 08 33 13.10									
0300 Scissors type folding gate, ptd. steel, single, 6-1/2' high, 5-1/2' wide	2 Sswk	4	4	Opng.	205	247		452	625
0350 6-1/2' wide		4	4		500	247		747	950
0400 7-1/2' wide		4	4		500	247		747	950
0600 Double gate, 8' high, 8' wide		2.50	6.400		400	395		795	1,075
0650 10' wide		2.50	6.400		435	395		830	1,125
0700 12' wide		2	8		530	495		1,025	1,375
0750 14' wide		2	8		530	495		1,025	1,375
0900 Door gate, folding steel, 4' wide, 61" high		4	4		167	247		414	585
1000 71" high		4	4		179	247		426	595
1200 81" high		4	4		194	247		441	615
1300 Window gates, 2' to 4' wide, 31" high		4	4		134	247		381	545
1500 55" high		3.75	4.267		245	264		509	695
1600 79" high		3.50	4.571		245	283		528	725

10 22 19 – Demountable Partitions

10 22 19.43 Demountable Composite Partitions

	Crew	Daily Output	Labor- Hours	Unit	Material	2022 Bare Costs Labor	Equipment	Total	Total Incl O&P
0010 **DEMOUNTABLE COMPOSITE PARTITIONS**, add for doors									
0100 Do not deduct door openings from total L.F.									
0900 Demountable gypsum system on 2" to 2-1/2"									
1000 Steel studs, 9' high, 3" to 3-3/4" thick									
1200 Vinyl-clad gypsum	2 Carp	48	.333	L.F.	72	18.75		90.75	109
1300 Fabric clad gypsum		44	.364		154	20.50		174.50	202
1500 Steel clad gypsum		40	.400		173	22.50		195.50	226
1600 1.75 system, aluminum framing, vinyl-clad hardboard,									
1800 Paper honeycomb core panel, 1-3/4" to 2-1/2" thick									
1900 9' high	2 Carp	48	.333	L.F.	104	18.75		122.75	144
2100 7' high		60	.267		93	15		108	126
2200 5' high		80	.200		78.50	11.25		89.75	104
2250 Unitized gypsum system									
2300 Unitized panel, 9' high, 2" to 2-1/2" thick									
2350 Vinyl-clad gypsum	2 Carp	48	.333	L.F.	134	18.75		152.75	177
2400 Fabric clad gypsum	"	44	.364	"	220	20.50		240.50	275
2500 Unitized mineral fiber system									
2510 Unitized panel, 9' high, 2-1/4" thick, aluminum frame									
7550 Vinyl-clad mineral fiber	2 Carp	48	.333	L.F.	133	18.75		151.75	176
2600 Fabric clad mineral fiber	"	44	.364	"	198	20.50		218.50	251
2800 Movable steel walls, modular system									
2900 Unitized panels, 9' high, 48" wide									
3100 Baked enamel, pre-finished	2 Carp	60	.267	L.F.	151	15		166	190
3200 Fabric clad steel	"	56	.286	"	218	16.10		234.10	265
5500 For acoustical partitions, add, unsealed				S.F.	2.40			2.40	2.64
5550 Sealed				"	11.20			11.20	12.30
5700 For doors, see Section 08 16									
5800 For door hardware, see Section 08 71									
6100 In-plant modular office system, w/prehung hollow core door									
6200 3" thick polystyrene core panels									
6250 12' x 12', 2 wall	2 Clab	3.80	4.211	Ea.	4,650	192		4,842	5,425
6300 4 wall		1.90	8.421		6,750	385		7,135	8,025
6350 16' x 16', 2 wall		3.60	4.444		6,425	203		6,628	7,400
6400 4 wall		1.80	8.889		9,425	405		9,830	11,000
9000 Minimum labor/equipment charge	2 Carp	3	5.333	Job		300		300	475

10 22 Partitions

10 22 23 – Portable Partitions, Screens, and Panels

10 22 23.13 Wall Screens

	10 22 23.13 Wall Screens	Crew	Daily Output	Labor-Hours	Unit	Material	2022 Bare Costs Labor	Equipment	Total	Total Incl O&P
0010	**WALL SCREENS**, divider panels, free standing, fiber core									
0020	Fabric face straight									
0100	3'-0" long, 4'-0" high	2 Carp	100	.160	L.F.	108	9		117	132
0200	5'-0" high		90	.178		87.50	10		97.50	112
0500	6'-0" high		75	.213		93.50	12		105.50	122
0900	5'-0" long, 4'-0" high		175	.091		87.50	5.15		92.65	105
1000	5'-0" high		150	.107		104	6		110	124
1500	6'-0" high	2 Carp	125	.128	L.F.	134	7.20		141.20	159
1600	6'-0" long, 5'-0" high		162	.099		134	5.55		139.55	157
3200	Economical panels, fabric face, 4'-0" long, 5'-0" high		132	.121		152	6.80		158.80	179
3250	6'-0" high		112	.143		152	8.05		160.05	181
3300	5'-0" long, 5'-0" high		150	.107		128	6		134	151
3350	6'-0" high		125	.128		128	7.20		135.20	152
3450	Acoustical panels, 60 to 90 NRC, 3'-0" long, 5'-0" high		90	.178		119	10		129	147
3550	6'-0" high		75	.213		113	12		125	144
3600	5'-0" long, 5'-0" high		150	.107		128	6		134	151
3650	6'-0" high		125	.128		86.50	7.20		93.70	106
3700	6'-0" long, 5'-0" high		162	.099		61	5.55		66.55	76
3750	6'-0" high		138	.116		110	6.55		116.55	131
3800	Economy acoustical panels, 40 NRC, 4'-0" long, 5'-0" high		132	.121		152	6.80		158.80	179
3850	6'-0" high		112	.143		152	8.05		160.05	181
3900	5'-0" long, 6'-0" high		125	.128		128	7.20		135.20	152
3950	6'-0" long, 5'-0" high		162	.099		112	5.55		117.55	132
9000	Minimum labor/equipment charge		3	5.333	Job		300		300	475

10 22 33 – Accordion Folding Partitions

10 22 33.10 Partitions, Accordion Folding

		Crew	Daily Output	Labor-Hours	Unit	Material	2022 Bare Costs Labor	Equipment	Total	Total Incl O&P
0010	**PARTITIONS, ACCORDION FOLDING**									
0100	Vinyl covered, over 150 S.F., frame not included									
0300	Residential, 1.25 lb./S.F., 8' maximum height	2 Carp	300	.053	S.F.	29	3		32	37
0400	Commercial, 1.75 lb./S.F., 8' maximum height		225	.071		23	4		27	31.50
0900	Acoustical, 3 lb./S.F., 17' maximum height		100	.160		36	9		45	54
1200	5 lb./S.F., 20' maximum height		95	.168		58	9.50		67.50	79
1500	Vinyl-clad wood or steel, electric operation, 5.0 psf		160	.100		73	5.65		78.65	89
1900	Wood, non-acoustic, birch or mahogany, to 10' high		300	.053		39.50	3		42.50	48
9000	Minimum labor/equipment charge		4	4	Job		225		225	360

10 22 39 – Folding Panel Partitions

10 22 39.10 Partitions, Folding Panel

		Crew	Daily Output	Labor-Hours	Unit	Material	2022 Bare Costs Labor	Equipment	Total	Total Incl O&P
0010	**PARTITIONS, FOLDING PANEL**, acoustic, wood									
0100	Vinyl faced, to 18' high, 6 psf, economy trim	2 Carp	60	.267	S.F.	84	15		99	117
0150	Standard trim		45	.356		101	20		121	143
0200	Premium trim		30	.533		130	30		160	191
0400	Plastic laminate or hardwood finish, standard trim		60	.267		86.50	15		101.50	120
0500	Premium trim		30	.533		91.50	30		121.50	149
0600	Wood, low acoustical type, 4.5 psf, to 14' high		50	.320		63	18		81	98
9000	Minimum labor/equipment charge		4	4	Job		225		225	360

10 26 Wall and Door Protection

10 26 23 - Protective Wall Covering

10 26 23.10 Wall Covering, Protective

	10 26 23.10 Wall Covering, Protective	Crew	Daily Output	Labor-Hours	Unit	Material	2022 Bare Costs Labor	Equipment	Total	Total Incl O&P
0010	**WALL COVERING, PROTECTIVE**									
1775	Wall end guard, stainless steel, 16 ga., 36" tall, screwed to studs	1 Skwk	30	.267	Ea.	44.50	15.70		60.20	74

10 26 23.13 Impact Resistant Wall Protection

		Crew	Daily Output	Labor-Hours	Unit	Material	2022 Bare Costs Labor	Equipment	Total	Total Incl O&P
0010	**IMPACT RESISTANT WALL PROTECTION**									
0100	Vinyl wall protection, complete instl. incl. panels, trim, adhesive									
0110	.040" thk., std. colors	2 Carp	320	.050	S.F.	10.75	2.82		13.57	16.25
0120	.040" thk., element patterns		300	.053		9.20	3		12.20	14.85
0130	.060" thk., std. colors		300	.053		11.30	3		14.30	17.15
0140	.060" thk., element patterns		280	.057		10.75	3.22		13.97	16.90
1775	Wall protection, stainless steel, 16 ga., 48" x 36" tall, screwed to studs	2 Skwk	500	.032		21	1.88		22.88	26

10 26 33 - Door and Frame Protection

10 26 33.10 Protection, Door and Frame

		Crew	Daily Output	Labor-Hours	Unit	Material	2022 Bare Costs Labor	Equipment	Total	Total Incl O&P
0010	**PROTECTION, DOOR AND FRAME**									
0100	Door frame guard, vinyl, 3" x 3" x 4'	1 Carp	30	.267	Ea.	98	15		113	132
0110	Door frame guard, vinyl, 3" x 3" x 8'		26	.308		128	17.30		145.30	168
0120	Door frame guard, stainless steel, 3" x 3" x 4'		30	.267		415	15		430	480
0500	Steel door track/wheel guards, 4'-0" high	E-4	22	1.455		48	90.50	6.70	145.20	206

10 28 Toilet, Bath, and Laundry Accessories

10 28 13 - Toilet Accessories

10 28 13.13 Commercial Toilet Accessories

		Crew	Daily Output	Labor-Hours	Unit	Material	2022 Bare Costs Labor	Equipment	Total	Total Incl O&P
0010	**COMMERCIAL TOILET ACCESSORIES**									
0200	Curtain rod, stainless steel, 5' long, 1" diameter	1 Carp	13	.615	Ea.	41	34.50		75.50	100
0300	1-1/4" diameter		13	.615		18.65	34.50		53.15	75.50
0350	Chrome, 1" diameter		13	.615		10.15	34.50		44.65	66
0360	For vinyl curtain, add		1950	.004	S.F.	134	.23		134.23	147
0400	Diaper changing station, horizontal, wall mounted, plastic		10	.800	Ea.	320	45		365	420
0420	Vertical		10	.800		212	45		257	305
0430	Oval shaped		10	.800		212	45		257	305
0440	Recessed, with stainless steel flange		6	1.333		800	75		875	1,000
0500	Dispenser units, combined soap & towel dispensers,									
0510	Mirror and shelf, flush mounted	1 Carp	10	.800	Ea.	320	45		365	425
0600	Towel dispenser and waste receptacle,									
0610	18 gallon capacity	1 Carp	10	.800	Ea.	445	45		490	560
0800	Grab bar, straight, 1-1/4" diameter, stainless steel, 18" long		24	.333		28.50	18.75		47.25	61.50
0900	24" long		23	.348		21.50	19.60		41.10	54.50
1000	30" long	1 Carp	22	.364	Ea.	33	20.50		53.50	69
1100	36" long		20	.400		18.90	22.50		41.40	57
1105	42" long		20	.400		32.50	22.50		55	72
1120	Corner, 36" long		20	.400		89	22.50		111.50	134
1200	1-1/2" diameter, 24" long		23	.348		22	19.60		41.60	55.50
1300	36" long		20	.400		35	22.50		57.50	74.50
1310	42" long		18	.444		38.50	25		63.50	82
1500	Tub bar, 1-1/4" diameter, 24" x 36"		14	.571		95.50	32		127.50	156
1600	Plus vertical arm		12	.667		153	37.50		190.50	228
1900	End tub bar, 1" diameter, 90° angle, 16" x 32"		12	.667		92.50	37.50		130	162
2010	Tub/shower/toilet, 2-wall, 36" x 24"		12	.667		92.50	37.50		130	162
2300	Hand dryer, surface mounted, electric, 115 volt, 20 amp		4	2		415	113		528	635
2400	230 volt, 10 amp		4	2		585	113		698	825
2450	Hand dryer, touch free, 1400 watt, 81,000 rpm		4	2		1,450	113		1,563	1,775

For customer support on your Commercial Renovation Costs with RSMeans data, call 800.448.8182.

359

10 28 13.13 Commercial Toilet Accessories		Crew	Daily Output	Labor-Hours	Unit	Material	2022 Bare Costs Labor	2022 Bare Costs Equipment	Total	Total Incl O&P
2600	Hat and coat strip, stainless steel, 4 hook, 36" long	1 Carp	24	.333	Ea.	39.50	18.75		58.25	73.50
2700	6 hook, 60" long		20	.400		92	22.50		114.50	137
3000	Mirror, with stainless steel 3/4" square frame, 18" x 24"		20	.400		43	22.50		65.50	83.50
3100	36" x 24"		15	.533		67	30		97	121
3200	48" x 24"		10	.800		98	45		143	180
3300	72" x 24"		6	1.333		121	75		196	253
3500	With 5" stainless steel shelf, 18" x 24"		20	.400		97	22.50		119.50	143
3600	36" x 24"		15	.533		122	30		152	182
3700	48" x 24"		10	.800		162	45		207	250
3800	72" x 24"		6	1.333		262	75		337	410
4100	Mop holder strip, stainless steel, 5 holders, 48" long		20	.400		78.50	22.50		101	122
4200	Napkin/tampon dispenser, recessed		15	.533		390	30		420	480
4220	Semi-recessed		6.50	1.231		375	69.50		444.50	525
4250	Napkin receptacle, recessed		6.50	1.231		177	69.50		246.50	305
4300	Robe hook, single, regular		96	.083		21.50	4.69		26.19	31.50
4400	Heavy duty, concealed mounting		56	.143		21.50	8.05		29.55	37
4600	Soap dispenser, chrome, surface mounted, liquid		20	.400		52.50	22.50		75	93.50
5000	Recessed stainless steel, liquid		10	.800		199	45		244	290
5600	Shelf, stainless steel, 5" wide, 18 ga., 24" long		24	.333		63.50	18.75		82.25	100
5700	48" long		16	.500		103	28		131	158
5800	8" wide shelf, 18 ga., 24" long		22	.364		55	20.50		75.50	93
5900	48" long		14	.571		158	32		190	225
6000	Toilet seat cover dispenser, stainless steel, recessed	1 Carp	20	.400	Ea.	197	22.50		219.50	253
6050	Surface mounted		15	.533		37	30		67	88
6100	Toilet tissue dispenser, surface mounted, SS, single roll		30	.267		14.60	15		29.60	40
6200	Double roll		24	.333		98	18.75		116.75	138
6240	Plastic, twin/jumbo dbl. roll		24	.333		33	18.75		51.75	66.50
6290	Toilet seat	1 Plum	40	.200		32.50	13.95		46.45	57
6400	Towel bar, stainless steel, 18" long	1 Carp	23	.348		34	19.60		53.60	68.50
6500	30" long		21	.381		57.50	21.50		79	97.50
6700	Towel dispenser, stainless steel, surface mounted		16	.500		67	28		95	118
6800	Flush mounted, recessed		10	.800		44	45		89	120
6900	Plastic, touchless, battery operated		16	.500		120	28		148	177
7000	Towel holder, hotel type, 2 guest size		20	.400		67.50	22.50		90	111
7200	Towel shelf, stainless steel, 24" long, 8" wide		20	.400		92.50	22.50		115	138
7400	Tumbler holder, for tumbler only		30	.267		18.80	15		33.80	44.50
7410	Tumbler holder, recessed		20	.400		8.95	22.50		31.45	46
7500	Soap, tumbler & toothbrush		30	.267		18.95	15		33.95	45
7510	Tumbler & toothbrush holder		20	.400		12.15	22.50		34.65	49.50
8000	Waste receptacles, stainless steel, with top, 13 gallon		10	.800		340	45		385	445
8100	36 gallon		8	1		430	56.50		486.50	565
9000	Minimum labor/equipment charge		5	1.600	Job		90		90	143
9996	Bathroom access., grab bar, straight, 1-1/2" diam., SS, 42" L install only		18	.444	Ea.		25		25	40

10 28 16 – Bath Accessories

10 28 16.20 Medicine Cabinets

0010	MEDICINE CABINETS									
0020	With mirror, sst frame, 16" x 22", unlighted	1 Carp	14	.571	Ea.	226	32		258	300
0100	Wood frame		14	.571		149	32		181	215
0300	Sliding mirror doors, 20" x 16" x 4-3/4", unlighted		7	1.143		145	64.50		209.50	262
0400	24" x 19" x 8-1/2", lighted		5	1.600		167	90		257	325
0600	Triple door, 30" x 32", unlighted, plywood body		7	1.143		239	64.50		303.50	365
0700	Steel body		7	1.143		410	64.50		474.50	550

10 28 Toilet, Bath, and Laundry Accessories

10 28 16 – Bath Accessories

10 28 16.20 Medicine Cabinets

	10 28 16.20 Medicine Cabinets	Crew	Daily Output	Labor-Hours	Unit	Material	2022 Bare Costs Labor	Equipment	Total	Total Incl O&P
0900	Oak door, wood body, beveled mirror, single door	1 Carp	7	1.143	Ea.	138	64.50		202.50	254
1000	Double door		6	1.333		450	75		525	615
1200	Hotel cabinets, stainless, with lower shelf, unlighted		10	.800		345	45		390	450
1300	Lighted		5	1.600		360	90		450	545
9000	Minimum labor/equipment charge		4	2	Job		113		113	179

10 28 19 – Tub and Shower Enclosures

10 28 19.10 Partitions, Shower

	10 28 19.10 Partitions, Shower	Crew	Daily Output	Labor-Hours	Unit	Material	2022 Bare Costs Labor	Equipment	Total	Total Incl O&P
0010	**PARTITIONS, SHOWER** floor mounted, no plumbing									
0400	Cabinet, one piece, fiberglass, 32" x 32"	2 Carp	5	3.200	Ea.	460	180		640	790
0420	36" x 36"		5	3.200		950	180		1,130	1,325
0440	36" x 48"		5	3.200		1,025	180		1,205	1,400
0460	Acrylic, 32" x 32"		5	3.200		197	180		377	505
0480	36" x 36"		5	3.200		935	180		1,115	1,300
0500	36" x 48"		5	3.200		815	180		995	1,175
0520	Shower door for above, clear plastic, 24" wide	1 Carp	8	1		111	56.50		167.50	212
0540	28" wide		8	1		187	56.50		243.50	296
0560	Tempered glass, 24" wide		8	1		135	56.50		191.50	239
0580	28" wide		8	1		160	56.50		216.50	266
4100	Shower doors, economy plastic, 24" wide	1 Shee	9	.889		131	60		191	238
4200	Tempered glass door, economy		8	1		273	67.50		340.50	405
4700	Deluxe, tempered glass, chrome on brass frame, 42" to 44"		8	1		320	67.50		387.50	455
4800	39" to 48" wide		1	8		600	540		1,140	1,500
9990	Minimum labor/equipment charge		2.50	3.200	Job		216		216	335

10 31 Manufactured Fireplaces

10 31 13 – Manufactured Fireplace Chimneys

10 31 13.10 Fireplace Chimneys

	10 31 13.10 Fireplace Chimneys	Crew	Daily Output	Labor-Hours	Unit	Material	2022 Bare Costs Labor	Equipment	Total	Total Incl O&P
0010	**FIREPLACE CHIMNEYS**									
0500	Chimney dbl. wall, all stainless, over 8'-6", 7" diam., add to fireplace	1 Carp	33	.242	V.L.F.	57	13.65		70.65	84
0600	10" diameter, add to fireplace		32	.250		116	14.10		130.10	151
0700	12" diameter, add to fireplace		31	.258		211	14.55		225.55	255
0800	14" diameter, add to fireplace		30	.267		219	15		234	265
1000	Simulated brick chimney top, 4' high, 16" x 16"		10	.800	Ea.	500	45		545	620
1100	24" x 24"		7	1.143	"	595	64.50		659.50	755

10 31 13.20 Chimney Accessories

	10 31 13.20 Chimney Accessories	Crew	Daily Output	Labor-Hours	Unit	Material	2022 Bare Costs Labor	Equipment	Total	Total Incl O&P
0010	**CHIMNEY ACCESSORIES**									
0020	Chimney screens, galv., 13" x 13" flue	1 Bric	8	1	Ea.	59	55.50		114.50	154
0050	24" x 24" flue		5	1.600		126	88.50		214.50	281
0200	Stainless steel, 13" x 13" flue		8	1		107	55.50		162.50	207
0250	20" x 20" flue		5	1.600		151	88.50		239.50	310
2400	Squirrel and bird screens, galvanized, 8" x 8" flue		16	.500		58	27.50		85.50	108
2450	13" x 13" flue		12	.667		151	37		188	225
9000	Minimum labor/equipment charge		3.50	2.286	Job		126		126	202

10 31 16 – Manufactured Fireplace Forms

10 31 16.10 Fireplace Forms

	10 31 16.10 Fireplace Forms	Crew	Daily Output	Labor-Hours	Unit	Material	2022 Bare Costs Labor	Equipment	Total	Total Incl O&P
0010	**FIREPLACE FORMS**									
1800	Fireplace forms, no accessories, 32" opening	1 Bric	3	2.667	Ea.	710	147		857	1,025
1900	36" opening		2.50	3.200		905	177		1,082	1,275
2000	40" opening		2	4		1,200	221		1,421	1,675
2100	78" opening		1.50	5.333		1,750	295		2,045	2,400

For customer support on your Commercial Renovation Costs with RSMeans data, call 800.448.8182.

10 31 Manufactured Fireplaces

10 31 23 – Prefabricated Fireplaces

10 31 23.10 Fireplace, Prefabricated	Crew	Daily Output	Labor-Hours	Unit	Material	2022 Bare Costs Labor	Equipment	Total	Total Incl O&P
0010 **FIREPLACE, PREFABRICATED**, free standing or wall hung									
0100 With hood & screen, painted	1 Carp	1.30	6.154	Ea.	1,700	345		2,045	2,425
0150 Average		1	8		1,575	450		2,025	2,475
0200 Stainless steel		.90	8.889	▼	2,900	500		3,400	3,975
1500 Simulated logs, gas fired, 40,000 BTU, 2' long, manual safety pilot		7	1.143	Set	790	64.50		854.50	970
1600 Adjustable flame remote pilot		6	1.333		2,525	75		2,600	2,900
1700 Electric, 1,500 BTU, 1'-6" long, incandescent flame		7	1.143		294	64.50		358.50	425
1800 1,500 BTU, LED flame		6	1.333	▼	355	75		430	510
2000 Fireplace, built-in, 36" hearth, radiant		1.30	6.154	Ea.	790	345		1,135	1,425
2100 Recirculating, small fan		1	8		945	450		1,395	1,775
2150 Large fan		.90	8.889		2,525	500		3,025	3,575
2200 42" hearth, radiant		1.20	6.667		1,100	375		1,475	1,825
2300 Recirculating, small fan		.90	8.889		1,275	500		1,775	2,200
2350 Large fan		.80	10		1,375	565		1,940	2,400
2400 48" hearth, radiant		1.10	7.273		2,175	410		2,585	3,050
2500 Recirculating, small fan		.80	10		2,775	565		3,340	3,950
2550 Large fan		.70	11.429		2,775	645		3,420	4,075
3000 See through, including doors		.80	10		2,200	565		2,765	3,325
3200 Corner (2 wall)	▼	1	8	▼	3,675	450		4,125	4,750

10 32 Fireplace Specialties

10 32 13 – Fireplace Dampers

10 32 13.10 Dampers

	Crew	Daily Output	Labor-Hours	Unit	Material	2022 Bare Costs Labor	Equipment	Total	Total Incl O&P
0010 **DAMPERS**									
0800 Damper, rotary control, steel, 30" opening	1 Bric	6	1.333	Ea.	127	73.50		200.50	258
0850 Cast iron, 30" opening		6	1.333		126	73.50		199.50	257
0880 36" opening		6	1.333		130	73.50		203.50	261
0900 48" opening		6	1.333		152	73.50		225.50	286
0920 60" opening		6	1.333		365	73.50		438.50	520
1000 84" opening, special order		5	1.600		945	88.50		1,033.50	1,175
1050 96" opening, special order		4	2		955	111		1,066	1,225
1200 Steel plate, poker control, 60" opening		8	1		335	55.50		390.50	455
1250 84" opening, special order	▼	5	1.600	▼	610	88.50		698.50	810
1400 "Universal" type, chain operated, 32" x 20" opening	1 Bric	8	1	Ea.	259	55.50		314.50	375
1450 48" x 24" opening	"	5	1.600	"	385	88.50		473.50	565

10 32 23 – Fireplace Doors

10 32 23.10 Doors

	Crew	Daily Output	Labor-Hours	Unit	Material	2022 Bare Costs Labor	Equipment	Total	Total Incl O&P
0010 **DOORS**									
0400 Cleanout doors and frames, cast iron, 8" x 8"	1 Bric	12	.667	Ea.	54	37		91	119
0450 12" x 12"		10	.800		105	44		149	186
0500 18" x 24"		8	1		156	55.50		211.50	260
0550 Cast iron frame, steel door, 24" x 30"		5	1.600		385	88.50		473.50	565
1600 Dutch oven door and frame, cast iron, 12" x 15" opening		13	.615		136	34		170	205
1650 Copper plated, 12" x 15" opening	▼	13	.615	▼	266	34		300	350

10 35 Stoves

10 35 13 – Heating Stoves

10 35 13.10 Wood Burning Stoves	Crew	Daily Output	Labor-Hours	Unit	Material	2022 Bare Costs Labor	Equipment	Total	Total Incl O&P
0010 **WOOD BURNING STOVES**									
0015 Cast iron, less than 1,500 S.F.	2 Carp	1.30	12.308	Ea.	1,700	695		2,395	2,975
0020 1,500 to 2,000 S.F.		1	16		2,800	900		3,700	4,500
0030 greater than 2,000 S.F.		.80	20		2,850	1,125		3,975	4,925
0050 For gas log lighter, add					33			33	36.50

10 43 Emergency Aid Specialties

10 43 13 – Defibrillator Cabinets

10 43 13.05 Defibrillator Cabinets	Crew	Daily Output	Labor-Hours	Unit	Material	Labor	Equipment	Total	Total Incl O&P
0010 **DEFIBRILLATOR CABINETS**, not equipped, stainless steel									
0050 Defibrillator cabinet, stainless steel with strobe & alarm 12" x 27"	1 Carp	10	.800	Ea.	325	45		370	425
0100 Automatic External Defibrillator	"	30	.267	"	1,125	15		1,140	1,275

10 44 Fire Protection Specialties

10 44 13 – Fire Protection Cabinets

10 44 13.53 Fire Equipment Cabinets	Crew	Daily Output	Labor-Hours	Unit	Material	Labor	Equipment	Total	Total Incl O&P
0010 **FIRE EQUIPMENT CABINETS**, not equipped, 20 ga. steel box									
0040 Recessed, D.S. glass in door, box size given									
1000 Portable extinguisher, single, 8" x 12" x 27", alum. door & frame	Q-12	8	2	Ea.	116	123		239	320
1100 Steel door and frame	"	8	2	"	93	123		216	292
3000 Hose rack assy., 1-1/2" valve & 100' hose, 24" x 40" x 5-1/2"									
3200 Steel door and frame	Q-12	6	2.667	Ea.	227	165		392	505
4000 Hose rack assy., 2-1/2" x 1-1/2" valve, 100' hose, 24" x 40" x 8"									
4200 Steel door and frame	Q-12	6	2.667	Ea.	840	165		1,005	1,175
5000 Hose rack assy., 2-1/2" x 1-1/2" valve, 100' hose									
5010 and extinguisher, 30" x 40" x 8"									
5200 Steel door and frame	Q-12	5	3.200	Ea.	280	197		477	610

10 44 16 – Fire Extinguishers

10 44 16.13 Portable Fire Extinguishers	Crew	Daily Output	Labor-Hours	Unit	Material	Labor	Equipment	Total	Total Incl O&P
0010 **PORTABLE FIRE EXTINGUISHERS**									
0140 CO_2, with hose and "H" horn, 10 lb.				Ea.	330			330	365
1000 Dry chemical, pressurized									
1040 Standard type, portable, painted, 2-1/2 lb.				Ea.	25.50			25.50	28
1080 10 lb.					62.50			62.50	68.50
1100 20 lb.					152			152	167
1120 30 lb.					870			870	955
2000 ABC all purpose type, portable, 2-1/2 lb.					17.45			17.45	19.20
2080 9-1/2 lb.					60			60	66
3500 Halotron 1, 2-1/2 lb.					110			110	121
3600 5 lb.					197			197	217
3700 60 lb.					400			400	440

10 51 13.10 Lockers		Crew	Daily Output	Labor-Hours	Unit	Material	2022 Bare Costs Labor	Equipment	Total	Total Incl O&P
0011	**LOCKERS** steel, baked enamel, knock down construction									
0012	Body- 24ga, door- 16 or 18ga									
0110	1-tier locker, 12" x 15" x 72"	1 Shee	20	.400	Ea.	156	27		183	214
0120	18" x 15" x 72"		20	.400		495	27		522	585
0130	12" x 18" x 72"		20	.400		240	27		267	305
0140	18" x 18" x 72"		20	.400		390	27		417	470
0410	2- tier, 12" x 15" x 36"		30	.267		390	18		408	460
0420	18" x 15" x 36"		30	.267		405	18		423	475
0430	12" x 18" x 36"		30	.267		405	18		423	475
0440	18" x 18" x 36"		30	.267		345	18		363	405
0450	3- tier 12" x 15" x 24"		30	.267		420	18		438	490
0455	12" x 18" x 24"		30	.267		455	18		473	530
0480	4- tier 12" x 15" x 18"		30	.267		184	18		202	230
0485	12" x 18" x 18"		30	.267		228	18		246	279
0500	Two person, 18" x 15" x 72"		20	.400		345	27		372	420
0510	18" x 18" x 72"		20	.400		345	27		372	420
0520	Duplex, 15" x 15" x 72"		20	.400		925	27		952	1,075
0530	15" x 21" x 72"		20	.400		445	27		472	530
1098	All welded, ventilated athletic lockers, body- 16 ga., door- 14 ga.									
1100	1-tier, 12" x 18" x 72"	1 Shee	7.50	1.067	Ea.	260	72		332	400
1110	18" x 15" x 72"		7.50	1.067		260	72		332	400
1115	18" x 18" x 72"		7.50	1.067		180	72		252	310
1130	2-tier, 12" x 15" x 36"	1 Shee	7.50	1.067	Ea.	885	72		957	1,075
1135	12" x 18" x 36"		7.50	1.067		885	72		957	1,075
1140	18" x 15" x 36"		7.50	1.067		245	72		317	380
1145	18" x 18" x 36"		7.50	1.067		256	72		328	395
1160	3-tier, 12" x 18" x 24"		7.50	1.067		169	72		241	298
1180	4-tier, 12" x 18" x 18"		7.50	1.067		209	72		281	340
1190	5H box, 12" x 18" x 14.4"		7.50	1.067		350	72		422	495
1195	18" x 18" x 14.4"		7.50	1.067		247	72		319	385
1196	6H box, 12" x 15" x 12"		7.50	1.067		415	72		487	565
1197	12" x 18" x 12"		7.50	1.067		460	72		532	615
1198	18" x 18" x 12"		7.50	1.067		625	72		697	800
2399	Standard duty lockers, body- 24 ga., doors- 16-18 ga.									
2400	16-person locker unit with clothing rack									
2500	72" wide x 15" deep x 72" high	1 Shee	15	.533	Ea.	750	36		786	880
2550	18" deep		15	.533	"	975	36		1,011	1,125
3250	Rack w/24 wire mesh baskets		1.50	5.333	Set	605	360		965	1,225
3260	30 baskets		1.25	6.400		510	430		940	1,250
3270	36 baskets		.95	8.421		600	570		1,170	1,550
3280	42 baskets		.80	10		615	675		1,290	1,725
3600	For hanger rods, add				Ea.	3.27			3.27	3.60
3650	For number plate kit, 100 plates #1 - #100, add	1 Shee	4	2		113	135		248	335
3700	For locker base, closed front panel		90	.089		7.75	6		13.75	17.90
3710	End panel, bolted		36	.222		26	15		41	52
3800	For sloping top, 12" wide		24	.333		41.50	22.50		64	80.50
3810	15" wide		24	.333		86.50	22.50		109	131
3820	18" wide		24	.333		67	22.50		89.50	109
3850	Sloping top end panel, 12" deep		72	.111		21	7.50		28.50	35
3860	15" deep		72	.111		22.50	7.50		30	36
3870	18" deep		72	.111		26	7.50		33.50	40
3900	For finish end panels, steel, 60" high, 15" deep		12	.667		110	45		155	191
3910	72" high, 12" deep		12	.667		35.50	45		80.50	109

10 51 Lockers

10 51 13 – Metal Lockers

10 51 13.10 Lockers

		Crew	Daily Output	Labor-Hours	Unit	Material	2022 Bare Costs Labor	Equipment	Total	Total Incl O&P
3920	18" deep	1 Shee	12	.667	Ea.	57	45		102	133
5000	For "ready to assemble" lockers,									
5010	Add to labor						75%			
5020	Deduct from material					20%				
6000	Heavy duty for detention facility, tamper proof, 14 ga. welded steel, solid									
6100	24" W x 24" D x 74" H, single tier	1 Shee	18	.444	Ea.	785	30		815	910
6110	Double tier	1 Shee	18	.444	Ea.	745	30		775	860
6120	Triple tier		18	.444	"	795	30		825	920
9000	Minimum labor/equipment charge	↓	2.50	3.200	Job		216		216	335

10 55 Postal Specialties

10 55 23 – Mail Boxes

10 55 23.10 Commercial Mail Boxes

		Crew	Daily Output	Labor-Hours	Unit	Material	2022 Bare Costs Labor	Equipment	Total	Total Incl O&P
0010	**COMMERCIAL MAIL BOXES**									
0020	Horiz., key lock, 5"H x 6"W x 15"D, alum., rear load	1 Carp	34	.235	Ea.	24.50	13.25		37.75	48
0100	Front loading		34	.235		48.50	13.25		61.75	74
0200	Double, 5"H x 12"W x 15"D, rear loading		26	.308		112	17.30		129.30	151
0300	Front loading		26	.308		90	17.30		107.30	127
0500	Quadruple, 10"H x 12"W x 15"D, rear loading		20	.400		81.50	22.50		104	126
0600	Front loading		20	.400		49	22.50		71.50	90
1600	Vault type, horizontal, for apartments, 4" x 5"		34	.235		48.50	13.25		61.75	74.50
1700	Alphabetical directories, 120 names		10	.800		510	45		555	630
1900	Letter slot, residential		20	.400		49	22.50		71.50	90
2000	Post office type		8	1	↓	213	56.50		269.50	325
9000	Minimum labor/equipment charge	↓	5	1.600	Job		90		90	143

10 56 Storage Assemblies

10 56 13 – Metal Storage Shelving

10 56 13.10 Shelving

		Crew	Daily Output	Labor-Hours	Unit	Material	2022 Bare Costs Labor	Equipment	Total	Total Incl O&P
0010	**SHELVING**									
0020	Metal, industrial, cross-braced, 3' W, 12" D	1 Sswk	175	.046	SF Shlf	17.25	2.83		20.08	23.50
0100	24" D		330	.024		11.25	1.50		12.75	14.75
2200	Wide span, 1600 lb. capacity per shelf, 6' W, 24" D		380	.021		30	1.30		31.30	35
2400	36" D	↓	440	.018	↓	11.80	1.12		12.92	14.75
9000	Minimum labor/equipment charge	1 Carp	4	2	Job		113		113	179

10 57 Wardrobe and Closet Specialties

10 57 23 – Closet and Utility Shelving

10 57 23.19 Wood Closet and Utility Shelving

		Crew	Daily Output	Labor-Hours	Unit	Material	2022 Bare Costs Labor	Equipment	Total	Total Incl O&P
0010	**WOOD CLOSET AND UTILITY SHELVING**									
0020	Pine, clear grade, no edge band, 1" x 8"	1 Carp	115	.070	L.F.	3.15	3.92		7.07	9.65
0100	1" x 10"		110	.073		3.93	4.09		8.02	10.80
0200	1" x 12"		105	.076		4.73	4.29		9.02	12
0600	Plywood, 3/4" thick with lumber edge, 12" wide		75	.107		3.47	6		9.47	13.35
0700	24" wide		70	.114	↓	6.55	6.45		13	17.40
0900	Bookcase, clear grade pine, shelves 12" OC, 8" deep, per S.F. shelf		70	.114	S.F.	10.25	6.45		16.70	21.50
1000	12" deep shelves	↓	65	.123	"	15.35	6.95		22.30	28

10 57 Wardrobe and Closet Specialties

10 57 23 – Closet and Utility Shelving

10 57 23.19 Wood Closet and Utility Shelving	Crew	Daily Output	Labor-Hours	Unit	Material	2022 Bare Costs Labor	Equipment	Total	Total Incl O&P
1200	Adjustable closet rod and shelf, 12" wide, 3' long	1 Carp	20	.400	Ea.	13.85	22.50		36.35	51.50
1300	8' long	1 Carp	15	.533	Ea.	25	30		55	75
1500	Prefinished shelves with supports, stock, 8" wide		75	.107	L.F.	10.05	6		16.05	20.50
1600	10" wide		70	.114	"	15.70	6.45		22.15	27.50
9000	Minimum labor/equipment charge		4	2	Job		113		113	179

10 73 Protective Covers

10 73 13 – Awnings

10 73 13.10 Awnings, Fabric

	10 73 13.10 Awnings, Fabric	Crew	Daily Output	Labor-Hours	Unit	Material	2022 Bare Costs Labor	Equipment	Total	Total Incl O&P
0010	**AWNINGS, FABRIC**									
0020	Including acrylic canvas and frame, standard design									
0100	Door and window, slope, 3' high, 4' wide	1 Carp	4.50	1.778	Ea.	940	100		1,040	1,175
0110	6' wide		3.50	2.286		1,200	129		1,329	1,525
0120	8' wide		3	2.667		1,475	150		1,625	1,875
0200	Quarter round convex, 4' wide		3	2.667		1,450	150		1,600	1,850
0210	6' wide		2.25	3.556		1,900	200		2,100	2,425
0220	8' wide		1.80	4.444		2,325	250		2,575	2,950
0300	Dome, 4' wide		7.50	1.067		565	60		625	715
0310	6' wide		3.50	2.286		1,275	129		1,404	1,600
0320	8' wide		2	4		2,250	225		2,475	2,825
0350	Elongated dome, 4' wide		1.33	6.015		2,125	340		2,465	2,875
0360	6' wide		1.11	7.207		2,525	405		2,930	3,425
0370	8' wide		1	8		2,950	450		3,400	3,975
1000	Entry or walkway, peak, 12' long, 4' wide	2 Carp	.90	17.778		10,600	1,000		11,600	13,300
1010	6' wide		.60	26.667		16,400	1,500		17,900	20,400
1020	8' wide		.40	40		22,600	2,250		24,850	28,400
1100	Radius with dome end, 4' wide		1.10	14.545		8,050	820		8,870	10,200
1110	6' wide		.70	22.857		12,900	1,275		14,175	16,300
1120	8' wide		.50	32		18,400	1,800		20,200	23,100
2000	Retractable lateral arm awning, manual									
2010	To 12' wide, 8'-6" projection	2 Carp	1.70	9.412	Ea.	2,275	530		2,805	3,350
2020	To 14' wide, 8'-6" projection		1.10	14.545		2,650	820		3,470	4,225
2030	To 19' wide, 8'-6" projection		.85	18.824		3,600	1,050		4,650	5,625
2040	To 24' wide, 8'-6" projection		.67	23.881		4,550	1,350		5,900	7,125
2050	Motor for above, add	1 Carp	2.67	3		655	169		824	990
3000	Patio/deck canopy with frame									
3010	12' wide, 12' projection	2 Carp	2	8	Ea.	3,200	450		3,650	4,250
3020	16' wide, 14' projection	"	1.20	13.333		4,975	750		5,725	6,675
9000	For fire retardant canvas, add					7%				
9010	For lettering or graphics, add					35%				
9020	For painted or coated acrylic canvas, deduct				Ea.	8%				
9030	For translucent or opaque vinyl canvas, add					10%				
9040	For 6 or more units, deduct					20%	15%			

10 73 16 – Canopies

10 73 16.10 Canopies, Residential

	10 73 16.10 Canopies, Residential	Crew	Daily Output	Labor-Hours	Unit	Material	2022 Bare Costs Labor	Equipment	Total	Total Incl O&P
0010	**CANOPIES, RESIDENTIAL** Prefabricated									
0500	Carport, free standing, baked enamel, alum., .032", 40 psf									
0520	16' x 8', 4 posts	2 Carp	3	5.333	Ea.	2,650	300		2,950	3,400
0600	20' x 10', 6 posts		2	8		3,425	450		3,875	4,500
0605	30' x 10', 8 posts		2	8		5,150	450		5,600	6,375

10 73 16 – Canopies

	10 73 16.10 Canopies, Residential	Crew	Daily Output	Labor-Hours	Unit	Material	2022 Bare Costs Labor	Equipment	Total	Total Incl O&P
1000	Door canopies, extruded alum., .032", 42" projection, 4' wide	1 Carp	8	1	Ea.	236	56.50		292.50	350
1020	6' wide	"	6	1.333		289	75		364	440
1040	8' wide	2 Carp	9	1.778		495	100		595	705
1060	10' wide		7	2.286		620	129		749	885
1080	12' wide		5	3.200		600	180		780	945
1200	54" projection, 4' wide	1 Carp	8	1		237	56.50		293.50	350
1220	6' wide	"	6	1.333		298	75		373	450
1240	8' wide	2 Carp	9	1.778		400	100		500	600
1260	10' wide		7	2.286		825	129		954	1,125
1280	12' wide		5	3.200		890	180		1,070	1,275
1300	Painted, add					20%				
1310	Bronze anodized, add					50%				
3000	Window awnings, aluminum, window 3' high, 4' wide	1 Carp	10	.800		115	45		160	199
3020	6' wide	"	8	1		187	56.50		243.50	296
3040	9' wide	2 Carp	9	1.778		420	100		520	620
3060	12' wide	"	5	3.200		530	180		710	870
3100	Window, 4' high, 4' wide	1 Carp	10	.800		284	45		329	380
3120	6' wide	"	8	1		335	56.50		391.50	460
3140	9' wide	2 Carp	9	1.778		715	100		815	945
3160	12' wide	"	5	3.200		765	180		945	1,125
3200	Window, 6' high, 4' wide	1 Carp	10	.800		380	45		425	485
3220	6' wide	"	8	1		390	56.50		446.50	520
3240	9' wide	2 Carp	9	1.778		970	100		1,070	1,225
3260	12' wide	"	5	3.200		1,475	180		1,655	1,900
3400	Roll-up aluminum, 2'-6" wide	1 Carp	14	.571		211	32		243	283
3420	3' wide		12	.667		259	37.50		296.50	345
3440	4' wide		10	.800		280	45		325	380
3460	6' wide		8	1		335	56.50		391.50	460
3480	9' wide	2 Carp	9	1.778		400	100		500	600
3500	12' wide	2 Carp	5	3.200	Ea.	535	180		715	875
3600	Window awnings, canvas, 24" drop, 3' wide	1 Carp	30	.267	L.F.	68	15		83	99
3620	4' wide		40	.200		60.50	11.25		71.75	84.50
3700	30" drop, 3' wide		30	.267		61	15		76	91
3720	4' wide		40	.200		56.50	11.25		67.75	80.50
3740	5' wide		45	.178		47.50	10		57.50	68.50
3760	6' wide		48	.167		56	9.40		65.40	77
3780	8' wide		48	.167		51	9.40		60.40	71
3800	10' wide		50	.160		43	9		52	62
3900	Repair canvas, minimum		8	1	Ea.		56.50		56.50	89.50
3920	Maximum		4	2	"		113		113	179
9000	Minimum labor/equipment charge		3	2.667	Job		150		150	239

10 73 16.20 Metal Canopies

	10 73 16.20 Metal Canopies	Crew	Daily Output	Labor-Hours	Unit	Material	2022 Bare Costs Labor	Equipment	Total	Total Incl O&P
0010	**METAL CANOPIES**									
0020	Wall hung, .032", aluminum, prefinished, 8' x 10'	K-2	1.30	18.462	Ea.	3,500	1,075	290	4,865	5,925
0300	8' x 20'		1.10	21.818		5,275	1,275	345	6,895	8,250
1000	12' x 20'		1	24		5,275	1,400	375	7,050	8,500
2300	Aluminum entrance canopies, flat soffit, .032"									
2500	3'-6" x 4'-0", clear anodized	2 Carp	4	4	Ea.	1,500	225		1,725	2,000
9000	Minimum labor/equipment charge	"	2	8	Job		450		450	715

10 74 Manufactured Exterior Specialties

10 74 23 – Cupolas

10 74 23.10 Wood Cupolas

		Crew	Daily Output	Labor-Hours	Unit	Material	2022 Bare Costs Labor	Equipment	Total	Total Incl O&P
0010	**WOOD CUPOLAS**									
0020	Stock units, pine, painted, 18" sq., 28" high, alum. roof	1 Carp	4.10	1.951	Ea.	249	110		359	450
0100	Copper roof		3.80	2.105		259	119		378	475
0300	23" square, 33" high, aluminum roof		3.70	2.162		385	122		507	615
0400	Copper roof		3.30	2.424		625	136		761	900
0600	30" square, 37" high, aluminum roof		3.70	2.162		610	122		732	865
0700	Copper roof		3.30	2.424		630	136		766	905
0900	Hexagonal, 31" wide, 46" high, copper roof		4	2		1,125	113		1,238	1,400
1000	36" wide, 50" high, copper roof		3.50	2.286		1,275	129		1,404	1,600
1200	For deluxe stock units, add to above					25%				
1400	For custom built units, add to above					50%	50%			
9000	Minimum labor/equipment charge	1 Carp	2.75	2.909	Job		164		164	260

10 74 29 – Steeples

10 74 29.10 Prefabricated Steeples

		Crew	Daily Output	Labor-Hours	Unit	Material	2022 Bare Costs Labor	Equipment	Total	Total Incl O&P
0010	**PREFABRICATED STEEPLES**									
4000	Steeples, translucent fiberglass, 30" square, 15' high	F-3	2	20	Ea.	1,525	1,150	990	3,665	4,600
4150	25' high	F-3	1.80	22.222	Ea.	5,600	1,275	1,100	7,975	9,400
4600	Aluminum, baked finish, 16" square, 14' high					6,100			6,100	6,700
4640	35' high, 8' base					38,000			38,000	41,800
4680	152' high, custom					711,000			711,000	782,000

10 74 33 – Weathervanes

10 74 33.10 Residential Weathervanes

		Crew	Daily Output	Labor-Hours	Unit	Material	2022 Bare Costs Labor	Equipment	Total	Total Incl O&P
0010	**RESIDENTIAL WEATHERVANES**									
0020	Residential types, 18" to 24"	1 Carp	8	1	Ea.	185	56.50		241.50	293
0100	24" to 48"		2	4	"	219	225		444	600
9000	Minimum labor/equipment charge		4	2	Job		113		113	179

10 74 46 – Window Wells

10 74 46.10 Area Window Wells

		Crew	Daily Output	Labor-Hours	Unit	Material	2022 Bare Costs Labor	Equipment	Total	Total Incl O&P
0010	**AREA WINDOW WELLS**, Galvanized steel									
0020	20 ga., 3'-2" wide, 1' deep	1 Sswk	29	.276	Ea.	17.80	17.05		34.85	47
0100	2' deep		23	.348		425	21.50		446.50	505
0300	16 ga., 3'-2" wide, 1' deep		29	.276		335	17.05		352.05	400
0400	3' deep		23	.348		50.50	21.50		72	90
0600	Welded grating for above, 15 lb., painted		45	.178		107	11		118	136
0700	Galvanized		45	.178		131	11		142	162
0900	Translucent plastic cap for above		60	.133		21.50	8.25		29.75	37

10 86 Security Mirrors and Domes

10 86 10 – Security Mirrors

10 86 10.10 Exterior Traffic Control Mirrors

		Crew	Daily Output	Labor-Hours	Unit	Material	2022 Bare Costs Labor	Equipment	Total	Total Incl O&P
0010	**EXTERIOR TRAFFIC CONTROL MIRRORS**									
0100	Convex, stainless steel, 20 ga., 26" diameter	1 Carp	12	.667	Ea.	305	37.50		342.50	395

10 86 20 – Security Domes

10 86 20.10 Domes

		Crew	Daily Output	Labor-Hours	Unit	Material	2022 Bare Costs Labor	Equipment	Total	Total Incl O&P
0010	**DOMES** for security cameras (CCTV)									
0100	Ceiling mounted, 10" diameter	1 Carp	30	.267	Ea.	54	15		69	83.50
0110	12" diameter	"	30	.267	"	54	15		69	83.50

11 12 Parking Control Equipment

11 12 33 – Parking Gates

11 12 33.13 Lift Arm Parking Gates	Crew	Daily Output	Labor-Hours	Unit	Material	2022 Bare Costs Labor	2022 Bare Costs Equipment	Total	Total Incl O&P
5800	Inductive loop	2 Elec	4	4	Ea.	135	265		400	555
5950	Vehicle detector, microprocessor based	1 Elec	3	2.667	↓	380	177		557	690
7100	Traffic spike unit, flush mount, spring loaded, 72" L	B-89	4	4	Ea.	1,625	215	157	1,997	2,300
7200	Surface mount, 72" L	2 Skwk	10	1.600	"	2,000	94		2,094	2,350

11 13 Loading Dock Equipment

11 13 13 – Loading Dock Bumpers

11 13 13.10 Dock Bumpers

		Crew	Daily Output	Labor-Hours	Unit	Material	Labor	Equipment	Total	Total Incl O&P
0010	**DOCK BUMPERS** Bolts not included									
0012	2" x 6" to 4" x 8", average	1 Carp	300	.027	B.F.	32	1.50		33.50	37.50
0050	Bumpers, lam. rubber blocks 4-1/2" thick, 10" high, 14" long		26	.308	Ea.	45.50	17.30		62.80	77.50
0200	24" long		22	.364		70.50	20.50		91	110
0300	36" long		17	.471		106	26.50		132.50	158
0500	12" high, 14" long	↓	25	.320	↓	55	18		73	89

11 13 16 – Loading Dock Seals and Shelters

11 13 16.10 Dock Seals and Shelters

		Crew	Daily Output	Labor-Hours	Unit	Material	Labor	Equipment	Total	Total Incl O&P
0010	**DOCK SEALS AND SHELTERS**									
6200	Shelters, fabric, for truck or train, scissor arms, minimum	1 Carp	1	8	Ea.	2,725	450		3,175	3,725
6300	Maximum	"	.50	16	"	2,525	900		3,425	4,200

11 13 19 – Stationary Loading Dock Equipment

11 13 19.10 Dock Equipment

		Crew	Daily Output	Labor-Hours	Unit	Material	Labor	Equipment	Total	Total Incl O&P
0010	**DOCK EQUIPMENT**									
2200	Dock boards, heavy duty, 60" x 60", aluminum, 5,000 lb. capacity				Ea.	1,075			1,075	1,175
4200	Platform lifter, 6' x 6', portable, 3,000 lb. capacity				"	8,150			8,150	8,950
6000	Dock leveler, 15 ton capacity									
6100	I-beam construction, mechanical, 6' x 8'	E-16	.50	32	Ea.	5,950	2,025	295	8,270	10,100
6150	Hydraulic		.50	32		6,650	2,025	295	8,970	10,900
6200	Formed beam deck construction, mechanical, 6' x 8'		.50	32		2,200	2,025	295	4,520	6,000
6250	Hydraulic		.50	32		4,200	2,025	295	6,520	8,200
6300	Edge of dock leveler, mechanical, 15 ton capacity		2	8		2,200	505	73.50	2,778.50	3,325
6301	Edge of dock leveler, mechanical, 10 ton capacity		2	8	↓	2,200	505	73.50	2,778.50	3,325
7000	22.5 ton capacity									
7100	Vertical storing dock leveler, hydraulic, 6' x 6'	E-16	.40	40	Ea.	8,625	2,525	370	11,520	14,000

11 21 Retail and Service Equipment

11 21 13 – Cash Registers and Checking Equipment

11 21 13.10 Checkout Counter

		Crew	Daily Output	Labor-Hours	Unit	Material	Labor	Equipment	Total	Total Incl O&P
0010	**CHECKOUT COUNTER**									
0020	Supermarket conveyor, single belt	2 Clab	10	1.600	Ea.	3,900	73		3,973	4,425
0100	Double belt, power take-away		9	1.778		5,375	81		5,456	6,050
0400	Double belt, power take-away, incl. side scanning		7	2.286		5,925	104		6,029	6,700
0800	Warehouse or bulk type	↓	6	2.667	↓	6,900	122		7,022	7,775
1000	Scanning system, 2 lanes, w/registers, scan gun & memory				System	19,600			19,600	21,600
1100	10 lanes, single processor, full scan, with scales				System	186,000			186,000	204,500
2000	Register, restaurant, minimum				Ea.	820			820	905
2100	Maximum					207			207	228
2150	Store, minimum					820			820	905
2200	Maximum				↓	3,150			3,150	3,450

For customer support on your Commercial Renovation Costs with RSMeans data, call 800.448.8182.

371

11 21 33 – Checkroom Equipment

11 21 33.10 Clothes Check Equipment	Crew	Daily Output	Labor-Hours	Unit	Material	2022 Bare Costs Labor	2022 Bare Costs Equipment	Total	Total Incl O&P
0010 **CLOTHES CHECK EQUIPMENT**									
0030 Clothes check rack, free standing, st. stl., 2-tier, 90 bag capacity				Ea.	1,600			1,600	1,750
0050 Wall mounted, 45 bag capacity	L-2	8	2		935	99.50		1,034.50	1,175
0100 Garment checking bag, green mesh fabric, 21" H x 17" W with 4.5" hook					29			29	31.50

11 21 53 – Barber and Beauty Shop Equipment

11 21 53.10 Barber Equipment

	Crew	Daily Output	Labor-Hours	Unit	Material	2022 Bare Costs Labor	2022 Bare Costs Equipment	Total	Total Incl O&P
0010 **BARBER EQUIPMENT**									
0020 Chair, hydraulic, movable, minimum	1 Carp	24	.333	Ea.	645	18.75		663.75	735
0050 Maximum	"	16	.500		4,000	28		4,028	4,425
0200 Wall hung styling station with mirrors, minimum	L-2	8	2		660	99.50		759.50	885
0300 Maximum	"	4	4		2,650	199		2,849	3,250
0500 Sink, hair washing basin, rough plumbing not incl.	1 Plum	8	1		249	70		319	380
1000 Sterilizer, liquid solution for tools					165			165	181
1100 Total equipment, rule of thumb, per chair, minimum	L-8	1	20		1,300	1,175		2,475	3,275
1150 Maximum	"	1	20		5,900	1,175		7,075	8,350

11 21 73 – Commercial Laundry and Dry Cleaning Equipment

11 21 73.13 Dry Cleaning Equipment

	Crew	Daily Output	Labor-Hours	Unit	Material	2022 Bare Costs Labor	2022 Bare Costs Equipment	Total	Total Incl O&P
0010 **DRY CLEANING EQUIPMENT**									
2000 Dry cleaners, electric, 20 lb. capacity, not incl. rough-in	L-1	.20	80	Ea.	35,800	5,450		41,250	47,800
2050 25 lb. capacity		.17	94.118		54,000	6,400		60,400	69,000
2100 30 lb. capacity		.15	107		56,500	7,250		63,750	73,500
2150 60 lb. capacity		.09	178		80,500	12,100		92,600	107,500

11 21 73.16 Drying and Conditioning Equipment

	Crew	Daily Output	Labor-Hours	Unit	Material	2022 Bare Costs Labor	2022 Bare Costs Equipment	Total	Total Incl O&P
0010 **DRYING AND CONDITIONING EQUIPMENT**									
0100 Dryers, not including rough-in									
1500 Industrial, 30 lb. capacity	1 Plum	2	4	Ea.	1,650	279		1,929	2,250
1600 50 lb. capacity	"	1.70	4.706		4,675	330		5,005	5,650
4700 Lint collector, ductwork not included, 8,000 to 10,000 CFM	Q-10	.30	80		11,500	5,050		16,550	20,600

11 21 73.19 Finishing Equipment

	Crew	Daily Output	Labor-Hours	Unit	Material	2022 Bare Costs Labor	2022 Bare Costs Equipment	Total	Total Incl O&P
0010 **FINISHING EQUIPMENT**									
3500 Folders, blankets & sheets, minimum	1 Elec	.17	47.059	Ea.	36,800	3,125		39,925	45,300
3700 King size with automatic stacker		.10	80		66,500	5,300		71,800	81,000
3800 For conveyor delivery, add		.45	17.778		16,900	1,175		18,075	20,400

11 21 73.23 Commercial Ironing Equipment

	Crew	Daily Output	Labor-Hours	Unit	Material	2022 Bare Costs Labor	2022 Bare Costs Equipment	Total	Total Incl O&P
0010 **COMMERCIAL IRONING EQUIPMENT**									
4500 Ironers, institutional, 110", single roll	1 Elec	.20	40	Ea.	35,600	2,650		38,250	43,300

11 21 73.26 Commercial Washers and Extractors

	Crew	Daily Output	Labor-Hours	Unit	Material	2022 Bare Costs Labor	2022 Bare Costs Equipment	Total	Total Incl O&P
0010 **COMMERCIAL WASHERS AND EXTRACTORS**, not including rough-in									
6000 Combination washer/extractor, 20 lb. capacity	L-6	1.50	8	Ea.	5,050	550		5,600	6,400
6100 30 lb. capacity		.80	15		11,200	1,025		12,225	13,900
6200 50 lb. capacity		.68	17.647		14,100	1,200		15,300	17,500
6300 75 lb. capacity		.30	40		19,700	2,750		22,450	25,800
6350 125 lb. capacity		.16	75		30,800	5,150		35,950	41,800

11 21 73.33 Coin-Operated Laundry Equipment

	Crew	Daily Output	Labor-Hours	Unit	Material	2022 Bare Costs Labor	2022 Bare Costs Equipment	Total	Total Incl O&P
0010 **COIN-OPERATED LAUNDRY EQUIPMENT**									
0990 Dryer, gas fired									
1000 Commercial, 30 lb. capacity, coin operated, single	1 Plum	3	2.667	Ea.	1,175	186		1,361	1,550
1100 Double stacked	"	2	4	"	3,000	279		3,279	3,725
5290 Clothes washer									
5300 Commercial, coin operated, average	1 Plum	3	2.667	Ea.	1,400	186		1,586	1,800

11 21 Retail and Service Equipment

11 21 83 – Photo Processing Equipment

11 21 83.13 Darkroom Equipment

		Crew	Daily Output	Labor-Hours	Unit	Material	2022 Bare Costs Labor	2022 Bare Costs Equipment	Total	Total Incl O&P
0010	**DARKROOM EQUIPMENT**									
0020	Developing sink, 5" deep, 24" x 48"	Q-1	2	8	Ea.	430	505		935	1,250
0050	48" x 52"		1.70	9.412		1,725	590		2,315	2,800
0200	10" deep, 24" x 48"		1.70	9.412		1,900	590		2,490	3,000
0250	24" x 108"		1.50	10.667		5,175	670		5,845	6,725
0500	Dryers, dehumidified filtered air, 36" x 25" x 68" high	L-7	6	4.667		585	255		840	1,050
3000	Viewing lights, 20" x 24"		6	4.667		320	255		575	755
3100	20" x 24" with color correction		6	4.667		990	255		1,245	1,500

11 22 Banking Equipment

11 22 13 – Vault Equipment

11 22 13.16 Safes

		Crew	Daily Output	Labor-Hours	Unit	Material	2022 Bare Costs Labor	2022 Bare Costs Equipment	Total	Total Incl O&P
0010	**SAFES**									
0800	Money, "B" label, 9" x 14" x 14"				Ea.	630			630	695
0900	Tool resistive, 24" x 24" x 20"					4,225			4,225	4,650
1050	Tool and torch resistive, 24" x 24" x 20"					8,925			8,925	9,800

11 22 16 – Teller and Service Equipment

11 22 16.13 Teller Equipment Systems

		Crew	Daily Output	Labor-Hours	Unit	Material	2022 Bare Costs Labor	2022 Bare Costs Equipment	Total	Total Incl O&P
0010	**TELLER EQUIPMENT SYSTEMS**									
0020	Alarm system, police	2 Elec	1.60	10	Ea.	6,600	660		7,260	8,300
0400	Bullet resistant teller window, 44" x 60"	1 Glaz	.60	13.333		6,050	720		6,770	7,800
0500	48" x 60"	"	.60	13.333		7,875	720		8,595	9,800
3000	Counters for banks, frontal only	2 Carp	1	16	Station	2,750	900		3,650	4,450
3100	Complete with steel undercounter	2 Carp	.50	32	Station	4,825	1,800		6,625	8,150
5400	Partitions, bullet-resistant, 1-3/16" glass, 8' high	"	10	1.600	L.F.	267	90		357	435
5600	Pass thru, bullet-res. window, painted steel, 24" x 36"	2 Sswk	1.60	10	Ea.	4,900	620		5,520	6,400

11 30 Residential Equipment

11 30 13 – Residential Appliances

11 30 13.15 Cooking Equipment

		Crew	Daily Output	Labor-Hours	Unit	Material	2022 Bare Costs Labor	2022 Bare Costs Equipment	Total	Total Incl O&P
0010	**COOKING EQUIPMENT**									
0020	Cooking range, 30" free standing, 1 oven, minimum	2 Clab	10	1.600	Ea.	630	73		703	805
0050	Maximum	"	4	4		900	182		1,082	1,275
0350	Built-in, 30" wide, 1 oven, minimum	1 Elec	6	1.333		1,200	88.50		1,288.50	1,450
0400	Maximum	2 Carp	2	8		1,900	450		2,350	2,825
0900	Countertop cooktops, 4 burner, standard, minimum	1 Elec	6	1.333		294	88.50		382.50	460
0950	Maximum		3	2.667		540	177		717	865
1250	Microwave oven, minimum		4	2		119	132		251	335
1300	Maximum		2	4		320	265		585	755
5380	Oven, built-in, standard		4	2		615	132		747	885
5390	Deluxe		2	4		1,600	265		1,865	2,150

11 30 13.16 Refrigeration Equipment

			Crew	Daily Output	Labor-Hours	Unit	Material	2022 Bare Costs Labor	2022 Bare Costs Equipment	Total	Total Incl O&P
0010	**REFRIGERATION EQUIPMENT**										
5450	Refrigerator, no frost, 6 C.F.		2 Clab	15	1.067	Ea.	1,000	48.50		1,048.50	1,175
5500	Refrigerator, no frost, 10 C.F. to 12 C.F., minimum			10	1.600		560	73		633	735
5600	Maximum			6	2.667		500	122		622	745
6790	Energy-star qualified, 18 C.F., minimum	G	2 Carp	4	4		1,800	225		2,025	2,325
6795	Maximum	G		2	8		670	450		1,120	1,450

11 30 13 – Residential Appliances

11 30 13.16 Refrigeration Equipment		Crew	Daily Output	Labor-Hours	Unit	Material	2022 Bare Costs Labor	Equipment	Total	Total Incl O&P
6797	21.7 C.F., minimum	2 Carp [G]	4	4	Ea.	1,550	225		1,775	2,050
6799	Maximum	[G]	4	4		3,200	225		3,425	3,875

11 30 13.17 Kitchen Cleaning Equipment

		Crew	Daily Output	Labor-Hours	Unit	Material	Labor	Equipment	Total	Total Incl O&P
0010	**KITCHEN CLEANING EQUIPMENT**									
2750	Dishwasher, built-in, 2 cycles, minimum	L-1	4	4	Ea.	380	272		652	835
2800	Maximum		2	8		835	545		1,380	1,750
3100	Energy-star qualified, minimum	[G]	4	4		530	272		802	1,000
3110	Maximum	[G]	2	8		1,200	545		1,745	2,150

11 30 13.18 Waste Disposal Equipment

		Crew	Daily Output	Labor-Hours	Unit	Material	Labor	Equipment	Total	Total Incl O&P
0010	**WASTE DISPOSAL EQUIPMENT**									
1750	Compactor, residential size, 4 to 1 compaction, minimum	1 Carp	5	1.600	Ea.	1,100	90		1,190	1,350
1800	Maximum	"	3	2.667		1,250	150		1,400	1,625
3300	Garbage disposal, sink type, minimum	L-1	10	1.600		99	109		208	276
3350	Maximum	"	10	1.600		239	109		348	430

11 30 13.19 Kitchen Ventilation Equipment

		Crew	Daily Output	Labor-Hours	Unit	Material	Labor	Equipment	Total	Total Incl O&P
0010	**KITCHEN VENTILATION EQUIPMENT**									
4150	Hood for range, 2 speed, vented, 30" wide, minimum	L-3	5	3.200	Ea.	116	197		313	440
4200	Maximum	L-3	3	5.333	Ea.	835	330		1,165	1,425
4300	42" wide, minimum		5	3.200		157	197		354	485
4330	Custom		5	3.200		129	197		326	450
4350	Maximum		3	5.333		157	330		487	690
4400	Ventless hood, 2 speed, 30" wide, minimum	1 Elec	8	1		895	66		961	1,075
4450	Maximum		7	1.143		625	75.50		700.50	800
4510	36" wide, minimum		7	1.143		85	75.50		160.50	210
4550	Maximum		6	1.333		875	88.50		963.50	1,100
4580	42" wide, minimum		6	1.333		108	88.50		196.50	255
4600	Maximum		5	1.600		680	106		786	915
4650	For vented 1 speed, deduct from maximum					63.50			63.50	69.50

11 30 13.24 Washers

		Crew	Daily Output	Labor-Hours	Unit	Material	Labor	Equipment	Total	Total Incl O&P
0010	**WASHERS**									
5000	Residential, 4 cycle, average	1 Plum	3	2.667	Ea.	1,025	186		1,211	1,400
6750	Energy star qualified, front loading, minimum	[G]	3	2.667		1,050	186		1,236	1,425
6760	Maximum	[G]	1	8		1,300	560		1,860	2,275
6764	Top loading, minimum	[G]	3	2.667		550	186		736	890
6766	Maximum	[G]	3	2.667		850	186		1,036	1,225

11 30 13.25 Dryers

		Crew	Daily Output	Labor-Hours	Unit	Material	Labor	Equipment	Total	Total Incl O&P
0010	**DRYERS**									
0500	Gas fired residential, 16 lb. capacity, average	1 Plum	3	2.667	Ea.	970	186		1,156	1,350
6770	Electric, front loading, energy-star qualified, minimum	[G] L-2	3	5.333		700	266		966	1,200
6780	Maximum	[G] "	2	8		900	400		1,300	1,625

11 30 15 – Miscellaneous Residential Appliances

11 30 15.13 Sump Pumps

		Crew	Daily Output	Labor-Hours	Unit	Material	Labor	Equipment	Total	Total Incl O&P
0010	**SUMP PUMPS**									
6400	Cellar drainer, pedestal, 1/3 HP, molded PVC base	1 Plum	3	2.667	Ea.	138	186		324	440
6450	Solid brass	"	2	4	"	250	279		529	705
6460	Sump pump, see also Section 22 14 29.16									

374

For customer support on your Commercial Renovation Costs with RSMeans data, call 800.448.8182.

11 30 Residential Equipment

11 30 33 – Retractable Stairs

11 30 33.10 Disappearing Stairway

		Crew	Daily Output	Labor-Hours	Unit	Material	2022 Bare Costs Labor	2022 Bare Costs Equipment	Total	Total Incl O&P
0010	**DISAPPEARING STAIRWAY** No trim included									
0020	One piece, yellow pine, 8'-0" ceiling	2 Carp	4	4	Ea.	415	225		640	815
0030	9'-0" ceiling		4	4		415	225		640	815
0040	10'-0" ceiling		3	5.333		415	300		715	930
0050	11'-0" ceiling		3	5.333		475	300		775	1,000
0060	12'-0" ceiling		3	5.333		475	300		775	995
0100	Custom grade, pine, 8'-6" ceiling, minimum	1 Carp	4	2		110	113		223	300
0150	Average		3.50	2.286		191	129		320	415
0200	Maximum		3	2.667		271	150		421	535
0500	Heavy duty, pivoted, from 7'-7" to 12'-10" floor to floor		3	2.667		490	150		640	780
0600	16'-0" ceiling	1 Carp	2	4	Ea.	2,575	225		2,800	3,200
0800	Economy folding, pine, 8'-6" ceiling		4	2		96.50	113		209.50	285
0900	9'-6" ceiling		4	2		102	113		215	291
1100	Automatic electric, aluminum, floor to floor height, 8' to 9'	2 Carp	1	16		3,925	900		4,825	5,750
1400	11' to 12'		.90	17.778		4,075	1,000		5,075	6,075
1700	14' to 15'		.70	22.857		8,425	1,275		9,700	11,300
9000	Minimum labor/equipment charge	1 Carp	2	4	Job		225		225	360

11 32 Unit Kitchens

11 32 13 – Metal Unit Kitchens

11 32 13.10 Commercial Unit Kitchens

		Crew	Daily Output	Labor-Hours	Unit	Material	2022 Bare Costs Labor	2022 Bare Costs Equipment	Total	Total Incl O&P
0010	**COMMERCIAL UNIT KITCHENS**									
1500	Combination range, refrigerator and sink, 30" wide, minimum	L-1	2	8	Ea.	1,275	545		1,820	2,225
1550	Maximum	"	1	16	"	1,275	1,100		2,375	3,075
1640	Combination range, refrigerator, sink, microwave									
1660	Oven and ice maker	L-1	.80	20	Ea.	5,725	1,350		7,075	8,400

11 41 Foodservice Storage Equipment

11 41 13 – Refrigerated Food Storage Cases

11 41 13.10 Refrigerated Food Cases

		Crew	Daily Output	Labor-Hours	Unit	Material	2022 Bare Costs Labor	2022 Bare Costs Equipment	Total	Total Incl O&P
0010	**REFRIGERATED FOOD CASES**									
0030	Dairy, multi-deck, 12' long	Q-5	3	5.333	Ea.	12,900	340		13,240	14,600
0100	For rear sliding doors, add					2,000			2,000	2,200
0200	Delicatessen case, service deli, 12' long, single deck	Q-5	3.90	4.103		13,000	260		13,260	14,700
0300	Multi-deck, 18 S.F. shelf display		3	5.333		8,425	340		8,765	9,800
0400	Freezer, self-contained, chest-type, 30 C.F.		3.90	4.103		5,075	260		5,335	5,975
0500	Glass door, upright, 78 C.F.		3.30	4.848		9,400	305		9,705	10,800
0600	Frozen food, chest type, 12' long		3.30	4.848		6,925	305		7,230	8,100
0700	Glass door, reach-in, 5 door		3	5.333		10,700	340		11,040	12,300
0800	Island case, 12' long, single deck		3.30	4.848		8,875	305		9,180	10,200
0900	Multi-deck		3	5.333		10,700	340		11,040	12,300
1000	Meat case, 12' long, single deck		3.30	4.848		7,525	305		7,830	8,750
1050	Multi-deck		3.10	5.161		10,300	325		10,625	11,800
1100	Produce, 12' long, single deck		3.30	4.848		6,675	305		6,980	7,800
1200	Multi-deck		3.10	5.161		8,625	325		8,950	9,975

11 41 13 – Refrigerated Food Storage Cases

11 41 13.20 Refrigerated Food Storage Equipment		Crew	Daily Output	Labor-Hours	Unit	Material	2022 Bare Costs Labor	Equipment	Total	Total Incl O&P
0010	**REFRIGERATED FOOD STORAGE EQUIPMENT**									
2350	Cooler, reach-in, beverage, 6' long	Q-1	6	2.667	Ea.	2,900	168		3,068	3,450
4300	Freezers, reach-in, 44 C.F.		4	4		4,475	251		4,726	5,300
4500	68 C.F.		3	5.333		5,500	335		5,835	6,575
4600	Freezer, pre-fab, 8' x 8' w/refrigeration	2 Carp	.45	35.556	Ea.	9,325	2,000		11,325	13,500
4620	8' x 12'		.35	45.714		11,300	2,575		13,875	16,500
4640	8' x 16'		.25	64		12,600	3,600		16,200	19,500
4660	8' x 20'		.17	94.118		15,400	5,300		20,700	25,300
4680	Reach-in, 1 compartment	Q-1	4	4		4,000	251		4,251	4,775
4685	Energy star rated Ⓖ	R-18	7.80	3.333		3,675	194		3,869	4,325
4700	2 compartment	Q-1	3	5.333		5,300	335		5,635	6,375
4705	Energy star rated Ⓖ	R-18	6.20	4.194		3,900	244		4,144	4,650
4710	3 compartment	Q-1	3	5.333		6,450	335		6,785	7,600
4715	Energy star rated Ⓖ	R-18	5.60	4.643		5,375	270		5,645	6,350
8320	Refrigerator, reach-in, 1 compartment		7.80	3.333		4,250	194		4,444	4,950
8325	Energy star rated Ⓖ		7.80	3.333		3,675	194		3,869	4,325
8330	2 compartment		6.20	4.194		3,525	244		3,769	4,250
8335	Energy star rated Ⓖ		6.20	4.194		3,900	244		4,144	4,650
8340	3 compartment		5.60	4.643		5,000	270		5,270	5,925
8345	Energy star rated Ⓖ		5.60	4.643		5,375	270		5,645	6,350
8350	Pre-fab, with refrigeration, 8' x 8'	2 Carp	.45	35.556		6,475	2,000		8,475	10,300
8360	8' x 12'		.35	45.714		9,925	2,575		12,500	15,000
8370	8' x 16'		.25	64		10,500	3,600		14,100	17,300
8380	8' x 20'		.17	94.118		13,500	5,300		18,800	23,200
8390	Pass-thru/roll-in, 1 compartment	R-18	7.80	3.333		4,350	194		4,544	5,075
8400	2 compartment		6.24	4.167		5,575	242		5,817	6,500
8410	3 compartment		5.60	4.643		23,900	270		24,170	26,700
8420	Walk-in, alum, door & floor only, no refrig, 6' x 6' x 7'-6"	2 Carp	1.40	11.429		6,100	645		6,745	7,750
8430	10' x 6' x 7'-6"		.55	29.091		5,650	1,650		7,300	8,800
8440	12' x 14' x 7'-6"		.25	64		11,900	3,600		15,500	18,800
8450	12' x 20' x 7'-6"		.17	94.118		15,100	5,300		20,400	25,100
8460	Refrigerated cabinets, mobile					7,700			7,700	8,475
8470	Refrigerator/freezer, reach-in, 1 compartment	R-18	5.60	4.643		2,925	270		3,195	3,625
8480	2 compartment	"	4.80	5.417		5,275	315		5,590	6,275

11 41 13.30 Wine Cellar

		Crew	Daily Output	Labor-Hours	Unit	Material	2022 Bare Costs Labor	Equipment	Total	Total Incl O&P
0010	**WINE CELLAR**, refrigerated, Redwood interior, carpeted, walk-in type									
0020	6'-8" high, including racks									
0200	80" W x 48" D for 900 bottles	2 Carp	1.50	10.667	Ea.	4,575	600		5,175	5,975
0250	80" W x 72" D for 1300 bottles		1.33	12.030		6,375	675		7,050	8,075
0300	80" W x 94" D for 1900 bottles		1.17	13.675		7,450	770		8,220	9,425
0400	80" W x 124" D for 2500 bottles		1	16		8,625	900		9,525	10,900

11 41 33 – Foodservice Shelving

11 41 33.20 Metal Food Storage Shelving

		Crew	Daily Output	Labor-Hours	Unit	Material	2022 Bare Costs Labor	Equipment	Total	Total Incl O&P
0010	**METAL FOOD STORAGE SHELVING**									
8600	Stainless steel shelving, louvered 4-tier, 20" x 3'	1 Clab	6	1.333	Ea.	1,225	61		1,286	1,450
8605	20" x 4'		6	1.333		298	61		359	420
8610	20" x 6'		6	1.333		345	61		406	475
8615	24" x 3'		6	1.333		1,900	61		1,961	2,200
8620	24" x 4'		6	1.333		2,375	61		2,436	2,700
8625	24" x 6'		6	1.333		3,225	61		3,286	3,650
8630	Flat 4-tier, 20" x 3'		6	1.333		1,225	61		1,286	1,450

11 41 Foodservice Storage Equipment

11 41 33 – Foodservice Shelving

11 41 33.20 Metal Food Storage Shelving	Crew	Daily Output	Labor-Hours	Unit	Material	2022 Bare Costs Labor	Equipment	Total	Total Incl O&P
8635	20" x 4'	1 Clab	6	1.333	Ea.	1,325	61		1,386	1,550
8640	20" x 5'		6	1.333		1,575	61		1,636	1,825
8645	24" x 3'		6	1.333		835	61		896	1,000
8650	24" x 4'		6	1.333		2,150	61		2,211	2,450
8655	24" x 6'		6	1.333		2,650	61		2,711	3,000
8700	Galvanized shelving, louvered 4-tier, 20" x 3'		6	1.333		735	61		796	905
8705	20" x 4'		6	1.333		815	61		876	990
8710	20" x 6'		6	1.333		860	61		921	1,050
8715	24" x 3'		6	1.333		835	61		896	1,025
8720	24" x 4'		6	1.333		880	61		941	1,075
8725	24" x 6'		6	1.333		345	61		406	475
8730	Flat 4-tier, 20" x 3'		6	1.333		585	61		646	735
8735	20" x 4'		6	1.333		635	61		696	795
8740	20" x 6'		6	1.333		915	61		976	1,100
8745	24" x 3'		6	1.333		740	61		801	910
8750	24" x 4'		6	1.333		515	61		576	660
8755	24" x 6'		6	1.333		169	61		230	283
8760	Stainless steel dunnage rack, 24" x 3'		8	1		1,225	45.50		1,270.50	1,425
8765	24" x 4'		8	1		340	45.50		385.50	450
8770	Galvanized dunnage rack, 24" x 3'		8	1		130	45.50		175.50	216
8775	24" x 4'	↓	8	1	↓	179	45.50		224.50	270

11 42 Food Preparation Equipment

11 42 10 – Commercial Food Preparation Equipment

11 42 10.10 Choppers, Mixers and Misc. Equipment

		Crew	Daily Output	Labor-Hours	Unit	Material	2022 Bare Costs Labor	Equipment	Total	Total Incl O&P
0010	**CHOPPERS, MIXERS AND MISC. EQUIPMENT**									
1700	Choppers, 5 pounds	R-18	7	3.714	Ea.	3,200	216		3,416	3,825
1720	16 pounds		5	5.200		2,075	300		2,375	2,775
1740	35 to 40 pounds	↓	4	6.500		3,250	380		3,630	4,150
1840	Coffee brewer, 5 burners	1 Plum	3	2.667		855	186		1,041	1,225
1850	Coffee urn, twin 6 gallon urns		2	4		4,225	279		4,504	5,075
1860	Single, 3 gallon	↓	3	2.667		2,800	186		2,986	3,350
3000	Fast food equipment, total package, minimum	6 Skwk	.08	600		206,500	35,300		241,800	283,500
3100	Maximum	"	.07	686		265,500	40,400		305,900	356,000
3800	Food mixers, bench type, 20 quarts	L-7	7	4		2,600	219		2,819	3,200
3850	40 quarts		5.40	5.185		5,425	283		5,708	6,425
3900	60 quarts		5	5.600		10,900	305		11,205	12,500
4040	80 quarts		3.90	7.179		17,300	390		17,690	19,700
4100	Floor type, 20 quarts		15	1.867		3,475	102		3,577	3,975
4120	60 quarts		14	2		9,850	109		9,959	11,000
4140	80 quarts		12	2.333		13,500	128		13,628	15,100
4160	140 quarts	↓	8.60	3.256		24,300	178		24,478	27,100
6700	Peelers, small	R-18	8	3.250		2.79	189		191.79	293
6720	Large	"	6	4.333		3,850	252		4,102	4,600
6800	Pulper/extractor, close coupled, 5 HP	1 Plum	1.90	4.211		2,500	294		2,794	3,200
8580	Slicer with table	R-18	9	2.889	↓	4,100	168		4,268	4,750

11 43 13 – Food Delivery Carts

11 43 13.10 Mobile Carts, Racks and Trays	Crew	Daily Output	Labor-Hours	Unit	Material	2022 Bare Costs Labor	Equipment	Total	Total Incl O&P
0010 **MOBILE CARTS, RACKS AND TRAYS**									
1650 Cabinet, heated, 1 compartment, reach-in	R-18	5.60	4.643	Ea.	3,150	270		3,420	3,900
1655 Pass-thru roll-in		5.60	4.643		3,125	270		3,395	3,875
1660 2 compartment, reach-in		4.80	5.417		11,700	315		12,015	13,400
1670 Mobile					2,575			2,575	2,850
2000 Hospital food cart, hot and cold service, 20 tray capacity					15,800			15,800	17,300
6850 Mobile rack w/pan slide					855			855	940
9180 Tray and silver dispenser, mobile	1 Clab	16	.500		4,850	23		4,873	5,375

11 44 Food Cooking Equipment

11 44 13 – Commercial Ranges

11 44 13.10 Cooking Equipment

	Crew	Daily Output	Labor-Hours	Unit	Material	2022 Bare Costs Labor	Equipment	Total	Total Incl O&P
0010 **COOKING EQUIPMENT**									
0020 Bake oven, gas, one section	Q-1	8	2	Ea.	3,825	126		3,951	4,425
0300 Two sections		7	2.286		6,850	144		6,994	7,750
0600 Three sections		6	2.667		8,225	168		8,393	9,300
0900 Electric convection, single deck	L-7	4	7		9,300	385		9,685	10,800
1300 Broiler, without oven, standard	Q-1	8	2		2,825	126		2,951	3,325
1550 Infrared	L-7	4	7		6,400	385		6,785	7,650
4750 Fryer, with twin baskets, modular model	Q-1	7	2.286		1,300	144		1,444	1,650
5000 Floor model, on 6" legs	"	5	3.200		3,750	201		3,951	4,425
5100 Extra single basket, large					62			62	68
5170 Energy star rated, 50 lb. capacity [G]	R-18	4	6.500		6,075	380		6,455	7,250
5175 85 lb. capacity [G]	R-18	4	6.500	Ea.	1,825	380		2,205	2,600
5300 Griddle, SS, 24" plate, w/4" legs, elec, 208 V, 3 phase, 3' long	Q-1	7	2.286		4,875	144		5,019	5,575
5550 4' long	"	6	2.667		2,875	168		3,043	3,425
6200 Iced tea brewer	1 Plum	3.44	2.326		705	162		867	1,025
6350 Kettle, w/steam jacket, tilting, w/positive lock, SS, 20 gallons	L-7	7	4		12,900	219		13,119	14,500
6600 60 gallons	"	6	4.667		15,900	255		16,155	17,900
6900 Range, restaurant type, 6 burners and 1 standard oven, 36" wide	Q-1	7	2.286		1,250	144		1,394	1,600
6950 Convection		7	2.286		5,450	144		5,594	6,225
7150 2 standard ovens, 24" griddle, 60" wide		6	2.667		4,600	168		4,768	5,300
7200 1 standard, 1 convection oven		6	2.667		13,300	168		13,468	14,900
7450 Heavy duty, single 34" standard oven, open top		5	3.200		7,175	201		7,376	8,200
7500 Convection oven		5	3.200		6,725	201		6,926	7,700
7700 Griddle top		6	2.667		2,675	168		2,843	3,200
7750 Convection oven		6	2.667		7,300	168		7,468	8,275
7760 Induction cooker, electric	L-7	7	4		2,350	219		2,569	2,950
8850 Steamer, electric 27 KW		7	4		31,700	219		31,919	35,200
9100 Electric, 10 KW or gas 100,000 BTU		5	5.600		22,500	305		22,805	25,300
9150 Toaster, conveyor type, 16-22 slices/minute					795			795	875
9160 Pop-up, 2 slot					108			108	119
9200 For deluxe models of above equipment, add					75%				
9400 Rule of thumb: Equipment cost based									
9410 on kitchen work area									
9420 Office buildings, minimum	L-7	77	.364	S.F.	119	19.90		138.90	163
9450 Maximum		58	.483		193	26.50		219.50	254
9550 Public eating facilities, minimum		77	.364		158	19.90		177.90	206
9600 Maximum		46	.609		253	33.50		286.50	330
9750 Hospitals, minimum		58	.483		154	26.50		180.50	211
9800 Maximum		39	.718		283	39		322	370

11 46 Food Dispensing Equipment

11 46 16 – Service Line Equipment

11 46 16.10 Commercial Food Dispensing Equipment

		Crew	Daily Output	Labor-Hours	Unit	Material	2022 Bare Costs Labor	2022 Bare Costs Equipment	Total	Total Incl O&P
0010	**COMMERCIAL FOOD DISPENSING EQUIPMENT**									
1050	Butter pat dispenser	1 Clab	13	.615	Ea.	795	28		823	915
1100	Bread dispenser, counter top		13	.615		206	28		234	272
1900	Cup and glass dispenser, drop in		4	2		620	91		711	830
1920	Disposable cup, drop in		16	.500		780	23		803	895
2650	Dish dispenser, drop in, 12"		11	.727		430	33		463	530
2660	Mobile	1 Clab	10	.800	Ea.	3,625	36.50		3,661.50	4,025
3300	Food warmer, counter, 1.2 KW					645			645	705
3550	1.6 KW					675			675	740
3600	Well, hot food, built-in, rectangular, 12" x 20"	R-30	10	2.600		335	140		475	590
3610	Circular, 7 qt.		10	2.600		405	140		545	665
3620	Refrigerated, 2 compartments		10	2.600		2,950	140		3,090	3,475
3630	3 compartments		9	2.889		5,375	155		5,530	6,175
3640	4 compartments		8	3.250		5,100	174		5,274	5,875
4720	Frost cold plate		9	2.889		23,500	155		23,655	26,100
5700	Hot chocolate dispenser	1 Plum	4	2		1,200	140		1,340	1,550
5750	Ice dispenser 567 pound	Q-1	6	2.667		6,125	168		6,293	6,975
6250	Jet spray dispenser	R-18	4.50	5.778		1,225	335		1,560	1,875
6300	Juice dispenser, concentrate	"	4.50	5.778		2,025	335		2,360	2,750
6690	Milk dispenser, bulk, 2 flavor	R-30	8	3.250		2,450	174		2,624	2,975
6695	3 flavor	"	8	3.250		3,025	174		3,199	3,600
8800	Serving counter, straight	1 Carp	40	.200	L.F.	2,025	11.25		2,036.25	2,275
8820	Curved section	"	30	.267	"	2,600	15		2,615	2,900
8825	Solid surface, see Section 12 36 61.16									
8860	Sneeze guard with lights, 60" L	1 Clab	16	.500	Ea.	340	23		363	410
8900	Sneeze guard, stainless steel and glass, single sided									
8910	Portable, 48" W				Ea.	405			405	445
8920	Portable, 72" W					465			465	510
8930	Adjustable, 36" W	1 Carp	24	.333		360	18.75		378.75	425
8940	Adjustable, 48" W	"	20	.400		325	22.50		347.50	395
9100	Soft serve ice cream machine, medium	R-18	11	2.364		9,675	137		9,812	10,800
9110	Large	"	9	2.889		25,400	168		25,568	28,300

11 46 83 – Ice Machines

11 46 83.10 Commercial Ice Equipment

		Crew	Daily Output	Labor-Hours	Unit	Material	2022 Bare Costs Labor	2022 Bare Costs Equipment	Total	Total Incl O&P
0010	**COMMERCIAL ICE EQUIPMENT**									
5800	Ice cube maker, 50 lbs./day	Q-1	6	2.667	Ea.	1,775	168		1,943	2,225
5810	65 lbs./day, energy star rated		6	2.667		3,075	168		3,243	3,625
5900	250 lbs./day		1.20	13.333		2,150	840		2,990	3,650
5950	300 lbs./day, remote condensing		1.20	13.333		2,475	840		3,315	4,025
6050	500 lbs./day		4	4		2,575	251		2,826	3,200
6060	With bin		1.20	13.333		3,875	840		4,715	5,550
6070	Modular, with bin and condenser		1.20	13.333		3,750	840		4,590	5,425
6090	1000 lbs./day, with bin		1	16		5,475	1,000		6,475	7,575
6100	Ice flakers, 300 lbs./day		1.60	10		3,700	630		4,330	5,025
6120	600 lbs./day	Q-1	.95	16.842	Ea.	5,525	1,050		6,575	7,700
6130	1000 lbs./day		.75	21.333		5,450	1,350		6,800	8,075
6140	2000 lbs./day		.65	24.615		22,100	1,550		23,650	26,700
6160	Ice storage bin, 500 pound capacity	Q-5	1	16		1,000	1,025		2,025	2,650
6180	1000 pound	"	.56	28.571		3,075	1,800		4,875	6,200

11 48 Foodservice Cleaning and Disposal Equipment

11 48 13 – Commercial Dishwashers

11 48 13.10 Dishwashers	Crew	Daily Output	Labor-Hours	Unit	Material	2022 Bare Costs Labor	Equipment	Total	Total Incl O&P
0010 **DISHWASHERS**									
2700 Dishwasher, commercial, rack type									
2720 10 to 12 racks/hour	Q-1	3.20	5	Ea.	27,000	315		27,315	30,200
2730 Energy star rated, 35 to 40 racks/hour [G]		1.30	12.308		6,425	775		7,200	8,275
2740 50 to 60 racks/hour [G]		1.30	12.308		10,700	775		11,475	13,000
2800 Automatic, 190 to 230 racks/hour	L-6	.35	34.286		25,600	2,350		27,950	31,700
2820 235 to 275 racks/hour		.25	48		33,900	3,300		37,200	42,400
2840 8,750 to 12,500 dishes/hour		.10	120		56,000	8,225		64,225	74,000
2950 Dishwasher hood, canopy type	L-3A	10	1.200	L.F.	565	73.50		638.50	735
2960 Pant leg type	"	2.50	4.800	Ea.	7,950	295		8,245	9,225
5200 Garbage disposal 1.5 HP, 100 GPH	L-1	4.80	3.333		2,200	227		2,427	2,750
5210 3 HP, 120 GPH		4.60	3.478		4,550	237		4,787	5,375
5220 5 HP, 250 GPH		4.50	3.556		4,550	242		4,792	5,400
6750 Pot sink, 3 compartment	1 Plum	7.25	1.103	L.F.	860	77		937	1,075
6760 Pot washer, low temp wash/rinse		1.60	5	Ea.	5,475	350		5,825	6,575
6770 High pressure wash, high temperature rinse		1.20	6.667		16,900	465		17,365	19,300
9170 Trash compactor, small, up to 125 lb. compacted weight	L-4	4	6		1,050	320		1,370	1,650
9175 Large, up to 175 lb. compacted weight	"	3	8		28,700	430		29,130	32,300

11 52 Audio-Visual Equipment

11 52 13 – Projection Screens

11 52 13.10 Projection Screens, Wall or Ceiling Hung

	Crew	Daily Output	Labor-Hours	Unit	Material	2022 Bare Costs Labor	Equipment	Total	Total Incl O&P
0010 **PROJECTION SCREENS, WALL OR CEILING HUNG**, matte white									
0100 Manually operated, economy	2 Carp	500	.032	S.F.	6.70	1.80		8.50	10.20
0300 Intermediate		450	.036		8	2		10	12
0400 Deluxe		400	.040		10.15	2.25		12.40	14.75
9000 Minimum labor/equipment charge		3	5.333	Job		300		300	475

11 52 16 – Projectors

11 52 16.10 Movie Equipment

	Crew	Daily Output	Labor-Hours	Unit	Material	2022 Bare Costs Labor	Equipment	Total	Total Incl O&P
0011 **MOVIE EQUIPMENT**									
0020 Changeover, minimum				Ea.	525			525	580
3000 Projection screens, rigid, in wall, acrylic, 1/4" thick	2 Glaz	195	.082	S.F.	49.50	4.43		53.93	61.50
3100 1/2" thick	2 Glaz	130	.123	S.F.	57	6.65		63.65	73
3700 Sound systems, incl. amplifier, mono, minimum	1 Elec	.90	8.889	Ea.	3,925	590		4,515	5,200
3800 Dolby/Super Sound, maximum		.40	20		21,100	1,325		22,425	25,200
4100 Dual system, 2 channel, front surround, minimum		.70	11.429		5,400	755		6,155	7,100
4200 Dolby/Super Sound, 4 channel, maximum		.40	20		20,200	1,325		21,525	24,300
5700 Seating, painted steel, upholstered, minimum	2 Carp	35	.457		176	25.50		201.50	235
5800 Maximum	"	28	.571		570	32		602	675

11 53 Laboratory Equipment

11 53 33 – Emergency Safety Appliances

11 53 33.13 Emergency Equipment	Crew	Daily Output	Labor-Hours	Unit	Material	2022 Bare Costs Labor	Equipment	Total	Total Incl O&P
0010 **EMERGENCY EQUIPMENT**									
1400 Safety equipment, eye wash, hand held				Ea.	440			440	485
1450 Deluge shower				"	275			275	300

11 53 43 – Service Fittings and Accessories

11 53 43.13 Fittings

11 53 43.13 Fittings	Crew	Daily Output	Labor-Hours	Unit	Material	2022 Bare Costs Labor	Equipment	Total	Total Incl O&P
0010 **FITTINGS**									
1600 Sink, one piece plastic, flask wash, hose, free standing	1 Plum	1.60	5	Ea.	286	350		636	855
1610 Epoxy resin sink, 25" x 16" x 10"	"	2	4	"	1,550	279		1,829	2,150
1630 Steel table, open underneath				L.F.	630			630	690
1640 Doors underneath				"	1,025			1,025	1,125
1900 Utility cabinet, stainless steel				Ea.	1,150			1,150	1,275
1950 Utility table, acid resistant top with drawers	2 Carp	30	.533	L.F.	199	30		229	267
8000 Alternate pricing method: as percent of lab furniture									
8050 Installation, not incl. plumbing & duct work				% Furn.				22%	22%
8100 Plumbing, final connections, simple system								10%	10%
8110 Moderately complex system								15%	15%
8120 Complex system								20%	20%
8150 Electrical, simple system								10%	10%
8160 Moderately complex system								20%	20%
8170 Complex system								35%	35%

11 53 53 – Biological Safety Cabinets

11 53 53.10 Pharmacy Cabinets

11 53 53.10 Pharmacy Cabinets	Crew	Daily Output	Labor-Hours	Unit	Material	2022 Bare Costs Labor	Equipment	Total	Total Incl O&P
0010 **PHARMACY CABINETS**, vertical flow									
0100 Class II, type B2, 6' L	2 Carp	1.50	10.667	Ea.	8,625	600		9,225	10,500

11 57 Vocational Shop Equipment

11 57 10 – Shop Equipment

11 57 10.10 Vocational School Shop Equipment

11 57 10.10 Vocational School Shop Equipment	Crew	Daily Output	Labor-Hours	Unit	Material	2022 Bare Costs Labor	Equipment	Total	Total Incl O&P
0010 **VOCATIONAL SCHOOL SHOP EQUIPMENT**									
0020 Benches, work, wood, average	2 Carp	5	3.200	Ea.	410	180		590	735
0100 Metal, average	2 Carp	5	3.200	Ea.	380	180		560	705
0400 Combination belt & disc sander, 6"		4	4		1,850	225		2,075	2,375
0700 Drill press, floor mounted, 12", 1/2 HP		4	4		1,325	225		1,550	1,800
0800 Dust collector, not incl. ductwork, 6" diameter	1 Shee	1.10	7.273		8,275	490		8,765	9,875
0810 Dust collector bag, 20" diameter	"	5	1.600		870	108		978	1,125
1000 Grinders, double wheel, 1/2 HP	2 Carp	5	3.200		159	180		339	460
1300 Jointer, 4", 3/4 HP		4	4		1,975	225		2,200	2,500
1600 Kilns, 16 C.F., to 2000°		4	4		6,100	225		6,325	7,050
1900 Lathe, woodworking, 10", 1/2 HP		4	4		550	225		775	965
2200 Planer, 13" x 6"		4	4		1,400	225		1,625	1,900
2500 Potter's wheel, motorized		4	4		1,825	225		2,050	2,375
2800 Saws, band, 14", 3/4 HP		4	4		900	225		1,125	1,350
3100 Metal cutting band saw, 14"		4	4		3,025	225		3,250	3,675
3400 Radial arm saw, 10", 2 HP		4	4		3,950	225		4,175	4,700
3700 Scroll saw, 24"		4	4		895	225		1,120	1,350
4000 Table saw, 10", 3 HP		4	4		1,500	225		1,725	2,000
4300 Welder AC arc, 30 amp capacity		4	4		5,475	225		5,700	6,375

11 61 Broadcast, Theater, and Stage Equipment

11 61 23 – Folding and Portable Stages

11 61 23.10 Portable Stages

	11 61 23.10 Portable Stages	Crew	Daily Output	Labor-Hours	Unit	Material	2022 Bare Costs Labor	Equipment	Total	Total Incl O&P
0010	**PORTABLE STAGES**									
5000	Stages, portable with steps, folding legs, stock, 8" high				SF Stg.	47.50			47.50	52.50
5100	16" high					59			59	65
5200	32" high					60.50			60.50	66.50
5300	40" high					65.50			65.50	72

11 61 33 – Rigging Systems and Controls

11 61 33.10 Controls

	11 61 33.10 Controls	Crew	Daily Output	Labor-Hours	Unit	Material	2022 Bare Costs Labor	Equipment	Total	Total Incl O&P
0010	**CONTROLS**									
0050	Control boards with dimmers and breakers, minimum	1 Elec	1	8	Ea.	15,600	530		16,130	18,000
0150	Maximum	"	.20	40	"	125,000	2,650		127,650	141,500

11 61 43 – Stage Curtains

11 61 43.10 Curtains

	11 61 43.10 Curtains	Crew	Daily Output	Labor-Hours	Unit	Material	2022 Bare Costs Labor	Equipment	Total	Total Incl O&P
0010	**CURTAINS**									
0500	Curtain track, straight, light duty	2 Carp	20	.800	L.F.	31.50	45		76.50	106
0700	Curved sections		12	1.333	"	24.50	75		99.50	146
1000	Curtains, velour, medium weight		600	.027	S.F.	8.75	1.50		10.25	12.05

11 66 Athletic Equipment

11 66 13 – Exercise Equipment

11 66 13.10 Physical Training Equipment

	11 66 13.10 Physical Training Equipment	Crew	Daily Output	Labor-Hours	Unit	Material	2022 Bare Costs Labor	Equipment	Total	Total Incl O&P
0010	**PHYSICAL TRAINING EQUIPMENT**									
0020	Abdominal rack, 2 board capacity				Ea.	470			470	520
0050	Abdominal board, upholstered					705			705	775
0200	Bicycle trainer, minimum					900			900	990
0300	Deluxe, electric					1,800			1,800	1,975
0400	Barbell set, chrome plated steel, 25 lb.					174			174	191
0420	100 lb.					169			169	186
0450	200 lb.					360			360	395
0500	Weight plates, cast iron, per lb.				Lb.	1.85			1.85	2.04
0520	Storage rack, 10 station				Ea.	850			850	935
0600	Circuit training apparatus, 12 machines minimum	2 Clab	1.25	12.800	Set	25,500	585		26,085	28,900
0700	Average		1	16		14,500	730		15,230	17,100
0800	Maximum		.75	21.333		14,500	975		15,475	17,500
0820	Dumbbell set, cast iron, with rack and 5 pair					1,350			1,350	1,500
0900	Squat racks	2 Clab	5	3.200	Ea.	800	146		946	1,100
4150	Exercise equipment, bicycle trainer					1,200			1,200	1,325
4180	Chinning bar, adjustable, wall mounted	1 Carp	5	1.600		241	90		331	410
4200	Exercise ladder, 16' x 1'-7", suspended	L-2	3	5.333		1,400	266		1,666	1,975
4210	High bar, floor plate attached	1 Carp	4	2		350	113		463	565
4240	Parallel bars, adjustable		4	2		1,625	113		1,738	1,975
4270	Uneven parallel bars, adjustable		4	2		4,050	113		4,163	4,625
4280	Wall mounted, adjustable	L-2	1.50	10.667	Set	975	530		1,505	1,925
4300	Rope, ceiling mounted, 18' long	1 Carp	3.66	2.186	Ea.	208	123		331	425
4330	Side horse, vaulting		5	1.600		1,700	90		1,790	2,025
4360	Treadmill, motorized, deluxe, training type		5	1.600		4,150	90		4,240	4,725
4390	Weight lifting multi-station, minimum	2 Clab	1	16		310	730		1,040	1,500

11 66 Athletic Equipment

11 66 23 – Gymnasium Equipment

11 66 23.13 Basketball Equipment

		Crew	Daily Output	Labor-Hours	Unit	Material	2022 Bare Costs Labor	Equipment	Total	Total Incl O&P
0010	**BASKETBALL EQUIPMENT**									
1000	Backstops, wall mtd., 6' extended, fixed, minimum	L-2	1	16	Ea.	2,000	795		2,795	3,475
1100	Maximum		1	16		2,100	795		2,895	3,575
1200	Swing up, minimum		1	16		1,575	795		2,370	3,025
1250	Maximum		1	16		3,675	795		4,470	5,325
1300	Portable, manual, heavy duty, spring operated		1.90	8.421		14,200	420		14,620	16,300
1400	Ceiling suspended, stationary, minimum		.78	20.513		3,925	1,025		4,950	5,925
1450	Fold up, with accessories, maximum		.40	40		6,225	2,000		8,225	10,000
1600	For electrically operated, add	1 Elec	1	8		2,600	530		3,130	3,675
5800	Wall pads, 1-1/2" thick, standard (not fire rated)	2 Carp	640	.025	S.F.	132	1.41		133.41	147

11 66 23.19 Boxing Ring

		Crew	Daily Output	Labor-Hours	Unit	Material	2022 Bare Costs Labor	Equipment	Total	Total Incl O&P
0010	**BOXING RING**									
4100	Elevated, 22' x 22'	L-4	.10	240	Ea.	6,375	12,900		19,275	27,500
4110	For cellular plastic foam padding, add		.10	240		1,275	12,900		14,175	21,900
4120	Floor level, including posts and ropes only, 20' x 20'		.80	30		5,000	1,600		6,600	8,050
4130	Canvas, 30' x 30'		5	4.800		1,525	258		1,783	2,075

11 66 23.47 Gym Mats

		Crew	Daily Output	Labor-Hours	Unit	Material	2022 Bare Costs Labor	Equipment	Total	Total Incl O&P
0010	**GYM MATS**									
5500	2" thick, naugahyde covered				S.F.	4.39			4.39	4.83
5600	Vinyl/nylon covered					9.05			9.05	9.95
6000	Wrestling mats, 1" thick, heavy duty					4.94			4.94	5.45

11 66 43 – Interior Scoreboards

11 66 43.10 Scoreboards

		Crew	Daily Output	Labor-Hours	Unit	Material	2022 Bare Costs Labor	Equipment	Total	Total Incl O&P
0010	**SCOREBOARDS**									
7000	Baseball, minimum	R-3	1.30	15.385	Ea.	4,800	1,025	133	5,958	6,975
7200	Maximum		.05	400		11,400	26,300	3,475	41,175	57,000
7300	Football, minimum		.86	23.256		5,200	1,525	202	6,927	8,275
7400	Maximum		.20	100		14,900	6,575	865	22,340	27,500
7500	Basketball (one side), minimum		2.07	9.662		2,400	635	84	3,119	3,725
7600	Maximum		.30	66.667		6,225	4,400	580	11,205	14,200
7700	Hockey-basketball (four sides), minimum		.25	80		10,300	5,275	695	16,270	20,300
7800	Maximum		.15	133		17,600	8,775	1,150	27,525	34,100

11 66 53 – Gymnasium Dividers

11 66 53.10 Divider Curtains

		Crew	Daily Output	Labor-Hours	Unit	Material	2022 Bare Costs Labor	Equipment	Total	Total Incl O&P
0010	**DIVIDER CURTAINS**									
4500	Gym divider curtain, mesh top, vinyl bottom, manual	L-4	500	.048	S.F.	9.90	2.58		12.48	14.95
4700	Electric roll up	L-7	400	.070	"	14.50	3.83		18.33	22

11 71 Medical Sterilizing Equipment

11 71 10 – Medical Sterilizers & Distillers

11 71 10.10 Sterilizers and Distillers

		Crew	Daily Output	Labor-Hours	Unit	Material	2022 Bare Costs Labor	Equipment	Total	Total Incl O&P
0010	**STERILIZERS AND DISTILLERS**									
3010	Portable, top loading, 105-135 degree C, 3 to 30 psi, 50 L chamber				Ea.	10,200			10,200	11,300
3020	Stainless steel basket, 10.7" diam. x 11.8" H					226			226	248
3025	Stainless steel pail, 10.7" diam. x 10.7" H					335			335	370
3050	85 L chamber					19,000			19,000	20,900
3060	Stainless steel basket, 15.3" diam. x 11.5" H					490			490	535
3065	Stainless steel pail, 15.3" diam. x 11" H					715			715	785

11 76 Operating Room Equipment

11 76 10 – Equipment for Operating Rooms

11 76 10.10 Surgical Equipment	Crew	Daily Output	Labor-Hours	Unit	Material	2022 Bare Costs Labor	Equipment	Total	Total Incl O&P
0010 **SURGICAL EQUIPMENT**									
6550 Major surgery table, minimum	1 Sswk	.50	16	Ea.	20,600	990		21,590	24,200
6570 Maximum		.50	16		18,800	990		19,790	22,300
6600 Hydraulic, hand-held control, general surgery		.60	13.333		38,800	825		39,625	44,000
6650 Stationary, universal		.50	16		48,900	990		49,890	55,500
6800 Surgical lights, major operating room, dual head, minimum	2 Elec	1	16		4,675	1,050		5,725	6,775
6850 Maximum		1	16		6,675	1,050		7,725	8,975
6900 Ceiling mount articulation, single arm		1	16		4,450	1,050		5,500	6,525

11 81 Facility Maintenance Equipment

11 81 19 – Vacuum Cleaning Systems

11 81 19.10 Vacuum Cleaning

11 81 19.10 Vacuum Cleaning	Crew	Daily Output	Labor-Hours	Unit	Material	2022 Bare Costs Labor	Equipment	Total	Total Incl O&P
0010 **VACUUM CLEANING**									
0020 Central, 3 inlet, residential	1 Skwk	.90	8.889	Total	1,150	525		1,675	2,100
0200 Commercial		.70	11.429		1,375	675		2,050	2,575
0400 5 inlet system, residential		.50	16		1,825	940		2,765	3,500
0600 7 inlet system, commercial		.40	20		1,200	1,175		2,375	3,175

11 82 Facility Solid Waste Handling Equipment

11 82 26 – Facility Waste Compactors

11 82 26.10 Compactors

11 82 26.10 Compactors	Crew	Daily Output	Labor-Hours	Unit	Material	2022 Bare Costs Labor	Equipment	Total	Total Incl O&P
0010 **COMPACTORS**									
0020 Compactors, 115 volt, 250 lbs./hr., chute fed	L-4	1	24	Ea.	12,300	1,300		13,600	15,600
0100 Hand fed		2.40	10		16,500	535		17,035	19,000
1000 Heavy duty industrial compactor, 0.5 C.Y. capacity		1	24		10,900	1,300		12,200	14,100
1050 1.0 C.Y. capacity		1	24		16,700	1,300		18,000	20,400
1400 For handling hazardous waste materials, 55 gallon drum packer, std.					21,400			21,400	23,600
1410 55 gallon drum packer w/HEPA filter					25,800			25,800	28,400
1420 55 gallon drum packer w/charcoal & HEPA filter					34,400			34,400	37,800
1430 All of the above made explosion proof, add					2,000			2,000	2,200

11 82 39 – Medical Waste Disposal Systems

11 82 39.10 Off-Site Disposal

11 82 39.10 Off-Site Disposal	Crew	Daily Output	Labor-Hours	Unit	Material	2022 Bare Costs Labor	Equipment	Total	Total Incl O&P
0010 **OFF-SITE DISPOSAL**									
0100 Medical waste disposal, Red Bag system, pick up & treat, 200 lbs./week				Week	213			213	234
0110 Per month				Month	770			770	845
0150 Red bags, 7-10 gal., 1.2 mil, pkg of 500				Ea.	67.50			67.50	74
0200 15 gal., package of 250					66			66	72.50
0250 33 gal., package of 250					74.50			74.50	82
0300 45 gal., package of 100					61			61	67.50

11 82 39.20 Disposal Carts

11 82 39.20 Disposal Carts	Crew	Daily Output	Labor-Hours	Unit	Material	2022 Bare Costs Labor	Equipment	Total	Total Incl O&P
0010 **DISPOSAL CARTS**									
2010 Medical waste disposal cart, HDPE, w/lid, 28 gal. capacity				Ea.	286			286	315
2020 96 gal. capacity					335			335	370
2030 150 gal. capacity, low profile					485			485	530
2040 200 gal. capacity					1,000			1,000	1,100

11 82 Facility Solid Waste Handling Equipment

11 82 39 – Medical Waste Disposal Systems

11 82 39.30 Medical Waste Sanitizers		Crew	Daily Output	Labor-Hours	Unit	Material	2022 Bare Costs Labor	Equipment	Total	Total Incl O&P
0010	**MEDICAL WASTE SANITIZERS**									
2010	Small, hand loaded, 1.5 C.Y., 225 lb. capacity				Ea.	82,000			82,000	90,000
2020	Medium, cart loaded, 6.25 C.Y., 938 lb. capacity					109,500			109,500	120,500
2030	Large, cart loaded, 15 C.Y., 2250 lb. capacity					131,500			131,500	145,000
3010	Cart, aluminum, 75 lb. capacity					2,175			2,175	2,400
3020	95 lb. capacity					2,375			2,375	2,600
4010	Stainless steel, 173 lb. capacity					3,100			3,100	3,425
4020	232 lb. capacity					3,475			3,475	3,825
4030	Cart lift, hydraulic scissor type					6,150			6,150	6,775
4040	Portable aluminum ramp					1,750			1,750	1,900
4050	Fold-down steel tracks					1,425			1,425	1,550
4060	Pull-out drawer, small					6,625			6,625	7,300
4070	Medium					9,575			9,575	10,500
4080	Large				↓	13,500			13,500	14,900
5000	Medical waste treatment, sanitize, on-site									
5010	Less than 15,000 lbs./month				Lb.	.20			.20	.22
5020	Over 15,000 lbs./month				"	.16			.16	.18

11 91 Religious Equipment

11 91 13 – Baptisteries

11 91 13.10 Baptistry

		Crew	Daily Output	Labor-Hours	Unit	Material	2022 Bare Costs Labor	Equipment	Total	Total Incl O&P
0010	**BAPTISTRY**									
0150	Fiberglass, 3'-6" deep, x 13'-7" long,									
0160	steps at both ends, incl. plumbing, minimum	L-8	1	20	Ea.	7,275	1,175		8,450	9,850
0200	Maximum	"	.70	28.571		11,900	1,675		13,575	15,700
0250	Add for filter, heater and lights				↓	1,625			1,625	1,800

11 91 23 – Sanctuary Equipment

11 91 23.10 Sanctuary Furnishings

		Crew	Daily Output	Labor-Hours	Unit	Material	2022 Bare Costs Labor	Equipment	Total	Total Incl O&P
0010	**SANCTUARY FURNISHINGS**									
0020	Altar, wood, custom design, plain	1 Carp	1.40	5.714	Ea.	2,675	320		2,995	3,450
0050	Deluxe		.20	40	"	12,800	2,250		15,050	17,700
5000	Wall cross, aluminum, extruded, 2" x 2" section		34	.235	L.F.	240	13.25		253.25	285
5150	4" x 4" section		29	.276		345	15.55		360.55	405
5300	Bronze, extruded, 1" x 2" section		31	.258		475	14.55		489.55	545
5350	2-1/2" x 2-1/2" section	↓	34	.235	↓	715	13.25		728.25	810

11 97 Security Equipment

11 97 30 – Security Drawers

11 97 30.10 Pass Through Drawer

		Crew	Daily Output	Labor-Hours	Unit	Material	2022 Bare Costs Labor	Equipment	Total	Total Incl O&P
0010	**PASS THROUGH DRAWER**									
0100	Pass-thru drawer for personal items, 18" x 15" x 24"	1 Skwk	2	4	Ea.	2,750	236		2,986	3,400
0110	Including speakers	"	1.50	5.333	"	3,100	315		3,415	3,900

11 98 Detention Equipment

11 98 21 – Detention Windows

11 98 21.13 Visitor Cubicle Windows	Crew	Daily Output	Labor-Hours	Unit	Material	2022 Bare Costs Labor	Equipment	Total	Total Incl O&P
0010 **VISITOR CUBICLE WINDOWS**									
4000 Visitor cubicle, vision panel, no intercom	E-4	2	16	Ea.	3,975	1,000	74	5,049	6,050

11 98 30 – Detention Cell Equipment

11 98 30.10 Cell Equipment

	Crew	Daily Output	Labor-Hours	Unit	Material	2022 Bare Costs Labor	Equipment	Total	Total Incl O&P
0010 **CELL EQUIPMENT**									
3000 Toilet apparatus including wash basin, average	L-8	1.50	13.333	Ea.	3,550	785		4,335	5,175

For customer support on your Commercial Renovation Costs with RSMeans data, call 800.448.8182.

Estimating Tips
General

- The items in this division are usually priced per square foot or each. Most of these items are purchased by the owner and installed by the contractor. Do not assume the items in Division 12 will be purchased and installed by the contractor. Check the specifications for responsibilities and include receiving, storage, installation, and mechanical and electrical hookups in the appropriate divisions.

- Some items in this division require some type of support system that is not usually furnished with the item. Examples of these systems include blocking for the attachment of casework and heavy drapery rods. The required blocking must be added to the estimate in the appropriate division.

Reference Numbers

Reference numbers are shown at the beginning of some major classifications. These numbers refer to related items in the Reference Section. The reference information may be an estimating procedure, an alternate pricing method, or technical information.

Same Data. Simplified.

Enjoy the convenience and efficiency of accessing your costs anywhere:

- **Skip the multiplier** by setting your location
- **Quickly search,** edit, favorite and share costs
- **Stay on top of price changes** with automatic updates

Discover more at rsmeans.com/online

12 21 Window Blinds

12 21 13 – Horizontal Louver Blinds

12 21 13.13 Metal Horizontal Louver Blinds	Crew	Daily Output	Labor-Hours	Unit	Material	2022 Bare Costs Labor	Equipment	Total	Total Incl O&P
0010 **METAL HORIZONTAL LOUVER BLINDS**									
0020 Horizontal, 1" aluminum slats, solid color, stock	1 Carp	590	.014	S.F.	5.25	.76		6.01	7

12 21 13.33 Vinyl Horizontal Louver Blinds

	Crew	Daily Output	Labor-Hours	Unit	Material	Labor	Equipment	Total	Total Incl O&P
0010 **VINYL HORIZONTAL LOUVER BLINDS**									
0100 2" composite, 48" wide, 48" high	1 Carp	30	.267	Ea.	119	15		134	155
0120 72" high		29	.276		161	15.55		176.55	202
0140 96" high		28	.286		219	16.10		235.10	267
0200 60" wide, 60" high		27	.296		246	16.70		262.70	297
0220 72" high		25	.320		283	18		301	340
0240 96" high		24	.333		425	18.75		443.75	495
0300 72" wide, 72" high		25	.320		355	18		373	420
0320 96" high		23	.348		485	19.60		504.60	565
0400 96" wide, 96" high		20	.400		615	22.50		637.50	710
1000 2" faux wood, 48" wide, 48" high		30	.267		80.50	15		95.50	113
1020 72" high		29	.276		104	15.55		119.55	140
1040 96" high		28	.286		124	16.10		140.10	163
1300 72" wide, 72" high		25	.320		222	18		240	273
1320 96" high		23	.348		330	19.60		349.60	395
1400 96" wide, 96" high		20	.400		335	22.50		357.50	405

12 23 Interior Shutters

12 23 13 – Wood Interior Shutters

12 23 13.10 Wood Interior Shutters

	Crew	Daily Output	Labor-Hours	Unit	Material	Labor	Equipment	Total	Total Incl O&P
0010 **WOOD INTERIOR SHUTTERS**, louvered									
0200 Two panel, 27" wide, 36" high	1 Carp	5	1.600	Set	212	90		302	375
0300 33" wide, 36" high	"	5	1.600	"	274	90		364	445
0500 47" wide, 36" high	1 Carp	5	1.600	Set	365	90		455	550
1000 Four panel, 27" wide, 36" high		5	1.600		282	90		372	455
1100 33" wide, 36" high		5	1.600		360	90		450	540
1300 47" wide, 36" high		5	1.600		480	90		570	675
1400 Plantation shutters, 16" x 48"		5	1.600	Ea.	198	90		288	360
1450 16" x 96"		4	2		325	113		438	535
1460 36" x 96"		3	2.667		715	150		865	1,025

12 23 13.13 Wood Panels

	Crew	Daily Output	Labor-Hours	Unit	Material	Labor	Equipment	Total	Total Incl O&P
0010 **WOOD PANELS**									
3000 Wood folding panels with movable louvers, 7" x 20" each	1 Carp	17	.471	Pr.	93.50	26.50		120	145
4000 Fixed louver type, stock units, 8" x 20" each		17	.471		123	26.50		149.50	177
4450 18" x 40" each		17	.471		176	26.50		202.50	236

12 24 Window Shades

12 24 13 – Roller Window Shades

12 24 13.10 Shades

		Crew	Daily Output	Labor-Hours	Unit	Material	2022 Bare Costs Labor	Equipment	Total	Total Incl O&P
0010	**SHADES**									
0020	Basswood, roll-up, stain finish, 3/8" slats	1 Carp	300	.027	S.F.	25	1.50		26.50	30
0200	7/8" slats		300	.027		61	1.50		62.50	70
0900	Mylar, single layer, non-heat reflective		685	.012		5.90	.66		6.56	7.55
0910	Mylar, single layer, heat reflective		685	.012		4.52	.66		5.18	6
1000	Double layered, heat reflective		685	.012		7.50	.66		8.16	9.30
1100	Triple layered, heat reflective		685	.012		11.30	.66		11.96	13.45
2000	Polyester, room darkening, with continuous cord, GEI									
2010	36" x 72"	1 Carp	38	.211	Ea.	242	11.85		253.85	285
2020	48" x 72"		28	.286		315	16.10		331.10	375
2030	60" x 72"		23	.348		425	19.60		444.60	500
2040	72" x 72"		19	.421		450	23.50		473.50	535
5011	Insulative shades		125	.064	S.F.	19.45	3.60		23.05	27
5030	R10.7		44	.182	"	23	10.25		33.25	42
5050	Magnetic clips, set of 20				Set	43			43	47.50
6011	[G] Solar screening, fiberglass	1 Carp	85	.094	S.F.	8.90	5.30		14.20	18.20

12 32 Manufactured Wood Casework

12 32 23 – Hardwood Casework

12 32 23.10 Manufactured Wood Casework, Stock Units

		Crew	Daily Output	Labor-Hours	Unit	Material	2022 Bare Costs Labor	Equipment	Total	Total Incl O&P
0010	**MANUFACTURED WOOD CASEWORK, STOCK UNITS**									
0300	Built-in drawer units, pine, 18" deep, 32" high, unfinished									
0400	Minimum	2 Carp	53	.302	L.F.	139	17		156	180
0500	Maximum	"	40	.400	"	164	22.50		186.50	217
0700	Kitchen base cabinets, hardwood, not incl. counter tops,									
0710	24" deep, 35" high, prefinished									
0800	One top drawer, one door below, 12" wide	2 Carp	24.80	.645	Ea.	390	36.50		426.50	485
0820	15" wide		24	.667		400	37.50		437.50	500
0840	18" wide		23.30	.687		415	38.50		453.50	515
0860	21" wide		22.70	.705		425	39.50		464.50	535
0880	24" wide		22.30	.717		470	40.50		510.50	580
1000	Four drawers, 12" wide		24.80	.645		735	36.50		771.50	870
1020	15" wide		24	.667		740	37.50		777.50	875
1040	18" wide		23.30	.687		770	38.50		808.50	910
1060	24" wide		22.30	.717		815	40.50		855.50	960
1200	Two top drawers, two doors below, 27" wide		22	.727		480	41		521	595
1220	30" wide		21.40	.748		510	42		552	625
1240	33" wide		20.90	.766		530	43		573	655
1260	36" wide		20.30	.788		540	44.50		584.50	665
1280	42" wide		19.80	.808		660	45.50		705.50	800
1300	48" wide		18.90	.847		680	47.50		727.50	825
1500	Range or sink base, two doors below, 30" wide		21.40	.748		365	42		407	470
1520	33" wide		20.90	.766		390	43		433	500
1540	36" wide		20.30	.788		405	44.50		449.50	515
1560	42" wide		19.80	.808		425	45.50		470.50	545
1580	48" wide		18.90	.847		445	47.50		492.50	565
1800	For sink front units, deduct					177			177	195
2000	Corner base cabinets, 36" wide, standard	2 Carp	18	.889		385	50		435	505
2100	Lazy Susan with revolving door	"	16.50	.970		510	54.50		564.50	645
4000	Kitchen wall cabinets, hardwood, 12" deep with two doors									
4050	12" high, 30" wide	2 Carp	24.80	.645	Ea.	222	36.50		258.50	300

12 32 23 – Hardwood Casework

12 32 23.10 Manufactured Wood Casework, Stock Units	Crew	Daily Output	Labor-Hours	Unit	Material	2022 Bare Costs Labor	Equipment	Total	Total Incl O&P
4400	15" high, 30" wide	2 Carp	24	.667	Ea.	330	37.50		367.50	420
4420	33" wide		23.30	.687		148	38.50		186.50	225
4440	36" wide		22.70	.705		155	39.50		194.50	234
4450	42" wide		22.70	.705		360	39.50		399.50	460
4700	24" high, 30" wide		23.30	.687		166	38.50		204.50	245
4720	36" wide		22.70	.705		475	39.50		514.50	590
4740	42" wide		22.30	.717		300	40.50		340.50	395
5000	30" high, one door, 12" wide		22	.727		108	41		149	184
5020	15" wide		21.40	.748		246	42		288	335
5040	18" wide		20.90	.766		270	43		313	365
5060	24" wide		20.30	.788		315	44.50		359.50	415
5300	Two doors, 27" wide		19.80	.808		181	45.50		226.50	273
5320	30" wide	2 Carp	19.30	.829	Ea.	405	46.50		451.50	520
5340	36" wide		18.80	.851		385	48		433	500
5360	42" wide		18.50	.865		455	48.50		503.50	580
5380	48" wide		18.40	.870		780	49		829	935
6000	Corner wall, 30" high, 24" wide		18	.889		201	50		251	300
6050	30" wide		17.20	.930		530	52.50		582.50	665
6100	36" wide		16.50	.970		470	54.50		524.50	600
6500	Revolving Lazy Susan		15.20	1.053		146	59.50		205.50	255
7000	Broom cabinet, 84" high, 24" deep, 18" wide		10	1.600		815	90		905	1,050
7500	Oven cabinets, 84" high, 24" deep, 27" wide		8	2		1,050	113		1,163	1,325
7750	Valance board trim		396	.040	L.F.	24.50	2.27		26.77	30.50
7780	Toe kick trim	1 Carp	256	.031	"	3.78	1.76		5.54	6.95
7790	Base cabinet corner filler		16	.500	Ea.	53	28		81	103
7800	Cabinet filler, 3" x 24"		20	.400		20.50	22.50		43	58.50
7810	3" x 30"		20	.400		25.50	22.50		48	64
7820	3" x 42"		18	.444		36	25		61	79.50
7830	3" x 80"		16	.500		68.50	28		96.50	120
7850	Cabinet panel		50	.160	S.F.	12	9		21	27.50
9000	For deluxe models of all cabinets, add					40%				
9500	For custom built in place, add					25%	10%			
9558	Rule of thumb, kitchen cabinets not including									
9560	appliances & counter top, minimum	2 Carp	30	.533	L.F.	208	30		238	277
9600	Maximum	"	25	.640	"	460	36		496	570
9700	Minimum labor/equipment charge	1 Carp	3	2.667	Job		150		150	239

12 32 23.30 Manufactured Wood Casework Vanities

		Crew	Daily Output	Labor-Hours	Unit	Material	2022 Bare Costs Labor	Equipment	Total	Total Incl O&P
0010	**MANUFACTURED WOOD CASEWORK VANITIES**									
8000	Vanity bases, 2 doors, 30" high, 21" deep, 24" wide	2 Carp	20	.800	Ea.	455	45		500	570
8050	30" wide		16	1		480	56.50		536.50	615
8100	36" wide		13.33	1.200		405	67.50		472.50	550
8150	48" wide		11.43	1.400		530	79		609	705
9000	For deluxe models of all vanities, add to above					40%				
9500	For custom built in place, add to above					25%	10%			

12 32 23.35 Manufactured Wood Casework Hardware

		Crew	Daily Output	Labor-Hours	Unit	Material	2022 Bare Costs Labor	Equipment	Total	Total Incl O&P
0010	**MANUFACTURED WOOD CASEWORK HARDWARE**									
1000	Catches, minimum	1 Carp	235	.034	Ea.	1.63	1.92		3.55	4.84
1040	Maximum	"	80	.100	"	9.40	5.65		15.05	19.30
2000	Door/drawer pulls, handles									
2200	Handles and pulls, projecting, metal, minimum	1 Carp	48	.167	Ea.	1.13	9.40		10.53	16.15
2240	Maximum	"	36	.222	"	12	12.50		24.50	33
2300	Wood, minimum	1 Carp	48	.167	Ea.	6.05	9.40		15.45	21.50

12 32 Manufactured Wood Casework

12 32 23 – Hardwood Casework

12 32 23.35 Manufactured Wood Casework Hardware	Crew	Daily Output	Labor-Hours	Unit	Material	2022 Bare Costs Labor	2022 Bare Costs Equipment	Total	Total Incl O&P
2340	Maximum	1 Carp	36	.222	Ea.	11.20	12.50		23.70	32.50
2400	Drawer pulls, antimicrobial copper alloy finish		50	.160		13.75	9		22.75	29.50
2600	Flush, metal, minimum		48	.167		6.05	9.40		15.45	21.50
2640	Maximum		36	.222		11.35	12.50		23.85	32.50
2900	Drawer knobs, antimicrobial copper alloy finish		50	.160	↓	7.85	9		16.85	23
3000	Drawer tracks/glides, minimum		48	.167	Pr.	10.10	9.40		19.50	26
3040	Maximum		24	.333		29.50	18.75		48.25	62
4000	Cabinet hinges, minimum		160	.050		1.48	2.82		4.30	6.10
4040	Maximum		68	.118	↓	10.60	6.60		17.20	22
7000	Appliance pulls, antimicrobial copper alloy finish	↓	50	.160	L.F.	77	9		86	99.50

12 34 Manufactured Plastic Casework

12 34 16 – Manufactured Solid-Plastic Casework

12 34 16.10 Outdoor Casework

		Crew	Daily Output	Labor-Hours	Unit	Material	Labor	Equipment	Total	Total Incl O&P
0010	**OUTDOOR CASEWORK**									
0020	Cabinet, base, sink/range, 36"	2 Carp	20.30	.788	Ea.	1,875	44.50		1,919.50	2,125
0100	Base, 36"		20.30	.788		1,875	44.50		1,919.50	2,125
0200	Filler strip, 1" x 30"		158	.101		32.50	5.70		38.20	45
0210	Filler strip, 2" x 30"	↓	158	.101	↓	54	5.70		59.70	68.50

12 35 Specialty Casework

12 35 53 – Laboratory Casework

12 35 53.13 Metal Laboratory Casework

		Crew	Daily Output	Labor-Hours	Unit	Material	Labor	Equipment	Total	Total Incl O&P
0010	**METAL LABORATORY CASEWORK**									
0020	Cabinets, base, door units, metal	2 Carp	18	.889	L.F.	2,375	50		2,425	2,675
0300	Drawer units		18	.889		1,675	50		1,725	1,900
0700	Tall storage cabinets, open, 7' high		20	.800		284	45		329	380
0900	With glazed doors		20	.800		730	45		775	870
1300	Wall cabinets, metal, 12-1/2" deep, open		20	.800		151	45		196	238
1500	With doors	↓	20	.800	↓	296	45		341	395

12 35 70 – Healthcare Casework

12 35 70.13 Hospital Casework

		Crew	Daily Output	Labor-Hours	Unit	Material	Labor	Equipment	Total	Total Incl O&P
0010	**HOSPITAL CASEWORK**									
3000	Hospital cabinets, stainless steel with glass door(s), lockable									
3010	One door, 24" W x 18" D x 60" H	2 Clab	18	.889	Ea.	3,400	40.50		3,440.50	3,825
3020	Two doors, 36" W x 18" D x 60" H		15	1.067		3,925	48.50		3,973.50	4,375
3030	36" W x 24" D x 67" H		15	1.067		5,025	48.50		5,073.50	5,625
3040	48" W x 24" D x 66" H		12	1.333		6,250	61		6,311	6,975
3050	48" W x 24" D x 72" H		12	1.333		5,125	61		5,186	5,725
3060	60" W x 24" D x 72" H	↓	9	1.778	↓	7,750	81		7,831	8,650

For customer support on your Commercial Renovation Costs with RSMeans data, call 800.448.8182.

391

12 36 Countertops

12 36 16 – Metal Countertops

12 36 16.10 Stainless Steel Countertops

12 36 16.10 Stainless Steel Countertops	Crew	Daily Output	Labor-Hours	Unit	Material	2022 Bare Costs Labor	Equipment	Total	Total Incl O&P
0010 **STAINLESS STEEL COUNTERTOPS**									
3200 Stainless steel, custom	1 Carp	24	.333	S.F.	200	18.75		218.75	250

12 36 19 – Wood Countertops

12 36 19.10 Maple Countertops

	Crew	Daily Output	Labor-Hours	Unit	Material	2022 Bare Costs Labor	Equipment	Total	Total Incl O&P
0010 **MAPLE COUNTERTOPS**									
2900 Solid, laminated, 1-1/2" thick, no splash	1 Carp	28	.286	L.F.	93.50	16.10		109.60	129
3000 With square splash		28	.286	"	111	16.10		127.10	148
3400 Recessed cutting block with trim, 16" x 20" x 1"		8	1	Ea.	144	56.50		200.50	248

12 36 23 – Plastic Countertops

12 36 23.13 Plastic-Laminate-Clad Countertops

	Crew	Daily Output	Labor-Hours	Unit	Material	2022 Bare Costs Labor	Equipment	Total	Total Incl O&P
0010 **PLASTIC-LAMINATE-CLAD COUNTERTOPS**									
0020 Stock, 24" wide w/backsplash, minimum	1 Carp	30	.267	L.F.	17.65	15		32.65	43.50
0100 Maximum		25	.320		42.50	18		60.50	75.50
0300 Custom plastic, 7/8" thick, aluminum molding, no splash		30	.267		39	15		54	67
0400 Cove splash		30	.267		50	15		65	79
0600 1-1/4" thick, no splash		28	.286		43	16.10		59.10	73
0700 Square splash		28	.286		47.50	16.10		63.60	78
0900 Square edge, plastic face, 7/8" thick, no splash		30	.267		37	15		52	65
1000 With splash		30	.267		57.50	15		72.50	87
1200 For stainless channel edge, 7/8" thick, add					4.06			4.06	4.47
1300 1-1/4" thick, add					4.66			4.66	5.15
1500 For solid color suede finish, add					6.10			6.10	6.70
1700 For end splash, add				Ea.	28			28	30.50
1900 For cut outs, standard, add, minimum	1 Carp	32	.250		17.75	14.10		31.85	42
2000 Maximum		8	1		9.20	56.50		65.70	99.50
2100 Postformed, including backsplash and front edge		30	.267	L.F.	13.30	15		28.30	38.50
2110 Mitred, add		12	.667	Ea.		37.50		37.50	59.50
2200 Built-in place, 25" wide, plastic laminate		25	.320	L.F.	70	18		88	106
9000 Minimum labor/equipment charge		3.75	2.133	Job		120		120	191

12 36 33 – Tile Countertops

12 36 33.10 Ceramic Tile Countertops

	Crew	Daily Output	Labor-Hours	Unit	Material	2022 Bare Costs Labor	Equipment	Total	Total Incl O&P
0010 **CERAMIC TILE COUNTERTOPS**									
2300 Ceramic tile mosaic	1 Carp	25	.320	L.F.	47	18		65	80

12 36 40 – Stone Countertops

12 36 40.10 Natural Stone Countertops

	Crew	Daily Output	Labor-Hours	Unit	Material	2022 Bare Costs Labor	Equipment	Total	Total Incl O&P
0010 **NATURAL STONE COUNTERTOPS**									
2500 Marble, stock, with splash, 1/2" thick, minimum	1 Bric	17	.471	L.F.	56	26		82	104
2700 3/4" thick, maximum	"	13	.615	"	137	34		171	205

12 36 53 – Laboratory Countertops

12 36 53.10 Laboratory Countertops and Sinks

	Crew	Daily Output	Labor-Hours	Unit	Material	2022 Bare Costs Labor	Equipment	Total	Total Incl O&P
0010 **LABORATORY COUNTERTOPS AND SINKS**									
0020 Countertops, epoxy resin, not incl. base cabinets, acid-proof, minimum	2 Carp	82	.195	S.F.	97	11		108	124
0030 Maximum		70	.229		550	12.85		562.85	625
0040 Stainless steel		82	.195		53	11		64	76

12 36 61 – Simulated Stone Countertops

12 36 61.16 Solid Surface Countertops	Crew	Daily Output	Labor-Hours	Unit	Material	2022 Bare Costs Labor	Equipment	Total	Total Incl O&P
0010 **SOLID SURFACE COUNTERTOPS**, Acrylic polymer									
0020 Pricing for orders of 100 L.F. or greater									
0100 25" wide, solid colors	2 Carp	28	.571	L.F.	58.50	32		90.50	116
0200 Patterned colors		28	.571		74.50	32		106.50	133
0300 Premium patterned colors		28	.571		99	32		131	160
0400 With silicone attached 4" backsplash, solid colors		27	.593		68.50	33.50		102	129
0500 Patterned colors		27	.593		87	33.50		120.50	149
0600 Premium patterned colors		27	.593		108	33.50		141.50	172
0700 With hard seam attached 4" backsplash, solid colors		23	.696		68.50	39		107.50	138
0800 Patterned colors		23	.696		87	39		126	158
0900 Premium patterned colors		23	.696		108	39		147	181
1000 Pricing for order of 51-99 L.F.									
1100 25" wide, solid colors	2 Carp	24	.667	L.F.	67	37.50		104.50	134
1200 Patterned colors		24	.667		85.50	37.50		123	154
1300 Premium patterned colors		24	.667		114	37.50		151.50	185
1400 With silicone attached 4" backsplash, solid colors		23	.696		79	39		118	149
1500 Patterned colors		23	.696		100	39		139	172
1600 Premium patterned colors		23	.696		125	39		164	199
1700 With hard seam attached 4" backsplash, solid colors		20	.800		79	45		124	158
1800 Patterned colors		20	.800		100	45		145	182
1900 Premium patterned colors		20	.800		125	45		170	209
2000 Pricing for order of 1-50 L.F.									
2100 25" wide, solid colors	2 Carp	20	.800	L.F.	79	45		124	159
2200 Patterned colors		20	.800		100	45		145	182
2300 Premium patterned colors		20	.800		134	45		179	219
2400 With silicone attached 4" backsplash, solid colors		19	.842		92.50	47.50		140	178
2500 Patterned colors		19	.842		117	47.50		164.50	205
2600 Premium patterned colors		19	.842		146	47.50		193.50	237
2700 With hard seam attached 4" backsplash, solid colors		15	1.067		92.50	60		152.50	198
2800 Patterned colors		15	1.067		117	60		177	225
2900 Premium patterned colors	2 Carp	15	1.067	L.F.	146	60		206	257
3000 Sinks, pricing for order of 100 or greater units									
3100 Single bowl, hard seamed, solid colors, 13" x 17"	1 Carp	3	2.667	Ea.	420	150		570	705
3200 10" x 15"		7	1.143		195	64.50		259.50	315
3300 Cutouts for sinks		8	1			56.50		56.50	89.50
3400 Sinks, pricing for order of 51-99 units									
3500 Single bowl, hard seamed, solid colors, 13" x 17"	1 Carp	2.55	3.137	Ea.	485	177		662	815
3600 10" x 15"		6	1.333		224	75		299	365
3700 Cutouts for sinks		7	1.143			64.50		64.50	102
3800 Sinks, pricing for order of 1-50 units									
3900 Single bowl, hard seamed, solid colors, 13" x 17"	1 Carp	2	4	Ea.	570	225		795	985
4000 10" x 15"		4.55	1.758		263	99		362	445
4100 Cutouts for sinks		5.25	1.524			86		86	136
4200 Cooktop cutouts, pricing for 100 or greater units		4	2		28.50	113		141.50	211
4300 51-99 units		3.40	2.353		33	132		165	246
4400 1-50 units		3	2.667		38.50	150		188.50	282

12 36 61.17 Solid Surface Vanity Tops

12 36 61.17 Solid Surface Vanity Tops	Crew	Daily Output	Labor-Hours	Unit	Material	2022 Bare Costs Labor	Equipment	Total	Total Incl O&P
0010 **SOLID SURFACE VANITY TOPS**									
0015 Solid surface, center bowl, 17" x 19"	1 Carp	12	.667	Ea.	300	37.50		337.50	390
0020 19" x 25"		12	.667		305	37.50		342.50	400
0030 19" x 31"		12	.667		360	37.50		397.50	455
0040 19" x 37"		12	.667		420	37.50		457.50	520

12 36 Countertops

12 36 61 – Simulated Stone Countertops

12 36 61.17 Solid Surface Vanity Tops

		Crew	Daily Output	Labor-Hours	Unit	Material	2022 Bare Costs Labor	2022 Bare Costs Equipment	Total	Total Incl O&P
0050	22" x 25"	1 Carp	10	.800	Ea.	305	45		350	410
0060	22" x 31"		10	.800		360	45		405	465
0070	22" x 37"		10	.800		420	45		465	530
0080	22" x 43"		10	.800		475	45		520	595
0090	22" x 49"		10	.800		530	45		575	650
0110	22" x 55"		8	1		600	56.50		656.50	750
0120	22" x 61"		8	1		685	56.50		741.50	840
0220	Double bowl, 22" x 61"		8	1		775	56.50		831.50	940
0230	Double bowl, 22" x 73"		8	1		1,500	56.50		1,556.50	1,750
0240	For aggregate colors, add					35%				
0250	For faucets and fittings, see Section 22 41 39.10									

12 36 61.19 Quartz Agglomerate Countertops

		Crew	Daily Output	Labor-Hours	Unit	Material	2022 Bare Costs Labor	2022 Bare Costs Equipment	Total	Total Incl O&P
0010	**QUARTZ AGGLOMERATE COUNTERTOPS**									
0100	25" wide, 4" backsplash, color group A, minimum	2 Carp	15	1.067	L.F.	64	60		124	166
0110	Maximum		15	1.067		105	60		165	212
0120	Color group B, minimum		15	1.067		74.50	60		134.50	178
0130	Maximum		15	1.067		118	60		178	226
0140	Color group C, minimum	2 Carp	15	1.067	L.F.	93	60		153	198
0150	Maximum		15	1.067		167	60		227	279
0160	Color group D, minimum		15	1.067		99.50	60		159.50	205
0170	Maximum		15	1.067		128	60		188	237

12 48 Rugs and Mats

12 48 13 – Entrance Floor Mats and Frames

12 48 13.13 Entrance Floor Mats

			Crew	Daily Output	Labor-Hours	Unit	Material	2022 Bare Costs Labor	2022 Bare Costs Equipment	Total	Total Incl O&P
0010	**ENTRANCE FLOOR MATS**										
2000	Recycled rubber tire tile, 12" x 12" x 3/8" thick	G	1 Clab	125	.064	S.F.	11.30	2.92		14.22	17.05
2510	Natural cocoa fiber, 1/2" thick	G		125	.064		8.40	2.92		11.32	13.90
2520	3/4" thick	G		125	.064		7.70	2.92		10.62	13.15
2530	1" thick	G		125	.064		8.40	2.92		11.32	13.90
3000	Hospital tacky mats, package of 30 with frame					Ea.	76			76	83.50
3010	4 packages of 30					"	75			75	82.50

12 51 Office Furniture

12 51 16 – Case Goods

12 51 16.13 Metal Case Goods

		Crew	Daily Output	Labor-Hours	Unit	Material	2022 Bare Costs Labor	2022 Bare Costs Equipment	Total	Total Incl O&P
0010	**METAL CASE GOODS**									
0020	Desks, 29" high, double pedestal, 30" x 60", metal, minimum				Ea.	1,425			1,425	1,550
0030	Maximum				"	1,650			1,650	1,825

12 54 Hospitality Furniture

12 54 13 – Hotel and Motel Furniture

12 54 13.10 Hotel Furniture

		Crew	Daily Output	Labor-Hours	Unit	Material	2022 Bare Costs Labor	Equipment	Total	Total Incl O&P
0010	**HOTEL FURNITURE**									
0020	Standard quality set, minimum				Room	2,675			2,675	2,925
0200	Maximum				"	8,500			8,500	9,350

12 54 16 – Restaurant Furniture

12 54 16.20 Furniture, Restaurant

		Crew	Daily Output	Labor-Hours	Unit	Material	2022 Bare Costs Labor	Equipment	Total	Total Incl O&P
0010	**FURNITURE, RESTAURANT**									
0020	Bars, built-in, front bar	1 Carp	5	1.600	L.F.	320	90		410	500
0200	Back bar		5	1.600	"	234	90		324	400
0500	Booth unit, molded plastic, stub wall and 2 seats, minimum		2	4	Set	355	225		580	755
0600	Maximum		1.50	5.333	"	1,775	300		2,075	2,425
0800	Booth seat, upholstered, foursome, single (end) minimum		5	1.600	Ea.	930	90		1,020	1,175
0900	Maximum		4	2		1,250	113		1,363	1,550
1000	Foursome, double, minimum		4	2		1,200	113		1,313	1,500
1100	Maximum		3	2.667		2,100	150		2,250	2,575
1300	Circle booth, upholstered, 1/4 circle, minimum		3	2.667		1,700	150		1,850	2,125
1400	Maximum		2	4		2,025	225		2,250	2,575
1500	3/4 circle, minimum		1.50	5.333		7,550	300		7,850	8,775
1600	Maximum		1	8		5,375	450		5,825	6,650

12 55 Detention Furniture

12 55 13 – Detention Bunks

12 55 13.13 Cots

		Crew	Daily Output	Labor-Hours	Unit	Material	2022 Bare Costs Labor	Equipment	Total	Total Incl O&P
0010	**COTS**									
2500	Bolted, single, painted steel	E-4	20	1.600	Ea.	355	100	7.40	462.40	560
2700	Stainless steel	"	20	1.600	"	1,025	100	7.40	1,132.40	1,300

12 56 Institutional Furniture

12 56 43 – Dormitory Furniture

12 56 43.10 Dormitory Furnishings

		Crew	Daily Output	Labor-Hours	Unit	Material	2022 Bare Costs Labor	Equipment	Total	Total Incl O&P
0010	**DORMITORY FURNISHINGS**									
0300	Bunkable bed, twin, minimum				Ea.	410			410	455
0320	Maximum				"	475			475	520

12 56 51 – Library Furniture

12 56 51.10 Library Furnishings

		Crew	Daily Output	Labor-Hours	Unit	Material	2022 Bare Costs Labor	Equipment	Total	Total Incl O&P
0010	**LIBRARY FURNISHINGS**									
1710	Carrels, hardwood, 36" x 24", minimum	1 Carp	5	1.600	Ea.	570	90		660	775
1720	Maximum		4	2		1,950	113		2,063	2,300
1730	Metal, minimum		5	1.600		187	90		277	350
1740	Maximum		4	2		470	113		583	695
6010	Bookshelf, metal, 90" high, 10" shelf, double face		11.50	.696	L.F.	213	39		252	297
6020	Single face		12	.667	"	119	37.50		156.50	191
6050	For 8" shelving, subtract from above					10%				
6070	For 42" high with countertop, subtract from above					20%				

12 56 Institutional Furniture

12 56 70 – Healthcare Furniture

12 56 70.10 Furniture, Hospital	Crew	Daily Output	Labor-Hours	Unit	Material	2022 Bare Costs Labor	Equipment	Total	Total Incl O&P
0010 **FURNITURE, HOSPITAL**									
0020 Beds, manual, minimum				Ea.	555			555	610
0100 Maximum				"	3,075			3,075	3,375
1100 Patient wall systems, not incl. plumbing, minimum				Room	1,375			1,375	1,500
1200 Maximum				"	2,475			2,475	2,725

12 63 Stadium and Arena Seating

12 63 13 – Stadium and Arena Bench Seating

12 63 13.13 Bleachers

	Crew	Daily Output	Labor-Hours	Unit	Material	2022 Bare Costs Labor	Equipment	Total	Total Incl O&P
0010 **BLEACHERS**									
3000 Telescoping, manual to 15 tier, minimum	F-5	65	.492	Seat	98	28		126	153
3100 Maximum		60	.533		211	30.50		241.50	280
3300 16 to 20 tier, minimum		60	.533		213	30.50		243.50	282
3400 Maximum		55	.582		415	33		448	515
3600 21 to 30 tier, minimum		50	.640		279	36.50		315.50	365
3700 Maximum		40	.800		445	45.50		490.50	560
3900 For integral power operation, add, minimum	2 Elec	300	.053		800	3.53		803.53	880
4000 Maximum	"	250	.064		1,475	4.24		1,479.24	1,625
5000 Benches, folding, in wall, 14' table, 2 benches	L-4	2	12	Set	2,500	645		3,145	3,775

12 67 Pews and Benches

12 67 13 – Pews

12 67 13.13 Sanctuary Pews

	Crew	Daily Output	Labor-Hours	Unit	Material	2022 Bare Costs Labor	Equipment	Total	Total Incl O&P
0010 **SANCTUARY PEWS**									
1500 Bench type, hardwood, minimum	1 Carp	20	.400	L.F.	135	22.50		157.50	185
1550 Maximum	"	15	.533		650	30		680	765
1570 For kneeler, add					20.50			20.50	22.50

12 93 Interior Public Space Furnishings

12 93 13 – Bicycle Racks

12 93 13.10 Bicycle Racks

	Crew	Daily Output	Labor-Hours	Unit	Material	2022 Bare Costs Labor	Equipment	Total	Total Incl O&P
0010 **BICYCLE RACKS**									
0020 Single side, grid, 1-5/8" OD stl. pipe, w/1/2" bars, galv, 5 bike cap	2 Clab	10	1.600	Ea.	345	73		418	495
0025 Powder coat finish		10	1.600		380	73		453	535
0030 Single side, grid, 1-5/8" OD stl. pipe, w/1/2" bars, galv, 9 bike cap		8	2		380	91		471	560
0035 Powder coat finish		8	2		600	91		691	805
0040 Single side, grid, 1-5/8" OD stl. pipe, w/1/2" bars, galv, 18 bike cap		4	4		710	182		892	1,075
0045 Powder coat finish		4	4		710	182		892	1,075
0050 S curve, 1-7/8" OD stl. pipe, 11 ga., galv, 5 bike cap		10	1.600		219	73		292	355
0055 Powder coat finish		10	1.600		187	73		260	320
0060 S curve, 1-7/8" OD stl. pipe, 11 ga., galv, 7 bike cap		9	1.778		330	81		411	495
0065 Powder coat finish		9	1.778		289	81		370	450
0070 S curve, 1-7/8" OD stl. pipe, 11 ga., galv, 9 bike cap		8	2		425	91		516	615
0075 Powder coat finish		8	2		340	91		431	520
0080 S curve, 1-7/8" OD stl. pipe, 11 ga., galv, 11 bike cap		6	2.667		540	122		662	790
0085 Powder coat finish		6	2.667		540	122		662	790

12 93 23 – Trash and Litter Receptacles

12 93 23.10 Trash Receptacles		Crew	Daily Output	Labor-Hours	Unit	Material	2022 Bare Costs Labor	Equipment	Total	Total Incl O&P
0010	**TRASH RECEPTACLES**									
0500	Recycled plastic, var colors, round, 32 gal., 28" x 38" high	G	2 Clab	5	3.200	Ea.	880	146		1,026	1,200
0510	32 gal., 31" x 32" high	G		5	3.200		945	146		1,091	1,275
9110	Plastic, with dome lid, 32 gal. capacity			35	.457		70.50	21		91.50	111
9120	Recycled plastic slats, plastic dome lid, 32 gal. capacity		↓	35	.457	↓	350	21		371	420

397

For customer support on your Commercial Renovation Costs with RSMeans data, call 800.448.8182.

Division Notes

	CREW	DAILY OUTPUT	LABOR-HOURS	UNIT	BARE COSTS				TOTAL INCL O&P
					MAT.	LABOR	EQUIP.	TOTAL	

Estimating Tips

General

- The items and systems in this division are usually estimated, purchased, supplied, and installed as a unit by one or more subcontractors. The estimator must ensure that all parties are operating from the same set of specifications and assumptions, and that all necessary items are estimated and will be provided. Many times the complex items and systems are covered, but the more common ones, such as excavation or a crane, are overlooked for the very reason that everyone assumes nobody could miss them. The estimator should be the central focus and be able to ensure that all systems are complete.

- It is important to consider factors such as site conditions, weather, shape and size of building, as well as labor availability as they may impact the overall cost of erecting special structures and systems included in this division.

- Another area where problems can develop in this division is at the interface between systems.

The estimator must ensure, for instance, that anchor bolts, nuts, and washers are estimated and included for the air-supported structures and pre-engineered buildings to be bolted to their foundations. Utility supply is a common area where essential items or pieces of equipment can be missed or overlooked because each subcontractor may feel it is another's responsibility. The estimator should also be aware of certain items which may be supplied as part of a package but installed by others, and ensure that the installing contractor's estimate includes the cost of installation. Conversely, the estimator must also ensure that items are not costed by two different subcontractors, resulting in an inflated overall estimate.

13 30 00 Special Structures

- The foundations and floor slab, as well as rough mechanical and electrical, should be estimated, as this work is required for the assembly and erection of the structure. Generally, as noted in the data set, the pre-engineered building comes as a shell. Pricing is based on the size and structural design parameters stated in the reference section. Additional features, such as windows and doors with their related structural framing, must also be included by the estimator. Here again, the estimator must have a clear understanding of the scope of each portion of the work and all the necessary interfaces.

Reference Numbers

Reference numbers are shown at the beginning of some major classifications. These numbers refer to related items in the Reference Section. The reference information may be an estimating procedure, an alternate pricing method, or technical information.

Note: Not all subdivisions listed here necessarily appear. ■

---

13 05 05 – Selective Demolition for Special Construction

13 05 05.10 Selective Demolition, Air Supported Structures		Crew	Daily Output	Labor-Hours	Unit	Material	2022 Bare Costs Labor	Equipment	Total	Total Incl O&P
0010	**SELECTIVE DEMOLITION, AIR SUPPORTED STRUCTURES**									
0020	Tank covers, scrim, dbl. layer, vinyl poly w/hdwe., blower & controls									
0050	Round and rectangular R024119-10	B-2	9000	.004	S.F.		.20		.20	.32
0100	Warehouse structures									
0120	Poly/vinyl fabric, 28 oz., incl. tension cables & inflation system	4 Clab	9000	.004	SF Flr.		.16		.16	.26
0150	Reinforced vinyl, 12 oz., 3,000 S.F.	"	5000	.006			.29		.29	.46
0200	12,000 to 24,000 S.F.	8 Clab	20000	.003			.15		.15	.23
0250	Tedlar vinyl fabric, 28 oz. w/liner, to 3,000 S.F.	4 Clab	5000	.006			.29		.29	.46
0300	12,000 to 24,000 S.F.	8 Clab	20000	.003	▼		.15		.15	.23
0350	Greenhouse/shelter, woven polyethylene with liner									
0400	3,000 S.F.	4 Clab	5000	.006	SF Flr.		.29		.29	.46
0450	12,000 to 24,000 S.F.	8 Clab	20000	.003			.15		.15	.23
0500	Tennis/gymnasium, poly/vinyl fabric, 28 oz., incl. thermal liner	4 Clab	9000	.004			.16		.16	.26
0600	Stadium/convention center, teflon coated fiberglass, incl. thermal liner	9 Clab	40000	.002	▼		.08		.08	.13
0700	Doors, air lock, 15' long, 10' x 10'	2 Carp	1.50	10.667	Ea.		600		600	955
0720	15' x 15'		.80	20			1,125		1,125	1,800
0750	Revolving personnel door, 6' diam. x 6'-6" high	▼	1.50	10.667	▼		600		600	955

13 05 05.20 Selective Demolition, Garden Houses

		Crew	Daily Output	Labor-Hours	Unit	Material	Labor	Equipment	Total	Total Incl O&P
0010	**SELECTIVE DEMOLITION, GARDEN HOUSES**									
0020	Prefab, wood, excl. foundation, average	2 Clab	400	.040	SF Flr.		1.82		1.82	2.90

13 05 05.25 Selective Demolition, Geodesic Domes

		Crew	Daily Output	Labor-Hours	Unit	Material	Labor	Equipment	Total	Total Incl O&P
0010	**SELECTIVE DEMOLITION, GEODESIC DOMES**									
0050	Shell only, interlocking plywood panels, 30' diameter	F-5	3.20	10	Ea.		570		570	900
0060	34' diameter		2.30	13.913			790		790	1,250
0070	39' diameter	▼	2	16			910		910	1,450
0080	45' diameter	F-3	2.20	18.182			1,050	900	1,950	2,650
0090	55' diameter		2	20			1,150	990	2,140	2,925
0100	60' diameter		2	20			1,150	990	2,140	2,925
0110	65' diameter	▼	1.60	25	▼		1,450	1,250	2,700	3,650

13 05 05.30 Selective Demolition, Greenhouses

		Crew	Daily Output	Labor-Hours	Unit	Material	Labor	Equipment	Total	Total Incl O&P
0010	**SELECTIVE DEMOLITION, GREENHOUSES**									
0020	Resi-type, free standing, excl. foundations, 9' long x 8' wide	2 Clab	160	.100	SF Flr.		4.56		4.56	7.25
0030	9' long x 11' wide		170	.094			4.29		4.29	6.80
0040	9' long x 14' wide		220	.073			3.32		3.32	5.25
0050	9' long x 17' wide		320	.050			2.28		2.28	3.62
0060	Lean-to type, 4' wide		64	.250			11.40		11.40	18.10
0070	7' wide		120	.133	▼		6.10		6.10	9.65
0080	Geodesic hemisphere, 1/8" plexiglass glazing, 8' diam.		4	4	Ea.		182		182	290
0090	24' diam.		.80	20			910		910	1,450
0100	48' diam.	▼	.40	40			1,825		1,825	2,900

13 05 05.35 Selective Demolition, Hangars

		Crew	Daily Output	Labor-Hours	Unit	Material	Labor	Equipment	Total	Total Incl O&P
0010	**SELECTIVE DEMOLITION, HANGARS**									
0020	T type hangars, prefab, steel, galv roof & walls, incl doors, excl fndtn	E-2	2550	.022	SF Flr.		1.34	.97	2.31	3.22
0030	Circular type, prefab, steel frame, plastic skin, incl foundation, 80' diam	E-2	.50	112	Total		6,850	4,950	11,800	16,400

13 05 05.45 Selective Demolition, Lightning Protection

		Crew	Daily Output	Labor-Hours	Unit	Material	Labor	Equipment	Total	Total Incl O&P
0010	**SELECTIVE DEMOLITION, LIGHTNING PROTECTION**									
0020	Air terminal & base, copper, 3/8" diam. x 10", to 75' H	1 Clab	16	.500	Ea.		23		23	36
0030	1/2" diam. x 12", over 75' H		16	.500			23		23	36
0050	Aluminum, 1/2" diam. x 12", to 75' H		16	.500			23		23	36
0060	5/8" diam. x 12", over 75' H		16	.500	▼		23		23	36
0070	Cable, copper, 220 lb. per thousand feet, to 75' H		640	.013	L.F.		.57		.57	.91

13 05 Common Work Results for Special Construction

13 05 05 – Selective Demolition for Special Construction

13 05 05.45 Selective Demolition, Lightning Protection		Crew	Daily Output	Labor-Hours	Unit	Material	2022 Bare Costs Labor	2022 Bare Costs Equipment	Total	Total Incl O&P
0080	375 lb. per thousand feet, over 75' H	1 Clab	460	.017	L.F.		.79		.79	1.26
0090	Aluminum, 101 lb. per thousand feet, to 75' H		560	.014			.65		.65	1.04
0100	199 lb. per thousand feet, over 75' H		480	.017			.76		.76	1.21
0110	Arrester, 175 V AC, to ground		16	.500	Ea.		23		23	36
0120	650 V AC, to ground		13	.615	"		28		28	44.50

13 05 05.50 Selective Demolition, Pre-Engineered Steel Buildings		Crew	Daily Output	Labor-Hours	Unit	Material	2022 Bare Costs Labor	2022 Bare Costs Equipment	Total	Total Incl O&P
0010	**SELECTIVE DEMOLITION, PRE-ENGINEERED STEEL BUILDINGS**									
0500	Pre-engd. steel bldgs., rigid frame, clear span & multi post, excl. salvage									
0550	3,500 to 7,500 S.F.	L-10	1000	.024	SF Flr.		1.51	1.98	3.49	4.59
0600	7,501 to 12,500 S.F.		1500	.016			1.01	1.32	2.33	3.06
0650	12,501 S.F. or greater		1650	.015			.92	1.20	2.12	2.78
0700	Pre-engd. steel building components									
0710	Entrance canopy, including frame 4' x 4'	E-24	8	4	Ea.		246	262	508	685
0720	4' x 8'	"	7	4.571			281	299	580	780
0730	HM doors, self framing, single leaf	2 Skwk	8	2			118		118	186
0740	Double leaf		5	3.200			188		188	298
0760	Gutter, eave type		600	.027	L.F.		1.57		1.57	2.49
0770	Sash, single slide, double slide or fixed		24	.667	Ea.		39.50		39.50	62
0780	Skylight, fiberglass, to 30 S.F.		16	1			59		59	93
0785	Roof vents, circular, 12" to 24" diameter		12	1.333			78.50		78.50	124
0790	Continuous, 10' long		8	2			118		118	186
0900	Shelters, aluminum frame									
0910	Acrylic glazing, 3' x 9' x 8' high	2 Skwk	2	8	Ea.		470		470	745
0920	9' x 12' x 8' high	"	1.50	10.667	"		630		630	995

13 05 05.60 Selective Demolition, Silos		Crew	Daily Output	Labor-Hours	Unit	Material	2022 Bare Costs Labor	2022 Bare Costs Equipment	Total	Total Incl O&P
0010	**SELECTIVE DEMOLITION, SILOS**									
0020	Conc stave, indstrl, conical/sloping bott, excl fndtn, 12' diam., 35' H	E-24	.18	178	Ea.		10,900	11,600	22,500	30,300
0030	16' diam., 45' H		.12	267			16,400	17,400	33,800	45,400
0040	25' diam., 75' H		.08	400			24,600	26,200	50,800	68,000
0050	Steel, factory fabricated, 30,000 gal. cap, painted or epoxy lined	L-5	2	28			1,750	1,050	2,800	3,950

13 05 05.65 Selective Demolition, Sound Control		Crew	Daily Output	Labor-Hours	Unit	Material	2022 Bare Costs Labor	2022 Bare Costs Equipment	Total	Total Incl O&P
0010	**SELECTIVE DEMOLITION, SOUND CONTROL**									
0120	Acoustical enclosure, 4" thick walls & ceiling panels, 8 lb./S.F.	3 Carp	144	.167	SF Surf		9.40		9.40	14.90
0130	10.5 lb./S.F.		128	.188			10.55		10.55	16.75
0140	Reverb chamber, parallel walls, 4" thick		120	.200			11.25		11.25	17.90
0150	Skewed walls, parallel roof, 4" thick		110	.218			12.30		12.30	19.50
0160	Skewed walls/roof, 4" layer/air space		96	.250			14.10		14.10	22.50
0170	Sound-absorbing panels, painted metal, 2'-6" x 8', under 1,000 S.F.		430	.056			3.14		3.14	4.99
0180	Over 1,000 S.F.		480	.050			2.82		2.82	4.47
0190	Flexible transparent curtain, clear	3 Shee	430	.056			3.77		3.77	5.85
0192	50% clear, 50% foam		430	.056			3.77		3.77	5.85
0194	25% clear, 75% foam		430	.056			3.77		3.77	5.85
0196	100% foam		430	.056			3.77		3.77	5.85
0200	Audio-masking sys., incl. speakers, amplfr., signal gnrtr.									
0205	Ceiling mounted, 5,000 S.F.	2 Elec	4800	.003	S.F.		.22		.22	.34
0210	10,000 S.F.		5600	.003			.19		.19	.29
0220	Plenum mounted, 5,000 S.F.		7600	.002			.14		.14	.21
0230	10,000 S.F.		8800	.002			.12		.12	.18

13 05 05.70 Selective Demolition, Special Purpose Rooms		Crew	Daily Output	Labor-Hours	Unit	Material	2022 Bare Costs Labor	2022 Bare Costs Equipment	Total	Total Incl O&P
0010	**SELECTIVE DEMOLITION, SPECIAL PURPOSE ROOMS**									
0100	Audiometric rooms, under 500 S.F. surface	4 Carp	200	.160	SF Surf		9		9	14.30
0110	Over 500 S.F. surface	"	240	.133	"		7.50		7.50	11.95

13 05 05.70 Selective Demolition, Special Purpose Rooms	Crew	Daily Output	Labor-Hours	Unit	Material	2022 Bare Costs Labor	Equipment	Total	Total Incl O&P
0200 Clean rooms, 12' x 12' soft wall, class 100	1 Carp	.30	26.667	Ea.		1,500		1,500	2,375
0210 Class 1,000		.30	26.667			1,500		1,500	2,375
0220 Class 10,000		.35	22.857			1,275		1,275	2,050
0230 Class 100,000		.35	22.857			1,275		1,275	2,050
0300 Darkrooms, shell complete, 8' high	2 Carp	220	.073	SF Flr.		4.09		4.09	6.50
0310 12' high		110	.145	"		8.20		8.20	13
0350 Darkrooms doors, mini-cylindrical, revolving		4	4	Ea.		225		225	360
0400 Music room, practice modular		140	.114	SF Surf		6.45		6.45	10.20
0500 Refrigeration structures and finishes									
0510 Wall finish, 2 coat Portland cement plaster, 1/2" thick	1 Clab	200	.040	S.F.		1.82		1.82	2.90
0522 Fiberglass panels, 1/8" thick		400	.020			.91		.91	1.45
0530 Ceiling finish, polystyrene plastic, 1" to 2" thick		500	.016			.73		.73	1.16
0540 4" thick		450	.018			.81		.81	1.29
0550 Refrigerator, prefab aluminum walk-in, 7'-6" high, 6' x 6' OD	2 Carp	100	.160	SF Flr.		9		9	14.30
0560 10' x 10' OD		160	.100			5.65		5.65	8.95
0570 Over 150 S.F.		200	.080			4.50		4.50	7.15
0600 Sauna, prefabricated, including heater & controls, 7' high, to 30 S.F.	2 Carp	120	.133	SF Flr.		7.50		7.50	11.95
0610 To 40 S.F.		140	.114			6.45		6.45	10.20
0620 To 60 S.F.		175	.091			5.15		5.15	8.20
0630 To 100 S.F.		220	.073			4.09		4.09	6.50
0640 To 130 S.F.		250	.064			3.60		3.60	5.70
0650 Steam bath, heater, timer, head, single, to 140 C.F.	1 Plum	2.20	3.636	Ea.		254		254	390
0660 To 300 C.F.		2.20	3.636			254		254	390
0670 Steam bath, comm. size, w/blow-down assembly, to 800 C.F.		1.80	4.444			310		310	480
0680 To 2,500 C.F.		1.60	5			350		350	540
0690 Steam bath, comm. size, multiple, for motels, apts, 500 C.F., 2 baths		2	4			279		279	430
0700 1,000 C.F., 4 baths		1.40	5.714			400		400	615

13 05 05.75 Selective Demolition, Storage Tanks	Crew	Daily Output	Labor-Hours	Unit	Material	2022 Bare Costs Labor	Equipment	Total	Total Incl O&P
0010 **SELECTIVE DEMOLITION, STORAGE TANKS**									
0500 Steel tank, single wall, above ground, not incl. fdn., pumps or piping									
0510 Single wall, 275 gallon	Q-1	3	5.333	Ea.		335		335	515
0520 550 thru 2,000 gallon	B-34P	2	12			725	355	1,080	1,525
0530 5,000 thru 10,000 gallon	B-34Q	2	12			735	1,325	2,060	2,600
0540 15,000 thru 30,000 gallon	B-34S	2	16			1,025	2,050	3,075	3,875
0600 Steel tank, double wall, above ground not incl. fdn., pumps & piping									
0620 500 thru 2,000 gallon	B-34P	2	12	Ea.		725	355	1,080	1,525

13 05 05.85 Selective Demolition, Swimming Pool Equip	Crew	Daily Output	Labor-Hours	Unit	Material	2022 Bare Costs Labor	Equipment	Total	Total Incl O&P
0010 **SELECTIVE DEMOLITION, SWIMMING POOL EQUIP**									
0020 Diving stand, stainless steel, 3 meter	2 Clab	3	5.333	Ea.		243		243	385
0030 1 meter		5	3.200			146		146	232
0040 Diving board, 16' long, aluminum		5.40	2.963			135		135	215
0050 Fiberglass		5.40	2.963			135		135	215
0070 Ladders, heavy duty, stainless steel, 2 tread		14	1.143			52		52	83
0080 4 tread		12	1.333			61		61	96.50
0090 Lifeguard chair, stainless steel, fixed		5	3.200			146		146	232
0100 Slide, tubular, fiberglass, aluminum handrails & ladder, 5', straight		4	4			182		182	290
0110 8', curved		6	2.667			122		122	193
0120 10', curved		3	5.333			243		243	385
0130 12' straight, with platform		2.50	6.400			292		292	465
0140 Removable access ramp, stainless steel		4	4			182		182	290
0150 Removable stairs, stainless steel, collapsible		4	4			182		182	290

13 05 Common Work Results for Special Construction

13 05 05 – Selective Demolition for Special Construction

13 05 05.90 Selective Demolition, Tension Structures

		Crew	Daily Output	Labor-Hours	Unit	Material	2022 Bare Costs Labor	2022 Bare Costs Equipment	Total	Total Incl O&P
0010	**SELECTIVE DEMOLITION, TENSION STRUCTURES**									
0020	Steel/alum. frame, fabric shell, 60' clear span, 6,000 S.F.	B-41	2000	.022	SF Flr.		1.04	.16	1.20	1.81
0030	12,000 S.F.		2200	.020			.94	.14	1.08	1.66
0040	80' clear span, 20,800 S.F.	↓	2440	.018			.85	.13	.98	1.49
0050	100' clear span, 10,000 S.F.	L-5	4350	.013			.80	.48	1.28	1.82
0060	26,000 S.F.		4600	.012			.76	.45	1.21	1.72
0070	36,000 S.F.	↓	5000	.011	↓		.70	.42	1.12	1.58

13 05 05.95 Selective Demo, X-Ray/Radio Freq Protection

		Crew	Daily Output	Labor-Hours	Unit	Material	2022 Bare Costs Labor	2022 Bare Costs Equipment	Total	Total Incl O&P
0010	**SELECTIVE DEMO, X-RAY/RADIO FREQ PROTECTION**									
0020	Shielding lead, lined door frame, excl. hdwe., 1/16" thick	1 Clab	4.80	1.667	Ea.		76		76	121
0030	Lead sheets, 1/16" thick	2 Clab	270	.059	S.F.		2.70		2.70	4.29
0040	1/8" thick		240	.067			3.04		3.04	4.83
0050	Lead shielding, 1/4" thick		270	.059			2.70		2.70	4.29
0060	1/2" thick	↓	240	.067	↓		3.04		3.04	4.83
0070	Lead glass, 1/4" thick, 2.0 mm LE, 12" x 16"	2 Glaz	16	1	Ea.		54		54	85.50
0080	24" x 36"		8	2			108		108	171
0090	36" x 60"		4	4			216		216	345
0100	Lead glass window frame, with 1/16" lead & voice passage, 36" x 60"		4	4			216		216	345
0110	Lead glass window frame, 24" x 36"	↓	8	2	↓		108		108	171
0120	Lead gypsum board, 5/8" thick with 1/16" lead	2 Clab	320	.050	S.F.		2.28		2.28	3.62
0130	1/8" lead		280	.057			2.61		2.61	4.14
0140	1/32" lead		400	.040	↓		1.82		1.82	2.90
0150	Butt joints, 1/8" lead or thicker, 2" x 7' long batten strip		480	.033	Ea.		1.52		1.52	2.42
0160	X-ray protection, average radiography room, up to 300 S.F., 1/16" lead, min		.50	32	Total		1,450		1,450	2,325
0170	Maximum		.30	53.333			2,425		2,425	3,875
0180	Deep therapy X-ray room, 250 kV cap, up to 300 S.F., 1/4" lead, min		.20	80			3,650		3,650	5,800
0190	Maximum		.12	133	↓		6,075		6,075	9,650
0880	Radio frequency shielding, prefab or screen-type copper or steel, minimum		360	.044	SF Surf		2.03		2.03	3.22
0890	Average		310	.052			2.35		2.35	3.74
0895	Maximum	↓	290	.055	↓		2.52		2.52	4

13 12 Fountains

13 12 13 – Exterior Fountains

13 12 13.10 Outdoor Fountains

		Crew	Daily Output	Labor-Hours	Unit	Material	2022 Bare Costs Labor	2022 Bare Costs Equipment	Total	Total Incl O&P
0010	**OUTDOOR FOUNTAINS**									
0100	Outdoor fountain, 48" high with bowl and figures	2 Clab	2	8	Ea.	415	365		780	1,025
0200	Commercial, concrete or cast stone, 40-60" H, simple		2	8		1,050	365		1,415	1,725
0220	Average		2	8		2,025	365		2,390	2,800
0240	Ornate		2	8		4,600	365		4,965	5,625
0260	Metal, 72" high		2	8		1,625	365		1,990	2,375
0280	90" high		2	8		2,350	365		2,715	3,175
0300	120" high		2	8		5,625	365		5,990	6,775
0320	Resin or fiberglass, 40-60" H, wall type		2	8		735	365		1,100	1,400
0340	Waterfall type		2	8	↓	1,250	365		1,615	1,950

13 12 23 – Interior Fountains

13 12 23.10 Indoor Fountains

		Crew	Daily Output	Labor-Hours	Unit	Material	2022 Bare Costs Labor	2022 Bare Costs Equipment	Total	Total Incl O&P
0010	**INDOOR FOUNTAINS**									
0100	Commercial, floor type, resin or fiberglass, lighted, cascade type	2 Clab	2	8	Ea.	460	365		825	1,100
0120	Tiered type		2	8		730	365		1,095	1,375
0140	Waterfall type		2	8	↓	515	365		880	1,150

13 17 Tubs and Pools

13 17 13 – Hot Tubs

13 17 13.10 Redwood Hot Tub System

		Crew	Daily Output	Labor-Hours	Unit	Material	2022 Bare Costs Labor	2022 Bare Costs Equipment	Total	Total Incl O&P
0010	**REDWOOD HOT TUB SYSTEM**									
7050	4' diameter x 4' deep	Q-1	1	16	Ea.	4,725	1,000		5,725	6,750
7150	6' diameter x 4' deep		.80	20		6,400	1,250		7,650	8,975
7200	8' diameter x 4' deep		.80	20		9,150	1,250		10,400	12,000

13 17 33 – Whirlpool Tubs

13 17 33.10 Whirlpool Bath

		Crew	Daily Output	Labor-Hours	Unit	Material	2022 Bare Costs Labor	2022 Bare Costs Equipment	Total	Total Incl O&P
0010	**WHIRLPOOL BATH**									
6000	Whirlpool, bath with vented overflow, molded fiberglass									
6100	66" x 36" x 24"	Q-1	1	16	Ea.	1,475	1,000		2,475	3,175
6400	72" x 36" x 21"		1	16		1,325	1,000		2,325	3,025
6500	60" x 34" x 21"		1	16		1,250	1,000		2,250	2,925
6600	72" x 42" x 23"		1	16		1,150	1,000		2,150	2,825
6710	For color, add					10%				
6711	For designer colors and trim, add					25%				

13 21 Controlled Environment Rooms

13 21 26 – Cold Storage Rooms

13 21 26.50 Refrigeration

		Crew	Daily Output	Labor-Hours	Unit	Material	2022 Bare Costs Labor	2022 Bare Costs Equipment	Total	Total Incl O&P
0010	**REFRIGERATION**									
0020	Curbs, 12" high, 4" thick, concrete	2 Carp	58	.276	L.F.	3.31	15.55		18.86	28
6300	Rule of thumb for complete units, w/o doors & refrigeration, cooler		146	.110	SF Flr.	176	6.15		182.15	203
6400	Freezer		109.60	.146	"	113	8.20		121.20	137

13 24 Special Activity Rooms

13 24 16 – Saunas

13 24 16.50 Saunas and Heaters

		Crew	Daily Output	Labor-Hours	Unit	Material	2022 Bare Costs Labor	2022 Bare Costs Equipment	Total	Total Incl O&P
0010	**SAUNAS AND HEATERS**									
0020	Prefabricated, incl. heater & controls, 7' high, 6' x 4', C/C	L-7	2.20	12.727	Ea.	6,875	695		7,570	8,650
1700	Door only, cedar, 2' x 6', w/ 1' x 4' tempered insulated glass window	2 Carp	3.40	4.706		635	265		900	1,125
1800	Prehung, incl. jambs, pulls & hardware	"	12	1.333		1,000	75		1,075	1,225
2500	Heaters only (incl. above), wall mounted, to 200 C.F.					1,100			1,100	1,200
4480	For additional equipment, see Section 11 66 13.10									

13 24 26 – Steam Baths

13 24 26.50 Steam Baths and Components

		Crew	Daily Output	Labor-Hours	Unit	Material	2022 Bare Costs Labor	2022 Bare Costs Equipment	Total	Total Incl O&P
0010	**STEAM BATHS AND COMPONENTS**									
0020	Heater, timer & head, single, to 140 C.F.	1 Plum	1.20	6.667	Ea.	2,300	465		2,765	3,250
0500	To 300 C.F.	"	1.10	7.273		2,600	510		3,110	3,625
2000	Multiple, motels, apts., 2 baths, w/blow-down assm., 500 C.F.	Q-1	1.30	12.308		7,225	775		8,000	9,150
2500	4 baths	"	.70	22.857		10,900	1,425		12,325	14,200

13 28 33 – Athletic and Recreational Court Walls

13 28 33.50 Sport Court	Crew	Daily Output	Labor-Hours	Unit	Material	2022 Bare Costs Labor	Equipment	Total	Total Incl O&P
0010 **SPORT COURT**									
0020 Sport court floors, no. 2 & better maple, 25/32" thick, see 09 64 66.10				SF Flr. Court					
0300 Squash, regulation court in existing building, minimum				Court	45,200			45,200	49,700
0400 Maximum				"	50,500			50,500	55,500
0450 Rule of thumb for components:									
0470 Walls	3 Carp	.15	160	Court	13,800	9,000		22,800	29,500
0500 Floor	"	.25	96		10,900	5,400		16,300	20,600
0550 Lighting	2 Elec	.60	26.667		2,575	1,775		4,350	5,550

13 34 Fabricated Engineered Structures

13 34 13 – Glazed Structures

13 34 13.13 Greenhouses

	Crew	Daily Output	Labor-Hours	Unit	Material	2022 Bare Costs Labor	Equipment	Total	Total Incl O&P
0010 **GREENHOUSES**, Shell only, stock units, not incl. 2' stub walls,									
0020 foundation, floors, heat or compartments									
0300 Residential type, free standing, 8'-6" long x 7'-6" wide	2 Carp	59	.271	SF Flr.	25.50	15.25		40.75	53
0400 10'-6" wide		85	.188		47	10.60		57.60	68.50
0600 13'-6" wide		108	.148		49.50	8.35		57.85	68
0700 17'-0" wide		160	.100		46	5.65		51.65	59.50
0900 Lean-to type, 3'-10" wide	2 Carp	34	.471	SF Flr.	46	26.50		72.50	92.50
1000 6'-10" wide		58	.276		28	15.55		43.55	55.50
1050 8'-0" wide		60	.267		81	15		96	113
1100 Wall mounted to existing window, 3' x 3'	1 Carp	4	2	Ea.	2,850	113		2,963	3,300
1120 4' x 5'	"	3	2.667		2,625	150		2,775	3,125
3900 Cooling, 1,200 CFM exhaust fan, add					350			350	385
4000 7,850 CFM					920			920	1,000
4200 For heaters, 10 MBH, add					265			265	291
4300 60 MBH, add					790			790	870
4500 For benches, 2' x 8', add					204			204	224
4600 4' x 10', add					535			535	590
4800 For ventilation & humidity control w/4 integrated outlets, add				Total	310			310	340
4900 For environmental controls and automation, 8 outputs, 9 stages, add				"	1,425			1,425	1,575
5100 For humidification equipment, add				Ea.	350			350	385
5200 For vinyl shading, add				S.F.	.33			.33	.36

13 34 13.19 Swimming Pool Enclosures

	Crew	Daily Output	Labor-Hours	Unit	Material	2022 Bare Costs Labor	Equipment	Total	Total Incl O&P
0010 **SWIMMING POOL ENCLOSURES** Translucent, free standing									
0020 not including foundations, heat or light									
0200 Economy	2 Carp	200	.080	SF Hor.	48	4.50		52.50	60
0600 Deluxe	"	70	.229	"	131	12.85		143.85	166

13 34 23 – Fabricated Structures

13 34 23.16 Fabricated Control Booths

	Crew	Daily Output	Labor-Hours	Unit	Material	2022 Bare Costs Labor	Equipment	Total	Total Incl O&P
0010 **FABRICATED CONTROL BOOTHS**									
0100 Guard House, prefab conc. w/bullet resistant doors & windows, roof & wiring									
0110 8' x 8', Level III	L-10	1	24	Ea.	59,500	1,525	1,975	63,000	70,000
0120 8' x 8', Level IV	"	1	24	"	76,500	1,525	1,975	80,000	89,000

13 34 23.45 Kiosks

	Crew	Daily Output	Labor-Hours	Unit	Material	2022 Bare Costs Labor	Equipment	Total	Total Incl O&P
0010 **KIOSKS**									
0020 Round, advertising type, 5' diameter, 7' high, aluminum wall, illuminated				Ea.	22,500			22,500	24,800
0100 Aluminum wall, non-illuminated					21,800			21,800	23,900
0500 Rectangular, 5' x 9', 7'-6" high, aluminum wall, illuminated					24,800			24,800	27,300
0600 Aluminum wall, non-illuminated					25,500			25,500	28,000

For customer support on your Commercial Renovation Costs with RSMeans data, call 800.448.8182.

405

13 34 Fabricated Engineered Structures

13 34 63 – Natural Fiber Construction

13 34 63.50 Straw Bale Construction

		Crew	Daily Output	Labor-Hours	Unit	Material	2022 Bare Costs Labor	Equipment	Total	Total Incl O&P
0010	**STRAW BALE CONSTRUCTION**									
2020	Straw bales in walls w/modified post and beam frame	G 2 Carp	320	.050	S.F.	7.10	2.82		9.92	12.25

13 42 Building Modules

13 42 63 – Detention Cell Modules

13 42 63.16 Steel Detention Cell Modules

		Crew	Daily Output	Labor-Hours	Unit	Material	Labor	Equipment	Total	Total Incl O&P
0010	**STEEL DETENTION CELL MODULES**									
2000	Cells, prefab., 5' to 6' wide, 7' to 8' high, 7' to 8' deep,									
2010	bar front, cot, not incl. plumbing	E-4	1.50	21.333	Ea.	9,750	1,325	98.50	11,173.50	13,000

13 48 Sound, Vibration, and Seismic Control

13 48 13 – Manufactured Sound and Vibration Control Components

13 48 13.50 Audio Masking

		Crew	Daily Output	Labor-Hours	Unit	Material	Labor	Equipment	Total	Total Incl O&P
0010	**AUDIO MASKING**, acoustical enclosure, 4" thick wall and ceiling									
0020	8 lb./S.F., up to 12' span	3 Carp	72	.333	SF Surf	33.50	18.75		52.25	67
0300	Better quality panels, 10.5 #/S.F		64	.375		38	21		59	75.50
0400	Reverb-chamber, 4" thick, parallel walls		60	.400		47.50	22.50		70	88.50

13 49 Radiation Protection

13 49 13 – Integrated X-Ray Shielding Assemblies

13 49 13.50 Lead Sheets

		Crew	Daily Output	Labor-Hours	Unit	Material	Labor	Equipment	Total	Total Incl O&P
0010	**LEAD SHEETS**									
0300	Lead sheets, 1/16" thick	2 Lath	135	.119	S.F.	10.65	6.65		17.30	22
0400	1/8" thick		120	.133		24	7.45		31.45	37.50
0500	Lead shielding, 1/4" thick		135	.119		45	6.65		51.65	60
0550	1/2" thick		120	.133		83.50	7.45		90.95	104
1200	X-ray protection, average radiography or fluoroscopy									
1210	room, up to 300 S.F. floor, 1/16" lead, economy	2 Lath	.25	64	Total	14,900	3,575		18,475	22,000
1500	7'-0" walls, deluxe	"	.15	107	"	17,900	5,975		23,875	29,000
1600	Deep therapy X-ray room, 250 kV capacity,									
1800	up to 300 S.F. floor, 1/4" lead, economy	2 Lath	.08	200	Total	41,700	11,200		52,900	63,500
1900	7'-0" walls, deluxe		.06	267	"	51,500	14,900		66,400	80,000
1999	Minimum labor/equipment charge		4.50	3.556	Job		199		199	310

13 49 21 – Lead Glazing

13 49 21.50 Lead Glazing

		Crew	Daily Output	Labor-Hours	Unit	Material	Labor	Equipment	Total	Total Incl O&P
0010	**LEAD GLAZING**									
0600	Lead glass, 1/4" thick, 2.0 mm LE, 12" x 16"	2 Glaz	13	1.231	Ea.	515	66.50		581.50	675
0700	24" x 36"	"	8	2	"	1,875	108		1,983	2,250
2000	X-ray viewing panels, clear lead plastic									
2010	7 mm thick, 0.3 mm LE, 2.3 lb./S.F.	H-3	139	.115	S.F.	288	5.60		293.60	325
2020	12 mm thick, 0.5 mm LE, 3.9 lb./S.F.		82	.195		385	9.50		394.50	440
2030	18 mm thick, 0.8 mm LE, 5.9 lb./S.F.		54	.296		470	14.40		484.40	540
2040	22 mm thick, 1.0 mm LE, 7.2 lb./S.F.		44	.364		775	17.70		792.70	885
2050	35 mm thick, 1.5 mm LE, 11.5 lb./S.F.		28	.571		1,025	28		1,053	1,175
2060	46 mm thick, 2.0 mm LE, 15.0 lb./S.F.		21	.762		1,350	37		1,387	1,525

13 49 Radiation Protection

13 49 23 – Integrated RFI/EMI Shielding Assemblies

13 49 23.50 Modular Shielding Partitions	Crew	Daily Output	Labor-Hours	Unit	Material	2022 Bare Costs Labor	Equipment	Total	Total Incl O&P
0010 **MODULAR SHIELDING PARTITIONS**									
4000 X-ray barriers, modular, panels mounted within framework for									
4002 attaching to floor, wall or ceiling, upper portion is clear lead									
4005 plastic window panels 48"H, lower portion is opaque leaded									
4008 steel panels 36"H, structural supports not incl.									
4010 1-section barrier, 36"W x 84"H overall									
4020 0.5 mm LE panels	H-3	6.40	2.500	Ea.	10,400	122		10,522	11,700
4030 0.8 mm LE panels		6.40	2.500		11,200	122		11,322	12,600
4040 1.0 mm LE panels		5.33	3.002		15,600	146		15,746	17,300
4050 1.5 mm LE panels		5.33	3.002		18,300	146		18,446	20,400
4060 2-section barrier, 72"W x 84"H overall									
4070 0.5 mm LE panels	H-3	4	4	Ea.	15,800	195		15,995	17,700
4080 0.8 mm LE panels		4	4		16,300	195		16,495	18,200
4090 1.0 mm LE panels		3.56	4.494		17,700	219		17,919	19,900
5000 1.5 mm LE panels		3.20	5		31,700	243		31,943	35,200
5010 3-section barrier, 108"W x 84"H overall									
5020 0.5 mm LE panels	H-3	3.20	5	Ea.	23,700	243		23,943	26,400
5030 0.8 mm LE panels		3.20	5		24,400	243		24,643	27,200
5040 1.0 mm LE panels		2.67	5.993		26,500	292		26,792	29,700
5050 1.5 mm LE panels		2.46	6.504		48,000	315		48,315	53,500
7000 X-ray barriers, mobile, mounted within framework w/casters on									
7005 bottom, clear lead plastic window panels on upper portion,									
7010 opaque on lower, 30"W x 75"H overall, incl. framework									
7020 24"H upper w/0.5 mm LE, 48"H lower w/0.8 mm LE	1 Carp	16	.500	Ea.	5,325	28		5,353	5,925
7030 48"W x 75"H overall, incl. framework									
7040 36"H upper w/0.5 mm LE, 36"H lower w/0.8 mm LE	1 Carp	16	.500	Ea.	7,575	28		7,603	8,375
7050 36"H upper w/1.0 mm LE, 36"H lower w/1.5 mm LE	"	16	.500	"	8,425	28		8,453	9,325
7060 72"W x 75"H overall, incl. framework									
7070 36"H upper w/0.5 mm LE, 36"H lower w/0.8 mm LE	1 Carp	16	.500	Ea.	8,425	28		8,453	9,325
7080 36"H upper w/1.0 mm LE, 36"H lower w/1.5 mm LE	"	16	.500	"	10,600	28		10,628	11,600

Division Notes

	CREW	DAILY OUTPUT	LABOR-HOURS	UNIT	BARE COSTS				TOTAL INCL O&P
					MAT.	LABOR	EQUIP.	TOTAL	

Estimating Tips
General

- Many products in Division 14 will require some type of support or blocking for installation not included with the item itself. Examples are supports for conveyors or tube systems, attachment points for lifts, and footings for hoists or cranes. Add these supports in the appropriate division.

14 10 00 Dumbwaiters
14 20 00 Elevators

- Dumbwaiters and elevators are estimated and purchased in a method similar to buying a car. The manufacturer has a base unit with standard features. Added to this base unit price will be whatever options the owner or specifications require. Increased load capacity, additional vertical travel, additional stops, higher speed, and cab finish options are items to be considered. When developing an estimate for dumbwaiters and elevators, remember that some items needed by the installers may have to be included as part of the general contract.

Examples are:

- ☐ shaftway
- ☐ rail support brackets
- ☐ machine room
- ☐ electrical supply
- ☐ sill angles
- ☐ electrical connections
- ☐ pits
- ☐ roof penthouses
- ☐ pit ladders

Check the job specifications and drawings before pricing.

- Installation of elevators and handicapped lifts in historic structures can require significant additional costs. The associated structural requirements may involve cutting into and repairing finishes, moldings, flooring, etc. The estimator must account for these special conditions.

14 30 00 Escalators and Moving Walks

- Escalators and moving walks are specialty items installed by specialty contractors. There are numerous options associated with these items. For specific options, contact a manufacturer or contractor. In a method similar to estimating dumbwaiters and elevators, you should verify the extent of general contract work and add items as necessary.

14 40 00 Lifts
14 90 00 Other Conveying Equipment

- Products such as correspondence lifts, chutes, and pneumatic tube systems, as well as other items specified in this subdivision, may require trained installers. The general contractor might not have any choice as to who will perform the installation or when it will be performed. Long lead times are often required for these products, making early decisions in scheduling necessary.

Reference Numbers

Reference numbers are shown at the beginning of some major classifications. These numbers refer to related items in the Reference Section. The reference information may be an estimating procedure, an alternate pricing method, or technical information.

Note: Not all subdivisions listed here necessarily appear. ■

Same Data. Simplified.

Enjoy the convenience and efficiency of accessing your costs anywhere:

- **Skip the multiplier** by setting your location
- **Quickly search,** edit, favorite and share costs
- **Stay on top of price changes** with automatic updates

Discover more at rsmeans.com/online

14 11 Manual Dumbwaiters

14 11 10 – Hand Operated Dumbwaiters

14 11 10.20 Manual Dumbwaiters	Crew	Daily Output	Labor-Hours	Unit	Material	2022 Bare Costs Labor	Equipment	Total	Total Incl O&P
0010 **MANUAL DUMBWAITERS**									
0020 2 stop, hand powered, up to 75 lb. capacity	2 Elev	.75	21.333	Ea.	3,200	2,000		5,200	6,575
0100 76 lb. capacity and up		.50	32	"	6,275	3,000		9,275	11,500
0300 For each additional stop, add		.75	21.333	Stop	1,025	2,000		3,025	4,175

14 12 Electric Dumbwaiters

14 12 10 – Dumbwaiters

14 12 10.10 Electric Dumbwaiters

	Crew	Daily Output	Labor-Hours	Unit	Material	2022 Bare Costs Labor	Equipment	Total	Total Incl O&P
0010 **ELECTRIC DUMBWAITERS**									
0020 2 stop, up to 75 lb. capacity	2 Elev	.13	123	Ea.	4,225	11,500		15,725	22,300
0100 76 lb. capacity and up		.11	145	"	10,600	13,600		24,200	32,500
0600 For each additional stop, add		.54	29.630	Stop	555	2,775		3,330	4,850

14 21 Electric Traction Elevators

14 21 13 – Electric Traction Freight Elevators

14 21 13.10 Electric Traction Freight Elevators and Options

		Crew	Daily Output	Labor-Hours	Unit	Material	2022 Bare Costs Labor	Equipment	Total	Total Incl O&P
0010 **ELECTRIC TRACTION FREIGHT ELEVATORS AND OPTIONS**	R142000-10									
0425 Electric freight, base unit, 4000 lb., 200 fpm, 4 stop, std. fin.	R142000-20	2 Elev	.05	320	Ea.	130,000	29,900		159,900	189,000
0450 For 5000 lb. capacity, add	R142000-30					7,400			7,400	8,125
0500 For 6000 lb. capacity, add	R142000-40					17,300			17,300	19,100
0525 For 7000 lb. capacity, add						23,700			23,700	26,100
0550 For 8000 lb. capacity, add						27,700			27,700	30,500
0575 For 10000 lb. capacity, add						38,600			38,600	42,500
0600 For 12000 lb. capacity, add						46,700			46,700	51,500
0625 For 16000 lb. capacity, add						57,000			57,000	62,500
0650 For 20000 lb. capacity, add						66,000			66,000	72,500
0675 For increased speed, 250 fpm, add						17,200			17,200	18,900
0700 300 fpm, geared electric, add						21,500			21,500	23,700
0725 350 fpm, geared electric, add						26,300			26,300	28,900
0750 400 fpm, geared electric, add						30,800			30,800	33,900
0775 500 fpm, gearless electric, add						42,900			42,900	47,200
0800 600 fpm, gearless electric, add						47,600			47,600	52,500
0825 700 fpm, gearless electric, add						56,500			56,500	62,000
0850 800 fpm, gearless electric, add						62,500			62,500	69,000
0875 For class "B" loading, add						3,500			3,500	3,850
0900 For class "C-1" loading, add						8,600			8,600	9,450
0925 For class "C-2" loading, add						10,100			10,100	11,100
0950 For class "C-3" loading, add						14,100			14,100	15,600
0975 For travel over 40 V.L.F., add		2 Elev	7.25	2.207	V.L.F.	720	206		926	1,100
1000 For number of stops over 4, add		"	.27	59.259	Stop	5,450	5,550		11,000	14,500

14 21 23 – Electric Traction Passenger Elevators

14 21 23.10 Electric Traction Passenger Elevators and Options

		Crew	Daily Output	Labor-Hours	Unit	Material	2022 Bare Costs Labor	Equipment	Total	Total Incl O&P
0010 **ELECTRIC TRACTION PASSENGER ELEVATORS AND OPTIONS**	R142000-10									
1625 Electric pass., base unit, 2,000 lb., 200 fpm, 4 stop, std. fin.	R142000-20	2 Elev	.05	320	Ea.	109,500	29,900		139,400	166,500
1650 For 2,500 lb. capacity, add	R142000-30				"	4,675			4,675	5,150
1675 For 3,000 lb. capacity, add	R142000-40				Ea.	5,125			5,125	5,650
1700 For 3,500 lb. capacity, add						6,525			6,525	7,175
1725 For 4,000 lb. capacity, add						7,650			7,650	8,425
1750 For 4,500 lb. capacity, add						10,600			10,600	11,600

14 21 Electric Traction Elevators

14 21 23 – Electric Traction Passenger Elevators

14 21 23.10 Electric Traction Passenger Elevators and Options	Crew	Daily Output	Labor-Hours	Unit	Material	2022 Bare Costs Labor	Equipment	Total	Total Incl O&P
1775	For 5000 lb. capacity, add				Ea.	13,600			13,600	14,900
1800	For increased speed, 250 fpm, geared electric, add					4,125			4,125	4,550
1825	300 fpm, geared electric, add					7,975			7,975	8,750
1850	350 fpm, geared electric, add					10,200			10,200	11,300
1875	400 fpm, geared electric, add					14,900			14,900	16,400
1900	500 fpm, gearless electric, add					36,000			36,000	39,600
1925	600 fpm, gearless electric, add					58,500			58,500	64,000
1950	700 fpm, gearless electric, add					65,500			65,500	72,000
1975	800 fpm, gearless electric, add					73,000			73,000	80,000
2000	For travel over 40 V.L.F., add	2 Elev	7.25	2.207	V.L.F.	775	206		981	1,175
2025	For number of stops over 4, add		.27	59.259	Stop	3,475	5,550		9,025	12,300
2400	Electric hospital, base unit, 4,000 lb., 200 fpm, 4 stop, std fin.		.05	320	Ea.	101,500	29,900		131,400	157,500
2425	For 4,500 lb. capacity, add					6,500			6,500	7,150
2450	For 5,000 lb. capacity, add					8,500			8,500	9,350
2475	For increased speed, 250 fpm, geared electric, add					4,600			4,600	5,050
2500	300 fpm, geared electric, add					8,800			8,800	9,700
2525	350 fpm, geared electric, add					10,600			10,600	11,700
2550	400 fpm, geared electric, add					15,400			15,400	16,900
2575	500 fpm, gearless electric, add					42,300			42,300	46,500
2600	600 fpm, gearless electric, add					63,500			63,500	70,000
2625	700 fpm, gearless electric, add					70,500			70,500	77,500
2650	800 fpm, gearless electric, add					79,500			79,500	87,000
2675	For travel over 40 V.L.F., add	2 Elev	7.25	2.207	V.L.F.	205	206		411	540
2700	For number of stops over 4, add	"	.27	59.259	Stop	5,175	5,550		10,725	14,200

14 21 33 – Electric Traction Residential Elevators

14 21 33.20 Residential Elevators

		Crew	Daily Output	Labor-Hours	Unit	Material	Labor	Equipment	Total	Total Incl O&P
0010	**RESIDENTIAL ELEVATORS**									
7000	Residential, cab type, 1 floor, 2 stop, economy model	2 Elev	.20	80	Ea.	12,800	7,475		20,275	25,600
7100	Custom model		.10	160		17,000	15,000		32,000	41,600
7200	2 floor, 3 stop, economy model		.12	133		13,700	12,500		26,200	34,200
7300	Custom model		.06	267		23,500	24,900		48,400	64,000

14 24 Hydraulic Elevators

14 24 13 – Hydraulic Freight Elevators

14 24 13.10 Hydraulic Freight Elevators and Options

			Crew	Daily Output	Labor-Hours	Unit	Material	Labor	Equipment	Total	Total Incl O&P
0010	**HYDRAULIC FREIGHT ELEVATORS AND OPTIONS**										
1025	Hydraulic freight, base unit, 2,000 lb., 50 fpm, 2 stop, std. fin.	R142000-10 R142000-20	2 Elev	.10	160	Ea.	96,000	15,000		111,000	129,000
1050	For 2,500 lb. capacity, add	R142000-30					4,550			4,550	5,000
1075	For 3,000 lb. capacity, add	R142000-40					6,000			6,000	6,600
1100	For 3,500 lb. capacity, add						9,825			9,825	10,800
1125	For 4,000 lb. capacity, add						11,300			11,300	12,400
1150	For 4,500 lb. capacity, add						13,000			13,000	14,300
1175	For 5,000 lb. capacity, add						17,700			17,700	19,500
1200	For 6,000 lb. capacity, add						20,000			20,000	22,000
1225	For 7,000 lb. capacity, add						30,800			30,800	33,900
1250	For 8,000 lb. capacity, add						37,300			37,300	41,000
1275	For 10,000 lb. capacity, add						39,400			39,400	43,400
1300	For 12,000 lb. capacity, add						48,000			48,000	53,000
1325	For 16,000 lb. capacity, add						64,500			64,500	71,000
1350	For 20,000 lb. capacity, add						71,500			71,500	78,500

411

14 24 Hydraulic Elevators

14 24 13 – Hydraulic Freight Elevators

14 24 13.10 Hydraulic Freight Elevators and Options	Crew	Daily Output	Labor-Hours	Unit	Material	2022 Bare Costs Labor	Equipment	Total	Total Incl O&P
1375	For increased speed, 100 fpm, add				Ea.	1,300			1,300	1,425
1400	125 fpm, add					3,400			3,400	3,750
1425	150 fpm, add					5,350			5,350	5,900
1450	175 fpm, add					7,675			7,675	8,425
1475	For class "B" loading, add					3,050			3,050	3,350
1500	For class "C-1" loading, add					7,675			7,675	8,450
1525	For class "C-2" loading, add					9,225			9,225	10,200
1550	For class "C-3" loading, add					12,700			12,700	13,900
1575	For travel over 20 V.L.F., add	2 Elev	7.25	2.207	V.L.F.	965	206		1,171	1,375
1600	For number of stops over 2, add	"	.27	59.259	Stop	2,575	5,550		8,125	11,300

14 24 23 – Hydraulic Passenger Elevators

14 24 23.10 Hydraulic Passenger Elevators and Options

		Crew	Daily Output	Labor-Hours	Unit	Material	2022 Bare Costs Labor	Equipment	Total	Total Incl O&P
0010	**HYDRAULIC PASSENGER ELEVATORS AND OPTIONS** R142000-10									
2050	Hyd. pass., base unit, 1,500 lb., 100 fpm, 2 stop, std. fin. R142000-20	2 Elev	.10	160	Ea.	46,900	15,000		61,900	74,500
2075	For 2,000 lb. capacity, add R142000-30					890			890	980
2100	For 2,500 lb. capacity, add R142000-40					3,325			3,325	3,675
2125	For 3,000 lb. capacity, add					4,800			4,800	5,275
2150	For 3,500 lb. capacity, add					8,000			8,000	8,800
2175	For 4,000 lb. capacity, add					9,625			9,625	10,600
2200	For 4,500 lb. capacity, add					12,700			12,700	14,000
2225	For 5,000 lb. capacity, add					17,900			17,900	19,700
2250	For increased speed, 125 fpm, add					1,300			1,300	1,425
2275	150 fpm, add				Ea.	2,875			2,875	3,175
2300	175 fpm, add					5,600			5,600	6,175
2325	200 fpm, add					10,800			10,800	11,900
2350	For travel over 12 V.L.F., add	2 Elev	7.25	2.207	V.L.F.	765	206		971	1,150
2375	For number of stops over 2, add		.27	59.259	Stop	1,175	5,550		6,725	9,750
2725	Hydraulic hospital, base unit, 4,000 lb., 100 fpm, 2 stop, std. fin.		.10	160	Ea.	77,000	15,000		92,000	108,000
2775	For 4,500 lb. capacity, add					8,125			8,125	8,950
2800	For 5,000 lb. capacity, add					11,800			11,800	13,000
2825	For increased speed, 125 fpm, add					2,425			2,425	2,675
2850	150 fpm, add					4,050			4,050	4,450
2875	175 fpm, add					6,825			6,825	7,500
2900	200 fpm, add					10,000			10,000	11,000
2925	For travel over 12 V.L.F., add	2 Elev	7.25	2.207	V.L.F.	445	206		651	805
2950	For number of stops over 2, add	"	.27	59.259	Stop	5,375	5,550		10,925	14,400

14 27 Custom Elevator Cabs and Doors

14 27 13 – Custom Elevator Cab Finishes

14 27 13.10 Cab Finishes

		Crew	Daily Output	Labor-Hours	Unit	Material	2022 Bare Costs Labor	Equipment	Total	Total Incl O&P
0010	**CAB FINISHES**									
3325	Passenger elevator cab finishes (based on 3,500 lb. cab size)									
3350	Acrylic panel ceiling				Ea.	825			825	910
3375	Aluminum eggcrate ceiling					745			745	820
3400	Stainless steel doors					4,300			4,300	4,725
3425	Carpet flooring					660			660	725
3450	Epoxy flooring					495			495	545
3475	Quarry tile flooring					960			960	1,050
3500	Slate flooring					1,725			1,725	1,900
3525	Textured rubber flooring					695			695	765

14 27 Custom Elevator Cabs and Doors

14 27 13 – Custom Elevator Cab Finishes

14 27 13.10 Cab Finishes	Crew	Daily Output	Labor-Hours	Unit	Material	2022 Bare Costs Labor	2022 Bare Costs Equipment	Total	Total Incl O&P
3550	Stainless steel walls				Ea.	4,325			4,325	4,750
3575	Stainless steel returns at door				↓	1,250			1,250	1,375
4450	Hospital elevator cab finishes (based on 3,500 lb. cab size)									
4475	Aluminum eggcrate ceiling				Ea.	755			755	830
4500	Stainless steel doors					4,300			4,300	4,725
4525	Epoxy flooring					495			495	545
4550	Quarry tile flooring					960			960	1,050
4575	Textured rubber flooring					690			690	755
4600	Stainless steel walls					4,900			4,900	5,375
4625	Stainless steel returns at door					965			965	1,075

14 28 Elevator Equipment and Controls

14 28 10 – Elevator Equipment and Control Options

14 28 10.10 Elevator Controls and Doors

		Crew	Daily Output	Labor-Hours	Unit	Material	2022 Bare Costs Labor	2022 Bare Costs Equipment	Total	Total Incl O&P
0010	**ELEVATOR CONTROLS AND DOORS**									
2975	Passenger elevator options									
3000	2 car group automatic controls	2 Elev	.66	24.242	Ea.	3,750	2,275		6,025	7,600
3025	3 car group automatic controls		.44	36.364		5,750	3,400		9,150	11,500
3050	4 car group automatic controls		.33	48.485		10,000	4,525		14,525	18,000
3075	5 car group automatic controls		.26	61.538		15,100	5,750		20,850	25,400
3100	6 car group automatic controls		.22	72.727		23,000	6,800		29,800	35,700
3125	Intercom service		3	5.333		605	500		1,105	1,425
3150	Duplex car selective collective		.66	24.242		4,350	2,275		6,625	8,275
3175	Center opening 1 speed doors		2	8		2,175	750		2,925	3,525
3200	Center opening 2 speed doors		2	8		3,050	750		3,800	4,500
3225	Rear opening doors (opposite front)		2	8		4,650	750		5,400	6,250
3250	Side opening 2 speed doors		2	8		7,975	750		8,725	9,925
3275	Automatic emergency power switching		.66	24.242		1,325	2,275		3,600	4,925
3300	Manual emergency power switching	↓	8	2		560	187		747	900
3625	Hall finishes, stainless steel doors					1,550			1,550	1,700
3650	Stainless steel frames					1,600			1,600	1,775
3675	12 month maintenance contract								4,425	4,425
3700	Signal devices, hall lanterns	2 Elev	8	2		565	187		752	905
3725	Position indicators, up to 3		9.40	1.702		365	159		524	645
3750	Position indicators, per each over 3	↓	32	.500		98.50	47		145.50	180
3775	High speed heavy duty door opener					2,875			2,875	3,175
3800	Variable voltage, O.H. gearless machine, min.	2 Elev	.16	100		36,000	9,350		45,350	54,000
3815	Maximum		.07	229		80,500	21,400		101,900	121,000
3825	Basement installed geared machine	↓	.33	48.485	↓	16,300	4,525		20,825	25,000
3850	Freight elevator options									
3875	Doors, bi-parting	2 Elev	.66	24.242	Ea.	6,875	2,275		9,150	11,100
3900	Power operated door and gate	"	.66	24.242		27,600	2,275		29,875	33,900
3925	Finishes, steel plate floor					1,250			1,250	1,375
3950	14 ga. 1/4" x 4' steel plate walls					2,275			2,275	2,500
3975	12 month maintenance contract								3,400	3,400
4000	Signal devices, hall lanterns	2 Elev	8	2		585	187		772	930
4025	Position indicators, up to 3		9.40	1.702		415	159		574	700
4050	Position indicators, per each over 3		32	.500		113	47		160	197
4075	Variable voltage basement installed geared machine	↓	.66	24.242	↓	23,100	2,275		25,375	28,900
4100	Hospital elevator options									
4125	2 car group automatic controls	2 Elev	.66	24.242	Ea.	3,750	2,275		6,025	7,600

14 28 Elevator Equipment and Controls

14 28 10 – Elevator Equipment and Control Options

14 28 10.10 Elevator Controls and Doors

		Crew	Daily Output	Labor-Hours	Unit	Material	2022 Bare Costs Labor	Equipment	Total	Total Incl O&P
4150	3 car group automatic controls	2 Elev	.44	36.364	Ea.	5,725	3,400		9,125	11,500
4175	4 car group automatic controls		.33	48.485		13,600	4,525		18,125	22,000
4200	5 car group automatic controls		.26	61.538		13,600	5,750		19,350	23,800
4225	6 car group automatic controls		.22	72.727		20,800	6,800		27,600	33,300
4250	Intercom service		3	5.333		575	500		1,075	1,400
4275	Duplex car selective collective		.66	24.242		4,400	2,275		6,675	8,325
4300	Center opening 1 speed doors		2	8		2,150	750		2,900	3,500
4325	Center opening 2 speed doors		2	8		2,850	750		3,600	4,275
4350	Rear opening doors (opposite front)		2	8		4,650	750		5,400	6,250
4375	Side opening 2 speed doors		2	8		6,900	750		7,650	8,750
4400	Automatic emergency power switching		.66	24.242		1,300	2,275		3,575	4,900
4425	Manual emergency power switching		8	2		545	187		732	885
4675	Hall finishes, stainless steel doors					1,600			1,600	1,775
4700	Stainless steel frames					1,625			1,625	1,775
4725	12 month maintenance contract								6,125	6,125
4750	Signal devices, hall lanterns	2 Elev	8	2		540	187		727	880
4775	Position indicators, up to 3		9.40	1.702		370	159		529	650
4800	Position indicators, per each over 3		32	.500		99	47		146	181
4825	High speed heavy duty door opener					2,925			2,925	3,200
4850	Variable voltage, O.H. gearless machine, min.	2 Elev	.16	100		37,800	9,350		47,150	56,000
4865	Maximum		.07	229		82,000	21,400		103,400	122,500
4875	Basement installed geared machine		.33	48.485		19,000	4,525		23,525	27,900
5000	Drilling for piston, casing included, 18" diameter	B-48	80	.700	V.L.F.	63.50	36	15.30	114.80	143

14 42 Wheelchair Lifts

14 42 13 – Inclined Wheelchair Lifts

14 42 13.10 Inclined Wheelchair Lifts and Stairclimbers

		Crew	Daily Output	Labor-Hours	Unit	Material	2022 Bare Costs Labor	Equipment	Total	Total Incl O&P
0010	**INCLINED WHEELCHAIR LIFTS AND STAIRCLIMBERS**									
7700	Stair climber (chair lift), single seat, minimum	2 Elev	1	16	Ea.	4,125	1,500		5,625	6,825
7800	Maximum		.20	80		7,125	7,475		14,600	19,400
8700	Stair lift, minimum		1	16		13,800	1,500		15,300	17,400
8900	Maximum		.20	80		19,900	7,475		27,375	33,400

14 42 16 – Vertical Wheelchair Lifts

14 42 16.10 Wheelchair Lifts

		Crew	Daily Output	Labor-Hours	Unit	Material	2022 Bare Costs Labor	Equipment	Total	Total Incl O&P
0010	**WHEELCHAIR LIFTS**									
8000	Wheelchair lift, minimum	2 Elev	1	16	Ea.	6,500	1,500		8,000	9,450
8500	Maximum	"	.50	32	"	15,700	3,000		18,700	21,900

14 45 Vehicle Lifts

14 45 10 – Hydraulic Vehicle Lifts

14 45 10.10 Hydraulic Lifts

		Crew	Daily Output	Labor-Hours	Unit	Material	2022 Bare Costs Labor	Equipment	Total	Total Incl O&P
0010	**HYDRAULIC LIFTS**									
2200	Single post, 8,000 lb. capacity	L-4	.40	60	Ea.	5,575	3,225		8,800	11,300
2810	Double post, 6,000 lb. capacity		2.67	8.989		2,775	485		3,260	3,825
2815	9,000 lb. capacity		2.29	10.480		3,650	565		4,215	4,925
2820	15,000 lb. capacity		2	12		7,625	645		8,270	9,400
2822	Four post, 26,000 lb. capacity		1.80	13.333		15,900	715		16,615	18,700
2825	30,000 lb. capacity		1.60	15		72,000	805		72,805	81,000
2830	Ramp style, 4 post, 25,000 lb. capacity		2	12		14,200	645		14,845	16,600

14 45 Vehicle Lifts

14 45 10 – Hydraulic Vehicle Lifts

14 45 10.10 Hydraulic Lifts		Crew	Daily Output	Labor-Hours	Unit	Material	2022 Bare Costs Labor	Equipment	Total	Total Incl O&P
2835	35,000 lb. capacity	L-4	1	24	Ea.	22,700	1,300		24,000	27,000
2840	50,000 lb. capacity		1	24		112,500	1,300		113,800	125,500
2845	75,000 lb. capacity	↓	1	24		110,000	1,300		111,300	123,000
2850	For drive thru tracks, add, minimum					1,275			1,275	1,400
2855	Maximum					2,200			2,200	2,400
2860	Ramp extensions, 3' (set of 2)					520			520	570
2865	Rolling jack platform					1,825			1,825	2,000
2870	Electric/hydraulic jacking beam					5,250			5,250	5,775
2880	Scissor lift, portable, 6,000 lb. capacity				↓	2,000			2,000	2,200

14 91 Facility Chutes

14 91 33 – Laundry and Linen Chutes

14 91 33.10 Chutes		Crew	Daily Output	Labor-Hours	Unit	Material	2022 Bare Costs Labor	Equipment	Total	Total Incl O&P
0011	CHUTES, linen, trash or refuse									
0050	Aluminized steel, 16 ga., 18" diameter	2 Shee	3.50	4.571	Floor	1,925	310		2,235	2,600
0100	24" diameter		3.20	5		2,150	340		2,490	2,875
0400	Galvanized steel, 16 ga., 18" diameter		3.50	4.571		1,150	310		1,460	1,750
0500	24" diameter		3.20	5		1,300	340		1,640	1,950
0800	Stainless steel, 18" diameter		3.50	4.571		3,725	310		4,035	4,575
0900	24" diameter	↓	3.20	5	↓	3,250	340		3,590	4,100

14 91 82 – Trash Chutes

14 91 82.10 Trash Chutes and Accessories		Crew	Daily Output	Labor-Hours	Unit	Material	2022 Bare Costs Labor	Equipment	Total	Total Incl O&P
0010	TRASH CHUTES AND ACCESSORIES									
2900	Package chutes, spiral type, minimum	2 Shee	4.50	3.556	Floor	2,750	240		2,990	3,400
3000	Maximum	"	1.50	10.667	"	7,200	720		7,920	9,050
9000	Minimum labor/equipment charge	1 Shee	1	8	Job		540		540	840

14 92 Pneumatic Tube Systems

14 92 10 – Conventional, Automatic and Computer Controlled Pneumatic Tube Systems

14 92 10.10 Pneumatic Tube Systems		Crew	Daily Output	Labor-Hours	Unit	Material	2022 Bare Costs Labor	Equipment	Total	Total Incl O&P
0010	PNEUMATIC TUBE SYSTEMS									
0020	100' long, single tube, 2 stations, stock									
0100	3" diameter	2 Stpi	.12	133	Total	5,625	9,375		15,000	20,700
0300	4" diameter	"	.09	178	"	6,675	12,500		19,175	26,700
0400	Twin tube, two stations or more, conventional system									
0700	3" round	2 Stpi	46	.348	L.F.	56.50	24.50		81	99.50
1050	Add for blower		2	8	System	6,025	565		6,590	7,500
1110	Plus for each round station, add		7.50	2.133	Ea.	1,850	150		2,000	2,250
1200	Alternate pricing method: base cost, economy model		.75	21.333	Total	5,625	1,500		7,125	8,525
1300	Custom model		.25	64	"	12,600	4,500		17,100	20,900
1500	Plus total system length, add, for economy model		93.40	.171	L.F.	8.15	12.05		20.20	27.50
1600	For custom model	↓	37.60	.426	"	17.90	30		47.90	65.50

Division Notes

		CREW	DAILY OUTPUT	LABOR-HOURS	UNIT	BARE COSTS				TOTAL INCL O&P
						MAT.	LABOR	EQUIP.	TOTAL	

Estimating Tips

Pipe for fire protection and all uses is located in Subdivisions 21 11 13 and 22 11 13.

The labor adjustment factors listed in Subdivision 22 01 02.20 also apply to Division 21.

Many, but not all, areas in the U.S. require backflow protection in the fire system. Insurance underwriters may have specific requirements for the type of materials to be installed or design requirements based on the hazard to be protected. Local jurisdictions may have requirements not covered by code. It is advisable to be aware of any special conditions.

For your reference, the following is a list of the most applicable Fire Codes and Standards, which may be purchased from the NFPA, 1 Batterymarch Park, Quincy, MA 02169-7471.

- NFPA 1: Uniform Fire Code
- NFPA 10: Portable Fire Extinguishers
- NFPA 11: Low-, Medium-, and High-Expansion Foam
- NFPA 12: Carbon Dioxide Extinguishing Systems (Also companion 12A)
- NFPA 13: Installation of Sprinkler Systems (Also companion 13D, 13E, and 13R)
- NFPA 14: Installation of Standpipe and Hose Systems
- NFPA 15: Water Spray Fixed Systems for Fire Protection
- NFPA 16: Installation of Foam-Water Sprinkler and Foam-Water Spray Systems
- NFPA 17: Dry Chemical Extinguishing Systems (Also companion 17A)
- NFPA 18: Wetting Agents
- NFPA 20: Installation of Stationary Pumps for Fire Protection
- NFPA 22: Water Tanks for Private Fire Protection
- NFPA 24: Installation of Private Fire Service Mains and their Appurtenances
- NFPA 25: Inspection, Testing and Maintenance of Water-Based Fire Protection

Reference Numbers

Reference numbers are shown at the beginning of some major classifications. These numbers refer to related items in the Reference Section. The reference information may be an estimating procedure, an alternate pricing method, or technical information.

Same Data. Simplified.

Enjoy the convenience and efficiency of accessing your costs anywhere:

- **Skip the multiplier** by setting your location
- **Quickly search,** edit, favorite and share costs
- **Stay on top of price changes** with automatic updates

Discover more at rsmeans.com/online

Note: "Powered in part by CINX™, based on licensed proprietary information of Harrison Publishing House, Inc."

Note: Trade Service, in part, has been used as a reference source for some of the material prices used in Division 21.

21 05 Common Work Results for Fire Suppression

21 05 23 – General-Duty Valves for Water-Based Fire-Suppression Piping

21 05 23.50 General-Duty Valves	Crew	Daily Output	Labor-Hours	Unit	Material	2022 Bare Costs Labor	Equipment	Total	Total Incl O&P
0010 **GENERAL-DUTY VALVES**, for water-based fire suppression									
6500 Check, swing, C.I. body, brass fittings, auto. ball drip									
6520 4" size	Q-12	3	5.333	Ea.	475	330		805	1,025
9990 Minimum labor/equipment charge	1 Spri	3	2.667	Job		183		183	282

21 11 Facility Fire-Suppression Water-Service Piping

21 11 16 – Facility Fire Hydrants

21 11 16.50 Fire Hydrants for Buildings	Crew	Daily Output	Labor-Hours	Unit	Material	2022 Bare Costs Labor	Equipment	Total	Total Incl O&P
0010 **FIRE HYDRANTS FOR BUILDINGS**									
3750 Hydrants, wall, w/caps, single, flush, polished brass									
3800 2-1/2" x 2-1/2"	Q-12	5	3.200	Ea.	236	197		433	565
4350 Double, projecting, polished brass									
4400 2-1/2" x 2-1/2" x 4"	Q-12	5	3.200	Ea.	282	197		479	615

21 11 19 – Fire-Department Connections

21 11 19.50 Connections for the Fire-Department	Crew	Daily Output	Labor-Hours	Unit	Material	2022 Bare Costs Labor	Equipment	Total	Total Incl O&P
0010 **CONNECTIONS FOR THE FIRE-DEPARTMENT**									
7140 Standpipe connections, wall, w/plugs & chains									
7280 Double, flush, polished brass									
7300 2-1/2" x 2-1/2" x 4"	Q-12	5	3.200	Ea.	575	197		772	935
9990 Minimum labor/equipment charge	1 Spri	4	2	Job		137		137	212

21 12 Fire-Suppression Standpipes

21 12 13 – Fire-Suppression Hoses and Nozzles

21 12 13.50 Fire Hoses and Nozzles	Crew	Daily Output	Labor-Hours	Unit	Material	2022 Bare Costs Labor	Equipment	Total	Total Incl O&P
0010 **FIRE HOSES AND NOZZLES**									
0200 Adapters, rough brass, straight hose threads									
0220 One piece, female to male, rocker lugs									
0240 1" x 1"				Ea.	60.50			60.50	66.50
0320 2" x 2"					70			70	77
0380 2-1/2" x 2-1/2"				↓	122			122	134
2200 Hose, less couplings									
2260 Synthetic jacket, lined, 300 lb. test, 1-1/2" diameter	Q-12	2600	.006	L.F.	3.31	.38		3.69	4.23
2270 2" diameter		2200	.007		4.63	.45		5.08	5.80
2280 2-1/2" diameter		2200	.007		5.80	.45		6.25	7.10
2290 3" diameter		2200	.007		9.10	.45		9.55	10.70
2360 High strength, 500 lb. test, 1-1/2" diameter		2600	.006		3.62	.38		4	4.57
2380 2-1/2" diameter	↓	2200	.007	↓	6.10	.45		6.55	7.40
5600 Nozzles, brass									
5620 Adjustable fog, 3/4" booster line				Ea.	154			154	170
5640 1-1/2" leader line					120			120	132
5660 2-1/2" direct connection					280			280	310
5680 2-1/2" playpipe nozzle				↓	335			335	370
5780 For chrome plated, add					8%				
9900 Minimum labor/equipment charge	1 Plum	2	4	Job		279		279	430

21 12 Fire-Suppression Standpipes

21 12 19 – Fire-Suppression Hose Racks

21 12 19.50 Fire Hose Racks

		Crew	Daily Output	Labor-Hours	Unit	Material	2022 Bare Costs Labor	Equipment	Total	Total Incl O&P
0010	**FIRE HOSE RACKS**									
2600	Hose rack, swinging, for 1-1/2" diameter hose,									
2620	Enameled steel, 50' and 75' lengths of hose	Q-12	20	.800	Ea.	132	49.50		181.50	221

21 12 23 – Fire-Suppression Hose Valves

21 12 23.70 Fire Hose Valves

		Crew	Daily Output	Labor-Hours	Unit	Material	2022 Bare Costs Labor	Equipment	Total	Total Incl O&P
0010	**FIRE HOSE VALVES**									
3000	Gate, hose, wheel handle, N.R.S., rough brass, 1-1/2"	1 Spri	12	.667	Ea.	149	45.50		194.50	235
3040	2-1/2", 300 lb.	"	7	1.143		210	78.50		288.50	350
3080	For polished brass, add					40%				
3090	For polished chrome, add			↓		50%				

21 13 Fire-Suppression Sprinkler Systems

21 13 13 – Wet-Pipe Sprinkler Systems

21 13 13.50 Wet-Pipe Sprinkler System Components

		Crew	Daily Output	Labor-Hours	Unit	Material	2022 Bare Costs Labor	Equipment	Total	Total Incl O&P
0010	**WET-PIPE SPRINKLER SYSTEM COMPONENTS**									
1220	Water motor gong	1 Spri	4	2	Ea.	365	137		502	610
1800	Firecycle system, controls, includes panel									
1900	Flexible sprinkler head connectors									
1910	Braided stainless steel hose with mounting bracket									
1920	1/2" and 3/4" outlet size									
1940	40" length	1 Spri	30	.267	Ea.	67.50	18.30		85.80	103
1960	60" length	1 Spri	22	.364	Ea.	79	25		104	125
1982	May replace hard-pipe armovers									
1984	for wet and pre-action systems.									
2600	Sprinkler heads, not including supply piping									
3700	Standard spray, pendent or upright, brass, 135°F to 286°F									
3740	1/2" NPT, K5.6	1 Spri	16	.500	Ea.	16.45	34.50		50.95	71
3790	For riser & feeder piping, see Section 22 11 13.00									

21 13 16 – Dry-Pipe Sprinkler Systems

21 13 16.50 Dry-Pipe Sprinkler System Components

		Crew	Daily Output	Labor-Hours	Unit	Material	2022 Bare Costs Labor	Equipment	Total	Total Incl O&P
0010	**DRY-PIPE SPRINKLER SYSTEM COMPONENTS**									
0600	Accelerator	1 Spri	8	1	Ea.	1,075	68.50		1,143.50	1,275
0800	Air compressor for dry pipe system, automatic, complete									
0820	30 gal. system capacity, 3/4 HP	1 Spri	1.30	6.154	Ea.	1,450	420		1,870	2,250
2600	Sprinkler heads, not including supply piping									
2640	Dry, pendent, 1/2" orifice, 3/4" or 1" NPT									
2660	3" to 6" length	1 Spri	14	.571	Ea.	211	39		250	293
2700	15-1/4" to 18" length	"	14	.571	"	279	39		318	365
6330	Valves and components									
8200	Dry pipe valve, incl. trim and gauges, 3" size	Q-12	2	8	Ea.	3,450	495		3,945	4,525
8220	4" size	"	1	16	"	3,700	985		4,685	5,575

21 13 39 – Foam-Water Systems

21 13 39.50 Foam-Water System Components

		Crew	Daily Output	Labor-Hours	Unit	Material	2022 Bare Costs Labor	Equipment	Total	Total Incl O&P
0010	**FOAM-WATER SYSTEM COMPONENTS**									
2600	Sprinkler heads, not including supply piping									
3600	Foam-water, pendent or upright, 1/2" NPT	1 Spri	12	.667	Ea.	268	45.50		313.50	365

Division Notes

	CREW	DAILY OUTPUT	LABOR-HOURS	UNIT	BARE COSTS				TOTAL INCL O&P
					MAT.	LABOR	EQUIP.	TOTAL	

Plumbing

Estimating Tips
22 10 00 Plumbing Piping and Pumps

This subdivision is primarily basic pipe and related materials. The pipe may be used by any of the mechanical disciplines, i.e., plumbing, fire protection, heating, and air conditioning.

Note: CPVC plastic piping approved for fire protection is located in 21 11 13.

- The labor adjustment factors listed in Subdivision 22 01 02.20 apply throughout Divisions 21, 22, and 23. CAUTION: the correct percentage may vary for the same items. For example, the percentage add for the basic pipe installation should be based on the maximum height that the installer must install for that particular section. If the pipe is to be located 14' above the floor but it is suspended on threaded rod from beams, the bottom flange of which is 18' high (4' rods), then the height is actually 18' and the add is 20%. The pipe cover, however, does not have to go above the 14' and so the add should be 10%.
- Most pipe is priced first as straight pipe with a joint (coupling, weld, etc.) every 10' and a hanger usually every 10'. There are exceptions with hanger spacing such as for cast iron pipe (5')

and plastic pipe (3 per 10'). Following each type of pipe there are several lines listing sizes and the amount to be subtracted to delete couplings and hangers. This is for pipe that is to be buried or supported together on trapeze hangers. The reason that the couplings are deleted is that these runs are usually long, and frequently longer lengths of pipe are used. By deleting the couplings, the estimator is expected to look up and add back the correct reduced number of couplings.

- When preparing an estimate, it may be necessary to approximate the fittings. Fittings usually run between 25% and 50% of the cost of the pipe. The lower percentage is for simpler runs, and the higher number is for complex areas, such as mechanical rooms.
- For historic restoration projects, the systems must be as invisible as possible, and pathways must be sought for pipes, conduit, and ductwork. While installations in accessible spaces (such as basements and attics) are relatively straightforward to estimate, labor costs may be more difficult to determine when delivery systems must be concealed.

22 40 00 Plumbing Fixtures

- Plumbing fixture costs usually require two lines: the fixture itself and its "rough-in, supply, and waste."
- In the Assemblies Section (Plumbing D2010) for the desired fixture, the System Components Group at the center of the page shows the fixture on the first line. The rest of the list (fittings, pipe, tubing, etc.) will total up to what we refer to in the Unit Price section as "Rough-in, supply, waste, and vent." Note that for most fixtures we allow a nominal 5' of tubing to reach from the fixture to a main or riser.
- Remember that gas- and oil-fired units need venting.

Reference Numbers

Reference numbers are shown at the beginning of some major classifications. These numbers refer to related items in the Reference Section. The reference information may be an estimating procedure, an alternate pricing method, or technical information.

Note: Not all subdivisions listed here necessarily appear. ∎

Same Data. Simplified.

Enjoy the convenience and efficiency of accessing your costs anywhere:

- **Skip the multiplier** by setting your location
- **Quickly search**, edit, favorite and share costs
- **Stay on top of price changes** with automatic updates

Discover more at rsmeans.com/online

22 01 02.20 Labor Adjustment Factors	Crew	Daily Output	Labor-Hours	Unit	Material	2022 Bare Costs Labor	Equipment	Total	Total Incl O&P
0010 **LABOR ADJUSTMENT FACTORS** (For Div. 21, 22 and 23) R220102-20									
0100 Labor factors: The below are reasonable suggestions, but									
0110 each project must be evaluated for its own peculiarities, and									
0120 the adjustments be increased or decreased depending on the									
0130 severity of the special conditions.									
1000 Add to labor for elevated installation (Above floor level)									
1080 10' to 14.5' high						10%			
1100 15' to 19.5' high						20%			
1120 20' to 24.5' high						25%			
1140 25' to 29.5' high						35%			
1160 30' to 34.5' high						40%			
1180 35' to 39.5' high						50%			
1200 40' and higher						55%			
2000 Add to labor for crawl space									
2100 3' high						40%			
2140 4' high						30%			
3000 Add to labor for multi-story building									
3010 For new construction (No elevator available)									
3100 Add for floors 3 thru 10						5%			
3110 Add for floors 11 thru 15						10%			
3120 Add for floors 16 thru 20						15%			
3130 Add for floors 21 thru 30						20%			
3140 Add for floors 31 and up						30%			
3170 For existing structure (Elevator available)									
3180 Add for work on floor 3 and above						2%			
4000 Add to labor for working in existing occupied buildings									
4100 Hospital						35%			
4140 Office building						25%			
4180 School						20%			
4220 Factory or warehouse						15%			
4260 Multi dwelling						15%			
5000 Add to labor, miscellaneous									
5100 Cramped shaft						35%			
5140 Congested area						15%			
5180 Excessive heat or cold						30%			
9000 Labor factors: The above are reasonable suggestions, but									
9010 each project should be evaluated for its own peculiarities.									
9100 Other factors to be considered are:									
9140 Movement of material and equipment through finished areas									
9180 Equipment room									
9220 Attic space									
9260 No service road									
9300 Poor unloading/storage area									
9340 Congested site area/heavy traffic									

22 05 Common Work Results for Plumbing

22 05 05 – Selective Demolition for Plumbing

22 05 05.10 Plumbing Demolition

		Crew	Daily Output	Labor-Hours	Unit	Material	2022 Bare Costs Labor	Equipment	Total	Total Incl O&P
0010	**PLUMBING DEMOLITION** R220105-10									
1020	Fixtures, including 10' piping									
1100	Bathtubs, cast iron R024119-10	1 Plum	4	2	Ea.		140		140	215
1120	Fiberglass		6	1.333			93		93	143
1140	Steel		5	1.600			112		112	172
1200	Lavatory, wall hung		10	.800			56		56	86
1220	Counter top		8	1			70		70	108
1300	Sink, single compartment		8	1			70		70	108
1320	Double compartment		7	1.143			80		80	123
1400	Water closet, floor mounted		8	1			70		70	108
1420	Wall mounted		7	1.143			80		80	123
1500	Urinal, floor mounted		4	2			140		140	215
1520	Wall mounted		7	1.143			80		80	123
1600	Water fountains, free standing		8	1			70		70	108
1620	Wall or deck mounted		6	1.333	▼		93		93	143
2000	Piping, metal, up thru 1-1/2" diameter		200	.040	L.F.		2.79		2.79	4.30
2050	2" thru 3-1/2" diameter	▼	150	.053			3.72		3.72	5.75
2100	4" thru 6" diameter	2 Plum	100	.160			11.15		11.15	17.20
2150	8" thru 14" diameter	"	60	.267			18.60		18.60	28.50
2153	16" thru 20" diameter	Q-18	70	.343			22.50	.89	23.39	35.50
2155	24" thru 26" diameter		55	.436			28.50	1.13	29.63	45
2156	30" thru 36" diameter	▼	40	.600			39.50	1.55	41.05	62
2160	Plastic pipe with fittings, up thru 1-1/2" diameter	1 Plum	250	.032			2.23		2.23	3.44
2162	2" thru 3" diameter	"	200	.040			2.79		2.79	4.30
2164	4" thru 6" diameter	Q-1	200	.080			5.05		5.05	7.75
2166	8" thru 14" diameter		150	.107			6.70		6.70	10.35
2168	16" diameter	▼	100	.160	▼		10.05		10.05	15.50
2250	Water heater, 40 gal.	1 Plum	6	1.333	Ea.		93		93	143
6000	Remove and reset fixtures, easy access		6	1.333			93		93	143
6100	Difficult access		4	2	▼		140		140	215
9000	Minimum labor/equipment charge	▼	2	4	Job		279		279	430

22 05 23 – General-Duty Valves for Plumbing Piping

22 05 23.20 Valves, Bronze

		Crew	Daily Output	Labor-Hours	Unit	Material	2022 Bare Costs Labor	Equipment	Total	Total Incl O&P
0010	**VALVES, BRONZE**									
1020	Angle, 150 lb., rising stem, threaded									
1070	3/4"	1 Plum	20	.400	Ea.	259	28		287	330
1080	1"	"	19	.421	"	360	29.50		389.50	440
1100	1-1/2"	1 Plum	13	.615	Ea.	605	43		648	730
1102	Soldered same price as threaded									
1110	2"	1 Plum	11	.727	Ea.	950	51		1,001	1,125
1300	Ball									
1398	Threaded, 150 psi									
1460	3/4"	1 Plum	20	.400	Ea.	31	28		59	77
1522	Solder the same price as threaded									
1750	Check, swing, class 150, regrinding disc, threaded									
1850	1/2"	1 Plum	24	.333	Ea.	86.50	23.50		110	131
1860	3/4"		20	.400		136	28		164	192
1870	1"		19	.421		190	29.50		219.50	255
1890	1-1/2"		13	.615		305	43		348	400
1900	2"	▼	11	.727	▼	445	51		496	570
2850	Gate, N.R.S., soldered, 125 psi									
2920	1/2"	1 Plum	24	.333	Ea.	87	23.50		110.50	132

For customer support on your Commercial Renovation Costs with RSMeans data, call 800.448.8182.

423

22 05 23.20 Valves, Bronze

		Crew	Daily Output	Labor-Hours	Unit	Material	2022 Bare Costs Labor	2022 Bare Costs Equipment	Total	Total Incl O&P
2940	3/4"	1 Plum	20	.400	Ea.	96.50	28		124.50	149
2950	1"		19	.421		139	29.50		168.50	198
2970	1-1/2"		13	.615		236	43		279	325
2980	2"		11	.727		335	51		386	445
4850	Globe, class 150, rising stem, threaded									
4950	1/2"	1 Plum	24	.333	Ea.	142	23.50		165.50	192
4960	3/4"	"	20	.400	"	190	28		218	251
5600	Relief, pressure & temperature, self-closing, ASME, threaded									
5640	3/4"	1 Plum	28	.286	Ea.	280	19.95		299.95	340
5650	1"		24	.333		450	23.50		473.50	530
5660	1-1/4"		20	.400		900	28		928	1,025
5670	1-1/2"		18	.444		1,725	31		1,756	1,950
5680	2"		16	.500		1,875	35		1,910	2,125
6400	Pressure, water, ASME, threaded									
6440	3/4"	1 Plum	28	.286	Ea.	59.50	19.95		79.45	96
6450	1"		24	.333		375	23.50		398.50	445
6470	1-1/2"		18	.444		820	31		851	955
6900	Reducing, water pressure									
6920	300 psi to 25-75 psi, threaded or sweat									
6940	1/2"	1 Plum	24	.333	Ea.	585	23.50		608.50	680
6950	3/4"		20	.400		615	28		643	725
6960	1"		19	.421		955	29.50		984.50	1,100
8350	Tempering, water, sweat connections									
8400	1/2"	1 Plum	24	.333	Ea.	159	23.50		182.50	211
8440	3/4"	"	20	.400	"	184	28		212	246
8650	Threaded connections									
8700	1/2"	1 Plum	24	.333	Ea.	176	23.50		199.50	229
8740	3/4"	"	20	.400	"	1,100	28		1,128	1,275
8800	Water heater water & gas safety shut off									
8810	Protection against a leaking water heater									
8814	Shut off valve	1 Plum	16	.500	Ea.	196	35		231	269
8818	Water heater dam		32	.250		52.50	17.45		69.95	85
8822	Gas control wiring harness		32	.250		25.50	17.45		42.95	55
8830	Whole house flood safety shut off									
8834	Connections									
8838	3/4" NPT	1 Plum	12	.667	Ea.	1,050	46.50		1,096.50	1,225
8842	1" NPT		11	.727		1,075	51		1,126	1,250
8846	1-1/4" NPT		10	.800		1,100	56		1,156	1,275
9000	Minimum labor/equipment charge		4	2	Job		140		140	215

22 05 23.60 Valves, Plastic

		Crew	Daily Output	Labor-Hours	Unit	Material	2022 Bare Costs Labor	2022 Bare Costs Equipment	Total	Total Incl O&P
0010	**VALVES, PLASTIC**									
1150	Ball, PVC, socket or threaded, true union									
1230	1/2"	1 Plum	26	.308	Ea.	39.50	21.50		61	76.50
1240	3/4"		25	.320		47	22.50		69.50	86
1250	1"		23	.348		56	24.50		80.50	99
1260	1-1/4"		21	.381		88.50	26.50		115	139
1270	1-1/2"		20	.400		88.50	28		116.50	141
1280	2"		17	.471		141	33		174	206
1650	CPVC, socket or threaded, single union									
1700	1/2"	1 Plum	26	.308	Ea.	61.50	21.50		83	101
1720	3/4"		25	.320		77.50	22.50		100	120
1730	1"		23	.348		93	24.50		117.50	140

424

For customer support on your Commercial Renovation Costs with RSMeans data, call 800.448.8182.

22 05 Common Work Results for Plumbing

22 05 23 – General-Duty Valves for Plumbing Piping

22 05 23.60 Valves, Plastic		Crew	Daily Output	Labor-Hours	Unit	Material	2022 Bare Costs Labor	2022 Bare Costs Equipment	Total	Total Incl O&P
1750	1-1/4"	1 Plum	21	.381	Ea.	148	26.50		174.50	204
1760	1-1/2"		20	.400		148	28		176	206
1770	2"	↓	17	.471	↓	204	33		237	275
3150	Ball check, PVC, socket or threaded									
3290	2"	1 Plum	17	.471	Ea.	154	33		187	221
9000	Minimum labor/equipment charge	"	3.50	2.286	Job		160		160	246

22 05 76 – Facility Drainage Piping Cleanouts

22 05 76.10 Cleanouts

22 05 76.10 Cleanouts		Crew	Daily Output	Labor-Hours	Unit	Material	2022 Bare Costs Labor	2022 Bare Costs Equipment	Total	Total Incl O&P
0010	**CLEANOUTS**									
0060	Floor type									
0080	Round or square, scoriated nickel bronze top									
0100	2" pipe size	1 Plum	10	.800	Ea.	460	56		516	590
0120	3" pipe size		8	1		485	70		555	645
0140	4" pipe size	↓	6	1.333	↓	650	93		743	860
0980	Round top, recessed for terrazzo									
1000	2" pipe size	1 Plum	9	.889	Ea.	705	62		767	870
1080	3" pipe size		6	1.333		755	93		848	975
1100	4" pipe size	↓	4	2		900	140		1,040	1,200
1120	5" pipe size	Q-1	6	2.667	↓	1,425	168		1,593	1,800
9000	Minimum labor/equipment charge	1 Plum	3	2.667	Job		186		186	287

22 05 76.20 Cleanout Tees

22 05 76.20 Cleanout Tees		Crew	Daily Output	Labor-Hours	Unit	Material	2022 Bare Costs Labor	2022 Bare Costs Equipment	Total	Total Incl O&P
0010	**CLEANOUT TEES**									
0100	Cast iron, B&S, with countersunk plug									
0220	3" pipe size	1 Plum	3.60	2.222	Ea.	229	155		384	490
0240	4" pipe size	"	3.30	2.424	"	345	169		514	640
0500	For round smooth access cover, same price									
4000	Plastic, tees and adapters. Add plugs									
4010	ABS, DWV									
4020	Cleanout tee, 1-1/2" pipe size	1 Plum	15	.533	Ea.	27.50	37		64.50	87.50
9000	Minimum labor/equipment charge	"	2.75	2.909	Job		203		203	315

22 07 Plumbing Insulation

22 07 16 – Plumbing Equipment Insulation

22 07 16.10 Insulation for Plumbing Equipment

22 07 16.10 Insulation for Plumbing Equipment			Crew	Daily Output	Labor-Hours	Unit	Material	2022 Bare Costs Labor	2022 Bare Costs Equipment	Total	Total Incl O&P
0010	**INSULATION FOR PLUMBING EQUIPMENT**										
2900	Domestic water heater wrap kit										
2920	1-1/2" with vinyl jacket, 20 to 60 gal.	G	1 Plum	8	1	Ea.	29.50	70		99.50	141
2925	50 to 80 gal.	G	"	8	1	"	40.50	70		110.50	153

22 07 19 – Plumbing Piping Insulation

22 07 19.10 Piping Insulation

22 07 19.10 Piping Insulation			Crew	Daily Output	Labor-Hours	Unit	Material	2022 Bare Costs Labor	2022 Bare Costs Equipment	Total	Total Incl O&P
0010	**PIPING INSULATION**										
0110	Insulation req'd. is based on the surface size/area to be covered										
0230	Insulated protectors (ADA)										
0235	For exposed piping under sinks or lavatories	♿									
0240	Vinyl coated foam, velcro tabs										
0245	P Trap, 1-1/4" or 1-1/2"		1 Plum	32	.250	Ea.	17.10	17.45		34.55	46
0260	Valve and supply cover										
0265	1/2", 3/8", and 7/16" pipe size		1 Plum	32	.250	Ea.	16.80	17.45		34.25	45.50
0280	Tailpiece offset (wheelchair)	♿									
0285	1-1/4" pipe size		1 Plum	32	.250	Ea.	13.95	17.45		31.40	42.50

For customer support on your Commercial Renovation Costs with RSMeans data, call 800.448.8182.

425

22 07 Plumbing Insulation

22 07 19 – Plumbing Piping Insulation

22 07 19.10 Piping Insulation

22 07 19.10 Piping Insulation		Crew	Daily Output	Labor-Hours	Unit	Material	2022 Bare Costs Labor	Equipment	Total	Total Incl O&P
0600	Pipe covering (price copper tube one size less than IPS)									
6120	2" wall, 1/2" iron pipe size	G Q-14	145	.110	L.F.	7.60	6.25		13.85	18.25
6210	4" iron pipe size	G "	125	.128	"	15.05	7.25		22.30	28
6600	Fiberglass, with all service jacket									
6840	1" wall, 1/2" iron pipe size	G Q-14	240	.067	L.F.	2.96	3.78		6.74	9.25
6860	3/4" iron pipe size	G	230	.070		1.09	3.95		5.04	7.45
6870	1" iron pipe size	G	220	.073		2.32	4.13		6.45	9.05
6880	1-1/4" iron pipe size	G	210	.076		3.72	4.32		8.04	10.95
6890	1-1/2" iron pipe size	G	210	.076		2.71	4.32		7.03	9.85
6900	2" iron pipe size	G	200	.080		2.92	4.54		7.46	10.35
7080	1-1/2" wall, 1/2" iron pipe size	G	230	.070		3.75	3.95		7.70	10.40
7140	2" iron pipe size	G	190	.084		5.05	4.78		9.83	13.15
7160	3" iron pipe size	G	170	.094		4.10	5.35		9.45	12.95
7879	Rubber tubing, flexible closed cell foam									
8100	1/2" wall, 1/4" iron pipe size	G 1 Asbe	90	.089	L.F.	1.14	5.60		6.74	10.10
8120	3/8" iron pipe size	G	90	.089		1.13	5.60		6.73	10.10
8130	1/2" iron pipe size	G	89	.090		1.25	5.65		6.90	10.35
8140	3/4" iron pipe size	G	89	.090		.22	5.65		5.87	9.20
8150	1" iron pipe size	G	88	.091		1.56	5.75		7.31	10.75
8170	1-1/2" iron pipe size	G	87	.092		1.97	5.80		7.77	11.30
8180	2" iron pipe size	G	86	.093		2.52	5.85		8.37	12
8200	3" iron pipe size	G	85	.094		3.71	5.95		9.66	13.45
8220	4" iron pipe size	G	80	.100		5.55	6.30		11.85	16.10
8300	3/4" wall, 1/4" iron pipe size	G	90	.089		.27	5.60		5.87	9.15
8330	1/2" iron pipe size	G	89	.090		1.90	5.65		7.55	11.05
8340	3/4" iron pipe size	G	89	.090		2.28	5.65		7.93	11.45
8350	1" iron pipe size	G	88	.091		.46	5.75		6.21	9.55
8360	1-1/4" iron pipe size	G	87	.092		3.17	5.80		8.97	12.65
8370	1-1/2" iron pipe size	G	87	.092		4.21	5.80		10.01	13.80
8380	2" iron pipe size	G	86	.093		5.10	5.85		10.95	14.90
8444	1" wall, 1/2" iron pipe size	G	86	.093		.57	5.85		6.42	9.90
8445	3/4" iron pipe size	G	84	.095		.70	6		6.70	10.25
8446	1" iron pipe size	G	84	.095		5.10	6		11.10	15.10
8447	1-1/4" iron pipe size	G	82	.098		5.90	6.15		12.05	16.20
8448	1-1/2" iron pipe size	G	82	.098		6.55	6.15		12.70	16.90
8449	2" iron pipe size	G	80	.100		7.85	6.30		14.15	18.60
8450	2-1/2" iron pipe size	G	80	.100		7.90	6.30		14.20	18.65
8456	Rubber insulation tape, 1/8" x 2" x 30'	G			Ea.	25			25	27.50
9600	Minimum labor/equipment charge	1 Plum	4	2	Job		140		140	215

22 11 Facility Water Distribution

22 11 13 – Facility Water Distribution Piping

22 11 13.14 Pipe, Brass

22 11 13.14 Pipe, Brass		Crew	Daily Output	Labor-Hours	Unit	Material	2022 Bare Costs Labor	Equipment	Total	Total Incl O&P
0010	**PIPE, BRASS**, Plain end									
0900	Field threaded, coupling & clevis hanger assembly 10' OC									
0920	Regular weight									
1120	1/2" diameter	1 Plum	48	.167	L.F.	9.15	11.65		20.80	28
1140	3/4" diameter		46	.174		11.85	12.15		24	32
1160	1" diameter		43	.186		6.40	13		19.40	27
1180	1-1/4" diameter	Q-1	72	.222		19.35	13.95		33.30	43
1200	1-1/2" diameter		65	.246		25	15.45		40.45	51.50

22 11 Facility Water Distribution

22 11 13 – Facility Water Distribution Piping

22 11 13.14 Pipe, Brass		Crew	Daily Output	Labor-Hours	Unit	Material	2022 Bare Costs Labor	2022 Bare Costs Equipment	Total	Total Incl O&P
1220	2" diameter	Q-1	53	.302	L.F.	31	18.95		49.95	63
9000	Minimum labor/equipment charge	1 Plum	4	2	Job		140		140	215

22 11 13.16 Pipe Fittings, Brass

		Crew	Daily Output	Labor-Hours	Unit	Material	2022 Bare Costs Labor	2022 Bare Costs Equipment	Total	Total Incl O&P
0010	**PIPE FITTINGS, BRASS**, Rough bronze, threaded, lead free.									
1000	Standard wt., 90° elbow									
1100	1/2"	1 Plum	12	.667	Ea.	28.50	46.50		75	103
1120	3/4"		11	.727		38	51		89	121
1140	1"	↓	10	.800		62	56		118	154
1160	1-1/4"	Q-1	17	.941		100	59		159	201
1180	1-1/2"	"	16	1		139	63		202	250
1500	45° elbow, 1/8"	1 Plum	13	.615		35	43		78	105
1580	1/2"		12	.667		35	46.50		81.50	110
1600	3/4"		11	.727		50	51		101	133
1620	1"	↓	10	.800		85	56		141	180
1640	1-1/4"	Q-1	17	.941		135	59		194	239
1660	1-1/2"	"	16	1		170	63		233	284
2000	Tee, 1/8"	1 Plum	9	.889		33.50	62		95.50	133
2080	1/2"		8	1		35	70		105	147
2100	3/4"		7	1.143		47.50	80		127.50	176
2120	1"	↓	6	1.333		99	93		192	252
2140	1-1/4"	Q-1	10	1.600		148	101		249	320
2160	1-1/2"	"	9	1.778		167	112		279	355
2500	Coupling, 1/8"	1 Plum	26	.308		24	21.50		45.50	59
2580	1/2"		15	.533		24	37		61	83.50
2600	3/4"		14	.571		33.50	40		73.50	98.50
2620	1"	↓	13	.615		57	43		100	129
2640	1-1/4"	Q-1	22	.727		95.50	45.50		141	176
2660	1-1/2"		20	.800		124	50.50		174.50	214
2680	2"	↓	18	.889	↓	205	56		261	310
9000	Minimum labor/equipment charge	1 Plum	4	2	Job		140		140	215

22 11 13.23 Pipe/Tube, Copper

		Crew	Daily Output	Labor-Hours	Unit	Material	2022 Bare Costs Labor	2022 Bare Costs Equipment	Total	Total Incl O&P
0010	**PIPE/TUBE, COPPER**, Solder joints									
1000	Type K tubing, couplings & clevis hanger assemblies 10' OC									
1100	1/4" diameter	1 Plum	84	.095	L.F.	6.75	6.65		13.40	17.65
1200	1" diameter		66	.121		15.70	8.45		24.15	30.50
1260	2" diameter	↓	40	.200	↓	39.50	13.95		53.45	65
2000	Type L tubing, couplings & clevis hanger assemblies 10' OC									
2100	1/4" diameter	1 Plum	88	.091	L.F.	3.01	6.35		9.36	13.10
2120	3/8" diameter		84	.095		3.85	6.65		10.50	14.50
2140	1/2" diameter		81	.099		4.42	6.90		11.32	15.50
2160	5/8" diameter		79	.101		5.90	7.05		12.95	17.40
2180	3/4" diameter		76	.105		6.45	7.35		13.80	18.45
2200	1" diameter		68	.118		8.35	8.20		16.55	22
2220	1-1/4" diameter		58	.138		11.45	9.65		21.10	27.50
2240	1-1/2" diameter		52	.154		14.75	10.75		25.50	33
2260	2" diameter	↓	42	.190		22.50	13.30		35.80	45.50
2280	2-1/2" diameter	Q-1	62	.258		35	16.20		51.20	63.50
2300	3" diameter		56	.286		47.50	17.95		65.45	79.50
2320	3-1/2" diameter		43	.372		69	23.50		92.50	112
2340	4" diameter		39	.410		80	26		106	128
2360	5" diameter	↓	34	.471		194	29.50		223.50	259
2380	6" diameter	Q-2	40	.600	↓	262	39		301	350

22 11 13.23 Pipe/Tube, Copper		Crew	Daily Output	Labor-Hours	Unit	Material	2022 Bare Costs Labor	Equipment	Total	Total Incl O&P
2410	For other than full hard temper, add				L.F.	21%				
2590	For silver solder, add						15%			
3000	Type M tubing, couplings & clevis hanger assemblies 10' OC									
3140	1/2" diameter	1 Plum	84	.095	L.F.	4.82	6.65		11.47	15.55
3180	3/4" diameter		78	.103		7.20	7.15		14.35	18.95
3200	1" diameter	↓	70	.114	↓	9.90	8		17.90	23
4000	Type DWV tubing, couplings & clevis hanger assemblies 10' OC									
4100	1-1/4" diameter	1 Plum	60	.133	L.F.	13.85	9.30		23.15	29.50
4120	1-1/2" diameter		54	.148		8.55	10.35		18.90	25.50
4140	2" diameter	↓	44	.182		12.10	12.70		24.80	33
4160	3" diameter	Q-1	58	.276		25.50	17.35		42.85	54.50
4180	4" diameter		40	.400		89.50	25		114.50	137
4200	5" diameter	↓	36	.444		180	28		208	241
4220	6" diameter	Q-2	42	.571	↓	270	37		307	355
9000	Minimum labor/equipment charge	1 Plum	4	2	Job		140		140	215

22 11 13.25 Pipe/Tube Fittings, Copper

		Crew	Daily Output	Labor-Hours	Unit	Material	2022 Bare Costs Labor	Equipment	Total	Total Incl O&P
0010	**PIPE/TUBE FITTINGS, COPPER**, Wrought unless otherwise noted									
0020	For silver solder, add						15%			
0040	Solder joints, copper x copper									
0070	90° elbow, 1/4"	1 Plum	22	.364	Ea.	7.40	25.50		32.90	47
0100	1/2"		20	.400		2.59	28		30.59	46
0120	3/4"		19	.421		5.70	29.50		35.20	52
0130	1"		16	.500		13.95	35		48.95	69.50
0250	45° elbow, 1/4"		22	.364		13.15	25.50		38.65	53.50
0270	3/8"		22	.364		11.25	25.50		36.75	51.50
0280	1/2"		20	.400		3.96	28		31.96	47.50
0290	5/8"		19	.421		19.75	29.50		49.25	67
0300	3/4"		19	.421		9.30	29.50		38.80	55.50
0310	1"		16	.500		23.50	35		58.50	79.50
0320	1-1/4"		15	.533		26.50	37		63.50	87
0330	1-1/2"		13	.615		32.50	43		75.50	102
0340	2"	↓	11	.727		54.50	51		105.50	139
0350	2-1/2"	Q-1	13	1.231		106	77.50		183.50	236
0360	3"		13	1.231		148	77.50		225.50	282
0370	3-1/2"		10	1.600		255	101		356	435
0380	4"		9	1.778		305	112		417	505
0390	5"	↓	6	2.667		1,100	168		1,268	1,475
0400	6"	Q-2	9	2.667		1,750	174		1,924	2,200
0450	Tee, 1/4"	1 Plum	14	.571		14.75	40		54.75	78
0470	3/8"		14	.571		11.85	40		51.85	74.50
0480	1/2"		13	.615		3.68	43		46.68	70
0490	5/8"		12	.667		27	46.50		73.50	101
0500	3/4"		12	.667		12.30	46.50		58.80	85
0510	1"		10	.800		14.85	56		70.85	102
0520	1-1/4"		9	.889		43	62		105	143
0530	1-1/2"		8	1		65.50	70		135.50	180
0540	2"	↓	7	1.143		104	80		184	237
0550	2-1/2"	Q-1	8	2		191	126		317	405
0560	3"		7	2.286		275	144		419	520
0570	3-1/2"		6	2.667		795	168		963	1,125
0580	4"		5	3.200		645	201		846	1,025
0590	5"	↓	4	4		1,975	251		2,226	2,550

22 11 13.25 Pipe/Tube Fittings, Copper		Crew	Daily Output	Labor-Hours	Unit	Material	2022 Bare Costs Labor	2022 Bare Costs Equipment	Total	Total Incl O&P
0600	6"	Q-2	6	4	Ea.	2,700	261		2,961	3,350
0612	Tee, reducing on the outlet, 1/4"	1 Plum	15	.533		42	37		79	104
0613	3/8"		15	.533		35.50	37		72.50	97
0614	1/2"		14	.571		33	40		73	98
0615	5/8"		13	.615		66.50	43		109.50	139
0616	3/4"		12	.667		15.20	46.50		61.70	88
0617	1"		11	.727		72	51		123	158
0618	1-1/4"		10	.800		71	56		127	165
0619	1-1/2"		9	.889		75.50	62		137.50	179
0620	2"	↓	8	1		123	70		193	244
0621	2-1/2"	Q-1	9	1.778		340	112		452	545
0622	3"		8	2		385	126		511	620
0623	4"		6	2.667		695	168		863	1,025
0624	5"	↓	5	3.200		3,000	201		3,201	3,600
0625	6"	Q-2	7	3.429		5,275	223		5,498	6,150
0626	8"	"	6	4		20,400	261		20,661	22,800
0630	Tee, reducing on the run, 1/4"	1 Plum	15	.533		47.50	37		84.50	110
0631	3/8"		15	.533		54	37		91	117
0632	1/2"		14	.571		45.50	40		85.50	112
0633	5/8"		13	.615		18.65	43		61.65	86.50
0634	3/4"		12	.667		37	46.50		83.50	112
0635	1"		11	.727		58	51		109	143
0636	1-1/4"		10	.800		93.50	56		149.50	189
0637	1-1/2"		9	.889		161	62		223	273
0638	2"	↓	8	1		206	70		276	335
0639	2-1/2"	Q-1	9	1.778		440	112		552	655
0640	3"		8	2		610	126		736	865
0641	4"		6	2.667		1,300	168		1,468	1,700
0642	5"	↓	5	3.200		3,675	201		3,876	4,350
0643	6"	Q-2	7	3.429		5,575	223		5,798	6,475
0644	8"	"	6	4		15,600	261		15,861	17,600
0650	Coupling, 1/4"	1 Plum	24	.333		2.71	23.50		26.21	39
0670	3/8"		24	.333		3.54	23.50		27.04	40
0680	1/2"		22	.364		2.92	25.50		28.42	42
0690	5/8"		21	.381		7.70	26.50		34.20	49.50
0700	3/4"		21	.381		5.85	26.50		32.35	47.50
0710	1"		18	.444		11.70	31		42.70	61
0715	1-1/4"		17	.471		20.50	33		53.50	73
0716	1-1/2"		15	.533		27	37		64	87
0718	2"		13	.615		45	43		88	116
0721	2-1/2"	Q-1	15	1.067		92.50	67		159.50	205
0722	3"		13	1.231		126	77.50		203.50	258
0724	3-1/2"		8	2		237	126		363	455
0726	4"		7	2.286		278	144		422	525
0728	5"	↓	6	2.667		580	168		748	895
0731	6"	Q-2	8	3	↓	960	195		1,155	1,350
2000	DWV, solder joints, copper x copper									
2030	90° elbow, 1-1/4"	1 Plum	13	.615	Ea.	41.50	43		84.50	112
2050	1-1/2"		12	.667		55	46.50		101.50	132
2070	2"	↓	10	.800		88	56		144	183
2090	3"	Q-1	10	1.600		212	101		313	390
2100	4"	"	9	1.778		1,025	112		1,137	1,300
2250	Tee, sanitary, 1-1/4"	1 Plum	9	.889		72.50	62		134.50	175

22 11 13 – Facility Water Distribution Piping

22 11 13.25 Pipe/Tube Fittings, Copper

		Crew	Daily Output	Labor-Hours	Unit	Material	2022 Bare Costs Labor	2022 Bare Costs Equipment	Total	Total Incl O&P
2270	1-1/2"	1 Plum	8	1	Ea.	90	70		160	207
2290	2"	↓	7	1.143		124	80		204	259
2310	3"	Q-1	7	2.286		520	144		664	790
2330	4"	"	6	2.667		1,150	168		1,318	1,500
2400	Coupling, 1-1/4"	1 Plum	14	.571		17.20	40		57.20	80.50
2420	1-1/2"		13	.615		21.50	43		64.50	89.50
2440	2"	↓	11	.727		29.50	51		80.50	111
2460	3"	Q-1	11	1.455		68.50	91.50		160	217
2480	4"	"	10	1.600	↓	151	101		252	320
9000	Minimum labor/equipment charge	1 Plum	4	2	Job		140		140	215

22 11 13.44 Pipe, Steel

		Crew	Daily Output	Labor-Hours	Unit	Material	2022 Bare Costs Labor	2022 Bare Costs Equipment	Total	Total Incl O&P
0010	**PIPE, STEEL** R221113-50									
0020	All pipe sizes are to Spec. A-53 unless noted otherwise									
0050	Schedule 40, threaded, with couplings, and clevis hanger									
0060	assemblies sized for covering, 10' OC									
0540	Black, 1/4" diameter	1 Plum	66	.121	L.F.	7.95	8.45		16.40	22
0560	1/2" diameter		63	.127		2.81	8.85		11.66	16.75
0570	3/4" diameter		61	.131		3.41	9.15		12.56	17.85
0580	1" diameter	↓	53	.151		3.89	10.55		14.44	20.50
0590	1-1/4" diameter	Q-1	89	.180		5.05	11.30		16.35	23
0600	1-1/2" diameter		80	.200		5.90	12.55		18.45	26
0610	2" diameter		64	.250		7.90	15.70		23.60	32.50
0620	2-1/2" diameter		50	.320		14.70	20		34.70	47
0630	3" diameter		43	.372		19.20	23.50		42.70	57
0640	3-1/2" diameter		40	.400		28.50	25		53.50	70
0650	4" diameter	↓	36	.444	↓	29	28		57	74.50
1280	All pipe sizes are to Spec. A-53 unless noted otherwise									
1281	Schedule 40, threaded, with couplings and clevis hanger									
1282	assemblies sized for covering, 10' OC									
1290	Galvanized, 1/4" diameter	1 Plum	66	.121	L.F.	10.95	8.45		19.40	25
1310	1/2" diameter		63	.127		3.19	8.85		12.04	17.15
1320	3/4" diameter		61	.131		3.92	9.15		13.07	18.40
1330	1" diameter	↓	53	.151		4.63	10.55		15.18	21.50
1350	1-1/2" diameter	Q-1	80	.200		7.10	12.55		19.65	27
1360	2" diameter		64	.250		9.55	15.70		25.25	34.50
1370	2-1/2" diameter		50	.320		16.50	20		36.50	49
1380	3" diameter		43	.372		21.50	23.50		45	60
1400	4" diameter	↓	36	.444	↓	33.50	28		61.50	80
2000	Welded, sch. 40, on yoke & roll hanger assy's, sized for covering, 10' OC									
2040	Black, 1" diameter	Q-15	93	.172	L.F.	4.65	10.80	.67	16.12	22.50
2070	2" diameter		61	.262		7.85	16.50	1.02	25.37	35.50
2090	3" diameter		43	.372		16.15	23.50	1.44	41.09	55.50
2110	4" diameter		37	.432		22.50	27	1.68	51.18	69
2120	5" diameter	↓	32	.500		25.50	31.50	1.94	58.94	78.50
2130	6" diameter	Q-16	36	.667	↓	27	43.50	1.73	72.23	98.50
9990	Minimum labor/equipment charge	1 Plum	3	2.667	Job		186		186	287

22 11 13.45 Pipe Fittings, Steel, Threaded	Crew	Daily Output	Labor-Hours	Unit	Material	2022 Bare Costs Labor	Equipment	Total	Total Incl O&P
0010 **PIPE FITTINGS, STEEL, THREADED** R221113-50									
0020 Cast iron									
0040 Standard weight, black									
0060 90° elbow, straight									
0070 1/4"	1 Plum	16	.500	Ea.	13.50	35		48.50	69
0080 3/8"		16	.500		19.55	35		54.55	75.50
0090 1/2"		15	.533		8.55	37		45.55	67
0100 3/4"		14	.571		8.90	40		48.90	71.50
0110 1"		13	.615		10.55	43		53.55	77.50
0130 1-1/2"	Q-1	20	.000		20.50	50.50		71	101
0140 2"		18	.889		32.50	56		88.50	122
0150 2-1/2"		14	1.143		77.50	72		149.50	197
0160 3"		10	1.600		127	101		228	295
0180 4"		6	2.667		235	168		403	515
0500 Tee, straight									
0510 1/4"	1 Plum	10	.800	Ea.	22	56		78	111
0520 3/8"		10	.800		20.50	56		76.50	109
0540 3/4"		9	.889		15.55	62		77.55	113
0550 1"		8	1		13.85	70		83.85	123
0570 1-1/2"	Q-1	13	1.231		33	77.50		110.50	155
0580 2"		11	1.455		45.50	91.50		137	191
0590 2-1/2"		9	1.778		119	112		231	305
0600 3"		6	2.667		182	168		350	460
0620 4"		4	4		355	251		606	775
0700 Standard weight, galvanized cast iron									
0720 90° elbow, straight									
0730 1/4"	1 Plum	16	.500	Ea.	18.65	35		53.65	74.50
0740 3/8"		16	.500		18.65	35		53.65	74.50
0750 1/2"		15	.533		23.50	37		60.50	83.50
0760 3/4"		14	.571		18.20	40		58.20	81.50
0770 1"		13	.615		24	43		67	92.50
0790 1-1/2"	Q-1	20	.800		51.50	50.50		102	134
0800 2"		18	.889		66	56		122	159
0810 2-1/2"		14	1.143		135	72		207	260
0820 3"		10	1.600		206	101		307	380
0840 4"		6	2.667		375	168		543	675
1100 Tee, straight									
1110 1/4"	1 Plum	10	.800	Ea.	22.50	56		78.50	111
1120 3/8"		10	.800		19.70	56		75.70	108
1140 3/4"		9	.889		30	62		92	129
1150 1"		8	1		28	70		98	139
1170 1-1/2"	Q-1	13	1.231		65.50	77.50		143	191
1180 2"		11	1.455		81.50	91.50		173	231
1190 2-1/2"		9	1.778		167	112		279	355
1200 3"		6	2.667		545	168		713	860
1220 4"		4	4		645	251		896	1,100
5000 Malleable iron, 150 lb.									
5020 Black									
5040 90° elbow, straight									
5060 1/4"	1 Plum	16	.500	Ea.	6.65	35		41.65	61.50
5070 3/8"		16	.500		6.65	35		41.65	61.50
5090 3/4"		14	.571		5.55	40		45.55	67.50

22 11 Facility Water Distribution

22 11 13 – Facility Water Distribution Piping

22 11 13.45 Pipe Fittings, Steel, Threaded

		Crew	Daily Output	Labor-Hours	Unit	Material	2022 Bare Costs Labor	Equipment	Total	Total Incl O&P
5100	1"	1 Plum	13	.615	Ea.	9.65	43		52.65	76.50
5120	1-1/2"	Q-1	20	.800		21	50.50		71.50	101
5130	2"		18	.889		36	56		92	126
5140	2-1/2"		14	1.143		80.50	72		152.50	200
5150	3"		10	1.600		118	101		219	284
5170	4"		6	2.667		224	168		392	505
5450	Tee, straight									
5470	1/4"	1 Plum	10	.800	Ea.	9.65	56		65.65	96.50
5480	3/8"		10	.800		9.65	56		65.65	96.50
5500	3/4"		9	.889		8.85	62		70.85	105
5510	1"		8	1		15.10	70		85.10	125
5520	1-1/4"	Q-1	14	1.143		24.50	72		96.50	138
5530	1-1/2"	Q-1	13	1.231	Ea.	30.50	77.50		108	153
5540	2"		11	1.455		52	91.50		143.50	198
5550	2-1/2"		9	1.778		112	112		224	295
5560	3"		6	2.667		165	168		333	440
5570	3-1/2"		5	3.200		400	201		601	750
5580	4"		4	4		400	251		651	825
5650	Coupling									
5670	1/4"	1 Plum	19	.421	Ea.	8.25	29.50		37.75	54.50
5680	3/8"		19	.421		8.25	29.50		37.75	54.50
5690	1/2"		19	.421		6.35	29.50		35.85	52.50
5700	3/4"		18	.444		7.45	31		38.45	56
5710	1"		15	.533		11.15	37		48.15	70
5720	1-1/4"	Q-1	26	.615		12.80	38.50		51.30	73.50
5730	1-1/2"		24	.667		17.35	42		59.35	83.50
5740	2"		21	.762		25.50	48		73.50	102
5750	2-1/2"		18	.889		71	56		127	164
5760	3"		14	1.143		95.50	72		167.50	216
5780	4"		10	1.600		193	101		294	365
6000	For galvanized elbows, tees, and couplings, add					20%				
9990	Minimum labor/equipment charge	1 Plum	4	2	Job		140		140	215

22 11 13.47 Pipe Fittings, Steel

		Crew	Daily Output	Labor-Hours	Unit	Material	2022 Bare Costs Labor	Equipment	Total	Total Incl O&P
0010	**PIPE FITTINGS, STEEL**, flanged, welded & special									
3000	Weld joint, butt, carbon steel, standard weight									
3040	90° elbow, long radius									
3050	1/2" pipe size	Q-15	16	1	Ea.	51	63	3.88	117.88	157
3060	3/4" pipe size		16	1		51	63	3.88	117.88	157
3070	1" pipe size		16	1		30.50	63	3.88	97.38	135
3080	1-1/4" pipe size		14	1.143		30.50	72	4.43	106.93	149
3090	1-1/2" pipe size		13	1.231		30.50	77.50	4.78	112.78	158
3100	2" pipe size		10	1.600		22	101	6.20	129.20	186
3110	2-1/2" pipe size		8	2		27	126	7.75	160.75	232
3120	3" pipe size		7	2.286		33	144	8.85	185.85	267
3130	4" pipe size		5	3.200		54	201	12.40	267.40	385
3136	5" pipe size		4	4		117	251	15.50	383.50	530
3140	6" pipe size	Q-16	5	4.800		119	315	12.45	446.45	625
3350	Tee, straight									
3360	1/2" pipe size	Q-15	10	1.600	Ea.	134	101	6.20	241.20	310
3370	3/4" pipe size		10	1.600		134	101	6.20	241.20	310
3380	1" pipe size		10	1.600		50	101	6.20	157.20	217
3390	1-1/4" pipe size	Q-15	9	1.778	Ea.	50	112	6.90	168.90	235

432

22 11 13 – Facility Water Distribution Piping

22 11 13.47 Pipe Fittings, Steel

		Crew	Daily Output	Labor-Hours	Unit	Material	2022 Bare Costs Labor	2022 Bare Costs Equipment	Total	Total Incl O&P
3400	1-1/2" pipe size	Q-15	8	2	Ea.	50	126	7.75	183.75	258
3410	2" pipe size		6	2.667		49	168	10.35	227.35	325
3420	2-1/2" pipe size		5	3.200		91.50	201	12.40	304.90	425
3430	3" pipe size		4	4		158	251	15.50	424.50	575
3440	4" pipe size		3	5.333		115	335	20.50	470.50	665
3446	5" pipe size		2.50	6.400		191	400	25	616	860
3450	6" pipe size	Q-16	3	8		198	520	20.50	738.50	1,050
9805	2" x 1"	1 Plum	50	.160		132	11.15		143.15	162
9810	2" x 1-1/4"		50	.160		154	11.15		165.15	187
9815	2" x 1-1/2"		50	.160		160	11.15		171.15	193
9820	2-1/2" x 1"	Q-1	80	.200		138	12.55		150.55	171
9825	2-1/2" x 1-1/4"		80	.200		203	12.55		215.55	242
9830	2-1/2" x 1-1/2"		80	.200		203	12.55		215.55	242
9835	2-1/2" x 2"		80	.200		138	12.55		150.55	171
9840	3" x 1"		80	.200		149	12.55		161.55	183
9845	3" x 1-1/4"		80	.200		194	12.55		206.55	233
9850	3" x 1-1/2"		80	.200		194	12.55		206.55	233
9852	3" x 2"		80	.200		237	12.55		249.55	280
9856	4" x 1"		67	.239		185	15		200	227
9858	4" x 1-1/4"		67	.239		243	15		258	290
9860	4" x 1-1/2"		67	.239		245	15		260	293
9862	4" x 2"		67	.239		254	15		269	305
9864	4" x 2-1/2"		67	.239		262	15		277	310
9866	4" x 3"		67	.239		262	15		277	310
9870	6" x 1-1/2"	Q-2	50	.480		305	31.50		336.50	390
9872	6" x 2"		50	.480		310	31.50		341.50	395
9874	6" x 2-1/2"		50	.480		310	31.50		341.50	390
9876	6" x 3"		50	.480		355	31.50		386.50	440
9878	6" x 4"		50	.480		395	31.50		426.50	485
9880	8" x 2"		42	.571		530	37		567	645
9882	8" x 2-1/2"		42	.571		530	37		567	645
9884	8" x 3"		42	.571		560	37		597	675
9886	8" x 4"		42	.571		1,150	37		1,187	1,300
9990	Minimum labor/equipment charge	Q-15	3	5.333	Job		335	20.50	355.50	540

22 11 13.48 Pipe, Fittings and Valves, Steel, Grooved-Joint

		Crew	Daily Output	Labor-Hours	Unit	Material	2022 Bare Costs Labor	2022 Bare Costs Equipment	Total	Total Incl O&P
0010	**PIPE, FITTINGS AND VALVES, STEEL, GROOVED-JOINT**									
0012	Fittings are ductile iron. Steel fittings noted.									
0020	Pipe includes coupling & clevis type hanger assemblies, 10' OC									
1000	Schedule 40, black									
1040	3/4" diameter	1 Plum	71	.113	L.F.	6.10	7.85		13.95	18.80
1050	1" diameter		63	.127		5.65	8.85		14.50	19.85
1060	1-1/4" diameter		58	.138		7.25	9.65		16.90	23
1070	1-1/2" diameter		51	.157		8.15	10.95		19.10	26
1080	2" diameter		40	.200		9.65	13.95		23.60	32
1090	2-1/2" diameter	Q-1	57	.281		15.30	17.65		32.95	44
1100	3" diameter		50	.320		18.75	20		38.75	51.50
1110	4" diameter		45	.356		26.50	22.50		49	64
1120	5" diameter		37	.432		32	27		59	77.50
3990	Fittings: coupling material required at joints not incl. in fitting price.									
3994	Add 1 selected coupling, material only, per joint for installed price.									
4000	Elbow, 90° or 45°, painted									
4030	3/4" diameter	1 Plum	50	.160	Ea.	101	11.15		112.15	128

22 11 13.48 Pipe, Fittings and Valves, Steel, Grooved-Joint	Crew	Daily Output	Labor-Hours	Unit	Material	2022 Bare Costs Labor	Equipment	Total	Total Incl O&P
4040	1" diameter	1 Plum	50	.160	Ea.	54	11.15		65.15	76.50
4050	1-1/4" diameter		40	.200		54	13.95		67.95	81
4060	1-1/2" diameter		33	.242		54	16.90		70.90	85.50
4070	2" diameter		25	.320		54	22.50		76.50	94
4080	2-1/2" diameter	Q-1	40	.400		54	25		79	98
4090	3" diameter		33	.485		95	30.50		125.50	151
4100	4" diameter		25	.640		103	40		143	175
4110	5" diameter		20	.800		244	50.50		294.50	345
4250	For galvanized elbows, add					26%				
4690	Tee, painted									
4700	3/4" diameter	1 Plum	38	.211	Ea.	108	14.70		122.70	142
4740	1" diameter		33	.242		83.50	16.90		100.40	118
4750	1-1/4" diameter		27	.296		83.50	20.50		104	124
4760	1-1/2" diameter		22	.364		83.50	25.50		109	131
4770	2" diameter		17	.471		83.50	33		116.50	142
4780	2-1/2" diameter	Q-1	27	.593		83.50	37		120.50	149
4790	3" diameter		22	.727		114	45.50		159.50	196
4800	4" diameter		17	.941		173	59		232	281
4810	5" diameter		13	1.231		400	77.50		477.50	560
4900	For galvanized tees, add					24%				
4933	20" diameter	Q-2	16	1.500		835	97.50		932.50	1,075
4940	Flexible, standard, painted									
4950	3/4" diameter	1 Plum	100	.080	Ea.	31	5.60		36.60	42.50
4960	1" diameter		100	.080		31	5.60		36.60	42.50
4970	1-1/4" diameter		80	.100		40	7		47	55
4980	1-1/2" diameter	1 Plum	67	.119	Ea.	43	8.35		51.35	60.50
4990	2" diameter	"	50	.160		46.50	11.15		57.65	68
5000	2-1/2" diameter	Q-1	80	.200		54	12.55		66.55	78.50
5010	3" diameter		67	.239		59	15		74	88
5020	3-1/2" diameter		57	.281		84.50	17.65		102.15	120
5030	4" diameter		50	.320		85.50	20		105.50	125
5040	5" diameter		40	.400		128	25		153	180
5050	6" diameter	Q-2	50	.480		152	31.50		183.50	215
5200	For galvanized couplings, add					33%				
7790	Valves: coupling material required at joints not incl. in valve price.									
7794	Add 1 selected coupling, material only, per joint for installed price.									
9990	Minimum labor/equipment charge	1 Plum	4	2	Job		140		140	215

22 11 13.74 Pipe, Plastic

		Crew	Daily Output	Labor-Hours	Unit	Material	2022 Bare Costs Labor	Equipment	Total	Total Incl O&P
0010	**PIPE, PLASTIC**									
1800	PVC, couplings 10' OC, clevis hanger assemblies, 3 per 10'									
1820	Schedule 40									
1860	1/2" diameter	1 Plum	54	.148	L.F.	6.35	10.35		16.70	23
1870	3/4" diameter		51	.157		6.90	10.95		17.85	24.50
1880	1" diameter		46	.174		6.25	12.15		18.40	25.50
1890	1-1/4" diameter		42	.190		7.55	13.30		20.85	29
1900	1-1/2" diameter		36	.222		8.95	15.50		24.45	34
1910	2" diameter	Q-1	59	.271		11.10	17.05		28.15	39
1920	2-1/2" diameter		56	.286		16.70	17.95		34.65	46
1930	3" diameter		53	.302		20.50	18.95		39.45	52
1940	4" diameter		48	.333		14.60	21		35.60	48.50
1950	5" diameter		43	.372		30	23.50		53.50	69
1960	6" diameter		39	.410		31.50	26		57.50	74.50

22 11 13.74 Pipe, Plastic

		Crew	Daily Output	Labor-Hours	Unit	Material	2022 Bare Costs Labor	Equipment	Total	Total Incl O&P
4100	DWV type, schedule 40, couplings 10' OC, clevis hanger assy's, 3 per 10'									
4210	ABS, schedule 40, foam core type									
4212	Plain end black									
4214	1-1/2" diameter	1 Plum	39	.205	L.F.	3.40	14.30		17.70	25.50
4216	2" diameter	Q-1	62	.258		4.02	16.20		20.22	29.50
4218	3" diameter		56	.286		7.20	17.95		25.15	35.50
4220	4" diameter		51	.314		10.45	19.70		30.15	42
4222	6" diameter		42	.381		27.50	24		51.50	67
4240	To delete coupling & hangers, subtract									
4244	1-1/2" diam. to 6" diam.					43%	48%			
4400	PVC									
4410	1-1/4" diameter	1 Plum	42	.190	L.F.	5.95	13.30		19.25	27
4460	2" diameter	Q-1	59	.271		6.25	17.05		23.30	33.50
4470	3" diameter		53	.302		8.30	18.95		27.25	38
4480	4" diameter		48	.333		11.55	21		32.55	45
4490	6" diameter		39	.410		25.50	26		51.50	67.50
5300	CPVC, socket joint, couplings 10' OC, clevis hanger assemblies, 3 per 10'									
5302	Schedule 40									
5304	1/2" diameter	1 Plum	54	.148	L.F.	6.25	10.35		16.60	23
5305	3/4" diameter		51	.157		7.75	10.95		18.70	25.50
5306	1" diameter		46	.174		11.85	12.15		24	32
5307	1-1/4" diameter		42	.190		8.15	13.30		21.45	29.50
5308	1-1/2" diameter		36	.222		8	15.50		23.50	33
5309	2" diameter	Q-1	59	.271		19.75	17.05		36.80	48.50
5310	2-1/2" diameter		56	.286		19.55	17.95		37.50	49
5311	3" diameter		53	.302		34	18.95		52.95	66
5360	CPVC, threaded, couplings 10' OC, clevis hanger assemblies, 3 per 10'									
5380	Schedule 40									
5460	1/2" diameter	1 Plum	54	.148	L.F.	7.45	10.35		17.80	24
5470	3/4" diameter		51	.157		9.85	10.95		20.80	28
5480	1" diameter		46	.174		14	12.15		26.15	34
5490	1-1/4" diameter		42	.190		9.85	13.30		23.15	31.50
5500	1-1/2" diameter		36	.222		9.45	15.50		24.95	34.50
5510	2" diameter	Q-1	59	.271		21.50	17.05		38.55	50
5520	2-1/2" diameter		56	.286		21.50	17.95		39.45	51
5530	3" diameter		53	.302		36.50	18.95		55.45	69
9900	Minimum labor/equipment charge	1 Plum	4	2	Job		140		140	215

22 11 13.76 Pipe Fittings, Plastic

		Crew	Daily Output	Labor-Hours	Unit	Material	2022 Bare Costs Labor	Equipment	Total	Total Incl O&P
0010	**PIPE FITTINGS, PLASTIC**									
2700	PVC (white), schedule 40, socket joints									
2760	90° elbow, 1/2"	1 Plum	33.30	.240	Ea.	.68	16.75		17.43	27
2810	2"	Q-1	36.40	.440	"	4.04	27.50		31.54	47
4500	DWV, ABS, non pressure, socket joints									
4540	1/4 bend, 1-1/4"	1 Plum	20.20	.396	Ea.	13.55	27.50		41.05	57.50
4560	1-1/2"	"	18.20	.440		10.50	30.50		41	59
4570	2"	Q-1	33.10	.483		10.35	30.50		40.85	58.50
4650	1/8 bend, same as 1/4 bend									
4800	Tee, sanitary									
4820	1-1/4"	1 Plum	13.50	.593	Ea.	18.05	41.50		59.55	84
4830	1-1/2"	"	12.10	.661	"	15.55	46		61.55	88
4840	2"	Q-1	20	.800	Ea.	24	50.50		74.50	104
5000	DWV, PVC, schedule 40, socket joints									

22 11 13 – Facility Water Distribution Piping

22 11 13.76 Pipe Fittings, Plastic		Crew	Daily Output	Labor-Hours	Unit	Material	2022 Bare Costs Labor	Equipment	Total	Total Incl O&P
5040	1/4 bend, 1-1/4"	1 Plum	20.20	.396	Ea.	27.50	27.50		55	72.50
5060	1-1/2"	"	18.20	.440		7.85	30.50		38.35	56
5070	2"	Q-1	33.10	.483		12.40	30.50		42.90	60.50
5080	3"		20.80	.769		36.50	48.50		85	115
5090	4"		16.50	.970		72	61		133	173
5100	6"		10.10	1.584		252	99.50		351.50	430
5105	8"	Q-2	9.30	2.581		325	168		493	620
5110	1/4 bend, long sweep, 1-1/2"	1 Plum	18.20	.440		18.25	30.50		48.75	67.50
5112	2"	Q-1	33.10	.483		20.50	30.50		51	69.50
5114	3"		20.80	.769		47	48.50		95.50	126
5116	4"		16.50	.970		88.50	61		149.50	192
5150	1/8 bend, 1-1/4"	1 Plum	20.20	.396		18.70	27.50		46.20	63
5215	8"	Q-2	9.30	2.581		108	168		276	380
5250	Tee, sanitary 1-1/4"	1 Plum	13.50	.593		29.50	41.50		71	96.50
5254	1-1/2"	"	12.10	.661		13.70	46		59.70	86
5255	2"	Q-1	20	.800		20	50.50		70.50	99.50
5256	3"		13.90	1.151		53	72.50		125.50	170
5257	4"		11	1.455		97	91.50		188.50	248
5259	6"		6.70	2.388		390	150		540	660
5261	8"	Q-2	6.20	3.871		865	252		1,117	1,350
5264	2" x 1-1/2"	Q-1	22	.727		17.85	45.50		63.35	90
5266	3" x 1-1/2"		15.50	1.032		38.50	65		103.50	143
5268	4" x 3"		12.10	1.322		114	83		197	253
5271	6" x 4"		6.90	2.319		375	146		521	640
5314	Combination Y & 1/8 bend, 1-1/2"	1 Plum	12.10	.661		33.50	46		79.50	108
5315	2"	Q-1	20	.800		41.50	50.50		92	123
5317	3"		13.90	1.151		91.50	72.50		164	212
5318	4"		11	1.455		180	91.50		271.50	340
5324	Combination Y & 1/8 bend, reducing									
5325	2" x 2" x 1-1/2"	Q-1	22	.727	Ea.	47	45.50		92.50	122
5327	3" x 3" x 1-1/2"		15.50	1.032		83.50	65		148.50	192
5328	3" x 3" x 2"		15.30	1.046		62.50	65.50		128	170
5329	4" x 4" x 2"		12.20	1.311		94	82.50		176.50	231
5331	Wye, 1-1/4"	1 Plum	13.50	.593		37.50	41.50		79	106
5332	1-1/2"	"	12.10	.661		25	46		71	98.50
5333	2"	Q-1	20	.800		24.50	50.50		75	105
5334	3"		13.90	1.151		66.50	72.50		139	184
5335	4"		11	1.455		121	91.50		212.50	274
5336	6"		6.70	2.388		350	150		500	615
5337	8"	Q-2	6.20	3.871		580	252		832	1,025
5341	2" x 1-1/2"	Q-1	22	.727		13.35	45.50		58.85	85
5342	3" x 1-1/2"		15.50	1.032		19.80	65		84.80	122
5343	4" x 3"		12.10	1.322		43.50	83		126.50	176
5344	6" x 4"		6.90	2.319		118	146		264	355
5345	8" x 6"	Q-2	6.40	3.750		255	244		499	655
5347	Double wye, 1-1/2"	1 Plum	9.10	.879		56.50	61.50		118	157
5348	2"	Q-1	16.60	.964		67	60.50		127.50	167
5349	3"		10.40	1.538		130	96.50		226.50	293
5350	4"		8.25	1.939		281	122		403	500
5353	Double wye, reducing									
5354	2" x 2" x 1-1/2" x 1-1/2"	Q-1	16.80	.952	Ea.	57.50	60		117.50	155
5355	3" x 3" x 2" x 2"		10.60	1.509		97.50	95		192.50	253
5356	4" x 4" x 3" x 3"		8.45	1.893		210	119		329	415

22 11 13.76 Pipe Fittings, Plastic	Crew	Daily Output	Labor-Hours	Unit	Material	2022 Bare Costs Labor	Equipment	Total	Total Incl O&P
5357	6" x 6" x 4" x 4"	Q-1	7.25	2.207	Ea.	735	139		874	1,025
5374	Coupling, 1-1/4"	1 Plum	20.20	.396		17.70	27.50		45.20	62
5376	1-1/2"	"	18.20	.440		3.70	30.50		34.20	51.50
5378	2"	Q-1	33.10	.483		5.05	30.50		35.55	52.50
5380	3"		20.80	.769		17.65	48.50		66.15	94
5390	4"		16.50	.970		30	61		91	127
5410	Reducer bushing, 2" x 1-1/4"		36.50	.438		4.64	27.50		32.14	47.50
5412	3" x 1-1/2"		27.30	.586		13.70	37		50.70	72
5414	4" x 2"		18.20	.879		53.50	55		108.50	144
5416	6" x 4"	▼	11.10	1.441		138	90.50		228.50	292
5418	8" x 6"	Q-2	10.20	2.353	▼	270	153		423	535
5500	CPVC, Schedule 80, threaded joints									
5540	90° elbow, 1/4"	1 Plum	32	.250	Ea.	18.35	17.45		35.80	47
5560	1/2"		30.30	.264		10.65	18.45		29.10	40.50
5570	3/4"		26	.308		15.95	21.50		37.45	50.50
5580	1"		22.70	.352		22.50	24.50		47	62.50
5590	1-1/4"		20.20	.396		43	27.50		70.50	90
5600	1-1/2"	▼	18.20	.440		46.50	30.50		77	98.50
5610	2"	Q-1	33.10	.483		62	30.50		92.50	116
5620	2-1/2"		24.20	.661		193	41.50		234.50	276
5630	3"	▼	20.80	.769		207	48.50		255.50	300
5730	Coupling, 1/4"	1 Plum	32	.250		23.50	17.45		40.95	52.50
5732	1/2"		30.30	.264		19.25	18.45		37.70	49.50
5734	3/4"		26	.308		31	21.50		52.50	67
5736	1"		22.70	.352		35.50	24.50		60	77
5738	1-1/4"		20.20	.396		37.50	27.50		65	83.50
5740	1-1/2"	▼	18.20	.440		40	30.50		70.50	91.50
5742	2"	Q-1	33.10	.483		47.50	30.50		78	99
5744	2-1/2"		24.20	.661		85	41.50		126.50	158
5746	3"	▼	20.80	.769	▼	99	48.50		147.50	184
5900	CPVC, Schedule 80, socket joints									
5904	90° elbow, 1/4"	1 Plum	32	.250	Ea.	17.50	17.45		34.95	46.50
5906	1/2"		30.30	.264		6.85	18.45		25.30	36
5908	3/4"		26	.308		8.75	21.50		30.25	42.50
5910	1"		22.70	.352		13.85	24.50		38.35	53.50
5912	1-1/4"		20.20	.396		30	27.50		57.50	75.50
5914	1-1/2"	▼	18.20	.440		33.50	30.50		64	84.50
5916	2"	Q-1	33.10	.483		40.50	30.50		71	91.50
5918	2-1/2"		24.20	.661		93	41.50		134.50	166
5920	3"	▼	20.80	.769		105	48.50		153.50	191
5930	45° elbow, 1/4"	1 Plum	32	.250		26	17.45		43.45	55.50
5932	1/2"		30.30	.264		8.40	18.45		26.85	37.50
5934	3/4"		26	.308		12.10	21.50		33.60	46.50
5936	1"		22.70	.352		19.25	24.50		43.75	59
5938	1-1/4"		20.20	.396		38	27.50		65.50	84
5940	1-1/2"	▼	18.20	.440		39	30.50		69.50	90
5942	2"	Q-1	33.10	.483		43.50	30.50		74	95
5944	2-1/2"		24.20	.661		89	41.50		130.50	162
5946	3"	▼	20.80	.769		114	48.50		162.50	201
5990	Coupling, 1/4"	1 Plum	32	.250		18.60	17.45		36.05	47.50
5992	1/2"		30.30	.264		7.20	18.45		25.65	36.50
5994	3/4"		26	.308		10.10	21.50		31.60	44
5996	1"	▼	22.70	.352	▼	13.65	24.50		38.15	53

22 11 13 – Facility Water Distribution Piping

22 11 13.76 Pipe Fittings, Plastic		Crew	Daily Output	Labor-Hours	Unit	Material	2022 Bare Costs Labor	Equipment	Total	Total Incl O&P
5998	1-1/4"	1 Plum	20.20	.396	Ea.	20.50	27.50		48	65
6000	1-1/2"	↓	18.20	.440		25.50	30.50		56	75.50
6002	2"	Q-1	33.10	.483		30	30.50		60.50	80
6004	2-1/2"		24.20	.661		66.50	41.50		108	137
6006	3"	↓	20.80	.769	↓	72	48.50		120.50	154
9990	Minimum labor/equipment charge	1 Plum	4	2	Job		140		140	215

22 11 19 – Domestic Water Piping Specialties

22 11 19.10 Flexible Connectors

22 11 19.10 Flexible Connectors		Crew	Daily Output	Labor-Hours	Unit	Material	2022 Bare Costs Labor	Equipment	Total	Total Incl O&P
0010	**FLEXIBLE CONNECTORS**, Corrugated, 5/8" OD, 3/4" ID									
0050	Gas, seamless brass, steel fittings									
0200	12" long	1 Plum	36	.222	Ea.	20.50	15.50		36	47
0220	18" long		36	.222		25.50	15.50		41	52.50
0240	24" long	↓	34	.235	↓	30.50	16.40		46.90	59
0260	30" long	1 Plum	34	.235	Ea.	33	16.40		49.40	61.50
9000	Minimum labor/equipment charge	"	4	2	Job		140		140	215

22 11 19.18 Mixing Valve

22 11 19.18 Mixing Valve		Crew	Daily Output	Labor-Hours	Unit	Material	2022 Bare Costs Labor	Equipment	Total	Total Incl O&P
0010	**MIXING VALVE**, Automatic, water tempering.									
0040	1/2" size	1 Stpi	19	.421	Ea.	490	29.50		519.50	585
0050	3/4" size		18	.444	"	440	31.50		471.50	530
9000	Minimum labor/equipment charge		5	1.600	Job		113		113	173

22 11 19.38 Water Supply Meters

22 11 19.38 Water Supply Meters		Crew	Daily Output	Labor-Hours	Unit	Material	2022 Bare Costs Labor	Equipment	Total	Total Incl O&P
0010	**WATER SUPPLY METERS**									
2000	Domestic/commercial, bronze									
2020	Threaded									
2060	5/8" diameter, to 20 GPM	1 Plum	16	.500	Ea.	148	35		183	217
2080	3/4" diameter, to 30 GPM		14	.571		370	40		410	465
2100	1" diameter, to 50 GPM	↓	12	.667	↓	500	46.50		546.50	620
2300	Threaded/flanged									
2340	1-1/2" diameter, to 100 GPM	1 Plum	8	1	Ea.	1,150	70		1,220	1,350
2600	Flanged, compound									
9000	Minimum labor/equipment charge	1 Plum	3.25	2.462	Job		172		172	265

22 11 19.42 Backflow Preventers

22 11 19.42 Backflow Preventers		Crew	Daily Output	Labor-Hours	Unit	Material	2022 Bare Costs Labor	Equipment	Total	Total Incl O&P
0010	**BACKFLOW PREVENTERS**, Includes valves									
0020	and four test cocks, corrosion resistant, automatic operation									
1000	Double check principle									
1010	Threaded, with ball valves									
1020	3/4" pipe size	1 Plum	16	.500	Ea.	203	35		238	277
1030	1" pipe size		14	.571		315	40		355	405
1040	1-1/2" pipe size		10	.800		655	56		711	805
1050	2" pipe size	↓	7	1.143	↓	900	80		980	1,125
1080	Threaded, with gate valves									
1100	3/4" pipe size	1 Plum	16	.500	Ea.	1,575	35		1,610	1,775
1120	1" pipe size		14	.571		1,600	40		1,640	1,825
1140	1-1/2" pipe size		10	.800		1,425	56		1,481	1,650
1160	2" pipe size	↓	7	1.143	↓	2,525	80		2,605	2,900
4000	Reduced pressure principle									
4100	Threaded, bronze, valves are ball									
4120	3/4" pipe size	1 Plum	16	.500	Ea.	560	35		595	670
4140	1" pipe size		14	.571		595	40		635	715
4150	1-1/4" pipe size		12	.667		1,450	46.50		1,496.50	1,675
4160	1-1/2" pipe size		10	.800		1,200	56		1,256	1,400

22 11 19 – Domestic Water Piping Specialties

22 11 19.42 Backflow Preventers

		Crew	Daily Output	Labor-Hours	Unit	Material	2022 Bare Costs Labor	2022 Bare Costs Equipment	Total	Total Incl O&P
4180	2" pipe size	1 Plum	7	1.143	Ea.	1,925	80		2,005	2,225
5000	Flanged, bronze, valves are OS&Y									
5060	2-1/2" pipe size	Q-1	5	3.200	Ea.	6,475	201		6,676	7,425
5080	3" pipe size		4.50	3.556		7,100	223		7,323	8,175
5100	4" pipe size		3	5.333		9,100	335		9,435	10,500
5120	6" pipe size	Q-2	3	8		12,900	520		13,420	14,900
5600	Flanged, iron, valves are OS&Y									
5660	2-1/2" pipe size	Q-1	5	3.200	Ea.	3,225	201		3,426	3,850
5680	3" pipe size		4.50	3.556		3,325	223		3,548	4,000
5700	4" pipe size		3	5.333		4,300	335		4,635	5,250
5720	6" pipe size	Q-2	3	8		6,350	520		6,870	7,800
5800	Rebuild 4" diameter reduced pressure	1 Plum	4	2		500	140		640	765
5810	6" diameter		2.66	3.008		530	210		740	905
5820	8" diameter		2	4		660	279		939	1,150
5830	10" diameter		1.60	5		770	350		1,120	1,375
9010	Minimum labor/equipment charge		2	4	Job		279		279	430

22 11 19.50 Vacuum Breakers

		Crew	Daily Output	Labor-Hours	Unit	Material	2022 Bare Costs Labor	2022 Bare Costs Equipment	Total	Total Incl O&P
0010	**VACUUM BREAKERS**									
0013	See also backflow preventers Section 22 11 19.42									
1000	Anti-siphon continuous pressure type									
1010	Max. 150 psi - 210°F									
1020	Bronze body									
1030	1/2" size	1 Stpi	24	.333	Ea.	274	23.50		297.50	335
1040	3/4" size		20	.400		274	28		302	345
1050	1" size		19	.421		280	29.50		309.50	355
1060	1-1/4" size		15	.533		555	37.50		592.50	675
1070	1-1/2" size		13	.615		675	43.50		718.50	805
1080	2" size		11	.727		440	51		491	560
1200	Max. 125 psi with atmospheric vent									
1210	Brass, in-line construction									
1220	1/4" size	1 Stpi	24	.333	Ea.	187	23.50		210.50	241
1230	3/8" size	"	24	.333		187	23.50		210.50	241
1260	For polished chrome finish, add					13%				
2000	Anti-siphon, non-continuous pressure type									
2010	Hot or cold water 125 psi - 210°F									
2020	Bronze body									
2030	1/4" size	1 Stpi	24	.333	Ea.	114	23.50		137.50	161
2040	3/8" size		24	.333		114	23.50		137.50	161
2050	1/2" size		24	.333		128	23.50		151.50	177
2060	3/4" size		20	.400		153	28		181	212
2070	1" size		19	.421		236	29.50		265.50	305
2080	1-1/4" size	1 Stpi	15	.533	Ea.	415	37.50		452.50	515
2090	1-1/2" size		13	.615		460	43.50		503.50	570
2100	2" size		11	.727		755	51		806	910
2110	2-1/2" size		8	1		2,175	70.50		2,245.50	2,475
2120	3" size		6	1.333		2,875	94		2,969	3,325
2150	For polished chrome finish, add					50%				
3000	Air gap fitting									
3020	1/2" NPT size	1 Plum	19	.421	Ea.	83.50	29.50		113	137
3030	1" NPT size	"	15	.533		83.50	37		120.50	149
3040	2" NPT size	Q-1	21	.762		105	48		153	190
3050	3" NPT size		14	1.143		340	72		412	485

439

For customer support on your Commercial Renovation Costs with RSMeans data, call 800.448.8182.

22 11 19 – Domestic Water Piping Specialties

22 11 19.50 Vacuum Breakers		Crew	Daily Output	Labor-Hours	Unit	Material	2022 Bare Costs Labor	Equipment	Total	Total Incl O&P
3060	4" NPT size	Q-1	10	1.600	Ea.	223	101		324	400

22 11 19.54 Water Hammer Arresters/Shock Absorbers

		Crew	Daily Output	Labor-Hours	Unit	Material	Labor	Equipment	Total	Total Incl O&P
0010	**WATER HAMMER ARRESTERS/SHOCK ABSORBERS**									
0490	Copper									
0500	3/4" male IPS for 1 to 11 fixtures	1 Plum	12	.667	Ea.	35.50	46.50		82	111
0600	1" male IPS for 12 to 32 fixtures		8	1		59.50	70		129.50	174
0700	1-1/4" male IPS for 33 to 60 fixtures		8	1		58	70		128	172
0800	1-1/2" male IPS for 61 to 113 fixtures		8	1		84	70		154	201
0900	2" male IPS for 114 to 154 fixtures		8	1		124	70		194	244
1000	2-1/2" male IPS for 155 to 330 fixtures		4	2		355	140		495	605
9000	Minimum labor/equipment charge		3.50	2.286	Job		160		160	246

22 11 19.64 Hydrants

		Crew	Daily Output	Labor-Hours	Unit	Material	Labor	Equipment	Total	Total Incl O&P
0010	**HYDRANTS**									
0050	Wall type, moderate climate, bronze, encased									
0200	3/4" IPS connection	1 Plum	16	.500	Ea.	1,375	35		1,410	1,550
1000	Non-freeze, bronze, exposed									
1100	3/4" IPS connection, 4" to 9" thick wall	1 Plum	14	.571	Ea.	655	40		695	780
1120	10" to 14" thick wall	"	12	.667		575	46.50		621.50	700
1280	For anti-siphon type, add					246			246	271
9000	Minimum labor/equipment charge	1 Plum	3	2.667	Job		186		186	287

22 13 16 – Sanitary Waste and Vent Piping

22 13 16.20 Pipe, Cast Iron

		Crew	Daily Output	Labor-Hours	Unit	Material	Labor	Equipment	Total	Total Incl O&P
0010	**PIPE, CAST IRON**, Soil, on clevis hanger assemblies, 5' OC　　R221113-50									
0020	Single hub, service wt., lead & oakum joints 10' OC									
2120	2" diameter	Q-1	63	.254	L.F.	17.15	15.95		33.10	43.50
2140	3" diameter		60	.267		22.50	16.75		39.25	50.50
2160	4" diameter		55	.291		29	18.30		47.30	60
2180	5" diameter	Q-2	76	.316	L.F.	31	20.50		51.50	65.50
2200	6" diameter	"	73	.329		49.50	21.50		71	87.50
2220	8" diameter	Q-3	59	.542		79.50	36		115.50	143
4000	No hub, couplings 10' OC									
4100	1-1/2" diameter	Q-1	71	.225	L.F.	8.65	14.15		22.80	31.50
4120	2" diameter		67	.239		9.10	15		24.10	33
4140	3" diameter		64	.250		11.70	15.70		27.40	37
4160	4" diameter		58	.276		15.45	17.35		32.80	43.50
4180	5" diameter	Q-2	83	.289		22.50	18.85		41.35	54
9000	Minimum labor/equipment charge	1 Plum	4	2	Job		140		140	215

22 13 16.30 Pipe Fittings, Cast Iron

		Crew	Daily Output	Labor-Hours	Unit	Material	Labor	Equipment	Total	Total Incl O&P
0010	**PIPE FITTINGS, CAST IRON**, Soil　　R221113-50									
0040	Hub and spigot, service weight, lead & oakum joints									
0080	1/4 bend, 2"	Q-1	16	1	Ea.	27	63		90	127
0120	3"		14	1.143		36	72		108	151
0140	4"		13	1.231		56.50	77.50		134	181
0160	5"	Q-2	18	1.333		78.50	87		165.50	220
0180	6"	"	17	1.412		98	92		190	250
0200	8"	Q-3	11	2.909		295	193		488	625
0340	1/8 bend, 2"	Q-1	16	1		19.10	63		82.10	118
0350	3"		14	1.143		30	72		102	144

22 13 16.30 Pipe Fittings, Cast Iron		Crew	Daily Output	Labor-Hours	Unit	Material	2022 Bare Costs Labor	Equipment	Total	Total Incl O&P
0360	4"	Q-1	13	1.231	Ea.	44	77.50		121.50	168
0380	5"	Q-2	18	1.333		61.50	87		148.50	202
0400	6"	"	17	1.412		74.50	92		166.50	224
0420	8"	Q-3	11	2.909		222	193		415	540
0500	Sanitary tee, 2"	Q-1	10	1.600		37.50	101		138.50	197
0540	3"		9	1.778		60.50	112		172.50	239
0620	4"		8	2		74.50	126		200.50	276
0700	5"	Q-2	12	2		148	130		278	365
0800	6"	"	11	2.182		168	142		310	405
0880	8"	Q-3	7	4.571		445	305		750	960
5990	No hub									
6000	Cplg. & labor required at joints not incl. in fitting									
6010	price. Add 1 coupling per joint for installed price									
6020	1/4 bend, 1-1/2"				Ea.	13.85			13.85	15.25
6060	2"					15.05			15.05	16.55
6080	3"					21			21	23
6120	4"					31			31	34.50
6184	1/4 bend, long sweep, 1-1/2"					35			35	38.50
6186	2"					33			33	36.50
6188	3"				Ea.	40			40	44
6189	4"					63.50			63.50	69.50
6190	5"					123			123	135
6191	6"					140			140	154
6192	8"					365			365	400
6193	10"					705			705	775
6200	1/8 bend, 1-1/2"					11.55			11.55	12.70
6210	2"					13			13	14.30
6212	3"					17.45			17.45	19.15
6214	4"					23			23	25
6380	Sanitary tee, tapped, 1-1/2"					27.50			27.50	30
6382	2" x 1-1/2"					24.50			24.50	27
6384	2"					26			26	28.50
6386	3" x 2"					39			39	42.50
6388	3"					67			67	73.50
6390	4" x 1-1/2"					34.50			34.50	38
6392	4" x 2"					39			39	43
6393	4"					39			39	43
6394	6" x 1-1/2"					92.50			92.50	102
6396	6" x 2"					98			98	108
6459	Sanitary tee, 1-1/2"					19.40			19.40	21.50
6460	2"					21			21	23
6470	3"					25.50			25.50	28
6472	4"					48.50			48.50	53.50
8000	Coupling, standard (by CISPI Mfrs.)									
8020	1-1/2"	Q-1	48	.333	Ea.	14.65	21		35.65	48.50
8040	2"		44	.364		16.35	23		39.35	53
8080	3"		38	.421		19.20	26.50		45.70	62
8120	4"		33	.485		25	30.50		55.50	74.50
8300	Coupling, cast iron clamp & neoprene gasket (by MG)									
8310	1-1/2"	Q-1	48	.333	Ea.	10.25	21		31.25	44
8320	2"		44	.364		13.75	23		36.75	50
8330	3"		38	.421		13.50	26.50		40	56
8340	4"		33	.485		20.50	30.50		51	69.50

22 13 16 – Sanitary Waste and Vent Piping

22 13 16.30 Pipe Fittings, Cast Iron

		Crew	Daily Output	Labor-Hours	Unit	Material	2022 Bare Costs Labor	Equipment	Total	Total Incl O&P
8600	Coupling, stainless steel, heavy duty									
8620	1-1/2"	Q-1	48	.333	Ea.	5.65	21		26.65	38.50
8630	2"		44	.364		6.10	23		29.10	41.50
8640	2" x 1-1/2"		44	.364		13.75	23		36.75	50
8650	3"		38	.421		6.40	26.50		32.90	48
8660	4"		33	.485		7.20	30.50		37.70	55
9000	Minimum labor/equipment charge	1 Plum	4	2	Job		140		140	215

22 13 16.60 Traps

		Crew	Daily Output	Labor-Hours	Unit	Material	2022 Bare Costs Labor	Equipment	Total	Total Incl O&P
0010	**TRAPS**									
0030	Cast iron, service weight									
0050	Running P trap, without vent									
1100	2"	Q-1	16	1	Ea.	225	63		288	345
1150	4"	"	13	1.231		194	77.50		271.50	330
1160	6"	Q-2	17	1.412		860	92		952	1,075
3000	P trap, B&S, 2" pipe size	Q-1	16	1		46	63		109	148
3040	3" pipe size	"	14	1.143		68.50	72		140.50	187
4700	Copper, drainage, drum trap									
4840	3" x 6" swivel, 1-1/2" pipe size	1 Plum	16	.500	Ea.	1,200	35		1,235	1,375
5100	P trap, standard pattern									
5200	1-1/4" pipe size	1 Plum	18	.444	Ea.	355	31		386	445
5240	1-1/2" pipe size		17	.471		272	33		305	350
5260	2" pipe size		15	.533		420	37		457	520
5280	3" pipe size		11	.727		1,200	51		1,251	1,400
9000	Minimum labor/equipment charge		3	2.667	Job		186		186	287

22 13 16.80 Vent Flashing and Caps

		Crew	Daily Output	Labor-Hours	Unit	Material	2022 Bare Costs Labor	Equipment	Total	Total Incl O&P
0010	**VENT FLASHING AND CAPS**									
0120	Vent caps									
0140	Cast iron									
0180	2-1/2" to 3-5/8" pipe	1 Plum	21	.381	Ea.	55.50	26.50		82	102
0190	4" to 4-1/8" pipe	"	19	.421	"	80	29.50		109.50	134
0900	Vent flashing									
1000	Aluminum with lead ring									
1050	3" pipe	1 Plum	17	.471	Ea.	6.30	33		39.30	57.50
1060	4" pipe	"	16	.500	"	7.60	35		42.60	62.50
1350	Copper with neoprene ring									
1440	2" pipe	1 Plum	18	.444	Ea.	78.50	31		109.50	134
1450	3" pipe		17	.471		94.50	33		127.50	155
1460	4" pipe		16	.500		94.50	35		129.50	158
9000	Minimum labor/equipment charge		4	2	Job		140		140	215

22 13 19 – Sanitary Waste Piping Specialties

22 13 19.13 Sanitary Drains

		Crew	Daily Output	Labor-Hours	Unit	Material	2022 Bare Costs Labor	Equipment	Total	Total Incl O&P
0010	**SANITARY DRAINS**									
0400	Deck, auto park, CI, 13" top									
0440	3", 4", 5", and 6" pipe size	Q-1	8	2	Ea.	2,275	126		2,401	2,700
0480	For galvanized body, add				"	1,400			1,400	1,550
2000	Floor, medium duty, CI, deep flange, 7" diam. top									
2040	2" and 3" pipe size	Q-1	12	1.333	Ea.	415	84		499	585
2080	For galvanized body, add					175			175	192
2120	With polished bronze top					475			475	525
2500	Heavy duty, cleanout & trap w/bucket, CI, 15" top									
2540	2", 3", and 4" pipe size	Q-1	6	2.667	Ea.	7,475	168		7,643	8,475
2560	For galvanized body, add					3,025			3,025	3,325

22 13 Facility Sanitary Sewerage

22 13 19 – Sanitary Waste Piping Specialties

22 13 19.13 Sanitary Drains

	Crew	Daily Output	Labor-Hours	Unit	Material	2022 Bare Costs Labor	Equipment	Total	Total Incl O&P
2580	With polished bronze top				Ea.	12,100			12,100	13,300

22 13 19.14 Floor Receptors

	Crew	Daily Output	Labor-Hours	Unit	Material	2022 Bare Costs Labor	Equipment	Total	Total Incl O&P
0010	**FLOOR RECEPTORS**, For connection to 2", 3" & 4" diameter pipe									
0200	12-1/2" square top, 25 sq. in. open area	Q-1	10	1.600	Ea.	1,825	101		1,926	2,175
0300	For grate with 4" diameter x 3-3/4" high funnel, add					2,300			2,300	2,525
0400	For grate with 6" diameter x 6" high funnel, add					264			264	290
0700	For acid-resisting bucket, add					335			335	370
0900	For stainless steel mesh bucket liner, add					340			340	375
2000	12-5/8" diameter top, 40 sq. in. open area	Q-1	10	1.600		880	101		981	1,125
2100	For options, add same prices as square top									
3000	8" x 4" rectangular top, 7.5 sq. in. open area	Q-1	14	1.143	Ea.	1,075	72		1,147	1,275
3100	For trap primer connections, add				"	106			106	117
9000	Minimum labor/equipment charge	Q-1	3	5.333	Job		335		335	515

22 13 19.15 Sink Waste Treatment

	Crew	Daily Output	Labor-Hours	Unit	Material	2022 Bare Costs Labor	Equipment	Total	Total Incl O&P
0010	**SINK WASTE TREATMENT**, System for commercial kitchens									
0100	includes clock timer & fittings									
0200	System less chemical, wall mounted cabinet	1 Plum	16	.500	Ea.	480	35		515	585
2000	Chemical, 1 gallon, add					45.50			45.50	50
2100	6 gallons, add					206			206	226
2200	15 gallons, add					560			560	615
2300	30 gallons, add					1,050			1,050	1,150
2400	55 gallons, add					1,775			1,775	1,950

22 13 23 – Sanitary Waste Interceptors

22 13 23.10 Interceptors

	Crew	Daily Output	Labor-Hours	Unit	Material	2022 Bare Costs Labor	Equipment	Total	Total Incl O&P
0010	**INTERCEPTORS**									
0150	Grease, fabricated steel, 4 GPM, 8 lb. fat capacity	1 Plum	4	2	Ea.	1,850	140		1,990	2,250
0200	7 GPM, 14 lb. fat capacity		4	2		2,475	140		2,615	2,950
1000	10 GPM, 20 lb. fat capacity		4	2		3,025	140		3,165	3,550
1120	50 GPM, 100 lb. fat capacity	Q-1	2	8		9,750	505		10,255	11,500
1160	100 GPM, 200 lb. fat capacity		2	8		22,800	505		23,305	25,900
1240	300 GPM, 600 lb. fat capacity		1	16		48,900	1,000		49,900	55,000
9000	Minimum labor/equipment charge	1 Plum	3	2.667	Job		186		186	287

22 13 29 – Sanitary Sewerage Pumps

22 13 29.14 Sewage Ejector Pumps

	Crew	Daily Output	Labor-Hours	Unit	Material	2022 Bare Costs Labor	Equipment	Total	Total Incl O&P
0010	**SEWAGE EJECTOR PUMPS**, With operating and level controls									
0100	Simplex system incl. tank, cover, pump 15' head									
0500	37 gal. PE tank, 12 GPM, 1/2 HP, 2" discharge	Q-1	3.20	5	Ea.	555	315		870	1,100
0510	3" discharge		3.10	5.161		590	325		915	1,150
0530	87 GPM, .7 HP, 2" discharge		3.20	5		855	315		1,170	1,425
0540	3" discharge		3.10	5.161		925	325		1,250	1,525
0600	45 gal. coated stl. tank, 12 GPM, 1/2 HP, 2" discharge		3	5.333		995	335		1,330	1,625
0610	3" discharge		2.90	5.517		1,025	345		1,370	1,675
0630	87 GPM, .7 HP, 2" discharge		3	5.333		1,275	335		1,610	1,925
0640	3" discharge		2.90	5.517		1,350	345		1,695	2,000
0660	134 GPM, 1 HP, 2" discharge		2.80	5.714		1,375	360		1,735	2,075
0680	3" discharge		2.70	5.926		1,450	370		1,820	2,175
0700	70 gal. PE tank, 12 GPM, 1/2 HP, 2" discharge		2.60	6.154		1,075	385		1,460	1,775
0710	3" discharge		2.40	6.667		1,150	420		1,570	1,925
0730	87 GPM, .7 HP, 2" discharge		2.50	6.400		1,425	400		1,825	2,200
0740	3" discharge		2.30	6.957		1,475	435		1,910	2,300
0760	134 GPM, 1 HP, 2" discharge		2.20	7.273		1,500	455		1,955	2,350

22 13 Facility Sanitary Sewerage

22 13 29 – Sanitary Sewerage Pumps

22 13 29.14 Sewage Ejector Pumps	Crew	Daily Output	Labor-Hours	Unit	Material	2022 Bare Costs Labor	Equipment	Total	Total Incl O&P
0770	3" discharge	Q-1	2	8	Ea.	1,600	505		2,105	2,550
9000	Minimum labor/equipment charge		2.50	6.400	Job		400		400	620

22 14 Facility Storm Drainage

22 14 26 – Facility Storm Drains

22 14 26.13 Roof Drains

		Crew	Daily Output	Labor-Hours	Unit	Material	2022 Bare Costs Labor	Equipment	Total	Total Incl O&P
0010	**ROOF DRAINS**									
0140	Cornice, CI, 45° or 90° outlet									
0200	3" and 4" pipe size	Q-1	12	1.333	Ea.	430	84		514	600
0260	For galvanized body, add					138			138	152
0280	For polished bronze dome, add					103			103	113
3860	Roof, flat metal deck, CI body, 12" CI dome									
3890	3" pipe size	Q-1	14	1.143	Ea.	505	72		577	665
3900	4" pipe size	"	13	1.231	"	710	77.50		787.50	900
4620	Main, all aluminum, 12" low profile dome									
4640	2", 3" and 4" pipe size	Q-1	14	1.143	Ea.	700	72		772	880
9000	Minimum labor/equipment charge	1 Plum	4	2	Job		140		140	215

22 14 29 – Sump Pumps

22 14 29.16 Submersible Sump Pumps

		Crew	Daily Output	Labor-Hours	Unit	Material	2022 Bare Costs Labor	Equipment	Total	Total Incl O&P
0010	**SUBMERSIBLE SUMP PUMPS**									
1000	Elevator sump pumps, automatic									
1010	Complete systems, pump, oil detector, controls and alarm									
1020	1-1/2" discharge, does not include the sump pit/tank									
1040	1/3 HP, 115 V	1 Plum	4.40	1.818	Ea.	1,900	127		2,027	2,300
1050	1/2 HP, 115 V		4	2		2,025	140		2,165	2,450
1060	1/2 HP, 230 V		4	2		2,000	140		2,140	2,425
1070	3/4 HP, 115 V		3.60	2.222		2,125	155		2,280	2,575
1080	3/4 HP, 230 V		3.60	2.222		2,100	155		2,255	2,550
1100	Sump pump only									
1110	1/3 HP, 115 V	1 Plum	6.40	1.250	Ea.	177	87.50		264.50	330
1120	1/2 HP, 115 V		5.80	1.379		253	96.50		349.50	425
1130	1/2 HP, 230 V		5.80	1.379		294	96.50		390.50	475
1140	3/4 HP, 115 V		5.40	1.481		335	103		438	530
1150	3/4 HP, 230 V		5.40	1.481		380	103		483	580
1200	Oil detector, control and alarm only									
1210	115 V	1 Plum	8	1	Ea.	1,825	70		1,895	2,125
1220	230 V	"	8	1	"	1,825	70		1,895	2,125
7000	Sump pump, automatic									
7100	Plastic, 1-1/4" discharge, 1/4 HP	1 Plum	6.40	1.250	Ea.	173	87.50		260.50	325
7140	1/3 HP		6	1.333		229	93		322	395
7180	1-1/2" discharge, 1/2 HP		5.20	1.538		350	107		457	550
7500	Cast iron, 1-1/4" discharge, 1/4 HP		6	1.333		285	93		378	460
7540	1/3 HP		6	1.333		286	93		379	460
7560	1/2 HP		5	1.600		345	112		457	550
9000	Minimum labor/equipment charge		4	2	Job		140		140	215

22 15 General Service Compressed-Air Systems

22 15 19 – General Service Packaged Air Compressors and Receivers

22 15 19.10 Air Compressors

		Crew	Daily Output	Labor-Hours	Unit	Material	2022 Bare Costs Labor	Equipment	Total	Total Incl O&P
0010	**AIR COMPRESSORS**									
5250	Air, reciprocating air cooled, splash lubricated, tank mounted									
5300	Single stage, 1 phase, 140 psi									
5303	1/2 HP, 17 gal. tank	1 Stpi	3	2.667	Ea.	2,100	188		2,288	2,600
5305	3/4 HP, 30 gal. tank		2.60	3.077		1,725	216		1,941	2,225
5307	1 HP, 30 gal. tank		2.20	3.636		2,550	256		2,806	3,200

22 31 Domestic Water Softeners

22 31 13 – Residential Domestic Water Softeners

22 31 13.10 Residential Water Softeners

		Crew	Daily Output	Labor-Hours	Unit	Material	2022 Bare Costs Labor	Equipment	Total	Total Incl O&P
0010	**RESIDENTIAL WATER SOFTENERS**									
7350	Water softener, automatic, to 30 grains per gallon	2 Plum	5	3.200	Ea.	575	223		798	975
7400	To 100 grains per gallon	"	4	4	"	945	279		1,224	1,475

22 33 Electric Domestic Water Heaters

22 33 13 – Instantaneous Electric Domestic Water Heaters

22 33 13.20 Instantaneous Elec. Point-Of-Use Water Heaters

			Crew	Daily Output	Labor-Hours	Unit	Material	2022 Bare Costs Labor	Equipment	Total	Total Incl O&P
0010	**INSTANTANEOUS ELECTRIC POINT-OF-USE WATER HEATERS**										
8965	Point of use, electric, glass lined										
8969	Energy saver										
8970	2.5 gal. single element	G	1 Plum	2.80	2.857	Ea.	227	199		426	555
8971	4 gal. single element	G		2.80	2.857		266	199		465	600
8974	6 gal. single element	G		2.50	3.200		320	223		543	695
8975	10 gal. single element	G		2.50	3.200		440	223		663	830
8976	15 gal. single element	G		2.40	3.333		495	233		728	905
8977	20 gal. single element	G		2.40	3.333		505	233		738	915
8978	30 gal. single element	G		2.30	3.478		640	243		883	1,075
8979	40 gal. single element	G		2.20	3.636		1,250	254		1,504	1,775
8988	Commercial (ASHRAE energy std. 90)										
8989	6 gallon	G	1 Plum	2.50	3.200	Ea.	945	223		1,168	1,400
8990	10 gallon	G		2.50	3.200		1,100	223		1,323	1,550
8991	15 gallon	G		2.40	3.333		1,100	233		1,333	1,550
8992	20 gallon	G		2.40	3.333		1,250	233		1,483	1,725
8993	30 gallon	G		2.30	3.478		2,650	243		2,893	3,300
8995	Under the sink, copper, w/bracket										
8996	2.5 gallon	G	1 Plum	4	2	Ea.	375	140		515	630
9000	Minimum labor/equipment charge		"	1.75	4.571	Job		320		320	490

22 33 30 – Residential, Electric Domestic Water Heaters

22 33 30.13 Residential, Small-Capacity Elec. Water Heaters

		Crew	Daily Output	Labor-Hours	Unit	Material	2022 Bare Costs Labor	Equipment	Total	Total Incl O&P
0010	**RESIDENTIAL, SMALL-CAPACITY ELECTRIC DOMESTIC WATER HEATERS**									
1000	Residential, electric, glass lined tank, 5 yr., 10 gal., single element	1 Plum	2.30	3.478	Ea.	420	243		663	840
1060	30 gallon, double element		2.20	3.636		800	254		1,054	1,275
1080	40 gallon, double element		2	4		830	279		1,109	1,350
1100	52 gallon, double element		2	4		1,450	279		1,729	2,000
1120	66 gallon, double element		1.80	4.444		1,950	310		2,260	2,625
1140	80 gallon, double element		1.60	5		2,225	350		2,575	3,000

For customer support on your Commercial Renovation Costs with RSMeans data, call 800.448.8182.

445

22 33 Electric Domestic Water Heaters

22 33 33 – Light-Commercial Electric Domestic Water Heaters

22 33 33.10 Commercial Electric Water Heaters

		Crew	Daily Output	Labor-Hours	Unit	Material	2022 Bare Costs Labor	Equipment	Total	Total Incl O&P
0010	**COMMERCIAL ELECTRIC WATER HEATERS**									
4000	Commercial, 100° rise. NOTE: for each size tank, a range of									
4010	heaters between the ones shown is available									
4020	Electric									
4100	5 gal., 3 kW, 12 GPH, 208 volt	1 Plum	2	4	Ea.	4,200	279		4,479	5,025
4160	50 gal., 36 kW, 148 GPH, 208 volt	"	1.80	4.444		9,700	310		10,010	11,200
4480	400 gal., 210 kW, 860 GPH, 480 volt	Q-1	1	16	↓	84,000	1,000		85,000	94,000

22 34 Fuel-Fired Domestic Water Heaters

22 34 13 – Instantaneous, Tankless, Gas Domestic Water Heaters

22 34 13.10 Instantaneous, Tankless, Gas Water Heaters

			Crew	Daily Output	Labor-Hours	Unit	Material	2022 Bare Costs Labor	Equipment	Total	Total Incl O&P
0010	**INSTANTANEOUS, TANKLESS, GAS WATER HEATERS**										
9410	Natural gas/propane, 3.2 GPM	G	1 Plum	2	4	Ea.	550	279		829	1,025
9420	6.4 GPM	G		1.90	4.211		600	294		894	1,125
9430	8.4 GPM	G		1.80	4.444		730	310		1,040	1,275
9440	9.5 GPM	G	↓	1.60	5	↓	930	350		1,280	1,575

22 34 30 – Residential Gas Domestic Water Heaters

22 34 30.13 Residential, Atmos, Gas Domestic Wtr Heaters

		Crew	Daily Output	Labor-Hours	Unit	Material	2022 Bare Costs Labor	Equipment	Total	Total Incl O&P
0010	**RESIDENTIAL, ATMOSPHERIC, GAS DOMESTIC WATER HEATERS**									
2000	Gas fired, foam lined tank, 10 yr., vent not incl.									
2040	30 gallon	1 Plum	2	4	Ea.	2,025	279		2,304	2,650
2060	40 gallon		1.90	4.211		695	294		989	1,225
2100	75 gallon	↓	1.50	5.333	↓	1,275	370		1,645	1,975
3000	Tank leak safety, water & gas shut off see 22 05 23.20 8800									

22 34 36 – Commercial Gas Domestic Water Heaters

22 34 36.13 Commercial, Atmos., Gas Domestic Water Htrs.

		Crew	Daily Output	Labor-Hours	Unit	Material	2022 Bare Costs Labor	Equipment	Total	Total Incl O&P
0010	**COMMERCIAL, ATMOSPHERIC, GAS DOMESTIC WATER HEATERS**									
6000	Gas fired, flush jacket, std. controls, vent not incl.									
6040	75 MBH input, 73 GPH	1 Plum	1.40	5.714	Ea.	3,375	400		3,775	4,325
6060	98 MBH input, 95 GPH		1.40	5.714		8,275	400		8,675	9,725
6180	200 MBH input, 192 GPH		.60	13.333		10,700	930		11,630	13,200
6900	For low water cutoff, add		8	1		370	70		440	515
6960	For bronze body hot water circulator, add	↓	4	2	↓	2,100	140		2,240	2,525

22 34 46 – Oil-Fired Domestic Water Heaters

22 34 46.10 Residential Oil-Fired Water Heaters

		Crew	Daily Output	Labor-Hours	Unit	Material	2022 Bare Costs Labor	Equipment	Total	Total Incl O&P
0010	**RESIDENTIAL OIL-FIRED WATER HEATERS**									
3000	Oil fired, glass lined tank, 5 yr., vent not included, 30 gallon	1 Plum	2	4	Ea.	1,525	279		1,804	2,100
3040	50 gallon	"	1.80	4.444	"	1,375	310		1,685	2,000

22 34 46.20 Commercial Oil-Fired Water Heaters

		Crew	Daily Output	Labor-Hours	Unit	Material	2022 Bare Costs Labor	Equipment	Total	Total Incl O&P
0010	**COMMERCIAL OIL-FIRED WATER HEATERS**									
8000	Oil fired, glass lined, UL listed, std. controls, vent not incl.									
8060	140 gal., 140 MBH input, 134 GPH	Q-1	2.13	7.512	Ea.	22,700	470		23,170	25,600
8080	140 gal., 199 MBH input, 191 GPH		2	8		23,500	505		24,005	26,600
8160	140 gal., 540 MBH input, 519 GPH	↓	.96	16.667	↓	32,000	1,050		33,050	36,800
8280	201 gal., 1,250 MBH input, 1,200 GPH	Q-2	1.22	19.672	Ea.	50,500	1,275		51,775	57,500

22 41 Residential Plumbing Fixtures

22 41 13 – Residential Water Closets, Urinals, and Bidets

22 41 13.13 Water Closets

		Crew	Daily Output	Labor-Hours	Unit	Material	2022 Bare Costs Labor	2022 Bare Costs Equipment	Total	Total Incl O&P
0010	**WATER CLOSETS**									
0032	For automatic flush, see Line 22 42 39.10 0972									
0150	Tank type, vitreous china, incl. seat, supply pipe w/stop, 1.6 gpf or noted									
0200	Wall hung									
0400	Two piece, close coupled	Q-1	5.30	3.019	Ea.	450	190		640	785
0960	For rough-in, supply, waste, vent and carrier	"	2.73	5.861	"	1,675	370		2,045	2,425
0999	Floor mounted									
1100	Two piece, close coupled	Q-1	5.30	3.019	Ea.	199	190		389	510
1102	Economy		5.30	3.019		114	190		304	420
1110	Two piece, close coupled, dual flush		5.30	3.019		284	190		474	600
1140	Two piece, close coupled, 1.28 gpf, ADA [G]		5.30	3.019		276	190		466	595
1960	For color, add					30%				
1980	For rough-in, supply, waste and vent	Q-1	3.05	5.246	Ea.	370	330		700	920

22 41 16 – Residential Lavatories and Sinks

22 41 16.13 Lavatories

		Crew	Daily Output	Labor-Hours	Unit	Material	2022 Bare Costs Labor	2022 Bare Costs Equipment	Total	Total Incl O&P
0010	**LAVATORIES**, With trim, white unless noted otherwise									
0500	Vanity top, porcelain enamel on cast iron									
0600	20" x 18"	Q-1	6.40	2.500	Ea.	375	157		532	655
0640	33" x 19" oval	"	6.40	2.500	"	815	157		972	1,125
0860	For color, add					25%				
1000	Cultured marble, 19" x 17", single bowl	Q-1	6.40	2.500	Ea.	162	157		319	420
1120	25" x 22", single bowl	"	6.40	2.500	"	182	157		339	440
1580	For color, same price									
1900	Stainless steel, self-rimming, 25" x 22", single bowl, ledge	Q-1	6.40	2.500	Ea.	480	157		637	765
1960	17" x 22", single bowl		6.40	2.500		465	157		622	750
2600	Steel, enameled, 20" x 17", single bowl		5.80	2.759		172	173		345	455
2900	Vitreous china, 20" x 16", single bowl		5.40	2.963		248	186		434	560
2960	20" x 17", single bowl		5.40	2.963		147	186		333	450
3580	Rough-in, supply, waste and vent for all above lavatories		2.30	6.957		590	435		1,025	1,325
4000	Wall hung									
4040	Porcelain enamel on cast iron, 16" x 14", single bowl	Q-1	8	2	Ea.	525	126		651	775
4180	20" x 18", single bowl	"	8	2	"	310	126		436	535
4580	For color, add					30%				
6000	Vitreous china, 18" x 15", single bowl with backsplash	Q-1	7	2.286	Ea.	196	144		340	435
6960	Rough-in, supply, waste and vent for above lavatories	"	1.66	9.639	"	815	605		1,420	1,825
7000	Pedestal type									
7600	Vitreous china, 27" x 21", white	Q-1	6.60	2.424	Ea.	750	152		902	1,075
7610	27" x 21", colored		6.60	2.424		950	152		1,102	1,275
7620	27" x 21", premium color		6.60	2.424		1,075	152		1,227	1,425
7660	26" x 20", white		6.60	2.424		730	152		882	1,050
7670	26" x 20", colored		6.60	2.424		925	152		1,077	1,250
7680	26" x 20", premium color		6.60	2.424		1,125	152		1,277	1,450
7700	24" x 20", white		6.60	2.424		500	152		652	790
7710	24" x 20", colored		6.60	2.424		610	152		762	905
7720	24" x 20", premium color		6.60	2.424		670	152		822	970
7760	21" x 18", white		6.60	2.424		300	152		452	565
7770	21" x 18", colored		6.60	2.424		340	152		492	610
7990	Rough-in, supply, waste and vent for pedestal lavatories		1.66	9.639		815	605		1,420	1,825
9000	Minimum labor/equipment charge	1 Plum	3	2.667	Job		186		186	287

447

For customer support on your Commercial Renovation Costs with RSMeans data, call 800.448.8182.

22 41 Residential Plumbing Fixtures

22 41 16 – Residential Lavatories and Sinks

22 41 16.16 Sinks

22 41 16.16 Sinks		Crew	Daily Output	Labor-Hours	Unit	Material	2022 Bare Costs Labor	Equipment	Total	Total Incl O&P
0010	**SINKS**, With faucets and drain									
2000	Kitchen, counter top style, PE on CI, 24" x 21" single bowl	Q-1	5.60	2.857	Ea.	435	180		615	755
2100	31" x 22" single bowl		5.60	2.857		1,050	180		1,230	1,425
2200	32" x 21" double bowl		4.80	3.333		400	209		609	765
3000	Stainless steel, self rimming, 19" x 18" single bowl		5.60	2.857		650	180		830	985
3100	25" x 22" single bowl		5.60	2.857		720	180		900	1,075
3200	33" x 22" double bowl		4.80	3.333		1,050	209		1,259	1,475
3300	43" x 22" double bowl		4.80	3.333		1,200	209		1,409	1,650
4000	Steel, enameled, with ledge, 24" x 21" single bowl		5.60	2.857		495	180		675	820
4100	32" x 21" double bowl		4.80	3.333		495	209		704	870
4960	For color sinks except stainless steel, add					10%				
4980	For rough-in, supply, waste and vent, counter top sinks	Q-1	2.14	7.477		520	470		990	1,300
5790	For rough-in, supply, waste & vent, sinks	"	1.85	8.649		520	545		1,065	1,400

22 41 19 – Residential Bathtubs

22 41 19.10 Baths

22 41 19.10 Baths		Crew	Daily Output	Labor-Hours	Unit	Material	2022 Bare Costs Labor	Equipment	Total	Total Incl O&P
0010	**BATHS**									
0100	Tubs, recessed porcelain enamel on cast iron, with trim									
0180	48" x 42"	Q-1	4	4	Ea.	3,350	251		3,601	4,050
0220	72" x 36"	"	3	5.333	"	3,275	335		3,610	4,125
0300	Mat bottom									
0380	5' long	Q-1	4.40	3.636	Ea.	1,450	228		1,678	1,950
0560	Corner 48" x 44"		4.40	3.636		3,050	228		3,278	3,700
2000	Enameled formed steel, 4'-6" long		5.80	2.759		560	173		733	880
4600	Module tub & showerwall surround, molded fiberglass									
4610	5' long x 34" wide x 76" high	Q-1	4	4	Ea.	860	251		1,111	1,325
4750	ADA compliant with 1-1/2" OD grab bar, antiskid bottom									
4760	60" x 32-3/4" x 72" high	Q-1	4	4	Ea.	620	251		871	1,075
4770	60" x 30" x 71" high with molded seat		3.50	4.571		780	287		1,067	1,300
9600	Rough-in, supply, waste and vent, for all above tubs, add		2.07	7.729		590	485		1,075	1,400
9900	Minimum labor/equipment charge		3	5.333	Job		335		335	515

22 41 23 – Residential Showers

22 41 23.20 Showers

22 41 23.20 Showers		Crew	Daily Output	Labor-Hours	Unit	Material	2022 Bare Costs Labor	Equipment	Total	Total Incl O&P
0011	**SHOWERS**, Stall, with drain only									
1520	32" square	Q-1	5	3.200	Ea.	1,200	201		1,401	1,625
1530	36" square		4.80	3.333		3,200	209		3,409	3,850
1540	Terrazzo receptor, 32" square		5	3.200		1,400	201		1,601	1,825
1580	36" corner angle		4.80	3.333		2,300	209		2,509	2,850
3000	Fiberglass, one piece, with 3 walls, 32" x 32" square		5.50	2.909		375	183		558	690
3100	36" x 36" square		5.50	2.909		420	183		603	740
4200	Rough-in, supply, waste and vent for above showers		2.05	7.805		625	490		1,115	1,450

22 41 23.40 Shower System Components

22 41 23.40 Shower System Components			Crew	Daily Output	Labor-Hours	Unit	Material	2022 Bare Costs Labor	Equipment	Total	Total Incl O&P
0010	**SHOWER SYSTEM COMPONENTS**										
5500	Head, water economizer, 1.6 GPM	G	1 Plum	24	.333	Ea.	55.50	23.50		79	97

22 41 36 – Residential Laundry Trays

22 41 36.10 Laundry Sinks

22 41 36.10 Laundry Sinks		Crew	Daily Output	Labor-Hours	Unit	Material	2022 Bare Costs Labor	Equipment	Total	Total Incl O&P
0010	**LAUNDRY SINKS**, With trim									
0020	Porcelain enamel on cast iron, black iron frame									
0050	24" x 21", single compartment	Q-1	6	2.667	Ea.	640	168		808	965
0100	26" x 21", single compartment	"	6	2.667	"	700	168		868	1,025
2000	Molded stone, on wall hanger or legs									

448

22 41 Residential Plumbing Fixtures

22 41 36 – Residential Laundry Trays

22 41 36.10 Laundry Sinks

		Crew	Daily Output	Labor-Hours	Unit	Material	2022 Bare Costs Labor	Equipment	Total	Total Incl O&P
2020	22" x 23", single compartment	Q-1	6	2.667	Ea.	183	168		351	460
2100	45" x 21", double compartment	"	5	3.200	"	375	201		576	725
3000	Plastic, on wall hanger or legs									
3020	18" x 23", single compartment	Q-1	6.50	2.462	Ea.	146	155		301	400
3100	20" x 24", single compartment		6.50	2.462		167	155		322	420
3200	36" x 23", double compartment		5.50	2.909		222	183		405	525
3300	40" x 24", double compartment		5.50	2.909		292	183		475	600
5000	Stainless steel, counter top, 22" x 17" single compartment		6	2.667		77.50	168		245.50	345
5200	33" x 22", double compartment		5	3.200		94	201		295	415
9600	Rough-in, supply, waste and vent, for all laundry sinks	Q-1	2.14	7.477	Ea.	520	470		990	1,300
9810	Minimum labor/equipment charge	1 Plum	3	2.667	Job		186		186	287

22 41 39 – Residential Faucets, Supplies and Trim

22 41 39.10 Faucets and Fittings

		Crew	Daily Output	Labor-Hours	Unit	Material	2022 Bare Costs Labor	Equipment	Total	Total Incl O&P
0010	**FAUCETS AND FITTINGS**									
0150	Bath, faucets, diverter spout combination, sweat	1 Plum	8	1	Ea.	86	70		156	203
0200	For integral stops, IPS unions, add					115			115	127
0420	Bath, press-bal mix valve w/diverter, spout, shower head, arm/flange	1 Plum	8	1		207	70		277	335
0500	Drain, central lift, 1-1/2" IPS male		20	.400		55	28		83	104
0600	Trip lever, 1-1/2" IPS male		20	.400		60	28		88	109
0810	Bidet									
0812	Fitting, over the rim, swivel spray/pop-up drain	1 Plum	8	1	Ea.	239	70		309	370
1000	Kitchen sink faucets, top mount, cast spout		10	.800		83.50	56		139.50	178
1100	For spray, add		24	.333		17.80	23.50		41.30	55.50
1200	Wall type, swing tube spout		10	.800		73.50	56		129.50	167
1300	Single control lever handle									
1310	With pull out spray									
1320	Polished chrome	1 Plum	10	.800	Ea.	233	56		289	345
2000	Laundry faucets, shelf type, IPS or copper unions		12	.667		61.50	46.50		108	139
2100	Lavatory faucet, centerset, without drain		10	.800		67	56		123	160
2120	With pop-up drain		6.66	1.201		86.50	84		170.50	224
2210	Porcelain cross handles and pop-up drain									
2220	Polished chrome	1 Plum	6.66	1.201	Ea.	222	84		306	375
2230	Polished brass	"	6.66	1.201	"	294	84		378	455
2260	Single lever handle and pop-up drain									
2280	Satin nickel	1 Plum	6.66	1.201	Ea.	370	84		454	535
2290	Polished chrome		6.66	1.201		262	84		346	415
2800	Self-closing, center set		10	.800		149	56		205	250
2810	Automatic sensor and operator, with faucet head [G]		6.15	1.301		615	91		706	815
4000	Shower by-pass valve with union		18	.444		57	31		88	111
4100	Shower arm with flange and head		22	.364		19.75	25.50		45.25	61
4140	Shower, hand held, pin mount, massage action, chrome		22	.364		76.50	25.50		102	123
4142	Polished brass		22	.364		166	25.50		191.50	221
4144	Shower, hand held, wall mtd, adj. spray, 2 wall mounts, chrome		20	.400		119	28		147	174
4146	Polished brass		20	.400		254	28		282	325
4148	Shower, hand held head, bar mounted 24", adj. spray, chrome		20	.400		190	28		218	252
4150	Polished brass	1 Plum	20	.400	Ea.	390	28		418	475
4200	Shower thermostatic mixing valve, concealed, with shower head trim kit	"	8	1	"	375	70		445	525
4220	Shower pressure balancing mixing valve									
4230	With shower head, arm, flange and diverter tub spout									
4240	Chrome	1 Plum	6.14	1.303	Ea.	435	91		526	620
4250	Satin nickel		6.14	1.303		570	91		661	765
4260	Polished graphite		6.14	1.303		555	91		646	750

22 41 Residential Plumbing Fixtures

22 41 39 – Residential Faucets, Supplies and Trim

22 41 39.10 Faucets and Fittings		Crew	Daily Output	Labor-Hours	Unit	Material	2022 Bare Costs Labor	Equipment	Total	Total Incl O&P
5000	Sillcock, compact, brass, IPS or copper to hose	1 Plum	24	.333	Ea.	15.85	23.50		39.35	53.50
6000	Stop and waste valves, bronze									
6100	Angle, solder end 1/2"	1 Plum	24	.333	Ea.	48	23.50		71.50	89
6110	3/4"		20	.400		56	28		84	105
6300	Straightway, solder end 3/8"		24	.333		20	23.50		43.50	58
6310	1/2"		24	.333		25.50	23.50		49	64
6410	Straightway, threaded 1/2"		24	.333		32	23.50		55.50	71.50
6420	3/4"		20	.400		36	28		64	82.50
6430	1"		19	.421		22.50	29.50		52	70.50
7800	Water closet, wax gasket		96	.083		1.30	5.80		7.10	10.40
7820	Gasket toilet tank to bowl		32	.250		2.99	17.45		20.44	30.50
7830	Replacement diaphragm washer assy for ballcock valve		12	.667		2.99	46.50		49.49	75
7850	Dual flush valve		12	.667		115	46.50		161.50	198
8000	Water supply stops, polished chrome plate									
8200	Angle, 3/8"	1 Plum	24	.333	Ea.	11.70	23.50		35.20	49
8300	1/2"		22	.364		13.40	25.50		38.90	54
8400	Straight, 3/8"		26	.308		16	21.50		37.50	50.50
8500	1/2"		24	.333		12.30	23.50		35.80	49.50
8600	Water closet, angle, w/flex riser, 3/8"		24	.333		41.50	23.50		65	82
9000	Minimum labor/equipment charge		4	2	Job		140		140	215

22 42 Commercial Plumbing Fixtures

22 42 13 – Commercial Water Closets, Urinals, and Bidets

22 42 13.13 Water Closets

		Crew	Daily Output	Labor-Hours	Unit	Material	2022 Bare Costs Labor	Equipment	Total	Total Incl O&P
0010	**WATER CLOSETS**									
3000	Bowl only, with flush valve, seat, 1.6 gpf unless noted									
3100	Wall hung	Q-1	5.80	2.759	Ea.	980	173		1,153	1,350
3200	For rough-in, supply, waste and vent, single WC		2.56	6.250		1,725	395		2,120	2,500
3360	With floor outlet, 1.28 gpf [G]		5.80	2.759		575	173		748	895
3362	With floor outlet, 1.28 gpf, ADA [G]		5.80	2.759		595	173		768	920
3370	For rough-in, supply, waste and vent, single WC		2.84	5.634		420	355		775	1,000
3390	Floor mounted children's size, 10-3/4" high									
3392	With automatic flush sensor, 1.6 gpf	Q-1	6.20	2.581	Ea.	680	162		842	995
3396	With automatic flush sensor, 1.28 gpf		6.20	2.581	"	625	162		787	940
9000	Minimum labor/equipment charge		4	4	Job		251		251	385

22 42 13.16 Urinals

		Crew	Daily Output	Labor-Hours	Unit	Material	2022 Bare Costs Labor	Equipment	Total	Total Incl O&P
0010	**URINALS**									
3000	Wall hung, vitreous china, with self-closing valve									
3100	Siphon jet type	Q-1	3	5.333	Ea.	340	335		675	890
3120	Blowout type		3	5.333		580	335		915	1,150
3140	Water saving .5 gpf [G]		3	5.333		600	335		935	1,175
3300	Rough-in, supply, waste & vent		2.83	5.654		810	355		1,165	1,450
5000	Stall type, vitreous china, includes valve		2.50	6.400		915	400		1,315	1,625
6980	Rough-in, supply, waste and vent		1.99	8.040		675	505		1,180	1,525
9000	Minimum labor/equipment charge		4	4	Job		251		251	385

22 42 16 – Commercial Lavatories and Sinks

22 42 16.13 Lavatories	Crew	Daily Output	Labor-Hours	Unit	Material	2022 Bare Costs Labor	Equipment	Total	Total Incl O&P
0010 **LAVATORIES**, With trim, white unless noted otherwise									
0020 Commercial lavatories same as residential. See Section 22 41 16									

22 42 16.40 Service Sinks

22 42 16.40 Service Sinks	Crew	Daily Output	Labor-Hours	Unit	Material	Labor	Equipment	Total	Total Incl O&P
0010 **SERVICE SINKS**									
6650 Service, floor, corner, PE on CI, 28" x 28"	Q-1	4.40	3.636	Ea.	1,500	228		1,728	2,000
6790 For rough-in, supply, waste & vent, floor service sinks		1.64	9.756	"	2,025	615		2,640	3,175
9000 Minimum labor/equipment charge		4	4	Job		251		251	385

22 42 23 – Commercial Showers

22 42 23.30 Group Showers

22 42 23.30 Group Showers	Crew	Daily Output	Labor-Hours	Unit	Material	Labor	Equipment	Total	Total Incl O&P
0010 **GROUP SHOWERS**									
6000 Group, w/pressure balancing valve, rough-in and rigging not included									
6800 Column, 6 heads, no receptors, less partitions	Q-1	3	5.333	Ea.	10,000	335		10,335	11,500
6900 With stainless steel partitions		1	16		13,300	1,000		14,300	16,200
7600 5 heads, no receptors, less partitions		3	5.333		6,875	335		7,210	8,100
7620 4 heads (1 ADA compliant) no receptors, less partitions		3	5.333		6,400	335		6,735	7,550
7700 With stainless steel partitions		1	16		6,900	1,000		7,900	9,125
9000 Minimum labor/equipment charge		4	4	Job		251		251	385

22 42 33 – Wash Fountains

22 42 33.20 Commercial Wash Fountains

22 42 33.20 Commercial Wash Fountains	Crew	Daily Output	Labor-Hours	Unit	Material	Labor	Equipment	Total	Total Incl O&P
0010 **COMMERCIAL WASH FOUNTAINS**									
1900 Group, foot control									
2000 Precast terrazzo, circular, 36" diam., 5 or 6 persons	Q-2	3	8	Ea.	7,425	520		7,945	8,950
2100 54" diam. for 8 or 10 persons		2.50	9.600		10,800	625		11,425	12,900
2400 Semi-circular, 36" diam. for 3 persons		3	8		6,050	520		6,570	7,450
2500 54" diam. for 4 or 5 persons	Q-2	2.50	9.600	Ea.	11,100	625		11,725	13,200
5610 Group, infrared control, barrier free									
5614 Precast terrazzo									
5620 Semi-circular 36" diam. for 3 persons	Q-2	3	8	Ea.	7,425	520		7,945	8,975
5630 46" diam. for 4 persons		2.80	8.571		8,075	560		8,635	9,725
5640 Circular, 54" diam. for 8 persons, button control		2.50	9.600		9,675	625		10,300	11,600
5700 Rough-in, supply, waste and vent for above wash fountains	Q-1	1.82	8.791		765	550		1,315	1,700
9000 Minimum labor/equipment charge	Q-2	3	8	Job		520		520	805

22 42 39 – Commercial Faucets, Supplies, and Trim

22 42 39.10 Faucets and Fittings

22 42 39.10 Faucets and Fittings	Crew	Daily Output	Labor-Hours	Unit	Material	Labor	Equipment	Total	Total Incl O&P
0010 **FAUCETS AND FITTINGS**									
0840 Flush valves, with vacuum breaker									
0850 Water closet									
0860 Exposed, rear spud	1 Plum	8	1	Ea.	157	70		227	281
0870 Top spud		8	1		205	70		275	335
0880 Concealed, rear spud		8	1		236	70		306	365
0890 Top spud		8	1		222	70		292	350
0900 Wall hung		8	1		235	70		305	365
0910 Dual flush flushometer		12	.667		246	46.50		292.50	345
0912 Flushometer retrofit kit		18	.444		18.35	31		49.35	68
0920 Urinal									
0930 Exposed, stall	1 Plum	8	1	Ea.	205	70		275	335
0940 Wall (washout)		8	1		166	70		236	291
0950 Pedestal, top spud		8	1		180	70		250	305
0960 Concealed, stall		8	1		175	70		245	300
0970 Wall (washout)		8	1		189	70		259	315

For customer support on your Commercial Renovation Costs with RSMeans data, call 800.448.8182.

451

22 42 Commercial Plumbing Fixtures

22 42 39 - Commercial Faucets, Supplies, and Trim

22 42 39.10 Faucets and Fittings

	22 42 39.10 Faucets and Fittings		Crew	Daily Output	Labor-Hours	Unit	Material	2022 Bare Costs Labor	Equipment	Total	Total Incl O&P
0971	Automatic flush sensor and operator for										
0972	urinals or water closets, standard	G	1 Plum	8	1	Ea.	580	70		650	750
0980	High efficiency water saving										
0984	Water closets, 1.28 gpf	G	1 Plum	8	1	Ea.	450	70		520	605
0988	Urinals, .5 gpf	G	"	8	1	"	450	70		520	605
2790	Faucets for lavatories										
2800	Self-closing, center set		1 Plum	10	.800	Ea.	149	56		205	250
2810	Automatic sensor and operator, with faucet head			6.15	1.301		615	91		706	815
3000	Service sink faucet, cast spout, pail hook, hose end			14	.571		76	40		116	145

22 42 39.30 Carriers and Supports

	22 42 39.30 Carriers and Supports	Crew	Daily Output	Labor-Hours	Unit	Material	2022 Bare Costs Labor	Equipment	Total	Total Incl O&P
0010	CARRIERS AND SUPPORTS, For plumbing fixtures									
3000	Lavatory, concealed arm									
3050	Floor mounted, single									
3100	High back fixture	1 Plum	6	1.333	Ea.	725	93		818	945
3200	Flat slab fixture		6	1.333		850	93		943	1,075
3220	ADA compliant		6	1.333		765	93		858	985
6980	Water closet, siphon jet									
7000	Horizontal, adjustable, caulk									
7040	Single, 4" pipe size	1 Plum	5.33	1.501	Ea.	1,525	105		1,630	1,825
7060	5" pipe size		5.33	1.501		1,125	105		1,230	1,375
7100	Double, 4" pipe size		5	1.600		2,850	112		2,962	3,325
7120	5" pipe size		5	1.600		1,625	112		1,737	1,975
8200	Water closet, residential									
8220	Vertical centerline, floor mount									
8240	Single, 3" caulk, 2" or 3" vent	1 Plum	6	1.333	Ea.	840	93		933	1,075
8260	4" caulk, 2" or 4" vent		6	1.333	"	1,175	93		1,268	1,450
9990	Minimum labor/equipment charge		3.50	2.286	Job		160		160	246

22 43 Healthcare Plumbing Fixtures

22 43 39 - Healthcare Faucets

22 43 39.10 Faucets and Fittings

	22 43 39.10 Faucets and Fittings	Crew	Daily Output	Labor-Hours	Unit	Material	2022 Bare Costs Labor	Equipment	Total	Total Incl O&P
0010	FAUCETS AND FITTINGS									
2850	Medical, bedpan cleanser, with pedal valve,	1 Plum	12	.667	Ea.	900	46.50		946.50	1,050
2860	With screwdriver stop valve		12	.667		465	46.50		511.50	585
2870	With self-closing spray valve		12	.667		340	46.50		386.50	445
2900	Faucet, gooseneck spout, wrist handles, grid drain		10	.800		191	56		247	296
2940	Mixing valve, knee action, screwdriver stops		4	2		505	140		645	770

22 45 Emergency Plumbing Fixtures

22 45 13 - Emergency Showers

22 45 13.10 Emergency Showers

	22 45 13.10 Emergency Showers	Crew	Daily Output	Labor-Hours	Unit	Material	2022 Bare Costs Labor	Equipment	Total	Total Incl O&P
0010	EMERGENCY SHOWERS, Rough-in not included									
5000	Shower, single head, drench, ball valve, pull, freestanding	Q-1	4	4	Ea.	430	251		681	860
5200	Horizontal or vertical supply		4	4		710	251		961	1,175
6000	Multi-nozzle, eye/face wash combination		4	4		945	251		1,196	1,425
6400	Multi-nozzle, 12 spray, shower only		4	4		2,300	251		2,551	2,900
6600	For freeze-proof, add		6	2.667		585	168		753	900
9000	Minimum labor/equipment charge		3	5.333	Job		335		335	515

22 45 Emergency Plumbing Fixtures

22 45 16 – Eyewash Equipment

22 45 16.10 Eyewash Safety Equipment	Crew	Daily Output	Labor-Hours	Unit	Material	2022 Bare Costs Labor	Equipment	Total	Total Incl O&P
0010 **EYEWASH SAFETY EQUIPMENT**, Rough-in not included									
1000 Eye wash fountain									
1400 Plastic bowl, pedestal mounted	Q-1	4	4	Ea.	565	251		816	1,000

22 47 Drinking Fountains and Water Coolers

22 47 13 – Drinking Fountains

22 47 13.10 Drinking Water Fountains

	Crew	Daily Output	Labor-Hours	Unit	Material	2022 Bare Costs Labor	Equipment	Total	Total Incl O&P
0010 **DRINKING WATER FOUNTAINS**, For connection to cold water supply									
1000 Wall mounted, non-recessed									
2700 Stainless steel, single bubbler, no back	1 Plum	4	2	Ea.	870	140		1,010	1,175
2740 With back		4	2		1,125	140		1,265	1,475
2780 Dual handle, ADA compliant		4	2		865	140		1,005	1,175
2820 Dual level, ADA compliant		3.20	2.500		1,875	175		2,050	2,350
3980 For rough-in, supply and waste, add		2.21	3.620		545	253		798	985
4000 Wall mounted, semi-recessed									
4200 Poly-marble, single bubbler	1 Plum	4	2	Ea.	1,150	140		1,290	1,475
4600 Stainless steel, satin finish, single bubbler	"	4	2	"	1,700	140		1,840	2,100
6000 Wall mounted, fully recessed									
6400 Poly-marble, single bubbler	1 Plum	4	2	Ea.	2,125	140		2,265	2,550
6800 Stainless steel, single bubbler		4	2		1,625	140		1,765	2,000
7580 For rough-in, supply and waste, add		1.83	4.372		545	305		850	1,075
7600 Floor mounted, pedestal type									
8600 Enameled iron, heavy duty service, 2 bubblers	1 Plum	2	4	Ea.	3,475	279		3,754	4,250
8880 For freeze-proof valve system, add		2	4		1,275	279		1,554	1,825
8900 For rough-in, supply and waste, add		1.83	4.372		545	305		850	1,075
9000 Minimum labor/equipment charge		2	4	Job		279		279	430

22 47 16 – Pressure Water Coolers

22 47 16.10 Electric Water Coolers

	Crew	Daily Output	Labor-Hours	Unit	Material	2022 Bare Costs Labor	Equipment	Total	Total Incl O&P
0010 **ELECTRIC WATER COOLERS**									
0100 Wall mounted, non-recessed									
0140 4 GPH	Q-1	4	4	Ea.	850	251		1,101	1,325
0160 8 GPH, barrier free, sensor operated		4	4		1,750	251		2,001	2,300
1000 Dual height, 8.2 GPH		3.80	4.211		2,175	265		2,440	2,800
1040 14.3 GPH		3.80	4.211		2,450	265		2,715	3,100
3300 Semi-recessed, 8.1 GPH		4	4		1,025	251		1,276	1,500
4600 Floor mounted, flush-to-wall									
4640 4 GPH	1 Plum	3	2.667	Ea.	980	186		1,166	1,350
4980 For stainless steel cabinet, add					170			170	187
5000 Dual height, 8.2 GPH	1 Plum	2	4		1,450	279		1,729	2,025
9000 Minimum labor/equipment charge	"	2	4	Job		279		279	430

22 66 53 – Laboratory Chemical-Waste and Vent Piping

22 66 53.30 Glass Pipe

		Crew	Daily Output	Labor-Hours	Unit	Material	2022 Bare Costs Labor	2022 Bare Costs Equipment	Total	Total Incl O&P
0010	**GLASS PIPE**, Borosilicate, couplings & clevis hanger assemblies, 10' OC									
0020	Drainage									
1100	1-1/2" diameter	Q-1	52	.308	L.F.	15	19.35		34.35	46.50
1120	2" diameter		44	.364		19.60	23		42.60	56.50
1140	3" diameter		39	.410		26	26		52	68.50
1160	4" diameter		30	.533		47	33.50		80.50	103
1180	6" diameter		26	.615		88.50	38.50		127	157
9000	Minimum labor/equipment charge	1 Plum	4	2	Job		140		140	215

22 66 53.40 Pipe Fittings, Glass

		Crew	Daily Output	Labor-Hours	Unit	Material	2022 Bare Costs Labor	2022 Bare Costs Equipment	Total	Total Incl O&P
0010	**PIPE FITTINGS, GLASS**									
0020	Drainage, beaded ends									
0040	Coupling & labor required at joints not incl. in fitting									
0050	price. Add 1 per joint for installed price									
0070	90° bend or sweep, 1-1/2"				Ea.	44			44	48
0090	2"					56			56	61.50
0100	3"					91.50			91.50	100
0110	4"					146			146	160
0120	6" (sweep only)					395			395	430
0350	Tee, single sanitary, 1-1/2"					71			71	78
0370	2"					71			71	78
0380	3"					102			102	113
0390	4"					189			189	208
0400	6"					505			505	555
0500	Coupling, stainless steel, TFE seal ring									
0520	1-1/2"	Q-1	32	.500	Ea.	43	31.50		74.50	95.50
0530	2"		30	.533		54	33.50		87.50	111
0540	3"		25	.640		72.50	40		112.50	142
0550	4"		23	.696		125	43.50		168.50	206
0560	6"		20	.800		281	50.50		331.50	390
9000	Minimum labor/equipment charge	1 Plum	4	2	Job		140		140	215

22 66 53.60 Corrosion Resistant Pipe

		Crew	Daily Output	Labor-Hours	Unit	Material	2022 Bare Costs Labor	2022 Bare Costs Equipment	Total	Total Incl O&P
0010	**CORROSION RESISTANT PIPE**, No couplings or hangers									
0020	Iron alloy, drain, mechanical joint									
1000	1-1/2" diameter	Q-1	70	.229	L.F.	93	14.35		107.35	124
1100	2" diameter		66	.242		96	15.25		111.25	129
1120	3" diameter		60	.267		105	16.75		121.75	141
1140	4" diameter		52	.308		133	19.35		152.35	176
2980	Plastic, epoxy, fiberglass filament wound, B&S joint									
3000	2" diameter	Q-1	62	.258	L.F.	14.80	16.20		31	41.50
3100	3" diameter		51	.314		17.35	19.70		37.05	49.50
3120	4" diameter		45	.356		24.50	22.50		47	61.50
3160	8" diameter	Q-2	38	.632		54.50	41		95.50	124
3200	12" diameter	"	28	.857		90.50	56		146.50	186
9800	Minimum labor/equipment charge	1 Plum	4	2	Job		140		140	215

22 66 53.70 Pipe Fittings, Corrosion Resistant

		Crew	Daily Output	Labor-Hours	Unit	Material	2022 Bare Costs Labor	2022 Bare Costs Equipment	Total	Total Incl O&P
0010	**PIPE FITTINGS, CORROSION RESISTANT**									
0030	Iron alloy									
0050	Mechanical joint									
0060	1/4 bend, 1-1/2"	Q-1	12	1.333	Ea.	154	84		238	298
0080	2"		10	1.600		252	101		353	430
0090	3"		9	1.778		300	112		412	500
0100	4"		8	2		345	126		471	575

For customer support on your Commercial Renovation Costs with RSMeans data, call 800.448.8182.

22 66 Chemical-Waste Systems for Lab. and Healthcare Facilities

22 66 53 – Laboratory Chemical-Waste and Vent Piping

22 66 53.70 Pipe Fittings, Corrosion Resistant	Crew	Daily Output	Labor-Hours	Unit	Material	2022 Bare Costs Labor	Equipment	Total	Total Incl O&P
0160 Tee and Y, sanitary, straight									
0170 1-1/2"	Q-1	8	2	Ea.	168	126		294	380
0180 2"		7	2.286		224	144		368	465
0190 3"		6	2.667		345	168		513	640
0200 4"		5	3.200		640	201		841	1,000
0360 Coupling, 1-1/2"		14	1.143		80.50	72		152.50	200
0380 2"		12	1.333		91	84		175	229
0390 3"		11	1.455		96	91.50		187.50	247
0400 4"		10	1.600		108	101		209	274
3000 Epoxy, filament wound									
3030 Quick-lock joint									
3040 90° elbow, 2"	Q-1	28	.571	Ea.	119	36		155	187
3060 3"		16	1		136	63		199	247
3070 4"		13	1.231		186	77.50		263.50	325
3190 Tee, 2"		19	.842		282	53		335	390
3200 3"		11	1.455		340	91.50		431.50	515
3210 4"		9	1.778		410	112		522	625
9000 Minimum labor/equipment charge	1 Plum	4	2	Job		140		140	215

Division Notes

	CREW	DAILY OUTPUT	LABOR-HOURS	UNIT	BARE COSTS				TOTAL INCL O&P
					MAT.	LABOR	EQUIP.	TOTAL	

Estimating Tips

The labor adjustment factors listed in Subdivision 22 01 02.20 also apply to Division 23.

23 10 00 Facility Fuel Systems

- The prices in this subdivision for above- and below-ground storage tanks do not include foundations or hold-down slabs, unless noted. The estimator should refer to Divisions 3 and 31 for foundation system pricing. In addition to the foundations, required tank accessories, such as tank gauges, leak detection devices, and additional manholes and piping, must be added to the tank prices.

23 50 00 Central Heating Equipment

- When estimating the cost of an HVAC system, check to see who is responsible for providing and installing the temperature control system. It is possible to overlook controls, assuming that they would be included in the electrical estimate.
- When looking up a boiler, be careful on specified capacity. Some manufacturers rate their products on output while others use input.
- Include HVAC insulation for pipe, boiler, and duct (wrap and liner).
- Be careful when looking up mechanical items to get the correct pressure rating and connection type (thread, weld, flange).

23 70 00 Central HVAC Equipment

- Combination heating and cooling units are sized by the air conditioning requirements. (See Reference No. R236000 20 for the preliminary sizing guide.)
- A ton of air conditioning is nominally 400 CFM.
- Rectangular duct is taken off by the linear foot for each size, but its cost is usually estimated by the pound. Remember that SMACNA standards now base duct on internal pressure.
- Prefabricated duct is estimated and purchased like pipe: straight sections and fittings.
- Note that cranes or other lifting equipment are not included on any lines in Division 23. For example, if a crane is required to lift a heavy piece of pipe into place high above a gym floor, or to put a rooftop unit on the roof of a four-story building, etc., it must be added. Due to the potential for extreme variation—from nothing additional required to a major crane or helicopter—we feel that including a nominal amount for "lifting contingency" would be useless and detract from the accuracy of the estimate. When using equipment rental cost data from RSMeans, do not forget to include the cost of the operator(s).

Reference Numbers

Reference numbers are shown at the beginning of some major classifications. These numbers refer to related items in the Reference Section. The reference information may be an estimating procedure, an alternate pricing method, or technical information.

Note: Not all subdivisions listed here necessarily appear. ∎

Same Data. Simplified.

Enjoy the convenience and efficiency of accessing your costs anywhere:

- **Skip the multiplier** by setting your location
- **Quickly search,** edit, favorite and share costs
- **Stay on top of price changes** with automatic updates

Discover more at rsmeans.com/online

---

23 05 Common Work Results for HVAC

23 05 05 – Selective Demolition for HVAC

23 05 05.10 HVAC Demolition

		Crew	Daily Output	Labor-Hours	Unit	Material	2022 Bare Costs Labor	Equipment	Total	Total Incl O&P
0010	**HVAC DEMOLITION** R220105-10									
0100	Air conditioner, split unit, 3 ton	Q-5	2	8	Ea.		505		505	780
0150	Package unit, 3 ton R024119-10	Q-6	3	8	"		525		525	810
0298	Boilers									
0300	Electric, up thru 148 kW	Q-19	2	12	Ea.		770		770	1,175
0310	150 thru 518 kW	"	1	24			1,550		1,550	2,375
0320	550 thru 2,000 kW	Q-21	.40	80			5,275		5,275	8,100
0330	2,070 kW and up	"	.30	107			7,025		7,025	10,800
0340	Gas and/or oil, up thru 150 MBH	Q-7	2.20	14.545			975		975	1,500
0350	160 thru 2,000 MBH		.80	40			2,675		2,675	4,125
0360	2,100 thru 4,500 MBH		.50	64			4,275		4,275	6,600
0370	4,600 thru 7,000 MBH		.30	107			7,150		7,150	11,000
0380	7,100 thru 12,000 MBH		.16	200			13,400		13,400	20,600
0390	12,200 thru 25,000 MBH		.12	267			17,900		17,900	27,500
1000	Ductwork, 4" high, 8" wide	1 Clab	200	.040	L.F.		1.82		1.82	2.90
1020	10" wide		190	.042			1.92		1.92	3.05
1040	14" wide		180	.044			2.03		2.03	3.22
1100	6" high, 8" wide		165	.048			2.21		2.21	3.51
1120	12" wide		150	.053			2.43		2.43	3.86
1140	18" wide		135	.059			2.70		2.70	4.29
1200	10" high, 12" wide		125	.064			2.92		2.92	4.64
1220	18" wide		115	.070			3.17		3.17	5.05
1240	24" wide		110	.073			3.32		3.32	5.25
1300	12"-14" high, 16"-18" wide		85	.094			4.29		4.29	6.80
1320	24" wide		75	.107			4.86		4.86	7.75
1340	48" wide		71	.113			5.15		5.15	8.15
1400	18" high, 24" wide		67	.119			5.45		5.45	8.65
1420	36" wide		63	.127			5.80		5.80	9.20
1440	48" wide		59	.136			6.20		6.20	9.80
1500	30" high, 36" wide		56	.143			6.50		6.50	10.35
1520	48" wide		53	.151			6.90		6.90	10.95
1540	72" wide		50	.160			7.30		7.30	11.60
1550	Duct heater, electric strip	1 Elec	8	1	Ea.		66		66	102
1850	Minimum labor/equipment charge	1 Clab	3	2.667	Job		122		122	193
2200	Furnace, electric	Q-20	2	10	Ea.		620		620	960
2300	Gas or oil, under 120 MBH	Q-9	4	4			243		243	380
2340	Over 120 MBH	"	3	5.333			325		325	505
2800	Heat pump, package unit, 3 ton	Q-5	2.40	6.667			420		420	650
2840	Split unit, 3 ton		2	8			505		505	780
3000	Mechanical equipment, light items. Unit is weight, not cooling.		.90	17.778	Ton		1,125		1,125	1,725
3600	Heavy items		1.10	14.545	"		920		920	1,425
5090	Remove refrigerant from system	1 Stpi	40	.200	Lb.		14.05		14.05	21.50
9000	Minimum labor/equipment charge	Q-6	3	8	Job		525		525	810

23 05 23 – General-Duty Valves for HVAC Piping

23 05 23.30 Valves, Iron Body

		Crew	Daily Output	Labor-Hours	Unit	Material	2022 Bare Costs Labor	Equipment	Total	Total Incl O&P
0010	**VALVES, IRON BODY**									
1020	Butterfly, wafer type, gear actuator, 200 lb.									
1030	2"	1 Plum	14	.571	Ea.	118	40		158	192
1060	4"	Q-1	5	3.200	"	138	201		339	460
1650	Gate, 125 lb., N.R.S.									
2150	Flanged									
2240	2-1/2"	Q-1	5	3.200	Ea.	985	201		1,186	1,375

23 05 Common Work Results for HVAC

23 05 23 – General-Duty Valves for HVAC Piping

23 05 23.30 Valves, Iron Body	Crew	Daily Output	Labor-Hours	Unit	Material	2022 Bare Costs Labor	Equipment	Total	Total Incl O&P
2260	3"	Q-1	4.50	3.556	Ea.	1,100	223		1,323	1,575
2280	4"	▼	3	5.333	▼	1,650	335		1,985	2,350
3550	OS&Y, 125 lb., flanged									
3680	4"	Q-1	3	5.333	Ea.	980	335		1,315	1,600
3690	5"	Q-2	3.40	7.059		1,875	460		2,335	2,750
3700	6"	"	3	8	▼	1,875	520		2,395	2,850
9000	Minimum labor/equipment charge	1 Plum	3	2.667	Job		186		186	287

23 05 93 – Testing, Adjusting, and Balancing for HVAC

23 05 93.10 Balancing, Air

0010	**BALANCING, AIR** (Subcontractor's quote incl. material and labor)								
0900	Heating and ventilating equipment								
1000	Centrifugal fans, utility sets				Ea.				420	420
1100	Heating and ventilating unit								630	630
1200	In-line fan								630	630
1300	Propeller and wall fan								119	119
1400	Roof exhaust fan								280	280
2000	Air conditioning equipment, central station								910	910
2100	Built-up low pressure unit								840	840
2200	Built-up high pressure unit								980	980
2300	Built-up high pressure dual duct								1,550	1,550
2400	Built-up variable volume								1,825	1,825
2500	Multi-zone A.C. and heating unit								630	630
2600	For each zone over one, add								140	140
2700	Package A.C. unit								350	350
2800	Rooftop heating and cooling unit								490	490
3000	Supply, return, exhaust, registers & diffusers, avg. height ceiling								84	84
3100	High ceiling								126	126
3200	Floor height								70	70
3300	Off mixing box				▼				56	56
3500	Induction unit				Ea.				91	91
3600	Lab fume hood								420	420
3700	Linear supply								210	210
3800	Linear supply high								245	245
4000	Linear return								70	70
4100	Light troffers								84	84
4200	Moduline - master								84	84
4300	Moduline - slaves								42	42
4400	Regenerators								560	560
4500	Taps into ceiling plenums								105	105
4600	Variable volume boxes				▼				84	84

23 05 93.20 Balancing, Water

0010	**BALANCING, WATER** (Subcontractor's quote incl. material and labor)								
0050	Air cooled condenser				Ea.				256	256
0080	Boiler								515	515
0100	Cabinet unit heater								88	88
0200	Chiller								620	620
0300	Convector								73	73
0400	Converter								365	365
0500	Cooling tower								475	475
0600	Fan coil unit, unit ventilator								132	132
0700	Fin tube and radiant panels								146	146
0800	Main and duct re-heat coils				▼				135	135

23 05 93 – Testing, Adjusting, and Balancing for HVAC

23 05 93.20 Balancing, Water		Crew	Daily Output	Labor-Hours	Unit	Material	2022 Bare Costs Labor	Equipment	Total	Total Incl O&P
0810	Heat exchanger				Ea.				135	135
0900	Main balancing cocks								110	110
1000	Pumps								320	320
1100	Unit heater				↓				102	102

23 07 HVAC Insulation

23 07 13 – Duct Insulation

23 07 13.10 Duct Thermal Insulation

			Crew	Daily Output	Labor-Hours	Unit	Material	2022 Bare Costs Labor	Equipment	Total	Total Incl O&P
0010	**DUCT THERMAL INSULATION**										
0110	Insulation req'd. is based on the surface size/area to be covered										
3000	Ductwork										
3020	Blanket type, fiberglass, flexible										
3030	Fire rated for grease and hazardous exhaust ducts										
3060	1-1/2" thick		Q-14	84	.190	S.F.	4.25	10.80		15.05	22
3090	Fire rated for plenums										
3100	1/2" x 24" x 25'		Q-14	1.94	8.247	Roll	191	470		661	950
3110	1/2" x 24" x 25'		"	98	.163	S.F.	3.81	9.25		13.06	18.85
3120	1/2" x 48" x 25'		Q-14	1.04	15.385	Roll	375	875		1,250	1,800
3126	1/2" x 48" x 25'		"	104	.154	S.F.	3.77	8.75		12.52	17.95
3140	FSK vapor barrier wrap, .75 lb. density										
3160	1" thick	G	Q-14	350	.046	S.F.	.22	2.59		2.81	4.34
3170	1-1/2" thick	G		320	.050		.55	2.84		3.39	5.10
3180	2" thick	G		300	.053		.77	3.03		3.80	5.65
3190	3" thick	G		260	.062		1.33	3.49		4.82	6.95
3200	4" thick	G		242	.066	↓	.70	3.75		4.45	6.70
3210	Vinyl jacket, same as FSK										
9600	Minimum labor/equipment charge		1 Stpi	4	2	Job		141		141	217

23 07 16 – HVAC Equipment Insulation

23 07 16.10 HVAC Equipment Thermal Insulation

		Crew	Daily Output	Labor-Hours	Unit	Material	2022 Bare Costs Labor	Equipment	Total	Total Incl O&P
0010	**HVAC EQUIPMENT THERMAL INSULATION**									
0110	Insulation req'd. is based on the surface size/area to be covered									
9610	Minimum labor/equipment charge	1 Stpi	4	2	Job		141		141	217

23 09 Instrumentation and Control for HVAC

23 09 23 – Direct-Digital Control System for HVAC

23 09 23.10 Control Components/DDC Systems

		Crew	Daily Output	Labor-Hours	Unit	Material	2022 Bare Costs Labor	Equipment	Total	Total Incl O&P
0010	**CONTROL COMPONENTS/DDC SYSTEMS** (Sub's quote incl. M & L)									
0100	Analog inputs									
0110	Sensors (avg. 50' run in 1/2" EMT)									
0120	Duct temperature				Ea.				465	465
0130	Space temperature								665	665
0140	Duct humidity, +/- 3%								595	595
0150	Space humidity, +/- 2%								1,150	1,150
0160	Duct static pressure								680	680
0170	CFM/transducer								920	920
0172	Water temperature								940	940
0174	Water flow								3,450	3,450
0176	Water pressure differential								975	975
0177	Steam flow				↓				2,400	2,400

23 09 23 – Direct-Digital Control System for HVAC

23 09 23.10 Control Components/DDC Systems	Crew	Daily Output	Labor-Hours	Unit	Material	2022 Bare Costs Labor	Equipment	Total	Total Incl O&P
0178	Steam pressure				Ea.				1,025	1,025
0180	KW/transducer								1,350	1,350
0182	KWH totalization (not incl. elec. meter pulse xmtr.)								625	625
0190	Space static pressure				▼				1,075	1,075
1000	Analog outputs (avg. 50' run in 1/2" EMT)									
1010	P/I transducer				Ea.				635	635
1020	Analog output, matl. in MUX								305	305
1030	Pneumatic (not incl. control device)				▼				645	645
1040	Electric (not incl. control device)				Ea.				380	380
2000	Status (alarms)									
2100	Digital inputs (avg. 50' run in 1/2" EMT)									
2110	Freeze				Ea.				435	435
2120	Fire								395	395
2130	Differential pressure (air)	2 Elec	3	5.333		355	355		710	930
2140	Differential pressure (water)								975	975
2150	Current sensor								435	435
2160	Duct high temperature thermostat								570	570
2170	Duct smoke detector				▼				705	705
2200	Digital output (avg. 50' run in 1/2" EMT)									
2210	Start/stop				Ea.				340	340
2220	On/off (maintained contact)				"				585	585
3000	Controller MUX panel, incl. function boards									
3100	48 point				Ea.				5,275	5,275
3110	128 point				"				7,225	7,225
3200	DDC controller (avg. 50' run in conduit)									
3210	Mechanical room									
3214	16 point controller (incl. 120 volt/1 phase power supply)				Ea.				3,275	3,275
3229	32 point controller (incl. 120 volt/1 phase power supply)				"				5,425	5,425
3230	Includes software programming and checkout									
3260	Space									
3266	VAV terminal box (incl. space temp. sensor)				Ea.				840	840
3280	Host computer (avg. 50' run in conduit)									
3281	Package complete with PC, keyboard,									
3282	printer, monitor, basic software				Ea.				3,150	3,150
4000	Front end costs									
4100	Computer (P.C.) with software program				Ea.				6,350	6,350
4200	Color graphics software								3,925	3,925
4300	Color graphics slides								490	490
4350	Additional printer				▼				980	980
4400	Communications trunk cable				L.F.				3.80	3.80
4500	Engineering labor (not incl. dftg.)				Point				94	94
4600	Calibration labor								120	120
4700	Start-up, checkout labor				▼				120	120
4800	Programming labor, as req'd									
5000	Communications bus (data transmission cable)									
5010	#18 twisted shielded pair in 1/2" EMT conduit				C.L.F.				380	380
8000	Applications software									
8050	Basic maintenance manager software (not incl. data base entry)				Ea.				1,950	1,950
8100	Time program				Point				6.85	6.85
8120	Duty cycle								13.65	13.65
8140	Optimum start/stop								41.50	41.50
8160	Demand limiting								20.50	20.50
8180	Enthalpy program				▼				41.50	41.50

23 09 Instrumentation and Control for HVAC

23 09 23 – Direct-Digital Control System for HVAC

23 09 23.10 Control Components/DDC Systems		Crew	Daily Output	Labor-Hours	Unit	Material	2022 Bare Costs Labor	Equipment	Total	Total Incl O&P
8200	Boiler optimization				Ea.				1,225	1,225
8220	Chiller optimization				"				1,625	1,625
8240	Custom applications									
8260	Cost varies with complexity									

23 09 53 – Pneumatic and Electric Control System for HVAC

23 09 53.10 Control Components

		Crew	Daily Output	Labor-Hours	Unit	Material	2022 Bare Costs Labor	Equipment	Total	Total Incl O&P
0010	**CONTROL COMPONENTS**									
5000	Thermostats									
5030	Manual	1 Stpi	8	1	Ea.	45.50	70.50		116	159
5040	1 set back, electric, timed [G]		8	1		101	70.50		171.50	219
5050	2 set back, electric, timed [G]		8	1		251	70.50		321.50	385
5200	24 hour, automatic, clock [G]	1 Shee	8	1		186	67.50		253.50	310
6000	Valves, motorized zone									
6100	Sweat connections, 1/2" C x C	1 Stpi	20	.400	Ea.	146	28		174	205
6110	3/4" C x C		20	.400		78.50	28		106.50	130
6120	1" C x C		19	.421		192	29.50		221.50	257
9000	Minimum labor/equipment charge	1 Plum	4	2	Job		140		140	215

23 21 Hydronic Piping and Pumps

23 21 20 – Hydronic HVAC Piping Specialties

23 21 20.34 Dielectric Unions

		Crew	Daily Output	Labor-Hours	Unit	Material	2022 Bare Costs Labor	Equipment	Total	Total Incl O&P
0010	**DIELECTRIC UNIONS**, Standard gaskets for water and air									
0020	250 psi maximum pressure									
0280	Female IPT to sweat, straight									
0340	3/4" pipe size	1 Plum	20	.400	Ea.	9.80	28		37.80	54
0780	Female IPT to female IPT, straight									
0800	1/2" pipe size	1 Plum	24	.333	Ea.	17.05	23.50		40.55	55
0840	3/4" pipe size		20	.400	"	24	28		52	69
9000	Minimum labor/equipment charge		4	2	Job		140		140	215

23 21 20.46 Expansion Tanks

		Crew	Daily Output	Labor-Hours	Unit	Material	2022 Bare Costs Labor	Equipment	Total	Total Incl O&P
0010	**EXPANSION TANKS**									
2000	Steel, liquid expansion, ASME, painted, 15 gallon capacity	Q-5	17	.941	Ea.	935	59.50		994.50	1,125
2040	30 gallon capacity	Q-5	12	1.333	Ea.	1,125	84.50		1,209.50	1,350
2080	60 gallon capacity		8	2		1,550	127		1,677	1,925
2120	100 gallon capacity		6	2.667		2,275	169		2,444	2,750
3000	Steel ASME expansion, rubber diaphragm, 19 gal. cap. accept.		12	1.333		3,550	84.50		3,634.50	4,025
3020	31 gallon capacity		8	2		3,950	127		4,077	4,550
3040	61 gallon capacity		6	2.667		5,925	169		6,094	6,775
3080	119 gallon capacity		4	4		5,975	253		6,228	6,975
9000	Minimum labor/equipment charge		4	4	Job		253		253	390

23 21 20.70 Steam Traps

		Crew	Daily Output	Labor-Hours	Unit	Material	2022 Bare Costs Labor	Equipment	Total	Total Incl O&P
0010	**STEAM TRAPS**									
0030	Cast iron body, threaded									
0040	Inverted bucket									
0050	1/2" pipe size	1 Stpi	12	.667	Ea.	170	47		217	260
0100	1" pipe size		9	.889		455	62.50		517.50	595
0120	1-1/4" pipe size		8	1		690	70.50		760.50	870
1000	Float & thermostatic, 15 psi									
1010	3/4" pipe size	1 Stpi	16	.500	Ea.	202	35		237	276
1020	1" pipe size		15	.533		194	37.50		231.50	272

23 21 Hydronic Piping and Pumps

23 21 20 − Hydronic HVAC Piping Specialties

23 21 20.70 Steam Traps

		Crew	Daily Output	Labor-Hours	Unit	Material	2022 Bare Costs Labor	2022 Bare Costs Equipment	Total	Total Incl O&P
1040	1-1/2" pipe size	1 Stpi	9	.889	Ea.	340	62.50		402.50	470
1060	2" pipe size	↓	6	1.333	↓	975	94		1,069	1,225
9000	Minimum labor/equipment charge	↓	4	2	Job		141		141	217

23 21 20.76 Strainers, Y Type, Bronze Body

		Crew	Daily Output	Labor-Hours	Unit	Material	2022 Bare Costs Labor	2022 Bare Costs Equipment	Total	Total Incl O&P
0010	**STRAINERS, Y TYPE, BRONZE BODY**									
0050	Screwed, 125 lb., 1/4" pipe size	1 Stpi	24	.333	Ea.	48	23.50		71.50	88.50
0100	1/2" pipe size		20	.400		46.50	28		74.50	94.50
0120	3/4" pipe size		19	.421		59.50	29.50		89	111
0140	1" pipe size		17	.471		87.50	33		120.50	147
0160	1-1/2" pipe size		14	.571		187	40		227	268
0180	2" pipe size		13	.615		246	43.50		289.50	340
0182	3" pipe size	↓	12	.667		1,650	47		1,697	1,875
0200	300 lb., 2-1/2" pipe size	Q-5	17	.941		930	59.50		989.50	1,125
0220	3" pipe size		16	1		1,350	63.50		1,413.50	1,575
0240	4" pipe size	↓	15	1.067		2,475	67.50		2,542.50	2,825
1000	Flanged, 150 lb., 1-1/2" pipe size	1 Stpi	11	.727		575	51		626	715
1020	2" pipe size	"	8	1		705	70.50		775.50	885
1030	2-1/2" pipe size	Q-5	5	3.200		1,175	203		1,378	1,600
1040	3" pipe size		4.50	3.556		1,450	225		1,675	1,950
1060	4" pipe size	↓	3	5.333		2,250	340		2,590	2,975
1100	6" pipe size	Q-6	3	8		4,325	525		4,850	5,550
1106	8" pipe size	"	2.60	9.231	↓	4,700	605		5,305	6,075
1500	For 300 lb. rating, add					40%				
9000	Minimum labor/equipment charge	1 Stpi	3.75	2.133	Job		150		150	231

23 21 23 − Hydronic Pumps

23 21 23.13 In-Line Centrifugal Hydronic Pumps

		Crew	Daily Output	Labor-Hours	Unit	Material	2022 Bare Costs Labor	2022 Bare Costs Equipment	Total	Total Incl O&P
0010	**IN-LINE CENTRIFUGAL HYDRONIC PUMPS**									
0600	Bronze, sweat connections, 1/40 HP, in line									
0640	3/4" size	Q-1	16	1	Ea.	257	63		320	380
1000	Flange connection, 3/4" to 1-1/2" size									
1040	1/12 HP	Q-1	6	2.667	Ea.	685	168		853	1,000
1060	1/8 HP		6	2.667		1,250	168		1,418	1,625
1100	1/3 HP		6	2.667		1,400	168		1,568	1,800
1140	2" size, 1/6 HP		5	3.200		1,800	201		2,001	2,275
1180	2-1/2" size, 1/4 HP		5	3.200		2,425	201		2,626	2,950
1220	3" size, 1/4 HP	↓	4	4	↓	2,675	251		2,926	3,325
9000	Minimum labor/equipment charge	↓	3.25	4.923	Job		310		310	475

23 23 Refrigerant Piping

23 23 16 − Refrigerant Piping Specialties

23 23 16.16 Refrigerant Line Sets

		Crew	Daily Output	Labor-Hours	Unit	Material	2022 Bare Costs Labor	2022 Bare Costs Equipment	Total	Total Incl O&P
0010	**REFRIGERANT LINE SETS**, Standard									
0100	Copper tube									
0110	1/2" insulation, both tubes									
0120	Combination 1/4" and 1/2" tubes									
0130	10' set	Q-5	42	.381	Ea.	53.50	24		77.50	96
0135	15' set		42	.381		77.50	24		101.50	122
0140	20' set		40	.400		74	25.50		99.50	121
0150	30' set		37	.432		105	27.50		132.50	157
0160	40' set	↓	35	.457		147	29		176	207

23 23 Refrigerant Piping

23 23 16 – Refrigerant Piping Specialties

23 23 16.16 Refrigerant Line Sets		Crew	Daily Output	Labor-Hours	Unit	Material	2022 Bare Costs Labor	Equipment	Total	Total Incl O&P
0170	50' set	Q-5	32	.500	Ea.	162	31.50		193.50	227
0180	100' set	↓	22	.727	↓	350	46		396	455
0300	Combination 1/4" and 3/4" tubes									
0310	10' set	Q-5	40	.400	Ea.	60	25.50		85.50	105
0320	20' set		38	.421		106	26.50		132.50	158
0330	30' set		35	.457		158	29		187	219
0340	40' set		33	.485		207	30.50		237.50	276
0350	50' set		30	.533		261	34		295	340
0380	100' set	↓	20	.800	↓	595	50.50		645.50	735
0500	Combination 3/8" & 3/4" tubes									
0510	10' set	Q-5	28	.571	Ea.	69	36		105	132
0520	20' set		36	.444		95	28		123	148
0530	30' set	↓	34	.471	↓	129	30		159	188
0540	40' set	Q-5	31	.516	Ea.	170	32.50		202.50	238
0550	50' set		28	.571		199	36		235	275
0580	100' set	↓	18	.889	↓	590	56.50		646.50	730
0700	Combination 3/8" & 1-1/8" tubes									
0710	10' set	Q-5	36	.444	Ea.	120	28		148	176
0720	20' set		33	.485		178	30.50		208.50	244
0730	30' set		31	.516		240	32.50		272.50	315
0740	40' set		28	.571		360	36		396	450
0750	50' set	↓	26	.615	↓	350	39		389	445
0900	Combination 1/2" & 3/4" tubes									
0910	10' set	Q-5	37	.432	Ea.	72.50	27.50		100	122
0920	20' set		35	.457		129	29		158	187
0930	30' set		33	.485		194	30.50		224.50	261
0940	40' set		30	.533		255	34		289	330
0950	50' set		27	.593		284	37.50		321.50	375
0980	100' set	↓	17	.941	↓	640	59.50		699.50	795
2100	Combination 1/2" & 1-1/8" tubes									
2110	10' set	Q-5	35	.457	Ea.	124	29		153	181
2120	20' set		31	.516		212	32.50		244.50	285
2130	30' set		29	.552		320	35		355	405
2140	40' set		25	.640		430	40.50		470.50	535
2150	50' set	↓	14	1.143	↓	470	72.50		542.50	625
2300	For 1" thick insulation add					30%	15%			
3000	Refrigerant line sets, min-split, flared									
3100	Combination 1/4" & 3/8" tubes									
3120	15' set	Q-5	41	.390	Ea.	76.50	24.50		101	123
3140	25' set		38.50	.416		108	26.50		134.50	160
3160	35' set		37	.432		137	27.50		164.50	192
3180	50' set	↓	30	.533	↓	186	34		220	256
3200	Combination 1/4" & 1/2" tubes									
3220	15' set	Q-5	41	.390	Ea.	82	24.50		106.50	128
3240	25' set		38.50	.416		111	26.50		137.50	163
3260	35' set		37	.432		143	27.50		170.50	199
3280	50' set	↓	30	.533	↓	193	34		227	265

23 31 HVAC Ducts and Casings

23 31 13 - Metal Ducts

23 31 13.13 Rectangular Metal Ducts

		Crew	Daily Output	Labor-Hours	Unit	Material	2022 Bare Costs Labor	2022 Bare Costs Equipment	Total	Total Incl O&P
0010	**RECTANGULAR METAL DUCTS**									
0020	Fabricated rectangular, includes fittings, joints, supports,									
0021	allowance for flexible connections and field sketches.									
0030	Does not include "as-built dwgs." or insulation.									
0031	NOTE: Fabrication and installation are combined									
0040	as LABOR cost. Approx. 25% fittings assumed.									
0042	Fabrication/Inst. is to commercial quality standards									
0043	(SMACNA or equiv.) for structure, sealing, leak testing, etc.									
0100	Aluminum, alloy 3003-H14, under 100 lb.	Q-10	75	.320	Lb.	3.18	20		23.18	35
0110	100 to 500 lb.		80	.300		1.80	18.90		20.70	31.50
0120	500 to 1,000 lb.		95	.253		2.04	15.95		17.99	27
0160	Over 5,000 lb.		145	.166		1.98	10.45		12.43	18.45
0500	Galvanized steel, under 200 lb.		235	.102		.63	6.45		7.08	10.70
0520	200 to 500 lb.		245	.098		.70	6.20		6.90	10.35
0540	500 to 1,000 lb.		255	.094		.66	5.95		6.61	10
0560	1,000 to 2,000 lb.		265	.091		.68	5.70		6.38	9.65
0580	Over 5,000 lb.		285	.084		.68	5.30		5.98	9
1000	Stainless steel, type 304, under 100 lb.		165	.145		5.20	9.15		14.35	20
1020	100 to 500 lb.		175	.137		4.17	8.65		12.82	18.05
1030	500 to 1,000 lb.		190	.126		3.03	7.95		10.98	15.75

23 31 13.16 Round and Flat-Oval Spiral Ducts

		Crew	Daily Output	Labor-Hours	Unit	Material	2022 Bare Costs Labor	2022 Bare Costs Equipment	Total	Total Incl O&P
0010	**ROUND AND FLAT-OVAL SPIRAL DUCTS**									
0020	Fabricated round and flat oval spiral, includes hangers,									
0021	supports and field sketches.									
5400	Spiral preformed, steel, galv., straight lengths, max 10" spwg.									
5410	4" diameter, 26 ga.	Q-9	360	.044	L.F.	7	2.70		9.70	11.90
5416	5" diameter, 26 ga.		320	.050		7.40	3.04		10.44	12.90
5420	6" diameter, 26 ga.		280	.057		7.80	3.47		11.27	13.95
5425	7" diameter, 26 ga.		240	.067		8.60	4.05		12.65	15.75
5430	8" diameter, 26 ga.		200	.080		9	4.86		13.86	17.45
5440	10" diameter, 26 ga.		160	.100		11	6.10		17.10	21.50
9990	Minimum labor/equipment charge	1 Shee	3	2.667	Job		180		180	280

23 31 16 - Nonmetal Ducts

23 31 16.13 Fibrous-Glass Ducts

		Crew	Daily Output	Labor-Hours	Unit	Material	2022 Bare Costs Labor	2022 Bare Costs Equipment	Total	Total Incl O&P
0010	**FIBROUS-GLASS DUCTS**									
3490	Rigid fiberglass duct board, foil reinf. kraft facing									
3500	Rectangular, 1" thick, alum. faced, (FRK), std. weight	Q-10	350	.069	SF Surf	.93	4.32		5.25	7.75
9990	Minimum labor/equipment charge	1 Shee	3	2.667	Job		180		180	280

23 31 16.19 PVC Ducts

		Crew	Daily Output	Labor-Hours	Unit	Material	2022 Bare Costs Labor	2022 Bare Costs Equipment	Total	Total Incl O&P
0010	**PVC DUCTS**									
4000	Rigid plastic, corrosive fume resistant PVC									
4020	Straight, 6" diameter	Q-9	220	.073	L.F.	15.10	4.42		19.52	23.50
4040	8" diameter		160	.100		22	6.10		28.10	33.50
4060	10" diameter		120	.133		27.50	8.10		35.60	43
4070	12" diameter		100	.160		32	9.75		41.75	50
4080	14" diameter		70	.229		40	13.90		53.90	65.50
4090	16" diameter		62	.258		46	15.70		61.70	75.50
4100	18" diameter		58	.276		63.50	16.75		80.25	96
4110	20" diameter	Q-10	75	.320		59	20		79	96.50
4130	24" diameter	"	55	.436		81	27.50		108.50	132
4250	Coupling, 6" diameter	Q-9	88	.182	Ea.	31	11.05		42.05	51
4270	8" diameter		78	.205		40	12.45		52.45	63.50

23 31 HVAC Ducts and Casings

23 31 16 – Nonmetal Ducts

23 31 16.19 PVC Ducts

		Crew	Daily Output	Labor-Hours	Unit	Material	2022 Bare Costs Labor	Equipment	Total	Total Incl O&P
4290	10" diameter	Q-9	70	.229	Ea.	46	13.90		59.90	72.50
4300	12" diameter		55	.291		50	17.70		67.70	83
4310	14" diameter		44	.364		65	22		87	106
4320	16" diameter		40	.400		68.50	24.50		93	114
4330	18" diameter		38	.421		88.50	25.50		114	138
4340	20" diameter	Q-10	55	.436		106	27.50		133.50	160
4360	24" diameter	"	45	.533		133	33.50		166.50	200
4470	Elbow, 90°, 6" diameter	Q-9	44	.364		140	22		162	189
4490	8" diameter		28	.571		157	34.50		191.50	227
4510	10" diameter		18	.889		191	54		245	294
4520	12" diameter		15	1.067		217	65		282	340
4540	16" diameter		10	1.600		325	97.50		422.50	510
4550	18" diameter	Q-10	15	1.600		450	101		551	650
4560	20" diameter		14	1.714		675	108		783	910
4580	24" diameter		12	2		830	126		956	1,100
4750	Elbow 45°, use 90° and deduct					25%				
9990	Minimum labor/equipment charge	1 Shee	3	2.667	Job		180		180	280

23 31 16.22 Duct Board

0010	**DUCT BOARD**									
3800	Rigid, resin bonded fibrous glass, FSK									
3810	Temperature, bacteria and fungi resistant									
3820	1" thick	Q-14	150	.107	S.F.	1.81	6.05		7.86	11.55
3830	1-1/2" thick		130	.123		2.39	7		9.39	13.70
3840	2" thick		120	.133		4.12	7.55		11.67	16.50

23 33 Air Duct Accessories

23 33 13 – Dampers

23 33 13.13 Volume-Control Dampers

0010	**VOLUME-CONTROL DAMPERS**									
5990	Multi-blade dampers, opposed blade, 8" x 6"	1 Shee	24	.333	Ea.	44	22.50		66.50	83
5994	8" x 8"	1 Shee	22	.364	Ea.	145	24.50		169.50	198
5996	10" x 10"		21	.381		124	25.50		149.50	176
6000	12" x 12"		21	.381		124	25.50		149.50	176
6020	12" x 18"		18	.444		189	30		219	255
6030	14" x 10"		20	.400		129	27		156	184
6031	14" x 14"		17	.471		148	32		180	212
6033	16" x 12"		17	.471		186	32		218	254
6035	16" x 16"		16	.500		148	34		182	215
6037	18" x 16"		15	.533		156	36		192	227
6038	18" x 18"		15	.533		186	36		222	261
6070	20" x 16"		14	.571		224	38.50		262.50	305
6072	20" x 20"		13	.615		186	41.50		227.50	270
6074	22" x 18"		14	.571		107	38.50		145.50	178
6076	24" x 16"		11	.727		233	49		282	335
6078	24" x 20"		8	1		198	67.50		265.50	320
6080	24" x 24"		8	1		214	67.50		281.50	340
6110	26" x 26"		6	1.333		248	90		338	415
6133	30" x 30"	Q-9	6.60	2.424		287	147		434	545
6135	32" x 32"		6.40	2.500		287	152		439	550
6180	48" x 36"		5.60	2.857		530	174		704	855
7500	Variable volume modulating motorized damper, incl. elect. mtr.									

23 33 Air Duct Accessories

23 33 13 – Dampers

23 33 13.13 Volume-Control Dampers

		Crew	Daily Output	Labor-Hours	Unit	Material	2022 Bare Costs Labor	Equipment	Total	Total Incl O&P
7504	8" x 6"	1 Shee	15	.533	Ea.	226	36		262	305
7506	10" x 6"		14	.571		226	38.50		264.50	310
7510	10" x 10"		13	.615		236	41.50		277.50	325
7520	12" x 12"		12	.667		243	45		288	335
7522	12" x 16"		11	.727		244	49		293	345
7524	16" x 10"		12	.667		240	45		285	335
7526	16" x 14"		10	.800		251	54		305	360
7528	16" x 18"		9	.889		259	60		319	380
7542	18" x 18"		8	1		263	67.50		330.50	395
7544	20" x 14"		8	1		273	67.50		340.50	405
7546	20" x 18"		7	1.143		282	77		359	430
7560	24" x 12"		8	1		275	67.50		342.50	410
7562	24" x 18"		7	1.143		296	77		373	445
7564	24" x 24"		6	1.333		310	90		400	480
7568	28" x 10"		7	1.143		275	77		352	425
7590	30" x 14"		5	1.600		283	108		391	480
7600	30" x 18"		4	2		365	135		500	610
7610	30" x 24"	↓	3.80	2.105	↓	485	142		627	755
7700	For thermostat, add	1 Shee	8	1	Ea.	54	67.50		121.50	164
8000	Multi-blade dampers, parallel blade									
8100	8" x 8"	1 Shee	24	.333	Ea.	136	22.50		158.50	185
8140	16" x 10"		20	.400		142	27		169	198
8160	18" x 12"		18	.444		145	30		175	206
8220	28" x 16"	↓	10	.800	↓	199	54		253	305

23 33 13.16 Fire Dampers

		Crew	Daily Output	Labor-Hours	Unit	Material	2022 Bare Costs Labor	Equipment	Total	Total Incl O&P
0010	**FIRE DAMPERS**									
3000	Fire damper, curtain type, 1-1/2 hr. rated, vertical, 6" x 6"	1 Shee	24	.333	Ea.	24	22.50		46.50	61.50
3020	8" x 6"		22	.364		32	24.50		56.50	73.50
3240	16" x 14"		18	.444		78	30		108	133
3400	24" x 20"	↓	8	1	↓	67.50	67.50		135	180

23 33 19 – Duct Silencers

23 33 19.10 Duct Silencers

		Crew	Daily Output	Labor-Hours	Unit	Material	2022 Bare Costs Labor	Equipment	Total	Total Incl O&P
0010	**DUCT SILENCERS**									
9000	Silencers, noise control for air flow, duct				MCFM	94.50			94.50	104

23 33 23 – Turning Vanes

23 33 23.13 Air Turning Vanes

		Crew	Daily Output	Labor-Hours	Unit	Material	2022 Bare Costs Labor	Equipment	Total	Total Incl O&P
0010	**AIR TURNING VANES**									
9400	Turning vane components									
9410	Turning vane rail	1 Shee	160	.050	L.F.	.96	3.38		4.34	6.30
9420	Double thick, factory fab. vane		300	.027		1.41	1.80		3.21	4.35
9428	12" high set		170	.047		2.16	3.18		5.34	7.35
9432	14" high set		160	.050	↓	2.52	3.38		5.90	8
9900	Minimum labor/equipment charge	↓	4	2	Job		135		135	210

23 33 46 – Flexible Ducts

23 33 46.10 Flexible Air Ducts

		Crew	Daily Output	Labor-Hours	Unit	Material	2022 Bare Costs Labor	Equipment	Total	Total Incl O&P
0010	**FLEXIBLE AIR DUCTS**									
1280	Add to labor for elevated installation									
1282	of prefabricated (purchased) ductwork									
1283	10' to 15' high						10%			
1284	15' to 20' high						20%			
1285	20' to 25' high						25%			

23 33 Air Duct Accessories

23 33 46 – Flexible Ducts

23 33 46.10 Flexible Air Ducts		Crew	Daily Output	Labor-Hours	Unit	Material	2022 Bare Costs Labor	Equipment	Total	Total Incl O&P
1286	25' to 30' high						35%			
1287	30' to 35' high						40%			
1288	35' to 40' high						50%			
1289	Over 40' high						55%			
1300	Flexible, coated fiberglass fabric on corr. resist. metal helix									
1400	pressure to 12" (WG) UL-181									
1500	Noninsulated, 3" diameter	Q-9	400	.040	L.F.	1.42	2.43		3.85	5.35
1540	5" diameter	"	320	.050	"	1.65	3.04		4.69	6.55
1560	6" diameter	Q-9	280	.057	L.F.	1.94	3.47		5.41	7.55
1580	7" diameter		240	.067		2.19	4.05		6.24	8.70
1600	8" diameter		200	.080		2.45	4.86		7.31	10.25
1640	10" diameter		160	.100		3.12	6.10		9.22	12.90
1660	12" diameter		120	.133		3.67	8.10		11.77	16.65
1900	Insulated, 1" thick, PE jacket, 3" diameter	G	380	.042		2.93	2.56		5.49	7.20
1910	4" diameter	G	340	.047		3.44	2.86		6.30	8.25
1920	5" diameter	G	300	.053		3.60	3.24		6.84	9
1940	6" diameter	G	260	.062		3.81	3.74		7.55	10
1960	7" diameter	G	220	.073		4.41	4.42		8.83	11.75
1980	8" diameter	G	180	.089		4.48	5.40		9.88	13.35
2020	10" diameter	G	140	.114		5.40	6.95		12.35	16.75
2040	12" diameter	G	100	.160		6.30	9.75		16.05	22
2060	14" diameter	G	80	.200		7.65	12.15		19.80	27.50
9990	Minimum labor/equipment charge	1 Shee	3	2.667	Job		180		180	280

23 33 53 – Duct Liners

23 33 53.10 Duct Liner Board

			Crew	Daily Output	Labor-Hours	Unit	Material	Labor	Equipment	Total	Total Incl O&P
0010	**DUCT LINER BOARD**										
3490	Board type, fiberglass liner, 3 lb. density										
3680	No finish										
3700	1" thick	G	Q-14	170	.094	S.F.	.77	5.35		6.12	9.30
3710	1-1/2" thick	G		140	.114		1.15	6.50		7.65	11.50
3720	2" thick	G		130	.123		1.51	7		8.51	12.70
3940	Board type, non-fibrous foam										
3950	Temperature, bacteria and fungi resistant										
3960	1" thick	G	Q-14	150	.107	S.F.	3.05	6.05		9.10	12.90
3970	1-1/2" thick	G		130	.123		5.30	7		12.30	16.90
3980	2" thick	G		120	.133		4.25	7.55		11.80	16.65

23 34 HVAC Fans

23 34 13 – Axial HVAC Fans

23 34 13.10 Axial Flow HVAC Fans

0010	**AXIAL FLOW HVAC FANS**									
0020	Air conditioning and process air handling									

23 34 14 – Blower HVAC Fans

23 34 14.10 Blower Type HVAC Fans

			Crew	Daily Output	Labor-Hours	Unit	Material	Labor	Equipment	Total	Total Incl O&P
0010	**BLOWER TYPE HVAC FANS**										
2500	Ceiling fan, right angle, extra quiet, 0.10" S.P.										
2540	210 CFM		Q-20	19	1.053	Ea.	335	65		400	470
2580	885 CFM			16	1.250		835	77.50		912.50	1,050
2620	2,960 CFM			11	1.818		1,550	113		1,663	1,875
2640	For wall or roof cap, add		1 Shee	16	.500		283	34		317	365

23 34 HVAC Fans

23 34 16 – Centrifugal HVAC Fans

23 34 16.10 Centrifugal Type HVAC Fans

		Crew	Daily Output	Labor-Hours	Unit	Material	2022 Bare Costs Labor	Equipment	Total	Total Incl O&P
0010	**CENTRIFUGAL TYPE HVAC FANS**									
0200	In-line centrifugal, supply/exhaust booster									
0220	aluminum wheel/hub, disconnect switch, 1/4" S.P.									
0240	500 CFM, 10" diameter connection	Q-20	3	6.667	Ea.	1,625	415		2,040	2,425
0280	1,520 CFM, 16" diameter connection		2	10		1,750	620		2,370	2,875
0320	3,480 CFM, 20" diameter connection		.80	25		2,250	1,550		3,800	4,850
0326	5,080 CFM, 20" diameter connection	↓	.75	26.667	↓	2,425	1,650		4,075	5,200
5000	Utility set, centrifugal, V belt drive, motor									
5020	1/4" S.P., 1,200 CFM, 1/4 HP	Q-20	6	3.333	Ea.	2,325	206		2,531	2,900
5040	1,520 CFM, 1/3 HP		5	4		2,675	248		2,923	3,325
5060	1,850 CFM, 1/2 HP		4	5		2,675	310		2,985	3,425
5080	2,180 CFM, 3/4 HP	↓	3	6.667	↓	3,050	415		3,465	4,025
7000	Roof exhauster, centrifugal, aluminum housing, 12" galvanized									
7020	curb, bird screen, back draft damper, 1/4" S.P.									
7100	Direct drive, 320 CFM, 11" sq. damper	Q-20	7	2.857	Ea.	770	177		947	1,125
7120	600 CFM, 11" sq. damper	"	6	3.333	"	985	206		1,191	1,400
9900	Minimum labor/equipment charge	1 Elec	4	2	Job		132		132	203

23 34 23 – HVAC Power Ventilators

23 34 23.10 HVAC Power Circulators and Ventilators

		Crew	Daily Output	Labor-Hours	Unit	Material	2022 Bare Costs Labor	Equipment	Total	Total Incl O&P
0010	**HVAC POWER CIRCULATORS AND VENTILATORS**									
6650	Residential, bath exhaust, grille, back draft damper									
6660	50 CFM	Q-20	24	.833	Ea.	47	51.50		98.50	132
6670	110 CFM		22	.909		94	56.50		150.50	192
6680	Light combination, squirrel cage, 100 watt, 70 CFM	↓	24	.833	↓	120	51.50		171.50	212
6700	Light/heater combination, ceiling mounted									
6710	70 CFM, 1,450 watt	Q-20	24	.833	Ea.	213	51.50		264.50	315
6800	Heater combination, recessed, 70 CFM		24	.833		66.50	51.50		118	153
6820	With 2 infrared bulbs		23	.870		94.50	54		148.50	188
6900	Kitchen exhaust, grille, complete, 160 CFM		22	.909		136	56.50		192.50	238
6910	180 CFM		20	1		103	62		165	210
6920	270 CFM	↓	18	1.111	↓	201	69		270	330
6940	Residential roof jacks and wall caps									
6944	Wall cap with back draft damper									
6946	3" & 4" diam. round duct	1 Shee	11	.727	Ea.	19.40	49		68.40	98
6948	6" diam. round duct	"	11	.727	"	53	49		102	135
6958	Roof jack with bird screen and back draft damper									
6960	3" & 4" diam. round duct	1 Shee	11	.727	Ea.	24.50	49		73.50	104
6962	3-1/4" x 10" rectangular duct	"	10	.800	"	38	54		92	126
6980	Transition									
6982	3-1/4" x 10" to 6" diam. round	1 Shee	20	.400	Ea.	25	27		52	69.50
8020	Attic, roof type									
8030	Aluminum dome, damper & curb									
8080	12" diameter, 1,000 CFM (gravity)	1 Elec	10	.800	Ea.	660	53		713	805
8090	16" diameter, 1,500 CFM (gravity)		9	.889		795	59		854	965
8100	20" diameter, 2,500 CFM (gravity)		8	1		975	66		1,041	1,175
8110	26" diameter, 4,000 CFM (gravity)		7	1.143		1,175	75.50		1,250.50	1,425
8120	32" diameter, 6,500 CFM (gravity)		6	1.333		1,625	88.50		1,713.50	1,900
8130	38" diameter, 8,000 CFM (gravity)		5	1.600		2,400	106		2,506	2,825
8140	50" diameter, 13,000 CFM (gravity)	↓	4	2	↓	3,475	132		3,607	4,025
8160	Plastic, ABS dome									
8180	1,050 CFM	1 Elec	14	.571	Ea.	193	38		231	271
8200	1,600 CFM	"	12	.667	"	290	44		334	390

23 34 23 – HVAC Power Ventilators

23 34 23.10 HVAC Power Circulators and Ventilators	Crew	Daily Output	Labor- Hours	Unit	Material	2022 Bare Costs Labor	2022 Bare Costs Equipment	Total	Total Incl O&P
8240	Attic, wall type, with shutter, one speed									
8250	12" diameter, 1,000 CFM	1 Elec	14	.571	Ea.	480	38		518	590
8260	14" diameter, 1,500 CFM		12	.667		520	44		564	640
8270	16" diameter, 2,000 CFM	↓	9	.889	↓	590	59		649	740
8290	Whole house, wall type, with shutter, one speed									
8300	30" diameter, 4,800 CFM	1 Elec	7	1.143	Ea.	1,250	75.50		1,325.50	1,500
8310	36" diameter, 7,000 CFM		6	1.333		1,375	88.50		1,463.50	1,625
8320	42" diameter, 10,000 CFM		5	1.600		1,525	106		1,631	1,875
8330	48" diameter, 16,000 CFM	↓	4	2		1,900	132		2,032	2,300
8340	For two speed, add				↓	115			115	126
8350	Whole house, lay-down type, with shutter, one speed									
8360	30" diameter, 4,500 CFM	1 Elec	8	1	Ea.	1,350	66		1,416	1,575
8370	36" diameter, 6,500 CFM		7	1.143		1,450	75.50		1,525.50	1,700
8380	42" diameter, 9,000 CFM		6	1.333		1,575	88.50		1,663.50	1,875
8390	48" diameter, 12,000 CFM	↓	5	1.600		1,800	106		1,906	2,150
8440	For two speed, add					86.50			86.50	95
8450	For 12 hour timer switch, add	1 Elec	32	.250	↓	86.50	16.55		103.05	121

23 36 Air Terminal Units

23 36 13 – Constant-Air-Volume Units

23 36 13.10 Constant Volume Mixing Boxes

		Crew	Daily Output	Labor- Hours	Unit	Material	Labor	Equipment	Total	Total Incl O&P
0010	CONSTANT VOLUME MIXING BOXES									
5180	Mixing box, includes electric or pneumatic motor									
5200	Constant volume, 150 to 270 CFM	Q-9	12	1.333	Ea.	1,275	81		1,356	1,525
5210	270 to 600 CFM	"	11	1.455	"	1,150	88.50		1,238.50	1,425

23 36 16 – Variable-Air-Volume Units

23 36 16.10 Variable Volume Mixing Boxes

		Crew	Daily Output	Labor- Hours	Unit	Material	Labor	Equipment	Total	Total Incl O&P
0010	VARIABLE VOLUME MIXING BOXES									
5180	Mixing box, includes electric or pneumatic motor									
5500	VAV cool only, pneumatic, pressure independent 300 to 600 CFM	Q-9	11	1.455	Ea.	920	88.50		1,008.50	1,150
5510	500 to 1,000 CFM		9	1.778	↓	945	108		1,053	1,225
5520	800 to 1,600 CFM		9	1.778		975	108		1,083	1,250
5530	1,100 to 2,000 CFM		8	2		995	122		1,117	1,300
5540	1,500 to 3,000 CFM		7	2.286		1,050	139		1,189	1,400
5550	2,000 to 4,000 CFM	↓	6	2.667		1,075	162		1,237	1,425

23 37 Air Outlets and Inlets

23 37 13 – Diffusers, Registers, and Grilles

23 37 13.10 Diffusers

		Crew	Daily Output	Labor- Hours	Unit	Material	Labor	Equipment	Total	Total Incl O&P
0010	DIFFUSERS, Aluminum, opposed blade damper unless noted									
0100	Ceiling, linear, also for sidewall									
0120	2" wide	1 Shee	32	.250	L.F.	26.50	16.90		43.40	56
0160	4" wide		26	.308		23	21		44	58
0180	6" wide		24	.333		28.50	22.50		51	66
0200	8" wide		22	.364	↓	51.50	24.50		76	95
0500	Perforated, 24" x 24" lay-in panel size, 6" x 6"		16	.500	Ea.	187	34		221	259
0520	8" x 8"		15	.533		96.50	36		132.50	162
0530	9" x 9"		14	.571		199	38.50		237.50	279
0540	10" x 10"	↓	14	.571	↓	199	38.50		237.50	279

470

For customer support on your Commercial Renovation Costs with RSMeans data, call 800.448.8182.

23 37 Air Outlets and Inlets

23 37 13 - Diffusers, Registers, and Grilles

23 37 13.10 Diffusers

		Crew	Daily Output	Labor-Hours	Unit	Material	2022 Bare Costs Labor	Equipment	Total	Total Incl O&P
0560	12" x 12"	1 Shee	12	.667	Ea.	208	45		253	298
0590	16" x 16"		11	.727		221	49		270	320
0600	18" x 18"		10	.800		237	54		291	345
0610	20" x 20"		10	.800		254	54		308	365
0620	24" x 24"		9	.889		298	60		358	425
1000	Rectangular, 1 to 4 way blow, 6" x 6"		16	.500		45.50	34		79.50	103
1010	8" x 8"		15	.533		62.50	36		98.50	125
1014	9" x 9"		15	.533		53	36		89	114
1016	10" x 10"		15	.533		92	36		128	157
1020	12" x 6"		15	.533		77	36		113	141
1040	12" x 9"		14	.571		82	38.50		120.50	150
1060	12" x 12"		12	.667		76.50	45		121.50	154
1070	14" x 6"	1 Shee	13	.615	Ea.	83.50	41.50		125	157
1074	14" x 14"		12	.667		136	45		181	220
1150	18" x 18"		9	.889		131	60		191	239
1170	24" x 12"		10	.800		182	54		236	284
2000	T-bar mounting, 24" x 24" lay-in frame, 6" x 6"		16	.500		72	34		106	132
2020	8" x 8"		14	.571		72.50	38.50		111	140
2040	12" x 12"		12	.667		88.50	45		133.50	168
2060	16" x 16"		11	.727		114	49		163	202
2080	18" x 18"		10	.800		112	54		166	207
6000	For steel diffusers instead of aluminum, deduct					10%				
9000	Minimum labor/equipment charge	1 Shee	4	2	Job		135		135	210

23 37 13.30 Grilles

		Crew	Daily Output	Labor-Hours	Unit	Material	2022 Bare Costs Labor	Equipment	Total	Total Incl O&P
0010	**GRILLES**									
0020	Aluminum, unless noted otherwise									
1000	Air return, steel, 6" x 6"	1 Shee	26	.308	Ea.	24	21		45	59
1020	10" x 6"		24	.333		24	22.50		46.50	61.50
1080	16" x 8"		22	.364		35	24.50		59.50	76.50
1100	12" x 12"		22	.364		35	24.50		59.50	76.50
1120	24" x 12"		18	.444		45.50	30		75.50	96.50
1300	48" x 24"		12	.667		164	45		209	251
9000	Minimum labor/equipment charge		4	2	Job		135		135	210

23 37 13.60 Registers

		Crew	Daily Output	Labor-Hours	Unit	Material	2022 Bare Costs Labor	Equipment	Total	Total Incl O&P
0010	**REGISTERS**									
0980	Air supply									
1000	Ceiling/wall, O.B. damper, anodized aluminum									
1010	One or two way deflection, adj. curved face bars									
1120	12" x 12"	1 Shee	18	.444	Ea.	56.50	30		86.50	109
1140	14" x 8"		17	.471		28.50	32		60.50	81
1240	20" x 6"		18	.444		48	30		78	99
1340	24" x 8"		13	.615		64.50	41.50		106	136
1350	24" x 18"		12	.667		119	45		164	201
2700	Above registers in steel instead of aluminum, deduct					10%				
3000	Baseboard, hand adj. damper, enameled steel									
3012	8" x 6"	1 Shee	26	.308	Ea.	10.80	21		31.80	44.50
3020	10" x 6"		24	.333		11.75	22.50		34.25	48
3040	12" x 5"		23	.348		9.80	23.50		33.30	47.50
3060	12" x 6"		23	.348		12.80	23.50		36.30	50.50
3080	12" x 8"		22	.364		18.50	24.50		43	58.50
3100	14" x 6"		20	.400		13.85	27		40.85	57.50
9000	Minimum labor/equipment charge		4	2	Job		135		135	210

23 37 Air Outlets and Inlets

23 37 23 – HVAC Gravity Ventilators

23 37 23.10 HVAC Gravity Air Ventilators	Crew	Daily Output	Labor-Hours	Unit	Material	2022 Bare Costs Labor	Equipment	Total	Total Incl O&P
0010 **HVAC GRAVITY AIR VENTILATORS**, Includes base									
1280 Rotary ventilators, wind driven, galvanized									
1300 4" neck diameter	Q-9	20	.800	Ea.	69	48.50		117.50	152
1400 12" neck diameter		10	1.600		97.50	97.50		195	258
1500 24" neck diameter	↓	8	2	↓	435	122		557	665
1600 For aluminum, add					30%				
9000 Minimum labor/equipment charge	1 Plum	2	4	Job		279		279	430

23 38 Ventilation Hoods

23 38 13 – Commercial-Kitchen Hoods

23 38 13.10 Hood and Ventilation Equipment

	Crew	Daily Output	Labor-Hours	Unit	Material	2022 Bare Costs Labor	Equipment	Total	Total Incl O&P
0010 **HOOD AND VENTILATION EQUIPMENT**									
2970 Exhaust hood, sst, gutter on all sides, 4' x 4' x 2'	1 Carp	1.80	4.444	Ea.	2,325	250		2,575	2,950
2980 4' x 4' x 7'	"	1.60	5	"	6,775	282		7,057	7,900

23 41 Particulate Air Filtration

23 41 13 – Panel Air Filters

23 41 13.10 Panel Type Air Filters

		Crew	Daily Output	Labor-Hours	Unit	Material	2022 Bare Costs Labor	Equipment	Total	Total Incl O&P
0010 **PANEL TYPE AIR FILTERS**										
2950 Mechanical media filtration units										
3000 High efficiency type, with frame, non-supported	G				MCFM	32.50			32.50	36
3100 Supported type	G				"	47			47	52
5500 Throwaway glass or paper media type, 12" x 36" x 1"					Ea.	2.29			2.29	2.52

23 41 16 – Renewable-Media Air Filters

23 41 16.10 Disposable Media Air Filters

	Crew	Daily Output	Labor-Hours	Unit	Material	2022 Bare Costs Labor	Equipment	Total	Total Incl O&P
0010 **DISPOSABLE MEDIA AIR FILTERS**									
5000 Renewable disposable roll				C.S.F.	5.10			5.10	5.60

23 41 19 – Washable Air Filters

23 41 19.10 Permanent Air Filters

		Crew	Daily Output	Labor-Hours	Unit	Material	2022 Bare Costs Labor	Equipment	Total	Total Incl O&P
0010 **PERMANENT AIR FILTERS**										
4500 Permanent washable	G				MCFM	24.50			24.50	26.50

23 41 23 – Extended Surface Filters

23 41 23.10 Expanded Surface Filters

		Crew	Daily Output	Labor-Hours	Unit	Material	2022 Bare Costs Labor	Equipment	Total	Total Incl O&P
0010 **EXPANDED SURFACE FILTERS**										
4000 Medium efficiency, extended surface	G				MCFM	5.70			5.70	6.30

23 42 Gas-Phase Air Filtration

23 42 13 – Activated-Carbon Air Filtration

23 42 13.10 Charcoal Type Air Filtration

		Crew	Daily Output	Labor-Hours	Unit	Material	2022 Bare Costs Labor	Equipment	Total	Total Incl O&P
0010	**CHARCOAL TYPE AIR FILTRATION**									
0050	Activated charcoal type, full flow				MCFM	635			635	700
0060	Full flow, impregnated media 12" deep					219			219	241
0070	HEPA filter & frame for field erection					490			490	540
0080	HEPA filter-diffuser, ceiling install.				MCFM	320			320	355

23 43 Electronic Air Cleaners

23 43 13 – Washable Electronic Air Cleaners

23 43 13.10 Electronic Air Cleaners

		Crew	Daily Output	Labor-Hours	Unit	Material	Labor	Equipment	Total	Total Incl O&P
0010	**ELECTRONIC AIR CLEANERS**									
2000	Electronic air cleaner, duct mounted									
2150	1,000 CFM	1 Shee	4	2	Ea.	430	135		565	685
2200	1,200 CFM		3.80	2.105		555	142		697	830
2250	1,400 CFM		3.60	2.222		565	150		715	860

23 51 Breechings, Chimneys, and Stacks

23 51 13 – Draft Control Devices

23 51 13.16 Vent Dampers

		Crew	Daily Output	Labor-Hours	Unit	Material	Labor	Equipment	Total	Total Incl O&P
0010	**VENT DAMPERS**									
5000	Vent damper, bi-metal, gas, 3" diameter	Q-9	24	.667	Ea.	94	40.50		134.50	166
5010	4" diameter		24	.667		73.50	40.50		114	144
5020	5" diameter		23	.696		94	42.50		136.50	169
5030	6" diameter		22	.727		73.50	44		117.50	150
5040	7" diameter		21	.762		75.50	46.50		122	155
5050	8" diameter		20	.800		81.50	48.50		130	165
9000	Minimum labor/equipment charge	1 Shee	4	2	Job		135		135	210

23 51 13.19 Barometric Dampers

		Crew	Daily Output	Labor-Hours	Unit	Material	Labor	Equipment	Total	Total Incl O&P
0010	**BAROMETRIC DAMPERS**									
1000	Barometric, gas fired system only, 6" size for 5" and 6" pipes	1 Shee	20	.400	Ea.	142	27		169	198
1040	8" size, for 7" and 8" pipes	"	18	.444	"	197	30		227	263
2000	All fuel, oil, oil/gas, coal									
2020	10" for 9" and 10" pipes	1 Shee	15	.533	Ea.	330	36		366	415
3260	For thermal switch for above, add	"	24	.333	"	129	22.50		151.50	177

23 51 23 – Gas Vents

23 51 23.10 Gas Chimney Vents

		Crew	Daily Output	Labor-Hours	Unit	Material	Labor	Equipment	Total	Total Incl O&P
0010	**GAS CHIMNEY VENTS**, Prefab metal, UL listed									
0020	Gas, double wall, galvanized steel									
0080	3" diameter	Q-9	72	.222	V.L.F.	8.45	13.50		21.95	30.50
0100	4" diameter		68	.235		10.45	14.30		24.75	34
0120	5" diameter		64	.250		11.85	15.20		27.05	36.50
0140	6" diameter		60	.267		14.10	16.20		30.30	40.50
0160	7" diameter		56	.286		27	17.35		44.35	56.50
0180	8" diameter		52	.308		29	18.70		47.70	60.50
0200	10" diameter		48	.333		55	20.50		75.50	92
0220	12" diameter		44	.364		65.50	22		87.50	107
0260	16" diameter		40	.400		157	24.50		181.50	211
0300	20" diameter	Q-10	36	.667		228	42		270	315
0340	24" diameter	Q-10	32	.750	V.L.F.	360	47.50		407.50	470

473

23 51 26.30 All-Fuel Vent Chimneys, Double Wall, St. Stl.		Crew	Daily Output	Labor-Hours	Unit	Material	2022 Bare Costs Labor	Equipment	Total	Total Incl O&P
0010	**ALL-FUEL VENT CHIMNEYS, DOUBLE WALL, STAINLESS STEEL**									
7780	All fuel, pressure tight, double wall, 4" insulation, UL listed, 1,400°F.									
7790	304 stainless steel liner, aluminized steel outer jacket									
7800	6" diameter	Q-9	60	.267	V.L.F.	58	16.20		74.20	89
7804	8" diameter		52	.308		84.50	18.70		103.20	122
7806	10" diameter		48	.333		96.50	20.50		117	138
7808	12" diameter		44	.364		112	22		134	159
7810	14" diameter		42	.381		126	23		149	175
7880	For 316 stainless steel liner, add				L.F.	30%				
8000	All fuel, double wall, stainless steel fittings									
8010	Roof support, 6" diameter	Q-9	30	.533	Ea.	130	32.50		162.50	194
8030	8" diameter		26	.615		137	37.50		174.50	209
8040	10" diameter		24	.667		163	40.50		203.50	242
8050	12" diameter		22	.727		171	44		215	258
8060	14" diameter		21	.762		184	46.50		230.50	274
8100	Elbow 45°, 6" diameter		30	.533		265	32.50		297.50	345
8140	8" diameter		26	.615		305	37.50		342.50	395
8160	10" diameter		24	.667		365	40.50		405.50	470
8180	12" diameter		22	.727		390	44		434	500
8200	14" diameter		21	.762		455	46.50		501.50	570
8300	Insulated tee, 6" diameter		30	.533		335	32.50		367.50	415
8360	8" diameter		26	.615		350	37.50		387.50	445
8380	10" diameter		24	.667		375	40.50		415.50	480
8400	12" diameter		22	.727		455	44		499	575
8420	14" diameter		20	.800		540	48.50		588.50	665
8500	Boot tee, 6" diameter		28	.571		760	34.50		794.50	890
8520	8" diameter		24	.667		810	40.50		850.50	955
8530	10" diameter		22	.727		915	44		959	1,075
8540	12" diameter		20	.800		1,075	48.50		1,123.50	1,250
8550	14" diameter		18	.889		1,100	54		1,154	1,300
8600	Rain cap with bird screen, 6" diameter		30	.533		340	32.50		372.50	425
8640	8" diameter		26	.615		370	37.50		407.50	470
8660	10" diameter		24	.667		380	40.50		420.50	480
8680	12" diameter		22	.727		535	44		579	655
8700	14" diameter		21	.762		560	46.50		606.50	685
8800	Flat roof flashing, 6" diameter	Q-9	30	.533	Ea.	125	32.50		157.50	188
8840	8" diameter		26	.615		127	37.50		164.50	199
8860	10" diameter		24	.667		137	40.50		177.50	213
8880	12" diameter		22	.727		152	44		196	237
8900	14" diameter		21	.762		160	46.50		206.50	248

23 52 Heating Boilers

23 52 13 – Electric Boilers

23 52 13.10 Electric Boilers, ASME

		Crew	Daily Output	Labor-Hours	Unit	Material	2022 Bare Costs Labor	Equipment	Total	Total Incl O&P
0010	**ELECTRIC BOILERS, ASME**, Standard controls and trim									
1000	Steam, 6 KW, 20.5 MBH	Q-19	1.20	20	Ea.	4,500	1,275		5,775	6,950
1160	60 KW, 205 MBH		1	24		7,150	1,550		8,700	10,200
1300	296 KW, 1,010 MBH	↓	.45	53.333		26,400	3,425		29,825	34,300
1400	592 KW, 2,020 MBH	Q-21	.34	94.118		35,700	6,200		41,900	48,800
1540	1,110 KW, 3,788 MBH	"	.19	168		50,000	11,100		61,100	72,000
2000	Hot water, 7.5 KW, 25.6 MBH	Q-19	1.30	18.462		5,150	1,175		6,325	7,475
2040	30 KW, 102 MBH		1.20	20		5,525	1,275		6,800	8,075
2060	45 KW, 164 MBH		1.20	20		5,400	1,275		6,675	7,925
2070	60 KW, 205 MBH		1.20	20		5,775	1,275		7,050	8,325
2080	75 KW, 256 MBH		1.10	21.818		6,100	1,400		7,500	8,875
2100	90 KW, 307 MBH	↓	1.10	21.818	↓	6,150	1,400		7,550	8,900
9000	Minimum labor/equipment charge	Q-20	1	20	Job		1,250		1,250	1,925

23 52 23 – Cast-Iron Boilers

23 52 23.20 Gas-Fired Boilers

		Crew	Daily Output	Labor-Hours	Unit	Material	2022 Bare Costs Labor	Equipment	Total	Total Incl O&P
0010	**GAS-FIRED BOILERS**, Natural or propane, standard controls, packaged									
1000	Cast iron, with insulated jacket									
2000	Steam, gross output, 81 MBH	Q-7	1.40	22.857	Ea.	3,150	1,525		4,675	5,800
2060	163 MBH		.90	35.556		3,425	2,375		5,800	7,450
2200	440 MBH		.51	62.500		6,250	4,175		10,425	13,300
2320	1,875 MBH		.30	107		29,000	7,150		36,150	42,900
2400	3,570 MBH		.18	182		40,300	12,200		52,500	63,000
2540	6,970 MBH		.10	320		96,500	21,400		117,900	139,500
3000	Hot water, gross output 80 MBH		1.46	21.918		2,850	1,475		4,325	5,400
3020	100 MBH		1.35	23.704		2,800	1,575		4,375	5,525
3040	122 MBH		1.10	29.091		3,025	1,950		4,975	6,325
3060	163 MBH		1	32		3,550	2,150		5,700	7,225
3200	440 MBH		.58	54.983		5,525	3,675		9,200	11,800
3320	2,000 MBH		.26	125		25,200	8,375		33,575	40,600
3540	6,970 MBH	↓	.09	360		129,000	24,100		153,100	179,000
7000	For tankless water heater, add				↓	10%				
7050	For additional zone valves up to 312 MBH, add				Ea.	199			199	219
9900	Minimum labor/equipment charge	Q-6	1	24	Job		1,575		1,575	2,425

23 52 23.30 Gas/Oil Fired Boilers

		Crew	Daily Output	Labor-Hours	Unit	Material	2022 Bare Costs Labor	Equipment	Total	Total Incl O&P
0010	**GAS/OIL FIRED BOILERS**, Combination with burners and controls, packaged									
1000	Cast iron with insulated jacket									
2000	Steam, gross output, 720 MBH	Q-7	.43	74.074	Ea.	16,100	4,950		21,050	25,400
2140	2,700 MBH		.19	166		26,200	11,100		37,300	45,900
2380	6,970 MBH	↓	.09	372	↓	112,000	24,900		136,900	162,000
2900	Hot water, gross output									
2910	200 MBH	Q-6	.62	39.024	Ea.	11,700	2,575		14,275	16,900
2920	300 MBH		.49	49.080		11,700	3,225		14,925	17,900
2930	400 MBH		.41	57.971		13,700	3,800		17,500	21,000
2940	500 MBH	↓	.36	67.039		14,800	4,400		19,200	23,100
3000	584 MBH	Q-7	.44	72.072		14,900	4,825		19,725	23,700
3060	1,460 MBH		.28	113		30,800	7,575		38,375	45,600
3300	13,500 MBH, 403.3 BHP	↓	.04	727	↓	212,500	48,700		261,200	308,500

23 52 23.40 Oil-Fired Boilers

		Crew	Daily Output	Labor-Hours	Unit	Material	2022 Bare Costs Labor	Equipment	Total	Total Incl O&P
0010	**OIL-FIRED BOILERS**, Standard controls, flame retention burner, packaged									
1000	Cast iron, with insulated flush jacket									
2000	Steam, gross output, 109 MBH	Q-7	1.20	26.667	Ea.	2,500	1,775		4,275	5,500
2060	207 MBH		.90	35.556		3,375	2,375		5,750	7,400

23 52 Heating Boilers

23 52 23 – Cast-Iron Boilers

23 52 23.40 Oil-Fired Boilers

		Crew	Daily Output	Labor-Hours	Unit	Material	2022 Bare Costs Labor	Equipment	Total	Total Incl O&P
2180	1,084 MBH	Q-7	.38	85.106	Ea.	11,700	5,700		17,400	21,700
2200	1,360 MBH		.33	98.160		14,300	6,575		20,875	25,900
2240	2,175 MBH		.24	134		17,400	8,975		26,375	33,000
2340	4,360 MBH		.15	215		39,400	14,400		53,800	65,500
2460	6,970 MBH	▼	.09	364	▼	108,500	24,300		132,800	157,000
3000	Hot water, same price as steam									
4000	For tankless coil in smaller sizes, add				Ea.	15%				

23 52 26 – Steel Boilers

23 52 26.40 Oil-Fired Boilers

		Crew	Daily Output	Labor-Hours	Unit	Material	2022 Bare Costs Labor	Equipment	Total	Total Incl O&P
0010	**OIL-FIRED BOILERS**, Standard controls, flame retention burner									
5000	Steel, with insulated flush jacket									
7000	Hot water, gross output, 103 MBH	Q-6	1.60	15	Ea.	2,225	985		3,210	3,975
7020	122 MBH		1.45	16.506		2,100	1,075		3,175	4,000
7040	137 MBH		1.36	17.595		2,850	1,150		4,000	4,925
7060	168 MBH		1.30	18.405		2,800	1,200		4,000	4,925
7080	225 MBH		1.22	19.704		4,300	1,300		5,600	6,725
7100	315 MBH		.96	25.105		5,625	1,650		7,275	8,750
7120	420 MBH		.70	34.483	▼	5,325	2,275		7,600	9,350
9000	Minimum labor/equipment charge	▼	1.75	13.714	Job		900		900	1,400

23 52 88 – Burners

23 52 88.10 Replacement Type Burners

		Crew	Daily Output	Labor-Hours	Unit	Material	2022 Bare Costs Labor	Equipment	Total	Total Incl O&P
0010	**REPLACEMENT TYPE BURNERS**									
0990	Residential, conversion, gas fired, LP or natural									
1000	Gun type, atmospheric input 50 to 225 MBH	Q-1	2.50	6.400	Ea.	1,725	400		2,125	2,525
1020	100 to 400 MBH		2	8		2,875	505		3,380	3,925
1040	300 to 1,000 MBH	▼	1.70	9.412	▼	8,700	590		9,290	10,500
3000	Flame retention oil fired assembly, input									
3020	.50 to 2.25 GPH	Q-1	2.40	6.667	Ea.	360	420		780	1,050
3040	2.0 to 5.0 GPH		2	8		335	505		840	1,150
3060	3.0 to 7.0 GPH		1.80	8.889		630	560		1,190	1,550
3080	6.0 to 12.0 GPH	▼	1.60	10	▼	1,025	630		1,655	2,100

23 54 Furnaces

23 54 13 – Electric-Resistance Furnaces

23 54 13.10 Electric Furnaces

		Crew	Daily Output	Labor-Hours	Unit	Material	2022 Bare Costs Labor	Equipment	Total	Total Incl O&P
0010	**ELECTRIC FURNACES**, Hot air, blowers, std. controls									
0011	not including gas, oil or flue piping									
1000	Electric, UL listed									
1070	10.2 MBH	Q-20	4.40	4.545	Ea.	345	281		626	810
1080	17.1 MBH		4.60	4.348		420	269		689	885
1100	34.1 MBH	▼	4.40	4.545	▼	530	281		811	1,025

23 54 16 – Fuel-Fired Furnaces

23 54 16.13 Gas-Fired Furnaces

		Crew	Daily Output	Labor-Hours	Unit	Material	2022 Bare Costs Labor	Equipment	Total	Total Incl O&P
0010	**GAS-FIRED FURNACES**									
3000	Gas, AGA certified, upflow, direct drive models									
3020	45 MBH input	Q-9	4	4	Ea.	715	243		958	1,175
3040	60 MBH input		3.80	4.211		645	256		901	1,100
3060	75 MBH input		3.60	4.444		880	270		1,150	1,400
3100	100 MBH input	▼	3.20	5	▼	840	305		1,145	1,400

23 54 Furnaces

23 54 16 – Fuel-Fired Furnaces

23 54 16.13 Gas-Fired Furnaces

		Crew	Daily Output	Labor-Hours	Unit	Material	2022 Bare Costs Labor	Equipment	Total	Total Incl O&P
3120	125 MBH input	Q-9	3	5.333	Ea.	875	325		1,200	1,475
3130	150 MBH input		2.80	5.714		1,750	345		2,095	2,475
3140	200 MBH input		2.60	6.154		3,425	375		3,800	4,325

23 54 16.16 Oil-Fired Furnaces

		Crew	Daily Output	Labor-Hours	Unit	Material	2022 Bare Costs Labor	Equipment	Total	Total Incl O&P
0010	**OIL-FIRED FURNACES**									
6000	Oil, UL listed, atomizing gun type burner									
6020	56 MBH output	Q-9	3.60	4.444	Ea.	3,375	270		3,645	4,150
6040	95 MBH output		3.40	4.706		3,600	286		3,886	4,400
6060	134 MBH output		3.20	5		3,500	305		3,805	4,325
6080	151 MBH output		3	5.333		3,900	325		4,225	4,775
6100	200 MBH input		2.60	6.154		4,800	375		5,175	5,850
9000	Minimum labor/equipment charge		2.75	5.818	Job		355		355	550

23 54 24 – Furnace Components for Cooling

23 54 24.10 Furnace Components and Combinations

		Crew	Daily Output	Labor-Hours	Unit	Material	2022 Bare Costs Labor	Equipment	Total	Total Incl O&P
0010	**FURNACE COMPONENTS AND COMBINATIONS**									
0080	Coils, A.C. evaporator, for gas or oil furnaces									
0090	Add-on, with holding charge									
0100	Upflow									
0120	1-1/2 ton cooling	Q-5	4	4	Ea.	243	253		496	655
0130	2 ton cooling		3.70	4.324		254	274		528	700
0140	3 ton cooling		3.30	4.848		395	305		700	910
0150	4 ton cooling		3	5.333		495	340		835	1,050
0160	5 ton cooling		2.70	5.926		555	375		930	1,200
0300	Downflow									
0330	2-1/2 ton cooling	Q-5	3	5.333	Ea.	310	340		650	860
0340	3-1/2 ton cooling		2.60	6.154		460	390		850	1,100
0350	5 ton cooling		2.20	7.273		555	460		1,015	1,325
0600	Horizontal									
0630	2 ton cooling	Q-5	3.90	4.103	Ea.	335	260		595	770
0640	3 ton cooling		3.50	4.571		420	290		710	910
0650	4 ton cooling		3.20	5		545	315		860	1,100
0660	5 ton cooling		2.90	5.517		595	350		945	1,200
2000	Cased evaporator coils for air handlers									
2100	1-1/2 ton cooling	Q-5	4.40	3.636	Ea.	276	230		506	660
2110	2 ton cooling		4.10	3.902		263	247		510	670
2120	2-1/2 ton cooling		3.90	4.103		310	260		570	740
2130	3 ton cooling		3.70	4.324		375	274		649	835
2140	3-1/2 ton cooling		3.50	4.571		515	290		805	1,000
2150	4 ton cooling		3.20	5		545	315		860	1,100
2160	5 ton cooling		2.90	5.517		540	350		890	1,125
3010	Air handler, modular									
3100	With cased evaporator cooling coil									
3120	1-1/2 ton cooling	Q-5	3.80	4.211	Ea.	635	267		902	1,100
3130	2 ton cooling		3.50	4.571		645	290		935	1,150
3140	2-1/2 ton cooling		3.30	4.848		755	305		1,060	1,300
3150	3 ton cooling		3.10	5.161		845	325		1,170	1,425
3160	3-1/2 ton cooling		2.90	5.517		845	350		1,195	1,475
3170	4 ton cooling		2.50	6.400		1,075	405		1,480	1,825
3180	5 ton cooling		2.10	7.619		1,150	485		1,635	2,000
3500	With no cooling coil									
3520	1-1/2 ton coil size	Q-5	12	1.333	Ea.	600	84.50		684.50	790
3530	2 ton coil size		10	1.600		630	101		731	850

23 54 Furnaces

23 54 24 – Furnace Components for Cooling

23 54 24.10 Furnace Components and Combinations

		Crew	Daily Output	Labor-Hours	Unit	Material	2022 Bare Costs Labor	Equipment	Total	Total Incl O&P
3540	2-1/2 ton coil size	Q-5	10	1.600	Ea.	710	101		811	935
3554	3 ton coil size		9	1.778		780	113		893	1,025
3560	3-1/2 ton coil size		9	1.778		845	113		958	1,100
3570	4 ton coil size		8.50	1.882		885	119		1,004	1,150
3580	5 ton coil size		8	2		955	127		1,082	1,250
4000	With heater									
4120	5 kW, 17.1 MBH	Q-5	16	1	Ea.	970	63.50		1,033.50	1,175
4130	7.5 kW, 25.6 MBH		15.60	1.026		855	65		920	1,050
4140	10 kW, 34.2 MBH		15.20	1.053		1,025	66.50		1,091.50	1,250
4150	12.5 kW, 42.7 MBH		14.80	1.081		1,225	68.50		1,293.50	1,450

23 55 Fuel-Fired Heaters

23 55 13 – Fuel-Fired Duct Heaters

23 55 13.16 Gas-Fired Duct Heaters

		Crew	Daily Output	Labor-Hours	Unit	Material	2022 Bare Costs Labor	Equipment	Total	Total Incl O&P
0010	**GAS-FIRED DUCT HEATERS**, Includes burner, controls, stainless steel									
0020	heat exchanger. Gas fired, electric ignition									
1000	Outdoor installation, with power venter									
1120	225 MBH output	Q-5	2.50	6.400	Ea.	5,850	405		6,255	7,050
1160	375 MBH output		1.60	10		6,675	635		7,310	8,300
1180	450 MBH output		1.40	11.429		7,550	725		8,275	9,425

23 55 33 – Fuel-Fired Unit Heaters

23 55 33.16 Gas-Fired Unit Heaters

		Crew	Daily Output	Labor-Hours	Unit	Material	2022 Bare Costs Labor	Equipment	Total	Total Incl O&P
0010	**GAS-FIRED UNIT HEATERS**, Cabinet, grilles, fan, ctrls., burner, no piping									
0022	thermostat, no piping. For flue see Section 23 51 23.10									
1000	Gas fired, floor mounted									
1100	60 MBH output	Q-5	10	1.600	Ea.	755	101		856	985
1180	180 MBH output		6	2.667		1,200	169		1,369	1,550
2000	Suspension mounted, propeller fan, 20 MBH output		8.50	1.882		775	119		894	1,050
2040	60 MBH output		7	2.286		1,850	145		1,995	2,275
2100	130 MBH output		5	3.200		1,300	203		1,503	1,750
2240	320 MBH output		2	8		2,550	505		3,055	3,575
2500	For powered venter and adapter, add					400			400	440
5000	Wall furnace, 17.5 MBH output	Q-5	6	2.667		640	169		809	965
5020	24 MBH output		5	3.200		640	203		843	1,025
5040	35 MBH output		4	4		680	253		933	1,150
9000	Minimum labor/equipment charge		3.50	4.571	Job		290		290	445

23 56 Solar Energy Heating Equipment

23 56 16 – Packaged Solar Heating Equipment

23 56 16.40 Solar Heating Systems

			Crew	Daily Output	Labor-Hours	Unit	Material	2022 Bare Costs Labor	Equipment	Total	Total Incl O&P
0010	**SOLAR HEATING SYSTEMS**										
0020	System/package prices, not including connecting										
0030	pipe, insulation, or special heating/plumbing fixtures										
0500	Hot water, standard package, low temperature										
0580	2 collectors, circulator, fittings, 120 gal. tank	G	Q-1	.40	40	Ea.	5,475	2,525		8,000	9,900

23 56 19.50 Solar Heating Ancillary

		Crew	Daily Output	Labor-Hours	Unit	Material	2022 Bare Costs Labor	2022 Bare Costs Equipment	Total	Total Incl O&P
0010	**SOLAR HEATING ANCILLARY**									
2300	Circulators, air									
2310	Blowers									
2550	Booster fan 6" diameter, 120 CFM	G	Q-9	16	1	Ea.	38	61		99	137
2580	8" diameter, 150 CFM	G		16	1		43	61		104	142
2660	Shutter/damper	G		12	1.333		59	81		140	191
2670	Shutter motor	G		16	1		147	61		208	257
2800	Circulators, liquid, 1/25 HP, 5.3 GPM	G	Q-1	14	1.143		209	72		281	340
2820	1/20 HP, 17 GPM	G	"	12	1.333		310	84		394	470
3000	Collector panels, air with aluminum absorber plate									
3010	Wall or roof mount									
3040	Flat black, plastic glazing									
3080	4' x 8'	G	Q-9	6	2.667	Ea.	690	162		852	1,000
3300	Collector panels, liquid with copper absorber plate									
3320	Black chrome, tempered glass glazing									
3330	Alum. frame, 4' x 8', 5/32" single glazing	G	Q-1	9.50	1.684	Ea.	1,050	106		1,156	1,325
3450	Flat black, alum. frame, 3.5' x 7.5'	G		9	1.778		765	112		877	1,000
3500	4' x 8'	G		5.50	2.909		930	183		1,113	1,300
3600	Liquid, full wetted, plastic, alum. frame, 4' x 10'	G		5	3.200		330	201		531	675
3650	Collector panel mounting, flat roof or ground rack	G		7	2.286		267	144		411	515
3670	Roof clamps	G		70	.229	Set	3.57	14.35		17.92	26
3700	Roof strap, teflon	G	1 Plum	205	.039	L.F.	25	2.72		27.72	31.50
3900	Differential controller with two sensors									
3930	Thermostat, hard wired	G	1 Plum	8	1	Ea.	115	70		185	235
4050	Pool valve system	G		2.50	3.200		149	223		372	510
4070	With 12 VAC actuator	G		2	4		330	279		609	795
4150	Sensors									
4220	Freeze prevention	G	1 Plum	32	.250	Ea.	26	17.45		43.45	55.50
4300	Heat exchanger									
4315	includes coil, blower, circulator									
4316	and controller for DHW and space hot air									
4330	Fluid to air coil, up flow, 45 MBH	G	Q-1	4	4	Ea.	335	251		586	755
4380	70 MBH	G		3.50	4.571		380	287		667	860
4400	80 MBH	G		3	5.333		500	335		835	1,075
4580	Fluid to fluid package includes two circulating pumps									
4590	expansion tank, check valve, relief valve									
4600	controller, high temperature cutoff and sensors	G	Q-1	2.50	6.400	Ea.	785	400		1,185	1,475
4650	Heat transfer fluid									
4700	Propylene glycol, inhibited anti-freeze	G	1 Plum	28	.286	Gal.	24.50	19.95		44.45	57.50
4800	Solar storage tanks, knocked down									
5190	6'-3" high, 7' x 7' = 306 C.F./2,000 gallons	G	Q-10	1.20	20	Ea.	14,000	1,250		15,250	17,400
5210	7' x 14' = 613 C.F./4,000 gallons	G		.60	40		22,100	2,525		24,625	28,200
5230	10'-6" x 14' = 919 C.F./6,000 gallons	G		.40	60		24,300	3,775		28,075	32,700
5250	14' x 17'-6" = 1,531 C.F./10,000 gallons	G	Q-11	.30	107		31,500	6,850		38,350	45,300
7000	Solar control valves and vents									
7070	Air eliminator, automatic 3/4" size	G	1 Plum	32	.250	Ea.	31	17.45		48.45	61
7090	Air vent, automatic, 1/8" fitting	G		32	.250		16	17.45		33.45	44.50
7100	Manual, 1/8" NPT	G		32	.250		3.31	17.45		20.76	30.50
7120	Backflow preventer, 1/2" pipe size	G		16	.500		89	35		124	152
7130	3/4" pipe size	G		16	.500		74.50	35		109.50	136
7150	Balancing valve, 3/4" pipe size	G		20	.400		56	28		84	105
7180	Draindown valve, 1/2" copper tube	G		9	.889		220	62		282	340

23 56 Solar Energy Heating Equipment

23 56 19 – Solar Heating Components

23 56 19.50 Solar Heating Ancillary		Crew	Daily Output	Labor-Hours	Unit	Material	2022 Bare Costs Labor	2022 Bare Costs Equipment	Total	Total Incl O&P
7200	Flow control valve, 1/2" pipe size	G 1 Plum	22	.364	Ea.	143	25.50		168.50	197
7220	Expansion tank, up to 5 gal.	G	32	.250		65.50	17.45		82.95	99
7400	Pressure gauge, 2" dial	G	32	.250		25.50	17.45		42.95	55
7450	Relief valve, temp. and pressure 3/4" pipe size	G	30	.267		26.50	18.60		45.10	58
7500	Solenoid valve, normally closed									
7520	Brass, 3/4" NPT, 24V	G 1 Plum	9	.889	Ea.	173	62		235	287
7530	1" NPT, 24V	G	9	.889		273	62		335	395
7750	Vacuum relief valve, 3/4" pipe size	G	32	.250		31	17.45		48.45	61
7800	Thermometers									
7820	Digital temperature monitoring, 4 locations	G 1 Plum	2.50	3.200	Ea.	168	223		391	530
7900	Upright, 1/2" NPT	G "	8	1	"	59.50	70		129.50	173
8250	Water storage tank with heat exchanger and electric element									
8270	66 gal. with 2" x 2 lb. density insulation	G 1 Plum	1.60	5	Ea.	2,075	350		2,425	2,850
8300	80 gal. with 2" x 2 lb. density insulation	G "	1.60	5	"	1,800	350		2,150	2,525
8500	Water storage module, plastic									
8600	Tubular, 12" diameter, 4' high	G 1 Carp	48	.167	Ea.	101	9.40		110.40	126
8610	12" diameter, 8' high	G "	40	.200		160	11.25		171.25	193
8650	Cap, 12" diameter	G				20			20	22.50
9000	Minimum labor/equipment charge	1 Plum	2	4	Job		279		279	430

23 61 Refrigerant Compressors

23 61 16 – Reciprocating Refrigerant Compressors

23 61 16.10 Reciprocating Compressors

		Crew	Daily Output	Labor-Hours	Unit	Material	2022 Bare Costs Labor	2022 Bare Costs Equipment	Total	Total Incl O&P
0010	**RECIPROCATING COMPRESSORS**									
0990	Refrigeration, recip. hermetic, switches & protective devices									
1000	10 ton	Q-5	1	16	Ea.	18,200	1,025		19,225	21,700
1100	20 ton	Q-6	.72	33.333		24,300	2,200		26,500	30,100
1400	50 ton	"	.20	120		36,000	7,875		43,875	51,500
1600	130 ton	Q-7	.21	152		45,400	10,200		55,600	65,500

23 62 Packaged Compressor and Condenser Units

23 62 13 – Packaged Air-Cooled Refrigerant Compressor and Condenser Units

23 62 13.10 Packaged Air-Cooled Refrig. Condensing Units

		Crew	Daily Output	Labor-Hours	Unit	Material	2022 Bare Costs Labor	2022 Bare Costs Equipment	Total	Total Incl O&P
0010	**PACKAGED AIR-COOLED REFRIGERANT CONDENSING UNITS**									
0020	Condensing unit									
0030	Air cooled, compressor, standard controls									
0050	1.5 ton	Q-5	2.50	6.400	Ea.	1,325	405		1,730	2,075
0200	2.5 ton		1.70	9.412		1,450	595		2,045	2,525
0300	3 ton		1.30	12.308		1,625	780		2,405	2,975
0400	4 ton		.90	17.778		2,000	1,125		3,125	3,925
0500	5 ton		.60	26.667		2,325	1,700		4,025	5,150

23 64 Packaged Water Chillers

23 64 19 – Reciprocating Water Chillers

23 64 19.10 Reciprocating Type Water Chillers

		Crew	Daily Output	Labor-Hours	Unit	Material	2022 Bare Costs Labor	Equipment	Total	Total Incl O&P
0010	**RECIPROCATING TYPE WATER CHILLERS**, With standard controls									
0494	Water chillers, integral air cooled condenser									
0980	Water cooled, multiple compressor, semi-hermetic, tower not incl.									
1090	45 ton cooling	Q-7	.29	111	Ea.	29,300	7,450		36,750	43,800
1160	100 ton cooling		.18	180		49,800	12,000		61,800	73,500
1200	145 ton cooling		.16	203		62,500	13,600		76,100	89,500

23 64 23 – Scroll Water Chillers

23 64 23.10 Scroll Water Chillers

		Crew	Daily Output	Labor-Hours	Unit	Material	2022 Bare Costs Labor	Equipment	Total	Total Incl O&P
0010	**SCROLL WATER CHILLERS**, With standard controls									
0480	Packaged w/integral air cooled condenser									
0482	10 ton cooling	Q-7	.34	94.118	Ea.	16,000	6,300		22,300	27,300
0490	15 ton cooling		.37	86.486		19,200	5,800		25,000	30,100
0500	20 ton cooling		.34	94.118		24,400	6,300		30,700	36,500
0520	40 ton cooling		.30	108		39,600	7,250		46,850	55,000
0680	Scroll water cooled, single compressor, hermetic, tower not incl.									
0760	10 ton cooling	Q-6	.36	67.039	Ea.	7,775	4,400		12,175	15,300
0900	Scroll water cooled, multiple compressor, hermetic, tower not incl.									

23 64 26 – Rotary-Screw Water Chillers

23 64 26.10 Rotary-Screw Type Water Chillers

		Crew	Daily Output	Labor-Hours	Unit	Material	2022 Bare Costs Labor	Equipment	Total	Total Incl O&P
0010	**ROTARY-SCREW TYPE WATER CHILLERS**, With standard controls									
0110	Screw, liquid chiller, air cooled, insulated evaporator									
0120	130 ton	Q-7	.14	229	Ea.	104,500	15,300		119,800	138,500
0124	160 ton	"	.13	246	"	135,000	16,500		151,500	173,500

23 64 33 – Modular Water Chillers

23 64 33.10 Direct Expansion Type Water Chillers

		Crew	Daily Output	Labor-Hours	Unit	Material	2022 Bare Costs Labor	Equipment	Total	Total Incl O&P
0010	**DIRECT EXPANSION TYPE WATER CHILLERS**, With standard controls									
8000	Direct expansion, shell and tube type, for built up systems									
8020	1 ton	Q-5	2	8	Ea.	8,750	505		9,255	10,400
8030	5 ton		1.90	8.421		14,500	535		15,035	16,800
8040	10 ton		1.70	9.412		18,000	595		18,595	20,700
9000	Minimum labor/equipment charge	Q-6	1	24	Job		1,575		1,575	2,425

23 65 Cooling Towers

23 65 13 – Forced-Draft Cooling Towers

23 65 13.10 Forced-Draft Type Cooling Towers

		Crew	Daily Output	Labor-Hours	Unit	Material	2022 Bare Costs Labor	Equipment	Total	Total Incl O&P
0010	**FORCED-DRAFT TYPE COOLING TOWERS**, Packaged units									
0070	Galvanized steel									
0080	Induced draft, crossflow									
0100	Vertical, belt drive, 61 tons	Q-6	90	.267	TonAC	218	17.50		235.50	267
0150	100 ton		100	.240		210	15.75		225.75	257
0200	115 ton		109	.220		183	14.45		197.45	224
0250	131 ton		120	.200		220	13.15		233.15	262
0260	162 ton		132	.182		178	11.95		189.95	214
1000	For higher capacities, use multiples									
1500	Induced air, double flow									
1900	Vertical, gear drive, 167 ton	Q-6	126	.190	TonAC	172	12.50		184.50	209
2000	297 ton		129	.186		106	12.20		118.20	136
2100	582 ton		132	.182		59	11.95		70.95	83.50
2200	1,016 ton		150	.160		80.50	10.50		91	105

23 65 Cooling Towers

23 65 13 – Forced-Draft Cooling Towers

23 65 13.10 Forced-Draft Type Cooling Towers	Crew	Daily Output	Labor-Hours	Unit	Material	2022 Bare Costs Labor	Equipment	Total	Total Incl O&P
3000	For higher capacities, use multiples								
3500	For pumps and piping, add	Q-6	38	.632	TonAC	109	41.50		150.50	184
4000	For absorption systems, add				TonAC	75%	75%			
4100	Cooling water chemical feeder	Q-5	3	5.333	Ea.	460	340		800	1,025
5000	Fiberglass tower on galvanized steel support structure									
5010	Draw thru									
5100	100 ton	Q-6	1.40	17.143	Ea.	14,600	1,125		15,725	17,700
5120	120 ton		1.20	20		17,500	1,325		18,825	21,200
5140	140 ton		1	24		17,800	1,575		19,375	22,000
5160	160 ton		.80	30		19,200	1,975		21,175	24,100
5180	180 ton		.65	36.923		22,500	2,425		24,925	28,500
5251	600 ton		.16	150		82,500	9,850		92,350	105,500
5252	1,000 ton		.10	250		137,500	16,400		153,900	176,500
5300	For stainless steel support structure, add					30%				
5360	For higher capacities, use multiples of each size									
6000	Stainless steel									
6010	Induced draft, crossflow, horizontal, belt drive									
6100	57 ton	Q-6	1.50	16	Ea.	29,000	1,050		30,050	33,500
6120	91 ton		.99	24.242		35,600	1,600		37,200	41,700
6140	111 ton		.43	55.814		53,500	3,675		57,175	64,500
6160	126 ton		.22	109		55,500	7,175		62,675	72,000
6170	Induced draft, crossflow, vertical, gear drive									
6172	167 ton	Q-6	.75	32	Ea.	65,000	2,100		67,100	75,000
6174	297 ton		.43	55.814		72,000	3,675		75,675	84,500
6176	582 ton		.23	104		115,500	6,850		122,350	137,500
6180	1,016 ton		.15	160		203,000	10,500		213,500	239,500
9000	Minimum labor/equipment charge		1	24	Job		1,575		1,575	2,425

23 72 Air-to-Air Energy Recovery Equipment

23 72 13 – Heat-Wheel Air-to-Air Energy-Recovery Equipment

23 72 13.10 Heat-Wheel Air-to-Air Energy Recovery Equip.

		Crew	Daily Output	Labor-Hours	Unit	Material	2022 Bare Costs Labor	Equipment	Total	Total Incl O&P
0010	**HEAT-WHEEL AIR-TO-AIR ENERGY RECOVERY EQUIPMENT**									
0100	Air to air									
9000	Minimum labor/equipment charge	1 Shee	4	2	Job		135		135	210

23 73 Indoor Central-Station Air-Handling Units

23 73 39 – Indoor, Direct Gas-Fired Heating and Ventilating Units

23 73 39.10 Make-Up Air Unit

		Crew	Daily Output	Labor-Hours	Unit	Material	2022 Bare Costs Labor	Equipment	Total	Total Incl O&P
0010	**MAKE-UP AIR UNIT**									
0020	Indoor suspension, natural/LP gas, direct fired,									
0032	standard control. For flue see Section 23 51 23.10									
0040	70°F temperature rise, MBH is input									
0100	75 MBH input	Q-6	3.60	6.667	Ea.	4,000	440		4,440	5,075
0220	225 MBH input		2.40	10		6,175	655		6,830	7,800
0300	400 MBH input		1.60	15		15,300	985		16,285	18,300

23 74 Packaged Outdoor HVAC Equipment

23 74 23 – Packaged, Outdoor, Heating-Only Makeup-Air Units

23 74 23.16 Packaged, Ind.-Fired, In/Outdoor

		Crew	Daily Output	Labor-Hours	Unit	Material	2022 Bare Costs Labor	2022 Bare Costs Equipment	Total	Total Incl O&P
0010	**PACKAGED, IND.-FIRED, IN/OUTDOOR**, Heat-Only Makeup-Air Units									
1000	Rooftop unit, natural gas, gravity vent, S.S. exchanger									
1010	70°F temperature rise, MBH is input									
1020	250 MBH	Q-6	4	6	Ea.	17,500	395		17,895	19,800
1040	400 MBH		3.60	6.667		20,000	440		20,440	22,600
1060	550 MBH		3.30	7.273		21,800	480		22,280	24,600
1600	For cleanable filters, add					5%				
1700	For electric modulating gas control, add					10%				
9000	Minimum labor/equipment charge	Q-6	2.75	8.727	Job		575		575	885

23 74 33 – Dedicated Outdoor-Air Units

23 74 33.10 Rooftop Air Conditioners

		Crew	Daily Output	Labor-Hours	Unit	Material	2022 Bare Costs Labor	2022 Bare Costs Equipment	Total	Total Incl O&P
0010	**ROOFTOP AIR CONDITIONERS**, Standard controls, curb, economizer									
1000	Single zone, electric cool, gas heat									
1100	3 ton cooling, 60 MBH heating	Q-5	.70	22.857	Ea.	1,800	1,450		3,250	4,200
1120	4 ton cooling, 95 MBH heating		.61	26.403		3,425	1,675		5,100	6,325
1150	7.5 ton cooling, 170 MBH heating		.50	32.258		6,425	2,050		8,475	10,200
1156	8.5 ton cooling, 170 MBH heating		.46	34.783		7,275	2,200		9,475	11,400
1160	10 ton cooling, 200 MBH heating	Q-6	.67	35.982		7,925	2,375		10,300	12,400
1220	30 ton cooling, 540 MBH heating	Q-7	.47	68.376		36,600	4,575		41,175	47,400
1240	40 ton cooling, 675 MBH heating	"	.35	91.168		44,700	6,100		50,800	58,500
2000	Multizone, electric cool, gas heat, economizer									
2120	20 ton cooling, 360 MBH heating	Q-7	.53	60.038	Ea.	63,000	4,025		67,025	75,500
2180	30 ton cooling, 540 MBH heating		.37	85.562		95,500	5,725		101,225	114,500
2210	50 ton cooling, 540 MBH heating		.23	142		140,000	9,525		149,525	168,500
2220	70 ton cooling, 1,500 MBH heating		.16	199		149,000	13,300		162,300	184,000
2240	80 ton cooling, 1,500 MBH heating		.14	229		170,000	15,300		185,300	210,500
2280	105 ton cooling, 1,500 MBH heating		.11	291		224,000	19,500		243,500	276,500
2400	For hot water heat coil, deduct					5%				
2500	For steam heat coil, deduct					2%				
2600	For electric heat, deduct					3%	5%			
9000	Minimum labor/equipment charge	Q-5	1.50	10.667	Job		675		675	1,050

23 81 Decentralized Unitary HVAC Equipment

23 81 19 – Self-Contained Air-Conditioners

23 81 19.10 Window Unit Air Conditioners

		Crew	Daily Output	Labor-Hours	Unit	Material	2022 Bare Costs Labor	2022 Bare Costs Equipment	Total	Total Incl O&P
0010	**WINDOW UNIT AIR CONDITIONERS**									
4000	Portable/window, 15 amp, 125 V grounded receptacle required									
4500	10,000 BTUH	1 Carp	6	1.333	Ea.	470	75		545	640
4520	12,000 BTUH	L-2	8	2	"	2,175	99.50		2,274.50	2,525
9000	Minimum labor/equipment charge	1 Carp	2	4	Job		225		225	360

23 81 19.20 Self-Contained Single Package

		Crew	Daily Output	Labor-Hours	Unit	Material	2022 Bare Costs Labor	2022 Bare Costs Equipment	Total	Total Incl O&P
0010	**SELF-CONTAINED SINGLE PACKAGE**									
0100	Air cooled, for free blow or duct, not incl. remote condenser									
0110	Constant volume									
0200	3 ton cooling	Q-5	1	16	Ea.	4,100	1,025		5,125	6,050
0210	4 ton cooling	"	.80	20		4,475	1,275		5,750	6,875
0240	10 ton cooling	Q-7	1	32		8,450	2,150		10,600	12,600
0280	30 ton cooling	"	.80	40		27,500	2,675		30,175	34,300
1000	Water cooled for free blow or duct, not including tower									
1010	Constant volume									

23 81 Decentralized Unitary HVAC Equipment

23 81 19 – Self-Contained Air-Conditioners

23 81 19.20 Self-Contained Single Package

		Crew	Daily Output	Labor-Hours	Unit	Material	2022 Bare Costs Labor	Equipment	Total	Total Incl O&P
1100	3 ton cooling	Q-6	1	24	Ea.	3,800	1,575		5,375	6,600
1120	5 ton cooling	"	1	24		4,975	1,575		6,550	7,900
1140	10 ton cooling	Q-7	.90	35.556		8,850	2,375		11,225	13,400
1180	30 ton cooling	"	.70	45.714		36,000	3,050		39,050	44,300
1240	60 ton cooling	Q-8	.30	107		57,000	7,150	207	64,357	73,500
1300	For hot water or steam heat coils, add					12%	10%			
9000	Minimum labor/equipment charge	Q-5	1	16	Job		1,025		1,025	1,550

23 81 23 – Computer-Room Air-Conditioners

23 81 23.10 Computer Room Units

		Crew	Daily Output	Labor-Hours	Unit	Material	2022 Bare Costs Labor	Equipment	Total	Total Incl O&P
0010	**COMPUTER ROOM UNITS**									
1000	Air cooled, includes remote condenser but not									
1020	interconnecting tubing or refrigerant									
1160	6 ton	Q-5	.30	53.333	Ea.	57,500	3,375		60,875	68,500
1240	10 ton	"	.25	64		62,500	4,050		66,550	75,500
1300	20 ton	Q-6	.26	92.308		82,500	6,050		88,550	100,000
1320	22 ton	"	.24	100		83,500	6,575		90,075	101,500

23 81 26 – Split-System Air-Conditioners

23 81 26.10 Split Ductless Systems

		Crew	Daily Output	Labor-Hours	Unit	Material	2022 Bare Costs Labor	Equipment	Total	Total Incl O&P
0010	**SPLIT DUCTLESS SYSTEMS**									
0100	Cooling only, single zone									
0110	Wall mount									
0120	3/4 ton cooling	Q-5	2	8	Ea.	1,100	505		1,605	2,000
0130	1 ton cooling	"	1.80	8.889	"	1,250	565		1,815	2,275
0140	1-1/2 ton cooling	Q-5	1.60	10	Ea.	1,975	635		2,610	3,150
0150	2 ton cooling		1.40	11.429		2,250	725		2,975	3,600
0160	3 ton cooling		1.20	13.333		2,250	845		3,095	3,775
1000	Ceiling mount									
1020	2 ton cooling	Q-5	1.40	11.429	Ea.	2,000	725		2,725	3,325
1030	3 ton cooling	"	1.20	13.333	"	2,550	845		3,395	4,100
3000	Multizone									
3010	Wall mount									
3020	2 @ 3/4 ton cooling	Q-5	1.80	8.889	Ea.	3,375	565		3,940	4,575
5000	Cooling/Heating									
5010	Wall mount									
5110	1 ton cooling	Q-5	1.70	9.412	Ea.	1,300	595		1,895	2,350
5120	1-1/2 ton cooling	"	1.50	10.667	"	1,975	675		2,650	3,225
7000	Accessories for all split ductless systems									
7010	Add for ambient frost control	Q-5	8	2	Ea.	259	127		386	480
7020	Add for tube/wiring kit (line sets)									
7030	15' kit	Q-5	32	.500	Ea.	73.50	31.50		105	130
7036	25' kit		28	.571		53	36		89	115
7040	35' kit		24	.667		127	42		169	205
7050	50' kit		20	.800		168	50.50		218.50	263

23 81 Decentralized Unitary HVAC Equipment

23 81 29 – Variable Refrigerant Flow HVAC Systems

23 81 29.10 Heat Pump, Gas Driven

		Crew	Daily Output	Labor-Hours	Unit	Material	2022 Bare Costs Labor	Equipment	Total	Total Incl O&P
0010	**HEAT PUMP, GAS DRIVEN**, Variable refrigerant volume (VRV) type									
0020	Not including interconnecting tubing or multi-zone controls									
1000	For indoor fan VRV type AHU see 23 82 19.40									
1010	Multi-zone split									
1020	Outdoor unit									
1100	8 tons cooling, for up to 17 zones	Q-5	1.30	12.308	Ea.	31,300	780		32,080	35,600
1110	Isolation rails		2.60	6.154	Pr.	1,100	390		1,490	1,800
1160	15 tons cooling, for up to 33 zones		1	16	Ea.	41,200	1,025		42,225	46,900
1170	Isolation rails	↓	2.60	6.154	Pr.	1,300	390		1,690	2,025
2000	Packaged unit									
2020	Outdoor unit									
2200	11 tons cooling	Q-5	1.30	12.308	Ea.	40,300	780		41,080	45,500
2210	Roof curb adapter	"	2.60	6.154	↓	700	390		1,090	1,375
2220	Thermostat	1 Stpi	1.30	6.154		420	435		855	1,125

23 81 43 – Air-Source Unitary Heat Pumps

23 81 43.10 Air-Source Heat Pumps

		Crew	Daily Output	Labor-Hours	Unit	Material	2022 Bare Costs Labor	Equipment	Total	Total Incl O&P
0010	**AIR-SOURCE HEAT PUMPS**, Not including interconnecting tubing									
1000	Air to air, split system, not including curbs, pads, fan coil and ductwork									
1012	Outside condensing unit only, for fan coil see Section 23 82 19.10									
1020	2 ton cooling, 8.5 MBH heat @ 0°F	Q-5	2	8	Ea.	1,625	505		2,130	2,575
1040	3 ton cooling, 13 MBH heat @ 0°F		1.20	13.333		2,000	845		2,845	3,500
1080	7.5 ton cooling, 33 MBH heat @ 0°F	↓	.45	35.556		3,725	2,250		5,975	7,575
1120	15 ton cooling, 64 MBH heat @ 0°F	Q-6	.50	48		10,700	3,150		13,850	16,700
1130	20 ton cooling, 85 MBH heat @ 0°F		.35	68.571		19,500	4,500		24,000	28,500
1140	25 ton cooling, 119 MBH heat @ 0°F	↓	.25	96	↓	22,400	6,300		28,700	34,300
1500	Single package, not including curbs, pads, or plenums									
1520	2 ton cooling, 6.5 MBH heat @ 0°F	Q-5	1.50	10.667	Ea.	3,400	675		4,075	4,800
1560	3 ton cooling, 10 MBH heat @ 0°F	"	1.20	13.333	"	3,800	845		4,645	5,475
6000	Air to water, single package, excluding storage tank and ductwork									
6010	Includes circulating water pump, air duct connections, digital temperature									
6020	controller with remote tank temp. probe and sensor for storage tank.									
6040	Water heating - air cooling capacity									
6110	35.5 MBH heat water, 2.3 ton cool air	Q-5	1.60	10	Ea.	17,100	635		17,735	19,800
6120	58 MBH heat water, 3.8 ton cool air		1.10	14.545		19,500	920		20,420	22,900
6130	76 MBH heat water, 4.9 ton cool air		.87	18.391		23,300	1,175		24,475	27,400
6140	98 MBH heat water, 6.5 ton cool air		.62	25.806		30,100	1,625		31,725	35,600
6150	113 MBH heat water, 7.4 ton cool air		.59	27.119		33,000	1,725		34,725	39,000
6160	142 MBH heat water, 9.2 ton cool air		.52	30.769		40,000	1,950		41,950	47,000
6170	171 MBH heat water, 11.1 ton cool air	↓	.49	32.653	↓	47,100	2,075		49,175	55,000

23 81 46 – Water-Source Unitary Heat Pumps

23 81 46.10 Water Source Heat Pumps

		Crew	Daily Output	Labor-Hours	Unit	Material	2022 Bare Costs Labor	Equipment	Total	Total Incl O&P
0010	**WATER SOURCE HEAT PUMPS**, Not incl. connecting tubing or water source									
2000	Water source to air, single package									
2100	1 ton cooling, 13 MBH heat @ 75°F	Q-5	2	8	Ea.	2,125	505		2,630	3,100
2140	2 ton cooling, 19 MBH heat @ 75°F		1.70	9.412		2,350	595		2,945	3,500
2220	5 ton cooling, 29 MBH heat @ 75°F		.90	17.778	↓	3,425	1,125		4,550	5,500
9000	Minimum labor/equipment charge	↓	1.75	9.143	Job		580		580	890

For customer support on your Commercial Renovation Costs with RSMeans data, call 800.448.8182.

485

23 82 16 – Air Coils

23 82 16.10 Flanged Coils		Crew	Daily Output	Labor-Hours	Unit	Material	2022 Bare Costs Labor	2022 Bare Costs Equipment	Total	Total Incl O&P
0010	**FLANGED COILS**									
1500	Hot water heating, 1 row, 24" x 48"	Q-5	4	4	Ea.	2,075	253		2,328	2,700
9000	Minimum labor/equipment charge	1 Plum	1	8	Job		560		560	860

23 82 16.20 Duct Heaters

		Crew	Daily Output	Labor-Hours	Unit	Material	2022 Bare Costs Labor	2022 Bare Costs Equipment	Total	Total Incl O&P
0010	**DUCT HEATERS**, Electric, 480 V, 3 Ph.									
0020	Finned tubular insert, 500°F									
0100	8" wide x 6" high, 4.0 kW	Q-20	16	1.250	Ea.	920	77.50		997.50	1,125
0120	12" high, 8.0 kW		15	1.333		1,525	82.50		1,607.50	1,800
0140	18" high, 12.0 kW		14	1.429		2,150	88.50		2,238.50	2,475
0160	24" high, 16.0 kW		13	1.538		2,750	95		2,845	3,175
0180	30" high, 20.0 kW		12	1.667		3,350	103		3,453	3,850
0300	12" wide x 6" high, 6.7 kW		15	1.333		975	82.50		1,057.50	1,200
0320	12" high, 13.3 kW		14	1.429		1,575	88.50		1,663.50	1,850
0340	18" high, 20.0 kW		13	1.538		2,200	95		2,295	2,575
8000	To obtain BTU multiply kW by 3413									
9900	Minimum labor/equipment charge	1 Shee	4	2	Job		135		135	210

23 82 19 – Fan Coil Units

23 82 19.10 Fan Coil Air Conditioning

		Crew	Daily Output	Labor-Hours	Unit	Material	2022 Bare Costs Labor	2022 Bare Costs Equipment	Total	Total Incl O&P
0010	**FAN COIL AIR CONDITIONING**									
0030	Fan coil AC, cabinet mounted, filters and controls									
0100	Chilled water, 1/2 ton cooling	Q-5	8	2	Ea.	750	127		877	1,025
0120	1 ton cooling		6	2.667		860	169		1,029	1,200
0160	2-1/2 ton cooling		5	3.200		2,400	203		2,603	2,925
0180	3 ton cooling		4	4		2,625	253		2,878	3,275
0262	For hot water coil, add					40%	10%			
0320	1 ton cooling	Q-5	6	2.667	Ea.	2,000	169		2,169	2,450
0940	Direct expansion, for use w/air cooled condensing unit, 1-1/2 ton cooling		5	3.200		770	203		973	1,150
1000	5 ton cooling		3	5.333		1,375	340		1,715	2,050
1040	10 ton cooling	Q-6	2.60	9.231		2,400	605		3,005	3,575
1060	20 ton cooling	"	.70	34.286		4,750	2,250		7,000	8,700
1500	For hot water coil, add					40%	10%			
1512	For condensing unit add see Section 23 62									
3000	Chilled water, horizontal unit, housing, 2 pipe, fan control, no valves									
3100	1/2 ton cooling	Q-5	8	2	Ea.	1,600	127		1,727	1,975
3110	1 ton cooling		6	2.667		1,875	169		2,044	2,300
3120	1-1/2 ton cooling		5.50	2.909		2,125	184		2,309	2,600
3130	2 ton cooling		5.25	3.048		2,500	193		2,693	3,050
3140	3 ton cooling		4	4		3,100	253		3,353	3,800
3150	3-1/2 ton cooling		3.80	4.211		3,475	267		3,742	4,225
3160	4 ton cooling		3.80	4.211		3,475	267		3,742	4,225
4000	With electric heat, 2 pipe									
4100	1/2 ton cooling	Q-5	8	2	Ea.	1,875	127		2,002	2,275
4105	3/4 ton cooling		7	2.286		2,075	145		2,220	2,525
4110	1 ton cooling		6	2.667		2,325	169		2,494	2,800
4120	1-1/2 ton cooling		5.50	2.909		2,600	184		2,784	3,125
4130	2 ton cooling		5.25	3.048		3,200	193		3,393	3,825
4135	2-1/2 ton cooling		4.80	3.333		4,500	211		4,711	5,275
4140	3 ton cooling		4	4		5,875	253		6,128	6,875
4150	3-1/2 ton cooling		3.80	4.211		7,025	267		7,292	8,125
4160	4 ton cooling		3.60	4.444		8,075	281		8,356	9,300
9000	Minimum labor/equipment charge		3.75	4.267	Job		270		270	415

23 82 19.40 Fan Coil Air Conditioning		Crew	Daily Output	Labor-Hours	Unit	Material	2022 Bare Costs Labor	Equipment	Total	Total Incl O&P
0010	**FAN COIL AIR CONDITIONING,** Variable refrigerant volume (VRV) type									
0020	Not including interconnecting tubing or multi-zone controls									
0030	For VRV condensing unit see section 23 81 29.10									
0050	Indoor type, ducted									
0100	Vertical concealed									
0130	1 ton cooling	G	Q-5	2.97	5.387	Ea.	2,950	340		3,290	3,775
0140	1.5 ton cooling	G		2.77	5.776		3,000	365		3,365	3,875
0150	2 ton cooling	G		2.70	5.926		3,350	375		3,725	4,275
0160	2.5 ton cooling	G		2.60	6.154		3,400	390		3,790	4,350
0170	3 ton cooling	G		2.50	6.400		3,575	405		3,980	4,575
0180	3.5 ton cooling	G		2.30	6.957		3,950	440		4,390	5,000
0190	4 ton cooling	G		2.19	7.306		3,800	465		4,265	4,900
0200	4.5 ton cooling	G	↓	2.08	7.692	↓	4,175	485		4,660	5,350
0230	Outside air connection possible									
1100	Ceiling concealed									
1130	0.6 ton cooling	G	Q-5	2.60	6.154	Ea.	2,075	390		2,465	2,875
1140	0.75 ton cooling	G		2.60	6.154		2,150	390		2,540	2,975
1150	1 ton cooling	G	↓	2.60	6.154		2,175	390		2,565	3,000
1152	Screening door	G	1 Stpi	5.20	1.538		121	108		229	300
1160	1.5 ton cooling	G	Q-5	2.60	6.154		2,250	390		2,640	3,100
1162	Screening door	G	1 Stpi	5.20	1.538		140	108		248	320
1170	2 ton cooling	G	Q-5	2.60	6.154		2,775	390		3,165	3,650
1172	Screening door	G	1 Stpi	5.20	1.538		121	108		229	300
1180	2.5 ton cooling	G	Q-5	2.60	6.154		3,075	390		3,465	3,975
1182	Screening door	G	1 Stpi	5.20	1.538		210	108		318	400
1190	3 ton cooling	G	Q-5	2.60	6.154		3,400	390		3,790	4,325
1192	Screening door	G	1 Stpi	5.20	1.538		210	108		318	400
1200	4 ton cooling	G	Q-5	2.30	6.957		3,475	440		3,915	4,500
1202	Screening door	G	1 Stpi	5.20	1.538		210	108		318	400
1220	6 ton cooling	G	Q-5	2.08	7.692		5,675	485		6,160	7,000
1230	8 ton cooling	G	"	1.89	8.466	↓	6,400	535		6,935	7,850
1260	Outside air connection possible									
4050	Indoor type, duct-free									
4100	Ceiling mounted cassette									
4130	0.75 ton cooling	G	Q-5	3.46	4.624	Ea.	2,125	293		2,418	2,800
4140	1 ton cooling	G		2.97	5.387		2,300	340		2,640	3,050
4150	1.5 ton cooling	G		2.77	5.776		2,350	365		2,715	3,175
4160	2 ton cooling	G		2.60	6.154		2,550	390		2,940	3,425
4170	2.5 ton cooling	G		2.50	6.400		2,850	405		3,255	3,750
4180	3 ton cooling	G		2.30	6.957		3,125	440		3,565	4,125
4190	4 ton cooling	G	↓	2.08	7.692	↓	3,400	485		3,885	4,500
4198	For ceiling cassette decoration panel, add	G	1 Stpi	5.20	1.538		365	108		473	565
4230	Outside air connection possible									
4500	Wall mounted									
4530	0.6 ton cooling	G	Q-5	5.20	3.077	Ea.	1,225	195		1,420	1,650
4540	0.75 ton cooling	G		5.20	3.077		1,275	195		1,470	1,700
4550	1 ton cooling	G		4.16	3.846		1,475	244		1,719	1,975
4560	1.5 ton cooling	G		3.77	4.244		1,600	269		1,869	2,200
4570	2 ton cooling	G	↓	3.46	4.624		1,825	293		2,118	2,450
4590	Condensate pump for the above air handlers	G	1 Stpi	6.46	1.238	↓	261	87		348	420
4700	Floor standing unit									
4730	1 ton cooling	G	Q-5	2.60	6.154	Ea.	2,050	390		2,440	2,850

23 82 19 – Fan Coil Units

23 82 19.40 Fan Coil Air Conditioning		Crew	Daily Output	Labor-Hours	Unit	Material	2022 Bare Costs Labor	Equipment	Total	Total Incl O&P
4740	1.5 ton cooling	G	Q-5	2.60	6.154	Ea.	2,325	390		2,715	3,175
4750	2 ton cooling	G	↓	2.60	6.154	↓	2,500	390		2,890	3,350
4800	Floor standing unit, concealed										
4830	1 ton cooling	G	Q-5	2.60	6.154	Ea.	1,975	390		2,365	2,775
4840	1.5 ton cooling	G		2.60	6.154		2,225	390		2,615	3,050
4850	2 ton cooling	G	↓	2.60	6.154	↓	2,425	390		2,815	3,250
4880	Outside air connection possible										
8000	Accessories										
8100	Branch divergence pipe fitting										
8110	Capacity under 76 MBH		1 Stpi	2.50	3.200	Ea.	190	225		415	555
8120	76 MBH to 112 MBH			2.50	3.200		218	225		443	585
8130	112 MBH to 234 MBH		↓	2.50	3.200	↓	615	225		840	1,025
8200	Header pipe fitting										
8210	Max 4 branches										
8220	Capacity under 76 MBH		1 Stpi	1.70	4.706	Ea.	315	330		645	855
8260	Max 8 branches										
8270	76 MBH to 112 MBH		1 Stpi	1.70	4.706	Ea.	550	330		880	1,125
8280	112 MBH to 234 MBH		"	1.30	6.154	"	875	435		1,310	1,625

23 82 29 – Radiators

23 82 29.10 Hydronic Heating

		Crew	Daily Output	Labor-Hours	Unit	Material	2022 Bare Costs Labor	Equipment	Total	Total Incl O&P
0010	**HYDRONIC HEATING**, Terminal units, not incl. main supply pipe									
1000	Radiation										
1100	Panel, baseboard, C.I., including supports, no covers	Q-5	46	.348	L.F.	53.50	22		75.50	93
3000	Radiators, cast iron										
3100	Free standing or wall hung, 6 tube, 25" high	Q-5	96	.167	Section	70.50	10.55		81.05	94
9000	Minimum labor/equipment charge	"	3.20	5	Job		315		315	490

23 82 36 – Finned-Tube Radiation Heaters

23 82 36.10 Finned Tube Radiation

		Crew	Daily Output	Labor-Hours	Unit	Material	2022 Bare Costs Labor	Equipment	Total	Total Incl O&P
0010	**FINNED TUBE RADIATION**, Terminal units, not incl. main supply pipe									
1150	Fin tube, wall hung, 14" slope top cover, with damper										
1200	1-1/4" copper tube, 4-1/4" alum. fin	Q-5	38	.421	L.F.	48.50	26.50		75	94
1250	1-1/4" steel tube, 4-1/4" steel fin		36	.444		42.50	28		70.50	90
1255	2" steel tube, 4-1/4" steel fin		32	.500		46.50	31.50		78	100
1310	Baseboard, pkgd, 1/2" copper tube, alum. fin, 7" high		60	.267		11.55	16.90		28.45	39
1320	3/4" copper tube, alum. fin, 7" high		58	.276		7.70	17.45		25.15	35.50
1340	1" copper tube, alum. fin, 8-7/8" high		56	.286		19.80	18.10		37.90	50
1360	1-1/4" copper tube, alum. fin, 8-7/8" high		54	.296	↓	29.50	18.75		48.25	61.50
1500	Note: fin tube may also require corners, caps, etc.										

23 82 39 – Unit Heaters

23 82 39.16 Propeller Unit Heaters

		Crew	Daily Output	Labor-Hours	Unit	Material	2022 Bare Costs Labor	Equipment	Total	Total Incl O&P
0010	**PROPELLER UNIT HEATERS**									
3950	Unit heaters, propeller, 115 V 2 psi steam, 60°F entering air										
4000	Horizontal, 12 MBH	Q-5	12	1.333	Ea.	425	84.50		509.50	600
4060	43.9 MBH		8	2		530	127		657	780
4240	286.9 MBH		2	8		1,800	505		2,305	2,750
4260	364 MBH		1.80	8.889		2,425	565		2,990	3,525
4270	404 MBH	↓	1.60	10	↓	2,650	635		3,285	3,900
4300	Vertical diffuser same price										
4310	Vertical flow, 40 MBH	Q-5	11	1.455	Ea.	690	92		782	900
4326	131 MBH	"	4	4		1,175	253		1,428	1,675
4354	420 MBH (460 V)	Q-6	1.80	13.333	↓	2,575	875		3,450	4,175

488

For customer support on your Commercial Renovation Costs with RSMeans data, call 800.448.8182.

23 82 Convection Heating and Cooling Units

23 82 39 – Unit Heaters

23 82 39.16 Propeller Unit Heaters	Crew	Daily Output	Labor-Hours	Unit	Material	2022 Bare Costs Labor	Equipment	Total	Total Incl O&P
4358	500 MBH (460 V)	Q-6	1.71	14.035	Ea.	3,275	920		4,195	5,025
4362	570 MBH (460 V)	↓	1.40	17.143	↓	4,075	1,125		5,200	6,200
4366	620 MBH (460 V)	Q-6	1.30	18.462	Ea.	4,350	1,200		5,550	6,675
4370	960 MBH (460 V)	"	1.10	21.818	"	8,300	1,425		9,725	11,300

23 83 Radiant Heating Units

23 83 33 – Electric Radiant Heaters

23 83 33.10 Electric Heating

		Crew	Daily Output	Labor-Hours	Unit	Material	Labor	Equipment	Total	Total Incl O&P
0010	**ELECTRIC HEATING**, not incl. conduit or feed wiring									
1100	Rule of thumb: Baseboard units, including control	1 Elec	4.40	1.818	kW	132	120		252	330
1300	Baseboard heaters, 2' long, 350 watt		8	1	Ea.	32.50	66		98.50	138
1400	3' long, 750 watt		8	1		37.50	66		103.50	144
1600	4' long, 1,000 watt		6.70	1.194		45.50	79		124.50	171
1800	5' long, 935 watt		5.70	1.404		54.50	93		147.50	203
2000	6' long, 1,500 watt		5	1.600		62.50	106		168.50	232
2400	8' long, 2,000 watt		4	2		76	132		208	287
2600	9' long, 1,680 watt		3.60	2.222		95	147		242	330
2800	10' long, 1,875 watt		3.30	2.424	↓	98.50	160		258.50	355
2950	Wall heaters with fan, 120 to 277 volt									
3170	1,000 watt	1 Elec	6	1.333	Ea.	370	88.50		458.50	540
3180	1,250 watt		5	1.600		126	106		232	300
3190	1,500 watt		4	2		400	132		532	645
3230	2,500 watt		3.50	2.286		405	151		556	675
3250	4,000 watt		2.70	2.963		365	196		561	700
3600	Thermostats, integral		16	.500		32.50	33		65.50	87
3800	Line voltage, 1 pole		8	1		34.50	66		100.50	140
3810	2 pole		8	1	↓	26	66		92	131
9990	Minimum labor/equipment charge	↓	4	2	Job		132		132	203

23 84 Humidity Control Equipment

23 84 13 – Humidifiers

23 84 13.10 Humidifier Units

		Crew	Daily Output	Labor-Hours	Unit	Material	Labor	Equipment	Total	Total Incl O&P
0010	**HUMIDIFIER UNITS**									
0520	Steam, room or duct, filter, regulators, auto. controls, 220 V									
0560	22 lb./hr.	Q-5	5	3.200	Ea.	3,125	203		3,328	3,750
0580	33 lb./hr.		4	4		3,200	253		3,453	3,925
0620	100 lb./hr.		3	5.333	↓	5,275	340		5,615	6,325
9000	Minimum labor/equipment charge	↓	3.50	4.571	Job		290		290	445

23 91 10.10 Prefabricated Pads and Stands	Crew	Daily Output	Labor-Hours	Unit	Material	2022 Bare Costs Labor	Equipment	Total	Total Incl O&P
0010 **PREFABRICATED PADS AND STANDS**									
6000 Pad, fiberglass reinforced concrete with polystyrene foam core									
6090 24" x 36"	1 Shee	12	.667	Ea.	66.50	45		111.50	143
6280 36" x 36"	Q-9	8	2		85	122		207	283
6300 36" x 40"		7	2.286		122	139		261	350
6320 36" x 48"		7	2.286		143	139		282	375

Estimating Tips

- When estimating material costs for special systems, it is always prudent to obtain manufacturers' quotations for equipment prices and special installation requirements that may affect the total cost.

- For cost modifications for elevated tray installation, add the percentages to labor according to the height of the installation and only to the quantities exceeding the different height levels, not to the total tray quantities. Refer to subdivision 26 01 02.20 for labor adjustment factors.

- Do not overlook the costs for equipment used in the installation.

Reference Numbers

Reference numbers are shown at the beginning of some major classifications. These numbers refer to related items in the Reference Section. The reference information may be an estimating procedure, an alternate pricing method, or technical information.

Same Data. Simplified.

Enjoy the convenience and efficiency of accessing your costs anywhere:

- **Skip the multiplier** by setting your location
- **Quickly search,** edit, favorite and share costs
- **Stay on top of price changes** with automatic updates

Discover more at rsmeans.com/online

Note: Trade Service, in part, has been used as a reference source for some of the material prices used in Division 25.

25 05 Common Work Results for Integrated Automation

25 05 28 – Pathways for Integrated Automation

25 05 28.39 Surface Raceways for Integrated Automation	Crew	Daily Output	Labor-Hours	Unit	Material	2022 Bare Costs Labor	Equipment	Total	Total Incl O&P
0010 **SURFACE RACEWAYS FOR INTEGRATED AUTOMATION**									
1000 Din rail, G type, 15 mm, steel, galvanized, 3', for termination blocks	1 Elec	48	.167	Ea.	12.75	11.05		23.80	31
1010 32 mm, steel, galvanized, 3', for termination blocks		48	.167		16.10	11.05		27.15	34.50
1020 35 mm, steel, galvanized, 3', for termination blocks		48	.167		17.50	11.05		28.55	36
1030 Din rail, rail mounted, 35 mm, aluminum, 3', for termination blocks		48	.167		22.50	11.05		33.55	42
1040 35 mm, aluminum, 6', for termination blocks		48	.167		40.50	11.05		51.55	61.50
2000 Terminal block, Din mounted, 35 mm, medium screw, connector end bracket		48	.167		2.63	11.05		13.68	19.85
2010 35 mm, large screw, connector end bracket		48	.167		5.25	11.05		16.30	22.50
2020 Large screw, connector end bracket		48	.167		5.25	11.05		16.30	23
3000 Terminal block, feed through, 35 mm, Din mounted, 600V, 20A		48	.167		.98	11.05		12.03	18.05
3010 35 mm, Din mounted, 600V, 30A		48	.167		1.36	11.05		12.41	18.45
3020 Terminal block, fuse holder, LED indicator, Din mounted, 110V, 6A		48	.167		8.05	11.05		19.10	26
3030 Din mounted, 220V, 6A		48	.167		8.35	11.05		19.40	26
3040 Din mounted, 300V, 6A		48	.167		8.45	11.05		19.50	26

25 36 Integrated Automation Instrumentation & Terminal Devices

25 36 19 – Integrated Automation Current Sensors

25 36 19.10 Integrated Automation Current Sensors

	Crew	Daily Output	Labor-Hours	Unit	Material	2022 Bare Costs Labor	Equipment	Total	Total Incl O&P
0010 **INTEGRATED AUTOMATION CURRENT SENSORS**									
1000 Current limiter inrush, 100 to 480 VAC, 10A, power supply	1 Elec	48	.167	Ea.	91.50	11.05		102.55	118

Estimating Tips
26 05 00 Common Work Results for Electrical

- Conduit should be taken off in three main categories—power distribution, branch power, and branch lighting—so the estimator can concentrate on systems and components, therefore making it easier to ensure all items have been accounted for.

- For cost modifications for elevated conduit installation, add the percentages to labor according to the height of installation and only to the quantities exceeding the different height levels, not to the total conduit quantities. Refer to subdivision 26 01 02.20 for labor adjustment factors.

- Remember that aluminum wiring of equal ampacity is larger in diameter than copper and may require larger conduit.

- If more than three wires at a time are being pulled, deduct percentages from the labor hours of that grouping of wires.

- When taking off grounding systems, identify separately the type and size of wire, and list each unique type of ground connection.

- The estimator should take the weights of materials into consideration when completing a takeoff. Topics to consider include: How will the materials be supported? What methods of support are available? How high will the support structure have to reach? Will the final support structure be able to withstand the total burden? Is the support material included or separate from the fixture, equipment, and material specified?

- Do not overlook the costs for equipment used in the installation. If scaffolding or highlifts are available in the field, contractors may use them in lieu of the proposed ladders and rolling staging.

26 20 00 Low-Voltage Electrical Transmission

- Supports and concrete pads may be shown on drawings for the larger equipment, or the support system may be only a piece of plywood for the back of a panelboard. In either case, they must be included in the costs.

26 40 00 Electrical and Cathodic Protection

- When taking off cathodic protection systems, identify the type and size of cable, and list each unique type of anode connection.

26 50 00 Lighting

- Fixtures should be taken off room by room using the fixture schedule, specifications, and the ceiling plan. For large concentrations of lighting fixtures in the same area, deduct the percentages from labor hours.

Reference Numbers

Reference numbers are shown at the beginning of some major classifications. These numbers refer to related items in the Reference Section. The reference information may be an estimating procedure, an alternate pricing method, or technical information.

Note: Not all subdivisions listed here necessarily appear. ■

Same Data. Simplified.

Enjoy the convenience and efficiency of accessing your costs anywhere:

- **Skip the multiplier** by setting your location
- **Quickly search,** edit, favorite and share costs
- **Stay on top of price changes** with automatic updates

Discover more at rsmeans.com/online

Note: "Powered in part by CINX™, based on licensed proprietary information of Harrison Publishing House, Inc."

Note: The following companies, in part, have been used as a reference source for some of the material prices used in Division 26:

Electriflex

Trade Service

26 01 02.20 Labor Adjustment Factors	Crew	Daily Output	Labor-Hours	Unit	Material	2022 Bare Costs Labor	Equipment	Total	Total Incl O&P
0010 **LABOR ADJUSTMENT FACTORS** (For Div. 26, 27 and 28) R260519-90									
0100 Subtract from labor for Economy of Scale for Wire									
0110 4-5 wires						25%			
0120 6-10 wires						30%			
0130 11-15 wires						35%			
0140 over 15 wires						40%			
0150 Labor adjustment factors (For Div. 26, 27, 28 and 48)									
0200 Labor factors: The below are reasonable suggestions, but									
0210 each project must be evaluated for its own peculiarities, and									
0220 the adjustments be increased or decreased depending on the									
0230 severity of the special conditions.									
1000 Add to labor for elevated installation (above floor level)									
1010 10' to 14.5' high						10%			
1020 15' to 19.5' high						20%			
1030 20' to 24.5' high						25%			
1040 25' to 29.5' high						35%			
1050 30' to 34.5' high						40%			
1060 35' to 39.5' high						50%			
1070 40' and higher						55%			
2000 Add to labor for crawl space									
2010 3' high						40%			
2020 4' high						30%			
3000 Add to labor for multi-story building									
3100 For new construction (No elevator available)									
3110 Add for floors 3 thru 10						5%			
3120 Add for floors 11 thru 15						10%			
3130 Add for floors 16 thru 20						15%			
3140 Add for floors 21 thru 30						20%			
3150 Add for floors 31 and up						30%			
3200 For existing structure (Elevator available)									
3210 Add for work on floor 3 and above						2%			
4000 Add to labor for working in existing occupied buildings									
4010 Hospital						35%			
4020 Office building						25%			
4030 School						20%			
4040 Factory or warehouse						15%			
4050 Multi-dwelling						15%			
5000 Add to labor, miscellaneous									
5010 Cramped shaft						35%			
5020 Congested area						15%			
5030 Excessive heat or cold						30%			
6000 Labor factors: the above are reasonable suggestions, but									
6100 each project should be evaluated for its own peculiarities									
6200 Other factors to be considered are:									
6210 Movement of material and equipment through finished areas						10%			
6220 Equipment room min security direct access w/authorization						15%			
6230 Attic space						25%			
6240 No service road						25%			
6250 Poor unloading/storage area, no hydraulic lifts or jacks						20%			
6260 Congested site area/heavy traffic						20%			
7000 Correctional facilities (no compounding division 1 adjustment factors)									
7010 Minimum security w/facilities escort						30%			
7020 Medium security w/facilities and correctional officer escort						40%			

26 01 Operation and Maintenance of Electrical Systems

26 01 02 – Labor Adjustment

26 01 02.20 Labor Adjustment Factors	Crew	Daily Output	Labor-Hours	Unit	Material	2022 Bare Costs Labor	Equipment	Total	Total Incl O&P
7030	Max security w/facilities & correctional officer escort (no inmate contact)						50%			

26 01 50 – Operation and Maintenance of Lighting

26 01 50.81 Luminaire Replacement

		Crew	Daily Output	Labor-Hours	Unit	Material	2022 Bare Costs Labor	Equipment	Total	Total Incl O&P
0010	**LUMINAIRE REPLACEMENT**									
0962	Remove and install new replacement lens, 2' x 2' troffer	1 Elec	16	.500	Ea.	18.45	33		51.45	71.50
0964	2' x 4' troffer		16	.500		15.10	33		48.10	67.50
3200	Remove and replace (reinstall), lighting fixture	▼	4	2	▼		132		132	203

26 05 Common Work Results for Electrical

26 05 05 – Selective Demolition for Electrical

26 05 05.10 Electrical Demolition

			Crew	Daily Output	Labor-Hours	Unit	Material	2022 Bare Costs Labor	Equipment	Total	Total Incl O&P
0010	**ELECTRICAL DEMOLITION**	R260105-30									
0020	Electrical demolition, conduit to 10' high, incl. fittings & hangers										
0100	Rigid galvanized steel, 1/2" to 1" diameter	R024119-10	1 Elec	242	.033	L.F.		2.19		2.19	3.36
0120	1-1/4" to 2"		"	200	.040			2.65		2.65	4.07
0140	2-1/2" to 3-1/2"		2 Elec	302	.053			3.51		3.51	5.40
0200	Electric metallic tubing (EMT), 1/2" to 1"		1 Elec	394	.020			1.34		1.34	2.06
0220	1-1/4" to 1-1/2"			326	.025			1.62		1.62	2.50
0240	2" to 3"		▼	236	.034	▼		2.24		2.24	3.45
0400	Wiremold raceway, including fittings & hangers										
0420	No. 3000		1 Elec	250	.032	L.F.		2.12		2.12	3.25
0440	No. 4000		"	217	.037	"		2.44		2.44	3.75
0500	Channels, steel, including fittings & hangers										
0520	3/4" x 1-1/2"		1 Elec	308	.026	L.F.		1.72		1.72	2.64
0540	1-1/2" x 1-1/2"			269	.030			1.97		1.97	3.02
0560	1-1/2" x 1-7/8"		▼	229	.035	▼		2.31		2.31	3.55
0600	Copper bus duct, indoor, 3 phase										
0610	Including hangers & supports										
0620	225 amp		2 Elec	135	.119	L.F.		7.85		7.85	12.05
0640	400 amp		"	106	.151	"		10		10	15.35
0660	600 amp		2 Elec	86	.186	L.F.		12.30		12.30	18.90
0680	1,000 amp			60	.267			17.65		17.65	27
0700	1,600 amp			40	.400			26.50		26.50	40.50
0720	3,000 amp		▼	10	1.600	▼		106		106	163
0800	Plug-in switches, 600 V 3 ph., incl. disconnecting										
0820	wire, conduit terminations, 30 amp		1 Elec	15.50	.516	Ea.		34		34	52.50
0840	60 amp			13.90	.576			38		38	58.50
0850	100 amp			10.40	.769			51		51	78
0860	200 amp		▼	6.20	1.290			85.50		85.50	131
0880	400 amp		2 Elec	5.40	2.963			196		196	300
0900	600 amp			3.40	4.706			310		310	480
0920	800 amp			2.60	6.154			405		405	625
0940	1,200 amp			2	8			530		530	815
0960	1,600 amp			1.70	9.412			625		625	955
0961	Remove and Replace, Voltage monitoring, 120/240 V		▼	1.70	9.412			625		625	955
1010	Safety switches, 250 or 600 V, incl. disconnection										
1050	of wire & conduit terminations										
1100	30 amp		1 Elec	12.30	.650	Ea.		43		43	66
1120	60 amp			8.80	.909			60		60	92.50
1140	100 amp		▼	7.30	1.096	▼		72.50		72.50	111

26 05 05.10 Electrical Demolition		Crew	Daily Output	Labor-Hours	Unit	Material	2022 Bare Costs Labor	2022 Bare Costs Equipment	Total	Total Incl O&P
1160	200 amp	1 Elec	5	1.600	Ea.		106		106	163
1210	Panel boards, incl. removal of all breakers,									
1220	conduit terminations & wire connections									
1230	3 wire, 120/240 V, 100A, to 20 circuits	1 Elec	2.60	3.077	Ea.		204		204	315
1240	200 amps, to 42 circuits	2 Elec	2.60	6.154			405		405	625
1241	225 amps, to 42 circuits		2.40	6.667			440		440	680
1250	400 amps, to 42 circuits	↓	2.20	7.273			480		480	740
1260	4 wire, 120/208 V, 125A, to 20 circuits	1 Elec	2.40	3.333			221		221	340
1270	200 amps, to 42 circuits	2 Elec	2.40	6.667	↓		440		440	680
1300	Transformer, dry type, 1 phase, incl. removal of									
1320	supports, wire & conduit terminations									
1340	1 kVA	1 Elec	7.70	1.039	Ea.		69		69	106
1360	5 kVA		4.70	1.702			113		113	173
1380	10 kVA	↓	3.60	2.222			147		147	226
1400	37.5 kVA	2 Elec	3	5.333			355		355	540
1420	75 kVA	"	2.50	6.400	↓		425		425	650
1440	3 phase to 600 V, primary									
1460	3 kVA	1 Elec	3.87	2.067	Ea.		137		137	210
1480	15 kVA	2 Elec	3.67	4.360	"		289		289	445
1481	15 kVA		40	.400	kVA		26.50		26.50	40.50
1490	25 kVA		3.51	4.558	Ea.		300		300	465
1500	30 kVA	↓	3.42	4.678			310		310	475
1530	112.5 kVA	R-3	2.90	6.897			455	60	515	765
1560	500 kVA		1.40	14.286			940	124	1,064	1,575
1570	750 kVA	↓	1.10	18.182	↓		1,200	158	1,358	2,025
1571	750 kVA	↓	700	.029	kVA		1.88	.25	2.13	3.17
1600	Pull boxes & cabinets, sheet metal, incl. removal									
1620	of supports and conduit terminations									
1640	6" x 6" x 4"	1 Elec	31.10	.257	Ea.		17.05		17.05	26
1660	12" x 12" x 4"		23.30	.343			22.50		22.50	35
1720	Junction boxes, 4" sq. & oct.		80	.100			6.60		6.60	10.15
1740	Handy box		107	.075			4.95		4.95	7.60
1760	Switch box		107	.075			4.95		4.95	7.60
1780	Receptacle & switch plates		257	.031	↓		2.06		2.06	3.17
1800	Wire, THW-THWN-THHN, removed from									
1810	in place conduit, to 10' high									
1830	#14	1 Elec	65	.123	C.L.F.		8.15		8.15	12.50
1840	#12		55	.145			9.65		9.65	14.80
1850	#10	↓	45.50	.176			11.65		11.65	17.90
1880	#4	2 Elec	53	.302			20		20	30.50
1890	#3		50	.320			21		21	32.50
1910	1/0		33.20	.482			32		32	49
1920	2/0		29.20	.548			36.50		36.50	55.50
1930	3/0		25	.640			42.50		42.50	65
1980	400 kcmil		17	.941			62.50		62.50	95.50
1990	500 kcmil	↓	16.20	.988	↓		65.50		65.50	100
2000	Interior fluorescent fixtures, incl. supports									
2010	& whips, to 10' high									
2100	Recessed drop-in 2' x 2', 2 lamp	2 Elec	35	.457	Ea.		30.50		30.50	46.50
2110	2' x 2', 4 lamp		30	.533			35.50		35.50	54
2120	2' x 4', 2 lamp		33	.485			32		32	49.50
2140	2' x 4', 4 lamp		30	.533			35.50		35.50	54
2160	4' x 4', 4 lamp	↓	20	.800	↓		53		53	81.50

26 05 05 – Selective Demolition for Electrical

26 05 05.10 Electrical Demolition

		Crew	Daily Output	Labor-Hours	Unit	Material	2022 Bare Costs Labor	Equipment	Total	Total Incl O&P
2180	Surface mount, acrylic lens & hinged frame									
2200	1' x 4', 2 lamp	2 Elec	44	.364	Ea.		24		24	37
2210	6" x 4', 2 lamp		44	.364			24		24	37
2220	2' x 2', 2 lamp	↓	44	.364	↓		24		24	37
2260	2' x 4', 4 lamp	2 Elec	33	.485	Ea.		32		32	49.50
2280	4' x 4', 4 lamp		23	.696			46		46	71
2281	4' x 4', 6 lamp	↓	23	.696	↓		46		46	71
2300	Strip fixtures, surface mount									
2320	4' long, 1 lamp	2 Elec	53	.302	Ea.		20		20	30.50
2340	4' long, 2 lamp		50	.320			21		21	32.50
2360	8' long, 1 lamp		42	.381			25		25	38.50
2380	8' long, 2 lamp	↓	40	.400	↓		26.50		26.50	40.50
2400	Pendant mount, industrial, incl. removal									
2410	of chain or rod hangers, to 10' high									
2420	4' long, 2 lamp	2 Elec	35	.457	Ea.		30.50		30.50	46.50
2421	4' long, 4 lamp		35	.457			30.50		30.50	46.50
2440	8' long, 2 lamp	↓	27	.593	↓		39		39	60.50
2460	Interior incandescent, surface, ceiling									
2470	or wall mount, to 10' high									
2480	Metal cylinder type, 75 Watt	2 Elec	62	.258	Ea.		17.10		17.10	26.50
2500	150 Watt	"	62	.258	"		17.10		17.10	26.50
2520	Metal halide, high bay									
2540	400 Watt	2 Elec	15	1.067	Ea.		70.50		70.50	108
2560	1,000 Watt		12	1.333			88.50		88.50	136
2580	150 Watt, low bay	↓	20	.800	↓		53		53	81.50
2600	Exterior fixtures, incandescent, wall mount									
2620	100 Watt	2 Elec	50	.320	Ea.		21		21	32.50
2640	Quartz, 500 Watt		33	.485			32		32	49.50
2660	1,500 Watt	↓	27	.593			39		39	60.50
2680	Wall pack, mercury vapor									
2700	175 Watt	2 Elec	25	.640	Ea.		42.50		42.50	65
2720	250 Watt	"	25	.640	"		42.50		42.50	65
9000	Minimum labor/equipment charge	1 Elec	4	2	Job		132		132	203
9900	Add to labor for higher elevated installation									
9905	10' to 14.5' high, add						10%			
9910	15' to 20' high, add						20%			
9920	20' to 25' high, add						25%			
9930	25' to 30' high, add						35%			
9940	30' to 35' high, add						40%			
9950	35' to 40' high, add						50%			
9960	Over 40' high, add						55%			

26 05 19 – Low-Voltage Electrical Power Conductors and Cables

26 05 19.20 Armored Cable

		Crew	Daily Output	Labor-Hours	Unit	Material	2022 Bare Costs Labor	Equipment	Total	Total Incl O&P
0010	**ARMORED CABLE**									
0050	600 volt, copper (BX), #14, 2 conductor, solid	1 Elec	2.40	3.333	C.L.F.	39.50	221		260.50	385
0100	3 conductor, solid		2.20	3.636		74	241		315	450
0150	#12, 2 conductor, solid		2.30	3.478		45	230		275	405
0200	3 conductor, solid		2	4		76	265		341	490
0250	#10, 2 conductor, solid		2	4		77.50	265		342.50	490
0300	3 conductor, solid		1.60	5		124	330		454	645
0340	#8, 2 conductor, stranded		1.50	5.333		173	355		528	730
0350	3 conductor, stranded	↓	1.30	6.154	↓	173	405		578	815

26 05 19.20 Armored Cable		Crew	Daily Output	Labor-Hours	Unit	Material	2022 Bare Costs Labor	Equipment	Total	Total Incl O&P
9010	600 volt, copper (MC) steel clad, #14, 2 wire	R-1A	8.16	1.961	C.L.F.	64.50	117		181.50	250
9020	3 wire		7.21	2.219		83.50	132		215.50	295
9030	4 wire		6.69	2.392		213	142		355	455
9040	#12, 2 wire		7.21	2.219		80	132		212	291
9050	3 wire		6.69	2.392		127	142		269	360
9070	#10, 2 wire		6.04	2.649		178	158		336	440
9080	3 wire		5.65	2.832		190	169		359	470
9100	#8, 2 wire, stranded		3.94	4.061		150	242		392	535
9110	3 wire, stranded		3.40	4.706		193	280		473	640
9900	Minimum labor/equipment charge	R-1C	4	8	Job		475	34	509	770

26 05 19.50 Mineral Insulated Cable

		Crew	Daily Output	Labor-Hours	Unit	Material	2022 Bare Costs Labor	Equipment	Total	Total Incl O&P
0010	**MINERAL INSULATED CABLE** 600 volt									
0100	1 conductor, #12	1 Elec	1.60	5	C.L.F.	475	330		805	1,025
0200	#10		1.60	5		710	330		1,040	1,300
0400	#8		1.50	5.333		800	355		1,155	1,425
0500	#6		1.40	5.714		955	380		1,335	1,625
0600	#4	2 Elec	2.40	6.667		1,275	440		1,715	2,075
0800	#2		2.20	7.273		1,825	480		2,305	2,750
0900	#1		2.10	7.619		2,100	505		2,605	3,075
1000	1/0		2	8		2,450	530		2,980	3,500
1100	2/0		1.90	8.421		2,925	555		3,480	4,075
1200	3/0		1.80	8.889		3,500	590		4,090	4,750
1400	4/0		1.60	10		4,050	660		4,710	5,475
1410	250 kcmil	3 Elec	2.40	10		4,550	660		5,210	6,050
1420	350 kcmil		1.95	12.308		5,350	815		6,165	7,150
1430	500 kcmil		1.95	12.308		6,700	815		7,515	8,625
1500	2 conductor, #12	1 Elec	1.40	5.714		1,175	380		1,555	1,850
1600	#10		1.20	6.667		1,425	440		1,865	2,250
1800	#8		1.10	7.273		1,550	480		2,030	2,450
2000	#6		1.05	7.619		2,275	505		2,780	3,275
2100	#4	2 Elec	2	8		3,050	530		3,580	4,200
2200	3 conductor, #12	1 Elec	1.20	6.667		1,450	440		1,890	2,275
2400	#10		1.10	7.273		1,650	480		2,130	2,575
2600	#8		1.05	7.619		2,075	505		2,580	3,050
2800	#6		1	8		2,725	530		3,255	3,825
3000	#4	2 Elec	1.80	8.889		3,425	590		4,015	4,650
3100	4 conductor, #12	1 Elec	1.20	6.667		1,525	440		1,965	2,350
3200	#10		1.10	7.273		1,825	480		2,305	2,775
3400	#8		1	8		2,425	530		2,955	3,475
3600	#6		.90	8.889		3,125	590		3,715	4,350
3620	7 conductor, #12		1.10	7.273		1,975	480		2,455	2,925
3640	#10		1	8		2,575	530		3,105	3,675
3800	Terminations, 600 volt, 1 conductor, #12		8	1	Ea.	24.50	66		90.50	129
5500	2 conductor, #12		6.70	1.194		24.50	79		103.50	148
5600	#10		6.40	1.250		37	83		120	168
5800	#8		6.20	1.290		37	85.50		122.50	172
6000	#6		5.70	1.404		37	93		130	184
6200	#4		5.30	1.509		82	100		182	244
6400	3 conductor, #12		5.70	1.404		37	93		130	184
6500	#10		5.50	1.455		37	96.50		133.50	189
6600	#8		5.20	1.538		37	102		139	197
6800	#6		4.80	1.667		37	110		147	211

26 05 Common Work Results for Electrical

26 05 19 – Low-Voltage Electrical Power Conductors and Cables

		Crew	Daily Output	Labor-Hours	Unit	Material	2022 Bare Costs Labor	2022 Bare Costs Equipment	Total	Total Incl O&P
26 05 19.50 Mineral Insulated Cable										
7200	#4	1 Elec	4.60	1.739	Ea.	82	115		197	267
7400	4 conductor, #12		4.60	1.739		41	115		156	222
7500	#10		4.40	1.818		41	120		161	230
7600	#8		4.20	1.905		41	126		167	239
8400	#6		4	2		86	132		218	298
8500	7 conductor, #12		3.50	2.286		41	151		192	277
8600	#10		3	2.667		89	177		266	370
8800	Crimping tool, plier type					90.50			90.50	99.50
9000	Stripping tool					340			340	375
9200	Hand vise					102			102	112
9500	Minimum labor/equipment charge	1 Elec	4	2	Job		132		132	203
26 05 19.55 Non-Metallic Sheathed Cable										
0010	**NON-METALLIC SHEATHED CABLE** 600 volt									
0100	Copper with ground wire (Romex)									
0150	#14, 2 conductor	1 Elec	2.70	2.963	C.L.F.	22.50	196		218.50	325
0200	3 conductor		2.40	3.333		31.50	221		252.50	375
0220	4 conductor		2.20	3.636		53.50	241		294.50	430
0250	#12, 2 conductor		2.50	3.200		29	212		241	355
0300	3 conductor		2.20	3.636		47	241		288	420
0320	4 conductor		2	4		78	265		343	490
0350	#10, 2 conductor		2.20	3.636		56.50	241		297.50	430
0400	3 conductor		1.80	4.444		77.50	294		371.50	535
0430	#8, 2 conductor		1.60	5		101	330		431	620
0450	3 conductor		1.50	5.333		144	355		499	700
0500	#6, 3 conductor		1.40	5.714		390	380		770	1,000
0550	SE type SER aluminum cable, 3 RHW and									
0600	1 bare neutral, 3 #8 & 1 #8	1 Elec	1.60	5	C.L.F.	72	330		402	590
0650	3 #6 & 1 #6	"	1.40	5.714		75	380		455	665
0700	3 #4 & 1 #6	2 Elec	2.40	6.667		82.50	440		522.50	770
0750	3 #2 & 1 #4		2.20	7.273		128	480		608	880
0800	3 #1/0 & 1 #2		2	8		174	530		704	1,000
0850	3 #2/0 & 1 #1		1.80	8.889		206	590		796	1,125
0900	3 #4/0 & 1 #2/0		1.60	10		305	660		965	1,375
6500	Service entrance cap for copper SEU									
6600	100 amp	1 Elec	12	.667	Ea.	6.40	44		50.40	75
6700	150 amp		10	.800		13.35	53		66.35	96
6800	200 amp		8	1		19.80	66		85.80	124
9000	Minimum labor/equipment charge		4	2	Job		132		132	203
26 05 19.90 Wire										
0010	**WIRE**, normal installation conditions in wireway, conduit, cable tray									
0020	600 volt, copper type THW, solid, #14 [R260533-22]	1 Elec	13	.615	C.L.F.	13.95	40.50		54.45	78
0030	#12		11	.727		16.95	48		64.95	92.50
0040	#10		10	.800		34.50	53		87.50	120
0050	Stranded, #14		13	.615		16.55	40.50		57.05	80.50
0100	#12		11	.727		23	48		71	99
0120	#10		10	.800		34.50	53		87.50	120
0140	#8		8	1		52.50	66		118.50	160
0160	#6		6.50	1.231		75	81.50		156.50	208
0180	#4	2 Elec	10.60	1.509		141	100		241	310
0200	#3		10	1.600		170	106		276	350
0220	#2		9	1.778		214	118		332	415
0240	#1		8	2		248	132		380	475

26 05 19.90 Wire		Crew	Daily Output	Labor-Hours	Unit	Material	2022 Bare Costs Labor	2022 Bare Costs Equipment	Total	Total Incl O&P
0260	1/0	2 Elec	6.60	2.424	C.L.F.	298	160		458	575
0280	2/0		5.80	2.759		410	183		593	730
0300	3/0		5	3.200		495	212		707	865
0350	4/0		4.40	3.636		590	241		831	1,025
0400	250 kcmil	3 Elec	6	4		715	265		980	1,200
0420	300 kcmil		5.70	4.211		940	279		1,219	1,450
0450	350 kcmil		5.40	4.444		1,075	294		1,369	1,650
0480	400 kcmil		5.10	4.706		1,225	310		1,535	1,825
0490	500 kcmil		4.80	5		1,550	330		1,880	2,200
0510	750 kcmil		3.30	7.273		2,800	480		3,280	3,825
0540	600 volt, aluminum type THHN, stranded, #6	1 Elec	8	1		15	66		81	119
0560	#4	2 Elec	13	1.231		20.50	81.50		102	148
0580	#2		10.60	1.509		27	100		127	184
0600	#1		9	1.778		39	118		157	224
0620	1/0		8	2		43.50	132		175.50	251
0640	2/0		7.20	2.222		51	147		198	282
0680	3/0		6.60	2.424		62.50	160		222.50	315
0700	4/0		6.20	2.581		72	171		243	340
0720	250 kcmil	3 Elec	8.70	2.759		85.50	183		268.50	375
0740	300 kcmil		8.10	2.963		112	196		308	425
0760	350 kcmil		7.50	3.200		120	212		332	455
0780	400 kcmil		6.90	3.478		135	230		365	505
0800	500 kcmil		6	4		142	265		407	560
0850	600 kcmil		5.70	4.211		194	279		473	645
0880	700 kcmil		5.10	4.706		214	310		524	715
9000	Minimum labor/equipment charge	1 Elec	4	2	Job		132		132	203

		Crew	Daily Output	Labor-Hours	Unit	Material	Labor	Equipment	Total	Total Incl O&P
0010	**GROUNDING**									
0030	Rod, copper clad, 8' long, 1/2" diameter	1 Elec	5.50	1.455	Ea.	18	96.50		114.50	168
0040	5/8" diameter		5.50	1.455		16.05	96.50		112.55	166
0050	3/4" diameter		5.30	1.509		30	100		130	187
0080	10' long, 1/2" diameter		4.80	1.667		22	110		132	194
0090	5/8" diameter		4.60	1.739		21	115		136	200
0100	3/4" diameter		4.40	1.818		60.50	120		180.50	252
0130	15' long, 3/4" diameter		4	2		65	132		197	275
0260	Wire ground bare armored, #8-1 conductor		2	4	C.L.F.	77.50	265		342.50	490
0280	#4-1 conductor		1.60	5		148	330		478	675
0390	Bare copper wire, stranded, #8		11	.727		34	48		82	111
0400	#6		10	.800		55	53		108	142
0600	#2	2 Elec	10	1.600		141	106		247	320
0800	3/0		6.60	2.424		267	160		427	540
1000	4/0		5.70	2.807		385	186		571	710
1200	250 kcmil	3 Elec	7.20	3.333		495	221		716	885
1800	Water pipe ground clamps, heavy duty									
2000	Bronze, 1/2" to 1" diameter	1 Elec	8	1	Ea.	29.50	66		95.50	135
2100	1-1/4" to 2" diameter		8	1		33	66		99	138
2200	2-1/2" to 3" diameter		6	1.333		64.50	88.50		153	207
2400	Grounding, exothermic welding									
2500	Exothermic welding reusable mold, cable to cable, parallel, vertical	1 Elec	8	1	Ea.	445	66		511	590
2505	Splice single		8	1		375	66		441	515
2510	Splice single to double		8	1		445	66		511	585

26 05 Common Work Results for Electrical

26 05 26 – Grounding and Bonding for Electrical Systems

26 05 26.80 Grounding

		Crew	Daily Output	Labor-Hours	Unit	Material	2022 Bare Costs Labor	Equipment	Total	Total Incl O&P
2520	Cable to cable, termination, Tee	1 Elec	8	1	Ea.	435	66		501	580
2530	Cable to rod, termination, Tee		8	1		340	66		406	475
2535	Cable to rebar over, termination, Tee		8	1		124	66		190	238
2540	Cable to rod, termination, 90 Deg		8	1		355	66		421	490
2550	Cable to vertical flat steel		8	1		440	66		506	585
2555	Cable to 45 Deg		8	1		445	66		511	590
2730	Exothermic weld, 4/0 wire to 1" ground rod		7	1.143		2.40	75.50		77.90	119
2800	Brazed connections, #6 wire		12	.667		18.60	44		62.60	88.50
3000	#2 wire		10	.800		25	53		78	109
3100	3/0 wire		8	1		37	66		103	143
3200	4/0 wire		7	1.143		42.50	75.50		118	163
3400	250 kcmil wire		5	1.600		49.50	106		155.50	218
3600	500 kcmil wire		4	2		61	132		193	271
7000	Exothermic welding kit, multi vertical		8	1		705	66		771	875
9000	Minimum labor/equipment charge		4	2	Job		132		132	203

26 05 29 – Hangers and Supports for Electrical Systems

26 05 29.30 Fittings and Channel Support

		Crew	Daily Output	Labor-Hours	Unit	Material	2022 Bare Costs Labor	Equipment	Total	Total Incl O&P
0010	**FITTINGS & CHANNEL SUPPORT**									
0020	Rooftop channel support									
0200	2-7/8" L x 1-5/8" W, 12 ga. pre galv., dbl. base	1 Elec	46	.174	Ea.	3.91	11.50		15.41	22
0210	2-7/8" L x 1-5/8" W, 12 ga. hot dip galv., dbl. base		46	.174		2.28	11.50		13.78	20
0220	2-7/16" L x 1-5/8" W, 12 ga. pre galv., sngl base		48	.167		1.88	11.05		12.93	19
0230	2-7/16" L x 1-5/8" W, 12 ga. hot dip galv., sngl base		48	.167		6.70	11.05		17.75	24.50
0240	13/16" L x 1-5/8" W, 14 ga. pre galv., sngl base		56	.143		15.15	9.45		24.60	31
0250	13/16" L x 1-5/8" W, 14 ga. hot dip galv., sngl base		56	.143		6.55	9.45		16	22
0260	1-5/8" L x 1-5/8" W, 12 ga. pre galv., sngl base		54	.148		17.35	9.80		27.15	34
0270	1-5/8" L x 1-5/8" W, 12 ga. hot dip galv., sngl base		54	.148		1.83	9.80		11.63	17.05
0280	1-5/8" x 28", 12 ga. pre galv., dbl. H block base		42	.190		3.97	12.60		16.57	23.50
0290	1-5/8" x 36", 12 ga. pre galv., dbl. H block base	1 Elec	40	.200	Ea.	4.20	13.25		17.45	25
0300	1-5/8" x 42", 12 ga. pre galv., dbl. H block base		34	.235		4.71	15.60		20.31	29
0310	1-5/8" x 50", 12 ga. pre galv., dbl. H block base		38	.211		5	13.95		18.95	27
0320	1-5/8" x 60", 12 ga. pre galv., dbl. H block base		36	.222		5.50	14.70		20.20	28.50
0330	13/16" x 8" H, threaded rod pre galv., H block base		35	.229		2.66	15.15		17.81	26.50
0340	13/16" x 12" H, threaded rod pre galv., H block base		34.60	.231		3.33	15.30		18.63	27
0350	1-5/8" x 16" H, threaded rod dbl. H block base		34	.235		4.97	15.60		20.57	29.50
0360	1-5/8" x 12" H, pre galv., dbl. H block base		34.60	.231		9.50	15.30		24.80	34
0370	1-5/8" x 24"H, pre galv., dbl. H block base		34	.235		10.05	15.60		25.65	35
0380	3-1/4" x 24" H, pre galv., dbl. H block base		32	.250		23	16.55		39.55	50.50
0390	3-1/4" x 36" H, pre galv., dbl. H block base		30	.267		25	17.65		42.65	54.50

26 05 33 – Raceway and Boxes for Electrical Systems

26 05 33.13 Conduit

		Crew	Daily Output	Labor-Hours	Unit	Material	2022 Bare Costs Labor	Equipment	Total	Total Incl O&P
0010	**CONDUIT** To 10' high, includes 2 terminations, 2 elbows,	R260533-22								
0020	11 beam clamps, and 11 couplings per 100 L.F.									
1161	Field bends, 45° to 90°, 1/2" diameter	1 Elec	53	.151	Ea.		10		10	15.35
1162	3/4" diameter		47	.170			11.25		11.25	17.30
1163	1" diameter		44	.182			12.05		12.05	18.50
1164	1-1/4" diameter		23	.348			23		23	35.50
1165	1-1/2" diameter		21	.381			25		25	38.50
1166	2" diameter		16	.500			33		33	51
1691	See note on line 26 05 33.13 9995									
1750	Rigid galvanized steel, 1/2" diameter	1 Elec	90	.089	L.F.	3.24	5.90		9.14	12.60
1770	3/4" diameter		80	.100		3.50	6.60		10.10	14

501

26 05 33.13 Conduit

		Crew	Daily Output	Labor-Hours	Unit	Material	2022 Bare Costs Labor	2022 Bare Costs Equipment	Total	Total Incl O&P
1800	1" diameter	1 Elec	65	.123	L.F.	5.05	8.15		13.20	18.05
1830	1-1/4" diameter		60	.133		7.25	8.85		16.10	21.50
1850	1-1/2" diameter		55	.145		8.25	9.65		17.90	24
1870	2" diameter		45	.178		10	11.75		21.75	29
1991	Field bends, 45° to 90°, 1/2" diameter		44	.182	Ea.		12.05		12.05	18.50
1992	3/4" diameter		40	.200			13.25		13.25	20.50
1993	1" diameter		36	.222			14.70		14.70	22.50
1994	1-1/4" diameter		19	.421			28		28	43
1995	1-1/2" diameter		18	.444			29.50		29.50	45
1996	2" diameter		13	.615			40.50		40.50	62.50
2500	Steel, intermediate conduit (IMC), 1/2" diameter		100	.080	L.F.	2.34	5.30		7.64	10.70
2530	3/4" diameter		90	.089	"	2.86	5.90		8.76	12.20
2550	1" diameter	1 Elec	70	.114	L.F.	3.93	7.55		11.48	15.95
2570	1-1/4" diameter		65	.123		5.45	8.15		13.60	18.50
2600	1-1/2" diameter		60	.133		6.70	8.85		15.55	21
2630	2" diameter		50	.160		7.95	10.60		18.55	25
2650	2-1/2" diameter		40	.200		14.65	13.25		27.90	36.50
2670	3" diameter	2 Elec	60	.267		19.85	17.65		37.50	49
2700	3-1/2" diameter		54	.296		24	19.60		43.60	56.50
2730	4" diameter		50	.320		27	21		48	62
2731	Field bends, 45° to 90°, 1/2" diameter	1 Elec	44	.182	Ea.		12.05		12.05	18.50
2732	3/4" diameter		40	.200			13.25		13.25	20.50
2733	1" diameter		36	.222			14.70		14.70	22.50
2734	1-1/4" diameter		19	.421			28		28	43
2735	1-1/2" diameter		18	.444			29.50		29.50	45
2736	2" diameter		13	.615			40.50		40.50	62.50
5000	Electric metallic tubing (EMT), 1/2" diameter		170	.047	L.F.	.82	3.12		3.94	5.70
5020	3/4" diameter		130	.062		1.23	4.07		5.30	7.60
5040	1" diameter		115	.070		2.07	4.61		6.68	9.30
5060	1-1/4" diameter		100	.080		3.39	5.30		8.69	11.90
5080	1-1/2" diameter		90	.089		4.11	5.90		10.01	13.55
5100	2" diameter		80	.100		4.82	6.60		11.42	15.45
5120	2-1/2" diameter		60	.133		6.80	8.85		15.65	21
5140	3" diameter	2 Elec	100	.160		8.60	10.60		19.20	25.50
5160	3-1/2" diameter		90	.178		11.10	11.75		22.85	30.50
5180	4" diameter		80	.200		11.65	13.25		24.90	33.50
5200	Field bends, 45° to 90°, 1/2" diameter	1 Elec	89	.090	Ea.		5.95		5.95	9.15
5220	3/4" diameter		80	.100			6.60		6.60	10.15
5240	1" diameter		73	.110			7.25		7.25	11.15
5260	1-1/4" diameter		38	.211			13.95		13.95	21.50
5280	1-1/2" diameter		36	.222			14.70		14.70	22.50
5300	2" diameter		26	.308			20.50		20.50	31.50
5320	Offsets, 1/2" diameter		65	.123			8.15		8.15	12.50
5340	3/4" diameter		62	.129			8.55		8.55	13.10
5360	1" diameter		53	.151			10		10	15.35
5380	1-1/4" diameter		30	.267			17.65		17.65	27
5400	1-1/2" diameter		28	.286			18.90		18.90	29
9990	Minimum labor/equipment charge		4	2	Job		132		132	203

26 05 33.16 Boxes for Electrical Systems

		Crew	Daily Output	Labor-Hours	Unit	Material	2022 Bare Costs Labor	2022 Bare Costs Equipment	Total	Total Incl O&P
0010	**BOXES FOR ELECTRICAL SYSTEMS**									
0020	Pressed steel, octagon, 4"	1 Elec	20	.400	Ea.	1.32	26.50		27.82	42
0060	Covers, blank	1 Elec	64	.125	Ea.	.68	8.30		8.98	13.45

26 05 Common Work Results for Electrical

26 05 33 – Raceway and Boxes for Electrical Systems

26 05 33.16 Boxes for Electrical Systems

		Crew	Daily Output	Labor-Hours	Unit	Material	2022 Bare Costs Labor	Equipment	Total	Total Incl O&P
0100	Extension rings	1 Elec	40	.200	Ea.	2.19	13.25		15.44	23
0150	Square, 4"		20	.400		2.06	26.50		28.56	43
0200	Extension rings		40	.200		2.65	13.25		15.90	23.50
0250	Covers, blank		64	.125		1.19	8.30		9.49	14
0300	Plaster rings		64	.125		1.71	8.30		10.01	14.60
0652	Switchbox		24	.333		4.77	22		26.77	39.50
1100	Concrete, floor, 1 gang		5.30	1.509		126	100		226	293
9000	Minimum labor/equipment charge		4	2	Job		132		132	203

26 05 33.17 Outlet Boxes, Plastic

		Crew	Daily Output	Labor-Hours	Unit	Material	2022 Bare Costs Labor	Equipment	Total	Total Incl O&P
0010	**OUTLET BOXES, PLASTIC**									
0050	4" diameter, round with 2 mounting nails	1 Elec	25	.320	Ea.	3.23	21		24.23	36
0100	Bar hanger mounted		25	.320		5	21		26	38
0200	4", square with 2 mounting nails		25	.320		5.35	21		26.35	38.50
0300	Plaster ring		64	.125		2.12	8.30		10.42	15.05
0400	Switch box with 2 mounting nails, 1 gang		30	.267		2.11	17.65		19.76	29.50
0500	2 gang		25	.320		3.50	21		24.50	36.50
0600	3 gang		20	.400		5.90	26.50		32.40	47
0700	Old work box		30	.267		2.31	17.65		19.96	29.50
9000	Minimum labor/equipment charge		4	2	Job		132		132	203

26 05 33.18 Pull Boxes

		Crew	Daily Output	Labor-Hours	Unit	Material	2022 Bare Costs Labor	Equipment	Total	Total Incl O&P
0010	**PULL BOXES**									
0100	Steel, pull box, NEMA 1, type SC, 6" W x 6" H x 4" D	1 Elec	8	1	Ea.	9.70	66		75.70	113
0200	8" W x 8" H x 4" D		8	1		13.25	66		79.25	117
0300	10" W x 12" H x 6" D		5.30	1.509		32	100		132	190
0400	16" W x 20" H x 8" D		4	2		84	132		216	296
0500	20" W x 24" H x 8" D		3.20	2.500		98	166		264	360
0600	24" W x 36" H x 8" D		2.70	2.963		152	196		348	465
0650	Pull box, hinged, NEMA 1, 6" W x 6" H x 4" D		8	1		12.85	66		78.85	116
0800	12" W x 16" H x 6" D		4.70	1.702		80	113		193	262
1000	20" W x 20" H x 6" D		3.60	2.222		271	147		418	525
1200	20" W x 20" H x 8" D		3.20	2.500		170	166		336	440
1400	24" W x 36" H x 8" D		2.70	2.963		249	196		445	575
1600	24" W x 42" H x 8" D		2	4		365	265		630	810

26 05 33.23 Wireway

		Crew	Daily Output	Labor-Hours	Unit	Material	2022 Bare Costs Labor	Equipment	Total	Total Incl O&P
0010	**WIREWAY** to 10' high									
0020	For higher elevations, see Section 26 05 36.40									
0100	NEMA 1, screw cover w/fittings and supports, 2-1/2" x 2-1/2"	1 Elec	45	.178	L.F.	15.85	11.75		27.60	35.50
0200	4" x 4"	"	40	.200		19.75	13.25		33	42
0400	6" x 6"	2 Elec	60	.267		24	17.65		41.65	53
0600	8" x 8"	2 Elec	40	.400	L.F.	23	26.50		49.50	65.50
4475	NEMA 3R, screw cover w/fittings and supports, 4" x 4"	1 Elec	36	.222		17.60	14.70		32.30	42
4480	6" x 6"	2 Elec	55	.291		19.55	19.25		38.80	51
4485	8" x 8"		36	.444		42	29.50		71.50	91.50
4490	12" x 12"		18	.889		92.50	59		151.50	193

26 05 33.95 Cutting and Drilling

		Crew	Daily Output	Labor-Hours	Unit	Material	2022 Bare Costs Labor	Equipment	Total	Total Incl O&P
0010	**CUTTING AND DRILLING**									
0100	Hole drilling to 10' high, concrete wall									
0110	8" thick, 1/2" pipe size	R-31	12	.667	Ea.	.29	44	5.85	50.14	75
0120	3/4" pipe size		12	.667		.29	44	5.85	50.14	75
0130	1" pipe size		9.50	.842		.40	56	7.40	63.80	94
0140	1-1/4" pipe size		9.50	.842		.40	56	7.40	63.80	94
0150	1-1/2" pipe size		9.50	.842		.40	56	7.40	63.80	94

26 05 33.95 Cutting and Drilling		Crew	Daily Output	Labor-Hours	Unit	Material	2022 Bare Costs Labor	Equipment	Total	Total Incl O&P
0160	2" pipe size	R-31	4.40	1.818	Ea.	.57	120	15.95	136.52	203
0170	2-1/2" pipe size		4.40	1.818		.57	120	15.95	136.52	203
0180	3" pipe size		4.40	1.818		.57	120	15.95	136.52	203
0190	3-1/2" pipe size		3.30	2.424		.75	160	21.50	182.25	271
0200	4" pipe size		3.30	2.424		.75	160	21.50	182.25	271
0500	12" thick, 1/2" pipe size		9.40	.851		.44	56.50	7.45	64.39	95
0520	3/4" pipe size		9.40	.851		.44	56.50	7.45	64.39	95
0540	1" pipe size		7.30	1.096		.60	72.50	9.60	82.70	122
0560	1-1/4" pipe size		7.30	1.096		.60	72.50	9.60	82.70	122
0570	1-1/2" pipe size		7.30	1.096		.60	72.50	9.60	82.70	122
0580	2" pipe size		3.60	2.222		.85	147	19.50	167.35	248
0590	2-1/2" pipe size		3.60	2.222		.85	147	19.50	167.35	248
0600	3" pipe size		3.60	2.222		.85	147	19.50	167.35	248
0610	3-1/2" pipe size		2.80	2.857		1.13	189	25	215.13	320
0630	4" pipe size		2.50	3.200		1.13	212	28	241.13	355
0650	16" thick, 1/2" pipe size		7.60	1.053		.58	69.50	9.25	79.33	118
0670	3/4" pipe size		7	1.143		.58	75.50	10	86.08	128
0690	1" pipe size		6	1.333		.80	88.50	11.70	101	150
0710	1-1/4" pipe size		5.50	1.455		.80	96.50	12.75	110.05	163
0730	1-1/2" pipe size		5.50	1.455		.80	96.50	12.75	110.05	163
0750	2" pipe size		3	2.667		1.13	177	23.50	201.63	298
0770	2-1/2" pipe size		2.70	2.963		1.13	196	26	223.13	330
0790	3" pipe size		2.50	3.200		1.13	212	28	241.13	355
0810	3-1/2" pipe size		2.30	3.478		1.51	230	30.50	262.01	390
0830	4" pipe size		2	4		1.51	265	35	301.51	445
0850	20" thick, 1/2" pipe size		6.40	1.250		.73	83	10.95	94.68	140
0870	3/4" pipe size		6	1.333		.73	88.50	11.70	100.93	150
0890	1" pipe size		5	1.600		1	106	14.05	121.05	180
0910	1-1/4" pipe size		4.80	1.667		1	110	14.60	125.60	187
0930	1-1/2" pipe size		4.60	1.739		1	115	15.25	131.25	195
0950	2" pipe size		2.70	2.963		1.41	196	26	223.41	330
0970	2-1/2" pipe size		2.40	3.333		1.41	221	29	251.41	375
0990	3" pipe size		2.20	3.636		1.41	241	32	274.41	405
1010	3-1/2" pipe size		2	4		1.88	265	35	301.88	445
1030	4" pipe size		1.70	4.706		1.88	310	41.50	353.38	530
1050	24" thick, 1/2" pipe size		5.50	1.455		.87	96.50	12.75	110.12	163
1070	3/4" pipe size		5.10	1.569		.87	104	13.75	118.62	176
1090	1" pipe size		4.30	1.860		1.20	123	16.30	140.50	208
1110	1-1/4" pipe size		4	2		1.20	132	17.55	150.75	224
1130	1-1/2" pipe size		4	2		1.20	132	17.55	150.75	224
1150	2" pipe size		2.40	3.333		1.70	221	29	251.70	375
1170	2-1/2" pipe size		2.20	3.636		1.70	241	32	274.70	405
1190	3" pipe size		2	4		1.70	265	35	301.70	445
1210	3-1/2" pipe size		1.80	4.444		2.26	294	39	335.26	495
1230	4" pipe size		1.50	5.333		2.26	355	47	404.26	595
1500	Brick wall, 8" thick, 1/2" pipe size		18	.444		.29	29.50	3.90	33.69	49.50
1520	3/4" pipe size		18	.444		.29	29.50	3.90	33.69	49.50
1540	1" pipe size		13.30	.602		.40	40	5.30	45.70	67
1560	1-1/4" pipe size		13.30	.602		.40	40	5.30	45.70	67
1580	1-1/2" pipe size		13.30	.602		.40	40	5.30	45.70	67
1600	2" pipe size		5.70	1.404		.57	93	12.30	105.87	157
1620	2-1/2" pipe size		5.70	1.404		.57	93	12.30	105.87	157
1640	3" pipe size		5.70	1.404		.57	93	12.30	105.87	157

26 05 33.95 Cutting and Drilling	Crew	Daily Output	Labor-Hours	Unit	Material	2022 Bare Costs Labor	2022 Bare Costs Equipment	Total	Total Incl O&P
1660	3-1/2" pipe size	R-31	4.40	1.818	Ea.	.75	120	15.95	136.70	203
1680	4" pipe size		4	2		.75	132	17.55	150.30	223
1700	12" thick, 1/2" pipe size		14.50	.552		.44	36.50	4.84	41.78	62
1720	3/4" pipe size		14.50	.552		.44	36.50	4.84	41.78	62
1740	1" pipe size		11	.727		.60	48	6.40	55	81.50
1760	1-1/4" pipe size		11	.727		.60	48	6.40	55	81.50
1780	1-1/2" pipe size		11	.727		.60	48	6.40	55	81.50
1800	2" pipe size		5	1.600		.85	106	14.05	120.90	179
1820	2-1/2" pipe size		5	1.600		.85	106	14.05	120.90	179
1840	3" pipe size		5	1.600		.85	106	14.05	120.90	179
1860	3-1/2" pipe size		3.80	2.105		1.13	139	18.45	158.58	236
1880	4" pipe size		3.30	2.424		1.13	160	21.50	182.63	272
1900	16" thick, 1/2" pipe size		12.30	.650		.58	43	5.70	49.28	73
1920	3/4" pipe size		12.30	.650		.58	43	5.70	49.28	73
1940	1" pipe size		9.30	.860		.80	57	7.55	65.35	96.50
1960	1-1/4" pipe size		9.30	.860		.80	57	7.55	65.35	96.50
1980	1-1/2" pipe size		9.30	.860		.80	57	7.55	65.35	96.50
2000	2" pipe size		4.40	1.818		1.13	120	15.95	137.08	204
2010	2-1/2" pipe size		4.40	1.818		1.13	120	15.95	137.08	204
2030	3" pipe size		4.40	1.818		1.13	120	15.95	137.08	204
2050	3-1/2" pipe size		3.30	2.424		1.51	160	21.50	183.01	272
2070	4" pipe size		3	2.667		1.51	177	23.50	202.01	298
2090	20" thick, 1/2" pipe size		10.70	.748		.73	49.50	6.55	56.78	84
2110	3/4" pipe size		10.70	.748		.73	49.50	6.55	56.78	84
2130	1" pipe size		8	1		1	66	8.75	75.75	113
2150	1-1/4" pipe size		8	1		1	66	8.75	75.75	113
2170	1-1/2" pipe size		8	1		1	66	8.75	75.75	113
2190	2" pipe size		4	2		1.41	132	17.55	150.96	224
2210	2-1/2" pipe size		4	2		1.41	132	17.55	150.96	224
2230	3" pipe size		4	2		1.41	132	17.55	150.96	224
2250	3-1/2" pipe size		3	2.667		1.88	177	23.50	202.38	299
2270	4" pipe size		2.70	2.963		1.88	196	26	223.88	330
2290	24" thick, 1/2" pipe size		9.40	.851		.87	56.50	7.45	64.82	95.50
2310	3/4" pipe size		9.40	.851		.87	56.50	7.45	64.82	95.50
2330	1" pipe size		7.10	1.127		1.20	74.50	9.90	85.60	127
2350	1-1/4" pipe size		7.10	1.127		1.20	74.50	9.90	85.60	127
2370	1-1/2" pipe size		7.10	1.127		1.20	74.50	9.90	85.60	127
2390	2" pipe size		3.60	2.222		1.70	147	19.50	168.20	249
2410	2-1/2" pipe size		3.60	2.222		1.70	147	19.50	168.20	249
2430	3" pipe size		3.60	2.222		1.70	147	19.50	168.20	249
2450	3-1/2" pipe size		2.80	2.857		2.26	189	25	216.26	320
2470	4" pipe size		2.50	3.200		2.26	212	28	242.26	360
3000	Knockouts to 8' high, metal boxes & enclosures									
3020	With hole saw, 1/2" pipe size	1 Elec	53	.151	Ea.		10		10	15.35
3040	3/4" pipe size		47	.170			11.25		11.25	17.30
3050	1" pipe size		40	.200			13.25		13.25	20.50
3060	1-1/4" pipe size		36	.222			14.70		14.70	22.50
3070	1-1/2" pipe size		32	.250			16.55		16.55	25.50
3080	2" pipe size		27	.296			19.60		19.60	30
3090	2-1/2" pipe size		20	.400			26.50		26.50	40.50
4010	3" pipe size		16	.500			33		33	51
4030	3-1/2" pipe size		13	.615			40.50		40.50	62.50
4050	4" pipe size		11	.727			48		48	74

26 05 39 – Underfloor Raceways for Electrical Systems

26 05 39.30 Conduit In Concrete Slab	Crew	Daily Output	Labor-Hours	Unit	Material	2022 Bare Costs Labor	Equipment	Total	Total Incl O&P
0010 **CONDUIT IN CONCRETE SLAB** Including terminations,									
0020 fittings and supports									
3230 PVC, schedule 40, 1/2" diameter	1 Elec	270	.030	L.F.	.62	1.96		2.58	3.70
3250 3/4" diameter		230	.035		.67	2.30		2.97	4.27
3270 1" diameter		200	.040		.87	2.65		3.52	5
3300 1-1/4" diameter		170	.047		1.31	3.12		4.43	6.25
3330 1-1/2" diameter		140	.057		1.52	3.78		5.30	7.45
3350 2" diameter		120	.067		1.86	4.41		6.27	8.85
4350 Rigid galvanized steel, 1/2" diameter		200	.040		2.85	2.65		5.50	7.20
4400 3/4" diameter		170	.047		2.94	3.12		6.06	8
4450 1" diameter		130	.062		4.45	4.07		8.52	11.15
4500 1-1/4" diameter		110	.073		6.45	4.81		11.26	14.45
4600 1-1/2" diameter		100	.080		7.20	5.30		12.50	16.10
4800 2" diameter		90	.089		8.85	5.90		14.75	18.80
9000 Minimum labor/equipment charge		4	2	Job		132		132	203

26 05 39.40 Conduit In Trench

	Crew	Daily Output	Labor-Hours	Unit	Material	2022 Bare Costs Labor	Equipment	Total	Total Incl O&P
0010 **CONDUIT IN TRENCH** Includes terminations and fittings									
0020 Does not include excavation or backfill, see Section 31 23 16									
0200 Rigid galvanized steel, 2" diameter	1 Elec	150	.053	L.F.	8.50	3.53		12.03	14.75
0400 2-1/2" diameter	"	100	.080		16.55	5.30		21.85	26.50
0600 3" diameter	2 Elec	160	.100		19	6.60		25.60	31
0800 3-1/2" diameter		140	.114		25.50	7.55		33.05	39.50
1000 4" diameter		100	.160		26.50	10.60		37.10	46
1200 5" diameter		80	.200		54	13.25		67.25	80
1400 6" diameter		60	.267		73	17.65		90.65	108
9000 Minimum labor/equipment charge	1 Elec	4	2	Job		132		132	203

26 05 83 – Wiring Connections

26 05 83.10 Motor Connections

	Crew	Daily Output	Labor-Hours	Unit	Material	2022 Bare Costs Labor	Equipment	Total	Total Incl O&P
0010 **MOTOR CONNECTIONS**									
0020 Flexible conduit and fittings, 115 volt, 1 phase, up to 1 HP motor	1 Elec	8	1	Ea.	5.20	66		71.20	108
9000 Minimum labor/equipment charge	"	4	2	Job		132		132	203

26 05 90 – Residential Applications

26 05 90.10 Residential Wiring

	Crew	Daily Output	Labor-Hours	Unit	Material	2022 Bare Costs Labor	Equipment	Total	Total Incl O&P
0010 **RESIDENTIAL WIRING**									
0020 20' avg. runs and #14/2 wiring incl. unless otherwise noted									
1000 Service & panel, includes 24' SE-AL cable, service eye, meter,									
1010 Socket, panel board, main bkr., ground rod, 15 or 20 amp									
1020 1-pole circuit breakers, and misc. hardware									
1100 100 amp, with 10 branch breakers	1 Elec	1.19	6.723	Ea.	415	445		860	1,150
1110 With PVC conduit and wire		.92	8.696		460	575		1,035	1,400
1120 With RGS conduit and wire		.73	10.959		655	725		1,380	1,850
1150 150 amp, with 14 branch breakers		1.03	7.767		585	515		1,100	1,425
1170 With PVC conduit and wire		.82	9.756		660	645		1,305	1,725
1180 With RGS conduit and wire		.67	11.940		1,050	790		1,840	2,400
1200 200 amp, with 18 branch breakers	2 Elec	1.80	8.889		1,250	590		1,840	2,275
1220 With PVC conduit and wire		1.46	10.959		1,350	725		2,075	2,600
1230 With RGS conduit and wire		1.24	12.903		1,800	855		2,655	3,300
1800 Lightning surge suppressor	1 Elec	32	.250		97.50	16.55		114.05	133
2000 Switch devices									
2100 Single pole, 15 amp, ivory, with a 1-gang box, cover plate,									
2110 Type NM (Romex) cable	1 Elec	17.10	.468	Ea.	7.85	31		38.85	56

26 05 90.10 Residential Wiring	Crew	Daily Output	Labor-Hours	Unit	Material	2022 Bare Costs Labor	Equipment	Total	Total Incl O&P
2120	Type MC cable	1 Elec	14.30	.559	Ea.	21.50	37		58.50	81
2130	EMT & wire		5.71	1.401		37	93		130	183
2150	3-way, #14/3, type NM cable		14.55	.550		9.95	36.50		46.45	67
2170	Type MC cable		12.31	.650		29	43		72	98
2180	EMT & wire		5	1.600		40.50	106		146.50	208
2200	4-way, #14/3, type NM cable		14.55	.550		17.15	36.50		53.65	75
2220	Type MC cable		12.31	.650		36	43		79	106
2230	EMT & wire		5	1.600		47.50	106		153.50	216
2250	S.P., 20 amp, #12/2, type NM cable		13.33	.600		12.30	39.50		51.80	74.50
2270	Type MC cable		11.43	.700		26	46.50		72.50	99.50
2280	EMT & wire		4.85	1.649		45	109		154	218
2290	S.P. rotary dimmer, 600 W, no wiring		17	.471		33.50	31		64.50	85
2300	S.P. rotary dimmer, 600 W, type NM cable		14.55	.550		38	36.50		74.50	98
2320	Type MC cable		12.31	.650		52	43		95	123
2330	EMT & wire		5	1.600		70	106		176	240
2350	3-way rotary dimmer, type NM cable		13.33	.600		25.50	39.50		65	89
2370	Type MC cable		11.43	.700		39.50	46.50		86	115
2380	EMT & wire		4.85	1.649		58	109		167	232
2400	Interval timer wall switch, 20 amp, 1-30 min., #12/2									
2410	Type NM cable	1 Elec	14.55	.550	Ea.	69	36.50		105.50	132
2420	Type MC cable	1 Elec	12.31	.650	Ea.	75	43		118	149
2430	EMT & wire	"	5	1.600	"	102	106		208	275
2500	Decorator style									
2510	S.P., 15 amp, type NM cable	1 Elec	17.10	.468	Ea.	13.40	31		44.40	62.50
2520	Type MC cable		14.30	.559		27.50	37		64.50	87
2530	EMT & wire		5.71	1.401		43	93		136	189
2550	3-way, #14/3, type NM cable		14.55	.550		15.50	36.50		52	73
2570	Type MC cable		12.31	.650		34.50	43		77.50	104
2580	EMT & wire		5	1.600		46	106		152	214
2600	4-way, #14/3, type NM cable		14.55	.550		22.50	36.50		59	81
2620	Type MC cable		12.31	.650		41.50	43		84.50	112
2630	EMT & wire		5	1.600		53	106		159	222
2650	S.P., 20 amp, #12/2, type NM cable		13.33	.600		17.80	39.50		57.30	80.50
2670	Type MC cable		11.43	.700		31.50	46.50		78	106
2680	EMT & wire		4.85	1.649		50.50	109		159.50	224
2700	S.P., slide dimmer, type NM cable		17.10	.468		39.50	31		70.50	90.50
2720	Type MC cable		14.30	.559		53	37		90	116
2730	EMT & wire		5.71	1.401		71.50	93		164.50	221
2750	S.P., touch dimmer, type NM cable		17.10	.468		32	31		63	83
2770	Type MC cable		14.30	.559		46	37		83	108
2780	EMT & wire		5.71	1.401		64.50	93		157.50	213
2800	3-way touch dimmer, type NM cable		13.33	.600		44.50	39.50		84	110
2820	Type MC cable		11.43	.700		58.50	46.50		105	136
2830	EMT & wire		4.85	1.649		76.50	109		185.50	253
3000	Combination devices									
3100	S.P. switch/15 amp recpt., ivory, 1-gang box, plate									
3110	Type NM cable	1 Elec	11.43	.700	Ea.	21	46.50		67.50	94.50
3120	Type MC cable		10	.800		35	53		88	120
3130	EMT & wire		4.40	1.818		53.50	120		173.50	244
3150	S.P. switch/pilot light, type NM cable		11.43	.700		22.50	46.50		69	96
3170	Type MC cable		10	.800		36.50	53		89.50	122
3180	EMT & wire		4.43	1.806		55	120		175	245
3190	2-S.P. switches, 2-#14/2, no wiring		14	.571		5.40	38		43.40	64

26 05 90.10 Residential Wiring	Crew	Daily Output	Labor-Hours	Unit	Material	2022 Bare Costs Labor	Equipment	Total	Total Incl O&P
3200 2-S.P. switches, 2-#14/2, type NM cables	1 Elec	10	.800	Ea.	23.50	53		76.50	107
3220 Type MC cable		8.89	.900		40.50	59.50		100	136
3230 EMT & wire		4.10	1.951		59.50	129		188.50	264
3250 3-way switch/15 amp recpt., #14/3, type NM cable		10	.800		29.50	53		82.50	114
3270 Type MC cable		8.89	.900		48.50	59.50		108	145
3280 EMT & wire		4.10	1.951		60	129		189	264
3300 2-3 way switches, 2-#14/3, type NM cables		8.89	.900		36.50	59.50		96	132
3320 Type MC cable		8	1		64	66		130	173
3330 EMT & wire		4	2		69	132		201	279
3350 S.P. switch/20 amp recpt., #12/2, type NM cable		10	.800		35	53		88	120
3370 Type MC cable		8.89	.900		41	59.50		100.50	137
3380 EMT & wire		4.10	1.951		67.50	129		196.50	273
3400 Decorator style									
3410 S.P. switch/15 amp recpt., type NM cable	1 Elec	11.43	.700	Ea.	27	46.50		73.50	101
3420 Type MC cable		10	.800		40.50	53		93.50	126
3430 EMT & wire		4.40	1.818		59	120		179	250
3450 S.P. switch/pilot light, type NM cable		11.43	.700		28.50	46.50		75	102
3470 Type MC cable		10	.800		42	53		95	128
3480 EMT & wire		4.40	1.818		60.50	120		180.50	252
3500 2-S.P. switches, 2-#14/2, type NM cables		10	.800		29	53		82	114
3520 Type MC cable		8.89	.900		46	59.50		105.50	143
3530 EMT & wire		4.10	1.951		65	129		194	270
3550 3-way/15 amp recpt., #14/3, type NM cable		10	.800		35.50	53		88.50	121
3570 Type MC cable		8.89	.900		54.50	59.50		114	151
3580 EMT & wire		4.10	1.951		65.50	129		194.50	270
3650 2-3 way switches, 2-#14/3, type NM cables		8.89	.900		42	59.50		101.50	138
3670 Type MC cable		8	1		69.50	66		135.50	179
3680 EMT & wire		4	2		74.50	132		206.50	285
3700 S.P. switch/20 amp recpt., #12/2, type NM cable		10	.800		40.50	53		93.50	126
3720 Type MC cable		8.89	.900		46.50	59.50		106	143
3730 EMT & wire		4.10	1.951		73	129		202	279
4000 Receptacle devices									
4010 Duplex outlet, 15 amp recpt., ivory, 1-gang box, plate									
4015 Type NM cable	1 Elec	14.55	.550	Ea.	8.45	36.50		44.95	65.50
4020 Type MC cable		12.31	.650		22.50	43		65.50	90.50
4030 EMT & wire		5.33	1.501		38	99.50		137.50	195
4050 With #12/2, type NM cable		12.31	.650		9.75	43		52.75	76.50
4070 Type MC cable		10.67	.750		23.50	49.50		73	102
4080 EMT & wire		4.71	1.699		42.50	112		154.50	220
4100 20 amp recpt., #12/2, type NM cable		12.31	.650		21.50	43		64.50	89.50
4120 Type MC cable		10.67	.750		35	49.50		84.50	115
4130 EMT & wire		4.71	1.699		54	112		166	233
4140 For GFI see Section 26 05 90.10 line 4300 below									
4150 Decorator style, 15 amp recpt., type NM cable	1 Elec	14.55	.550	Ea.	14	36.50		50.50	71.50
4170 Type MC cable		12.31	.650		28	43		71	96.50
4180 EMT & wire		5.33	1.501		43.50	99.50		143	201
4200 With #12/2, type NM cable		12.31	.650		15.25	43		58.25	83
4220 Type MC cable		10.67	.750		29	49.50		78.50	109
4230 EMT & wire		4.71	1.699		48	112		160	226
4250 20 amp recpt., #12/2, type NM cable		12.31	.650		27	43		70	95.50
4270 Type MC cable		10.67	.750		40.50	49.50		90	121
4280 EMT & wire		4.71	1.699		59.50	112		171.50	239
4300 GFI, 15 amp recpt., type NM cable		12.31	.650		25.50	43		68.50	94

26 05 90 – Residential Applications

26 05 90.10 Residential Wiring	Crew	Daily Output	Labor-Hours	Unit	Material	2022 Bare Costs Labor	Equipment	Total	Total Incl O&P
4320 Type MC cable	1 Elec	10.67	.750	Ea.	39.50	49.50		89	120
4330 EMT & wire		4.71	1.699		55	112		167	234
4350 GFI with #12/2, type NM cable		10.67	.750		27	49.50		76.50	106
4370 Type MC cable		9.20	.870		40.50	57.50		98	133
4380 EMT & wire		4.21	1.900		59.50	126		185.50	259
4400 20 amp recpt., #12/2, type NM cable		10.67	.750		38.50	49.50		88	119
4420 Type MC cable		9.20	.870		52	57.50		109.50	146
4430 EMT & wire		4.21	1.900		71	126		197	271
4500 Weather-proof cover for above receptacles, add	↓	32	.250	↓	4.02	16.55		20.57	30
4550 Air conditioner outlet, 20 amp-240 volt recpt.									
4560 30' of #12/2, 2 pole circuit breaker									
4570 Type NM cable	1 Elec	10	.800	Ea.	68.50	53		121.50	157
4580 Type MC cable		9	.889		84	59		143	183
4590 EMT & wire		4	2		103	132		235	315
4600 Decorator style, type NM cable		10	.800		75	53		128	164
4620 Type MC cable		9	.889		90.50	59		149.50	190
4630 EMT & wire	↓	4	2	↓	110	132		242	325
4650 Dryer outlet, 30 amp-240 volt recpt., 20' of #10/3									
4660 2 pole circuit breaker									
4670 Type NM cable	1 Elec	6.41	1.248	Ea.	59	82.50		141.50	192
4680 Type MC cable		5.71	1.401		64.50	93		157.50	213
4690 EMT & wire	↓	3.48	2.299	↓	96	152		248	340
4700 Range outlet, 50 amp-240 volt recpt., 30' of #8/3									
4710 Type NM cable	1 Elec	4.21	1.900	Ea.	93.50	126		219.50	296
4720 Type MC cable		4	2		112	132		244	325
4730 EMT & wire		2.96	2.703		128	179		307	415
4750 Central vacuum outlet, type NM cable		6.40	1.250		67	83		150	201
4770 Type MC cable		5.71	1.401		80	93		173	230
4780 EMT & wire	↓	3.48	2.299	↓	109	152		261	355
4800 30 amp-110 volt locking recpt., #10/2 circ. bkr.									
4810 Type NM cable	1 Elec	6.20	1.290	Ea.	80.50	85.50		166	220
4820 Type MC cable		5.40	1.481		90.50	98		188.50	251
4830 EMT & wire	↓	3.20	2.500	↓	137	166		303	405
4900 Low voltage outlets									
4910 Telephone recpt., 20' of 4/C phone wire	1 Elec	26	.308	Ea.	8.25	20.50		28.75	40.50
4920 TV recpt., 20' of RG59U coax wire, F type connector	"	16	.500	"	15.80	33		48.80	68.50
4950 Door bell chime, transformer, 2 buttons, 60' of bellwire									
4970 Economy model	1 Elec	11.50	.696	Ea.	56	46		102	133
4980 Custom model		11.50	.696		102	46		148	183
4990 Luxury model, 3 buttons	↓	9.50	.842	↓	192	56		248	298
6000 Lighting outlets									
6050 Wire only (for fixture), type NM cable	1 Elec	32	.250	Ea.	7.10	16.55		23.65	33.50
6070 Type MC cable		24	.333		11.85	22		33.85	47
6080 EMT & wire		10	.800		26.50	53		79.50	111
6100 Box (4") and wire (for fixture), type NM cable		25	.320		16.40	21		37.40	50.50
6120 Type MC cable		20	.400		21	26.50		47.50	64
6130 EMT & wire	↓	11	.727	↓	36	48		84	114
6200 Fixtures (use with line 6050 or 6100 above)									
6210 Canopy style, economy grade	1 Elec	40	.200	Ea.	22	13.25		35.25	45
6220 Custom grade		40	.200		57	13.25		70.25	83
6250 Dining room chandelier, economy grade		19	.421		87	28		115	139
6270 Luxury grade		15	.533		3,050	35.50		3,085.50	3,400
6310 Kitchen fixture (fluorescent), economy grade	↓	30	.267		81.50	17.65		99.15	117

26 05 90.10 Residential Wiring		Crew	Daily Output	Labor-Hours	Unit	Material	2022 Bare Costs Labor	2022 Bare Costs Equipment	Total	Total Incl O&P
6320	Custom grade	1 Elec	25	.320	Ea.	220	21		241	275
6350	Outdoor, wall mounted, economy grade		30	.267		33	17.65		50.65	63.50
6360	Custom grade		30	.267		125	17.65		142.65	164
6370	Luxury grade		25	.320		248	21		269	305
6410	Outdoor PAR floodlights, 1 lamp, 150 watt		20	.400		21	26.50		47.50	63.50
6420	2 lamp, 150 watt each		20	.400		28.50	26.50		55	72
6425	Motion sensing, 2 lamp, 150 watt each		20	.400		106	26.50		132.50	157
6430	For infrared security sensor, add		32	.250		114	16.55		130.55	151
6450	Outdoor, quartz-halogen, 300 watt flood		20	.400		43	26.50		69.50	88
6600	Recessed downlight, round, pre-wired, 50 or 75 watt trim		30	.267		48	17.65		65.65	79.50
6610	With shower light trim	1 Elec	30	.267	Ea.	74.50	17.65		92.15	109
6620	With wall washer trim		28	.286		75	18.90		93.90	112
6630	With eye-ball trim		28	.286		29.50	18.90		48.40	61.50
6700	Porcelain lamp holder		40	.200		2.78	13.25		16.03	23.50
6710	With pull switch		40	.200		12.15	13.25		25.40	34
6750	Fluorescent strip, 2-20 watt tube, wrap around diffuser, 24"		24	.333		46.50	22		68.50	85.50
6760	1-34 watt tube, 48"		24	.333		117	22		139	163
6770	2-34 watt tubes, 48"		20	.400		134	26.50		160.50	188
6800	Bathroom heat lamp, 1-250 watt		28	.286		33.50	18.90		52.40	66
6810	2-250 watt lamps		28	.286		60	18.90		78.90	95
6820	For timer switch, see Section 26 05 90.10 line 2400									
6900	Outdoor post lamp, incl. post, fixture, 35' of #14/2									
6910	Type NM cable	1 Elec	3.50	2.286	Ea.	385	151		536	655
6920	Photo-eye, add		27	.296		15.20	19.60		34.80	46.50
6950	Clock dial time switch, 24 hr., w/enclosure, type NM cable		11.43	.700		74	46.50		120.50	152
6970	Type MC cable		11	.727		87.50	48		135.50	171
6980	EMT & wire		4.85	1.649		103	109		212	281
7000	Alarm systems									
7050	Smoke detectors, box, #14/3, type NM cable	1 Elec	14.55	.550	Ea.	30.50	36.50		67	89.50
7070	Type MC cable		12.31	.650		42	43		85	112
7080	EMT & wire		5	1.600		53.50	106		159.50	222
7090	For relay output to security system, add					9.75			9.75	10.75
8000	Residential equipment									
8050	Disposal hook-up, incl. switch, outlet box, 3' of flex									
8060	20 amp-1 pole circ. bkr., and 25' of #12/2									
8070	Type NM cable	1 Elec	10	.800	Ea.	25.50	53		78.50	110
8080	Type MC cable		8	1		40	66		106	146
8090	EMT & wire		5	1.600		63.50	106		169.50	233
8100	Trash compactor or dishwasher hook-up, incl. outlet box,									
8110	3' of flex, 15 amp-1 pole circ. bkr., and 25' of #14/2									
8120	Type NM cable	1 Elec	10	.800	Ea.	21.50	53		74.50	106
8130	Type MC cable		8	1		37.50	66		103.50	144
8140	EMT & wire		5	1.600		57	106		163	226
8150	Hot water sink dispenser hook-up, use line 8100									
8200	Vent/exhaust fan hook-up, type NM cable	1 Elec	32	.250	Ea.	7.10	16.55		23.65	33.50
8220	Type MC cable	"	24	.333	"	11.85	22		33.85	47
8230	EMT & wire	1 Elec	10	.800	Ea.	26.50	53		79.50	111
8250	Bathroom vent fan, 50 CFM (use with above hook-up)									
8260	Economy model	1 Elec	15	.533	Ea.	18.60	35.50		54.10	74.50
8270	Low noise model		15	.533		50	35.50		85.50	109
8280	Custom model		12	.667		115	44		159	195
8300	Bathroom or kitchen vent fan, 110 CFM									
8310	Economy model	1 Elec	15	.533	Ea.	67.50	35.50		103	128

510

26 05 90 – Residential Applications

26 05 90.10 Residential Wiring	Crew	Daily Output	Labor-Hours	Unit	Material	2022 Bare Costs Labor	Equipment	Total	Total Incl O&P
8320 Low noise model	1 Elec	15	.533	Ea.	108	35.50		143.50	173
8350 Paddle fan, variable speed (w/o lights)									
8360 Economy model (AC motor)	1 Elec	10	.800	Ea.	141	53		194	237
8362 With light kit		10	.800		182	53		235	282
8370 Custom model (AC motor)		10	.800		345	53		398	460
8372 With light kit		10	.800		385	53		438	505
8380 Luxury model (DC motor)		8	1		277	66		343	405
8382 With light kit		8	1		320	66		386	450
8390 Remote speed switch for above, add		12	.667		43.50	44		87.50	116
8500 Whole house exhaust fan, ceiling mount, 36", variable speed									
8510 Remote switch, incl. shutters, 20 amp-1 pole circ. bkr.									
8520 30' of #12/2, type NM cable	1 Elec	4	2	Ea.	1,550	132		1,682	1,900
8530 Type MC cable		3.50	2.286		1,575	151		1,726	1,950
8540 EMT & wire		3	2.667		1,600	177		1,777	2,025
8600 Whirlpool tub hook-up, incl. timer switch, outlet box									
8610 3' of flex, 20 amp-1 pole GFI circ. bkr.									
8620 30' of #12/2, type NM cable	1 Elec	5	1.600	Ea.	247	106		353	435
8630 Type MC cable		4.20	1.905		257	126		383	475
8640 EMT & wire		3.40	2.353		276	156		432	545
8650 Hot water heater hook-up, incl. 1-2 pole circ. bkr., box;									
8660 3' of flex, 20' of #10/2, type NM cable	1 Elec	5	1.600	Ea.	31	106		137	197
8670 Type MC cable		4.20	1.905		44	126		170	243
8680 EMT & wire		3.40	2.353		58	156		214	305
9000 Heating/air conditioning									
9050 Furnace/boiler hook-up, incl. firestat, local on-off switch									
9060 Emergency switch, and 40' of type NM cable	1 Elec	4	2	Ea.	67.50	132		199.50	277
9070 Type MC cable		3.50	2.286		84.50	151		235.50	325
9080 EMT & wire		1.50	5.333		117	355		472	670
9100 Air conditioner hook-up, incl. local 60 amp disc. switch									
9110 3' sealtite, 40 amp, 2 pole circuit breaker									
9130 40' of #8/2, type NM cable	1 Elec	3.50	2.286	Ea.	158	151		309	405
9140 Type MC cable		3	2.667		186	177		363	475
9150 EMT & wire		1.30	6.154		221	405		626	870
9200 Heat pump hook-up, 1-40 & 1-100 amp 2 pole circ. bkr.									
9210 Local disconnect switch, 3' sealtite									
9220 40' of #8/2 & 30' of #3/2									
9230 Type NM cable	1 Elec	1.30	6.154	Ea.	500	405		905	1,175
9240 Type MC cable		1.08	7.407		510	490		1,000	1,325
9250 EMT & wire		.94	8.511		655	565		1,220	1,575
9500 Thermostat hook-up, using low voltage wire									
9520 Heating only, 25' of #18-3	1 Elec	24	.333	Ea.	9.75	22		31.75	44.50
9530 Heating/cooling, 25' of #18-4	"	20	.400	"	10.55	26.50		37.05	52

For customer support on your Commercial Renovation Costs with RSMeans data, call 800.448.8182.

511

26 22 Low-Voltage Transformers

26 22 13 – Low-Voltage Distribution Transformers

26 22 13.10 Transformer, Dry-Type

26 22 13.10 Transformer, Dry-Type	Crew	Daily Output	Labor-Hours	Unit	Material	2022 Bare Costs Labor	Equipment	Total	Total Incl O&P
0010 **TRANSFORMER, DRY-TYPE**									
0050 Single phase, 240/480 volt primary, 120/240 volt secondary									
0100 1 kVA	1 Elec	2	4	Ea.	430	265		695	875
0300 2 kVA		1.60	5		600	330		930	1,175
0500 3 kVA		1.40	5.714		685	380		1,065	1,325
0700 5 kVA		1.20	6.667		825	440		1,265	1,600
0900 7.5 kVA	2 Elec	2.20	7.273		1,225	480		1,705	2,100
1100 10 kVA		1.60	10		1,550	660		2,210	2,750
1300 15 kVA		1.20	13.333		2,275	885		3,160	3,875
2300 3 phase, 480 volt primary, 120/208 volt secondary									
2310 Ventilated, 3 kVA	1 Elec	1	8	Ea.	1,225	530		1,755	2,175
2700 6 kVA		.80	10		2,275	660		2,935	3,525
2900 9 kVA		.70	11.429		2,625	755		3,380	4,025
3100 15 kVA	2 Elec	1.10	14.545		2,325	965		3,290	4,025
3300 30 kVA		.90	17.778		6,175	1,175		7,350	8,575
3500 45 kVA		.80	20		8,050	1,325		9,375	10,900
9000 Minimum labor/equipment charge	1 Elec	1	8	Job		530		530	815
9300 Energy efficient transformer 3 ph									
9400 15 kVA, 480VAC delta, 208Y/120VAC	R-3	1.50	13.333	Ea.	2,900	880	116	3,896	4,650
9405 30 kVA, 480VAC delta, 208Y/120VAC		1.50	13.333		3,025	880	116	4,021	4,800
9410 45 kVA, 480VAC delta, 208Y/120VAC		1.50	13.333		2,375	880	116	3,371	4,075
9415 75 kVA, 480VAC delta, 208Y/120VAC		1.25	16		2,900	1,050	139	4,089	4,950
9420 112 kVA, 480VAC delta, 208Y/120VAC	R-3	1.25	16	Ea.	8,475	1,050	139	9,664	11,100
9425 150 kVA, 480VAC delta, 208Y/120VAC		1.25	16		5,350	1,050	139	6,539	7,650
9430 225 kVA, 480VAC delta, 208Y/120VAC		1.25	16		6,875	1,050	139	8,064	9,325
9435 300 kVA, 480VAC delta, 208Y/120VAC		1.25	16		9,550	1,050	139	10,739	12,300

26 24 Switchboards and Panelboards

26 24 16 – Panelboards

26 24 16.20 Panelboard and Load Center Circuit Breakers

26 24 16.20 Panelboard and Load Center Circuit Breakers	Crew	Daily Output	Labor-Hours	Unit	Material	2022 Bare Costs Labor	Equipment	Total	Total Incl O&P
0010 **PANELBOARD AND LOAD CENTER CIRCUIT BREAKERS**									
0050 Bolt-on, 10,000 amp I.C., 120 volt, 1 pole									
0100 15-50 amp	1 Elec	10	.800	Ea.	27	53		80	112
0200 60 amp		8	1		23.50	66		89.50	128
0300 70 amp		8	1		47.50	66		113.50	155
0350 240 volt, 2 pole									
0400 15-50 amp	1 Elec	8	1	Ea.	63.50	66		129.50	172
0500 60 amp		7.50	1.067		63	70.50		133.50	177
0600 80-100 amp		5	1.600		154	106		260	335
0700 3 pole, 15-60 amp		6.20	1.290		193	85.50		278.50	345
0800 70 amp		5	1.600		246	106		352	435
0900 80-100 amp		3.60	2.222		287	147		434	540
1000 22,000 amp I.C., 240 volt, 2 pole, 70-225 amp		2.70	2.963		117	196		313	430
1100 3 pole, 70-225 amp		2.30	3.478		385	230		615	775
1200 14,000 amp I.C., 277 volts, 1 pole, 15-30 amp		8	1		108	66		174	220
1300 22,000 amp I.C., 480 volts, 2 pole, 70-225 amp		2.70	2.963		130	196		326	445
1400 3 pole, 70-225 amp		2.30	3.478		1,275	230		1,505	1,750
2060 Plug-in tandem, 120/240 V, 2-15 A, 1 pole		11	.727		21.50	48		69.50	98
2070 1-15 A & 1-20 A		11	.727		22	48		70	98
2080 2-20 A		11	.727		23	48		71	99
2082 Arc fault circuit interrupter, 120/240 V, 1-15 A & 1-20 A, 1 pole		11	.727		51	48		99	130

26 24 Switchboards and Panelboards

26 24 16 – Panelboards

26 24 16.20 Panelboard and Load Center Circuit Breakers	Crew	Daily Output	Labor-Hours	Unit	Material	2022 Bare Costs Labor	Equipment	Total	Total Incl O&P
5110 NEMA 1 enclosure only, 600V, 3 p, 14k AIC, 100A	1 Elec	2.20	3.636	Ea.	112	241		353	495
9000 Minimum labor/equipment charge	↓	3	2.667	Job		177		177	271

26 24 16.30 Panelboards Commercial Applications

	Crew	Daily Output	Labor-Hours	Unit	Material	2022 Bare Costs Labor	Equipment	Total	Total Incl O&P
0010 **PANELBOARDS COMMERCIAL APPLICATIONS**									
0050 NQ, w/20 amp 1 pole bolt-on circuit breakers									
0100 3 wire, 120/240 volts, 100 amp main lugs									
0150 10 circuits	1 Elec	1	8	Ea.	1,250	530		1,780	2,200
0200 14 circuits		.88	9.091		1,400	600		2,000	2,450
0250 18 circuits		.75	10.667		850	705		1,555	2,000
0300 20 circuits	↓	.65	12.308	↓	1,650	815		2,465	3,075
0600 4 wire, 120/208 volts, 100 amp main lugs, 12 circuits	1 Elec	1	8	Ea.	1,350	530		1,880	2,300
0650 16 circuits		.75	10.667		945	705		1,650	2,125
0700 20 circuits		.65	12.308		1,725	815		2,540	3,150
0750 24 circuits		.60	13.333		1,175	885		2,060	2,650
0800 30 circuits	↓	.53	15.094		1,175	1,000		2,175	2,825
0850 225 amp main lugs, 32 circuits	2 Elec	.90	17.778		1,325	1,175		2,500	3,250
0900 34 circuits		.84	19.048		2,500	1,250		3,750	4,675
0950 36 circuits		.80	20		2,575	1,325		3,900	4,850
1000 42 circuits		.68	23.529		1,600	1,550		3,150	4,150
1010 400 amp main lugs, 42 circs	↓	.68	23.529	↓	1,600	1,550		3,150	4,150
1200 NF, w/20 amp, 1 pole bolt-on circuit breakers									
1250 4 wire, 277/480 volts, 100 amp main lugs, 12 circuits	1 Elec	.88	9.091	Ea.	1,825	600		2,425	2,925
1300 20 circuits	"	.60	13.333		2,825	885		3,710	4,450
1350 225 amp main lugs, 24 circuits	2 Elec	.90	17.778		3,300	1,175		4,475	5,425
1400 30 circuits		.80	20		2,625	1,325		3,950	4,900
1448 32 circuits		4.90	3.265		4,600	216		4,816	5,375
1450 36 circuits	↓	.72	22.222	↓	4,600	1,475		6,075	7,300
1600 NQ panel, w/20 amp, 1 pole, circuit breakers									
2000 4 wire, 120/208 volts with main circuit breaker									
2050 100 amp main, 24 circuits	1 Elec	.47	17.021	Ea.	2,275	1,125		3,400	4,225
2100 30 circuits	"	.40	20		3,550	1,325		4,875	5,925
2200 225 amp main, 32 circuits	2 Elec	.72	22.222		3,825	1,475		5,300	6,450
2250 42 circuits		.56	28.571		4,475	1,900		6,375	7,825
2300 400 amp main, 42 circuits		.48	33.333		6,050	2,200		8,250	10,100
2350 600 amp main, 42 circuits	↓	.40	40	↓	9,700	2,650		12,350	14,800
2400 NF, with 20 amp, 1 pole circuit breaker									
2450 4 wire, 277/480 volts with main circuit breaker									
2500 100 amp main, 24 circuits	1 Elec	.42	19.048	Ea.	3,800	1,250		5,050	6,100
2550 30 circuits	"	.38	21.053		7,825	1,400		9,225	10,800
2600 225 amp main, 30 circuits	2 Elec	.72	22.222		9,850	1,475		11,325	13,100
2650 42 circuits	"	.56	28.571	↓	11,900	1,900		13,800	15,900
9000 Minimum labor/equipment charge	1 Elec	1	8	Job		530		530	815

For customer support on your Commercial Renovation Costs with RSMeans data, call 800.448.8182.

513

26 27 13 – Electricity Metering

26 27 13.10 Meter Centers and Sockets

26 27 13.10 Meter Centers and Sockets	Crew	Daily Output	Labor-Hours	Unit	Material	2022 Bare Costs Labor	Equipment	Total	Total Incl O&P
0010 **METER CENTERS AND SOCKETS**									
0100 Sockets, single position, 4 terminal, 100 amp	1 Elec	3.20	2.500	Ea.	71	166		237	330
0200 150 amp	"	2.30	3.478	"	59	230		289	420
0300 200 amp	1 Elec	1.90	4.211	Ea.	134	279		413	575
0400 Transformer rated, 20 amp		3.20	2.500		222	166		388	500
0500 Double position, 4 terminal, 100 amp		2.80	2.857		295	189		484	615
0600 150 amp		2.10	3.810		345	252		597	765
0700 200 amp		1.70	4.706		625	310		935	1,175
1100 Meter centers and sockets, three phase, single pos, 7 terminal, 100 amp		2.80	2.857		138	189		327	445
1200 200 amp		2.10	3.810		279	252		531	690
1400 400 amp		1.70	4.706		2,825	310		3,135	3,575
9000 Minimum labor/equipment charge		3	2.667	Job		177		177	271

26 27 16 – Electrical Cabinets and Enclosures

26 27 16.10 Cabinets

	Crew	Daily Output	Labor-Hours	Unit	Material	2022 Bare Costs Labor	Equipment	Total	Total Incl O&P
0010 **CABINETS**									
7000 Cabinets, current transformer									
7050 Single door, 24" H x 24" W x 10" D	1 Elec	1.60	5	Ea.	143	330		473	670
7100 30" H x 24" W x 10" D		1.30	6.154		159	405		564	800
7150 36" H x 24" W x 10" D		1.10	7.273		500	480		980	1,300
7200 30" H x 30" W x 10" D		1	8		270	530		800	1,100
7250 36" H x 30" W x 10" D		.90	8.889		345	590		935	1,275
7300 36" H x 36" W x 10" D		.80	10		375	660		1,035	1,425
7500 Double door, 48" H x 36" W x 10" D		.60	13.333		670	885		1,555	2,075
7550 24" H x 24" W x 12" D		1	8		200	530		730	1,025
9990 Minimum labor/equipment charge		2	4	Job		265		265	405

26 27 23 – Indoor Service Poles

26 27 23.40 Surface Raceway

	Crew	Daily Output	Labor-Hours	Unit	Material	2022 Bare Costs Labor	Equipment	Total	Total Incl O&P
0010 **SURFACE RACEWAY**									
0090 Metal, straight section									
0100 No. 500	1 Elec	100	.080	L.F.	1.22	5.30		6.52	9.50
0110 No. 700		100	.080		1.35	5.30		6.65	9.65
0400 No. 1500, small pancake		90	.089		2.62	5.90		8.52	11.95
0600 No. 2000, base & cover, blank		90	.089		2.68	5.90		8.58	12
0800 No. 3000, base & cover, blank		75	.107		5.15	7.05		12.20	16.50
1000 No. 4000, base & cover, blank		65	.123		8.20	8.15		16.35	21.50
1200 No. 6000, base & cover, blank		50	.160		13.70	10.60		24.30	31.50
2400 Fittings, elbows, No. 500		40	.200	Ea.	2.24	13.25		15.49	23
2800 Elbow cover, No. 2000		40	.200		4.32	13.25		17.57	25.50
2880 Tee, No. 500		42	.190		4.28	12.60		16.88	24
2900 No. 2000		27	.296		14.40	19.60		34	46
3000 Switch box, No. 500		16	.500		12.85	33		45.85	65
3400 Telephone outlet, No. 1500		16	.500		18.20	33		51.20	71
3600 Junction box, No. 1500	1 Elec	16	.500	Ea.	11.50	33		44.50	63.50
3800 Plugmold wired sections, No. 2000									
4000 1 circuit, 6 outlets, 3' long	1 Elec	8	1	Ea.	43.50	66		109.50	150
4100 2 circuits, 8 outlets, 6' long	"	5.30	1.509	"	76	100		176	238
9300 Non-metallic, straight section									
9310 7/16" x 7/8", base & cover, blank	1 Elec	160	.050	L.F.	2.33	3.31		5.64	7.65
9320 Base & cover w/adhesive		160	.050		1.60	3.31		4.91	6.85
9340 7/16" x 1-5/16", base & cover, blank		145	.055		2.56	3.65		6.21	8.40
9350 Base & cover w/adhesive		145	.055		2.44	3.65		6.09	8.30
9370 11/16" x 2-1/4", base & cover, blank		130	.062		3.10	4.07		7.17	9.65

26 27 Low-Voltage Distribution Equipment

26 27 23 – Indoor Service Poles

26 27 23.40 Surface Raceway

		Crew	Daily Output	Labor-Hours	Unit	Material	2022 Bare Costs Labor	Equipment	Total	Total Incl O&P
9380	Base & cover w/adhesive	1 Elec	130	.062	L.F.	3.40	4.07		7.47	10
9385	1-11/16" x 5-1/4", two compartment base & cover w/screws		80	.100	↓	11.30	6.60		17.90	22.50
9400	Fittings, elbows, 7/16" x 7/8"		50	.160	Ea.	2.13	10.60		12.73	18.60
9410	7/16" x 1-5/16"		45	.178		2.20	11.75		13.95	20.50
9420	11/16" x 2-1/4"		40	.200		2.39	13.25		15.64	23
9425	1-11/16" x 5-1/4"		28	.286		12.55	18.90		31.45	43
9430	Tees, 7/16" x 7/8"		35	.229		2.75	15.15		17.90	26.50
9440	7/16" x 1-5/16"		32	.250		2.82	16.55		19.37	28.50
9450	11/16" x 2-1/4"		30	.267		2.90	17.65		20.55	30
9455	1-11/16" x 5-1/4"		24	.333		20	22		42	56
9460	Cover clip, 7/16" x 7/8"		80	.100		.56	6.60		7.16	10.75
9470	7/16" x 1-5/16"		72	.111		.51	7.35		7.86	11.85
9480	11/16" x 2-1/4"		64	.125		.85	8.30		9.15	13.65
9484	1-11/16" x 5-1/4"		42	.190		2.97	12.60		15.57	22.50
9486	Wire clip, 1-11/16" x 5-1/4"		68	.118		.52	7.80		8.32	12.50
9490	Blank end, 7/16" x 7/8"		50	.160		.77	10.60		11.37	17.10
9500	7/16" x 1-5/16"		45	.178		.89	11.75		12.64	19.10
9510	11/16" x 2-1/4"		40	.200		1.32	13.25		14.57	22
9515	1-11/16" x 5-1/4"		38	.211		6.15	13.95		20.10	28.50
9520	Round fixture box, 5.5" diam. x 1"		25	.320		12.55	21		33.55	46.50
9530	Device box, 1 gang		30	.267		5.70	17.65		23.35	33.50
9540	2 gang		25	.320	↓	8.25	21		29.25	41.50
9990	Minimum labor/equipment charge		5	1.600	Job		106		106	163

26 27 26 – Wiring Devices

26 27 26.20 Wiring Devices Elements

		Crew	Daily Output	Labor-Hours	Unit	Material	2022 Bare Costs Labor	Equipment	Total	Total Incl O&P
0010	**WIRING DEVICES ELEMENTS**									
0200	Toggle switch, quiet type, single pole, 15 amp	1 Elec	40	.200	Ea.	2.15	13.25		15.40	23
0500	20 amp	"	27	.296	"	4.03	19.60		23.63	34.50
0550	Rocker, 15 amp	1 Elec	40	.200	Ea.	3.36	13.25		16.61	24
0560	20 amp		27	.296		6.40	19.60		26	37
0600	3 way, 15 amp		23	.348		1.56	23		24.56	37
0850	Rocker, 15 amp		23	.348		9.75	23		32.75	46
0860	20 amp		18	.444		8.85	29.50		38.35	55
0900	4 way, 15 amp		15	.533		12.95	35.50		48.45	68.50
1030	Rocker, 15 amp		15	.533		11.35	35.50		46.85	66.50
1040	20 amp		11	.727		20.50	48		68.50	96.50
1650	Dimmer switch, 120 volt, incandescent, 600 watt, 1 pole [G]		16	.500		25.50	33		58.50	79
2460	Receptacle, duplex, 120 volt, grounded, 15 amp		40	.200		1.48	13.25		14.73	22
2470	20 amp		27	.296		13.10	19.60		32.70	44.50
2480	Ground fault interrupting, 15 amp		27	.296		18.50	19.60		38.10	50.50
2482	20 amp		27	.296		30	19.60		49.60	63
2490	Dryer, 30 amp		15	.533		4.77	35.50		40.27	59.50
2500	Range, 50 amp		11	.727		11.60	48		59.60	87
2600	Wall plates, stainless steel, 1 gang		80	.100		1.49	6.60		8.09	11.80
2800	2 gang		53	.151		14.80	10		24.80	31.50
3200	Lampholder, keyless		26	.308		23	20.50		43.50	57
3400	Pullchain with receptacle		22	.364		24.50	24		48.50	63.50
4870	Receptacle box assembly, cast aluminum, 60A, 4P5W, 3P, 120/208V, NEMA PR4		22	.364		840	24		864	960
4875	100A, 4P5W, 3P, 347/600V, NEMA PR4		22	.364		930	24		954	1,050
4880	200A, 4P5W, 3P, 347/600V, NEMA PR4		22	.364		2,025	24		2,049	2,250
8100	Pin/sleeve, 20A, 480V, DSN1, male inlet, 3 pole, 3 phase, 4 wire		16	.500		52	33		85	109
8110	20A, 125V, DSN20, male inlet, 3 pole, 3 phase, 3 wire		16	.500		47.50	33		80.50	104

26 27 26 – Wiring Devices

26 27 26.20 Wiring Devices Elements

		Crew	Daily Output	Labor-Hours	Unit	Material	2022 Bare Costs Labor	Equipment	Total	Total Incl O&P
8120	60A, 480V, DS60, male inlet, 3 pole, 3 phase, 3 wire	1 Elec	16	.500	Ea.	395	33		428	485
8130	Pin/sleeve, 20A, 480V, DSN10, female RCPT, 3P, 3 phase, 4 wire		16	.500		105	33		138	166
8140	30A, 480V, DS30, female RCPT, 3P, 3 phase, 4 wire		16	.500		365	33		398	450
8500	Wiring device terminal strip, 2 pole, 12-2AWG, 300VAC, CSA, 600VAC, 10A		36	.222		3.98	14.70		18.68	27
8505	4 pole, 12-2AWG, 300VAC, CSA, 600VAC, 10A		36	.222		5.60	14.70		20.30	28.50
8510	6 pole, 12-2AWG, 300VAC, CSA, 600VAC, 10A		36	.222		6.35	14.70		21.05	29.50
8515	8 pole, 12-2AWG, 300VAC, CSA, 600VAC, 10A		30	.267		7.10	17.65		24.75	35
8520	10 pole, 12-2AWG, 300VAC, CSA, 600VAC, 10A		30	.267		7.80	17.65		25.45	35.50
8525	12 pole, 12-2AWG, 300VAC, CSA, 600VAC, 10A		30	.267		8.60	17.65		26.25	36.50
8530	Wiring device terminal strip jumper, 10A		36	.222		.98	14.70		15.68	23.50
8535	Wiring device terminal strip, 2 pole, 12-22AWG, 300VAC, CSA, 600VAC, 20A		34	.235		4.32	15.60		19.92	29
8540	4 pole, 12-22AWG, 300VAC, CSA, 600VAC, 20A	1 Elec	34	.235	Ea.	5.90	15.60		21.50	30.50
8545	6 pole, 12-22AWG, 300VAC, CSA, 600VAC, 20A		34	.235		6.70	15.60		22.30	31.50
8550	8 pole, 12-22AWG, 300VAC, CSA, 600VAC, 20A		30	.267		7.45	17.65		25.10	35
8555	10 pole, 12-22AWG, 300VAC, CSA, 600VAC, 20A		30	.267		8.30	17.65		25.95	36
8560	12 pole, 12-22AWG, 300VAC, CSA, 600VAC, 20A		30	.267		9.10	17.65		26.75	37
8565	Wiring device terminal strip jumper, 20A		34	.235		1.21	15.60		16.81	25.50
8570	Wiring device terminal shorting type no cover, 4 pole, 6-10AWG, 600VAC, 50A		30	.267		23	17.65		40.65	52.50
8575	6 pole, 6-10AWG, 600VAC, 50A		30	.267		29.50	17.65		47.15	59.50
8580	8 pole, 6-10AWG, 600VAC, 50A		30	.267		34	17.65		51.65	64.50
8585	12 pole, 6-10AWG, 600VAC, 50A		30	.267		37.50	17.65		55.15	68.50
8590	Wiring device terminal 2 shorting pin 1/2 cover, 2 pole, 8-4AWG, 600V, 45A		30	.267		24	17.65		41.65	53.50
8595	4 pole, 8-4AWG, 600V, 45A		30	.267		30	17.65		47.65	60
8600	6 pole, 8-4AWG, 600V, 45A		30	.267		37	17.65		54.65	67.50
8605	8 pole, 8-4AWG, 600V, 45A		30	.267		53.50	17.65		71.15	86
8610	12 pole, 8-4AWG, 600V, 45A		30	.267		64	17.65		81.65	97.50
8615	Wiring device terminal 2 shorting pin W/cover, 4 pole, 8-4AWG, 600V, 45A		30	.267		40.50	17.65		58.15	71.50
8620	6 pole, 8-4AWG, 600V, 45A		30	.267		48	17.65		65.65	80
8625	8 pole, 8-4AWG, 600V, 45A		30	.267		64.50	17.65		82.15	98
8630	12 pole, 8-4AWG, 600V, 45A		30	.267		77	17.65		94.65	112
9000	Minimum labor/equipment charge		4	2	Job		132		132	203

26 27 33 – Power Distribution Units

26 27 33.10 Power Distribution Unit Cabinet

		Crew	Daily Output	Labor-Hours	Unit	Material	2022 Bare Costs Labor	Equipment	Total	Total Incl O&P
0010	**POWER DISTRIBUTION UNIT CABINET**									
0100	Power distribution unit, single cabinet, 50 kVA output, 480/208 input	1 Elec	2	4	Ea.	9,950	265		10,215	11,300
0110	75 kVA output, 480/208 input		2	4		10,800	265		11,065	12,300
0120	100 kVA output, 480/208 input		2	4		11,800	265		12,065	13,400
0130	125 kVA output, 480/208 input		1.50	5.333		12,900	355		13,255	14,700
0140	150 kVA output, 480/208 input		1.50	5.333		15,700	355		16,055	17,800
0150	200 kVA output, 480/208 input		1	8		17,300	530		17,830	19,800
0160	225 kVA output, 480/208 input		1	8		18,300	530		18,830	20,900

26 27 73 – Door Chimes

26 27 73.10 Doorbell System

		Crew	Daily Output	Labor-Hours	Unit	Material	2022 Bare Costs Labor	Equipment	Total	Total Incl O&P
0010	**DOORBELL SYSTEM**, incl. transformer, button & signal									
0100	6" bell	1 Elec	4	2	Ea.	165	132		297	385
0200	Buzzer		4	2		132	132		264	350
1000	Door chimes, 2 notes		16	.500		32.50	33		65.50	87
1020	with ambient light		12	.667		120	44		164	200
1100	Tube type, 3 tube system	1 Elec	12	.667	Ea.	237	44		281	330
1180	4 tube system		10	.800		475	53		528	600
1900	For transformer & button, add		5	1.600		17.25	106		123.25	182
3000	For push button only		24	.333		.99	22		22.99	35

For customer support on your Commercial Renovation Costs with RSMeans data, call 800.448.8182.

26 27 Low-Voltage Distribution Equipment

26 27 73 – Door Chimes

26 27 73.10 Doorbell System	Crew	Daily Output	Labor-Hours	Unit	Material	2022 Bare Costs Labor	Equipment	Total	Total Incl O&P
3200	Bell transformer	1 Elec	16	.500	Ea.	28.50	33		61.50	82.50
9000	Minimum labor/equipment charge		4	2	Job		132		132	203

26 28 Low-Voltage Circuit Protective Devices

26 28 16 – Enclosed Switches and Circuit Breakers

26 28 16.10 Circuit Breakers

		Crew	Daily Output	Labor-Hours	Unit	Material	2022 Bare Costs Labor	Equipment	Total	Total Incl O&P
0010	**CIRCUIT BREAKERS** (in enclosure)									
0100	Enclosed (NEMA 1), 600 volt, 3 pole, 30 amp	1 Elec	3.20	2.500	Ea.	645	166		811	965
0200	60 amp		2.80	2.057		645	189		834	1,000
0400	100 amp		2.30	3.478		760	230		990	1,200
0500	200 amp		1.50	5.333		1,150	355		1,505	1,825
0600	225 amp		1.50	5.333		1,150	355		1,505	1,825
0700	400 amp	2 Elec	1.60	10		4,525	660		5,185	6,025
9000	Minimum labor/equipment charge	1 Elec	4	2	Job		132		132	203

26 28 16.13 Circuit Breakers

		Crew	Daily Output	Labor-Hours	Unit	Material	2022 Bare Costs Labor	Equipment	Total	Total Incl O&P
0010	**CIRCUIT BREAKERS**									
0100	Circuit breaker current limiter, 225 to 400, ampere, 1 pole	1 Elec	48	.167	Ea.	845	11.05		856.05	940
0200	Circuit breaker current limiter, 200 to 500, ampere, 1 pole, thermal magnet		46	.174		510	11.50		521.50	585
0300	300 to 500, ampere, 1 pole, thermal magnet		46	.174		1,425	11.50		1,436.50	1,600
0400	Circuit breaker current limiter, 600 to 1000, ampere, 1 pole, tri pac		46	.174		1,850	11.50		1,861.50	2,050

26 28 16.20 Safety Switches

		Crew	Daily Output	Labor-Hours	Unit	Material	2022 Bare Costs Labor	Equipment	Total	Total Incl O&P
0010	**SAFETY SWITCHES**									
0100	General duty 240 volt, 3 pole NEMA 1, fusible, 30 amp	1 Elec	3.20	2.500	Ea.	65	166		231	325
0200	60 amp		2.30	3.478		110	230		340	475
0300	100 amp		1.90	4.211		188	279		467	635
0400	200 amp		1.30	6.154		410	405		815	1,075
0500	400 amp	2 Elec	1.80	8.889		995	590		1,585	2,000
1100	Heavy duty, 600 volt, 3 pole NEMA 1 nonfused									
1110	30 amp	1 Elec	3.20	2.500	Ea.	94.50	166		260.50	360
1500	60 amp		2.30	3.478		163	230		393	535
1700	100 amp		1.90	4.211		262	279		541	720
1900	200 amp		1.30	6.154		395	405		800	1,050
2100	400 amp	2 Elec	1.80	8.889		900	590		1,490	1,900
2300	600 amp		1.20	13.333		1,700	885		2,585	3,225
2500	800 amp		.94	17.021		3,175	1,125		4,300	5,225
2700	1,200 amp		.80	20		4,500	1,325		5,825	6,975
4350	600 volt, 3 pole, fusible, 30 amp	1 Elec	3.20	2.500		175	166		341	445
4380	60 amp		2.30	3.478		213	230		443	590
4400	100 amp		1.90	4.211		390	279		669	855
4420	200 amp		1.30	6.154		555	405		960	1,225
4440	400 amp	2 Elec	1.80	8.889		1,425	590		2,015	2,475
4450	600 amp		1.20	13.333		2,600	885		3,485	4,225
4460	800 amp		.94	17.021		4,950	1,125		6,075	7,175
4480	1,200 amp		.80	20		7,150	1,325		8,475	9,875
5510	Heavy duty, 600 volt, 3 pole 3 ph. NEMA 3R fusible, 30 amp	1 Elec	3.10	2.581		300	171		471	590
5520	60 amp		2.20	3.636		380	241		621	785
5530	100 amp		1.80	4.444		535	294		829	1,025
5540	200 amp		1.20	6.667		755	440		1,195	1,525
5550	400 amp	2 Elec	1.60	10		1,725	660		2,385	2,925
5700	600 volt, 3 pole NEMA 3R nonfused									

For customer support on your Commercial Renovation Costs with RSMeans data, call 800.448.8182.

517

26 28 16 – Enclosed Switches and Circuit Breakers

26 28 16.20 Safety Switches		Crew	Daily Output	Labor-Hours	Unit	Material	2022 Bare Costs Labor	Equipment	Total	Total Incl O&P
5710	30 amp	1 Elec	3.10	2.581	Ea.	170	171		341	450
5900	60 amp		2.20	3.636		297	241		538	695
6100	100 amp		1.80	4.444		420	294		714	910
6300	200 amp		1.20	6.667		490	440		930	1,225
6500	400 amp	2 Elec	1.60	10		1,225	660		1,885	2,375
6700	600 amp	"	1	16		2,475	1,050		3,525	4,350
6900	600 volt, 6 pole NEMA 3R nonfused, 30 amp	1 Elec	2.70	2.963		1,150	196		1,346	1,575
7100	60 amp		2	4		1,450	265		1,715	1,975
7300	100 amp		1.50	5.333		1,900	355		2,255	2,625
7500	200 amp		1.20	6.667		4,025	440		4,465	5,125
9990	Minimum labor/equipment charge		3	2.667	Job		177		177	271

26 28 16.40 Time Switches

		Crew	Daily Output	Labor-Hours	Unit	Material	2022 Bare Costs Labor	Equipment	Total	Total Incl O&P
0010	**TIME SWITCHES**									
0100	Single pole, single throw, 24 hour dial	1 Elec	4	2	Ea.	143	132		275	360
0200	24 hour dial with reserve power		3.60	2.222		770	147		917	1,075
0300	Astronomic dial		3.60	2.222		268	147		415	520
0400	Astronomic dial with reserve power		3.30	2.424		1,000	160		1,160	1,350
0500	7 day calendar dial		3.30	2.424		223	160		383	490
0600	7 day calendar dial with reserve power		3.20	2.500		191	166		357	465
0700	Photo cell 2,000 watt		8	1		23	66		89	128
1080	Load management device, 4 loads		2	4		1,200	265		1,465	1,725
1100	8 loads		1	8		2,700	530		3,230	3,800
9000	Minimum labor/equipment charge		3.50	2.286	Job		151		151	232

26 28 16.50 Meter Socket Entry Hub

		Crew	Daily Output	Labor-Hours	Unit	Material	2022 Bare Costs Labor	Equipment	Total	Total Incl O&P
0010	**METER SOCKET ENTRY HUB**									
0100	Meter socket entry hub closing cap, 3/4 to 4 inch	1 Elec	40	.200	Ea.	2.46	13.25		15.71	23
0110	Meter socket entry hub, 3/4 conduit		40	.200		14.65	13.25		27.90	36.50
0120	1 inch conduit		40	.200		15.65	13.25		28.90	37.50
0130	1.25 inch conduit		40	.200		14.65	13.25		27.90	36.50
0140	1.50 inch conduit		40	.200		14.50	13.25		27.75	36.50
0150	2 inch conduit		40	.200		28	13.25		41.25	51.50
0160	2.50 inch conduit		40	.200		41.50	13.25		54.75	66.50
0170	3 inch conduit		40	.200		72	13.25		85.25	99.50

26 29 Low-Voltage Controllers

26 29 13 – Enclosed Controllers

26 29 13.40 Relays

		Crew	Daily Output	Labor-Hours	Unit	Material	2022 Bare Costs Labor	Equipment	Total	Total Incl O&P
0010	**RELAYS** Enclosed (NEMA 1)									
0050	600 volt AC, 1 pole, 12 amp	1 Elec	5.30	1.509	Ea.	117	100		217	282
0100	2 pole, 12 amp		5	1.600		145	106		251	320
0200	4 pole, 10 amp		4.50	1.778		193	118		311	395

26 29 23 – Variable-Frequency Motor Controllers

26 29 23.10 Variable Frequency Drives/Adj. Frequency Drives

			Crew	Daily Output	Labor-Hours	Unit	Material	2022 Bare Costs Labor	Equipment	Total	Total Incl O&P
0010	**VARIABLE FREQUENCY DRIVES/ADJ. FREQUENCY DRIVES**										
0100	Enclosed (NEMA 1), 460 volt, for 3 HP motor size	G	1 Elec	.80	10	Ea.	1,875	660		2,535	3,075
0110	5 HP motor size	G		.80	10		2,125	660		2,785	3,375
0120	7.5 HP motor size	G		.67	11.940		2,575	790		3,365	4,075
0130	10 HP motor size	G		.67	11.940		2,825	790		3,615	4,350
0140	15 HP motor size	G	2 Elec	.89	17.978		3,650	1,200		4,850	5,850
0150	20 HP motor size	G		.89	17.978		4,600	1,200		5,800	6,875

26 29 Low-Voltage Controllers

26 29 23 – Variable-Frequency Motor Controllers

26 29 23.10 Variable Frequency Drives/Adj. Frequency Drives		Crew	Daily Output	Labor-Hours	Unit	Material	2022 Bare Costs Labor	Equipment	Total	Total Incl O&P
0160	25 HP motor size	G 2 Elec	.67	23.881	Ea.	5,450	1,575		7,025	8,425
0170	30 HP motor size	G	.67	23.881		6,625	1,575		8,200	9,700
0180	40 HP motor size	G	.67	23.881		7,650	1,575		9,225	10,900
0190	50 HP motor size	G	.53	30.189		10,100	2,000		12,100	14,200
0200	60 HP motor size	G R-3	.56	35.714		12,600	2,350	310	15,260	17,900
0210	75 HP motor size	G	.56	35.714		14,600	2,350	310	17,260	20,100
0220	100 HP motor size	G	.50	40		17,600	2,625	345	20,570	23,800
0230	125 HP motor size	G	.50	40		19,200	2,625	345	22,170	25,500
0240	150 HP motor size	G	.50	40		25,000	2,625	345	27,970	31,900
0250	200 HP motor size	G	.42	47.619		35,400	3,125	415	38,940	44,200
1100	Custom-engineered, 460 volt, for 3 HP motor size	G 1 Elec	.56	14.286		4,625	945		5,570	6,525
1110	5 HP motor size	G	.56	14.286		4,675	945		5,620	6,600
1120	7.5 HP motor size	G	.47	17.021		3,950	1,125		5,075	6,075
1130	10 HP motor size	G	.47	17.021		4,200	1,125		5,325	6,350
1140	15 HP motor size	G 2 Elec	.62	25.806		6,350	1,700		8,050	9,625
1150	20 HP motor size	G	.62	25.806		3,425	1,700		5,125	6,400
1160	25 HP motor size	G	.47	34.043		7,325	2,250		9,575	11,500
1170	30 HP motor size	G	.47	34.043		8,500	2,250		10,750	12,800
1180	40 HP motor size	G 2 Elec	.47	34.043	Ea.	9,575	2,250		11,825	14,000
1190	50 HP motor size	G "	.37	43.243		14,200	2,875		17,075	20,000
1200	60 HP motor size	G R-3	.39	51.282		21,300	3,375	445	25,120	29,100
1210	75 HP motor size	G	.39	51.282		17,800	3,375	445	21,620	25,300
1220	100 HP motor size	G	.35	57.143		25,000	3,750	495	29,245	33,800
1230	125 HP motor size	G	.35	57.143		26,800	3,750	495	31,045	35,700
1240	150 HP motor size	G	.35	57.143		30,300	3,750	495	34,545	39,600
1250	200 HP motor size	G	.29	68.966		34,800	4,550	600	39,950	46,000
2000	For complex & special design systems to meet specific									
2010	requirements, obtain quote from vendor.									

26 32 Packaged Generator Assemblies

26 32 13 – Engine Generators

26 32 13.16 Gas-Engine-Driven Generator Sets

		Crew	Daily Output	Labor-Hours	Unit	Material	Labor	Equipment	Total	Total Incl O&P
0010	**GAS-ENGINE-DRIVEN GENERATOR SETS**									
0020	Gas or gasoline operated, includes battery,									
0050	charger & muffler									
0200	7.5 kW	R-3	.83	24.096	Ea.	7,100	1,575	209	8,884	10,500
0300	11.5 kW		.71	28.169		10,100	1,850	244	12,194	14,200
0400	20 kW		.63	31.746		10,600	2,100	275	12,975	15,200
0500	35 kW		.55	36.364		15,600	2,400	315	18,315	21,200

26 33 Battery Equipment

26 33 19 – Battery Units

26 33 19.10 Battery Units

		Crew	Daily Output	Labor-Hours	Unit	Material	2022 Bare Costs Labor	2022 Bare Costs Equipment	Total	Total Incl O&P
0010	**BATTERY UNITS**									
0500	Salt water battery, 2.1 kWA, 24V, 750W, 30A, wired in crate, IP22 rated	1 Elec	16.25	.492	Ea.	920	32.50		952.50	1,075
0510	2.2 kWh, 48V, 800W, 17A, wired in crate, IP22 rated		16.25	.492		1,150	32.50		1,182.50	1,325
0520	25.9 kWh, 48V, 11700W, 240A, wired in crate, IP2X rated		16.25	.492		13,800	32.50		13,832.50	15,300
0700	Nickel iron nife battery, deep cycle, 100Ah, 12V, 10 cells		14.25	.561		990	37		1,027	1,125
0710	24V, 10 cells		14.25	.561		2,100	37		2,137	2,375
0720	48V, 10 cells		14.25	.561		3,950	37		3,987	4,400
0730	Nickel iron nife battery, deep cycle, 300Ah, 12V, 10 cells		12.25	.653		2,900	43		2,943	3,275
0740	24V, 10 cells		12.25	.653		5,825	43		5,868	6,475
0750	48V, 10 cells	↓	12.25	.653	↓	12,200	43		12,243	13,500

26 51 Interior Lighting

26 51 13 – Interior Lighting Fixtures, Lamps, and Ballasts

26 51 13.50 Interior Lighting Fixtures

		Crew	Daily Output	Labor-Hours	Unit	Material	2022 Bare Costs Labor	2022 Bare Costs Equipment	Total	Total Incl O&P
0010	**INTERIOR LIGHTING FIXTURES** Including lamps, mounting									
0030	hardware and connections									
0100	Fluorescent, C.W. lamps, troffer, recess mounted in grid, RS									
0130	Grid ceiling mount									
0200	Acrylic lens, 1' W x 4' L, two 40 watt	1 Elec	5.70	1.404	Ea.	55	93		148	203
0300	2' W x 2' L, two U40 watt		5.70	1.404		145	93		238	300
0600	2' W x 4' L, four 40 watt		4.70	1.702		66.50	113		179.50	246
0910	Acrylic lens, 1' W x 4' L, two 32 watt T8 [G]		5.70	1.404		111	93		204	266
0930	2' W x 2' L, two U32 watt T8 [G]		5.70	1.404		147	93		240	305
0940	2' W x 4' L, two 32 watt T8 [G]		5.30	1.509		102	100		202	266
0950	2' W x 4' L, three 32 watt T8 [G]		5	1.600		93.50	106		199.50	266
0960	2' W x 4' L, four 32 watt T8 [G]	↓	4.70	1.702	↓	156	113		269	345
1000	Surface mounted, RS									
1030	Acrylic lens with hinged & latched door frame									
1100	1' W x 4' L, two 40 watt	1 Elec	7	1.143	Ea.	71.50	75.50		147	195
1200	2' W x 2' L, two U40 watt		7	1.143		77	75.50		152.50	201
1500	2' W x 4' L, four 40 watt		5.30	1.509		89	100		189	252
1501	2' W x 4' L, six 40 watt T8	↓	5.20	1.538	↓	89	102		191	254
2100	Strip fixture									
2130	Surface mounted									
2200	4' long, one 40 watt, RS	1 Elec	8.50	.941	Ea.	108	62.50		170.50	215
2300	4' long, two 40 watt, RS	"	8	1		143	66		209	259
2600	8' long, one 75 watt, SL	2 Elec	13.40	1.194		106	79		185	238
2700	8' long, two 75 watt, SL	"	12.40	1.290	↓	118	85.50		203.50	260
3000	Strip, pendent mounted, industrial, white porcelain enamel									
3100	4' long, two 40 watt, RS	1 Elec	5.70	1.404	Ea.	54	93		147	203
3200	4' long, two 60 watt, HO	"	5	1.600		83.50	106		189.50	255
3300	8' long, two 75 watt, SL	2 Elec	8.80	1.818	↓	102	120		222	297
4220	Metal halide, integral ballast, ceiling, recess mounted									
4230	prismatic glass lens, floating door									
4240	2' W x 2' L, 250 watt	1 Elec	3.20	2.500	Ea.	375	166		541	665
4250	2' W x 2' L, 400 watt	2 Elec	5.80	2.759		380	183		563	700
4260	Surface mounted, 2' W x 2' L, 250 watt	1 Elec	2.70	2.963		345	196		541	680
4270	400 watt	2 Elec	4.80	3.333		410	221		631	790
4280	High bay, aluminum reflector,									
4290	Single unit, 400 watt	2 Elec	4.60	3.478	Ea.	430	230		660	825
4300	Single unit, 1,000 watt	↓	4	4	↓	820	265		1,085	1,300

26 51 Interior Lighting

26 51 13 – Interior Lighting Fixtures, Lamps, and Ballasts

26 51 13.50 Interior Lighting Fixtures

		Crew	Daily Output	Labor-Hours	Unit	Material	2022 Bare Costs Labor	Equipment	Total	Total Incl O&P
4310	Twin unit, 400 watt	2 Elec	3.20	5	Ea.	830	330		1,160	1,425
4320	Low bay, aluminum reflector, 250W DX lamp	1 Elec	3.20	2.500	↓	365	166		531	660
4340	High pressure sodium integral ballast ceiling, recess mounted									
4350	prismatic glass lens, floating door									
4360	2' W x 2' L, 150 watt lamp	1 Elec	3.20	2.500	Ea.	400	166		566	700
4370	2' W x 2' L, 400 watt lamp	2 Elec	5.80	2.759		480	183		663	810
4380	Surface mounted, 2' W x 2' L, 150 watt lamp	1 Elec	2.70	2.963		490	196		686	840
4390	400 watt lamp	2 Elec	4.80	3.333	↓	550	221		771	945
4400	High bay, aluminum reflector,									
4410	Single unit, 400 watt lamp	2 Elec	4.60	3.478	Ea.	395	230		625	790
4430	Single unit, 1,000 watt lamp	"	4	4	↓	555	265		820	1,025
4440	Low bay, aluminum reflector, 150 watt lamp	1 Elec	3.20	2.500	↓	340	166		506	630
4450	Incandescent, high hat can, round alzak reflector, prewired									
4470	100 watt	1 Elec	8	1	Ea.	69	66		135	178
4480	150 watt		8	1		105	66		171	218
4500	300 watt	↓	6.70	1.194	↓	244	79		323	390
4600	Square glass lens with metal trim, prewired									
4630	100 watt	1 Elec	6.70	1.194	Ea.	55.50	79		134.50	182
4680	150 watt		6.70	1.194		99	79		178	230
4700	200 watt		6.70	1.194		99	79		178	230
6010	Vapor tight, incandescent, ceiling mounted, 200 watt		6.20	1.290		71	85.50		156.50	209
6100	Fluorescent, surface mounted, 2 lamps, 4' L, RS, 40 watt		3.20	2.500		121	166		287	385
6850	Vandalproof, surface mounted, fluorescent, two 32 watt T8 ⒼG		3.20	2.500		271	166		437	550
6860	Incandescent, one 150 watt		8	1		101	66		167	213
6900	Mirror light, fluorescent, RS, acrylic enclosure, two 40 watt		8	1		118	66		184	232
6910	One 40 watt		8	1		102	66		168	214
6920	One 20 watt	↓	12	.667	↓	88.50	44		132.50	165
7500	Ballast replacement, by weight of ballast, to 15' high									
7520	Indoor fluorescent, less than 2 lb.	1 Elec	10	.800	Ea.	27.50	53		80.50	112
7540	Two 40W, watt reducer, 2 to 5 lb.		9.40	.851		69.50	56.50		126	163
7560	Two F96 slimline, over 5 lb.		8	1		110	66		176	223
7580	Vaportite ballast, less than 2 lb.		9.40	.851		27.50	56.50		84	117
7600	2 lb. to 5 lb.		8.90	.899		69.50	59.50		129	168
7620	Over 5 lb.		7.60	1.053		110	69.50		179.50	228
7630	Electronic ballast for two tubes		8	1		38	66		104	144
7640	Dimmable ballast one-lamp ⒼG		8	1		107	66		173	219
7650	Dimmable ballast two-lamp ⒼG	↓	7.60	1.053	↓	112	69.50		181.50	230
7660	Dimmable ballast three-lamp	1 Elec	6.90	1.159	Ea.	128	77		205	258
9000	Minimum labor/equipment charge	"	3	2.667	Job		177		177	271

26 51 13.70 Residential Fixtures

		Crew	Daily Output	Labor-Hours	Unit	Material	2022 Bare Costs Labor	Equipment	Total	Total Incl O&P
0010	**RESIDENTIAL FIXTURES**									
0400	Fluorescent, interior, surface, circline, 32 watt & 40 watt	1 Elec	20	.400	Ea.	183	26.50		209.50	242
0500	2' x 2', two U-tube 32 watt T8		8	1		127	66		193	242
0700	Shallow under cabinet, two 20 watt		16	.500		83	33		116	143
0900	Wall mounted, 4' L, two 32 watt T8, with baffle		10	.800		164	53		217	263
2000	Incandescent, exterior lantern, wall mounted, 60 watt		16	.500		61	33		94	118
2100	Post light, 150 W, with 7' post		4	2		330	132		462	570
2500	Lamp holder, weatherproof with 150 W PAR		16	.500		36	33		69	90.50
2550	With reflector and guard		12	.667		57.50	44		101.50	131
2600	Interior pendent, globe with shade, 150 W		20	.400	↓	182	26.50		208.50	241
9000	Minimum labor/equipment charge		4	2	Job		132		132	203

26 51 19.10 LED Interior Lighting	Crew	Daily Output	Labor-Hours	Unit	Material	2022 Bare Costs Labor	Equipment	Total	Total Incl O&P
0010 **LED INTERIOR LIGHTING**									
7000 Interior LED fixts, tape rope kit, 120V, 3' strip, 4W, w/5' cord, 3,000K	1 Elec	24	.333	Ea.	17.35	22		39.35	53
7010 5,000K		24	.333		11.05	22		33.05	46
7020 Interior LED fixts, tape rope kit, 120V, 6' strip, 8W, w/5' cord, 3,000K		23	.348		15.10	23		38.10	52
7030 5,000K		23	.348		22.50	23		45.50	60
7040 Interior LED fixts, tape rope kit, 120V, 13' strip, 15W, w/5' cord, 3,000K		22	.364		26	24		50	65.50
7050 5,000K		22	.364		28	24		52	67.50
7060 Interior LED fixts, tape rope kit, 120V, 20' strip, 22W, w/5' cord, 3,000K		21	.381		40.50	25		65.50	83
7070 5,000K		21	.381		40.50	25		65.50	83
7080 Interior LED fixts, tape rope kit, 120V, 33' strip, 37W, w/5' cord, 3,000K		20	.400		59.50	26.50		86	106
7090 5,000K		20	.400		59.50	26.50		86	106
7100 Interior LED fixts, tape rope kit, 120V, 50' strip, 55W, w/5' cord, 3,000K		19	.421		90	28		118	142
7110 5,000K		19	.421		129	28		157	185
7120 Interior LED tape light, starstrand, dimmable, 12V, super star, 2", 3,000K		25	.320		11.05	21		32.05	44.50
7130 Ultra star, 2", 3,000K		25	.320		14.10	21		35.10	48
7140 Rainbow star, 2", 3,000K		25	.320		11.30	21		32.30	45
7150 Interior LED tape light, starstrand, dimmable, 24V, elite star, 4", 35K		25	.320		5.05	21		26.05	38
7160 12", 35K		25	.320		19.35	21		40.35	54
7170 60", 35K		23	.348		54	23		77	95
7180 120", 35K		21	.381		101	25		126	150
7190 240", 35K		18	.444		310	29.50		339.50	385
7200 Interior LED tape lighting channel , 36", AL flex, starstrand, elite star	1 Elec	48	.167	Ea.	11.35	11.05		22.40	29.50

26 52 13.10 Emergency Lighting and Battery Units

	Crew	Daily Output	Labor-Hours	Unit	Material	Labor	Equipment	Total	Total Incl O&P
0010 **EMERGENCY LIGHTING AND BATTERY UNITS**									
0300 Emergency light units, battery operated									
0350 Twin sealed beam light, 25 W, 6 V each									
0500 Lead battery operated	1 Elec	4	2	Ea.	143	132		275	360
0700 Nickel cadmium battery operated		4	2	"	350	132		482	590
9000 Minimum labor/equipment charge		4	2	Job		132		132	203

26 52 13.16 Exit Signs

	Crew	Daily Output	Labor-Hours	Unit	Material	Labor	Equipment	Total	Total Incl O&P
0010 **EXIT SIGNS**									
0080 Exit light ceiling or wall mount, incandescent, single face	1 Elec	8	1	Ea.	60.50	66		126.50	169
0100 Double face	"	6.70	1.194		60.50	79		139.50	188
1500 Exit sign, 12 V, 1 face, remote end mounted (Type 602A1)	R-19	18	1.111		76.50	73.50		150	197
1780 With emergency battery, explosion proof	"	7.70	2.597		3,800	172		3,972	4,450
9000 Minimum labor/equipment charge	1 Elec	4	2	Job		132		132	203

26 55 33 – Hazard Warning Lighting

26 55 33.10 Warning Beacons	Crew	Daily Output	Labor-Hours	Unit	Material	2022 Bare Costs Labor	Equipment	Total	Total Incl O&P
0010 **WARNING BEACONS**									
0015 Surface mount with colored or clear lens									
0100 Rotating beacon									
0110 120V, 40 watt halogen	1 Elec	3.50	2.286	Ea.	235	151		386	490
0120 24V, 20 watt halogen	"	3.50	2.286	"	275	151		426	530
0200 Steady beacon									
0210 120V, 40 watt halogen	1 Elec	3.50	2.286	Ea.	267	151		418	525
0220 24V, 20 watt		3.50	2.286		275	151		426	530
0230 12V DC, incandescent	↓	3.50	2.286	↓	109	151		260	350
0300 Flashing beacon									
0310 120V, 40 watt halogen	1 Elec	3.50	2.286	Ea.	229	151		380	485
0320 24V, 20 watt halogen		3.50	2.286		254	151		405	510
0410 12V DC with two 6V lantern batteries	↓	7	1.143	↓	238	75.50		313.50	380

26 55 59 – Display Lighting

26 55 59.10 Track Lighting	Crew	Daily Output	Labor-Hours	Unit	Material	2022 Bare Costs Labor	Equipment	Total	Total Incl O&P
0010 **TRACK LIGHTING**									
0080 Track, 1 circuit, 4' section	1 Elec	6.70	1.194	Ea.	51	79		130	177
0100 8' section	2 Elec	10.60	1.509		71	100		171	232
0300 3 circuits, 4' section	1 Elec	6.70	1.194		135	79		214	270
0400 8' section	2 Elec	10.60	1.509	↓	176	100		276	350
9000 Minimum labor/equipment charge	1 Elec	3	2.667	Job		177		177	271

26 55 63 – Detention Lighting

26 55 63.10 Detention Lighting Fixtures	Crew	Daily Output	Labor-Hours	Unit	Material	2022 Bare Costs Labor	Equipment	Total	Total Incl O&P
0010 **DETENTION LIGHTING FIXTURES**									
1000 Surface mounted, cold rolled steel, 14 ga., 1' x 4'	1 Elec	5.40	1.481	Ea.	1,000	98		1,098	1,250
1010 2' x 2'		5.40	1.481		975	98		1,073	1,225
1020 2' x 4'		5	1.600		1,175	106		1,281	1,475
1100 12 ga., 1' x 4'		5.40	1.481		1,250	98		1,348	1,525
1110 2' x 2'		5.40	1.481		975	98		1,073	1,225
1120 2' x 4'		5	1.600		1,275	106		1,381	1,575
3000 Fluorescent vandal resistant light fixture, high abuse troffer		5.40	1.481		261	98		359	440
3010 50" L x 8" W x 4" H		5.40	1.481		216	98		314	390
3020 32W		5.40	1.481		330	98		428	515
3030 T8/32W		5.40	1.481		229	98		327	405
3040 No lamp		5.60	1.429		395	94.50		489.50	580
3050 Lamp included		5.40	1.481		455	98		553	650
3060 16 ga. cold rolled steel		5.40	1.481		450	98		548	650
3070 16 ga. cold rolled steel, ceiling/wall mount		5.40	1.481		380	98		478	565
3080 w/lamp, white		5.40	1.481		295	98		393	475
3090 wo/lamp, bronze		5.60	1.429		262	94.50		356.50	435
3100 w/lamp 6" L x 4-1/2" W x 8-1/2" H		5.40	1.481		42.50	98		140.50	198
3110 w/lamp 5-7/8" W x 4-1/2" D x 8-1/2" H clear		5.40	1.481		119	98		217	282
3120 w/lamp white		5.40	1.481		595	98		693	800
3130 Tall wallpack, w/lamp, bronze		5.40	1.481		139	98		237	305
3140 Tall wallpack, w/lamp, white	↓	5.40	1.481	↓	118	98		216	281

523

For customer support on your Commercial Renovation Costs with RSMeans data, call 800.448.8182.

26 56 Exterior Lighting

26 56 19 – LED Exterior Lighting

26 56 19.55 Roadway LED Luminaire

		Crew	Daily Output	Labor-Hours	Unit	Material	2022 Bare Costs Labor	Equipment	Total	Total Incl O&P
0010	**ROADWAY LED LUMINAIRE**									
2000	Exterior fixtures, LED roadway, type 2, 50W, neutral slipfitter bronze	2 Elec	5.20	3.077	Ea.	735	204		939	1,125
2010	3, 50W, neutral slipfitter bronze		5.20	3.077		730	204		934	1,125
2020	4, 50W, neutral slipfitter bronze		5.20	3.077		730	204		934	1,125
2030	Exterior fixtures, LED roadway, type 2, 78W, neutral slipfitter bronze		5.20	3.077		710	204		914	1,100
2040	3, 78W, neutral slipfitter bronze		5.20	3.077		715	204		919	1,100
2050	4, 78W, neutral slipfitter bronze		5.20	3.077		720	204		924	1,100
2060	Exterior fixtures, LED roadway, type 2, 105W, neutral slipfitter bronze		5.20	3.077		860	204		1,064	1,250
2070	3, 105W, neutral slipfitter bronze		5.20	3.077		860	204		1,064	1,250
2080	4, 105W, neutral slipfitter bronze		5.20	3.077		860	204		1,064	1,250
2090	Exterior fixtures, LED roadway, type 2, 125W, neutral slipfitter bronze		5.20	3.077		860	204		1,064	1,250

26 56 23 – Area Lighting

26 56 23.10 Exterior Fixtures

		Crew	Daily Output	Labor-Hours	Unit	Material	2022 Bare Costs Labor	Equipment	Total	Total Incl O&P
0010	**EXTERIOR FIXTURES** With lamps									
0200	Wall mounted, incandescent, 100 watt	1 Elec	8	1	Ea.	67	66		133	176
0400	Quartz, 500 watt		5.30	1.509		123	100		223	289
1100	Wall pack, low pressure sodium, 35 watt		4	2		182	132		314	405
1150	55 watt		4	2		214	132		346	440
1160	High pressure sodium, 70 watt		4	2		113	132		245	325
1170	150 watt		4	2		276	132		408	510
1175	High pressure sodium, 250 watt		4	2		276	132		408	510
1180	Metal halide, 175 watt		4	2		173	132		305	395
1190	250 watt		4	2		210	132		342	435
1195	400 watt		4	2		320	132		452	555
1250	Induction lamp, 40 watt		4	2		460	132		592	710
1260	80 watt		4	2		197	132		329	420
1278	LED, poly lens, 26 watt		4	2		145	132		277	365
1280	110 watt		4	2		765	132		897	1,050
1500	LED, glass lens, 13 watt		4	2		145	132		277	365

26 56 33 – Walkway Lighting

26 56 33.10 Walkway Luminaire

		Crew	Daily Output	Labor-Hours	Unit	Material	2022 Bare Costs Labor	Equipment	Total	Total Incl O&P
0010	**WALKWAY LUMINAIRE**									
9000	Minimum labor/equipment charge	1 Elec	3.75	2.133	Job		141		141	217

26 56 33.55 Walkway LED Luminaire

		Crew	Daily Output	Labor-Hours	Unit	Material	2022 Bare Costs Labor	Equipment	Total	Total Incl O&P
0010	**WALKWAY LED LUMINAIRE**									
0600	LED bollard pole, 6" diam. x 36" H, cast alum, surf. mtd., 8W, 6,000K, 120V	1 Elec	8.50	.941	Ea.	405	62.50		467.50	540
0610	6" diam. x 42" H, cast alum, surf. mtd., 12W, 5,100K,120V		8.50	.941		630	62.50		692.50	785
0620	6" diam. x 42" H, cast alum, surf. mtd., 18W, 5,100K,120V		8.50	.941		640	62.50		702.50	795
0630	6" diam. x 42" H, cast alum, surf. mtd., 24W, 5,100K,120V		8.50	.941		660	62.50		722.50	820
0640	9.5" square x 42" H, cast resin, surf. mtd., A19, 9W, 120V		8.50	.941		640	62.50		702.50	800
0650	9.75" diam. x 42" H, cast resin, side mount, dome, 9W, 120V		8.50	.941		730	62.50		792.50	895
0660	9.75" diam. x 43" H, cast resin, burial base, dome, 9W, 120V		8.50	.941		730	62.50		792.50	895
0670	8" diam. x 36" H, concrete security base, w/rebar, 8W, HID		8	1		1,075	66		1,141	1,300
0680	10" diam. x 49" H, concrete base, w/Sch 40 Steel pipe, 8W		6.75	1.185		1,225	78.50		1,303.50	1,475
0690	12" diam. x 48" H, concrete base, w/Sch 40 Steel pipe, 8W		6.75	1.185		1,400	78.50		1,478.50	1,675
0700	8" diam. x 36" H, concrete, cast aluminum dome, 90 degree, 4.1W		7	1.143		1,100	75.50		1,175.50	1,325
0710	8" diam. x 36" H, concrete, cast aluminum dome, 180 degree, 4.1W		7	1.143		1,325	75.50		1,400.50	1,600
0720	8" diam. x 48" H, concrete, cast aluminum dome, 360 degree, 4.1W	1 Elec	7	1.143	Ea.	1,650	75.50		1,725.50	1,925

524

For customer support on your Commercial Renovation Costs with RSMeans data, call 800.448.8182.

26 56 Exterior Lighting

26 56 36 – Flood Lighting

26 56 36.20 Floodlights	Crew	Daily Output	Labor-Hours	Unit	Material	2022 Bare Costs Labor	Equipment	Total	Total Incl O&P
0010 **FLOODLIGHTS** with ballast and lamp,									
1290 floor mtd, mount with swivel bracket									
1400 Pole mounted, pole not included									
2250 Low pressure sodium, 55 watt	1 Elec	2.70	2.963	Ea.	435	196		631	780
2270 90 watt	"	2	4	"	625	265		890	1,100

26 61 Lighting Systems and Accessories

26 61 23 – Lamps Applications

26 61 23.10 Lamps		Crew	Daily Output	Labor-Hours	Unit	Material	2022 Bare Costs Labor	Equipment	Total	Total Incl O&P
0010 **LAMPS**										
0080 Fluorescent, rapid start, cool white, 2' long, 20 watt		1 Elec	1	8	C	292	530		822	1,125
0100 4' long, 40 watt			.90	8.889		300	590		890	1,225
0170 4' long, 34 watt energy saver	G		.90	8.889		4.58	590		594.58	910
0176 2' long, T8, 17 watt energy saver	G		1	8		415	530		945	1,275
0178 3' long, T8, 25 watt energy saver	G		.90	8.889		415	590		1,005	1,375
0180 4' long, T8, 32 watt energy saver	G		.90	8.889		780	590		1,370	1,775
0560 Twin tube compact lamp	G		.90	8.889		340	590		930	1,275
0570 Double twin tube compact lamp	G		.80	10		745	660		1,405	1,850
0600 Mercury vapor, mogul base, deluxe white, 100 watt			.30	26.667		15,400	1,775		17,175	19,600
0800 400 watt			.30	26.667		11,300	1,775		13,075	15,100
1000 Metal halide, mogul base, 175 watt			.30	26.667		1,300	1,775		3,075	4,150
1100 250 watt			.30	26.667		3,175	1,775		4,950	6,200
1350 High pressure sodium, 70 watt			.30	26.667		1,825	1,775		3,600	4,700
1370 150 watt			.30	26.667		1,725	1,775		3,500	4,575
1500 Low pressure sodium, 35 watt			.30	26.667		10,400	1,775		12,175	14,200
1600 90 watt			.30	26.667		15,800	1,775		17,575	20,100
1800 Incandescent, interior, A21, 100 watt			1.60	5		450	330		780	1,000
1900 A21, 150 watt			1.60	5		172	330		502	700
2300 R30, 75 watt			1.30	6.154		860	405		1,265	1,575
2500 Exterior, PAR 38, 75 watt			1.30	6.154		1,600	405		2,005	2,400
2600 PAR 38, 150 watt			1.30	6.154		2,475	405		2,880	3,350
9000 Minimum labor/equipment charge			4	2	Job		132		132	203

Division Notes

	CREW	DAILY OUTPUT	LABOR-HOURS	UNIT	BARE COSTS				TOTAL INCL O&P
					MAT.	LABOR	EQUIP.	TOTAL	

Estimating Tips
27 20 00 Data Communications
27 30 00 Voice Communications
27 40 00 Audio-Video Communications

- When estimating material costs for special systems, it is always prudent to obtain manufacturers' quotations for equipment prices and special installation requirements that may affect the total cost.

- For cost modifications for elevated tray installation, add the percentages to labor according to the height of the installation and only to the quantities exceeding the different height levels, not to the total tray quantities. Refer to subdivision 26 01 02.20 for labor adjustment factors.

- Do not overlook the costs for equipment used in the installation. If scissor lifts and boom lifts are available in the field, contractors may use them in lieu of the proposed ladders and rolling staging.

Reference Numbers
Reference numbers are shown at the beginning of some major classifications. These numbers refer to related items in the Reference Section. The reference information may be an estimating procedure, an alternate pricing method, or technical information.

Note: Not all subdivisions listed here necessarily appear. ■

Same Data. Simplified.

Enjoy the convenience and efficiency of accessing your costs anywhere:

- **Skip the multiplier** by setting your location
- **Quickly search,** edit, favorite and share costs
- **Stay on top of price changes** with automatic updates

Discover more at rsmeans.com/online

27 13 23 – Communications Optical Fiber Backbone Cabling

27 13 23.13 Communications Optical Fiber	Crew	Daily Output	Labor-Hours	Unit	Material	2022 Bare Costs Labor	Equipment	Total	Total Incl O&P
0010 **COMMUNICATIONS OPTICAL FIBER**									
0040 Specialized tools & techniques cause installation costs to vary.									
0070 Fiber optic, cable, bulk simplex, single mode	1 Elec	8	1	C.L.F.	22.50	66		88.50	127
0080 Multi mode		8	1		29.50	66		95.50	135
0090 4 strand, single mode		7.34	1.090		38	72		110	153
0095 Multi mode		7.34	1.090		50.50	72		122.50	167
0100 12 strand, single mode		6.67	1.199		72.50	79.50		152	202
0105 Multi mode		6.67	1.199		96.50	79.50		176	228
0150 Jumper				Ea.	47			47	52
0200 Pigtail					34			34	37.50
0300 Connector	1 Elec	24	.333		28	22		50	64.50
0350 Finger splice		32	.250		41.50	16.55		58.05	71
0400 Transceiver (low cost bi-directional)		8	1		490	66		556	635
0450 Rack housing, 4 rack spaces, 12 panels (144 fibers)		2	4		1,200	265		1,465	1,700
0500 Patch panel, 12 ports		6	1.333		350	88.50		438.50	520
0600 Cable connector panel, 6 fiber		40	.200		61.50	13.25		74.75	88
1000 Cable, 62.5 microns, direct burial, 4 fiber	R-15	1200	.040	L.F.	.85	2.59	.27	3.71	5.25
1020 Indoor, 2 fiber	R-19	1000	.020		.54	1.33		1.87	2.63
1040 Outdoor, aerial/duct	"	1670	.012		.71	.79		1.50	2
1060 50 microns, direct burial, 8 fiber	R-22	4000	.009		1.42	.56		1.98	2.43
1080 12 fiber		4000	.009		1.68	.56		2.24	2.72
1100 Indoor, 12 fiber		759	.049		1.53	2.98		4.51	6.25
1120 Connectors, 62.5 micron cable, transmission	R-19	40	.500	Ea.	16.45	33		49.45	69
1140 Cable splice		40	.500		19.20	33		52.20	72
1160 125 micron cable, transmission		16	1.250		17.30	83		100.30	146
1180 Receiver, 1.2 mile range		20	1		375	66.50		441.50	515
1200 1.9 mile range		20	1		250	66.50		316.50	380
1220 6.2 mile range		5	4		415	265		680	860
1240 Transmitter, 1.2 mile range		20	1		415	66.50		481.50	560
1260 1.9 mile range		20	1		315	66.50		381.50	445
1280 6.2 mile range		5	4		1,300	265		1,565	1,825
1300 Modem, 1.2 mile range		5	4		221	265		486	650
1320 6.2 mile range		5	4		415	265		680	865
1340 1.9 mile range, 12 channel		5	4		2,325	265		2,590	2,950
1360 Repeater, 1.2 mile range		10	2		375	133		508	620
1380 1.9 mile range		10	2		550	133		683	810
1400 6.2 mile range		5	4		1,000	265		1,265	1,500
1420 1.2 mile range, digital		5	4		510	265		775	965
1510 1/4", black, non-metallic, flexible, tube, liquid tight	R-19	8	2.500	C.L.F.	211	166		377	485
1520 3/8"		8	2.500		133	166		299	400
1530 1/2"		7.34	2.725		132	181		313	425
1540 3/4"		7.34	2.725		235	181		416	535
1550 1"		7.34	2.725		370	181		551	685
1560 1-1/4"		6.67	2.999		630	199		829	995
1570 1-1/2"		6.67	2.999		685	199		884	1,050
1580 2"		6.35	3.150		1,150	209		1,359	1,575
2040 Fiber optic cable, 48 strand, single mode, armored, Plenum		1.67	11.976	M.L.F.	5,050	795		5,845	6,775
2042 72 strand		1.50	13.307		7,050	880		7,930	9,100
2044 96 strand		1.42	14.085		16,200	935		17,135	19,200
2046 144 strand		1.50	13.307		18,800	880		19,680	22,100

27 15 Communications Horizontal Cabling

27 15 01 – Communications Horizontal Cabling Applications

27 15 01.23 Audio-Video Communications Horizontal Cabling	Crew	Daily Output	Labor-Hours	Unit	Material	2022 Bare Costs Labor	Equipment	Total	Total Incl O&P
0010 **AUDIO-VIDEO COMMUNICATIONS HORIZONTAL CABLING**									
3000 S-video cable, MD4, 1M Pro interconnects, 3'	1 Elec	38	.211	Ea.	26.50	13.95		40.45	51
3010 6'		38	.211		37.50	13.95		51.45	63
3020 9'		38	.211		38	13.95		51.95	63.50
3030 16'		36	.222		63.50	14.70		78.20	92.50
3040 22'		36	.222		74.50	14.70		89.20	104
3300 HDMI cable, coaxial, HDMI/HDMI, Non plenum, 3', 24 kt terminations		40	.200		14.35	13.25		27.60	36.50
3310 6', 24 kt terminations		40	.200		16.95	13.25		30.20	39
3320 HDMI cable, coaxial, HDMI/HDMI metal, Non plenum, 3', 24 kt terminations		40	.200		16.40	13.25		29.65	38.50
3330 6', 24 kt terminations		40	.200		19.05	13.25		32.30	41.50
3340 15', 24 kt terminations		38	.211		26.50	13.95		40.45	50.50
3350 25', 24 kt terminations		38	.211		34.50	13.95		48.45	59.50

27 15 10 – Special Communications Cabling

27 15 10.23 Sound and Video Cables and Fittings

	Crew	Daily Output	Labor-Hours	Unit	Material	Labor	Equipment	Total	Total Incl O&P
0010 **SOUND AND VIDEO CABLES & FITTINGS**									
1250 Nonshielded, #22-2 conductor	1 Elec	10	.800	C.L.F.	14.25	53		67.25	97

27 15 13 – Communications Copper Horizontal Cabling

27 15 13.13 Communication Cables and Fittings

	Crew	Daily Output	Labor-Hours	Unit	Material	Labor	Equipment	Total	Total Incl O&P
0010 **COMMUNICATION CABLES AND FITTINGS**									
5000 High performance unshielded twisted pair (UTP)									
7000 Category 5, #24, 4 pair solid, PVC jacket	1 Elec	7	1.143	C.L.F.	67	75.50		142.50	190
7100 4 pair solid, plenum		7	1.143		21.50	75.50		97	140
7200 4 pair stranded, PVC jacket		7	1.143		35	75.50		110.50	155
7210 Category 5e, #24, 4 pair solid, PVC jacket		7	1.143		17.45	75.50		92.95	135
7212 4 pair solid, plenum		7	1.143		47.50	75.50		123	169
7214 4 pair stranded, PVC jacket		7	1.143		24.50	75.50		100	143
7240 Category 6, #24, 4 pair solid, PVC jacket		7	1.143		64	75.50		139.50	187
7242 4 pair solid, plenum		7	1.143		15.30	75.50		90.80	133
7244 4 pair stranded, PVC jacket		7	1.143		25.50	75.50		101	144
7300 Connector, RJ45, category 5		80	.100	Ea.	.95	6.60		7.55	11.20
7302 Shielded RJ45, category 5		72	.111		1.79	7.35		9.14	13.25
7310 Jack, UTP RJ45, category 3		72	.111		.69	7.35		8.04	12.05
7312 Category 5		65	.123		8.20	8.15		16.35	21.50
7314 Category 5e		65	.123		6.50	8.15		14.65	19.65
7316 Category 6		65	.123		4.83	8.15		12.98	17.80
7402 RJ45, green		16.50	.485		6.70	32		38.70	57
7408 RJ45, orange		16.50	.485		6.40	32		38.40	56.50
7420 5M, RJ45, 568B, cord set		16.50	.485		36.50	32		68.50	89.50
7422 3M, RJ45, 568B, cord set		16.50	.485		31.50	32		63.50	84.50
7424 1M, RJ45, 568B, cord set		16.50	.485		27	32		59	79
7450 M12 TO RJ45, 2M ultra-loch D-mode, cord set		15.50	.516		37	34		71	93.50

27 32 Voice Communications Terminal Equipment

27 32 26 – Ring-Down Emergency Telephones

27 32 26.10 Emergency Phone Stations	Crew	Daily Output	Labor-Hours	Unit	Material	2022 Bare Costs Labor	Equipment	Total	Total Incl O&P
0010 **EMERGENCY PHONE STATIONS**									
0100 Hands free emergency speaker phone w/LED strobe	1 Elec	7.22	1.108	Ea.	880	73.50		953.50	1,075
0110 Wall/pole mounted emergency phone station w/LED strobe	"	4.60	1.739		1,550	115		1,665	1,875
2000 Call station, 30", pole mounted emergency phone, 120V/ VoIP, blue beacon	2 Elec	5	3.200		1,375	212		1,587	1,850
2010 Call station, 36"		5	3.200		1,600	212		1,812	2,100
2020 Call station, 5'-8"		4.80	3.333		2,725	221		2,946	3,350
2030 Call station, 9'		4.80	3.333		2,750	221		2,971	3,375

27 41 Audio-Video Systems

27 41 33 – Master Antenna Television Systems

27 41 33.10 TV Systems

	Crew	Daily Output	Labor-Hours	Unit	Material	2022 Bare Costs Labor	Equipment	Total	Total Incl O&P
0010 **TV SYSTEMS,** not including rough-in wires, cables & conduits									
5000 Antenna, small	1 Elec	6	1.333	Ea.	54.50	88.50		143	196
5100 Large	"	4	2	"	230	132		362	455

27 51 Distributed Audio-Video Communications Systems

27 51 16 – Public Address Systems

27 51 16.10 Public Address System

	Crew	Daily Output	Labor-Hours	Unit	Material	2022 Bare Costs Labor	Equipment	Total	Total Incl O&P
0010 **PUBLIC ADDRESS SYSTEM**									
0100 Conventional, office	1 Elec	5.33	1.501	Speaker	265	99.50		364.50	445
0200 Industrial		2.70	2.963	"	239	196		435	565
9000 Minimum labor/equipment charge		3.50	2.286	Job		151		151	232

27 51 19 – Sound Masking Systems

27 51 19.10 Sound System

	Crew	Daily Output	Labor-Hours	Unit	Material	2022 Bare Costs Labor	Equipment	Total	Total Incl O&P
0010 **SOUND SYSTEM,** not including rough-in wires, cables & conduits									
2000 Intercom, 30 station capacity, master station	2 Elec	2	8	Ea.	1,775	530		2,305	2,775
2020 10 station capacity	"	4	4		1,475	265		1,740	2,025
3600 House telephone, talking station	1 Elec	1.60	5		560	330		890	1,125
3800 Press to talk, release to listen	"	5.30	1.509		237	100		337	415
4000 System-on button					78.50			78.50	86
4200 Door release	1 Elec	4	2		140	132		272	355
4400 Combination speaker and microphone		8	1		238	66		304	365
4600 Termination box		3.20	2.500		75	166		241	335
4800 Amplifier or power supply		5.30	1.509		860	100		960	1,100
5000 Vestibule door unit		16	.500	Name	165	33		198	233
5200 Strip cabinet		27	.296	Ea.	299	19.60		318.60	360
5400 Directory		16	.500	"	160	33		193	227
9000 Minimum labor/equipment charge		3.50	2.286	Job		151		151	232

Estimating Tips

- When estimating material costs for electronic safety and security systems, it is always prudent to obtain manufacturers' quotations for equipment prices and special installation requirements that may affect the total cost.

- Fire alarm systems consist of control panels, annunciator panels, batteries with rack, charger, and fire alarm actuating and indicating devices. Some fire alarm systems include speakers, telephone lines, door closer controls, and other components. Be careful not to overlook the costs related to installation for these items. Also be aware of costs for integrated automation instrumentation and terminal devices, control equipment, control wiring, and programming. Insurance underwriters may have specific requirements for the type of materials to be installed or design requirements based on the hazard to be protected. Local jurisdictions may have requirements not covered by code. It is advisable to be aware of any special conditions.

- Security equipment includes items such as CCTV, access control, and other detection and identification systems to perform alert and alarm functions. Be sure to consider the costs related to installation for this security equipment, such as for integrated automation instrumentation and terminal devices, control equipment, control wiring, and programming.

Reference Numbers

Reference numbers are shown at the beginning of some major classifications. These numbers refer to related items in the Reference Section. The reference information may be an estimating procedure, an alternate pricing method, or technical information.

Same Data. Simplified.

Enjoy the convenience and efficiency of accessing your costs anywhere:

- **Skip the multiplier** by setting your location
- **Quickly search,** edit, favorite and share costs
- **Stay on top of price changes** with automatic updates

Discover more at rsmeans.com/online

Note: Trade Service, in part, has been used as a reference source for some of the material prices used in Division 28.

28 15 Access Control Hardware Devices

28 15 11 – Integrated Credential Readers and Field Entry Management

28 15 11.11 Standard Card Readers

28 15 11.11 Standard Card Readers		Crew	Daily Output	Labor-Hours	Unit	Material	2022 Bare Costs Labor	Equipment	Total	Total Incl O&P
0010	**STANDARD CARD READERS**									
0015	Card key access									
0020	Computerized system, processor, proximity reader and cards									
0030	Does not include door hardware, lockset or wiring									
0040	Card key system for 1 door				Ea.	1,675			1,675	1,850
0060	Card key system for 2 doors					2,400			2,400	2,650
0080	Card key system for 4 doors					2,625			2,625	2,875
0100	Processor for card key access system					1,250			1,250	1,375
0160	Magnetic lock for electric access, 600 pound holding force					192			192	211
0170	Magnetic lock for electric access, 1200 pound holding force					192			192	211
0200	Proximity card reader					199			199	219

28 15 11.15 Biometric Identity Devices

		Crew	Daily Output	Labor-Hours	Unit	Material	Labor	Equipment	Total	Total Incl O&P
0010	**BIOMETRIC IDENTITY DEVICES**									
0220	Hand geometry scanner, mem of 512 users, excl. striker/power	1 Elec	3	2.667	Ea.	3,075	177		3,252	3,650
0230	Memory upgrade for, adds 9,700 user profiles		8	1		550	66		616	705
0240	Adds 32,500 user profiles		8	1		735	66		801	905
0250	Prison type, memory of 256 users, excl. striker/power		3	2.667		3,525	177		3,702	4,150
0260	Memory upgrade for, adds 3,300 user profiles		8	1		515	66		581	670
0270	Adds 9,700 user profiles		8	1		635	66		701	800
0280	Adds 27,900 user profiles	1 Elec	8	1	Ea.	970	66		1,036	1,175
0290	All weather, mem of 512 users, excl. striker/power		3	2.667		5,575	177		5,752	6,400
0300	Facial & fingerprint scanner, combination unit, excl. striker/power		3	2.667		1,175	177		1,352	1,575
0310	Access for, for initial setup, excl. striker/power		3	2.667		1,750	177		1,927	2,200

28 18 Security Access Detection Equipment

28 18 11 – Security Access Metal Detectors

28 18 11.13 Security Access Metal Detectors

		Crew	Daily Output	Labor-Hours	Unit	Material	Labor	Equipment	Total	Total Incl O&P
0010	**SECURITY ACCESS METAL DETECTORS**									
0240	Metal detector, hand-held, wand type, unit only				Ea.	129			129	138
0250	Metal detector, walk through portal type, single zone	1 Elec	2	4		1,100	265		1,365	1,600
0260	Multi-zone	"	2	4		2,350	265		2,615	2,975

28 18 13 – Security Access X-Ray Equipment

28 18 13.16 Security Access X-Ray Equipment

		Crew	Daily Output	Labor-Hours	Unit	Material	Labor	Equipment	Total	Total Incl O&P
0010	**SECURITY ACCESS X-RAY EQUIPMENT**									
0290	X-ray machine, desk top, for mail/small packages/letters	1 Elec	4	2	Ea.	3,425	132		3,557	3,975
0300	Conveyor type, incl. monitor		2	4		18,800	265		19,065	21,000
0310	Includes additional features		2	4		33,400	265		33,665	37,100
0320	X-ray machine, large unit, for airports, incl. monitor	2 Elec	1	16		46,900	1,050		47,950	53,000
0330	Full console	"	.50	32		80,500	2,125		82,625	92,000

28 18 15 – Security Access Explosive Detection Equipment

28 18 15.23 Security Access Explosive Detection Equipment

		Crew	Daily Output	Labor-Hours	Unit	Material	Labor	Equipment	Total	Total Incl O&P
0010	**SECURITY ACCESS EXPLOSIVE DETECTION EQUIPMENT**									
0270	Explosives detector, walk through portal type	1 Elec	2	4	Ea.	175,000	265		175,265	193,000
0280	Hand-held, battery operated				"				25,500	28,100

28 23 Video Management System

28 23 23 – Video Surveillance Systems Infrastructure

28 23 23.50 Video Surveillance Equipments	Crew	Daily Output	Labor-Hours	Unit	Material	2022 Bare Costs Labor	Equipment	Total	Total Incl O&P
0010 **VIDEO SURVEILLANCE EQUIPMENTS**									
0200 Video cameras, wireless, hidden in exit signs, clocks, etc., incl. receiver	1 Elec	3	2.667	Ea.	149	177		326	435
0210 Accessories for video recorder, single camera		3	2.667		178	177		355	465
0220 For multiple cameras		3	2.667		1,600	177		1,777	2,025
0230 Video cameras, wireless, for under vehicle searching, complete	↓	2	4	↓	14,200	265		14,465	16,000

28 31 Intrusion Detection

28 31 16 – Intrusion Detection Systems Infrastructure

28 31 16.50 Intrusion Detection

	Crew	Daily Output	Labor-Hours	Unit	Material	2022 Bare Costs Labor	Equipment	Total	Total Incl O&P
0010 **INTRUSION DETECTION**, not including wires & conduits									
0100 Burglar alarm, battery operated, mechanical trigger	1 Elec	4	2	Ea.	284	132		416	515
0200 Electrical trigger		4	2		340	132		472	575
0400 For outside key control, add	↓	8	1	↓	87.50	66		153.50	198
0600 For remote signaling circuitry, add	1 Elec	8	1	Ea.	144	66		210	261
0800 Card reader, flush type, standard		2.70	2.963		680	196		876	1,050
1000 Multi-code		2.70	2.963		1,225	196		1,421	1,650
1200 Door switches, hinge switch		5.30	1.509		66.50	100		166.50	227
1400 Magnetic switch		5.30	1.509		125	100		225	291
1600 Exit control locks, horn alarm		4	2		237	132		369	465
1800 Flashing light alarm		4	2		239	132		371	465
2000 Indicating panels, 1 channel	↓	2.70	2.963		218	196		414	540
2200 10 channel	2 Elec	3.20	5		1,150	330		1,480	1,775
2400 20 channel		2	8		2,450	530		2,980	3,525
2600 40 channel	↓	1.14	14.035		4,550	930		5,480	6,425
2800 Ultrasonic motion detector, 12 V	1 Elec	2.30	3.478		164	230		394	535
3000 Infrared photoelectric detector		4	2		120	132		252	335
3200 Passive infrared detector		4	2		54	132		186	263
3400 Glass break alarm switch		8	1		66	66		132	175
3420 Switchmats, 30" x 5'		5.30	1.509		118	100		218	284
3440 30" x 25'		4	2		207	132		339	430
3460 Police connect panel		4	2		286	132		418	520
3480 Telephone dialer		5.30	1.509		395	100		495	585
3500 Alarm bell		4	2		99	132		231	310
3520 Siren		4	2		149	132		281	365
3540 Microwave detector, 10' to 200'		2	4		545	265		810	1,000
3560 10' to 350'	↓	2	4		1,550	265		1,815	2,125

28 46 Fire Detection and Alarm

28 46 11 – Fire Sensors and Detectors

28 46 11.21 Carbon-Monoxide Detection Sensors

	Crew	Daily Output	Labor-Hours	Unit	Material	2022 Bare Costs Labor	Equipment	Total	Total Incl O&P
0010 **CARBON-MONOXIDE DETECTION SENSORS**									
8400 Smoke and carbon monoxide alarm battery operated photoelectric low profile	1 Elec	24	.333	Ea.	57.50	22		79.50	97.50
8410 low profile photoelectric battery powered		24	.333		45	22		67	83.50
8420 photoelectric low profile sealed lithium		24	.333		79	22		101	121
8430 Photoelectric low profile sealed lithium smoke and CO with voice combo		24	.333		55	22		77	94.50
8500 Carbon monoxide sensor, wall mount 1Mod 1 relay output smoke & heat		24	.333		460	22		482	545
8700 Carbon monoxide detector, battery operated, wall mounted		16	.500		45	33		78	101
8710 Hardwired, wall and ceiling mounted		8	1		77.50	66		143.50	187
8720 Duct mounted	↓	8	1		340	66		406	475

28 46 Fire Detection and Alarm

28 46 11 – Fire Sensors and Detectors

28 46 11.21 Carbon-Monoxide Detection Sensors	Crew	Daily Output	Labor-Hours	Unit	Material	2022 Bare Costs Labor	Equipment	Total	Total Incl O&P
8730	Continuous air monitoring system, PC based, incl. remote sensors, min.	2 Elec	.25	64	Ea.	33,300	4,225		37,525	43,200
8740	Maximum	"	.10	160	↓	60,000	10,600		70,600	82,500

28 46 11.27 Other Sensors

		Crew	Daily Output	Labor-Hours	Unit	Material	Labor	Equipment	Total	Total Incl O&P
0010	**OTHER SENSORS**									
5200	Smoke detector, ceiling type	1 Elec	6.20	1.290	Ea.	143	85.50		228.50	288
5240	Smoke detector, addressable type		6	1.333		226	88.50		314.50	385
5400	Duct type		3.20	2.500		310	166		476	600
5420	Duct addressable type		3.20	2.500		190	166		356	465
8300	Smoke alarm with integrated strobe light 120 V, 16DB 60 fpm flash rate		16	.500		133	33		166	197
8310	Photoelectric smoke detector with strobe 120 V, 90 DB ceiling mount		12	.667		237	44		281	330
8320	120 V, 90 DB wall mount	↓	12	.667	↓	229	44		273	320

28 46 11.50 Fire and Heat Detectors

		Crew	Daily Output	Labor-Hours	Unit	Material	Labor	Equipment	Total	Total Incl O&P
0010	**FIRE & HEAT DETECTORS**									
5000	Detector, rate of rise	1 Elec	8	1	Ea.	31	66		97	137
5100	Fixed temp fire alarm		7	1.143		41	75.50		116.50	162
5200	125 V SRF MT-135F		7.25	1.103		35.50	73		108.50	152
5300	MT-195F	↓	7.25	1.103	↓	36.50	73		109.50	152

28 46 20 – Fire Alarm

28 46 20.50 Alarm Panels and Devices

		Crew	Daily Output	Labor-Hours	Unit	Material	Labor	Equipment	Total	Total Incl O&P
0010	**ALARM PANELS AND DEVICES**, not including wires & conduits									
2200	Intercom remote station	1 Elec	8	1	Ea.	94.50	66		160.50	206
2400	Intercom outlet		8	1		66	66		132	175
2600	Sound system, intercom handset		8	1		490	66		556	635
3590	Signal device, beacon		8	1		227	66		293	350
3594	2 zone	↓	1.35	5.926		400	390		790	1,050
3598	3 zone	2 Elec	2	8		495	530		1,025	1,350
3600	4 zone		2	8		460	530		990	1,325
3700	5 zone		1.50	10.667		780	705		1,485	1,925
3800	8 zone		1	16		575	1,050		1,625	2,250
3900	10 zone		1.25	12.800		890	845		1,735	2,275
4000	12 zone	↓	.67	23.988		2,700	1,600		4,300	5,425
4020	Alarm device, tamper, flow	1 Elec	8	1		217	66		283	340
4025	Fire alarm, loop expander card		16	.500		555	33		588	660
4200	Battery and rack		4	2		490	132		622	745
4400	Automatic charger		8	1		555	66		621	710
4600	Signal bell		8	1		50.50	66		116.50	158
4610	Fire alarm signal bell 10" red 20-24 V P		8	1		168	66		234	287
4800	Trouble buzzer or manual station		8	1		80	66		146	190
5425	Duct smoke and heat detector 2 wire		8	1		148	66		214	265
5430	Fire alarm duct detector controller		3	2.667		186	177		363	475
5435	Fire alarm duct detector sensor kit		8	1		77	66		143	187
5440	Remote test station for smoke detector duct type		5.30	1.509		49.50	100		149.50	209
5460	Remote fire alarm indicator light		5.30	1.509		26	100		126	183
5600	Strobe and horn		5.30	1.509		120	100		220	286
5610	Strobe and horn (ADA type)		5.30	1.509		143	100		243	310
5620	Visual alarm (ADA type)		6.70	1.194		109	79		188	241
5800	electric bell		6.70	1.194		53	79		132	180
6000	Door holder, electro-magnetic		4	2		106	132		238	320
6200	Combination holder and closer		3.20	2.500		165	166		331	435
6600	Drill switch		8	1		510	66		576	660
6800	Master box		2.70	2.963		6,175	196		6,371	7,100
7000	Break glass station		8	1		64	66		130	173

28 46 Fire Detection and Alarm

28 46 20 – Fire Alarm

28 46 20.50 Alarm Panels and Devices	Crew	Daily Output	Labor-Hours	Unit	Material	2022 Bare Costs Labor	Equipment	Total	Total Incl O&P
7800	Remote annunciator, 8 zone lamp	1 Elec	1.80	4.444	Ea.	325	294		619	810
8000	12 zone lamp	2 Elec	2.60	6.154		485	405		890	1,150
8200	16 zone lamp	"	2.20	7.273		380	480		860	1,150

28 47 Mass Notification

28 47 12 – Notification Systems

28 47 12.10 Mass Notification System

		Crew	Daily Output	Labor-Hours	Unit	Material	2022 Bare Costs Labor	Equipment	Total	Total Incl O&P
0010	**MASS NOTIFICATION SYSTEM**									
0100	Wireless command center, 10,000 devices	2 Elec	1.33	12.030	Ea.	3,000	795		3,795	4,525
0200	Option, email notification					2,100			2,100	2,325
0210	Remote device supervision & monitor					2,625			2,625	2,900
0300	Antenna VHF or UHF, for medium range	1 Elec	4	2.		95	132		227	310
0310	For high-power transmitter		2	4		1,100	265		1,365	1,600
0400	Transmitter, 25 watt		4	2		1,725	132		1,857	2,100
0410	40 watt		2.66	3.008		2,000	199		2,199	2,500
0420	100 watt		1.33	6.015		6,475	400		6,875	7,725
0500	Wireless receiver/control module for speaker		8	1		285	66		351	415
0600	Desktop paging controller, stand alone		4	2		895	132		1,027	1,200
0700	Auxiliary level inputs		8.36	.957		103	63.50		166.50	211

Division Notes

	CREW	DAILY OUTPUT	LABOR-HOURS	UNIT	BARE COSTS				TOTAL INCL O&P
					MAT.	LABOR	EQUIP.	TOTAL	

Estimating Tips
31 05 00 Common Work Results for Earthwork

- Estimating the actual cost of performing earthwork requires careful consideration of the variables involved. This includes items such as type of soil, whether water will be encountered, dewatering, whether banks need bracing, disposal of excavated earth, and length of haul to fill or spoil sites, etc. If the project has large quantities of cut or fill, consider raising or lowering the site to reduce costs, while paying close attention to the effect on site drainage and utilities.

- If the project has large quantities of fill, creating a borrow pit on the site can significantly lower the costs.

- It is very important to consider what time of year the project is scheduled for completion. Bad weather can create large cost overruns from dewatering, site repair, and lost productivity from cold weather.

Reference Numbers

Reference numbers are shown at the beginning of some major classifications. These numbers refer to related items in the Reference Section. The reference information may be an estimating procedure, an alternate pricing method, or technical information.

Note: Not all subdivisions listed here necessarily appear. ■

Same Data. Simplified.

Enjoy the convenience and efficiency of accessing your costs anywhere:

- **Skip the multiplier** by setting your location
- **Quickly search,** edit, favorite and share costs
- **Stay on top of price changes** with automatic updates

Discover more at rsmeans.com/online

31 05 Common Work Results for Earthwork

31 05 13 – Soils for Earthwork

31 05 13.10 Borrow	Crew	Daily Output	Labor-Hours	Unit	Material	2022 Bare Costs Labor	Equipment	Total	Total Incl O&P
0010 **BORROW**									
0020 Spread, 200 HP dozer, no compaction, 2 mile RT haul									
0200 Common borrow	B-15	600	.047	C.Y.	12.55	2.52	4.81	19.88	23
0700 Screened loam		600	.047		34	2.52	4.81	41.33	47
0800 Topsoil, weed free		600	.047		20.50	2.52	4.81	27.83	32
0900 For 5 mile haul, add	B-34B	200	.040			2.12	3.84	5.96	7.60

31 05 16 – Aggregates for Earthwork

31 05 16.10 Borrow	Crew	Daily Output	Labor-Hours	Unit	Material	2022 Bare Costs Labor	Equipment	Total	Total Incl O&P
0010 **BORROW**									
0020 Spread, with 200 HP dozer, no compaction, 2 mile RT haul									
0100 Bank run gravel	B-15	600	.047	L.C.Y.	27.50	2.52	4.81	34.83	39.50
0300 Crushed stone (1.40 tons per C.Y.), 1-1/2"		600	.047		20.50	2.52	4.81	27.83	32
0320 3/4"		600	.047		20.50	2.52	4.81	27.83	32
0340 1/2"		600	.047		18.55	2.52	4.81	25.88	30
0360 3/8"		600	.047		23	2.52	4.81	30.33	34.50
0400 Sand, washed, concrete		600	.047		18.75	2.52	4.81	26.08	30
0500 Dead or bank sand		600	.047		13.70	2.52	4.81	21.03	24.50
0600 Select structural fill		600	.047		16.25	2.52	4.81	23.58	27
0900 For 5 mile haul, add	B-34B	200	.040			2.12	3.84	5.96	7.60

31 06 Schedules for Earthwork

31 06 60 – Schedules for Special Foundations and Load Bearing Elements

31 06 60.14 Piling Special Costs	Crew	Daily Output	Labor-Hours	Unit	Material	2022 Bare Costs Labor	Equipment	Total	Total Incl O&P
0010 **PILING SPECIAL COSTS**									
0011 Piling special costs, pile caps, see Section 03 30 53.40									
0500 Cutoffs, concrete piles, plain	1 Pile	5.50	1.455	Ea.		84.50		84.50	135
0600 With steel thin shell, add		38	.211			12.25		12.25	19.55
0700 Steel pile or "H" piles		19	.421			24.50		24.50	39
0800 Wood piles		38	.211			12.25		12.25	19.55
1000 Testing, any type piles, test load is twice the design load									
1050 50 ton design load, 100 ton test				Ea.				14,000	15,500
1100 100 ton design load, 200 ton test								20,000	22,000
1200 200 ton design load, 400 ton test								28,000	31,000

31 06 60.15 Mobilization	Crew	Daily Output	Labor-Hours	Unit	Material	2022 Bare Costs Labor	Equipment	Total	Total Incl O&P
0010 **MOBILIZATION**									
0020 Set up & remove, air compressor, 600 CFM	A-5	3.30	5.455	Ea.		252	26	278	430
0100 1,200 CFM	"	2.20	8.182			375	38.50	413.50	645
0200 Crane, with pile leads and pile hammer, 75 ton	B-19	.60	107			6,325	3,500	9,825	13,900
0300 150 ton	"	.36	178			10,500	5,850	16,350	23,100

31 14 Earth Stripping and Stockpiling

31 14 13 – Soil Stripping and Stockpiling

31 14 13.23 Topsoil Stripping and Stockpiling

		Crew	Daily Output	Labor-Hours	Unit	Material	2022 Bare Costs Labor	Equipment	Total	Total Incl O&P
0010	**TOPSOIL STRIPPING AND STOCKPILING**									
1500	Loam or topsoil, remove/stockpile on site									
1510	By hand, 6" deep, 50' haul, less than 100 S.Y.	B-1	100	.240	S.Y.		11.10		11.10	17.65
1520	By skid steer, 6" deep, 100' haul, 101-500 S.Y.	B-62	500	.048			2.37	.53	2.90	4.32
1530	100' haul, 501-900 S.Y.	"	900	.027			1.32	.29	1.61	2.40
1540	200' haul, 901-1,100 S.Y.	B-63	1000	.040	S.Y.		1.92	.26	2.18	3.32
1550	By dozer, 200' haul, 1,101-4,000 S.Y.	B-10B	4000	.003	"		.17	.34	.51	.63

31 23 Excavation and Fill

31 23 16 – Excavation

31 23 16.13 Excavating, Trench

		Crew	Daily Output	Labor-Hours	Unit	Material	2022 Bare Costs Labor	Equipment	Total	Total Incl O&P
0010	**EXCAVATING, TRENCH**									
0011	Or continuous footing									
0020	Common earth with no sheeting or dewatering included									
1400	By hand with pick and shovel 2' to 6' deep, light soil	1 Clab	8	1	B.C.Y.		45.50		45.50	72.50
1500	Heavy soil	"	4	2	"		91		91	145
1700	For tamping backfilled trenches, air tamp, add	A-1G	100	.080	E.C.Y.		3.65	.83	4.48	6.70
1900	Vibrating plate, add	B-18	180	.133	"		6.15	.76	6.91	10.65
2100	Trim sides and bottom for concrete pours, common earth		1500	.016	S.F.		.74	.09	.83	1.28
2300	Hardpan		600	.040	"		1.85	.23	2.08	3.19
9000	Minimum labor/equipment charge	1 Clab	4	2	Job		91		91	145

31 23 16.14 Excavating, Utility Trench

		Crew	Daily Output	Labor-Hours	Unit	Material	2022 Bare Costs Labor	Equipment	Total	Total Incl O&P
0010	**EXCAVATING, UTILITY TRENCH**									
0011	Common earth									
0050	Trenching with chain trencher, 12 HP, operator walking									
0100	4" wide trench, 12" deep	B-53	800	.010	L.F.		.57	.24	.81	1.15
0150	18" deep		750	.011			.61	.26	.87	1.23
0200	24" deep		700	.011			.65	.27	.92	1.31
0300	6" wide trench, 12" deep		650	.012			.70	.30	1	1.41
0350	18" deep		600	.013			.76	.32	1.08	1.53
0400	24" deep		550	.015			.83	.35	1.18	1.67
0450	36" deep		450	.018			1.01	.43	1.44	2.05
0600	8" wide trench, 12" deep		475	.017			.96	.40	1.36	1.94
0650	18" deep		400	.020			1.14	.48	1.62	2.31
0700	24" deep		350	.023			1.30	.55	1.85	2.63
0750	36" deep		300	.027			1.52	.64	2.16	3.07
0900	Minimum labor/equipment charge		2	4	Job		228	96	324	460
1000	Backfill by hand including compaction, add									
1050	4" wide trench, 12" deep	A-1G	800	.010	L.F.		.46	.10	.56	.83
1100	18" deep		530	.015			.69	.16	.85	1.26
1150	24" deep		400	.020			.91	.21	1.12	1.68
1300	6" wide trench, 12" deep		540	.015			.68	.15	.83	1.24
1350	18" deep		405	.020			.90	.20	1.10	1.65
1400	24" deep		270	.030			1.35	.31	1.66	2.49
1450	36" deep		180	.044			2.03	.46	2.49	3.72
1600	8" wide trench, 12" deep	A-1G	400	.020	L.F.		.91	.21	1.12	1.68
1650	18" deep		265	.030			1.38	.31	1.69	2.53
1700	24" deep		200	.040			1.82	.41	2.23	3.35
1750	36" deep		135	.059			2.70	.61	3.31	4.96
2000	Chain trencher, 40 HP operator riding									
2050	6" wide trench and backfill, 12" deep	B-54	1200	.007	L.F.		.38	.33	.71	.96

31 23 Excavation and Fill

31 23 16 – Excavation

31 23 16.14 Excavating, Utility Trench

		Crew	Daily Output	Labor-Hours	Unit	Material	2022 Bare Costs Labor	Equipment	Total	Total Incl O&P
2100	18" deep	B-54	1000	.008	L.F.		.46	.40	.86	1.15
2150	24" deep		975	.008			.47	.41	.88	1.18
2200	36" deep		900	.009			.51	.45	.96	1.28
2250	48" deep		750	.011			.61	.54	1.15	1.54
2300	60" deep		650	.012			.70	.62	1.32	1.77
2400	8" wide trench and backfill, 12" deep		1000	.008			.46	.40	.86	1.15
2450	18" deep		950	.008			.48	.42	.90	1.22
2500	24" deep		900	.009			.51	.45	.96	1.28
2550	36" deep		800	.010			.57	.50	1.07	1.44
2600	48" deep		650	.012			.70	.62	1.32	1.77
2700	12" wide trench and backfill, 12" deep		975	.008			.47	.41	.88	1.18
2750	18" deep		860	.009			.53	.47	1	1.34
2800	24" deep		800	.010			.57	.50	1.07	1.44
2850	36" deep		725	.011			.63	.55	1.18	1.59
3000	16" wide trench and backfill, 12" deep		835	.010			.55	.48	1.03	1.38
3050	18" deep		750	.011			.61	.54	1.15	1.54
3100	24" deep	▼	700	.011	▼		.65	.57	1.22	1.64
3200	Compaction with vibratory plate, add								35%	35%
5100	Hand excavate and trim for pipe bells after trench excavation									
5200	8" pipe	1 Clab	155	.052	L.F.		2.35		2.35	3.74
5300	18" pipe	"	130	.062	"		2.81		2.81	4.46
9000	Minimum labor/equipment charge	A-1G	4	2	Job		91	20.50	111.50	168

31 23 16.16 Structural Excavation for Minor Structures

		Crew	Daily Output	Labor-Hours	Unit	Material	2022 Bare Costs Labor	Equipment	Total	Total Incl O&P
0010	**STRUCTURAL EXCAVATION FOR MINOR STRUCTURES**									
0015	Hand, pits to 6' deep, sandy soil	1 Clab	8	1	B.C.Y.		45.50		45.50	72.50
0100	Heavy soil or clay		4	2			91		91	145
0300	Pits 6' to 12' deep, sandy soil		5	1.600			73		73	116
0500	Heavy soil or clay		3	2.667			122		122	193
0700	Pits 12' to 18' deep, sandy soil		4	2			91		91	145
0900	Heavy soil or clay		2	4			182		182	290
1100	Hand loading trucks from stock pile, sandy soil		12	.667			30.50		30.50	48.50
1300	Heavy soil or clay	▼	8	1	▼		45.50		45.50	72.50
1500	For wet or muck hand excavation, add to above								50%	50%
6000	Machine excavation, for spread and mat footings, elevator pits,									
6001	and small building foundations									
6030	Common earth, hydraulic backhoe, 1/2 C.Y. bucket	B-12E	55	.291	B.C.Y.		15.85	11.90	27.75	38
6035	3/4 C.Y. bucket	B-12F	90	.178			9.70	9.70	19.40	26
6040	1 C.Y. bucket	B-12A	108	.148			8.05	8.45	16.50	22
6050	1-1/2 C.Y. bucket	B-12B	144	.111			6.05	6.85	12.90	17.05
6060	2 C.Y. bucket	B-12C	200	.080			4.36	4.85	9.21	12.20
6070	Sand and gravel, 3/4 C.Y. bucket	B-12F	100	.160			8.70	8.75	17.45	23.50
6080	1 C.Y. bucket	B-12A	120	.133			7.25	7.60	14.85	19.75
6090	1-1/2 C.Y. bucket	B-12B	160	.100			5.45	6.20	11.65	15.35
6100	2 C.Y. bucket	B-12C	220	.073			3.96	4.41	8.37	11.10
6110	Clay, till, or blasted rock, 3/4 C.Y. bucket	B-12F	80	.200			10.90	10.95	21.85	29
6120	1 C.Y. bucket	B-12A	95	.168			9.20	9.60	18.80	25
6130	1-1/2 C.Y. bucket	B-12B	130	.123			6.70	7.60	14.30	18.90
6140	2 C.Y. bucket	B-12C	175	.091			4.98	5.55	10.53	13.95
6230	Sandy clay & loam, hydraulic backhoe, 1/2 C.Y. bucket	B-12E	60	.267			14.55	10.90	25.45	35
6235	3/4 C.Y. bucket	B-12F	98	.163			8.90	8.90	17.80	24
6240	1 C.Y. bucket	B-12A	116	.138			7.50	7.85	15.35	20.50
6250	1-1/2 C.Y. bucket	B-12B	156	.103	▼		5.60	6.35	11.95	15.75

31 23 Excavation and Fill

31 23 16 – Excavation

31 23 16.16 Structural Excavation for Minor Structures	Crew	Daily Output	Labor-Hours	Unit	Material	2022 Bare Costs Labor	2022 Bare Costs Equipment	Total	Total Incl O&P
9000	Minimum labor/equipment charge	1 Clab	4	2	Job		91		91	145

31 23 16.42 Excavating, Bulk Bank Measure

		Crew	Daily Output	Labor-Hours	Unit	Material	Labor	Equipment	Total	Total Incl O&P
0010	**EXCAVATING, BULK BANK MEASURE** R312316-40									
0011	Common earth piled									
0020	For loading onto trucks, add								15%	15%
0200	Excavator, hydraulic, crawler mtd., 1 C.Y. cap. = 100 C.Y./hr. R312316-45	B-12A	800	.020	B.C.Y.		1.09	1.14	2.23	2.96
1200	Front end loader, track mtd., 1-1/2 C.Y. cap. = 70 C.Y./hr.	B-10N	560	.021			1.19	1.05	2.24	3.02
1500	Wheel mounted, 3/4 C.Y. cap. = 45 C.Y./hr.	B-10R	360	.033	↓		1.85	.90	2.75	3.89
5000	Excavating, bulk bank measure, sandy clay & loam piled									
5020	For loading onto trucks, add								15%	15%
5100	Excavator, hydraulic, crawler mtd., 1 C.Y. cap. = 120 C.Y./hr.	B-12A	960	.017	B.C.Y.		.91	.95	1.86	2.48
9000	Minimum labor/equipment charge	B-10L	2	6	Job		335	268	603	815

31 23 16.46 Excavating, Bulk, Dozer

		Crew	Daily Output	Labor-Hours	Unit	Material	Labor	Equipment	Total	Total Incl O&P
0010	**EXCAVATING, BULK, DOZER**									
0011	Open site									
2000	80 HP, 50' haul, sand & gravel	B-10L	460	.026	B.C.Y.		1.45	1.16	2.61	3.55
2200	150' haul, sand & gravel		230	.052			2.89	2.33	5.22	7.10
2400	300' haul, sand & gravel	↓	120	.100	↓		5.55	4.46	10.01	13.60
3000	105 HP, 50' haul, sand & gravel	B-10W	700	.017	B.C.Y.		.95	.94	1.89	2.53
3200	150' haul, sand & gravel		310	.039			2.15	2.13	4.28	5.70
3300	300' haul, sand & gravel	↓	140	.086			4.75	4.71	9.46	12.65
4000	200 HP, 50' haul, sand & gravel	B-10B	1400	.009			.48	.96	1.44	1.80
4200	150' haul, sand & gravel		595	.020			1.12	2.27	3.39	4.24
4400	300' haul, sand & gravel	↓	310	.039	↓		2.15	4.35	6.50	8.15

31 23 23 – Fill

31 23 23.13 Backfill

		Crew	Daily Output	Labor-Hours	Unit	Material	Labor	Equipment	Total	Total Incl O&P
0010	**BACKFILL** R312323-30									
0015	By hand, no compaction, light soil	1 Clab	14	.571	L.C.Y.		26		26	41.50
0100	Heavy soil	↓	11	.727	"		33		33	52.50
0300	Compaction in 6" layers, hand tamp, add to above	↓	20.60	.388	E.C.Y.		17.70		17.70	28
0400	Roller compaction operator walking, add	B-10A	100	.120			6.65	1.37	8.02	11.95
0500	Air tamp, add	B-9D	190	.211			9.70	1.87	11.57	17.45
0600	Vibrating plate, add	A-1D	60	.133			6.10	1.88	7.98	11.70
0800	Compaction in 12" layers, hand tamp, add to above	1 Clab	34	.235			10.75		10.75	17.05
0900	Roller compaction operator walking, add	B-10A	150	.080			4.44	.91	5.35	7.95
1000	Air tamp, add	B-9	285	.140			6.45	1.16	7.61	11.55
1100	Vibrating plate, add	A-1E	90	.089	↓		4.05	1.52	5.57	8.15
1200	Trench, dozer, no compaction, 60 HP	B-10L	425	.028	L.C.Y.		1.57	1.26	2.83	3.84
1300	Dozer backfilling, bulk, up to 300' haul, no compaction	B-10B	1200	.010	"		.55	1.12	1.67	2.11
1400	Air tamped, add	B-11B	80	.200	E.C.Y.		10.25	5.60	15.85	22.50
1900	Dozer backfilling, trench, up to 300' haul, no compaction	B-10B	900	.013	L.C.Y.		.74	1.50	2.24	2.81
2000	Air tamped, add	B-11B	80	.200	E.C.Y.		10.25	5.60	15.85	22.50
2350	Spreading in 8" layers, small dozer	B-10B	1060	.011	L.C.Y.		.63	1.27	1.90	2.38
2450	Compacting with vibrating plate, 8" lifts	A-1D	73	.110	E.C.Y.		5	1.54	6.54	9.65

31 23 23.14 Backfill, Structural

		Crew	Daily Output	Labor-Hours	Unit	Material	Labor	Equipment	Total	Total Incl O&P
0010	**BACKFILL, STRUCTURAL**									
0011	Dozer or F.E. loader									
0020	From existing stockpile, no compaction									
1000	55 HP wheeled loader, 50' haul, common earth	B-11C	200	.080	L.C.Y.		4.24	1.64	5.88	8.45
2000	80 HP, 50' haul, sand & gravel	B-10L	1100	.011			.61	.49	1.10	1.49
2010	Sandy clay & loam	↓	1070	.011	↓		.62	.50	1.12	1.52

541

31 23 23.14 Backfill, Structural

		Crew	Daily Output	Labor-Hours	Unit	Material	2022 Bare Costs Labor	2022 Bare Costs Equipment	Total	Total Incl O&P
2020	Common earth	B-10L	975	.012	L.C.Y.		.68	.55	1.23	1.67
2040	Clay		850	.014			.78	.63	1.41	1.92
2400	300' haul, sand & gravel		370	.032			1.80	1.45	3.25	4.41
2410	Sandy clay & loam		360	.033			1.85	1.49	3.34	4.54
2420	Common earth		330	.036			2.02	1.62	3.64	4.95
2440	Clay	B-10L	290	.041	L.C.Y.		2.30	1.85	4.15	5.65
3000	105 HP, 50' haul, sand & gravel	B-10W	1350	.009			.49	.49	.98	1.31
3010	Sandy clay & loam		1325	.009			.50	.50	1	1.34
3020	Common earth		1225	.010			.54	.54	1.08	1.44
3040	Clay		1100	.011			.61	.60	1.21	1.61
3300	300' haul, sand & gravel		465	.026			1.43	1.42	2.85	3.80
3310	Sandy clay & loam		455	.026			1.46	1.45	2.91	3.89
3320	Common earth		415	.029			1.60	1.59	3.19	4.26
3340	Clay		370	.032			1.80	1.78	3.58	4.78

31 23 23.16 Fill By Borrow and Utility Bedding

		Crew	Daily Output	Labor-Hours	Unit	Material	2022 Bare Costs Labor	2022 Bare Costs Equipment	Total	Total Incl O&P
0010	**FILL BY BORROW AND UTILITY BEDDING**									
0049	Utility bedding, for pipe & conduit, not incl. compaction									
0050	Crushed or screened bank run gravel	B-6	150	.160	L.C.Y.	31	7.90	2.19	41.09	49
0100	Crushed stone 3/4" to 1/2"		150	.160		20.50	7.90	2.19	30.59	37.50
0200	Sand, dead or bank		150	.160		13.70	7.90	2.19	23.79	30
0500	Compacting bedding in trench	A-1D	90	.089	E.C.Y.		4.05	1.25	5.30	7.85
0600	If material source exceeds 2 miles, add for extra mileage.									
0610	See Section 31 23 23.20 for hauling mileage add.									

31 23 23.17 General Fill

		Crew	Daily Output	Labor-Hours	Unit	Material	2022 Bare Costs Labor	2022 Bare Costs Equipment	Total	Total Incl O&P
0010	**GENERAL FILL**									
0011	Spread dumped material, no compaction									
0020	By dozer	B-10B	1000	.012	L.C.Y.		.67	1.35	2.02	2.52
0100	By hand	1 Clab	12	.667	"		30.50		30.50	48.50
9000	Minimum labor/equipment charge	"	4	2	Job		91		91	145

31 23 23.20 Hauling

		Crew	Daily Output	Labor-Hours	Unit	Material	2022 Bare Costs Labor	2022 Bare Costs Equipment	Total	Total Incl O&P
0010	**HAULING**									
0011	Excavated or borrow, loose cubic yards									
0012	no loading equipment, including hauling, waiting, loading/dumping									
0013	time per cycle (wait, load, travel, unload or dump & return)									
0014	8 C.Y. truck, 15 MPH avg., cycle 0.5 miles, 10 min. wait/ld./uld.	B-34A	320	.025	L.C.Y.		1.33	1.23	2.56	3.46
0016	cycle 1 mile		272	.029			1.56	1.44	3	4.08
0018	cycle 2 miles		208	.038			2.04	1.88	3.92	5.30
0020	cycle 4 miles		144	.056			2.95	2.72	5.67	7.70
0022	cycle 6 miles		112	.071			3.79	3.50	7.29	9.90
0024	cycle 8 miles		88	.091			4.82	4.45	9.27	12.60
0026	20 MPH avg., cycle 0.5 mile		336	.024			1.26	1.17	2.43	3.29
0028	cycle 1 mile		296	.027			1.43	1.32	2.75	3.75
0030	cycle 2 miles		240	.033			1.77	1.63	3.40	4.62
0032	cycle 4 miles		176	.045			2.41	2.23	4.64	6.30
0034	cycle 6 miles		136	.059			3.12	2.88	6	8.15
0036	cycle 8 miles	B-34A	112	.071	L.C.Y.		3.79	3.50	7.29	9.90
0044	25 MPH avg., cycle 4 miles		192	.042			2.21	2.04	4.25	5.75
0046	cycle 6 miles		160	.050			2.65	2.45	5.10	6.95
0048	cycle 8 miles		128	.063			3.32	3.06	6.38	8.65
0050	30 MPH avg., cycle 4 miles		216	.037			1.96	1.81	3.77	5.15
0052	cycle 6 miles		176	.045			2.41	2.23	4.64	6.30
0054	cycle 8 miles		144	.056			2.95	2.72	5.67	7.70

31 23 Excavation and Fill

31 23 23 – Fill

31 23 23.20 Hauling		Crew	Daily Output	Labor-Hours	Unit	Material	2022 Bare Costs Labor	2022 Bare Costs Equipment	Total	Total Incl O&P
0114	15 MPH avg., cycle 0.5 mile, 15 min. wait/ld./uld.	B-34A	224	.036	L.C.Y.		1.89	1.75	3.64	4.95
0116	cycle 1 mile		200	.040			2.12	1.96	4.08	5.55
0118	cycle 2 miles		168	.048			2.53	2.33	4.86	6.60
0120	cycle 4 miles		120	.067			3.54	3.27	6.81	9.25
0122	cycle 6 miles		96	.083			4.42	4.08	8.50	11.55
0124	cycle 8 miles		80	.100			5.30	4.90	10.20	13.85
0126	20 MPH avg., cycle 0.5 mile		232	.034			1.83	1.69	3.52	4.78
0128	cycle 1 mile		208	.038			2.04	1.88	3.92	5.30
0130	cycle 2 miles		184	.043			2.31	2.13	4.44	6
0132	cycle 4 miles		144	.056			2.95	2.72	5.67	7.70
0134	cycle 6 miles		112	.071			3.79	3.50	7.29	9.90
0136	cycle 8 miles		96	.083			4.42	4.08	8.50	11.55
0144	25 MPH avg., cycle 4 miles		152	.053			2.79	2.58	5.37	7.30
0146	cycle 6 miles		128	.063			3.32	3.06	6.38	8.65
0148	cycle 8 miles		112	.071			3.79	3.50	7.29	9.90
0150	30 MPH avg., cycle 4 miles		168	.048			2.53	2.33	4.86	6.60
0152	cycle 6 miles		144	.056			2.95	2.72	5.67	7.70
0154	cycle 8 miles		120	.067			3.54	3.27	6.81	9.25
0214	15 MPH avg., cycle 0.5 mile, 20 min. wait/ld./uld.		176	.045			2.41	2.23	4.64	6.30
0216	cycle 1 mile		160	.050			2.65	2.45	5.10	6.95
0218	cycle 2 miles		136	.059			3.12	2.88	6	8.15
0220	cycle 4 miles		104	.077			4.08	3.77	7.85	10.65
0222	cycle 6 miles		88	.091			4.82	4.45	9.27	12.60
0224	cycle 8 miles		72	.111			5.90	5.45	11.35	15.40
0226	20 MPH avg., cycle 0.5 mile		176	.045			2.41	2.23	4.64	6.30
0228	cycle 1 mile		168	.048			2.53	2.33	4.86	6.60
0230	cycle 2 miles		144	.056			2.95	2.72	5.67	7.70
0232	cycle 4 miles		120	.067			3.54	3.27	6.81	9.25
0234	cycle 6 miles		96	.083			4.42	4.08	8.50	11.55
0236	cycle 8 miles		88	.091			4.82	4.45	9.27	12.60
0244	25 MPH avg., cycle 4 miles		128	.063			3.32	3.06	6.38	8.65
0246	cycle 6 miles	B-34A	112	.071	L.C.Y.		3.79	3.50	7.29	9.90
0248	cycle 8 miles		96	.083			4.42	4.08	8.50	11.55
0250	30 MPH avg., cycle 4 miles		136	.059			3.12	2.88	6	8.15
0252	cycle 6 miles		120	.067			3.54	3.27	6.81	9.25
0254	cycle 8 miles		104	.077			4.08	3.77	7.85	10.65
0314	15 MPH avg., cycle 0.5 mile, 25 min. wait/ld./uld.		144	.056			2.95	2.72	5.67	7.70
0316	cycle 1 mile		128	.063			3.32	3.06	6.38	8.65
0318	cycle 2 miles		112	.071			3.79	3.50	7.29	9.90
0320	cycle 4 miles		96	.083			4.42	4.08	8.50	11.55
0322	cycle 6 miles		80	.100			5.30	4.90	10.20	13.85
0324	cycle 8 miles		64	.125			6.65	6.15	12.80	17.30
0326	20 MPH avg., cycle 0.5 mile		144	.056			2.95	2.72	5.67	7.70
0328	cycle 1 mile		136	.059			3.12	2.88	6	8.15
0330	cycle 2 miles		120	.067			3.54	3.27	6.81	9.25
0332	cycle 4 miles		104	.077			4.08	3.77	7.85	10.65
0334	cycle 6 miles		88	.091			4.82	4.45	9.27	12.60
0336	cycle 8 miles		80	.100			5.30	4.90	10.20	13.85
0344	25 MPH avg., cycle 4 miles		112	.071			3.79	3.50	7.29	9.90
0346	cycle 6 miles		96	.083			4.42	4.08	8.50	11.55
0348	cycle 8 miles		88	.091			4.82	4.45	9.27	12.60
0350	30 MPH avg., cycle 4 miles		112	.071			3.79	3.50	7.29	9.90
0352	cycle 6 miles		104	.077			4.08	3.77	7.85	10.65

543

31 23 23.20 Hauling		Crew	Daily Output	Labor-Hours	Unit	Material	2022 Bare Costs		Total	Total Incl O&P
							Labor	Equipment		
0354	cycle 8 miles	B-34A	96	.083	L.C.Y.		4.42	4.08	8.50	11.55
0414	15 MPH avg., cycle 0.5 mile, 30 min. wait/ld./uld.		120	.067			3.54	3.27	6.81	9.25
0416	cycle 1 mile		112	.071			3.79	3.50	7.29	9.90
0418	cycle 2 miles		96	.083			4.42	4.08	8.50	11.55
0420	cycle 4 miles		80	.100			5.30	4.90	10.20	13.85
0422	cycle 6 miles		72	.111			5.90	5.45	11.35	15.40
0424	cycle 8 miles		64	.125			6.65	6.15	12.80	17.30
0426	20 MPH avg., cycle 0.5 mile		120	.067			3.54	3.27	6.81	9.25
0428	cycle 1 mile		112	.071			3.79	3.50	7.29	9.90
0430	cycle 2 miles		104	.077			4.08	3.77	7.85	10.65
0432	cycle 4 miles		88	.091			4.82	4.45	9.27	12.60
0434	cycle 6 miles		80	.100			5.30	4.90	10.20	13.85
0436	cycle 8 miles		72	.111			5.90	5.45	11.35	15.40
0444	25 MPH avg., cycle 4 miles		96	.083			4.42	4.08	8.50	11.55
0446	cycle 6 miles		88	.091			4.82	4.45	9.27	12.60
0448	cycle 8 miles		80	.100			5.30	4.90	10.20	13.85
0450	30 MPH avg., cycle 4 miles	B-34A	96	.083	L.C.Y.		4.42	4.08	8.50	11.55
0452	cycle 6 miles		88	.091			4.82	4.45	9.27	12.60
0454	cycle 8 miles		80	.100			5.30	4.90	10.20	13.85
0514	15 MPH avg., cycle 0.5 mile, 35 min. wait/ld./uld.		104	.077			4.08	3.77	7.85	10.65
0516	cycle 1 mile		96	.083			4.42	4.08	8.50	11.55
0518	cycle 2 miles		88	.091			4.82	4.45	9.27	12.60
0520	cycle 4 miles		72	.111			5.90	5.45	11.35	15.40
0522	cycle 6 miles		64	.125			6.65	6.15	12.80	17.30
0524	cycle 8 miles		56	.143			7.60	7	14.60	19.80
0526	20 MPH avg., cycle 0.5 mile		104	.077			4.08	3.77	7.85	10.65
0528	cycle 1 mile		96	.083			4.42	4.08	8.50	11.55
0530	cycle 2 miles		96	.083			4.42	4.08	8.50	11.55
0532	cycle 4 miles		80	.100			5.30	4.90	10.20	13.85
0534	cycle 6 miles		72	.111			5.90	5.45	11.35	15.40
0536	cycle 8 miles		64	.125			6.65	6.15	12.80	17.30
0544	25 MPH avg., cycle 4 miles		88	.091			4.82	4.45	9.27	12.60
0546	cycle 6 miles		80	.100			5.30	4.90	10.20	13.85
0548	cycle 8 miles		72	.111			5.90	5.45	11.35	15.40
0550	30 MPH avg., cycle 4 miles		88	.091			4.82	4.45	9.27	12.60
0552	cycle 6 miles		80	.100			5.30	4.90	10.20	13.85
0554	cycle 8 miles		72	.111			5.90	5.45	11.35	15.40
1014	12 C.Y. truck, cycle 0.5 mile, 15 MPH avg., 15 min. wait/ld./uld.	B-34B	336	.024			1.26	2.28	3.54	4.52
1016	cycle 1 mile		300	.027			1.41	2.56	3.97	5.05
1018	cycle 2 miles		252	.032			1.68	3.05	4.73	6.05
1020	cycle 4 miles		180	.044			2.36	4.26	6.62	8.45
1022	cycle 6 miles		144	.056			2.95	5.35	8.30	10.55
1024	cycle 8 miles		120	.067			3.54	6.40	9.94	12.70
1025	cycle 10 miles		96	.083			4.42	8	12.42	15.85
1026	20 MPH avg., cycle 0.5 mile		348	.023			1.22	2.21	3.43	4.37
1028	cycle 1 mile		312	.026			1.36	2.46	3.82	4.88
1030	cycle 2 miles		276	.029			1.54	2.78	4.32	5.50
1032	cycle 4 miles		216	.037			1.96	3.55	5.51	7.05
1034	cycle 6 miles		168	.048			2.53	4.57	7.10	9.10
1036	cycle 8 miles		144	.056			2.95	5.35	8.30	10.55
1038	cycle 10 miles		120	.067			3.54	6.40	9.94	12.70
1040	25 MPH avg., cycle 4 miles		228	.035			1.86	3.37	5.23	6.65
1042	cycle 6 miles		192	.042			2.21	4	6.21	7.90

31 23 23 – Fill

31 23 23.20 Hauling		Crew	Daily Output	Labor-Hours	Unit	Material	2022 Bare Costs Labor	2022 Bare Costs Equipment	Total	Total Incl O&P
1044	cycle 8 miles	B-34B	168	.048	L.C.Y.		2.53	4.57	7.10	9.10
1046	cycle 10 miles	B-34B	144	.056	L.C.Y.		2.95	5.35	8.30	10.55
1050	30 MPH avg., cycle 4 miles		252	.032			1.68	3.05	4.73	6.05
1052	cycle 6 miles		216	.037			1.96	3.55	5.51	7.05
1054	cycle 8 miles		180	.044			2.36	4.26	6.62	8.45
1056	cycle 10 miles		156	.051			2.72	4.92	7.64	9.75
1060	35 MPH avg., cycle 4 miles		264	.030			1.61	2.91	4.52	5.75
1062	cycle 6 miles		228	.035			1.86	3.37	5.23	6.65
1064	cycle 8 miles		204	.039			2.08	3.76	5.84	7.45
1066	cycle 10 miles		180	.044			2.36	4.26	6.62	8.45
1068	cycle 20 miles		120	.067			3.54	6.40	9.94	12.70
1069	cycle 30 miles		84	.095			5.05	9.15	14.20	18.10
1070	cycle 40 miles		72	.111			5.90	10.65	16.55	21
1072	40 MPH avg., cycle 6 miles		240	.033			1.77	3.20	4.97	6.35
1074	cycle 8 miles		216	.037			1.96	3.55	5.51	7.05
1076	cycle 10 miles		192	.042			2.21	4	6.21	7.90
1078	cycle 20 miles		120	.067			3.54	6.40	9.94	12.70
1080	cycle 30 miles		96	.083			4.42	8	12.42	15.85
1082	cycle 40 miles		72	.111			5.90	10.65	16.55	21
1084	cycle 50 miles		60	.133			7.05	12.80	19.85	25.50
1094	45 MPH avg., cycle 8 miles		216	.037			1.96	3.55	5.51	7.05
1096	cycle 10 miles		204	.039			2.08	3.76	5.84	7.45
1098	cycle 20 miles		132	.061			3.22	5.80	9.02	11.50
1100	cycle 30 miles		108	.074			3.93	7.10	11.03	14.05
1102	cycle 40 miles		84	.095			5.05	9.15	14.20	18.10
1104	cycle 50 miles		72	.111			5.90	10.65	16.55	21
1106	50 MPH avg., cycle 10 miles		216	.037			1.96	3.55	5.51	7.05
1108	cycle 20 miles		144	.056			2.95	5.35	8.30	10.55
1110	cycle 30 miles		108	.074			3.93	7.10	11.03	14.05
1112	cycle 40 miles		84	.095			5.05	9.15	14.20	18.10
1114	cycle 50 miles		72	.111			5.90	10.65	16.55	21
1214	15 MPH avg., cycle 0.5 mile, 20 min. wait/ld./uld.		264	.030			1.61	2.91	4.52	5.75
1216	cycle 1 mile		240	.033			1.77	3.20	4.97	6.35
1218	cycle 2 miles		204	.039			2.08	3.76	5.84	7.45
1220	cycle 4 miles		156	.051			2.72	4.92	7.64	9.75
1222	cycle 6 miles		132	.061			3.22	5.80	9.02	11.50
1224	cycle 8 miles		108	.074			3.93	7.10	11.03	14.05
1225	cycle 10 miles		96	.083			4.42	8	12.42	15.85
1226	20 MPH avg., cycle 0.5 mile		264	.030			1.61	2.91	4.52	5.75
1228	cycle 1 mile		252	.032			1.68	3.05	4.73	6.05
1230	cycle 2 miles	B-34B	216	.037	L.C.Y.		1.96	3.55	5.51	7.05
1232	cycle 4 miles		180	.044			2.36	4.26	6.62	8.45
1234	cycle 6 miles		144	.056			2.95	5.35	8.30	10.55
1236	cycle 8 miles		132	.061			3.22	5.80	9.02	11.50
1238	cycle 10 miles		108	.074			3.93	7.10	11.03	14.05
1240	25 MPH avg., cycle 4 miles		192	.042			2.21	4	6.21	7.90
1242	cycle 6 miles		168	.048			2.53	4.57	7.10	9.10
1244	cycle 8 miles		144	.056			2.95	5.35	8.30	10.55
1246	cycle 10 miles		132	.061			3.22	5.80	9.02	11.50
1250	30 MPH avg., cycle 4 miles		204	.039			2.08	3.76	5.84	7.45
1252	cycle 6 miles		180	.044			2.36	4.26	6.62	8.45
1254	cycle 8 miles		156	.051			2.72	4.92	7.64	9.75
1256	cycle 10 miles		144	.056			2.95	5.35	8.30	10.55

545

31 23 23.20 Hauling		Crew	Daily Output	Labor-Hours	Unit	Material	2022 Bare Costs Labor	2022 Bare Costs Equipment	Total	Total Incl O&P
1260	35 MPH avg., cycle 4 miles	B-34B	216	.037	L.C.Y.		1.96	3.55	5.51	7.05
1262	cycle 6 miles		192	.042			2.21	4	6.21	7.90
1264	cycle 8 miles		168	.048			2.53	4.57	7.10	9.10
1266	cycle 10 miles		156	.051			2.72	4.92	7.64	9.75
1268	cycle 20 miles		108	.074			3.93	7.10	11.03	14.05
1269	cycle 30 miles		72	.111			5.90	10.65	16.55	21
1270	cycle 40 miles		60	.133			7.05	12.80	19.85	25.50
1272	40 MPH avg., cycle 6 miles		192	.042			2.21	4	6.21	7.90
1274	cycle 8 miles		180	.044			2.36	4.26	6.62	8.45
1276	cycle 10 miles		156	.051			2.72	4.92	7.64	9.75
1278	cycle 20 miles		108	.074			3.93	7.10	11.03	14.05
1280	cycle 30 miles		84	.095			5.05	9.15	14.20	18.10
1282	cycle 40 miles		72	.111			5.90	10.65	16.55	21
1284	cycle 50 miles		60	.133			7.05	12.80	19.85	25.50
1294	45 MPH avg., cycle 8 miles		180	.044			2.36	4.26	6.62	8.45
1296	cycle 10 miles		168	.048			2.53	4.57	7.10	9.10
1298	cycle 20 miles		120	.067			3.54	6.40	9.94	12.70
1300	cycle 30 miles		96	.083			4.42	8	12.42	15.85
1302	cycle 40 miles		72	.111			5.90	10.65	16.55	21
1304	cycle 50 miles		60	.133			7.05	12.80	19.85	25.50
1306	50 MPH avg., cycle 10 miles		180	.044			2.36	4.26	6.62	8.45
1308	cycle 20 miles		132	.061			3.22	5.80	9.02	11.50
1310	cycle 30 miles		96	.083			4.42	8	12.42	15.85
1312	cycle 40 miles		84	.095			5.05	9.15	14.20	18.10
1314	cycle 50 miles		72	.111			5.90	10.65	16.55	21
1414	15 MPH avg., cycle 0.5 mile, 25 min. wait/ld./uld.		204	.039			2.08	3.76	5.84	7.45
1416	cycle 1 mile	B-34B	192	.042	L.C.Y.		2.21	4	6.21	7.90
1418	cycle 2 miles		168	.048			2.53	4.57	7.10	9.10
1420	cycle 4 miles		132	.061			3.22	5.80	9.02	11.50
1422	cycle 6 miles		120	.067			3.54	6.40	9.94	12.70
1424	cycle 8 miles		96	.083			4.42	8	12.42	15.85
1425	cycle 10 miles		84	.095			5.05	9.15	14.20	18.10
1426	20 MPH avg., cycle 0.5 mile		216	.037			1.96	3.55	5.51	7.05
1428	cycle 1 mile		204	.039			2.08	3.76	5.84	7.45
1430	cycle 2 miles		180	.044			2.36	4.26	6.62	8.45
1432	cycle 4 miles		156	.051			2.72	4.92	7.64	9.75
1434	cycle 6 miles		132	.061			3.22	5.80	9.02	11.50
1436	cycle 8 miles		120	.067			3.54	6.40	9.94	12.70
1438	cycle 10 miles		96	.083			4.42	8	12.42	15.85
1440	25 MPH avg., cycle 4 miles		168	.048			2.53	4.57	7.10	9.10
1442	cycle 6 miles		144	.056			2.95	5.35	8.30	10.55
1444	cycle 8 miles		132	.061			3.22	5.80	9.02	11.50
1446	cycle 10 miles		108	.074			3.93	7.10	11.03	14.05
1450	30 MPH avg., cycle 4 miles		168	.048			2.53	4.57	7.10	9.10
1452	cycle 6 miles		156	.051			2.72	4.92	7.64	9.75
1454	cycle 8 miles		132	.061			3.22	5.80	9.02	11.50
1456	cycle 10 miles		120	.067			3.54	6.40	9.94	12.70
1460	35 MPH avg., cycle 4 miles		180	.044			2.36	4.26	6.62	8.45
1462	cycle 6 miles		156	.051			2.72	4.92	7.64	9.75
1464	cycle 8 miles		144	.056			2.95	5.35	8.30	10.55
1466	cycle 10 miles		132	.061			3.22	5.80	9.02	11.50
1468	cycle 20 miles		96	.083			4.42	8	12.42	15.85
1469	cycle 30 miles		72	.111			5.90	10.65	16.55	21

31 23 23 – Fill

31 23 23.20 Hauling		Crew	Daily Output	Labor-Hours	Unit	Material	2022 Bare Costs Labor	Equipment	Total	Total Incl O&P
1470	cycle 40 miles	B-34B	60	.133	L.C.Y.		7.05	12.80	19.85	25.50
1472	40 MPH avg., cycle 6 miles		168	.048			2.53	4.57	7.10	9.10
1474	cycle 8 miles		156	.051			2.72	4.92	7.64	9.75
1476	cycle 10 miles		144	.056			2.95	5.35	8.30	10.55
1478	cycle 20 miles		96	.083			4.42	8	12.42	15.85
1480	cycle 30 miles		84	.095			5.05	9.15	14.20	18.10
1482	cycle 40 miles		60	.133			7.05	12.80	19.85	25.50
1484	cycle 50 miles		60	.133			7.05	12.80	19.85	25.50
1494	45 MPH avg., cycle 8 miles		156	.051			2.72	4.92	7.64	9.75
1496	cycle 10 miles		144	.056			2.95	5.35	8.30	10.55
1498	cycle 20 miles		108	.074			3.93	7.10	11.03	14.05
1500	cycle 30 miles	↓	84	.095	↓		5.05	9.15	14.20	18.10
1502	cycle 40 miles	B-34B	72	.111	L.C.Y.		5.90	10.65	16.55	21
1504	cycle 50 miles		60	.133			7.05	12.80	19.85	25.50
1506	50 MPH avg., cycle 10 miles		156	.051			2.72	4.92	7.64	9.75
1508	cycle 20 miles		120	.067			3.54	6.40	9.94	12.70
1510	cycle 30 miles		96	.083			4.42	8	12.42	15.85
1512	cycle 40 miles		72	.111			5.90	10.65	16.55	21
1514	cycle 50 miles		60	.133			7.05	12.80	19.85	25.50
1614	15 MPH avg., cycle 0.5 mile, 30 min. wait/ld./uld.		180	.044			2.36	4.26	6.62	8.45
1616	cycle 1 mile		168	.048			2.53	4.57	7.10	9.10
1618	cycle 2 miles		144	.056			2.95	5.35	8.30	10.55
1620	cycle 4 miles		120	.067			3.54	6.40	9.94	12.70
1622	cycle 6 miles		108	.074			3.93	7.10	11.03	14.05
1624	cycle 8 miles		84	.095			5.05	9.15	14.20	18.10
1625	cycle 10 miles		84	.095			5.05	9.15	14.20	18.10
1626	20 MPH avg., cycle 0.5 mile		180	.044			2.36	4.26	6.62	8.45
1628	cycle 1 mile		168	.048			2.53	4.57	7.10	9.10
1630	cycle 2 miles		156	.051			2.72	4.92	7.64	9.75
1632	cycle 4 miles		132	.061			3.22	5.80	9.02	11.50
1634	cycle 6 miles		120	.067			3.54	6.40	9.94	12.70
1636	cycle 8 miles		108	.074			3.93	7.10	11.03	14.05
1638	cycle 10 miles		96	.083			4.42	8	12.42	15.85
1640	25 MPH avg., cycle 4 miles		144	.056			2.95	5.35	8.30	10.55
1642	cycle 6 miles		132	.061			3.22	5.80	9.02	11.50
1644	cycle 8 miles		108	.074			3.93	7.10	11.03	14.05
1646	cycle 10 miles		108	.074			3.93	7.10	11.03	14.05
1650	30 MPH avg., cycle 4 miles		144	.056			2.95	5.35	8.30	10.55
1652	cycle 6 miles		132	.061			3.22	5.80	9.02	11.50
1654	cycle 8 miles		120	.067			3.54	6.40	9.94	12.70
1656	cycle 10 miles		108	.074			3.93	7.10	11.03	14.05
1660	35 MPH avg., cycle 4 miles		156	.051			2.72	4.92	7.64	9.75
1662	cycle 6 miles		144	.056			2.95	5.35	8.30	10.55
1664	cycle 8 miles		132	.061			3.22	5.80	9.02	11.50
1666	cycle 10 miles		120	.067			3.54	6.40	9.94	12.70
1668	cycle 20 miles		84	.095			5.05	9.15	14.20	18.10
1669	cycle 30 miles		72	.111			5.90	10.65	16.55	21
1670	cycle 40 miles		60	.133			7.05	12.80	19.85	25.50
1672	40 MPH avg., cycle 6 miles		144	.056			2.95	5.35	8.30	10.55
1674	cycle 8 miles		132	.061			3.22	5.80	9.02	11.50
1676	cycle 10 miles	↓	120	.067	↓		3.54	6.40	9.94	12.70
1678	cycle 20 miles	B-34B	96	.083	L.C.Y.		4.42	8	12.42	15.85
1680	cycle 30 miles	↓	72	.111			5.90	10.65	16.55	21

For customer support on your Commercial Renovation Costs with RSMeans data, call 800.448.8182.

547

31 23 23.20 Hauling		Crew	Daily Output	Labor-Hours	Unit	Material	2022 Bare Costs Labor	Equipment	Total	Total Incl O&P
1682	cycle 40 miles	B-34B	60	.133	L.C.Y.		7.05	12.80	19.85	25.50
1684	cycle 50 miles		48	.167			8.85	16	24.85	31.50
1694	45 MPH avg., cycle 8 miles		144	.056			2.95	5.35	8.30	10.55
1696	cycle 10 miles		132	.061			3.22	5.80	9.02	11.50
1698	cycle 20 miles		96	.083			4.42	8	12.42	15.85
1700	cycle 30 miles		84	.095			5.05	9.15	14.20	18.10
1702	cycle 40 miles		60	.133			7.05	12.80	19.85	25.50
1704	cycle 50 miles		60	.133			7.05	12.80	19.85	25.50
1706	50 MPH avg., cycle 10 miles		132	.061			3.22	5.80	9.02	11.50
1708	cycle 20 miles		108	.074			3.93	7.10	11.03	14.05
1710	cycle 30 miles		84	.095			5.05	9.15	14.20	18.10
1712	cycle 40 miles		72	.111			5.90	10.65	16.55	21
1714	cycle 50 miles		60	.133			7.05	12.80	19.85	25.50
2000	Hauling, 8 C.Y. truck, small project cost per hour	B-34A	8	1	Hr.		53	49	102	139
2100	12 C.Y. truck	B-34B	8	1			53	96	149	191
2150	16.5 C.Y. truck	B-34C	8	1			53	82	135	175
2175	18 C.Y. 8 wheel truck	B-34I	8	1			53	95	148	190
2200	20 C.Y. truck	B-34D	8	1			53	84	137	177
2300	Grading at dump, or embankment if required, by dozer	B-10B	1000	.012	L.C.Y.		.67	1.35	2.02	2.52
2310	Spotter at fill or cut, if required	1 Clab	8	1	Hr.		45.50		45.50	72.50
9014	18 C.Y. truck, 8 wheels,15 min. wait/ld./uld.,15 MPH, cycle 0.5 mi.	B-34I	504	.016	L.C.Y.		.84	1.51	2.35	3
9016	cycle 1 mile		450	.018			.94	1.69	2.63	3.36
9018	cycle 2 miles		378	.021			1.12	2.02	3.14	4.01
9020	cycle 4 miles		270	.030			1.57	2.82	4.39	5.60
9022	cycle 6 miles		216	.037			1.96	3.53	5.49	7
9024	cycle 8 miles		180	.044			2.36	4.23	6.59	8.40
9025	cycle 10 miles		144	.056			2.95	5.30	8.25	10.50
9026	20 MPH avg., cycle 0.5 mile		522	.015			.81	1.46	2.27	2.91
9028	cycle 1 mile		468	.017			.91	1.63	2.54	3.24
9030	cycle 2 miles		414	.019			1.03	1.84	2.87	3.65
9032	cycle 4 miles		324	.025			1.31	2.35	3.66	4.68
9034	cycle 6 miles		252	.032			1.68	3.02	4.70	6
9036	cycle 8 miles		216	.037			1.96	3.53	5.49	7
9038	cycle 10 miles		180	.044			2.36	4.23	6.59	8.40
9040	25 MPH avg., cycle 4 miles		342	.023			1.24	2.23	3.47	4.43
9042	cycle 6 miles		288	.028			1.47	2.64	4.11	5.25
9044	cycle 8 miles	B-34I	252	.032	L.C.Y.		1.68	3.02	4.70	6
9046	cycle 10 miles		216	.037			1.96	3.53	5.49	7
9050	30 MPH avg., cycle 4 miles		378	.021			1.12	2.02	3.14	4.01
9052	cycle 6 miles		324	.025			1.31	2.35	3.66	4.68
9054	cycle 8 miles		270	.030			1.57	2.82	4.39	5.60
9056	cycle 10 miles		234	.034			1.81	3.26	5.07	6.45
9060	35 MPH avg., cycle 4 miles		396	.020			1.07	1.92	2.99	3.83
9062	cycle 6 miles		342	.023			1.24	2.23	3.47	4.43
9064	cycle 8 miles		288	.028			1.47	2.64	4.11	5.25
9066	cycle 10 miles		270	.030			1.57	2.82	4.39	5.60
9068	cycle 20 miles		162	.049			2.62	4.70	7.32	9.35
9070	cycle 30 miles		126	.063			3.37	6.05	9.42	12
9072	cycle 40 miles		90	.089			4.72	8.45	13.17	16.80
9074	40 MPH avg., cycle 6 miles		360	.022			1.18	2.12	3.30	4.21
9076	cycle 8 miles		324	.025			1.31	2.35	3.66	4.68
9078	cycle 10 miles		288	.028			1.47	2.64	4.11	5.25
9080	cycle 20 miles		180	.044			2.36	4.23	6.59	8.40

31 23 23 – Fill

31 23 23.20 Hauling		Crew	Daily Output	Labor-Hours	Unit	Material	Labor	Equipment	Total	Total Incl O&P
9082	cycle 30 miles	B-34I	144	.056	L.C.Y.		2.95	5.30	8.25	10.50
9084	cycle 40 miles		108	.074			3.93	7.05	10.98	14
9086	cycle 50 miles		90	.089			4.72	8.45	13.17	16.80
9094	45 MPH avg., cycle 8 miles		324	.025			1.31	2.35	3.66	4.68
9096	cycle 10 miles		306	.026			1.39	2.49	3.88	4.95
9098	cycle 20 miles		198	.040			2.14	3.85	5.99	7.65
9100	cycle 30 miles		144	.056			2.95	5.30	8.25	10.50
9102	cycle 40 miles		126	.063			3.37	6.05	9.42	12
9104	cycle 50 miles		108	.074			3.93	7.05	10.98	14
9106	50 MPH avg., cycle 10 miles		324	.025			1.31	2.35	3.66	4.68
9108	cycle 20 miles		216	.037			1.96	3.53	5.49	7
9110	cycle 30 miles		162	.049			2.62	4.70	7.32	9.35
9112	cycle 40 miles		126	.063			3.37	6.05	9.42	12
9114	cycle 50 miles		108	.074			3.93	7.05	10.98	14
9214	20 min. wait/ld./uld.,15 MPH, cycle 0.5 mi.		396	.020			1.07	1.92	2.99	3.83
9216	cycle 1 mile		360	.022			1.18	2.12	3.30	4.21
9218	cycle 2 miles		306	.026			1.39	2.49	3.88	4.95
9220	cycle 4 miles		234	.034			1.81	3.26	5.07	6.45
9222	cycle 6 miles		198	.040			2.14	3.85	5.99	7.65
9224	cycle 8 miles		162	.049			2.62	4.70	7.32	9.35
9225	cycle 10 miles		144	.056			2.95	5.30	8.25	10.50
9226	20 MPH avg., cycle 0.5 mile		396	.020			1.07	1.92	2.99	3.83
9228	cycle 1 mile	B-34I	378	.021	L.C.Y.		1.12	2.02	3.14	4.01
9230	cycle 2 miles		324	.025			1.31	2.35	3.66	4.68
9232	cycle 4 miles		270	.030			1.57	2.82	4.39	5.60
9234	cycle 6 miles		216	.037			1.96	3.53	5.49	7
9236	cycle 8 miles		198	.040			2.14	3.85	5.99	7.65
9238	cycle 10 miles		162	.049			2.62	4.70	7.32	9.35
9240	25 MPH avg., cycle 4 miles		288	.028			1.47	2.64	4.11	5.25
9242	cycle 6 miles		252	.032			1.68	3.02	4.70	6
9244	cycle 8 miles		216	.037			1.96	3.53	5.49	7
9246	cycle 10 miles		198	.040			2.14	3.85	5.99	7.65
9250	30 MPH avg., cycle 4 miles		306	.026			1.39	2.49	3.88	4.95
9252	cycle 6 miles		270	.030			1.57	2.82	4.39	5.60
9254	cycle 8 miles		234	.034			1.81	3.26	5.07	6.45
9256	cycle 10 miles		216	.037			1.96	3.53	5.49	7
9260	35 MPH avg., cycle 4 miles		324	.025			1.31	2.35	3.66	4.68
9262	cycle 6 miles		288	.028			1.47	2.64	4.11	5.25
9264	cycle 8 miles		252	.032			1.68	3.02	4.70	6
9266	cycle 10 miles		234	.034			1.81	3.26	5.07	6.45
9268	cycle 20 miles		162	.049			2.62	4.70	7.32	9.35
9270	cycle 30 miles		108	.074			3.93	7.05	10.98	14
9272	cycle 40 miles		90	.089			4.72	8.45	13.17	16.80
9274	40 MPH avg., cycle 6 miles		288	.028			1.47	2.64	4.11	5.25
9276	cycle 8 miles		270	.030			1.57	2.82	4.39	5.60
9278	cycle 10 miles		234	.034			1.81	3.26	5.07	6.45
9280	cycle 20 miles		162	.049			2.62	4.70	7.32	9.35
9282	cycle 30 miles		126	.063			3.37	6.05	9.42	12
9284	cycle 40 miles		108	.074			3.93	7.05	10.98	14
9286	cycle 50 miles		90	.089			4.72	8.45	13.17	16.80
9294	45 MPH avg., cycle 8 miles		270	.030			1.57	2.82	4.39	5.60
9296	cycle 10 miles		252	.032			1.68	3.02	4.70	6
9298	cycle 20 miles		180	.044			2.36	4.23	6.59	8.40

31 23 23.20 Hauling		Crew	Daily Output	Labor-Hours	Unit	Material	2022 Bare Costs		Total	Total Incl O&P
							Labor	Equipment		
9300	cycle 30 miles	B-34I	144	.056	L.C.Y.		2.95	5.30	8.25	10.50
9302	cycle 40 miles		108	.074			3.93	7.05	10.98	14
9304	cycle 50 miles		90	.089			4.72	8.45	13.17	16.80
9306	50 MPH avg., cycle 10 miles		270	.030			1.57	2.82	4.39	5.60
9308	cycle 20 miles		198	.040			2.14	3.85	5.99	7.65
9310	cycle 30 miles		144	.056			2.95	5.30	8.25	10.50
9312	cycle 40 miles		126	.063			3.37	6.05	9.42	12
9314	cycle 50 miles		108	.074			3.93	7.05	10.98	14
9414	25 min. wait/ld./uld., 15 MPH, cycle 0.5 mi.	B-34I	306	.026	L.C.Y.		1.39	2.49	3.88	4.95
9416	cycle 1 mile		288	.028			1.47	2.64	4.11	5.25
9418	cycle 2 miles		252	.032			1.68	3.02	4.70	6
9420	cycle 4 miles		198	.040			2.14	3.85	5.99	7.65
9422	cycle 6 miles		180	.044			2.36	4.23	6.59	8.40
9424	cycle 8 miles		144	.056			2.95	5.30	8.25	10.50
9425	cycle 10 miles		126	.063			3.37	6.05	9.42	12
9426	20 MPH avg., cycle 0.5 mile		324	.025			1.31	2.35	3.66	4.68
9428	cycle 1 mile		306	.026			1.39	2.49	3.88	4.95
9430	cycle 2 miles		270	.030			1.57	2.82	4.39	5.60
9432	cycle 4 miles		234	.034			1.81	3.26	5.07	6.45
9434	cycle 6 miles		198	.040			2.14	3.85	5.99	7.65
9436	cycle 8 miles		180	.044			2.36	4.23	6.59	8.40
9438	cycle 10 miles		144	.056			2.95	5.30	8.25	10.50
9440	25 MPH avg., cycle 4 miles		252	.032			1.68	3.02	4.70	6
9442	cycle 6 miles		216	.037			1.96	3.53	5.49	7
9444	cycle 8 miles		198	.040			2.14	3.85	5.99	7.65
9446	cycle 10 miles		180	.044			2.36	4.23	6.59	8.40
9450	30 MPH avg., cycle 4 miles		252	.032			1.68	3.02	4.70	6
9452	cycle 6 miles		234	.034			1.81	3.26	5.07	6.45
9454	cycle 8 miles		198	.040			2.14	3.85	5.99	7.65
9456	cycle 10 miles		180	.044			2.36	4.23	6.59	8.40
9460	35 MPH avg., cycle 4 miles		270	.030			1.57	2.82	4.39	5.60
9462	cycle 6 miles		234	.034			1.81	3.26	5.07	6.45
9464	cycle 8 miles		216	.037			1.96	3.53	5.49	7
9466	cycle 10 miles		198	.040			2.14	3.85	5.99	7.65
9468	cycle 20 miles		144	.056			2.95	5.30	8.25	10.50
9470	cycle 30 miles		108	.074			3.93	7.05	10.98	14
9472	cycle 40 miles		90	.089			4.72	8.45	13.17	16.80
9474	40 MPH avg., cycle 6 miles		252	.032			1.68	3.02	4.70	6
9476	cycle 8 miles		234	.034			1.81	3.26	5.07	6.45
9478	cycle 10 miles		216	.037			1.96	3.53	5.49	7
9480	cycle 20 miles		144	.056			2.95	5.30	8.25	10.50
9482	cycle 30 miles		126	.063			3.37	6.05	9.42	12
9484	cycle 40 miles		90	.089			4.72	8.45	13.17	16.80
9486	cycle 50 miles		90	.089			4.72	8.45	13.17	16.80
9494	45 MPH avg., cycle 8 miles		234	.034			1.81	3.26	5.07	6.45
9496	cycle 10 miles		216	.037			1.96	3.53	5.49	7
9498	cycle 20 miles		162	.049			2.62	4.70	7.32	9.35
9500	cycle 30 miles	B-34I	126	.063	L.C.Y.		3.37	6.05	9.42	12
9502	cycle 40 miles		108	.074			3.93	7.05	10.98	14
9504	cycle 50 miles		90	.089			4.72	8.45	13.17	16.80
9506	50 MPH avg., cycle 10 miles		234	.034			1.81	3.26	5.07	6.45
9508	cycle 20 miles		180	.044			2.36	4.23	6.59	8.40
9510	cycle 30 miles		144	.056			2.95	5.30	8.25	10.50

31 23 23.20 Hauling		Crew	Daily Output	Labor-Hours	Unit	Material	2022 Bare Costs Labor	2022 Bare Costs Equipment	Total	Total Incl O&P
9512	cycle 40 miles	B-34I	108	.074	L.C.Y.		3.93	7.05	10.98	14
9514	cycle 50 miles		90	.089			4.72	8.45	13.17	16.80
9614	30 min. wait/ld./uld.,15 MPH, cycle 0.5 mi.		270	.030			1.57	2.82	4.39	5.60
9616	cycle 1 mile		252	.032			1.68	3.02	4.70	6
9618	cycle 2 miles		216	.037			1.96	3.53	5.49	7
9620	cycle 4 miles		180	.044			2.36	4.23	6.59	8.40
9622	cycle 6 miles		162	.049			2.62	4.70	7.32	9.35
9624	cycle 8 miles		126	.063			3.37	6.05	9.42	12
9625	cycle 10 miles		126	.063			3.37	6.05	9.42	12
9626	20 MPH avg., cycle 0.5 mile		270	.030			1.57	2.82	4.39	5.60
9628	cycle 1 mile		252	.032			1.68	3.02	4.70	6
9630	cycle 2 miles		234	.034			1.81	3.26	5.07	6.45
9632	cycle 4 miles		198	.040			2.14	3.85	5.99	7.65
9634	cycle 6 miles		180	.044			2.36	4.23	6.59	8.40
9636	cycle 8 miles		162	.049			2.62	4.70	7.32	9.35
9638	cycle 10 miles		144	.056			2.95	5.30	8.25	10.50
9640	25 MPH avg., cycle 4 miles		216	.037			1.96	3.53	5.49	7
9642	cycle 6 miles		198	.040			2.14	3.85	5.99	7.65
9644	cycle 8 miles		180	.044			2.36	4.23	6.59	8.40
9646	cycle 10 miles		162	.049			2.62	4.70	7.32	9.35
9650	30 MPH avg., cycle 4 miles		216	.037			1.96	3.53	5.49	7
9652	cycle 6 miles		198	.040			2.14	3.85	5.99	7.65
9654	cycle 8 miles		180	.044			2.36	4.23	6.59	8.40
9656	cycle 10 miles		162	.049			2.62	4.70	7.32	9.35
9660	35 MPH avg., cycle 4 miles		234	.034			1.81	3.26	5.07	6.45
9662	cycle 6 miles		216	.037			1.96	3.53	5.49	7
9664	cycle 8 miles		198	.040			2.14	3.85	5.99	7.65
9666	cycle 10 miles		180	.044			2.36	4.23	6.59	8.40
9668	cycle 20 miles		126	.063			3.37	6.05	9.42	12
9670	cycle 30 miles		108	.074			3.93	7.05	10.98	14
9672	cycle 40 miles		90	.089			4.72	8.45	13.17	16.80
9674	40 MPH avg., cycle 6 miles		216	.037			1.96	3.53	5.49	7
9676	cycle 8 miles		198	.040			2.14	3.85	5.99	7.65
9678	cycle 10 miles	B-34I	180	.044	L.C.Y.		2.36	4.23	6.59	8.40
9680	cycle 20 miles		144	.056			2.95	5.30	8.25	10.50
9682	cycle 30 miles		108	.074			3.93	7.05	10.98	14
9684	cycle 40 miles		90	.089			4.72	8.45	13.17	16.80
9686	cycle 50 miles		72	.111			5.90	10.60	16.50	21
9694	45 MPH avg., cycle 8 miles		216	.037			1.96	3.53	5.49	7
9696	cycle 10 miles		198	.040			2.14	3.85	5.99	7.65
9698	cycle 20 miles		144	.056			2.95	5.30	8.25	10.50
9700	cycle 30 miles		126	.063			3.37	6.05	9.42	12
9702	cycle 40 miles		108	.074			3.93	7.05	10.98	14
9704	cycle 50 miles		90	.089			4.72	8.45	13.17	16.80
9706	50 MPH avg., cycle 10 miles		198	.040			2.14	3.85	5.99	7.65
9708	cycle 20 miles		162	.049			2.62	4.70	7.32	9.35
9710	cycle 30 miles		126	.063			3.37	6.05	9.42	12
9712	cycle 40 miles		108	.074			3.93	7.05	10.98	14
9714	cycle 50 miles		90	.089			4.72	8.45	13.17	16.80

For customer support on your Commercial Renovation Costs with RSMeans data, call 800.448.8182.

551

31 23 23.23 Compaction

		Crew	Daily Output	Labor-Hours	Unit	Material	2022 Bare Costs Labor	2022 Bare Costs Equipment	Total	Total Incl O&P
0010	**COMPACTION**									
5000	Riding, vibrating roller, 6" lifts, 2 passes	B-10Y	3000	.004	E.C.Y.		.22	.16	.38	.52
5020	3 passes		2300	.005			.29	.20	.49	.67
5040	4 passes		1900	.006			.35	.25	.60	.82
5050	8" lifts, 2 passes		4100	.003			.16	.11	.27	.38
5060	12" lifts, 2 passes		5200	.002			.13	.09	.22	.30
5080	3 passes		3500	.003			.19	.13	.32	.45
5100	4 passes		2600	.005			.26	.18	.44	.60
5600	Sheepsfoot or wobbly wheel roller, 6" lifts, 2 passes	B-10G	2400	.005			.28	.31	.59	.77
5620	3 passes		1735	.007			.38	.42	.80	1.06
5640	4 passes		1300	.009			.51	.56	1.07	1.42
5680	12" lifts, 2 passes		5200	.002			.13	.14	.27	.35
5700	3 passes		3500	.003			.19	.21	.40	.53
5720	4 passes		2600	.005			.26	.28	.54	.71
7000	Walk behind, vibrating plate 18" wide, 6" lifts, 2 passes	A-1D	200	.040			1.82	.56	2.38	3.52
7020	3 passes		185	.043			1.97	.61	2.58	3.80
7040	4 passes		140	.057			2.61	.80	3.41	5.05
7200	12" lifts, 2 passes, 21" wide	A-1E	560	.014			.65	.24	.89	1.31
7220	3 passes		375	.021			.97	.37	1.34	1.95
7240	4 passes		280	.029			1.30	.49	1.79	2.61
7500	Vibrating roller 24" wide, 6" lifts, 2 passes	B-10A	420	.029			1.58	.33	1.91	2.84
7520	3 passes	"	280	.043			2.38	.49	2.87	4.26
7540	4 passes	B-10A	210	.057	E.C.Y.		3.17	.65	3.82	5.70
7600	12" lifts, 2 passes		840	.014			.79	.16	.95	1.42
7620	3 passes		560	.021			1.19	.24	1.43	2.13
7640	4 passes		420	.029			1.58	.33	1.91	2.84
8000	Rammer tamper, 6" to 11", 4" lifts, 2 passes	A-1F	130	.062			2.81	.58	3.39	5.10
8050	3 passes		97	.082			3.76	.78	4.54	6.85
8100	4 passes		65	.123			5.60	1.17	6.77	10.20
8200	8" lifts, 2 passes		260	.031			1.40	.29	1.69	2.55
8250	3 passes		195	.041			1.87	.39	2.26	3.40
8300	4 passes		130	.062			2.81	.58	3.39	5.10
8400	13" to 18", 4" lifts, 2 passes	A-1G	390	.021			.94	.21	1.15	1.72
8450	3 passes		290	.028			1.26	.28	1.54	2.31
8500	4 passes		195	.041			1.87	.42	2.29	3.44
8600	8" lifts, 2 passes		780	.010			.47	.11	.58	.86
8650	3 passes		585	.014			.62	.14	.76	1.15
8700	4 passes		390	.021			.94	.21	1.15	1.72
9000	Water, 3,000 gal. truck, 3 mile haul	B-45	1888	.008		1.22	.48	.45	2.15	2.60
9010	6 mile haul		1444	.011		1.22	.63	.59	2.44	2.98
9020	12 mile haul		1000	.016		1.22	.91	.86	2.99	3.71
9030	6,000 gal. wagon, 3 mile haul	B-59	2000	.004		1.22	.21	.24	1.67	1.94
9040	6 mile haul	"	1600	.005		1.22	.27	.30	1.79	2.09

31 23 23.24 Compaction, Structural

		Crew	Daily Output	Labor-Hours	Unit	Material	2022 Bare Costs Labor	2022 Bare Costs Equipment	Total	Total Incl O&P
0010	**COMPACTION, STRUCTURAL** R312323-30									
0020	Steel wheel tandem roller, 5 tons	B-10E	8	1.500	Hr.		83	33	116	166
0050	Air tamp, 6" to 8" lifts, common fill	B-9	250	.160	E.C.Y.		7.35	1.33	8.68	13.15
0060	Select fill	"	300	.133			6.15	1.11	7.26	10.95
0600	Vibratory plate, 8" lifts, common fill	A-1D	200	.040			1.82	.56	2.38	3.52
0700	Select fill	"	216	.037			1.69	.52	2.21	3.25
9000	Minimum labor/equipment charge	1 Clab	4	2	Job		91		91	145

31 25 14 – Stabilization Measures for Erosion and Sedimentation Control

31 25 14.16 Rolled Erosion Control Mats and Blankets		Crew	Daily Output	Labor-Hours	Unit	Material	2022 Bare Costs Labor	Equipment	Total	Total Incl O&P
0010	**ROLLED EROSION CONTROL MATS AND BLANKETS**									
0020	Jute mesh, 100 S.Y. per roll, 4' wide, stapled G	B-80A	2400	.010	S.Y.	.99	.46	.16	1.61	1.98
0070	Paper biodegradable mesh G	B-1	2500	.010		.20	.44		.64	.93
0100	Plastic netting, stapled, 2" x 1" mesh, 20 mil G	"	2500	.010		1.76	.44		2.20	2.65
0120	Revegetation mat, webbed G	2 Clab	1000	.016		2.81	.73		3.54	4.25
0200	Polypropylene mesh, stapled, 6.5 oz./S.Y. G	B-1	2500	.010		1.76	.44		2.20	2.65
0300	Tobacco netting, or jute mesh #2, stapled G	"	2500	.010		.25	.44		.69	.99
0600	Straw in polymeric netting, biodegradable log	A-2	1000	.024	L.F.	2.25	1.13	.34	3.72	4.65
0705	Sediment Log, Filter Sock, 9"		1000	.024		2.26	1.13	.34	3.73	4.66
0710	Sediment Log, Filter Sock, 12"		1000	.024		3.04	1.13	.34	4.51	5.50
1000	Silt fence, install and remove G	B-62	650	.037		.53	1.82	.40	2.75	3.91
1250	Straw bale, install and remove		450	.053		1.89	2.63	.58	5.10	6.85
1275	Straw bale, synthetic, install and remove		450	.053		3.83	2.63	.58	7.04	9
1625	Erosion control, Silt Sac, for Catch Basin, Single 2ft x 2ft		24	1		45	49.50	10.95	105.45	140
2000	Snow fence, install and remove		650	.037		.57	1.82	.40	2.79	3.96

31 31 Soil Treatment

31 31 16 – Termite Control

31 31 16.13 Chemical Termite Control

31 31 16.13 Chemical Termite Control		Crew	Daily Output	Labor-Hours	Unit	Material	2022 Bare Costs Labor	Equipment	Total	Total Incl O&P
0010	**CHEMICAL TERMITE CONTROL**									
0020	Slab and walls, residential	1 Skwk	1200	.007	SF Flr.	.35	.39		.74	1.01
0030	SS mesh, no chemicals, avg 1,400 S.F. home, min G		1000	.008		.35	.47		.82	1.14
0100	Commercial, minimum		2496	.003		.40	.19		.59	.74
0200	Maximum		1645	.005		.60	.29		.89	1.11
0390	Minimum labor/equipment charge		4	2	Job		118		118	186
0400	Insecticides for termite control, minimum		14	.571	Gal.	2.45	33.50		35.95	56
0500	Maximum		11	.727	"	4.91	43		47.91	73.50

31 41 Shoring

31 41 13 – Timber Shoring

31 41 13.10 Building Shoring

31 41 13.10 Building Shoring		Crew	Daily Output	Labor-Hours	Unit	Material	2022 Bare Costs Labor	Equipment	Total	Total Incl O&P
0010	**BUILDING SHORING**									
0020	Shoring, existing building, with timber, no salvage allowance	B-51	2.20	21.818	M.B.F.	1,375	1,025	154	2,554	3,300
1000	On cribbing with 35 ton screw jacks, per box and jack		3.60	13.333	Ea.	64.50	625	94.50	784	1,175
1090	Minimum labor/equipment charge		2	24	"		1,125	170	1,295	1,950
1100	Masonry openings in walls, see Section 02 41 19.16									

31 41 16 – Sheet Piling

31 41 16.10 Sheet Piling Systems

31 41 16.10 Sheet Piling Systems		Crew	Daily Output	Labor-Hours	Unit	Material	2022 Bare Costs Labor	Equipment	Total	Total Incl O&P
0010	**SHEET PILING SYSTEMS**									
0020	Sheet piling, 50,000 psi steel, not incl. wales, 22 psf, left in place	B-40	10.81	5.920	Ton	1,825	350	335	2,510	2,925
0100	Drive, extract & salvage R314116-40		6	10.667	"	510	630	605	1,745	2,225
1200	15' deep excavation, 22 psf, left in place		983	.065	S.F.	21	3.86	3.69	28.55	33.50
1300	Drive, extract & salvage		545	.117	"	5.70	6.95	6.65	19.30	24.50
2100	Rent steel sheet piling and wales, first month				Ton	320			320	350
2200	Per added month				"	32			32	35
3900	Wood, solid sheeting, incl. wales, braces and spacers,									
3910	drive, extract & salvage, 8' deep excavation	B-31	330	.121	S.F.	2.95	5.85	.80	9.60	13.40
4520	Left in place, 8' deep, 55 S.F./hr.	B-31	440	.091	S.F.	5.30	4.38	.60	10.28	13.45
4990	Minimum labor/equipment charge	"	2	20	Job		965	132	1,097	1,675

For customer support on your Commercial Renovation Costs with RSMeans data, call 800.448.8182.

553

31 41 Shoring

31 41 16 – Sheet Piling

31 41 16.10 Sheet Piling Systems	Crew	Daily Output	Labor-Hours	Unit	Material	2022 Bare Costs Labor	Equipment	Total	Total Incl O&P
5000	For treated lumber add cost of treatment to lumber									

31 43 Concrete Raising

31 43 13 – Pressure Grouting

31 43 13.13 Concrete Pressure Grouting

		Crew	Daily Output	Labor-Hours	Unit	Material	2022 Bare Costs Labor	Equipment	Total	Total Incl O&P
0010	**CONCRETE PRESSURE GROUTING**									
0020	Grouting, pressure, cement & sand, 1:1 mix, minimum	B-61	124	.323	Bag	24	15.55	2.59	42.14	54
0100	Maximum		51	.784	"	24	38	6.30	68.30	93.50
0200	Cement and sand, 1:1 mix, minimum		250	.160	C.F.	48	7.70	1.29	56.99	66.50
0300	Maximum		100	.400		72.50	19.30	3.22	95.02	114
0400	Epoxy cement grout, minimum		137	.292		935	14.10	2.35	951.45	1,050
0500	Maximum		57	.702		935	34	5.65	974.65	1,075
0700	Alternate pricing method: (Add for materials)									
0710	5 person crew and equipment	B-61	1	40	Day		1,925	320	2,245	3,400

31 46 Needle Beams

31 46 13 – Cantilever Needle Beams

31 46 13.10 Needle Beams

		Crew	Daily Output	Labor-Hours	Unit	Material	2022 Bare Costs Labor	Equipment	Total	Total Incl O&P
0010	**NEEDLE BEAMS**									
0011	Incl. wood shoring 10' x 10' opening									
0400	Block, concrete, 8" thick	B-9	7.10	5.634	Ea.	67	259	47	373	535
0420	12" thick		6.70	5.970		84	275	49.50	408.50	580
0800	Brick, 4" thick with 8" backup block		5.70	7.018		84	325	58.50	467.50	670
1000	Brick, solid, 8" thick		6.20	6.452		67	297	53.50	417.50	605
1040	12" thick		4.90	8.163		84	375	68	527	760
1080	16" thick		4.50	8.889		117	410	74	601	860
2000	Add for additional floors of shoring	B-1	6	4		67	185		252	370
9000	Minimum labor/equipment charge	"	2	12	Job		555		555	880

31 48 Underpinning

31 48 13 – Underpinning Piers

31 48 13.10 Underpinning Foundations

		Crew	Daily Output	Labor-Hours	Unit	Material	2022 Bare Costs Labor	Equipment	Total	Total Incl O&P
0010	**UNDERPINNING FOUNDATIONS**									
0011	Including excavation,									
0020	forming, reinforcing, concrete and equipment									
0100	5' to 16' below grade, 100 to 500 C.Y.	B-52	2.30	24.348	C.Y.	345	1,275	257	1,877	2,650
0200	Over 500 C.Y.		2.50	22.400		315	1,175	236	1,726	2,450
0400	16' to 25' below grade, 100 to 500 C.Y.		2	28		380	1,450	295	2,125	3,050
0500	Over 500 C.Y.		2.10	26.667		360	1,400	281	2,041	2,900
0700	26' to 40' below grade, 100 to 500 C.Y.		1.60	35		415	1,825	370	2,610	3,750
0800	Over 500 C.Y.	B-52	1.80	31.111	C.Y.	380	1,625	330	2,335	3,350
0900	For under 50 C.Y., add				"	10%	40%			

31 62 Driven Piles

31 62 13 – Concrete Piles

31 62 13.23 Prestressed Concrete Piles

		Crew	Daily Output	Labor-Hours	Unit	Material	2022 Bare Costs Labor	Equipment	Total	Total Incl O&P
0010	**PRESTRESSED CONCRETE PILES**, 200 piles									
0020	Unless specified otherwise, not incl. pile caps or mobilization									
3100	Precast, prestressed, 40' long, 10" thick, square	B-19	700	.091	V.L.F.	13.95	5.40	3	22.35	27
3200	12" thick, square		680	.094		25	5.55	3.09	33.64	39.50
3400	14" thick, square	↓	600	.107	↓	28	6.30	3.51	37.81	45

31 62 16 – Steel Piles

31 62 16.13 Steel Piles

		Crew	Daily Output	Labor-Hours	Unit	Material	2022 Bare Costs Labor	Equipment	Total	Total Incl O&P
0010	**STEEL PILES**									
0100	Step tapered, round, concrete filled									
0110	8" tip, 12" butt, 60 ton capacity, 30' depth	B-19	760	.004	V.L.F.	17.05	4.99	2.77	24.81	29.50
0120	60' depth with extension		740	.086		36.50	5.10	2.84	44.44	51
0130	80' depth with extensions		700	.091		56.50	5.40	3	64.90	74
0250	"H" Sections, 50' long, HP8 x 36	↓	640	.100	↓	17.05	5.90	3.29	26.24	31.50

31 62 19 – Timber Piles

31 62 19.10 Wood Piles

		Crew	Daily Output	Labor-Hours	Unit	Material	2022 Bare Costs Labor	Equipment	Total	Total Incl O&P
0010	**WOOD PILES**									
0011	Friction or end bearing, not including									
0050	mobilization or demobilization									
0100	ACZA treated piles, 1.0 lb./C.F., up to 30' long, 12" butts, 8" points	B-19	625	.102	V.L.F.	14.50	6.05	3.36	23.91	29.50
0200	30' to 39' long, 12" butts, 8" points	"	700	.091	"	15.30	5.40	3	23.70	28.50
0800	ACZA Treated piles, 1.5 lb./C.F.									
0810	friction or end bearing, ASTM class B									
1000	Up to 30' long, 12" butts, 8" points	B-19	625	.102	V.L.F.	17.20	6.05	3.36	26.61	32
1100	30' to 39' long, 12" butts, 8" points		700	.091		17.80	5.40	3	26.20	31.50
2700	Mobilization for 10,000 L.F. pile job, add		3300	.019			1.15	.64	1.79	2.52
2800	25,000 L.F. pile job, add	↓	8500	.008	↓		.45	.25	.70	.98

31 62 23 – Composite Piles

31 62 23.13 Concrete-Filled Steel Piles

		Crew	Daily Output	Labor-Hours	Unit	Material	2022 Bare Costs Labor	Equipment	Total	Total Incl O&P
0010	**CONCRETE-FILLED STEEL PILES** no mobilization or demobilization									
2600	Pipe piles, 50' L, 8" diam., 29 lb./L.F., no concrete	B-19	500	.128	V.L.F.	18.70	7.60	4.21	30.51	37
2700	Concrete filled		460	.139		23.50	8.25	4.57	36.32	44
2900	10" diameter, 34 lb./L.F., no concrete		500	.128		22.50	7.60	4.21	34.31	41.50
3000	Concrete filled		450	.142		28.50	8.40	4.67	41.57	49.50
3200	12" diameter, 44 lb./L.F., no concrete	↓	475	.135	↓	35	8	4.43	47.43	56
3300	Concrete filled	B-19	415	.154	V.L.F.	34.50	9.15	5.05	48.70	58
3500	14" diameter, 46 lb./L.F., no concrete		430	.149		31.50	8.80	4.89	45.19	54
3600	Concrete filled	↓	355	.180	↓	37.50	10.70	5.90	54.10	65

31 63 Bored Piles

31 63 26 – Drilled Caissons

31 63 26.13 Fixed End Caisson Piles

		Crew	Daily Output	Labor-Hours	Unit	Material	2022 Bare Costs Labor	Equipment	Total	Total Incl O&P
0010	**FIXED END CAISSON PILES**									
0015	Including excavation, concrete, 50 lb. reinforcing									
0020	per C.Y., not incl. mobilization, boulder removal, disposal									
0100	Open style, machine drilled, to 50' deep, in stable ground, no [R316326-60]									
0110	casings or ground water, 18" diam., 0.065 C.Y./L.F.	B-43	200	.240	V.L.F.	12.70	12.05	3.93	28.68	37.50
0200	24" diameter, 0.116 C.Y./L.F.		190	.253	"	22.50	12.70	4.14	39.34	49.50
0210	4' bell diameter, add		20	2.400	Ea.	264	121	39.50	424.50	525
0500	48" diameter, 0.465 C.Y./L.F.	↓	100	.480	V.L.F.	91	24	7.85	122.85	147

31 63 Bored Piles

31 63 26 – Drilled Caissons

31 63 26.13 Fixed End Caisson Piles

		Crew	Daily Output	Labor-Hours	Unit	Material	2022 Bare Costs Labor	Equipment	Total	Total Incl O&P
0510	9' bell diameter, add	B-43	2	24	Ea.	1,525	1,200	395	3,120	4,000
1200	Open style, machine drilled, to 50' deep, in wet ground, pulled									
1220	casing and pumping									
1400	24" diameter, 0.116 C.Y./L.F.	B-48	125	.448	V.L.F.	22.50	23	9.80	55.30	72
1410	4' bell diameter, add	"	19.80	2.828	Ea.	264	145	62	471	585
1700	48" diameter, 0.465 C.Y./L.F.	B-49	55	1.600	V.L.F.	91	86	60.50	237.50	305
1710	9' bell diameter, add	"	3.30	26.667	Ea.	1,525	1,425	1,000	3,950	5,050
2300	Open style, machine drilled, to 50' deep, in soft rocks and									
2320	medium hard shales									
2500	24" diameter, 0.116 C.Y./L.F.	B-49	30	2.933	V.L.F.	22.50	158	111	291.50	395
2510	4' bell diameter, add		10.90	8.073	Ea.	264	435	305	1,004	1,300
2800	48" diameter, 0.465 C.Y./L.F.		10	8.800	V.L.F.	91	475	330	896	1,200
2810	9' bell diameter, add		1.10	80	Ea.	1,150	4,300	3,025	8,475	11,400
3600	For rock excavation, sockets, add, minimum		120	.733	C.F.		39.50	27.50	67	92.50
3650	Average		95	.926			50	35	85	117
3700	Maximum		48	1.833			99	69	168	232
3900	For 50' to 100' deep, add				V.L.F.				7%	7%
4000	For 100' to 150' deep, add								25%	25%
4100	For 150' to 200' deep, add								30%	30%
4200	For casings left in place, add				Lb.	1.34			1.34	1.47
4300	For other than 50 lb. reinf. per C.Y., add or deduct				"	1.26			1.26	1.39
4400	For steel I-beam cores, add	B-49	8.30	10.602	Ton	2,625	570	400	3,595	4,225
4500	Load and haul excess excavation, 2 miles	B-34B	178	.045	L.C.Y.		2.38	4.31	6.69	8.55
4600	For mobilization, 50 mile radius, rig to 36"	B-43	2	24	Ea.		1,200	395	1,595	2,325
4650	Rig to 84"	B-48	1.75	32			1,650	700	2,350	3,350
4700	For low headroom, add								50%	50%
4750	For difficult access, add								25%	25%

31 63 29 – Drilled Concrete Piers and Shafts

31 63 29.13 Uncased Drilled Concrete Piers

		Crew	Daily Output	Labor-Hours	Unit	Material	2022 Bare Costs Labor	Equipment	Total	Total Incl O&P
0010	**UNCASED DRILLED CONCRETE PIERS**									
0020	Unless specified otherwise, not incl. pile caps or mobilization									
0800	Cast in place friction pile, 50' long, fluted,									
0810	tapered steel, 4,000 psi concrete, no reinforcing									
0900	12" diameter, 7 ga.	B-19	600	.107	V.L.F.	28.50	6.30	3.51	38.31	45
1200	18" diameter, 7 ga.	"	480	.133	"	42.50	7.90	4.38	54.78	64.50
1300	End bearing, fluted, constant diameter,									
1320	4,000 psi concrete, no reinforcing									
1340	12" diameter, 7 ga.	B-19	600	.107	V.L.F.	30	6.30	3.51	39.81	47
1400	18" diameter, 7 ga.	"	480	.133	"	47.50	7.90	4.38	59.78	70

Estimating Tips
32 01 00 Operations and Maintenance of Exterior Improvements

- Recycling of asphalt pavement is becoming very popular and is an alternative to removal and replacement. It can be a good value engineering proposal if removed pavement can be recycled, either at the project site or at another site that is reasonably close to the project site. Sections on repair of flexible and rigid pavement are included.

32 10 00 Bases, Ballasts, and Paving

- When estimating paving, keep in mind the project schedule. Also note that prices for asphalt and concrete are generally higher in the cold seasons. Lines for pavement markings, including tactile warning systems and fence lines, are included.

32 90 00 Planting

- The timing of planting and guarantee specifications often dictate the costs for establishing tree and shrub growth and a stand of grass or ground cover. Establish the work performance schedule to coincide with the local planting season. Maintenance and growth guarantees can add 20–100% to the total landscaping cost and can be contractually cumbersome. The cost to replace trees and shrubs can be as high as 5% of the total cost, depending on the planting zone, soil conditions, and time of year.

Reference Numbers

Reference numbers are shown at the beginning of some major classifications. These numbers refer to related items in the Reference Section. The reference information may be an estimating procedure, an alternate pricing method, or technical information.

Note: Not all subdivisions listed here necessarily appear. ■

Same Data. Simplified.

Enjoy the convenience and efficiency of accessing your costs anywhere:

- **Skip the multiplier** by setting your location
- **Quickly search,** edit, favorite and share costs
- **Stay on top of price changes** with automatic updates

Discover more at rsmeans.com/online

32 01 Operation and Maintenance of Exterior Improvements

32 01 11 – Paving Cleaning

32 01 11.51 Rubber and Paint Removal From Paving

32 01 11.51 Rubber and Paint Removal From Paving	Crew	Daily Output	Labor-Hours	Unit	Material	2022 Bare Costs Labor	Equipment	Total	Total Incl O&P
0010 **RUBBER AND PAINT REMOVAL FROM PAVING**									
0015 Does not include traffic control costs									
0020 See other items in Section 32 17 23.13									
0100 Remove permanent painted traffic lines and markings	B-78A	500	.016	C.L.F.		.91	3.25	4.16	4.99
0200 Temporary traffic line tape	2 Clab	1500	.011	L.F.		.49		.49	.77
0300 Thermoplastic traffic lines and markings	B-79A	500	.024	C.L.F.		1.37	4.91	6.28	7.55
0400 Painted pavement markings	B-78B	500	.036	S.F.		1.69	.89	2.58	3.64
1000 Prepare and clean, 25,000 S.Y.	B-91A	7.36	2.174	Mile		110	84.50	194.50	268
2000 Minimum labor/equipment charge	1 Clab	2	4	Job		182		182	290

32 01 13 – Flexible Paving Surface Treatment

32 01 13.61 Slurry Seal (Latex Modified)

0010 **SLURRY SEAL (LATEX MODIFIED)**									
0011 Chip seal, slurry seal, and microsurfacing, see section 32 12 36									

32 01 13.62 Asphalt Surface Treatment

	Crew	Daily Output	Labor-Hours	Unit	Material	Labor	Equipment	Total	Total Incl O&P
0010 **ASPHALT SURFACE TREATMENT**									
3000 Pavement overlay, polypropylene									
3040 6 oz./S.Y., ideal conditions	B-63	10000	.004	S.Y.	.97	.19	.03	1.19	1.40
3190 Prime coat, bituminous, 0.28 gallon/S.Y.	B-45	2400	.007	C.S.F.	14.80	.38	.36	15.54	17.25
3200 Tack coat, emulsion, 0.05 gal./S.Y., 1,000 S.Y.	"	2500	.006	S.Y.	.39	.36	.34	1.09	1.38
3240 10,000 S.Y.	B-45	10000	.002	S.Y.	.31	.09	.09	.49	.57
3275 10,000 S.Y.		10000	.002		.59	.09	.09	.77	.87
3280 0.15 gal./S.Y., 1,000 S.Y.		2500	.006		1.07	.36	.34	1.77	2.13
3320 10,000 S.Y.		10000	.002		.86	.09	.09	1.04	1.17
3780 Rubberized asphalt (latex) seal		5000	.003		1.42	.18	.17	1.77	2.04
5400 Thermoplastic coal-tar, Type I, small or irregular area	B-90	2400	.027		3	1.35	.96	5.31	6.50
5450 Roadway or large area		8000	.008		3	.40	.29	3.69	4.26
5500 Type II, small or irregular area		2400	.027		3.83	1.35	.96	6.14	7.40
6000 Gravel surfacing on asphalt, screened and rolled	B-11L	160	.100	C.Y.	16.25	5.30	4.82	26.37	31.50
7000 For subbase treatment, see Section 31 32 13									

32 01 13.64 Sand Seal

	Crew	Daily Output	Labor-Hours	Unit	Material	Labor	Equipment	Total	Total Incl O&P
0010 **SAND SEAL**									
2080 Sand sealing, sharp sand, asphalt emulsion, small area	B-91	10000	.006	S.Y.	1.56	.35	.24	2.15	2.54
2120 Roadway or large area	"	18000	.004	"	1.34	.19	.14	1.67	1.93
3000 Sealing random cracks, min. 1/2" wide, to 1-1/2", 1,000 L.F.	B-77	2800	.014	L.F.	2.67	.67	.29	3.63	4.33
3040 10,000 L.F.		4000	.010	"	1.85	.47	.20	2.52	3.01
3080 Alternate method, 1,000 L.F.		200	.200	Gal.	40	9.40	4.05	53.45	63.50
3120 10,000 L.F.		325	.123	"	32	5.80	2.49	40.29	47.50
3200 Multi-cracks (flooding), 1 coat, small area	B-92	460	.070	S.Y.	2.59	3.21	1.97	7.77	10.10
3320 Large area		1425	.022	"	16.15	1.04	.64	17.83	20
3360 Alternate method, small area		115	.278	Gal.	15.80	12.85	7.90	36.55	46.50
3400 Large area		715	.045	"	14.85	2.06	1.27	18.18	21

32 01 13.66 Fog Seal

	Crew	Daily Output	Labor-Hours	Unit	Material	Labor	Equipment	Total	Total Incl O&P
0010 **FOG SEAL**									
0012 Sealcoating, 2 coat coal tar pitch emulsion over 10,000 S.Y.	B-45	5000	.003	S.Y.	.89	.18	.17	1.24	1.46
0030 1,000 to 10,000 S.Y.	"	3000	.005		.89	.30	.29	1.48	1.77
0100 Under 1,000 S.Y.	B-1	1050	.023		.89	1.06		1.95	2.66
0300 Petroleum resistant, over 10,000 S.Y.	B-45	5000	.003		1.24	.18	.17	1.59	1.84
0320 1,000 to 10,000 S.Y.	"	3000	.005		1.24	.30	.29	1.83	2.15
0400 Under 1,000 S.Y.	B-1	1050	.023		1.24	1.06		2.30	3.04
0600 Non-skid pavement renewal, over 10,000 S.Y.	B-45	5000	.003		1.30	.18	.17	1.65	1.91
0620 1,000 to 10,000 S.Y.	"	3000	.005		1.30	.30	.29	1.89	2.22

32 01 13 – Flexible Paving Surface Treatment

32 01 13.66 Fog Seal

		Crew	Daily Output	Labor-Hours	Unit	Material	2022 Bare Costs Labor	Equipment	Total	Total Incl O&P
0700	Under 1,000 S.Y.	B-1	1050	.023	S.Y.	1.30	1.06		2.36	3.11
0800	Prepare and clean surface for above	A-2	8545	.003	↓		.13	.04	.17	.25
1000	Hand seal asphalt curbing	B-1	4420	.005	L.F.	.68	.25		.93	1.15
1900	Asphalt surface treatment, single course, small area									
1901	0.30 gal./S.Y. asphalt material, 20 lb./S.Y. aggregate	B-91	5000	.013	S.Y.	1.33	.69	.49	2.51	3.09
1910	Roadway or large area	B-91	10000	.006	S.Y.	1.22	.35	.24	1.81	2.17
1950	Asphalt surface treatment, dbl. course for small area		3000	.021		2.66	1.16	.81	4.63	5.65
1960	Roadway or large area		6000	.011		2.39	.58	.41	3.38	3.99
1980	Asphalt surface treatment, single course, for shoulders	↓	7500	.009	↓	1.30	.46	.33	2.09	2.52

32 01 16 – Flexible Paving Rehabilitation

32 01 16.71 Cold Milling Asphalt Paving

		Crew	Daily Output	Labor-Hours	Unit	Material	2022 Bare Costs Labor	Equipment	Total	Total Incl O&P
0010	**COLD MILLING ASPHALT PAVING**									
5200	Cold planing & cleaning, 1" to 3" asphalt pavmt., over 25,000 S.Y.	B-71	6000	.009	S.Y.		.49	.71	1.20	1.56
5280	5,000 S.Y. to 10,000 S.Y.		4000	.014	"		.73	1.07	1.80	2.33
5285	5,000 S.Y. to 10,000 S.Y.		36000	.002	S.F.		.08	.12	.20	.26
5300	Asphalt pavement removal from conc. base, no haul									
5320	Rip, load & sweep 1" to 3"	B-70	8000	.007	S.Y.		.37	.23	.60	.83
5330	3" to 6" deep	"	5000	.011			.58	.36	.94	1.32
5340	Profile grooving, asphalt pavement load & sweep, 1" deep	B-71	12500	.004			.23	.34	.57	.75
5350	3" deep		9000	.006			.32	.48	.80	1.03
5360	6" deep	↓	5000	.011	↓		.58	.86	1.44	1.86

32 01 16.73 In Place Cold Reused Asphalt Paving

		Crew	Daily Output	Labor-Hours	Unit	Material	2022 Bare Costs Labor	Equipment	Total	Total Incl O&P
0010	**IN PLACE COLD REUSED ASPHALT PAVING**									
5000	Reclamation, pulverizing and blending with existing base									
5040	Aggregate base, 4" thick pavement, over 15,000 S.Y.	B-73	2400	.027	S.Y.		1.47	1.86	3.33	4.35
5080	5,000 S.Y. to 15,000 S.Y.		2200	.029			1.60	2.03	3.63	4.74
5120	8" thick pavement, over 15,000 S.Y.		2200	.029			1.60	2.03	3.63	4.74
5160	5,000 S.Y. to 15,000 S.Y.	↓	2000	.032	↓		1.76	2.23	3.99	5.20

32 01 16.74 In Place Hot Reused Asphalt Paving

			Crew	Daily Output	Labor-Hours	Unit	Material	2022 Bare Costs Labor	Equipment	Total	Total Incl O&P
0010	**IN PLACE HOT REUSED ASPHALT PAVING**										
5500	Recycle asphalt pavement at site										
5520	Remove, rejuvenate and spread 4" deep	G	B-72	2500	.026	S.Y.	4.60	1.36	2.93	8.89	10.40
5521	6" deep	G	"	2000	.032	"	6.75	1.70	3.66	12.11	14.10

32 01 17 – Flexible Paving Repair

32 01 17.10 Repair of Asphalt Pavement Holes

		Crew	Daily Output	Labor-Hours	Unit	Material	2022 Bare Costs Labor	Equipment	Total	Total Incl O&P
0010	**REPAIR OF ASPHALT PAVEMENT HOLES** (cold patch)									
0100	Flexible pavement repair holes, roadway, light traffic, 1 C.F. size each	B-37A	24	1	Ea.	6.80	47	23	76.80	108
0150	Group of two, 1 C.F. size each		16	1.500	Set	13.60	71	35	119.60	166
0200	Group of three, 1 C.F. size each		12	2	"	20.50	94.50	46.50	161.50	224
0300	Medium traffic, 1 C.F. size each	B-37B	24	1.333	Ea.	6.80	62.50	23	92.30	132
0350	Group of two, 1 C.F. size each		16	2	Set	13.60	93.50	35	142.10	202
0400	Group of three, 1 C.F. size each	↓	12	2.667	"	20.50	125	46.50	192	272
0500	Highway/heavy traffic, 1 C.F. size each	B-37C	24	1.333	Ea.	6.80	64	37.50	108.30	151
0550	Group of two, 1 C.F. size each	B-37C	16	2	Set	13.60	96	56	165.60	230
0600	Group of three, 1 C.F. size each	"	12	2.667	"	20.50	128	74.50	223	310
0700	Add police officer and car for traffic control				Hr.	69			69	76
1000	Flexible pavement repair holes, parking lot, bag material, 1 C.F. size each	B-37D	18	.889	Ea.	67.50	42.50	6.40	116.40	149
1010	Economy bag material, 1 C.F. size each		18	.889	"	41.50	42.50	6.40	90.40	121
1100	Group of two, 1 C.F. size each		12	1.333	Set	135	64	9.60	208.60	261
1110	Group of two, economy bag, 1 C.F. size each		12	1.333		82.50	64	9.60	156.10	204
1200	Group of three, 1 C.F. size each	↓	10	1.600	↓	202	77	11.50	290.50	355

32 01 17.10 Repair of Asphalt Pavement Holes

		Crew	Daily Output	Labor-Hours	Unit	Material	2022 Bare Costs Labor	2022 Bare Costs Equipment	Total	Total Incl O&P
1210	Economy bag, group of three, 1 C.F. size each	B-37D	10	1.600	Set	124	77	11.50	212.50	271
1300	Flexible pavement repair holes, parking lot, 1 C.F. single hole	A-3A	4	2	Ea.	67.50	114	45	226.50	300
1310	Economy material, 1 C.F. single hole		4	2	"	41.50	114	45	200.50	273
1400	Flexible pavement repair holes, parking lot, 1 C.F. four holes		2	4	Set	269	228	90	587	750
1410	Economy material, 1 C.F. four holes	↓	2	4	"	165	228	90	483	635
1500	Flexible pavement repair holes, large parking lot, bulk matl, 1 C.F. size	B-37A	32	.750	Ea.	6.80	35.50	17.40	59.70	83

32 01 17.20 Repair of Asphalt Pavement Patches

		Crew	Daily Output	Labor-Hours	Unit	Material	2022 Bare Costs Labor	2022 Bare Costs Equipment	Total	Total Incl O&P
0010	**REPAIR OF ASPHALT PAVEMENT PATCHES**									
0100	Flexible pavement patches, roadway, light traffic, sawcut 10-25 S.F.	B-89	12	1.333	Ea.		71.50	52.50	124	171
0150	Sawcut 26-60 S.F.		10	1.600			86	62.50	148.50	204
0200	Sawcut 61-100 S.F.		8	2			107	78.50	185.50	255
0210	Sawcut groups of small size patches		24	.667			36	26	62	85
0220	Large size patches	↓	16	1			53.50	39	92.50	128
0300	Flexible pavement patches, roadway, light traffic, digout 10-25 S.F.	B-6	16	1.500			74	20.50	94.50	140
0350	Digout 26-60 S.F.		12	2			99	27.50	126.50	186
0400	Digout 61-100 S.F.	↓	8	3	↓		148	41	189	279
0450	Add 8 C.Y. truck, small project debris haulaway	B-34A	8	1	Hr.		53	49	102	139
0460	Add 12 C.Y. truck, small project debris haulaway	B-34B	8	1			53	96	149	191
0480	Add flagger for non-intersection medium traffic	1 Clab	8	1			45.50		45.50	72.50
0490	Add flasher truck for intersection medium traffic or heavy traffic	A-2B	8	1	↓		50.50	42.50	93	127
0500	Flexible pavement patches, roadway, repave, cold, 15 S.F., 4" D	B-37	8	6	Ea.	31.50	287	33	351.50	525
0510	6" depth		8	6		47.50	287	33	367.50	545
0520	Repave, cold, 2Q S.F., 4" depth		8	6		41.50	287	33	361.50	535
0530	6" depth		8	6		63	287	33	383	560
0540	Repave, cold, 25 S.F., 4" depth		8	6		52	287	33	372	550
0550	6" depth		8	6		78.50	287	33	398.50	580
0600	Repave, cold, 30 S.F., 4" depth		8	6		62.50	287	33	382.50	560
0610	6" depth		8	6		94.50	287	33	414.50	595
0640	Repave, cold, 40 S.F., 4" depth		8	6		83	287	33	403	585
0650	6" depth		8	6		126	287	33	446	630
0680	Repave, cold, 50 S.F., 4" depth		8	6		104	287	33	424	605
0690	6" depth		8	6		157	287	33	477	665
0720	Repave, cold, 60 S.F., 4" depth		8	6		125	287	33	445	630
0730	6" depth		8	6		189	287	33	509	700
0800	Repave, cold, 70 S.F., 4" depth		7	6.857		146	330	37.50	513.50	720
0810	6" depth		7	6.857		220	330	37.50	587.50	805
0820	Repave, cold, 75 S.F., 4" depth		7	6.857		155	330	37.50	522.50	735
0830	6" depth		7	6.857		236	330	37.50	603.50	820
0900	Repave, cold, 80 S.F., 4" depth		6	8		166	385	44	595	835
0910	6" depth		6	8		252	385	44	681	930
0940	Repave, cold, 90 S.F., 4" depth		6	8		187	385	44	616	860
0950	6" depth		6	8		284	385	44	713	965
0980	Repave, cold, 100 S.F., 4" depth		6	8		208	385	44	637	885
0990	6" depth	↓	6	8	↓	315	385	44	744	1,000
1000	Add flasher truck for paving operations in medium or heavy traffic	A-2B	8	1	Hr.		50.50	42.50	93	127
1100	Prime coat for repair 15-40 S.F., 25% overspray	B-37A	48	.500	Ea.	9.75	23.50	11.60	44.85	61
1150	41-60 S.F.		40	.600		19.50	28.50	13.90	61.90	82
1175	61-80 S.F.		32	.750		26.50	35.50	17.40	79.40	105
1200	81-100 S.F.		32	.750	↓	32.50	35.50	17.40	85.40	111
1210	Groups of patches w/25% overspray	↓	3600	.007	S.F.	.32	.31	.15	.78	1.03
1300	Flexible pavement repair patches, street repave, hot, 60 S.F., 4" D	B-37	8	6	Ea.	94.50	287	33	414.50	595
1310	6" depth	↓	8	6	↓	144	287	33	464	650

32 01 17 – Flexible Paving Repair

32 01 17.20 Repair of Asphalt Pavement Patches

		Crew	Daily Output	Labor-Hours	Unit	Material	2022 Bare Costs Labor	Equipment	Total	Total Incl O&P
1320	Repave, hot, 70 S.F., 4" depth	B-37	7	6.857	Ea.	110	330	37.50	477.50	685
1330	6" depth		7	6.857		167	330	37.50	534.50	745
1340	Repave, hot, 75 S.F., 4" depth		7	6.857		118	330	37.50	485.50	690
1350	6" depth		7	6.857		179	330	37.50	546.50	760
1360	Repave, hot, 80 S.F., 4" depth		6	8		126	385	44	555	795
1370	6" depth		6	8		191	385	44	620	865
1380	Repave, hot, 90 S.F., 4" depth		6	8		142	385	44	571	810
1390	6" depth		6	8		216	385	44	645	890
1400	Repave, hot, 100 S.F., 4" depth		6	8		158	385	44	587	830
1410	6" depth		6	8		239	385	44	668	915
1420	Pave hot groups of patches, 4" depth		900	.053	S.F.	1.57	2.55	.29	4.41	6.10
1430	6" depth		900	.053	"	2.39	2.55	.29	5.23	7
1500	Add 8 C.Y. truck for hot asphalt paving operations	B-34A	1	8	Day		425	390	815	1,100
1550	Add flasher truck for hot paving operations in med/heavy traffic	A-2B	8	1	Hr.		50.50	42.50	93	127
2000	Add police officer and car for traffic control				"	69			69	76

32 01 17.61 Sealing Cracks In Asphalt Paving

		Crew	Daily Output	Labor-Hours	Unit	Material	2022 Bare Costs Labor	Equipment	Total	Total Incl O&P
0010	**SEALING CRACKS IN ASPHALT PAVING**									
0100	Sealing cracks in asphalt paving, 1/8" wide x 1/2" depth, slow set	B-37A	2000	.012	L.F.	.15	.57	.28	1	1.37
0110	1/4" wide x 1/2" depth		2000	.012		.17	.57	.28	1.02	1.40
0130	3/8" wide x 1/2" depth		2000	.012		.19	.57	.28	1.04	1.42
0140	1/2" wide x 1/2" depth		1800	.013		.21	.63	.31	1.15	1.57
0150	3/4" wide x 1/2" depth		1600	.015		.25	.71	.35	1.31	1.79
0160	1" wide x 1/2" depth		1600	.015		.30	.71	.35	1.36	1.84
0165	1/8" wide x 1" depth, rapid set	B-37F	2000	.016		.12	.75	.33	1.20	1.69
0170	1/4" wide x 1" depth		2000	.016		.24	.75	.33	1.32	1.82
0175	3/8" wide x 1" depth		1800	.018		.36	.83	.37	1.56	2.13
0180	1/2" wide x 1" depth		1800	.018		.48	.83	.37	1.68	2.26
0185	5/8" wide x 1" depth		1800	.018		.60	.83	.37	1.80	2.39
0190	3/4" wide x 1" depth		1600	.020		.72	.94	.42	2.08	2.74
0195	1" wide x 1" depth		1600	.020		.96	.94	.42	2.32	3
0200	1/8" wide x 2" depth, rapid set		2000	.016		.24	.75	.33	1.32	1.82
0210	1/4" wide x 2" depth		2000	.016		.48	.75	.33	1.56	2.09
0220	3/8" wide x 2" depth		1800	.018		.72	.83	.37	1.92	2.52
0230	1/2" wide x 2" depth		1800	.018		.96	.83	.37	2.16	2.78
0240	5/8" wide x 2" depth		1800	.018		1.20	.83	.37	2.40	3.05
0250	3/4" wide x 2" depth		1600	.020		1.44	.94	.42	2.80	3.53
0260	1" wide x 2" depth		1600	.020		1.92	.94	.42	3.28	4.06
0300	Add flagger for non-intersection medium traffic	1 Clab	8	1	Hr.		45.50		45.50	72.50
0400	Add flasher truck for intersection medium traffic or heavy traffic	A-2B	8	1	"		50.50	42.50	93	127

32 01 19 – Rigid Paving Surface Treatment

32 01 19.61 Sealing of Joints In Rigid Paving

		Crew	Daily Output	Labor-Hours	Unit	Material	2022 Bare Costs Labor	Equipment	Total	Total Incl O&P
0010	**SEALING OF JOINTS IN RIGID PAVING**									
1000	Sealing, concrete pavement, preformed elastomeric	B-77	500	.080	L.F.	3.79	3.76	1.62	9.17	11.90
2000	Waterproofing, membrane, tar and fabric, small area	B-63	233	.172	S.Y.	16.80	8.20	1.13	26.13	32.50
2500	Large area		1435	.028		14.10	1.33	.18	15.61	17.80
3000	Preformed rubberized asphalt, small area		100	.400		20.50	19.15	2.63	42.28	56
3500	Large area		367	.109		18.75	5.20	.72	24.67	29.50
9000	Minimum labor/equipment charge	1 Clab	2	4	Job		182		182	290

For customer support on your Commercial Renovation Costs with RSMeans data, call 800.448.8182.

561

32 01 29.61 Partial Depth Patching of Rigid Pavement

	32 01 29.61 Partial Depth Patching of Rigid Pavement	Crew	Daily Output	Labor-Hours	Unit	Material	2022 Bare Costs Labor	Equipment	Total	Total Incl O&P
0010	**PARTIAL DEPTH PATCHING OF RIGID PAVEMENT**									
0100	Rigid pavement repair, roadway, light traffic, 25% pitting 15 S.F.	B-37F	16	2	Ea.	21.50	93.50	42	157	219
0110	Pitting 20 S.F.		16	2		29	93.50	42	164.50	227
0120	Pitting 25 S.F.		16	2		36	93.50	42	171.50	235
0130	Pitting 30 S.F.		16	2		43.50	93.50	42	179	243
0140	Pitting 40 S.F.		16	2		58	93.50	42	193.50	259
0150	Pitting 50 S.F.		12	2.667		72.50	125	56	253.50	340
0160	Pitting 60 S.F.		12	2.667		87	125	56	268	355
0170	Pitting 70 S.F.		12	2.667		101	125	56	282	370
0180	Pitting 75 S.F.		12	2.667		109	125	56	290	380
0190	Pitting 80 S.F.		8	4		116	187	83.50	386.50	515
0200	Pitting 90 S.F.		8	4		130	187	83.50	400.50	535
0210	Pitting 100 S.F.		8	4		145	187	83.50	415.50	550
0300	Roadway, light traffic, 50% pitting 15 S.F.		12	2.667		43.50	125	56	224.50	310
0310	Pitting 20 S.F.		12	2.667		58	125	56	239	325
0320	Pitting 25 S.F.		12	2.667		72.50	125	56	253.50	340
0330	Pitting 30 S.F.		12	2.667		87	125	56	268	355
0340	Pitting 40 S.F.		12	2.667		116	125	56	297	385
0350	Pitting 50 S.F.		8	4		145	187	83.50	415.50	550
0360	Pitting 60 S.F.		8	4		174	187	83.50	444.50	580
0370	Pitting 70 S.F.		8	4		203	187	83.50	473.50	615
0380	Pitting 75 S.F.		8	4		217	187	83.50	487.50	630
0390	Pitting 80 S.F.		6	5.333		232	250	112	594	775
0400	Pitting 90 S.F.		6	5.333		261	250	112	623	805
0410	Pitting 100 S.F.		6	5.333		290	250	112	652	840
1000	Rigid pavement repair, light traffic, surface patch, 2" deep, 15 S.F.		8	4		87	187	83.50	357.50	485
1010	Surface patch, 2" deep, 20 S.F.		8	4		116	187	83.50	386.50	515
1020	Surface patch, 2" deep, 25 S.F.		8	4		145	187	83.50	415.50	550
1030	Surface patch, 2" deep, 30 S.F.		8	4		174	187	83.50	444.50	580
1040	Surface patch, 2" deep, 40 S.F.		8	4		232	187	83.50	502.50	645
1050	Surface patch, 2" deep, 50 S.F.		6	5.333		290	250	112	652	840
1060	Surface patch, 2" deep, 60 S.F.		6	5.333		350	250	112	712	900
1070	Surface patch, 2" deep, 70 S.F.		6	5.333		405	250	112	767	965
1080	Surface patch, 2" deep, 75 S.F.		6	5.333		435	250	112	797	1,000
1090	Surface patch, 2" deep, 80 S.F.		5	6.400		465	300	134	899	1,125
1100	Surface patch, 2" deep, 90 S.F.		5	6.400		520	300	134	954	1,200
1200	Surface patch, 2" deep, 100 S.F.		5	6.400		580	300	134	1,014	1,250
2000	Add flagger for non-intersection medium traffic	1 Clab	8	1	Hr.		45.50		45.50	72.50
2100	Add flasher truck for intersection medium traffic or heavy traffic	A-2B	8	1			50.50	42.50	93	127
2200	Add police officer and car for traffic control					69			69	76

32 01 29.70 Full Depth Patching of Rigid Pavement

	32 01 29.70 Full Depth Patching of Rigid Pavement	Crew	Daily Output	Labor-Hours	Unit	Material	2022 Bare Costs Labor	Equipment	Total	Total Incl O&P
0010	**FULL DEPTH PATCHING OF RIGID PAVEMENT**									
0015	Pavement preparation includes sawcut, remove pavement and replace									
0020	6" of subbase and one layer of reinforcement									
0030	Trucking and haulaway of debris excluded									
0100	Rigid pavement replace, light traffic, prep, 15 S.F., 6" depth	B-37E	16	3.500	Ea.	25.50	177	88.50	291	405
0110	Replacement preparation 20 S.F., 6" depth		16	3.500		33.50	177	88.50	299	415
0115	25 S.F., 6" depth		16	3.500		42	177	88.50	307.50	425
0120	30 S.F., 6" depth		14	4		50.50	203	101	354.50	485
0125	35 S.F., 6" depth		14	4		59	203	101	363	495
0130	40 S.F., 6" depth		14	4		67.50	203	101	371.50	505
0135	45 S.F., 6" depth		14	4		76	203	101	380	515

32 01 29 – Rigid Paving Repair

32 01 29.70 Full Depth Patching of Rigid Pavement

		Crew	Daily Output	Labor-Hours	Unit	Material	2022 Bare Costs Labor	Equipment	Total	Total Incl O&P
0140	50 S.F., 6" depth	B-37E	12	4.667	Ea.	84.50	237	118	439.50	600
0145	55 S.F., 6" depth		12	4.667		92.50	237	118	447.50	605
0150	60 S.F., 6" depth		10	5.600		101	284	141	526	715
0155	65 S.F., 6" depth		10	5.600		110	284	141	535	725
0160	70 S.F., 6" depth		10	5.600		118	284	141	543	735
0165	75 S.F., 6" depth		10	5.600		126	284	141	551	745
0170	80 S.F., 6" depth		8	7		135	355	177	667	900
0175	85 S.F., 6" depth		8	7		143	355	177	675	910
0180	90 S.F., 6" depth		8	7		152	355	177	684	920
0185	95 S.F., 6" depth		8	7		160	355	177	692	930
0190	100 S.F., 6" depth	↓	8	7	↓	169	355	177	701	940
0200	Pavement preparation for 8" rigid paving same as 6"									
0290	Pavement preparation for 9" rigid paving same as 10"									
0300	Rigid pavement replace, light traffic, prep, 15 S.F., 10" depth	B-37E	16	3.500	Ea.	45	177	88.50	310.50	425
0302	Pavement preparation 10" includes sawcut, remove pavement and									
0304	replace 6" of subbase and two layers of reinforcement									
0306	Trucking and haulaway of debris excluded									
0310	Replacement preparation 20 S.F., 10" depth	B-37E	16	3.500	Ea.	60	177	88.50	325.50	445
0315	25 S.F., 10" depth		16	3.500		75	177	88.50	340.50	460
0320	30 S.F., 10" depth		14	4		90	203	101	394	530
0325	35 S.F., 10" depth		14	4		65.50	203	101	369.50	505
0330	40 S.F., 10" depth		14	4		120	203	101	424	565
0335	45 S.F., 10" depth		12	4.667	↓	135	237	118	490	655
0340	50 S.F., 10" depth	B-37E	12	4.667	Ea.	150	237	118	505	670
0345	55 S.F., 10" depth		12	4.667		165	237	118	520	685
0350	60 S.F., 10" depth		10	5.600		180	284	141	605	805
0355	65 S.F., 10" depth		10	5.600		195	284	141	620	820
0360	70 S.F., 10" depth		10	5.600		210	284	141	635	835
0365	75 S.F., 10" depth		8	7		225	355	177	757	1,000
0370	80 S.F., 10" depth		8	7		240	355	177	772	1,025
0375	85 S.F., 10" depth		8	7		254	355	177	786	1,025
0380	90 S.F., 10" depth		8	7		269	355	177	801	1,050
0385	95 S.F., 10" depth		8	7		284	355	177	816	1,075
0390	100 S.F., 10" depth	↓	8	7	↓	299	355	177	831	1,075
0500	Add 8 C.Y. truck, small project debris haulaway	B-34A	8	1	Hr.		53	49	102	139
0550	Add 12 C.Y. truck, small project debris haulaway	B-34B	8	1			53	96	149	191
0600	Add flagger for non-intersection medium traffic	1 Clab	8	1			45.50		45.50	72.50
0650	Add flasher truck for intersection medium traffic or heavy traffic	A-2B	8	1			50.50	42.50	93	127
0700	Add police officer and car for traffic control				↓	69			69	76
0990	Concrete will be replaced using quick set mix with aggregate									
1000	Rigid pavement replace, light traffic, repour, 15 S.F., 6" depth	B-37F	18	1.778	Ea.	125	83	37	245	310
1010	20 S.F., 6" depth		18	1.778		167	83	37	287	355
1015	25 S.F., 6" depth		18	1.778		209	83	37	329	405
1020	30 S.F., 6" depth		16	2		250	93.50	42	385.50	470
1025	35 S.F., 6" depth		16	2		292	93.50	42	427.50	515
1030	40 S.F., 6" depth		16	2		335	93.50	42	470.50	560
1035	45 S.F., 6" depth		16	2		375	93.50	42	510.50	610
1040	50 S.F., 6" depth		16	2		415	93.50	42	550.50	655
1045	55 S.F., 6" depth		12	2.667		460	125	56	641	765
1050	60 S.F., 6" depth		12	2.667		500	125	56	681	810
1055	65 S.F., 6" depth		12	2.667		545	125	56	726	855
1060	70 S.F., 6" depth		12	2.667		585	125	56	766	905
1065	75 S.F., 6" depth	↓	10	3.200	↓	625	150	67	842	1,000

32 01 29 – Rigid Paving Repair

32 01 29.70 Full Depth Patching of Rigid Pavement	Crew	Daily Output	Labor-Hours	Unit	Material	2022 Bare Costs Labor	Equipment	Total	Total Incl O&P
1070	80 S.F., 6" depth	B-37F	10	3.200	Ea.	670	150	67	887	1,050
1075	85 S.F., 6" depth		8	4		710	187	83.50	980.50	1,175
1080	90 S.F., 6" depth		8	4		750	187	83.50	1,020.50	1,225
1085	95 S.F., 6" depth		8	4		795	187	83.50	1,065.50	1,250
1090	100 S.F., 6" depth	▼	8	4	▼	835	187	83.50	1,105.50	1,300
1099	Concrete will be replaced using 4,500 psi concrete ready mix									
1100	Rigid pavement replace, light traffic, repour, 15 S.F., 6" depth	A-2	12	2	Ea.	395	94.50	28.50	518	615
1110	20-50 S.F., 6" depth		12	2		395	94.50	28.50	518	615
1120	55-65 S.F., 6" depth		10	2.400		435	113	34	582	700
1130	70-80 S.F., 6" depth		10	2.400		470	113	34	617	735
1140	85-100 S.F., 6" depth		8	3		540	142	42.50	724.50	865
1200	Repour 15-40 S.F., 8" depth		12	2		395	94.50	28.50	518	615
1210	45-50 S.F., 8" depth		12	2		435	94.50	28.50	558	660
1220	55-60 S.F., 8" depth		10	2.400		470	113	34	617	735
1230	65-80 S.F., 8" depth		10	2.400		540	113	34	687	815
1240	85-90 S.F., 8" depth		8	3		580	142	42.50	764.50	910
1250	95-100 S.F., 8" depth		8	3		615	142	42.50	799.50	945
1300	Repour 15-30 S.F., 10" depth		12	2		395	94.50	28.50	518	615
1310	35-40 S.F., 10" depth		12	2		435	94.50	28.50	558	660
1320	45 S.F., 10" depth		12	2		470	94.50	28.50	593	695
1330	50-65 S.F., 10" depth		10	2.400		540	113	34	687	815
1340	70 S.F., 10" depth		10	2.400		580	113	34	727	860
1350	75-80 S.F., 10" depth		10	2.400		615	113	34	762	895
1360	85-95 S.F., 10" depth		8	3		690	142	42.50	874.50	1,025
1370	100 S.F., 10" depth	▼	8	3	▼	730	142	42.50	914.50	1,075

32 06 Schedules for Exterior Improvements

32 06 10 – Schedules for Bases, Ballasts, and Paving

32 06 10.10 Sidewalks, Driveways and Patios

		Crew	Daily Output	Labor-Hours	Unit	Material	2022 Bare Costs Labor	Equipment	Total	Total Incl O&P
0010	**SIDEWALKS, DRIVEWAYS AND PATIOS** No base									
0020	Asphaltic concrete, 2" thick	B-37	720	.067	S.Y.	7.20	3.19	.37	10.76	13.40
0100	2-1/2" thick	"	660	.073	"	9.15	3.48	.40	13.03	16
0110	Bedding for brick or stone, mortar, 1" thick	D-1	300	.053	S.F.	.88	2.67		3.55	5.25
0120	2" thick	"	200	.080		2.21	4.01		6.22	8.85
0130	Sand, 2" thick	B-18	8000	.003		.47	.14	.02	.63	.76
0140	4" thick	"	4000	.006	▼	.94	.28	.03	1.25	1.52
0300	Concrete, 3,000 psi, CIP, 6 x 6 - W1.4 x W1.4 mesh,									
0310	broomed finish, no base, 4" thick	B-24	600	.040	S.F.	2.88	2.07		4.95	6.45
0350	5" thick		545	.044		3.54	2.28		5.82	7.50
0400	6" thick	▼	510	.047		4.13	2.43		6.56	8.40
0450	For bank run gravel base, 4" thick, add	B-18	2500	.010		.64	.44	.05	1.13	1.47
0520	8" thick, add	"	1600	.015		1.28	.69	.09	2.06	2.60
0550	Exposed aggregate finish, add to above, minimum	B-24	1875	.013		.09	.66		.75	1.15
0600	Maximum	"	455	.053	▼	.30	2.73		3.03	4.64
0950	Concrete tree grate, 5' square	B-6	25	.960	Ea.	475	47.50	13.15	535.65	615
0955	Tree well & cover, concrete, 3' square		25	.960		276	47.50	13.15	336.65	395
0960	Cast iron tree grate with frame, 2 piece, round, 5' diameter		25	.960		1,275	47.50	13.15	1,335.65	1,500
0980	Square, 5' side		25	.960	▼	1,275	47.50	13.15	1,335.65	1,500
1700	Redwood, prefabricated, 4' x 4' sections	2 Carp	316	.051	S.F.	13.10	2.85		15.95	18.95
1750	Redwood planks, 1" thick, on sleepers	"	240	.067	"	13.10	3.75		16.85	20.50
2250	Stone dust, 4" thick	B-62	900	.027	S.Y.	6.25	1.32	.29	7.86	9.30

For customer support on your Commercial Renovation Costs with RSMeans data, call 800.448.8182.

32 06 Schedules for Exterior Improvements

32 06 10 – Schedules for Bases, Ballasts, and Paving

32 06 10.10 Sidewalks, Driveways and Patios

		Crew	Daily Output	Labor-Hours	Unit	Material	2022 Bare Costs Labor	Equipment	Total	Total Incl O&P
9000	Minimum labor/equipment charge	D-1	2	8	Job		400		400	640

32 06 10.20 Steps

		Crew	Daily Output	Labor-Hours	Unit	Material	2022 Bare Costs Labor	Equipment	Total	Total Incl O&P
0010	**STEPS**									
0011	Incl. excav., borrow & concrete base as required									
0100	Brick steps	B-24	35	.686	LF Riser	27.50	35.50		63	86.50
0200	Railroad ties	2 Clab	25	.640		4.33	29		33.33	51.50
0300	Bluestone treads, 12" x 2" or 12" x 1-1/2"	B-24	30	.800	↓	54	41.50		95.50	125
0490	Minimum labor/equipment charge	D-1	2	8	Job		400		400	640
0500	Concrete, cast in place, see Section 03 30 53.40									
0600	Precast concrete, see Section 03 41 23.50									

32 11 Base Courses

32 11 23 – Aggregate Base Courses

32 11 23.23 Base Course Drainage Layers

		Crew	Daily Output	Labor-Hours	Unit	Material	2022 Bare Costs Labor	Equipment	Total	Total Incl O&P
0010	**BASE COURSE DRAINAGE LAYERS**									
0011	For Soil Stabilization, see Section 31 32									
0012	For roadways and large areas									
0050	Crushed 3/4" stone base, compacted, 3" deep	B-36C	5200	.008	S.Y.	1.99	.43	.71	3.13	3.64
0100	6" deep		5000	.008		3.98	.45	.74	5.17	5.90
0200	9" deep		4600	.009		5.95	.49	.80	7.24	8.20
0300	12" deep	↓	4200	.010		7.95	.54	.88	9.37	10.55
0301	Crushed 1-1/2" stone base, compacted to 4" deep	B-36B	6000	.011		5.50	.58	.73	6.81	7.75
0302	6" deep		5400	.012		8.25	.64	.81	9.70	10.95
0303	8" deep		4500	.014		11	.77	.97	12.74	14.40
0304	12" deep	↓	3800	.017	↓	16.50	.91	1.15	18.56	21
0310	Minimum labor/equipment charge	B-36	1	40	Job		2,075	1,750	3,825	5,200
0350	Bank run gravel, spread and compacted									
0370	6" deep	B-32	6000	.005	S.Y.	5.30	.30	.41	6.01	6.75
0390	9" deep		4900	.007		7.95	.37	.50	8.82	9.90
0400	12" deep	↓	4200	.008	↓	10.65	.43	.59	11.67	13
6900	For small and irregular areas, add						50%	50%		
6990	Minimum labor/equipment charge	1 Clab	3	2.667	Job		122		122	193
7000	Prepare and roll sub-base, small areas to 2,500 S.Y.	B-32A	1500	.016	S.Y.		.89	.85	1.74	2.32
8000	Large areas over 2,500 S.Y.	"	3500	.007	"		.38	.36	.74	1
8050	For roadways	B-32	4000	.008	S.Y.		.45	.61	1.06	1.39
9000	Minimum labor/equipment charge	1 Clab	4	2	Job		91		91	145

32 12 Flexible Paving

32 12 16 – Asphalt Paving

32 12 16.13 Plant-Mix Asphalt Paving

		Crew	Daily Output	Labor-Hours	Unit	Material	2022 Bare Costs Labor	Equipment	Total	Total Incl O&P
0010	**PLANT-MIX ASPHALT PAVING**									
0080	Binder course, 1-1/2" thick	B-25	7725	.011	S.Y.	5.55	.57	.38	6.50	7.45
0120	2" thick		6345	.014		7.40	.69	.46	8.55	9.75
0130	2-1/2" thick		5620	.016		9.25	.78	.52	10.55	12
0160	3" thick		4905	.018		11.10	.89	.59	12.58	14.25
0170	3-1/2" thick	↓	4520	.019		12.95	.97	.64	14.56	16.50
0300	Wearing course, 1" thick	B-25B	10575	.009		3.68	.46	.31	4.45	5.10
0340	1-1/2" thick		7725	.012		6.20	.63	.42	7.25	8.25
0380	2" thick		6345	.015	↓	8.30	.77	.51	9.58	10.90

32 12 16.13 Plant-Mix Asphalt Paving

		Crew	Daily Output	Labor-Hours	Unit	Material	2022 Bare Costs Labor	Equipment	Total	Total Incl O&P
0420	2-1/2" thick	B-25B	5480	.018	S.Y.	10.25	.89	.59	11.73	13.30
0460	3" thick		4900	.020		12.20	.99	.66	13.85	15.70
0470	3-1/2" thick		4520	.021		14.30	1.08	.72	16.10	18.25
0480	4" thick		4140	.023		16.35	1.18	.78	18.31	20.50
0600	Porous pavement, 1-1/2" open graded friction course	B-25	7725	.011		18.85	.57	.38	19.80	22
0800	Alternate method of figuring paving costs									
0810	Binder course, 1-1/2" thick	B-25	630	.140	Ton	68	6.95	4.62	79.57	91
0811	2" thick		690	.128		68	6.35	4.22	78.57	89.50
0812	3" thick		800	.110		68	5.50	3.64	77.14	87.50
0813	4" thick		900	.098		68	4.87	3.23	76.10	86.50
0850	Wearing course, 1" thick	B-25B	575	.167		75.50	8.45	5.65	89.60	103
0851	1-1/2" thick		630	.152		75.50	7.75	5.15	88.40	101
0852	2" thick		690	.139		75.50	7.05	4.71	87.26	99.50
0853	2-1/2" thick		765	.125		75.50	6.35	4.25	86.10	97.50
0854	3" thick		800	.120		75.50	6.10	4.06	85.66	97
1000	Pavement replacement over trench, 2" thick	1 B-17	90	.533	S.Y.	7.65	26.50	21	55.15	74
1050	4" thick		70	.686		15.15	34.50	27	76.65	101
1080	6" thick		55	.873		24	44	34.50	102.50	134
3000	Prime coat, emulsion, 0.30 gal./S.Y., 1000 S.Y.	B-45	2500	.006		2.34	.36	.34	3.04	3.52
3100	Tack coat, emulsion, 0.10 gal./S.Y., 1000 S.Y.	"	2500	.006		.78	.36	.34	1.48	1.81

32 12 16.14 Asphaltic Concrete Paving

		Crew	Daily Output	Labor-Hours	Unit	Material	2022 Bare Costs Labor	Equipment	Total	Total Incl O&P
0011	**ASPHALTIC CONCRETE PAVING**, parking lots & driveways									
0015	No asphalt hauling included									
0020	6" stone base, 2" binder course, 1" topping	B-25C	9000	.005	S.F.	2.02	.27	.28	2.57	2.96
0025	2" binder course, 2" topping	"	9000	.005	"	2.48	.27	.28	3.03	3.46
0030	3" binder course, 2" topping	B-25C	9000	.005	S.F.	2.89	.27	.28	3.44	3.92
0035	4" binder course, 2" topping		9000	.005		3.30	.27	.28	3.85	4.37
0040	1-1/2" binder course, 1" topping		9000	.005		1.81	.27	.28	2.36	2.73
0042	3" binder course, 1" topping		9000	.005		2.43	.27	.28	2.98	3.42
0045	3" binder course, 3" topping		9000	.005		3.35	.27	.28	3.90	4.43
0050	4" binder course, 3" topping		9000	.005		3.76	.27	.28	4.31	4.88
0055	4" binder course, 4" topping		9000	.005		4.21	.27	.28	4.76	5.40
0300	Binder course, 1-1/2" thick		35000	.001		.62	.07	.07	.76	.87
0400	2" thick		25000	.002		.80	.10	.10	1	1.14
0500	3" thick		15000	.003		1.24	.16	.17	1.57	1.81
0600	4" thick		10800	.004		1.62	.23	.24	2.09	2.40
0800	Sand finish course, 3/4" thick		41000	.001		.33	.06	.06	.45	.52
0900	1" thick		34000	.001		.40	.07	.07	.54	.63
1000	Fill pot holes, hot mix, 2" thick	B-16	4200	.008		.80	.37	.18	1.35	1.66
1100	4" thick		3500	.009		1.17	.44	.22	1.83	2.22
1120	6" thick		3100	.010		1.57	.50	.25	2.32	2.78
1140	Cold patch, 2" thick	B-51	3000	.016		1.06	.75	.11	1.92	2.48
1160	4" thick		2700	.018		2.02	.83	.13	2.98	3.69
1180	6" thick		1900	.025		3.15	1.18	.18	4.51	5.55
3000	Prime coat, emulsion, 0.30 gal./S.Y., 1000 S.Y.	B-45	2500	.006	S.Y.	2.34	.36	.34	3.04	3.52
3100	Tack coat, emulsion, 0.10 gal./S.Y., 1000 S.Y.	"	2500	.006	"	.78	.36	.34	1.48	1.81

32 12 Flexible Paving

32 12 36 – Seal Coats

32 12 36.13 Chip Seal

		Crew	Daily Output	Labor-Hours	Unit	Material	2022 Bare Costs Labor	2022 Bare Costs Equipment	Total	Total Incl O&P
0010	**CHIP SEAL**									
0011	Excludes crack repair and flush coat									
1000	Fine - PMCRS-2h (20lbs/sy, 1/4" (No.10), .30gal/sy app. rate)									
1010	Small, irregular areas	B-91	5000	.013	S.Y.	1.25	.69	.49	2.43	3.01
1020	Parking Lot	B-91D	15000	.007		1.25	.36	.23	1.84	2.21
1030	Roadway	"	30000	.003		1.25	.18	.12	1.55	1.80
1090	For Each .5% Latex Additive, Add				↓	.29			.29	.32
1100	Medium Fine - PMCRS-2h (25lbs/sy, 5/16" (No.8), .35gal/sy app. rate)									
1110	Small, irregular areas	B-91	4000	.016	S.Y.	1.55	.87	.61	3.03	3.74
1120	Parking Lot	B-91D	12000	.009		1.55	.46	.29	2.30	2.75
1130	Roadway	"	24000	.004		1.55	.23	.15	1.93	2.23
1190	For Each .5% Latex Additive, Add				↓	.36			.36	.40
1200	Medium - PMCRS-2h (30lbs/sy, 3/8" (No.6), .40gal/sy app. rate)									
1210	Small, irregular areas	B-91	3330	.019	S.Y.	1.60	1.04	.73	3.37	4.21
1220	Parking Lot	B-91D	10000	.010	"	1.60	.55	.35	2.50	3.01
1230	Roadway	B-91D	20000	.005	S.Y.	1.60	.27	.18	2.05	2.38
1290	For Each .5% Latex Additive, Add				"	.43			.43	.48
1300	Course - PMCRS-2h (30lbs/sy, 1/2" (No.4), .40gal/sy app. rate)									
1310	Small, irregular areas	B-91	2500	.026	S.Y.	1.56	1.39	.98	3.93	4.96
1320	Parking Lot	B-91D	7500	.014		1.56	.73	.47	2.76	3.37
1330	Roadway	"	15000	.007		1.56	.36	.23	2.15	2.54
1390	For Each .5% Latex Additive, Add				↓	.43			.43	.48
1400	Double - PMCRS-2h (Course Base with Fine Top)									
1410	Small, irregular areas	B-91	2000	.032	S.Y.	2.66	1.73	1.22	5.61	7
1420	Parking Lot	B-91D	6000	.017		2.66	.91	.58	4.15	5
1430	Roadway	"	12000	.009		2.66	.46	.29	3.41	3.96
1490	For Each .5% Latex Additive, Add				↓	.72			.72	.79

32 12 36.14 Flush Coat

		Crew	Daily Output	Labor-Hours	Unit	Material	2022 Bare Costs Labor	2022 Bare Costs Equipment	Total	Total Incl O&P
0010	**FLUSH COAT**									
0011	Fog Seal with Sand Cover (.18gal/sy, 6lbs/sy)									
1010	Small, irregular areas	B-91	2000	.032	S.Y.	1.14	1.73	1.22	4.09	5.30
1020	Parking lot		6000	.011		1.14	.58	.41	2.13	2.61
1030	Roadway	↓	12000	.005	↓	1.14	.29	.20	1.63	1.92

32 12 36.33 Slurry Seal

		Crew	Daily Output	Labor-Hours	Unit	Material	2022 Bare Costs Labor	2022 Bare Costs Equipment	Total	Total Incl O&P
0010	**SLURRY SEAL**									
0011	Includes sweeping and cleaning of area									
1000	Type I-PMCQS-1h-EAS (12lbs/sy, 1/8", 20% asphalt emulsion)									
1010	Small, irregular areas	B-90	8000	.008	S.Y.	1.98	.40	.29	2.67	3.14
1020	Parking Lot		25000	.003		1.98	.13	.09	2.20	2.48
1030	Roadway	↓	50000	.001		1.98	.06	.05	2.09	2.33
1090	For Each .5% Latex Additive, Add				↓	.19			.19	.21
1100	Type II-PMCQS-1h-EAS (15lbs/sy, 1/4", 18% asphalt emulsion)									
1110	Small, irregular areas	B-90	6000	.011	S.Y.	2.20	.54	.38	3.12	3.69
1120	Parking lot		20000	.003		2.20	.16	.11	2.47	2.81
1130	Roadway	↓	40000	.002		2.20	.08	.06	2.34	2.61
1190	For Each .5% Latex Additive, Add				↓	.23			.23	.25
1200	Type III-PMCQS-1h-EAS (25lbs/sy, 3/8", 15% asphalt emulsion)									
1210	Small, irregular areas	B-90	4000	.016	S.Y.	3.03	.81	.57	4.41	5.25
1220	Parking lot		12000	.005		3.03	.27	.19	3.49	3.97
1230	Roadway	↓	24000	.003		3.03	.13	.10	3.26	3.65
1290	For Each .5% Latex Additive, Add					.37			.37	.41

32 12 Flexible Paving

32 12 36 – Seal Coats

32 12 36.36 Microsurfacing

		Crew	Daily Output	Labor-Hours	Unit	Material	2022 Bare Costs Labor	2022 Bare Costs Equipment	Total	Total Incl O&P
0010	**MICROSURFACING**									
1100	Type II-MSE (20lbs/sy, 1/4", 18% microsurfacing emulsion)									
1110	Small, irregular areas	B-90	5000	.013	S.Y.	1.99	.65	.46	3.10	3.71
1120	Parking lot		15000	.004		1.99	.22	.15	2.36	2.70
1130	Roadway	↓	30000	.002	↓	1.99	.11	.08	2.18	2.44
1200	Type IIIa-MSE (32lbs/sy, 3/8", 15% microsurfacing emulsion)									
1210	Small, irregular areas	B-90	3000	.021	S.Y.	2.75	1.08	.76	4.59	5.55
1220	Parking lot		9000	.007		2.75	.36	.25	3.36	3.87
1230	Roadway	↓	18000	.004	↓	2.75	.18	.13	3.06	3.44

32 13 Rigid Paving

32 13 13 – Concrete Paving

32 13 13.25 Concrete Pavement, Highways

		Crew	Daily Output	Labor-Hours	Unit	Material	2022 Bare Costs Labor	2022 Bare Costs Equipment	Total	Total Incl O&P
0010	**CONCRETE PAVEMENT, HIGHWAYS**									
0015	Including joints, finishing and curing									
0020	Fixed form, 12' pass, unreinforced, 6" thick	B-26	3000	.029	S.Y.	27	1.48	1.12	29.60	33.50
0100	8" thick		2750	.032		37	1.62	1.22	39.84	45
0400	12" thick		1800	.049		53	2.47	1.86	57.33	64
0410	Conc. pavement, w/jt., fnsh.& curing, fix form, 24' pass, unreinforced, 6"T		6000	.015		25.50	.74	.56	26.80	30
0430	8" thick		5500	.016		35	.81	.61	36.42	40.50
0460	12" thick	↓	3600	.024		50.50	1.24	.93	52.67	58.50
0520	Welded wire fabric, sheets for rigid paving 2.33 lb./S.Y.	2 Rodm	389	.041		2	2.50		4.50	6.10
0530	Reinforcing steel for rigid paving 12 lb./S.Y.		666	.024		9.15	1.46		10.61	12.40
0540	Reinforcing steel for rigid paving 18 lb./S.Y.	↓	444	.036		13.75	2.19		15.94	18.50
0620	Slip form, 12' pass, unreinforced, 6" thick	B-26A	5600	.016		25.50	.79	.63	26.92	30.50
0624	8" thick		5300	.017		35.50	.84	.66	37	41
0630	12" thick		3470	.025		51	1.28	1.01	53.29	59
0632	15" thick		2890	.030		64.50	1.54	1.21	67.25	75
0640	Slip form, 24' pass, unreinforced, 6" thick		11200	.008		25.50	.40	.31	26.21	29
0644	8" thick		10600	.008		34	.42	.33	34.75	38.50
0650	12" thick		6940	.013		49.50	.64	.51	50.65	56
0652	15" thick	↓	5780	.015		62.50	.77	.61	63.88	70.50
0700	Finishing, broom finish small areas	2 Cefi	120	.133			7.10		7.10	11.05
1000	Curing, with sprayed membrane by hand	2 Clab	1500	.011	↓	1.22	.49		1.71	2.11

32 14 Unit Paving

32 14 13 – Precast Concrete Unit Paving

32 14 13.18 Precast Concrete Plantable Pavers

		Crew	Daily Output	Labor-Hours	Unit	Material	2022 Bare Costs Labor	2022 Bare Costs Equipment	Total	Total Incl O&P
0010	**PRECAST CONCRETE PLANTABLE PAVERS** (50% grass)									
0300	3/4" crushed stone base for plantable pavers, 6" depth	B-62	1000	.024	S.Y.	3.09	1.19	.26	4.54	5.55
0400	8" depth		900	.027		4.12	1.32	.29	5.73	6.95
0500	10" depth	↓	800	.030	↓	5.15	1.48	.33	6.96	8.35
0600	12" depth	B-62	700	.034	S.Y.	6.20	1.69	.38	8.27	9.90
0700	Hydro seeding plantable pavers	B-81A	20	.800	M.S.F.	33	38.50	35	106.50	136
0800	Apply fertilizer and seed to plantable pavers	1 Clab	8	1	"	50	45.50		95.50	128

32 14 Unit Paving

32 14 16 – Brick Unit Paving

32 14 16.10 Brick Paving

		Crew	Daily Output	Labor-Hours	Unit	Material	2022 Bare Costs Labor	Equipment	Total	Total Incl O&P
0010	**BRICK PAVING**									
0012	4" x 8" x 1-1/2", without joints (4.5 bricks/S.F.)	D-1	110	.145	S.F.	4.37	7.30		11.67	16.50
0100	Grouted, 3/8" joint (3.9 bricks/S.F.)		90	.178		3.98	8.90		12.88	18.70
0200	4" x 8" x 2-1/4", without joints (4.5 bricks/S.F.)		110	.145		4.59	7.30		11.89	16.75
0300	Grouted, 3/8" joint (3.9 bricks/S.F.)		90	.178		3.98	8.90		12.88	18.70
0455	Pervious brick paving, 4" x 8" x 3-1/4", without joints (4.5 bricks/S.F.)		110	.145		2.86	7.30		10.16	14.85
0500	Bedding, asphalt, 3/4" thick	B-25	5130	.017		.68	.85	.57	2.10	2.72
0540	Course washed sand bed, 1" thick	B-18	5000	.005		.53	.22	.03	.78	.96
0580	Mortar, 1" thick	D-1	300	.053		.74	2.67		3.41	5.10
0620	2" thick		200	.080		1.47	4.01		5.48	8
1500	Brick on 1" thick sand bed laid flat, 4.5/S.F.		100	.160		3.81	8		11.81	17.05
2000	Brick pavers, laid on edge, 7.2/S.F.		70	.229		7.55	11.45		19	26.50
2500	For 4" thick concrete bed and joints, add		595	.027		1.86	1.35		3.21	4.21
2800	For steam cleaning, add	A-1H	950	.008		.12	.38	.08	.58	.83
9000	Minimum labor/equipment charge	1 Bric	2	4	Job		221		221	355

32 14 23 – Asphalt Unit Paving

32 14 23.10 Asphalt Blocks

		Crew	Daily Output	Labor-Hours	Unit	Material	2022 Bare Costs Labor	Equipment	Total	Total Incl O&P
0010	**ASPHALT BLOCKS**									
0020	Rectangular, 6" x 12" x 1-1/4", w/bed & neopr. adhesive	D-1	135	.119	S.F.	10.95	5.95		16.90	21.50
0100	3" thick		130	.123		15.30	6.15		21.45	27
0300	Hexagonal tile, 8" wide, 1-1/4" thick		135	.119		10.95	5.95		16.90	21.50
0400	2" thick		130	.123		15.30	6.15		21.45	27
0500	Square, 8" x 8", 1-1/4" thick		135	.119		10.95	5.95		16.90	21.50
0600	2" thick		130	.123		15.30	6.15		21.45	27
9000	Minimum labor/equipment charge	1 Bric	2	4	Job		221		221	355
7010	Non-Vegetative Surface, Rubber, 1/2" Thick	D-1	130	.123	S.F.	3.68	6.15		9.83	13.95

32 14 40 – Stone Paving

32 14 40.10 Stone Pavers

		Crew	Daily Output	Labor-Hours	Unit	Material	2022 Bare Costs Labor	Equipment	Total	Total Incl O&P
0010	**STONE PAVERS**									
1300	Slate, natural cleft, irregular, 3/4" thick	D-1	92	.174	S.F.	9.30	8.70		18	24
1350	Random rectangular, gauged, 1/2" thick		105	.152		20	7.65		27.65	34.50
1400	Random rectangular, butt joint, gauged, 1/4" thick		150	.107		21.50	5.35		26.85	32.50
1450	For sand rubbed finish, add					11.35			11.35	12.45
1550	Granite blocks, 3-1/2" x 3-1/2" x 3-1/2"	D-1	92	.174	S.F.	27	8.70		35.70	43.50
1600	4" to 12" long, 3" to 5" wide, 3" to 5" thick	"	98	.163	"	22.50	8.20		30.70	37.50

32 16 Curbs, Gutters, Sidewalks, and Driveways

32 16 13 – Curbs and Gutters

32 16 13.13 Cast-in-Place Concrete Curbs and Gutters

		Crew	Daily Output	Labor-Hours	Unit	Material	2022 Bare Costs Labor	Equipment	Total	Total Incl O&P
0010	**CAST-IN-PLACE CONCRETE CURBS AND GUTTERS**									
0290	Forms only, no concrete									
0300	Concrete, wood forms, 6" x 18", straight	C-2	500	.096	L.F.	4.18	5.25		9.43	12.95
0400	6" x 18", radius	"	200	.240		4.37	13.15		17.52	26
0404	Concrete, wood forms, 6" x 18", straight & concrete	C-2A	500	.096		8.65	5.20		13.85	17.75
0406	6" x 18", radius	"	200	.240		8.85	13.05		21.90	30
0415	Machine formed, 6" x 18", straight	B-69A	2000	.024		6.50	1.19	.63	8.32	9.75
0416	6" x 18", radius	"	900	.053		6.60	2.65	1.41	10.66	13

For customer support on your Commercial Renovation Costs with RSMeans data, call 800.448.8182.

569

32 16 13 – Curbs and Gutters

32 16 13.23 Precast Concrete Curbs and Gutters

		Daily Output	Labor-Hours	Unit	Material	2022 Bare Costs Labor	Equipment	Total	Total Incl O&P
	32 16 13.23 Precast Concrete Curbs and Gutters	Crew							
0010	**PRECAST CONCRETE CURBS AND GUTTERS**								
0550	Precast, 6" x 18", straight	B-29	700	.080	L.F.	11.10	3.97	1.23	16.30	19.85
0600	6" x 18", radius	"	325	.172	"	11.85	8.55	2.64	23.04	29.50

32 16 13.33 Asphalt Curbs

0010	**ASPHALT CURBS**									
0012	Curbs, asphaltic, machine formed, 8" wide, 6" high, 40 L.F./ton	B-27	1000	.032	L.F.	2.07	1.48	.70	4.25	5.40
0100	8" wide, 8" high, 30 L.F./ton		900	.036		2.77	1.64	.78	5.19	6.50
0150	Asphaltic berm, 12" W, 3" to 6" H, 35 L.F./ton, before pavement	↓	700	.046		.05	2.11	1	3.16	4.51
0200	12" W, 1-1/2" to 4" H, 60 L.F./ton, laid with pavement	B-2	1050	.038	↓	.03	1.75		1.78	2.81

32 16 13.43 Stone Curbs

0010	**STONE CURBS**									
1000	Granite, split face, straight, 5" x 16"	D-13	275	.175	L.F.	13.85	9.35	2.08	25.28	32.50
1100	6" x 18"	"	250	.192		18.20	10.30	2.29	30.79	39
1300	Radius curbing, 6" x 18", over 10' radius	B-29	260	.215	↓	22.50	10.70	3.30	36.50	45
1400	Corners, 2' radius	"	80	.700	Ea.	75	34.50	10.70	120.20	149
1600	Edging, 4-1/2" x 12", straight	D-13	300	.160	L.F.	6.95	8.60	1.91	17.46	23.50
1800	Curb inlets (guttermouth) straight	B-29	41	1.366	Ea.	166	68	21	255	315
2000	Indian granite (Belgian block)									
2100	Jumbo, 10-1/2" x 7-1/2" x 4", grey	D-1	150	.107	L.F.	10.10	5.35		15.45	19.65
2150	Pink		150	.107		13.55	5.35		18.90	23.50
2200	Regular, 9" x 4-1/2" x 4-1/2", grey		160	.100		7.25	5		12.25	16.05
2250	Pink		160	.100		11.05	5		16.05	20
2300	Cubes, 4" x 4" x 4", grey		175	.091		6.15	4.58		10.73	14.10
2350	Pink		175	.091		7.35	4.58		11.93	15.45
2400	6" x 6" x 6", pink	↓	155	.103	↓	17.35	5.15		22.50	27.50
2500	Alternate pricing method for Indian granite									
2550	Jumbo, 10-1/2" x 7-1/2" x 4" (30 lb.), grey				Ton	575			575	635
2600	Pink					785			785	865
2650	Regular, 9" x 4-1/2" x 4-1/2" (20 lb.), grey					510			510	565
2700	Pink					770			770	850
2750	Cubes, 4" x 4" x 4" (5 lb.), grey					745			745	820
2800	Pink					945			945	1,050
2850	6" x 6" x 6" (25 lb.), pink					675			675	740
2900	For pallets, add				↓	22			22	24

32 17 Paving Specialties

32 17 13 – Parking Bumpers

32 17 13.13 Metal Parking Bumpers

0010	**METAL PARKING BUMPERS**									
0015	Bumper rails for garages, 12 ga. rail, 6" wide, with steel									
0020	posts 12'-6" OC, minimum	E-4	190	.168	L.F.	21	10.50	.78	32.28	41
0030	Average		165	.194		26	12.10	.89	38.99	49.50
0100	Maximum	↓	140	.229	↓	31.50	14.25	1.05	46.80	58.50
1300	Pipe bollards, conc. filled/paint, 8' L x 4' D hole, 6" diam.	B-6	20	1.200	Ea.	470	59.50	16.45	545.95	625
1400	8" diam.		15	1.600		515	79	22	616	715
1500	12" diam.	↓	12	2		600	99	27.50	726.50	845
8890	Security bollards, SS, lighted, hyd, incl. controls, Single Unit	L-7	.08	337		22,200	18,400		40,600	53,500
8900	Security bollards, SS, lighted, hyd., incl. controls, group of 3		.06	509		45,900	27,800		73,700	94,500
8910	Group of 5	↓	.04	683	↓	80,500	37,300		117,800	147,500
9000	Minimum labor/equipment charge	E-4	2	16	Job		1,000	74	1,074	1,675

570

For customer support on your **Commercial Renovation Costs** with RSMeans data, call 800.448.8182.

32 17 Paving Specialties

32 17 13 – Parking Bumpers

32 17 13.16 Plastic Parking Bumpers

		Crew	Daily Output	Labor-Hours	Unit	Material	2022 Bare Costs Labor	Equipment	Total	Total Incl O&P
0010	**PLASTIC PARKING BUMPERS**									
1600	Bollards, recycled plastic, 5" x 5" x 5' L ⒢	B-6	20	1.200	Ea.	155	59.50	16.45	230.95	283
1610	Wheel stops, recycled plastic, yellow, 4" x 6" x 6' L ⒢		20	1.200		57	59.50	16.45	132.95	174
1620	Speed bumps, recycled plastic, yellow, 3" H x 10" W x 6' L ⒢		20	1.200		179	59.50	16.45	254.95	310
1630	3" H x 10" W x 9' L ⒢		20	1.200		229	59.50	16.45	304.95	365

32 17 13.19 Precast Concrete Parking Bumpers

		Crew	Daily Output	Labor-Hours	Unit	Material	2022 Bare Costs Labor	Equipment	Total	Total Incl O&P
0010	**PRECAST CONCRETE PARKING BUMPERS**									
1000	Wheel stops, precast concrete incl. dowels, 6" x 10" x 6'-0"	B-2	120	.333	Ea.	46	15.35		61.35	75
1100	8" x 13" x 6'-0"	"	120	.333		90	15.35		105.35	124
1540	Bollards, precast concrete, free standing, round, 15" diam. x 36" H	B-6	14	1.714		184	84.50	23.50	292	365
1550	16" diam. x 30" H		14	1.714		237	84.50	23.50	345	420
1560	18" diam. x 34" H		14	1.714		370	84.50	23.50	478	565
1570	18" diam. x 72" H		14	1.714		585	84.50	23.50	693	805
1572	26" diam. x 18" H		12	2		335	99	27.50	461.50	550
1574	26" diam. x 30" H	B-6	12	2	Ea.	555	99	27.50	681.50	795
1578	26" diam. x 50" H		12	2		745	99	27.50	871.50	1,000
1580	Square, 22" x 32" H		12	2		350	99	27.50	476.50	575
1585	22" x 41" H		12	2		520	99	27.50	646.50	755
1590	Hexagon, 26" x 41" H		10	2.400		435	119	33	587	700

32 17 13.26 Wood Parking Bumpers

		Crew	Daily Output	Labor-Hours	Unit	Material	2022 Bare Costs Labor	Equipment	Total	Total Incl O&P
0010	**WOOD PARKING BUMPERS**									
0020	Parking barriers, timber w/saddles, treated type									
0100	4" x 4" for cars	B-2	520	.077	L.F.	5.20	3.54		8.74	11.35
0200	6" x 6" for trucks	"	520	.077	"	11	3.54		14.54	17.70

32 17 23 – Pavement Markings

32 17 23.13 Painted Pavement Markings

		Crew	Daily Output	Labor-Hours	Unit	Material	2022 Bare Costs Labor	Equipment	Total	Total Incl O&P
0010	**PAINTED PAVEMENT MARKINGS**									
0020	Acrylic waterborne, white or yellow, 4" wide, less than 3,000 L.F.	B-78	20000	.002	L.F.	.14	.11	.03	.28	.36
0200	6" wide, less than 3,000 L.F.		11000	.004		.21	.20	.05	.46	.61
0500	8" wide, less than 3,000 L.F.		10000	.005		.27	.22	.06	.55	.72
0600	12" wide, less than 3,000 L.F.		4000	.012		.41	.56	.14	1.11	1.49
0620	Arrows or gore lines		2300	.021	S.F.	.26	.98	.24	1.48	2.11
0640	Temporary paint, white or yellow, less than 3,000 L.F.		15000	.003	L.F.	.16	.15	.04	.35	.45
0660	Removal	1 Clab	300	.027			1.22		1.22	1.93
0680	Temporary tape	2 Clab	1500	.011		.53	.49		1.02	1.35
0710	Thermoplastic, white or yellow, 4" wide, less than 6,000 L.F.	B-79	15000	.003		.92	.13	.09	1.14	1.31
0730	6" wide, less than 6,000 L.F.		14000	.003		1.38	.13	.10	1.61	1.84
0740	8" wide, less than 6,000 L.F.		12000	.003		1.84	.16	.11	2.11	2.40
0750	12" wide, less than 6,000 L.F.		6000	.007		2.71	.31	.23	3.25	3.74
0760	Arrows		660	.061	S.F.	2.76	2.85	2.08	7.69	9.85
0770	Gore lines		2500	.016		2.76	.75	.55	4.06	4.83
0780	Letters		660	.061		.92	2.85	2.08	5.85	7.80

32 17 23.14 Pavement Parking Markings

		Crew	Daily Output	Labor-Hours	Unit	Material	2022 Bare Costs Labor	Equipment	Total	Total Incl O&P
0010	**PAVEMENT PARKING MARKINGS**									
0790	Layout of pavement marking	A-2	25000	.001	L.F.		.05	.01	.06	.08
0800	Lines on pvmt., parking stall, paint, white, 4" wide	B-78B	400	.045	Stall	6	2.11	1.11	9.22	11.15
0825	Parking stall, small quantities	2 Pord	80	.200		11.95	9.55		21.50	28.50
0830	Lines on pvmt., parking stall, thermoplastic, white, 4" wide	B-79	300	.133		21	6.25	4.57	31.82	38
1000	Street letters and numbers	B-78B	1600	.011	S.F.	.83	.53	.28	1.64	2.05

32 31 Fences and Gates

32 31 13 – Chain Link Fences and Gates

32 31 13.20 Fence, Chain Link Industrial

		Crew	Daily Output	Labor-Hours	Unit	Material	2022 Bare Costs Labor	Equipment	Total	Total Incl O&P
0010	**FENCE, CHAIN LINK INDUSTRIAL**									
0011	Schedule 40, including concrete									
0020	3 strands barb wire, 2" post @ 10' OC, set in concrete, 6' H									
0200	9 ga. wire, galv. steel, in concrete	B-80C	240	.100	L.F.	21.50	4.72	1.62	27.84	33
0250	Vinyl coated fabric		240	.100		.90	4.72	1.62	7.24	10.25
0300	Aluminized steel		240	.100		25.50	4.72	1.62	31.84	38
0301	Fence, wrought iron		240	.100		46.50	4.72	1.62	52.84	61
0303	Fence, commercial 4' high		240	.100		35.50	4.72	1.62	41.84	48.50
0304	Fence, commercial 6' high		240	.100		63	4.72	1.62	69.34	79
0500	6 ga. wire, galv. steel		240	.100		27.50	4.72	1.62	33.84	39.50
0600	Aluminized steel		240	.100		37	4.72	1.62	43.34	50
0800	6 ga. wire, 6' high but omit barbed wire, galv. steel		250	.096		24.50	4.53	1.56	30.59	36
0900	Aluminized steel, in concrete		250	.096		33.50	4.53	1.56	39.59	45.50
0920	8' H, 6 ga. wire, 2-1/2" line post, galv. steel, in concrete		180	.133		38.50	6.30	2.16	46.96	55
0940	Aluminized steel, in concrete		180	.133	↓	47	6.30	2.16	55.46	64.50
1100	Add for corner posts, 3" diam., galv. steel, in concrete		40	.600	Ea.	96	28.50	9.70	134.20	162
1200	Aluminized steel, in concrete		40	.600		107	28.50	9.70	145.20	174
1300	Add for braces, galv. steel		80	.300		53	14.15	4.86	72.01	86
1350	Aluminized steel		80	.300		56	14.15	4.86	75.01	90
1400	Gate for 6' high fence, 1-5/8" frame, 3' wide, galv. steel		10	2.400		225	113	39	377	470
1500	Aluminized steel, in concrete	↓	10	2.400	↓	242	113	39	394	490
2000	5'-0" high fence, 9 ga., no barbed wire, 2" line post, in concrete									
2010	10' OC, 1-5/8" top rail, in concrete									
2100	Galvanized steel, in concrete	B-80C	300	.080	L.F.	22.50	3.78	1.30	27.58	32
2200	Aluminized steel, in concrete		300	.080	"	22	3.78	1.30	27.08	32
2400	Gate, 4' wide, 5' high, 2" frame, galv. steel, in concrete		10	2.400	Ea.	247	113	39	399	495
2500	Aluminized steel, in concrete		10	2.400	"	198	113	39	350	440
3100	Overhead slide gate, chain link, 6' high, to 18' wide, in concrete	↓	38	.632	L.F.	105	30	10.25	145.25	175
3110	Cantilever type, in concrete	B-80	48	.667		205	33.50	11.30	249.80	290
3120	8' high, in concrete		24	1.333		197	67	22.50	286.50	350
3130	10' high, in concrete		18	1.778	↓	228	89	30	347	425
9000	Minimum labor/equipment charge	↓	2	16	Job		800	271	1,071	1,575

32 31 13.25 Fence, Chain Link Residential

		Crew	Daily Output	Labor-Hours	Unit	Material	2022 Bare Costs Labor	Equipment	Total	Total Incl O&P
0010	**FENCE, CHAIN LINK RESIDENTIAL**									
0011	Schedule 20, 11 ga. wire, 1-5/8" post									
3300	Residential, 11 ga. wire, 1-5/8" line post @ 10' OC									
3310	1-3/8" top rail									
3350	4' high, galvanized steel	B-80C	475	.051	L.F.	13.90	2.38	.82	17.10	20
3400	Aluminized		475	.051	"	18.50	2.38	.82	21.70	25
3600	Gate, 3' wide, 1-3/8" frame, galv. steel		10	2.400	Ea.	120	113	39	272	355
3700	Aluminized		10	2.400	"	182	113	39	334	425
3900	3' high, galvanized steel		620	.039	L.F.	145	1.83	.63	147.46	163
4000	Aluminized		620	.039	"	131	1.83	.63	133.46	148
4200	Gate, 3' wide, 1-3/8" frame, galv. steel		12	2	Ea.	101	94.50	32.50	228	297
4300	Aluminized	↓	12	2	"	107	94.50	32.50	234	305
9000	Minimum labor/equipment charge	B-1	2	12	Job		555		555	880

32 31 13.26 Tennis Court Fences and Gates

		Crew	Daily Output	Labor-Hours	Unit	Material	2022 Bare Costs Labor	Equipment	Total	Total Incl O&P
0010	**TENNIS COURT FENCES AND GATES**									
3150	Tennis courts, 11 ga. wire, 1-3/4" mesh, 2-1/2" line posts									
3170	1-5/8" top rail, 3" corner and gate posts									
3190	10' high, galvanized steel	B-80	190	.168	L.F.	34	8.45	2.85	45.30	54
3210	Vinyl covered 9 ga. wire	"	190	.168	"	42.50	8.45	2.85	53.80	63.50

For customer support on your Commercial Renovation Costs with RSMeans data, call 800.448.8182.

32 31 Fences and Gates

32 31 13 – Chain Link Fences and Gates

32 31 13.40 Fence, Fabric and Accessories

		Crew	Daily Output	Labor-Hours	Unit	Material	2022 Bare Costs Labor	2022 Bare Costs Equipment	Total	Total Incl O&P
0010	**FENCE, FABRIC & ACCESSORIES**									
8000	Components only, fabric, galvanized, 4' high	B-80C	585	.041	L.F.	4.18	1.94	.66	6.78	8.40
8040	6' high		430	.056	"	6.30	2.63	.90	9.83	12.15
8200	Posts, line post, 4' high		34	.706	Ea.	17.50	33.50	11.45	62.45	85
8240	6' high		26	.923	"	21.50	43.50	14.95	79.95	109
8400	Top rails, 1-3/8" OD		1000	.024	L.F.	1.85	1.13	.39	3.37	4.27
8440	1-5/8" OD		1000	.024	"	3.05	1.13	.39	4.57	5.60

32 31 13.80 Residential Chain Link Gate

		Crew	Daily Output	Labor-Hours	Unit	Material	2022 Bare Costs Labor	2022 Bare Costs Equipment	Total	Total Incl O&P
0010	**RESIDENTIAL CHAIN LINK GATE**									
0110	Residential 4' gate, single incl. hardware and concrete	B-80C	10	2.400	Ea.	214	113	39	366	460
0120	5'		10	2.400		225	113	39	377	470
0130	6'		10	2.400		233	113	39	385	480
0510	Residential 4' gate, double incl. hardware and concrete		10	2.400		320	113	39	472	580
0520	5'		10	2.400		340	113	39	492	595
0530	6'		10	2.400		380	113	39	532	640

32 31 13.82 Internal Chain Link Gate

		Crew	Daily Output	Labor-Hours	Unit	Material	2022 Bare Costs Labor	2022 Bare Costs Equipment	Total	Total Incl O&P
0010	**INTERNAL CHAIN LINK GATE**									
0110	Internal 6' gate, single incl. post flange, hardware and concrete	B-80C	10	2.400	Ea.	375	113	39	527	640
0120	8'		10	2.400		425	113	39	577	690
0130	10'		10	2.400		595	113	39	747	875
0510	Internal 6' gate, double incl. post flange, hardware and concrete		10	2.400		625	113	39	777	915
0520	8'		10	2.400		710	113	39	862	1,000
0530	10'		10	2.400		920	113	39	1,072	1,225

32 31 13.84 Industrial Chain Link Gate

		Crew	Daily Output	Labor-Hours	Unit	Material	2022 Bare Costs Labor	2022 Bare Costs Equipment	Total	Total Incl O&P
0010	**INDUSTRIAL CHAIN LINK GATE**									
0110	Industrial 8' gate, single incl. hardware and concrete	B-80C	10	2.400	Ea.	640	113	39	792	930
0120	10'		10	2.400		725	113	39	877	1,025
0510	Industrial 8' gate, double incl. hardware and concrete		10	2.400		930	113	39	1,082	1,250
0520	10'		10	2.400		1,050	113	39	1,202	1,375

32 31 13.88 Chain Link Transom

		Crew	Daily Output	Labor-Hours	Unit	Material	2022 Bare Costs Labor	2022 Bare Costs Equipment	Total	Total Incl O&P
0010	**CHAIN LINK TRANSOM**									
0110	Add for, single transom, 3' wide, incl. components & hardware	B-80C	10	2.400	Ea.	128	113	39	280	365
0120	Add for, double transom, 6' wide, incl. components & hardware	"	10	2.400	"	137	113	39	289	375

32 31 19 – Decorative Metal Fences and Gates

32 31 19.10 Decorative Fence

		Crew	Daily Output	Labor-Hours	Unit	Material	2022 Bare Costs Labor	2022 Bare Costs Equipment	Total	Total Incl O&P
0010	**DECORATIVE FENCE**									
5300	Tubular picket, steel, 6' sections, 1-9/16" posts, 4' high	B-80C	300	.080	L.F.	41.50	3.78	1.30	46.58	53
5400	2" posts, 5' high		240	.100		47	4.72	1.62	53.34	61.50
5600	2" posts, 6' high		200	.120		57	5.65	1.94	64.59	74
5700	Staggered picket, 1-9/16" posts, 4' high		300	.080		48.50	3.78	1.30	53.58	61
5800	2" posts, 5' high		240	.100		48.50	4.72	1.62	54.84	63
5900	2" posts, 6' high		200	.120		63.50	5.65	1.94	71.09	81
6200	Gates, 4' high, 3' wide	B-1	10	2.400	Ea.	360	111		471	575
6300	5' high, 3' wide		10	2.400		265	111		376	470
6400	6' high, 3' wide		10	2.400		400	111		511	615
6500	4' wide		10	2.400		355	111		466	565

For customer support on your Commercial Renovation Costs with RSMeans data, call 800.448.8182.

573

32 31 Fences and Gates

32 31 23 – Plastic Fences and Gates

32 31 23.20 Fence, Recycled Plastic

	32 31 23.20 Fence, Recycled Plastic	Crew	Daily Output	Labor-Hours	Unit	Material	2022 Bare Costs Labor	Equipment	Total	Total Incl O&P
0010	**FENCE, RECYCLED PLASTIC**									
9015	Fence rail, made from recycled plastic, various colors, 2 rail [G]	B-1	150	.160	L.F.	12.25	7.40		19.65	25
9018	3 rail [G]		150	.160		11.05	7.40		18.45	24
9020	4 rail [G]		150	.160	↓	17.05	7.40		24.45	30.50
9030	Fence pole, made from recycled plastic, various colors, 7' [G]	↓	96	.250	Ea.	65	11.55		76.55	90
9040	Stockade fence, made from recycled plastic, various colors, 4' high [G]	B-80C	160	.150	L.F.	35	7.10	2.43	44.53	52.50
9050	6' high [G]		160	.150	"	40	7.10	2.43	49.53	58
9060	6' pole [G]		96	.250	Ea.	46.50	11.80	4.05	62.35	74
9070	9' pole [G]		96	.250	"	46	11.80	4.05	61.85	73.50
9080	Picket fence, made from recycled plastic, various colors, 3' high [G]		160	.150	L.F.	25.50	7.10	2.43	35.03	42
9090	4' high [G]		160	.150	"	37.50	7.10	2.43	47.03	55.50
9100	3' high gate [G]		8	3	Ea.	26.50	142	48.50	217	310
9110	4' high gate [G]		8	3		34	142	48.50	224.50	315
9120	5' high pole [G]		96	.250		49	11.80	4.05	64.85	77
9130	6' high pole [G]	↓	96	.250		45.50	11.80	4.05	61.35	73
9140	Pole cap only [G]				↓	5.25			5.25	5.80
9150	Keeper pins only [G]				Ea.	.44			.44	.48

32 31 26 – Wire Fences and Gates

32 31 26.10 Fences, Misc. Metal

	32 31 26.10 Fences, Misc. Metal	Crew	Daily Output	Labor-Hours	Unit	Material	2022 Bare Costs Labor	Equipment	Total	Total Incl O&P
0010	**FENCES, MISC. METAL**									
0012	Chicken wire, posts @ 4', 1" mesh, 4' high	B-80C	410	.059	L.F.	4.66	2.76	.95	8.37	10.60
0100	2" mesh, 6' high		350	.069		4.04	3.24	1.11	8.39	10.80
0200	Galv. steel, 12 ga., 2" x 4" mesh, posts 5' OC, 3' high		300	.080		3.01	3.78	1.30	8.09	10.75
0300	5' high		300	.080		3.56	3.78	1.30	8.64	11.35
0400	14 ga., 1" x 2" mesh, 3' high		300	.080		3.34	3.78	1.30	8.42	11.10
0500	5' high	↓	300	.080	↓	4.68	3.78	1.30	9.76	12.60
1000	Kennel fencing, 1-1/2" mesh, 6' long, 3'-6" wide, 6'-2" high	2 Clab	4	4	Ea.	545	182		727	885
1050	12' long		4	4		770	182		952	1,125
1200	Top covers, 1-1/2" mesh, 6' long		15	1.067		158	48.50		206.50	251
1250	12' long	↓	12	1.333	↓	236	61		297	355
4492	Security fence, prison grade, barbed wire, set in concrete, 10' high	B-80	22	1.455	L.F.	69.50	73	24.50	167	219
4494	Security fence, prison grade, razor wire, set in concrete, 10' high		18	1.778		71	89	30	190	253
4500	Security fence, prison grade, set in concrete, 12' high		25	1.280		60	64	21.50	145.50	192
4600	16' high		20	1.600		94	80	27	201	260
4990	Security fence, prison grade, set in concrete, 10' high	↓	25	1.280	↓	60	64	21.50	145.50	192

32 31 29 – Wood Fences and Gates

32 31 29.20 Fence, Wood Rail

	32 31 29.20 Fence, Wood Rail	Crew	Daily Output	Labor-Hours	Unit	Material	2022 Bare Costs Labor	Equipment	Total	Total Incl O&P
0010	**FENCE, WOOD RAIL**									
0012	Picket, No. 2 cedar, Gothic, 2 rail, 3' high	B-1	160	.150	L.F.	10.95	6.95		17.90	23
0050	Gate, 3'-6" wide	B-80C	9	2.667	Ea.	124	126	43	293	385
0600	Open rail, rustic, No. 1 cedar, 2 rail, 3' high		160	.150	L.F.	15.30	7.10	2.43	24.83	30.50
0650	Gate, 3' wide		9	2.667	Ea.	118	126	43	287	380
1200	Stockade, No. 2 cedar, treated wood rails, 6' high		160	.150	L.F.	11.70	7.10	2.43	21.23	27
1250	Gate, 3' wide		9	2.667	Ea.	167	126	43	336	430
3300	Board, shadow box, 1" x 6", treated pine, 6' high		160	.150	L.F.	27.50	7.10	2.43	37.03	44.50
3400	No. 1 cedar, 6' high		150	.160		27.50	7.55	2.59	37.64	45.50
3900	Basket weave, No. 1 cedar, 6' high	↓	160	.150	↓	49.50	7.10	2.43	59.03	68.50
3950	Gate, 3'-6" wide	B-1	8	3	Ea.	370	139		509	630
4000	Treated pine, 6' high		150	.160	L.F.	23	7.40		30.40	37
4200	Gate, 3'-6" wide		9	2.667	Ea.	254	123		377	475
8000	Posts only, 4' high	↓	40	.600	↓	17.80	28		45.80	63.50

32 31 Fences and Gates

32 31 29 – Wood Fences and Gates

	32 31 29.20 Fence, Wood Rail	Crew	Daily Output	Labor-Hours	Unit	Material	2022 Bare Costs Labor	Equipment	Total	Total Incl O&P
8040	6' high	B-1	30	.800	Ea.	23	37		60	84.50
9000	Minimum labor/equipment charge	1 Clab	2	4	Job		182		182	290

32 32 Retaining Walls

32 32 29 – Timber Retaining Walls

32 32 29.10 Landscape Timber Retaining Walls

		Crew	Daily Output	Labor-Hours	Unit	Material	2022 Bare Costs Labor	Equipment	Total	Total Incl O&P
0010	**LANDSCAPE TIMBER RETAINING WALLS**									
0100	Treated timbers, 6" x 6"	1 Clab	265	.030	L.F.	2.68	1.38		4.06	5.15
0110	6" x 8'	"	200	.040	"	7	1.82		8.82	10.60
0120	Drilling holes in timbers for fastening, 1/2"	1 Carp	450	.018	Inch		1		1	1.59
0130	5/8"	"	450	.018	"		1		1	1.59
0140	Reinforcing rods for fastening, 1/2"	1 Clab	312	.026	L.F.	.53	1.17		1.70	2.45
0150	5/8"	"	312	.026	"	.83	1.17		2	2.78
0160	Reinforcing fabric	2 Clab	2500	.006	S.Y.	1.87	.29		2.16	2.52
0170	Gravel backfill	↓	28	.571	C.Y.	25	26		51	69
0180	Perforated pipe, 4" diameter with silt sock		1200	.013	L.F.	1.46	.61		2.07	2.58
0190	Galvanized 60d common nails	1 Clab	625	.013	Ea.	.15	.58		.73	1.09
0200	20d common nails	"	3800	.002	"	.03	.10		.13	.19

32 32 53 – Stone Retaining Walls

32 32 53.10 Retaining Walls, Stone

		Crew	Daily Output	Labor-Hours	Unit	Material	2022 Bare Costs Labor	Equipment	Total	Total Incl O&P
0010	**RETAINING WALLS, STONE**									
0015	Including excavation, concrete footing and									
0020	stone 3' below grade. Price is exposed face area.									
0200	Decorative random stone, to 6' high, 1'-6" thick, dry set	D-1	35	.457	S.F.	86	23		109	131
0300	Mortar set		40	.400		88.50	20		108.50	130
0500	Cut stone, to 6' high, 1'-6" thick, dry set		35	.457		89	23		112	135
0600	Mortar set		40	.400		89.50	20		109.50	131
0800	Random stone, 6' to 10' high, 2' thick, dry set		45	.356		91.50	17.80		109.30	130
0900	Mortar set		50	.320		109	16.05		125.05	146
1100	Cut stone, 6' to 10' high, 2' thick, dry set		45	.356		91.50	17.80		109.30	130
1200	Mortar set		50	.320	↓	109	16.05		125.05	146
9000	Minimum labor/equipment charge	↓	2	8	Job		400		400	640

32 33 Site Furnishings

32 33 33 – Site Manufactured Planters

32 33 33.10 Planters

		Crew	Daily Output	Labor-Hours	Unit	Material	2022 Bare Costs Labor	Equipment	Total	Total Incl O&P
0010	**PLANTERS**									
0012	Concrete, sandblasted, precast, 48" diameter, 24" high	2 Clab	15	1.067	Ea.	520	48.50		568.50	650
0100	Fluted, precast, 7' diameter, 36" high		10	1.600		1,600	73		1,673	1,875
0300	Fiberglass, circular, 36" diameter, 24" high		15	1.067		385	48.50		433.50	500
0400	60" diameter, 24" high	↓	10	1.600	↓	865	73		938	1,075
9000	Minimum labor/equipment charge	1 Clab	2	4	Job		182		182	290

32 33 43 – Site Seating and Tables

32 33 43.13 Site Seating

		Crew	Daily Output	Labor-Hours	Unit	Material	2022 Bare Costs Labor	Equipment	Total	Total Incl O&P
0010	**SITE SEATING**									
0012	Seating, benches, park, precast conc., w/backs, wood rails, 4' long	2 Clab	5	3.200	Ea.	820	146		966	1,125
0100	8' long		4	4		845	182		1,027	1,225
0300	Fiberglass, without back, one piece, 4' long	↓	10	1.600	↓	690	73		763	875

32 33 Site Furnishings

32 33 43 – Site Seating and Tables

32 33 43.13 Site Seating

		Crew	Daily Output	Labor-Hours	Unit	Material	2022 Bare Costs Labor	Equipment	Total	Total Incl O&P
0400	8' long	2 Clab	7	2.286	Ea.	505	104		609	720
0500	Steel barstock pedestals w/backs, 2" x 3" wood rails, 4' long		10	1.600		1,400	73		1,473	1,675
0510	8' long		7	2.286		520	104		624	735
0515	Powder coated steel, 4" x 4" plastic slats, 6' long		8	2		630	91		721	835
0520	3" x 8" wood plank, 4' long		10	1.600		1,775	73		1,848	2,075
0530	8' long		7	2.206		840	104		944	1,075
0540	Backless, 4" x 4" wood plank, 4' square		10	1.600		1,450	73		1,523	1,700
0550	8' long		7	2.286		1,475	104		1,579	1,800
0560	Powder coated steel, with back and 2 anti-vagrant dividers, 6' long		8	2		1,375	91		1,466	1,675
0600	Aluminum pedestals, with backs, aluminum slats, 8' long		8	2		1,525	91		1,616	1,825
0610	15' long		5	3.200		1,150	146		1,296	1,500
0620	Portable, aluminum slats, 8' long		8	2		515	91		606	710
0630	15' long		5	3.200		745	146		891	1,050
0800	Cast iron pedestals, back & arms, wood slats, 4' long		8	2		505	91		596	700
0820	8' long		5	3.200		1,300	146		1,446	1,650
0840	Backless, wood slats, 4' long		8	2		1,350	91		1,441	1,625
0860	8' long		5	3.200		1,475	146		1,621	1,850
1700	Steel frame, fir seat, 10' long		10	1.600		460	73		533	625
2000	Benches, park, with back, galv. stl. frame, 4" x 4" plastic slats, 6' long		7	2.286		630	104		734	855
9000	Minimum labor/equipment charge		2	8	Job		365		365	580

32 84 Planting Irrigation

32 84 23 – Underground Sprinklers

32 84 23.10 Sprinkler Irrigation System

		Crew	Daily Output	Labor-Hours	Unit	Material	2022 Bare Costs Labor	Equipment	Total	Total Incl O&P
0010	**SPRINKLER IRRIGATION SYSTEM**									
0011	For lawns									
0800	Residential system, custom, 1" supply	B-20	2000	.012	S.F.	.32	.61		.93	1.32
0900	1-1/2" supply	"	1800	.013	"	.47	.68		1.15	1.59

32 91 Planting Preparation

32 91 13 – Soil Preparation

32 91 13.23 Structural Soil Mixing

		Crew	Daily Output	Labor-Hours	Unit	Material	2022 Bare Costs Labor	Equipment	Total	Total Incl O&P
0010	**STRUCTURAL SOIL MIXING**									
0100	Rake topsoil, site material, harley rock rake, ideal	B-6	33	.727	M.S.F.		36	9.95	45.95	67.50
0200	Adverse	"	7	3.429			169	47	216	320
0300	Screened loam, york rake and finish, ideal	B-62	24	1			49.50	10.95	60.45	90
0400	Adverse	B-62	20	1.200	M.S.F.		59.50	13.15	72.65	108
1000	Remove topsoil & stock pile on site, 75 HP dozer, 6" deep, 50' haul	B-10L	30	.400			22	17.85	39.85	54.50
1050	300' haul		6.10	1.967			109	88	197	268
1100	12" deep, 50' haul		15.50	.774			43	34.50	77.50	106
1150	300' haul		3.10	3.871			215	173	388	525
1200	200 HP dozer, 6" deep, 50' haul	B-10B	125	.096			5.35	10.80	16.15	20
1250	300' haul		30.70	.391			21.50	44	65.50	82.50
1300	12" deep, 50' haul		62	.194			10.75	22	32.75	41
1350	300' haul		15.40	.779			43	87.50	130.50	164
1400	Alternate method, 75 HP dozer, 50' haul	B-10L	860	.014	C.Y.		.77	.62	1.39	1.90
1450	300' haul	"	114	.105			5.85	4.70	10.55	14.30
1500	200 HP dozer, 50' haul	B-10B	2660	.005			.25	.51	.76	.95
1600	300' haul	"	570	.021			1.17	2.37	3.54	4.43

32 91 Planting Preparation

32 91 13 – Soil Preparation

32 91 13.23 Structural Soil Mixing

		Crew	Daily Output	Labor-Hours	Unit	Material	2022 Bare Costs Labor	Equipment	Total	Total Incl O&P
1800	Rolling topsoil, hand push roller	1 Clab	3200	.003	S.F.		.11		.11	.18
1850	Tractor drawn roller	B-66	10666	.001	"		.04	.03	.07	.10
2000	Root raking and loading, residential, no boulders	B-6	53.30	.450	M.S.F.		22	6.15	28.15	42
2100	With boulders		32	.750			37	10.30	47.30	70
2200	Municipal, no boulders		200	.120			5.95	1.64	7.59	11.15
2300	With boulders		120	.200			9.90	2.74	12.64	18.60
2400	Large commercial, no boulders	B-10B	400	.030			1.66	3.37	5.03	6.30
2500	With boulders	"	240	.050			2.77	5.60	8.37	10.55
3000	Scarify subsoil, residential, skid steer loader w/scarifiers, 50 HP	B-66	32	.250			14.25	8.40	22.65	31.50
3050	Municipal, skid steer loader w/scarifiers, 50 HP	"	120	.067			3.80	2.24	6.04	8.35
3100	Large commercial, 75 HP, dozer w/scarifier	B-10L	240	.050			2.77	2.23	5	6.80
3500	Screen topsoil from stockpile, vibrating screen, wet material (organic)	B-10P	200	.060	C.Y.		3.33	5.90	9.23	11.70
3550	Dry material	"	300	.040			2.22	3.94	6.16	7.80
3600	Mixing with conditioners, manure and peat	B-10R	550	.022			1.21	.59	1.80	2.55
3650	Mobilization add for 2 days or less operation	B-34K	3	2.667	Job		141	297	438	550
3800	Spread conditioned topsoil, 6" deep, by hand	B-1	360	.067	S.Y.	4.50	3.08		7.58	9.85
3850	300 HP dozer	B-10M	27	.444	M.S.F.	485	24.50	60.50	570	640
4000	Spread soil conditioners, alum. sulfate, 1#/S.Y., hand push spreader	1 Clab	17500	.001	S.Y.	18.45	.02		18.47	20.50
4050	Tractor spreader	B-66	700	.011	M.S.F.	2,050	.65	.38	2,051.03	2,250
4100	Fertilizer, 0.2#/S.Y., push spreader	1 Clab	17500	.001	S.Y.	.07	.02		.09	.11
4150	Tractor spreader	B-66	700	.011	M.S.F.	8	.65	.38	9.03	10.25
4200	Ground limestone, 1#/S.Y., push spreader	1 Clab	17500	.001	S.Y.	.15	.02		.17	.20
4250	Tractor spreader	B-66	700	.011	M.S.F.	16.65	.65	.38	17.68	19.80
4400	Manure, 18#/S.Y., push spreader	1 Clab	2500	.003	S.Y.	7.15	.15		7.30	8.10
4450	Tractor spreader	B-66	280	.029	M.S.F.	795	1.63	.96	797.59	880
4500	Perlite, 1" deep, push spreader	1 Clab	17500	.001	S.Y.	10.65	.02		10.67	11.75
4550	Tractor spreader	B-66	700	.011	M.S.F.	1,175	.65	.38	1,176.03	1,300
4600	Vermiculite, push spreader	1 Clab	17500	.001	S.Y.	8.10	.02		8.12	9
4650	Tractor spreader	B-66	700	.011	M.S.F.	900	.65	.38	901.03	990
5000	Spread topsoil, skid steer loader and hand dress	B-62	270	.089	C.Y.	20.50	4.39	.97	25.86	30.50
5100	Articulated loader and hand dress	B-100	320	.038		20.50	2.08	2.92	25.50	29
5200	Articulated loader and 75 HP dozer	B-10M	500	.024		20.50	1.33	3.26	25.09	28
5300	Road grader and hand dress	B-11L	1000	.016		20.50	.85	.77	22.12	24.50
6000	Tilling topsoil, 20 HP tractor, disk harrow, 2" deep	B-66	450	.018	M.S.F.		1.01	.60	1.61	2.24
6050	4" deep		360	.022			1.27	.75	2.02	2.79
6100	6" deep		270	.030			1.69	1	2.69	3.72
6150	26" rototiller, 2" deep	A-1J	1250	.006	S.Y.		.29	.07	.36	.53
6200	4" deep		1000	.008			.36	.08	.44	.67
6250	6" deep		750	.011			.49	.11	.60	.89

32 91 13.26 Planting Beds

		Crew	Daily Output	Labor-Hours	Unit	Material	2022 Bare Costs Labor	Equipment	Total	Total Incl O&P
0010	**PLANTING BEDS**									
0100	Backfill planting pit, by hand, on site topsoil	2 Clab	18	.889	C.Y.		40.50		40.50	64.50
0200	Prepared planting mix, by hand	"	24	.667			30.50		30.50	48.50
0300	Skid steer loader, on site topsoil	B-62	340	.071			3.49	.77	4.26	6.35
0400	Prepared planting mix	"	410	.059			2.89	.64	3.53	5.25
1000	Excavate planting pit, by hand, sandy soil	2 Clab	16	1			45.50		45.50	72.50
1100	Heavy soil or clay	"	8	2			91		91	145
1200	1/2 C.Y. backhoe, sandy soil	B-11C	150	.107			5.65	2.19	7.84	11.30
1300	Heavy soil or clay	"	115	.139			7.35	2.86	10.21	14.75
2000	Mix planting soil, incl. loam, manure, peat, by hand	2 Clab	60	.267		35.50	12.15		47.65	58.50
2100	Skid steer loader	B-62	150	.160		35.50	7.90	1.75	45.15	53.50
3000	Pile sod, skid steer loader	"	2800	.009	S.Y.		.42	.09	.51	.77

577

32 91 Planting Preparation

32 91 13 – Soil Preparation

32 91 13.26 Planting Beds		Crew	Daily Output	Labor-Hours	Unit	Material	2022 Bare Costs Labor	Equipment	Total	Total Incl O&P
3100	By hand	2 Clab	400	.040	S.Y.		1.82		1.82	2.90
4000	Remove sod, F.E. loader	B-10S	2000	.006			.33	.22	.55	.76
4100	Sod cutter	B-12K	3200	.005			.27	.47	.74	.94
4200	By hand	2 Clab	240	.067	↓		3.04		3.04	4.83

32 91 19 – Landscape Grading

32 91 19.13 Topsoil Placement and Grading

		Crew	Daily Output	Labor-Hours	Unit	Material	2022 Bare Costs Labor	Equipment	Total	Total Incl O&P
0010	**TOPSOIL PLACEMENT AND GRADING**									
0700	Furnish and place, truck dumped, screened, 4" deep	B-10S	1300	.009	S.Y.	4.25	.51	.34	5.10	5.85
0800	6" deep	"	820	.015	"	5.45	.81	.54	6.80	7.85
0810	Minimum labor/equipment charge	B-10S	2	6	Ea.		335	222	557	765
0900	Fine grading and seeding, incl. lime, fertilizer & seed,									
1000	With equipment	B-14	1000	.048	S.Y.	.57	2.30	.33	3.20	4.61
2000	Minimum labor/equipment charge	1 Clab	4	2	Job		91		91	145

32 92 Turf and Grasses

32 92 19 – Seeding

32 92 19.13 Mechanical Seeding

		Crew	Daily Output	Labor-Hours	Unit	Material	2022 Bare Costs Labor	Equipment	Total	Total Incl O&P
0010	**MECHANICAL SEEDING**									
0020	Mechanical seeding, 215 lb./acre	B-66	1.50	5.333	Acre	565	305	179	1,049	1,300
0100	44 lb./M.S.Y.	"	2500	.003	S.Y.	.23	.18	.11	.52	.65
0101	44 lb./M.S.Y.	1 Clab	13950	.001	S.F.	.03	.03		.06	.07
0300	Fine grading and seeding incl. lime, fertilizer & seed,									
0310	with equipment	B-14	1000	.048	S.Y.	.45	2.30	.33	3.08	4.49
0400	Fertilizer hand push spreader, 35 lb./M.S.F.	1 Clab	200	.040	M.S.F.	24	1.82		25.82	29.50
0600	Limestone hand push spreader, 50 lb./M.S.F.		180	.044		7.15	2.03		9.18	11.05
0800	Grass seed hand push spreader, 4.5 lb./M.S.F.		180	.044	↓	25.50	2.03		27.53	31
1000	Hydro or air seeding for large areas, incl. seed and fertilizer	B-81	8900	.003	S.Y.	.58	.14	.11	.83	1
1100	With wood fiber mulch added	"	8900	.003	"	2.12	.14	.11	2.37	2.69
1300	Seed only, over 100 lb., field seed, minimum				Lb.	2.42			2.42	2.66
1400	Maximum					1.61			1.61	1.77
1500	Lawn seed, minimum					1.29			1.29	1.42
1600	Maximum				↓	2.89			2.89	3.18
1800	Aerial operations, seeding only, field seed	B-58	50	.480	Acre	785	23.50	103	911.50	1,025
1900	Lawn seed		50	.480		420	23.50	103	546.50	610
2100	Seed and liquid fertilizer, field seed		50	.480		860	23.50	103	986.50	1,100
2200	Lawn seed		50	.480	↓	490	23.50	103	616.50	690
9000	Minimum labor/equipment charge	1 Clab	4	2	Job		91		91	145

32 92 23 – Sodding

32 92 23.10 Sodding Systems

		Crew	Daily Output	Labor-Hours	Unit	Material	2022 Bare Costs Labor	Equipment	Total	Total Incl O&P
0010	**SODDING SYSTEMS**									
0020	Sodding, 1" deep, bluegrass sod, on level ground, over 8 M.S.F.	B-63	22	1.818	M.S.F.	570	87	11.95	668.95	775
0200	4 M.S.F.		17	2.353		335	113	15.50	463.50	560
0300	1,000 S.F.		13.50	2.963		570	142	19.50	731.50	870
0500	Sloped ground, over 8 M.S.F.		6	6.667		570	320	44	934	1,175
0600	4 M.S.F.		5	8		335	385	52.50	772.50	1,025
0700	1,000 S.F.		4	10		570	480	66	1,116	1,450
1000	Bent grass sod, on level ground, over 6 M.S.F.		20	2		570	96	13.15	679.15	790
1100	3 M.S.F.		18	2.222		850	106	14.60	970.60	1,125
1200	Sodding 1,000 S.F. or less	B-63	14	2.857	M.S.F.	850	137	18.80	1,005.80	1,175
1500	Sloped ground, over 6 M.S.F.	↓	15	2.667	↓	570	128	17.55	715.55	845

For customer support on your Commercial Renovation Costs with RSMeans data, call 800.448.8182.

32 92 Turf and Grasses

32 92 23 – Sodding

	32 92 23.10 Sodding Systems	Crew	Daily Output	Labor-Hours	Unit	Material	2022 Bare Costs Labor	Equipment	Total	Total Incl O&P
1600	3 M.S.F.	B-63	13.50	2.963	M.S.F.	850	142	19.50	1,011.50	1,175
1700	1,000 S.F.	↓	12	3.333	↓	850	160	22	1,032	1,200

32 93 Plants

32 93 13 – Ground Covers

32 93 13.10 Ground Cover Plants

	32 93 13.10 Ground Cover Plants	Crew	Daily Output	Labor-Hours	Unit	Material	2022 Bare Costs Labor	Equipment	Total	Total Incl O&P
0010	**GROUND COVER PLANTS**									
0012	Plants, pachysandra, in prepared beds	B-1	15	1.600	C	120	74		194	250
0200	Vinca minor, 1 yr., bare root, in prepared beds		12	2	"	110	92.50		202.50	268
0600	Stone chips, in 50 lb. bags, Georgia marble		520	.046	Bag	4.58	2.14		6.72	8.45
0700	Onyx gemstone		260	.092		44.50	4.27		48.77	55.50
0800	Quartz		260	.092	↓	19	4.27		23.27	28
0900	Pea gravel, truckload lots	↓	28	.857	Ton	17.35	39.50		56.85	82

32 93 33 – Shrubs

32 93 33.10 Shrubs and Trees

	32 93 33.10 Shrubs and Trees	Crew	Daily Output	Labor-Hours	Unit	Material	2022 Bare Costs Labor	Equipment	Total	Total Incl O&P
0010	**SHRUBS AND TREES**									
0011	Evergreen, in prepared beds, B&B									
0100	Arborvitae pyramidal, 4'-5'	B-17	30	1.067	Ea.	113	53.50	24	190.50	237
0150	Globe, 12"-15"	B-1	96	.250		16.55	11.55		28.10	36.50
0300	Cedar, blue, 8'-10'	B-17	18	1.778		215	89.50	40	344.50	420
0500	Hemlock, Canadian, 2-1/2'-3'	B-1	36	.667		19.95	31		50.95	71
0550	Holly, Savannah, 8'-10' H		9.68	2.479		275	115		390	485
0600	Juniper, andorra, 18"-24"		80	.300		60	13.90		73.90	88
0620	Wiltoni, 15"-18"	↓	80	.300		27.50	13.90		41.40	52.50
0640	Skyrocket, 4-1/2'-5'	B-17	55	.582		161	29.50	13.10	203.60	238
0660	Blue pfitzer, 2'-2-1/2'	B-1	44	.545		39.50	25		64.50	83.50
0680	Ketleerie, 2-1/2'-3'		50	.480		64	22		86	106
0700	Pine, black, 2-1/2'-3'		50	.480		56	22		78	97
0720	Mugo, 18"-24"	↓	60	.400		53	18.50		71.50	87.50
0740	White, 4'-5'	B-17	75	.427		56	21.50	9.60	87.10	106
0800	Spruce, blue, 18"-24"	B-1	60	.400		71	18.50		89.50	108
0840	Norway, 4'-5'	B-17	75	.427		80	21.50	9.60	111.10	133
0900	Yew, denisforma, 12"-15"	B-1	60	.400		39.50	18.50		58	73
1000	Capitata, 18"-24"		30	.800		35	37		72	97.50
1100	Hicksi, 2'-2-1/2'	↓	30	.800	↓	105	37		142	174

32 93 33.20 Shrubs

	32 93 33.20 Shrubs	Crew	Daily Output	Labor-Hours	Unit	Material	2022 Bare Costs Labor	Equipment	Total	Total Incl O&P
0010	**SHRUBS**									
0011	Broadleaf Evergreen, planted in prepared beds									
0100	Andromeda, 15"-18", cont	B-1	96	.250	Ea.	38.50	11.55		50.05	61
0200	Azalea, 15"-18", cont	B-1	96	.250	Ea.	26.50	11.55		38.05	48
0300	Barberry, 9"-12", cont		130	.185		22.50	8.55		31.05	38
0400	Boxwood, 15"-18", B&B		96	.250		60	11.55		71.55	84.50
0500	Euonymus, emerald gaiety, 12"-15", cont		115	.209		27.50	9.65		37.15	45.50
0600	Holly, 15"-18", B&B		96	.250		62	11.55		73.55	86.50
0900	Mount laurel, 18"-24", B&B		80	.300		64.50	13.90		78.40	93
1000	Paxistema, 9"-12" H		130	.185		28.50	8.55		37.05	45
1100	Rhododendron, 18"-24", cont		48	.500		59	23		82	102
1200	Rosemary, 1 gal. cont		600	.040		26	1.85		27.85	31.50
2000	Deciduous, planted in prepared beds, amelanchier, 2'-3', B&B		57	.421		117	19.50		136.50	160
2100	Azalea, 15"-18", B&B	↓	96	.250	↓	29	11.55		40.55	50.50

32 93 Plants

32 93 33 – Shrubs

32 93 33.20 Shrubs

		Crew	Daily Output	Labor-Hours	Unit	Material	2022 Bare Costs Labor	2022 Bare Costs Equipment	Total	Total Incl O&P
2300	Bayberry, 2'-3', B&B	B-1	57	.421	Ea.	40	19.50		59.50	75
2600	Cotoneaster, 15"-18", B&B	↓	80	.300		30	13.90		43.90	55
2800	Dogwood, 3'-4', B&B	B-17	40	.800		35	40.50	18	93.50	122
2900	Euonymus, alatus compacta, 15"-18", cont	B-1	80	.300		34.50	13.90		48.40	60
3200	Forsythia, 2'-3', cont	"	60	.400		15.85	18.50		34.35	47
3300	Hibiscus, 3'-4', B&B	B 17	75	.427		56	21.50	9.60	87.10	106
3400	Honeysuckle, 3'-4', B&B	B-1	60	.400		79.50	18.50		98	117
3500	Hydrangea, 2'-3', B&B	"	57	.421		29	19.50		48.50	63
3600	Lilac, 3'-4', B&B	B-17	40	.800		50	40.50	18	108.50	138
3900	Privet, bare root, 18"-24"	B-1	80	.300		34.50	13.90		48.40	60
4100	Quince, 2'-3', B&B	"	57	.421		38.50	19.50		58	73.50
4200	Russian olive, 3'-4', B&B	B-17	75	.427		32	21.50	9.60	63.10	79.50
4400	Spirea, 3'-4', B&B	B-1	70	.343		35	15.85		50.85	63.50
4500	Viburnum, 3'-4', B&B	B-17	40	.800		71	40.50	18	129.50	161

32 93 43 – Trees

32 93 43.20 Trees

			Crew	Daily Output	Labor-Hours	Unit	Material	2022 Bare Costs Labor	2022 Bare Costs Equipment	Total	Total Incl O&P
0010	**TREES**										
0011	Deciduous, in prep. beds, balled & burlapped (B&B)										
0100	Ash, 2" caliper	G	B-17	8	4	Ea.	222	201	90	513	665
0200	Beech, 5'-6'	G		50	.640		244	32	14.40	290.40	335
0300	Birch, 6'-8', 3 stems	G		20	1.600		105	80.50	36	221.50	283
0500	Crabapple, 6'-8'	G		20	1.600		90	80.50	36	206.50	266
0600	Dogwood, 4'-5'	G		40	.800		80	40.50	18	138.50	171
0700	Eastern redbud, 4'-5'	G		40	.800		108	40.50	18	166.50	202
0800	Elm, 8'-10'	G		20	1.600		420	80.50	36	536.50	630
0900	Ginkgo, 6'-7'	G		24	1.333		99.50	67	30	196.50	248
1000	Hawthorn, 8'-10', 1" caliper	G		20	1.600		645	80.50	36	761.50	875
1100	Honeylocust, 10'-12', 1-1/2" caliper	G	B-17	10	3.200	Ea.	595	161	72	828	990
1300	Larch, 8'	G		32	1		90	50.50	22.50	163	204
1400	Linden, 8'-10', 1" caliper	G		20	1.600		105	80.50	36	221.50	283
1500	Magnolia, 4'-5'	G		20	1.600		100	80.50	36	216.50	277
1600	Maple, red, 8'-10', 1-1/2" caliper	G		10	3.200		105	161	72	338	450
1700	Mountain ash, 8'-10', 1" caliper	G		16	2		227	101	45	373	460
1800	Oak, 2-1/2"-3" caliper	G		6	5.333		285	268	120	673	870
2100	Planetree, 9'-11', 1-1/4" caliper	G		10	3.200		340	161	72	573	710
2200	Plum, 6'-8', 1" caliper	G		20	1.600		148	80.50	36	264.50	330
2300	Poplar, 9'-11', 1-1/4" caliper	G		10	3.200		96	161	72	329	440
2500	Sumac, 2'-3'	G		75	.427		70	21.50	9.60	101.10	122
2700	Tulip, 5'-6'	G		40	.800		54.50	40.50	18	113	143
2800	Willow, 6'-8', 1" caliper	G		20	1.600		105	80.50	36	221.50	283
9000	Minimum labor/equipment charge		1 Clab	4	2	Job		91		91	145

32 94 Planting Accessories

32 94 13 – Landscape Edging

32 94 13.20 Edging

		Crew	Daily Output	Labor-Hours	Unit	Material	2022 Bare Costs Labor	Equipment	Total	Total Incl O&P
0010	**EDGING**									
0050	Aluminum alloy, including stakes, 1/8" x 4", mill finish	B-1	390	.062	L.F.	3.54	2.85		6.39	8.40
0051	Black paint		390	.062		4.11	2.85		6.96	9.05
0052	Black anodized		390	.062		4.74	2.85		7.59	9.70
0100	Brick, set horizontally, 1-1/2 bricks per L.F.	D-1	370	.043		1.77	2.17		3.94	5.40
0150	Set vertically, 3 bricks per L.F.	"	135	.119		4.20	5.95		10.15	14.10
0200	Corrugated aluminum, roll, 4" wide	1 Carp	650	.012		3.54	.69		4.23	4.99
0250	6" wide	"	550	.015		4.43	.82		5.25	6.15
0600	Railroad ties, 6" x 8"	2 Carp	170	.094		3.25	5.30		8.55	12
0650	7" x 9"		136	.118		3.61	6.60		10.21	14.45
0750	Redwood 2" x 4"		330	.048		6.25	2.73		8.98	11.20
0800	Steel edge strips, incl. stakes, 1/4" x 5"	B-1	390	.062		5.60	2.85		8.45	10.70
0850	3/16" x 4"	"	390	.062		4.44	2.85		7.29	9.40
9000	Minimum labor/equipment charge	1 Carp	4	2	Job		113		113	179

32 94 50 – Tree Guying

32 94 50.10 Tree Guying Systems

		Crew	Daily Output	Labor-Hours	Unit	Material	2022 Bare Costs Labor	Equipment	Total	Total Incl O&P
0010	**TREE GUYING SYSTEMS**									
0015	Tree guying including stakes, guy wire and wrap									
0100	Less than 3" caliper, 2 stakes	2 Clab	35	.457	Ea.	15.10	21		36.10	49.50
0200	3" to 4" caliper, 3 stakes	"	21	.762	"	21.50	34.50		56	79
1000	Including arrowhead anchor, cable, turnbuckles and wrap									
1100	Less than 3" caliper, 3" anchors	2 Clab	20	.800	Ea.	45.50	36.50		82	108
1200	3" to 6" caliper, 4" anchors	2 Clab	15	1.067	Ea.	36	48.50		84.50	118
1300	6" caliper, 6" anchors		12	1.333		45	61		106	146
1400	8" caliper, 8" anchors		9	1.778		147	81		228	291

32 96 Transplanting

32 96 23 – Plant and Bulb Transplanting

32 96 23.23 Planting

		Crew	Daily Output	Labor-Hours	Unit	Material	2022 Bare Costs Labor	Equipment	Total	Total Incl O&P
0010	**PLANTING**									
0012	Moving shrubs on site, 12" ball	B-62	28	.857	Ea.		42.50	9.40	51.90	77.50
0100	24" ball	"	22	1.091	"		54	11.95	65.95	98

32 96 23.43 Moving Trees

		Crew	Daily Output	Labor-Hours	Unit	Material	2022 Bare Costs Labor	Equipment	Total	Total Incl O&P
0010	**MOVING TREES**, On site									
0300	Moving trees on site, 36" ball	B-6	3.75	6.400	Ea.		315	87.50	402.50	595
0400	60" ball	"	1	24	"		1,175	330	1,505	2,225

32 96 43 – Tree Transplanting

32 96 43.20 Tree Removal

		Crew	Daily Output	Labor-Hours	Unit	Material	2022 Bare Costs Labor	Equipment	Total	Total Incl O&P
0010	**TREE REMOVAL**									
0100	Dig & lace, shrubs, broadleaf evergreen, 18"-24" high	B-1	55	.436	Ea.		20		20	32
0200	2'-3'	"	35	.686			31.50		31.50	50.50
0300	3'-4'	B-6	30	.800			39.50	10.95	50.45	74.50
0400	4'-5'	"	20	1.200			59.50	16.45	75.95	112
1000	Deciduous, 12"-15"	B-1	110	.218			10.10		10.10	16.05
1100	18"-24"		65	.369			17.10		17.10	27
1200	2'-3'		55	.436			20		20	32
1300	3'-4'	B-6	50	.480			23.50	6.60	30.10	45
2000	Evergreen, 18"-24"	B-1	55	.436			20		20	32
2100	2' to 2.5'		50	.480			22		22	35.50
2200	2.5' to 3'		35	.686			31.50		31.50	50.50

32 96 Transplanting

32 96 43 – Tree Transplanting

32 96 43.20 Tree Removal		Crew	Daily Output	Labor-Hours	Unit	Material	2022 Bare Costs Labor	2022 Bare Costs Equipment	Total	Total Incl O&P
2300	3' to 3.5'	B-1	20	1.200	Ea.		55.50		55.50	88
3000	Trees, deciduous, small, 2'-3'	↓	55	.436			20		20	32
3100	3'-4'	B-6	50	.480			23.50	6.60	30.10	45
3200	4'-5'		35	.686			34	9.40	43.40	64
3300	5'-6'		30	.800			39.50	10.95	50.45	74.50
4000	Shade, 5'-6'		50	.480			23.50	6.60	30.10	45
4100	6'-8'		35	.686			34	9.40	43.40	64
4200	8'-10'		25	.960			47.50	13.15	60.65	89.50
4300	2" caliper		12	2			99	27.50	126.50	186
5000	Evergreen, 4'-5'		35	.686			34	9.40	43.40	64
5100	5'-6'		25	.960			47.50	13.15	60.65	89.50
5200	6'-7'		19	1.263			62.50	17.30	79.80	118
5300	7'-8'		15	1.600			79	22	101	149
5400	8'-10'	↓	11	2.182	↓		108	30	138	203

Estimating Tips

33 10 00 Water Utilities
33 30 00 Sanitary Sewerage Utilities
33 40 00 Storm Drainage Utilities

- Never assume that the water, sewer, and drainage lines will go in at the early stages of the project. Consider the site access needs before dividing the site in half with open trenches, loose pipe, and machinery obstructions. Always inspect the site to establish that the site drawings are complete. Check off all existing utilities on your drawings as you locate them. Be especially careful with underground utilities because appurtenances are sometimes buried during regrading or repaving operations. If you find any discrepancies, mark up the site plan for further research. Differing site conditions can be very costly if discovered later in the project.

- See also Section 33 01 00 for restoration of pipe where removal/replacement may be undesirable. Use of new types of piping materials can reduce the overall project cost. Owners/design engineers should consider the installing contractor as a valuable source of current information on utility products and local conditions that could lead to significant cost savings.

Reference Numbers

Reference numbers are shown at the beginning of some major classifications. These numbers refer to related items in the Reference Section. The reference information may be an estimating procedure, an alternate pricing method, or technical information.

Note: Not all subdivisions listed here necessarily appear. ∎

Same Data. Simplified.

Enjoy the convenience and efficiency of accessing your costs anywhere:

- **Skip the multiplier** by setting your location
- **Quickly search,** edit, favorite and share costs
- **Stay on top of price changes** with automatic updates

Discover more at rsmeans.com/online

33 01 Operation and Maintenance of Utilities

33 01 30 – Operation and Maintenance of Sewer Utilities

33 01 30.11 Television Inspection of Sewers	Crew	Daily Output	Labor-Hours	Unit	Material	2022 Bare Costs Labor	2022 Bare Costs Equipment	Total	Total Incl O&P
0010 **TELEVISION INSPECTION OF SEWERS**									
0100 Pipe internal cleaning & inspection, cleaning, pressure pipe systems									
0120 Pig method, lengths 1000' to 10,000'									
0140 4" diameter thru 24" diameter, minimum				L.F.				3.60	4.14
0160 Maximum				"				18	21
6000 Sewage/sanitary systems									
6100 Power rodder with header & cutters									
6110 Mobilization charge, minimum				Total				695	800
6120 Mobilization charge, maximum				"				9,125	10,600
6140 Cleaning 4"-12" diameter				L.F.				6	6.60
6190 14"-24" diameter								8	8.80
6240 30" diameter								9.60	10.50
6250 36" diameter								12	13.20
6260 48" diameter								16	17.60
6270 60" diameter								24	26.50
6280 72" diameter								48	53
9000 Inspection, television camera with video									
9060 up to 500 linear feet				Total				715	820
9220 up to 500 linear feet				"					

33 01 30.72 Cured-In-Place Pipe Lining

	Crew	Daily Output	Labor-Hours	Unit	Material	2022 Bare Costs Labor	2022 Bare Costs Equipment	Total	Total Incl O&P
0010 **CURED-IN-PLACE PIPE LINING**									
0011 Not incl. bypass or cleaning									
0020 Less than 10,000 L.F., urban, 6" to 10"	C-17E	130	.615	L.F.	10.10	36.50	.29	46.89	69
0050 10" to 12"		125	.640		12.40	38	.30	50.70	74
0070 12" to 16"		115	.696		12.70	41.50	.33	54.53	80
0100 16" to 20"		95	.842		14.95	50	.40	65.35	96
0200 24" to 36"		90	.889		16.40	52.50	.42	69.32	102
0300 48" to 72"		80	1		25.50	59.50	.47	85.47	123
0500 Rural, 6" to 10"		180	.444		10.10	26.50	.21	36.81	53
0550 10" to 12"		175	.457		12.40	27	.21	39.61	57
0570 12" to 16"		160	.500		13.25	29.50	.24	42.99	62
0600 16" to 20"		135	.593		15.40	35	.28	50.68	73
0700 24" to 36"		125	.640		16.80	38	.30	55.10	79
0800 48" to 72"		100	.800		25.50	47.50	.38	73.38	103
1000 Greater than 10,000 L.F., urban, 6" to 10"		160	.500		10.10	29.50	.24	39.84	58.50
1050 10" to 12"		155	.516		12.40	30.50	.24	43.14	62.50
1070 12" to 16"		140	.571		12.70	34	.27	46.97	68
1100 16" to 20"	C-17E	120	.667	L.F.	15.40	39.50	.31	55.21	80
1200 24" to 36"		115	.696		16.80	41.50	.33	58.63	84.50
1300 48" to 72"		95	.842		25.50	50	.40	75.90	107
1500 Rural, 6" to 10"		215	.372		10.10	22	.17	32.27	46.50
1550 10" to 12"		210	.381		12.40	22.50	.18	35.08	49.50
1570 12" to 16"		185	.432		12.70	25.50	.20	38.40	54.50
1600 16" to 20"		150	.533		15.40	31.50	.25	47.15	67
1700 24" to 36"		140	.571		16.80	34	.27	51.07	72.50
1800 48" to 72"		120	.667		27	39.50	.31	66.81	92.50
2000 Cured in place pipe, non-pressure, flexible felt resin									
2100 6" diameter				L.F.				25.50	28
2200 8" diameter								26.50	29
2300 10" diameter								29	32
2400 12" diameter								34.50	38
2500 15" diameter								59	65

33 01 30 – Operation and Maintenance of Sewer Utilities

33 01 30.72 Cured-In-Place Pipe Lining	Crew	Daily Output	Labor-Hours	Unit	Material	2022 Bare Costs Labor	Equipment	Total	Total Incl O&P
2600	18" diameter				L.F.				82	90
2700	21" diameter								105	115
2800	24" diameter								177	195
2900	30" diameter								195	215
3000	36" diameter								205	225
3100	48" diameter				▼				218	240

33 01 30.75 Cured-In-place Pipe Lining

		Crew	Daily Output	Labor-Hours	Unit	Material	2022 Bare Costs Labor	Equipment	Total	Total Incl O&P
0010	**CURED-IN-PLACE PIPE LINING**									
0050	Bypass, Sewer	C-17E	500	.160	L.F.		9.50	.08	9.58	15.10
0090	Mobilize - (Any Size) 1 Mob per Job.		1	80	Job		4,750	37.50	4,787.50	7,550
0091	Relocate - (°12") (8hrs, 3 per day)		3	26.667	Ea.		1,575	12.55	1,587.55	2,525
0092	Relocate - (12" to 24") (8hrs, 2 per day)		2	40			2,375	18.80	2,393.80	3,775
0093	Relocate - (24" to 42") (8hrs, 1.5 per day)		1.50	53.333			3,175	25	3,200	5,025
0094	Relocate - (42") (8hrs, 1.5 per day)		1	80			4,750	37.50	4,787.50	7,550
0095	Additional Travel, $2.00 per Mile		3362.50	.024	Mile		1.41	.01	1.42	2.24
0096	Demobilize - (Any Size) 1 Demob per Job.		1	80	Job		4,750	37.50	4,787.50	7,550
0097	Refrigeration Truck (12", Cost per Each Location)		3	26.667	Ea.		1,575	12.55	1,587.55	2,525
0098	Refrigeration Truck (12"-24", Each Pipe / Location)		3	26.667			1,575	12.55	1,587.55	2,525
0099	Refrigeration Truck (24"-42", Each Pipe / Location)		1.50	53.333			3,175	25	3,200	5,025
0100	Refrigeration Truck (42", Each Pipe / Location)		1	80	▼		4,750	37.50	4,787.50	7,550
1210	12", 6mm, Epoxy, 250k, 2ft	▼	6	13.333	L.F.	54.50	790	6.25	850.75	1,325
1211	12", 6mm, Epoxy, 250k, 4ft	C-17E	12	6.667	L.F.	54.50	395	3.13	452.63	690
1212	12", 6mm, Epoxy, 250k, 10ft		30	2.667		54.50	158	1.25	213.75	310
1213	12", 6mm, Epoxy, 250k, 20ft		60	1.333		54.50	79	.63	134.13	186
1216	12", 6mm, Epoxy, 250k, 300ft		900	.089		54.50	5.25	.04	59.79	68.50
1217	12", 6mm, Epoxy, 250k, 328ft	▼	1000	.080		54.50	4.74	.04	59.28	67.50
1218	12", 6mm, Epoxy, 250k, Transitions, add					20%				
1219	12", 6mm, Epoxy, 250k, 45 and 90 deg, add					50%				
1230	12", 5mm, Epoxy, 400k, 2ft	C-17E	6	13.333		45.50	790	6.25	841.75	1,300
1231	12", 5mm, Epoxy, 400k, 4ft		12	6.667		45.50	395	3.13	443.63	680
1232	12", 5mm, Epoxy, 400k, 10ft		30	2.667		45.50	158	1.25	204.75	300
1233	12", 5mm, Epoxy, 400k, 20ft		60	1.333		45.50	79	.63	125.13	176
1234	12", 5mm, Epoxy, 400k, 50ft		150	.533		45.50	31.50	.25	77.25	100
1235	12", 5mm, Epoxy, 400k, 100ft		300	.267		45.50	15.80	.13	61.43	75
1236	12", 5mm, Epoxy, 400k, 300ft		900	.089		45.50	5.25	.04	50.79	58.50
1237	12", 5mm, Epoxy, 400k, 328ft	▼	1000	.080		45.50	4.74	.04	50.28	57.50
1238	12", 5mm, Epoxy, 400k, Transitions, add					20%				
1239	12", 5mm, Epoxy, 400k, 45 and 90 deg, add					50%				
2410	24", 12mm, Epoxy, 250k, 2ft	C-17E	4	20		218	1,175	9.40	1,402.40	2,125
2411	24", 12mm, Epoxy, 250k, 4ft		8	10		218	595	4.70	817.70	1,175
2412	24", 12mm, Epoxy, 250k, 10ft		20	4		218	237	1.88	456.88	615
2413	24", 12mm, Epoxy, 250k, 20ft		40	2		218	119	.94	337.94	430
2414	24", 12mm, Epoxy, 250k, 50ft		100	.800		218	47.50	.38	265.88	315
2415	24", 12mm, Epoxy, 250k, 100ft		200	.400		218	23.50	.19	241.69	278
2416	24", 12mm, Epoxy, 250k, 300ft		600	.133		218	7.90	.06	225.96	253
2417	24", 12mm, Epoxy, 250k, 328ft	▼	656	.122		218	7.25	.06	225.31	252
2418	24", 12mm, Epoxy, 250k, Transitions, add					20%				
2419	24", 12mm, Epoxy, 250k, 45 and 90 deg, add					50%				
2430	24", 10.5mm, Epoxy, 400k, 2ft	C-17E	4	20		191	1,175	9.40	1,375.40	2,100
2431	24", 10.5mm, Epoxy, 400k, 4ft		8	10		191	595	4.70	790.70	1,150
2432	24", 10.5mm, Epoxy, 400k, 10ft		20	4		191	237	1.88	429.88	585
2433	24", 10.5mm, Epoxy, 400k, 20ft	▼	40	2		191	119	.94	310.94	400

585

For customer support on your Commercial Renovation Costs with RSMeans data, call 800.448.8182.

33 01 30 – Operation and Maintenance of Sewer Utilities

33 01 30.75 Cured-In-place Pipe Lining	Crew	Daily Output	Labor-Hours	Unit	Material	2022 Bare Costs Labor	Equipment	Total	Total Incl O&P
2434	24", 10.5mm, Epoxy, 400k, 50ft	C-17E	100	.800	L.F.	191	47.50	.38	238.88	285
2435	24", 10.5mm, Epoxy, 400k, 100ft	C-17E	200	.400	L.F.	191	23.50	.19	214.69	248
2436	24", 10.5mm, Epoxy, 400k, 300ft		600	.133		191	7.90	.06	198.96	223
2437	24", 10.5mm, Epoxy, 400k, 328ft		656	.122		191	7.25	.06	198.31	222
2438	24", 10.5mm, Epoxy, 400k, Transitions, add					20%				
2439	24", 10.5mm, Epoxy, 400k, 45 and 90 deg, add					50%				

33 05 Common Work Results for Utilities

33 05 07 – Trenchless Installation of Utility Piping

33 05 07.23 Utility Boring and Jacking

		Crew	Daily Output	Labor-Hours	Unit	Material	2022 Bare Costs Labor	Equipment	Total	Total Incl O&P
0010	**UTILITY BORING AND JACKING**									
0011	Casing only, 100' minimum,									
0020	not incl. jacking pits or dewatering									
0100	Roadwork, 1/2" thick wall, 24" diameter casing	B-42	20	3.200	L.F.	154	164	135	453	575
0200	36" diameter		16	4		238	205	168	611	770
0300	48" diameter		15	4.267		330	218	179	727	905
0500	Railroad work, 24" diameter		15	4.267		154	218	179	551	710
0600	36" diameter		14	4.571		238	234	192	664	840
0700	48" diameter		12	5.333		330	273	224	827	1,050
0900	For ledge, add								20%	20%
1000	Small diameter boring, 3", sandy soil	B-82	900	.018		29.50	.91	.12	30.53	34
1040	Rocky soil	"	500	.032		29.50	1.64	.21	31.35	35.50

33 05 07.36 Microtunneling

		Crew	Daily Output	Labor-Hours	Unit	Material	2022 Bare Costs Labor	Equipment	Total	Total Incl O&P
0010	**MICROTUNNELING**									
0011	Not including excavation, backfill, shoring,									
0020	or dewatering, average 50'/day, slurry method									
0100	24" to 48" outside diameter, minimum				L.F.				965	965
0110	Adverse conditions, add				"				500	500
1000	Rent microtunneling machine, average monthly lease				Month				97,500	107,000
1010	Operating technician				Day				630	705
1100	Mobilization and demobilization, minimum				Job				41,200	45,900
1110	Maximum				"				445,500	490,500

33 05 61 – Concrete Manholes

33 05 61.10 Storm Drainage Manholes, Frames and Covers

		Crew	Daily Output	Labor-Hours	Unit	Material	2022 Bare Costs Labor	Equipment	Total	Total Incl O&P
0010	**STORM DRAINAGE MANHOLES, FRAMES & COVERS**									
0020	Excludes footing, excavation, backfill (See line items for frame & cover)									
0050	Brick, 4' inside diameter, 4' deep	D-1	1	16	Ea.	995	800		1,795	2,375
0100	6' deep		.70	22.857		1,425	1,150		2,575	3,400
0150	8' deep		.50	32		1,850	1,600		3,450	4,625
0200	Add for 1' depth increase		4	4	V.L.F.	250	201		451	595
1110	Precast, 4' ID, 4' deep	B-22	4.10	7.317	Ea.	1,025	390	63.50	1,478.50	1,800
1120	6' deep		3	10		1,200	530	86.50	1,816.50	2,250
1130	8' deep		2	15		2,125	800	130	3,055	3,725
1140	For depths over 8', add		16	1.875	V.L.F.	119	100	16.25	235.25	305

33 05 63 – Concrete Vaults and Chambers

33 05 63.13 Precast Concrete Utility Structures

		Crew	Daily Output	Labor-Hours	Unit	Material	2022 Bare Costs Labor	Equipment	Total	Total Incl O&P
0010	**PRECAST CONCRETE UTILITY STRUCTURES**, 6" thick									
0050	5' x 10' x 6' high, ID	B-13	2	28	Ea.	1,975	1,400	1,050	4,425	5,525
0350	Hand hole, precast concrete, 1-1/2" thick									
0400	1'-0" x 2'-0" x 1'-9", ID, light duty	B-1	4	6	Ea.	440	278		718	925

33 05 Common Work Results for Utilities

33 05 63 – Concrete Vaults and Chambers

33 05 63.13 Precast Concrete Utility Structures	Crew	Daily Output	Labor-Hours	Unit	Material	2022 Bare Costs Labor	Equipment	Total	Total Incl O&P
0450 4'-6" x 3'-2" x 2'-0", OD, heavy duty	B-6	3	8	Ea.	2,050	395	110	2,555	3,000

33 05 97 – Identification and Signage for Utilities

33 05 97.05 Utility Connection

	Crew	Daily Output	Labor-Hours	Unit	Material	2022 Bare Costs Labor	Equipment	Total	Total Incl O&P
0010 **UTILITY CONNECTION**									
0020 Water, sanitary, stormwater, gas, single connection	B-14	1	48	Ea.	3,050	2,300	330	5,680	7,325
0030 Telecommunication	"	3	16	"	415	765	110	1,290	1,775

33 05 97.10 Utility Accessories

		Crew	Daily Output	Labor-Hours	Unit	Material	2022 Bare Costs Labor	Equipment	Total	Total Incl O&P
0010 **UTILITY ACCESSORIES**	R312316-40									
0400 Underground tape, detectable, reinforced, alum. foil core, 2"		1 Clab	150	.053	C.L.F.	2	2.43		4.43	6.05
9000 Minimum labor/equipment charge		"	4	2	Job		91		91	145

33 11 Groundwater Sources

33 11 13 – Potable Water Supply Wells

33 11 13.10 Wells and Accessories

		Crew	Daily Output	Labor-Hours	Unit	Material	2022 Bare Costs Labor	Equipment	Total	Total Incl O&P
0010 **WELLS & ACCESSORIES**	R221113-50									
0011 Domestic										
0100 Drilled, 4" to 6" diameter		B-23	120	.333	L.F.		15.35	9.70	25.05	35
0200 8" diameter		"	95.20	.420	"		19.35	12.20	31.55	44
1400 Remove & reset pump, minimum		B-21	4	7	Ea.		370	43.50	413.50	630
1420 Maximum		"	2	14	"		735	86.50	821.50	1,275
1500 Pumps, installed in wells to 100' deep, 4" submersible										
1510 1/2 HP		Q-1	3.22	4.969	Ea.	575	310		885	1,125
1520 3/4 HP			2.66	6.015		795	380		1,175	1,450
1600 1 HP			2.29	6.987		935	440		1,375	1,700
1700 1-1/2 HP		Q-22	1.60	10		1,250	630	1,250	3,130	3,725
1800 2 HP			1.33	12.030		1,725	755	1,500	3,980	4,725
1900 3 HP			1.14	14.035		2,325	880	1,750	4,955	5,825
2000 5 HP			1.14	14.035		2,850	880	1,750	5,480	6,425
2050 Remove and install motor only, 4 HP			1.14	14.035		1,575	880	1,750	4,205	5,000
5000 Wells to 180' deep, 4" submersible, 1 HP		B-21	1.10	25.455		2,525	1,325	158	4,008	5,075
5500 2 HP		B-21	1.10	25.455	Ea.	4,000	1,325	158	5,483	6,700
6000 3 HP			1	28		4,625	1,475	173	6,273	7,625
7000 5 HP			.90	31.111		8,000	1,625	193	9,818	11,600
9000 Minimum labor/equipment charge			1.80	15.556	Job		815	96.50	911.50	1,400

33 14 Water Utility Transmission and Distribution

33 14 13 – Public Water Utility Distribution Piping

33 14 13.15 Water Supply, Ductile Iron Pipe

		Crew	Daily Output	Labor-Hours	Unit	Material	2022 Bare Costs Labor	Equipment	Total	Total Incl O&P
0010 **WATER SUPPLY, DUCTILE IRON PIPE**	R331113-80									
0020 Not including excavation or backfill										
2000 Pipe, class 50 water piping, 18' lengths										
2020 Mechanical joint, 4" diameter		B-21	200	.140	L.F.	57	7.35	.87	65.22	75
3000 Push-on joint, 4" diameter			400	.070		24	3.68	.43	28.11	33
3020 6" diameter			333.33	.084		23.50	4.41	.52	28.43	33.50
8000 Piping, fittings, mechanical joint, AWWA C110										
8006 90° bend, 4" diameter		B-20	16	1.500	Ea.	216	76		292	360
8020 6" diameter			12.80	1.875		365	95		460	550
8200 Wye or tee, 4" diameter			10.67	2.249		550	114		664	785

587

For customer support on your Commercial Renovation Costs with RSMeans data, call 800.448.8182.

33 14 Water Utility Transmission and Distribution

33 14 13 – Public Water Utility Distribution Piping

33 14 13.15 Water Supply, Ductile Iron Pipe

33 14 13.15 Water Supply, Ductile Iron Pipe	Crew	Daily Output	Labor-Hours	Unit	Material	2022 Bare Costs Labor	Equipment	Total	Total Incl O&P
8220 6" diameter	B-20	8.53	2.814	Ea.	745	143		888	1,050
8398 45° bend, 4" diameter		16	1.500		233	76		309	375
8400 6" diameter	↓	12.80	1.875		330	95		425	515
8405 8" diameter	B-21	10.67	2.624		490	138	16.25	644.25	770
8450 Decreaser, 6" x 4" diameter	B-20	14.22	1.688		268	85.50		353.50	430
8460 8" x 6" diameter	B-21	11.64	2.406	↓	400	126	14.90	540.90	655
8550 Piping, butterfly valves, cast iron									
8560 4" diameter	B-20	6	4	Ea.	640	203		843	1,025
9600 Steel sleeve with tap, 4" diameter		3	8		550	405		955	1,250
9620 6" diameter		2	12		590	610		1,200	1,625

33 14 17 – Site Water Utility Service Laterals

33 14 17.15 Tapping, Crosses and Sleeves

	Crew	Daily Output	Labor-Hours	Unit	Material	2022 Bare Costs Labor	Equipment	Total	Total Incl O&P
0010 **TAPPING, CROSSES AND SLEEVES**									
4000 Drill and tap pressurized main (labor only)									
4100 6" main, 1" to 2" service	Q-1	3	5.333	Ea.		335		335	515
4150 8" main, 1" to 2" service	"	2.75	5.818	"		365		365	565
4500 Tap and insert gate valve									
4600 8" main, 4" branch	B-21	3.20	8.750	Ea.		460	54	514	785
4650 6" branch		2.70	10.370			545	64	609	930
4800 12" main, 6" branch	↓	2.35	11.915	↓		625	74	699	1,075

33 31 Sanitary Sewerage Piping

33 31 11 – Public Sanitary Sewerage Gravity Piping

33 31 11.25 Sewage Collection, Polyvinyl Chloride Pipe

	Crew	Daily Output	Labor-Hours	Unit	Material	2022 Bare Costs Labor	Equipment	Total	Total Incl O&P
0010 **SEWAGE COLLECTION, POLYVINYL CHLORIDE PIPE**									
0020 Not including excavation or backfill									
2000 20' lengths, SDR 35, B&S, 4" diameter	B-20	375	.064	L.F.	2.21	3.24		5.45	7.60
2040 6" diameter		350	.069		4.26	3.48		7.74	10.20
2080 13' lengths, SDR 35, B&S, 8" diameter	↓	335	.072		12.70	3.63		16.33	19.70
2120 10" diameter	B-21	330	.085		12.05	4.46	.53	17.04	21
4000 Piping, DWV PVC, no exc./bkfill., 10' L, Sch 40, 4" diameter	B-20	375	.064		8.55	3.24		11.79	14.55
4010 6" diameter		350	.069		20.50	3.48		23.98	28
4020 8" diameter	↓	335	.072		32.50	3.63		36.13	42

33 34 Onsite Wastewater Disposal

33 34 13 – Septic Tanks

33 34 13.13 Concrete Septic Tanks

	Crew	Daily Output	Labor-Hours	Unit	Material	2022 Bare Costs Labor	Equipment	Total	Total Incl O&P
0010 **CONCRETE SEPTIC TANKS**									
0011 Not including excavation or piping									
0015 Septic tanks, precast, 1,000 gallon	B-21	8	3.500	Ea.	990	184	21.50	1,195.50	1,425
0020 1,250 gallon		8	3.500		1,200	184	21.50	1,405.50	1,650
0060 1,500 gallon		7	4		1,425	210	25	1,660	1,900
0100 2,000 gallon		5	5.600		2,400	294	34.50	2,728.50	3,150
0140 2,500 gallon		5	5.600		2,350	294	34.50	2,678.50	3,075
0180 4,000 gallon	↓	4	7		6,200	370	43.50	6,613.50	7,425
0220 5,000 gallon, 4 piece	B-13	3	18.667		9,075	925	700	10,700	12,200
0300 15,000 gallon, 4 piece	B-13B	1.70	32.941		20,300	1,625	1,325	23,250	26,400
0400 25,000 gallon, 4 piece		1.10	50.909		37,500	2,525	2,050	42,075	47,600
0500 40,000 gallon, 4 piece	↓	.80	70		60,000	3,475	2,825	66,300	74,500

33 34 Onsite Wastewater Disposal

33 34 13 – Septic Tanks

33 34 13.13 Concrete Septic Tanks

		Crew	Daily Output	Labor-Hours	Unit	Material	2022 Bare Costs Labor	Equipment	Total	Total Incl O&P
0520	50,000 gallon, 5 piece	B-13C	.60	93.333	Ea.	69,000	4,625	2,350	75,975	86,000
0640	75,000 gallon, cast in place	C-14C	.25	448		84,000	24,100	112	108,212	130,500
0660	100,000 gallon	"	.15	747		104,000	40,200	187	144,387	178,000
1150	Leaching field chambers, 13' x 3'-7" x 1'-4", standard	B-13	16	3.500		480	174	131	785	950
1200	Heavy duty, 8' x 4' x 1'-6"		14	4		284	199	149	632	790
1300	13' x 3'-9" x 1'-6"		12	4.667		1,400	232	174	1,806	2,100
1350	20' x 4' x 1'-6"		5	11.200		1,250	555	420	2,225	2,700
1420	Leaching pit, precast concrete, 6' diameter, 3' deep	B-21	4.70	5.957		1,125	315	37	1,477	1,750
1600	Leaching pit, 6'-6" diameter, 6' deep		5	5.600		885	294	34.50	1,213.50	1,475
1620	8' deep		4	7		1,175	370	43.50	1,588.50	1,900
1700	8' diameter, H-20 load, 6' deep		4	7		1,400	370	43.50	1,813.50	2,175
1720	8' deep		3	9.333		2,600	490	58	3,148	3,700
2000	Velocity reducing pit, precast conc., 6' diameter, 3' deep	B-21	4.70	5.957	Ea.	1,875	315	37	2,227	2,600

33 34 13.33 Polyethylene Septic Tanks

		Crew	Daily Output	Labor-Hours	Unit	Material	2022 Bare Costs Labor	Equipment	Total	Total Incl O&P
0010	**POLYETHYLENE SEPTIC TANKS**									
0011	High density polyethylene, 300 gallon	B-21	9.50	2.947	Ea.	725	155	18.25	898.25	1,075
0012	High density polyethylene, 500 gallon		9	3.111		1,150	163	19.25	1,332.25	1,550
0014	High density polyethylene, 750 gallon		8.50	3.294		1,225	173	20.50	1,418.50	1,650
0015	High density polyethylene, 1,000 gallon		8	3.500		1,325	184	21.50	1,530.50	1,775
0020	1,250 gallon		8	3.500		1,725	184	21.50	1,930.50	2,225
0025	1,500 gallon		7	4		2,300	210	25	2,535	2,875
0030	2,500 gallon		6	4.667		4,750	245	29	5,024	5,650
0035	3,000 gallon		5.50	5.091		6,500	267	31.50	6,798.50	7,600
0040	3,500 gallon		5	5.600		7,750	294	34.50	8,078.50	9,025

33 34 16 – Septic Tank Effluent Filters

33 34 16.13 Septic Tank Gravity Effluent Filters

		Crew	Daily Output	Labor-Hours	Unit	Material	2022 Bare Costs Labor	Equipment	Total	Total Incl O&P
0010	**SEPTIC TANK GRAVITY EFFLUENT FILTERS**									
3000	Effluent filter, 4" diameter	1 Skwk	8	1	Ea.	36	59		95	133
3020	6" diameter		7	1.143		44	67.50		111.50	155
3030	8" diameter		7	1.143		256	67.50		323.50	390
3040	8" diameter, very fine		7	1.143		510	67.50		577.50	665
3050	10" diameter, very fine		6	1.333		242	78.50		320.50	390
3060	10" diameter		6	1.333		276	78.50		354.50	430
3080	12" diameter		6	1.333		665	78.50		743.50	855
3090	15" diameter		5	1.600		1,100	94		1,194	1,350

33 34 51 – Drainage Field Systems

33 34 51.10 Drainage Field Excavation and Fill

		Crew	Daily Output	Labor-Hours	Unit	Material	2022 Bare Costs Labor	Equipment	Total	Total Incl O&P
0010	**DRAINAGE FIELD EXCAVATION AND FILL**									
2200	Septic tank & drainage field excavation with 3/4 C.Y. backhoe	B-12F	145	.110	C.Y.		6	6.05	12.05	16.10
2400	4' trench for disposal field, 3/4 C.Y. backhoe	"	335	.048	L.F.		2.60	2.61	5.21	6.95
2600	Gravel fill, run of bank	B-6	150	.160	C.Y.	25	7.90	2.19	35.09	42.50
2800	Crushed stone, 3/4"	"	150	.160	"	28.50	7.90	2.19	38.59	46

33 34 51.13 Utility Septic Tank Tile Drainage Field

		Crew	Daily Output	Labor-Hours	Unit	Material	2022 Bare Costs Labor	Equipment	Total	Total Incl O&P
0010	**UTILITY SEPTIC TANK TILE DRAINAGE FIELD**									
0015	Distribution box, concrete, 5 outlets	2 Clab	20	.800	Ea.	75.50	36.50		112	141
0020	7 outlets		16	1		94	45.50		139.50	176
0025	9 outlets		8	2		535	91		626	730
0115	Distribution boxes, HDPE, 5 outlets		20	.800		75.50	36.50		112	141
0117	6 outlets		15	1.067		75.50	48.50		124	161
0118	7 outlets		15	1.067		75	48.50		123.50	160
0120	8 outlets		10	1.600		79.50	73		152.50	204

33 34 Onsite Wastewater Disposal

33 34 51 – Drainage Field Systems

33 34 51.13 Utility Septic Tank Tile Drainage Field

		Crew	Daily Output	Labor-Hours	Unit	Material	2022 Bare Costs Labor	Equipment	Total	Total Incl O&P
0240	Distribution boxes, outlet flow leveler	1 Clab	50	.160	Ea.	3.01	7.30		10.31	14.90
0300	Precast concrete, galley, 4' x 4' x 4'	B-21	16	1.750	Ea.	248	92	10.85	350.85	430
0350	HDPE infiltration chamber 12" H x 15" W	2 Clab	300	.053	L.F.	7.10	2.43		9.53	11.65
0351	12" H x 15" W end cap	1 Clab	32	.250	Ea.	19.15	11.40		30.55	39
0355	chamber 12" H x 22" W	2 Clab	300	.053	L.F.	6.75	2.43		9.18	11.25
0356	12" H x 22" W end cap	1 Clab	32	.250	Ea.	23.50	11.40		34.90	44
0360	chamber 13" H x 34" W	2 Clab	300	.053	L.F.	14.55	2.43		16.98	19.85
0361	13" H x 34" W end cap	1 Clab	32	.250	Ea.	51.50	11.40		62.90	74.50
0365	chamber 16" H x 34" W	2 Clab	300	.053	L.F.	14.95	2.43		17.38	20.50
0366	16" H x 34" W end cap	1 Clab	32	.250	Ea.	17	11.40		28.40	37
0370	chamber 8" H x 16" W	2 Clab	300	.053	L.F.	11.40	2.43		13.83	16.40
0371	8" H x 16" W end cap	1 Clab	32	.250	Ea.	11.20	11.40		22.60	30.50

33 41 16.10 Piping, Subdrainage, Vitrified Clay

		Crew	Daily Output	Labor-Hours	Unit	Material	2022 Bare Costs Labor	Equipment	Total	Total Incl O&P
0010	**PIPING, SUBDRAINAGE, VITRIFIED CLAY**									
0020	Not including excavation and backfill									
1000	Foundation drain, 6" diameter	B-14	315	.152	L.F.	6.10	7.30	1.04	14.44	19.45
1010	8" diameter		290	.166		9.10	7.90	1.13	18.13	24
1020	12" diameter		275	.175		19.60	8.35	1.20	29.15	36
3000	Perforated, 5' lengths, C700, 4" diameter		400	.120		3.45	5.75	.82	10.02	13.80
3020	6" diameter		315	.152		6.60	7.30	1.04	14.94	19.95
3040	8" diameter		290	.166		7.25	7.90	1.13	16.28	22
3060	12" diameter		275	.175		19.95	8.35	1.20	29.50	36.50

33 42 Stormwater Conveyance

33 42 11 – Stormwater Gravity Piping

33 42 11.40 Piping, Storm Drainage, Corrugated Metal

		Crew	Daily Output	Labor-Hours	Unit	Material	2022 Bare Costs Labor	Equipment	Total	Total Incl O&P
0010	**PIPING, STORM DRAINAGE, CORRUGATED METAL** R221113-50									
0020	Not including excavation or backfill									
2000	Corrugated metal pipe, galvanized									
2040	8" diameter, 16 ga.	B-14	330	.145	L.F.	7.60	6.95	1	15.55	20.50
2080	12" diameter, 16 ga.	"	210	.229	"	12.55	10.95	1.57	25.07	33

33 42 11.60 Sewage/Drainage Collection, Concrete Pipe

		Crew	Daily Output	Labor-Hours	Unit	Material	2022 Bare Costs Labor	Equipment	Total	Total Incl O&P
0010	**SEWAGE/DRAINAGE COLLECTION, CONCRETE PIPE**									
0020	Not including excavation or backfill									
1000	Non-reinforced pipe, extra strength, B&S or T&G joints									
1010	6" diameter	B-14	265.04	.181	L.F.	10.80	8.65	1.24	20.69	27
1020	8" diameter		224	.214		11.90	10.25	1.47	23.62	31
1040	12" diameter		200	.240		13.70	11.50	1.64	26.84	35
2000	Reinforced culvert, class 3, no gaskets									
2010	12" diameter	B-14	150	.320	L.F.	19.95	15.30	2.19	37.44	48.50
2020	15" diameter	"	150	.320	"	24	15.30	2.19	41.49	52.50

33 42 33 – Stormwater Curbside Drains and Inlets

33 42 33.13 Catch Basins

		Crew	Daily Output	Labor-Hours	Unit	Material	2022 Bare Costs Labor	Equipment	Total	Total Incl O&P
0010	**CATCH BASINS**									
0011	Not including footing & excavation									
1600	Frames & grates, C.I., 24" square, 500 lb.	B-6	7.80	3.077	Ea.	410	152	42	604	735
1700	26" D shape, 600 lb.	"	7	3.429	"	760	169	47	976	1,150
3320	Frames and covers, existing, raised for paving, 2", including									
3340	row of brick, concrete collar, up to 12" wide frame	B-6	18	1.333	Ea.	54	66	18.25	138.25	184

33 42 Stormwater Conveyance

33 42 33 - Stormwater Curbside Drains and Inlets

33 42 33.15 Remove and Replace Catch Basin Cover	Crew	Daily Output	Labor-Hours	Unit	Material	2022 Bare Costs Labor	Equipment	Total	Total Incl O&P
0010 **REMOVE AND REPLACE CATCH BASIN COVER**									
0023 Remove catch basin cover	1 Clab	80	.100	Ea.		4.56		4.56	7.25
0033 Replace catch basin cover	"	80	.100	"		4.56		4.56	7.25

33 52 Hydrocarbon Transmission and Distribution

33 52 13 - Liquid Hydrocarbon Piping

33 52 13.16 Gasoline Piping

33 52 13.16 Gasoline Piping	Crew	Daily Output	Labor-Hours	Unit	Material	2022 Bare Costs Labor	Equipment	Total	Total Incl O&P
0010 **GASOLINE PIPING**									
0020 Primary containment pipe, fiberglass-reinforced									
0030 Plastic pipe 15' & 30' lengths									
0040 2" diameter	Q-6	425	.056	L.F.	6.60	3.71		10.31	12.95
0050 3" diameter		400	.060		8.95	3.94		12.89	15.90
0060 4" diameter		375	.064		11.60	4.20		15.80	19.25
0100 Fittings									
0110 Elbows, 90° & 45°, bell ends, 2"	Q-6	24	1	Ea.	39.50	65.50		105	145
0120 3" diameter		22	1.091		52	71.50		123.50	167
0130 4" diameter		20	1.200		65	79		144	193
0200 Tees, bell ends, 2"		21	1.143		52.50	75		127.50	174
0210 3" diameter		18	1.333		59.50	87.50		147	201
0220 4" diameter		15	1.600		73.50	105		178.50	243
0230 Flanges bell ends, 2"		24	1		28.50	65.50		94	132
0240 3" diameter		22	1.091		35	71.50		106.50	149
0250 4" diameter		20	1.200		39	79		118	164
0260 Sleeve couplings, 2"		21	1.143		10.85	75		85.85	128
0270 3" diameter		18	1.333		17.10	87.50		104.60	154
0280 4" diameter		15	1.600		19	105		124	183
0290 Threaded adapters, 2"		21	1.143		17.80	75		92.80	136
0300 3" diameter		18	1.333		23	87.50		110.50	161
0310 4" diameter		15	1.600		33	105		138	199
0320 Reducers, 2"		27	.889		24.50	58.50		83	117
0330 3" diameter		22	1.091		24	71.50		95.50	137
0340 4" diameter		20	1.200		32.50	79		111.50	157
1010 Gas station product line for secondary containment (double wall)									
1100 Fiberglass reinforced plastic pipe 25' lengths									
1120 Pipe, plain end, 3" diameter	Q-6	375	.064	L.F.	32.50	4.20		36.70	42.50
1130 4" diameter		350	.069		26.50	4.50		31	36.50
1140 5" diameter		325	.074		30	4.85		34.85	40.50
1150 6" diameter		300	.080		33	5.25		38.25	44.50
1200 Fittings									
1230 Elbows, 90° & 45°, 3" diameter	Q-6	18	1.333	Ea.	138	87.50		225.50	287
1240 4" diameter		16	1.500		166	98.50		264.50	335
1250 5" diameter		14	1.714		159	113		272	350
1260 6" diameter		12	2		198	131		329	420
1270 Tees, 3" diameter		15	1.600		172	105		277	350
1280 4" diameter		12	2		163	131		294	380
1290 5" diameter		9	2.667		214	175		389	505
1300 6" diameter		6	4		300	263		563	735
1310 Couplings, 3" diameter		18	1.333		49	87.50		136.50	189
1320 4" diameter		16	1.500		111	98.50		209.50	274
1330 5" diameter		14	1.714		198	113		311	390

591

33 52 Hydrocarbon Transmission and Distribution

33 52 13 – Liquid Hydrocarbon Piping

33 52 13.16 Gasoline Piping	Crew	Daily Output	Labor-Hours	Unit	Material	2022 Bare Costs Labor	Equipment	Total	Total Incl O&P
1340	6" diameter	Q-6	12	2	Ea.	310	131		441	540
1350	Cross-over nipples, 3" diameter		18	1.333		9.25	87.50		96.75	145
1360	4" diameter		16	1.500		11.30	98.50		109.80	164
1370	5" diameter		14	1.714		15.65	113		128.65	190
1380	6" diameter		12	2		16.35	131		147.35	220
1400	Telescoping, reducers, concentric 4" x 3"		18	1.333		39.50	87.50		127	179
1410	5" x 4"		17	1.412		84.50	92.50		177	236
1420	6" x 5"		16	1.500		237	98.50		335.50	415

33 52 16 – Gas Hydrocarbon Piping

33 52 16.20 Piping, Gas Service and Distribution, P.E.

		Crew	Daily Output	Labor-Hours	Unit	Material	2022 Bare Costs Labor	Equipment	Total	Total Incl O&P
0010	**PIPING, GAS SERVICE AND DISTRIBUTION, POLYETHYLENE**									
0020	Not including excavation or backfill									
1000	60 psi coils, compression coupling @ 100', 1/2" diameter, SDR 11	B-20A	608	.053	L.F.	.57	2.88		3.45	5.15
1010	1" diameter, SDR 11		544	.059		1.55	3.22		4.77	6.75
1040	1-1/4" diameter, SDR 11		544	.059		3.26	3.22		6.48	8.65
1100	2" diameter, SDR 11		488	.066		5.75	3.59		9.34	11.90
1160	3" diameter, SDR 11		408	.078		6.25	4.29		10.54	13.60
1500	60 psi 40' joints with coupling, 3" diameter, SDR 11	B-21A	408	.098		9.30	5.55	1.40	16.25	20.50
1540	4" diameter, SDR 11		352	.114		18	6.40	1.62	26.02	31.50
1600	6" diameter, SDR 11		328	.122		46	6.90	1.74	54.64	63
1640	8" diameter, SDR 11	B-21A	272	.147	L.F.	77	8.30	2.10	87.40	100
9000	Minimum labor/equipment charge	B-20	2	12	Job		610		610	965

33 71 Electrical Utility Transmission and Distribution

33 71 19 – Electrical Underground Ducts and Manholes

33 71 19.17 Electric and Telephone Underground

		Crew	Daily Output	Labor-Hours	Unit	Material	2022 Bare Costs Labor	Equipment	Total	Total Incl O&P
0010	**ELECTRIC AND TELEPHONE UNDERGROUND**									
0011	Not including excavation									
0200	backfill and cast in place concrete									
4200	Underground duct, banks ready for concrete fill, min. of 7.5"									
4400	between conduits, center to center									
4600	2 @ 2" diameter	2 Elec	240	.067	L.F.	2.24	4.41		6.65	9.25
4800	4 @ 2" diameter		120	.133		4.49	8.85		13.34	18.50
5600	4 @ 4" diameter		80	.200		8.80	13.25		22.05	30
6200	Rigid galvanized steel, 2 @ 2" diameter		180	.089		17.05	5.90		22.95	28
6400	4 @ 2" diameter		90	.178		34	11.75		45.75	55.50
7400	4 @ 4" diameter		34	.471		105	31		136	164
9990	Minimum labor/equipment charge	1 Elec	3.50	2.286	Job		151		151	232

Estimating Tips
34 11 00 Rail Tracks
This subdivision includes items that may involve either repair of existing or construction of new railroad tracks. Additional preparation work, such as the roadbed earthwork, would be found in Division 31. Additional new construction siding and turnouts are found in Subdivision 34 72. Maintenance of railroads is found under 34 01 23 Operation and Maintenance of Railways.

34 40 00 Traffic Signals
This subdivision includes traffic signal systems. Other traffic control devices such as traffic signs are found in Subdivision 10 14 53 Traffic Signage.

34 70 00 Vehicle Barriers
This subdivision includes security vehicle barriers, guide and guard rails, crash barriers, and delineators. The actual maintenance and construction of concrete and asphalt pavement are found in Division 32.

Reference Numbers
Reference numbers are shown at the beginning of some major classifications. These numbers refer to related items in the Reference Section. The reference information may be an estimating procedure, an alternate pricing method, or technical information.

Note: Not all subdivisions listed here necessarily appear. ■

34 71 13.26 Vehicle Guide Rails	Crew	Daily Output	Labor-Hours	Unit	Material	2022 Bare Costs Labor	Equipment	Total	Total Incl O&P
0010 **VEHICLE GUIDE RAILS**									
0012 Corrugated stl., galv. stl. posts, 6'-3" OC - "W-Beam Guiderail"	B-80	850	.038	L.F.	12.55	1.89	.64	15.08	17.50
0200 End sections, galvanized, flared		50	.640	Ea.	99.50	32	10.85	142.35	172
0300 Wrap around end		50	.640		142	32	10.85	184.85	219
0325 W-Beam, stl., galv. Terminal Connector, Concrete Mounted		8	4		264	201	67.50	532.50	680
0410 Thrie Beam, stl., galv. stl. posts, 6'-3" OC		850	.038	L.F.	24	1.89	.64	26.53	29.50
0440 Thrie Beam, stl., galv, Wrap Around End Section		50	.640	Ea.	94.50	32	10.85	137.35	167
0450 Thrie Beam, stl., galv. Terminal Connector, Concrete Mounted		8	4	"	289	201	67.50	557.50	710
0510 Rub Rail, stl, galv, steel posts (not included), 6'-3" OC		850	.038	L.F.	2.04	1.89	.64	4.57	5.95
1650 Timber guide rail, 4" x 8" with 6" x 8" wood posts, treated		960	.033	"	14.05	1.67	.56	16.28	18.70

Estimating Tips

Products such as conveyors, material handling cranes and hoists, and other items specified in this division require trained installers. The general contractor may not have any choice as to who will perform the installation or when it will be performed. Long lead times are often required for these products, making early decisions in purchasing and scheduling necessary. The installation of this type of equipment may require the embedment of mounting hardware during the construction of floors, structural walls, or interior walls/partitions. Electrical connections will require coordination with the electrical contractor.

Reference Numbers

Reference numbers are shown at the beginning of some major classifications. These numbers refer to related items in the Reference Section. The reference information may be an estimating procedure, an alternate pricing method, or technical information.

41 22 Cranes and Hoists

41 22 13 – Cranes

41 22 13.10 Crane Rail

		Crew	Daily Output	Labor-Hours	Unit	Material	2022 Bare Costs Labor	Equipment	Total	Total Incl O&P
0010	**CRANE RAIL**									
0020	Box beam bridge, no equipment included	E-4	3400	.009	Lb.	1.64	.59	.04	2.27	2.79
0200	Running track only, 104 lb. per yard		5600	.006	"	.82	.36	.03	1.21	1.50
0210	Running track only, 104 lb. per yard, 20' piece	↓	160	.200	L.F.	28.50	12.45	.92	41.87	52.50

41 22 13.13 Bridge Cranes

		Crew	Daily Output	Labor-Hours	Unit	Material	2022 Bare Costs Labor	Equipment	Total	Total Incl O&P
0010	**BRIDGE CRANES**									
0100	1 girder, 20' span, 3 ton	M-3	1	34	Ea.	25,600	2,375	172	28,147	32,100
0125	5 ton		1	34		29,100	2,375	172	31,647	35,900
0150	7.5 ton		1	34		30,600	2,375	172	33,147	37,600
0175	10 ton		.80	42.500		40,900	2,975	215	44,090	49,800
0176	12.5 ton		.80	42.500		49,100	2,975	215	52,290	59,000
0200	15 ton		.80	42.500		66,500	2,975	215	69,690	78,500
0225	30' span, 3 ton		1	34		25,800	2,375	172	28,347	32,300
0250	5 ton		1	34		29,400	2,375	172	31,947	36,200
0275	7.5 ton		1	34		33,000	2,375	172	35,547	40,200
0300	10 ton		.80	42.500		41,100	2,975	215	44,290	50,000
0325	15 ton	↓	.80	42.500		69,500	2,975	215	72,690	81,500
0350	2 girder, 40' span, 3 ton	M-4	.50	72		23,100	4,975	510	28,585	33,700
0375	5 ton		.50	72		45,900	4,975	510	51,385	59,000
0400	7.5 ton		.50	72		50,500	4,975	510	55,985	64,000
0425	10 ton		.40	90		43,800	6,225	640	50,665	58,500
0450	15 ton		.40	90		73,000	6,225	640	79,865	90,500
0475	25 ton		.30	120		82,500	8,300	855	91,655	104,000
0500	50' span, 3 ton		.50	72		44,500	4,975	510	49,985	57,500
0525	5 ton		.50	72		48,000	4,975	510	53,485	61,500
0550	7.5 ton		.50	72		54,000	4,975	510	59,485	68,000
0575	10 ton		.40	90		65,500	6,225	640	72,365	82,500
0600	15 ton		.40	90		78,000	6,225	640	84,865	96,000
0625	25 ton	↓	.30	120	↓	69,500	8,300	855	78,655	90,000

41 22 13.19 Jib Cranes

		Crew	Daily Output	Labor-Hours	Unit	Material	2022 Bare Costs Labor	Equipment	Total	Total Incl O&P
0010	**JIB CRANES**									
0020	Jib crane, wall cantilever, 500 lb. capacity, 8' span	2 Mill	1	16	Ea.	1,550	970		2,520	3,200
0040	12' span		1	16		1,700	970		2,670	3,375
0060	16' span		1	16		1,975	970		2,945	3,675
0080	20' span		1	16		2,400	970		3,370	4,125
0100	1,000 lb. capacity, 8' span		1	16		1,650	970		2,620	3,325
0120	12' span		1	16		1,800	970		2,770	3,500
0130	16' span		1	16		2,525	970		3,495	4,275
0150	20' span	↓	1	16	↓	2,800	970		3,770	4,600

Estimating Tips

- When estimating costs for the installation of electrical power generation equipment, factors to review include access to the job site, access and setting up at the installation site, required connections, uncrating pads, anchors, leveling, final assembly of the components, and temporary protection from physical damage, such as environmental exposure.

- Be aware of the costs of equipment supports, concrete pads, and vibration isolators. Cross-reference them against other trades' specifications. Also, review site and structural drawings for items that must be included in the estimates.

- It is important to include items that are not documented in the plans and specifications but must be priced. These items include, but are not limited to, testing, dust protection, roof penetration, core drilling concrete floors and walls, patching, cleanup, and final adjustments. Add a contingency or allowance for utility company fees for power hookups, if needed.

- The project size and scope of electrical power generation equipment will have a significant impact on cost. The intent of RSMeans cost data is to provide a benchmark cost so that owners, engineers, and electrical contractors will have a comfortable number with which to start a project. Additionally, there are many websites available to use for research and to obtain a vendor's quote to finalize costs.

Reference Numbers

Reference numbers are shown at the beginning of some major classifications. These numbers refer to related items in the Reference Section. The reference information may be an estimating procedure, an alternate pricing method, or technical information.

Same Data. Simplified.

Enjoy the convenience and efficiency of accessing your costs anywhere:

- **Skip the multiplier** by setting your location
- **Quickly search,** edit, favorite and share costs
- **Stay on top of price changes** with automatic updates

Discover more at rsmeans.com/online

Note: Trade Service, in part, has been used as a reference source for some of the material prices used in Division 48.

48 13 Hydroelectric Plant Electrical Power Generation Equipment

48 13 13 – Hydroelectric Power Plant Water Turbines

48 13 13.10 Water Turbines and Components	Crew	Daily Output	Labor-Hours	Unit	Material	2022 Bare Costs Labor	Equipment	Total	Total Incl O&P
0010 **WATER TURBINES & COMPONENTS**									
0500 Water turbine, 100-500 watt, 3P, 48VDC, 54"W x 24"H, 260LBS	2 Elec	2	8	Ea.	11,900	530		12,430	13,900
0600 Water turbine, 1500-2000 watt, 3P, 130-200VAC to DC, 95LBS	"	3	5.333	"	4,500	355		4,855	5,500

Assemblies Section

Table of Contents

Table of Contents

RSMeans data: Assemblies— How They Work

Assemblies estimating provides a fast and reasonably accurate way to develop construction costs. An assembly is the grouping of individual work items—with appropriate quantities— to provide a cost for a major construction component in a convenient unit of measure.

An assemblies estimate is often used during early stages of design development to compare

the cost impact of various design alternatives on total building cost.

Assemblies estimates are also used as an efficient tool to verify construction estimates.

Assemblies estimates do not require a completed design or detailed drawings. Instead, they are based on the general size of the structure and

other known parameters of the project. The degree of accuracy of an assemblies estimate is generally within +/- 15%.

Most assemblies consist of three major elements: a graphic, the system components, and the cost data itself. The **Graphic** is a visual representation showing the typical appearance of the assembly

① Unique 12-character Identifier

Our assemblies are identified by a **unique 12-character identifier**. The assemblies are numbered using UNIFORMAT II, ASTM Standard E1557. The first five characters represent this system to Level 3. The last seven characters represent further breakdown in order to arrange items in understandable groups of similar tasks. Line numbers are consistent across all of our publications, so a line number in any assemblies data set will always refer to the same item.

② Narrative Descriptions

Our assemblies descriptions appear in two formats: narrative and table. **Narrative descriptions** are shown in a hierarchical structure to make them readable. In order to read a complete description, read up through the indents to the top of the section. Include everything that is above and to the left that is not contradicted by information below.

A10 Foundations

A1010 Standard Foundations

This page illustrates and describes a spread footing system including concrete, forms, reinforcing and anchor bolts. Lines within System Components give the unit price and total price on a cost each basis for this system. Prices for spread footing systems are on Line Items A1010 230 1200 thru 2300. Both material quantities and labor costs have been adjusted for the system listed.

Factors: To adjust for job conditions other than normal working situations use Lines A1010 230 2900 thru 4000.

Example: You are to install the system in an existing occupied building. Access to the site and protection to the building are mandatory. Go to Lines A1010 230 3600 and 3800 and apply these percentages to the appropriate MAT. and INST. costs.

System Components	QUANTITY	UNIT	COST EACH MAT.	COST EACH INST.	COST EACH TOTAL
Interior column footing, 3' square 1' thick, 2000 psi concrete, Including forms, reinforcing, and anchor bolts.					
Concrete, 2000 psi	.330	C.Y.	55.11		55.11
Placing concrete	.330	C.Y.		21.82	21.82
Forms, footing, 4 uses	12.000	SFCA	13.44	79.20	92.64
Reinforcing	11.000	Lb.	8.80	7.92	16.72
Anchor bolts, 3/4" diameter	2.000	Ea.	4.92	10.70	15.62
TOTAL			82.27	119.64	201.91

A1010 230	Spread Footing	COST EACH MAT.	COST EACH INST.	COST EACH TOTAL
1200	3' square x 1' thick, 2000 psi concrete	82.50	120	202.50
1300	3000 psi concrete	85	120	205
1400	4000 psi concrete	86.50	120	206.50
1600	For alternate footing systems:			
1700	4' square x 1' thick, 2000 psi concrete	137	169	306
1800	3000 psi concrete	142	169	311
1900	4000 psi concrete	145	169	314
2100	5' square x 1'-3" thick, 2000 psi concrete	257	280	537
2200	3000 psi concrete	266	280	546
2300	4000 psi concrete	272	280	552
2600				
2700				
2900	Cut & patch to match existing construction, add, minimum	2%	3%	
3000	Maximum	5%	9%	
3100	Dust protection, add, minimum	1%	2%	
3200	Maximum	4%	11%	
3300	Equipment usage curtailment, add, minimum	1%	1%	
3400	Maximum	3%	10%	
3500	Material handling & storage limitation, add, minimum	1%	1%	
3600	Maximum	6%	7%	
3700	Protection of existing work, add, minimum	2%	2%	
3800	Maximum	5%	7%	
3900	Shift work requirements, add, minimum		5%	
4000	Maximum		30%	

For supplemental customizable square foot estimating forms, visit: **RSMeans.com/2022books** in question. It is frequently accompanied by additional explanatory technical information describing the class of items. The **System Components** is a listing of the individual tasks that make up the assembly, including the quantity and unit of measure for each item, along with the cost of material and installation. The **Assemblies**

data below lists prices for other similar systems with dimensional and/or size variations.

All of our assemblies costs represent the cost for the installing contractor. An allowance for profit has been added to all material, labor, and equipment rental costs. A markup for labor burdens, including workers' compensation, fixed

overhead, and business overhead, is included with installation costs.

The information in RSMeans cost data represents a "national average" cost. This data should be modified to the project location using the **City Cost Indexes** or **Location Factors** tables found in the Reference Section.

A10 Foundations

A1010   Standard Foundations

This page illustrates and describes a spread footing system including concrete, forms, reinforcing and anchor bolts. Lines within System Components give the unit price and total price on a cost each basis for this system. Prices for spread footing systems are on Line Items A1010 230 1200 thru 2300. Both material quantities and labor costs have been adjusted for the system listed.

Factors: To adjust for job conditions other than normal working situations use Lines A1010 230 2900 thru 4000.

Example: You are to install the system in an existing occupied building. Access to the site and protection to the building are mandatory. Go to Lines A1010 230 3600 and 3800 and apply these percentages to the appropriate MAT. and INST. costs.

3 Unit of Measure

All RSMeans data: Assemblies include a typical **Unit of Measure** used for estimating that item. For instance, for continuous footings or foundation walls the unit is linear feet (L.F.). For spread footings the unit is each (Ea.). The estimator needs to take special care that the unit in the data matches the unit in the takeoff. Abbreviations and unit conversions can be found in the Reference Section.

4 System Components

System components are listed separately to detail what is included in the development of the total system price.

System Components 4	QUANTITY	UNIT	COST EACH MAT.	COST EACH INST.	COST EACH TOTAL
Interior column footing, 3' square 1' thick, 2000 psi concrete, Including forms, reinforcing, and anchor bolts.					
Concrete, 2000 psi	.330	C.Y.	55.11		55.11
Placing concrete	.330	C.Y.		21.82	21.82
Forms, footing, 4 uses	12.000	SFCA	13.44	79.20	92.64
Reinforcing	11.000	Lb.	8.80	7.92	16.72
Anchor bolts, 3/4" diameter	2.000	Ea.	4.92	10.70	15.62
TOTAL			82.27	119.64	201.91

A1010 230	Spread Footing	MAT.	INST.	TOTAL
1200	3' square x 1' thick, 2000 psi concrete	82.50	120	202.50
1300	3000 psi concrete	85	120	205
1400	4000 psi concrete	86.50	120	206.50
1600	For alternate footing systems:			
1700	4' square x 1' thick, 2000 psi concrete	137	169	306
1800	3000 psi concrete	142	169	311
1900	4000 psi concrete	145	169	314
2100	5' square x 1'-3" thick, 2000 psi concrete	257	280	537
2200	3000 psi concrete	266	280	546
2300	4000 psi concrete	272	280	552
2600				
2700				
2900	Cut & patch to match existing construction, add, minimum	2%	3%	
3000	Maximum	5%	9%	
3100	Dust protection, add, minimum	1%	2%	
3200	Maximum	4%	11%	
3300	Equipment usage curtailment, add, minimum	1%	1%	
3400	Maximum	3%	10%	
3500	Material handling & storage limitation, add, minimum	1%	1%	
3600	Maximum	6%	7%	
3700	Protection of existing work, add, minimum	2%	2%	
3800	Maximum	5%	7%	
3900	Shift work requirements, add, minimum		5%	
4000	Maximum		30%	

Sample Estimate

This sample demonstrates the elements of an estimate, including a tally of the RSMeans data lines. Published assemblies costs include all markups for labor burden and profit for the installing contractor. This estimate adds a summary of the markups applied by a general contractor on the installing contractor's work. These figures represent the total cost to the owner. The location factor with RSMeans data is applied at the bottom of the estimate to adjust the cost of the work to a specific location.

Project Name:	Interior Fit-out, ABC Office			
Location:	**Anywhere, USA**	Date: 1/1/2022		**R+R**
Assembly Number	**Description**	**Qty.**	**Unit**	**Subtotal**
❶ C1010 124 1200	Wood partition, 2 x 4 @ 16" OC w/5/8" FR gypsum board	560	S.F.	$3,320.80
C1020 114 1800	Metal door & frame, flush hollow core, 3'-0" x 7'-0"	2	Ea.	$3,040.00
C3010 230 0080	Painting, brushwork, primer & 2 coats	1,120	S.F.	$1,534.40
C3020 410 0140	Carpet, tufted, nylon, roll goods, 12' wide, 26 oz	240	S.F.	$700.80
C3030 210 6000	Acoustic ceilings, 24" x 48" tile, tee grid suspension	200	S.F.	$1,908.00
D5020 125 0560	Receptacles incl plate, box, conduit, wire, 20 A duplex	8	Ea.	$2,676.00
D5020 125 0720	Light switch incl plate, box, conduit, wire, 20 A single pole	2	Ea.	$649.00
D5020 210 0560	Fluorescent fixtures, recess mounted, 20 per 1000 SF	200	S.F.	$2,940.00
	Assembly Subtotal			**$16,769.00**
	Sales Tax @ **❷**	5 %		$ 419.23
	General Requirements @ **❸**	7 %		$ 1,173.83
	Subtotal A			**$18,362.06**
	GC Overhead @ **❹**	5 %		$ 918.10
	Subtotal B			**$19,280.16**
	GC Profit @ **❺**	5 %		$ 964.01
	Subtotal C			**$20,244.17**
	Adjusted by Location Factor ❻	113.1		$ 22,896.15
	Architects Fee @ **❼**	8 %		$ 1,831.69
	Contingency @ **❽**	15 %		$ 3,434.42
	Project Total Cost			**$ 28,162.27**

This estimate is based on an interactive spreadsheet. You are free to download it and adjust it to your methodology.
A copy of this spreadsheet is available at **RSMeans.com/2022books**.

1 Work Performed

The body of the estimate shows the RSMeans data selected, including line numbers, a brief description of each item, its takeoff quantity and unit, and the total installed cost, including the installing contractor's overhead and profit.

2 Sales Tax

If the work is subject to state or local sales taxes, the amount must be added to the estimate. In a conceptual estimate, it can be assumed that one half of the total represents material costs. Therefore, apply the sales tax rate to 50% of the assembly subtotal.

3 General Requirements

This item covers project-wide needs provided by the general contractor. These items vary by project but may include temporary facilities and utilities, security, testing, project cleanup, etc. In assemblies estimates a percentage is used—typically between 5% and 15% of project cost.

4 General Contractor Overhead

This entry represents the general contractor's markup on all work to cover project administration costs.

5 General Contractor Profit

This entry represents the GC's profit on all work performed. The value included here can vary widely by project and is influenced by the GC's perception of the project's financial risk and market conditions.

6 Location Factor

RSMeans published data are based on national average costs. If necessary, adjust the total cost of the project using a location factor from the "Location Factor" table or the "City Cost Indexes" table found in the Reference Section. Use location factors if the work is general, covering the work of multiple trades. If the work is by a single trade (e.g., masonry), use the more specific data found in the City Cost Indexes.

To adjust costs by location factors, multiply the base cost by the factor and divide by 100.

7 Architect's Fee

If appropriate, add the design cost to the project estimate. These fees vary based on project complexity and size. Typical design and engineering fees can be found in the Reference Section.

8 Contingency

A factor for contingency may be added to any estimate to represent the cost of unknowns that may occur between the time that the estimate is performed and the time the project is constructed. The amount of the allowance will depend on the stage of design at which the estimate is done, as well as the contractor's assessment of the risk involved.

Same Data. Simplified.

Enjoy the convenience and efficiency of accessing your costs anywhere:

- **Skip the multiplier** by setting your location
- **Quickly search,** edit, favorite and share costs
- **Stay on top of price changes** with automatic updates

Discover more at rsmeans.com/online

A1010 Standard Foundations

This page illustrates and describes a strip footing system including concrete, forms, reinforcing, keyway and dowels. Lines within System Components give the unit price and total price per linear foot for this system. Prices for strip footing systems are on Line Items A1010 120 1400 thru 2500. Both material quantities and labor costs have been adjusted for the system listed.

Factors: To adjust for job conditions other than normal working situations use Lines A1010 120 2900 thru 4000.

Example: You are to install this footing, and due to a lack of accessibility, only hand tools can be used. Material handling is also a problem. Go to Lines A1010 120 3400 and 3600 and apply these percentages to the appropriate MAT. and INST. costs.

System Components	QUANTITY	UNIT	COST PER L.F.		
			MAT.	INST.	TOTAL
Strip footing, 2'-0" wide x 1'-0" thick, 2000 psi concrete including forms Reinforcing, keyway, and dowels.					
Concrete, 2000 psi	.074	C.Y.	12.36		12.36
Placing concrete	.074	C.Y.		2.26	2.26
Forms, footing, 4 uses	2.000	S.F.	5.64	11.20	16.84
Reinforcing	3.170	Lb.	2.54	2.28	4.82
Keyway, 2" x 4", 4 uses	1.000	L.F.	.38	1.35	1.73
Dowels, #4 bars, 2' long, 24" O.C.	.500	Ea.	.59	1.58	2.17
TOTAL			21.51	18.67	40.18

A1010 120	Strip Footing	COST PER L.F.		
		MAT.	INST.	TOTAL
1400	2'-0" wide x 1' thick, 2000 psi concrete	21.50	18.65	40.15
1500	3000 psi concrete	22	18.65	40.65
1600	4000 psi concrete	22.50	18.65	41.15
1800	For alternate footing systems:			
1900	2'-6" wide x 1' thick, 2000 psi concrete	25.50	19.85	45.35
2000	3000 psi concrete	26	19.85	45.85
2100	4000 psi concrete	26.50	19.85	46.35
2300	3'-0" wide x 1' thick, 2000 psi concrete	29	21	50
2400	3000 psi concrete	29.50	21	50.50
2500	4000 psi concrete	30.50	21	51.50
2700				
2800				
2900	Cut & patch to match existing construction, add, minimum	2%	3%	
3000	Maximum	5%	9%	
3100	Dust protection, add, minimum	1%	2%	
3200	Maximum	4%	11%	
3300	Equipment usage curtailment, add, minimum	1%	1%	
3400	Maximum	3%	10%	
3500	Material handling & storage limitation, add, minimum	1%	1%	
3600	Maximum	6%	7%	
3700	Protection of existing work, add, minimum	2%	2%	
3800	Maximum	5%	7%	
3900	Shift work requirements, add, minimum		5%	
4000	Maximum		30%	

A1010 Standard Foundations

This page illustrates and describes a spread footing system including concrete, forms, reinforcing and anchor bolts. Lines within System Components give the unit price and total price on a cost each basis for this system. Prices for spread footing systems are on Line Items A1010 230 1200 thru 2300. Both material quantities and labor costs have been adjusted for the system listed.

Factors: To adjust for job conditions other than normal working situations use Lines A1010 230 2900 thru 4000.

Example: You are to install the system in an existing occupied building. Access to the site and protection to the building are mandatory. Go to Lines A1010 230 3600 and 3800 and apply these percentages to the appropriate MAT. and INST. costs.

System Components	QUANTITY	UNIT	COST EACH		
			MAT.	INST.	TOTAL
Interior column footing, 3' square 1' thick, 2000 psi concrete, Including forms, reinforcing, and anchor bolts.					
Concrete, 2000 psi	.330	C.Y.	55.11		55.11
Placing concrete	.330	C.Y.		21.82	21.82
Forms, footing, 4 uses	12.000	SFCA	13.44	79.20	92.64
Reinforcing	11.000	Lb.	8.80	7.92	16.72
Anchor bolts, 3/4" diameter	2.000	Ea.	4.92	10.70	15.62
TOTAL			82.27	119.64	201.91

A1010 230	Spread Footing	COST EACH		
		MAT.	INST.	TOTAL
1200	3' square x 1' thick, 2000 psi concrete	82.50	120	202.50
1300	3000 psi concrete	85	120	205
1400	4000 psi concrete	86.50	120	206.50
1600	For alternate footing systems:			
1700	4' square x 1' thick, 2000 psi concrete	137	169	306
1800	3000 psi concrete	142	169	311
1900	4000 psi concrete	145	169	314
2100	5' square x 1'-3" thick, 2000 psi concrete	257	280	537
2200	3000 psi concrete	266	280	546
2300	4000 psi concrete	272	280	552
2600				
2700				
2900	Cut & patch to match existing construction, add, minimum	2%	3%	
3000	Maximum	5%	9%	
3100	Dust protection, add, minimum	1%	2%	
3200	Maximum	4%	11%	
3300	Equipment usage curtailment, add, minimum	1%	1%	
3400	Maximum	3%	10%	
3500	Material handling & storage limitation, add, minimum	1%	1%	
3600	Maximum	6%	7%	
3700	Protection of existing work, add, minimum	2%	2%	
3800	Maximum	5%	7%	
3900	Shift work requirements, add, minimum		5%	
4000	Maximum		30%	

A10 Foundations

A1010 Standard Foundations

General: Footing drains can be placed either inside or outside of foundation walls depending upon the source of water to be intercepted. If the source of subsurface water is principally from grade or a subsurface stream above the bottom of the footing, outside drains should be used. For high water tables, use inside drains or both inside and outside.

The effectiveness of underdrains depends on good waterproofing. This must be carefully installed and protected during construction.

Costs below include the labor and materials for the pipe and 6″ of crushed stone around pipe. Excavation and backfill are not included.

System Components		QUANTITY	UNIT	COST PER L.F.		
				MAT.	INST.	TOTAL
Foundation underdrain, outside, PVC 4″ diameter.						
	PVC pipe 4″ diam. S.D.R. 35	1.000	L.F.	2.43	5.15	7.58
	Crushed stone 3/4″ to 1/2″	.070	C.Y.	1.58	1.04	2.62
	TOTAL			4.01	6.19	10.20

A1010 310	Foundation Underdrain	COST PER L.F.		
		MAT.	INST.	TOTAL
1000	Foundation underdrain, outside only, PVC, 4″ diameter	4.01	6.20	10.21
1100	6″ diameter	6.70	6.85	13.55
1400	Perforated HDPE, 6″ diameter	4.54	2.63	7.17
1450	8″ diameter	9.20	3.29	12.49
1500	12″ diameter	11.05	8.10	19.15
1600	Corrugated metal, 16 ga. asphalt coated, 6″ diameter	10.15	6.45	16.60
1650	8″ diameter	15.80	6.85	22.65
1700	10″ diameter	19.55	8.90	28.45
3000	Outside and inside, PVC, 4″ diameter	8	12.40	20.40
3100	6″ diameter	13.45	13.65	27.10
3400	Perforated HDPE, 6″ diameter	9.05	5.25	14.30
3450	8″ diameter	18.35	6.60	24.95
3500	12″ diameter	22	16.15	38.15
3600	Corrugated metal, 16 ga., asphalt coated, 6″ diameter	20.50	12.85	33.35
3650	8″ diameter	31.50	13.65	45.15
3700	10″ diameter	39	17.80	56.80
4700	Cut & patch to match existing construction, add, minimum	2%	3%	
4800	Maximum	5%	9%	
4900	Dust protection, add, minimum	1%	2%	
5000	Maximum	4%	11%	
5100	Equipment usage curtailment, add, minimum	1%	1%	
5200	Maximum	3%	10%	
5300	Material handling & storage limitation, add, minimum	1%	1%	
5400	Maximum	6%	7%	
5500	Protection of existing work, add, minimum	2%	2%	
5600	Maximum	5%	7%	
5700	Shift work requirements, add, minimum		5%	
5800	Maximum		30%	
5900	Temporary shoring and bracing, add, minimum	2%	5%	
6000	Maximum	5%	12%	
8840	Waterproof exterior wall, 2 layers w/ protection board	2.61	5.05	7.66
8850	Waterproof roof structure, 2 layers w/ protection board	2.61	5.05	7.66

A1030 Slab on Grade

This page illustrates and describes a slab on grade system including slab, bank run gravel, bulkhead forms, placing concrete, welded wire fabric, vapor barrier, steel trowel finish and curing paper. Lines within System Components give the unit price and total price per square foot for this system. Prices for slab on grade systems are on Line Items A1030 110 1400 thru 2600. Both material quantities and labor costs have been adjusted for the system listed.

Factors: To adjust for job conditions other than normal working situations use Lines A1030 110 2900 thru 4000.

Example: You are to install the system at a site where protection of the existing building is required. Go to Line A1030 110 3800 and apply these percentages to the appropriate MAT. and INST. costs.

System Components	QUANTITY	UNIT	COST PER S.F.		
			MAT.	INST.	TOTAL
Ground slab, 4″ thick, 3000 psi concrete, 4″ granular base, vapor barrier					
Welded wire fabric, screed and steel trowel finish.					
Concrete, 4″ thick, 3000 psi concrete	.012	C.Y.	2.10		2.10
Bank run gravel, 4″ deep	.012	C.Y.	.36	.11	.47
Polyethylene vapor barrier, 10 mil.	.011	C.S.F.	.10	.21	.31
Bulkhead forms, expansion material	.100	L.F.	.05	.45	.50
Welded wire fabric, 6 x 6 - #10/10	.011	C.S.F.	.27	.48	.75
Place concrete	.012	C.Y.		.40	.40
Screed & steel trowel finish	1.000	S.F.		1.17	1.17
TOTAL			2.88	2.82	5.70

A1030 110	Interior Slab on Grade	COST PER S.F.		
		MAT.	INST.	TOTAL
1400	4″ thick slab, 3000 PSI concrete, 4″ deep bank run gravel	2.88	2.82	5.70
1500	6″ deep bank run gravel	2.62	2.73	5.35
1600	12″ deep bank run gravel	2.95	2.76	5.71
1700				
1800				
1900				
2000	For alternate slab systems:			
2100	5″ thick slab, 3000 psi concrete, 6″ deep bank run gravel	3.69	2.91	6.60
2200	12″ deep bank run gravel	4.33	2.96	7.29
2300				
2400				
2500	6″ thick slab, 3000 psi concrete, 6″ deep bank run gravel	4.39	3.04	7.43
2600	12″ deep bank run gravel	5.05	3.09	8.14
2700				
2900	Cut & patch to match existing construction, add, minimum	2%	3%	
3000	Maximum	5%	9%	
3100	Dust protection, add, minimum	1%	2%	
3200	Maximum	4%	11%	
3300	Equipment usage curtailment, add, minimum	1%	1%	
3400	Maximum	3%	10%	
3500	Material handling & storage limitation, add, minimum	1%	1%	
3600	Maximum	6%	7%	
3700	Protection of existing work, add, minimum	2%	2%	
3800	Maximum	5%	7%	
3900	Shift work requirements, add, minimum		5%	
4000	Maximum		30%	

A20 Basement Construction

A2010 Basement Excavation

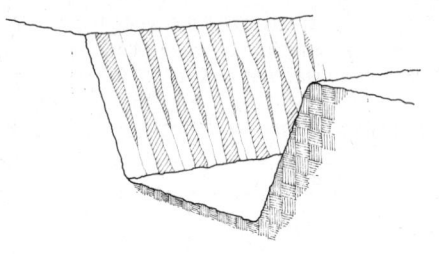

This page illustrates and describes continuous footing and trench excavation systems including a wheel mounted backhoe, operator, equipment rental, fuel, oil, mobilization, no hauling or backfill. Lines within System Components give the unit price and total price per linear foot for this system. Prices for alternate continuous footing and trench excavation systems are on Line Items A2010 130 1200 thru 2700. Both material quantities and labor costs have been adjusted for the system listed.

Factors: To adjust for job conditions other than normal working situations use Lines A2010 130 2900 thru 4000.

Example: You are to install the system and supply dust protection. Go to Line A2010 130 3000 and apply this percentage to the appropriate TOTAL costs.

System Components	QUANTITY	UNIT	COST PER L.F. EQUIP.	COST PER L.F. LABOR	COST PER L.F. TOTAL
Continuous footing or trench excav. w/ 3/4 C.Y. wheel mntd backhoe Including operator, equipment rental, fuel, oil and mobilization. Trench Is 4' wide at bottom, 3' deep w/sides sloped 1 to 2 and 200' long. Costs Are based on a production rate of 240 C.Y./day. No hauling/backfill incl.					
3/4 C.Y. backhoe, wheel mounted	.500	Day	180	665	845
Mobilization	1.000	Ea.	1,250	2,175	3,425
TOTAL			1,430	2,840	4,270
COST PER L.F.		L.F.	7.15	14.20	21.35

A2010 130	Excavation, Footings or Trench	COST PER L.F. EQUIP.	COST PER L.F. LABOR	COST PER L.F. TOTAL
1200	For alternate trench sizes, 4' bottom with sloped sides:			
1300	2' deep, 50' long	8.65	27	35.65
1400	100' long	4.33	13.60	17.93
1500	300' long	4.77	9.45	14.22
1600	4' deep, 50' long	8.65	27	35.65
1700	100' long	4.33	13.60	17.93
1800	300' long	2.08	6.90	8.98
1900	6' deep, 50' long	8.65	27	35.65
2000	100' long	4.87	15.60	20.47
2100	300' long	3.24	11.20	14.44
2200	8' deep, 50' long	8.65	27	35.65
2300	100' long	6.15	20.50	26.65
2400	300' long	4.44	15.60	20.04
2500	10' deep, 50' long	10.45	34	44.45
2600	100' long	7.70	26	33.70
2700	300' long	5.90	21	26.90
2800				
2900	Dust protection, add, minimum		2%	2%
3000	Maximum		11%	11%
3100	Equipment usage curtailment, add, minimum		1%	1%
3200	Maximum		10%	10%
3300	Material handling & storage limitation, add, minimum		1%	1%
3400	Maximum		7%	7%
3500	Protection of existing work, add, minimum		2%	2%
3600	Maximum		7%	7%
3700	Shift work requirements, add, minimum		5%	5%
3800	Maximum		30%	30%
3900	Temporary shoring and bracing, add, minimum		5%	5%
4000	Maximum		12%	12%

A2010 Basement Excavation

This page illustrates and describes foundation excavation systems including a backhoe-loader, operator, equipment rental, fuel, oil, mobilization, hauling material, no backfilling. Lines within System Components give the unit price and total price per cubic yard for this system. Prices for foundation excavation systems are on Line Items A2010 140 1600 thru 2200. Both material quantities and labor costs have been adjusted for the system listed.

Factors: To adjust for job conditions other than normal working situations use Lines A2010 140 2900 thru 4000.

Example: You are to install the system with the use of temporary shoring and bracing. Go to Line A2010 140 3900 and apply these percentages to the appropriate TOTAL costs.

System Components	QUANTITY	UNIT	COST PER C.Y. EQUIP.	COST PER C.Y. LABOR	COST PER C.Y. TOTAL
Foundation excav w/ 3/4 C.Y. backhoe-loader, incl. operator, equip Rental, fuel, oil, and mobilization. Hauling of excavated material is Included. Prices based on one day production of 360 C.Y. in medium soil Without backfilling.					
Backhoe-loader, 3/4 C.Y.	1.000	Day	360	1,330	1,690
Hauling, 12 C.Y. trucks, 1 mile round trip	360.000	L.C.Y.	1,206	964.80	2,170.80
Mobilization	1.000	Ea.	253	695	948
TOTAL			1,819	2,989.80	4,808.80
COST PER C.Y.		C.Y.	5.04	8.28	13.36

A2010 140	Excavation, Foundation	COST PER C.Y. EQUIP.	COST PER C.Y. LABOR	COST PER C.Y. TOTAL
1600	100 C.Y.	6.90	13.35	20.25
1700	200 C.Y.	5.60	9.90	15.50
1800	300 C.Y.	5.25	8.90	14.15
1850	360 C.Y.	5.07	8.33	13.40
1900	400 C.Y.	4.97	8.10	13.07
2000	500 C.Y.	4.86	7.80	12.66
2100	600 C.Y.	4.73	7.40	12.13
2200	700 C.Y.	4.69	7.30	11.99
2300				
2400				
2900	Dust protection, add, minimum		2%	2%
3000	Maximum		11%	11%
3100	Equipment usage curtailment, add, minimum		1%	1%
3200	Maximum		10%	10%
3300	Material handling & storage limitation, add, minimum		1%	1%
3400	Maximum		7%	7%
3500	Protection of existing work, add, minimum		2%	2%
3600	Maximum		7%	7%
3700	Shift work requirements, add, minimum		5%	5%
3800	Maximum		30%	30%
3900	Temporary shoring and bracing, add, minimum		5%	5%
4000	Maximum		12%	12%

613

A20 Basement Construction

A2020 Basement Walls

This page illustrates and describes a concrete wall system including concrete, placing concrete, forms, reinforcing, insulation, waterproofing and anchor bolts. Lines within System Components give the unit price and total price per linear foot for this system. Prices for concrete wall systems are on Line Items A2020 120 1400 thru 2600. Both material quantities and labor costs have been adjusted for the system listed.

Factors: To adjust for job conditions other than normal working situations use Lines A2020 120 2900 thru 4000.

Example: You are to install this wall system where delivery of material is difficult. Go to Line A2020 120 3600 and apply these percentages to the appropriate MAT. and INST. costs.

System Components	QUANTITY	UNIT	COST PER L.F.		
			MAT.	INST.	TOTAL
Cast in place concrete foundation wall, 8" thick, 3' high, 2500 psi					
Concrete including forms, reinforcing, waterproofing, and anchor bolts.					
Concrete, 2500 psi, 8" thick, 3' high	.070	C.Y.	11.90		11.90
Forms, wall, 4 uses	6.000	S.F.	10.38	49.80	60.18
Reinforcing	6.000	Lb.	4.80	3.06	7.86
Placing concrete	.070	C.Y.		4.05	4.05
Waterproofing	3.000	S.F.	1.41	4.11	5.52
Rigid insulaton, 1" polystyrene	3.000	S.F.	.93	3.15	4.08
Anchor bolts, 1/2" diameter, 4' O.C.	.250	Ea.	.68	1.35	2.03
TOTAL			30.10	65.52	95.62

A2020 120	Concrete Wall	COST PER L.F.		
		MAT.	INST.	TOTAL
1400	8" thick, 2500 psi concrete, 3' high	30	65.50	95.50
1500	4' high	41	87.50	128.50
1600	6' high	61	130	191
1700	8' high	81.50	173	254.50
1800	3500 psi concrete, 4' high	41.50	87.50	129
1900	6' high	62.50	130	192.50
2000	8' high	83	173	256
2100	12" thick, 2500 psi concrete, 4' high	52.50	92	144.50
2200	6' high	76.50	136	212.50
2300	8' high	102	181	283
2400	3500 psi concrete, 4' high	54	92	146
2500	8' high	104	181	285
2600	10' high	129	226	355
2700				
2900	Cut & patch to match existing construction, add, minimum	2%	3%	
3000	Maximum	5%	9%	
3100	Dust protection, add, minimum	1%	2%	
3200	Maximum	4%	11%	
3300	Equipment usage curtailment, add, minimum	1%	1%	
3400	Maximum	3%	10%	
3500	Material handling & storage limitation, add, minimum	1%	1%	
3600	Maximum	6%	7%	
3700	Protection of existing work, add, minimum	2%	2%	
3800	Maximum	5%	7%	
3900	Shift work requirements, add, minimum		5%	
4000	Maximum		30%	

A2020 Basement Walls

This page illustrates and describes a concrete block wall system including concrete block, masonry reinforcing, parging, waterproofing insulation and anchor bolts. Lines within System Components give the unit price and total price per linear foot for this system. Prices for concrete block wall systems are on Line Items A2020 140 1200 thru 2600. Both material quantities and labor costs have been adjusted for the system listed.

Factors: To adjust for job conditions other than normal working situations use Lines A2020 140 2900 thru 4000.

Example: You are to install the system to match an existing foundation wall. Go to Line A2020 140 3000 and apply these percentages to the appropriate MAT. and INST. costs.

System Components	QUANTITY	UNIT	COST PER L.F.		
			MAT.	INST.	TOTAL
Concrete block, 8″ thick, masonry reinforcing, parged and Waterproofed, insulation and anchor bolts, wall 2'-8″ high.					
Concrete block, 8″ x 8″ x 16″	2.670	S.F.	9.18	20.56	29.74
Masonry reinforcing	2.000	L.F.	.58	.48	1.06
Parging	.193	S.Y.	.78	3.13	3.91
Waterproofing	2.670	S.F.	1.25	3.66	4.91
Insulation, 1″ rigid polystyrene	2.670	S.F.	.83	2.80	3.63
Anchor bolts, 1/2″ diameter, 4' O.C.	.250	Ea.	.68	1.35	2.03
TOTAL			13.30	31.98	45.28

A2020 140	Concrete Block Wall	COST PER L.F.		
		MAT.	INST.	TOTAL
1200	8″ thick block, 2'-8″ high	13.30	32	45.30
1300	4' high	19.60	47.50	67.10
1400	6' high	29	70	99
1500	8' high	38.50	93.50	132
1600	Grouted solid, 4' high	25.50	63.50	89
1700	6' high	38	94	132
1800	8' high	50.50	125	175.50
2100	12″ thick block, 4' high	28	68	96
2200	6' high	41.50	101	142.50
2300	8' high	55	135	190
2400	Grouted solid, 4' high	38	85.50	123.50
2500	6' high	56	127	183
2600	8' high	75	169	244
2700				
2900	Cut & patch to match existing construction, add, minimum	2%	3%	
3000	Maximum	5%	9%	
3100	Dust protection, add, minimum	1%	2%	
3200	Maximum	4%	11%	
3300	Equipment usage curtailment, add, minimum	1%	1%	
3400	Maximum	3%	10%	
3500	Material handling & storage limitation, add, minimum	1%	1%	
3600	Maximum	6%	7%	
3700	Protection of existing work, add, minimum	2%	2%	
3800	Maximum	5%	7%	
3900	Shift work requirements, add, minimum		5%	
4000	Maximum		30%	

Same Data. Simplified.

Enjoy the convenience and efficiency of accessing your costs anywhere:

- **Skip the multiplier** by setting your location
- **Quickly search,** edit, favorite and share costs
- **Stay on top of price changes** with automatic updates

Discover more at rsmeans.com/online

This page illustrates and describes a reinforced concrete slab system including concrete, placing concrete, formwork, reinforcing, steel trowel finish, V.A. floor tile and acoustical spray ceiling finish. Lines within System Components give the unit price per square foot for this system. Prices for reinforced concrete slab systems are on Line Items B1010 225 1400 thru 1800. Both material quantities and labor costs have been adjusted for the system listed.

Factors: To adjust for job conditions other than normal working situations use Lines B1010 225 2700 thru 4000.

Example: You are to install the system to match an existing floor system. Go to Line B1010 225 2800 and apply these percentages to the appropriate MAT. and INST. costs.

System Components	QUANTITY	UNIT	COST PER S.F.		
			MAT.	INST.	TOTAL
Flat slab system, reinforced concrete with vinyl tile floor, sprayed					
Acoustical ceiling finish, not including columns.					
Concrete, 4000 psi, 6" thick	.020	C.Y.	3.60		3.60
Placing concrete	.020	C.Y.		.83	.83
Formwork, 4 use	1.000	S.F.	2.06	7.45	9.51
Edge form	.120	S.F.	.11	1.45	1.56
Reinforcing steel	3.000	Lb.	2.40	2.16	4.56
Steel trowel finish	1.000	S.F.		1.51	1.51
Vinyl floor tile, 1/8" thick, premium colors/patterns	1.000	S.F.	6.70	1.33	8.03
Acoustical spray ceiling finish, solvent based	1.000	S.F.	.32	.21	.53
TOTAL			15.19	14.94	30.13

B1010 225	Floor - Ceiling, Concrete Slab	COST PER S.F.		
		MAT.	INST.	TOTAL
1400	Concrete, 4000 psi, 6" thick	15.20	14.95	30.15
1500	7" thick	15.90	15.55	31.45
1600	8" thick	16.95	16.35	33.30
1700	9" thick	17.80	17	34.80
1800	10" thick	18.85	17.80	36.65
1900				
2000				
2100				
2200				
2300				
2400				
2500				
2700	Cut & patch to match existing construction, add, minimum	2%	3%	
2800	Maximum	5%	9%	
2900	Dust protection, add, minimum	1%	2%	
3000	Maximum	4%	11%	
3100	Equipment usage curtailment, add, minimum	1%	1%	
3200	Maximum	3%	10%	
3300	Material handling & storage limitation, add, minimum	1%	1%	
3400	Maximum	6%	7%	
3500	Protection of existing work, add, minimum	2%	2%	
3600	Maximum	5%	7%	
3700	Shift work requirements, add, minimum		5%	
3800	Maximum		30%	
3900	Temporary shoring and bracing, add, minimum	2%	5%	
4000	Maximum	5%	12%	

B1010 Floor Construction

This page illustrates and describes a hollow core prestressed concrete panel system including hollow core slab, grout, carpet, carpet padding, and sprayed ceiling. Lines within System Components give the unit price and total price per square foot for this system. Prices for hollow core prestressed concrete panel systems are on Line Items B1010 232 1200 thru 1600. Both material and labor costs have been adjusted for the system listed.

Factors: To adjust for job conditions other than normal working situations use Lines B1010 232 2700 thru 3800.

Example: You are to install the system where dust control is a major concern. Go to Line B1010 232 2800 and apply these percentages to the appropriate MAT. and INST. costs.

System Components	QUANTITY	UNIT	COST PER S.F. MAT.	COST PER S.F. INST.	COST PER S.F. TOTAL
Precast hollow core plank with carpeted floors, padding					
And sprayed textured ceiling.					
Hollow core plank, 4" thick with grout topping	1.000	S.F.	9.05	4.03	13.08
Nylon carpet, 26 oz. medium traffic	.110	S.Y.	5.01	.97	5.98
Carpet padding, 20 oz./sq. yd.	.110	S.Y.	.59	.49	1.08
Sprayed texture ceiling	1.000	S.F.	.06	.87	.93
TOTAL			14.71	6.36	21.07

B1010 232	Floor - Ceiling, Conc. Panel	COST PER S.F. MAT.	COST PER S.F. INST.	COST PER S.F. TOTAL
1200	Hollow core concrete plank, with grout, 4" thick	14.70	6.35	21.05
1300	6" thick	14.80	5.80	20.60
1400	8" thick	17.40	5.35	22.75
1500	10" thick	16.30	5	21.30
1600	12" thick	18.20	4.76	22.96
1700				
1800				
1900				
2000				
2100				
2200				
2300				
2400				
2500				
2700	Dust protection, add, minimum	1%	2%	
2800	Maximum	4%	11%	
2900	Equipment usage curtailment, add, minimum	1%	1%	
3000	Maximum	3%	10%	
3100	Material handling & storage limitation, add, minimum	1%	1%	
3200	Maximum	6%	7%	
3300	Protection of existing work, add, minimum	2%	2%	
3400	Maximum	5%	7%	
3500	Shift work requirements, add, minimum		5%	
3600	Maximum		30%	
3700	Temporary shoring and bracing, add, minimum	2%	5%	
3800	Maximum	5%	12%	

B1010 Floor Construction

This page illustrates and describes a hollow core plank system including precast hollow core plank, concrete topping, V.A. tile, concealed suspension system and ceiling tile. Lines within System Components give the unit price and total price per square foot for this system. Prices for hollow core plank systems are on Line Items B1010 233 1400 thru 1800. Both material quantities and labor costs have been adjusted for the system listed.

Factors: To adjust for job conditions other than normal working situations use Lines B1010 233 2900 thru 4000.

Example: You are to install the system where equipment usage is a problem. Go to Line B1010 233 3200 and apply these percentages to the appropriate MAT. and INST. costs.

System Components	QUANTITY	UNIT	COST PER S.F. MAT.	COST PER S.F. INST.	COST PER S.F. TOTAL
Precast hollow core plank with 2″ topping, vinyl composition floor tile					
And suspended acoustical tile ceiling.					
Hollow core concrete plank, 4″ thick	1.000	S.F.	9.05	4.03	13.08
Finishing floors, granolithic topping, 1:1:1-1/2 mix, 2″ thick	1.000	S.F.	1.68	6.84	8.52
Vinyl composition tile, .08″ thick	1.000	S.F.	1.42	1.33	2.75
Concealed Z bar suspension system, 12″ module	1.000	S.F.	1.03	1.38	2.41
Ceiling tile, mineral fiber, 3/4″ thick	1.000	S.F.	5.15	2.39	7.54
TOTAL			18.33	15.97	34.30

B1010 233	Floor - Ceiling, Conc. Plank	COST PER S.F. MAT.	COST PER S.F. INST.	COST PER S.F. TOTAL
1400	Hollow core concrete plank, 4″ thick	18.35	15.95	34.30
1500	6″ thick	18.45	15.40	33.85
1600	8″ thick	21	14.95	35.95
1700	10″ thick	19.95	14.65	34.60
1800	12″ thick	22	14.35	36.35
1900				
2000				
2100				
2200				
2300				
2400				
2500				
2600				
2700				
2900	Dust protection, add, minimum	1%	2%	
3000	Maximum	4%	11%	
3100	Equipment usage curtailment, add, minimum	1%	1%	
3200	Maximum	3%	10%	
3300	Material handling & storage limitation, add, minimum	1%	1%	
3400	Maximum	6%	7%	
3500	Protection of existing work, add, minimum	2%	2%	
3600	Maximum	5%	7%	
3700	Shift work requirements, add, minimum		5%	
3800	Maximum		30%	
3900	Temporary shoring and bracing, add, minimum	2%	5%	
4000	Maximum	5%	12%	

620

B1010 Floor Construction

This page illustrates and describes a structural steel w/metal decking and concrete system including steel beams, steel decking, shear studs, concrete, placing concrete, edge form, steel trowel finish, curing, wire fabric, fireproofing, beams and decking, tile floor, and suspended ceiling. Lines within System Components give the unit price and total price per square foot for this system. Prices for alternate structural steel w/metal decking and concrete systems are on Line Items B1010 243 1900 thru 2400. Both material quantities and labor costs have been adjusted for the system listed.

Factors: To adjust for job conditions other than normal working situations use Lines B1010 243 2900 thru 4000.

Example: You are to install the system where material handling and storage are a problem. Go to Line B1010 243 3400 and apply these percentages to the appropriate MAT. and INST. costs.

System Components	QUANTITY	UNIT	COST PER S.F. MAT.	COST PER S.F. INST.	COST PER S.F. TOTAL
Composite structural beams, 20 ga. 3" deep steel decking, shear studs, 3000 psi, concrete, placing concrete, edge forms, steel trowel finish, Welded wire fabric fireproofing, tile floor and suspended ceiling.					
Steel deck, 20 gage, 3" deep, galvanized	1.000	S.F.	3.87	1.12	4.99
Structural steel framing	.004	Ton	14.40	3.34	17.74
Shear studs, 3/4"	.250	Ea.	.28	.53	.81
Concrete, 3000 psi, 5" thick	.015	C.Y.	2.63		2.63
Place concrete	.015	C.Y.		.63	.63
Edge form	.140	L.F.	.05	.76	.81
Steel trowel finish	1.000	S.F.		1.51	1.51
Welded wire fabric 6 x 6 - #10/10	.011	C.S.F.	.27	.48	.75
Fireproofing, sprayed	1.880	S.F.	1.20	2.46	3.66
Vinyl composition floor tile	1.000	S.F.	.85	1.33	2.18
Suspended acoustical ceiling	1.000	S.F.	7.10	1.88	8.98
TOTAL			30.65	14.04	44.69

B1010 243	Floor - Ceiling, Struc. Steel	COST PER S.F. MAT.	COST PER S.F. INST.	COST PER S.F. TOTAL
1900	Composite deck, galvanized, 3" deep, 22 ga.	29.50	13.95	43.45
1950	20 ga.	30.50	14.05	44.55
2000	18 ga.	30	14.10	44.10
2100	16 ga.	32.50	14.15	46.65
2200	Composite deck, galvanized, 2" deep, 22 ga.	29.50	13.80	43.30
2300	20 ga.	30.50	13.85	44.35
2400	16 ga.	32	13.95	45.95
2500				
2900	Dust protection, add, minimum	1%	2%	
3000	Maximum	4%	11%	
3100	Equipment usage curtailment, add, minimum	1%	1%	
3200	Maximum	3%	10%	
3300	Material handling & storage limitation, add, minimum	1%	1%	
3400	Maximum	6%	7%	
3500	Protection of existing work, add, minimum	2%	2%	
3600	Maximum	5%	7%	
3700	Shift work requirements, add, minimum		5%	
3800	Maximum		30%	
3900	Temporary shoring and bracing, add, minimum	2%	5%	
4000	Maximum	5%	12%	

621

This page illustrates and describes an open web joist and steel slab-form system including open web steel joist, slab form, concrete, placing concrete, wire fabric, steel trowel finish, tile floor and plasterboard. Lines within System Components give the unit price and total price per square foot for this system. Prices for open web joists and steel slab-form systems are on Line Items B1010 245 1800 thru 2200. Both material quantities and labor costs have been adjusted for the systems listed.

Factors: To adjust for job conditions other than normal working situations use Lines B1010 245 2900 thru 4000.

Example: You are to install the system in a congested commercial area and most work will be done at night. Go to Line B1010 245 3800 and apply this percentage to the appropriate INST. cost.

System Components	QUANTITY	UNIT	COST PER S.F.		
			MAT.	INST.	TOTAL
Open web steel joists, 24″ O.C., slab form, 3000 psi concrete, welded					
Wire fabric, steel trowel finish, floor tile, 5/8″ drywall ceiling.					
Open web steel joists, 12″ deep, 5.2#/L.F., 24″ O.C.	2.630	Lb.	4.18	1.42	5.60
Slab form, 28 gage, 9/16″ deep, galvanized	1.000	S.F.	1.74	.84	2.58
Concrete, 3000 psi, 2-1/2″ thick	.008	C.Y.	1.40		1.40
Placing concrete	.008	C.Y.		.33	.33
Welded wire fabric 6 x 6 - #10/10	.011	C.S.F.	.27	.48	.75
Steel trowel finish	1.000	S.F.		1.51	1.51
Vinyl composition floor tile	1.000	S.F.	.85	1.33	2.18
Ceiling furring, 3/4″ channels, 24″ O.C.	1.000	S.F.	.29	1.66	1.95
Gypsum drywall, 5/8″ thick, finished	1.000	S.F.	.34	1.87	2.21
Paint ceiling	1.000	S.F.	.14	.80	.94
TOTAL			9.21	10.24	19.45

B1010 245	Floor - Ceiling, Steel Joists	COST PER S.F.		
		MAT.	INST.	TOTAL
1800	Open web joists, 12″ deep, 5.2#/L.F.	9.20	10.25	19.45
1900	16″ deep, 6.6#/L.F.	10.30	10.60	20.90
2000	20″ deep, 8.4# L.F.	11.70	11.10	22.80
2100	24″ deep, 11.5# L.F.	14.15	11.90	26.05
2200	26″ deep, 12.8# L.F.	15.20	12.30	27.50
2300				
2400				
2500				
2600				
2700				
2900	Dust protection, add, minimum	1%	2%	
3000	Maximum	4%	11%	
3100	Equipment usage curtailment, add, minimum	1%	1%	
3200	Maximum	3%	10%	
3300	Material handling & storage limitation, add, minimum	1%	1%	
3400	Maximum	6%	7%	
3500	Protection of existing work, add, minimum	2%	2%	
3600	Maximum	5%	7%	
3700	Shift work requirements, add, minimum		5%	
3800	Maximum		30%	
3900	Temporary shoring and bracing, add, minimum	2%	5%	
4000	Maximum	5%	12%	

B1010 Floor Construction

Joist Hangers

Existing Girder

Fiberglass Insulation, Kraft Faced, 9" Thick

Steel Cross Bridging

12" Open Web Wood Joists, 2'-0" o.c.

3/4" CDX Plywood Sub-floor

Existing Foundation

2" x 8" Sill

Anchor Bolts, 1/2" x 8"

Open Web Wood Joists

This page illustrates and describes floor framing systems including sills, joist hangers, joists, bridging, subfloor and insulation. Lines within systems components give the unit price and total price per square foot for this system. Prices for alternate joist types are on line items B1010 262 0100 through 0700. Both material quantities and labor costs have been adjusted for the system listed.

Factors: To adjust for job conditions other than normal working situations use lines B1010 262 2700 through 4000.

Example: You are to install the system where delivery of material is difficult. Go to line B1010 262 3400 and apply these percentages to the appropriate MAT. and INST. costs.

System Components	QUANTITY	UNIT	COST PER S.F.		
			MAT.	INST.	TOTAL
Open web wood joists, sill, butt to girder, 5/8" subfloor, insul.					
Anchor bolt, hook type, with nut and washer, 1/2" x 8" long	.025	Ea.	.06	.13	.19
Joist & beam hanger, 18 ga. galvanized	.075	Ea.	.16	.33	.49
Bridging, steel, compression, for 2" x 12" joists	.010	C.Pr.	.02	.04	.06
Framing, open web wood joists, 12" deep	.001	M.L.F.	8.35	1.63	9.98
Framing, sills, 2" x 8"	.001	M.B.F.	1.18	2.13	3.31
Sub-floor, plywood, CDX, 3/4" thick	1.000	S.F.	.90	1.06	1.96
Insulation, fiberglass, kraft face, 9" thick x 23" wide	1.000	S.F.	1.25	.53	1.78
TOTAL			11.92	5.85	17.77

B1010 262	Open Web Wood Joists	COST PER S.F.		
		MAT.	INST.	TOTAL
0100	12" open web joists, sill, butt to girder, 5/8" sub-floor	11.90	5.85	17.75
0200	14" joists	12.65	5.95	18.60
0300	16" joists	16.55	6.05	22.60
0400	18" joists	13.65	6.15	19.80
0450	Wood struc." I " joists, 24" O.C., to 24' span, 50 PSF LL, butt to girder	8.85	5.40	14.25
0500	55 PSF LL	9.15	5.50	14.65
0600	to 30' span, 45 PSF LL	9.95	5.30	15.25
0700	55 PSF LL	12.95	5.40	18.35
2700	Cut & patch to match existing construction, add, minimum	2%	3%	
2800	Maximum	5%	9%	
2900	Dust protection, add, minimum	1%	2%	
3000	Maximum	4%	11%	
3100	Equipment usage curtailment, add, minimum	1%	1%	
3200	Maximum	3%	10%	
3300	Material handling & storage limitation, add, minimum	1%	1%	
3400	Maximum	6%	7%	
3500	Protection of existing work, add, minimum	2%	2%	
3600	Maximum	5%	7%	
3700	Shift work requirements, add, minimum		5%	
3800	Maximum		30%	
3900	Temporary shoring and bracing, add, minimum	2%	5%	
4000	Maximum	5%	12%	

B10 Superstructure

B1010 Floor Construction

This page illustrates and describes a wood joist floor system including wood joist, oak floor, sub-floor, bridging, sand and finish floor, furring, plasterboard, taped, finished and painted ceiling. Lines within System Components give the unit price and total price per square foot for this system. Prices for wood joist floor systems are on Line Items B1010 263 1700 thru 2300. Both material quantities and labor costs have been adjusted for the system listed.

Factors: To adjust for job conditions other than normal working situations use Lines B1010 263 2700 thru 4000.

Example: You are to install the system during off peak hours, 6 P.M. to 2 A.M. Go to Line B1010 263 3800 and apply this percentage to the appropriate INST. costs.

System Components	QUANTITY	UNIT	COST PER S.F.		
			MAT.	INST.	TOTAL
Wood joists, 2" x 8", 16" O.C., oak floor (sanded & finished), 1/2" sub Floor, 1"x 3" bridging, furring, 5/8" drywall, taped, finished and painted.					
Wood joists, 2" x 8", 16" O.C.	1.000	L.F.	1.58	1.30	2.88
Subfloor plywood CDX 1/2"	1.000	SF Flr.	.84	.95	1.79
Bridging 1" x 3"	.150	Pr.	.13	.83	.96
Oak flooring, No. 1 common	1.000	S.F.	5.10	4.21	9.31
Sand and finish floor	1.000	S.F.	.27	1.96	2.23
Wood furring, 1" x 3", 16" O.C.	1.000	L.F.	.51	2.04	2.55
Gypsum drywall, 5/8" thick	1.000	S.F.	.29	.80	1.09
Tape and finishing	1.000	S.F.	.06	.72	.78
Paint ceiling	1.000	S.F.	.14	.80	.94
TOTAL			8.92	13.61	22.53

B1010 263	Floor-Ceiling, Wood Joists	COST PER S.F.		
		MAT.	INST.	TOTAL
1700	16" on center, 2" x 6" joists	8.40	13.45	21.85
1750	2" x 8"	8.90	13.60	22.50
1800	2" x 10"	9.50	13.90	23.40
1900	2" x 12"	10.40	13.95	24.35
2000	2" x 14"	13.15	14.15	27.30
2100	12" on center, 2" x 10" joists	10.05	14.30	24.35
2200	2" x 12"	11.15	14.35	25.50
2300	2" x 14"	14.60	14.65	29.25
2400				
2500				
2700	Cut & patch to match existing construction, add, minimum	2%	3%	
2800	Maximum	5%	9%	
2900	Dust protection, add, minimum	1%	2%	
3000	Maximum	4%	11%	
3100	Equipment usage curtailment, add, minimum	1%	1%	
3200	Maximum	3%	10%	
3300	Material handling & storage limitation, add, minimum	1%	1%	
3400	Maximum	6%	7%	
3500	Protection of existing work, add, minimum	2%	2%	
3600	Maximum	5%	7%	
3700	Shift work requirements, add, minimum		5%	
3800	Maximum		30%	
3900	Temporary shoring and bracing, add, minimum	2%	5%	
4000	Maximum	5%	12%	

B1020 Roof Construction

This page illustrates and describes a flat roof bar joist system including asphalt and gravel roof, lightweight concrete, metal decking, open web joists, and suspended acoustic ceiling. Lines within System Components give the unit price and total price per square foot for this system. Prices for flat roof bar joist systems are on Line Items B1020 106 1200 thru 1600. Both material quantities and labor costs have been adjusted for the system listed.

Factors: To adjust for job conditions other than normal working situations use Lines B1020 106 2700 thru 4000.

Example: You are to install the system where material handling and storage present some problem. Go to Line B1020 106 3300 and apply these percentages to the appropriate MAT. and INST. costs.

System Components	QUANTITY	UNIT	COST PER S.F.		
			MAT.	INST.	TOTAL
Three ply asphalt and gravel roof on lightweight concrete, metal decking on Web joist 4′ O.C., with suspended acoustic ceiling.					
Open web joist, 10″ deep, 4′ O.C.	1.250	Lb.	1.99	.68	2.67
Metal decking, 1-1/2″ deep, 22 Ga., galvanized	1.000	S.F.	2.79	.75	3.54
Gravel roofing, 3 ply asphalt	.010	C.S.F.	1.40	2.31	3.71
Perlite roof fill, 3″ thick	1.000	S.F.	3.04	.68	3.72
Suspended ceiling, 3/4″ mineral fiber on "Z" bar suspension	1.000	S.F.	2.29	4.77	7.06
TOTAL			11.51	9.19	20.70

B1020 106	Steel Joist Roof & Ceiling	COST PER S.F.		
		MAT.	INST.	TOTAL
1200	Open web bar joist, 10″ deep, 5#/L.F.	11.50	9.20	20.70
1300	16″ deep, 5#/L.F.	12.15	9.40	21.55
1400	18″ deep, 6.6#/L.F.	12.70	9.60	22.30
1500	20″ deep, 9.6#/L.F.	13.35	9.80	23.15
1600	24″ deep, 11.52#/L.F.	14.10	10.05	24.15
1700				
1800				
1900				
2000				
2100				
2200				
2300				
2400				
2500				
2700	Cut & patch to match existing construction, add, minimum	2%	3%	
2800	Maximum	5%	9%	
2900	Dust protection, add, minimum	1%	2%	
3000	Maximum	4%	11%	
3100	Equipment usage curtailment, add, minimum	1%	1%	
3200	Maximum	3%	10%	
3300	Material handling & storage limitation, add, minimum	1%	1%	
3400	Maximum	6%	7%	
3500	Protection of existing work, add, minimum	2%	2%	
3600	Maximum	5%	7%	
3700	Shift work requirements, add, minimum		5%	
3800	Maximum		30%	
3900	Temporary shoring and bracing, add, minimum	2%	5%	
4000	Maximum	5%	12%	

This page illustrates and describes a wood frame roof system including rafters, ceiling joists, sheathing, building paper, asphalt shingles, roof trim, furring, insulation, plaster and paint. Lines within System Components give the unit price and total price per square foot for this system. Prices for wood frame roof systems are on Line Items B1020 202 1800 thru 2700. Both material quantities and labor costs have been adjusted for the system listed.

Factors: To adjust for job conditions other than normal working situations use Lines B1020 202 3300 thru 4000.

Example: You are to install the system while protecting existing work. Go to Line B1020 202 3800 and apply these percentages to the appropriate MAT. and INST. costs.

System Components	QUANTITY	UNIT	COST PER S.F.		
			MAT.	INST.	TOTAL
Wood frame roof system, 4 in 12 pitch, including rafters, sheathing, Shingles, insulation, drywall, thin coat plaster, and painting.					
Rafters, 2" x 6", 16" O.C., 4 in 12 pitch	1.080	L.F.	1.16	1.54	2.70
Ceiling joists, 2" x 6", 16 O.C.	1.000	L.F.	1.07	1.14	2.21
Sheathing, 1/2" CDX	1.080	S.F.	.91	1.10	2.01
Building paper, 15# felt	.011	C.S.F.	.05	.21	.26
Asphalt, standard strip shingles, inorganic, class A, 30 year	.011	C.S.F.	1.53	1.51	3.04
Roof trim	.100	L.F.	.39	.31	.70
Furring, 1" x 3", 16" O.C.	1.000	L.F.	.51	2.04	2.55
Fiberglass insulation, 6" batts	1.000	S.F.	1.07	.53	1.60
Gypsum board, 1/2" thick	1.000	S.F.	.28	.80	1.08
Thin coat plaster	1.000	S.F.	.13	.90	1.03
Paint, roller, 2 coats	1.000	S.F.	.14	.80	.94
TOTAL			7.24	10.88	18.12

B1020 202	Wood Frame Roof & Ceiling	COST PER S.F.		
		MAT.	INST.	TOTAL
1800	Rafters 16" O.C., 2" x 6"	7.25	10.90	18.15
1900	2" x 8"	7.80	10.95	18.75
2000	2" x 10"	8.45	11.80	20.25
2100	2" x 12"	9.35	12.05	21.40
2200	Rafters 24" O.C., 2" x 6"	6.65	10.20	16.85
2300	2" x 8"	7.05	10.25	17.30
2400	2" x 10"	7.55	10.85	18.40
2500	2" x 12"	8.25	11.05	19.30
2600	Roof pitch, 6 in 12, add	3%	10%	
2700	8 in 12, add	5%	12%	
2800				
2900				
3000				
3100				
3300	Cut & patch to match existing construction, add, minimum	2%	3%	
3400	Maximum	5%	9%	
3500	Material handling & storage limitation, add, minimum	1%	1%	
3600	Maximum	6%	7%	
3700	Protection of existing work, add, minimum	2%	2%	
3800	Maximum	5%	7%	
3900	Shift work requirements, add, minimum		5%	
4000	Maximum		30%	

B1020 Roof Construction

5/8" Plywood Sheathing

Truss Anchors

Roof Trusses at 2'-0" o.c.

9" Fiberglass Insulation

1" x 8" Fascia Board

Wood Fabricated Roof Truss System

This page illustrates and describes a sloped roof truss framing system including trusses, truss anchors, sheathing, fascia and insulation. Lines B1020 210 1500 through 1800 are for a flat roof framing system including flat roof trusses, truss anchors, sheathing, fascia and insulation. Lines within system components give the unit price and total price per square foot for this system. Prices for alternate sizes and systems are given in line items B1020 210 0100 through 1800. Both material quantities and labor costs have been adjusted for the system listed.

Factors: To adjust for job conditions other than normal working situations use lines B1020 210 2700 through 4000.

Example: You are to install the system where crane placement and movement will be difficult. Go to line B1020 210 3200 and apply these percentages to the appropriate MAT. and INST. costs.

System Components	QUANTITY	UNIT	COST PER S.F.		
			MAT.	INST.	TOTAL
Roof truss, 2' O.C., 4/12 slope, 12' span, 5/8 sheath., fascia, insul.					
Timber connectors, rafter anchors, galv., 1 1/2" x 5 1/4"	.084	Ea.	.05	.41	.46
Plywood sheathing, CDX, 5/8" thick	1.054	S.F.	.95	1.16	2.11
Truss, 2' O.C., metal plate connectors, 12' span	.042	Ea.	2.33	2.21	4.54
Molding, fascia trim, 1" x 8"	.167	L.F.	.64	.52	1.16
Insulation, fiberglass, kraft face, 9" thick x 23" wide	1.000	S.F.	1.25	.53	1.78
TOTAL			5.22	4.83	10.05

B1020 210	Roof Truss, Wood	COST PER S.F.		
		MAT.	INST.	TOTAL
0100	Roof truss, 2'O.C., 4/12p, 1'ovhg, 12'span, 5/8"sheath, fascia, insul.	5.20	4.83	10.03
0200	20' span	6.25	4.65	10.90
0300	24' span	5.35	4.32	9.67
0400	26' span	8.95	6.05	15
0500	28' span	5.80	4.22	10.02
0600	30' span	6.10	4.17	10.27
0700	32' span	6.20	4.08	10.28
0800	34' span	6.15	4.03	10.18
0890	60' span	8.15	4.11	12.26
0900	8/12 pitch, 20' span	7.85	5.05	12.90
1000	24' span	7.65	4.65	12.30
1100	26' span	7.05	4.54	11.59
1200	28' span	8	4.53	12.53
1300	32' span	7.60	4.44	12.04
1400	36' span	7.70	4.34	12.04
1500	Flat roof frame, fab.strl.joists, 2' O.C., 15'-24' span, 50 PSF LL	7.85	3.27	11.12
1600	55 PSF LL	8.15	3.35	11.50
1700	24'-30'span, 45 PSF LL	8.90	3.13	12.03
1800	55 PSF LL	11.90	3.22	15.12
1900				
2700	Cut & patch to match existing construction, add, minimum	2%	3%	
2800	Maximum	5%	9%	
2900	Dust protection, add, minimum	1%	2%	
3000	Maximum	4%	11%	
3100	Equipment usage curtailment, add, minimum	1%	1%	
3200	Maximum	3%	10%	
3300	Material handling & storage limitation, add, minimum	1%	1%	
3400	Maximum	6%	7%	

B1020 Roof Construction

B1020 210	Roof Truss, Wood	COST PER S.F.		
		MAT.	INST.	TOTAL
3500	Protection of existing work, add, minimum	2%	2%	
3600	Maximum	5%	7%	
3700	Shift work requirements, add, minimum		5%	
3800	Maximum		30%	
3900	Temporary shoring and bracing, add, minimum	2%	5%	
4000	Maximum	5%	12%	

B2010 Exterior Walls

This page illustrates and describes a masonry concrete block system including concrete block wall, pointed, reinforcing, waterproofing, gypsum plaster on gypsum lath, and furring. Lines within System Components give the unit price and total price per square foot for this system. Prices for masonry concrete block wall systems are on Line Items B2010 108 1400 thru 2200. Both material quantities and labor costs have been adjusted for the system listed.

Factors: To adjust for job conditions other than normal working situations use Lines B2010 108 2700 thru 4000.

Example: You are to install the system and match existing construction at several locations. Go to Line B2010 108 2800 and apply these percentages to the appropriate MAT. and INST. costs.

System Components	QUANTITY	UNIT	COST PER S.F.		
			MAT.	INST.	TOTAL
Concrete block, 8″ thick, reinforced every 2 courses, waterproofing gypsum Plaster over gypsum lath on 1″ x 3″ furring, interior painting & baseboard.					
Concrete block, 8″ x 8″ x 16″, reinforced	1.000	S.F.	5.10	7.90	13
Silicone waterproofing, 2 coats	1.000	S.F.	.93	.23	1.16
Bituminous coating, 1/16″ thick	1.000	S.F.	.50	1.37	1.87
On concrete	1.000	L.F.	.56	2.75	3.31
Gypsum lath, 3/8″ thick	.110	S.Y.	.41	.90	1.31
Gypsum plaster, 2 coats	.110	S.Y.	.42	3.38	3.80
Painting, 2 coats	1.000	S.F.	.10	.62	.72
Baseboard wood, 9/16″ x 2-5/8″	.100	L.F.	.13	.30	.43
Paint baseboard, primer + 1 coat enamel	.100	L.F.	.01	.08	.09
TOTAL			8.16	17.53	25.69

B2010 108	Masonry Wall, Concrete Block	COST PER S.F.		
		MAT.	INST.	TOTAL
1400	8″ thick block, regular	8.15	17.50	25.65
1500	Fluted 2 sides	9.05	20.50	29.55
1600	Deep grooved	8.20	20.50	28.70
1700	Slump block	13.90	17.90	31.80
1800	Split rib	10.85	20.50	31.35
1900				
2000	12″ thick block, regular	7.85	21	28.85
2100	Slump block	18.30	19.50	37.80
2200	Split rib	11.95	23.50	35.45
2300				
2700	Cut & patch to match existing construction, add, minimum	2%	3%	
2800	Maximum	5%	9%	
2900	Dust protection, add, minimum	1%	2%	
3000	Maximum	4%	11%	
3100	Equipment usage curtailment, add, minimum	1%	1%	
3200	Maximum	3%	10%	
3300	Material handling & storage limitation, add, minimum	1%	1%	
3400	Maximum	6%	7%	
3500	Protection of existing work, add, minimum	2%	2%	
3600	Maximum	5%	7%	
3700	Shift work requirements, add, minimum		5%	
3800	Maximum		30%	
3900	Temporary shoring and bracing, add, minimum	2%	5%	
4000	Maximum	5%	12%	

B2010 Exterior Walls

This page illustrates and describes a masonry wall, brick-stone system including brick, concrete block, durawall, insulation, plasterboard, taped and finished, furring, baseboard and painting interior. Lines within System Components give the unit price and total price per square foot for this system. Prices for masonry wall, brick-stone systems are on Line Item B2010 129 1400 thru 2500. Both material quantities and labor costs have been adjusted for the system listed.

Factors: To adjust for job conditions other than normal working situations use Lines B2010 129 3100 thru 4200.

Example: You are to install the system without damaging the existing work. Go to Line B2010 129 3900 and apply these percentages to the appropriate MAT. and INST. costs.

System Components	QUANTITY	UNIT	COST PER S.F.		
			MAT.	INST.	TOTAL
Face brick, 4"thick, concrete block back-up, reinforce every second course, 3/4"insulation, furring, 1/2"drywall, taped, finish, and painted, baseboard					
Face brick, 4" brick	1.000	S.F.	5.60	14.90	20.50
Concrete back-up block, reinforced 8" thick	1.000	S.F.	3.05	8.30	11.35
3/4" rigid polystyrene insulation	1.000	S.F.	.80	.89	1.69
Furring, 1" x 3", wood, 16" O.C.	1.000	L.F.	.56	1.45	2.01
Drywall, 1/2" thick	1.000	S.F.	.28	.72	1
Taping & finishing	1.000	S.F.	.06	.72	.78
Painting, 2 coats	1.000	S.F.	.14	.80	.94
Baseboard, wood, 9/16" x 2-5/8"	.100	L.F.	.13	.30	.43
Paint baseboard, primer + 1 coat enamel	.100	L.F.	.01	.08	.09
TOTAL			10.63	28.16	38.79

B2010 131	Masonry Wall, Brick - Stone	COST PER S.F.		
		MAT.	INST.	TOTAL
1400	Face brick, standard, red, 4" x 2-2/3" x 8" (6.75/S.F.)	10.65	28	38.65
1500	Norman, 4" x 2-2/3" x 12" (4.5 per S.F.)	12.45	23.50	35.95
1600	Roman, 4" x 2" x 12" (6.0 per S.F.)	14	26.50	40.50
1700	Engineer, 4" x 3-1/5" x 8" (5.63 per S.F.)	9.40	26	35.40
1800	S.C.R., 6" x 2-2/3" x 12" (4.5 per S.F.)	11.80	24	35.80
1900	Jumbo, 6" x 4" x 12" (3.0 per S.F.)	11	21	32
2000	Norwegian, 6" x 3-1/5" x 12" (3.75 per S.F.)	13.55	22	35.55
2100				
2200				
2300	Stone, veneer, fieldstone, 6" thick	12.85	24	36.85
2400	Marble, 2" thick	56	42	98
2500	Limestone, 2" thick	38	25.50	63.50
3100	Cut & patch to match existing construction, add, minimum	2%	3%	
3200	Maximum	5%	9%	
3300	Dust protection, add, minimum	1%	2%	
3400	Maximum	4%	11%	
3500	Equipment usage curtailment, add, minimum	1%	1%	
3600	Maximum	3%	10%	
3700	Material handling & storage limitation, add, minimum	1%	1%	
3800	Maximum	6%	7%	
3900	Protection of existing work, add, minimum	2%	2%	
4000	Maximum	5%	7%	
4100	Shift work requirements, add, minimum		5%	
4200	Maximum		30%	

B2010 Exterior Walls

This page illustrates and describes a parapet wall system including wall, coping, flashing and cant strip. Lines within System Components give the unit price and cost per lineal foot for this system. Prices for parapet wall systems are on Lines B2010 142 1300 thru 2400. Both material quantities and labor costs have been adjusted for the system listed.

Factors: To adjust for job conditions other than normal working situations, use Lines B2010 142 2900 thru 4000.

Example: You are to install the system without damaging the adjacent property. Go to Line B2010 142 3600 and apply these percentages to the appropriate MAT. and INST. costs.

System Components	QUANTITY	UNIT	COST PER L.F.		
			MAT.	INST.	TOTAL
Concrete block parapet, incl. reinf., coping, flashing & cant strip, 2'high					
8" concrete block	2.000	S.F.	10.20	15.80	26
Masonry reinforcing	1.000	L.F.	.29	.24	.53
Coping, precast	1.000	L.F.	8.55	18.35	26.90
Roof cant	1.000	L.F.	.65	2.10	2.75
16 oz.	1.000	L.F.	9.55	3.18	12.73
Cap flashing	1.000	L.F.	15.55	5.95	21.50
TOTAL			44.79	45.62	90.41

B2010 142	Parapet Wall	COST PER L.F.		
		MAT.	INST.	TOTAL
1300	Concrete block, 8" thick	45	45.50	90.50
1400	12" thick	55	59.50	114.50
1500	Split rib block, 8" thick	50	51.50	101.50
1600	12" thick	56.50	61	117.50
1700	Brick, common brick, 8" wall	67	78.50	145.50
1800	12" wall	84.50	98.50	183
1900	4" brick, 4" backup block	50	75	125
2000	8" backup block	51.50	77	128.50
2100	Stucco on masonry	45	51	96
2200	On wood frame	31.50	41.50	73
2300	Wood, T 1-11 siding	49.50	49.50	99
2400	Boards, 1" x 6" cedar	60	55	115
2500				
2600				
2900	Dust protection, add, minimum	1%	2%	
3000	Maximum	4%	11%	
3100	Equipment usage curtailment, add, minimum	1%	1%	
3200	Maximum	3%	10%	
3300	Material handling & storage limitation, add, minimum	1%	1%	
3400	Maximum	6%	7%	
3500	Protection of existing work, add, minimum	2%	2%	
3600	Maximum	5%	7%	
3700	Shift work requirements, add, minimum		5%	
3800	Maximum		30%	
3900	Temporary shoring and bracing, add, minimum	2%	5%	
4000	Maximum	5%	12%	

This page illustrates and describes a masonry cleaning and restoration system, including staging, cleaning, repointing. Lines within System Components give the unit price and cost per square foot for this system. Prices for systems are on Lines B2010 143 1300 thru 2500. Both material quantities and labor costs have been adjusted for the system listed.

Factors: To adjust for conditions other than normal working situations, use Lines B2010 143 2900 thru 4200.

Example: You are to clean a wall and be concerned about dust control. Go to line B2010 143 3100 and apply these percentages to the appropriate MAT. and INST. costs.

System Components	QUANTITY	UNIT	COST PER S.F. MAT.	COST PER S.F. INST.	COST PER S.F. TOTAL
Repoint existing building, brick, common bond, high pressure cleaning, Water only, soft old mortar.					
Scaffolding, steel tubular, bldg ext wall face, labor only, 1 to 5 stories	.010	C.S.F.		2.68	2.68
Cleaning masonry, high pressure wash, water only, avg soil, bio staining	1.100	S.F.		2.04	2.04
Repoint, brick common bond	1.000	S.F.	.91	7.40	8.31
TOTAL			.91	12.12	13.03

B2010 143	Masonry Restoration - Cleaning	COST PER S.F. MAT.	COST PER S.F. INST.	COST PER S.F. TOTAL
1300	Brick, common bond	.91	12.10	13.01
1400	Flemish bond	.96	12.55	13.51
1500	English bond	.96	13.35	14.31
1700	Stone, 2' x 2' blocks	1.21	9.15	10.36
1800	2' x 4' blocks	.91	8.05	8.96
2000	Add to above prices for alternate cleaning systems:			
2100	Chemical brush and wash	.05	.81	.86
2200	High pressure chemical and water	.08	.69	.77
2300	Sandblasting, wet system	.47	1.94	2.41
2400	Dry system	.44	1.13	1.57
2500	Steam cleaning		.54	.54
2600				
2900	Cut & patch to match existing construction, add, minimum	2%	3%	
3000	Maximum	5%	9%	
3100	Dust protection, add, minimum	1%	2%	
3200	Maximum	4%	11%	
3300	Equipment usage curtailment, add, minimum	1%	1%	
3400	Maximum	3%	10%	
3500	Material handling & storage limitation, add, minimum	1%	1%	
3600	Maximum	6%	7%	
3900	Protection of existing work, add, minimum	2%	2%	
4000	Maximum	5%	7%	
4100	Shift work requirements, add, minimum		5%	
4200	Maximum		30%	

B2010 Exterior Walls

This page illustrates and describes a wood frame exterior wall system including wood studs, sheathing, felt, insulation, plasterboard, taped and finished, baseboard and painted interior. Lines within System Components give the unit price and total price per square foot for this system. Prices for wood frame exterior wall systems are on Line Items B2010 150 1600 thru 2700. Both material quantities and labor costs have been adjusted for the system listed.

Factors: To adjust for job conditions other than normal working situations use Lines B2010 150 3100 thru 4200.

Example: You are to install the system with need for complete temporary bracing. Go to Line B2010 150 4200 and apply these percentages to the appropriate MAT. and INST. costs.

System Components	QUANTITY	UNIT	COST PER S.F.		
			MAT.	INST.	TOTAL
Wood stud wall, cedar shingle siding, building paper, plywood sheathing, Insulation, 5/8″ drywall, taped, finished and painted, baseboard.					
2″ x 4″ wood studs, 16″ O.C.	.100	L.F.	.76	1.43	2.19
1/2″ CDX sheathing	1.000	S.F.	.84	1.02	1.86
18″ No. 1 red cedar shingles, 7-1/2″ exposure	.008	C.S.F.	3	2.56	5.56
15# felt paper	.010	C.S.F.	.05	.19	.24
3-1/2″ fiberglass insulation	1.000	S.F.	.77	.53	1.30
5/8″ drywall	1.000	S.F.	.29	.72	1.01
Drywall finishing adder	1.000	S.F.	.06	.72	.78
Baseboard trim, stock pine, 9/16″ x 3-1/2″, painted	.100	L.F.	.13	.30	.43
Paint baseboard, primer + 1 coat enamel	.100	L.F.	.01	.08	.09
Paint, 2 coats, interior	1.000	S.F.	.14	.80	.94
TOTAL			6.05	8.35	14.40

B2010 150	Wood Frame Exterior Wall	COST PER S.F.		
		MAT.	INST.	TOTAL
1600	Cedar shingle siding, painted	6.05	8.35	14.40
1700	Aluminum siding, horizontal clapboard	6.60	8.55	15.15
1800	Cedar bevel siding, 1/2″ x 6″, vertical , painted	8.10	8.20	16.30
1900	Redwood siding 1″ x 4″ to 1″ x 6″ vertical, T & G	8.75	9.05	17.80
2000	Board and batten	8.60	7.50	16.10
2100	Ship lap siding	9.05	7.95	17
2200	Plywood, grooved (T1-11) fir	4.67	7.90	12.57
2300	Redwood	6	7.90	13.90
2400	Southern yellow pine	4.88	7.90	12.78
2500	Masonry on stud wall, stucco, wire and plaster	4.12	12.65	16.77
2600	Stone veneer	10.85	16.55	27.40
2700	Brick veneer, brick $275 per M	8.65	20.50	29.15
3100	Cut & patch to match existing construction, add, minimum	2%	3%	
3200	Maximum	5%	9%	
3300	Dust protection, add, minimum	1%	2%	
3400	Maximum	4%	11%	
3500	Material handling & storage limitation, add, minimum	1%	1%	
3600	Maximum	6%	7%	
3700	Protection of existing work, add, minimum	2%	2%	
3800	Maximum	5%	7%	
3900	Shift work requirements, add, minimum		5%	
4000	Maximum		30%	
4100	Temporary shoring and bracing, add, minimum	2%	5%	
4200	Maximum	5%	12%	

B2010 Exterior Walls

B2010 190	Selective Price Sheet	COST PER S.F.		
		MAT.	INST.	TOTAL
0100	Exterior surface, masonry, concrete block, standard 4" thick	1.96	7.60	9.56
0200	6" thick	4.30	8.20	12.50
0300	8" thick	2.75	8.75	11.50
0400	12" thick	4.81	11.35	16.16
0500	Split rib, 4" thick	5.55	9.50	15.05
0600	8" thick	7.80	10.95	18.75
0700	Brick running bond, standard size, 6.75/S.F.	5.60	14.90	20.50
0800	Buff, 6.75/S.F.	5.90	14.90	20.80
0900	Stucco, on frame	1.08	6.95	8.03
1000	On masonry	.34	5.35	5.69
1100	Metal, aluminum, horizontal, plain	3.53	2.78	6.31
1200	Insulated	3.41	2.78	6.19
1300	Vertical, plain	2.97	2.78	5.75
1400	Insulated	3.10	2.90	6
1500	Wood, beveled siding, "A" grade cedar, 1/2" x 6"	5.05	2.43	7.48
1600	1/2" x 8"	8.60	2.17	10.77
1700	Shingles, 16" #1 red, 7-1/2" exposure	3.40	3.50	6.90
1800	18" perfections, 7-1/2" exposure	2.90	2.92	5.82
1900	Handsplit, 10" exposure	5.50	2.86	8.36
2000	White cedar, 7-1/2" exposure	1.26	3.60	4.86
2100	Vertical, board & batten, redwood	5.70	3.25	8.95
2200	White pine	2.76	2.17	4.93
2300	T. & G. boards, redwood, 1" x 4"	8	3.25	11.25
2400	1' x 8"	10.30	2.17	12.47
2500				
2600	Interior surface, drywall, taped & finished, standard, 1/2"	.33	1.48	1.81
2700	5/8" thick	.34	1.48	1.82
2800	Fire resistant, 1/2" thick	.56	1.48	2.04
2900	5/8" thick	.43	1.48	1.91
3000	Moisture resistant, 1/2" thick	.42	1.48	1.90
3100	5/8" thick	.53	1.48	2.01
3200	Core board, 1" thick	.80	2.98	3.78
3300	Plaster, gypsum, 2 coats	.42	3.41	3.83
3400	3 coats	.59	4.15	4.74
3500	Perlite or vermiculite, 2 coats	.58	3.92	4.50
3600	3 coats	.83	4.85	5.68
3700	Gypsum lath, standard, 3/8" thick	.42	.91	1.33
3800	1/2" thick	.36	.97	1.33
3900	Fire resistant, 3/8" thick	.31	1.11	1.42
4000	1/2" thick	.35	1.19	1.54
4100	Metal lath, diamond, 2.5 lb.	.62	.91	1.53
4200	Rib, 3.4 lb.	.62	1.11	1.73
4300	Framing metal studs including top and bottom			
4400	Runners, walls 10' high			
4500	24" O.C., non load bearing 20 gauge, 2-1/2" wide	.49	1.19	1.68
4600	3-5/8" wide	.92	1.21	2.13
4700	4" wide	.76	1.21	1.97
4800	6" wide	.82	1.23	2.05
4900	Load bearing 18 gauge, 2-1/2" wide	.63	1.34	1.97
5000	3-5/8" wide	.75	1.37	2.12
5100	4" wide	.81	1.74	2.55
5200	6" wide	.99	1.42	2.41
5300	16" O.C., non load bearing 20 gauge, 2-1/2" wide	.63	1.83	2.46
5400	3-5/8" wide	1.28	1.86	3.14
5500	4" wide	1.03	1.88	2.91
5600	6" wide	1.02	1.91	2.93
5700	Load bearing 18 gauge, 2-1/2" wide	.84	1.86	2.70
5800	3-5/8" wide	1	1.89	2.89

For customer support on your Commercial Renovation Costs with RSMeans data, call 800.448.8182.

B2010 Exterior Walls

B2010 190	Selective Price Sheet	COST PER S.F.		
		MAT.	**INST.**	**TOTAL**
5900	4" wide	1.02	1.94	2.96
6000	6" wide	1.33	1.96	3.29
6100	Framing wood studs incl. double top plate and			
6200	Single bottom plate, walls 10' high			
6300	24" O.C., 2" x 4"	.58	1.15	1.73
6400	2" x 6"	.85	1.25	2.10
6500	16" O.C., 2" x 4"	.76	1.43	2.19
6600	2" x 6"	1.12	1.59	2.71
6700	Sheathing, boards, 1" x 6"	2.32	2.20	4.52
6800	1" x 8"	2.37	1.87	4.24
6900	Plywood, 3/8" thick	.75	1.19	1.94
7000	1/2" thick	.84	1.27	2.11
7100	5/8" thick	.90	1.36	2.26
7200	3/4" thick	1.04	1.47	2.51
7300	Wood fiber, 5/8" thick	.76	1.19	1.95
7400	Gypsum weatherproof, 1/2" thick	.57	1.27	1.84
7500	Insulation, fiberglass batts, 3-1/2" thick, R13	.80	1.02	1.82
7600	6" thick, R19	1.09	1.19	2.28
7700	Poured 4" thick, fiberglass wool, R4/inch	.79	3.58	4.37
7800	Mineral wool, R3/inch	.74	3.58	4.32
7900	Polystyrene, R4/inch	1.96	3.58	5.54
8000	Perlite or vermiculite, R2.7/inch	3.47	3.58	7.05
8100	Rigid, fiberglass, R4.3/inch, 1" thick	.80	.72	1.52
8200	R8.7/inch, 2" thick	1.62	.80	2.42

B2020 Exterior Windows

| B2020 108 | | | Vinyl Clad Windows | | | | |

	MATERIAL	TYPE	GLAZING	SIZE	DETAIL	COST PER UNIT		
						MAT.	INST.	TOTAL
1000	Vinyl clad	casement	insul. glass	1'-4" x 4'-0"		465	165	630
1010				2'-0" x 3'-0"		480	165	645
1020				2'-0" x 4'-0"		560	173	733
1030				2'-0" x 5'-0"		640	183	823
1040				2'-0" x 6'-0"		770	183	953
1050				2'-4" x 4'-0"		555	183	738
1060				2'-6" x 5'-0"		705	178	883
1070				3'-0" x 5'-0"		600	183	783
1080				4'-0" x 3'-0"		1,100	183	1,283
1090				4'-0" x 4'-0"		925	183	1,108
1100				4'-8" x 4'-0"		920	183	1,103
1110				4'-8" x 5'-0"		1,050	213	1,263
1120				4'-8" x 6'-0"		1,125	213	1,338
1130				6'-0" x 4'-0"		1,125	213	1,338
1140				6'-0" x 5'-0"		1,175	213	1,388
3000		double-hung	insul. glass	2'-0" x 4'-0"		510	159	669
3050				2'-0" x 5'-0"		455	159	614
3100				2'-4" x 4'-0"		530	165	695
3150				2'-4" x 4'-8"		305	165	470
3200				2'-4" x 6'-0"		450	165	615
3250				2'-6" x 4'-0"		405	165	570
3300				2'-8" x 4'-0"		560	213	773
3350				2'-8" x 5'-0"		540	183	723
3375				2'-8" x 6'-0"		475	183	658
3400				3'-0 x 3'-6"		390	165	555
3450				3'-0 x 4'-0"		435	173	608
3500				3'-0 x 4'-8"		480	173	653
3550				3'-0 x 5'-0"		480	183	663
3600				3'-0 x 6'-0"		435	183	618
3700				4'-0 x 5'-0"		565	196	761
3800				4'-0 x 6'-0"		670	196	866

B2020 Exterior Windows

This page illustrates and describes an aluminum window system including double hung aluminum window, exterior and interior trim, hardware and insulating glass. Lines within System Components give the unit price and total price on a cost each basis for this system. Prices for aluminum window systems are on Line Items B2020 110 1000 thru 2400. Both material quantities and labor costs have been adjusted for the system listed.

Factors: To adjust for job conditions other than normal working situations use Lines B2020 110 3100 thru 4000.

Example: You are to install the system and cut and patch to match existing construction. Go to Line B2020 110 3200 and apply these percentages to the appropriate MAT. and INST. costs.

System Components	QUANTITY	UNIT	COST EACH MAT.	COST EACH INST.	COST EACH TOTAL
Double hung aluminum window, 2'-4" x 2'-6" exterior and interior trim, Hardware glazed with insulating glass.					
Double-hung aluminum window, 2'-4" x 2' 6"	1.000	Ea.	268.18	46.35	314.53
Interior Trim	1.000	Set	40	55	95
Hardware	1.000	Set	7.60	30	37.60
TOTAL			315.78	131.35	447.13

B2020 110	Windows - Aluminum	MAT.	INST.	TOTAL
1000	Aluminum, double hung, 2'-4" x 2'-6"	315	131	446
1100	2'-8" x 4'-6"	605	197	802
1200	3'- 4" x 5'-6"	945	295	1,240
1300	Casement, 3'-6" x 2'-4"	430	150	580
1400	4'-6" x 2'-4"	550	185	735
1500	5'-6" x 2'-4"	705	251	956
1600	Projected window, 2'-1" x 3'-0"	310	135	445
1700	3'-5" x 3'-0"	485	183	668
1800	4'-0" x 4'-0"	775	276	1,051
1900	Horizontal sliding 3'-0" x 3'-0"	282	157	439
2000	3'-6" x 4'-0"	420	213	633
2100	5'-0" x 6'-0"	880	390	1,270
2200	Picture window, 3'-8" x 3'-1"	290	175	465
2300	4'-4" x 4'-5"	465	254	719
2400	5'-8" x 4'-9"	680	365	1,045
2500				
2600				
2700				
2800				
2900				
3100	Cut & patch to match existing construction, add, minimum	2%	3%	
3200	Maximum	5%	9%	
3300	Dust protection, add, minimum	1%	2%	
3400	Maximum	4%	11%	
3500	Material handling & storage limitation, add, minimum	1%	1%	
3600	Maximum	6%	7%	
3700	Protection of existing work, add, minimum	2%	2%	
3800	Maximum	5%	7%	
3900	Shift work requirements, add, minimum		5%	
4000	Maximum		30%	

B2020 Exterior Windows

This page illustrates and describes a wood window system including wood picture window, exterior and interior trim, hardware and insulating glass. Lines within System Components give the unit price and total price on a cost each basis for this system. Prices for wood window systems are on Line Items B2020 112 1200 thru 2900. Both material quantities and labor costs have been adjusted for the system listed.

Factors: To adjust for job conditions other than normal working situations use Lines B2020 112 3100 thru 4000.

Example: You are to install the above system where dust control is vital. Go to Line B2020 112 3400 and apply these percentages to the appropriate MAT. and INST. costs.

System Components	QUANTITY	UNIT	COST EACH		
			MAT.	INST.	TOTAL
Wood picture window 4'-0" x 4'-6", exterior and interior trim, hardware, Glazed with insulating glass.					
4'-0" x 4'-6" wood picture window with insulating glass	1.000	Ea.	880	130	1,010
Oak interior trim	1.000	Set	94.50	119	213.50
Hardware	1.000	Set	7.60	30	37.60
TOTAL			982.10	279	1,261.10

B2020 112	Windows - Wood	COST EACH		
		MAT.	INST.	TOTAL
1200	Picture window, 4'-0" x 4'-6"	980	279	1,259
1300	5'-0" x 4'-0"	975	279	1,254
1400	6'-0" x 4'-6"	1,025	292	1,317
1500	Bow, bay window, 8'-0" x 5'-0", standard	2,925	292	3,217
1600	Deluxe	2,150	292	2,442
1700	Bow, bay window, 12'-0" x 6'-0" standard	4,250	390	4,640
1800	Deluxe	3,000	390	3,390
1900	Double hung, 3'-0" x 4'-0"	315	181	496
2000	4'-0" x 4'-6"	405	239	644
2100	Casement 2'-0" x 3'-0"	445	157	602
2200	2 leaf, 4'-0" x 4'-0"	985	261	1,246
2300	3 leaf, 6'-0" x 6'-0"	3,075	420	3,495
2400	Awning, 2'-10" x 1'-10"	525	157	682
2500	3'-6" x 2'-4"	560	181	741
2600	4'-0" x 3'-0"	830	239	1,069
2700	Horizontal sliding 3'-0" x 2'-0"	870	157	1,027
2800	4'-0" x 3'-6"	675	181	856
2900	6'-0" x 5'-0"	965	239	1,204
3100	Cut & patch to match existing construction, add, minimum	2%	3%	
3200	Maximum	5%	9%	
3300	Dust protection, add, minimum	1%	2%	
3400	Maximum	4%	11%	
3500	Material handling & storage limitation, add, minimum	1%	1%	
3600	Maximum	6%	7%	
3700	Protection of existing work, add, minimum	2%	2%	
3800	Maximum	5%	7%	
3900	Shift work requirements, add, minimum		5%	
4000	Maximum		30%	

This page illustrates and describes storm window and door systems based on a cost each price. Prices for storm window and door systems are on Line Items B2020 116 0400 thru 2400. Both material quantities and labor costs have been adjusted for the system listed.

Factors: To adjust for job conditions other than normal working situations use Lines B2020 116 3100 thru 4000 and apply these percentages to the appropriate MAT and INST. costs.

Example: You are to install the system and protect all existing construction. Go to Line B2020 116 3800 and apply these percentages to the appropriate MAT. and INST. costs.

B2020 116	Storm Windows & Doors	COST EACH		
		MAT.	INST.	TOTAL
0350	Storm Window and door systems, single glazed			
0400	Window, custom aluminum anodized, 2'-0" x 3'-5"	135	47.50	182.50
0500	2'-6" x 5'-0"	197	51	248
0600	4'-0" x 6'-0"	350	57.50	407.50
0700	White painted aluminum, 2'-0" x 3'-5"	165	47.50	212.50
0800	2'-6" x 5'-0"	280	51	331
0900	4'-0" x 6'-0"	430	57.50	487.50
1000	Average quality aluminum, anodized, 2'-0" x 3'-5"	150	47.50	197.50
1100	2'-6" x 5'-0"	186	51	237
1200	4'-0" x 6'-0"	191	57.50	248.50
1300	White painted aluminum, 2'-0" x 3'-5"	155	47.50	202.50
1400	2'-6" x 5'-0"	263	51	314
1500	4'-0" x 6'-0"	174	57.50	231.50
1600	Mill finish, 2'-0" x 3'-5"	124	47.50	171.50
1700	2'-6" x 5'-0"	156	51	207
1800	4'-0" x 6'-0"	197	57.50	254.50
1900				
2000	Door, aluminum anodized 3'-0" x 6'-8"	216	102	318
2100	White painted aluminum, 3'-0" x 6'-8"	330	102	432
2200	Mill finish, 3'-0" x 6'-8"	300	102	402
2300				
2400	Wood, storm and screen, painted	375	159	534
2500				
2600				
2700				
2800				
3100	Cut & patch to match existing construction, add, minimum	2%	3%	
3200	Maximum	5%	9%	
3300	Dust protection, add, minimum	1%	2%	
3400	Maximum	4%	11%	
3500	Material handling & storage limitation, add, minimum	1%	1%	
3600	Maximum	6%	7%	
3700	Protection of existing work, add, minimum	2%	2%	
3800	Maximum	5%	7%	
3900	Shift work requirements, add, minimum		5%	
4000	Maximum		30%	

B20 Exterior Enclosure

B2020 Exterior Windows

This page illustrates and describes a window wall system including aluminum tube framing, caulking, and glass. Lines within System Components give the unit price and total price per square foot for this system. Prices for window wall systems are on Line Items B2020 124 1300 thru 2600. Both material quantities and labor costs have been adjusted for the system listed.

Factors: To adjust for job conditions other than normal working situations use Lines B2020 124 3100 thru 4000.

Example: You are to install the system with need for complete dust protection. Go to Line B2020 124 3400 and apply these percentages to the appropriate MAT. and INST. costs.

System Components	QUANTITY	UNIT	COST PER S.F.		
			MAT.	INST.	TOTAL
Window wall, including aluminum header, sill, mullions, caulking And glass					
Header, mill finish, 1-3/4" x 4-1/2" deep	.167	L.F.	3.92	2.76	6.68
Sill, mill finish, 1-3/4" x 4-1/2" deep	.167	L.F.	3.92	2.69	6.61
Vertical mullion, 1-3/4" x 4-1/2" deep, 6' O.C.	.191	L.F.	10.02	6.71	16.73
Caulking	.381	L.F.	.16	.94	1.10
Glass, 1/4" plate	1.000	S.F.	11.10	11.40	22.50
TOTAL			29.12	24.50	53.62

B2020 124	Aluminum Frame, Window Wall	COST PER S.F.		
		MAT.	INST.	TOTAL
1300	Mill finish, 1-3/4" x 4-1/2" deep, 1/4" plate glass	29	24.50	53.50
1400	2" x 4-1/2" deep, insulating glass	43.50	28.50	72
1500	Thermo-break, 2-1/4" x 4-1/2" deep, insulating glass	45	29	74
1600	Bronze finish, 1-3/4" x 4-1/2" deep, 1/4" plate glass	32.50	26.50	59
1700	2" x 4-1/2" deep, insulating glass	46	30.50	76.50
1800	Thermo-break, 2-1/4" x 4-1/2" deep, insulating glass	47.50	31.50	79
2000	Black finish, 1-3/4" x 4-1/2" deep, 1/4" plate glass	34	28	62
2100	2" x 4-1/2" deep, insulating glass	47	32	79
2200	Thermo-break, 2-1/4" x 4-1/2" deep, insulating glass	48.50	32.50	81
2300				
2400	Stainless steel, 1-3/4" x 4-1/2" deep, 1/4" plate glass	42.50	33.50	76
2500	2" x 4-1/2" deep, insulating glass	53.50	38	91.50
2600	Thermo-break, 2-1/4" x 4-1/2" deep, insulating glass	55.50	39	94.50
2700				
3100	Cut & patch to match existing construction, add, minimum	2%	3%	
3200	Maximum	5%	9%	
3300	Dust protection, add, minimum	1%	2%	
3400	Maximum	4%	11%	
3500	Material handling, & storage limitation, add, minimum	1%	1%	
3600	Maximum	6%	7%	
3700	Protection of existing work, add, minimum	2%	2%	
3800	Maximum	5%	7%	
3900	Shift work requirements, add, minimum		5%	
4000	Maximum		30%	

B2030 Exterior Doors

This page illustrates and describes a commercial metal door system, including a single aluminum and glass door, narrow stiles, jamb, hardware weatherstripping, panic hardware and closer. Lines within System Components give the unit price and total price on a cost each basis for this system. Prices for commercial metal door systems are on Line Items B2030 125 1200 thru 2500. Both material quantities and labor costs have been adjusted for the system listed.

Factors: To adjust for job conditions other than normal working situations, use Lines B2030 125 3100 thru 4000.

Example: You are to install the system and cut and patch to match existing construction. Go to Line B2030 125 3200 and apply these percentages to the appropriate MAT. and INST. costs.

System Components	QUANTITY	UNIT	COST EACH MAT.	COST EACH INST.	COST EACH TOTAL
Single aluminum and glass door, 3'-0"x7'-0", with narrow stiles, ext. jamb. Weatherstripping, 1/2" tempered insul. glass, panic hardware, and closer.					
Aluminum door, 3'-0" x 7'-0" x 1-3/4", narrow stiles	1.000	Ea.	1,225	795	2,020
Tempered insulating glass, 1/2" thick	20.000	S.F.	710	500	1,210
Panic hardware	1.000	Set	1,150	119	1,269
Surface mount regular arm	1.000	Ea.	170	110	280
TOTAL			3,255	1,524	4,779

B2030 125	Doors, Metal - Commercial	COST EACH MAT.	COST EACH INST.	COST EACH TOTAL
1200	Single aluminum and glass, 3'-0" x 7'-0"	3,250	1,525	4,775
1300	With transom, 3'-0" x 10'-0"	3,825	1,825	5,650
1400	Anodized aluminum and glass, 3'-0" x 7'-0"	3,675	1,800	5,475
1500	With transom, 3'-0" x 10'-0"	4,350	2,125	6,475
1600	Steel, deluxe, hollow metal 3'-0" x 7'-0"	2,500	595	3,095
1700	With transom 3'-0" x 10'-0"	3,100	680	3,780
1800	Fire door, "A" label, 3'-0" x 7'-0"	2,800	595	3,395
1900	Double, aluminum and glass, 6'-0" x 7'-0"	4,725	2,625	7,350
2000	With transom, 6'-0" x 10'-0"	5,900	3,250	9,150
2100	Anodized aluminum and glass 6'-0" x 7'-0"	5,150	3,050	8,200
2200	With transom, 6'-0" x 10'-0"	6,525	3,750	10,275
2300	Steel, deluxe, hollow metal, 6'-0" x 7'-0"	4,375	1,025	5,400
2400	With transom, 6'-0" x 10'-0"	5,575	1,200	6,775
2500	Fire door, "A" label, 6'-0" x 7'-0"	4,625	1,025	5,650
2800				
2900				
3100	Cut & patch to match existing construction, add, minimum	2%	3%	
3200	Maximum	5%	9%	
3300	Dust protection, add, minimum	1%	2%	
3400	Maximum	4%	11%	
3500	Material handling & storage limitation, add, minimum	1%	1%	
3600	Maximum	6%	7%	
3700	Protection of existing work, add, minimum	2%	2%	
3800	Maximum	5%	7%	
3900	Shift work requirements, add, minimum		5%	
4000	Maximum		30%	

B2030 Exterior Doors

This page illustrates and describes sliding door systems including a sliding door, frame, interior and exterior trim with exterior staining. Lines within System Components give the unit price and total price on a cost each basis for this system. Prices for sliding door systems are on Line Items B2030 150 1000 thru 2400. Both material quantities and labor costs have been adjusted for the system listed.

Factors: To adjust for job conditions other than normal working situations use Lines B2030 150 2700 thru 4000.

Example: You are to install the system with temporary shoring and bracing. Go to Line B2030 150 3900 and apply these percentages to the appropriate MAT. and INST. costs.

System Components	QUANTITY	UNIT	COST EACH		
			MAT.	INST.	TOTAL
Sliding wood door, 6'-0" x 6'-8", with wood frame, interior and exterior Trim and exterior staining.					
Sliding wood door, standard, 6'-0" x 6'-8", insulated glass	1.000	Ea.	1,300	360	1,660
Interior & exterior trim	1.000	Set	31.80	57.20	89
Stain door	1.000	Ea.	4.01	35.50	39.51
Stain trim	1.000	Ea.	2.20	15.20	17.40
TOTAL			1,338.01	467.90	1,805.91

B2030 150	Doors, Sliding - Patio	COST EACH		
		MAT.	INST.	TOTAL
1000	Wood, standard, 6'-0" x 6'-8", insulated glass	1,350	470	1,820
1100	8'-0" x 6'-8"	1,600	590	2,190
1200	12'-0" x 6'-8"	2,900	715	3,615
1300	Vinyl coated, 6'-0" x 6'-8"	1,700	470	2,170
1400	8'-0" x 6'-8"	2,175	590	2,765
1500	12'-0" x 6'-8"	3,475	715	4,190
1700	Aluminum, standard, 6'-0" x 6'-8", insulated glass	1,050	505	1,555
1800	8'-0" x 6'-8"	1,775	635	2,410
1900	12'-0" x 6'-8"	1,800	755	2,555
2000	Anodized, 6'-0" x 6'-8"	1,750	505	2,255
2100	8'-0" x 6'-8"	2,000	635	2,635
2200	12'-0" x 6'-8"	3,425	755	4,180
2300				
2400	Deduct for single glazing	37.50		37.50
2500				
2700	Cut & patch to match existing construction, add, minimum	2%	3%	
2800	Maximum	5%	9%	
2900	Dust protection, add, minimum	1%	2%	
3000	Maximum	4%	11%	
3100	Equipment usage curtailment, add, minimum	1%	1%	
3200	Maximum	3%	10%	
3300	Material handling & storage limitation, add, minimum	1%	1%	
3400	Maximum	6%	7%	
3500	Protection of existing, work, add, minimum	2%	2%	
3600	Maximum	5%	7%	
3700	Shift work requirements, add, minimum		5%	
3800	Maximum		30%	
3900	Temporary shoring and bracing, add, minimum	2%	5%	
4000	Maximum	5%	12%	

B2030 Exterior Doors

This page illustrates and describes residential door systems including a door, frame, trim, hardware, weatherstripping, stained and finished. Lines within System Components give the unit price and total price on a cost each basis for this system. Prices for residential door systems are on Line Items B2030 240 1500 thru 2200. Both material quantities and labor costs have been adjusted for the system listed.

Factors: To adjust for job conditions other than normal working situations use Lines B2030 240 3100 thru 4000.

Example: You are to install the system with a material handling and storage limitation. Go to Line B2030 240 3600 and apply these percentages to the appropriate MAT. and INST. costs.

System Components	QUANTITY	UNIT	COST EACH MAT.	COST EACH INST.	COST EACH TOTAL
Single solid wood colonial door, 3' x 6'-8" with wood frame and trim,					
Stained and finished, including hardware and weatherstripping.					
Solid wood colonial door, fir 3' x 6'-8" x 1-3/4", hinges	1.000	Ea.	186	95.50	281.50
Hinges, ball bearing	1.000	Ea.	16		16
Exterior frame, with trim	1.000	Set	162.35	64.94	227.29
Interior trim	1.000	Set	26.50	59.50	86
Average quality	1.000	Set	160	51	211
Sill, oak 8" deep	3.500	Ea.	85.75	50.05	135.80
Weatherstripping	1.000	Set	24.50	94	118.50
Stained and finished	1.000	Ea.	8.55	75.50	84.05
TOTAL			669.65	490.49	1,160.14

B2030 240	Doors, Residential - Exterior	COST EACH MAT.	COST EACH INST.	COST EACH TOTAL
1400	Single doors, 3' x 6'-8"			
1500	Solid colonial door, fir	670	490	1,160
1600	Hollow metal exterior door, plain	1,100	460	1,560
1700	Solid core wood door, plain	1,050	470	1,520
1800				
1900	Double doors, 6' x 6'-8"			
2000	Solid colonial double doors, fir	965	650	1,615
2100	Hollow metal exterior doors, plain	1,800	610	2,410
2200	Hollow core wood doors, plain	1,700	630	2,330
2300				
2800				
2900				
3100	Cut & patch to match existing construction, add, minimum	2%	3%	
3200	Maximum	5%	9%	
3300	Dust protection, add, minimum	1%	2%	
3400	Maximum	4%	11%	
3500	Material handling & storage limitation, add, minimum	1%	1%	
3600	Maximum	6%	7%	
3700	Protection of existing work, add, minimum	2%	2%	
3800	Maximum	5%	7%	
3900	Shift work requirements, add, minimum		5%	
4000	Maximum		30%	

B2030 Exterior Doors

This page illustrates and describes overhead door systems including an overhead door, track, hardware, trim, and electric door opener. Lines within System Components give the unit price and total price on a cost each basis for this system. Prices for overhead door systems are on Line Items B2030 410 1200 thru 2300. Both material quantities and labor costs have been adjusted for the system listed.

Factors: To adjust for job conditions other than normal working situations use Lines B2030 410 3100 thru 4000.

Example: You are to install the system and match existing construction. Go to Line B2030 410 3200 and apply these percentages to the appropriate MAT. and INST. costs.

System Components	QUANTITY	UNIT	COST EACH		
			MAT.	INST.	TOTAL
Wood overhead door, commercial sectional, including track, door, hardware					
And trim, electrically operated.					
Commercial, heavy duty wood door, 8' x 8' x 1-3/4" thick	1.000	Ea.	1,325	715	2,040
Frame, 2 x 8, pressure treated	1.000	Set	42.48	90.48	132.96
Wood trim	1.000	Set	31.68	71.52	103.20
Painting, two coats	1.000	Ea.	41.76	241.92	283.68
Electric trolley operator	1.000	Ea.	955	360	1,315
TOTAL			2,395.92	1,478.92	3,874.84

B2030 410	Doors, Overhead	COST EACH		
		MAT.	INST.	TOTAL
1200	Commercial, wood, 1-3/4" thick, 8' x 8'	2,400	1,475	3,875
1300	12' x 12'	4,200	2,075	6,275
1400	14' x 14'	5,350	2,450	7,800
1500	Fiberglass & aluminum 12' x 12'	4,900	2,125	7,025
1600	20' x 20'	9,850	5,375	15,225
1700	Residential, wood, 9' x 7'	2,625	560	3,185
1800	16' x 7'	4,000	835	4,835
1900	Hardboard faced, 9' x 7'	900	560	1,460
2000	16' x 7'	1,625	835	2,460
2100	Fiberglass & aluminum, 9' x 7'	1,150	670	1,820
2200	16' x 7'	2,550	835	3,385
2300	For residential electric opener, add	705	89.50	794.50
2400				
2500				
2600				
2700				
2800				
2900				
3100	Cut & patch to match existing construction, add, minimum	2%	3%	
3200	Maximum	5%	9%	
3300	Dust protection, add, minimum	1%	2%	
3400	Maximum	4%	11%	
3500	Material handling & storage limitation, add, minimum	1%	1%	
3600	Maximum	6%	7%	
3700	Protection of existing work, add, minimum	2%	2%	
3800	Maximum	5%	7%	
3900	Shift work requirements, add, minimum		5%	
4000	Maximum		30%	

B2030 Exterior Doors

B2030 810	Selective Price Sheet	COST EACH		
		MAT.	INST.	TOTAL
0100	Door closer, rack and pinion	184	119	303
0200	Backcheck and adjustable power	184	119	303
0300	Regular, hinge face mount, all sizes, regular arm	184	119	303
0400	Hold open arm	112	119	231
0500	Top jamb mount, all sizes, regular arm	184	119	303
0600	Hold open arm	112	119	231
0700	Stop face mount, all sizes, regular arm	184	119	303
0800	Hold open arm	112	119	231
0900	Fusible link, hinge face mount, all sizes, regular arm	395	119	514
1000	Hold open arm	325	119	444
1100	Top jamb mount, all sizes, regular arm	395	119	514
1200	Hold open arm	325	119	444
1300	Stop face mount, all sizes, regular arm	395	119	514
1400	Hold open arm	325	119	444
1500				
1600				
1700	Door stops			
1800				
1900	Holder & bumper, floor or wall	35.50	22.50	58
2000	Wall bumper	6.20	22.50	28.70
2100	Floor bumper	6.80	22.50	29.30
2200	Plunger type, door mounted	43.50	22.50	66
2300	Hinges, full mortise, material only, per pair			
2400	Low frequency, 4-1/2″ x 4-1/2″, steel base, USP	3.77		3.77
2500	Brass base, US10	103		103
2600	Stainless steel base, US32	88.50		88.50
2700	Average frequency, 4-1/2″ x 4-1/2″, steel base, USP	38		38
2800	Brass base, US10	54.50		54.50
2900	Stainless steel base, US32	61		61
3000	High frequency, 4-1/2″ x 4-1/2″, steel base, USP	71.50		71.50
3100	Brass base, US10	66.50		66.50
3200	Stainless steel base, US32	61		61
3300	Kick plate			
3400				
3500	6″ high, for 3′-0″ door, aluminum	41	47.50	88.50
3600	Bronze	74	47.50	121.50
3700	Panic device			
3800				
3900	For rim locks, single door, exit	1,150	119	1,269
4000	Outside key and pull	1,575	143	1,718
4100	Bar and vertical rod, exit only	1,600	143	1,743
4200	Outside key and pull	1,800	179	1,979
4300	Lockset			
4400				
4500	Heavy duty, cylindrical, passage doors	325	59.50	384.50
4600	Classroom	525	89.50	614.50
4700	Bedroom, bathroom, and inner office doors	217	59.50	276.50
4800	Apartment, office, and corridor doors	288	71.50	359.50
4900	Standard duty, cylindrical, exit doors	196	71.50	267.50
5100	Passage doors	158	59.50	217.50
5200	Public restroom, classroom, & office doors	292	89.50	381.50
5300	Deadlock, mortise, heavy duty	266	79.50	345.50
5400	Double cylinder	172	79.50	251.50
5500	Entrance lock, cylinder, deadlocking latch	202	79.50	281.50
5600	Deadbolt	259	89.50	348.50
5700	Commercial, mortise, wrought knob, keyed, minimum	187	89.50	276.50
5800	Maximum	460	102	562
5900	Cast knob, keyed, minimum	300	79.50	379.50

B2030 Exterior Doors

B2030 810	Selective Price Sheet	COST EACH		
		MAT.	INST.	TOTAL
6000	Maximum	610	79.50	689.50
6100	Push-pull			
6200				
6300	Aluminum	10.95	59.50	70.45
6400	Bronze	25	59.50	84.50
6500	Door pull, designer style, minimum	25.50	59.50	85
6600	Maximum	49	130	179
6700	Threshold			
6800				
6900	3'-0" long door saddles, aluminum, minimum	13.30	14.90	28.20
7000	Maximum	66	59.50	125.50
7100	Bronze, minimum	40.50	11.95	52.45
7200	Maximum	151	59.50	210.50
7300	Rubber, 1/2" thick, 5-1/2" wide	39	36	75
7400	2-3/4" wide	28	36	64
7500	Weatherstripping, per set			
7600				
7700	Doors, wood frame, interlocking for 3' x 7' door, zinc	59	239	298
7800	Bronze	76.50	239	315.50
7900	Wood frame, spring type for 3' x 7' door, bronze	24.50	94	118.50
8000	Metal frame, spring type for 3' x 7' door, bronze	56	239	295
8100	For stainless steel, spring type, add	133%		
8200				
8300	Metal frame, extruded sections, 3' x 7' door, aluminum	36.50	239	275.50
8400	Bronze	105	239	344

B3010 Roof Coverings

B3010 160	Selective Price Sheet	COST PER S.F.		
		MAT.	INST.	TOTAL
0100	Roofing, built-up, asphalt roll roof, 3 ply organic/mineral surface	.79	1.88	2.67
0200	3 plies glass fiber felt type iv, 1 ply mineral surface	1.36	2.03	3.39
0300	Cold applied, 3 ply		.66	.66
0400	Coal tar pitch, 4 ply tarred felt	2.44	2.42	4.86
0500	Mopped, 3 ply glass fiber	2.01	2.67	4.68
0600	4 ply organic felt	2.44	2.42	4.86
0700	Elastomeric, hypalon, neoprene unreinforced	1.48	4.47	5.95
0800	Polyester reinforced	1.66	5.30	6.96
0900	Neoprene, 5 coats 60 mils	4.95	15.60	20.55
1000	Over 10,000 S.F.	4.95	8.10	13.05
1600				
1700	Asphalt, strip, 210-235#/sq.	1.06	1.24	2.30
1800	235-240#/sq.	1.39	1.37	2.76
1900	Class A laminated	1.35	1.52	2.87
2000	Class C laminated	1.35	1.71	3.06
2100	Slate, buckingham, 3/16" thick	7.60	3.90	11.50
2200	Black, 1/4" thick	7.60	3.90	11.50
2300	Wood, shingles, 16" no. 1, 5" exp.	5.10	2.86	7.96
2400	Red cedar, 18" perfections	5.10	2.60	7.70
2500	Shakes, 24", 10" exposure	5.50	2.86	8.36
2600	18", 8-1/2" exposure	4.55	3.60	8.15
2700	Insulation, ceiling batts, fiberglass, 3-1/2" thick, R13	.66	.53	1.19
2800	6" thick, R19	.95	.53	1.48
2900	9" thick, R30	1.25	.62	1.87
3000	12" thick, R38	1.58	.62	2.20
3100	Mineral fiber, 3-1/2" thick, R13	1.14	.45	1.59
3200	6" thick, R19	1.80	.45	2.25
3300	Roof deck, fiberboard, 1" thick, R2.78	.98	.88	1.86
3400	Mineral fiber, 2" thick, R5.26	1.36	.88	2.24
3500	Perlite boards, 3/4" thick, R2.08	.56	.88	1.44
3600	2" thick, R5.26	1.33	.97	2.30
3700	Polystyrene extruded, 15 PSI, 1" thick, R5.26	.76	.57	1.33
3800	2" thick, R10	.95	.64	1.59
3900	40 PSI, 1" thick, R5	1.09	.57	1.66
4000	Tapered for drainage	1.10	.60	1.70
4300	Ceiling, plaster, gypsum, 2 coats	.42	3.92	4.34
4400	3 coats	.59	4.62	5.21
4500	Perlite or vermiculite, 2 coats	.58	4.56	5.14
4600	3 coats	.83	5.70	6.53
4700	Gypsum lath, plain 3/8" thick	.42	.91	1.33
4800	1/2" thick	.36	.97	1.33
4900	Firestop, 3/8" thick	.31	1.11	1.42
5000	1/2" thick	.35	1.19	1.54
5100	Metal lath, rib, 2.75 lb.	.49	1.03	1.52
5200	3.40 lb.	.62	1.11	1.73
5300	Diamond, 2.50 lb.	.62	1.03	1.65
5400	3.40 lb.	.51	1.29	1.80
5500	Drywall, taped and finished standard, 1/2" thick	.33	1.87	2.20
5600	5/8" thick	.34	1.87	2.21
5700	Fire resistant, 1/2" thick	.56	1.87	2.43
5800	5/8" thick	.43	1.87	2.30
5900	Water resist., 1/2" thick	.42	1.87	2.29
6000	5/8" thick	.53	1.87	2.40
6100	Finish, instead of taping			
6200	For thin coat plaster, add	.13	.90	1.03
6300	Finish textured spray, add	.06	.87	.93
6500	Tile, stapled glued, mineral fiber plastic coated, 5/8" thick	4.75	2.39	7.14
6600	3/4" thick	5.15	2.39	7.54

B3010 Roof Coverings

B3010 160	Selective Price Sheet	COST PER S.F.		
		MAT.	INST.	TOTAL
6700	Wood fiber, 1/2" thick	1.25	1.79	3.04
6800	3/4" thick	1.77	1.79	3.56
6900	Suspended, fiberglass film faced, 5/8" thick	5.05	1.14	6.19
7000	3" thick	4.61	1.59	6.20
7100	Mineral fiber 5/8" thick, standard face	1.38	1.06	2.44
7200	Aluminum faced	4.22	1.19	5.41
7500	Ceiling suspension systems, for tile, "T" bar, class "A", 2' x 4' grid	1.66	.89	2.55
7600	2' x 2' grid	2.13	1.10	3.23
7700	Concealed "Z" bar, 12" module	1.03	1.38	2.41
7800				
7900	Plaster/drywall, 3/4" channels, steel furring, 16" O.C.	.43	2.41	2.84
8000	24" O.C.	.29	1.66	1.95
8100	1-1/2" channels, 16" O.C.	.58	2.69	3.27
8200	24" O.C.	.39	1.79	2.18
8300	Ceiling framing, 2" x 4" studs, 16" O.C.	.54	1.19	1.73
8400	24" O.C.	.36	.80	1.16

B30 Roofing

B3020 Roof Openings

This page illustrates and describes a roof hatch system. Lines within System Components give the unit price and total cost each for this system. Prices for systems are on Lines B3020 230 1400 thru 2400. Both material quantities and labor costs have been adjusted for the system listed.

Factors: To adjust for job conditions other than normal working situtations use Lines B3020 230 2900 thru 4000.

Example: You are to install the system and cut and patch to match existing construction. Use line B3020 230 3000 and apply these percentages to the appropriate MAT. and INST. costs.

System Components	QUANTITY	UNIT	COST EACH		
			MAT.	INST.	TOTAL
Roof hatch, 2'-6" x 3'-0", aluminum, curb and cover included.					
Through steel construction.					
Roof hatch, 2'-6" x 3'-0", aluminum	1.000	Ea.	1,250	284	1,534
Cutout decking	11.000	L.F.	12.54	28.82	41.36
Frame opening	44.000	L.F.	47.52	427.24	474.76
Flashing, 16 oz. copper	16.000	S.F.	248.80	95.20	344
Cant strip, 4 x 4, mineral fiber	12.000	L.F.	7.80	25.20	33
TOTAL			1,566.66	860.46	2,427.12

B3020 230	Roof Hatches, Skylights	COST EACH		
		MAT.	INST.	TOTAL
1400	Roof hatch, aluminum, curb and cover, 2'-6" x 3'-0"	1,575	860	2,435
1500	2'-6" x 4'-6"	1,925	1,050	2,975
1600	2'-6" x 8'-0"	3,025	1,500	4,525
1700	Skylight, plexiglass dome, with curb mounting, 30" x 32"	855	815	1,670
1800	30" x 45"	1,025	1,025	2,050
1900	40" x 45"	1,250	1,350	2,600
2000	Smoke hatch, unlabeled, 2'-6" x 3'-0"	1,875	930	2,805
2100	2'-6" x 4'-6"	2,300	1,125	3,425
2200	2'-6" x 8'-0"	3,275	1,550	4,825
2300				
2400	For steel ladder, add per vertical linear foot	59.50	40	99.50
2450				
2600				
2700				
2900	Cut & patch to match existing construction, add, minimum	2%	3%	
3000	Maximum	5%	9%	
3100	Dust protection, add, minimum	1%	2%	
3200	Maximum	4%	11%	
3300	Equipment usage curtailment, add, minimum	1%	1%	
3400	Maximum	3%	10%	
3500	Material handling & storage limitation, add, minimum	1%	1%	
3600	Maximum	6%	7%	
3700	Protection of existing work, add, minimum	2%	2%	
3800	Maximum	5%	7%	
3900	Shift work requirements, add, minimum		5%	
4000	Maximum		30%	

B3020 Roof Openings

B3020 240	Selective Price Sheet	COST PER UNIT		
		MAT.	INST.	TOTAL
0100	Downspouts per L.F., aluminum, enameled .024" thick, 2" x 3"	1.54	4.67	6.21
0200	3" x 4"	3.19	6	9.19
0300	Round .025" thick, 3" diam.	5.30	4.43	9.73
0400	4" diam.	9.20	6	15.20
0500	Copper, round 16 oz. stock, 2" diam.	10.65	4.43	15.08
0600	3" diam.	10.65	4.43	15.08
0700	4" diam.	14.35	5.80	20.15
0800	5" diam.	21	6.45	27.45
0900	Rectangular, 2" x 3"	9.85	4.43	14.28
1000	3" x 4"	10.90	5.80	16.70
1100	Lead coated copper, round, 2" diam.	38.50	4.43	42.93
1200	3" diam.	36.50	4.43	40.93
1300	4" diam.	41	5.80	46.80
1400	5" diam.	21	6.45	27.45
1500	Rectangular, 2" x 3"	38	4.43	42.43
1600	3" x 4"	60.50	5.80	66.30
1700	Steel galvanized, round 28 gauge, 3" diam.	3.80	4.43	8.23
1800	4" diam.	5.15	5.80	10.95
1900	5" diam.	5.65	6.45	12.10
2000	6" diam.	6.50	8	14.50
2100	Rectangular, 2" x 3"	3.41	4.43	7.84
2200	3" x 4"	4.18	5.80	9.98
2300	Elbows, aluminum, round, 3" diam.	5.05	8.40	13.45
2400	4" diam.	9.10	8.40	17.50
2500	Rectangular, 2" x 3"	1.55	8.40	9.95
2600	3" x 4"	1.55	8.40	9.95
2700	Copper, round 16 oz., 2" diam.	13.65	8.40	22.05
2800	3" diam.	15.50	8.40	23.90
2900	4" diam.	19.95	8.40	28.35
3100	Rectangular, 2" x 3"	12.05	8.40	20.45
3200	3" x 4"	18.55	8.40	26.95
3300	Drip edge per L.F., aluminum, 5" wide	.76	1.79	2.55
3400	8" wide	.92	1.79	2.71
3500	28" wide	6.80	7.15	13.95
3600				
3700	Steel galvanized, 5" wide	1.06	1.79	2.85
3800	8" wide	1.09	1.79	2.88
3900				
4000				
4100				
4200				
4300	Flashing 12" wide per S.F., aluminum, mill finish, .013" thick	1.46	4.71	6.17
4400	.019" thick	5.75	4.71	10.46
4500	.040" thick	4.92	4.71	9.63
4600	.050" thick	6.85	4.71	11.56
4700	Copper, mill finish, 16 oz.	15.55	5.95	21.50
4800	20 oz.	18.90	6.20	25.10
4900	24 oz.	22.50	6.50	29
5000	32 oz.	30	6.85	36.85
5100	Lead, 2.5 lb./S.F., 12" wide	7.10	5.05	12.15
5200	Over 12" wide	4.88	5.05	9.93
5900	Polyvinyl chloride, black, .010" thick	.31	2.40	2.71
6000	.020" thick	.30	2.40	2.70
6100	.030" thick	.37	2.40	2.77
6200	.056" thick	.99	2.40	3.39
6300	Steel, galvanized, 20 gauge	3.30	5.25	8.55
6400	30 gauge	1.83	4.27	6.10
6500	Stainless, 32 gauge, .010" thick	3.99	4.41	8.40

B3020 Roof Openings

B3020 240	Selective Price Sheet	COST PER UNIT		
		MAT.	INST.	TOTAL
6600	28 gauge, .015" thick	9.15	4.41	13.56
6700	26 gauge, .018" thick	7.10	4.41	11.51
6800	24 gauge, .025" thick	8.50	4.41	12.91
6900	Gutters per L.F., aluminum, 5" box, .027" thick	2.75	6.95	9.70
7000	.032" thick	2.97	6.75	9.72
7100	Copper, half round, 4" wide	9.10	6.75	15.85
7200	6" wide	12.25	7.15	19.40
7300	Steel, 26 gauge galvanized, 5" wide	3.99	6.75	10.74
7400	6" wide	5.40	6.75	12.15
7500	Wood, treated hem-fir, 3" x 4"	11.85	7.15	19
7600	4" x 5"	21	7.15	28.15
7700	Reglet per L.F., aluminum, .025" thick	2.48	3.18	5.66
7800	Copper, 10 oz.	9.55	3.18	12.73
7900	Steel, galvanized, 24 gauge	2.38	3.18	5.56
8000	Stainless, .020" thick	5.35	3.18	8.53
8100	Counter flashing 12" wide per L.F., aluminum, .032" thick	3.75	5.60	9.35
8200	Aluminum, .025" thick	3.75	5.60	9.35
8300	Steel, galvanized, 24 gauge	2.15	5.60	7.75
8400	Stainless, .020" thick	9.05	5.60	14.65

Same Data. Simplified.

Enjoy the convenience and efficiency of accessing your costs anywhere:

- **Skip the multiplier** by setting your location
- **Quickly search,** edit, favorite and share costs
- **Stay on top of price changes** with automatic updates

Discover more at rsmeans.com/online

This page illustrates and describes a concrete block wall system including concrete block, horizontal reinforcing alternate courses, mortar, and tooled joints both sides. Lines within System Components give the unit price and total price per square foot for this system. Prices for concrete block wall systems are on Line Items C1010 106 0800 thru 2100. Both material quantities and labor costs have been adjusted for the system listed.

Factors: To adjust for job conditions other than normal working situations use Lines C1010 106 2900 thru 4000.

Example: You are to install the system and protect all existing work. Go to Line C1010 106 3600 and apply these percentages to the appropriate MAT. and INST. costs.

System Components	QUANTITY	UNIT	COST PER S.F.		
			MAT.	INST.	TOTAL
Concrete block partition, including horizontal reinforcing every second Course, mortar, tooled joints both sides.					
8" x 16" concrete block, normal weight, 4" thick	1.000	S.F.	1.96	7.60	9.56
Horizontal reinforcing every second course	.750	L.F.	.22	.18	.40
TOTAL			2.18	7.78	9.96

C1010 106	Partitions, Concrete Block	COST PER S.F.		
		MAT.	INST.	TOTAL
0800	8" x 16" concrete block, normal weight, 4" thick	2.18	7.80	9.98
0900	6" thick	4.52	8.40	12.92
1000	8" thick	2.97	8.95	11.92
1100	10" thick	4.33	9.30	13.63
1200	12" thick	5.05	11.55	16.60
1300	8" x 16" concrete block, lightweight, 4" thick	2.36	7.65	10.01
1400	6" thick	2.14	8.20	10.34
1500	8" thick	3.69	8.70	12.39
1600	10" thick	3.67	9.05	12.72
1700	12" thick	4.40	11.20	15.60
1800	8" x 16" glazed concrete block, 4" thick	14.65	9.70	24.35
1900	8" thick	19.15	10.75	29.90
2000	Structural facing tile, 6T series, glazed 2 sides, 4" thick	19.60	17	36.60
2100	6" thick	25	17.90	42.90
2200				
2300				
2400				
2500				
2600				
2900	Cut & patch to match existing construction, add, minimum	2%	3%	
3000	Maximum	5%	9%	
3100	Dust protection, add, minimum	1%	2%	
3200	Maximum	4%	11%	
3300	Material handling & storage limitation, add, minimum	1%	1%	
3400	Maximum	6%	7%	
3500	Protection of existing work, add, minimum	2%	2%	
3600	Maximum	5%	7%	
3700	Shift work requirements, add, minimum		5%	
3800	Maximum		30%	
3900	Temporary shoring and bracing, add, minimum	2%	5%	
4000	Maximum	5%	12%	

This page illustrates and describes a wood stud partition system including wood studs with plates, gypsum plasterboard – taped and finished, insulation, baseboard and painting. Lines within System Components give the unit price and total price per square foot for this system. Prices for wood stud partition systems are on Line Items C1010 132 1300 thru 2700. Both material quantities and labor costs have been adjusted for the system listed.

Factors: To adjust for job conditions other than normal working situations use Lines C1010 132 2900 thru 4000.

Example: You are to install the system where material handling and storage present a serious problem. Go to Line C1010 132 3400 and apply these percentages to the appropriate MAT. and INST. costs.

System Components	QUANTITY	UNIT	COST PER S.F.		
			MAT.	INST.	TOTAL
Wood stud wall, 2"x4", 16" O.C., dbl. top plate, sngl bot. plate, 5/8" dwl. taped, finished and painted on 2 faces, insulation, baseboard, wall 8' high.					
Wood studs, 2" x 4", 16" O.C., 8' high	1.000	S.F.	.81	1.79	2.60
Gypsum drywall, 5/8" thick	2.000	S.F.	.58	1.44	2.02
Taping and finishing	2.000	S.F.	.12	1.44	1.56
Insulation, 3-1/2" fiberglass batts	1.000	S.F.	.77	.53	1.30
Baseboard	.250	L.F.	.33	.75	1.08
Paint baseboard, primer + 2 coats	.250	L.F.	.04	.45	.49
Painting, roller, 2 coats	2.000	S.F.	.28	1.60	1.88
TOTAL			2.93	8	10.93

C1010 132	Partitions, Wood Stud	COST PER S.F.		
		MAT.	INST.	TOTAL
1300	2" x 3" studs, 8' high, 16" O.C.	2.88	7.90	10.78
1400	24" O.C.	2.71	7.55	10.26
1500	10' high, 16" O.C.	2.84	7.55	10.39
1600	24" O.C.	2.67	7.30	9.97
1650	2" x 4" studs, 8' high, 16" O.C.	2.93	8	10.93
1700	24" O.C.	2.75	7.65	10.40
1800	10' high, 16" O.C.	2.88	7.65	10.53
1900	24" O.C.	2.70	7.35	10.05
2000	12' high, 16" O.C.	2.81	7.65	10.46
2100	24" O.C.	2.64	7.35	9.99
2200	2" x 6" studs, 8' high, 16" O.C.	3.32	8.20	11.52
2300	24" O.C.	3.05	7.75	10.80
2400	10' high, 16" O.C.	3.24	7.80	11.04
2500	24" O.C.	2.97	7.45	10.42
2600	12' high, 16" O.C.	3.14	7.85	10.99
2700	24" O.C.	2.88	7.45	10.33
2900	Cut & patch to match existing construction, add, minimum	2%	3%	
3000	Maximum	5%	9%	
3100	Dust protection, add, minimum	1%	2%	
3200	Maximum	4%	11%	
3300	Material handling & storage limitation, add, minimum	1%	1%	
3400	Maximum	6%	7%	
3500	Protection of existing work, add, minimum	2%	2%	
3600	Maximum	5%	7%	
3700	Shift work requirements, add, minimum		5%	
3800	Maximum		30%	
3900	Temporary shoring and bracing, add, minimum	2%	5%	
4000	Maximum	5%	12%	

C1010 Partitions

This page illustrates and describes a non-load bearing metal stud partition system including metal studs with runners, gypsum plasterboard, taped and finished, insulation, baseboard and painting. Lines within System Components give the unit price and total price per square foot for this system. Prices for non-load bearing metal stud partition systems are on Line Items C1010 134 1300 thru 2300. Both material quantities and labor costs have been adjusted for the system listed.

Factors: To adjust for job conditions other than normal working situations use Lines C1010 134 2900 thru 4000.

Example: You are to install the system and cut and patch to match existing construction. Go to Line C1010 134 3000 and apply these percentages to the appropriate MAT. and INST. costs.

System Components	QUANTITY	UNIT	COST PER S.F. MAT.	COST PER S.F. INST.	COST PER S.F. TOTAL
Non-load bearing metal studs, including top & bottom runners, 5/8″ drywall, Taped, finished and painted 2 faces, insulation, painted baseboard.					
Metal studs, 25 ga., 3-5/8″ wide, 24″ O.C.	1.000	S.F.	.48	.97	1.45
Gypsum drywall, 5/8″ thick	2.000	S.F.	.58	1.44	2.02
Taping & finishing	2.000	S.F.	.12	1.44	1.56
Insulation, 3-1/2″ fiberglass batts	1.000	S.F.	.77	.53	1.30
Baseboard	.200	L.F.	.26	.60	.86
Paint baseboard, primer + 2 coats	.200	L.F.	.03	.36	.39
Painting, roller work, 2 coats	2.000	S.F.	.28	1.60	1.88
TOTAL			2.52	6.94	9.46

C1010 134	Partitions, Metal Stud, NLB	COST PER S.F. MAT.	COST PER S.F. INST.	COST PER S.F. TOTAL
1300	Non-load bearing, 25 ga., 24″ O.C., 2-1/2″ wide	2.49	6.90	9.39
1350	3-5/8″ wide	2.52	6.95	9.47
1400	6″ wide	2.67	6.95	9.62
1500	16″ O.C., 2-1/2″ wide	2.60	7.15	9.75
1600	3-5/8″ wide	2.64	7.15	9.79
1700	6″ wide	2.83	7.20	10.03
1800	20 ga., 24″ O.C., 2-1/2″ wide	2.53	7.15	9.68
1900	3-5/8″ wide	2.80	7.15	9.95
2000	6″ wide	2.86	7.20	10.06
2100	16″ O.C., 2-1/2″ wide	2.65	7.45	10.10
2200	3-5/8″ wide	2.99	7.45	10.44
2300	6″ wide	3.07	7.50	10.57
2500				
2600				
2700				
2900	Cut & patch to match existing construction, add, minimum	2%	3%	
3000	Maximum	5%	9%	
3100	Dust protection, add, minimum	1%	2%	
3200	Maximum	4%	11%	
3300	Material handling & storage limitation, add, minimum	1%	1%	
3400	Maximum	6%	7%	
3500	Protection of existing work, add, minimum	2%	2%	
3600	Maximum	5%	7%	
3700	Shift work requirements, add, minimum		5%	
3800	Maximum		30%	
3900	Temporary shoring and bracing, add, minimum	2%	5%	
4000	Maximum	5%	12%	

This page illustrates and describes a drywall system including gypsum plasterboard, taped and finished, metal studs with runners, insulation, baseboard and painting. Lines within System Components give the unit price and total price per square foot for this system. Prices for drywall systems are on Line Items C1010 136 1200 thru 1900. Both material quantities and labor costs have been adjusted for the system listed.

Factors: To adjust for job conditions other than normal working situations use Lines C1010 136 2900 thru 4000.

Example: You are to install the system and control dust in the work area. Go to Line C1010 136 3100 and apply these percentages to the appropriate MAT. and INST. costs.

System Components	QUANTITY	UNIT	COST PER S.F. MAT.	COST PER S.F. INST.	COST PER S.F. TOTAL
Gypsum drywall, taped, finished and painted 2 faces, galvanized metal studs					
Including top & bottom runners, insulation, painted baseboard, wall 10'high					
Gypsum drywall, 5/8" thick, standard	2.000	S.F.	.58	1.44	2.02
Taping and finishing	2.000	S.F.	.12	1.44	1.56
Metal studs, 20 ga., 3-5/8" wide, 24" O.C.	1.000	S.F.	.92	1.21	2.13
Insulation, 3-1/2" fiberglass batts	1.000	S.F.	.77	.53	1.30
Baseboard	.200	L.F.	.26	.60	.86
Paint baseboard, primer + 2 coats	.200	L.F.	.02	.15	.17
Painting, roller 2 coats	2.000	S.F.	.80	1.52	2.32
TOTAL			3.47	6.89	10.36

C1010 136	Partitions, Drywall	COST PER S.F. MAT.	COST PER S.F. INST.	COST PER S.F. TOTAL
1200	Gypsum drywall, 5/8" thick, standard	3.48	6.90	10.38
1300	Fire resistant	3.20	6.65	9.85
1400	Water resistant	3.40	6.65	10.05
1500	1/2" thick, standard	3.02	6.65	9.67
1600	Fire resistant	3.48	6.65	10.13
1700	Water resistant	3.18	6.65	9.83
1800	3/8" thick, vinyl faced, standard	3.50	8.40	11.90
1900	5/8" thick, vinyl faced, fire resistant	4.64	8.40	13.04
2000				
2100				
2200				
2300				
2400				
2500				
2600				
2700				
2900	Cut & patch to match existing construction, add, minimum	2%	3%	
3000	Maximum	5%	9%	
3100	Dust protection, add, minimum	1%	2%	
3200	Maximum	4%	11%	
3300	Material handling & storage limitation, add, minimum	1%	1%	
3400	Maximum	6%	7%	
3500	Protection of existing work, add, minimum	2%	2%	
3600	Maximum	5%	7%	
3700	Shift work requirements, add, minimum		5%	
3800	Maximum		30%	
3900	Temporary shoring and bracing, add, minimum	2%	5%	
4000	Maximum	5%	12%	

C1010 Partitions

This page illustrates and describes a load bearing metal stud wall system including metal studs, sheetrock–taped and finished, insulation, baseboard and painting. Lines within System Components give the unit price and total price per square foot for this system. Prices for load bearing metal stud wall systems are on Line Items C1010 138 1500 thru 2500. Both material quantities and labor costs have been adjusted for the system listed.

Factors: To adjust for job conditions other than normal working situations use Lines C1010 138 2900 thru 4000.

Example: You are to install the system using temporary shoring and bracing. Go to Line C1010 138 3900 and apply these percentages to the appropriate MAT. and INST. costs.

System Components	QUANTITY	UNIT	COST PER S.F.		
			MAT.	INST.	TOTAL
Load bearing, 18 ga., 3-5/8", galvanized metal studs, 24" O.C.,including					
Top and bottom runners, 1/2" drywall, taped, finished and painted 2					
Faces, 3" insulation, and painted baseboard, wall 10' high.					
Metal studs, 24" OC, 18 ga., 3-5/8" wide, galvanized	1.000	S.F.	.75	1.37	2.12
Gypsum drywall, 1/2" thick	2.000	S.F.	.56	1.44	2
Taping and finishing	2.000	S.F.	.12	1.44	1.56
Insulation, 3-1/2" fiberglass batts	1.000	S.F.	.77	.53	1.30
Baseboard	.200	L.F.	.26	.60	.86
Paint baseboard, primer + 2 coats	.200	L.F.	.03	.36	.39
Painting, roller 2 coats	2.000	S.F.	.28	1.60	1.88
TOTAL			2.77	7.34	10.11

C1010 138	Partitions, Metal Stud, LB	COST PER S.F.		
		MAT.	INST.	TOTAL
1500	Load bearing, 18 ga., 24" O.C., 2-1/2" wide	2.65	7.30	9.95
1550	3-5/8" wide	2.77	7.35	10.12
1600	6" wide	3.01	7.40	10.41
1700	16" O.C. 2-1/2" wide	2.80	7.65	10.45
1800	3-5/8" wide	2.95	7.65	10.60
1900	6" wide	3.26	7.75	11.01
2000	16 ga., 24" O.C., 2-1/2" wide	2.82	7.50	10.32
2100	3-5/8" wide	2.94	7.50	10.44
2200	6" wide	3.27	7.60	10.87
2300	16" O.C., 2-1/2" wide	3.02	7.85	10.87
2400	3-5/8" wide	3.17	7.90	11.07
2500	6" wide	3.58	8	11.58
2600				
2700				
2900	Cut & patch to match existing construction, add, minimum	2%	3%	
3000	Maximum	5%	9%	
3100	Dust protection, add, minumum	1%	2%	
3200	Maximum	4%	11%	
3300	Material handling & storage limitation, add, minimum	1%	1%	
3400	Maximum	6%	7%	
3500	Protection of existing work, add, minimum	2%	2%	
3600	Maximum	5%	7%	
3700	Shift work requirements, add, minimum		5%	
3800	Maximum		30%	
3900	Temporary shoring and bracing, add, minimum	2%	5%	
4000	Maximum	5%	12%	

C10 Interior Construction

C1010 Partitions

This page illustrates and describes a plaster and lath system including gypsum plaster, gypsum lath, wood studs with plates, insulation, baseboard and painting. Lines within System Components give the unit price and price per square foot for this system. Prices for plaster and lath systems are on Line Items C1010 148 1400 thru 2500. Both material quantities and labor costs have been adjusted for the system listed.

Factors: To adjust for job conditions other than normal working situations use Lines C1010 148 2900 thru 4000.

Example: You are to install the system during evening hours only. Go to Line C1010 148 3800 and apply this percentage to the appropriate INST. costs.

System Components	QUANTITY	UNIT	COST PER S.F. MAT.	COST PER S.F. INST.	COST PER S.F. TOTAL
Gypsum plaster, 2 coats, over 3/8" lath, 2 faces, 2" x 4" wood Stud partition, 24"O.C. incl. double top plate, single bottom plate, 3-1/2" Insulation, baseboard and painting.					
Gypsum plaster, 2 coats	.220	S.Y.	.84	6.75	7.59
Lath, gypsum, 3/8" thick	.220	S.Y.	.83	1.80	2.63
Wood studs, 2" x 4", 24" O.C.	1.000	S.F.	.58	1.15	1.73
Insulation, 3-1/2" fiberglass batts	1.000	S.F.	.77	.53	1.30
Baseboard, 9/16" x 3-1/2"	.200	L.F.	.26	.60	.86
Paint baseboard, primer + 2 coats	.200	L.F.	.02	.15	.17
Paint, 2 coats	2.000	S.F.	.28	1.60	1.88
TOTAL			3.58	12.58	16.16

C1010 148	Partitions, Plaster & Lath		COST PER S.F. MAT.	COST PER S.F. INST.	COST PER S.F. TOTAL
1400	Gypsum plaster, 2 coats		3.58	12.60	16.18
1500	3 coats		3.92	14.05	17.97
1600	Perlite plaster, 2 coats		3.90	13.60	17.50
1700	3 coats		4.39	15.45	19.84
1800					
1900					
2000	For alternate lath systems:				
2100	Gypsum lath, 1/2" thick		3.79	13.75	17.54
2200	Foil back, 3/8" thick		3.90	13.85	17.75
2300	1/2" thick		3.88	14	17.88
2400	Metal lath, 2.5 Lb. diamond		4.30	13.60	17.90
2500	3.4 Lb. diamond		4.09	13.85	17.94
2600					
2700					
2900	Cut & patch to match existing construction, add, minimum		2%	3%	
3000	Maximum		5%	9%	
3100	Dust protection, add, minimum		1%	2%	
3200	Maximum		4%	11%	
3300	Material handling & storage limitation, add, minimum		1%	1%	
3400	Maximum		6%	7%	
3500	Protection of existing work, add, minimum		2%	2%	
3600	Maximum		5%	7%	
3700	Shift work requirements, add, minimum			5%	
3800	Maximum			30%	
3900	Temporary shoring and bracing, add, minimum		2%	5%	
4000	Maximum		5%	12%	

For customer support on your Commercial Renovation Costs with RSMeans data, call 800.448.8182.

C1010 Partitions

C1010 170	Selective Price Sheet	COST PER S.F.		
		MAT.	INST.	TOTAL
0100	Studs			
0200				
0300	24" O.C. metal, 10' high wall, including			
0400	Top and bottom runners			
0500	Non load bearing, galvanized 25 Ga., 1-5/8" wide	.45	.94	1.39
0600	2-1/2" wide	.45	.95	1.40
0700	3-5/8" wide	.48	.97	1.45
0800	4" wide	.58	.97	1.55
0900	6" wide	.63	.99	1.62
1000				
1100	Galvanized 20 Ga., 2-1/2" wide	.49	1.19	1.68
1200	3-5/8" wide	.92	1.21	2.13
1300	4" wide	.76	1.21	1.97
1400	6" wide	.82	1.23	2.05
1500	Load bearing, painted 18 Ga., 2-1/2" wide	.63	1.34	1.97
1600	3-5/8" wide	.75	1.37	2.12
1700	4" wide	.65	1.39	2.04
1800	6" wide	.99	1.42	2.41
1900	Galvanized 18 Ga., 2-1/2" wide	.63	1.34	1.97
2000	3-5/8" wide	.75	1.37	2.12
2100	4" wide	.65	1.39	2.04
2200	6" wide	.99	1.42	2.41
2300	Galvanized 16 Ga., 2-1/2" wide	.80	1.53	2.33
2400	3-5/8" wide	.92	1.56	2.48
2500	4" wide	.97	1.59	2.56
2600	6" wide	1.25	1.63	2.88
2700				
2800				
2900	24" O.C. wood, 10' high wall, including			
3000	Double top plate and shoe			
3100	2" x 4"	.68	1.35	2.03
3200	2" x 6"	.99	1.44	2.43
3300				
3400				
3500	Furring, 24" O.C. 10' high wall, metal, 3/4" channels	.29	1.66	1.95
3600	1-1/2" channels	.39	1.79	2.18
3700	Wood, on wood, 1" x 2" strips	.17	.65	.82
3800	1" x 3" strips	.26	.65	.91
3900	On masonry, 1" x 2" strips	.19	.73	.92
4000	1" x 3" strips	.28	.73	1.01
4100	On concrete, 1" x 2" strips	.19	1.38	1.57
4200	1" x 3" strips	.28	1.38	1.66
4300	Studs			
4400				
4500	16" O.C., metal, 10' high wall, including			
4600	Top and bottom runners			
4700	Non load bearing, galvanized 25 Ga., 1-5/8" wide	.62	1.45	2.07
4800	2-1/2" wide	.61	1.46	2.07
4900	3-5/8" wide	.65	1.49	2.14
5000	4" wide	.81	1.51	2.32
5100	6" wide	.86	1.52	2.38
5200				
5300	Galvanized 20 Ga., 2-1/2" wide	.63	1.83	2.46
5400	3-5/8" wide	1.28	1.86	3.14
5500	4" wide	1.03	1.88	2.91
5600	6" wide	1.02	1.91	2.93
5700	Load bearing, painted 18 Ga., 2-1/2" wide	.84	1.86	2.70
5800	3-5/8" wide	1	1.89	2.89

C1010 Partitions

C1010 170	Selective Price Sheet	COST PER S.F.		
		MAT.	INST.	TOTAL
5900	4" wide	1.02	1.94	2.96
6000	6" wide	1.33	1.96	3.29
6100	Galvanized 18 Ga., 2-1/2" wide	.84	1.86	2.70
6200	3-5/8" wide	1	1.89	2.89
6300	4" wide	1.02	1.94	2.96
6400	6" wide	1.33	1.96	3.29
6500	Galvanized 16 Ga., 2-1/2" wide	1.10	2.10	3.20
6600	3-5/8" wide	1.26	2.15	3.41
6700	4" wide	1.35	2.20	3.55
6800	6" wide	1.72	2.25	3.97
6900				
7000				
7100	16" O.C., wood, 10' high wall, including			
7200	Double top plate and shoe			
7300	2" x 4"	.83	1.56	2.39
7400	2" x 6"	.99	1.40	2.39
7500				
7600				
7700	Furring, 16" O.C. 10' high wall, metal, 3/4" channels	.43	2.41	2.84
7800	1-1/2" channels	.58	2.69	3.27
7900	Wood, on wood, 1" x 2" strips	.25	.98	1.23
8000	1" x 3" strips	.38	.98	1.36
8100	On masonry, 1" x 2" strips	.28	1.09	1.37
8200	1" x 3" strips	.42	1.09	1.51
8300	On concrete, 1" x 2" strips	.28	2.06	2.34
8400	1" x 3" strips	.42	2.06	2.48

661

C1010 Partitions

C1010 180	Selective Price Sheet, Drywall & Plaster	COST PER S.F.		
		MAT.	INST.	TOTAL
0100	Lath, gypsum perforated			
0200				
0300	Lath, gypsum, perforated, regular, 3/8" thick	.42	.91	1.33
0400	1/2" thick	.36	1.11	1.47
0500	Fire resistant, 3/8" thick	.31	1.11	1.42
0600	1/2" thick	.35	1.19	1.54
0700	Foil back, 3/8" thick	.42	1.03	1.45
0800	1/2" thick	.41	1.11	1.52
0900				
1000	Metal lath			
1100	Metal lath, diamond, painted, 2.5 lb.	.62	.91	1.53
1200	3.4 lb.	.51	1.03	1.54
1300	Rib painted, 2.75 lb	.49	1.03	1.52
1400	3.40 lb	.62	1.11	1.73
1500				
1600				
1700	Plaster, gypsum, 2 coats	.42	3.41	3.83
1800	3 coats	.59	4.15	4.74
1900	Perlite/vermiculite, 2 coats	.58	3.92	4.50
2000	3 coats	.83	4.85	5.68
2100	Bondcrete, 1 coat	.45	1.80	2.25
2200				
2500				
2600				
2700	Drywall, standard, 3/8" thick, no finish included	.28	.72	1
2800	1/2" thick, no finish included	.28	.72	1
2900	Taped and finished	.33	1.48	1.81
3000	5/8" thick, no finish included	.29	.72	1.01
3100	Taped and finished	.34	1.48	1.82
3200	Fire resistant, 1/2" thick, no finish included	.51	.72	1.23
3300	Taped and finished	.56	1.48	2.04
3400	5/8" thick, no finish included	.37	.72	1.09
3500	Taped and finished	.43	1.48	1.91
3600	Water resistant, 1/2" thick, no finish included	.36	.72	1.08
3700	Taped and finished	.42	1.48	1.90
3800	5/8" thick, no finish included	.47	.72	1.19
3900	Taped and finished	.53	1.48	2.01
4000	Finish, instead of taping			
4100	For thin coat plaster, add	.13	.90	1.03
4200	Finish, textured spray, add	.06	.87	.93

C1010 Partitions

This page illustrates and describes a movable office partition system including demountable office partitions of various styles and sizes based on a cost per square foot basis. Prices for movable office partition systems are on Line Items C1010 210 0300 thru 3600.

C1010 210	Partitions, Movable Office	COST PER S.F.		
		MAT.	INST.	TOTAL
0250	Office partition, demountable, no deduction for door opening, add for doors			
0300	Air wall, cork finish, semi acoustic, 1-5/8" thick, minimum	46.50	4.40	50.90
0400	Maximum	55.50	7.55	63.05
0500	Acoustic, 2" thick, minimum	47.50	4.69	52.19
0600	Maximum	66	6.35	72.35
0700				
0800				
0900	In-plant modular office system, w/prehung steel door			
1000	3" thick honeycomb core panels			
1100	12' x 12', 2 wall	10.25	.61	10.86
1200	4 wall	22.50	1.83	24.33
1300				
1400				
1500	Gypsum, demountable, 3" to 3-3/4" thick x 9' high, vinyl clad	8.75	3.33	12.08
1600	Fabric clad	18.75	3.61	22.36
1700	1.75 system, vinyl clad hardboard, paper honeycomb core panel			
1800	1-3/4" to 2-1/2" thick x 9' high	12.65	3.33	15.98
1900	Unitized gypsum panel, 2" to 2-1/2" thick x 9' high, vinyl clad	16.30	3.33	19.63
2000	Fabric clad	27	3.61	30.61
2100				
2200				
2300	Unitized mineral fiber panel system, 2-1/4" thick x 9' high			
2400	Vinyl clad mineral fiber	16.20	3.33	19.53
2500	Fabric clad mineral fiber	24	3.61	27.61
2600				
2700	Movable steel walls, modular system			
2800	Unitized panels, 48" wide x 9' high			
2900	Baked enamel, pre-finished	18.45	2.66	21.11
3000	Fabric clad	26.50	2.83	29.33
3100	For acoustical partitions, add, unsealed	2.64		2.64
3200	Sealed	12.30		12.30
3300				
3400				
3500	Note: For door prices, see Div. 08			
3600	For door hardware prices, see Div. 08			
3700				
3800				
3900				
4000				

663

C1020 Interior Doors

This page illustrates and describes flush interior door systems including hollow core door, jamb, header and trim with hardware. Lines within System Components give the unit price and total price on a cost each basis for this system. Prices for flush interior door systems are on Line items C1020 106 1000 thru 2400. Both material quantities and labor costs have been adjusted for the system listed.

Factors: To adjust for job conditions other than normal working situations use Lines C1020 106 2700 thru 3800.

Example: You are to install the system in an area where dust protection is vital. Go to Line C1020 106 3000 and apply these percentages to the appropriate MAT. and INST. costs.

System Components			COST EACH		
	QUANTITY	UNIT	MAT.	INST.	TOTAL
Single hollow core door, include jamb, header, trim and hardware, painted.					
Hollow core Lauan, 1-3/8" thick, 2'-0" x 6'-8", painted	1.000	Ea.	81	79.50	160.50
Paint door and frame, 1 coat	1.000	Ea.	4.01	35.50	39.51
Wood jamb, 4-9/16" deep	1.000	Set	118.40	61.12	179.52
Trim, casing	1.000	Set	50.88	91.52	142.40
Hardware, hinges	1.000	Set	3.77		3.77
Hardware, lockset	1.000	Set	61.50	44.50	106
TOTAL			319.56	312.14	631.70

C1020 106	Doors, Interior Flush, Wood	COST EACH		
		MAT.	INST.	TOTAL
1000	Lauan (Mahogany) hollow core, 1-3/8" x 2'-0" x 6'-8"	320	310	630
1100	2'-6" x 6'-8"	335	315	650
1200	2'-8" x 6'-8"	355	320	675
1300	3'-0" x 6'-8"	330	330	660
1500	Birch, hollow core, 1-3/8" x 2'-0" x 6'-8"	365	310	675
1600	2'-6" x 6'-8"	365	315	680
1700	2'-8" x 6'-8"	385	320	705
1800	3'-0" x 6'-8"	380	330	710
1900	Solid core, pre-hung, 1-3/8" x 2'-6" x 6'-8"	590	320	910
2000	2'-8" x 6'-8"	515	325	840
2100	3'-0" x 6'-8"	630	335	965
2200				
2400	For metal frame instead of wood, add	50%	20%	
2600				
2700	Cut & patch to match existing construction, add, minimum	2%	3%	
2800	Maximum	5%	9%	
2900	Dust protection, add, minimum	1%	2%	
3000	Maximum	4%	11%	
3100	Equipment usage curtailment, add, minimum	1%	1%	
3200	Maximum	3%	10%	
3300	Material handling & storage limitation, add, minimum	1%	1%	
3400	Maximum	6%	7%	
3500	Protection of existing work, add, minimum	2%	2%	
3600	Maximum	5%	7%	
3700	Shift work requirements, add, minimum		5%	
3800	Maximum		30%	
3900				

C1020   Interior Doors

This page illustrates and describes interior, solid and louvered door systems including a pine panel door, wood jambs, header, and trim with hardware. Lines within System Components give the unit price and total price on a cost each basis for this system. Prices for interior, solid and louvered systems are on Line Items C1020 108 1100 thru 2400. Both material quantities and labor costs have been adjusted for the system listed.

Factors: To adjust for job conditions other than normal working situations use Lines C1020 108 2900 thru 4000.

Example: You are to install the system during night hours only. Go to Line C1020 108 4000 and apply these percentages to the appropriate INST. costs.

System Components	QUANTITY	UNIT	COST EACH		
			MAT.	INST.	TOTAL
Single interior door, including jamb, header, trim and hardware, painted.					
Solid pine panel door, 1-3/8" thick, 2'-0" x 6'-8"	1.000	Ea.	350	79.50	429.50
Paint door and frame, 1 coat	1.000	Ea.	4.01	35.50	39.51
Wooden jamb, 4-5/8" deep	1.000	Set	118.40	61.12	179.52
Trim, casing	1.000	Set	50.88	91.52	142.40
Hardware, hinges	1.000	Set	3.77		3.77
Hardware, lockset	1.000	Set	61.50	44.50	106
TOTAL			588.56	312.14	900.70

C1020 108	Doors, Interior Solid & Louvered	COST EACH		
		MAT.	INST.	TOTAL
1000				
1100	Solid pine, painted raised panel, 1-3/8" x 2'-0" x 6'-8"	590	310	900
1200	2'-6" x 6'-8"	625	315	940
1300	2'-8" x 6'-8"	620	320	940
1400	3'-0" x 6'-8"	660	330	990
1500				
1600	Louvered pine, painted 1'-6" x 6'-8"	460	310	770
1700	2'-0" x 6'-8"	490	315	805
1800	2'-6" x 6'-8"	510	320	830
1900	3'-0" x 6'-8"	520	330	850
2200	For prehung door, deduct	5%	30%	
2300				
2400	For metal frame instead of wood, add	50%	20%	
2500				
2900	Cut & patch to match existing construction, add, minimum	2%	3%	
3000	Maximum	5%	9%	
3100	Dust protection, add, minimum	1%	2%	
3200	Maximum	4%	11%	
3300	Equipment usage curtailment, add, minimum	1%	1%	
3400	Maximum	3%	10%	
3500	Material handling & storage limitation, add, minimum	1%	1%	
3600	Maximum	6%	7%	
3700	Protection of existing work, add, minimum	2%	2%	
3800	Maximum	5%	7%	
3900	Shift work requirements, add, minimum		5%	
4000	Maximum		30%	

C1020 Interior Doors

This page illustrates and describes interior metal door systems including a metal door, metal frame and hardware. Lines within System Components give the unit price and total price on a cost each for this system. Prices for interior metal door systems are on Line Items C1020 110 1000 thru 2000. Both material quantities and labor costs have been adjusted for the system listed.

Factors: To adjust for job conditions other than normal working situations use Lines C1020 110 2900 thru 4000.

Example: You are to install the system while protecting existing construction. Go to Line C1020 110 3700 and apply these percentages to the appropriate MAT. and INST. costs.

System Components	QUANTITY	UNIT	COST EACH MAT.	COST EACH INST.	COST EACH TOTAL
Single metal door, including frame and hardware.					
Hollow metal door, 1-3/8" thick, 2'-6" x 6'-8", painted	1.000	Ea.	700	95.50	795.50
Metal frame, 6-1/4" deep	1.000	Set	234	89.50	323.50
Paint door and frame, 1 coat	1.000	Ea.	4.01	35.50	39.51
Hardware, hinges	1.000	Set	63.50		63.50
Hardware, passage lockset	1.000	Set	61.50	44.50	106
TOTAL			1,063.01	265	1,328.01

C1020 110	Doors, Interior Flush, Metal	COST EACH MAT.	COST EACH INST.	COST EACH TOTAL
1000	Hollow metal doors, 1-3/8" thick, 2'-6" x 6'-8"	1,075	265	1,340
1100	1-3/8" thick, 2'-8" x 6'-8"	1,075	265	1,340
1200	3'-0" x 7'-0"	1,075	278	1,353
1300				
1400	Interior fire door, 1-3/8" thick, 2'-6" x 6'-8"	1,150	265	1,415
1500	2'-8" x 6'-8"	1,050	265	1,315
1600	3'-0" x 7'-0"	1,325	278	1,603
1700				
1800	Add to fire doors:			
1900	Baked enamel finish	30%	15%	
2000	Galvanizing	20%		
2200				
2300				
2400				
2900	Cut & patch to match existing construction, add, minimum	2%	3%	
3000	Maximum	5%	9%	
3100	Dust protection, add, minimum	1%	2%	
3200	Maximum	4%	11%	
3300	Equipment usage curtailment, add, minimum	1%	1%	
3400	Maximum	3%	10%	
3500	Material handling & storage limitation, add, minimum	1%	1%	
3600	Maximum	6%	7%	
3700	Protection of existing work, add, minimum	2%	2%	
3800	Maximum	5%	7%	
3900	Shift work requirements, add, minimum		5%	
4000	Maximum		30%	

C1020 Interior Doors

This page illustrates and describes an interior closet door system including an interior closet door, painted, with trim and hardware. Prices for interior closet door systems are on Line Items C1020 112 0500 thru 2200. Both material quantities and labor costs have been adjusted for the system listed.

Factors: To adjust for job conditions other than normal working situations use Lines C1020 112 2900 thru 4000.

Example: You are to install the system and match the existing construction. Go to Line C1020 112 2900 and apply these percentages to the appropriate MAT. and INST. costs.

C1020 112	Doors, Closet	COST PER SET		
		MAT.	INST.	TOTAL
0350	Interior closet door painted, including frame, trim and hardware, prehung.			
0400	Bi-fold doors			
0500	Pine paneled, 3'-0" x 6'-8"	775	259	1,034
0600	6'-0" x 6'-8"	680	350	1,030
0700	Birch, hollow core, 3'-0" x 6'-8"	390	281	671
0800	6'-0" x 6'-8"	650	375	1,025
0900	Lauan, hollow core, 3'-0" x 6'-8"	350	259	609
1000	6'-0" x 6'-8"	570	350	920
1100	Louvered pine, 3'-0" x 6'-8"	435	259	694
1200	6'-0" x 6'-8"	540	350	890
1300				
1400	Sliding, bi-passing closet doors			
1500	Pine paneled, 4'-0" x 6'-8"	750	275	1,025
1600	6'-0" x 6'-8"	1,275	350	1,625
1700	Birch, hollow core, 4'-0" x 6'-8"	465	275	740
1800	6'-0" x 6'-8"	610	350	960
1900	Lauan, hollow core, 4'-0" x 6'-8"	360	275	635
2000	6'-0" x 6'-8"	380	350	730
2100	Louvered pine, 4'-0" x 6'-8"	690	275	965
2200	6'-0" x 6'-8"	1,100	350	1,450
2300				
2400				
2500				
2600				
2700				
2800				
2900	Cut & patch to match existing construction, add, minimum	2%	3%	
3000	Maximum	5%	9%	
3100	Dust protection, add, minimum	1%	2%	
3200	Maximum	4%	11%	
3300	Equipment usage curtailment, add, minimum	1%	1%	
3400	Maximum	3%	10%	
3500	Material handling & storage limitation, add, minimum	1%	1%	
3600	Maximum	6%	7%	
3700	Protection of existing work, add, minimum	2%	2%	
3800	Maximum	5%	7%	
3900	Shift work requirements, add, minimum		5%	
4000	Maximum		30%	

C2010 Stair Construction

This page illustrates and describes a stair system based on a cost per flight price. Prices for various stair systems are on Line Items C2010 130 0700 thru 3200. Both material quantities and labor costs have been adjusted for the system listed.

Factors: To adjust for job conditions other than normal working situations use Lines C2010 130 3500 thru 4200.

Example: You are to install the system during evenings only. Go to Line C2010 130 4200 and apply this percentage to the appropriate MAT. and INST. costs.

System Components	QUANTITY	UNIT	COST PER FLIGHT		
			MAT.	INST.	TOTAL
Below are various stair systems based on cost per flight of stairs, no side Walls. Stairs are 4'-0" wide, railings are included unless otherwise noted.					

C2010 130	Stairs	COST PER FLIGHT		
		MAT.	INST.	TOTAL
0700	Concrete, cast in place, no nosings, no railings, 12 risers	455	2,450	2,905
0800	24 risers	910	4,875	5,785
0900	Add for 1 intermediate landing	243	675	918
1000	Concrete, cast in place, with nosings, no railings, 12 risers	1,350	2,700	4,050
1100	24 risers	2,675	5,400	8,075
1200	Add for 1 intermediate landing	315	700	1,015
1300	Steel, grating tread, safety nosing, 12 risers	5,400	1,275	6,675
1400	24 risers	10,800	2,550	13,350
1500	Add for intermediate landing	1,325	335	1,660
1600	Steel, metal pan tread for concrete in-fill, 12 risers	8,325	1,275	9,600
1700	24 risers	16,600	2,550	19,150
1800	Add for intermediate landing	1,325	335	1,660
1900	Spiral, industrial, 4' - 6" diameter, 12 risers	7,800	900	8,700
2000	24 risers	15,600	1,800	17,400
2100	Wood, box stairs, oak treads, 12 risers	31,600	545	32,145
2200	24 risers	63,000	1,100	64,100
2300	Add for 1 intermediate landing	276	183	459
2400	Wood, basement stairs, no risers, 12 steps	1,100	27.50	1,127.50
2500	24 steps	2,200	55	2,255
2600	Add for 1 intermediate landing	107	35	142
2700	Wood, open, rough sawn cedar, 12 steps	26,600	655	27,255
2800	24 steps	53,000	1,300	54,300
2900	Add for 1 intermediate landing	38	36	74
3000	Wood, residential, oak treads, 12 risers	3,300	3,575	6,875
3100	24 risers	6,600	7,150	13,750
3200	Add for 1 intermediate landing	223	166	389
3500	Dust protection, add, minimum	1%	2%	
3600	Maximum	4%	11%	
3700	Material handling & storage limitation, add, minimum	1%	1%	
3800	Maximum	6%	7%	
3900	Protection of existing work, add, minimum	2%	2%	
4000	Maximum	5%	7%	
4100	Shift work requirements, add, minimum		5%	
4200	Maximum		30%	

C3010 Wall Finishes

C3010 210	Selective Price Sheet	COST PER S.F.		
		MAT.	INST.	TOTAL
0100	Painting, on plaster or drywall, brushwork, primer and 1 ct.	.40	.89	1.29
0200	Primer and 2 ct.	.47	1.42	1.89
0300	Rollerwork, primer and 1 ct.	.40	.76	1.16
0400	Primer and 2 ct.	.47	1.21	1.68
0500	Woodwork incl. puttying, brushwork, primer and 1 ct.	.10	1.34	1.44
0600	Primer and 2 ct.	.14	1.78	1.92
0700	Wood trim to 6" wide, enamel, primer and 1 ct.	.11	.76	.87
0800	Primer and 2 ct.	.15	.96	1.11
0900	Cabinets and casework, enamel, primer and 1 ct.	.11	1.51	1.62
1000	Primer and 2 ct.	.15	1.86	2.01
1100	On masonry or concrete, latex, brushwork, primer and 1 ct.	.22	1.26	1.48
1200	Primer and 2 ct.	.27	1.81	2.08
1300	For block filler, add	.20	1.55	1.75
1400				
1500	Varnish, wood trim, sealer 1 ct., sanding, puttying, quality work	.40	2.72	3.12
1600	Medium work	.31	2.11	2.42
1700	Without sanding	.17	.32	.49
1800				
1900	Wall coverings, wall paper, at low price per double roll, avg workmanship	.66	.95	1.61
2000	At medium price per double roll, average workmanship	1.49	1.14	2.63
2100	At high price per double roll, quality workmanship	2.57	1.40	3.97
2200				
2300	Grass cloths with lining paper	1.39	1.52	2.91
2400	Premium texture/color	3.82	1.74	5.56
2500	Vinyl, fabric backed, light weight	1.73	.95	2.68
2600	Medium weight	1.35	1.27	2.62
2700	Heavy weight	2.06	1.40	3.46
2800				
2900	Cork tiles, 12" x 12", 3/16" thick	4.44	2.53	6.97
3000	5/16" thick	3.72	2.59	6.31
3100	Granular surface, 12" x 36", 1/2" thick	1.49	1.58	3.07
3200	1" thick	1.94	1.64	3.58
3300	Aluminum foil	1.18	2.21	3.39
3400				
3500	Tile, ceramic, adhesive set, 4-1/4" x 4-1/4"	1.50	6.30	7.80
3600	6" x 6"	2.02	6.85	8.87
3700	Decorated, 4-1/4" x 4-1/4", minimum	9.20	4.44	13.64
3800	Color group 4	68.50	6.65	75.15
3900	For epoxy grout, add	.43	1.50	1.93
4000	Pregrouted sheets	5.70	4.99	10.69
4100	Glass mosaics, 3/4" tile on 12" sheets, minimum	8.60	16.40	25
4200	Color group 8	18.60	18.70	37.30
4300	Metal, tile pattern, 4' x 4' sheet, 24 ga., nailed			
4400	Stainless steel	28	2.80	30.80
4500	Aluminized steel	12.85	2.80	15.65
4600				
4700	Brick, interior veneer, 4" face brick, running bond, minimum	5.50	15.25	20.75
4800	Maximum	5.50	15.25	20.75
4900	Simulated, urethane pieces, set in mastic	9.20	4.77	13.97
5000	Fiberglass panels	11	3.58	14.58
5100	Wall coating, on drywall, thin coat, plain	.13	.90	1.03
5200	Stipple	.13	.90	1.03
5300	Textured spray	.06	.87	.93
5400				
5500	Paneling not incl. furring or trim, hardboard, tempered, 1/8" thick	.35	2.86	3.21
5600	1/4" thick	.77	2.86	3.63
5700	Plastic faced, 1/8" thick	.59	2.86	3.45
5800	1/4" thick	1.16	2.86	4.02

669

C3010 Wall Finishes

C3010 210	Selective Price Sheet	COST PER S.F.		
		MAT.	INST.	TOTAL
5900	Woodgrained, 1/4" thick, minimum	.99	2.86	3.85
6000	Maximum	1.87	3.37	5.24
6100	Plywood, 4' x 8' shts. 1/4" thick, prefin., birch faced, min.	.97	2.86	3.83
6200	Maximum	2.20	4.09	6.29
6300	Walnut, minimum	3.72	2.86	6.58
6400	Maximum	3.44	3.58	7.02
6500	Mahogany, African	3.63	3.58	7.21
6600	Philippine	.58	2.86	3.44
6700	Chestnut	7.55	3.82	11.37
6800	Pecan	3.52	3.58	7.10
6900	Rosewood	4.31	4.47	8.78
7000	Teak	4.58	3.58	8.16
7100	Aromatic cedar, plywood	3.23	3.58	6.81
7200	Particle board	1.63	3.58	5.21
7300	Wood board, 3/4" thick, knotty pine	2.56	4.77	7.33
7400	Rough sawn cedar	5.65	4.77	10.42
7500	Redwood, clear	15.45	4.77	20.22
7600	Aromatic cedar	2.73	5.20	7.93

C30 Interior Finishes

C3020 Floor Finishes

C3020 430	Selective Price Sheet	COST PER S.F.		
		MAT.	INST.	TOTAL
0100	Flooring, carpet, acrylic, 26 oz. light traffic	1.94	.98	2.92
0200	35 oz. heavy traffic	6.85	.98	7.83
0300	Nylon anti-static, 15 oz. light traffic	2.89	.98	3.87
0400	22 oz. medium traffic	4.72	.98	5.70
0500	26 oz. heavy traffic	6.55	1.05	7.60
0600	28 oz. heavy traffic	3.94	1.05	4.99
0700	Tile, foamed back, needle punch	4.66	1.16	5.82
0800	Tufted loop	3.86	1.16	5.02
0900	Wool, 36 oz. medium traffic	12.10	1.05	13.15
1000	42 oz. heavy traffic	13.30	1.05	14.35
1100	Composition, epoxy, with colored chips, minimum	3.89	5.40	9.29
1200	Maximum	5.25	7.45	12.70
1300	Trowelled, minimum	4.18	6.50	10.68
1400	Maximum	7.25	7.60	14.85
1500	Terrazzo, 1/4" thick, chemical resistant, minimum	7.05	7.45	14.50
1600	Maximum	12	9.95	21.95
1700	Resilient, asphalt tile, 1/8" thick	1.49	1.66	3.15
1800	Conductive flooring, rubber, 1/8" thick	16.85	2.11	18.96
1900	Cork tile 1/8" thick, standard finish	3.86	2.11	5.97
2000	Urethane finish	6.30	2.11	8.41
2100	PVC sheet goods for gyms, 1/4" thick	6.75	8.30	15.05
2200	3/8" thick	9.85	11.05	20.90
2300	Vinyl composition 12" x 12" tile, plain, 1/16" thick	.85	1.33	2.18
2400	1/8" thick	1.21	1.33	2.54
2500	Vinyl tile, 12" x 12" x 1/8" thick, minimum	6.70	1.33	8.03
2600	Premium colors/pattern	3.58	1.33	4.91
2700	Vinyl sheet goods, backed, .080" thick, plain pattern/colors	3.07	2.89	5.96
2900	Slate, random rectangular, 1/4" thick	24	8.55	32.55
3000	1/2" thick	22	12.25	34.25
3100	Natural cleft, irregular, 3/4" thick	10.20	13.95	24.15
3200	For sand rubbed finish, add	12.45		12.45
3300	Terrazzo, cast in place, bonded 1-3/4" thick, gray cement	4.50	19.85	24.35
3400	White cement	5.30	19.85	25.15
3500	Not bonded 3" thick, gray cement	5.55	21.50	27.05
3600	White cement	6.40	21.50	27.90
3700	Precast, 12" x 12" x 1" thick	32	21.50	53.50
3800	1-1/4" thick	30	21.50	51.50
3900	16" x 16" x 1-1/4" thick	33	25.50	58.50
4000	1-1/2" thick	30	28.50	58.50
4100	Marble travertine, standard, 12" x 12" x 3/4" thick	9.10	19.95	29.05
4200				
4300	Tile, ceramic, natural clay, thin set	1.93	6.55	8.48
4400	Porcelain, thin set	1.25	6.30	7.55
4500	Specialty, decorator finish	13	6.55	19.55
4600				
4700	Quarry, red, mud set, 4" x 4" x 1/2" thick	8.50	10	18.50
4800	6" x 6" x 1/2" thick	9.30	8.55	17.85
4900	Brown, imported, 6" x 6" x 3/4" thick	8.20	10	18.20
5000	8" x 8" x 1" thick	9.35	10.90	20.25
5100	Slate, Vermont, thin set, 6" x 6" x 1/4" thick	8.55	6.65	15.20
5200				
5300	Wood, maple strip, 25/32" x 2-1/4", finished, select	6	4.53	10.53
5400	2nd and better	5.70	4.53	10.23
5500	Oak, 25/32" x 2-1/4" finished, clear	4.75	4.53	9.28
5600	No. 1 common	5.25	4.53	9.78
5700	Parquet, standard, 5/16", finished, minimum	12.20	4.79	16.99
5800	Maximum	22	7.45	29.45
5900	Custom, finished, plain pattern	25.50	7.15	32.65

C3020 Floor Finishes

C3020 430	Selective Price Sheet	COST PER S.F.		
		MAT.	INST.	TOTAL
6000	Intricate pattern	35	14.30	49.30
6100	Prefinished, oak, 2-1/4" wide	4.94	4.21	9.15
6200	Ranch plank	6.50	4.94	11.44
6300	Sleepers on concrete, treated, 24" O.C., 1" x 2"	.40	.47	.87
6400	1" x 3"	.45	.61	1.06
6500	2" x 4"	.89	.84	1.73
6600	2" x 6"	1.28	.97	2.25
6700	Refinish old floors, soft wood	.27	1.45	1.72
6800	Hard wood	.27	4.46	4.73
6900	Subflooring, plywood, CDX, 1/2" thick	.84	.95	1.79
7000	5/8" thick	.90	1.06	1.96
7100	3/4" thick	1.04	1.14	2.18
7200				
7300	1" x 10" boards, S4S, laid regular	2.55	1.30	3.85
7400	Laid diagonal	2.55	1.59	4.14
7500	1" x 8" boards, S4S, laid regular	2.37	1.43	3.80
7600	Laid diagonal	2.37	1.68	4.05
7700	Underlayment, plywood, underlayment grade, 3/8" thick	2.27	.95	3.22
7800	1/2" thick	3.03	.99	4.02
7900	5/8" thick	3.09	1.02	4.11
8000	3/4" thick	3.39	1.10	4.49
8100	Particle board, 3/8" thick	.64	.95	1.59
8200	1/2" thick	1.10	.99	2.09
8300	5/8" thick	1.13	1.02	2.15
8400	3/4" thick	1.17	1.10	2.27

C3030 Ceiling Finishes

This page illustrates suspended acoustical board systems including acoustic ceiling board, hangers, and T bar suspension. Lines within System Components give the unit price and total price per square foot for this system. Prices for suspended acoustical board systems are on Line Items C3030 210 1000 thru 2400. Both material quantities and labor costs have been adjusted for the system listed.

Factors: To adjust for job conditions other than normal working situations use Lines C3030 210 2900 thru 4000.

Example: You are to install the system and protect existing construction. Go to Line C3030 210 3800 and apply these percentages to the appropriate MAT. and INST. costs.

System Components	QUANTITY	UNIT	COST PER S.F. MAT.	COST PER S.F. INST.	COST PER S.F. TOTAL
Suspended acoustical ceiling board installed on exposed grid system.					
Fiberglass boards, film faced, 2' x 4', 5/8" thick	1.000	S.F.	5.05	1.14	6.19
Hangers, #12 wire	1.000	S.F.	.08	.69	.77
T bar suspension system, 2' x 4' grid	1.000	S.F.	1.66	.89	2.55
TOTAL			6.79	2.72	9.51

C3030 210	Ceiling, Suspended Acoustical	COST PER S.F. MAT.	COST PER S.F. INST.	COST PER S.F. TOTAL
1000	2' x 4' grid, fiberglass board, film faced, 5/8" thick	6.80	2.72	9.52
1100	Mineral fiber board, aluminum faced	6.30	2.68	8.98
1200	Standard faced	3.12	2.64	5.76
1300	Plastic faced	3.19	3.37	6.56
1400	Fiberglass, film faced, 3" thick, R11	6.35	3.17	9.52
1500	Grass cloth faced, 3/4" thick	5.45	3.01	8.46
1600	1" thick	9.75	3.06	12.81
1700	1-1/2" thick, nubby face	4.88	3.09	7.97
2200				
2300				
2400	Add for 2' x 2' grid system	.43	.22	.65
2500				
2600				
2700				
2900	Cut & patch to match existing construction, add, minimum	2%	3%	
3000	Maximum	5%	9%	
3100	Dust protection, add, minimum	1%	2%	
3200	Maximum	4%	11%	
3300	Equipment usage curtailment, add, minimum	1%	1%	
3400	Maximum	3%	10%	
3500	Material handling & storage limitation, add, minimum	1%	1%	
3600	Maximum	6%	7%	
3700	Protection of existing work, add, minimum	2%	2%	
3800	Maximum	5%	7%	
3900	Shift work requirements, add, minimum		5%	
4000	Maximum		30%	

C3030 Ceiling Finishes

This page illustrates and describes suspended gypsum board systems including gypsum board, metal furring, taping, finished and painted. Lines within System Components give the unit price and total price per square foot for this system. Prices for suspended gypsum board systems are on Line Items C3030 220 1400 thru 1700. Both material quantities and labor costs have been adjusted for the system listed.

Factors: To adjust for job conditions other than normal working situations use Lines C3030 220 2900 thru 4000.

Example: You are to install the system and control dust in the working area. Go to Line C3030 220 3200 and apply these percentages to the appropriate MAT. and INST. costs.

System Components	QUANTITY	UNIT	COST PER S.F.		
			MAT.	INST.	TOTAL
Suspended ceiling gypsum board,4' x 8' x 5/8" thick,					
On metal furring, taped, finished and painted.					
Gypsum drywall, 4' x 8', 5/8" thick, screwed	1.000	S.F.	.29	.80	1.09
Main runners, 1-1/2" C.R.C., 4' O.C.	.500	S.F.	.20	.90	1.10
25 ga., channels, 2' O.C.	1.000	S.F.	.29	1.66	1.95
Taped and finished	1.000	S.F.	.06	.72	.78
Paint, 2 coats, roller work	1.000	S.F.	.14	.80	.94
TOTAL			.98	4.88	5.86

C3030 220	Ceilings, Suspended Gypsum Board	COST PER S.F.		
		MAT.	INST.	TOTAL
1400	4' x 8' x 5/8", gypsum drywall, 2 coats paint	.98	4.88	5.86
1500	Thin coat plaster, 2 coats paint	.91	4.26	5.17
1600	Spray-on sand finish, no paint	.84	4.23	5.07
1700	12" x 12" x 3/4" acoustical wood fiber tile	2.55	5.15	7.70
1800				
1900				
2000				
2100				
2200				
2300				
2400				
2500				
2600				
2700				
2900	Cut & patch to match existing construction, add, minimum	2%	3%	
3000	Maximum	5%	9%	
3100	Dust protection, add, minimum	1%	2%	
3200	Maximum	4%	11%	
3300	Equipment usage curtailment, add, minimum	1%	1%	
3400	Maximum	3%	10%	
3500	Material handling & storage limitation, add, minimum	1%	1%	
3600	Maximum	6%	7%	
3700	Protection of existing work, add, minimum	2%	2%	
3800	Maximum	5%	7%	
3900	Shift work requirements, add, minimum		5%	
4000	Maximum		30%	

C3030 Ceiling Finishes

This page illustrates and describes suspended plaster and lath systems including gypsum plaster, lath, furring and runners with ceiling painted. Lines within System Components give the unit price and total price per square foot for this system. Prices for suspended plaster and lath systems are on Line Items C3030 230 1200 thru 2300. Both material quantities and labor costs have been adjusted for the system listed.

Factors: To adjust for job conditions other than normal working situations use Lines C3030 230 2900 thru 4000.

Example: You are to install the system to match existing construction. Go to Line C3030 230 2900 and apply these percentages to the appropriate MAT. and INST. costs.

System Components	QUANTITY	UNIT	COST PER S.F.		
			MAT.	INST.	TOTAL
Gypsum plaster, 3 coats, on 3.4# rib lath, on 3/4″ C.R.C. furring on 1-1/2″ main runners, ceiling painted.					
Gypsum plaster, 3 coats	.110	S.Y.	.59	4.57	5.16
3.4# rib lath	.110	S.Y.	.62	1.10	1.72
Main runners, 1-1/2″ C.R.C. 24″ O.C.	1.000	S.F.	.39	1.79	2.18
Furring 3/4″ C.R.C. 16″ OC	1.000	S.F.	.43	2.41	2.84
Painting, 2 coats, roller work	1.000	S.F.	.14	.80	.94
TOTAL			2.17	10.67	12.84

C3030 230	Ceiling, Suspended Plaster	COST PER S.F.		
		MAT.	INST.	TOTAL
1200	Gypsum plaster, 3 coats, on 3.4# rib lath	2.16	10.65	12.81
1300	On 2.5# diamond lath	2.16	10.60	12.76
1400	On 3/8″ gypsum lath	1.96	10.60	12.56
1500	2 coats, on 3.4# rib lath	1.91	10.65	12.56
1600	On 2.5# diamond lath	1.99	9.90	11.89
1700	On 3/8″ gypsum lath	1.79	9.90	11.69
1800	Perlite plaster, 3 coats, on 3.4# rib lath	2.40	11.75	14.15
1900	On 2.5# diamond lath	2.40	11.70	14.10
2000	On 3/8″ gypsum lath	2.20	11.70	13.90
2100	2 coats, on 3.4# rib lath	2.15	10.60	12.75
2200	On 2.5# diamond lath	2.15	10.55	12.70
2300	On 3/8″ gypsum lath	1.95	10.55	12.50
2400				
2500				
2600				
2700				
2900	Cut & patch to match existing construction, add, minimum	2%	3%	
3000	Maximum	5%	9%	
3100	Dust protection, add, minimum	1%	2%	
3200	Maximum	4%	11%	
3300	Equipment usage curtailment, add, minimum	1%	1%	
3400	Maximum	3%	10%	
3500	Material handling & storage limitation, add, minimum	1%	1%	
3600	Maximum	6%	7%	
3700	Protection of existing work, add, minimum	2%	2%	
3800	Maximum	5%	7%	
3900	Shift work requirements, add, minimum		5%	
4000	Maximum		30%	

C3030 Ceiling Finishes

C3030 250	Selective Price Sheet	COST PER S.F.		
		MAT.	INST.	TOTAL
0100	Ceiling, plaster, gypsum, 2 coats	.42	3.92	4.34
0200	3 coats	.59	4.62	5.21
0300	Perlite or vermiculite, 2 coats	.58	4.56	5.14
0400	3 coats	.83	5.70	6.53
0500	Gypsum lath, plain, 3/8" thick	.42	.91	1.33
0600	1/2" thick	.36	.97	1.33
0700	Firestop, 3/8" thick	.31	1.11	1.42
0800	1/2" thick	.35	1.19	1.54
0900	Metal lath, rib, 2.75 lb.	.49	1.03	1.52
1000	3.40 lb.	.62	1.11	1.73
1100	Diamond, 2.50 lb.	.62	1.03	1.65
1200	3.40 lb.	.51	1.29	1.80
1500	Drywall, standard, 1/2" thick, no finish included	.28	.80	1.08
1600	Taped and finished	.33	1.87	2.20
1700	5/8" thick, no finish included	.29	.80	1.09
1800	Taped and finished	.34	1.87	2.21
1900	Fire resistant, 1/2" thick, no finish included	.51	.80	1.31
2000	Taped and finished	.56	1.87	2.43
2100	5/8" thick, no finish included	.37	.80	1.17
2200	Taped and finished	.43	1.87	2.30
2300	Water resistant, 1/2" thick, no finish included	.36	.80	1.16
2400	Taped and finished	.42	1.87	2.29
2500	5/8" thick, no finish included	.47	.80	1.27
2600	Taped and finished	.53	1.87	2.40
2700	Finish, instead of taping			
2800	For thin coat plaster, add	.13	.90	1.03
2900	Finish, textured spray, add	.06	.87	.93
3000				
3100				
3200				
3300	Tile, stapled or glued, mineral fiber plastic coated, 5/8" thick	4.75	2.39	7.14
3400	3/4" thick	5.15	2.39	7.54
3500	Wood fiber, 1/2" thick	1.25	1.79	3.04
3600	3/4" thick	1.77	1.79	3.56
3700	Suspended, fiberglass, film faced, 5/8" thick	5.05	1.14	6.19
3800	3" thick	4.61	1.59	6.20
3900	Mineral fiber, 5/8" thick, standard	1.38	1.06	2.44
4000	Aluminum	4.22	1.19	5.41
4100	Wood fiber, reveal edge, 1" thick			
4200	3" thick			
4300	Framing, metal furring, 3/4" channels, 12" O.C.	.48	3.33	3.81
4400	16" O.C.	.43	2.41	2.84
4500	24" O.C.	.29	1.66	1.95
4600				
4700	1-1/2" channels, 12" O.C.	.65	3.68	4.33
4800	16" O.C.	.58	2.69	3.27
4900	24" O.C.	.39	1.79	2.18
5000				
5100				
5200				
5300	Ceiling suspension systems, for tile,			
5400	Concealed "Z" bar, 12" module	1.03	1.38	2.41
5500	Class A, "T" bar 2'-0" x 4'-0" grid	1.66	.89	2.55
5600	2'-0" x 2'-0" grid	2.13	1.10	3.23
5700	Carrier channels for lighting fixtures, add	.24	1.56	1.80
5800				
5900				
6000				

C3030 Ceiling Finishes

C3030 250	Selective Price Sheet	COST PER S.F.		
		MAT.	INST.	TOTAL
6100	Wood, furring 1" x 3", on wood, 12" O.C.	.51	2.04	2.55
6200	16" O.C..	.38	1.53	1.91
6300	24" O.C.	.26	1.02	1.28
6400				
6500	On concrete, 12" O.C.	.56	3.41	3.97
6600	16" O.C.	.42	2.56	2.98
6700	24" O.C.	.28	1.71	1.99
6800				
6900	Joists, 2" x 4", 12" O.C.	.93	1.47	2.40
7000	16" O.C.	.78	1.23	2.01
7100	24" O.C.	.63	1	1.63
7200	32" O.C.	.48	.76	1.24
7300	2" x 6", 12" O.C.	1.40	1.49	2.89
7400	16" O.C.	1.16	1.23	2.39
7500	24" O.C.	.92	.98	1.90
7600	32" O.C.	.68	.73	1.41

D1010 Elevators and Lifts

This page illustrates and describes oil hydraulic elevator systems. Prices for various oil hydraulic elevator systems are on Line Items D1010 130 0400 thru 2000. Both material quantities and labor costs have been adjusted for the system listed.

Factors: To adjust for job conditions other than normal working situations use Lines D1010 130 3100 thru 4000.

Example: You are to install the system with limited equipment usage. Go to Line D1010 130 3400 and apply this percentage to the appropriate TOTAL costs.

D1010 130	Oil Hydraulic Elevators	COST EACH		
		MAT.	INST.	TOTAL
0310	Oil-hydraulic elevator systems			
0320	Including piston and piston shaft			
0330				
0400	1500 Lb. passenger, 2 floors	51,500	22,900	74,400
0500	3 floors	63,000	35,400	98,400
0600	4 floors	89,000	49,700	138,700
0700	5 floors	100,000	62,000	162,000
0800	6 floors	110,000	74,000	184,000
0900				
1000	2500 Lb. passenger, 2 floors	55,000	22,900	77,900
1100	3 floors	66,500	35,400	101,900
1200	4 floors	87,500	49,700	137,200
1300	5 floors	98,500	62,000	160,500
1400	6 floors	114,000	74,000	188,000
1500				
1600	4000 Lb. passenger, 2 floors	62,000	22,900	84,900
1700	3 floors	75,000	35,400	110,400
1800	4 floors	99,500	49,700	149,200
1900	5 floors	110,500	62,000	172,500
2000	6 floors	120,500	74,000	194,500
2100				
2200				
2300				
2400				
2500				
2600				
2700				
2800				
2900				
3000				
3100	Dust protection, add, minimum	1%	2%	
3200	Maximum	4%	11%	
3300	Equipment usage curtailment, add, minimum	1%	1%	
3400	Maximum	3%	10%	
3500	Material handling & storage limitation, add, minimum	1%	1%	
3600	Maximum	6%	7%	
3700	Protection of existing work, add, minimum	2%	2%	
3800	Maximum	5%	7%	
3900	Shift work requirements, add, minimum		5%	
4000	Maximum		30%	

D2010 Plumbing Fixtures

This page illustrates and describes a women's public restroom system including a water closet, lavatory, accessories, and service piping. Lines within System Components give the unit price and total price on a cost each basis for this system. Prices for women's public restroom systems are on Line Items D2010 956 1300 thru 1700. Both material quantities and labor costs have been adjusted for the system listed.

Factors: To adjust for job conditions other than normal working situations use Lines D2010 956 2900 thru 4000.

Example: You are to install the system and protect surrounding area from dust. Go to Line D2010 956 3100 and apply these percentages to the MAT. and INST. costs.

System Components	QUANTITY	UNIT	COST EACH MAT.	COST EACH INST.	COST EACH TOTAL
Public women's restroom incl. water closet, lavatory, accessories and Necessary service piping to install this system in one wall.					
Water closet, wall mounted, one piece	1.000	Ea.	1,075	267	1,342
Rough-in waste and vent for water closet	1.000	Set	1,900	605	2,505
Lavatory, 20" x 18" P.E. cast iron with accessories	1.000	Ea.	340	194	534
Rough-in waste and vent for lavatory	1.000	Set	895	935	1,830
Toilet partition, painted metal between walls, floor mounted	1.000	Ea.	525	239	764
For handicap unit, add	1.000	Ea.	365		365
Grab bar, 36" long	2.000	Ea.	42	72	114
Mirror, 18" x 24", with stainless steel shelf	1.000	Ea.	107	36	143
Napkin/tampon dispenser, recessed	1.000	Ea.	430	47.50	477.50
Soap Dispenser, chrome, surface mounted, liquid	1.000	Ea.	57.50	36	93.50
Toilet tissue dispenser, surface mounted, stainless steel	1.000	Ea.	16.05	24	40.05
Towel dispenser, surface mounted, stainless steel	1.000	Ea.	73.50	44.50	118
TOTAL			5,826.05	2,500	8,326.05

D2010 956	Plumbing - Public Restroom	COST EACH MAT.	COST EACH INST.	COST EACH TOTAL
1200				
1300	Public women's restroom, one water closet, one lavatory	5,825	2,500	8,325
1400	Two water closets, two lavatories	11,900	4,775	16,675
1500				
1600	For each additional water closet over 2, add	4,100	1,100	5,200
1700	For each additional lavatory over 2, add	1,400	1,200	2,600
1800				
1900				
2400	NOTE: PLUMBING APPROXIMATIONS			
2500	WATER CONTROL: water meter, backflow preventer,			
2600	Shock absorbers, vacuum breakers, mixer....10 to 15% of fixtures			
2700	PIPE AND FITTINGS: .... 30 to 60% of fixtures			
2800				
2900	Cut & patch to match existing construction, add, minimum	2%	3%	
3000	Maximum	5%	9%	
3100	Dust protection, add, minimum	1%	2%	
3200	Maximum	4%	11%	
3300	Equipment usage curtailment, add, minimum	1%	1%	
3400	Maximum	3%	10%	
3500	Material handling & storage limitation, add, minimum	1%	1%	

D2010 Plumbing Fixtures

D2010 956	Plumbing - Public Restroom	COST EACH		
		MAT.	INST.	TOTAL
3600	Maximum	6%	7%	
3700	Protection of existing work, add, minimum	2%	2%	
3800	Maximum	5%	7%	
3900	Shift work requirements, add, minimum		5%	
4000	Maximum		30%	

D20 Plumbing

D2010 Plumbing Fixtures

This page illustrates and describes a men's public restroom system including a water closet, urinal, lavatory, accessories and service piping. Lines within System Components give the unit price and total price on a cost each basis for this system. Prices for men's public restroom systems are on Line Items D2010 957 1700 thru 2200. Both material quantities and labor costs have been adjusted for the system listed.

Factors: To adjust for job conditions other than normal working situations use Lines D2010 957 2900 thru 4000.

Example: You are to install the system and match existing construction. Go to Line D2010 957 3000 and apply these percentages to the appropriate MAT. and INST. costs.

System Components	QUANTITY	UNIT	COST EACH		
			MAT.	INST.	TOTAL
Public men's restroom incl. water closet, urinal, lavatory, accessories, And necessary service piping to install this system in one wall.					
Water closet, wall mounted, one piece	1.000	Ea.	1,075	267	1,342
Rough-in waste & vent for water closet	1.000	Set	1,900	605	2,505
Urinal, wall hung	1.000	Ea.	375	515	890
Rough-in waste & vent for urinal	1.000	Set	890	550	1,440
Lavatory, 20" x 18", P.E. cast iron with accessories	1.000	Ea.	340	194	534
Rough-in waste & vent for lavatory	1.000	Set	895	935	1,830
Partition, painted mtl., between walls, floor mntd.	1.000	Ea.	525	239	764
For handicap unit, add	1.000	Ea.	365		365
Grab bars	2.000	Ea.	42	72	114
Urinal screen, powder coated steel, wall mounted	1.000	Ea.	116	143	259
Mirror, 18" x 24" with stainless steel shelf	1.000	Ea.	107	36	143
Soap dispenser, surface mounted, liquid	1.000	Ea.	57.50	36	93.50
Toilet tissue dispenser, surface mounted, stainless steel	1.000	Ea.	16.05	24	40.05
Towel dispenser, surface mounted, stainless steel	1.000	Ea.	73.50	44.50	118
TOTAL			6,777.05	3,660.50	10,437.55

D2010 957	Plumbing - Public Restroom	COST EACH		
		MAT.	INST.	TOTAL
1600				
1700	Public men's restroom, one water closet, one urinal, one lavatory	6,775	3,650	10,425
1800	Two water closets, two urinals, two lavatories	14,200	7,125	21,325
1900				
2000	For each additional water closet over 2, add	4,100	1,100	5,200
2100	For each additional urinal over 2, add	1,375	1,200	2,575
2200	For each additional lavatory over 2, add	1,400	1,200	2,600
2300				
2400	NOTE: PLUMBING APPROXIMATIONS			
2500	WATER CONTROL: water meter, backflow preventer,			
2600	Shock absorbers, vacuum breakers, mixer....10 to 15% of fixtures			
2700	PIPE AND FITTINGS: .... 30 to 60% of fixtures			
2800				
2900	Cut & patch to match existing construction, add, minimum	2%	3%	
3000	Maximum	5%	9%	
3100	Dust protection, add, minimum	1%	2%	
3200	Maximum	4%	11%	
3300	Equipment usage curtailment, add, minimum	1%	1%	

D2010 Plumbing Fixtures

D2010 957	Plumbing - Public Restroom	COST EACH		
		MAT.	INST.	TOTAL
3400	Maximum	3%	10%	
3500	Material handling, & storage limitation, add, minimum	1%	1%	
3600	Maximum	6%	7%	
3700	Protection of existing work, add, minimum	.2%	2%	
3800	Maximum	5%	7%	
3900	Shift work requirements, add, minimum		5%	
4000	Maximum		30%	

D20 Plumbing

D2010 Plumbing Fixtures

This page illustrates and describes a two fixture lavatory system including a water closet, lavatory, accessories and all service piping. Lines within System Components give the unit price and total price on a cost each basis for this system. Prices for two fixture lavatory systems are on Line Items D2010 958 1800 and 1900. Both material quantities and labor costs have been adjusted for the system listed.

Factors: To adjust for job conditions other than normal working situations use Lines D2010 958 2900 thru 4000.

Example: You are to install the system while controlling dust in the work area. Go to Line D2010 958 3200 and apply these percentages to the appropriate MAT. and INST. costs.

System Components	QUANTITY	UNIT	COST EACH		
			MAT.	INST.	TOTAL
Two fixture bathroom incl. water closet, lavatory, accessories and Necessary service piping to install this system in 2 walls.					
Water closet, floor mounted, 2 piece, close coupled	1.000	Ea.	219	292	511
Rough in waste & vent for water closet	1.000	Set	410	510	920
Lavatory, 20" x 18", P.E. cast iron with accessories	1.000	Ea.	340	194	534
Rough in waste & vent for lavatory	1.000	Set	895	935	1,830
Additional service piping					
1/2" copper pipe with sweat solder joints	10.000	L.F.	48.60	106.50	155.10
2" black schedule 40 steel pipe with threaded couplings	12.000	L.F.	104.40	288	392.40
4" cast iron soil pipe with lead and oakum joints	7.000	L.F.	224	196	420
Accessories					
Toilet tissue dispenser, chrome, single roll	1.000	Ea.	16.05	24	40.05
18" long stainless steel towel bar	1.000	Ea.	37.50	31	68.50
Medicine cabinet with mirror, 20" x 16", unlighted	1.000	Ea.	249	51	300
TOTAL			2,543.55	2,627.50	5,171.05

D2010 958	Plumbing - Two Fixture Bathroom	COST EACH		
		MAT.	INST.	TOTAL
1800	Water closet, lavavatory & accessories, in 2 walls with all service piping	2,550	2,625	5,175
1900	Above system installed in 1 wall	2,400	2,275	4,675
2000				
2400	NOTE: PLUMBING APPROXIMATIONS			
2500	WATER CONTROL: water meter, backflow preventer,			
2600	Shock absorbers, vacuum breakers, mixer....10 to 15% of fixtures			
2700	PIPE AND FITTINGS: .... 30 to 60% of fixtures			
2800				
2900	Cut & patch to match existing construction, add, minimum	2%	3%	
3000	Maximum	5%	9%	
3100	Dust protection, add, minimum	1%	2%	
3200	Maximum	4%	11%	
3300	Equipment usage curtailment, add, minimum	1%	1%	
3400	Maximum	3%	10%	
3500	Material handling & storage limitation, add, minimum	1%	1%	
3600	Maximum	6%	7%	
3700	Protection of existing work, add, minimum	2%	2%	
3800	Maximum	5%	7%	
3900	Shift work requirements, add, minimum		5%	
4000	Maximum		30%	

D20 Plumbing

D2010 Plumbing Fixtures

This page illustrates and describes a three fixture bathroom system including a water closet, tub, lavatory, accessories and service piping. Lines within System Components give the unit price and total price on a cost each basis for this system. Prices for a three fixture bathroom system is on Line Item D2010 959 1700. Both material quantities and labor costs have been adjusted for the system listed.

Factors: To adjust for job conditions other than normal working situations use Lines D2010 959 2900 thru 4000.

Example: You are to install the system and protect all existing work. Go to Line D2010 959 3800 and apply these percentages to the appropriate MAT. and INST. costs.

System Components	QUANTITY	UNIT	COST EACH		
			MAT.	INST.	TOTAL
Three fixture bathroom incl. water closet, bathtub, lavatory, accessories, And necessary service piping to install this system in 1 wall.					
Water closet, tank vit china flr mtd close cpld, 2 pc. w/seat supply & stop	1.000	Ea.	219	292	511
For rough in, supply, waste and vent	.930	Set	381.30	474.30	855.60
Bath tub, P.E., cast iron, 5′ long, with accessories	1.000	Ea.	1,600	350	1,950
Rough-in, supply, waste and vent, for all above tubs, add	.950	Set	617.50	712.50	1,330
Lavatory, 20″ x 18″ P.E. cast iron with accessories	1.000	Ea.	340	194	534
Rough-in, supply, waste and vent for above lavatories	1.000	Set	895	935	1,830
Accessories					
Toilet tissue dispenser, surface mounted, SS, single roll	1.000	Ea.	16.05	24	40.05
Towel bar, stainless steel, 18″ long	2.000	Ea.	75	62	137
Medicine cabinets with mirror, sst. frame, 16″ x 22″, unlighted	1.000	Ea.	249	51	300
TOTAL			4,392.85	3,094.80	7,487.65

D2010 959	Plumbing - Three Fixture Bathroom	COST EACH		
		MAT.	INST.	TOTAL
1600				
1700	Water closet, lavatory, bathtub, in 1 wall with all service piping	4,400	3,100	7,500
2400	NOTE: PLUMBING APPROXIMATIONS			
2500	WATER CONTROL: water meter, backflow preventer,			
2600	Shock absorbers, vacuum breakers, mixer....10 to 15% of fixtures			
2700	PIPE AND FITTINGS: .... 30 to 60% of fixtures			
2800				
2900	Cut & patch to match existing construction, add, minimum	2%	3%	
3000	Maximum	5%	9%	
3100	Dust protection, add, minimum	1%	2%	
3200	Maximum	4%	11%	
3300	Equipment usage curtailment, add, minimum	1%	1%	
3400	Maximum	3%	10%	
3500	Material handling & storage limitation, add, minimum	1%	1%	
3600	Maximum	6%	7%	
3700	Protection of existing work, add, minimum	2%	2%	
3800	Maximum	5%	7%	
3900	Shift work requirements, add, minimum		5%	
4000	Maximum		30%	

D2010 Plumbing Fixtures

This page illustrates and describes a three fixture bathroom system including a water closet, tub, lavatory, accessories, and service piping. Lines within System Components give the unit price and total price on a cost each basis for this system. Prices for a three fixture bathroom system are on Line Item D2010 960 1900 thru 2000. Both material quantities and labor costs have been adjusted for the system listed.

Factors: To adjust for job conditions other than normal working situations use Lines D2010 960 2900 thru 4000.

Example: You are to install the system and protect the surrounding area from dust. Go to Line D2010 960 3100 and apply these percentages to the appropriate MAT. and INST. costs.

System Components	QUANTITY	UNIT	COST EACH		
			MAT.	INST.	TOTAL
Three fixture bathroom incl. water closet, bathtub, lavatory, accessories, And necessary service piping to install this system in 2 walls.					
Water closet, floor mounted, 2 piece, close coupled	1.000	Ea.	219	292	511
Rough-in waste & vent for water closet	1.000	Set	410	510	920
Bathtub, P.E. cast iron, 5' long with accessories	1.000	Ea.	1,600	350	1,950
Rough-in waste & vent for bathtub	1.000	Set	650	750	1,400
Lavatory, 20" x 18" P.E. cast iron with accessories	1.000	Ea.	340	194	534
Rough-in waste & vent for lavatory	1.000	Set	895	935	1,830
Additional service piping					
1-1/4" copper DWV type tubing, sweat solder joints	6.000	L.F.	91.50	86.10	177.60
2" black schedule 40 steel pipe with threaded couplings	12.000	L.F.	104.40	288	392.40
Accessories					
Toilet tissue dispenser, chrome, single roll	1.000	Ea.	16.05	24	40.05
18" long stainless steel towel bar	2.000	Ea.	75	62	137
Medicine cabinet with mirror, 20" x 16", unlighted	1.000	Ea.	249	51	300
TOTAL			4,649.95	3,542.10	8,192.05

D2010 960	Plumbing - Three Fixture Bathroom	COST EACH		
		MAT.	INST.	TOTAL
1900	Water closet, lavatory, bathtub, in 2 walls with all service piping	4,650	3,550	8,200
2000	Above system with corner contour tub, P.E. cast iron	6,400	3,550	9,950
2300				
2400	NOTE: PLUMBING APPROXIMATIONS			
2500	WATER CONTROL: water meter, backflow preventer,			
2600	Shock absorbers, vacuum breakers, mixer....10 to 15% of fixtures			
2700	PIPE AND FITTINGS: .... 30 to 60% of fixtures			
2800				
2900	Cut & patch to match existing construction, add, minimum	2%	3%	
3000	Maximum	5%	9%	
3100	Dust protection, add, minimum	1%	2%	
3200	Maximum	4%	11%	
3300	Equipment usage curtailment, add, minimum	1%	1%	
3400	Maximum	3%	10%	
3500	Material handling & storage limitation, add, minimum	1%	1%	
3600	Maximum	6%	7%	
3700	Protection of existing work, add, minimum	2%	2%	
3800	Maximum	5%	7%	
3900	Shift work requirements, add, minimum		5%	
4000	Maximum		30%	

This page illustrates and describes a three fixture bathroom system including a water closet, shower, lavatory, accessories, and service piping, Lines within System Components give the unit price and total price on a cost each basis for this system. Prices for three fixture bathroom systems are on Line Items D2010 961 2100 thru 2200. Both material quantities and labor costs have been adjusted for the system listed.

Factors: To adjust for job conditions other than normal working situations use Lines D2010 961 2900 thru 4000.

Example: You are to install the system and protect existing construction. Go to Line D2010 961 3700 and apply these percentages to the appropriate MAT. and INST. costs.

System Components	QUANTITY	UNIT	COST EACH MAT.	COST EACH INST.	COST EACH TOTAL
Three fixture bathroom incl. water closet, shower, lavatory, accessories,					
And necessary service piping to install this system in 2 walls.					
Water closet, floor mounted, 2 piece, close coupled	1.000	Ea.	219	292	511
Rough-in waste & vent for water closet	1.000	Set	410	510	920
32" shower, enameled stall, molded stone receptor	1.000	Ea.	1,325	310	1,635
Rough-in waste & vent for shower	1.000	Set	690	755	1,445
Lavatory, 20" x 18" P.E. cast iron with accessories	1.000	Ea.	340	194	534
Rough-in waste & vent for lavatory	1.000	Set	895	935	1,830
Additional service piping					
1-1/4" Copper DWV type tubing, sweat solder joints	4.000	L.F.	61	57.40	118.40
2" black schedule 40 steel pipe with threaded couplings	6.000	L.F.	52.20	144	196.20
4" cast iron soil with lead and oakum joints	7.000	L.F.	224	196	420
Accessories					
Toilet tissue dispenser, chrome, single roll	1.000	Ea.	16.05	24	40.05
18" long stainless steel towel bar	2.000	Ea.	75	62	137
Medicine cabinet with mirror, 20" x 16", unlighted	1.000	Ea.	249	51	300
TOTAL			4,556.25	3,530.40	8,086.65

D2010 961	Plumbing - Three Fixture Bathroom	COST EACH MAT.	COST EACH INST.	COST EACH TOTAL
2100	Water closet, 32" shower, lavavatory, in 2 walls with all service piping	4,550	3,525	8,075
2200	Above system with 36" corner angle shower	5,750	3,550	9,300
2300				
2400	NOTE: PLUMBING APPROXIMATIONS			
2500	WATER CONTROL: water meter, backflow preventer,			
2600	Shock absorbers, vacuum breakers, mixer....10 to 15% of fixtures			
2700	PIPE AND FITTINGS: .... 30 to 60% of fixtures			
2800				
2900	Cut & patch to match existing construction, add, minimum	2%	3%	
3000	Maximum	5%	9%	
3100	Dust protection, add, minimum	1%	2%	
3200	Maximum	4%	11%	
3300	Equipment usage curtailment, add, minimum	1%	1%	
3400	Maximum	3%	10%	
3500	Material handling & storage limitation, add, minimum	1%	1%	
3600	Maximum	6%	7%	
3700	Protection of existing work, add, minimum	2%	2%	
3800	Maximum	5%	7%	
3900	Shift work requirements, add, minimum		5%	
4000	Maximum		30%	

For customer support on your Commercial Renovation Costs with RSMeans data, call 800.448.8182.

D2010   Plumbing Fixtures

This page illustrates and describes a four fixture bathroom system including a water closet, shower, bathtub, lavatory, accessories, and service piping. Lines within System Components give the unit price and total price on a cost each basis for this system. Prices for four fixture bathroom systems are on Line Items D2010 962 1800 thru 1900. Both material quantities and labor costs have been adjusted for the system listed.

Factors: To adjust for job conditions other than normal working situations use Lines D2010 962 2900 thru 4000.

Example: You are to install the system during weekends and evenings. Go to Line D2010 962 4000 and apply these percentages to the appropriate MAT. and INST. costs.

System Components	QUANTITY	UNIT	COST EACH		
			MAT.	INST.	TOTAL
Four fixture bathroom incl. water closet, shower, bathtub, lavatory					
Accessories and necessary service piping to install this system in 2 walls.					
Water closet, floor mounted, 2 piece, close coupled	1.000	Ea.	219	292	511
Rough-in waste & vent for water closet	1.000	Set	410	510	920
32" shower, enameled steel stall, molded stone receptor	1.000	Ea.	1,325	310	1,635
Rough-in waste & vent for shower	1.000	Set	690	755	1,445
Bathtub, P.E. cast iron, 5' long with accessories	1.000	Ea.	1,600	350	1,950
Rough-in waste & vent for bathtub	1.000	Set	650	750	1,400
Lavatory, 20" x 18" P.E. cast iron with accessories	1.000	Ea.	340	194	534
Rough-in waste & vent for lavatory	1.000	Set	895	935	1,830
Accessories					
Toilet tissue dispenser, chrome, single roll	1.000	Ea.	16.05	24	40.05
18" long stainless steel towel bar	2.000	Ea.	75	62	137
Medicine cabinet with mirror, 20" x 16" unlighted	1.000	Ea.	249	51	300
TOTAL			6,469.05	4,233	10,702.05

D2010 962	Plumbing - Four Fixture Bathroom	COST EACH		
		MAT.	INST.	TOTAL
1800	Water closet, 32" shower, bathtub, lavavatory, in 2 walls	6,475	4,225	10,700
1900	Above system with 36" corner angle shower, plumbing in 3 walls	7,825	4,650	12,475
2000				
2300				
2400	NOTE: PLUMBING APPROXIMATIONS			
2500	WATER CONTROL: water meter, backflow preventer,			
2600	Shock absorbers, vacuum breakers, mixer....10 to 15% of fixtures			
2700	PIPE AND FITTINGS: .... 30 to 60% of fixtures			
2800				
2900	Cut & patch to match existing construction, add, minimum	2%	3%	
3000	Maximum	5%	9%	
3100	Dust protection, add, minimum	1%	2%	
3200	Maximum	4%	11%	
3300	Equipment usage curtailment, add, minimum	1%	1%	
3400	Maximum	3%	10%	
3500	Material handling & storage limitation, add, minimum	1%	1%	
3600	Maximum	6%	7%	
3700	Protection of existing work, add, minimum	2%	2%	
3800	Maximum	5%	7%	
3900	Shift work requirements, add, minimum		5%	
4000	Maximum		30%	

D2010 Plumbing Fixtures

This page illustrates and describes a five fixture bathroom system including a water closet, shower, bathtub, lavatories, accessories, and service piping. Lines within System Components give the unit price and total price on a cost each basis for this system. Prices for five fixture bathroom systems are on Line Items D2010 963 1800 thru 1900. Both material quantities and labor costs have been adjusted for the system listed.

Factors: To adjust for job conditions other than normal working situations use Lines D2010 963 2900 thru 4000.

Example: You are to install and match any existing construction. Go to Line D2010 963 2900 and apply these percentages to the appropriate MAT. and INST. costs.

System Components	QUANTITY	UNIT	COST EACH MAT.	COST EACH INST.	COST EACH TOTAL
Five fixture bathroom incl. water closet, shower, bathtub, 2 lavatories, Accessories and necessary service piping to install this system in 1 wall.					
Water closet, floor mounted, 2 piece, close coupled	1.000	Ea.	219	292	511
Rough-in waste & vent for water closet	1.000	Set	410	510	920
36" corner angle shower, enameled steel stall, molded stone receptor	1.000	Ea.	2,525	325	2,850
Rough-in waste & vent for shower	1.000	Set	690	755	1,445
Bathtub, P.E. cast iron, 5' long with accessories	1.000	Ea.	1,600	350	1,950
Rough-in waste & vent for bathtub	1.000	Set	650	750	1,400
Countertop, plastic laminate, with backsplash	2.000	Ea.	252	96	348
Vanity base, 2 doors, 24" wide	2.000	Ea.	860	260	1,120
Lavatories, 20" x 18" cabinet mntd, PECI	2.000	Ea.	830	484	1,314
Rough-in waste & vent for lavatories	1.600	Set	1,040	1,080	2,120
Accessories					
Toilet tissue dispenser, chrome, single roll	1.000	Ea.	16.05	24	40.05
18" long stainless steel towel bars	2.000	Ea.	75	62	137
Medicine cabinet with mirror, 20" x 16", unlighted	2.000	Ea.	498	102	600
TOTAL			9,665.05	5,090	14,755.05

D2010 963	Plumbing - Five Fixture Bathroom	COST EACH MAT.	COST EACH INST.	COST EACH TOTAL
1800	Water closet, shower, bathtub, 2 lavatories, on 1 wall	9,675	5,100	14,775
1900	Above system installed in 2 walls with all necessary service piping	9,725	5,275	15,000
2000				
2100				
2200				
2300				
2400	NOTE: PLUMBING APPROXIMATIONS			
2500	WATER CONTROL: water meter, backflow preventer,			
2600	Shock absorbers, vacuum breakers, mixer....10 to 15% of fixtures			
2700	PIPE AND FITTINGS: .... 30 to 60% of fixtures			
2800				
2900	Cut & patch to match existing construction, add, minimum	2%	3%	
3000	Maximum	5%	9%	
3100	Dust protection, add, minimum	1%	2%	
3200	Maximum	4%	11%	
3300	Equipment usage curtailment, add, minimum	1%	1%	
3400	Maximum	3%	10%	
3500	Material handling & storage limitation, add, minimum	1%	1%	

For customer support on your Commercial Renovation Costs with RSMeans data, call 800.448.8182.

D2010 Plumbing Fixtures

D2010 963	Plumbing - Five Fixture Bathroom	COST EACH		
		MAT.	INST.	TOTAL
3600	Maximum	6%	7%	
3700	Protection of existing work, add, minimum	2%	2%	
3800	Maximum	5%	7%	
3900	Shift work requirements, add, minimum		5%	
4000	Maximum		30%	

This page illustrates and describes an oil fired hot water baseboard system including an oil fired boiler, fin tube radiation and all fittings and piping. Lines within System Components give the unit price and total price per square foot for this system. Prices for oil fired hot water baseboard systems are on Line Items D3010 540 1600 thru 2100. Both material quantities and labor costs have been adjusted for the system listed.

Factors: To adjust for job conditions other than normal working situations use Lines D3010 540 2900 thru 4000.

Example: You are to install the system while protecting all existing work. Go to Line D3010 540 3700 and apply these percentages to the appropriate MAT. and INST. costs.

System Components	QUANTITY	UNIT	COST PER S.F.		
			MAT.	INST.	TOTAL
Oil fired hot water baseboard system including boiler, fin tube radiation					
And all necessary fittings and piping.					
Area to 800 S.F.					
Boiler, cast iron, w/oil piping, 97 MBH	1.000	Ea.	3,062.50	1,906.25	4,968.75
Copper piping	130.000	L.F.	1,638	1,930.50	3,568.50
Fin tube radiation	68.000	L.F.	578	1,836	2,414
Circulator	1.000	Ea.	750	258	1,008
Oil tank	1.000	Ea.	730	310	1,040
Expansion tank, ASME	1.000	Ea.	840	112	952
TOTAL			7,598.50	6,352.75	13,951.25
COST PER S.F.		S.F.	9.50	7.94	17.44

D3010 540	Heating - Oil Fired Hot Water	COST PER S.F.		
		MAT.	INST.	TOTAL
1600	Cast iron boiler, area to 800 S.F.	9.50	7.94	17.44
1700	To 1000 S.F.	8.05	6.85	14.90
1800	To 1200 S.F.	6.95	6.20	13.15
1900	To 1600 S.F.	8.80	6.20	15
2000	To 2000 S.F.	7.30	5.50	12.80
2100	To 3000 S.F.	5.35	4.63	9.98
2200				
2300				
2700				
2800				
2900	Cut & patch to match existing construction, add, minimum	2%	3%	
3000	Maximum	5%	9%	
3100	Dust protection, add, minimum	1%	2%	
3200	Maximum	4%	11%	
3300	Equipment usage curtailment, add, minimum	1%	1%	
3400	Maximum	3%	10%	
3500	Material handling & storage limitation, add, minimum	1%	1%	
3600	Maximum	6%	7%	
3700	Protection of existing work, add, minimum	2%	2%	
3800	Maximum	5%	7%	
3900	Shift work requirements, add, minimum		5%	
4000	Maximum		30%	

D3010 Energy Supply

This page illustrates and describes a gas fired hot water baseboard system including a gas fired boiler, fin tube radiation, fittings and piping. Lines within System Components give the unit price and total price per square foot for this system. Prices for gas fired hot water baseboard systems are on Line Items D3010 550 1400 thru 1900. Both material quantities and labor costs have been adjusted for the system listed.

Factors: To adjust for job conditions other than normal working situations use Lines D3010 550 2900 thru 4000.

Example: You are to install the system with minimal equipment usage. Go to Line D3010 550 3400 and apply these percentages to the appropriate MAT. and INST. costs.

System Components	QUANTITY	UNIT	COST PER S.F. MAT.	COST PER S.F. INST.	COST PER S.F. TOTAL
Gas fired hot water baseboard system including boiler, fin tube radiation,					
All necessary fittings and piping.					
Area to 800 S.F.					
Cast iron boiler, insulating jacket, gas piping, 80 MBH	1.000	Ea.	3,937.50	2,812.50	6,750
Copper piping	130.000	L.F.	1,638	1,930.50	3,568.50
Fin tube radiation	68.000	L.F.	578	1,836	2,414
Circulator, flange connection	1.000	Ea.	750	258	1,008
Expansion tank, ASME	1.000	Ea.	840	112	952
TOTAL			7,743.50	6,949	14,692.50
COST PER S.F.		S.F.	9.68	8.69	18.37

D3010 550	Heating - Gas Fired Hot Water	COST PER S.F. MAT.	COST PER S.F. INST.	COST PER S.F. TOTAL
1400	Cast iron boiler, area to 800 S.F.	9.69	8.69	18.38
1500	To 1000 S.F.	8.10	7.65	15.75
1600	To 1200 S.F.	6.95	6.85	13.80
1700	To 1600 S.F.	8.85	6.70	15.55
1800	To 2000 S.F.	7.30	6	13.30
1900	To 3000 S.F.	5.60	5.25	10.85
2000				
2100				
2900	Cut & patch to match existing construction, add, minimum	2%	3%	
3000	Maximum	5%	9%	
3100	Dust protection, add, minimum	1%	2%	
3200	Maximum	4%	11%	
3300	Equipment usage curtailment, add, minimum	1%	1%	
3400	Maximum	3%	10%	
3500	Material handling & storage limitation, add, minimum	1%	1%	
3600	Maximum	6%	7%	
3700	Protection of existing work, add, minimum	2%	2%	
3800	Maximum	5%	7%	
3900	Shift work requirements, add, minimum		5%	
4000	Maximum		30%	

D3010 Energy Supply

This page illustrates and describes a wood burning stove system including a free standing stove, preformed hearth, masonry chimney, and necessary piping and fittings. Lines within System Components give the unit price and total price on a cost each basis for this system. Prices for wood burning stove systems are on Line Items D3010 991 1400 thru 2000. Both material quantities and labor costs have been adjusted for the system listed.

Factors: To adjust for job conditions other than normal working situations use Lines D3010 991 2900 thru 4000.

Example: You are to install the system and protect existing construction. Go to Line D3010 991 3500 and apply these percentages to the appropriate MAT., and INST. costs.

System Components	QUANTITY	UNIT	COST EACH MAT.	INST.	TOTAL
Cast iron, free standing, wood burning stove with preformed hearth, masonry Chimney, and all necessary piping and fittings to install in chimney.					
Cast iron wood burning stove, stove pipe	1.000	Ea.	1,875	1,100	2,975
Wall panel or hearth, non-combustible	1.000	Ea.	978.50	203.30	1,181.80
16" x 16" brick chimney, 8" x 8" flue	20.000	V.L.F.	1,187.50	1,762.50	2,950
Concrete in place, chimney foundations	.500	C.Y.	106	148.47	254.47
TOTAL			4,147	3,214.27	7,361.27

D3010 991	Wood Burning Stoves	COST EACH MAT.	INST.	TOTAL
1400	Cast iron, free standing with preformed hearth	4,150	3,225	7,375
1500				
1600	Installed in existing fireplace	2,400	1,200	3,600
1700				
1800	Installed with insulated metal chimney			
1900	System including ceiling package,			
2000	Metal chimney to 10 L.F.	4,200	1,725	5,925
2100				
2200				
2300				
2400				
2500				
2600				
2700				
2900	Dust protection, add, minimum	1%	2%	
3000	Maximum	4%	11%	
3100	Equipment usage curtailment, add, minimum	1%	1%	
3200	Maximum	3%	10%	
3300	Material handling & storage limitation, add, minimum	1%	1%	
3400	Maximum	6%	7%	
3500	Protection of existing work, add, minimum	2%	2%	
3600	Maximum	5%	7%	
3700	Shift work requirements, add, minimum		5%	
3800	Maximum		30%	
3900	Temporary shoring and bracing, add, minimum	2%	5%	
4000	Maximum	5%	12%	

D3010 Energy Supply

This page illustrates and describes masonry fireplace systems including a brick fireplace, footing, foundation, hearth, firebox, chimney and flue. Lines within System Components give the unit price and total price on a cost each basis for this system. Prices for masonry fireplace systems are on Line Items D3010 992 1200 thru 1500. Both material quantities and labor costs have been adjusted for the system listed.

Factors: To adjust for job conditions other than normal working situations use Lines D3010 992 2700 thru 4000.

Example: You are to install the system with some temporary shoring and bracing. Go to Line D3010 992 3900 and apply these percentages to the appropriate MAT. and INST. costs.

System Components	QUANTITY	UNIT	COST EACH MAT.	COST EACH INST.	COST EACH TOTAL
Brick masonry fireplace, including footing, foundation, hearth, firebox, Chimney and flue, chimney 12′ above firebox.					
Footing, 4′ x 7′ x 12″ thick, 3000 psi concrete	1.040	C.Y.	292.24	354.72	646.96
Foundation, 12″ concrete block	180.000	S.F.	999	2,313	3,312
Fireplace, brick faced, 6′-0″ wide x 5′-0″ high	1.000	Ea.	665	3,200	3,865
Hearth	1.000	Ea.	226	640	866
Chimney, 20″ x 20″, one 12″ x 12″ flue, 12 V.L.F.	12.000	V.L.F.	996	1,128	2,124
Mantle, wood	7.000	L.F.	88.20	139.30	227.50
TOTAL			3,266.44	7,775.02	11,041.46

D3010 992	Masonry Fireplace	COST EACH MAT.	COST EACH INST.	COST EACH TOTAL
1200	Chimney, 20″ x 20″, one 12″ x 12″ flue, 12 V.L.F.	3,275	7,775	11,050
1300	18 V.L.F.	3,775	8,350	12,125
1400	24 V.L.F.	4,250	8,900	13,150
1500	Fieldstone face instead of brick	3,550	7,775	11,325
1600				
1700				
1800				
1900				
2000				
2100				
2200				
2300				
2700	Cut & patch to match existing construction, add, minimum	2%	3%	
2800	Maximum	5%	9%	
2900	Dust protection, add, minimum	1%	2%	
3000	Maximum	4%	11%	
3100	Equipment usage curtailment, add, minimum	1%	1%	
3200	Maximum	3%	10%	
3300	Material handling & storage limitation, add, minimum	1%	1%	
3400	Maximum	6%	7%	
3500	Protection of existing work, add, minimum	2%	2%	
3600	Maximum	5%	7%	
3700	Shift work requirements, add, minimum		5%	
3800	Maximum		30%	
3900	Temporary shoring and bracing, add, minimum	2%	5%	
4000	Maximum	5%	12%	

For customer support on your Commercial Renovation Costs with RSMeans data, call 800.448.8182.

D3020 Heat Generating Systems

This page illustrates and describes an oil fired forced air system including an oil fired furnace, ductwork, registers and hookups. Lines within System Components give the unit price and total price per square foot for this system. Prices for oil fired forced air systems are on Line Items D3020 122 1600 thru 2800. Both material quantities and labor costs have been adjusted for the system listed.

Factors: To adjust for job conditions other than normal working situations use Lines D3020 122 3100 thru 4200.

Example: You are to install the system during evenings and weekends. Go to Line D3020 122 4200 and apply this percentage to the appropriate INST. cost.

System Components			COST PER S.F.		
	QUANTITY	UNIT	MAT.	INST.	TOTAL
Oil fired hot air heating system including furnace, ductwork, registers **And all necessary hookups.**					
Area to 800 S.F., heat only					
Furnace, oil, atomizing gun type burner, w/oil piping	1.000	Ea.	4,656.25	525	5,181.25
Oil tank, steel, 275 gallon	1.000	Ea.	730	310	1,040
Duct, galvanized, steel	312.000	Lb.	215.28	3,120	3,335.28
Insulation, blanket type, duct work	270.000	S.F.	1,263.60	4,617	5,880.60
Flexible duct, 6" diameter, insulated	100.000	L.F.	419	580	999
Registers, baseboard, gravity, 12" x 6"	8.000	Ea.	112.40	292	404.40
Return, damper, 18" x 36"	1.000	Ea.	202	84	286
TOTAL			7,598.53	9,528	17,126.53
COST PER S.F.		S.F.	9.50	11.91	21.41

D3020 122	Heating-Cooling, Oil, Forced Air	COST PER S.F.		
		MAT.	INST.	TOTAL
1600	Oil fired, area to 800 S.F.	9.50	11.91	21.41
1700	To 1000 S.F.	7.65	9.60	17.25
1800	To 1200 S.F.	6.55	8.30	14.85
1900	To 1600 S.F.	5.20	6.70	11.90
2000	To 2000 S.F.	5.35	9.15	14.50
2100	To 3000 S.F.	4.04	7.30	11.34
2200	For combined heating and cooling systems:			
2300	Oil fired, heating and cooling, area to 800 S.F.	12.45	13.35	25.80
2400	To 1000 S.F.	9.75	10.80	20.55
2500	To 1200 S.F.	8.35	9.30	17.65
2600	To 1600 S.F.	7.05	7.45	14.50
2700	To 2000 S.F.	6.75	9.85	16.60
2800	To 3000 S.F.	5	7.80	12.80
3100	Cut & patch to match existing construction, add, minimum	2%	3%	
3200	Maximum	5%	9%	
3300	Dust protection, add, minimum	1%	2%	
3400	Maximum	4%	11%	
3500	Equipment usage curtailment, add, minimum	1%	1%	
3600	Maximum	3%	10%	
3700	Material handling & storage limitation, add, minimum	1%	1%	
3800	Maximum	6%	7%	
3900	Protection of existing work, add, minimum	2%	2%	
4000	Maximum	5%	7%	
4100	Shift work requirements, add, minimum		5%	
4200	Maximum		30%	

D3020 Heat Generating Systems

This page illustrates and describes a gas fired forced air system including a gas fired furnace, ductwork, registers and hookups. Lines within System Components give the unit price and total price per square foot for this system. Prices for gas fired forced air systems are on Line Items D3020 124 1400 thru 2600. Both material quantities and labor costs have been adjusted for the system listed.

Factors: To adjust for job conditions other than normal working situations use Lines D3020 124 2900 thru 4000.

Example: You are to install the system with material handling and storage limitations. Go to Line D3020 124 3500 and apply these percentages to the appropriate MAT. and INST. costs.

System Components	QUANTITY	UNIT	COST PER S.F. MAT.	COST PER S.F. INST.	COST PER S.F. TOTAL
Gas fired hot air heating system including furnace, ductwork, registers					
And all necessary hookups.					
Area to 800 S.F., heat only					
Furnace, gas, AGA certified, direct drive, w/gas piping, 44 MBH	1.000	Ea.	981.25	475	1,456.25
Duct, galvanized steel	312.000	Lb.	215.28	3,120	3,335.28
Insulation, blanket type, ductwork	270.000	S.F.	1,263.60	4,617	5,880.60
Flexible duct, 6" diameter, insulated	100.000	L.F.	419	580	999
Registers, baseboard, gravity, 12" x 6"	8.000	Ea.	112.40	292	404.40
Return, damper, 18" x 36"	1.000	Ea.	202	84	286
TOTAL			3,193.53	9,168	12,361.53
COST PER S.F.		S.F.	3.99	11.46	15.45

D3020 124	Heating-Cooling, Gas, Forced Air	COST PER S.F. MAT.	COST PER S.F. INST.	COST PER S.F. TOTAL
1400	Gas fired, area to 800 S.F.	4	11.47	15.47
1500	To 1000 S.F.	3.15	9.25	12.40
1600	To 1200 S.F.	2.82	8	10.82
1700	To 1600 S.F.	2.85	7.30	10.15
1800	To 2000 S.F.	3.10	9	12.10
1900	To 3000 S.F.	2.45	7.20	9.65
2000	For combined heating and cooling systems:			
2100	Gas fired, heating and cooling, area to 800 S.F.	6.95	12.90	19.85
2200	To 1000 S.F.	5.30	10.45	15.75
2300	To 1200 S.F.	4.60	9	13.60
2400	To 1600 S.F.	4.66	8.10	12.76
2500	To 2000 S.F.	4.49	9.70	14.19
2600	To 3000 S.F.	3.41	7.70	11.11
2900	Cut & patch to match existing construction, add, minimum	2%	3%	
3000	Maximum	5%	9%	
3100	Dust protection, add, minimum	1%	2%	
3200	Maximum	4%	11%	
3300	Equipment usage curtailment, add, minimum	1%	1%	
3400	Maximum	3%	10%	
3500	Material handling & storage limitation, add, minimum	1%	1%	
3600	Maximum	6%	7%	
3700	Protection of existing work, add, minimum	2%	2%	
3800	Maximum	5%	7%	
3900	Shift work requirements, add, minimum		5%	
4000	Maximum		30%	

D3020 Heat Generating Systems

Cast Iron Boiler, Hot Water, Gas Fired

This page illustrates and describes boilers including expansion tank, circulating pump and all service piping. Lines within Systems Components give the material and installation price on a cost each basis for the components. Prices for boiler systems are on Line Items D3020 130 1010 thru D3020 138 1040. Material quantities and labor costs have been adjusted for the system listed.

Factors: To adjust for job conditions other than normal working situations use Line D3020 138 2700 thru 4000.

System Components	QUANTITY	UNIT	COST EACH		
			MAT.	INST.	TOTAL
Cast iron boiler, including piping, chimney and accessories					
Boilers, gas fired, std controls, CI, insulated, HW, gross output 100 MBH	1.000	Ea.	3,075	2,450	5,525
Pipe, black steel, Sch 40, threaded, W/coupling & hangers, 10' OC, 3/4" dia	22.500	L.F.	84.38	317.25	401.63
Pipe, black steel, Sch 40, threaded, W/coupling & hangers, 10' OC, 1" dia.	21.500	L.F.	91.81	349.38	441.19
Elbow, 90°, black, straight, 3/4" dia.	9.000	Ea.	54.90	553.50	608.40
Elbow, 90°, black, straight, 1" dia.	6.000	Ea.	63.60	396	459.60
Tee, black, straight, 3/4" dia.	2.000	Ea.	19.40	191	210.40
Tee, black, straight, 1" dia.	2.000	Ea.	33.20	216	249.20
Tee, black, reducing, 1" dia.	2.000	Ea.	37.90	216	253.90
Pipe cap, black, 3/4" dia.	1.000	Ea.	6.95	27	33.95
Union, black with brass seat, 3/4" dia.	2.000	Ea.	43	132	175
Union, black with brass seat, 1" dia.	2.000	Ea.	56	143	199
Pipe nipples, black, 3/4" dia.	5.000	Ea.	9.38	35.25	44.63
Pipe nipples, black, 1" dia.	3.000	Ea.	6.41	24.38	30.79
Valves, bronze, gate, N.R.S., threaded, class 150, 3/4" size	2.000	Ea.	266	86	352
Valves, bronze, gate, N.R.S., threaded, class 150, 1" size	2.000	Ea.	392	91	483
Gas cock, brass, 3/4" size	1.000	Ea.	42.50	39	81.50
Thermometer, stem type, 9" case, 8" stem, 3/4" NPT	2.000	Ea.	422	62	484
Tank, steel, liquid expansion, ASME, painted, 15 gallon capacity	1.000	Ea.	1,025	92	1,117
Pump, circulating, bronze, flange connection, 3/4" to 1-1/2" size, 1/8 HP	1.000	Ea.	1,375	258	1,633
Vent chimney, all fuel, pressure tight, double wall, SS, 6" dia.	20.000	L.F.	1,540	500	2,040
Vent chimney, elbow, 90° fixed, 6" dia.	2.000	Ea.	980	101	1,081
Vent chimney, Tee, 6" dia.	2.000	Ea.	588	126	714
Vent chimney, ventilated roof thimble, 6" dia.	1.000	Ea.	315	58.50	373.50
Vent chimney, flat roof flashing, 6" dia.	1.000	Ea.	137	50.50	187.50
Vent chimney, stack cap, 6" diameter	1.000	Ea.	350	33	383
Insulation, fiberglass pipe covering, 1" wall, 1" IPS	20.000	L.F.	51	130	181
TOTAL			11,065.43	6,677.76	17,743.19

D3020 130	Boiler, Cast Iron, Hot Water, Gas	COST EACH		
		MAT.	INST.	TOTAL
1000	For alternate boilers:			
1010	Boiler, cast iron, gas, hot water, 100 MBH	11,100	6,675	17,775
1020	200 MBH	12,800	7,625	20,425
1030	320 MBH	11,000	8,300	19,300
1040	440 MBH	15,800	10,700	26,500
1050	544 MBH	26,100	16,700	42,800
1060	765 MBH	33,600	18,000	51,600
1070	1088 MBH	37,600	19,200	56,800

D30 HVAC

D3020 Heat Generating Systems

D3020 130	Boiler, Cast Iron, Hot Water, Gas	COST EACH		
		MAT.	INST.	TOTAL
1080	1530 MBH	44,600	25,400	70,000
1090	2312 MBH	54,500	29,900	84,400

D3020 134	Boiler, Cast Iron, Steam, Gas	COST EACH		
		MAT.	INST.	TOTAL
1010	Boiler, cast iron, gas, steam, 100 MBH	8,600	6,475	15,075
1020	200 MBH	19,800	12,200	32,000
1030	320 MBH	20,700	13,700	34,400
1040	544 MBH	28,900	21,300	50,200
1050	765 MBH	35,100	22,200	57,300
1060	1275 MBH	42,000	24,400	66,400
1070	2675 MBH	53,500	35,500	89,000

D3020 136	Boiler, Cast Iron, Hot Water, Gas/Oil	COST EACH		
		MAT.	INST.	TOTAL
1010	Boiler, cast iron, gas & oil, hot water, 584 MBH	33,200	18,600	51,800
1020	876 MBH	45,600	19,900	65,500
1030	1168 MBH	53,500	22,500	76,000
1040	1460 MBH	56,500	27,100	83,600
1050	2044 MBH	68,500	28,000	96,500
1060	2628 MBH	87,500	33,900	121,400

D3020 138	Boiler, Cast Iron, Steam, Gas/Oil	COST EACH		
		MAT.	INST.	TOTAL
1010	Boiler, cast iron, gas & oil, steam, 810 MBH	40,500	24,100	64,600
1020	1360 MBH	46,100	25,700	71,800
1030	2040 MBH	50,500	33,000	83,500
1040	2700 MBH	53,500	36,700	90,200
2700	Cut & patch to match existing construction, add, minimum	2%	3%	
2800	Maximum	5%	9%	
2900	Dust protection, add, minimum	1%	2%	
3000	Maximum	4%	11%	
3100	Equipment usage curtailment, add, minimum	1%	1%	
3200	Maximum	3%	10%	
3300	Material handling & storage limitation, add, minimum	1%	1%	
3400	Maximum	6%	7%	
3500	Protection of existing work, add, minimum	2%	2%	
3600	Maximum	5%	7%	
3700	Shift work requirements, add, minimum		5%	
3800	Maximum		30%	
3900	Temporary shoring and bracing, add, minimum	2%	5%	
4000	Maximum	5%	12%	

D3020 Heat Generating Systems

Flow ↓ ↑ Flow

Gate Valve →
Y Strainer →
Pressure Gauge →
Expansion Joint →
← Combination Valve
← Pressure Gauge
← Expansion Joint

Base Mounted End–Suction Pump

This page illustrates and describes end suction base mounted circulating pumps, including strainer and piping. Lines within system components give the material and installation price on a cost each basis for the components. Prices for circulating pumps are on Line items D3020 330 1010 thru 1050. Material quantities and labor costs have been adjusting for the system listed.

Factors: To adjust for job conditions other than normal working situations uses Lines D3020 330 2700 thru 4000.

System Components	QUANTITY	UNIT	COST EACH		
			MAT.	INST.	TOTAL
Circulator pump, end suction, including pipe, Valves, fittings and guages.					
Pump, circulating, CI, base mounted, 2-1/2" size, 3 HP, to 150 GPM	1.000	Ea.	11,100	860	11,960
Pipe, black steel, Sch. 40, on yoke & roll hangers, 10' O.C., 2-1/2" dia	12.000	L.F.	175.20	413.40	588.60
Elbow, 90°, weld joint, steel, 2-1/2" pipe size	1.000	Ea.	29.50	202.55	232.05
Flange, weld neck, 150 LB, 2-1/2" pipe size	8.000	Ea.	524	810.16	1,334.16
Valve, iron body, gate, 125 lb., N.R.S., flanged, 2-1/2" size	1.000	Ea.	1,075	310	1,385
Strainer, Y type, iron body, flanged, 125 lb., 2-1/2" pipe size	1.000	Ea.	192	310	502
Multipurpose valve, CI body, 2" size	1.000	Ea.	675	108	783
Expansion joint, flanged spool, 6" F to F, 2-1/2" dia	2.000	Ea.	800	252	1,052
T-O-L, weld joint, socket, 1/4" pipe size, nozzle	2.000	Ea.	22.40	140.94	163.34
T-O-L, weld joint, socket, 1/2" pipe size, nozzle	2.000	Ea.	22.40	147.22	169.62
Control gauges, pressure or vacuum, 3-1/2" diameter dial	2.000	Ea.	56	54	110
Pressure/temperature relief plug, 316 SS, 3/4" OD, 7-1/2" insertion	2.000	Ea.	194	54	248
Insulation, fiberglass pipe covering, 1-1/2" wall, 2-1/2" IPS	12.000	L.F.	71.40	95.40	166.80
Water flow	1.000	Ea.	1,625	1,325	2,950
Pump balancing	1.000	Ea.		380	380
TOTAL			16,561.90	5,462.67	22,024.57

D3020 330	Circulating Pump Systems, End Suction	COST EACH		
		MAT.	INST.	TOTAL
1000	For alternate pump sizes:			
1010	Pump, base mtd with motor, end-suction, 2-1/2" size, 3 HP, to 150 GPM	16,600	5,475	22,075
1020	3" size, 5 HP, to 225 GPM	20,000	6,175	26,175
1030	4" size, 7-1/2 HP, to 350 GPM	20,900	7,325	28,225
1040	5" size, 15 HP, to 1000 GPM	28,500	10,800	39,300
1050	6" size, 25 HP, to 1550 GPM	40,500	13,100	53,600
2700	Cut & patch to match existing construction, add, minimum	2%	3%	
2800	Maximum	5%	9%	
2900	Dust protection, add, minimum	1%	2%	
3000	Maximum	4%	11%	
3100	Equipment usage curtailment, add, minimum	1%	1%	
3200	Maximum	3%	10%	
3300	Material handling & storage limitation, add, minimum	1%	1%	
3400	Maximum	6%	7%	
3500	Protection of existing work, add, minimum	2%	2%	
3600	Maximum	5%	7%	
3700	Shift work requirements, add, minimum		5%	
3800	Maximum		30%	

D30 HVAC

D3020 Heat Generating Systems

D3020 330	Circulating Pump Systems, End Suction	COST EACH		
		MAT.	INST.	TOTAL
3900	Temporary shoring and bracing, add, minimum	2%	5%	
4000	Maximum	5%	12%	

701

D3040 Distribution Systems

Galvanized Steel Duct

Fiberglass Duct Insulation

Flexible Fiberglass Duct

Horizontal Fan Coil Air Conditioning System

This page illustrates and describes fan coil air conditioners including duct work, duct installation, piping and diffusers. Lines within Systems Components give the material and installation price on a cost each basis for the components. Prices for fan coil A/C unit systems are on Line Items D3040 118 1010 thru D3040 126 1120. Material quantities and labor costs have been adjusted for the system listed.

Factors: To adjust for job conditions other than normal working situations use Lines D3040 126 2700 thru 4000.

System Components	QUANTITY	UNIT	COST EACH		
			MAT.	INST.	TOTAL
Fan coil A/C system, horizontal, with housing, controls, 2 pipe, 1/2 ton					
Fan coil A/C, horizontal housing, filters, chilled water, 1/2 ton cooling	1.000	Ea.	1,775	195	1,970
Pipe, black steel, Sch 40, threaded, W/cplgs & hangers, 10' OC, 3/4" dia.	21.500	L.F.	80.63	303.15	383.78
Elbow, 90°, black, straight, 3/4" dia.	6.000	Ea.	36.60	369	405.60
Tee, black, straight, 3/4" dia.	2.000	Ea.	19.40	191	210.40
Union, black with brass seat, 3/4" dia.	2.000	Ea.	43	132	175
Pipe nipples, 3/4" diam	3.000	Ea.	5.63	21.15	26.78
Valves, bronze, gate, N.R.S., threaded, class 150, 3/4" size	2.000	Ea.	266	86	352
Circuit setter, balance valve, bronze body, threaded, 3/4" pipe size	1.000	Ea.	99	43.50	142.50
Insulation, fiberglass pipe covering, 1" wall, 3/4" IPS	20.000	L.F.	24	125	149
Ductwork, 12" x 8" fabricated, galvanized steel, 12 LF	55.000	Lb.	37.95	550	587.95
Insulation, ductwork, blanket type, fiberglass, 1" thk, 1-1/2 LB density	40.000	S.F.	187.20	684	871.20
Diffusers, aluminum, OB damper, ceiling, perf, 24"x24" panel size, 6"x6"	2.000	Ea.	412	105	517
Ductwork, flexible, fiberglass fabric, insulated, 1"thk, PE jacket, 6" dia	16.000	L.F.	67.04	92.80	159.84
Round volume control damper 6" dia.	2.000	Ea.	95	76	171
Fan coil unit balancing	1.000	Ea.		109	109
Re-heat coil balancing	1.000	Ea.		160	160
Diffuser/register, high, balancing	2.000	Ea.		302	302
TOTAL			3,148.45	3,544.60	6,693.05

D3040 118	Fan Coil A/C Unit, Two Pipe	COST EACH		
		MAT.	INST.	TOTAL
1000	For alternate A/C units:			
1010	Fan coil A/C system, cabinet mounted, controls, 2 pipe, 1/2 Ton	1,400	1,625	3,025
1020	1 Ton	1,775	1,825	3,600
1030	1-1/2 Ton	1,825	1,850	3,675
1040	2 Ton	2,950	2,250	5,200
1050	3 Ton	4,475	2,325	6,800

D3040 120	Fan Coil A/C Unit, Two Pipe, Electric Heat	COST EACH		
		MAT.	INST.	TOTAL
1010	Fan coil A/C system, cabinet mntd, elect. ht, controls, 2 pipe, 1/2 Ton	2,425	1,625	4,050
1020	1 Ton	3,025	1,825	4,850
1030	1-1/2 Ton	3,475	1,850	5,325
1040	2 Ton	5,100	2,275	7,375
1050	3 Ton	8,350	2,325	10,675

D3040 Distribution Systems

D3040 122	Fan Coil A/C Unit, Four Pipe	COST EACH		
		MAT.	INST.	TOTAL
1010	Fan coil A/C system, cabinet mounted, controls, 4 pipe, 1/2 Ton	2,325	3,150	5,475
1020	1 Ton	2,725	3,350	6,075
1030	1-1/2 Ton	2,800	3,375	6,175
1040	2 Ton	3,575	3,500	7,075
1050	3 Ton	5,950	3,750	9,700

D3040 124	Fan Coil A/C, Horizontal, Duct Mount, 2 Pipe	COST EACH		
		MAT.	INST.	TOTAL
1010	Fan coil A/C system, horizontal w/housing, controls, 2 pipe, 1/2 Ton	3,150	3,550	6,700
1020	1 Ton	4,225	5,650	9,875
1030	1-1/2 Ton	4,725	7,575	12,300
1040	2 Ton	6,225	9,425	15,650
1050	3 Ton	7,900	13,100	21,000
1060	3-1/2 Ton	8,325	13,400	21,725
1070	4 Ton	8,750	15,600	24,350
1080	5 Ton	10,400	20,100	30,500
1090	6 Ton	10,600	21,400	32,000
1100	7 Ton	10,900	22,000	32,900
1110	8 Ton	10,900	22,300	33,200
1120	10 Ton	11,500	22,900	34,400

D3040 126	Fan Coil A/C, Horiz., Duct Mount, 2 Pipe, Elec. Ht.	COST EACH		
		MAT.	INST.	TOTAL
1010	Fan coil A/C system, horiz. hsng, elect. ht, ctrls, 2 pipe, 1/2 Ton	3,450	3,550	7,000
1020	1 Ton	4,725	5,650	10,375
1030	1-1/2 Ton	5,250	7,575	12,825
1040	2 Ton	7,000	9,425	16,425
1050	3 Ton	11,000	13,400	24,400
1060	3-1/2 Ton	12,200	13,400	25,600
1070	4 Ton	13,800	15,600	29,400
1080	5 Ton	16,400	20,100	36,500
1090	6 Ton	16,600	21,500	38,100
1100	7 Ton	17,700	22,300	40,000
1110	8 Ton	18,000	23,000	41,000
1120	10 Ton	19,200	22,000	41,200
2700	Cut & patch to match existing construction, add, minimum	2%	3%	
2800	Maximum	5%	9%	
2900	Dust protection, add, minimum	1%	2%	
3000	Maximum	4%	11%	
3100	Equipment usage curtailment, add, minimum	1%	1%	
3200	Maximum	3%	10%	
3300	Material handling & storage limitation, add, minimum	1%	1%	
3400	Maximum	6%	7%	
3500	Protection of existing work, add, minimum	2%	2%	
3600	Maximum	5%	7%	
3700	Shift work requirements, add, minimum		5%	
3800	Maximum		30%	
3900	Temporary shoring and bracing, add, minimum	2%	5%	
4000	Maximum	5%	12%	

D3040 Distribution Systems

Plate Heat Exchanger   Shell and Tube Heat Exchanger

This page illustrates and describes plate and shell and tube type heat exchangers including pressure gauge and exchanger control system. Lines within Systems Components give the material and installation price on a cost each basis for the components. Prices for heat exchanger systems are on Line Items D3040 610 1010 thru D3040 620 1040. Material quantities and labor costs have been adjusted for the system listed.

Factors To adjust for job conditions other then normal working situations use Lines D3040 620 2700 thru 4000.

System Components	QUANTITY	UNIT	COST EACH		
			MAT.	INST.	TOTAL
Shell and tube type heat exchanger					
with related valves and piping.					
Heat exchanger, 4 pass, 3/4″ O.D. copper tubes, by steam at 10 psi, 40 GPM	1.000	Ea.	7,600	390	7,990
Pipe, black steel, Sch 40, threaded, w/couplings & hangers, 10′ OC, 2″ dia.	50.500	L.F.	439.35	1,212	1,651.35
Elbow, 90°, straight, 2″ dia.	16.000	Ea.	632	1,376	2,008
Reducer, black steel, concentric, 2″ dia.	2.000	Ea.	67	148	215
Valves, bronze, gate, rising stem, threaded, class 150, 2″ size	6.000	Ea.	2,220	471	2,691
Valves, bronze, globe, class 150, rising stem, threaded, 2″ size	2.000	Ea.	1,900	157	2,057
Strainers, Y type, bronze body, screwed, 125 lb, 2″ pipe size	2.000	Ea.	542	133	675
Valve, electric motor actuated, brass, 2 way, screwed, 1-1/2″ pipe size	1.000	Ea.	745	66.50	811.50
Union, black with brass seat, 2″ dia.	8.000	Ea.	588	728	1,316
Tee, black, straight, 2″ dia.	9.000	Ea.	450	1,269	1,719
Tee, black, reducing run and outlet, 2″ dia.	4.000	Ea.	240	564	804
Pipe nipple, black, 2″ dia.	2.000	Ea.	4.35	12	16.35
Thermometers, stem type, 9″ case, 8″ stem, 3/4″ NPT	2.000	Ea.	422	62	484
Gauges, pressure or vacuum, 3-1/2″ diameter dial	2.000	Ea.	56	54	110
Insulation, fiberglass pipe covering, 1″ wall, 2″ IPS	40.000	L.F.	128.40	286	414.40
Heat exchanger control system	1.000	Ea.	3,525	2,750	6,275
Coil balancing	1.000	Ea.		160	160
TOTAL			19,559.10	9,838.50	29,397.60

D3040 610	Heat Exchanger, Plate Type	COST EACH		
		MAT.	INST.	TOTAL
1000	For alternate heat exchangers:			
1010	Plate heat exchanger, 400 GPM	75,500	19,800	95,300
1020	800 GPM	123,000	25,300	148,300
1030	1200 GPM	187,500	34,200	221,700
1040	1800 GPM	256,000	42,600	298,600

D3040 620	Heat Exchanger, Shell & Tube	COST EACH		
		MAT.	INST.	TOTAL
1010	Shell & tube heat exchanger, 40 GPM	19,600	9,850	29,450
1020	96 GPM	74,500	17,400	91,900
1030	240 GPM	168,500	23,600	192,100
1040	600 GPM	121,000	33,300	154,300
2700	Cut & patch to match existing construction, add, minimum	2%	3%	
2800	Maximum	5%	9%	
2900	Dust protection, add, minimum	1%	2%	
3000	Maximum	4%	11%	

D3040 Distribution Systems

D3040 620	Heat Exchanger, Shell & Tube	COST EACH		
		MAT.	INST.	TOTAL
3100	Equipment usage curtailment, add, minimum	1%	1%	
3200	Maximum	3%	10%	
3300	Material handling & storage limitation, add, minimum	1%	1%	
3400	Maximum	6%	7%	
3500	Protection of existing work, add, minimum	2%	2%	
3600	Maximum	5%	7%	
3700	Shift work requirements, add, minimum		5%	
3800	Maximum		30%	
3900	Temporary shoring and bracing, add, minimum	2%	5%	
4000	Maximum	5%	12%	

D3050 Terminal & Package Units

Weather Cap Roof Vent →

Roof Flashing →

Gas Cock →

Vent Pipe

90° Elbow

Tee

Tee Cap

Gas Unit Heater

Hydronic Unit Heater

This page illustrates and describes unit heaters including piping and vents. Lines within Systems Components give the material and installation price on a cost each basis for this system. Prices for unit heater systems are on Line items D3050 120 1010 thru D3050 140 1040. Material quantities and labor costs have been adjusted for the system listed.

Factors: To adjust for job conditions other than normal working situations use Lines D3050 140 2700 thru 4000.

System Components			COST EACH		
	QUANTITY	UNIT	MAT.	INST.	TOTAL
Unit heaters complete with piping, thermostats and chimney as required					
Space heater, propeller fan, 20 MBH output	1.000	Ea.	855	184	1,039
Pipe, black steel, Sch 40, threaded, w/coupling & hangers, 10' OC, 3/4" dia	21.000	L.F.	78.75	296.10	374.85
Elbow, 90°, black steel, straight, 1/2" dia.	3.000	Ea.	15.15	172.50	187.65
Elbow, 90°, black steel, straight, 3/4" dia.	3.000	Ea.	18.30	184.50	202.80
Tee, black steel, reducing, 3/4" dia.	1.000	Ea.	15	95.50	110.50
Union, black with brass seat, 3/4" dia.	1.000	Ea.	21.50	66	87.50
Pipe nipples, black, 1/2" dia	2.000	Ea.	6.18	27.30	33.48
Pipe nipples, black, 3/4" dia	2.000	Ea.	3.75	14.10	17.85
Pipe cap, black, 3/4" dia.	1.000	Ea.	6.95	27	33.95
Gas cock, brass, 3/4" size	1.000	Ea.	42.50	39	81.50
Thermostat, 1 set back, electric, timed	1.000	Ea.	111	108	219
Wiring, thermostat hook-up, 25' of #18-3	1.000	Ea.	10.70	34	44.70
Vent chimney, prefab metal, U.L. listed, gas, double wall, galv. st, 4" dia	12.000	L.F.	137.40	270	407.40
Vent chimney, gas, double wall, galv. steel, elbow 90°, 4" dia.	2.000	Ea.	65	89	154
Vent chimney, gas, double wall, galv. steel, Tee, 4" dia.	1.000	Ea.	46	58.50	104.50
Vent chimney, gas, double wall, galv. steel, T cap, 4" dia.	1.000	Ea.	3.08	36	39.08
Vent chimney, gas, double wall, galv. steel, roof flashing, 4" dia.	1.000	Ea.	11.15	44.50	55.65
Vent chimney, gas, double wall, galv. steel, top, 4" dia.	1.000	Ea.	23	34.50	57.50
TOTAL			1,470.41	1,780.50	3,250.91

D3050 120	Unit Heaters, Gas	COST EACH		
		MAT.	INST.	TOTAL
1000	For alternate unit heaters:			
1010	Space heater, suspended, gas fired, propeller fan, 20 MBH	1,475	1,775	3,250
1020	60 MBH	2,725	1,850	4,575
1030	100 MBH	2,025	1,950	3,975
1040	160 MBH	2,350	2,150	4,500
1050	200 MBH	2,975	2,375	5,350
1060	280 MBH	5,225	2,550	7,775
1070	320 MBH	5,025	2,875	7,900

D3050 130	Unit Heaters, Hydronic	COST EACH		
		MAT.	INST.	TOTAL
1010	Space heater, suspended, horiz. mount, HW, prop. fan, 20 MBH	2,450	1,900	4,350
1020	60 MBH	3,550	2,125	5,675
1030	100 MBH	4,175	2,325	6,500
1040	150 MBH	4,650	2,575	7,225

706

D30 HVAC

D3050 Terminal & Package Units

D3050 130	Unit Heaters, Hydronic	COST EACH		
		MAT.	INST.	TOTAL
1050	200 MBH	5,250	2,900	8,150
1060	300 MBH	5,125	3,400	8,525

D3050 140	Cabinet Unit Heaters, Hydronic	COST EACH		
		MAT.	INST.	TOTAL
1010	Unit heater, cabinet type, horizontal blower, hot water, 20 MBH	3,300	1,850	5,150
1020	60 MBH	4,800	2,050	6,850
1030	100 MBH	5,575	2,250	7,825
1040	120 MBH	5,075	2,325	7,400
2700	Cut & patch to match existing construction, add, minimum	2%	3%	
2800	Maximum	5%	9%	
2900	Dust protection, add, minimum	1%	2%	
3000	Maximum	4%	11%	
3100	Equipment usage curtailment, add, minimum	1%	1%	
3200	Maximum	3%	10%	
3300	Material handling & storage limitation, add, minimum	1%	1%	
3400	Maximum	6%	7%	
3500	Protection of existing work, add, minimum	2%	2%	
3600	Maximum	5%	7%	
3700	Shift work requirements, add, minimum		5%	
3800	Maximum		30%	
3900	Temporary shoring and bracing, add, minimum	2%	5%	
4000	Maximum	5%	12%	

D30 HVAC

D3050 Terminal & Package Units

Gas Cock ←

Roof Curb ←

Rooftop Air Conditioner Unit

Gas Piping

This page illustrates and describes packaged air conditioners including rooftop DX unit, pipes and fittings. Lines within Systems Components give the material and installation price on a cost each basis for the components. Prices for A/C systems are on Line Items D3050 175 1010 thru D3050 180 1060. Material quantities and labor costs have been adjusted for the system listed.

Factors: To adjust for job conditions other than normal working situations use Lines D3050 175 2700 thru 4000.

System Components	QUANTITY	UNIT	COST EACH		
			MAT.	INST.	TOTAL
Packaged rooftop DX unit, gas, with related piping and valves.					
Roof top A/C, curb, economizer, sgl zone, elec cool, gas ht, 5 ton, 112 MBH	1.000	Ea.	4,625	2,775	7,400
Pipe, black steel, Sch 40, threaded, w/cplgs & hangers, 10' OC, 1" dia	21.000	L.F.	89.67	341.25	430.92
Elbow, 90°, black, straight, 3/4" dia.	3.000	Ea.	18.30	184.50	202.80
Elbow, 90°, black, straight, 1" dia.	3.000	Ea.	31.80	198	229.80
Tee, black, reducing, 1" dia.	1.000	Ea.	18.95	108	126.95
Union, black with brass seat, 1" dia.	1.000	Ea.	28	71.50	99.50
Pipe nipple, black, 3/4" dia	2.000	Ea.	7.50	28.20	35.70
Pipe nipple, black, 1" dia	2.000	Ea.	4.27	16.25	20.52
Cap, black, 1" dia.	1.000	Ea.	8.35	28.50	36.85
Gas cock, brass, 1" size	1.000	Ea.	50.50	45.50	96
Control system, pneumatic, Rooftop A/C unit	1.000	Ea.	3,625	3,050	6,675
Rooftop unit heat/cool balancing	1.000	Ea.		590	590
TOTAL			8,507.34	7,436.70	15,944.04

D3050 175	Rooftop Air Conditioner, Const. Volume	COST EACH		
		MAT.	INST.	TOTAL
1000	For alternate A/C systems:			
1010	A/C, rooftop, DX cool, gas heat, curb, economizer, fltrs, 5 Ton	8,500	7,425	15,925
1020	7-1/2 Ton	11,000	7,750	18,750
1030	12-1/2 Ton	13,800	8,525	22,325
1040	18 Ton	18,200	9,325	27,525
1050	25 Ton	43,900	10,600	54,500
1060	40 Ton	53,000	14,200	67,200

D3050 180	Rooftop Air Conditioner, Variable Air Volume	COST EACH		
		MAT.	INST.	TOTAL
1010	A/C, rooftop, DX cool, gas heat, curb, ecmizr, fltrs, VAV, 12-1/2 Ton	17,000	9,100	26,100
1020	18 Ton	22,200	10,000	32,200
1030	25 Ton	49,000	11,500	60,500
1040	40 Ton	61,500	15,600	77,100
1050	60 Ton	87,500	21,300	108,800
1060	80 Ton	111,000	26,600	137,600
2700	Cut & patch to match existing construction, add, minimum	2%	3%	
2800	Maximum	5%	9%	
2900	Dust protection, add, minimum	1%	2%	
3000	Maximum	4%	11%	
3100	Equipment usage curtailment, add, minimum	1%	1%	
3200	Maximum	3%	10%	

D3050 Terminal & Package Units

D3050 180	Rooftop Air Conditioner, Variable Air Volume	COST EACH		
		MAT.	INST.	TOTAL
3300	Material handling & storage limitation, add, minimum	1%	1%	
3400	Maximum	6%	7%	
3500	Protection of existing work, add, minimum	2%	2%	
3600	Maximum	5%	7%	
3700	Shift work requirements, add, minimum		5%	
3800	Maximum		30%	
3900	Temporary shoring and bracing, add, minimum	2%	5%	
4000	Maximum	5%	12%	

D30 HVAC

D3050 Terminal & Package Units

Condenser Supply Water Piping

Self—Contained Air Conditioner Unit, Water Cooled

Condenser Return Water Piping

This page illustrates and describes packaged electrical air conditioning units including multizone control and related piping. Lines within Systems Components give the material and installation price on a cost each basis for the components. Prices for packaged electrical A/C unit systems are on Line Items D3050 210 1010 thru D3050 225 1060. Material quantities and labor costs have been adjusted for the system listed.

Factors: To adjust for job conditions other than normal working situations use Lines D3050 225 2700 thru 4000.

System Components	QUANTITY	UNIT	COST EACH		
			MAT.	INST.	TOTAL
Packaged water cooled electric air conditioning unit with related piping and valves.					
Self-contained, water cooled, elect. heat, not inc tower, 5 ton, const vol	1.000	Ea.	10,100	2,025	12,125
Pipe, black steel, Sch 40, threaded, w/cplg & hangers, 10' OC, 1-1/4" dia.	21.500	L.F.	119.33	374.10	493.43
Elbow, 90°, black, straight, 1-1/4" dia.	6.000	Ea.	105	423	528
Tee, black, straight, 1-1/4" dia.	2.000	Ea.	54	222	276
Tee, black, reducing, 1-1/4" dia.	2.000	Ea.	66	222	288
Thermometers, stem type, 9" case, 8" stem, 3/4" NPT	2.000	Ea.	422	62	484
Union, black with brass seat, 1-1/4" dia.	2.000	Ea.	102	148	250
Pipe nipple, black, 1-1/4" dia	3.000	Ea.	8.33	26.10	34.43
Valves, bronze, gate, N.R.S., threaded, class 150, 1-1/4" size	2.000	Ea.	452	115	567
Circuit setter, bal valve, bronze body, threaded, 1-1/4" pipe size	1.000	Ea.	184	58	242
Insulation, fiberglass pipe covering, 1" wall, 1-1/4" IPS	20.000	L.F.	81.80	137	218.80
Control system, pneumatic, A/C Unit with heat	1.000	Ea.	3,625	3,050	6,675
Re-heat coil balancing	1.000	Ea.		160	160
Rooftop unit heat/cool balancing	1.000	Ea.		590	590
TOTAL			15,319.46	7,612.20	22,931.66

D3050 210	AC Unit, Package, Elec. Ht., Water Cooled	COST EACH		
		MAT.	INST.	TOTAL
1000	For alternate A/C systems:			
1010	A/C, self contained, single pkg., water cooled, elect. heat, 5 Ton	15,300	7,600	22,900
1020	10 Ton	22,900	9,375	32,275
1030	20 Ton	30,800	11,900	42,700
1040	30 Ton	54,000	12,900	66,900
1050	40 Ton	69,000	14,300	83,300
1060	50 Ton	90,500	17,100	107,600

D3050 215	AC Unit, Package, Elec. Ht., Water Cooled, VAV	COST EACH		
		MAT.	INST.	TOTAL
1010	A/C, self contained, single pkg., water cooled, elect. ht, VAV, 10 Ton	25,800	9,375	35,175
1020	20 Ton	35,900	11,900	47,800
1030	30 Ton	61,500	12,900	74,400
1040	40 Ton	77,500	14,300	91,800
1050	50 Ton	101,500	17,100	118,600
1060	60 Ton	115,000	20,700	135,700

D30 HVAC

D3050 Terminal & Package Units

D3050 220	AC Unit, Package, Hot Water Coil, Water Cooled	COST EACH		
		MAT.	INST.	TOTAL
1010	A/C, self contn'd, single pkg., water cool, H/W ht, const. vol, 5 Ton	13,400	10,600	24,000
1020	10 Ton	18,600	12,600	31,200
1030	20 Ton	43,200	15,700	58,900
1040	30 Ton	56,000	19,900	75,900
1050	40 Ton	65,500	24,200	89,700
1060	50 Ton	76,000	29,700	105,700

D3050 225	AC Unit, Package, HW Coil, Water Cooled, VAV	COST EACH		
		MAT.	INST.	TOTAL
1010	A/C, self contn'd, single pkg., water cool, H/W ht, VAV, 10 Ton	21,400	11,800	33,200
1020	20 Ton	30,800	15,300	46,100
1030	30 Ton	53,000	19,000	72,000
1040	40 Ton	69,000	20,800	89,800
1050	50 Ton	89,500	24,500	114,000
1060	60 Ton	104,000	26,500	130,500
2700	Cut & patch to match existing construction, add, minimum	2%	3%	
2800	Maximum	5%	9%	
2900	Dust protection, add, minimum	1%	2%	
3000	Maximum	4%	11%	
3100	Equipment usage curtailment, add, minimum	1%	1%	
3200	Maximum	3%	10%	
3300	Material handling & storage limitation, add, minimum	1%	1%	
3400	Maximum	6%	7%	
3500	Protection of existing work, add, minimum	2%	2%	
3600	Maximum	5%	7%	
3700	Shift work requirements, add, minimum		5%	
3800	Maximum		30%	
3900	Temporary shoring and bracing, add, minimum	2%	5%	
4000	Maximum	5%	12%	

D3050 Terminal & Package Units

Thru—Wall Heat Pump   Thru—Wall Air Conditioner

This page illustrates and describes thru-wall air conditioning and heat pump units. Lines within Systems Components give the material and installation price on a cost each basis for the components. Prices for thru-wall unit systems are on Line Items D3050 255 1010 thru D3050 260 1050. Material quantities and labor costs have been adjusted for the system listed.

Factors: To adjust for job conditions other than normal working conditions use Lines D3050 255 2700 thru 4000.

System Components		QUANTITY	UNIT	COST EACH		
				MAT.	INST.	TOTAL
Electric thru-wall air conditioning unit.						
A/C unit, thru-wall, electric heat., cabinet, louver, 1/2 ton		1.000	Ea.	1,050	260	1,310
	TOTAL			1,050	260	1,310

D3050 255	Thru-Wall A/C Unit	COST EACH		
		MAT.	INST.	TOTAL
1000	For alternate A/C units:			
1010	A/C unit, thru-the-wall, supplemental elect. heat, cabinet, louver, 1/2 Ton	1,050	260	1,310
1020	3/4 Ton	2,800	310	3,110
1030	1 Ton	2,950	390	3,340
1040	1-1/2 Ton	3,725	650	4,375
1050	2 Ton	3,900	820	4,720

D3050 260	Thru-Wall Heat Pump	COST EACH		
		MAT.	INST.	TOTAL
1010	Heat pump, thru-the-wall, cabinet, louver, 1/2 Ton	3,100	195	3,295
1020	3/4 Ton	3,250	260	3,510
1030	1 Ton	4,225	390	4,615
1040	Supplemental. elect. heat, 1-1/2 Ton	3,700	1,000	4,700
1050	2 Ton	3,750	1,050	4,800
2700	Cut & patch to match existing construction, add, minimum	2%	3%	
2800	Maximum	5%	9%	
2900	Dust protection, add, minimum	1%	2%	
3000	Maximum	4%	11%	
3100	Equipment usage curtailment, add, minimum	1%	1%	
3200	Maximum	3%	10%	
3300	Material handling & storage limitation, add, minimum	1%	1%	
3400	Maximum	6%	7%	
3500	Protection of existing work, add, minimum	2%	2%	
3600	Maximum	5%	7%	
3700	Shift work requirements, add, minimum		5%	
3800	Maximum		30%	
3900	Temporary shoring and bracing, add, minimum	2%	5%	
4000	Maximum	5%	12%	

D4010 Sprinklers

This page illustrates and describes a dry type fire sprinkler system. Lines within System Components give the unit cost of a system for a 2,000 square foot building. Lines D4010 320 1700 thru 2400 give the square foot costs for alternate systems. Both material quantities and labor costs have been adjusted for the system listed.

Factors: To adjust for conditions other than normal working conditions, use Lines D4010 320 2900 thru 4000.

System Components	QUANTITY	UNIT	COST EACH MAT.	COST EACH INST.	COST EACH TOTAL
Pre-action fire sprinkler system, ordinary hazard, open area to 2000 S.F.					
On one floor.					
Air compressor	1.000	Ea.	1,600	650	2,250
4" OS & Y valve	1.000	Ea.	1,075	515	1,590
Pipe riser 4" diameter	10.000	L.F.	250	438.50	688.50
Sprinkler head supply piping, 1"	163.000	L.F.	320.25	1,218.75	1,539
Sprinkler head supply piping, 1-1/4"	163.000	L.F.	305.25	957	1,262.25
Sprinkler head supply piping, 2-1/2"	163.000	L.F.	615.60	1,178	1,793.60
Pipe fittings, 1"	8.000	Ea.	132.80	864	996.80
Pipe fittings, 1-1/4"	8.000	Ea.	216	888	1,104
Pipe fittings, 2-1/2"	4.000	Ea.	492	688	1,180
Pipe fittings, 4"	2.000	Ea.	880	770	1,650
Detectors	2.000	Ea.	2,100	106	2,206
Sprinkler heads	16.000	Ea.	289.60	848	1,137.60
Fire department connection	1.000	Ea.	630	305	935
TOTAL			**8,906.50**	**9,426.25**	**18,332.75**

D4010 320	Fire Sprinkler Systems, Dry	COST PER S.F. MAT.	COST PER S.F. INST.	COST PER S.F. TOTAL
1700	Ordinary hazard, one floor, area to 2000 S.F./floor	4.45	4.71	9.16
1800	For each additional floor, add per floor	1.75	3.93	5.68
1900	Area to 3200 S.F./ floor	3.43	4.67	8.10
2000	For each additional floor, add per floor	1.74	4.18	5.92
2100	Area to 5000 S.F./ floor	3.65	4.22	7.87
2200	For each additional floor, add per floor	2.42	4.42	6.84
2300	Area to 8000 S.F./ floor	2.90	4	6.90
2400	For each additional floor, add per floor	1.84	3.78	5.62
2500				
2600				
2700				
2800				
2900	Cut & patch to match existing construction, add, minimum	2%	3%	
3000	Maximum	5%	9%	
3100	Dust protection, add, minimum	1%	2%	
3200	Maximum	4%	11%	
3300	Equipment usage curtailment, add, minimum	1%	1%	
3400	Maximum	3%	10%	

D40 Fire Protection

D4010 Sprinklers

D4010 320	Fire Sprinkler Systems, Dry	COST PER S.F.		
		MAT.	INST.	TOTAL
3500	Material handling & storage limitation, add, minimum	1%	1%	
3600	Maximum	6%	7%	
3700	Protection of existing work, add, minimum	2%	2%	
3800	Maximum	5%	7%	
3900	Shift work requirements, add, minimum		5%	
4000	Maximum		30%	

D40 Fire Protection

D4010 Sprinklers

This page illustrates and describes a wet type fire sprinkler system. Lines within System Components give the unit cost of a system for a 2,000 square foot building. Lines D4010 420 1900 thru 2600 give the square foot costs for alternate systems. Both material quantities and labor costs have been adjusted for the system listed.

Factors: To adjust for conditions other than normal working conditions, use Lines D4010 420 2900 thru 4000.

System Components	QUANTITY	UNIT	COST EACH MAT.	COST EACH INST.	COST EACH TOTAL
Wet pipe fire sprinkler system, ordinary hazard, open area to 2000 S.F.					
On one floor.					
4" pipe size	1.000	Ea.	5,025	515	5,540
Tamper switch (valve supervisory switch)	3.000	Ea.	5,025	515	5,540
4" OS & Y valve	1.000	Ea.	1,075	515	1,590
Wet pipe alarm valve	1.000	Ea.	2,975	760	3,735
Water motor alarm	1.000	Ea.	400	212	612
3" check valve	1.000	Ea.	520	510	1,030
Pipe riser, 4" diameter	10.000	L.F.	250	438.50	688.50
Water gauges and trim	1.000	Set	4,050	1,525	5,575
Electric fire horn	1.000	Ea.	58.50	121	179.50
Sprinkler head supply piping, 1" diameter	168.000	L.F.	320.25	1,218.75	1,539
Sprinkler head supply piping, 1-1/2" diameter	168.000	L.F.	357.50	1,064.25	1,421.75
Sprinkler head supply piping, 2-1/2" diameter	168.000	L.F.	615.60	1,178	1,793.60
Pipe fittings, 1"	8.000	Ea.	132.80	864	996.80
Pipe fittings, 1-1/2"	8.000	Ea.	268	952	1,220
Pipe fittings, 2-1/2"	4.000	Ea.	492	688	1,180
Pipe fittings, 4"	2.000	Ea.	880	770	1,650
Sprinkler heads	16.000	Ea.	289.60	848	1,137.60
Fire department connection	1.000	Ea.	630	305	935
TOTAL			19,254.25	12,643.50	31,897.75

D4010 420	Fire Sprinkler Systems, Wet	COST PER S.F. MAT.	COST PER S.F. INST.	COST PER S.F. TOTAL
1900	Ordinary hazard, one floor, area to 2000 S.F./floor	9.65	6.30	15.95
2000	For each additional floor, add per floor	1.80	4.01	5.81
2100	Area to 3200 S.F./floor	6.65	5.65	12.30
2200	For each additional floor, add per floor	1.74	4.18	5.92
2300	Area to 5000 S.F./ floor	5.55	5.35	10.90
2400	For each additional floor, add per floor	2.42	4.42	6.84
2500	Area to 8000 S.F./ floor	3.79	4.36	8.15
2600	For each additional floor, add per floor	1.97	4.02	5.99
2700				
2900	Cut & patch to match existing construction, add minimum	2%	3%	
3000	Maximum	5%	9%	
3100	Dust protection, add, minimum	1%	2%	
3200	Maximum	4%	11%	
3300	Equipment usage curtailment, add, minimum	1%	1%	
3400	Maximum	3%	10%	
3500	Material handling & storage limitation, add, minimum	1%	1%	
3600	Maximum	6%	7%	

715

For customer support on your Commercial Renovation Costs with RSMeans data, call 800.448.8182.

This page illustrates and describes commercial service systems including a meter socket, service head and cable, entrance switch, steel conduit, copper wire, panel board, ground rod, wire, and conduit. Lines within System Components give the unit price and total price on a cost each basis for this system. Prices for commercial service systems are on Line Items D5010 210 1500 thru 1700. Both material quantities and labor costs have been adjusted for the system listed.

Factors: To adjust for job conditions other than normal working situations use Lines D5010 210 2900 thru 4000.

Example: You are to install the system with maximum equipment usage curtailment. Go to Line D5010 210 3400 and apply these percentages to the appropriate MAT. and INST. costs.

System Components	QUANTITY	UNIT	COST EACH		
			MAT.	INST.	TOTAL
Commercial electric service including service breakers, metering 120/208 Volt, 3 phase, 4 wire, feeder, and panel board.					
100 Amp Service					
Meter socket	1.000	Ea.	78	254	332
Service head	.200	C.L.F.	28.20	163	191.20
Service entrance switch	1.000	Ea.	207	430	637
Rigid steel conduit	20.000	L.F.	120	250	370
600 volt copper wire #3	1.000	C.L.F.	187	163	350
Panel board, 24 circuits, 20 Amp breakers	1.000	Ea.	1,300	1,350	2,650
Ground rod	1.000	Ea.	19.80	148	167.80
TOTAL			1,940	2,758	4,698

D5010 210	Commercial Service - 3 Phase	COST EACH		
		MAT.	INST.	TOTAL
1500	120/208 Volt, 3 phase, 4 wire service, 60 Amp	1,850	2,150	4,000
1550	100 Amp	1,950	2,750	4,700
1600	200 Amp	3,550	4,875	8,425
1700	400 Amp	6,800	6,100	12,900
1800				
1900				
2000				
2100				
2200				
2300				
2400				
2500				
2900	Cut & patch to match existing construction, add, minimum	2%	3%	
3000	Maximum	5%	9%	
3100	Dust protection, add, minimum	1%	2%	
3200	Maximum	4%	11%	
3300	Equipment usage curtailment, add, minimum	1%	1%	
3400	Maximum	3%	10%	
3500	Material handling & storage limitation, add, minimum	1%	1%	
3600	Maximum	6%	7%	
3700	Protection of existing work, add, minimum	2%	2%	
3800	Maximum	5%	7%	
3900	Shift work requirements, add, minimum		5%	
4000	Maximum		30%	

D50 Electrical

D5010 Electrical Service/Distribution

This page illustrates and describes a residential, single phase system including a weather cap, service entrance cable, meter socket, entrance switch, ground rod, ground cable, EMT, and panelboard. Lines within System Components give the unit price and total price on a cost each basis for this system. Prices for a residential, single phase system are also given. Both material quantities and labor costs have been adjusted for the system listed.

Factors: To adjust for job conditions other than normal working situations use Lines D5010 220 2900 thru 4000.

Example: You are to install the system with a minimum equipment usage curtailment. Go to Line D5010 220 3300 and apply these percentages to the appropriate MAT. and INST. costs.

System Components	QUANTITY	UNIT	COST EACH MAT.	COST EACH INST.	COST EACH TOTAL
100 Amp Service, single phase					
Weathercap	1.000	Ea.	7.05	68	75.05
Service entrance cable	.200	C.L.F.	28.20	163	191.20
Meter socket	1.000	Ea.	78	254	332
Entrance disconnect switch	1.000	Ea.	207	430	637
Ground rod, with clamp	1.000	Ea.	24	170	194
Ground cable	.100	C.L.F.	16.30	51	67.30
Panelboard, 12 circuit	1.000	Ea.	247	680	927
TOTAL			607.55	1,816	2,423.55
200 Amp Service , single phase					
Weathercap	1.000	Ea.	22	102	124
Service entrance cable	.200	C.L.F.	68	230	298
Meter socket	1.000	Ea.	147	430	577
Entrance disconnect switch	1.000	Ea.	450	625	1,075
Ground rod, with clamp	1.000	Ea.	66.50	185	251.50
Ground cable	.100	C.L.F.	45	28.10	73.10
3/4" EMT	10.000	L.F.	13.50	62.50	76
Panelboard, 24 circuit	1.000	Ea.	495	1,125	1,620
TOTAL			1,307	2,787.60	4,094.60

D5010 220	Residential Service - Single Phase	COST EACH MAT.	COST EACH INST.	COST EACH TOTAL
2800				
2810	100 Amp Service, single phase	610	1,825	2,435
2820	200 Amp Service, single phase	1,300	2,800	4,100
2900	Cut & patch to match existing construction, add, minimum	2%	3%	
3000	Maximum	5%	9%	
3100	Dust protection, add, minimum	1%	2%	
3200	Maximum	4%	11%	
3300	Equipment usage curtailment, add, minimum	1%	1%	
3400	Maximum	3%	10%	
3500	Material handling & storage limitation, add, minimum	1%	1%	
3600	Maximum	6%	7%	
3700	Protection of existing work, add, minimum	2%	2%	
3800	Maximum	5%	7%	
3900	Shift work requirements, add, minimum		5%	
4000	Maximum		30%	

D5020 Lighting and Branch Wiring

D5020 180	Selective Price Sheet	COST EACH		
		MAT.	INST.	TOTAL
0100	Using non-metallic sheathed, cable, air conditioning receptacle	23.50	81.50	105
0200	Disposal wiring	11.75	90.50	102.25
0300	Dryer circuit	36	148	184
0400	Duplex receptacle	23.50	62.50	86
0500	Fire alarm or smoke detector	90	81.50	171.50
0600	Furnace circuit & switch	38.50	136	174.50
0700	Ground fault receptacle	46	102	148
0800	Heater circuit	22	102	124
0900	Lighting wiring	25	51	76
1000	Range circuit	87.50	203	290.50
1100	Switches single pole	13.50	51	64.50
1200	3-way	19.05	68	87.05
1300	Water heater circuit	30	163	193
1400	Weatherproof receptacle	275	136	411
1500	Using BX cable, air conditioning receptacle	38.50	98	136.50
1600	Disposal wiring	26	108	134
1700	Dryer circuit	42.50	177	219.50
1800	Duplex receptacle	38.50	75.50	114
1900	Fire alarm or smoke detector	90	81.50	171.50
2000	Furnace circuit & switch	54	163	217
2100	Ground fault receptacle	56	123	179
2200	Heater circuit	28.50	123	151.50
2300	Lighting wiring	30	61	91
2400	Range circuit	104	247	351
2500	Switches, single pole	28.50	61	89.50
2600	3-way	25.50	81.50	107
2700	Water heater circuit	44.50	194	238.50
2800	Weatherproof receptacle	282	163	445
2900	Using EMT conduit, air conditioning receptacle	59.50	121	180.50
3000	Disposal wiring	50	136	186
3100	Dryer circuit	76.50	220	296.50
3200	Duplex receptacle	59.50	93.50	153
3300	Fire alarm or smoke detector	116	121	237
3400	Furnace circuit & switch	71	203	274
3500	Ground fault receptacle	81.50	151	232.50
3600	Heater circuit	49.50	151	200.50
3700	Lighting wiring	49	76	125
3800	Range circuit	113	300	413
3900	Switches, single pole	49.50	76	125.50
4000	3-way	43	102	145
4100	Water heater circuit	60	239	299
4200	Weatherproof receptacle	305	203	508
4300	Using aluminum conduit, air conditioning receptacle	56.50	163	219.50
4400	Disposal wiring	47	181	228
4500	Dryer circuit	79.50	291	370.50
4600	Duplex receptacle	53	125	178
4700	Fire alarm or smoke detector	119	163	282
4800	Furnace circuit & switch	68.50	271	339.50
4900	Ground fault receptacle	82	203	285
5000	Heater circuit	44.50	203	247.50
5100	Lighting wiring	54	102	156
5200	Range circuit	125	405	530
5300	Switches, single pole	50	102	152
5400	3-way	43.50	136	179.50
5500	Water heater circuit	61.50	325	386.50
5600	Weatherproof receptacle	315	271	586
5700	Using galvanized steel conduit	68	173	241
5800	Disposal wiring	53	194	247

D50 Electrical

D5020 Lighting and Branch Wiring

D5020 180	Selective Price Sheet	COST EACH		
		MAT.	INST.	TOTAL
5900	Dryer circuit	100	315	415
6000	Duplex receptacle	68	133	201
6100	Fire alarm or smoke detector	125	173	298
6200	Furnace circuit & switch	74.50	291	365.50
6300	Ground fault receptacle	101	214	315
6400	Heater circuit	58.50	214	272.50
6500	Lighting wiring	62.50	108	170.50
6600	Range circuit	145	430	575
6700	Switches, single pole	69.50	108	177.50
6800	3-way	63.50	140	203.50
6900	Water heater circuit	78.50	340	418.50
7000	Weatherproof receptacle	325	291	616
7100				
7200				
7300				
7400				
7500				
7600				
7700				
7800				
7900				
8000				
8100				
8200				
8300				
8400				

This page illustrates and describes fluorescent lighting systems including a fixture, lamp, outlet box and wiring. Lines within System Components give the unit price and total price on a cost each basis for this system. Prices for fluorescent lighting systems are on Line Items D5020 248 1200 thru 1500. Both material quantities and labor costs have been adjusted for the system listed.

Factors: To adjust for job conditions other than normal working situations use Lines D5020 248 2900 thru 4000.

Example: You are to install the system during evening hours. Go to Line D5020 248 3900 and apply this percentage to the appropriate INST. cost.

System Components				COST EACH		
	QUANTITY	UNIT	MAT.	INST.	TOTAL
Fluorescent lighting, including fixture, lamp, outlet box and wiring.						
Recessed lighting fixture, on suspended system	1.000	Ea.	73	173	246
Outlet box	1.000	Ea.	2.27	45	47.27
#12 wire	.660	C.L.F.	9.41	48.84	58.25
Conduit, EMT, 1/2" conduit	20.000	L.F.	18	95.80	113.80
TOTAL			102.68	362.64	465.32

D5020 248	Lighting, Fluorescent	COST EACH		
		MAT.	INST.	TOTAL
1200	Recessed lighting fixture, on suspended system	103	365	468
1300	Surface mounted, 2' x 4', acrylic prismatic diffuser	126	380	506
1400	Strip fixture, 8' long, two 8' lamps	157	360	517
1500	Pendant mounted, industrial, 8' long, with reflectors	140	415	555
1600				
1700				
1800				
1900				
2000				
2100				
2200				
2300				
2900	Cut & patch to match existing construction, add, minimum	2%	3%	
3000	Maximum	5%	9%	
3100	Dust protection, add, minimum	1%	2%	
3200	Maximum	4%	11%	
3300	Equipment usage curtailment, add, minimum	1%	1%	
3400	Maximum	3%	10%	
3500	Material handling & storage limitation, add, minimum	1%	1%	
3600	Maximum	6%	7%	
3700	Protection of existing work, add, minimum	2%	2%	
3800	Maximum	5%	7%	
3900	Shift work requirements, add, minimum		5%	
4000	Maximum		30%	

D5020 Lighting and Branch Wiring

This page illustrates and describes incandescent lighting systems including a fixture, lamp, outlet box, conduit and wiring. Lines within System Components give the unit price and total price on a cost each basis for this system. Prices for an alternate incandescent lighting system are also given. Both material quantities and labor costs have been adjusted for the system listed.

Factors: To adjust for conditions other than normal working situations use Lines D5020 250 2900 thru 4000.

Example: You are to install the system and cut and match existing construction. Go to Line D5020 250 3000 and apply this percentage to the appropriate INST. costs.

System Components	QUANTITY	UNIT	COST EACH		
			MAT.	INST.	TOTAL
Incandescent light fixture, including lamp, outlet box, conduit and wiring.					
Recessed wide reflector with flat glass lens	1.000	Ea.	109	121	230
Outlet box	1.000	Ea.	2.27	45	47.27
Armored cable, 3 wire	.200	C.L.F.	16.30	74	90.30
TOTAL			127.57	240	367.57
Recessed flood light fixture, including lamp, outlet box, conduit & wire					
Recessed, R-40 flood lamp with refractor	1.000	Ea.	116	102	218
150 watt R-40 flood lamp	.010	Ea.	22.50	6.25	28.75
Outlet box	1.000	Ea.	2.27	45	47.27
Outlet box cover	1.000	Ea.	1.31	12.70	14.01
Romex, 12-2 with ground	.200	C.L.F.	6.40	74	80.40
Conduit, 1/2" EMT	20.000	L.F.	18	95.80	113.80
TOTAL			166.48	335.75	502.23

D5020 250	Lighting, Incandescent	COST EACH		
		MAT.	INST.	TOTAL
2810	Incandescent light fixture, including lamp, outlet box, conduit and wiring.	128	240	368
2820	Recessed flood light fixture, including lamp, outlet box, conduit & wire	166	335	501
2900	Cut & patch to match existing construction, add, minimum	2%	3%	
3000	Maximum	5%	9%	
3100	Dust protection, add, minimum	1%	2%	
3200	Maximum	4%	11%	
3300	Equipment usage curtailment, add, minimum	1%	1%	
3400	Maximum	3%	10%	
3500	Material handling & storage limitation, add, minimum	1%	1%	
3600	Maximum	6%	7%	
3700	Protection of existing work, add, minimum	2%	2%	
3800	Maximum	5%	7%	
3900	Shift work requirements, add, minimum		5%	
4000	Maximum		30%	

D5020 Lighting and Branch Wiring

This page illustrates and describes high intensity lighting systems including a lamp, EMT conduit, EMT "T" fitting with cover, and wire. Lines within System Components give the unit price and total price on a cost each basis for this system. Prices for high intensity lighting systems are on Line Items D5020 252 1100 thru 1800. Both material quantities and labor costs have been adjusted for the system listed.

Factors: To adjust for job conditions other than normal working situations use Lines D5020 252 2900 thru 4000.

Example: You are to install the system and protect existing construction. Go to Line D5020 252 3700 and apply these percentages to the appropriate MAT. and INST. costs.

System Components	QUANTITY	UNIT	COST EACH MAT.	COST EACH INST.	COST EACH TOTAL
High intensity lighting system consisting of 400 watt metal halide fixture And lamp with 1/2" EMT conduit and fittings using #12 wire, high bay.					
Single unit, 400 watt	1.000	Ea.	470	355	825
Electric metallic tubing (EMT), 1/2" diameter	30.000	Ea.	27	143.70	170.70
1/2" EMT "T" fitting with cover	1.000	Ea.	5.70	51	56.70
Wire, 600 volt, type THWN-THHN, copper, solid, #12	.600	Ea.	8.55	44.40	52.95
TOTAL			511.25	594.10	1,105.35

D5020 252	Lighting, High Intensity		MAT.	INST.	TOTAL
1100	High bay: 400 watt, metal halide fixture and lamp		510	595	1,105
1200	High pressure sodium and lamp		475	595	1,070
1400	1000 watt, metal halide		940	645	1,585
1500	High pressure sodium		650	645	1,295
1700	Low bay: 250 watt, metal halide		445	495	940
1800	150 watt, high pressure sodium		415	495	910
1900					
2000					
2100					
2200					
2300					
2400					
2500					
2600					
2700					
2900	Cut & patch to match existing construction, add, minimum		2%	3%	
3000	Maximum		5%	9%	
3100	Dust protection, add, minimum		1%	2%	
3200	Maximum		4%	11%	
3300	Equipment usage curtailment, add, minimum		1%	1%	
3400	Maximum		3%	10%	
3500	Material handling & storage limitation, add, minimum		1%	1%	
3600	Maximum		6%	7%	
3700	Protection of existing work, add, minimum		2%	2%	
3800	Maximum		5%	7%	
3900	Shift work requirements, add, minimum			5%	
4000	Maximum			30%	

D5090 Other Electrical Systems

This page illustrates and describes baseboard heat systems including a thermostat, outlet box, breaker, and feed. Lines within System Components give the unit price and total price on a cost each basis for this system. Prices for baseboard heat systems are on Line Items D5090 430 1500 thru 1800. Both material quantities and labor costs have been adjusted for the system listed.

Factors: To adjust for job conditions other than normal working situations use Lines D5090 430 2900 thru 4000.

Example: You are to install the system during evenings and weekends only. Go to Line D5090 430 4000 and apply this percentage to the appropriate INST. cost.

System Components	QUANTITY	UNIT	COST EACH MAT.	COST EACH INST.	COST EACH TOTAL
Baseboard heat including thermostat, outlet box, breaker and feed.					
Electric baseboard heater, 4' long	1.000	Ea.	50	121	171
Thermostat, integral	1.000	Ea.	36	51	87
Romex, 12-3 with ground	.400	C.L.F.	20.80	162	182.80
Panel board breaker, 15-50 amp	1.000	Ea.	70	102	172
Outlet box	1.000	Ea.	2.27	45	47.27
TOTAL			179.07	481	660.07

D5090 530	Heat, Baseboard	COST EACH MAT.	COST EACH INST.	COST EACH TOTAL
1400	Heat, electric baseboard			
1500	2' long	165	460	625
1550	4' long	179	480	659
1552	5 long	189	505	694
1600	6' long	198	525	723
1700	8' long	213	565	778
1800	10' long	237	605	842
1900				
2000				
2100				
2200				
2300				
2400				
2500				
2600				
2700				
2900	Cut & patch to match existing construction, add, minimum	2%	3%	
3000	Maximum	5%	9%	
3100	Dust protection, add, minimum	1%	2%	
3200	Maximum	4%	11%	
3300	Equipment usage curtailment, add, minimum	1%	1%	
3400	Maximum	3%	10%	
3500	Material handling & storage limitation, add, minimum	1%	1%	
3600	Maximum	6%	7%	
3700	Protection of existing work, add, minimum	2%	2%	
3800	Maximum	5%	7%	
3900	Shift work requirements, add, minimum		5%	
4000	Maximum		30%	

Same Data. Simplified.

Enjoy the convenience and efficiency of accessing your costs anywhere:

- **Skip the multiplier** by setting your location
- **Quickly search,** edit, favorite and share costs
- **Stay on top of price changes** with automatic updates

Discover more at rsmeans.com/online

This page illustrates and describes kitchen systems including top and bottom cabinets, custom laminated plastic top, single bowl sink, and appliances. Lines within System Components give the unit price and total price on a cost each basis for this system. Prices for kitchen systems are on Line Items E1090 310 1400 thru 1600. Both material quantities and labor costs have been adjusted for the system listed.

Factors: To adjust for job conditions other than normal working situations use Lines E1090 310 2900 thru 4000.

Example: You are to install the system and protect the work area from dust. Go to Line E1090 310 3200 and apply these percentages to the appropriate MAT. and INST. costs.

System Components	QUANTITY	UNIT	COST EACH MAT.	COST EACH INST.	COST EACH TOTAL
Kitchen cabinets including wall and base cabinets, custom laminated					
Plastic top, sink & appliances, no plumbing or electrical rough-in included.					
Prefinished wood cabinets, average quality, wall and base	20.000	L.F.	4,580	950	5,530
Custom laminated plastic counter top	20.000	L.F.	388	480	868
Stainless steel sink, 22" x 25"	1.000	Ea.	790	277	1,067
Faucet, top mount	1.000	Ea.	91.50	86	177.50
Dishwasher, built-in	1.000	Ea.	415	420	835
Compactor, built-in	1.000	Ea.	1,200	143	1,343
Range hood, vented, 30" wide	1.000	Ea.	128	310	438
TOTAL			7,592.50	2,666	10,258.50

E1090 310	Kitchens	COST EACH MAT.	COST EACH INST.	COST EACH TOTAL
1400	Prefinished wood cabinets, average quality	7,600	2,675	10,275
1500	Prefinished wood cabinets, high quality	14,700	3,050	17,750
1600	Custom cabinets, built in place, high quality	19,400	3,525	22,925
1700				
1800				
1900				
2000				
2100				
2200	NOTE: No plumbing or electric rough-ins are included in the above			
2300	Prices, for plumbing see Division 22, for electric see Division 26.			
2400				
2500				
2600				
2700				
2900	Cut & patch to match existing construction, add, minimum	2%	3%	
3000	Maximum	5%	9%	
3100	Dust protection, add, minimum	1%	2%	
3200	Maximum	4%	11%	
3300	Equipment usage curtailment, add, minimum	1%	1%	
3400	Maximum	3%	10%	
3500	Material handling & storage limitation, add, minimum	1%	1%	
3600	Maximum	6%	7%	
3700	Protection of existing work, add, minimum	2%	2%	
3800	Maximum	5%	7%	
3900	Shift work requirements, add, minimum		5%	
4000	Maximum		30%	

E1090 Other Equipment

E1090 315	Selective Price Sheet	COST EACH		
		MAT.	INST.	TOTAL
0100	Cabinets standard wood, base, one drawer one door, 12" wide	425	57.50	482.50
0200	15" wide	440	59.50	499.50
0300	18" wide	455	61.50	516.50
0400	21" wide	470	63	533
0500	24" wide	515	64	579
0600				
0700	Two drawers two doors, 27" wide	530	65	595
0800	30" wide	560	67	627
0900	33" wide	585	68.50	653.50
1000	36" wide	595	70.50	665.50
1100	42" wide	725	72.50	797.50
1200	48" wide	750	75.50	825.50
1300	Drawer base (4 drawers), 12" wide	810	57.50	867.50
1400	15" wide	815	59.50	874.50
1500	18" wide	850	61.50	911.50
1600	24" wide	895	64	959
1700	Sink or range base, 30" wide	405	67	472
1800	33" wide	430	68.50	498.50
1900	36" wide	445	70.50	515.50
2000	42" wide	470	72.50	542.50
2100	Corner base, 36" wide	425	79.50	504.50
2200	Lazy Susan with revolving door	560	86.50	646.50
2300	Cabinets standard wood, wall two doors, 12" high, 30" wide	244	57.50	301.50
2500	15" high, 30" high	360	59.50	419.50
2600	36" wide	171	63	234
2700	24" high, 30" wide	183	61.50	244.50
2800	36" wide	525	63	588
2900	30" high, 30" wide	445	74	519
3000	36" wide	425	76	501
3100	42" wide	500	77.50	577.50
3200	48" wide	855	78	933
3300	One door, 30" high, 12" wide	119	65	184
3400	15" wide	270	67	337
3500	18" wide	297	68.50	365.50
3600	24" wide	345	70.50	415.50
3700	Corner, 30" high, 24" wide	221	79.50	300.50
3800	36" wide	515	86.50	601.50
3900	Broom, 84" high, 24" deep, 18" wide	895	143	1,038
4000	Oven, 84" high, 24" deep, 27" wide	1,150	179	1,329
4100	Valance board, 4' long	108	14.45	122.45
4200	6" long	162	21.50	183.50
4300	Counter tops, laminated plastic, stock 25" wide w/backsplash, min.	19.40	24	43.40
4400	Maximum	47	28.50	75.50
4500	Custom, 7/8" thick, no splash	43	24	67
4600	Cove splash	55	24	79
4700	1-1/4" thick, no splash	47.50	25.50	73
4800	Square splash	52.50	25.50	78
4900	Post formed	14.65	24	38.65
5000				
5100	Maple laminated 1-1/2" thick, no splash	103	25.50	128.50
5200	Square splash	122	25.50	147.50
5300				
5400				
5500	Appliances, range, free standing, minimum	690	116	806
5600	Maximum	990	290	1,280
5700	Built-in, minimum	1,325	136	1,461
5800	Maximum	2,100	715	2,815
5900	Counter top range 4 burner, minimum	325	136	461

E1090 Other Equipment

E1090 315	Selective Price Sheet	COST EACH		
		MAT.	INST.	TOTAL
6000	Maximum	595	271	866
6100	Compactor, built-in, minimum	1,200	143	1,343
6200	Maximum	1,375	239	1,614
6300	Dishwasher, built-in, minimum	415	420	835
6400	Maximum	920	835	1,755
6500	Garbage disposer, sink-pipe, minimum	109	167	276
6600	Maximum	263	167	430
6700	Range hood, 30" wide, 2 speed, minimum	128	310	438
6800	Maximum	920	515	1,435
6900	Refrigerator, no frost, 12 cu. ft.	620	116	736
7000	20 cu. ft.	550	193	743
7100	Plumb. not incl. rough-ins, sinks porc. C.I., single bowl, 21" x 24"	480	277	757
7200	21" x 30"	1,150	277	1,427
7300	Double bowl, 20" x 32"	440	325	765
7400				
7500	Stainless steel, single bowl, 19" x 18"	710	277	987
7600	22" x 25"	790	277	1,067
7700				
7800				
7900				
8000				
8100				
8200				
8300				
8400				

Same Data. Simplified.

Enjoy the convenience and efficiency of accessing your costs anywhere:

■ **Skip the multiplier** by setting your location

■ **Quickly search,** edit, favorite and share costs

■ **Stay on top of price changes** with automatic updates

Discover more at rsmeans.com/online

G10 Site Preparation

G1030 Site Earthwork

This page illustrates and describes utility trench excavation with backfill systems including a trench, backfill, excavated material, utility pipe or ductwork not included. Lines within System Components give the unit price and total price on a linear foot basis for this system. Prices for utility trench excavation with backfill systems are on Line Items G1030 810 1100 thru 2500. Both material

quantities and labor costs have been adjusted for the system listed.

Factors: To adjust for job conditions other than normal working situations use Lines G1030 810 3100 thru 4000.

Example: You are to install the system and protect existing construction. Go to Line G1030 810 3600 and apply these percentages to the appropriate TOTAL cost.

System Components			COST PER L.F.		
	QUANTITY	UNIT	MAT.	INST.	TOTAL
Cont. utility trench excav. with 3/8 C.Y. wheel mtd. backhoe, Backfilling with excav. mat. and comp. in 12″ lifts. Trench is 2′ wide by 4′ dp. Cost of utility piping or ductwork is not incl. Cost based on an Excavation production rate of 150 C.Y. daily no hauling included.					
Machine excavate trench, 2′ wide by 4′ deep	.296	C.Y.	.71	2.63	3.34
Dozer backfill with excavated material	.296	C.Y.	.41	.73	1.14
Compact in 12″ lifts, vibrating plate	.296	C.Y.	.25	2.90	3.15
TOTAL			1.37	6.26	7.63

G1030 810	Excavation, Utility Trench	COST PER L.F.		
		MAT.	INST.	TOTAL
1100	2′ wide x 2′ deep	.69	3.13	3.82
1200	3′ deep	1.03	4.70	5.75
1250	4′ deep	1.37	6.25	7.65
1300	With sloping sides, 2′ wide x 5′ deep	3.87	17.60	21.50
1400	6′ deep	5.15	23.50	28.50
1500	7′ deep	6.60	30	37
1600	8′ deep	8.25	37.50	46
1700	9′ deep	10.05	46	56
1800	10′ deep	12.05	55	67
2000	For hauling excavated material up to 2 miles & backfilling w/ gravel			
2100	Gravel, 2′ wide by 2′ deep, add	5.05	4.25	9.30
2200	4′ deep, add	10.05	8.50	18.55
2300	6′ deep, add	38	32	70
2400	8′ deep, add	60.50	51	111.50
2500	10′ deep, add	88	74.50	162.50
2600				
2700				
2800	For shallow, hand excavated trenches, no backfill, to 6′ dp, light soil		72.50	72.50
2900	Heavy soil		145	145
3000				
3100	Equipment usage curtailment, add, minimum	1%	1%	
3200	Maximum	3%	10%	
3300	Material handling & storage limitation, add, minimum	1%	1%	
3400	Maximum	6%	7%	
3500	Protection of existing work, add, minimum	2%	2%	
3600	Maximum	5%	7%	
3700	Shift work requirements, add, minimum		5%	
3800	Maximum		30%	
3900	Temporary shoring and bracing, add, minimum	2%	5%	
4000	Maximum	5%	12%	

G20 Site Improvements

G2010 Roadways

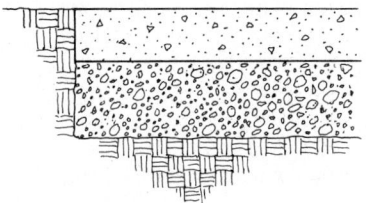

This page illustrates and describes driveway systems including concrete slab, gravel base, broom finish, compaction, and joists. Lines within System Components give the unit price and total price on a cost each basis for this system. Prices for driveway systems are on Line Items G2010 240 1200 thru 2700. Both material quantities and labor costs have been adjusted for the system listed.

Factors: To adjust for job conditions other than normal working situations use Lines G2010 240 2900 thru 4000.

Example: You are to install the system and match existing construction. Go to Line G2010 240 3000 and apply these percentages to the appropriate MAT. and INST. costs.

System Components	QUANTITY	UNIT	COST EACH		
			MAT.	INST.	TOTAL
Complete driveway, 10' x 30', 4" concrete slab, 3000 psi, on 6" crushed Stone base, concrete broom finished and cured with 10' x 10' joints.					
Grade and compact subgrade	33.000	S.Y.		76.56	76.56
Place and compact 6" crushed stone base	33.000	S.Y.	144.54	50.16	194.70
Place and remove edgeforms (4 use)	80.000	L.F.	40.80	363.20	404
Concrete for slab, 3000 psi	4.000	C.Y.	700		700
Place concrete, 4" thick slab, direct chute	4.000	C.Y.		132.24	132.24
Screed, float and broom finish slab	4.000	C.Y.		309	309
Spray on membrane curing compound	3.000	S.F.	44.70	36.60	81.30
Saw cut 3" deep joints	60.000	L.F.	7.20	124.80	132
TOTAL			**937.24**	**1,092.56**	**2,029.80**

G2010 240	Driveways	COST EACH		
		MAT.	INST.	TOTAL
1200	Concrete: 10' x 30' with 4" concrete on 6" crushed stone	935	1,100	2,035
1300	With 6" concrete	1,275	1,150	2,425
1400	20' x 30' with 4" concrete	1,850	2,100	3,950
1500	With 6" concrete	2,550	2,225	4,775
1600	10' wide, for each additional 10' length over 30', 4" thick, add	296	350	646
1700	6" thick, add	425	375	800
1800	20' wide, for each additional 10' length over 30', 4" thick, add	585	675	1,260
1900	6" thick, add	805	720	1,525
2000	Asphalt: 10' x 30' with 2" binder, 1" topping on 6" cr. st., sealed	630	1,225	1,855
2100	3" binder, 1" topping	765	1,250	2,015
2200	20' x 30' with 2" binder, 1" topping	1,250	1,500	2,750
2300	3" binding, 1" topping	1,525	1,550	3,075
2400	10' wide for each add'l. 10' length over 30', 2" b + 1" t, add	210	93.50	303.50
2500	3" binder, 1" topping	254	98.50	352.50
2600	20' wide for each add'l. 10' length over 30', 2" b + 1" t, add	420	187	607
2700	3" binder, 1" topping	510	197	707
2800				
2900	Cut & patch to match existing construction, add, minimum	2%	3%	
3000	Maximum	5%	9%	
3100	Dust protection, add, minimum	1%	2%	
3200	Maximum	4%	11%	
3300	Equipment usage curtailment, add, minimum	1%	1%	
3400	Maximum	3%	10%	
3500	Material handling & storage limitation, add, minimum	1%	1%	
3600	Maximum	6%	7%	
3700	Protection of existing work, add, minimum	2%	2%	
3800	Maximum	5%	7%	
3900	Shift work requirements, add, minimum		5%	
4000	Maximum		30%	

G2020 Parking Lots

This page illustrates and describes asphalt parking lot systems including asphalt binder, topping, crushed stone base, painted parking stripes and concrete parking blocks. Lines within System Components give the unit price and total price per square yard for this system. Prices for asphalt parking lot systems are on Line Items G2020 230 1400 thru 2500. Both material quantities and labor costs have been adjusted for the system listed.

Factors: To adjust for job conditions other than normal working situations use Lines G2020 230 2900 thru 4000.

Example: You are to install the system and match existing construction. Go to Line G2020 230 2900 and apply these percentages to the appropriate MAT. and INST. costs.

System Components	QUANTITY	UNIT	COST PER S.Y.		
			MAT.	INST.	TOTAL
Parking lot consisting of 2″ asphalt binder and 1″ topping on 6″ Crushed stone base with painted parking stripes and concrete parking blocks.					
Fine grade and compact subgrade	1.000	S.Y.		2.32	2.32
6″ crushed stone base, stone	.320	Ton	1.40	.49	1.89
2″ asphalt binder	1.000	S.Y.	8.15	1.59	9.74
1″ asphalt topping	1.000	S.Y.	4.04	1.07	5.11
Paint parking stripes	.500	L.F.	.08	.11	.19
6″ x 10″ x 6′ precast concrete parking blocks	.020	Ea.	1.01	.49	1.50
Mobilization of equipment	.005	Ea.		4.75	4.75
TOTAL			14.68	10.82	25.50

G2020 230	Parking Lots, Asphalt	COST PER S.Y.		
		MAT.	INST.	TOTAL
1400	2″ binder plus 1″ topping on 6″ crushed stone	14.70	10.80	25.50
1500	9″ crushed stone	19.85	11.95	31.80
1600	12″ crushed stone	22	12.10	34.10
1700	On bank run gravel, 6″ deep	19.15	11.25	30.40
1800	9″ deep	22	11.45	33.45
1900	12″ deep	25	11.65	36.65
2000	3″ binder plus 1″ topping on 6″ crushed stone	21.50	12.30	33.80
2100	9″ deep crushed stone	24	12.45	36.45
2200	12″ deep crushed stone	26	12.60	38.60
2300	On bank run gravel, 6″ deep	23	11.70	34.70
2400	9″ deep	26	11.90	37.90
2500	12″ deep	29	12.10	41.10
2600				
2700				
2900	Cut & patch to match existing construction, add, minimum	2%	3%	
3000	Maximum	5%	9%	
3100	Dust protection, add, minimum	1%	2%	
3200	Maximum	4%	11%	
3300	Equipment usage curtailment, add, minimum	1%	1%	
3400	Maximum	3%	10%	
3500	Material handling & storage limitation, add, minimum	1%	1%	
3600	Maximum	6%	7%	
3700	Protection of existing work, add, minimum	2%	2%	
3800	Maximum	5%	7%	
3900	Shift work requirements, add, minimum		5%	
4000	Maximum		30%	

G2020 Parking Lots

This page illustrates and describes concrete parking lot systems including a concrete slab, unreinforced, compacted gravel base, broom finished, cured, joints, painted parking stripes, and parking blocks. Lines within System Components give the unit price and total price per square yard for this system. Prices for concrete parking lot systems are on Line Items G2020 240 1500 thru 2500. All materials have been adjusted according to the system listed.

Factors: To adjust for job conditions other than normal working situations use Lines G2020 240 2900 thru 4000.

Example: You are to install the system and protect existing construction. Go to Line G2020 240 3800 and apply these percentages to the appropriate MAT. and INST. costs.

System Components	QUANTITY	UNIT	COST PER S.Y.		
			MAT.	INST.	TOTAL
Parking lot, 4" slab, 3000 psi concrete on 6" crushed stone base					
With 12' x 12' joints, painted parking strips, conc. parking blocks.					
Grade and compact subgrade	1.000	S.Y.		2.32	2.32
6" crushed stone base, stone	.320	Ton	1.40	.49	1.89
Place and remove edge forms, 4 use	.250	L.F.	.13	1.14	1.27
Concrete for slab, 3000 psi	1.000	S.F.	19.25		19.25
Place concrete, 4" thick slab, direct chute	1.000	S.F.		3.64	3.64
Spray on membrane curing compound	1.000	S.F.	1.34	.77	2.11
Saw cut 3" deep joints	2.250	L.F.	.27	4.69	4.96
Paint parking stripes	.500	L.F.	.08	.11	.19
6" x 10" x 6' precast concrete parking blocks	.020	Ea.	1.01	.49	1.50
TOTAL			23.48	13.65	37.13

G2020 240	Parking Lots, Concrete	COST PER S.Y.		
		MAT.	INST.	TOTAL
1500	4" concrete slab, 3000 psi on 3" crushed stone	24.50	14.60	39.10
1550	6" crushed stone	23.50	13.60	37.10
1600	9" crushed stone	28.50	14.80	43.30
1700	On bank run gravel, 6" deep	28	14.05	42.05
1800	9" deep	31	14.25	45.25
1900	On compacted subgrade only	22	13.15	35.15
2000	6" concrete slab, 3000 psi on 3" crushed stone	33.50	16.35	49.85
2100	6" crushed stone	35.50	16.40	51.90
2200	9" crushed stone	38	16.55	54.55
2300	On bank run gravel, 6" deep	37	15.80	52.80
2400	9" deep	40	16	56
2500	On compacted subgrade only	35.50	16.40	51.90
2600				
2700				
2900	Cut & patch to match existing construction, add, minimum	2%	3%	
3000	Maximum	5%	9%	
3100	Dust protection, add, minimum	1%	2%	
3200	Maximum	4%	11%	
3300	Equipment usage curtailment, add, minimum	1%	1%	
3400	Maximum	3%	10%	
3500	Material handling & storage limitation, add, minimum	1%	1%	
3600	Maximum	6%	7%	
3700	Protection of existing work, add, minimum	2%	2%	
3800	Maximum	5%	7%	
3900	Shift work requirements, add, minimum		5%	
4000	Maximum		30%	

733

G2030 Pedestrian Paving

This page illustrates and describes sidewalk systems including concrete, welded wire and broom finish. Lines within System Components give unit price and total price per square foot for this system. Prices for sidewalk systems are on Line Items G2030 105 1700 thru 1900. Both material quantities and labor costs have been adjusted for the system listed.

Factors: To adjust for job conditions other than normal working situations use Lines G2030 105 2900 thru 4000.

Example: You are to install the system and match existing construction. Go to Line G2030 105 2900 and apply these percentages to the appropriate MAT. and INST. costs.

System Components	QUANTITY	UNIT	MAT.	INST.	TOTAL
4″ thick concrete sidewalk with welded wire fabric					
3000 psi air entrained concrete, broom finish.					
Gravel fill, 4″ deep	.012	C.Y.	.36	.11	.47
Compact fill	.012	C.Y.		.04	.04
Hand grade	1.000	S.F.		2.74	2.74
Edge form	.250	L.F.	.13	1.14	1.27
Welded wire fabric	.011	S.F.	.27	.48	.75
Concrete, 3000 psi air entrained	.012	C.Y.	2.10		2.10
Place concrete	.012	C.Y.		.40	.40
Broom finish	1.000	S.F.		1.03	1.03
TOTAL			2.86	5.94	8.80

G2030 105	Sidewalks	MAT.	INST.	TOTAL
1700	Asphalt (bituminous), 2″ thick	1.23	3.49	4.72
1750	4″ thick	2.86	5.95	8.81
1800	Brick, on sand, bed, 4.5 brick per S.F.	4.19	12.85	17.04
1900	Flagstone, slate, 1″ thick, rectangular	17.70	13.95	31.65
2000				
2100				
2200				
2300				
2400				
2500				
2600				
2700				
2900	Cut & patch to match existing construction, add, minimum	2%	3%	
3000	Maximum	5%	9%	
3100	Dust protection, add, minimum	1%	2%	
3200	Maximum	4%	11%	
3300	Equipment usage curtailment, add, minimum	1%	1%	
3400	Maximum	3%	10%	
3500	Material handling & storage limitation, add, minimum	1%	1%	
3600	Maximum	6%	7%	
3700	Protection of existing work, add, minimum	2%	2%	
3800	Maximum	5%	7%	
3900	Shift work requirements, add, minimum		5%	
4000	Maximum		30%	

G2050 Landscaping

This page describes landscaping—lawn establishment systems including loam, lime, fertilizer, side and top mulching. Lines within system components give the unit price and total price per square yard for this system. Prices for landscaping—lawn establishment systems are on Line Items G2050 420 1000 thru 1200. Both material quantities and labor costs have been adjusted for the system listed.

Factors: To adjust for job conditions other than normal working situations use Lines G2050 420 2900 thru 4000.

Example: You are to install the system and provide dust protection. Go to Line G2050 420 3200 and apply these percentages to the appropriate MAT. and INST. costs.

System Components	QUANTITY	UNIT	COST PER S.Y.		
			MAT.	INST.	TOTAL
Establishing lawns with loam, lime, fertilizer, seed and top mulching					
On rough graded areas.					
Furnish and place loam 4" deep	.110	C.Y.	4.67	1.18	5.85
Fine grade, lime, fertilize and seed	1.000	S.Y.	.62	3.99	4.61
Hay mulch, 1 bale/M.S.F.	1.000	S.Y.	.28	.71	.99
Rolling with hand roller	1.000	S.Y.		1.20	1.20
TOTAL			5.57	7.08	12.65

G2050 420	Landscaping - Lawn Establishment	COST PER S.Y.		
		MAT.	INST.	TOTAL
1000	Establishing lawns with loam, lime, fertilizer, seed and hay mulch	5.55	7.10	12.65
1100	Above system with jute mesh in place of hay mulch	6.40	7.25	13.65
1200	Above system with sod in place of seed	10.60	2.60	13.20
1300				
1400				
1500				
1600				
1700				
1800				
1900				
2000				
2100				
2200				
2300				
2400				
2500				
2600				
2700				
2900	Cut & patch to match existing construction, add, minimum	2%	3%	
3000	Maximum	5%	9%	
3100	Dust protection, add, minimum	1%	2%	
3200	Maximum	4%	11%	
3300	Equipment usage curtailment, add, minimum	1%	1%	
3400	Maximum	3%	10%	
3500	Material handling & storage limitation, add, minimum	1%	1%	
3600	Maximum	6%	7%	
3700	Protection of existing work, add, minimum	2%	2%	
3800	Maximum	5%	7%	
3900	Shift work requirements, add, minimum		5%	
4000	Maximum		30%	

735

G2050 Landscaping

G2050 510	Selective Price Sheet	COST PER UNIT		
		MAT.	INST.	TOTAL
0100	Shrubs and trees, Evergreen, in prepared beds			
0200				
0300	Arborvitae pyramidal, 4'-5'	125	112	237
0400	Globe, 12"-15"	18.20	18.40	36.60
0500	Cedar, blue, 8'-10'	237	185	422
0600	Hemlock, Canadian, 2-1/2'-3'	22	49	71
0700	Juniper, andora, 18"-24"	66	22	88
0800	Wiltoni, 15"-18"	30.50	22	52.50
0900	Skyrocket, 4-1/2'-5'	177	61	238
1000	Blue pfitzer, 2'-2-1/2'	43.50	40	83.50
1100	Ketleerie, 2-1/2'-3'	70	35.50	105.50
1200	Pine, black, 2-1/2'-3'	61.50	35.50	97
1300	Mugo, 18"-24"	58	29.50	87.50
1400	White, 4'-5'	61.50	44.50	106
1500	Spruce, blue, 18"-24"	78	29.50	107.50
1600	Norway, 4'-5'	88	44.50	132.50
1700	Yew, denisforma, 12"-15"	43.50	29.50	73
1800	Capitata, 18"-24"	38.50	59	97.50
1900	Hicksi, 2'-2-1/2'	115	59	174
2000				
2100	Trees, Deciduous, in prepared beds			
2200				
2300	Beech, 5'-6'	268	67	335
2400	Dogwood, 4'-5'	88	83.50	171.50
2500	Elm, 8'-10'	465	167	632
2600	Magnolia, 4'-5'	110	167	277
2700	Maple, red, 8'-10', 1-1/2" caliper	116	335	451
2800	Oak, 2-1/2"-3" caliper	315	555	870
2900	Willow, 6'-8', 1" caliper	116	167	283
3000				
3100	Shrubs, Broadleaf Evergreen, in prepared beds			
3200				
3300	Andromeda, 15"-18", container	42.50	18.40	60.90
3400	Azalea, 15"-18", container	29.50	18.40	47.90
3500	Barberry, 9"-12", container	24.50	13.55	38.05
3600	Boxwood, 15"-18", B & B	66	18.40	84.40
3700	Euonymus, emerald gaiety, 12"-15", container	30	15.35	45.35
3800	Holly, 15"-18", B & B	68	18.40	86.40
3900	Mount laurel, 18"-24", B & B	71	22	93
4000	Privet, 18"-24", B & B	31.50	13.55	45.05
4100	Rhododendron, 18"-24", container	65	37	102
4200	Rosemary, 1 gal. container	28.50	2.94	31.44
4300	Deciduous, amalanchier, 2'-3', B & B	129	31	160
4400	Azalea, 15"-18", B & B	32	18.40	50.40
4500	Bayberry, 2'-3', B & B	44	31	75
4600	Cotoneaster, 15"-18", B & B	33	22	55
4700	Dogwood, 3'-4', B & B	38.50	83.50	122
4800	Euonymus, alatus compacta, 15"-18", container	38	22	60
4900	Forsythia, 2'-3', container	17.45	29.50	46.95
5000	Honeysuckle, 3'-4', B & B	87.50	29.50	117
5100	Hydrangea, 2'-3', B & B	32	31	63
5200	Lilac, 3'-4', B & B	55	83.50	138.50
5300	Quince, 2'-3', B & B	42.50	31	73.50
5400				
5500	Ground Cover			
5600				
5700	Plants, Pachysandra, prepared beds, per hundred	132	118	250
5800	Vinca minor or English ivy, per hundred	121	147	268

G20 Site Improvements

G2050 Landscaping

G2050 510	Selective Price Sheet	COST PER UNIT		
		MAT.	INST.	TOTAL
5900	Plant bed prep. 18" dp., mach., per square foot	2.17	.54	2.71
6000	By hand, per square foot	2.17	3.74	5.91
6100	Stone chips, Georgia marble, 50# bags, per bag	5.05	3.39	8.44
6200	Onyx gemstone, per bag	48.50	6.80	55.30
6300	Quartz, per bag	21	6.80	27.80
6400	Pea gravel, truck load lots, per cubic yard	19.10	63	82.10
6500	Mulch, polyethylene mulch, per square yard	.73	.58	1.31
6600	Wood chips, 2" deep, per square yard	1.96	2.63	4.59
6700	Peat moss, 1" deep, per square yard	5.45	.64	6.09
6800	Erosion control per square yard, Jute mesh stapled	1.09	.89	1.98
6900	Plastic netting, stapled	1.94	.71	2.65
7000	Polypropylene mesh, stapled	1.94	.71	2.65
7100	Tobacco netting, stapled	.28	.71	.99
7200	Lawns per square yard, seeding incl. fine grade, limestone, fert. and seed	.62	3.99	4.61
7300	Sodding, incl. fine grade, level ground	5.65	1.36	7.01
7500	Edging, per linear foot, Redwood, untreated, 1" x 4"	3.61	2.86	6.47
7600	2" x 4"	6.85	4.34	11.19
7700	Stl. edge strips, 1/4" x 5" inc. stakes	4.72	4.34	9.06
7800	3/16" x 4"	4.66	4.34	9
7900	Brick edging, set on edge	3.90	9.50	13.40
8000	Set flat	1.95	3.47	5.42
8100				
8200				
8300				
8400				

737

G3020 Sanitary Sewer

This page illustrates and describes drainage and utilities – septic system including a tank, distribution box, excavation, piping, crushed stone and backfill. Lines within System Components give the unit price and total price on a cost each basis for this system. Prices for drainage and utilities – septic systems are on Line Items G3020 610 1600 thru 2000. Both material quantities and labor costs have been adjusted for the system listed.

Factors: To adjust for job conditions other than normal working situations use Lines G3020 610 3100 thru 4000.

Example: You are to install the system with a material handling limitation. Go to Line G3020 610 3500 and apply these percentages to the appropriate MAT. and INST. costs.

System Components	QUANTITY	UNIT	COST EACH MAT.	COST EACH INST.	COST EACH TOTAL
Septic system including tank, distribution box, excavation, piping, crushed Stone and backfill for a 1000 S.F. leaching field.					
Precast concrete septic tank, 1000 gal. capacity	1.000	Ea.	1,100	315	1,415
Concrete distribution box	1.000	Ea.	103	72.50	175.50
Sch. 40 sewer pipe, PVC, 4″ diameter	170.000	L.F.	413.10	875.50	1,288.60
Sch. 40 sewer pipe fittings, PVC, 4″ diameter	8.000	Ea.	140	744	884
Excavation for septic tank	18.000	C.Y.		289.80	289.80
Excavation for disposal field	101.000	C.Y.		702.96	702.96
Crushed stone backfill	76.000	C.Y.	2,090	1,129.36	3,219.36
Backfill with excavated material	26.000	C.Y.		54.86	54.86
Asphalt felt paper, #15	6.000	Sq.	29.22	116.10	145.32
TOTAL			3,875.32	4,300.08	8,175.40

G3020 610	Septic Systems	COST EACH MAT.	COST EACH INST.	COST EACH TOTAL
1600	1000 gal. tank with 1000 S.F. field	3,875	4,300	8,175
1700	1000 gal. tank with 2000 S.F. field	5,650	6,325	11,975
1800	With leaching pits	3,825	2,550	6,375
1900	2000 gal. tank with 2000 S.F. field	7,200	6,525	13,725
2000	With leaching pits	5,500	3,275	8,775
2100				
2200				
2300				
2400				
2500				
2600				
2700				
2800				
2900				
3100	Dust protection, add, minimum	1%	2%	
3200	Maximum	4%	11%	
3300	Equipment usage curtailment, add, minimum	1%	1%	
3400	Maximum	3%	10%	
3500	Material handling & storage limitation, add, minimum	1%	1%	
3600	Maximum	6%	7%	
3700	Protection of existing work, add, minimum	2%	2%	
3800	Maximum	5%	7%	
3900	Shift work requirements, add, minimum		5%	
4000	Maximum		30%	

G3060 Fuel Distribution

2" Vent Cap
3" Fill Box
Manway Extension
Supply
Return
Pea Gravel Backfill
Hold Down Straps
Concrete Slab w/ Reinforcing Steel

Fiberglass Underground Fuel Storage Tank, Single Wall

This page illustrates and describes fiberglass tanks including hold down slab, reinforcing steel and peastone gravel. Lines within Systems Components give the material and installation price on a cost each basis for the components. Prices for fiberglass tank systems are on Line Items G3060 305 1010 thru G3060 310 1100. Material quantities and labor costs have been adjusted for the system listed.

Factors: To adjust for job conditions other than normal working situations use Lines G3060 310 2700 thru 4000.

System Components	QUANTITY	UNIT	COST EACH		
			MAT.	INST.	TOTAL
Double wall underground fiberglass storage tanks including, excavation, hold down slab, backfill, manway extension and related piping.					
Excavating trench, no sheeting or dewatering, 6'-10 ' D, 3/4 CY hyd backhoe	18.000	Ea.		186.66	186.66
Corrosion resistance wrap & coat, small diam. pipe, 1" diam., add	120.000	Ea.	112.80		112.80
Fiberglass-reinforced elbows 90 & 45 degree, 2 inch	7.000	Ea.	304.50	707	1,011.50
Fiberglass-reinforced threaded adapters, 2 inch	2.000	Ea.	39.10	232	271.10
Remote tank gauging system, 30', 5" pointer travel	1.000	Ea.	7,275	435	7,710
Corrosion resistance wrap & coat, small diam. pipe. 4 " diam. add	30.000	Ea.	92.70		92.70
Pea gravel	9.000	Ea.	1,197	567	1,764
Reinforcing in hold down pad, #3 to #7	120.000	Ea.	96	86.40	182.40
Tank leak detector system, 8 channel, external monitoring	1.000	Ea.	1,850		1,850
Tubbing copper, type L, 3/4" diam.	120.000	Ea.	852	1,362	2,214
Elbow 90 degree ,wrought cu x cu 3/4 " diam.	4.000	Ea.	25	182	207
Pipe, black steel welded, sch 40, 3" diam.	30.000	Ea.	532.50	1,127.70	1,660.20
Elbow 90 degree, steel welded butt joint, 3 " diam.	3.000	Ea.	108	692.25	800.25
Fiberglass-reinforced plastic pipe, 2 inch	156.000	Ea.	1,131	889.20	2,020.20
Fiberglass-reinforced plastic pipe, 3 inch	36.000	Ea.	1,296	234	1,530
Fiberglass-reinforced elbows 90 & 45 degree, 3 inch	3.000	Ea.	456	405	861
Tank leak detection system, probes, well monitoring, liquid phase detection	2.000	Ea.	1,680		1,680
Double wall fiberglass annular space	1.000	Ea.	525		525
Flange, weld neck, 150 lb, 3 " pipe size	1.000	Ea.	69	115.88	184.88
Vent protector/breather, 2" diam.	1.000	Ea.	68	27	95
Corrosion resistance wrap & coat, small diam. pipe, 2" diam., add	36.000	Ea.	46.44		46.44
Concrete hold down pad, 6' x 5' x 8" thick	30.000	Ea.	139.50	57.60	197.10
600 gallon capacity, underground storage tank, double wall fiberglass	1.000	Ea.	11,600	645	12,245
Tank, for hold-downs 500-4,000 gal., add	1.000	Ea.	565	206	771
Foot valve, single poppet, 3/4" diam.	1.000	Ea.	149	48	197
Fuel fill box, locking inner cover, 3" diam.	1.000	Ea.	146	173	319
Fuel oil specialties, valve,ball chk, globe type fusible, 3/4" diam.	2.000	Ea.	374	87	461
TOTAL			30,729.54	8,465.69	39,195.23

G3060 310	Fiberglass Fuel Tank, Double Wall	COST EACH		
		MAT.	INST.	TOTAL
1010	Storage tank, fuel, underground, double wall fiberglass, 600 Gal.	30,700	8,475	39,175
1020	2500 Gal.	46,800	11,700	58,500
1030	4000 Gal.	56,500	14,500	71,000
1040	6000 Gal.	64,500	16,400	80,900
1050	8000 Gal.	80,000	19,800	99,800
1060	10,000 Gal.	92,500	22,800	115,300

739

G3060 Fuel Distribution

G3060 310	Fiberglass Fuel Tank, Double Wall	COST EACH		
		MAT.	INST.	TOTAL
1070	15,000 Gal.	139,500	26,300	165,800
1080	20,000 Gal.	159,000	31,800	190,800
1090	25,000 Gal.	183,000	37,000	220,000
1100	30,000 Gal.	208,500	39,300	247,800
2700	Cut & patch to match existing construction, add, minimum	2%	3%	
2800	Maximum	5%	9%	
2900	Dust protection, add, minimum	1%	2%	
3000	Maximum	4%	11%	
3100	Equipment usage curtailment, add, minimum	1%	1%	
3200	Maximum	3%	10%	
3300	Material handling & storage limitation, add, minimum	1%	1%	
3400	Maximum	6%	7%	
3500	Protection of existing work, add, minimum	2%	2%	
3600	Maximum	5%	7%	
3700	Shift work requirements, add, minimum		5%	
3800	Maximum		30%	
3900	Temporary shoring and bracing, add, minimum	2%	5%	
4000	Maximum	5%	12%	

G3060 Fuel Distribution

Steel Underground Fuel Storage Tank, Double Wall

System Components	QUANTITY	UNIT	COST EACH MAT.	COST EACH INST.	COST EACH TOTAL
Double wall underground steel storage tank including, excavation, hold down slab, backfill, manway extension and related piping.					
Fiberglass-reinforced plastic pipe, 2 inch	156.000	Ea.	1,131	889.20	2,020.20
Fiberglass-reinforced plastic pipe, 3 inch	36.000	Ea.	1,296	234	1,530
Fiberglass-reinforced elbows 90 & 45 degree, 3 inch	3.000	Ea.	456	405	861
Tank leak detection system, probes, well monitoring, liquid phase detection	2.000	Ea.	1,680		1,680
Flange, weld neck, 150 lb, 3" pipe size	1.000	Ea.	69	115.88	184.88
Vent protector/breather, 2" diam.	1.000	Ea.	68	27	95
Concrete hold down pad, 6' x 5' x 8" thick	30.000	Ea.	139.50	57.60	197.10
Tanks, for hold-downs, 500-2,000 gal, add	1.000	Ea.	269	206	475
Double wall steel tank annular space	1.000	Ea.	315		315
Foot valve, single poppet, 3/4" diam.	1.000	Ea.	149	48	197
Fuel fill box, locking inner cover, 3" diam.	1.000	Ea.	146	173	319
Fuel oil specialties, valve, ball chk, globe type fusible, 3/4" diam.	2.000	Ea.	374	87	461
Pea gravel	9.000	Ea.	1,197	567	1,764
Reinforcing in hold down pad, #3 to #7	120.000	Ea.	96	86.40	182.40
Tank, for manways, add	1.000	Ea.	1,500		1,500
Tank leak detector system, 8 channel, external monitoring	1.000	Ea.	1,850		1,850
Tubbing copper, type L, 3/4" diam.	120.000	Ea.	852	1,362	2,214
Elbow 90 degree, wrought cu x cu 3/4" diam.	8.000	Ea.	50	364	414
Pipe, black steel welded, sch 40, 3" diam.	30.000	Ea.	532.50	1,127.70	1,660.20
Elbow 90 degree, steel welded butt joint, 3" diam.	3.000	Ea.	108	692.25	800.25
Excavating trench, no sheeting or dewatering, 6'-10' D, 3/4 CY hyd backhoe	18.000	Ea.		186.66	186.66
Fiberglass-reinforced elbows 90 & 45 degree, 2 inch	7.000	Ea.	304.50	707	1,011.50
Fiberglass-reinforced threaded adapters, 2 inch	2.000	Ea.	39.10	232	271.10
500 gallon capacity, underground storage tank, double wall steel	1.000	Ea.	4,450	650	5,100
Remote tank gauging system, 30', 5" pointer travel	1.000	Ea.	7,275	435	7,710
TOTAL			**24,346.60**	**8,652.69**	**32,999.29**

G3060 320	Steel Fuel Tank, Double Wall	COST EACH MAT.	COST EACH INST.	COST EACH TOTAL
1010	Storage tank, fuel, underground, double wall steel, 500 Gal.	24,300	8,650	32,950
1020	2000 Gal.	31,000	11,300	42,300
1030	4000 Gal.	42,300	14,700	57,000
1040	6000 Gal.	48,000	16,300	64,300
1050	8000 Gal.	54,500	20,100	74,600
1060	10,000 Gal.	62,500	23,100	85,600

G3060 Fuel Distribution

G3060 320	Steel Fuel Tank, Double Wall	COST EACH		
		MAT.	INST.	TOTAL
1070	15,000 Gal.	76,500	26,900	103,400
1080	20,000 Gal.	107,000	32,300	139,300
1090	25,000 Gal.	122,000	37,000	159,000
1100	30,000 Gal.	121,500	39,500	161,000
1110	40,000 Gal.	255,000	49,600	304,600
1120	50,000 Gal.	307,500	59,500	367,000

G3060 325	Steel Tank, Above Ground, Single Wall	COST EACH		
		MAT.	INST.	TOTAL
1010	Storage tank, fuel, above ground, single wall steel, 550 Gal.	19,700	6,050	25,750
1020	2000 Gal.	30,400	6,175	36,575
1030	5000 Gal.	41,000	9,450	50,450
1040	10,000 Gal.	54,500	10,100	64,600
1050	15,000 Gal.	68,500	10,400	78,900
1060	20,000 Gal.	79,500	10,700	90,200
1070	25,000 Gal.	133,000	11,000	144,000
1080	30,000 Gal.	103,000	11,400	114,400

G3060 330	Steel Tank, Above Ground, Double Wall	COST EACH		
		MAT.	INST.	TOTAL
1010	Storage tank, fuel, above ground, double wall steel, 500 Gal.	18,300	6,100	24,400
1020	2000 Gal.	21,600	6,250	27,850
1030	4000 Gal.	29,900	6,375	36,275
1040	6000 Gal.	59,000	9,800	68,800
1050	8000 Gal.	67,000	10,100	77,100
1060	10,000 Gal.	75,000	10,300	85,300
1070	15,000 Gal.	97,500	10,600	108,100
1080	20,000 Gal.	114,000	11,000	125,000
1090	25,000 Gal.	108,500	11,300	119,800
1100	30,000 Gal.	141,500	11,700	153,200
2700	Cut & patch to match existing construction, add, minimum	2%	3%	
2800	Maximum	5%	9%	
2900	Dust protection, add, minimum	1%	2%	
3000	Maximum	4%	11%	
3100	Equipment usage curtailment, add, minimum	1%	1%	
3200	Maximum	3%	10%	
3300	Material handling & storage limitation, add, minimum	1%	1%	
3400	Maximum	6%	7%	
3500	Protection of existing work, add, minimum	2%	2%	
3600	Maximum	5%	7%	
3700	Shift work requirements, add, minimum		5%	
3800	Maximum		30%	
3900	Temporary shoring and bracing, add, minimum	2%	5%	
4000	Maximum	5%	12%	

G4020 Site Lighting

This page illustrates and describes light poles for parking or walkway area lighting. Included are aluminum or steel light poles, single or multiple fixture bracket arms, excavation, concrete footing, backfill and compaction. Lines within system components give the unit price and total price on a cost per each basis. Prices for systems are shown on lines G4020 210 0200 through 1440. Both material quantities and labor costs have been adjusted for the system listed.

Factors: To adjust for job conditions other than normal working situations use lines G4020 210 2700 through 3600.

Example: You are to install the system where you must be careful not to damage existing walks or parking areas. Go to line G4020 210 330 and apply these percentages to the appropriate MAT. and INST. costs.

System Components	QUANTITY	UNIT	COST EACH		
			MAT.	INST.	TOTAL
Light poles, aluminum, 20' high, 1 arm bracket					
Aluminum light pole, 20', no concrete base	1.000	Ea.	960	766	1,726
Bracket arm for aluminum light pole	1.000	Ea.	495	102	597
Excavation by hand, pits to 6' deep, heavy soil or clay	2.368	C.Y.		343.36	343.36
Footing, concrete incl forms, reinforcing, spread, under 1 C.Y.	.465	C.Y.	130.67	158.60	289.27
Backfill by hand	1.903	C.Y.		99.91	99.91
Compaction vibrating plate	1.903	C.Y.		15.47	15.47
TOTAL			1,585.67	1,485.34	3,071.01

G4020 210	Light Pole (Installed)	COST EACH		
		MAT.	INST.	TOTAL
0200	Light pole, aluminum, 20' high, 1 arm bracket	1,575	1,475	3,050
0240	2 arm brackets	2,075	1,475	3,550
0280	3 arm brackets	1,550	1,525	3,075
0320	4 arm brackets	1,725	1,550	3,275
0360	30' high, 1 arm bracket	2,425	1,875	4,300
0400	2 arm brackets	2,925	1,875	4,800
0440	3 arm brackets	2,400	1,925	4,325
0480	4 arm brackets	2,575	1,950	4,525
0680	40' high, 1 arm bracket	3,475	2,525	6,000
0720	2 arm brackets	3,950	2,525	6,475
0760	3 arm brackets	3,450	2,575	6,025
0800	4 arm brackets	3,600	2,575	6,175
0840	Steel, 20' high, 1 arm bracket	1,725	1,575	3,300
0880	2 arm brackets	2,025	1,575	3,600
0920	3 arm brackets	2,125	1,625	3,750
0960	4 arm brackets	2,300	1,625	3,925
1000	30' high, 1 arm bracket	1,975	2,000	3,975
1040	2 arm brackets	2,300	2,000	4,300
1080	3 arm brackets	2,375	2,050	4,425
1120	4 arm brackets	2,550	2,050	4,600
1320	40' high, 1 arm bracket	2,300	2,700	5,000
1360	2 arm brackets	2,600	2,700	5,300
1400	3 arm brackets	2,700	2,750	5,450
1440	4 arm brackets	2,850	2,750	5,600
2700	Cut & patch to match existing construction, add, minimum	2%	3%	
2800	Maximum	5%	9%	
2900	Equipment usage curtailment, add, minimum	1%	1%	
3000	Maximum	3%	10%	

743

G4020 Site Lighting

G4020 210	Light Pole (Installed)	COST EACH		
		MAT.	INST.	TOTAL
3100	Material handling & storage limitation, add, minimum	1%	1%	
3200	Maximum	6%	7%	
3300	Protection of existing work, add, minimum	2%	2%	
3400	Maximum	5%	7%	
3500	Shift work requirements, add, minimum		5%	
3600	Maximum		30%	

Reference Section

All the reference information is in one section, making it easy to find what you need to know . . . and easy to use the data set on a daily basis. This section is visually identified by a vertical black bar on the page edges.

In this Reference Section, we've included Equipment Rental Costs, a listing of rental and operating costs; Crew Listings, a full listing of all crews and equipment, and their costs; Historical Cost Indexes for cost comparisons over time; City Cost Indexes and Location Factors for adjusting costs to the region you are in; Reference Tables, where you will find explanations, estimating information and procedures, or technical data; Change Orders, information on pricing changes to contract documents; and an explanation of all the Abbreviations in the data set.

Table of Contents

Reference Tables (cont.)

Estimating Tips

- This section contains the average costs to rent and operate hundreds of pieces of construction equipment. This is useful information when one is estimating the time and material requirements of any particular operation in order to establish a unit or total cost. Bare equipment costs shown on a unit cost line include, not only rental, but also operating costs for equipment under normal use.

Rental Costs

- Equipment rental rates are obtained from the following industry sources throughout North America: contractors, suppliers, dealers, manufacturers, and distributors.

- Rental rates vary throughout the country, with larger cities generally having lower rates. Lease plans for new equipment are available for periods in excess of six months, with a percentage of payments applying toward purchase.

- Monthly rental rates vary from 2% to 5% of the purchase price of the equipment depending on the anticipated life of the equipment and its wearing parts.

- Rental rates can also be treated as reimbursement costs for contractor-owned equipment. Owned equipment costs include depreciation, loan payments, interest, taxes, insurance, storage, and major repairs.

Operating Costs

- The operating costs include parts and labor for routine servicing, such as the repair and replacement of pumps, filters, and worn lines. Normal operating expendables, such as fuel, lubricants, tires, and electricity (where applicable), are also included.

- Extraordinary operating expendables with highly variable wear patterns, such as diamond bits and blades, are excluded. These costs can be found as material costs in the Unit Price section.

- The hourly operating costs listed do not include the operator's wages.

Equipment Cost/Day

- Any power equipment required by a crew is shown in the Crew Listings with a daily cost.

- This daily cost of equipment needed by a crew includes both the rental cost and the operating cost and is based on dividing the weekly rental rate by 5 (the number of working days in the week), then adding the hourly operating cost multiplied by 8 (the number of hours in a day). This "Equipment Cost/Day" is shown in the far right column of the Equipment Rental section.

- If equipment is needed for only one or two days, it is best to develop your own cost by including components for daily rent and hourly operating costs. This is important when the listed Crew for a task does not contain the equipment needed, such as a crane for lifting mechanical heating/cooling equipment up onto a roof.

- If the quantity of work is less than the crew's Daily Output shown for a Unit Price line item that includes a bare unit equipment cost, the recommendation is to estimate one day's rental cost and operating cost for equipment shown in the Crew Listing for that line item.

- Please note, in some cases the equipment description in the crew is followed by a time period in parenthesis. For example: (daily) or (monthly). In these cases the equipment cost/day is calculated by adding the rental cost per time period to the hourly operating cost multiplied by 8.

Mobilization, Demobilization Costs

- The cost to move construction equipment from an equipment yard or rental company to the job site and back again is not included in equipment rental costs listed in the Reference Section. It is also not included in the bare equipment cost of any Unit Price line item or in any equipment costs shown in the Crew Listings.

- Mobilization (to the site) and demobilization (from the site) costs can be found in the Unit Price section.

- If a piece of equipment is already at the job site, it is not appropriate to utilize mobilization or demobilization costs again in an estimate. ■

Same Data. Simplified.

Enjoy the convenience and efficiency of accessing your costs anywhere:

- **Skip the multiplier** by setting your location
- **Quickly search,** edit, favorite and share costs
- **Stay on top of price changes** with automatic updates

Discover more at rsmeans.com/online

01 54 33 | Equipment Rental

		UNIT	HOURLY OPER. COST	RENT PER DAY	RENT PER WEEK	RENT PER MONTH	EQUIPMENT COST/DAY	
10	**0010**	**CONCRETE EQUIPMENT RENTAL** without operators [R015433 -10]							**10**
	0200	Bucket, concrete lightweight, 1/2 C.Y.	Ea.	.78	55	165	495	39.20	
	0300	1 C.Y.		1.01	63.33	190	570	46.08	
	0400	1-1/2 C.Y.		1.32	85	255	765	61.53	
	0500	2 C.Y.		1.43	99.83	299.50	898.50	71.32	
	0580	8 C.Y.		7.15	126.67	380	1,140	133.20	
	0600	Cart, concrete, self-propelled, operator walking, 10 C.F.		2.94	159	477	1,431	118.93	
	0700	Operator riding, 18 C.F.		4.95	166.83	500.50	1,501.50	139.72	
	0800	Conveyer for concrete, portable, gas, 16" wide, 26' long		10.93	103.33	310	930	149.42	
	0900	46' long		11.32	126.67	380	1,140	166.53	
	1000	56' long		11.48	151.67	455	1,365	182.87	
	1100	Core drill, electric, 2-1/2 H.P., 1" to 8" bit diameter		1.61	99.89	286.11	671.89	70.13	
	1150	11 H.P., 8" to 18" cores		5.56	147.07	373.33	924.05	119.12	
	1200	Finisher, concrete floor, gas, riding trowel, 96" wide		9.95	250.41	779.18	1,999.08	235.40	
	1300	Gas, walk-behind, 3 blade, 36" trowel		2.10	101	313.69	910	79.52	
	1400	4 blade, 48" trowel		3.16	109.76	307.50	724.64	86.80	
	1500	Float, hand-operated (Bull float), 48" wide		.08	13	40.44	117	8.76	
	1570	Curb builder, 14 H.P., gas, single screw		14.42	141.33	424	1,272	200.15	
	1590	Double screw		12.75	333.33	1,000	3,000	302	
	1600	Floor grinder, concrete and terrazzo, electric, 22" path		3.13	130.88	382.38	898.33	101.51	
	1700	Edger, concrete, electric, 7" path		1.22	80	198.33	530	49.42	
	1750	Vacuum pick-up system for floor grinders, wet/dry		1.67	103.07	329.44	879.78	79.22	
	1800	Mixer, powered, mortar and concrete, gas, 6 C.F., 18 H.P.		7.63	100.17	260.71	678.57	113.21	
	1900	10 C.F., 25 H.P.		9.26	114.71	332.67	806.90	140.63	
	2000	16 C.F.		9.63	119.68	304.88	746.10	138.02	
	2100	Concrete, stationary, tilt drum, 2 C.Y.		8.14	98.94	258.57	673.67	116.83	
	2120	Pump, concrete, truck mounted, 4" line, 80' boom		31.86	833	2,499	7,497	754.72	
	2140	5" line, 110' boom		40.79	1,066.33	3,199	9,597	966.12	
	2160	Mud jack, 50 C.F. per hr.		6.63	142.67	428	1,284	138.60	
	2180	225 C.F. per hr.		8.78	192	576	1,728	185.44	
	2190	Shotcrete pump rig, 12 C.Y./hr.		17.06	843	2,529	7,587	642.32	
	2200	35 C.Y./hr.		16.23	1,098.33	3,295	9,885	788.84	
	2600	Saw, concrete, manual, gas, 18 H.P.		5.70	116.85	323	767.46	110.19	
	2650	Self-propelled, gas, 30 H.P.		8.13	174.56	468	1,252.06	158.60	
	2675	V-groove crack chaser, manual, gas, 6 H.P.		1.69	130.44	339.11	960.33	81.33	
	2700	Vibrators, concrete, electric, 60 cycle, 2 H.P.		.48	71.15	206.93	524.98	45.25	
	2800	3 H.P.		.58	69.68	199.76	485.24	44.58	
	2900	Gas engine, 5 H.P.		1.59	17.05	53.11	154	23.33	
	3000	8 H.P.		2.14	17.25	53.65	155	27.87	
	3050	Vibrating screed, gas engine, 8 H.P.		2.89	119.38	315.63	768	86.24	
	3120	Concrete transit mixer, 6 x 4, 250 H.P., 8 C.Y., rear discharge		52.10	157.67	473	1,419	511.44	
	3200	Front discharge		60.49	95.33	286	858	541.09	
	3300	6 x 6, 285 H.P., 12 C.Y., rear discharge		59.73	333.33	1,000	3,000	677.86	
	3400	Front discharge		62.24	333.33	1,000	3,000	697.90	
20	**0010**	**EARTHWORK EQUIPMENT RENTAL** without operators [R015433 -10]							**20**
	0040	Aggregate spreader, push type, 8' to 12' wide	Ea.	2.67	70	210	630	63.37	
	0045	Tailgate type, 8' wide		2.62	52.67	158	474	52.52	
	0055	Earth auger, truck mounted, for fence & sign posts, utility poles		14.23	84.17	252.50	757.50	164.35	
	0060	For borings and monitoring wells		43.81	63.17	189.50	568.50	388.41	
	0070	Portable, trailer mounted		2.37	109.76	299.85	713.44	78.91	
	0075	Truck mounted, for caissons, water wells		87.72	139.83	419.50	1,258.50	785.64	
	0080	Horizontal boring machine, 12" to 36" diameter, 45 H.P.		23.39	275	825	2,475	352.12	
	0090	12" to 48" diameter, 65 H.P.		32.10	324	972	2,916	451.20	
	0095	Auger, for fence posts, gas engine, hand held		.46	73.04	226.70	542.87	49.02	
	0100	Excavator, diesel hydraulic, crawler mounted, 1/2 C.Y. cap.		22.35	831.63	2,375.88	5,349.75	654	
	0120	5/8 C.Y. capacity		29.88	918.63	2,450.75	5,696	729.23	
	0140	3/4 C.Y. capacity		33.61	1,004.63	3,025.73	6,854.27	874.03	
	0150	1 C.Y. capacity		42.41	1,018.33	2,863.33	6,600	911.97	

For customer support on your Commercial Renovation Costs with RSMeans data, call 800.448.8182.

			HOURLY OPER. COST	RENT PER DAY	RENT PER WEEK	RENT PER MONTH	EQUIPMENT COST/DAY	
	01 54 33 | Equipment Rental	UNIT						
20	0200	1-1/2 C.Y. capacity	Ea.	50.00	962.78	2,942.78	7,273.33	988.59	20
	0300	2 C.Y. capacity		58.23	810	2,519.91	7,300	969.84	
	0320	2-1/2 C.Y. capacity		85.05	952.22	3,301.67	9,303.33	1,340.70	
	0325	3-1/2 C.Y. capacity		123.62	2,025	6,296.66	18,200	2,248.30	
	0330	4-1/2 C.Y. capacity		156.05	3,700	11,543.88	33,400	3,557.16	
	0335	6 C.Y. capacity		197.99	3,250	10,127.13	29,300	3,609.36	
	0340	7 C.Y. capacity		180.31	3,350	13,350	40,000	4,112.45	
	0342	Excavator attachments, bucket thumbs		3.50	261	812.90	2,350	190.57	
	0345	Grapples		3.23	225	699.09	2,025	165.68	
	0346	Hydraulic hammer for boom mounting, 4000 ft lb.		13.87	900	2,802.02	8,100	671.39	
	0347	5000 ft lb.		16.42	960	2,990.92	8,650	729.52	
	0348	8000 ft lb.		24.23	1,225	3,778	10,900	949.42	
	0349	12,000 ft lb.		26.47	1,125	3,497.80	10,100	911.33	
	0350	Gradall type, truck mounted, 3 ton @ 15' radius, 5/8 C.Y.		44.71	833.33	2,500	7,500	857.66	
	0370	1 C.Y. capacity		61.13	1,250	5,000	15,000	1,489.06	
	0400	Backhoe-loader, 40 to 45 H.P., 5/8 C.Y. capacity		12.25	261.30	853.80	1,853.04	268.74	
	0450	45 H.P. to 60 H.P., 3/4 C.Y. capacity		18.55	272.22	902.22	1,852.78	328.84	
	0460	80 H.P., 1-1/4 C.Y. capacity		20.95	442.22	1,173.89	2,595	402.39	
	0470	112 H.P., 1-1/2 C.Y. capacity		33.95	620	1,923.86	5,575	656.34	
	0482	Backhoe-loader attachment, compactor, 20,000 lb.		6.63	390.56	1,001.67	2,963.33	253.34	
	0485	Hydraulic hammer, 750 ft lb.		3.79	108	336	970	97.51	
	0486	Hydraulic hammer, 1200 ft lb.		6.74	207	644.90	1,875	182.89	
	0500	Brush chipper, gas engine, 6" cutter head, 35 H.P.		9.43	263.78	809.63	1,936.46	237.40	
	0550	Diesel engine, 12" cutter head, 130 H.P.		24.36	480.78	1,564.44	3,971.89	507.78	
	0600	15" cutter head, 165 H.P.		27.36	515	1,602.17	4,625	539.32	
	0750	Bucket, clamshell, general purpose, 3/8 C.Y.		1.44	36.67	110	330	33.50	
	0800	1/2 C.Y.		1.56	36.67	110	330	34.47	
	0850	3/4 C.Y.		1.69	40	120	360	37.50	
	0900	1 C.Y.		1.75	48.33	145	435	42.98	
	0950	1-1/2 C.Y.		2.87	48.33	145	435	51.93	
	1000	2 C.Y.		3.00	190	570	1,710	138.02	
	1010	Bucket, dragline, medium duty, 1/2 C.Y.		.84	32.50	97.50	292.50	26.24	
	1020	3/4 C.Y.		.80	35.17	105.50	316.50	27.52	
	1030	1 C.Y.		.82	40.17	120.50	361.50	30.66	
	1040	1-1/2 C.Y.		1.29	41	123	369	34.95	
	1050	2 C.Y.		1.33	46.67	140	420	38.62	
	1070	3 C.Y.		2.14	52	156	468	48.30	
	1200	Compactor, manually guided 2-drum vibratory smooth roller, 7.5 H.P.		7.43	142.24	388.05	959.15	137.04	
	1250	Rammer/tamper, gas, 8"		2.27	87.39	288.37	683.46	75.84	
	1260	15"		2.71	98	305	709	82.65	
	1300	Vibratory plate, gas, 18" plate, 3000 lb. blow		2.19	173	475.61	1,169.66	112.64	
	1350	21" plate, 5000 lb. blow		2.69	206.56	577.80	1,380.61	137.11	
	1370	Curb builder/extruder, 14 H.P., gas, single screw		12.75	1,000	3,000	9,000	702	
	1390	Double screw		12.75	1,000	3,000	9,000	702	
	1500	Disc harrow attachment, for tractor		.49	350	1,100	3,300	223.90	
	1810	Feller buncher, shearing & accumulating trees, 100 H.P.		44.63	1,266.67	3,800	11,400	1,117	
	1860	Grader, self-propelled, 25,000 lb.		34.32	1,075	3,175	9,500	909.56	
	1910	30,000 lb.		33.81	850	2,500	5,500	770.49	
	1920	40,000 lb.		53.40	1,000	4,000	12,000	1,227.20	
	1930	55,000 lb.		68.88	1,250	5,000	15,000	1,551.05	
	1950	Hammer, pavement breaker, self-propelled, diesel, 1000 to 1250 lb.		29.23	450	1,800	4,500	593.82	
	2000	1300 to 1500 lb.		44.05	450	1,800	4,500	712.39	
	2050	Pile driving hammer, steam or air, 4150 ft lb. @ 225 bpm		12.51	505	1,574.17	4,550	414.88	
	2100	8750 ft lb. @ 145 bpm		14.76	710	2,203.83	6,375	558.88	
	2150	15,000 ft lb. @ 60 bpm		15.10	845	2,623.61	7,600	645.54	
	2200	24,450 ft lb. @ 111 bpm		16.15	980	3,043.39	8,800	737.84	
	2250	Leads, 60' high for pile driving hammers up to 20,000 ft lb.		3.78	305	944.50	2,725	219.12	
	2300	90' high for hammers over 20,000 ft lb.		5.61	545	1,700.10	4,925	384.88	

For customer support on your Commercial Renovation Costs with RSMeans data, call 800.448.8182.

749

01 54 33 | Equipment Rental

		UNIT	HOURLY OPER. COST	RENT PER DAY	RENT PER WEEK	RENT PER MONTH	EQUIPMENT COST/DAY	
20	2350	Diesel type hammer, 22,400 ft lb.	Ea.	18.34	495	1,544.51	4,475	455.60	20
	2400	41,300 ft lb.		26.43	625	1,950.96	5,650	601.64	
	2450	141,000 ft lb.		42.53	995	3,089.02	8,925	958.07	
	2500	Vib. elec. hammer/extractor, 200 kW diesel generator, 34 H.P.		42.58	725	2,249.02	6,500	790.42	
	2550	80 H.P.		75.15	1,050	3,251.60	9,400	1,251.55	
	2600	150 H.P.		139.08	2,000	6,259.32	18,100	2,364.52	
	2700	Hydro Excavator w/EXT boom 12 C.Y., 1200 gallons		38.92	1,625	5,037.33	14,600	1,318.79	
	2800	Log chipper, up to 22" diameter, 600 H.P.		45.90	345	1,035	3,105	574.20	
	2850	Logger, for skidding & stacking logs, 150 H.P.		44.71	86.67	260	780	409.71	
	2860	Mulcher, diesel powered, trailer mounted		20.46	747.50	2,242.50	6,727.50	612.19	
	2900	Rake, spring tooth, with tractor		15.15	50.78	133.33	352.22	147.84	
	3000	Roller, vibratory, tandem, smooth drum, 20 H.P.		8.03	331.67	996.67	2,348.89	263.60	
	3050	35 H.P.		10.43	424.44	1,277.78	2,786.67	338.98	
	3100	Towed type vibratory compactor, smooth drum, 50 H.P.		26.01	575	2,300	6,900	668.06	
	3150	Sheepsfoot, 50 H.P.		26.39	389	1,100	2,893	431.11	
	3170	Landfill compactor, 220 H.P.		71.92	1,778.33	5,335	16,005	1,642.32	
	3200	Pneumatic tire roller, 80 H.P.		13.30	410	1,273.54	3,675	361.08	
	3250	120 H.P.		19.95	570	1,700	5,100	499.63	
	3300	Sheepsfoot vibratory roller, 240 H.P.		64.02	389	1,100	2,893	732.18	
	3320	340 H.P.		86.27	1,500	4,500	13,500	1,590.14	
	3350	Smooth drum vibratory roller, 75 H.P.		24.02	411.63	1,382.38	3,014.50	468.61	
	3400	125 H.P.		28.42	450	1,350	3,950	497.35	
	3410	Rotary mower, brush, 60", with tractor		19.34	119.11	342.22	897.22	223.13	
	3420	Rototiller, walk-behind, gas, 5 H.P.		2.20	109.24	324.52	777	82.48	
	3422	8 H.P.		2.89	117.93	335.33	853.73	90.20	
	3440	Scrapers, towed type, 7 C.Y. capacity		6.63	129	401.03	1,150	133.22	
	3450	10 C.Y. capacity		7.41	172	536.51	1,550	166.62	
	3500	15 C.Y. capacity		7.61	199	617.80	1,775	184.47	
	3525	Self-propelled, single engine, 14 C.Y. capacity		137.17	1,666.67	5,000	15,000	2,097.38	
	3550	Dual engine, 21 C.Y. capacity		145.49	1,650	5,000	15,000	2,163.94	
	3600	31 C.Y. capacity		193.32	1,325	6,625	26,500	2,871.57	
	3640	44 C.Y. capacity		239.46	1,400	7,000	28,000	3,315.67	
	3650	Elevating type, single engine, 11 C.Y. capacity		63.67	800	2,400	7,200	989.35	
	3700	22 C.Y. capacity		117.96	1,250	5,000	15,000	1,943.69	
	3710	Screening plant, 110 H.P. w/5' x 10' screen		21.75	650	2,024.81	5,850	578.94	
	3720	5' x 16' screen		27.46	1,222.22	3,666.67	11,000	952.98	
	3850	Shovel, crawler-mounted, front-loading, 7 C.Y. capacity		225.02	3,975	12,356.07	35,700	4,271.41	
	3855	12 C.Y. capacity		346.72	5,500	17,125.08	49,500	6,198.74	
	3860	Shovel/backhoe bucket, 1/2 C.Y.		2.77	74	229.78	665	68.13	
	3870	3/4 C.Y.		2.74	83	257.96	745	73.54	
	3880	1 C.Y.		2.84	43.45	137.25	386	50.15	
	3890	1-1/2 C.Y.		3.04	108	336	970	91.52	
	3910	3 C.Y.		3.54	146	455.22	1,325	119.34	
	3950	Stump chipper, 18" deep, 30 H.P.		7.09	321.22	1,015.56	2,516.67	259.82	
	4110	Dozer, crawler, torque converter, diesel 80 H.P.		25.99	619.44	1,638.89	4,514.44	535.68	
	4150	105 H.P.		35.34	605	1,885.49	5,450	659.79	
	4200	140 H.P.		42.47	660	1,980	5,940	735.80	
	4260	200 H.P.		65.00	1,381	4,143	12,843	1,348.56	
	4310	300 H.P.		83.08	1,611.11	4,833.33	14,500	1,631.32	
	4360	410 H.P.		109.85	2,000	8,000	24,000	2,478.84	
	4370	500 H.P.		137.27	2,333.33	7,000	21,000	2,498.17	
	4380	700 H.P.		236.87	2,600	10,400	31,200	3,974.98	
	4400	Loader, crawler, torque conv., diesel, 1-1/2 C.Y., 80 H.P.		30.40	555	1,729.41	5,000	589.09	
	4450	1-1/2 to 1-3/4 C.Y., 95 H.P.		31.15	705	2,198.44	6,350	688.92	
	4510	1-3/4 to 2-1/4 C.Y., 130 H.P.		49.15	675	2,700	8,100	933.20	
	4530	2-1/2 to 3-1/4 C.Y., 190 H.P.		59.47	1,125	3,525.80	10,200	1,180.92	
	4560	3-1/2 to 5 C.Y., 275 H.P.		73.49	1,100	4,400	13,200	1,467.90	
	4610	Front end loader, 4WD, articulated frame, diesel, 1 to 1-1/4 C.Y., 70 H.P.		17.11	335.11	931.67	2,722.33	323.21	

01 54 33 | Equipment Rental

		UNIT	HOURLY OPER. COST	RENT PER DAY	RENT PER WEEK	RENT PER MONTH	EQUIPMENT COST/DAY	
20	4620	1-1/2 to 1-3/4 C.Y., 95 H.P.	Ea.	20.58	350	1,400	4,200	444.62	**20**
	4650	1-3/4 to 2 C.Y., 130 H.P.		21.68	491	1,588.56	4,320.89	491.15	
	4710	2-1/2 to 3-1/2 C.Y., 145 H.P.		30.39	500	2,000	6,000	643.09	
	4730	3 to 4-1/2 C.Y., 185 H.P.		33.03	825	2,501.67	4,682.22	764.54	
	4760	5-1/4 to 5-3/4 C.Y., 270 H.P.		54.74	900	3,600	10,800	1,157.96	
	4810	7 to 9 C.Y., 475 H.P.		93.83	1,500	6,000	18,000	1,950.65	
	4870	9 to 11 C.Y., 620 H.P.		135.76	2,725	8,508.35	24,600	2,787.78	
	4880	Skid-steer loader, wheeled, 10 C.F., 30 H.P. gas		9.85	325	921.67	1,936.67	263.12	
	4890	1 C.Y., 78 H.P., diesel		18.97	430.56	1,349.44	3,163.89	421.68	
	4892	Skid-steer attachment, auger		.77	115.13	319.49	819.49	70.04	
	4893	Backhoe		.76	99.67	321.67	816.67	70.42	
	4894	Broom		.73	143.13	378.75	882	81.57	
	4895	Forks		.16	40.33	107.44	289.67	22.75	
	4896	Grapple		.74	81.25	221.25	613	50.21	
	4897	Concrete hammer		1.08	203.25	579.38	1,516	124.54	
	4898	Tree spade		.62	315	945	2,268	193.94	
	4899	Trencher		.67	115.40	313.13	806.38	68	
	4900	Trencher, chain, boom type, gas, operator walking, 12 H.P.		4.30	249.39	787.07	1,735.85	191.80	
	4910	Operator riding, 40 H.P.		17.17	474.74	1,321.03	3,023.33	401.58	
	5000	Wheel type, diesel, 4' deep, 12" wide		70.71	975	3,034.82	8,775	1,172.66	
	5100	6' deep, 20" wide		90.13	838.41	2,364.76	7,865.24	1,194.02	
	5150	Chain type, diesel, 5' deep, 8" wide		16.78	365	1,138.06	3,300	361.81	
	5200	Diesel, 8' deep, 16" wide		92.28	1,950	6,069.65	17,600	1,952.14	
	5202	Rock trencher, wheel type, 6" wide x 18" deep		48.41	838.11	2,366.56	7,865.22	860.58	
	5206	Chain type, 18" wide x 7' deep		107.55	243.33	730	2,190	1,006.38	
	5210	Tree spade, self-propelled		14.07	374.17	1,122.50	3,367.50	337.03	
	5250	Truck, dump, 2-axle, 12 ton, 8 C.Y. payload, 220 H.P.		24.65	344.46	974.12	2,604.20	392.01	
	5300	Three axle dump, 16 ton, 12 C.Y. payload, 400 H.P.		45.93	666.67	2,000	6,000	767.47	
	5310	Four axle dump, 25 ton, 18 C.Y. payload, 450 H.P.		51.46	350	1,750	7,580	761.66	
	5350	Dump trailer only, rear dump, 16-1/2 C.Y.		5.91	153	476.90	1,375	142.67	
	5400	20 C.Y.		6.38	172	536.51	1,550	158.36	
	5450	Flatbed, single axle, 1-1/2 ton rating		19.61	255	915	2,400	339.86	
	5500	3 ton rating		26.47	275.83	827.50	2,482.50	377.25	
	5550	Off highway rear dump, 25 ton capacity		64.69	1,150	4,600	13,700	1,437.52	
	5600	35 ton capacity		69.06	1,150	4,600	13,800	1,472.45	
	5610	50 ton capacity		86.57	1,625	6,500	19,450	1,992.59	
	5620	65 ton capacity		92.45	2,025	6,286.42	18,200	1,996.90	
	5630	100 ton capacity		125.15	2,975	9,267.05	26,800	2,854.63	
	6000	Vibratory plow, 25 H.P., walking		6.98	323.33	970	2,910	249.84	
40	0010	**GENERAL EQUIPMENT RENTAL** without operators [R015433-10]							**40**
	0020	Aerial lift, scissor type, to 20' high, 1200 lb. capacity, electric	Ea.	3.59	183.89	375.56	699.44	103.86	
	0030	To 30' high, 1200 lb. capacity		3.90	271.92	690.51	1,362.69	169.26	
	0040	Over 30' high, 1500 lb. capacity		5.30	338.25	781.50	1,649.38	198.70	
	0070	Articulating boom, to 45' high, 500 lb. capacity, diesel [R015433-15]		10.24	2,390	4,693.89	9,760.56	1,020.68	
	0075	To 60' high, 500 lb. capacity		14.10	647.78	1,628.89	3,541.11	438.57	
	0080	To 80' high, 500 lb. capacity		16.57	1,089.44	2,707.22	5,666.11	674	
	0085	To 125' high, 500 lb. capacity		18.94	1,986.59	5,280.45	13,543.86	1,207.58	
	0100	Telescoping boom to 40' high, 500 lb. capacity, diesel		11.60	491.54	1,168.46	2,436.41	326.48	
	0105	To 45' high, 500 lb. capacity		12.91	512.22	1,161.11	2,505	335.51	
	0110	To 60' high, 500 lb. capacity		16.89	512.22	1,161.11	2,505	367.32	
	0115	To 80' high, 500 lb. capacity		21.96	972.45	2,372.87	4,948.40	650.22	
	0120	To 100' high, 500 lb. capacity		29.64	1,089.44	2,707.22	5,666.11	778.56	
	0125	To 120' high, 500 lb. capacity		30.10	1,601.91	4,231.41	11,028.77	1,087.10	
	0195	Air compressor, portable, 6.5 CFM, electric		.93	48.56	143.92	342.45	36.24	
	0196	Gasoline		.67	59.51	166.79	401.79	38.74	
	0200	Towed type, gas engine, 60 CFM		9.74	129	401.84	1,175	158.26	
	0300	160 CFM		10.81	212.57	592.03	1,180.81	204.90	

For customer support on your Commercial Renovation Costs with RSMeans data, call 800.448.8182.

751

01 54 33 | Equipment Rental

		UNIT	HOURLY OPER. COST	RENT PER DAY	RENT PER WEEK	RENT PER MONTH	EQUIPMENT COST/DAY	
40	0400	Diesel engine, rotary screw, 250 CFM	Ea.	12.47	162.43	552.03	1,122.70	210.20	40
	0500	365 CFM		16.51	355	1,109.59	3,200	354.03	
	0550	450 CFM		20.59	345.95	949.59	2,692.97	354.64	
	0600	600 CFM		35.20	251	780.38	2,250	437.66	
	0700	750 CFM		35.74	639.38	1,612.60	3,901.88	608.41	
	0915	Air purifier		.72	9.51	22.05	62.95	10.18	
	0917	Air scrubber, 500 - 700 CFM, 110 V		.71	81.50	254.07	735	56.53	
	0919	200 - 1200 CFM, 110 V		.71	96.50	300.73	870	65.86	
	0930	Air tools, breaker, pavement, 60 lb.		.58	58.72	164.58	402.22	37.59	
	0940	80 lb.		.58	63.68	181.22	419.32	40.90	
	0950	Drills, hand (jackhammer), 65 lb.		.69	64.62	178.11	483.92	41.15	
	0960	Track or wagon, swing boom, 4" drifter		63.75	1,000	3,000	9,000	1,110	
	0970	5" drifter		63.75	1,000	3,000	9,000	1,110	
	0975	Track mounted quarry drill, 6" diameter drill		105.00	738.67	2,216	6,648	1,283.21	
	0980	Dust control per drill		1.07	9.33	28	84	14.18	
	0990	Hammer, chipping, 12 lb.		.62	44.67	121.78	293.11	29.30	
	1000	Hose, air with couplings, 50' long, 3/4" diameter		.07	13.79	36.86	90.47	7.93	
	1100	1" diameter		.08	15.78	40.56	116.67	8.75	
	1200	1-1/2" diameter		.23	42.54	107.03	281.22	23.22	
	1300	2" diameter		.25	65.89	157.22	406.11	33.42	
	1400	2-1/2" diameter		.37	70	168.57	507.86	36.67	
	1410	3" diameter		.43	82.67	207.78	620	44.99	
	1450	Drill, steel, 7/8" x 2'		.09	9.91	22.94	58.17	5.29	
	1460	7/8" x 6'		.12	19.85	61.78	179	13.31	
	1520	Moil points		.03	10.14	22.72	60.03	4.77	
	1525	Pneumatic nailer w/accessories		.49	31.33	81.67	213.67	20.27	
	1530	Sheeting driver for 60 lb. breaker		.05	7.85	24.39	70.50	5.24	
	1540	For 90 lb. breaker		.13	10.65	33.06	95.50	7.69	
	1550	Spade, 25 lb.		.52	7.50	23.30	67.50	8.80	
	1560	Tamper, single, 35 lb.		.61	72.67	222.22	626.67	49.31	
	1570	Triple, 140 lb.		.91	57.67	166.11	399.67	40.54	
	1580	Wrenches, impact, air powered, up to 3/4" bolt		.44	45.50	129.11	333	29.33	
	1590	Up to 1-1/4" bolt		.59	86.37	228	591.29	50.35	
	1600	Barricades, barrels, reflectorized, 1 to 99 barrels		.03	5.11	12.11	37.39	2.69	
	1610	100 to 200 barrels		.03	6.11	15.44	41.22	3.29	
	1620	Barrels with flashers, 1 to 99 barrels		.04	6.11	15.44	41.22	3.38	
	1630	100 to 200 barrels		.03	6.11	15.44	41.22	3.32	
	1640	Barrels with steady burn type C lights		.05	6.11	15.44	41.22	3.48	
	1650	Illuminated board, trailer mounted, with generator		3.38	99.11	260.56	574.22	79.16	
	1670	Portable barricade, stock, with flashers, 1 to 6 units		.04	4.78	12.89	41.78	2.86	
	1680	25 to 50 units		.03	4.78	12.89	41.78	2.85	
	1685	Butt fusion machine, wheeled, 1.5 HP electric, 2" - 8" diameter pipe		2.71	379	1,137	3,411	249.07	
	1690	Tracked, 20 HP diesel, 4"-12" diameter pipe		10.24	373.33	1,120	3,360	305.93	
	1695	83 HP diesel, 8" - 24" diameter pipe		52.88	513.33	1,540	4,620	731.01	
	1700	Carts, brick, gas engine, 1000 lb. capacity		3.04	30.25	79.36	202.50	40.16	
	1800	1500 lb., 7-1/2' lift		3.01	70	218.28	630	67.73	
	1822	Dehumidifier, medium, 6 lb./hr., 150 CFM		1.23	53	170	533.33	43.81	
	1824	Large, 18 lb./hr., 600 CFM		2.26	250	1,750	5,250	368.11	
	1830	Distributor, asphalt, trailer mounted, 2000 gal., 38 H.P. diesel		11.35	355	1,110.96	3,225	312.98	
	1840	3000 gal., 38 H.P. diesel		13.28	385	1,192.25	3,450	344.69	
	1850	Drill, rotary hammer, electric		1.15	54.78	142.11	325.89	37.61	
	1860	Carbide bit, 1-1/2" diameter, add to electric rotary hammer		.03	48.08	117.44	335.28	23.75	
	1865	Rotary, crawler, 250 H.P.		140.15	2,458.33	7,375	22,125	2,596.20	
	1870	Emulsion sprayer, 65 gal., 5 H.P. gas engine		2.86	108	336	970	90.05	
	1880	200 gal., 5 H.P. engine		7.46	181	563.61	1,625	172.38	
	1900	Floor auto-scrubbing machine, walk-behind, 28" path		5.81	219.72	719.03	1,476.94	190.25	
	1930	Floodlight, mercury vapor, or quartz, on tripod, 1000 watt		.47	40.71	110.40	275.31	25.85	
	1940	2000 watt		.61	42.28	115.69	324.72	28.01	

For customer support on your Commercial Renovation Costs with RSMeans data, call 800.448.8182.

01 54 33 | Equipment Rental

		UNIT	HOURLY OPER. COST	RENT PER DAY	RENT PER WEEK	RENT PER MONTH	EQUIPMENT COST/DAY	
40	1950	Floodlights, trailer mounted with generator, 1 - 300 watt light	Ea.	3.66	168.89	433.33	850	115.94	40
	1960	2 - 1000 watt lights		4.63	168.89	433.33	850	123.71	
	2000	4 - 300 watt lights		4.38	168.89	433.33	850	121.71	
	2005	Foam spray rig, incl. box trailer, compressor, generator, proportioner		26.28	540	1,679.99	4,850	546.24	
	2015	Forklift, pneumatic tire, rough terr, straight mast, 5000 lb, 12' lift, gas		19.19	504.44	1,248.89	2,715.56	403.33	
	2025	8000 lb, 12' lift		23.41	593.89	1,561.67	3,292.78	499.64	
	2030	5000 lb, 12' lift, diesel		15.90	392.08	1,024.86	2,278.75	332.20	
	2035	8000 lb, 12' lift, diesel		17.24	564.89	1,519.43	3,262.39	441.79	
	2045	All terrain, telescoping boom, diesel, 5000 lb, 10' reach, 19' lift		17.75	466.14	1,199.89	2,626.93	382	
	2055	6600 lb, 29' reach, 42' lift		21.71	495.91	1,332.95	2,889.43	440.31	
	2065	10,000 lb, 31' reach, 45' lift		23.77	851.67	2,190.56	4,532.78	628.27	
	2070	Cushion tire, smooth floor, gas, 5000 lb capacity		8.49	327.22	815	1,971.11	230.92	
	2075	8000 lb capacity		11.70	381.53	951.25	2,282.22	283.82	
	2085	Diesel, 5000 lb capacity		7.98	257	798.49	2,300	223.51	
	2090	12,000 lb capacity		12.40	639.09	1,706.70	3,981.59	440.55	
	2095	20,000 lb capacity		17.76	813.89	2,310.14	6,005.83	604.09	
	2100	Generator, electric, gas engine, 1.5 kW to 3 kW		2.65	59.78	169.78	422.22	55.16	
	2200	5 kW		3.31	83.94	257.81	605.72	78.05	
	2300	10 kW		6.10	128.89	356.25	907.64	120.06	
	2400	25 kW		7.62	321.39	795.14	1,969.17	220.01	
	2500	Diesel engine, 20 kW		9.48	251.25	642.78	1,526.53	204.37	
	2600	50 kW		16.41	375.89	961.67	2,492.22	323.62	
	2700	100 kW		29.43	546.25	1,334.03	3,446.39	502.26	
	2800	250 kW		55.94	830.71	2,100.71	5,479	867.65	
	2850	Hammer, hydraulic, for mounting on boom, to 500 ft lb.		2.98	94.50	293.73	850	82.63	
	2860	1000 ft lb.		4.74	141	438.97	1,275	125.71	
	2900	Heaters, space, oil or electric, 50 MBH		1.51	87.78	251.22	686.44	62.29	
	3000	100 MBH		2.80	131	330.83	771.25	88.55	
	3100	300 MBH		8.16	675.56	1,791.11	5,030	423.47	
	3150	500 MBH		13.54	240.57	581.29	1,405.57	224.60	
	3200	Hose, water, suction with coupling, 20' long, 2" diameter		.02	7.59	18.74	52.41	3.91	
	3210	3" diameter		.03	17.18	47.44	130.88	9.74	
	3220	4" diameter		.03	28.50	88.15	255	17.88	
	3230	6" diameter		.11	41	127.51	370	26.41	
	3240	8" diameter		.28	54	167.91	485	35.80	
	3250	Discharge hose with coupling, 50' long, 2" diameter		.01	8.66	22.77	55.29	4.64	
	3260	3" diameter		.01	10.60	25.77	69.89	5.24	
	3270	4" diameter		.02	21	65.59	190	13.28	
	3280	6" diameter		.06	29.50	91.30	264	18.77	
	3290	8" diameter		.25	38	118.06	340	25.58	
	3295	Insulation blower		.86	118	366.62	1,050	80.17	
	3300	Ladders, extension type, 16' to 36' long		.19	41.17	103.26	282.71	22.14	
	3400	40' to 60' long		.66	136.74	345.11	851.94	74.28	
	3405	Lance for cutting concrete		2.27	133.33	400	1,200	98.15	
	3407	Lawn mower, rotary, 22", 5 H.P.		1.09	36.33	117.56	305	32.20	
	3408	48" self-propelled		2.98	175.02	461.56	1,251.88	116.17	
	3410	Level, electronic, automatic, with tripod and leveling rod		1.08	56.13	138.83	376.77	36.43	
	3430	Laser type, for pipe and sewer line and grade		2.24	152.86	401.43	1,101.14	98.20	
	3440	Rotating beam for interior control		.93	87.57	218.54	657.57	51.14	
	3460	Builder's optical transit, with tripod and rod		.10	55.63	138.69	372.71	28.55	
	3500	Light towers, towable, with diesel generator, 2000 watt		4.39	168.89	433.33	850	121.80	
	3600	4000 watt		4.64	161.53	430.28	859.31	123.21	
	3700	Mixer, powered, plaster and mortar, 6 C.F., 7 H.P.		2.12	101.17	246.71	678.86	66.29	
	3800	10 C.F., 9 H.P.		2.31	117.96	301.49	761.06	78.76	
	3850	Nailer, pneumatic		.49	31.33	81.67	213.67	20.29	
	3900	Paint sprayers complete, 8 CFM		.88	143.21	418.50	1,082.76	90.71	
	4000	17 CFM		1.65	169.55	405.45	1,055	94.27	
	4020	Pavers, bituminous, rubber tires, 8' wide, 50 H.P., diesel		32.96	575	1,788.38	5,175	621.32	

For customer support on your Commercial Renovation Costs with RSMeans data, call 800.448.8182.

753

01 54 33 | Equipment Rental

		UNIT	HOURLY OPER. COST	RENT PER DAY	RENT PER WEEK	RENT PER MONTH	EQUIPMENT COST/DAY	
40	4030	10' wide, 150 H.P.	Ea.	98.70	1,975	6,123.84	17,700	2,014.40	**40**
	4050	Crawler, 8' wide, 100 H.P., diesel		90.42	2,075	6,476.10	18,700	2,018.54	
	4060	10' wide, 150 H.P.		107.32	2,375	7,397.38	21,400	2,338.01	
	4070	Concrete paver, 12' to 24' wide, 250 H.P.		90.45	1,700	5,310.94	15,400	1,785.75	
	4080	Placer-spreader-trimmer, 24' wide, 300 H.P.		121.30	2,575	8,047.70	23,300	2,579.95	
	4100	Pump, centrifugal gas pump, 1-1/2" diam., 65 GPM		4.05	54.50	170.17	490	66.43	
	4200	2" diameter, 130 GPM		5.14	49.45	131.82	358.79	67.48	
	4300	3" diameter, 250 GPM		5.28	59.55	171.82	488.33	76.61	
	4400	6" diameter, 1500 GPM		22.91	395	1,045	2,605.56	392.30	
	4500	Submersible electric pump, 1-1/4" diameter, 55 GPM		.41	47.79	133.06	345.29	29.90	
	4600	1-1/2" diameter, 83 GPM		.46	51.85	155.47	384.12	34.76	
	4700	2" diameter, 120 GPM		1.70	55	171.11	495	47.78	
	4800	3" diameter, 300 GPM		3.13	124.04	326.57	917.21	90.35	
	4900	4" diameter, 560 GPM		15.20	75.83	227.50	682.50	167.10	
	5000	6" diameter, 1590 GPM		22.75	211.67	635	1,905	308.97	
	5100	Diaphragm pump, gas, single, 1-1/2" diameter		1.16	40	120	360	33.28	
	5200	2" diameter		4.10	45	135	405	59.79	
	5300	3" diameter		4.17	74.21	220.64	561.21	77.51	
	5400	Double, 4" diameter		6.22	204.87	458.94	1,170.81	141.52	
	5450	Pressure washer 5 GPM, 3000 psi		3.99	99.65	344.79	919.59	100.91	
	5460	7 GPM, 3000 psi		5.10	99.49	343.24	920.71	109.42	
	5470	High Pressure Water Jet 10 KSI		40.83	730	2,266.80	6,550	779.96	
	5480	40 KSI		28.78	990	3,085.37	8,925	847.30	
	5500	Trash pump, self-priming, gas, 2" diameter		3.93	77.33	232	696	77.86	
	5600	Diesel, 4" diameter		6.88	285.78	723.33	1,912.33	199.70	
	5650	Diesel, 6" diameter		17.36	395	1,045	2,605.56	347.88	
	5655	Grout Pump		19.30	284	883.35	2,550	331.07	
	5700	Salamanders, L.P. gas fired, 100,000 Btu		2.98	60.06	166.82	440.45	57.17	
	5705	50,000 Btu		1.72	23.82	68.09	178.03	27.34	
	5720	Sandblaster, portable, open top, 3 C.F. capacity		.62	145	395	945	83.94	
	5730	6 C.F. capacity		1.04	330	790	2,035	166.28	
	5740	Accessories for above		.14	24	74.79	216	16.09	
	5750	Sander, floor		.79	71.06	201.88	489.66	46.72	
	5760	Edger		.53	41	117.78	288.44	27.83	
	5800	Saw, chain, gas engine, 18" long		1.81	71.22	170	477.78	48.47	
	5900	Hydraulic powered, 36" long		.81	73.17	219.50	658.50	50.34	
	5950	60" long		.80	81.67	245	735	55.44	
	6000	Masonry, table mounted, 14" diameter, 5 H.P.		1.36	81.57	279.48	679.70	66.79	
	6050	Portable cut-off, 8 H.P.		1.87	77.85	257.24	625.79	66.39	
	6100	Circular, hand held, electric, 7-1/4" diameter		.23	15.45	40.48	102.21	9.97	
	6200	12" diameter		.24	47.55	128.79	334.36	27.71	
	6250	Wall saw, w/hydraulic power, 10 H.P.		3.39	129.33	388	1,164	104.72	
	6275	Shot blaster, walk-behind, 20" wide		4.89	284	883.35	2,550	215.77	
	6280	Sidewalk broom, walk-behind		2.31	83.50	260.13	755	70.52	
	6300	Steam cleaner, 100 gallons per hour		3.45	83.50	260.13	755	79.60	
	6310	200 gallons per hour		4.47	101	314.32	910	98.64	
	6340	Tar Kettle/Pot, 400 gallons		16.98	126.67	380	1,140	211.82	
	6350	Torch, cutting, acetylene-oxygen, 150' hose, excludes gases		.47	15.50	48.23	140	13.37	
	6360	Hourly operating cost includes tips and gas		21.59	7.10	22.02	63.50	177.16	
	6410	Toilet, portable chemical		.14	23.50	72.62	210	15.61	
	6420	Recycle flush type		.17	29	89.96	260	19.33	
	6430	Toilet, fresh water flush, garden hose,		.20	34.50	107.30	310	23.06	
	6440	Hoisted, non-flush, for high rise		.16	28	87.79	254	18.86	
	6465	Tractor, farm with attachment		17.93	405.65	1,359.02	2,927.28	415.21	
	6480	Trailers, platform, flush deck, 2 axle, 3 ton capacity		1.74	96	298.04	860	73.56	
	6500	25 ton capacity		6.43	145	449.80	1,300	141.42	
	6600	40 ton capacity		8.30	206	639.48	1,850	194.29	
	6700	3 axle, 50 ton capacity		9.00	500	1,300	2,975	332.01	

01 54 33 | Equipment Rental

		UNIT	HOURLY OPER. COST	RENT PER DAY	RENT PER WEEK	RENT PER MONTH	EQUIPMENT COST/DAY	
40	6800	75 ton capacity	Ea.	11.43	305	941.21	2,725	279.72	40
	6810	Trailer mounted cable reel for high voltage line work		6.06	252.67	758	2,274	200.11	
	6820	Trailer mounted cable tensioning rig		12.03	83.33	250	750	146.21	
	6830	Cable pulling rig		76.00	142.50	427.50	1,282.50	693.54	
	6850	Portable cable/wire puller, 8000 lb max pulling capacity		3.81	189.35	524.76	1,258.33	135.46	
	6900	Water tank trailer, engine driven discharge, 5000 gallons		7.39	160	498.58	1,450	158.82	
	6925	10,000 gallons		10.06	218	677.42	1,950	216	
	6950	Water truck, off highway, 6000 gallons		74.06	1,333.33	4,000	12,000	1,392.52	
	7010	Tram car for high voltage line work, powered, 2 conductor		7.09	29.33	88	264	74.32	
	7020	Transit (builder's level) with tripod		.10	51.78	129.89	345.56	26.79	
	7030	Trench box, 3000 lb., 6' x 8'		.58	97.50	304	880	65.42	
	7040	7200 lb., 6' x 20'		.74	189	587.69	1,700	123.46	
	7050	8000 lb., 8' x 16'		1.11	205	637.76	1,850	136.44	
	7060	9500 lb., 8' x 20'		1.24	235	731.61	2,125	156.25	
	7065	11,000 lb., 8' x 24'		1.30	221	688.25	2,000	148.07	
	7070	12,000 lb., 10' x 20'		1.54	267	829.16	2,400	178.14	
	7100	Truck, pickup, 3/4 ton, 2 wheel drive		9.53	62.50	194.01	560	115.08	
	7200	4 wheel drive		9.79	174.39	508.64	1,505.15	180.03	
	7250	Crew carrier, 9 passenger		13.07	109	340.44	985	172.61	
	7290	Flat bed truck, 20,000 lb. GVW		15.75	134	417.29	1,200	209.49	
	7300	Tractor, 4 x 2, 220 H.P.		22.97	218	677.42	1,950	319.23	
	7410	330 H.P.		33.37	298	926.71	2,675	452.31	
	7500	6 x 4, 380 H.P.		37.26	345	1,073.03	3,100	512.66	
	7600	450 H.P.		45.66	420	1,300.64	3,775	625.39	
	7610	Tractor, with A frame, boom and winch, 225 H.P.		25.54	296	919.57	2,650	388.20	
	7620	Vacuum truck, hazardous material, 2500 gallons		13.20	315	975.48	2,825	300.72	
	7625	5,000 gallons		13.44	445	1,381.93	4,000	383.94	
	7650	Vacuum, HEPA, 16 gallon, wet/dry		.88	134.09	386.21	1,005.30	84.26	
	7655	55 gallon, wet/dry		.80	33.91	86.79	228.64	23.77	
	7660	Water tank, portable		.75	147.40	425.74	1,083.40	91.17	
	7690	Sewer/catch basin vacuum, 14 C.Y., 1500 gallons		17.87	670	2,086.44	6,025	560.27	
	7700	Welder, electric, 200 amp		3.94	36.22	80.22	226.25	47.54	
	7800	300 amp		5.73	32.08	81.30	214.09	62.07	
	7900	Gas engine, 200 amp		9.23	102.73	283.48	666.52	130.57	
	8000	300 amp		10.45	112.44	318.78	711.33	147.37	
	8100	Wheelbarrow, any size		.07	13.94	34.91	98.91	7.51	
	8200	Wrecking ball, 4000 lb.		2.58	62	186	558	57.84	
50	0010	**HIGHWAY EQUIPMENT RENTAL** without operators							50
	0050	Asphalt batch plant, portable drum mixer, 100 ton/hr.	Ea.	91.26	1,550	4,850.30	14,000	1,700.12	
	0060	200 ton/hr.		105.27	1,675	5,175.46	15,000	1,877.29	
	0070	300 ton/hr.		123.72	1,950	6,069.65	17,600	2,203.69	
	0100	Backhoe attachment, long stick, up to 185 H.P., 10.5' long		.38	26	80.21	232	19.09	
	0140	Up to 250 H.P., 12' long		.42	29	89.96	260	21.39	
	0180	Over 250 H.P., 15' long		.58	39.50	122.48	355	29.16	
	0200	Special dipper arm, up to 100 H.P., 32' long		1.19	80.50	250.37	725	59.63	
	0240	Over 100 H.P., 33' long		1.49	101	314.32	910	74.79	
	0280	Catch basin/sewer cleaning truck, 3 ton, 9 C.Y., 1000 gal.		36.53	425	1,327.74	3,850	557.79	
	0300	Concrete batch plant, portable, electric, 200 C.Y./hr.		24.96	565	1,761.28	5,100	551.97	
	0520	Grader/dozer attachment, ripper/scarifier, rear mounted, up to 135 H.P.		3.25	64	199.43	575	65.90	
	0540	Up to 180 H.P.		4.27	97	301.31	870	94.41	
	0580	Up to 250 H.P.		6.04	155	482.32	1,400	144.78	
	0700	Pvmt. removal bucket, for hyd. excavator, up to 90 H.P.		2.23	59	183.17	530	54.46	
	0740	Up to 200 H.P.		2.38	75.50	234.11	675	65.86	
	0780	Over 200 H.P.		2.60	92.50	287.22	830	78.24	
	0900	Aggregate spreader, self-propelled, 187 H.P.		52.23	750	2,330.31	6,750	883.90	
	1000	Chemical spreader, 3 C.Y.		3.26	70	210	630	68.11	
	1900	Hammermill, traveling, 250 H.P.		69.39	600	1,800	5,400	915.14	

R015433 -10

For customer support on your Commercial Renovation Costs with RSMeans data, call 800.448.8182.

755

01 54 33 | Equipment Rental

			UNIT	HOURLY OPER. COST	RENT PER DAY	RENT PER WEEK	RENT PER MONTH	EQUIPMENT COST/DAY	
50	2000	Horizontal borer, 3" diameter, 13 H.P. gas driven	Ea.	5.58	103.33	310	930	106.64	50
	2150	Horizontal directional drill, 20,000 lb. thrust, 78 H.P. diesel		28.42	738.67	2,216	6,648	670.52	
	2160	30,000 lb. thrust, 115 H.P.		34.93	738.67	2,216	6,648	722.60	
	2170	50,000 lb. thrust, 170 H.P.		50.07	738.67	2,216	6,648	843.77	
	2190	Mud trailer for HDD, 1500 gallons, 175 H.P., gas		26.28	659.83	1,979.50	5,938.50	606.11	
	2200	Hydromulcher, diesel, 3000 gallon, for truck mounting		17.96	926.67	2,780	8,340	699.68	
	2300	Gas, 600 gallon		7.72	433.33	1,300	3,900	321.77	
	2400	Joint & crack cleaner, walk behind, 25 H.P.		3.25	217.33	652	1,956	156.44	
	2500	Filler, trailer mounted, 400 gallons, 20 H.P.		8.60	345	1,035	3,105	275.79	
	3000	Paint striper, self-propelled, 40 gallon, 22 H.P.		6.97	20	60	180	67.72	
	3100	120 gallon, 120 H.P.		18.18	133.33	400	1,200	225.45	
	3200	Post drivers, 6" I-Beam frame, for truck mounting		12.79	75	225	675	147.31	
	3400	Road sweeper, self-propelled, 8' wide, 90 H.P.		37.06	372.78	1,043.33	2,660	505.17	
	3450	Road sweeper, vacuum assisted, 4 C.Y., 220 gallons		60.16	3,035	6,515	12,545	1,784.25	
	4000	Road mixer, self-propelled, 130 H.P.		47.73	835	2,601.28	7,525	902.06	
	4100	310 H.P.		77.42	2,175	6,801.26	19,700	1,979.64	
	4220	Cold mix paver, incl. pug mill and bitumen tank, 165 H.P.		98.03	750	2,150	6,400	1,214.25	
	4240	Pavement brush, towed		3.54	101	314.32	910	91.15	
	4250	Paver, asphalt, wheel or crawler, 130 H.P., diesel		97.27	2,300	7,153.51	20,700	2,208.89	
	4300	Paver, road widener, gas, 1' to 6', 67 H.P.		48.16	985	3,061.92	8,850	997.70	
	4400	Diesel, 2' to 14', 88 H.P.		58.20	1,175	3,630.95	10,500	1,191.78	
	4600	Slipform pavers, curb and gutter, 2 track, 75 H.P.		59.70	1,275	3,956.11	11,400	1,268.80	
	4700	4 track, 165 H.P.		36.84	855	2,655.47	7,675	825.78	
	4800	Median barrier, 215 H.P.		60.31	1,350	4,227.08	12,200	1,327.93	
	4901	Trailer, low bed, 75 ton capacity		11.05	286	888.77	2,575	266.17	
	5000	Road planer, walk behind, 10" cutting width, 10 H.P.		2.53	353.33	991.78	2,268.78	218.58	
	5100	Self-propelled, 12" cutting width, 64 H.P.		8.51	326.11	877.78	2,430	243.60	
	5120	Traffic line remover, metal ball blaster, truck mounted, 115 H.P.		47.97	2,066.67	6,200	18,600	1,623.77	
	5140	Grinder, truck mounted, 115 H.P.		52.43	2,066.67	6,200	18,600	1,659.47	
	5160	Walk-behind, 11 H.P.		3.66	286.67	860	2,580	201.31	
	5200	Pavement profiler, 4' to 6' wide, 450 H.P.		223.15	1,000	3,000	9,000	2,385.17	
	5300	8' to 10' wide, 750 H.P.		341.64	1,000	3,000	9,000	3,333.09	
	5400	Roadway plate, steel, 1" x 8' x 20'		.09	61.50	191.52	555	39.01	
	5600	Stabilizer, self-propelled, 150 H.P.		42.38	233.33	700	2,100	479.06	
	5700	310 H.P.		78.49	291.67	875	2,625	802.91	
	5800	Striper, truck mounted, 120 gallon paint, 460 H.P.		50.22	348.67	1,046	3,138	610.97	
	5900	Thermal paint heating kettle, 115 gallons		7.94	150	450	1,350	153.51	
	6000	Tar kettle, 330 gallon, trailer mounted		12.65	191.67	575	1,725	216.16	
	7000	Tunnel locomotive, diesel, 8 to 12 ton		30.72	625	1,950.96	5,650	635.92	
	7005	Electric, 10 ton		30.19	715	2,221.92	6,425	685.90	
	7010	Muck cars, 1/2 C.Y. capacity		2.37	27	84	243	35.77	
	7020	1 C.Y. capacity		2.59	35	109.47	315	42.64	
	7030	2 C.Y. capacity		2.74	39.50	122.48	355	46.42	
	7040	Side dump, 2 C.Y. capacity		2.96	49	151.74	440	54.05	
	7050	3 C.Y. capacity		3.98	53.50	166.92	485	65.20	
	7060	5 C.Y. capacity		5.80	69.50	215.69	625	89.57	
	7100	Ventilating blower for tunnel, 7-1/2 H.P.		2.21	53.50	165.83	480	50.82	
	7110	10 H.P.		2.50	55.50	173.42	500	54.67	
	7120	20 H.P.		3.66	72.50	225.44	650	74.33	
	7140	40 H.P.		6.34	96	298.06	860	110.30	
	7160	60 H.P.		8.97	103	319.74	925	135.70	
	7175	75 H.P.		10.70	160	498.58	1,450	185.34	
	7180	200 H.P.		21.45	315	980.90	2,850	367.81	
	7800	Windrow loader, elevating		55.57	1,759.58	5,278.75	15,836.25	1,500.35	
60	0010	**LIFTING AND HOISTING EQUIPMENT RENTAL** without operators (R015433 -10)							60
	0150	Crane, flatbed mounted, 3 ton capacity	Ea.	14.88	849.70	2,252.27	5,480.30	569.46	
	0200	Crane, climbing, 106' jib, 6000 lb. capacity, 410 fpm (R312316 -45)		40.93	2,758.34	8,275.02	24,825.06	1,982.43	
	0300	101' jib, 10,250 lb. capacity, 270 fpm		47.84	2,904.26	8,712.77	26,138.31	2,125.28	

01 54 33 | Equipment Rental

		UNIT	HOURLY OPER. COST	RENT PER DAY	RENT PER WEEK	RENT PER MONTH	EQUIPMENT COST/DAY
0500	Tower, static, 130' high, 106' jib, 6200 lb. capacity at 400 fpm	Ea.	46.53	2,374.49	7,123.48	21,370.44	1,796.91
0520	Mini crawler spider crane, up to 24" wide, 1990 lb. lifting capacity		12.91	555	1,734.19	5,025	450.08
0525	Up to 30" wide, 6450 lb. lifting capacity		14.99	660	2,059.34	5,950	531.80
0530	Up to 52" wide, 6680 lb. lifting capacity		23.85	810	2,519.99	7,300	694.78
0535	Up to 55" wide, 8920 lb. lifting capacity		26.63	895	2,790.95	8,075	771.20
0540	Up to 66" wide, 13,350 lb. lifting capacity		36.05	1,400	4,335.46	12,500	1,155.46
0600	Crawler mounted, lattice boom, 1/2 C.Y., 15 tons at 12' radius		38.08	878.25	2,634.74	7,904.22	831.61
0700	3/4 C.Y., 20 tons at 12' radius		55.81	986.74	2,960.22	8,880.66	1,038.56
0800	1 C.Y., 25 tons at 12' radius		69.47	1,043.05	3,129.14	9,387.42	1,181.56
0900	1-1/2 C.Y., 40 tons at 12' radius		68.32	1,193.43	3,580.28	10,740.84	1,262.65
1000	2 C.Y., 50 tons at 12' radius		91.46	1,414.53	4,243.60	12,730.80	1,580.38
1100	3 C.Y., 75 tons at 12' radius		66.96	2,475.43	7,426.30	22,278.90	2,020.96
1200	100 ton capacity, 60' boom		88.51	1,166.67	3,500	10,500	1,408.10
1300	165 ton capacity, 60' boom		109.32	1,400	4,200	12,600	1,714.56
1400	200 ton capacity, 70' boom		142.40	1,556.67	4,670	14,010	2,073.21
1500	350 ton capacity, 80' boom		187.72	3,266.67	9,800	29,400	3,461.78
1600	Truck mounted, lattice boom, 6 x 4, 20 tons at 10' radius		40.96	963.62	2,437.13	7,195.32	815.12
1700	25 tons at 10' radius		44.02	2,475.43	7,426.30	22,278.90	1,837.44
1800	8 x 4, 30 tons at 10' radius		55.80	2,652.25	7,956.75	23,870.25	2,037.78
1900	40 tons at 12' radius		55.80	2,899.79	8,699.38	26,098.14	2,186.31
2000	60 tons at 15' radius		55.31	3,041.25	9,123.74	27,371.22	2,267.25
2050	82 tons at 15' radius		61.23	3,129.83	9,389.48	28,168.44	2,367.72
2100	90 tons at 15' radius		68.40	3,221.50	9,664.49	28,993.47	2,480.12
2200	115 tons at 15' radius		77.17	3,316.94	9,950.83	29,852.49	2,607.51
2300	150 tons at 18' radius		83.55	2,864.43	8,593.29	25,779.87	2,387.05
2350	165 tons at 18' radius		89.67	2,917.65	8,752.94	26,258.82	2,467.97
2400	Truck mounted, hydraulic, 12 ton capacity		30.40	2,899.79	8,699.38	26,098.14	1,983.05
2500	25 ton capacity		37.46	2,988.03	8,964.09	26,892.27	2,092.48
2550	33 ton capacity		55.80	2,988.03	8,964.09	26,892.27	2,239.25
2560	40 ton capacity		55.80	3,014.47	9,043.40	27,130.20	2,255.11
2600	55 ton capacity		55.41	3,041.25	9,123.74	27,371.22	2,268.02
2700	80 ton capacity		72.55	3,094.12	9,282.36	27,847.08	2,436.89
2720	100 ton capacity		77.23	3,226.65	9,679.94	29,039.82	2,553.84
2740	120 ton capacity		108.47	3,271.28	9,813.84	29,441.52	2,830.50
2760	150 ton capacity		116.08	3,478.32	10,434.96	31,304.88	3,015.60
2800	Self-propelled, 4 x 4, with telescoping boom, 5 ton		15.60	370	1,110	3,330	346.78
2900	12-1/2 ton capacity		22.06	658.33	1,975	5,925	571.52
3000	15 ton capacity		35.46	750	2,250	6,750	733.70
3050	20 ton capacity		23.00	841.67	2,525	7,575	689.01
3100	25 ton capacity		37.80	900	2,700	8,100	842.41
3150	40 ton capacity		46.26	1,089.33	3,268	9,804	1,023.68
3200	Derricks, guy, 20 ton capacity, 60' boom, 75' mast		23.43	1,503.11	4,509.34	13,528.02	1,089.34
3300	100' boom, 115' mast		37.13	2,121.80	6,365.40	19,096.20	1,570.15
3400	Stiffleg, 20 ton capacity, 70' boom, 37' mast		26.18	654.39	1,963.18	5,889.54	602.05
3500	100' boom, 47' mast		40.51	707.27	2,121.80	6,365.40	748.43
3550	Helicopter, small, lift to 1250 lb. maximum, w/pilot		102.15	6,666.67	20,000	60,000	4,817.17
3600	Hoists, chain type, overhead, manual, 3/4 ton		.15	15	38	75	8.79
3900	10 ton		.81	124	310	620	68.50
4000	Hoist and tower, 5000 lb. cap., portable electric, 40' high		4.92	150.38	451.14	1,353.42	129.56
4100	For each added 10' section, add		.12	33.65	100.94	302.82	21.14
4200	Hoist and single tubular tower, 5000 lb. electric, 100' high		7.17	111.58	334.75	1,004.25	124.28
4300	For each added 6'-6" section, add		.21	40.51	121.54	364.62	26.01
4400	Hoist and double tubular tower, 5000 lb., 100' high		7.80	111.58	334.75	1,004.25	129.36
4500	For each added 6'-6" section, add		.23	43.95	131.84	395.52	28.24
4550	Hoist and tower, mast type, 6000 lb., 100' high		9.51	487.53	1,462.60	4,387.80	368.57
4570	For each added 10' section, add		.14	33.65	100.94	302.82	21.27
4600	Hoist and tower, personnel, electric, 2000 lb., 100' @ 125 fpm		18.03	528.73	1,586.20	4,758.60	461.46
4700	3000 lb., 100' @ 200 fpm		20.63	555.17	1,665.51	4,996.53	498.13

01 54 33 | Equipment Rental

		UNIT	HOURLY OPER. COST	RENT PER DAY	RENT PER WEEK	RENT PER MONTH	EQUIPMENT COST/DAY	
60	4800	3000 lb., 150' @ 300 fpm	Ea.	22.89	749.84	2,249.52	6,748.56	633.04	60
	4900	4000 lb., 100' @ 300 fpm		23.68	775.93	2,327.80	6,983.40	655.01	
	5000	6000 lb., 100' @ 275 fpm	↓	25.45	802.71	2,408.14	7,224.42	685.22	
	5100	For added heights up to 500', add	L.F.	.01	3.43	10.30	30.90	2.14	
	5200	Jacks, hydraulic, 20 ton	Ea.	.05	21.28	55.98	101.51	11.61	
	5500	100 ton		.41	26.06	64.70	133.48	16.23	
	6100	Jacks, hydraulic, climbing w/50' jackrods, control console, 30 ton cap.		2.23	32.96	98.88	296.64	37.63	
	6150	For each added 10' jackrod section, add		.05	5.15	15.45	46.35	3.50	
	6300	50 ton capacity		3.59	35.36	106.09	318.27	49.92	
	6350	For each added 10' jackrod section, add		.06	5.15	15.45	46.35	3.58	
	6500	125 ton capacity		9.38	54.93	164.80	494.40	108.01	
	6550	For each added 10' jackrod section, add		.63	5.15	15.45	46.35	8.17	
	6600	Cable jack, 10 ton capacity with 200' cable		1.63	37.77	113.30	339.90	35.72	
	6650	For each added 50' of cable, add	↓	.23	15.79	47.38	142.14	11.30	
70	0010	**WELLPOINT EQUIPMENT RENTAL** without operators							70
	0020	Based on 2 months rental							
	0100	Combination jetting & wellpoint pump, 60 H.P. diesel	Ea.	16.15	316.55	949.66	2,848.98	319.10	
	0200	High pressure gas jet pump, 200 H.P., 300 psi	"	34.86	165	495	1,485	377.85	
	0300	Discharge pipe, 8" diameter	L.F.	.01	50	150	450	30.07	
	0350	12" diameter		.01	53.33	160	480	32.10	
	0400	Header pipe, flows up to 150 GPM, 4" diameter		.01	25.33	76	228	15.26	
	0500	400 GPM, 6" diameter		.01	39.33	118	354	23.68	
	0600	800 GPM, 8" diameter		.01	46.67	140	420	28.10	
	0700	1500 GPM, 10" diameter		.01	54	162	486	32.51	
	0800	2500 GPM, 12" diameter		.03	61.33	184	552	37.01	
	0900	4500 GPM, 16" diameter		.03	50	150	450	30.26	
	0950	For quick coupling aluminum and plastic pipe, add	↓	.03	9.90	29.71	89.12	6.22	
	1100	Wellpoint, 25' long, with fittings & riser pipe, 1-1/2" or 2" diameter	Ea.	.07	77.25	231.75	695.25	46.90	
	1200	Wellpoint pump, diesel powered, 4" suction, 20 H.P.		7.21	166.67	500	1,500	157.71	
	1300	6" suction, 30 H.P.		9.68	280	840	2,520	245.42	
	1400	8" suction, 40 H.P.		13.11	250	750	2,250	254.89	
	1500	10" suction, 75 H.P.		19.13	320	960	2,880	345	
	1600	12" suction, 100 H.P.		28.06	336.67	1,010	3,030	426.50	
	1700	12" suction, 175 H.P.	↓	40.16	298.33	895	2,685	500.27	
80	0010	**MARINE EQUIPMENT RENTAL** without operators							80
	0200	Barge, 400 Ton, 30' wide x 90' long	Ea.	18.20	1,200	3,766.43	10,900	898.90	
	0240	800 Ton, 45' wide x 90' long		22.85	1,500	4,633.53	13,400	1,109.54	
	2000	Tugboat, diesel, 100 H.P.		30.52	240	747.87	2,175	393.75	
	2040	250 H.P.		59.26	435	1,354.83	3,925	745.07	
	2080	380 H.P.		129.02	1,300	4,064.50	11,800	1,845.03	
	3000	Small work boat, gas, 16-foot, 50 H.P.		11.71	48.50	150.38	435	123.78	
	4000	Large, diesel, 48-foot, 200 H.P.	↓	77.08	1,375	4,308.37	12,500	1,478.34	

The R015433-10 reference box appears next to lines 0010/0020 (item 70) and line 0010/0200 (item 80).

Crew A-1

Crew A-1	Hr.	Daily	Hr.	Daily	Bare Costs	Incl. O&P
1 Building Laborer	$45.60	$364.80	$72.45	$579.60	$45.60	$72.45
1 Concrete Saw, Gas Manual		117.14		128.86	14.64	16.11
8 L.H., Daily Totals		$481.94		$708.46	$60.24	$88.56

Crew A-1A

Crew A-1A	Hr.	Daily	Hr.	Daily	Bare Costs	Incl. O&P
1 Skilled Worker	$58.90	$471.20	$93.20	$745.60	$58.90	$93.20
1 Shot Blaster, 20"		215.77		237.34	26.97	29.67
8 L.H., Daily Totals		$686.97		$982.94	$85.87	$122.87

Crew A-1B

Crew A-1B	Hr.	Daily	Hr.	Daily	Bare Costs	Incl. O&P
1 Building Laborer	$45.60	$364.80	$72.45	$579.60	$45.60	$72.45
1 Concrete Saw		115.94		127.54	14.49	15.94
8 L.H., Daily Totals		$480.74		$707.14	$60.09	$88.39

Crew A-1C

Crew A-1C	Hr.	Daily	Hr.	Daily	Bare Costs	Incl. O&P
1 Building Laborer	$45.60	$364.80	$72.45	$579.60	$45.60	$72.45
1 Chain Saw, Gas, 18"		53.87		59.26	6.73	7.41
8 L.H., Daily Totals		$418.67		$638.86	$52.33	$79.86

Crew A-1D

Crew A-1D	Hr.	Daily	Hr.	Daily	Bare Costs	Incl. O&P
1 Building Laborer	$45.60	$364.80	$72.45	$579.60	$45.60	$72.45
1 Vibrating Plate, Gas, 18"		32.77		36.05	4.10	4.51
8 L.H., Daily Totals		$397.57		$615.65	$49.70	$76.96

Crew A-1E

Crew A-1E	Hr.	Daily	Hr.	Daily	Bare Costs	Incl. O&P
1 Building Laborer	$45.60	$364.80	$72.45	$579.60	$45.60	$72.45
1 Vibrating Plate, Gas, 21"		171.39		188.53	21.42	23.57
8 L.H., Daily Totals		$536.19		$768.13	$67.02	$96.02

Crew A-1F

Crew A-1F	Hr.	Daily	Hr.	Daily	Bare Costs	Incl. O&P
1 Building Laborer	$45.60	$364.80	$72.45	$579.60	$45.60	$72.45
1 Rammer/Tamper, Gas, 8"		48.45		53.30	6.06	6.66
8 L.H., Daily Totals		$413.25		$632.90	$51.66	$79.11

Crew A-1G

Crew A-1G	Hr.	Daily	Hr.	Daily	Bare Costs	Incl. O&P
1 Building Laborer	$45.60	$364.80	$72.45	$579.60	$45.60	$72.45
1 Rammer/Tamper, Gas, 15"		56.33		61.97	7.04	7.75
8 L.H., Daily Totals		$421.13		$641.57	$52.64	$80.20

Crew A-1H

Crew A-1H	Hr.	Daily	Hr.	Daily	Bare Costs	Incl. O&P
1 Building Laborer	$45.60	$364.80	$72.45	$579.60	$45.60	$72.45
1 Exterior Steam Cleaner		79.60		87.56	9.95	10.94
8 L.H., Daily Totals		$444.40		$667.16	$55.55	$83.39

Crew A-1J

Crew A-1J	Hr.	Daily	Hr.	Daily	Bare Costs	Incl. O&P
1 Building Laborer	$45.60	$364.80	$72.45	$579.60	$45.60	$72.45
1 Cultivator, Walk-Behind, 5 H.P.		54.91		60.40	6.86	7.55
8 L.H., Daily Totals		$419.71		$640.00	$52.46	$80.00

Crew A-1K

Crew A-1K	Hr.	Daily	Hr.	Daily	Bare Costs	Incl. O&P
1 Building Laborer	$45.60	$364.80	$72.45	$579.60	$45.60	$72.45
1 Cultivator, Walk-Behind, 8 H.P.		95.21		104.73	11.90	13.09
8 L.H., Daily Totals		$460.01		$684.33	$57.50	$85.54

Crew A-1M

Crew A-1M	Hr.	Daily	Hr.	Daily	Bare Costs	Incl. O&P
1 Building Laborer	$45.60	$364.80	$72.45	$579.60	$45.60	$72.45
1 Snow Blower, Walk-Behind		70.52		77.57	8.81	9.70
8 L.H., Daily Totals		$435.32		$657.17	$54.41	$82.15

Crew A-2

Crew A-2	Hr.	Daily	Hr.	Daily	Bare Costs	Incl. O&P
2 Laborers	$45.60	$729.60	$72.45	$1159.20	$47.20	$75.07
1 Truck Driver (light)	50.40	403.20	80.30	642.40		
1 Flatbed Truck, Gas, 1.5 Ton		203.25		223.57	8.47	9.32
24 L.H., Daily Totals		$1336.05		$2025.17	$55.67	$84.38

Crew A-2A

Crew A-2A	Hr.	Daily	Hr.	Daily	Bare Costs	Incl. O&P
2 Laborers	$45.60	$729.60	$72.45	$1159.20	$47.20	$75.07
1 Truck Driver (light)	50.40	403.20	80.30	642.40		
1 Flatbed Truck, Gas, 1.5 Ton		203.25		223.57		
1 Concrete Saw		115.94		127.54	13.30	14.63
24 L.H., Daily Totals		$1451.99		$2152.71	$60.50	$89.70

Crew A-2B

Crew A-2B	Hr.	Daily	Hr.	Daily	Bare Costs	Incl. O&P
1 Truck Driver (light)	$50.40	$403.20	$80.30	$642.40	$50.40	$80.30
1 Flatbed Truck, Gas, 1.5 Ton		203.25		223.57	25.41	27.95
8 L.H., Daily Totals		$606.45		$865.97	$75.81	$108.25

Crew A-3A

Crew A-3A	Hr.	Daily	Hr.	Daily	Bare Costs	Incl. O&P
1 Equip. Oper. (light)	$57.00	$456.00	$88.80	$710.40	$57.00	$88.80
1 Pickup Truck, 4x4, 3/4 Ton		182.00		200.20	22.75	25.03
8 L.H., Daily Totals		$638.00		$910.60	$79.75	$113.83

Crew A-3B

Crew A-3B	Hr.	Daily	Hr.	Daily	Bare Costs	Incl. O&P
1 Equip. Oper. (medium)	$60.40	$483.20	$94.10	$752.80	$56.73	$89.33
1 Truck Driver (heavy)	53.05	424.40	84.55	676.40		
1 Dump Truck, 12 C.Y., 400 H.P.		594.66		654.12		
1 F.E. Loader, W.M., 2.5 C.Y.		657.89		723.68	78.28	86.11
16 L.H., Daily Totals		$2160.15		$2807.00	$135.01	$175.44

Crew A-3C

Crew A-3C	Hr.	Daily	Hr.	Daily	Bare Costs	Incl. O&P
1 Equip. Oper. (light)	$57.00	$456.00	$88.80	$710.40	$57.00	$88.80
1 Loader, Skid Steer, 78 H.P.		460.30		506.33	57.54	63.29
8 L.H., Daily Totals		$916.30		$1216.73	$114.54	$152.09

Crew A-3D

Crew A-3D	Hr.	Daily	Hr.	Daily	Bare Costs	Incl. O&P
1 Truck Driver (light)	$50.40	$403.20	$80.30	$642.40	$50.40	$80.30
1 Pickup Truck, 4x4, 3/4 Ton		182.00		200.20		
1 Flatbed Trailer, 25 Ton		141.42		155.56	40.43	44.47
8 L.H., Daily Totals		$726.62		$998.16	$90.83	$124.77

Crew A-3E

Crew A-3E	Hr.	Daily	Hr.	Daily	Bare Costs	Incl. O&P
1 Equip. Oper. (crane)	$63.40	$507.20	$98.75	$790.00	$58.23	$91.65
1 Truck Driver (heavy)	53.05	424.40	84.55	676.40		
1 Pickup Truck, 4x4, 3/4 Ton		182.00		200.20	11.38	12.51
16 L.H., Daily Totals		$1113.60		$1666.60	$69.60	$104.16

Crew A-3F

Crew A-3F	Hr.	Daily	Hr.	Daily	Bare Costs	Incl. O&P
1 Equip. Oper. (crane)	$63.40	$507.20	$98.75	$790.00	$58.23	$91.65
1 Truck Driver (heavy)	53.05	424.40	84.55	676.40		
1 Pickup Truck, 4x4, 3/4 Ton		182.00		200.20		
1 Truck Tractor, 6x4, 380 H.P.		512.66		563.93		
1 Lowbed Trailer, 75 Ton		266.17		292.79	60.05	66.06
16 L.H., Daily Totals		$1892.43		$2523.31	$118.28	$157.71

Crew A-3G

Crew No.	Bare Costs Hr.	Bare Costs Daily	Incl. Subs O&P Hr.	Incl. Subs O&P Daily	Cost Per Labor-Hour Bare Costs	Cost Per Labor-Hour Incl. O&P
1 Equip. Oper. (crane)	$63.40	$507.20	$98.75	$790.00	$58.23	$91.65
1 Truck Driver (heavy)	53.05	424.40	84.55	676.40		
1 Pickup Truck, 4x4, 3/4 Ton		182.00		200.20		
1 Truck Tractor, 6x4, 450 H.P.		625.39		687.93		
1 Lowbed Trailer, 75 Ton		266.17		292.79	67.10	73.81
16 L.H., Daily Totals		$2005.16		$2647.32	$125.32	$165.46

Crew A-3H

Crew No.	Bare Costs Hr.	Bare Costs Daily	Incl. Subs O&P Hr.	Incl. Subs O&P Daily	Cost Per Labor-Hour Bare Costs	Cost Per Labor-Hour Incl. O&P
1 Equip. Oper. (crane)	$63.40	$507.20	$98.75	$790.00	$63.40	$98.75
1 Hyd. Crane, 12 Ton (Daily)		756.78		832.46	94.60	104.06
8 L.H., Daily Totals		$1263.98		$1622.46	$158.00	$202.81

Crew A-3I

Crew No.	Bare Costs Hr.	Bare Costs Daily	Incl. Subs O&P Hr.	Incl. Subs O&P Daily	Cost Per Labor-Hour Bare Costs	Cost Per Labor-Hour Incl. O&P
1 Equip. Oper. (crane)	$63.40	$507.20	$98.75	$790.00	$63.40	$98.75
1 Hyd. Crane, 25 Ton (Daily)		835.58		919.14	104.45	114.89
8 L.H., Daily Totals		$1342.78		$1709.14	$167.85	$213.64

Crew A-3J

Crew No.	Bare Costs Hr.	Bare Costs Daily	Incl. Subs O&P Hr.	Incl. Subs O&P Daily	Cost Per Labor-Hour Bare Costs	Cost Per Labor-Hour Incl. O&P
1 Equip. Oper. (crane)	$63.40	$507.20	$98.75	$790.00	$63.40	$98.75
1 Hyd. Crane, 40 Ton (Daily)		1327.41		1460.15	165.93	182.52
8 L.H., Daily Totals		$1834.61		$2250.15	$229.33	$281.27

Crew A-3K

Crew No.	Bare Costs Hr.	Bare Costs Daily	Incl. Subs O&P Hr.	Incl. Subs O&P Daily	Cost Per Labor-Hour Bare Costs	Cost Per Labor-Hour Incl. O&P
1 Equip. Oper. (crane)	$63.40	$507.20	$98.75	$790.00	$58.70	$91.45
1 Equip. Oper. (oiler)	54.00	432.00	84.15	673.20		
1 Hyd. Crane, 55 Ton (Daily)		1420.54		1562.59		
1 P/U Truck, 3/4 Ton (Daily)		147.80		162.58	98.02	107.82
16 L.H., Daily Totals		$2507.54		$3188.37	$156.72	$199.27

Crew A-3L

Crew No.	Bare Costs Hr.	Bare Costs Daily	Incl. Subs O&P Hr.	Incl. Subs O&P Daily	Cost Per Labor-Hour Bare Costs	Cost Per Labor-Hour Incl. O&P
1 Equip. Oper. (crane)	$63.40	$507.20	$98.75	$790.00	$58.70	$91.45
1 Equip. Oper. (oiler)	54.00	432.00	84.15	673.20		
1 Hyd. Crane, 80 Ton (Daily)		2124.93		2337.42		
1 P/U Truck, 3/4 Ton (Daily)		147.80		162.58	142.05	156.25
16 L.H., Daily Totals		$3211.93		$3963.20	$200.75	$247.70

Crew A-3M

Crew No.	Bare Costs Hr.	Bare Costs Daily	Incl. Subs O&P Hr.	Incl. Subs O&P Daily	Cost Per Labor-Hour Bare Costs	Cost Per Labor-Hour Incl. O&P
1 Equip. Oper. (crane)	$63.40	$507.20	$98.75	$790.00	$58.70	$91.45
1 Equip. Oper. (oiler)	54.00	432.00	84.15	673.20		
1 Hyd. Crane, 100 Ton (Daily)		2324.92		2557.42		
1 P/U Truck, 3/4 Ton (Daily)		147.80		162.58	154.55	170.00
16 L.H., Daily Totals		$3411.92		$4183.20	$213.25	$261.45

Crew A-3N

Crew No.	Bare Costs Hr.	Bare Costs Daily	Incl. Subs O&P Hr.	Incl. Subs O&P Daily	Cost Per Labor-Hour Bare Costs	Cost Per Labor-Hour Incl. O&P
1 Equip. Oper. (crane)	$63.40	$507.20	$98.75	$790.00	$63.40	$98.75
1 Tower Crane (monthly)		1794.74		1974.21	224.34	246.78
8 L.H., Daily Totals		$2301.94		$2764.21	$287.74	$345.53

Crew A-3P

Crew No.	Bare Costs Hr.	Bare Costs Daily	Incl. Subs O&P Hr.	Incl. Subs O&P Daily	Cost Per Labor-Hour Bare Costs	Cost Per Labor-Hour Incl. O&P
1 Equip. Oper. (light)	$57.00	$456.00	$88.80	$710.40	$57.00	$88.80
1 A.T. Forklift, 31' reach, 45' lift		356.07		391.68	44.51	48.96
8 L.H., Daily Totals		$812.07		$1102.08	$101.51	$137.76

Crew A-3Q

Crew No.	Bare Costs Hr.	Bare Costs Daily	Incl. Subs O&P Hr.	Incl. Subs O&P Daily	Cost Per Labor-Hour Bare Costs	Cost Per Labor-Hour Incl. O&P
1 Equip. Oper. (light)	$57.00	$456.00	$88.80	$710.40	$57.00	$88.80
1 Pickup Truck, 4x4, 3/4 Ton		182.00		200.20		
1 Flatbed Trailer, 3 Ton		73.56		80.91	31.94	35.14
8 L.H., Daily Totals		$711.56		$991.51	$88.94	$123.94

Crew A-3R

Crew No.	Bare Costs Hr.	Bare Costs Daily	Incl. Subs O&P Hr.	Incl. Subs O&P Daily	Cost Per Labor-Hour Bare Costs	Cost Per Labor-Hour Incl. O&P
1 Equip. Oper. (light)	$57.00	$456.00	$88.80	$710.40	$57.00	$88.80
1 Forklift, Smooth Floor, 8,000 Lb.		292.16		321.38	36.52	40.17
8 L.H., Daily Totals		$748.16		$1031.78	$93.52	$128.97

Crew A-4

Crew No.	Bare Costs Hr.	Bare Costs Daily	Incl. Subs O&P Hr.	Incl. Subs O&P Daily	Cost Per Labor-Hour Bare Costs	Cost Per Labor-Hour Incl. O&P
2 Carpenters	$56.30	$900.80	$89.45	$1431.20	$53.47	$84.83
1 Painter, Ordinary	47.80	382.40	75.60	604.80		
24 L.H., Daily Totals		$1283.20		$2036.00	$53.47	$84.83

Crew A-5

Crew No.	Bare Costs Hr.	Bare Costs Daily	Incl. Subs O&P Hr.	Incl. Subs O&P Daily	Cost Per Labor-Hour Bare Costs	Cost Per Labor-Hour Incl. O&P
2 Laborers	$45.60	$729.60	$72.45	$1159.20	$46.13	$73.32
.25 Truck Driver (light)	50.40	100.80	80.30	160.60		
.25 Flatbed Truck, Gas, 1.5 Ton		50.81		55.89	2.82	3.11
18 L.H., Daily Totals		$881.21		$1375.69	$48.96	$76.43

Crew A-6

Crew No.	Bare Costs Hr.	Bare Costs Daily	Incl. Subs O&P Hr.	Incl. Subs O&P Daily	Cost Per Labor-Hour Bare Costs	Cost Per Labor-Hour Incl. O&P
1 Instrument Man	$58.90	$471.20	$93.20	$745.60	$56.45	$89.42
1 Rodman/Chainman	54.00	432.00	85.65	685.20		
1 Level, Electronic		35.52		39.08	2.22	2.44
16 L.H., Daily Totals		$938.73		$1469.88	$58.67	$91.87

Crew A-7

Crew No.	Bare Costs Hr.	Bare Costs Daily	Incl. Subs O&P Hr.	Incl. Subs O&P Daily	Cost Per Labor-Hour Bare Costs	Cost Per Labor-Hour Incl. O&P
1 Chief of Party	$70.50	$564.00	$108.75	$870.00	$61.13	$95.87
1 Instrument Man	58.90	471.20	93.20	745.60		
1 Rodman/Chainman	54.00	432.00	85.65	685.20		
1 Level, Electronic		35.52		39.08	1.48	1.63
24 L.H., Daily Totals		$1502.72		$2339.88	$62.61	$97.49

Crew A-8

Crew No.	Bare Costs Hr.	Bare Costs Daily	Incl. Subs O&P Hr.	Incl. Subs O&P Daily	Cost Per Labor-Hour Bare Costs	Cost Per Labor-Hour Incl. O&P
1 Chief of Party	$70.50	$564.00	$108.75	$870.00	$59.35	$93.31
1 Instrument Man	58.90	471.20	93.20	745.60		
2 Rodmen/Chainmen	54.00	864.00	85.65	1370.40		
1 Level, Electronic		35.52		39.08	1.11	1.22
32 L.H., Daily Totals		$1934.72		$3025.08	$60.46	$94.53

Crew A-9

Crew No.	Bare Costs Hr.	Bare Costs Daily	Incl. Subs O&P Hr.	Incl. Subs O&P Daily	Cost Per Labor-Hour Bare Costs	Cost Per Labor-Hour Incl. O&P
1 Asbestos Foreman	$63.55	$508.40	$100.40	$803.20	$63.11	$99.70
7 Asbestos Workers	63.05	3530.80	99.60	5577.60		
64 L.H., Daily Totals		$4039.20		$6380.80	$63.11	$99.70

Crew A-10A

Crew No.	Bare Costs Hr.	Bare Costs Daily	Incl. Subs O&P Hr.	Incl. Subs O&P Daily	Cost Per Labor-Hour Bare Costs	Cost Per Labor-Hour Incl. O&P
1 Asbestos Foreman	$63.55	$508.40	$100.40	$803.20	$63.22	$99.87
2 Asbestos Workers	63.05	1008.80	99.60	1593.60		
24 L.H., Daily Totals		$1517.20		$2396.80	$63.22	$99.87

Crew A-10B

Crew No.	Bare Costs Hr.	Bare Costs Daily	Incl. Subs O&P Hr.	Incl. Subs O&P Daily	Cost Per Labor-Hour Bare Costs	Cost Per Labor-Hour Incl. O&P
1 Asbestos Foreman	$63.55	$508.40	$100.40	$803.20	$63.17	$99.80
3 Asbestos Workers	63.05	1513.20	99.60	2390.40		
32 L.H., Daily Totals		$2021.60		$3193.60	$63.17	$99.80

Crew A-10C

Crew No.	Bare Costs Hr.	Bare Costs Daily	Incl. Subs O&P Hr.	Incl. Subs O&P Daily	Cost Per Labor-Hour Bare Costs	Cost Per Labor-Hour Incl. O&P
3 Asbestos Workers	$63.05	$1513.20	$99.60	$2390.40	$63.05	$99.60
1 Flatbed Truck, Gas, 1.5 Ton		203.25		223.57	8.47	9.32
24 L.H., Daily Totals		$1716.45		$2613.97	$71.52	$108.92

Crews - Renovation

Crew No.	Bare Costs Hr.	Bare Costs Daily	Incl. Subs O&P Hr.	Incl. Subs O&P Daily	Cost Per Labor-Hour Bare Costs	Cost Per Labor-Hour Incl. O&P
Crew A-10D	Hr.	Daily	Hr.	Daily	Bare Costs	Incl. O&P
2 Asbestos Workers	$63.05	$1008.80	$99.60	$1593.60	$60.88	$95.53
1 Equip. Oper. (crane)	63.40	507.20	98.75	790.00		
1 Equip. Oper. (oiler)	54.00	432.00	84.15	673.20		
1 Hydraulic Crane, 33 Ton		1012.08		1113.29	31.63	34.79
32 L.H., Daily Totals		$2960.08		$4170.09	$92.50	$130.32
Crew A-11	Hr.	Daily	Hr.	Daily	Bare Costs	Incl. O&P
1 Asbestos Foreman	$63.55	$508.40	$100.40	$803.20	$63.11	$99.70
7 Asbestos Workers	63.05	3530.80	99.60	5577.60		
2 Chip. Hammers, 12 Lb., Elec.		67.95		74.75	1.06	1.17
64 L.H., Daily Totals		$4107.15		$6455.55	$64.17	$100.87
Crew A-12	Hr.	Daily	Hr.	Daily	Bare Costs	Incl. O&P
1 Asbestos Foreman	$63.55	$508.40	$100.40	$803.20	$63.11	$99.70
7 Asbestos Workers	63.05	3530.80	99.60	5577.60		
1 Trk-Mtd Vac, 14 CY, 1500 Gal.		560.27		616.30		
1 Flatbed Truck, 20,000 GVW		209.49		230.44	12.03	13.23
64 L.H., Daily Totals		$4808.96		$7227.53	$75.14	$112.93
Crew A-13	Hr.	Daily	Hr.	Daily	Bare Costs	Incl. O&P
1 Equip. Oper. (light)	$57.00	$456.00	$88.80	$710.40	$57.00	$88.80
1 Trk-Mtd Vac, 14 CY, 1500 Gal.		560.27		616.30		
1 Flatbed Truck, 20,000 GVW		209.49		230.44	96.22	105.84
8 L.H., Daily Totals		$1225.76		$1557.13	$153.22	$194.64
Crew B-1	Hr.	Daily	Hr.	Daily	Bare Costs	Incl. O&P
1 Labor Foreman (outside)	$47.60	$380.80	$75.60	$604.80	$46.27	$73.50
2 Laborers	45.60	729.60	72.45	1159.20		
24 L.H., Daily Totals		$1110.40		$1764.00	$46.27	$73.50
Crew B-1A	Hr.	Daily	Hr.	Daily	Bare Costs	Incl. O&P
1 Labor Foreman (outside)	$47.60	$380.80	$75.60	$604.80	$46.27	$73.50
2 Laborers	45.60	729.60	72.45	1159.20		
2 Cutting Torches		26.74		29.41		
2 Sets of Gases		354.31		389.75	15.88	17.47
24 L.H., Daily Totals		$1491.45		$2183.16	$62.14	$90.97
Crew B-1B	Hr.	Daily	Hr.	Daily	Bare Costs	Incl. O&P
1 Labor Foreman (outside)	$47.60	$380.80	$75.60	$604.80	$50.55	$79.81
2 Laborers	45.60	729.60	72.45	1159.20		
1 Equip. Oper. (crane)	63.40	507.20	98.75	790.00		
2 Cutting Torches		26.74		29.41		
2 Sets of Gases		354.31		389.75		
1 Hyd. Crane, 12 Ton		489.34		538.27	27.20	29.92
32 L.H., Daily Totals		$2487.99		$3511.43	$77.75	$109.73
Crew B-1C	Hr.	Daily	Hr.	Daily	Bare Costs	Incl. O&P
1 Labor Foreman (outside)	$47.60	$380.80	$75.60	$604.80	$46.27	$73.50
2 Laborers	45.60	729.60	72.45	1159.20		
1 Telescoping Boom Lift, to 60'		301.02		331.12	12.54	13.80
24 L.H., Daily Totals		$1411.42		$2095.12	$58.81	$87.30
Crew B-1D	Hr.	Daily	Hr.	Daily	Bare Costs	Incl. O&P
2 Laborers	$45.60	$729.60	$72.45	$1159.20	$45.60	$72.45
1 Small Work Boat, Gas, 50 H.P.		123.78		136.15		
1 Pressure Washer, 7 GPM		96.77		106.44	13.78	15.16
16 L.H., Daily Totals		$950.14		$1401.80	$59.38	$87.61
Crew B-1E	Hr.	Daily	Hr.	Daily	Bare Costs	Incl. O&P
1 Labor Foreman (outside)	$47.60	$380.80	$75.60	$604.80	$46.10	$73.24
3 Laborers	45.60	1094.40	72.45	1738.80		
1 Work Boat, Diesel, 200 H.P.		1478.34		1626.17		
2 Pressure Washers, 7 GPM		193.53		212.89	52.25	57.47
32 L.H., Daily Totals		$3147.07		$4182.66	$98.35	$130.71
Crew B-1F	Hr.	Daily	Hr.	Daily	Bare Costs	Incl. O&P
2 Skilled Workers	$58.90	$942.40	$93.20	$1491.20	$54.47	$86.28
1 Laborer	45.60	364.80	72.45	579.60		
1 Small Work Boat, Gas, 50 H.P.		123.78		136.15		
1 Pressure Washer, 7 GPM		96.77		106.44	9.19	10.11
24 L.H., Daily Totals		$1527.74		$2313.40	$63.66	$96.39
Crew B-1G	Hr.	Daily	Hr.	Daily	Bare Costs	Incl. O&P
2 Laborers	$45.60	$729.60	$72.45	$1159.20	$45.60	$72.45
1 Small Work Boat, Gas, 50 H.P.		123.78		136.15	7.74	8.51
16 L.H., Daily Totals		$853.38		$1295.35	$53.34	$80.96
Crew B-1H	Hr.	Daily	Hr.	Daily	Bare Costs	Incl. O&P
2 Skilled Workers	$58.90	$942.40	$93.20	$1491.20	$54.47	$86.28
1 Laborer	45.60	364.80	72.45	579.60		
1 Small Work Boat, Gas, 50 H.P.		123.78		136.15	5.16	5.67
24 L.H., Daily Totals		$1430.98		$2206.95	$59.62	$91.96
Crew B-1J	Hr.	Daily	Hr.	Daily	Bare Costs	Incl. O&P
1 Labor Foreman (inside)	$46.10	$368.80	$73.25	$586.00	$45.85	$72.85
1 Laborer	45.60	364.80	72.45	579.60		
16 L.H., Daily Totals		$733.60		$1165.60	$45.85	$72.85
Crew B-1K	Hr.	Daily	Hr.	Daily	Bare Costs	Incl. O&P
1 Carpenter Foreman (inside)	$56.80	$454.40	$90.25	$722.00	$56.55	$89.85
1 Carpenter	56.30	450.40	89.45	715.60		
16 L.H., Daily Totals		$904.80		$1437.60	$56.55	$89.85
Crew B-2	Hr.	Daily	Hr.	Daily	Bare Costs	Incl. O&P
1 Labor Foreman (outside)	$47.60	$380.80	$75.60	$604.80	$46.00	$73.08
4 Laborers	45.60	1459.20	72.45	2318.40		
40 L.H., Daily Totals		$1840.00		$2923.20	$46.00	$73.08
Crew B-2A	Hr.	Daily	Hr.	Daily	Bare Costs	Incl. O&P
1 Labor Foreman (outside)	$47.60	$380.80	$75.60	$604.80	$46.27	$73.50
2 Laborers	45.60	729.60	72.45	1159.20		
1 Telescoping Boom Lift, to 60'		301.02		331.12	12.54	13.80
24 L.H., Daily Totals		$1411.42		$2095.12	$58.81	$87.30
Crew B-3	Hr.	Daily	Hr.	Daily	Bare Costs	Incl. O&P
1 Labor Foreman (outside)	$47.60	$380.80	$75.60	$604.80	$50.88	$80.62
2 Laborers	45.60	729.60	72.45	1159.20		
1 Equip. Oper. (medium)	60.40	483.20	94.10	752.80		
2 Truck Drivers (heavy)	53.05	848.80	84.55	1352.80		
1 Crawler Loader, 3 C.Y.		1180.92		1299.01		
2 Dump Trucks, 12 C.Y., 400 H.P.		1189.32		1308.25	49.38	54.32
48 L.H., Daily Totals		$4812.63		$6476.86	$100.26	$134.93
Crew B-3A	Hr.	Daily	Hr.	Daily	Bare Costs	Incl. O&P
4 Laborers	$45.60	$1459.20	$72.45	$2318.40	$48.56	$76.78
1 Equip. Oper. (medium)	60.40	483.20	94.10	752.80		
1 Hyd. Excavator, 1.5 C.Y.		714.87		786.36	17.87	19.66
40 L.H., Daily Totals		$2657.27		$3857.56	$66.43	$96.44

Crew No.	Bare Costs Hr.	Daily	Incl. Subs O&P Hr.	Daily	Cost Per Labor-Hour Bare Costs	Incl. O&P
Crew B-3B	Hr.	Daily	Hr.	Daily	Bare Costs	Incl. O&P
2 Laborers	$45.60	$729.60	$72.45	$1159.20	$51.16	$80.89
1 Equip. Oper. (medium)	60.40	483.20	94.10	752.80		
1 Truck Driver (heavy)	53.05	424.40	84.55	676.40		
1 Backhoe Loader, 80 H.P.		240.94		265.04		
1 Dump Truck, 12 C.Y., 400 H.P.		594.66		654.12	26.11	28.72
32 L.H., Daily Totals		$2472.80		$3507.56	$77.28	$109.61
Crew B-3C	Hr.	Daily	Hr.	Daily	Bare Costs	Incl. O&P
3 Laborers	$45.60	$1094.40	$72.45	$1738.80	$49.30	$77.86
1 Equip. Oper. (medium)	60.40	483.20	94.10	752.80		
1 Crawler Loader, 4 C.Y.		1500.46		1650.51	46.89	51.58
32 L.H., Daily Totals		$3078.06		$4142.11	$96.19	$129.44
Crew B-4	Hr.	Daily	Hr.	Daily	Bare Costs	Incl. O&P
1 Labor Foreman (outside)	$47.60	$380.80	$75.60	$604.80	$47.17	$74.99
4 Laborers	45.60	1459.20	72.45	2318.40		
1 Truck Driver (heavy)	53.05	424.40	84.55	676.40		
1 Truck Tractor, 220 H.P.		319.23		351.15		
1 Flatbed Trailer, 40 Ton		194.29		213.72	10.70	11.77
48 L.H., Daily Totals		$2777.92		$4164.47	$57.87	$86.76
Crew B-5	Hr.	Daily	Hr.	Daily	Bare Costs	Incl. O&P
1 Labor Foreman (outside)	$47.60	$380.80	$75.60	$604.80	$50.11	$79.09
4 Laborers	45.60	1459.20	72.45	2318.40		
2 Equip. Oper. (medium)	60.40	966.40	94.10	1505.60		
1 Air Compressor, 250 cfm		208.68		229.55		
2 Breakers, Pavement, 60 lb.		110.96		122.06		
2 -50' Air Hoses, 1.5"		47.19		51.91		
1 Crawler Loader, 3 C.Y.		1180.92		1299.01	27.64	30.40
56 L.H., Daily Totals		$4354.15		$6131.33	$77.75	$109.49
Crew B-5D	Hr.	Daily	Hr.	Daily	Bare Costs	Incl. O&P
1 Labor Foreman (outside)	$47.60	$380.80	$75.60	$604.80	$50.48	$79.77
4 Laborers	45.60	1459.20	72.45	2318.40		
2 Equip. Oper. (medium)	60.40	966.40	94.10	1505.60		
1 Truck Driver (heavy)	53.05	424.40	84.55	676.40		
1 Air Compressor, 250 cfm		208.68		229.55		
2 Breakers, Pavement, 60 lb.		110.96		122.06		
2 -50' Air Hoses, 1.5"		47.19		51.91		
1 Crawler Loader, 3 C.Y.		1180.92		1299.01		
1 Dump Truck, 12 C.Y., 400 H.P.		594.66		654.12	33.48	36.82
64 L.H., Daily Totals		$5373.21		$7461.85	$83.96	$116.59
Crew B-5E	Hr.	Daily	Hr.	Daily	Bare Costs	Incl. O&P
1 Labor Foreman (outside)	$47.60	$380.80	$75.60	$604.80	$50.48	$79.77
4 Laborers	45.60	1459.20	72.45	2318.40		
2 Equip. Oper. (medium)	60.40	966.40	94.10	1505.60		
1 Truck Driver (heavy)	53.05	424.40	84.55	676.40		
1 Water Tank Trailer, 5000 Gal.		158.82		174.70		
1 High Pressure Water Jet 40 KSI		847.30		932.03		
2 -50' Air Hoses, 1.5"		47.19		51.91		
1 Crawler Loader, 3 C.Y.		1180.92		1299.01		
1 Dump Truck, 12 C.Y., 400 H.P.		594.66		654.12	44.20	48.62
64 L.H., Daily Totals		$6059.69		$8216.98	$94.68	$128.39
Crew B-6	Hr.	Daily	Hr.	Daily	Bare Costs	Incl. O&P
2 Laborers	$45.60	$729.60	$72.45	$1159.20	$49.40	$77.90
1 Equip. Oper. (light)	57.00	456.00	88.80	710.40		
1 Backhoe Loader, 48 H.P.		221.72		243.89	9.24	10.16
24 L.H., Daily Totals		$1407.32		$2113.49	$58.64	$88.06

Crew No.	Bare Costs Hr.	Daily	Incl. Subs O&P Hr.	Daily	Cost Per Labor-Hour Bare Costs	Incl. O&P
Crew B-6A	Hr.	Daily	Hr.	Daily	Bare Costs	Incl. O&P
.5 Labor Foreman (outside)	$47.60	$190.40	$75.60	$302.40	$51.92	$81.74
1 Laborer	45.60	364.80	72.45	579.60		
1 Equip. Oper. (medium)	60.40	483.20	94.10	752.80		
1 Vacuum Truck, 5000 Gal.		383.94		422.34	19.20	21.12
20 L.H., Daily Totals		$1422.34		$2057.14	$71.12	$102.86
Crew B-6B	Hr.	Daily	Hr.	Daily	Bare Costs	Incl. O&P
2 Labor Foremen (outside)	$47.60	$761.60	$75.60	$1209.60	$46.27	$73.50
4 Laborers	45.60	1459.20	72.45	2318.40		
1 S.P. Crane, 4x4, 5 Ton		393.98		433.38		
1 Flatbed Truck, Gas, 1.5 Ton		203.25		223.57		
1 Butt Fusion Mach., 4"-12" diam.		434.93		478.43	21.50	23.65
48 L.H., Daily Totals		$3252.96		$4663.38	$67.77	$97.15
Crew B-6C	Hr.	Daily	Hr.	Daily	Bare Costs	Incl. O&P
2 Labor Foremen (outside)	$47.60	$761.60	$75.60	$1209.60	$46.27	$73.50
4 Laborers	45.60	1459.20	72.45	2318.40		
1 S.P. Crane, 4x4, 12 Ton		445.72		490.30		
1 Flatbed Truck, Gas, 3 Ton		877.96		965.75		
1 Butt Fusion Mach., 8"-24" diam.		1119.59		1231.55	50.90	55.99
48 L.H., Daily Totals		$4664.07		$6215.60	$97.17	$129.49
Crew B-6D	Hr.	Daily	Hr.	Daily	Bare Costs	Incl. O&P
.5 Labor Foreman (outside)	$47.60	$190.40	$75.60	$302.40	$51.92	$81.74
1 Laborer	45.60	364.80	72.45	579.60		
1 Equip. Oper. (medium)	60.40	483.20	94.10	752.80		
1 Hydro Excavator, 12 C.Y.		1318.79		1450.67	65.94	72.53
20 L.H., Daily Totals		$2357.19		$3085.47	$117.86	$154.27
Crew B-7	Hr.	Daily	Hr.	Daily	Bare Costs	Incl. O&P
1 Labor Foreman (outside)	$47.60	$380.80	$75.60	$604.80	$48.40	$76.58
4 Laborers	45.60	1459.20	72.45	2318.40		
1 Equip. Oper. (medium)	60.40	483.20	94.10	752.80		
1 Brush Chipper, 12", 130 H.P.		376.37		414.01		
1 Crawler Loader, 3 C.Y.		1180.92		1299.01		
2 Chain Saws, Gas, 36" Long		86.21		94.83	34.24	37.66
48 L.H., Daily Totals		$3966.70		$5483.84	$82.64	$114.25
Crew B-7A	Hr.	Daily	Hr.	Daily	Bare Costs	Incl. O&P
2 Laborers	$45.60	$729.60	$72.45	$1159.20	$49.40	$77.90
1 Equip. Oper. (light)	57.00	456.00	88.80	710.40		
1 Rake w/Tractor		354.21		389.63		
2 Chain Saws, Gas, 18"		107.75		118.52	19.25	21.17
24 L.H., Daily Totals		$1647.55		$2377.75	$68.65	$99.07
Crew B-7B	Hr.	Daily	Hr.	Daily	Bare Costs	Incl. O&P
1 Labor Foreman (outside)	$47.60	$380.80	$75.60	$604.80	$49.06	$77.72
4 Laborers	45.60	1459.20	72.45	2318.40		
1 Equip. Oper. (medium)	60.40	483.20	94.10	752.80		
1 Truck Driver (heavy)	53.05	424.40	84.55	676.40		
1 Brush Chipper, 12", 130 H.P.		376.37		414.01		
1 Crawler Loader, 3 C.Y.		1180.92		1299.01		
2 Chain Saws, Gas, 36" Long		86.21		94.83		
1 Dump Truck, 8 C.Y., 220 H.P.		419.41		461.35	36.84	40.52
56 L.H., Daily Totals		$4810.51		$6621.60	$85.90	$118.24

Crews - Renovation

Crew No.	Hr.	Daily	Hr.	Daily	Bare Costs	Incl. O&P
Crew B-7C						
1 Labor Foreman (outside)	$47.60	$380.80	$75.60	$604.80	$49.06	$77.72
4 Laborers	45.60	1459.20	72.45	2318.40		
1 Equip. Oper. (medium)	60.40	483.20	94.10	752.80		
1 Truck Driver (heavy)	53.05	424.40	84.55	676.40		
1 Brush Chipper, 12", 130 H.P.		376.37		414.01		
1 Crawler Loader, 3 C.Y.		1180.92		1299.01		
2 Chain Saws, Gas, 36" Long		86.21		94.83		
1 Dump Truck, 12 C.Y., 400 H.P.		594.66		654.12	39.97	43.96
56 L.H., Daily Totals		$4985.75		$6814.37	$89.03	$121.69

Crew No.	Hr.	Daily	Hr.	Daily	Bare Costs	Incl. O&P
Crew B-8						
1 Labor Foreman (outside)	$47.60	$380.80	$75.60	$604.80	$52.46	$82.74
2 Laborers	45.60	729.60	72.45	1159.20		
2 Equip. Oper. (medium)	60.40	966.40	94.10	1505.60		
1 Equip. Oper. (oiler)	54.00	432.00	84.15	673.20		
2 Truck Drivers (heavy)	53.05	848.80	84.55	1352.80		
1 Hyd. Crane, 25 Ton		603.44		663.78		
1 Crawler Loader, 3 C.Y.		1180.92		1299.01		
2 Dump Trucks, 12 C.Y., 400 H.P.		1189.32		1308.25	46.46	51.11
64 L.H., Daily Totals		$6331.27		$8566.64	$98.93	$133.85

Crew No.	Hr.	Daily	Hr.	Daily	Bare Costs	Incl. O&P
Crew B-9						
1 Labor Foreman (outside)	$47.60	$380.80	$75.60	$604.80	$46.00	$73.08
4 Laborers	45.60	1459.20	72.45	2318.40		
1 Air Compressor, 250 cfm		208.68		229.55		
2 Breakers, Pavement, 60 lb.		110.96		122.06		
2 -50' Air Hoses, 1.5"		47.19		51.91	9.17	10.09
40 L.H., Daily Totals		$2206.83		$3326.72	$55.17	$83.17

Crew No.	Hr.	Daily	Hr.	Daily	Bare Costs	Incl. O&P
Crew B-9A						
2 Laborers	$45.60	$729.60	$72.45	$1159.20	$48.08	$76.48
1 Truck Driver (heavy)	53.05	424.40	84.55	676.40		
1 Water Tank Trailer, 5000 Gal.		158.82		174.70		
1 Truck Tractor, 220 H.P.		319.23		351.15		
2 -50' Discharge Hoses, 3"		9.30		10.23	20.31	22.34
24 L.H., Daily Totals		$1641.35		$2371.69	$68.39	$98.82

Crew No.	Hr.	Daily	Hr.	Daily	Bare Costs	Incl. O&P
Crew B-9B						
2 Laborers	$45.60	$729.60	$72.45	$1159.20	$48.08	$76.48
1 Truck Driver (heavy)	53.05	424.40	84.55	676.40		
2 -50' Discharge Hoses, 3"		9.30		10.23		
1 Water Tank Trailer, 5000 Gal.		158.82		174.70		
1 Truck Tractor, 220 H.P.		319.23		351.15		
1 Pressure Washer		100.40		110.44	24.49	26.94
24 L.H., Daily Totals		$1741.75		$2482.12	$72.57	$103.42

Crew No.	Hr.	Daily	Hr.	Daily	Bare Costs	Incl. O&P
Crew B-9D						
1 Labor Foreman (outside)	$47.60	$380.80	$75.60	$604.80	$46.00	$73.08
4 Common Laborers	45.60	1459.20	72.45	2318.40		
1 Air Compressor, 250 cfm		208.68		229.55		
2 -50' Air Hoses, 1.5"		47.19		51.91		
2 Air Powered Tampers		82.32		90.55	8.45	9.30
40 L.H., Daily Totals		$2178.19		$3295.21	$54.45	$82.38

Crew No.	Hr.	Daily	Hr.	Daily	Bare Costs	Incl. O&P
Crew B-9E						
1 Cement Finisher	$53.15	$425.20	$83.05	$664.40	$49.38	$77.75
1 Laborer	45.60	364.80	72.45	579.60		
1 Chip. Hammers, 12 Lb., Elec.		33.98		37.37	2.12	2.34
16 L.H., Daily Totals		$823.98		$1281.37	$51.50	$80.09

Crew No.	Hr.	Daily	Hr.	Daily	Bare Costs	Incl. O&P
Crew B-10						
1 Equip. Oper. (medium)	$60.40	$483.20	$94.10	$752.80	$55.47	$86.88
.5 Laborer	45.60	182.40	72.45	289.80		
12 L.H., Daily Totals		$665.60		$1042.60	$55.47	$86.88

Crew No.	Hr.	Daily	Hr.	Daily	Bare Costs	Incl. O&P
Crew B-10A						
1 Equip. Oper. (medium)	$60.40	$483.20	$94.10	$752.80	$55.47	$86.88
.5 Laborer	45.60	182.40	72.45	289.80		
1 Roller, 2-Drum, W.B., 7.5 H.P.		171.95		189.14	14.33	15.76
12 L.H., Daily Totals		$837.55		$1231.74	$69.80	$102.65

Crew No.	Hr.	Daily	Hr.	Daily	Bare Costs	Incl. O&P
Crew B-10B						
1 Equip. Oper. (medium)	$60.40	$483.20	$94.10	$752.80	$55.47	$86.88
.5 Laborer	45.60	182.40	72.45	289.80		
1 Dozer, 200 H.P.		1567.46		1724.21	130.62	143.68
12 L.H., Daily Totals		$2233.06		$2766.81	$186.09	$230.57

Crew No.	Hr.	Daily	Hr.	Daily	Bare Costs	Incl. O&P
Crew B-10C						
1 Equip. Oper. (medium)	$60.40	$483.20	$94.10	$752.80	$55.47	$86.88
.5 Laborer	45.60	182.40	72.45	289.80		
1 Dozer, 200 H.P.		1567.46		1724.21		
1 Vibratory Roller, Towed, 23 Ton		536.14		589.75	175.30	192.83
12 L.H., Daily Totals		$2769.20		$3356.56	$230.77	$279.71

Crew No.	Hr.	Daily	Hr.	Daily	Bare Costs	Incl. O&P
Crew B-10D						
1 Equip. Oper. (medium)	$60.40	$483.20	$94.10	$752.80	$55.47	$86.88
.5 Laborer	45.60	182.40	72.45	289.80		
1 Dozer, 200 H.P.		1567.46		1724.21		
1 Sheepsft. Roller, Towed		439.25		483.18	167.23	183.95
12 L.H., Daily Totals		$2672.31		$3249.98	$222.69	$270.83

Crew No.	Hr.	Daily	Hr.	Daily	Bare Costs	Incl. O&P
Crew B-10E						
1 Equip. Oper. (medium)	$60.40	$483.20	$94.10	$752.80	$55.47	$86.88
.5 Laborer	45.60	182.40	72.45	289.80		
1 Tandem Roller, 5 Ton		267.23		293.96	22.27	24.50
12 L.H., Daily Totals		$932.83		$1336.56	$77.74	$111.38

Crew No.	Hr.	Daily	Hr.	Daily	Bare Costs	Incl. O&P
Crew B-10F						
1 Equip. Oper. (medium)	$60.40	$483.20	$94.10	$752.80	$55.47	$86.88
.5 Laborer	45.60	182.40	72.45	289.80		
1 Tandem Roller, 10 Ton		254.53		279.98	21.21	23.33
12 L.H., Daily Totals		$920.13		$1322.58	$76.68	$110.21

Crew No.	Hr.	Daily	Hr.	Daily	Bare Costs	Incl. O&P
Crew B-10G						
1 Equip. Oper. (medium)	$60.40	$483.20	$94.10	$752.80	$55.47	$86.88
.5 Laborer	45.60	182.40	72.45	289.80		
1 Sheepsfoot Roller, 240 H.P.		1404.71		1545.18	117.06	128.76
12 L.H., Daily Totals		$2070.31		$2587.78	$172.53	$215.65

Crew No.	Hr.	Daily	Hr.	Daily	Bare Costs	Incl. O&P
Crew B-10H						
1 Equip. Oper. (medium)	$60.40	$483.20	$94.10	$752.80	$55.47	$86.88
.5 Laborer	45.60	182.40	72.45	289.80		
1 Diaphragm Water Pump, 2"		90.40		99.44		
1 -20' Suction Hose, 2"		3.69		4.06		
2 -50' Discharge Hoses, 2"		8.25		9.08	8.53	9.38
12 L.H., Daily Totals		$767.95		$1155.18	$64.00	$96.27

For customer support on your **Commercial Renovation Costs** with RSMeans data, call 800.448.8182.

763

Crew No.	Bare Costs		Incl. Subs O&P		Cost Per Labor-Hour	
Crew B-10I	Hr.	Daily	Hr.	Daily	Bare Costs	Incl. O&P
1 Equip. Oper. (medium)	$60.40	$483.20	$94.10	$752.80	$55.47	$86.88
.5 Laborer	45.60	182.40	72.45	289.80		
1 Diaphragm Water Pump, 4"		109.44		120.39		
1 -20' Suction Hose, 4"		17.88		19.67		
2 -50' Discharge Hoses, 4"		26.56		29.22	12.82	14.11
12 L.H., Daily Totals		$819.49		$1211.87	$68.29	$100.99

Crew No.	Bare Costs		Incl. Subs O&P		Cost Per Labor-Hour	
Crew B-10J	Hr.	Daily	Hr.	Daily	Bare Costs	Incl. O&P
1 Equip. Oper. (medium)	$60.40	$483.20	$94.10	$752.80	$55.47	$86.88
.5 Laborer	45.60	182.40	72.45	289.80		
1 Centrifugal Water Pump, 3"		76.47		84.12		
1 -20' Suction Hose, 3"		9.06		9.97		
2 -50' Discharge Hoses, 3"		9.30		10.23	7.90	8.69
12 L.H., Daily Totals		$760.43		$1146.92	$63.37	$95.58

Crew No.	Bare Costs		Incl. Subs O&P		Cost Per Labor-Hour	
Crew B-10K	Hr.	Daily	Hr.	Daily	Bare Costs	Incl. O&P
1 Equip. Oper. (medium)	$60.40	$483.20	$94.10	$752.80	$55.47	$86.88
.5 Laborer	45.60	182.40	72.45	289.80		
1 Centr. Water Pump, 6"		240.92		265.01		
1 -20' Suction Hose, 6"		26.41		29.05		
2 -50' Discharge Hoses, 6"		37.54		41.29	25.41	27.95
12 L.H., Daily Totals		$970.46		$1377.95	$80.87	$114.83

Crew No.	Bare Costs		Incl. Subs O&P		Cost Per Labor-Hour	
Crew B-10L	Hr.	Daily	Hr.	Daily	Bare Costs	Incl. O&P
1 Equip. Oper. (medium)	$60.40	$483.20	$94.10	$752.80	$55.47	$86.88
.5 Laborer	45.60	182.40	72.45	289.80		
1 Dozer, 80 H.P.		417.40		459.14	34.78	38.26
12 L.H., Daily Totals		$1083.00		$1501.74	$90.25	$125.15

Crew No.	Bare Costs		Incl. Subs O&P		Cost Per Labor-Hour	
Crew B-10M	Hr.	Daily	Hr.	Daily	Bare Costs	Incl. O&P
1 Equip. Oper. (medium)	$60.40	$483.20	$94.10	$752.80	$55.47	$86.88
.5 Laborer	45.60	182.40	72.45	289.80		
1 Dozer, 300 H.P.		1840.03		2024.04	153.34	168.67
12 L.H., Daily Totals		$2505.63		$3066.64	$208.80	$255.55

Crew No.	Bare Costs		Incl. Subs O&P		Cost Per Labor-Hour	
Crew B-10N	Hr.	Daily	Hr.	Daily	Bare Costs	Incl. O&P
1 Equip. Oper. (medium)	$60.40	$483.20	$94.10	$752.80	$55.47	$86.88
.5 Laborer	45.60	182.40	72.45	289.80		
1 F.E. Loader, T.M., 1.5 C.Y.		589.09		648.00	49.09	54.00
12 L.H., Daily Totals		$1254.69		$1690.60	$104.56	$140.88

Crew No.	Bare Costs		Incl. Subs O&P		Cost Per Labor-Hour	
Crew B-10O	Hr.	Daily	Hr.	Daily	Bare Costs	Incl. O&P
1 Equip. Oper. (medium)	$60.40	$483.20	$94.10	$752.80	$55.47	$86.88
.5 Laborer	45.60	182.40	72.45	289.80		
1 F.E. Loader, T.M., 2.25 C.Y.		953.18		1048.50	79.43	87.37
12 L.H., Daily Totals		$1618.78		$2091.10	$134.90	$174.26

Crew No.	Bare Costs		Incl. Subs O&P		Cost Per Labor-Hour	
Crew B-10P	Hr.	Daily	Hr.	Daily	Bare Costs	Incl. O&P
1 Equip. Oper. (medium)	$60.40	$483.20	$94.10	$752.80	$55.47	$86.88
.5 Laborer	45.60	182.40	72.45	289.80		
1 Crawler Loader, 3 C.Y.		1180.92		1299.01	98.41	108.25
12 L.H., Daily Totals		$1846.52		$2341.61	$153.88	$195.13

Crew No.	Bare Costs		Incl. Subs O&P		Cost Per Labor-Hour	
Crew B-10Q	Hr.	Daily	Hr.	Daily	Bare Costs	Incl. O&P
1 Equip. Oper. (medium)	$60.40	$483.20	$94.10	$752.80	$55.47	$86.88
.5 Laborer	45.60	182.40	72.45	289.80		
1 Crawler Loader, 4 C.Y.		1500.46		1650.51	125.04	137.54
12 L.H., Daily Totals		$2166.06		$2693.11	$180.51	$224.43

Crew No.	Bare Costs		Incl. Subs O&P		Cost Per Labor-Hour	
Crew B-10R	Hr.	Daily	Hr.	Daily	Bare Costs	Incl. O&P
1 Equip. Oper. (medium)	$60.40	$483.20	$94.10	$752.80	$55.47	$86.88
.5 Laborer	45.60	182.40	72.45	289.80		
1 F.E. Loader, W.M., 1 C.Y.		314.63		346.09	26.22	28.84
12 L.H., Daily Totals		$980.23		$1388.69	$81.69	$115.72

Crew No.	Bare Costs		Incl. Subs O&P		Cost Per Labor-Hour	
Crew B-10S	Hr.	Daily	Hr.	Daily	Bare Costs	Incl. O&P
1 Equip. Oper. (medium)	$60.40	$483.20	$94.10	$752.80	$55.47	$86.88
.5 Laborer	45.60	182.40	72.45	289.80		
1 F.E. Loader, W.M., 1.5 C.Y.		454.98		500.48	37.92	41.71
12 L.H., Daily Totals		$1120.58		$1543.08	$93.38	$128.59

Crew No.	Bare Costs		Incl. Subs O&P		Cost Per Labor-Hour	
Crew B-10T	Hr.	Daily	Hr.	Daily	Bare Costs	Incl. O&P
1 Equip. Oper. (medium)	$60.40	$483.20	$94.10	$752.80	$55.47	$86.88
.5 Laborer	45.60	182.40	72.45	289.80		
1 F.E. Loader, W.M., 2.5 C.Y.		657.89		723.68	54.82	60.31
12 L.H., Daily Totals		$1323.49		$1766.28	$110.29	$147.19

Crew No.	Bare Costs		Incl. Subs O&P		Cost Per Labor-Hour	
Crew B-10U	Hr.	Daily	Hr.	Daily	Bare Costs	Incl. O&P
1 Equip. Oper. (medium)	$60.40	$483.20	$94.10	$752.80	$55.47	$86.88
.5 Laborer	45.60	182.40	72.45	289.80		
1 F.E. Loader, W.M., 5.5 C.Y.		996.48		1096.13	83.04	91.34
12 L.H., Daily Totals		$1662.08		$2138.73	$138.51	$178.23

Crew No.	Bare Costs		Incl. Subs O&P		Cost Per Labor-Hour	
Crew B-10V	Hr.	Daily	Hr.	Daily	Bare Costs	Incl. O&P
1 Equip. Oper. (medium)	$60.40	$483.20	$94.10	$752.80	$55.47	$86.88
.5 Laborer	45.60	182.40	72.45	289.80		
1 Dozer, 700 H.P.		5335.27		5868.80	444.61	489.07
12 L.H., Daily Totals		$6000.87		$6911.40	$500.07	$575.95

Crew No.	Bare Costs		Incl. Subs O&P		Cost Per Labor-Hour	
Crew B-10W	Hr.	Daily	Hr.	Daily	Bare Costs	Incl. O&P
1 Equip. Oper. (medium)	$60.40	$483.20	$94.10	$752.80	$55.47	$86.88
.5 Laborer	45.60	182.40	72.45	289.80		
1 Dozer, 105 H.P.		659.79		725.77	54.98	60.48
12 L.H., Daily Totals		$1325.39		$1768.37	$110.45	$147.36

Crew No.	Bare Costs		Incl. Subs O&P		Cost Per Labor-Hour	
Crew B-10X	Hr.	Daily	Hr.	Daily	Bare Costs	Incl. O&P
1 Equip. Oper. (medium)	$60.40	$483.20	$94.10	$752.80	$55.47	$86.88
.5 Laborer	45.60	182.40	72.45	289.80		
1 Dozer, 410 H.P.		2896.31		3185.94	241.36	265.50
12 L.H., Daily Totals		$3561.91		$4228.54	$296.83	$352.38

Crew No.	Bare Costs		Incl. Subs O&P		Cost Per Labor-Hour	
Crew B-10Y	Hr.	Daily	Hr.	Daily	Bare Costs	Incl. O&P
1 Equip. Oper. (medium)	$60.40	$483.20	$94.10	$752.80	$55.47	$86.88
.5 Laborer	45.60	182.40	72.45	289.80		
1 Vibr. Roller, Towed, 12 Ton		603.24		663.57	50.27	55.30
12 L.H., Daily Totals		$1268.84		$1706.17	$105.74	$142.18

Crew No.	Bare Costs		Incl. Subs O&P		Cost Per Labor-Hour	
Crew B-11A	Hr.	Daily	Hr.	Daily	Bare Costs	Incl. O&P
1 Equipment Oper. (med.)	$60.40	$483.20	$94.10	$752.80	$53.00	$83.28
1 Laborer	45.60	364.80	72.45	579.60		
1 Dozer, 200 H.P.		1567.46		1724.21	97.97	107.76
16 L.H., Daily Totals		$2415.46		$3056.61	$150.97	$191.04

Crew No.	Bare Costs		Incl. Subs O&P		Cost Per Labor-Hour	
Crew B-11B	Hr.	Daily	Hr.	Daily	Bare Costs	Incl. O&P
1 Equipment Oper. (light)	$57.00	$456.00	$88.80	$710.40	$51.30	$80.63
1 Laborer	45.60	364.80	72.45	579.60		
1 Air Powered Tamper		41.16		45.28		
1 Air Compressor, 365 cfm		354.03		389.44		
2 -50' Air Hoses, 1.5"		47.19		51.91	27.65	30.41
16 L.H., Daily Totals		$1263.18		$1776.62	$78.95	$111.04

For customer support on your Commercial Renovation Costs with RSMeans data, call 800.448.8182.

Crew No.	Bare Costs		Incl. Subs O&P		Cost Per Labor-Hour	

Crew B-11C	Hr.	Daily	Hr.	Daily	Bare Costs	Incl. O&P
1 Equipment Oper. (med.)	$60.40	$483.20	$94.10	$752.80	$53.00	$83.28
1 Laborer	45.60	364.80	72.45	579.60		
1 Backhoe Loader, 48 H.P.		221.72		243.89	13.86	15.24
16 L.H., Daily Totals		$1069.72		$1576.29	$66.86	$98.52

Crew B-11K	Hr.	Daily	Hr.	Daily	Bare Costs	Incl. O&P
1 Equipment Oper. (med.)	$60.40	$483.20	$94.10	$752.80	$53.00	$83.28
1 Laborer	45.60	364.80	72.45	579.60		
1 Trencher, Chain Type, 8' D		1952.14		2147.36	122.01	134.21
16 L.H., Daily Totals		$2800.14		$3479.76	$175.01	$217.48

Crew B-11L	Hr.	Daily	Hr.	Daily	Bare Costs	Incl. O&P
1 Equipment Oper. (med.)	$60.40	$483.20	$94.10	$752.80	$53.00	$83.28
1 Laborer	45.60	364.80	72.45	579.60		
1 Grader, 30,000 Lbs.		1108.49		1219.34	69.28	76.21
16 L.H., Daily Totals		$1956.49		$2551.74	$122.28	$159.48

Crew B-11M	Hr.	Daily	Hr.	Daily	Bare Costs	Incl. O&P
1 Equipment Oper. (med.)	$60.40	$483.20	$94.10	$752.80	$53.00	$83.28
1 Laborer	45.60	364.80	72.45	579.60		
1 Backhoe Loader, 80 H.P.		240.94		265.04	15.06	16.56
16 L.H., Daily Totals		$1088.94		$1597.44	$68.06	$99.84

Crew B-11S	Hr.	Daily	Hr.	Daily	Bare Costs	Incl. O&P
1 Equipment Operator (med.)	$60.40	$483.20	$94.10	$752.80	$55.47	$86.88
.5 Laborer	45.60	182.40	72.45	289.80		
1 Dozer, 300 H.P.		1840.03		2024.04		
1 Ripper, Beam & 1 Shank		94.41		103.85	161.20	177.32
12 L.H., Daily Totals		$2600.04		$3170.49	$216.67	$264.21

Crew B-11T	Hr.	Daily	Hr.	Daily	Bare Costs	Incl. O&P
1 Equipment Operator (med.)	$60.40	$483.20	$94.10	$752.80	$55.47	$86.88
.5 Laborer	45.60	182.40	72.45	289.80		
1 Dozer, 410 H.P.		2896.31		3185.94		
1 Ripper, Beam & 2 Shanks		144.78		159.26	253.42	278.77
12 L.H., Daily Totals		$3706.69		$4387.80	$308.89	$365.65

Crew B-11W	Hr.	Daily	Hr.	Daily	Bare Costs	Incl. O&P
1 Equipment Operator (med.)	$60.40	$483.20	$94.10	$752.80	$53.04	$84.34
1 Common Laborer	45.60	364.80	72.45	579.60		
10 Truck Drivers (heavy)	53.05	4244.00	84.55	6764.00		
1 Dozer, 200 H.P.		1567.46		1724.21		
1 Vibratory Roller, Towed, 23 Ton		536.14		589.75		
10 Dump Trucks, 8 C.Y., 220 H.P.		4194.13		4613.54	65.60	72.16
96 L.H., Daily Totals		$11389.73		$15023.90	$118.64	$156.50

Crew B-11Y	Hr.	Daily	Hr.	Daily	Bare Costs	Incl. O&P
1 Labor Foreman (outside)	$47.60	$380.80	$75.60	$604.80	$50.76	$80.02
5 Common Laborers	45.60	1824.00	72.45	2898.00		
3 Equipment Operators (med.)	60.40	1449.60	94.10	2258.40		
1 Dozer, 80 H.P.		417.40		459.14		
2 Rollers, 2-Drum, W.B., 7.5 H.P.		343.89		378.28		
4 Vibrating Plates, Gas, 21"		685.57		754.13	20.10	22.10
72 L.H., Daily Totals		$5101.27		$7352.75	$70.85	$102.12

Crew B-12A	Hr.	Daily	Hr.	Daily	Bare Costs	Incl. O&P
1 Equip. Oper. (crane)	$63.40	$507.20	$98.75	$790.00	$54.50	$85.60
1 Laborer	45.60	364.80	72.45	579.60		
1 Hyd. Excavator, 1 C.Y.		857.81		943.59	53.61	58.97
16 L.H., Daily Totals		$1729.81		$2313.19	$108.11	$144.57

Crew B-12B	Hr.	Daily	Hr.	Daily	Bare Costs	Incl. O&P
1 Equip. Oper. (crane)	$63.40	$507.20	$98.75	$790.00	$54.50	$85.60
1 Laborer	45.60	364.80	72.45	579.60		
1 Hyd. Excavator, 1.5 C.Y.		714.87		786.36	44.68	49.15
16 L.H., Daily Totals		$1586.87		$2155.96	$99.18	$134.75

Crew B-12C	Hr.	Daily	Hr.	Daily	Bare Costs	Incl. O&P
1 Equip. Oper. (crane)	$63.40	$507.20	$98.75	$790.00	$54.50	$85.60
1 Laborer	45.60	364.80	72.45	579.60		
1 Hyd. Excavator, 2 C.Y.		969.84		1066.82	60.61	66.68
16 L.H., Daily Totals		$1841.84		$2436.42	$115.11	$152.28

Crew B-12D	Hr.	Daily	Hr.	Daily	Bare Costs	Incl. O&P
1 Equip. Oper. (crane)	$63.40	$507.20	$98.75	$790.00	$54.50	$85.60
1 Laborer	45.60	364.80	72.45	579.60		
1 Hyd. Excavator, 3.5 C.Y.		2248.30		2473.13	140.52	154.57
16 L.H., Daily Totals		$3120.30		$3842.73	$195.02	$240.17

Crew B-12E	Hr.	Daily	Hr.	Daily	Bare Costs	Incl. O&P
1 Equip. Oper. (crane)	$63.40	$507.20	$98.75	$790.00	$54.50	$85.60
1 Laborer	45.60	364.80	72.45	579.60		
1 Hyd. Excavator, .5 C.Y.		470.93		518.02	29.43	32.38
16 L.H., Daily Totals		$1342.93		$1887.62	$83.93	$117.98

Crew B-12F	Hr.	Daily	Hr.	Daily	Bare Costs	Incl. O&P
1 Equip. Oper. (crane)	$63.40	$507.20	$98.75	$790.00	$54.50	$85.60
1 Laborer	45.60	364.80	72.45	579.60		
1 Hyd. Excavator, .75 C.Y.		723.26		795.59	45.20	49.72
16 L.H., Daily Totals		$1595.26		$2165.19	$99.70	$135.32

Crew B-12G	Hr.	Daily	Hr.	Daily	Bare Costs	Incl. O&P
1 Equip. Oper. (crane)	$63.40	$507.20	$98.75	$790.00	$54.50	$85.60
1 Laborer	45.60	364.80	72.45	579.60		
1 Crawler Crane, 15 Ton		835.19		918.71		
1 Clamshell Bucket, .5 C.Y.		70.08		77.09	56.58	62.24
16 L.H., Daily Totals		$1777.28		$2365.41	$111.08	$147.84

Crew B-12H	Hr.	Daily	Hr.	Daily	Bare Costs	Incl. O&P
1 Equip. Oper. (crane)	$63.40	$507.20	$98.75	$790.00	$54.50	$85.60
1 Laborer	45.60	364.80	72.45	579.60		
1 Crawler Crane, 25 Ton		1185.81		1304.39		
1 Clamshell Bucket, 1 C.Y.		71.59		78.75	78.59	86.45
16 L.H., Daily Totals		$2129.40		$2752.74	$133.09	$172.05

Crew B-12I	Hr.	Daily	Hr.	Daily	Bare Costs	Incl. O&P
1 Equip. Oper. (crane)	$63.40	$507.20	$98.75	$790.00	$54.50	$85.60
1 Laborer	45.60	364.80	72.45	579.60		
1 Crawler Crane, 20 Ton		1042.58		1146.84		
1 Dragline Bucket, .75 C.Y.		64.03		70.44	69.16	76.08
16 L.H., Daily Totals		$1978.62		$2586.88	$123.66	$161.68

Crew B-12J	Hr.	Daily	Hr.	Daily	Bare Costs	Incl. O&P
1 Equip. Oper. (crane)	$63.40	$507.20	$98.75	$790.00	$54.50	$85.60
1 Laborer	45.60	364.80	72.45	579.60		
1 Gradall, 5/8 C.Y.		876.16		963.78	54.76	60.24
16 L.H., Daily Totals		$1748.16		$2333.38	$109.26	$145.84

For customer support on your Commercial Renovation Costs with RSMeans data, call 800.448.8182.

765

Crews - Renovation

Crew B-12K

Crew No.	Bare Costs Hr.	Bare Costs Daily	Incl. Subs O&P Hr.	Incl. Subs O&P Daily	Cost Per Labor-Hour Bare Costs	Cost Per Labor-Hour Incl. O&P
1 Equip. Oper. (crane)	$63.40	$507.20	$98.75	$790.00	$54.50	$85.60
1 Laborer	45.60	364.80	72.45	579.60		
1 Gradall, 3 Ton, 1 C.Y.		1012.80		1114.08	63.30	69.63
16 L.H., Daily Totals		$1884.80		$2483.68	$117.80	$155.23

Crew B-12L

Crew No.	Bare Costs Hr.	Bare Costs Daily	Incl. Subs O&P Hr.	Incl. Subs O&P Daily	Cost Per Labor-Hour Bare Costs	Cost Per Labor-Hour Incl. O&P
1 Equip. Oper. (crane)	$63.40	$507.20	$98.75	$790.00	$54.50	$85.60
1 Laborer	45.60	364.80	72.45	579.60		
1 Crawler Crane, 15 Ton		835.19		918.71		
1 F.E. Attachment, .5 C.Y.		68.13		74.94	56.46	62.10
16 L.H., Daily Totals		$1775.32		$2363.25	$110.96	$147.70

Crew B-12M

Crew No.	Bare Costs Hr.	Bare Costs Daily	Incl. Subs O&P Hr.	Incl. Subs O&P Daily	Cost Per Labor-Hour Bare Costs	Cost Per Labor-Hour Incl. O&P
1 Equip. Oper. (crane)	$63.40	$507.20	$98.75	$790.00	$54.50	$85.60
1 Laborer	45.60	364.80	72.45	579.60		
1 Crawler Crane, 20 Ton		1042.58		1146.84		
1 F.E. Attachment, .75 C.Y.		73.54		80.89	69.76	76.73
16 L.H., Daily Totals		$1988.12		$2597.33	$124.26	$162.33

Crew B-12N

Crew No.	Bare Costs Hr.	Bare Costs Daily	Incl. Subs O&P Hr.	Incl. Subs O&P Daily	Cost Per Labor-Hour Bare Costs	Cost Per Labor-Hour Incl. O&P
1 Equip. Oper. (crane)	$63.40	$507.20	$98.75	$790.00	$54.50	$85.60
1 Laborer	45.60	364.80	72.45	579.60		
1 Crawler Crane, 25 Ton		1185.81		1304.39		
1 F.E. Attachment, 1 C.Y.		79.82		87.80	79.10	87.01
16 L.H., Daily Totals		$2137.63		$2761.79	$133.60	$172.61

Crew B-12O

Crew No.	Bare Costs Hr.	Bare Costs Daily	Incl. Subs O&P Hr.	Incl. Subs O&P Daily	Cost Per Labor-Hour Bare Costs	Cost Per Labor-Hour Incl. O&P
1 Equip. Oper. (crane)	$63.40	$507.20	$98.75	$790.00	$54.50	$85.60
1 Laborer	45.60	364.80	72.45	579.60		
1 Crawler Crane, 40 Ton		1267.52		1394.27		
1 F.E. Attachment, 1.5 C.Y.		91.52		100.67	84.94	93.43
16 L.H., Daily Totals		$2231.04		$2864.54	$139.44	$179.03

Crew B-12P

Crew No.	Bare Costs Hr.	Bare Costs Daily	Incl. Subs O&P Hr.	Incl. Subs O&P Daily	Cost Per Labor-Hour Bare Costs	Cost Per Labor-Hour Incl. O&P
1 Equip. Oper. (crane)	$63.40	$507.20	$98.75	$790.00	$54.50	$85.60
1 Laborer	45.60	364.80	72.45	579.60		
1 Crawler Crane, 40 Ton		1267.52		1394.27		
1 Dragline Bucket, 1.5 C.Y.		67.96		74.76	83.47	91.81
16 L.H., Daily Totals		$2207.48		$2838.63	$137.97	$177.41

Crew B-12Q

Crew No.	Bare Costs Hr.	Bare Costs Daily	Incl. Subs O&P Hr.	Incl. Subs O&P Daily	Cost Per Labor-Hour Bare Costs	Cost Per Labor-Hour Incl. O&P
1 Equip. Oper. (crane)	$63.40	$507.20	$98.75	$790.00	$54.50	$85.60
1 Laborer	45.60	364.80	72.45	579.60		
1 Hyd. Excavator, 5/8 C.Y.		623.13		685.45	38.95	42.84
16 L.H., Daily Totals		$1495.13		$2055.05	$93.45	$128.44

Crew B-12S

Crew No.	Bare Costs Hr.	Bare Costs Daily	Incl. Subs O&P Hr.	Incl. Subs O&P Daily	Cost Per Labor-Hour Bare Costs	Cost Per Labor-Hour Incl. O&P
1 Equip. Oper. (crane)	$63.40	$507.20	$98.75	$790.00	$54.50	$85.60
1 Laborer	45.60	364.80	72.45	579.60		
1 Hyd. Excavator, 2.5 C.Y.		1613.66		1775.03	100.85	110.94
16 L.H., Daily Totals		$2485.66		$3144.63	$155.35	$196.54

Crew B-12T

Crew No.	Bare Costs Hr.	Bare Costs Daily	Incl. Subs O&P Hr.	Incl. Subs O&P Daily	Cost Per Labor-Hour Bare Costs	Cost Per Labor-Hour Incl. O&P
1 Equip. Oper. (crane)	$63.40	$507.20	$98.75	$790.00	$54.50	$85.60
1 Laborer	45.60	364.80	72.45	579.60		
1 Crawler Crane, 75 Ton		2031.06		2234.16		
1 F.E. Attachment, 3 C.Y.		119.34		131.28	134.40	147.84
16 L.H., Daily Totals		$3022.40		$3735.04	$188.90	$233.44

Crew B-12V

Crew No.	Bare Costs Hr.	Bare Costs Daily	Incl. Subs O&P Hr.	Incl. Subs O&P Daily	Cost Per Labor-Hour Bare Costs	Cost Per Labor-Hour Incl. O&P
1 Equip. Oper. (crane)	$63.40	$507.20	$98.75	$790.00	$54.50	$85.60
1 Laborer	45.60	364.80	72.45	579.60		
1 Crawler Crane, 75 Ton		2031.06		2234.16		
1 Dragline Bucket, 3 C.Y.		74.71		82.18	131.61	144.77
16 L.H., Daily Totals		$2977.77		$3685.95	$186.11	$230.37

Crew B-12Y

Crew No.	Bare Costs Hr.	Bare Costs Daily	Incl. Subs O&P Hr.	Incl. Subs O&P Daily	Cost Per Labor-Hour Bare Costs	Cost Per Labor-Hour Incl. O&P
1 Equip. Oper. (crane)	$63.40	$507.20	$98.75	$790.00	$51.53	$81.22
2 Laborers	45.60	729.60	72.45	1159.20		
1 Hyd. Excavator, 3.5 C.Y.		2248.30		2473.13	93.68	103.05
24 L.H., Daily Totals		$3485.10		$4422.33	$145.21	$184.26

Crew B-12Z

Crew No.	Bare Costs Hr.	Bare Costs Daily	Incl. Subs O&P Hr.	Incl. Subs O&P Daily	Cost Per Labor-Hour Bare Costs	Cost Per Labor-Hour Incl. O&P
1 Equip. Oper. (crane)	$63.40	$507.20	$98.75	$790.00	$51.53	$81.22
2 Laborers	45.60	729.60	72.45	1159.20		
1 Hyd. Excavator, 2.5 C.Y.		1613.66		1775.03	67.24	73.96
24 L.H., Daily Totals		$2850.46		$3724.23	$118.77	$155.18

Crew B-13

Crew No.	Bare Costs Hr.	Bare Costs Daily	Incl. Subs O&P Hr.	Incl. Subs O&P Daily	Cost Per Labor-Hour Bare Costs	Cost Per Labor-Hour Incl. O&P
1 Labor Foreman (outside)	$47.60	$380.80	$75.60	$604.80	$49.63	$78.33
4 Laborers	45.60	1459.20	72.45	2318.40		
1 Equip. Oper. (crane)	63.40	507.20	98.75	790.00		
1 Equip. Oper. (oiler)	54.00	432.00	84.15	673.20		
1 Hyd. Crane, 25 Ton		603.44		663.78	10.78	11.85
56 L.H., Daily Totals		$3382.64		$5050.18	$60.40	$90.18

Crew B-13A

Crew No.	Bare Costs Hr.	Bare Costs Daily	Incl. Subs O&P Hr.	Incl. Subs O&P Daily	Cost Per Labor-Hour Bare Costs	Cost Per Labor-Hour Incl. O&P
1 Labor Foreman (outside)	$47.60	$380.80	$75.60	$604.80	$52.24	$82.54
2 Laborers	45.60	729.60	72.45	1159.20		
2 Equipment Operators (med.)	60.40	966.40	94.10	1505.60		
2 Truck Drivers (heavy)	53.05	848.80	84.55	1352.80		
1 Crawler Crane, 75 Ton		2031.06		2234.16		
1 Crawler Loader, 4 C.Y.		1500.46		1650.51		
2 Dump Trucks, 8 C.Y., 220 H.P.		838.83		922.71	78.04	85.85
56 L.H., Daily Totals		$7295.94		$9429.78	$130.28	$168.39

Crew B-13B

Crew No.	Bare Costs Hr.	Bare Costs Daily	Incl. Subs O&P Hr.	Incl. Subs O&P Daily	Cost Per Labor-Hour Bare Costs	Cost Per Labor-Hour Incl. O&P
1 Labor Foreman (outside)	$47.60	$380.80	$75.60	$604.80	$49.63	$78.33
4 Laborers	45.60	1459.20	72.45	2318.40		
1 Equip. Oper. (crane)	63.40	507.20	98.75	790.00		
1 Equip. Oper. (oiler)	54.00	432.00	84.15	673.20		
1 Hyd. Crane, 55 Ton		1019.39		1121.33	18.20	20.02
56 L.H., Daily Totals		$3798.59		$5507.73	$67.83	$98.35

Crew B-13C

Crew No.	Bare Costs Hr.	Bare Costs Daily	Incl. Subs O&P Hr.	Incl. Subs O&P Daily	Cost Per Labor-Hour Bare Costs	Cost Per Labor-Hour Incl. O&P
1 Labor Foreman (outside)	$47.60	$380.80	$75.60	$604.80	$49.63	$78.33
4 Laborers	45.60	1459.20	72.45	2318.40		
1 Equip. Oper. (crane)	63.40	507.20	98.75	790.00		
1 Equip. Oper. (oiler)	54.00	432.00	84.15	673.20		
1 Crawler Crane, 100 Ton		2384.09		2622.50	42.57	46.83
56 L.H., Daily Totals		$5163.29		$7008.90	$92.20	$125.16

Crew B-13D

Crew No.	Bare Costs Hr.	Bare Costs Daily	Incl. Subs O&P Hr.	Incl. Subs O&P Daily	Cost Per Labor-Hour Bare Costs	Cost Per Labor-Hour Incl. O&P
1 Laborer	$45.60	$364.80	$72.45	$579.60	$54.50	$85.60
1 Equip. Oper. (crane)	63.40	507.20	98.75	790.00		
1 Hyd. Excavator, 1 C.Y.		857.81		943.59		
1 Trench Box		123.46		135.81	61.33	67.46
16 L.H., Daily Totals		$1853.27		$2449.00	$115.83	$153.06

For customer support on your Commercial Renovation Costs with RSMeans data, call 800.448.8182.

Crews - Renovation

Crew B-13E

Crew B-13E	Hr.	Daily	Hr.	Daily	Bare Costs	Incl. O&P
1 Laborer	$45.60	$364.80	$72.45	$579.60	$54.50	$85.60
1 Equip. Oper. (crane)	63.40	507.20	98.75	790.00		
1 Hyd. Excavator, 1.5 C.Y.		714.87		786.36		
1 Trench Box		123.46		135.81	52.40	57.64
16 L.H., Daily Totals		$1710.33		$2291.76	$106.90	$143.24

Crew B-13F

Crew B-13F	Hr.	Daily	Hr.	Daily	Bare Costs	Incl. O&P
1 Laborer	$45.60	$364.80	$72.45	$579.60	$54.50	$85.60
1 Equip. Oper. (crane)	63.40	507.20	98.75	790.00		
1 Hyd. Excavator, 3.5 C.Y.		2248.30		2473.13		
1 Trench Box		123.46		135.81	148.24	163.06
16 L.H., Daily Totals		$3243.76		$3978.54	$202.74	$248.66

Crew B-13G

Crew B-13G	Hr.	Daily	Hr.	Daily	Bare Costs	Incl. O&P
1 Laborer	$45.60	$364.80	$72.45	$579.60	$54.50	$85.60
1 Equip. Oper. (crane)	63.40	507.20	98.75	790.00		
1 Hyd. Excavator, .75 C.Y.		723.26		795.59		
1 Trench Box		123.46		135.81	52.92	58.21
16 L.H., Daily Totals		$1718.73		$2301.00	$107.42	$143.81

Crew B-13H

Crew B-13H	Hr.	Daily	Hr.	Daily	Bare Costs	Incl. O&P
1 Laborer	$45.60	$364.80	$72.45	$579.60	$54.50	$85.60
1 Equip. Oper. (crane)	63.40	507.20	98.75	790.00		
1 Gradall, 5/8 C.Y.		876.16		963.78		
1 Trench Box		123.46		135.81	62.48	68.72
16 L.H., Daily Totals		$1871.62		$2469.19	$116.98	$154.32

Crew B-13I

Crew B-13I	Hr.	Daily	Hr.	Daily	Bare Costs	Incl. O&P
1 Laborer	$45.60	$364.80	$72.45	$579.60	$54.50	$85.60
1 Equip. Oper. (crane)	63.40	507.20	98.75	790.00		
1 Gradall, 3 Ton, 1 C.Y.		1012.80		1114.08		
1 Trench Box		123.46		135.81	71.02	78.12
16 L.H., Daily Totals		$2008.27		$2619.49	$125.52	$163.72

Crew B-13J

Crew B-13J	Hr.	Daily	Hr.	Daily	Bare Costs	Incl. O&P
1 Laborer	$45.60	$364.80	$72.45	$579.60	$54.50	$85.60
1 Equip. Oper. (crane)	63.40	507.20	98.75	790.00		
1 Hyd. Excavator, 2.5 C.Y.		1613.66		1775.03		
1 Trench Box		123.46		135.81	108.57	119.43
16 L.H., Daily Totals		$2609.13		$3280.44	$163.07	$205.03

Crew B-13K

Crew B-13K	Hr.	Daily	Hr.	Daily	Bare Costs	Incl. O&P
2 Equip. Opers. (crane)	$63.40	$1014.40	$98.75	$1580.00	$63.40	$98.75
1 Hyd. Excavator, .75 C.Y.		723.26		795.59		
1 Hyd. Hammer, 4000 ft-lb		671.39		738.52		
1 Hyd. Excavator, .75 C.Y.		723.26		795.59	132.37	145.61
16 L.H., Daily Totals		$3132.31		$3909.70	$195.77	$244.36

Crew B-13L

Crew B-13L	Hr.	Daily	Hr.	Daily	Bare Costs	Incl. O&P
2 Equip. Opers. (crane)	$63.40	$1014.40	$98.75	$1580.00	$63.40	$98.75
1 Hyd. Excavator, 1.5 C.Y.		714.87		786.36		
1 Hyd. Hammer, 5000 ft-lb		729.52		802.47		
1 Hyd. Excavator, .75 C.Y.		723.26		795.59	135.48	149.03
16 L.H., Daily Totals		$3182.05		$3964.42	$198.88	$247.78

Crew B-13M

Crew B-13M	Hr.	Daily	Hr.	Daily	Bare Costs	Incl. O&P
2 Equip. Opers. (crane)	$63.40	$1014.40	$98.75	$1580.00	$63.40	$98.75
1 Hyd. Excavator, 2.5 C.Y.		1613.66		1775.03		
1 Hyd. Hammer, 8000 ft-lb		949.42		1044.36		
1 Hyd. Excavator, 1.5 C.Y.		714.87		786.36	204.87	225.36
16 L.H., Daily Totals		$4292.35		$5185.74	$268.27	$324.11

Crew B-13N

Crew B-13N	Hr.	Daily	Hr.	Daily	Bare Costs	Incl. O&P
2 Equip. Opers. (crane)	$63.40	$1014.40	$98.75	$1580.00	$63.40	$98.75
1 Hyd. Excavator, 3.5 C.Y.		2248.30		2473.13		
1 Hyd. Hammer, 12,000 ft-lb		911.33		1002.46		
1 Hyd. Excavator, 1.5 C.Y.		714.87		786.36	242.16	266.37
16 L.H., Daily Totals		$4888.90		$5841.94	$305.56	$365.12

Crew B-14

Crew B-14	Hr.	Daily	Hr.	Daily	Bare Costs	Incl. O&P
1 Labor Foreman (outside)	$47.60	$380.80	$75.60	$604.80	$47.83	$75.70
4 Laborers	45.60	1459.20	72.45	2318.40		
1 Equip. Oper. (light)	57.00	456.00	88.80	710.40		
1 Backhoe Loader, 48 H.P.		221.72		243.89	4.62	5.08
48 L.H., Daily Totals		$2517.72		$3877.49	$52.45	$80.78

Crew B-14A

Crew B-14A	Hr.	Daily	Hr.	Daily	Bare Costs	Incl. O&P
1 Equip. Oper. (crane)	$63.40	$507.20	$98.75	$790.00	$57.47	$89.98
.5 Laborer	45.60	182.40	72.45	289.80		
1 Hyd. Excavator, 4.5 C.Y.		3557.16		3912.87	296.43	326.07
12 L.H., Daily Totals		$4246.76		$4992.67	$353.90	$416.06

Crew B-14B

Crew B-14B	Hr.	Daily	Hr.	Daily	Bare Costs	Incl. O&P
1 Equip. Oper. (crane)	$63.40	$507.20	$98.75	$790.00	$57.47	$89.98
.5 Laborer	45.60	182.40	72.45	289.80		
1 Hyd. Excavator, 6 C.Y.		3609.36		3970.30	300.78	330.86
12 L.H., Daily Totals		$4298.96		$5050.10	$358.25	$420.84

Crew B-14C

Crew B-14C	Hr.	Daily	Hr.	Daily	Bare Costs	Incl. O&P
1 Equip. Oper. (crane)	$63.40	$507.20	$98.75	$790.00	$57.47	$89.98
.5 Laborer	45.60	182.40	72.45	289.80		
1 Hyd. Excavator, 7 C.Y.		3579.34		3937.27	298.28	328.11
12 L.H., Daily Totals		$4268.94		$5017.07	$355.74	$418.09

Crew B-14F

Crew B-14F	Hr.	Daily	Hr.	Daily	Bare Costs	Incl. O&P
1 Equip. Oper. (crane)	$63.40	$507.20	$98.75	$790.00	$57.47	$89.98
.5 Laborer	45.60	182.40	72.45	289.80		
1 Hyd. Shovel, 7 C.Y.		4271.41		4698.55	355.95	391.55
12 L.H., Daily Totals		$4961.01		$5778.35	$413.42	$481.53

Crew B-14G

Crew B-14G	Hr.	Daily	Hr.	Daily	Bare Costs	Incl. O&P
1 Equip. Oper. (crane)	$63.40	$507.20	$98.75	$790.00	$57.47	$89.98
.5 Laborer	45.60	182.40	72.45	289.80		
1 Hyd. Shovel, 12 C.Y.		6198.74		6818.61	516.56	568.22
12 L.H., Daily Totals		$6888.34		$7898.41	$574.03	$658.20

Crew B-14J

Crew B-14J	Hr.	Daily	Hr.	Daily	Bare Costs	Incl. O&P
1 Equip. Oper. (medium)	$60.40	$483.20	$94.10	$752.80	$55.47	$86.88
.5 Laborer	45.60	182.40	72.45	289.80		
1 F.E. Loader, 8 C.Y.		2356.88		2592.56	196.41	216.05
12 L.H., Daily Totals		$3022.48		$3635.16	$251.87	$302.93

For customer support on your Commercial Renovation Costs with RSMeans data, call 800.448.8182.

767

Crew B-14K

	Bare Costs Hr.	Bare Costs Daily	Incl. Subs O&P Hr.	Incl. Subs O&P Daily	Cost Per Labor-Hour Bare Costs	Cost Per Labor-Hour Incl. O&P
1 Equip. Oper. (medium)	$60.40	$483.20	$94.10	$752.80	$55.47	$86.88
.5 Laborer	45.60	182.40	72.45	289.80		
1 F.E. Loader, 10 C.Y.		2787.78		3066.56	232.32	255.55
12 L.H., Daily Totals		$3453.38		$4109.16	$287.78	$342.43

Crew B-15

	Bare Costs Hr.	Bare Costs Daily	Incl. Subs O&P Hr.	Incl. Subs O&P Daily	Cost Per Labor-Hour Bare Costs	Cost Per Labor-Hour Incl. O&P
1 Equipment Oper. (med.)	$60.40	$483.20	$94.10	$752.80	$54.09	$85.55
.5 Laborer	45.60	182.40	72.45	289.80		
2 Truck Drivers (heavy)	53.05	848.80	84.55	1352.80		
2 Dump Trucks, 12 C.Y., 400 H.P.		1189.32		1308.25		
1 Dozer, 200 H.P.		1567.46		1724.21	98.46	108.30
28 L.H., Daily Totals		$4271.18		$5427.85	$152.54	$193.85

Crew B-16

	Bare Costs Hr.	Bare Costs Daily	Incl. Subs O&P Hr.	Incl. Subs O&P Daily	Cost Per Labor-Hour Bare Costs	Cost Per Labor-Hour Incl. O&P
1 Labor Foreman (outside)	$47.60	$380.80	$75.60	$604.80	$47.96	$76.26
2 Laborers	45.60	729.60	72.45	1159.20		
1 Truck Driver (heavy)	53.05	424.40	84.55	676.40		
1 Dump Truck, 12 C.Y., 400 H.P.		594.66		654.12	18.58	20.44
32 L.H., Daily Totals		$2129.46		$3094.52	$66.55	$96.70

Crew B-17

	Bare Costs Hr.	Bare Costs Daily	Incl. Subs O&P Hr.	Incl. Subs O&P Daily	Cost Per Labor-Hour Bare Costs	Cost Per Labor-Hour Incl. O&P
2 Laborers	$45.60	$729.60	$72.45	$1159.20	$50.31	$79.56
1 Equip. Oper. (light)	57.00	456.00	88.80	710.40		
1 Truck Driver (heavy)	53.05	424.40	84.55	676.40		
1 Backhoe Loader, 48 H.P.		221.72		243.89		
1 Dump Truck, 8 C.Y., 220 H.P.		419.41		461.35	20.04	22.04
32 L.H., Daily Totals		$2251.13		$3251.24	$70.35	$101.60

Crew B-17A

	Bare Costs Hr.	Bare Costs Daily	Incl. Subs O&P Hr.	Incl. Subs O&P Daily	Cost Per Labor-Hour Bare Costs	Cost Per Labor-Hour Incl. O&P
2 Labor Foremen (outside)	$47.60	$761.60	$75.60	$1209.60	$48.86	$77.55
6 Laborers	45.60	2188.80	72.45	3477.60		
1 Skilled Worker Foreman (out)	60.90	487.20	96.35	770.80		
1 Skilled Worker	58.90	471.20	93.20	745.60		
80 L.H., Daily Totals		$3908.80		$6203.60	$48.86	$77.55

Crew B-17B

	Bare Costs Hr.	Bare Costs Daily	Incl. Subs O&P Hr.	Incl. Subs O&P Daily	Cost Per Labor-Hour Bare Costs	Cost Per Labor-Hour Incl. O&P
2 Laborers	$45.60	$729.60	$72.45	$1159.20	$50.31	$79.56
1 Equip. Oper. (light)	57.00	456.00	88.80	710.40		
1 Truck Driver (heavy)	53.05	424.40	84.55	676.40		
1 Backhoe Loader, 48 H.P.		221.72		243.89		
1 Dump Truck, 12 C.Y., 400 H.P.		594.66		654.12	25.51	28.06
32 L.H., Daily Totals		$2426.37		$3444.01	$75.82	$107.63

Crew B-18

	Bare Costs Hr.	Bare Costs Daily	Incl. Subs O&P Hr.	Incl. Subs O&P Daily	Cost Per Labor-Hour Bare Costs	Cost Per Labor-Hour Incl. O&P
1 Labor Foreman (outside)	$47.60	$380.80	$75.60	$604.80	$46.27	$73.50
2 Laborers	45.60	729.60	72.45	1159.20		
1 Vibrating Plate, Gas, 21"		171.39		188.53	7.14	7.86
24 L.H., Daily Totals		$1281.79		$1952.53	$53.41	$81.36

Crew B-19

	Bare Costs Hr.	Bare Costs Daily	Incl. Subs O&P Hr.	Incl. Subs O&P Daily	Cost Per Labor-Hour Bare Costs	Cost Per Labor-Hour Incl. O&P
1 Pile Driver Foreman (outside)	$60.20	$481.60	$96.15	$769.20	$59.23	$93.70
4 Pile Drivers	58.20	1862.40	92.95	2974.40		
2 Equip. Oper. (crane)	63.40	1014.40	98.75	1580.00		
1 Equip. Oper. (oiler)	54.00	432.00	84.15	673.20		
1 Crawler Crane, 40 Ton		1267.52		1394.27		
1 Lead, 90' High		384.88		423.37		
1 Hammer, Diesel, 22k ft-lb		455.60		501.16	32.94	36.23
64 L.H., Daily Totals		$5898.40		$8315.59	$92.16	$129.93

Crew B-19A

	Bare Costs Hr.	Bare Costs Daily	Incl. Subs O&P Hr.	Incl. Subs O&P Daily	Cost Per Labor-Hour Bare Costs	Cost Per Labor-Hour Incl. O&P
1 Pile Driver Foreman (outside)	$60.20	$481.60	$96.15	$769.20	$59.23	$93.70
4 Pile Drivers	58.20	1862.40	92.95	2974.40		
2 Equip. Oper. (crane)	63.40	1014.40	98.75	1580.00		
1 Equip. Oper. (oiler)	54.00	432.00	84.15	673.20		
1 Crawler Crane, 75 Ton		2031.06		2234.16		
1 Lead, 90' High		384.88		423.37		
1 Hammer, Diesel, 41k ft-lb		601.64		661.80	47.15	51.86
64 L.H., Daily Totals		$6807.97		$9316.13	$106.37	$145.56

Crew B-19B

	Bare Costs Hr.	Bare Costs Daily	Incl. Subs O&P Hr.	Incl. Subs O&P Daily	Cost Per Labor-Hour Bare Costs	Cost Per Labor-Hour Incl. O&P
1 Pile Driver Foreman (outside)	$60.20	$481.60	$96.15	$769.20	$59.23	$93.70
4 Pile Drivers	58.20	1862.40	92.95	2974.40		
2 Equip. Oper. (crane)	63.40	1014.40	98.75	1580.00		
1 Equip. Oper. (oiler)	54.00	432.00	84.15	673.20		
1 Crawler Crane, 40 Ton		1267.52		1394.27		
1 Lead, 90' High		384.88		423.37		
1 Hammer, Diesel, 22k ft-lb		455.60		501.16		
1 Barge, 400 Ton		898.90		988.79	46.98	51.68
64 L.H., Daily Totals		$6797.30		$9304.39	$106.21	$145.38

Crew B-19C

	Bare Costs Hr.	Bare Costs Daily	Incl. Subs O&P Hr.	Incl. Subs O&P Daily	Cost Per Labor-Hour Bare Costs	Cost Per Labor-Hour Incl. O&P
1 Pile Driver Foreman (outside)	$60.20	$481.60	$96.15	$769.20	$59.23	$93.70
4 Pile Drivers	58.20	1862.40	92.95	2974.40		
2 Equip. Oper. (crane)	63.40	1014.40	98.75	1580.00		
1 Equip. Oper. (oiler)	54.00	432.00	84.15	673.20		
1 Crawler Crane, 75 Ton		2031.06		2234.16		
1 Lead, 90' High		384.88		423.37		
1 Hammer, Diesel, 41k ft-lb		601.64		661.80		
1 Barge, 400 Ton		898.90		988.79	61.19	67.31
64 L.H., Daily Totals		$7706.88		$10304.92	$120.42	$161.01

Crew B-20

	Bare Costs Hr.	Bare Costs Daily	Incl. Subs O&P Hr.	Incl. Subs O&P Daily	Cost Per Labor-Hour Bare Costs	Cost Per Labor-Hour Incl. O&P
1 Labor Foreman (outside)	$47.60	$380.80	$75.60	$604.80	$50.70	$80.42
1 Skilled Worker	58.90	471.20	93.20	745.60		
1 Laborer	45.60	364.80	72.45	579.60		
24 L.H., Daily Totals		$1216.80		$1930.00	$50.70	$80.42

Crew B-20A

	Bare Costs Hr.	Bare Costs Daily	Incl. Subs O&P Hr.	Incl. Subs O&P Daily	Cost Per Labor-Hour Bare Costs	Cost Per Labor-Hour Incl. O&P
1 Labor Foreman (outside)	$47.60	$380.80	$75.60	$604.80	$54.71	$85.44
1 Laborer	45.60	364.80	72.45	579.60		
1 Plumber	69.80	558.40	107.60	860.80		
1 Plumber Apprentice	55.85	446.80	86.10	688.80		
32 L.H., Daily Totals		$1750.80		$2734.00	$54.71	$85.44

Crew B-21

	Bare Costs Hr.	Bare Costs Daily	Incl. Subs O&P Hr.	Incl. Subs O&P Daily	Cost Per Labor-Hour Bare Costs	Cost Per Labor-Hour Incl. O&P
1 Labor Foreman (outside)	$47.60	$380.80	$75.60	$604.80	$52.51	$83.04
1 Skilled Worker	58.90	471.20	93.20	745.60		
1 Laborer	45.60	364.80	72.45	579.60		
.5 Equip. Oper. (crane)	63.40	253.60	98.75	395.00		
.5 S.P. Crane, 4x4, 5 Ton		196.99		216.69	7.04	7.74
28 L.H., Daily Totals		$1667.39		$2541.69	$59.55	$90.77

Crew B-21A

	Bare Costs Hr.	Bare Costs Daily	Incl. Subs O&P Hr.	Incl. Subs O&P Daily	Cost Per Labor-Hour Bare Costs	Cost Per Labor-Hour Incl. O&P
1 Labor Foreman (outside)	$47.60	$380.80	$75.60	$604.80	$56.45	$88.10
1 Laborer	45.60	364.80	72.45	579.60		
1 Plumber	69.80	558.40	107.60	860.80		
1 Plumber Apprentice	55.85	446.80	86.10	688.80		
1 Equip. Oper. (crane)	63.40	507.20	98.75	790.00		
1 S.P. Crane, 4x4, 12 Ton		445.72		490.30	11.14	12.26
40 L.H., Daily Totals		$2703.72		$4014.30	$67.59	$100.36

Crews - Renovation

Crew B-21B	Hr.	Daily	Hr.	Daily	Bare Costs	Incl. O&P
1 Labor Foreman (outside)	$47.60	$380.80	$75.60	$604.80	$49.56	$78.34
3 Laborers	45.60	1094.40	72.45	1738.80		
1 Equip. Oper. (crane)	63.40	507.20	98.75	790.00		
1 Hyd. Crane, 12 Ton		489.34		538.27	12.23	13.46
40 L.H., Daily Totals		$2471.74		$3671.87	$61.79	$91.80

Crew B-21C	Hr.	Daily	Hr.	Daily	Bare Costs	Incl. O&P
1 Labor Foreman (outside)	$47.60	$380.80	$75.60	$604.80	$49.63	$78.33
4 Laborers	45.60	1459.20	72.45	2318.40		
1 Equip. Oper. (crane)	63.40	507.20	98.75	790.00		
1 Equip. Oper. (oiler)	54.00	432.00	84.15	673.20		
2 Cutting Torches		26.74		29.41		
2 Sets of Gases		354.31		389.75		
1 Lattice Boom Crane, 90 Ton		1767.55		1944.30	38.37	42.20
56 L.H., Daily Totals		$4927.80		$6749.86	$88.00	$120.53

Crew B-22	Hr.	Daily	Hr.	Daily	Bare Costs	Incl. O&P
1 Labor Foreman (outside)	$47.60	$380.80	$75.60	$604.80	$53.24	$84.08
1 Skilled Worker	58.90	471.20	93.20	745.60		
1 Laborer	45.60	364.80	72.45	579.60		
.75 Equip. Oper. (crane)	63.40	380.40	98.75	592.50		
.75 S.P. Crane, 4x4, 5 Ton		295.49		325.03	9.85	10.83
30 L.H., Daily Totals		$1892.69		$2847.53	$63.09	$94.92

Crew B-22A	Hr.	Daily	Hr.	Daily	Bare Costs	Incl. O&P
1 Labor Foreman (outside)	$47.60	$380.80	$75.60	$604.80	$52.22	$82.49
1 Skilled Worker	58.90	471.20	93.20	745.60		
2 Laborers	45.60	729.60	72.45	1159.20		
1 Equipment Operator, Crane	63.40	507.20	98.75	790.00		
1 S.P. Crane, 4x4, 5 Ton		393.98		433.38		
1 Butt Fusion Mach., 4"-12" diam.		434.93		478.43	20.72	22.80
40 L.H., Daily Totals		$2917.71		$4211.40	$72.94	$105.29

Crew B-22B	Hr.	Daily	Hr.	Daily	Bare Costs	Incl. O&P
1 Labor Foreman (outside)	$47.60	$380.80	$75.60	$604.80	$52.22	$82.49
1 Skilled Worker	58.90	471.20	93.20	745.60		
2 Laborers	45.60	729.60	72.45	1159.20		
1 Equip. Oper. (crane)	63.40	507.20	98.75	790.00		
1 S.P. Crane, 4x4, 5 Ton		393.98		433.38		
1 Butt Fusion Mach., 8"-24" diam.		1119.59		1231.55	37.84	41.62
40 L.H., Daily Totals		$3602.37		$4964.53	$90.06	$124.11

Crew B-22C	Hr.	Daily	Hr.	Daily	Bare Costs	Incl. O&P
1 Skilled Worker	$58.90	$471.20	$93.20	$745.60	$52.25	$82.83
1 Laborer	45.60	364.80	72.45	579.60		
1 Butt Fusion Mach., 2"-8" diam.		139.58		153.54	8.72	9.60
16 L.H., Daily Totals		$975.58		$1478.74	$60.97	$92.42

Crew B-23	Hr.	Daily	Hr.	Daily	Bare Costs	Incl. O&P
1 Labor Foreman (outside)	$47.60	$380.80	$75.60	$604.80	$46.00	$73.08
4 Laborers	45.60	1459.20	72.45	2318.40		
1 Drill Rig, Truck-Mounted		785.11		863.62		
1 Flatbed Truck, Gas, 3 Ton		877.96		965.75	41.58	45.73
40 L.H., Daily Totals		$3503.07		$4752.58	$87.58	$118.81

Crew B-23A	Hr.	Daily	Hr.	Daily	Bare Costs	Incl. O&P
1 Labor Foreman (outside)	$47.60	$380.80	$75.60	$604.80	$51.20	$80.72
1 Laborer	45.60	364.80	72.45	579.60		
1 Equip. Oper. (medium)	60.40	483.20	94.10	752.80		
1 Drill Rig, Truck-Mounted		785.11		863.62		
1 Pickup Truck, 3/4 Ton		115.08		126.59	37.51	41.26
24 L.H., Daily Totals		$2128.99		$2927.41	$88.71	$121.98

Crew B-23B	Hr.	Daily	Hr.	Daily	Bare Costs	Incl. O&P
1 Labor Foreman (outside)	$47.60	$380.80	$75.60	$604.80	$51.20	$80.72
1 Laborer	45.60	364.80	72.45	579.60		
1 Equip. Oper. (medium)	60.40	483.20	94.10	752.80		
1 Drill Rig, Truck-Mounted		785.11		863.62		
1 Pickup Truck, 3/4 Ton		115.08		126.59		
1 Centr. Water Pump, 6"		240.92		265.01	47.55	52.30
24 L.H., Daily Totals		$2369.91		$3192.42	$98.75	$133.02

Crew B-24	Hr.	Daily	Hr.	Daily	Bare Costs	Incl. O&P
1 Cement Finisher	$53.15	$425.20	$83.05	$664.40	$51.68	$81.65
1 Laborer	45.60	364.80	72.45	579.60		
1 Carpenter	56.30	450.40	89.45	715.60		
24 L.H., Daily Totals		$1240.40		$1959.60	$51.68	$81.65

Crew B-25	Hr.	Daily	Hr.	Daily	Bare Costs	Incl. O&P
1 Labor Foreman (outside)	$47.60	$380.80	$75.60	$604.80	$49.82	$78.64
7 Laborers	45.60	2553.60	72.45	4057.20		
3 Equip. Oper. (medium)	60.40	1449.60	94.10	2258.40		
1 Asphalt Paver, 130 H.P.		2208.89		2429.77		
1 Tandem Roller, 10 Ton		254.53		279.98		
1 Roller, Pneum. Whl., 12 Ton		361.08		397.18	32.10	35.31
88 L.H., Daily Totals		$7208.49		$10027.34	$81.91	$113.95

Crew B-25B	Hr.	Daily	Hr.	Daily	Bare Costs	Incl. O&P
1 Labor Foreman (outside)	$47.60	$380.80	$75.60	$604.80	$50.70	$79.93
7 Laborers	45.60	2553.60	72.45	4057.20		
4 Equip. Oper. (medium)	60.40	1932.80	94.10	3011.20		
1 Asphalt Paver, 130 H.P.		2208.89		2429.77		
2 Tandem Rollers, 10 Ton		509.05		559.96		
1 Roller, Pneum. Whl., 12 Ton		361.08		397.18	32.07	35.28
96 L.H., Daily Totals		$7946.21		$11060.11	$82.77	$115.21

Crew B-25C	Hr.	Daily	Hr.	Daily	Bare Costs	Incl. O&P
1 Labor Foreman (outside)	$47.60	$380.80	$75.60	$604.80	$50.87	$80.19
3 Laborers	45.60	1094.40	72.45	1738.80		
2 Equip. Oper. (medium)	60.40	966.40	94.10	1505.60		
1 Asphalt Paver, 130 H.P.		2208.89		2429.77		
1 Tandem Roller, 10 Ton		254.53		279.98	51.32	56.45
48 L.H., Daily Totals		$4905.01		$6558.95	$102.19	$136.64

Crew B-25D	Hr.	Daily	Hr.	Daily	Bare Costs	Incl. O&P
1 Labor Foreman (outside)	$47.60	$380.80	$75.60	$604.80	$51.10	$80.56
3 Laborers	45.60	1094.40	72.45	1738.80		
2.125 Equip. Oper. (medium)	60.40	1026.80	94.10	1599.70		
.125 Truck Driver (heavy)	53.05	53.05	84.55	84.55		
.125 Truck Tractor, 6x4, 380 H.P.		64.08		70.49		
.125 Dist. Tanker, 3000 Gallon		43.09		47.40		
1 Asphalt Paver, 130 H.P.		2208.89		2429.77		
1 Tandem Roller, 10 Ton		254.53		279.98	51.41	56.55
50 L.H., Daily Totals		$5125.63		$6855.49	$102.51	$137.11

Crew No.	Bare Costs Hr.	Daily	Incl. Subs O&P Hr.	Daily	Cost Per Labor-Hour Bare Costs	Incl. O&P
Crew B-25E	Hr.	Daily	Hr.	Daily	Bare Costs	Incl. O&P
1 Labor Foreman (outside)	$47.60	$380.80	$75.60	$604.80	$51.32	$80.89
3 Laborers	45.60	1094.40	72.45	1738.80		
2.250 Equip. Oper. (medium)	60.40	1087.20	94.10	1693.80		
.25 Truck Driver (heavy)	53.05	106.10	84.55	169.10		
.25 Truck Tractor, 6x4, 380 H.P.		128.17		140.98		
.25 Dist. Tanker, 3000 Gallon		86.17		94.79		
1 Asphalt Paver, 130 H.P.		2208.89		2429.77		
1 Tandem Roller, 10 Ton		254.53		279.98	51.50	56.64
52 L.H., Daily Totals		$5346.25		$7152.02	$102.81	$137.54

Crew B-26	Hr.	Daily	Hr.	Daily	Bare Costs	Incl. O&P
1 Labor Foreman (outside)	$47.60	$380.80	$75.60	$604.80	$50.53	$79.68
6 Laborers	45.60	2188.80	72.45	3477.60		
2 Equip. Oper. (medium)	60.40	966.40	94.10	1505.60		
1 Rodman (reinf.)	60.70	485.60	94.90	759.20		
1 Cement Finisher	53.15	425.20	83.05	664.40		
1 Grader, 30,000 Lbs.		1108.49		1219.34		
1 Paving Mach. & Equip.		2579.95		2837.94	41.91	46.11
88 L.H., Daily Totals		$8135.23		$11068.88	$92.45	$125.78

Crew B-26A	Hr.	Daily	Hr.	Daily	Bare Costs	Incl. O&P
1 Labor Foreman (outside)	$47.60	$380.80	$75.60	$604.80	$50.53	$79.68
6 Laborers	45.60	2188.80	72.45	3477.60		
2 Equip. Oper. (medium)	60.40	966.40	94.10	1505.60		
1 Rodman (reinf.)	60.70	485.60	94.90	759.20		
1 Cement Finisher	53.15	425.20	83.05	664.40		
1 Grader, 30,000 Lbs.		1108.49		1219.34		
1 Paving Mach. & Equip.		2579.95		2837.94		
1 Concrete Saw		115.94		127.54	43.23	47.55
88 L.H., Daily Totals		$8251.18		$11196.41	$93.76	$127.23

Crew B-26B	Hr.	Daily	Hr.	Daily	Bare Costs	Incl. O&P
1 Labor Foreman (outside)	$47.60	$380.80	$75.60	$604.80	$51.35	$80.88
6 Laborers	45.60	2188.80	72.45	3477.60		
3 Equip. Oper. (medium)	60.40	1449.60	94.10	2258.40		
1 Rodman (reinf.)	60.70	485.60	94.90	759.20		
1 Cement Finisher	53.15	425.20	83.05	664.40		
1 Grader, 30,000 Lbs.		1108.49		1219.34		
1 Paving Mach. & Equip.		2579.95		2837.94		
1 Concrete Pump, 110' Boom		506.49		557.14	43.70	48.07
96 L.H., Daily Totals		$9124.92		$12378.81	$95.05	$128.95

Crew B-26C	Hr.	Daily	Hr.	Daily	Bare Costs	Incl. O&P
1 Labor Foreman (outside)	$47.60	$380.80	$75.60	$604.80	$49.55	$78.23
6 Laborers	45.60	2188.80	72.45	3477.60		
1 Equip. Oper. (medium)	60.40	483.20	94.10	752.80		
1 Rodman (reinf.)	60.70	485.60	94.90	759.20		
1 Cement Finisher	53.15	425.20	83.05	664.40		
1 Paving Mach. & Equip.		2579.95		2837.94		
1 Concrete Saw		115.94		127.54	33.70	37.07
80 L.H., Daily Totals		$6659.49		$9224.28	$83.24	$115.30

Crew B-27	Hr.	Daily	Hr.	Daily	Bare Costs	Incl. O&P
1 Labor Foreman (outside)	$47.60	$380.80	$75.60	$604.80	$46.10	$73.24
3 Laborers	45.60	1094.40	72.45	1738.80		
1 Berm Machine		261.22		287.34	8.16	8.98
32 L.H., Daily Totals		$1736.42		$2630.94	$54.26	$82.22

Crew B-28	Hr.	Daily	Hr.	Daily	Bare Costs	Incl. O&P
2 Carpenters	$56.30	$900.80	$89.45	$1431.20	$52.73	$83.78
1 Laborer	45.60	364.80	72.45	579.60		
24 L.H., Daily Totals		$1265.60		$2010.80	$52.73	$83.78

Crew B-29	Hr.	Daily	Hr.	Daily	Bare Costs	Incl. O&P
1 Labor Foreman (outside)	$47.60	$380.80	$75.60	$604.80	$49.63	$78.33
4 Laborers	45.60	1459.20	72.45	2318.40		
1 Equip. Oper. (crane)	63.40	507.20	98.75	790.00		
1 Equip. Oper. (oiler)	54.00	432.00	84.15	673.20		
1 Gradall, 5/8 C.Y.		876.16		963.78	15.65	17.21
56 L.H., Daily Totals		$3655.36		$5350.18	$65.27	$95.54

Crew B-30	Hr.	Daily	Hr.	Daily	Bare Costs	Incl. O&P
1 Equip. Oper. (medium)	$60.40	$483.20	$94.10	$752.80	$55.50	$87.73
2 Truck Drivers (heavy)	53.05	848.80	84.55	1352.80		
1 Hyd. Excavator, 1.5 C.Y.		714.87		786.36		
2 Dump Trucks, 12 C.Y., 400 H.P.		1189.32		1308.25	79.34	87.28
24 L.H., Daily Totals		$3236.19		$4200.20	$134.84	$175.01

Crew B-31	Hr.	Daily	Hr.	Daily	Bare Costs	Incl. O&P
1 Labor Foreman (outside)	$47.60	$380.80	$75.60	$604.80	$48.14	$76.48
3 Laborers	45.60	1094.40	72.45	1738.80		
1 Carpenter	56.30	450.40	89.45	715.60		
1 Air Compressor, 250 cfm		208.68		229.55		
1 Sheeting Driver		7.69		8.46		
2 -50' Air Hoses, 1.5"		47.19		51.91	6.59	7.25
40 L.H., Daily Totals		$2189.16		$3349.11	$54.73	$83.73

Crew B-32	Hr.	Daily	Hr.	Daily	Bare Costs	Incl. O&P
1 Laborer	$45.60	$364.80	$72.45	$579.60	$56.70	$88.69
3 Equip. Oper. (medium)	60.40	1449.60	94.10	2258.40		
1 Grader, 30,000 Lbs.		1108.49		1219.34		
1 Tandem Roller, 10 Ton		254.53		279.98		
1 Dozer, 200 H.P.		1567.46		1724.21	91.58	100.74
32 L.H., Daily Totals		$4744.87		$6061.52	$148.28	$189.42

Crew B-32A	Hr.	Daily	Hr.	Daily	Bare Costs	Incl. O&P
1 Laborer	$45.60	$364.80	$72.45	$579.60	$55.47	$86.88
2 Equip. Oper. (medium)	60.40	966.40	94.10	1505.60		
1 Grader, 30,000 Lbs.		1108.49		1219.34		
1 Roller, Vibratory, 25 Ton		693.41		762.76	75.08	82.59
24 L.H., Daily Totals		$3133.10		$4067.29	$130.55	$169.47

Crew B-32B	Hr.	Daily	Hr.	Daily	Bare Costs	Incl. O&P
1 Laborer	$45.60	$364.80	$72.45	$579.60	$55.47	$86.88
2 Equip. Oper. (medium)	60.40	966.40	94.10	1505.60		
1 Dozer, 200 H.P.		1567.46		1724.21		
1 Roller, Vibratory, 25 Ton		693.41		762.76	94.20	103.62
24 L.H., Daily Totals		$3592.07		$4572.16	$149.67	$190.51

Crew B-32C	Hr.	Daily	Hr.	Daily	Bare Costs	Incl. O&P
1 Labor Foreman (outside)	$47.60	$380.80	$75.60	$604.80	$53.33	$83.80
2 Laborers	45.60	729.60	72.45	1159.20		
3 Equip. Oper. (medium)	60.40	1449.60	94.10	2258.40		
1 Grader, 30,000 Lbs.		1108.49		1219.34		
1 Tandem Roller, 10 Ton		254.53		279.98		
1 Dozer, 200 H.P.		1567.46		1724.21	61.05	67.16
48 L.H., Daily Totals		$5490.47		$7245.92	$114.38	$150.96

Crew No.	Bare Costs		Incl. Subs O&P		Cost Per Labor-Hour	
	Hr.	Daily	Hr.	Daily	Bare Costs	Incl. O&P
Crew B-33A						
1 Equip. Oper. (medium)	$60.40	$483.20	$94.10	$752.80	$56.17	$87.91
.5 Laborer	45.60	182.40	72.45	289.80		
.25 Equip. Oper. (medium)	60.40	120.80	94.10	188.20		
1 Scraper, Towed, 7 C.Y.		133.22		146.54		
1.25 Dozers, 300 H.P.		2300.04		2530.05	173.80	191.19
14 L.H., Daily Totals		$3219.66		$3907.39	$229.98	$279.10
Crew B-33B	Hr.	Daily	Hr.	Daily	Bare Costs	Incl. O&P
1 Equip. Oper. (medium)	$60.40	$483.20	$94.10	$752.80	$56.17	$87.91
.5 Laborer	45.60	182.40	72.45	289.80		
.25 Equip. Oper. (medium)	60.40	120.80	94.10	188.20		
1 Scraper, Towed, 10 C.Y.		166.62		183.28		
1.25 Dozers, 300 H.P.		2300.04		2530.05	176.19	193.81
14 L.H., Daily Totals		$3253.06		$3944.13	$232.36	$281.72
Crew B-33C	Hr.	Daily	Hr.	Daily	Bare Costs	Incl. O&P
1 Equip. Oper. (medium)	$60.40	$483.20	$94.10	$752.80	$56.17	$87.91
.5 Laborer	45.60	182.40	72.45	289.80		
.25 Equip. Oper. (medium)	60.40	120.80	94.10	188.20		
1 Scraper, Towed, 15 C.Y.		184.47		202.92		
1.25 Dozers, 300 H.P.		2300.04		2530.05	177.47	195.21
14 L.H., Daily Totals		$3270.91		$3963.76	$233.64	$283.13
Crew B-33D	Hr.	Daily	Hr.	Daily	Bare Costs	Incl. O&P
1 Equip. Oper. (medium)	$60.40	$483.20	$94.10	$752.80	$56.17	$87.91
.5 Laborer	45.60	182.40	72.45	289.80		
.25 Equip. Oper. (medium)	60.40	120.80	94.10	188.20		
1 S.P. Scraper, 14 C.Y.		2495.24		2744.76		
.25 Dozer, 300 H.P.		460.01		506.01	211.09	232.20
14 L.H., Daily Totals		$3741.65		$4481.57	$267.26	$320.11
Crew B-33E	Hr.	Daily	Hr.	Daily	Bare Costs	Incl. O&P
1 Equip. Oper. (medium)	$60.40	$483.20	$94.10	$752.80	$56.17	$87.91
.5 Laborer	45.60	182.40	72.45	289.80		
.25 Equip. Oper. (medium)	60.40	120.80	94.10	188.20		
1 S.P. Scraper, 21 C.Y.		2735.18		3008.70		
.25 Dozer, 300 H.P.		460.01		506.01	228.23	251.05
14 L.H., Daily Totals		$3981.59		$4745.51	$284.40	$338.97
Crew B-33F	Hr.	Daily	Hr.	Daily	Bare Costs	Incl. O&P
1 Equip. Oper. (medium)	$60.40	$483.20	$94.10	$752.80	$56.17	$87.91
.5 Laborer	45.60	182.40	72.45	289.80		
.25 Equip. Oper. (medium)	60.40	120.80	94.10	188.20		
1 Elev. Scraper, 11 C.Y.		1090.07		1199.08		
.25 Dozer, 300 H.P.		460.01		506.01	110.72	121.79
14 L.H., Daily Totals		$2336.48		$2935.89	$166.89	$209.71
Crew B-33G	Hr.	Daily	Hr.	Daily	Bare Costs	Incl. O&P
1 Equip. Oper. (medium)	$60.40	$483.20	$94.10	$752.80	$56.17	$87.91
.5 Laborer	45.60	182.40	72.45	289.80		
.25 Equip. Oper. (medium)	60.40	120.80	94.10	188.20		
1 Elev. Scraper, 22 C.Y.		1949.58		2144.54		
.25 Dozer, 300 H.P.		460.01		506.01	172.11	189.33
14 L.H., Daily Totals		$3195.99		$3881.35	$228.29	$277.24

Crew No.	Bare Costs		Incl. Subs O&P		Cost Per Labor-Hour	
	Hr.	Daily	Hr.	Daily	Bare Costs	Incl. O&P
Crew B-33K						
1 Equipment Operator (med.)	$60.40	$483.20	$94.10	$752.80	$56.17	$87.91
.25 Equipment Operator (med.)	60.40	120.80	94.10	188.20		
.5 Laborer	45.60	182.40	72.45	289.80		
1 S.P. Scraper, 31 C.Y.		3818.46		4200.30		
.25 Dozer, 410 H.P.		724.08		796.49	324.47	356.91
14 L.H., Daily Totals		$5328.93		$6227.59	$380.64	$444.83
Crew B-34A	Hr.	Daily	Hr.	Daily	Bare Costs	Incl. O&P
1 Truck Driver (heavy)	$53.05	$424.40	$84.55	$676.40	$53.05	$84.55
1 Dump Truck, 8 C.Y., 220 H.P.		419.41		461.35	52.43	57.67
8 L.H., Daily Totals		$843.81		$1137.75	$105.48	$142.22
Crew B-34B	Hr.	Daily	Hr.	Daily	Bare Costs	Incl. O&P
1 Truck Driver (heavy)	$53.05	$424.40	$84.55	$676.40	$53.05	$84.55
1 Dump Truck, 12 C.Y., 400 H.P.		594.66		654.12	74.33	81.77
8 L.H., Daily Totals		$1019.06		$1330.52	$127.38	$166.32
Crew B-34C	Hr.	Daily	Hr.	Daily	Bare Costs	Incl. O&P
1 Truck Driver (heavy)	$53.05	$424.40	$84.55	$676.40	$53.05	$84.55
1 Truck Tractor, 6x4, 380 H.P.		512.66		563.93		
1 Dump Trailer, 16.5 C.Y.		142.67		156.94	81.92	90.11
8 L.H., Daily Totals		$1079.73		$1397.26	$134.97	$174.66
Crew B-34D	Hr.	Daily	Hr.	Daily	Bare Costs	Incl. O&P
1 Truck Driver (heavy)	$53.05	$424.40	$84.55	$676.40	$53.05	$84.55
1 Truck Tractor, 6x4, 380 H.P.		512.66		563.93		
1 Dump Trailer, 20 C.Y.		158.36		174.20	83.88	92.27
8 L.H., Daily Totals		$1095.42		$1414.53	$136.93	$176.82
Crew B-34E	Hr.	Daily	Hr.	Daily	Bare Costs	Incl. O&P
1 Truck Driver (heavy)	$53.05	$424.40	$84.55	$676.40	$53.05	$84.55
1 Dump Truck, Off Hwy., 25 Ton		1471.56		1618.72	183.95	202.34
8 L.H., Daily Totals		$1895.96		$2295.12	$237.00	$286.89
Crew B-34F	Hr.	Daily	Hr.	Daily	Bare Costs	Incl. O&P
1 Truck Driver (heavy)	$53.05	$424.40	$84.55	$676.40	$53.05	$84.55
1 Dump Truck, Off Hwy., 35 Ton		971.45		1068.60	121.43	133.57
8 L.H., Daily Totals		$1395.85		$1745.00	$174.48	$218.12
Crew B-34G	Hr.	Daily	Hr.	Daily	Bare Costs	Incl. O&P
1 Truck Driver (heavy)	$53.05	$424.40	$84.55	$676.40	$53.05	$84.55
1 Dump Truck, Off Hwy., 50 Ton		2040.69		2244.75	255.09	280.59
8 L.H., Daily Totals		$2465.09		$2921.15	$308.14	$365.14
Crew B-34H	Hr.	Daily	Hr.	Daily	Bare Costs	Incl. O&P
1 Truck Driver (heavy)	$53.05	$424.40	$84.55	$676.40	$53.05	$84.55
1 Dump Truck, Off Hwy., 65 Ton		1996.90		2196.59	249.61	274.57
8 L.H., Daily Totals		$2421.30		$2872.99	$302.66	$359.12
Crew B-34I	Hr.	Daily	Hr.	Daily	Bare Costs	Incl. O&P
1 Truck Driver (heavy)	$53.05	$424.40	$84.55	$676.40	$53.05	$84.55
1 Dump Truck, 18 C.Y., 450 H.P.		774.61		852.07	96.83	106.51
8 L.H., Daily Totals		$1199.01		$1528.47	$149.88	$191.06
Crew B-34J	Hr.	Daily	Hr.	Daily	Bare Costs	Incl. O&P
1 Truck Driver (heavy)	$53.05	$424.40	$84.55	$676.40	$53.05	$84.55
1 Dump Truck, Off Hwy., 100 Ton		2854.63		3140.09	356.83	392.51
8 L.H., Daily Totals		$3279.03		$3816.49	$409.88	$477.06

Crew B-34K

Crew No.	Bare Costs Hr.	Bare Costs Daily	Incl. Subs O&P Hr.	Incl. Subs O&P Daily	Cost Per Labor-Hour Bare Costs	Cost Per Labor-Hour Incl. O&P
1 Truck Driver (heavy)	$53.05	$424.40	$84.55	$676.40	$53.05	$84.55
1 Truck Tractor, 6x4, 450 H.P.		625.39		687.93		
1 Lowbed Trailer, 75 Ton		266.17		292.79	111.44	122.59
8 L.H., Daily Totals		$1315.96		$1657.12	$164.50	$207.14

Crew B-34L

Crew No.	Bare Costs Hr.	Bare Costs Daily	Incl. Subs O&P Hr.	Incl. Subs O&P Daily	Cost Per Labor-Hour Bare Costs	Cost Per Labor-Hour Incl. O&P
1 Equip. Oper. (light)	$57.00	$456.00	$88.80	$710.40	$57.00	$88.80
1 Flatbed Truck, Gas, 1.5 Ton		203.25		223.57	25.41	27.95
8 L.H., Daily Totals		$659.25		$933.97	$82.41	$116.75

Crew B-34M

Crew No.	Bare Costs Hr.	Bare Costs Daily	Incl. Subs O&P Hr.	Incl. Subs O&P Daily	Cost Per Labor-Hour Bare Costs	Cost Per Labor-Hour Incl. O&P
1 Equip. Oper. (light)	$57.00	$456.00	$88.80	$710.40	$57.00	$88.80
1 Flatbed Truck, Gas, 3 Ton		877.96		965.75	109.74	120.72
8 L.H., Daily Totals		$1333.96		$1676.15	$166.74	$209.52

Crew B-34N

Crew No.	Bare Costs Hr.	Bare Costs Daily	Incl. Subs O&P Hr.	Incl. Subs O&P Daily	Cost Per Labor-Hour Bare Costs	Cost Per Labor-Hour Incl. O&P
1 Truck Driver (heavy)	$53.05	$424.40	$84.55	$676.40	$56.73	$89.33
1 Equip. Oper. (medium)	60.40	483.20	94.10	752.80		
1 Truck Tractor, 6x4, 380 H.P.		512.66		563.93		
1 Flatbed Trailer, 40 Ton		194.29		213.72	44.18	48.60
16 L.H., Daily Totals		$1614.55		$2206.85	$100.91	$137.93

Crew B-34P

Crew No.	Bare Costs Hr.	Bare Costs Daily	Incl. Subs O&P Hr.	Incl. Subs O&P Daily	Cost Per Labor-Hour Bare Costs	Cost Per Labor-Hour Incl. O&P
1 Pipe Fitter	$70.35	$562.80	$108.40	$867.20	$60.38	$94.27
1 Truck Driver (light)	50.40	403.20	80.30	642.40		
1 Equip. Oper. (medium)	60.40	483.20	94.10	752.80		
1 Flatbed Truck, Gas, 3 Ton		877.96		965.75		
1 Backhoe Loader, 48 H.P.		221.72		243.89	45.82	50.40
24 L.H., Daily Totals		$2548.87		$3472.04	$106.20	$144.67

Crew B-34Q

Crew No.	Bare Costs Hr.	Bare Costs Daily	Incl. Subs O&P Hr.	Incl. Subs O&P Daily	Cost Per Labor-Hour Bare Costs	Cost Per Labor-Hour Incl. O&P
1 Pipe Fitter	$70.35	$562.80	$108.40	$867.20	$61.38	$95.82
1 Truck Driver (light)	50.40	403.20	80.30	642.40		
1 Equip. Oper. (crane)	63.40	507.20	98.75	790.00		
1 Flatbed Trailer, 25 Ton		141.42		155.56		
1 Dump Truck, 8 C.Y., 220 H.P.		419.41		461.35		
1 Hyd. Crane, 25 Ton		603.44		663.78	48.51	53.36
24 L.H., Daily Totals		$2637.47		$3580.29	$109.89	$149.18

Crew B-34R

Crew No.	Bare Costs Hr.	Bare Costs Daily	Incl. Subs O&P Hr.	Incl. Subs O&P Daily	Cost Per Labor-Hour Bare Costs	Cost Per Labor-Hour Incl. O&P
1 Pipe Fitter	$70.35	$562.80	$108.40	$867.20	$61.38	$95.82
1 Truck Driver (light)	50.40	403.20	80.30	642.40		
1 Equip. Oper. (crane)	63.40	507.20	98.75	790.00		
1 Flatbed Trailer, 25 Ton		141.42		155.56		
1 Dump Truck, 8 C.Y., 220 H.P.		419.41		461.35		
1 Hyd. Crane, 25 Ton		603.44		663.78		
1 Hyd. Excavator, 1 C.Y.		857.81		943.59	84.25	92.68
24 L.H., Daily Totals		$3495.27		$4523.88	$145.64	$188.50

Crew B-34S

Crew No.	Bare Costs Hr.	Bare Costs Daily	Incl. Subs O&P Hr.	Incl. Subs O&P Daily	Cost Per Labor-Hour Bare Costs	Cost Per Labor-Hour Incl. O&P
2 Pipe Fitters	$70.35	$1125.60	$108.40	$1734.40	$64.29	$100.03
1 Truck Driver (heavy)	53.05	424.40	84.55	676.40		
1 Equip. Oper. (crane)	63.40	507.20	98.75	790.00		
1 Flatbed Trailer, 40 Ton		194.29		213.72		
1 Truck Tractor, 6x4, 380 H.P.		512.66		563.93		
1 Hyd. Crane, 80 Ton		1502.22		1652.44		
1 Hyd. Excavator, 2 C.Y.		969.84		1066.82	99.34	109.28
32 L.H., Daily Totals		$5236.21		$6697.71	$163.63	$209.30

Crew B-34T

Crew No.	Bare Costs Hr.	Bare Costs Daily	Incl. Subs O&P Hr.	Incl. Subs O&P Daily	Cost Per Labor-Hour Bare Costs	Cost Per Labor-Hour Incl. O&P
2 Pipe Fitters	$70.35	$1125.60	$108.40	$1734.40	$64.29	$100.03
1 Truck Driver (heavy)	53.05	424.40	84.55	676.40		
1 Equip. Oper. (crane)	63.40	507.20	98.75	790.00		
1 Flatbed Trailer, 40 Ton		194.29		213.72		
1 Truck Tractor, 6x4, 380 H.P.		512.66		563.93		
1 Hyd. Crane, 80 Ton		1502.22		1652.44	69.04	75.94
32 L.H., Daily Totals		$4266.37		$5630.88	$133.32	$175.97

Crew B-34U

Crew No.	Bare Costs Hr.	Bare Costs Daily	Incl. Subs O&P Hr.	Incl. Subs O&P Daily	Cost Per Labor-Hour Bare Costs	Cost Per Labor-Hour Incl. O&P
1 Truck Driver (heavy)	$53.05	$424.40	$84.55	$676.40	$55.02	$86.67
1 Equip. Oper. (light)	57.00	456.00	88.80	710.40		
1 Truck Tractor, 220 H.P.		319.23		351.15		
1 Flatbed Trailer, 25 Ton		141.42		155.56	28.79	31.67
16 L.H., Daily Totals		$1341.05		$1893.51	$83.82	$118.34

Crew B-34V

Crew No.	Bare Costs Hr.	Bare Costs Daily	Incl. Subs O&P Hr.	Incl. Subs O&P Daily	Cost Per Labor-Hour Bare Costs	Cost Per Labor-Hour Incl. O&P
1 Truck Driver (heavy)	$53.05	$424.40	$84.55	$676.40	$57.82	$90.70
1 Equip. Oper. (crane)	63.40	507.20	98.75	790.00		
1 Equip. Oper. (light)	57.00	456.00	88.80	710.40		
1 Truck Tractor, 6x4, 450 H.P.		625.39		687.93		
1 Equipment Trailer, 50 Ton		213.73		235.10		
1 Pickup Truck, 4x4, 3/4 Ton		182.00		200.20	42.55	46.80
24 L.H., Daily Totals		$2408.72		$3300.03	$100.36	$137.50

Crew B-34W

Crew No.	Bare Costs Hr.	Bare Costs Daily	Incl. Subs O&P Hr.	Incl. Subs O&P Daily	Cost Per Labor-Hour Bare Costs	Cost Per Labor-Hour Incl. O&P
5 Truck Drivers (heavy)	$53.05	$2122.00	$84.55	$3382.00	$55.68	$87.96
2 Equip. Opers. (crane)	63.40	1014.40	98.75	1580.00		
1 Equip. Oper. (mechanic)	63.50	508.00	98.95	791.60		
1 Laborer	45.60	364.80	72.45	579.60		
4 Truck Tractors, 6x4, 380 H.P.		2050.64		2255.71		
2 Equipment Trailers, 50 Ton		427.46		470.20		
2 Flatbed Trailers, 40 Ton		388.58		427.44		
1 Pickup Truck, 4x4, 3/4 Ton		182.00		200.20		
1 S.P. Crane, 4x4, 20 Ton		592.53		651.78	50.57	55.63
72 L.H., Daily Totals		$7650.41		$10338.54	$106.26	$143.59

Crew B-35

Crew No.	Bare Costs Hr.	Bare Costs Daily	Incl. Subs O&P Hr.	Incl. Subs O&P Daily	Cost Per Labor-Hour Bare Costs	Cost Per Labor-Hour Incl. O&P
1 Labor Foreman (outside)	$47.60	$380.80	$75.60	$604.80	$58.44	$91.34
1 Skilled Worker	58.90	471.20	93.20	745.60		
2 Welders	69.80	1116.80	107.60	1721.60		
1 Laborer	45.60	364.80	72.45	579.60		
1 Equip. Oper. (crane)	63.40	507.20	98.75	790.00		
1 Equip. Oper. (oiler)	54.00	432.00	84.15	673.20		
2 Welder, Electric, 300 amp		221.51		243.66		
1 Hyd. Excavator, .75 C.Y.		723.26		795.59	16.87	18.56
56 L.H., Daily Totals		$4217.57		$6154.05	$75.31	$109.89

Crew B-35A

Crew No.	Bare Costs Hr.	Bare Costs Daily	Incl. Subs O&P Hr.	Incl. Subs O&P Daily	Cost Per Labor-Hour Bare Costs	Cost Per Labor-Hour Incl. O&P
1 Labor Foreman (outside)	$47.60	$380.80	$75.60	$604.80	$54.99	$86.31
2 Laborers	45.60	729.60	72.45	1159.20		
1 Skilled Worker	58.90	471.20	93.20	745.60		
1 Welder (plumber)	69.80	558.40	107.60	860.80		
1 Equip. Oper. (crane)	63.40	507.20	98.75	790.00		
1 Equip. Oper. (oiler)	54.00	432.00	84.15	673.20		
1 Welder, Gas Engine, 300 amp		152.88		168.16		
1 Crawler Crane, 75 Ton		2031.06		2234.16	39.00	42.90
56 L.H., Daily Totals		$5263.13		$7235.93	$93.98	$129.21

For customer support on your Commercial Renovation Costs with RSMeans data, call 800.448.8182.

Crews - Renovation

Crew B-36	Hr.	Daily	Hr.	Daily	Bare Costs	Incl. O&P
1 Labor Foreman (outside)	$47.60	$380.80	$75.60	$604.80	$51.92	$81.74
2 Laborers	45.60	729.60	72.45	1159.20		
2 Equip. Oper. (medium)	60.40	966.40	94.10	1505.60		
1 Dozer, 200 H.P.		1567.46		1724.21		
1 Aggregate Spreader		61.81		67.99		
1 Tandem Roller, 10 Ton		254.53		279.98	47.09	51.80
40 L.H., Daily Totals		$3960.59		$5341.77	$99.01	$133.54

Crew B-36A	Hr.	Daily	Hr.	Daily	Bare Costs	Incl. O&P
1 Labor Foreman (outside)	$47.60	$380.80	$75.60	$604.80	$54.34	$85.27
2 Laborers	45.60	729.60	72.45	1159.20		
4 Equip. Oper. (medium)	60.40	1932.80	94.10	3011.20		
1 Dozer, 200 H.P.		1567.46		1724.21		
1 Aggregate Spreader		61.81		67.99		
1 Tandem Roller, 10 Ton		254.53		279.98		
1 Roller, Pneum. Whl., 12 Ton		361.08		397.18	40.09	44.10
56 L.H., Daily Totals		$5288.07		$7244.56	$94.43	$129.37

Crew B-36B	Hr.	Daily	Hr.	Daily	Bare Costs	Incl. O&P
1 Labor Foreman (outside)	$47.60	$380.80	$75.60	$604.80	$54.18	$85.18
2 Laborers	45.60	729.60	72.45	1159.20		
4 Equip. Oper. (medium)	60.40	1932.80	94.10	3011.20		
1 Truck Driver (heavy)	53.05	424.40	84.55	676.40		
1 Grader, 30,000 Lbs.		1108.49		1219.34		
1 F.E. Loader, Crl, 1.5 C.Y.		688.92		757.81		
1 Dozer, 300 H.P.		1840.03		2024.04		
1 Roller, Vibratory, 25 Ton		693.41		762.76		
1 Truck Tractor, 6x4, 450 H.P.		625.39		687.93		
1 Water Tank Trailer, 5000 Gal.		158.82		174.70	79.92	87.92
64 L.H., Daily Totals		$8582.66		$11078.17	$134.10	$173.10

Crew B-36C	Hr.	Daily	Hr.	Daily	Bare Costs	Incl. O&P
1 Labor Foreman (outside)	$47.60	$380.80	$75.60	$604.80	$56.37	$88.49
3 Equip. Oper. (medium)	60.40	1449.60	94.10	2258.40		
1 Truck Driver (heavy)	53.05	424.40	84.55	676.40		
1 Grader, 30,000 Lbs.		1108.49		1219.34		
1 Dozer, 300 H.P.		1840.03		2024.04		
1 Roller, Vibratory, 25 Ton		693.41		762.76		
1 Truck Tractor, 6x4, 450 H.P.		625.39		687.93		
1 Water Tank Trailer, 5000 Gal.		158.82		174.70	110.65	121.72
40 L.H., Daily Totals		$6680.94		$8408.36	$167.02	$210.21

Crew B-37	Hr.	Daily	Hr.	Daily	Bare Costs	Incl. O&P
1 Labor Foreman (outside)	$47.60	$380.80	$75.60	$604.80	$47.83	$75.70
4 Laborers	45.60	1459.20	72.45	2318.40		
1 Equip. Oper. (light)	57.00	456.00	88.80	710.40		
1 Tandem Roller, 5 Ton		267.23		293.96	5.57	6.12
48 L.H., Daily Totals		$2563.23		$3927.56	$53.40	$81.82

Crew B-37A	Hr.	Daily	Hr.	Daily	Bare Costs	Incl. O&P
2 Laborers	$45.60	$729.60	$72.45	$1159.20	$47.20	$75.07
1 Truck Driver (light)	50.40	403.20	80.30	642.40		
1 Flatbed Truck, Gas, 1.5 Ton		203.25		223.57		
1 Tar Kettle, T.M.		160.79		176.87	15.17	16.69
24 L.H., Daily Totals		$1496.84		$2202.04	$62.37	$91.75

Crew B-37B	Hr.	Daily	Hr.	Daily	Bare Costs	Incl. O&P
3 Laborers	$45.60	$1094.40	$72.45	$1738.80	$46.80	$74.41
1 Truck Driver (light)	50.40	403.20	80.30	642.40		
1 Flatbed Truck, Gas, 1.5 Ton		203.25		223.57		
1 Tar Kettle, T.M.		160.79		176.87	11.38	12.51
32 L.H., Daily Totals		$1861.64		$2781.64	$58.18	$86.93

Crew B-37C	Hr.	Daily	Hr.	Daily	Bare Costs	Incl. O&P
2 Laborers	$45.60	$729.60	$72.45	$1159.20	$48.00	$76.38
2 Truck Drivers (light)	50.40	806.40	80.30	1284.80		
2 Flatbed Trucks, Gas, 1.5 Ton		406.50		447.15		
1 Tar Kettle, T.M.		160.79		176.87	17.73	19.50
32 L.H., Daily Totals		$2103.29		$3068.02	$65.73	$95.88

Crew B-37D	Hr.	Daily	Hr.	Daily	Bare Costs	Incl. O&P
1 Laborer	$45.60	$364.80	$72.45	$579.60	$48.00	$76.38
1 Truck Driver (light)	50.40	403.20	80.30	642.40		
1 Pickup Truck, 3/4 Ton		115.08		126.59	7.19	7.91
16 L.H., Daily Totals		$883.08		$1348.59	$55.19	$84.29

Crew B-37E	Hr.	Daily	Hr.	Daily	Bare Costs	Incl. O&P
3 Laborers	$45.60	$1094.40	$72.45	$1738.80	$50.71	$80.12
1 Equip. Oper. (light)	57.00	456.00	88.80	710.40		
1 Equip. Oper. (medium)	60.40	483.20	94.10	752.80		
2 Truck Drivers (light)	50.40	806.40	80.30	1284.80		
4 Barrels w/ Flasher		17.23		18.95		
1 Concrete Saw		115.94		127.54		
1 Rotary Hammer Drill		54.02		59.42		
1 Hammer Drill Bit		26.18		28.80		
1 Loader, Skid Steer, 30 H.P.		184.81		203.29		
1 Conc. Hammer Attach.		122.73		135.00		
1 Vibrating Plate, Gas, 18"		32.77		36.05		
2 Flatbed Trucks, Gas, 1.5 Ton		406.50		447.15	17.15	18.86
56 L.H., Daily Totals		$3800.18		$5543.00	$67.86	$98.98

Crew B-37F	Hr.	Daily	Hr.	Daily	Bare Costs	Incl. O&P
3 Laborers	$45.60	$1094.40	$72.45	$1738.80	$46.80	$74.41
1 Truck Driver (light)	50.40	403.20	80.30	642.40		
4 Barrels w/ Flasher		17.23		18.95		
1 Concrete Mixer, 10 C.F.		151.35		166.49		
1 Air Compressor, 60 cfm		158.26		174.09		
1 -50' Air Hose, 3/4"		7.40		8.15		
1 Spade (Chipper)		8.80		9.68		
1 Flatbed Truck, Gas, 1.5 Ton		203.25		223.57	17.07	18.78
32 L.H., Daily Totals		$2043.90		$2982.13	$63.87	$93.19

Crew B-37G	Hr.	Daily	Hr.	Daily	Bare Costs	Incl. O&P
1 Labor Foreman (outside)	$47.60	$380.80	$75.60	$604.80	$47.83	$75.70
4 Laborers	45.60	1459.20	72.45	2318.40		
1 Equip. Oper. (light)	57.00	456.00	88.80	710.40		
1 Berm Machine		261.22		287.34		
1 Tandem Roller, 5 Ton		267.23		293.96	11.01	12.11
48 L.H., Daily Totals		$2824.45		$4214.90	$58.84	$87.81

Crew B-37H	Hr.	Daily	Hr.	Daily	Bare Costs	Incl. O&P
1 Labor Foreman (outside)	$47.60	$380.80	$75.60	$604.80	$47.83	$75.70
4 Laborers	45.60	1459.20	72.45	2318.40		
1 Equip. Oper. (light)	57.00	456.00	88.80	710.40		
1 Tandem Roller, 5 Ton		267.23		293.96		
1 Flatbed Truck, Gas, 1.5 Ton		203.25		223.57		
1 Tar Kettle, T.M.		160.79		176.87	13.15	14.47
48 L.H., Daily Totals		$2927.27		$4328.00	$60.98	$90.17

For customer support on your Commercial Renovation Costs with RSMeans data, call 800.448.8182.

773

Crew No.	Bare Costs		Incl. Subs O&P		Cost Per Labor-Hour	
Crew B-37I	Hr.	Daily	Hr.	Daily	Bare Costs	Incl. O&P
3 Laborers	$45.60	$1094.40	$72.45	$1738.80	$50.71	$80.12
1 Equip. Oper. (light)	57.00	456.00	88.80	710.40		
1 Equip. Oper. (medium)	60.40	483.20	94.10	752.80		
2 Truck Drivers (light)	50.40	806.40	80.30	1284.80		
4 Barrels w/ Flasher		17.23		18.95		
1 Concrete Saw		115.94		127.54		
1 Rotary Hammer Drill		54.02		59.42		
1 Hammer Drill Bit		26.18		28.80		
1 Air Compressor, 60 cfm		158.26		174.09		
1 -50' Air Hose, 3/4"		7.40		8.15		
1 Spade (Chipper)		8.80		9.68		
1 Loader, Skid Steer, 30 H.P.		184.81		203.29		
1 Conc. Hammer Attach.		122.73		135.00		
1 Concrete Mixer, 10 C.F.		151.35		166.49		
1 Vibrating Plate, Gas, 18"		32.77		36.05		
2 Flatbed Trucks, Gas, 1.5 Ton		406.50		447.15	22.96	25.26
56 L.H., Daily Totals		$4126.00		$5901.40	$73.68	$105.38
Crew B-37J	Hr.	Daily	Hr.	Daily	Bare Costs	Incl. O&P
1 Labor Foreman (outside)	$47.60	$380.80	$75.60	$604.80	$47.83	$75.70
4 Laborers	45.60	1459.20	72.45	2318.40		
1 Equip. Oper. (light)	57.00	456.00	88.80	710.40		
1 Air Compressor, 60 cfm		158.26		174.09		
1 -50' Air Hose, 3/4"		7.40		8.15		
2 Concrete Mixers, 10 C.F.		302.71		332.98		
2 Flatbed Trucks, Gas, 1.5 Ton		406.50		447.15		
1 Shot Blaster, 20"		215.77		237.34	22.72	24.99
48 L.H., Daily Totals		$3386.64		$4833.30	$70.55	$100.69
Crew B-37K	Hr.	Daily	Hr.	Daily	Bare Costs	Incl. O&P
1 Labor Foreman (outside)	$47.60	$380.80	$75.60	$604.80	$47.83	$75.70
4 Laborers	45.60	1459.20	72.45	2318.40		
1 Equip. Oper. (light)	57.00	456.00	88.80	710.40		
1 Air Compressor, 60 cfm		158.26		174.09		
1 -50' Air Hose, 3/4"		7.40		8.15		
2 Flatbed Trucks, Gas, 1.5 Ton		406.50		447.15		
1 Shot Blaster, 20"		215.77		237.34	16.42	18.06
48 L.H., Daily Totals		$3083.93		$4500.33	$64.25	$93.76
Crew B-38	Hr.	Daily	Hr.	Daily	Bare Costs	Incl. O&P
1 Labor Foreman (outside)	$47.60	$380.80	$75.60	$604.80	$51.24	$80.68
2 Laborers	45.60	729.60	72.45	1159.20		
1 Equip. Oper. (light)	57.00	456.00	88.80	710.40		
1 Equip. Oper. (medium)	60.40	483.20	94.10	752.80		
1 Backhoe Loader, 48 H.P.		221.72		243.89		
1 Hyd. Hammer (1200 lb.)		182.89		201.18		
1 F.E. Loader, W.M., 4 C.Y.		782.70		860.97		
1 Pvmt. Rem. Bucket		65.86		72.45	31.33	34.46
40 L.H., Daily Totals		$3302.77		$4605.69	$82.57	$115.14
Crew B-39	Hr.	Daily	Hr.	Daily	Bare Costs	Incl. O&P
1 Labor Foreman (outside)	$47.60	$380.80	$75.60	$604.80	$47.83	$75.70
4 Laborers	45.60	1459.20	72.45	2318.40		
1 Equip. Oper. (light)	57.00	456.00	88.80	710.40		
1 Air Compressor, 250 cfm		208.68		229.55		
2 Breakers, Pavement, 60 lb.		110.96		122.06		
2 -50' Air Hoses, 1.5"		47.19		51.91	7.64	8.41
48 L.H., Daily Totals		$2662.83		$4037.12	$55.48	$84.11
Crew B-40	Hr.	Daily	Hr.	Daily	Bare Costs	Incl. O&P
1 Pile Driver Foreman (outside)	$60.20	$481.60	$96.15	$769.20	$59.23	$93.70
4 Pile Drivers	58.20	1862.40	92.95	2974.40		
2 Equip. Oper. (crane)	63.40	1014.40	98.75	1580.00		
1 Equip. Oper. (oiler)	54.00	432.00	84.15	673.20		
1 Crawler Crane, 40 Ton		1267.52		1394.27		
1 Vibratory Hammer & Gen.		2364.52		2600.97	56.75	62.43
64 L.H., Daily Totals		$7422.44		$9992.04	$115.98	$156.13
Crew B-40B	Hr.	Daily	Hr.	Daily	Bare Costs	Incl. O&P
1 Labor Foreman (outside)	$47.60	$380.80	$75.60	$604.80	$50.30	$79.31
3 Laborers	45.60	1094.40	72.45	1738.80		
1 Equip. Oper. (crane)	63.40	507.20	98.75	790.00		
1 Equip. Oper. (oiler)	54.00	432.00	84.15	673.20		
1 Lattice Boom Crane, 40 Ton		2198.13		2417.95	45.79	50.37
48 L.H., Daily Totals		$4612.53		$6224.75	$96.09	$129.68
Crew B-41	Hr.	Daily	Hr.	Daily	Bare Costs	Incl. O&P
1 Labor Foreman (outside)	$47.60	$380.80	$75.60	$604.80	$47.15	$74.75
4 Laborers	45.60	1459.20	72.45	2318.40		
.25 Equip. Oper. (crane)	63.40	126.80	98.75	197.50		
.25 Equip. Oper. (oiler)	54.00	108.00	84.15	168.30		
.25 Crawler Crane, 40 Ton		316.88		348.57	7.20	7.92
44 L.H., Daily Totals		$2391.68		$3637.57	$54.36	$82.67
Crew B-42	Hr.	Daily	Hr.	Daily	Bare Costs	Incl. O&P
1 Labor Foreman (outside)	$47.60	$380.80	$75.60	$604.80	$51.17	$81.01
4 Laborers	45.60	1459.20	72.45	2318.40		
1 Equip. Oper. (crane)	63.40	507.20	98.75	790.00		
1 Equip. Oper. (oiler)	54.00	432.00	84.15	673.20		
1 Welder	61.95	495.60	99.75	798.00		
1 Hyd. Crane, 25 Ton		603.44		663.78		
1 Welder, Gas Engine, 300 amp		152.88		168.16		
1 Horz. Boring Csg. Mch.		337.69		371.46	17.09	18.80
64 L.H., Daily Totals		$4368.80		$6387.80	$68.26	$99.81
Crew B-43	Hr.	Daily	Hr.	Daily	Bare Costs	Incl. O&P
1 Labor Foreman (outside)	$47.60	$380.80	$75.60	$604.80	$50.30	$79.31
3 Laborers	45.60	1094.40	72.45	1738.80		
1 Equip. Oper. (crane)	63.40	507.20	98.75	790.00		
1 Equip. Oper. (oiler)	54.00	432.00	84.15	673.20		
1 Drill Rig, Truck-Mounted		785.11		863.62	16.36	17.99
48 L.H., Daily Totals		$3199.51		$4670.42	$66.66	$97.30
Crew B-44	Hr.	Daily	Hr.	Daily	Bare Costs	Incl. O&P
1 Pile Driver Foreman (outside)	$60.20	$481.60	$96.15	$769.20	$58.17	$92.24
4 Pile Drivers	58.20	1862.40	92.95	2974.40		
2 Equip. Oper. (crane)	63.40	1014.40	98.75	1580.00		
1 Laborer	45.60	364.80	72.45	579.60		
1 Crawler Crane, 40 Ton		1267.52		1394.27		
1 Lead, 60' High		219.12		241.03		
1 Hammer, Diesel, 15K ft.-lbs.		645.54		710.10	33.32	36.65
64 L.H., Daily Totals		$5855.38		$8248.60	$91.49	$128.88
Crew B-45	Hr.	Daily	Hr.	Daily	Bare Costs	Incl. O&P
1 Equip. Oper. (medium)	$60.40	$483.20	$94.10	$752.80	$56.73	$89.33
1 Truck Driver (heavy)	53.05	424.40	84.55	676.40		
1 Dist. Tanker, 3000 Gallon		344.69		379.16		
1 Truck Tractor, 6x4, 380 H.P.		512.66		563.93	53.58	58.94
16 L.H., Daily Totals		$1764.95		$2372.29	$110.31	$148.27

For customer support on your Commercial Renovation Costs with RSMeans data, call 800.448.8182.

Crew No.	Bare Costs		Incl. Subs O&P		Cost Per Labor-Hour	

Crew B-46	Hr.	Daily	Hr.	Daily	Bare Costs	Incl. O&P
1 Pile Driver Foreman (outside)	$60.20	$481.60	$96.15	$769.20	$52.23	$83.23
2 Pile Drivers	58.20	931.20	92.95	1487.20		
3 Laborers	45.60	1094.40	72.45	1738.80		
1 Chain Saw, Gas, 36" Long		43.10		47.42	.90	.99
48 L.H., Daily Totals		$2550.30		$4042.62	$53.13	$84.22

Crew B-47	Hr.	Daily	Hr.	Daily	Bare Costs	Incl. O&P
1 Blast Foreman (outside)	$47.60	$380.80	$75.60	$604.80	$50.07	$78.95
1 Driller	45.60	364.80	72.45	579.60		
1 Equip. Oper. (light)	57.00	456.00	88.80	710.40		
1 Air Track Drill, 4"		1160.28		1276.31		
1 Air Compressor, 600 cfm		437.66		481.43		
2 -50' Air Hoses, 3"		79.46		87.40	69.89	76.88
24 L.H., Daily Totals		$2879.01		$3739.95	$119.96	$155.83

Crew B-47A	Hr.	Daily	Hr.	Daily	Bare Costs	Incl. O&P
1 Drilling Foreman (outside)	$47.60	$380.80	$75.60	$604.80	$55.00	$86.17
1 Equip. Oper. (heavy)	63.40	507.20	98.75	790.00		
1 Equip. Oper. (oiler)	54.00	432.00	84.15	673.20		
1 Air Track Drill, 5"		1160.28		1276.31	48.35	53.18
24 L.H., Daily Totals		$2480.28		$3344.31	$103.35	$139.35

Crew B-47C	Hr.	Daily	Hr.	Daily	Bare Costs	Incl. O&P
1 Laborer	$45.60	$364.80	$72.45	$579.60	$51.30	$80.63
1 Equip. Oper. (light)	57.00	456.00	88.80	710.40		
1 Air Compressor, 750 C.F.M.		613.58		674.94		
2 -50' Air Hoses, 3"		79.46		87.40		
1 Air Track Drill, 4"		1160.28		1276.31	115.83	127.42
16 L.H., Daily Totals		$2674.12		$3328.66	$167.13	$208.04

Crew B-47E	Hr.	Daily	Hr.	Daily	Bare Costs	Incl. O&P
1 Labor Foreman (outside)	$47.60	$380.80	$75.60	$604.80	$46.10	$73.24
3 Laborers	45.60	1094.40	72.45	1738.80		
1 Flatbed Truck, Gas, 3 Ton		877.96		965.75	27.44	30.18
32 L.H., Daily Totals		$2353.16		$3309.35	$73.54	$103.42

Crew B-47G	Hr.	Daily	Hr.	Daily	Bare Costs	Incl. O&P
1 Labor Foreman (outside)	$47.60	$380.80	$75.60	$604.80	$48.95	$77.33
2 Laborers	45.60	729.60	72.45	1159.20		
1 Equip. Oper. (light)	57.00	456.00	88.80	710.40		
1 Air Track Drill, 4"		1160.28		1276.31		
1 Air Compressor, 600 cfm		437.66		481.43		
2 -50' Air Hoses, 3"		79.46		87.40		
1 Gunite Pump Rig		331.07		364.17	62.76	69.04
32 L.H., Daily Totals		$3574.87		$4683.72	$111.71	$146.37

Crew B-47H	Hr.	Daily	Hr.	Daily	Bare Costs	Incl. O&P
1 Skilled Worker Foreman (out)	$60.90	$487.20	$96.35	$770.80	$59.40	$93.99
3 Skilled Workers	58.90	1413.60	93.20	2236.80		
1 Flatbed Truck, Gas, 3 Ton		877.96		965.75	27.44	30.18
32 L.H., Daily Totals		$2778.76		$3973.35	$86.84	$124.17

Crew B-48	Hr.	Daily	Hr.	Daily	Bare Costs	Incl. O&P
1 Labor Foreman (outside)	$47.60	$380.80	$75.60	$604.80	$51.26	$80.66
3 Laborers	45.60	1094.40	72.45	1738.80		
1 Equip. Oper. (crane)	63.40	507.20	98.75	790.00		
1 Equip. Oper. (oiler)	54.00	432.00	84.15	673.20		
1 Equip. Oper. (light)	57.00	456.00	88.80	710.40		
1 Centr. Water Pump, 6"		240.92		265.01		
1 -20' Suction Hose, 6"		26.41		29.05		
1 -50' Discharge Hose, 6"		18.77		20.65		
1 Drill Rig, Truck-Mounted		785.11		863.62	19.13	21.04
56 L.H., Daily Totals		$3941.60		$5695.52	$70.39	$101.71

Crew B-49	Hr.	Daily	Hr.	Daily	Bare Costs	Incl. O&P
1 Labor Foreman (outside)	$47.60	$380.80	$75.60	$604.80	$53.87	$84.86
3 Laborers	45.60	1094.40	72.45	1738.80		
2 Equip. Oper. (crane)	63.40	1014.40	98.75	1580.00		
2 Equip. Oper. (oilers)	54.00	864.00	84.15	1346.40		
1 Equip. Oper. (light)	57.00	456.00	88.80	710.40		
2 Pile Drivers	58.20	931.20	92.95	1487.20		
1 Hyd. Crane, 25 Ton		603.44		663.78		
1 Centr. Water Pump, 6"		240.92		265.01		
1 -20' Suction Hose, 6"		26.41		29.05		
1 -50' Discharge Hose, 6"		18.77		20.65		
1 Drill Rig, Truck-Mounted		785.11		863.62	19.03	20.93
88 L.H., Daily Totals		$6415.44		$9309.70	$72.90	$105.79

Crew B-50	Hr.	Daily	Hr.	Daily	Bare Costs	Incl. O&P
2 Pile Driver Foremen (outside)	$60.20	$963.20	$96.15	$1538.40	$56.23	$89.21
6 Pile Drivers	58.20	2793.60	92.95	4461.60		
2 Equip. Oper. (crane)	63.40	1014.40	98.75	1580.00		
1 Equip. Oper. (oiler)	54.00	432.00	84.15	673.20		
3 Laborers	45.60	1094.40	72.45	1738.80		
1 Crawler Crane, 40 Ton		1267.52		1394.27		
1 Lead, 60' High		219.12		241.03		
1 Hammer, Diesel, 15K ft.-lbs.		645.54		710.10		
1 Air Compressor, 600 cfm		437.66		481.43		
2 -50' Air Hoses, 3"		79.46		87.40		
1 Chain Saw, Gas, 36" Long		43.10		47.42	24.04	26.44
112 L.H., Daily Totals		$8990.01		$12953.65	$80.27	$115.66

Crew B-51	Hr.	Daily	Hr.	Daily	Bare Costs	Incl. O&P
1 Labor Foreman (outside)	$47.60	$380.80	$75.60	$604.80	$46.73	$74.28
4 Laborers	45.60	1459.20	72.45	2318.40		
1 Truck Driver (light)	50.40	403.20	80.30	642.40		
1 Flatbed Truck, Gas, 1.5 Ton		203.25		223.57	4.23	4.66
48 L.H., Daily Totals		$2446.45		$3789.17	$50.97	$78.94

Crew B-52	Hr.	Daily	Hr.	Daily	Bare Costs	Incl. O&P
1 Carpenter Foreman (outside)	$58.30	$466.40	$92.60	$740.80	$52.16	$82.42
1 Carpenter	56.30	450.40	89.45	715.60		
3 Laborers	45.60	1094.40	72.45	1738.80		
1 Cement Finisher	53.15	425.20	83.05	664.40		
.5 Rodman (reinf.)	60.70	242.80	94.90	379.60		
.5 Equip. Oper. (medium)	60.40	241.60	94.10	376.40		
.5 Crawler Loader, 3 C.Y.		590.46		649.50	10.54	11.60
56 L.H., Daily Totals		$3511.26		$5265.10	$62.70	$94.02

Crew B-53	Hr.	Daily	Hr.	Daily	Bare Costs	Incl. O&P
1 Equip. Oper. (light)	$57.00	$456.00	$88.80	$710.40	$57.00	$88.80
1 Trencher, Chain, 12 H.P.		163.91		180.30	20.49	22.54
8 L.H., Daily Totals		$619.91		$890.70	$77.49	$111.34

Crew No.	Bare Costs Hr.	Bare Costs Daily	Incl. Subs O & P Hr.	Incl. Subs O & P Daily	Cost Per Labor-Hour Bare Costs	Cost Per Labor-Hour Incl. O&P
Crew B-54	Hr.	Daily	Hr.	Daily	Bare Costs	Incl. O&P
1 Equip. Oper. (light)	$57.00	$456.00	$88.80	$710.40	$57.00	$88.80
1 Trencher, Chain, 40 H.P.		465.06		511.57	58.13	63.95
8 L.H., Daily Totals		$921.06		$1221.97	$115.13	$152.75
Crew B-54A	Hr.	Daily	Hr.	Daily	Bare Costs	Incl. O&P
.17 Labor Foreman (outside)	$47.60	$64.74	$75.60	$102.82	$58.54	$91.41
1 Equipment Operator (med.)	60.40	483.20	94.10	752.80		
1 Wheel Trencher, 67 H.P.		1172.66		1289.93	125.28	137.81
9.36 L.H., Daily Totals		$1720.60		$2145.55	$183.82	$229.23
Crew B-54B	Hr.	Daily	Hr.	Daily	Bare Costs	Incl. O&P
.25 Labor Foreman (outside)	$47.60	$95.20	$75.60	$151.20	$57.84	$90.40
1 Equipment Operator (med.)	60.40	483.20	94.10	752.80		
1 Wheel Trencher, 150 H.P.		1271.72		1398.89	127.17	139.89
10 L.H., Daily Totals		$1850.12		$2302.89	$185.01	$230.29
Crew B-54D	Hr.	Daily	Hr.	Daily	Bare Costs	Incl. O&P
1 Laborer	$45.60	$364.80	$72.45	$579.60	$53.00	$83.28
1 Equipment Operator (med.)	60.40	483.20	94.10	752.80		
1 Rock Trencher, 6" Width		443.83		488.21	27.74	30.51
16 L.H., Daily Totals		$1291.83		$1820.61	$80.74	$113.79
Crew B-54E	Hr.	Daily	Hr.	Daily	Bare Costs	Incl. O&P
1 Laborer	$45.60	$364.80	$72.45	$579.60	$53.00	$83.28
1 Equipment Operator (med.)	60.40	483.20	94.10	752.80		
1 Rock Trencher, 18" Width		1038.46		1142.30	64.90	71.39
16 L.H., Daily Totals		$1886.46		$2474.70	$117.90	$154.67
Crew B-55	Hr.	Daily	Hr.	Daily	Bare Costs	Incl. O&P
2 Laborers	$45.60	$729.60	$72.45	$1159.20	$47.20	$75.07
1 Truck Driver (light)	50.40	403.20	80.30	642.40		
1 Truck-Mounted Earth Auger		402.88		443.17		
1 Flatbed Truck, Gas, 3 Ton		877.96		965.75	53.37	58.71
24 L.H., Daily Totals		$2413.64		$3210.52	$100.57	$133.77
Crew B-56	Hr.	Daily	Hr.	Daily	Bare Costs	Incl. O&P
1 Laborer	$45.60	$364.80	$72.45	$579.60	$51.30	$80.63
1 Equip. Oper. (light)	57.00	456.00	88.80	710.40		
1 Air Track Drill, 4"		1160.28		1276.31		
1 Air Compressor, 600 cfm		437.66		481.43		
1 -50' Air Hose, 3"		39.73		43.70	102.35	112.59
16 L.H., Daily Totals		$2458.48		$3091.44	$153.65	$193.22
Crew B-57	Hr.	Daily	Hr.	Daily	Bare Costs	Incl. O&P
1 Labor Foreman (outside)	$47.60	$380.80	$75.60	$604.80	$52.20	$82.03
2 Laborers	45.60	729.60	72.45	1159.20		
1 Equip. Oper. (crane)	63.40	507.20	98.75	790.00		
1 Equip. Oper. (light)	57.00	456.00	88.80	710.40		
1 Equip. Oper. (oiler)	54.00	432.00	84.15	673.20		
1 Crawler Crane, 25 Ton		1185.81		1304.39		
1 Clamshell Bucket, 1 C.Y.		71.59		78.75		
1 Centr. Water Pump, 6"		240.92		265.01		
1 -20' Suction Hose, 6"		26.41		29.05		
20 -50' Discharge Hoses, 6"		375.41		412.95	39.59	43.54
48 L.H., Daily Totals		$4405.73		$6027.74	$91.79	$125.58

Crew No.	Bare Costs Hr.	Bare Costs Daily	Incl. Subs O & P Hr.	Incl. Subs O & P Daily	Cost Per Labor-Hour Bare Costs	Cost Per Labor-Hour Incl. O&P
Crew B-58	Hr.	Daily	Hr.	Daily	Bare Costs	Incl. O&P
2 Laborers	$45.60	$729.60	$72.45	$1159.20	$49.40	$77.90
1 Equip. Oper. (light)	57.00	456.00	88.80	710.40		
1 Backhoe Loader, 48 H.P.		221.72		243.89		
1 Small Helicopter, w/Pilot		2165.30		2381.83	99.46	109.40
24 L.H., Daily Totals		$3572.61		$4495.31	$148.86	$187.30
Crew B-59	Hr.	Daily	Hr.	Daily	Bare Costs	Incl. O&P
1 Truck Driver (heavy)	$53.05	$424.40	$84.55	$676.40	$53.05	$84.55
1 Truck Tractor, 220 H.P.		319.23		351.15		
1 Water Tank Trailer, 5000 Gal.		158.82		174.70	59.76	65.73
8 L.H., Daily Totals		$902.45		$1202.25	$112.81	$150.28
Crew B-59A	Hr.	Daily	Hr.	Daily	Bare Costs	Incl. O&P
2 Laborers	$45.60	$729.60	$72.45	$1159.20	$48.08	$76.48
1 Truck Driver (heavy)	53.05	424.40	84.55	676.40		
1 Water Tank Trailer, 5000 Gal.		158.82		174.70		
1 Truck Tractor, 220 H.P.		319.23		351.15	19.92	21.91
24 L.H., Daily Totals		$1632.05		$2361.45	$68.00	$98.39
Crew B-60	Hr.	Daily	Hr.	Daily	Bare Costs	Incl. O&P
1 Labor Foreman (outside)	$47.60	$380.80	$75.60	$604.80	$52.89	$83.00
2 Laborers	45.60	729.60	72.45	1159.20		
1 Equip. Oper. (crane)	63.40	507.20	98.75	790.00		
2 Equip. Oper. (light)	57.00	912.00	88.80	1420.80		
1 Equip. Oper. (oiler)	54.00	432.00	84.15	673.20		
1 Crawler Crane, 40 Ton		1267.52		1394.27		
1 Lead, 60' High		219.12		241.03		
1 Hammer, Diesel, 15K ft.-lbs.		645.54		710.10		
1 Backhoe Loader, 48 H.P.		221.72		243.89	42.03	46.24
56 L.H., Daily Totals		$5315.50		$7237.29	$94.92	$129.24
Crew B-61	Hr.	Daily	Hr.	Daily	Bare Costs	Incl. O&P
1 Labor Foreman (outside)	$47.60	$380.80	$75.60	$604.80	$48.28	$76.35
3 Laborers	45.60	1094.40	72.45	1738.80		
1 Equip. Oper. (light)	57.00	456.00	88.80	710.40		
1 Cement Mixer, 2 C.Y.		115.40		126.94		
1 Air Compressor, 160 cfm		218.71		240.59	8.35	9.19
40 L.H., Daily Totals		$2265.31		$3421.52	$56.63	$85.54
Crew B-62	Hr.	Daily	Hr.	Daily	Bare Costs	Incl. O&P
2 Laborers	$45.60	$729.60	$72.45	$1159.20	$49.40	$77.90
1 Equip. Oper. (light)	57.00	456.00	88.80	710.40		
1 Loader, Skid Steer, 30 H.P.		184.81		203.29	7.70	8.47
24 L.H., Daily Totals		$1370.41		$2072.89	$57.10	$86.37
Crew B-62A	Hr.	Daily	Hr.	Daily	Bare Costs	Incl. O&P
2 Laborers	$45.60	$729.60	$72.45	$1159.20	$49.40	$77.90
1 Equip. Oper. (light)	57.00	456.00	88.80	710.40		
1 Loader, Skid Steer, 30 H.P.		184.81		203.29		
1 Trencher Attachment		68.63		75.49	10.56	11.62
24 L.H., Daily Totals		$1439.04		$2148.38	$59.96	$89.52
Crew B-63	Hr.	Daily	Hr.	Daily	Bare Costs	Incl. O&P
4 Laborers	$45.60	$1459.20	$72.45	$2318.40	$47.88	$75.72
1 Equip. Oper. (light)	57.00	456.00	88.80	710.40		
1 Loader, Skid Steer, 30 H.P.		184.81		203.29	4.62	5.08
40 L.H., Daily Totals		$2100.01		$3232.09	$52.50	$80.80

Crews - Renovation

Crew B-63B	Hr.	Daily	Hr.	Daily	Bare Costs	Incl. O&P
1 Labor Foreman (inside)	$46.10	$368.80	$73.25	$586.00	$48.58	$76.74
2 Laborers	45.60	729.60	72.45	1159.20		
1 Equip. Oper. (light)	57.00	456.00	88.80	710.40		
1 Loader, Skid Steer, 78 H.P.		460.30		506.33	14.38	15.82
32 L.H., Daily Totals		$2014.70		$2961.93	$62.96	$92.56

Crew B-64	Hr.	Daily	Hr.	Daily	Bare Costs	Incl. O&P
1 Laborer	$45.60	$364.80	$72.45	$579.60	$48.00	$76.38
1 Truck Driver (light)	50.40	403.20	80.30	642.40		
1 Power Mulcher (small)		207.17		227.89		
1 Flatbed Truck, Gas, 1.5 Ton		203.25		223.57	25.65	28.22
16 L.H., Daily Totals		$1178.42		$1673.47	$73.65	$104.59

Crew B-65	Hr.	Daily	Hr.	Daily	Bare Costs	Incl. O&P
1 Laborer	$45.60	$364.80	$72.45	$579.60	$48.00	$76.38
1 Truck Driver (light)	50.40	403.20	80.30	642.40		
1 Power Mulcher (Large)		355.38		390.92		
1 Flatbed Truck, Gas, 1.5 Ton		203.25		223.57	34.91	38.41
16 L.H., Daily Totals		$1326.63		$1836.49	$82.91	$114.78

Crew B-66	Hr.	Daily	Hr.	Daily	Bare Costs	Incl. O&P
1 Equip. Oper. (light)	$57.00	$456.00	$88.80	$710.40	$57.00	$88.80
1 Loader-Backhoe, 40 H.P.		275.82		303.40	34.48	37.93
8 L.H., Daily Totals		$731.82		$1013.80	$91.48	$126.73

Crew B-67	Hr.	Daily	Hr.	Daily	Bare Costs	Incl. O&P
1 Millwright	$60.55	$484.40	$93.55	$748.40	$58.77	$91.17
1 Equip. Oper. (light)	57.00	456.00	88.80	710.40		
1 R.T. Forklift, 5,000 Lb., diesel		280.85		308.94	17.55	19.31
16 L.H., Daily Totals		$1221.25		$1767.74	$76.33	$110.48

Crew B-67B	Hr.	Daily	Hr.	Daily	Bare Costs	Incl. O&P
1 Millwright Foreman (inside)	$61.05	$488.40	$94.30	$754.40	$60.80	$93.92
1 Millwright	60.55	484.40	93.55	748.40		
16 L.H., Daily Totals		$972.80		$1502.80	$60.80	$93.92

Crew B-68	Hr.	Daily	Hr.	Daily	Bare Costs	Incl. O&P
2 Millwrights	$60.55	$968.80	$93.55	$1496.80	$59.37	$91.97
1 Equip. Oper. (light)	57.00	456.00	88.80	710.40		
1 R.T. Forklift, 5,000 Lb., diesel		280.85		308.94	11.70	12.87
24 L.H., Daily Totals		$1705.65		$2516.14	$71.07	$104.84

Crew B-68A	Hr.	Daily	Hr.	Daily	Bare Costs	Incl. O&P
1 Millwright Foreman (inside)	$61.05	$488.40	$94.30	$754.40	$60.72	$93.80
2 Millwrights	60.55	968.80	93.55	1496.80		
1 Forklift, Smooth Floor, 8,000 Lb.		292.16		321.38	12.17	13.39
24 L.H., Daily Totals		$1749.36		$2572.58	$72.89	$107.19

Crew B-68B	Hr.	Daily	Hr.	Daily	Bare Costs	Incl. O&P
1 Millwright Foreman (inside)	$61.05	$488.40	$94.30	$754.40	$64.88	$100.00
2 Millwrights	60.55	968.80	93.55	1496.80		
2 Electricians	66.20	1059.20	101.70	1627.20		
2 Plumbers	69.80	1116.80	107.60	1721.60		
1 R.T. Forklift, 5,000 Lb., gas		291.20		320.32	5.20	5.72
56 L.H., Daily Totals		$3924.40		$5920.32	$70.08	$105.72

Crew B-68C	Hr.	Daily	Hr.	Daily	Bare Costs	Incl. O&P
1 Millwright Foreman (inside)	$61.05	$488.40	$94.30	$754.40	$64.40	$99.29
1 Millwright	60.55	484.40	93.55	748.40		
1 Electrician	66.20	529.60	101.70	813.60		
1 Plumber	69.80	558.40	107.60	860.80		
1 R.T. Forklift, 5,000 Lb., gas		291.20		320.32	9.10	10.01
32 L.H., Daily Totals		$2352.00		$3497.52	$73.50	$109.30

Crew B-68D	Hr.	Daily	Hr.	Daily	Bare Costs	Incl. O&P
1 Labor Foreman (inside)	$46.10	$368.80	$73.25	$586.00	$49.57	$78.17
1 Laborer	45.60	364.80	72.45	579.60		
1 Equip. Oper. (light)	57.00	456.00	88.80	710.40		
1 R.T. Forklift, 5,000 Lb., gas		291.20		320.32	12.13	13.35
24 L.H., Daily Totals		$1480.80		$2196.32	$61.70	$91.51

Crew B-68E	Hr.	Daily	Hr.	Daily	Bare Costs	Incl. O&P
1 Struc. Steel Foreman (inside)	$62.35	$498.80	$100.40	$803.20	$61.97	$99.79
3 Struc. Steel Workers	61.85	1484.40	99.60	2390.40		
1 Welder	61.95	495.60	99.75	798.00		
1 Forklift, Smooth Floor, 8,000 Lb.		292.16		321.38	7.30	8.03
40 L.H., Daily Totals		$2770.96		$4312.98	$69.27	$107.82

Crew B-68F	Hr.	Daily	Hr.	Daily	Bare Costs	Incl. O&P
1 Skilled Worker Foreman (out)	$60.90	$487.20	$96.35	$770.80	$59.57	$94.25
2 Skilled Workers	58.90	942.40	93.20	1491.20		
1 R.T. Forklift, 5,000 Lb., gas		291.20		320.32	12.13	13.35
24 L.H., Daily Totals		$1720.80		$2582.32	$71.70	$107.60

Crew B-68G	Hr.	Daily	Hr.	Daily	Bare Costs	Incl. O&P
2 Structural Steel Workers	$61.85	$989.60	$99.60	$1593.60	$61.85	$99.60
1 R.T. Forklift, 5,000 Lb., gas		291.20		320.32	18.20	20.02
16 L.H., Daily Totals		$1280.80		$1913.92	$80.05	$119.62

Crew B-69	Hr.	Daily	Hr.	Daily	Bare Costs	Incl. O&P
1 Labor Foreman (outside)	$47.60	$380.80	$75.60	$604.80	$50.30	$79.31
3 Laborers	45.60	1094.40	72.45	1738.80		
1 Equip. Oper. (crane)	63.40	507.20	98.75	790.00		
1 Equip. Oper. (oiler)	54.00	432.00	84.15	673.20		
1 Hyd. Crane, 80 Ton		1502.22		1652.44	31.30	34.43
48 L.H., Daily Totals		$3916.62		$5459.24	$81.60	$113.73

Crew B-69A	Hr.	Daily	Hr.	Daily	Bare Costs	Incl. O&P
1 Labor Foreman (outside)	$47.60	$380.80	$75.60	$604.80	$49.66	$78.35
3 Laborers	45.60	1094.40	72.45	1738.80		
1 Equip. Oper. (medium)	60.40	483.20	94.10	752.80		
1 Concrete Finisher	53.15	425.20	83.05	664.40		
1 Curb/Gutter Paver, 2-Track		1268.80		1395.68	26.43	29.08
48 L.H., Daily Totals		$3652.40		$5156.48	$76.09	$107.43

Crew B-69B	Hr.	Daily	Hr.	Daily	Bare Costs	Incl. O&P
1 Labor Foreman (outside)	$47.60	$380.80	$75.60	$604.80	$49.66	$78.35
3 Laborers	45.60	1094.40	72.45	1738.80		
1 Equip. Oper. (medium)	60.40	483.20	94.10	752.80		
1 Cement Finisher	53.15	425.20	83.05	664.40		
1 Curb/Gutter Paver, 4-Track		825.78		908.36	17.20	18.92
48 L.H., Daily Totals		$3209.38		$4669.16	$66.86	$97.27

For customer support on your Commercial Renovation Costs with RSMeans data, call 800.448.8182.

777

Crew No.		Bare Costs		Incl. Subs O&P		Cost Per Labor-Hour	
Crew B-70	Hr.	Daily	Hr.	Daily	Bare Costs	Incl. O&P
1 Labor Foreman (outside)	$47.60	$380.80	$75.60	$604.80	$52.23	$82.18
3 Laborers	45.60	1094.40	72.45	1738.80		
3 Equip. Oper. (medium)	60.40	1449.60	94.10	2258.40		
1 Grader, 30,000 Lbs.		1108.49		1219.34		
1 Ripper, Beam & 1 Shank		94.41		103.85		
1 Road Sweeper, S.P., 8' wide		745.48		820.02		
1 F.E. Loader, W.M., 1.5 C.Y.		454.98		500.48	42.92	47.21
56 L.H., Daily Totals		$5328.15		$7245.69	$95.15	$129.39
Crew B-71	Hr.	Daily	Hr.	Daily	Bare Costs	Incl. O&P
1 Labor Foreman (outside)	$47.60	$380.80	$75.60	$604.80	$52.23	$82.18
3 Laborers	45.60	1094.40	72.45	1738.80		
3 Equip. Oper. (medium)	60.40	1449.60	94.10	2258.40		
1 Pvmt. Profiler, 750 H.P.		3565.85		3922.43		
1 Road Sweeper, S.P., 8' wide		745.48		820.02		
1 F.E. Loader, W.M., 1.5 C.Y.		454.98		500.48	85.11	93.62
56 L.H., Daily Totals		$7691.10		$9844.93	$137.34	$175.80
Crew B-72	Hr.	Daily	Hr.	Daily	Bare Costs	Incl. O&P
1 Labor Foreman (outside)	$47.60	$380.80	$75.60	$604.80	$53.25	$83.67
3 Laborers	45.60	1094.40	72.45	1738.80		
4 Equip. Oper. (medium)	60.40	1932.80	94.10	3011.20		
1 Pvmt. Profiler, 750 H.P.		3565.85		3922.43		
1 Hammermill, 250 H.P.		879.87		967.85		
1 Windrow Loader		1507.52		1658.28		
1 Mix Paver, 165 H.P.		2239.34		2463.27		
1 Roller, Pneum. Whl., 12 Ton		361.08		397.18	133.65	147.02
64 L.H., Daily Totals		$11961.65		$14763.81	$186.90	$230.68
Crew B-73	Hr.	Daily	Hr.	Daily	Bare Costs	Incl. O&P
1 Labor Foreman (outside)	$47.60	$380.80	$75.60	$604.80	$55.10	$86.38
2 Laborers	45.60	729.60	72.45	1159.20		
5 Equip. Oper. (medium)	60.40	2416.00	94.10	3764.00		
1 Road Mixer, 310 H.P.		1979.64		2177.61		
1 Tandem Roller, 10 Ton		254.53		279.98		
1 Hammermill, 250 H.P.		879.87		967.85		
1 Grader, 30,000 Lbs.		1108.49		1219.34		
.5 F.E. Loader, W.M., 1.5 C.Y.		227.49		250.24		
.5 Truck Tractor, 220 H.P.		159.61		175.58		
.5 Water Tank Trailer, 5000 Gal.		79.41		87.35	73.27	80.59
64 L.H., Daily Totals		$8215.44		$10685.94	$128.37	$166.97
Crew B-74	Hr.	Daily	Hr.	Daily	Bare Costs	Incl. O&P
1 Labor Foreman (outside)	$47.60	$380.80	$75.60	$604.80	$55.11	$86.69
1 Laborer	45.60	364.80	72.45	579.60		
4 Equip. Oper. (medium)	60.40	1932.80	94.10	3011.20		
2 Truck Drivers (heavy)	53.05	848.80	84.55	1352.80		
1 Grader, 30,000 Lbs.		1108.49		1219.34		
1 Ripper, Beam & 1 Shank		94.41		103.85		
2 Stabilizers, 310 H.P.		2889.92		3178.91		
1 Flatbed Truck, Gas, 3 Ton		877.96		965.75		
1 Chem. Spreader, Towed		88.12		96.93		
1 Roller, Vibratory, 25 Ton		693.41		762.76		
1 Water Tank Trailer, 5000 Gal.		158.82		174.70		
1 Truck Tractor, 220 H.P.		319.23		351.15	97.35	107.08
64 L.H., Daily Totals		$9757.55		$12401.79	$152.46	$193.78
Crew B-75	Hr.	Daily	Hr.	Daily	Bare Costs	Incl. O&P
1 Labor Foreman (outside)	$47.60	$380.80	$75.60	$604.80	$55.41	$87.00
1 Laborer	45.60	364.80	72.45	579.60		
4 Equip. Oper. (medium)	60.40	1932.80	94.10	3011.20		
1 Truck Driver (heavy)	53.05	424.40	84.55	676.40		
1 Grader, 30,000 Lbs.		1108.49		1219.34		
1 Ripper, Beam & 1 Shank		94.41		103.85		
2 Stabilizers, 310 H.P.		2889.92		3178.91		
1 Dist. Tanker, 3000 Gallon		344.69		379.16		
1 Truck Tractor, 6x4, 380 H.P.		512.66		563.93		
1 Roller, Vibratory, 25 Ton		693.41		762.76	100.78	110.86
56 L.H., Daily Totals		$8746.38		$11079.94	$156.19	$197.86
Crew B-76	Hr.	Daily	Hr.	Daily	Bare Costs	Incl. O&P
1 Dock Builder Foreman (outside)	$60.20	$481.60	$96.15	$769.20	$59.11	$93.62
5 Dock Builders	58.20	2328.00	92.95	3718.00		
2 Equip. Oper. (crane)	63.40	1014.40	98.75	1580.00		
1 Equip. Oper. (oiler)	54.00	432.00	84.15	673.20		
1 Crawler Crane, 50 Ton		1586.15		1744.77		
1 Barge, 400 Ton		898.90		988.79		
1 Hammer, Diesel, 15K ft.-lbs.		645.54		710.10		
1 Lead, 60' High		219.12		241.03		
1 Air Compressor, 600 cfm		437.66		481.43		
2 -50' Air Hoses, 3"		79.46		87.40	53.71	59.08
72 L.H., Daily Totals		$8122.84		$10993.92	$112.82	$152.69
Crew B-76A	Hr.	Daily	Hr.	Daily	Bare Costs	Incl. O&P
1 Labor Foreman (outside)	$47.60	$380.80	$75.60	$604.80	$49.13	$77.59
5 Laborers	45.60	1824.00	72.45	2898.00		
1 Equip. Oper. (crane)	63.40	507.20	98.75	790.00		
1 Equip. Oper. (oiler)	54.00	432.00	84.15	673.20		
1 Crawler Crane, 50 Ton		1586.15		1744.77		
1 Barge, 400 Ton		898.90		988.79	38.83	42.71
64 L.H., Daily Totals		$5629.05		$7699.56	$87.95	$120.31
Crew B-77	Hr.	Daily	Hr.	Daily	Bare Costs	Incl. O&P
1 Labor Foreman (outside)	$47.60	$380.80	$75.60	$604.80	$46.96	$74.65
3 Laborers	45.60	1094.40	72.45	1738.80		
1 Truck Driver (light)	50.40	403.20	80.30	642.40		
1 Crack Cleaner, 25 H.P.		54.53		59.98		
1 Crack Filler, Trailer Mtd.		176.12		193.73		
1 Flatbed Truck, Gas, 3 Ton		877.96		965.75	27.72	30.49
40 L.H., Daily Totals		$2987.01		$4205.47	$74.68	$105.14
Crew B-78	Hr.	Daily	Hr.	Daily	Bare Costs	Incl. O&P
1 Labor Foreman (outside)	$47.60	$380.80	$75.60	$604.80	$46.73	$74.28
4 Laborers	45.60	1459.20	72.45	2318.40		
1 Truck Driver (light)	50.40	403.20	80.30	642.40		
1 Paint Striper, S.P., 40 Gallon		132.19		145.41		
1 Flatbed Truck, Gas, 3 Ton		877.96		965.75		
1 Pickup Truck, 3/4 Ton		115.08		126.59	23.44	25.79
48 L.H., Daily Totals		$3368.43		$4803.35	$70.18	$100.07
Crew B-78A	Hr.	Daily	Hr.	Daily	Bare Costs	Incl. O&P
1 Equip. Oper. (light)	$57.00	$456.00	$88.80	$710.40	$57.00	$88.80
1 Line Rem. (Metal Balls) 115 H.P.		1026.71		1129.39	128.34	141.17
8 L.H., Daily Totals		$1482.71		$1839.79	$185.34	$229.97

For customer support on your Commercial Renovation Costs with RSMeans data, call 800.448.8182.

Crews - Renovation

Crew B-78B

Crew No.	Hr.	Daily	Hr.	Daily	Bare Costs	Incl. O&P
2 Laborers	$45.60	$729.60	$72.45	$1159.20	$46.87	$74.27
.25 Equip. Oper. (light)	57.00	114.00	88.80	177.60		
1 Pickup Truck, 3/4 Ton		115.08		126.59		
1 Line Rem.,11 H.P.,Walk Behind		118.49		130.34		
.25 Road Sweeper, S.P., 8' wide		186.37		205.01	23.33	25.66
18 L.H., Daily Totals		$1263.54		$1798.74	$70.20	$99.93

Crew B-78C

Crew No.	Hr.	Daily	Hr.	Daily	Bare Costs	Incl. O&P
1 Labor Foreman (outside)	$47.60	$380.80	$75.60	$604.80	$46.73	$74.28
4 Laborers	45.60	1459.20	72.45	2318.40		
1 Truck Driver (light)	50.40	403.20	80.30	642.40		
1 Paint Striper, T.M., 120 Gal.		618.71		680.58		
1 Flatbed Truck, Gas, 3 Ton		877.96		965.75		
1 Pickup Truck, 3/4 Ton		115.08		126.59	33.58	36.94
48 L.H., Daily Totals		$3854.95		$5338.52	$80.31	$111.22

Crew B-78D

Crew No.	Hr.	Daily	Hr.	Daily	Bare Costs	Incl. O&P
2 Labor Foremen (outside)	$47.60	$761.60	$75.60	$1209.60	$46.48	$73.86
7 Laborers	45.60	2553.60	72.45	4057.20		
1 Truck Driver (light)	50.40	403.20	80.30	642.40		
1 Paint Striper, T.M., 120 Gal.		618.71		680.58		
1 Flatbed Truck, Gas, 3 Ton		877.96		965.75		
3 Pickup Trucks, 3/4 Ton		345.24		379.77		
1 Air Compressor, 60 cfm		158.26		174.09		
1 -50' Air Hose, 3/4"		7.40		8.15		
1 Breaker, Pavement, 60 lb.		55.48		61.03	25.79	28.37
80 L.H., Daily Totals		$5781.46		$8178.56	$72.27	$102.23

Crew B-78E

Crew No.	Hr.	Daily	Hr.	Daily	Bare Costs	Incl. O&P
2 Labor Foremen (outside)	$47.60	$761.60	$75.60	$1209.60	$46.33	$73.63
9 Laborers	45.60	3283.20	72.45	5216.40		
1 Truck Driver (light)	50.40	403.20	80.30	642.40		
1 Paint Striper, T.M., 120 Gal.		618.71		680.58		
1 Flatbed Truck, Gas, 3 Ton		877.96		965.75		
4 Pickup Trucks, 3/4 Ton		460.33		506.36		
2 Air Compressors, 60 cfm		316.52		348.18		
2 -50' Air Hoses, 3/4"		14.81		16.29		
2 Breakers, Pavement, 60 lb.		110.96		122.06	24.99	27.49
96 L.H., Daily Totals		$6847.29		$9707.62	$71.33	$101.12

Crew B-78F

Crew No.	Hr.	Daily	Hr.	Daily	Bare Costs	Incl. O&P
2 Labor Foremen (outside)	$47.60	$761.60	$75.60	$1209.60	$46.23	$73.46
11 Laborers	45.60	4012.80	72.45	6375.60		
1 Truck Driver (light)	50.40	403.20	80.30	642.40		
1 Paint Striper, T.M., 120 Gal.		618.71		680.58		
1 Flatbed Truck, Gas, 3 Ton		877.96		965.75		
7 Pickup Trucks, 3/4 Ton		805.57		886.13		
3 Air Compressors, 60 cfm		474.79		522.26		
3 -50' Air Hoses, 3/4"		22.21		24.44		
3 Breakers, Pavement, 60 lb.		166.45		183.09	26.48	29.13
112 L.H., Daily Totals		$8143.28		$11489.85	$72.71	$102.59

Crew B-79

Crew No.	Hr.	Daily	Hr.	Daily	Bare Costs	Incl. O&P
1 Labor Foreman (outside)	$47.60	$380.80	$75.60	$604.80	$46.96	$74.65
3 Laborers	45.60	1094.40	72.45	1738.80		
1 Truck Driver (light)	50.40	403.20	80.30	642.40		
1 Paint Striper, T.M., 120 Gal.		618.71		680.58		
1 Heating Kettle, 115 Gallon		110.18		121.20		
1 Flatbed Truck, Gas, 3 Ton		877.96		965.75		
2 Pickup Trucks, 3/4 Ton		230.16		253.18	45.93	50.52
40 L.H., Daily Totals		$3715.41		$5006.71	$92.89	$125.17

Crew B-79A

Crew No.	Hr.	Daily	Hr.	Daily	Bare Costs	Incl. O&P
1.5 Equip. Oper. (light)	$57.00	$684.00	$88.80	$1065.60	$57.00	$88.80
.5 Line Remov. (Grinder) 115 H.P.		531.21		584.33		
1 Line Rem. (Metal Balls) 115 H.P.		1026.71		1129.39	129.83	142.81
12 L.H., Daily Totals		$2241.92		$2779.31	$186.83	$231.61

Crew B-79B

Crew No.	Hr.	Daily	Hr.	Daily	Bare Costs	Incl. O&P
1 Laborer	$45.60	$364.80	$72.45	$579.60	$45.60	$72.45
1 Set of Gases		177.16		194.87	22.14	24.36
8 L.H., Daily Totals		$541.96		$774.47	$67.74	$96.81

Crew B-79C

Crew No.	Hr.	Daily	Hr.	Daily	Bare Costs	Incl. O&P
1 Labor Foreman (outside)	$47.60	$380.80	$75.60	$604.80	$46.57	$74.02
5 Laborers	45.60	1824.00	72.45	2898.00		
1 Truck Driver (light)	50.40	403.20	80.30	642.40		
1 Paint Striper, T.M., 120 Gal.		618.71		680.58		
1 Heating Kettle, 115 Gallon		110.18		121.20		
1 Flatbed Truck, Gas, 3 Ton		877.96		965.75		
3 Pickup Trucks, 3/4 Ton		345.24		379.77		
1 Air Compressor, 60 cfm		158.26		174.09		
1 -50' Air Hose, 3/4"		7.40		8.15		
1 Breaker, Pavement, 60 lb.		55.48		61.03	38.81	42.69
56 L.H., Daily Totals		$4781.24		$6535.76	$85.38	$116.71

Crew B-79D

Crew No.	Hr.	Daily	Hr.	Daily	Bare Costs	Incl. O&P
2 Labor Foremen (outside)	$47.60	$761.60	$75.60	$1209.60	$46.70	$74.22
5 Laborers	45.60	1824.00	72.45	2898.00		
1 Truck Driver (light)	50.40	403.20	80.30	642.40		
1 Paint Striper, T.M., 120 Gal.		618.71		680.58		
1 Heating Kettle, 115 Gallon		110.18		121.20		
1 Flatbed Truck, Gas, 3 Ton		877.96		965.75		
4 Pickup Trucks, 3/4 Ton		460.33		506.36		
1 Air Compressor, 60 cfm		158.26		174.09		
1 -50' Air Hose, 3/4"		7.40		8.15		
1 Breaker, Pavement, 60 lb.		55.48		61.03	35.76	39.33
64 L.H., Daily Totals		$5277.12		$7267.15	$82.45	$113.55

Crew B-79E

Crew No.	Hr.	Daily	Hr.	Daily	Bare Costs	Incl. O&P
2 Labor Foremen (outside)	$47.60	$761.60	$75.60	$1209.60	$46.48	$73.86
7 Laborers	45.60	2553.60	72.45	4057.20		
1 Truck Driver (light)	50.40	403.20	80.30	642.40		
1 Paint Striper, T.M., 120 Gal.		618.71		680.58		
1 Heating Kettle, 115 Gallon		110.18		121.20		
1 Flatbed Truck, Gas, 3 Ton		877.96		965.75		
5 Pickup Trucks, 3/4 Ton		575.41		632.95		
2 Air Compressors, 60 cfm		316.52		348.18		
2 -50' Air Hoses, 3/4"		14.81		16.29		
2 Breakers, Pavement, 60 lb.		110.96		122.06	32.81	36.09
80 L.H., Daily Totals		$6342.95		$8796.20	$79.29	$109.95

Crew B-80

Crew No.	Hr.	Daily	Hr.	Daily	Bare Costs	Incl. O&P
1 Labor Foreman (outside)	$47.60	$380.80	$75.60	$604.80	$50.15	$79.29
1 Laborer	45.60	364.80	72.45	579.60		
1 Truck Driver (light)	50.40	403.20	80.30	642.40		
1 Equip. Oper. (light)	57.00	456.00	88.80	710.40		
1 Flatbed Truck, Gas, 3 Ton		877.96		965.75		
1 Earth Auger, Truck-Mtd.		208.12		228.94	33.94	37.33
32 L.H., Daily Totals		$2690.88		$3731.89	$84.09	$116.62

For customer support on your Commercial Renovation Costs with RSMeans data, call 800.448.8182.

779

Crew No.	Bare Costs		Incl. Subs O&P		Cost Per Labor-Hour	
	Hr.	Daily	Hr.	Daily	Bare Costs	Incl. O&P
Crew B-80A					Bare Costs	Incl. O&P
3 Laborers	$45.60	$1094.40	$72.45	$1738.80	$45.60	$72.45
1 Flatbed Truck, Gas, 3 Ton		877.96		965.75	36.58	40.24
24 L.H., Daily Totals		$1972.36		$2704.55	$82.18	$112.69
Crew B-80B	Hr.	Daily	Hr.	Daily	Bare Costs	Incl. O&P
3 Laborers	$45.60	$1094.40	$72.45	$1738.80	$48.45	$76.54
1 Equip. Oper. (light)	57.00	456.00	88.80	710.40		
1 Crane, Flatbed Mounted, 3 Ton		245.58		270.14	7.67	8.44
32 L.H., Daily Totals		$1795.98		$2719.34	$56.12	$84.98
Crew B-80C	Hr.	Daily	Hr.	Daily	Bare Costs	Incl. O&P
2 Laborers	$45.60	$729.60	$72.45	$1159.20	$47.20	$75.07
1 Truck Driver (light)	50.40	403.20	80.30	642.40		
1 Flatbed Truck, Gas, 1.5 Ton		203.25		223.57		
1 Manual Fence Post Auger, Gas		56.37		62.01	10.82	11.90
24 L.H., Daily Totals		$1392.42		$2087.18	$58.02	$86.97
Crew B-81	Hr.	Daily	Hr.	Daily	Bare Costs	Incl. O&P
1 Laborer	$45.60	$364.80	$72.45	$579.60	$53.02	$83.70
1 Equip. Oper. (medium)	60.40	483.20	94.10	752.80		
1 Truck Driver (heavy)	53.05	424.40	84.55	676.40		
1 Hydromulcher, T.M., 3000 Gal.		263.98		290.37		
1 Truck Tractor, 220 H.P.		319.23		351.15	24.30	26.73
24 L.H., Daily Totals		$1855.61		$2650.33	$77.32	$110.43
Crew B-81A	Hr.	Daily	Hr.	Daily	Bare Costs	Incl. O&P
1 Laborer	$45.60	$364.80	$72.45	$579.60	$48.00	$76.38
1 Truck Driver (light)	50.40	403.20	80.30	642.40		
1 Hydromulcher, T.M., 600 Gal.		121.48		133.62		
1 Flatbed Truck, Gas, 3 Ton		877.96		965.75	62.46	68.71
16 L.H., Daily Totals		$1767.43		$2321.38	$110.46	$145.09
Crew B-82	Hr.	Daily	Hr.	Daily	Bare Costs	Incl. O&P
1 Laborer	$45.60	$364.80	$72.45	$579.60	$51.30	$80.63
1 Equip. Oper. (light)	57.00	456.00	88.80	710.40		
1 Horiz. Borer, 6 H.P.		190.24		209.27	11.89	13.08
16 L.H., Daily Totals		$1011.04		$1499.27	$63.19	$93.70
Crew B-82A	Hr.	Daily	Hr.	Daily	Bare Costs	Incl. O&P
2 Laborers	$45.60	$729.60	$72.45	$1159.20	$51.30	$80.63
2 Equip. Opers. (light)	57.00	912.00	88.80	1420.80		
2 Dump Truck, 8 C.Y., 220 H.P.		838.83		922.71		
1 Flatbed Trailer, 25 Ton		141.42		155.56		
1 Horiz. Dir. Drill, 20k lb. Thrust		560.43		616.47		
1 Mud Trailer for HDD, 1500 Gal.		320.19		352.21		
1 Pickup Truck, 4x4, 3/4 Ton		182.00		200.20		
1 Flatbed Trailer, 3 Ton		73.56		80.91		
1 Loader, Skid Steer, 78 H.P.		460.30		506.33	80.52	88.57
32 L.H., Daily Totals		$4218.32		$5414.40	$131.82	$169.20
Crew B-82B	Hr.	Daily	Hr.	Daily	Bare Costs	Incl. O&P
2 Laborers	$45.60	$729.60	$72.45	$1159.20	$51.30	$80.63
2 Equip. Opers. (light)	57.00	912.00	88.80	1420.80		
2 Dump Truck, 8 C.Y., 220 H.P.		838.83		922.71		
1 Flatbed Trailer, 25 Ton		141.42		155.56		
1 Horiz. Dir. Drill, 30k lb. Thrust		666.98		733.67		
1 Mud Trailer for HDD, 1500 Gal.		320.19		352.21		
1 Pickup Truck, 4x4, 3/4 Ton		182.00		200.20		
1 Flatbed Trailer, 3 Ton		73.56		80.91		
1 Loader, Skid Steer, 78 H.P.		460.30		506.33	83.85	92.24
32 L.H., Daily Totals		$4324.87		$5531.60	$135.15	$172.86

Crew No.	Bare Costs		Incl. Subs O&P		Cost Per Labor-Hour	
	Hr.	Daily	Hr.	Daily	Bare Costs	Incl. O&P
Crew B-82C					Bare Costs	Incl. O&P
2 Laborers	$45.60	$729.60	$72.45	$1159.20	$51.30	$80.63
2 Equip. Opers. (light)	57.00	912.00	88.80	1420.80		
2 Dump Truck, 8 C.Y., 220 H.P.		838.83		922.71		
1 Flatbed Trailer, 25 Ton		141.42		155.56		
1 Horiz. Dir. Drill, 50k lb. Thrust		847.85		932.64		
1 Mud Trailer for HDD, 1500 Gal.		320.19		352.21		
1 Pickup Truck, 4x4, 3/4 Ton		182.00		200.20		
1 Flatbed Trailer, 3 Ton		73.56		80.91		
1 Loader, Skid Steer, 78 H.P.		460.30		506.33	89.50	98.45
32 L.H., Daily Totals		$4505.75		$5730.56	$140.80	$179.08
Crew B-82D	Hr.	Daily	Hr.	Daily	Bare Costs	Incl. O&P
1 Equip. Oper. (light)	$57.00	$456.00	$88.80	$710.40	$57.00	$88.80
1 Mud Trailer for HDD, 1500 Gal.		320.19		352.21	40.02	44.03
8 L.H., Daily Totals		$776.19		$1062.61	$97.02	$132.83
Crew B-83	Hr.	Daily	Hr.	Daily	Bare Costs	Incl. O&P
1 Tugboat Captain	$60.40	$483.20	$94.10	$752.80	$53.00	$83.28
1 Tugboat Hand	45.60	364.80	72.45	579.60		
1 Tugboat, 250 H.P.		745.07		819.58	46.57	51.22
16 L.H., Daily Totals		$1593.07		$2151.98	$99.57	$134.50
Crew B-84	Hr.	Daily	Hr.	Daily	Bare Costs	Incl. O&P
1 Equip. Oper. (medium)	$60.40	$483.20	$94.10	$752.80	$60.40	$94.10
1 Rotary Mower/Tractor		382.30		420.53	47.79	52.57
8 L.H., Daily Totals		$865.50		$1173.33	$108.19	$146.67
Crew B-85	Hr.	Daily	Hr.	Daily	Bare Costs	Incl. O&P
3 Laborers	$45.60	$1094.40	$72.45	$1738.80	$50.05	$79.20
1 Equip. Oper. (medium)	60.40	483.20	94.10	752.80		
1 Truck Driver (heavy)	53.05	424.40	84.55	676.40		
1 Telescoping Boom Lift, to 80'		399.18		439.10		
1 Brush Chipper, 12", 130 H.P.		376.37		414.01		
1 Pruning Saw, Rotary		27.36		30.10	20.07	22.08
40 L.H., Daily Totals		$2804.91		$4051.20	$70.12	$101.28
Crew B-86	Hr.	Daily	Hr.	Daily	Bare Costs	Incl. O&P
1 Equip. Oper. (medium)	$60.40	$483.20	$94.10	$752.80	$60.40	$94.10
1 Stump Chipper, S.P.		195.25		214.77	24.41	26.85
8 L.H., Daily Totals		$678.45		$967.57	$84.81	$120.95
Crew B-86A	Hr.	Daily	Hr.	Daily	Bare Costs	Incl. O&P
1 Equip. Oper. (medium)	$60.40	$483.20	$94.10	$752.80	$60.40	$94.10
1 Grader, 30,000 Lbs.		1108.49		1219.34	138.56	152.42
8 L.H., Daily Totals		$1591.69		$1972.14	$198.96	$246.52
Crew B-86B	Hr.	Daily	Hr.	Daily	Bare Costs	Incl. O&P
1 Equip. Oper. (medium)	$60.40	$483.20	$94.10	$752.80	$60.40	$94.10
1 Dozer, 200 H.P.		1567.46		1724.21	195.93	215.53
8 L.H., Daily Totals		$2050.66		$2477.01	$256.33	$309.63
Crew B-87	Hr.	Daily	Hr.	Daily	Bare Costs	Incl. O&P
1 Laborer	$45.60	$364.80	$72.45	$579.60	$57.44	$89.77
4 Equip. Oper. (medium)	60.40	1932.80	94.10	3011.20		
2 Feller Bunchers, 100 H.P.		1292.22		1421.44		
1 Log Chipper, 22" Tree		569.41		626.36		
1 Dozer, 105 H.P.		659.79		725.77		
1 Chain Saw, Gas, 36" Long		43.10		47.42	64.11	70.52
40 L.H., Daily Totals		$4862.13		$6411.78	$121.55	$160.29

For customer support on your Commercial Renovation Costs with RSMeans data, call 800.448.8182.

Crew No.	Bare Costs		Incl. Subs O&P		Cost Per Labor-Hour	
	Hr.	Daily	Hr.	Daily	Bare Costs	Incl. O&P
Crew B-88					$58.29	$91.01
1 Laborer	$45.60	$364.80	$72.45	$579.60		
6 Equip. Oper. (medium)	60.40	2899.20	94.10	4516.80		
2 Feller Bunchers, 100 H.P.		1292.22		1421.44		
1 Log Chipper, 22" Tree		569.41		626.36		
2 Log Skidders, 50 H.P.		1882.34		2070.57		
1 Dozer, 105 H.P.		659.79		725.77		
1 Chain Saw, Gas, 36" Long		43.10		47.42	79.41	87.35
56 L.H., Daily Totals		$7710.86		$9987.95	$137.69	$178.36
Crew B-89	Hr.	Daily	Hr.	Daily	Bare Costs	Incl. O&P
1 Equip. Oper. (light)	$57.00	$456.00	$88.80	$710.40	$53.70	$84.55
1 Truck Driver (light)	50.40	403.20	80.30	642.40		
1 Flatbed Truck, Gas, 3 Ton		877.96		965.75		
1 Concrete Saw		115.94		127.54		
1 Water Tank, 65 Gal.		106.63		117.29	68.78	75.66
16 L.H., Daily Totals		$1959.73		$2563.38	$122.48	$160.21
Crew B-89A	Hr.	Daily	Hr.	Daily	Bare Costs	Incl. O&P
1 Skilled Worker	$58.90	$471.20	$93.20	$745.60	$52.25	$82.83
1 Laborer	45.60	364.80	72.45	579.60		
1 Core Drill (Large)		125.34		137.87	7.83	8.62
16 L.H., Daily Totals		$961.34		$1463.07	$60.08	$91.44
Crew B-89B	Hr.	Daily	Hr.	Daily	Bare Costs	Incl. O&P
1 Equip. Oper. (light)	$57.00	$456.00	$88.80	$710.40	$53.70	$84.55
1 Truck Driver (light)	50.40	403.20	80.30	642.40		
1 Wall Saw, Hydraulic, 10 H.P.		89.13		98.04		
1 Generator, Diesel, 100 kW		537.22		590.94		
1 Water Tank, 65 Gal.		106.63		117.29		
1 Flatbed Truck, Gas, 3 Ton		877.96		965.75	100.68	110.75
16 L.H., Daily Totals		$2470.14		$3124.83	$154.38	$195.30
Crew B-89C	Hr.	Daily	Hr.	Daily	Bare Costs	Incl. O&P
1 Cement Finisher	$53.15	$425.20	$83.05	$664.40	$53.15	$83.05
1 Masonry cut-off saw, gas		60.05		66.06	7.51	8.26
8 L.H., Daily Totals		$485.25		$730.46	$60.66	$91.31
Crew B-90	Hr.	Daily	Hr.	Daily	Bare Costs	Incl. O&P
1 Labor Foreman (outside)	$47.60	$380.80	$75.60	$604.80	$50.56	$79.96
3 Laborers	45.60	1094.40	72.45	1738.80		
2 Equip. Oper. (light)	57.00	912.00	88.80	1420.80		
2 Truck Drivers (heavy)	53.05	848.80	84.55	1352.80		
1 Road Mixer, 310 H.P.		1979.64		2177.61		
1 Dist. Truck, 2000 Gal.		312.98		344.27	35.82	39.40
64 L.H., Daily Totals		$5528.62		$7639.08	$86.38	$119.36
Crew B-90A	Hr.	Daily	Hr.	Daily	Bare Costs	Incl. O&P
1 Labor Foreman (outside)	$47.60	$380.80	$75.60	$604.80	$54.34	$85.27
2 Laborers	45.60	729.60	72.45	1159.20		
4 Equip. Oper. (medium)	60.40	1932.80	94.10	3011.20		
2 Graders, 30,000 Lbs.		2216.97		2438.67		
1 Tandem Roller, 10 Ton		254.53		279.98		
1 Roller, Pneum. Whl., 12 Ton		361.08		397.18	50.58	55.64
56 L.H., Daily Totals		$5875.78		$7891.03	$104.92	$140.91

Crew No.	Bare Costs		Incl. Subs O&P		Cost Per Labor-Hour	
	Hr.	Daily	Hr.	Daily	Bare Costs	Incl. O&P
Crew B-90B					$53.33	$83.80
1 Labor Foreman (outside)	$47.60	$380.80	$75.60	$604.80		
2 Laborers	45.60	729.60	72.45	1159.20		
3 Equip. Oper. (medium)	60.40	1449.60	94.10	2258.40		
1 Roller, Pneum. Whl., 12 Ton		361.08		397.18		
1 Road Mixer, 310 H.P.		1979.64		2177.61	48.77	53.64
48 L.H., Daily Totals		$4900.72		$6597.19	$102.10	$137.44
Crew B-90C	Hr.	Daily	Hr.	Daily	Bare Costs	Incl. O&P
1 Labor Foreman (outside)	$47.60	$380.80	$75.60	$604.80	$51.85	$81.94
4 Laborers	45.60	1459.20	72.45	2318.40		
3 Equip. Oper. (medium)	60.40	1449.60	94.10	2258.40		
3 Truck Drivers (heavy)	53.05	1273.20	84.55	2029.20		
3 Road Mixers, 310 H.P.		5938.93		6532.82	67.49	74.24
88 L.H., Daily Totals		$10501.73		$13743.62	$119.34	$156.18
Crew B-90D	Hr.	Daily	Hr.	Daily	Bare Costs	Incl. O&P
1 Labor Foreman (outside)	$47.60	$380.80	$75.60	$604.80	$50.89	$80.48
6 Laborers	45.60	2188.80	72.45	3477.60		
3 Equip. Oper. (medium)	60.40	1449.60	94.10	2258.40		
3 Truck Drivers (heavy)	53.05	1273.20	84.55	2029.20		
3 Road Mixers, 310 H.P.		5938.93		6532.82	57.11	62.82
104 L.H., Daily Totals		$11231.33		$14902.82	$107.99	$143.30
Crew B-90E	Hr.	Daily	Hr.	Daily	Bare Costs	Incl. O&P
1 Labor Foreman (outside)	$47.60	$380.80	$75.60	$604.80	$51.58	$81.36
4 Laborers	45.60	1459.20	72.45	2318.40		
3 Equip. Oper. (medium)	60.40	1449.60	94.10	2258.40		
1 Truck Driver (heavy)	53.05	424.40	84.55	676.40		
1 Road Mixer, 310 H.P.		1979.64		2177.61	27.50	30.24
72 L.H., Daily Totals		$5693.64		$8035.61	$79.08	$111.61
Crew B-91	Hr.	Daily	Hr.	Daily	Bare Costs	Incl. O&P
1 Labor Foreman (outside)	$47.60	$380.80	$75.60	$604.80	$54.18	$85.18
2 Laborers	45.60	729.60	72.45	1159.20		
4 Equip. Oper. (medium)	60.40	1932.80	94.10	3011.20		
1 Truck Driver (heavy)	53.05	424.40	84.55	676.40		
1 Dist. Tanker, 3000 Gallon		344.69		379.16		
1 Truck Tractor, 6x4, 380 H.P.		512.66		563.93		
1 Aggreg. Spreader, S.P.		883.90		972.29		
1 Roller, Pneum. Whl., 12 Ton		361.08		397.18		
1 Tandem Roller, 10 Ton		254.53		279.98	36.83	40.51
64 L.H., Daily Totals		$5824.46		$8044.14	$91.01	$125.69
Crew B-91A	Hr.	Daily	Hr.	Daily	Bare Costs	Incl. O&P
2 Truck Drivers (light)	$50.40	$806.40	$80.30	$1284.80	$50.40	$80.30
1 Road Sweeper, S.P., 8' wide		745.48		820.02		
1 Pickup Truck, 3/4 Ton		115.08		126.59	53.78	59.16
16 L.H., Daily Totals		$1666.96		$2231.41	$104.18	$139.46
Crew B-91B	Hr.	Daily	Hr.	Daily	Bare Costs	Incl. O&P
1 Laborer	$45.60	$364.80	$72.45	$579.60	$53.00	$83.28
1 Equipment Oper. (med.)	60.40	483.20	94.10	752.80		
1 Road Sweeper, Vac. Assist.		903.96		994.35	56.50	62.15
16 L.H., Daily Totals		$1751.96		$2326.75	$109.50	$145.42
Crew B-91C	Hr.	Daily	Hr.	Daily	Bare Costs	Incl. O&P
1 Laborer	$45.60	$364.80	$72.45	$579.60	$48.00	$76.38
1 Truck Driver (light)	50.40	403.20	80.30	642.40		
1 Catch Basin Cleaning Truck		557.79		613.57	34.86	38.35
16 L.H., Daily Totals		$1325.79		$1835.57	$82.86	$114.72

For customer support on your Commercial Renovation Costs with RSMeans data, call 800.448.8182.

781

Crews - Renovation

Crew B-91D

Crew No.	Bare Costs Hr.	Daily	Incl. Subs O&P Hr.	Daily	Cost Per Labor-Hour Bare Costs	Incl. O&P
1 Labor Foreman (outside)	$47.60	$380.80	$75.60	$604.80	$52.59	$82.88
5 Laborers	45.60	1824.00	72.45	2898.00		
5 Equip. Oper. (medium)	60.40	2416.00	94.10	3764.00		
2 Truck Drivers (heavy)	53.05	848.80	84.55	1352.80		
1 Aggreg. Spreader, S.P.		883.90		972.29		
2 Truck Tractors, 6x4, 380 H.P.		1025.32		1127.85		
2 Dist. Tankers, 3000 Gallon		689.38		758.32		
2 Pavement Brushes, Towed		182.31		200.54		
2 Rollers Pneum. Whl., 12 Ton		722.15		794.37	33.68	37.05
104 L.H., Daily Totals		$8972.66		$12472.97	$86.28	$119.93

Crew B-92

Crew No.	Bare Costs Hr.	Daily	Incl. Subs O&P Hr.	Daily	Cost Per Labor-Hour Bare Costs	Incl. O&P
1 Labor Foreman (outside)	$47.60	$380.80	$75.60	$604.80	$46.10	$73.24
3 Laborers	45.60	1094.40	72.45	1738.80		
1 Crack Cleaner, 25 H.P.		54.53		59.98		
1 Air Compressor, 60 cfm		158.26		174.09		
1 Tar Kettle, T.M.		160.79		176.87		
1 Flatbed Truck, Gas, 3 Ton		877.96		965.75	39.11	43.02
32 L.H., Daily Totals		$2726.74		$3720.29	$85.21	$116.26

Crew B-93

Crew No.	Bare Costs Hr.	Daily	Incl. Subs O&P Hr.	Daily	Cost Per Labor-Hour Bare Costs	Incl. O&P
1 Equip. Oper. (medium)	$60.40	$483.20	$94.10	$752.80	$60.40	$94.10
1 Feller Buncher, 100 H.P.		646.11		710.72	80.76	88.84
8 L.H., Daily Totals		$1129.31		$1463.52	$141.16	$182.94

Crew B-94A

Crew No.	Bare Costs Hr.	Daily	Incl. Subs O&P Hr.	Daily	Cost Per Labor-Hour Bare Costs	Incl. O&P
1 Laborer	$45.60	$364.80	$72.45	$579.60	$45.60	$72.45
1 Diaphragm Water Pump, 2"		90.40		99.44		
1 -20' Suction Hose, 2"		3.69		4.06		
2 -50' Discharge Hoses, 2"		8.25		9.08	12.79	14.07
8 L.H., Daily Totals		$467.15		$692.18	$58.39	$86.52

Crew B-94B

Crew No.	Bare Costs Hr.	Daily	Incl. Subs O&P Hr.	Daily	Cost Per Labor-Hour Bare Costs	Incl. O&P
1 Laborer	$45.60	$364.80	$72.45	$579.60	$45.60	$72.45
1 Diaphragm Water Pump, 4"		109.44		120.39		
1 -20' Suction Hose, 4"		17.88		19.67		
2 -50' Discharge Hoses, 4"		26.56		29.22	19.24	21.16
8 L.H., Daily Totals		$518.69		$748.87	$64.84	$93.61

Crew B-94C

Crew No.	Bare Costs Hr.	Daily	Incl. Subs O&P Hr.	Daily	Cost Per Labor-Hour Bare Costs	Incl. O&P
1 Laborer	$45.60	$364.80	$72.45	$579.60	$45.60	$72.45
1 Centrifugal Water Pump, 3"		76.47		84.12		
1 -20' Suction Hose, 3"		9.06		9.97		
2 -50' Discharge Hoses, 3"		9.30		10.23	11.85	13.04
8 L.H., Daily Totals		$459.63		$683.92	$57.45	$85.49

Crew B-94D

Crew No.	Bare Costs Hr.	Daily	Incl. Subs O&P Hr.	Daily	Cost Per Labor-Hour Bare Costs	Incl. O&P
1 Laborer	$45.60	$364.80	$72.45	$579.60	$45.60	$72.45
1 Centr. Water Pump, 6"		240.92		265.01		
1 -20' Suction Hose, 6"		26.41		29.05		
2 -50' Discharge Hoses, 6"		37.54		41.29	38.11	41.92
8 L.H., Daily Totals		$669.66		$914.95	$83.71	$114.37

Crew C-1

Crew No.	Bare Costs Hr.	Daily	Incl. Subs O&P Hr.	Daily	Cost Per Labor-Hour Bare Costs	Incl. O&P
3 Carpenters	$56.30	$1351.20	$89.45	$2146.80	$53.63	$85.20
1 Laborer	45.60	364.80	72.45	579.60		
32 L.H., Daily Totals		$1716.00		$2726.40	$53.63	$85.20

Crew C-2

Crew No.	Bare Costs Hr.	Daily	Incl. Subs O&P Hr.	Daily	Cost Per Labor-Hour Bare Costs	Incl. O&P
1 Carpenter Foreman (outside)	$58.30	$466.40	$92.60	$740.80	$54.85	$87.14
4 Carpenters	56.30	1801.60	89.45	2862.40		
1 Laborer	45.60	364.80	72.45	579.60		
48 L.H., Daily Totals		$2632.80		$4182.80	$54.85	$87.14

Crew C-2A

Crew No.	Bare Costs Hr.	Daily	Incl. Subs O&P Hr.	Daily	Cost Per Labor-Hour Bare Costs	Incl. O&P
1 Carpenter Foreman (outside)	$58.30	$466.40	$92.60	$740.80	$54.33	$86.08
3 Carpenters	56.30	1351.20	89.45	2146.80		
1 Cement Finisher	53.15	425.20	83.05	664.40		
1 Laborer	45.60	364.80	72.45	579.60		
48 L.H., Daily Totals		$2607.60		$4131.60	$54.33	$86.08

Crew C-3

Crew No.	Bare Costs Hr.	Daily	Incl. Subs O&P Hr.	Daily	Cost Per Labor-Hour Bare Costs	Incl. O&P
1 Rodman Foreman (outside)	$62.70	$501.60	$98.05	$784.40	$56.71	$88.92
4 Rodmen (reinf.)	60.70	1942.40	94.90	3036.80		
1 Equip. Oper. (light)	57.00	456.00	88.80	710.40		
2 Laborers	45.60	729.60	72.45	1159.20		
3 Stressing Equipment		58.72		64.60		
.5 Grouting Equipment		127.30		140.03	2.91	3.20
64 L.H., Daily Totals		$3815.62		$5895.43	$59.62	$92.12

Crew C-4

Crew No.	Bare Costs Hr.	Daily	Incl. Subs O&P Hr.	Daily	Cost Per Labor-Hour Bare Costs	Incl. O&P
1 Rodman Foreman (outside)	$62.70	$501.60	$98.05	$784.40	$61.20	$95.69
3 Rodmen (reinf.)	60.70	1456.80	94.90	2277.60		
3 Stressing Equipment		58.72		64.60	1.84	2.02
32 L.H., Daily Totals		$2017.12		$3126.60	$63.04	$97.71

Crew C-4A

Crew No.	Bare Costs Hr.	Daily	Incl. Subs O&P Hr.	Daily	Cost Per Labor-Hour Bare Costs	Incl. O&P
2 Rodmen (reinf.)	$60.70	$971.20	$94.90	$1518.40	$60.70	$94.90
4 Stressing Equipment		78.30		86.13	4.89	5.38
16 L.H., Daily Totals		$1049.50		$1604.53	$65.59	$100.28

Crew C-5

Crew No.	Bare Costs Hr.	Daily	Incl. Subs O&P Hr.	Daily	Cost Per Labor-Hour Bare Costs	Incl. O&P
1 Rodman Foreman (outside)	$62.70	$501.60	$98.05	$784.40	$60.41	$94.36
4 Rodmen (reinf.)	60.70	1942.40	94.90	3036.80		
1 Equip. Oper. (crane)	63.40	507.20	98.75	790.00		
1 Equip. Oper. (oiler)	54.00	432.00	84.15	673.20		
1 Hyd. Crane, 25 Ton		603.44		663.78	10.78	11.85
56 L.H., Daily Totals		$3986.64		$5948.18	$71.19	$106.22

Crew C-6

Crew No.	Bare Costs Hr.	Daily	Incl. Subs O&P Hr.	Daily	Cost Per Labor-Hour Bare Costs	Incl. O&P
1 Labor Foreman (outside)	$47.60	$380.80	$75.60	$604.80	$47.19	$74.74
4 Laborers	45.60	1459.20	72.45	2318.40		
1 Cement Finisher	53.15	425.20	83.05	664.40		
2 Gas Engine Vibrators		55.74		61.31	1.16	1.28
48 L.H., Daily Totals		$2320.94		$3648.91	$48.35	$76.02

Crew C-6A

Crew No.	Bare Costs Hr.	Daily	Incl. Subs O&P Hr.	Daily	Cost Per Labor-Hour Bare Costs	Incl. O&P
2 Cement Finishers	$53.15	$850.40	$83.05	$1328.80	$53.15	$83.05
1 Concrete Vibrator, Elec, 2 HP		47.42		52.16	2.96	3.26
16 L.H., Daily Totals		$897.82		$1380.96	$56.11	$86.31

For customer support on your Commercial Renovation Costs with RSMeans data, call 800.448.8182.

Crews - Renovation

Crew No.	Bare Costs		Incl. Subs O&P		Cost Per Labor-Hour	
Crew C-7	Hr.	Daily	Hr.	Daily	Bare Costs	Incl. O&P
1 Labor Foreman (outside)	$47.60	$380.80	$75.60	$604.80	$49.24	$77.68
5 Laborers	45.60	1824.00	72.45	2898.00		
1 Cement Finisher	53.15	425.20	83.05	664.40		
1 Equip. Oper. (medium)	60.40	483.20	94.10	752.80		
1 Equip. Oper. (oiler)	54.00	432.00	84.15	673.20		
2 Gas Engine Vibrators		55.74		61.31		
1 Concrete Bucket, 1 C.Y.		47.48		52.23		
1 Hyd. Crane, 55 Ton		1019.39		1121.33	15.59	17.15
72 L.H., Daily Totals		$4667.81		$6828.07	$64.83	$94.83
Crew C-8	Hr.	Daily	Hr.	Daily	Bare Costs	Incl. O&P
1 Labor Foreman (outside)	$47.60	$380.80	$75.60	$604.80	$50.16	$79.02
3 Laborers	45.60	1094.40	72.45	1738.80		
2 Cement Finishers	53.15	850.40	83.05	1328.80		
1 Equip. Oper. (medium)	60.40	483.20	94.10	752.80		
1 Concrete Pump (Small)		435.09		478.60	7.77	8.55
56 L.H., Daily Totals		$3243.89		$4903.80	$57.93	$87.57
Crew C-8A	Hr.	Daily	Hr.	Daily	Bare Costs	Incl. O&P
1 Labor Foreman (outside)	$47.60	$380.80	$75.60	$604.80	$48.45	$76.51
3 Laborers	45.60	1094.40	72.45	1738.80		
2 Cement Finishers	53.15	850.40	83.05	1328.80		
48 L.H., Daily Totals		$2325.60		$3672.40	$48.45	$76.51
Crew C-8B	Hr.	Daily	Hr.	Daily	Bare Costs	Incl. O&P
1 Labor Foreman (outside)	$47.60	$380.80	$75.60	$604.80	$48.96	$77.41
3 Laborers	45.60	1094.40	72.45	1738.80		
1 Equip. Oper. (medium)	60.40	483.20	94.10	752.80		
1 Vibrating Power Screed		90.52		99.57		
1 Roller, Vibratory, 25 Ton		693.41		762.76		
1 Dozer, 200 H.P.		1567.46		1724.21	58.78	64.66
40 L.H., Daily Totals		$4309.79		$5682.93	$107.74	$142.07
Crew C-8C	Hr.	Daily	Hr.	Daily	Bare Costs	Incl. O&P
1 Labor Foreman (outside)	$47.60	$380.80	$75.60	$604.80	$49.66	$78.35
3 Laborers	45.60	1094.40	72.45	1738.80		
1 Cement Finisher	53.15	425.20	83.05	664.40		
1 Equip. Oper. (medium)	60.40	483.20	94.10	752.80		
1 Shotcrete Pump Rig, 12 C.Y./hr		276.88		304.57		
1 Air Compressor, 160 cfm		218.71		240.59		
4 -50' Air Hoses, 1"		33.27		36.59		
4 -50' Air Hoses, 2"		119.90		131.89	13.52	14.87
48 L.H., Daily Totals		$3032.36		$4474.44	$63.17	$93.22
Crew C-8D	Hr.	Daily	Hr.	Daily	Bare Costs	Incl. O&P
1 Labor Foreman (outside)	$47.60	$380.80	$75.60	$604.80	$50.84	$79.97
1 Laborer	45.60	364.80	72.45	579.60		
1 Cement Finisher	53.15	425.20	83.05	664.40		
1 Equipment Oper. (light)	57.00	456.00	88.80	710.40		
1 Air Compressor, 250 cfm		208.68		229.55		
2 -50' Air Hoses, 1"		16.63		18.30	7.04	7.75
32 L.H., Daily Totals		$1852.11		$2807.04	$57.88	$87.72

Crew No.	Bare Costs		Incl. Subs O&P		Cost Per Labor-Hour	
Crew C-8E	Hr.	Daily	Hr.	Daily	Bare Costs	Incl. O&P
1 Labor Foreman (outside)	$47.60	$380.80	$75.60	$604.80	$49.09	$77.47
3 Laborers	45.60	1094.40	72.45	1738.80		
1 Cement Finisher	53.15	425.20	83.05	664.40		
1 Equipment Oper. (light)	57.00	456.00	88.80	710.40		
1 Shotcrete Rig, 35 C.Y./hr		310.01		341.01		
1 Air Compressor, 250 cfm		208.68		229.55		
4 -50' Air Hoses, 1"		33.27		36.59		
4 -50' Air Hoses, 2"		119.90		131.89	14.00	15.40
48 L.H., Daily Totals		$3028.26		$4457.44	$63.09	$92.86
Crew C-9	Hr.	Daily	Hr.	Daily	Bare Costs	Incl. O&P
1 Cement Finisher	$53.15	$425.20	$83.05	$664.40	$50.34	$79.19
2 Laborers	45.60	729.60	72.45	1159.20		
1 Equipment Oper. (light)	57.00	456.00	88.80	710.40		
1 Grout Pump, 50 C.F./hr.		196.51		216.16		
1 Air Compressor, 160 cfm		218.71		240.59		
2 -50' Air Hoses, 1"		16.63		18.30		
2 -50' Air Hoses, 2"		59.95		65.95	15.37	16.91
32 L.H., Daily Totals		$2102.61		$3074.99	$65.71	$96.09
Crew C-10	Hr.	Daily	Hr.	Daily	Bare Costs	Incl. O&P
1 Laborer	$45.60	$364.80	$72.45	$579.60	$50.63	$79.52
2 Cement Finishers	53.15	850.40	83.05	1328.80		
24 L.H., Daily Totals		$1215.20		$1908.40	$50.63	$79.52
Crew C-10B	Hr.	Daily	Hr.	Daily	Bare Costs	Incl. O&P
3 Laborers	$45.60	$1094.40	$72.45	$1738.80	$48.62	$76.69
2 Cement Finishers	53.15	850.40	83.05	1328.80		
1 Concrete Mixer, 10 C.F.		151.35		166.49		
2 Trowels, 48" Walk-Behind		194.75		214.23	8.65	9.52
40 L.H., Daily Totals		$2290.91		$3448.32	$57.27	$86.21
Crew C-10C	Hr.	Daily	Hr.	Daily	Bare Costs	Incl. O&P
1 Laborer	$45.60	$364.80	$72.45	$579.60	$50.63	$79.52
2 Cement Finishers	53.15	850.40	83.05	1328.80		
1 Trowel, 48" Walk-Behind		97.38		107.11	4.06	4.46
24 L.H., Daily Totals		$1312.58		$2015.51	$54.69	$83.98
Crew C-10D	Hr.	Daily	Hr.	Daily	Bare Costs	Incl. O&P
1 Laborer	$45.60	$364.80	$72.45	$579.60	$50.63	$79.52
2 Cement Finishers	53.15	850.40	83.05	1328.80		
1 Vibrating Power Screed		90.52		99.57		
1 Trowel, 48" Walk-Behind		97.38		107.11	7.83	8.61
24 L.H., Daily Totals		$1403.10		$2115.09	$58.46	$88.13
Crew C-10E	Hr.	Daily	Hr.	Daily	Bare Costs	Incl. O&P
1 Laborer	$45.60	$364.80	$72.45	$579.60	$50.63	$79.52
2 Cement Finishers	53.15	850.40	83.05	1328.80		
1 Vibrating Power Screed		90.52		99.57		
1 Cement Trowel, 96" Ride-On		176.03		193.63	11.11	12.22
24 L.H., Daily Totals		$1481.75		$2201.61	$61.74	$91.73
Crew C-10F	Hr.	Daily	Hr.	Daily	Bare Costs	Incl. O&P
1 Laborer	$45.60	$364.80	$72.45	$579.60	$50.63	$79.52
2 Cement Finishers	53.15	850.40	83.05	1328.80		
1 Telescoping Boom Lift, to 60'		301.02		331.12	12.54	13.80
24 L.H., Daily Totals		$1516.22		$2239.52	$63.18	$93.31

For customer support on your Commercial Renovation Costs with RSMeans data, call 800.448.8182.

783

Crew No.	Bare Costs		Incl. Subs O&P		Cost Per Labor-Hour	
Crew C-11	**Hr.**	**Daily**	**Hr.**	**Daily**	**Bare Costs**	**Incl. O&P**
1 Struc. Steel Foreman (outside)	$63.85	$510.80	$102.80	$822.40	$61.37	$98.14
6 Struc. Steel Workers	61.85	2968.80	99.60	4780.80		
1 Equip. Oper. (crane)	63.40	507.20	98.75	790.00		
1 Equip. Oper. (oiler)	54.00	432.00	84.15	673.20		
1 Lattice Boom Crane, 150 Ton		2398.73		2638.61	33.32	36.65
72 L.H., Daily Totals		$6817.53		$9705.01	$94.69	$134.79
Crew C-12	**Hr.**	**Daily**	**Hr.**	**Daily**	**Bare Costs**	**Incl. O&P**
1 Carpenter Foreman (outside)	$58.30	$466.40	$92.60	$740.80	$56.03	$88.69
3 Carpenters	56.30	1351.20	89.45	2146.80		
1 Laborer	45.60	364.80	72.45	579.60		
1 Equip. Oper. (crane)	63.40	507.20	98.75	790.00		
1 Hyd. Crane, 12 Ton		489.34		538.27	10.19	11.21
48 L.H., Daily Totals		$3178.94		$4795.47	$66.23	$99.91
Crew C-13	**Hr.**	**Daily**	**Hr.**	**Daily**	**Bare Costs**	**Incl. O&P**
1 Struc. Steel Worker	$61.85	$494.80	$99.60	$796.80	$60.03	$96.27
1 Welder	61.95	495.60	99.75	798.00		
1 Carpenter	56.30	450.40	89.45	715.60		
1 Welder, Gas Engine, 300 amp		152.88		168.16	6.37	7.01
24 L.H., Daily Totals		$1593.68		$2478.56	$66.40	$103.27
Crew C-14	**Hr.**	**Daily**	**Hr.**	**Daily**	**Bare Costs**	**Incl. O&P**
1 Carpenter Foreman (outside)	$58.30	$466.40	$92.60	$740.80	$54.93	$86.57
5 Carpenters	56.30	2252.00	89.45	3578.00		
4 Laborers	45.60	1459.20	72.45	2318.40		
4 Rodmen (reinf.)	60.70	1942.40	94.90	3036.80		
2 Cement Finishers	53.15	850.40	83.05	1328.80		
1 Equip. Oper. (crane)	63.40	507.20	98.75	790.00		
1 Equip. Oper. (oiler)	54.00	432.00	84.15	673.20		
1 Hyd. Crane, 80 Ton		1502.22		1652.44	10.43	11.48
144 L.H., Daily Totals		$9411.82		$14118.44	$65.36	$98.04
Crew C-14A	**Hr.**	**Daily**	**Hr.**	**Daily**	**Bare Costs**	**Incl. O&P**
1 Carpenter Foreman (outside)	$58.30	$466.40	$92.60	$740.80	$56.27	$89.02
16 Carpenters	56.30	7206.40	89.45	11449.60		
4 Rodmen (reinf.)	60.70	1942.40	94.90	3036.80		
2 Laborers	45.60	729.60	72.45	1159.20		
1 Cement Finisher	53.15	425.20	83.05	664.40		
1 Equip. Oper. (medium)	60.40	483.20	94.10	752.80		
1 Gas Engine Vibrator		27.87		30.65		
1 Concrete Pump (Small)		435.09		478.60	2.31	2.55
200 L.H., Daily Totals		$11716.16		$18312.85	$58.58	$91.56
Crew C-14B	**Hr.**	**Daily**	**Hr.**	**Daily**	**Bare Costs**	**Incl. O&P**
1 Carpenter Foreman (outside)	$58.30	$466.40	$92.60	$740.80	$56.15	$88.79
16 Carpenters	56.30	7206.40	89.45	11449.60		
4 Rodmen (reinf.)	60.70	1942.40	94.90	3036.80		
2 Laborers	45.60	729.60	72.45	1159.20		
2 Cement Finishers	53.15	850.40	83.05	1328.80		
1 Equip. Oper. (medium)	60.40	483.20	94.10	752.80		
1 Gas Engine Vibrator		27.87		30.65		
1 Concrete Pump (Small)		435.09		478.60	2.23	2.45
208 L.H., Daily Totals		$12141.36		$18977.25	$58.37	$91.24

Crew No.	Bare Costs		Incl. Subs O&P		Cost Per Labor-Hour	
Crew C-14C	**Hr.**	**Daily**	**Hr.**	**Daily**	**Bare Costs**	**Incl. O&P**
1 Carpenter Foreman (outside)	$58.30	$466.40	$92.60	$740.80	$53.79	$85.14
6 Carpenters	56.30	2702.40	89.45	4293.60		
2 Rodmen (reinf.)	60.70	971.20	94.90	1518.40		
4 Laborers	45.60	1459.20	72.45	2318.40		
1 Cement Finisher	53.15	425.20	83.05	664.40		
1 Gas Engine Vibrator		27.87		30.65	.25	.27
112 L.H., Daily Totals		$6052.27		$9566.25	$54.04	$85.41
Crew C-14D	**Hr.**	**Daily**	**Hr.**	**Daily**	**Bare Costs**	**Incl. O&P**
1 Carpenter Foreman (outside)	$58.30	$466.40	$92.60	$740.80	$55.91	$88.58
18 Carpenters	56.30	8107.20	89.45	12880.80		
2 Rodmen (reinf.)	60.70	971.20	94.90	1518.40		
2 Laborers	45.60	729.60	72.45	1159.20		
1 Cement Finisher	53.15	425.20	83.05	664.40		
1 Equip. Oper. (medium)	60.40	483.20	94.10	752.80		
1 Gas Engine Vibrator		27.87		30.65		
1 Concrete Pump (Small)		435.09		478.60	2.31	2.55
200 L.H., Daily Totals		$11645.76		$18225.65	$58.23	$91.13
Crew C-14E	**Hr.**	**Daily**	**Hr.**	**Daily**	**Bare Costs**	**Incl. O&P**
1 Carpenter Foreman (outside)	$58.30	$466.40	$92.60	$740.80	$54.88	$86.50
2 Carpenters	56.30	900.80	89.45	1431.20		
4 Rodmen (reinf.)	60.70	1942.40	94.90	3036.80		
3 Laborers	45.60	1094.40	72.45	1738.80		
1 Cement Finisher	53.15	425.20	83.05	664.40		
1 Gas Engine Vibrator		27.87		30.65	.32	.35
88 L.H., Daily Totals		$4857.07		$7642.65	$55.19	$86.85
Crew C-14F	**Hr.**	**Daily**	**Hr.**	**Daily**	**Bare Costs**	**Incl. O&P**
1 Labor Foreman (outside)	$47.60	$380.80	$75.60	$604.80	$50.86	$79.87
2 Laborers	45.60	729.60	72.45	1159.20		
6 Cement Finishers	53.15	2551.20	83.05	3986.40		
1 Gas Engine Vibrator		27.87		30.65	.39	.43
72 L.H., Daily Totals		$3689.47		$5781.05	$51.24	$80.29
Crew C-14G	**Hr.**	**Daily**	**Hr.**	**Daily**	**Bare Costs**	**Incl. O&P**
1 Labor Foreman (outside)	$47.60	$380.80	$75.60	$604.80	$50.20	$78.96
2 Laborers	45.60	729.60	72.45	1159.20		
4 Cement Finishers	53.15	1700.80	83.05	2657.60		
1 Gas Engine Vibrator		27.87		30.65	.50	.55
56 L.H., Daily Totals		$2839.07		$4452.25	$50.70	$79.50
Crew C-14H	**Hr.**	**Daily**	**Hr.**	**Daily**	**Bare Costs**	**Incl. O&P**
1 Carpenter Foreman (outside)	$58.30	$466.40	$92.60	$740.80	$55.06	$86.98
2 Carpenters	56.30	900.80	89.45	1431.20		
1 Rodman (reinf.)	60.70	485.60	94.90	759.20		
1 Laborer	45.60	364.80	72.45	579.60		
1 Cement Finisher	53.15	425.20	83.05	664.40		
1 Gas Engine Vibrator		27.87		30.65	.58	.64
48 L.H., Daily Totals		$2670.67		$4205.85	$55.64	$87.62
Crew C-14L	**Hr.**	**Daily**	**Hr.**	**Daily**	**Bare Costs**	**Incl. O&P**
1 Carpenter Foreman (outside)	$58.30	$466.40	$92.60	$740.80	$52.64	$83.51
6 Carpenters	56.30	2702.40	89.45	4293.60		
4 Laborers	45.60	1459.20	72.45	2318.40		
1 Cement Finisher	53.15	425.20	83.05	664.40		
1 Gas Engine Vibrator		27.87		30.65	.29	.32
96 L.H., Daily Totals		$5081.07		$8047.85	$52.93	$83.83

For customer support on your Commercial Renovation Costs with RSMeans data, call 800.448.8182.

Crew C-14M

Crew No.	Bare Costs Hr.	Daily	Incl. Subs O&P Hr.	Daily	Cost Per Labor-Hour Bare Costs	Incl. O&P
1 Carpenter Foreman (outside)	$58.30	$466.40	$92.60	$740.80	$54.54	$86.06
2 Carpenters	56.30	900.80	89.45	1431.20		
1 Rodman (reinf.)	60.70	485.60	94.90	759.20		
2 Laborers	45.60	729.60	72.45	1159.20		
1 Cement Finisher	53.15	425.20	83.05	664.40		
1 Equip. Oper. (medium)	60.40	483.20	94.10	752.80		
1 Gas Engine Vibrator		27.87		30.65		
1 Concrete Pump (Small)		435.09		478.60	7.23	7.96
64 L.H., Daily Totals		$3953.76		$6016.85	$61.78	$94.01

Crew C-15

Crew No.	Bare Costs Hr.	Daily	Incl. Subs O&P Hr.	Daily	Cost Per Labor-Hour Bare Costs	Incl. O&P
1 Carpenter Foreman (outside)	$58.30	$466.40	$92.60	$740.80	$52.74	$83.32
2 Carpenters	56.30	900.80	89.45	1431.20		
3 Laborers	45.60	1094.40	72.45	1738.80		
2 Cement Finishers	53.15	850.40	83.05	1328.80		
1 Rodman (reinf.)	60.70	485.60	94.90	759.20		
72 L.H., Daily Totals		$3797.60		$5998.80	$52.74	$83.32

Crew C-16

Crew No.	Bare Costs Hr.	Daily	Incl. Subs O&P Hr.	Daily	Cost Per Labor-Hour Bare Costs	Incl. O&P
1 Labor Foreman (outside)	$47.60	$380.80	$75.60	$604.80	$50.16	$79.02
3 Laborers	45.60	1094.40	72.45	1738.80		
2 Cement Finishers	53.15	850.40	83.05	1328.80		
1 Equip. Oper. (medium)	60.40	483.20	94.10	752.80		
1 Gunite Pump Rig		331.07		364.17		
2 -50' Air Hoses, 3/4"		14.81		16.29		
2 -50' Air Hoses, 2"		59.95		65.95	7.25	7.97
56 L.H., Daily Totals		$3214.63		$4871.61	$57.40	$86.99

Crew C-16A

Crew No.	Bare Costs Hr.	Daily	Incl. Subs O&P Hr.	Daily	Cost Per Labor-Hour Bare Costs	Incl. O&P
1 Laborer	$45.60	$364.80	$72.45	$579.60	$53.08	$83.16
2 Cement Finishers	53.15	850.40	83.05	1328.80		
1 Equip. Oper. (medium)	60.40	483.20	94.10	752.80		
1 Gunite Pump Rig		331.07		364.17		
2 -50' Air Hoses, 3/4"		14.81		16.29		
2 -50' Air Hoses, 2"		59.95		65.95		
1 Telescoping Boom Lift, to 60'		301.02		331.12	22.09	24.30
32 L.H., Daily Totals		$2405.24		$3438.73	$75.16	$107.46

Crew C-17

Crew No.	Bare Costs Hr.	Daily	Incl. Subs O&P Hr.	Daily	Cost Per Labor-Hour Bare Costs	Incl. O&P
2 Skilled Worker Foremen (out)	$60.90	$974.40	$96.35	$1541.60	$59.30	$93.83
8 Skilled Workers	58.90	3769.60	93.20	5964.80		
80 L.H., Daily Totals		$4744.00		$7506.40	$59.30	$93.83

Crew C-17A

Crew No.	Bare Costs Hr.	Daily	Incl. Subs O&P Hr.	Daily	Cost Per Labor-Hour Bare Costs	Incl. O&P
2 Skilled Worker Foremen (out)	$60.90	$974.40	$96.35	$1541.60	$59.35	$93.89
8 Skilled Workers	58.90	3769.60	93.20	5964.80		
.125 Equip. Oper. (crane)	63.40	63.40	98.75	98.75		
.125 Hyd. Crane, 80 Ton		187.78		206.55	2.32	2.55
81 L.H., Daily Totals		$4995.18		$7811.70	$61.67	$96.44

Crew C-17B

Crew No.	Bare Costs Hr.	Daily	Incl. Subs O&P Hr.	Daily	Cost Per Labor-Hour Bare Costs	Incl. O&P
2 Skilled Worker Foremen (out)	$60.90	$974.40	$96.35	$1541.60	$59.40	$93.95
8 Skilled Workers	58.90	3769.60	93.20	5964.80		
.25 Equip. Oper. (crane)	63.40	126.80	98.75	197.50		
.25 Hyd. Crane, 80 Ton		375.55		413.11		
.25 Trowel, 48" Walk-Behind		24.34		26.78	4.88	5.36
82 L.H., Daily Totals		$5270.70		$8143.79	$64.28	$99.31

Crew C-17C

Crew No.	Bare Costs Hr.	Daily	Incl. Subs O&P Hr.	Daily	Cost Per Labor-Hour Bare Costs	Incl. O&P
2 Skilled Worker Foremen (out)	$60.90	$974.40	$96.35	$1541.60	$59.45	$94.01
8 Skilled Workers	58.90	3769.60	93.20	5964.80		
.375 Equip. Oper. (crane)	63.40	190.20	98.75	296.25		
.375 Hyd. Crane, 80 Ton		563.33		619.66	6.79	7.47
83 L.H., Daily Totals		$5497.53		$8422.31	$66.24	$101.47

Crew C-17D

Crew No.	Bare Costs Hr.	Daily	Incl. Subs O&P Hr.	Daily	Cost Per Labor-Hour Bare Costs	Incl. O&P
2 Skilled Worker Foremen (out)	$60.90	$974.40	$96.35	$1541.60	$59.50	$94.06
8 Skilled Workers	58.90	3769.60	93.20	5964.80		
.5 Equip. Oper. (crane)	63.40	253.60	98.75	395.00		
.5 Hyd. Crane, 80 Ton		751.11		826.22	8.94	9.84
84 L.H., Daily Totals		$5748.71		$8727.62	$68.44	$103.90

Crew C-17E

Crew No.	Bare Costs Hr.	Daily	Incl. Subs O&P Hr.	Daily	Cost Per Labor-Hour Bare Costs	Incl. O&P
2 Skilled Worker Foremen (out)	$60.90	$974.40	$96.35	$1541.60	$59.30	$93.83
8 Skilled Workers	58.90	3769.60	93.20	5964.80		
1 Hyd. Jack with Rods		37.76		41.54	.47	.52
80 L.H., Daily Totals		$4781.76		$7547.94	$59.77	$94.35

Crew C-18

Crew No.	Bare Costs Hr.	Daily	Incl. Subs O&P Hr.	Daily	Cost Per Labor-Hour Bare Costs	Incl. O&P
.125 Labor Foreman (outside)	$47.60	$47.60	$75.60	$75.60	$45.82	$72.80
1 Laborer	45.60	364.80	72.45	579.60		
1 Concrete Cart, 10 C.F.		120.90		132.99	13.43	14.78
9 L.H., Daily Totals		$533.30		$788.19	$59.26	$87.58

Crew C-19

Crew No.	Bare Costs Hr.	Daily	Incl. Subs O&P Hr.	Daily	Cost Per Labor-Hour Bare Costs	Incl. O&P
.125 Labor Foreman (outside)	$47.60	$47.60	$75.60	$75.60	$45.82	$72.80
1 Laborer	45.60	364.80	72.45	579.60		
1 Concrete Cart, 18 C.F.		143.42		157.76	15.94	17.53
9 L.H., Daily Totals		$555.82		$812.96	$61.76	$90.33

Crew C-20

Crew No.	Bare Costs Hr.	Daily	Incl. Subs O&P Hr.	Daily	Cost Per Labor-Hour Bare Costs	Incl. O&P
1 Labor Foreman (outside)	$47.60	$380.80	$75.60	$604.80	$48.64	$76.88
5 Laborers	45.60	1824.00	72.45	2898.00		
1 Cement Finisher	53.15	425.20	83.05	664.40		
1 Equip. Oper. (medium)	60.40	483.20	94.10	752.80		
2 Gas Engine Vibrators		55.74		61.31		
1 Concrete Pump (Small)		435.09		478.60	7.67	8.44
64 L.H., Daily Totals		$3604.02		$5459.91	$56.31	$85.31

Crew C-21

Crew No.	Bare Costs Hr.	Daily	Incl. Subs O&P Hr.	Daily	Cost Per Labor-Hour Bare Costs	Incl. O&P
1 Labor Foreman (outside)	$47.60	$380.80	$75.60	$604.80	$48.64	$76.88
5 Laborers	45.60	1824.00	72.45	2898.00		
1 Cement Finisher	53.15	425.20	83.05	664.40		
1 Equip. Oper. (medium)	60.40	483.20	94.10	752.80		
2 Gas Engine Vibrators		55.74		61.31		
1 Concrete Conveyer		212.33		233.56	4.19	4.61
64 L.H., Daily Totals		$3381.27		$5214.87	$52.83	$81.48

Crew C-22

Crew No.	Bare Costs Hr.	Daily	Incl. Subs O&P Hr.	Daily	Cost Per Labor-Hour Bare Costs	Incl. O&P
1 Rodman Foreman (outside)	$62.70	$501.60	$98.05	$784.40	$60.99	$95.34
4 Rodmen (reinf.)	60.70	1942.40	94.90	3036.80		
.125 Equip. Oper. (crane)	63.40	63.40	98.75	98.75		
.125 Equip. Oper. (oiler)	54.00	54.00	84.15	84.15		
.125 Hyd. Crane, 25 Ton		75.43		82.97	1.80	1.98
42 L.H., Daily Totals		$2636.83		$4087.07	$62.78	$97.31

For customer support on your Commercial Renovation Costs with RSMeans data, call 800.448.8182.

785

Crew No.	Bare Costs		Incl. Subs O&P		Cost Per Labor-Hour	
Crew C-23	Hr.	Daily	Hr.	Daily	Bare Costs	Incl. O&P
2 Skilled Worker Foremen (out)	$60.90	$974.40	$96.35	$1541.60	$59.26	$93.48
6 Skilled Workers	58.90	2827.20	93.20	4473.60		
1 Equip. Oper. (crane)	63.40	507.20	98.75	790.00		
1 Equip. Oper. (oiler)	54.00	432.00	84.15	673.20		
1 Lattice Boom Crane, 90 Ton		1767.55		1944.30	22.09	24.30
80 L.H., Daily Totals		$6508.35		$9422.70	$81.35	$117.78
Crew C-24	Hr.	Daily	Hr.	Daily	Bare Costs	Incl. O&P
2 Skilled Worker Foremen (out)	$60.90	$974.40	$96.35	$1541.60	$59.26	$93.48
6 Skilled Workers	58.90	2827.20	93.20	4473.60		
1 Equip. Oper. (crane)	63.40	507.20	98.75	790.00		
1 Equip. Oper. (oiler)	54.00	432.00	84.15	673.20		
1 Lattice Boom Crane, 150 Ton		2398.73		2638.61	29.98	32.98
80 L.H., Daily Totals		$7139.53		$10117.01	$89.24	$126.46
Crew C-25	Hr.	Daily	Hr.	Daily	Bare Costs	Incl. O&P
2 Rodmen (reinf.)	$60.70	$971.20	$94.90	$1518.40	$48.98	$79.50
2 Rodmen Helpers	37.25	596.00	64.10	1025.60		
32 L.H., Daily Totals		$1567.20		$2544.00	$48.98	$79.50
Crew C-27	Hr.	Daily	Hr.	Daily	Bare Costs	Incl. O&P
2 Cement Finishers	$53.15	$850.40	$83.05	$1328.80	$53.15	$83.05
1 Concrete Saw		115.94		127.54	7.25	7.97
16 L.H., Daily Totals		$966.34		$1456.34	$60.40	$91.02
Crew C-28	Hr.	Daily	Hr.	Daily	Bare Costs	Incl. O&P
1 Cement Finisher	$53.15	$425.20	$83.05	$664.40	$53.15	$83.05
1 Portable Air Compressor, Gas		40.12		44.13	5.01	5.52
8 L.H., Daily Totals		$465.32		$708.53	$58.16	$88.57
Crew C-29	Hr.	Daily	Hr.	Daily	Bare Costs	Incl. O&P
1 Laborer	$45.60	$364.80	$72.45	$579.60	$45.60	$72.45
1 Pressure Washer		100.40		110.44	12.55	13.80
8 L.H., Daily Totals		$465.20		$690.04	$58.15	$86.25
Crew C-30	Hr.	Daily	Hr.	Daily	Bare Costs	Incl. O&P
1 Laborer	$45.60	$364.80	$72.45	$579.60	$45.60	$72.45
1 Concrete Mixer, 10 C.F.		151.35		166.49	18.92	20.81
8 L.H., Daily Totals		$516.15		$746.09	$64.52	$93.26
Crew C-31	Hr.	Daily	Hr.	Daily	Bare Costs	Incl. O&P
1 Cement Finisher	$53.15	$425.20	$83.05	$664.40	$53.15	$83.05
1 Grout Pump		331.07		364.17	41.38	45.52
8 L.H., Daily Totals		$756.27		$1028.57	$94.53	$128.57
Crew C-32	Hr.	Daily	Hr.	Daily	Bare Costs	Incl. O&P
1 Cement Finisher	$53.15	$425.20	$83.05	$664.40	$49.38	$77.75
1 Laborer	45.60	364.80	72.45	579.60		
1 Crack Chaser Saw, Gas, 6 H.P.		75.72		83.30		
1 Vacuum Pick-Up System		77.50		85.25	9.58	10.53
16 L.H., Daily Totals		$943.22		$1412.54	$58.95	$88.28
Crew D-1	Hr.	Daily	Hr.	Daily	Bare Costs	Incl. O&P
1 Bricklayer	$55.25	$442.00	$88.50	$708.00	$50.13	$80.30
1 Bricklayer Helper	45.00	360.00	72.10	576.80		
16 L.H., Daily Totals		$802.00		$1284.80	$50.13	$80.30

Crew No.	Bare Costs		Incl. Subs O&P		Cost Per Labor-Hour	
Crew D-2	Hr.	Daily	Hr.	Daily	Bare Costs	Incl. O&P
3 Bricklayers	$55.25	$1326.00	$88.50	$2124.00	$51.62	$82.62
2 Bricklayer Helpers	45.00	720.00	72.10	1153.60		
.5 Carpenter	56.30	225.20	89.45	357.80		
44 L.H., Daily Totals		$2271.20		$3635.40	$51.62	$82.62
Crew D-3	Hr.	Daily	Hr.	Daily	Bare Costs	Incl. O&P
3 Bricklayers	$55.25	$1326.00	$88.50	$2124.00	$51.40	$82.30
2 Bricklayer Helpers	45.00	720.00	72.10	1153.60		
.25 Carpenter	56.30	112.60	89.45	178.90		
42 L.H., Daily Totals		$2158.60		$3456.50	$51.40	$82.30
Crew D-4	Hr.	Daily	Hr.	Daily	Bare Costs	Incl. O&P
1 Bricklayer	$55.25	$442.00	$88.50	$708.00	$50.56	$80.38
2 Bricklayer Helpers	45.00	720.00	72.10	1153.60		
1 Equip. Oper. (light)	57.00	456.00	88.80	710.40		
1 Grout Pump, 50 C.F./hr.		196.51		216.16	6.14	6.76
32 L.H., Daily Totals		$1814.51		$2788.16	$56.70	$87.13
Crew D-5	Hr.	Daily	Hr.	Daily	Bare Costs	Incl. O&P
1 Bricklayer	55.25	442.00	88.50	708.00	55.25	88.50
8 L.H., Daily Totals		$442.00		$708.00	$55.25	$88.50
Crew D-6	Hr.	Daily	Hr.	Daily	Bare Costs	Incl. O&P
3 Bricklayers	$55.25	$1326.00	$88.50	$2124.00	$50.37	$80.67
3 Bricklayer Helpers	45.00	1080.00	72.10	1730.40		
.25 Carpenter	56.30	112.60	89.45	178.90		
50 L.H., Daily Totals		$2518.60		$4033.30	$50.37	$80.67
Crew D-7	Hr.	Daily	Hr.	Daily	Bare Costs	Incl. O&P
1 Tile Layer	$53.15	$425.20	$82.95	$663.60	$47.98	$74.88
1 Tile Layer Helper	42.80	342.40	66.80	534.40		
16 L.H., Daily Totals		$767.60		$1198.00	$47.98	$74.88
Crew D-8	Hr.	Daily	Hr.	Daily	Bare Costs	Incl. O&P
3 Bricklayers	$55.25	$1326.00	$88.50	$2124.00	$51.15	$81.94
2 Bricklayer Helpers	45.00	720.00	72.10	1153.60		
40 L.H., Daily Totals		$2046.00		$3277.60	$51.15	$81.94
Crew D-9	Hr.	Daily	Hr.	Daily	Bare Costs	Incl. O&P
3 Bricklayers	$55.25	$1326.00	$88.50	$2124.00	$50.13	$80.30
3 Bricklayer Helpers	45.00	1080.00	72.10	1730.40		
48 L.H., Daily Totals		$2406.00		$3854.40	$50.13	$80.30
Crew D-10	Hr.	Daily	Hr.	Daily	Bare Costs	Incl. O&P
1 Bricklayer Foreman (outside)	$57.25	$458.00	$91.70	$733.60	$55.23	$87.76
1 Bricklayer	55.25	442.00	88.50	708.00		
1 Bricklayer Helper	45.00	360.00	72.10	576.80		
1 Equip. Oper. (crane)	63.40	507.20	98.75	790.00		
1 S.P. Crane, 4x4, 12 Ton		445.72		490.30	13.93	15.32
32 L.H., Daily Totals		$2212.92		$3298.70	$69.15	$103.08
Crew D-11	Hr.	Daily	Hr.	Daily	Bare Costs	Incl. O&P
1 Bricklayer Foreman (outside)	$57.25	$458.00	$91.70	$733.60	$52.50	$84.10
1 Bricklayer	55.25	442.00	88.50	708.00		
1 Bricklayer Helper	45.00	360.00	72.10	576.80		
24 L.H., Daily Totals		$1260.00		$2018.40	$52.50	$84.10

For customer support on your Commercial Renovation Costs with RSMeans data, call 800.448.8182.

Crew D-12

Crew No.	Bare Costs Hr.	Daily	Incl. Subs O&P Hr.	Daily	Cost Per Labor-Hour Bare Costs	Incl. O&P
1 Bricklayer Foreman (outside)	$57.25	$458.00	$91.70	$733.60	$50.63	$81.10
1 Bricklayer	55.25	442.00	88.50	708.00		
2 Bricklayer Helpers	45.00	720.00	72.10	1153.60		
32 L.H., Daily Totals		$1620.00		$2595.20	$50.63	$81.10

Crew D-13

Crew No.	Bare Costs Hr.	Daily	Incl. Subs O&P Hr.	Daily	Cost Per Labor-Hour Bare Costs	Incl. O&P
1 Bricklayer Foreman (outside)	$57.25	$458.00	$91.70	$733.60	$53.70	$85.43
1 Bricklayer	55.25	442.00	88.50	708.00		
2 Bricklayer Helpers	45.00	720.00	72.10	1153.60		
1 Carpenter	56.30	450.40	89.45	715.60		
1 Equip. Oper. (crane)	63.40	507.20	98.75	790.00		
1 S.P. Crane, 4x4, 12 Ton		445.72		490.30	9.29	10.21
48 L.H., Daily Totals		$3023.32		$4591.10	$62.99	$95.65

Crew D-14

Crew No.	Bare Costs Hr.	Daily	Incl. Subs O&P Hr.	Daily	Cost Per Labor-Hour Bare Costs	Incl. O&P
3 Bricklayers	$55.25	$1326.00	$88.50	$2124.00	$52.69	$84.40
1 Bricklayer Helper	45.00	360.00	72.10	576.80		
32 L.H., Daily Totals		$1686.00		$2700.80	$52.69	$84.40

Crew E-1

Crew No.	Bare Costs Hr.	Daily	Incl. Subs O&P Hr.	Daily	Cost Per Labor-Hour Bare Costs	Incl. O&P
1 Welder Foreman (outside)	$63.95	$511.60	$102.95	$823.60	$60.97	$97.17
1 Welder	61.95	495.60	99.75	798.00		
1 Equip. Oper. (light)	57.00	456.00	88.80	710.40		
1 Welder, Gas Engine, 300 amp		152.88		168.16	6.37	7.01
24 L.H., Daily Totals		$1616.08		$2500.16	$67.34	$104.17

Crew E-2

Crew No.	Bare Costs Hr.	Daily	Incl. Subs O&P Hr.	Daily	Cost Per Labor-Hour Bare Costs	Incl. O&P
1 Struc. Steel Foreman (outside)	$63.85	$510.80	$102.80	$822.40	$61.24	$97.73
4 Struc. Steel Workers	61.85	1979.20	99.60	3187.20		
1 Equip. Oper. (crane)	63.40	507.20	98.75	790.00		
1 Equip. Oper. (oiler)	54.00	432.00	84.15	673.20		
1 Lattice Boom Crane, 90 Ton		1767.55		1944.30	31.56	34.72
56 L.H., Daily Totals		$5196.75		$7417.10	$92.80	$132.45

Crew E-3

Crew No.	Bare Costs Hr.	Daily	Incl. Subs O&P Hr.	Daily	Cost Per Labor-Hour Bare Costs	Incl. O&P
1 Struc. Steel Foreman (outside)	$63.85	$510.80	$102.80	$822.40	$62.55	$100.72
1 Struc. Steel Worker	61.85	494.80	99.60	796.80		
1 Welder	61.95	495.60	99.75	798.00		
1 Welder, Gas Engine, 300 amp		152.88		168.16	6.37	7.01
24 L.H., Daily Totals		$1654.08		$2585.36	$68.92	$107.72

Crew E-3A

Crew No.	Bare Costs Hr.	Daily	Incl. Subs O&P Hr.	Daily	Cost Per Labor-Hour Bare Costs	Incl. O&P
1 Struc. Steel Foreman (outside)	$63.85	$510.80	$102.80	$822.40	$62.55	$100.72
1 Struc. Steel Worker	61.85	494.80	99.60	796.80		
1 Welder	61.95	495.60	99.75	798.00		
1 Welder, Gas Engine, 300 amp		152.88		168.16		
1 Telescoping Boom Lift, to 40'		290.76		319.84	18.48	20.33
24 L.H., Daily Totals		$1944.84		$2905.20	$81.03	$121.05

Crew E-4

Crew No.	Bare Costs Hr.	Daily	Incl. Subs O&P Hr.	Daily	Cost Per Labor-Hour Bare Costs	Incl. O&P
1 Struc. Steel Foreman (outside)	$63.85	$510.80	$102.80	$822.40	$62.35	$100.40
3 Struc. Steel Workers	61.85	1484.40	99.60	2390.40		
1 Welder, Gas Engine, 300 amp		152.88		168.16	4.78	5.26
32 L.H., Daily Totals		$2148.08		$3380.96	$67.13	$105.66

Crew E-5

Crew No.	Bare Costs Hr.	Daily	Incl. Subs O&P Hr.	Daily	Cost Per Labor-Hour Bare Costs	Incl. O&P
2 Struc. Steel Foremen (outside)	$63.85	$1021.60	$102.80	$1644.80	$61.63	$98.63
5 Struc. Steel Workers	61.85	2474.00	99.60	3984.00		
1 Equip. Oper. (crane)	63.40	507.20	98.75	790.00		
1 Welder	61.95	495.60	99.75	798.00		
1 Equip. Oper. (oiler)	54.00	432.00	84.15	673.20		
1 Lattice Boom Crane, 90 Ton		1767.55		1944.30		
1 Welder, Gas Engine, 300 amp		152.88		168.16	24.01	26.41
80 L.H., Daily Totals		$6850.83		$10002.47	$85.64	$125.03

Crew E-6

Crew No.	Bare Costs Hr.	Daily	Incl. Subs O&P Hr.	Daily	Cost Per Labor-Hour Bare Costs	Incl. O&P
3 Struc. Steel Foremen (outside)	$63.85	$1532.40	$102.80	$2467.20	$61.53	$98.52
9 Struc. Steel Workers	61.85	4453.20	99.60	7171.20		
1 Equip. Oper. (crane)	63.40	507.20	98.75	790.00		
1 Welder	61.95	495.60	99.75	798.00		
1 Equip. Oper. (oiler)	54.00	432.00	84.15	673.20		
1 Equip. Oper. (light)	57.00	456.00	88.80	710.40		
1 Lattice Boom Crane, 90 Ton		1767.55		1944.30		
1 Welder, Gas Engine, 300 amp		152.88		168.16		
1 Air Compressor, 160 cfm		218.71		240.59		
2 Impact Wrenches		93.50		102.84	17.44	19.19
128 L.H., Daily Totals		$10109.04		$15065.90	$78.98	$117.70

Crew E-7

Crew No.	Bare Costs Hr.	Daily	Incl. Subs O&P Hr.	Daily	Cost Per Labor-Hour Bare Costs	Incl. O&P
1 Struc. Steel Foreman (outside)	$63.85	$510.80	$102.80	$822.40	$61.65	$98.66
4 Struc. Steel Workers	61.85	1979.20	99.60	3187.20		
1 Equip. Oper. (crane)	63.40	507.20	98.75	790.00		
1 Equip. Oper. (oiler)	54.00	432.00	84.15	673.20		
1 Welder Foreman (outside)	63.95	511.60	102.95	823.60		
2 Welders	61.95	991.20	99.75	1596.00		
1 Lattice Boom Crane, 90 Ton		1767.55		1944.30		
2 Welder, Gas Engine, 300 amp		305.75		336.33	25.92	28.51
80 L.H., Daily Totals		$7005.30		$10173.03	$87.57	$127.16

Crew E-8

Crew No.	Bare Costs Hr.	Daily	Incl. Subs O&P Hr.	Daily	Cost Per Labor-Hour Bare Costs	Incl. O&P
1 Struc. Steel Foreman (outside)	$63.85	$510.80	$102.80	$822.40	$61.34	$98.07
4 Struc. Steel Workers	61.85	1979.20	99.60	3187.20		
1 Welder Foreman (outside)	63.95	511.60	102.95	823.60		
4 Welders	61.95	1982.40	99.75	3192.00		
1 Equip. Oper. (crane)	63.40	507.20	98.75	790.00		
1 Equip. Oper. (oiler)	54.00	432.00	84.15	673.20		
1 Equip. Oper. (light)	57.00	456.00	88.80	710.40		
1 Lattice Boom Crane, 90 Ton		1767.55		1944.30		
4 Welder, Gas Engine, 300 amp		611.51		672.66	22.88	25.16
104 L.H., Daily Totals		$8758.26		$12815.76	$84.21	$123.23

Crew E-9

Crew No.	Bare Costs Hr.	Daily	Incl. Subs O&P Hr.	Daily	Cost Per Labor-Hour Bare Costs	Incl. O&P
2 Struc. Steel Foremen (outside)	$63.85	$1021.60	$102.80	$1644.80	$61.57	$98.56
5 Struc. Steel Workers	61.85	2474.00	99.60	3984.00		
1 Welder Foreman (outside)	63.95	511.60	102.95	823.60		
5 Welders	61.95	2478.00	99.75	3990.00		
1 Equip. Oper. (crane)	63.40	507.20	98.75	790.00		
1 Equip. Oper. (oiler)	54.00	432.00	84.15	673.20		
1 Equip. Oper. (light)	57.00	456.00	88.80	710.40		
1 Lattice Boom Crane, 90 Ton		1767.55		1944.30		
5 Welder, Gas Engine, 300 amp		764.38		840.82	19.78	21.76
128 L.H., Daily Totals		$10412.33		$15401.12	$81.35	$120.32

Crew E-10

Crew No.	Bare Costs Hr.	Daily	Incl. Subs O&P Hr.	Daily	Cost Per Labor-Hour Bare Costs	Incl. O&P
1 Welder Foreman (outside)	$63.95	$511.60	$102.95	$823.60	$62.95	$101.35
1 Welder	61.95	495.60	99.75	798.00		
1 Welder, Gas Engine, 300 amp		152.88		168.16		
1 Flatbed Truck, Gas, 3 Ton		877.96		965.75	64.43	70.87
16 L.H., Daily Totals		$2038.03		$2755.52	$127.38	$172.22

Crew E-11

Crew No.	Bare Costs Hr.	Daily	Incl. Subs O&P Hr.	Daily	Cost Per Labor-Hour Bare Costs	Incl. O&P
2 Painters, Struc. Steel	$48.90	$782.40	$82.95	$1327.20	$50.10	$81.79
1 Building Laborer	45.60	364.80	72.45	579.60		
1 Equip. Oper. (light)	57.00	456.00	88.80	710.40		
1 Air Compressor, 250 cfm		208.68		229.55		
1 Sandblaster, Portable, 3 C.F.		86.86		95.55		
1 Set Sand Blasting Accessories		16.09		17.70	9.74	10.71
32 L.H., Daily Totals		$1914.84		$2960.00	$59.84	$92.50

Crew E-11A

Crew No.	Bare Costs Hr.	Daily	Incl. Subs O&P Hr.	Daily	Cost Per Labor-Hour Bare Costs	Incl. O&P
2 Painters, Struc. Steel	$48.90	$782.40	$82.95	$1327.20	$50.10	$81.79
1 Building Laborer	45.60	364.80	72.45	579.60		
1 Equip. Oper. (light)	57.00	456.00	88.80	710.40		
1 Air Compressor, 250 cfm		208.68		229.55		
1 Sandblaster, Portable, 3 C.F.		86.86		95.55		
1 Set Sand Blasting Accessories		16.09		17.70		
1 Telescoping Boom Lift, to 60'		301.02		331.12	19.15	21.06
32 L.H., Daily Totals		$2215.85		$3291.12	$69.25	$102.85

Crew E-11B

Crew No.	Bare Costs Hr.	Daily	Incl. Subs O&P Hr.	Daily	Cost Per Labor-Hour Bare Costs	Incl. O&P
2 Painters, Struc. Steel	$48.90	$782.40	$82.95	$1327.20	$47.80	$79.45
1 Building Laborer	45.60	364.80	72.45	579.60		
2 Paint Sprayer, 8 C.F.M.		91.47		100.62		
1 Telescoping Boom Lift, to 60'		301.02		331.12	16.35	17.99
24 L.H., Daily Totals		$1539.69		$2338.54	$64.15	$97.44

Crew E-12

Crew No.	Bare Costs Hr.	Daily	Incl. Subs O&P Hr.	Daily	Cost Per Labor-Hour Bare Costs	Incl. O&P
1 Welder Foreman (outside)	$63.95	$511.60	$102.95	$823.60	$60.48	$95.88
1 Equip. Oper. (light)	57.00	456.00	88.80	710.40		
1 Welder, Gas Engine, 300 amp		152.88		168.16	9.55	10.51
16 L.H., Daily Totals		$1120.48		$1702.16	$70.03	$106.39

Crew E-13

Crew No.	Bare Costs Hr.	Daily	Incl. Subs O&P Hr.	Daily	Cost Per Labor-Hour Bare Costs	Incl. O&P
1 Welder Foreman (outside)	$63.95	$511.60	$102.95	$823.60	$61.63	$98.23
.5 Equip. Oper. (light)	57.00	228.00	88.80	355.20		
1 Welder, Gas Engine, 300 amp		152.88		168.16	12.74	14.01
12 L.H., Daily Totals		$892.48		$1346.96	$74.37	$112.25

Crew E-14

Crew No.	Bare Costs Hr.	Daily	Incl. Subs O&P Hr.	Daily	Cost Per Labor-Hour Bare Costs	Incl. O&P
1 Welder Foreman (outside)	$63.95	$511.60	$102.95	$823.60	$63.95	$102.95
1 Welder, Gas Engine, 300 amp		152.88		168.16	19.11	21.02
8 L.H., Daily Totals		$664.48		$991.76	$83.06	$123.97

Crew E-16

Crew No.	Bare Costs Hr.	Daily	Incl. Subs O&P Hr.	Daily	Cost Per Labor-Hour Bare Costs	Incl. O&P
1 Welder Foreman (outside)	$63.95	$511.60	$102.95	$823.60	$62.95	$101.35
1 Welder	61.95	495.60	99.75	798.00		
1 Welder, Gas Engine, 300 amp		152.88		168.16	9.55	10.51
16 L.H., Daily Totals		$1160.08		$1789.76	$72.50	$111.86

Crew E-17

Crew No.	Bare Costs Hr.	Daily	Incl. Subs O&P Hr.	Daily	Cost Per Labor-Hour Bare Costs	Incl. O&P
1 Struc. Steel Foreman (outside)	$63.85	$510.80	$102.80	$822.40	$62.85	$101.20
1 Structural Steel Worker	61.85	494.80	99.60	796.80		
16 L.H., Daily Totals		$1005.60		$1619.20	$62.85	$101.20

Crew E-18

Crew No.	Bare Costs Hr.	Daily	Incl. Subs O&P Hr.	Daily	Cost Per Labor-Hour Bare Costs	Incl. O&P
1 Struc. Steel Foreman (outside)	$63.85	$510.80	$102.80	$822.40	$61.96	$99.14
3 Structural Steel Workers	61.85	1484.40	99.60	2390.40		
1 Equipment Operator (med.)	60.40	483.20	94.10	752.80		
1 Lattice Boom Crane, 20 Ton		1577.49		1735.24	39.44	43.38
40 L.H., Daily Totals		$4055.89		$5700.84	$101.40	$142.52

Crew E-19

Crew No.	Bare Costs Hr.	Daily	Incl. Subs O&P Hr.	Daily	Cost Per Labor-Hour Bare Costs	Incl. O&P
1 Struc. Steel Foreman (outside)	$63.85	$510.80	$102.80	$822.40	$60.90	$97.07
1 Structural Steel Worker	61.85	494.80	99.60	796.80		
1 Equip. Oper. (light)	57.00	456.00	88.80	710.40		
1 Lattice Boom Crane, 20 Ton		1577.49		1735.24	65.73	72.30
24 L.H., Daily Totals		$3039.09		$4064.84	$126.63	$169.37

Crew E-20

Crew No.	Bare Costs Hr.	Daily	Incl. Subs O&P Hr.	Daily	Cost Per Labor-Hour Bare Costs	Incl. O&P
1 Struc. Steel Foreman (outside)	$63.85	$510.80	$102.80	$822.40	$61.31	$97.96
5 Structural Steel Workers	61.85	2474.00	99.60	3984.00		
1 Equip. Oper. (crane)	63.40	507.20	98.75	790.00		
1 Equip. Oper. (oiler)	54.00	432.00	84.15	673.20		
1 Lattice Boom Crane, 40 Ton		2198.13		2417.95	34.35	37.78
64 L.H., Daily Totals		$6122.13		$8687.55	$95.66	$135.74

Crew E-22

Crew No.	Bare Costs Hr.	Daily	Incl. Subs O&P Hr.	Daily	Cost Per Labor-Hour Bare Costs	Incl. O&P
1 Skilled Worker Foreman (out)	$60.90	$487.20	$96.35	$770.80	$59.57	$94.25
2 Skilled Workers	58.90	942.40	93.20	1491.20		
24 L.H., Daily Totals		$1429.60		$2262.00	$59.57	$94.25

Crew E-24

Crew No.	Bare Costs Hr.	Daily	Incl. Subs O&P Hr.	Daily	Cost Per Labor-Hour Bare Costs	Incl. O&P
3 Structural Steel Workers	$61.85	$1484.40	$99.60	$2390.40	$61.49	$98.22
1 Equipment Operator (med.)	60.40	483.20	94.10	752.80		
1 Hyd. Crane, 25 Ton		603.44		663.78	18.86	20.74
32 L.H., Daily Totals		$2571.04		$3806.98	$80.34	$118.97

Crew E-25

Crew No.	Bare Costs Hr.	Daily	Incl. Subs O&P Hr.	Daily	Cost Per Labor-Hour Bare Costs	Incl. O&P
1 Welder Foreman (outside)	$63.95	$511.60	$102.95	$823.60	$63.95	$102.95
1 Cutting Torch		13.37		14.71	1.67	1.84
8 L.H., Daily Totals		$524.97		$838.31	$65.62	$104.79

Crew E-26

Crew No.	Bare Costs Hr.	Daily	Incl. Subs O&P Hr.	Daily	Cost Per Labor-Hour Bare Costs	Incl. O&P
1 Struc. Steel Foreman (outside)	$63.85	$510.80	$102.80	$822.40	$63.33	$101.28
1 Struc. Steel Worker	61.85	494.80	99.60	796.80		
1 Welder	61.95	495.60	99.75	798.00		
.25 Electrician	66.20	132.40	101.70	203.40		
.25 Plumber	69.80	139.60	107.60	215.20		
1 Welder, Gas Engine, 300 amp		152.88		168.16	5.46	6.01
28 L.H., Daily Totals		$1926.08		$3003.96	$68.79	$107.28

Crew E-27

Crew No.	Bare Costs Hr.	Daily	Incl. Subs O&P Hr.	Daily	Cost Per Labor-Hour Bare Costs	Incl. O&P
1 Struc. Steel Foreman (outside)	$63.85	$510.80	$102.80	$822.40	$61.31	$97.96
5 Struc. Steel Workers	61.85	2474.00	99.60	3984.00		
1 Equip. Oper. (crane)	63.40	507.20	98.75	790.00		
1 Equip. Oper. (oiler)	54.00	432.00	84.15	673.20		
1 Hyd. Crane, 12 Ton		489.34		538.27		
1 Hyd. Crane, 80 Ton		1502.22		1652.44	31.12	34.23
64 L.H., Daily Totals		$5915.56		$8460.31	$92.43	$132.19

Crew F-3

Crew No.	Bare Costs Hr.	Daily	Incl. Subs O&P Hr.	Daily	Cost Per Labor-Hour Bare Costs	Incl. O&P
4 Carpenters	$56.30	$1801.60	$89.45	$2862.40	$57.72	$91.31
1 Equip. Oper. (crane)	63.40	507.20	98.75	790.00		
1 Hyd. Crane, 12 Ton		489.34		538.27	12.23	13.46
40 L.H., Daily Totals		$2798.14		$4190.67	$69.95	$104.77

For customer support on your Commercial Renovation Costs with RSMeans data, call 800.448.8182.

Crew No.	Bare Costs		Incl. Subs O&P		Cost Per Labor-Hour	
Crew F-4	Hr.	Daily	Hr.	Daily	Bare Costs	Incl. O&P
4 Carpenters	$56.30	$1801.60	$89.45	$2862.40	$57.10	$90.12
1 Equip. Oper. (crane)	63.40	507.20	98.75	790.00		
1 Equip. Oper. (oiler)	54.00	432.00	84.15	673.20		
1 Hyd. Crane, 55 Ton		1019.39		1121.33	21.24	23.36
48 L.H., Daily Totals		$3760.19		$5446.93	$78.34	$113.48
Crew F-5	Hr.	Daily	Hr.	Daily	Bare Costs	Incl. O&P
1 Carpenter Foreman (outside)	$58.30	$466.40	$92.60	$740.80	$56.80	$90.24
3 Carpenters	56.30	1351.20	89.45	2146.80		
32 L.H., Daily Totals		$1817.60		$2887.60	$56.80	$90.24
Crew F-6	Hr.	Daily	Hr.	Daily	Bare Costs	Incl. O&P
2 Carpenters	$56.30	$900.80	$89.45	$1431.20	$53.44	$84.51
2 Building Laborers	45.60	729.60	72.45	1159.20		
1 Equip. Oper. (crane)	63.40	507.20	98.75	790.00		
1 Hyd. Crane, 12 Ton		489.34		538.27	12.23	13.46
40 L.H., Daily Totals		$2626.94		$3918.67	$65.67	$97.97
Crew F-7	Hr.	Daily	Hr.	Daily	Bare Costs	Incl. O&P
2 Carpenters	$56.30	$900.80	$89.45	$1431.20	$50.95	$80.95
2 Building Laborers	45.60	729.60	72.45	1159.20		
32 L.H., Daily Totals		$1630.40		$2590.40	$50.95	$80.95
Crew G-1	Hr.	Daily	Hr.	Daily	Bare Costs	Incl. O&P
1 Roofer Foreman (outside)	$51.60	$412.80	$88.80	$710.40	$46.36	$79.77
4 Roofers Composition	49.60	1587.20	85.35	2731.20		
2 Roofer Helpers	37.25	596.00	64.10	1025.60		
1 Application Equipment		200.52		220.57		
1 Tar Kettle/Pot		215.43		236.98		
1 Crew Truck		172.61		189.87	10.51	11.56
56 L.H., Daily Totals		$3184.56		$5114.61	$56.87	$91.33
Crew G-2	Hr.	Daily	Hr.	Daily	Bare Costs	Incl. O&P
1 Plasterer	$51.40	$411.20	$81.55	$652.40	$47.62	$75.58
1 Plasterer Helper	45.85	366.80	72.75	582.00		
1 Building Laborer	45.60	364.80	72.45	579.60		
1 Grout Pump, 50 C.F./hr.		196.51		216.16	8.19	9.01
24 L.H., Daily Totals		$1339.31		$2030.16	$55.80	$84.59
Crew G-2A	Hr.	Daily	Hr.	Daily	Bare Costs	Incl. O&P
1 Roofer Composition	$49.60	$396.80	$85.35	$682.80	$44.15	$73.97
1 Roofer Helper	37.25	298.00	64.10	512.80		
1 Building Laborer	45.60	364.80	72.45	579.60		
1 Foam Spray Rig, Trailer-Mtd.		546.24		600.86		
1 Pickup Truck, 3/4 Ton		115.08		126.59	27.56	30.31
24 L.H., Daily Totals		$1720.92		$2502.65	$71.71	$104.28
Crew G-3	Hr.	Daily	Hr.	Daily	Bare Costs	Incl. O&P
2 Sheet Metal Workers	$67.55	$1080.80	$105.15	$1682.40	$56.58	$88.80
2 Building Laborers	45.60	729.60	72.45	1159.20		
32 L.H., Daily Totals		$1810.40		$2841.60	$56.58	$88.80
Crew G-4	Hr.	Daily	Hr.	Daily	Bare Costs	Incl. O&P
1 Labor Foreman (outside)	$47.60	$380.80	$75.60	$604.80	$46.27	$73.50
2 Building Laborers	45.60	729.60	72.45	1159.20		
1 Flatbed Truck, Gas, 1.5 Ton		203.25		223.57		
1 Air Compressor, 160 cfm		218.71		240.59	17.58	19.34
24 L.H., Daily Totals		$1532.36		$2228.16	$63.85	$92.84

Crew No.	Bare Costs		Incl. Subs O&P		Cost Per Labor-Hour	
Crew G-5	Hr.	Daily	Hr.	Daily	Bare Costs	Incl. O&P
1 Roofer Foreman (outside)	$51.60	$412.80	$88.80	$710.40	$45.06	$77.54
2 Roofers Composition	49.60	793.60	85.35	1365.60		
2 Roofer Helpers	37.25	596.00	64.10	1025.60		
1 Application Equipment		200.52		220.57	5.01	5.51
40 L.H., Daily Totals		$2002.92		$3322.17	$50.07	$83.05
Crew G-6A	Hr.	Daily	Hr.	Daily	Bare Costs	Incl. O&P
2 Roofers Composition	$49.60	$793.60	$85.35	$1365.60	$49.60	$85.35
1 Small Compressor, Electric		40.64		44.70		
2 Pneumatic Nailers		57.32		63.05	6.12	6.73
16 L.H., Daily Totals		$891.56		$1473.35	$55.72	$92.08
Crew G-7	Hr.	Daily	Hr.	Daily	Bare Costs	Incl. O&P
1 Carpenter	$56.30	$450.40	$89.45	$715.60	$56.30	$89.45
1 Small Compressor, Electric		40.64		44.70		
1 Pneumatic Nailer		28.66		31.53	8.66	9.53
8 L.H., Daily Totals		$519.70		$791.83	$64.96	$98.98
Crew H-1	Hr.	Daily	Hr.	Daily	Bare Costs	Incl. O&P
2 Glaziers	$54.00	$864.00	$85.65	$1370.40	$57.92	$92.63
2 Struc. Steel Workers	61.85	989.60	99.60	1593.60		
32 L.H., Daily Totals		$1853.60		$2964.00	$57.92	$92.63
Crew H-2	Hr.	Daily	Hr.	Daily	Bare Costs	Incl. O&P
2 Glaziers	$54.00	$864.00	$85.65	$1370.40	$51.20	$81.25
1 Building Laborer	45.60	364.80	72.45	579.60		
24 L.H., Daily Totals		$1228.80		$1950.00	$51.20	$81.25
Crew H-3	Hr.	Daily	Hr.	Daily	Bare Costs	Incl. O&P
1 Glazier	$54.00	$432.00	$85.65	$685.20	$48.65	$77.72
1 Helper	43.30	346.40	69.80	558.40		
16 L.H., Daily Totals		$778.40		$1243.60	$48.65	$77.72
Crew H-4	Hr.	Daily	Hr.	Daily	Bare Costs	Incl. O&P
1 Carpenter	$56.30	$450.40	$89.45	$715.60	$53.08	$84.04
1 Carpenter Helper	43.30	346.40	69.80	558.40		
.5 Electrician	66.20	264.80	101.70	406.80		
20 L.H., Daily Totals		$1061.60		$1680.80	$53.08	$84.04
Crew J-1	Hr.	Daily	Hr.	Daily	Bare Costs	Incl. O&P
3 Plasterers	$51.40	$1233.60	$81.55	$1957.20	$49.18	$78.03
2 Plasterer Helpers	45.85	733.60	72.75	1164.00		
1 Mixing Machine, 6 C.F.		116.55		128.20	2.91	3.21
40 L.H., Daily Totals		$2083.75		$3249.40	$52.09	$81.23
Crew J-2	Hr.	Daily	Hr.	Daily	Bare Costs	Incl. O&P
3 Plasterers	$51.40	$1233.60	$81.55	$1957.20	$50.32	$79.58
2 Plasterer Helpers	45.85	733.60	72.75	1164.00		
1 Lather	56.00	448.00	87.35	698.80		
1 Mixing Machine, 6 C.F.		116.55		128.20	2.43	2.67
48 L.H., Daily Totals		$2531.75		$3948.20	$52.74	$82.25
Crew J-3	Hr.	Daily	Hr.	Daily	Bare Costs	Incl. O&P
1 Terrazzo Worker	$53.40	$427.20	$83.35	$666.80	$49.10	$76.65
1 Terrazzo Helper	44.80	358.40	69.95	559.60		
1 Floor Grinder, 22" Path		99.18		109.09		
1 Terrazzo Mixer		167.65		184.42	16.68	18.34
16 L.H., Daily Totals		$1052.43		$1519.91	$65.78	$94.99

Crew J-4

Crew No.	Bare Costs Hr.	Daily	Incl. Subs O&P Hr.	Daily	Cost Per Labor-Hour Bare Costs	Incl. O&P
2 Cement Finishers	$53.15	$850.40	$83.05	$1328.80	$50.63	$79.52
1 Laborer	45.60	364.80	72.45	579.60		
1 Floor Grinder, 22" Path		99.18		109.09		
1 Floor Edger, 7" Path		45.53		50.08		
1 Vacuum Pick-Up System		77.50		85.25	9.26	10.18
24 L.H., Daily Totals		$1437.40		$2152.82	$59.89	$89.70

Crew J-4A

Crew No.	Bare Costs Hr.	Daily	Incl. Subs O&P Hr.	Daily	Cost Per Labor-Hour Bare Costs	Incl. O&P
2 Cement Finishers	$53.15	$850.40	$83.05	$1328.80	$49.38	$77.75
2 Laborers	45.60	729.60	72.45	1159.20		
1 Floor Grinder, 22" Path		99.18		109.09		
1 Floor Edger, 7" Path		45.53		50.08		
1 Vacuum Pick-Up System		77.50		85.25		
1 Floor Auto Scrubber		185.41		203.95	12.74	14.01
32 L.H., Daily Totals		$1987.61		$2936.37	$62.11	$91.76

Crew J-4B

Crew No.	Bare Costs Hr.	Daily	Incl. Subs O&P Hr.	Daily	Cost Per Labor-Hour Bare Costs	Incl. O&P
1 Laborer	$45.60	$364.80	$72.45	$579.60	$45.60	$72.45
1 Floor Auto Scrubber		185.41		203.95	23.18	25.49
8 L.H., Daily Totals		$550.21		$783.55	$68.78	$97.94

Crew J-6

Crew No.	Bare Costs Hr.	Daily	Incl. Subs O&P Hr.	Daily	Cost Per Labor-Hour Bare Costs	Incl. O&P
2 Painters	$47.80	$764.80	$75.60	$1209.60	$49.55	$78.11
1 Building Laborer	45.60	364.80	72.45	579.60		
1 Equip. Oper. (light)	57.00	456.00	88.80	710.40		
1 Air Compressor, 250 cfm		208.68		229.55		
1 Sandblaster, Portable, 3 C.F.		86.86		95.55		
1 Set Sand Blasting Accessories		16.09		17.70	9.74	10.71
32 L.H., Daily Totals		$1897.24		$2842.40	$59.29	$88.83

Crew J-7

Crew No.	Bare Costs Hr.	Daily	Incl. Subs O&P Hr.	Daily	Cost Per Labor-Hour Bare Costs	Incl. O&P
2 Painters	$47.80	$764.80	$75.60	$1209.60	$47.80	$75.60
1 Floor Belt Sander		51.97		57.17		
1 Floor Sanding Edger		26.05		28.65	4.88	5.36
16 L.H., Daily Totals		$842.82		$1295.42	$52.68	$80.96

Crew K-1

Crew No.	Bare Costs Hr.	Daily	Incl. Subs O&P Hr.	Daily	Cost Per Labor-Hour Bare Costs	Incl. O&P
1 Carpenter	$56.30	$450.40	$89.45	$715.60	$53.35	$84.88
1 Truck Driver (light)	50.40	403.20	80.30	642.40		
1 Flatbed Truck, Gas, 3 Ton		877.96		965.75	54.87	60.36
16 L.H., Daily Totals		$1731.56		$2323.75	$108.22	$145.23

Crew K-2

Crew No.	Bare Costs Hr.	Daily	Incl. Subs O&P Hr.	Daily	Cost Per Labor-Hour Bare Costs	Incl. O&P
1 Struc. Steel Foreman (outside)	$63.85	$510.80	$102.80	$822.40	$58.70	$94.23
1 Struc. Steel Worker	61.85	494.80	99.60	796.80		
1 Truck Driver (light)	50.40	403.20	80.30	642.40		
1 Flatbed Truck, Gas, 3 Ton		877.96		965.75	36.58	40.24
24 L.H., Daily Totals		$2286.76		$3227.35	$95.28	$134.47

Crew L-1

Crew No.	Bare Costs Hr.	Daily	Incl. Subs O&P Hr.	Daily	Cost Per Labor-Hour Bare Costs	Incl. O&P
1 Electrician	$66.20	$529.60	$101.70	$813.60	$68.00	$104.65
1 Plumber	69.80	558.40	107.60	860.80		
16 L.H., Daily Totals		$1088.00		$1674.40	$68.00	$104.65

Crew L-2

Crew No.	Bare Costs Hr.	Daily	Incl. Subs O&P Hr.	Daily	Cost Per Labor-Hour Bare Costs	Incl. O&P
1 Carpenter	$56.30	$450.40	$89.45	$715.60	$49.80	$79.63
1 Carpenter Helper	43.30	346.40	69.80	558.40		
16 L.H., Daily Totals		$796.80		$1274.00	$49.80	$79.63

Crew L-3

Crew No.	Bare Costs Hr.	Daily	Incl. Subs O&P Hr.	Daily	Cost Per Labor-Hour Bare Costs	Incl. O&P
1 Carpenter	$56.30	$450.40	$89.45	$715.60	$61.59	$96.44
.5 Electrician	66.20	264.80	101.70	406.80		
.5 Sheet Metal Worker	67.55	270.20	105.15	420.60		
16 L.H., Daily Totals		$985.40		$1543.00	$61.59	$96.44

Crew L-3A

Crew No.	Bare Costs Hr.	Daily	Incl. Subs O&P Hr.	Daily	Cost Per Labor-Hour Bare Costs	Incl. O&P
1 Carpenter Foreman (outside)	$58.30	$466.40	$92.60	$740.80	$61.38	$96.78
.5 Sheet Metal Worker	67.55	270.20	105.15	420.60		
12 L.H., Daily Totals		$736.60		$1161.40	$61.38	$96.78

Crew L-4

Crew No.	Bare Costs Hr.	Daily	Incl. Subs O&P Hr.	Daily	Cost Per Labor-Hour Bare Costs	Incl. O&P
2 Skilled Workers	$58.90	$942.40	$93.20	$1491.20	$53.70	$85.40
1 Helper	43.30	346.40	69.80	558.40		
24 L.H., Daily Totals		$1288.80		$2049.60	$53.70	$85.40

Crew L-5

Crew No.	Bare Costs Hr.	Daily	Incl. Subs O&P Hr.	Daily	Cost Per Labor-Hour Bare Costs	Incl. O&P
1 Struc. Steel Foreman (outside)	$63.85	$510.80	$102.80	$822.40	$62.36	$99.94
5 Struc. Steel Workers	61.85	2474.00	99.60	3984.00		
1 Equip. Oper. (crane)	63.40	507.20	98.75	790.00		
1 Hyd. Crane, 25 Ton		603.44		663.78	10.78	11.85
56 L.H., Daily Totals		$4095.44		$6260.18	$73.13	$111.79

Crew L-5A

Crew No.	Bare Costs Hr.	Daily	Incl. Subs O&P Hr.	Daily	Cost Per Labor-Hour Bare Costs	Incl. O&P
1 Struc. Steel Foreman (outside)	$63.85	$510.80	$102.80	$822.40	$62.74	$100.19
2 Structural Steel Workers	61.85	989.60	99.60	1593.60		
1 Equip. Oper. (crane)	63.40	507.20	98.75	790.00		
1 S.P. Crane, 4x4, 25 Ton		1192.78		1312.06	37.27	41.00
32 L.H., Daily Totals		$3200.38		$4518.06	$100.01	$141.19

Crew L-5B

Crew No.	Bare Costs Hr.	Daily	Incl. Subs O&P Hr.	Daily	Cost Per Labor-Hour Bare Costs	Incl. O&P
1 Struc. Steel Foreman (outside)	$63.85	$510.80	$102.80	$822.40	$64.23	$100.57
2 Structural Steel Workers	61.85	989.60	99.60	1593.60		
2 Electricians	66.20	1059.20	101.70	1627.20		
2 Steamfitters/Pipefitters	70.35	1125.60	108.40	1734.40		
1 Equip. Oper. (crane)	63.40	507.20	98.75	790.00		
1 Equip. Oper. (oiler)	54.00	432.00	84.15	673.20		
1 Hyd. Crane, 80 Ton		1502.22		1652.44	20.86	22.95
72 L.H., Daily Totals		$6126.62		$8893.24	$85.09	$123.52

Crew L-6

Crew No.	Bare Costs Hr.	Daily	Incl. Subs O&P Hr.	Daily	Cost Per Labor-Hour Bare Costs	Incl. O&P
1 Plumber	$69.80	$558.40	$107.60	$860.80	$68.60	$105.63
.5 Electrician	66.20	264.80	101.70	406.80		
12 L.H., Daily Totals		$823.20		$1267.60	$68.60	$105.63

Crew L-7

Crew No.	Bare Costs Hr.	Daily	Incl. Subs O&P Hr.	Daily	Cost Per Labor-Hour Bare Costs	Incl. O&P
2 Carpenters	$56.30	$900.80	$89.45	$1431.20	$54.66	$86.34
1 Building Laborer	45.60	364.80	72.45	579.60		
.5 Electrician	66.20	264.80	101.70	406.80		
28 L.H., Daily Totals		$1530.40		$2417.60	$54.66	$86.34

Crew L-8

Crew No.	Bare Costs Hr.	Daily	Incl. Subs O&P Hr.	Daily	Cost Per Labor-Hour Bare Costs	Incl. O&P
2 Carpenters	$56.30	$900.80	$89.45	$1431.20	$59.00	$93.08
.5 Plumber	69.80	279.20	107.60	430.40		
20 L.H., Daily Totals		$1180.00		$1861.60	$59.00	$93.08

For customer support on your Commercial Renovation Costs with RSMeans data, call 800.448.8182.

Crew No.	Bare Costs		Incl. Subs O&P		Cost Per Labor-Hour	

Crew L-9	Hr.	Daily	Hr.	Daily	Bare Costs	Incl. O&P
1 Labor Foreman (inside)	$46.10	$368.80	$73.25	$586.00	$51.61	$81.91
2 Building Laborers	45.60	729.60	72.45	1159.20		
1 Struc. Steel Worker	61.85	494.80	99.60	796.80		
.5 Electrician	66.20	264.80	101.70	406.80		
36 L.H., Daily Totals		$1858.00		$2948.80	$51.61	$81.91

Crew L-10	Hr.	Daily	Hr.	Daily	Bare Costs	Incl. O&P
1 Struc. Steel Foreman (outside)	$63.85	$510.80	$102.80	$822.40	$63.03	$100.38
1 Structural Steel Worker	61.85	494.80	99.60	796.80		
1 Equip. Oper. (crane)	63.40	507.20	98.75	790.00		
1 Hyd. Crane, 12 Ton		489.34		538.27	20.39	22.43
24 L.H., Daily Totals		$2002.14		$2947.47	$83.42	$122.81

Crew L-11	Hr.	Daily	Hr.	Daily	Bare Costs	Incl. O&P
2 Wreckers	$45.60	$729.60	$73.45	$1175.20	$52.90	$83.61
1 Equip. Oper. (crane)	63.40	507.20	98.75	790.00		
1 Equip. Oper. (light)	57.00	456.00	88.80	710.40		
1 Hyd. Excavator, 2.5 C.Y.		1613.66		1775.03		
1 Loader, Skid Steer, 78 H.P.		460.30		506.33	64.81	71.29
32 L.H., Daily Totals		$3766.76		$4956.96	$117.71	$154.90

Crew M-1	Hr.	Daily	Hr.	Daily	Bare Costs	Incl. O&P
3 Elevator Constructors	$93.55	$2245.20	$143.15	$3435.60	$88.88	$136.00
1 Elevator Apprentice	74.85	598.80	114.55	916.40		
5 Hand Tools		51.95		57.15	1.62	1.79
32 L.H., Daily Totals		$2895.95		$4409.15	$90.50	$137.79

Crew M-3	Hr.	Daily	Hr.	Daily	Bare Costs	Incl. O&P
1 Electrician Foreman (outside)	$68.20	$545.60	$104.80	$838.40	$69.95	$107.88
1 Common Laborer	45.60	364.80	72.45	579.60		
.25 Equipment Operator (med.)	60.40	120.80	94.10	188.20		
1 Elevator Constructor	93.55	748.40	143.15	1145.20		
1 Elevator Apprentice	74.85	598.80	114.55	916.40		
.25 S.P. Crane, 4x4, 20 Ton		148.13		162.95	4.36	4.79
34 L.H., Daily Totals		$2526.53		$3830.75	$74.31	$112.67

Crew M-4	Hr.	Daily	Hr.	Daily	Bare Costs	Incl. O&P
1 Electrician Foreman (outside)	$68.20	$545.60	$104.80	$838.40	$69.23	$106.82
1 Common Laborer	45.60	364.80	72.45	579.60		
.25 Equipment Operator, Crane	63.40	126.80	98.75	197.50		
.25 Equip. Oper. (oiler)	54.00	108.00	84.15	168.30		
1 Elevator Constructor	93.55	748.40	143.15	1145.20		
1 Elevator Apprentice	74.85	598.80	114.55	916.40		
.25 S.P. Crane, 4x4, 40 Ton		195.96		215.56	5.44	5.99
36 L.H., Daily Totals		$2688.36		$4060.96	$74.68	$112.80

Crew Q-1	Hr.	Daily	Hr.	Daily	Bare Costs	Incl. O&P
1 Plumber	$69.80	$558.40	$107.60	$860.80	$62.83	$96.85
1 Plumber Apprentice	55.85	446.80	86.10	688.80		
16 L.H., Daily Totals		$1005.20		$1549.60	$62.83	$96.85

Crew Q-1A	Hr.	Daily	Hr.	Daily	Bare Costs	Incl. O&P
.25 Plumber Foreman (outside)	$71.80	$143.60	$110.65	$221.30	$70.20	$108.21
1 Plumber	69.80	558.40	107.60	860.80		
10 L.H., Daily Totals		$702.00		$1082.10	$70.20	$108.21

Crew Q-1C	Hr.	Daily	Hr.	Daily	Bare Costs	Incl. O&P
1 Plumber	$69.80	$558.40	$107.60	$860.80	$62.02	$95.93
1 Plumber Apprentice	55.85	446.80	86.10	688.80		
1 Equip. Oper. (medium)	60.40	483.20	94.10	752.80		
1 Trencher, Chain Type, 8' D		1952.14		2147.36	81.34	89.47
24 L.H., Daily Totals		$3440.54		$4449.76	$143.36	$185.41

Crew Q-2	Hr.	Daily	Hr.	Daily	Bare Costs	Incl. O&P
2 Plumbers	$69.80	$1116.80	$107.60	$1721.60	$65.15	$100.43
1 Plumber Apprentice	55.85	446.80	86.10	688.80		
24 L.H., Daily Totals		$1563.60		$2410.40	$65.15	$100.43

Crew Q-3	Hr.	Daily	Hr.	Daily	Bare Costs	Incl. O&P
1 Plumber Foreman (inside)	$70.30	$562.40	$108.35	$866.80	$66.44	$102.41
2 Plumbers	69.80	1116.80	107.60	1721.60		
1 Plumber Apprentice	55.85	446.80	86.10	688.80		
32 L.H., Daily Totals		$2126.00		$3277.20	$66.44	$102.41

Crew Q-4	Hr.	Daily	Hr.	Daily	Bare Costs	Incl. O&P
1 Plumber Foreman (inside)	$70.30	$562.40	$108.35	$866.80	$66.44	$102.41
1 Plumber	69.80	558.40	107.60	860.80		
1 Welder (plumber)	69.80	558.40	107.60	860.80		
1 Plumber Apprentice	55.85	446.80	86.10	688.80		
1 Welder, Electric, 300 amp		110.75		121.83	3.46	3.81
32 L.H., Daily Totals		$2236.75		$3399.03	$69.90	$106.22

Crew Q-5	Hr.	Daily	Hr.	Daily	Bare Costs	Incl. O&P
1 Steamfitter	$70.35	$562.80	$108.40	$867.20	$63.33	$97.58
1 Steamfitter Apprentice	56.30	450.40	86.75	694.00		
16 L.H., Daily Totals		$1013.20		$1561.20	$63.33	$97.58

Crew Q-6	Hr.	Daily	Hr.	Daily	Bare Costs	Incl. O&P
2 Steamfitters	$70.35	$1125.60	$108.40	$1734.40	$65.67	$101.18
1 Steamfitter Apprentice	56.30	450.40	86.75	694.00		
24 L.H., Daily Totals		$1576.00		$2428.40	$65.67	$101.18

Crew Q-7	Hr.	Daily	Hr.	Daily	Bare Costs	Incl. O&P
1 Steamfitter Foreman (inside)	$70.85	$566.80	$109.20	$873.60	$66.96	$103.19
2 Steamfitters	70.35	1125.60	108.40	1734.40		
1 Steamfitter Apprentice	56.30	450.40	86.75	694.00		
32 L.H., Daily Totals		$2142.80		$3302.00	$66.96	$103.19

Crew Q-8	Hr.	Daily	Hr.	Daily	Bare Costs	Incl. O&P
1 Steamfitter Foreman (inside)	$70.85	$566.80	$109.20	$873.60	$66.96	$103.19
1 Steamfitter	70.35	562.80	108.40	867.20		
1 Welder (steamfitter)	70.35	562.80	108.40	867.20		
1 Steamfitter Apprentice	56.30	450.40	86.75	694.00		
1 Welder, Electric, 300 amp		110.75		121.83	3.46	3.81
32 L.H., Daily Totals		$2253.55		$3423.83	$70.42	$106.99

Crew Q-9	Hr.	Daily	Hr.	Daily	Bare Costs	Incl. O&P
1 Sheet Metal Worker	$67.55	$540.40	$105.15	$841.20	$60.80	$94.65
1 Sheet Metal Apprentice	54.05	432.40	84.15	673.20		
16 L.H., Daily Totals		$972.80		$1514.40	$60.80	$94.65

Crew Q-10	Hr.	Daily	Hr.	Daily	Bare Costs	Incl. O&P
2 Sheet Metal Workers	$67.55	$1080.80	$105.15	$1682.40	$63.05	$98.15
1 Sheet Metal Apprentice	54.05	432.40	84.15	673.20		
24 L.H., Daily Totals		$1513.20		$2355.60	$63.05	$98.15

For customer support on your Commercial Renovation Costs with RSMeans data, call 800.448.8182.

Crew No.	Bare Costs Hr.	Daily	Incl. Subs O&P Hr.	Daily	Cost Per Labor-Hour Bare Costs	Incl. O&P
Crew Q-11	Hr.	Daily	Hr.	Daily	Bare Costs	Incl. O&P
1 Sheet Metal Foreman (inside)	$68.05	$544.40	$105.95	$847.60	$64.30	$100.10
2 Sheet Metal Workers	67.55	1080.80	105.15	1682.40		
1 Sheet Metal Apprentice	54.05	432.40	84.15	673.20		
32 L.H., Daily Totals		$2057.60		$3203.20	$64.30	$100.10
Crew Q-12	Hr.	Daily	Hr.	Daily	Bare Costs	Incl. O&P
1 Sprinkler Installer	$68.55	$548.40	$105.75	$846.00	$61.70	$95.17
1 Sprinkler Apprentice	54.85	438.80	84.60	676.80		
16 L.H., Daily Totals		$987.20		$1522.80	$61.70	$95.17
Crew Q-13	Hr.	Daily	Hr.	Daily	Bare Costs	Incl. O&P
1 Sprinkler Foreman (inside)	$69.05	$552.40	$106.50	$852.00	$65.25	$100.65
2 Sprinkler Installers	68.55	1096.80	105.75	1692.00		
1 Sprinkler Apprentice	54.85	438.80	84.60	676.80		
32 L.H., Daily Totals		$2088.00		$3220.80	$65.25	$100.65
Crew Q-14	Hr.	Daily	Hr.	Daily	Bare Costs	Incl. O&P
1 Asbestos Worker	$63.05	$504.40	$99.60	$796.80	$56.75	$89.65
1 Asbestos Apprentice	50.45	403.60	79.70	637.60		
16 L.H., Daily Totals		$908.00		$1434.40	$56.75	$89.65
Crew Q-15	Hr.	Daily	Hr.	Daily	Bare Costs	Incl. O&P
1 Plumber	$69.80	$558.40	$107.60	$860.80	$62.83	$96.85
1 Plumber Apprentice	55.85	446.80	86.10	688.80		
1 Welder, Electric, 300 amp		110.75		121.83	6.92	7.61
16 L.H., Daily Totals		$1115.95		$1671.43	$69.75	$104.46
Crew Q-16	Hr.	Daily	Hr.	Daily	Bare Costs	Incl. O&P
2 Plumbers	$69.80	$1116.80	$107.60	$1721.60	$65.15	$100.43
1 Plumber Apprentice	55.85	446.80	86.10	688.80		
1 Welder, Electric, 300 amp		110.75		121.83	4.61	5.08
24 L.H., Daily Totals		$1674.35		$2532.23	$69.76	$105.51
Crew Q-17	Hr.	Daily	Hr.	Daily	Bare Costs	Incl. O&P
1 Steamfitter	$70.35	$562.80	$108.40	$867.20	$63.33	$97.58
1 Steamfitter Apprentice	56.30	450.40	86.75	694.00		
1 Welder, Electric, 300 amp		110.75		121.83	6.92	7.61
16 L.H., Daily Totals		$1123.95		$1683.03	$70.25	$105.19
Crew Q-17A	Hr.	Daily	Hr.	Daily	Bare Costs	Incl. O&P
1 Steamfitter	$70.35	$562.80	$108.40	$867.20	$63.35	$97.97
1 Steamfitter Apprentice	56.30	450.40	86.75	694.00		
1 Equip. Oper. (crane)	63.40	507.20	98.75	790.00		
1 Hyd. Crane, 12 Ton		489.34		538.27		
1 Welder, Electric, 300 amp		110.75		121.83	25.00	27.50
24 L.H., Daily Totals		$2120.49		$3011.30	$88.35	$125.47
Crew Q-18	Hr.	Daily	Hr.	Daily	Bare Costs	Incl. O&P
2 Steamfitters	$70.35	$1125.60	$108.40	$1734.40	$65.67	$101.18
1 Steamfitter Apprentice	56.30	450.40	86.75	694.00		
1 Welder, Electric, 300 amp		110.75		121.83	4.61	5.08
24 L.H., Daily Totals		$1686.75		$2550.23	$70.28	$106.26
Crew Q-19	Hr.	Daily	Hr.	Daily	Bare Costs	Incl. O&P
1 Steamfitter	$70.35	$562.80	$108.40	$867.20	$64.28	$98.95
1 Steamfitter Apprentice	56.30	450.40	86.75	694.00		
1 Electrician	66.20	529.60	101.70	813.60		
24 L.H., Daily Totals		$1542.80		$2374.80	$64.28	$98.95

Crew No.	Bare Costs Hr.	Daily	Incl. Subs O&P Hr.	Daily	Cost Per Labor-Hour Bare Costs	Incl. O&P
Crew Q-20	Hr.	Daily	Hr.	Daily	Bare Costs	Incl. O&P
1 Sheet Metal Worker	$67.55	$540.40	$105.15	$841.20	$61.88	$96.06
1 Sheet Metal Apprentice	54.05	432.40	84.15	673.20		
.5 Electrician	66.20	264.80	101.70	406.80		
20 L.H., Daily Totals		$1237.60		$1921.20	$61.88	$96.06
Crew Q-21	Hr.	Daily	Hr.	Daily	Bare Costs	Incl. O&P
2 Steamfitters	$70.35	$1125.60	$108.40	$1734.40	$65.80	$101.31
1 Steamfitter Apprentice	56.30	450.40	86.75	694.00		
1 Electrician	66.20	529.60	101.70	813.60		
32 L.H., Daily Totals		$2105.60		$3242.00	$65.80	$101.31
Crew Q-22	Hr.	Daily	Hr.	Daily	Bare Costs	Incl. O&P
1 Plumber	$69.80	$558.40	$107.60	$860.80	$62.83	$96.85
1 Plumber Apprentice	55.85	446.80	86.10	688.80		
1 Hyd. Crane, 12 Ton		489.34		538.27	30.58	33.64
16 L.H., Daily Totals		$1494.54		$2087.87	$93.41	$130.49
Crew Q-22A	Hr.	Daily	Hr.	Daily	Bare Costs	Incl. O&P
1 Plumber	$69.80	$558.40	$107.60	$860.80	$58.66	$91.22
1 Plumber Apprentice	55.85	446.80	86.10	688.80		
1 Laborer	45.60	364.80	72.45	579.60		
1 Equip. Oper. (crane)	63.40	507.20	98.75	790.00		
1 Hyd. Crane, 12 Ton		489.34		538.27	15.29	16.82
32 L.H., Daily Totals		$2366.54		$3457.47	$73.95	$108.05
Crew Q-23	Hr.	Daily	Hr.	Daily	Bare Costs	Incl. O&P
1 Plumber Foreman (outside)	$71.80	$574.40	$110.65	$885.20	$67.33	$104.12
1 Plumber	69.80	558.40	107.60	860.80		
1 Equip. Oper. (medium)	60.40	483.20	94.10	752.80		
1 Lattice Boom Crane, 20 Ton		1577.49		1735.24	65.73	72.30
24 L.H., Daily Totals		$3193.49		$4234.04	$133.06	$176.42
Crew R-1	Hr.	Daily	Hr.	Daily	Bare Costs	Incl. O&P
1 Electrician Foreman	$66.70	$533.60	$102.45	$819.60	$61.87	$95.04
3 Electricians	66.20	1588.80	101.70	2440.80		
2 Electrician Apprentices	52.95	847.20	81.35	1301.60		
48 L.H., Daily Totals		$2969.60		$4562.00	$61.87	$95.04
Crew R-1A	Hr.	Daily	Hr.	Daily	Bare Costs	Incl. O&P
1 Electrician	$66.20	$529.60	$101.70	$813.60	$59.58	$91.53
1 Electrician Apprentice	52.95	423.60	81.35	650.80		
16 L.H., Daily Totals		$953.20		$1464.40	$59.58	$91.53
Crew R-1B	Hr.	Daily	Hr.	Daily	Bare Costs	Incl. O&P
1 Electrician	$66.20	$529.60	$101.70	$813.60	$57.37	$88.13
2 Electrician Apprentices	52.95	847.20	81.35	1301.60		
24 L.H., Daily Totals		$1376.80		$2115.20	$57.37	$88.13
Crew R-1C	Hr.	Daily	Hr.	Daily	Bare Costs	Incl. O&P
2 Electricians	$66.20	$1059.20	$101.70	$1627.20	$59.58	$91.53
2 Electrician Apprentices	52.95	847.20	81.35	1301.60		
1 Portable cable puller, 8000 lb.		105.93		116.52	3.31	3.64
32 L.H., Daily Totals		$2012.33		$3045.32	$62.89	$95.17

Crew No.	Bare Costs Hr.	Daily	Incl. Subs O&P Hr.	Daily	Cost Per Labor-Hour Bare Costs	Incl. O&P
Crew R-2	Hr.	Daily	Hr.	Daily	Bare Costs	Incl. O&P
1 Electrician Foreman	$66.70	$533.60	$102.45	$819.60	$62.09	$95.57
3 Electricians	66.20	1588.80	101.70	2440.80		
2 Electrician Apprentices	52.95	847.20	81.35	1301.60		
1 Equip. Oper. (crane)	63.40	507.20	98.75	790.00		
1 S.P. Crane, 4x4, 5 Ton		393.98		433.38	7.04	7.74
56 L.H., Daily Totals		$3870.78		$5785.38	$69.12	$103.31
Crew R-3	Hr.	Daily	Hr.	Daily	Bare Costs	Incl. O&P
1 Electrician Foreman	$66.70	$533.60	$102.45	$819.60	$65.84	$101.41
1 Electrician	66.20	529.60	101.70	813.60		
.5 Equip. Oper. (crane)	63.40	253.60	98.75	395.00		
.5 S.P. Crane, 4x4, 5 Ton		196.99		216.69	9.85	10.83
20 L.H., Daily Totals		$1513.79		$2244.89	$75.69	$112.24
Crew R-4	Hr.	Daily	Hr.	Daily	Bare Costs	Incl. O&P
1 Struc. Steel Foreman (outside)	$63.85	$510.80	$102.80	$822.40	$63.12	$100.66
3 Struc. Steel Workers	61.85	1484.40	99.60	2390.40		
1 Electrician	66.20	529.60	101.70	813.60		
1 Welder, Gas Engine, 300 amp		152.88		168.16	3.82	4.20
40 L.H., Daily Totals		$2677.68		$4194.56	$66.94	$104.86
Crew R-5	Hr.	Daily	Hr.	Daily	Bare Costs	Incl. O&P
1 Electrician Foreman	$66.70	$533.60	$102.45	$819.60	$57.92	$90.17
4 Electrician Linemen	66.20	2118.40	101.70	3254.40		
2 Electrician Operators	66.20	1059.20	101.70	1627.20		
4 Electrician Groundmen	43.30	1385.60	69.80	2233.60		
1 Crew Truck		172.61		189.87		
1 Flatbed Truck, 20,000 GVW		209.49		230.44		
1 Pickup Truck, 3/4 Ton		115.08		126.59		
.2 Hyd. Crane, 55 Ton		203.88		224.27		
.2 Hyd. Crane, 12 Ton		97.87		107.65		
.2 Earth Auger, Truck-Mtd.		41.62		45.79		
1 Tractor w/Winch		388.20		427.02	13.96	15.36
88 L.H., Daily Totals		$6325.55		$9286.42	$71.88	$105.53
Crew R-6	Hr.	Daily	Hr.	Daily	Bare Costs	Incl. O&P
1 Electrician Foreman	$66.70	$533.60	$102.45	$819.60	$57.92	$90.17
4 Electrician Linemen	66.20	2118.40	101.70	3254.40		
2 Electrician Operators	66.20	1059.20	101.70	1627.20		
4 Electrician Groundmen	43.30	1385.60	69.80	2233.60		
1 Crew Truck		172.61		189.87		
1 Flatbed Truck, 20,000 GVW		209.49		230.44		
1 Pickup Truck, 3/4 Ton		115.08		126.59		
.2 Hyd. Crane, 55 Ton		203.88		224.27		
.2 Hyd. Crane, 12 Ton		97.87		107.65		
.2 Earth Auger, Truck-Mtd.		41.62		45.79		
1 Tractor w/Winch		388.20		427.02		
3 Cable Trailers		198.95		218.85		
.5 Tensioning Rig		57.01		62.71		
.5 Cable Pulling Rig		312.92		344.21	20.43	22.47
88 L.H., Daily Totals		$6894.43		$9912.20	$78.35	$112.64
Crew R-7	Hr.	Daily	Hr.	Daily	Bare Costs	Incl. O&P
1 Electrician Foreman	$66.70	$533.60	$102.45	$819.60	$47.20	$75.24
5 Electrician Groundmen	43.30	1732.00	69.80	2792.00		
1 Crew Truck		172.61		189.87	3.60	3.96
48 L.H., Daily Totals		$2438.21		$3801.47	$50.80	$79.20

Crew No.	Bare Costs Hr.	Daily	Incl. Subs O&P Hr.	Daily	Cost Per Labor-Hour Bare Costs	Incl. O&P
Crew R-8	Hr.	Daily	Hr.	Daily	Bare Costs	Incl. O&P
1 Electrician Foreman	$66.70	$533.60	$102.45	$819.60	$58.65	$91.19
3 Electrician Linemen	66.20	1588.80	101.70	2440.80		
2 Electrician Groundmen	43.30	692.80	69.80	1116.80		
1 Pickup Truck, 3/4 Ton		115.08		126.59		
1 Crew Truck		172.61		189.87	5.99	6.59
48 L.H., Daily Totals		$3102.89		$4693.66	$64.64	$97.78
Crew R-9	Hr.	Daily	Hr.	Daily	Bare Costs	Incl. O&P
1 Electrician Foreman	$66.70	$533.60	$102.45	$819.60	$54.81	$85.84
1 Electrician Lineman	66.20	529.60	101.70	813.60		
2 Electrician Operators	66.20	1059.20	101.70	1627.20		
4 Electrician Groundmen	43.30	1385.60	69.80	2233.60		
1 Pickup Truck, 3/4 Ton		115.08		126.59		
1 Crew Truck		172.61		189.87	4.50	4.94
64 L.H., Daily Totals		$3795.69		$5810.46	$59.31	$90.79
Crew R-10	Hr.	Daily	Hr.	Daily	Bare Costs	Incl. O&P
1 Electrician Foreman	$66.70	$533.60	$102.45	$819.60	$62.47	$96.51
4 Electrician Linemen	66.20	2118.40	101.70	3254.40		
1 Electrician Groundman	43.30	346.40	69.80	558.40		
1 Crew Truck		172.61		189.87		
3 Tram Cars		224.91		247.41	8.28	9.11
48 L.H., Daily Totals		$3395.92		$5069.67	$70.75	$105.62
Crew R-11	Hr.	Daily	Hr.	Daily	Bare Costs	Incl. O&P
1 Electrician Foreman	$66.70	$533.60	$102.45	$819.60	$62.93	$97.21
4 Electricians	66.20	2118.40	101.70	3254.40		
1 Equip. Oper. (crane)	63.40	507.20	98.75	790.00		
1 Common Laborer	45.60	364.80	72.45	579.60		
1 Crew Truck		172.61		189.87		
1 Hyd. Crane, 12 Ton		489.34		538.27	11.82	13.00
56 L.H., Daily Totals		$4185.95		$6171.74	$74.75	$110.21
Crew R-12	Hr.	Daily	Hr.	Daily	Bare Costs	Incl. O&P
1 Carpenter Foreman (inside)	$56.80	$454.40	$90.25	$722.00	$53.33	$84.69
4 Carpenters	56.30	1801.60	89.45	2862.40		
4 Common Laborers	45.60	1459.20	72.45	2318.40		
1 Equip. Oper. (medium)	60.40	483.20	94.10	752.80		
1 Steel Worker	61.85	494.80	99.60	796.80		
1 Dozer, 200 H.P.		1567.46		1724.21		
1 Pickup Truck, 3/4 Ton		115.08		126.59	19.12	21.03
88 L.H., Daily Totals		$6375.74		$9303.19	$72.45	$105.72
Crew R-13	Hr.	Daily	Hr.	Daily	Bare Costs	Incl. O&P
1 Electrician Foreman	$66.70	$533.60	$102.45	$819.60	$63.84	$98.36
3 Electricians	66.20	1588.80	101.70	2440.80		
.25 Equip. Oper. (crane)	63.40	126.80	98.75	197.50		
1 Equipment Oiler	54.00	432.00	84.15	673.20		
.25 Hydraulic Crane, 33 Ton		253.02		278.32	6.02	6.63
42 L.H., Daily Totals		$2934.22		$4409.42	$69.86	$104.99
Crew R-15	Hr.	Daily	Hr.	Daily	Bare Costs	Incl. O&P
1 Electrician Foreman	$66.70	$533.60	$102.45	$819.60	$64.75	$99.67
4 Electricians	66.20	2118.40	101.70	3254.40		
1 Equipment Oper. (light)	57.00	456.00	88.80	710.40		
1 Telescoping Boom Lift, to 40'		290.76		319.84	6.06	6.66
48 L.H., Daily Totals		$3398.76		$5104.24	$70.81	$106.34

Crew No.	Bare Costs		Incl. Subs O & P		Cost Per Labor-Hour	
Crew R-18	Hr.	Daily	Hr.	Daily	Bare Costs	Incl. O&P
.25 Electrician Foreman	$66.70	$133.40	$102.45	$204.90	$58.08	$89.23
1 Electrician	66.20	529.60	101.70	813.60		
2 Electrician Apprentices	52.95	847.20	81.35	1301.60		
26 L.H., Daily Totals		$1510.20		$2320.10	$58.08	$89.23
Crew R-19	Hr.	Daily	Hr.	Daily	Bare Costs	Incl. O&P
.5 Electrician Foreman	$66.70	$266.80	$102.45	$409.80	$66.30	$101.85
2 Electricians	66.20	1059.20	101.70	1627.20		
20 L.H., Daily Totals		$1326.00		$2037.00	$66.30	$101.85
Crew R-21	Hr.	Daily	Hr.	Daily	Bare Costs	Incl. O&P
1 Electrician Foreman	$66.70	$533.60	$102.45	$819.60	$66.18	$101.70
3 Electricians	66.20	1588.80	101.70	2440.80		
.1 Equip. Oper. (medium)	60.40	48.32	94.10	75.28		
.1 S.P. Crane, 4x4, 25 Ton		119.28		131.21	3.64	4.00
32.8 L.H., Daily Totals		$2290.00		$3466.89	$69.82	$105.70
Crew R-22	Hr.	Daily	Hr.	Daily	Bare Costs	Incl. O&P
.66 Electrician Foreman	$66.70	$352.18	$102.45	$540.94	$60.58	$93.07
2 Electricians	66.20	1059.20	101.70	1627.20		
2 Electrician Apprentices	52.95	847.20	81.35	1301.60		
37.28 L.H., Daily Totals		$2258.58		$3469.74	$60.58	$93.07
Crew R-30	Hr.	Daily	Hr.	Daily	Bare Costs	Incl. O&P
.25 Electrician Foreman (outside)	$68.20	$136.40	$104.80	$209.60	$53.68	$83.94
1 Electrician	66.20	529.60	101.70	813.60		
2 Laborers (Semi-Skilled)	45.60	729.60	72.45	1159.20		
26 L.H., Daily Totals		$1395.60		$2182.40	$53.68	$83.94
Crew R-31	Hr.	Daily	Hr.	Daily	Bare Costs	Incl. O&P
1 Electrician	$66.20	$529.60	$101.70	$813.60	$66.20	$101.70
1 Core Drill, Electric, 2.5 H.P.		64.76		71.23	8.09	8.90
8 L.H., Daily Totals		$594.36		$884.84	$74.29	$110.60

For customer support on your Commercial Renovation Costs with RSMeans data, call 800.448.8182.

Historical Cost Indexes

The table below lists both the RSMeans® historical cost index based on Jan. 1, 1993 = 100 as well as the computed value of an index based on Jan. 1, 2022 costs. Since the Jan. 1, 2022 figure is estimated, space is left to write in the actual index figures as they become available through the quarterly *RSMeans Construction Cost Indexes*.

To compute the actual index based on Jan. 1, 2022 = 100, divide the historical cost index for a particular year by the actual Jan. 1, 2022 construction cost index. Space has been left to advance the index figures as the year progresses.

Year	Historical Cost Index Jan. 1, 1993 = 100		Current Index Based on Jan. 1, 2022 = 100		Year	Historical Cost Index Jan. 1, 1993 = 100	Current Index Based on Jan. 1, 2022 = 100		Year	Historical Cost Index Jan. 1, 1993 = 100	Current Index Based on Jan. 1, 2022 = 100	
	Est.	Actual	Est.	Actual		Actual	Est.	Actual		Actual	Est.	Actual
Oct 2022					July 2005	151.6	60.7		1985	82.6	34.6	
July 2022					2004	143.7	55.8		1984	82.0	33.9	
April 2022					2003	132.0	54.4		1983	80.2	32.2	
Jan 2022	261.6		100.0		2002	128.7	52.9		1982	76.1	29.6	
July 2021		257.5	99.1		2001	125.1	51.1		1981	70.0	26.6	
2020		234.6	98.1		2000	120.9	49.7		1980	62.9	24.4	
2019		232.2	94.2		1999	117.6	48.6		1979	57.8	22.6	
2018		222.9	90.2		1998	115.1	47.7		1978	53.5	20.9	
2017		213.6	87.6		1997	112.8	46.6		1977	49.5	19.8	
2016		207.3	87.1		1996	110.2	45.5		1976	46.9	18.9	
2015		206.2	86.6		1995	107.6	44.1		1975	44.8	17.5	
2014		204.9	85.0		1994	104.4	43.0		1974	41.4	15.9	
2013		201.2	82.2		1993	101.7	42.0		1973	37.7	14.7	
2012		194.6	80.8		1992	99.4	40.9		1972	34.8	13.6	
2011		191.2	77.5		1991	96.8	39.8		1971	32.1	12.1	
2010		183.5	76.1		1990	94.3	38.9		1970	28.7	11.4	
2009		180.1	76.2		1989	92.1	38.0					
2008		180.4	71.6		1988	89.9	37.0					
2007		169.4	68.4		1987	87.7	35.6					
2006		162.0	64.0		1986	84.2	34.9					

Adjustments to Costs

The "Historical Cost Index" can be used to convert national average building costs at a particular time to the approximate building costs for some other time.

Example:

Estimate and compare construction costs for different years in the same city.

To estimate the national average construction cost of a building in 1970, knowing that it cost $900,000 in 2022:

INDEX in 1970 = 28.7

INDEX in 2022 = 261.6

Note: The city cost indexes for Canada can be used to convert U.S. national averages to local costs in Canadian dollars.

Example:

To estimate and compare the cost of a building in Toronto, ON in 2022 with the known cost of $600,000 (US$) in New York, NY in 2022:

INDEX Toronto = 118.4

INDEX New York = 141.1

$$\frac{\text{INDEX Toronto}}{\text{INDEX New York}} \times \text{Cost New York} = \text{Cost Toronto}$$

$$\frac{118.4}{141.1} \times \$600,000 = .839 \times \$600,000 = \$503,400$$

The construction cost of the building in Toronto is $503,400 (CN$).

Time Adjustment Using the Historical Cost Indexes:

$$\frac{\text{Index for Year A}}{\text{Index for Year B}} \times \text{Cost in Year B} = \text{Cost in Year A}$$

$$\frac{\text{INDEX 1970}}{\text{INDEX 2022}} \times \text{Cost 2022} = \text{Cost 1970}$$

$$\frac{28.7}{261.6} \times \$900,000 = .109 \times \$900,000 = \$98,739$$

The construction cost of the building in 1970 was $98,739.

*Historical Cost Index updates and other resources are provided on the following website: **rsmeans.com/2022books**

For customer support on your Commercial Renovation Costs with RSMeans data, call 800.448.8182.

795

How to Use the City Cost Indexes

What you should know before you begin

RSMeans City Cost Indexes (CCI) are an extremely useful tool for when you want to compare costs from city to city and region to region.

This publication contains average construction cost indexes for 731 U.S. and Canadian cities covering over 930 three-digit zip code locations, as listed directly under each city.

Keep in mind that a City Cost Index number is a percentage ratio of a specific city's cost to the national average cost of the same item at a stated time period.

In other words, these index figures represent relative construction factors (or, if you prefer, multipliers) for material and installation costs, as well as the weighted average for Total In Place costs for each CSI MasterFormat division. Installation costs include both labor and equipment rental costs. When estimating equipment rental rates only for a specific location, use 01 54 33 EQUIPMENT RENTAL COSTS in the Reference Section.

The 30 City Average Index is the average of 30 major U.S. cities and serves as a national average. This national average represents the baseline from which all locations are calculated.

Index figures for both material and installation are based on the 30 major city average of 100 and represent the cost relationship as of July 1, 2021. The index for each division is computed from representative material and labor quantities for that division. The weighted average for each city is a weighted total of the components listed above it. It does not include relative productivity between trades or cities.

As changes occur in local material prices, labor rates, and equipment rental rates (including fuel costs), the impact of these changes should be accurately measured by the change in the City Cost Index for each particular city (as compared to the 30 city average).

Therefore, if you know (or have estimated) building costs in one city today, you can easily convert those costs to expected building costs in another city.

In addition, by using the Historical Cost Index, you can easily convert national average building costs at a particular time to the approximate building costs for some other time. The City Cost Indexes can then be applied to calculate the costs for a particular city.

Quick calculations

Location Adjustment Using the City Cost Indexes:

$$\frac{\text{Index for City A}}{\text{Index for City B}} \times \text{Cost in City B} = \text{Cost in City A}$$

Time Adjustment for the National Average Using the Historical Cost Index:

$$\frac{\text{Index for Year A}}{\text{Index for Year B}} \times \text{Cost in Year B} = \text{Cost in Year A}$$

Adjustment from the National Average:

$$\frac{\text{Index for City A}}{100} \times \text{National Average Cost} = \text{Cost in City A}$$

Since each of the other RSMeans data sets contains many different items, any *one* item multiplied by the particular city index may give incorrect results. However, the larger the number of items compiled, the closer the results should be to actual costs for that particular city.

The City Cost Indexes for Canadian cities are calculated using Canadian material and equipment prices and labor rates in Canadian dollars. When compared to the baseline mentioned previously, the resulting index enables the user to convert baseline costs to local costs. Therefore, indexes for Canadian cities can be used to convert U.S. national average prices to local costs in Canadian dollars.

How to use this section

1. Compare costs from city to city.

In using the RSMeans Indexes, remember that an index number is not a fixed number but a ratio: It's a percentage ratio of a building component's cost at any stated time to the national average cost of that same component at the same time period. Put in the form of an equation:

$$\frac{\text{Specific City Cost}}{\text{National Average Cost}} \times 100 = \text{City Index Number}$$

Therefore, when making cost comparisons between cities, do not subtract one city's index number from the index number of another city and read the result as a percentage difference. Instead, divide one city's index number by that of the other city. The resulting number may then be used as a multiplier to calculate cost differences from city to city.

The formula used to find cost differences between cities for the purpose of comparison is as follows:

$$\frac{\text{City A Index}}{\text{City B Index}} \times \text{City B Cost (Known)} = \text{City A Cost (Unknown)}$$

In addition, you can use RSMeans CCI to calculate and compare costs division by division between cities using the same basic formula. (Just be sure that you're comparing similar divisions.)

2. Compare a specific city's construction costs with the national average.

When you're studying construction location feasibility, it's advisable to compare a prospective project's cost index with an index of the national average cost.

For example, divide the weighted average index of construction costs of a specific city by that of the 30 City Average, which = 100.

$$\frac{\text{City Index}}{100} = \text{\% of National Average}$$

As a result, you get a ratio that indicates the relative cost of construction in that city in comparison with the national average.

3. Convert U.S. national average to actual costs in Canadian City.

$$\frac{\text{Index for Canadian City}}{100} \times \text{National Average Cost} = \text{Cost in Canadian City in \$ CAN}$$

4. Adjust construction cost data based on a national average.

When you use a source of construction cost data which is based on a national average (such as RSMeans cost data), it is necessary to adjust those costs to a specific location.

$$\frac{\text{City Index}}{100} \times \frac{\text{Cost Based on}}{\text{National Average Costs}} = \frac{\text{City Cost}}{\text{(Unknown)}}$$

5. When applying the City Cost Indexes to demolition projects, use the appropriate division installation index. For example, for removal of existing doors and windows, use the Division 8 (Openings) index.

What you might like to know about how we developed the Indexes

The information presented in the CCI is organized according to the Construction Specifications Institute (CSI) MasterFormat 2018 classification system.

To create a reliable index, RSMeans researched the building type most often constructed in the United States and Canada. Because it was concluded that no one type of building completely represented the building construction industry, nine different types of buildings were combined to create a composite model.

The exact material, labor, and equipment quantities are based on detailed analyses of these nine building types, and then each quantity is weighted in proportion to expected usage. These various material items, labor hours, and equipment rental rates are thus combined to form a composite building representing as closely as possible the actual usage of materials, labor, and equipment in the North American building construction industry.

The following structures were chosen to make up that composite model:

1. Factory, 1 story
2. Office, 2–4 stories
3. Store, Retail
4. Town Hall, 2–3 stories
5. High School, 2–3 stories
6. Hospital, 4–8 stories
7. Garage, Parking
8. Apartment, 1–3 stories
9. Hotel/Motel, 2–3 stories

For the purposes of ensuring the timeliness of the data, the components of the index for the composite model have been streamlined. They currently consist of:

- specific quantities of 66 commonly used construction materials;
- specific labor-hours for 21 building construction trades; and
- specific days of equipment rental for 6 types of construction equipment (normally used to install the 66 material items by the 21 trades.) Fuel costs and routine maintenance costs are included in the equipment cost.

Material and equipment price quotations are gathered quarterly from cities in the United States and Canada. These prices and the latest negotiated labor wage rates for 21 different building trades are used to compile the quarterly update of the City Cost Index.

The 30 major U.S. cities used to calculate the national average are:

Atlanta, GA	Memphis, TN
Baltimore, MD	Milwaukee, WI
Boston, MA	Minneapolis, MN
Buffalo, NY	Nashville, TN
Chicago, IL	New Orleans, LA
Cincinnati, OH	New York, NY
Cleveland, OH	Philadelphia, PA
Columbus, OH	Phoenix, AZ
Dallas, TX	Pittsburgh, PA
Denver, CO	St. Louis, MO
Detroit, MI	San Antonio, TX
Houston, TX	San Diego, CA
Indianapolis, IN	San Francisco, CA
Kansas City, MO	Seattle, WA
Los Angeles, CA	Washington, DC

What the CCI does not indicate

The weighted average for each city is a total of the divisional components weighted to reflect typical usage. It does not include the productivity variations between trades or cities.

In addition, the CCI does not take into consideration factors such as the following:

- managerial efficiency
- competitive conditions
- automation
- restrictive union practices
- unique local requirements
- regional variations due to specific building codes

DIVISION		UNITED STATES 30 CITY AVERAGE			ALABAMA															
					ANNISTON 362			BIRMINGHAM 350 - 352			BUTLER 369			DECATUR 356			DOTHAN 363		
		MAT.	INST.	TOTAL	MAT.	INST.	TOTAL	MAT.	INST.	TOTAL	MAT.	INST.	TOTAL	MAT.	INST.	TOTAL	MAT.	INST.	TOTAL
015433	CONTRACTOR EQUIPMENT		100.0	100.0		99.3	99.3		104.4	104.4		97.0	97.0		99.3	99.3		97.0	97.0
0241, 31 - 34	SITE & INFRASTRUCTURE, DEMOLITION	100.0	100.0	100.0	100.7	85.6	90.4	99.5	94.2	95.9	116.3	81.8	92.8	90.4	85.2	86.9	113.4	81.8	91.9
0310	Concrete Forming & Accessories	100.0	100.0	100.0	86.2	65.3	69.4	99.6	68.3	74.3	81.1	67.6	70.2	99.0	64.2	70.9	93.1	67.8	72.7
0320	Concrete Reinforcing	100.0	100.0	100.0	88.4	69.0	82.2	99.0	69.2	89.4	93.2	69.9	85.7	92.2	66.8	84.1	93.2	69.3	85.5
0330	Cast-in-Place Concrete	100.0	100.0	100.0	88.0	66.6	80.6	106.8	69.9	94.0	85.8	66.8	79.2	99.1	66.5	87.8	85.8	67.0	79.3
03	CONCRETE	100.0	100.0	100.0	92.1	68.2	82.5	100.0	70.4	88.2	93.1	69.4	83.6	93.6	67.2	83.0	92.9	69.5	83.5
04	MASONRY	100.0	100.0	100.0	98.2	61.2	74.8	98.0	62.9	75.9	103.1	61.2	76.7	92.3	61.0	72.5	104.7	62.7	78.2
05	METALS	100.0	100.0	100.0	101.3	94.2	99.6	99.3	91.7	97.5	100.2	95.1	99.0	101.1	92.9	99.2	100.3	95.3	99.1
06	WOOD, PLASTICS & COMPOSITES	100.0	100.0	100.0	80.7	65.5	74.6	95.8	68.5	84.6	74.9	68.5	72.2	99.1	64.0	84.6	88.7	68.3	80.3
07	THERMAL & MOISTURE PROTECTION	100.0	100.0	100.0	98.2	60.7	81.8	99.5	66.5	85.0	98.2	64.3	83.3	96.5	63.8	82.1	98.3	63.9	83.2
08	OPENINGS	100.0	100.0	100.0	91.0	66.9	85.2	98.2	68.9	91.1	91.0	68.9	85.7	102.8	65.7	93.9	91.1	68.7	85.7
0920	Plaster & Gypsum Board	100.0	100.0	100.0	89.6	65.6	74.6	99.7	68.0	79.9	85.5	68.2	74.7	99.1	63.7	76.9	103.6	68.1	81.4
0950, 0980	Ceilings & Acoustic Treatment	100.0	100.0	100.0	90.1	65.6	74.2	104.0	68.0	80.7	90.1	68.2	75.9	101.3	63.7	76.9	90.1	68.1	75.8
0960	Flooring	100.0	100.0	100.0	85.9	66.7	80.0	99.4	68.1	89.8	89.6	66.7	82.6	93.1	66.7	85.0	93.5	69.4	86.1
0970, 0990	Wall Finishes & Painting/Coating	100.0	100.0	100.0	96.3	52.3	69.5	101.8	52.3	71.6	96.3	44.3	64.6	90.3	58.9	71.2	96.3	80.7	86.8
09	FINISHES	100.0	100.0	100.0	89.6	64.2	75.8	100.7	66.4	82.0	92.4	64.8	77.4	95.3	63.8	78.2	96.1	69.3	81.5
COVERS	DIVS. 10 - 14, 25, 28, 41, 43, 44, 46	100.0	83.7	96.5	100.0	85.2	96.8	100.0	85.2	96.8	100.0	84.8	96.7	100.0	81.5	96.0	100.0	84.9	96.7
21, 22, 23	FIRE SUPPRESSION, PLUMBING & HVAC	100.0	100.0	100.0	102.8	50.6	81.4	100.8	63.9	85.7	98.5	62.8	83.8	100.9	63.2	85.4	98.5	62.8	83.9
26, 27, 3370	ELECTRICAL, COMMUNICATIONS & UTIL.	100.0	100.0	100.0	96.9	55.5	77.0	96.4	64.9	81.2	98.3	55.7	77.7	93.5	62.7	78.7	97.5	74.8	86.6
MF2018	WEIGHTED AVERAGE	100.0	100.0	100.0	97.5	65.3	84.4	99.4	71.0	87.8	97.4	68.2	85.5	98.1	68.6	86.1	97.8	71.7	87.1

DIVISION		ALABAMA																	
		EVERGREEN 364			GADSDEN 359			HUNTSVILLE 357 - 358			JASPER 355			MOBILE 365 - 366			MONTGOMERY 360 - 361		
		MAT.	INST.	TOTAL	MAT.	INST.	TOTAL	MAT.	INST.	TOTAL	MAT.	INST.	TOTAL	MAT.	INST.	TOTAL	MAT.	INST.	TOTAL
015433	CONTRACTOR EQUIPMENT		97.0	97.0		99.3	99.3		99.3	99.3		99.3	99.3		97.0	97.0		105.1	105.1
0241, 31 - 34	SITE & INFRASTRUCTURE, DEMOLITION	116.8	81.7	93.0	97.5	85.6	89.4	90.1	85.5	87.0	97.4	85.6	89.4	107.6	81.8	90.1	105.9	95.5	98.8
0310	Concrete Forming & Accessories	77.0	65.5	67.7	87.7	65.8	70.0	99.0	66.5	72.8	95.2	65.7	71.4	91.8	67.5	72.2	98.0	68.0	73.8
0320	Concrete Reinforcing	93.3	69.9	85.7	97.4	69.2	88.3	92.2	77.2	84.8	92.2	69.1	84.8	91.5	69.9	84.6	100.5	69.2	90.5
0330	Cast-in-Place Concrete	85.8	66.8	79.2	99.1	66.8	88.0	96.5	66.7	86.2	109.9	66.8	95.0	90.0	66.6	82.0	89.9	69.5	82.8
03	CONCRETE	93.2	68.5	83.3	97.3	68.5	85.7	92.5	70.1	83.5	100.6	68.4	87.7	90.0	69.4	81.7	92.8	70.2	83.7
04	MASONRY	103.1	61.2	76.7	90.9	61.9	72.6	93.7	61.8	73.6	88.9	61.2	71.4	101.9	62.1	76.8	101.9	62.2	76.9
05	METALS	100.2	94.9	99.0	98.7	94.7	97.8	101.1	97.7	100.3	98.6	94.6	97.7	102.6	95.0	100.8	101.8	91.7	99.4
06	WOOD, PLASTICS & COMPOSITES	70.9	65.9	68.9	87.0	65.9	78.3	99.1	66.9	85.8	95.5	65.9	83.3	86.8	68.3	79.2	90.8	68.7	81.6
07	THERMAL & MOISTURE PROTECTION	98.1	63.3	82.8	96.6	64.2	82.4	96.4	64.3	82.3	96.7	63.5	82.2	97.9	63.8	82.9	98.9	66.2	84.6
08	OPENINGS	91.0	67.5	85.4	99.7	67.3	91.9	102.6	69.9	94.7	99.7	67.3	91.9	93.4	68.9	87.5	92.1	68.9	86.5
0920	Plaster & Gypsum Board	84.5	65.6	72.7	86.2	65.6	73.3	99.1	66.6	78.8	93.4	65.6	76.0	97.1	68.1	78.9	99.1	68.0	79.7
0950, 0980	Ceilings & Acoustic Treatment	90.1	65.6	74.2	93.8	65.6	75.5	102.8	66.6	79.3	93.8	65.6	75.5	98.5	68.1	78.7	100.5	68.0	79.5
0960	Flooring	87.9	66.7	81.4	89.8	68.1	83.1	93.1	68.1	85.4	91.6	66.7	84.0	93.5	68.1	85.7	93.0	68.1	85.4
0970, 0990	Wall Finishes & Painting/Coating	96.3	44.3	64.6	90.3	52.3	67.1	90.3	60.2	72.0	90.3	52.3	67.1	99.4	44.3	65.8	101.8	52.3	71.6
09	FINISHES	91.8	63.3	76.3	91.1	64.5	76.6	95.6	65.9	79.5	92.7	64.2	77.2	96.3	65.0	79.3	99.0	66.2	81.2
COVERS	DIVS. 10 - 14, 25, 28, 41, 43, 44, 46	100.0	84.6	96.6	100.0	83.8	96.5	100.0	81.8	96.0	100.0	83.7	96.5	100.0	84.9	96.7	100.0	85.0	96.7
21, 22, 23	FIRE SUPPRESSION, PLUMBING & HVAC	98.5	56.7	81.3	104.1	63.9	87.6	100.8	63.6	85.6	104.1	63.3	87.4	100.8	59.9	84.0	100.8	62.9	85.3
26, 27, 3370	ELECTRICAL, COMMUNICATIONS & UTIL.	96.5	55.7	76.8	93.7	64.9	79.8	94.3	62.6	79.0	93.3	55.5	75.1	99.0	59.2	79.8	99.2	71.9	86.1
MF2018	WEIGHTED AVERAGE	97.2	66.4	84.7	98.2	69.8	86.6	98.1	70.1	86.7	98.7	68.2	86.3	98.4	68.2	86.1	98.8	71.7	87.8

DIVISION		ALABAMA									ALASKA								
		PHENIX CITY 368			SELMA 367			TUSCALOOSA 354			ANCHORAGE 995 - 996			FAIRBANKS 997			JUNEAU 998		
		MAT.	INST.	TOTAL	MAT.	INST.	TOTAL	MAT.	INST.	TOTAL	MAT.	INST.	TOTAL	MAT.	INST.	TOTAL	MAT.	INST.	TOTAL
015433	CONTRACTOR EQUIPMENT		97.0	97.0		97.0	97.0		99.3	99.3		105.0	105.0		109.7	109.7		105.0	105.0
0241, 31 - 34	SITE & INFRASTRUCTURE, DEMOLITION	121.3	81.8	94.5	113.1	81.8	91.9	90.7	85.6	87.2	138.8	112.8	121.2	136.6	119.0	124.7	153.4	112.8	125.9
0310	Concrete Forming & Accessories	86.1	67.3	71.0	82.9	65.8	69.1	99.0	67.6	73.7	103.0	111.1	109.5	108.9	110.7	110.4	105.0	111.1	109.9
0320	Concrete Reinforcing	93.2	66.6	84.7	93.2	69.1	85.5	92.2	69.2	84.8	124.6	118.4	122.6	120.1	118.4	119.5	124.6	118.4	122.6
0330	Cast-in-Place Concrete	85.8	66.8	79.2	85.8	66.8	79.2	100.5	66.9	88.9	131.5	113.6	125.3	130.3	110.8	123.6	131.4	113.6	125.2
03	CONCRETE	95.9	68.7	85.0	92.1	68.5	82.7	94.2	69.3	84.3	117.6	112.7	115.6	110.3	111.7	110.9	122.4	112.7	118.5
04	MASONRY	103.1	61.2	76.7	107.6	61.2	78.3	92.6	62.1	73.4	202.3	114.5	147.0	202.5	113.4	146.4	182.0	114.5	139.5
05	METALS	100.2	93.7	98.7	100.2	94.7	98.9	100.4	94.8	99.1	127.5	103.2	121.8	122.6	105.7	118.7	122.7	103.2	118.1
06	WOOD, PLASTICS & COMPOSITES	80.5	68.0	75.3	77.0	65.9	72.4	99.1	68.3	86.4	100.5	108.4	103.8	113.4	108.3	111.3	105.0	108.4	106.4
07	THERMAL & MOISTURE PROTECTION	98.6	64.2	83.6	98.0	64.0	83.1	96.5	64.6	82.5	134.6	111.6	124.5	143.8	110.2	129.1	145.8	111.6	130.8
08	OPENINGS	91.0	67.8	85.4	91.0	67.3	85.3	102.6	68.6	94.4	125.9	112.2	122.6	128.4	111.6	124.4	126.8	112.2	123.3
0920	Plaster & Gypsum Board	92.0	67.7	76.8	88.6	65.6	74.2	99.1	68.0	79.7	116.5	108.6	111.6	167.2	108.6	130.6	168.8	108.6	131.1
0950, 0980	Ceilings & Acoustic Treatment	90.1	67.7	75.6	90.1	65.6	74.2	102.8	68.0	80.3	119.2	108.6	112.3	103.7	108.6	106.9	112.8	108.6	110.1
0960	Flooring	91.0	66.7	83.6	89.8	66.7	82.8	93.1	68.1	85.4	104.8	113.7	107.5	106.3	113.7	108.6	120.5	113.7	118.4
0970, 0990	Wall Finishes & Painting/Coating	96.3	77.7	85.0	96.3	52.3	69.5	90.3	52.3	67.1	101.8	118.6	112.1	97.0	114.5	107.7	98.7	118.6	110.8
09	FINISHES	94.3	68.1	80.1	92.6	64.2	77.2	95.6	65.9	79.5	111.4	111.7	111.9	116.3	111.6	113.7	121.4	112.3	116.4
COVERS	DIVS. 10 - 14, 25, 28, 41, 43, 44, 46	100.0	82.6	96.2	100.0	83.8	96.5	100.0	84.1	96.5	100.0	110.7	102.3	100.0	110.7	102.2	100.0	110.7	102.3
21, 22, 23	FIRE SUPPRESSION, PLUMBING & HVAC	98.5	62.7	83.8	98.5	62.2	83.6	100.9	63.8	85.7	100.8	103.3	101.8	100.7	104.3	102.2	101.0	103.3	101.9
26, 27, 3370	ELECTRICAL, COMMUNICATIONS & UTIL.	97.9	65.6	82.3	97.3	70.5	84.4	94.0	64.9	80.0	97.9	105.0	101.3	114.9	105.0	110.1	104.0	105.0	104.5
MF2018	WEIGHTED AVERAGE	98.1	69.7	86.5	97.3	69.8	86.1	98.1	70.2	86.8	117.1	108.7	113.7	117.8	109.2	114.3	118.2	108.7	114.3

DIVISION		ALASKA KETCHIKAN 999			ARIZONA CHAMBERS 865			FLAGSTAFF 860			GLOBE 855			KINGMAN 864			MESA/TEMPE 852		
		MAT.	INST.	TOTAL	MAT.	INST.	TOTAL	MAT.	INST.	TOTAL	MAT.	INST.	TOTAL	MAT.	INST.	TOTAL	MAT.	INST.	TOTAL
015433	CONTRACTOR EQUIPMENT		109.7	109.7		85.9	85.9		85.9	85.9		85.9	85.9		86.9	86.9		86.9	86.9
0241, 31 - 34	SITE & INFRASTRUCTURE, DEMOLITION	202.1	119.1	145.7	69.9	86.5	81.2	89.1	86.8	87.5	107.0	87.3	93.6	69.9	86.8	81.3	96.9	87.5	90.5
0310	Concrete Forming & Accessories	98.7	111.1	108.7	100.9	67.5	74.0	108.6	73.7	80.4	91.3	67.6	72.2	98.3	63.7	70.4	95.3	69.8	74.7
0320	Concrete Reinforcing	94.6	118.4	102.3	89.8	73.3	84.5	89.7	73.4	84.5	83.4	73.3	80.1	89.9	73.3	84.5	80.8	80.5	80.5
0330	Cast-in-Place Concrete	264.1	112.4	211.7	80.8	67.1	76.1	80.9	67.4	76.2	78.6	66.8	74.5	80.6	67.3	76.0	79.2	67.0	75.0
03	CONCRETE	180.9	112.4	153.5	88.4	68.4	80.4	105.2	71.4	91.6	92.2	68.4	82.7	88.0	66.8	79.6	85.8	69.5	79.3
04	MASONRY	213.6	114.4	151.1	96.5	58.9	72.8	96.7	59.6	73.3	105.1	58.8	76.0	96.5	59.0	72.9	105.4	58.9	76.1
05	METALS	122.8	105.7	118.8	95.9	70.9	90.0	96.4	71.8	90.6	95.4	71.7	89.8	96.6	71.6	90.8	95.7	72.5	90.3
06	WOOD, PLASTICS & COMPOSITES	103.3	108.3	105.4	100.0	68.8	87.1	107.6	76.9	94.9	86.0	69.0	79.0	94.0	63.5	81.4	90.4	71.7	82.7
07	THERMAL & MOISTURE PROTECTION	151.1	109.9	133.0	101.8	68.1	87.0	103.9	69.5	88.8	99.7	66.8	85.3	101.7	67.9	86.9	99.4	67.5	85.4
08	OPENINGS	124.3	112.2	121.4	101.0	65.6	92.5	101.1	73.5	94.4	91.1	65.7	85.0	101.2	64.3	92.2	91.2	67.7	85.5
0920	Plaster & Gypsum Board	148.8	108.6	123.7	105.5	68.4	82.3	111.4	76.6	89.7	90.7	68.4	76.7	92.6	62.9	74.0	96.0	71.2	80.5
0950, 0980	Ceilings & Acoustic Treatment	98.3	108.6	105.0	109.4	68.4	82.8	110.1	76.6	88.4	95.8	68.4	78.0	110.1	62.9	79.5	95.8	71.2	79.8
0960	Flooring	105.6	113.7	108.1	87.3	60.0	78.9	89.2	63.3	81.3	96.9	60.0	85.6	86.2	60.0	78.2	98.1	62.2	87.1
0970, 0990	Wall Finishes & Painting/Coating	97.0	118.6	110.2	86.7	56.5	68.3	86.7	56.5	68.3	90.7	56.5	69.9	86.7	56.5	68.3	90.7	56.5	69.9
09	FINISHES	117.4	112.3	114.6	95.5	64.7	78.8	99.0	70.1	83.3	96.9	64.8	79.5	93.3	61.6	76.1	96.9	66.9	80.6
COVERS	DIVS. 10 - 14, 25, 28, 41, 43, 44, 46	100.0	110.6	102.3	100.0	83.8	96.5	100.0	84.6	96.7	100.0	84.1	96.5	100.0	83.2	96.3	100.0	84.4	96.6
21, 22, 23	FIRE SUPPRESSION, PLUMBING & HVAC	97.7	103.3	100.0	98.0	76.0	89.0	101.3	76.0	90.9	95.9	76.0	87.8	98.0	76.0	89.0	101.2	76.0	90.9
26, 27, 3370	ELECTRICAL, COMMUNICATIONS & UTIL.	114.8	105.0	110.1	103.9	66.4	85.8	103.3	62.2	83.5	99.5	62.2	81.5	103.9	62.2	83.8	97.6	62.2	80.5
MF2018	WEIGHTED AVERAGE	128.9	109.3	120.9	96.5	70.3	85.8	100.4	71.5	88.6	96.3	69.8	85.5	96.3	69.0	85.2	96.2	70.5	85.7

DIVISION		ARIZONA PHOENIX 850,853			PRESCOTT 863			SHOW LOW 859			TUCSON 856 - 857			ARKANSAS BATESVILLE 725			CAMDEN 717		
		MAT.	INST.	TOTAL	MAT.	INST.	TOTAL	MAT.	INST.	TOTAL	MAT.	INST.	TOTAL	MAT.	INST.	TOTAL	MAT.	INST.	TOTAL
015433	CONTRACTOR EQUIPMENT		96.7	96.7		85.9	85.9		86.9	86.9		86.9	86.9		86.5	86.5		86.5	86.5
0241, 31 - 34	SITE & INFRASTRUCTURE, DEMOLITION	96.5	96.2	96.3	77.0	86.7	83.6	109.4	87.3	94.4	92.0	87.5	89.0	73.5	81.7	79.1	79.5	81.7	81.0
0310	Concrete Forming & Accessories	101.0	72.0	77.6	103.1	69.6	76.1	100.9	67.7	74.1	96.3	78.3	81.8	78.4	58.7	62.5	76.7	59.1	62.5
0320	Concrete Reinforcing	80.1	73.4	77.9	89.7	73.3	84.4	83.9	73.3	80.5	68.9	73.4	70.3	97.1	65.9	87.1	96.6	67.0	87.1
0330	Cast-in-Place Concrete	83.4	70.8	79.1	80.9	67.3	76.2	78.6	66.8	74.5	81.5	67.1	76.5	61.7	72.9	65.6	67.1	72.8	69.1
03	CONCRETE	87.6	72.0	81.3	93.0	69.5	83.6	94.4	68.4	84.0	83.5	73.4	79.5	71.9	65.7	69.4	74.3	66.0	71.0
04	MASONRY	98.5	60.6	74.6	96.7	59.0	72.9	105.1	58.8	76.0	91.7	59.5	71.4	114.2	59.8	79.9	95.2	60.2	74.5
05	METALS	103.2	75.0	96.6	96.4	71.6	90.6	95.1	71.7	89.7	96.5	72.6	90.9	89.0	75.6	85.8	95.2	75.8	90.6
06	WOOD, PLASTICS & COMPOSITES	98.1	73.7	88.0	101.9	71.5	89.3	95.7	69.0	84.6	90.9	83.1	87.7	84.0	59.9	74.0	84.7	60.4	74.6
07	THERMAL & MOISTURE PROTECTION	100.3	69.8	86.9	102.4	68.8	87.7	100.1	66.8	85.5	100.5	68.9	86.6	103.2	59.7	84.1	98.5	60.3	81.7
08	OPENINGS	101.5	71.3	94.2	101.1	67.3	93.0	90.5	67.3	84.9	87.9	77.0	85.3	96.2	58.1	87.0	99.3	58.7	89.5
0920	Plaster & Gypsum Board	105.1	72.8	84.9	106.3	71.1	84.3	99.7	68.4	80.1	100.9	82.9	89.7	90.7	59.4	71.1	87.4	59.9	70.2
0950, 0980	Ceilings & Acoustic Treatment	109.9	72.8	85.8	107.4	71.1	83.8	95.8	68.4	78.0	97.9	82.9	88.2	98.6	59.4	73.2	91.9	59.9	71.1
0960	Flooring	103.1	62.4	90.7	88.0	62.4	80.2	99.5	60.0	87.4	89.3	63.6	81.4	86.5	68.6	81.0	88.8	70.0	83.1
0970, 0990	Wall Finishes & Painting/Coating	98.7	56.3	72.8	86.7	56.5	68.3	90.7	56.5	69.9	91.8	56.5	70.3	95.9	51.0	68.5	98.7	52.6	70.6
09	FINISHES	103.2	68.5	84.4	95.9	66.8	80.1	99.3	64.8	80.6	95.7	73.9	83.8	87.9	59.6	72.5	87.9	60.2	72.9
COVERS	DIVS. 10 - 14, 25, 28, 41, 43, 44, 46	100.0	86.0	97.0	100.0	84.1	96.5	100.0	84.1	96.5	100.0	85.6	96.9	100.0	78.6	95.4	100.0	78.5	95.3
21, 22, 23	FIRE SUPPRESSION, PLUMBING & HVAC	97.6	77.0	89.1	101.3	76.0	90.9	95.9	76.0	87.8	101.2	74.5	90.2	96.9	50.5	77.8	96.8	55.7	79.9
26, 27, 3370	ELECTRICAL, COMMUNICATIONS & UTIL.	101.6	62.2	82.6	103.1	62.2	83.4	97.8	62.2	80.6	99.0	59.5	80.0	93.4	56.0	75.4	92.7	53.9	74.0
MF2018	WEIGHTED AVERAGE	98.8	72.6	88.1	98.1	70.3	86.8	96.7	69.9	85.8	95.1	71.9	85.7	91.1	61.7	79.1	92.3	62.7	80.3

DIVISION		ARKANSAS FAYETTEVILLE 727			FORT SMITH 729			HARRISON 726			HOT SPRINGS 719			JONESBORO 724			LITTLE ROCK 720 - 722		
		MAT.	INST.	TOTAL	MAT.	INST.	TOTAL	MAT.	INST.	TOTAL	MAT.	INST.	TOTAL	MAT.	INST.	TOTAL	MAT.	INST.	TOTAL
015433	CONTRACTOR EQUIPMENT		86.5	86.5		86.5	86.5		86.5	86.5		86.5	86.5		108.9	108.9		93.6	93.6
0241, 31 - 34	SITE & INFRASTRUCTURE, DEMOLITION	73.0	81.7	78.9	78.5	81.7	80.7	78.1	81.7	80.5	82.6	81.7	82.0	97.7	97.9	97.9	84.6	92.3	89.8
0310	Concrete Forming & Accessories	72.6	59.1	61.7	98.5	59.0	66.6	84.6	58.9	63.9	72.7	59.2	61.8	82.6	59.4	63.9	98.2	59.5	67.0
0320	Concrete Reinforcing	97.1	63.7	86.4	98.3	64.7	87.5	96.7	67.0	87.1	94.9	66.8	85.8	94.5	66.7	85.6	99.6	67.2	89.2
0330	Cast-in-Place Concrete	61.7	72.8	65.5	70.5	72.9	71.3	68.4	72.7	69.9	68.7	73.3	70.3	67.1	74.2	69.6	68.4	75.4	70.8
03	CONCRETE	71.4	65.5	69.1	78.3	65.6	73.2	77.1	65.9	72.7	76.4	66.2	72.3	75.0	67.5	72.0	81.3	67.0	75.6
04	MASONRY	104.4	60.2	76.6	118.2	60.6	81.9	114.4	60.2	80.3	74.6	58.9	64.7	106.5	60.6	77.6	114.1	60.7	80.4
05	METALS	89.0	74.7	85.6	91.4	74.8	87.5	90.3	75.7	86.9	95.1	75.4	90.5	86.1	90.2	87.0	91.4	73.9	87.3
06	WOOD, PLASTICS & COMPOSITES	79.6	60.4	71.6	103.8	60.4	85.8	91.1	60.4	78.4	81.2	60.4	72.6	88.2	60.7	76.8	98.1	60.7	82.6
07	THERMAL & MOISTURE PROTECTION	104.1	60.3	84.9	104.8	60.1	85.2	103.6	60.3	84.6	98.7	59.3	81.4	108.5	60.1	87.3	98.7	61.7	82.5
08	OPENINGS	96.2	57.8	86.9	98.1	57.8	88.3	96.9	58.6	87.6	99.3	58.1	89.4	101.4	58.8	91.1	91.8	58.1	83.7
0920	Plaster & Gypsum Board	88.8	59.9	70.7	101.3	59.9	75.4	99.9	59.9	74.9	84.7	59.9	69.2	103.5	59.9	76.2	112.8	59.9	79.7
0950, 0980	Ceilings & Acoustic Treatment	98.6	59.9	73.5	100.7	59.9	74.2	100.7	59.9	74.2	91.9	59.9	71.1	101.3	59.9	74.4	103.8	59.9	75.3
0960	Flooring	83.9	70.0	79.6	92.4	75.3	87.2	88.6	70.0	82.9	87.8	66.0	81.1	61.4	75.3	65.6	94.5	78.4	89.6
0970, 0990	Wall Finishes & Painting/Coating	95.9	52.6	69.5	95.9	51.3	68.7	95.9	52.6	69.5	98.7	50.6	69.4	85.3	52.6	65.3	100.4	51.6	70.6
09	FINISHES	86.9	60.2	72.4	91.9	61.2	75.3	90.8	60.2	74.2	87.5	59.4	72.3	87.0	61.6	73.2	101.0	62.2	79.9
COVERS	DIVS. 10 - 14, 25, 28, 41, 43, 44, 46	100.0	78.5	95.3	100.0	78.7	95.4	100.0	78.5	95.3	100.0	78.9	95.4	100.0	79.4	95.5	100.0	79.4	95.5
21, 22, 23	FIRE SUPPRESSION, PLUMBING & HVAC	97.0	59.9	81.8	101.2	48.1	79.4	96.9	48.2	76.9	96.8	48.8	77.1	101.7	49.8	80.4	100.8	48.1	79.2
26, 27, 3370	ELECTRICAL, COMMUNICATIONS & UTIL.	89.7	49.9	70.5	91.8	57.4	75.2	92.5	53.3	73.6	93.9	60.4	77.7	98.8	58.9	79.6	99.6	57.4	79.3
MF2018	WEIGHTED AVERAGE	90.2	62.9	79.1	94.3	61.6	81.0	92.5	61.0	79.7	91.7	61.9	79.6	93.4	65.2	81.9	95.3	62.7	82.0

City Cost Indexes

ARKANSAS / CALIFORNIA

DIVISION		PINE BLUFF 716 MAT.	INST.	TOTAL	RUSSELLVILLE 728 MAT.	INST.	TOTAL	TEXARKANA 718 MAT.	INST.	TOTAL	WEST MEMPHIS 723 MAT.	INST.	TOTAL	ALHAMBRA 917-918 MAT.	INST.	TOTAL	ANAHEIM 928 MAT.	INST.	TOTAL
015433	CONTRACTOR EQUIPMENT		86.5	86.5		86.5	86.5		87.7	87.7		108.9	108.9		92.8	92.8		96.5	96.5
0241, 31 - 34	SITE & INFRASTRUCTURE, DEMOLITION	85.2	81.7	82.8	74.8	81.7	79.5	95.3	83.6	87.4	103.9	97.8	99.7	118.5	101.1	106.7	111.2	101.2	104.4
0310	Concrete Forming & Accessories	72.4	59.2	61.7	79.6	59.1	63.0	81.6	59.1	63.5	89.9	59.5	65.4	113.2	140.1	134.9	101.5	140.1	132.6
0320	Concrete Reinforcing	96.6	67.2	87.2	97.7	66.8	87.8	96.1	67.3	86.8	94.5	59.0	83.1	103.6	130.4	112.2	83.9	130.1	98.7
0330	Cast-in-Place Concrete	68.7	73.0	70.2	64.6	73.2	67.6	74.9	72.8	74.2	70.4	74.3	71.8	94.4	126.7	105.5	99.2	129.6	109.7
03	CONCRETE	77.1	66.2	72.8	74.2	66.1	71.0	76.4	66.1	72.3	80.3	66.3	74.7	98.9	132.6	112.4	98.5	133.7	112.5
04	MASONRY	107.4	60.6	77.9	117.4	58.9	80.5	85.5	60.6	69.8	98.0	58.8	73.3	124.8	139.1	133.8	80.7	135.5	115.2
05	METALS	95.7	75.9	91.1	89.0	75.2	85.8	88.0	76.0	85.2	85.4	87.0	85.8	81.0	114.3	88.8	106.8	115.0	108.8
06	WOOD, PLASTICS & COMPOSITES	80.7	60.4	72.3	85.5	60.4	75.1	90.1	60.4	77.8	95.3	60.7	81.0	97.9	139.1	115.0	95.5	139.3	113.7
07	THERMAL & MOISTURE PROTECTION	98.8	60.4	82.0	104.4	59.3	84.6	99.7	60.1	82.3	109.2	59.2	87.3	93.5	130.2	109.6	96.3	132.4	112.1
08	OPENINGS	100.5	58.3	90.3	96.2	58.1	87.0	104.9	58.7	93.8	99.0	56.5	88.8	85.1	136.6	97.6	99.7	135.3	108.3
0920	Plaster & Gypsum Board	84.2	59.9	69.0	90.7	59.9	71.4	88.8	59.9	70.7	108.1	59.9	78.0	101.8	140.6	126.1	118.0	140.6	132.1
0950, 0980	Ceilings & Acoustic Treatment	91.9	59.9	71.1	98.6	59.9	73.5	98.7	59.9	73.5	100.6	59.9	74.2	107.5	140.6	129.0	106.4	140.6	128.6
0960	Flooring	87.7	75.3	83.9	86.0	66.0	79.9	89.5	76.4	85.5	63.4	66.0	64.2	99.7	119.8	105.9	95.3	119.8	102.8
0970, 0990	Wall Finishes & Painting/Coating	98.7	51.3	69.8	95.9	50.6	68.3	98.7	52.6	70.6	85.3	50.6	64.1	97.4	121.8	112.3	85.3	116.7	104.5
09	FINISHES	87.5	61.2	73.2	88.0	59.4	72.5	90.9	61.5	74.9	89.0	59.7	73.1	101.0	134.4	119.2	99.9	133.9	118.4
COVERS	DIVS. 10 - 14, 25, 28, 41, 43, 44, 46	100.0	78.7	95.4	100.0	78.9	95.4	100.0	78.5	95.3	100.0	79.9	95.6	100.0	117.3	103.8	100.0	117.8	103.9
21, 22, 23	FIRE SUPPRESSION, PLUMBING & HVAC	101.1	51.5	80.8	97.0	48.3	77.0	101.1	54.9	82.1	97.5	60.9	82.5	96.4	128.4	109.6	100.6	128.5	112.1
26, 27, 3370	ELECTRICAL, COMMUNICATIONS & UTIL.	92.7	56.6	75.3	91.8	53.3	73.2	93.8	55.5	75.3	99.8	60.5	80.8	115.9	130.0	122.7	85.1	110.4	97.3
MF2018	WEIGHTED AVERAGE	94.2	62.4	81.3	91.5	60.8	79.0	92.8	63.2	80.8	93.0	66.8	82.4	97.0	127.8	109.5	99.1	124.9	109.6

CALIFORNIA

DIVISION		BAKERSFIELD 932 - 933 MAT.	INST.	TOTAL	BERKELEY 947 MAT.	INST.	TOTAL	EUREKA 955 MAT.	INST.	TOTAL	FRESNO 936 - 938 MAT.	INST.	TOTAL	INGLEWOOD 903 - 905 MAT.	INST.	TOTAL	LONG BEACH 906 - 908 MAT.	INST.	TOTAL
015433	CONTRACTOR EQUIPMENT		99.3	99.3		97.3	97.3		94.4	94.4		94.6	94.6		94.2	94.2		94.2	94.2
0241, 31 - 34	SITE & INFRASTRUCTURE, DEMOLITION	111.7	107.6	109.0	119.4	101.7	107.4	126.9	98.6	107.7	115.0	99.0	104.2	99.8	97.9	98.5	109.3	97.9	101.6
0310	Concrete Forming & Accessories	104.9	139.1	132.4	122.4	169.4	160.3	110.3	156.1	147.3	106.4	155.4	145.9	101.5	140.4	132.9	94.4	140.4	131.5
0320	Concrete Reinforcing	100.4	130.2	110.0	91.3	131.8	104.3	92.5	131.6	105.1	83.1	130.9	98.5	102.2	130.4	111.3	101.4	130.4	110.7
0330	Cast-in-Place Concrete	95.2	130.5	107.4	109.1	133.0	117.3	107.7	129.3	115.2	103.2	129.0	112.1	107.3	129.0	114.8	122.1	129.0	124.5
03	CONCRETE	98.8	133.3	112.6	106.1	148.2	123.0	109.0	141.0	121.8	98.0	140.5	115.0	102.4	133.7	114.9	112.0	133.7	120.7
04	MASONRY	100.7	138.0	124.2	107.4	152.1	135.6	103.3	153.9	135.2	105.7	143.1	129.3	79.7	139.2	117.2	89.5	139.2	120.8
05	METALS	99.4	112.4	102.4	100.1	117.8	104.2	106.5	117.6	109.1	100.0	116.7	103.9	89.7	115.9	95.9	89.6	115.9	95.8
06	WOOD, PLASTICS & COMPOSITES	94.5	138.3	112.6	111.0	175.7	137.8	109.5	160.5	130.6	101.5	160.5	125.9	95.5	139.4	113.7	87.9	139.4	109.3
07	THERMAL & MOISTURE PROTECTION	100.3	123.6	110.5	102.3	154.3	125.1	99.2	148.7	120.9	89.9	131.8	108.3	94.1	131.7	110.6	94.3	131.7	110.7
08	OPENINGS	89.8	134.4	100.6	87.4	162.7	105.6	99.2	143.4	109.9	91.7	147.1	105.1	83.8	136.7	96.6	83.7	136.7	96.5
0920	Plaster & Gypsum Board	102.2	139.4	125.4	116.2	177.6	154.5	125.6	162.2	148.5	96.5	162.2	137.6	112.8	140.6	130.2	104.8	140.6	127.2
0950, 0980	Ceilings & Acoustic Treatment	111.4	139.4	129.6	101.6	177.6	150.9	111.2	162.2	144.3	99.9	162.2	140.4	102.4	140.6	127.2	102.4	140.6	127.2
0960	Flooring	102.8	119.8	108.0	110.4	145.3	121.0	98.8	145.3	113.0	101.9	125.0	108.9	108.6	119.8	112.0	106.1	119.8	110.3
0970, 0990	Wall Finishes & Painting/Coating	98.7	110.9	106.1	97.1	168.4	140.6	87.6	132.1	114.8	101.6	120.3	113.0	99.1	121.8	112.9	99.1	121.8	112.9
09	FINISHES	103.4	132.9	119.4	105.9	166.5	138.8	106.0	153.3	131.7	100.0	148.5	126.3	104.6	134.7	121.0	103.2	134.7	120.3
COVERS	DIVS. 10 - 14, 25, 28, 41, 43, 44, 46	100.0	118.0	103.9	100.0	131.3	106.8	100.0	128.4	106.2	100.0	128.3	106.2	100.0	118.1	103.9	100.0	118.1	103.9
21, 22, 23	FIRE SUPPRESSION, PLUMBING & HVAC	100.8	127.7	111.9	96.8	167.1	125.7	96.4	129.6	110.0	100.8	129.6	112.7	95.9	128.5	109.3	95.9	128.5	109.3
26, 27, 3370	ELECTRICAL, COMMUNICATIONS & UTIL.	91.0	105.6	98.0	95.9	164.1	128.8	91.5	125.9	108.1	82.9	104.5	93.3	95.7	130.0	112.2	95.5	130.0	112.1
MF2018	WEIGHTED AVERAGE	98.5	124.1	108.9	100.1	151.5	121.0	102.2	134.1	115.2	97.7	128.9	110.4	94.8	128.0	108.3	96.5	128.0	109.3

CALIFORNIA

DIVISION		LOS ANGELES 900 - 902 MAT.	INST.	TOTAL	MARYSVILLE 959 MAT.	INST.	TOTAL	MODESTO 953 MAT.	INST.	TOTAL	MOJAVE 935 MAT.	INST.	TOTAL	OAKLAND 946 MAT.	INST.	TOTAL	OXNARD 930 MAT.	INST.	TOTAL
015433	CONTRACTOR EQUIPMENT		106.5	106.5		94.4	94.4		94.4	94.4		94.6	94.6		97.3	97.3		93.6	93.6
0241, 31 - 34	SITE & INFRASTRUCTURE, DEMOLITION	108.4	111.3	110.4	122.3	98.5	106.2	115.9	98.5	104.1	106.2	99.3	101.5	125.9	101.7	109.5	115.4	97.5	103.2
0310	Concrete Forming & Accessories	99.2	140.9	132.8	97.1	155.7	144.4	92.0	155.8	143.5	116.7	139.1	134.8	106.2	169.5	157.2	106.6	140.2	133.7
0320	Concrete Reinforcing	101.1	130.5	110.5	92.5	131.0	104.9	95.9	131.0	107.2	101.6	130.3	110.8	93.4	134.3	106.5	99.7	130.1	109.5
0330	Cast-in-Place Concrete	110.9	131.3	117.9	120.4	129.1	123.4	107.7	129.1	115.1	89.2	128.9	102.9	103.5	133.0	113.7	103.9	130.1	112.6
03	CONCRETE	109.6	134.8	119.7	110.5	140.7	122.6	102.1	140.7	117.5	92.7	132.9	108.8	106.4	148.7	123.3	100.0	133.5	113.4
04	MASONRY	96.2	139.3	123.4	104.6	144.1	129.5	103.7	144.1	129.1	103.3	137.9	125.1	116.1	152.1	138.8	106.7	135.1	124.6
05	METALS	104.5	118.5	107.8	105.9	116.3	108.4	100.0	116.4	103.8	98.8	114.7	102.5	95.4	118.7	100.8	95.3	114.8	99.9
06	WOOD, PLASTICS & COMPOSITES	100.6	140.0	116.9	94.6	160.5	121.9	89.6	160.5	118.9	102.4	138.2	117.2	95.7	175.7	128.8	96.2	139.4	114.1
07	THERMAL & MOISTURE PROTECTION	91.6	133.2	109.8	98.5	139.5	116.5	97.9	138.6	115.7	94.9	121.6	106.6	100.7	154.3	124.2	97.6	131.2	112.3
08	OPENINGS	97.8	137.1	107.3	98.5	148.7	110.6	97.6	149.3	110.1	87.0	134.3	98.4	83.8	163.3	105.8	89.3	135.4	100.5
0920	Plaster & Gypsum Board	105.1	140.9	127.4	113.3	162.2	143.9	118.0	162.2	145.7	110.6	139.4	128.6	105.2	177.6	150.4	100.7	140.6	125.7
0950, 0980	Ceilings & Acoustic Treatment	112.0	140.9	130.7	110.5	162.2	144.1	106.4	162.2	142.6	103.9	139.4	126.9	103.8	177.6	151.7	102.2	140.6	127.1
0960	Flooring	110.6	119.8	113.4	95.7	123.6	104.2	95.9	135.6	108.1	102.7	119.8	107.9	105.5	145.3	117.7	95.6	119.8	103.0
0970, 0990	Wall Finishes & Painting/Coating	98.7	121.8	112.8	87.6	132.1	114.8	87.6	136.1	117.2	88.7	110.9	102.3	97.1	168.4	140.6	88.7	116.7	105.8
09	FINISHES	106.8	135.1	122.2	102.3	149.5	128.0	101.6	152.1	129.0	101.9	132.8	118.7	103.8	166.5	137.9	98.4	134.1	117.8
COVERS	DIVS. 10 - 14, 25, 28, 41, 43, 44, 46	100.0	118.8	104.1	100.0	128.4	106.2	100.0	128.4	106.2	100.0	126.8	105.8	100.0	131.3	106.8	100.0	118.0	103.9
21, 22, 23	FIRE SUPPRESSION, PLUMBING & HVAC	100.6	128.6	112.1	96.4	129.7	110.0	100.6	131.7	113.4	96.5	127.7	109.3	101.1	167.2	128.2	100.7	128.5	112.1
26, 27, 3370	ELECTRICAL, COMMUNICATIONS & UTIL.	94.4	130.0	111.6	89.6	119.1	103.9	91.3	110.1	100.4	81.3	105.6	93.0	95.4	164.1	128.6	87.5	116.3	101.3
MF2018	WEIGHTED AVERAGE	102.0	129.6	113.2	101.5	131.4	113.7	99.8	130.9	112.5	95.2	123.8	106.9	100.1	151.6	121.1	97.4	125.4	108.8

For CCI updates and additional guidance please go to www.rsmeans.com/2022books

City Cost Indexes

CALIFORNIA

DIVISION		PALM SPRINGS 922			PALO ALTO 943			PASADENA 910 - 912			REDDING 960			RICHMOND 948			RIVERSIDE 925		
		MAT.	INST.	TOTAL	MAT.	INST.	TOTAL	MAT.	INST.	TOTAL	MAT.	INST.	TOTAL	MAT.	INST.	TOTAL	MAT.	INST.	TOTAL
015433	CONTRACTOR EQUIPMENT		95.4	95.4		97.3	97.3		92.8	92.8		94.4	94.4		97.3	97.3		95.4	95.4
0241, 31 - 34	SITE & INFRASTRUCTURE, DEMOLITION	100.9	99.4	99.9	115.2	101.7	106.0	114.2	101.1	105.3	140.3	98.5	111.9	124.6	101.6	109.0	109.3	99.4	102.6
0310	Concrete Forming & Accessories	97.5	140.2	131.9	103.7	169.6	156.9	98.5	140.1	132.0	91.6	155.7	143.4	126.6	169.1	160.9	102.5	140.1	132.8
0320	Concrete Reinforcing	97.1	130.3	107.8	91.3	134.3	105.1	104.5	130.4	112.8	90.3	131.0	103.4	91.3	134.2	105.1	94.4	130.1	105.8
0330	Cast-in-Place Concrete	94.6	129.7	106.7	92.3	133.1	106.4	89.5	126.7	102.4	131.4	129.1	130.6	106.1	132.8	115.4	102.9	129.6	112.1
03	CONCRETE	94.8	133.7	110.4	96.4	148.8	117.3	94.6	132.6	109.8	115.5	140.7	125.6	108.6	148.4	124.5	100.4	133.7	113.7
04	MASONRY	78.4	138.2	116.0	94.0	152.1	130.6	108.2	139.1	127.6	124.3	144.1	136.8	107.1	152.1	135.5	79.6	135.2	114.6
05	METALS	107.3	115.3	109.1	93.4	118.8	99.4	81.1	114.3	88.8	101.5	116.3	105.0	93.5	118.3	99.3	105.5	115.0	107.7
06	WOOD, PLASTICS & COMPOSITES	89.1	139.3	109.9	92.6	175.7	127.0	82.3	139.1	105.8	91.0	160.5	119.7	115.0	175.7	140.1	95.5	139.3	113.7
07	THERMAL & MOISTURE PROTECTION	95.5	132.2	111.6	99.9	155.7	124.4	92.9	130.2	109.2	115.9	139.5	126.3	100.9	154.3	124.3	96.4	132.0	112.0
08	OPENINGS	96.1	136.7	105.9	87.5	162.6	105.6	85.2	136.6	97.6	112.0	148.7	120.8	87.5	162.6	105.6	98.5	135.3	107.4
0920	Plaster & Gypsum Board	108.7	140.6	128.6	102.5	177.6	149.4	92.6	140.6	122.6	123.3	162.2	147.6	116.8	177.6	154.8	117.2	140.6	131.8
0950, 0980	Ceilings & Acoustic Treatment	102.3	140.6	127.2	102.3	177.6	151.1	107.5	140.6	129.0	129.2	162.2	150.6	102.3	177.6	151.1	109.2	140.6	129.6
0960	Flooring	98.5	119.8	105.0	104.5	145.3	117.0	95.0	119.8	102.6	85.7	123.6	97.2	112.2	145.3	122.3	99.5	119.8	105.7
0970, 0990	Wall Finishes & Painting/Coating	84.3	121.8	107.2	97.1	168.4	140.6	97.4	121.8	112.3	90.5	132.1	115.9	97.1	168.4	140.6	84.3	116.7	104.1
09	FINISHES	97.6	134.6	117.7	102.0	166.5	137.1	98.0	134.4	117.8	107.5	149.5	130.3	107.2	166.5	139.4	101.1	133.9	119.0
COVERS	DIVS. 10 - 14, 25, 28, 41, 43, 44, 46	100.0	117.8	103.9	100.0	131.4	106.8	100.0	117.3	103.8	100.0	128.4	106.2	100.0	131.2	106.8	100.0	117.8	103.9
21, 22, 23	FIRE SUPPRESSION, PLUMBING & HVAC	96.4	128.5	109.6	96.8	168.7	126.3	96.4	128.4	109.6	100.6	129.7	112.5	96.8	158.2	122.0	100.6	128.5	112.1
26, 27, 3370	ELECTRICAL, COMMUNICATIONS & UTIL.	86.7	111.7	98.7	95.3	181.7	137.0	114.0	130.0	121.7	92.7	119.1	105.5	95.6	135.0	114.6	84.8	111.7	97.8
MF2018	WEIGHTED AVERAGE	97.0	125.4	108.5	96.1	154.5	119.8	95.0	127.8	108.3	105.7	131.4	116.2	99.3	145.5	118.1	98.9	124.9	109.5

CALIFORNIA

DIVISION		SACRAMENTO 942,956 - 958			SALINAS 939			SAN BERNARDINO 923 - 924			SAN DIEGO 919 - 921			SAN FRANCISCO 940 - 941			SAN JOSE 951		
		MAT.	INST.	TOTAL	MAT.	INST.	TOTAL	MAT.	INST.	TOTAL	MAT.	INST.	TOTAL	MAT.	INST.	TOTAL	MAT.	INST.	TOTAL
015433	CONTRACTOR EQUIPMENT		96.1	96.1		94.6	94.6		95.4	95.4		105.7	105.7		112.2	112.2		96.0	96.0
0241, 31 - 34	SITE & INFRASTRUCTURE, DEMOLITION	102.9	106.8	105.5	132.2	99.1	109.7	84.0	99.4	94.5	123.5	112.8	116.2	125.3	117.7	120.1	156.7	94.1	114.2
0310	Concrete Forming & Accessories	104.8	158.2	147.9	112.0	158.6	149.6	107.6	140.2	133.9	104.7	130.6	125.6	108.4	169.9	158.1	99.5	169.4	155.9
0320	Concrete Reinforcing	85.6	131.0	100.2	100.4	131.5	110.4	94.4	130.3	105.9	101.1	130.2	110.4	101.8	134.5	112.3	83.8	134.3	100.0
0330	Cast-in-Place Concrete	87.5	130.1	102.2	102.6	129.5	111.9	71.0	129.7	91.3	102.7	127.9	111.4	115.6	134.7	122.2	125.3	132.2	127.6
03	CONCRETE	97.5	141.8	115.2	108.1	142.1	121.7	76.9	133.8	99.7	105.4	128.8	114.8	116.5	149.9	129.9	109.7	148.7	124.1
04	MASONRY	94.7	144.1	125.8	103.7	149.4	132.5	85.7	138.2	118.8	100.8	134.0	121.7	118.3	155.1	141.5	132.7	152.1	144.9
05	METALS	91.0	111.5	95.8	101.8	117.5	105.5	105.5	115.4	107.8	103.8	115.4	106.5	105.0	124.6	109.6	99.1	124.1	104.9
06	WOOD, PLASTICS & COMPOSITES	91.5	163.5	121.3	112.1	163.2	127.4	99.9	139.3	116.2	94.0	128.1	108.1	95.9	175.7	128.9	100.6	175.5	131.6
07	THERMAL & MOISTURE PROTECTION	106.5	141.2	121.7	95.5	145.8	117.5	94.7	133.2	111.6	98.6	122.2	108.9	109.1	157.3	130.3	95.3	155.7	121.8
08	OPENINGS	99.5	151.0	111.9	90.3	155.9	106.2	96.1	136.7	105.9	100.4	126.6	106.7	99.2	163.3	114.7	89.4	163.2	107.2
0920	Plaster & Gypsum Board	100.4	165.1	140.9	101.9	165.1	141.4	119.5	140.6	132.7	102.8	128.6	119.0	101.7	177.6	149.1	111.7	177.6	152.9
0950, 0980	Ceilings & Acoustic Treatment	102.3	165.1	143.0	103.9	165.1	143.6	106.4	140.6	128.6	120.1	128.6	125.6	119.8	177.6	157.3	105.1	177.6	152.1
0960	Flooring	103.9	129.8	111.8	97.7	145.3	112.2	101.0	119.8	106.8	104.3	119.8	109.1	102.9	145.3	115.8	89.7	145.3	106.7
0970, 0990	Wall Finishes & Painting/Coating	95.0	136.1	120.1	89.8	168.4	137.7	84.3	128.4	111.2	98.7	121.8	112.8	98.7	177.0	146.4	87.3	168.4	136.8
09	FINISHES	100.4	152.9	128.9	101.7	158.9	132.8	99.4	135.3	118.9	106.6	127.6	118.0	105.9	167.5	139.4	100.7	166.4	136.4
COVERS	DIVS. 10 - 14, 25, 28, 41, 43, 44, 46	100.0	129.2	106.4	100.0	128.8	106.3	100.0	117.8	103.9	100.0	116.5	103.6	100.0	131.5	106.9	100.0	130.5	106.7
21, 22, 23	FIRE SUPPRESSION, PLUMBING & HVAC	101.0	130.2	113.0	96.5	137.2	113.2	96.4	128.5	109.6	100.6	127.2	111.5	101.0	179.6	133.2	100.6	168.6	128.5
26, 27, 3370	ELECTRICAL, COMMUNICATIONS & UTIL.	91.1	120.8	105.4	82.3	132.9	106.7	86.7	113.9	99.8	96.9	103.9	100.3	95.6	188.8	140.5	95.1	181.7	136.9
MF2018	WEIGHTED AVERAGE	97.0	132.8	111.6	99.0	137.6	114.7	94.2	125.8	107.1	102.4	122.2	110.4	105.0	160.3	127.5	102.3	154.3	123.5

CALIFORNIA

DIVISION		SAN LUIS OBISPO 934			SAN MATEO 944			SAN RAFAEL 949			SANTA ANA 926 - 927			SANTA BARBARA 931			SANTA CRUZ 950		
		MAT.	INST.	TOTAL	MAT.	INST.	TOTAL	MAT.	INST.	TOTAL	MAT.	INST.	TOTAL	MAT.	INST.	TOTAL	MAT.	INST.	TOTAL
015433	CONTRACTOR EQUIPMENT		94.6	94.6		97.3	97.3		96.5	96.5		95.4	95.4		94.6	94.6		96.0	96.0
0241, 31 - 34	SITE & INFRASTRUCTURE, DEMOLITION	122.0	99.3	106.6	122.1	101.7	108.2	116.1	106.7	109.7	98.8	99.4	99.2	115.4	99.3	104.4	156.0	93.9	113.9
0310	Concrete Forming & Accessories	119.0	140.3	136.2	112.7	169.5	158.5	123.9	169.2	160.4	108.0	140.2	133.9	107.1	140.2	133.8	99.5	158.7	147.3
0320	Concrete Reinforcing	101.6	130.3	110.8	91.3	134.4	105.2	91.6	131.9	104.6	97.4	130.3	108.0	99.7	130.3	109.5	104.3	131.5	113.0
0330	Cast-in-Place Concrete	110.3	129.0	116.8	102.7	133.0	113.2	119.6	132.0	123.9	90.7	129.7	104.2	103.5	129.0	112.3	124.4	131.1	126.7
03	CONCRETE	107.4	133.5	117.9	105.0	148.7	122.5	125.8	147.6	134.5	92.8	133.7	109.2	99.9	133.5	113.3	111.8	143.0	124.3
04	MASONRY	105.5	137.1	125.4	106.8	155.0	137.2	87.1	154.9	129.9	75.7	138.4	115.3	103.9	137.1	124.8	136.6	149.6	144.8
05	METALS	100.0	115.0	103.5	93.3	118.9	99.3	95.5	114.1	99.9	105.6	115.4	107.8	96.7	114.9	101.0	106.0	121.6	109.6
06	WOOD, PLASTICS & COMPOSITES	105.0	139.4	119.2	102.4	175.7	132.8	104.2	175.5	133.7	101.9	139.3	117.4	96.2	139.4	114.1	100.6	163.4	126.6
07	THERMAL & MOISTURE PROTECTION	95.7	130.3	110.9	100.5	155.2	124.5	105.0	154.8	126.8	96.0	132.3	111.9	95.0	131.3	110.9	95.9	147.6	118.0
08	OPENINGS	88.7	135.0	99.9	87.4	163.3	105.8	97.1	161.9	112.8	95.4	136.7	105.4	90.2	136.7	101.4	90.4	155.9	106.2
0920	Plaster & Gypsum Board	111.8	140.6	129.8	112.7	177.6	153.2	117.0	177.6	154.9	121.2	140.6	133.3	100.7	140.6	125.7	118.4	165.1	147.6
0950, 0980	Ceilings & Acoustic Treatment	103.9	140.6	127.7	102.3	177.6	151.1	113.3	177.6	155.0	106.4	140.6	128.6	102.2	140.6	127.1	107.1	165.1	144.7
0960	Flooring	103.4	119.8	108.4	107.3	145.3	118.9	117.3	145.3	125.9	101.5	119.8	107.1	96.9	119.8	103.9	93.6	145.3	109.4
0970, 0990	Wall Finishes & Painting/Coating	88.7	116.7	105.8	97.1	168.4	140.6	94.0	163.7	136.5	84.3	121.8	107.2	88.7	116.7	105.8	87.4	168.4	136.8
09	FINISHES	103.6	134.1	120.2	105.0	166.5	138.4	108.7	165.9	139.8	101.2	134.6	119.4	98.9	134.1	118.0	103.3	159.0	133.6
COVERS	DIVS. 10 - 14, 25, 28, 41, 43, 44, 46	100.0	127.0	105.9	100.0	131.3	106.8	100.0	130.7	106.7	100.0	117.8	103.9	100.0	117.9	103.9	100.0	129.0	106.3
21, 22, 23	FIRE SUPPRESSION, PLUMBING & HVAC	96.5	128.5	109.6	96.8	164.7	124.7	96.8	179.5	130.7	96.4	128.5	109.6	100.6	128.5	112.1	100.6	137.2	115.6
26, 27, 3370	ELECTRICAL, COMMUNICATIONS & UTIL.	81.3	109.3	94.8	95.3	172.1	132.3	90.0	125.9	107.3	86.7	113.3	99.5	80.7	107.5	93.6	94.8	132.9	113.2
MF2018	WEIGHTED AVERAGE	98.3	125.0	109.2	98.3	152.6	120.4	101.4	148.8	120.7	96.6	125.6	108.4	96.9	124.5	108.1	104.6	137.8	118.1

CALIFORNIA / COLORADO

DIVISION		SANTA ROSA 954			STOCKTON 952			SUSANVILLE 961			VALLEJO 945			VAN NUYS 913 - 916			ALAMOSA 811		
		MAT.	INST.	TOTAL	MAT.	INST.	TOTAL	MAT.	INST.	TOTAL	MAT.	INST.	TOTAL	MAT.	INST.	TOTAL	MAT.	INST.	TOTAL
015433	CONTRACTOR EQUIPMENT		94.9	94.9		94.4	94.4		94.4	94.4		96.5	96.5		92.8	92.8		87.8	87.8
0241, 31 - 34	SITE & INFRASTRUCTURE, DEMOLITION	115.9	98.5	104.0	115.6	98.5	104.0	149.6	98.5	114.9	103.3	106.6	105.5	138.2	101.1	113.0	137.5	80.6	98.8
0310	Concrete Forming & Accessories	98.8	168.4	154.9	97.6	157.8	146.1	93.1	155.7	143.6	109.8	168.2	156.9	107.2	140.1	133.7	100.6	69.0	75.1
0320	Concrete Reinforcing	93.2	131.9	105.6	95.9	131.0	107.2	90.3	131.0	103.4	92.8	131.8	105.3	104.5	130.4	112.8	97.9	65.1	87.4
0330	Cast-in-Place Concrete	118.3	130.5	122.5	105.0	129.1	113.3	119.5	129.1	122.8	95.2	131.2	107.6	94.4	126.7	105.6	90.1	70.1	83.2
03	CONCRETE	110.4	147.0	125.1	101.3	141.6	117.4	116.1	140.7	125.9	102.9	146.8	120.5	106.5	132.6	117.0	104.6	69.3	90.5
04	MASONRY	102.7	154.0	135.0	103.5	144.1	129.1	122.3	144.1	136.0	70.8	154.0	123.2	124.8	139.1	133.8	130.5	58.0	84.8
05	METALS	105.9	119.1	109.0	100.6	116.3	104.3	99.6	116.3	103.5	95.4	113.5	99.6	80.2	114.3	88.2	97.0	77.9	92.6
06	WOOD, PLASTICS & COMPOSITES	91.1	175.3	125.9	96.0	163.2	123.9	92.6	160.5	120.7	90.6	175.5	125.7	92.0	139.1	111.5	94.7	71.3	85.0
07	THERMAL & MOISTURE PROTECTION	96.4	153.7	121.5	98.4	137.9	115.7	118.6	139.5	127.8	102.7	154.0	125.2	94.3	130.2	110.0	108.7	65.4	89.7
08	OPENINGS	96.9	162.5	112.7	97.6	150.2	110.3	112.9	148.7	121.6	98.7	161.9	113.9	85.0	136.6	97.5	92.3	69.4	86.8
0920	Plaster & Gypsum Board	112.5	177.6	153.2	118.0	165.1	147.5	124.2	162.2	148.0	106.4	177.6	150.9	99.5	140.6	125.2	87.1	70.7	76.8
0950, 0980	Ceilings & Acoustic Treatment	106.4	177.6	152.6	113.3	165.1	146.9	121.7	162.2	148.0	114.8	177.6	155.5	104.8	140.6	128.0	104.6	70.7	82.6
0960	Flooring	98.9	132.9	109.3	95.9	135.6	108.1	86.0	123.6	97.5	112.8	145.3	122.7	97.5	119.8	104.3	106.5	80.3	98.5
0970, 0990	Wall Finishes & Painting/Coating	84.3	163.7	132.7	87.6	132.1	114.8	90.5	132.1	115.9	95.0	163.7	136.9	97.4	121.8	112.3	97.8	74.2	83.4
09	FINISHES	100.3	163.3	134.5	103.1	153.3	130.4	107.5	149.5	130.3	104.6	165.6	137.7	100.9	134.4	119.1	101.8	71.8	85.5
COVERS	DIVS. 10 - 14, 25, 28, 41, 43, 44, 46	100.0	129.8	106.5	100.0	128.7	106.2	100.0	128.4	106.2	100.0	130.3	106.6	100.0	117.3	103.8	100.0	84.9	96.7
21, 22, 23	FIRE SUPPRESSION, PLUMBING & HVAC	96.4	174.9	128.6	100.6	131.7	113.4	96.4	129.7	110.0	101.1	146.9	119.9	96.4	128.4	109.6	96.6	68.5	85.1
26, 27, 3370	ELECTRICAL, COMMUNICATIONS & UTIL.	86.9	121.0	103.3	91.3	112.5	101.5	92.9	119.1	105.6	87.7	126.1	106.2	114.0	130.0	121.7	95.3	58.6	77.6
MF2018	WEIGHTED AVERAGE	100.6	146.4	119.2	100.0	131.6	112.8	104.9	131.4	115.7	97.4	141.6	115.4	98.1	127.8	110.2	100.6	68.8	87.7

COLORADO

DIVISION		BOULDER 803			COLORADO SPRINGS 808 - 809			DENVER 800 - 802			DURANGO 813			FORT COLLINS 805			FORT MORGAN 807		
		MAT.	INST.	TOTAL	MAT.	INST.	TOTAL	MAT.	INST.	TOTAL	MAT.	INST.	TOTAL	MAT.	INST.	TOTAL	MAT.	INST.	TOTAL
015433	CONTRACTOR EQUIPMENT		89.9	89.9		87.9	87.9		95.0	95.0		87.8	87.8		89.9	89.9		89.9	89.9
0241, 31 - 34	SITE & INFRASTRUCTURE, DEMOLITION	92.3	88.1	89.5	94.7	83.1	86.8	99.1	94.8	96.2	131.1	80.6	96.8	104.2	87.4	92.8	94.4	86.8	89.2
0310	Concrete Forming & Accessories	99.9	72.5	77.8	87.4	68.8	72.4	99.3	69.0	74.8	108.9	68.7	76.5	96.9	69.0	74.4	100.6	68.7	74.9
0320	Concrete Reinforcing	97.4	69.7	88.5	97.0	69.4	88.1	101.4	69.5	91.1	97.9	65.1	87.4	97.5	66.8	87.6	97.6	65.6	87.3
0330	Cast-in-Place Concrete	97.8	77.3	90.8	100.5	72.3	90.7	101.8	73.5	92.0	103.7	70.0	92.1	110.5	72.1	97.2	96.0	69.4	86.8
03	CONCRETE	99.0	74.2	89.0	101.5	70.7	89.2	108.3	71.1	93.4	108.0	69.2	91.7	108.0	70.3	92.9	97.7	69.0	86.2
04	MASONRY	111.4	75.7	88.9	115.9	61.7	81.7	123.2	62.4	84.9	118.1	59.0	80.8	128.5	66.5	89.4	127.8	66.5	89.2
05	METALS	89.6	79.3	87.2	92.7	79.0	89.5	101.9	77.9	96.3	97.0	77.7	92.5	90.5	78.2	87.6	89.3	77.4	86.6
06	WOOD, PLASTICS & COMPOSITES	97.7	70.9	86.6	85.5	71.1	79.5	99.4	71.1	87.7	106.1	71.3	91.7	94.3	70.9	84.6	97.7	70.9	86.6
07	THERMAL & MOISTURE PROTECTION	106.2	74.7	92.4	107.1	69.4	90.6	106.7	72.3	91.6	108.7	65.6	89.8	106.6	70.8	90.9	106.1	69.0	89.8
08	OPENINGS	91.2	70.5	86.2	95.0	70.6	89.1	100.4	70.6	93.2	98.6	69.4	91.6	91.2	69.8	86.0	91.2	69.2	85.9
0920	Plaster & Gypsum Board	111.4	70.7	85.9	89.8	70.7	77.9	102.8	70.7	82.7	110.3	70.7	85.5	99.5	70.7	81.5	111.4	70.7	85.9
0950, 0980	Ceilings & Acoustic Treatment	92.0	70.7	78.2	103.6	70.7	82.3	106.5	70.7	83.3	104.6	70.7	82.6	92.0	70.7	78.2	92.0	70.7	78.2
0960	Flooring	106.2	80.3	98.3	97.7	80.3	92.4	102.5	80.3	95.7	110.8	80.3	101.5	102.7	80.3	95.9	106.6	80.3	98.6
0970, 0990	Wall Finishes & Painting/Coating	88.9	74.2	79.9	88.6	78.6	82.5	98.7	74.2	83.7	97.8	74.2	83.4	88.9	74.2	79.9	88.9	74.2	79.9
09	FINISHES	99.9	73.9	85.8	96.7	72.0	83.3	101.4	71.5	85.2	105.9	71.8	87.4	97.6	71.5	83.4	100.0	71.5	84.5
COVERS	DIVS. 10 - 14, 25, 28, 41, 43, 44, 46	100.0	87.2	97.2	100.0	84.2	96.6	100.0	84.2	96.6	100.0	84.9	96.7	100.0	84.1	96.5	100.0	84.1	96.5
21, 22, 23	FIRE SUPPRESSION, PLUMBING & HVAC	96.7	76.5	88.4	101.1	68.4	87.7	100.9	75.0	90.3	96.6	55.8	79.9	101.0	70.1	88.3	96.7	69.5	85.5
26, 27, 3370	ELECTRICAL, COMMUNICATIONS & UTIL.	98.7	80.1	89.7	100.5	72.9	87.2	101.6	80.1	91.2	94.9	51.5	74.0	98.7	80.1	89.7	99.0	72.4	86.2
MF2018	WEIGHTED AVERAGE	96.5	77.4	88.8	98.8	71.8	87.9	103.1	75.3	91.8	101.2	65.2	86.6	99.7	73.9	89.2	97.1	72.3	87.0

COLORADO

DIVISION		GLENWOOD SPRINGS 816			GOLDEN 804			GRAND JUNCTION 815			GREELEY 806			MONTROSE 814			PUEBLO 810		
		MAT.	INST.	TOTAL	MAT.	INST.	TOTAL	MAT.	INST.	TOTAL	MAT.	INST.	TOTAL	MAT.	INST.	TOTAL	MAT.	INST.	TOTAL
015433	CONTRACTOR EQUIPMENT		90.8	90.8		89.9	89.9		90.8	90.8		89.9	89.9		89.3	89.3		87.8	87.8
0241, 31 - 34	SITE & INFRASTRUCTURE, DEMOLITION	146.5	87.6	106.5	104.0	87.0	92.5	131.0	88.1	101.9	91.8	87.4	88.8	140.1	84.4	102.3	122.9	80.9	94.4
0310	Concrete Forming & Accessories	96.1	68.5	73.8	90.5	68.8	73.0	107.3	69.5	76.8	94.0	69.0	73.8	95.6	68.7	73.9	103.8	69.1	75.8
0320	Concrete Reinforcing	97.3	65.5	87.1	97.6	65.6	87.3	97.5	66.7	87.6	97.4	66.8	87.6	97.3	65.3	87.0	94.0	69.5	86.1
0330	Cast-in-Place Concrete	90.1	69.0	82.8	96.0	72.0	87.7	99.8	72.6	90.4	92.3	72.1	85.3	90.1	69.5	83.0	89.5	72.5	83.6
03	CONCRETE	108.8	68.8	92.8	106.1	69.9	91.6	103.7	70.7	90.5	94.6	70.3	84.9	101.2	69.0	88.3	95.1	71.0	85.4
04	MASONRY	104.0	66.4	80.3	131.2	66.3	90.3	138.5	64.9	92.1	121.1	66.5	86.7	110.4	62.9	80.5	100.4	62.9	76.8
05	METALS	96.7	77.9	92.4	89.5	77.6	86.7	98.0	78.6	93.5	90.5	78.2	87.6	95.5	77.9	91.3	100.3	80.1	95.6
06	WOOD, PLASTICS & COMPOSITES	89.5	71.1	81.9	88.0	70.9	81.0	104.2	71.1	90.5	91.3	70.9	82.9	90.0	71.2	82.2	98.0	71.3	87.0
07	THERMAL & MOISTURE PROTECTION	108.5	67.7	90.6	107.3	70.5	91.1	107.6	69.3	90.4	105.8	70.8	90.4	108.6	66.7	90.3	107.1	68.4	90.1
08	OPENINGS	97.6	69.3	90.8	91.2	69.2	85.9	98.3	69.9	91.4	91.2	69.8	86.0	98.7	69.4	91.7	93.8	70.7	88.3
0920	Plaster & Gypsum Board	118.6	70.7	88.7	94.0	70.7	79.4	141.8	70.7	97.4	97.4	70.7	80.7	85.4	70.7	76.2	93.2	70.7	79.1
0950, 0980	Ceilings & Acoustic Treatment	103.9	70.7	82.4	92.0	70.7	78.2	103.9	70.7	82.4	92.0	70.7	78.2	104.6	70.7	82.6	112.8	70.7	85.5
0960	Flooring	105.7	80.3	97.9	100.5	80.3	94.4	110.2	80.3	101.1	101.6	80.3	95.1	108.6	80.3	99.9	107.5	80.3	99.2
0970, 0990	Wall Finishes & Painting/Coating	97.8	74.2	83.4	88.9	74.2	79.9	97.8	74.2	83.4	88.9	74.2	79.9	97.8	74.2	83.4	97.8	74.2	83.4
09	FINISHES	107.1	71.6	87.8	96.8	71.5	83.1	110.2	72.0	89.4	96.2	71.5	82.8	102.1	71.7	85.6	103.0	71.7	86.0
COVERS	DIVS. 10 - 14, 25, 28, 41, 43, 44, 46	100.0	84.3	96.6	100.0	84.0	96.5	100.0	84.8	96.7	100.0	84.1	96.5	100.0	84.6	96.7	100.0	84.8	96.7
21, 22, 23	FIRE SUPPRESSION, PLUMBING & HVAC	96.6	55.7	79.8	96.7	69.4	85.5	100.8	71.4	88.7	101.0	70.1	88.5	96.6	55.7	79.8	100.8	68.4	87.5
26, 27, 3370	ELECTRICAL, COMMUNICATIONS & UTIL.	93.4	51.5	73.2	99.0	74.7	87.3	94.8	54.5	75.4	98.7	80.1	89.7	94.8	51.5	73.9	95.3	63.4	79.9
MF2018	WEIGHTED AVERAGE	100.9	66.5	86.9	98.3	72.8	87.9	102.9	70.6	89.8	97.0	73.9	87.6	99.6	65.9	85.9	99.5	70.6	87.7

For CCI updates and additional guidance please go to www.rsmeans.com/2022books

COLORADO / CONNECTICUT

DIVISION		COLORADO SALIDA 812 MAT.	INST.	TOTAL	BRIDGEPORT 066 MAT.	INST.	TOTAL	BRISTOL 060 MAT.	INST.	TOTAL	HARTFORD 061 MAT.	INST.	TOTAL	MERIDEN 064 MAT.	INST.	TOTAL	NEW BRITAIN 060 MAT.	INST.	TOTAL
015433	CONTRACTOR EQUIPMENT		89.3	89.3		92.3	92.3		92.3	92.3		97.1	97.1		92.7	92.7		92.3	92.3
0241, 31 - 34	SITE & INFRASTRUCTURE, DEMOLITION	130.8	84.1	99.1	101.3	93.2	95.8	100.5	93.2	95.6	97.5	101.2	100.0	97.8	93.9	95.2	100.7	93.2	95.6
0310	Concrete Forming & Accessories	107.2	68.9	76.3	99.2	114.5	111.5	99.2	114.4	111.4	100.7	114.6	111.9	98.9	114.3	111.3	100.1	114.4	111.6
0320	Concrete Reinforcing	97.1	65.1	86.9	104.6	141.5	116.5	104.6	141.5	116.5	99.8	141.5	113.2	104.6	141.5	116.5	104.6	141.5	116.5
0330	Cast-in-Place Concrete	103.3	69.6	91.6	81.7	125.0	96.6	76.5	125.0	93.3	78.0	126.7	94.8	73.6	124.9	91.3	77.7	125.0	94.1
03	CONCRETE	102.5	69.1	89.2	92.2	122.3	104.3	90.0	122.2	102.9	90.8	122.8	103.6	88.7	122.2	102.1	90.6	122.2	103.2
04	MASONRY	138.8	59.0	88.5	105.0	129.2	120.3	97.8	129.2	117.6	100.7	129.3	118.8	97.4	129.2	117.5	99.4	129.2	118.2
05	METALS	95.1	77.8	91.1	97.6	117.6	102.3	97.6	117.5	102.2	101.3	115.5	104.6	94.4	117.4	99.8	93.7	117.5	99.2
06	WOOD, PLASTICS & COMPOSITES	99.2	71.2	87.6	99.6	111.8	104.6	99.6	111.8	104.6	92.0	112.0	100.2	99.6	111.8	104.6	99.6	111.8	104.6
07	THERMAL & MOISTURE PROTECTION	107.6	65.6	89.2	103.1	120.4	110.7	103.3	117.7	109.6	110.8	119.3	114.5	103.3	117.7	109.6	103.3	117.7	109.6
08	OPENINGS	92.4	69.4	86.8	92.8	118.8	99.0	92.8	118.8	99.0	94.2	118.9	100.1	94.5	118.8	100.4	92.8	118.8	99.0
0920	Plaster & Gypsum Board	85.9	70.7	76.4	128.4	112.0	118.2	128.4	112.0	118.2	117.2	112.0	114.0	130.6	112.0	119.0	128.4	112.0	118.2
0950, 0980	Ceilings & Acoustic Treatment	104.6	70.7	82.6	103.5	112.0	109.1	103.5	112.0	109.1	104.8	112.0	109.5	113.7	112.0	112.6	103.5	112.0	109.1
0960	Flooring	113.1	80.3	103.1	97.7	122.9	105.4	97.7	122.9	105.4	105.3	125.2	111.4	97.7	115.8	103.3	97.7	122.9	105.4
0970, 0990	Wall Finishes & Painting/Coating	97.8	74.2	83.4	90.1	127.4	112.9	90.1	127.4	112.9	98.7	127.4	116.2	90.1	127.4	112.9	90.1	127.4	112.9
09	FINISHES	102.3	71.7	85.7	102.1	117.0	110.2	102.2	117.0	110.2	105.0	117.5	111.8	104.8	115.8	110.7	102.2	117.0	110.2
COVERS	DIVS. 10 - 14, 25, 28, 41, 43, 44, 46	100.0	84.6	96.7	100.0	111.4	102.5	100.0	111.4	102.5	100.0	111.9	102.6	100.0	111.4	102.5	100.0	111.4	102.5
21, 22, 23	FIRE SUPPRESSION, PLUMBING & HVAC	96.6	69.5	85.5	101.0	115.7	107.1	101.0	115.7	107.1	100.9	115.8	107.0	96.8	115.7	104.6	101.0	115.7	107.1
26, 27, 3370	ELECTRICAL, COMMUNICATIONS & UTIL.	94.9	58.6	77.4	98.2	105.5	101.7	98.2	107.5	102.7	101.2	106.6	103.8	98.1	98.1	98.1	98.2	107.5	102.7
MF2018	WEIGHTED AVERAGE	100.2	69.4	87.6	98.3	115.2	105.2	97.7	115.4	104.9	99.3	115.9	106.1	96.2	113.9	103.4	97.0	115.2	104.5

CONNECTICUT

DIVISION		NEW HAVEN 065 MAT.	INST.	TOTAL	NEW LONDON 063 MAT.	INST.	TOTAL	NORWALK 068 MAT.	INST.	TOTAL	STAMFORD 069 MAT.	INST.	TOTAL	WATERBURY 067 MAT.	INST.	TOTAL	WILLIMANTIC 062 MAT.	INST.	TOTAL
015433	CONTRACTOR EQUIPMENT		92.7	92.7		92.7	92.7		92.3	92.3		92.3	92.3		92.3	92.3		92.3	92.3
0241, 31 - 34	SITE & INFRASTRUCTURE, DEMOLITION	100.5	93.9	96.0	92.8	93.9	93.5	101.1	93.2	95.8	101.7	93.2	96.0	101.0	93.2	95.7	101.1	93.1	95.7
0310	Concrete Forming & Accessories	98.9	114.3	111.3	98.9	114.2	111.2	99.2	114.5	111.5	99.1	114.5	111.5	99.2	114.3	111.4	99.1	114.1	111.2
0320	Concrete Reinforcing	104.6	141.5	116.5	82.0	141.5	101.1	104.6	141.6	116.5	104.6	141.6	116.5	104.6	141.5	116.5	104.6	141.5	116.5
0330	Cast-in-Place Concrete	79.1	123.1	94.3	67.4	123.1	86.6	80.3	126.1	96.1	81.7	126.1	97.0	81.7	125.0	96.6	76.3	121.2	91.8
03	CONCRETE	102.1	121.6	109.9	78.7	121.5	95.8	91.6	122.7	104.1	92.2	122.7	104.4	92.2	122.2	104.2	89.8	120.8	102.2
04	MASONRY	98.0	129.2	117.7	96.2	129.2	117.0	97.5	130.4	118.2	98.3	130.4	118.5	97.6	129.2	117.8	97.6	129.2	117.5
05	METALS	93.9	117.4	99.4	93.6	117.3	99.1	97.6	117.7	102.3	97.6	117.7	102.3	97.6	117.5	102.2	97.3	117.2	102.0
06	WOOD, PLASTICS & COMPOSITES	99.6	111.8	104.6	99.6	111.8	104.6	99.6	111.8	104.6	99.6	111.8	104.6	99.6	111.8	104.6	99.6	111.8	104.6
07	THERMAL & MOISTURE PROTECTION	103.4	117.4	109.5	103.1	117.4	109.4	103.3	120.9	111.0	103.3	120.9	111.0	103.3	117.7	109.6	103.4	116.9	109.3
08	OPENINGS	92.8	118.8	99.0	94.9	117.3	100.3	92.8	118.8	99.0	92.8	118.8	99.0	92.8	118.8	99.0	95.0	118.8	100.7
0920	Plaster & Gypsum Board	128.4	112.0	118.2	128.4	112.0	118.2	128.4	112.0	118.2	128.4	112.0	118.2	128.4	112.0	118.2	128.4	112.0	118.2
0950, 0980	Ceilings & Acoustic Treatment	103.5	112.0	109.1	102.8	112.0	108.8	103.5	112.0	109.1	103.5	112.0	109.1	103.5	112.0	109.1	102.8	112.0	108.8
0960	Flooring	97.7	122.9	105.4	97.7	118.1	104.0	97.7	122.9	105.4	97.7	122.9	105.4	97.7	122.9	105.4	97.7	112.7	102.3
0970, 0990	Wall Finishes & Painting/Coating	90.1	127.4	112.9	90.1	123.6	110.5	90.1	127.4	112.9	90.1	127.4	112.9	90.1	127.4	112.9	90.1	127.4	112.9
09	FINISHES	102.2	117.0	110.2	101.5	115.7	109.2	102.2	117.0	110.2	102.2	117.0	110.2	102.0	117.0	110.2	102.1	115.2	109.2
COVERS	DIVS. 10 - 14, 25, 28, 41, 43, 44, 46	100.0	111.4	102.5	100.0	111.4	102.5	100.0	111.4	102.5	100.0	111.4	102.5	100.0	111.4	102.5	100.0	111.4	102.5
21, 22, 23	FIRE SUPPRESSION, PLUMBING & HVAC	101.0	115.7	107.1	96.8	115.7	104.6	101.0	115.7	107.1	101.0	115.7	107.1	101.0	115.7	107.1	101.0	115.7	107.0
26, 27, 3370	ELECTRICAL, COMMUNICATIONS & UTIL.	98.1	105.8	101.8	96.3	100.0	98.1	98.2	105.5	101.7	98.2	108.2	103.0	97.9	105.8	101.7	98.2	103.1	100.6
MF2018	WEIGHTED AVERAGE	98.6	115.1	105.3	94.1	114.0	102.2	97.9	115.4	105.0	98.0	115.8	105.3	98.0	115.1	105.0	97.8	114.3	104.5

D.C. / DELAWARE / FLORIDA

DIVISION		D.C. WASHINGTON 200 - 205 MAT.	INST.	TOTAL	DOVER 199 MAT.	INST.	TOTAL	NEWARK 197 MAT.	INST.	TOTAL	WILMINGTON 198 MAT.	INST.	TOTAL	DAYTONA BEACH 321 MAT.	INST.	TOTAL	FORT LAUDERDALE 333 MAT.	INST.	TOTAL
015433	CONTRACTOR EQUIPMENT		104.6	104.6		114.6	114.6		114.6	114.6		114.8	114.8		97.0	97.0		90.5	90.5
0241, 31 - 34	SITE & INFRASTRUCTURE, DEMOLITION	111.5	98.7	102.8	111.9	104.1	106.6	110.5	103.5	105.7	107.9	105.3	106.1	133.7	81.3	98.1	107.8	70.5	82.5
0310	Concrete Forming & Accessories	102.1	73.4	78.9	102.1	98.7	99.4	98.2	98.4	98.4	101.7	99.0	99.5	95.7	62.3	68.7	91.8	60.6	66.6
0320	Concrete Reinforcing	101.0	90.6	97.7	99.6	118.7	105.7	98.4	118.7	104.9	99.4	118.7	105.6	90.9	60.2	81.0	89.8	61.5	80.7
0330	Cast-in-Place Concrete	129.4	79.3	112.1	119.7	107.9	115.7	99.8	105.4	101.7	116.4	108.0	113.5	90.3	65.4	81.7	93.0	63.4	82.8
03	CONCRETE	114.0	79.3	100.1	105.8	106.4	106.0	97.4	105.6	100.7	104.3	106.5	105.2	90.6	64.7	80.2	92.3	63.5	80.7
04	MASONRY	95.8	85.6	89.4	106.2	102.7	104.0	101.4	102.6	102.1	98.8	102.7	101.3	88.3	59.9	70.4	95.7	56.9	71.2
05	METALS	100.7	95.4	99.5	95.5	123.0	101.9	97.9	126.4	104.5	95.9	123.2	102.3	101.2	88.7	98.3	98.0	89.2	95.9
06	WOOD, PLASTICS & COMPOSITES	100.2	71.8	88.4	92.5	95.9	93.9	92.0	95.6	93.5	89.9	95.9	92.4	90.0	61.2	78.1	79.6	60.8	71.9
07	THERMAL & MOISTURE PROTECTION	104.9	86.9	97.0	104.5	114.1	108.7	107.7	112.3	109.7	103.2	114.1	108.0	101.2	63.8	84.8	106.3	60.0	86.0
08	OPENINGS	101.0	75.7	94.9	88.2	108.1	93.0	88.0	107.9	92.8	85.7	108.1	91.1	89.5	59.9	82.4	90.6	60.0	83.2
0920	Plaster & Gypsum Board	116.0	71.1	87.9	116.5	95.7	103.5	113.9	95.7	102.5	116.5	95.7	103.5	100.9	60.8	75.8	119.7	60.4	82.6
0950, 0980	Ceilings & Acoustic Treatment	112.5	71.1	85.6	109.4	95.7	100.5	105.7	95.7	99.2	101.7	95.7	97.8	83.2	60.8	68.7	96.9	60.4	73.2
0960	Flooring	93.3	77.1	88.3	95.4	110.4	100.0	92.6	110.4	98.1	102.6	110.4	105.0	105.8	59.5	91.7	101.7	61.9	89.5
0970, 0990	Wall Finishes & Painting/Coating	98.7	70.7	81.6	98.7	119.6	111.4	87.6	119.6	107.1	98.7	119.6	111.4	99.5	60.5	75.7	91.2	58.1	71.0
09	FINISHES	103.7	72.7	86.9	103.5	102.0	102.7	97.5	101.7	99.8	104.1	102.0	103.0	98.8	61.1	78.3	101.1	60.0	78.8
COVERS	DIVS. 10 - 14, 25, 28, 41, 43, 44, 46	100.0	95.3	99.0	100.0	97.5	99.5	100.0	96.7	99.3	100.0	108.5	101.8	100.0	82.1	96.1	100.0	82.3	96.2
21, 22, 23	FIRE SUPPRESSION, PLUMBING & HVAC	100.6	92.5	97.3	99.1	122.1	108.5	99.4	122.0	108.7	100.6	122.1	109.4	100.5	75.3	90.1	100.5	68.8	87.5
26, 27, 3370	ELECTRICAL, COMMUNICATIONS & UTIL.	99.3	101.3	100.3	99.2	110.5	104.7	98.7	110.6	104.4	98.9	110.5	104.5	96.5	60.2	79.0	95.8	71.0	83.8
MF2018	WEIGHTED AVERAGE	102.7	88.5	96.9	99.4	110.6	103.9	98.1	110.6	103.2	98.9	111.1	103.8	97.9	69.1	86.2	97.3	67.8	85.3

City Cost Indexes

FLORIDA

DIVISION		FORT MYERS 339,341			GAINESVILLE 326,344			JACKSONVILLE 320,322			LAKELAND 338			MELBOURNE 329			MIAMI 330 - 332,340		
		MAT.	INST.	TOTAL	MAT.	INST.	TOTAL	MAT.	INST.	TOTAL	MAT.	INST.	TOTAL	MAT.	INST.	TOTAL	MAT.	INST.	TOTAL
015433	CONTRACTOR EQUIPMENT		97.0	97.0		97.0	97.0		97.0	97.0		97.0	97.0		97.0	97.0		97.1	97.1
0241, 31 - 34	SITE & INFRASTRUCTURE, DEMOLITION	123.2	81.4	94.8	144.3	81.1	101.4	133.6	81.1	98.0	125.4	81.4	95.5	143.0	81.4	101.1	108.6	83.4	91.5
0310	Concrete Forming & Accessories	86.2	60.7	65.6	89.1	54.6	61.2	95.3	62.5	68.9	81.5	61.8	65.6	90.7	62.4	67.9	102.5	63.3	70.9
0320	Concrete Reinforcing	90.8	73.3	85.1	96.3	59.6	84.5	90.9	59.6	80.8	93.0	76.3	87.6	91.9	67.2	83.9	100.4	61.5	87.9
0330	Cast-in-Place Concrete	96.9	65.0	85.9	103.7	65.0	90.3	91.2	65.0	82.1	99.2	65.0	87.4	108.9	65.4	93.9	90.0	69.2	82.8
03	CONCRETE	92.9	66.1	82.2	100.8	61.0	84.8	91.0	64.6	80.4	94.5	67.2	83.5	101.0	66.0	87.0	94.7	66.5	83.4
04	MASONRY	90.7	55.8	68.7	102.1	58.4	74.6	88.3	57.7	69.0	107.1	59.2	76.9	87.6	59.9	70.1	99.7	62.2	76.1
05	METALS	100.0	93.0	98.4	99.9	87.3	97.0	99.8	87.3	96.9	99.9	94.5	98.6	109.1	91.3	104.9	97.9	86.0	95.1
06	WOOD, PLASTICS & COMPOSITES	75.7	59.6	69.0	83.0	51.5	70.0	90.0	62.2	78.5	70.6	61.2	66.7	84.9	61.2	75.1	94.4	61.2	80.6
07	THERMAL & MOISTURE PROTECTION	106.0	60.6	86.1	101.5	60.9	83.7	101.5	61.9	84.1	105.9	61.6	86.5	101.7	62.9	84.7	105.2	64.6	87.4
08	OPENINGS	92.2	62.1	84.9	89.4	54.5	80.9	89.5	60.3	82.5	92.1	63.6	85.2	88.9	61.5	82.3	93.0	60.2	85.1
0920	Plaster & Gypsum Board	111.9	59.1	78.9	95.5	50.8	67.5	100.9	61.7	76.4	107.6	60.8	78.3	95.5	60.8	73.8	111.9	60.4	79.7
0950, 0980	Ceilings & Acoustic Treatment	90.6	59.1	70.2	78.7	50.8	60.6	83.2	61.7	69.3	90.6	60.8	71.3	82.6	60.8	68.4	97.6	60.4	73.4
0960	Flooring	98.9	71.8	90.7	103.6	58.3	89.8	105.8	61.3	92.2	97.3	60.7	86.1	103.9	60.7	90.7	101.4	61.3	89.1
0970, 0990	Wall Finishes & Painting/Coating	96.0	62.0	75.3	99.5	62.0	76.6	99.5	62.0	76.6	96.0	62.0	75.3	99.5	58.6	86.7	98.7	58.1	73.9
09	FINISHES	100.6	62.5	79.9	97.6	55.2	74.5	98.8	62.0	78.8	99.6	61.2	78.7	98.0	63.2	79.1	102.0	61.7	80.1
COVERS	DIVS. 10 - 14, 25, 28, 41, 43, 44, 46	100.0	80.4	95.7	100.0	79.2	95.5	100.0	80.4	95.7	100.0	81.8	96.0	100.0	82.1	96.1	100.0	85.2	96.8
21, 22, 23	FIRE SUPPRESSION, PLUMBING & HVAC	97.2	55.5	80.1	97.5	59.4	81.9	100.5	64.5	85.7	97.2	58.0	81.1	100.5	75.4	90.2	96.1	67.0	84.2
26, 27, 3370	ELECTRICAL, COMMUNICATIONS & UTIL.	96.6	61.8	79.8	96.5	54.3	76.1	96.4	63.3	80.4	95.5	54.3	75.6	97.5	66.6	82.6	97.3	75.7	86.9
MF2018	WEIGHTED AVERAGE	97.5	65.4	84.4	99.1	62.8	84.3	97.7	66.9	85.2	98.1	65.5	84.8	101.1	70.8	88.8	97.5	70.1	86.4

FLORIDA

DIVISION		ORLANDO 327 - 328,347			PANAMA CITY 324			PENSACOLA 325			SARASOTA 342			ST. PETERSBURG 337			TALLAHASSEE 323		
		MAT.	INST.	TOTAL	MAT.	INST.	TOTAL	MAT.	INST.	TOTAL	MAT.	INST.	TOTAL	MAT.	INST.	TOTAL	MAT.	INST.	TOTAL
015433	CONTRACTOR EQUIPMENT		105.1	105.1		97.0	97.0		97.0	97.0		97.0	97.0		97.0	97.0		105.1	105.1
0241, 31 - 34	SITE & INFRASTRUCTURE, DEMOLITION	131.1	95.0	106.6	150.0	81.3	103.4	149.5	81.2	103.1	134.5	81.4	98.5	127.9	81.4	96.3	126.4	95.0	105.0
0310	Concrete Forming & Accessories	102.2	62.4	70.1	94.4	65.0	70.7	91.3	62.0	67.7	95.3	61.6	68.1	91.2	62.0	67.6	101.5	63.0	70.4
0320	Concrete Reinforcing	101.0	63.6	89.0	94.9	64.4	85.1	97.2	65.3	87.0	92.9	71.0	85.9	93.0	73.4	86.7	101.2	59.7	87.9
0330	Cast-in-Place Concrete	111.7	67.8	96.5	95.8	65.3	85.3	118.4	64.4	99.8	106.9	65.0	92.4	100.2	65.3	88.1	93.6	67.8	84.7
03	CONCRETE	104.4	65.9	89.0	99.0	66.6	86.0	108.4	65.1	91.1	97.4	66.1	84.9	96.2	66.8	84.4	97.5	65.6	84.7
04	MASONRY	102.1	60.0	75.5	92.9	60.0	72.1	110.8	59.0	78.1	92.6	59.2	71.5	147.1	58.4	91.2	88.5	60.1	70.6
05	METALS	96.4	86.5	94.1	100.4	90.0	98.0	101.4	90.2	98.8	102.2	91.9	99.8	100.5	93.2	98.8	100.4	85.1	96.8
06	WOOD, PLASTICS & COMPOSITES	90.6	61.6	78.6	88.6	65.2	79.0	86.4	62.2	76.4	92.5	61.2	79.6	81.1	61.2	72.9	95.9	62.5	82.1
07	THERMAL & MOISTURE PROTECTION	108.0	65.5	89.4	101.8	62.1	84.4	101.7	60.9	83.8	101.1	61.6	83.8	106.2	60.5	86.2	99.4	63.8	83.8
08	OPENINGS	93.9	60.9	86.0	87.9	63.1	81.9	87.9	61.6	81.6	94.2	62.0	86.4	91.0	63.0	84.2	94.1	60.5	86.0
0920	Plaster & Gypsum Board	113.5	60.8	80.5	99.6	64.9	77.9	105.2	61.7	78.0	107.0	60.8	78.1	116.5	60.8	81.7	107.3	61.7	78.8
0950, 0980	Ceilings & Acoustic Treatment	97.6	60.8	73.7	82.6	64.9	71.1	82.6	61.7	69.0	92.1	60.8	71.8	92.1	60.8	71.8	97.6	61.7	74.3
0960	Flooring	105.6	59.5	91.5	105.5	70.2	94.7	101.9	59.5	88.9	104.8	51.0	88.4	100.7	56.0	87.0	105.6	58.6	91.2
0970, 0990	Wall Finishes & Painting/Coating	98.7	60.5	75.4	99.5	62.0	76.6	99.5	62.0	76.6	100.6	62.0	77.0	96.0	62.0	75.3	98.7	62.0	76.3
09	FINISHES	105.8	61.4	81.6	100.1	65.4	81.3	99.6	61.4	78.8	103.0	59.5	79.4	102.4	60.5	79.6	103.9	61.9	81.1
COVERS	DIVS. 10 - 14, 25, 28, 41, 43, 44, 46	100.0	82.9	96.3	100.0	81.5	96.0	100.0	80.6	95.8	100.0	80.5	95.8	100.0	80.8	95.8	100.0	81.2	95.9
21, 22, 23	FIRE SUPPRESSION, PLUMBING & HVAC	96.1	56.5	79.9	100.5	64.6	85.8	100.5	60.5	84.1	100.1	54.7	81.5	100.5	57.2	82.7	97.2	63.0	83.2
26, 27, 3370	ELECTRICAL, COMMUNICATIONS & UTIL.	98.1	62.0	80.7	95.7	54.3	75.7	98.2	50.3	75.1	96.7	54.3	76.2	95.9	59.3	78.3	99.7	56.4	78.8
MF2018	WEIGHTED AVERAGE	99.7	66.5	86.2	99.4	67.0	86.2	101.8	64.6	86.7	100.0	64.1	85.3	101.2	65.6	86.7	99.0	66.9	86.0

FLORIDA / GEORGIA

DIVISION		TAMPA 335 - 336,346			WEST PALM BEACH 334,349			ALBANY 317,398			ATHENS 306			ATLANTA 300 - 303,399			AUGUSTA 308 - 309		
		MAT.	INST.	TOTAL	MAT.	INST.	TOTAL	MAT.	INST.	TOTAL	MAT.	INST.	TOTAL	MAT.	INST.	TOTAL	MAT.	INST.	TOTAL
015433	CONTRACTOR EQUIPMENT		97.0	97.0		90.5	90.5		91.3	91.3		90.3	90.3		99.1	99.1		90.3	90.3
0241, 31 - 34	SITE & INFRASTRUCTURE, DEMOLITION	128.1	83.2	97.6	103.5	70.4	81.1	119.7	73.4	88.3	148.7	87.3	107.0	143.0	99.1	113.2	136.2	87.9	103.4
0310	Concrete Forming & Accessories	95.1	62.4	68.7	96.9	60.1	67.2	86.9	66.3	70.3	90.0	41.1	50.6	97.4	74.2	78.7	91.4	71.9	75.7
0320	Concrete Reinforcing	89.8	76.3	85.5	92.3	53.6	79.8	92.5	70.1	85.3	92.3	58.8	81.6	101.0	71.6	91.5	92.5	69.3	85.1
0330	Cast-in-Place Concrete	98.0	65.6	86.8	88.4	63.2	79.7	91.0	68.3	83.2	125.9	67.8	105.9	130.0	71.6	109.8	118.9	69.1	101.7
03	CONCRETE	94.7	67.6	83.9	89.8	61.9	78.6	87.0	69.5	80.0	106.8	55.0	86.0	110.5	73.4	95.7	100.4	71.2	88.7
04	MASONRY	96.6	59.9	73.5	95.0	52.9	68.4	93.6	68.1	77.5	89.9	66.5	75.2	103.2	68.4	81.3	105.0	68.5	82.0
05	METALS	99.9	94.8	98.7	96.7	85.7	94.1	106.1	97.4	104.1	92.4	77.0	88.8	98.5	82.8	94.8	92.1	83.0	90.0
06	WOOD, PLASTICS & COMPOSITES	85.9	61.2	75.7	85.7	60.8	75.4	79.4	66.4	74.0	89.7	33.8	66.6	99.1	76.9	89.9	91.6	74.4	84.5
07	THERMAL & MOISTURE PROTECTION	106.6	62.0	87.0	106.0	59.4	85.6	102.8	67.0	87.1	92.7	63.6	79.9	93.7	72.4	84.4	92.4	69.2	82.2
08	OPENINGS	92.0	63.6	85.1	90.3	58.2	82.6	82.2	68.7	78.9	87.8	48.4	78.3	99.8	75.4	93.9	87.9	73.4	84.4
0920	Plaster & Gypsum Board	121.4	60.8	83.5	128.3	60.4	85.8	108.1	66.1	81.8	100.2	32.5	57.9	107.6	76.4	88.1	102.9	74.2	85.0
0950, 0980	Ceilings & Acoustic Treatment	96.6	60.8	73.5	90.6	60.4	71.0	79.1	66.1	70.7	100.0	32.5	56.2	91.5	76.4	81.7	100.8	74.2	83.6
0960	Flooring	101.7	60.7	89.2	103.3	51.0	87.3	100.8	69.3	91.2	89.9	69.3	83.6	95.3	70.7	87.8	89.9	70.7	84.1
0970, 0990	Wall Finishes & Painting/Coating	96.0	62.0	75.3	91.2	57.8	70.8	94.7	92.5	93.4	91.2	87.2	88.8	98.7	94.2	95.9	91.2	83.7	86.6
09	FINISHES	104.6	61.4	81.1	101.3	57.8	77.6	96.6	69.2	81.7	99.6	49.1	72.1	100.7	75.7	87.1	99.0	73.0	84.9
COVERS	DIVS. 10 - 14, 25, 28, 41, 43, 44, 46	100.0	80.8	95.8	100.0	82.3	96.1	100.0	84.1	96.5	100.0	80.2	95.7	100.0	86.1	97.0	100.0	85.0	96.7
21, 22, 23	FIRE SUPPRESSION, PLUMBING & HVAC	100.5	58.4	83.2	97.2	56.4	80.4	100.5	67.9	87.1	96.4	61.4	82.1	100.6	70.1	88.1	100.7	66.1	86.5
26, 27, 3370	ELECTRICAL, COMMUNICATIONS & UTIL.	95.4	65.1	80.8	96.3	62.5	80.0	98.3	59.7	79.7	98.1	59.2	79.4	98.3	70.4	84.8	98.5	65.9	82.8
MF2018	WEIGHTED AVERAGE	99.0	67.4	86.2	96.0	62.6	82.4	97.6	70.7	86.6	97.8	62.4	83.4	102.2	75.4	91.3	98.1	72.2	87.6

City Cost Indexes

GEORGIA

DIVISION		COLUMBUS 318 - 319 MAT.	INST.	TOTAL	DALTON 307 MAT.	INST.	TOTAL	GAINESVILLE 305 MAT.	INST.	TOTAL	MACON 310 - 312 MAT.	INST.	TOTAL	SAVANNAH 313 - 314 MAT.	INST.	TOTAL	STATESBORO 304 MAT.	INST.	TOTAL
015433	CONTRACTOR EQUIPMENT		91.3	91.3		104.9	104.9		90.3	90.3		100.5	100.5		99.2	99.2		93.2	93.2
0241, 31 - 34	SITE & INFRASTRUCTURE, DEMOLITION	119.6	73.7	88.4	148.8	92.1	110.3	146.8	87.2	106.4	121.8	87.2	98.3	122.0	86.7	98.1	149.5	74.0	98.2
0310	Concrete Forming & Accessories	86.9	68.9	72.4	79.1	60.7	64.3	95.3	39.0	49.9	87.0	67.6	71.3	95.6	71.4	76.1	72.5	70.2	70.6
0320	Concrete Reinforcing	93.3	70.2	85.9	92.1	56.4	80.6	92.3	58.8	81.5	94.5	70.2	86.7	100.8	69.3	90.7	91.9	66.0	83.6
0330	Cast-in-Place Concrete	90.6	68.7	83.0	122.3	66.4	103.0	132.4	67.4	109.9	89.4	69.1	82.4	91.9	70.7	84.6	125.7	67.1	105.5
03	CONCRETE	87.0	70.8	80.5	105.3	63.9	88.7	109.3	53.9	87.1	86.7	70.4	80.2	88.3	72.2	81.9	105.8	70.1	91.5
04	MASONRY	93.6	69.1	78.1	91.0	64.9	74.5	99.4	65.5	78.1	105.5	67.1	81.3	88.5	69.1	76.3	93.4	66.3	76.3
05	METALS	105.9	98.2	104.1	93.8	91.4	93.2	91.8	77.0	88.3	100.2	98.2	99.8	102.1	93.9	100.2	96.9	96.3	96.8
06	WOOD, PLASTICS & COMPOSITES	79.4	69.8	75.4	74.6	61.5	69.2	94.3	31.4	68.3	82.0	68.5	76.4	90.2	73.3	83.2	66.6	73.0	69.3
07	THERMAL & MOISTURE PROTECTION	102.6	68.5	87.7	94.0	65.2	81.3	92.7	62.2	79.6	101.4	69.1	87.2	99.7	69.4	86.4	92.7	65.5	80.7
08	OPENINGS	82.2	71.1	79.5	88.8	63.0	82.6	87.8	47.0	78.0	82.3	70.4	79.4	91.3	72.9	86.8	89.8	71.8	85.4
0920	Plaster & Gypsum Board	108.1	69.6	84.0	93.4	61.0	73.1	104.6	30.0	57.9	104.0	68.2	81.6	109.8	72.9	86.7	87.5	72.9	78.3
0950, 0980	Ceilings & Acoustic Treatment	79.1	69.6	72.9	111.7	61.0	78.8	100.0	30.0	54.6	75.2	68.2	70.7	94.9	72.9	80.6	108.5	72.9	85.4
0960	Flooring	100.8	70.7	91.6	90.0	69.3	83.7	91.0	69.3	84.4	80.7	69.3	77.2	94.9	70.7	87.5	108.3	69.3	96.4
0970, 0990	Wall Finishes & Painting/Coating	94.7	86.2	89.5	81.8	65.0	71.6	91.2	87.2	88.8	97.3	92.5	94.4	98.7	83.1	89.2	89.3	65.0	74.5
09	FINISHES	96.5	70.7	82.5	108.3	62.5	83.4	100.3	47.4	71.6	85.6	70.1	77.2	99.6	72.5	84.9	110.4	69.6	88.2
COVERS	DIVS. 10 - 14, 25, 28, 41, 43, 44, 46	100.0	84.5	96.6	100.0	78.6	95.4	100.0	79.7	95.6	100.0	84.0	96.5	100.0	85.6	96.9	100.0	84.3	96.6
21, 22, 23	FIRE SUPPRESSION, PLUMBING & HVAC	100.5	64.9	85.9	96.5	56.3	80.0	96.4	60.9	81.8	100.5	70.0	88.0	100.6	62.9	85.1	97.3	64.5	83.8
26, 27, 3370	ELECTRICAL, COMMUNICATIONS & UTIL.	98.2	69.5	84.4	110.6	57.7	85.1	98.1	68.1	83.6	96.1	60.4	78.9	101.2	61.4	82.0	99.4	61.0	80.9
MF2018	WEIGHTED AVERAGE	97.5	72.2	87.2	99.8	66.4	86.2	98.5	62.9	84.0	95.9	72.7	86.4	98.2	71.9	87.5	99.8	70.3	87.8

GEORGIA / HAWAII / IDAHO

DIVISION		VALDOSTA 316 MAT.	INST.	TOTAL	WAYCROSS 315 MAT.	INST.	TOTAL	HILO 967 MAT.	INST.	TOTAL	HONOLULU 968 MAT.	INST.	TOTAL	STATES & POSS., GUAM 969 MAT.	INST.	TOTAL	BOISE 836 - 837 MAT.	INST.	TOTAL
015433	CONTRACTOR EQUIPMENT		91.3	91.3		91.3	91.3		95.1	95.1		101.3	101.3		155.9	155.9		92.1	92.1
0241, 31 - 34	SITE & INFRASTRUCTURE, DEMOLITION	132.2	73.4	92.3	127.7	73.0	90.6	166.6	98.8	120.6	174.0	110.1	130.6	225.3	95.7	137.3	85.4	87.8	87.1
0310	Concrete Forming & Accessories	74.5	70.9	71.6	76.9	69.8	71.2	95.6	121.2	116.3	108.9	121.3	118.9	98.1	61.9	68.9	98.7	79.5	83.2
0320	Concrete Reinforcing	94.9	59.3	83.5	94.9	55.3	82.2	93.9	119.3	102.1	107.2	119.3	111.1	157.3	42.9	120.5	92.8	74.5	86.9
0330	Cast-in-Place Concrete	89.0	68.2	81.9	100.6	66.9	89.0	221.5	121.8	187.1	172.8	123.1	155.6	191.9	101.6	160.7	81.6	96.2	86.6
03	CONCRETE	90.1	69.6	81.9	93.6	67.8	83.2	153.5	120.4	140.2	143.4	120.7	134.3	154.3	73.2	121.9	92.1	84.7	89.1
04	MASONRY	98.5	68.1	79.3	99.5	66.4	78.6	147.8	121.3	131.1	140.5	121.4	128.5	208.4	47.6	107.0	126.8	85.5	100.8
05	METALS	104.7	92.8	102.0	104.1	86.2	99.9	107.6	107.1	107.5	122.2	104.8	118.2	138.3	80.8	124.9	103.2	81.0	98.0
06	WOOD, PLASTICS & COMPOSITES	65.7	73.0	68.7	67.5	73.0	69.7	94.4	121.7	105.7	111.4	121.7	115.7	107.4	62.2	88.7	90.9	78.1	85.6
07	THERMAL & MOISTURE PROTECTION	102.9	67.7	87.5	102.6	65.7	86.4	117.3	115.8	116.6	133.5	117.7	126.6	136.8	68.0	106.6	99.8	85.3	93.5
08	OPENINGS	79.2	70.1	77.0	79.5	66.5	76.4	110.6	120.3	113.0	123.8	120.3	123.0	114.5	51.6	99.3	92.7	73.3	88.0
0920	Plaster & Gypsum Board	94.0	72.9	80.8	94.0	72.9	80.8	127.7	122.4	124.4	169.5	122.4	140.0	259.3	51.7	129.5	93.9	77.7	83.8
0950, 0980	Ceilings & Acoustic Treatment	77.8	72.9	74.6	76.2	72.9	74.0	117.6	122.4	120.7	127.4	122.4	124.1	230.9	51.7	114.6	107.5	77.7	88.2
0960	Flooring	96.4	69.3	88.2	97.3	69.3	88.8	98.1	136.9	109.9	120.3	136.9	125.4	116.1	46.1	94.7	90.0	82.8	87.8
0970, 0990	Wall Finishes & Painting/Coating	94.7	90.7	92.3	94.7	65.0	76.6	86.1	134.5	115.6	98.7	134.5	120.5	90.5	38.6	58.8	90.7	60.5	72.3
09	FINISHES	94.0	72.8	82.5	93.4	69.6	80.5	110.4	126.0	118.9	127.1	126.0	126.5	192.6	56.8	118.8	95.2	78.1	85.9
COVERS	DIVS. 10 - 14, 25, 28, 41, 43, 44, 46	100.0	84.7	96.7	100.0	80.2	95.7	100.0	111.1	102.4	100.0	111.2	102.4	100.0	90.2	97.9	100.0	89.2	97.7
21, 22, 23	FIRE SUPPRESSION, PLUMBING & HVAC	100.5	67.6	87.0	97.9	58.9	81.9	100.6	107.2	103.3	100.6	107.2	103.3	102.7	51.0	81.5	101.1	71.0	88.7
26, 27, 3370	ELECTRICAL, COMMUNICATIONS & UTIL.	97.0	54.6	76.6	99.6	61.0	81.0	100.3	119.6	109.6	100.8	119.6	109.9	139.6	47.4	95.1	101.4	69.1	85.8
MF2018	WEIGHTED AVERAGE	97.5	70.2	86.4	97.5	67.4	85.2	114.8	115.0	114.9	119.3	115.8	117.9	136.6	61.9	106.2	99.5	78.3	90.9

IDAHO / ILLINOIS

DIVISION		COEUR D'ALENE 838 MAT.	INST.	TOTAL	IDAHO FALLS 834 MAT.	INST.	TOTAL	LEWISTON 835 MAT.	INST.	TOTAL	POCATELLO 832 MAT.	INST.	TOTAL	TWIN FALLS 833 MAT.	INST.	TOTAL	BLOOMINGTON 617 MAT.	INST.	TOTAL
015433	CONTRACTOR EQUIPMENT		87.3	87.3		92.1	92.1		87.3	87.3		92.1	92.1		92.1	92.1		98.9	98.9
0241, 31 - 34	SITE & INFRASTRUCTURE, DEMOLITION	83.7	85.2	84.7	84.2	88.2	86.9	90.5	83.5	85.8	87.0	87.8	87.5	93.7	87.7	89.7	90.5	91.3	91.1
0310	Concrete Forming & Accessories	97.4	85.0	87.4	90.1	79.6	81.6	104.5	84.7	88.5	98.9	79.1	82.9	100.4	79.5	83.5	77.3	109.2	103.0
0320	Concrete Reinforcing	97.0	100.7	98.2	93.9	74.6	87.7	97.0	100.5	98.1	93.0	74.5	87.0	94.1	74.3	87.7	99.9	95.3	98.4
0330	Cast-in-Place Concrete	88.2	83.1	86.5	77.7	96.4	84.1	91.7	84.7	89.3	83.9	96.1	88.1	86.1	96.4	89.6	88.3	106.7	94.6
03	CONCRETE	96.8	87.1	92.9	85.2	84.8	85.0	100.1	87.5	95.0	91.4	84.5	88.6	97.8	84.7	92.6	89.7	106.9	96.6
04	MASONRY	128.8	87.9	103.0	122.2	86.8	99.8	129.3	86.3	102.2	124.5	86.5	100.5	127.4	82.0	98.8	121.0	113.7	116.4
05	METALS	96.0	90.7	94.8	110.3	81.0	103.5	95.5	90.4	94.3	110.7	80.7	103.7	110.4	80.6	103.5	90.4	120.5	97.4
06	WOOD, PLASTICS & COMPOSITES	89.9	84.2	87.5	83.4	78.1	81.2	96.7	84.2	91.5	90.9	78.1	85.6	92.4	78.1	86.5	77.4	106.6	89.5
07	THERMAL & MOISTURE PROTECTION	148.8	83.5	120.2	98.5	79.6	90.2	149.2	82.4	119.9	99.2	76.1	89.1	100.0	82.7	92.5	96.4	105.3	100.3
08	OPENINGS	108.9	78.8	101.6	95.4	72.1	89.8	102.0	82.5	97.3	93.3	66.9	86.9	96.1	73.3	90.6	88.2	111.7	93.9
0920	Plaster & Gypsum Board	151.6	84.1	109.4	85.7	77.7	80.7	154.0	84.1	110.3	87.6	77.7	81.4	92.4	77.7	83.2	77.5	107.1	96.0
0950, 0980	Ceilings & Acoustic Treatment	132.7	84.1	101.2	106.0	77.7	87.6	132.7	84.1	101.2	112.8	77.7	90.1	109.4	77.7	88.9	86.5	107.1	99.9
0960	Flooring	131.7	93.5	120.0	90.2	79.1	86.8	134.5	77.6	117.1	93.1	77.6	88.4	94.1	82.8	90.7	86.5	112.4	94.4
0970, 0990	Wall Finishes & Painting/Coating	104.6	71.2	84.3	90.7	42.5	61.3	104.6	60.5	77.7	90.6	60.5	72.2	90.7	60.5	72.3	88.0	125.0	110.6
09	FINISHES	158.6	85.1	118.7	93.1	75.5	83.5	159.8	80.6	116.8	96.1	77.0	85.8	97.2	78.2	86.8	85.8	111.5	99.8
COVERS	DIVS. 10 - 14, 25, 28, 41, 43, 44, 46	100.0	98.4	99.6	100.0	89.4	97.7	100.0	98.2	99.6	100.0	89.2	97.7	100.0	89.4	97.7	100.0	104.7	101.0
21, 22, 23	FIRE SUPPRESSION, PLUMBING & HVAC	100.1	86.1	94.4	102.4	79.0	92.8	102.1	85.7	95.4	100.8	72.4	89.2	100.8	71.1	88.6	96.4	97.6	96.9
26, 27, 3370	ELECTRICAL, COMMUNICATIONS & UTIL.	86.8	81.6	84.3	86.6	70.7	79.0	85.7	84.6	85.2	97.1	70.7	84.4	87.4	70.7	79.4	94.1	83.7	89.1
MF2018	WEIGHTED AVERAGE	104.3	86.0	96.9	98.5	79.9	90.9	104.7	85.6	97.0	100.5	78.2	91.4	101.0	78.1	91.7	93.4	103.0	97.3

City Cost Indexes

DIVISION		CARBONDALE 629 MAT.	INST.	TOTAL	CENTRALIA 628 MAT.	INST.	TOTAL	CHAMPAIGN 618 - 619 MAT.	INST.	TOTAL	CHICAGO 606 - 608 MAT.	INST.	TOTAL	DECATUR 625 MAT.	INST.	TOTAL	EAST ST. LOUIS 620 - 622 MAT.	INST.	TOTAL
015433	CONTRACTOR EQUIPMENT		107.0	107.0		107.0	107.0		99.8	99.8		100.0	100.0		99.8	99.8		107.0	107.0
0241, 31 - 34	SITE & INFRASTRUCTURE, DEMOLITION	98.6	91.9	94.0	98.9	93.6	95.3	98.4	92.2	94.2	91.8	104.1	100.2	94.6	92.5	93.2	101.5	93.5	96.1
0310	Concrete Forming & Accessories	88.6	102.3	99.6	91.0	107.3	104.2	86.1	109.7	105.1	93.5	160.2	147.3	93.1	116.4	111.9	85.1	113.1	107.7
0320	Concrete Reinforcing	102.4	96.2	100.4	102.4	96.8	100.6	99.9	94.6	98.2	99.0	151.6	115.9	100.1	99.5	99.9	102.4	103.0	102.6
0330	Cast-in-Place Concrete	86.0	96.1	89.5	86.5	111.9	95.2	102.4	103.7	102.8	107.9	155.0	124.2	94.0	112.0	100.2	87.9	116.9	97.9
03	CONCRETE	83.7	100.7	90.5	84.2	108.7	94.0	100.2	105.8	102.4	103.4	156.3	124.6	92.3	112.7	100.4	84.9	114.2	96.6
04	MASONRY	88.2	104.1	98.2	88.3	114.8	105.0	148.3	117.1	128.6	99.6	165.2	140.9	83.5	120.4	106.8	88.6	123.2	110.4
05	METALS	95.5	126.8	102.8	95.6	129.6	103.6	90.4	117.1	96.6	97.6	148.1	109.4	100.4	121.2	105.3	96.4	134.0	105.2
06	WOOD, PLASTICS & COMPOSITES	88.0	99.9	92.9	91.2	104.9	96.8	85.6	107.5	94.6	101.2	159.6	125.4	91.2	115.9	101.4	84.6	110.1	95.2
07	THERMAL & MOISTURE PROTECTION	93.4	93.8	93.6	93.5	103.4	97.8	97.3	107.1	101.6	88.7	149.6	115.4	98.9	111.0	104.2	93.4	110.7	101.0
08	OPENINGS	84.2	109.4	90.3	84.2	112.2	90.9	88.7	110.9	94.1	101.6	169.7	118.0	94.3	116.9	99.8	84.2	114.8	91.6
0920	Plaster & Gypsum Board	83.8	100.2	94.1	85.5	105.3	97.9	80.4	108.0	97.7	85.6	161.2	132.9	87.4	116.7	105.7	81.9	110.7	99.9
0950, 0980	Ceilings & Acoustic Treatment	87.4	100.2	95.7	87.4	105.3	99.0	86.5	108.0	100.5	79.8	161.2	132.6	96.3	116.7	109.5	87.4	110.7	102.5
0960	Flooring	118.5	108.5	115.5	119.3	108.5	116.0	89.0	108.5	94.9	92.9	162.2	114.1	106.7	113.2	108.7	117.6	102.9	113.1
0970, 0990	Wall Finishes & Painting/Coating	98.7	97.4	97.9	98.7	103.6	101.7	88.0	106.6	99.3	98.7	164.7	138.9	92.2	109.3	102.6	98.7	110.3	105.8
09	FINISHES	93.0	103.7	98.8	93.5	107.2	100.9	87.6	109.5	99.5	89.3	161.9	128.7	94.2	115.7	105.9	92.5	110.7	102.4
COVERS	DIVS. 10 - 14, 25, 28, 41, 43, 44, 46	100.0	98.4	99.7	100.0	97.4	99.4	100.0	105.3	101.2	100.0	127.5	106.0	100.0	106.2	101.3	100.0	113.2	102.9
21, 22, 23	FIRE SUPPRESSION, PLUMBING & HVAC	96.4	98.9	97.5	96.4	90.6	94.0	96.4	99.1	97.5	100.5	133.4	114.0	100.7	96.0	98.8	100.7	95.6	98.6
26, 27, 3370	ELECTRICAL, COMMUNICATIONS & UTIL.	94.0	99.9	96.8	94.7	98.5	96.5	96.0	88.4	92.4	95.1	135.3	114.5	96.4	96.2	96.3	94.5	101.7	98.0
MF2018	WEIGHTED AVERAGE	92.7	102.7	96.8	93.0	104.1	97.5	96.7	103.8	99.6	98.5	144.8	117.3	96.9	107.1	101.1	94.0	109.0	100.1

DIVISION		EFFINGHAM 624 MAT.	INST.	TOTAL	GALESBURG 614 MAT.	INST.	TOTAL	JOLIET 604 MAT.	INST.	TOTAL	KANKAKEE 609 MAT.	INST.	TOTAL	LA SALLE 613 MAT.	INST.	TOTAL	NORTH SUBURBAN 600 - 603 MAT.	INST.	TOTAL
015433	CONTRACTOR EQUIPMENT		99.8	99.8		98.9	98.9		90.6	90.6		90.6	90.6		98.9	98.9		90.6	90.6
0241, 31 - 34	SITE & INFRASTRUCTURE, DEMOLITION	97.6	91.1	93.1	92.6	91.2	91.6	87.6	94.2	92.1	82.2	92.0	88.9	92.2	91.9	92.0	87.1	92.8	91.0
0310	Concrete Forming & Accessories	99.0	107.7	106.0	85.9	110.1	105.4	91.1	162.9	149.1	83.0	136.5	126.2	104.9	117.8	115.3	90.2	149.0	137.6
0320	Concrete Reinforcing	103.0	87.3	98.0	99.6	98.3	99.2	109.3	140.8	119.4	109.7	125.7	114.8	99.7	123.7	107.4	109.3	138.7	118.8
0330	Cast-in-Place Concrete	93.6	102.8	96.8	90.9	102.4	94.9	98.1	152.5	116.9	91.4	126.9	103.7	90.8	114.0	98.8	98.1	143.0	113.6
03	CONCRETE	93.3	103.3	97.3	92.4	106.5	98.0	98.7	154.8	121.1	93.4	131.1	108.5	93.7	118.5	103.6	98.6	144.7	117.1
04	MASONRY	92.7	107.7	102.1	121.3	114.4	116.9	98.0	165.1	140.3	95.0	137.3	121.7	121.3	121.2	121.2	95.6	148.3	128.8
05	METALS	97.5	111.8	100.9	90.4	124.2	98.3	89.6	142.3	101.9	89.6	129.5	98.9	90.5	138.1	101.6	90.7	137.7	101.6
06	WOOD, PLASTICS & COMPOSITES	94.4	107.5	99.8	85.4	108.8	95.1	91.5	163.6	121.4	84.3	136.5	105.9	104.4	116.2	109.3	91.2	149.3	115.2
07	THERMAL & MOISTURE PROTECTION	98.4	101.1	99.6	96.7	102.0	99.0	94.4	147.1	117.5	93.4	129.2	109.1	97.2	113.6	104.4	94.7	136.8	113.2
08	OPENINGS	89.8	107.8	94.1	88.2	111.2	93.8	95.4	168.5	113.0	89.0	147.0	103.0	88.2	125.4	97.2	95.4	158.0	110.5
0920	Plaster & Gypsum Board	88.4	108.0	100.7	80.4	109.4	98.5	76.3	165.7	132.2	75.6	137.9	114.5	93.0	117.0	108.0	83.6	151.0	125.7
0950, 0980	Ceilings & Acoustic Treatment	87.4	108.0	100.8	86.5	109.4	101.4	80.3	165.7	135.7	80.3	137.9	117.6	86.5	117.0	106.3	80.3	151.0	126.2
0960	Flooring	107.8	108.5	108.0	88.8	112.4	96.0	91.5	162.0	113.1	89.3	140.2	104.9	93.9	116.1	100.7	91.8	148.4	109.1
0970, 0990	Wall Finishes & Painting/Coating	92.2	101.4	97.8	88.0	91.4	90.0	87.7	164.7	134.7	87.7	125.9	111.0	88.0	123.4	109.6	89.7	152.1	127.8
09	FINISHES	92.8	107.9	101.0	87.0	109.0	99.0	85.3	164.2	128.1	84.3	136.6	112.7	90.5	116.9	104.8	86.7	150.0	121.1
COVERS	DIVS. 10 - 14, 25, 28, 41, 43, 44, 46	100.0	103.9	100.9	100.0	104.9	101.1	100.0	127.0	105.9	100.0	110.6	102.3	100.0	105.9	101.3	100.0	123.4	105.1
21, 22, 23	FIRE SUPPRESSION, PLUMBING & HVAC	96.5	96.8	96.6	96.4	98.0	97.1	100.6	135.2	114.8	96.3	120.0	106.1	96.4	119.0	105.7	100.5	126.2	111.0
26, 27, 3370	ELECTRICAL, COMMUNICATIONS & UTIL.	95.2	99.9	97.4	94.6	80.1	87.6	94.1	135.8	114.2	90.8	127.0	108.2	92.9	127.0	109.4	94.3	125.6	109.4
MF2018	WEIGHTED AVERAGE	95.4	102.5	98.3	94.1	102.5	97.5	95.1	143.9	115.0	92.1	126.3	106.0	94.6	119.3	104.7	95.3	134.1	111.1

DIVISION		PEORIA 615 - 616 MAT.	INST.	TOTAL	QUINCY 623 MAT.	INST.	TOTAL	ROCK ISLAND 612 MAT.	INST.	TOTAL	ROCKFORD 610 - 611 MAT.	INST.	TOTAL	SOUTH SUBURBAN 605 MAT.	INST.	TOTAL	SPRINGFIELD 626 - 627 MAT.	INST.	TOTAL
015433	CONTRACTOR EQUIPMENT		98.9	98.9		99.8	99.8		98.9	98.9		98.9	98.9		90.6	90.6		106.0	106.0
0241, 31 - 34	SITE & INFRASTRUCTURE, DEMOLITION	93.7	92.0	92.5	96.3	91.5	93.0	91.0	90.0	90.3	93.3	93.6	93.5	87.1	92.7	90.9	98.5	102.9	101.5
0310	Concrete Forming & Accessories	90.8	118.9	113.5	95.7	111.1	108.2	88.0	93.1	92.1	95.8	134.6	127.1	90.2	156.3	143.6	94.3	117.1	112.7
0320	Concrete Reinforcing	97.2	104.5	99.5	102.8	80.2	95.6	99.6	92.1	97.2	91.4	136.5	105.9	109.3	138.7	118.8	99.9	99.6	99.8
0330	Cast-in-Place Concrete	88.2	116.0	97.8	93.8	99.7	95.8	89.0	92.0	90.0	90.3	132.4	104.8	98.1	142.9	113.6	89.3	110.9	96.8
03	CONCRETE	90.0	116.2	100.5	93.0	102.5	96.8	90.7	93.8	92.0	90.0	134.8	108.0	98.6	148.0	118.4	91.0	112.4	99.6
04	MASONRY	119.5	123.9	122.3	117.5	105.8	110.1	121.0	93.6	103.8	93.4	146.3	126.8	95.6	148.2	128.8	96.1	121.1	111.9
05	METALS	93.0	126.2	100.7	97.6	109.6	100.4	90.4	115.2	96.2	93.0	147.1	105.6	90.7	137.5	101.6	98.2	118.5	102.9
06	WOOD, PLASTICS & COMPOSITES	95.4	117.2	104.4	91.0	113.4	100.3	87.4	92.0	89.3	95.4	131.9	110.5	91.2	159.2	119.3	89.6	116.2	100.6
07	THERMAL & MOISTURE PROTECTION	97.6	112.7	104.2	98.3	99.6	98.9	96.7	91.2	94.3	100.6	133.5	115.0	94.7	137.9	113.6	104.2	113.5	108.3
08	OPENINGS	93.1	121.2	99.9	90.4	108.7	94.9	88.2	97.2	90.4	93.1	141.5	104.8	95.4	163.5	111.9	95.0	117.1	100.3
0920	Plaster & Gypsum Board	87.5	118.0	106.6	85.5	114.1	103.4	80.4	92.1	87.7	87.5	133.1	116.0	83.6	161.2	132.1	87.1	116.7	105.6
0950, 0980	Ceilings & Acoustic Treatment	92.7	118.0	109.1	87.4	114.1	104.7	86.5	92.1	90.1	92.7	133.1	118.9	80.3	161.2	132.8	99.0	116.7	110.5
0960	Flooring	91.5	126.1	102.1	106.7	105.6	106.4	89.6	88.0	89.1	91.5	130.1	103.3	91.8	148.4	109.1	107.1	125.5	112.7
0970, 0990	Wall Finishes & Painting/Coating	88.0	131.9	114.8	92.2	103.4	99.0	88.0	88.6	88.4	88.0	142.0	120.9	89.7	152.1	127.8	98.7	111.0	106.2
09	FINISHES	90.3	122.0	107.5	92.0	110.4	102.0	87.2	91.4	89.5	90.3	134.4	114.3	86.7	155.9	124.3	98.2	118.8	109.4
COVERS	DIVS. 10 - 14, 25, 28, 41, 43, 44, 46	100.0	106.9	101.5	100.0	103.7	100.8	100.0	99.7	99.8	100.0	112.0	102.6	100.0	124.5	105.3	100.0	107.1	101.6
21, 22, 23	FIRE SUPPRESSION, PLUMBING & HVAC	100.6	99.5	100.2	96.5	95.3	96.0	96.4	90.3	93.9	100.8	118.9	108.2	100.5	126.2	111.0	100.7	99.7	100.3
26, 27, 3370	ELECTRICAL, COMMUNICATIONS & UTIL.	95.2	93.6	94.4	93.8	76.2	85.3	89.9	86.3	88.2	95.5	132.0	113.1	94.3	125.6	109.4	102.9	88.0	95.7
MF2018	WEIGHTED AVERAGE	96.0	109.8	101.6	96.2	98.7	97.2	93.4	93.2	93.3	95.1	129.5	109.1	95.3	135.7	111.7	98.0	107.9	102.0

City Cost Indexes

INDIANA

DIVISION		ANDERSON 460			BLOOMINGTON 474			COLUMBUS 472			EVANSVILLE 476 - 477			FORT WAYNE 467 - 468			GARY 463 - 464		
		MAT.	INST.	TOTAL	MAT.	INST.	TOTAL	MAT.	INST.	TOTAL	MAT.	INST.	TOTAL	MAT.	INST.	TOTAL	MAT.	INST.	TOTAL
015433	CONTRACTOR EQUIPMENT		91.8	91.8		79.3	79.3		79.3	79.3		107.4	107.4		91.8	91.8		91.8	91.8
0241, 31 - 34	SITE & INFRASTRUCTURE, DEMOLITION	91.5	85.0	87.1	77.4	84.0	81.9	73.8	83.7	80.5	82.0	110.6	101.4	93.1	85.2	87.8	92.1	88.6	89.7
0310	Concrete Forming & Accessories	90.1	77.1	79.6	101.5	81.6	85.4	93.6	77.6	80.7	92.1	81.9	83.9	88.0	81.1	82.4	90.2	106.5	103.3
0320	Concrete Reinforcing	101.2	82.0	95.0	92.0	85.1	89.8	92.5	83.0	89.4	100.6	97.4	99.6	101.2	79.7	94.3	101.2	109.9	104.0
0330	Cast-in-Place Concrete	89.4	73.6	84.0	97.0	74.7	89.3	96.5	71.6	87.9	92.7	83.8	89.6	95.0	73.1	87.4	93.5	106.9	98.1
03	CONCRETE	89.7	77.4	84.8	97.1	79.5	90.1	96.3	76.2	88.3	97.7	85.5	92.8	92.1	78.6	86.7	91.5	107.2	97.8
04	MASONRY	84.5	74.7	78.3	91.2	72.2	79.2	91.0	71.2	78.5	86.7	81.0	83.1	89.5	72.9	79.0	85.9	105.1	98.0
05	METALS	95.8	88.9	94.2	94.1	76.5	90.0	94.2	74.8	89.6	87.8	91.9	88.8	95.8	88.2	94.0	95.8	105.8	98.1
06	WOOD, PLASTICS & COMPOSITES	89.6	77.5	84.6	108.3	82.7	97.7	101.5	78.2	91.9	90.5	81.0	86.6	89.4	84.1	87.2	86.5	104.9	94.1
07	THERMAL & MOISTURE PROTECTION	111.3	73.5	94.7	97.3	76.4	88.1	96.7	74.3	86.9	101.8	82.8	93.5	111.0	78.0	96.6	110.2	100.4	105.9
08	OPENINGS	88.9	75.6	85.7	93.5	79.5	90.1	90.1	76.4	86.8	88.2	82.0	86.7	88.9	77.9	86.2	88.9	110.0	94.0
0920	Plaster & Gypsum Board	97.6	77.4	84.9	100.3	83.1	89.5	94.7	78.5	84.6	92.1	80.4	84.7	96.6	84.1	88.8	86.0	105.6	98.2
0950, 0980	Ceilings & Acoustic Treatment	90.5	77.4	82.0	80.6	83.1	82.2	80.6	78.5	79.2	84.4	80.4	81.8	90.5	84.1	86.4	90.5	105.6	100.3
0960	Flooring	93.0	74.1	87.2	99.6	79.8	93.6	95.3	79.8	90.6	95.7	74.1	89.1	93.0	74.8	87.4	93.0	104.9	96.6
0970, 0990	Wall Finishes & Painting/Coating	90.8	66.1	75.8	87.4	80.1	82.9	87.4	77.7	81.5	92.6	86.0	88.5	90.8	71.2	78.9	90.8	114.2	105.1
09	FINISHES	90.8	75.3	82.4	93.1	81.2	86.7	91.0	78.1	84.0	91.6	80.5	85.6	90.5	79.5	84.6	88.8	106.8	98.6
COVERS	DIVS. 10 - 14, 25, 28, 41, 43, 44, 46	100.0	89.3	97.7	100.0	89.4	97.7	100.0	88.7	97.5	100.0	92.2	98.3	100.0	90.6	98.0	100.0	99.9	100.0
21, 22, 23	FIRE SUPPRESSION, PLUMBING & HVAC	100.9	74.6	90.1	100.4	77.4	91.0	96.2	74.8	87.4	100.7	80.3	92.3	100.9	73.4	89.6	100.9	100.3	100.7
26, 27, 3370	ELECTRICAL, COMMUNICATIONS & UTIL.	85.0	81.8	83.5	101.9	84.4	93.5	101.4	83.9	92.9	96.4	83.3	90.1	86.1	77.8	82.1	95.8	104.0	99.8
MF2018	WEIGHTED AVERAGE	94.0	78.7	87.8	96.7	79.6	89.7	95.0	77.6	87.9	94.2	85.4	90.7	94.6	78.7	88.2	95.2	103.1	98.4

INDIANA

DIVISION		INDIANAPOLIS 461 - 462			KOKOMO 469			LAFAYETTE 479			LAWRENCEBURG 470			MUNCIE 473			NEW ALBANY 471		
		MAT.	INST.	TOTAL	MAT.	INST.	TOTAL	MAT.	INST.	TOTAL	MAT.	INST.	TOTAL	MAT.	INST.	TOTAL	MAT.	INST.	TOTAL
015433	CONTRACTOR EQUIPMENT		90.5	90.5		91.8	91.8		79.3	79.3		97.9	97.9		90.3	90.3		88.2	88.2
0241, 31 - 34	SITE & INFRASTRUCTURE, DEMOLITION	91.4	96.3	94.8	87.7	85.0	85.9	74.4	83.7	80.7	74.4	97.6	90.1	78.1	84.1	82.2	70.5	85.9	80.9
0310	Concrete Forming & Accessories	92.7	85.3	86.7	93.9	75.5	79.0	90.1	79.9	81.9	88.9	75.7	78.3	88.7	76.6	79.0	89.9	74.4	77.4
0320	Concrete Reinforcing	100.6	85.5	95.8	91.5	83.3	88.9	92.0	83.2	89.2	91.7	76.2	86.8	101.8	82.0	95.5	92.6	78.6	88.1
0330	Cast-in-Place Concrete	86.9	85.9	86.6	88.5	78.9	85.2	97.1	76.5	90.0	90.8	71.0	83.9	101.9	72.8	91.9	93.7	70.9	85.8
03	CONCRETE	93.9	85.0	90.3	86.3	78.7	83.2	96.3	79.0	89.4	90.5	74.9	84.2	96.7	76.9	88.8	95.1	74.3	86.8
04	MASONRY	93.6	79.4	84.7	84.1	74.0	77.8	96.0	74.2	82.3	75.8	71.1	72.9	93.3	74.7	81.6	82.4	68.5	73.7
05	METALS	96.8	76.1	92.0	92.0	89.2	91.4	92.5	75.4	88.5	89.4	85.5	88.5	95.8	88.8	94.2	91.2	81.6	88.9
06	WOOD, PLASTICS & COMPOSITES	95.0	85.9	91.2	93.1	74.9	85.6	97.9	81.2	91.0	87.8	76.0	82.9	98.2	77.1	89.5	90.8	75.3	84.4
07	THERMAL & MOISTURE PROTECTION	101.2	81.4	92.5	110.5	74.2	94.6	96.6	76.6	87.8	100.1	73.7	88.5	99.6	74.8	88.7	91.7	71.2	82.7
08	OPENINGS	102.3	81.9	97.4	84.5	74.6	82.1	88.9	78.0	86.3	89.9	73.1	85.8	87.3	75.4	84.4	87.7	71.6	83.9
0920	Plaster & Gypsum Board	86.6	85.5	85.9	105.4	74.7	86.2	91.2	81.6	85.2	73.6	76.2	75.2	92.1	77.4	82.9	87.5	75.3	79.8
0950, 0980	Ceilings & Acoustic Treatment	88.7	85.5	86.6	91.2	74.7	80.5	76.5	81.6	79.8	89.9	76.2	81.0	80.6	77.4	78.5	84.4	75.3	78.5
0960	Flooring	93.2	82.2	89.9	96.3	85.7	93.0	94.3	79.8	89.9	70.1	79.8	73.1	95.8	74.1	89.2	92.5	50.3	79.6
0970, 0990	Wall Finishes & Painting/Coating	98.7	80.1	87.3	90.8	65.5	75.4	87.4	80.1	83.0	87.8	69.9	76.9	87.4	66.1	74.4	92.6	62.8	74.4
09	FINISHES	91.6	84.4	87.7	93.0	76.2	83.9	89.3	80.1	84.3	82.7	76.0	79.1	90.4	75.0	82.0	90.2	68.5	78.4
COVERS	DIVS. 10 - 14, 25, 28, 41, 43, 44, 46	100.0	92.6	98.4	100.0	89.2	97.6	100.0	88.9	97.6	100.0	86.2	97.0	100.0	88.4	97.5	100.0	86.0	97.0
21, 22, 23	FIRE SUPPRESSION, PLUMBING & HVAC	100.8	78.1	91.5	96.7	75.0	87.8	96.2	75.7	87.8	97.3	71.7	86.8	100.4	74.5	89.7	96.5	74.6	87.5
26, 27, 3370	ELECTRICAL, COMMUNICATIONS & UTIL.	101.6	86.6	94.3	87.5	74.6	81.3	101.1	77.3	89.6	94.4	71.2	83.2	92.2	72.8	82.9	95.6	72.0	84.2
MF2018	WEIGHTED AVERAGE	97.9	83.2	91.9	91.8	78.0	86.2	94.6	78.0	87.8	91.4	76.4	85.3	95.3	77.2	87.9	92.8	74.5	85.3

INDIANA / IOWA

DIVISION		SOUTH BEND 465 - 466			TERRE HAUTE 478			WASHINGTON 475			BURLINGTON 526			CARROLL 514			CEDAR RAPIDS 522 - 524		
		MAT.	INST.	TOTAL	MAT.	INST.	TOTAL	MAT.	INST.	TOTAL	MAT.	INST.	TOTAL	MAT.	INST.	TOTAL	MAT.	INST.	TOTAL
015433	CONTRACTOR EQUIPMENT		109.6	109.6		107.4	107.4		107.4	107.4		95.6	95.6		95.6	95.6		92.6	92.6
0241, 31 - 34	SITE & INFRASTRUCTURE, DEMOLITION	91.3	97.7	95.6	83.2	110.5	101.7	82.1	110.7	101.5	86.5	87.0	86.9	77.8	87.8	84.6	88.0	87.3	87.6
0310	Concrete Forming & Accessories	95.9	76.1	79.9	93.5	76.8	80.0	94.6	79.7	82.6	93.3	88.5	89.4	76.3	78.9	78.4	101.3	83.8	87.2
0320	Concrete Reinforcing	99.7	81.0	93.7	100.6	82.1	94.7	93.1	83.0	89.9	95.1	91.0	93.8	95.8	79.9	90.7	95.8	97.1	96.2
0330	Cast-in-Place Concrete	89.5	79.0	85.9	89.7	77.0	85.3	97.6	83.7	92.8	118.2	51.3	95.1	118.2	76.8	103.9	118.5	82.2	106.0
03	CONCRETE	96.8	79.4	89.8	99.9	78.2	91.2	104.1	82.0	95.2	102.1	76.6	91.9	100.6	79.2	92.0	102.5	86.4	96.0
04	MASONRY	85.5	73.5	77.9	94.3	73.5	81.2	86.9	76.3	80.2	101.7	67.2	79.9	103.0	67.7	80.7	107.3	80.6	90.5
05	METALS	99.6	100.5	99.8	88.5	85.5	87.8	92.4	86.5	83.4	82.8	96.3	85.9	82.8	92.7	85.1	85.2	101.3	89.0
06	WOOD, PLASTICS & COMPOSITES	93.7	75.4	86.2	93.2	76.6	86.3	93.6	79.0	87.6	90.9	92.6	91.6	74.4	83.4	78.1	99.6	83.6	93.0
07	THERMAL & MOISTURE PROTECTION	109.0	77.8	95.3	101.9	77.8	91.3	102.0	80.2	92.4	100.4	71.4	87.7	100.5	73.0	88.4	101.6	79.0	91.7
08	OPENINGS	89.2	74.8	85.7	88.6	74.8	85.3	86.0	78.5	84.2	91.5	90.2	91.2	95.5	79.3	91.6	95.8	85.8	93.4
0920	Plaster & Gypsum Board	90.0	74.8	80.5	92.1	75.8	81.9	93.2	78.3	83.9	97.3	92.8	94.4	90.5	83.3	86.0	106.0	83.7	92.1
0950, 0980	Ceilings & Acoustic Treatment	93.5	74.8	81.4	84.4	75.8	78.9	79.6	78.3	78.8	102.2	92.8	96.1	102.2	83.3	90.0	105.0	83.7	91.2
0960	Flooring	96.8	85.1	93.2	95.7	74.7	89.3	96.5	79.8	91.4	97.2	64.3	87.1	92.2	74.0	86.6	111.6	85.7	103.7
0970, 0990	Wall Finishes & Painting/Coating	98.7	81.2	88.0	92.6	77.0	83.1	92.6	84.3	87.5	93.6	78.5	84.4	93.6	78.5	84.4	95.2	70.4	80.1
09	FINISHES	94.2	78.0	85.4	91.6	76.2	83.2	91.2	79.9	85.1	95.9	84.1	89.5	92.2	78.4	84.7	101.9	82.7	91.5
COVERS	DIVS. 10 - 14, 25, 28, 41, 43, 44, 46	100.0	91.4	98.1	100.0	90.6	97.9	100.0	91.6	98.2	100.0	93.2	98.5	100.0	89.2	97.6	100.0	92.4	98.3
21, 22, 23	FIRE SUPPRESSION, PLUMBING & HVAC	100.8	73.8	89.8	100.7	76.0	90.6	96.5	77.8	88.8	96.6	77.6	88.8	96.6	72.1	86.6	100.9	82.1	93.2
26, 27, 3370	ELECTRICAL, COMMUNICATIONS & UTIL.	98.2	85.5	92.1	95.4	82.1	89.0	95.7	80.6	88.4	98.9	67.0	83.5	99.3	71.3	85.8	97.5	77.2	87.7
MF2018	WEIGHTED AVERAGE	97.4	81.7	91.0	95.0	80.9	89.3	92.9	82.8	88.8	94.5	79.2	88.3	94.1	77.4	87.3	97.1	84.4	91.9

City Cost Indexes

IOWA

	DIVISION	COUNCIL BLUFFS 515			CRESTON 508			DAVENPORT 527 - 528			DECORAH 521			DES MOINES 500 - 503,509			DUBUQUE 520		
		MAT.	INST.	TOTAL	MAT.	INST.	TOTAL	MAT.	INST.	TOTAL	MAT.	INST.	TOTAL	MAT.	INST.	TOTAL	MAT.	INST.	TOTAL
015433	CONTRACTOR EQUIPMENT		91.9	91.9		95.6	95.6		95.6	95.6		95.6	95.6		102.1	102.1		91.5	91.5
0241, 31 - 34	SITE & INFRASTRUCTURE, DEMOLITION	89.0	84.4	85.9	82.5	88.7	86.7	87.5	90.8	89.7	85.3	86.8	86.3	86.4	99.8	95.5	86.2	84.8	85.2
0310	Concrete Forming & Accessories	75.2	74.6	74.7	69.9	81.5	79.3	100.6	98.1	98.6	89.8	67.5	71.8	93.1	88.0	89.0	77.4	81.2	80.5
0320	Concrete Reinforcing	97.7	82.6	92.9	90.1	79.9	86.8	95.8	97.6	96.4	95.1	79.3	90.0	99.3	100.2	99.6	94.4	78.9	89.5
0330	Cast-in-Place Concrete	123.2	85.8	110.3	121.4	80.1	107.1	114.1	94.0	107.2	115.0	71.1	99.8	103.4	91.6	99.3	116.0	81.6	104.2
03	CONCRETE	104.1	80.9	94.8	100.2	81.5	92.7	100.5	97.2	99.2	100.0	72.0	88.8	99.2	91.9	96.3	98.7	81.8	91.9
04	MASONRY	106.8	76.0	87.4	101.5	74.6	84.5	102.6	92.7	96.4	122.4	64.6	86.0	86.9	86.0	86.3	108.3	68.2	83.0
05	METALS	89.1	90.5	90.5	82.1	93.3	84.7	85.2	107.1	90.3	82.9	91.6	84.9	87.3	102.2	90.8	83.6	92.8	85.7
06	WOOD, PLASTICS & COMPOSITES	72.9	73.4	73.1	66.7	83.4	73.6	99.6	98.0	98.9	87.2	66.9	78.8	88.3	86.6	87.6	75.0	82.2	78.0
07	THERMAL & MOISTURE PROTECTION	100.5	73.4	88.6	109.8	76.1	94.9	101.0	91.0	96.6	100.6	64.8	84.9	101.8	86.2	94.9	100.9	74.6	89.3
08	OPENINGS	95.0	78.2	91.0	104.6	80.9	98.9	95.8	99.6	96.7	94.2	73.3	89.1	97.2	93.3	96.3	95.1	81.2	92.0
0920	Plaster & Gypsum Board	90.5	73.2	79.7	93.2	83.3	87.0	106.0	98.3	101.2	95.6	66.3	77.3	91.7	86.4	88.4	90.5	82.3	85.3
0950, 0980	Ceilings & Acoustic Treatment	102.2	73.2	83.4	90.8	83.3	85.9	105.0	98.3	100.6	102.2	66.3	78.9	94.2	86.4	89.1	102.2	82.3	89.3
0960	Flooring	91.2	83.2	88.8	80.2	64.3	75.3	99.2	88.3	95.9	96.9	64.3	86.9	90.0	90.5	90.1	103.8	69.6	93.4
0970, 0990	Wall Finishes & Painting/Coating	90.9	55.5	69.3	86.5	78.5	81.6	93.6	90.6	91.8	93.6	78.5	84.4	98.7	85.7	90.8	94.8	77.3	84.1
09	FINISHES	93.0	73.7	82.5	85.7	78.1	81.6	98.5	95.5	96.9	95.5	67.8	80.5	92.7	88.0	90.1	96.5	78.6	86.8
COVERS	DIVS. 10 - 14, 25, 28, 41, 43, 44, 46	100.0	89.9	97.8	100.0	91.5	98.1	100.0	96.5	99.2	100.0	88.7	97.5	100.0	94.8	98.9	100.0	91.3	98.1
21, 22, 23	FIRE SUPPRESSION, PLUMBING & HVAC	100.9	77.5	91.3	96.6	73.9	87.3	100.9	93.0	97.6	96.6	69.0	85.3	100.7	88.5	95.7	100.9	74.2	89.9
26, 27, 3370	ELECTRICAL, COMMUNICATIONS & UTIL.	100.6	79.0	90.2	92.1	71.3	82.1	94.9	84.8	90.0	97.4	43.8	71.6	101.5	87.9	95.0	99.7	74.1	87.4
MF2018	WEIGHTED AVERAGE	97.3	79.9	90.2	93.7	79.1	87.7	96.1	94.2	95.3	95.1	69.3	84.6	95.9	91.0	93.9	95.7	78.7	88.7

IOWA

	DIVISION	FORT DODGE 505			MASON CITY 504			OTTUMWA 525			SHENANDOAH 516			SIBLEY 512			SIOUX CITY 510 - 511		
		MAT.	INST.	TOTAL	MAT.	INST.	TOTAL	MAT.	INST.	TOTAL	MAT.	INST.	TOTAL	MAT.	INST.	TOTAL	MAT.	INST.	TOTAL
015433	CONTRACTOR EQUIPMENT		95.6	95.6		95.6	95.6		91.5	91.5		91.9	91.9		95.6	95.6		95.6	95.6
0241, 31 - 34	SITE & INFRASTRUCTURE, DEMOLITION	88.0	85.7	86.5	88.1	86.7	87.2	85.7	82.3	83.4	87.1	84.2	85.1	94.4	86.8	89.2	96.8	88.9	91.4
0310	Concrete Forming & Accessories	70.7	73.3	72.8	75.9	67.1	68.8	86.9	82.0	82.9	77.6	72.6	73.6	78.4	35.6	43.9	101.3	73.4	78.8
0320	Concrete Reinforcing	90.1	79.3	86.6	91.0	79.3	86.5	95.1	91.2	93.9	97.7	79.9	92.0	95.8	95.7	95.8	101.3	97.5	95.8
0330	Cast-in-Place Concrete	113.9	39.1	88.1	113.9	66.6	97.6	119.0	59.9	98.6	119.1	77.5	104.8	116.7	51.9	94.3	117.5	85.0	106.3
03	CONCRETE	95.8	63.4	82.9	96.2	70.2	85.8	101.5	76.7	91.6	101.7	76.7	91.7	100.7	50.8	80.7	101.9	82.4	94.1
04	MASONRY	100.5	48.5	67.7	113.3	63.5	81.9	104.3	51.6	71.1	106.3	69.6	83.2	125.1	48.7	77.0	101.0	66.7	79.4
05	METALS	82.1	91.3	84.3	82.2	91.6	84.4	82.7	97.6	86.2	88.4	93.6	89.6	83.0	90.7	84.8	85.2	100.1	88.7
06	WOOD, PLASTICS & COMPOSITES	67.2	83.4	73.9	71.9	66.6	69.8	83.7	92.4	87.3	75.0	74.0	74.6	76.0	32.2	57.9	99.6	72.9	88.5
07	THERMAL & MOISTURE PROTECTION	109.1	60.4	87.8	108.7	65.3	89.7	101.2	65.1	85.4	99.9	67.9	85.9	100.1	51.1	78.6	101.0	69.4	87.2
08	OPENINGS	99.0	71.5	92.3	91.8	73.3	87.3	95.5	87.3	93.5	87.5	71.9	83.7	92.5	43.3	80.6	95.8	79.0	91.8
0920	Plaster & Gypsum Board	93.2	83.3	87.0	93.2	66.3	76.4	91.0	92.8	92.1	90.5	73.9	80.1	90.5	30.7	53.1	106.0	72.5	85.1
0950, 0980	Ceilings & Acoustic Treatment	90.8	83.3	85.9	90.8	66.3	74.9	102.2	92.8	96.1	102.2	73.9	83.8	102.2	30.7	55.8	105.0	72.5	83.9
0960	Flooring	81.3	64.3	76.1	82.8	64.3	77.1	106.5	64.3	93.6	91.8	68.8	84.8	93.0	64.3	84.2	99.3	69.6	90.2
0970, 0990	Wall Finishes & Painting/Coating	86.5	75.4	79.8	86.5	78.5	81.6	94.8	78.5	84.9	90.9	78.5	83.3	93.6	77.6	83.8	93.6	66.0	76.8
09	FINISHES	87.2	73.1	79.5	87.5	67.5	76.7	97.4	79.5	87.7	93.0	72.3	81.7	94.9	43.4	66.9	99.8	71.4	84.3
COVERS	DIVS. 10 - 14, 25, 28, 41, 43, 44, 46	100.0	85.2	96.8	100.0	88.3	97.5	100.0	86.8	97.1	100.0	88.1	97.4	100.0	79.8	95.6	100.0	89.9	97.8
21, 22, 23	FIRE SUPPRESSION, PLUMBING & HVAC	96.6	66.2	84.1	96.6	72.0	86.5	96.6	67.7	84.8	96.6	77.7	88.9	96.6	66.2	84.2	100.9	76.7	91.0
26, 27, 3370	ELECTRICAL, COMMUNICATIONS & UTIL.	95.8	64.3	80.6	95.3	43.8	70.4	98.8	65.8	82.9	97.4	72.7	85.5	97.5	43.8	71.6	97.5	71.3	84.9
MF2018	WEIGHTED AVERAGE	93.2	69.3	83.4	93.1	69.5	83.5	94.9	74.0	86.4	94.9	77.0	87.6	95.2	58.7	80.4	96.8	78.3	89.3

	DIVISION	IOWA SPENCER 513			WATERLOO 506 - 507			KANSAS BELLEVILLE 669			COLBY 677			DODGE CITY 678			EMPORIA 668		
		MAT.	INST.	TOTAL	MAT.	INST.	TOTAL	MAT.	INST.	TOTAL	MAT.	INST.	TOTAL	MAT.	INST.	TOTAL	MAT.	INST.	TOTAL
015433	CONTRACTOR EQUIPMENT		95.6	95.6		95.6	95.6		100.6	100.6		100.6	100.6		100.6	100.6		98.9	98.9
0241, 31 - 34	SITE & INFRASTRUCTURE, DEMOLITION	94.5	85.6	88.5	93.4	88.5	90.1	104.8	87.5	93.1	98.3	88.0	91.3	101.2	88.3	92.4	97.9	85.2	89.3
0310	Concrete Forming & Accessories	86.8	35.3	45.3	90.8	68.4	72.8	93.0	51.9	59.9	98.2	57.7	65.6	88.7	59.4	65.1	80.9	65.9	68.8
0320	Concrete Reinforcing	97.7	79.2	91.8	90.7	79.3	87.0	102.8	97.4	101.0	111.5	97.4	106.9	108.7	97.2	105.0	101.3	97.5	100.1
0330	Cast-in-Place Concrete	116.7	61.3	97.6	122.0	82.6	108.4	106.0	79.3	96.8	101.7	81.6	94.8	103.5	85.9	97.4	102.5	81.7	95.3
03	CONCRETE	101.3	53.9	82.3	102.1	76.4	91.8	104.5	70.9	91.0	103.7	74.4	92.0	104.1	76.6	93.1	97.9	78.1	90.0
04	MASONRY	125.1	48.7	77.0	101.5	74.5	84.5	95.0	56.6	70.8	97.7	60.3	74.1	113.8	58.3	78.8	100.8	61.7	76.1
05	METALS	83.0	90.4	84.7	84.3	93.6	86.5	91.0	97.3	92.5	88.5	98.1	90.7	89.5	97.8	91.4	90.7	98.3	92.5
06	WOOD, PLASTICS & COMPOSITES	83.5	32.2	62.3	87.7	64.3	78.0	90.9	48.3	73.3	97.7	54.1	79.7	87.5	54.9	74.0	79.9	65.0	73.7
07	THERMAL & MOISTURE PROTECTION	101.1	51.7	79.5	109.2	75.4	94.4	99.1	59.1	81.6	105.1	61.3	85.9	104.9	65.1	87.5	97.0	69.8	85.1
08	OPENINGS	102.8	43.3	88.4	92.0	72.8	87.4	90.4	59.2	82.9	94.3	62.4	86.6	94.2	62.9	86.7	88.6	70.4	84.2
0920	Plaster & Gypsum Board	91.0	30.7	53.3	106.2	63.6	79.6	84.1	47.1	61.0	96.9	53.1	69.5	85.5	54.0	65.8	78.3	64.3	69.6
0950, 0980	Ceilings & Acoustic Treatment	102.2	30.7	55.8	92.8	63.6	73.9	86.7	47.1	61.0	81.1	53.1	63.0	81.1	54.0	63.5	86.7	64.3	72.2
0960	Flooring	95.3	64.3	85.8	86.8	77.9	84.1	88.9	63.6	81.2	84.0	63.6	77.7	81.3	64.9	76.2	85.6	63.6	78.9
0970, 0990	Wall Finishes & Painting/Coating	93.6	53.1	68.9	86.5	82.4	84.0	91.7	95.6	94.1	97.1	95.6	96.2	97.1	95.6	96.2	91.7	95.6	94.1
09	FINISHES	95.6	39.6	65.2	91.4	70.6	80.1	87.7	57.3	71.1	86.9	61.6	73.1	84.4	62.8	72.7	84.8	68.0	75.7
COVERS	DIVS. 10 - 14, 25, 28, 41, 43, 44, 46	100.0	79.7	95.6	100.0	90.0	97.8	100.0	77.3	95.1	100.0	79.2	95.5	100.0	83.1	96.3	100.0	79.1	95.4
21, 22, 23	FIRE SUPPRESSION, PLUMBING & HVAC	96.6	66.2	84.2	100.8	78.8	91.8	96.1	66.1	83.8	96.5	65.9	83.9	100.7	74.2	89.9	96.1	66.1	84.6
26, 27, 3370	ELECTRICAL, COMMUNICATIONS & UTIL.	98.5	43.8	72.1	93.3	61.8	78.1	95.9	58.1	77.7	100.5	62.9	82.4	98.8	68.5	84.2	94.5	62.0	78.8
MF2018	WEIGHTED AVERAGE	96.5	58.5	81.1	95.2	76.5	87.6	95.6	67.6	84.2	96.0	70.1	85.5	97.4	73.2	87.5	94.0	72.4	85.2

City Cost Indexes

KANSAS

DIVISION		FORT SCOTT 667 MAT.	INST.	TOTAL	HAYS 676 MAT.	INST.	TOTAL	HUTCHINSON 675 MAT.	INST.	TOTAL	INDEPENDENCE 673 MAT.	INST.	TOTAL	KANSAS CITY 660-662 MAT.	INST.	TOTAL	LIBERAL 679 MAT.	INST.	TOTAL
015433	CONTRACTOR EQUIPMENT		99.7	99.7		100.6	100.6		100.6	100.6		100.6	100.6		97.5	97.5		100.6	100.6
0241, 31 - 34	SITE & INFRASTRUCTURE, DEMOLITION	95.1	85.3	88.5	102.1	87.8	92.4	84.5	88.1	87.0	101.4	88.2	92.4	91.7	86.5	88.1	101.8	87.8	92.3
0310	Concrete Forming & Accessories	104.6	78.0	83.1	94.8	56.3	63.8	81.0	53.5	58.8	110.1	63.9	72.8	99.6	99.8	99.8	89.2	55.6	62.1
0320	Concrete Reinforcing	100.6	93.8	98.4	108.7	97.4	105.1	108.7	97.4	105.1	108.1	93.8	103.5	97.8	103.8	99.7	110.3	97.1	106.0
0330	Cast-in-Place Concrete	95.1	77.2	88.9	80.2	79.6	80.0	74.3	81.5	76.8	103.9	81.6	96.2	81.4	97.0	86.8	80.2	78.3	79.6
03	CONCRETE	94.8	81.4	89.4	96.2	73.1	86.9	82.3	72.4	78.3	105.7	76.5	94.0	87.8	100.0	92.7	97.6	72.2	87.4
04	MASONRY	103.7	52.5	71.4	109.8	56.7	76.3	97.2	60.3	73.9	95.2	60.3	73.2	101.8	99.2	100.2	111.3	50.1	72.7
05	METALS	90.7	96.7	92.1	88.0	98.1	90.4	87.8	97.6	90.1	87.7	96.7	89.8	99.4	106.7	101.1	88.3	97.3	90.4
06	WOOD, PLASTICS & COMPOSITES	103.5	85.4	96.0	94.0	54.1	77.5	81.4	48.5	67.8	111.1	62.1	90.9	98.7	99.9	99.2	88.3	54.1	74.1
07	THERMAL & MOISTURE PROTECTION	98.4	67.5	84.8	105.4	59.8	85.4	103.6	60.3	84.6	105.3	69.4	89.6	97.1	98.4	97.7	105.5	57.1	84.3
08	OPENINGS	88.6	80.9	86.7	94.2	62.4	86.5	94.2	59.3	85.8	92.4	66.0	86.0	89.5	98.0	91.5	94.2	62.4	86.6
0920	Plaster & Gypsum Board	89.0	85.3	86.7	92.7	53.1	68.0	83.6	47.3	60.9	112.3	61.4	80.5	81.4	100.2	93.2	86.7	53.1	65.7
0950, 0980	Ceilings & Acoustic Treatment	86.7	85.3	85.8	81.1	53.1	63.0	81.1	47.3	59.2	81.1	61.4	68.3	86.7	100.2	95.5	81.1	53.1	63.0
0960	Flooring	97.6	62.2	86.7	83.1	63.6	77.1	79.2	63.6	74.4	87.0	62.2	79.4	77.8	93.6	82.6	81.5	63.6	76.0
0970, 0990	Wall Finishes & Painting/Coating	92.9	71.9	80.1	97.1	95.6	96.2	97.1	95.6	96.2	97.1	95.6	96.2	99.2	103.7	101.9	97.1	95.6	96.2
09	FINISHES	90.0	75.6	82.2	86.4	60.7	72.4	82.1	58.3	69.2	90.3	66.5	77.4	85.7	99.1	93.0	85.2	60.2	71.6
COVERS	DIVS. 10 - 14, 25, 28, 41, 43, 44, 46	100.0	79.9	95.6	100.0	78.0	95.2	100.0	78.6	95.3	100.0	80.1	95.7	100.0	95.9	99.1	100.0	77.3	95.1
21, 22, 23	FIRE SUPPRESSION, PLUMBING & HVAC	96.1	62.4	82.3	96.5	63.5	83.0	96.5	65.9	83.9	96.5	65.3	83.7	100.3	101.4	100.8	96.5	64.3	83.3
26, 27, 3370	ELECTRICAL, COMMUNICATIONS & UTIL.	94.2	62.9	79.1	100.0	62.9	82.1	97.3	56.5	77.6	98.4	67.5	83.5	96.7	96.1	96.4	98.8	62.9	81.5
MF2018	WEIGHTED AVERAGE	94.2	72.3	85.3	95.3	68.8	84.5	91.6	68.2	82.1	96.1	71.9	86.3	95.7	98.8	97.0	95.4	68.0	84.2

KANSAS / KENTUCKY

DIVISION		SALINA 674 MAT.	INST.	TOTAL	TOPEKA 664-666 MAT.	INST.	TOTAL	WICHITA 670-672 MAT.	INST.	TOTAL	ASHLAND 411-412 MAT.	INST.	TOTAL	BOWLING GREEN 421-422 MAT.	INST.	TOTAL	CAMPTON 413-414 MAT.	INST.	TOTAL
015433	CONTRACTOR EQUIPMENT		100.6	100.6		106.0	106.0		106.1	106.1		94.8	94.8		88.2	88.2		94.2	94.2
0241, 31 - 34	SITE & INFRASTRUCTURE, DEMOLITION	93.0	87.9	89.5	94.8	97.5	96.7	91.0	97.6	95.5	102.4	75.0	83.8	71.2	86.0	81.3	79.0	86.3	83.9
0310	Concrete Forming & Accessories	83.8	60.2	64.8	94.8	66.6	72.1	95.3	57.2	64.6	81.0	84.3	83.6	84.9	79.5	80.5	85.3	76.6	78.2
0320	Concrete Reinforcing	108.1	97.3	104.6	99.4	98.8	99.2	100.3	101.0	100.5	93.2	89.2	91.9	91.3	77.4	86.8	92.1	86.6	90.4
0330	Cast-in-Place Concrete	89.8	82.4	87.3	90.2	87.9	89.4	84.1	77.3	81.7	85.3	88.6	86.4	84.6	66.7	78.4	94.4	63.9	83.9
03	CONCRETE	93.4	75.8	86.4	92.2	80.6	87.6	89.7	73.0	83.2	90.8	87.9	89.6	89.7	75.1	83.9	93.1	74.4	85.6
04	MASONRY	128.7	53.7	81.4	97.1	63.4	75.9	95.6	54.4	69.6	93.3	85.9	88.7	95.5	67.8	78.0	92.3	51.1	66.3
05	METALS	89.3	98.4	91.4	94.6	97.5	95.3	91.6	96.9	92.8	90.5	105.5	94.0	91.7	84.6	90.0	91.2	88.0	90.4
06	WOOD, PLASTICS & COMPOSITES	83.2	59.3	73.3	95.7	65.4	83.1	100.1	54.8	81.4	72.5	82.7	76.7	84.1	81.9	83.2	82.2	83.3	82.7
07	THERMAL & MOISTURE PROTECTION	104.3	60.6	85.1	103.2	82.4	94.1	102.6	61.4	84.5	91.6	82.6	87.7	91.6	75.6	84.6	102.0	62.9	84.8
08	OPENINGS	94.2	65.8	87.3	99.3	73.2	93.0	97.2	64.1	89.2	86.8	84.1	86.1	87.8	77.7	85.4	89.0	82.2	87.3
0920	Plaster & Gypsum Board	83.6	58.5	67.9	90.4	64.3	74.1	89.0	53.4	66.8	59.8	82.7	74.1	80.0	82.0	81.2	80.0	82.7	81.7
0950, 0980	Ceilings & Acoustic Treatment	81.1	58.5	66.4	103.1	64.3	77.9	96.3	53.4	68.5	80.6	82.7	82.0	84.4	82.0	82.8	84.4	82.7	83.3
0960	Flooring	80.2	64.9	75.5	91.0	64.9	83.0	90.3	68.9	83.8	74.5	75.6	74.9	90.7	58.8	81.0	93.6	59.8	83.2
0970, 0990	Wall Finishes & Painting/Coating	97.1	95.6	96.2	98.7	95.6	96.8	98.7	55.8	72.5	92.5	83.9	87.2	92.6	65.2	75.9	92.6	50.0	66.6
09	FINISHES	83.1	63.8	72.6	95.3	68.0	80.5	93.2	57.8	74.0	77.9	82.5	80.4	88.4	74.2	80.7	89.3	71.1	79.4
COVERS	DIVS. 10 - 14, 25, 28, 41, 43, 44, 46	100.0	85.2	96.8	100.0	83.8	96.5	100.0	85.9	96.9	100.0	81.2	95.9	100.0	88.7	97.6	100.0	87.1	97.2
21, 22, 23	FIRE SUPPRESSION, PLUMBING & HVAC	100.7	67.5	87.1	100.5	72.5	89.0	100.5	69.2	87.7	96.2	78.8	89.0	100.7	75.7	90.5	96.5	70.9	86.0
26, 27, 3370	ELECTRICAL, COMMUNICATIONS & UTIL.	98.7	72.6	86.1	98.3	69.3	84.3	101.6	72.6	87.6	93.5	81.8	87.8	95.9	73.6	85.2	94.3	81.7	88.3
MF2018	WEIGHTED AVERAGE	96.1	72.0	86.3	97.1	76.4	88.7	96.1	71.9	86.2	91.8	84.2	88.7	93.4	76.5	86.5	93.1	74.6	85.6

KENTUCKY

DIVISION		CORBIN 407-409 MAT.	INST.	TOTAL	COVINGTON 410 MAT.	INST.	TOTAL	ELIZABETHTOWN 427 MAT.	INST.	TOTAL	FRANKFORT 406 MAT.	INST.	TOTAL	HAZARD 417-418 MAT.	INST.	TOTAL	HENDERSON 424 MAT.	INST.	TOTAL
015433	CONTRACTOR EQUIPMENT		94.2	94.2		97.9	97.9		88.2	88.2		99.7	99.7		94.2	94.2		107.4	107.4
0241, 31 - 34	SITE & INFRASTRUCTURE, DEMOLITION	82.5	86.7	85.4	75.5	96.9	90.0	65.8	85.1	78.9	79.3	98.1	92.1	77.2	87.3	84.1	72.6	109.0	97.3
0310	Concrete Forming & Accessories	79.6	71.4	73.0	79.7	66.7	69.2	77.6	68.1	69.9	99.7	74.1	79.0	80.4	77.4	78.0	89.4	72.4	75.7
0320	Concrete Reinforcing	90.5	85.9	89.0	91.4	73.2	85.5	91.7	77.9	87.3	99.1	78.3	92.4	92.6	86.1	90.5	91.4	78.5	87.2
0330	Cast-in-Place Concrete	87.7	67.9	80.9	90.3	76.3	85.5	76.5	63.4	72.0	87.9	72.5	82.6	90.7	65.3	82.0	74.7	79.1	76.2
03	CONCRETE	84.0	73.3	79.7	91.3	72.2	83.7	82.7	68.9	77.2	89.9	74.7	83.8	90.2	75.2	84.2	86.9	76.4	82.7
04	MASONRY	92.6	56.0	69.5	106.2	67.2	81.6	79.6	57.7	65.8	93.4	69.4	78.3	91.0	53.0	67.1	98.5	73.5	82.8
05	METALS	86.5	87.1	86.6	89.3	85.9	88.5	91.1	83.5	89.3	89.0	82.8	87.6	91.2	87.9	90.4	82.2	84.8	82.8
06	WOOD, PLASTICS & COMPOSITES	69.2	73.2	70.9	78.7	64.4	72.8	77.7	69.8	74.4	94.4	73.2	85.6	78.1	83.3	80.2	87.4	71.5	80.8
07	THERMAL & MOISTURE PROTECTION	96.6	64.4	82.5	100.2	67.5	85.9	91.0	63.5	78.9	95.9	71.3	85.1	101.8	64.5	85.4	101.1	76.7	90.4
08	OPENINGS	83.7	66.3	79.5	90.7	68.6	85.4	87.8	68.7	83.2	94.8	74.6	89.9	89.3	82.1	87.5	86.3	73.8	83.3
0920	Plaster & Gypsum Board	91.8	72.3	79.6	67.3	64.2	65.4	77.3	69.5	72.4	103.4	72.3	83.9	77.3	82.7	80.7	86.5	70.6	76.5
0950, 0980	Ceilings & Acoustic Treatment	80.9	72.3	75.3	89.9	64.2	73.2	84.4	69.5	74.8	104.0	72.3	83.4	84.4	82.7	83.3	79.6	70.6	73.8
0960	Flooring	89.9	59.8	80.7	68.0	77.7	71.0	88.4	68.7	82.4	97.7	73.1	90.2	92.1	59.8	82.2	94.9	70.7	87.5
0970, 0990	Wall Finishes & Painting/Coating	89.5	57.5	70.0	87.8	66.0	74.5	92.6	63.2	74.7	101.8	85.1	91.6	92.6	50.0	66.6	92.6	80.5	85.2
09	FINISHES	87.1	67.9	76.7	81.3	68.1	74.1	87.0	67.1	76.2	99.9	75.1	86.4	88.3	71.6	79.2	89.0	73.0	80.3
COVERS	DIVS. 10 - 14, 25, 28, 41, 43, 44, 46	100.0	88.5	97.5	100.0	87.3	97.2	100.0	85.7	96.9	100.0	90.4	97.9	100.0	85.7	96.9	100.0	51.4	89.4
21, 22, 23	FIRE SUPPRESSION, PLUMBING & HVAC	96.7	68.7	85.2	97.3	70.4	86.3	96.7	72.2	86.6	100.8	75.1	90.2	96.5	71.9	86.4	96.7	73.5	87.2
26, 27, 3370	ELECTRICAL, COMMUNICATIONS & UTIL.	92.2	81.7	87.1	96.0	67.1	82.1	94.2	72.2	83.6	97.5	72.2	85.3	94.3	81.7	88.3	95.6	72.2	84.3
MF2018	WEIGHTED AVERAGE	89.9	73.3	83.2	92.8	73.3	84.9	90.3	71.7	82.7	94.8	76.9	87.5	92.5	75.2	85.5	90.5	76.8	84.9

City Cost Indexes

KENTUCKY

	DIVISION	LEXINGTON 403 - 405			LOUISVILLE 400 - 402			OWENSBORO 423			PADUCAH 420			PIKEVILLE 415 - 416			SOMERSET 425 - 426		
		MAT.	INST.	TOTAL	MAT.	INST.	TOTAL	MAT.	INST.	TOTAL	MAT.	INST.	TOTAL	MAT.	INST.	TOTAL	MAT.	INST.	TOTAL
015433	CONTRACTOR EQUIPMENT		94.2	94.2		96.5	96.5		107.4	107.4		107.4	107.4		94.8	94.8		94.2	94.2
0241, 31 - 34	SITE & INFRASTRUCTURE, DEMOLITION	85.2	89.1	87.9	77.8	98.1	91.6	81.5	110.9	101.3	74.6	109.2	98.1	110.9	74.2	86.0	71.1	86.8	81.8
0310	Concrete Forming & Accessories	96.3	73.2	77.7	98.7	79.7	83.4	87.1	76.7	78.7	84.4	76.3	77.9	93.4	80.5	83.0	81.9	72.0	73.9
0320	Concrete Reinforcing	99.4	84.8	94.7	98.6	84.8	94.1	91.4	78.0	87.1	91.9	76.7	87.1	93.7	89.1	92.2	91.7	86.1	89.9
0330	Cast-in-Place Concrete	89.8	81.8	87.0	84.5	69.0	79.1	87.2	81.9	85.4	79.7	75.9	78.4	93.8	83.7	90.3	74.7	82.3	77.4
03	CONCRETE	88.0	78.8	84.3	88.2	77.1	83.7	97.0	79.3	89.9	90.6	76.8	85.1	102.5	84.5	95.3	78.8	78.7	78.7
04	MASONRY	91.3	69.9	77.8	85.4	68.9	75.0	91.1	78.0	82.8	94.0	74.0	81.3	90.3	77.7	82.4	85.6	59.0	68.8
05	METALS	88.9	88.1	88.7	90.7	84.9	89.3	83.3	86.8	84.2	81.1	85.3	82.1	90.4	105.3	93.9	91.1	87.7	90.3
06	WOOD, PLASTICS & COMPOSITES	87.6	71.5	80.9	92.2	82.2	88.1	84.7	75.7	81.0	81.9	76.8	79.8	84.9	82.7	84.0	78.7	73.2	76.4
07	THERMAL & MOISTURE PROTECTION	97.1	75.7	87.7	94.3	73.4	85.2	101.8	79.6	92.1	101.2	76.6	90.4	92.7	74.6	84.8	101.0	66.7	86.0
08	OPENINGS	83.8	74.8	81.7	85.2	77.3	83.3	86.3	77.6	84.2	85.7	76.3	83.4	87.3	81.0	85.8	88.4	72.5	84.6
0920	Plaster & Gypsum Board	109.7	70.6	85.2	103.2	82.0	89.9	83.8	74.9	78.2	82.6	76.0	78.5	67.1	82.7	76.8	77.3	72.3	74.2
0950, 0980	Ceilings & Acoustic Treatment	83.6	70.6	75.1	101.1	82.0	88.7	79.6	74.9	76.6	79.6	76.0	77.3	80.6	82.7	82.0	84.4	72.3	76.6
0960	Flooring	93.9	62.2	84.2	98.2	58.6	86.1	94.4	58.8	83.5	93.5	70.7	86.5	78.1	59.8	72.5	92.3	59.8	82.4
0970, 0990	Wall Finishes & Painting/Coating	89.5	78.4	82.8	98.7	65.1	78.2	92.6	88.4	90.0	92.6	67.9	77.5	92.5	85.1	88.0	92.6	63.2	74.7
09	FINISHES	91.7	71.0	80.5	98.2	74.4	85.2	88.9	74.0	80.8	88.2	74.4	80.7	80.8	77.4	78.9	87.7	68.7	77.4
COVERS	DIVS. 10 - 14, 25, 28, 41, 43, 44, 46	100.0	90.1	97.9	100.0	89.6	97.7	100.0	93.2	98.5	100.0	83.9	96.5	100.0	44.4	87.9	100.0	88.7	97.5
21, 22, 23	FIRE SUPPRESSION, PLUMBING & HVAC	100.9	74.8	90.2	100.6	76.3	90.6	100.7	75.8	90.5	96.7	72.5	86.8	96.2	76.2	88.0	96.7	68.5	85.1
26, 27, 3370	ELECTRICAL, COMMUNICATIONS & UTIL.	93.8	73.4	84.0	101.5	73.6	88.1	95.6	71.1	83.8	97.1	71.6	84.8	95.3	81.8	88.8	94.6	81.7	88.4
MF2018	WEIGHTED AVERAGE	92.6	76.9	86.2	93.9	77.9	87.4	92.9	80.0	87.7	90.7	78.0	85.5	94.1	80.1	88.4	90.5	74.8	84.1

LOUISIANA

	DIVISION	ALEXANDRIA 713 - 714			BATON ROUGE 707 - 708			HAMMOND 704			LAFAYETTE 705			LAKE CHARLES 706			MONROE 712		
		MAT.	INST.	TOTAL	MAT.	INST.	TOTAL	MAT.	INST.	TOTAL	MAT.	INST.	TOTAL	MAT.	INST.	TOTAL	MAT.	INST.	TOTAL
015433	CONTRACTOR EQUIPMENT		87.7	87.7		94.7	94.7		86.0	86.0		86.0	86.0		85.6	85.6		87.7	87.7
0241, 31 - 34	SITE & INFRASTRUCTURE, DEMOLITION	101.2	84.3	89.7	104.1	96.1	98.7	101.4	83.5	89.2	103.0	83.7	89.9	103.8	83.0	89.7	101.2	84.2	89.7
0310	Concrete Forming & Accessories	75.4	59.8	62.8	98.6	70.0	75.6	75.3	53.3	57.6	99.1	64.1	70.8	98.4	67.5	73.5	74.8	59.3	62.3
0320	Concrete Reinforcing	98.0	52.6	83.4	99.6	52.7	84.5	94.5	50.7	80.4	95.9	50.7	81.4	95.9	52.6	82.0	96.9	52.6	82.7
0330	Cast-in-Place Concrete	78.2	65.7	73.9	87.8	68.9	81.3	82.9	64.3	76.5	82.5	66.0	76.8	86.8	66.9	80.0	78.2	65.0	73.7
03	CONCRETE	80.2	61.4	72.7	89.3	67.2	80.5	84.3	57.6	73.6	86.1	63.0	76.9	88.0	65.3	78.9	79.9	61.0	72.3
04	MASONRY	101.7	62.9	77.3	85.0	60.8	69.7	88.9	60.6	71.1	88.8	63.7	73.0	88.4	65.2	73.7	97.4	61.6	74.9
05	METALS	86.5	70.6	82.8	91.8	73.0	87.4	84.6	68.3	80.8	83.9	68.4	80.3	83.9	69.2	80.5	86.5	70.5	82.7
06	WOOD, PLASTICS & COMPOSITES	84.6	58.7	73.9	98.8	73.3	88.3	77.7	51.2	66.7	102.8	64.1	86.8	100.7	67.9	87.1	83.8	58.7	73.4
07	THERMAL & MOISTURE PROTECTION	100.1	64.2	84.4	95.5	66.7	82.9	94.6	61.5	80.1	95.4	65.1	82.1	95.1	66.2	82.4	100.0	63.7	84.1
08	OPENINGS	106.4	58.2	94.8	93.6	70.3	88.0	91.7	51.6	82.0	95.0	58.5	86.2	95.0	61.7	87.0	106.4	56.7	94.4
0920	Plaster & Gypsum Board	85.0	58.2	68.2	104.3	72.8	84.6	95.4	50.4	67.3	113.5	63.8	82.4	113.5	67.7	84.9	84.5	58.2	68.1
0950, 0980	Ceilings & Acoustic Treatment	93.2	58.2	70.5	103.4	72.8	83.5	99.6	50.4	67.7	96.1	63.8	75.1	96.9	67.7	77.9	93.2	58.2	70.5
0960	Flooring	87.8	65.9	81.1	96.5	65.9	87.2	96.4	65.9	87.0	104.2	65.9	92.5	104.2	65.9	92.5	87.5	65.9	80.9
0970, 0990	Wall Finishes & Painting/Coating	98.7	58.8	74.4	101.8	58.8	75.6	99.9	55.5	72.8	99.9	55.4	72.8	99.9	58.8	74.8	98.7	60.6	75.5
09	FINISHES	89.2	60.5	73.6	102.3	68.5	84.0	99.5	55.0	75.4	103.9	63.5	81.9	104.0	66.4	83.6	89.0	60.4	73.4
COVERS	DIVS. 10 - 14, 25, 28, 41, 43, 44, 46	100.0	80.3	95.7	100.0	85.1	96.8	100.0	82.0	96.1	100.0	84.3	96.6	100.0	85.2	96.8	100.0	79.9	95.6
21, 22, 23	FIRE SUPPRESSION, PLUMBING & HVAC	101.1	61.3	84.8	100.9	59.9	84.1	96.8	58.3	81.0	101.0	59.6	84.0	101.0	62.5	85.2	101.1	60.2	84.3
26, 27, 3370	ELECTRICAL, COMMUNICATIONS & UTIL.	90.6	52.9	72.4	97.5	57.1	78.0	93.9	64.5	79.7	94.6	59.1	77.5	94.3	63.6	79.5	91.6	56.3	74.6
MF2018	WEIGHTED AVERAGE	93.5	63.3	81.2	95.8	67.2	84.2	91.8	62.2	79.8	93.8	64.5	81.9	94.0	66.9	83.0	93.3	63.3	81.1

LOUISIANA / MAINE

	DIVISION	LOUISIANA									MAINE								
		NEW ORLEANS 700 - 701			SHREVEPORT 710 - 711			THIBODAUX 703			AUGUSTA 043			BANGOR 044			BATH 045		
		MAT.	INST.	TOTAL	MAT.	INST.	TOTAL	MAT.	INST.	TOTAL	MAT.	INST.	TOTAL	MAT.	INST.	TOTAL	MAT.	INST.	TOTAL
015433	CONTRACTOR EQUIPMENT		90.2	90.2		96.9	96.9		86.0	86.0		98.4	98.4		92.3	92.3		92.3	92.3
0241, 31 - 34	SITE & INFRASTRUCTURE, DEMOLITION	103.7	96.4	98.8	103.5	97.7	99.6	103.8	83.5	90.0	88.3	102.4	97.9	90.1	91.0	90.7	87.4	89.4	88.8
0310	Concrete Forming & Accessories	103.6	69.9	76.4	99.2	61.2	68.6	87.8	60.9	66.1	97.7	77.1	81.1	90.8	78.4	80.8	85.3	76.7	78.4
0320	Concrete Reinforcing	100.0	52.9	84.9	99.6	51.7	84.2	94.5	50.7	80.4	99.9	81.2	93.9	85.2	81.4	84.0	84.4	81.0	83.3
0330	Cast-in-Place Concrete	79.8	71.8	77.0	80.9	67.9	76.4	89.0	64.0	80.4	74.1	109.2	86.2	58.6	111.5	76.9	58.6	107.9	75.7
03	CONCRETE	89.1	64.9	80.5	86.3	62.5	76.8	88.7	60.9	77.6	88.9	89.5	89.1	79.7	91.1	84.3	79.5	89.0	83.3
04	MASONRY	91.7	62.9	73.5	84.8	62.1	70.5	111.2	60.2	79.0	86.2	92.0	89.8	104.6	93.2	97.4	110.6	89.5	97.3
05	METALS	100.2	61.9	91.3	88.7	68.4	83.9	84.6	68.3	80.8	106.7	91.4	103.1	91.9	94.1	92.4	90.7	93.1	91.3
06	WOOD, PLASTICS & COMPOSITES	103.8	72.3	90.8	99.7	61.0	83.7	86.3	61.8	76.2	92.2	74.1	84.7	88.2	75.5	82.9	81.2	74.1	78.2
07	THERMAL & MOISTURE PROTECTION	93.6	68.8	82.7	98.1	65.8	84.0	95.0	62.4	80.7	106.1	98.7	102.8	104.0	100.2	102.3	103.8	96.5	100.6
08	OPENINGS	97.9	64.9	90.0	100.5	56.4	89.8	95.8	53.6	85.6	98.7	77.9	93.6	92.8	81.5	90.0	92.7	76.9	88.9
0920	Plaster & Gypsum Board	101.0	71.7	82.7	107.0	60.1	77.7	98.3	61.4	75.2	105.5	73.3	85.3	104.3	74.7	85.8	95.6	73.3	81.6
0950, 0980	Ceilings & Acoustic Treatment	90.1	71.7	78.2	102.7	60.1	75.1	99.6	61.4	74.8	96.9	73.3	81.6	80.4	74.7	76.7	80.4	73.3	75.8
0960	Flooring	115.4	67.2	100.6	95.5	65.9	86.4	101.7	65.3	90.6	91.7	107.1	96.4	86.1	107.1	92.5	84.7	102.9	90.3
0970, 0990	Wall Finishes & Painting/Coating	104.5	60.8	77.8	101.8	60.6	76.7	100.3	55.5	73.0	102.0	87.5	93.2	95.5	95.9	95.7	95.5	82.5	87.6
09	FINISHES	106.8	68.7	86.1	101.1	61.8	79.8	101.7	61.1	79.7	98.6	82.8	90.0	90.9	84.5	87.4	89.1	81.4	84.9
COVERS	DIVS. 10 - 14, 25, 28, 41, 43, 44, 46	100.0	85.5	96.8	100.0	81.1	95.9	100.0	82.9	96.3	100.0	97.8	99.5	100.0	100.3	100.1	100.0	93.6	98.6
21, 22, 23	FIRE SUPPRESSION, PLUMBING & HVAC	100.6	62.6	85.0	100.9	60.9	84.5	96.8	58.0	80.9	101.2	73.5	89.8	101.4	75.7	90.8	97.1	73.5	87.4
26, 27, 3370	ELECTRICAL, COMMUNICATIONS & UTIL.	101.9	68.6	85.9	98.5	63.1	81.4	93.1	64.5	79.3	105.5	75.5	91.0	100.8	68.2	85.1	99.4	75.5	87.9
MF2018	WEIGHTED AVERAGE	98.8	68.6	86.5	95.4	65.8	83.3	93.9	63.6	81.6	99.6	84.5	93.5	94.5	84.2	90.3	93.2	82.9	89.0

For CCI updates and additional guidance please go to www.rsmeans.com/2022books

MAINE

DIVISION		HOULTON 047			KITTERY 039			LEWISTON 042			MACHIAS 046			PORTLAND 040 - 041			ROCKLAND 048		
		MAT.	INST.	TOTAL	MAT.	INST.	TOTAL	MAT.	INST.	TOTAL	MAT.	INST.	TOTAL	MAT.	INST.	TOTAL	MAT.	INST.	TOTAL
015433	CONTRACTOR EQUIPMENT		92.3	92.3		92.3	92.3		92.3	92.3		92.3	92.3		95.5	95.5		92.3	92.3
0241, 31 - 34	SITE & INFRASTRUCTURE, DEMOLITION	89.2	89.4	89.3	76.2	89.4	85.2	88.0	91.0	90.0	88.6	89.4	89.1	88.9	97.7	94.9	85.6	89.4	88.2
0310	Concrete Forming & Accessories	96.0	76.6	80.4	87.2	76.8	78.8	98.1	78.4	82.2	91.9	76.7	79.6	101.1	79.1	83.3	93.4	76.7	79.9
0320	Concrete Reinforcing	85.2	81.0	83.9	85.3	81.1	83.9	104.7	81.4	97.2	85.2	81.0	83.9	98.8	81.5	93.2	85.2	81.0	83.9
0330	Cast-in-Place Concrete	58.6	106.9	75.3	58.9	108.0	75.8	59.8	111.6	77.7	58.6	107.9	75.7	71.1	112.9	85.5	59.8	107.9	76.5
03	CONCRETE	80.7	88.6	83.9	75.4	89.1	80.9	81.9	91.1	85.6	80.1	89.0	83.7	87.6	91.8	89.2	78.0	89.0	82.4
04	MASONRY	88.3	89.5	89.0	106.6	89.5	95.8	89.0	93.2	91.7	88.3	89.5	89.0	99.1	93.3	95.4	83.4	89.5	87.2
05	METALS	91.0	93.1	91.5	86.0	93.3	87.7	96.2	94.1	95.7	91.0	93.1	91.5	101.8	92.6	99.6	90.8	93.1	91.4
06	WOOD, PLASTICS & COMPOSITES	92.8	74.1	81.6	86.8	74.1	81.5	95.1	75.5	87.0	89.4	74.1	83.0	95.2	75.5	87.1	90.4	74.1	83.6
07	THERMAL & MOISTURE PROTECTION	104.1	96.5	100.8	106.7	96.5	102.2	103.8	100.2	102.2	104.0	96.5	100.7	109.0	101.8	105.8	103.7	96.5	100.5
08	OPENINGS	92.8	76.9	89.0	92.9	80.2	89.8	95.3	81.5	92.0	92.8	76.9	89.0	93.0	81.5	90.2	92.8	76.9	88.9
0920	Plaster & Gypsum Board	107.7	73.3	86.2	96.0	73.3	81.8	111.6	74.7	88.5	106.0	73.3	85.5	106.5	74.7	86.6	106.0	73.3	85.5
0950, 0980	Ceilings & Acoustic Treatment	80.4	73.3	75.8	89.7	73.3	79.0	92.1	74.7	80.8	80.4	73.3	75.8	100.1	74.7	83.6	80.4	73.3	75.8
0960	Flooring	87.2	102.9	92.0	90.7	102.9	94.4	88.5	107.1	94.2	86.6	102.9	91.6	91.7	167.9	115.0	86.9	102.9	91.8
0970, 0990	Wall Finishes & Painting/Coating	95.5	82.5	87.6	81.6	94.8	89.6	95.5	95.9	95.7	95.5	82.5	87.6	101.8	95.9	98.2	95.5	82.5	87.6
09	FINISHES	91.8	81.4	86.1	92.4	82.7	87.1	94.9	84.5	89.3	91.3	81.4	85.9	98.4	97.0	97.6	91.1	81.4	85.8
COVERS	DIVS. 10 - 14, 25, 28, 41, 43, 44, 46	100.0	90.6	98.0	100.0	93.6	98.6	100.0	100.3	100.1	100.0	90.6	98.0	100.0	100.4	100.1	100.0	93.6	98.6
21, 22, 23	FIRE SUPPRESSION, PLUMBING & HVAC	97.1	73.5	87.4	96.9	73.5	87.3	101.4	75.7	90.8	97.1	73.5	87.4	100.8	75.8	90.5	97.1	73.5	87.4
26, 27, 3370	ELECTRICAL, COMMUNICATIONS & UTIL.	102.3	75.5	89.4	99.9	75.5	88.1	102.3	75.6	89.4	102.3	75.5	89.4	101.6	75.6	89.1	102.2	75.5	89.3
MF2018	WEIGHTED AVERAGE	93.2	82.8	88.9	91.6	83.3	88.2	95.8	85.2	91.5	93.0	82.8	88.9	98.1	87.3	93.7	92.4	82.9	88.5

DIVISION		MAINE			MARYLAND														
		WATERVILLE 049			ANNAPOLIS 214			BALTIMORE 210 - 212			COLLEGE PARK 207 - 208			CUMBERLAND 215			EASTON 216		
		MAT.	INST.	TOTAL	MAT.	INST.	TOTAL	MAT.	INST.	TOTAL	MAT.	INST.	TOTAL	MAT.	INST.	TOTAL	MAT.	INST.	TOTAL
015433	CONTRACTOR EQUIPMENT		92.3	92.3		106.4	106.4		104.5	104.5		104.2	104.2		99.5	99.5		99.5	99.5
0241, 31 - 34	SITE & INFRASTRUCTURE, DEMOLITION	88.9	89.4	89.2	109.8	96.4	100.7	110.0	97.8	101.7	108.5	88.8	95.2	99.7	86.0	90.4	107.2	84.3	91.7
0310	Concrete Forming & Accessories	84.6	76.7	78.2	103.3	73.8	79.5	103.2	74.3	79.9	80.2	71.9	73.5	91.2	78.7	81.1	88.7	69.7	73.4
0320	Concrete Reinforcing	85.2	81.0	83.9	99.1	88.4	95.7	101.0	91.3	97.9	98.4	98.1	98.3	88.3	87.0	87.9	87.5	81.2	85.5
0330	Cast-in-Place Concrete	58.6	107.9	75.7	103.7	76.3	94.3	111.4	79.2	100.3	136.6	71.7	114.2	86.1	83.2	85.1	95.6	61.8	83.9
03	CONCRETE	80.8	89.0	84.1	97.2	78.4	89.7	104.6	79.9	94.7	114.8	78.0	100.1	85.3	83.1	84.4	91.8	70.5	83.3
04	MASONRY	98.5	89.5	92.8	97.7	71.6	81.2	96.9	76.0	83.7	109.5	69.8	84.5	96.5	86.7	90.4	109.7	55.1	75.3
05	METALS	90.9	93.1	91.4	97.9	100.6	98.5	99.6	98.0	99.3	82.9	110.2	89.3	93.6	103.6	95.9	93.8	100.1	95.3
06	WOOD, PLASTICS & COMPOSITES	80.5	74.1	77.8	94.1	73.1	85.4	102.1	73.3	90.2	73.6	71.4	72.7	83.7	76.4	80.7	81.2	76.3	79.1
07	THERMAL & MOISTURE PROTECTION	104.0	96.5	100.7	105.2	79.0	93.8	105.5	83.6	95.9	103.9	76.9	92.0	101.2	78.9	91.4	101.4	69.6	87.4
08	OPENINGS	92.8	76.9	89.0	98.3	79.3	93.7	101.5	80.0	96.3	98.6	80.9	87.5	94.1	81.1	90.9	92.6	79.1	89.3
0920	Plaster & Gypsum Board	95.6	73.3	81.6	115.0	72.4	88.4	115.0	72.4	88.4	101.4	70.9	82.4	112.7	76.2	89.9	112.7	76.1	89.8
0950, 0980	Ceilings & Acoustic Treatment	80.4	73.3	75.8	110.0	72.4	85.6	110.0	72.4	85.6	117.6	70.9	87.3	116.8	76.2	90.5	116.8	76.1	90.4
0960	Flooring	84.5	102.9	90.1	96.2	74.0	89.4	100.5	75.7	92.9	87.9	73.0	83.3	92.6	91.9	92.4	91.8	70.4	85.3
0970, 0990	Wall Finishes & Painting/Coating	95.5	82.5	87.6	101.8	67.7	81.0	105.0	69.0	83.0	97.4	67.7	79.3	97.2	78.6	85.8	97.2	67.7	79.2
09	FINISHES	89.2	81.4	84.9	105.0	72.6	87.4	106.2	73.3	88.3	99.9	70.9	84.1	105.8	81.0	92.4	106.2	69.8	86.4
COVERS	DIVS. 10 - 14, 25, 28, 41, 43, 44, 46	100.0	90.5	97.9	100.0	89.3	97.7	100.0	89.6	97.7	100.0	79.8	95.6	100.0	90.9	98.0	100.0	81.3	95.9
21, 22, 23	FIRE SUPPRESSION, PLUMBING & HVAC	97.1	73.5	87.4	100.7	80.7	92.5	100.7	83.2	93.5	96.5	81.2	90.2	96.3	68.9	85.0	96.3	67.6	84.5
26, 27, 3370	ELECTRICAL, COMMUNICATIONS & UTIL.	102.3	75.5	89.4	98.1	82.3	90.5	99.2	86.7	93.1	99.8	101.4	100.6	97.9	79.0	88.8	97.6	57.8	78.4
MF2018	WEIGHTED AVERAGE	93.2	82.8	89.0	99.6	81.8	92.3	101.5	83.7	94.2	97.3	84.0	91.9	95.2	81.6	89.7	96.8	74.9	86.2

DIVISION		MARYLAND															MASSACHUSETTS		
		ELKTON 219			HAGERSTOWN 217			SALISBURY 218			SILVER SPRING 209			WALDORF 206			BOSTON 020 - 022, 024		
		MAT.	INST.	TOTAL	MAT.	INST.	TOTAL	MAT.	INST.	TOTAL	MAT.	INST.	TOTAL	MAT.	INST.	TOTAL	MAT.	INST.	TOTAL
015433	CONTRACTOR EQUIPMENT		99.5	99.5		99.5	99.5		99.5	99.5		96.4	96.4		96.4	96.4		103.1	103.1
0241, 31 - 34	SITE & INFRASTRUCTURE, DEMOLITION	93.3	84.8	87.5	98.3	86.3	90.1	107.2	84.1	91.5	93.7	81.3	85.3	100.4	81.1	87.3	87.7	103.4	98.4
0310	Concrete Forming & Accessories	96.9	86.9	88.9	90.0	83.3	84.6	108.7	47.6	59.4	89.6	71.2	74.7	99.6	71.1	76.6	101.2	135.4	128.8
0320	Concrete Reinforcing	87.5	109.2	94.5	88.3	87.1	87.9	87.5	60.7	78.9	97.8	98.0	97.9	98.5	98.0	98.3	101.6	151.6	117.6
0330	Cast-in-Place Concrete	77.4	67.9	74.1	82.0	83.4	82.5	95.6	59.8	83.2	139.8	72.0	116.4	156.7	71.8	127.4	80.3	141.6	101.5
03	CONCRETE	79.5	85.3	81.8	82.3	85.2	83.5	93.3	56.3	78.5	113.8	77.6	99.3	125.0	77.5	106.0	92.2	139.9	111.3
04	MASONRY	95.1	62.7	74.7	101.2	86.7	92.1	109.2	51.8	73.0	109.7	70.1	84.7	95.2	70.0	79.3	112.1	147.5	134.4
05	METALS	93.9	111.7	98.0	93.7	104.0	96.1	93.9	91.0	93.2	87.5	106.3	91.8	87.5	105.7	91.7	100.2	134.8	108.3
06	WOOD, PLASTICS & COMPOSITES	89.4	95.6	91.9	82.7	82.2	82.5	102.8	48.5	80.3	81.6	70.7	77.1	89.9	70.7	82.0	99.9	133.9	113.9
07	THERMAL & MOISTURE PROTECTION	100.9	75.2	89.6	101.4	83.4	93.5	101.9	65.1	85.8	106.1	81.4	95.2	106.8	81.4	95.6	106.5	138.9	120.7
08	OPENINGS	92.6	97.2	93.7	92.5	84.2	90.5	92.8	58.3	84.4	82.2	80.5	81.8	82.7	80.5	82.2	97.9	142.1	108.6
0920	Plaster & Gypsum Board	116.6	95.9	103.7	112.7	82.1	93.6	131.6	47.5	79.0	112.8	70.9	86.6	117.7	70.9	88.4	112.6	134.7	126.4
0950, 0980	Ceilings & Acoustic Treatment	116.8	95.9	103.3	119.1	82.1	95.1	116.8	47.5	71.8	125.8	70.9	90.2	125.8	70.9	90.2	105.3	134.7	124.4
0960	Flooring	93.9	70.4	86.7	92.2	91.9	92.1	97.1	70.4	88.9	93.1	73.0	87.0	95.8	73.0	88.9	97.5	161.9	117.2
0970, 0990	Wall Finishes & Painting/Coating	97.2	67.7	79.2	97.2	70.2	80.7	97.2	67.7	79.2	105.3	67.7	82.4	105.3	67.7	82.4	105.0	155.2	135.6
09	FINISHES	106.3	83.0	93.6	105.9	83.5	93.7	110.7	52.7	79.2	100.5	70.3	84.1	102.3	70.4	85.0	102.5	142.7	124.3
COVERS	DIVS. 10 - 14, 25, 28, 41, 43, 44, 46	100.0	52.2	89.6	100.0	91.5	98.2	100.0	77.3	95.1	100.0	78.4	95.3	100.0	76.5	94.9	100.0	116.7	103.6
21, 22, 23	FIRE SUPPRESSION, PLUMBING & HVAC	96.3	73.0	86.7	100.5	86.0	94.6	96.3	65.6	83.7	96.5	81.6	90.4	96.5	81.5	90.3	96.4	132.9	111.4
26, 27, 3370	ELECTRICAL, COMMUNICATIONS & UTIL.	98.7	80.5	90.0	97.8	79.0	88.7	96.8	55.7	77.0	95.9	101.3	98.5	94.4	101.3	97.8	99.4	132.7	115.4
MF2018	WEIGHTED AVERAGE	94.3	81.0	88.9	95.8	86.3	91.9	97.5	63.5	83.6	96.8	83.2	91.2	98.1	83.0	91.9	98.5	134.5	113.1

City Cost Indexes

MASSACHUSETTS

DIVISION		BROCKTON 023 MAT.	INST.	TOTAL	BUZZARDS BAY 025 MAT.	INST.	TOTAL	FALL RIVER 027 MAT.	INST.	TOTAL	FITCHBURG 014 MAT.	INST.	TOTAL	FRAMINGHAM 017 MAT.	INST.	TOTAL	GREENFIELD 013 MAT.	INST.	TOTAL
015433	CONTRACTOR EQUIPMENT		95.2	95.2		95.2	95.2		96.1	96.1		92.3	92.3		94.0	94.0		92.3	92.3
0241, 31 - 34	SITE & INFRASTRUCTURE, DEMOLITION	87.2	94.1	91.9	78.0	93.7	88.6	86.3	94.1	91.6	80.4	93.6	89.3	77.5	93.0	88.0	83.4	92.1	89.3
0310	Concrete Forming & Accessories	98.0	119.8	115.6	95.3	119.1	114.5	98.0	119.6	115.4	91.8	112.4	108.4	101.7	119.8	116.3	89.5	116.1	111.0
0320	Concrete Reinforcing	99.4	140.9	112.7	79.7	118.5	92.1	99.4	118.6	105.5	82.0	132.7	98.3	82.0	140.7	100.8	83.8	118.1	94.8
0330	Cast-in-Place Concrete	69.2	129.3	89.9	57.5	129.0	82.2	66.9	129.5	88.5	63.2	128.9	85.9	63.3	129.2	86.0	64.9	114.5	82.0
03	CONCRETE	84.2	126.6	101.2	71.2	122.3	91.6	82.2	122.7	98.4	72.3	121.5	92.0	74.7	126.5	95.5	74.9	115.5	91.1
04	MASONRY	105.1	132.9	122.6	97.5	132.8	119.8	105.7	132.8	122.8	95.2	124.9	113.9	101.4	128.8	118.7	99.7	115.6	109.7
05	METALS	94.6	127.8	102.4	88.6	117.7	95.4	94.6	118.3	100.1	93.4	120.1	99.6	93.4	127.0	101.3	96.7	111.4	100.2
06	WOOD, PLASTICS & COMPOSITES	95.8	119.0	105.5	91.6	119.3	103.1	95.8	119.5	105.6	91.5	109.9	99.1	99.6	119.0	107.6	89.1	119.6	101.7
07	THERMAL & MOISTURE PROTECTION	102.5	123.5	111.7	101.3	120.3	109.6	102.2	120.2	110.1	101.7	116.3	108.1	101.9	121.3	110.4	101.7	106.6	103.8
08	OPENINGS	93.5	127.4	101.7	89.9	116.6	96.4	93.5	117.7	99.3	95.6	120.2	101.5	86.5	127.3	96.3	95.6	118.2	101.1
0920	Plaster & Gypsum Board	104.3	119.7	113.9	95.9	119.7	110.8	104.3	119.7	113.9	104.2	110.1	107.9	110.8	119.7	116.4	104.3	120.1	114.2
0950, 0980	Ceilings & Acoustic Treatment	103.7	119.7	114.1	85.7	119.7	107.8	103.7	119.7	114.1	87.9	110.1	102.3	87.9	119.7	108.6	97.5	120.1	112.2
0960	Flooring	89.8	154.8	109.7	87.2	154.8	107.9	88.2	154.8	108.6	90.3	154.8	110.0	91.7	154.8	111.0	89.7	133.3	103.0
0970, 0990	Wall Finishes & Painting/Coating	89.9	135.3	117.6	89.9	135.3	117.6	89.9	135.3	117.6	87.5	135.3	116.7	88.4	135.3	117.0	87.5	112.0	102.4
09	FINISHES	95.8	128.2	113.4	89.2	128.2	110.4	95.4	128.4	113.3	91.3	122.7	108.4	92.6	128.0	111.8	93.5	119.9	107.9
COVERS	DIVS. 10 - 14, 25, 28, 41, 43, 44, 46	100.0	111.8	102.6	100.0	111.8	102.6	100.0	112.3	102.7	100.0	103.0	100.7	100.0	111.4	102.5	100.0	104.4	101.0
21, 22, 23	FIRE SUPPRESSION, PLUMBING & HVAC	101.4	101.5	101.5	96.5	96.8	96.8	101.4	101.2	101.3	97.0	97.8	97.4	97.0	113.3	103.7	97.0	95.5	96.4
26, 27, 3370	ELECTRICAL, COMMUNICATIONS & UTIL.	100.7	96.1	98.5	98.6	92.8	95.8	100.3	97.4	98.9	100.7	96.5	98.7	98.5	115.3	106.6	100.7	97.7	99.3
MF2018	WEIGHTED AVERAGE	96.1	114.7	103.7	90.4	111.2	98.9	95.7	113.0	102.8	92.5	110.2	99.7	92.2	119.3	103.2	93.9	106.7	99.1

MASSACHUSETTS

DIVISION		HYANNIS 026 MAT.	INST.	TOTAL	LAWRENCE 019 MAT.	INST.	TOTAL	LOWELL 018 MAT.	INST.	TOTAL	NEW BEDFORD 027 MAT.	INST.	TOTAL	PITTSFIELD 012 MAT.	INST.	TOTAL	SPRINGFIELD 010 - 011 MAT.	INST.	TOTAL
015433	CONTRACTOR EQUIPMENT		95.2	95.2		94.6	94.6		92.3	92.3		96.1	96.1		92.3	92.3		92.3	92.3
0241, 31 - 34	SITE & INFRASTRUCTURE, DEMOLITION	84.2	93.9	90.8	88.7	93.7	92.1	87.8	94.0	92.0	85.1	94.1	91.2	88.7	92.2	91.1	88.2	92.6	91.2
0310	Concrete Forming & Accessories	87.2	119.5	113.3	103.4	120.3	117.0	98.8	122.2	117.7	98.0	119.7	115.5	98.8	104.0	103.0	99.0	117.4	113.8
0320	Concrete Reinforcing	79.7	118.6	92.1	101.4	138.6	113.3	102.3	138.3	113.9	99.4	118.5	105.5	83.2	114.8	93.4	102.3	118.1	107.4
0330	Cast-in-Place Concrete	63.1	129.2	85.9	73.1	129.4	92.5	66.4	132.2	89.2	59.0	129.6	83.4	72.5	112.7	86.4	69.2	116.3	85.4
03	CONCRETE	75.3	122.6	94.2	87.6	126.4	103.1	81.3	128.1	100.0	79.7	122.8	97.0	79.8	108.9	91.4	82.5	116.7	96.2
04	MASONRY	104.1	132.8	122.2	106.9	135.3	124.8	94.4	136.4	120.9	104.1	131.4	121.3	95.1	110.4	104.7	94.7	118.8	109.9
05	METALS	89.7	118.1	96.3	97.2	127.1	104.2	97.1	123.9	103.4	94.6	118.3	100.1	96.9	109.7	99.9	100.0	111.5	102.7
06	WOOD, PLASTICS & COMPOSITES	83.3	119.3	98.5	100.6	119.2	108.3	99.3	119.2	107.6	95.8	119.5	105.6	99.3	104.5	101.5	99.3	119.6	107.7
07	THERMAL & MOISTURE PROTECTION	101.6	122.7	110.9	102.5	124.2	112.0	102.3	125.3	112.4	102.2	119.8	109.9	102.3	103.1	102.6	102.3	108.0	104.8
08	OPENINGS	90.5	117.8	97.1	89.6	126.7	98.6	96.6	126.7	103.9	93.5	121.9	100.3	96.6	109.0	99.6	96.6	118.2	101.8
0920	Plaster & Gypsum Board	88.9	119.7	108.2	113.7	119.7	117.5	113.7	119.7	117.5	104.3	119.7	113.9	113.7	104.5	108.0	113.7	120.1	117.7
0950, 0980	Ceilings & Acoustic Treatment	93.4	119.7	110.5	98.3	119.7	112.2	98.3	119.7	112.2	103.7	119.7	114.1	98.3	104.5	102.3	98.3	120.1	112.4
0960	Flooring	85.3	154.8	106.5	92.0	154.8	111.2	92.0	154.8	111.2	88.2	154.8	108.6	92.3	128.2	103.3	91.5	133.3	104.3
0970, 0990	Wall Finishes & Painting/Coating	89.9	135.3	117.6	87.6	135.3	116.7	87.5	135.3	116.7	89.9	135.3	117.6	87.5	112.0	102.4	88.3	112.0	102.7
09	FINISHES	89.5	128.2	110.5	96.1	128.2	113.5	96.0	129.5	114.2	95.3	128.4	113.2	96.1	109.5	103.4	96.0	120.7	109.4
COVERS	DIVS. 10 - 14, 25, 28, 41, 43, 44, 46	100.0	111.8	102.6	100.0	111.8	102.6	100.0	113.5	102.9	100.0	112.3	102.7	100.0	102.0	100.4	100.0	105.5	101.2
21, 22, 23	FIRE SUPPRESSION, PLUMBING & HVAC	101.4	105.6	103.1	100.9	117.0	107.5	100.9	123.1	110.0	101.4	101.2	101.3	100.9	92.6	97.5	100.9	97.2	99.4
26, 27, 3370	ELECTRICAL, COMMUNICATIONS & UTIL.	99.5	97.4	98.5	100.1	120.9	110.1	100.4	119.7	109.7	101.6	99.4	100.5	100.4	97.7	99.1	100.4	99.0	99.8
MF2018	WEIGHTED AVERAGE	92.8	113.9	101.4	96.8	121.7	106.9	96.0	123.1	107.0	95.4	113.3	102.7	95.8	102.4	98.5	96.8	107.9	101.3

MASSACHUSETTS / MICHIGAN

DIVISION		MASSACHUSETTS WORCESTER 015 - 016 MAT.	INST.	TOTAL	ANN ARBOR 481 MAT.	INST.	TOTAL	BATTLE CREEK 490 MAT.	INST.	TOTAL	BAY CITY 487 MAT.	INST.	TOTAL	DEARBORN 481 MAT.	INST.	TOTAL	DETROIT 482 MAT.	INST.	TOTAL
015433	CONTRACTOR EQUIPMENT		92.3	92.3		105.9	105.9		93.6	93.6		105.9	105.9		105.9	105.9		95.5	95.5
0241, 31 - 34	SITE & INFRASTRUCTURE, DEMOLITION	88.1	93.6	91.8	78.3	88.8	85.4	82.6	77.8	79.3	71.8	87.1	82.2	78.0	88.9	85.4	92.6	99.2	97.1
0310	Concrete Forming & Accessories	99.5	121.0	116.8	98.6	102.9	102.0	99.0	79.5	83.2	98.7	77.6	81.7	98.5	103.6	102.6	100.4	103.7	103.1
0320	Concrete Reinforcing	102.3	145.3	116.1	97.9	102.4	99.3	95.3	79.7	90.3	97.9	101.7	99.1	97.9	103.7	99.7	101.3	102.1	101.6
0330	Cast-in-Place Concrete	68.8	129.1	89.6	91.3	96.6	93.1	80.8	90.7	84.2	87.3	83.3	85.9	89.3	97.5	92.1	108.7	98.2	105.1
03	CONCRETE	82.3	127.6	100.5	91.4	101.6	95.5	84.9	83.3	84.3	89.7	85.3	87.9	90.5	102.5	95.3	102.6	101.0	102.0
04	MASONRY	94.3	131.3	117.6	91.4	97.5	95.3	89.2	75.3	80.4	90.9	80.0	84.0	91.2	120.3	109.6	95.9	98.7	97.7
05	METALS	100.1	126.2	106.2	92.9	117.7	98.7	98.9	82.5	95.1	93.5	113.3	98.1	93.0	119.6	99.2	97.5	93.0	96.4
06	WOOD, PLASTICS & COMPOSITES	99.9	121.1	108.6	94.7	104.5	98.8	95.4	79.1	88.7	94.7	77.3	87.5	94.7	104.5	98.8	100.1	105.7	102.5
07	THERMAL & MOISTURE PROTECTION	102.3	119.3	109.7	111.5	97.1	105.2	103.8	76.7	91.9	109.2	81.6	97.1	110.3	104.6	107.8	112.9	103.7	108.9
08	OPENINGS	96.6	129.6	104.6	90.3	99.8	92.6	83.7	75.1	81.6	90.3	81.7	88.2	90.3	99.7	92.6	97.0	101.1	98.0
0920	Plaster & Gypsum Board	113.7	121.6	118.6	94.8	104.6	100.9	85.4	75.7	79.3	94.8	76.5	83.4	94.8	104.6	100.9	90.3	105.9	100.0
0950, 0980	Ceilings & Acoustic Treatment	98.3	121.6	113.4	94.1	104.6	100.9	92.9	75.7	81.7	94.9	76.5	83.0	94.1	104.6	100.9	100.5	105.9	104.0
0960	Flooring	92.0	156.0	111.5	93.1	108.0	97.7	93.2	65.5	84.7	93.1	83.3	90.1	92.0	102.3	95.1	95.3	104.2	98.0
0970, 0990	Wall Finishes & Painting/Coating	87.5	135.3	116.7	88.4	97.9	94.2	90.1	70.8	78.3	88.4	80.3	83.4	88.4	96.1	93.1	102.2	97.8	99.5
09	FINISHES	96.0	129.5	114.2	96.2	103.5	100.2	91.0	75.6	82.7	96.0	78.2	86.3	95.9	102.6	99.6	99.6	103.6	101.7
COVERS	DIVS. 10 - 14, 25, 28, 41, 43, 44, 46	100.0	106.6	101.4	100.0	101.0	100.2	100.0	101.3	100.3	100.0	94.8	98.9	100.0	101.5	100.3	100.0	100.7	100.2
21, 22, 23	FIRE SUPPRESSION, PLUMBING & HVAC	100.9	102.2	101.4	101.0	91.5	97.1	100.9	77.8	91.4	101.0	80.3	92.5	101.0	100.3	100.7	100.8	100.6	100.8
26, 27, 3370	ELECTRICAL, COMMUNICATIONS & UTIL.	100.4	102.4	101.4	98.7	101.3	100.0	96.5	72.0	84.7	97.6	76.5	87.4	98.7	96.4	97.6	102.7	101.4	102.1
MF2018	WEIGHTED AVERAGE	96.7	115.6	104.4	95.6	99.5	97.2	94.6	78.2	87.9	95.1	84.1	90.7	95.5	103.4	98.7	100.0	100.3	100.1

MICHIGAN

DIVISION		FLINT 484 - 485			GAYLORD 497			GRAND RAPIDS 493,495			IRON MOUNTAIN 498 - 499			JACKSON 492			KALAMAZOO 491		
		MAT.	INST.	TOTAL	MAT.	INST.	TOTAL	MAT.	INST.	TOTAL	MAT.	INST.	TOTAL	MAT.	INST.	TOTAL	MAT.	INST.	TOTAL
015433	CONTRACTOR EQUIPMENT		105.9	105.9		100.8	100.8		99.2	99.2		88.7	88.7		100.8	100.8		93.6	93.6
0241, 31 - 34	SITE & INFRASTRUCTURE, DEMOLITION	69.7	87.5	81.8	80.3	75.7	77.2	81.6	90.6	87.7	84.9	84.2	84.4	94.8	77.2	82.8	82.9	77.8	79.4
0310	Concrete Forming & Accessories	103.0	84.8	88.3	98.0	69.7	75.2	96.5	76.2	80.1	91.8	74.1	77.5	93.7	77.5	80.6	99.0	79.7	83.4
0320	Concrete Reinforcing	97.9	102.0	99.2	89.0	86.6	88.2	99.6	79.6	93.2	88.0	79.1	85.2	86.4	101.5	91.3	95.3	78.3	89.9
0330	Cast-in-Place Concrete	91.9	86.3	90.0	80.6	73.5	78.2	85.0	88.4	86.2	95.5	63.4	84.4	80.5	87.5	82.9	82.4	90.8	85.3
03	CONCRETE	92.0	89.7	91.1	81.7	75.7	79.3	92.1	80.9	87.6	88.9	72.0	82.2	77.6	86.5	81.2	87.5	83.2	85.8
04	MASONRY	91.5	89.1	90.0	98.8	67.9	79.3	87.3	74.8	79.4	86.2	74.4	78.8	80.3	81.1	80.8	87.9	82.0	84.1
05	METALS	93.0	113.9	97.9	100.7	108.1	102.5	95.7	81.0	92.3	100.1	88.9	97.5	101.0	111.1	103.4	98.9	82.7	95.1
06	WOOD, PLASTICS & COMPOSITES	99.4	84.6	93.3	89.2	69.3	80.9	95.5	74.8	86.9	87.5	74.5	82.1	87.7	75.3	82.5	95.4	79.1	88.7
07	THERMAL & MOISTURE PROTECTION	109.5	87.2	99.7	102.6	69.3	88.0	105.9	71.8	91.0	106.5	69.3	90.2	101.8	82.1	93.2	103.8	78.8	92.9
08	OPENINGS	90.3	84.9	89.0	82.7	74.1	80.6	97.9	74.4	92.2	88.6	65.7	83.1	82.0	81.8	81.9	83.7	76.3	81.9
0920	Plaster & Gypsum Board	98.0	84.1	89.3	85.0	67.7	74.2	95.0	71.7	80.4	54.3	74.5	66.9	82.8	73.9	77.2	85.4	75.7	79.3
0950, 0980	Ceilings & Acoustic Treatment	94.1	84.1	87.6	90.9	67.7	75.9	104.0	71.7	83.0	91.4	74.5	80.4	90.9	73.9	79.8	92.9	75.7	81.7
0960	Flooring	93.1	88.5	91.7	86.1	79.6	84.1	95.5	85.8	92.5	105.3	84.7	99.0	84.9	73.2	81.3	93.2	73.4	87.1
0970, 0990	Wall Finishes & Painting/Coating	88.4	78.9	82.6	86.2	75.7	79.8	101.8	75.7	85.9	101.3	64.6	78.9	86.2	87.1	86.7	90.1	76.6	81.9
09	FINISHES	95.9	84.4	89.6	90.8	71.4	80.3	98.3	77.5	87.0	91.4	75.0	82.5	91.4	76.7	83.5	91.0	77.8	83.9
COVERS	DIVS. 10 - 14, 25, 28, 41, 43, 44, 46	100.0	96.9	99.3	100.0	93.8	98.7	100.0	99.8	100.0	100.0	92.5	98.4	100.0	97.1	99.4	100.0	101.3	100.3
21, 22, 23	FIRE SUPPRESSION, PLUMBING & HVAC	101.0	83.9	94.0	96.9	73.2	87.2	100.8	79.6	92.1	96.8	77.8	89.0	96.9	79.5	89.8	100.9	79.4	92.1
26, 27, 3370	ELECTRICAL, COMMUNICATIONS & UTIL.	98.7	91.8	95.4	95.4	70.7	83.5	101.9	83.2	92.9	100.1	74.7	87.9	97.9	95.4	96.7	96.2	77.0	87.0
MF2018	WEIGHTED AVERAGE	95.5	89.9	93.2	93.7	76.3	86.6	97.2	80.7	90.5	95.2	77.1	87.8	92.9	85.9	90.1	94.8	80.3	88.9

MICHIGAN / MINNESOTA

DIVISION		LANSING 488 - 489			MUSKEGON 494			ROYAL OAK 480,483			SAGINAW 486			TRAVERSE CITY 496			BEMIDJI 566		
		MAT.	INST.	TOTAL	MAT.	INST.	TOTAL	MAT.	INST.	TOTAL	MAT.	INST.	TOTAL	MAT.	INST.	TOTAL	MAT.	INST.	TOTAL
015433	CONTRACTOR EQUIPMENT		110.7	110.7		93.6	93.6		87.1	87.1		105.9	105.9		88.7	88.7		94.5	94.5
0241, 31 - 34	SITE & INFRASTRUCTURE, DEMOLITION	86.6	99.7	95.5	80.9	77.8	78.8	79.7	87.3	84.9	72.9	87.1	82.5	73.1	83.3	80.0	83.3	90.6	88.3
0310	Concrete Forming & Accessories	96.5	81.4	84.3	99.7	78.7	82.8	93.2	101.7	100.1	98.6	82.2	85.4	91.8	68.2	72.7	81.6	82.9	82.6
0320	Concrete Reinforcing	100.4	101.8	100.8	96.2	80.0	91.0	88.7	97.5	91.5	97.9	101.6	99.1	89.4	74.7	84.7	97.6	101.2	98.8
0330	Cast-in-Place Concrete	100.6	86.7	95.8	80.5	89.7	83.7	79.9	91.2	83.8	90.2	83.2	87.8	74.5	75.1	74.7	112.2	92.5	105.4
03	CONCRETE	96.8	88.0	93.3	83.9	82.7	83.4	79.4	96.7	86.3	90.9	87.4	89.5	75.3	72.7	74.2	97.4	90.9	94.8
04	MASONRY	86.9	88.5	87.9	86.8	80.9	83.1	85.8	93.0	90.3	92.9	80.0	84.8	84.1	69.9	75.1	100.4	100.4	100.4
05	METALS	91.8	109.3	95.8	97.3	83.2	94.0	95.8	89.6	94.4	93.0	113.2	97.7	100.0	86.8	96.9	85.8	118.9	93.5
06	WOOD, PLASTICS & COMPOSITES	93.0	79.8	87.6	93.8	77.5	87.0	89.9	104.5	95.9	92.3	83.7	88.7	87.5	67.9	79.4	68.3	78.2	72.4
07	THERMAL & MOISTURE PROTECTION	105.8	85.6	97.0	102.7	73.7	90.0	107.3	94.0	101.4	110.0	82.3	97.9	105.4	69.0	89.5	102.7	87.3	96.0
08	OPENINGS	98.1	82.1	94.3	82.9	76.0	81.2	90.0	97.6	91.8	88.6	85.2	87.8	88.6	61.6	82.1	96.0	102.2	97.5
0920	Plaster & Gypsum Board	94.0	79.2	84.8	70.5	74.0	72.7	90.0	104.6	99.1	94.8	83.1	87.5	54.3	67.7	62.7	92.4	78.2	83.5
0950, 0980	Ceilings & Acoustic Treatment	99.2	79.2	86.2	92.9	74.0	80.7	93.5	104.6	100.7	94.1	83.1	87.0	91.4	67.7	76.1	120.9	78.2	93.2
0960	Flooring	97.3	85.8	93.7	91.9	85.8	90.0	90.7	99.2	93.3	93.1	83.3	90.1	105.3	79.6	97.4	94.5	101.5	96.7
0970, 0990	Wall Finishes & Painting/Coating	101.8	75.7	85.9	88.3	76.9	81.4	90.0	87.1	88.2	88.4	80.3	83.4	101.3	36.6	61.9	90.9	94.1	92.8
09	FINISHES	98.5	81.1	89.0	87.5	79.5	83.2	94.9	100.0	97.7	95.9	82.0	88.3	90.4	66.6	77.5	98.1	86.9	92.0
COVERS	DIVS. 10 - 14, 25, 28, 41, 43, 44, 46	100.0	96.7	99.3	100.0	101.3	100.3	100.0	100.3	100.1	100.0	95.5	99.0	100.0	91.2	98.1	100.0	96.8	99.3
21, 22, 23	FIRE SUPPRESSION, PLUMBING & HVAC	100.8	85.2	94.4	100.7	81.9	93.0	97.0	93.9	95.8	101.0	79.9	92.3	96.8	73.1	87.1	96.9	81.4	90.6
26, 27, 3370	ELECTRICAL, COMMUNICATIONS & UTIL.	101.9	90.9	96.6	96.8	73.9	85.8	100.1	95.6	97.9	97.2	84.1	90.9	97.0	70.7	84.3	102.9	93.3	98.3
MF2018	WEIGHTED AVERAGE	97.2	89.7	94.1	93.6	80.3	88.2	93.4	94.8	94.0	95.1	86.1	91.5	92.5	73.5	84.8	95.2	92.6	94.1

MINNESOTA

DIVISION		BRAINERD 564			DETROIT LAKES 565			DULUTH 556 - 558			MANKATO 560			MINNEAPOLIS 553 - 555			ROCHESTER 559		
		MAT.	INST.	TOTAL	MAT.	INST.	TOTAL	MAT.	INST.	TOTAL	MAT.	INST.	TOTAL	MAT.	INST.	TOTAL	MAT.	INST.	TOTAL
015433	CONTRACTOR EQUIPMENT		97.0	97.0		94.5	94.5		103.1	103.1		97.0	97.0		103.7	103.7		97.3	97.3
0241, 31 - 34	SITE & INFRASTRUCTURE, DEMOLITION	86.8	95.2	92.5	81.7	90.9	88.0	92.6	102.5	99.3	83.9	96.0	92.1	87.4	102.6	97.7	90.1	93.7	92.6
0310	Concrete Forming & Accessories	80.7	83.5	83.0	77.4	78.6	78.4	99.9	100.8	100.7	91.9	99.3	97.8	97.9	119.4	115.2	101.0	104.9	104.1
0320	Concrete Reinforcing	97.0	101.3	98.4	97.6	101.1	98.7	99.9	101.8	100.5	97.0	107.7	100.4	94.3	111.7	99.9	92.6	108.2	97.6
0330	Cast-in-Place Concrete	122.0	97.1	113.4	108.9	95.5	104.3	141.1	103.4	128.1	112.3	107.4	110.6	124.9	119.6	123.1	134.7	101.8	123.3
03	CONCRETE	101.4	92.8	97.9	95.0	90.0	93.0	119.7	102.8	112.9	97.3	104.6	100.2	112.0	118.7	114.7	108.7	105.4	107.4
04	MASONRY	126.4	109.5	115.7	124.5	106.1	112.9	104.5	122.5	115.9	114.3	122.7	119.6	100.1	124.4	115.4	98.9	113.7	108.2
05	METALS	86.8	118.9	94.3	85.7	118.0	93.3	93.9	119.5	99.9	86.6	122.2	95.0	98.2	127.0	104.9	93.3	125.3	100.8
06	WOOD, PLASTICS & COMPOSITES	80.6	75.3	78.4	65.0	70.2	67.1	98.0	99.2	98.5	92.0	90.0	91.2	101.2	117.5	107.9	102.7	104.0	103.3
07	THERMAL & MOISTURE PROTECTION	99.4	95.2	97.6	102.4	91.2	97.5	94.9	105.7	99.6	100.0	97.4	98.9	91.7	119.5	103.9	97.5	98.3	97.9
08	OPENINGS	83.8	100.6	87.9	95.9	97.8	96.4	99.4	109.8	101.9	88.1	109.2	93.2	99.9	127.2	106.5	94.7	118.6	100.5
0920	Plaster & Gypsum Board	81.3	75.3	77.6	91.9	69.9	78.2	92.6	99.4	96.9	88.1	90.5	89.6	91.9	118.1	108.3	100.8	104.8	103.3
0950, 0980	Ceilings & Acoustic Treatment	60.3	75.3	70.1	120.9	69.9	87.8	100.9	99.4	100.0	60.3	90.5	79.9	100.5	118.1	112.0	98.1	104.8	102.4
0960	Flooring	93.4	101.5	95.9	93.4	79.6	89.1	98.2	105.8	100.5	94.9	86.6	92.4	108.5	126.4	114.0	98.2	94.0	96.9
0970, 0990	Wall Finishes & Painting/Coating	84.4	94.1	90.3	90.9	77.4	82.6	101.7	110.1	106.8	97.1	96.2	96.6	105.0	128.7	119.5	92.0	103.1	98.8
09	FINISHES	82.3	87.0	84.8	97.6	78.0	87.0	98.5	102.8	100.8	84.4	96.4	90.9	101.3	121.7	112.4	96.3	103.1	100.0
COVERS	DIVS. 10 - 14, 25, 28, 41, 43, 44, 46	100.0	98.6	99.7	100.0	97.8	99.5	100.0	101.1	100.2	100.0	104.5	101.0	100.0	106.7	101.5	100.0	101.4	100.3
21, 22, 23	FIRE SUPPRESSION, PLUMBING & HVAC	95.9	85.1	91.5	96.9	84.0	91.6	100.8	98.2	99.8	95.9	93.0	94.7	100.7	114.6	106.4	100.8	98.9	100.0
26, 27, 3370	ELECTRICAL, COMMUNICATIONS & UTIL.	101.8	101.5	101.6	102.6	65.9	84.9	99.7	101.5	100.6	105.7	95.7	100.9	101.7	115.4	108.3	100.5	95.7	98.2
MF2018	WEIGHTED AVERAGE	94.6	96.2	95.3	95.7	88.4	92.7	101.3	104.4	102.6	94.6	102.4	97.7	101.2	117.8	108.0	99.0	104.3	101.2

DIVISION		MINNESOTA														MISSISSIPPI		
		SAINT PAUL 550 - 551			ST. CLOUD 563			THIEF RIVER FALLS 567			WILLMAR 562			WINDOM 561			BILOXI 395		
		MAT.	INST.	TOTAL	MAT.	INST.	TOTAL	MAT.	INST.	TOTAL	MAT.	INST.	TOTAL	MAT.	INST.	TOTAL	MAT.	INST.	TOTAL
015433	CONTRACTOR EQUIPMENT		103.1	103.1		97.0	97.0		94.5	94.5		97.0	97.0		97.0	97.0		97.5	97.5
0241, 31 - 34	SITE & INFRASTRUCTURE, DEMOLITION	90.1	103.2	99.0	83.4	96.4	92.2	82.5	90.5	87.9	82.1	95.1	91.0	76.8	94.4	88.8	115.0	82.4	92.9
0310	Concrete Forming & Accessories	97.9	117.7	113.9	77.1	117.0	109.3	82.7	82.2	82.3	76.8	83.3	82.0	82.6	79.3	80.0	94.9	66.4	71.9
0320	Concrete Reinforcing	94.7	108.4	99.1	97.1	108.1	100.6	97.8	101.0	98.9	96.9	107.8	100.4	96.9	106.7	100.0	88.2	63.1	80.1
0330	Cast-in-Place Concrete	141.6	117.2	133.2	107.6	114.3	109.9	111.2	89.6	103.7	109.3	76.5	98.0	94.5	80.7	89.7	128.8	66.3	107.2
03	CONCRETE	118.4	116.5	117.6	93.0	115.2	101.9	96.3	89.5	93.6	93.1	86.7	90.5	84.8	86.1	85.3	102.3	67.4	88.3
04	MASONRY	96.7	122.3	112.8	109.5	118.6	115.3	100.4	100.3	100.4	112.5	111.7	112.0	123.8	86.8	100.5	85.0	62.3	70.7
05	METALS	93.8	124.1	100.9	87.3	124.3	95.9	89.5	117.9	93.4	86.6	122.1	94.9	86.5	120.0	94.3	89.3	89.9	89.5
06	WOOD, PLASTICS & COMPOSITES	98.1	115.5	105.3	77.2	115.0	92.9	69.7	78.2	73.2	76.9	75.4	76.2	82.2	75.4	79.4	94.3	67.0	83.0
07	THERMAL & MOISTURE PROTECTION	91.5	118.4	103.3	99.6	114.9	106.3	103.7	85.3	95.6	99.3	95.5	97.7	99.4	79.0	90.5	100.9	61.8	83.7
08	OPENINGS	94.4	124.9	101.8	88.5	124.7	97.2	96.0	102.2	97.5	85.7	98.8	88.8	88.9	98.8	91.3	92.6	61.7	85.1
0920	Plaster & Gypsum Board	91.9	116.2	107.1	81.3	116.2	103.1	92.4	78.2	83.5	81.3	75.4	77.6	81.3	75.4	77.6	125.3	66.7	88.6
0950, 0980	Ceilings & Acoustic Treatment	100.5	116.2	110.7	60.3	116.2	96.6	120.9	78.2	93.2	60.3	75.4	70.1	60.3	75.4	70.1	94.9	66.7	76.6
0960	Flooring	97.4	122.4	105.0	90.5	91.5	90.8	94.2	79.6	89.7	91.9	79.6	88.1	94.1	79.6	89.7	93.0	62.7	83.7
0970, 0990	Wall Finishes & Painting/Coating	101.8	127.9	117.8	97.1	128.7	116.4	90.9	77.4	82.6	90.9	77.4	82.6	90.9	96.2	94.1	90.7	47.1	64.1
09	FINISHES	98.0	119.7	109.8	82.0	114.0	99.4	98.0	81.3	88.9	82.0	81.2	81.6	82.1	80.6	81.3	98.3	63.9	79.6
COVERS	DIVS. 10 - 14, 25, 28, 41, 43, 44, 46	100.0	106.2	101.4	100.0	105.2	101.1	100.0	96.7	99.3	100.0	98.5	99.7	100.0	95.6	99.0	100.0	82.4	96.2
21, 22, 23	FIRE SUPPRESSION, PLUMBING & HVAC	100.7	114.7	106.4	100.2	111.8	104.9	96.9	81.0	90.4	95.9	101.3	98.2	95.9	78.5	88.8	100.6	55.1	81.9
26, 27, 3370	ELECTRICAL, COMMUNICATIONS & UTIL.	102.0	115.4	108.5	101.8	115.4	108.4	101.3	65.9	84.2	101.8	77.6	90.1	105.7	95.7	100.9	96.9	54.0	76.2
MF2018	WEIGHTED AVERAGE	100.4	116.7	107.0	94.0	114.1	102.2	94.9	87.5	91.9	92.8	95.2	93.8	92.8	89.4	91.4	96.9	65.2	84.0

DIVISION		MISSISSIPPI																
		CLARKSDALE 386			COLUMBUS 397			GREENVILLE 387			GREENWOOD 389			JACKSON 390 - 392			LAUREL 394		
		MAT.	INST.	TOTAL	MAT.	INST.	TOTAL	MAT.	INST.	TOTAL	MAT.	INST.	TOTAL	MAT.	INST.	TOTAL	MAT.	INST.	TOTAL
015433	CONTRACTOR EQUIPMENT		97.5	97.5		97.5	97.5		97.5	97.5		97.5	97.5		102.9	102.9		97.5	97.5
0241, 31 - 34	SITE & INFRASTRUCTURE, DEMOLITION	112.7	80.8	91.0	113.5	81.5	91.8	120.7	82.4	94.7	116.9	80.5	92.2	110.0	91.5	97.4	119.9	80.8	93.4
0310	Concrete Forming & Accessories	80.5	41.6	49.1	78.9	43.7	50.5	76.1	60.5	63.6	93.5	41.8	51.8	98.6	66.0	72.3	79.2	56.3	60.7
0320	Concrete Reinforcing	100.1	62.0	87.9	94.3	62.1	84.0	100.7	62.2	88.3	100.1	62.0	87.9	99.5	66.1	88.7	94.9	30.5	74.3
0330	Cast-in-Place Concrete	105.9	56.1	88.7	131.4	58.2	106.1	109.1	64.6	93.7	113.9	55.9	93.9	111.0	68.6	96.4	128.5	57.4	103.9
03	CONCRETE	96.4	52.3	78.8	103.8	54.2	83.9	100.9	64.0	86.1	102.5	52.3	82.4	99.5	68.3	87.0	105.2	54.1	84.7
04	MASONRY	93.1	48.8	65.2	109.8	51.3	73.0	138.8	61.9	90.3	93.9	48.7	65.4	91.6	62.7	73.4	105.6	51.3	71.4
05	METALS	92.3	86.0	90.9	86.3	88.4	86.8	93.1	89.5	92.3	92.3	85.8	90.8	95.6	88.1	93.8	86.4	74.9	83.7
06	WOOD, PLASTICS & COMPOSITES	76.3	41.9	62.1	76.2	42.7	62.3	72.3	59.3	66.9	92.1	41.9	71.3	98.6	66.3	85.2	77.5	60.9	70.6
07	THERMAL & MOISTURE PROTECTION	98.3	48.7	76.5	100.6	51.9	79.3	98.7	60.9	82.1	98.8	50.7	77.7	98.4	63.7	83.2	100.9	53.9	80.3
08	OPENINGS	91.5	45.3	80.4	92.3	45.8	81.1	91.3	55.0	82.5	91.5	45.3	80.4	95.6	59.8	86.9	89.8	48.7	79.9
0920	Plaster & Gypsum Board	93.5	40.9	60.6	106.0	41.7	65.8	92.6	58.8	71.4	114.6	40.9	68.5	107.0	65.6	81.1	106.0	60.4	77.5
0950, 0980	Ceilings & Acoustic Treatment	90.1	40.9	58.2	87.3	41.7	57.7	91.6	58.8	70.3	90.1	40.9	58.2	99.0	65.6	77.3	87.3	60.4	69.8
0960	Flooring	100.3	62.7	88.8	88.2	62.7	80.4	99.0	62.7	87.9	104.8	62.7	92.0	90.8	64.0	82.6	87.2	62.7	79.7
0970, 0990	Wall Finishes & Painting/Coating	95.6	43.5	63.8	90.7	43.5	61.9	95.6	54.7	70.7	95.6	43.5	63.8	98.7	54.7	71.9	90.7	43.5	61.9
09	FINISHES	95.6	45.4	68.3	92.3	46.5	67.4	96.2	60.1	76.6	100.7	45.4	70.7	98.3	64.5	80.0	92.7	57.2	73.4
COVERS	DIVS. 10 - 14, 25, 28, 41, 43, 44, 46	100.0	44.2	87.9	100.0	45.1	88.1	100.0	81.4	96.0	100.0	44.2	87.9	100.0	83.0	96.3	100.0	31.6	85.1
21, 22, 23	FIRE SUPPRESSION, PLUMBING & HVAC	98.7	47.5	77.7	98.2	49.4	78.2	100.7	54.9	81.9	98.7	47.9	77.9	100.6	58.4	83.3	98.3	44.1	76.0
26, 27, 3370	ELECTRICAL, COMMUNICATIONS & UTIL.	96.9	38.6	68.8	95.3	51.0	73.9	96.9	55.0	76.7	96.9	36.1	67.6	99.0	55.0	77.8	96.2	53.7	75.7
MF2018	WEIGHTED AVERAGE	96.0	52.6	78.3	96.1	55.8	79.7	99.2	63.8	84.8	97.5	52.3	79.1	98.4	66.8	85.5	96.2	55.1	79.5

DIVISION		MISSISSIPPI									MISSOURI								
		MCCOMB 396			MERIDIAN 393			TUPELO 388			BOWLING GREEN 633			CAPE GIRARDEAU 637			CHILLICOTHE 646		
		MAT.	INST.	TOTAL	MAT.	INST.	TOTAL	MAT.	INST.	TOTAL	MAT.	INST.	TOTAL	MAT.	INST.	TOTAL	MAT.	INST.	TOTAL
015433	CONTRACTOR EQUIPMENT		97.5	97.5		97.5	97.5		97.5	97.5		104.0	104.0		104.0	104.0		98.7	98.7
0241, 31 - 34	SITE & INFRASTRUCTURE, DEMOLITION	104.6	80.6	88.3	109.8	82.4	91.2	110.5	80.7	90.3	86.6	85.5	85.9	88.5	85.6	86.6	105.2	84.2	90.9
0310	Concrete Forming & Accessories	78.9	42.9	49.8	75.2	66.6	68.3	76.8	43.1	49.6	89.7	88.6	88.8	80.2	79.5	79.7	82.4	90.2	88.7
0320	Concrete Reinforcing	95.5	32.0	75.1	94.3	66.0	85.2	98.0	61.9	86.4	103.7	89.9	99.3	105.0	83.0	97.9	102.3	95.5	100.1
0330	Cast-in-Place Concrete	114.0	55.7	93.9	122.0	66.3	102.8	105.9	57.0	89.0	79.6	89.4	83.0	78.7	82.9	80.2	93.4	80.6	89.0
03	CONCRETE	94.0	47.6	75.4	98.0	68.0	86.0	95.6	53.3	78.7	87.9	90.7	89.0	87.1	83.2	85.5	95.4	88.8	92.8
04	MASONRY	109.6	48.3	70.9	84.4	62.3	70.5	128.6	49.7	78.9	129.1	91.5	105.4	125.3	78.1	95.5	101.2	86.1	91.7
05	METALS	86.5	73.9	83.6	87.1	91.2	88.1	92.2	85.6	90.7	85.8	114.7	92.5	86.6	111.0	92.3	82.0	108.1	88.1
06	WOOD, PLASTICS & COMPOSITES	76.2	44.2	62.9	73.0	67.0	70.5	73.0	42.7	60.5	93.5	89.1	91.7	84.5	77.8	81.7	88.8	91.6	90.0
07	THERMAL & MOISTURE PROTECTION	100.1	50.8	78.5	100.2	62.1	83.5	98.2	49.0	76.6	95.3	91.9	93.8	94.7	80.6	88.5	90.3	85.9	88.3
08	OPENINGS	92.4	38.8	79.4	92.1	63.4	85.1	91.5	48.4	81.1	95.1	92.4	94.4	95.1	76.2	90.5	83.7	90.5	85.3
0920	Plaster & Gypsum Board	106.0	43.2	66.7	106.0	66.7	81.4	92.6	41.7	60.8	82.0	89.3	86.6	81.0	77.7	78.9	78.4	91.4	86.6
0950, 0980	Ceilings & Acoustic Treatment	87.3	43.2	58.7	88.0	66.7	74.2	90.1	41.7	58.7	98.0	89.3	92.3	98.0	77.7	84.8	94.6	91.4	92.5
0960	Flooring	88.2	62.7	80.4	87.2	62.7	79.7	99.2	62.7	88.0	97.9	90.1	95.5	95.1	82.5	91.2	93.8	92.7	93.5
0970, 0990	Wall Finishes & Painting/Coating	90.7	43.5	61.9	90.7	54.8	68.8	95.6	42.0	62.9	100.0	98.6	99.1	100.0	74.8	84.6	98.1	95.8	96.7
09	FINISHES	91.6	46.6	67.2	91.8	64.7	77.1	95.2	46.2	68.6	95.8	89.7	92.5	94.8	78.9	86.1	94.6	91.5	92.9
COVERS	DIVS. 10 - 14, 25, 28, 41, 43, 44, 46	100.0	46.9	88.5	100.0	82.4	96.2	100.0	45.1	88.1	100.0	96.3	99.2	100.0	95.1	98.9	100.0	96.1	99.2
21, 22, 23	FIRE SUPPRESSION, PLUMBING & HVAC	98.2	47.0	77.2	100.6	58.5	83.3	98.8	48.8	78.3	96.5	92.2	94.8	100.7	95.6	98.6	96.5	92.6	94.9
26, 27, 3370	ELECTRICAL, COMMUNICATIONS & UTIL.	94.3	51.9	73.9	96.2	55.9	76.8	96.8	51.0	74.7	98.4	71.2	85.3	98.4	91.3	94.9	95.6	70.3	83.4
MF2018	WEIGHTED AVERAGE	94.4	52.6	77.4	94.8	66.5	83.3	97.2	55.1	80.0	94.5	90.2	92.7	95.1	88.5	92.4	92.5	88.7	90.9

City Cost Indexes

MISSOURI

DIVISION		COLUMBIA 652 MAT.	INST.	TOTAL	FLAT RIVER 636 MAT.	INST.	TOTAL	HANNIBAL 634 MAT.	INST.	TOTAL	HARRISONVILLE 647 MAT.	INST.	TOTAL	JEFFERSON CITY 650-651 MAT.	INST.	TOTAL	JOPLIN 648 MAT.	INST.	TOTAL
015433	CONTRACTOR EQUIPMENT		107.0	107.0		104.0	104.0		104.0	104.0		98.7	98.7		114.1	114.1		102.0	102.0
0241, 31 - 34	SITE & INFRASTRUCTURE, DEMOLITION	88.7	90.2	89.7	89.0	85.2	86.4	84.4	85.3	85.0	96.7	85.2	88.9	89.2	101.3	97.4	106.8	88.9	94.6
0310	Concrete Forming & Accessories	76.2	85.3	83.6	98.6	83.0	86.0	87.6	77.8	79.7	78.5	93.3	90.4	96.9	75.2	79.4	97.5	74.0	78.6
0320	Concrete Reinforcing	95.6	97.7	96.3	105.0	95.8	102.1	103.1	89.8	98.9	102.1	103.3	102.5	99.9	87.1	95.8	105.7	100.0	103.9
0330	Cast-in-Place Concrete	78.3	82.5	79.8	82.2	84.7	83.0	75.3	87.5	79.5	95.7	93.4	94.9	83.1	82.2	82.8	101.2	73.1	91.5
03	CONCRETE	79.6	88.3	83.1	90.8	87.6	89.5	84.9	85.1	85.0	92.6	96.0	93.9	87.7	81.5	85.2	96.4	79.5	89.6
04	MASONRY	151.5	84.6	109.3	125.8	71.3	91.5	119.4	88.3	99.8	95.2	92.9	93.7	100.7	84.7	90.7	95.3	78.5	84.7
05	METALS	91.0	118.8	97.5	85.7	116.0	92.7	85.8	114.2	92.4	82.3	112.5	89.4	91.1	110.4	95.6	84.1	106.1	89.3
06	WOOD, PLASTICS & COMPOSITES	76.8	84.7	80.1	103.7	84.5	95.7	91.3	76.6	85.2	84.9	93.2	88.3	98.6	71.6	87.4	103.3	73.5	90.9
07	THERMAL & MOISTURE PROTECTION	89.5	85.0	87.6	95.7	84.3	90.7	95.1	86.6	91.4	89.3	93.4	91.1	95.9	84.9	91.1	90.3	78.4	85.1
08	OPENINGS	91.1	86.7	90.0	95.0	91.7	94.2	95.1	78.8	91.1	83.4	97.2	86.7	91.1	76.7	87.6	84.5	76.7	82.6
0920	Plaster & Gypsum Board	73.7	84.6	80.5	91.6	84.6	87.2	81.5	76.4	78.3	76.8	93.1	87.0	85.8	70.7	76.4	89.8	72.8	79.2
0950, 0980	Ceilings & Acoustic Treatment	83.2	84.6	84.1	98.0	84.6	89.3	98.0	76.4	84.0	94.6	93.1	93.6	88.7	70.7	77.0	96.0	72.8	80.9
0960	Flooring	88.7	93.5	90.2	100.6	78.6	93.9	97.3	90.1	95.1	89.3	93.7	90.7	93.7	66.5	85.4	117.5	69.7	102.9
0970, 0990	Wall Finishes & Painting/Coating	98.9	79.9	87.3	100.0	66.6	79.6	100.0	87.3	92.3	102.8	100.0	101.1	100.2	79.9	87.8	97.8	70.4	81.1
09	FINISHES	83.7	85.9	84.8	98.2	79.8	88.2	95.3	80.0	87.0	92.7	93.9	93.3	92.3	73.3	82.0	101.6	72.9	86.0
COVERS	DIVS. 10 - 14, 25, 28, 41, 43, 44, 46	100.0	94.9	98.9	100.0	92.1	98.3	100.0	94.0	98.7	100.0	97.4	99.4	100.0	95.9	99.1	100.0	92.9	98.5
21, 22, 23	FIRE SUPPRESSION, PLUMBING & HVAC	100.6	96.1	98.8	96.5	89.7	93.7	96.5	90.6	94.1	96.5	93.4	95.2	100.7	90.7	96.6	100.8	68.0	87.3
26, 27, 3370	ELECTRICAL, COMMUNICATIONS & UTIL.	96.0	79.4	88.0	101.2	91.3	96.4	97.6	71.2	84.9	99.7	93.8	96.9	101.6	79.4	90.9	94.2	71.9	83.5
MF2018	WEIGHTED AVERAGE	94.4	90.9	93.0	95.4	88.5	92.6	93.4	86.6	90.6	91.9	95.2	93.2	95.1	86.8	91.7	94.4	78.4	87.9

MISSOURI

DIVISION		KANSAS CITY 640-641 MAT.	INST.	TOTAL	KIRKSVILLE 635 MAT.	INST.	TOTAL	POPLAR BLUFF 639 MAT.	INST.	TOTAL	ROLLA 654-655 MAT.	INST.	TOTAL	SEDALIA 653 MAT.	INST.	TOTAL	SIKESTON 638 MAT.	INST.	TOTAL
015433	CONTRACTOR EQUIPMENT		104.8	104.8		94.5	94.5		96.7	96.7		107.0	107.0		97.5	97.5		96.7	96.7
0241, 31 - 34	SITE & INFRASTRUCTURE, DEMOLITION	98.4	98.5	98.5	86.7	80.6	82.6	74.3	84.0	80.9	87.0	90.0	89.0	86.4	84.9	85.3	77.5	84.6	82.3
0310	Concrete Forming & Accessories	97.6	102.5	101.6	78.6	73.9	74.8	78.7	74.6	75.4	86.5	95.7	93.9	84.7	73.7	75.8	79.7	74.6	75.6
0320	Concrete Reinforcing	100.4	108.5	103.0	103.1	77.6	94.9	106.4	70.7	94.9	96.1	98.9	97.0	94.3	102.7	97.0	105.7	70.7	94.5
0330	Cast-in-Place Concrete	99.3	102.1	100.2	82.1	77.0	80.3	63.2	78.1	68.3	80.1	94.5	85.1	83.7	76.2	81.1	67.5	78.1	71.2
03	CONCRETE	96.6	103.8	99.5	99.3	77.0	90.3	79.8	76.5	78.5	81.5	97.3	87.8	91.7	81.0	87.4	82.6	76.5	80.2
04	MASONRY	100.8	100.1	100.4	133.5	79.2	99.3	123.6	69.2	89.3	120.6	86.5	99.1	128.1	76.3	95.4	123.3	69.2	89.2
05	METALS	97.7	110.4	100.7	86.1	98.5	89.0	86.6	95.2	88.6	90.7	119.3	97.4	91.2	110.0	95.6	87.4	95.2	89.3
06	WOOD, PLASTICS & COMPOSITES	101.2	103.1	102.0	77.8	72.2	75.5	76.7	75.5	76.2	85.9	97.8	90.8	80.5	72.4	77.1	78.9	75.5	77.5
07	THERMAL & MOISTURE PROTECTION	91.0	102.6	96.1	99.9	84.2	93.0	98.8	76.9	89.2	89.9	92.1	90.9	94.1	81.3	88.5	99.0	76.1	89.0
08	OPENINGS	97.0	102.0	98.2	100.3	74.2	94.0	101.1	70.8	93.8	91.1	94.3	91.9	96.0	81.9	92.6	101.1	70.8	93.8
0920	Plaster & Gypsum Board	89.4	103.1	98.0	73.2	71.9	72.4	73.5	75.3	74.6	76.1	98.1	89.8	70.5	71.9	71.4	77.4	75.3	76.1
0950, 0980	Ceilings & Acoustic Treatment	99.9	103.1	102.0	96.6	71.9	80.6	98.0	75.3	83.3	83.2	98.1	92.9	83.2	71.9	75.9	98.0	75.3	83.3
0960	Flooring	93.2	100.5	95.5	76.2	89.7	80.4	90.0	78.6	86.5	91.6	94.2	92.4	71.2	67.8	70.2	90.5	78.6	86.9
0970, 0990	Wall Finishes & Painting/Coating	105.0	110.7	108.5	95.5	72.7	81.6	94.7	62.2	74.9	98.9	91.1	94.1	98.9	95.8	97.0	94.7	62.2	74.9
09	FINISHES	96.6	103.1	100.2	94.0	76.0	84.2	93.7	73.5	82.7	84.9	94.9	90.3	83.4	74.1	78.3	94.7	73.8	83.3
COVERS	DIVS. 10 - 14, 25, 28, 41, 43, 44, 46	100.0	100.0	100.0	100.0	93.5	98.6	100.0	92.6	98.4	100.0	97.9	99.5	100.0	91.6	98.2	100.0	90.4	97.9
21, 22, 23	FIRE SUPPRESSION, PLUMBING & HVAC	100.6	101.0	100.8	96.5	90.0	93.9	96.5	88.0	93.0	96.4	98.4	97.2	96.4	88.0	92.9	96.5	88.0	93.0
26, 27, 3370	ELECTRICAL, COMMUNICATIONS & UTIL.	101.8	96.2	99.1	96.8	71.1	84.5	97.4	91.2	94.4	95.1	73.9	84.9	95.2	93.8	94.5	96.9	91.2	94.1
MF2018	WEIGHTED AVERAGE	98.6	101.6	99.8	96.4	81.8	90.5	93.2	82.4	88.8	92.5	94.0	93.1	94.7	86.1	91.2	93.8	82.4	89.2

MISSOURI / MONTANA

DIVISION		SPRINGFIELD 656-658 MAT.	INST.	TOTAL	ST. JOSEPH 644-645 MAT.	INST.	TOTAL	ST. LOUIS 630-631 MAT.	INST.	TOTAL	BILLINGS 590-591 MAT.	INST.	TOTAL	BUTTE 597 MAT.	INST.	TOTAL	GREAT FALLS 594 MAT.	INST.	TOTAL
015433	CONTRACTOR EQUIPMENT		99.7	99.7		98.7	98.7		108.7	108.7		94.7	94.7		94.5	94.5		94.5	94.5
0241, 31 - 34	SITE & INFRASTRUCTURE, DEMOLITION	88.1	88.5	88.3	100.6	83.4	88.9	95.0	99.3	97.9	83.4	89.3	87.4	86.7	88.7	88.1	90.0	88.9	89.3
0310	Concrete Forming & Accessories	95.3	76.6	80.2	96.1	91.2	92.1	96.1	103.6	102.2	102.0	68.1	74.7	84.1	67.6	70.8	102.0	67.7	74.4
0320	Concrete Reinforcing	91.5	97.3	93.4	99.1	106.2	101.4	100.7	110.0	103.7	91.8	82.2	88.7	99.7	81.6	93.9	91.9	82.0	88.8
0330	Cast-in-Place Concrete	85.0	75.3	81.7	94.0	94.2	94.1	88.9	103.9	94.1	126.4	70.1	106.9	139.1	68.6	114.8	147.0	69.6	120.3
03	CONCRETE	88.9	80.9	85.7	92.1	95.8	93.6	91.6	105.7	97.3	102.9	72.3	90.6	107.2	71.4	92.9	111.8	71.8	95.8
04	MASONRY	95.0	82.3	87.0	95.7	90.9	92.7	101.6	104.4	103.4	152.8	83.1	108.9	146.1	74.3	100.8	152.3	78.3	105.7
05	METALS	97.3	104.5	98.9	88.8	113.3	94.5	97.4	120.2	102.7	100.7	90.6	98.4	94.7	89.4	93.4	98.1	90.3	96.3
06	WOOD, PLASTICS & COMPOSITES	89.4	74.6	83.3	103.3	91.3	98.3	98.7	102.7	100.3	89.7	65.0	79.5	73.8	65.0	70.2	91.3	65.0	80.4
07	THERMAL & MOISTURE PROTECTION	93.9	76.2	86.2	90.1	90.0	90.0	97.2	104.6	100.4	128.2	72.3	103.7	127.6	67.7	101.3	128.4	69.9	102.8
08	OPENINGS	97.9	86.2	95.1	87.2	97.7	89.7	101.3	108.1	102.9	95.3	66.0	88.3	93.7	65.6	86.9	96.4	70.0	90.0
0920	Plaster & Gypsum Board	78.0	74.2	75.6	91.4	91.2	91.3	87.4	103.1	97.2	116.6	64.6	84.1	105.8	64.6	80.0	123.0	64.6	86.4
0950, 0980	Ceilings & Acoustic Treatment	83.2	74.2	77.4	103.5	91.2	95.5	94.4	103.1	100.0	101.7	64.6	77.6	107.3	64.6	79.6	109.3	64.6	80.3
0960	Flooring	90.3	69.7	84.0	97.7	97.4	97.6	99.5	97.6	98.9	93.1	78.2	88.6	91.0	77.4	86.9	97.3	75.2	90.5
0970, 0990	Wall Finishes & Painting/Coating	92.9	101.8	98.4	98.1	100.4	99.5	105.0	107.1	106.3	91.0	91.4	91.3	89.9	62.9	73.4	89.8	88.7	89.2
09	FINISHES	87.4	78.1	82.4	99.0	93.4	96.0	96.9	102.6	100.0	96.3	71.7	82.9	94.9	68.4	80.5	100.1	70.6	84.1
COVERS	DIVS. 10 - 14, 25, 28, 41, 43, 44, 46	100.0	93.8	98.7	100.0	96.7	99.3	100.0	101.6	100.4	100.0	94.0	98.7	100.0	94.0	98.7	100.0	93.7	98.6
21, 22, 23	FIRE SUPPRESSION, PLUMBING & HVAC	100.7	72.0	88.9	100.8	88.2	95.6	100.8	104.6	102.4	100.7	75.0	90.2	100.7	66.6	86.7	100.7	72.0	89.0
26, 27, 3370	ELECTRICAL, COMMUNICATIONS & UTIL.	98.1	64.6	82.0	99.8	77.0	88.8	101.9	103.9	102.9	97.5	73.7	86.0	101.3	67.8	85.1	96.9	70.6	84.2
MF2018	WEIGHTED AVERAGE	96.0	79.6	89.3	95.2	91.2	93.6	98.4	105.4	101.3	102.1	77.4	92.0	101.2	73.0	89.7	103.3	75.7	92.0

MONTANA

DIVISION		HAVRE 595 MAT.	INST.	TOTAL	HELENA 596 MAT.	INST.	TOTAL	KALISPELL 599 MAT.	INST.	TOTAL	MILES CITY 593 MAT.	INST.	TOTAL	MISSOULA 598 MAT.	INST.	TOTAL	WOLF POINT 592 MAT.	INST.	TOTAL
015433	CONTRACTOR EQUIPMENT		94.5	94.5		101.9	101.9		94.5	94.5		94.5	94.5		94.5	94.5		94.5	94.5
0241, 31 - 34	SITE & INFRASTRUCTURE, DEMOLITION	91.6	88.3	89.3	80.7	100.1	93.9	77.6	88.4	84.9	82.7	88.3	86.5	72.2	88.5	83.2	95.9	88.3	90.7
0310	Concrete Forming & Accessories	74.5	66.2	67.8	104.8	67.4	74.6	88.8	67.1	71.3	99.7	66.3	72.8	88.0	67.6	71.7	89.6	66.3	70.8
0320	Concrete Reinforcing	100.5	81.8	94.5	101.8	81.6	95.3	102.3	81.8	95.8	99.9	81.8	94.1	101.4	81.6	95.1	101.4	81.8	95.1
0330	Cast-in-Place Concrete	149.9	66.6	121.1	107.2	70.1	94.4	120.7	67.8	102.5	132.2	66.7	109.6	102.5	68.3	90.7	148.2	65.7	119.7
03	CONCRETE	114.3	70.0	96.6	107.8	71.6	89.7	97.8	70.9	87.0	104.7	70.1	90.8	87.1	71.2	80.7	117.5	69.8	98.4
04	MASONRY	147.4	75.2	101.9	135.9	73.2	96.4	144.4	77.2	102.0	155.1	75.2	104.8	174.2	74.0	111.0	156.7	75.2	105.4
05	METALS	91.2	88.7	90.6	96.5	86.1	94.0	91.0	88.9	90.5	90.5	88.8	90.1	91.4	88.7	90.8	90.6	88.8	90.2
06	WOOD, PLASTICS & COMPOSITES	63.8	65.0	64.3	94.4	65.3	82.4	77.9	65.0	72.5	87.7	65.0	78.3	77.9	65.0	72.5	77.2	65.0	72.1
07	THERMAL & MOISTURE PROTECTION	127.9	61.8	98.9	131.6	68.9	104.1	127.2	68.6	101.5	127.7	63.1	99.4	126.7	69.8	101.8	128.6	63.0	99.8
08	OPENINGS	94.2	65.6	87.3	93.3	65.7	86.6	94.2	65.6	87.3	93.8	65.6	87.0	93.8	65.6	87.0	93.8	65.6	87.0
0920	Plaster & Gypsum Board	98.3	64.6	77.2	111.3	64.6	82.1	105.8	64.6	80.0	121.7	64.6	86.0	105.8	64.6	80.0	111.3	64.6	82.1
0950, 0980	Ceilings & Acoustic Treatment	107.3	64.6	79.6	109.9	64.6	80.5	107.3	64.6	79.6	103.2	64.6	78.1	107.3	64.6	79.6	103.2	64.6	78.1
0960	Flooring	88.9	84.6	87.6	103.0	77.4	95.2	92.7	84.6	90.2	97.2	84.6	93.3	92.7	77.4	88.0	93.9	84.6	91.1
0970, 0990	Wall Finishes & Painting/Coating	89.9	62.9	73.4	98.7	55.7	72.5	89.9	84.5	86.6	89.9	62.9	73.4	89.9	84.5	86.6	89.9	62.9	73.4
09	FINISHES	93.7	69.2	80.4	103.1	67.6	83.8	94.9	72.0	82.5	98.1	69.2	82.4	94.4	70.8	81.5	96.8	69.2	81.8
COVERS	DIVS. 10 - 14, 25, 28, 41, 43, 44, 46	100.0	82.2	96.1	100.0	89.9	97.8	100.0	82.8	96.3	100.0	82.2	96.1	100.0	89.5	97.7	100.0	82.2	96.1
21, 22, 23	FIRE SUPPRESSION, PLUMBING & HVAC	96.5	63.9	83.1	100.0	66.1	86.6	96.5	64.3	83.3	96.5	69.8	85.5	100.7	66.5	86.7	96.5	69.8	85.5
26, 27, 3370	ELECTRICAL, COMMUNICATIONS & UTIL.	96.8	64.2	81.1	103.3	68.2	86.4	99.3	63.1	81.9	96.8	69.0	83.4	100.0	66.5	83.8	96.8	69.0	83.4
MF2018	WEIGHTED AVERAGE	100.1	71.3	88.4	101.4	73.3	90.0	97.8	72.2	87.4	99.3	73.3	88.7	98.4	72.9	88.0	101.3	73.3	89.9

NEBRASKA

DIVISION		ALLIANCE 693 MAT.	INST.	TOTAL	COLUMBUS 686 MAT.	INST.	TOTAL	GRAND ISLAND 688 MAT.	INST.	TOTAL	HASTINGS 689 MAT.	INST.	TOTAL	LINCOLN 683 - 685 MAT.	INST.	TOTAL	MCCOOK 690 MAT.	INST.	TOTAL
015433	CONTRACTOR EQUIPMENT		92.5	92.5		98.9	98.9		98.9	98.9		98.9	98.9		105.5	105.5		98.9	98.9
0241, 31 - 34	SITE & INFRASTRUCTURE, DEMOLITION	95.8	91.4	92.8	97.0	86.1	89.6	102.0	86.8	91.7	100.2	86.1	90.6	92.0	97.7	95.9	96.4	86.0	89.4
0310	Concrete Forming & Accessories	82.8	52.1	58.0	95.9	70.6	75.5	95.3	79.2	82.3	83.3	74.8	74.4	96.1	72.4	77.0	89.1	52.6	59.6
0320	Concrete Reinforcing	111.8	81.3	102.0	101.3	80.0	94.5	100.8	78.6	93.6	100.8	70.9	91.2	99.5	78.9	92.9	107.8	71.0	96.0
0330	Cast-in-Place Concrete	97.4	76.6	90.2	97.8	76.6	90.5	103.5	74.3	93.5	103.5	69.3	91.7	81.9	80.8	81.5	105.3	69.2	92.9
03	CONCRETE	109.1	66.6	92.1	97.4	75.6	88.7	101.2	78.5	92.1	101.5	70.4	89.1	90.8	77.5	85.5	101.1	63.3	86.0
04	MASONRY	114.2	72.3	87.8	121.9	72.3	90.6	115.0	71.2	87.4	125.9	69.1	90.1	102.5	75.4	85.4	107.7	72.3	85.4
05	METALS	94.0	83.1	91.5	86.0	95.5	88.3	87.4	95.8	89.3	88.4	91.4	89.1	95.2	93.1	94.7	89.1	91.4	89.6
06	WOOD, PLASTICS & COMPOSITES	80.9	46.0	66.4	92.7	70.6	83.6	91.8	80.2	87.0	96.3	68.0	84.6	98.3	70.9	87.0	88.5	46.9	71.2
07	THERMAL & MOISTURE PROTECTION	103.3	70.2	88.8	102.7	74.5	90.3	102.9	76.9	91.5	103.0	72.0	89.4	102.1	77.4	91.3	97.1	70.5	85.5
08	OPENINGS	87.8	54.7	79.8	88.7	69.4	84.0	88.7	78.0	86.1	88.7	65.7	83.1	101.9	66.8	93.4	88.2	52.5	79.6
0920	Plaster & Gypsum Board	73.5	44.8	55.6	88.0	70.1	76.8	86.8	79.9	82.5	89.2	67.4	75.6	97.5	70.1	80.4	80.4	45.7	58.7
0950, 0980	Ceilings & Acoustic Treatment	89.7	44.8	60.6	82.9	70.1	74.6	82.9	79.9	81.0	82.9	67.4	72.9	109.3	70.1	83.9	86.5	45.7	60.0
0960	Flooring	92.4	81.3	89.0	81.3	81.3	81.3	81.1	73.4	78.7	82.1	75.6	80.1	92.1	80.5	88.5	89.5	81.3	87.0
0970, 0990	Wall Finishes & Painting/Coating	165.9	47.9	94.0	72.9	55.7	62.4	72.9	75.0	74.2	72.9	55.7	62.4	98.7	73.6	83.4	88.0	41.8	59.8
09	FINISHES	91.3	55.5	71.8	84.0	70.8	76.8	84.0	77.7	80.6	84.7	68.1	75.7	98.4	73.7	84.9	87.8	55.3	70.2
COVERS	DIVS. 10 - 14, 25, 28, 41, 43, 44, 46	100.0	84.8	96.7	100.0	87.6	97.3	100.0	89.6	97.7	100.0	87.3	97.2	100.0	89.3	97.7	100.0	85.0	96.7
21, 22, 23	FIRE SUPPRESSION, PLUMBING & HVAC	96.8	69.5	85.6	96.7	69.8	85.7	101.0	79.8	92.3	96.7	69.1	85.4	100.7	79.9	92.2	96.4	69.7	85.4
26, 27, 3370	ELECTRICAL, COMMUNICATIONS & UTIL.	92.0	60.8	77.0	94.3	75.5	85.2	93.5	63.6	79.1	93.1	73.6	83.7	102.3	63.6	83.6	93.2	60.9	77.6
MF2018	WEIGHTED AVERAGE	97.1	68.9	85.6	94.2	76.0	86.8	95.6	78.3	88.6	95.4	73.6	86.5	98.1	78.2	90.0	94.5	68.8	84.1

NEBRASKA / NEVADA

DIVISION		NORFOLK 687 MAT.	INST.	TOTAL	NORTH PLATTE 691 MAT.	INST.	TOTAL	OMAHA 680 - 681 MAT.	INST.	TOTAL	VALENTINE 692 MAT.	INST.	TOTAL	CARSON CITY 897 MAT.	INST.	TOTAL	ELKO 898 MAT.	INST.	TOTAL
015433	CONTRACTOR EQUIPMENT		89.0	89.0		98.9	98.9		96.9	96.9		92.2	92.2		97.6	97.6		92.1	92.1
0241, 31 - 34	SITE & INFRASTRUCTURE, DEMOLITION	78.7	85.2	83.1	98.3	85.9	89.9	86.6	97.8	94.2	82.4	90.4	87.8	81.0	96.9	91.8	66.1	86.9	80.2
0310	Concrete Forming & Accessories	79.7	69.7	71.7	92.7	71.6	75.7	95.5	76.7	80.3	79.2	50.3	55.9	110.3	82.6	88.0	115.6	90.3	95.2
0320	Concrete Reinforcing	101.4	61.9	88.7	107.3	76.1	97.3	100.7	79.1	93.7	107.9	61.8	93.1	102.3	115.0	106.4	103.2	88.2	98.4
0330	Cast-in-Place Concrete	98.4	66.7	87.5	105.3	63.8	91.0	80.9	80.7	80.8	92.9	50.7	78.4	88.9	82.6	86.7	83.2	67.7	77.9
03	CONCRETE	95.8	68.0	84.7	101.2	70.9	89.1	90.6	78.9	85.9	92.9	53.7	80.5	102.9	88.2	97.0	93.1	82.0	88.7
04	MASONRY	130.7	72.2	93.9	97.6	71.8	81.3	103.3	79.6	88.4	109.1	71.8	85.6	109.1	65.6	81.7	116.7	63.2	83.0
05	METALS	89.2	78.3	86.6	88.1	93.7	89.4	95.2	85.0	92.8	99.6	77.7	94.5	103.2	94.0	101.1	107.3	84.8	102.1
06	WOOD, PLASTICS & COMPOSITES	77.4	70.2	74.4	90.9	72.6	83.3	95.4	76.4	87.5	76.5	44.8	63.4	96.3	83.6	91.0	108.9	96.5	103.8
07	THERMAL & MOISTURE PROTECTION	101.1	72.5	88.6	97.1	74.1	87.0	99.8	80.8	91.5	96.6	68.5	84.3	119.2	79.2	101.7	114.2	68.2	94.0
08	OPENINGS	89.4	64.6	83.4	87.6	72.4	83.9	96.0	77.3	91.5	89.1	50.3	79.8	96.6	81.8	93.1	97.7	83.2	94.2
0920	Plaster & Gypsum Board	87.5	70.1	76.6	80.4	72.1	75.2	95.2	76.1	83.3	81.8	43.9	58.1	110.2	83.1	93.3	130.6	96.7	109.4
0950, 0980	Ceilings & Acoustic Treatment	96.3	70.1	79.3	86.5	72.1	77.2	90.8	76.1	81.2	101.0	43.9	64.0	101.3	83.1	89.5	96.3	96.7	96.6
0960	Flooring	102.3	81.3	95.8	90.4	73.4	85.2	91.9	88.1	90.7	116.7	81.3	105.8	99.1	62.4	87.9	98.2	62.4	87.2
0970, 0990	Wall Finishes & Painting/Coating	135.4	55.7	86.8	88.0	70.7	77.4	98.7	56.6	73.0	162.2	56.6	97.8	98.7	83.0	89.1	93.6	73.6	81.4
09	FINISHES	100.8	70.5	84.3	88.1	71.8	79.2	93.5	76.4	84.2	107.9	55.5	79.4	100.1	78.5	88.3	98.5	84.7	91.0
COVERS	DIVS. 10 - 14, 25, 28, 41, 43, 44, 46	100.0	86.6	97.1	100.0	87.6	97.3	100.0	89.1	97.6	100.0	83.6	96.4	100.0	103.0	100.7	100.0	83.8	96.5
21, 22, 23	FIRE SUPPRESSION, PLUMBING & HVAC	96.3	68.9	85.1	100.6	71.7	88.8	100.7	76.8	90.9	95.9	68.2	84.5	100.9	79.2	92.0	98.8	71.1	87.4
26, 27, 3370	ELECTRICAL, COMMUNICATIONS & UTIL.	93.5	75.4	84.8	92.1	63.6	78.4	102.5	80.0	91.6	90.2	60.9	76.0	102.2	88.5	95.6	98.8	81.8	90.6
MF2018	WEIGHTED AVERAGE	95.4	72.9	86.2	94.8	74.2	86.4	97.0	80.7	90.4	97.0	66.0	84.4	101.5	83.9	94.3	100.1	78.6	91.4

City Cost Indexes

NEVADA / NEW HAMPSHIRE

DIVISION		ELY 893 MAT.	INST.	TOTAL	LAS VEGAS 889-891 MAT.	INST.	TOTAL	RENO 894-895 MAT.	INST.	TOTAL	CHARLESTON 036 MAT.	INST.	TOTAL	CLAREMONT 037 MAT.	INST.	TOTAL	CONCORD 032-033 MAT.	INST.	TOTAL
015433	CONTRACTOR EQUIPMENT		92.1	92.1		92.1	92.1		92.1	92.1		92.3	92.3		92.3	92.3		96.8	96.8
0241, 31 - 34	SITE & INFRASTRUCTURE, DEMOLITION	71.4	88.0	82.6	74.9	91.9	86.4	71.6	88.6	83.2	77.9	91.0	86.8	72.6	91.0	85.1	87.2	101.3	96.7
0310	Concrete Forming & Accessories	104.9	95.2	97.1	105.9	111.1	110.1	100.2	82.5	85.9	83.9	81.0	81.5	92.3	81.0	83.2	102.2	94.0	95.6
0320	Concrete Reinforcing	102.0	88.4	97.6	94.5	122.9	103.6	96.5	121.0	104.3	85.3	85.0	85.2	85.3	85.0	85.2	98.7	85.3	94.4
0330	Cast-in-Place Concrete	89.3	89.0	89.2	86.5	107.8	93.8	94.5	80.3	89.6	70.5	108.6	83.6	64.7	108.6	79.9	85.8	115.4	96.0
03	CONCRETE	99.8	91.8	96.6	95.1	111.5	101.7	98.6	88.5	94.6	81.8	91.6	85.7	76.3	91.6	82.4	94.1	99.8	96.4
04	MASONRY	121.7	69.9	89.0	108.7	107.1	107.7	115.8	65.4	84.0	87.9	92.9	91.1	88.8	92.9	91.4	99.6	98.0	98.6
05	METALS	107.3	87.6	102.7	116.2	105.6	113.7	108.9	98.1	106.4	92.7	90.6	92.2	92.7	90.6	92.2	100.3	89.4	97.7
06	WOOD, PLASTICS & COMPOSITES	96.5	98.5	97.3	94.1	109.3	100.4	88.9	83.3	86.6	84.6	78.6	82.1	92.1	78.6	86.6	97.1	93.8	95.7
07	THERMAL & MOISTURE PROTECTION	114.5	85.4	101.8	130.1	102.3	117.9	114.0	76.9	97.7	106.1	100.1	103.5	106.0	100.1	103.4	113.7	107.7	111.1
08	OPENINGS	97.7	84.3	94.4	96.7	113.6	100.8	95.6	83.0	92.5	92.8	78.6	89.4	93.9	78.6	90.2	93.5	90.2	92.7
0920	Plaster & Gypsum Board	121.7	98.7	107.3	105.5	109.9	108.2	101.6	83.1	90.1	96.0	78.0	84.7	97.0	78.0	85.1	97.3	93.6	95.0
0950, 0980	Ceilings & Acoustic Treatment	96.3	98.7	97.9	103.2	109.9	107.5	99.8	83.1	89.0	89.7	78.0	82.1	89.7	78.0	82.1	88.4	93.6	91.8
0960	Flooring	96.1	62.4	85.8	88.7	102.5	93.0	93.4	62.4	83.9	89.5	105.2	94.3	91.4	105.2	95.7	98.1	109.5	101.6
0970, 0990	Wall Finishes & Painting/Coating	93.6	103.9	99.9	96.6	122.7	112.5	93.6	83.0	87.1	81.6	83.5	82.8	81.6	83.5	82.8	98.7	101.6	100.4
09	FINISHES	97.0	90.9	93.7	94.0	110.8	103.1	93.5	78.3	85.3	91.3	85.1	87.9	91.5	85.1	88.0	95.4	97.4	96.5
COVERS	DIVS. 10 - 14, 25, 28, 41, 43, 44, 46	100.0	65.1	92.4	100.0	104.7	101.0	100.0	102.5	100.5	100.0	81.8	96.0	100.0	81.8	96.0	100.0	106.6	101.4
21, 22, 23	FIRE SUPPRESSION, PLUMBING & HVAC	98.8	88.9	94.7	101.1	109.3	104.5	100.9	79.1	92.0	96.9	76.2	88.4	96.9	76.2	88.4	100.9	85.3	94.5
26, 27, 3370	ELECTRICAL, COMMUNICATIONS & UTIL.	99.1	86.2	92.8	104.7	105.8	105.2	99.1	88.6	94.0	100.6	69.7	85.7	100.6	69.7	85.7	101.6	77.8	90.1
MF2018	WEIGHTED AVERAGE	101.1	86.1	95.0	103.0	107.2	104.7	101.0	83.6	93.9	93.0	83.6	89.2	92.4	83.6	88.8	98.6	92.3	96.0

NEW HAMPSHIRE / NEW JERSEY

DIVISION		KEENE 034 MAT.	INST.	TOTAL	LITTLETON 035 MAT.	INST.	TOTAL	MANCHESTER 031 MAT.	INST.	TOTAL	NASHUA 030 MAT.	INST.	TOTAL	PORTSMOUTH 038 MAT.	INST.	TOTAL	ATLANTIC CITY 082,084 MAT.	INST.	TOTAL
015433	CONTRACTOR EQUIPMENT		92.3	92.3		92.3	92.3		97.1	97.1		92.3	92.3		92.3	92.3		90.0	90.0
0241, 31 - 34	SITE & INFRASTRUCTURE, DEMOLITION	84.4	91.0	88.9	72.9	90.1	84.5	84.2	101.0	95.6	86.4	91.4	89.8	80.8	92.2	88.5	118.1	142.0	137.4
0310	Concrete Forming & Accessories	90.3	81.2	83.0	106.2	75.5	81.4	101.9	94.2	95.7	101.6	94.2	95.6	85.9	93.5	92.0	77.2	134.6	95.6
0320	Concrete Reinforcing	85.3	85.0	85.2	86.0	85.0	85.7	100.3	85.3	95.5	106.5	85.3	99.7	85.3	85.3	85.3	74.4	152.1	99.4
0330	Cast-in-Place Concrete	70.8	108.7	83.9	63.5	100.5	76.3	80.8	115.8	92.9	67.0	114.8	83.5	63.5	114.9	81.3	65.4	131.8	88.3
03	CONCRETE	81.8	91.8	85.8	76.4	86.3	80.4	92.2	100.1	95.4	84.6	99.8	90.7	75.7	99.7	85.3	79.9	135.8	102.3
04	MASONRY	90.2	92.9	91.9	100.2	78.5	86.5	97.0	98.0	97.6	93.2	97.9	96.1	88.8	97.0	94.0	102.7	136.5	124.0
05	METALS	93.2	91.0	92.7	93.2	90.6	92.6	100.6	89.7	98.0	100.0	91.9	98.1	94.4	93.5	94.2	97.0	114.8	101.1
06	WOOD, PLASTICS & COMPOSITES	90.3	78.6	85.5	104.6	78.6	93.8	98.2	93.9	96.4	102.7	93.9	99.0	86.2	93.9	89.4	119.1	143.5	129.2
07	THERMAL & MOISTURE PROTECTION	106.9	100.1	103.9	106.2	93.6	100.7	113.6	107.7	111.0	107.5	106.2	106.9	106.7	103.7	105.4	104.4	133.1	117.0
08	OPENINGS	91.6	81.8	89.3	94.7	78.6	90.8	92.3	91.9	92.2	95.4	89.6	94.0	96.1	80.8	92.4	93.0	138.6	104.0
0920	Plaster & Gypsum Board	96.5	78.0	84.9	121.9	78.0	94.4	101.8	93.6	96.7	111.2	93.6	100.2	96.0	93.6	94.5	128.7	144.7	138.7
0950, 0980	Ceilings & Acoustic Treatment	89.7	78.0	82.1	89.7	78.0	82.1	93.2	93.6	93.5	100.8	93.6	96.1	90.5	93.6	92.5	95.6	144.7	127.4
0960	Flooring	91.2	105.2	95.5	99.8	105.2	101.5	94.5	111.7	99.8	94.2	107.3	98.2	89.6	107.3	95.0	101.7	155.6	118.2
0970, 0990	Wall Finishes & Painting/Coating	81.6	95.9	90.3	81.6	83.5	82.8	98.7	115.0	108.6	81.6	115.0	101.9	81.6	94.9	89.7	81.4	143.5	119.2
09	FINISHES	93.0	86.5	89.4	97.8	81.5	88.9	96.0	99.4	97.9	98.5	98.5	98.5	92.1	96.0	94.3	99.6	146.2	124.9
COVERS	DIVS. 10 - 14, 25, 28, 41, 43, 44, 46	100.0	87.5	97.3	100.0	88.3	97.5	100.0	106.7	101.5	100.0	106.8	101.5	100.0	106.3	101.4	100.0	115.3	103.3
21, 22, 23	FIRE SUPPRESSION, PLUMBING & HVAC	96.9	76.2	88.4	96.9	68.7	85.3	100.9	85.4	94.5	101.1	85.5	94.7	101.1	84.7	94.4	100.4	126.5	111.1
26, 27, 3370	ELECTRICAL, COMMUNICATIONS & UTIL.	100.6	69.7	85.7	101.3	45.5	74.4	101.6	77.8	90.1	102.1	77.8	90.4	100.9	74.7	88.3	92.4	132.9	111.9
MF2018	WEIGHTED AVERAGE	93.5	84.2	89.7	93.7	75.9	86.5	98.2	92.6	95.1	93.7	91.8	95.1	94.0	90.5	92.6	95.4	129.2	109.1

NEW JERSEY

DIVISION		CAMDEN 081 MAT.	INST.	TOTAL	DOVER 078 MAT.	INST.	TOTAL	ELIZABETH 072 MAT.	INST.	TOTAL	HACKENSACK 076 MAT.	INST.	TOTAL	JERSEY CITY 073 MAT.	INST.	TOTAL	LONG BRANCH 077 MAT.	INST.	TOTAL
015433	CONTRACTOR EQUIPMENT		90.0	90.0		92.3	92.3		92.3	92.3		92.3	92.3		90.0	90.0		89.6	89.6
0241, 31 - 34	SITE & INFRASTRUCTURE, DEMOLITION	87.9	95.6	93.1	102.3	96.6	98.4	106.9	96.6	99.9	103.5	96.6	98.8	93.5	96.5	95.5	97.7	96.0	96.5
0310	Concrete Forming & Accessories	105.0	144.9	137.2	94.0	145.3	135.4	109.6	145.4	138.5	94.0	145.3	135.4	99.0	145.4	136.4	99.8	131.4	125.3
0320	Concrete Reinforcing	101.3	137.2	112.8	74.4	152.1	99.4	74.4	152.1	99.4	74.4	152.1	99.4	96.6	152.1	114.5	74.4	151.5	99.2
0330	Cast-in-Place Concrete	63.4	135.9	88.5	67.7	134.9	90.9	58.2	136.2	85.1	66.2	136.1	90.3	52.9	134.9	81.2	58.7	125.5	81.8
03	CONCRETE	82.6	139.0	105.2	77.9	141.6	103.4	76.1	142.1	102.5	76.6	142.0	102.8	76.5	141.4	102.5	77.1	131.7	99.0
04	MASONRY	92.5	140.8	122.9	94.7	140.8	123.8	111.5	140.8	130.0	98.8	140.8	125.3	89.4	140.8	121.8	103.8	126.6	118.2
05	METALS	104.0	116.4	106.9	94.8	125.2	101.9	95.9	125.3	102.8	94.9	125.2	102.0	102.0	122.3	106.7	94.9	120.8	101.0
06	WOOD, PLASTICS & COMPOSITES	104.9	146.1	121.9	91.7	146.1	114.2	108.9	146.1	124.3	91.7	146.1	114.2	93.0	146.1	115.0	94.2	131.4	109.6
07	THERMAL & MOISTURE PROTECTION	104.1	135.9	118.0	94.1	134.2	111.7	94.5	134.4	112.0	93.9	136.5	112.6	93.5	134.2	111.4	93.8	122.6	106.4
08	OPENINGS	94.8	141.1	105.9	98.4	142.8	109.1	96.9	142.8	108.0	96.3	142.8	107.6	94.5	142.8	106.2	91.1	134.7	101.6
0920	Plaster & Gypsum Board	120.0	147.4	137.1	104.8	147.4	131.4	118.1	147.4	136.4	104.8	147.4	131.4	108.5	147.4	132.8	107.9	132.3	123.2
0950, 0980	Ceilings & Acoustic Treatment	108.0	147.4	133.6	81.1	147.4	124.1	81.8	147.4	124.4	81.1	147.4	124.1	90.1	147.4	127.3	81.1	132.3	114.3
0960	Flooring	98.9	155.6	116.2	82.0	182.4	112.7	85.7	182.4	115.3	82.0	182.4	112.7	82.7	182.4	113.2	82.9	164.6	107.9
0970, 0990	Wall Finishes & Painting/Coating	81.4	143.5	119.2	83.7	143.5	120.2	83.7	143.5	120.2	83.7	143.5	120.2	83.8	143.5	120.2	83.8	143.5	120.2
09	FINISHES	100.0	148.1	126.1	89.0	152.2	123.3	92.8	152.8	125.4	88.8	152.2	123.3	91.5	152.9	124.8	90.0	139.4	116.8
COVERS	DIVS. 10 - 14, 25, 28, 41, 43, 44, 46	100.0	116.0	103.5	100.0	127.9	106.1	100.0	127.9	106.1	100.0	127.9	106.1	100.0	127.9	106.1	100.0	112.1	102.6
21, 22, 23	FIRE SUPPRESSION, PLUMBING & HVAC	100.8	133.9	114.4	100.3	135.0	114.5	100.6	135.1	114.8	100.3	135.0	114.5	100.6	135.0	114.7	100.3	130.3	112.6
26, 27, 3370	ELECTRICAL, COMMUNICATIONS & UTIL.	95.2	139.6	116.6	92.7	142.2	116.6	93.1	142.2	116.8	92.7	137.7	114.4	95.6	137.7	115.9	92.5	120.4	106.0
MF2018	WEIGHTED AVERAGE	97.1	133.3	111.9	93.8	136.0	111.0	95.1	136.2	111.8	93.7	135.5	110.7	94.8	135.2	111.3	93.4	125.7	106.6

NEW JERSEY

DIVISION		NEW BRUNSWICK 088 - 089			NEWARK 070 - 071			PATERSON 074 - 075			POINT PLEASANT 087			SUMMIT 079			TRENTON 085 - 086		
		MAT.	INST.	TOTAL	MAT.	INST.	TOTAL	MAT.	INST.	TOTAL	MAT.	INST.	TOTAL	MAT.	INST.	TOTAL	MAT.	INST.	TOTAL
015433	CONTRACTOR EQUIPMENT		89.6	89.6		97.7	97.7		92.3	92.3		89.6	89.6		92.3	92.3		95.7	95.7
0241, 31 - 34	SITE & INFRASTRUCTURE, DEMOLITION	98.6	96.4	97.1	109.5	104.7	106.3	105.0	96.6	99.3	99.9	96.0	97.2	104.5	96.6	99.1	87.0	105.9	99.8
0310	Concrete Forming & Accessories	109.6	144.7	137.9	100.1	145.7	136.9	96.5	145.5	136.0	101.5	131.3	125.5	97.5	145.2	136.0	103.6	144.9	136.9
0320	Concrete Reinforcing	78.0	151.6	101.7	100.4	152.2	117.0	96.6	152.2	114.5	78.0	151.5	101.6	74.4	152.1	99.4	100.7	135.6	111.9
0330	Cast-in-Place Concrete	80.8	129.2	97.5	76.6	138.3	97.9	67.3	136.2	91.1	80.8	127.4	96.9	56.3	136.1	83.8	77.5	136.9	98.0
03	CONCRETE	92.6	139.0	111.2	88.9	142.8	110.5	82.7	142.1	106.5	92.0	132.3	108.1	73.6	142.0	100.9	89.6	139.1	109.4
04	MASONRY	100.9	132.9	121.1	99.2	140.9	125.5	95.3	140.8	124.0	89.7	126.6	112.9	97.6	140.8	124.8	99.4	140.9	125.6
05	METALS	97.1	121.0	102.6	103.2	123.4	107.9	97.6	125.4	104.1	97.1	120.7	102.6	94.8	125.1	101.9	103.3	115.6	106.2
06	WOOD, PLASTICS & COMPOSITES	111.4	146.1	125.7	95.3	146.4	116.4	94.6	146.1	115.9	101.8	131.4	114.1	95.9	146.1	116.7	97.0	146.1	117.3
07	THERMAL & MOISTURE PROTECTION	104.7	128.5	115.1	97.7	135.9	114.5	94.2	136.5	112.8	104.7	124.7	113.4	94.5	134.4	112.0	106.9	137.4	120.3
08	OPENINGS	88.5	142.8	101.6	96.6	143.0	107.8	100.9	142.8	111.1	90.2	136.0	101.3	102.2	142.8	112.0	93.0	139.3	104.2
0920	Plaster & Gypsum Board	125.5	147.4	139.2	101.3	147.4	130.1	108.5	147.4	132.8	116.1	132.3	126.2	107.9	147.4	132.6	112.9	147.4	134.5
0950, 0980	Ceilings & Acoustic Treatment	95.6	147.4	129.2	90.0	147.4	127.2	90.1	147.4	127.3	95.6	132.3	119.4	81.1	147.4	124.1	102.3	147.4	131.6
0960	Flooring	100.0	182.4	125.2	97.2	182.4	123.2	82.7	182.4	113.2	98.0	155.6	115.6	82.9	182.4	113.3	103.8	180.7	127.3
0970, 0990	Wall Finishes & Painting/Coating	81.4	143.5	119.2	98.7	143.5	126.0	83.7	143.5	120.2	81.4	143.5	119.2	83.7	143.5	120.2	98.7	143.5	126.0
09	FINISHES	99.8	152.2	128.2	96.7	153.0	127.3	91.6	152.8	124.9	97.7	137.8	119.5	90.0	152.2	123.8	102.6	152.4	129.7
COVERS	DIVS. 10 - 14, 25, 28, 41, 43, 44, 46	100.0	127.8	106.0	100.0	128.6	106.2	100.0	127.9	106.1	100.0	102.7	100.6	100.0	127.9	106.1	100.0	115.9	103.5
21, 22, 23	FIRE SUPPRESSION, PLUMBING & HVAC	100.4	133.5	114.0	100.7	135.2	114.8	100.6	135.0	114.7	100.4	130.3	112.7	100.3	131.2	113.0	100.8	133.5	114.2
26, 27, 3370	ELECTRICAL, COMMUNICATIONS & UTIL.	92.8	123.9	107.8	100.4	137.7	118.4	95.6	142.2	118.0	92.4	120.4	105.9	93.1	142.2	116.8	99.4	128.9	113.6
MF2018	WEIGHTED AVERAGE	96.9	131.4	111.0	98.8	136.3	114.1	96.0	136.3	112.4	96.3	125.4	108.1	96.3	135.3	110.8	98.7	133.0	112.6

DIVISION		NEW JERSEY VINELAND 080,083			NEW MEXICO ALBUQUERQUE 870 - 872			CARRIZOZO 883			CLOVIS 881			FARMINGTON 874			GALLUP 873		
		MAT.	INST.	TOTAL	MAT.	INST.	TOTAL	MAT.	INST.	TOTAL	MAT.	INST.	TOTAL	MAT.	INST.	TOTAL	MAT.	INST.	TOTAL
015433	CONTRACTOR EQUIPMENT		90.0	90.0		104.6	104.6		104.6	104.6		104.6	104.6		104.6	104.6		104.6	104.6
0241, 31 - 34	SITE & INFRASTRUCTURE, DEMOLITION	90.9	95.1	93.8	89.8	93.2	92.1	106.7	93.2	97.5	94.3	93.2	93.6	95.9	93.2	94.1	104.2	93.2	96.7
0310	Concrete Forming & Accessories	97.2	130.0	123.6	101.3	66.3	73.1	96.4	66.3	72.1	96.4	66.1	72.0	101.5	66.3	73.1	101.5	66.3	73.1
0320	Concrete Reinforcing	77.2	122.2	91.6	94.7	69.8	86.7	99.9	69.8	90.2	101.0	69.8	91.0	103.5	69.8	92.7	99.1	69.8	89.7
0330	Cast-in-Place Concrete	70.5	125.6	89.5	82.6	67.4	77.4	85.4	67.4	79.2	85.3	67.3	79.1	83.4	67.4	77.9	78.5	67.4	74.7
03	CONCRETE	82.7	126.1	100.1	90.1	68.7	81.5	107.9	68.7	92.2	98.2	68.6	86.3	93.9	68.7	83.8	98.7	68.7	86.7
04	MASONRY	91.7	126.6	113.7	103.5	59.1	75.5	109.1	59.1	77.5	109.1	59.1	77.6	110.7	59.1	78.2	98.7	59.1	73.7
05	METALS	96.9	110.4	100.1	100.5	91.3	98.4	98.9	91.3	97.2	98.6	91.2	96.8	97.5	91.3	96.0	96.8	91.3	95.6
06	WOOD, PLASTICS & COMPOSITES	97.7	130.0	111.1	100.3	67.8	86.8	90.9	67.8	81.4	90.9	67.8	81.4	100.5	67.8	86.9	100.5	67.8	86.9
07	THERMAL & MOISTURE PROTECTION	104.0	128.0	114.5	106.5	70.5	90.7	105.7	70.5	90.3	104.4	70.5	89.5	106.8	70.5	90.9	108.0	70.5	91.6
08	OPENINGS	89.8	128.8	99.2	94.3	67.0	87.7	92.3	67.0	86.2	92.5	67.0	86.3	96.6	67.0	89.4	96.6	67.0	89.5
0920	Plaster & Gypsum Board	112.7	130.8	124.0	121.6	66.9	85.9	85.9	66.9	74.0	85.9	66.9	74.0	108.9	66.9	82.6	108.9	66.9	82.6
0950, 0980	Ceilings & Acoustic Treatment	95.6	130.8	118.5	109.8	66.9	82.0	104.6	66.9	80.1	104.6	66.9	80.1	104.6	66.9	80.1	104.6	66.9	80.1
0960	Flooring	97.4	155.6	115.2	87.5	67.3	81.3	94.9	67.3	86.4	94.9	67.3	86.4	88.9	67.3	82.3	88.9	67.3	82.3
0970, 0990	Wall Finishes & Painting/Coating	81.4	143.5	119.2	93.2	51.7	67.9	90.7	51.7	66.9	90.7	51.7	66.9	87.5	51.7	65.7	87.5	51.7	65.7
09	FINISHES	96.2	137.0	118.4	98.7	64.8	80.3	96.4	64.8	79.2	95.0	64.8	78.6	95.8	64.8	79.0	97.1	64.8	79.6
COVERS	DIVS. 10 - 14, 25, 28, 41, 43, 44, 46	100.0	112.1	102.6	100.0	85.8	96.9	100.0	85.8	96.9	100.0	85.8	96.9	100.0	85.8	96.9	100.0	85.8	96.9
21, 22, 23	FIRE SUPPRESSION, PLUMBING & HVAC	100.4	122.4	109.5	101.0	65.3	86.4	98.3	65.3	84.8	98.3	64.9	84.6	100.9	65.3	86.3	98.3	65.3	84.8
26, 27, 3370	ELECTRICAL, COMMUNICATIONS & UTIL.	92.4	132.9	111.9	85.0	69.3	77.4	88.4	69.3	79.2	87.0	69.3	78.5	83.5	69.3	76.6	83.1	69.3	76.4
MF2018	WEIGHTED AVERAGE	94.6	123.7	106.4	96.8	71.0	86.3	98.9	71.0	87.6	96.9	70.9	86.4	97.0	71.0	86.4	96.8	71.0	86.3

NEW MEXICO

DIVISION		LAS CRUCES 880			LAS VEGAS 877			ROSWELL 882			SANTA FE 875			SOCORRO 878			TRUTH/CONSEQUENCES 879		
		MAT.	INST.	TOTAL	MAT.	INST.	TOTAL	MAT.	INST.	TOTAL	MAT.	INST.	TOTAL	MAT.	INST.	TOTAL	MAT.	INST.	TOTAL
015433	CONTRACTOR EQUIPMENT		81.3	81.3		104.6	104.6		104.6	104.6		108.7	108.7		104.6	104.6		81.3	81.3
0241, 31 - 34	SITE & INFRASTRUCTURE, DEMOLITION	95.5	74.2	81.0	94.5	93.2	93.6	96.9	93.2	94.4	96.9	100.0	99.0	91.3	93.2	92.6	109.2	74.2	85.4
0310	Concrete Forming & Accessories	92.7	65.3	70.6	101.5	66.3	73.1	96.4	66.3	72.1	102.5	66.5	73.5	101.5	66.3	73.1	99.5	65.3	71.9
0320	Concrete Reinforcing	99.5	69.6	89.9	100.8	69.8	90.9	101.0	69.8	91.0	101.0	69.9	91.0	102.7	69.8	92.2	98.8	69.7	89.5
0330	Cast-in-Place Concrete	80.6	60.8	73.8	81.1	67.4	76.4	85.3	67.4	79.1	86.2	69.8	80.5	79.5	67.4	75.3	87.1	60.8	78.0
03	CONCRETE	80.3	65.5	74.4	91.6	68.7	82.4	98.9	68.7	86.8	97.2	69.4	86.1	90.9	68.7	82.0	85.5	65.5	77.5
04	MASONRY	104.5	57.3	74.8	99.1	59.1	73.8	120.0	59.1	81.6	96.4	59.2	73.0	98.9	59.1	73.8	95.9	58.8	72.5
05	METALS	96.7	83.8	93.7	96.5	91.3	95.3	99.6	91.3	97.7	95.1	88.6	93.6	96.8	91.3	95.6	95.9	83.9	93.1
06	WOOD, PLASTICS & COMPOSITES	81.9	66.8	75.6	100.5	67.8	86.9	90.9	67.8	81.4	96.9	68.0	84.9	100.5	67.8	86.9	93.7	66.8	82.6
07	THERMAL & MOISTURE PROTECTION	92.7	65.8	80.9	106.4	70.5	90.6	104.5	70.5	89.6	110.7	72.8	94.1	106.3	70.5	90.6	94.3	66.2	82.0
08	OPENINGS	87.8	66.5	82.7	93.4	67.0	87.0	92.3	67.0	86.2	94.7	67.1	88.1	93.2	67.0	86.9	86.7	66.5	81.8
0920	Plaster & Gypsum Board	84.3	66.9	73.4	108.9	66.9	82.6	85.9	66.9	74.0	118.7	66.9	86.3	108.9	66.9	82.6	110.4	66.9	83.2
0950, 0980	Ceilings & Acoustic Treatment	95.2	66.9	76.8	104.6	66.9	80.1	104.6	66.9	80.1	104.0	66.9	79.9	104.6	66.9	80.1	99.8	66.9	78.5
0960	Flooring	123.4	67.3	106.2	88.9	67.3	82.3	94.9	67.3	86.4	97.7	67.3	88.4	88.9	67.3	82.3	115.8	67.3	101.0
0970, 0990	Wall Finishes & Painting/Coating	78.4	51.7	62.1	87.5	51.7	65.7	90.7	51.7	66.9	98.7	51.7	70.0	87.5	51.7	65.7	78.8	51.7	62.3
09	FINISHES	104.4	64.0	82.5	95.7	64.8	78.9	95.1	64.8	78.7	102.5	65.0	82.1	95.6	64.8	78.9	107.9	64.1	84.1
COVERS	DIVS. 10 - 14, 25, 28, 41, 43, 44, 46	100.0	83.6	96.4	100.0	85.8	96.9	100.0	85.8	96.9	100.0	86.4	97.0	100.0	85.8	96.9	100.0	83.6	96.4
21, 22, 23	FIRE SUPPRESSION, PLUMBING & HVAC	101.5	65.0	86.6	98.3	65.3	84.8	100.8	65.3	86.3	100.9	65.3	86.3	98.3	65.3	84.8	98.0	65.0	84.5
26, 27, 3370	ELECTRICAL, COMMUNICATIONS & UTIL.	90.3	77.7	84.2	84.4	69.3	77.1	87.8	69.3	78.9	100.9	69.3	85.7	83.3	69.3	76.5	86.9	69.2	78.4
MF2018	WEIGHTED AVERAGE	94.8	69.1	84.3	95.2	71.0	85.4	98.4	71.0	87.2	98.6	71.5	87.6	95.0	71.0	85.2	94.6	68.0	83.8

NEW MEXICO / NEW YORK

DIVISION		NEW MEXICO TUCUMCARI 884			NEW YORK ALBANY 120 - 122			BINGHAMTON 137 - 139			BRONX 104			BROOKLYN 112			BUFFALO 140 - 142		
		MAT.	INST.	TOTAL	MAT.	INST.	TOTAL	MAT.	INST.	TOTAL	MAT.	INST.	TOTAL	MAT.	INST.	TOTAL	MAT.	INST.	TOTAL
015433	CONTRACTOR EQUIPMENT		104.6	104.6		112.3	112.3		113.5	113.5		101.6	101.6		105.5	105.5		95.7	95.7
0241, 31 - 34	SITE & INFRASTRUCTURE, DEMOLITION	93.9	93.2	93.4	75.8	103.0	94.3	90.7	85.3	87.0	94.1	104.8	101.4	113.1	114.6	114.1	89.9	96.9	94.6
0310	Concrete Forming & Accessories	96.4	66.1	72.0	101.7	103.9	103.5	100.3	90.2	92.1	87.8	178.9	161.3	107.5	178.9	165.1	100.7	112.5	110.2
0320	Concrete Reinforcing	99.0	69.8	89.7	100.0	111.8	103.8	97.1	104.3	99.4	92.6	169.6	117.3	97.3	225.9	138.6	100.4	112.9	104.4
0330	Cast-in-Place Concrete	85.3	67.3	79.1	76.3	115.0	89.7	105.5	102.0	104.3	69.6	162.4	101.6	107.4	160.7	125.8	126.3	118.6	123.7
03	CONCRETE	97.4	68.6	85.9	92.1	110.1	99.3	95.2	99.1	96.8	80.2	171.2	116.6	104.6	179.3	134.5	112.9	114.3	113.5
04	MASONRY	120.7	59.1	81.9	102.5	116.1	111.1	110.1	101.0	104.3	86.4	181.3	146.2	118.2	181.3	158.0	114.7	120.4	118.3
05	METALS	98.6	91.2	96.8	97.3	122.2	103.1	90.1	133.6	100.2	88.3	170.8	107.6	97.5	170.8	114.6	99.6	105.3	100.9
06	WOOD, PLASTICS & COMPOSITES	90.9	67.8	81.4	96.6	100.1	98.1	100.3	86.9	94.8	89.2	178.0	126.0	104.3	177.7	134.7	97.6	111.2	103.2
07	THERMAL & MOISTURE PROTECTION	104.3	70.5	89.5	113.7	109.4	111.8	111.1	91.7	102.6	105.1	160.6	129.5	112.9	160.0	133.5	108.0	109.2	108.5
08	OPENINGS	92.2	67.0	86.2	93.1	101.8	95.2	87.7	92.4	88.8	90.2	187.6	113.7	85.6	187.4	110.1	100.7	107.9	102.4
0920	Plaster & Gypsum Board	85.9	66.9	74.0	102.7	100.0	101.0	118.2	86.5	98.4	100.1	180.2	150.2	113.6	180.2	155.2	107.4	111.5	110.0
0950, 0980	Ceilings & Acoustic Treatment	104.6	66.9	80.1	104.8	100.0	101.7	105.3	86.5	93.1	76.0	180.2	143.6	99.2	180.2	151.8	107.1	111.5	110.0
0960	Flooring	94.9	67.3	86.4	94.5	107.6	98.5	104.5	97.2	102.3	101.2	179.1	125.0	111.1	179.1	131.9	91.7	116.1	99.2
0970, 0990	Wall Finishes & Painting/Coating	90.7	51.7	66.9	98.7	100.2	99.6	84.0	98.9	93.1	99.3	164.7	139.2	111.9	164.7	144.1	98.7	113.6	107.8
09	FINISHES	95.0	64.8	78.6	98.2	103.5	101.1	98.6	91.5	94.8	95.3	178.1	140.3	111.0	177.9	147.3	98.5	113.7	106.7
COVERS	DIVS. 10 - 14, 25, 28, 41, 43, 44, 46	100.0	85.8	96.9	100.0	100.2	100.1	100.0	97.2	99.4	100.0	133.7	107.3	100.0	132.9	107.2	100.0	106.3	101.4
21, 22, 23	FIRE SUPPRESSION, PLUMBING & HVAC	98.3	64.9	84.6	100.9	107.4	103.6	101.4	97.9	100.0	100.4	167.3	127.8	100.5	167.2	127.8	100.9	99.2	100.2
26, 27, 3370	ELECTRICAL, COMMUNICATIONS & UTIL.	88.4	69.3	79.2	101.9	106.2	104.0	100.9	97.5	99.3	96.4	176.3	135.0	103.1	176.3	138.4	101.9	100.4	101.2
MF2018	WEIGHTED AVERAGE	97.4	70.9	86.6	97.8	108.5	102.1	97.0	99.2	97.9	92.7	167.1	122.9	101.5	168.8	128.9	102.6	106.8	104.3

NEW YORK

DIVISION		ELMIRA 148 - 149			FAR ROCKAWAY 116			FLUSHING 113			GLENS FALLS 128			HICKSVILLE 115,117,118			JAMAICA 114		
		MAT.	INST.	TOTAL	MAT.	INST.	TOTAL	MAT.	INST.	TOTAL	MAT.	INST.	TOTAL	MAT.	INST.	TOTAL	MAT.	INST.	TOTAL
015433	CONTRACTOR EQUIPMENT		115.5	115.5		105.5	105.5		105.5	105.5		108.7	108.7		105.5	105.5		105.5	105.5
0241, 31 - 34	SITE & INFRASTRUCTURE, DEMOLITION	88.8	84.6	85.9	115.5	114.6	114.9	115.6	114.6	114.9	68.1	94.2	85.8	106.8	113.2	111.2	110.3	114.6	113.2
0310	Concrete Forming & Accessories	76.6	92.1	89.1	89.3	178.9	161.6	94.4	178.9	162.5	79.1	93.9	91.0	84.4	152.1	139.0	94.4	178.9	162.5
0320	Concrete Reinforcing	100.4	103.1	101.2	97.3	225.9	138.8	99.0	225.9	139.8	97.3	104.9	99.7	97.3	171.1	121.0	97.3	225.9	138.6
0330	Cast-in-Place Concrete	111.1	99.9	107.3	116.2	160.8	131.6	116.2	160.8	131.6	73.5	104.3	84.1	98.7	152.3	117.2	107.4	160.8	125.8
03	CONCRETE	97.1	99.1	97.9	109.5	179.3	137.5	110.3	179.3	137.9	80.3	100.8	88.5	96.7	154.8	120.0	103.7	179.3	133.9
04	MASONRY	108.6	100.5	103.5	124.2	181.3	160.2	117.2	181.3	157.6	107.3	105.2	106.0	114.1	167.6	147.8	121.8	181.3	159.3
05	METALS	91.8	133.8	101.6	97.5	170.8	114.6	97.5	170.8	114.6	91.1	120.7	98.0	98.6	147.8	110.1	97.5	170.8	114.6
06	WOOD, PLASTICS & COMPOSITES	75.7	90.5	81.8	84.1	177.7	122.8	89.9	177.7	126.2	80.5	90.7	84.7	80.1	149.7	108.9	89.9	177.7	126.2
07	THERMAL & MOISTURE PROTECTION	110.8	90.6	101.9	112.6	160.2	133.4	112.7	160.2	133.5	104.0	99.2	101.9	112.1	151.8	129.5	112.4	160.2	133.4
08	OPENINGS	93.5	94.1	93.7	84.4	187.4	109.3	84.4	187.4	109.3	87.4	93.2	88.8	84.8	158.5	102.6	84.4	187.4	109.3
0920	Plaster & Gypsum Board	99.0	90.4	93.6	92.6	180.2	147.4	97.9	180.2	149.3	84.1	90.6	88.2	92.1	151.4	129.2	97.9	180.2	149.3
0950, 0980	Ceilings & Acoustic Treatment	113.9	90.4	98.6	84.1	180.2	146.5	84.1	180.2	146.5	94.6	90.6	92.0	83.3	151.4	127.5	84.1	180.2	146.5
0960	Flooring	88.4	95.3	90.5	106.7	179.1	128.8	108.0	179.1	129.7	83.9	100.2	88.9	105.8	179.1	128.3	108.0	179.1	129.7
0970, 0990	Wall Finishes & Painting/Coating	94.7	86.6	89.7	111.9	164.7	144.1	111.9	164.7	144.1	87.4	96.9	93.2	111.9	164.7	144.1	111.9	164.7	144.1
09	FINISHES	95.4	91.8	93.4	103.5	177.9	143.9	104.7	177.9	144.4	88.5	94.7	91.8	102.1	158.1	132.5	104.2	177.9	144.2
COVERS	DIVS. 10 - 14, 25, 28, 41, 43, 44, 46	100.0	95.1	98.9	100.0	132.9	107.2	100.0	132.9	107.2	100.0	89.1	97.6	100.0	128.1	106.1	100.0	132.9	107.2
21, 22, 23	FIRE SUPPRESSION, PLUMBING & HVAC	96.8	91.9	94.8	96.1	167.2	125.3	96.1	167.2	125.3	96.7	101.2	98.6	100.5	154.0	122.4	96.1	167.2	125.3
26, 27, 3370	ELECTRICAL, COMMUNICATIONS & UTIL.	99.4	96.5	98.0	107.4	176.3	140.6	107.4	176.3	140.6	96.9	97.3	97.1	102.4	140.3	120.7	101.8	176.3	137.7
MF2018	WEIGHTED AVERAGE	96.3	97.7	96.9	101.1	168.8	128.7	101.1	168.8	128.7	91.8	100.5	95.4	99.2	149.6	119.7	99.6	168.8	127.8

NEW YORK

DIVISION		JAMESTOWN 147			KINGSTON 124			LONG ISLAND CITY 111			MONTICELLO 127			MOUNT VERNON 105			NEW ROCHELLE 108		
		MAT.	INST.	TOTAL	MAT.	INST.	TOTAL	MAT.	INST.	TOTAL	MAT.	INST.	TOTAL	MAT.	INST.	TOTAL	MAT.	INST.	TOTAL
015433	CONTRACTOR EQUIPMENT		87.0	87.0		105.5	105.5		105.5	105.5		105.5	105.5		101.6	101.6		101.6	101.6
0241, 31 - 34	SITE & INFRASTRUCTURE, DEMOLITION	90.3	84.7	86.5	132.4	109.6	117.0	114.3	114.6	114.5	128.3	109.6	115.6	99.1	100.8	100.2	98.3	100.8	100.0
0310	Concrete Forming & Accessories	76.6	84.7	83.1	80.6	122.9	114.7	100.3	178.9	163.7	90.6	122.9	116.6	76.8	131.2	120.7	93.4	127.9	121.3
0320	Concrete Reinforcing	100.4	103.5	101.4	97.9	149.3	114.4	97.3	225.9	138.6	97.1	149.3	113.9	92.7	169.9	117.5	92.8	169.8	117.5
0330	Cast-in-Place Concrete	115.3	98.2	109.4	99.2	136.7	112.2	110.9	160.8	128.1	92.9	136.8	108.0	77.8	141.0	99.6	77.8	140.7	99.5
03	CONCRETE	100.0	92.9	97.2	98.2	132.2	111.8	106.6	179.3	135.7	94.2	132.2	109.4	87.0	142.2	109.1	87.0	140.6	108.5
04	MASONRY	118.6	97.7	105.4	124.6	145.1	137.5	114.7	181.3	156.7	117.2	145.1	134.8	92.7	147.3	127.1	92.7	147.3	127.1
05	METALS	89.5	98.7	91.7	98.3	131.9	106.2	97.5	170.8	114.6	98.3	132.0	106.2	88.0	167.9	106.7	88.3	167.2	106.7
06	WOOD, PLASTICS & COMPOSITES	74.3	81.3	77.2	81.9	117.0	96.4	97.3	177.7	130.6	90.6	117.0	101.5	78.8	125.0	97.9	97.7	121.4	107.5
07	THERMAL & MOISTURE PROTECTION	110.2	88.4	100.6	126.8	135.4	130.6	112.7	160.2	133.5	126.6	135.4	130.4	105.9	138.4	120.2	106.2	136.1	119.3
08	OPENINGS	93.4	87.5	91.9	90.4	135.5	101.3	84.4	187.4	109.3	86.6	135.5	98.4	90.2	161.5	107.4	90.3	159.5	107.0
0920	Plaster & Gypsum Board	87.0	80.9	83.2	84.2	117.8	105.2	107.3	180.2	152.9	84.9	117.8	105.5	93.2	125.7	113.5	116.4	122.0	119.9
0950, 0980	Ceilings & Acoustic Treatment	109.1	80.9	90.8	79.5	117.8	104.4	84.1	180.2	146.5	79.5	117.8	104.4	74.0	125.7	107.5	74.0	122.0	105.1
0960	Flooring	90.8	95.3	92.2	104.3	149.4	118.1	109.4	179.1	130.7	106.1	149.4	119.3	92.8	167.1	115.6	100.3	152.6	116.3
0970, 0990	Wall Finishes & Painting/Coating	96.3	93.4	94.5	115.8	121.5	119.2	111.9	164.7	144.1	115.8	114.5	115.0	97.7	164.7	138.6	97.7	164.7	138.6
09	FINISHES	93.2	86.7	89.6	101.5	126.9	115.3	106.3	177.9	145.2	101.8	126.1	115.0	92.2	140.1	118.2	97.8	135.0	118.0
COVERS	DIVS. 10 - 14, 25, 28, 41, 43, 44, 46	100.0	98.7	99.7	100.0	111.5	102.5	100.0	132.9	107.2	100.0	111.5	102.5	100.0	119.4	104.2	100.0	104.5	101.0
21, 22, 23	FIRE SUPPRESSION, PLUMBING & HVAC	96.6	86.8	92.6	96.6	124.7	108.2	100.5	167.2	127.8	96.6	128.7	109.8	96.3	142.3	115.2	96.3	142.1	115.1
26, 27, 3370	ELECTRICAL, COMMUNICATIONS & UTIL.	98.1	87.9	93.2	98.6	107.8	103.0	102.1	176.3	137.9	98.6	107.8	103.0	93.9	158.1	124.8	93.9	131.3	111.9
MF2018	WEIGHTED AVERAGE	96.3	90.2	93.8	100.1	125.4	110.4	101.0	168.8	128.6	98.9	126.1	110.0	92.5	143.6	113.3	93.2	138.1	111.5

City Cost Indexes

NEW YORK

DIVISION		NEW YORK 100-102 MAT.	INST.	TOTAL	NIAGARA FALLS 143 MAT.	INST.	TOTAL	PLATTSBURGH 129 MAT.	INST.	TOTAL	POUGHKEEPSIE 125-126 MAT.	INST.	TOTAL	QUEENS 110 MAT.	INST.	TOTAL	RIVERHEAD 119 MAT.	INST.	TOTAL
015433	CONTRACTOR EQUIPMENT		103.8	103.8		87.0	87.0		91.1	91.1		105.5	105.5		105.5	105.5		105.5	105.5
0241, 31 - 34	SITE & INFRASTRUCTURE, DEMOLITION	102.0	108.8	106.6	92.1	85.6	87.7	107.2	91.4	96.5	129.3	108.6	115.3	109.6	114.6	113.0	107.6	113.2	111.4
0310	Concrete Forming & Accessories	97.7	182.8	166.4	76.6	106.3	100.6	87.1	85.8	86.1	80.6	157.4	142.6	84.7	178.9	160.7	90.7	152.8	140.8
0320	Concrete Reinforcing	101.4	171.3	123.8	99.1	104.2	100.7	99.9	104.2	101.3	97.9	148.9	114.3	99.0	225.9	139.8	99.2	238.5	143.9
0330	Cast-in-Place Concrete	80.6	166.1	110.1	119.3	115.9	118.1	89.9	94.0	91.3	96.1	126.5	106.6	102.0	160.8	122.3	100.3	152.7	118.4
03	CONCRETE	92.1	174.5	125.0	102.1	109.0	104.8	91.8	91.9	91.8	95.9	144.2	115.2	99.8	179.3	131.6	97.8	166.7	125.4
04	MASONRY	98.0	183.7	152.0	125.3	116.8	120.0	102.3	90.9	95.1	116.3	130.3	125.1	108.2	181.3	154.3	119.0	167.7	149.7
05	METALS	106.1	171.4	121.3	92.0	99.2	93.7	94.3	96.4	94.8	98.4	131.6	106.1	97.5	170.8	114.6	99.3	173.1	116.5
06	WOOD, PLASTICS & COMPOSITES	98.0	181.8	132.7	74.2	102.1	85.7	89.2	83.2	86.7	81.9	169.1	118.0	80.3	177.7	120.6	86.1	149.7	112.4
07	THERMAL & MOISTURE PROTECTION	110.5	165.2	134.5	110.4	101.9	106.6	118.6	90.0	106.1	126.7	134.6	130.2	112.1	160.2	133.2	113.5	153.3	130.9
08	OPENINGS	97.7	192.8	120.6	93.4	98.5	94.6	94.7	89.1	93.3	90.4	160.5	107.3	84.4	187.4	109.3	84.8	175.1	106.6
0920	Plaster & Gypsum Board	109.6	183.8	156.0	87.0	102.3	96.6	115.1	82.5	94.7	84.2	171.4	138.7	92.1	180.2	147.2	93.9	151.4	129.9
0950, 0980	Ceilings & Acoustic Treatment	94.2	183.8	152.4	109.1	102.3	104.7	110.7	82.5	92.4	79.5	171.4	139.2	84.1	180.2	146.5	84.0	151.4	127.8
0960	Flooring	101.6	181.5	126.0	90.8	104.3	94.9	105.7	97.7	103.3	104.3	147.6	117.5	105.8	179.1	128.3	106.7	179.1	128.8
0970, 0990	Wall Finishes & Painting/Coating	98.7	164.7	138.9	96.3	103.4	100.6	113.4	88.3	98.1	115.8	114.5	115.0	111.9	164.7	144.1	111.9	164.7	144.1
09	FINISHES	101.5	181.4	144.0	93.3	105.8	100.1	101.0	87.6	93.7	101.4	153.9	129.9	102.4	177.9	143.4	102.8	158.1	132.8
COVERS	DIVS. 10 - 14, 25, 28, 41, 43, 44, 46	100.0	141.2	108.9	100.0	106.2	101.4	100.0	85.3	96.8	100.0	107.1	101.6	100.0	132.9	107.2	100.0	128.1	106.1
21, 22, 23	FIRE SUPPRESSION, PLUMBING & HVAC	99.3	172.4	129.3	96.6	97.4	97.0	96.6	93.0	95.1	96.6	111.5	102.7	100.5	167.2	127.8	100.7	154.6	122.8
26, 27, 3370	ELECTRICAL, COMMUNICATIONS & UTIL.	100.2	183.8	140.5	97.2	91.6	94.5	91.7	85.0	88.5	98.6	112.4	105.2	102.4	176.3	138.1	103.4	140.6	121.3
MF2018	WEIGHTED AVERAGE	100.1	171.3	129.1	97.4	100.9	98.8	96.3	90.4	93.9	99.3	128.1	111.1	99.2	168.8	127.5	100.0	154.4	122.1

NEW YORK

DIVISION		ROCHESTER 144-146 MAT.	INST.	TOTAL	SCHENECTADY 123 MAT.	INST.	TOTAL	STATEN ISLAND 103 MAT.	INST.	TOTAL	SUFFERN 109 MAT.	INST.	TOTAL	SYRACUSE 130-132 MAT.	INST.	TOTAL	UTICA 133-135 MAT.	INST.	TOTAL
015433	CONTRACTOR EQUIPMENT		111.7	111.7		108.7	108.7		101.6	101.6		101.6	101.6		108.7	108.7		108.7	108.7
0241, 31 - 34	SITE & INFRASTRUCTURE, DEMOLITION	81.9	99.5	93.9	77.7	95.5	89.8	103.5	104.8	104.4	95.8	98.7	97.7	89.2	94.5	92.8	69.4	94.0	86.1
0310	Concrete Forming & Accessories	101.7	98.0	98.7	102.2	103.6	103.3	76.2	179.1	159.2	86.3	132.3	123.4	99.5	92.4	93.8	100.1	90.7	92.5
0320	Concrete Reinforcing	100.4	103.4	101.3	96.3	111.8	101.3	92.6	201.3	127.5	92.8	141.5	108.4	98.1	104.5	100.1	98.1	103.4	99.8
0330	Cast-in-Place Concrete	106.7	106.2	106.5	88.8	112.9	97.1	77.8	162.4	107.0	75.4	130.0	94.3	97.9	106.4	100.8	89.6	105.2	94.9
03	CONCRETE	103.7	102.9	103.4	92.5	109.4	99.3	88.3	176.2	123.4	84.8	132.7	104.0	97.1	100.8	98.6	95.0	99.4	96.7
04	MASONRY	104.8	106.5	105.9	105.0	116.0	111.9	97.9	181.3	150.5	92.1	132.8	117.7	100.2	107.9	105.1	92.6	106.3	101.2
05	METALS	100.8	118.3	104.9	95.5	125.8	102.6	86.5	171.0	106.3	86.5	128.4	96.3	94.5	120.4	100.6	92.7	119.9	99.1
06	WOOD, PLASTICS & COMPOSITES	97.4	96.0	96.8	103.8	99.9	102.2	77.6	178.0	119.1	88.6	134.4	107.6	97.5	88.3	93.7	97.5	86.3	92.9
07	THERMAL & MOISTURE PROTECTION	114.1	101.0	108.4	106.0	107.6	106.7	105.4	160.6	129.6	106.0	131.7	117.2	107.0	98.2	103.1	97.0	97.5	97.2
08	OPENINGS	97.1	97.4	97.1	92.2	101.7	94.5	90.2	187.6	113.7	90.3	143.3	103.1	89.3	91.4	89.8	91.6	90.0	91.2
0920	Plaster & Gypsum Board	109.0	95.9	100.8	100.4	100.0	100.2	93.1	180.2	147.6	99.2	135.4	121.8	108.7	88.1	95.8	108.7	86.0	94.5
0950, 0980	Ceilings & Acoustic Treatment	109.4	95.9	100.6	106.9	100.0	102.4	76.0	180.2	143.6	74.0	135.4	113.8	105.3	88.1	94.1	105.3	86.0	92.8
0960	Flooring	91.4	107.0	96.1	89.4	107.6	95.0	96.7	179.1	121.9	96.3	159.6	115.7	91.4	98.2	93.5	88.4	98.3	91.4
0970, 0990	Wall Finishes & Painting/Coating	98.7	99.4	99.1	87.4	100.2	95.2	99.3	164.7	139.2	97.7	117.8	110.0	88.5	101.8	96.6	81.6	101.8	93.9
09	FINISHES	98.9	99.7	99.3	95.9	103.4	100.0	94.1	178.1	139.7	93.7	136.6	117.0	97.2	93.7	95.3	95.4	92.3	93.7
COVERS	DIVS. 10 - 14, 25, 28, 41, 43, 44, 46	100.0	98.3	99.6	100.0	99.7	99.9	100.0	133.7	107.3	100.0	105.6	101.2	100.0	98.2	99.6	100.0	95.1	98.9
21, 22, 23	FIRE SUPPRESSION, PLUMBING & HVAC	100.8	90.9	96.8	101.0	105.4	102.8	100.4	167.3	127.8	96.3	114.6	103.8	101.1	93.8	98.1	101.1	101.0	101.0
26, 27, 3370	ELECTRICAL, COMMUNICATIONS & UTIL.	102.3	92.9	97.8	99.7	106.2	102.8	96.4	176.3	135.0	98.1	106.2	102.0	100.8	102.8	101.7	98.0	102.8	100.3
MF2018	WEIGHTED AVERAGE	100.8	99.5	100.3	97.1	107.6	101.4	93.9	167.8	124.0	92.4	122.1	104.5	97.5	100.0	98.5	95.6	100.7	97.7

NEW YORK / NORTH CAROLINA

DIVISION		WATERTOWN 136 MAT.	INST.	TOTAL	WHITE PLAINS 106 MAT.	INST.	TOTAL	YONKERS 107 MAT.	INST.	TOTAL	ASHEVILLE 287-288 MAT.	INST.	TOTAL	CHARLOTTE 281-282 MAT.	INST.	TOTAL	DURHAM 277 MAT.	INST.	TOTAL
015433	CONTRACTOR EQUIPMENT		108.7	108.7		101.6	101.6		101.6	101.6		95.9	95.9		101.4	101.4		100.9	100.9
0241, 31 - 34	SITE & INFRASTRUCTURE, DEMOLITION	76.5	94.3	88.6	94.3	100.8	98.7	101.4	100.8	101.0	104.8	74.9	84.5	107.0	85.8	92.6	111.0	83.1	92.0
0310	Concrete Forming & Accessories	79.7	95.1	92.1	92.1	142.0	132.3	92.3	142.1	132.5	92.1	61.5	67.4	101.4	63.7	71.0	98.2	61.9	68.9
0320	Concrete Reinforcing	98.7	104.5	100.5	92.8	170.0	117.6	96.5	170.0	120.1	98.2	61.6	86.4	100.4	66.8	89.6	99.2	62.3	87.3
0330	Cast-in-Place Concrete	104.3	108.1	105.6	69.1	141.2	94.0	77.3	141.2	99.3	111.5	70.4	97.3	113.7	72.4	99.5	126.1	70.4	106.9
03	CONCRETE	105.3	102.6	104.3	80.2	147.2	107.0	87.5	147.2	111.4	95.7	66.3	83.9	98.3	68.6	86.4	104.7	66.6	89.4
04	MASONRY	94.1	110.7	104.6	91.6	147.3	126.7	93.2	147.3	127.3	88.6	63.6	72.8	90.7	64.4	74.1	91.5	63.6	73.9
05	METALS	92.8	120.3	99.2	87.7	168.3	106.5	97.7	168.4	114.2	93.3	88.7	92.2	94.5	87.4	92.9	109.4	88.9	104.6
06	WOOD, PLASTICS & COMPOSITES	76.0	91.1	82.2	94.9	139.1	113.2	94.7	139.1	113.1	89.4	59.5	77.0	92.3	61.9	79.8	97.6	60.1	82.0
07	THERMAL & MOISTURE PROTECTION	97.1	99.9	98.4	105.8	141.5	121.5	106.2	141.5	121.7	104.2	61.1	85.3	97.4	65.3	83.3	101.5	61.2	83.8
08	OPENINGS	91.6	94.5	92.3	90.3	169.3	109.3	93.3	169.3	111.6	86.3	58.4	79.6	95.7	60.2	87.1	92.6	58.9	84.5
0920	Plaster & Gypsum Board	91.5	91.0	91.2	105.7	140.2	127.3	110.1	140.2	128.9	114.2	58.4	79.3	112.4	60.8	80.1	112.0	59.1	78.9
0950, 0980	Ceilings & Acoustic Treatment	105.3	91.0	96.0	74.0	140.2	116.9	97.3	140.2	125.1	101.4	58.4	73.5	101.7	60.8	75.2	88.4	59.1	69.4
0960	Flooring	82.8	98.3	87.5	98.6	179.1	123.3	98.0	179.1	122.8	97.9	62.1	87.0	99.6	66.7	89.5	108.9	62.1	94.6
0970, 0990	Wall Finishes & Painting/Coating	81.6	101.0	93.4	97.7	164.7	138.6	97.7	164.7	138.6	102.2	53.0	72.2	98.7	65.1	78.2	103.0	53.0	72.5
09	FINISHES	91.7	95.8	93.9	95.0	150.9	125.3	101.3	150.9	128.2	102.1	60.4	79.4	102.4	64.1	81.6	100.5	60.7	78.9
COVERS	DIVS. 10 - 14, 25, 28, 41, 43, 44, 46	100.0	99.1	99.8	100.0	123.6	105.1	100.0	130.6	106.7	100.0	84.4	96.6	100.0	85.1	96.8	100.0	84.5	96.6
21, 22, 23	FIRE SUPPRESSION, PLUMBING & HVAC	101.1	89.6	96.4	101.0	142.5	118.0	101.0	142.5	118.0	101.6	62.0	85.4	100.8	62.2	84.9	101.7	62.0	85.4
26, 27, 3370	ELECTRICAL, COMMUNICATIONS & UTIL.	100.8	90.9	96.0	93.9	160.0	125.7	98.2	160.0	128.0	100.1	53.6	77.7	100.9	81.1	91.3	95.7	53.2	75.2
MF2018	WEIGHTED AVERAGE	97.0	98.4	97.6	92.7	146.7	114.7	97.2	146.9	117.4	96.9	65.3	84.0	98.3	71.0	87.2	101.7	66.0	87.2

For CCI updates and additional guidance please go to www.rsmeans.com/2022books

NORTH CAROLINA

DIVISION		ELIZABETH CITY 279			FAYETTEVILLE 283			GASTONIA 280			GREENSBORO 270,272 - 274			HICKORY 286			KINSTON 285		
		MAT.	INST.	TOTAL	MAT.	INST.	TOTAL	MAT.	INST.	TOTAL	MAT.	INST.	TOTAL	MAT.	INST.	TOTAL	MAT.	INST.	TOTAL
015433	CONTRACTOR EQUIPMENT		104.8	104.8		100.9	100.9		95.9	95.9		100.9	100.9		100.9	100.9		100.9	100.9
0241, 31 - 34	SITE & INFRASTRUCTURE, DEMOLITION	115.4	84.3	94.3	104.0	82.8	89.6	104.3	74.9	84.3	110.8	83.1	92.0	103.1	82.8	89.3	101.6	82.8	88.8
0310	Concrete Forming & Accessories	78.2	60.2	63.7	91.5	60.1	66.2	101.1	62.1	69.6	97.9	61.9	68.9	86.5	62.0	66.8	81.3	60.1	64.2
0320	Concrete Reinforcing	97.7	65.1	87.2	101.4	62.3	88.9	98.5	62.4	86.9	98.6	62.3	86.9	98.2	62.3	86.6	97.6	62.3	86.3
0330	Cast-in-Place Concrete	126.3	69.9	106.9	116.9	67.7	99.9	109.0	70.5	95.7	125.1	70.4	106.2	111.4	70.2	97.2	107.6	67.7	93.8
03	CONCRETE	103.8	66.2	88.8	98.1	64.9	84.8	94.9	66.7	83.6	104.1	66.6	89.1	95.3	66.6	83.8	92.4	64.9	81.4
04	MASONRY	103.9	58.8	75.5	91.6	58.8	70.9	92.5	63.6	74.3	88.1	63.6	72.7	78.3	63.5	69.0	84.5	58.8	68.3
05	METALS	95.5	91.0	94.5	111.9	88.9	106.5	94.2	89.1	93.0	101.1	88.9	98.3	93.6	88.3	92.3	92.5	88.9	91.7
06	WOOD, PLASTICS & COMPOSITES	76.2	59.8	69.4	87.8	60.1	76.3	99.5	60.1	83.2	97.1	60.1	81.8	82.8	60.1	73.4	78.5	60.1	70.9
07	THERMAL & MOISTURE PROTECTION	100.2	58.8	82.0	103.6	59.1	84.1	104.5	61.2	85.5	101.0	61.2	83.6	104.6	61.2	85.6	104.3	59.1	84.5
08	OPENINGS	90.1	59.3	82.7	86.4	58.9	79.8	89.3	58.9	82.0	92.6	58.9	84.5	86.4	58.9	79.7	86.5	58.9	79.8
0920	Plaster & Gypsum Board	95.7	58.2	72.2	117.6	59.1	81.0	124.8	59.1	83.7	111.9	59.1	78.9	114.2	59.1	79.7	112.4	59.1	79.1
0950, 0980	Ceilings & Acoustic Treatment	88.4	58.2	68.8	105.5	59.1	75.4	108.2	59.1	76.3	88.4	59.1	69.4	101.4	59.1	73.9	108.2	59.1	76.3
0960	Flooring	101.2	62.1	89.2	98.0	62.1	87.1	100.4	62.1	88.7	108.9	62.1	94.6	97.9	62.1	87.0	95.7	62.1	85.5
0970, 0990	Wall Finishes & Painting/Coating	103.0	53.0	72.5	102.2	53.0	72.2	102.2	53.0	72.2	103.0	53.0	72.5	102.2	53.0	72.2	102.2	53.0	72.2
09	FINISHES	96.2	59.4	76.2	103.5	59.5	79.6	105.9	60.7	81.3	100.5	60.7	78.9	102.3	60.7	79.7	102.7	59.5	79.2
COVERS	DIVS. 10 - 14, 25, 28, 41, 43, 44, 46	100.0	84.8	96.7	100.0	82.9	96.3	100.0	84.6	96.7	100.0	84.5	96.6	100.0	84.6	96.7	100.0	82.9	96.3
21, 22, 23	FIRE SUPPRESSION, PLUMBING & HVAC	97.3	59.8	81.9	101.4	59.6	84.2	101.6	62.1	85.4	101.5	62.0	85.3	97.4	62.1	82.9	97.4	59.6	81.8
26, 27, 3370	ELECTRICAL, COMMUNICATIONS & UTIL.	95.1	64.1	80.1	101.6	53.2	78.2	99.8	81.1	90.7	95.0	53.6	75.0	98.7	81.1	90.2	98.5	51.7	76.0
MF2018	WEIGHTED AVERAGE	97.6	66.6	85.0	101.3	64.4	86.3	97.8	69.3	86.2	99.7	66.0	86.0	95.4	69.8	85.0	95.0	64.2	82.5

NORTH CAROLINA / NORTH DAKOTA

DIVISION		MURPHY 289			RALEIGH 275 - 276			ROCKY MOUNT 278			WILMINGTON 284			WINSTON-SALEM 271			BISMARCK 585		
		MAT.	INST.	TOTAL	MAT.	INST.	TOTAL	MAT.	INST.	TOTAL	MAT.	INST.	TOTAL	MAT.	INST.	TOTAL	MAT.	INST.	TOTAL
015433	CONTRACTOR EQUIPMENT		95.9	95.9		107.5	107.5		100.9	100.9		95.9	95.9		100.9	100.9		101.2	101.2
0241, 31 - 34	SITE & INFRASTRUCTURE, DEMOLITION	105.6	74.6	84.5	109.7	95.7	100.2	113.4	82.7	92.6	105.7	74.6	84.6	111.3	83.1	92.1	88.1	99.6	95.9
0310	Concrete Forming & Accessories	102.1	56.7	65.5	98.5	61.6	68.8	87.1	61.4	66.4	93.9	60.1	66.6	100.1	61.9	69.3	115.2	97.9	109.6
0320	Concrete Reinforcing	97.6	60.1	85.6	99.8	62.3	87.8	97.7	62.3	86.3	98.9	62.3	87.2	98.6	62.3	86.9	111.9	87.6	103.5
0330	Cast-in-Place Concrete	115.3	67.7	98.9	129.8	71.5	109.7	123.6	69.5	104.9	111.0	67.7	96.1	128.3	70.4	108.3	107.6	84.8	98.5
03	CONCRETE	98.6	62.9	84.3	105.0	66.7	89.7	104.2	66.0	88.9	95.8	64.9	83.4	105.6	66.6	90.0	107.6	84.8	98.5
04	MASONRY	81.8	58.8	67.3	89.7	62.5	72.5	81.2	62.3	69.3	79.3	58.8	66.4	88.5	63.6	72.8	97.8	82.0	87.8
05	METALS	91.5	88.1	90.7	94.6	85.5	92.5	94.8	88.2	93.2	93.0	88.9	92.1	98.6	88.9	96.4	90.3	92.8	90.9
06	WOOD, PLASTICS & COMPOSITES	100.4	55.6	81.8	94.2	60.3	80.1	85.5	60.1	75.0	91.2	60.1	78.3	97.1	60.1	81.8	94.7	73.7	86.0
07	THERMAL & MOISTURE PROTECTION	104.6	58.3	84.3	96.8	62.4	81.7	100.9	60.7	83.3	104.2	59.1	84.4	101.0	61.2	83.6	98.2	84.4	92.1
08	OPENINGS	86.3	55.9	79.0	94.5	59.0	85.9	89.5	58.9	82.1	86.5	58.9	79.8	92.6	58.9	84.5	100.1	82.7	95.9
0920	Plaster & Gypsum Board	123.3	54.4	80.2	107.6	59.1	77.3	99.5	59.1	74.2	116.0	59.1	80.4	111.9	59.1	78.9	103.9	73.1	84.7
0950, 0980	Ceilings & Acoustic Treatment	101.4	54.4	70.9	84.6	59.1	68.0	85.7	59.1	68.4	105.5	59.1	75.4	88.4	59.1	69.4	107.4	73.1	85.2
0960	Flooring	100.7	62.1	88.9	99.7	62.1	88.2	104.6	62.1	91.6	98.4	62.1	87.3	108.9	62.1	94.6	92.2	53.6	80.4
0970, 0990	Wall Finishes & Painting/Coating	102.2	53.0	72.2	100.1	53.0	71.4	103.0	53.0	72.5	102.2	53.0	72.2	103.0	53.0	72.5	98.7	59.5	74.7
09	FINISHES	104.5	56.9	78.6	98.3	60.5	77.8	97.3	60.4	77.2	103.4	59.5	79.5	100.5	60.7	78.9	98.8	71.5	84.0
COVERS	DIVS. 10 - 14, 25, 28, 41, 43, 44, 46	100.0	82.4	96.2	100.0	84.5	96.6	100.0	84.1	96.5	100.0	82.9	96.3	100.0	84.5	96.6	100.0	91.7	98.2
21, 22, 23	FIRE SUPPRESSION, PLUMBING & HVAC	97.4	59.5	81.8	100.8	61.5	84.7	97.3	61.4	82.6	101.6	59.6	84.3	101.5	62.0	85.3	100.6	76.0	90.5
26, 27, 3370	ELECTRICAL, COMMUNICATIONS & UTIL.	100.6	51.7	77.0	100.9	52.2	77.4	96.3	53.2	75.5	100.5	51.7	77.0	95.0	53.6	75.0	102.3	73.4	88.4
MF2018	WEIGHTED AVERAGE	96.2	62.7	82.6	98.9	66.3	85.6	96.8	65.5	84.1	96.7	63.6	83.2	99.5	66.0	85.9	98.9	81.2	91.7

NORTH DAKOTA

DIVISION		DEVILS LAKE 583			DICKINSON 586			FARGO 580 - 581			GRAND FORKS 582			JAMESTOWN 584			MINOT 587		
		MAT.	INST.	TOTAL	MAT.	INST.	TOTAL	MAT.	INST.	TOTAL	MAT.	INST.	TOTAL	MAT.	INST.	TOTAL	MAT.	INST.	TOTAL
015433	CONTRACTOR EQUIPMENT		94.5	94.5		94.5	94.5		101.5	101.5		94.5	94.5		94.5	94.5		94.5	94.5
0241, 31 - 34	SITE & INFRASTRUCTURE, DEMOLITION	91.3	88.8	89.6	96.4	88.7	91.2	88.2	99.8	96.0	94.3	89.5	91.1	90.4	88.8	89.3	92.4	89.3	90.3
0310	Concrete Forming & Accessories	102.9	67.6	74.4	88.3	67.5	71.5	97.2	68.3	73.9	93.9	74.4	78.1	90.5	67.5	72.0	87.8	67.7	71.6
0320	Concrete Reinforcing	111.4	97.9	107.1	112.5	97.8	107.8	99.8	98.5	99.4	109.7	98.0	105.9	112.1	98.4	107.7	113.6	97.8	108.5
0330	Cast-in-Place Concrete	131.3	76.4	112.3	118.7	76.3	104.1	108.7	83.6	100.0	118.7	82.7	106.3	129.6	76.4	111.3	118.7	76.6	104.2
03	CONCRETE	111.4	76.8	97.6	109.2	76.7	96.2	105.4	79.5	95.0	107.0	82.1	97.0	109.8	76.8	96.6	106.1	76.9	94.4
04	MASONRY	112.4	80.3	92.2	111.5	77.7	90.2	95.6	89.0	91.4	105.4	83.4	91.5	123.9	88.8	101.8	106.1	77.1	87.8
05	METALS	89.1	94.3	90.3	89.1	93.7	90.1	94.4	93.4	94.2	89.1	95.1	90.5	89.1	94.9	90.4	89.4	94.6	90.6
06	WOOD, PLASTICS & COMPOSITES	97.2	63.7	83.3	81.9	63.7	74.3	94.4	64.1	81.9	87.3	70.9	80.5	84.2	63.7	75.7	81.5	63.7	74.1
07	THERMAL & MOISTURE PROTECTION	97.8	78.6	89.3	98.1	79.0	89.7	93.1	83.9	89.1	97.9	81.7	90.8	97.4	81.3	90.4	97.5	79.4	89.6
08	OPENINGS	97.4	77.2	92.5	97.4	77.2	92.5	97.5	77.4	92.6	96.1	81.1	92.5	97.4	77.2	92.5	96.2	77.2	91.6
0920	Plaster & Gypsum Board	111.1	63.2	81.2	95.6	63.2	75.4	91.1	63.2	73.7	97.8	70.6	80.8	97.3	63.2	76.0	95.6	63.2	75.4
0950, 0980	Ceilings & Acoustic Treatment	99.2	63.2	75.9	99.2	63.2	75.9	99.0	63.2	75.8	99.2	70.6	80.7	99.2	63.2	75.9	99.2	63.2	75.9
0960	Flooring	96.4	53.6	83.3	90.8	53.6	79.4	97.9	53.6	84.4	92.4	53.6	80.5	91.5	53.6	79.9	90.6	53.6	79.3
0970, 0990	Wall Finishes & Painting/Coating	89.0	52.3	66.6	89.0	52.3	66.6	98.7	62.5	76.6	89.0	61.3	72.1	89.0	52.3	66.6	89.0	53.3	67.2
09	FINISHES	95.6	63.6	78.3	92.4	63.6	76.8	96.5	65.1	79.4	92.7	69.6	80.1	92.0	63.6	76.6	91.7	63.8	76.5
COVERS	DIVS. 10 - 14, 25, 28, 41, 43, 44, 46	100.0	80.3	95.7	100.0	80.3	95.7	100.0	89.3	97.7	100.0	88.1	97.4	100.0	80.3	95.7	100.0	88.4	97.5
21, 22, 23	FIRE SUPPRESSION, PLUMBING & HVAC	96.5	72.7	86.8	96.5	68.0	84.8	100.7	68.8	87.6	100.8	75.9	90.6	96.5	67.0	84.4	100.8	66.9	86.9
26, 27, 3370	ELECTRICAL, COMMUNICATIONS & UTIL.	96.1	67.4	82.3	101.3	67.1	84.9	102.2	64.3	83.9	98.2	67.4	83.4	96.1	63.3	80.3	100.1	67.9	84.6
MF2018	WEIGHTED AVERAGE	97.9	75.8	88.9	97.8	74.4	88.3	98.8	77.2	90.0	97.7	79.0	90.1	97.7	75.0	88.5	97.7	74.7	88.3

City Cost Indexes

NORTH DAKOTA / OHIO

DIVISION		WILLISTON 588 MAT.	INST.	TOTAL	AKRON 442-443 MAT.	INST.	TOTAL	ATHENS 457 MAT.	INST.	TOTAL	CANTON 446-447 MAT.	INST.	TOTAL	CHILLICOTHE 456 MAT.	INST.	TOTAL	CINCINNATI 451-452 MAT.	INST.	TOTAL
015433	CONTRACTOR EQUIPMENT		94.5	94.5		86.5	86.5		82.6	82.6		86.5	86.5		92.8	92.8		97.2	97.2
0241, 31 - 34	SITE & INFRASTRUCTURE, DEMOLITION	91.8	89.6	90.3	83.5	90.4	88.2	89.8	81.1	83.9	83.6	90.0	87.9	83.1	90.3	88.0	83.2	99.0	93.9
0310	Concrete Forming & Accessories	96.3	74.3	78.6	98.5	83.7	86.6	88.2	74.3	77.0	98.5	76.7	80.9	92.3	76.7	79.7	96.7	73.8	78.2
0320	Concrete Reinforcing	114.6	97.8	109.2	102.9	88.5	98.3	97.7	83.0	93.0	102.9	75.9	94.3	96.4	82.7	92.0	100.0	83.6	94.8
0330	Cast-in-Place Concrete	118.7	82.7	106.3	101.0	87.3	96.2	111.7	89.7	104.1	102.0	85.3	96.2	101.3	86.5	96.2	96.8	77.3	90.0
03	CONCRETE	107.6	82.0	97.4	99.7	85.4	94.0	102.9	81.1	94.2	100.2	79.4	91.9	98.0	81.7	91.5	96.5	76.9	88.7
04	MASONRY	100.4	80.1	87.6	103.9	90.9	95.7	86.3	90.8	89.1	104.5	83.1	91.0	93.7	84.5	87.9	99.3	81.6	88.2
05	METALS	89.2	94.8	90.5	91.0	81.0	88.7	95.8	78.5	91.8	91.0	75.5	87.4	88.7	87.6	88.4	97.0	82.1	93.5
06	WOOD, PLASTICS & COMPOSITES	89.2	70.9	81.6	102.5	82.3	94.2	82.4	69.8	77.2	103.3	75.3	91.7	93.9	73.6	85.5	98.5	72.1	87.6
07	THERMAL & MOISTURE PROTECTION	97.8	80.8	90.4	102.7	89.4	96.9	100.2	84.8	93.5	103.4	85.9	95.8	101.3	81.9	92.8	101.7	81.0	92.3
08	OPENINGS	97.4	81.1	93.5	103.6	82.4	98.5	92.9	71.4	87.7	97.8	72.2	91.6	84.8	73.4	82.1	98.6	72.4	92.3
0920	Plaster & Gypsum Board	97.8	70.6	80.8	91.8	81.9	85.7	81.2	69.1	73.6	94.7	74.7	82.2	87.7	73.4	78.8	86.6	71.4	77.1
0950, 0980	Ceilings & Acoustic Treatment	99.2	70.6	80.7	93.0	81.9	85.8	103.9	69.1	81.3	93.0	74.7	81.1	97.8	73.4	82.0	88.7	71.4	77.5
0960	Flooring	93.2	53.6	81.1	93.0	76.6	88.0	127.8	70.0	110.1	93.4	70.6	86.4	103.4	70.0	93.2	107.6	80.0	99.2
0970, 0990	Wall Finishes & Painting/Coating	89.0	58.9	70.6	96.4	88.7	91.7	101.7	82.7	90.1	96.4	75.3	83.5	99.4	82.7	89.2	98.7	73.3	83.2
09	FINISHES	92.8	69.4	80.1	93.4	82.9	87.7	98.7	73.7	85.1	94.0	74.7	83.5	96.9	75.3	85.2	95.4	74.4	84.0
COVERS	DIVS. 10 - 14, 25, 28, 41, 43, 44, 46	100.0	90.2	97.9	100.0	92.6	98.4	100.0	89.6	97.7	100.0	90.8	98.0	100.0	88.7	97.6	100.0	88.1	97.4
21, 22, 23	FIRE SUPPRESSION, PLUMBING & HVAC	96.5	72.7	86.7	100.5	87.7	95.3	96.5	75.2	87.8	100.5	78.2	91.4	97.1	85.6	92.4	100.6	78.5	91.6
26, 27, 3370	ELECTRICAL, COMMUNICATIONS & UTIL.	98.5	67.9	83.7	102.3	82.5	92.7	104.0	83.1	94.0	101.9	84.0	93.2	101.3	83.2	92.5	101.0	76.8	89.3
MF2018	WEIGHTED AVERAGE	96.9	78.0	89.2	98.2	85.9	93.2	97.5	79.8	90.3	97.8	80.3	90.6	94.6	83.2	90.0	98.3	79.8	90.8

OHIO

DIVISION		CLEVELAND 441 MAT.	INST.	TOTAL	COLUMBUS 430-432 MAT.	INST.	TOTAL	DAYTON 453-454 MAT.	INST.	TOTAL	HAMILTON 450 MAT.	INST.	TOTAL	LIMA 458 MAT.	INST.	TOTAL	LORAIN 440 MAT.	INST.	TOTAL
015433	CONTRACTOR EQUIPMENT		92.9	92.9		95.6	95.6		86.4	86.4		92.8	92.8		85.8	85.8		86.5	86.5
0241, 31 - 34	SITE & INFRASTRUCTURE, DEMOLITION	83.0	97.5	92.3	80.4	96.6	91.4	80.6	89.9	86.9	80.6	89.9	87.0	85.2	80.9	82.3	82.7	90.1	87.8
0310	Concrete Forming & Accessories	98.5	89.9	91.6	99.2	78.0	82.1	94.8	77.6	80.9	94.8	73.0	77.2	88.2	72.1	75.2	98.6	79.1	82.9
0320	Concrete Reinforcing	101.1	92.3	98.3	101.4	79.3	94.3	102.2	78.6	94.6	102.2	71.9	92.5	97.7	78.7	91.6	102.9	88.6	98.3
0330	Cast-in-Place Concrete	100.7	97.6	99.6	99.6	90.1	96.3	86.8	80.9	84.8	93.0	77.6	87.7	102.6	84.0	96.2	96.1	90.6	94.2
03	CONCRETE	98.9	92.6	96.4	98.7	82.5	92.2	90.9	78.9	86.1	93.6	75.1	86.2	96.6	77.7	89.0	97.6	84.6	92.4
04	MASONRY	108.6	99.2	102.7	90.4	86.9	88.2	92.5	74.1	80.9	92.9	75.9	82.2	116.3	73.0	89.0	100.5	91.1	94.6
05	METALS	97.3	84.6	94.3	97.2	80.1	93.2	90.4	77.4	87.4	90.3	83.9	88.8	95.9	80.5	92.3	91.6	81.2	89.2
06	WOOD, PLASTICS & COMPOSITES	99.8	87.9	94.9	100.3	75.7	90.1	97.7	78.2	89.6	96.7	72.8	86.8	82.3	70.9	77.6	102.5	77.3	92.1
07	THERMAL & MOISTURE PROTECTION	101.7	97.4	99.8	110.0	84.5	98.8	104.8	77.0	92.6	101.3	76.0	90.3	99.9	80.7	91.5	103.3	87.4	96.4
08	OPENINGS	101.0	86.3	97.5	98.7	73.7	92.6	91.3	75.3	87.4	89.2	71.3	84.9	92.9	70.5	87.5	97.8	79.7	93.4
0920	Plaster & Gypsum Board	90.6	87.6	88.7	92.0	74.9	81.3	89.6	78.2	82.5	89.6	72.6	79.0	81.2	70.1	74.3	91.8	76.8	82.4
0950, 0980	Ceilings & Acoustic Treatment	90.8	87.6	88.7	92.1	74.9	81.0	97.8	78.2	85.1	97.8	72.6	81.4	103.9	70.1	82.0	93.0	76.8	82.5
0960	Flooring	97.5	89.0	94.9	98.3	75.0	91.2	107.2	71.8	96.4	104.4	73.0	94.8	127.4	71.8	110.4	93.4	87.0	91.4
0970, 0990	Wall Finishes & Painting/Coating	100.1	88.7	93.1	98.7	78.4	86.3	99.4	68.6	80.6	99.4	66.2	79.1	101.8	71.0	83.0	96.4	88.7	91.7
09	FINISHES	94.0	89.5	91.6	94.4	76.9	84.9	98.0	75.5	85.8	97.2	72.1	83.6	98.2	71.2	83.6	93.5	81.2	86.8
COVERS	DIVS. 10 - 14, 25, 28, 41, 43, 44, 46	100.0	97.1	99.4	100.0	90.0	97.8	100.0	87.4	97.3	100.0	86.6	97.1	100.0	90.6	98.0	100.0	93.6	98.6
21, 22, 23	FIRE SUPPRESSION, PLUMBING & HVAC	100.5	92.4	97.2	100.6	91.0	96.7	101.5	77.8	91.8	101.3	69.5	88.2	96.5	85.1	91.8	100.5	79.0	91.7
26, 27, 3370	ELECTRICAL, COMMUNICATIONS & UTIL.	102.2	89.3	96.0	102.3	81.2	92.2	99.5	74.1	87.3	99.8	69.9	85.4	104.3	70.1	87.8	102.0	76.9	89.9
MF2018	WEIGHTED AVERAGE	99.3	92.0	96.3	98.5	84.6	92.8	95.4	77.8	88.3	95.4	75.0	87.1	97.7	77.5	89.5	97.3	82.7	91.4

OHIO

DIVISION		MANSFIELD 448-449 MAT.	INST.	TOTAL	MARION 433 MAT.	INST.	TOTAL	SPRINGFIELD 455 MAT.	INST.	TOTAL	STEUBENVILLE 439 MAT.	INST.	TOTAL	TOLEDO 434-436 MAT.	INST.	TOTAL	YOUNGSTOWN 444-445 MAT.	INST.	TOTAL
015433	CONTRACTOR EQUIPMENT		86.5	86.5		86.8	86.8		86.4	86.4		90.3	90.3		90.3	90.3		86.5	86.5
0241, 31 - 34	SITE & INFRASTRUCTURE, DEMOLITION	79.3	89.7	86.4	74.3	86.5	82.6	80.9	90.0	87.1	101.9	93.9	96.5	77.5	87.2	84.1	83.3	89.9	87.8
0310	Concrete Forming & Accessories	84.2	77.6	78.9	95.7	74.7	78.8	94.8	78.1	81.3	94.4	74.6	78.4	100.7	89.5	91.7	98.5	78.0	82.0
0320	Concrete Reinforcing	93.8	73.6	87.3	97.9	73.6	90.1	102.2	78.7	94.6	96.4	95.2	96.0	106.3	82.9	98.8	102.9	84.2	96.9
0330	Cast-in-Place Concrete	93.5	82.5	89.7	86.1	82.6	84.9	89.1	81.6	86.5	93.6	84.0	90.3	94.3	89.5	92.7	100.0	86.4	95.3
03	CONCRETE	91.5	78.5	86.3	87.6	77.4	83.5	91.9	79.4	86.9	90.9	81.3	87.1	95.5	88.4	92.7	99.3	81.8	92.3
04	MASONRY	102.7	83.1	90.3	92.9	83.9	87.3	92.7	75.2	81.7	81.0	85.1	83.6	99.2	90.8	93.9	104.2	86.7	93.2
05	METALS	92.1	74.9	88.1	92.3	77.2	88.8	90.4	77.4	87.3	89.4	83.1	87.9	92.9	86.6	91.5	91.0	78.9	88.2
06	WOOD, PLASTICS & COMPOSITES	87.1	77.3	83.0	92.7	73.2	84.6	98.7	78.2	90.2	87.1	72.1	80.9	97.8	90.8	94.9	102.5	76.4	91.7
07	THERMAL & MOISTURE PROTECTION	101.8	84.9	94.4	104.1	82.4	94.6	104.7	77.5	92.8	117.9	81.7	102.1	105.7	90.1	98.9	103.6	86.8	96.2
08	OPENINGS	98.8	73.0	92.5	87.5	70.7	83.4	89.7	75.7	86.3	87.4	75.4	84.5	89.9	84.5	88.6	97.8	79.6	93.4
0920	Plaster & Gypsum Board	79.7	76.8	77.9	89.9	72.7	79.1	89.6	78.2	82.5	83.7	71.1	75.8	93.1	90.8	91.6	91.8	75.9	81.9
0950, 0980	Ceilings & Acoustic Treatment	93.7	76.8	82.7	98.0	72.7	81.6	97.8	78.2	85.1	95.6	71.1	79.7	98.0	90.8	93.3	93.0	75.9	81.9
0960	Flooring	88.5	90.8	89.2	95.0	90.8	93.7	107.2	71.8	96.4	125.8	88.5	114.4	95.4	93.3	94.7	93.4	89.4	92.1
0970, 0990	Wall Finishes & Painting/Coating	96.4	72.2	81.6	99.8	72.2	83.0	99.4	68.6	80.6	111.3	80.8	92.7	99.9	85.6	91.2	96.4	78.6	85.5
09	FINISHES	90.1	79.3	84.2	95.0	77.1	85.3	98.0	75.7	85.9	111.2	76.9	92.6	95.9	90.0	92.7	93.6	79.6	86.0
COVERS	DIVS. 10 - 14, 25, 28, 41, 43, 44, 46	100.0	90.4	97.9	100.0	87.9	97.4	100.0	87.8	97.3	100.0	89.0	97.6	100.0	94.0	98.7	100.0	90.9	98.0
21, 22, 23	FIRE SUPPRESSION, PLUMBING & HVAC	96.3	77.5	88.6	96.4	86.2	92.2	101.5	78.4	92.0	96.9	84.8	91.9	100.6	92.5	97.3	100.5	82.3	93.0
26, 27, 3370	ELECTRICAL, COMMUNICATIONS & UTIL.	100.5	84.1	92.6	99.8	84.1	92.2	99.5	80.6	90.4	93.1	98.7	95.8	104.0	93.4	98.9	102.0	84.0	93.3
MF2018	WEIGHTED AVERAGE	95.1	80.5	89.2	93.5	81.7	88.7	95.4	79.1	88.8	94.5	85.4	90.8	96.8	90.3	94.2	97.6	83.1	91.7

For CCI updates and additional guidance please go to www.rsmeans.com/2022books

City Cost Indexes

		OHIO			OKLAHOMA														
		ZANESVILLE			ARDMORE			CLINTON			DURANT			ENID			GUYMON		
	DIVISION	437 - 438			734			736			747			737			739		
		MAT.	INST.	TOTAL	MAT.	INST.	TOTAL	MAT.	INST.	TOTAL	MAT.	INST.	TOTAL	MAT.	INST.	TOTAL	MAT.	INST.	TOTAL
015433	CONTRACTOR EQUIPMENT		86.8	86.8		77.8	77.8		77.0	77.0		77.0	77.0		77.0	77.0		77.0	77.0
0241, 31 - 34	SITE & INFRASTRUCTURE, DEMOLITION	76.4	86.3	83.1	97.0	88.2	91.0	98.5	86.8	90.5	95.4	86.5	89.4	100.8	86.8	91.3	103.0	86.0	91.5
0310	Concrete Forming & Accessories	91.5	73.8	77.2	88.2	53.8	60.4	86.2	53.8	60.0	80.3	52.2	57.6	91.3	54.0	61.2	95.3	52.1	60.5
0320	Concrete Reinforcing	97.4	86.2	93.8	88.7	66.1	81.5	89.2	66.1	81.8	97.7	59.3	85.4	88.7	66.1	81.4	89.2	58.9	79.5
0330	Cast-in-Place Concrete	90.8	81.1	87.4	83.7	71.3	79.4	80.9	71.3	77.6	80.9	69.1	76.8	80.9	71.4	77.6	80.9	68.6	76.7
03	CONCRETE	90.6	78.7	85.8	82.6	62.1	74.4	82.1	62.1	74.1	83.0	59.5	73.6	82.6	62.2	74.5	84.9	59.1	74.6
04	MASONRY	91.3	80.9	84.7	88.8	60.2	70.8	115.0	60.2	80.5	88.9	58.1	69.5	97.9	60.2	74.2	90.1	51.8	65.9
05	METALS	93.6	84.8	91.5	90.6	62.2	84.0	90.7	62.2	84.0	86.7	59.6	80.4	91.7	62.4	84.9	91.2	57.8	83.4
06	WOOD, PLASTICS & COMPOSITES	86.8	73.2	81.2	93.9	51.3	76.3	92.8	51.3	75.7	83.5	51.3	70.2	97.1	51.3	78.1	101.1	51.3	80.5
07	THERMAL & MOISTURE PROTECTION	104.2	78.8	93.1	103.2	62.7	85.5	103.3	62.7	85.5	98.3	61.4	82.1	103.5	62.7	85.6	103.9	57.9	83.7
08	OPENINGS	87.5	76.0	84.7	99.5	52.4	88.1	99.5	52.4	88.2	92.3	50.8	82.2	100.7	53.8	89.4	99.7	50.8	87.9
0920	Plaster & Gypsum Board	82.0	72.7	76.2	95.9	50.7	67.6	95.4	50.7	67.5	79.5	50.7	61.5	97.8	50.7	68.4	97.8	50.7	68.4
0950, 0980	Ceilings & Acoustic Treatment	98.0	72.7	81.6	93.6	50.7	65.7	93.6	50.7	65.7	85.5	50.7	62.9	93.6	50.7	65.7	93.6	50.7	65.7
0960	Flooring	93.7	70.0	86.4	90.2	53.3	78.9	89.1	53.3	78.2	93.8	48.1	79.8	90.7	69.8	84.3	92.1	53.3	80.3
0970, 0990	Wall Finishes & Painting/Coating	99.8	82.7	89.4	91.9	43.3	62.2	91.9	43.3	62.2	92.5	43.3	62.5	91.9	43.3	62.2	91.9	39.2	59.7
09	FINISHES	93.6	73.3	82.5	92.4	51.2	70.0	92.3	51.2	70.0	88.0	49.2	66.9	93.2	54.6	72.2	94.0	51.0	70.7
COVERS	DIVS. 10 - 14, 25, 28, 41, 43, 44, 46	100.0	87.0	97.2	100.0	78.4	95.3	100.0	78.4	95.3	100.0	77.1	95.0	100.0	78.4	95.3	100.0	77.1	95.0
21, 22, 23	FIRE SUPPRESSION, PLUMBING & HVAC	96.4	83.3	91.0	97.0	66.9	84.6	97.0	66.9	84.6	96.9	64.9	83.8	101.2	66.9	87.1	97.0	63.7	83.3
26, 27, 3370	ELECTRICAL, COMMUNICATIONS & UTIL.	100.3	83.1	92.0	94.2	73.7	84.3	94.8	73.7	84.6	96.2	64.2	80.7	94.8	73.7	84.6	95.8	60.8	79.0
MF2018	WEIGHTED AVERAGE	94.1	81.1	88.8	93.3	65.2	81.9	94.4	65.1	82.5	91.5	62.1	79.6	95.2	65.7	83.2	94.4	60.6	80.6

		OKLAHOMA																	
		LAWTON			MCALESTER			MIAMI			MUSKOGEE			OKLAHOMA CITY			PONCA CITY		
	DIVISION	735			745			743			744			730 - 731			746		
		MAT.	INST.	TOTAL	MAT.	INST.	TOTAL	MAT.	INST.	TOTAL	MAT.	INST.	TOTAL	MAT.	INST.	TOTAL	MAT.	INST.	TOTAL
015433	CONTRACTOR EQUIPMENT		77.8	77.8		77.0	77.0		87.7	87.7		87.7	87.7		90.4	90.4		77.0	77.0
0241, 31 - 34	SITE & INFRASTRUCTURE, DEMOLITION	96.4	88.2	90.8	88.5	86.5	87.1	89.2	83.9	85.6	90.5	83.8	86.0	96.1	102.0	100.1	96.1	86.5	89.6
0310	Concrete Forming & Accessories	95.0	57.0	64.3	77.8	54.8	59.3	95.7	52.2	60.6	101.9	55.9	64.8	100.6	62.5	69.9	89.4	52.2	59.4
0320	Concrete Reinforcing	88.9	66.1	81.6	97.4	64.9	86.9	95.9	66.1	86.3	96.8	65.0	86.6	100.4	66.2	89.4	96.8	66.1	86.9
0330	Cast-in-Place Concrete	78.2	71.4	75.9	71.0	69.6	70.5	74.4	70.1	72.9	75.3	70.7	73.7	82.2	72.4	78.8	83.1	69.3	78.3
03	CONCRETE	79.9	63.6	73.4	75.6	61.8	70.0	79.6	61.9	72.5	81.3	63.6	74.2	88.1	66.5	79.5	84.9	60.7	75.2
04	MASONRY	90.0	60.2	71.2	106.6	57.5	75.7	91.5	56.6	69.5	108.9	59.0	77.4	88.4	57.7	69.0	84.4	56.5	66.8
05	METALS	96.0	62.4	88.1	86.6	61.3	80.7	86.6	76.6	84.3	87.7	76.3	85.0	90.2	64.8	84.3	86.6	62.0	80.9
06	WOOD, PLASTICS & COMPOSITES	100.0	55.4	81.6	80.8	54.3	69.8	99.0	51.5	79.3	104.3	55.5	84.1	97.0	64.1	83.4	93.3	51.3	75.9
07	THERMAL & MOISTURE PROTECTION	103.3	63.1	85.7	97.8	61.8	82.0	98.6	61.8	82.4	98.9	61.9	82.7	95.6	64.4	81.9	98.6	61.0	82.1
08	OPENINGS	102.2	56.0	91.0	92.2	54.7	83.2	92.2	52.5	82.6	93.4	54.4	84.0	95.6	60.8	87.2	92.2	52.4	82.6
0920	Plaster & Gypsum Board	100.3	54.9	71.9	78.8	53.7	63.1	90.7	50.7	65.7	93.4	54.9	69.3	101.0	63.4	77.5	88.0	50.7	64.7
0950, 0980	Ceilings & Acoustic Treatment	105.2	54.9	72.5	85.5	53.7	64.9	85.5	50.7	62.9	98.6	54.9	70.2	94.2	63.4	74.2	85.5	50.7	62.9
0960	Flooring	92.3	69.8	85.5	92.9	62.4	83.5	99.2	48.1	83.6	101.4	61.4	89.2	95.6	69.8	87.8	96.6	53.3	83.4
0970, 0990	Wall Finishes & Painting/Coating	91.9	43.3	62.2	92.5	45.4	63.8	92.5	39.2	60.0	92.5	39.2	60.0	98.7	43.3	64.9	92.5	43.3	62.5
09	FINISHES	95.9	57.0	74.8	87.0	54.4	69.3	90.5	48.9	67.9	94.4	54.2	72.6	97.5	61.5	77.9	90.3	51.2	69.0
COVERS	DIVS. 10 - 14, 25, 28, 41, 43, 44, 46	100.0	78.9	95.4	100.0	77.8	95.2	100.0	77.5	95.1	100.0	78.4	95.3	100.0	79.7	95.6	100.0	77.1	95.0
21, 22, 23	FIRE SUPPRESSION, PLUMBING & HVAC	101.2	66.9	87.1	96.9	61.5	82.4	96.9	59.0	81.3	101.1	61.6	84.9	101.1	65.6	86.5	96.9	59.0	81.3
26, 27, 3370	ELECTRICAL, COMMUNICATIONS & UTIL.	95.9	73.7	85.2	95.2	68.5	82.3	96.1	68.5	82.8	94.9	83.5	89.4	100.8	73.7	87.8	95.0	60.8	78.5
MF2018	WEIGHTED AVERAGE	95.7	66.4	83.8	90.8	63.3	79.6	91.4	63.0	79.8	93.8	67.0	82.9	95.4	68.5	84.5	91.8	60.9	79.2

		OKLAHOMA												OREGON					
		POTEAU			SHAWNEE			TULSA			WOODWARD			BEND			EUGENE		
	DIVISION	749			748			740 - 741			738			977			974		
		MAT.	INST.	TOTAL	MAT.	INST.	TOTAL	MAT.	INST.	TOTAL	MAT.	INST.	TOTAL	MAT.	INST.	TOTAL	MAT.	INST.	TOTAL
015433	CONTRACTOR EQUIPMENT		86.5	86.5		77.0	77.0		87.7	87.7		77.0	77.0		94.6	94.6		94.6	94.6
0241, 31 - 34	SITE & INFRASTRUCTURE, DEMOLITION	76.2	81.9	80.1	98.9	86.5	90.5	96.3	84.0	87.9	98.9	86.8	90.7	127.8	94.0	104.9	115.4	94.0	100.9
0310	Concrete Forming & Accessories	86.6	52.2	58.9	80.1	52.3	57.7	101.7	55.1	64.1	86.3	57.1	62.8	109.9	99.4	101.4	104.6	99.7	100.6
0320	Concrete Reinforcing	97.9	66.1	87.7	96.8	61.1	85.3	97.0	66.2	87.1	88.7	66.1	81.4	90.7	111.7	97.5	94.1	111.7	99.8
0330	Cast-in-Place Concrete	74.4	70.1	72.9	85.8	69.3	80.1	82.0	70.9	78.2	80.9	71.4	77.6	133.2	98.4	121.2	129.0	98.5	118.4
03	CONCRETE	81.2	61.9	73.5	85.9	59.8	75.5	85.4	63.5	76.7	82.3	63.6	74.8	115.8	100.8	109.6	108.3	101.0	105.4
04	MASONRY	91.8	56.6	69.6	108.2	56.5	75.6	92.6	57.6	70.5	85.4	60.2	69.5	107.5	100.1	102.8	104.2	100.1	101.6
05	METALS	86.6	76.7	84.3	86.5	60.3	80.4	91.6	77.0	88.2	90.7	62.4	84.1	106.2	97.1	104.1	106.8	97.5	104.6
06	WOOD, PLASTICS & COMPOSITES	89.3	51.5	73.6	83.3	51.3	70.1	103.3	54.4	83.1	93.0	55.6	77.5	106.5	99.7	103.7	100.9	99.7	100.4
07	THERMAL & MOISTURE PROTECTION	98.6	61.8	82.4	98.5	59.7	81.5	98.9	63.7	83.5	103.4	63.2	85.8	113.1	99.8	107.2	112.1	102.4	107.8
08	OPENINGS	92.2	52.5	82.6	92.2	51.2	82.3	94.8	55.3	85.2	99.6	55.8	89.0	93.5	102.6	95.7	93.7	102.6	95.9
0920	Plaster & Gypsum Board	84.9	50.7	63.5	79.5	50.7	61.5	93.4	53.7	68.6	95.4	55.1	70.2	123.9	99.7	108.8	119.8	99.7	107.3
0950, 0980	Ceilings & Acoustic Treatment	85.5	50.7	62.9	85.5	50.7	62.9	98.6	53.7	69.5	93.6	55.1	68.6	91.3	99.7	96.8	91.3	99.7	96.8
0960	Flooring	96.0	48.1	81.4	93.8	53.3	81.4	100.3	60.8	88.2	89.1	67.1	82.4	98.8	103.8	100.3	97.7	103.8	99.5
0970, 0990	Wall Finishes & Painting/Coating	92.5	43.3	62.5	92.5	39.2	60.0	92.5	53.5	68.7	91.9	43.3	62.2	95.9	77.5	84.7	95.9	64.9	77.0
09	FINISHES	87.7	49.3	66.8	88.3	49.8	67.4	94.3	55.0	73.0	92.4	56.6	72.9	104.4	97.8	100.8	102.0	96.4	99.0
COVERS	DIVS. 10 - 14, 25, 28, 41, 43, 44, 46	100.0	77.4	95.1	100.0	77.2	95.0	100.0	78.2	95.3	100.0	78.9	95.4	100.0	101.8	100.4	100.0	101.8	100.4
21, 22, 23	FIRE SUPPRESSION, PLUMBING & HVAC	96.9	59.0	81.3	96.9	64.9	83.8	101.1	61.5	84.9	97.0	66.9	84.6	96.5	99.6	97.8	101.0	114.2	106.4
26, 27, 3370	ELECTRICAL, COMMUNICATIONS & UTIL.	95.0	68.5	82.2	96.2	73.7	85.4	96.2	68.8	83.0	95.8	73.7	85.1	93.4	92.5	93.0	93.0	92.5	92.8
MF2018	WEIGHTED AVERAGE	90.8	62.9	79.5	92.8	63.5	80.9	94.9	65.0	82.7	93.4	66.3	82.3	103.2	98.1	101.1	102.5	101.2	101.9

OREGON

DIVISION		KLAMATH FALLS 976			MEDFORD 975			PENDLETON 978			PORTLAND 970 - 972			SALEM 973			VALE 979		
		MAT.	INST.	TOTAL	MAT.	INST.	TOTAL	MAT.	INST.	TOTAL	MAT.	INST.	TOTAL	MAT.	INST.	TOTAL	MAT.	INST.	TOTAL
015433	CONTRACTOR EQUIPMENT		94.6	94.6		94.6	94.6		92.2	92.2		94.6	94.6		99.4	99.4		92.2	92.2
0241, 31 - 34	SITE & INFRASTRUCTURE, DEMOLITION	133.1	94.0	106.5	125.3	94.0	104.0	122.2	87.8	98.8	118.1	94.0	101.8	109.3	101.5	104.0	106.1	87.7	93.6
0310	Concrete Forming & Accessories	99.3	99.1	99.1	97.7	99.2	98.9	103.1	99.3	100.0	106.5	99.6	101.0	104.6	99.7	100.7	112.9	98.0	100.9
0320	Concrete Reinforcing	90.7	111.6	97.4	92.2	111.7	98.5	91.1	111.8	97.7	95.0	111.9	100.4	101.1	111.9	104.6	88.9	111.6	96.1
0330	Cast-in-Place Concrete	133.2	93.3	119.4	133.2	98.3	121.1	134.1	93.7	120.2	132.5	98.5	120.8	121.9	100.6	114.5	105.4	93.9	101.5
03	CONCRETE	117.7	98.8	110.1	113.4	100.6	108.3	103.0	99.2	101.4	110.3	101.0	106.5	108.2	101.6	105.5	87.2	98.6	91.7
04	MASONRY	120.9	100.1	107.8	100.9	100.1	100.4	111.6	102.2	105.7	106.4	102.1	103.7	106.9	102.3	104.0	108.6	102.2	104.6
05	METALS	106.2	96.8	104.0	106.4	96.9	104.2	114.1	97.6	110.2	108.6	97.6	106.0	114.9	95.5	110.4	113.9	96.3	109.8
06	WOOD, PLASTICS & COMPOSITES	93.9	99.7	96.3	92.5	99.7	95.5	98.0	99.8	98.7	102.2	99.7	101.2	94.4	99.9	96.7	109.5	99.8	105.5
07	THERMAL & MOISTURE PROTECTION	113.2	95.5	105.4	112.7	95.1	105.0	104.3	92.8	99.3	112.1	100.3	106.9	107.2	102.4	105.1	103.7	89.1	97.3
08	OPENINGS	93.5	102.6	95.7	96.0	102.6	97.6	90.0	102.6	93.1	91.9	102.6	94.4	98.9	102.7	99.8	90.0	90.2	90.0
0920	Plaster & Gypsum Board	110.5	99.7	103.8	109.6	99.7	103.4	101.5	99.7	100.4	116.8	99.7	106.1	113.3	99.7	104.8	114.6	99.7	105.3
0950, 0980	Ceilings & Acoustic Treatment	100.2	99.7	99.9	104.0	99.7	101.3	67.8	99.7	88.5	93.6	99.7	97.6	110.5	99.7	103.5	67.8	99.7	88.5
0960	Flooring	96.7	103.8	98.8	96.3	103.8	98.6	64.7	103.8	76.6	95.4	103.8	97.9	97.5	103.8	99.4	66.2	103.8	77.7
0970, 0990	Wall Finishes & Painting/Coating	95.9	74.8	83.1	95.9	74.8	83.1	91.0	74.8	81.1	95.9	74.8	83.0	98.7	72.5	82.7	91.0	77.5	82.8
09	FINISHES	104.2	97.5	100.6	103.7	97.5	100.3	74.2	97.6	86.9	101.5	97.5	99.3	103.5	97.4	100.2	75.5	97.9	87.7
COVERS	DIVS. 10 - 14, 25, 28, 41, 43, 44, 46	100.0	101.7	100.4	100.0	101.7	100.4	100.0	102.1	100.5	100.0	101.8	100.4	100.0	102.3	100.5	100.0	101.9	100.4
21, 22, 23	FIRE SUPPRESSION, PLUMBING & HVAC	96.5	99.5	97.7	101.0	105.2	102.7	99.7	105.9	102.2	101.0	108.1	103.9	99.9	108.2	103.3	99.7	67.9	86.6
26, 27, 3370	ELECTRICAL, COMMUNICATIONS & UTIL.	92.7	76.6	84.9	95.1	76.6	86.2	86.5	89.8	88.1	93.6	106.3	99.7	98.4	92.5	95.5	86.5	63.1	75.2
MF2018	WEIGHTED AVERAGE	103.9	95.4	100.4	103.6	96.9	100.9	100.1	98.4	99.4	103.1	102.2	102.7	104.7	100.8	103.1	97.6	85.7	92.8

PENNSYLVANIA

DIVISION		ALLENTOWN 181			ALTOONA 166			BEDFORD 155			BRADFORD 167			BUTLER 160			CHAMBERSBURG 172		
		MAT.	INST.	TOTAL	MAT.	INST.	TOTAL	MAT.	INST.	TOTAL	MAT.	INST.	TOTAL	MAT.	INST.	TOTAL	MAT.	INST.	TOTAL
015433	CONTRACTOR EQUIPMENT		108.7	108.7		108.7	108.7		106.5	106.5		108.7	108.7		108.7	108.7		107.9	107.9
0241, 31 - 34	SITE & INFRASTRUCTURE, DEMOLITION	87.8	93.1	91.4	90.2	92.9	92.0	90.8	89.5	89.9	86.3	90.8	89.3	81.5	92.6	89.1	84.0	90.1	88.1
0310	Concrete Forming & Accessories	98.7	108.4	106.6	78.2	85.6	84.2	81.8	76.5	77.6	81.6	90.9	89.1	79.9	89.8	87.9	85.9	72.2	74.8
0320	Concrete Reinforcing	98.1	112.3	102.6	95.0	107.2	98.9	96.3	98.4	97.0	97.1	98.5	97.5	95.6	118.4	103.0	95.4	104.4	98.3
0330	Cast-in-Place Concrete	88.7	102.8	93.6	98.8	89.2	95.5	101.9	80.3	94.4	94.4	84.7	91.1	87.3	94.4	89.7	94.0	87.6	91.8
03	CONCRETE	92.2	108.1	98.6	88.3	92.0	89.8	98.2	83.1	92.1	93.4	91.2	92.5	81.4	97.6	87.9	96.1	84.9	91.6
04	MASONRY	96.4	98.4	97.7	100.7	89.2	93.5	113.3	74.9	89.1	98.6	76.6	84.7	102.8	89.5	94.4	99.5	73.9	83.4
05	METALS	94.2	122.0	100.7	88.8	116.8	95.4	90.9	110.1	95.4	92.0	110.1	96.2	88.6	121.3	96.2	93.9	115.3	98.9
06	WOOD, PLASTICS & COMPOSITES	96.8	109.8	102.2	73.0	84.1	77.6	79.8	77.0	78.7	78.7	95.5	85.6	74.6	88.7	80.5	80.2	71.3	76.5
07	THERMAL & MOISTURE PROTECTION	107.0	110.8	108.7	105.0	92.9	99.7	103.2	82.5	94.2	106.6	83.3	96.4	104.5	91.7	98.8	104.4	77.0	92.3
08	OPENINGS	89.3	107.9	93.8	83.5	88.2	84.6	90.1	82.4	88.3	89.4	91.5	89.9	83.5	97.0	86.7	86.1	76.6	83.8
0920	Plaster & Gypsum Board	106.5	110.3	108.9	90.1	83.8	86.2	100.6	76.7	85.6	91.4	95.5	94.0	90.1	88.6	89.1	101.5	70.7	82.2
0950, 0980	Ceilings & Acoustic Treatment	95.0	110.3	104.9	98.4	83.8	88.9	102.2	76.7	85.6	98.5	95.5	96.6	99.2	88.6	92.3	94.5	70.7	79.0
0960	Flooring	91.4	97.2	93.2	85.0	104.0	90.8	89.6	96.5	91.7	86.7	96.5	89.7	85.8	98.0	89.5	93.9	72.5	87.4
0970, 0990	Wall Finishes & Painting/Coating	88.5	108.1	100.5	84.9	109.1	99.7	86.7	98.0	93.6	88.5	98.0	94.3	84.9	98.5	93.2	90.1	93.4	92.1
09	FINISHES	94.6	106.4	101.0	91.3	90.8	91.0	99.0	81.8	89.7	91.7	93.1	92.4	91.2	91.7	91.4	93.1	73.8	82.6
COVERS	DIVS. 10 - 14, 25, 28, 41, 43, 44, 46	100.0	103.2	100.7	100.0	97.9	99.5	100.0	94.3	98.8	100.0	97.1	99.4	100.0	99.0	99.8	100.0	92.0	98.3
21, 22, 23	FIRE SUPPRESSION, PLUMBING & HVAC	101.1	117.0	107.6	100.5	87.7	95.2	92.6	78.3	86.7	96.8	84.0	91.5	96.2	90.0	93.7	94.7	83.1	89.9
26, 27, 3370	ELECTRICAL, COMMUNICATIONS & UTIL.	100.0	98.1	99.1	91.6	110.1	100.5	101.4	110.1	105.6	96.0	110.1	102.8	91.9	99.8	95.7	93.4	79.6	86.7
MF2018	WEIGHTED AVERAGE	96.3	107.4	100.8	92.8	95.5	93.9	95.7	88.1	92.6	94.3	92.8	93.7	90.8	96.2	93.0	94.1	83.9	89.9

PENNSYLVANIA

DIVISION		DOYLESTOWN 189			DUBOIS 158			ERIE 164 - 165			GREENSBURG 156			HARRISBURG 170 - 171			HAZLETON 182		
		MAT.	INST.	TOTAL	MAT.	INST.	TOTAL	MAT.	INST.	TOTAL	MAT.	INST.	TOTAL	MAT.	INST.	TOTAL	MAT.	INST.	TOTAL
015433	CONTRACTOR EQUIPMENT		88.2	88.2		106.5	106.5		108.7	108.7		106.5	106.5		112.1	112.1		108.7	108.7
0241, 31 - 34	SITE & INFRASTRUCTURE, DEMOLITION	100.3	81.1	87.3	94.6	89.8	91.4	87.5	93.0	91.2	87.3	91.4	90.1	85.9	99.9	95.4	81.1	91.7	88.3
0310	Concrete Forming & Accessories	77.2	119.2	111.1	81.2	78.9	79.3	97.3	87.8	89.6	91.1	84.6	85.9	102.3	90.9	93.1	74.2	83.5	81.7
0320	Concrete Reinforcing	94.6	140.4	109.3	95.6	111.1	100.6	97.1	110.5	101.4	95.6	118.1	102.8	99.2	112.2	103.3	95.2	105.2	98.4
0330	Cast-in-Place Concrete	83.8	121.2	96.7	98.2	90.8	95.7	97.2	91.0	95.0	94.5	93.9	94.3	92.7	101.5	95.8	83.8	89.3	85.7
03	CONCRETE	87.3	123.1	101.6	99.4	90.1	90.7	88.4	94.1	90.7	94.2	95.0	94.5	92.3	99.4	95.1	85.0	90.8	87.3
04	MASONRY	100.2	126.0	116.5	113.7	87.2	97.0	87.9	92.4	90.7	122.4	85.8	99.3	93.3	94.4	94.0	110.3	83.6	93.4
05	METALS	91.9	120.2	98.6	90.9	115.4	96.6	89.1	117.6	95.8	90.9	119.3	97.5	100.6	117.4	104.5	93.9	116.2	99.1
06	WOOD, PLASTICS & COMPOSITES	73.9	118.6	92.4	78.4	77.0	77.8	93.8	85.8	90.5	87.7	82.1	85.4	96.6	88.5	93.2	72.1	82.1	76.3
07	THERMAL & MOISTURE PROTECTION	103.8	122.0	111.8	103.6	89.0	97.2	105.4	91.1	99.1	103.2	89.6	97.2	109.2	107.7	108.5	105.9	93.3	100.4
08	OPENINGS	91.4	125.3	99.6	90.1	85.3	89.0	83.6	90.9	85.4	90.1	93.4	90.9	97.1	89.1	95.2	89.8	87.8	89.3
0920	Plaster & Gypsum Board	89.1	119.3	108.0	98.2	76.7	84.7	106.5	85.5	93.4	101.6	81.9	89.3	111.7	88.1	96.9	89.4	81.8	84.6
0950, 0980	Ceilings & Acoustic Treatment	94.4	119.3	110.6	102.2	76.7	85.6	95.0	85.5	88.8	102.2	81.9	89.0	101.7	88.1	92.9	95.7	81.8	86.7
0960	Flooring	76.7	129.6	92.9	89.4	96.5	91.6	89.4	104.0	93.9	92.6	72.5	86.5	101.7	96.0	100.0	84.4	83.1	84.0
0970, 0990	Wall Finishes & Painting/Coating	88.5	133.3	115.8	86.7	98.5	93.9	93.6	94.3	94.0	86.7	98.5	93.9	98.7	90.5	93.7	88.5	96.6	93.4
09	FINISHES	85.3	122.3	105.4	98.9	83.2	90.4	94.9	90.9	92.7	99.6	83.2	90.7	101.0	91.2	95.7	89.6	84.5	86.9
COVERS	DIVS. 10 - 14, 25, 28, 41, 43, 44, 46	100.0	113.9	103.0	100.0	96.1	99.2	100.0	99.0	99.8	100.0	98.1	99.6	100.0	98.5	99.7	100.0	97.8	99.5
21, 22, 23	FIRE SUPPRESSION, PLUMBING & HVAC	96.2	122.8	107.2	92.6	81.2	87.9	100.5	94.7	98.1	92.6	82.4	88.4	96.3	97.6	96.8	96.8	88.5	93.4
26, 27, 3370	ELECTRICAL, COMMUNICATIONS & UTIL.	94.9	122.7	108.3	101.8	110.1	105.8	92.8	90.2	91.6	101.8	110.1	105.8	99.4	91.4	95.6	96.5	83.4	90.2
MF2018	WEIGHTED AVERAGE	93.2	119.4	103.9	96.0	91.9	94.3	93.0	95.0	93.8	95.6	93.6	94.8	97.6	97.6	97.6	93.7	90.1	92.3

		PENNSYLVANIA																
	DIVISION	INDIANA 157			JOHNSTOWN 159			KITTANNING 162			LANCASTER 175 - 176			LEHIGH VALLEY 180			MONTROSE 188		
		MAT.	INST.	TOTAL	MAT.	INST.	TOTAL	MAT.	INST.	TOTAL	MAT.	INST.	TOTAL	MAT.	INST.	TOTAL	MAT.	INST.	TOTAL
015433	CONTRACTOR EQUIPMENT		106.5	106.5		106.5	106.5		108.7	108.7		107.9	107.9		108.7	108.7		108.7	108.7
0241, 31 - 34	SITE & INFRASTRUCTURE, DEMOLITION	85.7	90.4	88.9	91.2	91.7	91.5	83.8	92.5	89.7	76.7	92.4	87.4	84.7	92.1	89.8	83.5	91.9	89.2
0310	Concrete Forming & Accessories	82.7	87.7	86.7	81.1	85.3	84.5	79.9	84.7	83.8	88.5	89.1	89.0	90.0	103.7	101.0	75.5	84.0	82.4
0320	Concrete Reinforcing	94.9	118.5	102.5	96.3	118.4	103.4	95.6	118.3	102.9	95.1	110.0	99.9	95.2	100.5	96.9	99.7	108.2	102.4
0330	Cast-in-Place Concrete	92.7	91.2	92.1	102.8	88.5	97.9	90.7	91.5	91.0	80.0	95.7	85.4	90.7	92.7	91.4	88.9	86.9	88.2
03	CONCRETE	91.8	95.4	93.2	98.9	93.4	96.7	83.6	94.2	87.9	85.8	96.5	90.1	91.0	100.3	94.7	89.9	90.7	90.2
04	MASONRY	109.1	87.3	95.4	110.2	86.9	95.5	105.1	82.9	91.1	104.6	88.6	94.5	96.4	90.1	92.5	96.4	85.5	89.5
05	METALS	91.0	119.2	97.6	90.9	119.3	97.6	88.7	120.1	96.0	93.9	120.9	100.2	93.9	115.5	98.9	92.1	116.9	97.9
06	WOOD, PLASTICS & COMPOSITES	80.7	88.6	84.0	78.4	84.0	80.7	74.6	84.3	78.6	83.9	88.2	85.7	86.3	106.2	94.5	73.0	82.1	76.8
07	THERMAL & MOISTURE PROTECTION	103.0	90.3	97.4	103.2	89.6	97.3	104.6	88.6	97.6	103.7	103.0	103.4	106.6	103.7	105.3	106.0	84.3	96.5
08	OPENINGS	90.1	93.4	90.9	90.1	90.8	90.3	83.5	94.5	86.1	86.1	88.6	86.7	89.8	100.3	92.3	87.0	86.4	86.9
0920	Plaster & Gypsum Board	101.1	88.6	93.3	98.2	83.8	89.2	90.1	84.0	86.3	107.0	88.1	95.2	93.0	106.5	101.4	90.0	81.8	84.8
0950, 0980	Ceilings & Acoustic Treatment	102.2	88.6	93.3	102.2	83.8	90.3	99.2	84.0	89.3	94.5	88.1	90.3	95.7	106.5	102.7	98.5	81.8	87.6
0960	Flooring	90.0	96.5	92.0	89.4	101.1	93.0	85.8	96.5	89.1	94.8	94.0	94.6	89.0	87.6	88.5	85.0	96.5	88.5
0970, 0990	Wall Finishes & Painting/Coating	86.7	98.5	93.9	86.7	109.1	100.4	84.9	98.5	93.2	90.1	87.2	88.3	88.5	93.7	91.7	88.5	96.6	93.4
09	FINISHES	98.8	90.0	94.0	98.5	90.0	93.9	91.3	87.5	89.2	93.6	89.2	91.2	91.8	99.8	96.1	90.7	86.9	88.6
COVERS	DIVS. 10 - 14, 25, 28, 41, 43, 44, 46	100.0	97.3	99.4	100.0	97.5	99.5	100.0	97.1	99.4	100.0	96.6	99.3	100.0	101.9	100.4	100.0	98.4	99.7
21, 22, 23	FIRE SUPPRESSION, PLUMBING & HVAC	92.6	79.3	87.2	92.6	87.4	90.5	96.2	84.2	91.3	94.7	95.4	95.0	96.8	108.9	101.8	96.8	87.8	93.1
26, 27, 3370	ELECTRICAL, COMMUNICATIONS & UTIL.	101.8	110.1	105.8	101.8	110.1	105.8	91.6	110.1	100.5	94.3	92.8	93.6	96.5	119.9	107.8	96.0	88.2	92.3
MF2018	WEIGHTED AVERAGE	94.6	94.0	94.3	95.7	95.4	95.6	91.3	94.3	92.5	92.9	95.6	94.0	94.4	104.8	98.6	93.3	90.9	92.4

		PENNSYLVANIA																	
	DIVISION	NEW CASTLE 161			NORRISTOWN 194			OIL CITY 163			PHILADELPHIA 190 - 191			PITTSBURGH 150 - 152			POTTSVILLE 179		
		MAT.	INST.	TOTAL	MAT.	INST.	TOTAL	MAT.	INST.	TOTAL	MAT.	INST.	TOTAL	MAT.	INST.	TOTAL	MAT.	INST.	TOTAL
015433	CONTRACTOR EQUIPMENT		108.7	108.7		93.7	93.7		108.7	108.7		101.1	101.1		97.3	97.3		107.9	107.9
0241, 31 - 34	SITE & INFRASTRUCTURE, DEMOLITION	82.0	93.1	89.5	102.4	89.8	93.8	80.6	91.0	87.7	106.1	103.8	104.6	91.8	94.9	93.9	79.3	90.6	87.0
0310	Concrete Forming & Accessories	79.9	93.7	91.0	79.1	117.8	110.3	79.9	87.3	85.9	103.1	140.2	133.0	107.4	99.1	100.7	77.1	82.4	81.4
0320	Concrete Reinforcing	94.6	99.0	96.0	97.2	140.4	111.0	95.6	88.3	93.3	101.5	150.1	117.1	101.0	123.0	108.1	94.3	106.2	98.2
0330	Cast-in-Place Concrete	88.1	94.3	90.2	94.2	119.0	102.8	85.6	88.2	86.5	105.7	135.1	115.9	101.4	99.1	100.6	85.2	90.0	86.9
03	CONCRETE	81.6	96.0	87.4	91.6	121.7	103.7	80.3	89.3	83.9	103.8	139.0	117.9	100.5	103.1	101.6	88.4	90.7	89.4
04	MASONRY	103.4	92.3	96.4	110.0	122.5	117.9	101.1	82.9	89.6	96.9	135.9	121.5	96.3	99.9	98.5	98.0	81.8	87.8
05	METALS	88.7	114.3	94.7	93.1	120.2	99.4	88.7	111.9	94.1	99.7	123.9	105.4	99.0	107.3	101.0	94.2	116.9	99.5
06	WOOD, PLASTICS & COMPOSITES	74.6	94.1	82.7	73.7	118.6	92.3	74.6	88.7	80.5	102.2	141.3	118.4	107.7	99.5	104.3	72.1	80.3	75.5
07	THERMAL & MOISTURE PROTECTION	104.5	91.1	98.6	106.5	120.5	112.6	104.3	85.4	96.1	104.6	137.8	119.2	106.2	98.9	103.0	103.7	92.3	98.7
08	OPENINGS	83.5	95.2	86.3	83.7	125.3	93.7	83.5	90.6	85.2	98.9	142.7	109.5	97.9	104.7	99.5	86.1	87.5	86.5
0920	Plaster & Gypsum Board	90.1	94.1	92.6	95.6	119.3	110.4	90.1	88.6	89.1	116.5	142.4	132.7	110.1	99.4	103.4	96.6	79.9	86.2
0950, 0980	Ceilings & Acoustic Treatment	99.2	94.1	95.9	99.5	119.3	112.4	99.2	88.6	92.3	107.1	142.4	130.0	96.3	99.4	98.3	94.5	79.9	85.0
0960	Flooring	85.8	100.8	90.4	89.3	129.6	101.6	85.8	96.5	89.1	101.0	148.3	115.5	96.0	105.7	98.9	91.1	96.5	92.7
0970, 0990	Wall Finishes & Painting/Coating	84.9	106.4	98.0	88.8	133.3	115.9	84.9	98.5	93.2	98.7	168.6	141.3	93.3	109.1	102.9	90.1	96.6	94.1
09	FINISHES	91.2	96.3	94.0	91.8	121.4	107.8	91.1	90.1	90.5	104.5	145.0	126.5	101.1	101.0	101.1	91.2	85.5	88.1
COVERS	DIVS. 10 - 14, 25, 28, 41, 43, 44, 46	100.0	99.5	99.9	100.0	112.6	102.7	100.0	97.5	99.5	100.0	119.2	104.2	100.0	101.1	100.2	100.0	97.6	99.5
21, 22, 23	FIRE SUPPRESSION, PLUMBING & HVAC	96.2	93.5	95.1	95.1	121.0	105.7	96.2	87.3	92.6	96.3	141.3	114.8	96.4	96.5	96.4	94.7	88.7	92.2
26, 27, 3370	ELECTRICAL, COMMUNICATIONS & UTIL.	91.9	97.0	94.4	95.5	131.6	112.9	93.3	99.8	96.4	97.7	155.1	125.4	102.2	112.1	107.0	93.2	84.6	89.0
MF2018	WEIGHTED AVERAGE	90.9	96.6	93.2	94.2	120.2	104.8	90.7	92.2	91.3	99.9	137.6	115.2	99.1	102.0	100.3	92.6	90.2	91.6

		PENNSYLVANIA																	
	DIVISION	READING 195 - 196			SCRANTON 184 - 185			STATE COLLEGE 168			STROUDSBURG 183			SUNBURY 178			UNIONTOWN 154		
		MAT.	INST.	TOTAL	MAT.	INST.	TOTAL	MAT.	INST.	TOTAL	MAT.	INST.	TOTAL	MAT.	INST.	TOTAL	MAT.	INST.	TOTAL
015433	CONTRACTOR EQUIPMENT		114.6	114.6		108.7	108.7		107.9	107.9		108.7	108.7		108.7	108.7		106.5	106.5
0241, 31 - 34	SITE & INFRASTRUCTURE, DEMOLITION	107.8	102.7	104.3	88.4	92.8	91.4	78.7	91.7	87.5	82.7	92.0	89.0	90.5	91.4	91.1	86.1	91.3	89.6
0310	Concrete Forming & Accessories	100.8	88.7	91.1	98.9	89.7	91.5	82.9	85.9	85.3	82.5	85.1	84.6	88.9	79.7	81.5	73.0	89.7	86.5
0320	Concrete Reinforcing	98.4	112.2	102.9	98.1	112.1	102.6	96.3	107.4	99.9	98.4	108.3	101.6	97.0	104.5	99.4	95.6	118.5	103.0
0330	Cast-in-Place Concrete	84.4	96.2	88.5	92.6	92.8	92.7	89.3	88.3	89.0	87.3	88.3	87.6	93.2	88.6	91.6	92.7	93.8	93.1
03	CONCRETE	90.8	96.8	93.2	94.0	96.0	94.8	93.3	91.8	92.7	88.8	91.7	90.0	92.4	88.7	90.9	91.2	97.3	93.6
04	MASONRY	98.2	91.6	94.1	96.9	93.5	94.7	102.6	90.5	95.0	94.6	89.9	91.6	99.5	75.6	84.5	126.8	91.7	104.7
05	METALS	93.4	121.8	100.1	96.1	120.6	101.8	91.8	117.2	97.8	93.9	117.3	99.4	93.9	115.8	99.0	90.7	119.8	97.5
06	WOOD, PLASTICS & COMPOSITES	95.2	86.1	91.4	96.8	88.1	93.2	80.8	84.1	82.2	79.7	82.1	80.7	81.1	80.3	80.8	70.6	88.6	78.0
07	THERMAL & MOISTURE PROTECTION	107.7	104.5	106.3	106.8	91.7	100.2	105.9	102.7	104.5	106.3	84.6	96.8	104.8	85.9	96.5	102.7	92.3	98.1
08	OPENINGS	87.8	94.8	89.5	89.3	92.4	90.1	86.8	88.2	87.2	89.8	89.0	89.6	86.2	84.6	85.8	90.1	95.1	91.3
0920	Plaster & Gypsum Board	110.6	85.9	95.1	108.7	87.9	95.7	92.9	83.8	87.2	91.0	81.8	85.2	95.8	79.9	85.9	93.8	88.6	90.5
0950, 0980	Ceilings & Acoustic Treatment	89.1	85.9	87.0	105.3	87.9	94.0	95.1	83.8	87.8	94.4	81.8	86.2	90.3	79.9	83.6	102.2	88.6	93.3
0960	Flooring	92.6	94.0	93.0	91.4	96.5	93.0	89.6	94.0	90.9	87.2	87.5	87.3	91.7	96.5	93.2	86.9	102.4	91.6
0970, 0990	Wall Finishes & Painting/Coating	87.6	108.1	100.1	88.5	108.1	100.5	88.5	109.1	101.1	88.5	96.6	93.4	90.1	96.6	94.1	86.7	106.4	98.8
09	FINISHES	93.2	90.7	91.8	97.2	92.3	94.5	91.0	89.1	89.9	90.5	86.0	88.1	91.6	84.0	87.5	96.7	93.0	94.7
COVERS	DIVS. 10 - 14, 25, 28, 41, 43, 44, 46	100.0	98.7	99.7	100.0	99.5	99.9	100.0	96.2	99.2	100.0	99.4	99.9	100.0	93.6	98.6	100.0	98.8	99.7
21, 22, 23	FIRE SUPPRESSION, PLUMBING & HVAC	101.0	114.6	106.5	101.1	96.0	99.0	96.8	95.3	96.2	96.8	90.7	94.3	94.7	80.9	89.0	92.6	86.5	90.1
26, 27, 3370	ELECTRICAL, COMMUNICATIONS & UTIL.	100.7	95.2	98.0	100.0	93.5	96.8	95.1	110.1	102.4	96.5	131.4	113.3	93.8	85.2	89.6	99.8	110.1	104.8
MF2018	WEIGHTED AVERAGE	96.3	101.8	98.6	97.1	96.6	96.9	93.9	97.1	95.2	93.8	98.4	95.7	93.7	87.1	91.0	94.7	96.9	95.6

PENNSYLVANIA

DIVISION		WASHINGTON 153 MAT.	INST.	TOTAL	WELLSBORO 169 MAT.	INST.	TOTAL	WESTCHESTER 193 MAT.	INST.	TOTAL	WILKES-BARRE 186-187 MAT.	INST.	TOTAL	WILLIAMSPORT 177 MAT.	INST.	TOTAL	YORK 173-174 MAT.	INST.	TOTAL
015433	CONTRACTOR EQUIPMENT		106.5	106.5		108.7	108.7		93.7	93.7		108.7	108.7		108.7	108.7		107.9	107.9
0241, 31 - 34	SITE & INFRASTRUCTURE, DEMOLITION	86.2	92.1	90.2	89.3	91.3	90.7	109.3	90.7	96.6	80.8	92.5	88.7	82.2	92.4	89.1	81.6	92.4	89.0
0310	Concrete Forming & Accessories	83.0	95.8	93.3	80.6	79.1	79.4	88.1	119.0	113.0	86.4	87.8	87.5	84.4	84.9	84.8	81.6	87.6	86.4
0320	Concrete Reinforcing	95.6	118.7	103.0	96.3	108.1	100.1	96.3	140.4	110.5	97.1	109.9	101.2	96.2	110.0	100.7	97.0	112.0	101.8
0330	Cast-in-Place Concrete	92.7	94.4	93.3	93.6	82.4	89.7	104.4	120.9	110.1	83.8	90.9	86.2	78.8	89.3	82.4	85.6	95.7	89.1
03	CONCRETE	91.9	100.3	95.3	95.2	86.8	91.8	99.1	123.0	108.6	86.3	94.1	89.4	81.7	92.3	86.0	89.8	96.1	92.3
04	MASONRY	106.8	94.2	98.8	101.9	76.0	85.6	105.0	126.0	118.2	110.6	91.7	98.7	90.6	89.6	90.0	100.7	88.6	93.1
05	METALS	90.7	121.0	97.8	91.9	116.1	97.6	93.1	120.2	99.4	92.1	119.3	98.4	93.9	119.6	99.9	95.1	120.9	101.1
06	WOOD, PLASTICS & COMPOSITES	80.8	96.4	87.3	78.0	79.6	78.7	81.6	118.6	96.9	82.5	86.4	84.1	77.3	84.1	80.1	75.8	86.1	80.1
07	THERMAL & MOISTURE PROTECTION	103.0	94.5	99.3	106.9	79.6	95.0	107.1	117.7	111.8	106.1	90.1	99.1	104.0	88.7	97.3	103.9	103.5	103.7
08	OPENINGS	90.0	101.2	92.7	89.4	84.7	88.3	83.7	125.3	93.7	87.0	90.5	87.8	86.2	89.3	87.0	86.1	87.8	86.5
0920	Plaster & Gypsum Board	101.1	96.6	98.3	90.7	79.2	83.5	96.6	119.3	110.8	91.9	86.1	88.3	96.6	83.8	88.6	96.6	85.9	89.9
0950, 0980	Ceilings & Acoustic Treatment	102.2	96.6	98.5	95.1	79.2	84.8	99.5	119.3	112.4	98.5	86.1	90.5	94.5	83.8	87.6	93.7	85.9	88.6
0960	Flooring	90.1	102.4	93.9	86.4	96.5	89.5	91.7	129.6	103.3	87.9	94.5	89.9	90.7	92.1	91.1	92.4	94.0	92.9
0970, 0990	Wall Finishes & Painting/Coating	86.7	106.4	98.8	88.5	96.6	93.4	88.8	133.3	115.9	88.5	105.4	98.8	90.1	105.4	99.4	90.1	90.5	90.3
09	FINISHES	98.8	97.9	98.3	91.1	84.0	87.2	93.1	122.2	108.9	91.6	90.3	90.9	91.6	87.8	89.6	91.4	88.3	89.7
COVERS	DIVS. 10 - 14, 25, 28, 41, 43, 44, 46	100.0	99.7	99.9	100.0	93.5	98.6	100.0	113.7	103.0	100.0	98.9	99.8	100.0	95.6	99.1	100.0	96.3	99.2
21, 22, 23	FIRE SUPPRESSION, PLUMBING & HVAC	92.6	89.9	91.5	96.8	80.9	90.3	95.1	122.8	106.5	96.8	94.6	95.9	94.7	90.9	93.1	100.9	95.4	98.7
26, 27, 3370	ELECTRICAL, COMMUNICATIONS & UTIL.	101.4	110.1	105.6	96.0	88.2	92.3	95.5	122.7	108.6	96.5	88.6	92.7	94.1	82.0	88.2	95.1	82.4	89.0
MF2018	WEIGHTED AVERAGE	94.4	99.6	96.5	94.7	87.1	91.6	95.4	120.0	105.4	93.6	94.6	94.0	91.6	91.9	91.8	94.8	93.9	94.4

PUERTO RICO / RHODE ISLAND / SOUTH CAROLINA

DIVISION		SAN JUAN 009 MAT.	INST.	TOTAL	NEWPORT 028 MAT.	INST.	TOTAL	PROVIDENCE 029 MAT.	INST.	TOTAL	AIKEN 298 MAT.	INST.	TOTAL	BEAUFORT 299 MAT.	INST.	TOTAL	CHARLESTON 294 MAT.	INST.	TOTAL
015433	CONTRACTOR EQUIPMENT		88.8	88.8		95.0	95.0		100.5	100.5		100.5	100.5		100.5	100.5		100.5	100.5
0241, 31 - 34	SITE & INFRASTRUCTURE, DEMOLITION	123.1	87.2	98.7	82.5	93.5	89.9	86.5	103.5	98.1	144.9	81.9	102.1	139.3	81.9	100.3	121.8	82.0	94.8
0310	Concrete Forming & Accessories	72.8	20.9	31.0	97.9	118.1	114.2	101.6	118.3	115.0	94.7	61.1	67.6	93.0	62.5	68.4	91.8	61.4	67.3
0320	Concrete Reinforcing	118.5	17.9	86.2	99.4	118.8	105.6	99.6	118.8	105.8	98.9	56.2	85.2	98.0	61.9	86.4	97.8	62.7	86.5
0330	Cast-in-Place Concrete	81.0	32.0	64.1	56.3	113.7	76.1	77.2	115.1	90.3	97.6	65.2	86.4	97.6	65.3	86.4	114.6	65.5	97.6
03	CONCRETE	85.0	25.4	61.1	78.6	116.5	93.7	90.5	116.9	101.0	101.8	63.4	86.4	99.6	65.0	85.8	98.8	64.8	85.2
04	MASONRY	96.0	22.9	49.9	98.3	122.7	113.7	104.4	122.8	116.0	87.6	55.9	67.6	103.4	55.9	73.5	105.4	63.9	79.2
05	METALS	92.1	40.8	80.1	94.5	115.1	99.3	98.4	112.8	101.7	94.5	86.6	92.6	94.5	88.7	93.1	96.5	90.1	95.0
06	WOOD, PLASTICS & COMPOSITES	62.1	20.0	44.7	95.7	116.9	104.5	99.0	117.1	106.5	92.5	63.0	80.3	91.1	64.8	80.2	89.4	63.0	78.5
07	THERMAL & MOISTURE PROTECTION	140.4	26.7	90.5	101.9	114.8	107.5	111.4	116.2	113.5	102.2	58.7	83.1	101.7	59.5	83.2	100.5	61.9	83.6
08	OPENINGS	151.5	18.5	119.4	93.5	118.9	99.6	95.3	119.0	101.0	91.0	58.8	83.2	91.0	61.0	83.8	94.2	61.3	86.3
0920	Plaster & Gypsum Board	136.7	17.9	62.4	103.0	117.3	112.0	114.3	117.3	116.2	102.8	62.1	77.3	110.6	63.9	81.4	109.8	62.1	80.0
0950, 0980	Ceilings & Acoustic Treatment	285.1	17.9	111.7	89.8	117.3	107.7	101.0	117.3	111.6	86.2	62.1	70.6	93.8	63.9	74.4	93.8	62.1	73.2
0960	Flooring	220.6	27.6	161.6	88.2	131.2	101.3	88.2	131.2	101.4	105.7	83.5	98.9	107.2	70.5	96.0	106.9	78.0	98.0
0970, 0990	Wall Finishes & Painting/Coating	187.5	21.5	86.2	89.9	119.7	108.1	98.7	119.7	111.5	97.8	61.3	75.5	97.8	52.8	70.3	97.8	66.2	78.5
09	FINISHES	211.4	22.1	108.6	92.1	121.2	107.9	98.1	121.3	110.7	101.5	65.3	81.8	104.3	63.1	81.9	101.4	65.1	81.7
COVERS	DIVS. 10 - 14, 25, 28, 41, 43, 44, 46	98.6	24.9	82.6	100.0	106.7	101.5	100.0	107.0	101.5	100.0	82.9	96.3	100.0	77.5	95.1	100.0	82.9	96.3
21, 22, 23	FIRE SUPPRESSION, PLUMBING & HVAC	122.4	17.7	79.5	100.6	108.4	103.8	101.1	108.4	104.1	97.1	51.3	78.3	97.1	56.6	80.5	101.3	56.6	83.0
26, 27, 3370	ELECTRICAL, COMMUNICATIONS & UTIL.	122.5	23.2	74.7	101.3	98.7	100.0	100.5	98.7	99.6	93.4	57.3	76.0	95.6	61.5	79.2	94.7	54.6	75.4
MF2018	WEIGHTED AVERAGE	117.0	28.6	81.0	94.5	111.3	101.3	98.2	112.0	103.8	97.7	63.2	83.7	98.3	65.1	84.8	99.1	65.5	85.4

SOUTH CAROLINA / SOUTH DAKOTA

DIVISION		COLUMBIA 290-292 MAT.	INST.	TOTAL	FLORENCE 295 MAT.	INST.	TOTAL	GREENVILLE 296 MAT.	INST.	TOTAL	ROCK HILL 297 MAT.	INST.	TOTAL	SPARTANBURG 293 MAT.	INST.	TOTAL	ABERDEEN 574 MAT.	INST.	TOTAL
015433	CONTRACTOR EQUIPMENT		106.1	106.1		100.5	100.5		100.5	100.5		100.5	100.5		100.5	100.5		94.5	94.5
0241, 31 - 34	SITE & INFRASTRUCTURE, DEMOLITION	121.0	95.2	103.5	132.8	81.8	98.2	127.2	82.1	96.5	123.7	81.8	95.3	127.0	82.1	96.5	88.5	89.1	88.9
0310	Concrete Forming & Accessories	97.6	58.5	66.1	75.6	58.5	61.8	91.5	58.7	65.0	88.9	60.8	66.2	95.3	58.7	65.8	95.4	66.3	71.9
0320	Concrete Reinforcing	100.6	62.7	88.5	97.4	62.7	86.2	97.3	60.1	85.3	98.1	61.0	86.2	97.3	62.7	86.2	98.5	69.0	89.1
0330	Cast-in-Place Concrete	116.0	66.2	98.8	97.5	65.3	86.4	97.5	65.5	86.5	97.5	64.8	86.2	97.5	65.5	86.5	125.1	74.2	107.5
03	CONCRETE	99.0	63.5	84.8	95.0	63.3	82.3	94.7	63.1	82.0	93.3	63.9	81.5	95.0	63.5	82.4	106.7	70.5	92.2
04	MASONRY	98.1	64.0	76.6	87.7	63.8	72.6	85.3	63.9	71.8	110.6	55.2	75.7	87.7	63.9	72.7	117.9	73.2	89.7
05	METALS	94.1	86.4	92.3	95.1	89.3	93.7	95.1	89.7	93.8	94.5	88.2	93.0	95.1	90.2	93.9	87.4	82.7	86.3
06	WOOD, PLASTICS & COMPOSITES	94.2	59.1	79.7	72.8	59.2	67.1	88.7	59.2	76.5	86.8	63.0	77.0	93.5	59.2	79.3	95.3	61.2	81.2
07	THERMAL & MOISTURE PROTECTION	96.9	62.8	82.0	100.8	61.5	83.6	100.9	61.5	83.6	100.7	54.0	80.2	101.0	61.5	83.7	98.8	72.8	87.4
08	OPENINGS	96.3	59.1	87.3	91.0	59.1	83.3	91.0	58.9	83.2	91.0	59.9	83.5	91.0	59.1	83.3	93.8	72.4	88.6
0920	Plaster & Gypsum Board	109.5	58.1	77.4	91.2	58.1	70.5	100.1	58.1	73.9	99.2	62.1	76.0	104.2	58.1	75.4	107.2	60.7	78.1
0950, 0980	Ceilings & Acoustic Treatment	93.5	58.1	70.5	87.6	58.1	68.5	86.2	58.1	68.0	86.2	62.1	70.6	86.2	58.1	68.0	94.8	60.7	72.7
0960	Flooring	99.8	78.0	93.2	98.7	78.0	92.4	104.6	78.0	96.5	103.6	70.5	93.5	105.9	78.0	97.4	95.9	65.6	86.6
0970, 0990	Wall Finishes & Painting/Coating	100.1	66.2	79.4	97.8	66.2	78.5	97.8	66.2	78.5	97.8	61.3	75.5	97.8	66.2	78.5	86.4	76.1	80.1
09	FINISHES	101.1	62.8	80.3	96.5	62.8	78.2	98.7	62.8	79.2	97.9	62.8	78.9	99.7	62.8	79.7	93.5	67.4	79.3
COVERS	DIVS. 10 - 14, 25, 28, 41, 43, 44, 46	100.0	82.4	96.2	100.0	82.5	96.2	100.0	82.6	96.2	100.0	82.7	96.2	100.0	82.6	96.2	100.0	83.5	96.4
21, 22, 23	FIRE SUPPRESSION, PLUMBING & HVAC	100.8	55.0	82.0	101.3	55.0	82.3	101.3	54.0	81.9	97.1	50.9	78.2	101.3	54.0	81.9	100.8	79.5	92.1
26, 27, 3370	ELECTRICAL, COMMUNICATIONS & UTIL.	98.9	58.2	79.3	93.4	58.2	76.4	94.7	80.4	87.8	94.7	80.4	87.8	94.7	80.4	87.8	95.7	62.3	79.6
MF2018	WEIGHTED AVERAGE	98.8	65.7	85.3	96.8	64.9	83.8	97.1	67.9	85.2	96.7	66.2	84.3	97.3	68.0	85.4	97.4	74.1	88.0

For CCI updates and additional guidance please go to www.rsmeans.com/2022books

SOUTH DAKOTA

DIVISION		MITCHELL 573			MOBRIDGE 576			PIERRE 575			RAPID CITY 577			SIOUX FALLS 570 - 571			WATERTOWN 572		
		MAT.	INST.	TOTAL	MAT.	INST.	TOTAL	MAT.	INST.	TOTAL	MAT.	INST.	TOTAL	MAT.	INST.	TOTAL	MAT.	INST.	TOTAL
015433	CONTRACTOR EQUIPMENT		94.5	94.5		94.5	94.5		101.2	101.2		94.5	94.5		101.1	101.1		94.5	94.5
0241, 31 - 34	SITE & INFRASTRUCTURE, DEMOLITION	85.0	88.4	87.3	84.9	88.4	87.3	88.2	98.7	95.4	87.0	88.6	88.1	82.4	99.5	94.0	84.8	88.4	87.2
0310	Concrete Forming & Accessories	94.1	42.5	52.5	81.3	42.9	50.3	96.3	82.7	85.3	105.3	54.5	64.3	96.2	83.0	85.5	76.6	70.5	71.7
0320	Concrete Reinforcing	98.0	65.9	87.7	100.6	65.8	89.4	100.7	94.0	98.5	92.1	93.6	92.6	100.9	97.6	99.9	95.3	65.7	85.8
0330	Cast-in-Place Concrete	121.6	49.2	96.6	121.6	72.6	104.7	126.9	77.0	109.7	120.6	73.4	104.3	99.9	81.4	93.6	121.6	72.3	104.6
03	CONCRETE	104.4	50.5	82.8	104.1	58.8	85.9	110.5	82.8	99.4	103.5	69.0	89.7	98.6	85.2	93.3	102.6	71.2	90.0
04	MASONRY	104.6	72.3	84.3	112.2	68.0	84.3	104.7	77.1	87.3	111.6	70.9	85.9	106.8	77.2	88.1	138.4	68.8	94.6
05	METALS	86.7	80.4	85.2	86.7	80.7	85.3	90.0	87.5	89.4	89.3	89.2	89.3	90.5	91.4	90.7	86.7	80.1	85.2
06	WOOD, PLASTICS & COMPOSITES	94.1	31.8	68.3	80.3	31.5	60.1	99.7	82.3	92.5	100.7	45.2	77.7	93.9	82.1	89.0	75.5	70.0	73.2
07	THERMAL & MOISTURE PROTECTION	98.6	68.0	85.2	98.5	69.6	85.8	99.8	80.8	91.5	99.4	74.9	88.6	99.9	84.1	93.0	98.2	71.0	86.3
08	OPENINGS	93.0	55.5	83.9	95.4	55.4	85.7	96.5	89.3	94.7	97.2	68.9	90.4	100.2	89.9	97.7	93.0	57.9	84.5
0920	Plaster & Gypsum Board	106.0	30.4	58.7	95.5	30.2	54.7	100.7	82.0	89.0	106.6	44.3	67.6	100.7	81.8	88.9	92.7	69.7	78.3
0950, 0980	Ceilings & Acoustic Treatment	91.4	30.4	51.8	94.8	30.2	52.8	104.4	82.0	89.8	96.9	44.3	62.7	99.0	81.8	87.8	91.4	69.7	77.3
0960	Flooring	95.6	73.0	88.7	91.6	43.2	76.8	105.5	70.2	94.7	95.2	67.5	86.8	101.0	75.2	93.1	90.3	40.8	75.2
0970, 0990	Wall Finishes & Painting/Coating	86.4	76.1	80.1	86.4	76.1	80.1	98.7	115.9	109.2	86.4	115.9	104.4	98.7	115.9	109.2	86.4	76.1	80.1
09	FINISHES	92.3	49.7	69.2	90.3	45.6	66.0	100.9	85.3	92.4	93.6	62.7	76.8	98.3	85.0	91.1	88.7	67.7	77.3
COVERS	DIVS. 10 - 14, 25, 28, 41, 43, 44, 46	100.0	75.5	94.7	100.0	75.4	94.6	100.0	88.9	97.6	100.0	83.5	96.4	100.0	88.9	97.6	100.0	78.9	95.4
21, 22, 23	FIRE SUPPRESSION, PLUMBING & HVAC	96.6	45.5	75.6	96.6	64.0	83.2	100.7	73.1	89.4	100.8	72.2	89.1	100.7	70.3	88.3	96.6	48.3	76.8
26, 27, 3370	ELECTRICAL, COMMUNICATIONS & UTIL.	94.6	58.8	77.3	95.7	36.9	67.3	101.1	44.2	73.7	93.6	44.2	69.8	101.1	64.2	83.3	94.1	58.8	77.1
MF2018	WEIGHTED AVERAGE	95.2	59.6	80.7	95.4	60.7	81.3	99.3	77.1	90.3	97.2	69.2	85.8	97.8	80.1	90.6	95.7	65.7	83.5

TENNESSEE

DIVISION		CHATTANOOGA 373 - 374			COLUMBIA 384			COOKEVILLE 385			JACKSON 383			JOHNSON CITY 376			KNOXVILLE 377 - 379		
		MAT.	INST.	TOTAL	MAT.	INST.	TOTAL	MAT.	INST.	TOTAL	MAT.	INST.	TOTAL	MAT.	INST.	TOTAL	MAT.	INST.	TOTAL
015433	CONTRACTOR EQUIPMENT		101.9	101.9		97.0	97.0		97.0	97.0		102.8	102.8		96.2	96.2		96.2	96.2
0241, 31 - 34	SITE & INFRASTRUCTURE, DEMOLITION	103.8	90.1	94.5	101.1	81.0	87.4	108.4	80.1	89.2	112.9	90.9	98.0	109.5	80.0	89.5	90.5	80.8	83.9
0310	Concrete Forming & Accessories	100.3	60.6	68.3	76.3	58.9	62.3	76.5	30.9	39.7	85.8	68.8	72.1	81.0	58.5	62.8	98.4	64.3	70.9
0320	Concrete Reinforcing	93.6	66.8	85.0	87.1	60.5	78.6	87.1	60.3	78.5	87.1	67.7	80.9	93.9	62.9	83.9	93.6	67.9	85.3
0330	Cast-in-Place Concrete	94.9	62.2	83.6	94.4	61.7	83.1	107.1	55.0	89.1	104.5	64.6	90.7	76.3	56.6	69.5	89.2	64.6	80.7
03	CONCRETE	93.3	63.9	81.6	92.6	61.9	80.3	101.6	46.9	79.7	94.4	68.6	84.1	99.3	60.4	83.7	91.0	66.6	81.2
04	MASONRY	97.4	55.4	70.9	118.0	51.8	76.3	113.7	36.8	65.3	118.5	56.9	79.6	114.2	42.8	69.2	77.6	56.4	64.2
05	METALS	95.0	89.8	93.8	92.8	87.0	91.4	92.9	86.2	91.4	95.8	90.0	94.4	91.3	87.9	90.5	95.1	90.0	93.9
06	WOOD, PLASTICS & COMPOSITES	104.6	61.3	86.7	66.4	59.8	63.7	66.6	27.8	50.6	80.7	70.4	76.4	75.8	63.8	70.8	93.8	65.1	81.9
07	THERMAL & MOISTURE PROTECTION	100.4	60.6	82.9	95.9	58.5	79.5	96.4	49.0	75.6	97.8	63.3	82.7	95.8	55.5	78.1	93.7	63.4	80.4
08	OPENINGS	98.1	59.9	88.9	87.6	50.2	78.6	87.6	33.4	74.5	93.8	63.8	86.6	94.7	60.7	86.5	92.1	58.2	83.9
0920	Plaster & Gypsum Board	94.5	60.8	73.5	89.6	59.3	70.7	89.6	26.4	50.1	95.7	70.2	79.7	96.7	63.4	75.9	111.7	64.8	82.4
0950, 0980	Ceilings & Acoustic Treatment	97.7	60.8	73.8	80.5	59.3	66.8	80.5	26.4	45.4	87.3	70.2	76.2	94.0	63.4	74.2	95.3	64.8	75.5
0960	Flooring	95.3	59.9	84.5	85.3	50.2	74.5	85.3	46.5	73.5	82.7	54.7	74.2	93.6	54.7	81.7	97.7	57.0	85.2
0970, 0990	Wall Finishes & Painting/Coating	96.3	55.3	71.3	86.0	52.4	65.5	86.0	52.4	65.5	89.0	59.0	70.7	92.6	55.3	69.9	92.6	55.3	69.9
09	FINISHES	95.1	59.5	75.8	91.2	56.4	72.3	91.8	33.9	60.4	91.3	65.4	77.2	97.5	57.5	75.8	92.1	61.8	75.7
COVERS	DIVS. 10 - 14, 25, 28, 41, 43, 44, 46	100.0	80.9	95.9	100.0	80.6	95.8	100.0	73.4	94.2	100.0	83.3	96.4	100.0	77.8	95.2	100.0	82.1	96.1
21, 22, 23	FIRE SUPPRESSION, PLUMBING & HVAC	100.9	55.2	82.1	97.8	68.3	85.7	97.8	62.2	83.2	100.8	64.6	85.9	100.7	51.6	80.6	100.7	61.8	84.7
26, 27, 3370	ELECTRICAL, COMMUNICATIONS & UTIL.	102.1	76.8	89.9	94.7	46.9	71.6	95.7	56.8	77.0	102.1	60.9	82.2	92.4	48.7	71.3	96.0	57.4	77.4
MF2018	WEIGHTED AVERAGE	98.0	67.1	85.4	95.0	63.1	82.0	96.5	55.1	79.6	98.4	68.8	86.4	97.3	59.1	81.7	94.8	65.9	83.0

TENNESSEE / TEXAS

DIVISION		MCKENZIE 382			MEMPHIS 375,380 - 381			NASHVILLE 370 - 372			ABILENE 795 - 796			AMARILLO 790 - 791			AUSTIN 786 - 787		
		MAT.	INST.	TOTAL	MAT.	INST.	TOTAL	MAT.	INST.	TOTAL	MAT.	INST.	TOTAL	MAT.	INST.	TOTAL	MAT.	INST.	TOTAL
015433	CONTRACTOR EQUIPMENT		97.0	97.0		97.3	97.3		105.0	105.0		87.7	87.7		95.2	95.2		95.3	95.3
0241, 31 - 34	SITE & INFRASTRUCTURE, DEMOLITION	107.7	80.3	89.1	89.5	89.9	89.8	101.0	98.5	99.3	95.6	83.9	87.7	95.7	94.7	95.1	90.0	94.0	92.7
0310	Concrete Forming & Accessories	87.3	33.0	43.5	99.6	66.8	73.1	101.8	71.8	77.6	98.9	60.5	67.9	100.1	53.8	62.8	99.5	53.8	62.7
0320	Concrete Reinforcing	87.2	61.1	78.8	100.7	71.2	91.2	101.5	68.8	91.0	99.7	50.6	83.9	100.0	49.3	83.7	100.4	44.3	82.3
0330	Cast-in-Place Concrete	104.8	56.1	88.0	92.0	69.3	84.1	86.9	71.1	81.4	77.1	65.2	73.0	75.3	68.3	72.9	87.1	57.5	75.2
03	CONCRETE	100.8	48.4	79.8	96.8	69.3	85.8	94.8	71.9	85.6	84.2	61.3	75.0	89.2	59.0	77.1	87.1	57.5	75.2
04	MASONRY	117.2	42.1	69.9	88.5	59.0	69.9	93.1	63.6	74.5	98.6	61.7	75.4	95.0	62.3	74.4	102.1	61.7	76.7
05	METALS	92.9	86.8	91.5	99.1	82.2	95.1	99.0	84.9	95.7	101.4	70.8	94.3	97.2	68.3	90.4	94.2	65.3	87.4
06	WOOD, PLASTICS & COMPOSITES	77.1	29.9	57.6	99.9	65.9	85.8	101.9	72.2	89.6	99.7	62.6	84.4	101.3	52.2	81.0	96.4	53.7	78.7
07	THERMAL & MOISTURE PROTECTION	96.5	49.9	76.1	95.9	66.7	83.1	100.4	69.5	86.8	102.0	63.3	85.0	100.1	64.0	84.3	95.0	65.4	82.0
08	OPENINGS	87.6	35.0	74.9	97.3	62.8	88.9	101.3	68.8	93.4	97.5	57.9	88.0	98.9	51.9	87.6	97.3	48.9	85.6
0920	Plaster & Gypsum Board	95.7	28.6	53.7	99.1	65.2	77.9	102.8	71.5	83.2	105.8	62.1	78.5	111.2	51.0	73.6	100.3	52.6	70.5
0950, 0980	Ceilings & Acoustic Treatment	80.5	28.6	46.8	99.0	65.2	77.0	91.5	71.5	78.5	104.5	62.1	77.0	103.8	51.0	69.5	98.3	52.6	68.7
0960	Flooring	87.6	50.2	76.1	100.8	55.8	87.0	93.8	62.4	84.2	99.0	69.5	90.0	98.2	66.1	88.4	96.8	59.3	85.3
0970, 0990	Wall Finishes & Painting/Coating	86.0	41.0	58.5	98.7	75.1	84.3	98.7	67.5	79.7	93.1	55.6	70.2	98.7	52.9	70.8	98.7	41.0	63.5
09	FINISHES	93.3	35.2	61.7	98.8	65.2	80.5	96.2	69.4	81.6	96.6	61.8	77.7	102.1	55.4	76.7	97.2	53.0	73.2
COVERS	DIVS. 10 - 14, 25, 28, 41, 43, 44, 46	100.0	74.2	94.4	100.0	84.2	96.6	100.0	85.2	96.8	100.0	79.5	95.5	100.0	80.1	95.7	100.0	79.2	95.5
21, 22, 23	FIRE SUPPRESSION, PLUMBING & HVAC	97.8	54.4	80.0	100.6	73.6	89.5	100.8	81.4	92.9	101.1	48.6	79.5	100.9	53.0	81.2	101.0	58.2	83.4
26, 27, 3370	ELECTRICAL, COMMUNICATIONS & UTIL.	95.6	51.3	74.2	99.0	64.6	82.5	101.0	64.9	83.6	96.5	53.1	75.6	98.8	59.5	79.8	100.1	57.2	79.4
MF2018	WEIGHTED AVERAGE	96.7	53.8	79.2	98.2	70.8	87.1	99.0	75.2	89.3	97.4	60.6	82.4	97.8	61.7	83.1	96.5	61.5	82.2

City Cost Indexes

TEXAS

DIVISION		BEAUMONT 776-777 MAT.	INST.	TOTAL	BROWNWOOD 768 MAT.	INST.	TOTAL	BRYAN 778 MAT.	INST.	TOTAL	CHILDRESS 792 MAT.	INST.	TOTAL	CORPUS CHRISTI 783-784 MAT.	INST.	TOTAL	DALLAS 752-753 MAT.	INST.	TOTAL
015433	CONTRACTOR EQUIPMENT		91.2	91.2		87.7	87.7		91.2	91.2		87.7	87.7		97.6	97.6		109.3	109.3
0241, 31 - 34	SITE & INFRASTRUCTURE, DEMOLITION	92.3	87.3	88.9	104.2	83.8	90.3	82.2	87.0	85.5	106.6	83.8	91.1	129.8	79.2	95.4	115.5	100.9	105.6
0310	Concrete Forming & Accessories	105.1	56.5	65.9	97.9	57.3	65.1	79.6	54.9	59.6	97.0	57.4	65.0	105.2	52.6	62.8	100.5	62.6	69.9
0320	Concrete Reinforcing	101.0	63.6	89.0	90.4	47.6	76.7	102.2	47.6	84.7	99.8	47.6	83.1	88.3	50.6	76.2	100.9	51.1	84.9
0330	Cast-in-Place Concrete	71.8	65.8	69.7	96.5	55.3	82.3	57.5	55.5	56.9	79.1	55.3	70.9	101.8	65.4	89.3	101.6	70.1	90.7
03	CONCRETE	85.3	61.8	76.0	91.3	55.8	77.1	73.0	54.9	65.8	91.0	55.9	77.0	91.6	58.7	78.4	97.5	64.5	84.3
04	MASONRY	98.7	63.1	76.3	127.5	56.8	82.9	139.3	56.6	87.2	102.4	56.3	73.4	84.6	60.0	69.1	95.2	58.5	72.1
05	METALS	92.2	76.7	88.6	96.7	68.9	90.2	92.0	69.7	86.8	98.5	69.0	91.6	91.3	84.5	89.7	99.9	80.5	95.4
06	WOOD, PLASTICS & COMPOSITES	112.3	56.0	89.0	102.8	59.5	84.9	76.2	55.9	67.8	99.0	59.5	82.7	116.6	52.4	90.0	101.3	64.6	86.1
07	THERMAL & MOISTURE PROTECTION	94.0	65.5	81.5	94.6	59.3	79.1	87.4	60.3	75.5	102.7	59.2	83.6	99.9	61.5	83.1	93.4	66.0	81.4
08	OPENINGS	91.1	57.7	83.0	96.6	54.4	86.4	91.7	52.6	82.3	92.3	54.4	83.1	97.7	50.7	86.4	96.6	59.5	87.7
0920	Plaster & Gypsum Board	110.5	55.5	76.1	93.0	59.0	71.7	84.1	55.4	66.1	104.8	59.0	76.2	98.6	51.6	69.2	95.1	63.4	75.2
0950, 0980	Ceilings & Acoustic Treatment	108.8	55.5	74.2	88.4	59.0	69.3	98.1	55.4	70.4	99.7	59.0	73.3	105.6	51.6	70.5	94.9	63.4	74.4
0960	Flooring	114.5	75.2	102.4	77.3	50.8	69.2	87.3	62.3	79.7	97.4	50.8	83.2	112.3	73.3	100.4	92.6	68.1	85.1
0970, 0990	Wall Finishes & Painting/Coating	88.6	58.0	69.9	86.5	47.9	63.0	86.2	53.3	66.1	93.1	47.9	65.5	105.5	44.4	68.2	98.7	52.9	70.8
09	FINISHES	98.6	59.7	77.5	84.7	55.1	68.6	84.3	55.6	68.7	96.5	55.1	74.0	103.1	55.2	77.1	96.5	62.6	78.1
COVERS	DIVS. 10 - 14, 25, 28, 41, 43, 44, 46	100.0	78.2	95.3	100.0	78.8	95.4	100.0	77.5	95.1	100.0	78.8	95.4	100.0	77.8	95.2	100.0	81.5	96.0
21, 22, 23	FIRE SUPPRESSION, PLUMBING & HVAC	101.0	63.3	85.5	96.8	46.1	76.0	96.7	60.0	81.6	96.8	50.2	77.7	101.0	58.7	83.6	100.9	61.5	84.7
26, 27, 3370	ELECTRICAL, COMMUNICATIONS & UTIL.	92.6	63.9	78.8	98.7	42.8	71.7	93.3	58.7	76.6	96.5	51.7	74.9	98.0	63.6	81.4	99.7	61.1	81.1
MF2018	WEIGHTED AVERAGE	94.7	66.0	83.0	97.0	56.1	80.4	91.9	61.4	79.5	96.8	58.2	81.1	97.4	63.3	83.5	99.3	67.1	86.2

TEXAS

DIVISION		DEL RIO 788 MAT.	INST.	TOTAL	DENTON 762 MAT.	INST.	TOTAL	EASTLAND 764 MAT.	INST.	TOTAL	EL PASO 798-799,885 MAT.	INST.	TOTAL	FORT WORTH 760-761 MAT.	INST.	TOTAL	GALVESTON 775 MAT.	INST.	TOTAL
015433	CONTRACTOR EQUIPMENT		87.2	87.2		96.7	96.7		87.7	87.7		95.3	95.3		96.9	96.9		103.5	103.5
0241, 31 - 34	SITE & INFRASTRUCTURE, DEMOLITION	111.2	82.0	91.3	104.8	78.4	86.9	107.0	83.8	91.2	90.8	94.3	93.1	99.3	97.4	98.0	112.3	85.7	94.3
0310	Concrete Forming & Accessories	100.5	48.2	58.3	108.9	57.5	67.5	99.2	57.3	65.4	102.9	58.8	67.3	98.9	61.1	68.4	91.1	52.6	60.0
0320	Concrete Reinforcing	88.6	44.3	74.4	91.1	47.8	77.2	90.5	47.6	76.7	101.3	50.7	85.0	100.3	50.3	84.3	101.9	55.6	87.0
0330	Cast-in-Place Concrete	109.6	54.7	90.6	74.6	56.4	68.3	102.0	55.3	85.9	68.1	66.4	67.5	86.8	67.0	79.9	76.4	56.5	69.5
03	CONCRETE	107.3	50.8	84.7	74.9	57.3	67.9	95.3	55.9	79.5	87.4	60.7	76.7	88.6	62.0	77.9	85.7	56.6	74.1
04	MASONRY	99.6	56.5	72.5	135.8	58.6	87.2	95.4	56.8	71.1	94.1	64.2	75.3	93.9	59.7	72.4	95.7	56.7	71.1
05	METALS	90.3	65.4	84.6	96.0	84.0	93.2	96.5	69.0	90.0	97.6	68.2	90.7	99.0	69.0	92.0	93.6	88.4	92.4
06	WOOD, PLASTICS & COMPOSITES	99.2	47.5	77.8	116.8	59.7	93.2	109.1	59.5	88.6	92.7	59.4	78.9	96.9	63.0	82.9	92.2	52.7	75.8
07	THERMAL & MOISTURE PROTECTION	97.2	58.7	80.3	93.0	60.6	78.8	95.0	59.3	79.4	103.1	65.6	86.7	92.1	65.8	80.6	87.0	60.6	75.4
08	OPENINGS	93.1	44.9	81.5	114.2	54.1	99.7	67.8	54.4	64.6	92.1	54.3	83.0	96.8	58.1	87.5	95.7	52.7	85.3
0920	Plaster & Gypsum Board	91.7	46.7	63.6	101.1	59.0	74.8	93.0	59.0	71.7	113.8	58.5	79.2	98.7	62.1	75.8	97.1	51.9	68.8
0950, 0980	Ceilings & Acoustic Treatment	99.9	46.7	65.4	92.6	59.0	70.8	88.4	59.0	69.3	102.6	58.5	74.0	98.3	62.1	74.8	103.6	51.9	70.0
0960	Flooring	96.2	50.8	82.3	72.5	50.8	65.9	96.9	50.8	82.8	104.1	77.0	95.8	96.6	69.5	88.3	101.8	62.3	89.7
0970, 0990	Wall Finishes & Painting/Coating	95.2	39.5	61.2	96.0	45.4	65.1	88.0	47.9	63.5	98.7	47.9	67.7	98.7	53.0	70.8	97.3	49.8	68.3
09	FINISHES	94.9	47.1	69.0	83.5	54.9	68.0	90.4	55.1	71.2	103.3	60.8	80.3	98.2	61.8	78.4	95.7	53.3	72.7
COVERS	DIVS. 10 - 14, 25, 28, 41, 43, 44, 46	100.0	75.0	94.6	100.0	79.1	95.5	100.0	78.8	95.4	100.0	81.4	96.0	100.0	80.4	95.7	100.0	77.4	95.1
21, 22, 23	FIRE SUPPRESSION, PLUMBING & HVAC	96.6	55.1	79.6	96.8	51.9	78.3	96.8	46.1	76.0	100.7	62.4	85.0	101.0	56.9	82.9	96.7	60.1	81.7
26, 27, 3370	ELECTRICAL, COMMUNICATIONS & UTIL.	99.3	55.4	78.2	101.6	54.4	78.8	98.5	54.3	77.2	99.6	50.6	76.0	98.5	57.6	78.8	93.7	58.8	76.9
MF2018	WEIGHTED AVERAGE	97.5	56.9	81.0	96.9	60.3	82.0	94.2	57.7	79.3	96.9	63.8	83.5	97.3	63.9	83.8	94.4	62.9	81.6

TEXAS

DIVISION		GIDDINGS 789 MAT.	INST.	TOTAL	GREENVILLE 754 MAT.	INST.	TOTAL	HOUSTON 770-772 MAT.	INST.	TOTAL	HUNTSVILLE 773 MAT.	INST.	TOTAL	LAREDO 780 MAT.	INST.	TOTAL	LONGVIEW 756 MAT.	INST.	TOTAL
015433	CONTRACTOR EQUIPMENT		87.2	87.2		97.4	97.4		101.8	101.8		91.2	91.2		87.2	87.2		89.0	89.0
0241, 31 - 34	SITE & INFRASTRUCTURE, DEMOLITION	97.5	82.2	87.1	103.7	81.1	88.4	111.9	94.2	99.9	97.6	87.0	90.4	92.2	82.4	85.6	100.2	86.9	91.2
0310	Concrete Forming & Accessories	97.5	48.4	57.9	87.6	54.3	60.7	98.8	59.4	67.0	89.8	51.1	58.6	100.5	52.9	62.1	81.4	57.4	62.1
0320	Concrete Reinforcing	88.9	45.1	74.9	101.7	47.8	84.4	100.4	50.7	84.4	102.3	47.6	84.8	88.6	50.5	76.4	101.5	47.3	84.1
0330	Cast-in-Place Concrete	92.8	55.0	79.8	104.0	56.4	87.6	71.0	61.9	75.0	79.1	55.5	70.9	78.4	65.2	73.8	121.1	55.3	98.4
03	CONCRETE	88.9	51.3	73.8	94.9	55.7	79.2	83.8	61.9	75.0	90.7	52.4	75.7	84.2	57.8	73.6	109.0	55.8	87.7
04	MASONRY	107.3	56.6	75.4	154.7	58.6	94.1	94.0	62.1	73.9	136.7	56.6	86.2	92.5	62.9	72.6	150.8	55.4	90.7
05	METALS	89.8	66.7	84.4	93.1	82.6	90.6	99.9	74.3	94.0	91.9	69.5	86.6	93.1	69.8	87.6	85.8	68.0	81.7
06	WOOD, PLASTICS & COMPOSITES	98.4	47.5	77.3	84.0	55.1	72.1	100.6	60.1	83.8	87.2	51.0	72.2	99.2	52.3	79.8	77.2	59.6	69.9
07	THERMAL & MOISTURE PROTECTION	97.6	59.3	80.8	92.1	59.6	77.9	88.9	68.1	79.8	88.7	59.7	76.0	96.3	63.8	82.1	93.3	58.5	78.1
08	OPENINGS	92.3	45.8	81.1	88.6	52.1	79.8	104.2	56.9	92.8	91.7	49.4	81.5	93.2	50.2	82.8	79.8	54.1	73.6
0920	Plaster & Gypsum Board	90.0	46.7	62.9	77.3	54.3	62.9	98.7	58.9	73.8	91.4	50.2	65.7	93.0	51.6	67.1	76.0	59.0	65.4
0950, 0980	Ceilings & Acoustic Treatment	99.9	46.7	65.4	88.6	54.3	66.3	98.3	58.9	72.8	98.1	50.2	67.1	106.1	51.6	70.7	82.4	59.0	67.2
0960	Flooring	96.5	50.8	82.5	85.0	50.8	74.6	105.8	73.3	95.9	90.8	50.8	78.6	96.1	65.6	86.8	89.6	50.8	77.7
0970, 0990	Wall Finishes & Painting/Coating	95.2	41.0	62.2	88.8	47.9	63.9	98.7	56.5	73.0	86.2	53.3	66.1	95.2	44.4	64.2	81.8	45.4	59.6
09	FINISHES	93.7	47.3	68.5	88.8	52.5	69.1	102.0	61.6	80.0	87.5	50.6	67.5	94.9	53.7	72.5	93.2	54.9	72.4
COVERS	DIVS. 10 - 14, 25, 28, 41, 43, 44, 46	100.0	77.3	95.1	100.0	78.8	95.4	100.0	80.1	95.7	100.0	76.9	95.0	100.0	77.8	95.2	100.0	79.0	95.4
21, 22, 23	FIRE SUPPRESSION, PLUMBING & HVAC	96.7	59.1	81.3	96.7	53.9	79.1	101.0	63.2	85.5	96.7	59.9	81.6	100.9	59.9	84.0	96.6	53.5	78.9
26, 27, 3370	ELECTRICAL, COMMUNICATIONS & UTIL.	97.3	51.5	75.2	97.8	54.4	76.8	95.2	63.9	80.1	93.3	54.8	74.7	99.4	58.0	79.5	96.7	47.2	72.8
MF2018	WEIGHTED AVERAGE	94.4	57.6	79.4	97.1	60.1	82.0	97.8	66.6	85.1	95.0	59.7	80.7	95.0	61.5	81.3	96.7	58.2	81.0

For CCI updates and additional guidance please go to www.rsmeans.com/2022books

City Cost Indexes

TEXAS

DIVISION		LUBBOCK 793-794 MAT.	INST.	TOTAL	LUFKIN 759 MAT.	INST.	TOTAL	MCALLEN 785 MAT.	INST.	TOTAL	MCKINNEY 750 MAT.	INST.	TOTAL	MIDLAND 797 MAT.	INST.	TOTAL	ODESSA 797 MAT.	INST.	TOTAL
015433	CONTRACTOR EQUIPMENT		99.3	99.3		89.0	89.0		97.7	97.7		97.4	97.4		99.3	99.3		87.7	87.7
0241, 31 - 34	SITE & INFRASTRUCTURE, DEMOLITION	122.8	82.9	95.7	94.7	86.2	88.9	133.3	79.3	96.6	99.8	81.1	87.1	125.5	82.9	96.6	95.8	84.0	87.8
0310	Concrete Forming & Accessories	100.1	54.2	63.1	85.5	51.2	57.9	104.5	48.9	59.6	86.3	54.3	60.5	105.5	61.1	69.7	98.9	60.9	68.2
0320	Concrete Reinforcing	100.4	50.5	84.4	103.3	62.1	90.1	88.4	50.5	76.2	101.7	47.8	84.4	100.9	50.6	84.7	99.7	50.5	83.9
0330	Cast-in-Place Concrete	77.2	66.6	73.5	108.3	54.7	89.8	110.6	56.7	92.0	97.6	56.4	83.3	82.1	66.8	76.8	77.1	64.8	72.9
03	CONCRETE	83.3	59.9	73.9	101.9	55.2	83.2	97.7	53.9	80.2	90.4	55.8	76.6	86.9	63.0	77.4	84.2	61.3	75.0
04	MASONRY	98.0	63.1	76.0	113.4	56.5	77.5	100.2	60.9	75.5	166.8	58.6	98.6	114.8	61.7	81.3	98.6	61.6	75.3
05	METALS	104.8	85.4	100.3	93.3	71.6	88.3	90.2	83.9	88.7	93.0	82.7	90.6	103.2	85.6	99.1	100.7	70.7	93.7
06	WOOD, PLASTICS & COMPOSITES	101.3	52.9	81.3	84.0	51.1	70.4	114.9	47.7	87.1	82.7	55.1	71.3	106.8	62.7	88.6	99.7	62.6	84.4
07	THERMAL & MOISTURE PROTECTION	92.1	63.2	79.4	93.2	57.9	77.7	99.8	61.3	82.9	91.9	59.6	77.7	92.5	63.8	79.9	102.0	62.8	84.8
08	OPENINGS	104.1	52.7	91.7	60.7	53.2	58.9	97.2	46.4	84.9	88.6	52.0	79.7	105.6	58.3	94.2	97.5	58.3	88.1
0920	Plaster & Gypsum Board	106.0	52.1	72.3	74.9	50.2	59.5	100.0	46.7	66.7	77.3	54.3	62.9	107.8	62.1	79.2	105.8	62.1	78.5
0950, 0980	Ceilings & Acoustic Treatment	105.2	52.1	70.7	76.9	50.2	59.6	106.1	46.7	67.5	88.6	54.3	66.3	101.1	62.1	75.8	104.5	62.1	77.0
0960	Flooring	91.9	66.0	84.0	119.1	50.8	98.2	111.9	73.3	100.1	84.7	50.8	74.3	92.9	67.4	85.1	99.0	64.9	88.6
0970, 0990	Wall Finishes & Painting/Coating	103.4	52.9	72.6	81.8	47.9	61.1	105.5	39.5	65.2	88.8	47.9	63.9	103.4	52.9	72.6	93.1	52.9	68.6
09	FINISHES	99.8	55.9	75.9	99.4	50.5	72.9	103.8	52.1	75.7	88.4	52.5	68.9	99.9	61.4	79.0	96.6	60.8	77.2
COVERS	DIVS. 10 - 14, 25, 28, 41, 43, 44, 46	100.0	79.5	95.5	100.0	77.2	95.1	100.0	77.5	95.1	100.0	78.8	95.4	100.0	80.1	95.7	100.0	79.9	95.6
21, 22, 23	FIRE SUPPRESSION, PLUMBING & HVAC	100.4	50.1	79.8	96.6	54.0	79.1	96.7	45.9	75.8	96.7	56.0	80.0	96.2	53.6	78.7	101.1	49.6	80.0
26, 27, 3370	ELECTRICAL, COMMUNICATIONS & UTIL.	95.2	51.5	74.2	97.5	56.9	77.9	97.9	29.3	64.8	97.9	54.4	76.9	95.2	59.0	77.8	96.6	59.0	78.5
MF2018	WEIGHTED AVERAGE	98.9	60.8	83.4	94.4	59.2	80.1	97.8	54.4	80.2	96.8	60.6	82.0	99.2	64.0	84.9	97.2	61.6	82.7

TEXAS

DIVISION		PALESTINE 758 MAT.	INST.	TOTAL	SAN ANGELO 769 MAT.	INST.	TOTAL	SAN ANTONIO 781-782 MAT.	INST.	TOTAL	TEMPLE 765 MAT.	INST.	TOTAL	TEXARKANA 755 MAT.	INST.	TOTAL	TYLER 757 MAT.	INST.	TOTAL
015433	CONTRACTOR EQUIPMENT		89.0	89.0		87.7	87.7		94.5	94.5		87.7	87.7		89.0	89.0		89.0	89.0
0241, 31 - 34	SITE & INFRASTRUCTURE, DEMOLITION	100.4	86.9	91.2	100.5	83.8	89.1	90.3	94.2	93.0	89.1	83.3	85.1	89.0	86.9	87.6	99.5	86.9	90.9
0310	Concrete Forming & Accessories	73.6	57.4	60.6	98.4	48.7	58.3	98.9	54.7	63.3	103.1	48.3	58.9	95.9	57.8	65.1	88.0	57.5	63.4
0320	Concrete Reinforcing	100.8	47.6	83.7	90.3	50.3	77.5	102.0	50.6	85.5	90.4	47.3	76.6	100.6	50.0	84.3	101.5	50.0	85.0
0330	Cast-in-Place Concrete	98.9	55.3	83.8	91.1	64.4	81.8	74.7	68.6	72.6	74.6	54.8	67.7	99.6	63.6	87.2	118.9	55.4	97.0
03	CONCRETE	101.3	55.9	83.1	87.7	55.6	74.8	84.4	59.5	74.4	76.7	51.5	66.6	95.4	59.3	81.0	108.6	56.3	87.7
04	MASONRY	108.5	55.4	75.1	123.8	56.8	81.6	92.6	62.8	73.8	134.6	56.6	85.4	171.3	55.4	98.3	161.4	55.4	94.6
05	METALS	93.1	68.1	87.2	96.9	70.1	90.6	100.1	65.8	92.1	96.5	67.3	89.7	85.7	69.3	81.9	92.9	69.0	87.4
06	WOOD, PLASTICS & COMPOSITES	73.4	59.6	67.7	103.2	47.6	80.1	99.8	53.7	80.7	112.3	47.6	85.5	92.5	59.6	78.9	86.1	59.6	75.1
07	THERMAL & MOISTURE PROTECTION	93.5	58.7	78.2	94.3	59.7	79.1	90.1	67.0	80.0	93.9	58.3	78.3	93.1	60.1	78.7	93.6	58.7	78.3
08	OPENINGS	60.6	54.5	59.2	96.6	48.5	85.0	102.6	51.8	90.3	64.6	47.8	60.5	79.8	55.1	73.8	60.6	55.1	59.3
0920	Plaster & Gypsum Board	74.1	59.0	64.7	93.0	46.7	64.0	91.1	52.6	67.0	93.0	46.7	64.0	85.9	59.0	69.1	74.9	59.0	65.0
0950, 0980	Ceilings & Acoustic Treatment	76.9	59.0	65.3	88.4	46.7	61.3	99.0	52.6	68.9	88.4	46.7	61.3	82.4	59.0	67.2	76.9	59.0	65.3
0960	Flooring	113.2	50.8	94.1	77.5	50.8	69.3	108.9	74.2	98.3	98.4	50.8	83.8	95.1	59.3	84.1	120.7	50.8	99.3
0970, 0990	Wall Finishes & Painting/Coating	81.8	47.9	61.1	86.5	47.9	63.0	98.7	44.4	65.6	88.0	41.0	59.3	81.8	47.9	61.1	81.8	47.9	61.1
09	FINISHES	98.4	55.2	74.9	84.4	48.1	64.7	99.6	56.8	76.3	89.2	47.3	66.5	95.6	56.9	74.6	100.2	55.2	75.7
COVERS	DIVS. 10 - 14, 25, 28, 41, 43, 44, 46	100.0	78.9	95.4	100.0	77.3	95.1	100.0	79.5	95.5	100.0	77.5	95.1	100.0	79.0	95.4	100.0	79.0	95.4
21, 22, 23	FIRE SUPPRESSION, PLUMBING & HVAC	96.6	52.7	78.6	96.8	48.0	76.7	100.7	60.9	84.4	96.8	49.4	77.3	96.6	51.6	78.2	96.6	54.1	79.2
26, 27, 3370	ELECTRICAL, COMMUNICATIONS & UTIL.	95.0	44.9	70.8	101.4	50.6	76.9	100.1	60.4	81.0	99.4	48.2	74.7	97.5	50.6	74.8	96.7	49.8	74.1
MF2018	WEIGHTED AVERAGE	93.8	57.8	79.1	96.6	56.3	80.2	97.4	63.6	83.7	92.5	55.3	77.4	95.8	59.2	80.9	97.4	59.0	81.8

TEXAS / UTAH

DIVISION		VICTORIA 779 MAT.	INST.	TOTAL	WACO 766-767 MAT.	INST.	TOTAL	WAXAHACHIE 751 MAT.	INST.	TOTAL	WHARTON 774 MAT.	INST.	TOTAL	WICHITA FALLS 763 MAT.	INST.	TOTAL	LOGAN 843 MAT.	INST.	TOTAL
015433	CONTRACTOR EQUIPMENT		102.2	102.2		87.7	87.7		97.4	97.4		103.5	103.5		87.7	87.7		91.4	91.4
0241, 31 - 34	SITE & INFRASTRUCTURE, DEMOLITION	117.2	83.5	94.3	97.9	83.5	88.1	101.9	80.9	87.7	123.1	85.5	97.5	98.7	83.9	88.6	97.7	85.7	89.6
0310	Concrete Forming & Accessories	90.1	48.5	56.6	101.1	60.5	68.4	86.3	57.6	63.2	83.7	50.1	56.6	101.1	60.6	68.4	100.7	69.8	75.8
0320	Concrete Reinforcing	97.9	47.4	81.7	90.2	50.5	77.5	101.7	47.8	84.4	101.9	47.4	84.4	90.2	50.5	77.5	93.0	83.3	89.9
0330	Cast-in-Place Concrete	85.8	56.4	75.6	80.7	64.3	75.0	103.0	56.1	86.8	88.1	55.5	76.8	86.2	69.6	80.5	77.8	73.3	76.2
03	CONCRETE	90.9	53.3	75.9	81.9	60.9	73.5	93.9	57.2	79.2	94.2	53.7	78.0	84.4	62.8	75.7	98.0	73.9	88.4
04	MASONRY	113.5	56.7	77.7	93.3	60.9	72.9	155.3	58.5	94.3	96.9	56.6	71.5	93.9	60.5	72.8	110.2	61.5	79.5
05	METALS	91.9	85.1	90.3	99.2	69.7	92.3	93.1	82.3	90.5	93.6	84.6	91.5	99.1	70.9	92.5	101.9	83.2	97.5
06	WOOD, PLASTICS & COMPOSITES	93.8	47.6	74.7	110.2	62.6	90.5	82.7	59.8	73.2	83.5	49.7	69.5	110.2	62.6	90.5	85.3	70.4	79.1
07	THERMAL & MOISTURE PROTECTION	89.9	57.8	75.8	94.7	63.5	81.0	92.0	60.0	78.0	87.3	60.2	75.4	94.7	62.5	80.6	103.2	67.3	87.5
08	OPENINGS	95.6	48.0	84.1	75.0	58.2	70.9	88.6	54.6	80.4	95.7	49.1	84.5	75.0	57.9	70.9	88.1	71.6	84.1
0920	Plaster & Gypsum Board	87.7	46.7	62.1	93.1	62.1	73.7	77.7	59.0	66.0	87.4	48.8	63.3	93.1	62.1	73.7	86.2	69.9	76.0
0950, 0980	Ceilings & Acoustic Treatment	105.0	46.7	67.2	89.1	62.1	71.6	90.6	59.0	70.1	103.6	48.8	68.1	89.1	62.1	71.6	106.0	69.9	82.5
0960	Flooring	101.5	50.8	86.0	97.9	68.2	88.8	84.7	50.8	74.3	99.6	62.0	88.1	98.4	72.4	90.4	94.9	61.3	84.6
0970, 0990	Wall Finishes & Painting/Coating	97.5	53.3	70.5	88.0	55.6	68.2	88.8	47.9	63.9	97.3	51.7	69.5	89.5	52.9	67.2	90.7	61.1	72.6
09	FINISHES	92.8	48.7	68.9	89.8	61.5	74.4	89.2	55.2	70.8	94.5	51.7	71.2	90.1	62.1	74.9	96.0	67.0	80.3
COVERS	DIVS. 10 - 14, 25, 28, 41, 43, 44, 46	100.0	76.8	95.0	100.0	79.5	95.5	100.0	79.3	95.5	100.0	77.0	95.0	100.0	79.5	95.5	100.0	85.3	96.8
21, 22, 23	FIRE SUPPRESSION, PLUMBING & HVAC	96.7	60.0	81.6	101.0	59.7	84.0	96.7	53.9	79.1	96.7	57.4	80.6	101.0	53.7	81.6	100.8	72.5	89.2
26, 27, 3370	ELECTRICAL, COMMUNICATIONS & UTIL.	96.9	50.1	74.3	101.5	54.6	78.9	97.8	54.4	76.9	95.7	54.8	76.0	104.2	53.9	79.9	95.4	71.5	83.9
MF2018	WEIGHTED AVERAGE	95.9	59.8	81.2	94.4	62.9	81.6	96.9	60.8	82.2	95.9	60.6	81.6	95.1	62.0	81.6	98.7	72.9	88.2

City Cost Indexes

UTAH / VERMONT

DIVISION		OGDEN 842,844 MAT.	INST.	TOTAL	PRICE 845 MAT.	INST.	TOTAL	PROVO 846-847 MAT.	INST.	TOTAL	SALT LAKE CITY 840-841 MAT.	INST.	TOTAL	BELLOWS FALLS 051 MAT.	INST.	TOTAL	BENNINGTON 052 MAT.	INST.	TOTAL
015433	CONTRACTOR EQUIPMENT		91.4	91.4		90.5	90.5		90.5	90.5		91.3	91.3		92.3	92.3		92.3	92.3
0241, 31 - 34	SITE & INFRASTRUCTURE, DEMOLITION	86.0	85.7	85.8	94.9	83.4	87.1	94.2	84.3	87.4	85.7	85.6	85.7	84.8	91.1	89.1	84.2	91.1	88.9
0310	Concrete Forming & Accessories	100.7	69.8	75.8	104.3	62.0	70.1	103.3	69.8	76.3	103.3	69.8	76.3	99.8	96.1	96.9	96.3	96.0	96.1
0320	Concrete Reinforcing	92.8	83.3	89.7	99.9	83.4	94.6	100.7	83.3	95.1	95.2	83.3	91.4	81.0	81.0	81.0	81.0	81.0	81.0
0330	Cast-in-Place Concrete	79.0	73.3	77.0	77.9	69.2	74.9	77.9	73.3	76.3	86.3	73.3	81.8	70.9	111.5	84.9	70.9	111.4	84.9
03	CONCRETE	89.3	73.9	83.1	99.8	68.9	87.4	98.6	73.9	88.7	105.7	73.9	93.0	79.2	98.9	87.1	79.0	98.8	86.9
04	MASONRY	104.3	61.5	77.3	115.8	64.6	83.6	116.0	61.5	81.6	118.6	61.5	82.6	91.6	90.8	91.1	101.9	90.8	94.9
05	METALS	102.4	83.2	97.9	99.0	82.4	95.2	99.8	83.2	95.9	106.2	83.2	100.9	93.5	90.2	92.7	93.4	90.1	92.6
06	WOOD, PLASTICS & COMPOSITES	85.3	70.4	79.1	89.4	60.4	77.4	87.2	70.4	80.3	87.6	70.4	80.5	103.7	100.7	102.5	100.5	100.7	100.6
07	THERMAL & MOISTURE PROTECTION	101.9	67.3	86.8	105.1	65.0	87.5	105.2	67.3	88.6	109.1	67.3	90.8	101.7	85.8	94.7	101.6	85.8	94.7
08	OPENINGS	88.1	71.6	84.1	91.8	73.0	87.2	91.8	71.6	86.9	89.7	71.6	85.3	96.8	91.8	95.6	96.8	91.8	95.6
0920	Plaster & Gypsum Board	86.2	69.9	76.0	91.7	59.5	71.6	87.4	69.9	76.4	92.5	69.9	78.3	96.0	100.6	98.9	94.8	100.6	98.4
0950, 0980	Ceilings & Acoustic Treatment	106.0	69.9	82.5	106.0	59.5	75.8	106.0	69.9	82.5	100.6	69.9	80.7	84.8	100.6	95.1	84.8	100.6	95.1
0960	Flooring	92.8	61.3	83.2	95.8	52.2	82.4	95.6	61.3	85.1	97.6	61.3	86.5	91.3	110.1	97.0	90.5	110.1	96.5
0970, 0990	Wall Finishes & Painting/Coating	90.7	61.1	72.6	90.7	56.3	69.7	90.7	61.1	72.6	94.1	60.5	73.6	84.0	94.2	90.2	84.0	94.2	90.2
09	FINISHES	94.1	67.0	79.4	97.6	58.8	76.5	96.7	67.0	80.6	96.0	67.0	80.2	87.9	99.1	94.0	87.5	99.1	93.8
COVERS	DIVS. 10 - 14, 25, 28, 41, 43, 44, 46	100.0	85.3	96.8	100.0	84.2	96.6	100.0	85.3	96.8	100.0	85.3	96.8	100.0	99.7	99.9	100.0	99.6	99.9
21, 22, 23	FIRE SUPPRESSION, PLUMBING & HVAC	100.8	72.5	89.2	98.6	60.5	83.0	100.8	72.5	89.2	101.1	72.5	89.4	96.9	81.3	90.5	96.9	81.3	90.5
26, 27, 3370	ELECTRICAL, COMMUNICATIONS & UTIL.	95.9	71.5	84.2	98.5	63.7	81.8	95.8	68.9	82.8	98.8	71.5	85.6	101.5	73.7	88.1	101.5	50.5	76.9
MF2018	WEIGHTED AVERAGE	96.9	72.9	87.2	98.9	67.5	86.1	99.0	72.4	88.2	101.4	72.9	89.8	93.5	88.8	91.6	93.8	85.4	90.4

VERMONT

DIVISION		BRATTLEBORO 053 MAT.	INST.	TOTAL	BURLINGTON 054 MAT.	INST.	TOTAL	GUILDHALL 059 MAT.	INST.	TOTAL	MONTPELIER 056 MAT.	INST.	TOTAL	RUTLAND 057 MAT.	INST.	TOTAL	ST. JOHNSBURY 058 MAT.	INST.	TOTAL
015433	CONTRACTOR EQUIPMENT		92.3	92.3		97.3	97.3		92.3	92.3		97.3	97.3		92.3	92.3		92.3	92.3
0241, 31 - 34	SITE & INFRASTRUCTURE, DEMOLITION	85.5	91.1	89.3	90.1	101.9	98.1	84.0	90.8	88.6	88.0	101.0	96.8	88.3	92.2	90.9	84.1	90.2	88.2
0310	Concrete Forming & Accessories	100.2	96.2	96.9	102.0	83.7	87.2	95.9	90.7	91.7	102.0	97.4	98.3	100.5	83.7	86.9	93.1	90.6	91.1
0320	Concrete Reinforcing	80.0	81.0	80.4	99.6	83.5	94.4	81.7	81.0	81.4	98.9	83.5	94.0	101.2	83.5	95.5	80.0	81.0	80.3
0330	Cast-in-Place Concrete	73.1	111.5	86.4	87.9	114.5	97.1	68.7	103.7	80.8	81.5	114.5	92.9	69.5	113.4	84.6	68.7	103.6	80.8
03	CONCRETE	80.7	98.9	87.9	95.3	94.6	95.0	77.2	93.6	83.8	92.4	100.7	95.7	83.6	94.3	87.9	76.7	93.6	83.4
04	MASONRY	101.7	90.8	94.8	103.7	94.3	97.8	102.0	77.1	86.3	100.6	94.3	96.6	83.7	94.2	90.3	128.2	77.1	96.0
05	METALS	93.4	90.2	92.7	101.5	88.8	98.5	93.5	89.9	92.6	100.5	88.8	97.7	100.3	91.1	98.2	93.4	89.8	92.6
06	WOOD, PLASTICS & COMPOSITES	104.1	100.7	102.7	100.4	82.1	92.8	99.4	100.7	100.0	98.0	100.7	99.1	104.3	82.0	95.1	93.6	100.7	96.5
07	THERMAL & MOISTURE PROTECTION	101.8	85.8	94.8	108.8	96.8	103.5	101.3	79.6	91.8	110.0	98.8	105.1	101.9	95.2	99.0	101.1	79.6	91.7
08	OPENINGS	96.8	92.0	95.6	98.9	82.5	94.9	96.8	92.2	95.7	99.2	92.8	97.7	99.5	82.5	95.4	96.8	92.2	95.7
0920	Plaster & Gypsum Board	96.0	100.6	98.9	105.0	81.5	90.3	108.3	100.6	103.5	105.0	100.6	102.3	96.4	81.5	87.1	112.4	100.6	105.1
0950, 0980	Ceilings & Acoustic Treatment	84.8	100.6	95.1	94.7	81.5	86.1	84.8	100.6	95.1	90.9	100.6	97.2	90.7	81.5	84.7	84.8	100.6	95.1
0960	Flooring	91.4	110.1	97.1	100.8	110.1	103.6	93.7	110.1	98.7	96.5	110.1	100.7	91.3	110.1	97.0	96.3	110.1	100.5
0970, 0990	Wall Finishes & Painting/Coating	84.0	94.2	90.2	98.7	101.9	100.6	84.0	101.9	94.9	98.7	101.9	100.6	84.0	101.9	94.9	84.0	101.9	94.9
09	FINISHES	88.1	99.1	94.1	98.2	89.9	93.7	90.5	96.5	93.8	96.2	100.8	98.7	89.4	89.9	89.7	91.9	96.5	94.4
COVERS	DIVS. 10 - 14, 25, 28, 41, 43, 44, 46	100.0	99.7	99.9	100.0	98.9	99.8	100.0	95.1	98.9	100.0	100.8	100.2	100.0	98.8	99.7	100.0	95.1	98.9
21, 22, 23	FIRE SUPPRESSION, PLUMBING & HVAC	96.9	81.3	90.5	100.9	67.8	87.3	96.9	58.9	81.3	96.7	67.8	84.8	101.1	67.8	87.4	96.9	58.9	81.3
26, 27, 3370	ELECTRICAL, COMMUNICATIONS & UTIL.	101.5	73.7	88.1	100.8	56.0	79.2	101.5	50.5	76.9	100.8	56.0	79.2	101.6	56.0	79.6	101.5	50.5	76.9
MF2018	WEIGHTED AVERAGE	94.1	88.8	92.0	99.8	82.4	92.7	93.8	77.9	87.3	98.0	85.3	92.8	96.5	81.7	90.5	94.8	77.8	87.9

VERMONT / VIRGINIA

DIVISION		WHITE RIVER JCT. 050 MAT.	INST.	TOTAL	ALEXANDRIA 223 MAT.	INST.	TOTAL	ARLINGTON 222 MAT.	INST.	TOTAL	BRISTOL 242 MAT.	INST.	TOTAL	CHARLOTTESVILLE 229 MAT.	INST.	TOTAL	CULPEPER 227 MAT.	INST.	TOTAL
015433	CONTRACTOR EQUIPMENT		92.3	92.3		102.1	102.1		100.9	100.9		100.9	100.9		104.8	104.8		100.9	100.9
0241, 31 - 34	SITE & INFRASTRUCTURE, DEMOLITION	87.6	90.3	89.4	123.9	84.6	97.2	135.5	82.5	99.6	118.4	82.7	94.1	123.1	84.3	96.8	121.3	82.2	94.7
0310	Concrete Forming & Accessories	92.5	91.1	91.4	92.7	71.2	75.3	89.9	60.9	66.5	83.9	58.1	63.1	81.7	59.8	64.0	77.8	67.2	69.2
0320	Concrete Reinforcing	81.0	81.0	81.0	84.0	88.9	85.6	94.9	86.7	92.2	95.2	68.8	86.7	94.5	70.2	86.7	94.9	88.8	92.9
0330	Cast-in-Place Concrete	73.1	104.3	83.9	116.7	74.8	102.2	113.5	74.6	100.1	113.1	48.6	90.9	117.7	71.9	101.9	116.2	72.7	101.2
03	CONCRETE	81.9	94.1	86.8	101.6	77.0	91.7	105.8	71.9	92.2	102.8	58.5	85.1	103.6	67.7	89.2	101.0	74.3	90.4
04	MASONRY	114.6	78.4	91.8	92.0	72.5	79.7	106.5	71.6	84.5	95.8	54.4	69.7	120.2	54.6	78.9	107.4	68.6	82.9
05	METALS	93.5	89.8	92.6	98.6	102.2	99.4	96.8	101.5	97.9	95.8	92.6	95.1	96.1	95.5	95.9	96.1	101.0	97.3
06	WOOD, PLASTICS & COMPOSITES	96.5	100.7	98.2	93.0	70.0	83.5	88.6	56.5	75.3	79.1	56.4	69.7	77.3	56.9	68.9	75.7	66.1	71.7
07	THERMAL & MOISTURE PROTECTION	101.8	80.2	92.3	104.5	77.9	92.9	106.4	75.9	93.0	105.5	59.2	85.2	105.0	66.5	88.1	105.2	75.3	92.1
08	OPENINGS	96.8	92.2	95.7	92.2	74.9	88.0	90.6	67.3	85.0	93.2	63.3	85.9	91.6	63.9	84.9	91.9	72.7	87.3
0920	Plaster & Gypsum Board	92.4	100.6	97.5	115.5	69.3	86.6	109.1	55.4	75.5	101.5	55.2	72.6	101.5	55.2	72.6	101.6	65.3	78.9
0950, 0980	Ceilings & Acoustic Treatment	84.8	100.6	95.1	105.1	69.3	81.9	102.4	55.4	71.9	101.7	55.2	71.6	101.7	55.2	71.6	102.4	65.3	78.3
0960	Flooring	89.8	110.1	96.0	101.5	73.4	93.0	100.2	75.5	92.6	97.3	53.7	83.9	96.0	53.7	83.0	96.0	69.2	87.8
0970, 0990	Wall Finishes & Painting/Coating	84.0	101.9	94.9	117.1	68.3	87.3	117.1	71.1	89.0	102.1	73.0	84.4	102.1	57.4	74.8	117.1	68.1	87.2
09	FINISHES	87.3	96.8	92.5	105.3	70.5	86.4	104.8	63.1	82.1	100.8	58.6	77.9	100.5	57.0	76.9	101.1	66.8	82.4
COVERS	DIVS. 10 - 14, 25, 28, 41, 43, 44, 46	100.0	95.5	99.0	100.0	87.1	97.2	100.0	86.6	97.1	100.0	82.0	96.1	100.0	83.1	96.3	100.0	87.1	97.2
21, 22, 23	FIRE SUPPRESSION, PLUMBING & HVAC	96.9	59.6	81.6	101.2	83.8	94.1	101.2	85.6	94.8	97.0	48.7	77.2	97.0	84.1	91.7	97.0	81.0	90.4
26, 27, 3370	ELECTRICAL, COMMUNICATIONS & UTIL.	101.5	50.4	76.9	96.0	97.0	96.5	94.3	99.7	96.9	95.6	33.7	65.7	95.5	69.3	82.9	97.1	93.7	95.4
MF2018	WEIGHTED AVERAGE	94.7	78.2	88.0	99.8	83.0	93.0	100.5	81.3	92.7	98.1	58.4	81.9	99.2	72.7	88.4	98.4	80.2	91.0

VIRGINIA

DIVISION		FAIRFAX 220-221 MAT.	INST.	TOTAL	FARMVILLE 239 MAT.	INST.	TOTAL	FREDERICKSBURG 224-225 MAT.	INST.	TOTAL	GRUNDY 246 MAT.	INST.	TOTAL	HARRISONBURG 228 MAT.	INST.	TOTAL	LYNCHBURG 245 MAT.	INST.	TOTAL
015433	CONTRACTOR EQUIPMENT		100.9	100.9		104.8	104.8		100.9	100.9		100.9	100.9		100.9	100.9		100.9	100.9
0241, 31 - 34	SITE & INFRASTRUCTURE, DEMOLITION	133.6	82.4	98.8	119.3	83.2	94.8	120.8	82.1	94.5	116.0	81.2	92.4	130.1	82.5	97.8	116.8	82.7	93.6
0310	Concrete Forming & Accessories	81.9	60.3	64.5	102.7	59.1	67.5	81.9	61.8	65.7	88.5	57.2	63.3	76.3	60.2	63.3	83.9	59.5	64.2
0320	Concrete Reinforcing	94.9	88.9	93.0	90.8	63.5	82.0	95.5	75.7	89.2	93.8	41.2	76.9	94.9	78.7	89.7	94.5	69.1	86.4
0330	Cast-in-Place Concrete	113.5	73.5	99.7	117.2	82.1	105.1	115.2	70.8	99.9	113.1	48.7	90.9	113.5	74.1	99.9	113.1	72.8	99.2
03	CONCRETE	105.3	71.6	91.8	105.0	69.5	90.8	100.9	69.0	88.2	101.8	53.0	82.2	103.3	70.0	90.0	101.6	67.6	88.0
04	MASONRY	106.3	71.7	84.5	101.4	55.4	72.4	106.6	79.5	89.5	96.4	55.3	70.5	104.8	70.4	83.1	111.4	54.6	75.6
05	METALS	96.2	101.8	97.5	94.2	88.9	92.9	96.2	97.0	96.4	95.3	73.8	90.7	96.1	98.1	96.6	96.0	94.5	95.6
06	WOOD, PLASTICS & COMPOSITES	79.1	55.8	69.5	96.7	56.9	80.2	79.1	61.0	71.6	82.7	56.4	71.8	74.4	56.5	67.0	79.1	56.5	69.8
07	THERMAL & MOISTURE PROTECTION	106.0	75.3	92.6	105.2	68.0	88.9	105.3	77.1	92.9	105.5	57.7	84.5	105.7	77.4	93.3	105.3	66.5	88.3
08	OPENINGS	90.6	67.0	84.9	91.1	52.0	81.6	91.7	65.9	85.4	93.2	43.5	81.2	91.9	65.4	85.5	91.9	63.4	85.0
0920	Plaster & Gypsum Board	101.6	54.7	72.3	119.5	55.2	79.3	101.6	60.0	75.6	101.5	55.2	72.6	101.5	55.4	72.7	101.5	55.4	72.7
0950, 0980	Ceilings & Acoustic Treatment	102.4	54.7	71.4	98.1	55.2	70.3	102.4	60.0	74.9	101.7	55.2	71.6	101.7	55.4	71.7	101.7	55.4	71.7
0960	Flooring	97.5	74.5	90.5	98.3	53.7	84.6	97.5	67.4	88.3	98.3	53.7	84.7	95.7	74.9	89.4	97.3	53.7	83.9
0970, 0990	Wall Finishes & Painting/Coating	117.1	68.3	87.3	96.8	53.1	70.1	117.1	52.9	77.9	102.1	29.0	57.5	117.1	57.4	80.7	102.1	73.0	84.4
09	FINISHES	102.8	62.1	80.7	101.8	56.8	77.4	101.4	60.9	79.4	100.9	53.8	75.3	101.7	61.3	79.8	100.6	58.4	77.7
COVERS	DIVS. 10 - 14, 25, 28, 41, 43, 44, 46	100.0	86.5	97.1	100.0	81.5	96.0	100.0	72.1	93.9	100.0	80.2	95.7	100.0	82.2	96.1	100.0	81.8	96.0
21, 22, 23	FIRE SUPPRESSION, PLUMBING & HVAC	97.0	81.8	90.8	97.2	51.9	78.6	97.0	74.1	87.6	97.0	61.7	82.5	97.0	73.4	87.3	97.0	84.0	91.7
26, 27, 3370	ELECTRICAL, COMMUNICATIONS & UTIL.	96.2	99.7	97.9	92.2	67.2	80.2	94.5	99.7	97.0	95.6	67.2	81.9	95.7	75.4	85.9	96.2	62.8	80.1
MF2018	WEIGHTED AVERAGE	99.3	80.3	91.6	98.1	64.6	84.5	98.2	78.1	90.0	97.9	61.9	83.3	98.8	74.2	88.8	98.5	71.6	87.5

VIRGINIA

DIVISION		NEWPORT NEWS 236 MAT.	INST.	TOTAL	NORFOLK 233-235 MAT.	INST.	TOTAL	PETERSBURG 238 MAT.	INST.	TOTAL	PORTSMOUTH 237 MAT.	INST.	TOTAL	PULASKI 243 MAT.	INST.	TOTAL	RICHMOND 230-232 MAT.	INST.	TOTAL
015433	CONTRACTOR EQUIPMENT		104.9	104.9		104.9	104.9		104.8	104.8		104.8	104.8		100.9	100.9		104.9	104.9
0241, 31 - 34	SITE & INFRASTRUCTURE, DEMOLITION	118.3	84.4	95.2	118.1	89.5	98.7	122.2	84.3	96.5	116.8	84.0	94.6	115.3	81.4	92.3	109.3	89.5	95.9
0310	Concrete Forming & Accessories	101.5	59.4	67.5	102.0	59.0	67.3	92.2	59.9	66.1	86.0	59.1	64.3	88.5	57.7	63.6	102.0	59.8	67.9
0320	Concrete Reinforcing	90.6	65.3	82.5	99.7	65.2	88.6	90.3	70.3	83.9	90.3	65.3	82.2	93.8	78.2	88.8	99.6	70.3	90.2
0330	Cast-in-Place Concrete	113.9	74.6	100.3	126.5	75.6	108.9	121.0	74.9	105.1	112.8	63.6	95.8	113.1	78.2	101.1	96.3	76.7	89.5
03	CONCRETE	102.3	67.6	88.4	109.7	67.5	92.8	107.0	68.7	91.7	100.6	63.6	85.2	101.8	70.1	89.1	96.4	69.1	85.5
04	MASONRY	95.5	54.6	69.7	99.7	54.7	71.4	110.0	55.8	75.9	101.8	54.6	72.0	91.9	52.1	66.8	98.7	56.0	71.8
05	METALS	96.8	93.2	95.9	98.0	88.9	95.9	94.3	95.7	94.6	95.7	92.8	95.0	95.8	94.0	95.4	98.6	91.6	97.0
06	WOOD, PLASTICS & COMPOSITES	95.2	57.1	79.4	95.1	56.9	79.3	83.9	56.9	72.7	79.5	56.9	70.1	82.7	56.4	71.8	96.3	56.9	80.0
07	THERMAL & MOISTURE PROTECTION	105.2	64.7	87.4	104.9	67.8	88.7	105.1	67.2	88.5	105.0	65.0	87.5	105.5	66.6	88.4	101.1	69.0	87.0
08	OPENINGS	91.3	62.4	84.3	92.0	62.7	84.9	90.8	63.9	84.3	91.4	62.3	84.4	93.2	54.7	83.9	98.0	63.8	89.8
0920	Plaster & Gypsum Board	121.1	54.5	80.0	111.6	55.4	76.5	106.9	55.2	74.6	108.0	55.2	75.0	101.5	55.2	72.6	117.2	55.4	78.6
0950, 0980	Ceilings & Acoustic Treatment	105.7	54.5	73.0	96.3	55.4	69.7	100.2	55.2	71.0	105.7	55.2	72.9	101.7	55.2	71.6	101.7	55.4	71.7
0960	Flooring	98.3	53.7	84.6	103.7	53.7	88.4	94.9	55.8	82.9	92.4	53.7	80.5	98.3	53.7	84.7	91.5	55.8	80.6
0970, 0990	Wall Finishes & Painting/Coating	96.8	56.5	72.2	98.7	57.4	73.5	96.8	57.4	72.7	96.8	57.4	72.7	102.1	44.8	67.1	98.7	57.4	73.5
09	FINISHES	103.5	57.0	78.3	102.4	56.9	77.7	99.6	57.5	76.7	99.8	57.0	76.6	100.9	54.7	75.8	101.2	57.4	77.4
COVERS	DIVS. 10 - 14, 25, 28, 41, 43, 44, 46	100.0	83.6	96.4	100.0	83.1	96.3	100.0	83.1	96.3	100.0	83.5	96.4	100.0	81.1	95.9	100.0	83.1	96.3
21, 22, 23	FIRE SUPPRESSION, PLUMBING & HVAC	101.4	64.1	86.1	100.8	63.6	85.5	97.2	84.1	91.8	101.4	64.1	86.1	97.0	62.2	82.7	100.8	84.0	93.9
26, 27, 3370	ELECTRICAL, COMMUNICATIONS & UTIL.	93.9	67.0	80.9	99.7	60.0	80.6	94.0	69.3	82.1	92.9	60.0	77.0	95.6	77.2	86.7	99.8	69.3	85.1
MF2018	WEIGHTED AVERAGE	99.1	67.8	86.4	101.0	67.6	87.1	98.6	73.0	88.2	98.3	66.2	85.3	97.7	64.1	85.7	99.4	73.2	88.7

VIRGINIA / WASHINGTON

DIVISION		ROANOKE 240-241 MAT.	INST.	TOTAL	STAUNTON 244 MAT.	INST.	TOTAL	WINCHESTER 226 MAT.	INST.	TOTAL	CLARKSTON 994 MAT.	INST.	TOTAL	EVERETT 982 MAT.	INST.	TOTAL	OLYMPIA 985 MAT.	INST.	TOTAL
015433	CONTRACTOR EQUIPMENT		100.9	100.9		104.8	104.8		100.9	100.9		87.7	87.7		96.5	96.5		99.5	99.5
0241, 31 - 34	SITE & INFRASTRUCTURE, DEMOLITION	116.2	82.8	93.5	119.4	83.1	94.7	128.7	82.1	97.0	110.5	81.9	91.1	106.1	101.6	103.1	107.9	105.6	106.3
0310	Concrete Forming & Accessories	97.7	59.7	67.1	88.0	59.0	64.6	79.7	62.5	65.8	87.0	60.6	65.7	119.6	106.9	109.4	101.8	106.9	105.9
0320	Concrete Reinforcing	94.9	69.1	86.6	94.5	67.7	85.9	94.2	74.7	88.0	89.4	99.8	92.8	95.6	118.4	102.9	102.5	118.0	107.5
0330	Cast-in-Place Concrete	128.6	82.1	112.5	117.7	81.9	105.3	113.5	59.8	95.0	97.6	77.7	90.7	111.6	109.7	110.9	98.3	111.7	102.9
03	CONCRETE	107.2	71.0	92.7	103.4	70.2	90.1	103.0	65.3	87.9	91.4	73.9	84.4	99.0	109.4	103.1	97.9	109.8	102.6
04	MASONRY	98.2	56.4	71.8	106.8	55.4	74.4	102.2	65.1	78.8	109.6	85.6	94.5	104.1	104.7	104.5	107.5	99.8	103.0
05	METALS	98.5	94.4	97.5	96.0	90.0	94.6	96.2	95.3	96.0	95.6	88.7	94.0	106.4	101.7	105.3	107.5	99.0	105.5
06	WOOD, PLASTICS & COMPOSITES	94.1	56.5	78.6	82.7	56.9	72.0	77.3	61.1	70.6	90.6	55.2	75.9	115.9	106.6	112.1	97.7	106.8	101.5
07	THERMAL & MOISTURE PROTECTION	105.3	68.7	89.3	105.1	67.9	88.8	105.8	72.0	91.0	110.5	76.8	95.7	103.7	107.8	105.5	101.8	106.7	103.9
08	OPENINGS	92.2	63.4	85.2	91.9	52.9	82.5	93.2	64.3	86.2	112.5	63.5	100.7	102.4	110.1	104.3	105.9	110.3	107.0
0920	Plaster & Gypsum Board	115.5	55.4	77.9	101.5	55.2	72.6	101.6	60.2	75.7	142.0	54.0	87.0	125.0	107.1	113.8	111.8	107.1	108.9
0950, 0980	Ceilings & Acoustic Treatment	105.1	55.4	72.9	101.7	55.2	71.6	102.4	60.2	75.0	88.0	54.0	66.0	105.9	107.1	106.7	116.8	107.1	110.5
0960	Flooring	101.5	55.3	87.4	97.9	53.7	84.4	97.0	69.2	88.5	84.8	72.1	80.9	102.5	98.9	101.4	100.0	89.4	96.8
0970, 0990	Wall Finishes & Painting/Coating	102.1	73.0	84.4	102.1	28.5	57.2	117.1	72.0	89.6	78.9	66.9	71.6	92.2	94.4	93.5	101.8	94.9	97.6
09	FINISHES	104.2	59.0	79.7	100.8	54.1	75.4	102.1	64.0	81.4	99.2	61.6	78.8	107.6	103.6	105.4	106.7	101.9	104.1
COVERS	DIVS. 10 - 14, 25, 28, 41, 43, 44, 46	100.0	82.1	96.1	100.0	81.5	96.0	100.0	76.1	94.8	100.0	93.0	98.5	100.0	105.2	101.1	100.0	103.1	100.7
21, 22, 23	FIRE SUPPRESSION, PLUMBING & HVAC	101.2	65.7	86.7	97.0	55.7	80.0	97.0	75.5	88.2	96.5	74.7	87.6	101.0	107.3	103.6	100.9	105.2	102.7
26, 27, 3370	ELECTRICAL, COMMUNICATIONS & UTIL.	95.6	57.4	77.2	94.9	75.7	85.6	94.7	85.1	90.1	82.8	88.6	85.6	108.0	107.7	107.9	105.3	104.6	105.0
MF2018	WEIGHTED AVERAGE	100.5	67.7	87.2	98.5	66.5	85.5	98.7	74.5	88.9	97.2	77.8	89.3	103.6	106.0	104.6	103.3	104.5	103.8

City Cost Indexes

WASHINGTON

DIVISION		RICHLAND 993			SEATTLE 980 - 981,987			SPOKANE 990 - 992			TACOMA 983 - 984			VANCOUVER 986			WENATCHEE 988		
		MAT.	INST.	TOTAL	MAT.	INST.	TOTAL	MAT.	INST.	TOTAL	MAT.	INST.	TOTAL	MAT.	INST.	TOTAL	MAT.	INST.	TOTAL
015433	CONTRACTOR EQUIPMENT		87.7	87.7		99.0	99.0		87.7	87.7		96.5	96.5		93.1	93.1		96.5	96.5
0241, 31 - 34	SITE & INFRASTRUCTURE, DEMOLITION	113.1	82.9	92.6	111.8	102.7	105.6	112.8	82.9	92.5	109.6	101.5	104.1	123.0	89.2	100.1	121.4	98.8	106.1
0310	Concrete Forming & Accessories	87.1	83.4	84.1	103.4	107.4	106.7	94.5	83.0	85.2	106.2	106.9	106.7	105.5	98.5	99.9	108.3	77.5	83.4
0320	Concrete Reinforcing	85.8	99.7	90.3	102.0	115.8	106.4	86.4	100.5	91.0	94.9	118.2	102.4	95.4	118.4	102.8	95.4	89.4	93.4
0330	Cast-in-Place Concrete	97.9	86.1	93.8	117.6	111.4	115.5	101.7	86.0	96.3	114.8	109.6	113.0	128.2	97.7	117.7	117.1	86.2	106.4
03	CONCRETE	90.8	87.2	89.3	107.8	109.8	108.6	93.2	87.1	90.8	100.3	109.3	103.9	109.6	101.5	106.3	106.9	82.9	97.3
04	MASONRY	110.7	85.7	94.9	109.5	105.0	106.7	111.4	100.0	104.2	106.0	103.0	104.1	106.0	104.2	104.9	109.5	88.2	96.1
05	METALS	95.9	90.0	94.5	105.7	101.6	104.7	97.3	90.8	95.7	108.7	101.4	107.0	108.2	101.9	106.8	105.7	88.4	101.7
06	WOOD, PLASTICS & COMPOSITES	90.7	81.7	87.0	102.3	106.8	104.1	100.0	81.7	92.4	102.7	106.6	104.3	97.1	98.5	97.7	105.3	76.1	93.2
07	THERMAL & MOISTURE PROTECTION	112.6	85.8	100.8	102.5	109.4	105.5	110.7	87.3	100.4	103.4	107.0	105.0	104.6	98.9	102.1	103.3	82.6	94.2
08	OPENINGS	110.7	78.6	102.9	105.1	109.6	106.2	111.2	78.2	103.3	103.0	110.2	104.8	98.4	104.0	99.7	102.6	72.5	95.4
0920	Plaster & Gypsum Board	142.0	81.2	104.0	110.4	107.1	108.3	135.6	81.2	101.6	113.7	107.1	109.6	111.5	99.0	103.7	121.9	75.8	93.1
0950, 0980	Ceilings & Acoustic Treatment	93.4	81.2	85.5	107.4	107.1	107.2	90.3	81.2	84.4	110.0	107.1	108.1	106.4	99.0	101.6	101.6	75.8	84.8
0960	Flooring	85.1	79.7	83.4	96.3	101.9	98.0	83.4	98.9	88.1	97.2	98.9	97.7	104.9	102.5	104.2	99.4	72.1	91.0
0970, 0990	Wall Finishes & Painting/Coating	78.9	63.6	69.6	98.7	94.4	96.0	79.1	77.1	77.9	92.2	94.9	93.9	93.3	84.5	87.9	92.2	65.8	76.1
09	FINISHES	100.6	79.7	89.3	102.8	104.4	103.6	98.5	85.1	91.2	105.4	103.6	104.5	103.6	97.5	100.3	106.9	74.2	89.1
COVERS	DIVS. 10 - 14, 25, 28, 41, 43, 44, 46	100.0	97.7	99.5	100.0	103.1	100.7	100.0	97.6	99.5	100.0	102.6	100.6	100.0	97.8	99.5	100.0	95.5	99.0
21, 22, 23	FIRE SUPPRESSION, PLUMBING & HVAC	100.8	109.5	104.4	101.0	120.2	108.9	100.8	84.3	94.0	101.1	105.2	102.8	101.1	107.5	103.7	96.9	85.3	92.1
26, 27, 3370	ELECTRICAL, COMMUNICATIONS & UTIL.	79.4	102.0	90.3	104.0	120.7	112.0	80.2	83.3	81.7	107.9	104.6	106.4	117.3	105.0	111.4	108.4	87.6	98.4
MF2018	WEIGHTED AVERAGE	97.9	92.7	95.8	104.3	110.8	106.9	98.6	86.9	93.8	104.1	104.8	104.4	106.1	102.0	104.4	104.1	85.1	96.4

WASHINGTON / WEST VIRGINIA

DIVISION		YAKIMA 989			BECKLEY 258 - 259			BLUEFIELD 247 - 248			BUCKHANNON 262			CHARLESTON 250 - 253			CLARKSBURG 263 - 264		
		MAT.	INST.	TOTAL	MAT.	INST.	TOTAL	MAT.	INST.	TOTAL	MAT.	INST.	TOTAL	MAT.	INST.	TOTAL	MAT.	INST.	TOTAL
015433	CONTRACTOR EQUIPMENT		96.5	96.5		100.9	100.9		100.9	100.9		100.9	100.9		103.8	103.8		100.9	100.9
0241, 31 - 34	SITE & INFRASTRUCTURE, DEMOLITION	113.0	100.9	104.8	108.8	83.1	91.4	109.2	83.1	91.5	116.0	83.5	93.9	106.9	91.5	96.4	116.9	83.5	94.2
0310	Concrete Forming & Accessories	106.7	80.8	85.8	83.9	82.3	82.6	84.0	82.1	82.4	83.1	80.8	81.3	103.5	87.4	90.5	79.6	81.1	80.8
0320	Concrete Reinforcing	95.2	100.7	96.9	94.4	83.2	90.8	93.5	76.8	88.2	94.1	81.5	90.1	99.9	83.5	94.6	94.1	95.8	94.7
0330	Cast-in-Place Concrete	122.6	88.0	110.6	108.9	86.9	101.3	110.6	86.8	102.4	110.3	89.6	103.2	105.4	88.5	99.5	120.9	85.7	108.7
03	CONCRETE	104.9	86.9	97.7	96.0	85.3	91.7	98.1	84.1	92.5	100.4	85.3	94.3	98.0	88.0	94.0	104.5	86.5	97.3
04	MASONRY	100.0	100.4	100.3	96.4	83.2	88.1	93.5	83.2	87.0	104.6	88.1	94.2	90.8	87.7	88.8	109.1	88.1	95.9
05	METALS	106.2	94.3	103.4	91.3	101.9	93.8	96.0	99.5	96.8	96.2	101.5	97.5	90.0	99.1	92.1	96.2	106.8	98.7
06	WOOD, PLASTICS & COMPOSITES	103.4	76.1	92.1	80.6	82.5	81.4	81.2	82.5	81.7	80.3	77.7	79.2	95.7	88.0	92.5	76.0	77.7	76.7
07	THERMAL & MOISTURE PROTECTION	103.6	90.3	97.8	114.7	81.1	100.0	105.2	81.1	94.6	105.6	84.3	96.3	110.5	86.1	99.8	105.4	84.1	96.0
08	OPENINGS	102.6	86.8	98.8	91.3	78.8	88.3	93.6	77.4	89.7	93.8	75.8	89.3	91.6	81.9	89.2	93.6	81.4	90.7
0920	Plaster & Gypsum Board	114.9	75.8	90.4	98.1	82.1	88.1	100.0	82.1	88.8	100.8	77.2	86.0	114.8	87.7	97.8	96.4	77.2	84.4
0950, 0980	Ceilings & Acoustic Treatment	103.2	75.8	85.4	88.0	82.1	84.2	98.3	82.1	87.8	101.7	77.2	85.8	98.6	87.7	91.5	101.7	77.2	85.8
0960	Flooring	97.8	79.7	92.3	94.7	97.1	95.4	95.2	97.1	95.8	95.0	92.3	94.2	97.1	97.1	97.1	94.1	92.3	93.6
0970, 0990	Wall Finishes & Painting/Coating	92.2	77.1	83.0	98.9	85.5	90.7	102.1	83.3	90.6	102.1	86.7	92.7	101.8	90.0	94.6	102.1	86.7	92.7
09	FINISHES	104.5	78.2	90.2	94.6	85.6	89.7	98.5	85.3	91.3	100.0	83.1	90.8	102.1	89.7	95.4	98.9	83.1	90.3
COVERS	DIVS. 10 - 14, 25, 28, 41, 43, 44, 46	100.0	97.1	99.4	100.0	88.9	97.6	100.0	91.0	98.0	100.0	92.1	98.3	100.0	92.4	98.3	100.0	92.1	98.3
21, 22, 23	FIRE SUPPRESSION, PLUMBING & HVAC	101.1	110.7	105.0	97.2	87.6	93.3	97.0	77.9	89.2	97.0	92.4	95.1	100.8	88.7	95.9	97.0	92.7	95.2
26, 27, 3370	ELECTRICAL, COMMUNICATIONS & UTIL.	109.7	102.0	106.0	92.9	78.0	85.7	94.9	78.0	86.7	95.8	92.5	94.2	99.6	85.1	92.6	95.8	92.5	94.2
MF2018	WEIGHTED AVERAGE	104.2	96.4	101.0	95.4	85.6	91.4	96.9	83.1	91.3	98.1	88.9	94.4	97.2	89.0	93.9	98.8	89.8	95.1

WEST VIRGINIA

DIVISION		GASSAWAY 266			HUNTINGTON 255 - 257			LEWISBURG 249			MARTINSBURG 254			MORGANTOWN 265			PARKERSBURG 261		
		MAT.	INST.	TOTAL	MAT.	INST.	TOTAL	MAT.	INST.	TOTAL	MAT.	INST.	TOTAL	MAT.	INST.	TOTAL	MAT.	INST.	TOTAL
015433	CONTRACTOR EQUIPMENT		100.9	100.9		100.9	100.9		100.9	100.9		100.9	100.9		100.9	100.9		100.9	100.9
0241, 31 - 34	SITE & INFRASTRUCTURE, DEMOLITION	113.2	83.5	93.0	114.2	84.5	94.0	126.6	83.1	97.1	113.0	83.5	93.0	110.2	84.7	92.9	119.8	84.4	95.7
0310	Concrete Forming & Accessories	82.4	83.3	83.1	100.6	86.1	88.9	80.1	82.1	81.7	83.9	74.7	76.5	80.1	81.4	81.2	86.0	87.5	87.2
0320	Concrete Reinforcing	94.1	85.3	91.3	95.8	88.3	93.4	94.1	76.9	88.6	94.4	88.5	92.5	94.1	95.9	94.7	93.5	81.4	89.6
0330	Cast-in-Place Concrete	115.5	88.5	106.2	118.7	91.8	109.4	110.7	86.8	102.4	114.1	82.6	103.2	110.3	94.8	104.9	112.7	90.0	104.9
03	CONCRETE	101.2	86.7	95.4	101.8	89.6	96.9	105.8	84.1	96.9	99.2	81.2	92.0	97.8	89.8	94.6	102.3	88.5	96.7
04	MASONRY	109.5	88.0	96.0	94.8	89.8	91.7	95.7	83.2	87.8	97.6	78.4	85.5	126.0	88.1	102.1	83.2	81.2	82.0
05	METALS	96.2	102.7	97.7	94.0	104.4	96.4	96.1	99.5	96.9	91.7	101.5	94.0	96.3	107.1	98.8	96.7	101.7	97.9
06	WOOD, PLASTICS & COMPOSITES	78.9	81.3	79.9	95.2	85.3	91.1	76.3	82.5	78.9	80.6	74.2	77.9	76.3	77.7	76.9	80.8	88.8	84.1
07	THERMAL & MOISTURE PROTECTION	105.2	85.0	96.4	115.1	84.8	101.8	106.2	81.1	95.2	114.9	74.9	97.3	105.2	85.4	96.5	105.5	83.6	95.9
08	OPENINGS	92.1	78.6	88.8	90.5	81.4	88.3	93.6	77.4	89.7	92.9	69.5	87.2	94.7	81.4	91.5	92.6	81.9	90.0
0920	Plaster & Gypsum Board	98.3	80.9	87.4	110.4	85.0	94.5	96.4	82.1	87.4	98.1	73.5	82.7	96.4	77.2	84.4	101.5	88.6	93.4
0950, 0980	Ceilings & Acoustic Treatment	101.7	80.9	88.2	88.0	85.0	86.0	101.7	82.1	89.0	88.0	73.5	78.6	101.7	77.2	85.8	101.7	88.6	93.2
0960	Flooring	94.8	79.7	91.1	101.7	97.1	100.5	94.2	97.1	95.1	94.7	92.0	93.9	94.2	92.3	93.6	97.6	97.9	97.7
0970, 0990	Wall Finishes & Painting/Coating	102.1	90.0	94.7	98.9	86.9	91.6	102.1	59.8	76.3	102.1	77.7	86.0	102.1	86.7	92.7	102.1	89.0	94.1
09	FINISHES	99.2	86.6	92.3	98.5	88.3	93.0	100.3	82.7	90.8	94.9	77.8	85.6	98.4	83.1	90.1	100.9	89.8	94.9
COVERS	DIVS. 10 - 14, 25, 28, 41, 43, 44, 46	100.0	92.5	98.4	100.0	92.4	98.4	100.0	88.9	97.6	100.0	74.9	94.5	100.0	92.2	98.3	100.0	91.8	98.2
21, 22, 23	FIRE SUPPRESSION, PLUMBING & HVAC	97.0	80.3	90.2	101.4	89.1	96.4	97.0	87.5	93.1	97.2	75.0	88.1	97.0	92.7	95.2	101.2	88.1	95.8
26, 27, 3370	ELECTRICAL, COMMUNICATIONS & UTIL.	95.8	85.1	90.6	95.3	86.4	91.0	93.4	78.0	85.9	96.5	71.2	84.3	95.9	92.5	94.3	95.9	87.8	92.0
MF2018	WEIGHTED AVERAGE	98.1	86.1	93.3	98.3	89.4	94.7	98.5	84.8	92.9	96.6	78.8	89.4	98.4	90.4	95.2	98.6	88.3	94.4

City Cost Indexes

DIVISION		WEST VIRGINIA PETERSBURG 268			ROMNEY 267			WHEELING 260			WISCONSIN BELOIT 535			EAU CLAIRE 547			GREEN BAY 541 - 543		
		MAT.	INST.	TOTAL	MAT.	INST.	TOTAL	MAT.	INST.	TOTAL	MAT.	INST.	TOTAL	MAT.	INST.	TOTAL	MAT.	INST.	TOTAL
015433	CONTRACTOR EQUIPMENT		100.9	100.9		100.9	100.9		100.9	100.9		96.0	96.0		96.7	96.7		94.5	94.5
0241, 31 - 34	SITE & INFRASTRUCTURE, DEMOLITION	109.5	84.4	92.4	112.7	84.4	93.5	120.5	84.4	96.0	83.3	96.8	92.5	84.9	95.9	92.4	87.5	92.1	90.6
0310	Concrete Forming & Accessories	84.6	84.1	84.2	79.1	83.9	83.0	88.6	80.3	81.9	90.0	91.0	90.8	94.9	96.4	96.1	111.4	106.7	107.6
0320	Concrete Reinforcing	93.5	85.3	90.9	94.1	88.8	92.4	92.9	95.5	93.9	101.3	139.9	113.7	93.3	113.3	99.7	92.4	110.2	98.1
0330	Cast-in-Place Concrete	110.3	89.6	103.1	115.5	81.6	103.8	112.7	91.2	105.3	89.7	94.2	91.3	110.0	102.3	107.4	113.8	101.9	109.7
03	CONCRETE	97.9	87.4	93.7	101.0	85.1	94.7	102.3	88.1	96.6	92.9	100.8	96.1	98.7	101.7	99.9	102.3	105.7	103.6
04	MASONRY	99.6	88.1	92.4	96.2	88.0	91.0	107.7	85.5	93.7	92.2	107.0	101.5	92.2	108.2	102.3	121.0	102.1	109.1
05	METALS	96.3	102.7	97.8	96.4	103.7	98.1	96.9	107.0	99.2	92.3	115.4	97.7	86.4	108.3	91.6	90.7	107.3	94.6
06	WOOD, PLASTICS & COMPOSITES	81.5	82.5	81.9	75.3	82.5	78.3	82.7	77.7	80.7	90.8	88.8	90.0	98.1	94.2	96.5	111.1	108.0	109.8
07	THERMAL & MOISTURE PROTECTION	105.3	80.4	94.4	105.3	80.3	94.3	105.7	83.1	96.0	94.5	90.7	92.8	100.1	103.3	101.5	103.0	102.7	102.9
08	OPENINGS	94.7	78.7	90.8	94.6	79.5	91.0	93.3	81.4	90.4	95.5	104.5	97.7	100.0	95.3	98.8	95.6	107.4	98.5
0920	Plaster & Gypsum Board	100.8	82.1	89.1	95.9	82.1	87.3	101.5	77.2	86.3	86.7	89.1	88.2	96.9	94.6	95.5	102.7	108.8	106.5
0950, 0980	Ceilings & Acoustic Treatment	101.7	82.1	89.0	101.7	82.1	89.0	101.7	77.2	85.8	87.8	89.1	88.7	90.6	94.6	93.2	83.7	108.8	100.0
0960	Flooring	95.7	92.3	94.7	94.0	92.0	93.4	98.3	92.3	96.5	94.9	110.2	99.6	83.8	119.6	94.8	99.1	121.2	105.9
0970, 0990	Wall Finishes & Painting/Coating	102.1	86.7	92.7	102.1	86.7	92.7	102.1	86.7	92.7	94.5	96.1	95.5	86.1	83.1	84.3	92.8	123.1	111.3
09	FINISHES	99.5	85.9	92.1	98.5	85.8	91.6	101.1	82.5	91.0	91.9	94.4	93.3	88.9	98.9	94.4	93.4	111.6	103.2
COVERS	DIVS. 10 - 14, 25, 28, 41, 43, 44, 46	100.0	92.6	98.4	100.0	92.6	98.4	100.0	88.5	97.5	100.0	97.6	99.5	100.0	96.4	99.2	100.0	98.6	99.7
21, 22, 23	FIRE SUPPRESSION, PLUMBING & HVAC	97.0	80.8	90.3	97.0	80.5	90.2	101.2	91.3	97.2	98.7	88.8	94.6	100.8	88.5	95.8	101.1	95.5	98.8
26, 27, 3370	ELECTRICAL, COMMUNICATIONS & UTIL.	97.8	71.3	85.0	97.3	71.3	84.8	94.2	92.5	93.4	101.7	78.1	90.3	100.5	102.4	101.4	97.7	104.4	100.9
MF2018	WEIGHTED AVERAGE	97.7	84.3	92.2	97.9	84.0	92.2	99.6	89.3	95.4	95.4	95.5	95.4	95.8	99.0	97.1	98.3	102.6	100.1

DIVISION		WISCONSIN KENOSHA 531			LA CROSSE 546			LANCASTER 538			MADISON 537			MILWAUKEE 530,532			NEW RICHMOND 540		
		MAT.	INST.	TOTAL	MAT.	INST.	TOTAL	MAT.	INST.	TOTAL	MAT.	INST.	TOTAL	MAT.	INST.	TOTAL	MAT.	INST.	TOTAL
015433	CONTRACTOR EQUIPMENT		94.1	94.1		96.7	96.7		96.0	96.0		100.2	100.2		88.6	88.6		97.0	97.0
0241, 31 - 34	SITE & INFRASTRUCTURE, DEMOLITION	88.3	95.3	93.0	79.6	95.9	90.7	81.9	96.3	91.7	82.7	104.2	97.3	78.1	93.8	88.7	84.5	94.7	91.4
0310	Concrete Forming & Accessories	104.7	111.9	110.5	76.6	96.5	92.6	89.1	89.9	89.7	95.2	105.9	103.8	95.3	116.6	112.5	87.3	87.1	87.1
0320	Concrete Reinforcing	101.2	113.8	105.3	93.2	108.9	98.2	102.0	99.8	101.2	99.6	109.1	102.6	99.6	113.9	104.2	91.9	107.2	96.8
0330	Cast-in-Place Concrete	97.7	106.1	100.6	98.9	102.3	100.1	89.2	93.7	90.8	82.6	107.4	91.2	83.2	111.0	92.8	114.5	96.4	108.3
03	CONCRETE	97.4	110.0	102.5	90.0	101.0	94.4	92.6	93.3	92.9	93.3	106.8	98.7	93.5	113.2	101.4	97.0	94.4	96.0
04	MASONRY	90.5	113.0	104.7	91.1	102.7	98.4	92.3	107.0	101.6	91.6	105.6	100.4	94.7	117.4	109.0	115.8	107.1	110.3
05	METALS	94.2	107.1	97.2	86.3	106.5	91.1	89.4	99.8	91.8	96.1	102.5	97.6	97.4	97.5	97.5	88.2	104.6	92.1
06	WOOD, PLASTICS & COMPOSITES	101.5	112.2	106.0	79.8	94.2	85.8	89.9	88.8	89.5	93.3	105.9	98.5	99.4	115.9	106.2	85.9	85.1	85.6
07	THERMAL & MOISTURE PROTECTION	95.6	107.7	100.9	99.2	101.1	100.1	94.2	90.5	92.6	92.9	105.7	98.5	92.9	112.5	101.5	99.8	97.0	98.6
08	OPENINGS	89.9	111.2	95.0	99.9	94.0	98.5	91.8	86.5	90.5	99.4	106.4	101.1	102.8	113.2	105.3	85.9	85.0	85.7
0920	Plaster & Gypsum Board	82.5	113.2	101.7	89.4	94.6	92.6	85.4	89.1	87.7	89.8	106.3	100.1	89.1	116.4	106.2	81.5	85.4	83.9
0950, 0980	Ceilings & Acoustic Treatment	87.8	113.2	104.3	89.2	94.6	92.7	81.6	89.1	86.5	95.4	106.3	102.5	92.0	116.4	107.8	58.9	85.4	76.1
0960	Flooring	112.9	115.3	113.7	78.8	121.2	91.8	94.7	103.2	97.3	95.6	119.3	102.9	100.2	115.1	104.7	93.4	106.5	97.4
0970, 0990	Wall Finishes & Painting/Coating	104.0	119.3	113.3	86.1	122.7	108.4	94.5	90.6	92.2	101.8	102.0	101.9	101.8	122.7	114.6	97.1	75.8	84.1
09	FINISHES	96.7	113.9	106.0	85.7	103.6	95.4	90.3	92.2	91.3	95.0	108.3	102.2	95.6	117.5	107.5	82.7	89.2	86.2
COVERS	DIVS. 10 - 14, 25, 28, 41, 43, 44, 46	100.0	100.6	100.1	100.0	99.6	99.9	100.0	97.6	99.5	100.0	101.8	100.4	100.0	103.7	100.8	100.0	87.5	97.3
21, 22, 23	FIRE SUPPRESSION, PLUMBING & HVAC	101.1	99.5	100.5	100.8	88.5	95.8	94.6	82.7	89.7	96.4	97.6	96.9	96.4	106.7	100.6	95.9	82.4	90.4
26, 27, 3370	ELECTRICAL, COMMUNICATIONS & UTIL.	102.6	99.1	100.9	100.7	104.4	102.5	101.3	78.7	90.4	100.2	102.3	101.2	100.5	102.3	101.4	99.6	78.7	89.5
MF2018	WEIGHTED AVERAGE	97.0	105.3	100.4	93.9	99.0	96.0	93.3	90.8	92.3	96.1	103.5	99.1	96.9	108.0	101.4	93.9	90.5	92.5

DIVISION		WISCONSIN OSHKOSH 549			PORTAGE 539			RACINE 534			RHINELANDER 545			SUPERIOR 548			WAUSAU 544		
		MAT.	INST.	TOTAL	MAT.	INST.	TOTAL	MAT.	INST.	TOTAL	MAT.	INST.	TOTAL	MAT.	INST.	TOTAL	MAT.	INST.	TOTAL
015433	CONTRACTOR EQUIPMENT		94.5	94.5		96.0	96.0		96.0	96.0		94.5	94.5		97.0	97.0		94.5	94.5
0241, 31 - 34	SITE & INFRASTRUCTURE, DEMOLITION	79.9	91.0	87.5	74.9	97.0	89.9	83.9	98.7	94.0	89.6	90.6	90.3	81.6	94.5	90.3	76.5	91.7	86.8
0310	Concrete Forming & Accessories	87.9	90.4	89.9	79.4	90.5	88.4	90.4	111.9	107.8	84.7	89.3	88.4	84.6	83.6	83.8	87.2	92.1	91.1
0320	Concrete Reinforcing	92.5	103.2	95.9	102.0	99.9	101.3	101.3	113.8	105.3	92.5	100.0	94.9	91.9	110.0	97.7	92.5	109.2	97.9
0330	Cast-in-Place Concrete	105.5	93.3	101.3	76.6	92.5	82.1	88.0	105.9	94.2	119.7	92.0	110.1	108.0	92.3	102.6	98.3	87.0	94.4
03	CONCRETE	92.6	94.1	93.2	82.3	93.3	86.7	92.1	110.0	99.3	103.4	92.6	99.1	92.0	91.9	92.0	87.9	93.7	90.2
04	MASONRY	104.9	108.5	107.1	91.0	107.0	101.1	92.4	113.0	105.4	125.1	106.4	113.3	114.8	104.9	108.5	104.2	106.4	105.6
05	METALS	88.5	103.4	92.0	90.0	103.8	93.3	93.8	107.1	96.9	88.0	101.9	91.3	89.2	106.0	93.1	87.9	105.9	92.1
06	WOOD, PLASTICS & COMPOSITES	86.1	88.9	87.2	78.4	88.8	82.7	91.1	112.2	99.9	83.0	88.9	85.4	84.0	80.9	82.7	85.4	91.7	88.0
07	THERMAL & MOISTURE PROTECTION	101.7	82.8	93.4	93.5	96.5	94.8	94.6	107.9	100.4	102.6	81.1	93.2	99.3	91.9	96.1	101.5	92.3	97.5
08	OPENINGS	92.4	90.6	92.0	92.0	93.8	92.4	95.5	111.2	99.3	92.4	86.5	91.0	85.3	90.3	86.5	92.6	93.6	92.8
0920	Plaster & Gypsum Board	81.7	89.1	86.4	73.0	89.1	83.1	86.7	113.2	103.3	81.7	89.1	86.4	81.8	81.1	81.4	81.7	92.0	88.2
0950, 0980	Ceilings & Acoustic Treatment	83.7	89.1	87.2	85.7	89.1	87.9	87.8	113.2	104.3	83.7	89.1	87.2	60.3	81.1	73.8	83.7	92.0	89.1
0960	Flooring	91.9	106.8	96.5	91.6	106.5	96.1	94.9	115.3	101.2	91.4	106.5	96.0	94.6	114.5	100.7	91.8	106.5	96.3
0970, 0990	Wall Finishes & Painting/Coating	90.9	103.5	98.6	94.5	90.6	92.2	94.5	121.6	111.0	90.9	74.2	80.7	84.4	98.1	92.8	90.9	77.2	82.6
09	FINISHES	87.4	94.9	91.5	87.8	93.1	90.7	91.9	114.1	104.0	88.1	91.1	89.8	82.3	90.3	86.6	87.1	93.1	90.4
COVERS	DIVS. 10 - 14, 25, 28, 41, 43, 44, 46	100.0	96.1	99.1	100.0	97.7	99.5	100.0	100.6	100.1	100.0	95.3	99.0	100.0	86.4	97.0	100.0	95.7	99.1
21, 22, 23	FIRE SUPPRESSION, PLUMBING & HVAC	96.9	82.1	90.8	94.6	88.8	92.2	100.9	99.5	100.3	96.9	82.1	90.8	95.9	84.5	91.3	96.9	82.7	91.1
26, 27, 3370	ELECTRICAL, COMMUNICATIONS & UTIL.	99.9	78.1	89.4	103.9	87.2	95.8	102.5	96.1	99.4	99.5	72.4	86.4	102.6	91.3	97.2	100.6	72.4	87.0
MF2018	WEIGHTED AVERAGE	94.0	91.0	92.7	91.8	94.3	92.8	96.2	105.2	99.8	96.4	88.9	93.4	93.5	92.5	93.0	93.1	90.5	92.1

WYOMING

DIVISION		CASPER 826 MAT.	INST.	TOTAL	CHEYENNE 820 MAT.	INST.	TOTAL	NEWCASTLE 827 MAT.	INST.	TOTAL	RAWLINS 823 MAT.	INST.	TOTAL	RIVERTON 825 MAT.	INST.	TOTAL	ROCK SPRINGS 829 - 831 MAT.	INST.	TOTAL
015433	CONTRACTOR EQUIPMENT		99.2	99.2		92.1	92.1		92.1	92.1		92.1	92.1		92.1	92.1		92.1	92.1
0241, 31 - 34	SITE & INFRASTRUCTURE, DEMOLITION	100.0	98.2	98.8	92.5	86.2	88.2	83.9	86.2	85.4	97.3	86.2	89.8	91.2	86.2	87.8	87.5	86.2	86.6
0310	Concrete Forming & Accessories	102.2	64.3	71.6	105.1	63.5	71.5	91.4	62.8	68.3	97.3	62.8	69.5	89.9	62.7	67.9	100.0	64.9	71.7
0320	Concrete Reinforcing	101.7	82.2	95.4	92.8	82.3	89.4	100.2	82.3	94.5	100.0	82.3	94.3	100.8	82.3	94.9	100.8	81.9	94.8
0330	Cast-in-Place Concrete	95.7	79.5	90.1	90.2	77.9	86.0	91.1	77.3	86.4	91.2	77.3	86.4	91.2	77.3	86.4	91.1	77.9	86.6
03	CONCRETE	102.4	73.1	90.7	96.2	72.4	86.7	96.6	71.9	86.7	108.7	71.9	93.9	103.8	71.8	91.0	97.4	72.9	87.6
04	MASONRY	102.6	65.2	79.0	104.7	66.8	80.8	101.6	66.3	79.4	101.6	66.3	79.3	101.6	66.3	79.4	161.5	61.3	98.3
05	METALS	97.0	79.2	92.9	99.0	81.5	94.9	95.5	81.5	92.2	95.5	81.5	92.2	95.6	81.3	92.3	96.2	80.9	92.6
06	WOOD, PLASTICS & COMPOSITES	96.6	61.9	82.2	97.5	60.7	82.3	84.9	60.5	74.8	89.9	60.5	77.8	83.5	60.5	74.0	96.0	62.9	82.3
07	THERMAL & MOISTURE PROTECTION	107.4	67.5	89.9	102.0	66.7	86.5	102.6	65.2	86.2	104.3	65.2	87.2	103.6	68.4	88.1	102.8	68.0	87.6
08	OPENINGS	104.1	66.8	95.1	102.0	66.1	93.4	106.1	64.6	96.0	105.7	64.6	95.8	105.9	64.6	95.9	106.4	67.0	96.9
0920	Plaster & Gypsum Board	101.4	60.9	76.0	92.8	59.9	72.2	89.5	59.7	70.8	89.8	59.7	70.9	89.5	59.7	70.8	110.3	62.1	80.1
0950, 0980	Ceilings & Acoustic Treatment	112.7	60.9	79.0	100.0	59.9	74.0	102.3	59.7	74.6	102.3	59.7	74.6	102.3	59.7	74.6	102.3	62.1	76.2
0960	Flooring	101.5	73.7	93.0	100.8	73.7	92.5	95.2	66.0	86.3	97.5	66.0	87.8	94.7	66.0	85.9	99.4	58.0	86.7
0970, 0990	Wall Finishes & Painting/Coating	101.8	56.6	74.3	101.0	56.6	74.0	97.2	72.7	82.3	97.2	72.7	82.3	97.2	56.1	72.2	97.2	74.2	83.2
09	FINISHES	103.8	64.6	82.5	100.0	63.9	80.4	95.2	63.6	78.1	97.3	63.6	79.0	96.0	61.8	77.4	99.8	63.8	80.3
COVERS	DIVS. 10 - 14, 25, 28, 41, 43, 44, 46	100.0	86.4	97.0	100.0	87.9	97.4	100.0	95.5	99.0	100.0	95.5	99.0	100.0	84.7	96.7	100.0	87.3	97.2
21, 22, 23	FIRE SUPPRESSION, PLUMBING & HVAC	100.9	73.2	89.5	101.2	75.5	90.7	98.8	72.1	87.9	98.8	72.1	87.9	98.8	72.0	87.8	101.1	72.6	89.4
26, 27, 3370	ELECTRICAL, COMMUNICATIONS & UTIL.	102.1	61.8	82.6	99.8	67.8	84.4	97.9	57.8	78.6	97.9	57.8	78.6	97.9	61.8	80.5	96.9	64.9	81.5
MF2018	WEIGHTED AVERAGE	101.0	72.0	89.2	99.7	72.6	88.7	97.9	70.4	86.7	100.1	70.4	88.0	99.1	70.4	87.5	101.6	71.1	89.2

WYOMING / CANADA

DIVISION		SHERIDAN 828 MAT.	INST.	TOTAL	WHEATLAND 822 MAT.	INST.	TOTAL	WORLAND 824 MAT.	INST.	TOTAL	YELLOWSTONE NAT'L PA 821 MAT.	INST.	TOTAL	BARRIE, ONTARIO MAT.	INST.	TOTAL	BATHURST, NEW BRUNSWICK MAT.	INST.	TOTAL
015433	CONTRACTOR EQUIPMENT		92.1	92.1		92.1	92.1		92.1	92.1		92.1	92.1		96.9	96.9		95.9	95.9
0241, 31 - 34	SITE & INFRASTRUCTURE, DEMOLITION	91.7	86.2	88.0	88.1	86.2	86.8	85.4	86.2	85.9	85.5	86.2	86.0	120.3	90.0	99.8	113.9	85.6	94.7
0310	Concrete Forming & Accessories	102.6	63.1	70.7	94.0	62.5	68.6	94.1	62.8	68.9	94.2	62.8	68.8	126.4	79.7	88.7	99.1	55.9	64.3
0320	Concrete Reinforcing	100.8	82.3	94.9	100.2	81.6	94.3	100.8	82.3	94.9	102.6	82.3	96.1	152.2	83.4	130.1	70.5	55.2	65.6
0330	Cast-in-Place Concrete	94.2	77.9	88.6	95.1	77.2	88.9	91.1	77.3	86.4	91.1	77.3	86.4	165.2	78.7	135.3	115.6	53.6	94.2
03	CONCRETE	104.3	72.2	91.5	101.0	71.6	89.3	97.0	71.9	86.9	97.4	71.8	87.1	145.1	80.6	119.3	104.4	56.0	85.0
04	MASONRY	101.9	67.4	80.1	102.0	54.8	72.3	101.6	66.3	79.4	101.6	66.3	79.4	170.8	86.0	117.4	165.2	55.3	96.0
05	METALS	99.0	81.3	94.9	95.4	80.3	91.9	95.6	81.5	92.3	96.2	81.4	92.8	95.9	90.3	94.6	82.3	71.9	79.9
06	WOOD, PLASTICS & COMPOSITES	99.5	60.5	83.4	87.3	60.5	76.2	87.4	60.5	76.3	87.4	60.5	76.3	118.5	78.6	102.0	93.4	56.2	78.0
07	THERMAL & MOISTURE PROTECTION	104.6	66.8	88.0	102.9	62.1	85.0	102.7	65.2	86.3	102.2	65.2	86.0	113.6	82.0	99.8	110.0	55.6	86.1
08	OPENINGS	106.6	64.6	96.4	104.7	64.6	95.0	106.3	64.6	96.2	99.6	64.6	91.1	87.0	78.5	85.0	81.3	49.6	73.6
0920	Plaster & Gypsum Board	110.7	59.7	78.8	89.5	59.7	70.8	89.5	59.7	70.8	89.7	59.7	70.9	168.4	78.2	112.0	123.7	55.2	80.9
0950, 0980	Ceilings & Acoustic Treatment	105.4	59.7	75.7	102.3	59.7	74.6	102.3	59.7	74.6	103.0	59.7	74.9	98.1	78.2	85.2	116.4	55.2	76.7
0960	Flooring	100.8	56.1	87.1	96.4	66.0	87.1	96.4	66.0	87.1	96.4	66.0	87.1	119.0	83.8	108.2	101.5	39.8	82.6
0970, 0990	Wall Finishes & Painting/Coating	99.8	56.6	73.5	97.2	54.2	71.0	97.2	72.7	82.3	97.2	56.1	72.2	106.5	80.3	90.5	112.4	45.9	71.8
09	FINISHES	103.9	60.7	80.4	95.9	61.6	77.3	95.5	63.6	78.2	95.7	61.8	77.3	117.4	80.7	97.5	110.9	51.9	78.9
COVERS	DIVS. 10 - 14, 25, 28, 41, 43, 44, 46	100.0	85.8	96.9	100.0	85.4	96.8	100.0	95.6	99.0	100.0	95.6	99.0	139.3	61.5	122.4	131.1	54.7	114.5
21, 22, 23	FIRE SUPPRESSION, PLUMBING & HVAC	98.8	72.6	88.1	98.8	72.1	87.9	98.8	72.1	87.9	98.8	72.1	87.9	106.5	89.4	99.5	106.4	61.9	88.2
26, 27, 3370	ELECTRICAL, COMMUNICATIONS & UTIL.	99.9	61.8	81.5	97.9	61.8	80.5	97.9	61.8	80.5	97.3	61.8	80.2	120.4	79.3	100.6	120.3	53.4	88.0
MF2018	WEIGHTED AVERAGE	101.0	70.6	88.6	98.6	69.0	86.5	98.1	71.0	87.1	97.6	70.7	86.7	116.1	83.8	102.9	105.3	59.7	86.7

CANADA

DIVISION		BRANDON, MANITOBA MAT.	INST.	TOTAL	BRANTFORD, ONTARIO MAT.	INST.	TOTAL	BRIDGEWATER, NOVA SCOTIA MAT.	INST.	TOTAL	CALGARY, ALBERTA MAT.	INST.	TOTAL	CAP-DE-LA-MADELEINE, QUEBEC MAT.	INST.	TOTAL	CHARLESBOURG, QUEBEC MAT.	INST.	TOTAL
015433	CONTRACTOR EQUIPMENT		98.7	98.7		96.4	96.4		96.1	96.1		120.3	120.3		96.9	96.9		96.9	96.9
0241, 31 - 34	SITE & INFRASTRUCTURE, DEMOLITION	123.5	88.6	99.8	125.1	90.0	101.3	106.1	87.6	93.5	115.8	109.0	111.2	101.1	88.9	92.8	101.1	88.9	92.8
0310	Concrete Forming & Accessories	153.0	64.0	81.2	128.8	86.3	94.5	89.1	65.7	70.2	123.1	89.7	96.2	137.4	76.6	88.4	137.4	76.6	88.4
0320	Concrete Reinforcing	85.3	52.4	74.7	82.0	82.2	82.1	69.9	45.7	62.1	66.4	76.8	69.7	69.9	69.6	69.8	69.9	69.6	69.8
0330	Cast-in-Place Concrete	124.7	67.3	104.9	149.0	97.7	131.3	153.2	63.9	122.4	145.8	99.5	129.8	120.2	83.9	107.7	120.2	83.9	107.7
03	CONCRETE	118.9	64.2	97.0	123.4	90.0	110.0	121.1	62.7	97.7	128.8	91.6	113.9	108.7	78.6	96.6	108.7	78.6	96.6
04	MASONRY	221.0	58.5	118.5	175.0	90.0	121.4	170.1	63.2	102.7	226.6	83.1	136.2	170.3	73.6	109.3	170.3	73.6	109.3
05	METALS	91.4	77.2	88.1	80.8	90.7	83.1	80.2	74.7	78.9	91.2	95.0	92.1	79.6	83.7	80.5	79.6	83.7	80.5
06	WOOD, PLASTICS & COMPOSITES	150.9	65.0	115.3	120.8	85.3	106.1	83.7	65.5	76.2	101.7	89.8	96.7	134.5	76.8	110.6	134.5	76.8	110.6
07	THERMAL & MOISTURE PROTECTION	126.1	65.4	99.5	119.7	87.1	105.4	114.2	64.4	92.4	139.9	92.6	119.1	114.1	80.2	99.2	114.1	80.2	99.2
08	OPENINGS	96.9	58.1	87.5	84.6	84.1	84.5	79.9	59.5	74.9	84.1	77.9	82.6	86.1	70.2	82.3	86.1	70.2	82.3
0920	Plaster & Gypsum Board	138.7	63.9	91.9	131.9	85.1	102.7	117.3	64.8	84.5	134.8	88.7	106.0	160.6	76.3	107.9	160.6	76.3	107.9
0950, 0980	Ceilings & Acoustic Treatment	122.9	63.9	84.6	103.7	85.1	91.7	103.7	64.8	78.5	145.8	88.7	108.8	103.7	76.3	85.9	103.7	76.3	85.9
0960	Flooring	129.3	59.4	107.9	112.9	83.8	104.0	98.1	56.6	85.4	126.6	80.0	112.3	112.9	82.5	103.6	112.9	82.5	103.6
0970, 0990	Wall Finishes & Painting/Coating	117.6	51.8	77.5	110.9	88.1	97.0	110.9	56.8	77.9	110.1	98.7	103.1	110.9	80.1	92.1	110.9	80.1	92.1
09	FINISHES	127.6	62.1	92.0	113.0	86.3	98.5	106.0	63.3	82.8	126.6	89.4	106.4	117.0	78.4	96.0	117.0	78.4	96.0
COVERS	DIVS. 10 - 14, 25, 28, 41, 43, 44, 46	131.1	56.9	115.0	131.1	63.3	116.4	131.1	57.5	115.1	131.1	86.3	121.4	131.1	71.6	118.2	131.1	71.6	118.2
21, 22, 23	FIRE SUPPRESSION, PLUMBING & HVAC	106.7	75.4	93.8	106.5	92.1	100.6	106.5	76.0	94.0	107.5	83.4	97.6	106.9	80.5	96.1	106.9	80.5	96.1
26, 27, 3370	ELECTRICAL, COMMUNICATIONS & UTIL.	118.8	60.3	90.6	117.0	78.9	98.6	120.4	56.1	89.4	113.7	87.0	100.8	115.1	62.4	89.7	115.1	62.4	89.7
MF2018	WEIGHTED AVERAGE	115.3	67.8	96.0	109.0	87.2	100.1	106.7	67.5	90.8	114.8	89.0	104.3	106.2	76.9	94.3	106.2	76.9	94.3

		CANADA															
	DIVISION	CHARLOTTETOWN, PRINCE EDWARD ISLAND			CHICOUTIMI, QUEBEC			CORNER BROOK, NEWFOUNDLAND			CORNWALL, ONTARIO			DALHOUSIE, NEW BRUNSWICK			DARTMOUTH, NOVA SCOTIA		
		MAT.	INST.	TOTAL	MAT.	INST.	TOTAL	MAT.	INST.	TOTAL	MAT.	INST.	TOTAL	MAT.	INST.	TOTAL	MAT.	INST.	TOTAL
015433	CONTRACTOR EQUIPMENT		118.0	118.0		96.5	96.5		97.4	97.4		96.4	96.4		96.5	96.5		96.3	96.3
0241, 31 - 34	SITE & INFRASTRUCTURE, DEMOLITION	149.2	104.7	119.0	110.5	89.0	95.9	128.1	86.3	99.7	122.9	89.5	100.2	104.1	86.1	91.8	119.0	87.7	97.8
0310	Concrete Forming & Accessories	126.5	51.5	66.0	140.8	89.4	99.3	119.2	73.0	82.0	125.7	79.8	88.7	96.8	56.1	64.0	104.7	65.7	73.3
0320	Concrete Reinforcing	76.2	44.7	66.1	53.1	93.4	66.1	78.2	46.8	68.1	82.0	81.9	82.0	71.4	55.3	66.2	81.8	45.7	70.2
0330	Cast-in-Place Concrete	136.8	56.7	109.1	108.8	93.1	103.4	145.2	60.9	116.1	134.0	89.0	118.5	129.4	53.8	103.3	140.2	64.0	113.9
03	CONCRETE	128.0	53.8	98.3	97.9	91.6	95.4	148.0	65.1	114.8	116.6	83.9	103.5	114.3	56.2	91.1	131.2	62.7	103.8
04	MASONRY	186.6	53.0	102.4	167.6	89.1	118.1	216.6	71.5	125.1	173.5	82.5	116.1	171.5	55.3	98.3	232.3	63.2	125.7
05	METALS	94.9	75.8	90.5	82.1	92.0	84.4	91.7	74.2	87.6	80.6	89.5	82.7	75.8	72.5	75.0	91.9	74.8	87.9
06	WOOD, PLASTICS & COMPOSITES	104.7	51.2	82.6	134.8	90.3	116.3	123.8	79.0	105.3	118.9	79.3	102.5	91.2	56.3	76.7	109.2	65.6	91.1
07	THERMAL & MOISTURE PROTECTION	141.2	55.9	103.8	110.9	94.3	103.6	129.6	62.7	100.3	119.4	82.1	103.1	118.8	55.6	91.1	126.6	64.4	99.3
08	OPENINGS	86.7	44.5	76.5	84.9	76.0	82.8	102.5	64.5	93.3	86.1	78.2	84.2	82.7	49.7	74.7	88.6	59.5	81.6
0920	Plaster & Gypsum Board	133.9	49.2	80.9	159.4	90.1	116.1	153.6	78.5	106.6	178.7	79.0	116.4	127.5	55.2	82.3	145.1	64.8	94.9
0950, 0980	Ceilings & Acoustic Treatment	134.7	49.2	79.2	115.0	90.1	98.9	123.6	78.5	94.4	107.9	79.0	89.1	107.7	55.2	73.6	133.9	64.8	89.1
0960	Flooring	123.0	53.1	101.6	114.7	82.5	104.9	114.0	47.6	93.7	112.9	82.6	103.7	104.3	60.9	91.1	109.8	56.6	93.5
0970, 0990	Wall Finishes & Painting/Coating	112.3	38.2	67.1	112.4	102.7	106.5	117.5	54.1	78.8	110.9	82.0	93.3	114.1	45.9	72.5	117.5	56.8	80.5
09	FINISHES	129.1	50.7	86.5	119.7	90.6	103.9	126.5	67.7	94.5	121.5	80.7	99.4	111.8	56.2	81.6	124.8	63.3	91.4
COVERS	DIVS. 10 - 14, 25, 28, 41, 43, 44, 46	131.1	55.7	114.8	131.1	76.8	119.3	131.1	57.4	115.1	131.1	61.1	115.9	131.1	54.7	114.5	131.1	57.6	115.1
21, 22, 23	FIRE SUPPRESSION, PLUMBING & HVAC	106.9	56.5	86.2	106.4	79.7	95.5	106.7	63.0	88.8	106.9	90.2	100.0	106.5	61.9	88.2	106.7	76.0	94.1
26, 27, 3370	ELECTRICAL, COMMUNICATIONS & UTIL.	122.5	45.4	85.3	115.9	82.8	99.9	114.8	49.3	83.2	117.0	79.7	99.0	120.5	50.5	86.7	118.1	56.1	88.2
MF2018	WEIGHTED AVERAGE	115.9	58.3	92.5	105.4	86.2	97.6	119.1	65.7	97.3	108.7	84.0	98.6	105.8	60.0	87.1	115.9	67.6	96.2

		CANADA															
	DIVISION	EDMONTON, ALBERTA			FORT MCMURRAY, ALBERTA			FREDERICTON, NEW BRUNSWICK			GATINEAU, QUEBEC			GRANBY, QUEBEC			HALIFAX, NOVA SCOTIA		
		MAT.	INST.	TOTAL	MAT.	INST.	TOTAL	MAT.	INST.	TOTAL	MAT.	INST.	TOTAL	MAT.	INST.	TOTAL	MAT.	INST.	TOTAL
015433	CONTRACTOR EQUIPMENT		121.1	121.1		98.6	98.6		118.2	118.2		96.9	96.9		96.9	96.9		115.5	115.5
0241, 31 - 34	SITE & INFRASTRUCTURE, DEMOLITION	123.0	110.5	114.5	130.0	90.8	103.4	154.9	105.8	121.6	100.8	88.9	92.8	101.4	88.9	92.9	112.7	105.5	107.8
0310	Concrete Forming & Accessories	126.4	89.7	96.8	125.3	84.6	92.4	127.5	57.1	70.7	137.4	76.5	88.3	137.4	76.5	88.3	123.9	84.1	91.8
0320	Concrete Reinforcing	66.2	76.8	69.6	75.9	76.7	76.2	70.5	55.5	65.7	73.9	69.5	72.5	73.9	69.5	72.5	76.0	74.8	75.6
0330	Cast-in-Place Concrete	163.8	99.5	141.6	198.0	93.0	161.7	155.3	57.4	121.5	118.6	83.8	106.6	122.5	83.8	109.2	143.9	77.0	120.8
03	CONCRETE	136.9	91.6	118.8	143.4	86.5	120.7	135.1	58.4	104.4	108.8	78.5	96.7	110.6	78.5	97.7	130.2	81.0	110.5
04	MASONRY	226.8	83.1	136.2	217.3	80.8	131.2	193.7	56.7	107.3	170.1	73.6	109.2	170.5	73.6	109.4	186.8	82.4	121.0
05	METALS	93.7	94.9	94.0	100.4	86.4	97.1	95.1	81.1	91.8	79.6	83.6	80.5	79.7	83.5	80.6	94.8	95.3	94.9
06	WOOD, PLASTICS & COMPOSITES	100.7	89.8	96.2	113.6	84.3	101.5	107.1	57.0	86.3	134.5	76.8	110.6	134.5	76.8	110.6	104.1	84.6	96.0
07	THERMAL & MOISTURE PROTECTION	142.5	92.6	120.6	128.6	86.7	110.3	140.8	57.8	104.4	114.1	80.2	99.2	114.1	78.7	98.6	142.7	84.7	117.2
08	OPENINGS	82.7	77.9	81.5	86.1	74.9	83.4	89.9	49.0	80.0	86.1	65.9	81.2	86.1	65.9	81.2	90.9	76.2	87.3
0920	Plaster & Gypsum Board	133.0	88.7	105.3	125.5	83.8	99.4	136.8	55.2	85.8	138.3	76.3	99.6	138.3	76.3	99.6	135.7	83.7	103.2
0950, 0980	Ceilings & Acoustic Treatment	146.5	88.7	109.0	114.0	83.8	94.4	136.1	55.2	83.6	103.7	76.3	85.9	103.7	76.3	85.9	132.2	83.7	100.7
0960	Flooring	128.7	80.0	113.8	112.9	80.0	102.9	123.1	63.6	104.9	112.9	82.5	103.6	112.9	82.5	103.6	116.7	76.9	104.5
0970, 0990	Wall Finishes & Painting/Coating	113.1	98.7	104.3	111.0	84.6	94.9	112.9	58.5	79.7	110.9	80.1	92.1	110.9	80.1	92.1	112.6	86.2	96.5
09	FINISHES	126.4	89.4	106.3	115.5	84.0	98.4	133.0	58.7	92.6	113.4	78.4	94.4	113.4	78.4	94.4	126.9	83.6	103.3
COVERS	DIVS. 10 - 14, 25, 28, 41, 43, 44, 46	131.1	87.5	121.6	131.1	84.6	121.0	131.1	56.3	114.9	131.1	71.6	118.2	131.1	71.6	118.2	131.1	64.8	116.7
21, 22, 23	FIRE SUPPRESSION, PLUMBING & HVAC	107.4	83.6	97.6	106.9	89.0	99.6	107.0	70.5	92.0	106.9	80.5	96.1	106.5	80.5	95.8	107.3	77.4	95.0
26, 27, 3370	ELECTRICAL, COMMUNICATIONS & UTIL.	125.6	87.0	107.0	110.4	74.1	92.9	124.2	66.6	96.4	115.1	62.4	89.7	116.0	62.4	90.2	127.2	84.3	106.5
MF2018	WEIGHTED AVERAGE	117.7	89.2	106.1	117.4	84.1	103.9	118.1	67.3	97.4	106.0	76.7	94.1	106.3	76.6	94.2	116.1	83.8	103.0

		CANADA															
	DIVISION	HAMILTON, ONTARIO			HULL, QUEBEC			JOLIETTE, QUEBEC			KAMLOOPS, BRITISH COLUMBIA			KINGSTON, ONTARIO			KITCHENER, ONTARIO		
		MAT.	INST.	TOTAL	MAT.	INST.	TOTAL	MAT.	INST.	TOTAL	MAT.	INST.	TOTAL	MAT.	INST.	TOTAL	MAT.	INST.	TOTAL
015433	CONTRACTOR EQUIPMENT		112.6	112.6		96.9	96.9		96.9	96.9		99.9	99.9		98.6	98.6		98.5	98.5
0241, 31 - 34	SITE & INFRASTRUCTURE, DEMOLITION	115.3	103.9	107.6	100.8	88.9	92.8	101.6	88.9	93.0	127.2	92.3	103.5	122.9	93.0	102.6	102.0	94.9	97.2
0310	Concrete Forming & Accessories	133.2	90.3	98.6	137.4	76.5	88.3	137.4	76.6	88.4	116.5	79.6	86.8	125.8	79.8	88.7	116.6	85.9	91.9
0320	Concrete Reinforcing	67.1	100.2	77.7	73.9	69.5	72.5	69.9	69.6	69.8	54.8	72.8	60.6	82.0	81.9	82.0	48.4	100.0	65.0
0330	Cast-in-Place Concrete	151.6	100.3	133.9	118.6	83.8	106.6	123.5	83.9	109.8	106.9	88.0	100.3	134.0	89.0	118.5	141.2	93.9	124.9
03	CONCRETE	128.8	96.1	115.7	108.8	78.5	96.7	110.2	78.6	97.5	116.0	81.9	102.3	118.3	84.0	104.6	105.5	91.6	99.9
04	MASONRY	189.0	98.3	131.8	170.1	73.6	109.2	170.6	73.6	109.4	178.1	80.9	116.8	181.3	82.5	119.0	153.1	95.8	117.0
05	METALS	93.7	104.0	96.1	79.7	83.6	80.6	79.7	83.7	80.6	81.2	85.8	82.3	81.7	89.4	83.5	89.0	97.9	91.1
06	WOOD, PLASTICS & COMPOSITES	113.7	89.2	103.5	134.5	76.8	110.6	134.5	76.8	110.6	98.8	78.4	90.4	118.9	79.4	102.6	109.9	84.4	99.4
07	THERMAL & MOISTURE PROTECTION	132.5	97.2	117.0	114.1	80.2	99.2	114.1	80.2	99.2	130.1	77.8	107.2	119.4	83.2	103.5	115.7	93.3	105.9
08	OPENINGS	87.3	89.2	87.8	86.1	65.9	81.2	86.1	70.2	82.3	83.3	76.2	81.6	86.1	77.9	84.1	77.5	85.0	79.3
0920	Plaster & Gypsum Board	135.6	88.3	106.0	138.3	76.3	99.6	160.6	76.3	107.9	107.1	77.7	88.7	181.9	79.1	117.7	119.2	84.3	97.3
0950, 0980	Ceilings & Acoustic Treatment	141.9	88.3	107.1	103.7	76.3	85.9	103.7	76.3	85.9	103.7	77.7	86.8	123.6	79.1	94.7	110.2	84.3	93.4
0960	Flooring	123.4	89.6	113.1	112.9	82.5	103.6	112.9	82.5	103.6	114.8	47.2	94.2	112.9	82.6	103.7	102.2	89.7	98.4
0970, 0990	Wall Finishes & Painting/Coating	109.4	96.6	101.6	110.9	80.1	92.1	110.9	80.1	92.1	110.9	73.2	87.9	110.9	75.8	89.5	105.3	87.5	94.4
09	FINISHES	125.5	90.6	106.5	113.4	78.4	94.4	117.0	78.4	96.0	111.3	73.2	90.6	125.5	80.1	100.9	106.8	86.3	95.7
COVERS	DIVS. 10 - 14, 25, 28, 41, 43, 44, 46	131.1	86.7	121.5	131.1	71.6	118.2	131.1	71.6	118.2	131.1	79.9	120.0	131.1	61.2	115.9	131.1	84.1	120.9
21, 22, 23	FIRE SUPPRESSION, PLUMBING & HVAC	107.6	86.3	98.9	106.5	80.5	95.8	106.5	80.5	95.8	106.5	83.7	97.1	106.9	90.3	100.1	106.4	86.0	98.0
26, 27, 3370	ELECTRICAL, COMMUNICATIONS & UTIL.	117.7	100.0	109.2	117.9	62.4	91.1	116.0	62.4	90.2	120.0	70.9	96.3	117.0	78.6	98.5	113.2	97.9	105.8
MF2018	WEIGHTED AVERAGE	114.4	94.8	106.4	106.2	76.7	94.2	106.5	76.9	94.4	108.3	80.2	96.8	109.8	84.0	99.3	104.9	91.3	99.4

CANADA

DIVISION		LAVAL, QUEBEC			LETHBRIDGE, ALBERTA			LLOYDMINSTER, ALBERTA			LONDON, ONTARIO			MEDICINE HAT, ALBERTA			MONCTON, NEW BRUNSWICK		
		MAT.	INST.	TOTAL	MAT.	INST.	TOTAL	MAT.	INST.	TOTAL	MAT.	INST.	TOTAL	MAT.	INST.	TOTAL	MAT.	INST.	TOTAL
015433	CONTRACTOR EQUIPMENT		96.9	96.9		98.6	98.6		98.6	98.6		114.2	114.2		98.6	98.6		95.9	95.9
0241, 31 - 34	SITE & INFRASTRUCTURE, DEMOLITION	101.4	89.0	93.0	121.5	91.3	101.0	121.1	90.8	100.5	117.2	103.8	108.1	119.9	90.8	100.2	113.3	88.0	96.1
0310	Concrete Forming & Accessories	137.6	77.2	88.9	127.7	84.6	92.9	125.0	75.6	85.2	133.6	86.7	95.8	127.6	75.6	85.6	99.1	63.5	70.4
0320	Concrete Reinforcing	73.9	70.2	72.7	75.9	76.7	76.2	75.9	76.7	76.2	66.8	99.1	77.1	75.9	76.7	76.2	70.5	67.8	69.6
0330	Cast-in-Place Concrete	122.5	84.4	109.4	148.4	93.0	129.3	137.7	89.6	121.1	151.7	98.7	133.4	137.7	89.6	121.1	111.3	65.6	95.5
03	CONCRETE	110.6	79.1	98.0	121.7	86.6	107.7	116.8	81.3	102.6	128.7	93.8	114.8	117.0	81.3	102.7	102.5	66.1	87.9
04	MASONRY	170.4	74.2	109.8	190.2	80.8	121.2	169.8	77.6	109.9	198.6	96.8	134.4	169.8	74.8	109.9	164.8	73.2	107.0
05	METALS	79.6	83.9	80.6	96.3	86.5	94.0	81.2	86.3	82.4	93.7	104.7	96.3	81.3	86.3	82.5	82.3	85.5	83.0
06	WOOD, PLASTICS & COMPOSITES	134.5	77.5	110.9	117.8	84.3	103.9	113.6	75.3	97.8	116.5	85.0	103.5	117.8	75.3	100.2	93.4	61.3	80.1
07	THERMAL & MOISTURE PROTECTION	115.1	80.6	100.0	125.8	86.7	108.7	122.6	82.6	105.0	129.9	95.0	114.6	127.7	82.6	107.9	113.4	69.6	94.2
08	OPENINGS	86.1	66.5	81.4	86.1	74.9	83.4	86.1	69.9	82.2	82.5	85.9	83.3	86.1	69.9	82.2	81.3	58.7	75.8
0920	Plaster & Gypsum Board	138.6	77.0	100.1	121.4	83.8	97.9	115.4	74.5	89.8	136.3	84.0	103.6	119.2	74.5	91.3	123.7	60.4	84.1
0950, 0980	Ceilings & Acoustic Treatment	103.7	77.0	86.4	114.0	83.8	94.4	103.7	74.5	84.8	139.3	84.0	103.4	103.7	74.5	84.8	116.4	60.4	80.1
0960	Flooring	112.9	83.4	103.9	112.9	80.0	102.9	112.9	80.0	102.9	117.6	89.7	109.0	112.9	80.0	102.9	101.5	62.3	89.5
0970, 0990	Wall Finishes & Painting/Coating	110.9	80.8	92.5	110.8	92.3	99.5	111.0	71.9	87.2	112.2	93.6	100.9	110.8	71.9	87.1	112.4	77.5	91.1
09	FINISHES	113.4	79.0	94.7	113.9	84.9	98.1	110.8	75.9	91.8	127.5	87.7	105.9	111.3	75.9	92.0	110.9	64.2	85.5
COVERS	DIVS. 10 - 14, 25, 28, 41, 43, 44, 46	131.1	72.2	118.3	131.1	83.5	120.8	131.1	81.7	120.4	131.1	86.1	121.4	131.1	80.5	120.1	131.1	57.8	115.2
21, 22, 23	FIRE SUPPRESSION, PLUMBING & HVAC	106.7	81.2	96.2	106.8	86.0	98.2	106.9	85.9	98.3	107.6	84.7	98.2	106.5	82.9	96.8	106.4	76.5	94.1
26, 27, 3370	ELECTRICAL, COMMUNICATIONS & UTIL.	116.6	62.9	90.7	112.6	74.1	94.1	109.4	74.1	92.4	109.4	97.7	103.8	109.4	74.1	92.4	123.1	86.5	105.5
MF2018	WEIGHTED AVERAGE	106.4	77.2	94.5	112.3	83.6	100.6	107.1	80.6	96.3	113.6	93.0	105.2	107.2	79.9	96.1	105.3	74.6	92.8

CANADA

DIVISION		MONTREAL, QUEBEC			MOOSE JAW, SASKATCHEWAN			NEW GLASGOW, NOVA SCOTIA			NEWCASTLE, NEW BRUNSWICK			NORTH BAY, ONTARIO			OSHAWA, ONTARIO		
		MAT.	INST.	TOTAL	MAT.	INST.	TOTAL	MAT.	INST.	TOTAL	MAT.	INST.	TOTAL	MAT.	INST.	TOTAL	MAT.	INST.	TOTAL
015433	CONTRACTOR EQUIPMENT		115.2	115.2		95.1	95.1		96.3	96.3		95.9	95.9		96.6	96.6		98.5	98.5
0241, 31 - 34	SITE & INFRASTRUCTURE, DEMOLITION	127.9	103.1	111.1	120.7	85.4	96.7	110.0	87.7	94.9	113.9	85.6	94.7	126.0	89.3	101.1	113.4	95.3	101.1
0310	Concrete Forming & Accessories	131.5	90.2	98.2	103.4	52.8	62.6	104.6	65.7	73.2	99.1	56.1	64.4	154.3	77.5	92.3	124.6	89.2	96.1
0320	Concrete Reinforcing	66.7	93.5	75.3	53.6	58.9	55.3	78.2	45.7	67.8	70.5	55.3	65.6	92.5	81.5	89.0	76.6	100.6	84.3
0330	Cast-in-Place Concrete	117.6	97.4	110.6	133.6	60.9	108.5	140.2	64.0	113.9	115.6	53.7	94.2	135.0	76.9	114.9	163.4	103.4	142.7
03	CONCRETE	113.6	93.9	105.7	103.1	57.8	85.0	130.4	62.7	103.3	104.4	56.2	85.1	133.0	78.6	111.3	127.2	96.4	114.9
04	MASONRY	189.4	89.2	126.3	168.3	54.3	96.5	216.1	63.2	119.7	165.2	55.3	96.0	222.6	78.9	132.0	156.9	99.1	120.5
05	METALS	101.8	100.2	101.4	78.8	73.7	77.6	90.2	74.8	86.6	82.3	72.2	79.9	90.8	89.3	90.5	82.0	98.5	85.9
06	WOOD, PLASTICS & COMPOSITES	109.9	91.0	102.1	95.5	51.7	77.4	109.2	65.6	91.1	93.4	56.2	78.0	154.7	78.1	123.0	118.8	87.9	106.0
07	THERMAL & MOISTURE PROTECTION	130.3	96.4	115.5	113.0	57.1	88.5	126.6	64.4	99.3	113.4	55.6	88.0	132.8	78.8	109.1	116.7	99.7	109.3
08	OPENINGS	84.7	78.2	83.2	82.6	49.8	74.7	88.6	59.5	81.6	81.3	49.6	73.6	95.1	75.9	90.5	82.1	88.0	83.5
0920	Plaster & Gypsum Board	136.7	90.1	107.6	106.1	50.5	71.4	142.8	64.8	94.0	123.7	55.2	80.9	161.4	77.7	109.1	126.7	87.8	102.4
0950, 0980	Ceilings & Acoustic Treatment	144.4	90.1	109.2	103.7	50.5	69.2	122.9	64.8	85.2	116.4	55.2	76.7	122.9	77.7	93.6	105.4	87.8	94.0
0960	Flooring	120.7	85.4	109.9	104.7	51.5	88.4	109.8	56.6	93.5	101.5	60.9	89.1	129.3	82.6	115.0	104.6	91.7	100.6
0970, 0990	Wall Finishes & Painting/Coating	118.9	102.7	109.0	110.9	58.4	78.9	117.5	56.8	80.5	112.4	45.9	71.8	117.5	81.4	95.5	105.3	100.3	102.3
09	FINISHES	129.8	91.6	109.0	106.9	52.9	77.5	122.0	63.3	90.1	110.9	56.2	81.2	131.1	79.2	102.9	108.2	90.4	98.6
COVERS	DIVS. 10 - 14, 25, 28, 41, 43, 44, 46	131.1	78.5	119.7	131.1	54.4	114.5	131.1	57.6	115.1	131.1	54.7	114.5	131.1	60.0	115.7	131.1	84.8	121.1
21, 22, 23	FIRE SUPPRESSION, PLUMBING & HVAC	107.6	79.9	96.2	106.9	67.2	90.6	106.7	76.0	94.1	106.4	61.9	88.2	106.7	88.3	99.2	106.4	85.9	98.0
26, 27, 3370	ELECTRICAL, COMMUNICATIONS & UTIL.	119.3	82.8	101.7	119.3	53.9	87.8	115.4	56.1	86.8	119.8	53.4	87.8	117.3	79.6	99.1	113.9	100.3	107.3
MF2018	WEIGHTED AVERAGE	114.4	88.7	103.9	104.6	61.3	87.0	114.1	67.6	95.1	105.3	60.3	87.0	117.4	82.0	103.0	107.7	93.6	102.0

CANADA

DIVISION		OTTAWA, ONTARIO			OWEN SOUND, ONTARIO			PETERBOROUGH, ONTARIO			PORTAGE LA PRAIRIE, MANITOBA			PRINCE ALBERT, SASKATCHEWAN			PRINCE GEORGE, BRITISH COLUMBIA		
		MAT.	INST.	TOTAL	MAT.	INST.	TOTAL	MAT.	INST.	TOTAL	MAT.	INST.	TOTAL	MAT.	INST.	TOTAL	MAT.	INST.	TOTAL
015433	CONTRACTOR EQUIPMENT		114.5	114.5		96.9	96.9		96.4	96.4		98.6	98.6		95.1	95.1		99.9	99.9
0241, 31 - 34	SITE & INFRASTRUCTURE, DEMOLITION	117.5	103.9	108.2	120.3	89.9	99.7	125.1	89.5	100.9	121.3	88.4	99.0	116.4	85.5	95.4	131.3	92.3	104.8
0310	Concrete Forming & Accessories	126.6	90.9	97.8	126.4	76.2	85.9	128.8	78.4	88.2	127.9	63.6	76.0	103.4	52.6	62.4	111.3	75.0	82.0
0320	Concrete Reinforcing	66.6	99.1	77.0	152.2	83.4	130.1	82.0	82.0	82.0	75.9	52.3	68.4	56.0	58.9	56.9	54.8	72.8	60.6
0330	Cast-in-Place Concrete	137.9	101.4	125.3	165.2	73.4	133.5	149.0	78.3	124.6	137.7	66.7	113.2	121.0	60.8	100.2	134.0	88.0	118.1
03	CONCRETE	122.1	96.7	111.9	145.1	77.1	117.9	123.4	79.6	105.9	111.3	63.8	92.3	98.1	57.7	81.9	127.6	79.9	108.5
04	MASONRY	186.1	97.5	130.3	170.8	83.9	116.0	175.0	84.8	118.2	173.4	57.5	100.4	167.1	54.3	96.0	180.8	80.9	117.8
05	METALS	94.2	105.9	96.9	95.9	90.1	94.6	80.8	89.6	82.8	81.3	77.1	80.3	78.8	73.5	77.6	81.2	85.9	82.3
06	WOOD, PLASTICS & COMPOSITES	108.8	90.2	101.1	118.5	75.1	100.5	120.0	76.9	102.6	117.8	64.9	95.9	95.5	51.7	77.4	98.8	72.0	87.7
07	THERMAL & MOISTURE PROTECTION	142.6	97.4	122.8	113.6	79.5	98.7	119.7	83.9	104.0	113.7	65.0	92.3	112.9	56.1	88.0	124.4	77.1	103.7
08	OPENINGS	92.4	88.7	91.5	87.0	75.3	84.2	84.6	77.6	82.9	86.1	58.1	79.4	81.7	49.8	74.0	83.3	72.7	80.7
0920	Plaster & Gypsum Board	134.3	89.4	106.2	168.4	74.7	109.8	131.9	76.5	97.2	118.8	63.9	84.4	106.1	50.5	71.4	107.1	71.1	84.6
0950, 0980	Ceilings & Acoustic Treatment	150.8	89.4	110.9	98.1	74.7	82.9	103.7	76.5	86.0	103.7	63.9	77.9	103.7	50.5	69.2	103.7	71.1	82.6
0960	Flooring	116.6	85.5	107.1	119.0	83.8	108.2	112.9	82.6	103.7	112.9	59.4	96.6	104.7	51.5	88.4	109.1	64.7	95.5
0970, 0990	Wall Finishes & Painting/Coating	112.7	88.9	98.2	106.5	80.3	90.5	110.9	83.5	94.2	110.1	51.8	74.9	110.9	49.8	73.6	110.9	73.2	87.9
09	FINISHES	128.3	89.6	107.3	117.4	78.1	96.1	113.0	79.8	95.0	111.2	61.9	84.4	106.9	51.9	77.0	109.7	72.5	89.5
COVERS	DIVS. 10 - 14, 25, 28, 41, 43, 44, 46	131.1	84.9	121.1	139.3	60.5	122.2	131.1	61.3	116.0	131.1	56.6	114.9	131.1	54.4	114.5	131.1	79.2	119.9
21, 22, 23	FIRE SUPPRESSION, PLUMBING & HVAC	107.7	87.8	99.5	106.5	88.3	99.0	106.5	91.6	100.3	106.5	74.9	93.5	106.9	60.6	87.9	106.5	83.7	97.1
26, 27, 3370	ELECTRICAL, COMMUNICATIONS & UTIL.	111.1	98.5	105.0	121.9	78.6	101.0	117.0	79.3	98.8	118.9	52.1	86.7	119.3	53.9	87.8	117.0	70.9	94.8
MF2018	WEIGHTED AVERAGE	113.6	94.8	106.0	116.3	82.1	102.4	109.0	83.8	98.7	107.3	66.3	90.6	103.7	59.7	85.8	109.5	79.6	97.3

For CCI updates and additional guidance please go to www.rsmeans.com/2022books

City Cost Indexes

CANADA

DIVISION		QUEBEC CITY, QUEBEC			RED DEER, ALBERTA			REGINA, SASKATCHEWAN			RIMOUSKI, QUEBEC			ROUYN-NORANDA, QUEBEC			SAINT HYACINTHE, QUEBEC		
		MAT.	INST.	TOTAL	MAT.	INST.	TOTAL	MAT.	INST.	TOTAL	MAT.	INST.	TOTAL	MAT.	INST.	TOTAL	MAT.	INST.	TOTAL
015433	CONTRACTOR EQUIPMENT		116.6	116.6		98.6	98.6		125.3	125.3		96.9	96.9		96.9	96.9		96.9	96.9
0241, 31 - 34	SITE & INFRASTRUCTURE, DEMOLITION	127.9	103.3	111.2	119.9	90.8	100.2	139.2	116.7	123.9	101.3	89.3	93.2	100.8	88.9	92.8	101.4	88.9	92.9
0310	Concrete Forming & Accessories	128.6	90.4	97.8	135.3	75.6	87.1	129.2	86.3	94.6	137.4	89.4	98.7	137.4	76.5	88.3	137.4	76.5	88.3
0320	Concrete Reinforcing	65.4	93.5	74.4	75.9	76.7	76.2	64.5	82.7	70.3	52.7	93.4	65.8	73.9	69.5	72.5	73.9	69.5	72.5
0330	Cast-in-Place Concrete	117.6	97.5	110.7	137.7	89.6	121.1	176.1	93.9	147.7	124.8	93.2	113.9	118.6	83.8	106.6	122.5	83.8	109.2
03	CONCRETE	113.1	94.1	105.5	117.6	81.3	103.0	140.3	89.2	119.9	107.0	91.6	100.8	108.8	78.5	96.7	110.6	78.5	97.8
04	MASONRY	192.2	89.2	127.3	169.8	74.8	109.9	218.5	81.5	132.1	170.0	89.1	119.0	170.1	73.6	109.4	170.4	73.6	109.4
05	METALS	97.5	102.0	98.5	81.3	86.3	82.5	97.6	98.3	97.8	79.3	92.2	82.3	79.7	83.6	80.6	79.7	83.6	80.6
06	WOOD, PLASTICS & COMPOSITES	112.8	91.0	103.8	117.8	75.3	100.2	108.0	87.1	99.3	134.5	90.3	116.2	134.5	76.8	110.6	134.5	76.8	110.6
07	THERMAL & MOISTURE PROTECTION	127.7	96.5	114.0	136.3	82.6	112.8	148.7	81.9	119.4	114.1	94.3	105.4	114.1	80.2	99.2	114.8	80.2	99.6
08	OPENINGS	85.5	85.3	85.5	86.1	69.9	82.2	89.9	75.3	86.4	85.8	76.0	83.4	86.1	65.9	81.2	86.1	65.9	81.2
0920	Plaster & Gypsum Board	137.1	90.1	107.7	119.2	74.5	91.3	143.1	86.0	107.4	160.4	90.1	116.5	138.2	76.3	99.5	138.2	76.3	99.5
0950, 0980	Ceilings & Acoustic Treatment	138.4	90.1	107.1	103.7	74.5	84.8	162.9	86.0	113.0	103.1	90.1	94.7	103.1	76.3	85.7	103.1	76.3	85.7
0960	Flooring	117.0	85.4	107.4	117.1	80.0	105.7	128.2	89.7	116.4	114.1	82.5	104.4	112.9	82.5	103.6	112.9	82.5	103.6
0970, 0990	Wall Finishes & Painting/Coating	119.2	102.7	109.1	110.8	71.9	87.1	112.0	82.6	94.1	113.6	102.7	107.0	110.9	80.1	92.1	110.9	80.1	92.1
09	FINISHES	129.4	91.6	108.9	112.4	75.9	92.6	136.4	87.4	109.8	117.4	90.6	102.8	113.2	78.4	94.3	113.2	78.4	94.3
COVERS	DIVS. 10 - 14, 25, 28, 41, 43, 44, 46	131.1	78.5	119.7	131.1	80.5	120.1	131.1	66.0	117.0	131.1	76.9	119.3	131.1	71.6	118.2	131.1	71.6	118.2
21, 22, 23	FIRE SUPPRESSION, PLUMBING & HVAC	107.3	79.9	96.1	106.5	82.9	96.8	107.2	80.6	96.3	106.5	79.7	95.5	106.5	80.5	95.8	102.5	80.5	93.5
26, 27, 3370	ELECTRICAL, COMMUNICATIONS & UTIL.	126.3	82.8	105.3	109.4	74.1	92.4	123.3	84.8	104.7	116.0	82.8	100.0	116.0	62.4	90.2	116.7	62.4	90.5
MF2018	WEIGHTED AVERAGE	114.2	89.2	104.1	107.6	79.9	96.3	120.4	87.1	106.9	106.0	86.2	97.9	106.0	76.7	94.1	105.5	76.7	93.8

CANADA

DIVISION		SAINT JOHN, NEW BRUNSWICK			SARNIA, ONTARIO			SASKATOON, SASKATCHEWAN			SAULT STE MARIE, ONTARIO			SHERBROOKE, QUEBEC			SOREL, QUEBEC		
		MAT.	INST.	TOTAL	MAT.	INST.	TOTAL	MAT.	INST.	TOTAL	MAT.	INST.	TOTAL	MAT.	INST.	TOTAL	MAT.	INST.	TOTAL
015433	CONTRACTOR EQUIPMENT		95.9	95.9		96.4	96.4		95.3	95.3		96.4	96.4		96.9	96.9		96.9	96.9
0241, 31 - 34	SITE & INFRASTRUCTURE, DEMOLITION	114.1	88.0	96.4	123.4	89.6	100.4	117.1	88.9	97.9	112.9	89.2	96.8	101.4	88.9	92.9	101.6	88.9	93.0
0310	Concrete Forming & Accessories	128.2	64.3	76.7	127.0	84.8	93.0	102.0	85.4	88.6	111.4	84.1	89.4	137.4	76.5	88.3	137.4	76.6	88.4
0320	Concrete Reinforcing	70.5	67.8	69.6	58.5	83.3	66.5	57.2	82.6	65.4	52.6	82.2	62.1	73.9	69.5	72.5	69.9	69.6	69.8
0330	Cast-in-Place Concrete	113.8	65.6	97.2	137.5	90.4	121.2	132.9	88.2	117.5	123.5	77.1	107.5	122.5	83.8	109.2	123.5	83.9	109.8
03	CONCRETE	105.8	66.5	90.1	113.1	86.9	102.6	104.2	86.3	97.0	99.1	82.0	92.2	110.6	78.5	97.8	110.2	78.6	97.5
04	MASONRY	188.9	73.2	116.0	187.7	87.2	124.4	181.4	81.2	118.3	171.0	89.6	119.7	170.5	73.6	109.4	170.6	73.6	109.4
05	METALS	82.2	85.5	83.0	80.7	90.0	82.9	76.4	87.9	79.1	80.1	93.9	83.4	79.6	83.6	80.5	79.7	83.7	80.6
06	WOOD, PLASTICS & COMPOSITES	121.4	62.4	97.0	119.8	84.4	105.1	92.3	86.2	89.8	104.2	86.3	96.8	134.5	76.8	110.6	134.5	76.8	110.6
07	THERMAL & MOISTURE PROTECTION	114.1	69.6	94.6	119.7	87.2	105.5	116.7	78.5	100.0	118.3	84.1	103.3	114.1	80.2	99.2	114.1	80.2	99.2
08	OPENINGS	81.2	58.3	75.7	87.5	81.1	86.0	82.5	74.8	80.6	79.7	83.9	80.7	86.1	65.9	81.2	86.1	70.2	82.3
0920	Plaster & Gypsum Board	148.0	61.6	94.0	151.5	84.2	109.4	121.0	86.0	99.1	118.2	86.2	98.2	138.2	76.3	99.5	160.4	76.3	107.8
0950, 0980	Ceilings & Acoustic Treatment	121.9	61.6	82.7	109.9	84.2	93.2	130.3	86.0	101.6	103.7	86.2	92.4	103.1	76.3	85.7	103.1	76.3	85.7
0960	Flooring	111.2	62.3	96.2	112.9	90.3	106.0	108.2	89.7	102.5	107.1	87.4	101.0	112.9	82.5	103.6	112.9	82.5	103.6
0970, 0990	Wall Finishes & Painting/Coating	112.4	77.5	91.1	110.9	94.4	100.8	114.1	82.6	94.9	110.9	87.4	96.6	110.9	80.1	92.1	110.9	80.1	92.1
09	FINISHES	118.8	64.9	89.5	117.6	87.2	101.1	117.5	86.7	100.8	108.5	85.6	96.0	113.2	78.4	94.3	116.9	78.4	96.0
COVERS	DIVS. 10 - 14, 25, 28, 41, 43, 44, 46	131.1	58.2	115.3	131.1	62.6	116.2	131.1	64.3	116.6	131.1	82.3	120.5	131.1	71.6	118.2	131.1	71.6	118.2
21, 22, 23	FIRE SUPPRESSION, PLUMBING & HVAC	106.4	76.5	94.1	106.5	97.1	102.6	106.8	80.3	96.0	106.5	85.6	97.9	106.9	80.5	96.1	106.5	80.5	95.8
26, 27, 3370	ELECTRICAL, COMMUNICATIONS & UTIL.	125.1	86.5	106.5	121.3	81.3	102.0	121.8	84.8	104.0	119.3	79.6	100.2	116.0	62.4	90.2	116.0	62.4	90.2
MF2018	WEIGHTED AVERAGE	108.0	74.7	94.4	109.1	87.8	100.4	105.8	83.3	96.6	104.2	85.5	96.6	106.3	76.7	94.3	106.5	76.9	94.4

CANADA

DIVISION		ST CATHARINES, ONTARIO			ST JEROME, QUEBEC			ST. JOHN'S, NEWFOUNDLAND			SUDBURY, ONTARIO			SUMMERSIDE, PRINCE EDWARD ISLAND			SYDNEY, NOVA SCOTIA		
		MAT.	INST.	TOTAL	MAT.	INST.	TOTAL	MAT.	INST.	TOTAL	MAT.	INST.	TOTAL	MAT.	INST.	TOTAL	MAT.	INST.	TOTAL
015433	CONTRACTOR EQUIPMENT		96.5	96.5		96.9	96.9		118.8	118.8		96.5	96.5		96.2	96.2		96.3	96.3
0241, 31 - 34	SITE & INFRASTRUCTURE, DEMOLITION	101.8	91.7	94.9	100.8	88.9	92.8	113.4	109.8	111.0	101.9	91.2	94.6	119.6	85.2	96.3	105.3	87.7	93.4
0310	Concrete Forming & Accessories	113.6	91.1	95.5	137.4	76.5	88.3	128.1	81.9	90.8	107.7	86.7	90.8	104.9	50.8	61.3	104.6	65.7	73.2
0320	Concrete Reinforcing	48.8	100.1	65.2	73.9	69.5	72.5	83.6	79.8	82.4	49.2	96.7	64.4	77.2	44.6	66.7	78.2	45.7	67.8
0330	Cast-in-Place Concrete	134.9	95.5	121.3	118.6	83.8	106.6	136.7	95.9	122.6	136.0	92.1	120.8	133.0	52.4	105.2	108.1	64.0	92.8
03	CONCRETE	102.6	94.5	99.3	108.8	78.5	96.7	130.5	87.3	113.2	102.7	90.7	97.9	137.5	51.5	103.1	116.3	62.7	94.9
04	MASONRY	152.5	98.0	118.1	170.1	73.6	109.2	208.1	87.5	132.1	152.6	94.6	116.0	215.1	52.9	112.8	212.8	63.2	118.5
05	METALS	81.9	97.8	85.6	79.7	83.6	80.6	99.4	94.9	98.3	81.5	96.3	85.0	90.2	68.2	85.1	90.2	74.8	86.6
06	WOOD, PLASTICS & COMPOSITES	107.0	91.1	100.4	134.5	76.8	110.6	107.6	80.2	96.3	101.8	86.3	95.4	110.1	50.4	85.4	109.2	65.6	91.1
07	THERMAL & MOISTURE PROTECTION	115.6	96.4	107.2	114.1	80.2	99.2	144.1	91.0	120.8	114.5	91.7	104.5	126.6	54.5	95.0	126.6	64.4	99.3
08	OPENINGS	77.1	89.0	79.9	86.1	65.9	81.2	87.2	71.6	83.4	77.6	84.3	79.3	98.9	44.1	85.7	88.6	59.5	81.6
0920	Plaster & Gypsum Board	112.8	91.1	99.2	138.2	76.3	99.5	151.0	79.1	106.0	110.0	86.2	95.1	145.0	49.2	85.1	142.8	64.8	94.0
0950, 0980	Ceilings & Acoustic Treatment	105.4	91.1	96.1	103.1	76.3	85.7	148.9	79.1	103.6	99.2	86.2	90.8	122.9	49.2	75.1	122.9	64.8	85.2
0960	Flooring	101.2	86.7	96.8	112.9	82.5	103.6	121.8	49.3	99.6	99.7	87.4	95.9	109.8	53.1	92.4	109.8	56.6	93.5
0970, 0990	Wall Finishes & Painting/Coating	105.3	96.6	100.0	110.9	80.1	92.1	114.6	94.9	102.6	105.3	88.4	95.0	117.5	38.2	69.1	117.5	56.8	80.5
09	FINISHES	104.5	91.1	97.2	113.2	78.4	94.3	135.5	77.1	103.8	102.2	86.9	93.9	123.4	50.1	83.6	122.0	63.3	90.1
COVERS	DIVS. 10 - 14, 25, 28, 41, 43, 44, 46	131.1	65.4	116.8	131.1	71.6	118.2	131.1	64.2	116.6	131.1	84.0	120.9	131.1	54.0	114.4	131.1	57.6	115.1
21, 22, 23	FIRE SUPPRESSION, PLUMBING & HVAC	106.4	85.0	97.6	106.5	80.5	95.8	107.1	79.8	95.9	106.2	83.4	96.8	106.7	56.3	86.0	106.7	76.0	94.1
26, 27, 3370	ELECTRICAL, COMMUNICATIONS & UTIL.	115.4	94.4	105.3	116.9	62.4	90.6	126.1	75.4	101.6	112.5	94.7	103.9	113.9	45.4	80.8	115.4	56.1	86.8
MF2018	WEIGHTED AVERAGE	103.1	91.3	98.3	106.1	76.7	94.1	118.2	83.8	104.2	102.5	89.6	97.3	116.1	55.6	91.5	111.9	67.6	93.8

		CANADA																
	DIVISION	THUNDER BAY, ONTARIO			TIMMINS, ONTARIO			TORONTO, ONTARIO			TROIS RIVIERES, QUEBEC			TRURO, NOVA SCOTIA			VANCOUVER, BRITISH COLUMBIA		
		MAT.	INST.	TOTAL	MAT.	INST.	TOTAL	MAT.	INST.	TOTAL	MAT.	INST.	TOTAL	MAT.	INST.	TOTAL	MAT.	INST.	TOTAL
015433	CONTRACTOR EQUIPMENT		96.5	96.5		96.4	96.4		114.6	114.6		97.0	97.0		96.1	96.1		131.4	131.4
0241, 31 - 34	SITE & INFRASTRUCTURE, DEMOLITION	107.4	91.4	96.5	125.1	89.1	100.7	115.0	104.9	108.1	105.7	89.1	94.4	106.4	87.6	93.6	121.6	119.2	120.0
0310	Concrete Forming & Accessories	124.6	89.0	95.9	128.8	77.5	87.4	131.1	100.2	106.1	165.5	76.7	93.8	89.1	65.7	70.2	131.2	88.0	96.3
0320	Concrete Reinforcing	43.6	99.5	61.6	82.0	81.4	81.8	67.1	100.8	77.9	78.2	69.6	75.4	69.9	45.7	62.1	69.7	91.4	76.7
0330	Cast-in-Place Concrete	148.5	94.5	129.9	149.0	76.9	124.1	137.8	111.7	128.8	111.9	83.9	102.2	154.8	63.9	123.4	142.1	98.0	126.9
03	CONCRETE	111.1	93.0	103.9	123.4	78.6	105.5	122.5	104.8	115.4	120.6	78.6	103.8	121.8	62.7	98.1	127.8	93.2	114.0
04	MASONRY	153.5	96.2	117.4	175.0	78.9	114.4	187.6	105.9	136.1	217.3	73.6	126.7	170.3	63.2	102.8	187.2	90.2	126.0
05	METALS	81.9	96.6	85.4	80.8	89.2	82.7	94.1	107.0	97.1	89.7	83.8	88.3	80.2	74.7	78.9	94.2	107.5	97.3
06	WOOD, PLASTICS & COMPOSITES	118.8	88.8	106.4	120.8	78.1	103.1	116.5	98.8	109.2	172.5	76.9	132.9	83.7	65.5	76.2	110.3	87.2	100.8
07	THERMAL & MOISTURE PROTECTION	116.1	93.3	106.1	119.7	78.8	101.7	140.3	106.2	125.3	126.5	80.2	106.2	114.2	64.4	92.4	145.7	88.9	120.8
08	OPENINGS	76.4	87.2	79.0	84.6	75.9	82.5	86.0	96.4	88.5	96.9	70.2	90.5	79.9	59.5	74.9	85.5	84.9	85.3
0920	Plaster & Gypsum Board	138.1	88.7	107.2	131.9	77.7	98.0	135.3	98.2	112.1	196.7	76.3	121.4	117.3	64.8	84.5	135.9	86.0	104.7
0950, 0980	Ceilings & Acoustic Treatment	99.2	88.7	92.4	103.7	77.7	86.8	148.8	98.2	115.9	120.9	76.3	92.0	103.7	64.8	78.5	154.1	86.0	109.9
0960	Flooring	104.6	92.9	101.0	112.9	82.6	103.7	117.9	95.1	110.9	129.3	82.5	115.0	98.1	56.6	85.4	132.8	83.5	117.7
0970, 0990	Wall Finishes & Painting/Coating	105.3	89.3	95.5	110.9	81.4	92.9	108.6	100.3	103.6	117.5	80.1	94.7	110.9	56.8	77.9	108.5	95.9	100.8
09	FINISHES	108.5	89.8	98.3	113.0	79.2	94.6	125.8	99.3	111.4	135.6	78.4	104.5	106.0	63.3	82.8	133.1	87.7	108.5
COVERS	DIVS. 10 - 14, 25, 28, 41, 43, 44, 46	131.1	64.9	116.7	131.1	60.0	115.7	131.1	89.8	122.2	131.1	71.6	118.2	131.1	57.5	115.1	131.1	84.8	121.1
21, 22, 23	FIRE SUPPRESSION, PLUMBING & HVAC	106.4	84.5	97.4	106.5	88.3	99.0	107.7	95.4	102.7	106.7	80.5	95.9	106.5	76.0	94.0	107.6	85.7	98.6
26, 27, 3370	ELECTRICAL, COMMUNICATIONS & UTIL.	113.2	97.8	105.7	119.3	79.6	100.2	115.2	100.3	108.0	115.8	62.4	90.1	115.5	56.1	86.9	120.3	78.5	100.2
MF2018	WEIGHTED AVERAGE	104.6	90.8	99.0	109.2	82.0	98.1	113.4	100.9	108.3	115.1	76.9	99.6	106.3	67.5	90.5	115.3	91.0	105.4

		CANADA																	
	DIVISION	VICTORIA, BRITISH COLUMBIA			WHITEHORSE, YUKON			WINDSOR, ONTARIO			WINNIPEG, MANITOBA			YARMOUTH, NOVA SCOTIA			YELLOWKNIFE, NWT		
		MAT.	INST.	TOTAL	MAT.	INST.	TOTAL	MAT.	INST.	TOTAL	MAT.	INST.	TOTAL	MAT.	INST.	TOTAL	MAT.	INST.	TOTAL
015433	CONTRACTOR EQUIPMENT		103.0	103.0		127.3	127.3		96.5	96.5		123.9	123.9		96.3	96.3		123.8	123.8
0241, 31 - 34	SITE & INFRASTRUCTURE, DEMOLITION	130.0	97.4	107.9	132.9	114.4	120.3	98.8	91.5	93.9	124.0	112.9	116.5	109.8	87.7	94.8	148.7	112.9	124.4
0310	Concrete Forming & Accessories	109.4	85.3	89.9	134.4	54.8	70.2	124.6	87.7	94.8	130.2	62.4	75.5	104.6	65.7	73.2	139.4	73.6	86.3
0320	Concrete Reinforcing	56.0	91.2	67.3	81.5	60.7	74.9	47.8	98.9	64.2	65.2	57.6	62.8	78.2	45.7	67.8	73.1	63.3	70.0
0330	Cast-in-Place Concrete	137.7	93.1	122.3	178.7	69.0	140.8	138.2	96.1	123.6	172.1	69.4	136.6	138.7	64.0	112.9	199.6	85.5	160.2
03	CONCRETE	130.8	89.4	114.2	150.1	62.4	115.0	104.6	92.9	99.9	138.7	65.4	109.4	129.8	62.7	103.0	157.9	77.2	125.6
04	MASONRY	184.7	90.0	125.0	261.7	56.0	132.1	152.7	97.2	117.7	213.3	61.8	117.8	215.9	63.2	119.6	250.8	66.6	134.7
05	METALS	77.4	92.7	81.0	102.4	85.4	98.4	81.9	97.1	85.5	99.5	84.3	95.9	90.2	74.8	86.6	99.0	89.4	96.8
06	WOOD, PLASTICS & COMPOSITES	96.4	84.6	91.5	119.5	53.8	92.3	118.8	86.5	105.5	107.2	63.3	89.0	109.2	65.6	91.1	129.6	75.1	107.1
07	THERMAL & MOISTURE PROTECTION	127.1	84.6	108.5	143.9	61.8	107.9	115.9	93.6	106.1	141.5	67.3	109.0	126.6	64.4	99.3	145.2	76.8	115.2
08	OPENINGS	83.9	79.6	82.9	95.9	51.4	85.1	76.3	86.4	78.7	88.2	57.0	80.7	88.6	59.5	81.6	90.5	63.7	84.0
0920	Plaster & Gypsum Board	112.9	84.1	94.9	174.7	51.6	97.8	127.0	86.4	101.6	138.1	61.4	90.2	142.8	64.8	94.0	186.0	73.6	115.7
0950, 0980	Ceilings & Acoustic Treatment	106.3	84.1	91.9	176.7	51.6	95.5	99.2	86.4	90.9	154.0	61.4	93.9	122.9	64.8	85.2	175.0	73.6	109.2
0960	Flooring	112.7	64.7	98.0	127.2	53.0	104.5	104.6	90.4	100.2	122.9	64.6	105.0	109.8	56.6	93.5	126.1	79.6	111.9
0970, 0990	Wall Finishes & Painting/Coating	114.1	95.9	103.0	119.8	50.9	77.8	105.3	90.1	96.0	108.7	50.3	73.1	117.5	56.8	80.5	125.4	72.7	93.3
09	FINISHES	113.6	82.8	96.9	152.7	53.8	99.0	106.3	88.4	96.6	132.9	61.8	94.3	122.0	63.3	90.1	153.7	74.1	110.5
COVERS	DIVS. 10 - 14, 25, 28, 41, 43, 44, 46	131.1	61.8	116.1	131.1	57.3	115.1	131.1	64.9	116.7	131.1	59.8	115.6	131.1	57.6	115.1	131.1	60.8	115.8
21, 22, 23	FIRE SUPPRESSION, PLUMBING & HVAC	106.5	82.5	96.7	107.0	68.5	91.2	106.4	85.9	98.0	107.7	60.7	88.4	106.7	76.0	94.1	107.6	84.7	98.2
26, 27, 3370	ELECTRICAL, COMMUNICATIONS & UTIL.	118.8	79.3	99.8	145.3	54.6	101.6	119.7	98.5	109.5	124.9	59.9	93.6	115.4	56.1	86.8	137.9	74.1	107.2
MF2018	WEIGHTED AVERAGE	109.9	85.2	99.8	128.4	66.2	103.1	103.9	91.1	98.7	119.7	67.7	98.5	114.0	67.6	95.1	127.8	79.6	108.2

Location Factors - Commercial

Costs shown in RSMeans cost data publications are based on national averages for materials and installation. To adjust these costs to a specific location, simply multiply the base cost by the factor and divide by 100 for that city. The data is arranged alphabetically by state and postal zip code numbers. For a city not listed, use the factor for a nearby city with similar economic characteristics.

STATE/ZIP	CITY	MAT.	INST.	TOTAL
ALABAMA				
350-352	Birmingham	99.4	71.0	87.8
354	Tuscaloosa	98.1	70.2	86.8
355	Jasper	98.7	68.2	86.3
356	Decatur	98.1	68.6	86.1
357-358	Huntsville	98.1	70.1	86.7
359	Gadsden	98.2	69.8	86.6
360-361	Montgomery	98.8	71.7	87.8
362	Anniston	97.5	65.3	84.4
363	Dothan	97.8	71.7	87.1
364	Evergreen	97.2	66.4	84.7
365-366	Mobile	98.4	68.2	86.1
367	Selma	97.3	69.8	86.1
368	Phenix City	98.1	69.7	86.5
369	Butler	97.4	68.2	85.5
ALASKA				
995-996	Anchorage	117.1	108.7	113.7
997	Fairbanks	117.8	109.2	114.3
998	Juneau	118.2	108.7	114.3
999	Ketchikan	128.9	109.3	120.9
ARIZONA				
850,853	Phoenix	98.8	72.6	88.1
851,852	Mesa/Tempe	96.2	70.5	85.7
855	Globe	96.3	69.8	85.5
856-857	Tucson	95.1	71.9	85.7
859	Show Low	96.7	69.9	85.8
860	Flagstaff	100.4	71.5	88.6
863	Prescott	98.1	70.3	86.8
864	Kingman	96.3	69.0	85.2
865	Chambers	96.5	70.3	85.8
ARKANSAS				
716	Pine Bluff	94.2	62.4	81.3
717	Camden	92.3	62.7	80.3
718	Texarkana	92.8	63.2	80.8
719	Hot Springs	91.7	61.9	79.6
720-722	Little Rock	95.3	62.7	82.0
723	West Memphis	93.0	66.8	82.4
724	Jonesboro	93.4	65.2	81.9
725	Batesville	91.1	61.7	79.1
726	Harrison	92.5	61.0	79.7
727	Fayetteville	90.2	62.9	79.1
728	Russellville	91.5	60.8	79.0
729	Fort Smith	94.3	61.6	81.0
CALIFORNIA				
900-902	Los Angeles	102.0	129.6	113.2
903-905	Inglewood	94.8	128.0	108.3
906-908	Long Beach	96.5	128.0	109.3
910-912	Pasadena	95.0	127.8	108.3
913-916	Van Nuys	98.1	127.8	110.2
917-918	Alhambra	97.0	127.8	109.5
919-921	San Diego	102.4	122.2	110.4
922	Palm Springs	97.0	125.4	108.5
923-924	San Bernardino	94.2	125.8	107.1
925	Riverside	98.9	124.9	109.5
926-927	Santa Ana	96.6	125.6	108.4
928	Anaheim	99.1	124.9	109.6
930	Oxnard	97.4	125.4	108.8
931	Santa Barbara	96.9	124.5	108.1
932-933	Bakersfield	98.5	124.1	108.9
934	San Luis Obispo	98.3	125.0	109.2
935	Mojave	95.2	123.8	106.9
936-938	Fresno	97.7	128.9	110.4
939	Salinas	99.0	137.6	114.7
940-941	San Francisco	105.0	160.3	127.5
942,956-958	Sacramento	97.0	132.8	111.6
943	Palo Alto	96.1	154.5	119.8
944	San Mateo	98.3	152.6	120.4
945	Vallejo	97.4	141.6	115.4
946	Oakland	100.1	151.6	121.1
947	Berkeley	100.1	151.5	121.0
948	Richmond	99.3	145.5	118.1
949	San Rafael	101.4	148.8	120.7
950	Santa Cruz	104.6	137.8	118.1

STATE/ZIP	CITY	MAT.	INST.	TOTAL
CALIFORNIA (CONT'D)				
951	San Jose	102.3	154.3	123.5
952	Stockton	100.0	131.6	112.8
953	Modesto	99.8	130.9	112.5
954	Santa Rosa	100.6	146.4	119.2
955	Eureka	102.2	134.1	115.2
959	Marysville	101.5	131.4	113.7
960	Redding	105.7	131.4	116.2
961	Susanville	104.9	131.4	115.7
COLORADO				
800-802	Denver	103.1	75.3	91.8
803	Boulder	96.5	77.4	88.8
804	Golden	98.3	72.8	87.9
805	Fort Collins	99.7	73.9	89.2
806	Greeley	97.0	73.9	87.6
807	Fort Morgan	97.1	72.3	87.0
808-809	Colorado Springs	98.8	71.8	87.9
810	Pueblo	99.5	70.6	87.7
811	Alamosa	100.6	68.8	87.7
812	Salida	100.2	69.4	87.6
813	Durango	101.2	65.2	86.6
814	Montrose	99.6	65.9	85.9
815	Grand Junction	102.9	70.6	89.8
816	Glenwood Springs	100.9	66.5	86.9
CONNECTICUT				
060	New Britain	97.0	115.4	104.5
061	Hartford	99.3	115.9	106.1
062	Willimantic	97.8	114.3	104.5
063	New London	94.1	114.0	102.2
064	Meriden	96.2	113.9	103.4
065	New Haven	98.6	115.1	105.3
066	Bridgeport	98.3	115.2	105.2
067	Waterbury	98.0	115.1	105.0
068	Norwalk	97.9	115.4	105.0
069	Stamford	98.0	115.8	105.3
D.C.				
200-205	Washington	102.7	88.5	96.9
DELAWARE				
197	Newark	98.1	110.6	103.2
198	Wilmington	98.9	111.1	103.8
199	Dover	99.4	110.6	103.9
FLORIDA				
320,322	Jacksonville	97.7	66.9	85.2
321	Daytona Beach	97.9	69.1	86.2
323	Tallahassee	99.0	66.9	86.0
324	Panama City	99.4	67.0	86.2
325	Pensacola	101.8	64.6	86.7
326,344	Gainesville	99.1	62.8	84.3
327-328,347	Orlando	99.7	66.5	86.2
329	Melbourne	101.1	70.8	88.8
330-332,340	Miami	97.5	70.1	86.4
333	Fort Lauderdale	97.3	67.8	85.3
334,349	West Palm Beach	96.0	62.6	82.4
335-336,346	Tampa	99.0	67.4	86.2
337	St. Petersburg	101.2	65.6	86.7
338	Lakeland	98.1	65.5	84.8
339,341	Fort Myers	97.5	65.4	84.4
342	Sarasota	100.0	64.1	85.3
GEORGIA				
300-303,399	Atlanta	102.2	75.4	91.3
304	Statesboro	99.8	70.3	87.8
305	Gainesville	98.5	62.9	84.0
306	Athens	97.8	62.4	83.4
307	Dalton	99.8	66.4	86.2
308-309	Augusta	98.1	72.2	87.6
310-312	Macon	95.9	72.7	86.4
313-314	Savannah	98.2	71.9	87.5
315	Waycross	97.5	67.4	85.2
316	Valdosta	97.5	70.2	86.4
317,398	Albany	97.6	70.7	86.6
318-319	Columbus	97.5	72.2	87.2

STATE/ZIP	CITY	MAT.	INST.	TOTAL
HAWAII				
967	Hilo	114.8	115.0	114.9
968	Honolulu	119.3	115.8	117.9
STATES & POSS.				
969	Guam	136.6	61.9	106.2
IDAHO				
832	Pocatello	100.5	78.2	91.4
833	Twin Falls	101.0	78.1	91.7
834	Idaho Falls	98.5	79.9	90.9
835	Lewiston	104.7	85.6	97.0
836-837	Boise	99.5	78.3	90.9
838	Coeur d'Alene	104.3	86.0	96.9
ILLINOIS				
600-603	North Suburban	95.3	134.1	111.1
604	Joliet	95.1	143.9	115.0
605	South Suburban	95.3	135.7	111.7
606-608	Chicago	98.5	144.8	117.3
609	Kankakee	92.1	126.3	106.0
610-611	Rockford	95.1	129.5	109.1
612	Rock Island	93.4	93.2	93.3
613	La Salle	94.6	119.3	104.7
614	Galesburg	94.1	102.5	97.5
615-616	Peoria	96.0	109.8	101.6
617	Bloomington	93.4	103.0	97.3
618-619	Champaign	96.7	103.8	99.6
620-622	East St. Louis	94.0	109.0	100.1
623	Quincy	96.2	98.7	97.2
624	Effingham	95.4	102.5	98.3
625	Decatur	96.9	107.1	101.1
626-627	Springfield	98.0	107.9	102.0
628	Centralia	93.0	104.1	97.5
629	Carbondale	92.7	102.7	96.8
INDIANA				
460	Anderson	94.0	78.7	87.8
461-462	Indianapolis	97.9	83.2	91.9
463-464	Gary	95.2	103.1	98.4
465-466	South Bend	97.4	81.7	91.0
467-468	Fort Wayne	94.6	78.7	88.2
469	Kokomo	91.8	78.0	86.2
470	Lawrenceburg	91.4	76.4	85.3
471	New Albany	92.8	74.5	85.3
472	Columbus	95.0	77.6	87.9
473	Muncie	95.3	77.2	87.9
474	Bloomington	96.7	79.6	89.7
475	Washington	92.9	82.8	88.8
476-477	Evansville	94.2	85.4	90.7
478	Terre Haute	95.0	80.9	89.3
479	Lafayette	94.6	78.0	87.8
IOWA				
500-503,509	Des Moines	95.9	91.0	93.9
504	Mason City	93.1	69.5	83.5
505	Fort Dodge	93.2	69.3	83.4
506-507	Waterloo	95.2	76.5	87.6
508	Creston	93.7	79.1	87.7
510-511	Sioux City	96.8	78.3	89.3
512	Sibley	95.2	58.7	80.4
513	Spencer	96.5	58.5	81.1
514	Carroll	94.1	77.4	87.3
515	Council Bluffs	97.3	79.9	90.2
516	Shenandoah	94.9	77.0	87.6
520	Dubuque	95.7	78.7	88.7
521	Decorah	95.1	69.3	84.6
522-524	Cedar Rapids	97.1	84.4	91.9
525	Ottumwa	94.9	74.0	86.4
526	Burlington	94.5	79.2	88.3
527-528	Davenport	96.1	94.2	95.3
KANSAS				
660-662	Kansas City	95.7	98.8	97.0
664-666	Topeka	97.1	76.4	88.7
667	Fort Scott	94.2	72.3	85.3
668	Emporia	94.0	72.4	85.2
669	Belleville	95.6	67.6	84.2
670-672	Wichita	96.1	71.9	86.2
673	Independence	96.1	71.9	86.3
674	Salina	96.1	72.0	86.3
675	Hutchinson	91.6	68.2	82.1
676	Hays	95.3	68.8	84.5
677	Colby	96.0	70.1	85.5

STATE/ZIP	CITY	MAT.	INST.	TOTAL
KANSAS (CONT'D)				
678	Dodge City	97.4	73.2	87.5
679	Liberal	95.4	68.0	84.2
KENTUCKY				
400-402	Louisville	93.9	77.9	87.4
403-405	Lexington	92.6	76.9	86.2
406	Frankfort	94.8	76.9	87.5
407-409	Corbin	89.9	73.3	83.2
410	Covington	92.8	73.3	84.9
411-412	Ashland	91.8	84.2	88.7
413-414	Campton	93.1	74.6	85.6
415-416	Pikeville	94.1	80.1	88.4
417-418	Hazard	92.5	75.2	85.5
420	Paducah	90.7	78.0	85.5
421-422	Bowling Green	93.4	76.5	86.5
423	Owensboro	92.9	80.0	87.7
424	Henderson	90.5	76.8	84.9
425-426	Somerset	90.5	74.8	84.1
427	Elizabethtown	90.3	71.7	82.7
LOUISIANA				
700-701	New Orleans	98.8	68.6	86.5
703	Thibodaux	93.9	63.6	81.6
704	Hammond	91.8	62.2	79.8
705	Lafayette	93.8	64.5	81.9
706	Lake Charles	94.0	66.9	83.0
707-708	Baton Rouge	95.8	67.2	84.2
710-711	Shreveport	95.4	65.8	83.3
712	Monroe	93.3	63.3	81.1
713-714	Alexandria	93.5	63.3	81.2
MAINE				
039	Kittery	91.6	83.3	88.2
040-041	Portland	98.1	87.3	93.7
042	Lewiston	95.8	85.2	91.5
043	Augusta	99.6	84.5	93.5
044	Bangor	94.5	84.2	90.3
045	Bath	93.2	82.9	89.0
046	Machias	93.0	82.8	88.9
047	Houlton	93.2	82.8	88.9
048	Rockland	92.4	82.9	88.5
049	Waterville	93.2	82.8	89.0
MARYLAND				
206	Waldorf	98.1	83.0	91.9
207-208	College Park	97.3	84.0	91.9
209	Silver Spring	96.8	83.2	91.2
210-212	Baltimore	101.5	83.7	94.2
214	Annapolis	99.6	81.8	92.3
215	Cumberland	95.2	81.6	89.7
216	Easton	96.8	70.9	86.2
217	Hagerstown	95.8	86.3	91.9
218	Salisbury	97.5	63.5	83.6
219	Elkton	94.3	81.0	88.9
MASSACHUSETTS				
010-011	Springfield	96.8	107.9	101.3
012	Pittsfield	95.8	102.4	98.5
013	Greenfield	93.9	106.7	99.1
014	Fitchburg	92.5	110.2	99.7
015-016	Worcester	96.7	115.6	104.4
017	Framingham	92.2	119.3	103.2
018	Lowell	96.0	123.1	107.0
019	Lawrence	96.8	121.7	106.9
020-022, 024	Boston	98.5	134.5	113.1
023	Brockton	96.1	114.7	103.7
025	Buzzards Bay	90.4	111.2	98.9
026	Hyannis	92.8	113.9	101.4
027	New Bedford	95.4	113.3	102.7
MICHIGAN				
480,483	Royal Oak	93.4	94.8	94.0
481	Ann Arbor	95.6	99.5	97.2
482	Detroit	100.0	100.3	100.1
484-485	Flint	95.5	89.9	93.2
486	Saginaw	95.1	86.1	91.5
487	Bay City	95.1	84.1	90.7
488-489	Lansing	97.2	89.7	94.1
490	Battle Creek	94.6	78.2	87.9
491	Kalamazoo	94.8	80.3	88.9
492	Jackson	92.9	85.9	90.1
493,495	Grand Rapids	97.2	80.7	90.5
494	Muskegon	93.6	80.3	88.2

Location Factors - Commercial

STATE/ZIP	CITY	MAT.	INST.	TOTAL
MICHIGAN (CONT'D)				
496	Traverse City	92.5	73.5	84.8
497	Gaylord	93.7	76.3	86.6
498-499	Iron Mountain	95.2	77.1	87.8
MINNESOTA				
550-551	Saint Paul	100.4	116.7	107.0
553-555	Minneapolis	101.2	117.8	108.0
556-558	Duluth	101.3	104.4	102.6
559	Rochester	99.0	104.3	101.2
560	Mankato	94.6	102.4	97.7
561	Windom	92.8	89.4	91.4
562	Willmar	92.8	95.2	93.8
563	St. Cloud	94.0	114.1	102.2
564	Brainerd	94.6	96.2	95.3
565	Detroit Lakes	95.7	88.4	92.7
566	Bemidji	95.2	92.6	94.1
567	Thief River Falls	94.9	87.5	91.9
MISSISSIPPI				
386	Clarksdale	96.0	52.6	78.3
387	Greenville	99.2	63.8	84.8
388	Tupelo	97.2	55.1	80.0
389	Greenwood	97.5	52.3	79.1
390-392	Jackson	98.4	66.8	85.5
393	Meridian	94.8	66.5	83.3
394	Laurel	96.2	55.1	79.5
395	Biloxi	96.9	65.2	84.0
396	McComb	94.4	52.6	77.4
397	Columbus	96.1	55.8	79.7
MISSOURI				
630-631	St. Louis	98.4	105.4	101.3
633	Bowling Green	94.5	90.2	92.7
634	Hannibal	93.4	86.6	90.6
635	Kirksville	96.4	81.8	90.5
636	Flat River	95.4	88.5	92.6
637	Cape Girardeau	95.1	88.5	92.4
638	Sikeston	93.8	82.4	89.2
639	Poplar Bluff	93.2	82.4	88.8
640-641	Kansas City	98.6	101.6	99.8
644-645	St. Joseph	95.2	91.2	93.6
646	Chillicothe	92.5	88.7	90.9
647	Harrisonville	91.9	95.2	93.2
648	Joplin	94.4	78.4	87.9
650-651	Jefferson City	95.1	86.8	91.7
652	Columbia	94.4	90.9	93.0
653	Sedalia	94.7	86.1	91.2
654-655	Rolla	92.5	94.0	93.1
656-658	Springfield	96.0	79.6	89.3
MONTANA				
590-591	Billings	102.1	77.4	92.0
592	Wolf Point	101.3	73.3	89.9
593	Miles City	99.3	73.3	88.7
594	Great Falls	103.3	75.7	92.0
595	Havre	100.1	71.3	88.4
596	Helena	101.4	73.3	90.0
597	Butte	101.2	73.0	89.7
598	Missoula	98.4	72.9	88.0
599	Kalispell	97.8	72.2	87.4
NEBRASKA				
680-681	Omaha	97.0	80.7	90.4
683-685	Lincoln	98.1	78.2	90.0
686	Columbus	94.2	76.0	86.8
687	Norfolk	95.4	72.9	86.2
688	Grand Island	95.6	78.3	88.6
689	Hastings	95.4	73.6	86.5
690	McCook	94.5	68.8	84.1
691	North Platte	94.8	74.2	86.4
692	Valentine	97.0	66.0	84.4
693	Alliance	97.1	68.9	85.6
NEVADA				
889-891	Las Vegas	103.0	107.2	104.7
893	Ely	101.1	86.1	95.0
894-895	Reno	101.0	83.6	93.9
897	Carson City	101.5	83.9	94.3
898	Elko	100.1	78.6	91.4
NEW HAMPSHIRE				
030	Nashua	97.4	91.8	95.1
031	Manchester	98.2	92.6	95.9

STATE/ZIP	CITY	MAT.	INST.	TOTAL
NEW HAMPSHIRE (CONT'D)				
032-033	Concord	98.6	92.3	96.0
034	Keene	93.5	84.2	89.7
035	Littleton	93.7	75.9	86.5
036	Charleston	93.0	83.6	89.2
037	Claremont	92.4	83.6	88.8
038	Portsmouth	94.0	90.5	92.6
NEW JERSEY				
070-071	Newark	98.8	136.3	114.1
072	Elizabeth	95.1	136.2	111.8
073	Jersey City	94.8	135.2	111.3
074-075	Paterson	96.0	136.3	112.4
076	Hackensack	93.7	135.5	110.7
077	Long Branch	93.4	125.7	106.6
078	Dover	93.8	136.0	111.0
079	Summit	93.9	135.3	110.8
080,083	Vineland	94.6	123.7	106.4
081	Camden	97.1	133.3	111.9
082,084	Atlantic City	95.4	129.2	109.1
085-086	Trenton	98.7	133.0	112.6
087	Point Pleasant	96.3	125.4	108.1
088-089	New Brunswick	96.9	131.4	111.0
NEW MEXICO				
870-872	Albuquerque	96.8	71.0	86.3
873	Gallup	96.8	71.0	86.3
874	Farmington	97.0	71.0	86.4
875	Santa Fe	98.6	71.5	87.6
877	Las Vegas	95.2	71.0	85.4
878	Socorro	95.0	71.0	85.2
879	Truth/Consequences	94.6	68.0	83.8
880	Las Cruces	94.8	69.1	84.3
881	Clovis	96.9	70.9	86.4
882	Roswell	98.4	71.0	87.2
883	Carrizozo	98.9	71.0	87.6
884	Tucumcari	97.4	70.9	86.6
NEW YORK				
100-102	New York	100.1	171.3	129.1
103	Staten Island	93.9	167.8	124.0
104	Bronx	92.7	167.1	122.9
105	Mount Vernon	92.5	143.6	113.3
106	White Plains	92.7	146.7	114.7
107	Yonkers	97.2	146.9	117.4
108	New Rochelle	93.2	138.1	111.5
109	Suffern	92.4	122.1	104.5
110	Queens	99.2	168.8	127.5
111	Long Island City	101.0	168.8	128.6
112	Brooklyn	101.5	168.8	128.9
113	Flushing	101.1	168.8	128.7
114	Jamaica	99.6	168.8	127.8
115,117,118	Hicksville	99.2	149.6	119.7
116	Far Rockaway	101.1	168.8	128.7
119	Riverhead	100.0	154.4	122.1
120-122	Albany	97.8	108.5	102.1
123	Schenectady	97.1	107.6	101.4
124	Kingston	100.1	125.4	110.4
125-126	Poughkeepsie	99.3	128.1	111.1
127	Monticello	98.9	126.1	110.0
128	Glens Falls	91.8	100.5	95.4
129	Plattsburgh	96.3	90.4	93.9
130-132	Syracuse	97.5	100.0	98.5
133-135	Utica	95.6	100.7	97.7
136	Watertown	97.0	98.4	97.6
137-139	Binghamton	97.0	99.2	97.9
140-142	Buffalo	102.6	106.8	104.3
143	Niagara Falls	97.4	100.9	98.8
144-146	Rochester	100.8	99.5	100.3
147	Jamestown	96.3	90.2	93.8
148-149	Elmira	96.3	97.7	96.9
NORTH CAROLINA				
270,272-274	Greensboro	99.7	66.0	86.0
271	Winston-Salem	99.5	66.0	85.9
275-276	Raleigh	98.9	66.3	85.6
277	Durham	101.7	66.0	87.2
278	Rocky Mount	96.8	65.5	84.1
279	Elizabeth City	97.6	66.6	85.0
280	Gastonia	97.8	69.3	86.2
281-282	Charlotte	98.3	71.0	87.2
283	Fayetteville	101.3	64.4	86.3
284	Wilmington	96.7	63.6	83.2
285	Kinston	95.0	64.2	82.5

For customer support on your Commercial Renovation Costs with RSMeans data, call 800.448.8182.

841

Location Factors - Commercial

STATE/ZIP	CITY	MAT.	INST.	TOTAL	STATE/ZIP	CITY	MAT.	INST.	TOTAL
NORTH CAROLINA (CONT'D)					**PENNSYLVANIA (CONT'D)**				
286	Hickory	95.4	69.8	85.0	177	Williamsport	91.6	91.9	91.8
287-288	Asheville	96.9	65.3	84.0	178	Sunbury	93.7	87.1	91.0
289	Murphy	96.2	62.7	82.6	179	Pottsville	92.6	90.2	91.6
					180	Lehigh Valley	94.4	104.8	98.6
NORTH DAKOTA					181	Allentown	96.3	107.4	100.8
580-581	Fargo	98.8	77.2	90.0	182	Hazleton	93.7	90.1	92.3
582	Grand Forks	97.7	79.0	90.1	183	Stroudsburg	93.8	98.4	95.7
583	Devils Lake	97.9	75.8	88.9	184-185	Scranton	97.1	96.6	96.9
584	Jamestown	97.7	75.0	88.5	186-187	Wilkes-Barre	93.6	94.6	94.0
585	Bismarck	98.9	81.2	91.7	188	Montrose	93.3	90.9	92.4
586	Dickinson	97.8	74.4	88.3	189	Doylestown	93.2	119.4	103.9
587	Minot	97.7	74.7	88.3	190-191	Philadelphia	99.9	137.6	115.2
588	Williston	96.9	78.0	89.2	193	Westchester	95.4	120.0	105.4
					194	Norristown	94.2	120.2	104.8
OHIO					195-196	Reading	96.3	101.8	98.6
430-432	Columbus	98.5	84.6	92.8					
433	Marion	93.5	81.7	88.7	**PUERTO RICO**				
434-436	Toledo	96.8	90.3	94.2	009	San Juan	117.0	28.6	81.0
437-438	Zanesville	94.1	81.1	88.8					
439	Steubenville	94.5	85.4	90.8	**RHODE ISLAND**				
440	Lorain	97.3	82.7	91.4	028	Newport	94.5	111.3	101.3
441	Cleveland	99.3	92.0	96.3	029	Providence	98.2	112.0	103.8
442-443	Akron	98.2	85.9	93.2					
444-445	Youngstown	97.6	83.1	91.7	**SOUTH CAROLINA**				
446-447	Canton	97.8	80.3	90.6	290-292	Columbia	98.8	65.7	85.3
448-449	Mansfield	95.1	80.5	89.2	293	Spartanburg	97.3	68.0	85.4
450	Hamilton	95.4	75.0	87.1	294	Charleston	99.1	65.5	85.4
451-452	Cincinnati	98.3	79.8	90.8	295	Florence	96.8	64.9	83.8
453-454	Dayton	95.4	77.8	88.3	296	Greenville	97.1	67.9	85.2
455	Springfield	95.4	79.1	88.8	297	Rock Hill	96.7	66.2	84.3
456	Chillicothe	94.6	83.2	90.0	298	Aiken	97.7	63.2	83.7
457	Athens	97.5	79.8	90.3	299	Beaufort	98.3	65.1	84.8
458	Lima	97.7	77.5	89.5					
					SOUTH DAKOTA				
OKLAHOMA					570-571	Sioux Falls	97.8	80.1	90.6
730-731	Oklahoma City	95.4	68.5	84.5	572	Watertown	95.7	65.7	83.5
734	Ardmore	93.3	65.2	81.9	573	Mitchell	95.2	59.6	80.7
735	Lawton	95.7	66.4	83.8	574	Aberdeen	97.4	74.1	88.0
736	Clinton	94.4	65.1	82.5	575	Pierre	99.3	77.1	90.3
737	Enid	95.2	65.7	83.2	576	Mobridge	95.4	60.7	81.3
738	Woodward	93.4	66.3	82.3	577	Rapid City	97.2	69.2	85.8
739	Guymon	94.4	60.6	80.6					
740-741	Tulsa	94.9	65.0	82.7	**TENNESSEE**				
743	Miami	91.4	63.0	79.8	370-372	Nashville	99.0	75.2	89.3
744	Muskogee	93.8	67.0	82.9	373-374	Chattanooga	98.0	67.1	85.4
745	McAlester	90.8	63.3	79.6	375,380-381	Memphis	98.2	70.8	87.1
746	Ponca City	91.8	60.9	79.2	376	Johnson City	97.3	59.1	81.7
747	Durant	91.5	62.1	79.6	377-379	Knoxville	94.8	65.9	83.0
748	Shawnee	92.8	63.5	80.9	382	McKenzie	96.7	53.8	79.2
749	Poteau	90.8	62.9	79.5	383	Jackson	98.4	68.8	86.4
					384	Columbia	95.0	63.1	82.0
OREGON					385	Cookeville	96.5	55.1	79.6
970-972	Portland	103.1	102.2	102.7					
973	Salem	104.7	100.8	103.1	**TEXAS**				
974	Eugene	102.5	101.2	101.9	750	McKinney	96.8	60.6	82.0
975	Medford	103.6	96.9	100.9	751	Waxahachie	96.9	60.8	82.2
976	Klamath Falls	103.9	95.4	100.4	752-753	Dallas	99.3	67.1	86.2
977	Bend	103.2	98.1	101.1	754	Greenville	97.1	60.1	82.0
978	Pendleton	100.1	98.4	99.4	755	Texarkana	95.8	59.2	80.9
979	Vale	97.6	85.7	92.8	756	Longview	96.7	58.2	81.0
					757	Tyler	97.4	59.0	81.8
PENNSYLVANIA					758	Palestine	93.8	57.8	79.1
150-152	Pittsburgh	99.1	102.0	100.3	759	Lufkin	94.4	59.2	80.1
153	Washington	94.4	99.6	96.5	760-761	Fort Worth	97.3	63.9	83.8
154	Uniontown	94.7	96.9	95.6	762	Denton	96.9	60.3	82.0
155	Bedford	95.7	88.1	92.6	763	Wichita Falls	95.1	62.0	81.6
156	Greensburg	95.6	93.6	94.8	764	Eastland	94.2	57.7	79.3
157	Indiana	94.6	94.0	94.3	765	Temple	92.5	55.3	77.4
158	Dubois	96.0	91.9	94.3	766-767	Waco	94.4	62.9	81.6
159	Johnstown	95.7	95.4	95.6	768	Brownwood	97.0	56.1	80.4
160	Butler	90.8	96.2	93.0	769	San Angelo	96.6	56.3	80.2
161	New Castle	90.9	96.6	93.2	770-772	Houston	97.8	66.6	85.1
162	Kittanning	91.3	94.3	92.5	773	Huntsville	95.0	59.7	80.7
163	Oil City	90.7	92.2	91.3	774	Wharton	95.9	60.6	81.6
164-165	Erie	93.0	95.0	93.8	775	Galveston	94.4	62.9	81.6
166	Altoona	92.8	95.5	93.9	776-777	Beaumont	94.7	66.0	83.0
167	Bradford	94.3	92.8	93.7	778	Bryan	91.9	61.4	79.5
168	State College	93.9	97.1	95.2	779	Victoria	95.9	59.8	81.2
169	Wellsboro	94.7	87.1	91.6	780	Laredo	95.0	61.5	81.3
170-171	Harrisburg	97.6	97.6	97.6	781-782	San Antonio	97.4	63.6	83.7
172	Chambersburg	94.1	83.9	89.9	783-784	Corpus Christi	97.4	63.3	83.5
173-174	York	94.8	93.9	94.4	785	McAllen	97.8	54.4	80.2
175-176	Lancaster	92.9	95.6	94.0	786-787	Austin	96.5	61.5	82.2

For customer support on your Commercial Renovation Costs with RSMeans data, call 800.448.8182.

Location Factors - Commercial

STATE/ZIP	CITY	MAT.	INST.	TOTAL
TEXAS (CONT'D)				
788	Del Rio	97.5	56.9	81.0
789	Giddings	94.4	57.6	79.4
790-791	Amarillo	97.8	61.7	83.1
792	Childress	96.8	58.2	81.1
793-794	Lubbock	98.9	60.8	83.4
795-796	Abilene	97.4	60.6	82.4
797	Midland	99.2	64.0	84.9
798-799,885	El Paso	96.9	63.8	83.5
UTAH				
840-841	Salt Lake City	101.4	72.9	89.8
842,844	Ogden	96.9	72.9	87.2
843	Logan	98.7	72.9	88.2
845	Price	98.9	67.5	86.1
846-847	Provo	99.0	72.4	88.2
VERMONT				
050	White River Jct.	94.7	78.2	88.0
051	Bellows Falls	93.5	88.8	91.6
052	Bennington	93.8	85.4	90.4
053	Brattleboro	94.1	88.8	92.0
054	Burlington	99.8	82.4	92.7
056	Montpelier	98.0	85.3	92.8
057	Rutland	96.5	81.7	90.5
058	St. Johnsbury	94.8	77.8	87.9
059	Guildhall	93.8	77.9	87.3
VIRGINIA				
220-221	Fairfax	99.3	80.3	91.6
222	Arlington	100.5	81.3	92.7
223	Alexandria	99.8	83.0	93.0
224-225	Fredericksburg	98.2	78.1	90.0
226	Winchester	98.7	74.5	88.9
227	Culpeper	98.4	80.2	91.0
228	Harrisonburg	98.8	74.2	88.8
229	Charlottesville	99.2	72.7	88.4
230-232	Richmond	99.4	73.2	88.7
233-235	Norfolk	101.0	66.7	87.1
236	Newport News	99.1	67.8	86.4
237	Portsmouth	98.3	66.2	85.3
238	Petersburg	98.6	73.0	88.2
239	Farmville	98.1	64.6	84.5
240-241	Roanoke	100.5	67.7	87.2
242	Bristol	98.1	58.4	81.9
243	Pulaski	97.7	68.1	85.7
244	Staunton	98.5	66.5	85.5
245	Lynchburg	98.5	71.6	87.5
246	Grundy	97.9	61.9	83.3
WASHINGTON				
980-981,987	Seattle	104.3	110.8	106.9
982	Everett	103.6	106.0	104.6
983-984	Tacoma	104.1	104.8	104.4
985	Olympia	103.3	104.5	103.8
986	Vancouver	106.1	102.0	104.4
988	Wenatchee	104.1	85.1	96.4
989	Yakima	104.2	96.4	101.0
990-992	Spokane	98.6	86.9	93.8
993	Richland	97.9	92.7	95.8
994	Clarkston	97.2	77.8	89.3
WEST VIRGINIA				
247-248	Bluefield	96.9	83.1	91.3
249	Lewisburg	98.5	84.8	92.9
250-253	Charleston	97.2	89.0	93.9
254	Martinsburg	96.6	78.8	89.4
255-257	Huntington	98.3	89.4	94.7
258-259	Beckley	95.4	85.6	91.4
260	Wheeling	99.6	89.3	95.4
261	Parkersburg	98.6	88.3	94.4
262	Buckhannon	98.1	88.9	94.4
263-264	Clarksburg	98.8	89.8	95.1
265	Morgantown	98.4	90.4	95.2
266	Gassaway	98.1	86.1	93.3
267	Romney	97.9	84.0	92.2
268	Petersburg	97.7	84.3	92.2
WISCONSIN				
530,532	Milwaukee	96.9	108.0	101.4
531	Kenosha	97.0	105.3	100.4
534	Racine	96.2	105.2	99.8
535	Beloit	95.4	95.5	95.4
537	Madison	96.1	103.5	99.1

STATE/ZIP	CITY	MAT.	INST.	TOTAL
WISCONSIN (CONT'D)				
538	Lancaster	93.3	90.8	92.3
539	Portage	91.8	94.3	92.8
540	New Richmond	93.9	90.5	92.5
541-543	Green Bay	98.3	102.6	100.1
544	Wausau	93.1	90.5	92.1
545	Rhinelander	96.4	88.9	93.4
546	La Crosse	93.9	99.0	96.0
547	Eau Claire	95.8	99.0	97.1
548	Superior	93.5	92.5	93.0
549	Oshkosh	94.0	91.0	92.7
WYOMING				
820	Cheyenne	99.7	72.6	88.7
821	Yellowstone Nat'l Park	97.6	70.7	86.7
822	Wheatland	98.6	69.0	86.5
823	Rawlins	100.1	70.4	88.0
824	Worland	98.1	71.0	87.1
825	Riverton	99.1	70.4	87.5
826	Casper	101.0	72.0	89.2
827	Newcastle	97.9	70.4	86.7
828	Sheridan	101.0	70.6	88.6
829-831	Rock Springs	101.6	71.1	89.2
CANADIAN FACTORS (reflect Canadian currency)				
ALBERTA				
	Calgary	114.8	89.0	104.3
	Edmonton	117.7	89.2	106.1
	Fort McMurray	117.4	84.1	103.9
	Lethbridge	112.3	83.6	100.6
	Lloydminster	107.1	80.6	96.3
	Medicine Hat	107.2	79.9	96.1
	Red Deer	107.6	79.9	96.3
BRITISH COLUMBIA				
	Kamloops	108.3	80.2	96.8
	Prince George	109.5	79.6	97.3
	Vancouver	115.3	91.0	105.4
	Victoria	109.9	85.2	99.8
MANITOBA				
	Brandon	115.3	67.8	96.0
	Portage la Prairie	107.3	66.3	90.6
	Winnipeg	119.7	67.7	98.5
NEW BRUNSWICK				
	Bathurst	105.3	59.7	86.7
	Dalhousie	105.8	60.0	87.1
	Fredericton	118.1	67.3	97.4
	Moncton	105.3	74.6	92.8
	Newcastle	105.3	60.3	87.0
	St. John	108.0	74.7	94.4
NEWFOUNDLAND				
	Corner Brook	119.1	65.7	97.3
	St. Johns	118.2	83.8	104.2
NORTHWEST TERRITORIES				
	Yellowknife	127.8	79.6	108.2
NOVA SCOTIA				
	Bridgewater	106.7	67.5	90.8
	Dartmouth	115.9	67.6	96.2
	Halifax	116.1	83.8	103.0
	New Glasgow	114.1	67.6	95.1
	Sydney	111.9	67.6	93.8
	Truro	106.3	67.5	90.5
	Yarmouth	114.0	67.6	95.1
ONTARIO				
	Barrie	116.1	83.8	102.9
	Brantford	109.0	87.2	100.1
	Cornwall	108.7	84.0	98.6
	Hamilton	114.4	94.8	106.4
	Kingston	109.8	84.0	99.3
	Kitchener	104.9	91.3	99.4
	London	113.6	93.0	105.2
	North Bay	117.4	82.0	103.0
	Oshawa	107.7	93.6	102.0
	Ottawa	113.6	94.8	106.0
	Owen Sound	116.3	82.1	102.4
	Peterborough	109.0	83.8	98.7
	Sarnia	109.1	87.8	100.4

STATE/ZIP	CITY	MAT.	INST.	TOTAL
ONTARIO (CONT'D)				
	Sault Ste. Marie	104.2	85.5	96.6
	St. Catharines	103.1	91.3	98.3
	Sudbury	102.5	89.6	97.3
	Thunder Bay	104.6	90.8	99.0
	Timmins	109.2	82.0	98.1
	Toronto	113.4	100.9	108.3
	Windsor	103.9	91.1	98.7
PRINCE EDWARD ISLAND				
	Charlottetown	115.9	58.3	92.5
	Summerside	116.1	55.6	91.5
QUEBEC				
	Cap-de-la-Madeleine	106.2	76.9	94.3
	Charlesbourg	106.2	76.9	94.3
	Chicoutimi	105.4	86.2	97.6
	Gatineau	106.0	76.7	94.1
	Granby	106.3	76.6	94.2
	Hull	106.2	76.7	94.2
	Joliette	106.5	76.9	94.4
	Laval	106.4	77.2	94.5
	Montreal	114.4	88.7	103.9
	Quebec City	114.2	89.2	104.1
	Rimouski	106.0	86.2	97.9
	Rouyn-Noranda	106.0	76.7	94.1
	Saint-Hyacinthe	105.5	76.7	93.8
	Sherbrooke	106.3	76.7	94.3
	Sorel	106.5	76.9	94.4
	Saint-Jerome	106.1	76.7	94.1
	Trois-Rivieres	115.1	76.9	99.6
SASKATCHEWAN				
	Moose Jaw	104.6	61.3	87.0
	Prince Albert	103.7	59.7	85.8
	Regina	120.4	87.1	106.9
	Saskatoon	105.8	83.3	96.6
YUKON				
	Whitehorse	128.4	66.2	103.1

R011105-05 Tips for Accurate Estimating

1. Use pre-printed or columnar forms for orderly sequence of dimensions and locations and for recording telephone quotations.

2. Use only the front side of each paper or form except for certain pre-printed summary forms.

3. Be consistent in listing dimensions: For example, length x width x height. This helps in rechecking to ensure that, the total length of partitions is appropriate for the building area.

4. Use printed (rather than measured) dimensions where given.

5. Add up multiple printed dimensions for a single entry where possible.

6. Measure all other dimensions carefully.

7. Use each set of dimensions to calculate multiple related quantities.

8. Convert foot and inch measurements to decimal feet when listing. Memorize decimal equivalents to .01 parts of a foot (1/8″ equals approximately .01′).

9. Do not "round off" quantities until the final summary.

10. Mark drawings with different colors as items are taken off.

11. Keep similar items together, different items separate.

12. Identify location and drawing numbers to aid in future checking for completeness.

13. Measure or list everything on the drawings or mentioned in the specifications.

14. It may be necessary to list items not called for to make the job complete.

15. Be alert for: Notes on plans such as N.T.S. (not to scale); changes in scale throughout the drawings; reduced size drawings; discrepancies between the specifications and the drawings.

16. Develop a consistent pattern of performing an estimate. For example:
    a. Start the quantity takeoff at the lower floor and move to the next higher floor.
    b. Proceed from the main section of the building to the wings.
    c. Proceed from south to north or vice versa, clockwise or counterclockwise.
    d. Take off floor plan quantities first, elevations next, then detail drawings.

17. List all gross dimensions that can be either used again for different quantities, or used as a rough check of other quantities for verification (exterior perimeter, gross floor area, individual floor areas, etc.).

18. Utilize design symmetry or repetition (repetitive floors, repetitive wings, symmetrical design around a center line, similar room layouts, etc.). Note: Extreme caution is needed here so as not to omit or duplicate an area.

19. Do not convert units until the final total is obtained. For instance, when estimating concrete work, keep all units to the nearest cubic foot, then summarize and convert to cubic yards.

20. When figuring alternatives, it is best to total all items involved in the basic system, then total all items involved in the alternates. Therefore you work with positive numbers in all cases. When adds and deducts are used, it is often confusing whether to add or subtract a portion of an item; especially on a complicated or involved alternate.

R011105-50   Metric Conversion Factors

Description: This table is primarily for converting customary U.S. units in the left hand column to SI metric units in the right hand column. In addition, conversion factors for some commonly encountered Canadian and non-SI metric units are included.

	If You Know		Multiply By		To Find
Length	Inches	x	25.4[a]	=	Millimeters
	Feet	x	0.3048[a]	=	Meters
	Yards	x	0.9144[a]	=	Meters
	Miles (statute)	x	1.609	=	Kilometers
Area	Square inches	x	645.2	=	Square millimeters
	Square feet	x	0.0929	=	Square meters
	Square yards	x	0.8361	=	Square meters
Volume (Capacity)	Cubic inches	x	16,387	=	Cubic millimeters
	Cubic feet	x	0.02832	=	Cubic meters
	Cubic yards	x	0.7646	=	Cubic meters
	Gallons (U.S. liquids)[b]	x	0.003785	=	Cubic meters[c]
	Gallons (Canadian liquid)[b]	x	0.004546	=	Cubic meters[c]
	Ounces (U.S. liquid)[b]	x	29.57	=	Milliliters[c, d]
	Quarts (U.S. liquid)[b]	x	0.9464	=	Liters[c, d]
	Gallons (U.S. liquid)[b]	x	3.785	=	Liters[c, d]
Force	Kilograms force[d]	x	9.807	=	Newtons
	Pounds force	x	4.448	=	Newtons
	Pounds force	x	0.4536	=	Kilograms force[d]
	Kips	x	4448	=	Newtons
	Kips	x	453.6	=	Kilograms force[d]
Pressure, Stress, Strength (Force per unit area)	Kilograms force per square centimeter[d]	x	0.09807	=	Megapascals
	Pounds force per square inch (psi)	x	0.006895	=	Megapascals
	Kips per square inch	x	6.895	=	Megapascals
	Pounds force per square inch (psi)	x	0.07031	=	Kilograms force per square centimeter[d]
	Pounds force per square foot	x	47.88	=	Pascals
	Pounds force per square foot	x	4.882	=	Kilograms force per square meter[d]
Flow	Cubic feet per minute	x	0.4719	=	Liters per second
	Gallons per minute	x	0.0631	=	Liters per second
	Gallons per hour	x	1.05	=	Milliliters per second
Bending Moment Or Torque	Inch-pounds force	x	0.01152	=	Meter-kilograms force[d]
	Inch-pounds force	x	0.1130	=	Newton-meters
	Foot-pounds force	x	0.1383	=	Meter-kilograms force[d]
	Foot-pounds force	x	1.356	=	Newton-meters
	Meter-kilograms force[d]	x	9.807	=	Newton-meters
Mass	Ounces (avoirdupois)	x	28.35	=	Grams
	Pounds (avoirdupois)	x	0.4536	=	Kilograms
	Tons (metric)	x	1000	=	Kilograms
	Tons, short (2000 pounds)	x	907.2	=	Kilograms
	Tons, short (2000 pounds)	x	0.9072	=	Megagrams[e]
Mass per Unit Volume	Pounds mass per cubic foot	x	16.02	=	Kilograms per cubic meter
	Pounds mass per cubic yard	x	0.5933	=	Kilograms per cubic meter
	Pounds mass per gallon (U.S. liquid)[b]	x	119.8	=	Kilograms per cubic meter
	Pounds mass per gallon (Canadian liquid)[b]	x	99.78	=	Kilograms per cubic meter
Temperature	Degrees Fahrenheit	(F-32)/1.8		=	Degrees Celsius
	Degrees Fahrenheit	(F+459.67)/1.8		=	Degrees Kelvin
	Degrees Celsius	C+273.15		=	Degrees Kelvin

[a] The factor given is exact
[b] One U.S. gallon = 0.8327 Canadian gallon
[c] 1 liter = 1000 milliliters = 1000 cubic centimeters
1 cubic decimeter = 0.001 cubic meter
[d] Metric but not SI unit
[e] Called "tonne" in England and "metric ton" in other metric countries

R011105-60 Weights and Measures

Measures of Length
1 Mile = 1760 Yards = 5280 Feet
1 Yard = 3 Feet = 36 inches
1 Foot = 12 Inches
1 Mil = 0.001 Inch
1 Fathom = 2 Yards = 6 Feet
1 Rod = 5.5 Yards = 16.5 Feet
1 Hand = 4 Inches
1 Span = 9 Inches
1 Micro-inch = One Millionth Inch or 0.000001 Inch
1 Micron = One Millionth Meter + 0.00003937 Inch

Surveyor's Measure
1 Mile = 8 Furlongs = 80 Chains
1 Furlong = 10 Chains = 220 Yards
1 Chain = 4 Rods = 22 Yards = 66 Feet = 100 Links
1 Link = 7.92 Inches

Square Measure
1 Square Mile = 640 Acres = 6400 Square Chains
1 Acre = 10 Square Chains = 4840 Square Yards = 43,560 Sq. Ft.
1 Square Chain = 16 Square Rods = 484 Square Yards = 4356 Sq. Ft.
1 Square Rod = 30.25 Square Yards = 272.25 Square Feet = 625 Square Lines
1 Square Yard = 9 Square Feet
1 Square Foot = 144 Square Inches
An Acre equals a Square 208.7 Feet per Side

Cubic Measure
1 Cubic Yard = 27 Cubic Feet
1 Cubic Foot = 1728 Cubic Inches
1 Cord of Wood = 4 x 4 x 8 Feet = 128 Cubic Feet
1 Perch of Masonry = 16½ x 1½ x 1 Foot = 24.75 Cubic Feet

Avoirdupois or Commercial Weight
1 Gross or Long Ton = 2240 Pounds
1 Net or Short Ton = 2000 Pounds
1 Pound = 16 Ounces = 7000 Grains
1 Ounce = 16 Drachms = 437.5 Grains
1 Stone = 14 Pounds

Power
1 British Thermal Unit per Hour = 0.2931 Watts
1 Ton (Refrigeration) = 3.517 Kilowatts
1 Horsepower (Boiler) = 9.81 Kilowatts
1 Horsepower (550 ft-lb/s) = 0.746 Kilowatts

Shipping Measure
For Measuring Internal Capacity of a Vessel:
    1 Register Ton = 100 Cubic Feet

For Measurement of Cargo:
    Approximately 40 Cubic Feet of Merchandise is considered a Shipping Ton, unless that bulk would weigh more than 2000 Pounds, in which case Freight Charge may be based upon weight.

40 Cubic Feet = 32.143 U.S. Bushels = 31.16 Imp. Bushels

Liquid Measure
1 Imperial Gallon = 1.2009 U.S. Gallon = 277.42 Cu. In.
1 Cubic Foot = 7.48 U.S. Gallons

R011110-10 Architectural Fees

Tabulated below are typical percentage fees by project size, for good professional architectural service. Fees may vary from those listed depending upon degree of design difficulty and economic conditions in any particular area.

Rates can be interpolated horizontally and vertically. Various portions of the same project requiring different rates should be adjusted proportionately. For alterations, add 50% to the fee for the first $500,000 of project cost and add 25% to the fee for project cost over $500,000.

Architectural fees tabulated below include Structural, Mechanical and Electrical Engineering Fees. They do not include the fees for special consultants such as kitchen planning, security, acoustical, interior design, etc.

Civil Engineering fees are included in the Architectural fee for project sites requiring minimal design such as city sites. However, separate Civil Engineering fees must be added when utility connections require design, drainage calculations are needed, stepped foundations are required, or provisions are required to protect adjacent wetlands.

Building Types	Total Project Size in Thousands of Dollars						
	100	250	500	1,000	5,000	10,000	50,000
Factories, garages, warehouses, repetitive housing	9.0%	8.0%	7.0%	6.2%	5.3%	4.9%	4.5%
Apartments, banks, schools, libraries, offices, municipal buildings	12.2	12.3	9.2	8.0	7.0	6.6	6.2
Churches, hospitals, homes, laboratories, museums, research	15.0	13.6	12.7	11.9	9.5	8.8	8.0
Memorials, monumental work, decorative furnishings	—	16.0	14.5	13.1	10.0	9.0	8.3

R011110-30   Engineering Fees

Typical **Structural Engineering Fees** based on type of construction and total project size. These fees are included in Architectural Fees.

Type of Construction	Total Project Size (in thousands of dollars)			
	$500	$500-$1,000	$1,000-$5,000	Over $5000
Industrial buildings, factories & warehouses	Technical payroll times 2.0 to 2.5	1.60%	1.25%	1.00%
Hotels, apartments, offices, dormitories, hospitals, public buildings, food stores		2.00%	1.70%	1.20%
Museums, banks, churches and cathedrals		2.00%	1.75%	1.25%
Thin shells, prestressed concrete, earthquake resistive		2.00%	1.75%	1.50%
Parking ramps, auditoriums, stadiums, convention halls, hangars & boiler houses		2.50%	2.00%	1.75%
Special buildings, major alterations, underpinning & future expansion		Add to above 0.5%	Add to above 0.5%	Add to above 0.5%

For complex reinforced concrete or unusually complicated structures, add 20% to 50%.

Typical **Mechanical and Electrical Engineering Fees** are based on the size of the subcontract. The fee structure for both is shown below. These fees are included in Architectural Fees.

Type of Construction	Subcontract Size							
	$25,000	$50,000	$100,000	$225,000	$350,000	$500,000	$750,000	$1,000,000
Simple structures	6.4%	5.7%	4.8%	4.5%	4.4%	4.3%	4.2%	4.1%
Intermediate structures	8.0	7.3	6.5	5.6	5.1	5.0	4.9	4.8
Complex structures	10.1	9.0	9.0	8.0	7.5	7.5	7.0	7.0

For renovations, add 15% to 25% to applicable fee.

R012153-10   Repair and Remodeling

Cost figures are based on new construction utilizing the most cost-effective combination of labor, equipment and material with the work scheduled in proper sequence to allow the various trades to accomplish their work in an efficient manner.

The costs for repair and remodeling work must be modified due to the following factors that may be present in any given repair and remodeling project.

1. Equipment usage curtailment due to the physical limitations of the project, with only hand-operated equipment being used.

2. Increased requirement for shoring and bracing to hold up the building while structural changes are being made and to allow for temporary storage of construction materials on above-grade floors.

3. Material handling becomes more costly due to having to move within the confines of an enclosed building. For multi-story construction, low capacity elevators and stairwells may be the only access to the upper floors.

4. Large amount of cutting and patching and attempting to match the existing construction is required. It is often more economical to remove entire walls rather than create many new door and window openings. This sort of trade-off has to be carefully analyzed.

5. Cost of protection of completed work is increased since the usual sequence of construction usually cannot be accomplished.

6. Economies of scale usually associated with new construction may not be present. If small quantities of components must be custom fabricated due to job requirements, unit costs will naturally increase. Also, if only small work areas are available at a given time, job scheduling between trades becomes difficult and subcontractor quotations may reflect the excessive start-up and shut-down phases of the job.

7. Work may have to be done on other than normal shifts and may have to be done around an existing production facility which has to stay in production during the course of the repair and remodeling.

8. Dust and noise protection of adjoining non-construction areas can involve substantial special protection and alter usual construction methods.

9. Job may be delayed due to unexpected conditions discovered during demolition or removal. These delays ultimately increase construction costs.

10. Piping and ductwork runs may not be as simple as for new construction. Wiring may have to be snaked through walls and floors.

11. Matching "existing construction" may be impossible because materials may no longer be manufactured. Substitutions may be expensive.

12. Weather protection of existing structure requires additional temporary structures to protect building at openings.

13. On small projects, because of local conditions, it may be necessary to pay a tradesman for a minimum of four hours for a task that is completed in one hour.

All of the above areas can contribute to increased costs for a repair and remodeling project. Each of the above factors should be considered in the planning, bidding and construction stage in order to minimize the increased costs associated with repair and remodeling jobs.

R012153-60 Security Factors

Contractors entering, working in, and exiting secure facilities often lose productive time during a normal workday. The recommended allowances in this section are intended to provide for the loss of productivity by increasing labor costs. Note that different costs are associated with searches upon entry only and searches upon entry and exit. Time spent in a queue is unpredictable and not part of these allowances. Contractors should plan ahead for this situation.

Security checkpoints are designed to reflect the level of security required to gain access or egress. An extreme example is when contractors, along with any materials, tools, equipment, and vehicles, must be physically searched and have all materials, tools, equipment, and vehicles inventoried and documented prior to both entry and exit.

Physical searches without going through the documentation process represent the next level and take up less time.

Electronic searches—passing through a detector or x-ray machine with no documentation of materials, tools, equipment, and vehicles—take less time than physical searches.

Visual searches of materials, tools, equipment, and vehicles represent the next level of security.

Finally, access by means of an ID card or displayed sticker takes the least amount of time.

Another consideration is if the searches described above are performed each and every day, or if they are performed only on the first day with access granted by ID card or displayed sticker for the remainder of the project. The figures for this situation have been calculated to represent the initial check-in as described and subsequent entry by ID card or displayed sticker for up to 20 days on site. For the situation described above, where the time period is beyond 20 days, the impact on labor cost is negligible.

There are situations where tradespeople must be accompanied by an escort and observed during the work day. The loss of freedom of movement will slow down productivity for the tradesperson. Costs for the observer have not been included. Those costs are normally born by the owner.

R012153-65 Infectious Disease Precautions

Contractors entering and working on job sites may be required to take precautions to prevent the spread of infectious diseases. Those precautions will reduce the amount of productive time during a normal workday. The recommended allowances in this section are intended to provide for the loss of productive time by increasing labor costs.

The estimator should be aware that one, many or none of these precautions apply to the line items in an estimate. Job site requirements and sound judgement must be applied.

Labor cost implications are based upon:

Temperature checks	Once per day
Donning/doffing masks and gloves	Four times per day (2 times each)
Washing hands (additional over normal)	Eight times per day
Informational meetings	Once per day
Maintaining social distance	Throughout the day
Disinfecting tools or equipment	Throughout the day

R012157-20 Construction Time Requirements

The table below lists the construction durations for various building types along with their respective project sizes and project values. Design time runs 25% to 40% of construction time.

Building Type	Size S.F.	Project Value	Construction Duration
Industrial/Warehouse	100,000	$8,000,000	14 months
	500,000	$32,000,000	19 months
	1,000,000	$75,000,000	21 months
Offices/Retail	50,000	$7,000,000	15 months
	250,000	$28,000,000	23 months
	500,000	$58,000,000	34 months
Institutional/Hospitals/Laboratory	200,000	$45,000,000	31 months
	500,000	$110,000,000	52 months
	750,000	$160,000,000	55 months
	1,000,000	$210,000,000	60 months

For customer support on your Commercial Renovation Costs with RSMeans data, call 800.448.8182.

849

R012909-80   Sales Tax by State

State sales tax on materials is tabulated below (5 states have no sales tax). Many states allow local jurisdictions, such as a county or city, to levy additional sales tax.

Some projects may be sales tax exempt, particularly those constructed with public funds.

State	Tax (%)	State	Tax (%)	State	Tax (%)	State	Tax (%)
Alabama	4	Illinois	6.25	Montana	0	Rhode Island	7
Alaska	0	Indiana	7	Nebraska	5.5	South Carolina	6
Arizona	5.6	Iowa	6	Nevada	6.85	South Dakota	4.5
Arkansas	6.5	Kansas	6.5	New Hampshire	0	Tennessee	7
California	7.25	Kentucky	6	New Jersey	6.625	Texas	6.25
Colorado	2.9	Louisiana	4.45	New Mexico	5.125	Utah	6.10
Connecticut	6.35	Maine	5.5	New York	4	Vermont	6
Delaware	0	Maryland	6	North Carolina	4.75	Virginia	5.3
District of Columbia	6	Massachusetts	6.25	North Dakota	5	Washington	6.5
Florida	6	Michigan	6	Ohio	5.75	West Virginia	6
Georgia	4	Minnesota	6.875	Oklahoma	4.5	Wisconsin	5
Hawaii	4	Mississippi	7	Oregon	0	Wyoming	4
Idaho	6	Missouri	4.225	Pennsylvania	6	Average	5.11 %

Sales Tax by Province (Canada)

GST - a value-added tax, which the government imposes on most goods and services provided in or imported into Canada. PST - a retail sales tax, which five of the provinces impose on the prices of most goods and some

services. QST - a value-added tax, similar to the federal GST, which Quebec imposes. HST - Three provinces have combined their retail sales taxes with the federal GST into one harmonized tax.

Province	PST (%)	QST(%)	GST(%)	HST(%)
Alberta	0	0	5	0
British Columbia	7	0	5	0
Manitoba	7	0	5	0
New Brunswick	0	0	0	15
Newfoundland	0	0	0	15
Northwest Territories	0	0	5	0
Nova Scotia	0	0	0	15
Ontario	0	0	0	13
Prince Edward Island	0	0	0	15
Quebec	9.975	0	5	0
Saskatchewan	6	0	5	0
Yukon	0	0	5	0

R012909-85   Unemployment Taxes and Social Security Taxes

State unemployment tax rates vary not only from state to state, but also with the experience rating of the contractor. The federal unemployment tax rate is 6.2% of the first $7,000 of wages. This is reduced by a credit of up to 5.4% for timely payment to the state. The minimum federal unemployment tax is 0.6% after all credits.

Social security (FICA) for 2022 is estimated at time of publication to be 7.65% of wages up to $142,800.

R012909-86   Unemployment Tax by State

Information is from the U.S. Department of Labor, state unemployment tax rates.

State	Tax (%)	State	Tax (%)	State	Tax (%)	State	Tax (%)
Alabama	6.80	Illinois	6.40	Montana	6.12	Rhode Island	9.80
Alaska	5.9	Indiana	7.4	Nebraska	5.4	South Carolina	5.46
Arizona	20.6	Iowa	7.5	Nevada	5.4	South Dakota	9.5
Arkansas	14.2	Kansas	7.6	New Hampshire	7.5	Tennessee	10.0
California	6.2	Kentucky	10.0	New Jersey	5.4	Texas	6.5
Colorado	9.64	Louisiana	6.2	New Mexico	5.4	Utah	7.2
Connecticut	6.8	Maine	6.01	New York	7.9	Vermont	5.4
Delaware	8.20	Maryland	13.50	North Carolina	5.76	Virginia	6.20
District of Columbia	2.7	Massachusetts	12.65	North Dakota	9.69	Washington	6.86
Florida	5.4	Michigan	8.1	Ohio	9.3	West Virginia	8.5
Georgia	7.6	Minnesota	0.5	Oklahoma	7.5	Wisconsin	12.0
Hawaii	6.6	Mississippi	5.4	Oregon	5.4	Wyoming	9.8
Idaho	5.4	Missouri	6	Pennsylvania	9.93	Median	6.86%

R012909-90   Overtime

One way to improve the completion date of a project or eliminate negative float from a schedule is to compress activity duration times. This can be achieved by increasing the crew size or working overtime with the proposed crew.

To determine the costs of working overtime to compress activity duration times, consider the following examples. Below is an overtime efficiency and

cost chart based on a five, six, or seven day week with an eight through twelve hour day. Payroll percentage increases for time and one half and double times are shown for the various working days.

Days per Week	Hours per Day	Production Efficiency					Payroll Cost Factors	
		1st Week	2nd Week	3rd Week	4th Week	Average 4 Weeks	@ 1-1/2 Times	@ 2 Times
	8	100%	100%	100%	100%	100%	1.000	1.000
	9	100	100	95	90	96	1.056	1.111
5	10	100	95	90	85	93	1.100	1.200
	11	95	90	75	65	81	1.136	1.273
	12	90	85	70	60	76	1.167	1.333
	8	100	100	95	90	96	1.083	1.167
	9	100	95	90	85	93	1.130	1.259
6	10	95	90	85	80	88	1.167	1.333
	11	95	85	70	65	79	1.197	1.394
	12	90	80	65	60	74	1.222	1.444
	8	100	95	85	75	89	1.143	1.286
	9	95	90	80	70	84	1.183	1.365
7	10	90	85	75	65	79	1.214	1.429
	11	85	80	65	60	73	1.240	1.481
	12	85	75	60	55	69	1.262	1.524

For customer support on your Commercial Renovation Costs with RSMeans data, call 800.448.8182.

851

Reference Tables

R013113-40 Builder's Risk Insurance

Builder's risk insurance is insurance on a building during construction.
Premiums are paid by the owner or the contractor. Blasting, collapse and
underground insurance would raise total insurance costs.

R013113-50 General Contractor's Overhead

There are two distinct types of overhead on a construction project: Project
overhead and main office overhead. Project overhead includes those
costs at a construction site not directly associated with the installation of
construction materials. Examples of project overhead costs include the
following:

1. Superintendent
2. Construction office and storage trailers
3. Temporary sanitary facilities
4. Temporary utilities
5. Security fencing
6. Photographs
7. Cleanup
8. Performance and payment bonds

The above project overhead items are also referred to as general requirements
and therefore are estimated in Division 1. Division 1 is the first division
listed in the CSI MasterFormat but it is usually the last division estimated. The
sum of the costs in Divisions 1 through 49 is referred to as the sum of
the direct costs.

All construction projects also include indirect costs. The primary components
of indirect costs are the contractor's main office overhead and profit. The
amount of the main office overhead expense varies depending on the
following:

1. Owner's compensation
2. Project managers' and estimators' wages
3. Clerical support wages
4. Office rent and utilities
5. Corporate legal and accounting costs
6. Advertising
7. Automobile expenses
8. Association dues
9. Travel and entertainment expenses

These costs are usually calculated as a percentage of annual sales volume.
This percentage can range from 35% for a small contractor doing less than
$500,000 to 5% for a large contractor with sales in excess of $100 million.

R013113-55 Installing Contractor's Overhead

Installing contractors (subcontractors) also incur costs for general
requirements and main office overhead.

Included within the total incl. overhead and profit costs is a percent mark-up
for overhead that includes:

1. Compensation and benefits for office staff and project managers
2. Office rent, utilities, business equipment, and maintenance
3. Corporate legal and accounting costs

4. Advertising
5. Vehicle expenses (for office staff and project managers)
6. Association dues
7. Travel, entertainment
8. Insurance
9. Small tools and equipment

For customer support on your Commercial Renovation Costs with RSMeans data, call 800.448.8182.

R013113-60   Workers' Compensation Insurance Rates by Trade

The table below tabulates the national averages for Workers' Compensation insurance rates by trade and type of building. The average "Insurance Rate" is multiplied by the "% of Building Cost" for each trade. This produces

the "Workers' Compensation Cost" by % of total labor cost, to be added for each trade by building type to determine the weighted average Workers' Compensation rate for the building types analyzed.

Trade	Insurance Rate (% Labor Cost)			% of Building Cost			Workers' Compensation		
	Range		Average	Office Bldgs.	Schools & Apts.	Mfg.	Office Bldgs.	Schools & Apts.	Mfg.
Excavation, Grading, etc.	1.9 % to	14.7%	8.3%	4.8%	4.9%	4.5%	0.40%	0.41%	0.37%
Piles & Foundations	2.2 to	33.0	17.6	7.1	5.2	8.7	1.25	0.92	1.53
Concrete	2.7 to	25.1	13.9	5.0	14.8	3.7	0.70	2.06	0.51
Masonry	2.6 to	48.0	25.3	6.9	7.5	1.9	1.75	1.90	0.48
Structural Steel	2.3 to	30.9	16.6	10.7	3.9	17.6	1.78	0.65	2.92
Miscellaneous & Ornamental Metals	2.2 to	22.6	12.4	2.8	4.0	3.6	0.35	0.50	0.45
Carpentry & Millwork	2.6 to	26.8	14.7	3.7	4.0	0.5	0.54	0.59	0.07
Metal or Composition Siding	4.4 to	129.7	67.0	2.3	0.3	4.3	1.54	0.20	2.88
Roofing	4.4 to	101.9	53.2	2.3	2.6	3.1	1.22	1.38	1.65
Doors & Hardware	3.0 to	26.8	14.9	0.9	1.4	0.4	0.13	0.21	0.06
Sash & Glazing	3.3 to	20.1	11.7	3.5	4.0	1.0	0.41	0.47	0.12
Lath & Plaster	2.1 to	32.4	17.3	3.3	6.9	0.8	0.57	1.19	0.14
Tile, Marble & Floors	2.1 to	19.2	10.6	2.6	3.0	0.5	0.28	0.32	0.05
Acoustical Ceilings	1.7 to	21.5	11.6	2.4	0.2	0.3	0.28	0.02	0.03
Painting	2.7 to	35.8	19.2	1.5	1.6	1.6	0.29	0.31	0.31
Interior Partitions	2.6 to	26.8	14.7	3.9	4.3	4.4	0.57	0.63	0.65
Miscellaneous Items	1.4 to	97.7	49.6	5.2	3.7	9.7	2.58	1.84	4.81
Elevators	1.2 to	8.3	4.8	2.1	1.1	2.2	0.10	0.05	0.11
Sprinklers	1.6 to	13.2	7.4	0.5	—	2.0	0.04	—	0.15
Plumbing	1.3 to	12.3	6.8	4.9	7.2	5.2	0.33	0.49	0.35
Heat., Vent., Air Conditioning	2.3 to	14.7	8.5	13.5	11.0	12.9	1.15	0.94	1.10
Electrical	1.5 to	10.6	6.1	10.1	8.4	11.1	0.62	0.51	0.68
Total	1.2 % to	129.7%	65.5	100.0%	100.0%	100.0%	16.88%	15.59%	19.42%
		Overall Weighted Average	17.30%						

Workers' Compensation Insurance Rates by States

The table below lists the weighted average Workers' Compensation base rate for each state with a factor comparing this with the national average of 15.1%.

State	Weighted Average	Factor	State	Weighted Average	Factor	State	Weighted Average	Factor
Alabama	20.5%	174	Kentucky	20.4%	174	North Dakota	7.4%	63
Alaska	12.9	110	Louisiana	30.1	256	Ohio	7.5	64
Arizona	12.2	104	Maine	15.4	131	Oklahoma	14.2	121
Arkansas	8.1	69	Maryland	16.9	144	Oregon	12.2	104
California	35.7	304	Massachusetts	14.0	119	Pennsylvania	33.9	289
Colorado	7.0	60	Michigan	10.4	89	Rhode Island	12.8	109
Connecticut	23.3	198	Minnesota	23.6	201	South Carolina	28.7	244
Delaware	19.2	163	Mississippi	14.7	125	South Dakota	13.1	111
District of Columbia	12.5	106	Missouri	22.4	191	Tennessee	10.3	88
Florida	13.9	118	Montana	14.2	121	Texas	9.4	80
Georgia	53.7	457	Nebraska	17.6	150	Utah	10.7	91
Hawaii	16.3	139	Nevada	13.1	111	Vermont	15.8	134
Idaho	14.9	127	New Hampshire	14.5	123	Virginia	11.5	98
Illinois	33.5	285	New Jersey	25.0	213	Washington	12.2	104
Indiana	5.6	48	New Mexico	23.0	196	West Virginia	7.8	66
Iowa	16.8	143	New York	26.0	221	Wisconsin	18.6	158
Kansas	10.0	85	North Carolina	30.1	256	Wyoming	8.2	70
			Weighted Average for U.S. is	17.3% of payroll = 100%				

The weighted average skilled worker rate for 35 trades is 17.3%. For bidding purposes, apply the full value of Workers' Compensation directly to total labor costs, or if labor is 38%, materials 42% and overhead and profit 20% of total cost, carry 38/80 x 17.3% = 8.22% of cost (before overhead and

profit) into overhead. Rates vary not only from state to state but also with the experience rating of the contractor.

Rates are the most current available at the time of publication.

R013113-80 Performance Bond

This table shows the cost of a Performance Bond for a construction job scheduled to be completed in 12 months. Add 1% of the premium cost per month for jobs requiring more than 12 months to complete. The rates are "standard" rates offered to contractors that the bonding company considers financially sound and capable of doing the work. Preferred rates are offered by some bonding companies based upon financial strength of the contractor. Actual rates vary from contractor to contractor and from bonding company to bonding company. Contractors should prequalify through a bonding agency before submitting a bid on a contract that requires a bond.

Contract Amount	Building Construction Class B Projects		Highways & Bridges Class A New Construction		Highways & Bridges Class A-1 Highway Resurfacing	
First $ 100,000 bid	$25.00 per M		$15.00 per M		$9.40 per M	
Next 400,000 bid	$ 2,500 plus	$15.00 per M	$ 1,500 plus	$10.00 per M	$ 940 plus	$7.20 per M
Next 2,000,000 bid	8,500 plus	10.00 per M	5,500 plus	7.00 per M	3,820 plus	5.00 per M
Next 2,500,000 bid	28,500 plus	7.50 per M	19,500 plus	5.50 per M	15,820 plus	4.50 per M
Next 2,500,000 bid	47,250 plus	7.00 per M	33,250 plus	5.00 per M	28,320 plus	4.50 per M
Over 7,500,000 bid	64,750 plus	6.00 per M	45,750 plus	4.50 per M	39,570 plus	4.00 per M

R015423-10 Steel Tubular Scaffolding

On new construction, tubular scaffolding is efficient up to 60' high or five stories. Above this it is usually better to use a hung scaffolding if construction permits. Swing scaffolding operations may interfere with tenants. In this case, the tubular is more practical at all heights.

In repairing or cleaning the front of an existing building the cost of tubular scaffolding per S.F. of building front increases as the height increases above the first tier. The first tier cost is relatively high due to leveling and alignment.

The minimum efficient crew for erecting and dismantling is three workers. They can set up and remove 18 frame sections per day up to 5 stories high. For 6 to 12 stories high, a crew of four is most efficient. Use two or more on top and two on the bottom for handing up or hoisting. They can also set up and remove 18 frame sections per day. At 7' horizontal spacing, this will run about 800 S.F. per day of erecting and dismantling. Time for placing and removing planks must be added to the above. A crew of three can place and remove 72 planks per day up to 5 stories. For over 5 stories, a crew of four can place and remove 80 planks per day.

The table below shows the number of pieces required to erect tubular steel scaffolding for 1000 S.F. of building frontage. This area is made up of a scaffolding system that is 12 frames (11 bays) long by 2 frames high.

For jobs under twenty-five frames, add 50% to rental cost. Rental rates will be lower for jobs over three months duration. Large quantities for long periods can reduce rental rates by 20%.

Description of Component	Number of Pieces for 1000 S.F. of Building Front	Unit
5' Wide Standard Frame, 6'-4" High	24	Ea.
Leveling Jack & Plate	24	
Cross Brace	44	
Side Arm Bracket, 21"	12	
Guardrail Post	12	
Guardrail, 7' section	22	
Stairway Section	2	
Stairway Starter Bar	1	
Stairway Inside Handrail	2	
Stairway Outside Handrail	2	
Walk-Thru Frame Guardrail	2	

Scaffolding is often used as falsework over 15' high during construction of cast-in-place concrete beams and slabs. Two foot wide scaffolding is generally used for heavy beam construction. The span between frames depends upon the load to be carried with a maximum span of 5'.

Heavy duty shoring frames with a capacity of 10,000#/leg can be spaced up to 10' O.C. depending upon form support design and loading.

Scaffolding used as horizontal shoring requires less than half the material required with conventional shoring.

On new construction, erection is done by carpenters.

Rolling towers supporting horizontal shores can reduce labor and speed the job. For maintenance work, catwalks with spans up to 70' can be supported by the rolling towers.

R015423-20   Pump Staging

Pump staging is generally not available for rent. The table below shows the number of pieces required to erect pump staging for 2400 S.F. of building frontage. This area is made up of a pump jack system that is 3 poles (2 bays) wide by 2 poles high.

Item	Number of Pieces for 2400 S.F. of Building Front	Unit
Aluminum pole section, 24′ long	6	Ea.
Aluminum splice joint, 6′ long	3	
Aluminum foldable brace	3	
Aluminum pump jack	3	
Aluminum support for workbench/back safety rail	3	
Aluminum scaffold plank/workbench, 14″ wide x 24′ long	4	
Safety net, 22′ long	2	
Aluminum plank end safety rail	2	↓

The cost in place for this 2400 S.F. will depend on how many uses are realized during the life of the equipment.

R015433-10   Contractor Equipment

Rental Rates shown elsewhere in the data set pertain to late model high quality machines in excellent working condition, rented from equipment dealers. Rental rates from contractors may be substantially lower than the rental rates from equipment dealers depending upon economic conditions; for older, less productive machines, reduce rates by a maximum of 15%. Any overtime must be added to the base rates. For shift work, rates are lower. Usual rule of thumb is 150% of one shift rate for two shifts; 200% for three shifts.

For periods of less than one week, operated equipment is usually more economical to rent than renting bare equipment and hiring an operator.

Costs to move equipment to a job site (mobilization) or from a job site (demobilization) are not included in rental rates, nor in any Equipment costs on any Unit Price line items or crew listings. These costs can be found elsewhere. If a piece of equipment is already at a job site, it is not appropriate to utilize mob/demob costs in an estimate again.

Rental rates vary throughout the country with larger cities generally having lower rates. Lease plans for new equipment are available for periods in excess of six months with a percentage of payments applying toward purchase.

Rental rates can also be treated as reimbursement costs for contractor-owned equipment. Owned equipment costs include depreciation, loan payments, interest, taxes, insurance, storage, and major repairs.

Monthly rental rates vary from 2% to 5% of the cost of the equipment depending on the anticipated life of the equipment and its wearing parts. Weekly rates are about 1/3 the monthly rates and daily rental rates are about 1/3 the weekly rates.

The hourly operating costs for each piece of equipment include costs to the user such as fuel, oil, lubrication, normal expendables for the equipment, and a percentage of the mechanic's wages chargeable to maintenance. The hourly operating costs listed do not include the operator's wages.

The daily cost for equipment used in the standard crews is figured by dividing the weekly rate by five, then adding eight times the hourly operating cost to give the total daily equipment cost, not including the operator. This figure is in the right hand column of the Equipment listings under Equipment Cost/Day.

Pile Driving rates shown for the pile hammer and extractor do not include leads, cranes, boilers or compressors. Vibratory pile driving requires an added field specialist during set-up and pile driving operation for the electric model. The hydraulic model requires a field specialist for set-up only. Up to 125 reuses of sheet piling are possible using vibratory drivers. For normal conditions, crane capacity for hammer type and size is as follows.

Crane Capacity	Hammer Type and Size		
	Air or Steam	Diesel	Vibratory
25 ton	to 8,750 ft.-lb.		70 H.P.
40 ton	15,000 ft.-lb.	to 32,000 ft.-lb.	170 H.P.
60 ton	25,000 ft.-lb.		300 H.P.
100 ton		112,000 ft.-lb.	

Cranes should be specified for the job by size, building and site characteristics, availability, performance characteristics, and duration of time required.

Backhoes & Shovels rent for about the same as equivalent size cranes but maintenance and operating expenses are higher. The crane operator's rate must be adjusted for high boom heights. Average adjustments: for 150' boom add 2% per hour; over 185', add 4% per hour; over 210', add 6% per hour; over 250', add 8% per hour and over 295', add 12% per hour.

Tower Cranes of the climbing or static type have jibs from 50' to 200' and capacities at maximum reach range from 4,000 to 14,000 pounds. Lifting capacities increase up to maximum load as the hook radius decreases.

Typical rental rates, based on purchase price, are about 2% to 3% per month.

Erection and dismantling run between 500 and 2000 labor hours. Climbing operation takes 10 labor hours per 20' climb. Crane dead time is about 5 hours per 40' climb. If crane is bolted to side of the building add cost of ties and extra mast sections. Climbing cranes have from 80' to 180' of mast while static cranes have 80' to 800' of mast.

Truck Cranes can be converted to tower cranes by using tower attachments. Mast heights over 400' have been used.

A single 100' high material **Hoist and Tower** can be erected and dismantled in about 400 labor hours; a double 100' high hoist and tower in about 600 labor hours. Erection times for additional heights are 3 and 4 labor hours

per vertical foot respectively up to 150', and 4 to 5 labor hours per vertical foot over 150' high. A 40' high portable Buck hoist takes about 160 labor hours to erect and dismantle. Additional heights take 2 labor hours per vertical foot to 80' and 3 labor hours per vertical foot for the next 100'. Most material hoists do not meet local code requirements for carrying personnel.

A 150' high **Personnel Hoist** requires about 500 to 800 labor hours to erect and dismantle. Budget erection time at 5 labor hours per vertical foot for all trades. Local code requirements or labor scarcity requiring overtime can add up to 50% to any of the above erection costs.

Earthmoving Equipment: The selection of earthmoving equipment depends upon the type and quantity of material, moisture content, haul distance, haul road, time available, and equipment available. Short haul cut and fill operations may require dozers only, while another operation may require excavators, a fleet of trucks, and spreading and compaction equipment. Stockpiled material and granular material are easily excavated with front end loaders. Scrapers are most economically used with hauls between 300' and 1-1/2 miles if adequate haul roads can be maintained. Shovels are often used for blasted rock and any material where a vertical face of 8' or more can be excavated. Special conditions may dictate the use of draglines, clamshells, or backhoes. Spreading and compaction equipment must be matched to the soil characteristics, the compaction required and the rate the fill is being supplied.

R015433-15   Heavy Lifting

Hydraulic Climbing Jacks

The use of hydraulic heavy lift systems is an alternative to conventional type crane equipment. The lifting, lowering, pushing, or pulling mechanism is a hydraulic climbing jack moving on a square steel jackrod from 1-5/8" to 4" square, or a steel cable. The jackrod or cable can be vertical or horizontal, stationary or movable, depending on the individual application. When the jackrod is stationary, the climbing jack will climb the rod and push or pull the load along with itself. When the climbing jack is stationary, the jackrod is movable with the load attached to the end and the climbing jack will lift or lower the jackrod with the attached load. The heavy lift system is normally operated by a single control lever located at the hydraulic pump.

The system is flexible in that one or more climbing jacks can be applied wherever a load support point is required, and the rate of lift synchronized.

Economic benefits have been demonstrated on projects such as: erection of ground assembled roofs and floors, complete bridge spans, girders and trusses, towers, chimney liners and steel vessels, storage tanks, and heavy machinery. Other uses are raising and lowering offshore work platforms, caissons, tunnel sections and pipelines.

R015436-50 Mobilization

Costs to move rented construction equipment to a job site from an equipment dealer's or contractor's yard (mobilization) or off the job site (demobilization) are not included in the rental or operating rates, nor in the equipment cost on a unit price line or in a crew listing. These costs can be found consolidated in the Mobilization section of the data and elsewhere in particular site work sections. If a piece of equipment is already on the job site, it is not appropriate to include mob/demob costs in a new estimate that requires use of that equipment. The following table identifies approximate sizes of rented construction equipment that would be hauled on a towed trailer. Because this listing is not all-encompassing, the user can infer as to what size trailer might be required for a piece of equipment not listed.

3-ton Trailer	20-ton Trailer	40-ton Trailer	50-ton Trailer
20 H.P. Excavator	110 H.P. Excavator	200 H.P. Excavator	270 H.P. Excavator
50 H.P. Skid Steer	165 H.P. Dozer	300 H.P. Dozer	Small Crawler Crane
35 H.P. Roller	150 H.P. Roller	400 H.P. Scraper	500 H.P. Scraper
40 H.P. Trencher	Backhoe	450 H.P. Art. Dump Truck	500 H.P. Art. Dump Truck

Reference Tables

R024119-10 Demolition Defined

Whole Building Demolition - Demolition of the whole building with no concern for any particular building element, component, or material type being demolished. This type of demolition is accomplished with large pieces of construction equipment that break up the structure, load it into trucks and haul it to a disposal site, but disposal or dump fees are not included. Demolition of below-grade foundation elements, such as footings, foundation walls, grade beams, slabs on grade, etc., is not included. Certain mechanical equipment containing flammable liquids or ozone-depleting refrigerants, electric lighting elements, communication equipment components, and other building elements may contain hazardous waste, and must be removed, either selectively or carefully, as hazardous waste before the building can be demolished.

Foundation Demolition - Demolition of below-grade foundation footings, foundation walls, grade beams, and slabs on grade. This type of demolition is accomplished by hand or pneumatic hand tools, and does not include saw cutting, or handling, loading, hauling, or disposal of the debris.

Gutting - Removal of building interior finishes and electrical/mechanical systems down to the load-bearing and sub-floor elements of the rough building frame, with no concern for any particular building element, component, or material type being demolished. This type of demolition is accomplished by hand or pneumatic hand tools, and includes loading into trucks, but not hauling, disposal or dump fees, scaffolding, or shoring. Certain mechanical equipment containing flammable liquids or ozone-depleting refrigerants, electric lighting elements, communication equipment components, and other building elements may contain hazardous waste, and must be removed, either selectively or carefully, as hazardous waste, before the building is gutted.

Selective Demolition - Demolition of a selected building element, component, or finish, with some concern for surrounding or adjacent elements, components, or finishes (see the first Subdivision (s) at the beginning of appropriate Divisions). This type of demolition is accomplished by hand or pneumatic hand tools, and does not include handling, loading,

storing, hauling, or disposal of the debris, scaffolding, or shoring. "Gutting" methods may be used in order to save time, but damage that is caused to surrounding or adjacent elements, components, or finishes may have to be repaired at a later time.

Careful Removal - Removal of a piece of service equipment, building element or component, or material type, with great concern for both the removed item and surrounding or adjacent elements, components or finishes. The purpose of careful removal may be to protect the removed item for later re-use, preserve a higher salvage value of the removed item, or replace an item while taking care to protect surrounding or adjacent elements, components, connections, or finishes from cosmetic and/or structural damage. An approximation of the time required to perform this type of removal is 1/3 to 1/2 the time it would take to install a new item of like kind (see Reference Number R220105-10). This type of removal is accomplished by hand or pneumatic hand tools, and does not include loading, hauling, or storing the removed item, scaffolding, shoring, or lifting equipment.

Cutout Demolition - Demolition of a small quantity of floor, wall, roof, or other assembly, with concern for the appearance and structural integrity of the surrounding materials. This type of demolition is accomplished by hand or pneumatic hand tools, and does not include saw cutting, handling, loading, hauling, or disposal of debris, scaffolding, or shoring.

Rubbish Handling - Work activities that involve handling, loading or hauling of debris. Generally, the cost of rubbish handling must be added to the cost of all types of demolition, with the exception of whole building demolition.

Minor Site Demolition - Demolition of site elements outside the footprint of a building. This type of demolition is accomplished by hand or pneumatic hand tools, or with larger pieces of construction equipment, and may include loading a removed item onto a truck (check the Crew for equipment used). It does not include saw cutting, hauling or disposal of debris, and, sometimes, handling or loading.

R024119-20 Dumpsters

Dumpster rental costs on construction sites are presented in two ways.

The cost per week rental includes the delivery of the dumpster; its pulling or emptying once per week, and its final removal. The assumption is made that the dumpster contractor could choose to empty a dumpster by simply bringing in an empty unit and removing the full one. These costs also include the disposal of the materials in the dumpster.

The Alternate Pricing can be used when actual planned conditions are not approximated by the weekly numbers. For example, these lines can be used when a dumpster is needed for 4 weeks and will need to be emptied 2 or 3 times per week. Conversely the Alternate Pricing lines can be used when a dumpster will be rented for several weeks or months but needs to be emptied only a few times over this period.

R024119-30 Rubbish Handling Chutes

To correctly estimate the cost of rubbish handling chute systems, the individual components must be priced separately. First choose the size of the system; a 30-inch diameter chute is quite common, but the sizes range from 18 to 36 inches in diameter. The 30-inch chute comes in a standard weight and two thinner weights. The thinner weight chutes are sometimes chosen for cost savings, but they are more easily damaged.

There are several types of major chute pieces that make up the chute system. The first component to consider is the top chute section (top intake hopper) where the material is dropped into the chute at the highest point. After determining the top chute, the intermediate chute pieces called the regular chute sections are priced. Next, the number of chute control door sections (intermediate intake hoppers) must be determined. In the more complex systems, a chute control door section is provided at each floor level. The last major component to consider is bolt down frames; these are usually provided at every other floor level.

There are a number of accessories to consider for safe operation and control. There are covers for the top chute and the chute door sections. The top

chute can have a trough that allows for better loading of the chute. For the safest operation, a chute warning light system can be added that will warn the other chute intake locations not to load while another is being used. There are dust control devices that spray a water mist to keep down the dust as the debris is loaded into a Dumpster. There are special breakaway cords that are used to prevent damage to the chute if the dumpster is removed without disconnecting from the chute. There are chute liners that can be installed to protect the chute structure from physical damage from rough abrasive materials. Warning signs can be posted at each floor level that is provided with a chute control door section.

In summary, a complete rubbish handling chute system will include one top section, several intermediate regular sections, several intermediate control door (intake hopper) sections and bolt down frames at every other floor level starting with the top floor. If so desired, the system can also include covers and a light warning system for a safer operation. The bottom of the chute should always be above the Dumpster and should be tied off with a breakaway cord to the Dumpster.

Existing Conditions   R0265 Underground Storage Tank Removal

R026510-20   Underground Storage Tank Removal

Underground storage tank removal can be divided into two categories: non-leaking and leaking. Prior to removing an underground storage tank, tests should be made, with the proper authorities present, to determine whether a tank has been leaking or the surrounding soil has been contaminated.

To safely remove liquid underground storage tanks:
1. Excavate to the top of the tank.
2. Disconnect all piping.
3. Open all tank vents and access ports.
4. Remove all liquids and/or sludge.

5. Purge the tank with an inert gas.
6. Provide access to the inside of the tank and clean out the interior using proper personal protective equipment (PPE).
7. Excavate soil surrounding the tank using proper PPE for on-site personnel.
8. Pull and properly dispose of the tank.
9. Clean up the site of all contaminated material.
10. Install new tanks or close the excavation.

Existing Conditions   R0282 Asbestos Remediation

R028213-20   Asbestos Removal Process

Asbestos removal is accomplished by a specialty contractor who understands the federal and state regulations regarding the handling and disposal of the material. The process of asbestos removal is divided into many individual steps. An accurate estimate can be calculated only after all the steps have been priced.

The steps are generally as follows:
1. Obtain an asbestos abatement plan from an industrial hygienist.
2. Monitor the air quality in and around the removal area and along the path of travel between the removal area and transport area. This establishes the background contamination.
3. Construct a two part decontamination chamber at entrance to removal area.
4. Install a HEPA filter to create a negative pressure in the removal area.
5. Install wall, floor and ceiling protection as required by the plan, usually 2 layers of fireproof 6 mil polyethylene.

6. Industrial hygienist visually inspects work area to verify compliance with plan.
7. Provide temporary supports for conduit and piping affected by the removal process.
8. Proceed with asbestos removal and bagging process. Monitor air quality as described in Step #2. Discontinue operations when contaminate levels exceed applicable standards.
9. Document the legal disposal of materials in accordance with EPA standards.
10. Thoroughly clean removal area including all ledges, crevices and surfaces.
11. Post abatement inspection by industrial hygienist to verify plan compliance.
12. Provide a certificate from a licensed industrial hygienist attesting that contaminate levels are within acceptable standards before returning area to regular use.

Reference Tables

R028319-60   Lead Paint Remediation Methods

Lead paint remediation can be accomplished by the following methods.
1. Abrasive blast
2. Chemical stripping
3. Power tool cleaning with vacuum collection system
4. Encapsulation
5. Remove and replace
6. Enclosure

Each of these methods has strengths and weaknesses depending on the specific circumstances of the project. The following is an overview of each method.

1. **Abrasive blasting** is usually accomplished with sand or recyclable metallic blast. Before work can begin, the area must be contained to ensure the blast material with lead does not escape to the atmosphere. The use of vacuum blast greatly reduces the containment requirements. Lead abatement equipment that may be associated with this work includes a negative air machine. In addition, it is necessary to have an industrial hygienist monitor the project on a continual basis. When the work is complete, the spent blast sand with lead must be disposed of as a hazardous material. If metallic shot was used, the lead is separated from the shot and disposed of as hazardous material. Worker protection includes disposable clothing and respiratory protection.

2. **Chemical stripping** requires strong chemicals to be applied to the surface to remove the lead paint. Before the work can begin, the area under/adjacent to the work area must be covered to catch the chemical and removed lead. After the chemical is applied to the painted surface it is usually covered with paper. The chemical is left in place for the specified period, then the paper with lead paint is pulled or scraped off. The process may require several chemical applications. The paper with chemicals and lead paint adhered to it, plus the containment and loose scrapings collected by a HEPA (High Efficiency Particulate Air Filter) vac, must be disposed of as hazardous material. The chemical stripping process usually requires a neutralizing agent and several wash downs after the paint is removed. Worker protection includes a neoprene or other compatible protective clothing and respiratory

protection with face shield. An industrial hygienist is required intermittently during the process.

3. **Power tool cleaning** is accomplished using shrouded needle blasting guns. The shrouding with different end configurations is held up against the surface to be cleaned. The area is blasted with hardened needles and the shroud captures the lead with a HEPA vac and deposits it in a holding tank. An industrial hygienist monitors the project. Protective clothing and a respirator are required until air samples prove otherwise. When the work is complete the lead must be disposed of as hazardous material.

4. **Encapsulation** is a method that leaves the well bonded lead paint in place after the peeling paint has been removed. Before the work can begin, the area under/adjacent to the work must be covered to catch the scrapings. The scraped surface is then washed with a detergent and rinsed. The prepared surface is covered with approximately 10 mils of paint. A reinforcing fabric can also be embedded in the paint covering. The scraped paint and containment must be disposed of as hazardous material. Workers must wear protective clothing and respirators.

5. **Removing and replacing** are effective ways to remove lead paint from windows, gypsum walls and concrete masonry surfaces. The painted materials are removed and new materials are installed. Workers should wear a respirator and tyvek suit. The demolished materials must be disposed of as hazardous waste if it fails the TCLP (Toxicity Characteristic Leachate Process) test.

6. **Enclosure** is the process that permanently seals lead painted materials in place. This process has many applications such as covering lead painted drywall with new drywall, covering exterior construction with tyvek paper then residing, or covering lead painted structural members with aluminum or plastic. The seams on all enclosing materials must be securely sealed. An industrial hygienist monitors the project, and protective clothing and a respirator are required until air samples prove otherwise.

All the processes require clearance monitoring and wipe testing as required by the hygienist.

R031113-10   Wall Form Materials

Aluminum Forms
Approximate weight is 3 lbs. per S.F.C.A. Standard widths are available from 4″ to 36″ with 36″ most common. Standard lengths of 2′, 4′, 6′ to 8′ are available. Forms are lightweight and fewer ties are needed with the wider widths. The form face is either smooth or textured.

Metal Framed Plywood Forms
Manufacturers claim over 75 reuses of plywood and over 300 reuses of steel frames. Many specials such as corners, fillers, pilasters, etc. are available. Monthly rental is generally about 15% of purchase price for first month and 9% per month thereafter with 90% of rental applied to purchase for the first month and decreasing percentages thereafter. Aluminum framed forms cost 25% to 30% more than steel framed.

After the first month, extra days may be prorated from the monthly charge. Rental rates do not include ties, accessories, cleaning, loss of hardware or freight in and out. Approximate weight is 5 lbs. per S.F. for steel; 3 lbs. per S.F. for aluminum.

Forms can be rented with option to buy.

Plywood Forms, Job Fabricated
There are two types of plywood used for concrete forms.
1. Exterior plyform which is completely waterproof. This is face oiled to facilitate stripping. Ten reuses can be expected with this type with 25 reuses possible.
2. An overlaid type consists of a resin fiber fused to exterior plyform. No oiling is required except to facilitate cleaning. This is available in both high density (HDO) and medium density overlaid (MDO). Using HDO, 50 reuses can be expected with 200 possible.

Plyform is available in 5/8″ and 3/4″ thickness. High density overlaid is available in 3/8″, 1/2″, 5/8″ and 3/4″ thickness.

5/8″ thick is sufficient for most building forms, while 3/4″ is best on heavy construction.

Plywood Forms, Modular, Prefabricated
There are many plywood forming systems without frames. Most of these are manufactured from 1-1/8″ (HDO) plywood and have some hardware attached. These are used principally for foundation walls 8′ or less high. With care and maintenance, 100 reuses can be attained with decreasing quality of surface finish.

Steel Forms
Approximate weight is 6-1/2 lbs. per S.F.C.A. including accessories. Standard widths are available from 2″ to 24″, with 24″ most common. Standard lengths are from 2′ to 8′, with 4′ the most common. Forms are easily ganged into modular units.

Forms are usually leased for 15% of the purchase price per month prorated daily over 30 days.

Rental may be applied to sale price, and usually rental forms are bought. With careful handling and cleaning 200 to 400 reuses are possible.

Straight wall gang forms up to 12′ x 20′ or 8′ x 30′ can be fabricated. These crane handled forms usually lease for approx. 9% per month.

Individual job analysis is available from the manufacturer at no charge.

R031113-40   Forms for Reinforced Concrete

Design Economy
Avoid many sizes in proportioning beams and columns.

From story to story avoid changing column dimensions. Gain strength by adding steel or using a richer mix. If a change in size of column is necessary, vary one dimension only to minimize form alterations. Keep beams and columns the same width.

From floor to floor in a multi-story building vary beam depth, not width, as that will leave the slab panel form unchanged. It is cheaper to vary the strength of a beam from floor to floor by means of a steel area than by 2″ changes in either width or depth.

Cost Factors
Material includes the cost of lumber, cost of rent for metal pans or forms if used, nails, form ties, form oil, bolts and accessories.

Labor includes the cost of carpenters to make up, erect, remove and repair, plus common labor to clean and move. Having carpenters remove forms minimizes repairs.

Improper alignment and condition of forms will increase finishing cost. When forms are heavily oiled, concrete surfaces must be neutralized before finishing. Special curing compounds will cause spillages to spall off in first frost. Gang forming methods will reduce costs on large projects.

Materials Used
Boards are seldom used unless their architectural finish is required. Generally, steel, fiberglass and plywood are used for contact surfaces. Labor on plywood is 10% less than with boards. The plywood is backed up with

2 x 4's at 12″ to 32″ O.C. Walers are generally 2 - 2 x 4's. Column forms are held together with steel yokes or bands. Shoring is with adjustable shoring or scaffolding for high ceilings.

Reuse
Floor and column forms can be reused four or possibly five times without excessive repair. Remember to allow for 10% waste on each reuse.

When modular sized wall forms are made, up to twenty uses can be expected with exterior plyform.

When forms are reused, the cost to erect, strip, clean and move will not be affected. 10% replacement of lumber should be included and about one hour of carpenter time for repairs on each reuse per 100 S.F.

The reuse cost for certain accessory items normally rented on a monthly basis will be lower than the cost for the first use.

After the fifth use, new material required plus time needed for repair prevent the form cost from dropping further; it may go up. Much depends on care in stripping, the number of special bays, changes in beam or column sizes and other factors.

Costs for multiple use of formwork may be developed as follows:

2 Uses	3 Uses	4 Uses
$\dfrac{(\text{1st Use + Reuse})}{2} = \text{avg. cost/2 uses}$	$\dfrac{(\text{1st Use + 2 Reuses})}{3} = \text{avg. cost/3 uses}$	$\dfrac{(\text{1st use + 3 Reuses})}{4} = \text{avg. cost/4 uses}$

R031113-60   Formwork Labor-Hours

Item	Unit	Hours Required			Total Hours	Multiple Use		
		Fabricate	Erect & Strip	Clean & Move	1 Use	2 Use	3 Use	4 Use
Beam and Girder, interior beams, 12" wide	100 S.F.	6.4	8.3	1.3	16.0	13.3	12.4	12.0
Hung from steel beams		5.8	7.7	1.3	14.8	12.4	11.6	11.2
Beam sides only, 36" high		5.8	7.2	1.3	14.3	11.9	11.1	10.7
Beam bottoms only, 24" wide		6.6	13.0	1.3	20.9	18.1	17.2	16.7
Box out for openings		9.9	10.0	1.1	21.0	16.6	15.1	14.3
Buttress forms, to 8' high		6.0	6.5	1.2	13.7	11.2	10.4	10.0
Centering, steel, 3/4" rib lath			1.0		1.0			
3/8" rib lath or slab form			0.9		0.9			
Chamfer strip or keyway	100 L.F.		1.5		1.5	1.5	1.5	1.5
Columns, fiber tube 8" diameter			20.6		20.6			
12"			21.3		21.3			
16"			22.9		22.9			
20"			23.7		23.7			
24"			24.6		24.6			
30"			25.6		25.6			
Columns, round steel, 12" diameter			22.0		22.0	22.0	22.0	22.0
16"			25.6		25.6	25.6	25.6	25.6
20"			30.5		30.5	30.5	30.5	30.5
24"			37.7		37.7	37.7	37.7	37.7
Columns, plywood 8" x 8"	100 S.F.	7.0	11.0	1.2	19.2	16.2	15.2	14.7
12" x 12"		6.0	10.5	1.2	17.7	15.2	14.4	14.0
16" x 16"		5.9	10.0	1.2	17.1	14.7	13.8	13.4
24" x 24"		5.8	9.8	1.2	16.8	14.4	13.6	13.2
Columns, steel framed plywood 8" x 8"			10.0	1.0	11.0	11.0	11.0	11.0
12" x 12"			9.3	1.0	10.3	10.3	10.3	10.3
16" x 16"			8.5	1.0	9.5	9.5	9.5	9.5
24" x 24"			7.8	1.0	8.8	8.8	8.8	8.8
Drop head forms, plywood		9.0	12.5	1.5	23.0	19.0	17.7	17.0
Coping forms		8.5	15.0	1.5	25.0	21.3	20.0	19.4
Culvert, box			14.5	4.3	18.8	18.8	18.8	18.8
Curb forms, 6" to 12" high, on grade		5.0	8.5	1.2	14.7	12.7	12.1	11.7
On elevated slabs		6.0	10.8	1.2	18.0	15.5	14.7	14.3
Edge forms to 6" high, on grade	100 L.F.	2.0	3.5	0.6	6.1	5.6	5.4	5.3
7" to 12" high	100 S.F.	2.5	5.0	1.0	8.5	7.8	7.5	7.4
Equipment foundations		10.0	18.0	2.0	30.0	25.5	24.0	23.3
Flat slabs, including drops		3.5	6.0	1.2	10.7	9.5	9.0	8.8
Hung from steel		3.0	5.5	1.2	9.7	8.7	8.4	8.2
Closed deck for domes		3.0	5.8	1.2	10.0	9.0	8.7	8.5
Open deck for pans		2.2	5.3	1.0	8.5	7.9	7.7	7.6
Footings, continuous, 12" high		3.5	3.5	1.5	8.5	7.3	6.8	6.6
Spread, 12" high		4.7	4.2	1.6	10.5	8.7	8.0	7.7
Pile caps, square or rectangular		4.5	5.0	1.5	11.0	9.3	8.7	8.4
Grade beams, 24" deep		2.5	5.3	1.2	9.0	8.3	8.0	7.9
Lintel or Sill forms		8.0	17.0	2.0	27.0	23.5	22.3	21.8
Spandrel beams, 12" wide		9.0	11.2	1.3	21.5	17.5	16.2	15.5
Stairs			25.0	4.0	29.0	29.0	29.0	29.0
Trench forms in floor		4.5	14.0	1.5	20.0	18.3	17.7	17.4
Walls, Plywood, at grade, to 8' high		5.0	6.5	1.5	13.0	11.0	9.7	9.5
8' to 16'		7.5	8.0	1.5	17.0	13.8	12.7	12.1
16' to 20'		9.0	10.0	1.5	20.5	16.5	15.2	14.5
Foundation walls, to 8' high		4.5	6.5	1.0	12.0	10.3	9.7	9.4
8' to 16' high		5.5	7.5	1.0	14.0	11.8	11.0	10.6
Retaining wall to 12' high, battered		6.0	8.5	1.5	16.0	13.5	12.7	12.3
Radial walls to 12' high, smooth		8.0	9.5	2.0	19.5	16.0	14.8	14.3
2' chords		7.0	8.0	1.5	16.5	13.5	12.5	12.0
Prefabricated modular, to 8' high		—	4.3	1.0	5.3	5.3	5.3	5.3
Steel, to 8' high		—	6.8	1.2	8.0	8.0	8.0	8.0
8' to 16' high		—	9.1	1.5	10.6	10.3	10.2	10.2
Steel framed plywood to 8' high		—	6.8	1.2	8.0	7.5	7.3	7.2
8' to 16' high		—	9.3	1.2	10.5	9.5	9.2	9.0

R032110-10   Reinforcing Steel Weights and Measures

Bar Designation No.**	Nominal Weight Lb./Ft.	U.S. Customary Units			SI Units			
		Nominal Dimensions*			Nominal Weight kg/m	Nominal Dimensions*		
		Diameter in.	Cross Sectional Area, in.^{2}	Perimeter in.		Diameter mm	Cross Sectional Area, cm^2	Perimeter mm
3	.376	.375	.11	1.178	.560	9.52	.71	29.9
4	.668	.500	.20	1.571	.994	12.70	1.29	39.9
5	1.043	.625	.31	1.963	1.552	15.88	2.00	49.9
6	1.502	.750	.44	2.356	2.235	19.05	2.84	59.8
7	2.044	.875	.60	2.749	3.042	22.22	3.87	69.8
8	2.670	1.000	.79	3.142	3.973	25.40	5.10	79.8
9	3.400	1.128	1.00	3.544	5.059	28.65	6.45	90.0
10	4.303	1.270	1.27	3.990	6.403	32.26	8.19	101.4
11	5.313	1.410	1.56	4.430	7.906	35.81	10.06	112.5
14	7.650	1.693	2.25	5.320	11.384	43.00	14.52	135.1
18	13.600	2.257	4.00	7.090	20.238	57.33	25.81	180.1

* The nominal dimensions of a deformed bar are equivalent to those of a plain round bar having the same weight per foot as the deformed bar.

** Bar numbers are based on the number of eighths of an inch included in the nominal diameter of the bars.

R032110-80   Shop-Fabricated Reinforcing Steel

The material prices for reinforcing, shown in the unit cost sections of the data set, are for 50 tons or more of shop-fabricated reinforcing steel and include:

1. Mill base price of reinforcing steel
2. Mill grade/size/length extras
3. Mill delivery to the fabrication shop
4. Shop storage and handling
5. Shop drafting/detailing

6. Shop shearing and bending
7. Shop listing
8. Shop delivery to the job site

Both material and installation costs can be considerably higher for small jobs consisting primarily of smaller bars, while material costs may be slightly lower for larger jobs.

For customer support on your Commercial Renovation Costs with RSMeans data, call 800.448.8182.

863

Concrete   R0322 Welded Wire Fabric Reinforcing

R032205-30   Common Stock Styles of Welded Wire Fabric

This table provides some of the basic specifications, sizes, and weights of welded wire fabric used for reinforcing concrete.

	New Designation	Old Designation		Steel Area per Foot				Approximate Weight per 100 S.F.	
	Spacing — Cross Sectional Area (in.) — (Sq. in. 100)	Spacing — Wire Gauge (in.) — (AS & W)		Longitudinal		Transverse			
				in.	cm	in.	cm	lbs	kg
Rolls	6 x 6 — W1.4 x W1.4	6 x 6 — 10 x 10		.028	.071	.028	.071	21	9.53
	6 x 6 — W2.0 x W2.0	6 x 6 — 8 x 8	1	.040	.102	.040	.102	29	13.15
	6 x 6 — W2.9 x W2.9	6 x 6 — 6 x 6		.058	.147	.058	.147	42	19.05
	6 x 6 — W4.0 x W4.0	6 x 6 — 4 x 4		.080	.203	.080	.203	58	26.91
	4 x 4 — W1.4 x W1.4	4 x 4 — 10 x 10		.042	.107	.042	.107	31	14.06
	4 x 4 — W2.0 x W2.0	4 x 4 — 8 x 8	1	.060	.152	.060	.152	43	19.50
	4 x 4 — W2.9 x W2.9	4 x 4 — 6 x 6		.087	.227	.087	.227	62	28.12
	4 x 4 — W4.0 x W4.0	4 x 4 — 4 x 4		.120	.305	.120	.305	85	38.56
Sheets	6 x 6 — W2.9 x W2.9	6 x 6 — 6 x 6		.058	.147	.058	.147	42	19.05
	6 x 6 — W4.0 x W4.0	6 x 6 — 4 x 4		.080	.203	.080	.203	58	26.31
	6 x 6 — W5.5 x W5.5	6 x 6 — 2 x 2	2	.110	.279	.110	.279	80	36.29
	4 x 4 — W1.4 x W1.4	4 x 4 — 4 x 4		.120	.305	.120	.305	85	38.56

NOTES: 1. Exact W—number size for 8 gauge is W2.1
2. Exact W—number size for 2 gauge is W5.4

The above table was compiled with the following excerpts from the WRI Manual of Standard Practices, 7th Edition, Copyright 2006. Reproduced with permission of the Wire Reinforcement Institute, Inc.:

1. Chapter 3, page 7, Table 1 Common Styles of Metric Wire Reinforcement (WWR) With Equivalent US Customary Units
2. Chapter 6, page 19, Table 5 Customary Units
3. Chapter 6, Page 23, Table 7 Customary Units (in.) Welded Plain Wire Reinforcement
4. Chapter 6, Page 25 Table 8 Wire Size Comparison
5. Chapter 9, Page 30, Table 9 Weight of Longitudinal Wires Weight (Mass) Estimating Tables
6. Chapter 9, Page 31, Table 9M Weight of Longitudinal Wires Weight (Mass) Estimating Tables
7. Chapter 9, Page 32, Table 10 Weight of Transverse Wires Based on 62″ lengths of transverse wire (60″ width plus 1″ overhand each side)
8. Chapter 9, Page 33, Table 10M Weight of Transverse Wires

R033105-10   Proportionate Quantities

The tables below show both quantities per S.F. of floor areas as well as form and reinforcing quantities per C.Y. Unusual structural requirements would increase the ratios below. High strength reinforcing would reduce the steel weights. Figures are for 3000 psi concrete and 60,000 psi reinforcing unless specified otherwise.

Type of Construction	Live Load	Span	Per S.F. of Floor Area				Per C.Y. of Concrete		
			Concrete	Forms	Reinf.	Pans	Forms	Reinf.	Pans
Flat Plate	50 psf	15 Ft.	.46 C.F.	1.06 S.F.	1.71 lb.		62 S.F.	101 lb.	
		20	.63	1.02	2.40		44	104	
		25	.79	1.02	3.03		35	104	
	100	15	.46	1.04	2.14		61	126	
		20	.71	1.02	2.72		39	104	
		25	.83	1.01	3.47		33	113	
Flat Plate (waffle construction) 20" domes	50	20	.43	1.00	2.10	.84 S.F.	63	135	53 S.F.
		25	.52	1.00	2.90	.89	52	150	46
		30	.64	1.00	3.70	.87	42	155	37
	100	20	.51	1.00	2.30	.84	53	125	45
		25	.64	1.00	3.20	.83	42	135	35
		30	.76	1.00	4.40	.81	36	160	29
Waffle Construction 30" domes	50	25	.69	1.06	1.83	.68	42	72	40
		30	.74	1.06	2.39	.69	39	87	39
		35	.86	1.05	2.71	.69	33	85	39
		40	.78	1.00	4.80	.68	35	165	40
Flat Slab (two way with drop panels)	50	20	.62	1.03	2.34		45	102	
		25	.77	1.03	2.99		36	105	
		30	.95	1.03	4.09		29	116	
	100	20	.64	1.03	2.83		43	119	
		25	.79	1.03	3.88		35	133	
		30	.96	1.03	4.66		29	131	
	200	20	.73	1.03	3.03		38	112	
		25	.86	1.03	4.23		32	133	
		30	1.06	1.03	5.30		26	135	
One Way Joists 20" Pans	50	15	.36	1.04	1.40	.93	78	105	70
		20	.42	1.05	1.80	.94	67	120	60
		25	.47	1.05	2.60	.94	60	150	54
	100	15	.38	1.07	1.90	.93	77	140	66
		20	.44	1.08	2.40	.94	67	150	58
		25	.52	1.07	3.50	.94	55	185	49
One Way Joists 8" x 16" filler blocks	50	15	.34	1.06	1.80	.81 Ea.	84	145	64 Ea.
		20	.40	1.08	2.20	.82	73	145	55
		25	.46	1.07	3.20	.83	63	190	49
	100	15	.39	1.07	1.90	.81	74	130	56
		20	.46	1.09	2.80	.82	64	160	48
		25	.53	1.10	3.60	.83	56	190	42
One Way Beam & Slab	50	15	.42	1.30	1.73		84	111	
		20	.51	1.28	2.61		68	138	
		25	.64	1.25	2.78		53	117	
	100	15	.42	1.30	1.90		84	122	
		20	.54	1.35	2.69		68	154	
		25	.69	1.37	3.93		54	145	
	200	15	.44	1.31	2.24		80	137	
		20	.58	1.40	3.30		65	163	
		25	.69	1.42	4.89		53	183	
Two Way Beam & Slab	100	15	.47	1.20	2.26		69	130	
		20	.63	1.29	3.06		55	131	
		25	.83	1.33	3.79		43	123	
	200	15	.49	1.25	2.70		41	149	
		20	.66	1.32	4.04		54	165	
		25	.88	1.32	6.08		41	187	

R033105-10 Proportionate Quantities (cont.)

4000 psi Concrete and 60,000 psi Reinforcing—Form and Reinforcing Quantities per C.Y.

Item	Size	Forms	Reinforcing	Minimum	Maximum
Columns (square tied)	10″ x 10″	130 S.F.C.A.	#5 to #11	220 lbs.	875 lbs.
	12″ x 12″	108	#6 to #14	200	955
	14″ x 14″	92	#7 to #14	190	900
	16″ x 16″	81	#6 to #14	187	1082
	18″ x 18″	72	#6 to #14	170	906
	20″ x 20″	65	#7 to #18	150	1080
	22″ x 22″	59	#8 to #18	153	902
	24″ x 24″	54	#8 to #18	164	884
	26″ x 26″	50	#9 to #18	169	994
	28″ x 28″	46	#9 to #18	147	864
	30″ x 30″	43	#10 to #18	146	983
	32″ x 32″	40	#10 to #18	175	866
	34″ x 34″	38	#10 to #18	157	772
	36″ x 36″	36	#10 to #18	175	852
	38″ x 38″	34	#10 to #18	158	765
	40″ x 40″	32 ▼	#10 to #18	143	692

Item	Size	Form	Spiral	Reinforcing	Minimum	Maximum
Columns (spirally reinforced)	12″ diameter	34.5 L.F.	190 lbs.	#4 to #11	165 lbs.	1505 lb.
		34.5	190	#14 & #18	—	1100
	14″	25	170	#4 to #11	150	970
		25	170	#14 & #18	800	1000
	16″	19	160	#4 to #11	160	950
		19	160	#14 & #18	605	1080
	18″	15	150	#4 to #11	160	915
		15	150	#14 & #18	480	1075
	20″	12	130	#4 to #11	155	865
		12	130	#14 & #18	385	1020
	22″	10	125	#4 to #11	165	775
		10	125	#14 & #18	320	995
	24″	9	120	#4 to #11	195	800
		9	120	#14 & #18	290	1150
	26″	7.3	100	#4 to #11	200	729
		7.3	100	#14 & #18	235	1035
	28″	6.3	95	#4 to #11	175	700
		6.3	95	#14 & #18	200	1075
	30″	5.5	90	#4 to #11	180	670
		5.5	90	#14 & #18	175	1015
	32″	4.8	85	#4 to #11	185	615
		4.8	85	#14 & #18	155	955
	34″	4.3	80	#4 to #11	180	600
		4.3	80	#14 & #18	170	855
	36″	3.8	75	#4 to #11	165	570
		3.8	75	#14 & #18	155	865
	40″	3.0	70	#4 to #11	165	500
		3.0	70	#14 & #18	145	765

R033105-10 Proportionate Quantities (cont.)

		3000 psi Concrete and 60,000 psi Reinforcing—Form and Reinforcing Quantities per C.Y.				
Item	Type	Loading	Height	C.Y./L.F.	Forms/C.Y.	Reinf./C.Y.
Retaining Walls	Cantilever	Level Backfill	4 Ft.	0.2 C.Y.	49 S.F.	35 lbs.
			8	0.5	42	45
			12	0.8	35	70
			16	1.1	32	85
			20	1.6	28	105
		Highway Surcharge	4	0.3	41	35
			8	0.5	36	55
			12	0.8	33	90
			16	1.2	30	120
			20	1.7	27	155
		Railroad Surcharge	4	0.4	28	45
			8	0.8	25	65
			12	1.3	22	90
			16	1.9	20	100
			20	2.6	18	120
	Gravity, with Vertical Face	Level Backfill	4	0.4	37	None
			7	0.6	27	↓
			10	1.2	20	
		Sloping Backfill	4	0.3	31	
			7	0.8	21	
			10	1.6	15	↓

		Live Load in Kips per Linear Foot							
	Span	Under 1 Kip		2 to 3 Kips		4 to 5 Kips		6 to 7 Kips	
		Forms	Reinf.	Forms	Reinf.	Forms	Reinf.	Forms	Reinf.
Beams	10 Ft.	—	—	90 S.F.	170 #	85 S.F.	175 #	75 S.F.	185 #
	16	130 S.F.	165 #	85	180	75	180	65	225
	20	110	170	75	185	62	200	51	200
	26	90	170	65	215	62	215	—	—
	30	85	175	60	200	—	—	—	—

Item	Size	Type	Forms per C.Y.	Reinforcing per C.Y.
Spread Footings	Under 1 C.Y.	1,000 psf soil	24 S.F.	44 lbs.
		5,000	24	42
		10,000	24	52
	1 C.Y. to 5 C.Y.	1,000	14	49
		5,000	14	50
		10,000	14	50
	Over 5 C.Y.	1,000	9	54
		5,000	9	52
		10,000	9	56
Pile Caps (30 Ton Concrete Piles)	Under 5 C.Y.	shallow caps	20	65
		medium	20	50
		deep	20	40
	5 C.Y. to 10 C.Y.	shallow	14	55
		medium	15	45
		deep	15	40
	10 C.Y. to 20 C.Y.	shallow	11	60
		medium	11	45
		deep	12	35
	Over 20 C.Y.	shallow	9	60
		medium	9	45
		deep	10	40

For customer support on your Commercial Renovation Costs with RSMeans data, call 800.448.8182.

867

R033105-10 Proportionate Quantities (cont.)

Item	Size		Pile Spacing	50 T Pile	100 T Pile	50 T Pile	100 T Pile
Pile Caps (Steel H Piles)	Under 5 C.Y.		24" O.C.	24 S.F.	24 S.F.	75 lbs.	90 lbs.
			30"	25	25	80	100
			36"	24	24	80	110
	5 C.Y. to 10 C.Y.		24"	15	15	80	110
			30"	15	15	85	110
			36"	15	15	75	90
	Over 10 C.Y.		24"	13	13	85	90
			30"	11	11	85	95
			36"	10	10	85	90

3000 psi Concrete and 60,000 psi Reinforcing — Form and Reinforcing Quantities per C.Y.

		8" Thick		10" Thick		12" Thick		15" Thick	
	Height	Forms	Reinf.	Forms	Reinf.	Forms	Reinf.	Forms	Reinf.
Basement Walls	7 Ft.	81 S.F.	44 lbs.	65 S.F.	45 lbs.	54 S.F.	44 lbs.	41 S.F.	43 lbs.
	8		44		45		44		43
	9		46		45		44		43
	10		57		45		44		43
	12		83		50		52		43
	14		116		65		64		51
	16				86		90		65
	18						106		70

R033105-20 Materials for One C.Y. of Concrete

This is an approximate method of figuring quantities of cement, sand and coarse aggregate for a field mix with waste allowance included.

With crushed gravel as coarse aggregate, to determine barrels of cement required, divide 10 by total mix; that is, for 1:2:4 mix, 10 divided by 7 = 1-3/7 barrels. If the coarse aggregate is crushed stone, use 10-1/2 instead of 10 as given for gravel.

To determine tons of sand required, multiply barrels of cement by parts of sand and then by 0.2; that is, for the 1:2:4 mix, as above, 1-3/7 x 2 x .2 = .57 tons.

Tons of crushed gravel are in the same ratio to tons of sand as parts in the mix, or 4/2 x .57 = 1.14 tons.

1 bag cement = 94#	1 C.Y. sand or crushed gravel = 2700#	1 C.Y. crushed stone = 2575#
4 bags = 1 barrel	1 ton sand or crushed gravel = 20 C.F.	1 ton crushed stone = 21 C.F.

Average carload of cement is 692 bags; of sand or gravel is 56 tons.

Do not stack stored cement over 10 bags high.

R033105-70 Placing Ready-Mixed Concrete

For ground pours allow for 5% waste when figuring quantities.

Prices in the front of the data set assume normal deliveries. If deliveries are made before 8 A.M. or after 5 P.M. or on Saturday afternoons add 30%. Negotiated discounts for large volumes are not included in prices in front of the data set.

For the lower floors without truck access, concrete may be wheeled in rubber-tired buggies, conveyer handled, crane handled or pumped. Pumping is economical if there is top steel. Conveyers are more efficient for thick slabs.

At higher floors the rubber-tired buggies may be hoisted by a hoisting tower and wheeled to the location. Placement by a conveyer is limited to three floors and is best for high-volume pours. Pumped concrete is best when the building has no crane access. Concrete may be pumped directly as high as thirty-six stories using special pumping techniques. Normal maximum height is about fifteen stories.

The best pumping aggregate is screened and graded bank gravel rather than crushed stone.

Pumping downward is more difficult than pumping upward. The horizontal distance from pump to pour may increase preparation time prior to pour. Placing by cranes, either mobile, climbing or tower types, continues as the most efficient method for high-rise concrete buildings.

R033105-85   Lift Slabs

The cost advantage of the lift slab method is due to placing all concrete, reinforcing steel, inserts and electrical conduit at ground level and in reduction of formwork. Minimum economical project size is about 30,000 S.F. Slabs may be tilted for parking garage ramps.

It is now used in all types of buildings and has gone up to 22 stories high in apartment buildings. Current trend is to use post-tensioned flat plate slabs with spans from 22' to 35'. Cylindrical void forms are used when deep slabs are required. One pound of prestressing steel is about equal to seven pounds of conventional reinforcing.

To be considered cured for stressing and lifting, a slab must have attained 75% of design strength. Seven days are usually sufficient with four to five days possible if high early strength cement is used. Slabs can be stacked using two coats of a non-bonding agent to insure that slabs do not stick to each other. Lifting is done by companies specializing in this work. Lift rate is 5' to 15' per hour with an average of 10' per hour. Total areas up to 33,000 S.F. have been lifted at one time. 24 to 36 jacking columns are common. Most economical bay sizes are 24' to 28' with four to fourteen stories most efficient. Continuous design reduces reinforcing steel cost. Use of post-tensioned slabs allows larger bay sizes.

R033543-10   Polished Concrete Floors

A polished concrete floor has a glossy mirror-like appearance and is created by grinding the concrete floor with finer and finer diamond grits, similar to sanding wood, until the desired level of reflective clarity and sheen are achieved. The technical term for this type of polished concrete is bonded abrasive polished concrete. The basic piece of equipment used in the polishing process is a walk-behind planetary grinder for working large floor areas. This grinder drives diamond-impregnated abrasive discs, which progress from coarse- to fine-grit discs.

The process begins with the use of very coarse diamond segments or discs bonded in a metallic matrix. These segments are coarse enough to allow the removal of pits, blemishes, stains, and light coatings from the floor surface in preparation for final smoothing. The condition of the original concrete surface will dictate the grit coarseness of the initial grinding step which will generally end up being a three- to four-step process using ever finer grits. The purpose of this initial grinding step is to remove surface coatings and blemishes and to cut down into the cream for very fine aggregate exposure, or deeper into the fine aggregate layer just below the cream layer, or even deeper into the coarse aggregate layer. These initial grinding steps will progress up to the 100/120 grit. If wet grinding is done, a waste slurry is produced that must be removed between grit changes and disposed of properly. If dry grinding is done, a high performance vacuum will pick up the dust during grinding and collect it in bags which must be disposed of properly.

The process continues with honing the floor in a series of steps that progresses from 100-grit to 400-grit diamond abrasive discs embedded in a plastic or resin matrix. At some point during, or just prior to, the honing step, one or two coats of stain or dye can be sprayed onto the surface to give color to the concrete, and two coats of densifier/hardener must be applied to the floor surface and allowed to dry. This sprayed-on densifier/hardener will penetrate about 1/8" into the concrete to make the surface harder, denser and more abrasion-resistant.

The process ends with polishing the floor surface in a series of steps that progresses from resin-impregnated 800-grit (medium polish) to 1500-grit (high polish) to 3000-grit (very high polish), depending on the desired level of reflective clarity and sheen.

The Concrete Polishing Association of America (CPAA) has defined the flooring options available when processing concrete to a desired finish. The first category is aggregate exposure, the grinding of a concrete surface with bonded abrasives, in as many abrasive grits necessary, to achieve one of the following classes:

©Concrete Polishing Association of America "Glossary."

A. Cream – very little surface cut depth; little aggregate exposure

B. Fine aggregate (salt and pepper) – surface cut depth of 1/16"; fine aggregate exposure with little or no medium aggregate exposure at random locations

C. Medium aggregate – surface cut depth of 1/8"; medium aggregate exposure with little or no large aggregate exposure at random locations

D. Large aggregate – surface cut depth of 1/4"; large aggregate exposure with little or no fine aggregate exposure at random locations

The second CPAA defined category is reflective clarity and sheen, the polishing of a concrete surface with the minimum number of bonded abrasives as indicated to achieve one of the following levels:

1. Ground – flat appearance with none to very slight diffused reflection; none to very low reflective sheen; using a minimum total of 4 grit levels up to 100-grit

2. Honed – matte appearance with or without slight diffused reflection; low to medium reflective sheen; using a minimum total of 5 grit levels up to 400-grit

3. Semi-polished – objects being reflected are not quite sharp and crisp but can be easily identified; medium to high reflective sheen; using a minimum total of 6 grit levels up to 800-grit

4. Highly-polished – objects being reflected are sharp and crisp as would be seen in a mirror-like reflection; high to highest reflective sheen; using a minimum total of up to 8 grit levels up to 1500-grit or 3000-grit

The CPAA defines reflective clarity as the degree of sharpness and crispness of the reflection of overhead objects when viewed 5' above and perpendicular to the floor surface. Reflective sheen is the degree of gloss reflected from a surface when viewed at least 20' from and at an angle to the floor surface. These terms are relatively subjective. The final outcome depends on the internal makeup and surface condition of the original concrete floor, the experience of the floor polishing crew, and the expectations of the owner. Before the grinding, honing, and polishing work commences on the main floor area, it might be beneficial to do a mock-up panel in the same floor but in an out of the way place to demonstrate the sequence of steps with increasingly fine abrasive grits and to demonstrate the final reflective clarity and reflective sheen. This mock-up panel will be within the area of, and part of, the final work.

R034105-30 Prestressed Precast Concrete Structural Units

Type	Location	Depth	Span in Ft.		Live Load Lb. per S.F.
Double Tee	Floor	28" to 34"	60 to 80		50 to 80
	Roof	12" to 24"	30 to 50		40
	Wall	Width 8'	Up to 55' high		Wind
Multiple Tee	Roof	8" to 12"	15 to 40		40
	Floor	8" to 12"	15 to 30		100
Plank or or	Roof or Floor		Roof	Floor	
		4"	13	12	40 for Roof
		6"	22	18	
		8"	26	25	
		10"	33	29	100 for Floor
		12"	42	32	
Single Tee	Roof	28"	40		40
		32"	80		
		36"	100		
		48"	120		
AASHO Girder	Bridges	Type 4	100		Highway
		5	110		
		6	125		
Box Beam	Bridges	15"	40		Highway
		27"	to		
		33"	100		

The majority of precast projects today utilize double tees rather than single tees because of speed and ease of installation. As a result casting beds at manufacturing plants are normally formed for double tees. Single tee projects will therefore require an initial set up charge to be spread over the individual single tee costs.

For floors, a 2" to 3" topping is field cast over the shapes. For roofs, insulating concrete or rigid insulation is placed over the shapes.

Member lengths up to 40' are standard haul, 40' to 60' require special permits and lengths over 60' must be escorted. Excessive width and/or length can add up to 100% on hauling costs.

Large heavy members may require two cranes for lifting which would increase erection costs by about 45%. An eight man crew can install 12 to 20 double tees, or 45 to 70 quad tees or planks per day.

Grouting of connections must also be included.

Several system buildings utilizing precast members are available. Heights can go up to 22 stories for apartment buildings. The optimum design ratio is 3 S.F. of surface to 1 S.F. of floor area.

R035216-10   Lightweight Concrete

Lightweight aggregate concrete is usually purchased ready mixed, but it can also be field mixed.

Vermiculite or Perlite comes in bags of 4 C.F. under various trade names. The weight is about 8 lbs. per C.F. For insulating roof fill use 1:6 mix. For a structural deck use 1:4 mix over gypsum boards, steeltex, steel centering, etc., supported by closely spaced joists or bulb trees. For structural slabs use 1:3:2 vermiculite sand concrete over steeltex, metal lath, steel centering, etc., on joists spaced 2'-0" O.C. for maximum L.L. of 80 P.S.F. Use same mix

for slab base fill over steel flooring or regular reinforced concrete slab when tile, terrazzo or other finish is to be laid over.

For slabs on grade use 1:3:2 mix when tile, etc., finish is to be laid over. If radiant heating units are installed use a 1:6 mix for a base. After coils are in place, cover with a regular granolithic finish (mix 1:3:2) to a minimum depth of 1-1/2" over top of units.

Reinforce all slabs with 6 x 6 or 10 x 10 welded wire mesh.

Masonry   R0421 Clay Unit Masonry

R042110-20   Common and Face Brick

Common building brick manufactured according to ASTM C62 and facing brick manufactured according to ASTM C216 are the two standard bricks available for general building use.

Building brick is made in three grades: SW, where high resistance to damage caused by cyclic freezing is required; MW, where moderate resistance to cyclic freezing is needed; and NW, where little resistance to cyclic freezing is needed. Facing brick is made in only the two grades SW and MW. Additionally, facing brick is available in three types: FBS, for general use; FBX, for general use where a higher degree of precision and lower permissible variation in size than FBS are needed; and FBA, for general use to produce characteristic architectural effects resulting from non-uniformity in size and texture of the units.

In figuring the material cost of brickwork, an allowance of 25% mortar waste and 3% brick breakage was included. If bricks are delivered palletized

with 280 to 300 per pallet, or packaged, allow only 1-1/2% for breakage. Packaged or palletized delivery is practical when a job is big enough to have a crane or other equipment available to handle a package of brick. This is so on all industrial work but not always true on small commercial buildings.

The use of buff and gray face is increasing, and there is a continuing trend to the Norman, Roman, Jumbo and SCR brick.

Common red clay brick for backup is not used that often. Concrete block is the most usual backup material with occasional use of sand lime or cement brick. Building brick is commonly used in solid walls for strength and as a fire stop.

Brick panels built on the ground and then crane erected to the upper floors have proven to be economical. This allows the work to be done under cover and without scaffolding.

R042110-50   Brick, Block & Mortar Quantities

Running Bond					C.F. of Mortar per M Bricks, Waste Included		For Other Bonds Standard Size Add to S.F. Quantities in Table to Left		
Number of Brick per S.F. of Wall - Single Wythe with 3/8" Joints									
Type Brick	Nominal Size (incl. mortar) L H W			Modular Coursing	Number of Brick per S.F.	3/8" Joint	1/2" Joint	Bond Type	Description	Factor
Standard	8 x 2-2/3 x 4			3C=8"	6.75	8.1	10.3	Common	full header every fifth course	+20%
Economy	8 x 4 x 4			1C=4"	4.50	9.1	11.6		full header every sixth course	+16.7%
Engineer	8 x 3-1/5 x 4			5C=16"	5.63	8.5	10.8	English	full header every second course	+50%
Fire	9 x 2-1/2 x 4-1/2			2C=5"	6.40	550 # Fireclay	—	Flemish	alternate headers every course	+33.3%
Jumbo	12 x 4 x 6 or 8			1C=4"	3.00	22.5	29.2		every sixth course	+5.6%
Norman	12 x 2-2/3 x 4			3C=8"	4.50	11.2	14.3	Header = W x H exposed		+100%
Norwegian	12 x 3-1/5 x 4			5C=16"	3.75	11.7	14.9	Rowlock = H x W exposed		+100%
Roman	12 x 2 x 4			2C=4"	6.00	10.7	13.7	Rowlock stretcher = L x W exposed		+33.3%
SCR	12 x 2-2/3 x 6			3C=8"	4.50	21.8	28.0	Soldier = H x L exposed		—
Utility	12 x 4 x 4			1C=4"	3.00	12.3	15.7	Sailor = W x L exposed		-33.3%

Concrete Blocks Nominal Size		Approximate Weight per S.F.		Blocks per 100 S.F.	Mortar per M block, waste included	
		Standard	Lightweight		Partitions	Back up
2"	x 8" x 16"	20 PSF	15 PSF	113	27 C.F.	36 C.F.
4"		30	20		41	51
6"		42	30		56	66
8"		55	38		72	82
10"		70	47		87	97
12"		85	55		102	112

Brick & Mortar Quantities
©Brick Industry Association. 2009 Feb. Technical Notes on Brick Construction 10:
    Dimensioning and Estimating Brick Masonry. Reston (VA): BIA. Table 1 Modular Brick Sizes and Table 4 Quantity Estimates for Brick Masonry.

Reference Tables *(vertical left margin)*

R050521-20   Welded Structural Steel

Usual weight reductions with welded design run 10% to 20% compared with bolted or riveted connections. This amounts to about the same total cost compared with bolted structures since field welding is more expensive than bolts. For normal spans of 18' to 24' figure 6 to 7 connections per ton.

Trusses — For welded trusses add 4% to weight of main members for connections. Up to 15% less steel can be expected in a welded truss compared to one that is shop bolted. Cost of erection is the same whether shop bolted or welded.

General — Typical electrodes for structural steel welding are E6010, E6011, E60T and E70T. Typical buildings vary between 2# to 8# of weld rod per

ton of steel. Buildings utilizing continuous design require about three times as much welding as conventional welded structures. In estimating field erection by welding, it is best to use the average linear feet of weld per ton to arrive at the welding cost per ton. The type, size and position of the weld will have a direct bearing on the cost per linear foot. A typical field welder will deposit 1.8# to 2# of weld rod per hour manually. Using semiautomatic methods can increase production by as much as 50% to 75%.

R051223-10   Structural Steel

The bare material prices for structural steel, shown in the unit cost sections of the data set, are for 100 tons of shop-fabricated structural steel and include:
1. Mill base price of structural steel
2. Mill scrap/grade/size/length extras
3. Mill delivery to a metals service center (warehouse)
4. Service center storage and handling
5. Service center delivery to a fabrication shop
6. Shop storage and handling
7. Shop drafting/detailing
8. Shop fabrication

9. Shop coat of primer paint
10. Shop listing
11. Shop delivery to the job site

In unit cost sections of the data set that contain items for field fabrication of steel components, the bare material cost of steel includes:
1. Mill base price of structural steel
2. Mill scrap/grade/size/length extras
3. Mill delivery to a metals service center (warehouse)
4. Service center storage and handling
5. Service center delivery to the job site

R051223-15   Structural Steel Estimating for Repair and Remodeling Projects

The correct approach to estimating structural steel is dependent upon the amount of steel required for the particular project. If the project is a sizable addition to a building, the data can be used directly from the data set. This is not the case however if the project requires a small amount of steel, for instance to reinforce existing roof or floor structural systems. To better understand this, please refer to the unit price line for a W16x31 beam with bolted connections. Assume your project requires the reinforcement of the structural members of the roof system of a 3 story building required for the installation a new piece of HVAC equipment. The project will need 4 pieces of W16x31 that are 30' long. After pricing this 120 L.F. job using the unit prices for a W16x31, an analysis will reveal that the price is wholly inadequate.

The first problem is apparent if you examine the amount of steel that can be installed per day. The unit price line indicates 900 linear feet per day can be installed by a 5 man crew with an 90 ton crane. This productivity is correct for new construction but certainly is not for repair and remodeling work. Installation of new structural steel considers that each member is installed from the foundation to the roof of the structure in a planned and systematic manner with a crane having unrestricted access to all parts of the project. Additionally each connection is planned and detailed with full field access for fit-up and final bolting. The erection is planned and progresses such that interferences and conflicts with other structural members are minimized, if not completely eliminated. All of these assumptions are clearly not the case with a repair and remodeling job, and a significant decrease in the stated productivity will be observed.

A crane will certainly be needed to lift the members into the general area of the project but in most cases will not be able to place the beams into

their final position. An opening in the existing roof may not be large enough to permit the beams to pass through and it may be necessary to bring them into the building through existing windows or doors. Moving the beams to the actual area where they will be installed may involve hand labor and the use of dollies. Finally, hoists and/or jacks may be needed for final positioning.

The connection of new members to existing can often be accomplished by field bolting with accurate field measurements and good planning but in many cases access to both sides of existing members is not possible and field welding becomes the only alternative. In addition to the cost of the actual welding, protection of existing finishes, systems and structure and fire protection must be considered.

New beams can never be installed tight to the existing decks or floors which they must support and the use of shims and tack welding becomes necessary. Additionally, further planning and cost are involved in assuring that existing loads are minimized during the installation and shimming process.

It is apparent that installation of structural steel as part of a repair and remodeling project involves more than simply installing the members and estimating the cost in the same manner as new construction. The best procedure for estimating the total cost are adequate planning and coordination of each process and activity that will be needed. Unit costs for the materials, labor and equipment can then be attached to each needed activity and a final, complete price can be determined.

R053100-10 Decking Descriptions

General - All Deck Products

A steel deck is made by cold forming structural grade sheet steel into a repeating pattern of parallel ribs. The strength and stiffness of the panels are the result of the ribs and the material properties of the steel. Deck lengths can be varied to suit job conditions, but because of shipping considerations, are usually less than 40 feet. Standard deck width varies with the product used but full sheets are usually 12″, 18″, 24″, 30″, or 36″. The deck is typically furnished in a standard width with the ends cut square. Any cutting for width, such as at openings or for angular fit, is done at the job site.

The deck is typically attached to the building frame with arc puddle welds, self-drilling screws, or powder or pneumatically driven pins. Sheet to sheet fastening is done with screws, button punching (crimping), or welds.

Composite Floor Deck

After installation and adequate fastening, a floor deck serves several purposes. It (a) acts as a working platform, (b) stabilizes the frame, (c) serves as a concrete form for the slab, and (d) reinforces the slab to carry the design loads applied during the life of the building. Composite decks are distinguished by the presence of shear connector devices as part of the deck. These devices are designed to mechanically lock the concrete and deck together so that the concrete and the deck work together to carry subsequent floor loads. These shear connector devices can be rolled-in embossments, lugs, holes, or wires welded to the panels. The deck profile can also be used to interlock concrete and steel.

Composite deck finishes are either galvanized (zinc coated) or phosphatized/painted. Galvanized deck has a zinc coating on both the top and bottom surfaces. The phosphatized/painted deck has a bare (phosphatized) top surface that will come into contact with the concrete. This bare top surface can be expected to develop rust before the concrete is placed. The bottom side of the deck has a primer coat of paint.

A composite floor deck is normally installed so the panel ends do not overlap on the supporting beams. Shear lugs or panel profile shapes often prevent a tight metal to metal fit if the panel ends overlap; the air gap caused by overlapping will prevent proper fusion with the structural steel supports when the panel end laps are shear stud welded.

Adequate end bearing of the deck must be obtained as shown on the drawings. If bearing is actually less in the field than shown on the drawings, further investigation is required.

Roof Deck

A roof deck is not designed to act compositely with other materials. A roof deck acts alone in transferring horizontal and vertical loads into the building frame. Roof deck rib openings are usually narrower than floor deck rib openings. This provides adequate support of the rigid thermal insulation board.

A roof deck is typically installed to endlap approximately 2″ over supports. However, it can be butted (or lapped more than 2″) to solve field fit problems. Since designers frequently use the installed deck system as part of the horizontal bracing system (the deck as a diaphragm), any fastening substitution or change should be approved by the designer. Continuous perimeter support of the deck is necessary to limit edge deflection in the finished roof and may be required for diaphragm shear transfer.

Standard roof deck finishes are galvanized or primer painted. The standard factory applied paint for roof decks is a primer paint and is not intended to weather for extended periods of time. Field painting or touching up of abrasions and deterioration of the primer coat or other protective finishes is the responsibility of the contractor.

Cellular Deck

A cellular deck is made by attaching a bottom steel sheet to a roof deck or composite floor deck panel. A cellular deck can be used in the same manner as a floor deck. Electrical, telephone, and data wires are easily run through the chase created between the deck panel and the bottom sheet.

When used as part of the electrical distribution system, the cellular deck must be installed so that the ribs line up and create a smooth cell transition at abutting ends. The joint that occurs at butting cell ends must be taped or otherwise sealed to prevent wet concrete from seeping into the cell. Cell interiors must be free of welding burrs, or other sharp intrusions, to prevent damage to wires.

When used as a roof deck, the bottom flat plate is usually left exposed to view. Care must be maintained during erection to keep good alignment and prevent damage.

A cellular deck is sometimes used with the flat plate on the top side to provide a flat working surface. Installation of the deck for this purpose requires special methods for attachment to the frame because the flat plate, now on the top, can prevent direct access to the deck material that is bearing on the structural steel. It may be advisable to treat the flat top surface to prevent slipping.

A cellular deck is always furnished galvanized or painted over galvanized.

Form Deck

A form deck can be any floor or roof deck product used as a concrete form. Connections to the frame are by the same methods used to anchor floor and roof decks. Welding washers are recommended when welding a deck that is less than 20 gauge thickness.

A form deck is furnished galvanized, prime painted, or uncoated. Galvanized deck must be used for those roof deck systems where a form deck is used to carry a lightweight insulating concrete fill.

For customer support on your Commercial Renovation Costs with RSMeans data, call 800.448.8182.

873

Wood, Plastics & Comp. | R0611 Wood Framing

R061110-30 Lumber Product Material Prices

The price of forest products fluctuates widely from location to location and from season to season depending upon economic conditions. The bare material prices in the unit cost sections of the data set show the National Average material prices in effect Jan. 1 of this data year. It must be noted that lumber prices in general may change significantly during the year.

Availability of certain items depends upon geographic location and must be checked prior to firm-price bidding.

Wood, Plastics & Comp. | R0616 Sheathing

R061636-20 Plywood

There are two types of plywood used in construction: interior, which is moisture-resistant but not waterproofed, and exterior, which is waterproofed.

The grade of the exterior surface of the plywood sheets is designated by the first letter: A, for smooth surface with patches allowed; B, for solid surface with patches and plugs allowed; C, which may be surface plugged or may have knot holes up to 1″ wide; and D, which is used only for interior type plywood and may have knot holes up to 2-1/2″ wide. "Structural Grade" is specifically designed for engineered applications such as box beams. All CC & DD grades have roof and floor spans marked on them.

Underlayment-grade plywood runs from 1/4″ to 1-1/4″ thick. Thicknesses 5/8″ and over have optional tongue and groove joints which eliminate the need for blocking the edges. Underlayment 19/32″ and over may be referred to as Sturd-i-Floor.

The price of plywood can fluctuate widely due to geographic and economic conditions.

Typical uses for various plywood grades are as follows:

AA-AD Interior — cupboards, shelving, paneling, furniture

BB Plyform — concrete form plywood

CDX — wall and roof sheathing

Structural — box beams, girders, stressed skin panels

AA-AC Exterior — fences, signs, siding, soffits, etc.

Underlayment — base for resilient floor coverings

Overlaid HDO — high density for concrete forms & highway signs

Overlaid MDO — medium density for painting, siding, soffits & signs

303 Siding — exterior siding, textured, striated, embossed, etc.

Thermal & Moist. Protec. | R0751 Built-Up Bituminous Roofing

R075113-20 Built-Up Roofing

Asphalt is available in kegs of 100 lbs. each; coal tar pitch in 560 lb. kegs. Prepared roofing felts are available in a wide range of sizes, weights and characteristics. However, the most commonly used are #15 (432 S.F. per roll, 13 lbs. per square) and #30 (216 S.F. per roll, 27 lbs. per square).

Inter-ply bitumen varies from 24 lbs. per sq. (asphalt) to 30 lbs. per sq. (coal tar) per ply, MF4@ 25%. Flood coat bitumen also varies from 60 lbs. per sq. (asphalt) to 75 lbs. per sq. (coal tar), MF4@ 25%. Expendable equipment (mops, brooms, screeds, etc.) runs about 16% of the bitumen cost. For new, inexperienced crews this factor may be much higher.

A rigid insulation board is typically applied in two layers. The first is mechanically attached to nailable decks or spot or solid mopped to non-nailable decks; the second layer is then spot or solid mopped to the first layer. Membrane application follows the insulation, except in protected membrane roofs, where the membrane goes down first and the insulation on top, followed with ballast (stone or concrete pavers). Insulation and related labor costs are NOT included in prices for built-up roofing.

Thermal & Moist. Protec.   R0752 Modified Bituminous Membrane Roofing

Reference Tables

R075213-30   Modified Bitumen Roofing

The cost of modified bitumen roofing is highly dependent on the type of installation that is planned. Installation is based on the type of modifier used in the bitumen. The two most popular modifiers are atactic polypropylene (APP) and styrene butadiene styrene (SBS). The modifiers are added to heated bitumen during the manufacturing process to change its characteristics. A polyethylene, polyester or fiberglass reinforcing sheet is then sandwiched between layers of this bitumen. When completed, the result is a pre-assembled, built-up roof that has increased elasticity and weatherability. Some manufacturers include a surfacing material such as ceramic or mineral granules, metal particles or sand.

The preferred method of adhering SBS-modified bitumen roofing to the substrate is with hot-mopped asphalt (much the same as built-up roofing). This installation method requires a tar kettle/pot to heat the asphalt, as well as the labor, tools and equipment necessary to distribute and spread the hot asphalt.

The alternative method for applying APP and SBS modified bitumen is as follows. A skilled installer uses a torch to melt a small pool of bitumen off the membrane. This pool must form across the entire roll for proper adhesion. The installer must unroll the roofing at a pace slow enough to melt the bitumen, but fast enough to prevent damage to the rest of the membrane.

Modified bitumen roofing provides the advantages of both built-up and single-ply roofing. Labor costs are reduced over those of built-up roofing because only a single ply is necessary. The elasticity of single-ply roofing is attained with the reinforcing sheet and polymer modifiers. Modifieds have some self-healing characteristics and because of their multi-layer construction, they offer the reliability and safety of built-up roofing.

R078413-30   Firestopping

Firestopping is the sealing of structural, mechanical, electrical and other penetrations through fire-rated assemblies. The basic components of firestop systems are safing insulation and firestop sealant on both sides of wall penetrations and the top side of floor penetrations.

Pipe penetrations are assumed to be through concrete, grout, or joint compound and can be sleeved or unsleeved. Costs for the penetrations and sleeves are not included. An annular space of 1" is assumed. Escutcheons are not included.

A metallic pipe is assumed to be copper, aluminum, cast iron or similar metallic material. An insulated metallic pipe is assumed to be covered with a thermal insulating jacket of varying thickness and materials.

A non-metallic pipe is assumed to be PVC, CPVC, FR Polypropylene or similar plastic piping material. Intumescent firestop sealants or wrap strips are included. Collars on both sides of wall penetrations and a sheet metal plate on the underside of floor penetrations are included.

Ductwork is assumed to be sheet metal, stainless steel or similar metallic material. Duct penetrations are assumed to be through concrete, grout or joint compound. Costs for penetrations and sleeves are not included. An annular space of 1/2" is assumed.

Multi-trade openings include costs for sheet metal forms, firestop mortar, wrap strips, collars and sealants as necessary.

Structural penetrations joints are assumed to be 1/2" or less. CMU walls are assumed to be within 1-1/2" of the metal deck. Drywall walls are assumed to be tight to the underside of metal decking.

Metal panel, glass or curtain wall systems include a spandrel area of 5' filled with mineral wool foil-faced insulation. Fasteners and stiffeners are included.

Openings · R0813 Metal Doors

R081313-20   Steel Door Selection Guide

Standard steel doors are classified into four levels, as recommended by the Steel Door Institute in the chart below. Each of the four levels offers a range of construction models and designs to meet architectural requirements for preference and appearance, including full flush, seamless, and stile & rail. Recommended minimum gauge requirements are also included.

For complete standard steel door construction specifications and available sizes, refer to the Steel Door Institute Technical Data Series, ANSI A250.8-98 (SDI-100), and ANSI A250.4-94 Test Procedure and Acceptance Criteria for Physical Endurance of Steel Door and Hardware Reinforcements.

Level		Model	Construction	For Full Flush or Seamless		
				Min. Gauge	Thickness (in)	Thickness (mm)
I	Standard Duty	1	Full Flush			
		2	Seamless	20	0.032	0.8
II	Heavy Duty	1	Full Flush			
		2	Seamless	18	0.042	1.0
III	Extra Heavy Duty	1	Full Flush			
		2	Seamless			
		3	*Stile & Rail	16	0.053	1.3
IV	Maximum Duty	1	Full Flush			
		2	Seamless	14	0.067	1.6

*Stiles & rails are 16 gauge; flush panels, when specified, are 18 gauge

Openings · R0851 Metal Windows

R085123-10   Steel Sash

An ironworker crew will erect 25 S.F. or 1.3 sash unit per hour, whichever is less.

A mechanic will point 30 L.F. per hour.

A painter will paint 90 S.F. per coat per hour.

A Glazier production depends on light size.

Allow 1 lb. special steel sash putty per 16″ x 20″ light.

Openings · R0853 Plastic Windows

R085313-20   Replacement Windows

Replacement windows are typically measured per United Inch.

United Inches are calculated by rounding the width and height of the window opening up to the nearest inch, then adding the two figures.

The labor cost for replacement windows includes removal of sash, existing sash balance or weights, parting bead where necessary and installation of new window.

Debris hauling and dump fees are not included.

R087110-10   Hardware Finishes

This table describes hardware finishes used throughout the industry. It also shows the base metal and the respective symbols in the three predominate systems of identification. Many of these are used in pricing descriptions in Division Eight.

US″	BMHA*	CDN	Base	Description
US P	600	CP	Steel	Primed for Painting
US 1B	601	C1B	Steel	Bright Black Japanned
US 2C	602	C2C	Steel	Zinc Plated
US 2G	603	C2G	Steel	Zinc Plated
US 3	605	C3	Brass	Bright Brass, Clear Coated
US 4	606	C4	Brass	Satin Brass, Clear Coated
US 5	609	C5	Brass	Satin Brass, Blackened, Satin Relieved, Clear Coated
US 7	610	C7	Brass	Satin Brass, Blackened, Bright Relieved, Clear Coated
US 9	611	C9	Bronze	Bright Bronze, Clear Coated
US 10	612	C10	Bronze	Satin Bronze, Clear Coated
US 10A	641	C10A	Steel	Antiqued Bronze, Oiled and Lacquered
US 10B	613	C10B	Bronze	Antiqued Bronze, Oiled
US 11	616	C11	Bronze	Satin Bronze, Blackened, Satin Relieved, Clear Coated
US 14	618	C14	Brass/Bronze	Bright Nickel Plated, Clear Coated
US 15	619	C15	Brass/Bronze	Satin Nickel, Clear Coated
US 15A	620	C15A	Brass/Bronze	Satin Nickel Plated, Blackened, Satin Relieved, Clear Coated
US 17A	621	C17A	Brass/Bronze	Nickel Plated, Blackened, Relieved, Clear Coated
US 19	622	C19	Brass/Bronze	Flat Black Coated
US 20	623	C20	Brass/Bronze	Statuary Bronze, Light
US 20A	624	C20A	Brass/Bronze	Statuary Bronze, Dark
US 26	625	C26	Brass/Bronze	Bright Chromium
US 26D	626	C26D	Brass/Bronze	Satin Chromium
US 20	627	C27	Aluminum	Satin Aluminum Clear
US 28	628	C28	Aluminum	Anodized Dull Aluminum
US 32	629	C32	Stainless Steel	Bright Stainless Steel
US 32D	630	C32D	Stainless Steel	Stainless Steel
US 3	632	C3	Steel	Bright Brass Plated, Clear Coated
US 4	633	C4	Steel	Satin Brass, Clear Coated
US 7	636	C7	Steel	Satin Brass Plated, Blackened, Bright Relieved, Clear Coated
US 9	637	C9	Steel	Bright Bronze Plated, Clear Coated
US 5	638	C5	Steel	Satin Brass Plated, Blackened, Bright Relieved, Clear Coated
US 10	639	C10	Steel	Satin Bronze Plated, Clear Coated
US 10B	640	C10B	Steel	Antique Bronze, Oiled
US 10A	641	C10A	Steel	Antiqued Bronze, Oiled and Lacquered
US 11	643	C11	Steel	Satin Bronze Plated, Blackened, Bright Relieved, Clear Coated
US 14	645	C14	Steel	Bright Nickel Plated, Clear Coated
US 15	646	C15	Steel	Satin Nickel Plated, Clear Coated
US 15A	647	C15A	Steel	Nickel Plated, Blackened, Bright Relieved, Clear Coated
US 17A	648	C17A	Steel	Nickel Plated, Blackened, Relieved, Clear Coated
US 20	649	C20	Steel	Statuary Bronze, Light
US 20A	650	C20A	Steel	Statuary Bronze, Dark
US 26	651	C26	Steel	Bright Chromium Plated
US 26D	652	C26D	Steel	Satin Chromium Plated

* - BMHA Builders Hardware Manufacturing Association
″ - US Equivalent
^ - Canadian Equivalent
Japanning is imitating Asian lacquer work

Reference Tables

R088110-10 Glazing Productivity

Some glass sizes are estimated by the "united inch" (height + width). The table below shows the number of lights glazed in an eight-hour period by the crew size indicated, for glass up to 1/4″ thick. Square or nearly square lights are more economical on a S.F. basis. Long slender lights will have a high S.F. installation cost. For insulated glass reduce production by 33%. For 1/2″ float glass reduce production by 50%. Production time for glazing with two glaziers per day averages: 1/4″ float glass 120 S.F.; 1/2″ float glass 55 S.F.; 1/2″ insulated glass 95 S.F.; 3/4″ insulated glass 75 S.F.

Glazing Method	United Inches per Light							
	40″	60″	80″	100″	135″	165″	200″	240″
Number of Men in Crew	1	1	1	1	2	3	3	4
Industrial sash, putty	60	45	24	15	18	—	—	—
With stops, putty bed	50	36	21	12	16	8	4	3
Wood stops, rubber	40	27	15	9	11	6	3	2
Metal stops, rubber	30	24	14	9	9	6	3	2
Structural glass	10	7	4	3	—	—	—	—
Corrugated glass	12	9	7	4	4	4	3	—
Storefronts	16	15	13	11	7	6	4	4
Skylights, putty glass	60	36	21	12	16	—	—	—
Thiokol set	15	15	11	9	9	6	3	2
Vinyl set, snap on	18	18	13	12	12	7	5	4
Maximum area per light	2.8 S.F.	6.3 S.F.	11.1 S.F.	17.4 S.F.	31.6 S.F.	47 S.F.	69 S.F.	100 S.F.

R092000-50   Lath, Plaster and Gypsum Board

Gypsum board lath is available in 3/8" thick x 16" wide x 4' long sheets as a base material for multi-layer plaster applications. It is also available as a base for either multi-layer or veneer plaster applications in 1/2" and 5/8" thick–4' wide x 8', 10' or 12' long sheets. Fasteners are screws or blued ring shank nails for wood framing and screws for metal framing.

Metal lath is available in diamond mesh patterns with flat or self-furring profiles. Paper backing is available for applications where excessive plaster waste needs to be avoided. A slotted mesh ribbed lath should be used in areas where the span between structural supports is greater than normal. Most metal lath comes in 27" x 96" sheets. Diamond mesh weighs 1.75, 2.5 or 3.4 pounds per square yard, slotted mesh lath weighs 2.75 or 3.4 pounds per square yard. Metal lath can be nailed, screwed or tied in place.

Many **accessories** are available. Corner beads, flat reinforcing strips, casing beads, control and expansion joints, furring brackets and channels are some examples. Note that accessories are not included in plaster or stucco line items.

Plaster is defined as a material or combination of materials that when mixed with a suitable amount of water, forms a plastic mass or paste. When applied to a surface, the paste adheres to it and subsequently hardens, preserving in a rigid state the form or texture imposed during the period of elasticity.

Gypsum plaster is made from ground calcined gypsum. It is mixed with aggregates and water for use as a base coat plaster.

Vermiculite plaster is a fire-retardant plaster covering used on steel beams, concrete slabs and other heavy construction materials. Vermiculite is a group name for certain clay minerals, hydrous silicates or aluminum, magnesium and iron that have been expanded by heat.

Perlite plaster is a plaster using perlite as an aggregate instead of sand. Perlite is a volcanic glass that has been expanded by heat.

Gauging plaster is a mix of gypsum plaster and lime putty that when applied produces a quick drying finish coat.

Veneer plaster is a one or two component gypsum plaster used as a thin finish coat over special gypsum board.

Keenes cement is a white cementitious material manufactured from gypsum that has been burned at a high temperature and ground to a fine powder. Alum is added to accelerate the set. The resulting plaster is hard and strong and accepts and maintains a high polish, hence it is used as a finishing plaster.

Stucco is a Portland cement based plaster used primarily as an exterior finish.

Plaster is used on both interior and exterior surfaces. Generally it is applied in multiple-coat systems. A three-coat system uses the terms scratch, brown and finish to identify each coat. A two-coat system uses base and finish to describe each coat. Each type of plaster and application system has attributes that are chosen by the designer to best fit the intended use.

Gypsum Plaster	2 Coat, 5/8" Thick		3 Coat, 3/4" Thick		
	Base	Finish	Scratch	Brown	Finish
Quantities for 100 S.Y.	1:3 Mix	2:1 Mix	1:2 Mix	1:3 Mix	2:1 Mix
Gypsum plaster	1,300 lb.		1,350 lb.	650 lb.	
Sand	1.75 C.Y.		1.85 C.Y.	1.35 C.Y.	
Finish hydrated lime		340 lb.			340 lb.
Gauging plaster		170 lb.			170 lb.

Vermiculite or Perlite Plaster	2 Coat, 5/8" Thick		3 Coat, 3/4" Thick		
Quantities for 100 S.Y.	Base	Finish	Scratch	Brown	Finish
Gypsum plaster	1,250 lb.		1,450 lb.	800 lb.	
Vermiculite or perlite	7.8 bags		8.0 bags	3.3 bags	
Finish hydrated lime		340 lb.			340 lb.
Gauging plaster		170 lb.			170 lb.

Stucco–Three-Coat System Quantities for 100 S.Y.	On Wood Frame	On Masonry
Portland cement	29 bags	21 bags
Sand	2.6 C.Y.	2.0 C.Y.
Hydrated lime	180 lb.	120 lb.

R092910-10   Levels of Gypsum Board Finish

In the past, contract documents often used phrases such as "industry standard" and "workmanlike finish" to specify the expected quality of gypsum board wall and ceiling installations. The vagueness of these descriptions led to unacceptable work and disputes.

In order to resolve this problem, four major trade associations concerned with the manufacture, erection, finish and decoration of gypsum board wall and ceiling systems developed an industry-wide *Recommended Levels of Gypsum Board Finish.*

The finish of gypsum board walls and ceilings for specific final decoration is dependent on a number of factors. A primary consideration is the location of the surface and the degree of decorative treatment desired. Painted and unpainted surfaces in warehouses and other areas where appearance is normally not critical may simply require the taping of wallboard joints and 'spotting' of fastener heads. Blemish-free, smooth, monolithic surfaces often intended for painted and decorated walls and ceilings in habitated structures, ranging from single-family dwellings through monumental buildings, require additional finishing prior to the application of the final decoration.

Other factors to be considered in determining the level of finish of the gypsum board surface are (1) the type of angle of surface illumination (both natural and artificial lighting), and (2) the paint and method of application or the type and finish of wallcovering specified as the final decoration. Critical lighting conditions, gloss paints, and thin wall coverings require a higher level of gypsum board finish than heavily textured surfaces which are subsequently painted or surfaces which are to be decorated with heavy grade wall coverings.

The following descriptions were developed by the Association of the Wall and Ceiling Industries-International (AWCI), Ceiling & Interior Systems Construction Association (CISCA), Gypsum Association (GA), and Painting and Decorating Contractors of America (PDCA) as a guide.

Level 0: Used in temporary construction or wherever the final decoration has not been determined. Unfinished. No taping, finishing or corner beads are required. Also could be used where non-predecorated panels will be used in demountable-type partitions that are to be painted as a final finish.

Level 1: Frequently used in plenum areas above ceilings, in attics, in areas where the assembly would generally be concealed, or in building service corridors and other areas not normally open to public view. Some degree of sound and smoke control is provided; in some geographic areas, this level is referred to as "fire-taping," although this level of finish does not typically meet fire-resistant assembly requirements. Where a fire resistance rating is required for the gypsum board assembly, details of construction should be in accordance with reports of fire tests of assemblies that have met the requirements of the fire rating acceptable.

All joints and interior angles shall have tape embedded in joint compound. Accessories are optional at specifier discretion in corridors and other areas with pedestrian traffic. Tape and fastener heads need not be covered with joint compound. Surface shall be free of excess joint compound. Tool marks and ridges are acceptable.

Level 2: It may be specified for standard gypsum board surfaces in garages, warehouse storage, or other similar areas where surface appearance is not of primary importance.

All joints and interior angles shall have tape embedded in joint compound and shall be immediately wiped with a joint knife or trowel, leaving a thin coating of joint compound over all joints and interior angles. Fastener heads and accessories shall be covered with a coat of joint compound. Surface shall be free of excess joint compound. Tool marks and ridges are acceptable.

Level 3: Typically used in areas receiving heavy texture (spray or hand applied) finishes before final painting, or where commercial-grade (heavy duty) wall coverings are to be applied as the final decoration. This level of finish should not be used where smooth painted surfaces or where lighter weight wall coverings are specified. The prepared surface shall be coated with a drywall primer prior to the application of final finishes.

All joints and interior angles shall have tape embedded in joint compound and shall be immediately wiped with a joint knife or trowel, leaving a thin coating of joint compound over all joints and interior angles. One additional coat of joint compound shall be applied over all joints and interior angles. Fastener heads and accessories shall be covered with two separate coats of joint compound. All joint compounds shall be smooth and free of tool marks and ridges. The prepared surface shall be covered with a drywall primer prior to the application of the final decoration.

Level 4: This level should be used where residential grade (light duty) wall coverings, flat paints, or light textures are to be applied. The prepared surface shall be coated with a drywall primer prior to the application of final finishes. Release agents for wall coverings are specifically formulated to minimize damage if coverings are subsequently removed.

The weight, texture, and sheen level of the wall covering material selected should be taken into consideration when specifying wall coverings over this level of drywall treatment. Joints and fasteners must be sufficiently concealed if the wall covering material is lightweight, contains limited pattern, has a glossy finish, or has any combination of these features. In critical lighting areas, flat paints applied over light textures tend to reduce joint photographing. Gloss, semi-gloss, and enamel paints are not recommended over this level of finish.

All joints and interior angles shall have tape embedded in joint compound and shall be immediately wiped with a joint knife or trowel, leaving a thin coating of joint compound over all joints and interior angles. In addition, two separate coats of joint compound shall be applied over all flat joints and one separate coat of joint compound applied over interior angles. Fastener heads and accessories shall be covered with three separate coats of joint compound. All joint compounds shall be smooth and free of tool marks and ridges. The prepared surface shall be covered with a drywall primer like Sheetrock first coat prior to the application of the final decoration.

Level 5: The highest quality finish is the most effective method to provide a uniform surface and minimize the possibility of joint photographing and of fasteners showing through the final decoration. This level of finish is required where gloss, semi-gloss, or enamel is specified; when flat joints are specified over an untextured surface; or where critical lighting conditions occur. The prepared surface shall be coated with a drywall primer prior to the application of the final decoration.

All joints and interior angles shall have tape embedded in joint compound and be immediately wiped with a joint knife or trowel, leaving a thin coating of joint compound over all joints and interior angles. Two separate coats of joint compound shall be applied over all flat joints and one separate coat of joint compound applied over interior angles. Fastener heads and accessories shall be covered with three separate coats of joint compound.

A thin skim coat of joint compound shall be trowel applied to the entire surface. Excess compound is immediately troweled off, leaving a film or skim coating of compound completely covering the paper. As an alternative to a skim coat, a material manufactured especially for this purpose may be applied such as Sheetrock Tuff-Hide primer surfacer. The surface must be smooth and free of tool marks and ridges. The prepared surface shall be covered with a drywall primer prior to the application of the final decoration.

R142000-10   Freight Elevators

Capacities run from 2,000 lbs. to over 100,000 lbs. with 3,000 lbs. to 10,000 lbs. most common. Travel speeds are generally lower and the control is less intricate than on passenger elevators. Costs in the unit price sections are for hydraulic and geared elevators.

R142000-20   Elevator Selective Costs   See R142000-40 for cost development.

A. Base Unit	Passenger		Freight		Hospital	
	Hydraulic	Electric	Hydraulic	Electric	Hydraulic	Electric
Capacity	1,500 lb.	2,000 lb.	2,000 lb.	4,000 lb.	4,000 lb.	4,000 lb.
Speed	100 F.P.M.	200 F.P.M.	100 F.P.M.	200 F.P.M.	100 F.P.M.	200 F.P.M.
#Stops/Travel Ft.	2/12	4/40	2/20	4/40	2/20	4/40
Push Button Oper.	Yes	Yes	Yes	Yes	Yes	Yes
Telephone Box & Wire	"	"	"	"	"	"
Emergency Lighting	"	"	No	No	"	"
Cab	Plastic Lam. Walls	Plastic Lam. Walls	Painted Steel	Painted Steel	Plastic Lam. Walls	Plastic Lam. Walls
Cove Lighting	Yes	Yes	No	No	Yes	Yes
Floor	V.C.T.	V.C.T.	Wood w/Safety Treads	Wood w/Safety Treads	V.C.T.	V.C.T.
Doors, & Speedside Slide	Yes	Yes	Yes	Yes	Yes	Yes
Gates, Manual	No	No	No	No	No	No
Signals, Lighted Buttons	Car and Hall	Car and Hall	Car and Hall	Car and Hall	Car and Hall	Car and Hall
O.H. Geared Machine	N.A.	Yes	N.A.	Yes	N.A.	Yes
Variable Voltage Contr.	"	"	N.A.	"	"	"
Emergency Alarm	Yes	"	Yes	" •	Yes	"
Class "A" Loading	N.A.	N.A.	"	"	N.A.	N.A.

R142000-30   Passenger Elevators

Electric elevators are used generally but hydraulic elevators can be used for lifts up to 70' and where large capacities are required. Hydraulic speeds are limited to 200 F.P.M. but cars are self leveling at the stops. On low rises, hydraulic installation runs about 15% less than standard electric types but on higher rises this installation cost advantage is reduced. Maintenance of hydraulic elevators is about the same as the electric type but the underground portion is not included in the maintenance contract.

In electric elevators there are several control systems available, the choice of which will be based upon elevator use, size, speed and cost criteria. The two types of drives are geared for low speeds and gearless for 450 F.P.M. and over.

The tables on the preceding pages illustrate typical installed costs of the various types of elevators available.

R142000-40   Elevator Cost Development

To price a new car or truck from the factory, you must start with the manufacturer's basic model, then add or exchange optional equipment and features. The same is true for pricing elevators.

Requirement: One-passenger elevator, five-story hydraulic, 2,500 lb. capacity, 12' floor to floor, speed 150 F.P.M., emergency power switching and maintenance contract.

Example:

Description	Adjustment
A. Base Elevator: Hydraulic Passenger, 1500 lb. Capacity, 100 fpm, 2 Stops, Standard Finish	1 Ea.
B. Capacity Adjustment (2,500 lb.)	1 Ea.
C. Excess Travel Adjustment: 48' Total Travel (4 x 12') minus 12' Base Unit Travel =	36 V.L.F.
D. Stops Adjustment: 5 Total Stops minus 2 Stops (Base Unit) =	3 Stops
E. Speed Adjustment (150 F.P.M.)	1 Ea.
F. Options:	
1. Intercom Service	1 Ea.
2. Emergency Power Switching, Automatic	1 Ea.
3. Stainless Steel Entrance Doors	5 Ea.
4. Maintenance Contract (12 Months)	1 Ea.
5. Position Indicator for main floor level (none indicated in Base Unit)	1 Ea.

For customer support on your Commercial Renovation Costs with RSMeans data, call 800.448.8182.

881

R220102-20   Labor Adjustment Factors

Labor Adjustment Factors are provided for Divisions 21, 22, and 23 to assist the mechanical estimator account for the various complexities and special conditions of any particular project. While a single percentage has been entered on each line of Division 22 01 02.20, it should be understood that these are just suggested midpoints of ranges of values commonly used by mechanical estimators. They may be increased or decreased depending on the severity of the special conditions.

The group for "existing occupied buildings" has been the subject of requests for explanation. Actually there are two stages to this group: buildings that are existing and "finished" but unoccupied, and those that also are occupied. Buildings that are "finished" may result in higher labor costs due to the

workers having to be more careful not to damage finished walls, ceilings, floors, etc. and may necessitate special protective coverings and barriers. Also corridor bends and doorways may not accommodate long pieces of pipe or larger pieces of equipment. Work above an already hung ceiling can be very time consuming. The addition of occupants may force the work to be done on premium time (nights and/or weekends), eliminate the possible use of some preferred tools such as pneumatic drivers, powder charged drivers, etc. The estimator should evaluate the access to the work area and just how the work is going to be accomplished to arrive at an increase in labor costs over "normal" new construction productivity.

R220105-10   Demolition (Selective vs. Removal for Replacement)

Demolition can be divided into two basic categories.

One type of demolition involves the removal of material with no concern for its replacement. The labor-hours to estimate this work are found under "Selective Demolition" in the Fire Protection, Plumbing and HVAC Divisions. It is selective in that individual items or all the material installed as a system or trade grouping such as plumbing or heating systems are removed. This may be accomplished by the easiest way possible, such as sawing, torch cutting, or sledge hammering as well as simple unbolting.

The second type of demolition is the removal of some items for repair or replacement. This removal may involve careful draining, opening of unions,

disconnecting and tagging electrical connections, capping pipes/ducts to prevent entry of debris or leakage of the material contained as well as transporting the item away from its in-place location to a truck/dumpster. An approximation of the time required to accomplish this type of demolition is to use half of the time indicated as necessary to install a new unit. For example: installation of a new pump might be listed as requiring 6 labor-hours so if we had to estimate the removal of the old pump we would allow an additional 3 hours for a total of 9 hours. That is, the complete replacement of a defective pump with a new pump would be estimated to take 9 labor-hours.

R221113-50   Pipe Material Considerations

1. Malleable fittings should be used for gas service.
2. Malleable fittings are used where there are stresses/strains due to expansion and vibration.
3. Cast fittings may be broken as an aid to disassembling heating lines frozen by long use, temperature and minerals.
4. A cast iron pipe is extensively used for underground and submerged service.
5. Type M (light wall) copper tubing is available in hard temper only and is used for nonpressure and less severe applications than K and L.

6. Type L (medium wall) copper tubing, available hard or soft for interior service.
7. Type K (heavy wall) copper tubing, available in hard or soft temper for use where conditions are severe. For underground and interior service.
8. Hard drawn tubing requires fewer hangers or supports but should not be bent. Silver brazed fittings are recommended, but soft solder is normally used.
9. Type DMV (very light wall) copper tubing designed for drainage, waste and vent plus other non-critical pressure services.

Domestic/Imported Pipe and Fittings Costs

The prices shown in this publication for steel/cast iron pipe and steel, cast iron, and malleable iron fittings are based on domestic production sold at the normal trade discounts. The above listed items of foreign manufacture may be available at prices 1/3 to 1/2 of those shown. Some imported items after minor machining or finishing operations are being sold as domestic to further complicate the system.

Caution: Most pipe prices in this data set also include a coupling and pipe hangers which for the larger sizes can add significantly to the per foot cost and should be taken into account when comparing "book cost" with the quoted supplier's cost.

R260105-30 Electrical Demolition (Removal for Replacement)

The purpose of this reference number is to provide a guide to users for electrical "removal for replacement" by applying the rule of thumb: 1/3 of new installation time (typical range from 20% to 50%) for removal. Remember to use reasonable judgment when applying the suggested percentage factor. For example:

Contractors have been requested to remove an existing fluorescent lighting fixture and replace with a new fixture utilizing energy saver lamps and electronic ballast:

In order to fully understand the extent of the project, contractors should visit the job site and estimate the time to perform the renovation work in accordance with applicable national, state and local regulation and codes.

The contractor may need to add extra labor hours to his estimate if he discovers unknown concealed conditions such as: contaminated asbestos ceiling, broken acoustical ceiling tile and need to repair, patch and touch-up paint all the damaged or disturbed areas, tasks normally assigned to general contractors. In addition, the owner could request that the contractors salvage the materials removed and turn over the materials to the owner or dispose of the materials to a reclamation station. The normal removal item is 0.5 labor hour for a lighting fixture and 1.5 labor-hours for new installation time. Revise the estimate times from 2 labor-hours work up to a minimum 4 labor-hours work for just fluorescent lighting fixture.

For removal of large concentrations of lighting fixtures in the same area, apply an "economy of scale" to reduce estimating labor hours.

Electrical | **R2605 Common Work Results for Electrical**

R260519-90 Wire

Wire quantities are taken off by either measuring each cable run or by extending the conduit and raceway quantities times the number of conductors in the raceway. Ten percent should be added for waste and tie-ins. Keep in mind that the unit of measure of wire is C.L.F. not L.F. as in raceways so the formula would read:

$$\frac{(\text{L.F. Raceway x No. of Conductors) x 1.10}}{100} = \text{C.L.F.}$$

Price per C.L.F. of wire includes:
1. Setting up wire coils or spools on racks
2. Attaching wire to pull in means
3. Measuring and cutting wire
4. Pulling wire into a raceway
5. Identifying and tagging

Price does not include:
1. Connections to breakers, panelboards, or equipment
2. Splices

Job Conditions: Productivity is based on new construction to a height of 15' using rolling staging in an unobstructed area. Material staging is assumed to be within 100' of work being performed.

Economy of Scale: If more than three wires at a time are being pulled, deduct the following percentages from the labor of that grouping:

4-5 wires	25%
6-10 wires	30%
11-15 wires	35%
over 15	40%

If a wire pull is less than 100' in length and is interrupted several times by boxes, lighting outlets, etc., it may be necessary to add the following lengths to each wire being pulled:

Junction box to junction box	2 L.F.
Lighting panel to junction box	6 L.F.
Distribution panel to sub panel	8 L.F.
Switchboard to distribution panel	12 L.F.
Switchboard to motor control center	20 L.F.
Switchboard to cable tray	40 L.F.

Measure of Drops and Riser: It is important when taking off wire quantities to include the wire for drops to electrical equipment. If heights of electrical equipment are not clearly stated, use the following guide:

	Bottom A.F.F.	Top A.F.F.	Inside Cabinet
Safety switch to 100A	5'	6'	2'
Safety switch 400 to 600A	4'	6'	3'
100A panel 12 to 30 circuit	4'	6'	3'
42 circuit panel	3'	6'	4'
Switch box	3'	3'6"	1'
Switchgear	0'	8'	8'
Motor control centers	0'	8'	8'
Transformers - wall mount	4'	8'	2'
Transformers - floor mount	0'	12'	4'

R260533-22　Conductors in Conduit

The table below lists the maximum number of conductors for various sized conduit using THW, TW or THWN insulations.

Copper Wire Size	1/2"			3/4"			1"			1-1/4"			1-1/2"			2"			2-1/2"			3"		3-1/2"		4"	
	TW	THW	THWN	TW	THW	THWN	TW	THW	THWN	TW	THW	THWN	TW	THW	THWN	TW	THW	THWN	TW	THW	THWN	THW	THWN	THW	THWN	THW	THWN
#14	9	6	13	15	10	24	25	16	39	44	29	69	60	40	94	99	65	154	142	93		143		192			
#12	7	4	10	12	8	18	19	13	29	35	24	51	47	32	70	78	53	114	111	76	164	117		157			
#10	5	4	6	9	6	11	15	11	18	26	19	32	36	26	44	60	43	73	85	61	104	95	160	127		163	
#8	2	1	3	4	3	5	7	5	9	12	10	16	17	13	22	28	22	36	40	32	51	49	79	66	106	85	136
#6		1	1		2	4		4	6		7	11		10	15		16	26		23	37	36	57	48	76	62	98
#4		1	1		1	2		3	4		5	7		7	9		12	16		17	22	27	35	36	47	47	60
#3		1	1		1	1		2	3		4	6		6	8		10	13		15	19	23	29	31	39	40	51
#2		1	1		1	1		2	3		4	5		5	7		9	11		13	16	20	25	27	33	34	43
#1					1	1		1	1		3	3		4	5		6	8		9	12	14	18	19	25	25	32
1/0					1	1		1	1		2	3		3	4		5	7		8	10	12	15	16	21	21	27
2/0					1	1		1	1		1	2		3	3		5	6		7	8	10	13	14	17	18	22
3/0					1	1		1	1		1	1		2	3		4	5		6	7	9	11	12	14	15	18
4/0						1		1	1		1	1		1	2		3	4		5	6	7	9	10	12	13	15
250 kcmil								1	1		1	1		1	1		2	3		4	4	6	7	8	10	10	12
300								1	1		1	1		1	1		2	3		3	4	5	6	7	8	9	11
350									1		1	1		1	1		1	2		3	3	4	5	6	7	8	9
400											1	1		1	1		1	1		2	3	4	5	5	6	7	8
500											1	1		1	1		1	1		1	2	3	4	4	5	6	7
600												1		1	1		1	1		1	1	3	3	4	4	5	5
700														1	1		1	1		1	1	2	3	3	4	4	5
750														1	1		1	1		1	1	2	2	3	3	4	4

For customer support on your Commercial Renovation Costs with RSMeans data, call 800.448.8182.

R312316-40   Excavating

The selection of equipment used for structural excavation and bulk excavation or for grading is determined by the following factors.

1. Quantity of material
2. Type of material
3. Depth or height of cut
4. Length of haul
5. Condition of haul road
6. Accessibility of site
7. Moisture content and dewatering requirements
8. Availability of excavating and hauling equipment

Some additional costs must be allowed for hand trimming the sides and bottom of concrete pours and other excavation below the general excavation.

Number of B.C.Y. per truck = 1.5 C.Y. bucket x 8 passes = 12 loose C.Y.

$$= 12 \times \frac{100}{118} = 10.2 \text{ B.C.Y. per truck}$$

Truck Haul Cycle:

Load truck, 8 passes	=	4 minutes
Haul distance, 1 mile	=	9 minutes
Dump time	=	2 minutes
Return, 1 mile	=	7 minutes
Spot under machine	=	1 minute
		23 minute cycle

Add the mobilization and demobilization costs to the total excavation costs. When equipment is rented for more than three days, there is often no mobilization charge by the equipment dealer. On larger jobs outside of urban areas, scrapers can move earth economically provided a dump site or fill area and adequate haul roads are available. Excavation within sheeting bracing or cofferdam bracing is usually done with a clamshell and production

When planning excavation and fill, the following should also be considered.

1. Swell factor
2. Compaction factor
3. Moisture content
4. Density requirements

A typical example for scheduling and estimating the cost of excavation of a 15′ deep basement on a dry site when the material must be hauled off the site is outlined below.

Assumptions:

1. Swell factor, 18%
2. No mobilization or demobilization
3. Allowance included for idle time and moving on job
4. No dewatering, sheeting, or bracing
5. No truck spotter or hand trimming

Fleet Haul Production per day in B.C.Y.

$$4 \text{ trucks} \times \frac{50 \text{ min. hour}}{23 \text{ min. haul cycle}} \times 8 \text{ hrs.} \times 10.2 \text{ B.C.Y.}$$

$$= 4 \times 2.2 \times 8 \times 10.2 = 718 \text{ B.C.Y./day}$$

is low, since the clamshell may have to be guided by hand between the bracing. When excavating or filling an area enclosed with a wellpoint system, add 10% to 15% to the cost to allow for restricted access. When estimating earth excavation quantities for structures, allow work space outside the building footprint for construction of the foundation and a slope of 1:1 unless sheeting is used.

For customer support on your Commercial Renovation Costs with RSMeans data, call 800.448.8182.

885

Reference Tables

R312316-45 Excavating Equipment

The table below lists theoretical hourly production in C.Y./hr. bank measure for some typical excavation equipment. Figures assume 50 minute hours, 83% job efficiency, 100% operator efficiency, 90° swing and properly sized hauling units, which must be modified for adverse digging and loading conditions. Actual production costs in the front of the data set average about 50% of the theoretical values listed here.

Equipment	Soil Type	B.C.Y. Weight	% Swell	1 C.Y.	1-1/2 C.Y.	2 C.Y.	2-1/2 C.Y.	3 C.Y.	3-1/2 C.Y.	4 C.Y.
Hydraulic Excavator "Backhoe" 15' Deep Cut	Moist loam, sandy clay	3400 lb.	40%	165	195	200	275	330	385	440
	Sand and gravel	3100	18	140	170	225	240	285	330	380
	Common earth	2800	30	150	180	230	250	300	350	400
	Clay, hard, dense	3000	33	120	140	190	200	240	260	320
Power Shovel Optimum Cut (Ft.)	Moist loam, sandy clay	3400	40	170 (6.0)	245 (7.0)	295 (7.8)	335 (8.4)	385 (8.8)	435 (9.1)	475 (9.4)
	Sand and gravel	3100	18	165 (6.0)	225 (7.0)	275 (7.8)	325 (8.4)	375 (8.8)	420 (9.1)	460 (9.4)
	Common earth	2800	30	145 (7.8)	200 (9.2)	250 (10.2)	295 (11.2)	335 (12.1)	375 (13.0)	425 (13.8)
	Clay, hard, dense	3000	33	120 (9.0)	175 (10.7)	220 (12.2)	255 (13.3)	300 (14.2)	335 (15.1)	375 (16.0)
Drag Line Optimum Cut (Ft.)	Moist loam, sandy clay	3400	40	130 (6.6)	180 (7.4)	220 (8.0)	250 (8.5)	290 (9.0)	325 (9.5)	385 (10.0)
	Sand and gravel	3100	18	130 (6.6)	175 (7.4)	210 (8.0)	245 (8.5)	280 (9.0)	315 (9.5)	375 (10.0)
	Common earth	2800	30	110 (8.0)	160 (9.0)	190 (9.9)	220 (10.5)	250 (11.0)	280 (11.5)	310 (12.0)
	Clay, hard, dense	3000	33	90 (9.3)	130 (10.7)	160 (11.8)	190 (12.3)	225 (12.8)	250 (13.3)	280 (12.0)

				Wheel Loaders				Track Loaders		
				3 C.Y.	4 C.Y.	6 C.Y.	8 C.Y.	2-1/4 C.Y.	3 C.Y.	4 C.Y.
Loading Tractors	Moist loam, sandy clay	3400	40	260	340	510	690	135	180	250
	Sand and gravel	3100	18	245	320	480	650	130	170	235
	Common earth	2800	30	230	300	460	620	120	155	220
	Clay, hard, dense	3000	33	200	270	415	560	110	145	200
	Rock, well-blasted	4000	50	180	245	380	520	100	130	180

R312323-30 Compacting Backfill

Compaction of fill in embankments, around structures, in trenches, and under slabs is important to control settlement. Factors affecting compaction are:
1. Soil gradation
2. Moisture content
3. Equipment used
4. Depth of fill per lift
5. Density required

Example:

Compact granular fill around a building foundation using a 21" wide x 24" vibratory plate in 8" lifts. Operator moves at 50 F.P.M. working a 50 minute hour to develop 95% Modified Proctor Density with 4 passes.

Production Rate:

$$\frac{1.75' \text{ plate width x 50 F.P.M. x 50 min./hr. x .67' lift}}{27 \text{ C.F. per C.Y.}} = 108.5 \text{ C.Y./hr.}$$

Production Rate for 4 Passes:

$$\frac{108.5 \text{ C.Y.}}{4 \text{ passes}} = 27.125 \text{ C.Y./hr. x 8 hrs.} = 217 \text{ C.Y./day}$$

Earthwork | R3141 Shoring

R314116-40   Wood Sheet Piling

Wood sheet piling may be used for depths to 20' where there is no ground water. If moderate ground water is encountered Tongue & Groove sheeting will help to keep it out. When considerable ground water is present, steel sheeting must be used.

For estimating purposes on trench excavation, sizes are as follows:

Depth	Sheeting	Wales	Braces	B.F. per S.F.
To 8'	3 x 12's	6 x 8's, 2 line	6 x 8's, @ 10'	4.0 @ 8'
8' x 12'	3 x 12's	10 x 10's, 2 line	10 x 10's, @ 9'	5.0 average
12' to 20'	3 x 12's	12 x 12's, 3 line	12 x 12's, @ 8'	7.0 average

Sheeting to be toed in at least 2' depending upon soil conditions. A five person crew with an air compressor and sheeting driver can drive and brace 440 SF/day at 8' deep, 360 SF/day at 12' deep, and 320 SF/day at 16' deep.

For normal soils, piling can be pulled in 1/3 the time to install. Pulling difficulty increases with the time in the ground. Production can be increased by high pressure jetting.

Earthwork | R3163 Drilled Caissons

R316326-60   Caissons

The three principal types of caissons are:

(1) Belled Caissons, which except for shallow depths and poor soil conditions, are generally recommended. They provide more bearing than shaft area. Because of its conical shape, no horizontal reinforcement of the bell is required.

(2) Straight Shaft Caissons are used where relatively light loads are to be supported by caissons that rest on high value bearing strata. While the shaft is larger in diameter than for belled types this is more than offset by the savings in time and labor.

(3) Keyed Caissons are used when extremely heavy loads are to be carried. A keyed or socketed caisson transfers its load into rock by a combination of end-bearing and shear reinforcing of the shaft. The most economical shaft often consists of a steel casing, a steel wide flange core and concrete. Allowable compressive stresses of .225 f'c for concrete, 16,000 psi for the wide flange core, and 9,000 psi for the steel casing are commonly used. The usual range of shaft diameter is 18" to 84". The number of sizes specified for any one project should be limited due to the problems of casing and auger storage. When hand work is to be performed, shaft diameters should not be less than 32". When inspection of borings is required a minimum shaft diameter of 30" is recommended. Concrete caissons are intended to be poured against earth excavation so permanent forms, which add to cost, should not be used if the excavation is clean and the earth is sufficiently impervious to prevent excessive loss of concrete.

Soil Conditions for Belling		
Good	Requires Handwork	Not Recommended
Clay	Hard Shale	Silt
Sandy Clay	Limestone	Sand
Silty Clay	Sandstone	Gravel
Clayey Silt	Weathered Mica	Igneous Rock
Hard-pan		
Soft Shale		
Decomposed Rock		

Exterior Improvements | R3292 Turf & Grasses

R329219-50   Seeding

The type of grass is determined by light, shade and moisture content of soil plus intended use. Fertilizer should be disked 4" before seeding. For steep slopes disk five tons of mulch and lay two tons of hay or straw on surface per acre after seeding. Surface mulch can be staked, lightly disked or tar emulsion sprayed. Material for mulch can be wood chips, peat moss, partially rotted hay or straw, wood fibers and sprayed emulsions. Hemp seed blankets with fertilizer are also available. For spring seeding, watering is necessary. Late fall seeding may have to be reseeded in the spring. Hydraulic seeding, power mulching, and aerial seeding can be used on large areas.

For customer support on your Commercial Renovation Costs with RSMeans data, call 800.448.8182.

887

Reference Tables

R331113-80 Piping Designations

There are several systems currently in use to describe pipe and fittings. The following paragraphs will help to identify and clarify classifications of piping systems used for water distribution.

Piping may be classified by schedule. Piping schedules include 5S, 10S, 10, 20, 30, Standard, 40, 60, Extra Strong, 80, 100, 120, 140, 160 and Double Extra Strong. These schedules are dependent upon the pipe wall thickness. The wall thickness of a particular schedule may vary with pipe size.

Ductile iron pipe for water distribution is classified by Pressure Classes such as Class 150, 200, 250, 300 and 350. These classes are actually the rated water working pressure of the pipe in pounds per square inch (psi). The pipe in these pressure classes is designed to withstand the rated water working pressure plus a surge allowance of 100 psi.

The American Water Works Association (AWWA) provides standards for various types of **plastic pipe.** C-900 is the specification for polyvinyl chloride (PVC) piping used for water distribution in sizes ranging from 4″ through 12″. C-901 is the specification for polyethylene (PE) pressure pipe, tubing and fittings used for water distribution in sizes ranging from 1/2″ through 3″. C-905 is the specification for PVC piping sizes 14″ and greater.

PVC pressure-rated pipe is identified using the standard dimensional ratio (SDR) method. This method is defined by the American Society for Testing and Materials (ASTM) Standard D 2241. This pipe is available in SDR numbers 64, 41, 32.5, 26, 21, 17, and 13.5. A pipe with an SDR of 64 will have the thinnest wall while a pipe with an SDR of 13.5 will have the thickest wall. When the pressure rating (PR) of a pipe is given in psi, it is based on a line supplying water at 73 degrees F.

The National Sanitation Foundation (NSF) seal of approval is applied to products that can be used with potable water. These products have been tested to ANSI/NSF Standard 14.

Valves and strainers are classified by American National Standards Institute (ANSI) Classes. These Classes are 125, 150, 200, 250, 300, 400, 600, 900, 1500 and 2500. Within each class there is an operating pressure range dependent upon temperature. Design parameters should be compared to the appropriate material dependent, pressure-temperature rating chart for accurate valve selection.

Change Orders

Change Order Considerations

A change order is a written document usually prepared by the design professional and signed by the owner, the architect/engineer, and the contractor. A change order states the agreement of the parties to: an addition, deletion, or revision in the work; an adjustment in the contract sum, if any; or an adjustment in the contract time, if any. Change orders, or "extras", in the construction process occur after execution of the construction contract and impact architects/ engineers, contractors, and owners.

Change orders that are properly recognized and managed can ensure orderly, professional, and profitable progress for everyone involved in the project. There are many causes for change orders and change order requests. In all cases, change orders or change order requests should be addressed promptly and in a precise and prescribed manner. The following paragraphs include information regarding change order pricing and procedures.

The Causes of Change Orders

Reasons for issuing change orders include:

- Unforeseen field conditions that require a change in the work
- Correction of design discrepancies, errors, or omissions in the contract documents
- Owner-requested changes, either by design criteria, scope of work, or project objectives
- Completion date changes for reasons unrelated to the construction process
- Changes in building code interpretations, or other public authority requirements that require a change in the work
- Changes in availability of existing or new materials and products

Procedures

Properly written contract documents must include the correct change order procedures for all parties—owners, design professionals, and contractors—to follow in order to avoid costly delays and litigation.

Being "in the right" is not always a sufficient or acceptable defense. The contract provisions requiring notification and documentation must be adhered to within a defined or reasonable time frame.

The appropriate method of handling change orders is by a written proposal and acceptance by all parties involved. Prior to starting work on a project, all parties should identify their authorized agents who may sign and accept change orders, as well as any limits placed on their authority.

Time may be a critical factor when the need for a change arises. For such cases, the contractor might be directed to proceed on a "time and materials" basis, rather than wait for all paperwork to be processed—a delay that could impede progress. In this situation, the contractor must still follow the prescribed change order procedures including, but not limited to, notification and documentation.

Lack of documentation can be very costly, especially if legal judgments are to be made, and if certain field personnel are no longer available. For time and material change orders, the contractor should keep accurate daily records of all labor and material allocated to the change.

Owners or awarding authorities who do considerable and continual building construction (such as the federal government) realize the inevitability of change orders for numerous reasons, both predictable and unpredictable. As a result, the federal government, the American Institute of Architects (AIA), the Engineers Joint Contract Documents Committee (EJCDC), and other contractor, legal, and technical organizations have developed standards and procedures to be followed by all parties to achieve contract continuance and timely completion, while being financially fair to all concerned.

Pricing Change Orders

When pricing change orders, regardless of their cause, the most significant factor is when the change occurs. The need for a change may be perceived in the field or requested by the architect/engineer *before* any of the actual installation has begun, or may evolve or appear *during* construction when the item of work in question is partially installed. In the latter cases, the original sequence of construction is disrupted, along with all contiguous and supporting systems. Change orders cause the greatest impact when they occur *after* the installation has been completed and must be uncovered, or even replaced. Post-completion changes may be caused by necessary design changes, product failure, or changes in the owner's requirements that are not discovered until the building or the systems begin to function.

Specified procedures of notification and record keeping must be adhered to and enforced regardless of the stage of construction: *before*, *during*, or *after* installation. Some bidding documents anticipate change orders by requiring that unit prices including overhead and profit percentages—for additional as well as deductible changes—be listed. Generally these unit prices do not fully take into account the ripple effect, or impact on other trades, and should be used for general guidance only.

When pricing change orders, it is important to classify the time frame in which the change occurs. There are two basic time frames for change orders: *pre-installation change orders*, which occur before the start of construction, and *post-installation change orders*, which involve reworking after the original installation. Change orders that occur between these stages may be priced according to the extent of work completed using a combination of techniques developed for pricing *pre-* and *post-installation* changes.

Factors To Consider When Pricing Change Orders

As an estimator begins to prepare a change order, the following questions should be reviewed to determine their impact on the final price.

General

- *Is the change order work* pre-installation *or* post-installation?

  Change order work costs vary according to how much of the installation has been completed. Once workers have the project scoped in their minds, even though they have not started, it can be difficult to refocus. Consequently they may spend more than the normal amount of time understanding the change. Also, modifications to work in place, such as trimming or refitting, usually take more time than was initially estimated. The greater the amount of work in place, the more reluctant workers are to change it. Psychologically they may resent the change and as a result the rework takes longer than normal. Post-installation change order estimates must include demolition of existing work as required to accomplish the change. If the work is performed at a later time, additional obstacles, such as building finishes, may be present which must be protected. Regardless of whether the change occurs

For customer support on your Commercial Renovation Costs with RSMeans data, call 800.448.8182.

889

pre-installation or post-installation, attempt to isolate the identifiable factors and price them separately. For example, add shipping costs that may be required pre-installation or any demolition required post-installation. Then analyze the potential impact on productivity of psychological and/or learning curve factors and adjust the output rates accordingly. One approach is to break down the typical workday into segments and quantify the impact on each segment.

Change Order Installation Efficiency

The labor-hours expressed (for new construction) are based on average installation time, using an efficiency level. For change order situations, adjustments to this efficiency level should reflect the daily labor-hour allocation for that particular occurrence.

- *Will the change substantially delay the original completion date?*

A significant change in the project may cause the original completion date to be extended. The extended schedule may subject the contractor to new wage rates dictated by relevant labor contracts. Project supervision and other project overhead must also be extended beyond the original completion date. The schedule extension may also put installation into a new weather season. For example, underground piping scheduled for October installation was delayed until January. As a result, frost penetrated the trench area, thereby changing the degree of difficulty of the task. Changes and delays may have a ripple effect throughout the project. This effect must be analyzed and negotiated with the owner.

- *What is the net effect of a deduct change order?*

In most cases, change orders resulting in a deduction or credit reflect only bare costs. The contractor may retain the overhead and profit based on the original bid.

Materials

- *Will you have to pay more or less for the new material, required by the change order, than you paid for the original purchase?*

The same material prices or discounts will usually apply to materials purchased for change orders as new construction. In some instances, however, the contractor may forfeit the advantages of competitive pricing for change orders. Consider the following example:

A contractor purchased over $20,000 worth of fan coil units for an installation and obtained the maximum discount. Some time later it was determined the project required an additional matching unit. The contractor has to purchase this unit from the original supplier to ensure a match. The supplier at this time may not discount the unit because of the small quantity, and he is no longer in a competitive situation. The impact of quantity on purchase can add between 0% and 25% to material prices and/or subcontractor quotes.

- *If materials have been ordered or delivered to the job site, will they be subject to a cancellation charge or restocking fee?*

Check with the supplier to determine if ordered materials are subject to a cancellation charge. Delivered materials not used as a result of a change order may be subject to a restocking fee if returned to the supplier. Common restocking charges run between 20% and 40%. Also, delivery charges to return the goods to the supplier must be added.

Labor

- *How efficient is the existing crew at the actual installation?*

Is the same crew that performed the initial work going to do the change order? Possibly the change consists of the installation of a unit identical to one already installed; therefore, the change should take less time. Be sure to consider this potential productivity increase and modify the productivity rates accordingly.

- *If the crew size is increased, what impact will that have on supervision requirements?*

Under most bargaining agreements or management practices, there is a point at which a working foreman is replaced by a nonworking foreman. This replacement increases project overhead by adding a nonproductive worker. If additional workers are added to accelerate the project or to perform changes while maintaining the schedule, be sure to add additional supervision time if warranted. Calculate the hours involved and the additional cost directly if possible.

- *What are the other impacts of increased crew size?*

The larger the crew, the greater the potential for productivity to decrease. Some of the factors that cause this productivity loss are: overcrowding (producing restrictive conditions in the working space) and possibly a shortage of any special tools and equipment required. Such factors affect not only the crew working on the elements directly involved in the change order, but other crews whose movements may also be hampered. As the crew increases, check its basic composition for changes by the addition or deletion of apprentices or nonworking foreman, and quantify the potential effects of equipment shortages or other logistical factors.

- *As new crews, unfamiliar with the project, are brought onto the site, how long will it take them to become oriented to the project requirements?*

The orientation time for a new crew to become 100% effective varies with the site and type of project. Orientation is easiest at a new construction site and most difficult at existing, very restrictive renovation sites. The type of work also affects orientation time. When all elements of the work are exposed, such as concrete or masonry work, orientation is decreased. When the work is concealed or less visible, such as existing electrical systems, orientation takes longer. Usually orientation can be accomplished in one day or less. Costs for added orientation should be itemized and added to the total estimated cost.

- *How much actual production can be gained by working overtime?*

Short term overtime can be used effectively to accomplish more work in a day. However, as overtime is scheduled to run beyond several weeks, studies have shown marked decreases in output. The following chart shows the effect of long term overtime on worker efficiency. If the anticipated change requires extended overtime to keep the job on schedule, these factors can be used as a guide to predict the impact on time and cost. Add project overhead, particularly supervision, that may also be incurred.

Days per Week	Hours per Day	Production Efficiency					Payroll Cost Factors	
		1st Week	2nd Week	3rd Week	4th Week	Average 4 Weeks	@ 1-1/2 Times	@ 2 Times
5	8	100%	100%	100%	100%	100%	100%	100%
	9	100	100	95	90	96	1.056	1.111
	10	100	95	90	85	93	1.100	1.200
	11	95	90	75	65	81	1.136	1.273
	12	90	85	70	60	76	1.167	1.333
6	8	100	100	95	90	96	1.083	1.167
	9	100	95	90	85	93	1.130	1.259
	10	95	90	85	80	88	1.167	1.333
	11	95	85	70	65	79	1.197	1.394
	12	90	80	65	60	74	1.222	1.444
7	8	100	95	85	75	89	1.143	1.286
	9	95	90	80	70	84	1.183	1.365
	10	90	85	75	65	79	1.214	1.429
	11	85	80	65	60	73	1.240	1.481
	12	85	75	60	55	69	1.262	1.524

Effects of Overtime

Caution: Under many labor agreements, Sundays and holidays are paid at a higher premium than the normal overtime rate.

The use of long-term overtime is counterproductive on almost any construction job; that is, the longer the period of overtime, the lower the actual production rate. Numerous studies have been conducted, and while they have resulted in slightly different numbers, all reach the same conclusion. The figure above tabulates the effects of overtime work on efficiency.

As illustrated, there can be a difference between the *actual* payroll cost per hour and the *effective* cost per hour for overtime work. This is due to the reduced production efficiency with the increase in weekly hours beyond 40. This difference between actual and effective cost results from overtime work over a prolonged period. Short-term overtime work does not result in as great a reduction in efficiency and, in such cases, effective cost may not vary significantly from the actual payroll cost. As the total hours per week are increased on a regular basis, more time is lost due to fatigue, lowered morale, and an increased accident rate.

As an example, assume a project where workers are working 6 days a week, 10 hours per day. From the figure above (based on productivity studies), the average effective productive hours over a 4-week period are:

$$0.875 \times 60 = 52.5$$

Depending upon the locale and day of week, overtime hours may be paid at time and a half or double time. For time and a half, the overall (average) *actual* payroll cost (including regular and overtime hours) is determined as follows:

$$\frac{40 \text{ reg. hrs.} + (20 \text{ overtime hrs.} \times 1.5)}{60 \text{ hrs.}} = 1.167$$

Based on 60 hours, the payroll cost per hour will be 116.7% of the normal rate at 40 hours per week. However, because the effective production (efficiency) for 60 hours is reduced to the equivalent of 52.5 hours, the effective cost of overtime is calculated as follows:

For time and a half:

$$\frac{40 \text{ reg. hrs.} + (20 \text{ overtime hrs.} \times 1.5)}{52.5 \text{ hrs.}} = 1.33$$

The installed cost will be 133% of the normal rate (for labor).

Thus, when figuring overtime, the actual cost per unit of work will be higher than the apparent overtime payroll dollar increase, due to the reduced productivity of the longer work week. These efficiency calculations are true only for those cost factors determined by hours worked. Costs that are applied weekly or monthly, such as equipment rentals, will not be similarly affected.

Equipment

- *What equipment is required to complete the change order?*

Change orders may require extending the rental period of equipment already on the job site, or the addition of special equipment brought in to accomplish the change work. In either case, the additional rental charges and operator labor charges must be added.

Summary

The preceding considerations and others you deem appropriate should be analyzed and applied to a change order estimate. The impact of each should be quantified and listed on the estimate to form an audit trail.

Change orders that are properly identified, documented, and managed help to ensure the orderly, professional, and profitable progress of the work. They also minimize potential claims or disputes at the end of the project.

Back by customer demand!

You asked and we listened. For customer convenience and estimating ease, we have made the 2022 Project Costs available for download at **RSMeans.com/2022books**. You will also find sample estimates, an RSMeans data overview video, and a book registration form to receive quarterly data updates throughout 2022.

Estimating Tips

- The cost figures available in the download were derived from hundreds of projects contained in the RSMeans database of completed construction projects. They include the contractor's overhead and profit. The figures have been adjusted to January of the current year.

- These projects were located throughout the U.S. and reflect a tremendous variation in square foot (S.F.) costs. This is due to differences, not only in labor and material costs, but also in individual owners' requirements. For instance, a bank in a large city would have different features than one in a rural area. This is true of all the different types of buildings analyzed. Therefore, caution should be exercised when using these Project Costs. For example, for courthouses, costs in the database are local courthouse costs and will not apply to the larger, more elaborate federal courthouses.

- None of the figures "go with" any others. All individual cost items were computed and tabulated separately. Thus, the sum of the median figures for plumbing, HVAC, and electrical will not normally total up to the total mechanical and electrical costs arrived at by separate analysis and tabulation of the projects.

- Each building was analyzed as to total and component costs and percentages. The figures were arranged in ascending order with the results tabulated as shown. The 1/4 column shows that 25% of the projects had lower costs and 75% had higher. The 3/4 column shows that 75% of the projects had lower costs and 25% had higher. The median column shows that 50% of the projects had lower costs and 50% had higher.

- Project Costs are useful in the conceptual stage when no details are available. As soon as details become available in the project design, the square foot approach should be discontinued and the project should be priced as to its particular components. When more precision is required, or for estimating the replacement cost of specific buildings, the current edition of *Square Foot Costs with RSMeans data* should be used.

- In using the figures in this section, it is recommended that the median column be used for preliminary figures if no additional information is available. The median figures, when multiplied by the total city construction cost index figures (see City Cost Indexes) and then multiplied by the project size modifier at the end of this section, should present a fairly accurate base figure, which would then have to be adjusted in view of the estimator's experience, local economic conditions, code requirements, and the owner's particular requirements. There is no need to factor in the percentage figures, as these should remain constant from city to city.

- The editors of this data would greatly appreciate receiving cost figures on one or more of your recent projects, which would then be included in the averages for next year. All cost figures received will be kept confidential, except that they will be averaged with other similar projects to arrive at square foot cost figures for next year.

See the website above for details and the discount available for submitting one or more of your projects.

Same Data. Simplified.

Enjoy the convenience and efficiency of accessing your costs anywhere:

- **Skip the multiplier** by setting your location
- **Quickly search,** edit, favorite and share costs
- **Stay on top of price changes** with automatic updates

Discover more at rsmeans.com/online

50 17 00 | Project Costs

			UNIT	UNIT COSTS			% OF TOTAL			
				1/4	MEDIAN	3/4	1/4	MEDIAN	3/4	
01	0000	**Auto Sales with Repair**	S.F.							01
	0100	Architectural		119	134	145	58%	64%	67%	
	0200	Plumbing		10	10.45	14	4.84%	5.20%	6.80%	
	0300	Mechanical		13.40	17.95	19.80	6.40%	8.70%	10.15%	
	0400	Electrical		20.50	25.50	32	9.05%	11.70%	15.90%	
	0500	Total Project Costs		201	209	215				
02	0000	**Banking Institutions**	S.F.							02
	0100	Architectural		180	222	270	59%	65%	69%	
	0200	Plumbing		7.25	10.15	14.10	2.12%	3.39%	4.19%	
	0300	Mechanical		14.45	19.95	23.50	4.41%	5.10%	10.75%	
	0400	Electrical		35.50	42.50	65.50	10.45%	13.05%	15.90%	
	0500	Total Project Costs		300	335	415				
03	0000	**Court House**	S.F.							03
	0100	Architectural		95	187	261	54.50%	58.50%	61%	
	0200	Plumbing		3.59	13.40	13.40	2.07%	3.13%	3.13%	
	0300	Mechanical		22.50	27	27	6.25%	12.95%	12.95%	
	0400	Electrical		29	32.50	32.50	7.65%	16.60%	16.60%	
	0500	Total Project Costs		173	320	425				
04	0000	**Data Centers**	S.F.							04
	0100	Architectural		215	215	215	68%	68%	68%	
	0200	Plumbing		11.75	11.75	11.75	3.71%	3.71%	3.71%	
	0300	Mechanical		30	30	30	9.45%	9.45%	9.45%	
	0400	Electrical		28.50	28.50	28.50	9%	9%	9%	
	0500	Total Project Costs		315	315	315				
05	0000	**Detention Centers**	S.F.							05
	0100	Architectural		199	210	223	52%	53%	60.50%	
	0200	Plumbing		21	25.50	31	5.15%	7.10%	7.25%	
	0300	Mechanical		26.50	38	45.50	7.55%	9.50%	13.80%	
	0400	Electrical		43.50	51.50	67.50	10.90%	14.85%	17.95%	
	0500	Total Project Costs		335	355	420				
06	0000	**Fire Stations**	S.F.							06
	0100	Architectural		113	145	212	46%	54.50%	61.50%	
	0200	Plumbing		11.45	15.70	20.50	4.59%	5.60%	6.30%	
	0300	Mechanical		16.85	24.50	33	6.05%	8.25%	10.20%	
	0400	Electrical		26	33.50	45	10.75%	12.55%	14.95%	
	0500	Total Project Costs		236	269	375				
07	0000	**Gymnasium**	S.F.							07
	0100	Architectural		99.50	131	131	57%	64.50%	64.50%	
	0200	Plumbing		2.44	8	8	1.58%	3.48%	3.48%	
	0300	Mechanical		3.74	33.50	33.50	2.42%	14.65%	14.65%	
	0400	Electrical		12.25	24	24	7.95%	10.35%	10.35%	
	0500	Total Project Costs		155	230	230				
08	0000	**Hospitals**	S.F.							08
	0100	Architectural		121	198	215	41.50%	47.50%	48%	
	0200	Plumbing		8.85	36.50	43.50	6%	7.65%	10.35%	
	0300	Mechanical		53.50	66	86	12.55%	17.95%	23.50%	
	0400	Electrical		26.50	60	69.50	10.95%	14.10%	16.85%	
	0500	Total Project Costs		282	425	455				
09	0000	**Industrial Buildings**	S.F.							09
	0100	Architectural		66	130	262	52%	56%	82%	
	0200	Plumbing		1.90	5.15	14.90	1.57%	2.11%	6.30%	
	0300	Mechanical		5.40	10.30	49	4.77%	5.55%	14.80%	
	0400	Electrical		8.25	9.45	78.50	7.85%	13.55%	16.20%	
	0500	Total Project Costs		98	149	485				
10	0000	**Medical Clinics & Offices**	S.F.							10
	0100	Architectural		101	133	179	48.50%	54%	62%	
	0200	Plumbing		10.05	15.05	25	4.47%	6.65%	8.75%	
	0300	Mechanical		16.30	25	46.50	7.80%	10.90%	16.05%	
	0400	Electrical		22	27.50	42	9%	11.40%	13.85%	
	0500	Total Project Costs		188	250	330				

For customer support on your Commercial Renovation Costs with RSMeans data, call 800.448.8182.

50 17 00 | Project Costs	UNIT	UNIT COSTS			% OF TOTAL			
		1/4	MEDIAN	3/4	1/4	MEDIAN	3/4	
11 0000 Mixed Use	S.F.							11
0100 Architectural		124	171	234	45.50%	52.50%	61.50%	
0200 Plumbing		7	12.75	13.60	3.31%	3.47%	4.18%	
0300 Mechanical		18.50	40	53.50	6.10%	13.60%	17.05%	
0400 Electrical		24.50	41.50	59.50	8.60%	11.65%	15.65%	
0500 Total Project Costs		216	375	385				
12 0000 Multi-Family Housing	S.F.							12
0100 Architectural		86.50	128	191	54.50%	62%	68.50%	
0200 Plumbing		7.70	12.40	16.85	5.25%	6.65%	8%	
0300 Mechanical		8	10.65	30.50	4.92%	6.05%	10.25%	
0400 Electrical		11.20	17.50	25.50	6%	7.90%	10.25%	
0500 Total Project Costs		143	238	305				
13 0000 Nursing Home & Assisted Living	S.F.							13
0100 Architectural		81	106	133	51.50%	55.50%	63.50%	
0200 Plumbing		8.70	13.10	14.40	6.25%	7.40%	8.80%	
0300 Mechanical		7.15	10.55	20.50	4.04%	6.70%	9.55%	
0400 Electrical		11.80	18.65	26	7%	10.75%	13.10%	
0500 Total Project Costs		138	180	217				
14 0000 Office Buildings	S.F.							14
0100 Architectural		104	145	200	54.50%	61%	69%	
0200 Plumbing		5.75	9.05	16.95	2.70%	3.78%	5.85%	
0300 Mechanical		11.25	19.15	29.50	5.60%	8.20%	11.10%	
0400 Electrical		14.40	24.50	38	7.75%	10%	12.70%	
0500 Total Project Costs		178	225	320				
15 0000 Parking Garage	S.F.							15
0100 Architectural		35.50	43.50	45.50	70%	79%	88%	
0200 Plumbing		1.18	1.23	2.30	2.05%	2.70%	2.83%	
0300 Mechanical		.91	1.40	5.30	2.11%	3.62%	3.81%	
0400 Electrical		3.13	3.44	7.20	5.30%	6.35%	7.95%	
0500 Total Project Costs		43.50	53	57.50				
16 0000 Parking Garage/Mixed Use	S.F.							16
0100 Architectural		116	126	129	61%	62%	65.50%	
0200 Plumbing		3.71	4.86	7.45	2.47%	2.72%	3.66%	
0300 Mechanical		15.90	17.85	25.50	7.80%	13.10%	13.60%	
0400 Electrical		16.65	24	25	8.20%	12.65%	18.15%	
0500 Total Project Costs		189	197	203				
17 0000 Police Stations	S.F.							17
0100 Architectural		131	185	227	49%	61%	61.50%	
0200 Plumbing		17.25	20.50	21	5.05%	5.55%	7.70%	
0300 Mechanical		39	54.50	56.50	12.35%	14.55%	16.55%	
0400 Electrical		29.50	34	50.50	9.15%	12.10%	14%	
0500 Total Project Costs		244	340	375				
18 0000 Police/Fire	S.F.							18
0100 Architectural		127	127	390	55.50%	66%	68%	
0200 Plumbing		10.20	10.55	39	5.45%	5.50%	5.55%	
0300 Mechanical		15.60	24.50	89	8.35%	12.70%	12.80%	
0400 Electrical		17.75	22.50	102	9.50%	11.75%	14.55%	
0500 Total Project Costs		187	193	700				
19 0000 Public Assembly Buildings	S.F.							19
0100 Architectural		129	179	268	57.50%	61.50%	66%	
0200 Plumbing		6.85	10.05	15.10	2.60%	3.36%	4.79%	
0300 Mechanical		14.45	26	40	6.55%	8.95%	12.45%	
0400 Electrical		21.50	29	46.50	8.60%	10.75%	13%	
0500 Total Project Costs		208	285	420				
20 0000 Recreational	S.F.							20
0100 Architectural		125	196	266	49.50%	59%	64.50%	
0200 Plumbing		9.30	15.65	24.50	3.08%	4.67%	6.85%	
0300 Mechanical		14.85	22.50	33.50	5.15%	6.65%	11.35%	
0400 Electrical		21.50	31	47.50	7.15%	9.05%	10.95%	
0500 Total Project Costs		226	330	500				

For customer support on your Commercial Renovation Costs with RSMeans data, call 800.448.8182.

895

		50 17 00 | Project Costs	UNIT	UNIT COSTS			% OF TOTAL		
				1/4	MEDIAN	3/4	1/4	MEDIAN	3/4
21	0000	**Restaurants**	S.F.						
	0100	Architectural		141	221	280	57.50%	60%	63.50%
	0200	Plumbing		15.55	31	41	7.35%	7.70%	8.75%
	0300	Mechanical		16.75	22	41.50	6.50%	8.15%	11.15%
	0400	Electrical		16.65	27	54.50	6.75%	10.30%	11.60%
	0500	Total Project Costs		234	385	470			
22	0000	**Retail**	S.F.						
	0100	Architectural		63	98.50	203	54.50%	60%	64.50%
	0200	Plumbing		7.90	11	13.65	4.18%	5.45%	8.45%
	0300	Mechanical		7.40	10.45	19.20	4.98%	6.15%	7.05%
	0400	Electrical		11.75	23.50	35.50	7.90%	11.25%	12.45%
	0500	Total Project Costs		96	171	325			
23	0000	**Schools**	S.F.						
	0100	Architectural		109	138	185	50.50%	55.50%	60.50%
	0200	Plumbing		8.70	11.95	17.45	3.66%	4.57%	6.95%
	0300	Mechanical		20.50	29.50	43	8.90%	11.95%	14.55%
	0400	Electrical		20.50	28	36.50	9.45%	11.25%	13.30%
	0500	Total Project Costs		199	254	340			
24	0000	**University, College & Private School Classroom & Admin Buildings**	S.F.						
	0100	Architectural		140	173	203	50.50%	55%	59%
	0200	Plumbing		9.45	12.35	17.40	3.02%	4.46%	6.35%
	0300	Mechanical		28	42	52	9.95%	11.95%	14.70%
	0400	Electrical		22.50	31.50	38.50	7.70%	9.90%	11.50%
	0500	Total Project Costs		231	320	405			
25	0000	**University, College & Private School Dormitories**	S.F.						
	0100	Architectural		91	160	170	53%	61.50%	68%
	0200	Plumbing		12.05	17.05	19.55	6.35%	6.65%	8.95%
	0300	Mechanical		5.40	23	36.50	4.13%	9%	11.80%
	0400	Electrical		6.40	22.50	34	4.75%	7.35%	10.60%
	0500	Total Project Costs		135	256	300			
26	0000	**University, College & Private School Science, Eng. & Lab Buildings**	S.F.						
	0100	Architectural		157	185	274	48%	54.50%	58%
	0200	Plumbing		10.80	16.35	27.50	3.29%	3.77%	5%
	0300	Mechanical		41.50	77.50	79	11.70%	19.40%	23.50%
	0400	Electrical		32.50	42.50	43.50	9%	12.05%	13.15%
	0500	Total Project Costs		325	355	480			
27	0000	**University, College & Private School Student Union Buildings**	S.F.						
	0100	Architectural		124	325	325	54.50%	54.50%	59.50%
	0200	Plumbing		18.75	18.75	28	3.13%	4.27%	11.45%
	0300	Mechanical		35.50	57.50	57.50	9.60%	9.60%	14.55%
	0400	Electrical		31	54.50	54.50	9.05%	12.80%	13.15%
	0500	Total Project Costs		244	600	600			
28	0000	**Warehouses**	S.F.						
	0100	Architectural		53	81	148	60.50%	67%	71.50%
	0200	Plumbing		2.76	5.90	11.40	2.82%	3.72%	5%
	0300	Mechanical		3.27	18.65	29	4.56%	8.15%	10.70%
	0400	Electrical		6.90	22.50	37	7.75%	10.10%	18.30%
	0500	Total Project Costs		79	126	254			

Square Foot Project Size Modifier

One factor that affects the S.F. cost of a particular building is the size. In general, for buildings built to the same specifications in the same locality, the larger building will have the lower S.F. cost. This is due mainly to the decreasing contribution of the exterior walls plus the economy of scale usually achievable in larger buildings. The Area Conversion Scale shown below will give a factor to convert costs for the typical size building to an adjusted cost for the particular project.

The Square Foot Base Size lists the median costs, most typical project size in our accumulated data, and the range in size of the projects.

The Size Factor for your project is determined by dividing your project area in S.F. by the typical project size for the particular Building Type. With this factor, enter the Area Conversion Scale at the appropriate Size Factor and determine the appropriate Cost Multiplier for your building size.

Example: Determine the cost per S.F. for a 103,600 S.F. Multi-family housing.

$$\frac{\text{Proposed building area} = 103,600 \text{ S.F.}}{\text{Typical size from below} = 51,800 \text{ S.F.}} = 2.00$$

Enter Area Conversion Scale at 2.0, intersect curve, read horizontally the appropriate cost multiplier of .94. Size adjusted cost becomes .94 x $238.00 = $223.72 based on national average costs.

Note: For Size Factors less than .50, the Cost Multiplier is 1.1
For Size Factors greater than 3.5, the Cost Multiplier is .90

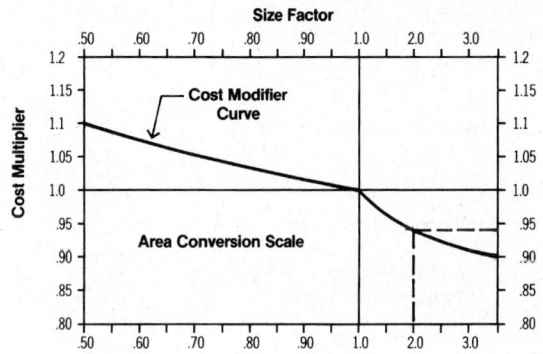

System	Median Cost (Total Project Costs)	Typical Size Gross S.F. (Median of Projects)	Typical Range (Low – High) (Projects)
Auto Sales with Repair	$209.00	24,900	4,700 – 29,300
Banking Institutions	335.00	9,300	3,300 – 38,100
Court House	320.00	56,000	24,700 – 70,500
Data Centers	315.00	14,400	14,400 – 14,400
Detention Centers	355.00	37,800	12,300 – 183,330
Fire Stations	269.00	12,600	6,300 – 49,600
Gymnasium	230.00	52,400	22,800 – 82,000
Hospitals	425.00	73,300	12,700 – 410,300
Industrial Buildings	149.00	21,100	5,100 – 200,600
Medical Clinics & Offices	250.00	24,100	2,300 – 327,000
Mixed Use	375.00	31,800	7,200 – 188,900
Multi-Family Housing	238.00	51,800	2,500 – 1,161,500
Nursing Home & Assisted Living	180.00	38,200	1,500 – 242,600
Office Buildings	225.00	20,600	1,100 – 930,000
Parking Garage	53.00	151,800	99,900 – 287,000
Parking Garage/Mixed Use	197.00	254,200	5,300 – 318,000
Police Stations	340.00	25,200	12,700 – 88,600
Police/Fire	193.00	44,300	8,600 – 50,300
Public Assembly Buildings	285.00	21,000	2,200 – 235,300
Recreational	330.00	25,300	1,000 – 223,800
Restaurants	385.00	5,900	2,100 – 42,000
Retail	171.00	26,000	4,000 – 84,300
Schools	254.00	69,400	1,300 – 410,800
University, College & Private School Classroom & Admin Buildings	320.00	48,000	2,400 – 196,200
University, College & Private School Dormitories	256.00	39,200	1,500 – 126,900
University, College & Private School Science, Eng. & Lab Buildings	355.00	60,000	5,300 – 117,600
University, College & Private School Student Union Buildings	600.00	48,700	42,100 – 50,000
Warehouses	126.00	10,900	600 – 303,800

Abbreviations

A	Area Square Feet; Ampere
AAFES	Army and Air Force Exchange Service
ABS	Acrylonitrile Butadiene Stryrene; Asbestos Bonded Steel
A.C., AC	Alternating Current; Air-Conditioning; Asbestos Cement; Plywood Grade A & C
ACI	American Concrete Institute
ACR	Air Conditioning Refrigeration
ADA	Americans with Disabilities Act
AD	Plywood, Grade A & D
Addit.	Additional
Adh.	Adhesive
Adj.	Adjustable
af	Audio-frequency
AFFF	Aqueous Film Forming Foam
AFUE	Annual Fuel Utilization Efficiency
AGA	American Gas Association
Agg.	Aggregate
A.H., Ah	Ampere Hours
A hr.	Ampere-hour
A.H.U., AHU	Air Handling Unit
A.I.A.	American Institute of Architects
AIC	Ampere Interrupting Capacity
Allow.	Allowance
alt., alt	Alternate
Alum.	Aluminum
a.m.	Ante Meridiem
Amp.	Ampere
Anod.	Anodized
ANSI	American National Standards Institute
APA	American Plywood Association
Approx.	Approximate
Apt.	Apartment
Asb.	Asbestos
A.S.B.C.	American Standard Building Code
Asbe.	Asbestos Worker
ASCE	American Society of Civil Engineers
A.S.H.R.A.E.	American Society of Heating, Refrig. & AC Engineers
ASME	American Society of Mechanical Engineers
ASTM	American Society for Testing and Materials
Attchmt.	Attachment
Avg., Ave.	Average
AWG	American Wire Gauge
AWWA	American Water Works Assoc.
Bbl.	Barrel
B&B, BB	Grade B and Better; Balled & Burlapped
B&S	Bell and Spigot
B.&W.	Black and White
b.c.c.	Body-centered Cubic
B.C.Y.	Bank Cubic Yards
BE	Bevel End
B.F.	Board Feet
Bg. cem.	Bag of Cement
BHP	Boiler Horsepower; Brake Horsepower
B.I.	Black Iron
bidir.	bidirectional
Bit., Bitum.	Bituminous
Bit., Conc.	Bituminous Concrete
Bk.	Backed
Bkrs.	Breakers
Bldg., bldg	Building
Blk.	Block
Bm.	Beam
Boil.	Boilermaker
bpm	Blows per Minute
BR	Bedroom
Brg., brng.	Bearing
Brhe.	Bricklayer Helper
Bric.	Bricklayer

Brk., brk	Brick
brkt	Bracket
Brs.	Brass
Brz.	Bronze
Bsn.	Basin
Btr.	Better
BTU	British Thermal Unit
BTUH	BTU per Hour
Bu.	Bushels
BUR	Built-up Roofing
BX	Interlocked Armored Cable
°C	Degree Centigrade
c	Conductivity, Copper Sweat
C	Hundred; Centigrade
C/C	Center to Center, Cedar on Cedar
C-C	Center to Center
Cab	Cabinet
Cair.	Air Tool Laborer
Cal.	Caliper
Calc	Calculated
Cap.	Capacity
Carp.	Carpenter
C.B.	Circuit Breaker
C.C.A.	Chromate Copper Arsenate
C.C.F.	Hundred Cubic Feet
cd	Candela
cd/sf	Candela per Square Foot
CD	Grade of Plywood Face & Back
CDX	Plywood, Grade C & D, exterior glue
Cefi.	Cement Finisher
Cem.	Cement
CF	Hundred Feet
C.F.	Cubic Feet
CFM	Cubic Feet per Minute
CFRP	Carbon Fiber Reinforced Plastic
c.g.	Center of Gravity
CHW	Chilled Water; Commercial Hot Water
C.I., CI	Cast Iron
C.I.P., CIP	Cast in Place
Circ.	Circuit
C.L.	Carload Lot
CL	Chain Link
Clab.	Common Laborer
Clam	Common Maintenance Laborer
C.L.F.	Hundred Linear Feet
CLF	Current Limiting Fuse
CLP	Cross Linked Polyethylene
cm	Centimeter
CMP	Corr. Metal Pipe
CMU	Concrete Masonry Unit
CN	Change Notice
Col.	Column
CO₂	Carbon Dioxide
Comb.	Combination
comm.	Commercial, Communication
Compr.	Compressor
Conc.	Concrete
Cont., cont	Continuous; Continued, Container
Corkbd.	Cork Board
Corr.	Corrugated
Cos	Cosine
Cot	Cotangent
Cov.	Cover
C/P	Cedar on Paneling
CPA	Control Point Adjustment
Cplg.	Coupling
CPM	Critical Path Method
CPVC	Chlorinated Polyvinyl Chloride
C.Pr.	Hundred Pair
CRC	Cold Rolled Channel
Creos.	Creosote
Crpt.	Carpet & Linoleum Layer
CRT	Cathode-ray Tube
CS	Carbon Steel, Constant Shear Bar Joist

Csc	Cosecant
C.S.F.	Hundred Square Feet
CSI	Construction Specifications Institute
CT	Current Transformer
CTS	Copper Tube Size
Cu	Copper, Cubic
Cu. Ft.	Cubic Foot
cw	Continuous Wave
C.W.	Cool White; Cold Water
Cwt.	100 Pounds
C.W.X.	Cool White Deluxe
C.Y.	Cubic Yard (27 cubic feet)
C.Y./Hr.	Cubic Yard per Hour
Cyl.	Cylinder
d	Penny (nail size)
D	Deep; Depth; Discharge
Dis., Disch.	Discharge
Db	Decibel
Dbl.	Double
DC	Direct Current
DDC	Direct Digital Control
Demob.	Demobilization
d.f.t.	Dry Film Thickness
d.f.u.	Drainage Fixture Units
D.H.	Double Hung
DHW	Domestic Hot Water
DI	Ductile Iron
Diag.	Diagonal
Diam., Dia	Diameter
Distrib.	Distribution
Div.	Division
Dk.	Deck
D.L.	Dead Load; Diesel
DLH	Deep Long Span Bar Joist
dlx	Deluxe
Do.	Ditto
DOP	Dioctyl Phthalate Penetration Test (Air Filters)
Dp., dp	Depth
D.P.S.T.	Double Pole, Single Throw
Dr.	Drive
DR	Dimension Ratio
Drink.	Drinking
D.S.	Double Strength
D.S.A.	Double Strength A Grade
D.S.B.	Double Strength B Grade
Dty.	Duty
DWV	Drain Waste Vent
DX	Deluxe White, Direct Expansion
dyn	Dyne
e	Eccentricity
E	Equipment Only; East; Emissivity
Ea.	Each
EB	Encased Burial
Econ.	Economy
E.C.Y	Embankment Cubic Yards
EDP	Electronic Data Processing
EIFS	Exterior Insulation Finish System
E.D.R.	Equiv. Direct Radiation
Eq.	Equation
EL	Elevation
Elec.	Electrician; Electrical
Elev.	Elevator; Elevating
EMT	Electrical Metallic Conduit; Thin Wall Conduit
Eng.	Engine, Engineered
EPDM	Ethylene Propylene Diene Monomer
EPS	Expanded Polystyrene
Eqhv.	Equip. Oper., Heavy
Eqlt.	Equip. Oper., Light
Eqmd.	Equip. Oper., Medium
Eqmm.	Equip. Oper., Master Mechanic
Eqol.	Equip. Oper., Oilers
Equip.	Equipment
ERW	Electric Resistance Welded

For customer support on your Commercial Renovation Costs with RSMeans data, call 800.448.8182.

Abbreviation	Meaning
E.S.	Energy Saver
Est.	Estimated
esu	Electrostatic Units
E.W.	Each Way
EWT	Entering Water Temperature
Excav.	Excavation
excl	Excluding
Exp., exp	Expansion, Exposure
Ext., ext	Exterior; Extension
Extru.	Extrusion
f.	Fiber Stress
F	Fahrenheit; Female; Fill
Fab., fab	Fabricated; Fabric
FBGS	Fiberglass
F.C.	Footcandles
f.c.c.	Face-centered Cubic
f'c.	Compressive Stress in Concrete; Extreme Compressive Stress
F.E.	Front End
FEP	Fluorinated Ethylene Propylene (Teflon)
F.G.	Flat Grain
F.H.A.	Federal Housing Administration
Fig.	Figure
Fin.	Finished
FIPS	Female Iron Pipe Size
Fixt.	Fixture
FJP	Finger jointed and primed
Fl. Oz.	Fluid Ounces
Flr.	Floor
Flrs.	Floors
FM	Frequency Modulation; Factory Mutual
Fmg.	Framing
FM/UL	Factory Mutual/Underwriters Labs
Fdn.	Foundation
FNPT	Female National Pipe Thread
Fori.	Foreman, Inside
Foro.	Foreman, Outside
Fount.	Fountain
fpm	Feet per Minute
FPT	Female Pipe Thread
Fr	Frame
F.R.	Fire Rating
FRK	Foil Reinforced Kraft
FSK	Foil/Scrim/Kraft
FRP	Fiberglass Reinforced Plastic
FS	Forged Steel
FSC	Cast Body; Cast Switch Box
Ft., ft	Foot; Feet
Ftng.	Fitting
Ftg.	Footing
Ft lb.	Foot Pound
Furn.	Furniture
FVNR	Full Voltage Non-Reversing
FVR	Full Voltage Reversing
FXM	Female by Male
Fy.	Minimum Yield Stress of Steel
g	Gram
G	Gauss
Ga.	Gauge
Gal., gal.	Gallon
Galv., galv	Galvanized
GC/MS	Gas Chromatograph/Mass Spectrometer
Gen.	General
GFI	Ground Fault Interrupter
GFRC	Glass Fiber Reinforced Concrete
Glaz.	Glazier
GPD	Gallons per Day
gpf	Gallon per Flush
GPH	Gallons per Hour
gpm, GPM	Gallons per Minute
GR	Grade
Gran.	Granular
Grnd.	Ground
GVW	Gross Vehicle Weight
GWB	Gypsum Wall Board
H	High Henry
HC	High Capacity
H.D., HD	Heavy Duty; High Density
H.D.O.	High Density Overlaid
HDPE	High Density Polyethylene Plastic
Hdr.	Header
Hdwe.	Hardware
H.I.D., HID	High Intensity Discharge
Help.	Helper Average
HEPA	High Efficiency Particulate Air Filter
Hg	Mercury
HIC	High Interrupting Capacity
HM	Hollow Metal
HMWPE	High Molecular Weight Polyethylene
HO	High Output
Horiz.	Horizontal
H.P., HP	Horsepower; High Pressure
H.P.F.	High Power Factor
Hr.	Hour
Hrs./Day	Hours per Day
HSC	High Short Circuit
Ht.	Height
Htg.	Heating
Htrs.	Heaters
HVAC	Heating, Ventilation & Air-Conditioning
Hvy.	Heavy
HW	Hot Water
Hyd.; Hydr.	Hydraulic
Hz	Hertz (cycles)
I.	Moment of Inertia
IBC	International Building Code
I.C.	Interrupting Capacity
ID	Inside Diameter
I.D.	Inside Dimension; Identification
I.F.	Inside Frosted
I.M.C.	Intermediate Metal Conduit
In.	Inch
Incan.	Incandescent
Incl.	Included; Including
Int.	Interior
Inst.	Installation
Insul., insul	Insulation/Insulated
I.P.	Iron Pipe
I.P.S., IPS	Iron Pipe Size
IPT	Iron Pipe Threaded
I.W.	Indirect Waste
J	Joule
J.I.C.	Joint Industrial Council
K	Thousand; Thousand Pounds; Heavy Wall Copper Tubing, Kelvin
K.A.H.	Thousand Amp. Hours
kcmil	Thousand Circular Mils
KD	Knock Down
K.D.A.T.	Kiln Dried After Treatment
kg	Kilogram
kG	Kilogauss
kgf	Kilogram Force
kHz	Kilohertz
Kip	1000 Pounds
KJ	Kilojoule
K.L.	Effective Length Factor
K.L.F.	Kips per Linear Foot
Km	Kilometer
KO	Knock Out
K.S.F.	Kips per Square Foot
K.S.I.	Kips per Square Inch
kV	Kilovolt
kVA	Kilovolt Ampere
kVAR	Kilovar (Reactance)
KW	Kilowatt
KWh	Kilowatt-hour
L	Labor Only; Length; Long; Medium Wall Copper Tubing
Lab.	Labor
lat	Latitude
Lath.	Lather
Lav.	Lavatory
lb.; #	Pound
L.B., LB	Load Bearing; L Conduit Body
L. & E.	Labor & Equipment
lb./hr.	Pounds per Hour
lb./L.F.	Pounds per Linear Foot
lbf/sq.in.	Pound-force per Square Inch
L.C.L.	Less than Carload Lot
L.C.Y.	Loose Cubic Yard
Ld.	Load
LE	Lead Equivalent
LED	Light Emitting Diode
L.F.	Linear Foot
L.F. Hdr	Linear Feet of Header
L.F. Nose	Linear Foot of Stair Nosing
L.F. Rsr	Linear Foot of Stair Riser
Lg.	Long; Length; Large
L & H	Light and Heat
LH	Long Span Bar Joist
L.H.	Labor Hours
L.L., LL	Live Load
L.L.D.	Lamp Lumen Depreciation
lm	Lumen
lm/sf	Lumen per Square Foot
lm/W	Lumen per Watt
LOA	Length Over All
log	Logarithm
L-O-L	Lateralolet
long.	Longitude
L.P., LP	Liquefied Petroleum; Low Pressure
L.P.F.	Low Power Factor
LR	Long Radius
L.S.	Lump Sum
Lt.	Light
Lt. Ga.	Light Gauge
L.T.L.	Less than Truckload Lot
Lt. Wt.	Lightweight
L.V.	Low Voltage
M	Thousand; Material; Male; Light Wall Copper Tubing
M²CA	Meters Squared Contact Area
m/hr.; M.H.	Man-hour
mA	Milliampere
Mach.	Machine
Mag. Str.	Magnetic Starter
Maint.	Maintenance
Marb.	Marble Setter
Mat; Mat'l.	Material
Max.	Maximum
MBF	Thousand Board Feet
MBH	Thousand BTU's per hr.
MC	Metal Clad Cable
MCC	Motor Control Center
M.C.F.	Thousand Cubic Feet
MCFM	Thousand Cubic Feet per Minute
M.C.M.	Thousand Circular Mils
MCP	Motor Circuit Protector
MD	Medium Duty
MDF	Medium-density fibreboard
M.D.O.	Medium Density Overlaid
Med.	Medium
MF	Thousand Feet
M.F.B.M.	Thousand Feet Board Measure
Mfg.	Manufacturing
Mfrs.	Manufacturers
mg	Milligram
MGD	Million Gallons per Day
MGPH	Million Gallons per Hour
MH, M.H.	Manhole; Metal Halide; Man-Hour
MHz	Megahertz
Mi.	Mile
MI	Malleable Iron; Mineral Insulated
MIPS	Male Iron Pipe Size
mj	Mechanical Joint
m	Meter
mm	Millimeter
Mill.	Millwright
Min., min.	Minimum, Minute

Abbreviations

Misc.	Miscellaneous	PCM	Phase Contrast Microscopy	SBS	Styrene Butadiere Styrene
ml	Milliliter, Mainline	PDCA	Painting and Decorating	SC	Screw Cover
M.L.F.	Thousand Linear Feet		Contractors of America	SCFM	Standard Cubic Feet per Minute
Mo.	Month	P.E., PE	Professional Engineer;	Scaf.	Scaffold
Mobil.	Mobilization		Porcelain Enamel;	Sch., Sched.	Schedule
Mog.	Mogul Base		Polyethylene; Plain End	S.C.R.	Modular Brick
MPH	Miles per Hour	P.E.C.I.	Porcelain Enamel on Cast Iron	S.D.	Sound Deadening
MPT	Male Pipe Thread	Perf.	Perforated	SDR	Standard Dimension Ratio
MRGWB	Moisture Resistant Gypsum	PEX	Cross Linked Polyethylene	S.E.	Surfaced Edge
	Wallboard	Ph.	Phase	Sel.	Select
MRT	Mile Round Trip	P.I.	Pressure Injected	SER, SEU	Service Entrance Cable
ms	Millisecond	Pile.	Pile Driver	S.F.	Square Foot
M.S.F.	Thousand Square Feet	Pkg.	Package	S.F.C.A.	Square Foot Contact Area
Mstz.	Mosaic & Terrazzo Worker	Pl.	Plate	S.F. Flr.	Square Foot of Floor
M.S.Y.	Thousand Square Yards	Plah.	Plasterer Helper	S.F.G.	Square Foot of Ground
Mtd., mtd., mtd	Mounted	Plas.	Plasterer	S.F. Hor.	Square Foot Horizontal
Mthe.	Mosaic & Terrazzo Helper	plf	Pounds Per Linear Foot	SFR	Square Feet of Radiation
Mtng.	Mounting	Pluh.	Plumber Helper	S.F. Shlf.	Square Foot of Shelf
Mult.	Multi; Multiply	Plum.	Plumber	S4S	Surface 4 Sides
MUTCD	Manual on Uniform Traffic Control	Ply.	Plywood	Shee.	Sheet Metal Worker
	Devices	p.m.	Post Meridiem	Sin.	Sine
M.V.A.	Million Volt Amperes	Pntd.	Painted	Skwk.	Skilled Worker
M.V.A.R.	Million Volt Amperes Reactance	Pord.	Painter, Ordinary	SL	Saran Lined
MV	Megavolt	pp	Pages	S.L.	Slimline
MW	Megawatt	PP, PPL	Polypropylene	Sldr.	Solder
MXM	Male by Male	P.P.M.	Parts per Million	SLH	Super Long Span Bar Joist
MYD	Thousand Yards	Pr.	Pair	S.N.	Solid Neutral
N	Natural; North	P.E.S.B.	Pre-engineered Steel Building	SO	Stranded with oil resistant inside
nA	Nanoampere	Prefab.	Prefabricated		insulation
NA	Not Available; Not Applicable	Prefin.	Prefinished	S-O-L	Socketolet
N.B.C.	National Building Code	Prop.	Propelled	sp	Standpipe
NC	Normally Closed	PSF, psf	Pounds per Square Foot	S.P.	Static Pressure; Single Pole; Self-
NEMA	National Electrical Manufacturers	PSI, psi	Pounds per Square Inch		Propelled
	Assoc.	PSIG	Pounds per Square Inch Gauge	Spri.	Sprinkler Installer
NEHB	Bolted Circuit Breaker to 600V.	PSP	Plastic Sewer Pipe	spwg	Static Pressure Water Gauge
NFPA	National Fire Protection Association	Pspr.	Painter, Spray	S.P.D.T.	Single Pole, Double Throw
NLB	Non-Load-Bearing	Psst.	Painter, Structural Steel	SPF	Spruce Pine Fir; Sprayed
NM	Non-Metallic Cable	P.T.	Potential Transformer		Polyurethane Foam
nm	Nanometer	P. & T.	Pressure & Temperature	S.P.S.T.	Single Pole, Single Throw
No.	Number	Ptd.	Painted	SPT	Standard Pipe Thread
NO	Normally Open	Ptns.	Partitions	Sq.	Square; 100 Square Feet
N.O.C.	Not Otherwise Classified	Pu	Ultimate Load	Sq. Hd.	Square Head
Nose.	Nosing	PVC	Polyvinyl Chloride	Sq. In.	Square Inch
NPT	National Pipe Thread	Pvmt.	Pavement	S.S.	Single Strength; Stainless Steel
NQOD	Combination Plug-on/Bolt on	PRV	Pressure Relief Valve	S.S.B.	Single Strength B Grade
	Circuit Breaker to 240V.	Pwr.	Power	sst, ss	Stainless Steel
N.R.C., NRC	Noise Reduction Coefficient/	Q	Quantity Heat Flow	Sswk.	Structural Steel Worker
	Nuclear Regulator Commission	Qt.	Quart	Sswl.	Structural Steel Welder
N.R.S.	Non Rising Stem	Quan., Qty.	Quantity	St.; Stl.	Steel
ns	Nanosecond	Q.C.	Quick Coupling	STC	Sound Transmission Coefficient
NTP	Notice to Proceed	r	Radius of Gyration	Std.	Standard
nW	Nanowatt	R	Resistance	Stg.	Staging
OB	Opposing Blade	R.C.P.	Reinforced Concrete Pipe	STK	Select Tight Knot
OC	On Center	Rect.	Rectangle	STP	Standard Temperature & Pressure
OD	Outside Diameter	recpt.	Receptacle	Stpi.	Steamfitter, Pipefitter
O.D.	Outside Dimension	Reg.	Regular	Str.	Strength; Starter; Straight
ODS	Overhead Distribution System	Reinf.	Reinforced	Strd.	Stranded
O.G.	Ogee	Req'd.	Required	Struct.	Structural
O.H.	Overhead	Res.	Resistant	Sty.	Story
O&P	Overhead and Profit	Resi.	Residential	Subj.	Subject
Oper.	Operator	RF	Radio Frequency	Subs.	Subcontractors
Opng.	Opening	RFID	Radio-frequency Identification	Surf.	Surface
Orna.	Ornamental	Rgh.	Rough	Sw.	Switch
OSB	Oriented Strand Board	RGS	Rigid Galvanized Steel	Swbd.	Switchboard
OS&Y	Outside Screw and Yoke	RHW	Rubber, Heat & Water Resistant;	S.Y.	Square Yard
OSHA	Occupational Safety and Health		Residential Hot Water	Syn.	Synthetic
	Act	rms	Root Mean Square	S.Y.P.	Southern Yellow Pine
Ovhd.	Overhead	Rnd.	Round	Sys.	System
OWG	Oil, Water or Gas	Rodm.	Rodman	t.	Thickness
Oz.	Ounce	Rofc.	Roofer, Composition	T	Temperature; Ton
P.	Pole; Applied Load; Projection	Rofp.	Roofer, Precast	Tan	Tangent
p.	Page	Rohe.	Roofer Helpers (Composition)	T.C.	Terra Cotta
Pape.	Paperhanger	Rots.	Roofer, Tile & Slate	T & C	Threaded and Coupled
P.A.P.R.	Powered Air Purifying Respirator	R.O.W.	Right of Way	T.D.	Temperature Difference
PAR	Parabolic Reflector	RPM	Revolutions per Minute	TDD	Telecommunications Device for
P.B., PB	Push Button	R.S.	Rapid Start		the Deaf
Pc., Pcs.	Piece, Pieces	Rsr	Riser	T.E.M.	Transmission Electron Microscopy
P.C.	Portland Cement; Power Connector	RT	Round Trip	temp	Temperature, Tempered, Temporary
P.C.F.	Pounds per Cubic Foot	S.	Suction; Single Entrance; South	TFFN	Nylon Jacketed Wire

Abbreviations

TFE	Tetrafluoroethylene (Teflon)	U.L., UL	Underwriters Laboratory	w/	With
T. & G.	Tongue & Groove;	Uld.	Unloading	W.C., WC	Water Column; Water Closet
	Tar & Gravel	Unfin.	Unfinished	W.F.	Wide Flange
Th., Thk.	Thick	UPS	Uninterruptible Power Supply	W.G.	Water Gauge
Thn.	Thin	URD	Underground Residential	Wldg.	Welding
Thrded	Threaded		Distribution	W. Mile	Wire Mile
Tilf.	Tile Layer, Floor	US	United States	W-O-L	Weldolet
Tilh.	Tile Layer, Helper	USGBC	U.S. Green Building Council	W.R.	Water Resistant
THHN	Nylon Jacketed Wire	USP	United States Primed	Wrck.	Wrecker
THW.	Insulated Strand Wire	UTMCD	Uniform Traffic Manual For Control	WSFU	Water Supply Fixture Unit
THWN	Nylon Jacketed Wire		Devices	W.S.P.	Water, Steam, Petroleum
T.L., TL	Truckload	UTP	Unshielded Twisted Pair	WT., Wt.	Weight
T.M.	Track Mounted	V	Volt	WWF	Welded Wire Fabric
Tot.	Total	VA	Volt Amperes	XFER	Transfer
T-O-L	Threadolet	VAT	Vinyl Asbestos Tile	XFMR	Transformer
tmpd	Tempered	V.C.T.	Vinyl Composition Tile	XHD	Extra Heavy Duty
TPO	Thermoplastic Polyolefin	VAV	Variable Air Volume	XHHW	Cross-Linked Polyethylene Wire
T.S.	Trigger Start	VC	Veneer Core	XLPE	Insulation
Tr.	Trade	VDC	Volts Direct Current	XLP	Cross-linked Polyethylene
Transf.	Transformer	Vent.	Ventilation	Xport	Transport
Trhv.	Truck Driver, Heavy	Vert.	Vertical	Y	Wye
Trlr	Trailer	V.F.	Vinyl Faced	yd	Yard
Trlt.	Truck Driver, Light	V.G.	Vertical Grain	yr	Year
TTY	Teletypewriter	VHF	Very High Frequency	Δ	Delta
TV	Television	VHO	Very High Output	%	Percent
T.W.	Thermoplastic Water Resistant	Vib.	Vibrating	~	Approximately
	Wire	VLF	Vertical Linear Foot	Ø	Phase; diameter
UCI	Uniform Construction Index	VOC	Volatile Organic Compound	@	At
UF	Underground Feeder	Vol.	Volume	#	Pound; Number
UGND	Underground Feeder	VRP	Vinyl Reinforced Polyester	<	Less Than
UHF	Ultra High Frequency	W	Wire; Watt; Wide; West	>	Greater Than
U.I.	United Inch			Z	Zone

Index

For customer support on your Commercial Renovation Costs with RSMeans data, call 800.448.8182.

For customer support on your Commercial Renovation Costs with RSMeans data, call 800.448.8182.

903

For customer support on your Commercial Renovation Costs with RSMeans data, call 800.448.8182.

Index

For customer support on your Commercial Renovation Costs with RSMeans data, call 800.448.8182.

For customer support on your Commercial Renovation Costs with RSMeans data, call 800.448.8182.

For customer support on your Commercial Renovation Costs with RSMeans data, call 800.448.8182.

909

Index

For customer support on your Commercial Renovation Costs with RSMeans data, call 800.448.8182.

For customer support on your Commercial Renovation Costs with RSMeans data, call 800.448.8182.

Index

Index

Virtual classes offered monthly. Call for the schedule. 📞 877.620.6245

RSMeans offers training classes in a variety of formats—eLearning training modules, instructor-led classes, on-site training, and virtual training classes, as well as custom solutions. Due to our current COVID-19 situation, RSMeans has adjusted our current training offerings and instructor-led classes for 2022 have not been determined. Please visit our website at rsmeans.com/products/training/seminars for our current schedule of class offerings.

Training classes that are offered in one or more of our formats are: Building Systems and the Construction Process; Construction Cost Estimating: Concepts & Practice; Introduction to Estimating; Scope of Work for Facilities Estimating; Facilities Construction Estimating; Fundamentals to Construction Cost Estimating; Maintenance & Repair Estimating for Facilities; Mechanical & Electrical Estimating; RSMeans CostWorks CD; and RSMeans Online Training.

Registration Information

How to register

By Phone

Register by phone at 877.620.6245

Online

Register online at RSMeans.com/products/services/training

Note: Purchase orders or credits cards are required to register.

Instructor-led Government Pricing

All federal government employees save off the regular seminar price. Other promotional discounts cannot be combined with the government discount. Call 781.422.5115 for government pricing.

If you have any questions or you would like to register for any class or purchase a training module, call us at 877.620.6245.

Construction Cost Estimating: Concepts and Practice

This one- or two-day introductory course to improve estimating skills and effectiveness starts with the details of interpreting bid documents and ends with the summary of the estimate and bid submission.

Some of what you'll learn:
- Using the plans and specifications to create estimates
- The takeoff process—deriving all tasks with correct quantities
- Summarizing the estimate to arrive at the final number
- Formulas for area and cubic measure, adding waste and adjusting productivity to specific projects
- Evaluating subcontractors' proposals and prices
- Adding insurance and bonds
- Understanding how labor costs are calculated
- Submitting bids and proposals

Who should attend: project managers, architects, engineers, owners' representatives, contractors, and anyone who's responsible for budgeting or estimating construction projects.

Fundamentals to Construction Cost Estimating

This self-paced course focuses on unit price estimating for facilities renovation, repair, and remodeling. Combining hands-on skill building with a discussion of best estimating practices and real-life problems, this informative online course will boost your estimating skills, bring you up to date with key concepts, and provide tips and guidelines that can save you time and help you avoid costly estimating oversights and errors.

Some of what you'll learn:
- The three phases of estimating
- Bid document review
- How to perform quantity takeoffs
- How to price the work

Who should attend: facilities professionals and anyone involved in facilities planning, design, and construction. For novice estimators, this course can serve as a primer to Facilities Construction Estimating Self-Paced Training or the following live training courses: Facilities Construction Estimating, Mechanical & Electrical Estimating, Assessing Scope of Work for Facilities Estimating, or Maintenance & Repair Estimating.

Mechanical & Electrical Estimating

This two-day course teaches attendees how to prepare more accurate and complete mechanical/electrical estimates, avoid the pitfalls of omission and double-counting, and understand the composition and rationale within the RSMeans mechanical/electrical database.

Some of what you'll learn:
- The unique way mechanical and electrical systems are interrelated
- M&E estimates—conceptual, planning, budgeting, and bidding stages
- Order of magnitude, square foot, assemblies, and unit price estimating
- Comparative cost analysis of equipment and design alternatives

Who should attend: architects, engineers, facilities managers, mechanical and electrical contractors, and others who need a highly reliable method for developing, understanding, and evaluating mechanical and electrical contracts.

Facilities Construction Estimating

In this two-day course, professionals working in facilities management can get help with their daily challenges to establish budgets for all phases of a project.

Some of what you'll learn:
- Determining the full scope of a project
- Understanding of RSMeans data and what is included in prices
- Identifying appropriate factors to be included in your estimate
- Creative solutions to estimating issues
- Organizing estimates for presentation and discussion
- Special estimating techniques for repair/remodel and maintenance projects
- Appropriate use of contingency, city cost indexes, and reference notes
- Techniques to get to the correct estimate quickly

Who should attend: facility managers, engineers, contractors, facility tradespeople, planners, and project managers.

Assessing Scope of Work for Facilities Construction Estimating

This two-day practical training program addresses the vital importance of understanding the scope of projects in order to produce accurate cost estimates for facilities repair and remodeling.

Some of what you'll learn:
- Discussions of site visits, plans/specs, record drawings of facilities, and site-specific lists
- Review of CSI divisions, including means, methods, materials, and the challenges of scoping each topic
- Exercises in scope identification and scope writing for accurate estimating of projects
- Hands-on exercises that require scope, takeoff, and pricing

Who should attend: corporate and government estimators, planners, facility managers, and others who need to produce accurate project estimates.

CostWorks CD Self-Paced Training Course

This self-paced course helps users become more familiar with the functionality of the CD. Each menu, icon, screen, and function found in the program is explained in depth. Time is devoted to hands-on estimating exercises.

Some of what you'll learn:
- Searching the database using all navigation methods
- Exporting RSMeans data to your preferred spreadsheet format
- Viewing crews, assembly components, and much more
- Automatically regionalizing the database

This training session requires users to bring a laptop computer to class and to follow along using a purchased copy of RSMeans CostWorks CD.

When you register for this course you will receive an outline for your laptop requirements.

Maintenance & Repair Estimating for Facilities

This two-day course teaches attendees how to plan, budget, and estimate the cost of ongoing and preventive maintenance and repair for existing buildings and grounds.

Some of what you'll learn:
- The most financially favorable maintenance, repair, and replacement scheduling and estimating
- Preventive planning and facilities upgrading
- Determining both in-house and contract-out service costs
- Annual, asset-protecting M&R plan

Who should attend: facility managers, maintenance supervisors, buildings and grounds superintendents, plant managers, planners, estimators, and others involved in facilities planning and budgeting.

RSMeans Online Training

Construction estimating is vital to the decision-making process at each state of every project. Our online solution works the way you do. It's systematic, flexible, and intuitive. In this self-paced course you will see how you can estimate any phase of any project faster and better.

Some of what you'll learn:
- Customizing our online estimating solution
- Making the most of RSMeans "Circle Reference" numbers
- How to integrate your cost data
- Generating reports, exporting estimates to MS Excel, sharing, collaborating, and more

Also offered as a self-paced or on-site training program!

RSMeans data Training

RSMeans Training Cancellation Policy

JAN - - 2022

Virtual Training (Individual)

If you are unable to attend a training class, a colleague may substitute at any time before the session starts by notifying the training registrar at 1.781.422.5115 or your sales representative. Gordian is not responsible for sending duplicate materials to the substitute colleague.

If you are unable to find a substitute and cancel fewer than ten (10) business days prior to the training class, you will either forfeit the registration fee or be allowed to apply your registration fee to the same class only once within the same calendar year. No-shows will forfeit their registration fees and no rescheduling is permitted. In the unfortunate circumstance that Gordian must cancel your event, Gordian will work with you to reschedule your attendance in the same training class at a later date or will fully refund your registration fee.

Any on-demand (self-paced) training modules are not eligible for cancellation, substitution, transfer, return, or refund.

Virtual Training (Private Group)

If you provide notice more than ten (10) business days prior to the training class, Gordian will work to reschedule your event. If you cancel your event fewer than ten (10) business days prior to the training class, you are responsible for paying for the trainers' time for the cancelled training, as well as the full cost of materials and shipping. In the unfortunate circumstance that Gordian must cancel your event, Gordian will work with you to reschedule your attendance in the same training class at a later date or will fully refund your registration fee.

Public Training Events (Individual)

If you are unable to attend a training conference, a colleague may substitute no fewer than ten (10) business days before the session starts by notifying the training registrar at 1.781.422.5115 or your sales representative. Gordian is not responsible for sending duplicate materials to the substitute colleague.

If you are unable to find a substitute and cancel fewer than ten (10) business days prior to the conference, you will forfeit the registration fee. No-shows will forfeit their registration fees and no rescheduling is permitted. In the unfortunate circumstance that Gordian must cancel your event, Gordian will fully refund your registration fee.

On-Site Training (Private Group)

If you provide notice more than ten (10) business days prior to the training class, Gordian will work with you to reschedule your event, but you are responsible for rescheduling fees associated with the trainer's time, airfare, and lodging. If you cancel fewer than ten (10) business days before your event, you are responsible for the full cost of the event. In the unfortunate event that Gordian must cancel your event, Gordian will work with you to reschedule your attendance at a later date or will fully refund your registration fee.

AACE Approved Courses

The AACE recognizes .1 CEUs per hour of training attended for all of Gordian's virtual training courses.

AIA Continuing Education

We are registered with the AIA Continuing Education System (AIA/CES) and are committed to developing quality learning activities in accordance with the CES criteria. Many seminars meet the AIA/CES criteria for Quality Level 2. AIA members may receive 14 learning units (LUs) for each two-day RSMeans course.

Daily Course Schedule

The first day of each seminar session begins at 8:30 a.m. and ends at 4:30 p.m. The second day begins at 8:00 a.m. and ends at 4:00 p.m. Participants are urged to bring a hand-held calculator since many actual problems will be worked out in each session.

Continental Breakfast

Your registration includes the cost of a continental breakfast and a morning and afternoon refreshment break. These informal segments allow you to discuss topics of mutual interest with other seminar attendees. You are free to make your own lunch and dinner arrangements.

Hotel/Transportation Arrangements

We arrange to hold a block of rooms at most host hotels. To take advantage of special group rates when making your reservation, be sure to mention that you are attending the RSMeans Institute data seminar. You are, of course, free to stay at the lodging place of your choice. Hotel reservations and transportation arrangements should be made directly by seminar attendees.

Important

Class sizes are limited, so please register as soon as possible.

RSMeans data Training